Toxic Plants of North America

Toxic Plants of North America

George E. Burrows
Ronald J. Tyrl

Iowa State University Press / Ames

George E. Burrows, DVM, PhD, is emeritus professor of toxicology, College of Veterinary Medicine, Oklahoma State University, Stillwater.

Ronald J. Tyrl, PhD, is professor of botany and curator of the herbarium at Oklahoma State University, Stillwater. He is coordinating editor of the Flora of Oklahoma Project.

Iowa State University Press
2121 South State Avenue, Ames, Iowa 50014

Orders: 1-800-862-6657
Office: 1-515-292-0140
Fax: 1-515-292-3348
Web site: www.isupress.com

♾ Printed on acid-free paper in the United States of America

First edition, 2001

Library of Congress Cataloging-in-Publication Data

Burrows, George E. (George Edward), 1935-
Toxic plants of North America / George E. Burrows, Ronald J. Tyrl.
p. cm.
Includes bibliographical references (p.).
ISBN 0-8138-2266-1 (alk. paper)
1. Poisonous plants—North America. I. Tyrl, Ronald J. II. Title

QK100.N6 B87 2001
581.6'59'097—dc21 00-053985

The last digit is the print number: 9 8 7 6 5 4 3 2 1

Contents

Contents

Color plates follow page 728

Toxic Plants of North America

Introduction 1

We humans have an intimate relationship with the plants that surround us. We take them for granted as we use them for food, clothes, and shelter. We use them medicinally; indeed, more than one-third of our modern pharmacopoeia has its origins in plant products. We please our senses, decorate our living spaces, and express our feelings for one another with them. Plants are an essential part of many of our religious and social rites. Paradoxically, some of the plants we prize for these varied uses also may pose a threat to us or to our domesticated animals. Toxic plants are very much a part of our environment. Until their effects, ranging from mild irritation or discomfort to rapid death, become apparent, they are often ignored or simply overlooked. Because of their ubiquity, there is a need for a comprehensive treatment of toxic plants likely to be encountered in North America north of the Tropic of Cancer, growing wild or cultivated. This book is written in response to that need.

Objective and Scope—The objective of this undertaking was to write a comprehensive treatment of toxic plants that brings together the currently available information on (1) their morphology and distribution, (2) the disease problem or problems associated with them, (3) their toxicants and mechanisms of action, (4) the clinical signs and pathologic changes associated with their toxicity, and (5) the principal aspects of treatment. Although the perspective of this book is primarily veterinary science, both human and animal intoxications are included because the information acquired about both often is interrelated and supportive.

Compilation of the information presented here began in the 1980s as a series of articles for the *Oklahoma Veterinarian* and an agricultural extension publication, *Poisonous Plants of Oklahoma and the Southern Plains.* Well received, these publications dealt primarily with native plants and their toxicity for livestock. Initially, the present book was anticipated to do the same for the United States. Gradually, however, its scope and depth of coverage evolved—larger area, more plant families, and greater detail than first envisioned. These changes came about in part because of the increasing popularity of ornamental plants for both house and garden. There has been a corresponding increase in awareness of toxicity problems associated with some of them.

The information presented is based upon reports extracted from the toxicological, veterinary, agronomy, chemical, biochemical, and physiological literature and from our personal observations. References are numerous. In the past, descriptions of intoxication problems were sometimes poorly documented, and a large amount of unsubstantiated anecdotal information was incorporated in earlier publications in such a form that it eventually became accepted as fact. Experimental studies have since confirmed or rejected much of this information. An effort has been made to document each point selectively to avoid being excessive, but it is anticipated that the incorporation of many references provides starting points for readers to delve more deeply into any topic.

The information presented is intended to be of interest to veterinarians, agricultural extension agents, horticulturists, animal scientists, botanists, personnel at poison control centers, physicians, pharmacists, agronomists, range scientists, toxicologists, wildlife biologists, ecologists, farmers, ranchers, students, and the general public. The book may be used as a textbook for graduate-level courses or as a general reference. The incorporation of tables associating the clinical signs and pathology of intoxications with specific plant genera and species permits its use in applied situations.

As always with a book such as this one, the caveat that it is not complete must be stated. As investigations of plants progress, there will be the discovery of new toxic species and the reassessment of the intoxication problems caused by known ones.

Organization and Format—The plant family serves as the organizational unit for the wealth of toxicological data compiled in this treatise. Each chapter is devoted to the toxic taxa of one family. To facilitate access and review, the material is organized into two columns. One column presents detailed textual information divided into seven sections: Taxonomy and Morphology, Distribution and Habitat, Disease Problems, Disease Genesis, Clinical Signs, Pathology, and Treatment. The second column offers the salient points of this information—key words, common names, line drawings, distribution maps, and chemical structures.

The descriptions given for each family describe the range of morphological variation for only its North American species. When a range of values is given for the numbers of genera and species in a family, differences in opinion among taxonomists are indicated. Unless otherwise attributed, information about the taxonomy and biology of each family was compiled primarily from Cronquist (1981), Heywood (1978), and Lawrence (1951).

To avoid repetition and conserve space, morphological features of the genus that are the same as those given for the family are generally not repeated; rather those features that are characteristic of or unique to the taxon are used. If a genus is monotypic or represented in North America by a single species, its morphological description is based on the species' appearance. The morphological descriptions of the genera and species are composites of those appearing in state and regional floras encompassing the distributional ranges of the taxa. Principal sources are listed in the references.

Concepts of families, genera, and species are based on current classifications. When significant changes in classification and/or nomenclature have occurred, the older names are given as synonyms in parentheses below the currently accepted name. When appropriate, brief explanations of the reasons for the name changes are given. The taxonomic keys for identification of the toxic species of a genus are not intended to be comprehensive nor to distinguish the toxic taxon from all other species of the genus. Rather they are intended to provide a quick, easy way of discriminating among the toxic taxa. It is anticipated that the reader will use floras specific for his or her locale to confirm identification of the suspected toxic plant.

Line drawings, maps, and chemical structures are based in part upon those appearing in the references cited below. Original line drawings are primarily the work of Bellamy Parks Jansen and Sheryl Holesko. Other drawings were obtained from the government publications listed in the references and were prepared by Regina Hughes and numerous other artists. Drawings have also been used with permission from *Flora of Missouri,* by J. A. Steyermark (1975). The maps and chemical structures are composites of the information available in both the references cited and the general literature.

In addition to the 75 chapters presenting the toxicological problems associated with each plant family, a chapter is included describing 37 families of questionable risk, a glossary, diagnostic synopses of the most important families, and tables cross-referencing disease syndromes and clinical signs.

Historical Perspective—We would be remiss in this endeavor if we did not recognize those who have gone before us and whose work has served as a foundation for this book. There are many individuals who should be recognized, and it is with some trepidation that we list them, because many who will not be included also have made substantial contributions to our understanding of toxic plants. Certainly L. H. Pammel and J. M. Kingsbury have been instrumental in providing a foundation and model upon which to write a book on toxic plants. Their efforts contributed greatly to our understanding of the effects of plants on livestock. Their work is especially significant because of the meager information they had in many instances upon which to base their conclusions about toxicity. Also of great importance were the efforts of early investigators and observers such as V. K. Chesnut and C. D. Marsh. The remarkable, astute observations of Marsh continue to be the basis for our understanding of the effects of many toxic plants as will be illustrated by the number of literature citations to his work throughout this book.

When reviewing those who have had great impact on our present state of knowledge of plant-caused problems, we cannot fail to recognize the personnel of the U.S. Department of Agriculture/Agricultural Research Service (ARS) Poisonous Plants Research Laboratory at Logan, Utah. These ARS scientists, both past and present, have had an immense impact on our understanding and ability to deal with the ever-present problems of plant intoxications in livestock. Many individuals have been involved in the lab's work, and the references throughout the book attest to their extensive efforts.

We are also indebted to the many personnel at state experiment stations who have contributed to the body of knowledge on the toxicity of plants, especially those in the western states. Worthy of particular note is the exceptional work conducted in Texas. Names that appear repeatedly in the toxicological literature and our references include I. B. Boughton, W. T. Hardy, and F. P. Mathews. Mathews was instrumental in opening the Locoweed Research Laboratory in Alpine, Texas, in 1930 and was responsible for many years for investigating the plant-related livestock problems in West Texas and surrounding areas.

Dedication—Following in the footsteps of Dr. Mathews was Dr. James W. Dollahite, a young veterinarian from west central Texas, and an individual who had a profound influence on the discipline of toxicology. His life and contributions were eloquently summarized by E. M. Bailey (1998) and are excerpted here with permission. Born in 1911, Dr. Dollahite was raised near Johnson City, Texas. He received his D.V.M. in 1933 from the Agricultural and Mechanical College of Texas. He worked for the U.S. government and practiced until World War II, when he served as an army veterinarian, later retiring as a lieutenant colonel in the Air Force Reserve. Following the war, he went back into veterinary practice in Marfa, Texas, but developed an interest in toxicology. Dr. Dollahite combined his practice and a part-time position with the Texas Agricultural Experiment Station in Alpine to further his interests in plant toxicology. He also worked for a time at the USDA research facility in Beltsville, Maryland. In 1956 he started a full-time experiment station position and was responsible for moving the Alpine Research Station, begun by Dr. Mathews, to Marfa, where it became the Marfa Toxic Plant Research Station. During this time he drove many

Dr. J.W. Dollahite

miles over West Texas and southern New Mexico, investigating toxic plant problems and conducting his toxic plant research. He closed the Marfa station in 1958 and moved his research endeavors to College Station, where he was a member of the veterinary research section of the College of Veterinary Medicine. Because there was no formal toxicology program at the time, he received his M.S. in veterinary physiology in 1961. He became an associate professor of pathology in 1964 and a professor in 1965. In 1968 he transferred to the Department of Veterinary Physiology and Pharmacology, where he was instrumental in establishing a doctoral program in toxicology in 1969.

Dr. Dollahite was a charter and founding diplomate of the American Board of Veterinary Toxicology (1966–1967). He continued his research until his retirement from Texas A&M in 1975. He continued to work on toxic plants at the USDA/ARS Veterinary Toxicology and Entomology Research Laboratory until his full retirement in 1980. He died in 1984.

Dr. Dollahite played a very important role in the development of veterinary toxicology in Texas, especially toxic plant research, and in the development of veterinary toxicology as a specialty within the American Veterinary Medical Association. However, these facts, dates, and accomplishments are but one aspect of the real man. One of us (GEB) had the opportunity to spend a week in 1979 traveling with him in a review of the toxic plants of Texas. It was this time that provided a glimpse of the person of whom others had long been aware. The respect paid to him by those with whom he had been associated in the field was impressive. He was truly a remarkable individual, not only for his powers of observation of clinical signs in diseased animals and contributions to our knowledge of toxic plants but also for his personal attributes. The legacy of his life was much more than professional success. He was an exemplary individual in many ways. We are sure that he would like to be remembered as a man of great faith in God, who made every effort to deal with others with respect, kindness, and gentleness. He had great integrity and was a gentleman in every sense of the word. He is truly a worthy role model.

It is with this in mind that we dedicate this book to Dr. J. W. Dollahite.

Acknowledgments—The writing of this book began as a traditional academic endeavor, that is, a review of the literature and an attempt to synthesize in a readable fashion the wealth of information accumulated. Initially the effort involved just the two of us, but as the project progressed, more and more individuals volunteered encouragement, support, time, and expertise. It is therefore necessary and certainly most appropriate to recognize formally their contributions at this point.

Thanks are expressed to Gayman Helman for critically reviewing all aspects of each chapter and making valued suggestions as to what additional information might be included, especially as pertains to the pathologic descriptions. Thanks also to Loyd Rowe for reviewing the large Fabaceae chapter.

To our wives, Connie Burrows and Lynda Tyrl, special thanks are given. Their tireless assistance with the odious and tedious editorial tasks, especially in the later stages of writing, was invaluable. Their words of encouragement represented vital contributions.

Likewise, completion of the work was facilitated by the technical assistance of Sheryl Holesko and Paula Shryock, staff members of the Department of Anatomy, Pathology, and Pharmacology and the Department of Botany at Oklahoma State University. Thanks to each are extended.

The financial assistance provided by the College of Veterinary Medicine via its long-term support of Burrows's research on toxic plants is gratefully acknowledged.

Long-term access to the library and herbarium collection at the Royal Botanic Gardens Kew for Tyrl also is treasured.

Finally, the individuals responsible for the transition of our manuscript to finished book certainly must be recognized. Our profound thanks to Gretchen Van Houten and Judi Brown of Iowa State University Press for their patience and ability to understand our vision of the book's final form. A special thanks to Rosemary Wetherold, our editor, whose careful work ensured accuracy, consistency, and clarity throughout the book. We also gratefully acknowledge the efforts of Nanette Cardon, our indexer who organized in a most logical fashion the plethora of names and terms that appear in this book. A final thanks to Fred Thompson, our production editor, whose editorial and organization skills facilitated the entire production process.

References

Bailey EM Jr: A tribute to Dr James W. Dollahite. In *Toxic Plants and Other Natural Toxicants,* Garland T, Barr AC eds, CAB International, New York, pp17–18, 1998

Belcher RO: A revision of the genus *Erechtites* (Compositae), with inquiries into *Senecio* and *Arrhenechtites. Ann Mo Bot Gard* 43;1–85, 1956

Brooks AB: *West Virginia Trees.* W Va Agric Exp Stn Bull 175, 1920

Burns GP, Otis CH: *The Trees of Vermont.* Vt Agric Exp Stn Bull 194, 1916

Clark RC: The woody plants of Alabama. *Ann Mo Bot Gard* 58;99-242, 1971

Connolly JD, Hill RA: *Dictionary of Terpenoids.* Vol 2, *Di- and Higher Terpenoids.* Chapman & Hall, London, pp1277–1279, 1991

Correll DS, Johnston MC: *Manual of the Vascular Plants of Texas,* 2d printing. Univ Texas at Dallas, Richardson, Tex, 1979

Crawford HS, Kucera CL, Ehrenreich JH: *Ozark Range and Wildlife Plants.* USDA Agric Handb 356, 1969

Cronquist A: *An Integrated System of Classification of Flowering Plants.* Columbia Univ Press, New York, 1981

Cronquist A, Holmgren AH, Holmgren NH, Reveal JL: *Intermountain Flora.* New York Botanical Garden, New York. Vol 1, 1972; vol 3, part B, Fabales, RC Barneby, 1989; vol 4, 1984; vol 5, Asterales, A Cronquist, 1994; vol 6, Monocotyledons, 1977

Dayton WA: *Important Western Browse Plants.* USDA Misc Publ 101, 1931

Dayton WA: *Notes on Western Range Forbs: Equisetaceae through Fumariaceae.* USDA Forest Service, Agric Handb 161, 1960

Dorn RD: *Vascular Plants of Wyoming.* Mountain West Publ, Cheyenne, Wyo, 1988

Flora of North America Committee: *Flora of North America.* Oxford Press, New York. Vol 1, 1993; vol 2, 1997

Frankton C, Mulligan GA: *Weeds of Canada.* Agric Can, Publ 948. New Canada Press, Toronto, 1987

Gates FC: *Weeds in Kansas.* Rep Kans State Board of Agric, vol 60, no 243, 1941

Gleason HA, Cronquist A: *Manual of Vascular Plants of Northeastern United States and Adjacent Canada,* 2d ed. New York Botanical Garden, Bronx, 1991

Great Plains Flora Association: *Atlas of the Flora of the Great Plains.* Iowa State Univ Press, Ames, 1977

Gunn CR, Seldin MJ: *Seeds and Fruits of North American Papaveraceae.* USDA Tech Bull 1517, 1976

Heywood VH ed: *Flowering Plants of the World.* Mayflower Books, New York, 1978

Hickman JC: *The Jepson Manual Higher Plants of California.* Univ California Press, Berkeley, 1993

Hitchcock AS, Chase A: *Manual of Grasses of the United States,* 2d ed. USDA Misc Publ 200, 1950

Hitchcock CL, Cronquist A: *Flora of the Pacific Northwest*. Univ Washington Press, Seattle, 1973
Hulten E: *Flora of Alaska and Neighboring Territories*. Stanford Univ Press, Stanford, Calif, 1968
Isely D: Leguminosae of the United States. I. Subfamily Mimosoideae. *Mem N Y Bot Gard* 25(1);1–152, 1973
Isely D: Leguminosae of the United States. II. Subfamily Caesalpinioideae. *Mem N Y Bot Gard* 25(2);1–228, 1975
Isely D: Leguminosae of the United States. II. Subfamily Papilionoidae. *Mem N Y Bot Gard* 25(2);1–264, 1981
Johnston IM: Plants of Coahuila, eastern Chihuahua, and adjoining Zacatecas and Durango, parts 1 & 2. *J Arnold Arbor* 24;306-339, 1943; 24;375–421, 1943
Jones GN: *Flora of Illinois,* 3d ed. Univ Notre Dame Press, Notre Dame, Ind, 1971
Jones GN, Fuller GD: *Vascular Plants of Illinois*. Univ Illinois Press, Urbana, 1955
Judd BI: *Principal Forage Plants of Southwestern Ranges*. USDA Rocky Mt Forest & Range Exp Stn Pap 69, 1962
Kearney TH, Peebles RH: *Arizona Flora,* 2d ed. Univ California Press, Berkeley, 1960
Kruschke EP: Preliminary reports on the flora of Wisconsin—31, Boraginaceae. *Trans Wis Acad Sci Arts Lett* 36;273–290, 1944
Lawrence GHM: *Taxonomy of Vascular Plants*. Macmillan, New York, 1951
Little EL Jr: To know the trees. *Yearb Agric* 1949;763–814, 1949
Little EL Jr: Fifty trees from foreign lands. *Yearb Agric* 1949;815–832, 1949
Little EL Jr: *Atlas of United States, Trees*. Vol 5, *Florida*. USDA Forest Service, Misc Publ 1361, 1978
Martin WC, Hutchins CR: *A Flora of New Mexico,* vols 1 & 2. J Cramer, Vaduz, Germany, 1980
McGregor RL, Barkley TM, Barker WT, Brooks RE, Churchill SP, Kaul RB, Sutherland DM, Van Bruggen T, Weedon RR, Wilson JS: *Atlas of the Flora of the Great Plains*. Iowa State Univ Press, Ames, 1977
Muller CH: The holacanthoid plants of North America. *Madrono* 6;128–137, 1941
Nelson EW, Burnside O: *Nebraska Weeds*. Nebraska Department of Agriculture, 1979; original as Bull 101-R, Weed and Seed Div, Nebr Dept Agric
Raup HM: The botany of southwestern Mackenzie. *Sargentia* 6;1–275, 1947
Reed CF, Hughes RO: *Selected Weeds of the United States*. USDA Agric Handb 366, 1970
Robbins WW, Bellue MK, Ball WS: *Weeds of California*. Calif Dept Agric Bull, 1941
Series on "The Biology of Canadian Weeds." *Can J Plant Sci;* reprinted in part as Contrib 33–61, Publ 1765, Agriculture Canada, Ottawa, 1984
Shetler SG, Skog LE: *Checklist of Species for Flora North America,* rev. Missouri Botanical Garden, 1978
Shreve F, Wiggins IL: *Vegetation and Flora of the Sonoran Desert,* vols 1 & 2. Stanford Univ Press, Stanford, Calif, 1964
Small JK: *Manual of the Southeastern Flora*. JK Small, New York, 1554pp, 1933
Smiley FJ: Weeds of California and methods of control. *Mon Bull Calif Dept Agric* 11;73–360, 1922
Southon IW: *Phytochemical Dictionary of the Leguminosae*. Vol 1, *Plants and Their Constituents*. Chapman & Hall, London, 1994
Southon IW, Buckingham J: *Dictionary of Alkaloids*. Chapman & Hall, London, 1989
Standley PC: Trees and Shrubs of Mexico, parts 1-5. *Contrib U S Natl Herb, Washington, DC* 23(1);1–171, 1920; 23(2);171–516, 1922; 23(3);517–848, 1923; 23(4);849–1312, 1924; 23(5);1313–1721, 1926
Stebbins GL Jr: The genetic approach to problems of rare and endemic species. *Madrono* 6;241–258, 1942
Steyermark JA: *Flora of Missouri*. Iowa State Univ Press, Ames, 1975
Svenson HK: Effects of post-Pleistocene submergence in eastern North America. *Rhodora* 29;105–114, 1927

Taylor RL, MacBryde B: *Vascular Plants of British Columbia.* Tech Bull 4, Bot Gard, Univ British Columbia, Vancouver BC, 1977

Turner BL: *The Legumes of Texas.* Univ Texas Press, Austin, 1959

Turner RM, Bowers JE, Burgess TL: *Sonoran Desert Plants: An Ecological Atlas.* Univ Arizona Press, Tucson, 1995

USDA, Div Plant Exploration and Introduction: *Contributions toward a Flora of Nevada,* nos 1-25, 1940

Van Bruggen T: *The Vascular Plants of South Dakota.* Iowa State Univ Press, Ames, 1976

Waterfall UT: *Keys to the Flora of Oklahoma,* 6th ed. UT Waterfall, Stillwater, Okla, 1979

Wax LM, Fawcett RS, Isely D: *Weeds of the North Central States.* North Cent Reg Res Publ 281, 1954, rev 1960

Weber WA: *Colorado Flora: Western Slope, Colorado.* Assoc Univ Press, Boulder, Colo, 1987

Weber WA: *Colorado Flora: Eastern Slope.* Univ Press of Colorado, Niwot, Colo, 1990

Welsh SL, Atwood ND, Goodrich S, Higgins LC: *A Utah Flora.* Great Basin Nat Mem 9, Brigham Young Univ, Provo, Utah, 1987

Wiggins IL: *Flora of Baja California.* Stanford Univ Press, Stanford, Calif, 1980

2 Aceraceae Juss.

■ Maple Family

Acer

Commonly known as the maple family, the Aceraceae comprises 2 genera, *Acer* and *Dipteronia,* and 100–150 species. Its distribution is cosmopolitan, with greatest diversity in China and northern temperate regions. In North America, only *Acer* is present. Prized for its wood—used in furniture, paneling, veneer, and flooring—the family also is economically important for its ornamentals. Wildlife consume buds, twigs, and fruits. Intoxication problems have been associated with only one species.

■ trees or shrubs; leaves opposite

Acer rubrum

Plants trees or rarely shrubs; polygamodioecious or dioecious. **Leaves** simple or 1-pinnately compound; opposite; stipules absent. **Simple Leaves:** venation palmate, margins palmately lobed. **Compound Leaves:** terminal leaflets present, venation pinnate. **Inflorescences** corymbs or panicles or racemes; terminal or axillary on short leafy branches. **Flowers** imperfect or perfect; perfect and imperfect similar; perianths in 1-series or 2-series; radially symmetrical. **Sepals** 5 or rarely 4 or 6; fused or free; green to greenish yellow or red. **Petals** 5 or rarely 4 or 6 or 0; free; red. **Stamens** 4–12. **Nectaries** present; may form disk. **Pistils** 1; compound; carpels 2; stigmas 2; styles 2; ovaries superior; placentation axile; ovules 2 per locule. **Fruits** schizocarps of 2 samaras. **Seeds** 2.

Acer L.

■ *A. rubrum*

red maple, swamp maple

■ trees; leaves palmately lobed with red petioles; fruits samaras

Taxonomy and Morphology—*Acer* is represented in North America by 13 native species. Numerous species from Eurasia have been introduced as ornamentals. The native species occupy a variety of upland and bottomland habitats primarily in the eastern half of the continent. Most commonly encountered are *A. saccharum* (sugar maple), a dominant of the northeastern deciduous forests; *A. saccharinum* (silver maple); and *A. negundo* (boxelder), which is found across the continent. Only one species of the genus in North America *has* confirmed toxicity potential:

A. rubrum L. — red maple, scarlet maple, swamp maple, water maple, soft maple, Carolina maple

Plants medium-sized trees; bark gray, smooth. **Branches** glabrous; red when young. **Leaves** simple; palmately 3-to-6-lobed; upper surfaces dark green; lower surfaces pale green to white-green; margins coarsely serrate or crenate; bases truncate to subcordate; petioles red. **Flowers** appearing before the leaves; clustered on pendulous peduncles; bright red. **Samaras** 15–25 mm long; bright red; angle of wing divergence acute.

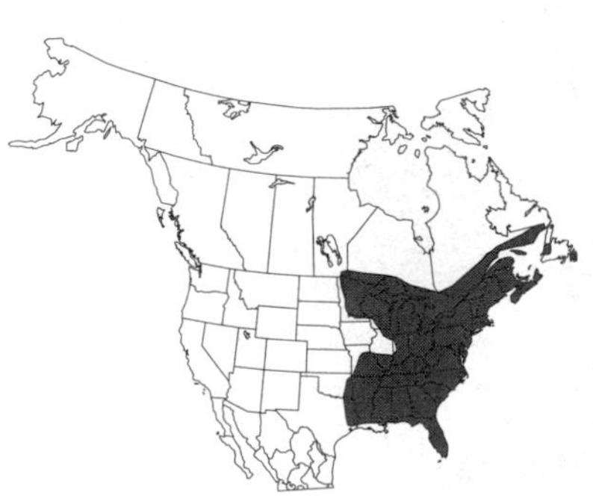

Acer rubrum

Distribution and Habitat—Exhibiting the greatest north-south distributional range of all tree species along the East Coast, *A. rubrum* occurs throughout the eastern half of North America. It typically is found in alluvial woods, along waterways, or on moist shady slopes.

Disease Problems—Acute lysis of red blood cells following consumption of *A. rubrum* leaves has been confirmed only in horses, but there is little reason to suspect that this disease is limited to this animal species, and it may not be limited to this species of *Acer*. With scattered occurrences in the summer and fall throughout the range of *A. rubrum,* the disease follows consumption of wilted or partly dried leaves from fallen or pruned branches (McConnico & Browne 1992). Enough leaves may be consumed in one day to produce intoxication similar to but typically more acute and severe than that produced by species of *Allium*. Fresh leaves are apparently nontoxic (Tennant et al. 1981; Divers et al. 1982; George et al. 1982).

- horses, abrupt onset of RBC lysis, 1–2 days after eating wilted maple leaves

Although it has been suggested that leaves are more toxic in the fall (George et al. 1982), severe intoxications have been reported as early as June (Tennant et al. 1981). Ingestion of 0.5–0.8% b.w. of dried but green leaves may cause intoxication. Typically horses will eat the leaves from a fallen limb for a day or two, with signs of intoxication appearing a day later. The leaves remain toxic for several weeks or more (Tennant et al. 1981; George et al. 1982). Experimentally, George et al. (1982) were unable to cause disease with fresh leaves. However, dried leaves (to 50% fresh weight), stored at room temperature and fed to horses at 1.5–3 g/kg b.w., produced the typical hemolytic disease up to 4 weeks after drying. Leaves collected after 15 September produced a more acute disease, with horses dying within 18 hours after ingestion.

An apparently unrelated disease problem has been reported in ducklings that ate seedlings of an unidentified European species of maple (Ludwig & Ludwig 1975). More than 300 of 3,900 ducklings died suddenly. The disease was reproduced experimentally by feeding seedlings to ducklings, with signs such as incoordination, weakness of the legs, and paralysis observed. Pathologic changes were limited to congestion of the liver and spleen and pale cardiac musculature.

Disease Genesis—Little is known of the identity of the specific toxicants, although *A. rubrum* is unusual among the maples in its pattern of galloyl constituents (tannins) and the type of ester binding in its leaves, the toxin may not be specific for this species (Bate-Smith 1978). These hydrolyzable tannins have numerous potentially reactive hydroxyl groups that conceivably could be involved in formation of reactants similar to the highly condensed quinones that play a role in photosensitization due to consumption of *Hypericum*. Although the toxicants and their mechanism of action are unknown, there appears to be formation of highly reactive oxidants in sufficient quantities to overwhelm the glutathione-reducing system in RBCs. As with the oxidant effects of species of *Allium* and *Brassica,* the result is increased RBC fragility, precipitation of hemoglobin, and formation of Heinz bodies in RBCs. However, in the situation with *Acer,* the formation of methemoglobinemia is much more prominent, affecting as much as 50% of the circulating hemoglobin. In this instance there may be a direct link between methemoglobin (MHb) formation and the appearance of Heinz bodies, in contrast to the situation with *Allium* and *Brassica*. There are a variety of types of MHb-forming agents, typically phenyl rings with quinones, phenolics, amines, or nitro moieties, some of which require bioactivation in the body (Kiese 1974; Smith 1993). Thiols such as cysteine and cysteamine are of only low potency. This would suggest that the causes of hemolysis in *Acer* are more likely to be nitrogenous compounds or polyphenolics, as indicated previously. Regardless of the cause, there is progressive intravascular and extravascular hemolysis due to RBC membrane perturbation and osmotic lysis, and removal by mononuclear phagocytic cells. The presence of severe hemolysis probably worsens the methemoglobinemia, because these conditions inhibit MHb reductase activity in red blood cells (Smith 1993). Death occurs because of the severe lack of oxygen delivery to vital cells due

- toxin exerts oxidizing action on RBCs, intra- and extravascular hemolysis, cause unknown but possibly polyphenolic or nitrogenous compounds

to anemia and to the oxidation of hemoglobin to MHb, which is incapable of oxygen transport. Renal effects due to hemoglobinuric nephrosis and shock-related tissue perfusion deficits may be serious complicating factors.

The cause of the duckling deaths following ingestion of the European maple seedlings also is unknown. Several glycine and alanine derivatives were suggested to be the toxins (Ludwig & Ludwig 1975). Maples, including *A. rubrum,* also contain the indole alkaloid gramine (Pachter 1959; Pachter et al. 1959). This alkaloid, however, produces antiepinephrine effects on blood pressure and smooth muscle in cats, and convulsant actions at high dosage parenterally in mice, rather than hematologic changes (Powell & Chen 1945).

■ horses, abrupt onset, depression, lethargy, anorexia, dark brown urine

severe anemia, Heinz bodies, methemoglobinemia

Clinical Signs—Rarely does peracute disease occur. When it does, there are few signs other than severe depression and cyanosis, terminating in death within 18 hours (George et al. 1982). More typically, the onset of signs is progressive over several days. Early signs—depression, lethargy, anorexia, and dark red to brown urine—will be seen 1 or 2 days after ingestion of the leaves. The following day may bring increased severity, with weakness, increased heart and respiratory rates, and cyanosis (Divers et al. 1982; Long & Payne 1984; Crowell 1988). At this time, hematologic changes will be marked: anemia, numerous Heinz bodies in RBCs, and MHb levels of up to 50% or more (Tennant et al. 1981; Tvedten 1988). The anemia may develop quickly or be progressive, reaching a hematocrit of 8–10% and hemoglobin concentration of 5 g/dl within 1 day or up to 7 days after ingestion. Heinz bodies may be apparent for several weeks. Because of the progressive nature of the anemia, death may be delayed until a week after ingestion of the leaves, when the anemia becomes most severe. During this time, there will be moderate elevation of various serum enzymes, including LDH, CK, SDH, and AST. There also may be elevated serum bilirubin and creatinine, and a marked decrease in the concentration of reduced glutathione in erythrocytes (Tennant et al. 1981; George et al. 1982). In some cases, oliguria and icterus may be the primary presenting signs, accompanied by severe anemia and indications of renal failure (Plumlee 1991).

■ gross pathology, spleen enlarged, kidneys swollen, dark colored

microscopic, kidney tubular necrosis, Hb casts

Pathology—Grossly, the spleen may be enlarged and congested with blood, and the kidneys swollen and dark reddish black. Microscopically, the most prominent change is tubular nephrosis, with hemoglobin casts and epithelial cellular debris in dilated tubules. There also may be moderate centrilobular hepatic degeneration and necrosis, and foci of erythrophagocytosis in splenic macrophages and hepatic Kupffer cells (Divers et al. 1982; Long & Payne 1984).

■ activated charcoal, fluids, whole blood, ascorbic acid, nonsteroidal antiinflammatory drugs

Treatment—Supportive care is very important, as well as avoidance of stresses that would complicate the existing problems of oxygen delivery. Fluids and whole blood may be given, but because of the continued presence of oxidants in the circulation, hemolysis may attenuate their beneficial effects. Early in the course of the disease, activated charcoal given orally may be of benefit. Treatment to reduce MHb to functional hemoglobin is not readily accomplished, because the horse responds poorly to the reducing effects of methylene blue (Kiese 1974). As an alternative, although slow in effect, ascorbic acid, 30 mg/kg b.w. i.v., may be given every 12 hours. McConnico and Browne (1992) reported some success with the use of 125 mg/kg ascorbic acid orally, 50 mg/kg s.c., and 30 mg/kg i.v. every 12 hours in concert with blood transfusion, nonsteroidal anti-inflammatory drugs, corticosteroids, and fluids.

References

Bate-Smith EC: Systematic aspects of the astringent tannins of *Acer* species. *Phytochemistry* 17;1945–1948, 1978

Crowell WA: A botanical hazard. *Vet Med* 83;651, 1988

Divers TJ, George LW, George JW: Hemolytic anemia in horses after the ingestion of red maple leaves. *J Am Vet Med Assoc* 180;300–302, 1982

George LW, Divers TJ, Mahaffey EA, Suarez MJH: Heinz body anemia and methemoglobinemia in ponies given red maple (*Acer rubrum* L.) leaves. *Vet Pathol* 19;521–533, 1982

Kiese M: *Methemoglobinemia: A Comprehensive Treatise.* CRC Press, Cleveland, 1974

Long PH, Payne JW: Red maple–associated pulmonary thrombosis in a horse. *J Am Vet Med Assoc* 184;977–978, 1984

Ludwig E, Ludwig G: Loss of ducklings following intake of maple seedlings. *Monatsh Veterinarmed* 30;766–768, 1975

McConnico RS, Browne CF: The use of ascorbic acid in the treatment of 2 cases of red maple *(Acer rubrum)*–poisoned horses. *Cornell Vet* 82;293–300, 1992

Pachter IJ: A note on the occurrence of gramine in *Acer rubrum* L. *J Am Pharm Assoc* 48;670, 1959

Pachter IJ, Zacharias DE, Ribeiro O: Indole alkaloids of *Acer saccharinum* (the silver maple), *Dictyoloma incanescens, Piptadenia colubrina,* and *Mimosa hostilis. J Org Chem* 24;1285–1287, 1959

Plumlee KH: Red maple toxicity in a horse. *Vet Hum Toxicol* 33;66–67, 1991

Powell CE, Chen KK: The action of gramine. *Proc Soc Exp Biol Med* 58;1–4, 1945

Smith RP: Toxic responses of the blood. In *Casarett and Doull's Toxicology,* 4th ed, Amdur MO, Doull J, Klaassen CD eds, Pergamon Press, New York, pp257–281, 1993

Tennant B, Dill SG, Glickman LT, Mirro EJ, King JM, Polak DM, Smith MC, Kradel DC: Acute hemolytic anemia, methemoglobinemia, and Heinz body formation associated with ingestion of red maple leaves by horses. *J Am Vet Med Assoc* 179;143–150, 1981

Tvedten HW: Red maple leaf toxicity warning. *J Equine Vet Sci* 8;77, 1988

3 Agavaceae Endl.

■ Agave Family

Agave
Nolina

Comprising 18–20 genera and 580–700 species native to warm, mostly arid regions of both the Old World and the New World, the Agavaceae is commonly known as the century plant or sisal family. The first common name reflects the monocarpic habit of some of the *Agave*. Because of the harsh growing conditions occupied by most species, plants grow vegetatively for many years or even decades. They are acaulescent, with a rosette of fleshy-firm leaves that may become quite massive. When flowering does occur, a flowering stem bearing a massive terminal inflorescence is quickly produced. The plant subsequently dies as the seeds mature. The second common name, *sisal,* reflects the family's economic importance. Strong, durable fibers for cordage and matting are extracted from the leaves of a number of species. Species of both *Agave* and *Yucca* are frequently used in landscaping, especially in xeric sites.

There are differences in taxonomic opinion about whether the family should be recognized. Although it was originally described by Endlicher in 1841, it has not been accepted as a distinct family by many taxonomists. Its members have been positioned in the families Liliaceae and Amaryllidaceae. However, accumulation of cytogenetic information and recognition of the distinctive adaptations for xeric habitats of the Agavaceae has led to its acceptance as a distinct family in modern classifications.

■ subshrubs or herbs; leaves simple and typically long

Plants subshrubs or herbs; perennials; from caudices or crowns; evergreen; caulescent or acaulescent; succulent or not succulent; bearing perfect flowers or polygamodioecious. **Leaves** simple; alternate; basal or cauline and crowded; sessile; spreading or reflexed; fibrous or fleshy; blades linear or lanceolate or oblong; venation parallel; apices acute, often spine tipped; margins entire or serrate; stipules absent. **Inflorescences** spikes or racemes or panicles; bracts absent or present. **Flowers** perfect or imperfect, similar; perianths in 1-series or 2-series; radially or slightly bilaterally symmetrical; campanulate or tubular or funnelform. **Perianth Parts** 6; all alike; petaloid; in 1 or 2 whorls; free or fused; greenish white to white to cream or yellow to orange. **Stamens** 6. **Pistils** 1; compound, carpels 3; stigmas 3; styles 1 or 0; ovaries superior or inferior; locules 3 or appearing to be 6; placentation axile. **Fruits** capsules or berries. **Seeds** numerous or 3; flattened or globose.

Because they are found mainly in dry desert-type ranges, members of the Agavaceae are well recognized for their value as emergency stock feeds (Wooton 1918; Forsling 1919). Especially valued are species of *Yucca,* commonly known as Spanish bayonet or soapweed. Members of the genus *Yucca* also are known as sources of steroidal saponins, which are composed of two groups, varying mainly in the glycosidic ether linkages. The spirostanols (monodesmosidic) have spirostan aglycones with mainly C-3–linked sugars, whereas the furostanols (mono-, di-, or tridesmosidic) are typically 26-C aglycones with glycosidic linkages at C-3 and C-26 and are composed of 2–5 or even up to 11 sugars (Hostettmann & Marston 1995). Although saponins are generally considered to be irritants of the digestive tract, the use of these forages for feed is not accompanied by noteworthy digestive disturbances (Wooton 1918; Forsling 1919). Even when chopped and fed to cattle at up to 9 kg/day, *Yucca* produced only mild diarrhea. Bloat was a more serious problem. Best results were obtained when cottonseed meal was given in addition to the chopped *Yucca*. *Nolina* and *Agave lechеguilla* were not as useful for feed.

- used as emergency stock feed, saponins in the leaves and seeds

Agave L.

Taxonomy and Morphology—Comprising some 300 species, *Agave*, commonly known as agave or maguey, is the largest genus of its family and certainly the most important. Its name comes from the Greek *agavos*, meaning "admirable," and presumably refers to the showy appearance of the plants in flower. In addition to being a source of fiber, *Agave* is the source of popular Mexican beverages and food (Gentry 1982). The sap, consumed fresh, is known as aguamiel; fermented, it is the source of pulque, and of mescal or tequila when distilled. Archaeological evidence indicates that species of the genus have been used for food for at least 9,000 years. Humans consumed, both raw and boiled, the soft, starchy, white meristems of the short stems; the white bases of the leaves; the immature flowering shoots; and the flowers of some species. In the 1960s, thousands of tons of leaves were fed to herds of dairy cattle in northeastern Mexico; and in Baja California, panicles of flowers were cut and fed to range cattle (Gentry 1982). Various species of the genus also are grown as ornamentals, especially for architectural effect, and are now propagated worldwide. In North America, some 30 species are present. Only one is of toxicologic importance:

- *A. lecheguilla*
  lechuguilla

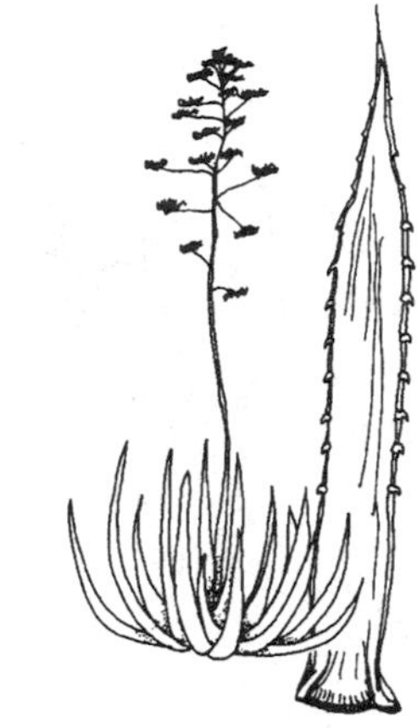

Agave lecheguilla

A. lecheguilla Torr. lechuguilla

Plants succulent herbs; perennials; from small, suckering, few-leaved rosettes, 30–50 cm in diameter and 40–60 cm high. **Leaves** 30–50 cm long; ascending to erect; light green to yellow green; stiff; blades linear-lanceolate; adaxial surfaces concave; abaxial surfaces convex; apices spine tipped; margins easily separated from blade when dry, coarsely serrate, teeth retrorse. **Inflorescences** spikes or racemes or rarely panicles; flowers borne in 2s or 3s; peduncles 2.5–3.5 m long; bracts present. **Flowers** perfect; funnelform. **Perianth Parts** yellow or tinged with red or purple; linear; ascending. **Stamens** clasped by perianth parts after anthesis. **Pistils** 1; ovaries inferior, fusiform. **Capsules** oblong to pyriform; short beaked. **Seeds** flattened; black.

- succulent perennial herb; leaves long and spiny, basal rosette; flowers yellow, borne on elongated stalk

Distribution and Habitat—All species of *Agave* are native to the Americas and generally occupy open, arid sites and a variety of soil types. One of the most abundant species in terms of numbers of rosettes, *A. lecheguilla* also has one of the most extensive ranges (Gentry 1982). A common component of different desert communities, it is found in rocky sites, especially limestone, often as the dominant plant. Because it often is locally very abundant, it may provide a captivating sight of desert beauty when numerous plants are in bloom.

- open, rocky, desert sites

Agave lecheguilla

Disease Problems—The spines on the leaf margins and the tips appear so menacing that it is difficult to comprehend that *A. lecheguilla* is eaten. As is so often the case in arid environments, problems due to ingestion usually occur in late winter or spring when there is little other forage available. Affecting primarily sheep and goats, the disease is a type of hepatogenous photosensitization called swellhead, goat fever, or lechuguilla fever. Cattle are affected much less commonly. In years when the plant is browsed extensively, morbidity rates may be as high as 30% in sheep and goats.

- abrupt onset, liver disease with photosensitization after eating leaves for several weeks, mainly sheep and goats

- saponins, crystalloid cholangiohepatopathy, calcium salts of a sapogenin

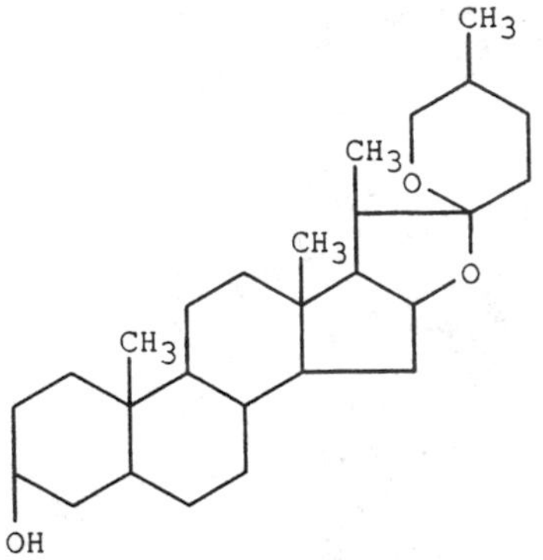

Smilagenin

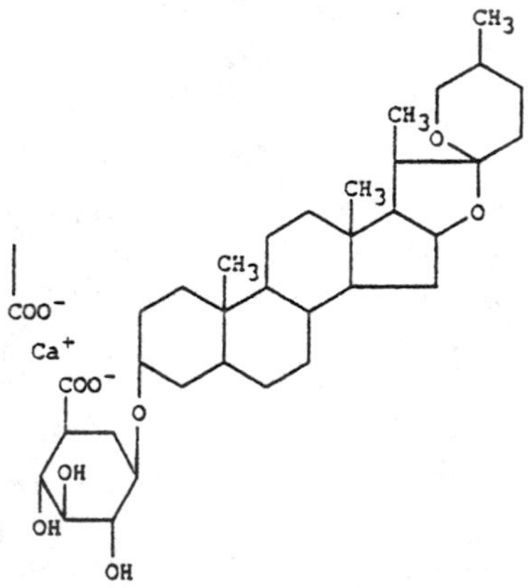

Ca^{2+} salt of epismilagenin ß-D-glucuronide

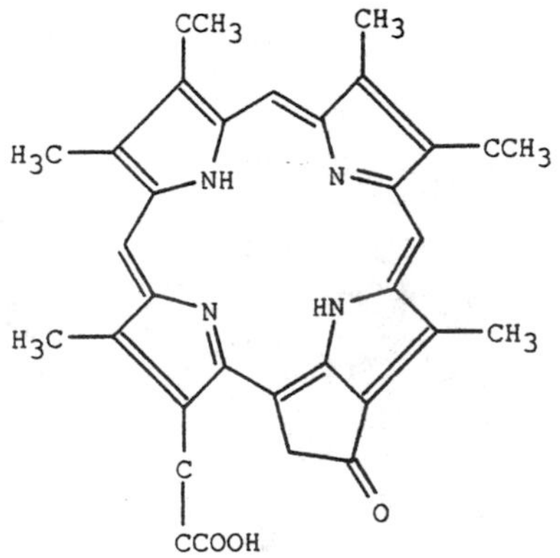

Phylloerythrin

Interestingly, during the same winter–spring period, mule deer may subsist extensively on lechuguilla without apparent ill effects (Brownlee 1981). The toxic potential of other species of *Agave* is unknown; they may be mechanically injurious and/or contain irritants.

Disease Genesis—Early studies indicated the presence of two toxicants, a photodynamic agent and a hepatotoxic saponin (Mathews 1937, 1938b). It is now clear that sapogenins such as smilagenin also are capable of causing hepatogenous photosensitization (Kellerman et al. 1991). The destructive effects of the toxins appear to affect the liver in a manner that renders it incapable of eliminating a photodynamic agent, presumably phylloerythrin. Whether an additional photodynamic factor is present is not resolved, but it is probably of only academic interest, because the action of smilagenin can account for all the observed disease effects. Similar-appearing bile duct crystal structures and the accompanying lesions are now recognized to be present in hepatogenous photosensitization caused by taxa from other families such as *Panicum* (see chapter 58) and *Tribulus* (see chapter 76). Interestingly, although these are diverse genera, they share a potential to cause disease through saponins (Kellerman et al. 1991). For *A. lecheguilla,* not only is smilagenin considered the cause, but the biliary crystals are also now known to be calcium salts of a sapogenin, probably smilagenin rather than cholesterol (Camp et al. 1988).

In addition, smilagenin has been shown to have abortifacient potential when given intravenously (Dollahite et al. 1962). Saponins are much like cardenolides; they are composed of a steroid sapogenin nucleus (the genin or aglycone) and one or more sugars attached at C-3 (Shoppee 1964). Thus smilagenin may be found with several different combinations of sugars to give various saponins. Lechuguilla leaves have about 1% (up to 2%) sapogenin, almost exclusively smilagenin (Wall et al. 1962). The concentrations are similar in dead leaves, but up to 2-fold higher in the plant's center, generally known as the heart or cajolla, which is selectively eaten by deer and livestock. The fruits and seeds contain lesser amounts of other sapogenins. Concentrations are quite consistent from site to site but may vary slightly during the year (highest in September and October) (Wall et al. 1962). Smilagenin also is found in *A. vilmoriniana* of Mexico, the consequences of which are not known (Wall et al. 1954). *Agave sisalana* contains high concentrations of the sapogenins, hecogenin, and tigogenin (Teixeira et al. 1989). Limited feeding studies on sisal residues following extraction revealed few indications of any toxicity potential, although the toxicants may have been leached out (Figueiredo 1975). Species of *Yucca* also contain similar concentrations of sapogenins but mainly sarsasapogenin rather than smilagenin (Wall 1954). *Agave americana* is reported to accumulate oxalates (Kellerman et al. 1988).

It should be pointed out that as with other disease problems involving the liver, animals need not be eating agave plants at the time of appearance of clinical signs. If hepatic changes are severe, regeneration and recovery may require an extended period. During this time, animals otherwise appearing normal will be susceptible to stresses that may precipitate the onset of signs of intoxication (Burrows & Stair 1990).

Clinical Signs—After feeding upon *A. lecheguilla* for several weeks or more, the animal may be listless and not inclined to keep up with the flock. A stringy, thick mucoid discharge may hang from the eyes and nose. Close examination will reveal icterus of the sclera and visible mucous membranes. In some cases the head and ears will be swollen, and when the head is handled, edema of the face, lips, and underside of the jaw may be readily detected. The edema may cause the animal to rub and scratch its head for several days. These are manifestations of photosensitivity and likely will be accompanied by purplish discoloration of the coronary bands. There will be progressive debilitation, weakness, and emaciation, and the urine may be a clear dark yellow or brown. Death may occur several days to a week or more after onset of signs. Cattle exhibit more diffuse skin changes.

Clinicopathologic changes during the course of the disease include marked elevation of serum bilirubin and hepatic enzymes.

- mucoid discharge from eyes and nose; swelling of head and ears; edema of face, lips, jaw; pruritis, weakness, emaciation
- elevation of serum bilirubin, hepatic enzymes

Pathology—In instances where photosensitization occurs, the most obvious changes will be in the skin of the head. There may be marked thickening of the skin and ears, with a gelatinous appearance extending into the deeper corium. Sloughing of patches, cracking, and even sloughing of an ear are features occasionally observed. Crusty and ulcerative or proliferative areas may be present, especially around the lips, eyes, and nose. The kidneys may be swollen and greenish black, with numerous gray foci. Typically the liver is brownish yellow.

Microscopically, edema of the skin will be accompanied by necrosis and a polymorphonuclear infiltrate in the deeper corium. The renal tubules will be distended with albumin and casts of pigment and cellular debris. Individual epithelial cells may show fatty degeneration and necrosis, with some tubules dilated and lined by a flattened epithelium. In the liver, fatty change, zonal necrosis, and bile pigment in macrophages are readily apparent, but the most distinctive lesions are the crystals or clefts in bile ducts. They may be surrounded by a brownish amorphous material filling the ducts or granuloma formation with necrosis of the biliary epithelium. Crystals may also be present in bile canaliculi and Kupffer cells. Originally the birefringent, acicular crystals were thought to be cholesterol, which they closely resemble (Mathews 1937, 1938a,b). They are now known to be calcium salts of a sapogenin. The crystals are best retained when acetone is substituted for alcohol to dehydrate tissues during processing for microscopic examination.

- gross pathology, thickened skin, head, especially ears; crusty, ulcerative areas

  microscopic, liver moderate necrosis, bile duct clefts or crystals; renal tubules dilated, with flattened epithelium

Treatment—Recovery from the disease is based upon general supportive care to allow the animal to regain adequate liver and kidney function. The animal may be protected from sunlight, but this is not generally necessary for survival. A discussion of the use of zinc salts for prevention of intoxication is presented in the following treatment of *Nolina*.

- supportive care, provide shade

Nolina Michx.

Taxonomy and Morphology—Native to the southwestern United States and Mexico, *Nolina* comprises 24–30 species and is closely related to *Yucca* and *Dasylirion*. Its name honors P. C. Nolin, an eighteenth-century French agriculturalist. Three species are of toxicologic concern:

N. bigelovii (Torr.) S.Watson	
N. microcarpa S.Watson	sacahuista, smallseed nolina
N. texana S.Watson (= *N. affinis* Trel.)	sacahuista, bunchgrass

- *N. bigelovii*
  N. microcarpa
  N. texana

  sacahuista
  nolina

Plants subshrubs or herbs; from large, woody, subterranean or aerial caudices; tussock-appearing. **Stems** present or absent. **Leaves** simple; alternate; numerous; basally clustered; sessile; spreading or arching; thick or thin; fibrous or fleshy; blades narrowly linear; apices often spine tipped; margins entire or serrate; stipules absent.

- subshrubs or herbs; leaves simple, long, narrow

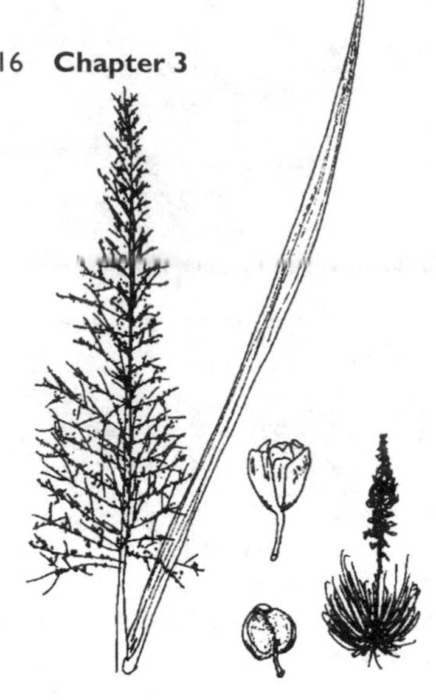

Nolina bigelovii

Inflorescences panicles; pedicels jointed; bracts present. **Flowers** small; numerous; perfect or imperfect, similar; campanulate or funnelform. **Perianth Parts** 6; free; white. **Stamens** 6. **Pistils** 1; ovaries inferior. **Capsules** ovoid; 3-lobed or 3-winged. **Seeds** 1–3; globose; grayish white or brown to blackish.

1. Plants caulescent. Leaves 15–35 mm wide. Capsules 10–15 mm in diameter; 8–10 mm long . . . . *N. bigelovii*
1. Plants acaulescent. Leaves 2–12 mm wide. Capsules 5–8 mm in diameter; 3–4 mm long.
   2. Leaf margins serrulate with numerous teeth. Capsules inflated; 7–8 mm in diameter . . . . *N. microcarpa*
   2. Leaf margins entire or sparsely serrate with few teeth. Capsules not inflated; 5–7 mm in diameter . . . . *N. texana*

■ open deserts

Distribution and Habitat—All species of *Nolina* are native to the southwestern United States and the Sonoran and Chihuahuan deserts of northern Mexico. They occupy a variety of soil types and habitats. Some species may be used occasionally as ornamentals in the southern states.

■ abrupt onset, liver disease with photosensitization, sheep and goats eating buds for several weeks

Disease Problems—Although animals are seldom in situations where they are forced to eat *Nolina* species, these plants are considered useful forage in some parts of their ranges (Mathews 1940). The value of *Nolina* as forage is probably reflected in its crude protein concentrations: 19% in the buds but only 5–6% in the leaves (Huston et al. 1981). Only the flower buds, open flowers, and ripe fruit present a significant hazard of the occurrence of photosensitization similar to that from *Agave lecheguilla*. This property of *Nolina* species minimizes the problem, because large numbers of flowers are not produced each year, but rather are produced once every 5 or 6 years, depending on the plant species (Mathews 1940). In years when flowers are not so abundant, plants may indeed be useful as forage. However, when flowers are profuse, *Nolina* lives up to its reputation as the most common cause of photosensitization in New Mexico (Hershey et al. 1945). Sheep and goats relish the buds and flowers but disdain the leaves. In contrast, cattle eat the leaves in the winter but seldom consume the toxic buds and flowers. Thus, sheep and goats are at considerable risk, whereas cattle are typically seldom affected (Mathews 1940; Hershey et al. 1945; Norris 1954). In some localities the leaves are harvested for making brooms, and the trimmings are fed to cattle (Nabhan & Burns 1985). When the outer leaves are removed, they are replaced with succulent new growth, which is readily grazed by cattle.

Nolina bigelovii

Nolina microcarpa

■ crystalloid cholangiohepatopathy, sapogenins suspected but not confirmed as cause of disease

Disease Genesis—The toxicant in *Nolina* is unknown, but the similarities in effects produced to those caused by other photosensitizing plants favor sapogenins. *Nolina* and several other genera cause a hepatotoxicity characterized by the presence of crystalloid material in the bile ducts. It now appears that most of these other genera contain toxic sapogenins (Kellerman et al. 1991). Thus, it is likely that similar toxins account for the effects produced by *Nolina*. Early studies failed to show sapogenins in the genus, but only leaves and whole plants were evaluated (Wall et al. 1954). Because it is the buds and flowers that are toxic, the negative results of the leaf analyses may have been misleading. An additional observation of interest is that the toxicant seems to be somewhat volatile, given that oven-dried plant material, originally used in toxicity experiments, was reported to be considerably less toxic after storage for 2 years (Samford et al. 1991). Whatever the identity of the toxin, it appears to affect the liver in a manner that renders it incapable of eliminating a photodynamic agent, presumably phylloerythrin. Photosensitization is reported to be worse when animals eat *Nolina* in pastures of green grass rather than in hay (Mathews 1940). At present it seems reasonable to consider *Nolina* intoxication from eating buds and flowers as hepatogenous photosensitization. A more detailed discussion of photosensitization is given in the treatment of Poaceae, in chapter 58. As little as 0.5% b.w. of dry *N. microcarpa* can cause serious disease (Rankins et al. 1989). *Nolina texana* is similarly toxic but may require a slightly higher dosage, in excess of 1% of b.w. (Mathews 1940). Although we know little of the toxicity potential of the many other species in this genus, it seems prudent to regard them with suspicion.

Nolina texana

In contrast to the liver derangement produced by ingestion of the buds and flowers, a wasting disease developed in rats and partridges that were fed seeds (Rankins et al. 1986; Smith et al. 1992). This may be indicative of a different toxin in the seeds or a difference in absorption and site of action of the toxin.

Clinical Signs—The disease appears sporadically in sheep and goats that have been eating *Nolina* for a week or more. Initially their appetite is much reduced, and then in another day or two the animals become depressed and reluctant to move about. Pruritis, with rubbing of the head and ears, may be noted for several days, and the urine may be dark yellowish brown. Upon close inspection, there is very obvious icterus; the sclera and mucous membranes are intensely yellow. The ruminal contents become quite dry because of dehydration, and there is a marked increase in fluid passage through the digestive tract (Rankins et al. 1989). By this time the head is quite swollen and the ears, lips, and face reddened and edematous—hence the name *bighead.* Most of the time, the animals stand in any shade available. Debilitation is progressive, with a copious sticky discharge from the eyes and nose and with death following in 1–2 weeks (Mathews 1940). The swelling and sloughing of the skin may be severe enough that the ears may be lost, even in animals that eventually recover (Hershey et al. 1945). In some instances, animals referred to as "fevered," may become ill and die without signs of photosensitization. It is of interest to note that *Nolina* given to rats readily produced severe illness and death, but there was no evidence of liver involvement. The disease appeared to be of a metabolic wasting type without elevation of serum hepatic enzymes, e.g., GGT and AST, and was readily reversible by eliminating *Nolina* from the feed (Rankins et al. 1986). This implies considerable variation among species in the signs of intoxication.

- head swollen; edema of ears, lips, and face; pruritis, icterus, dark brown urine

Clinicopathologic changes during the course of the disease include elevation of serum calcium, decrease of potassium and phosphorus, and marked elevation of bilirubin and serum hepatic enzymes. Alkaline phosphatase is elevated during the early phases (Rankins et al. 1988).

- elevation of serum bilirubin and hepatic enzymes

Pathology—The gross lesions, which are essentially the same as those produced by *A. lecheguilla,* are limited to the skin of the head and to the liver and kidney. The skin of the head may be thickened with a gelatinous, subcutaneous edema and areas of shallow ulceration. The kidneys may be swollen and dark greenish brown to black, whereas the liver will be a light yellowish brown, perhaps with a greenish sheen. Microscopically, there will be distortion of the hepatic cords and fatty degeneration of centrilobular cells. The characteristic lesions are thickened bile ducts filled with debris and crystalloid material or with clefts apparently left from previous crystals (Mathews 1940). The last change seems typical for several of the hepatogenous photosensitizing plant genera.

- gross pathology, edema, skin of head

  microscopic, liver, moderate necrosis with bile duct clefts or crystals

■ supportive care, provide shade; possible preventive effects of zinc, mixed grazing with sheep and cattle

Treatment—The most obvious approach to alleviating distress of the disease is to remove the animals from sunlight. However, even where this is possible, it does little for the underlying and much more serious problem of liver disease. Typically, the case fatality rate is high, especially if browsing of *Nolina* continues after signs of disease are manifested. Good nursing care to allow the liver to recover, often difficult to do under range conditions, is the primary approach.

A more specific approach is suggested by observations of the beneficial effects of zinc in preventing facial eczema due to the fungus *Pithomyces chartarum* (Smith & Towers 1985). Because facial eczema is somewhat similar to *Nolina* intoxication, the preventives might be expected to be interchangable. The value of zinc oxide for therapeutic use after disease onset is questionable, but when given intraruminally at a daily dose of 30 mg/kg b.w., it seems to provide some benefit (Rankins et al. 1988, 1993). Additional studies are needed to clearly delineate treatment conditions and the response. However, zinc supplementation appears to be of promise, at least as a preventive if not therapeutically. If it is used, care must be taken to avoid toxic amounts; 1 part zinc oxide to 3 or more parts water (w/v) to provide 20–30 mg/kg b.w. daily by oral drench is protective against facial eczema (Smith & Towers 1985). It may be given less often, provided the dose is increased proportionately. The foregoing recommendations are for zinc oxide; zinc sulfate is more toxic. In addition to drenching, zinc may also be effective when given in the drinking water or by spraying it on the pasture. It must be emphasized that zinc is not a proven treatment for *Nolina* intoxication at present.

Grazing systems employing a mixture of cattle, sheep, and goats together is an effective means of reducing disease losses, especially when combined with a pasture ration program (Merrill & Schuster 1978).

References

Brownlee S: *Desert Mule Deer Nutrition on the Black Gap Wildlife Management Area of Western Texas.* Tex Parks and Wildl Dept Rep W-109-R-3, Big Game Invest, 1981

Burrows GE, Stair EL: Apparent *Agave lecheguilla* intoxication in Angora goats [letter]. *Vet Hum Toxicol* 32;259–260, 1990

Camp BJ, Bridges CH, Hill DW, Patamali B, Wilson S: Isolation of a steroidal sapogenin from the bile of a sheep fed *Agave lecheguilla. Vet Hum Toxicol* 30;533–535, 1988

Dollahite JW, Shaver T, Camp BJ: Injected saponins as abortifacients. *Am J Vet Res* 23;1261–1263, 1962

Figueiredo LJC: Experimental studies of the toxicity of residues of *Agave sisalana* perene in cattle. *Arq Esc Vet Univ Fed Minas Gerais* 27;391–392, 1975

Forsling CL: *Chopped Soapweed as Emergency Feed for Cattle on Southwestern Ranges.* USDA Bull 745, 1919

Gentry HS: *Agaves of Continental North America.* Univ Arizona Press, Tucson, 1982

Hershey AL: Some poisonous plant problems of New Mexico. *N M Agric Exp Stn Bull* 322;10–12, 1945

Hostettmann K, Marston A: *Saponins.* Cambridge Univ Press, Cambridge, 1995

Huston JE, Rector BS, Merrill LB, Engdahl BS: *Nutritional Value of Range Plants in the Edwards Plateau Region of Texas.* Tex Agric Exp Stn Bull B-1357, 1981

Kellerman TS, Coetzer JAW, Naude TW: *Plant Poisonings and Mycotoxicoses of Livestock in Southern Africa.* Oxford Univ Press, Cape Town, 1988

Kellerman TS, Erasmus GL, Coetzer JAW, Brown JMM, Maartens BP: Photosensitivity in South Africa. 6. The experimental induction of geeldikkop in sheep with crude steroidal saponins from *Tribulus terrestris. Onderstepoort J Vet Res* 58;47–53, 1991

Mathews FP: *Lechuguilla* (Agave lecheguilla) *Poisoning in Sheep, Goats, and Laboratory Animals.* Tex Agric Exp Stn Bull 554, 1937

Mathews FP: An experimental investigation of lechuguilla poisoning. *Arch Pathol* 25;661–683, 1938a

Mathews FP: Lechuguilla *(Agave lecheguilla)* poisoning in sheep and goats. *J Am Vet Med Assoc* 93;168–175, 1938b

Mathews FP: *Poisoning in Sheep and Goats by Sacahuiste* (Nolina texana) *Buds and Blooms.* Tex Agric Exp Stn Bull 585, 1940

Merrill LB, Schuster JL: Grazing management practices affect livestock losses from poisonous plants. *J Range Manage* 31;351–354, 1978

Nabhan GP, Burns BT: Palmilla *(Nolina)* fiber: A native plant industry in the arid and semi-arid U.S./Mexico borderlands. *J Arid Environ* 9;97–103, 1985

Norris JJ, Valentine KA: Principal livestock-poisoning plants of New Mexico. *N M Agric Exp Stn Bull* 390;66–67, 1954

Rankins DL Jr, Smith GS, Ross TT: Rat study of beargrass (sacahuiste) toxicity. *Proc West Sec Am Soc Anim Sci* 37;224–226, 1986

Rankins DL Jr, Smith GS, Ross TT, Caton JS, Miller PR, Khan MF: *Nolina microcarpa* toxicosis in sheep. *Proc West Sec Am Soc Anim Sci* 39;218–221, 1988

Rankins DL Jr, Ross TT, Kloppenburg PB, Smith GS: Effects of intra-ruminally dosed sacahuista inflorescences on serum bilirubin and digesta kinetics of lambs fed alfalfa hay. *Proc West Sec Am Soc Anim Sci* 40;461–463, 1989

Rankins DL, Smith GS, Ross TT, Caton JS, Kloppenburg P: Characterization of toxicosis in sheep dosed with blossoms of sacahuiste *(Nolina microcarpa). J Anim Sci* 71;2489–2498, 1993

Samford MD, Ross TT, Edrington TS, Gentry PC: Toxicity of beargrass blossoms diminished during storage. *Proc West Sec Am Soc Anim Sci* 42;54–57, 1991

Shoppee CW: *Chemistry of the Steroids,* 2d ed. Butterworths, Washington, DC, pp398–432, 1964

Smith BL, Towers NR: Pithomycotoxicosis (facial eczema) in New Zealand and the use of zinc salts for its prevention. In *Plant Toxicology, Proceedings of the Australia/NUSA Poisonous Plants Symposium,* Seawright AA, Hegarty MP, James LF, Keeler RF eds, Queensland Poisonous Plants Committee, Yeerongpilly, Australia, pp70–79, 1985

Smith GS, Zornes ML, Schemnitz SD, Edrington TS, Stavanja M, Velastegui W: Toxicity of seeds of *Nolina microcarpa* (sacahuiste, beargrass) to chukar partridge *(Alectoris chukar). Proc West Sec Am Soc Anim Sci* 43;394–396, 1992

Teixeira Z, Marco A, Azzini A, Salgado ALB, Ciaramello D: Steroidal sapogenins in sisal. *Bragantia* 48;21–25, 1989

Wall ME, Krider MM, Krewson CF, Eddy CR, Willaman JJ, Correll DS, Gentry HS: *Steroidal Sapogenins 13. Supplementary Table of Data for Steroidal Sapogenins 7.* US Dept Agric Res Serv AIC-363, 1954

Wall ME, Warnock BH, Willaman JJ: Steroidal sapogenins. 48. Their occurrence in *Agave lecheguilla. Econ Bot* 16;266–269, 1962

Wooton EO: *Certain Desert Plants as Emergency Stock Feed.* USDA Bull 728, 1918

4 Aloaceae Batsch

■ Aloe Family

Aloe

Native to Africa, Madagascar, and Arabia, with greatest diversity in southern Africa, the Aloaceae comprises 5 genera and 700 species, many of which have been brought into cultivation. As with the Agavaceae, there have been differences in taxonomic opinion as to whether the family should be recognized. Described in 1802 by Batsch, it is clearly derived from the Liliaceae, and many taxonomists prefer to position its members in that family. However, classification of the Agavaceae and several other small liliaceous families as distinct requires that Aloaceae be recognized as well. Only *Aloe,* the largest genus, is associated with problems of intoxication.

■ succulent herbs or shrubs

Plants herbs or shrubs or small trees; perennials; evergreen; from caudices or rhizomes or fibrous root systems; caulescent or acaulescent; succulent or not succulent. **Leaves** simple; alternate; sessile; crowded in basal rosettes or dense rosettes at stem ends; blades of various shapes; venation parallel; stipules absent. **Inflorescences** terminal or lateral spikes or racemes or panicles; bracts present. **Flowers** perfect; perianths in 2-series; radially or sometimes slightly bilaterally symmetrical. **Perianth Parts** 6; all alike; petaloid; in 2 whorls; free or fused; of various colors. **Stamens** 6. **Pistils** 1; compound, carpels 3; stigmas 3-lobed; styles 1; ovaries superior; locules 3; placentation axile. **Capsules** dry or fleshy. **Seeds** numerous; flattened; usually winged.

Aloe L.

■ *A. arborescens*
A. vera

aloe
aloe vera
candelabra plant
octopus plant
torch plant

Taxonomy and Morphology—With greatest diversity in Africa, *Aloe* comprises 325–380 species. Its generic name is derived from the Greek *allal* or Arab *alloch,* meaning *bitter*. Exhibiting extraordinary diversity of form, many species of the genus are cultivated as ornamentals. In addition, they are prized for their medicinal properties, the best-known species being *A. vera,* the aloe vera plant, which is also used in cosmetics and shampoos. Two species are the most commonly encountered as ornamentals and medicinals:

A. arborescens Mill.	candelabra plant, torch plant, candelabra aloe, octopus plant
A. vera (L.) Burm.f. (= *A. barbadensis* Mill.)	aloe vera, Barbados aloe

■ succulent herbs and shrubs

Plants herbs or shrubs or small trees; succulent. **Leaves** crowded in basal rosettes or dense rosettes at stem ends; blades of various shapes; adaxial and abaxial surfaces smooth or scabrous; margins typically spinose or coarsely serrate. **Flowers** tubular. **Perianth Parts** red or orange or yellow. **Stamens** in 2 whorls. **Capsules** oblong or globose; papery or woody.

1. Plants acaulescent. Stolons present. Perianth parts yellow . . . . *A. vera*
1. Plants caulescent, 1–5 m tall. Stolons absent.
   Perianth parts red . . . . *A. arborescens*

Aloe arborescens

Distribution and Habitat—*Aloe arborescens* is native along the eastern coast of Africa from Mozambique to the Cape. Easily propagated, it is possibly the most widely cultivated of the genus. *Aloe vera,* sometimes still known by the illegitimate name *Aloe barbadensis,* is native to North Africa or Arabia but is now widely cultivated in many parts of the world, especially the West Indies (van Wyk & Smith 1996). In North America, both species are cultivated as houseplants or tub plants and are grown outdoors in selected warm areas. *Aloe vera* may escape cultivation in southern Florida.

■ cultivated ornamental in warm areas

Disease Problems—Exudates from the leaves of aloes have been used for centuries for medicinal purposes. Alexander the Great reputedly conquered the island of Socotra off the horn of Africa to gain control of the main supply of aloetic medicine (van Wyk & Smith 1996). Although aloes have been used for a great many problems, the plants are perhaps most well known for treatment of skin problems, from burns to frostbite (Ship 1977; Grindlay & Reynolds 1986; McCauley et al. 1990). Both *A. arborescens* and *A. vera* have been used for burns (Sato et al. 1990). At present, aloe vera juice remains popular for treatment of a variety of problems, through oral, i.v., topical, and intramammary routes of administration (Anderson 1983). Morton (1961) offers an extensive account of the uses, medicinal and otherwise, and the effects of the aloes, including *A. vera.* It is the possible purgative effects of oral exposure that is of concern as a toxicologic risk.

■ irritation, digestive disturbance, severe diarrhea

Disease Genesis—The component responsible for the reputed beneficial effects of *Aloe* is not known, but various compounds and their activity have been described—for example, bradykinase activity and modulation of host defense mechanisms (Fujita et al. 1976; 't Hart et al. 1988). Species of the genus also contain anthraquinone or anthracene glycosides, which act as potent purgatives (Fairbairn 1949). The crude extract is called aloin, which is a complex mixture of anthracene glycosides, the most important of which are the potent purgatives barbaloin and homonataloin. *Aloe ferox* and *A. candelabrum* have been the usual sources of the purgative of commerce known as bitter aloes or cape aloes, but many species, including all of the foregoing, contain barbaloin in significant concentrations (Watt & Breyer-Brandwijk 1962; Groom & Reynolds 1987). The concentrations of anthracenes are greatest in the young leaves, reaching levels as high as 30% d.w. of the exudate in some species (Groom & Reynolds 1987).

■ purgative anthracene glycosides in leaves; crude extract aloin, containing barbaloin, homonataloin

Barbaloin

Homonataloin

Barbaloin itself has only limited purgative activity; however, it is metabolized by intestinal bacterial glycosidases to the more potent genin or aglycone anthrone or further to the anthraquinone form, aloe-emodin (Thomson 1971; Hattori et al. 1988; Robinson 1991). The considerable variation among animals in their capacity to metabolize barbaloin is possibly a basis for the considerable interspecies variation in the observed toxicity potential of aloes. The purgative effect has been attributed partially to stimulation of prostaglandin production and increased activity of colonic mucosal adenyl cyclase (Capasso et al. 1983). Purgation is related to an increase in mucous secretion and water content of the colon; the increase in water precedes an increase in peristalsis (Ishii et al. 1994a,b). Consumption of the leaves of any of the species may result in a severe purgative effect. *Aloe* also has a reputation as an abortifacient (Watt & Breyer-Brandwijk 1962), and in vitro studies suggest that some of the anthracenes may be carcinogens (Wolfle et al. 1990).

■ increased mucous and water secretion into colon

■ diarrhea, colic; fluids and electrolytes

Anthraquinone

Anthrone

Anthranol

Clinical Signs, Pathology, and Treatment—Intoxication by species of *Aloe* will result in abrupt onset of moderate to severe diarrhea and possibly abdominal pain. In some instances there may be vomiting. The signs will be of moderate duration and are unlikely to be lethal. Treatment is aimed at relief of the increased intestinal activity and fluid loss.

References

Anderson BC: *Aloe vera* juice: a veterinary medicament. *Compend Contin Educ* 5;S364–S368, 1983

Capasso F, Mascolo N, Aulore G, Duraccio MR: Effect of indomethacin on aloin and 1,8-dioxianthraquinone–induced production of prostaglandins in rat isolated colon. *Prostaglandins* 26;557–562, 1983

Fairbairn JW: The active constituents of the vegetable purgatives containing anthracene derivatives. *J Pharm Pharmacol* 1;683–694, 1949

Fujita K, Teradaira R, Nagatsu T: Bradykinase activity of *Aloe* extract. *Biochem Pharmacol* 25;205, 1976

Grindlay D, Reynolds T: The *Aloe vera* phenomenon: a review of the properties and modern uses of the leaf parenchyma gel. *J Ethnopharmacol* 16;117–151, 1986

Groom QJ, Reynolds T: Barbaloin in *Aloe* species. *Planta Med* 53;345–348, 1987

Hattori M, Kanda T, Shu YZ, Akao T, Kobashi K, Namba T: Metabolism of barbaloin by intestinal bacteria. *Chem Pharm Bull* 36;4462–4466, 1988

Ishii Y, Tanizawa H, Takino Y: Studies of aloe. 4. Mechanism of cathartic effect. *Biol Pharm Bull* 17;495–497, 1994a

Ishii Y, Tanizawa H, Takino Y: Studies of aloe. 5. Mechanism of cathartic effect. *Biol Pharm Bull* 17;651–653, 1994b

McCauley RL, Heggers JP, Robson MC: Frostbite. Methods to minimize tissue loss. *Postgrad Med* 88;67–77, 1990

Morton JF: Folk uses and commercial exploitation of aloe leaf pulp. *Econ Bot* 15;311–319, 1961

Robinson T: *The Organic Constituents of Higher Plants.* Cordus Press, North Amherst, Mass, 1991

Sato Y, Ohta S, Shinoda M: Studies on chemical protectors against radiation. 31. Protection effects of *Aloe arborescens* on skin injury induced by x-irradiation. *Yakugaku-Zasshi* 110;876–884, 1990

Ship AG: Is topical aloe vera plant mucus helpful in burn treatment. *J Am Med Assoc* 238;1770, 1977

't Hart LA, van Enckevort PH, van Dijk H, Zaat R, de Silva KTD Labadie RP: Two functionally and chemically distinct immunomodulatory compounds in the gel of aloe vera. *J Ethnopharmacol* 23;61–71, 1988

Thomson RH: *Naturally Occurring Quinones,* 2d ed. Academic Press, London, pp399–404, 1971

van Wyk BE, Smith G: *Guide to the Aloes of South Africa.* Briza Publications, Pretoria, South Africa, 1996.

Watt JM, Breyer-Brandwijk MG: *The Medicinal and Poisonous Plants of Southern and Eastern Africa,* 2d ed. E & S Livingston, Edinburgh, 1962

Wolfle D, Schmutte C, Westendorf J, Marquardt H: Hydroxyanthraquinones as tumor promoters: enhancement of malignant transformation of mouse fibroblasts and growth stimulation of primary rat hepatocytes. *Cancer Res* 50;6540–6544, 1990

Amaranthaceae Juss.

5

Commonly known as the amaranth family, the Amaranthaceae comprises some 65 genera and 900 species, with greatest abundance in the tropical and subtropical regions of the world, in particular the Americas and Africa. In North America the family is represented by 17–19 genera and 110–120 species, both native and introduced. *Amaranthus* (pigweed) and *Alternanthera* (chaff flower) are the largest genera; most North American genera are relatively small, with 1–5 species each. Economically, the family is noted for both its weeds and its ornamentals. Prolific seed producers, many *Amaranthus* species are considered to be among the most noxious of weeds. In contrast, species of *Celosia* (cockscomb), *Gomphrena* (globe amaranth), and *Iresine* (bloodleaf) are prized for their beauty as pot or bedding plants. A few members of the family are eaten as potherbs or vegetables, and at one time the edible seeds of *Amaranthus* were an important component of the diets of pre-Columbian South and Central Americans.

■ Amaranth Family

Amaranthus

Plants herbs or rarely shrubs; annuals or perennials; armed or not armed with spines; bearing perfect flowers or dioecious or monoecious or polygamomonoecious. **Stems** prostrate to erect. **Leaves** herbaceous; simple; opposite or alternate; blades linear to ovate or obovate; venation pinnate; indumentum pubescent or stellate or absent; margins entire or sinuate; stipules absent. **Inflorescences** glomerules or spikes or panicles or heads or solitary flowers; terminal or axillary; bracts present, membranous or scarious; bracteoles present. **Flowers** perfect or imperfect, perfect and imperfect similar; perianths in 1-series. **Calyces** radially or rarely bilaterally symmetrical. **Sepals** 2–5; in 1 or rarely 2 whorls; free or fused; reddish green to white; scarious or membranous. **Petals** absent. **Stamens** 5 or 1–3; free or fused by filaments. **Pistils** 1; compound; carpels 2 or 3; stigmas 1–3, linear or capitate; styles 1; ovaries superior; locules 1; placentation basal; ovules 1–several. **Fruits** capsules or utricles; circumscissile. **Seeds** 1–several.

Intoxication problems associated with this family are related primarily to the ingestion of *Amaranthus* and are of three types: myocardial degeneration, renal disease, and nitrate intoxication.

■ herbs or shrubs; leaves simple; petals absent

Amaranthus L.

Taxonomy and Morphology—A cosmopolitan genus of some 60 species, *Amaranthus* is represented in North America by 41 taxa. The most commonly encountered are these:

A. albus L.	tumble pigweed, white pigweed
A. blitoides S.Watson	prostrate pigweed
A. hybridus L.	smooth pigweed, green pigweed, carelessweed, quelite morado

■ *A. albus*, *A. blitoides*, *A. hybridus*, *A. palmeri*, *A. powellii*, *A. retroflexus*, *A. spinosus*, *A. tuberculatus*

amaranth, carelessweed, hogweed, pigweed, quelite, stickerweed, tall water hemp

Amaranthus retroflexus

- annual herbs; leaves simple, alternate; spikes terminal or axillary; flowers imperfect

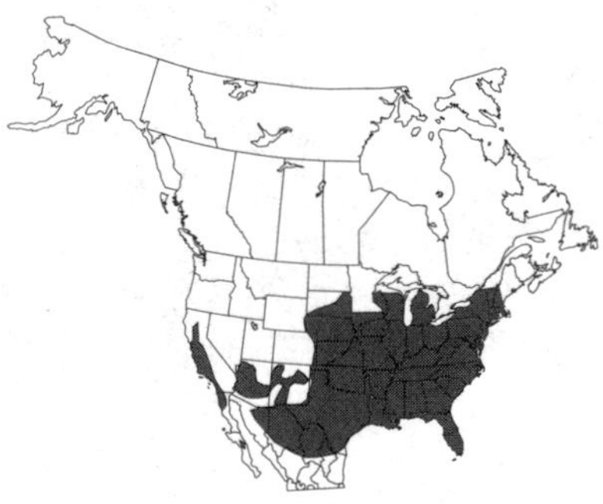

Amaranthus hybridus

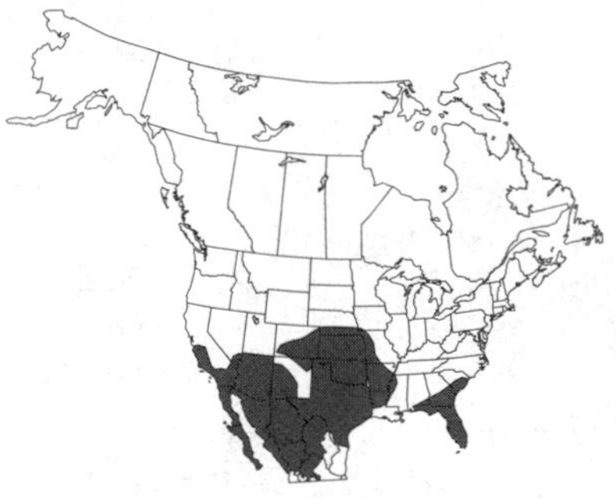

Amaranthus palmeri

A. palmeri S.Watson	Palmer pigweed, Palmer amaranth, carelessweed
A. powellii S.Watson	Powell pigweed, Powell amaranth
A. retroflexus L.	redroot pigweed, rough pigweed, quelite
A. spinosus L.	spiny pigweed, quelite espinoso, hogweed, stickerweed
A. tuberculatus (Moq.) J.D.Sauer	tall water hemp

Plants herbs; annuals; armed or not armed with spines; monoecious or dioecious or rarely bearing perfect flowers. **Root Systems** with a central taproot. **Stems** prostrate or ascending; branched at bases to form spheres; often strongly striated and turning red-brown or purple-red at maturity. **Leaves** simple; alternate; blades lanceolate or ovate or obovate; indumentum absent or occasionally puberulent; apices obtuse or acute or acuminate or mucronate; margins entire or sinuate; bases cuneate or attenuate; petioles long or short. **Inflorescences** glomerules or spikes; terminal and/or axillary. **Flowers** imperfect or rarely perfect, staminate and pistillate alike; each flower subtended by 3 conspicuous bracts. **Staminate Flowers:** Sepals 3 or 5; all alike; free; appressed or ascending; membranous; lanceolate or ovate or oblong or spathulate; venation a single vein; glabrous; apices acute or acuminate or obtuse or mucronate. Petals absent. Stamens 3 or 5; fused by filaments at bases. **Pistillate Flowers:** Sepals 3 or 5; all alike; free; appressed or ascending; membranous; lanceolate or ovate or oblong or spathulate; venation a single vein; glabrous; apices acute or acuminate or obtuse or mucronate. Petals absent. Pistils 1; compound; stigmas 3 or 2; styles 1; ovaries superior, carpels 2 or 3; locules 1; placentation basal; ovules 1. **Utricles** dehiscent; circumscissile; subglobose; glabrous. **Seeds** 1; reddish brown to black; smooth; lustrous; ovoid or lenticular or discoid.

Identification of species is complicated by their morphological similarity and frequent hybridization. Likewise, their ecology, reproductive biology, and potential toxicity are similar; thus a taxonomic key is not presented here.

Distribution and Habitat—Considered one of the most noxious weeds of row crops, plants of the genus quickly colonize disturbed sites with raw soil, e.g., tilled fields, gardens, farm lots, and waste areas. Seed production is prolific, and the seeds long-lived. *Amaranthus albus, A. blitoides,* and *A. retroflexus* occur across the continent; *A. hybridus* and *A. spinosus* occur in the eastern half from Canada to Mexico; *A. palmeri* occurs across the southern half; *A. powellii* is in the southwestern quarter from Oregon to southern Minnesota and south to Texas and Mexico; and *A. tuberculatus* also is in the eastern half, with the exception of the southeastern coastal states from Virginia to Florida.

Disease Problems—Species of *Amaranthus* generally are prolific seed producers; some are capable of producing 1 kg of seeds per plant. The seeds are highly nutritious and readily digested when cooked. They have protein levels of 12–15% and exhibit an excellent balance of amino acids (Pedersen et al. 1987). Especially rich in lysine, the seeds of *Amaranthus* were used by prehistoric humans 5,000 years ago in Central and South America and were a staple of the diet of the sixteenth-century Aztec and Mayan civilizations (Feine et al. 1979). When Cortez entered Mexico in 1519, he found warehouses with tons of stored amaranth seed. Spanish priests advocated suppression of its use in favor of corn and beans because the Aztecs mixed it with human blood to make idols in their religious ceremonies (Tucker 1986). Thereafter, the use of amaranth as a food crop declined. Recognition of its high lysine content, high productivity, and drought tolerance has led to an increase in the popularity of *Amaranthus* as a food staple, especially in developing countries. Its cultivation in Pennsylvania yielded 1,680 kg per hectare. The foliage also has considerable protein, commonly in excess of 20%, and concentrated leaf meal may have 16–24% (Feine et al. 1979; Cheeke et al. 1981).

■ nutrient grains

Although *Amaranthus* has desirable properties in terms of nutrition, toxicity problems are associated with it. Experimental feeding of concentrated leaf meal to rats was associated with poor feed intake and weight gain, an effect attributed to the presence of phenolics or saponins (Cheeke & Carlsson 1978; Cheeke et al. 1981). Both the foliage and the grain have caused more-severe problems in livestock. Sudden death due to myocardial degeneration has been reported for pigs fed large amounts of the grain of *A. caudatus* (tassel flower) for 5–7 weeks (Takken & Connor 1985).

■ pigs fed grain, sudden death, myocardial degeneration

The pigweeds also cause a disease of pigs, cattle, and sheep referred to as perirenal edema, because of characteristic pathologic changes. It develops in animals that consume large quantities of fresh plants for 5–10 days in the summer and fall, generally July until the first frost (Buck et al. 1965; Osweiler 1966; Brown 1974; Sanko 1975; Stuart et al. 1975; Wohlgemuth et al. 1987; Casteel et al. 1994; Rae & Binnington 1995). In one instance, cattle had been placed in a small pasture dominated by *A. retroflexus* in the seeding stage (Torres et al. 1997). The *Amaranthus* was grazed extensively, and after 3 days the cattle were moved to a large grassy pasture free of the weed. However, between 8 and 14 days after being put in the grassy pasture, 14 of 45 cows died, with signs and pathology consistent with the typical renal disease.

■ renal disease in pigs, cattle, sheep on pasture in late summer

■ nitrates, especially with plants in hay

Amaranthus retroflexus

Amaranthus spinosus

Amaranthus retroflexus is the species most commonly associated with the disease, but several others have been implicated as well, and all members of the genus probably should be considered toxic (Salles et al. 1991; Reit-Correa et al. 1994). The disease has been reported to occur in pigs, cattle, and sheep, and there is little reason to exclude horses and other species as susceptible, although Schamber and Misek (1985) were unable to produce symptoms in rabbits fed *A. retroflexus*. Intoxication may occur when animals are introduced into a weedy pasture with an abundance of *Amaranthus* or when dense local populations of the plants occur in an otherwise good pasture. Some animals seem predisposed to seek out and readily eat the plant (Jeppesen 1966). Appearance of the disease may occur with ingestion of mature plants or regrowth after cutting. Most intoxications are associated with fresh green plants. It is not known whether dried plants in hay represent a risk (Spearman 1989). There are a number of changes in serum chemistry consistent with renal failure, and death appears to be due to excess serum potassium and its effects on cardiac function as reflected in electrocardiographic changes (Osweiler et al. 1969).

In addition to the problems of myocardial degeneration and perirenal edema, species of *Amaranthus* accumulate oxalates and nitrates (Hill & Rawate 1982). Heretofore, oxalate accumulation has not been a significant problem, based on the absence of reports of abrupt-onset disease typical of these toxicants. Deaths due to nitrate intoxication are generally limited to ruminants, especially cattle (Egyed & Miller 1963; Duckworth 1975; Abbitt 1982). Nitrate intoxication may be seen when animals eat fresh plant material, but as with the grasses, it is more likely when dried plants of *Amaranthus* occur in hay. An additional factor contributing to the disease problem is that *Amaranthus* may be found as an abundant contaminant in hay composed of nitrate-accumulating grasses (Hibbs et al. 1978).

Disease Genesis—Knowledge of the potential toxicants in *Amaranthus* is limited, and the causes of the grain-associated myocardial degeneration and renal toxicity are

■ renal disease cause unknown; not due to oxalate salts

not known. Species of the genus are known to accumulate oxalates, and levels of 2–5% oxalic acid are reported for *A. retroflexus* (Hill & Rawate 1982). Much higher levels of 12–30% or more are sometimes encountered (Marshall et al. 1967; Mejia et al. 1985; Libert & Franceschi 1987). Although oxalates are suspect as causing the renal effects, the clinical signs and pathology of the disease are not consistent with this cause. For example, there are no hypocalcemic induced tremors or tetany, and few, if any, oxalate crystals are present in the tubules (Osweiler et al. 1969; Casteel et al. 1994; Torres et al. 1997). There may be an apparent initial response to calcium therapy, but this is only of short duration (Stuart et al. 1975). Further, the disease can be produced experimentally when animals are fed *Amaranthus,* but not when fed other oxalate accumulators such as *Chenopodium* or *Kochia.* The course of the disease is also much different from that produced by typical oxalate accumulators, which cause tubular injury within 24 hours of plant consumption (Sizelove et al. 1988). This is in contrast to the type produced by *Amaranthus* or the polyphenolic-induced renal injury caused by oak, both of which require ingestion for several days. When oxalate ingestion is spread over several days, the ruminal microflora adapt by degrading the oxalates to usable, nontoxic carbonates (Allison et al. 1977). Another point that must be considered when evaluating the relationship between *Amaranthus* and this disease is that perirenal edema is not specific for the genus but also may be produced by ingestion of other genera that cause kidney disease.

■ nitrate, nitrite; Hb to MHb, decreased oxygen transport

Under certain conditions, some plants are predisposed to accumulate nitrates into their tissues from the soil. Once concentrations in plants exceed 10,000 ppm, they are considered to be dangerous. Thus, species of *Amaranthus,* which may have concentrations of nitrate in the range of 15,000 to 40,000 ppm, present a risk under some conditions (Wilson 1943; Berg & McElroy 1953; Hill & Rawate 1982). Nitrate concentrations and risk may be temporarily increased by application of herbicides such as 2,4-D (Berg & McElroy 1953). While pigs ingesting high-nitrate plants may develop digestive tract problems, the lethal problem of nitrate intoxication is limited to ruminants, especially cattle, because of the reduction of nitrate to nitrite by ruminal microorganisms and the potential for accumulation and absorption of nitrite from the rumen. Absorbed nitrite may then oxidize iron in hemoglobin to the ferric form, resulting in formation of methemoglobin (MHb) from hemoglobin. The resulting MHb is incapable of oxygen transport, and the animal literally starves for oxygen. Problems with *Amaranthus* related to nitrate and nitrite are similar to those with other forages, and there is a higher risk with stems. Signs are accompanied by an increased incidence of abortions in cattle (Duckworth 1975; Abbitt 1982). Dried plant presents a greater danger because reduced moisture content allows more rapid consumption, on a dry-matter basis, of large amounts of nitrate. For a more complete discussion of nitrates, see under *Sorghum* in the Poaceae family (chapter 58).

■ renal, late summer, pasture, onset of signs over 1–2 days, weakness, incoordination, recumbency, death; increased BUN and serum K^+

Clinical Signs—After ingestion of *Amaranthus* for several days or more, the signs of renal disease appear over a period of several hours and progressively include weakness, trembling, unsteadiness when standing, incoordination, knuckling of the pasterns, recumbency, almost flaccid paralysis, coma, and then death in 1–2 days. In some instances there may be diarrhea with or without blood. The case fatality rate is likely to be in excess of 50%. Changes in serum chemistry include a marked increase in urea nitrogen (BUN, 90–200+ mg/dl) and elevation of potassium.

■ ruminants, nitrate, especially with plants in hay, abrupt onset, weakness, dark mucous membranes, death in min to hours

The signs of nitrate intoxication occur only in ruminants, with abrupt onset several hours after consumption of the forage, and may be fatal within minutes or a few hours. The primary effect is weakness and possibly dark-colored mucous membranes.

Pathology—At necropsy, there is widespread edema, especially around the kidneys and rectum and the omentum in general, and there is excess pale yellow, free fluid (perhaps blood-tinged) in the abdominal cavity and other areas. The kidneys are pale and either normal in size or swollen. There also may be edema, reddening, and ulceration of the mucosa of segments of the digestive tract. Microscopically, there are distinctive changes in the kidneys, including interstitial edema and scattered hemorrhages, plus tubular epithelial degeneration and necrosis. The proximal tubules are more affected than the distal. Some tubules will be dilated and filled with casts or proteinaceous material. In animals that do not die quickly, interstitial fibrosis of the kidneys may be present.

- renal, gross pathology, widespread edema, especially perirenal

microscopic, renal tubule dilation, flattened epithelium

With nitrate intoxication, there are few distinctive lesions at necropsy. The blood and tissues may be dark or chocolate colored, but this is not consistent. There also may be scattered, petechial to ecchymotic hemorrhages and increased foamy fluid in the terminal airways of the lungs, but these changes are not usually extensive.

- nitrate, no lesions, tissues dark

Treatment—As with other causes of renal failure, the basic goal of treatment is to assist the affected animal by symptomatic treatment and general nursing care in order to allow recovery of tubular function. Recovery will require several weeks or more, and the animal's prognosis generally must be considered poor because of the possibility of developing chronic renal failure.

- renal, nursing care to allow recovery of kidney function

With respect to nitrate intoxication, treatment is similar to that generally prescribed for other causes of the disease. The onset of signs requires the immediate use of 20 ml/100 kg b.w. of a 2% solution of methylene blue i.v.

- nitrate, methylene blue

References

Abbitt B: A case of nitrate-induced abortion in cattle. *Southwest Vet* 35;12, 1982

Allison MJ, Littledike ET, James LE: Changes in ruminal oxalate degradation rates associated with adaptation to oxalate ingestion. *J Anim Sci* 45;1173–1179, 1977

Berg RT, McElroy LW: Effect of 2,4-D on the nitrate content of forage crops and weeds. *Can J Agric Sci* 33;354–358, 1953

Brown CM: Chronic *Amaranthus* (pigweed) toxicity in cattle. *Vet Med/Small Anim Clin* 69;1551–1553, 1974

Buck WM, Preston KS, Abel M: Common weeds as a cause of perirenal edema in swine. *Iowa State Univ Vet* 1965;105–108, 1965

Casteel SW, Johnson GC, Miller MA, Chudomelka HJ, Cupps DE, Haskins HE, Gosser HS: *Amaranthus retroflexus* (redroot pigweed) poisoning in cattle. *J Am Vet Med Assoc* 204;1068–1070, 1994

Cheeke PR, Carlsson R: Evaluation of several crops as sources of leaf meal: composition, effect of drying procedure, and rat growth response. *Nutr Rep Int* 18;465–473, 1978

Cheeke PR, Carlsson R, Kohler GO: Nutritive value of leaf protein concentrates prepared from *Amaranthus* species. *Can J Anim Sci* 61;199–204, 1981

Duckworth RH: Poisoning of cattle by *Amaranthus. N Z Vet J* 23;154–155, 1975

Egyed M, Miller A: Nitrate poisoning in cattle due to feeding on *Amaranthus retroflexus. Refu Vet* 20;167–169, 1963

Feine LB, Harwood RR, Kauffman CS, Senft JP: Amaranth: gentle giant of the past and future. In *New Agricultural Crops,* Ritchie GA ed, Westview Press, Boulder, Colo, pp41–63, 1979

Hibbs CM, Stencil EL, Hill RM: Nitrate toxicosis in cattle. *Vet Hum Toxicol* 20;1–2, 1978

Hill RM, Rawate PD: Evaluation of food potential, some toxicological aspects, and preparation of a protein isolate from the aerial part of amaranth (pigweed). *J Agric Food Chem* 30;465–469, 1982

Jeppesen QE: Bovine perirenal disease associated with pigweed. *J Am Vet Med Assoc* 149;22, 1966
Libert B, Franceschi VR: Oxalate in crop plants. *J Agric Food Chem* 35;926–938, 1987
Marshall VL, Buck WB, Bell GL: Pigweed *(Amaranthus retroflexus):* an oxalate-containing plant. *Am J Vet Res* 28;888–889, 1967
Mejia MR, Montalvo VR, Martinez RR: Levels of oxalates in wild forages coming from the states of Hildalgo, Guanajuato, Mexico, Tlaxcala, and the Federal District. *Vet Mex* 16;21–25, 1985
Osweiler G: Swine toxicoses. *Iowa State Univ Vet* 1966;127–135, 1966
Osweiler GD, Buck WB, Bicknell EJ: Production of perirenal edema in swine with *Amaranthus retroflexus. Am J Vet Res* 30;557–566, 1969
Pedersen B, Kalinowski LS, Eggum BO: The nutritive value of amaranth grain *(Amaranthus caudatus)*. 1. Protein and minerals of raw and processed grain. *Plant Foods Hum Nutr* 36;309–324, 1987
Rae CA, Binnington BD: *Amaranthus retroflexus* (redroot pigweed) poisoning in lambs. *Can Vet J* 36;446, 1995
Reit-Correa F, Mendez MC, Barros CSL, Gava A: Poisonous plants of Rio Grande do Sul. In *Plant Associated Toxins—Agricultural, Phytochemical, and Ecological Aspects,* Colegate SM, Dorling PR eds, CAB International, Wallingford, UK, pp13–18, 1994
Salles MS, Lombardo de Barros CS, Lemos RA, Pilati C: Perirenal edema associated with *Amaranthus* spp poisoning in Brazilian swine. *Vet Hum Toxicol* 33;616–617, 1991
Sanko RE: Perirenal edema in swine caused by ingestion of *Amaranthus retroflexus* (pigweed). *Vet Med/Sm Anim Clin* 70;42–43, 1975
Schamber GJ, Misek AR: *Amaranthus retroflexus* (redroot pigweed): inability to cause renal toxicosis in rabbits. *Am J Vet Res* 46;266–267, 1985
Sizelove W, Hays T, Johnson BJ, Burrows GE: Perirenal edema in a calf [letter]. *Vet Hum Toxicol* 30;265–266, 1988
Spearman G: Redroot pigweed toxicosis in cattle. *Can Vet J* 30;255–256, 1989
Stuart BP, Nicholson SS, Smith JB: Perirenal edema and toxic nephrosis in cattle, associated with ingestion of pigweed. *J Am Vet Med Assoc* 167;949–950, 1975
Takken A, Connor JK: Some toxicological aspects of grain amaranth for pigs. In *Plant Toxicology—Proceedings of the Australia-USA Poisonous Plants Symposium,* Animal Research Institute, Yeerongpilly, Australia, pp170–177, 1985
Torres MB, Kommers GD, Dantas AFM, de Barros L: Redroot pigweed *(Amaranthus retroflexus)* poisoning of cattle in southern Brazil). *Vet Hum Toxicol* 39;94–96, 1997
Tucker JB: Amaranth: the once and future crop. *BioSci* 36;9–13, 1986
Wilson JK: Nitrate in plants: its relation to fertilizer injury, changes during silage making, and indirect toxicity to animals. *J Am Soc Agron* 35;279–290, 1943
Wohlgemuth K, Schamber GJ, Misek AR, Crenshaw JD: Pigweed is toxic to pigs. *N D Farm Res Bull* 44;21–22, 1987

Anacardiaceae Lindl.

6

Commonly known as the sumac or cashew family, the Anacardiaceae comprises some 60–80 genera and 600 species of trees, shrubs, and vines. Distribution is primarily tropical and subtropical, with equal representation in the Americas, Africa, and Asia. A few genera are native to temperate North America and Eurasia. In North America the family is represented by 12 genera and about 40 species, both native and introduced. The largest genus is *Rhus* (sumac), which is found in forests, grasslands, and deserts across the continent. Prized for their edible fruits are species of *Mangifera* (mango), *Anacardium* (cashew), and *Pistacia* (pistachio). Commonly grown as ornamentals are species of *Rhus* (sumac), *Cotinus* (smoke tree), and *Schinus* (pepper tree). Other species of the family are sources of tannins for the leather industry, and the resin of a Chinese species of *Rhus* is the basis of lacquer.

■ Cashew Family

Anacardium
Mangifera
Metopium
Rhus
Schinus
Toxicodendron

Plants trees or shrubs or woody vines; evergreen or deciduous; solitary or colonial and thicket-forming; polygamous or dioecious; sap viscous. **Leaves** 1-pinnately compound or simple; alternate or rarely opposite; venation pinnate; resin canals present; stipules absent. **Inflorescences** panicles or compound cymes; terminal or axillary. **Flowers** perfect or imperfect, perfect and imperfect similar; perianths in 2-series. **Sepals** 5, rarely 3 or 7; free or fused. **Corollas** radially symmetrical; imbricate. **Petals** 5, rarely 3 or 7; free; yellow or green or greenish white or red. **Stamens** 5 or 10, or 1 with 6–9 staminodia. **Pistils** 1–5; compound or simple; carpels typically 5; stigmas 1 or 3; styles 1 or 3; free or fused; ovaries superior; locules 1; placentation basal. **Drupes** fleshy or dry; often resinous or waxy. **Seeds** 1.

■ shrubs, vines, and trees; petals 5, yellow, green, greenish white, or red; fruits drupes

The Anacardiaceae is perhaps the most infamous of all plant families as the cause of allergic dermatitis in susceptible individuals coming in contact with the sap. The allergic manifestations are almost exclusively a problem in humans, although experimentally some laboratory animal species, such as guinea pigs, can be sensitized. Most notorious and most commonly encountered in North America are the species of *Toxicodendron,* commonly known as poison ivy, poison oak, or poison sumac. Other genera, such as *Metopium* (poisonwood), are even more potent. Pistachio and cashew nuts are likewise capable of producing dermatitis but are made edible by roasting. Susceptible individuals eating mangos may develop a "Florida grin" by inadvertently touching the rind to their cheeks and chin and subsequently developing a broad U-shaped blistering and reddening.

■ humans, allergic dermatitis

The dermatitis-causing genera of the family contain mixtures of mono- or dihydric phenols, resorcinols, or catechols, each with an unbranched and unsaturated side chain of 15, 17, or 19 carbons (Mitchell & Rook 1979). The potency of the compounds increases with the extent of unsaturation. These compounds may be distributed throughout the plant body, as in poison ivy, or concentrated in a particular organ, as in the fruit wall of the cashew or the rind of the mango. The resin canals that contain the allergens do not open to the surface; thus sensitization occurs only

■ resorcinols or catechols with long, unbranched side chains; cross-reactivity between genera of the family and with *Ginkgo*

when the surface of the plant is disrupted. Cross-reactivity between the allergic responses solicited by various genera is common (Keil et al. 1944; ElSohly et al. 1986). There may also be cross-reactivity with the allergen of *Ginkgo*. In addition to the allergenic compounds present, members of the family also contain various terpenoids that may act as immediate contact irritants.

Anacardium L.

■ *A. occidentale*
acajou
cashew
maranon
pajuil

Taxonomy and Morphology—Comprising 8 species native to the Americas, *Anacardium* is naturalized throughout the tropics of the world. It is the source of cashews, and an oil used in lubricants also is extracted from the fruit. A gum from the wood is used in inks, dyes, and varnishes. In North America the genus is represented by a single species:

A. occidentale L. cashew, maranon, acajou, pajuil

■ evergreen trees; leaves simple; fruits dry drupes

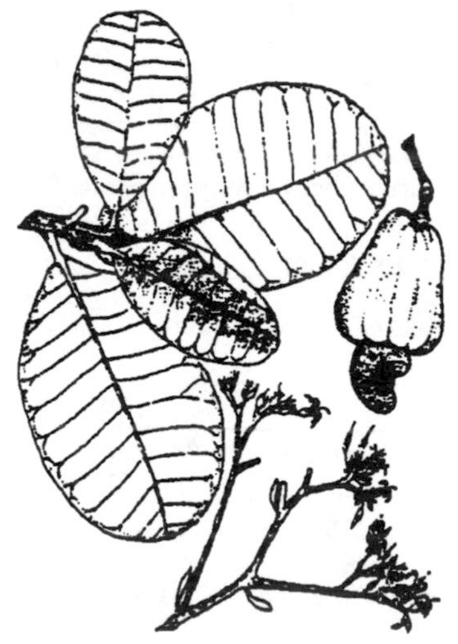

Anacardium occidentale

Plants trees; evergreen; trunk rarely straight; polygamomonoecious; sap white. **Leaves** simple; obovate to broadly elliptic; coriaceous; glabrous. **Inflorescences** panicles; bracts present. **Flowers** perfect or staminate. **Sepals** 5; fused. **Petals** 5; pale green, striped with red, becoming entirely red or pink. **Stamens** 10; 1 much longer than others. **Pistils** 1; compound. **Drupes** dry; reniform; gray-brown. **Seeds** 1; reniform.

Although commonly called a nut, the cashew is actually the seed of a dry drupe. As the drupe matures, the receptacle at its base enlarges rapidly to form a fleshy, club-shaped structure popularly known as the fruit. This enlarged receptacle is thin-skinned, juicy, spongy-textured, slightly acidic, and edible. Albeit very astringent when green, the flesh is yellow when mature and can be eaten raw or in preserves.

■ ornamental, warm areas

Distribution and Habitat—Commonly planted as a shade and ornamental tree, *A. occidentale* is restricted in distribution to the warmest portions of North America, southern Florida, southern California, and Mexico, where the temperature rarely falls below 15°C. It also is planted as a crop for the fruits, but it does not produce as well as it does farther south.

■ irritation and allergic dermatitis from handling raw fruit; affecting skin, eyes, and digestive tract

Disease Problems—Although the seed, or cashew, and the enlarged, fleshy receptacle, or "fruit," are edible raw, the tissue surrounding the cashew may cause problems. Cardol oil, the thick, black, very acrid oil present in the shell of the drupe is caustic and can be irritating and allergenic to the skin (Allen 1943). The delayed allergenic reactions are most troublesome. Problems are experienced mainly by agricultural workers exposed to the raw nuts and kernels. Less common are episodes of contact dermatitis associated with contamination of cashews with shell fragments (Aber et al. 1983). Very rarely, the nut-shell oil may contaminate other products such as cashew nut butter and cause a severe dermatitis (Rosen & Fordice 1994).

■ caustic and allergenic oil with 15-C side chains

Disease Genesis—The caustic oil of *Anacardium* is composed of approximately 82% anacardic acid, 13.8% cardol, 2.6% 2-methyl cardol, and 1.6% cardanol (Tyman 1976). The irritant and allergenic potency ranges in descending order from cardol to anacardic acid and anacardols (Keil et al. 1945a; Wasserman & Dawson 1948). Cardol is a resorcinol derivative, and anacardic acid and the anacardols are phenolic-type derivatives. All have 15 carbon monoene or diene side chains, which, as is typical of the family, are closely related chemically and allergenically to the catechol-derived urushiols in *Toxicodendron* (poison ivy) (Keil et al. 1945b; Mitchell & Rook

1979). Anacardol is a mixture of several phenol derivatives, more appropriately termed cardanols, that vary in the degree of unsaturation of the side chain (Wasserman & Dawson 1948). These derivatives are referred to as cardanol 15:0, 15:1, and 15:2, based upon the number of side chain carbons and double bonds. Species of *Anacardium, Schinus,* and *Toxicodendron* share similar toxicants, and there is considerable allergic cross-reactivity among them (Keil et al. 1945a; ElSohly et al. 1986). Thus individuals sensitive to one can be expected, more often than not, to be sensitive to the others. Contact with this combination of the oil's constituents, especially cardol, may cause immediate or direct irritation of the skin, eyes, and digestive tract, as well as a delayed or indirect allergic skin response.

The direct vesicant action of the oil is demonstrated by the natural protection that it appears to provide against insect damage to the plant and also by its reported antiseptic, anthelminthic, and molluscicidal properties (Sullivan et al. 1982; Moraes 1993). As with the effects on the skin, the antiparasitic potency increases with the degree of unsaturation of the side chains (Sullivan 1982). The oil has also been used for treatment of other skin diseases.

HO, OH

$(CH_2)_7CH{=}CHCH_2CH{=}CH(CH_2)CH_3$

or

$(CH_2)_7CH{=}CHCH_2CH{=}CHCH_2CH{=}CH_2$

Cardol

OH, COOH

$(CH_2)_7CH{=}CHCH_2CH{=}CH(CH_2)_2CH_3$

Anacardic acid (15:2)

OH

$(CH_2)_7CH{=}CHCH_2CH{=}CH(CH_2)_2CH_3$

Cardanol (15:2)

Clinical Signs—There is gradual development of reddening, blistering, roughness, and cracking of the skin following contact with the caustic oil of *Anacardium.* Although clinical signs may appear immediately, they often are delayed for 1–2 days in sensitized individuals or up to 1–2 weeks in those exposed for the first time.

- skin, reddened, blistered, and cracked

Treatment—Direct-contact irritant effects on the skin, eyes, or digestive tract may be treated with agents aimed at relief of the pain, itching, or inflammation. In many instances, treatment may not be required.

- topical, protective lotions before exposure, washing skin after exposure; drying agents after lesions develop

Prevention or treatment of the more troublesome allergic dermatitis caused by *Anacardium* is essentially the same as that recommended for *Toxicodendron;* application of protective lotions to prevent penetration of the oil, prompt removal of the oil following contact to reduce penetration, and application of drying agents to pustules and blisters formed. Severe cases may require the use of topical and/or systemic corticosteroids.

Mangifera L.

Taxonomy and Morphology—Comprising about 35 species native to tropical India, southern China, and islands of the southwestern Pacific, *Mangifera* is represented in North America by a single species:

M. indica L. mango

- *M. indica*

  mango

- large evergreen trees; leaves simple; petals 5, greenish white, yellow, pink, or red; fruits fleshy drupes

Plants large trees, to 20 m tall; evergreen; trunk stout; polygamomonoecious; sap white. **Leaves** simple; oblong-elliptic or oblong-lanceolate; subcoriaceous or papery; glabrous; red or yellowish when young. **Inflorescences** panicles; bracts present. **Flowers** perfect or staminate; small; fragrant. **Sepals** 5; fused. **Petals** 5; usually free; greenish white to yellow or pink to red. **Stamens** 5; 1 longer and fertile, 4 shorter and sterile. **Pistils** 1; compound. **Drupes** fleshy; ovoid-oblong; yellow or green or red; flesh yellow, juicy. **Seeds** 1; large; flattened; enclosed in thick and fibrous endocarp.

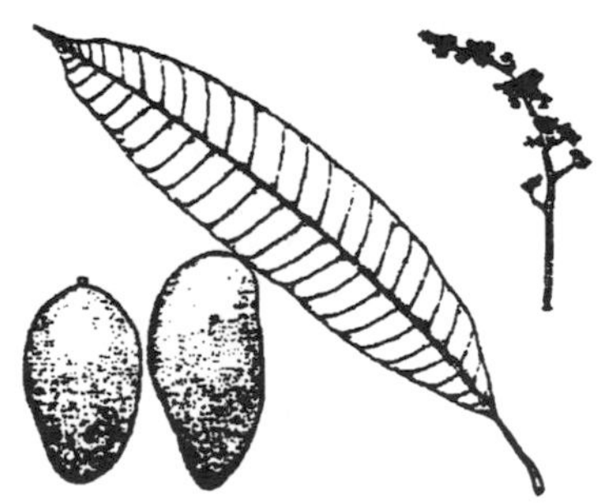

Mangifera indica

Distribution and Habitat—Possibly native to Burma, *M. indica* has been grown in India for 4,000 years and subsequently carried throughout tropical Southeast Asia, Africa, and South America. With hundreds of cultivars, it is considered to be the most

- cultivated for fruit and shade, warm areas

popular fruit of the Orient. Widely cultivated as a fruit and shade tree, *M. indica* is restricted in distribution to the warmest portions of North America, Mexico, southern California, and southern Florida. Mexico is one of the top two exporters of its fruits.

- weakly allergenic sap, reddening of lips and cheeks

Disease Problems—Although the mango is one of the most popular fruits consumed in the tropics and its flesh and juice do not cause a problem even in susceptible individuals, other parts of the plant are capable of causing allergic dermatitis (Mitchell & Rook 1979). The sap from the stems, leaves, and fruit exocarp (peel) and base is allergenic. Green fruits and even pollen are also allergenic for sensitive individuals. Thus care in preparation of the fruit before eating is advisable to avoid contaminating the fruit flesh. Inadvertent contact with the sap from the peel while eating the fruit is a concern. Reddening and blistering of the lips, cheeks, and chin produces the conspicuous condition known as the Florida grin, a condition experienced by many first-time visitors to the tropics.

- allergenic urushiol-type compounds

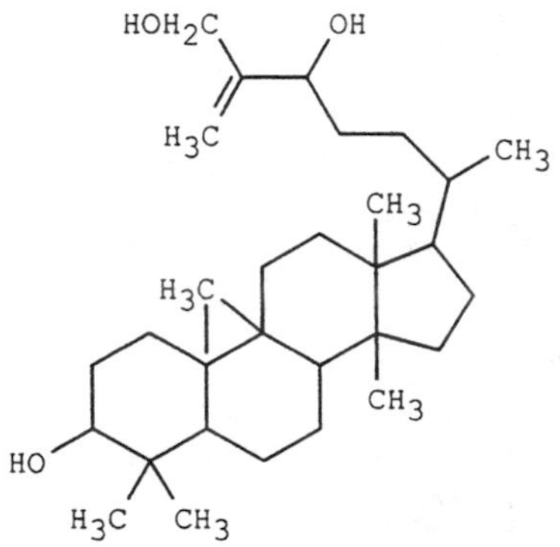

HOH_2C OH H_3C CH_3 CH_3 H_3C CH_3 HO H_3C CH_3

Terpenoid cycloartene

Disease Genesis—The sap of *M. indica* contains compounds similar to, but less potent, than urushiol, the allergen of *Toxicodendron* or poison ivy. The compounds may also be cross-reactive with urushiol. The plants additionally contain numerous terpenoid cycloartenes, which may contribute to a direct-contact irritant effect, but a complete understanding of their toxicologic significance is not clear (Anjaneyulu et al. 1985, 1989; Connolly & Hill 1991a,b). Some people develop severe reactions to the sap, but allergic responses are not as common as with poison ivy (Morton 1982).

- mouth and cheeks reddened, blistered

Clinical Signs—*Mangifera indica* is a weak allergenic sensitizer, but contact with the mouth, cheeks, and chin may result in reddening and formation of blisters.

- treatment generally not needed

Treatment—The dermatitis caused by *M. indica* typically does not require treatment, but when needed, treatment is essentially the same as that recommended for *Toxicodendron:* application of drying agents. Likewise, prevention is much the same as for *Toxicodendron:* avoidance of the sap and prompt removal of it following contact, to reduce the possibility of penetration. Unusually severe cases may require the use of topical and/or systemic corticosteroids.

Metopium P.D.Br.

- *M. toxiferum*

  coral sumac
  doctor-gum
  poisontree
  poisonwood

Taxonomy and Morphology—Comprising 3 species native to the West Indies, southern Mexico, and southern Florida,, *Metopium* is represented in North America by a single species:

M. toxiferum (L.) Krug & Urban	poisonwood, doctor-gum, coral sumac, poisontree

- evergreen trees; leaves pinnately compound; petals 5, greenish white or yellowish; fruits fleshy drupes

Plants trees; evergreen; dioecious or rarely polygamodioecious; sap watery; turning black when drying. **Leaves** 1-pinnately compound; leaflets 3 or 5 or 7, ovate, coriaceous, apices acuminate or obtuse, margins entire, bases obtuse or subcordate; petioles terete. **Inflorescences** panicles; axillary; bracts present, minute. **Flowers** imperfect; small. **Sepals** 5; fused. **Petals** 5; greenish white or yellowish, with darker markings. **Stamens** 5. **Pistils** 1; compound. **Drupes** fleshy and resinous; oblong or oblong-ovoid; apiculate; glabrous; dull orange or scarlet. **Seeds** 1; enclosed in parchmentlike endocarp.

A distinctive feature of poisonwood is that there are often tar-like black streaks, patches, or dots on the bark and/or leaves of injured trees caused by the sticky sap's turning black on contact with the air.

- black tarlike spots on leaves and bark from sticky sap

Distribution and Habitat—Native to and common in the pinelands and hammocks along the east coast of southern Florida, *M. toxiferum* also occurs in the Keys, the Bahamas, Cuba, Hispaniola, Mona, Puerto Rico, and Anguilla in the Leeward Islands. Typically growing in sandy soils, the trees are sometimes maintained as ornamentals in Florida when they have survived removal of the original forest.

- southern Florida

Disease Problems—The common names *poisonwood* and *poisontree* are derived from the dermatitis that follows contact with the tree's wood (Morton 1982). The dermatitis is similar to that caused by other members of the family such as *Anacardium, Mangifera,* and *Toxicodendron.*

- irritation and allergic skin reaction following contact with wood or sap

Disease Genesis—The allergens responsible for the dermatitis produced by *Metopium* are a complex mixture of 15-carbon mono- and dienes and a few 17-carbon mono-, di-, and triene catechols or urushiols similar to those of *Toxicodendron* (poison ivy) (Gross et al. 1975).

- allergenic catechols and urushiols

OH

OH

$(CH_2)_7C{=}C(CH_2)_5CH_3$

or

$(CH_2)_7C{=}CCC{=}C(CH_2)_2CH_3$

15-C urushiol catechols

Metopium toxiferum

Clinical Signs—Following contact with *M. toxiferum,* the clinical signs are those characteristic of the other allergenic members of the Anacardiaceae: reddening and blistering of the skin. The sap also stains the skin black.

- skin reddened, blistered, sometimes with black stain

Treatment—The direct-contact irritant effects on the skin, eyes, or digestive tract may be treated with agents aimed at relief of the pain, itching, or inflammation. Treatment may not be required in some instances.

Prevention or treatment of the more troublesome allergic dermatitis caused by *Anacardium* is essentially the same as that recommended for *Toxicodendron:* application of protective lotions to prevent penetration of the oil, prompt removal of the oil following contact to reduce penetration, and application of drying agents to pustules and blisters. Severe cases may require the use of topical and/or systemic corticosteroids.

- relief of pain and itching; often not needed

Rhus L.

Taxonomy and Morphology—Distributed in temperate and subtropical North America, South Africa, and eastern Asia, *Rhus* comprises some 200 species. In North America the genus is represented by 13 species and 2 named hybrids. Most commonly encountered are the following:

- *Rhus*
  polecat bush
  squawbush
  sumac

R. aromatica Aiton	lemon sumac, polecat bush, fragrant sumac, squawbush
R. copallina L.	winged sumac, shining sumac, dwarf sumac, mountain sumac
R. glabra L.	smooth sumac, scarlet sumac
R. typhina L.	staghorn sumac, velvet sumac, Virginia sumac

- shrubs or small trees; leaves 1-pinnately compound; fruits red drupes

Plants shrubs or small trees, often forming thickets; deciduous; polygamodioecious; sap viscous, white. **Stems** glabrous or densely indumented. **Leaves** 1-pinnately compound; leaflets 3 or 7–31; margins entire or variously toothed; petioles winged or not winged. **Inflorescences** dense panicles; typically terminal; bracts absent. **Flowers** perfect or imperfect; small. **Sepals** 5; fused. **Petals** 5; small; imbricate; greenish white or yellow. **Stamens** 5. **Pistils** 1; compound. **Drupes** fleshy; red; puberulent. **Seeds** 1.

- not generally a toxicity problem but confused with troublesome taxa

As presently circumscribed by taxonomists, *Rhus* does not include those species that cause allergic contact dermatitis, such as poison ivy, poison oak, and poison sumac, which are classified in the genus *Toxicodendron.* The alternative spelling *sumach* is encountered in some books.

- species are present across the continent

Distribution and Habitat—These four species of the genus occupy a variety of habitats—open woods, forest margins, grasslands, old fields, sand dunes, and roadsides—throughout their ranges. *Rhus copallina, R. glabra,* and *R. typhina* are found in the eastern half of North America, whereas *R. aromatica* occurs across the continent. Prized for the brilliant fall leaf colors of deep reds and purples or yellow, all are cultivated in gardens, with numerous named cultivars being sold.

- nonallergenic but rarely a contact dermatitis

Disease Problems—Species of *Rhus* normally do not cause the allergic contact dermatitis that is characteristic of other members of the family, in particular species of *Toxicodendron* which at one time were classified in the genus. Some individuals do, however, experience a short-term rash or reddening of the skin upon contact with the milky sap of *Rhus.*

Schinus L.

- *S. longifolius*
  S. molle
  S. polygamus
  S. terebinthifolius

  Christmas berry, false pepper, Florida holly, huigen, mastic tree, molle, peppertree, false pepper

Taxonomy and Morphology—Comprising about 30 species native to south Central America, *Schinus* is represented in North America by 4 introduced species:

S. longifolius (Lindl.) Speg.	
S. molle L.	Peruvian peppertree, molle, Peruvian mastic tree, California peppertree
S. polygamus (Cav.) Cabrera	huigen
S. terebinthifolius Raddi	Brazilian peppertree, Florida holly, false pepper, Christmas berry

- evergreen shrubs or trees; leaves pinnately compound, bad odor when crushed; flowers small; petals 4 or 5, white or yellow

Plants trees or shrubs; evergreen; dioecious; sap resinous. **Leaves** malodorous when crushed; 1-pinnately compound; leaflets 5–15 or more, coriaceous, margins entire or dentate; petioles with prominent abscission ring. **Inflorescences** panicles or racemes; axillary or terminal; bracts present. **Flowers** imperfect; small. **Sepals** 4 or 5; fused. **Petals** 4 or 5; white or yellow. **Stamens** 8–10; unequal in length. **Pistils** 1; compound; stigmas 3; styles 3. **Drupes** fleshy; pea-size; lavender to deep red. **Seeds** 1; enclosed in hard endocarps.

The two species most commonly encountered are *S. molle* and *S. terebinthifolius*. They are distinguished as follows:

1. Leaflets 15 or more; linear-lanceolate. Petals yellow . . . . *S. molle*
1. Leaflets 7; oblong. Petals white . . . . *S. terebinthifolius*

Schinus molle

Schinus terebinthifolius

Distribution and Habitat—Native to South America, species of *Schinus* have been naturalized through much of Mexico. The same has happened in southern Florida. Originally introduced in the early 1900s, plants of *S. terebinthifolius* have been dispersed by birds, and the species has become alarmingly abundant in various localities (Morton 1978). Its aggressive growth and allelopathic effects have facilitated spread. Plants of all four species are grown as ornamentals and street trees in Florida, southern Arizona, and southern California.

■ ornamentals in warm areas

Disease Problems—Exposure to the sap or trimmings of *S. terebinthifolius* causes severe irritation of the skin and eyes or produces an asthmalike condition (Morton 1982). The irritant effects range from a persistent rash and mild itching to burnlike lesions with exudation of serum and severe itching. Ingestion of the fruits produces a hot taste; hence the common name *pink pepper,* which has sometimes led to their use as a pepper substitute in small amounts. However, ingestion of large amounts of the fruits and seeds may result in severe irritation of the digestive tract, with subsequent vomiting and abdominal pain in both animals and humans (Watt & Breyer-Brandwijk 1962; Morton 1978; Fuller & McClintock 1986). The adverse effects of *S. molle* and other less common species are less severe than those due to *S. terebinthifolius*. Ingestion of plant material may result in severe digestive tract irritation. Colic in horses and bird kills have been related to the genus (Morton 1978).

In addition to the contact-type reaction, aerosolized components of *Schinus,* like those of *Toxicodendron,* may cause a severe asthmalike problem. There may be swelling of the eyelids, headache, and respiratory distress (Morton 1978).

■ contact irritation, mildly allergenic

■ irritant terpenoids

Disease Genesis—There appear to be two major classes of toxic compounds present in *Schinus*—terpenoids and phenol/catechols. Numerous monoterpenes and several triterpenes are present in the sap, leaves, and fruits (Paivo Campello & Marsaioli 1974; Lloyd et al. 1977). The immediate contact irritation that is produced may be related to the array of these terpenoids and especially to the monoterpenoid hydroperoxides. α-Phellandrene may pose a problem because it may form a stable toxic peroxide after contact with the skin (Stahl et al. 1983).

Schinus terebinthifolius

CH_3 ... H_3C CH_3

α-Phellandrene

OH ... $(CH_2)_7CH{=}CH(CH_2)_5CH_3$

Cardanol (15:1)

In addition to the direct irritant action, there are delayed effects similar to those caused by exposure to *Toxicodendron* (poison ivy) and *Anacardium* (cashew). The toxicants of *Schinus* are of the same type as those in these genera, and thus it is not surprising that susceptible individuals exhibit considerable cross-reactivity (Keil et al.

■ allergenic catechols

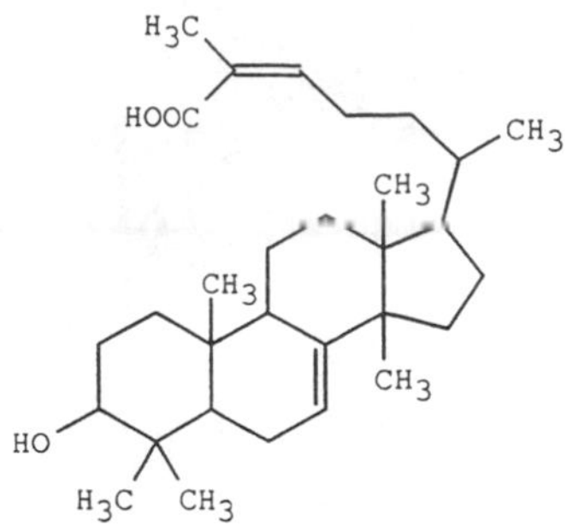

Schinol

■ skin rash, itching, vomiting, colic

■ topical relief of itching

1945a; ElSohly et al. 1986). The specific toxicants are phenol or catechol derivatives with unsaturated 15-C side chains at C-3. Cardanol 15:1 (15 carbons and 1 double bond) seems to be the toxicant of most concern because it may be present at concentrations up to 0.05% d.w. of the fruit of *S. terebinthifolius* (Stahl et al. 1983). It is irritating but less potent than the cardol and anacardic acid found in cashew nut oil. Thus the delayed allergic response to *Schinus,* while of some concern, is typically less severe than that produced by *Anacardium* and *Toxicodendron.*

Clinical Signs—Contact with the sap or foliage of *Schinus* may result in a delayed but persistent rash with itching, reddening, and swelling at the points of contact. Considerable discomfort and distress may result in both animals and people. Ingestion of the fruits or foliage produces more immediate effects, including retching and vomiting, colic, and less commonly diarrhea, which may last a day or more.

Treatment—When significant, the effects of direct-contact irritation on the skin, eyes, or digestive tract may be treated with agents aimed at relief of the pain, itching, and inflammation. The delayed allergic-type reaction may be treated in a manner similar to that recommended for *Toxicodendron.* This includes the application of drying agents to the pustules and blisters and, in severe cases, the use of steroidal and nonsteroidal anti-inflammatory drugs.

Toxicodendron Mill.

■ *T. diversilobum*
T. pubescens
T. radicans
T. rydbergii
T. vernix

poison ash
poison ivy
poison mercury
poison oak
poison sumac
three-leaved ivy

Taxonomy and Morphology—Comprising 6–10 species native to North America and eastern Asia, *Toxicodendron* is represented in North America by 5 species:

T. diversilobum (Torr. & A.Gray) Greene	poison oak, western poison oak
T. pubescens Mill.	eastern poison oak, three-leaved ivy
T. radicans (L.) Kuntze	poison ivy, poison mercury
T. rydbergii (Small ex Rydb.) Greene	western poison ivy
T. vernix (L.) Kuntze	poison sumac, poison elder, poison ash

The common names *poison ivy, poison oak,* and, to a lesser extent, *poison sumac* are applied almost interchangeably for the five species. Each individual person often has his or her own concept of what these poisonous plants are, depending upon where he or she has lived or has read or heard. The reason for this ambiguity is that species of the genus are morphologically quite variable and very similar. Plants can grow as herbs arising from rhizomes, climbing woody vines with aerial rootlets, thicket-forming shrubs, or small ill-shaped trees. Their leaflets can be large or small, toothed or lobed, and glossy dark green or pale yellow-green. Further complicating identification is the tendency for hybridization between some species. Taxonomists, however, have related the common names to specific species and their geographical distribution. An excellent comprehensive treatise on recognition of the various species of *Toxicodendron* has been written by Guin and his co-workers (1981).

In terms of botanical nomenclature, a plethora of names for the North American species have been published. Synonyms of *T. pubescens,* for example, include *T. quercifolium, T. toxicarium,* and *T. toxicodendron.* In addition, members of the genus were originally classified in the genus *Rhus,* and thus there is another suite of names in the older literature for these taxa.

Plants herbs or shrubs or trees or woody vines; deciduous; dioecious. **Stems** of woody vines bearing masses of adventitious roots. **Leaves** 1-pinnately compound; leaflets 3 or 7–13; margins entire or variously toothed or lobed; young leaves red or reddish green; petioles typically red. **Inflorescences** panicles or racemes; axillary; bracts absent. **Flowers** imperfect; small. **Sepals** 5. **Petals** 5; greenish white. **Stamens** 5; equal in length. **Pistils** 1; compound. **Drupes** fleshy; white or gray-white. **Seeds** 1.

- herbs, shrubs, vines, or trees; leaves pinnately compound, 3 leaflets or more for sumac, turn bright red or yellow-red in fall; flowers small; petals 5, greenish white; fruits drupes, white or gray-white

Toxicodendron diversilobum

Toxicodendron radicans

Toxicodendron vernix

Recognition of species of *Toxicodendron* in their many forms is not difficult for the wary individual. The distinguishing characteristics—three leaflets, with the central one having the longest petiolule or stalk; one leaf per node; and white drupes (popularly but incorrectly referred to as berries) arising laterally along the stems—should prompt immediate attention. In the fall, plants undergo color change earlier than other species and typically become conspicuously bright red or yellow-red. Color also is useful in the early spring when plants first begin to produce leaves; the branchlets and young leaves then are typically a shiny red or reddish brown.

The nontoxic *Acer negundo* (box elder) and *Parthenocissus quinquefolia* (Virginia creeper) are sometimes mistaken for poison ivy or poison oak. Observation of the box elder's two leaves per node or Virginia creeper's palmate arrangement of leaflets, however, allows one to decide quickly whether to touch them or to run for water with which to wash.

- key features—leaflets 3, central one with longer stalk
- differentiate from box elder and Virginia creeper by leaves

Distribution and Habitat—The true, or western, poison oak, *T. diversilobum* occurs in open woodlands only along the Pacific coast. In contrast, the two species of poison ivy, *T. radicans* and *T. rydbergii,* occur over much of the remainder of North America. In Oregon and Washington, *T. diversilobium* and *T. rydbergii* come together and apparently hybridize. The two species of poison ivy are found in a diversity of habitats: woods, thickets, bluffs, ridgetops, gravel bars in streams, and especially disturbed areas.

The shrubby eastern poison oak, *T. pubescens,* has a distribution similar to that of viny *T. radicans* but occurs only as far west as the eastern edge of the Great Plains. Despite its common name of *western poison ivy, T. rydbergii* likewise has a similar range. However, it typically is found at higher latitudes and occurs across Canada in sunny open areas.

Toxicodendron diversilobum

■ allergic reaction, reddened skin, watery blisters, severe itch; increased risk for sensitization in late childhood

Toxicodendron pubescens

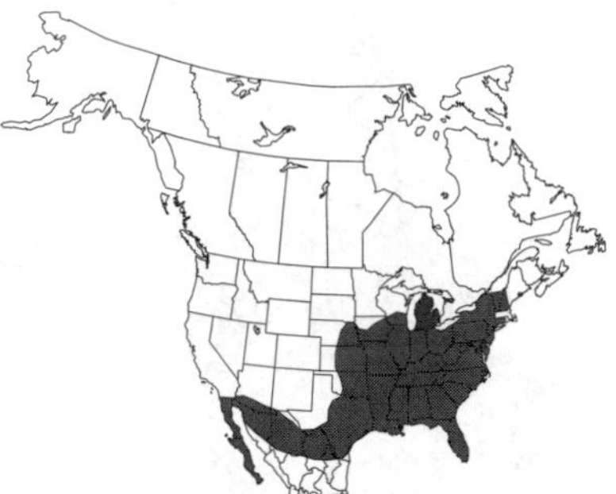

Toxicodendron radicans

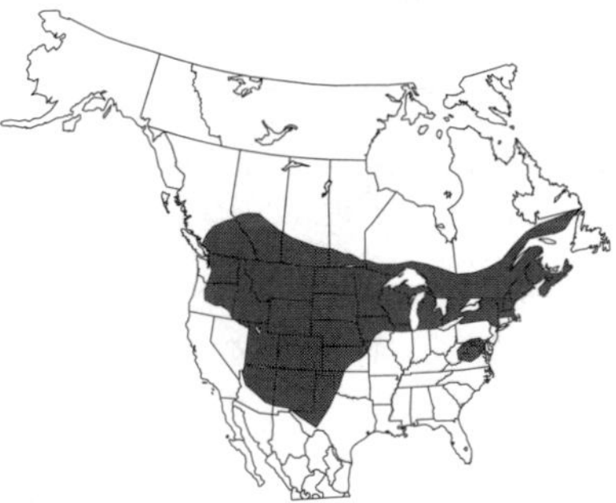

Toxicodendron rydbergii

Toxicodendron vernix

The distribution of poison sumac, *T. vernix*, is essentially east of the Mississippi River. Greatest abundance is in the southeastern quarter of the United States, where it is characteristically found in swamps under shade.

Disease Problems—The appearance of reddened skin and watery blisters accompanied by itching and an almost overpowering urge to scratch the inflamed region typically heralds the onset of a case of poison ivy, poison oak, or poison sumac. These three species are responsible for more cases of allergic contact dermatitis than all other plants, household chemicals, and industrial compounds combined. The dermatitis is a very significant cause of occupational injury, especially in forestry workers and fire fighters (Epstein 1994). The mere mention of poison ivy may evoke apprehension in some individuals because this is a problem that has affected a great many people (Benedix 1944). Approximately 80% of the population of North America is sensitive to poison ivy or poison oak, and an appreciable number (18–20%) of those afflicted require a physician's care or bed rest (Turner 1947; Vietmeyer 1986). Most individuals become sensitized in late childhood, and once they are sensitized, there is considerable cross-reactivity to contact with other members of the Anacardiaceae, as well as to *Ginkgo* (Epstein 1994).

Disease Genesis—The allergic reaction occurs when a susceptible individual comes in contact with the plant's oily resin, which contains a mixture of 3-alk(en)ylcatechols such as mono-, di-, and trienes (Symes & Dawson 1954; Gross et al. 1975). As with other genera in the family, this mixture of compounds is generally referred to as urushiol. The different species of *Toxicodendron* contain different mixtures of these catechol derivatives. For example, the 3-position side chains are mainly 17 carbons long, and a few 15 carbons long, in *T. diversilobum* and *T. pubescens.* In contrast, *T. radicans* and *T. rydbergii* have mainly 15-carbon chains and only a few chains with 17 carbons (Corbett & Billets 1975; Gross et al. 1975). *T. vernix* contains a slightly different mixture of 3-*n*-pentadec(en)ylcatechols.

The oil is present in the leaves, stems, roots, and immature fruits of all species, and readily adheres to humans, pets, and objects upon contact. Thus, exposure can be via direct contact with the plant or indirect contact when one strokes the fur of a pet, wears tainted clothing, or touches contaminated tools, toys, athletic equipment, and so on (Turner 1947). Urushiol may be aerosolized or carried on fragments when plants are burned and subsequently may cause severe problems when inhaled. In the winter, when the foliage is absent, the twigs may unwittingly be collected for fires.

Sensitization to poison ivy and its relatives requires one or more exposures to urushiol or the catechol compounds of other members of the Anacardiaceae (Keil et al. 1944; Vietmeyer 1986). Initial exposure results in a response within 5 days (Mitchell & Rook 1977). The components of urushiol differ in their allergenic potency. Decreased saturation of the side chain of the pentadecylcatechol is associated with increased antigenicity (Johnson et al. 1972). The allergen penetrates the outer skin layers and, after oxidation to a quinone form, chemically binds with ε-amino and thiol groups in cell proteins to form an antigen that elicits a cell-mediated immune reaction (Dunn et al. 1986; ElSohly et al. 1986). This likely occurs via macrophages or Langerhans cells as presenter cells for the T-lymphocytes, resulting in skin damage (Byers et al. 1979; Dunn et al. 1984). Reddening, swelling, blisters, and welts are typical lesions.

Individuals differ genetically in their sensitivity to urushiol. Some react to contact with amounts as small as 0.001 mg, whereas others are "immune" as long as the intensity of exposure does not exceed their higher levels of tolerance. Anyone can develop a sensitivity. The severity of the response varies with the individual, the extent of exposure to urushiol, the size of the exposed area, and the thickness of the outer layer of skin. There are many horror stories and almost as many myths related to unique exposures to these plants. Perhaps one of the best, or worst, is the episode in which a student participating in a Shakespearean play, draped himself, clad only in a loin cloth, with vines from a local woodland. Two days later he was hospitalized with a debilitating localized rash (Jaret 1988).

There have been occasional instances in which skin problems appear to be accompanied by systemic signs related to effects on the kidneys (Rytand 1948; Devich et al. 1975). However, experimental studies in rats and guinea pigs indicate that the systemic effects of urushiol are low, and even oral dosage of 4 mg urushiol/kg b.w. given 3 times per week was without apparent adverse consequences other than sensitization to skin testing (Murphy et al. 1983). The estimated LD_{50} in mice of urushiol given i.p. was quite high, approximately 74 mg/kg b.w. Thus any apparent systemic effects are likely to be secondary to and dependent upon the severity of the skin manifestations.

There are many myths that have been disseminated over the years about poison ivy, poison oak, and poison sumac—for example, scratching and breaking the blisters spreads the rash; an "immune" person will never get poison ivy or poison oak; or a sensitive person can get poison ivy simply by standing near a plant (Turner 1947; Vietmeyer 1986). Contrary to popular belief, the fluid from ruptured blisters does not spread the dermatitis; it is simply serous fluid from the tissue. Appearance of a rash away from the primary site of inflammation may be due to unsuspected contact with contaminated clothing or other articles. In addition, although the signs typically begin 24–72 hours after exposure, new lesions may develop over the next 2 weeks, giving the appearance of spread from the initial eruptions. One folk remedy that seems to have some validity is the belief that drinking the milk of goats or cows that have eaten poison ivy will desensitize a person. A variation of this approach has been used with limited success, with repeated i.m. or oral administration of urushiol over a prolonged period to reduce the allergic reactions of highly sensitive people (Abramowicz 1981; Vietmeyer 1986).

There has been considerable research effort to prevent or modify the allergic response (Epstein 1974; Watson et al. 1981; Vietmeyer 1986; Williford & Sheretz 1994). This has included the development of vaccines or immune modulants. Extracts of *Toxicodendron* have been used in homeopathic medicine to attenuate the effects of osteoarthritis, with questionable results (Gibson et al. 1980; Shipley et al. 1983).

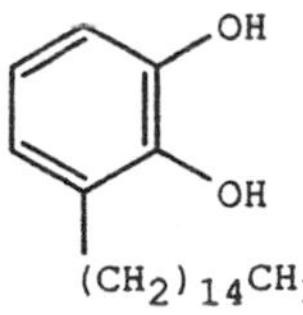

OH
OH
$(CH_2)_{14}CH_3$
or
$(CH_2)_7C{=}C(CH_2)_5CH_3$
or
$(CH_2)_7C{=}CCC{=}C(CH_2)_2CH_3$
or
$(CH_2)_7C{=}CCC{=}CCC{=}CH_2$

15-C urushiol catechols

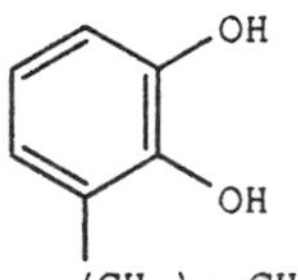

OH
OH
$(CH_2)_{16}CH_3$
or
$(CH_2)_7C{=}C(CH_2)_7CH_3$
or
$(CH_2)_7C{=}CCC{=}C(CH_2)_4CH_3$
or
$(CH_2)_7C{=}CCC{=}CCC{=}CCCH_3$

17-C urushiol catechols

- oily mixture of catechols (urushiol) composed of 15-C and 17-C side chains, present in all plant parts
- may present severe risk when sap is aerosolized
- allergic reaction requires direct contact with plant or contact with oily deposits on animal fur or inanimate objects
- blister fluid does not spread rash

Clinical Signs—In humans and primates, the signs develop in about 5 days on first exposure or in 24–72 hours in sensitized individuals. Initially, the skin shows mild reddening and swelling at the contact site accompanied by intense and intolerable itching. These are followed by more intense reddening and formation of vesicles that rupture and lead to fluid exudation, crusting, and scaling of the skin. New lesions may continue to develop for several days. In severe cases, there may be systemic signs such as vomiting, diarrhea, and abdominal pain. Without treatment, the dermatitis normally peaks in 4–5 days and resolves in 10–21 days.

- reaction 1–5 days after contact; skin reddened, watery vesicles, blisters, intense itching, peaking at 4–5 days, resolution in 2–3 weeks

Treatment—Recognition and avoidance of species of *Toxicodendron* are the best prevention. However, if contact is made, prompt removal of the urushiol may reduce the chances of developing dermatitis. About 10 minutes are required for urushiol to penetrate the skin. Thus it is possible to prevent the reaction by washing the affected area well with water. Care must be taken not to use solutions that remove protective

- immediate washing of site of contact

skin oils and, thus, actually enhance urushiol penetration. It is important to decontaminate tools and avoid contact with pets that have traversed an area infested with these plants.

- application of drying agents to affected areas; severe reactions may require corticosteroids

The vesicles are typically treated by application of drying agents such as calamine lotion, which may provide some relief. Numerous other palliative or folk remedies have been used at one time or another, including dripping the juice of a green tomato over the area, applying a combination of buttermilk or vinegar and salt to the rash, rubbing the inflamed area with the inner surface of a banana peel, rubbing wild touch-me-not *(Impatiens)* on the blisters, or applying hair spray, mouse-ear herb *(Cerastium)*, horse urine, or crushed plantain leaf to provide relief from the itching and weeping dermatitis (Duckett 1980; Jaret 1988; Zarrow et al. 1991). In some instances, allergic responses to topical medications complicates the overall disease manifestations (Williford & Sheretz 1994). Severe swelling and distress may require the use of corticosteroids topically or systemically. Except for the most potent types, topical corticosteroids are of limited value. Systemic corticosteroid therapy is much more effective. Typically, in severe cases these are given orally, in tapered dosages over a 2-to-3-week period (Williford & Sheretz 1994). In exceptionally severe cases, i.m. administration of potent corticosteroids may be necessary.

- prevention with protective barrier creams and lotions; desensitization of questionable value

Prevention of allergen penetration of the skin and subsequent oxidation and binding after contact with the plants of the genus have been attempted in a variety of ways. A number of protective or barrier creams or lotions have been formulated. Topical application of polyamine salts of a linoleic acid dimer provides a protective barrier in the majority of instances, as long as the skin is washed within 8–12 hours after contact (Orchard et al. 1986). A protective lotion of quaternium-18 bentonite is effective experimentally in guinea pigs (Marks et al. 1995). Prophylactic desensitization is possible for some individuals through i.m. injections or tablets of dilute urushiol extracts, but the effectiveness of this approach is questionable (Abramowicz 1981; Marks et al. 1987). Furthermore, it can take a prolonged time to accomplish and may be of short-lived effectiveness, i.e., only a few months. The use of urushiol analogs, such as diacetate esters or other forms, seems to hold some promise for immunization (Walker et al. 1995). Skin patch test kits are available for testing sensitivity to urushiol.

References

Aber R, Marks J, DeMelfi T, McCarthy MA, Witte E, Moore E, Hays CW: Dermatitis associated with cashew nut consumption—Pennsylvania. *Morb Mortal Wkly Rep* 32;129–130, 1983

Abramowicz M: Desensitization to poison ivy. *Med Lett* 23;40, 1981

Allen PH: Poisonous and injurious plants of Panama. *Am J Trop Med* 23 (Suppl);1–76, 1943

Anjaneyulu V, Prasad KH, Ravi K, Connolly JD: Terpenoids from *Mangifera indica*. *Phytochemistry* 24;2359–2367, 1985

Anjaneyulu V, Ravi K, Prasad KH, Connolly JD: Terpenoids from *Mangifera indica*. *Phytochemistry* 28;1471–1477, 1989

Benedix JA: The gremlin of the woods. *Appalachia* 10;127–130, 1944

Byers VS, Epstein WL, Castagnoli N, Baer H: In vitro studies of poison oak immunity. 1. In vitro reaction of human lymphocytes to urushiol. *J Clin Invest* 64;1437–1448, 1979

Connolly JD, Hill RA: *Dictionary of Terpenoids*. Vol 1, *Mono Terpenoids*. Chapman & Hall, London, 1991a

Connolly JD, Hill RA: *Dictionary of Terpenoids*. Vol 2, *Di- and Higher Terpenoids*. Chapman & Hall, London, 1991b

Corbett MD, Billets S: Characterization of poison oak urushiol. *J Pharm Sci* 64;1715–1718, 1975

Devich KB, Lee JC, Epstein WL, Spitler LE, Hopper J: Renal lesions accompanying poison oak dermatitis. *Clin Nephrol* 3;106, 1975

Duckett S: Plantain leaf for poison ivy [letter]. *N Engl J Med* 303;583, 1980

Dunn IS, Liberato DJ, Castagnoli N, Byers VS: Induction of suppressor T cells for lymph node cell proliferation after contact sensitization of mice with a poison oak urishiol component. *Immunology* 51;773–781, 1984

Dunn IS, Liberato DJ, Castagnoli N Jr, Byers V: Influence of chemical reactivity of urushiol-type haptens on sensitization and the induction of tolerance. *Cell Immunol* 97;189–196, 1986

ElSohly MA, Adawadkar PD, Benigni DA, Watson ES, Little TL Jr: Analogues of poison ivy urushiol: synthesis and biological activity of disubstituted *n*-alkylbenzenes. *J Med Chem* 29;606–611, 1986

Epstein WL: Poison oak and poison ivy dermatitis: an occupational problem. *Cutis* 13;544–548, 1974

Epstein WL: Occupational poison ivy and oak dermatitis. *Dermatol Clin* 12;511–516, 1994

Fuller TC, McClintock E: *Poisonous Plants of California.* Univ California Press, Berkeley, 1986

Gibson RG, Gibson SLM, MacNeill AD, Buchanan WW: Homeopathic therapy in rheumatoid arthritis: evaluation by a double blind clinical therapeutic trial. *Br J Clin Pharmacol* 9;453–459, 1980

Gross M, Baer H, Fales HM: Urushiols of poisonous Anacardiaceae. *Phytochemistry* 14;2263–2266, 1975

Guin JD, Gillis WT, Beaman JH: Recognizing the toxicodendrons (poison ivy, poison oak, and poison sumac). *J Am Acad Dermatol* 4;99–114, 1981

Jaret P: The big itch. *Hippocrates* 1988;76–81, 1988

Johnson RA, Baer H, Kirkpatrick CH, Dawson CR, Khurana RG: Comparison of the contact allergenicity of the four pentadecylcatechols derived from poison ivy urushiol in human subjects. *J Allergy Clin Immunol* 49;27–35, 1972

Keil H, Wasserman D, Dawson CR: The relation of chemical structure in catechol compounds and derivatives to poison ivy hypersensitiveness in man as shown by the patch test. *J Exp Med* 80;275–287, 1944

Keil H, Wasserman D, Dawson CR: The relation of hypersensitiveness to poison ivy and to cashew nut shell liquid. *Science* 102;279–280, 1945a

Keil H, Wasserman D, Dawson CR: The relation of hypersensitiveness to poison ivy and to the pure ingredients in cashew nut shell liquid and related substances. *Ind Med* 14;825–830, 1945b

Lloyd HA, Jaouni TM, Evans SL, Morton JF: Terpenes of *Schinus terebinthifolius.* *Phytochemistry* 16;1301–1302, 1977

Marks JG Jr, Fowler JF Jr, Sheretz EF, Rietschel RL: Prevention of poison ivy and poison oak allergic contact dermatitis by quaternium-18 bentonite. *J Am Acad Dermatol* 33;212–216, 1995

Marks JG, Trautlein JJ, Epstein WL, Laws DM, Sicard GR: Oral hyposensitization to poison ivy and poison oak. *Arch Dermatol* 123;476–478, 1987

Mitchell J, Rook AJ: Diagnosis of contact dermatitis from plants. *Int J Dermatol* 16;257–266, 1977

Mitchell J, Rook AJ: *Botanical Dermatology: Plants and Plant Products Injurious to the Skin.* Greengrass, Vancouver, BC, pp63–97, 1979

Moraes SM: Properties and pharmacological activities of cashew nut shell liquid, its constituents, and derivatives. *Rev Bras Farm* 74;87–90, 1993

Morton JF: Brazilian pepper—its impact on people, animals, and the environment. *Econ Bot* 32;353–359, 1978

Morton JF: *Plants Poisonous to People in Florida and Other Warm Areas.* Morton, Miami, pp117–125, 1982

Murphy JC, Watson ES, Harland EC: Toxicological evaluation of poison oak urushiol and its esterified derivative. *Toxicology* 26;135–142, 1983

Orchard S, Fellman JH, Storrs FJ: Poison ivy/oak dermatitis. *Arch Dermatol* 122;783–789, 1986

Paiva Campello J, Marsaioli AJ: Triterpenes of *Schinus terebinthifolius*. *Phytochemistry* 13;659–660, 1974

Rosen T, Fordice DB: Cashew nut dermatitis. *South Med J* 87;543–546, 1994

Rytand DA: Fatal anuria, the nephrotic syndrome, and glomerular nephritis as sequels of the dermatitis of poison oak. *Am J Med* 5;548, 1948

Shipley M, Berry H, Broster G, Jenkins M, Clover A, Williams I: Controlled trial of homeopathic treatment of osteoarthritis. *Lancet* 1;97–98, 1983

Stahl E, Keller K, Blinn C: Cardanol, a skin irritant in pink pepper. *Planta Med* 48;5–9, 1983

Sullivan JT, Richards CS, Lloyd HA, Krishna G: Anacardic acid: molluscicide in cashew nut shell liquid. *Planta Med* 44;175–177, 1982

Symes WF, Dawson CR: Poison ivy "urushiol." *J Am Chem Soc* 76;2959, 1954

Turner CE: *Rhus* dermatitis as a public health and health education problem. *Am J Public Health* 37;7–12, 1947

Tyman JH: Determination of the component phenols in natural and technical cashew nut-shell liquid by gas-liquid chromatography. *Anal Chem* 48;30–34, 1976

Vietmeyer N: Science has got its hands on poison-ivy, poison-oak, and poison-sumac. *USDA Forest Serv Fire Manage Notes* 47;23–28, 1986

Walker LA, Watson ES, ElSohly MA: Single dose parenteral hyposensitization to poison ivy urushiol in guinea pigs. *Immunopharmacol Immunotoxicol* 17;565–576, 1995

Wasserman D, Dawson CR: Cashew nut shell liquid. 3. The cardol component of Indian cashew nut shell liquid with reference to the liquid's vesicant activity. *J Am Chem Soc* 70;3675–3679, 1948

Watson ES, Murphy JC, Wirth PW, ElSohly MA, Skierkowski P: Immunological studies of poisonous Anacardiaceae: production of tolerance in guinea pigs using 3-*n*-pentadecylcatechol–"modified" autologous blood cells. *J Pharm Sci* 70;785–789, 1981

Watt JM, Breyer-Brandwijk MG: *The Medicinal and Poisonous Plants of Southern and Eastern Africa*. E&S Livingston, Edinburgh, 1962

Williford PM, Sheretz EF: Poison ivy dermatitis: nuances in treatment. *Arch Fam Med* 3;184–188, 1994

Zarrow S, McVeigh G, Rao L: Reader-tested home remedies. *Prevention* 1991;33–35,100–105, 1991

Apiaceae Lindl.

7

Commonly known as the carrot, parsley, or celery family, the Apiaceae comprises 200–300 genera and 2,500–3,000 species and is among the best-known families of flowering plants because of the economic importance of its genera and species as food, condiments, ornamentals, and as poisonous plants. Other familiar important members of the family include dill, chervil, caraway, coriander, anise, fennel, angelica, eryngo, water hemlock, and poison hemlock. Although members of the family are widely appreciated for their economic importance, two members are among the most hazardous of all plants for humans. Water hemlock and poison hemlock accounted for nearly 40% of human fatalities related to plants in North America between 1985 and 1994 (Krenzelok et al. 1996). Although early explorers were quite familiar with the usefulness of many plants, especially the pleasant-flavored roots of water parsnip *(Sium suave)*, Hudson Bay Company employees and other explorers were well advised by members of local Indian tribes to take great care in identifying the plants before eating the roots so as to avoid the deadly effects of the very similar water hemlock (Fernald and Kinsey 1958). Widespread in distribution, the family is most common in northern temperate regions. In North America it is represented by 75–94 genera and 380–440 species, both native and introduced. An alternative family name permitted by the rules of botanical nomenclature is Umbelliferae, typically encountered in older publications.

- Carrot Family

  Aethusa
  Ammi
  Cicuta
  Conium
  Cymopterus
  Heracleum
  Sphenosciadium

  QUESTIONABLE
  Daucus

- source of many spices—anise, caraway, coriander, dill, and fennel

Plants herbs; annuals, biennials, or perennials; foliage typically aromatic. **Stems** usually hollow or with soft pith. **Leaves** variable, but usually deeply lobed or dissected; herbaceous or cartilaginous; cauline or forming a basal rosette; 1-pinnately, 2-pinnately, or palmately compound or rarely simple; alternate; petioles typically sheathing; surfaces rugose, smooth, or punctate; terminal leaflets of compound leaves present; venation pinnate or a single vein; stipules absent. **Inflorescences** umbels or heads; umbels compound or simple; terminal or axillary; bracts present, subtending umbels and forming involucres; bracteoles present, subtending umbellets and forming involucels. **Flowers** perfect, perianths in 2-series or 1-series. **Calyces** radially symmetrical. **Sepals** 5 or 0; free. **Corollas** radially symmetrical or occasionally outermost flowers of umbels or umbellets bilaterally symmetrical. **Petals** 5; small; caducous; free; white, yellow, green, red, purple, or blue; spreading; apices usually inflexed. **Stamens** 5. **Pistils** 1; compound; carpels 2; stigmas 2; styles 2; stylopodia present; ovaries inferior, lobed in cross section, lobes 2; locules 2; placentation apical; ovules 1 per locule. **Nectaries** forming disk at bases of styles. **Fruits** schizocarps; mericarps with at least 5 primary ribs. **Seeds** 1 per mericarp.

- herbs; flowers small, in umbrella-shaped clusters; petals 5

Genera of Apiaceae are causes of photosensitization, neurotoxicosis, and teratogenic effects. Photosensitization is caused by several genera (table 7.1), and because its genesis, pathology, and treatment for all taxa are essentially the same, it will be described in a general sense. Aspects of the disease unique to particular genera will

- photosensitization, neurotoxic, teratogenic
- numerous taxa contain acetylenic compounds of unknown toxicity risk

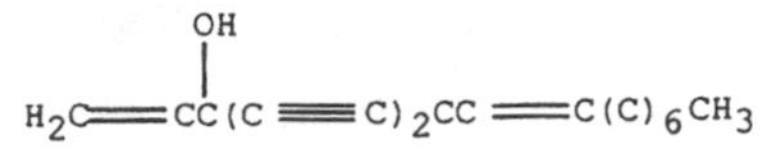

$H_2C{=}CC(OH)(C{\equiv}C)_2CC{=}C(C)_6CH_3$

Falcarinol

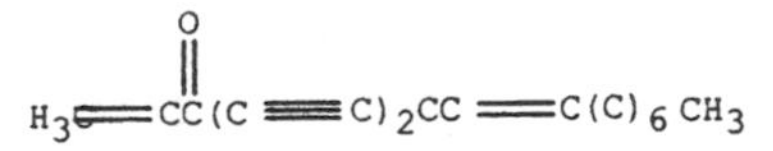

$H_3C{=}CC(O)(C{\equiv}C)_2CC{=}C(C)_6CH_3$

Falcarinone

be elaborated upon in the genus treatments. Most of the other effects, including the neurotoxicoses, seem to be associated with acetylenic compounds, except for a few instances in which alkaloids, generally sparsely represented in the family, prevail. It is estimated that at least 75% of the species of Apiaceae contain acetylenic compounds, especially falcarinol, falcarinone, and/or derivatives, some of which are neurotoxic (Crosby and Aharonson 1967; Bohlmann 1971; Yates and England 1982; Avalos et al. 1995). Even the common *Apium* (celery), *Daucus carota* (wild carrot), *Berula erecta* (cutleaf or narrow-leaved water parsnip), and species of *Torilis* (the hedge parsleys) contain measurable falcarinol or falcarinone. These compounds are known as potent irritants and contact allergens from the Araliaceae, *Hedera helix* (Hausen et al. 1987). *Berula* is also suspect as a cause of death losses in cattle (Kingsbury 1964).

Table 7.1. Genera known to cause photosensitization

Ammi	bishop's weed
Cymopterus	spring parsley
Heracleum	cow parsnip
Sphenosciadium	white-heads

Although intoxication problems in livestock are not reported, species of *Angelica* that contain acetylenic compounds have been used for suicidal purposes by members of American Indian tribes of Canada (Millspaugh 1974). Only the fresh root was used. When dried, other effects such as diuresis prevailed. An Old World species *Chaerophyllum temulum* (rough chervil) contains falcarinone and a glycoalkaloid, chaerophylline, possibly similar in effect to coniine (Raffauf 1970; Bartic and Piskac 1981). Experimentally, aqueous extracts of the seeds of *Petroselinum crispum* (common parsley) caused resorption of one-half of the embryos of 5 rats and minor developmental effects in the surviving fetuses (Nath et al. 1992). As its common name implies, *Sanicula bipinnata* (poison sanicle) has a reputation of toxicity of unknown origin. *Sium suave* (water parsnip), which is common through much of North America, contains falcarinone and other toxins, but reports of intoxication are questionable (Pammel 1911; Kingsbury 1964; Bohlmann 1971). As reported for a European species, signs would be limited to mild neurologic signs and possibly digestive tract irritation (Bartic and Piskac 1981). Species of *Thaspium* (meadow parsnips), also widely distributed in North America, contain acetylenic compounds and are reported to be similar to *Aethusa cynapium* in toxicity (Bohlmann 1971; Millspaugh 1974). A number of species in the family contain the hallucinogen myristicin, but in such low concentrations that there is little risk (Hallstrom and Thuvander 1997).

Conium is probably the only member of the family in which alkaloids are of toxicologic significance (Fairbairn 1971). The various genera, including *Cicuta* and *Conium,* are discussed individually. *Conium* has also been associated with teratogenic effects.

■ primary photosensitization, limited severity

Disease Problems—Sensitization to light, which may affect any animal species, including birds, has been related to the formation of transient reactive radicals in the peripheral circulation and their subsequent effect on surrounding tissues. In a now-considered-classic treatment on the disease, Clare (1952) described three types: type I—primary photosensitivity, type II—photosensitivity due to aberrant pigment synthesis, and type III—secondary, or hepatogenous, photosensitivity. Only types I and III are caused by plants. Type I is caused by several genera and many species of the Apiaceae (table 7.1)

The effects of photosensitization are seldom lethal, but large numbers of animals may be affected and the economic losses serious. Disease from these taxa is more akin to sunburn than is the disease caused by other photosensitizing genera. Systemic effects also may be noted, even in the absence of sunlight and in dark-skinned animals (Shlosberg et al. 1974; Egyed et al. 1977; Johnson 1983). Some genera of the Rutaceae produce a similar intoxication, as do a few taxa of the Leguminosae, Moraceae, and Orchidaceae (Pathak et al. 1962; Scott et al. 1976).

There is considerable variation in toxicity of plants, with a wide range of seasonal, climatic, insect predation, and other conditions affecting the toxicant levels in the leaves (Zangerl 1990; Zobel 1991; Zobel and Brown 1990a,b; Trumble 1992). Because of this variation, many of the plants of the family are toxic only under unusual circumstances or following ingestion for several days. There seems to be an increased likelihood of sensitization following contact with wet plants.

■ toxicity variable by season and other factors

Fungal and bacterial infections of plants also may be involved in the occurrence of photosensitization. The leaves of *Pastinaca* (common garden parsnip) and *Apium* (celery) may become toxic or be increased in toxicity under certain conditions, for celery most notably when it is subjected to fungal infection *(Sclerotina sclerotiorum)* or bacterial infection (Scheel et al. 1963; Karasawa et al. 1990; Avalos et al. 1995). The blistering action of celery is well known among those handling it, from harvesters to grocery store workers. A vesicular disease in white-skinned pigs also has been observed in the field and reproduced experimentally by feeding or rubbing parsnip or celery leaves on the snouts and feet (Montgomery et al. 1987a,b).

■ plant fungal infection increases the toxicity risk

Disease Genesis—The type I photosensitizing genera of the family contain members of a series of furanocoumarins and furanochromones and are able to elicit disease through topical or oral exposure and by single or repeated doses. The primary offenders are a series of linear furanocoumarins (psoralens), the most noteworthy of which are trimethylpsoralen, psoralen, xanthotoxin (8-methoxypsoralen), and bergapten (5-methoxypsoralen). Angular furanocoumarins such as angelicin also may be present, but these are of little consequence as photosensitizers (Murray et al. 1982). In addition, there are about two dozen other compounds of lesser importance (Pathak et al. 1962; Abu-mustafa et al. 1975). The toxins are in highest concentrations in the seeds, but the leaves also may have appreciable concentrations under some conditions (Berenbaum 1981). These toxins are present in plants presumably as deterrents to insect feeding (Downum and Rodriguez 1986).

■ photosensitization, extensive series of furanocoumarins

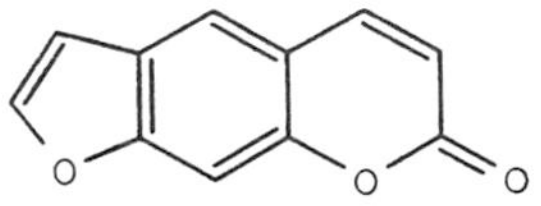

Psoralen

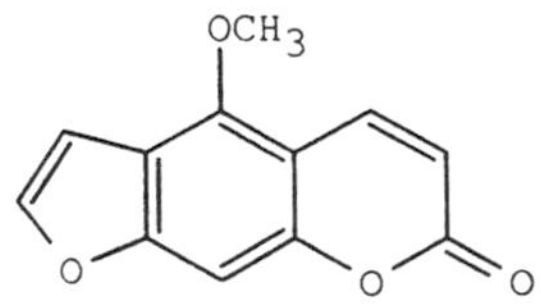

OCH_3

Bergapten

OCH_3

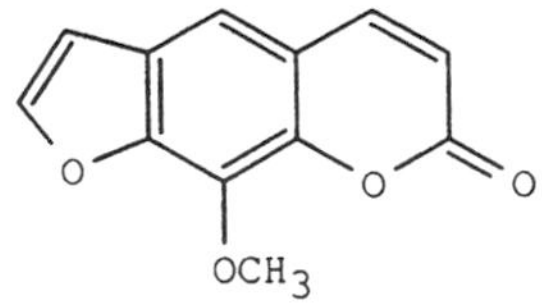

Xanthotoxin

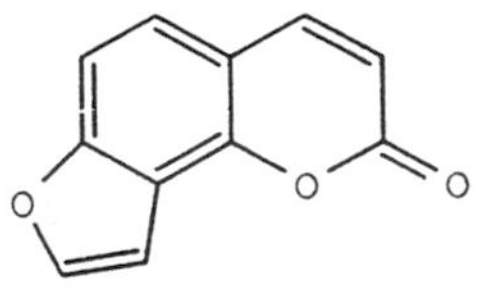

Angelicin

Furanocoumarins are activated by long-wavelength UV light (320–380 nm) and undergo type I anaerobic reactions with excitement of electrons to a transient higher-orbital energy state and subsequent interaction of the product directly with DNA, without oxygen interaction. There is cross-linking of DNA strands by linear furanocoumarins and monoadducts formed with angular types because of stearic misalignments (Scott et al. 1976; Song and Tapley 1979; Dodge and Knox 1986; Ivie 1987). Additional systemic effects occur by similar adduct formation with RNA and various cellular sites. The typical tissue injury accompanying photoactivation may in some measure be due to formation of covalent adducts with fatty acids followed by lipid-membrane alterations (Specht et al. 1988). Furanocoumarin transfer to the skin may occur by topical contact with oil ducts of the leaf surface or following ingestion of the plant. Sunlight subsequently initiates the reaction, with the skin remaining sensitive for up to a week following exposure to the plant. The term *phytophotodermatitis* has been suggested to describe the topical contact disease (Klaber 1942).

■ furanocoumarins activated by UV light (320–380 nm), contact topical injury, or secondary to ingestion

Some taxa have long been recognized for their therapeutic value, most notably to stimulate pigmentation in cases of vitiligo, a disease of patchy depigmentation of the skin. Some species, such as *Ammi majus*, are used as a source of 8-methoxypsoralen for photochemotherapy of psoriasis (deWolff and Thomas 1986).

■ minor therapeutic uses for some taxa

In cattle and sheep, the ingested furanocoumarins undergo ruminal O-demethylation, and hepatic oxidation and ring cleavage with glucuronide and sulfate conjugation. As a result of apparent metabolism upon first passage through the liver, the half-lives of these compounds are quite short; however, binding to cellular constituents such as DNA extends their biological effects for a much longer period (de Wolff and Thomas 1986). As shown in dogs, methoxsalen, a representative compound, is metabolized by opening of the furan ring and excreted almost equally in urine and bile. However, there is persistence for several weeks of the compound via protein binding (Kolis et al. 1979). There is very little public health risk, because there is little of the toxicants in milk or eggs (Ivie et al. 1986; Pangilinan et al. 1992).

Other genera that are known to contain linear furanocoumarins have not as yet been associated with disease (table 7.2). Additional genera containing a variety of furanocoumarin types that are of unknown toxicologic significance include *Ligusticum* (lovage), *Lomatium* (wild parsley), *Petroselinum* (parsley), *Glehnia* (= *Phellopterus,* wavewing), *Pimpinella* (burnet saxifrage), and *Seseli* (seseli).

Table 7.2. Genera containing linear furanocoumarins and potentially photosensitizing

Angelica	angelica
Apium	cultivated celery
Cicuta	water hemlock
Conium	poison hemlock
Sium	water parsnip

■ livestock, skin, erythema, edema, especially the muzzle, ears, and udder, moderate itch; eye, reddening of cornea, conjunctiva

Clinical Signs—Similar signs are produced by all the photosensitizing taxa. In livestock, these signs include erythema and edema of the skin of the muzzle, ears, udder, teats, and vulva and scratching of affected areas. In some cases the lesions may be limited to areas of contact with the plants, or they may be more widely distributed in less protected and minimally pigmented skin areas. Blisters may form in the skin, with exudation and subsequent ulceration. Involvement of the eyes is common, with sensitivity to light and inflammation of the cornea and conjunctiva, sometimes with corneal opacities. In poultry, similar effects occur in unfeathered areas, leading to serious deformities of the beak, legs, and feet; loss of comb, wattles, and feathers of the head; and increased pigmentation of the skin. Pigmentary retinopathy may accompany the other ocular effects.

■ humans, skin, erythema, blistering, 1–2 days after topical exposure

In humans, whether on first or subsequent exposures, reddening and blistering of the skin occurs about 48 hours after exposure (Mitchell and Rook 1977). Residual hyperpigmentation of the skin may develop and persist for several months.

■ in vitro assay available to aid diagnosis

Diagnosis or incrimination of a specific genus or species may be aided by the use of an in vitro microbiological assay of photosensitivity that utilizes the fungus *Candida albicans* and is especially useful with the furanocoumarins (Rowe and Norman 1989).

■ gross pathology, skin lesions;

microscopic, mild renal tubular and liver lesions

Pathology—The skin and eye changes as described above constitute the most prominent lesions. In addition, in some cases systemic lesions may be apparent, albeit usually only minor in degree. These include mild renal tubular dilation with flattened epithelia, and mild hepatic fatty change and necrosis (Egyed et al. 1977; Witzel et al. 1978).

■ skin lesions are usually mild and require no specific treatment

Treatment—The photosensitization is typically mild, and treatment generally is not required. Prevention of access to the offending plants is sufficient to allow recovery. In severe cases, recovery may require several weeks. Nursing offspring may require supplemental feeding.

Aethusa L.

Taxonomy and Morphology—The genus *Aethusa* is represented in North America by a single species:

A. cynapium L. — fool's parsley, dog's parsley, dog poison, garden hemlock, lesser hemlock, small hemlock

- *A. cynapium*

dog poison	garden hemlock
dog's parsley	lesser hemlock
fool's parsley	small hemlock

Plants annuals. **Stems** erect; freely branching; 20–70 cm tall. **Leaves** 2- or 3-pinnately dissected, the terminal segments narrow. **Umbels** compound; open; bracts of involucres absent or few; bracteoles of involucels 2–5. **Sepals** absent. **Petals** white. **Schizocarps** broadly ovate; nearly terete in cross section; with prominent corky ribs.

- erect, freely branching annuals; leaves pinnately dissected; petals white

Aethusa cynapium

Distribution and Habitat—A native of Eurasia, *A. cynapium* occurs as a weed in disturbed sites, primarily in the northeastern quarter. Populations have been reported in Alabama. Flowering is from June to August.

Disease Problems—Little is known of the toxicity of *A. cynapium;* the rarity of problems is perhaps related to its weedy nature, unpalatability, and occurrence in habitats not typically grazed by livestock. There is considerable disagreement on the toxicity of the species, and some workers view it as an innocuous plant. This may be an indication of markedly differing toxicity among varieties in diverse geographical locations (Millspaugh 1974). Extracts of the leaves, when administered to a dog in which vomiting was prevented by esophageal ligation, caused dilated pupils, decreased arousal and heart rate, and seizures (Millspaugh 1974). The animal died in 1 hour. In a situation where a sow and her pigs were abruptly allowed access to a weedy pen dominated by *A. cynapium,* there was rapid onset of incoordination of the pelvic limbs and death of the sow and 3 piglets in 2–3 hours (Barr and Davies 1963). A few hours later, 3 more piglets died. It was suspected that in this situation, the plants were eaten because the animals were allowed access to green forage after being indoors for a prolonged period. Swart (1975) describes a case involving goats, in which there was ataxia and digestive problems. *Aethusa cynapium* is also reputed to have produced numerous intoxications in humans when mistaken for parsley, with death in some instances (Kingsbury 1964; Millspaugh 1974).

- generally unpalatable, neurotoxic, vomiting, seizures

Aethusa cynapium

Disease Genesis—*Aethusa cynapium* is reputed to contain cynapine, a coniine-like alkaloid, but this is questionable (Bohlmann 1971; Bartic and Piskac 1981). This species does contain large amounts of numerous (at least 25) polyacetylenic compounds, including aethusin; most have 13-C side chains, and a few have 10-C and 11-C chains (Bohlmann 1971). The persistent vomiting and clonic-tonic picrotoxin-like seizures are consistent with polyacetylenic-type toxins (Frohne and Pfander 1984).

- neurotoxic, 10-to-13-C polyacetylenic compounds

$$H_3CC{\equiv}C(C{\equiv}C)_2(C{\equiv}C)_2CCH_3$$

$$HO{-}CC{\equiv}C(C{\equiv}C)_2(C{\equiv}C)_2CCH_3$$

$$H_3CC{\equiv}C(C{\equiv}C)_2C{\equiv}C\underset{\displaystyle OH}{C}CCCH_3$$

Representative acetylenic compounds

Clinical Signs—Ingestion of large amounts of the tops of the plants is said to cause stupor, digestive tract upset, pupillary dilation, seizures, and paralysis in horses, cattle, and sheep (Cary et al. 1924; Bartic and Piskac 1981). In children there is nausea, stomach cramps, headache, vomiting, diarrhea, and, in a few instances, seizures.

- livestock, digestive upset, seizures, paralysis
  humans, nausea, cramps, headache, vomiting, diarrhea

Treatment—Activated charcoal may be beneficial if given early in the course of the disease; otherwise treatment is directed toward relief of the specific signs. Rarely, control of seizures with diazepam may be indicated.

- activated charcoal

Ammi L.

- *A. majus*
  A. visnaga

  ammi
  bishop's weed
  bisnaga
  chellah
  khella

Taxonomy and Morphology—An Old World genus of 6 species, chiefly of the Mediterranean region, the genus name was used by Dioscorides. *Ammi* is represented in North America by two species:

A. majus L.	bishop's weed, greater ammi
A. visnaga (L.) Lam.	bisnaga, bishop's weed, toothpick ammi, khella, chellah

- erect, branching annuals; leaves pinnately compound or dissected; petals white

Plants annuals or biennials. **Stems** erect; branching; 20–80 cm tall. **Leaves** 3-pinnately compound or pinnately dissected, the terminal leaflets linear to lanceolate. **Umbels** compound; open or compact; bracts of involucres numerous; entire or divided; bracteoles of involucels entire. **Sepals** minute teeth. **Petals** white; wide; apices 2-lobed. **Schizocarps** oblong to ovate.

1. Plants biennials. Terminal leaf segments lanceolate. Inflorescences open; rays 20–60, spreading in fruit . . . . *A. majus*
1. Plants annuals. Terminal leaf segments linear. Inflorescences compact; rays 60–100, curving inward in fruit . . . . *A. visnaga*

Ammi majus

Ammi majus

Ammi visnaga

- waste areas

Distribution and Habitat—Of Eurasian origin, both species have inadvertently been introduced into the southeastern part of the continent. Occurring sporadically, plants are typically found in disturbed sites such as rights-of-way and waste areas. Seed is now available from wildflower seed producers.

- photosensitization, skin, eyes, in all livestock, including poultry; furanocoumarins

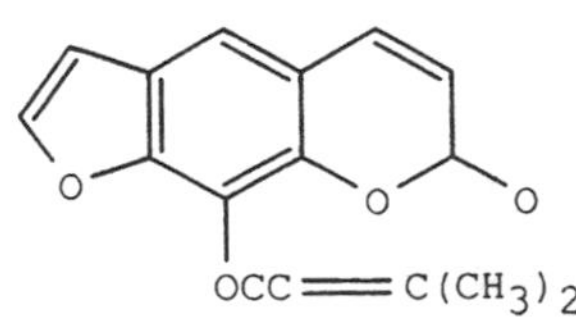

Imperatorin

Disease Problems and Genesis—Species of *Ammi* cause photosensitization; *A. majus* is the more toxic of the two, and its seeds are more toxic than its foliage. Furthermore, it is readily eaten, whereas *A. visnaga* is quite unpalatable (Shlosberg et al. 1974). Disease problems have been reported for cattle, sheep, and poultry (Egyed et al. 1977; Dollahite et al. 1978; Witzel et al. 1978; Mendez et al. 1991). Among poultry, ducks seem most susceptible, geese intermediate, and chickens and turkeys least sensitive (Egyed et al. 1975a). Ocular effects in ducks are severe and may be accompanied by fundic pigmentation (Egyed et al. 1975b). The photosensitizing furanocoumarins bergapten (0.2–0.3%), imperatorin (0.7–0.8%), and xanthotoxin (1–1.5%) have been isolated from *A. majus* seeds (Karawya et al. 1970). In one instance, eight 100-

kg calves grazing wheat stubble infested with *A. majus* developed typical signs (Lopez and Odriozola 1987). Total furocoumarin content was 1.4% in plants with ripe seeds. In another situation, Yeruham and co-workers (1988) observed three phases of signs in merino lambs eating *Vicia* hay contaminated with *A. majus*. Initially there was s.c. edema of the head and neck and then vesicles and erosions of the mouth and tongue. Eventually crusts appeared around the nostrils and mouth.

In addition to photosensitization, there are systemic effects, even in the absence of sunlight and in dark-skinned animals (Shlosberg et al. 1974; Egyed et al. 1977). The systemic effects, including kidney and liver degeneration, appear to be due to the furanochromones khellin and visnagin and to the pyranocoumarin visnadin, which have been identified in *A. visnaga*. Visnadin, a dilator of smooth muscle, appears to be the most potent constituent causing systemic effects, especially of coronary vessels and possibly bronchioles but having little effect on other systemic vessels (Anrep et al. 1949; Bagouri 1949). Some of its effects appear to be due possibly to Ca^{2+} channel inhibition (Rauwald et al. 1994). Because of these effects, the dried fruits of *A. visnaga* have been used for medicinal purposes in Middle Eastern countries (Anrep et al. 1949).

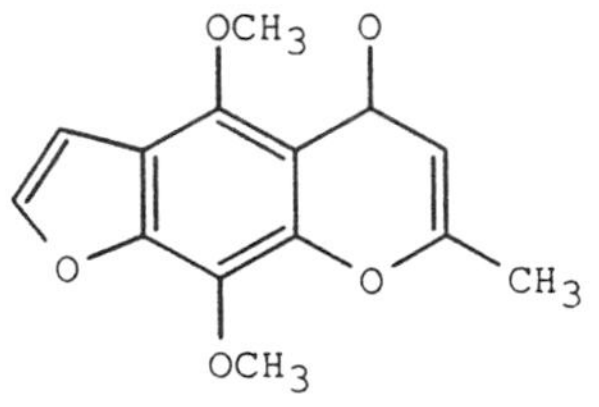

■ effects on liver and kidney, furanochromones, pyranocoumarin

OCH_3 O O O CH_3 OCH_3

Khellin

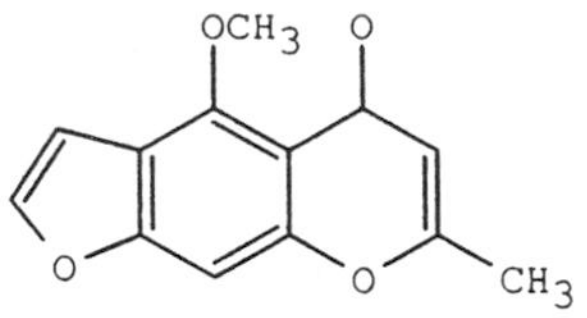

OCH_3 O O O CH_3

Visnagin

O O O H_3C H_3C CH_3 $OOCH_3$ $OCCCCH_3$ O

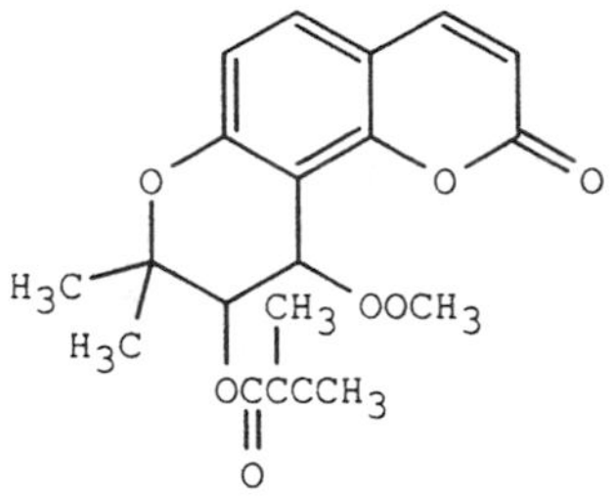

Visnadin

Cicuta L.

Taxonomy and Morphology—The genus *Cicuta*, traditionally known as water hemlock, is now considered to comprise 4 species in North America (see plate 7.1):

C. bulbifera L.	bulbous water hemlock
C. douglasii (DC.) J.M.Coult. & Rose	Douglas water hemlock
C. maculata L.	water hemlock, beaver poison, snakeweed, spotted cowbane, children's bane, death-of-man, musquash root, poison parsley
C. virosa L.	Mackenzie's water hemlock

■ *C. bulbifera*
C. douglasii
C. maculata
C. virosa

beaver poison
children's bane
cowbane
death-of-man
musquash root
poison hemlock
snakeweed
water hemlock

Other species names—e.g., *C. bolanderi* S.Watson (salt-marsh hemlock)—are encountered in the older literature; these taxa are now recognized to be only varieties of the morphologically variable, widespread *C. maculata*.

Plants perennials; from thickened tuberous bases. **Roots** clustered; fleshy-tuberous. **Stems** erect; stout; smooth; hollow; sometimes purplish streaked; to 2.5 m tall; primary rootstocks and lower portion of stems chambered, the chambers containing an oily yellowish liquid that turns reddish brown upon exposure to air. **Leaves** 1- or 2- or 3-pinnately compound; leaflets lanceolate, margins serrate, veins extending to sinuses between teeth. **Umbels** compound; large; open; bracts of involucres absent or few, narrow; bracteoles of involucels few or several, narrow. **Sepals** well developed;

■ erect perennials, stout hollow stems; lower stem and roots chambered with oily yellowish liquid; leaves pinnately compound, veins extending to sinuses or notches between teeth; umbels; petals white

■ in or near water

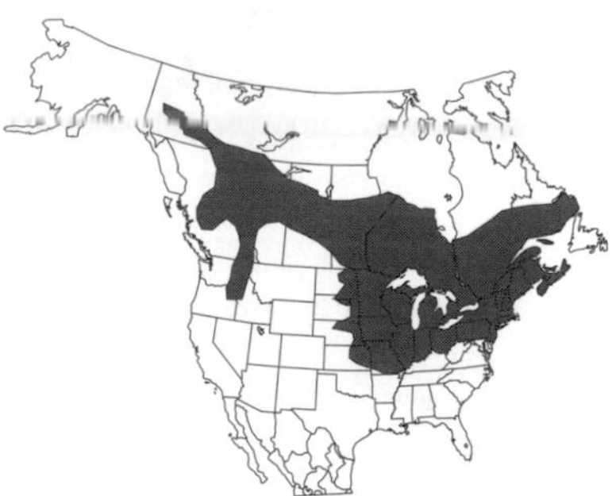

Cicuta bulbifera

Cicuta douglasii

Cicuta maculata

Cicuta virosa

triangular. **Petals** white. **Schizocarps** oval to orbicular; pale brown; with conspicuous ribs.

Cicuta maculata, or common water hemlock, is the most commonly encountered species. It is readily recognized by its typically solitary occurrence along waterways, tall stature, and large hemispheric inflorescence. Plants have a strong parsniplike odor, and the extensions of the veins to the sinuses between the teeth of the leaflets are different from those of most other genera of the family, such as *Sium,* whose veins extend to the apices of the teeth. The other species differ in minor features and are difficult to distinguish.

Distribution and Habitat—Of the 4 species, *C. maculata* has the greatest geographical distribution, occurring across the continent. *Cicuta bulbifera* also extends across the continent. *Cicuta douglasii* is a species of the northwestern corner of the continent, whereas *C. virosa* is found in the far north. All species are typically restricted to wet habitats along streams, near ponds, or in marshy areas, and plants sometimes may be found growing in water. Unlike *Conium, Cicuta* seldom grows in dense stands, more often occurring as scattered, solitary plants.

Cicuta maculata

Disease Problems—Species of *Cicuta* are well recognized to be extremely potent toxic plants that quickly act on the central nervous system. There are numerous recorded cases of intoxications in children and adults who have mistakenly eaten plants of the genus, thinking them to be wild carrots or wild parsnips (Starreveld and Hope 1975). Intoxications in humans are typified by vomiting and severe intermittent seizures. Nausea, sweating, and copious salivation also are common (Starreveld and Hope 1975; Mitchell and Routledge 1978; Knutsen and Paszkowski 1984). Death, which occurs within a few hours after eating the roots, occurs in the midst of violent muscular activity. It may take only a few bites to produce severe signs. A classic example is the case of six college students and two guides who were white-water rafting the Owyhee River in eastern Oregon (Landers et al. 1985). Six of the men tasted roots of what was called wild parsnip but in reality was *C. douglasii.* Four experienced only discomfort; however, one individual who ate two roots and another who ate one root developed grand mal–type seizures within 1 hour. The man who had eaten two roots experienced six severe intermittent seizures and died of cardiopulmonary arrest 2.5 hours later. The other individual, who had eaten one root, had five seizures of 2 minutes duration each over a period of 3 hours. He subsequently recovered without any lasting effects. A second illustration of the potency of only a few bites of the roots is the tale of two brothers who, while searching for wild ginseng, mistakenly ate the roots of *C. maculata.* One brother ate three bites, and the second brother ate one bite (Sweeney et al. 1994). Within 30 minutes, the first brother vomited and had a seizure and within another 15 minutes was unresponsive and cyanotic. Despite resuscitation efforts, he died about 3 hours after ingestion of the three bites of root. The main signs were profuse salivation and severe tonic-clonic seizures with apnea. The second brother developed seizures and delirium 2 hours after his single bite of the root but subsequently recovered with conservative treatment. It is of interest that humans who recover from *Cicuta* ingestion may have no memory of the intoxication events, despite the violence of the seizures and muscular activity (Gompertz 1926; Starreveld and Hope 1975; Landers et al. 1985).

- neurotoxic, humans and animals, vomiting, seizures, minutes after eating

The toxicity of the rootstocks is somewhat variable but probably can be considered lethal at dosages of 0.5% b.w. or less (Kingsbury 1964; Panter et al. 1988a). Dosage of 2 g/kg b.w. of the tubers to sheep resulted in death in 1–2 hours, whereas 1–1.5 g/kg b.w. caused subacute intoxication, but the sheep recovered (Panter et al. 1996). Consumption of only a few rootstocks by a cow may quickly produce clinical signs (Hedrick, 1897; Fleming et al. 1920; Pammel 1917, 1921). There are occasional episodes of intoxication in livestock eating the roots while feeding in particularly wet areas. Typically, few animals will be involved, although on occasion, circumstances will prevail that result in larger losses, as is typified by a report of illness in 17 bull calves in a group of 62; 10 of the 17 eventually died (Volker et al. 1983). In most instances, animals are found dead, sometimes with indications of a violent terminal struggle.

- highly lethal

Disease Genesis—There are two genera, *Cicuta* in North America and *Oenanthe* in Europe and several species, that contain similar toxins and produce nearly identical effects. An indication of the respect given these plants lies with the common names *dead-men's-fingers* and *five-finger death,* which refer to the appearance of the roots of *O. crocata.* The poisonous principals are acetylenic alcohols present in the oily yellowish liquid that accumulates in high concentrations in the rootstock chambers and lower stem (Anet et al. 1953). *Cicuta* produces cicutoxin and cicutol, and the similarly toxic *Oenanthe* yields oenanthotoxin, oenanthetol, and oenanthetone (Hill et al. 1955; Konoshima and Lee 1986). All are similar 17-C complex linear structures

- oily liquid in roots contains acetylenic alcohols

$$HOCCC(C\equiv C)_2(C=C)_3\overset{OH}{\overset{|}{C}}CCCH_3$$

Cicutoxin

$$HOCCC(C\equiv C)_2(C=C)_3CCCCH_3$$

Cicutol

$$HOCC=C(C\equiv C)_2(C=C)_2CC\overset{|}{C}CCCCH_3$$

Oenanthotoxin

containing two trienes and three dienes in differing orders. Cicutoxin and oenanthotoxin are diols, oenanthetone is a ketone, and the others are alcohols. *Cicuta virosa* has been shown to contain at least 11 acetylenic compounds, some novel, but no keto types (Wittstock et al. 1995).

- mechanism of action? possibly ion channel block or GABA inhibition

The specific mechanism of action of these acetylenic toxicants is uncertain. They are known, however, to inhibit electron transport in plants (Roshchina et al. 1980). Cicutoxin also is cytotoxic, with potent antileukemic activity (Konoshima and Lee 1986). Possibly of greater significance in animals are the toxin effects on Na^+ and K^+ channels, depending on the specific neuronal preparation. In early studies using oenanthotoxin, there was reversible blockade of Na^+ and/or K^+ currents mainly via action on closed channels (Dubois and Schneider 1982; Louvel et al. 1982). The outward current was most strongly affected. With the patch clamp technique and proliferation assays, cicutoxin produces reversible K^+ channel block in lymphocytes (Strauss et al. 1996). In contrast to other polyacetylenes of the family, cicutoxin causes an increase up to 6-fold in the duration of repolarization of giant cells of snail parietal ganglion (Wittstock et al. 1997). Various workers have suggested that prolongation of the action potential at excitatory synapses would promote increased excitatory activity (Louvel et al. 1982; Wittstock et al. 1997). The effects of cicutoxin also are noted to be very similar to those produced by picrotoxin, an inhibitor of GABA (Mitchell and Routledge 1978). This suggests that an alternative pathway may be present. The neurologic signs may be due to depression of the inhibitory modulation of activity in the central nervous system. This possibility is given further credibility by the similarity of some of the signs to those of strychnine, an antagonist of glycine spinal inhibitory pathways, and the effectiveness of diazepam, a GABA enhancer, in controlling the neurologic effects (Gompertz 1926; Nelson et al. 1978).

- neurologic effects, severe but generally of only a few hours' duration

Cicutoxin not only is a convulsant but also seems to produce cardiopulmonary effects. Early observations suggested effects on the brain stem, with direct cardiac actions (Grundy and Howarth 1956). This is consistent with clinical observations on the disparity in duration of the convulsant effects and those involving blood flow and pressure, such as the flushing of the skin, which lasts 1–2 days (Robson 1965; Withers et al. 1969). The duration of neurologic effects is usually limited to 3–4 hours because the toxins are rapidly eliminated or neutralized.

- toxicity variable by season and geographical location

Toxin concentrations vary considerably with season and geographical location (Fleming et al. 1920; Anet 1953). For the European species *C. virosa,* there appear to be quantitative but not qualitative variations in acetylenic profiles among plants from different locations and seasons (Wittstock et al. 1995). It has been our observation that the amount of oily yellowish liquid exuding from the root when cut is an indication of toxin concentration and toxicity of the plants. Plants with little liquid have been of low toxicity. Plants are said to be most toxic when dormant and least toxic when actively growing (Pammel 1911). The first new leaves and stems in the early spring are particularly dangerous, because they may be avidly eaten (Fleming et al. 1920). However, other studies indicate that cicutoxin concentrations in the plant may be high throughout the growing season (Payonk and Segelman 1980). The tuberous roots are apparently not too distasteful—cattle will sometimes eat them readily (Pammel 1917). The toxin levels, which are lowest in the upper vegetative parts, decline as the leaves mature and dry. Because of this, there is generally little problem if some mature plants are cut and included in hay. However, the concentration of toxin in the roots is sufficient to pose a serious hazard throughout the year. This presents a particular problem if plants are pulled up and allowed to remain lying on the ground where livestock can eat the roots or when the ground becomes wet enough to allow the entire plants to be easily pulled up and the roots eaten.

Clinical Signs—The signs of *Cicuta* intoxication are very consistent among all animal species, including humans. The predominant sign is increasingly violent muscular activity. The toxin is a potent, rapidly fatal convulsant that seems to act both on the central nervous system and on skeletal muscle. Three stages of intoxication may be recognized (Jacobson 1915; Fleming et al. 1920). During the first hour after consumption, there is uneasiness and twitching of the muscles of the lips, nose, face, and ears. Thereafter, seizures become increasingly evident, with champing of the jaws, grinding of the teeth, throwing of the head and neck grotesquely backwards, and sometimes frenzied activity. Peculiar, spasmodic diaphragmatic contractions accompanying the incipient seizures also have been noted (Marsh and Clawson 1914). Subsequently, the seizures become prolonged, so that the animal falls with kicking-running, tonic-clonic seizures. There may be accompanying bellowing, frothy bloody salivation, and bloat, and the tongue may be severely lacerated by the jaw's champing. Pigs may vomit and in particular may be prone to run about haphazardly (Hansen 1928).

- all animals, violent muscular activity; sequence of signs:
  (1) uneasiness, muscle twitching
  (2) jaw champing, frenzied activity
  (3) bellowing, violent seizures

In humans, some of the more distinctive signs are nausea, sweating, excess salivation, chewing movements, slightly bulging eyes, a clamped jaw, grand mal tonic-clonic seizures, and bizarre twitching of the muscles of the face, neck, and shoulders (Jacobson 1915; Fleming et al. 1920; Gompertz 1926; Starreveld and Hope 1976; Mitchell and Routledge 1978; Knutsen and Paszkowski 1984). The signs may begin within 30 minutes after ingestion.

- humans, nausea, sweating, grand mal seizures with bizarre twitching of shoulder muscles

Death may occur within 1–8 hours of ingestion and generally is probably due to cardiopulmonary arrest. Survival beyond 8 hours is a good indicator for eventual recovery if no complications from the convulsions ensue. In those who survive, there may be significant increases in serum LDH, AST, and CK enzymes 1–2 days after ingestion. Complete recovery may require several days. In some instances, animals may eat only a small amount of plant material, sufficient to cause muscular twitching but not the violent convulsions. Many of the signs have been likened to those of acute grass tetany in cattle (Dijkstra and Falkema 1981).

- death due to cardiopulmonary arrest; prognosis improved with survival more than a few hours

Pathology—There are no consistent gross lesions with this rapidly fatal disease. Bruises and lacerations may be evident on the skin, in the muscles, and on the tongue and lips. Scattered small hemorrhages may be apparent in various tissues, especially in the heart. Microscopically, there may be granular myocardial fiber cytoplasmic degeneration even in those animals dying within a few hours after ingestion of the roots (Panter et al. 1996). Diagnosis may be aided by careful examination of the stomach (rumen, reticulum, and omasum in the case of ruminants) for root fragments of *Cicuta*. In addition, extracts of the encountered fragments and/or the entire contents of the stomachs may be analyzed for cicutoxin and cicutol using gas chromatography/mass spectroscopy (Smith and Lewis 1987).

- gross pathology, bruises on skin from muscular activity; plant fragments in stomach; analysis of extracts of stomach contents

Treatment—There is no specific therapy, but sedation to control the convulsions is indicated. Barbiturates have been most effective, and diazepam also may be of value. Thiopentone infusion i.v. has provided effective control of the muscular activity in a human intoxication (Starreveld and Hope 1975). Experimentally, when administered early, treatment with 30 mg/kg b.w. of pentobarbital and 75 mg of atropine prevented the development of signs, lesions, and death in sheep given an otherwise fatal dose of the rootstocks (Panter et al. 1996). Activated charcoal may be helpful, but inserting a stomach tube for administration of the slurry can precipitate or worsen the convulsions. Hemodialysis and hemoperfusion, in combination with diuresis, a muscle relaxant, and mechanical ventilation, have also been used with success in an

- sedation, barbiturates, or diazepam to control seizures; atropine to control the cardiac effects

intoxication in a man (Knutsen and Paszkowski 1984). In some instances, rumenotomy and flushing with water to remove the root fragments may be appropriate (Wilcox 1899).

Conium L.

- *C. maculatum*

  carrot fern
  fool's parsley
  poison hemlock
  poison parsley
  spotted hemlock
  St. Bennett's weed
  stinkweed
  winter fern

- erect annuals or biennials, stout hollow stems, large white taproot; leaves 2-to-4-pinnately compound, finely dissected leaflets; compound umbels; petals white

Conium maculatum

- neurotoxic, early mild stimulation, later depression

Taxonomy and Morphology—An Old World genus comprising 2 or possibly 4 species, *Conium* is represented in North America by a single species:

C. maculatum L. poison hemlock, spotted hemlock, stinkweed, St. Bennett's weed, poison parsley, fool's parsley, carrot fern, winter fern

Plants annuals or biennials; from large, white taproots. **Stems** erect; stout; hollow; finely longitudinally ridged; to 3 m tall; often profusely splotched with purple and, when crushed, have a strong disagreeable odor sometimes described as "mousy." **Leaves** large; broadly triangular; 2-to-4-pinnately compound; leaflets ovate to oblong; pinnately incised. **Umbels** compound; large; open; bracts of involucres few, ovate to lanceolate; bracteoles of involucels few, resembling bracts but smaller. **Sepals** absent. **Petals** white. **Schizocarps** oval; pale brown; with conspicuous wavy longitudinal ribs.

The leaves of poison hemlock resemble those of carrot, parsley, and dill, and identification of immature plants is thus sometimes difficult. The species flowers from late April through June.

Distribution and Habitat—*Conium maculatum* is widely distributed throughout North America and absent only in the southwestern deserts. It is particularly abundant in the Great Plains and less common northward. Although plants may be found along stream banks, they are not restricted to such habitats and often occur in roadside borrow ditches and other disturbed areas with sufficient soil moisture. Often, large dense stands are established.

Disease Problems—*Conium maculatum* produces two types of intoxications: depression of the central nervous system in a variety of animals and teratogenicity primarily in cattle and pigs. The neurotoxic effects associated with this species are of classical interest because it is believed to be the plant from which a fresh juice decoction was prepared as a Grecian state poison. It is reputed that old men often gave one another these decoctions at "farewell" banquets, and it seems probable that Socrates drank such a potion (Millspaugh 1974). The decoctions were apparently employed to provide a regal exit for ancient Greek philosophers such as Theramenes and Phocion (Baskin 1967). Several biblical incidents have been interpreted by some scholars as possibly related to *Conium*. The toxicity of European migratory quail, known since ancient times and referred to in the Bible (Numbers 11:32–34), has been ascribed to consumption of *C. maculatum* seeds by the birds. This has now been largely discounted because the seeds are known to be quite toxic to quail (Kennedy and Grivetti 1980; Frank and Reed 1990). Davies and Davies (1994) have suggested that *Conium* may have been used to contaminate the incense involved in the deaths of Aaron's two oldest sons while they were offering "unholy fire" (Leviticus 10:1–2). Intoxications in humans are uncommon, although there are occasional occurrences. Rizzi et al. (1991) describe 17 cases, including 4 deaths, that occurred over a number of years. The most common finding was a curare-like effect resulting in reduction of muscle tone and leading to flaccid paralysis, including paralysis of the respiratory muscles. Fatalities also were associated with muscle degeneration and renal tubular necrosis secondary to myoglobin deposition. In other cases, the victims, sometimes children, became sleepy after ingestion and could not be aroused after eating the green plant tops (Drummer et al. 1995; Frank et al. 1995). Cases are reported from Italy in which people became intoxicated apparently after eating migratory birds on their return in late spring. Again the signs were mainly of a narcotic type, with muscle paralysis, muscle degeneration, and myoglobin-induced renal tubular necrosis (Scatizzi et al. 1993). In most cases there also is some vomiting, excess salivation, and other signs related to the digestive tract.

Conium maculatum

Acute neurotoxic disease in livestock rarely occurs and then usually only when animals are forced to consume large amounts of plant material. Such unusual circumstances have on occasion resulted in the loss of a large number of animals—e.g., several dozen deaths among cows given *Conium* in green-chop alfalfa, cows given heavily contaminated hay, and elk forced to subsist on *Conium* in late winter (Kubic et al. 1980; Jessup et al. 1986; Galey et al. 1992). Fresh young leaves are apparently not distasteful to pigs and perhaps sheep but are usually highly unpalatable when fully mature. There also are indications of an acquired taste for the plant (Gress 1935; Copithorne 1937; Jessup et al. 1986). Cases involving swine appear to be most frequent during the winter when plant growth is just beginning, with the root and young foliage being readily eaten. Because the plant forms dense stands along the edges of cultivated fields, intoxication problems may occur when *Conium* contaminates freshly cut (green-chop) alfalfa or sudan grass used for feeding penned livestock. Ingestion of fresh green plant material may quickly produce intoxication, with signs developing within an hour and lasting for several hours. Experimental results indicate that cows are most susceptible, and pigs, ponies, and sheep less so (Keeler et al. 1980; Panter et al. 1988a). Lethal doses are in the range of 0.2–0.5% of b.w. for cattle and 0.6–0.8% b.w. for pigs and sheep (Keeler and Balls 1978; Panter et al. 1983; Tokarnia et al. 1985; Panter et al. 1988a). Seeds are more toxic. Generally, dried plants are much reduced in toxicity (Keeler and Balls 1978; Tokarnia et al. 1985), but caution is warranted when *Conium* is found in particularly high proportions in hay (Galey et al. 1992).

- generally unpalatable, immature leaves are most likely to be eaten; green chopped feed poses a risk, hay less so

Field intoxications have been described for a broad range of species, including cattle, elk, goats, horses, pigs, poultry, and rabbits. Although *Conium* is of limited toxicity, the plants were present and used as fish poisons by members of western American Indian tribes by the time of settlement of the West (Ebeling 1986). Although acute disease has been the primary concern, an equally serious problem is teratogenicity, the effects involving bone deformation in piglets and calves. These appear to be the only animal species at substantial risk for teratogenic intoxication. Even laboratory animals such as rats and rabbits appear to be little affected, based upon limited experimental studies (Forsyth and Frank 1993).

- teratogenic, bone deformities in pigs and calves

Disease Genesis—*Conium* contains a series of volatile pyridine alkaloids of which coniine, *N*-methylconiine, and γ-coniceine are predominant (Cromwell 1956). Others of lesser importance include conhydrine, pseudoconhydrine, conhydrinone, *N*-methylpseudoconhydrine and 2-methylpiperidine (Panter et al. 1988c). The relative proportions of these alkaloids change as the plant grows, with a general increase in total concentrations and toxicity as it matures. The alkaloids are synthesized in the vegetative portions of the plant and translocated to be stored in the developing fruits (Fairbairn 1971). γ-Coniceine, which is considerably more toxic than coniine, appears to be the precursor alkaloid, being readily and rapidly reduced to coniine. Caution is needed when evaluating toxicity based simply on alkaloid levels, because their total concentrations as well as their relative proportions may change markedly with growth,

- toxicants, pyridine alkaloids

N CCCH$_3$

Coniine

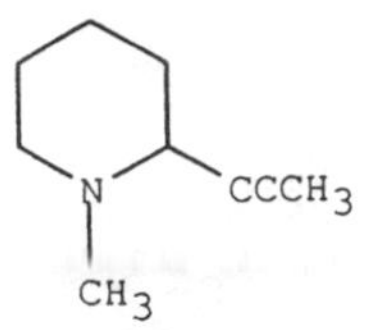

N-Methylconiine

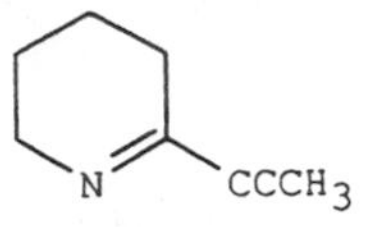

γ-Coniceine

season, moisture, temperature, and even time of day (Cromwell 1956; Fairbairn and Suwal 1961). γ-Coniceine may be markedly increased under wet conditions (Fairbairn 1971). Perhaps of even greater concern are the striking differences in toxicity, perhaps due to changes in alkaloid ratios, among varieties from different geographical locations (Panter et al. 1985b). This variation may account for the reputed edibility of the roots in some of the colder regions of the world (Millspaugh 1974).

- mechanism not entirely clear, appears to block nicotinic cholinergic receptors with mild stimulation due to initial depolarization

The mechanism of action of the alkaloids has not been clearly elucidated, but based on preliminary studies in laboratory animals, chemical structural similarities, and the clinical signs observed, the toxins appear to act in a manner similar to nicotine. Based on studies in mice, γ-coniceine is the most toxic of the three major alkaloids, with coniine one-eighth and *N*-methylconiine one-sixteenth as toxic (Bowman and Sanghvi 1963). In small doses, the alkaloids block spinal reflexes in a manner antagonistic to strychnine, without stimulant actions. They also stimulate and block autonomic ganglia, the predominant action varying with the specific alkaloid; γ-coniceine is mainly stimulatory, coniine is mixed, and *N*-methylconiine is primarily a blocker. At larger doses, nondepolarizing neuromuscular blockade is seen. Autonomic ganglia stimulation leads to some early signs of vomiting, increased salivation, and perhaps diarrhea and later depression of gastrointestinal motility and pupillary dilation. The narcotizing effects and neuromuscular blockade are preceded by only mild, if any, early signs of stimulation except for the muscular weakness. Death occurs due to respiratory failure.

- teratogenic effects occur with ingestion of plants during the first trimester of pregnancy

Coniine and γ-coniceine also are teratogenic, with their effects most likely to be manifested in cattle and pigs eating the plant for an extended period during the first trimester of pregnancy. The susceptible period for pigs ranges from days 30–60, with days 43–53 the period for severe skeletal malformations and the earlier period for cleft palate (Panter et al. 1985a,b). The susceptible period for cattle ranges from 50 to 75 days of gestation (Keeler and Balls 1978). Sheep may be affected, but the deformities are likely to be minor and probably outgrown by several months of age (Panter et al. 1988a). Structural requirements for the piperidine alkaloids to cause teratogenesis have been hypothesized, and they include a propyl or longer side chain—α to the nitrogen-containing ring—on a saturated or partially unsaturated ring (Keeler and Balls 1978; Keeler and Crowe 1985). In addition to or rather than actual malformation, fetal deformities may be due to the depressant effects of the alkaloids on the fetus during or after tissue formation as demonstrated by the strong correlation between impairment of movement of the fetus, and increased severity and permanence of the deformities (Panter et al. 1988b, 1990a; Bunch et al. 1992). The unusually long period of fetal susceptibility also supports this hypothesis.

- livestock, signs occur shortly after ingestion; nervousness, tremors, weakness, depression, paresis, recumbency, respiratory depression; low dosage may cause milk taint

Clinical Signs—The early signs that may occur within an hour after ingestion of the plant are, sequentially, slight nervousness and tremors, muscular weakness, incoordination, excess salivation and lacrimation, groaning, increased frequency of urination, and colic. If sufficient plant material has been eaten, the early stimulation phase is followed by severe depression and/or narcosis, progressive paresis leading to recumbency, bradycardia, hypothermia, and respiratory depression. Interestingly, horses seem not to fight the loss of locomotor control, going down relatively quietly (Macdonald 1937). Typically, the signs will subside, and the animal recovers in 6–8 hours, but in some instances the animal may become comatose and remain so for several hours or days, before death occurs from respiratory paralysis. Animals surviving can be depressed and appear "tucked up in the flanks" for several days. Because of the variability in plant toxicity, severe signs or death are rare; tainting of milk may be a more likely threat (Penny 1953). Although seizures are not typical, they may occasionally be seen, especially in poultry (Frank and Reed 1990). Of particular interest are the bulging eyes in mice and temporary blindness in cattle and pigs due to the third eyelid's extending across the eye (Bowman and Sanghvi 1963; Panter et al. 1983; Tokarnia et al. 1985).

In humans, intoxication produces initial mild signs of excess salivation, nausea, vomiting, dry mucous membranes, abdominal pain, and diarrhea in addition to the more pronounced muscular weakness, myoglobinuria, and flaccid paralysis. There may be significant increases in serum creatine phosphokinase and lactic dehydrogenase and stiff and swollen muscles in survivors.

■ humans salivation, vomiting, colic, diarrhea, paralysis

Teratogenic effects of *Conium* have been observed in pigs, goats, and cattle; horses and sheep are quite resistant. In most instances, teratogenic effects are preceded by systemic disease in the dam, but this is not invariably true (Edmonds et al. 1972; Widmer 1984; Hannam 1985; Markham 1985). Exposure early in gestation results in deformities such as severe carpal flexure that eventually may prevent sufficient extension for weight support, forelimb rotation with joint malalignment, elbow joint rigidity, rib cage anomalies, and vertebral curvature (Panter et al. 1990b). Arthrogryposis and forelimb deformities may become progressively worse with age, eventually becoming of such severity that the animal is forced to walk on its carpal joints (Keeler and Balls 1978).

■ fetal effects, carpal flexure, joint rotation, vertebral curvature, rib cage anomalies

Pathology—At necropsy there may be reddening of the gastric and duodenal mucosa; however, these effects are not consistently produced. Nonspecific vascular congestion of the abdominal viscera and pulmonary emphysema are more consistently seen. A "mousy" odor is commonly reported upon opening of the rumen.

■ no lesions

Treatment—Acute intoxications may be alleviated through administration of stimulants and activated charcoal, which is an effective absorbant of the toxic alkaloids. Although the signs are suggestive of organophosphate depression of acetylcholine esterase, atropine alleviates only the excess salivation and lacrimation, with little overall benefit (Jessup et al. 1986; Galey et al. 1992). Public health is a concern when dealing with intoxicated animals because of the possibility of alkaloid residues in meat. Although the alkaloids are volatile and thought to be cleared rapidly from the body, significant residues have been detected in the liver and muscle of poultry dying from intoxication and for up to 7 days in survivors (Frank and Reed 1990).

■ activated charcoal; stimulants if severe

Cymopterus Raf.

■ *C. ibapensis*

spring parsley

Taxonomy and Morphology—A western North American genus of 32–455 species, *Cymopterus* has only 1 species clearly recognized to be toxic:

C. ibapensis M.E. Jones spring parsley

■ perennials; leaves 2-to-4-pinnately dissected, basal rosettes; flowers white

A synonym that may be encountered in the older literature is *C. watsonii*.

Plants perennials; nearly acaulescent; from deep stout taproots. **Stems** absent or short. **Leaves** forming a basal rosette; 2-to-4-pinnately parted or dissected; segments oblong or linear. **Umbels** compound; small; dense; bracts of involucre absent; bracteoles of involucels large. **Sepals** small or absent. **Petals** white. **Schizocarps** large; oblong-ovate; flattened; ribs conspicuously winged.

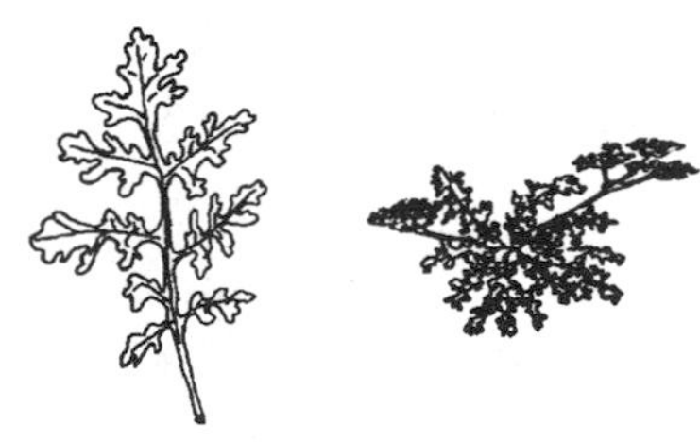

Cymopterus ibapensis

- dry brush sites
- photosensitization, mainly sheep

Cymopterus ibapensis

Distribution and Habitat—*Cymopterus ibapensis* is common in dry soils of the brush and juniper areas of southern Utah, and its distribution extends westward.

Disease Problems and Genesis—Although *C. ibapensis* is the only member of the genus clearly recognized as a cause of photosensitization, others also may be causes under appropriate conditions. As the common name implies, this is a parsleylike plant, relished by sheep and flourishing in the spring when other forage is limited. Up to 30% of a flock may be affected, but deaths are few, except for lambs that starve to death because ewes refuse to allow them to suckle (Binns et al. 1964). The disease also has been produced experimentally in chickens and is identical with that produced by *Ammi majus* (Williams and Binns 1968; Van Kampen et al. 1969). The specific toxin or toxins are unknown but presumably are furanocoumarins.

Heracleum L.

- *H. mantegazzianum*
  H. lanatum
  H. sphondylium

  hogweed
  cow parsnip

Taxonomy and Morphology—Primarily an Old World genus of some 60 species, *Heracleum* is represented in North America by 3 species:

H. mantegazzianum Sommier & Levier	giant hogweed
H. lanatum Michx.	cow parsnip
H. sphondylium L.	hogweed, common cow parsnip

Taxonomists differ in their interpretations of species boundaries in the genus and the binomials to be used for the taxa recognized. For example, some treat *H. lanatum* and *H. sphondylium* as conspecific, while others use *H. maximum* Bartr. instead of *H. lanatum*.

- erect, unbranched perennials from large taproots; leaves large, pinnately compound; flowers white
- meadows and brushlands

Plants biennials and perennials; from large taproots. **Stems** erect; unbranched; stout; deeply furrowed hollow; to 2.5 m tall. **Leaves** large; pinnately compound; leaflets 3 or 5–9, coarsely toothed or cleft. **Umbels** compound; open; large. **Sepals** minute or absent. **Petals** white. **Schizocarps** elliptic to obovate; strongly flattened; ribs broadly winged.

1. Rays of umbels 50–150 per peduncle . . . . *H. mantegazzianum*
1. Rays of umbels 15–45 per peduncle.
   2. Leaflets 3 . . . . *H. lanatum*
   2. Leaflets 5–9 . . . . *H. sphondylium*

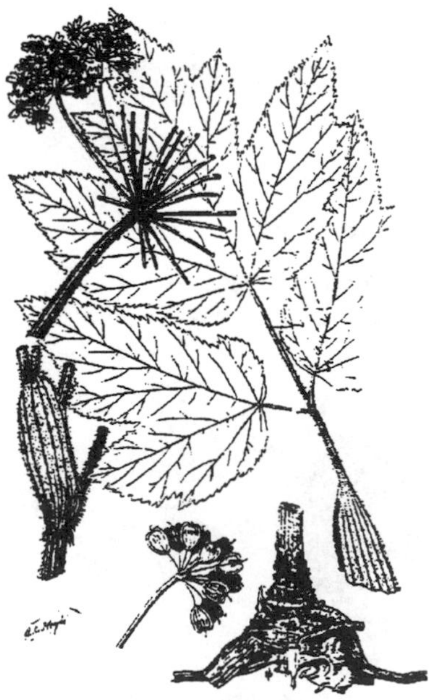

Heracleum lanatum

Distribution and Habitat—Native to North America, *H. lanatum* occurs across the continent in wet meadows, brushlands, and open woodlands. Greatest abundance is in the north. In contrast, both *H. mantegazzianum* and *H. sphondylium* are introduced from southwest Asia and Eurasia and have limited distribution in the Northeast in waste sites.

Disease Problems and Genesis—Very closely related to the garden parsnip, *H. lanatum* has long been suspected of causing photosensitization and otherwise being toxic to livestock but has never clearly been shown to be so (Irish 1889; Gress 1935; Reynard and Norton 1942). In contrast, photosensitization in people due to *H. mantegazzianum* is a well-recognized but variable problem, especially in Europe (Gunby 1980; Frohne and Pfander 1984; Andrews et al. 1985; Harwood 1985). In people handling the plant, especially when it is wet, skin reddening is seen within 24 hours after exposure and blistering thereafter at 48 hours. The stems and foliage also may cause contact irritation of the skin and oral ulcerations (Andrews et al. 1985). Many subspecies of *H. sphondylium* have been found to be positive when tested by the *Candida albicans* in vitro fungal assay (Weimarck and Nilsson 1980). The fruits, both mature and immature, are especially potent, but leaves and roots may also be hazardous. Species of *Heracleum* contain toxic furanocoumarins, mainly xanthotoxin and lesser amounts of others, including psoralen, bergapten, and imperatorin in the leaf tissue and on the surfaces (Zobel and Brown 1990a,b).

- photosensitization due to furanocoumarins

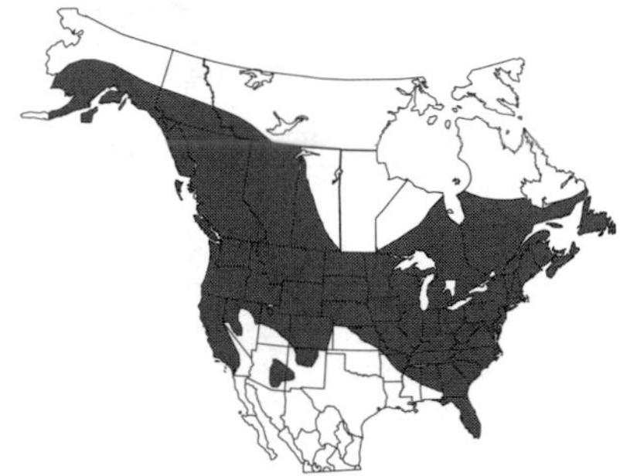

Heracleum lanatum

Heracleum mantegazzianum

O O O
OCH_3

Xanthotoxin

O O O

Psoralen

Sphenosciadium A.Gray

Taxonomy and Morphology—A monotypic genus of western North America, *Sphenosciadium* was described by Asa Gray in 1865. The single species is the following:

S. capitellatum A.Gray	white-heads, swamp white-heads, ranger's button, woolly-head parsnips

Plants perennials; from tuberous, strong-scented roots. **Stems** erect; 50–180 cm tall; branched; surfaces generally scabrous. **Leaves** 1- or 2-pinnately compound; leaflets lanceolate, margins toothed or cut or lobed; sheaths conspicuously inflated. **Umbels** compound; tomentose; umbellets headlike, spherical; bracts of involucre absent; bracteoles of involucels many, linear or bristles. **Sepals** absent. **Petals** white or purplish; obovate. **Schizocarps** obovate; flattened; ribbed; primary ribs 3, winged.

- *S. capitellatum*

  white-heads
  ranger's button
  woolly-head parsnips

- erect perennials, tuberous roots; leaves pinnately compound, margins toothed or lobed; flowers white

- moist areas

Distribution and Habitat—*Sphenosciadium capitellatum* is found along streams, in swampy areas, and in meadows from 900 to 3,000 m elevation, in the coastal

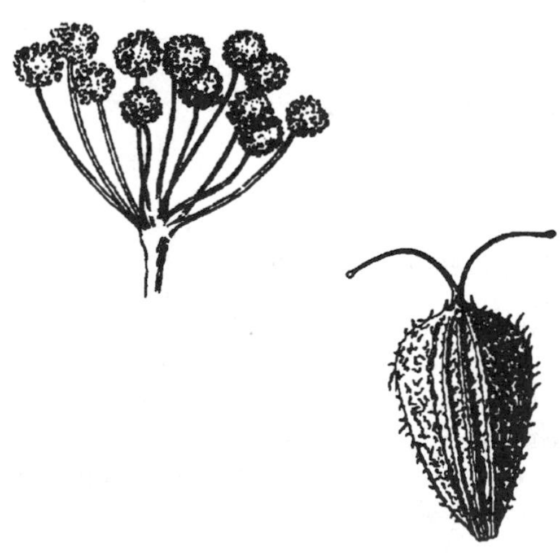

Sphenosciadium capitellatum

Sphenosciadium capitellatum

mountains of Oregon, California, and Baja California. Populations also occur in the Sierra Nevada and the Great Basin area of California, Nevada, and Idaho.

- cattle, severe pulmonary edema; horses, mild respiratory disease of slow onset; photosensitization

Disease Problems and Genesis—*Sphenosciadium capitellatum* is rarely a cause of intoxication, because it is seldom eaten. However, in some circumstances it may cause severe pulmonary edema and respiratory distress in cattle grazing mountain meadows (Fowler et al. 1970). Experimentally, the disease appears within a few hours after ingestion of 10 mg of plant material, either fresh or dried, per kilogram b.w. Death may occur within a few hours to a day or more. In contrast to cattle, horses develop only transient signs of respiratory disease 4–6 days after ingestion of the plants. Signs of primary photosensitization also are seen. Generally these signs disappear in several weeks to months without serious complications.

The specific cause of the toxic effects of *S. capitellatum* is not known, but it may be surmised that the photosensitizing effects are due to furanocoumarins similar to those present in other members of the family.

- cattle, labored open-mouth respiration with neck extended; weakness

Clinical Signs—The signs in cattle are mainly indicative of respiratory distress. The neck is extended, with open-mouth breathing, and respiration is labored. There is a slight increase in salivation, evidence of abdominal pain, coughing, expiratory grunt, cyanosis, weakness, bulging eyes, and incoordination of the pelvic limbs. In severe cases, death may occur within a few hours.

- horses, mild and transient increase in respiration rate and depth; photosensitization

In horses, respiratory signs are mild and transient, with the animals exhibiting some sweating. Several days later there is evidence of photosensitization, including erythema of the eyelids, nose, and other areas with thin or light-colored hair. Ocular discharge, corneal opacity, photophobia, and blindness also occur. The skin changes are predominantly in the less pigmented areas and are not accompanied by evidence of liver involvement.

- gross pathology, froth in airways

  microscopic, pulmonary alveolar and interstitial edema

Pathology—Grossly, there is abundant froth in the airways and scattered splotchy hemorrhages on abdominal visceral surfaces. Microscopically, there is extensive pulmonary alveolar and interstitial edema.

- generally ineffective; avoid stress

Treatment—Because the disease appears and progresses so rapidly primarily in grazing animals, it may not be recognized soon enough to allow treatment. Otherwise, therapy is directed toward relief of the respiratory distress.

Genera of Questionable Toxicity

Daucus L.

- *D. carota*
  D. pusillus

  Queen Anne's lace
  wild carrot

Comprising some 60 species around the world, *Daucus* is represented in North America by *D. carota* L. (Queen Anne's lace, or wild carrot) and *D. pusillus* Michx. (small wild carrot, or rattlesnake weed). Derived from the Greek work *daukos,* the genus name is one of the oldest, being used by Theophrastus in the third century B.C. Plants are annuals or biennials and are recognized by their stout or slender taproots, 2-pinnately compound leaflets, and large open compound umbels. A native of Eurasia, *D. carota* is now a naturalized weed in disturbed sites throughout most of North America. The cultivated carrot, with its massive orange taproot, is a variety or race. *Daucus pusillus* is native to North America and widespread in dry, open habitats, especially those somewhat disturbed.

Daucus is reputed to be mildly neurotoxic to horses and cattle in Europe, but there are few actual reports to indicate a risk in North America (Kingsbury 1964). There are, however, anecdotal reports of neurologic problems when pigs are allowed to forage on the plants. Ingestion of large amounts of the leaves may cause irritant effects on the digestive tract (Pammel 1911). There also may be some potential for reproductive problems, because aqueous extracts of seeds of *D. carota* have been shown to prevent implantation and to have abortifacient effects in rats (Garg and Garg 1970).

Similar to other members of the family, *Daucus* contains falcarinol (carotatoxin) in low concentrations: 10–20 ppm (2 mg toxin in 1 kg of carrot root). However, there is considerable variation in acetylene content among carrot cultivars (Mercier et al. 1993). Myristicin is also present in the roots but in concentrations too low to cause problems (Hallstrom and Thuvander 1997). In spite of the presence of toxins, the risk of intoxication is generally minimal because the LD_{50} in mice, even when given parenterally, is only 100 mg/kg b.w. (Crosby and Aharonson 1967). In instances of intoxication there may be neurologic signs and possibly indications of digestive tract irritation. Treatment would be based upon relief of the signs.

- slight neurotoxic effects especially in pigs, possibly due to falcarinol

Daucus carota

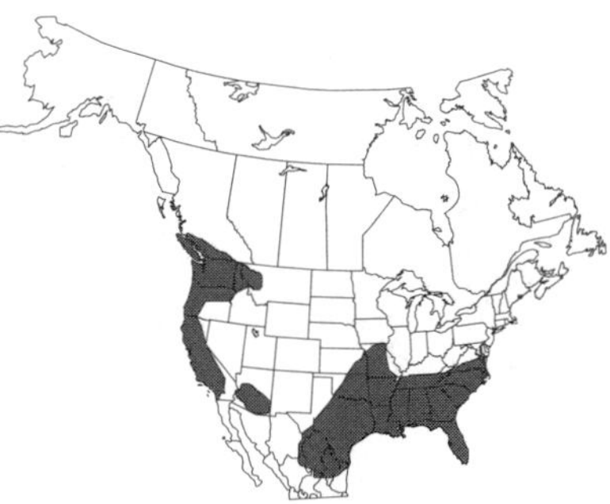

Daucus pusillus

Daucus carota

References

Abu-mustafa EA, El-bay FKA, Fayez MBE: Ammirin, a new coumarin constituent from *Ammi majus* L. fruits. *Naturwissenschaften* 62;39–40, 1975

Andrews AH, Giles CJ, Thomsett LR: Suspected poisoning of a goat by giant hogweed. *Vet Rec* 116;205–207, 1985

Anet EFLJ, Lythgoe B, Silk MH, Trippett S: Oenanthotoxin and cicutoxin. Isolation and structures. *J Chem Soc (London)* 1953;309–322, 1953

Anrep GV, Barsoum GS, Kenawy MR: The pharmacological actions of the crystalline principles of *Ammi visnaga* Linn. *J Pharm Pharmacol* 1;164–176, 1949

Avalos J, Fontan GP, Rodriguez E: Simultaneous HPLC quantification of two dermatotoxins, 5-methoxypsoralen and falcarinol, in healthy celery. *J Liq Chromatogr* 18;2069–2076, 1995

Bagouri MM: The coronary vasodilator action of the crystalline principles of *Ammi visnaga* Linn. *J Pharm Pharmacol* 1;177–180, 1949

Barr AG, Davies CS: An unusual case of poisoning in a sow and litter. *Vet Rec* 75;457, 1963

Bartik M, Piskac A: *Veterinary Toxicology*. Elsevier Scientific, Amsterdam, 1981

Baskin E: *The Poppy and Other Deadly Plants*. Delacorte Press, New York, 1967

Berenbaum M: Patterns of furanocoumarin distribution and insect herbivory in the Umbelliferae: plant chemistry and community structure. *Ecology* 62;1254–1266, 1981

Binns W, James LF, Brooksby W: *Cymopterus watsonii:* a photosensitizing plant for sheep. *Vet Med/Small Anim Clin* 59;375–379, 1964

Bohlmann F: Acetylenic compounds in the Umbelliferae. In *The Biology and Chemistry of the Umbelliferae,* Heywood VH, ed, Academic Press, London, pp279–291, 1971

Bowman WC, Sanghvi IS: Pharmacological actions of hemlock *(Conium maculatum)* alkaloids. *J Pharm Pharmacol* 15;1–25, 1963

Bunch TD, Panter KE, James LF: Ultrasound studies of the effects of certain poisonous plants on uterine function and fetal development in livestock. *J Anim Sci* 70;1639–1643, 1992

Cary CA, Miller ER, Johnstone GR: *Poisonous Plants of Alabama.* Ala Polytech Inst Ext Serv Circ 71, 1924

Clare NT: *Photosensitization in Diseases of Domestic Animals.* Commonwealth Agriculture Bureaux, Farnham Royal, Bucks, England, 1952

Cooper MR, Johnson AW: *Poisonous Plants in Britain and Their Effects on Animals and Man.* Ref book 161. Her Majesty's Stationery Office, London, 1984

Copithorne B: Suspected poisoning of goats by hemlock *(Conium maculatum). Vet Rec* 49;1018–1019, 1937
Cromwell BT: The separation, micro-estimation, and distribution of the alkaloids of hemlock (*Conium maculatum* L.). *Biochem J* 64;259–266, 1956
Crosby DG, Aharonson N: The structure of carotatoxin, a natural toxicant from carrot. *Tetrahedron* 23;465–472, 1967
Davies ML, Davies TAL: Hemlock: murder before the Lord. *Med Sci Law* 34;331–333, 1994
de Wolff FA, Thomas TV: Clinical pharmacokinetics of methoxsalen and other psoralens. *Clin Pharmacokin* 11;62–75, 1986
Dijkstra RG, Falkema R: Cicutoxin poisoning in two ponies (*Cicuta virosa,* cowbane). *Tijdschr Diergeneeskd* 106;1037–1039, 1981
Dodge AD, Knox JP: Photosensitizers from plants. *Pestic Sci* 17;579–586, 1986
Dollahite JW, Younger RL, Hoffman GO: Photosensitization in cattle and sheep caused by feeding *Ammi majus* (greater ammi; bishop's-weed). *Am J Vet Res* 39;193–197, 1978
Downum KR, Rodriguez E: Toxicological action and ecological importance of plant photosensitizers. *J Chem Ecol* 12;823–834, 1986
Drummer OH, Roberts AN, Bedford PJ, Crump KL, Phelan MH: Three deaths from hemlock poisoning. *Med J Aust* 162;592–593, 1995
Dubois JM, Schneider MF: Block of ionic and gating currents in node of Ranvier with oenanthotoxin. *Toxicon* 20;49–55, 1982
Dyson DA, Wrathall AE: Congenital deformities in pigs possibly associated with exposure to hemlock *(Conium maculatum). Vet Rec* 100;241–242, 1977
Ebeling W: *Handbook of Indian Foods and Fibers of Arid America.* Univ California Press, Berkeley, 1986
Edmonds LD, Selby LA, Case AA: Poisoning and congenital malformations associated with consumption of poison hemlock by sows. *J Am Vet Med Assoc* 160;1319–1324, 1972
Egyed MN, Shlosberg A, Eilat A, Malkinson M: Acute and chronic manifestations of *Ammi majus* induced photosensitization in ducks. *Vet Rec* 97;198–199, 1975a
Egyed MN, Singer L, Eilat A, Shlosberg A: Eye lesions in ducklings fed *Ammi majus* seeds. *Zentralbl Vet Med* A 22;764–768, 1975b
Egyed MN, Shlosberg A, Eilat A: Photosensitizing effects of furanocoumarin containing weeds on domestic birds. *Vet Hum Toxicol* 19;11–13, 1977
Fairbairn JW: The alkaloids of hemlock (*Conium maculatum* L.) (or *Conium maculatum* L: the odd man out). In *The Biology and Chemistry of the Umbelliferae,* Heywood VH ed, Academic Press, London, pp279–291, 1971
Fairbairn JW, Suwal PN: The alkaloids of hemlock (*Conium maculatum* L.). 2. Evidence for a rapid turnover of the major alkaloids. *Phytochemistry* 1;38–46, 1961
Fernald ML, Kinsey AL: *Edible Wild Plants of Eastern North America.* Harper and Row, New York, 1958
Fiske JG: *Some Poisonous Plants of New Jersey.* N J Agric Exp Sta Circ 261, 1932
Fleming CE, Peterson NF, Miller MR, Wright LH, Louck RC: *The Poison Parsnip or Water Hemlock* (Cicuta occidentalis). Nev Agric Exp Sta Bull 100, 1920
Forsyth CS, Frank AA: Evaluation of developmental toxicity of coniine to rats and rabbits. *Teratology* 48;59–64, 1993
Fowler ME, Berry LJ, Bushnell R, Hinkley HS: *Sphenosciadium capitellatum* (white-heads) toxicosis of cattle and horses. *J Am Vet Med Assoc* 157;1187–1192, 1970
Frank AA, Reed WM: *Conium maculatum* (poison hemlock) toxicosis in a flock of range turkeys. *Avian Dis* 31;386–391, 1987
Frank AA, Reed WM: Comparative toxicity of coniine, an alkaloid of *Conium maculatum* (poison hemlock), in chickens, quails, and turkeys. *Avian Dis* 34;433–437, 1990
Frank BS, Michelson WB, Panter KE, Gardner DR: Ingestion of poison hemlock *(Conium maculatum). West J Med* 163;573–574, 1995

Frohne D, Pfander HJ: *A Color Atlas of Poisonous Plants,* 2d ed. Wolfe Science, London, 1984

Fuller TC, McClintock E: *Poisonous Plants of California.* Univ California Press, Berkeley, 1986

Galey FD, Holstege DM, Fisher EG: Toxicosis in dairy cattle exposed to poison hemlock *(Conium maculatum)* in hay: isolation of *Conium* alkaloids in plants, hay, and urine. *J Vet Diagn Invest* 4;60–64, 1992

Garg SK, Garg GP: Antifertility screening of plants. Part 8. Effect of five indigenous plants on early pregnancy in albino rats. *Indian J Med Res* 59;302–306, 1970

Gompertz LM: Poisoning with water hemlock *(Cicuta maculata). J Am Med Assoc* 87;1277–1278, 1926

Gress EM: *Poisonous Plants of Pennsylvania.* Pa Dept Agric Gen Bull 531, 1935

Grundy HF, Howarth F: Pharmacological studies on hemlock water dropwort. *Br J Pharmacol* 11;225–230, 1956

Gunby P: Keep away from that "tree," folks. *J Am Med Assoc* 244;2596, 1980

Hallstrom H, Thuvander A: Toxicological evaluation of myristicin. *Nat Toxins* 5;186–192, 1997

Hannam DAR: Hemlock *(Conium maculatum)* poisoning in the pig. *Vet Rec* 116;322, 1985

Hansen AA: *Cicuta* or water hemlock poisoning. *North Am Vet* 9(6);34–39, 1928

Harwood DG: Giant hogweed and ducklings [letter]. *Vet Rec* 116;300, 1985

Hausen BM, Brohan J, Bruhn G, Faasch H, Hahn H, Konig WA: Allergic and contact dermatitis from falcarinol and dihydrofalcarinol in common ivy (*Hedera helix* L.). *Contact Dermatitis* 17;1–9, 1987

Hedrick VP: *Cicuta.* Oreg Agric Exp Sta Bull 46, 1897

Hill BE, Lythgoe B, Mirvish S, Trippett S: Oenanthotoxin and cicutoxin. Part 2. The synthesis of (+) cicutoxin and of oenanthetol. *J Chem Soc* 1955;1770–1775, 1955

Irish PH: *Plants Poisonous to Stock.* Oreg Agric Exp Sta Bull 3, 1889

Ivie GW: Biological actions and metabolic transformations of furanocoumarins. In *Light Activated Pesticides,* Heitz JR, Downum KR eds, American Chemical Society, Washington DC, pp217–230, 1987

Ivie GW, Beir RC, Bull DL, Oertli EH: Fate of [14C]xanthotoxin (8-methoxypsoralen) in a goat and in bovine ruminal fluid. *Am J Vet Res* 47;799–803, 1986

Jacobson CA: *Water Hemlock* (Cicuta). Nev Agric Exp Sta Tech Bull 81, 1915

Jessup DA, Boermans HJ, Kock ND: Toxicosis in tule elk caused by ingestion of poison hemlock. *J Am Vet Med Assoc* 189;1173–1175, 1986

Johnson AE: Photosensitizing toxins from plants and their biological effects. In *Plant and Fungal Toxins,* Keeler RF, Tu AT eds, Dekker, New York, pp345–359, 1983

Karasawa D, Shibata H, Horiuchi N, Andou Y, Sinada M: Photoactive furocoumarins in diseased celery *(Apium graveolens). Agric Biol Chem* 54;2141–2142, 1990

Karawya MS, Khayyal SE, Youssef GF: Estimation of xanthotoxin, imperitorin, and bergapten in *Ammi majus* fruits and formulations. *Planta Med* 18;195–200, 1970

Keeler RF: Coniine, a teratogenic principle from *Conium maculatum* producing congenital malformations in calves. *Clin Toxicol* 7;195–206, 1974

Keeler RF, Balls LD: Teratogenic effects in cattle of *Conium maculatum* and *Conium* alkaloids and analogs. *Clin Toxicol* 12;49–64, 1978

Keeler RF, Crowe MW: Anabasine, a teratogen from the *Nicotiana* genus. In *Plant Toxicology,* Seawright AA, Hegarty MP, James LF, Keeler RF eds, Animal Research Institute, Yeerongpilly, Queensland, Australia, pp324–333, 1985

Keeler RF, Balls LD, Shupe JL, Crowe MW: Teratogenicity and toxicity of coniine in cows, ewes, and mares. *Cornell Vet* 70;19–26, 1980

Kennedy BW, Grivetti LE: Toxic quail: a cultural-ecological investigation of coturnism. *Ecol Food Nutr* 9;15–42, 1980

Kingsbury JM: *Poisonous Plants of the United States and Canada.* Prentice-Hall, Englewood Cliffs, NJ, pp373–379, 1964

Klaber R: Phyto-photo-dermatitis. *Br J Dermatol Syph* 54;193–211, 1942

Knutsen OH, Paszkowski P: New aspects in the treatment of water hemlock poisoning. *Clin Toxicol* 22;157–166, 1984

Kolis SJ, Williams TH, Postma EJ, Sasso GJ, Confalone PN, Schwartz MA: The metabolism of 14C methoxsalen by the dog. *Drug Metab Dispos* 7;220–226, 1979

Konoshima T, Lee K-H: Antitumor agents, 85. Cicutoxin, an antileukemic principle from *Cicuta maculata,* and the cytotoxicity of the related derivatives. *J Nat Prod* 49;1117–1121, 1986

Krenzelok EP, Jacobsen TD, Aronis JM: Hemlock ingestions: the most deadly plant exposures [abstract]. *J Toxicol Clin Toxicol* 34;601, 1996

Kubic M, Rejholec J, Zachoval J: Outbreak of hemlock poisoning in cattle. *Veterinarstvi* 30;157–158, 1980; *Vet Bull* 50;6834, 1980

Landers D, Seppi K, Blauer W: Seizures and death on a white river float trip. *West J Med* 142;637–640, 1985

Lopez TA, Odriozola ER: Photosensitization risk in cattle grazing wheat-stubble infested with *Ammi majus.* Rev Med *Vet Argent* 68;98–101, 1987

Louvel J, Aldenhoff JB, Hofmeier G, Heinemann U: Effects of the convulsant drug oenanthotoxin on snail neurons and on cat cortex. In *Physiology and Pharmacology of epileptogenic phenomena,* Klee MR, Lux HD, Speckmann EJ eds, Raven Press, New York, pp47–52, 1982

Macdonald H: Hemlock poisoning in horses. *Vet Rec* 38;1211–1212, 1937

Markham K: Hemlock poisoning in piglets [letter]. *Vet Rec* 116;27, 1985

Marsh CD, Clawson AB: Cicuta, *or Water Hemlock.* US Dept Agric Bull 69, 1914

Mendez MC, Riet-Correa F, Schild AL, Ferreira JL, Pimentel MA: Photosensitization in cattle caused by *Ammi majus* (Umbelliferae), in southern Brazil. *Pesq Vet Bras* 11;17–19, 1991

Mercier J, Ponnampalam R, Berard LS, Arul J: Polyacetylene content and UV-induced 6-methoxymellein accumulation in carrot cultivars. *J Sci Food Agric* 63;313–317, 1993

Millspaugh CF: *American Medicinal Plants.* Dover, New York, pp241–270, 1974

Mitchell JC, Rook AJ: Diagnosis of contact dermatitis from plants. *Int J Dermatol* 16;257–266, 1977

Mitchell MI, Routledge PA: Hemlock water dropwort poisoning—a review. *Clin Toxicol* 12;417–426, 1978

Montgomery JF, Oliver RE, Poole WSH: A vesiculo-bullous disease in pigs resembling foot and mouth disease. 1. Field cases. *N Z Vet J* 35;21–26, 1987a

Montgomery JF, Oliver RE, Poole WSH, Julian AF: A vesiculo-bullous disease in pigs resembling foot and mouth disease. 2. Experimental reproduction of the lesion. *N Z Vet J* 35;27–30, 1987b

Mulligan GA: The genus *Cicuta* in North America. *Can J Bot* 58;1755–1767, 1980

Murray RDH, Mendez J, Brown SA: *The Natural Coumarins—Occurrence, Chemistry, and Biochemistry..* J Wiley and Sons, Chichester, pp291–311, 1982

Nath D, Sethi N, Singh RK, Jain AK: Commonly used Indian abortifacient plants with special reference to their teratologic effects in rats. *J Ethnopharmacol* 36;147–154, 1992

Nelson RB, North DS, Kaneriya M, Fletcher CV: The influence of biperiden, benztropine, physostigmine, and diazepam on the convulsive effects of *Cicuta douglasii. Proc West Pharmacol Soc* 21;137–139, 1978

Pammel LH: *A Manual of Poisonous Plants.* Torch Press, Cedar Rapids, Iowa, 1911

Pammel LH: Poisonous plants in a pasture. *Am J Vet Med* 12;235–236, 1917

Pammel LH: Western poison cowbane. *Vet Med* 16(11);33, 1921

Pangilinan NC, Ivie GW, Clement BA, Beier RC, Uwayjan M: Fate of [14C]xanthotoxin (8-methoxypsoralen) in laying hens and a lactating dairy goat. *J Chem Ecol* 18;253–270, 1992

Panter KE, Keeler RF, Buck WB, Shupe JL: Toxicity and teratogenicity of *Conium maculatum* in swine. *Toxicon* (Suppl) 3;333–336, 1983

Panter KE, Keeler RF, Buck WB: Induction of cleft palate in newborn pigs by maternal ingestion of poison hemlock *(Conium maculatum)*. Am J Vet Res 46;1368–1371, 1985a

Panter KE, Keeler RF, Buck WB: Congenital skeletal malformations induced by maternal ingestion of *Conium maculatum* (poison hemlock) in newborn pigs. *Am J Vet Res* 46;2064–2066, 1985b

Panter KE, Bunch TD, Keeler RF: Maternal and fetal toxicity of poison hemlock *(Conium maculatum)* in sheep. Am J Vet Res 49;281–283, 1988a

Panter KE, Bunch TD, Keeler RF, Sisson DV: Radio ultrasound observations of the fetotoxic effects in sheep from ingestion of *Conium maculatum* (poison-hemlock). Clin Toxicol 26;175–187, 1988b

Panter KE, Keeler RF, Baker DC: Toxicoses in livestock from the hemlocks (*Conium* and *Cicuta* spp). *J Anim Sci* 66;2407–2413, 1988c

Panter KE, Bunch TD, Keeler RF, Sisson DV, Callan RJ: Multiple congenital contractures (MCC) and cleft palate induced in goats by ingestion of piperidine alkaloid–containing plants: reduction in fetal movement as the probable cause. *Clin Toxicol* 28;69–83, 1990a

Panter KE, Keeler RF, Bunch TD, Callan RJ: Congenital skeletal malformations and cleft palate induced in goats by ingestion of *Lupinus, Conium,* and *Nicotiana* species. *Toxicon* 28;1377–1385, 1990b

Panter KE, Baker DC, Kechele PO: Water hemlock *(Cicuta douglasii)* toxicoses in sheep: pathologic description and prevention of lesions and death. *J Vet Diagn Invest* 8;474–480, 1996

Pathak MA, Daniels F Jr, Fitzpatrick TB: The presently known distribution of furanocoumarins (psoralens) in plants. *J Invest Dermatol* 39;225–239, 1962

Payonk GS, Segelman AB: Analytical toxicology and phytochemistry of the American water hemlock, *Cicuta maculata* L. (Umbelliferae) [abstract]. *Vet Hum Toxicol* 22;367, 1980

Penny RHC: Hemlock poisoning in cattle. *Vet Rec* 65;669–670, 1953

Raffauf RF: *A Handbook of Alkaloids and Alkaloid Containing Plants.* Wiley-Interscience, New York, 1970

Rauwald HW, Brehm O, Odenthal K-P: The involvement of a Ca2+ channel blocking mode of action in the pharmacology of *Ammi visnaga* fruits. *Planta Med* 60;101–105, 1994

Reynard GB, Norton JBS: *Poisonous Plants of Maryland in Relation to Livestock.* Md Agric Exp Sta Bull A10;249–312c, 1942

Rizzi D, Basile C, DiMaggio A, Sebastio A, Introna F Jr, Rizzi R, Scatizzi A, De Marco S, Smialek JE: Clinical spectrum of accidental hemlock poisoning: neurotoxic manifestations, rhabdomyolysis, and acute tubular necrosis. *Nephrol Dial Transplant* 6;939–943, 1991

Robson P: Water hemlock poisoning. *Lancet* 2;1274–1275, 1965

Roshchina VV, Solomatkin VP, Roshchina VD: Cicutoxin as an inhibitor of electron transport in photosynthesis. *Faziol Rast* (Moscow) 27;704–705, 1980; *Biol Abstr* 72;12707,1 1981

Rowe LD, Norman JO: Detection of phototoxic activity in plant specimens associated with primary photosensitization in livestock using a simple microbiological test. *J Vet Diagn Invest* 1;269–270, 1989

Scatizzi A, Di Maggio A, Rizzi D, Sebastio AM, Basile C: Acute renal failure due to tubular necrosis caused by wildfowl-mediated hemlock poisoning. *Renal Failure* 15;93–96, 1993

Scheel LD, Perone VB, Larkin RL, Kupel RE: The isolation and characterization of two phototoxic furanocoumarins (psoralens) from diseased celery. *Biochemistry* 2;1127–1131, 1963

Scott BR, Pathak MA, Mohn GR: Molecular and genetic basis of furanocoumarin reactions. *Mutat Res* 39;29–74, 1976

Shlosberg A, Egyed MN, Eilat A: The comparative photosensitizing properties of *Ammi majus* and *Ammi visnaga* in goslings. *Avian Dis* 18;544–550, 1974

Short SB, Edwards WC: Accidental *Conium maculata* poisoning in the rabbit. *Vet Hum Toxicol* 31;54–57, 1989

Smith RA, Lewis D. *Cicuta* toxicosis in cattle: a case history and simplified analytical method. *Vet Hum Toxicol* 29;240–241, 1987

Song PS, Tapley KJ jr: Photochemistry and photobiology of psoralens. *Photochem Photobiol* 29;1177–1197, 1979

Specht KG, Kittler L, Midden WR: A new biological target of furanocoumarins: photochemical formation of covalent adducts with unsaturated fatty acids. *Photochem Photobiol* 47;537–541, 1988

Starreveld E, Hope ChE: *Cicuta* poisoning (water hemlock). *Neurology* 25;730–738, 1975

Strauss U, Wittstock U, Schubert R, Teuscher E, Jung S, Mix E: Cicutoxin from *Cicuta virosa*—a new and potent potassium channel blocker in T lymphocytes. *Biochem Biophys Res Commun* 219;332–336, 1996

Swart FWJ: Poisoning of goats with fool's parsley, *Aethusa cynapium. Tijdschr Diergeneesk* 100;989–990, 1975; *Vet Bull* 46;942, 1976

Sweeney K, Gensheimer KF, Knowlton-Field J, Smith RA: Water hemlock poisoning—Maine, 1992. *J Am Med Assoc* 271;1475, 1994

Tokarnia CH, Dobrereiner J, Peixoto PV: Experimental intoxication with *Conium maculatum* (Umbelliferae) in cattle and sheep. *Pesq Vet Bras* 5;15–25, 1985

Trumble JT, Millar JG, Ott DE, Carson WC: Seasonal patterns and pesticidal effects on the phototoxic linear furanocoumarins in celery, *Apium graveolens* L. *J Agric Food Chem* 40;1501–1506, 1992

Van Kampen KR, Williams MC, Binns W: Deformities in chickens photosensitized by feeding spring parsley *(Cymopterus watsonii). Am J Vet Res* 30;1663–1665, 1969

Volker H, Schulz O, Albrecht K, Siering W: Poisoning of fattening bulls by cowbane *(Cicuta virosa). Monatsh Veterinarmed* 38;11–13, 1983

Weimarck G, Nilsson E: Phototoxicity of *Heracleum sphondylium. Planta Med* 38;97–111, 1980

Widmer WR: Poison hemlock toxicosis in swine. *Vet Med/Small Anim Clin* 79;405–408, 1984

Wilcox EW: *Poisonous Plants of Montana.* Mont Agric Exp Sta Bull 22, 1899

Williams MC: Xanthotoxin and bergapten in spring parsley *(Cymopterus watsonii). Weed Sci* 18;479–480, 1970

Williams MC, Binns W: Experimental photosensitization by spring parsley *(Cymopterus watsonii)* in chicks. *Am J Vet Res* 29;111–115, 1968

Withers LM, Cole FR, Nelson RB: Water-hemlock poisoning [letter]. *N Engl J Med* 281;566, 1969

Wittstock U, Hadacek F, Wurz G, Teuscher E, Greger H: Polyacetylenes from water hemlock, *Cicuta virosa. Planta Med* 61;439–445, 1995

Wittstock U, Lichtnow K-H, Teuscher E: Effects of cicutoxin and related polyacetylenes from *Cicuta virosa* on neuronal action potentials: a comparative study on the mechanism of the convulsive action. *Planta Med* 63;120–124, 1997

Witzel DA, Dollahite JW, Jones LP: Photosensitization in sheep fed *Ammi majus* (bishop's weed) seed. *Am J Vet Res* 39;319–320, 1978

Yajima T, Munakata K: Phloroglucinol-type furocoumarins, a group of potent naturally-occurring insect antifeedants. *Agric Biol Chem* 43;1701–1706, 1979

Yates SG, England RE: Isolation and identification of carrot *(Daucus carota)* constituents: myristicin, falcarinol, and falcarindiol. *J Agric Food Chem* 30;317–320, 1982

Yeruham I, Lemberg D, Natan A, Egyed MN: A case of photosensitization in sheep due to ingestion of feed contaminated with bishop's weed *(Ammi majus)* [abstract]. *Isr J Vet Med* 44;147, 1988

Zangerl AR: Furanocoumarin induction in wild parsnip: evidence for an induced defense against herbivores. *Ecology* 71;1926–1932, 1990

Zobel AM: Comparison of furanocoumarin concentrations of greenhouse-grown *Ruta chalepensis* with outdoor grown plants transferred to a greenhouse. *J Chem Ecol* 17;21–27, 1991

Zobel AM, Brown SA: Dermatitis-inducing furanocoumarins on leaf surfaces of eight species of rutaceous and umbelliferous plants. *J Chem Ecol* 16;693–700, 1990a

Zobel AM, Brown SA: Seasonal changes of furanocoumarin concentrations in leaves of *Heracleum lanatum*. *J Chem Ecol* 16;1623–1634, 1990b

8 Apocynaceae Juss.

- Dogbane Family

Acokanthera
Adenium
Allamanda
Angadenia
Apocynum
Catharanthus
Nerium
Pentalinon
Plumeria
Rauvolfia
Strophanthus
Thevetia
Urechites

QUESTIONABLE
Alstonia
Echites
Haplophyton
Ochrosia
Prestonia
Rhabdadenia
Tabernaemontana
Vinca

The Apocynaceae comprises 180–200 genera and 1,500–2,000 species of tall rainforest trees, shrubs, many vines, and a few temperate perennial herbs. Greatest abundance is in the tropics. The family is noted for its many showy ornamentals, including frangipani, allamanda, periwinkle, and oleander, which has been recognized as deadly toxic since antiquity. Edible species such as Natal plum are in this family, and several species are medicinally significant. The drugs reserpine and rescinnamine are produced by a genus of the family; Madagascar periwinkle produces alkaloids that are used to treat leukemia, Hodgkin's disease, and other cancers; and as described by Omino and Kokwaro (1993), many species of the family are used medicinally in Kenya. Some appear to be of value as mosquito larvacides (Evans & Kaleysa Raj 1988).

Closely related to the Asclepiadaceae, or milkweed family, most species of the Apocynaceae have a white latex sap similar to that produced by the milkweeds. Although numerous toxic species in the family have the milky sap, it is not always associated with toxicity; some latex-producing taxa are not toxic and are even purposefully used for their nutritional value (Morton 1982).

In North America, the Apocynaceae is represented by six native genera, of which *Apocynum*, or dogbane, is the most common. The introduced taxa are encountered primarily in the warmer, frost-free, southern portions of the continent, with oleander being the most common. Farther north, they are typically grown as tub plants. A few taxa, such as periwinkle, have escaped cultivation and are established in the wild.

- herbs, shrubs, vines, trees; leaves simple, opposite; petals 5, fused; many species with white latexlike sap

Plants trees or shrubs or woody vines or herbs; perennials; sap typically white. **Leaves** simple; opposite or rarely whorled; venation pinnate; margins entire; stipules absent or present, inconspicuous. **Inflorescences** solitary flowers or compound cymes; terminal or axillary; bracteoles present. **Flowers** perfect; often large, showy, and fragrant; perianths in 2-series. **Calyces** radially symmetrical. **Sepals** 5; fused. **Corollas** radially symmetrical; salverform or campanulate or funnelform; convolute. **Petals** 5; fused; of various colors. **Stamens** 5; epipetalous. **Pistils** 1; exhibiting various degrees of carpel fusion; compound; carpels 2; stigmas 1; styles 1 or 2; ovaries 1 or 2; superior or partly inferior; locules 1 or 2; placentation parietal. **Fruits** follicles, berries, capsules, or drupes. **Seeds** numerous; usually comose.

- 3 types of effects: cardiotoxic, neurotoxic, digestive disturbance

There are three general types of intoxications caused by members of the Apocynaceae: cardiotoxic, neurotoxic, and digestive tract irritant. Two, and possibly three, distinct types of toxicants are present in the family (table 8.1).

- cardiotoxic glycosides, cardenolides

Many of the most toxic species produce steroidal cardiotoxic glycosides or cardenolides (Ode et al. 1976). Interestingly, these compounds are the generally more toxic 5-β-series typical of the glycosides in *Digitalis* in the distantly related Scrophulariaceae rather than the 5-α-series of the closely related Asclepiadaceae (Shoppee 1964; Singh & Rastogi 1970). These glycosides are composed of a genin or aglycone

Table 8.1 Types of toxic compounds in the Apocynaceae and their effects

Genus	Type of Toxin	Predominant Effect
Acokanthera	cardenolide	cardiotoxic
Adenium	cardenolide	cardiotoxic
Allamanda	terpenoid	digestive tract irritation
Alstonia	alkaloid	neurotoxic, cardiovascular
Angadenia	cardenolide	digestive tract irritation, cardiotoxic
Apocynum	cardenolide	cardiotoxic
Catharanthus	alkaloid	neurotoxic
Echites	?	digestive tract irritation
Haplophyton	alkaloid	neurotoxic
Nerium	cardenolide	cardiotoxic
Ochrosia	alkaloid	neurotoxic
Pentalinon	cardenolide	digestive tract irritation, cardiotoxic
Plumeria	terpenoid	digestive tract irritation
Prestonia	?	digestive tract irritation
Rauvolfia	alkaloid	neurotoxic, cardiovascular
Rhabdadenia	?	digestive tract irritation
Strophanthus	cardenolide	cardiotoxic
Tabernaemontana	alkaloid	digestive tract irritation
Thevetia	cardenolide	cardiotoxic
Urechites	cardenolide	digestive tract irritation, cardiotoxic
Vinca	alkaloid	neurotoxic

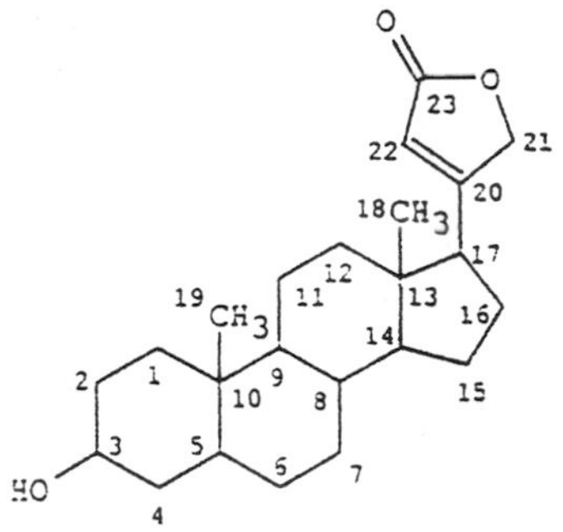

O
23
22
21
20
18CH3
17
12
11
13
16
19CH3
14
2
1
10
9
8
15
3
5
6
7
HO
4

Cardenolide

portion, in which resides the cardiac activity, and one or more sugars, responsible in large part for the potency of the toxin. Several genera may have the same aglycone but differ in the sugars present and, therefore, differ in toxicity.

■ mechanism, inhibition of Na^+,K^+-ATPase, block of cardiac conduction

The cardioactivity of the glycosides apparently is due to their ability to inhibit transport ATPase and increase myocardial contractility in the same manner as the digitalis glycosides (Kupchan et al. 1967; Singh & Rastogi 1970; Natochin & Lavrova 1972; Jortani et al. 1996). The primary effect is the result of cardenolide binding to the cell membrane, resulting in a conformational change in the Na^+,K^+-ATPase on the internal surface and subsequent alteration of Na^+ and K^+ movement and concentration balance across the cell membrane. This promotes an increase in intracellular Ca^{2+} (Hoffman & Bigger 1990). The toxic effects are primarily attributable to cardiac conduction alterations and other electrical activity changes. A-V block is one of the most common manifestations of toxicity, and death is typically due to ventricular fibrillation (Hoffman & Bigger 1990). In addition, cardenolides are cytotoxic against human nasopharyngeal carcinoma cells (Kupchan et al. 1964, 1967; Kelly et al. 1965). There appears to be good correlation between cardiac and cytotoxic potencies. Ruminants are reported to be somewhat resistant, presumably because of degradation of the glycosides by ruminal microorganisms (Craig & Kehoe 1925; Brander et al. 1991). For a more detailed discussion of the mechanisms of cardiac activity, see the discussion of *Digitalis* in chapter 68. Members of this family that produce these cardenolides have been used for a variety of purposes: arrow poisons, "trial-by-ordeal" potions, emetics, and cathartics.

■ neurotoxic alkaloids

A second group of genera in the Apocynaceae produce an extensive array of biologically active, indole iridoid-type alkaloids. Nearly 1,000 alkaloids have been isolated from species in the family (Raffauf 1996). Examples include the well-known reserpine, rescinnamine, and yohimbine. These compounds have neurologic, cardiovascular, and other systemic actions. *Rauvolfia* species native to India, containing reserpine-type alkaloids, have long been used for medicinal purposes (Sen & Bose 1931). Reserpine causes loss of adrenergic neuronal catecholamine storage vesicles, thus inducing a prolonged decrease in sympathetic activity in the central nervous system and elsewhere (Gerber & Nies 1990). The result is sedation, depression, and decreased heart rate and blood pressure. Medicinally, reserpine has been used primarily to lower blood pressure. Yohimbine acts to produce somewhat opposite effects, causing selective and competitive α_2-adrenergic receptor antagonism and thereby increasing heart rate, blood pressure, and motor activity in the brain, resulting in generalized tremors (Hoffman & Lefkowitz 1990).

■ irritant terpenoids

The third group of toxins present in the Apocynaceae are potential irritant-type compounds. They are triterpenoids, mainly pentacyclic alkyliridoid and ursane types, which are poorly absorbed and typically produce irritation of the digestive tract. In some instances they may be found in combination with cardenolides, and the problems observed may be due to the effects of both types of toxicants. Overall, the effects are seen primarily as a diarrhea.

A fourth group of plants, represented by *Echites, Prestonia,* and *Rhabdadenia* are of unknown toxin type. However, these three genera are closely akin taxonomically with *Angadenia* and *Pentalinon,* being included in the same subtribe Echitinae of the tribe Echiteae, subfamily Apocynoideae (Leeuwenberg 1994). Because of this close taxonomic kinship, it may be reasonable to assume some similarity in chemotaxonomy and the presence of cardenolides in all these species. The complex indole alkaloids typical of other members of the family are not a feature of members of this subfamily and tribe (Kisakurek et al. 1983; Raffauf 1996). Only a simple indole has been reported for a South American species of *Prestonia* (Snieckus 1968). Various compounds such as terpenes, sterol glycosides, and others have been identified from *Echites,* but their toxicological roles are not clear (Chien et al. 1979).

Other compounds identified as being present in the family include steroidal alkaloids, aminoglycosteroids, and nonpeptide kinin antagonists (Atta-ur-Rahman & Muzaffar 1988). Their significance with respect to intoxication is unknown. It is of interest that a few tropical African species of Apocynaceae produce alkaloids of the coronaridine group—e.g., ibogaine from *Tabernanthe iboga*—that have potent CNS activity (Bisset 1985, 1989a).

Acokanthera G.Don

■ *A. oblongifolia*
A. oppositifolia

bushman's poison
poison bush
poison arrow plant
wintersweet

Taxonomy and Morphology—An Old World genus of 5 species, *Acokanthera* is represented in North America by 2 cultivated species:

A. oblongifolia (Hochst.) Codd	bushman's poison, wintersweet, poison arrow plant
A. oppositifolia (Lam.) Codd	bushman's poison, poison bush

Derived from the Greek roots *akoke* and *anthera,* the name *Acokanthera* indicates that the anthers are pointed. The plants also are called poison bush and poison tree. The common names of the two species are used interchangeably.

■ evergreen shrubs or trees; leaves simple, opposite; flowers fragrant; petals white to pink; berries purple, plumlike

Plants evergreen shrubs or small trees; perennials; branchlets ribbed. **Leaves** simple; opposite; leathery; elliptic or obovate; decussate; glossy dark green. **Inflorescences** dense compound cymes; terminal or axillary. **Flowers** sweetly fragrant. **Sepals** fused only at base. **Corollas** salverform. **Petals** white to pink. **Stamens** attached at mouths of corolla tubes. **Pistils** 1; stigmas 1, hairy; styles 1; ovaries 1. **Fruits** berries; purple or purple-black; plumlike. **Seeds** 1 or 2.

1. Leaves elliptic. Corolla tubes 1.4–2 cm long . . . . *A. oblongifolia*
1. Leaves obovate. Corolla tubes 6.5–19.5 cm long . . . . *A. oppositifolia*

Distribution and Habitat—Native to tropical East Africa and South Africa, both species are cultivated to a limited extent as attractive hedges and specimen plants in the frost-free areas of North America. They may be encountered as tub plants elsewhere.

- ornamentals; frost-free areas

Disease Problems—Taxa of *Acokanthera* are probably the most infamous of the plant species that produce arrow poisons; they are widely used in eastern and southern Africa (Watt & Breyer-Brandwijk 1962). African arrow poisons most often contain cardenolides common to the genus, and the effects following their ingestion are typical of those with other cardiotoxic plants (Neuwinger 1974). With severe intoxication, there may be digestive tract problems, marked cardiac insufficiency, and possibly death due to ventricular fibrillation. All plant parts are toxic, with the leaves most potent, followed by the fruits and then the bark (Steyn 1934). The young leaves seem to be especially toxic (Joubert 1989). Problems occur when house pets or livestock eat the plants or discarded trimmings of the plants.

- digestive disturbance, diarrhea; cardiac insufficiency

Disease Genesis—The chemistry of the compounds of *Acokanthera* has been described in detail by Reichstein (1965). Plants of the genus contain several potent cardenolides of various genins, including gitoxigenin (Shoppee 1964; Singh & Rastogi 1970). Reichstein reported appreciable quantities of ouabain in some species. As demonstrated using immunoassay, estimated cardenolide concentrations are quite high, 8.7 μg digoxin eq/g of plant (Radford et al. 1994).

- steroidal cardenolides

Gitoxigenin

Ouabain

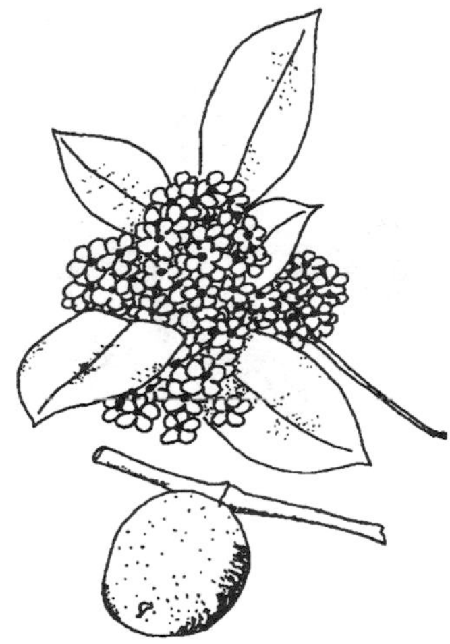

Acokanthera oblongifolia

Clinical Signs—In humans, clinical signs generally involve the digestive tract and include nausea, excess salivation, retching, vomiting, diarrhea, and extreme body weakness. Death from ingestion of this plant is unlikely. These signs are similar in pets when amounts of plant material ingested are small. In unusual situations in which livestock are exposed, the signs are similar except for the greater likelihood of severe cardiac effects. Signs include weakness, decreased appetite, excess salivation, colic, diarrhea with or without blood, cardiac arrhythmias, and cyanosis. Death is due to ventricular fibrillation or failure.

- nausea, vomiting, excess salivation, diarrhea, weakness, cardiac arrhythmias

Pathology—Distinctive changes due to intoxication by *Acokanthera* are minimal. There may be frothy material in the airways, indicative of pulmonary edema; reddening of the mucosa of the stomach and small intestine; visceral splanchnic congestion; and scattered serosal hemorrhages.

- gross pathology, reddened gut mucosa

Treatment—Symptomatic treatment as for other cardiotoxins is indicated (see *Nerium*, in this chapter) and may be augmented by administration of activated charcoal,

- activated charcoal, atropine

2 g/kg b.w. (Joubert & Schultz 1982a,b). Digoxin-specific Fab antibody fragments such as those used in treating intoxications caused by *Nerium* may be used if available (Bania et al. 1993).

Adenium Roem. & Shult.

- *A. obesum*

  desert azalea
  desert rose
  impala lily
  kudu lily
  mock azalea
  sabi star

Taxonomy and Morphology—Native to Arabia and tropical and southern Africa, *Adenium* is now believed to comprise a single, morphologically variable species with several subspecies:

A. obesum (Forssk.) Roem. & Schult.	impala lily, desert rose, mock azalea, kudu lily, sabi star, desert azalea

The binomial *A. multiflorum* Klotzsch, now considered to be a synonym, has been used in the toxicological literature.

- succulent shrubs; leaves simple, leathery; flowers showy, fragrant; petals red, pink, or white; milky sap

Plants succulent shrubs or small trees to 2 m tall; perennials; sap viscous, milky white. **Stems** fleshy; enlarged or swollen; sometimes almost entirely subterranean; tapering above; dividing irregularly into crown of thick branches. **Leaves** simple; spirally clustered at branch apices; leathery; lanceolate to obovate; fleshy. **Inflorescences** small corymbs; terminal. **Flowers** showy; 2–5 cm in diameter and 4–6 cm long; fragrant. **Sepals** 5. **Corollas** tubular. **Petals** red, sometimes with white centers, or pink or white. **Stamens** clustered together forming tassel-topped cone fused to stigma. **Pistils** 1; stigmas 1, capitate; styles 2; ovaries 2. **Fruits** follicles. **Seeds** numerous.

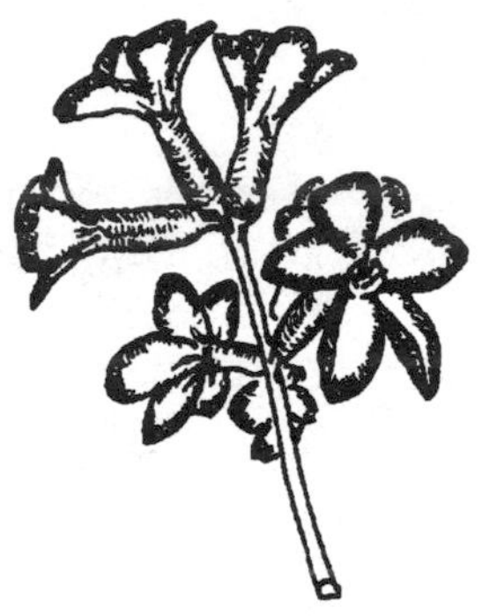

Adenium obesum

- ornamental; frost-free areas

Distribution and Habitat—*Adenium obesum* is a beautiful, novelty-type plant that is most likely to be encountered as a tub plant or houseplant. Outside, it survives only in the frost-free areas of North America.

- digestive disturbance; cardiac insufficiency; cardenolides, glycosides of digitoxigenin and oleandrigenin

Disease Problems and Genesis—Although rarely a cause of disease, *A. obesum* is capable of producing cardiotoxic effects—cardiac insufficiency and ventricular fibrillation—similar to those of species of *Acokanthera*. Some 30 glycosides produced by *A. obesum* have been identified (Yamauchi & Abe 1990). Many are found in other members of the family, and several are moderately potent derivatives of digitoxigenin and oleandrigenin (Shoppee 1964; Singh & Rastogi 1970). The most abundant is oleandrigenin-β-gentiobiosyl-β-D-thevetoside (Yamauchi & Abe 1990). This array of glycosides is the basis for the use of African plants of the species as arrow and trial-by-ordeal toxins (Bisset 1989b).

- vomiting, diarrhea, cardiac arrhythmias; treat with activated charcoal, atropine

Clinical Signs, Pathology, and Treatment—The clinical signs and pathological changes appearing as a result of *Adenium* intoxication are similar to those produced by other cardiotoxic plants of the family, such as *Nerium*. Symptomatic treatment as for other cardiotoxins is appropriate.

Allamanda L.

- *A. blanchetii*
  A. cathartica

  allamanda
  golden trumpet
  trompeta de oro

Taxonomy and Morphology—Native to tropical America, *Allamanda* comprises 12 species. It is represented in North America by 2 cultivated taxa:

A. blanchetii A.DC.	purple allamanda, violet allamanda
A. cathartica L.	yellow allamanda, golden trumpet, trompeta de oro

The binomial *A. violacea* formerly was used for purple allamanda. Several varieties and cultivars of *A. cathartica* have been described.

Plants evergreen shrubs; perennials; typically climbing. **Leaves** simple; whorled; leathery; oblong or obovate; margins entire or undulate; bases with axillary glands. **Inflorescences** few-flowered cymes; terminal. **Flowers** large and showy. **Sepals** fused; leafy. **Corollas** funnelform. **Petals** golden yellow or rose-purple. **Stamens** included within throats of corolla tubes. **Pistils** 1; stigmas 1; styles 1; ovaries 1. **Fruits** capsules; spiny. **Seeds** numerous; winged.

■ evergreen shrubs; leaves simple, whorled, leathery; flowers showy; petals golden yellow or rose-purple

1. Leaves obovate; apices obtuse or subacute; adaxial surfaces glabrous except for midribs. Flowers golden yellow . . . . . *A. cathartica*
1. Leaves oblong; apices acuminate; adaxial surfaces villous. Flowers rose-purple . . . . *A. blanchetii*

Allamanda cathartica

Distribution and Habitat—Introduced from tropical South America, *A. cathartica* is frost sensitive and grown mainly as a tub plant except in the frost-free areas of North America, where it is planted as a specimen plant. *Allamanda blanchetii* also is from South American and cultivated as an ornamental. It is often grafted to *A. cathartica.*

■ ornamental; frost-free areas

Disease Problems—*Allamanda cathartica* is widely used medicinally in the tropics (West 1957). All parts are toxic, especially the fruits. The principal effect following ingestion is severe purgation (Thorp & Watson 1953; Blohm 1962). At one time, the plant's milky latex sap was used as a purgative; however, the effects were so drastic that it is now little used (Morton 1982). The lethal dose of fresh foliage for cattle is approximately 30 g/kg b.w. (Tokarnia et al. 1996). However, the limited toxicity of *Allamanda* is shown by the difficulty in producing intoxication in sheep, in which a dosage of 15–20 mg/kg caused only slight to moderate effects of several days' duration (Armien & Tokarnia 1994). There is very little cardiotoxic effect.

■ digestive disturbance, purgation

Disease Genesis—Based on analyses of *A. cathartica* (Kupchan et al. 1974; Connolly & Hill 1991), species of *Allamanda* contain a series of alkyliridoid-type terpenoids such as allamdin, allamandin, and allamandicin. Concentrations are highest in the roots (Coppen & Cobb 1983). The purgative iridoids plumericin and plumieride also are present, the former in the roots and the latter throughout the plant. Alkaloids are present but of unknown significance (Raffauf 1996). Cardiotoxin concentrations are negligible (Radford et al. 1994).

■ terpenoids

Allamdin | Allamandin | Allamandicin

Clinical Signs—Following ingestion of plant material, there may be indications of irritation of the digestive tract with signs such as nausea, excess salivation, ruminal atony, diarrhea, and indications of pain. These effects usually are of short duration

■ nausea, excess salivation, diarrhea, colic, dehydration

and limited severity. However, as noted above, the sap produces severe purgation. Typically, cardiotoxic effects are not observed.

■ treat with fluids, electrolytes

Pathology and Treatment—Ingestion of *Allamanda* is highly unlikely to cause an intoxication problem leading to death. In those rare instances when a very large amount is eaten, there may be reddening and edema of the mucosa of the stomach and small intestine. Treatment is not likely to be necessary, but conservative use of antidiarrheal medications may be useful.

Angadenia Miers

■ *A. berterii*
pineland allamanda

Taxonomy and Morphology—Comprising 2 species, *Angadenia* is represented in North America by a single species:

A. berterii (A.DC.) Miers pineland allamanda

An alternative spelling of the specific epithet *berterii* sometimes encountered is *berteroi*. *Angadenia* is one member of a complex of closely related genera of shrubby vines, including *Echites*, *Prestonia*, and *Rhabdadenia*. Generic boundaries are tenuous, and individual species typically have been repeatedly moved from genus to genus by different taxonomists. A synonym of *A. berterii*, for example, is *Rhabdadenia corallicola* Small.

■ evergreen, perennial vines; leaves simple, opposite; flowers showy; petals yellow; follicles elongated

Plants evergreen shrubby vines; perennials. **Leaves** simple; opposite; elliptic to oblong-lanceolate; leathery; glossy above; apices acute or acuminate; bases rounded; petioles absent or short. **Inflorescences** solitary flowers or few-flowered cymes; axillary. **Flowers** large; showy. **Sepals** fused. **Corollas** funnelform. **Petals** yellow. **Stamens** attached within throats of corolla tubes. **Pistils** 1; stigmas 1; styles 1; ovaries 2. **Fruits** follicles; elongate. **Seeds** numerous; comose.

■ frost-free areas of Florida

Distribution and Habitat—Distributed in the frost-free pinelands of southern Florida and the Keys, *A. berterii* also occurs in the West Indies. Plants are sometimes cultivated in tubs.

■ irritant digestive disturbance, cardenolides

Disease Problems and Genesis—The viscid, milky sap of *Angadenia* is irritating, causes reddening and increased mucous production by the mucosa of the digestive tract when ingested, and as is characteristic of other species of the Apocynaceae, most likely causes reddening of the skin upon contact. Other intoxication problems have not been reported. Plants of *Angadenia* contain oleandrigenin cardenolides such as urechitin and the bioside urechitoxin (Chen & Henderson 1965; Pina Luis & Fanghaenel 1972).

Angadenia berterii

O, O, CH_3, CH_3, $OCOCH_3$, OH, O

Urechitin

O, O, CH_3, CH_3, $OCOCH_3$, OH, O, Oleandrin-glucose

Urechitoxin

Clinical Signs—Ingestion of *A. berterii* causes irritation of the mucosa of the digestive tract, and clinical signs include nausea, excess salivation, vomiting, and diarrhea. Any one animal, however, may not exhibit all of these signs. It is unlikely that sufficient plant material will be consumed to cause arrhythmias and/or other systemic effects.

■ nausea, excess salivation, vomiting, diarrhea

Pathology and Treatment—Because ingestion of *A. berterii* does not result in death, pathological changes have not been described. Treatment is unlikely to be necessary but could include the conservative use of antidiarrheal medications.

■ antidiarrheal drugs

Apocynum L.

Taxonomy and Morphology—Occurring in both temperate and tropical regions of both the Old World and the New World, *Apocynum* comprises some 70 species. In North America, it is represented by 2 species and 1 hybrid, albeit as many as 7 species were recognized at one time because of the complex variation pattern the genus exhibits. The North American taxa are these:

■ *A. androsaemifolium*
A. cannabinum
A. ×floribundum

American ipecac
bitterroot
bowman's root
colicroot
dogbane
honey bloom
Indian hemp
rheumatism root
wild cotton
wild ipecac

A. androsaemifolium L.	spreading dogbane, bitterroot, colicroot, flytrap dogbane, American ipecac, wild ipecac, honey bloom, rheumatism wood
A. cannabinum L.	Indian hemp, Indian dogbane, glabrous hemp, bowman's root, hemp dogbane, American hemp, wild cotton, rheumatism root, prairie dogbane, clasping-leaved dogbane
A. ×floribundum Greene	western dogbane

Now treated as synonyms of these names, the binomials *A. sibiricum, A. medium, A. pumilum, A. scopulorum, A. suksdorfii, A. jonesii,* and *A. milleri* appear in older literature.

■ perennial herbs; leaves simple, opposite; flowers small, white, pinkish, or greenish; follicles long, curved

Plants herbs; perennials; often forming colonies by root sprouts. **Leaves** simple; opposite or rarely whorled; margins entire; stipules absent. **Inflorescences** compound cymes; terminal or axillary. **Flowers** small. **Sepals** fused only at bases. **Corollas** campanulate or urceolate or tubular. **Petals** white or pinkish or greenish. **Stamens** borne at bases of petals; connivent and attached to stigmas. **Pistils** 1; stigmas 1; styles 2; ovaries 2, superior. **Fruits** follicles; elongate; curved; held together at stigma or rarely separate. **Seeds** numerous; comose.

1. Uppermost inflorescences surrounded or exceeded by foliage. Corollas urceolate . . . . *A. cannabinum*
1. Uppermost inflorescences borne above foliage. Corollas campanulate or tubular.
    2. Leaves drooping. Corolla lobes strongly spreading or recurved . . . . *A. androsaemifolium*
    2. Leaves spreading or ascending. Corolla lobes erect or slightly spreading . . . . *A. ×floribundum*

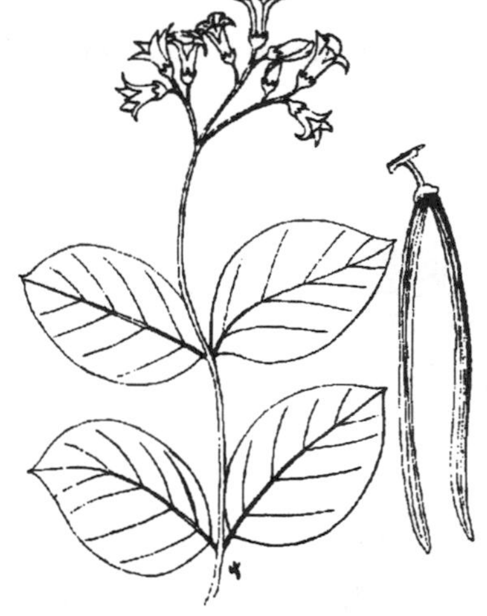

Apocynum androsaemifolium

Distribution and Habitat—With the exception of *A. ×floribundum,* which is restricted to the western half of the continent, plants of *Apocynum* are found throughout North America. *Apocynum androsaemifolium* is typically found in dry thickets and along the edges of woods, whereas *A. cannabinum* prefers moist soils in ditches, along waterways, and in open woodlands and fields. *Apocynum ×floribundum* may be

■ in dense stands

■ digestive disturbance, diarrhea, impaction

found in a variety of dry or moist sites, usually rocky. Because they spread prolifically by horizontal roots, plants of all three taxa may form dense stands.

Disease Problems—As is implied by the common name hemp, the mature stems are tough and fibrous. They were used by American Indians to make fishing nets and lines and for twine used in basket making. David Douglas (1959) recorded in his journal the use of fishing nets made of the stems of species of *Apocynum* for salmon fishing in the Pacific Northwest of the United States. There is little doubt that this genus is toxic, but the hazard is probably much exaggerated. Besides being very fibrous, the foliage is apparently quite distasteful and is generally ignored by livestock. As a result, although the species are common and often very abundant, they are seldom associated with disease. Early reports attributing high toxicity potential to *Apocynum* are false because of the misapplication of the results of initial work conducted on *Nerium oleander* (oleander) (Wilson 1909; Johnson & Archer 1922). There is little if any experimental work to verify the toxicity potential of *Apocynum* and confirm it as the likely cause of the few cases reported (Durrell & Newsom 1939). Marsh and co-workers (1928) were unable to produce adverse effects in cattle given 1% b.w. daily or sheep given up to 5.6% b.w. of the leaves of *Apocynum*. In a few instances adverse effects seem to be present. For example, a cooked concoction of Indian hemp eaten by an 84-year-old woman produced digestive disturbances and marked ECG changes, including prolongation of the PR interval (Bania et al. 1993). In another instance, two horses were reported to have died from hay containing *Apocynum* (Fuller & McClintock 1986). However, we have given 2.5 kg of wilted *A. cannabinum* via ruminal fistula to a 48-kg sheep over a 54-hour period and observed only diarrhea and a rumen distended with fibrous, stemmy plant material. There were no obvious effects on general demeanor or discernible effects on the heart. The same animal later given 75 g of partially dried oleander leaves died in less than 16 hours.

Apocynum cannabinum

- cardenolides, typically low toxicity potential

Disease Genesis—Toxicity of *Apocynum* is due to cardiotonic glycosides such as cymarin, the glycoside of apocynamarin or cynotoxin, and glycosides of strophanthidin (Finnemore 1908; Moore 1909; Couch 1937; Kupchan et al. 1964). These cardenolides are also known from other sources, e.g., cymarin from *Strophanthus* (corkscrew flower) in this family (Shoppee 1964). Several other compounds have been reported to be present in *Apocynum*, but they may simply be artifacts of the isolation procedures, i.e., breakdown products of the glycosides.

- risk increased when plants contaminate hay fed to horses

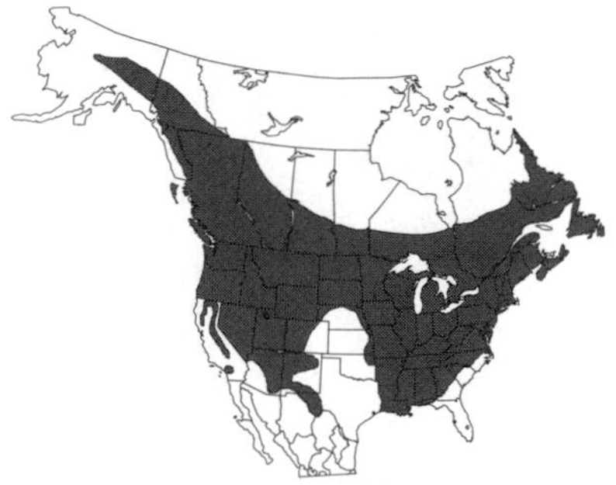

Apocynum androsaemifolium

D-Cymarose

Cymarin

Because the cardiotoxins are similar in structure to the digitalis glycosides, the toxic effects may in severe cases, include various arrhythmias as well as inhibition of cardiac conduction and function. In addition to being cardioactive, these glycosides, some of which are reported to be intensely bitter, have emetic and diuretic activity (Moore 1909). All parts of the plants are probably toxic but are likely to be a hazard only when the tender new shoots are eaten or when dried and baled with hay. Because of their low toxicity potential, an appreciable amount must be eaten, thus posing a problem only to herbivores. They are probably most hazardous for horses, inasmuch as the low toxicity potential would represent a threat only to the most susceptible species. Ruminants are reported to be somewhat resistant, presumably due to degradation of the glycosides by ruminal microorganisms (Brander et al. 1991; Craig & Kehoe 1925).

Apocynum cannabinum

Clinical Signs—The earliest and most consistent sign is diarrhea, with blood in some cases. Other less consistent but nevertheless important effects are manifestations of cardiac insufficiency and include weakness, cold extremities, bradycardia, and arrhythmias. There may be mydriasis, sweating, and a compensatory increase in heart rate in some animals. Because of the very fibrous nature of the plants, impaction and/or constipation may on occasion occur.

- diarrhea, weakness

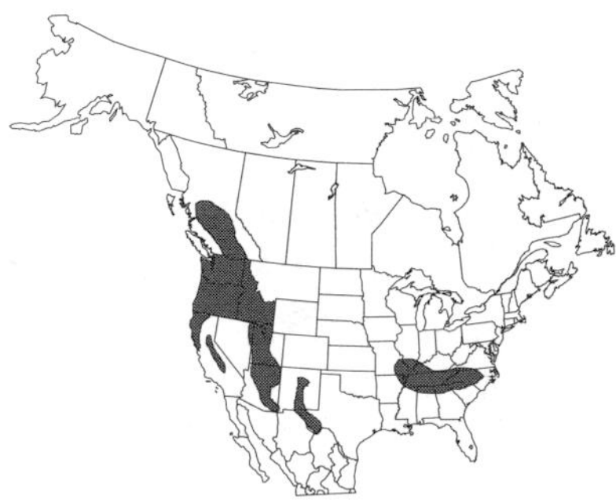

Apocynum xfloribundum

Pathology—There are unlikely to be any distinctive lesions at necropsy. There may be scattered ecchymotic or petechial hemorrhages on the surfaces of the heart and digestive tract and perhaps some pale streaks in the myocardium in a few instances. Because of the relatively large amount of plant material required for intoxication, it is likely that the fibrous stems and leaves will be identifiable in the stomach.

- no distinctive lesions; plants in stomach

Treatment—Symptomatic treatment as for other cardiotoxins is indicated and may be augmented by oral administration of activated charcoal, 2 g/kg b.w. (Joubert & Schultz 1982a,b). Digoxin-specific Fab antibody fragments such as those used in treating intoxications caused by *Nerium* may be used if available (Bania et al. 1993). Because the most likely source of plant material is as a contaminant in hay, hay should be examined before feeding and discarded if *Apocynum* is present.

- activated charcoal

Catharanthus G.Don

Taxonomy and Morphology—Native to Madagascar, *Catharanthus* comprises 8 species. In North America, it is represented by 1 cultivated taxon:

C. roseus (L.) G.Don	Madagascar periwinkle, pink periwinkle, old maid, cayenne jasmine, rose periwinkle

- *C. roseus*

  cayenne jasmine
  old maid
  pink periwinkle
  rose periwinkle

The binomials *Vinca rosea* and *Lochnera rosea* are encountered in the older literature.

Plants herbs or shrubs; perennials or annuals; erect; foliage dark green. **Leaves** simple; opposite; oblong; margins entire; apices mucronate; bases tapered; petioles short. **Inflorescences** solitary flowers or cymes of 2 or 3 flowers, borne in axils of leaves. **Flowers** large, showy. **Sepals** fused; narrow. **Corollas** salverform. **Petals** rosy purple, pink, red, or white, typically with red centers. **Stamens** borne just inside throats of corolla tubes. **Pistils** 1; stigmas 1; styles 2; ovaries 2, superior. **Fruits** follicles; short; indumented; held together at stigma or rarely separate. **Seeds** numerous.

- herbs; leaves simple, opposite; flowers large, showy; petals rosy purple, pink, red, or white, with red centers

■ ornamental, naturalized in warm areas

Distribution and Habitat—Although *C. roseus* is endemic to Madagascar, it has been carried throughout the tropics because of its horticultural use as a ground cover or bedding plant. It has naturalized in these areas. In temperate regions, it is generally cultivated as an annual from seed or cuttings. In frost-free areas of North America, it is perennial and has escaped to grow wild in a few areas of California and Florida and along the Gulf coast. It is naturalized in Sinaloa and scattered elsewhere in Mexico.

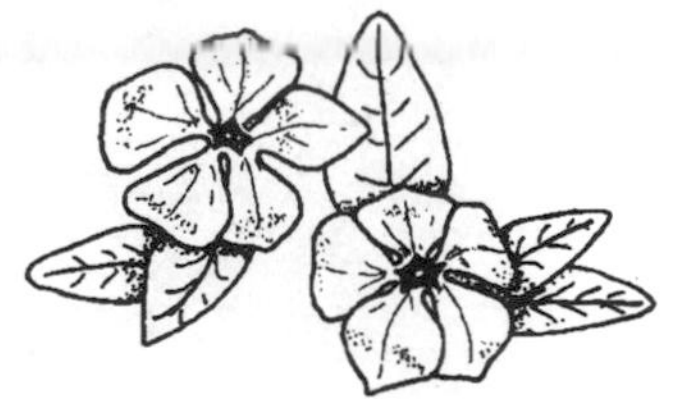

Catharanthus roseus

■ hypotensive, neurotoxic

Disease Problems—*Catharanthus rosea* has long been used in folk remedies, including as an anthelmintic, a diuretic, an emetic, an eyewash, and a laxative (Farnsworth 1961; Morton 1977). With respect to toxicity, the species appears to produce hypotensive and neurotoxic effects with continued exposure (Farnsworth 1961; Selva Raj & Ganapathy 1967; Morton 1977). Selva Raj & Ganapathy (1967) described distinctive neurologic signs for intoxication of cattle and sheep. Experimentally, ingestion of 0.1% of b.w. for several days caused similar signs in both species, including ataxia and tremors and hematologic changes.

■ reserpine-type alkaloids

Disease Genesis—*Catharanthus roseus* is noted for its array of dozens of alkaloids, including alstonine, reserpine, vinblastine, vincristine, yohimbine, and others of yohimbinoid and strychnoid bases (Taylor 1965). For many of these alkaloids, such as alstonine and reserpine, the predominant effects are blood pressure reduction, sedation, and slowed heart rate (Wakim & Chen 1947; Schlittler 1965).

Alstonine

Reserpine

Catharanthus roseus was originally studied because of its reputed efficacy as an herbal treatment for diabetes, but the benefits of its hypoglycemic effects were overshadowed by the discovery of its effects on bone marrow (Noble 1990). Some of its alkaloids are potent antineoplastic agents (Farnsworth 1961; Farnsworth et al. 1962). Particularly noteworthy are vinblastine, used for treatment of lymphomas, and vincristine, for treatment of leukemias (Noble 1990). Both bind tubulin protein, preventing microtubule assembly and resulting in metaphase arrest of mitosis (Sartorelli & Creasey 1969). In addition, these alkaloids produce other effects, including hypoglycemia and bone marrow hypoplasia (with severe leukopenia). They also are neurotoxic and teratogenic in laboratory rodents (Johnson et al. 1963; Svoboda et al. 1964; Keeler 1983; Zimmermann et al. 1991). Neurotoxicity is associated with inhibition of axoplasmic neuronal transport due to axon microtubular disruption (Gerzon 1980; Anthony & Graham 1991). This causes axonal degeneration and a dying-back axonopathy (Riopelle et al. 1984). These effects, however, are of concern only with medicinal use of these alkaloids.

■ antineoplastic, neurotoxic alkaloids

■ ataxia, tremors, seizures

■ activated charcoal; sedation

Clinical Signs, Pathology, and Treatment—Following continued ingestion of the foliage of *C. roseus,* animals exhibit a loss of appetite with abrupt onset of incoordination, lateral flexure of the neck, tremors, and seizures, followed by coma and death in 1–2 days. There also may be anemia. Few pathologic changes occur other than visceral congestion. Treatment is nonspecific and includes sedation and perhaps oral administration of activated charcoal.

Nerium L.

■ *N. oleander*

laurel blanco
laurel colorado
laurel rosa
oleander
rosa laurel

Taxonomy and Morphology—Native to the Old World, *Nerium* is now believed to comprise 1 polymorphic species with more than 400 cultivars (see plate 8.1):

N. oleander L. oleander, laurel rosa, laurel blanco, laurel colorado, rosa laurel

The binomials *N. indicum* and *N. odorum* are used in the older literature.

- evergreen shrubs; leaves simple, whorled or opposite, leathery; flowers fragrant, funnel-like, 5-lobed; petals white, pink, purple, or yellow; follicles long

Plants evergreen shrubs or small trees; perennials; 2–6 m tall; branches spreading to erect. **Leaves** simple; in whorls of 3 or 4 or rarely opposite; oblong or lanceolate; glossy dark green above, sometimes variegated; pale green below; with prominent midrib and secondary veins; apices acuminate; margins revolute; bases tapered; petioles short. **Inflorescences** dense compound cymes; terminal. **Flowers** fragrant. **Sepals** fused. **Corollas** funnelform; deeply 5-lobed; tube slender; throats dilated, with 5 cleft or ragged segments at apex. **Petals** white to pink or purple or yellow. **Stamens** attached within corolla tubes; anthers connivent and adhering to stigmas, with long apical appendages. **Pistils** 1; stigmas 1, styles 1; ovaries 2. **Fruits** follicles; elongate. **Seeds** numerous; comose.

Nerium oleander

- ornamental, naturalized in warmer areas of North America

Distribution and Habitat—Native from the Mediterranean region to western China, oleander is now widely naturalized throughout the warmer parts of the world, including North America. It is widely planted as a drought-tolerant ornamental in southern states and Mexico. It is cold-hardy to temperatures as low as 15–20°F. It is widely used in roadside plantings, for hedges, and for specimen yard plants.

Nerium oleander

- rarely eaten, digestive tract irritation, cardiotoxicity

Disease Problems—Because *N. oleander* was well known in antiquity as a medicinal and toxic species, there is little doubt concerning its deadly toxicity. It is mentioned in the writings of early Greeks and Romans such as Theophrastus, Pliny, and Galen. It produces cardiotoxic problems and, as with *Digitalis,* disturbances of vision and color perception (Langford & Boor 1996). The potency of oleander is shown by the death of two budgerigars given 250 mg of leaves in comparison with the lack of adverse effects in other birds given up to 1 g of *Digitalis* leaves (Shropshire et al. 1992). Even though oleander typically is not eaten or only reluctantly so, there are numerous reports of intoxications in a variety of animals, including humans. Species affected include dog, cat, sloth, monkey, bear, ducks, geese, goat, cattle, sheep, llama, ratites, budgerigar, mule, and horse (Wilson 1909; Szabuniewicz et al. 1972; Miller 1973; Schwartz et al. 1974; Alfonso et al. 1991; Shropshire et al. 1992; Langford & Boor 1996; Galey et al. 1998). Rats, mice, and chickens are somewhat resistant to the cardiotoxic effects of the species but display neurotoxic effects at high dosage (Szabuniewicz et al. 1972; Glendinning 1992). This disparity in effects has been reported for other cardenolides (Glendinning 1992) and may, in some way, be related to preferential effects on brain rather than heart ATPase.

- risk in humans via exposure to honey, teas, smoke, wood

In humans, intoxications have been ascribed to individuals eating honey derived from oleander blossoms, inhaling smoke from its burning branches, holding flowers or stems in the mouth, and using the plant as a medicament or as an ingredient in herbal tea (Shaw & Pearn 1979; Morton 1982; Haynes et al. 1985; Fuller & McClintock 1986; Blum & Rieders 1987). When oleander branches were used to stir soup or for roasting meat, considerable loss of life occurred in several instances (Wilson 1909). The leaves are also used for suicidal purposes but in spite of its reputation, relatively few deaths are reported (Shumaik et al. 1988; Langford & Boor 1996). The greatest risk seems to lie with children but even there, severe signs and death have been the exception rather than the rule, as shown for intoxications in children in Queensland, Australia, where oleander has escaped cultivation and become well established (Shaw & Pearn 1979). In instances of intoxication, the prognosis with prompt medical attention is favorable.

- livestock, risk from dried leaves, especially horses, greater than 50 mg/kg b.w. of leaves may be lethal; ruminants, lower risk

Typically the dried leaves, discarded in a manner allowing access to them by livestock, are the greatest hazard, because fresh, green foliage is, for the most part, quite disagreeable and rarely eaten. Galey et al. (1996) reported that of 37 cases of *Nerium* intoxication in livestock in California over a 7-year period, most were due to ingestion of clippings or dried leaves. Increased intoxication problems following unusually severe cold weather are presumably the result of ingestion of the more palatable dried leaves (Reagor 1985). In experimental studies (Wilson 1909), 20–30 g of leaves was sufficient to cause death in horses, while 18 g in a cow and 1–4 g in sheep was lethal. In sheep, 250 mg/kg b.w. was deadly (Kellerman et al. 1988). Dosage of 250–500 mg fresh leaves/kg b.w. was lethal to cattle (Tokarnia et al. 1996). A single dose of 50 mg/kg was lethal in 3 calves, as was the same dose divided into 3 potions given 24 hours apart (Oryan et al. 1996). Galey et al. (1998) reported minimum toxic doses in horses and cattle of 26 and 45 mg/kg, respectively. Steep response curves were apparent, since doses of 20 and 40 mg/kg, respectively, were without obvious adverse effects.

■ dogs at risk

Dogs are also at considerable risk because even a few dried leaves can cause problems (Reagor 1985). It is of interest that, as with digitalis glycosides, the female animal appears to be less sensitive to the toxic effects of oleander than the male (Grinnell & Smith 1957; Schwartz et al. 1974).

Disease Genesis—*Nerium oleander* contains numerous steroidal glycosidic cardenolides composed of several genins linked with various sugars or combinations of sugars (Jager et al. 1959; Kingsbury 1964; Shoppee 1964; Chen & Henderson 1965; Coffey 1970; Yamauchi et al. 1975; Joubert 1989; Paper & Franz 1989). Identified glycosides and their genins include the following:

adigoside	adigenin
adynerin	adynerigenin
deacetylnerigoside	gitoxigenin
neriantin	neriantogenin
nerigoside	oleandrigenin
neritaloside	oleandrigenin
odorosides A, D, F, G, H	digitoxigenin
odorosides B, K	uzarigenin
oleandrin (= oleandroside)	oleandrigenin

■ high levels of potent steroidal glycosidic cardenolides in all plant parts

All of these steroidal genins are 23-C except for oleandrigenin, which is acetylated gitoxigenin. Other similar glycosides such as odorobiosides and odorotriosides may be present. They contain sugars such as diginose, digitalose, glucose, oleandrose, and sarmentose.

Plants of *N. oleander* should be considered to have high cardiotoxic potential because they have high concentrations of these cardenolides—about 0.5%, as measured by a TLC-colorimetric assay or estimated as the equivalent of 3.4 µg of digoxin/g of plant as measured by immunoassay (Karawya et al. 1973; Radford et al. 1994). All plant parts have appreciable concentrations of cardenolides, especially the root, fruit, and seeds (Karawya et al. 1973). Differences in the array of the cardenolides represented in the various parts may indicate a difference in risk. The leaf and flower have been shown to contain considerable oleandrin, whereas the roots have much less. Although peak concentrations were seen at flowering, the differences at other times were not marked. In one evaluation, a red-flowering cultivar had slightly higher cardenolide concentrations than a white-flowered type (Karawya et al. 1973).

Adynerin

Numerous other cardenolides have been reported to be present in *Nerium*. In some instances, these compounds are chemically the same as those listed above, but because of nomenclatural confusion, they have been given different names. In other instances, they are artifacts of the isolation procedure, i.e., degradation products of the listed compounds mainly due to cleavage of sugar linkages. Even though artifacts, they may be significant because their parent compounds may undergo similar degradative processes in the digestive system of the animal and yield the same products. To some extent, degradation increases potency; monosides are typically more potent than the biosides and triosides (Chen & Henderson 1965). Not all of the glycosides are active or potent, for example, adigoside, adynerin, neriantin, and odoroside K have low toxicity (Chen & Henderson 1965; Hoffman et al. 1966). Compounds lacking the 5-membered lactone ring at C-17 are not active as cardiotoxins but may irritate the digestive tract and exhibit hemolytic activity typical of saponins (Shoppee 1964).

Even though some of the cardenolides are of low potency, the cardenolides of *Nerium* as a group are probably among the most toxic of the cardiac glycosides; some are almost as potent as ouabain (Chen & Henderson 1965; Coffey 1970). Other biological effects of these compounds include cytotoxicity of nasopharyngeal carcinoma and inhibition of brain transport ATPase (Kupchan et al. 1967).

Nerium also produces several ursane-type pentacyclic terpenoids such as kaneric and oleanderolic acids (Connolly & Hill 1991). They may contribute to the digestive disturbances associated with the cardenolides.

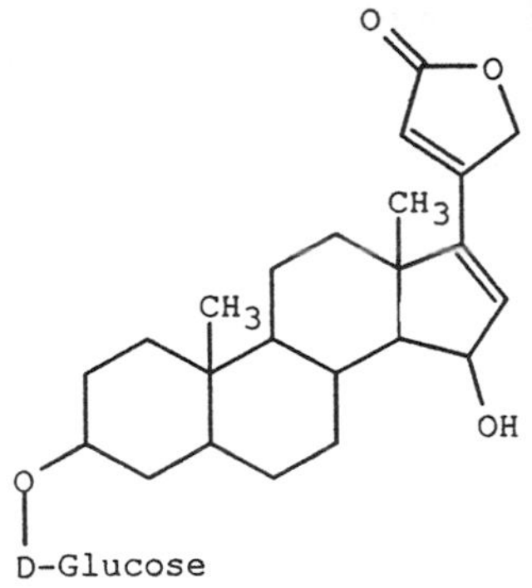

O
O
CH_3
CH_3
OH
O
D-Glucose

Neriantin

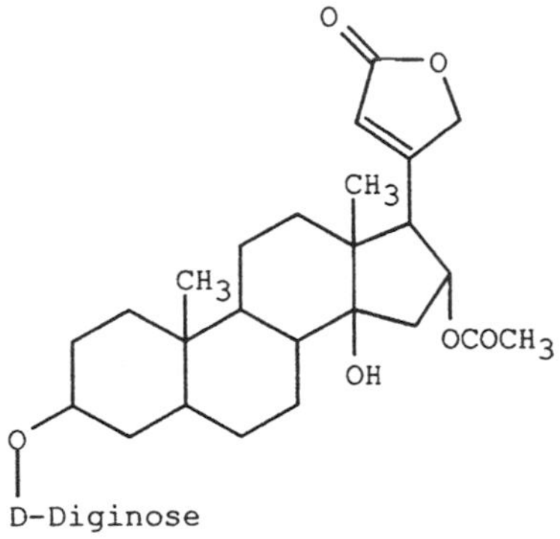

O
O
CH_3
CH_3
$OCOCH_3$
OH
O
D-Diginose

Nerigoside

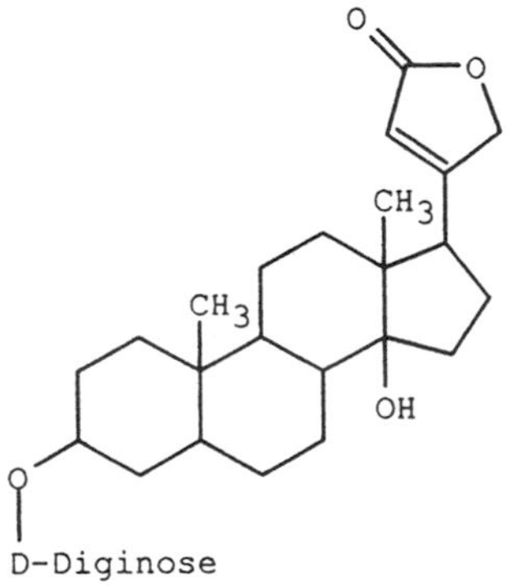

O
O
CH_3
CH_3
OH
O
D-Diginose

Odoroside A

Clinical Signs—The signs of *Nerium* intoxication vary somewhat depending on the species involved. Typically there is depression, anorexia, excess salivation, and varying degrees of diarrhea. Vomiting is common in dogs and cats. Diarrhea (± blood) is a consistent sign in all animals, including cattle. As the disease progresses, direct evidence of cardiac involvement emerges, e.g., bradycardia (in many but not all), paroxysmal tachycardia, ST-segment depression, and various arrhythmias, including AV block, ectopic beats, and gallop rhythm with dropped beats. These signs may be accompanied by cold extremities, irregular weak pulse, increased apical beat, tremors, mydriasis, polypnea, and a marked increase in heart rate with mild exertion (West 1957; Szabuniewicz et al. 1972; Miller 1973; Schwartz et al. 1974; Mahin et al. 1984). An accompanying sign on occasion is flaring and soreness around the mouth (Wilson 1909; Ansford & Morris 1983). Colic-type signs may be apparent in horses.

Signs are reported to be most severe or most likely to occur in males (Grinnell & Smith 1957; Schwartz et al. 1974). In some instances the disease progresses gradually, with clinical signs appearing several hours or more after ingestion of plant material. In other cases, however, the intoxication is so severe and the disease progresses so rapidly that premonitory signs are not observed and the animal is found dead. In cattle and sheep, the onset of signs may be delayed for 12 hours or more and then be prolonged for several days or more before death or recovery ensues.

- all animals, depression, excess salivation, vomiting, diarrhea, weakness, bradycardia, cardiac arrhythmias
- males at increased risk

Pathology—In instances when animals die quickly after ingestion of *Nerium*, there are few distinctive changes observable at necropsy. There may be subepicardial and subendocardial hemorrhages, a pale myocardium, increased pleural fluid, hemorrhages on the abomasal mucosa, pulmonary congestion and edema, congestion and hemorrhages of the abdominal viscera, and reddening of the gastroenteric mucosa. The changes are seldom severe and are not specific for *Nerium* (Thimnah 1972; Schwartz et al. 1974; Mahin et al. 1984; Oryan et al. 1996). Lesions, which are more likely in animals that live for several days, include a soft, flabby heart; scattered foci of myocardial necrosis with mononuclear inflammatory cell infiltrates, edema, and some fibrosis; and congestion, hemorrhages, and emphysema of the lungs (Newsholme & Coetzer 1984; Galey et al. 1996; Oryan et al. 1996). Serum chemistry changes are limited to inconsistent hyperkalemia, which occasionally is marked.

- gross pathology, cardiac hemorrhages, reddened gut mucosa

  microscopic, myocardial necrosis and fibrosis

- no distinctive lesions; may find the tough distinctively veined leaves in stomach or cardenolides in gut content extracts

Nerium's tough, leathery leaves and their venation are particularly distinctive and useful when examining stomach contents to determine if the plant has been ingested. The secondary veins are almost perpendicular to the midvein, are uniformly spaced, and become indistinct near the curled margin. Tertiary veins are indistinct. The leaves thus have a conspicuous ribbed appearance reminiscent of a trout skeleton.

Diagnosis may be facilitated by two-dimensional TLC of digestive tract contents extracted with dichloromethane as a sensitive and more reliable assay than radioimmunoassay (Galey et al. 1996). Although extracts of urine can be used, those from the rumen and the small and large bowel are preferred. An HPLC method using fluorescent detection on extracts from the digestive tract has also been developed (Tor et al. 1996).

- activated charcoal, atropine, propranolol

Treatment—Because cardenolide intoxication is typically acute, treatment may be directed toward prevention of further absorption of the toxicants from the digestive tract, reduction of serum concentrations of the toxins, and/or therapy for their deleterious effects. As has been demonstrated for other cardenolide-producing plants, oral administration of 2 g of activated charcoal/kg b.w. soon after the onset of clinical signs of intoxication may be effective in limiting absorption and disease severity (Joubert & Schultz 1982a,b). This treatment may also enhance biliary excretion and reduce enterohepatic recycling (Langford & Boor 1996).

Absorbed cardenolides may be rendered inactive by complexing them with antibody fragments. Oleander cardenolides are cross-reactive with digoxin-specific antibodies (Shumaik et al. 1988; Cheung et al. 1989; Jortani et al. 1996). Thus, digoxin-specific Fab antibody fragments are effective in reversing the cardiotoxic effects in a manner akin to that for digoxin toxicity (Schmidt & Butler 1971; Smith et al. 1976; Zucker et al. 1982; Shumaik et al. 1988).

Therapy for the deleterious cardiac effects seems to be best accomplished with a combination of atropine and propranolol, repeated several times. The dosage will vary somewhat, depending on the animal species; in dogs, 0.04 mg/kg of atropine and 1 mg/kg of propanalol i.m. or i.v. are suggested (Szabuniewicz et al. 1972). With such treatment, recovery may take several days. The deleterious effects of intracellular calcium in the heart have been attenuated by using dipotassium EDTA i.v. to replace some of the calcium in blood with potassium (Langford & Boor 1996). Flunixin is useful against the colic-type signs in horses (Galey et al. 1998).

Pentalinon Voigt

- *P. luteum*

  nightsage
  wild allamanda
  yellow allamanda

Taxonomy and Morphology—Native to Central America, the West Indies, and southern Florida, the genus *Pentalinon* comprises a single species:

P. luteum (L.) B.F.Hansen & Wunderlin	wild allamanda, nightsage, yellow nightshade

Wild allamanda has long been known as *Urechites lutea,* and toxicological references use this binomial. On the basis of recent studies, however, taxonomists now combine the genera *Pentalinon* and *Urechites.* The latter lacks nomenclatural priority and should no longer be used. In addition, *U. pinetorum* Small is now considered to be conspecific with *P. luteum.*

- evergreen shrubby vines; leaves simple, opposite; flowers showy, funnelform; petals yellow; follicles long

Plants evergreen shrubby vines; perennials; to 3 m tall. **Leaves** simple; opposite; oblong to obovate or suborbicular; thin-textured or leathery; glossy above; apices obtuse or cuspidate; bases rounded or cordate; petioles short. **Inflorescences** few-flowered cymes; axillary or terminal. **Flowers** showy. **Sepals** fused. **Corollas** funnelform. **Petals** yellow, occasionally marked with red in throats. **Stamens** attached within throats of corolla tubes. **Pistils** 1; stigmas 1; styles 1; ovaries 2. **Fruits** follicles; elongate; slightly curved. **Seeds** numerous; comose.

The yellow, goblet-shaped flowers of *Pentalinon* are very similar to those of *Allamanda,* but smaller.

Pentalinon luteum

Distribution and Habitat—In addition to occurring in Central America and the West Indies, plants of *P. luteum* occur in the coastal hammocks, pinelands, and mangrove thickets of southern Florida and the Keys.

■ southern Florida

Disease Problems—In Cuba, where the plants of *P. luteum* are locally abundant, two disease syndromes, sudden death and hemorrhagic diarrhea, have been described in cattle (Marrero et al. 1983; Infante et al. 1984; Faz 1996). In the sudden death syndrome, repeated ingestion of plant material in small amounts seems to cause cardiac damage, and when the animal subsequently is stressed, fatal ventricular fibrillation occurs. In contrast, hemorrhagic diarrhea, which occurs when a large amount of plant material is consumed in a short time, is associated with the feeding of chopped hay with *P. luteum* as a contaminant.

■ irritant digestive disturbance, diarrhea; cardiotoxicity

Disease Genesis—Plants of *P. luteum* contain oleandrigenin cardenolides, e.g., the biosides urechitoxin (L-oleandrose + D-glucose) and urechitin (Chen & Henderson 1965; Pina Luis & Fanghaenel 1973). The disease syndromes have been reproduced using cardiotoxic extracts (Angulo 1942; Marrero et al. 1982).

■ steroidal glycosidic cardenolides

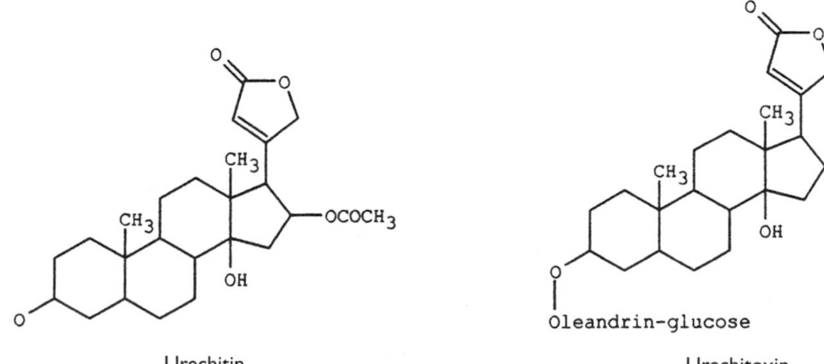

Urechitin

Urechitoxin

Clinical Signs—A day or two after the animals begin to eat *P. luteum,* there is onset of mucoid diarrhea that changes to a bloody diarrhea. This is accompanied by depression and labored respiration. Following these changes, typical cardiotoxic signs may develop, including decreased heart rate, arrhythmias, lethargy, and intolerance to exercise (Marrero et al. 1983; Faz 1996). With ingestion of small amounts of plant material, there may be few signs of digestive disturbance, while cardiotoxic signs, including cardiac failure, may be elicited with exercise.

■ diarrhea, mucoid to bloody; weakness, cardiac arrhythmias

Pathology—Consistent with the changes effected by other cardenolides, there are few distinctive pathologic changes other than reddening, ulceration, and hemorrhages of the mucosa of the stomach and small intestine. Hemorrhages also may be apparent on the surfaces of the myocardium. Microscopically, mild degenerative changes in the heart, liver, and renal tubular epithelium (Infante et al. 1984).

■ gross pathology, reddened and hemorrhagic gut mucosa

■ activated charcoal; atropine

Treatment—Symptomatic treatment as for other cardiotoxins is indicated and may be augmented by administration of activated charcoal, 2g/kg b.w. (Joubert & Schultz 1982a,b). Digoxin-specific Fab fragments, as for *Nerium,* may be used if available (Bania et al. 1993).

Plumeria L.

■ *P. alba*
P. obtusa
P. rubra

flor de mayo
frangipani
nosegay
pagoda tree
plumeria
temple tree
West Indian jasmine

Taxonomy and Morphology—Comprising 8 species, *Plumeria* is a genus of tropical America. In North America, 3 species are present:

P. alba L.	West Indian jasmine, white frangipani
P. obtusa L.	Singapore plumeria
P. rubra L.	frangipani, graveyard plumeria, nosegay, pagoda tree, temple tree, flor de mayo, rose-pink frangipani

These species are quite showy, and *P. rubra* was highly prized by the Aztecs. Some taxonomists recognize populations of *P. rubra* in northern Mexico as a distinct species and use the binomial *P. acutifolia* Poir.

■ small deciduous shrubs or trees; leaves simple, alternate, at ends of branches; flowers showy, sweet scent, white, yellow, or pink, with yellow centers; follicles elongated, paired

Plants deciduous small trees or shrubs; perennials; branches candelabriform, cylindrical, thick, fleshy, gray-green; leaf scars conspicuous. **Leaves** clustered at ends of branches, simple; alternate; oblong or oblanceolate-elliptic; glossy dark red-green; conspicuously pinnately veined, the secondary veins joined to a nerve running parallel to margins; petioles long. **Inflorescences** large cymes; terminal, borne on bare branches. **Flowers** sweetly fragrant; showy. **Sepals** fused; tips with glands. **Corollas** funnelform or salverform. **Petals** white or yellow or pink or mixtures of these colors; centers often yellow. **Stamens** attached near bases of corolla tubes. **Pistils** 1; stigmas 1; styles 1; ovaries 2. **Fruits** follicles; elongate; typically paired; leathery. **Seeds** numerous; winged.

1. Leaves elliptic or oblong-lanceolate. Leaf margins not revolute.
   Follicles to 30 cm long . . . . *P. rubra*
1. Leaves lanceolate to obovate. Leaf margins revolute.
   Follicles to 22 cm long.
   2. Leaves lanceolate. Flowers yellow with white centers.
      Follicles to 14 cm long . . . . *P. alba*
   2. Leaves obovate. Flowers white with yellow centers.
      Follicles to 22 cm long . . . . *P. obtusa*

Plumeria obtusa

■ ornamental, frost-free areas

Distribution and Habitat—Because they are frost sensitive, these introductions from tropical America are grown as specimen plants only in the southern part of the continent, areas with temperatures usually above 25°F. Elsewhere, they are grown as tub plants. *Plumeria rubra* is native to the Mexican states of Baja, Sonora, and Chihuahua. Both *P. alba* and *P. rubra* may be encountered as escapes in southern Florida.

■ contact irritation, skin, digestive tract

Disease Problems—The viscid, milky sap of *Plumeria* is quite irritating and causes reddening of the skin and digestive tract upon contact or ingestion, respectively (Blohm 1962).

■ irritant iridoids, terpenoids

Disease Genesis—Similar to *Allamanda,* species of the *Plumeria* genus produce several purgative iridoids such as allamcin, allamandin, plumericin, isoplumericin, and plumieride (Blohm 1962; Watt & Breyer-Brandwijk 1962; Coppen & Cobb 1983; Kardono et al. 1990; Connolly & Hill 1991a). Although all parts of the plant are considered toxic, these compounds are found primarily in the bark; very little is present in the leaves and flowers. Plumieride given orally to pregnant rats caused postimplantation embryotoxicity (Gunawardana et al. 1998). Stem-bark extracts caused diarrhea, decreased appetite, and lowered body weights. In addition to the iridoids present, several ursane-type pentacyclic terpenoids such as obtusin, obtusidin, obtusilin, and obtusinidin are present (Connolly & Hill 1991b). Cardenolide-type cardiotoxic activity is quite low (Watt & Breyer-Brandwijk 1962; Radford et al. 1986, 1994).

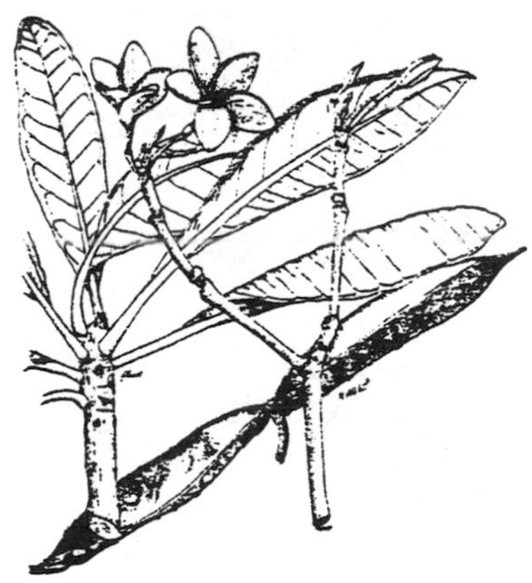

Plumeria rubra

Plumericin

Plumieride

Obtusin

Clinical Signs, Pathology, and Treatment—Contact with the sap may cause pain and reddening of the skin or mucous membranes of the mouth and eyes. Ingestion of foliage or flowers causes nausea and diarrhea, which are not likely to be severe unless large amounts of plant material are consumed. The disease is not severe enough to be fatal, and treatment is not generally needed.

Plumeria rubra

- pain, reddening of skin, eyes, and mouth; diarrhea

Rauvolfia L.

Taxonomy and Morphology—Comprising some 60 species of shrubs and small trees, *Rauvolfia* is represented in North America by 4 species:

R. nitida Jacq.	palo amargo, bitter ash
R. serpentina Benth. ex Kunz	serpentwood, snakeroot
R. tetraphylla L.	American serpentwood, devil pepper
R. viridis Willd. ex Roem. & Schult.	bitterbush

The genus name also has been spelled *Rauwolfia*.

- *R. nitida*
  R. serpentina
  R. tetraphylla
  R. viridis

  bitter ash
  bitterbush
  devil pepper
  serpentwood
  snakeroot

Plants evergreen small trees or shrubs; perennials. **Leaves** simple; whorled or opposite; ovate to elliptic to oblanceolate; margins entire. **Inflorescences** few-flowered cymes; terminal or axillary. **Flowers** small. **Sepals** fused. **Corollas** salverform; lobes typically twisted. **Petals** white to ivory or pink. **Stamens** attached near throats of corolla tubes. **Pistils** 1; stigmas 1; styles 1; ovaries 2. **Fruits** drupes; typically paired or sometimes united; black or red. **Seeds** 1 or 2.

- evergreen shrubs or trees; leaves simple, whorled or opposite; flowers small; petals white, ivory, or pink; drupes black or red

Distribution and Habitat—*Rauvolfia* is a genus of both Old and New World tropics and subtropics. As with other topical members of the family, species of the genus are cultivated in the frost-free parts of North America. *Rauvolfia tetraphylla* is the most

- ornamental, frost-free areas

Rauvolfia tetraphylla

commonly encountered. Other species occur on the southern edge of our range near the Tropic of Cancer in Mexico.

■ contact irritation skin, digestive tract

Disease Problems—The viscid, milky sap of *Rauvolfia* is irritating and causes reddening of the skin upon contact or inflammation of the mucosa of the digestive tract when ingested. Other intoxication problems have not been reported. The genus is perhaps best known as the source of reserpine, initially extracted from the Asian species *R. serpentina* and used to reduce blood pressure and treat mental disorders. Other preparations and decoctions of plants of *Rauvolfia* have been used in herbal preparations as emetics, anthelmintics, diuretics, expectorants, laxatives, and sedatives and for treating problems such as high blood pressure, epilepsy, and snakebite (Sen & Bose 1931; Morton 1977).

■ unknown irritant compounds; alkaloids

Disease Genesis—*Rauvolfia* is well known for its production of many alkaloids, including reserpine, rescinnamine, reserpinine, serpentine, and yohimbine (Schlittler 1965; Morton 1977). These alkaloids are found throughout the plants. Reserpine-type effects predominate, causing blood pressure reduction, sedation, and slowed heart rate (Schlittler 1965). The alkaloids have been suggested as a possible cause of teratogenic changes as well (Keeler 1983). Of the New World species, the tropical *R. tetraphylla* is the principal source of reserpine (Blohm 1962).

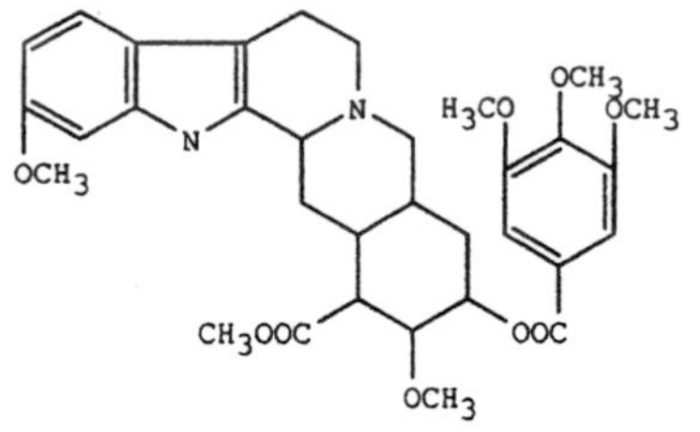

Reserpine

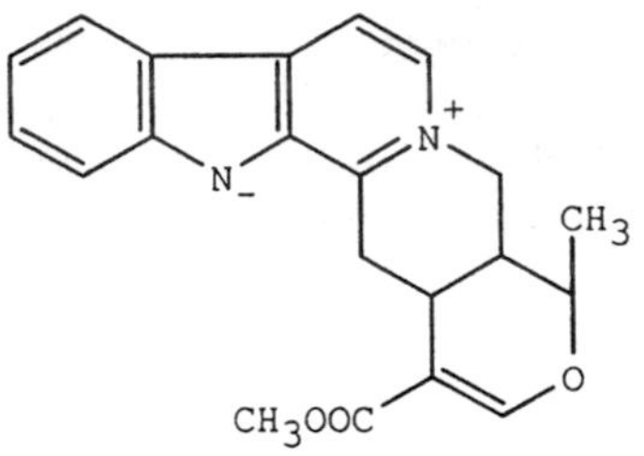

Serpentine

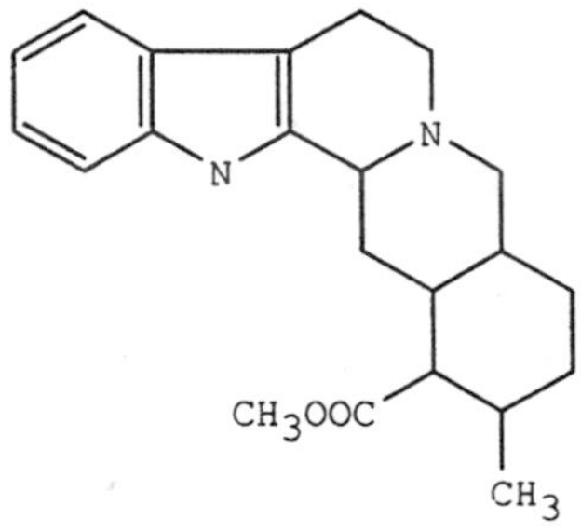

Yohimbine

■ pain, reddening of skin, eyes, and mouth; diarrhea

Clinical Signs, Pathology, and Treatment—Contact with the sap may cause pain and reddening of the skin or mucous membranes of the mouth and eyes. Ingestion of foliage or flowers may cause nausea and diarrhea, which are not likely to be severe unless large amounts of plant material are consumed. Because other signs of disease have not been described for *Rauvolfia,* the possible toxic effects of the alkaloids present can only be surmised. The disease is not severe enough to be fatal, and treatment generally is not needed.

Strophanthus DC.

■ *S. speciosus*
corkscrew flower

Taxonomy and Morphology—An Old World genus of some 38 species, *Strophanthus* is represented in North America by 1 cultivated species:

S. speciosus (Ward & Harv.) Reber   corkscrew flower

■ shrubs; leaves simple, whorled; flowers showy; petals cream yellow, red-spotted; follicles long

Plants erect or rambling shrubs; perennials; branches olive-green; glabrous. **Leaves** simple; whorled, 3 or 4 per node; oblong-lanceolate to lanceolate; leathery; apices acute or obtuse; bases cuneate; petioles short. **Inflorescences** few to many flowered cymes; terminal or axillary. **Flowers** showy. **Sepals** fused. **Corollas** funnelform or salverform; lobes typically twisted; with 10 clawlike scales in throats of tubes. **Petals** cream yellow; red-spotted. **Stamens** attached near throats of corolla tubes; converging into cone around stigma. **Pistils** 1; stigmas 1; styles 1; ovaries 2. **Fruits** follicles; elongate. **Seeds** numerous; spindle-shaped; comose at both ends.

Strophanthus speciosus

- ornamental, frost-free areas
- digestive disturbance, diarrhea; cardiac insufficiency
- steroidal glycosidic cardenolides

Distribution and Habitat—Cultivated in the frost-free areas of North America, *S. speciosus,* and possibly other species of the genus, may be encountered as tub plants. Although the plants are strikingly attractive ornamentals, they have never gained popularity in North America.

Disease Problems and Genesis—Plants of *Strophanthus* produce cardiotoxic problems similar to those caused by *Nerium oleander.* In Africa, various species of the genus have been favored as a source of toxicant for trial-by-ordeal potions (Watt & Breyer-Brandwijk 1962). Although quite toxic, plants are apparently not palatable and thus seldom eaten. All parts of the plant are toxic (Joubert 1989).

Species of *Strophanthus* contain a series of very potent cardenolides from the genin strophanthidin, e.g., *S. gratus* produces the deadly G-strophanthin (ouabain) and *S. kombe* contains K-strophanthin (Shoppee 1964; Singh & Rastogi 1970).

Strophanthidin

- nausea, vomiting, excess salivation, diarrhea, weakness, cardiac arrhythmias

Clinical Signs—In humans, clinical signs generally involve the digestive tract and include nausea, excess salivation, retching, vomiting, diarrhea, and extreme weakness. Death, however, is unlikely from ingestion of plant material. The same signs are observed in household pets when amounts ingested are small. In unusual situations in which livestock gain access to plants of *Strophanthus,* the signs are again similar except for the greater likelihood of cardiac effects. Signs observed in livestock include weakness, decreased appetite, excess salivation, colic, diarrhea with or without blood, cardiac arrhythmias, and cyanosis. Death in all animals is due to ventricular fibrillation or failure.

- gross pathology, reddened gut mucosa

Pathology—Distinctive changes due to ingestion of *Strophanthus* are minimal. There may be frothy edema fluid in the airways, reddening of the mucosa of the digestive tract, and visceral splanchnic congestion with scattered splotchy hemorrhages.

- activated charcoal, atropine

Treatment—Symptomatic treatment as for other cardiotoxins is indicated and may be augmented by administration of activated charcoal, 2 g/kg b.w. (Joubert & Schultz 1982a,b). Digoxin-specific Fab antibody fragments as for *Nerium* may be used if available (Bania et al. 1993).

Thevetia Adans.

- *T. ovata*
  T. peruviana
  T. thevetioides

  be-still tree
  codo de fraile
  friar's elbow
  lucky nut
  narciso amarillo
  thevetia
  tiger-apple
  trumpet flower
  yellow oleander
  yoyote

Taxonomy and Morphology—Comprising 8 species of evergreen trees and shrubs, the genus *Thevetia* is represented in North America by 3 species:

T. ovata (Cav.) A.DC.	narciso amarillo
T. peruviana (Pers.) K.Schum.	yellow oleander, lucky nut, be-still tree, tiger-apple, trumpet flower, friar's elbow, codo de fraile
T. thevetioides (Humb. Bonpl. & Kunth) K.Schum.	giant thevetia, yoyote

Synonyms of *T. peruviana* that appear in the older literature are *T. neriifolia* and *Cerbera thevetia*.

- evergreen shrubs or trees; leaves simple, alternate, leathery; flowers showy, fragrant, funnelform; petals saffron yellow, orange, or yellow-pink; drupes black

Plants evergreen shrubs or small trees; perennials. **Leaves** simple; alternate; linear or elliptic-oblong to oblanceolate; leathery; glossy dark green above; venation pinnate, midnerve conspicuous; apices acute or rounded; margins revolute; bases cuneate; petioles short. **Inflorescences** few-flowered compound cymes; terminal. **Flowers** large; showy; fragrant. **Sepals** fused at base; bearing glands inside. **Corollas** funnelform; lobes twisted; with downy scales in throats of tubes. **Petals** saffron yellow to orange or yellow-pink. **Stamens** attached at throats of corolla tubes. **Pistils** 1; stigmas 1; styles 1; ovaries 1. **Fruits** drupes; angular; red and then black. **Seeds** 1.

- ornamental, warmer areas of North America

Distribution and Habitat—*Thevetia* is a genus of tropical America and the West Indies. *Thevetia peruviana,* native to the Gulf coastal area of Mexico from Tamaulipas south through San Luis Potosí, is the most commonly encountered species and is used sparingly as an ornamental in the southern part of the continent. *Thevetia ovata* is found along the west coast of Mexico from Sinaloa south. The genus is not as cold-hardy as *Nerium* and thus is not widely cultivated.

- rarely eaten, digestive tract irritation; cardiotoxicity

Disease Problems—Ingestion of *Thevetia* produces both gastrointestinal and cardiotoxic effects. The intoxication signs are similar to those of *Nerium,* except that gastrointestinal effects predominate and the cardiac effects are reduced (Everist 1981; Kakrani et al. 1981; Chopra et al. 1984). The seeds are sometimes eaten by people attempting suicide (Saraswat et al. 1992; Ahlawat et al. 1994). In the reported cases, ingestion of more than four seeds or delay of treatment for more than 4 hours resulted in death. However, when four or fewer seeds were eaten and/or treatment was immediate, the prognosis for survival was good. As with *Nerium,* the risk of eating *Thevetia* may be exaggerated. A 7-year study in Australia revealed few serious problems of intoxication despite the common presence of the plants (Shaw & Pearn 1979). Studies in Brazil indicated that a dose of 14 g fresh foliage/kg b.w. of *T. peruviana* was lethal to cattle (Tokarnia et al. 1996). However, in sheep up to 20 g/kg caused only minor signs of intoxications (Armien & Tokarnia 1994). It is interesting that, as with *Nerium,* neurotoxicity in addition to or sometimes rather than cardiotoxicity appears when seeds of *Thevetia* are given to rats (Szabuniewicz et al. 1972; Pahwa & Chatterjee 1990; Oji et al. 1993). The ground seeds have potential for use as rodenticides. Significant hepatic and renal disease is apparent in rats after ingestion of *Thevetia.*

Thevetia peruviana

Disease Genesis—Species of *Thevetia* contain several cardenolides composed of digitoxigenin linked with thevetose alone, neriifolin, or thevetose plus glucose and thevetins A and B (Shoppee 1964). As with other members of the Apocynaceae, various isomers and breakdown products have been identified, complicating the nomenclature of the compounds present. All of these digitoxigenin glycosides are very similar in action to the cardenolides of *Nerium,* albeit probably less potent (Chen & Chen 1934a; Elderfield 1936; Joubert 1989; Radford et al. 1994). Although the toxins of *Thevetia* are, overall, considerably less potent than ouabain, the concentrations in the seeds are high enough to pose a serious hazard (Chen & Chen 1934a,b). Toxin concentrations are considerably lower in other plant parts, with a reduced risk (Thorp & Watson 1953). The cardenolides impart a bitter taste to the plants, rendering them unpalatable especially when green (Chopra et al. 1984).

- steroidal glycosidic cardenolides

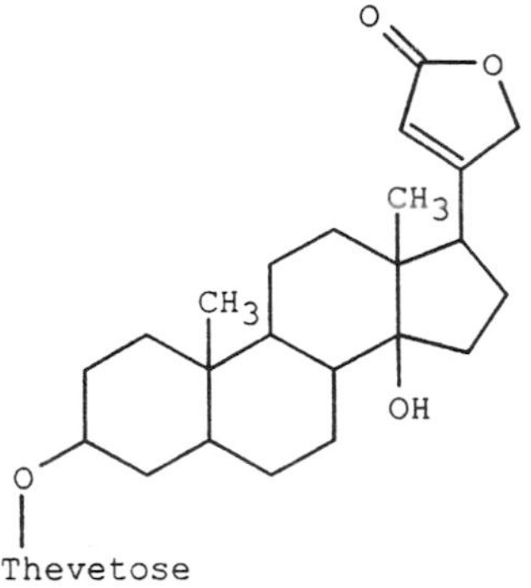

Neriifolin

Clinical Signs, Pathology, and Treatment—Ingestion of *Thevetia* produces clinical signs and pathologic changes as a consequence of both its gastrointestinal and cardiotoxic effects, similar to those produced by *Nerium* (Everist 1981; Chopra et al. 1984). Thus, diarrhea and perhaps cardiac arrhythmias are the primary signs. Aspects of treatment are the same as for *Nerium,* including the use of activated charcoal per os, atropine, and perhaps propranolol. In some instances, the use of digoxin-specific Fab antibody fragments may be appropriate.

- all animals, depression, excess salivation, vomiting, diarrhea, weakness, bradycardia, cardiac arrhythmias
- activated charcoal, atropine

Urechites

See treatment of *Pentalinon.*

Genera of Questionable Toxicity

Alstonia R.Br.

Comprising approximately 45 species in Africa, Central America, southeast Asia, Australia, and islands of the western Pacific, *Alstonia* is represented in North America by 2 species: *A. macrophylla* Wallich ex G.Don and *A. scholaris* (L.) R.Br. (scholar tree, devil's potato). These are evergreen shrubs or trees, with simple, whorled or occasionally opposite leaves. Indigenous to Australasia and India, plants of *Alstonia* are cultivated in the nearly frost-free areas North America for their fragrant flowers.

- *A. macrophylla*
  A. scholaris

  devil's potato
  scholar tree
- evergreen shrubs or trees; leaves simple, whorled or opposite, leathery; flowers fragrant

Although intoxication problems have not been reported for *A. macrophylla* or *A. scholaris,* the potential for such should be considered. An Australian species, *A. constricta,* has been associated with a neurologic disease in cattle and sheep (Everist 1981). Ingestion of large amounts of plant material for 2–3 weeks precipitates strychninelike tonic seizures. Species of *Alstonia* contain a number of alkaloids of unknown toxicity risk. Alstonine and echitamine are predominant, but there are many others, including macralstonine, nareline ethers, picrinine, reserpine, and villalstonine (Saxton 1965a; Govindachari 1968; Kam et al. 1997). Ajmaline, a similar alkaloid from a European species, inhibits atrioventricular conduction and has been widely used to treat cardiac arrhythmias, even though it has a slow onset and a short duration of action (Creasey 1983; Ban et al. 1988). Alstonine, which is also known from *Rauvolfia* and *Catharanthus roseus,* is hypotensive and adrenolytic and inhibits gut peristalsis (Wakim & Chen 1947). At high doses it causes a variety of effects, including cardiac arrhythmias. Echitamine is sympathomimetic at low dosage and antagonistic and somewhat curare-like at higher dosage (Creasey 1983). Likewise, the North American species of *Alstonia* contain many of these complex indole alkaloids. Some are tryptamine-like (5-OH-tryptamine) and have been used extensively for medicinal

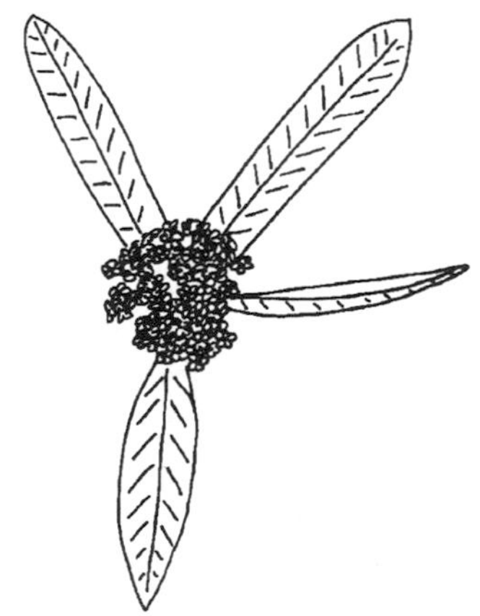

Alstonia scholaris

- ornamental, frost-free areas
- alkaloids with cardiac activity; neurotoxic

purposes. Also used by some societies, the bark of *A. scholaris* contains estrogenic substances, as determined in immature female mice (Biswas & Bhattacharyya 1996).

Echitamine

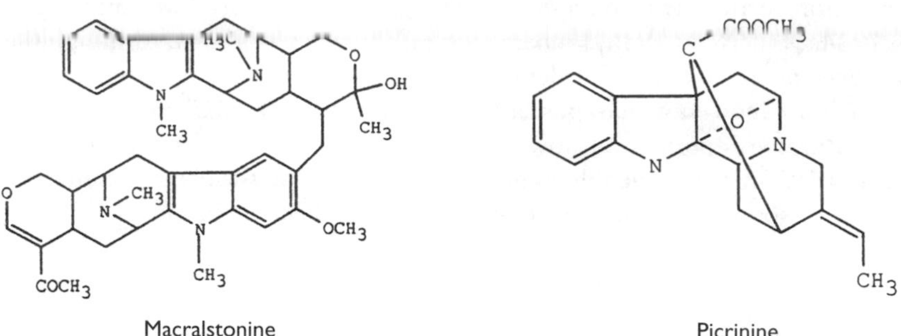

Macralstonine

Picrinine

Echites P.Browne

- *E. coulteri*
  E. umbellata
  rubber vine
- evergreen shrubby vines; southern Florida and Mexico

Comprising 6 tropical American species, *Echites* is represented in North America by *E. coulteri* S.Watson (rubber vine) and *E. umbellata* Jacq. (rubber vine, devil's potato). *Echites* is one member of a complex of closely related genera of shrubby vines, including *Angadenia, Prestonia,* and *Rhabdadenia.* Generic boundaries are tenuous, and individual species typically have been repeatedly moved from genus to genus by different taxonomists. Taxa originally described as species of *Echites,* for example, are now placed in *Angadenia* and *Rhabdadenia.* Evergreen shrubby vines, *Echites* are infrequent in the frost-free pinelands of southern and central Florida. *Echites coulteri* occurs from Coahuila to San Luis Potosí. Plants are sometimes cultivated in tubs.

Echites umbellata

- possibly irritation of skin and gut mucosa

Other than anecdotal reports, little is known about the toxic potential of *E. umbellata* and other members of this complex of shrubby vines. But because of their apparent close relationship, it has been suggested that we may be able to extrapolate somewhat from one species to another (Hardin & Arena 1974; Perkins & Payne 1978). Possessing the viscid, milky sap characteristic of the family and the closely related *Angadenia, E. umbellata* probably causes inflammation of the mucosa of the digestive tract when ingested and reddening of the skin upon contact. The toxins are not known; however, oleandrigenin cardenolides such as those isolated from *Angadenia* may be present. Alkaloids seem unlikely because they are poorly or not at all represented in the genus (Raffauf 1996). Clinical signs following ingestion of *E. umbellata* probably would be those associated with mucosal irritation. Because ingestion of *Echites* does not result in death, pathological changes are not likely to be observed. Treatment is probably not necessary but could include the conservative use of antidiarrheal medications.

Haplophyton A.DC.

- *H. cimicidum*
  cockroach plant

Native to the New World, *Haplophyton* comprises 3 species, one of which is found in North America, *H. cimicidum* A.DC. (cockroach plant, Mexican cockroach plant). The binomial *H. crooksii* is used in the older literature. Much-branched perennial shrubs, *H. cimicidum* occurs in populations on dry rocky slopes of the U.S. Southwest.

- bisindole alkaloids; digestive disturbance? neurotoxic?

Although the plant has not been reported to cause disease, extracts of the dried leaves of *H. cimicidum* are mixed with molasses or cornmeal to kill flies, cockroaches, mosquitoes, and other insects—hence the common name (Correll & Johnston 1979). Extracts also are used in lotions for parasite control. The species, therefore, possibly has the capacity to cause adverse effects. Closely related taxonomically to the *Alsto-*

nia, H. cimicidum contains many alkaloids of similar structure, including an extensive series of a dozen or more indole or unique bisindole alkaloids such as akuammicine, akuammidine, cimicidine, cimicine, crooksidine, crooksiine, eburnamine, haplocine, haplophytine, lanceomigine, and tubotaurine (Saxton 1965a; Adesomoju et al. 1983; Mroue & Alam 1991; Mroue et al. 1993, 1996). Most of these have mild to moderate acetylcholinesterase activity; the most potent (such as cimicidine, crooksidine, and tubotaurine) are about 10-to-50-fold less potent than are the medicinally used esterase inhibitors. Haplophytine, the major alkaloid, and probably others are active as insecticides; however, specific toxic effects, if any, are not known (Cordell 1983). Yohimbine, β-yohimbine, and crooksiine, a unique bisindole alkaloid, have also been isolated from *H. crooksii* (Mroue & Alam 1991). Crooksiine has moderate anticholinesterase activity, equal to about one-tenth that of neostigmine. Intoxication problems have not been reported. Because indole alkaloids are present, one might expect that if effects are ever observed, they will most likely be related to irritation of the digestive tract and/or neurologic dysfunction. Any treatment necessary should be based upon relief of the signs.

Haplophyton cimicidum

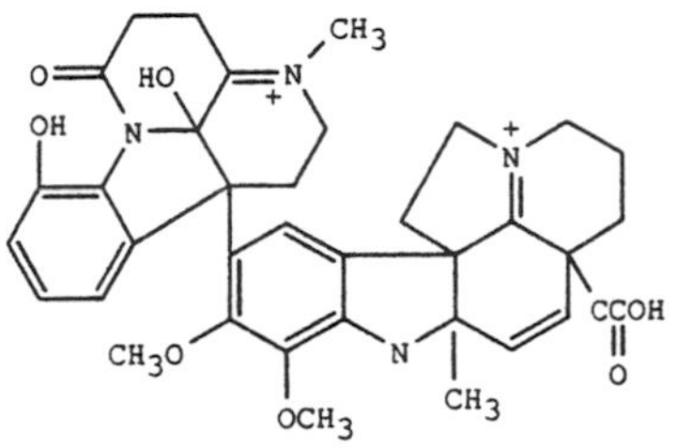

Haplophytine (in acid)

Ochrosia Juss.

With greatest diversity in Australasia, *Ochrosia* comprises some 23 species of trees or shrubs. In North America, the genus is represented by 2 introduced species, *O. compta* K.Schum. and *O. elliptica* Labill. (kopsia, ochrosia plum). Of the species, only *O. elliptica* is typically encountered. The genus name is derived from the Greek *ochros,* referring to the pale yellow color of the flowers. Small trees with sweetly scented flowers, *O. elliptica* is planted in southern Florida, where temperatures are normally above 25°F. Likewise, *O. compta* is found only in the warmest parts of the continent.

Although intoxication problems have not been reported, the attractive red drupes of *Ochrosia* are considered hazardous (Thorp & Watson 1953; Morton 1974; Perkins & Payne 1978). In contrast, the foliage is of little concern. Based on the compounds present, it is most likely that problems would involve irritation of the digestive tract or neurologic dysfunction. *Ochrosia* contains several monoterpenoid indole alkaloids of the heteroyohimbine-, strychnos-, and apparacine-types (Pawelka & Stoeckigt 1986). They include ellipticine, holeinine, powerine, reserpine, reserpiline, and ochrelline (Raffauf 1970). Ellipticine is capable of causing a decrease in heart rate and blood pressure, of β-adrenergic origin, because its effects can be antagonized by propranolol (Creasey 1983). At present, there are no descriptions of intoxications caused by species of *Ochrosia.* It is anticipated that clinical signs and pathology would reflect the genus's potential for irritation of the digestive tract or neurological dysfunction. Treatment would be based upon relief of the signs exhibited.

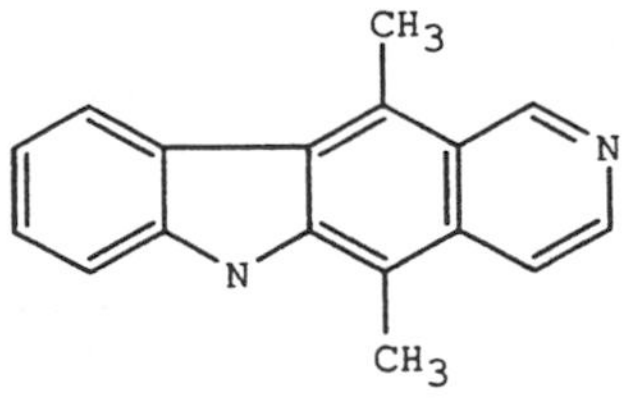

- *O. compta*
  O. elliptica

  kopsia
  ochrosia plum

- shrubs or trees; flowers sweet scented; petals pale yellow

- indole alkaloids; irritant or neurotoxic effects

Ellipticine

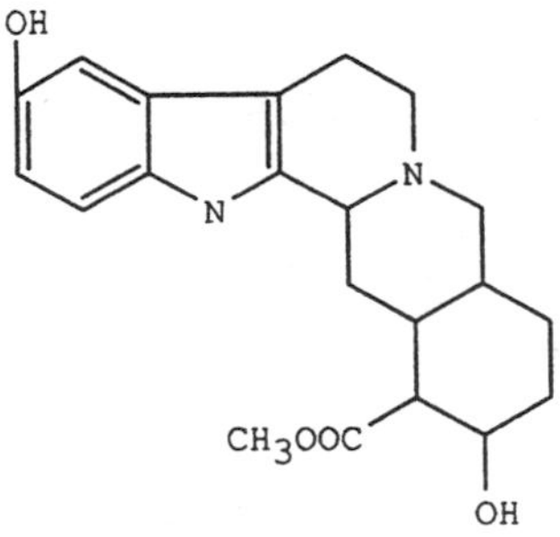

Powerine

Prestonia R.Br.

- *P. agglutinata*

  rubber vine
- evergreen shrubby vines, southern Florida and Mexico

Comprising some 66 tropical American species, *Prestonia* is represented in North America by *P. agglutinata* (Jacq.) Woods (rubber vine). *Prestonia* is one member of a complex of closely related genera of shrubby vines, including *Angadenia, Echites,* and *Rhabdadenia*. Generic boundaries are tenuous, and individual species typically have been repeatedly moved from genus to genus by different taxonomists. *Prestonia agglutinata,* for example, was originally described as a species of *Echites*. Occurring in the frost-free areas of southern Florida, plants of *P. agglutinata* and other species also occur on the southern edge of our range near the tropic of Cancer in Mexico.

- possibly irritation of skin and gut mucosa

Other than anecdotal reports, little is known about the toxins or toxic potential of *P. agglutinata* and other members of this complex of shrubby vines. Because of their apparent close relationship, it has been suggested that we may be able to extrapolate somewhat from one species to another (Hardin & Arena 1974; Perkins & Payne 1978). Possessing the viscid, milky sap characteristic of the family and the closely related *Angadenia, P. agglutinata* probably causes irritation and reddening of the mucosa of the digestive tract when ingested and reddening of the skin upon contact. Alkaloids are not reported for this genus except for a simple indole that has been isolated from a South American species (Snieckus 1968; Kisakurek et al. 1983; Raffauf 1996). Oleandrigenin cardenolides such as those isolated from other members of the complex, *Angadenia* and *Pentalinon,* possibly may be present, but this has not as yet been shown. Members of the genus have some local reputation for toxicity, including hallucinogenic activity, in tropical America (Schultes 1969). Signs following ingestion of *P. agglutinata* probably would be those associated with irritation of mucosal surfaces. Because ingestion of *Prestonia* is not likely to result in death, actual gross lesions have not been described. Treatment is probably not necessary but could include the conservative use of antidiarrheal medications.

Rhabdadenia Müll.Arg.

- *R. biflora*

  R. cordata

  R. paludosa

  rubber vine
- evergreen shrubby vines; flowers showy; southern Florida and Mexico

Rhabdadenia biflora

Comprising 4 tropical American species, *Rhabdadenia* is represented in North America by *R. biflora* (Jacq.) Müll.Arg. (rubber vine), *R. cordata* (Mill.) Miers, and *R. paludosa* (Vahl) Miers. *Rhabdadenia* is another member of the complex of closely related genera of shrubby rubber vines with showy flowers, including *Angadenia, Prestonia,* and *Echites*. Generic boundaries are tenuous, and individual species typically have been repeatedly moved from genus to genus by different taxonomists. *Rhabdadenia biflora,* for example, was originally described as a species of *Echites*. Common at the margins of mangrove swamps and in thickets in coastal hammocks of southern Florida and the Keys, plants of *R. biflora* also occur in the West Indies, Central America, and South America. Plants are sometimes cultivated in tubs. *Rhabdadenia cordata* occurs in Tamaulipas, Mexico, and *R. paludosa* is found in southern Florida.

Other than anecdotal reports, little is known about the toxic potential of *R. biflora* and other members of this complex of shrubby vines. But because of their apparent close relationship, it has been suggested that we may be able to extrapolate somewhat from one species to another (Hardin & Arena 1974; Perkins & Payne 1978). Possessing the viscid, milky sap characteristic of the family and the closely related *Angadenia, R. biflora* probably causes irritation and reddening of the mucosa of the digestive tract when ingested and of the skin upon contact. The toxins are not known with certainty, but alkaloids seem not to be present in the genus (Raffauf 1996). Oleandrigenin cardenolides such as those isolated from *Angadenia* may be present. Clinical signs following ingestion of *R. biflora* probably would be those associated with irritation of the mucosa. Because ingestion of *Rhabdadenia* does not result in death, gross lesions have not been described. Treatment is probably not necessary but could include the conservative use of antidiarrheal medications.

■ possibly irritation of skin and gut mucosa

Tabernaemontana L.

■ *T. citrifolia*
T. divaricata

crepe jasmine
rosebay
waxflower plant
paper gardenia

■ ornamental, warm areas

Comprising some 110 species of shrubs and small trees, *T. divaricata* (L.) R.Br. ex Roem. & Schult. is represented in North America by 2 species, *T. citrifolia* L. (palo de San Diego) and *T. divaricata* (L.) R.Br. ex Roem. & Schult. (crepe jasmine, rosebay, East Indian rosebay, Adam's apple, coffee rose, waxflower plant, paper gardenia, crepe gardenia, Nero's crown). Appearing in the older literature, the binomials *T. coronaria, Ervatamia coronaria,* and *Nerium divaricatum* are synonyms of *T. divaricata.* These evergreen shrubs bear showy, nocturnally fragrant flowers with doubled petal number and have become popular ornamentals in recent years. Because they are frost sensitive, these introductions from tropical Southeast Asia and the West Indies are restricted in distribution to areas with temperatures above 25°F. They are grown elsewhere as tub plants.

Tabernaemontana citrifolia

Specific intoxication problems caused by species of *Tabernaemontana* have not been reported. However, the seeds of *T. divaricata* are said to be narcotic and to cause delirium, and the leaves are purgative (Chopra et al. 1984). Studies by Ratnagiriswaran & Venkatachalam (1939) and Thorp & Watson (1953) indicated that the species has limited cardiotoxicity. Experiments with other species of the genus failed to show toxic effects in calves at oral doses up to 9 g/kg and 13.2 g/kg b.w. (Thorp & Watson 1953; Tokarnia et al. 1979). The juice from foliage has been applied to warts for their removal. *Tabernaemontana* members contain a large number of alkaloids, such as apparicine, vaophylline, and O-acetylvallesamine. Plants of the genus also contain two dozen or more alkaloids at lower concentrations, including affinine, coronarine, dregamine, montanine, tabernaemontane, vincamine, and the bisindole voacamine (Raffauf 1970; Southon & Buckingham 1989; Clivio et al. 1991; Dagnino et al. 1991). In one species, *T. corymbosa,* 33 alkaloids have been isolated from the leaves and root bark (Clivio et al. 1991). Although at least 2 of these alkaloids, coronarine and dregamine, are reported to be cardiotoxic (Chopra et al. 1984), their effects generally are not clinically manifested. Experimentally, apparicine at low dosage causes seizures in mice, and coronaridine-types are tremorgenic in addition to being weakly estrogenic (Creasey 1983). There is some interest in coronaridine-type alkaloids because of their affinity for opiate receptors and possible use as antiaddictive drugs (Deecher et al. 1992; O'Hearn & Molliver 1993). They also have some activity features similar to harmaline (as in Zygophyllaceae), including causing Purkinje cell degeneration in the cerebellar vermis. Plants of closely related genera, such as *Tabernanthe iboga* from West Africa, are of special interest because of their content of coronaridine-type alkaloids. These plants are widely used in counteracting fatigue and in mystical and religious rites (Bisset 1989). The alkaloids (ibogaine) are neurostimulants

Tabernaemontana citrifolia

■ numerous alkaloids

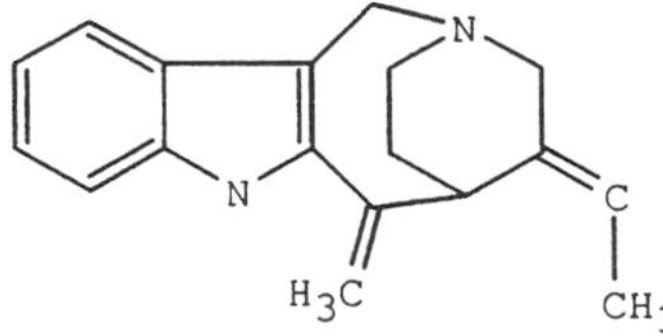

Apparicine

Coronaridine

and hallucinogens. They have anticholinesterase effects while concurrently decreasing catabolism of serotonin and increasing that of norepinephrine.

■ possibly irritation of the gut mucosa

Although the species of *Tabernaemontana* are of unknown toxicologic significance, it may be anticipated that effects related to the digestive tract will be foremost. If sufficient plant material is eaten, neurologic signs also may develop. Treatment will be symptomatic, that is, based upon relief of the signs exhibited.

Vinca L.

■ *V. major*
V. minor

periwinkle
running myrtle

■ trailing evergreen vines

An Old World genus, *Vinca* comprises 6 species, with 2 commonly encountered in North America: *V. major* L. (greater periwinkle, blue periwinkle) and *V. minor* L. (common periwinkle, running myrtle). Two other species, *V. herbacea* and *V. difformis,* are occasionally cultivated. Noted as a source of alkaloids used in cancer treatment, the Madagascar periwinkle, at one time known as *Vinca rosea,* belongs to the related genus *Catharanthus.* Native to southern Europe, both *V. major* and *V. minor* are widely cultivated as trailing evergreen ground covers in North America. They have become naturalized in woodlands in many areas, especially the eastern and southeastern states and as far west as central Texas.

Vinca minor

■ indole alkaloids; digestive disturbance; vomiting, diarrhea; neurotoxic?

Like the closely related *Catharanthus rosea* (Madagascar periwinkle), species of *Vinca* have a long history of medicinal use (Farnsworth 1961). However, in spite of their widespread cultivation, there have been practically no reports of intoxication associated with their ingestion by livestock or pets. When they are ingested, a digestive disturbance may be anticipated, and the alkaloids when given medicinally as antineoplastics for a prolonged time are neurotoxic. Plants of *Vinca* contain an extensive array of alkaloids, especially indole types (Farnsworth 1961). They also contain saponins and various other compounds but appear to be devoid of constituents with cardenolide activity (Farnsworth et al. 1962). Their alkaloid content seems to vary from region to region, especially between areas where they grow as annuals rather than perennials (Farnsworth 1961). *Vinca minor* mainly contains alkaloids with tryptophan bases such as vincadine, vincamine, and vincorine (Mokry & Kompis 1964; Taylor 1965). It also is reported to contain very low concentrations of reserpine (Taylor 1965). The overall effects of these compounds are a reduction in blood pressure and a possible curare-like action (Farnsworth 1961; Taylor 1965). Some of the alkaloids such as vincamine are convulsive at higher dosage (Farnsworth 1961). *Vinca major* also contains numerous alkaloids, several of which are common to *V. minor,* i.e., vincamine and vincadifformine (Taylor 1965). Many of the alkaloids of *V. major,* including akuammine, perivincine, reserpinine, and vinine, are hypotensive, although of variable duration. Their overall effect is a transient hypotension and possibly a ganglioplegic and sympatholytic action as well (Farnsworth et al. 1960; Farnsworth 1961; Taylor 1965). Akuammine also is a weak cholinesterase inhibitor and may, under some circumstances, potentiate sympathetic activity (Creasey 1983).

References

Adesomoju AA, Lakshmikantham MV, Cava MP: Norisohaplophytine and haplocidiphytine—new bisindole alkaloids from *Haplophyton cimicidum* (Apocynaceae). *Heterocycles* 20;1511–1517, 1983

Ahlawat SK, Agarwal AK, Wadhwa S: Rare poisoning with *Cerebra thevetia* (yellow oleander): a report of three cases. *Trop Doct* 24;37–38, 1994

Alfonso HA, Sanchez LM, Merino N, Gomez BC: *Nerium oleander* intoxication in geese. *Rev Salud Anim* 13;183–184, 1991

Angulo JJ: Poisoning by the toxic principle of "curamaguey." *Rev Med Trop Parasitol* (Havana) 8;47–49, 1942
Ansford AJ, Morris H: Oleander poisoning. *Toxicon* (Suppl) 3;15–16, 1983
Anthony DC, Graham DG: Toxic responses of the nervous system. In *Casarett and Doull's Toxicology,* 4th ed, Amdur MO, Doull J, Klaassen CD eds, Pergamon Press, New York, pp407–429, 1991
Armien AG, Tokarnia CH: Experiments on the toxicity of some ornamental plants in sheep. *Pesq Vet Bras* 14;69–73, 1994
Atta-ur-Rahman, Muzaffar A: Steroidal alkaloids of Apocynaceae and Buxaceae. In *The Alkaloids,* Brossi R ed, Academic Press, San Diego, vol 32, pp79–239, 1988
Ban Y, Murakami Y, Iwasawa Y, Tsuchiya M, Takano N: Indole alkaloids in medicine. *Med Res Rev* 8;231–308, 1988
Bania T, Hoffman RS, Howland MA, Goldfrank LR: Accidental Indian hemp *(Apocynum cannabinum)* cardiac glycoside toxicity [abstract]. *Vet Hum Toxicol* 35;328, 1993
Bisset NG: Phytochemistry and pharmacology of *Voacanya* species. In *Series of Revisions of Apocynaceae* 15, Leeuwenberg AJM ed, Wageningen Agric Univ Pap 85–3, The Netherlands, pp81–122, 1985
Bisset NG: Arrow and dart poisons. *J Ethnopharmacol* 25;1–41, 1989a
Bisset NG: *Tabernanthe:* uses, phytochemistry, and pharmacology. In *Series of Revisions of Apocynaceae* 29, 30, Leeuwenberg AJM ed, Wageningen Agric Univ Pap 89–4, The Netherlands, pp19–26, 1989b
Biswas D, Bhattacharyya B: Oestrogenic potency of petroleum ether extract of some indigenous plant materials. *Indian J Anim Sci* 66;152–153, 1996
Blohm H: *Poisonous Plants of Venezuela.* Harvard Univ Press, Cambridge, 1962
Blum LM, Rieders F: Oleandrin distribution in a fatality from rectal and oral *Nerium oleander* extract administration. *J Anal Toxicol* 11;219–221, 1987
Brander GC, Pugh DM, Bywater RJ, Jenkins WL: *Veterinary Applied Pharmacology and Therapeutics,* 5th ed. Bailliere Tindall, London, 1991
Chen KK, Chen AL: The action of crystalline thevetin, a cardiac glycoside of *Thevetia neriifolia. J Pharmacol Exp Ther* 51;23–34, 1934a
Chen KK, Chen AL: The constituents of be-still nuts, *Thevetia neriifolia. J Biol Chem* 105;231–240, 1934b
Chen KK, Henderson FG: Pharmacology of sixty-four cardiac glycosides and aglycones. *J Pharmacol Exp Ther* 111;365–383, 1965
Cheung K, Hinds JA, Duffy P: Detection of poisoning by plant-origin cardiac glycoside with the Abbott TDx analyser. *Clin Chem* 35;295–297, 1989
Chien MM, Svoboda GH, Schiff PL Jr, Slatkin DJ, Knapp JE: Chemical constituents of *Echites hirsuta* (Apocynaceae). *J Pharm Sci* 68;247–249, 1979
Chopra RN, Badhwar RL, Ghosh S: *Poisonous Plants of India.* Academic Publ, Jaipur, India, pp670–690, 1984
Clivio P, Richard B, Deverre J-R, Sevenet T, Zeches M, Le Men-Oliver L: Alkaloids form the leaves and root bark of *Ervatamia hirta. Phytochemistry* 30;3785–3792, 1991
Coffey S: *Rodd's Chemistry of Carbon Compounds,* 2d ed. Vol 2, part D, *Alicyclic Compounds.* Elsevier, Amsterdam, pp360–421, 1970
Connolly JD, Hill RA: *Dictionary of Terpenoids.* Vol 1, *Mono Terpenoids.* Chapman & Hall, London, 1991a
Connolly JD, Hill RA: *Dictionary of Terpenoids.* Vol 2, *Di- and Higher Terpenoids.* Chapman & Hall, London, 1991b
Coppen JJ, Cobb AL: The occurrence of iridoids in *Plumeria* and *Allamanda. Phytochemistry* 22;125–128, 1983
Cordell GA: The bisindole alkaloids. In *Indoles,* part 4, *The Monoterpenoid Indole Alkaloids,* Saxton JE ed, Wiley & Sons, New York, pp539–728, 1983
Correll DS, Johnston MC: *Manual of the Vascular Plants of Texas.* Univ Texas at Dallas, Richardson, Tex, 1979
Couch JF: The chemistry of stock-poisoning plants. *J Chem Educ* 14;16–30, 1937

Craig JF, Kehoe D: Plant poisoning. *Vet Rec* 5;795–816, 1925

Creasey WA: Pharmacology, biochemistry, and clinical applications of the monoterpenoid alkaloids. In *Indoles,* part 4, *The Monoterpenoid Indole Alkaloids,* Saxton JE ed, J Wiley & Sons, New York, pp783–829, 1983

Dagnino D, Schripsema J, Peltenburg A, Verpoorte R, Teunis K: Capillary gas chromatographic analysis of indole alkaloids: investigation of the indole alkaloids present in *Tabernaemontana divaricata* cell suspension culture. *J Nat Prod* 54;1558–1563, 1991

Deecher DC, Teitler M, Soderlund DM, Bornmann WG, Kuehne M, Glick SD: Mechanisms of action of ibogaine and harmaline congeners based on radioligand binding studies. *Brain Res* 571;242–247, 1992

Douglas D: *Journal Kept by David Douglas during His Travels in North America, 1823–1827, Together with a Particular Description of Thirty-three Species of American Oaks and Eighteen Species of* Pinus. Antiquarian Press, New York, 1959

Durrell LW, Newsom IE: Colorado's poisonous and injurious plants. *Colo Exp Sta Bull* 455;32, 1939

Elderfield RC: On thevetin. J Biol Chem 115;247–252, 1936

Evans DA, Kaleysa Raj R: Extracts of Indian plants as mosquito larvacides. *Indian J Med Res* 88;38–41, 1988

Everist SL: *Poisonous Plants of Australia,* 2d ed. Angus & Robertson, Sydney, Australia, pp77–89, 1981

Farnsworth NR: The pharmacognosy of the periwinkles: *Vinca* and *Catharanthus. Lloydia* 24;105–138, 1961

Farnsworth NR, Draus FJ, Sager RW, Bianculli JA: Studies on *Vinca major* L. (Apocynaceae). 1. Isolation of perivincine. *J Am Pharm Assoc* 49;589–591, 1960

Farnsworth NR, Fong HHS, Blomster RN, Draus FJ: Studies on *Vinca major* (Apocynaceae). 2. Phytochemical investigation. *J Pharm Sci* 51;217–224, 1962

Fax EM: *Urechites lutea* (L.) Britton toxicity in cattle. *Vet Hum Toxicol* 38;313–314, 1996

Finnemore H: The constituents of Canadian hemp. Part 1, Apocynin. *J Chem Soc* (London) 93;1513–1519, 1908

Fuller TC, McClintock E: *Poisonous Plants of California.* Univ California Press, Berkeley, 1986

Galey FD, Holstege DM, Plumlee KH, Tor E, Johnson B, Anderson ML, Blanchard PC, Brown F: Diagnosis of oleander poisoning in livestock. *J Vet Diagn Invest* 8;358–364, 1996

Galey FD, Holstege DM, Johnson BJ, Siemans L: Toxicity and diagnosis of oleander *(Nerium oleander)* poisoning in livestock. In *Toxic Plants and Other Natural Toxicants,* Garland T, Barr AC, eds, New York, pp215–219, 1998

Gerber JG, Nies AS: Antihypertensive agents and the drug therapy of hypertension. In *The Pharmacological Basis of Therapeutics,* 8th ed, Gilman AG, Rall TW, Nies AS, Taylor P eds, McGraw-Hill, New York, pp784–813, 1990

Gerzon K: Dimeric *Catharanthus* alkaloids. In *Anticancer Agents Based on Natural Product Models,* Cassady JM, Douros JD eds, Academic Press, New York, pp271–317, 1980

Glendinning JI: Effectiveness of cardenolides as feeding deterrents to *Peromyscus* mice. *J Chem Ecol* 18;1559–1575, 1992

Govindachari TR: Studies in alkaloids. *J Ind Chem Soc* 45;945–957, 1968

Grinnell EH, Smith PW: Effect of estrogens on myocardial sensitivity to toxic effects of digoxin. *Proc Soc Exp Biol Med* 94;524–526, 1957

Gunawardana VK, Goonesekera MM, Gunaherath GMKB: Embryotoxic effect of *Plumeria rubra.* In *Toxic Plants and Other Natural Toxicants,* Garland T, Barr AC eds, New York, pp317–322, 1998

Hardin JW, Arena JM: *Plant poisoning from native and cultivated plants.* Duke Univ Press, Durham, NC, pp132–133, 1974

Haynes BE, Bessen HA, Wightman WD: Oleander tea: herbal draught of death. *Ann Emerg Med* 14;350–353, 1985

Hoffman BB, Lefkowitz RJ: Adrenergic receptor antagonists. In *The Pharmacological Basis of Therapeutics,* 8th ed, Gilman AG, Rall TW, Nies AS, Taylor P eds, McGraw-Hill, New York, pp221–243, 1990

Hoffman BF, Bigger JT Jr: Digitalis and allied cardiac glycosides. In *The Pharmacological Basis of Therapeutics,* 8th ed, Gilman AG, Rall TW, Nies AS, Taylor P eds, McGraw-Hill, New York, pp814–839, 1990

Hoffman VS, Weiss E, Reichstein T: Adigosid, strukturbestimmung. *Helv Chim Acta* 49;1855–1872, 1966

Infante JF, Perdomo J, Merino N, Joa R: Estudio morfopatologico en bovinos intoxicados experimentalmente con bejuco marrullero *(Urechites lutea)* en la provincia Granma. *Rev Salud Anim* 6;193–204, 1984

Jager H, Schindler O, Reichstein T: Die glykoside der samen von *Nerium oleander* L. *Helv Chim Acta* 42;977–1014, 1959

Johnson EP, Archer WA: The principal stock-poisoning plants of New Mexico. *N M Agric Ext Serv Circ* 71, 1922

Johnson IS, Armstrong JG, Gorman M, Burnett JP Jr: The *Vinca* alkaloids: a new class of oncolytic agents. *Cancer Res* 23;1390–1427, 1963

Jortani SA, Helm RA, Valdes R jr: Inhibition of Na,K-Atpase by oleandrin and oleandrigenin, and their detection by digoxin immunoassays. *Clin Chem* 42;1654–1658, 1996

Joubert JPJ: Cardiac glycosides. In *Toxicants of Plant Origin,* vol 2, *Glycosides,* Cheeke PR ed, CRC Press, Boca Raton, FL, pp61–96, 1989

Joubert JPJ, Schultz RA: The minimal effective dose of activated charcoal in the treatment of sheep poisoned with cardiac glycoside containing plant *Moraea polystachya* (Thunb.) Ker Gawl. *J S Afr Vet Assoc* 53;265–266, 1982a

Joubert JPJ, Schultz RA: The treatment of *Moraea polystachya* (Thunb) ker-gawl (cardiac glycoside) poisoning in sheep and cattle with activated charcoal potassium chloride. *J S Afr Vet Assoc* 53;249–253, 1982b

Kakrani AL, Rajput CS, Khandare SK, Redkar VE: Yellow oleander seed poisoning with cardiotoxicity. *Indian Heart J* 33;31–33, 1981

Kam T-S, Nyeoh K-T, Sim K-M, Yoganathan K: Alkaloids from *Alstonia scholaris. Phytochemistry* 45;1303–1305, 1997

Karawya MS, Balbaa SI, Khayyal SE: Estimation of cardenolides in *Nerium oleander. Planta Med* 23;70–73, 1973

Kardono LBS, Tsauri S, Padmawinata K, Pezzuto JM, Kinghorn AD: Cytotoxic constituents of the bark of *Plumeria rubra* collected in Indonesia. *J Nat Prod* 53;1447–1455, 1990

Keeler RF: Naturally occurring teratogens from plants. In *Handbook of Natural Toxins,* vol 1, *Plant and Fungal Toxins,* Keeler RF, Tu AT eds, Marcel Dekker, New York, pp161–199, 1983

Kellerman TS, Coetzer JAW, Naude TW: *Plant Poisonings and Mycotoxicoses of Livestock in Southern Africa.* Oxford Univ Press, Cape Town, South Africa, pp84–87, 1988

Kelly RB, Daniels EG, Spaulding LB: Cytotoxicity of cardiac principles. *J Med Chem* 8;547–548, 1965

Kingsbury JM: *Poisonous Plants of the United States and Canada.* Prentice Hall, Englewood Cliffs, NJ, pp262–267, 1964

Kisakurek MV, Leeuwenberg AJM, Hesse M: A chemotaxonomic investigation of the plant families of Apocynaceae, Loganiaceae, and Rubiaceae by their indole alkaloid content. In *Alkaloids, Chemical and Biological Perspectives,* vol 1, Pelletier SW ed, J Wiley & Sons, New York, pp221–376, 1983

Krochmal A, Krochmal C: *A Guide To Medicinal Plants of the United States.* Quadrangle, New York, pp38–39, 1973

Kupchan SM, Hemingway RJ, Doskotch RW: Tumor inhibitors IV. Apocannoside and cymarin, the cytotoxic principles of *Apocynum cannabinum* L. *J Med Chem* 7;803–804, 1964

Kupchan SM, Mokotoff M, Sandhu RS, Hokin LE: The chemistry and biological activity of derivations of strophanthidin. *J Med Chem* 10:1025–1033, 1967

Kupchan SM, Dessertine AL, Blaylock BT, Bryan RF: Isolation and structural elucidation of allamandin, an antileukemic iridoid lactone from *Allamanda cathartica. J Org Chem* 39;2477–2482, 1974

Langford SD, Boor PJ: Oleander toxicity: an examination of human and animal toxic exposures. *Toxicology* 109;1–13, 1996

Leeuwenberg AJM: Taxa of the Apocynaceae above the genus level. In *Series of Revisions of Apocynaceae* 38, Wageningen Agric Univ Pap 94–3, The Netherlands, pp45–60, 1994

Mahin L, Marzou A, Huart A : A case report of *Nerium oleander* in cattle. *Vet Hum Toxicol* 26:303–304, 1984

Marrero E, Colome H, Alfonso HA, Aguilera JM, Mella C, Fajardo M: Efecto cardiaco de extractos acuosos de plantas sapuestamente toxicas procedentes de pastos de una region de la provincia Granma. *Rev Salud Anim* 4;127–140, 1982

Marrero E, Colome H, Infante JF, Valerino P, Fajardo M: Reproduccion experimental del sindrome hemorrhagico y muerte subita en terneros por consumo directo y decoccion de *Urechites lutea* (bejuco marrullero). *Rev Salud Anim* 5;345–352, 1983

Marsh CD, Clawson AB, Roe GC: *Four Species of Range Plants Not Poisonous to Livestock*. USDA Tech Bull 93, 1928

Miller R: Oleander poisoning in a two-toed sloth. *J Zoo Anim Med* 4(2);14, 1973

Mokry J, Kompis I: On the alkaloids from *Vinca minor. Lloydia* 27;428–434, 1964

Moore CW: The constituents of the rhizome of *Apocynum androsaemifolium. J Chem Soc* (London) 95;734–751, 1909

Morton JF: *500 Plants of South Florida.* EA Seemann, Miami, 1974

Morton JF: *Major Medicinal Plants: Botany, Culture, and Uses.* CC Thomas, Springfield, IL, pp233–267, 1977

Morton JF: *Plants Poisonous to People in Florida,* 2d ed. JF Morton, Southeastern Printing, Stuart, FL, 1982

Mroue M, Alam M: Crooksiine, a bisindole alkaloid from *Haplophyton crooksii. Phytochemistry* 30;1741–1744, 1991

Mroue MA, Ghuman MA, Alam: Crooksidine, an indole alkaloid from *Haplophyton crooksii. Phytochemistry* 33;217–219, 1993

Mroue MA, Euler KL, Ghuman MA, Alam M: Indole alkaloids of *Haplophyton crooksii. J Nat Prod* 59;890–893, 1996

Natochin YV, Lavrova EA: Inhibition of sodium transport in frog skin by cardiac glycosides of different molecular structure. *Experientia* 28;942–943, 1972

Neuwinger HD: Afrikanische pfeilgifte. *Naturwiss Rundsch* 27;340–359, 385–402, 1974

Newsholme SJ, Coetzer JAW: Myocardial pathology of domestic ruminants in southern Africa. *J S Afr Vet Assoc* 55;89–96, 1984

Noble RL: The discovery of the vinca alkaloids—chemotherapeutic agents against cancer. *Biochem Cell Biol* 68;1344–1351, 1990

Ode RH, Pettit GR, Kamano Y: Cardenolides and bufadienolides. In Int Rev Sci, *Org Chem,* ser 2, vol 8, *Steroids,* Johns WF ed, Butterworths, London, pp145–171, 1976

O'Hearn E, Molliver ME: Degeneration of Purkinje cells in parasagittal zones of the cerebellar vermis after treatment with ibogaine or harmaline. *Neuroscience* 55;303–310, 1993

Oji O, Madubuike FN, Nwalozie MC: Mortality in rats following dietary inclusion of *Thevetia neriifolia* seeds. *Fitoterapia* 64;137–139, 1993

Omino EA, Kokwaro JO: Ethnobotany of Apocynaceae species in Kenya. *J Ethnopharmacol* 40;167–180, 1993

Oryan A, Maham M, Rezakhani A, Maleki M: Morphological studies on experimental oleander poisoning in cattle. *J Vet Med A* 43;625–634, 1996

Pahwa R, Chatterjee VC: The toxicity of yellow oleander (*Thevetia neriifolia* Juss) seed kernels to rats. *Vet Hum Toxicol* 32;561–564, 1990

Paper D, Franz G: Glycosylation of cardenolide aglycones in the leaves of *Nerium oleander. Planta Medica* 55;30–34, 1989

Pawelka KH, Stoeckigt J: Indole alkaloids from *Ochrosia elliptica* plant cell suspension cultures. *Z Naturforsch C Biosci* 41;381–384, 1986

Perkins KD, Payne WW: *Guide to the Poisonous and Irritant Plants of Florida.* Fla Coop Ext Serv Circ 441, 1978

Pina Luis M delC, Fanghaenel E: Isolation of oleandrigenin from *Angadenia berterii. Ciencias,* ser 3, 14, 1972; *Chem Abstr* 78;94880, 1973

Radford DJ, Gillies AD, Hinds JA, Duffy P: Naturally occurring cardiac glycosides. *Med J Aust* 144;540–544, 1986

Radford DJ, Cheung K, Urech R, Gollogly JR, Duffy P: Immunological detection of cardiac glycosides in plants. *Aust Vet J* 71;236–238, 1994

Raffauf RF: *A Handbook of Alkaloids and Alkaloid-Containing Plants.* Wiley-Interscience, New York, 1970

Raffauf RF: *Plant Alkaloids: A Guide to Their Discovery and Distribution.* Haworth Press, New York, 1996

Ratnagiriswaran AN, Venkatachalam K: The phytochemistry of the bark of *Tabernaemontana coronaria* Br. *Q J Pharm Pharmacol* 12:174–181, 1939

Reagor JC: Increased oleander poisoning after extensive freezes in south/southeast Texas [report]. *Southwest Vet* 36;95, 1985

Reichstein T: Chemische rassen in *Acokanthera. Planta Med* 13;382–399, 1965

Riopelle RJ, Dow KE, Eisenhauer E: Some observations on vinca alkaloid and ethanol neurotoxicity using dissected neuronal cultures. *Can J Physiol Pharmacol* 62;1032–1036, 1984

Saraswat DK, Garg PK, Saraswat M: Rare poisoning with *Cerebra thevetia* (yellow oleander): review of 13 cases of suicidal attempt. *J Assoc Physicians India* 40;628–629, 1992

Sartorelli AC, Creasey WA: Cancer chemotherapy. *Ann Rev Pharmacol* 9;51–72, 1969

Saxton JE: Alkaloids of *Alstonia* species. In *The Alkaloids,* vol 8, Manske RHF ed, Academic Press, New York, pp159–202, 1965a

Saxton JE: Alkaloids of *Haplophyton cimicidum.* In *The Alkaloids,* vol 8, Manske RHF ed, Academic Press, New York, pp673–678, 1965b

Schlittler E: *Rauwolfia* alkaloids with special reference to the chemistry of reserpine. In *The Alkaloids,* vol 8, Manske RHF ed, Academic Press, New York, pp287–334, 1965

Schmidt DH, Butler VP Jr: Reversal of digoxin toxicity with specific antibodies. *J Clin Invest* 50;1738–1744, 1971

Schultes RE: Hallucinogens of plant origin. Science 163;245–254, 1969

Schwartz WL, Bay WW, Dollahite JW, Storts RW, Russell LH: Toxicity of *Nerium oleander* in the monkey *(Cebus apella). Vet Pathol* 11;259–277, 1974

Selva Raj VB, Ganapathy MS: Studies on the toxicity of *Lochnera pusilla* K Schum. *Indian Vet J* 44;871–876, 1967

Sen G, Bose KC: *Rauwolfia serpentina,* a new Indian drug for insanity and high blood pressure. *Indian J Med World* 2;194–201, 1931

Shaw D, Pearn J: Oleander poisoning. *Med J Aust* 2;267–269, 1979.

Shoppee CW: *Chemistry of the Steroids,* 2d ed. Butterworths, Washington, DC, pp326–397, 1964

Shropshire CM, Stauber E, Arai M: Evaluation of selected plants for acute toxicosis in budgerigars. *J Am Vet Med Assoc* 200;936–939, 1992

Shumaik GM, Wu AW, Ping AC: Oleander poisoning: treatment with digoxin-specific Fab antibody fragments. *Ann Emerg Med* 17;732–735, 1988

Singh B, Rastogi RP: Cardenolides—glycosides and genins. *Phytochemistry* 9;315–331, 1970

Smith TW, Haber E, Yeatman L, Butler VP Jr: Reversal of advanced digoxin intoxication with Fab Fragments of digoxin-specific antibodies. *N Engl J Med* 294;797–800, 1976

Snieckus V: The distribution of indole alkaloids in plants. In *The Alkaloids,* vol 11, Manske RHF ed, Academic Press, New York, pp1–40, 1968

Southon IW, Buckingham J: *Dictionary of Alkaloids.* Chapman & Hall, London, 1989

Steyn DG: *The Toxicology of Plants in South Africa.* Central News Agency, South Africa, 1934

Svoboda GH, Gorman M, Root MA: Alkaloids of *Vinca rosea (Catharanthus roseus).* 28. A preliminary report on hypoglycemic activity. *Lloydia* 27;361–363, 1964

Szabuniewicz M, Schwartz WL, McCrady JD, Russell LH, Camp BJ: Experimental oleander poisoning and treatment. *Southwest Vet* 25;105–114, 1972

Taylor WI: The *Vinca* alkaloids. In *The Alkaloids,* vol 8, Manske RHF ed, Academic Press, New York, pp269–285, 1965

Thimnah K: Nerium poisoning in cattle. *Ind Vet J* 49;942–946, 1972

Thorp RH, Watson TR: A survey of the occurrence of cardio-active constituents in plants growing wild in Australia. *Aust J Exp Biol* 31;529–532, 1953

Tokarnia CH, Dobereiner J, da Silva MF: *Plantas toxicas da Amazonia a bovinos e outros herbivoros.* Instituto Nacional de Pesquisas da Amazonia, Manaus, Brazil, 1979

Tokarnia CH, Armien AG, Peixoto PV, Barbosa JD: Experiments on the toxicity of some ornamental plants in cattle. *Pesq Vet Bras* 16;5–20, 1996

Tor ER, Holstege DM, Galey FD: Determination of oleander glycosides in biological matrices by high-performance liquid chromatography. *J Agric Food Chem* 44;2716–2719, 1996

Wakim KG, Chen KK: The action of alstonine. *J Pharm Exp Ther* 90;57–67, 1947

Watt JM, Breyer-Brandwijk MG: *The Medicinal and Poisonous Plants of Southern and Eastern Africa,* 2d ed. E & S Livingston, Edinburgh, 1962

West E: *Poisonous Plants around the Home. Fla Agric Exp Sta Circ* S-100, 1957

Westermarck H: Digitalis inactivation in vitro due to rumen microbes from sheep. *Acta Vet Scand* 1;67–73, 1959

Wilson FW: Oleander poisoning of livestock. *Ariz Agric Exp Sta Bull* 59;381–397, 1909

Yamauchi T, Abe F: Studies on the constituents of *Adenium.* 1. cardiac glycosides and pregnanes from *Adenium obesum. Chem Pharm Bull* 38;669–672, 1990

Yamauchi T, Takata N, Mimura T: Cardiac glycosides of the leaves of *Nerium odorum. Phytochemistry* 14;1379–1382, 1975

Zimmermann JL, Todd GC, Tamura RN: Target organ toxicity and leukopenia in Fischer 344 rats given intravenous doses of vinblastine and desacetyl vinblastine. *Fundam Appl Toxicol* 17;482–493, 1991

Zucker AR, Lacina SJ, Das Gupta DS, Fozzard HA, Mehlman D, Butler VP, Haber E, Smith TW: Fab fragments of digoxin-specific antibodies used to reverse ventricular fibrillation induced by digoxin in a child. *Pediatrics* 70;468–471, 1982

Aquifoliaceae Bartl. 9

Commonly known as the holly family, the Aquifoliaceae comprises 4 genera and 410–420 species, the majority of which are in the genus *Ilex*. Widely distributed in both temperate and tropical regions, the family is especially common in eastern North America and eastern Asia but is poorly represented in Africa and Australia. The Aquifoliaceae is economically important for wood and cultivated ornamentals. In North America, 2 genera are present: *Nemopanthus,* with 2 species commonly known as the mountain hollies, and *Ilex,* with 29 species. Reports of intoxication problems involve only *Ilex*.

■ Holly Family
Ilex

Plants trees or shrubs; deciduous or evergreen; dioecious or rarely polygamodioecious. **Stems** with terminal buds absent. **Leaves** simple; alternate; coriaceous or herbaceous; venation pinnate; apices acute or obtuse or spinescent; margins crenate or spinose; stipules present, caducous. **Inflorescences** of 2 types, staminate and pistillate different; axillary. **Staminate Inflorescences** compound cymes. **Pistillate Inflorescences** small clusters or solitary flowers. **Flowers** imperfect or perfect, staminate and pistillate similar; perianths in 2-series; radially symmetrical; rotate. **Sepals** 4–6; fused. **Petals** 4–6; free or fused; greenish white to yellow. **Stamens** 4–6; androecial rudiments present in pistillate flowers. **Pistils** 1; compound, carpels 4–6; stigmas 4–6, capitate; styles absent; ovaries superior; placentation axile; gynoecial rudiments present in staminate flowers. **Fruits** drupes; falsely resembling berries because of multiple stones. **Seeds** 1–6.

■ shrubs or trees; leaves simple, alternate; petals 4–6, greenish white to yellow; drupes with multiple stones

Ilex L.

Taxonomy and Morphology—Comprising about 400 species, *Ilex* is prized for its wood, foliage, and fruits. Its hard, white wood is used for carving, inlays, veneer, and paneling. Many of its species are cultivated as ornamental shrubs for their attractive foliage and colorful yellow, red, or orange fruits, which are very conspicuous in the fall and winter. Holly has long been symbolic of Christmas, and the common name is believed by some to be a corruption of *holy* (Galle 1997). Numerous cultivars, e.g., yaupon and Burford holly, have been developed. In South America, yerba mate, a popular tea, is made from *I. paraguariensis*. Because of the resemblance of the leaves to those of the Mediterranean holly oak, the generic name is believed derived from the ancient Latin name for *Quercus ilex*. In North America, 29 native and introduced species are present, of which the following are representative:

I. aquifolium L.	English holly, Christmas holly
I. cornuta Lindl. & Paxton	Chinese holly
I. crenata Thunb.	Japanese holly
I. glabra (L.) A.Gray	inkberry, gallberry

■ *I. aquifolium*
I. cornuta
I. crenata
I. glabra
I. opaca
I. verticillata
I. vomitoria

black alder
cassena
feverbush
gallberry
holly
inkberry
winterberry
yaupon

I. opaca Aiton	American holly, white holly
I. verticillata (L.) A.Gray	winterberry, black alder, feverbush
I. vomitoria Aiton	yaupon, cassena

■ shrubs or trees; petals white, greenish white, or yellow; drupes black, yellow, red, or orange

Plants trees or shrubs; deciduous or evergreen. **Stems** erect or ascending; 1–15 m tall. **Leaves** cartilaginous or coriaceous or herbaceous; blades of various shapes. **Flowers** imperfect or rarely perfect. **Sepals** fused. **Petals** free or fused; white or greenish white to yellow. **Stamens** inserted on corolla tubes. **Drupes** subglobose; black or yellow or red to orange.

Ilex opaca

1. Leaves deciduous; herbaceous . . . . *I. verticillata*
1. Leaves evergreen; coriaceous.
   2. Leaf margins spinose.
      3. Flowers and drupes borne on current season's wood. Plants native . . . . *I. opaca*
      3. Flowers and drupes borne on previous season's wood. Plants introduced ornamentals.
         4. Leaves elliptic or ovate or oblong-ovate . . . . *I. aquifolium*
         4. Leaves rhombic to rectangular . . . . *I. cornuta*
   2. Leaf margins serrate or crenate or entire.
      5. Drupes red or yellow. Stones ribbed . . . . *I. vomitoria*
      5. Drupes black. Stones smooth.
         6. Leaf margins serrate from apex to base. Petals and stamens 4 . . . . *I. crenata*
         6. Leaf margins serrate only at apex. Petals and stamens 5–8 . . . . *I. glabra*

Ilex verticillata

■ ornamentals

Distribution and Habitat—With an apparent center of distribution in Central and South America, *Ilex* is common in the eastern forests of North America and Asia. Native North American species have similar geographic ranges. *Ilex glabra* occurs in low sandy pinelands, swamps, and bogs of the Atlantic and Gulf coastal plain. *Ilex opaca* inhabits sandy soils of moist woods, stream banks, and swamps, especially in coastal areas of the Southeast. *Ilex verticillata* is present in swamps and wet woods of the Northeast. More tolerant of alkaline soils than the other species, *I. vomitoria* occurs in low woods and sandy pinelands in the southeastern states.

Ilex aquifolium, I. cornuta, and I. crenata are ornamental introductions from Europe, China, and Japan. *Ilex aquifolium* has naturalized in coastal California from San Francisco northward. Many of varieties of these 3 species are hardy only in the southern United States.

Ilex glabra

■ source of teas and medicinal preparations

Disease Problems—The foliage and drupes of *Ilex* are eaten by a variety of wildlife species (Gill & Healy 1973). Humans likewise have used the genus for food and medicine, as Galle (1997) described in extensive detail in his monograph. In addition to all-purpose yerba mate, or Paraguay tea, which is drunk by millions of people in South America, a drink known as guayusa is used as a coffee or tea substitute by Indian tribes in the Andes. The "black drink," or "cassena," brewed from the roasted leaves of *I. vomitoria* was used in ceremonies by American Indians of the southern United States to restore lost appetites, enhance health, and promote courage and agility in war. The medicinal value of *Ilex* also has long been recognized. In the Northeast, cases of intermittent fevers were treated with the fruits and bark of *I. verticillata*. Its bark also was used to treat gangrene and skin eruptions. *Ilex cassine,* a southeastern species, was mixed with lard as an ointment for smallpox. It was additionally employed to render coastal swamp water fit to drink.

Since the 1700s the drupes of various species of *Ilex* have been recognized to possess emetic and laxative properties and have been used medicinally as emetics. Millspaugh (1974) recorded an incident in which an individual ate 25 berries, became nauseous, vomited, and had diarrhea but did not exhibit colic. There have been numerous reports, mainly in the eastern United States, of children eating fruits (O'Leary 1964). The disease is usually mild to moderate in severity and limited to vomiting and diarrhea; rarely is there narcosis (Arena 1979). Rodrigues and co-workers (1984) described a case of 2-year-old twins eating a handful of drupes of *I. opaca*. After treatment with syrup of ipecac, vomiting persisted for several hours. Later one child had green, watery diarrhea plus several more episodes of vomiting. Rarely, digestive tract problems are reported for horses or other animal species (Finance 1987).

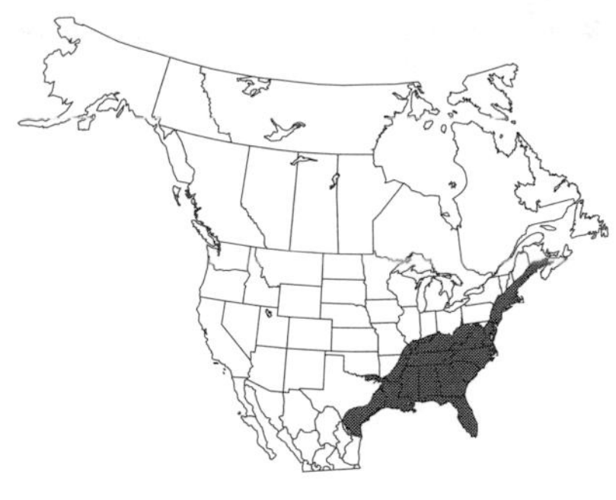

Ilex opaca

- ingestion of drupes, digestive disturbance, vomiting, diarrhea

- leaves, caffeine, theobromine, saponins; drupes, main risk, saponins

Disease Genesis—*Ilex* contains an array of potential toxicants, including glucosidic saponins in the leaves and fruit (West et al. 1977); the methylxanthines caffeine and possibly theobromine and theophylline in the leaves (Willaman & Schubert 1961); lactones (Thomas & Budzikiewicz 1980); and in some instances, a novel cyanogenic diglucoside, dihydromandelonitrile, in the leaves of *I. aquifolium, I. glabra,* and *I. verticillata* (Gibbs 1974; Willems 1988). However, even though this array of toxicants is present, most of the adverse effects appear to be due to the irritant effects of the saponins. These compounds, a series of glycosides of oleanolic acid and α-ilexanolic acid genins, also exert other actions, such as inhibition of histamine release from mast cells and deterrence of insect feeding on the leaves (Connolly & Hill 1991).

Ilex verticillata

Caffeine

Oleanolic acid

Cyanogenic potential seems to be quite variable in these species. It was reported to be highest in the fruits of *I. aquifolium*—0.28% fresh weight and 0.71% d.w. (Willems 1989). In another report, the seeds were negative and the fruit pulp positive for cyanogenesis (Barnea et al. 1993). Using the picrate paper test (see Rosaceae, in chapter 64), we have found the foliage of a number of species and cultivars to be generally negative for cyanogenesis. In addition to the methylxanthine stimulatory effects, digitalis-like strengthening of cardiac contraction and even ventricular arrest in frog and rabbit hearts in vitro have been shown, using water and/or alcohol extracts (Waud 1932).

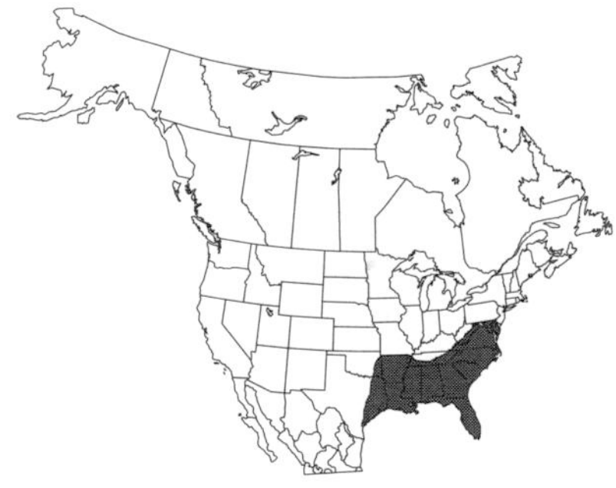

Ilex vomitoria

- weakly cyanogenic, little risk

Clinical Signs and Pathology—In most instances of *Ilex* intoxication, the signs will be indicative of mild to moderate dysfunction of the digestive tract and include vomiting and diarrhea, which last only a few hours. In more-severe cases there may be some degree of narcosis. A lethal outcome is not likely.

- transient vomiting, diarrhea

■ activated charcoal, fluids, electrolytes

Treatment—In the event that signs are severe enough to warrant intervention, treatment is primarily relief of the digestive effects, especially chemical control of the persistent vomiting and replacement of fluids and/or electrolytes. Activated charcoal given orally may provide adequate treatment in many instances.

References

Arena JM: Are holly berries toxic? *J Am Med Assoc* 242;2341, 1979

Barnea A, Harborne JB, Pannell C: What part of fleshy fruits contain secondary compounds toxic to birds and why? *Biochem Syst Ecol* 21;421–429, 1993

Connolly JD, Hill RA: *Dictionary of Terpenoids.* Vol 2, *Di- and Higher Terpenoids.* Chapman & Hall, London, pp654–1460, 1991

Finance B: Poisoning by leaves of holly *(Ilex aquifolium)* in two Comtois foals. *Pferdeheilkunde* 3;43–44, 1987

Galle FC: Hollies: *The Genus* Ilex. Timber Press, Portland, OR, 1997.

Gibbs RD: *Chemotaxonomy of Flowering Plants.* McGill-Queens Univ Press, Montreal, 1974

Gill JD, Healy WM: *Shrubs and Vines for Northeastern Wildlife.* USDA Forest Service, Gen Tech Rep NE-9, 1973

Millspaugh CF: *American Medicinal Plants.* Dover, New York, pp415–418, 1974 (reprinted from 1892)

O'Leary SB: Poisoning in man from eating poisonous plants. *Arch Environ Health* 9;216–242, 1964

Rodrigues TD, Johnson PN, Jeffrey LP: Holly berry ingestion: case report. *Vet Hum Toxicol* 26;157–158, 1984

Thomas H, Budzikiewicz H: Ilex-Lacton, ein Bisnormonoterpen neuartiger Struktur aus *Ilex aquifolium. Phytochemistry* 19;1866–1868, 1980

Waud RA: A digitalis-like action of extracts made from holly. *J Pharmacol Exp Ther* 45;279, 1932

West LG, McLaughlin JL, Eisenbeiss GK: Saponins and triterpenes from *Ilex opaca. Phytochemistry* 16;1846–1847, 1977

Willaman JJ, Schubert BG: Alkaloid-bearing plants and their contained alkaloids. USDA Tech Bull 1234, pp216–230, 1961

Willems M: A cyanogenic glycoside from *Ilex aquifolium. Phytochemistry* 27;1852–1853, 1988

Willems M: Quantitative determination and distribution of a cyanogenic glucoside in *Ilex aquifolium. Planta Med* 55;195, 1989

Araceae Juss. 10

Commonly known as the arum family, the Araceae comprises about 110 genera and 1,800–2,000 species, the vast majority of them occurring in tropical and subtropical regions. Many are cultivated in North America, and some are commonly recognized as toxic. In regions with warm winter climates, they may be used as outdoor ornamentals; however, in most parts of North America, they are grown as indoor ornamentals or used as cut flowers. Familiar ornamentals include *Philodendron, Caladium, Zantedeschia* (calla or arum lily), *Spathiphyllum* (white sails), *Anthurium* (painter's palette), and *Dieffenbachia* (dumb cane). Flowering in the early spring, species of *Arisaema* such as *A. dracontium* (green dragon) and *A. triphyllum* (jack-in-the-pulpit) are encountered in the moist soils of our eastern deciduous forests. Although most of the genera grown as houseplants rarely flower, all have the characteristic inflorescence of the family, which consists of numerous tiny flowers borne on a prominent fleshy axis (spadix) and subtended by a large bract (spathe). Typically quite conspicuous, the spathe is usually white, red, yellow, or pink.

In addition to their use as ornamentals, members of the Araceae are economically quite important for food. *Colocasia* (taro, dasheen) and *Xanthosoma* (tanier) are commercially cultivated in some parts of the world for their starchy, tuberous corms. The starch grains of *Colocasia* are small and easily digested, thus making it a good food for infants and individuals with digestive problems. Also eaten are the corms of species of *Alocasia, Amorphophallus,* and *Cyrtosperma.* An unusual, free-floating aquatic, *Pistia,* is used as an aquarium or fish pond ornamental but readily escapes and becomes so abundant as to clog irrigation canals and slow-moving streams.

- Arum Family

Alocasia	*Lysichiton*
Arisaema	*Monstera*
Arum	*Philodendron*
Caladium	*Pistia*
Calla	*Spathiphyllum*
Colocasia	*Symplocarpus*
Dieffenbachia	*Xanthosoma*
Epipremnum	*Zantedeschia*
arum	golden pothos
arum lily	ivy arum
blue tannia	jack-in-the-pulpit
caladium	peace lily
calla lily	philodendron
dasheen	skunk cabbage
devil's ivy	taro
dumb cane	white sails
giant elephant's ear	

- some species used for food, taro, tanier

Plants herbs or vines; perennials; typically from rhizomes or corms or tubers; foliage typically fleshy. **Leaves** simple or compound; alternate; often all basal; venation parallel or pinnate or palmate; petioles sheathing at bases; stipules absent. **Inflorescences** spadices almost always subtended by conspicuous, often brightly colored spathes. **Flowers** perfect or imperfect; when imperfect, staminate flowers borne above pistillate; small; numerous; perianths in 1-series or absent. **Perianth Parts** 4 or 6 or 8 or 0; free or fused. **Stamens** 4 or 6, or rarely 1 or 8; free or fused by filaments. **Pistils** 1; compound; carpels 3; stigmas 1; styles 1; ovaries superior, typically not lobed; locules 3 or 1; placentation axile or parietal. **Fruits** berries; fleshy or sometimes leathery; entire spadix ripening as a multiple fruit in some taxa.

- herbs or vines; leaves alternate; flowers in spadices; fruits berries

Disease Problems—The most commonly reported intoxication problem associated with genera of the Araceae is swelling and edema of the oral mucous membranes following chewing of the leaves or stems. An intense burning sensation occurs immediately after the first bite. Species with the fleshy, starchy rhizomes, corms, or tubers used for food are dried or heated to destroy their burning and toxic properties. Typically, the disease is self-limiting and is associated with acute, temporary painful dis-

- irritation of mucous membranes, mouth, eyes, digestive tract, skin; rarely internal systemic effects

Spathiphyllum

■ representative arum; with a spadix subtended by a spathe

comfort (Mrvos et al. 1991). However, in some instances, secondary complications may cause more-serious problems and rarely even death (McIntire et al. 1990). *Dieffenbachia* and *Philodendron* are the two genera most commonly reported to be ingested in both Europe and North America (Jaspersen-Schib et al. 1996; Krenzelok et al. 1996), in part because they are so favored as houseplants. *Dieffenbachia* is the genus most often associated with occasional severe oral or corneal injury.

Other problems also may be caused by aroids. Contact with wet foliage of *Philodendron* and perhaps other genera, for example, may cause contact dermatitis in sensitive individuals (Dorsey 1958; Hammershoy & Verdich 1980; Knight 1991). Neurotoxicity and nephrotoxicity are associated with ingestion of *Philodendron*, whereas the spadices or berries of other genera when eaten may cause different problems, as indicated for the specific genera. Because several species are employed as contraceptives or abortifacients by folk societies, systemic effects appear also to be factors of some concern (Plowman 1969).

Commonly encountered taxa of the family that cause similar toxicological problems are listed in table 10.1. Plants of these genera are undoubtedly hazardous for pets and other animals but seem to be a much greater problem for humans, especially small children (Krenzelok et al. 1996).

Table 10.1 Commonly encountered Araceae taxa causing toxicologic problems

Alocasia macrorrhiza	giant elephant's ear, giant taro
Arisaema triphyllum	jack-in-the-pulpit
Arum italicum	arum
Caladium bicolor	caladium
Calla palustris	wild calla
Colocasia esculenta	taro, dasheen, cocoyam
Dieffenbachia spp.	dumb cane, American arum
Lysichiton americanum	yellow skunk cabbage,
Philodendron spp.	philodendron
Pistia stratiotes	water lettuce, shellflower
Spathiphyllum spp.	white sails, peace lily
Symplocarpus foetidus	skunk cabbage
Xanthosoma sagittifolium	tannia, yautia
X. violaceum	blue tannia, blue taro
Zantedeschia aethiopica	calla lily, arum lily

■ calcium oxalate crystals, druses and raphides

Disease Genesis—The toxicants produced by members of the Araceae have not been completely characterized. However, the primary toxic effects, known to be associated with a heat-labile fraction, have long been thought to be due to calcium oxalates in the plant tissues. Raphides, aggregations of elongated crystals of what we now know are calcium oxalate, have been a recognized feature of the family since the first observations of cell components via microscopy by Leeuwenhoek in 1675 (Arnott & Pautard 1970). Later observations disclosed that raphides were one of several forms of intracellular, highly birefringent, solitary or compound, monohydrate crystals (Frey-Wyssling 1981). There are five types of crystal forms; spherical aggregates or clusters variously known as druses, sphaerraphides, sphaerocrystals, or sphaerites are the most common. Solitary crystals also are common, whereas microcrystalline forms (crystal dust or crystal sand), styloids, and raphides are less so (Haberlandt 1914; Arnott & Pautard 1970; Franceschi & Horner 1980). Crystals are intracellular for the most part, and typically the cells remain functional at maturity, although perhaps somewhat degenerated (Franceshi & Horner 1980). Although calcium oxalate crystals are common among many plant families, the only types associated with toxicity are raphides and less commonly druses (Doaigey 1991).

Crystals of calcium oxalate typically occur in idioblasts, specialized cells that differ markedly in appearance from surrounding cells. This is true of both raphides and druses. In the case of raphides, the idioblasts undergo a specialized, complex development, with an abundance of endoplasmic reticulum being formed, and may either be degenerating or exhibit low metabolic activity at maturity (Esau 1965; Mollenhauer & Larson 1966; Sakai & Hanson 1974; Sunell & Healey 1979). Raphides are long, individually distinct, needlelike crystals of equal size tightly bundled in an orderly parallel arrangement and held together by protoplasmic threads. They often occur in large idioblasts that are filled with mucilage that swells when it comes in contact with water, for example, as when the plant material is chewed. The ends of the cell wall are thin, and when the mucilage swells, the wall ruptures, and the raphides are forcibly extruded en masse into the mouth of the animal. In other instances, the raphides may be extruded by external pressure, either en masse or a few at a time. Raphides not contained in idioblasts seem to be less predisposed to cause irritant effects, as shown for those of *Impatiens,* which are found floating free or in large thin-walled cells (Rauber 1985). Druses, or sphaerocrystals, are large irregular crystalloid aggregations of an indistinct nature typically contained within a vacuole in a cell that is alive and functional at maturity (Price 1970).

- idioblast cells containing packets of grooved, needlelike oxalate crystals, which in some taxa are barbed

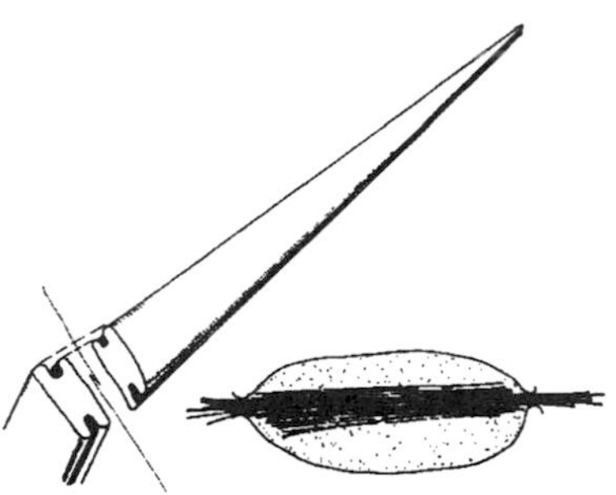

Idioblast with calcium oxalate crystals

- raphides extruded from idioblasts when mucilage in cell swells or when external pressure is applied
- calcium oxalate crystals, 60–130 µm long

Both druses and raphides are present in aroids, located mainly in the subepidermal mesophyll layer in both the photosynthetic palisade and loosely arranged spongy layers (Genua & Hillson 1985; Doaigey 1991). However, in a study of 14 genera of the Araceae, idioblasts were present in only 7, and these were filled exclusively with raphides rather than druses (Genua & Hillson 1985). The types of crystals, the numbers of idioblasts, and their distribution in the plant vary from taxon to taxon and even among varieties and/or individuals of a species. Studies of the raphide idioblasts of several aroids revealed that *Colocasia* and *Xanthosoma* each had a single type of raphide and idioblast throughout the plant (Sakai et al. 1972; Sakai & Hanson 1974). The two genera differed further in the size of the idioblasts. In contrast, three types of idioblasts and two sizes of raphide crystals were present in *Alocasia.* Only one type of idioblast and the larger crystal were present in rhizomes, whereas two types of idioblast and two crystal sizes were present in the petioles. Differences among the three genera in the mode of ejection of the raphides from the idioblast also were apparent; only those of *Colocasia* were forcefully released. The raphides of all three genera shared common features of general shape, the presence of material surrounding the crystals within the idioblasts that stains positive with periodic acid–Schiff base, and the presence of barbs and grooves on the crystals. The length of the crystals ranges from 60 to 130 µm, with one end tapered to a point and the other end blunt. Each crystal has two longitudinal grooves on opposite sides, thus giving it an H shape in cross section. This appearance is characteristic of the raphides in members of the Araceae (Franceschi & Horner 1980). Along the sides of the grooves are numerous barbs, 0.75 µm long, with their tips oriented away from the tapered end. It is hypothesized that the morphology of crystals enhances penetration of the tissues of the mouth and prevents subsequent dislodgment once embedded. It has been suggested that the crystals may serve as a defense against herbivory (Haberlandt 1914; Sakai et al. 1972).

Toxicity is believed to be due, at least in part, to the insoluble crystals of calcium oxalate, and in particular the raphides, acting as mechanical irritants upon contact with the mucous membranes (Ladeira et al. 1975; Kuballa et al. 1981). The crystals also are responsible for the acrid taste of the plants (Black 1918). When the crystals are washed free of all other plant materials, the contact irritation is not reduced (Black 1918). Similarly, conjunctivitis may be induced in rabbits from juice pressed from the stem of *Dieffenbachia* (Ellis et al. 1973). When the crystals are filtered out of the juice, inflammation is reduced. Persistence of the crystals in the cornea long after the clinical signs have disappeared is further evidence of their involvement.

Studies of raphide-containing taxa in other families confirm the importance of the crystals. For example, it has also been shown with the fishtail palm (*Caryota mitis*: Arecaceae) that contact irritation of short duration may be elicited by raphides alone (Snyder et al. 1979). Studies with kiwifruit (*Actinidia deliciosa*: Actinidiaceae), in which extraction procedures were used to eliminate other sources of activity, also show that raphides alone are capable of causing mild mechanical irritation (Perera et al. 1990).

- possible additional effects from chemical mediators of inflammation

The barbs and grooves of the raphides are unique to members of the Araceae (Franceschi & Horner 1980). The grooves may serve as channels to conduct compounds into the wound that act in concert with the mechanical irritation (Sakai & Hanson 1972; Sakai et al. 1974; Gardner 1994). Numerous studies support this hypothesis. Although the irritant effects are typically lost with boiling of the plant material, distillates retain some toxicity, indicating that full expression of aroid toxicity does not lie solely with the crystals (Osisiogu et al. 1974). In addition, in other studies the severity of the irritant effects did not appear to be directly proportional to raphide or calcium oxalate content (Choudhury & Hussain 1979; Saha & Hussain 1983). Other compounds possibly involved include prostaglandins, histamine, trypsinlike proteolytic enzymes that might cause the release of other mediators of inflammation, and glycosides such as of 3,4-dihydroxybenzaldehyde (Walter & Khanna 1972; Suzuki 1980; Saha & Hussain 1983). The other compounds have not been conclusively identified, and it is possible that no one factor is dominant.

A role for histamine has been proposed; however, reports are contradictory. Rizzini and Occhioni (1957) proposed that a toxic proteinaceous substance produced by aroids caused histamine release in the tissues of the animal. The presence of this additional compound was indicated by effects on heart and respiration rates, the presence of tonic seizures, and the development of tolerance to the effects of the plant that could not be accounted for by mechanical damage alone (Barnes & Fox 1955). Fochtman et al. (1969) observed a 2-fold increase in histamine levels in rats given the juice of *Dieffenbachia*. Pretreatment with diphenhydramine, an antihistamine, attenuated the rats' reaction to the juice. They and Walter (1967) suggested that the juice may have trypsinlike activity, causing noxious effects directly and/or releasing other agents such as endogenous histamines in the animal. This interpretation is supported by the isolation of L-asparagine and a proteolytic trypsinlike enzyme from the genus (Walter & Khanna 1972). In some studies, however, there has been a lack of protection afforded by pretreatment with either antihistamines such as chlorpheniramine or antihistamine/antiserotonins such as cyproheptadine (Ladeira et al. 1975; Kuballa et al. 1981). Yet, an antiprostaglandin, indomethacin, was protective (Kuballa et al. 1981). The occurrence of systemic reactions is also shown by the lethal effects caused by oral administration of material from several aroid genera to mice at 100 mg/35 g b.w. These effects were attributed to alkaloids, glycosides, biogenic amines, and/or saponins identified as being present by screening tests (Der Marderosian et al. 1976; Willaman & Schubert 1961).

- contact irritation, lips, tongue, mouth; swelling, pain; rarely lower digestive tract effects, vomiting, diarrhea

Clinical Signs—Primary signs are those associated with contact irritation of the tissues of the upper digestive tract; immediate swelling of the lips, tongue, and oral cavity, and excess salivation. In most instances, the signs are mild, but there will be a few severe cases in which the swelling and intense pain will render the animal unable to eat for several days. Contact with the plant juices may produce a severe conjunctivitis. In most instances, involvement will be restricted to the areas of the mouth and eyes, but in severe cases it may extend to the esophagus and stomach. The effects, though painful, are usually temporary. More serious intoxication and even death are possible but very unlikely, in large measure because of the rapid onset of pain and swelling of the lips and tongue, which promptly terminate interest in eating more of the plant.

Treatment—The most important considerations in treatment of aroid poisoning are alleviation of pain and reduction in swelling. Antihistamines are of limited value. Sedation and administration of nonsteroidal anti-inflammatory drugs such as aspirin or phenylbutazone are appropriate in most cases. Soft palatable foods are also desirable. Topical application of ointments or salves is probably of limited value, although lime juice is reported to be more effective than milk, water, or ice in relieving the swelling and pain (Lawrence & Schneider 1987). A steroid ointment may provide relief of inflammation of the eyes.

■ treatment generally not needed but control of pain and inflammation may be useful

Alocasia **(Schott) G.Don**

Taxonomy and Morphology—Comprising approximately 70 species, with greatest abundance in tropical Southeast Asia, *Alocasia* is represented in North America by cultivated ornamentals, most commonly as houseplants, although 1 species is apparently naturalized in Florida:

■ *A. macrorrhiza*

giant elephant's ear
giant taro

A. macrorrhiza (L.) G.Don giant elephant's ear, giant taro

The genus name is a variant of the closely related genus *Colocasia.*

Plants herbs; from rhizomes or tubers; stems erect or ascending; thick. **Leaves** simple; to 100 cm long; dark green or strikingly colored and/or variegated; blades erect or deflexed, cordate to sagittate, venation conspicuous; petioles long, cylindric. **Inflorescences** solitary or clustered; spathes yellow or green, margins overlapping to form tubes; spadices shorter than spathes; terminating in sterile appendage. **Flowers** imperfect; staminate flowers separated from pistillate by band of rudimentary sterile flowers. **Perianth Parts** absent. **Stamens** 3–8; fused. **Pistils** 1; compound; stigmas 1, 3-to-5-lobed; locules 1. **Berries** red or orange.

■ herbs from tubers or rhizomes; leaves simple, very large, on long petioles

Distribution and Habitat—Native to tropical Southeast Asia, *A. macrorrhiza* has escaped cultivation and occurs occasionally in wet disturbed thickets in southern Florida.

Disease Problems—As is characteristic of the family, species of *Alocasia* cause contact irritation. Based on poison control center reports, ingestions of one or more mouthfuls of the leaves or tubers of *A. macrorrhiza* causes oral numbness, pain, excess salivation, and rarely ulcers. These effects last for a few days, except for the numbness, which may persist for a week (Lin et al. 1998). A different disease has been reported to be caused by the ingestion of the fruit of an Asian species, *A. cucullata,* commonly known as Chinese taro (Goonasekera et al. 1993). After eating the berries, one child died 20 hours later, and another child, 5 days afterward. Cyanide was thought to be involved, but the presence of zonal hepatocellular necrosis and the protracted nature of the signs are not consistent with cyanide toxicosis.

■ contact irritation of mouth, numbness, pain, excess salivation

Arisaema Mart.

- *A. dracontium*
  A. triphyllum
  dragon root
  green dragon
  Indian turnip
  jack-in-the-pulpit
  parson-in-the-pulpit

Taxonomy and Morphology—Comprising approximately 150 species of tropical and temperate regions of both the Old World and the New World, *Arisaema* is represented in North America by 2 species:

A. dracontium (L.) Schott	green dragon, dragon root
A. triphyllum (L.) Torr.	Indian turnip, jack-in-the-pulpit, parson-in-the-pulpit

Arisaema triphyllum is morphologically quite variable, and 4 subspecies are now recognized. Previously described species such as *A. acuminatum, A. pusillum, A. quinatum, A. atrorubens,* and *A. stewardsonii* are now submerged in *A. triphyllum.*

- herbs; leaves 1-pinnately compound; spathes, outer surfaces green or white, inner bronze or purple; berries orange-red, clustered

Plants herbs; from rhizomes or tubers; acaulescent. **Leaves** palmately compound or pedately divided; leaflets 3–20; petioles long, emerging from several bracts. **Inflorescences** solitary; arising on elongate peduncles adjacent to leaves; spathes green or white, typically purple or bronze within, margins overlapping to form tubes, apices erect or forming a hood over spadix; spadices shorter or longer than spathes; terminating in an elongate sterile appendage. **Flowers** imperfect; borne at bases of spadices. **Perianth Parts** absent. **Stamens** 1–5; fused. **Pistils** 1; compound; stigmas 1, peltate or capitate; locules 1. **Berries** orange-red; borne in conspicuous clusters 1–2 cm in diameter.

Arisaema dracontium

Arisaema triphyllum

Arisaema triphyllum

1. Leaflets 5–13. Apices of spathes not forming hoods over spadices. Spadices exserted from spathes . . . . *A. dracontium*
1. Leaflets 3. Apices of spathes forming hoods over spadices. Spadices not exserted from spathes . . . . *A. triphyllum*

Distribution and Habitat—Both *A. triphyllum* and *A. dracontium* are understory plants of the moist woodlands, thickets, swamps, and bogs of the eastern deciduous forests. Flowering in the early spring, the fruits of *A. dracontium* are especially conspicuous in the fall.

■ moist or shaded sites

Disease Problems—As is characteristic of the family, both species of *Arisaema* cause contact irritation, and the corms are a favorite of Boy Scouts and field biologists who offer portions to the unsuspecting. The plants apparently are seldom eaten by livestock and are of little toxicologic importance. Sheep are probably the most likely species to eat the plants; however, they are also less likely to show adverse effects, based on their response to experimental administration of *Dieffenbachia* (Armien & Tokarnia 1994).

■ contact irritation

Calla L.

Taxonomy and Morphology—A monotypic genus, *Calla* is native to wetland habitats in temperate North America, Europe, and Asia. Recognized by the Roman natural historian Pliny, the genus derives its name from the Greek root *kallos,* which means *beauty.* Care must be taken not to confuse *Calla* with calla lilies, which are species of the genus *Zantedeschia,* or with black calla, which is a species of *Arum.* The genus consists of a single species:

■ *C. palustris*

bog arum
water arum
water dragon
wild calla

C. palustris L water arum, bog arum, wild calla, water dragon

Plants herbs; evergreen or deciduous; acaulescent; from rhizomes. **Rhizomes** horizontal; sparsely branched; leaf scars prominent. **Leaves** simple; basal; 2-ranked; glossy dark green; long petiolate; blades broadly ovate to reniform; venation pinnate; apices acute; bases cordate. **Inflorescences** solitary; spathes white; apices involute; margins overlapping to form tubes; spadices shorter than spathes; not enclosed; not terminating in sterile appendage. **Flowers** perfect or uppermost staminate. **Perianth Parts** absent. **Stamens** 6. **Pistils** 1; stigmas 1, broad; locules 1. **Berries** bright red; seeds 4–10, surrounded by gelatinous material.

■ herbs, prominent leaf scars; leaves simple, basal; berries bright red

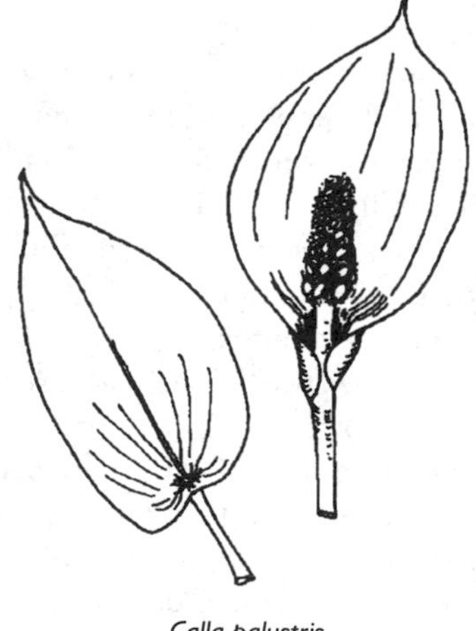

Calla palustris

Calla palustris

Distribution and Habitat—A vigorous and rapidly growing aquatic, *C. palustris* occupies a variety of wetland habitats—marshes, swamps, alder bogs, seeps, and lakesides.

■ moist sites; bogs, swamps, seeps

It is occasionally grown for its glossy dark green foliage and white spathes as an ornamental in aquatic gardens.

■ rarely, contact irritation

Disease Problems—As is characteristic of the family, *C. palustris* is capable of causing contact irritation; however, it is seldom eaten by livestock and thus is of limited toxicologic importance.

Dieffenbachia Schott

■ *D. maculata*
D. seguine
American arum
dumb cane
poison cane

Taxonomy and Morphology—Comprising some 25–30 species with greatest abundance in tropical America, *Dieffenbachia* is represented in North America only by cultivated ornamentals, most commonly as houseplants. Grown for their lush, tropical-appearing foliage, numerous named cultivars are in commercial trade. Especially prized for the mottled or herringbone variegation of their leaves, the 2 species most commonly encountered are morphologically similar:

D. maculata (Lodd.) G.Don	dumb cane
D. seguine (Jacq.) Schott	dumb cane, American arum, poison arum

Unfortunately, some horticulturalists apply the common name *mother-in-law's-tongue* to *Dieffenbachia,* a name usually associated with the genus *Sansevieria* in the Agavaceae. The binomial *D. picta,* which appears in the older literature, is a synonym of *D. maculata.*

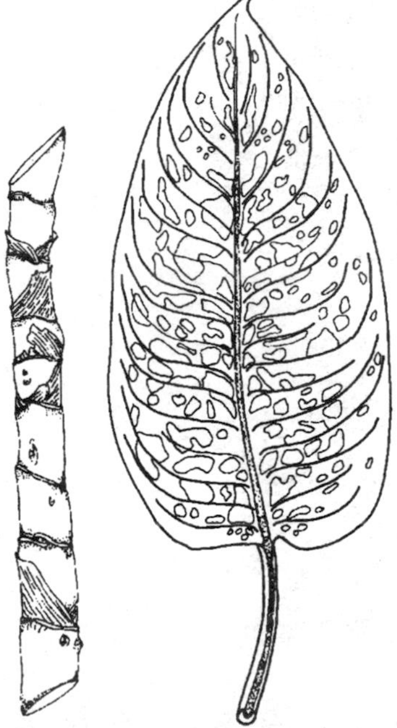

Dieffenbachia maculata

■ ornamental herbs, stout erect stems; leaves simple, often variegated; seldom flowers indoors

Plants herbs; stems erect; to 2.5 m tall; thick; bearing leaves to apices; leaf scars prominent. **Leaves** simple; green, but typically mottled or variegated in various shades of yellow, red, green, or white, the lighter colors occupying the spaces between veins; blades oblong to ovate, midrib and venation conspicuous; petioles long, sheathing stems. **Inflorescences** solitary; spathes green, margins overlapping to form tubes; spadices shorter or longer than spathes; terminating in sterile appendage. **Flowers** imperfect; staminate flowers separate from pistillate by band of rudimentary sterile flowers. **Perianth Parts** absent. **Stamens** 3–8; fused. **Pistils** 1; compound; stigmas 1, 2-or-3-lobed; locules 2 or 3. **Berries** red or orange-red.

■ contact irritation from chewing stem or leaf; may be severe; pets and humans

Disease Problems—As is characteristic of the family, species of *Dieffenbachia* cause severe contact irritation and are the subject of many inquiries to poison control centers, especially for cases involving small children (Schilling et al. 1980). In the Caribbean, *D. seguine* has been used since the 1600s to render people incapable of speaking by means of the severe swelling and edema that follows the chewing of its leaves and stems—hence the common name *dumb cane* (Barnes & Fox 1955). The juice of the stems was rubbed in the mouths of slaves as punishment, the severe reaction lasting up to several days. The species also had a reputation among Caribbean and Pacific Islanders as a contraceptive that lasted for 1–2 days (Barnes & Fox 1955; Plowman 1970). Experiments with animals and humans in Nazi Germany during World War II also indicated such an effect. Atrophy of the gonads and permanent sterility in rats was noted after daily oral administration of the juice for 1–2 months. In addition, female rats given extracts for 4 days experienced decreased ovulation and hormonal disruptions (Costa de Pascuale et al. 1984).

The stem is the most potent portion of the plant, followed by the petiole, and the blade of the leaf (Ladeira et al. 1975). *Dieffenbachia* differs from other members of the family in having raphides present in epidermal as well as mesophyll layers, which may account in part for its greater irritant potency (Genua & Hillson 1985). Although in the great majority of exposures, no signs or only mild signs observed, there are a few reports of individuals biting the stems or leaves and subsequently suffering more-severe injury (Mrvos et al. 1990; Pamies et al. 1992; Krenzelok & Provost 1995; Krenzelok et al. 1996). Inflammation and edema of the lips and tongue or cornea lasts for several hours to days (Schilling et al. 1980; Gardner 1994; Seet et al. 1995). There are instances of intense pain, inflammation of the mouth, and laryngeal edema with subsequent superficial necrosis of the tongue and cheeks, which may give the appearance of a caustic burn (Drach & Maloney 1963; Wiese et al. 1996). Injury may be severe even when the plant is spit out and not swallowed. Similarly, severe inflammation of the surface of the eye may follow exposure to the juice of the stems and leaves. Recovery may take several days, and the crystals may be present in the cornea for an even longer time (Ellis et al. 1973).

- irritation of mouth and tongue, swelling, pain, excess salivation, may last hours or rarely days; may appear like a burn

Livestock are at some risk, although in North America there is little opportunity for exposure because plants typically are grown inside or in gardens. Cattle administered 3.7–4.8 gm of plant material/kg b.w. developed moderate signs of intoxication, including increased salivation, and sublingual and submandibular edema (Tokarnia et al. 1996). Sheep exhibit similar responses to 2.9–10 g/kg b.w. dosage (Armien & Tokarnia 1994).

Epipremnum Schott

Taxonomy and Morphology—Comprising 8 species with greatest abundance in tropical Southeast Asia and the western Pacific, *Epipremnum* is represented in North America only by cultivated ornamentals. The following is perhaps the most common of all houseplants:

- *E. aureum*

  devil's ivy
  golden pothos
  hunter's robe
  ivy arum

E. aureum (Lindl. & André) Bunting	ivy arum, devil's ivy, golden pothos, hunter's-robe

Formerly known as *Scindapsus aureus*, this species is morphologically very similar to the viny philodendrons and is often mistakenly called one.

Plants herbaceous vines; evergreen; with adventitious roots arising from stems. **Leaves** simple; dimorphic, with juvenile and adult forms; dark green or variegated; blades ovate to lanceolate or oblong, coriaceous, margins entire or pinnately lobed, venation parallel-convergent; petioles long, sheathing at bases. **Inflorescences** solitary; spathes yellow to green or purple, margins not overlapping to form tubes; spadices shorter than spathes. **Flowers** perfect. **Perianth Parts** absent. **Stamens** 4; fused. **Pistils** 1; compound; stigmas 1, locules 1. **Berries** red or orange.

- herbaceous vines with adventitious roots from stems; leaves simple, often variegated; seldom flowers indoors

Epipremnum aureum

■ contact irritation

Disease Problems—As is characteristic of the family, *Epipremnum* is capable of causing contact irritation. In our experience this common ornamental has occasionally caused vomiting and diarrhea in dogs. However, experimentally, even very large doses in the range of 32–40 g/kg b.w. failed to cause adverse effects in cattle and sheep (Tokarnia et al. 1996).

Lysichiton Schott

■ *L. americanus*

skunk cabbage
yellow skunk cabbage

Taxonomy and Morphology—A monotypic genus, *Lysichiton* is native to western North America. The single species is the following:

L. americanus Hulten & St. John	yellow skunk cabbage, skunk cabbage

Unfortunately, the common name *skunk cabbage* is frequently applied to two species of the family. One is *L. americanus,* which occurs on the West Coast, and the other is *Symplocarpus foetidus,* which occurs in the northeastern quarter of the continent. In the older literature, the genus name is often incorrectly spelled *Lysichitum.*

■ perennial, deciduous herbs; leaves simple, ill-scented when crushed; spathes yellow or white; berries green

Plants herbs; perennials; deciduous; from short, stout, vertical rhizomes. **Leaves** simple; borne in large rosettelike clumps after flowering; dark green; malodorous when bruised or crushed; blades lanceolate to oblong-ovate, semierect or spreading, to 1.5 m long, margins entire, venation parallel-convergent; petioles short or absent. **Inflorescences** solitary; spathes yellow or white, conspicuous, margins overlapping to form tubes; spadices at first shorter than spathes and then elongating. **Flowers** perfect. **Perianth Parts** 4; erect; connivent. **Stamens** 4. **Pistils** 1; compound; ovaries embedded in spadix; locules 2. **Berries** green.

Lysichiton americanus

Lysichiton americanus

■ bogs

Distribution and Habitat—Plants of *L. americanus* are found in bogs primarily west of the Cascade Mountains.

■ rarely contact irritation

Disease Problems—As is characteristic of the family, *L. americanus* is capable of causing contact irritation; however, it is seldom eaten by livestock and thus is of limited toxicologic importance.

Monstera Adans.

Taxonomy and Morphology—Comprising some 22 species of the America tropics, *Monstera* is represented in North America only by cultivated ornamentals, most commonly as houseplants. Morphologically very similar to the viny philodendrons, species of the genus are often mistakenly called philodendrons. Most commonly encountered is this species:

■ *M. deliciosa*

ceriman
split-leaf philodendron
Swiss-cheese plant
windowleaf

M. deliciosa Liebm.	split-leaf philodendron, ceriman, Swiss cheese plant, windowleaf

This species is one of the most popular houseplants and at one time was cultivated for its spadix of berries, which are said to have the taste of pineapple and banana combined. When completely ripe, they were used in salads, drinks, and ices (Plowman 1969).

■ vines, long adventitious roots; leaves simple; seldom flowers indoors

Plants herbaceous vines; evergreen; stems ascending or sprawling, stout, with long adventitious roots. **Leaves** simple; dimorphic, with juvenile and adult forms; dark green or variegated; blades ovate to lanceolate or oblong, coriaceous, to 1 m long and 0.75 m wide, margins entire or typically pinnatifid, venation pinnate; petioles long, sheathing at bases. **Inflorescences** solitary; spathes white or cream; margins not overlapping to form tubes; spadices shorter than spathes. **Flowers** perfect. **Perianth Parts** absent. **Stamens** 4; fused. **Pistils** 1; compound; stigmas 1, locules 2. **Berries** white; fused together to form conelike structures; aromatic; seeds soft.

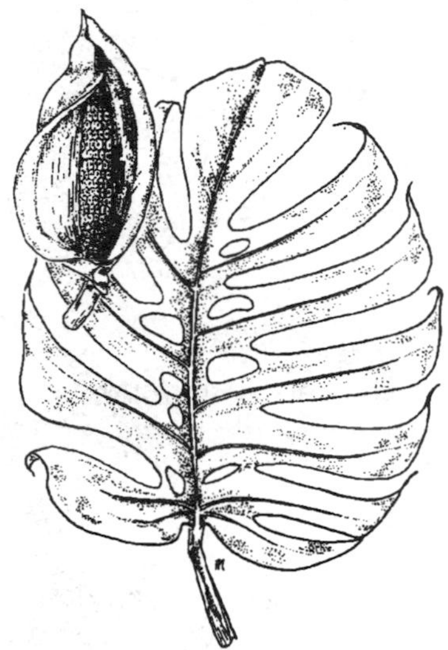

Monstera delicosa

■ contact irritation

Disease Problems—As is characteristic of the family, *Monstera* is capable of causing contact irritation. However, it does not appear to be a significant risk, because dosage of 16.5–26 g/kg b.w. fresh foliage failed to produce adverse effects in cattle, and only mild to moderate signs were seen with up to 20 g/kg in sheep (Armien & Tokarnia 1994; Tokarnia et al. 1996).

Philodendron Schott

■ *Philodendron*

philodendron

Taxonomy and Morphology—Comprising approximately 350 species of the American tropics, *Philodendron* is represented in North America only by cultivated ornamentals, most commonly as houseplants or in glasshouses. The genus name comes from the Greek roots *phileo* and *dendron*, which allude to the climbing or epiphytic habitat of many of the species of the genus. Morphologically quite diverse, *Philodendron* comprises a variety of forms; however, some of the cultivated plants called philodendron have never flowered and their classification is uncertain. Adding to the ambiguity is that commonly cultivated species of other genera, e.g., *Epipremnum* and *Monstera,* also are referred to as philodendrons. Because of the diversity of cultivated taxa in the genus and the attendant problems of identification and classification, a list of species is not presented here. If identification is needed, horticultural treatises should be consulted.

■ ornamental climbing herbs or shrubs; leaves simple, many shapes; seldom flowers indoors

Plants climbing herbs, shrubs, or small trees; evergreen; stems stout, internodes short, with adventitious roots. **Leaves** simple; usually dark green; blades cordate or sagittate or ovate to oblong, coriaceous, to 1 m long, margins entire or pinnately lobed, venation pinnate, petioles long, sheathing at bases. **Inflorescences** solitary or fascicled; axillary; spathes white to green and yellow, often marked with red and purple, margins overlapping to form tubes; spadices shorter than spathes. **Flowers** imperfect,

Philodendron

staminate and pistillate contiguous. **Perianth Parts** absent. **Stamens** 2–5; fused. **Pistils** 1; compound; locules 2-many. **Berries** white to orange or red.

- contact irritation mouth, tongue; rarely renal disease; possibly neurotoxic

Disease Problems—As is characteristic of the family, species of *Philodendron* cause contact irritation. Dermatitis is a continuing problem for nursery workers who come in repeated contact with wet leaves (Dorsey 1958). Despite the detrimental effects, a variety of medicinal uses have been found for various species of the genus. One is used as a contraceptive by the indigenous tribes of Columbia, another is used as an abortifacient, and yet another is used topically to treat testicular inflammation. Interestingly however, if used for too long, the testicles shrink. In Brazil, the species' common name is *capa homen,* which means "castrated man" in Portuguese (Plowman 1970).

In addition to the same irritant effects described for *Dieffenbachia,* philodendrons are purported to be nephrotoxic and possibly neurotoxic, the latter problem confined to cats (Greer 1969; Brogger 1970; Pierce 1970). Based upon a few retrospective case evaluations prompted by lesions of severe tubular nephrosis, signs of nephrotoxicity do not begin to appear until several days after ingestion of plant material by the animal. Although these two syndromes have been described clinically, experimental efforts to produce them have not been successful (Sellers et al. 1978). Furthermore, few instances of neurotoxicity have been reported. Experimentally, the effects have been variable in sheep and cattle given 6–20 g/kg of plant material, no signs in some individuals, and moderate in others (Armien & Tokarnia 1994; Tokarnia et al. 1996).

- irritation mouth, tongue swelling, pain, excess salivation, may last a few hours

Clinical Signs—The signs associated with the contact irritation are the same as those produced by *Dieffenbachia* and other members of the family: immediate swelling of the lips, tongue, and oral cavity, and excess salivation. In most instances, the signs are mild, but there may be a few severe cases in which the swelling and pain will be intense. Contact with the plant's juices or sap may produce conjunctivitis.

- renal failure a week after ingestion; increased frequency of urination, depression, marked increase in BUN

When nephrotoxicosis occurs, an increase in urine production and frequent urination are observed. They may be accompanied by severe depression and incoordination, with death a likely sequela. Renal disease can be monitored or verified by urinalysis, determination of blood urea nitrogen levels, and histopathologic evidence of tubular nephrosis.

Symplocarpus Salisb. ex Nutt.

- *S. foetidus*

  fetid pothos
  Midas's ears
  skunk cabbage

Taxonomy and Morphology—A monotypic genus, *Symplocarpus* is native to North America and Asia. The following is its single species:

S. foetidus (L.) Salisb. ex Nutt.	skunk cabbage, fetid pothos, Midas's ears, poison-in-the-pillory

A binomial for skunk cabbage that is encountered in the older literature is *Spathyema foetida.* Unfortunately, the common name *skunk cabbage* is frequently applied to two species of the family. One is *Symplocarpus foetidus,* which occurs in the northeastern portion of the continent, and the other is *Lysichiton americanus,* which occurs along the west coast from Alaska to northern California. *Lysichiton americanus* is more properly called yellow skunk cabbage.

- perennial deciduous herbs; leaves simple, ill-scented when crushed; spathes yellow-green, spotted or striped with red and purple

Plants herbs; perennials; deciduous; from short, stout, vertical rhizomes. **Leaves** simple; borne in large rosettelike clumps after flowering; dark green; malodorous when bruised or crushed; blades ovate-cordate, semierect or spreading, to 0.5 m long, margins entire, venation parallel-convergent; petioles long, sheathing at bases. **Inflorescences** solitary; spathes yellow-green, spotted and/or striped with red and purple, margins overlapping to form tubes, apices pointed; spadices shorter than spathes. **Flowers** perfect. **Perianth Parts** 4; erect; connivent. **Stamens** 4. **Pistils** 1; compound; stigmas 1, 4-lobed; ovaries embedded in spadix; locules 1. **Berries** fusing together to form a compound fruit.

Symplocarpus foetidus

Distribution and Habitat—Plants of *S. foetidus* occupy swamps, bogs, and low, wet soils of woodlands throughout the northeastern quarter of the continent. Because of its ability to flower when the soil is frozen and snow-covered, it is one of the first plants to flower in the spring.

■ swamps, bogs, wet woodlands

Disease Problems—As is characteristic of the family, *S. foetidus* is capable of causing contact irritation; however, it is seldom eaten by livestock and thus is of limited toxicologic importance. Sheep are probably the most likely species to eat the plants, but they are also less likely to show adverse effects.

■ contact irritation

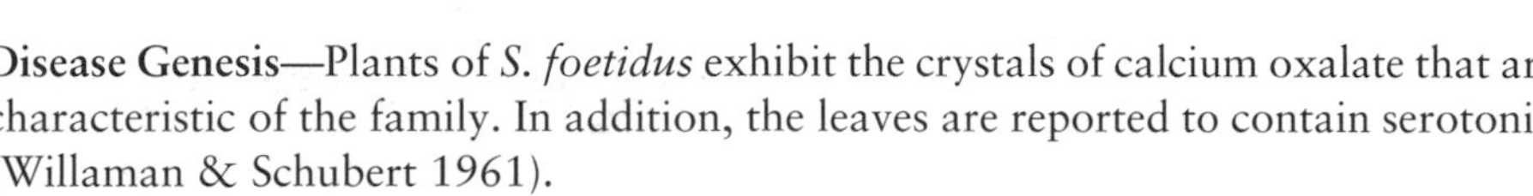

Disease Genesis—Plants of *S. foetidus* exhibit the crystals of calcium oxalate that are characteristic of the family. In addition, the leaves are reported to contain serotonin (Willaman & Schubert 1961).

Zantedeschia Spreng.

Taxonomy and Morphology—Comprising 6 species native to South Africa and naturalized elsewhere, *Zantedeschia* is represented in North America only by cultivated ornamentals, most commonly as houseplants. Most commonly encountered is this species:

■ *Z. aethiopica*
calla
calla lily
arum lily

Z. aethiopica (L.) Spreng. calla lily, calla, arum lily

Plants herbs; perennials; evergreen or deciduous; from fleshy, branched rhizomes. **Leaves** simple; dark green; coriaceous; blades lanceolate to ovate-cordate or hastate, erect or spreading, margins undulate, venation pinnate; petioles long, sheathing at bases. **Inflorescences** solitary; spathes white or yellow, margins overlapping to form tubes; spadices shorter than spathes. **Flowers** imperfect, staminate and pistillate flowers contiguous. **Perianth Parts** absent. **Stamens** 2 or 3; free. **Pistils** 1; compound; locules 3. **Berries** orange or red-orange; beaked.

■ herbs; leaves simple, on long stalks or petioles

■ contact irritation

Disease Problems—As is characteristic of the family, *Zantedeschia* is capable of causing contact irritation. In addition, Ladeira and co-workers (1975) reported that eating the spathes or spadices of *Z. aethiopica* was the cause of death in children in Brazil. In one instance, an 8-year-old girl was comatose for 12 hours after eating a spadix; she recovered after 24 hours.

Zantedeschia

References

Armien AG, Tokarnia CH: Experiments on the toxicity of some ornamental plants in sheep. *Pesq Vet Bras* 14;69–73, 1994

Arnott HJ, Pautard FGE: Calcification in plants. In *Biological Calcification: Cellular and Molecular Aspects,* Schafer H ed, Appleton-Century-Crofts, New York, 1970

Barnes BA, Fox LE: Poisoning with *Dieffenbachia*. *J Hist Med Allied Sci* 10;173–180, 1955

Black OF: Calcium oxalate in the dasheen. *Am J Bot* 5;447–451, 1918

Brogger JN: Renal failure from philodendron [letter]. *Mod Vet Pract* 51;46, 1970

Choudhury B, Hussain M: Chemical composition of the edible parts of aroids grown in Bangladesh. *Indian J Agric Sci* 49;110–115, 1979

Costa de Pascuale R, Ragusa S, Circosta C, Forestieri AM: Investigations on *Dieffenbachia amoena* Gentil. Part 1. Endocrine effects and contraceptive activity. *J Ethnopharmacol* 12;293–303, 1984

Der Marderosian AH, Giller FB, Roia FC Jr: Phytochemical and toxicological screening of household ornamental plants potentially toxic to humans. 1. *J Toxicol Environ Health* 1;939–953, 1976

Doaigey AR: Occurrence, type, and location of calcium oxalate crystals in leaves and stems of 16 species of poisonous plants. *Am J Bot* 78;1608–1616, 1991

Dorsey C: *Philodendron* dermatitis. *Calif Med* 88;329–330, 1958

Drach G, Maloney WH: Toxicity of the common houseplant *Dieffenbachia*. *J Am Med Assoc* 184;113–114, 1963

Ellis W, Barfort P, Mastman GJ: Keratoconjunctivitis with corneal crystals caused by the dieffenbachia plant. *Am J Ophthalmol* 76; 143–147, 1973

Esau K: *Plant Anatomy,* 2d ed. J Wiley, New York, 1965

Fochtman FW, Manno JE, Winek CL, Cooper JA: Toxicity of the genus *Dieffenbachia*. *Toxicol Appl Pharmacol* 15;38–45, 1969

Franceschi VR, Horner HT Jr: Calcium oxalate crystals in plants. *Bot Rev* 46;361–427, 1980

Frey-Wyssling A: Crystallography of the two hydrates of crystalline calcium oxalate in plants. *Am J Bot* 68;130–141, 1981

Gardner DG: Injury to the oral mucous membranes caused by the common houseplant, dieffenbachia: a review. *Oral Surg Oral Med Oral Pathol* 78;631–633, 1994

Genua JM, Hillson CJ: The occurrence, type, and location of calcium oxalate crystals in the leaves of fourteen species of Araceae. *Ann Bot* 56;351–361, 1985

Goonasekera CDA, Vasanthathilake VWJK, Ratnatunga N, Seneviratne CAS: Is nai habarala *(Alocasia cucullata)* a poisonous plant? *Toxicon* 31;813–816, 1993

Greer MJ: Plant poisoning in cats [letter]. *Mod Vet Pract* 42;62, 1969

Haberlandt G: *Physiological Plant Anatomy*. Macmillan, London, pp530–536, 1914

Hammershoy O, Verdich J: Allergic contact dermatitis from *Philodendron scandens oxycardium*. *Contact Derm* 6;95–99, 1980

Jaspersen-Schib R, Theus L, Guirguis-Oeschger M, Gossweiler B, Meier-Abt PJ: Serious plant poisonings in Switzerland 1966–1994. Case analysis from the Swiss Toxicology Information Center. *Schweiz Med Wochenschr* 126;1085–1098, 1996

Knight TE: Philodendron-induced dermatitis: report of cases and review of the literature. *Cutis* 48;375–378, 1991

Krenzelok E, Provost FJ: The ten most common plant exposures reported to poison information centers in the United States. *J Nat Toxins* 4;195–202, 1995

Krenzelok EP, Jacobsen TD, Aronis JM: A review of 96,659 dieffenbachia and philodendron exposures [abstract]. *J Toxicol Clin Toxicol* 34;601, 1996

Kuballa B, Lugnier AAJ, Anton R: Study of *Dieffenbachia*-induced edema in mouse and rat hindpaw: respective role of oxalate needles and trypsin-like protease. *Toxicol Appl Pharmacol* 58;444–451, 1981

Ladeira AM, Andrade SO, Sawaya P: Studies on *Dieffenbachia picta* Schott: toxic effects in guinea pigs. *Toxicol Appl Pharmacol* 34;363–373, 1975

Lawrence RA, Schneider MF: The toxicity of household plants: a twenty-five-year prospective study [abstract]. *Vet Hum Toxicol* 29;467, 1987

Lin T-J, Hung D-Z, Hu W-H, Yang DY, Wu T-C, Deng T-C: Calcium oxalate is the main toxic component in clinical presentations of *Alocasia macrorrhiza* (L) Schott & Endl poisonings. *Vet Hum Toxicol* 40;93–95, 1998

McIntire MS, Guest JR, Porterfield JF: Philodendron—an infant death.. *J Toxicol Clin Toxicol* 28;177–183, 1990

Mollenhauer HH, Larson DA: Developmental changes in raphide-forming cells of *Vanilla planifolia* and *Monstera deliciosa. J Ultrastruct Res* 16;55–70, 1966

Mrvos R, Dean BS, Krenzelok EP: *Philodendron/Dieffenbachia* ingestions: are they a problem? *Clin Toxicol* 29;485–491, 1991

Osisiogu IUW, Uzo JO, Ugochukwu EN: The irritant effects of cocoyams. *Planta Med* 26;166–169, 1974

Pamies RJ, Powell R, Herold AH, Martinez J: The dieffenbachia plant. Case history. *J Fla Med Assoc* 79;760–761, 1992

Perera CO, Hallett IC, Nguyen TT, Charles JC: Calcium oxalate crystals: the irritant factor in kiwifruit. *J Food Sci* 55;1066–1069, 1990

Pierce JH: Encephalitis signs from philodendron leaf [letter]. *Mod Vet Pract* 51;42, 1970

Plowman T: Folk uses of New World aroids. *Econ Bot* 23;97–122, 1969

Price JL: Ultrastructure of druse crystal idioblasts in leaves of *Cercidium floridum. Am J Bot* 57;1004–1009, 1970

Rauber A: Observations on the idioblasts of *Dieffenbachia. J Toxicol Clin Toxicol* 23;79–90, 1985

Rizzini CT, Occhioni P: Acao toxica das *Dieffenbachia picta* e *D. seguine. Rev Rodriguesia* 20;5–19, 1957

Saha BP, Hussain M: A study of the irritating principle of aroids. *Indian J Agric Sci* 53;833–836, 1983

Sakai WS, Hanson M: Mature raphid and raphid idioblast structure in plants of the edible aroid genera *Colocasia, Alocasia,* and *Xanthosoma. Ann Bot* 38;739–748, 1974

Sakai WS, Hanson M, Jones RC: Raphides with barbs and grooves in *Xanthosoma sagittifolium* (Araceae). *Science* 178;314–315, 1972

Schilling R, Der Marderosian A, Speaker J: Incidence of plant poisonings in Philadelphia noted as poison information calls. *Vet Hum Toxicol* 22;148–150, 1980

Seet B, Chan WK, Ang CL: Crystalline keratopathy from *Dieffenbachia* plant sap. *Br J Ophthalmol* 79;98–99, 1995

Sellers SJ, King M, Aronson CE, Der Marderosian A: Toxicologic assessment of *Philodendron oxycardium* Schott (Araceae) in domestic cats. *Vet Hum Toxicol* 20;92–96, 1978

Snyder DS, Hatfield GM, Lampe KF: Examination of the itch response from the raphides of the fishtail palm *(Caryota mitis). Toxicol Appl Pharmacol* 48;287–292, 1979

Sunell LA, Healey PL: Distribution of calcium oxalate crystal idioblasts in corms of taro *(Colocasia esculenta). Am J Bot* 66;1029–1032, 1979

Suzuki M: 3,4-Dihydroxybenzaldehyde-D-glucoside, the irritant substance of konnyaku. *Food Sci* 45;1075–1077, 1980

Tokarnia CH, Armien AG, Peixoto PV, Barbosa JD: Experiments on the toxicity of some ornamental plants in cattle. *Pesq Vet Bras* 16;5–20, 1996

Walter WG: *Dieffenbachia* toxicity. *J Am Med Assoc* 201;140–141, 1967

Walter WG, Khanna PN: Chemistry of the aroids I. *Dieffenbachia seguine, amoena,* and *picta. Econ Bot* 26;364–372, 1972

Wiese M, Kruszewska S, Kolacinski Z: Acute poisoning with *Dieffenbachia picta. Vet Hum Toxicol* 38;356–358, 1996

Willaman JJ, Schubert BG: *Alkaloid-Bearing Plants and Their Contained Alkaloids.* US Dept Agric Tech Bull 1234, 1961

11 Araliaceae Juss.

- Ginseng Family

  Aralia
  Hedera

Commonly known as the ginseng family, the Araliaceae comprises 55–70 genera and 700–800 species of tropical and temperate trees, shrubs, vines, and perennial herbs. Distribution is cosmopolitan, with greatest abundance in the tropics of Indomalaysia and tropical America. In North America, the family is represented by 15 genera and numerous species, both native and introduced. Familiar species include the cultivated *Hedera helix* (English ivy), *Schefflera,* and *Polyscias.* Species of *Panax* have been used medicinally in Asia for thousands of years. Considered of value for a wide variety of afflictions from nervous conditions, fevers, headaches, and rheumatism and as a female contraceptive, they are most sought after for their perceived qualities as aphrodisiacs. Exploitation by collectors has extirpated populations of the North American *Panax quinquefolius* in parts of its range in the continent's eastern deciduous forests. Medicinal extracts also have been obtained from species of *Aralia.*

- herbs, shrubs, vines, trees; flowers in umbels or heads; petals 5 or 3, white or greenish white; fruits berries or drupes

Plants trees, shrubs, woody vines, or herbs; perennials; from tubers or with perennating organs not apparent; deciduous or evergreen; armed or not armed with prickles. **Leaves** simple or palmately or 1-pinnately or 2-pinnately or 3-pinnately compound; alternate or whorled; stipules present. **Inflorescences** umbels or heads; terminal or axillary. **Flowers** perfect or imperfect, similar; perianths in 1-series or 2-series. **Sepals** 5 or 0; small, fused at bases. **Corollas** radially symmetrical; valvate. **Petals** 5 or rarely 3; caducous; free; white or greenish white. **Stamens** 5. **Pistils** 1; compound; carpels 2–5; stigmas 1; styles 2–5, free, stylopodia present; ovaries inferior; locules 2–5; placentation axile; ovules 1 per locule. **Nectaries** present; forming a disk at bases of styles. **Fruits** berries or drupes. **Seeds** 2–5.

Evidence that the Araliaceae and Apiaceae, the carrot family, form a monophyletic taxon has been offered, suggesting that the two families should be merged.

- many taxa contain compounds capable of irritant and other effects

The toxicological potential of members of the Araliaceae is not fully understood. Though some members have long been considered to be toxic, actual reported cases of intoxication are few and involve primarily species of *Hedera.* Their primary effect appears to be irritation of the digestive tract; however, respiratory depression and paralysis, cytotoxicity, suppression of hepatic biotransformation enzymes, cyanogenesis, and allergic contact dermatitis also have been observed, either in experimental studies or in rare clinical cases. Compounds similar to those found in *Hedera* also have been identified in other genera, including *Aralia*; species of *Polyscias* (wild aralia), and *Schefflera actinophylla* (umbrella tree). *Polyscias* and *Schefflera* are species introduced for cultivation as ornamentals.

Aralia L.

- *A. spinosa*

  Hercules'-club
  angelica tree

Taxonomy and Morphology—Commonly known as spikenard or sarsaparilla, *Aralia* comprises some 35 species in both the Old and New Worlds. In North

America, the genus is represented by 7 species, with 1 species the most commonly encountered:

A. spinosa L. Hercules'-club, angelica tree

Plants trees or shrubs; perennials; deciduous; armed or not armed with stout prickles. **Leaves** 2-pinnately or 3-pinnately compound; alternate or basal; leaflets lanceolate to ovate or obovate, apices acute or acuminate, margins serrate, bases acute or broadly rounded to cordate. **Inflorescences** numerous umbels; terminal. **Flowers** perfect; perianths in 2-series. **Sepals** 5. **Petals** 5. **Stamens** 5. **Pistils** 1; compound; styles 4 or 5, free or fused below. **Berries** purple or black. **Seeds** 4 or 5.

■ shrubs or trees; leaves pinnately compound; petals 5; dark berries

Distribution and Habitat—In eastern North America, plants of *Aralia* are typically found scattered through deciduous forests. Some species occupy dry, sandy soils, while others are found in moist loamy or silty soils of stream terraces. In the West, plants of *Aralia* are found in shade along streams in canyon bottoms.

■ forest or shady sites

Aralia spinosa

Aralia spinosa

Disease Problems and Genesis—The toxicologic potential of this genus is not clear, but the seeds and foliage of *A. spinosa* produce severe irritant effects (Reynard & Norton 1942). Plants contain several potential irritants, including terpenoids and acetylenic compounds (Hansen & Boll 1986). This indicates the potential for problems similar to those noted for species of *Hedera*. The acetylenic compounds include falcarinone, falcarinolone, falcarinol, and falcarindiol. The main pentacyclic terpenoids are a series of bidesmosidic aralosides (or elatosides), derivatives of oleanolic acid (Connolly & Hill 1991). Because of the presence of glycosidases in the plants, an array of monodesmosides and aglycones may also be present under varying circumstances. Experimentally, the aralosides cause a lowering of blood sugar when given orally to rats (Yoshikawa et al. 1996). This effect is suggested as the probable basis for use of the plants in Chinese traditional medicine.

■ digestive disturbance; terpenoids; acetylenic compounds also present

$$H_2C{=}CC(OH)(C{\equiv}C)_2CC{=}C(C)_6CH_3$$

Falcarinol

Clinical Signs and Treatment—Because of the presence of the same types of toxicants, the signs and treatment will be similar to those for *Hedera*.

■ vomiting, excess salivation, diarrhea; treatment not generally needed

Hedera L.

Taxonomy and Morphology—Comprising 11 species native to the Old World and numerous horticultural varieties, *Hedera* is the genus perhaps most widely used for ground cover. Frequently encountered taxa include the following:

■ *H. canariensis* *H. hibernica*
H. colchica *H. nepalensis*
H. helix

Atlantic ivy Irish ivy
Canary Island ivy Nepal ivy
common ivy Persian ivy
English ivy

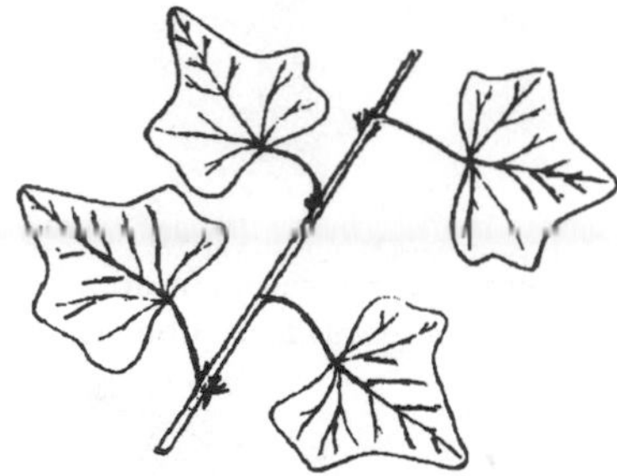

Hedera helix

H. canariensis Willd.	Canary Island ivy
H. colchica (K.Koch) Hibberd	Persian ivy, colchis ivy
H. helix L.	English ivy, common ivy, Irish ivy, poet's ivy
H. hibernica (Kirchn.) Bean	Atlantic ivy
H. nepalensis K.Koch	Nepal ivy

■ climbing woody vines; leaves simple, alternate; petals 5; berries

Plants woody vines, climbing or creeping; perennials; evergreen; adventitious roots present or absent. **Leaves** simple; alternate; of 2 forms, juvenile and adult; blades ovate or cordate; venation palmate; apices acute or obtuse; margins lobed or entire; bases rounded or cordate or cuneate; upper surfaces glabrous; lower surfaces stellate or scaly; acute or broadly rounded to cordate. **Inflorescences** umbels; solitary or clustered; terminal. **Flowers** perfect; perianths in 2-series. **Sepals** 5. **Petals** 5. **Stamens** 5. **Pistils** 1; compound; styles 1. **Berries** black or orange or yellow or cream. **Seeds** 3–5.

■ ornamental climbing vines

Distribution and Habitat—Species and varieties of *Hedera* are planted extensively throughout North America to cover ground, walls, sheds, and tree stumps. They are capable of growth in dense shade as well as sunny habitats. In addition, short-stemmed individuals are used as houseplants.

■ irritation, digestive disturbance

Disease Problems—Members of this genus have long been considered to be toxic, although there are few reported cases of intoxication. Their primary effect seems to be irritation of the digestive tract to produce vomiting and/or diarrhea. An individual who intentionally chewed one berry experienced a burning sensation in the throat that lasted about 15 minutes (Frohn & Pfander 1984). There may also be systemic effects. Bromel and Zettl (1986) reported an instance in which neurologic signs were observed in a roe deer that died and subsequently was discovered to have ivy leaves in its rumen.

■ irritant pentacyclic terpenoids, especially in fruits

Disease Genesis—*Hedera helix* contains a variety of potentially noxious compounds, the foremost of which are pentacyclic terpenoids. These terpenoids can be grouped under three classifications: the genins or aglycones, hederagenin and oleanolic acid; their respective monodesmosides, α-hederin and β-hederin, which are glycosides at C-3; and their respective bidesmosides, hederacoside C (hederasaponin C) and B, which are further glycosidic via esterification at C-20 (Connolly & Hill 1991; Quentin-Leclercq et al. 1992; Hostettmann & Marston 1995). Other glycosides of hederagenin include the monodesmosides caulosides and the hederacosides. α-Hederin is sometimes referred to as sapindoside A. Sometimes the mixture of genins, monodesmosides, and bidesmosides is referred to generally as saponins or simply as hederin. The confusion in nomenclature complicates understanding of the toxicologic role of the various types of compounds. The terpenoids are most commonly associated with the fruits. Barnea and co-workers (1993) found the unripe fruits to be negative for α-hederin and positive for hederagenin-3-α-arabinoside. In contrast, α-hederin was present in both the seed and the pulp of mature fruit.

ß-Hederin

Hederacoside B

Very early, Moore (1913) isolated a compound called hederin from the leaves of *H. helix* that was quite bitter and seemed to limit ingestion by animals. It was very irritating to the mucous membranes and, when administered parenterally, caused marked depression of respiration and eventually respiratory paralysis. It also caused hemolysis and was considered to be a saponin. Harshberger (1920) reported signs of neurointoxication in addition to the effects on the digestive tract. It is now apparent that the hederin originally referred to was most likely a mixture of many of the terpenoids. The specific compounds causing the toxic effects are not clear, but the presence of glycosidase activity in the plants and digestive tracts assures the exposure of the animal to the full range of terpenoids. α-Hederin, which has greater hemolytic activity than the bidesmosides, appears to of limited concern, given that oral dosage of 500–800 mg/kg of b.w. failed to cause adverse effects in sheep (Julien et al. 1985). However, it does appear to be the most cytotoxic (Quetin-Leclercq et al. 1992). α-Hederin binds to proteins such as bovine serum albumen, which may account for some of its effects (Danloy et al. 1994). Although direct toxicity seems to be limited for α-hederin, it is absorbed into the systemic circulation when given orally, as evidenced by its potent effects on liver flukes of sheep (Julien et al. 1985). Given parenterally, it also causes marked suppression of cytochrome P450 and a number of P-450–associated hepatic biotransformation enzymes (Liu et al. 1995). In vitro, α-hederin and hederagenin have significant smooth muscle relaxant activity (Trute et al. 1997).

■ irritant terpenoid mixture called hederin; also other actions

In addition to the terpenoids, *H. helix,* at least in some cultivars, also produces acetylenic compounds, cyanogenic precursors and other toxins. Acetylenic compounds such as falcarinone, falcarinolone, falcarinol, and falcarindiol are found consistently in various parts of the plant (Hansen & Boll 1986). These compounds, such as the neurotoxic falcarinol, may play an additive role, but this is not clear at present. Falcarinol and other acetylenics also are likely to be factors in causing the allergic contact dermatitis associated with *H. helix* (Garcia et al. 1995). In some instances, the fruit pulp, especially when unripe, has shown cyanogenic potential (Barnea et al. 1993). Some varieties of *H. helix* in Egypt also appear to contain the alkaloid emetine (Mahran et al. 1975).

■ other compounds of unknown significance, acetylenic and cyanogenic compounds

Clinical Signs and Treatment—The main clinical signs of intoxication are related to irritation of the digestive tract: vomiting, excess salivation, abdominal pain, and perhaps diarrhea. Except for depression, neurologic signs such as seizures are generally unlikely. Treatment generally is not required, although symptomatic relief of the irritant effects may be desirable in some instances.

■ vomiting, excess salivation, diarrhea; treatment not generally needed

References

Barnea A, Harborne JB, Pannell C: What parts of the fleshy fruits contain secondary compounds toxic to birds and why? *Biochem Syst Ecol* 21;421–429, 1993

Bromel J, Zettl K: Ivy poisoning in a roe deer. *Prakt Tierarzt* 67;967–968, 1986

Connolly JD, Hill RA: *Dictionary of Terpenoids.* Vol 2, *Di- and Higher Terpenoids.* Chapman & Hall, London, pp655–1460, 1991

Danloy S, Quetin-Leclercq J, Coucke P, De Pauw-Gillet M-Cl, Elias R, Balansard G, Angenot L, Bassleer R: Effects of α-hederin, a saponin extracted from *Hedera helix,* on cells cultured in vitro. *Planta Med* 60;45–49, 1994

Frohne D, Pfander HJ: *A Colour Atlas of Poisonous Plants.* Wolfe Science, London, pp61–62, 1984

Garcia M, Fernandez E, Navarro JA, del Pozo MD, Fernandez de Corres L: Allergic contact dermatitis from *Hedera helix* L. *Contact Derm* 33;133–134, 1995

Hansen L, Boll PM: Polyacetylenes in Araliaceae: their chemistry, biosynthesis, and biological significance. *Phytochemistry* 25;285–293, 1986

Harshberger JW: *Pastoral and Agricultural Botany.* Blakiston's Son & Co, Philadelphia, p87, 1920

Hostettmann K, Marston A: *Saponins.* Cambridge Univ Press, Cambridge, 1995

Julien J, Gasquet M, Maillard C, Balansard G, Timon-David P: Extracts of the ivy plant, *Hedera helix,* and their antihelminthic activity on liver flukes. *Planta Med* 53;205–208, 1985

Liu J, Liu Y, Bullock P, Klaassen CD: Suppression of liver cytochrome P450 by α-hederin: relevance to hepatoprotection. *Toxicol Appl Pharmacol* 134;124–131, 1995

Mahran GH, Hilal SH, El-Alfy TS: The isolation and characterization of emetine alkaloid from *Hedera helix*. *Planta Med* 27;127–132, 1975

Moore B: The chemical and pharmacological properties of hederin, sapo-glycoside contained in the leaves of common ivy *(Hedera helix)*. *J Pharmacol Exp Ther* 4;263–275, 1913

Quentin-Leclercq J, Elias R, Balansard G, Bassleer R, Angenot L: Cytotoxic activity of some triterpenoid saponins. *Planta Med* 58;279–281, 1992

Reynard GB, Norton JBS: *Poisonous Plants of Maryland in Relation to Livestock*. Md Agric Exp Sta Bull A10, pp249–312, 1942

Trute A, Gross J, Mutschler E, Nahrstedt A: In vitro antispasmodic compounds of the dry extract obtained from *Hedera helix*. *Planta Med* 63;125–129, 1997

Yoshikawa M, Murakami T, Harada E, Murakami N, Yamahara J, Matsuda H: Bioactive saponins and glycosides. 7. On the hypoglycemic principles from the root cortex of *Aralia elata* Seem.: structure related hypoglycemic activity of oleanolic acid oligoglycoside. *Chem Pharm Bull* 44;1923–1927, 1996

Asclepiadaceae R.Br.

12

Numerous plant species are characterized by a milky-white, viscid sap that exudes from broken leaves or stems. Although many such plants often are called milkweeds, taxonomists typically recognize only members of the Asclepiadaceae as the true milkweeds. Widespread in tropical, subtropical, and warm-temperate regions of the world, and especially abundant in Africa, the family comprises 250–350 genera and 2,000–3,000 species. It is closely related to the Apocynaceae, or dogbane family, and some taxonomists unite the two. However, the two families differ significantly in the nature of their stamens and pollen.

Of slight economic importance, the family contains numerous ornamentals, succulents, and attractors of butterflies. The popular wax plant *(Hoya carnosa),* stapelia (*Stapelia* sp.), and butterfly milkweed *(Asclepias tuberosa)* are representative. Species of the family also have been a source of fiber and oils. Medicinal emetics and purgatives were prepared by American Indians from species of *Asclepias.*

■ Milkweed Family

Asclepias
Calotropis
Cryptostegia

QUESTIONABLE
Araujia
Cynanchum
Hoya
Periploca
Sarcostemma

Plants herbs or herbaceous vines or shrubs or rarely small trees; perennials; sap viscous or thin, white or colorless. **Leaves** simple; opposite or subopposite or whorled; petiolate or sessile; herbaceous or fleshy; venation pinnate; margins entire; stipules absent. **Inflorescences** cymes or umbels or racemes or solitary or paired flowers; terminal or axillary. **Flowers** perfect; perianths in 2-series. **Sepals** 5; fused. **Corollas** radially symmetrical; rotate or campanulate; valvate. **Petals** 5; fused; of various colors; spreading or reflexed. **Coronas** present; petaloid; arising from filament bases and/or petals. **Stamens** 5; epipetalous; anthers fused to style heads; filaments free or fused; thecae 2; pollen in pollinia (1 pollinium per theca, attached via translator to pollinium of adjacent anther). **Pistils** 1; compound, carpels 2; stigmas 1, 5-lobed, discoid; styles 2, free; ovaries 2, free, superior; locules 2; placentation parietal. **Fruits** follicles. **Seeds** numerous; comose.

■ herbs, shrubs, vines, or small trees; leaves simple, opposite or whorled, margins entire; petals 5, fused; fruits follicles

Fusion of the anthers, and in some taxa the filaments, to the style forms a central column called the gynostegium, a structure specialized for mass transfer of pollen by insects and unique to the Asclepiadaceae.

The family contains at least 10 genera known to be toxic, most of which are native to the tropics. Those of some toxicologic importance in North America include *Asclepias, Cynanchum,* and *Sarcostemma.* Of these, only the species of *Asclepias* are important as livestock hazards. The ornamentals *Calotropis* and *Cryptostegia* may pose a slight risk in the South.

Asclepias L.

Taxonomy and Morphology—Primarily a New World genus, *Asclepias* comprises 120–150 species, of which 108 occur in North America (Woodson 1954). Its name is derived from *Asklepios,* the name of the Greek god of healing (Morse 1985). Of

■ *Asclepias*

butterfly weed	orangeroot
chigger flower	pleurisy root
Indian nosy	rubber root
milkweed	silkweed

the numerous species in North America, several are particularly important toxicologically. The following are representative of the two types of disease problems characteristic of *Asclepias* (see plates 12.1–12.5):

Asclepias flower

- herbs, sap white, latexlike; leaves opposite or whorled; flowers very characteristic; fruits follicles

A. asperula (Decne.) Woodson	antelope horn
A. brachystephana Engelm. ex Torr.	short-crown milkweed
A. eriocarpa Benth.	woolly-pod milkweed
A. fascicularis Decne.	narrow-leaved milkweed, Mexican milkweed
A. labriformis M.E.Jones	labriform milkweed
A. latifolia (Torr.) Raf.	broadleaf milkweed
A. mexicana Cav.	Mexican whorled milkweed
A. pumila (A.Gray) Vail	plains whorled milkweed
A. speciosa Torr.	showy milkweed
A. subulata Decne.	desert milkweed, yamate, ajamete
A. subverticillata (A.Gray) Vail	western whorled milkweed, horsetail milkweed
A. syriaca L.	common milkweed
A. tuberosa L.	butterfly milkweed, silkweed, orangeroot, pleurisy root, chigger flower, Indian nosy, orange swallowwort, rubber root, windroot, butterfly weed
A. verticillata L.	eastern whorled milkweed, spider milkweed
A. vestita Hook. & Arn.	woolly milkweed
A. viridis Walter	green milkweed, spider milkweed

Asclepias fascicularis

Plants herbs; sap white and viscous, rarely thin and colorless. **Stems** decumbent to erect; 4–150 cm tall; branched or not branched. **Leaves** opposite or whorled; herbaceous; decussate; sessile; blades of various shapes. **Inflorescences** umbellate clusters or solitary; terminal or extra-axillary. **Flowers** rotate. **Sepals** fused at bases. **Petals** fused; of various colors; spreading or reflexed. **Coronas** conspicuous; petaloid; hood shaped; erect or incurved; horns present or absent. **Follicles** globose to fusiform or ovoid; smooth or spiny; pedicels erect or recurved. **Seeds** large; flat.

Woodson, in his 1954 monograph of the North American species of *Asclepias*, recognized 9 subgenera, 108 species, and numerous subspecies. Identification is based primarily on features of the gynostegium and corona and is somewhat complicated because of the potential for hybridization (Kephart et al. 1988). In addition, some species form complexes of morphologically intergrading taxa. Because of the large number of species in the genus and also of toxicologic interest, a taxonomic key is not presented here; rather the reader is encouraged to use floristic treatments specific for his or her area.

It is important to note that plants identified as *A. galioides* and *A. mexicana* in early American toxicological publications are now considered to be *A. subverticillata* and *A. fascicularis*, respectively. The taxon *A. galioides* is a narrow-leaved form of *A. mexicana* that is strictly central Mexican in distribution (Woodson 1954).

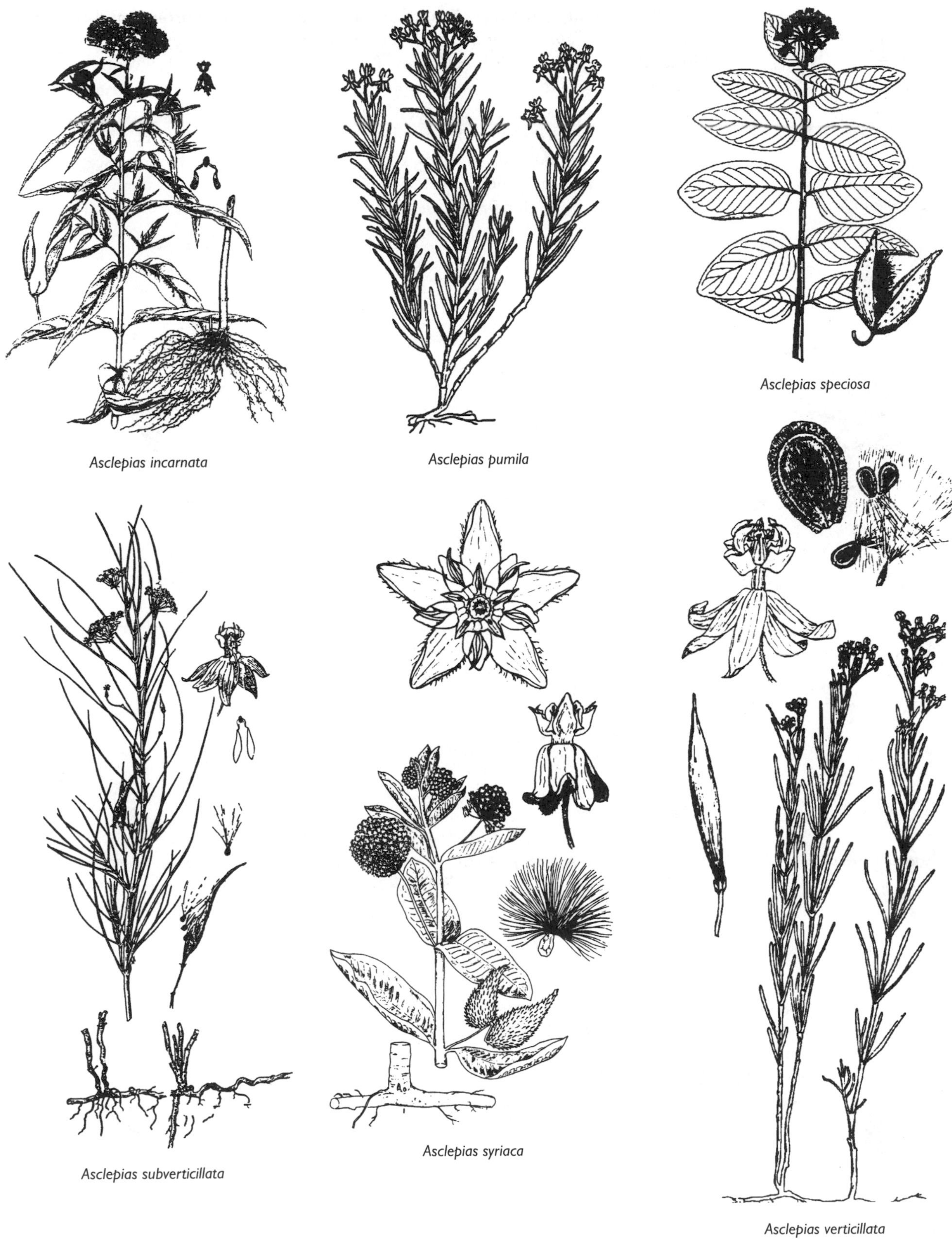

Asclepias incarnata

Asclepias pumila

Asclepias speciosa

Asclepias subverticillata

Asclepias syriaca

Asclepias verticillata

Asclepias viridis

■ wide variety of habitats

Distribution and Habitat—Adapted to a broad range of environmental conditions, species of *Asclepias* occur across the continent in almost all vegetation types. Greatest abundance is southward; relatively few are found in Canada. Woodson (1954) described six centers of species concentration, which he designated as Antillean, Floridian, Appalachian, Ozarkian, Mexican, and Californian. Illustrative of this pattern are the distributional maps of the representative species. Habitats occupied include open woodlands, prairies, rocky plains, sand dunes, marshes, seeps, canyon bottoms, dry washes, and desert flats. Some species are weedy and form small or large dense populations along roadsides or in overgrazed pastures.

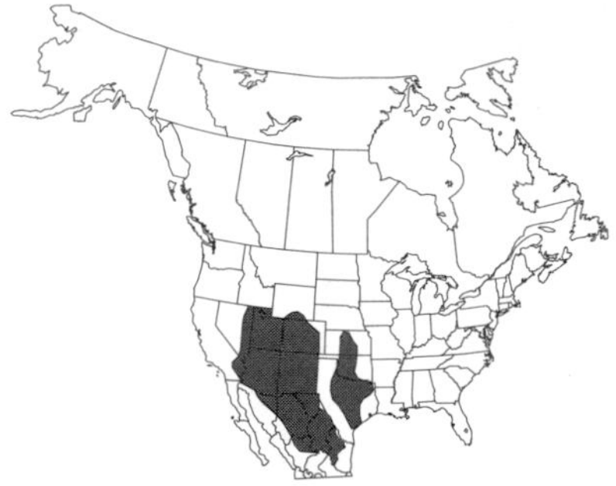

Asclepias asperula

Asclepias brachystephana

Asclepias eriocarpa

Asclepias fascicularis

Asclepias labriformis

Asclepias latifolia

Asclepias mexicana

Asclepias pumila

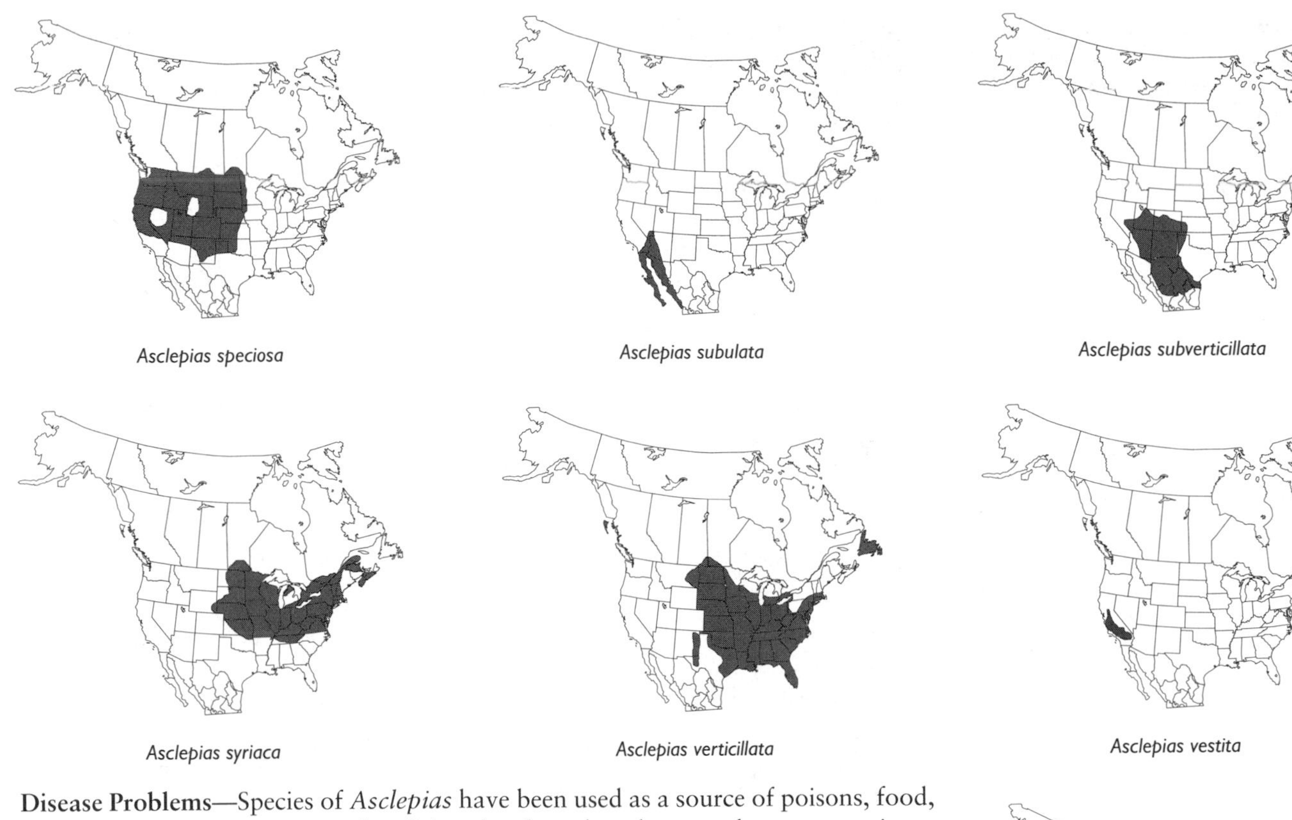
Asclepias speciosa

Asclepias subulata

Asclepias subverticillata

Asclepias syriaca

Asclepias verticillata

Asclepias vestita

Disease Problems—Species of *Asclepias* have been used as a source of poisons, food, and medicines. Containing cardenolides, they have long been used as arrow poisons in Africa. In North America, the roots of *A. tuberosa* were used extensively for ceremonial and medicinal purposes by members of American Indian tribes and early settlers (Kindscher 1992). Infusions or teas of the plant, which is sometimes called pleurisy root, were employed as a dewormer and a diuretic, although in excess they might cause nausea and vomiting. These uses and others were listed in the U.S. Pharmacopoeia published in the late 1800s. More recently, other uses have been proposed for some of the more common species. The potential for a favorable comparison of extractable yields of liquid fuels with coal for energy was explored (McLaughlin & Hoffman 1982; Adams et al. 1984). In addition when extracted with methanol to remove toxins, the residue may represent a useful livestock feed. Species of *Asclepias* also may be used as a source for the manufacture of paper (Adams et al. 1984).

Asclepias viridis

■ source of arrow poisons, food and medicines

In the early 1900s, devastating losses of hundreds of sheep in the western portions of the United States focused attention on *Asclepias* as one of the most toxic plants in North America (May 1920; Couch 1937). Illustrative are losses of 350 and 750 ewes from flocks of 1,700 and 1,400 animals held for 24 hours or less in fields containing large populations of what is now known to have been *A. subverticillata* (May 1920). Because of such losses and subsequent experimental work, this species came to be recognized as one of the most toxic. Its infamy led to the study of other species of the genus and an eventual elucidation of their toxicity potential, an effort greatly aided by information generated by entomologists interested in the ecological relationships between plants and insects.

■ two types of effects, all animal species including birds:
(1) digestive disturbance and cardiac
(2) neurologic only with whorled or verticillate-leaved species

The milkweeds appear to be toxic to all animals, although some species such as mice and quail are reported to be much less sensitive. In general the clinical signs that appear are similar in all animals, including birds. There are two distinct types of signs associated with species of *Asclepias:* a digestive tract/cardiac type and a neurologic

■ verticillate-leaved species, somewhat palatable, difficult to avoid when dried in hay, toxicity not diminished by drying

type (Fleming et al. 1920; Marsh & Clawson 1921a,b; Campbell 1931). Despite the well-known toxicity potential of many of the species cited above, only two verticillate-leaved taxa—*A. fascicularis* and *A. subverticillata*—appear to be responsible for most of the problems in livestock, including chickens. There are several apparent reasons for this disparity in hazard among species of the genus. Foremost is the likelihood of plants being eaten. Animals typically are very reluctant to eat milkweeds, especially the cardiotoxic species, many of which are quite fibrous in addition to being unpalatable (Nelson & Dexter 1945; Ogden et al. 1992b). In contrast, the verticillate-leaved species seem to be eaten when green by some animals and are even more readily accepted when dried and mixed in hay. Their toxicity is not appreciably reduced by drying. Typically the leaves of cardiotoxic species are more readily distinguished in hay and are easier for the animals to avoid in contrast to the fine leaves of the verticillate-leaved species. Thus the hazard is greater with the verticillate-leaved species because of increased likelihood of consumption.

■ neurotoxic; *A. fascicularis and A. subverticillata* form dense populations, greatest risk

An additional factor enhancing the hazard is the tendency of *A. fascicularis* and *A. subverticillata* to form very dense and extensive stands. Once a population is established, plants typically spread via an extensive network of rhizomes. Thus, not only are they likely to be eaten, but there also is often an abundance of plant biomass available. This is of particular importance with *A. subverticillata* because it is often found in great abundance along ditch banks of irrigated pastures and may be incorporated in large quantities in hay from such sites. Because of its fine stems and leaves, it is not conspicuous in either grass or alfalfa hay. None of the other verticillate-leaved species typically grow in sufficient abundance to be as serious problems as *A. fascicularis* and *A. subverticillata*. It appears that for verticillate-leaved species, all stages of growth and all plant parts are toxic, with the leaves of greatest hazard.

■ cardiotoxic species may be quite toxic but are generally unpalatable

The cardiotoxic species, some of which may also form dense stands—e.g., *A. speciosa* and *A. syriaca*—are unlikely to be eaten, fresh or in hay, unless the animals are forced to do so in unusual or dire circumstances. Although all animals are probably at risk, most reports involve sheep or cattle (Fleming et al. 1920; Marsh & Clawson 1924; Tunniclif & Cory 1930; Vail 1942; Rowe et al. 1970; Sperry et al. 1977; Ogden et al. 1992a,b). Similar signs are produced by all species, although the disease is generally more severe when the more toxic *A. labriformis, A. eriocarpa,* and *A. latifolia* are eaten. Typically, there are no convulsive episodes. Examples of the occasional episodes of intoxication include the death of 250 sheep from hay contaminated with *A. eriocarpa* in California and of 11 heifers over a 6-day period following ingestion of large amounts of *A. syriaca* in Hungary (Seiber et al. 1985; Salyi & Petri 1987). With respect to the cardiotoxic milkweeds, all plant parts are toxic; the fruits and seeds are the most toxic, and the lower leaves and roots the least.

Initial toxicity studies of *Asclepias* led toxicologists to divide the species into a "narrow-leaved" group and a "broad-leaved" group with blades wider than 3.5 cm. Based on clinical observations, it was originally believed that the broad-leaved group was cardiotoxic and the narrow-leaved group neurotoxic (Pammel 1917; Fleming et al. 1920; Radeleff 1970). However, it was later discovered (Ogden 1989; Ogden et al. 1992a,b) that this delineation is inappropriate. Several narrow-leaved species produce cardiotoxic effects. Furthermore, among the narrow-leaved taxa, only the verticillate-leaved species produce the neurotoxic effects.

■ verticillate-leaved species

A. fascicularis	*A. verticillata*
A. mexicana	and possibly
A. pumila	*A. incarnata*
A. subverticillata	

The verticillate-leaved species were classified by Woodson (1954) in the subgenus *Asclepias* and series *Incarnatae*. The series comprises 16 species, 5 of which are distinctly verticillate-leaved—*A. subverticillata, A. fascicularis, A. verticillata, A. pumila,* and *A. mexicana*—and 1 of which appears nearly so, *A. incarnata*. These 6 taxa form a complex of intergrading species across the western half of the United States and

Mexico and were hypothesized by Woodson to be derived from the easternmost *A. incarnata*. The neurotoxic effects of *A. subverticillata* have been described in sheep, cattle, horses, and poultry (Marsh et al. 1920; Stiles 1942; Sperry et al. 1977; Clark 1979; Sprowls 1982; Ogden et al. 1992a,b). The effects produced by *A. fascicularis, A. pumila,* and *A. verticillata* are similar (Glover et al. 1918; Fleming et al. 1920; Marsh & Clawson 1921a; Tehon et al. 1946; Sperry et al. 1977; Ogden et al. 1992a,b).

Disease Genesis—As early as the mid-1800s, naturalists had observed that birds avoided eating certain butterflies, such as monarchs *(Danaus plexippus),* whose larvae fed primarily on *Asclepias* species. It was proposed and later confirmed that the feeding larvae sequestered, by active regulation, emetic cardenolides that were retained and in some cases concentrated, in all stages of development, as a defense against vertebrate predators (Brower 1969; Rothschild & Ford 1970; Seiber et al. 1986). These cardenolides were recognized as very potent emetics (Brower et al. 1982). Birds eating butterflies containing them vomited shortly thereafter. Observations of this behavior led to the so-called gourmand-gourmet hypothesis, which proposes that a "naive" bird, initially accepting a wide range of foods, learns by conditioning to exercise judgment in selection on the basis of taste and eventually by the more efficient recognition of the visual characteristics of the insects. This in turn led to extensive studies on butterfly mimicry (Sheppard 1959; Rothschild 1967). Additionally, work was conducted to identify plant species containing emetic cardenolides and to characterize the compounds chemically. At present, 27 species of milkweed are recognized to be fed upon by butterfly larvae; the more toxic milkweed species are generally preferred (Malcolm & Brower 1986; Malcolm et al. 1989). Larvae accumulate much higher cardenolide concentrations when feeding on the more toxic species (Cohen & Brower 1983). The milkweeds are hosts for many other insects, herbivores, nectar seekers, and predators (Isman 1977; Cohen & Brower 1983; Morse 1985).

■ cardiotoxic species have emetic activity; exploited by butterflies; larval feeding on plants protects butterflies from bird predation

The digestive tract/cardiotoxic effects are typical of those produced by cardenolides and have been extensively studied (Roeske et al. 1976; Benson et al. 1978, 1979; Seiber et al. 1983; Seiber et al. 1985; Joubert 1989). The basic cardenolide aglycone is a 23-C steroidal genin with a 5-membered, singly unsaturated lactone ring at C-17, a hydroxyl group at C-14, and methyl groups at C-10 and C-13. Glycosidic linkage usually occurs at C-3 to one or more sugar moieties but can also involve C-2, creating a cyclic bridge to a single sugar. The sugars present include glucose and others such as rhamnose and thevetose. Additional methyl, hydroxyl, and carbonyl groups can be attached at other carbons of the genin, their presence further influencing lipid solubility as well as protein binding. These cardenolides inhibit Na^+,K^+-ATPase, and the structure-function relationship is believed to reside in the unsaturated lactone ring at C-17 and the hydroxyl group at C-14 (Joubert 1989).

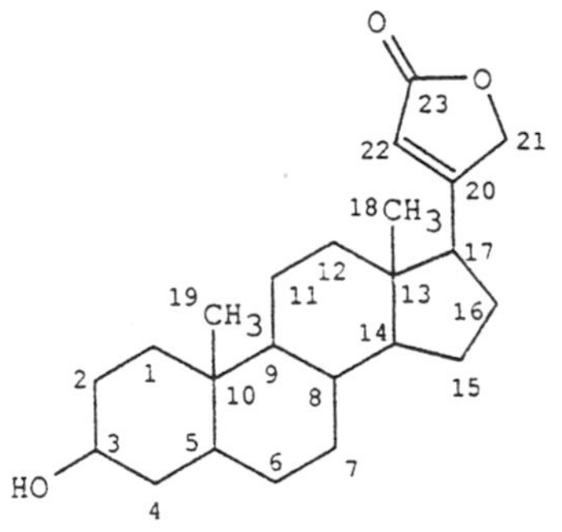

■ steroidal glycosidic cardenolides

Cardenolide

■ cardenolides inhibit Na^+,K^+-ATPase, increase calcium in cells

In the Asclepiadaceae, the cardenolides are steroids with 5–chair/chair *trans*-A/B ring fusion conformation, in contrast to the 5–boat/chair *cis*-A/B ring fusion of most other cardiotoxic genera. This may portend a slightly different activity potential, which may be the reason for the limited use of species of *Asclepias* for medicinal purposes. Differences among the various cardenolides of the *Asclepias* include a double bridge at C-2 and C-3 between genin and sugar, an epoxide rather than a double bond at C-7/C-8, and methyl derivatives at C-10 (Seiber et al. 1983). Cardenolides of more than one type may be present in the same plant (Seiber et al. 1985). Except for the configurational changes, the glycosides are similar to the medically more useful cardenolides from Apocynaceae and Scrophulariaceae (Seiber et al. 1983). Some of the asclepiad cardiotoxins are quite potent; labriformidin is 2 times more potent than the well-known ouabain (Benson et al. 1978). Even the minimally toxic

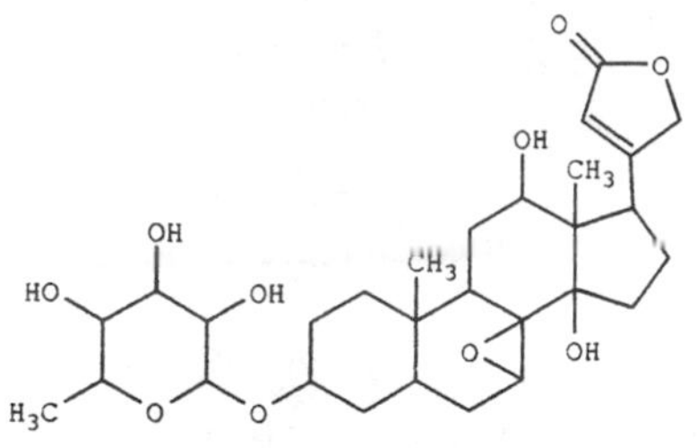

Aspecioside

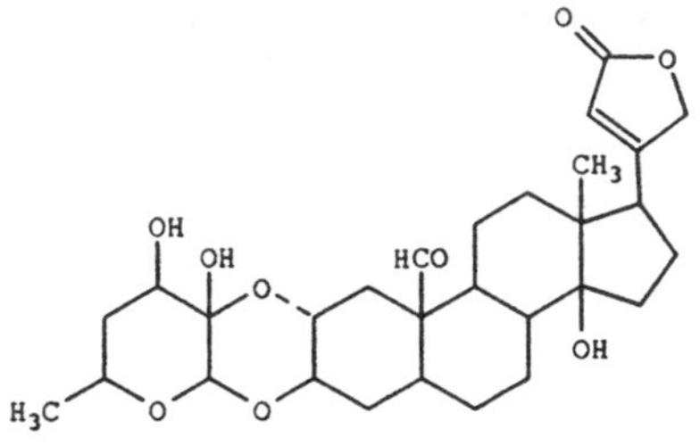

Calotropin

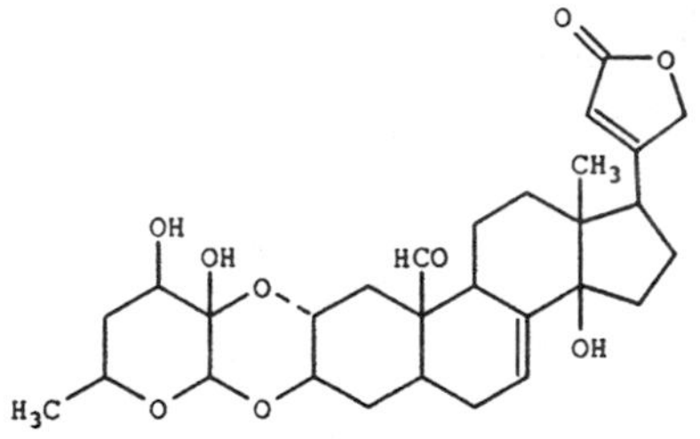

Humistratin

■ cardiotoxic risk highest in early growth

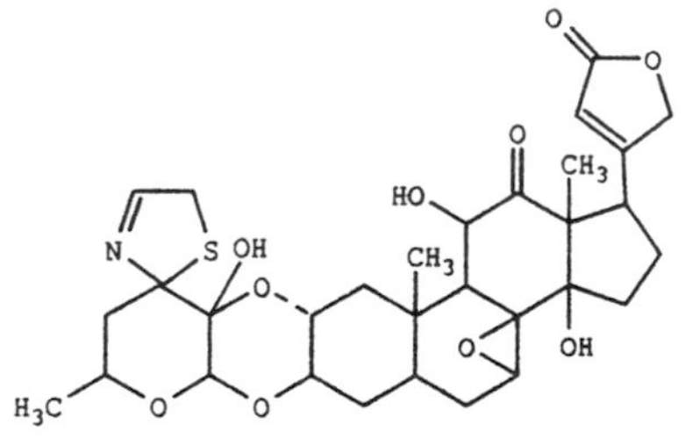

Labriformin

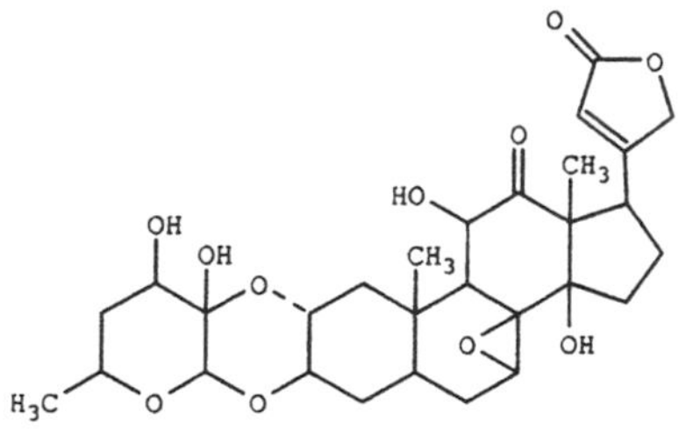

Syrioside

A. tuberosa contains cardenolides such as a diglucoside of coroglaucigenin. The Asclepiadaceae cardenolides affect the heart in a manner similar to that of the digitalis glycosides; there is inhibition of Na^+,K^+-ATPase and increased movement of sodium and calcium into the cells (Benson et al. 1978). Additional cytotoxic and neurotoxic effects have been ascribed to some of them (Kupchan et al. 1967; Piatak et al. 1985).

Cardenolide content may vary quantitatively and qualitatively with plant part, geographical location, and growing season (Roeske et al. 1976; Nelson et al. 1981; Brower et al. 1984; Moore & Scudder 1985). Concentrations of total cardenolides as digitoxin equivalents in the leaves and other organs are typically 0.01–0.5%, but in some extreme instances they may be up to 14.7% (Roeske et al. 1976). The most potent types are not necessarily present at these higher levels. Concentrations in the latex sap may be as much as 10 times higher (Nelson et al. 1981). Even species such as *A. asperula, A. speciosa,* and *A. viridis,* which have rarely been associated with intoxications, may have high cardenolide concentrations of 0.3–0.8% (Brower et al. 1984; Lynch & Martin 1987, 1993). Cardenolide concentrations in *A. eriocarpa,* which may exceed 0.5%, are highest in the sap, less in the pods and seeds, lower in the stems, lower still in the leaves, and lowest in the roots (Duffey et al. 1978; Nelson et al. 1981; Brower et al. 1982). The stem concentrations are highest in the early stages of growth and then decline steadily during the growing season to low levels in the fall. In the leaves, concentrations are highest in midsummer and decline markedly between October and December after they die (Nelson et al. 1981). This may account for the reputed decrease in toxicity of *A. latifolia* after the first frosts in the fall (Sperry et al. 1977). Typically, there seems to be an inverse relationship between polarity of the cardenolide and the toxicity (Seiber et al. 1986; Lynch & Martin 1987). A single species may contain more than a dozen similar but different cardenolides; a few may be rather novel ones, and the remainder the same as those present in other species of the genus (Brower et al. 1982; Jolad et al. 1986). Several species—e.g., *A. speciosa* and *A. syriaca*—may share an almost identical array of cardenolides (Seiber et al. 1986).

Cardenolides are biotransformed by hepatic microsomal mixed-function oxidases, hydroxylated at various sites to form more-polar derivatives, and excreted as glucuronide conjugates (Singh & Rastogi 1970). Larvae feeding on *Asclepias* containing high amounts of these toxins have increased activity of mixed-function oxidases in the midgut and fat body (Marty et al. 1982; Marty & Krieger 1984). This may portend a similar response in vertebrates, inducing biotransformation by hepatic mixed-function oxidases and hastening xenobiotic disposition.

The neurotoxic effects produced by the verticillate-leaved species of *Asclepias* appear to be caused by toxins other than the typical cardenolides. As noted above, Woodson (1954) classified these taxa into the series *Incarnatae* in the subgenus *Asclepias.* Toxicologically, they appear distinct from other species of *Asclepias* in that they contain little or no cardenolides and produce neurotoxic rather than cardiotoxic effects (Seiber et al. 1983, 1985; Burrows et al. 1990, 1992; Ogden et al. 1992a,b). Surprisingly, although these species are considered to possess negligible cardenolide content, they are reported to produce some cardiac effects, albeit of a minor nature (Ogden et al. 1992b).

Early research indicated that *A. subverticillata,* thought to be one of the most toxic plants in North America, contained an alcohol-extractable, glucosidal, resinoid neurotoxin called galitoxin (Couch 1937). However, these observations remain unconfirmed. Although neurotoxins have yet to be identified in *Asclepias,* studies in China and South Africa on *Cynanchum* and *Sarcostemma,* other genera of the family

Asclepiadaceae, have identified specific neurotoxicants. These genera lack cardenolides (Ritland 1991) and instead contain neurotoxic pregnane glycosides, e.g., the cynafoside series (Tsukamoto et al. 1985a,b, 1986; Zhang et al. 1985; Qui et al. 1989; Steyn et al. 1989; Lou et al. 1993). There is also the possibility that the neurotoxicants are structurally very similar to the cardiotoxins. Known to cause both cardiac and neurotoxic effects, the action of bufotoxins from toads can be altered to increase spastic activity at the expense of cardiac activity by removal of the C-14 hydroxy group or by converting it to an oxido group. There is evidence to indicate that the neurotoxicants are cinnamic ester pregnane glycosides (Robinson et al. 1998). Based on mass spectrophotometry data, the existence one such compound—verticenolide—has been proposed (Robinson et al. 1998).

Substantial information (table 12.1.) has been accumulated about the toxicity of the verticillate-leaved species and several of the cardiotoxic species, including *A. brachystephana, A. eriocarpa, A. labriformis, A. latifolia,* and *A. speciosa.* For most of the remainder, estimates of toxicity are based on concentrations of cardenolides or anecdotal reports. A wide variety of animal species are reported to be affected, including chickens, turkeys, rabbits, horses, cattle, and sheep. Chickens appear to be less susceptible than mammals to the cardiotoxic species (Sperry et al. 1977; Ogden et al. 1992a). Mice and quail are also reported to be little affected by the cardenolides (Rothschild & Ford 1970). In contrast, chickens and turkeys as well as many mammalian species are quite susceptible to the neurotoxins.

Additional toxicants in *A. speciosa* and *A. syriaca* include α- and β-amyrin (Brower et al. 1982).

Table 12.1 Toxicity and cardenolides content of species of *Asclepias*

Asclepias Species	Leaf Arrangement & Width	Cardenolide Content[a]	Toxicity Estimate[b]
A. amplexicaulis	opposite, broad	very low	—
A. asperula	opposite, broad	very high	>1%
A. brachystephana	opposite, narrow	very high	0.5%
A. californica	opposite, broad	high	—
A. cordifolia	opposite, broad	low[d]	>2%
A. curassavica	opposite, medium	very high	—
A. eriocarpa	opposite, broad	very high	0.2%
A. erosa	opposite, broad	intermediate	—
A. exaltata	opposite, broad	low	—
A. fascicularis	whorled, narrow	very low	0.5–1%
A. hirtella	opposite, medium	high[c]	1%
A. humistrata	opposite, broad	very high	—
A. incarnata	opposite, medium	very low[d]	2% or >
A. labriformis	opposite, narrow	very high	0.05–0.2%
A. latifolia	opposite, broad	high[c]	0.2–0.4%
A. pumila	whorled, narrow	very low[c]	1%
A. speciosa	opposite, broad	intermediate	1–2%
A. subulata	opposite, broad	very high	—
A. subverticillata	whorled, narrow	very low	0.2–0.4%
A. syriaca	opposite, broad	intermediate	1–2%
A. tuberosa	opposite, medium	very low	>2%
A. verticillata	whorled, narrow	very low	0.4–1%
A. vestita	opposite, medium	very high	—
A. viridiflora	opposite, broad	very low	—
A. viridis	opposite, broad	high	1%

a. very low = <0.25, low = 0.25–1, intermediate = 1–2, high = 2–4, very high = >4 mg/g plant.
b. estimate of the amount of the plant (green or dry) causing death, expressed as percentage of b.w. of the animal.
c. estimated, based on toxicity data.
d. Field test results indicate high concentrations.

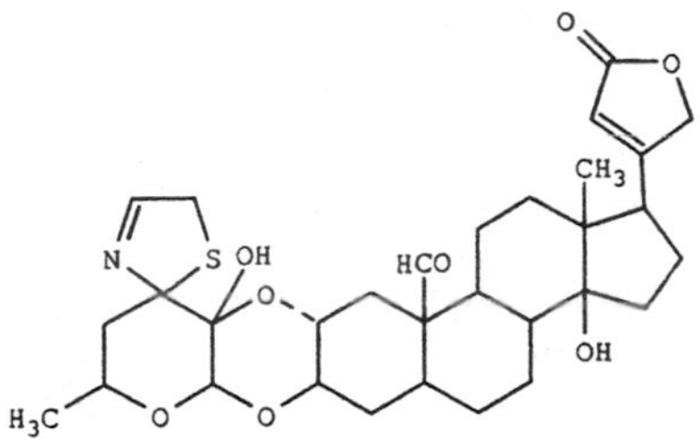

H3C O O S OH N HCO O CH3 OH O O

Uscharin

- neurotoxins isolated from other family members similar to cardenolides

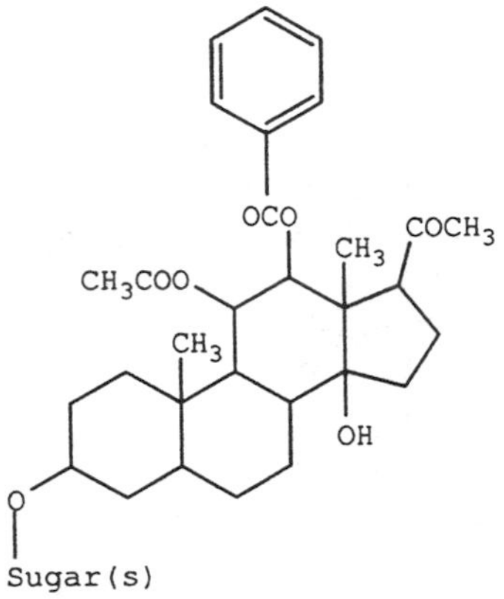

OCO CH3COO CH3 COCH3 CH3 OH O Sugar(s)

Cynafosides

- a pregnane glycoside structure has been proposed for the neurotoxin in verticillate-leaved species, verticenolide

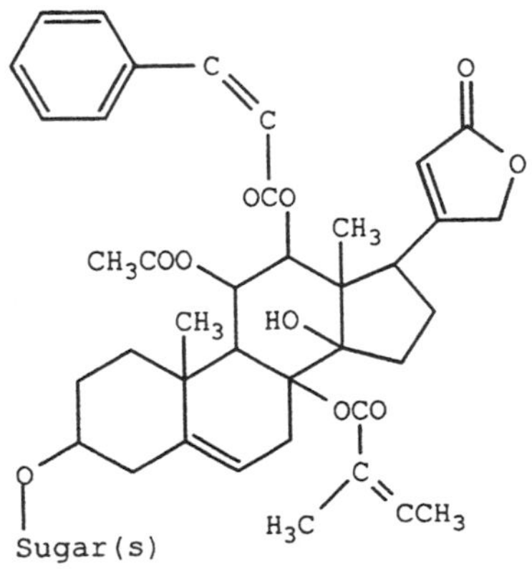

C C OCO CH3COO CH3 O O CH3 HO OCO C H3C CCH3 O Sugar(s)

Verticenolide (proposed structure)

Clinical Signs—As would be expected by the presence of two apparent classes of toxicants in *Asclepias,* the clinical signs appearing after ingestion of plants are quite different.

- cardiotoxic, depression, weakness, reluctance to stand, diarrhea, labored respiration, possible cardiac arrhythmias

As discussed above, disease due to consumption of the cardiotoxic species occasionally may occur, typically only in unusual circumstances. The animal becomes very depressed, often lying down and reluctant to stand. Abdominal pain with odontoprisis and diarrhea may also be evident. The most distinct and characteristic change occurs with the animal's respiration; it is often irregular, but typically difficult and with a peculiar and distinct groaning grunt, almost as if it were painful. It is important to note that even though the cardiotoxic species contain cardiac glycosides, arrhythmias as conspicuous as those produced by the glycosides of *Digitalis* are seldom seen. Death usually occurs without obvious signs of struggle. There are no convulsive episodes.

A field test with a detection threshold of 0.057% is available for detecting cardenolides in those instances when species of unknown toxicity are encountered (Sady & Seiber 1991).

- neurotoxic, especially horses, depression, trembling, weakness, ataxia, seizures with flexed thoracic limbs when down, paddling leg movements; death due to respiratory failure

The neurotoxic effects are cumulative, and signs may occur immediately after a single feeding of 1% b.w. or more of plant material, although there may be a delay of 12 or more hours between ingestion and the onset of obvious signs. Horses display signs of colic and marked uneasiness; they periodically lie down. Incoordination and weakness of the pelvic limbs almost to the extent of paresis follows. Because of this weakness, they become even more apprehensive and try to move about or fight their weakness. This leads to trembling; falls, some of which may produce serious injury; mydriasis; and profuse sweating. The animal's temperature often increases during this time. Finally, the weakness and incoordination progress into actual seizures, which may result in an inability to stand. Initially the seizures are tetanic and appear to be very agonizing; the animal falls with the neck bowed and thoracic limbs flexed. The tetanic seizures may give way to clonic types with paddling movements that occasionally become violent. The occurrence of violent seizures, while of great concern, does not necessarily signal impending death, because the animal may recover even after such severe signs (Marsh 1929). Death is apparently due to respiratory failure. Intoxication of an adult horse (ca. 450 kg) may occur with ingestion of as little as 1 kg of plant material. Recovery from the acute disease may be followed by the presence of some neurologic signs for several or more days.

Sheep are intoxicated by doses similar to those for horses, whereas cattle require substantially more, i.e., approximately two times as much. The signs in cattle are similar to those in horses and commence 12 or more hours after ingestion. They begin with depression, muscle fasciculation, and progressive weakness. They progress to incoordination and then to convulsions, which cause collapse, opisthotonus, and flexed thoracic limbs. Between convulsive episodes, bloat, groaning, and excess salivation may be prominent. Initially, the convulsive episodes are sufficiently separated to allow periods of uncoordinated standing, but they eventually become almost continuous and terminate in death in several hours. If the animal survives, it will be weak for several or more days.

In sheep, as in cattle and horses, neurotoxic signs are prominent. Weakness and pelvic limb incoordination are marked. The animal stands with a humped-up appearance and with the head held in an elevated position. When the animal is forced to move, an exaggerated jackrabbit-like gait with the head up has been described (Clark 1979). Convulsive episodes with opisthotonus, champing of the jaws, twitching of the eyelids, ears, and lips, head shaking, head pressing, and eventually death are common progressive stages of the disease.

Intoxication in poultry seems to be a particular problem when chicks or poults gain access to young green plants. Within a few hours of ingestion there are obvious signs, including loss of muscular control, excitement, bizarre posturing, and periodic violent convulsions with backward flips. After several more hours these subside to less dramatic tremors, torticollis, inability to stand, and running movements. With exhaustion and gasping dyspnea, death eventually supervenes.

■ neurotoxic, poultry, bizarre posturing

Pathology—Although the changes produced by the cardiotoxic species are not conspicuous, they are consistent and typically associated with the gastrointestinal tract. There is reddening, with or without hemorrhage, of the mucosa of the abomasum and small intestine, sometimes extending into the omasum and/orcecum. Mild to moderate pulmonary congestion, edema, and emphysema are not uncommon. Microscopically, acute ischemic-type glomerulopathy with tubular nephrosis, and pulmonary atelectasis with mucoid plugs, may be seen. Interestingly, mild nonsuppurative myocardial cellular degeneration and mononuclear infiltration may be observed with both the neurotoxic and the cardiotoxic species. Similarly, increases in serum creatinine kinase may be seen with any of the *Asclepias*.

■ cardiotoxic; gross pathology, reddened gut mucosa

microscopic, mild cardiac lesions

Few specific pathological changes occur with ingestion of the neurotoxic species. There are often hemorrhages in the trachea, in the lungs, and on the various heart surfaces. The urinary bladder is usually empty. Externally there may be obvious evidence of traumatic damage to the skin due to the sometimes violent seizures.

■ neurotoxic; gross pathology, few changes, may have mild cardiac lesions

Treatment and Control—Treatment of intoxications caused by the cardiotoxic species may be approached in a manner akin to that for the similar digitalis glycoside intoxications—activated charcoal, atropine for AV block, and/or antiarrhythmic drugs. There are no specific, effective means of treatment for the neurologic effects, although symptomatic therapy, including the use of sedatives for control of the convulsions, has been used with limited success. Chloral hydrate administered orally has been reported to be effective in sheep prior to onset of severe seizures (Clark 1979).

■ activated charcoal; cardiotoxic, atropine; neurotoxic, sedation

In most instances, prevention of plant consumption is the only effective means to reduce animal losses. It is very important to guard against feeding hay contaminated with milkweed. This is especially important when feeding alfalfa hay produced in areas within the range of *A. subverticillata*. Very close scrutiny is required to detect this species, even when the hay contains a high proportion of the plants.

Calotropis R.Br.

Taxonomy and Morphology—Comprising 4–6 species native to the Old World, *Calotropis* is closely related to *Asclepias* and often is called giant milkweed or crown flower. Its name is derived from the Greek roots *kalos* and *tropis*, meaning "beautiful" and "keel" and alluding to the flowers, which are made into leis in Hawaii and considered sacred to the Hindu god Shiva. Fibers are extracted from the bark and seeds of some species. In North America, only 1 introduced species is of toxicologic concern:

■ *C. procera*

calotrope
king's crown
rubber tree
St. Thomas bush
wild cotton
wild down

C. procera (Ait.) Ait.f.	calotrope, rubber tree, king's crown, St. Thomas bush, wild down, wild cotton

Plants shrubs; evergreen; sap viscous and white. **Stems** 100–250 cm tall; freely branched. **Leaves** opposite; sessile or subsessile; thick; gray-green; blades oblong. **Inflorescences** umbellate clusters borne in axils of upper leaves. **Flowers** campanulate;

■ evergreen shrubs with white, latexlike sap; leaves opposite; petals purple or white with purple tinge

glaucous. **Sepals** fused. **Petals** purple or white with purple-tinged corona; erect. **Coronas** of 5 fleshy scales. **Follicles** usually paired; gray-green; apices acute to acuminate.

■ ornamental houseplant

Distribution and Habitat—*Calotropis procera* is native to southwestern Asia and northern Africa. Capable of withstanding drought without stress, it is grown as a houseplant and sometimes as a planted ornamental in Mexico, California, and Florida. It has apparently naturalized in the West Indies and South America.

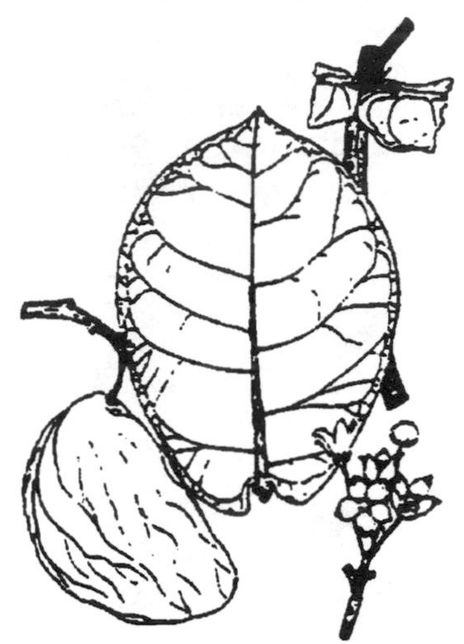

Calotropis procera

■ mild digestive tract irritation; possible cardiotoxic effects

Disease Problems—Decoctions and other preparations made from species of *Calotropis* have been widely used for medicinal purposes, mainly as purgatives and caustics (Watt & Breyer-Brandwijk 1962). The milky sap may cause contact irritation of skin in addition to cardiotoxic and digestive tract effects similar to the potent cardiotoxic species of *Asclepias* (Perkins & Payne 1978; Seiber et al. 1985). Even though these plants are recognized as cardiotoxic, there is considerable disagreement as to the risk they represent. The foliage is generally considered to be unpalatable. In experimental studies in the Sudan, sheep and goats that were fed the leaves of *C. procera* at 1% of b.w. became intoxicated but only after several or more weeks of daily ingestion (Mahmoud et al. 1979a). Administration of the milky sap produced similar effects in a much shorter period of time. It was lethal in a few hours at a dose of 1.2–1.6 ml/kg b.w. (Mahmoud et al. 1979b; El Sheikh et al. 1991). In rats, the sap from *C. procera* also has been shown to produce high mortality or fetal loss when fed for several days (Bhima Rao et al. 1974; Pahwa & Chatterjee 1988). In contrast, because of its high crude protein (19%) and general availability during unfavorable times, *C. procera* is used as dry-season livestock feed at up to 50% of the diet for prolonged periods in the Sudan without apparent adverse effects (Abbas et al. 1992). Furthermore, experimental results from northern Australia indicated no adverse effects when sheep were fed the species at 0.5–1% b.w. for 3 months and cattle at up to 1% b.w. for 84 days (Radunz et al. 1984). This further substantiates earlier reports of cattle eating the plants without harm (Watt & Breyer-Brandwijk 1962; Everist 1981). The presence of other plants in the diet that might stimulate processes of hepatic biotransformation may play a role in attenuating the toxicity, because dieldrin given experimentally to goats reduced toxicity (El Sheikh et al. 1991). Inadvertent ingestion of *C. procera* poses some risk for horses and perhaps pets but is unlikely to be a problem in ruminants.

■ steroidal cardenolides

Disease Genesis—In an extensive series of studies, Hesse and colleagues (1939, 1950, 1960) chronicled the numerous cardenolides of *Calotropis*. Some have unique sulfur-containing sugars (Crout et al. 1964). These potent cardenolides include calactin, calotoxin, calotropin, uscharidin, uscharin, and voruscharin, all glycosides of the unique, cyclic-bridged calotropagenin. This array of cardenolides is shared with some of the potent cardiotoxic species of *Asclepias* such as *A. curassavica* (Seiber et al. 1985). As measured immunologically and expressed as apparent digoxin concentration, cardenolide activity/content is estimated to be as high as 4.9µg/g of plant material (Radford et al. 1994).

Calotropin

■ depression, weakness, diarrhea; possibly signs of cardiac insufficiency

Clinical Signs—Typical of cardenolide effects, the signs include colic and diarrhea, depression, weakness, loss of appetite, slow heart rate, exercise intolerance, and collapse. Careful auscultation may reveal cardiac arrhythmias.

Pathology—The gross changes also are typical of cardenolides—mild to moderate reddening of the mucosa of the stomach and small intestine; scattered splotchy hemorrhages, especially on the heart surfaces; and excess fluid in the chest and around the heart. Pale and friable foci may be present in heart muscle, and microscopic examination may reveal myocardial degeneration and necrosis.

■ gross pathology, mild reddening of gut mucosa

Treatment—Treatment will generally be limited to controlling the digestive tract disturbance, and only rarely is intervention for cardiac disturbances required. Activated charcoal given orally at 2 gm/kg b.w. may be beneficial in limiting expression of the disease.

■ activated charcoal

Cryptostegia R.Br.

Taxonomy and Morphology—Native to the tropics of Africa, Madagascar, and India, *Cryptostegia* comprises 3 species. Its name is derived from the Greek *kryptos* and *stego,* meaning "hidden" and "scale," which refer to the corona scales covering the anthers. Two species have been introduced in North America:

■ *C. grandiflora*
C. madagascariensis
rubber vine
pink allamanda

C. grandiflora R.Br.	rubber vine, palay rubber vine, pink allamanda
C. madagascariensis Bojer ex Decne.	rubber vine

Plants shrubs or woody climbers; sap viscous and white. **Stems** arching or trailing; 100–200 cm tall. **Leaves** opposite; petiolate; thick; glossy dark green above; blades oblong-elliptic; apices obtuse. **Inflorescences** terminal cymes; branches typically trichotomously forked. **Flowers** large; 5–8 cm in diameter; funnelform-campanulate. **Sepals** large. **Petals** pink or rose-purple or rarely white. **Coronas** of 5 scales. **Follicles** usually paired; 3-angled or winged.

■ shrubs and woody climbers; sap latexlike; leaves opposite; petals pink or rose-purple

1. Calyces 10 mm long. Margins of crown scales deeply divided into filiform segments .... *C. grandiflora*
1. Calyces 5 mm long. Margins of crown scales entire . . . *C. madagascariensis*

Distribution and Habitat—Both species are cultivated in the warmer, frost-free areas of North America and have escaped in Florida, but they are relatively rare. *Cryptostegia grandiflora* also is naturalized in Sinaloa, Mexico.

■ ornamental, frost-free areas

Cryptostegia grandiflora

Cryptostegia grandiflora

■ digestive disturbance; cardiotoxic

Disease Problems—Although cardiotoxic plants are typically considered to be unpalatable, species of *Cryptostegia* are said to be palatable under some circumstances (Seawright 1989). Occasional intoxications are seen in horses—e.g., death following brief exercise of a pony and a donkey in Florida due to ingestion of *C. madagascariensis,* and death of several horses in Australia after eating the dried leaves of *C. grandiflora* (Morton 1982; Cook et al. 1990). Everist (1981) estimated the lethal dosage to be 0.03–0.06% of b.w. in horses and 0.1–0.5% or higher in ruminants in Australia where *Cryptostegia* is of some concern. The young leaves appear to be most toxic. The milky sap may cause contact irritation of the skin in addition to its cardiotoxic and digestive tract effects (Perkins & Payne 1978). Inadvertent ingestion of *Cryptostegia,* like that of *Calotropis procera,* may be a risk for horses and perhaps pets but is unlikely to be a problem in ruminants.

■ steroidal cardenolides

Disease Genesis—The very potent cardenolides present in the genus are glycosides, mainly rhamnosides, of digitoxigenin, oleandrigenin, and other genins (Sanduja et al. 1984). These include cryptograndosides A and B, plus a saponin (Morton 1982). As estimated from immunologic determination, cardenolide content is high—0.6µg and 2.1µg digoxin equivalents per g of plant material of *C. grandiflora* and *C. madagascariensis,* respectively (Radford et al. 1994).

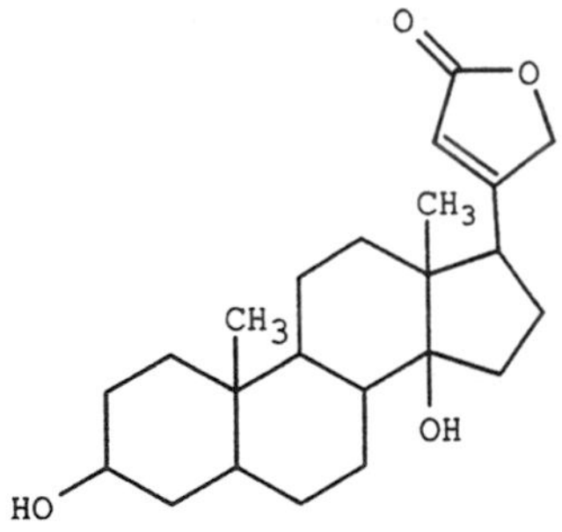

Digitoxigenin

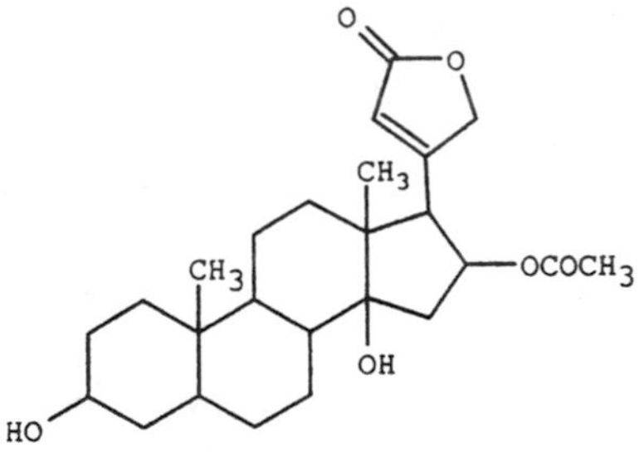

Oleandrigenin

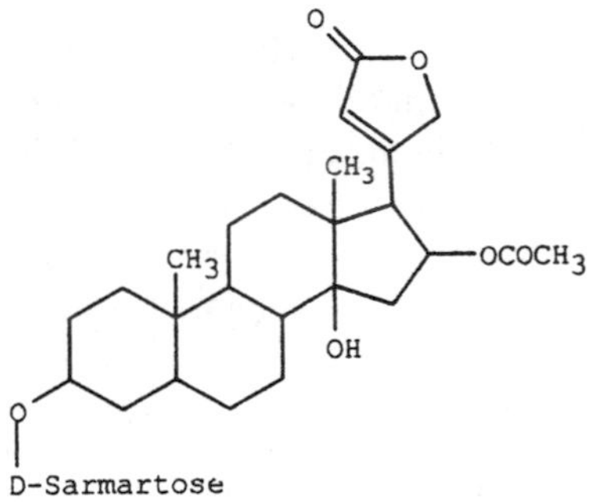

Cryptograndoside *A*

■ digestive disturbance, diarrhea; possible mild cardiac signs, slow heart rate, arrhythmias, weakness

Clinical Signs—Typically there are mild to moderate cardenolide effects; the signs including colic and diarrhea, depression, weakness, loss of appetite, slow heart rate, labored respiration, exercise intolerance, cyanosis, and eventual collapse. Careful auscultation may reveal cardiac arrhythmias. Death, which is due to ventricular fibrillation, rarely occurs.

■ gross pathology, moderate reddening of gut mucosa

microscopic, possible myocardial lesions

Pathology—The gross changes are typical of those caused by cardenolides, with mild to moderate reddening of the mucosa of the stomach and small intestine; scattered splotchy hemorrhages, especially on the heart surfaces; and excess fluid in the thorax and around the heart. Pale and friable foci may be present in heart muscle, and microscopic examination may reveal myocardial degeneration and necrosis.

■ activated charcoal; atropine

Treatment—Treatment will generally be limited to controlling the digestive tract disturbance and only rarely will intervention for cardiac disturbances be required. Activated charcoal given orally at 2 gm/kg b.w. may be beneficial in limiting expression of the disease.

Genera of Questionable Toxicity

Araujia Brot.

A South American genus of 4 or 5 species, *Araujia* is represented in North America by *A. sericifera* Brot., commonly known as bladder flower or cruel plant. Introduced from Brazil, it can withstand light frosts and has become a noxious weed in citrus groves and disturbed sites in California. It is an evergreen, twining, woody climber with stems to 12 m long. Its ovate-oblong leaves are glossy, dark green above, and white pubescent below, with acute apices and cordate to hastate bases. Borne in short, axillary racemes, the fragrant, waxy, campanulate or funnelform, white or pink flowers bear anthers with small, inflexed horns and a pistil head with 2 erect lobes. The follicles are coriaceous, deeply grooved, and 10–12 cm long.

- *A. sericifera*

  bladder flower
  cruel plant
- introduced weedy species in California

Although it has not been a problem in North America, *A. sericifera* is reported to cause neurologic and digestive tract problems elsewhere (Watt & Breyer-Brandwijk 1962). Its leaves and stems contain up to 0.1% serotonin and 0.13% 7-O-D-glucoluteolin. Smaller amounts are in the fruits, and none is in the seeds (Federici et al. 1988). Cardenolides do not seem to be important constituents of the genus, as is indicated by the negligible digoxin equivalents estimated for the Australian species *A. hortorum* E.Fourn., using immunologic methodology (Radford et al. 1994). This species, although rarely eaten, causes neurologic disease in poultry (Everist 1981). Dosage of 0.3–0.6% b.w. of the mature seeds produced ataxia and seizures.

- digestive system and neurologic disturbances; unknown toxins

If sufficient amounts of *A. sericifera* are eaten, abrupt onset of nausea, uneasiness, diarrhea, and incoordination may be expected, as well as seizures in severe cases. Treatment of this nonfatal disease should be directed toward relief of the digestive tract problems. Rarely would sedation for the neurologic effects be required.

- uneasiness, diarrhea; treatment may not be needed; activated charcoal

Cynanchum L.

Comprising 150–200 species in both the Old World and the New World, *Cynanchum*, commonly known as sand vine or vine milkweed, is represented in North America by approximately 24 species. Representative is *C. laeve* (Michx.) Pers., which ranges through the southeastern quarter of the United States in low, moist sites in woodlands, thickets, floodplains, and fields. Also known as blue vine or climbing milkweed, the species is a perennial vine with branched or unbranched stems from thickened rootstocks. The sap is milky, and the petiolate leaves are triangular to broadly ovate. Borne in axillary umbels or cymes, the flowers have green or purple-tinged sepal lobes and whitish to cream-colored petals and corona lobes. The prominent follicles are fusiform and smooth.

- *C. laeve*

  blue vine
  climbing milkweed
- perennial vines; milky viscous sap; follicles prominent and fusiform
- moist sites

Cynanchum laeve

Species of *Cynanchum* are not generally considered hazardous in North America, but Australian and African taxa are toxic (Everist 1981; Kellerman et al. 1988). The disease in Africa, called krimsiekte or cynanchosis, follows ingestion of *C. africanum* and other species of the genus. As little as 0.2% b.w. causes problems in horses, while 5–10 times that amount is required in ruminants (Steyn 1934). The signs of this disease are very similar to those produced by neurotoxic species of *Asclepias* in North America. Steroidal pregnane and modified pregnane glycosides have been isolated from Asian and African species of *Cynanchum* (Tsukamoto et al. 1985a,b, 1986; Zhang et al. 1985; Steyn et al. 1989; Konda et al. 1990). In some, called cynafosides (A–H), the lactone ring is absent at C-17 and occurs in the linkage of a phenyl group at C-12. The cynafosides are composed of cynafogenin and different combinations of sugars, mainly D- and L-cymaroses (Tsukamoto et al. 1985a, 1986; Zhang et al.

- toxins unknown; possible neurotoxic effects

1985; Steyn et al. 1989; Konda et al. 1990). Other compounds with different substituents at C-12 and C-17 (wilfosides) and glycosides of glaucogenin are also represented in some Asian species of *Cynanchum* (Nakagawa et al. 1983; Tsukamoto et al. 1985b; Qiu et al. 1989). The comparative toxicologic significance of these compounds is not clear, but the cynafosides are neurotoxic, causing paddling, tetanic seizures, and opisthotonus, often of several days' duration and eventually resulting in terminal paralysis (Kellerman et al. 1988).

The neurologic signs reported for the African species are ataxia, trembling, and seizures beginning as clonic types but later becoming tonic, with the head held backward (opisthotonus). There is an increase in respiratory and heart rates, and residual effects may be apparent for a few days after an intoxication episode. However, it is not certain whether the same signs are produced by the North American species of *Cynanchum*. There are few gross lesions and no specific treatment other than administration of sedatives.

Hoya R.Br.

■ *H. carnosa*
wax plant

Native to eastern Asia and Australasia, *Hoya,* commonly known as waxflower or porcelain flower, comprises 200–230 species. Representative of the genus is the popular ornamental *H. carnosa* (L.f.) R.Br. or wax plant. A succulent, evergreen, woody-stemmed vining climber to 4 m long, it has opposite, glabrous leaves with ovate-oblong, short-petioled, dark green, thick blades that are red when immature. Borne in short-peduncled umbels, the fragrant flowers are creamy white, with pink-tinged centers, and produce glabrous, acuminate follicles. Native to India, Burma, and southern China, *H. carnosa* and most of the other introduced species are frost sensitive and are grown as houseplants. Only in the extreme southern parts of Florida, California, and Mexico are they grown outside.

Although species of *Hoya* are generally not a problem in North America, an Australian species, *H. australis* R.Br. ex Traill, causes a neurologic disease with tetanic seizures in cattle (Everist 1981). The toxic dosage of green plant material is 0.2–0.8% b.w. for calves and sheep. Cardenolide concentrations, estimated as digoxin equivalents from immunologic data, are minimal for the species (Radford et al. 1994).

■ toxicity unknown; possible neurotoxic effects

Clinical signs produced by *H. australis* include incoordination, especially of the pelvic limbs, which causes the animals to knuckle over or squat. There are tremors, collapse, and struggling, with primarily tetanic-type seizures. There also may be constipation or diarrhea, but no indications of cardiotoxicosis. The gross changes are of a general nature, including scattered splotchy hemorrhages, especially on the heart surfaces, and mild pulmonary emphysema. There may also be reddening of the mucosa of the stomach and small intestine. Sedation is the primary treatment.

Periploca L.

■ *P. graeca*
silk vine

■ trailing vines or shrubs with milky sap; leaves simple, opposite; flowers ill scented

■ open woods, along streams

An Old World genus of 10–12 species, *Periploca* is quite different from other members of the Asclepiadaceae in the appearance of its pollinia and thus is segregated by some taxonomists in its own family, the Periplocaceae. The genus is represented in North America by one naturalized species, *P. graeca* L., commonly known as silk vine. It is a perennial vine or trailing shrub with milky sap; woody, glabrous, branched or unbranched stems; and simple, opposite, lanceolate to ovate, glossy dark green leaves with acuminate apices and entire margins. Its flowers are borne in axillary, long-peduncled corymbs and are malodorous, rotate, and 2.2–2.7 cm in diameter. The petals are adaxially maroon to dark brown and adaxially yellow-green, and the corona is a ring of five pairs of broad lobes. The narrowly cylindrical, acuminate follicles lack tubercles and indumentum. A native of southern Europe, *P. graeca* occurs in open woods or thickets along streams in the eastern half of the continent.

There is little doubt that the species of *Periploca* are cardiotoxic, but the risk of intoxication appears to be quite low. Various cardenolides are present, mainly glycosides of periplogenin in *P. graeca* and strophanthidin in *P. nigrescens* Afzel, an African species (Abisch & Reichstein 1962; Berthold et al. 1965a,b; Komissarenko & Bagirov 1969; Marks et al. 1975). Disease problems caused by these species is likely to be limited to mild to moderate digestive tract disturbance, such as nausea, increased salivation, and diarrhea, with little need for treatment.

Periploca graeca

- mild digestive disturbance, diarrhea; steroidal cardenolides

Periplogenin

Sarcostemma R.Br.

- *S. cynanchoides*

  climbing milkweed

- climbing vine; follicles narrowly fusiform

Commonly known as climbing milkweed, *Sarcostemma* comprises 30–60 species of worldwide distribution. Of the 5 species present in North America, primarily the Southwest, the most commonly encountered is *S. cynanchoides* Decne. (Holm 1950). It is a trailing or climbing, suffrutescent, profusely branched vine with petiolate, ovate to lanceolate, hastate or cordate leaves. Borne in axillary umbels, the rotate-campanulate flowers are greenish white to pinkish or purplish and bear a short, white corona. The follicles are narrowly fusiform and puberulent. Typically occurring in sandy or rocky soils and trailing across or climbing in shrubs, the species is found along watercourses, in desert washes, in thickets, and at the edges of woodlands throughout the southwest quarter of the continent from Texas and Oklahoma west to Utah, Nevada, and California, and south into central Mexico. It is reported to be a pest in gardens and cultivated fields in Arizona.

Sarcostemma cynanchoides

Although not generally considered to be hazardous in North America, species of *Sarcostemma* in Australia and Africa are toxic (van der Walt & Steyn 1939; Everist 1981; Kellerman et al. 1988). Livestock, however, are said to become somewhat tolerant of them (Watt & Breyer-Brandwijk 1962). The toxicity of some species appears to be variable, and plants may be eaten at times with apparent impunity. In some areas of western and southern Australia, *S. australe* R.Br. is used as forage, while in eastern Australia it causes neurologic disease in livestock during the dry season (Everist 1981). The lethal dose of green plants is 0.1–0.2% b.w. The milky sap is irritating to the skin and mucous membranes. The toxic effects of *Sarcostemma* appear to be due to glycosides that are similar to the cardenolides of *Asclepias* but lack the lactone ring at C-17. These glycosides are neurotoxic rather than cardiotoxic (Kellerman et al. 1988). Cardenolide activity is negligible as measured immunologically in *S. australe* (Radford et al. 1994).

Sarcostemma cynanchoides

When intoxication does occur, species of *Sarcostemma* produce an abrupt onset

- toxicity in North America not clear; possible neurotoxins similar in structure to cardenolides but identity unknown

of restlessness or anxiety, staggering incoordination, and collapse, followed by violent paddling seizures with champing or clamping of the jaws and rigidity of the neck muscles. Respiration and heart rates are increased. There may be constipation or diarrhea, and the seizures may be followed by paralysis. There are few changes visible at necropsy other than fluid accumulation in the chest and heart sac. The effects of irritation of the digestive tract are not typically prominent. In terms of treatment, there is little that can be done other than administration of sedatives to provide comfort.

References

Abbas B, El Tayeb AE, Sulleiman YR: *Calotropis procera:* feed potential for arid zones [letter]. *Vet Rec* 131;132, 1992

Abisch E, Reichstein T: Orientierende chemische untersuchung einiger Asclepiadaceae und Periplocaceen. *Helv Chim Acta* 45;2090–2116, 1962

Adams RP, Balandrin MF, Martineau JR: The showy milkweed, *Asclepias speciosa:* a potential new semi-arid land crop for energy and chemicals. *Biomass* 4;81–104, 1984

Benson JM, Seiber JN, Keeler RF, Johnson: Studies on the toxic principle of *Asclepias eriocarpa* and *Asclepias labriformis.* In *Effects of Poisonous Plants on Livestock,* Keeler RF, Van Kampen KR, James LF eds, Academic Press, New York, pp273–284, 1978

Benson JM, Seiber JN, Bagley CV, Keeler RF, Johnson AE, Young S: Effects on sheep of the milkweeds *Asclepias eriocarpa* and *A. labriformis* and of cardiac glycoside containing derivative material. *Toxicon* 17;155–165, 1979

Berthold R, Wehrli W, Reichstein T: Die Cardenolides von *Parquetina nigrescens* (Afzel.) Bullock. *Helv Chim Acta* 48;1634–1658, 1965a

Berthold R, Wehrli W, Reichstein T: Die Cardenolides von *Parquetina nigrescens* (Afzel.) Bullock. 3. Die konstitution 5 neuer Cardenolides (16-dehydrostrophanthidin und verwandt Substanzen). *Helv Chim Acta* 48;1659–1665, 1965b

Bhima Rao BS, Devaraj Sarkar HB, Sheshadri HS: Effect of latex of *Calotropis gigantea* on pregnancy in the albino rat [abstract]. *J Reprod Fert* 38;234, 1974

Brower LP: Ecological chemistry. *Sci Am* 220;22–29, 1969

Brower LP, Seiber JN, Nelson CJ, Lynch SP, Tuskes PM: Plant-determined variation in the cardenolide content, thin-layer chromatography profiles, and emetic potency of monarch butterflies, *Danaus plexippus,* reared on the milkweed *Asclepias eriocarpa* in California. *J Ecol Chem* 8;579–633, 1982

Brower LP, Seiber JN, Nelson CJ, Lynch SP, Hoggard MP, Cohen JA: Plant-determined variation in cardenolide content and thin-layer chromatography profiles of monarch butterflies, *Danaus plexippus,* reared on milkweed plants in California. 3. *Asclepias californica. J Chem Ecol* 10;1823–1857, 1984

Burrows GE, Ogden L, Tyrl RJ: Intoxication due to *Asclepias* spp. milkweeds. Third Pan-American Symposium, International Society on Toxinology, Oaxtepec, Mexico, January 1990

Burrows GE, Tyrl RJ, Jones L, Chisum A: Toxicity of verticillate leaved *Asclepias* (milkweeds) in chickens. Fourth Pan-American Symposium, International Society on Toxinology, Campinas, Brazil, July 1992

Campbell HW: Poisoning in chickens with whorled milkweed. *J Am Vet Med Assoc* 79;102–104, 1931

Clark JG: Whorled milkweed poisoning. *Vet Hum Toxicol* 21;431, 1979

Cohen JA, Brower LP: Cardenolide sequestration by the dogbane tiger moth. *J Chem Ecol* 9;521–532, 1983

Cook DR, Campbell GW, Meldrum AR: Suspected *Cryptostegia grandiflora* (rubber vine) poisoning in horses. *Aust Vet J* 67;344, 1990

Couch JF: The chemistry of stock-poisoning plants. *J Chem Educ* 14;16–30, 1937

Crout DHG, Hassall CH, Jones TL: Cardenolides. Part 6. Uscharidin, calotropin, and calotoxin. *J Chem Soc* 1964;2187–2194, 1964

Duffey SS, Blum MS, Isman MB, Scudder GGE: Cardiac glycosides: a physical system for their sequestration by the milkweed bug. *J Insect Physiol* 24;639–645, 1978

El Sheikh HA, Ali BH, Homeida AM, Hassa T, Odris OF, Hapke HJ: The activities of drug-metabolizing enzymes in goats treated orally with the latex of *Calotropis procera* and the influence of dieldrin pretreatment. *J Comp Pathol* 104;257–268, 1991

Everist SL: *Poisonous Plants of Australia,* 2d ed. Angus & Robertson, Sydney, Australia, pp 94–109, 1981

Federici E, Galeffi C, Nicoletti M: Constituents of *Araujia sericifera. J Nat Prod* 51;189–190, 1988

Fleming CE, Peterson NF, Miller MR, Vawter LR, Wright LH: *The Narrow-Leaved Milkweed and the Broad-Leaved or Showy Milkweed.* Nev Agric Exp Stn Bull 99, 1920

Glover GH, Newsom IE, Robbins WW: *A New Poisonous Plant, the Whorled Milkweed* (Asclepias verticillata). Colo Agric Exp Stn Bull 246, 1918

Hesse G, Ludwig G: African arrow poisons. 14. Voruscharine, a second sulfur-containing heart poison from *Calotropis procera* L. *Annalen* 632;158–171, 1960

Hesse G, Reicheneder F, Eysenbach H: African arrow poisons. II. Heart poisons in *Calotropis* latex. *Annalen* 537;67–86, 1939

Hesse G, Heuser LJ, Hutz E, Reicheneder F: African arrow poisons. 5. Relationships between the most important poisons of *Calotropis procera. Annalen* 566;130–139, 1950

Holm RW: American species of *Sarcostemma* R.Br. (Asclepiadaceae). *Ann Mo Bot Garden* 37;477–560, 1950

Isman MB: Dietary influence of cardenolides on larval growth and development of the milkweed bug *Oncopeltus fasciatus. J Insect Physiol* 23;1183–1187, 1977

Jolad SD, Bates RB, Cole JR, Hoffmann JJ, Siahaan TJ, Timmerman BN: Cardenolides and a lignan from *Asclepias subulata. Phytochemistry* 25;2581–2590, 1986

Joubert JPJ: Cardiac glycosides. In *Toxicants of Plant Origin,* vol 2, *Glycosides,* Cheeke PR ed, CRC Press, Boca Raton, Fla, pp 61–96, 1989

Kellerman TS, Coetzer JAW, Naude TW: *Plant Poisonings and Mycotoxicoses of Livestock in Southern Africa.* Oxford Univ Press, Cape Town, pp51–52, 1988

Kephart SR, Wyatt R, Parrella D: Hybridization in North American *Asclepias.* 1. Morphological evidence. *Syst Bot* 13;456–473, 1988

Kindscher K: *Medicinal Wild Plants of the Prairie.* Univ Press Kansas, Lawrence, 1992

Komissarenko NF, Bagirov RB: Chemical study of *Periploca graeca. Izv Akad Nauk Azerb S S R, Ser Biol Nauk* 1969;122–127, 1969 (*Chem Abstr* 73;123456t, 1970)

Konda Y, Urano M, Harigaya Y, Takayangai H, Ogura H, Li X, Low H, Onda M: Novel triterpenes, hancolupenone and hancolupenol, from *Cynanchum hancockianum. Chem Pharm Bull* 38;2899–2901, 1990

Kupchan SM, Mokotoff M, Sandhu RS, Hokin LE: The chemistry and biological activity of derivatives of strophanthidin. *Med Chem* 10;1025–1033, 1967

Lou H, Li X, Onda M, Konda Y, Machida T, Toda Y, Harigaya Y: Further isolation of glycosides from *Cynanchum hancockianum. J Nat Prod* 56;1437–1443, 1993

Lynch SP, Martin RA: Cardenolide content and thin-layer chromatography profiles of monarch butterflies, *Danaus plexippus* L., and their larval host-plant milkweed, *Asclepias viridis* Walt., in northwestern Louisiana. *J Chem Ecol* 13;47–70, 1987

Lynch SP, Martin RA: Milkweed host plant utilization and cardenolide sequestration by monarch butterflies in Louisiana and Texas. *Nat Hist Mus Los Ang Cty Sci Ser* 0;107–123, 1993

Mahmoud OM, Adam SEI, Tartour G: The effects of *Calotropis procera* on small ruminants. 1. Effects of feeding sheep with the plant. *J Comp Pathol* 89;241–250, 1979a

Mahmoud OM, Adam SEI, Tartour G: The effects of *Calotropis procera* on small ruminants. 2. Effects of administration of the latex to sheep and goats. *J Comp Pathol* 89;251–263, 1979b

Malcolm SB, Brower LP: Selective oviposition by monarch butterflies (*Danaus plexippus* L.) in a mixed stand of *Asclepias curassavica* L. and *A. incarnata* L. in south Florida. *Lepid Soc J* 40;255–263, 1986

Malcolm SB, Cockrell BJ, Brower LP: Cardenolide fingerprint of monarch butterflies reared on common milkweed, *Asclepias syriaca* L. *J Chem Ecol* 15;819–853, 1989

Marks WH, Fong HHS, Tin-Wa M, Farnsworth NR: Cytotoxic principles of *Parquetina nigrescens* (Afzel.) Bullock (Asclepiadaceae). *J Pharm Sci* 64;1674–1676, 1975

Marsh CD: *Stock Poisoning Plants of the Range.* USDA Bull 1245, 1929

Marsh CD, Clawson AB: *The Mexican Whorled Milkweed* (Asclepias mexicana) *as a Poisonous Plant.* USDA Bull 969, 1921a

Marsh CD, Clawson AB: *Poisonous Properties of the Whorled Milkweeds* Asclepias pumila *and* A. verticillata *var.* geyeri. USDA Bull 942, 1921b

Marsh CD, Clawson AB: *The Woolly-Pod Milkweed* (Asclepias eriocarpa) *as a Poisonous Plant.* USDA Bull 1212, 1924

Marsh CD, Clawson AB, Couch JF, Eggleston WW: *The Whorled Milkweed* (Asclepias galioides) *as a Poisonous Plant.* USDA Bull 800, 1920

Marty MA, Krieger RI: Metabolism of uscharirin, a milkweed cardenolide, by tissue homogenates of monarch butterfly larvae, *Danaus plexippus* L. *J Chem Ecol* 10;945–956, 1984

Marty MA, Gee SJ, Krieger RI: Monooxygenase activities of fat body and gut homogenates of monarch butterfly larvae, *Danaus plexippus,* fed four cardenolide-containing milkweeds, *Asclepias* spp. *J Chem Ecol* 8;797–805, 1982

May WL: *Whorled Milkweed—the Worst Stock-Poisoning Plant in Colorado.* Colo Agric Exp Stn Bull 255, 1920

McLaughlin SP, Hoffman JJ: Survey of biocrude-producing plants from the Southwest. *Econ Bot* 36;323–339, 1982

Moore LV, Scudder CGE: Selective sequestration of milkweed (*Asclepias* sp.) cardenolides in *Oncopeltus fasciatus* (Dallas) (Hemiptera: Lygaeidae). *J Chem Ecol* 11;667–687, 1985

Morse DH: Milkweeds and their visitors. *Sci Am* 253:112–119, 1985

Morton JF: *Plants Poisonous to People in Florida,* 2d ed. JF Morton, Southeastern Printing, Stuart, Fla, 1982

Nakagawa T, Hayashi K, Wada K, Mitsuhashi H: Studies on the constituents of Asclepiadaceae plants. 52. The structure of five glycosides, glaucoside-A, -B, -C, -D, and -E from Chinese drug "pai-ch'ien" *Cynanchum glaucescens* Hand-Mazz. *Tetrahedron* 39;607–612, 1983

Nelson CJ, Seiber JN, Brower LP: Seasonal and intraplant variation of cardenolide content in the California milkweed, *Asclepias eriocarpa,* and implications for plant defense. *J Chem Ecol* 7;981–1010, 1981

Nelson EG, Dexter ST: Fiber from the stems of common and swamp milkweed. *Q Bull Mich Agric Exp Stn* 28;20–28, 1945

Ogden L, Burrows GE, Tyrl RJ: Comparison of *Asclepias* species based on their toxic effects on chickens. In *Poisonous Plants: Proceedings of the Third International Symposium,* James LF, Keeler RF, Bailey EM Jr, Cheeke PR, Hegarty MP eds, Iowa State Univ Press, Ames, pp500–505, 1992a

Ogden L, Burrows GE, Tyrl RJ, Ely RW: Experimental intoxication in sheep by *Asclepias.* In *Poisonous Plants: Proceedings of the Third International Symposium,* James LF, Keeler RF, Bailey EM Jr, Cheeke PR, Hegarty MP eds, Iowa State Univ Press, Ames, pp495–499, 1992b

Pahwa R, Chatterjee VC: The toxicity of Indian *Calotropis procera* R.Br. latex in the black rat, *Rattus rattus* Linn. *Vet Hum Toxicol* 30;305–306, 1988

Pammel LH: Poisonous plants in a pasture. *Am J Vet Med* 12;235–236, 1917

Perkins KD, Payne WW: *Guide to the Poisonous and Irritant Plants of Florida.* Fla Coop Ext Serv Circ 441, 1978

Piatak DM, Patel J, Totten CE, Swenson RP, Brown P, Pettit GR: Cell growth inhibitory glycosides from *Asclepias amplexicaulis. J Nat Prod* 48;470–471, 1985

Qui S-X, Zhang Z-X, Zhou J: Steroidal glycosides from the root of *Cynanchum versicolor. Phytochemistry* 28;3175–3178, 1989

Radeleff RD: *Veterinary Toxicology,* 2d ed. Lea & Febiger, Philadelphia, pp118–121, 1970

Radford DJ, Cheung K, Urech R, Gollogly JR, Duffy P: Immunological detection of cardiac glycosides in plants. *Aust Vet J* 71;236–238, 1994

Radunz BL, Wilson G, Beere G: Feeding rubberbrush *(Calotropis procera)* to cattle and sheep. *Aust Vet J* 61;243–244, 1984

Ritland DJ: Palatability of aposematic queen butterflies *(Danaus gilippus)* feeding on *Sarcostemma clausum* (Asclepiadaceae) in Florida. *J Chem Ecol* 17;1593–1610, 1991

Robinson GH, Burrows GE, Holt EM, Tyrl RJ, Jones AD: Investigation of the neurotoxic compounds in *Asclepias subverticillata* (western whorled milkweed). In *Toxic Plants and Other Natural Toxicants,* Garland T, Barr AC eds, New York, pp435–439, 1998

Roeske CN, Seiber JN, Brower LP, Moffitt CM: Milkweed cardenolides and their comparative processing by monarch butterflies (*Danaus plexippus* L.). In *Biochemical Interaction between Plants and Insects, Recent Advances in Phytochemistry,* vol 10, Wallace JW, Mansell RL eds, Plenum Press, New York, pp93–167, 1976

Rothschild M: Mimicry: the deceptive way of life. *Nat Hist N Y* 76;44–51, 1967

Rothschild M, Ford R: Heart poisons and the monarch. *Nat Hist N Y* 79;36–37, 1970

Rowe LD, Ohlenbusch PD, Dollahite JW: Toxicity of *Asclepias brachystephana* (shortcrown milkweed) for sheep. *Southwest Vet* 23;219–224, 1970

Sady MB, Seiber JN: Field test for screening milkweed latex for cardenolides. *J Nat Prod* 54;1105–1107, 1991

Salyi G, Petri A: *Asclepias syriaca* poisoning of cattle. *Magy Allatorv Lapja* 42;56–58, 1987

Sanduja R, Lo WYR, Euler KL, Alam M: Cardenolides of *Cryptostegia madagascariensis. J Nat Prod* 47;260–265, 1984

Seawright AA: *Animal Health in Australia,* vol 2, 2d ed, *Chemical and Plant Poisons.* Australian Government Publ Serv, Canberra, 1989

Seiber JN, Lee SM, Benson JM: Cardiac glycosides (cardenolides) in species of *Asclepias* (Asclepiadaceae). In *Plant and Fungal Toxins, Handbook of Natural Toxins,* vol 1, Keeler RF, Tu AT, M Dekker eds, New York, pp43–83, 1983

Seiber JN, Lee SM, McChesney MM, Watson TR, Nelson CJ, Brower LP: New cardiac glycosides (cardenolides) from *Asclepias* species. In *Plant Toxicology—Proceedings of the Australia-USA Poisonous Plants Symposium,* Seawright AA, Hegarty MP, James LF, Keeler RF eds, Animal Research Institute, Yeerongpilly, Brisbane, Aust, pp427–437, 1985

Seiber JN, Brower LP, Lee SM, McChesney MM, Cheung HTA, Nelson CJ, Watson TR: Cardenolide connection between overwintering monarch butterflies from Mexico and their larval food plant, *Asclepias syriaca. J Chem Ecol* 12;1157–1170, 1986

Sheppard PM: The evolution of mimicry: a problem in ecology and genetics. *Cold Spring Harbor Symp Quant Biol* 24;131–140, 1959

Shoppee CW: *Chemistry of the Steroids,* 2d ed. Butterworths, Washington, DC, pp14–17, 1964

Singh B, Rastogi RP: Cardenolides—glycosides and genins. *Phytochemistry* 9;315–331, 1970

Sperry OE, Dollahite JW, Hoffman GO, Camp BJ: *Texas Plants Poisonous to Livestock.* Tex Agric Ext Serv reprint B-1028, 1977

Sprowls RW: Horsetail milkweed intoxications in horses and cattle. *Southwest Vet* 35;15, 1982

Steyn DG: *The Toxicology of Plants in South Africa.* Central News Agency, South Africa, 1934

Steyn PS, van Heerden FR, Vleggaar R, Erasmus GL, Anderson LAP: Toxic constituents of the Asclepiadaceae: structure elucidation of the cynafosides, toxic pregnane glycosides of *Cynanchum africanum* R.Br. *S Afr J Chem* 42;29–37, 1989

Stiles GW: Poisoning of turkey poults from whorled milkweed *(Asclepias galioides)*. *Poult Sci* 21;263–270, 1942

Tehon LR, Morrill CC, Graham R: *Illinois Plants Poisonous to Livestock*. Univ Ill Coll Agric Ext Serv Circ 599, 1946

Tsukamoto S, Hayashi K, Mitsuhashi H: Studies on the constituents of Asclepiadaceae plants. 40. Further studies on glycosides with a novel sugar chain containing a pair of optically isomeric sugars D- and L-cymarose from *Cynanchum wilfordi*. *Chem Pharm Bull* 33;2294–2304, 1985a

Tsukamoto S, Hayashi K, Mitsuhashi H, Snyckers FO: Studies on the constituents of Asclepiadaceae plants. 42. The structures of two glycosides, cynafoside-A and -B with a novel sugar chain containing a pair of optically isomeric sugars, D- and L-cymaroses, from *Cynanchum africanum* R.Br. *Chem Pharm Bull* 33;4807–4814, 1985b

Tsukamoto S, Hayashi K, Kaneko K, Mitsuhashi H, Snyckers FO: Studies on the constituents of Asclepiadaceae plants. 44. The structure elucidation of cynfogenin. *Chem Pharm Bull* 34;1337–1339, 1986

Tunniclif EA, Cory VL: Broad-leafed milkweed *(Asclepias latifolia)* poisonous for sheep and goats. *J Am Vet Med Assoc* 77;165–168, 1930

Vail EL: Woolly-pod or broad-leafed milkweed *(Ascepias eriocarpa)* poisoning of rabbits. *North Am Vet* 23;539–542, 1942

van der Walt SJ, Steyn DG: Recent investigations into the toxicity of known and unknown poisonous plants in the Union of South Africa, 9. *Onderstepoort J Vet Res Anim Ind* 12;335–360, 1939

Watt JM, Breyer-Brandwijk MG: *The Medicinal and Poisonous Plants of Southern Africa*. E & S Livingston, Edinburgh, pp197–305, 1962

Woodson RE: The North American species of Asclepias L. *Ann Mo Bot Gard* 41;1–211, 1954

Zhang ZX, Zhou J, Hayashi K, Mitsuhashi H: Studies on the constituents of Asclepiadaceae plants. 41. The structure of cynatratoside-F from Chinese drug "Pai-Wei," dried root of *Cynanchum atratum* Bunge. *Chem Pharm Bull* 33; 4188–4192, 1985

Asteraceae Dumort.

13

Generally considered to be the largest family of flowering plants, the Asteraceae, commonly known as the sunflower or aster family, comprises 1,300–1,530 genera and 21,000–25,000 species. Cosmopolitan in distribution with the exception of the Antarctic, the family exhibits greatest diversity in the semiarid regions of the tropics and subtropics, in temperate regions, and at higher altitudes in the tropics. Only in tropical rain forests is it poorly represented. As the native vegetation of most areas is cleared, its members typically appear and dominate the early stages of succession. Most species of the family are herbaceous—only about 2% are woody—and exhibit considerable diversity in life-form, mode of pollination, and seed dispersal. Unifying them is their inflorescence; all composites, as these plants are commonly called, possess a head—a structure consisting of one to many small, sessile flowers borne on a common receptacle and subtended by one to several series of bracts. This dense aggregation of flowers is sometimes mistaken by the novice for a single flower.

■ Sunflower Family

Artemisia	*Parthenium*
Baccharis	*Psathyrotes*
Baileya	*Psilostrophe*
Centaurea	*Rudbeckia*
Erechtites	*Sartwellia*
Eupatorium	*Senecio*
Flourensia	*Solidago*
Gutierrezia	*Tetradymia*
Helenium	*Verbesina*
Hymenoxys	*Viguiera*
Hypochaeris	*Xanthium*
Isocoma	*Xylorhiza*
Iva	

QUESTIONABLE

Achillea	*Conyza*
Ageratina	*Florestina*
Ageratum	*Helianthus*
Ambrosia	*Lactuca*
Anthemis	*Silybum*
Chrysothamnus	*Tanacetum*
Cichorium	*Vernonia*

Ecologically, the Asteraceae is considered to be relatively important. It is a major contributor to the diversity, and thus the stability and productivity, of grasslands, shrublands, and woodlands throughout the world. In addition, its fruits are an important component of the diet of many birds and small mammals. Economically, however, it is of relatively little importance when compared with other large families such as Poaceae, the grass family, or Fabaceae, the pea or legume family. It includes a few food and seasoning plants; a few medicinal plants, especially those used in folk remedies; several noxious weeds; and numerous garden ornamentals. Lettuce, endive, artichoke, tarragon, sunflower, and safflower are consumed by humans. Yarrow, coltsfoot, arnica, and chamomile are used medicinally. Widespread noxious weeds include Canada thistle, horseweed, ragweed, groundsel, dandelion, cocklebur, and star thistle. Representative ornamentals are marigold, zinnia, dahlia, daisie, and chrysanthemum. As described in detail below, other species of the family are toxic to both humans and livestock.

Recognized as a distinct plant group since the time of Theophrastus in the third century B.C., the Asteraceae is divided into 2–4 subfamilies and 13–21 tribes based on head morphology, biochemistry, pollen morphology, cytology, and anatomy (table 13.1.). The family's classification is in transition as taxonomists incorporate new data and reevaluate relationships. An alternative family name permitted by the rules of botanical nomenclature and commonly used is Compositae.

A rather specialized terminology has evolved to describe the distinctive morphology of the composite head. Many of the terms are those used for other families of flowering plants, but several important differences do exist. Most notable is the use of *florets* rather than *flowers* and *phyllaries* rather than *bracts*.

■ specialized terminology, florets, phyllaries

Plants herbs or rarely shrubs; perennials or annuals or biennials; from rhizomes or tubers or caudices or crowns or fleshy roots; terrestrial or emergent aquatics; caulescent

Table 13.1. Major subfamilies and tribes of the Asteraceae in North America

Subfamily	Tribe	Common Name(s)
Asteroideae	Anthemideae	yarrows, sagebrushes
	Astereae	asters, daisies, goldenrods
	Eupatorieae	eupatoriums, bonesets
	Gnaphalieae	everlastings, cudweeds
	Helenieae	sneezeweeds
	Heliantheae	sunflowers, ragweeds
	Inuleae	pussytoes, cudweeds
	Plucheeae	plucheas, stinkweeds
	Senecioneae	groundsels, ragworts
Cichorioideae	Cardueae	thistles
	Lactuceae	dandelions, hawkweeds
	Vernonieae	ironweeds

Source: Bremer 1994.

■ herbs or shrubs; flowers in heads, most commonly yellow, white, blue, or reddish purple, fruits achenes

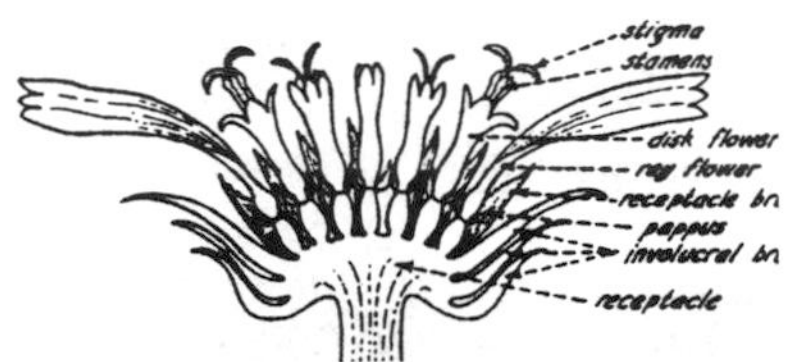

Composite head

or acaulescent; strongly aromatic or not aromatic; bearing perfect flowers or monoecious or dioecious or polygamous. **Stems** armed or not armed with spines or thorns or prickles. **Leaves** cauline or basal or forming basal rosettes; simple or 1-to-3-pinnately compound; alternate or opposite or alternate above and opposite below; venation pinnate or palmate or a single vein; stipules absent. **Inflorescences** heads; solitary or borne in racemes, corymbs, panicles, or clusters; terminal or axillary; bracts present; of 2 types—phyllaries and receptacular bracts. **Phyllaries** subtending heads, 4 to numerous, borne in 1 to several rows or fused into a hardened structure surrounding flowers. **Receptacular Bracts** (pales) present or absent, 1 per floret. **Heads** with disk florets only or both disk and ray florets or ligulate florets only. **Flowers** termed florets; fragrant or not fragrant; perfect or imperfect or neutral; of 3 forms—disk, ray, and ligulate. **Disk Florets** 2 to many per head, borne in center of receptacles, perfect or functionally staminate or pistillate, corollas radially symmetrical. **Ray Florets** 3 to many per head, borne at periphery of receptacles, pistillate or neutral; corollas bilaterally symmetrical, typically reflexed. **Ligulate Florets** 3 to many per head, borne both in center and at periphery of receptacles, perfect, corollas bilaterally symmetrical; perianths in 2-series or 1-series or absent. **Calyces** present or absent; termed the pappus; highly modified; consisting of bristles or scales or awns or teeth. **Corollas** radially or bilaterally symmetrical; tubular or ligulate or bilabiate. **Petals** 5 or 0; fused; of various colors, most commonly yellow or white or blue or reddish purple. **Stamens** 5; epipetalous; fused by anthers; anthers basifixed. **Pistils** 1; compound, carpels 2; stigmas 2, borne on inner surfaces of style branches, variously shaped; styles 1, divided, typically emerging laden with pollen from ring of fused anthers as florets mature, exserted beyond perianths; ovaries inferior; locules 1; placentation basal. **Fruits** achenes; wings absent or present. **Seeds** 1.

■ evolutionary success due to defensive chemicals, bitter sesquiterpene lactones and polyacetylenes in most taxa

As noted above, the Asteraceae is the largest family of flowering plants, with tremendous evolutionary radiation of genera and species in a variety of habitats. Its success throughout the world has been attributed by Cronquist (1981) to the presence of a defensive chemical combination of bitter sesquiterpene lactones and polyacetylenes in almost all taxa. In different tribes and genera, they are augmented by other compounds (Hegnauer 1977; Heywood et al. 1977). As might be expected, the family has a large number of toxic and potentially toxic genera that possess a variety of toxicants. An overview of these compounds and their distribution in the family is presented in the following paragraphs.

■ sesquiterpene lactones, pseudoguaianolides

Of greatest interest are the sesquiterpene lactones that are represented in all tribes of the family (Hegnauer 1977; Herz 1978). These compounds, of which there are sev-

Table 13.2. Species of unknown risk, that contain sesquiterpene lactones of possible toxicologic significance

Achillea spp.	yarrows, milfoils
Ambrosia spp.	ragweeds
Arnica chamissonis	leafy arnica, chamisso
A. longifolia	seep-spring arnica
A. montana	mountain arnica
Balduina angustifolia	
Gaillardia aristata	blanket flower
G. xgrandiflora	
G. pulchella	Indian blanket, Mexican firewheel, rose-ring gaillardia
Helianthus spp.	sunflowers
Inula spp.	elecampanes
Picradeniopsis spp. (= *Bahia*)	bahias
Rudbeckia spp.	coneflowers, brown-eyed Susans
Viguiera spp.	golden-eyes

Sources: Romo & deVivar 1967; Herz 1968; Gill et al. 1980; Krisken & Willuhn 1982; Seaman 1982.

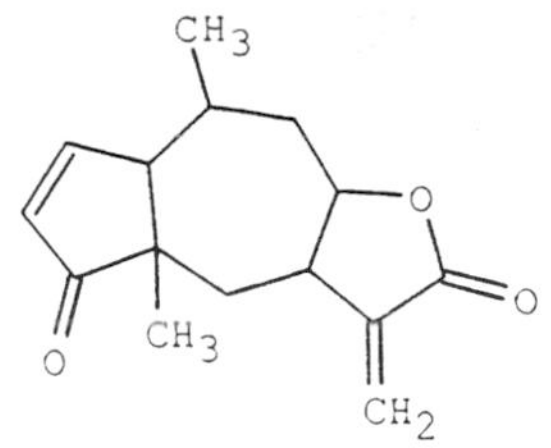

Helenanolide, a pseudoguaianolide

eral skeletal types, are known to produce an array of biological effects: antimicrobial, antineoplastic, cytotoxic, insect deterring, phytotoxic, and toxic to livestock (Rodriguez et al. 1976; Picman 1986). Although identification of actual intoxication problems caused by these lactones is limited to a few genera—*Baileya, Helenium, Hymenoxys,* and *Psilostrophe*—future work may reveal the toxic potential of other species that contain the same or similar compounds (table 13.2.). This possibility is particularly true for genera containing pseudoguaianolides. Intoxication by them may simply be a matter of unusual circumstances when sufficient plant material is eaten.

■ acetylenic compounds, some may be phototoxic

Acetylenic compounds with 9–17 carbons are distributed throughout the family except for the tribes Senecioneae and Lactuceae (Hegnauer 1977). They are structurally similar to those present in the Apiaceae (carrot family) and Araliaceae (ginseng family), and some contain sulfur (Sorensen 1968). Phototoxicity due to polyacetylene and thiophene derivatives is a suggested but not confirmed possibility for a number of genera in the Asteraceae, including *Artemisia, Bidens, Centaurea, Dahlia, Echinops, Solidago,* and *Tagetes* (McLachlan et al. 1986). In vitro evaluation revealed the presence of photoactive antimicrobial compounds in genera from the southwestern deserts such as *Dyssodia, Nicolletia (N. trifida), Pectis (P. canescens), Porophyllum (P. scoparium),* and *Thymophylla* (Downum et al. 1989).

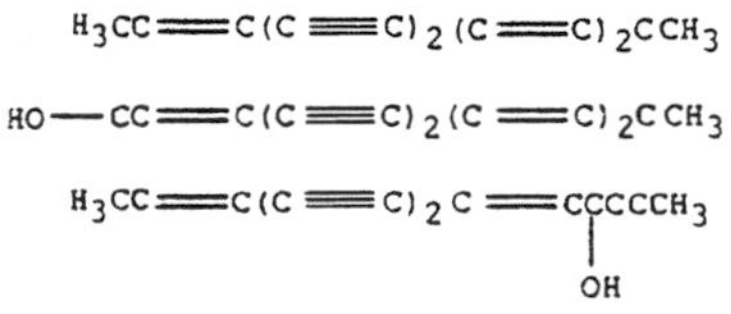

$H_3CC{\equiv}C(C{\equiv}C)_2(C{\equiv}C)_2CCH_3$

$HO{-}CC{\equiv}C(C{\equiv}C)_2(C{\equiv}C)_2CCH_3$

$H_3CC{\equiv}C(C{\equiv}C)_2C{=}CCCCCH_3$ (OH)

Representative acetylenic compounds

■ nitrate accumulation

A few genera of the family are nitrate accumulators and in some circumstances pose a toxicologic risk for livestock (table 13.3.). This risk is especially true for the thistles that are likely to be sprayed with herbicides, some of which—e.g., 2,4-D—cause the plants to accumulate higher levels of nitrates for a few days (Berg & McElroy 1953; Harradine 1990).

A variety of other toxicants have been identified in various genera of the Asteraceae. For example, species of *Machaeranthera* are secondary selenium accumulators and

Table 13.3. Genera and species of confirmed or potential concern as nitrate accumulators

Ambrosia spp.	ragweeds
Carduus nutans	musk thistle
Cirsium arvense	Canada thistle (to 4.8% NO_3)
Gnaphalium spp.	cudweeds
Helianthus annuus	annual sunflower
Lygodesmia spp.	skeleton-plants
Silybum marianum	milk thistle (to 10% NO_3)
Viguiera spp.	golden-eyes

Source: Clawson et at. 1935.

■ other toxicants, selenium, cyanogenic glycosides, terpenoids, bitter principles

may contain dangerously high concentrations in seleniferous areas. Other species, such as those of the genera *Osteospermum* (African daisies) and *Picradeniopsis*, contain cyanogenic glycosides and have caused animal losses (Deem et al. 1939; Watt & Breyer-Brandwijk 1962). Although *Balsamorhiza sagittata* is known to contain a series of tricyclic cycloartane terpenoids and suspected to be toxic, large amounts fed to horses and sheep failed to cause adverse effects (Chesnut & Wilcox 1901; Connolly & Hill 1991). Numerous genera also have been considered as imparting bitterness to milk (Pammel 1911; Watt & Breyer-Brandwijk 1962). These genera include *Achillea, Anthemis, Bidens, Cichorium, Dyssodia, Erigeron, Helenium, Lactuca,* and *Lygodesmia.*

Artemisia L.

■ *A. filifolia*
A. ludoviciana
A. spinescens
A. tridentata
mugwort
sage
sagebrush
sagewort
wormwood

Taxonomy and Morphology—Comprising 100–300 species distributed throughout the Northern Hemisphere, *Artemisia,* a member of the tribe Anthemideae, is commonly known as sagebrush, wormwood, or mugwort and is one of the genera most often associated with "the West." Shrubby *A. tridentata* is the sagebrush that is one of the dominants of the cold desert of the Great Basin. The generic name commemorates Artemisia, a botanist and medical scholar, who was queen of ancient Caria, a part of modern Turkey, in about 400 B.C. The genus is important for herbalists; its various aromatic species are used in a variety of medicinal infusions and are the source of absinthe and tarragon. In the Middle Ages, branches were typically spread on floors to keep fleas and other insects at bay. A number of species are cultivated as ornamentals for their foliage, particularly those with white, gray, or silvery dissected leaves.

In North America, approximately 60 species and innumerable subspecies and varieties are present. Some species are distinct and easily identified, whereas others intergrade and form complexes. Polyploidy and hybridization have been important aspects of the genus's evolution. A few taxonomists have proposed segregating several sections of the genus as distinct genera, but most employ a conservative approach, as is taken here. The following are representative species and species of toxicologic significance:

A. filifolia Torr.	sand sage, sand sagebrush, old-man sagebrush
A. ludoviciana Nutt.	white sage, western mugwort, prairie sage, sagewort, Louisiana wormwood
A. spinescens D.C.Eaton (= *Picrothamnus desertorum* Nutt.)	bud sage, button sage, spring sage
A. tridentata Nutt. (= *Seriphidium tridentatum* [Nutt.] W.A.Weber)	sagebrush, big sagebrush

■ herbs or shrubs; herbage aromatic; heads small, numerous, tan or yellowish white; disk florets only

Plants herbs or shrubs; annuals or perennials; from rhizomes or woody caudices or taproots; herbage aromatic; glabrous or glandular or indumented, often tomentose or sericeous-tomentose. **Stems** erect or ascending; 30–200 cm tall. **Leaves** simple; alternate; blades of various shapes; margins entire or lobed or pinnately dissected. **Heads** numerous; small; borne in spikes or racemes or spicate panicles; urceolate; phyllaries in 2 to several series, linear-lanceolate, densely tomentose; receptacles flat to hemispheric; receptacular bracts absent. **Ray Florets** absent. **Disk Florets** 3–20; perfect or functionally staminate or marginal ones pistillate; corollas tan or yellowish-

Artemisia filifolia

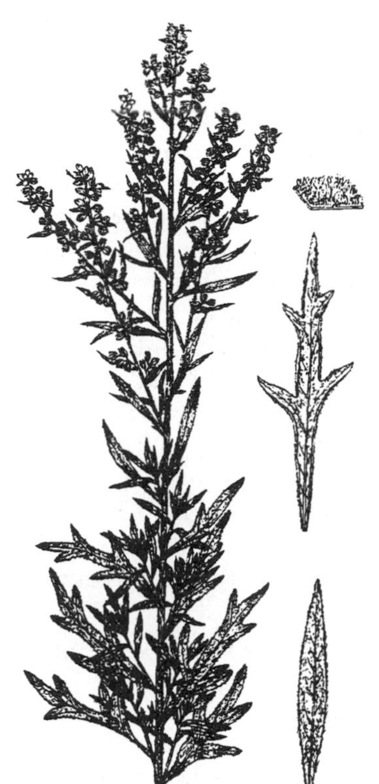
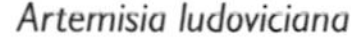

Artemisia ludoviciana

Artemisia tridentata

white. **Pappus** absent or rarely short coroniform. **Achenes** small; glabrous or resinous-glandular or indumented.

1. Plants herbs. Rhizomes present . . . . *A. ludoviciana*
1. Plants shrubs. Rhizomes absent.
   2. Thorns present. Corollas and achenes pilose-villous. Plants subshrubs; flowering in spring . . . . *A. spinescens*
   2. Thorns absent. Corollas and achenes glabrous or puberulent. Plants shrubs; flowering in late summer and fall.
      3. Leaves oblanceolate; apices 3-toothed. All disk florets perfect; fertile . . . . *A. tridentata*
      3. Leaves linear or filiform; apices rounded. Central disk florets functionally staminate, sterile; marginal disk florets pistillate, fertile . . . . *A. filifolia*

Distribution and Habitat—Species of *Artemisia* are characteristically encountered in a variety of soil types in drier habitats in the Northern Hemisphere. A few species are present in South America. Morphologically distinct and not readily confused with other taxa, *A. filifolia* occurs in sandy soils of the Southwest. *Artemisia ludoviciana* is widely distributed throughout much of North America. *Artemisia spinescens* is a species of mesas, plains, and dry alkaline sites throughout the intermountain west. *Artemisia tridentata* favors open dry plains, hills, and valleys from the Great Plains westward. It is very common and abundant, often forming extensive dense communities.

■ mainly open, dry soils; sometimes abundant

- valued winter browse when animals adapted to it; abrupt subsistence on large amounts, digestive disturbance; neurologic disturbance, sage sickness, horses

Disease Problems—Many species of *Artemisia* are valued browse for wildlife and livestock (Parker 1936). New growth of *A. spinescens* is relished and extensively browsed by range livestock (Cook et al. 1954). *Artemisia filifolia,* readily eaten by sheep in the winter, is also of some forage value for horses and more so for deer and antelope (Looman 1983). *Artemisia tridentata* may be of lesser palatability, but nevertheless it is a highly valued winter range plant for mule deer and livestock, especially sheep and goats (Cook et al. 1954; Welch & Pederson 1981; Wood et al. 1995). However, adverse effects may occur when livestock are abruptly forced to subsist on or otherwise eat large amounts of these species. When sheep or calves were given 4.3–5.7 kg of green *A. spinescens* in daily divided doses over a week or more, no adverse effects were noted (Fleming 1920). A smaller amount, 2 kg or more, in a single dose was lethal to yearling calves. Abrupt ingestion of 0.75% b.w./day of *A. tridentata* leaves by sheep was lethal in 2–4 days (Johnson et al. 1976). The same dosage given after preconditioning did not produce adverse effects. A similar effect appeared in horses introduced into a range containing *A. filifolia* and suddenly eating large amounts of the species (Beath et al. 1939). The disease, called sage sickness, causes alarming signs, but the animals are able to continue to eat and drink normally. The nervousness and foreleg incoordination gradually subside, and the animals seem to acquire a tolerance of the plant. Studies of the effects of *A. filifolia* and *A. tridentata* on reproduction indicate that there is little reason to associate the genus with reproductive abnormalities (Johnson et al. 1976). There is little information about the toxicity of *A. ludoviciana* and its numerous subspecies.

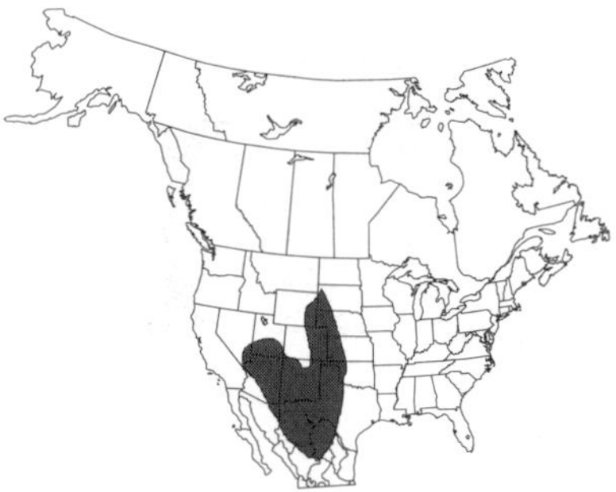

Artemisia filifolia

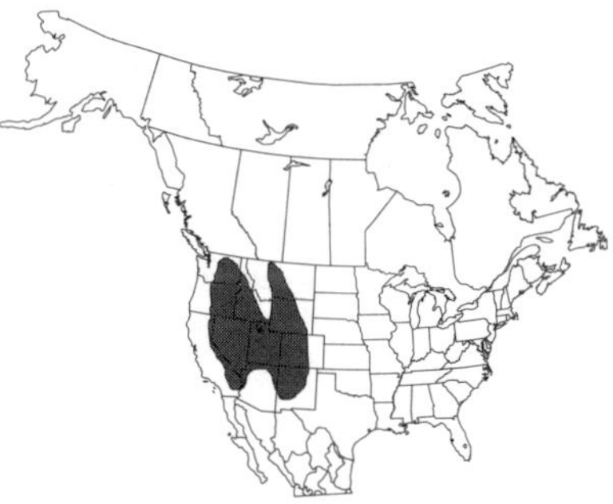

Artemisia spinescens

Artemisia tridentata

- variety of volatile aromatic monoterpene compounds and sesquiterpene lactones

Disease Genesis—*Artemisia* possesses an array of irritant, aromatic monoterpenes. *Artemisia tridentata* contains 1–3% d.w. monoterpenes, mainly camphor and thujones, but also others such as the pinenes and camphene (Welch & Pederson 1981). Terpenes, such as the pinenes and geraniol, have been associated with irritant effects when large amounts of a North African species, *A. campestris,* were eaten by goats and camels (El Bahri et al. 1997). Digestive disturbances associated with the genus appear to be the result of the presence of these aromatic components and volatile oils, which exert antibacterial effects on ruminal bacteria (Nagy & Tengerdy 1968). These effects can be significant with high dosage of plant material because adaptation of the ruminal flora is minimal at high concentrations of the oils. Also present and of significance are sesquiterpene lactones; scintonin is common to many species. These compounds exert a variety of effects. Crude extracts of the Brazilian *A. verlotorum* produced sedative/analgesic effects in mice (de Lima et al. 1993). Various species of *Artemisia* also seem to exert influences on the processes of hepatic biotransformation, inhibiting microsomal dealkylating activity and possibly inducing other types of activity (Jennings et al. 1978; Eissa et al. 1995, 1996). These effects may be a consequence of the wide array of sesquiterpene lactones present (Seaman 1982).

CH_3, O, C, $(CH_3)_2$

Camphor

CH_3, O, C, H_3C, CH_3

Thujone

Other types of bioactivity may be produced by the compounds present in *Artemisia*. Illustrative of this is the historical use of the classical dewormer santonin extracted from the seeds of the Old World *A. pauciflora*, commonly known as santonica or Levant wormseed (Hall 1921).

Clinical Signs and Treatment—Because of the irritant effects on the digestive tract, the animal exhibits loss of appetite, ruminal inactivity, and depression. Such signs are to be expected mainly when sheep are forced to consume large amounts of plant material abruptly. The neurologic effects—apprehension, nervousness, and incoordination—also resolve in a few days as the animals adapt to eating *Artemisia*. Treatment in either instance is generally not necessary.

■ irritant digestive disturbance, depression, anorexia, ruminal stasis; neurologic effects, apprehension, ataxia; treatment not needed

Baccharis L.

Taxonomy and Morphology—Native only to the New World, *Baccharis*, a member of the tribe Astereae, comprises about 400 species distributed primarily in tropical and warm-temperate regions. The generic name reputedly refers to the god Bacchus and alludes to the spicy odor of the roots. In North America, 22 species are present. The following are representative of the genus and implicated in toxicologic problems:

■ *B. angustifolia*
B. glomeruliflora
B. halimifolia
B. pteronioides

baccharis
consumption weed
false willow
groundsel bush
groundsel tree
yerba de pasmo
sea myrtle
silverling
yerba manza

B. angustifolia Michx.	baccharis, false willow, silverling, narrowleaf baccharis
B. glomeruliflora Pers.	baccharis, silverling, groundsel tree, southern baccharis, groundsel bush
B. halimifolia L.	eastern baccharis, sea myrtle, consumption weed, groundsel tree
B. pteronioides DC. (= *B. ramulosa* [DC.] A.Gray)	yerba manza, yerba del pasmo, yerba de pasmo, hierba del pasmo

Plants shrubs or subshrubs; perennials; herbage sometimes aromatic, glabrous or glandular-resinous; dioecious. **Stems** 50–400 cm tall; branchlets striate, angled or subterete. **Leaves** simple; alternate; reduced to bracts above; deciduous or evergreen; sessile or petioled; blades subulate to obovate; nerves 1 or 3; margins entire or serrate or dentate. **Heads** numerous; small; borne in racemes or panicles or corymbs; hemispheric to cylindrical; phyllaries imbricate, in several series, linear to lanceolate or ovate; receptacles flat; receptacular bracts absent or rarely present. **Ray Florets** absent. **Disk Florets** numerous; staminate tubular to funnelform; pistillate filiform; corollas white or pinkish or yellowish brown. **Pappus** of long capillary bristles; barbed or plumose. **Achenes** cylindrical; 4–10 ribbed; glabrous or hispid.

■ subshrubs or shrubs; leaves simple, alternate; heads numerous; florets disk only, white, pinkish, or yellowish brown

Baccharis pteronioides

1. Leaves linear or spathulate; 1–5 mm wide.
   2. Leaves 20–45 mm long; not fascicled. Heads in glomerules . . . . *B. angustifolia*
   2. Leaves 5–15 mm long; fascicled. Heads in racemes . . . . *B. pteronioides*
1. Leaves ovate or obovate or elliptic; 6–40 mm wide.
   3. Leaf punctations conspicuous. Heads solitary or in peduncled glomerules . . . . *B. halimifolia*
   3. Leaf punctations absent or inconspicuous. Heads in sessile glomerules . . . . *B. glomeruliflora*

■ moist sites, often saline or alkaline soils

Distribution and Habitat—Although occupying a variety of habitats and soil types, *Baccharis* is commonly associated with wet sites—marshes, washes, stream beds, stream banks, and ditches. Many species are tolerant of saline or alkaline soils and salt spray. *Baccharis angustifolia* and *B. glomeruliflora* are uncommon species of salt marshes and swamps along the southeastern coasts. *Baccharis halimifolia* occupies open, sandy coastal sites of the East and Southeast. Populations also occur inland in Oklahoma. *Baccharis pteronioides* is locally abundant between 1,300 and 2,300 m elevation in the Southwest.

Baccharis angustifolia

Baccharis pteronioides

Baccharis glomeruliflora

Baccharis halimifolia

■ digestive disturbance; stiffness, trembling, and other neurologic signs

Disease Problems—In the Southwest, *B. pteronioides* causes problems in cattle when they are forced to subsist on it in autumn and early winter (Huffman & Couch 1942). Affected cattle walk as if sore and stiff. They lie down, and if forced to move, they tremble and exhibit seizures (Marsh et al. 1920). The acute lethal dose in sheep is about 1% b.w. Plants are not palatable, but local abundance and absence of other forage lead to potential problems. In the eastern United States, *B. halimifolia* similarly causes problems in cattle, producing digestive disturbances, weakness, tremors, and death (Manley et al. 1982). There are few lesions other than scattered hemorrhages on the surfaces of the heart and digestive tract. Two Brazilian species, *B. cordifolia* and *B. megapotamica,* cause severe digestive disturbance, weakness, tremors, constipation, and hepatic necrosis in cattle (Tokarnia et al. 1992).

■ toxins unknown, although cardiotoxic saponins may play a role

Disease Genesis—*Baccharis* contains digitalis-like, cardiotoxic glycosidal saponins that cause both digestive and cardiac adverse effects (Duncan et al. 1957; *Poisonous Plants* 1980). *Baccharis glomeruliflora* is reported to be the most toxic, and *B. angustifolia* the least (Duncan et al. 1957). The signs reported for *B. cordifolia* and *B. megapotamica* also are somewhat consistent with cardiotoxic saponins. These Brazilian species, however, contain a variety of trichothecene mycotoxins, which may, in large measure, account for their toxicity (Busam & Habermehl 1982; Habermehl 1994). The concentrations are especially high in pistillate plants, 5-to-10-fold higher at flowering (Jarvis et al. 1996). The plants are seemingly resistant to these toxins, almost as if they had synthesized them. Also present in *B. pteronioides* are pyrrolizidine alkaloids of unknown toxicologic significance.

■ anorexia, diarrhea, stiffness, trembling, seizures

Clinical Signs and Pathology—The clinical signs produced by ingestion of *Baccharis* include ruminal stasis, loss of appetite, excess salivation, diarrhea, weakness, trembling, staggering, and prostration. There are few distinctive lesions other than reddening and edema of the mucosa of the digestive tract, which is most pronounced in the abomasum, where there may be ulceration. Seldom required, treatment may be directed toward relief of the diarrhea.

Baileya Harv. & A.Gray

Taxonomy and Morphology—Named for J. W. Bailey, an early American microscopist, *Baileya* is a member of the tribe Heliantheae and is endemic to the deserts of North America. The 3 species present are as follows:

- *B. multiradiata*
- *B. pauciradiata*
- *B. pleniradiata*

baileya
desert baileya
desert marigold
paperflower

B. multiradiata Harv. & A.Gray (= *B. perennis* [A.Nelson] Rydb.)	desert marigold, desert baileya, many-rayed baileya, paperflower
B. pauciradiata Harv. & A.Gray	baileya
B. pleniradiata Harv. & A.Gray	baileya

Plants herbs; annuals or perennials; from taproots or branching caudices; herbage floccose or tomentose. **Stems** erect or ascending; 10–50 cm tall; sparingly branched. **Leaves** simple; alternate; primarily basal or mostly cauline; petioled; blades oblong to oblanceolate; margins entire to pinnatifid. **Heads** showy; 1–3 per long or short peduncle; campanulate to hemispheric; phyllaries in 2 or 3 subequal series, linear-lanceolate, woolly; receptacles flat or slightly convex; receptacular bracts absent. **Ray Florets** 4–60; pistillate; fertile; corollas yellow; becoming reflexed and papery; persistent on achenes. **Disk Florets** 8 to many; perfect; fertile; yellow; corollas gland dotted. **Pappus** absent. **Achenes** cylindrical or clavate; terete or slightly angled; ribbed; glabrous or hispidulous; gland-dotted.

- herbs; leaves simple, alternate; heads showy; disk and ray florets yellow

Baileya multiradiata

1. Heads 2 or 3 per peduncle; 4–6 mm in diameter. Ray florets 4–8; pale yellow . . . . *B. pauciradiata*
1. Heads 1 per peduncle; 10–25 mm in diameter. Ray florets 20–60; bright yellow.
   2. Plants annuals. Branching caudices absent. Leaves primarily cauline. Corollas of ray florets 7–11 mm long . . . . *B. pleniradiata*
   2. Plants perennials. Branching caudices present. Leaves primarily basal. Corollas of ray florets 12–20 mm long . . . . *B. multiradiata*

Distribution and Habitat—Quite showy, the three species of *Baileya* are encountered in dry, open, often sandy or gravelly sites in the southwestern deserts. All are weedy, with populations appearing along roadsides and in severely overgrazed ranges. *Baileya multiradiata* is locally abundant in sandy valleys, desert flats, washes, and plains. It is one of the dominant spring flowers of Chihuahua and Coahuila in northern Mexico.

- dry, open, sandy or gravelly desert sites

Baileya multiradiata

Disease Problems—First noted as a problem in Texas in the early 1930s, extensive animal losses were associated with ranges where little forage other than *B. multiradiata* was available (Mathews 1933a). Death rates were as high as 25% in some flocks of sheep (Dollahite 1960). Palatability seems to vary. Normally the plants are not eaten, but in some instances, sheep relish the flowering and fruit heads, even when other forages are plentiful (Dollahite 1960; Hakkila et al. 1987). Sheep are much more prone to eat plants of this genus than are goats or cattle. Because the heads are more toxic than the leaves, the risk is greater with increased plant maturity. Experimentally, about 1% b.w. of dried plant material fed to sheep for 1 week or 5–6 weeks produced severe digestive and metabolic disease and death (Dollahite 1960).

- digestive disturbance, gradually increasing in severity; less commonly regurgitation of ruminal contents; death in a few days to a week or more

Disease Genesis—Paucin, a pseudoguaianolide glucoside, is found in concentrations as high as 0.04% in *B. pauciradiata* and *B. pleniradiata* (Waddell & Geissman 1969).

■ pseudoguaianolide sesquiterpene lactones

Hymenoxon and several other pseudoguaianolides have been identified in *B. multiradiata* (Hill et al. 1977; Seaman 1982). The presence of these compounds suggests that toxicity due to ingestion of species of *Baileya* is most likely another expression of the disease so commonly associated with species of *Helenium* and *Hymenoxys*.

Baileya pauciradiata

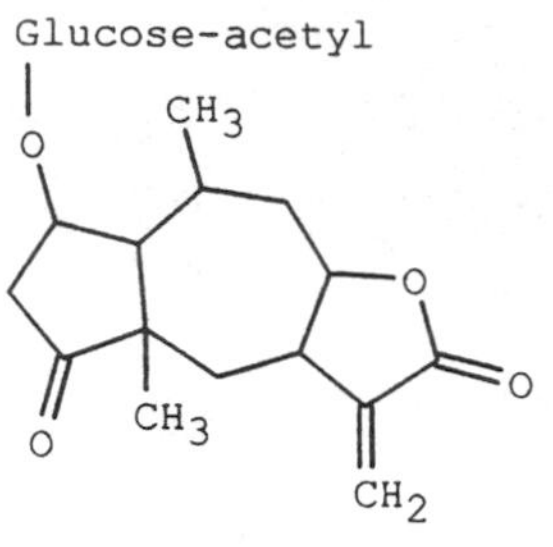

Paucin

Baileya pleniradiata

Clinical Signs—The signs of *Baileya* intoxication are very similar to those produced by *Helenium* and *Hymenoxys*. After the animal has consumed plants for several weeks, there is decreased appetite, stiff gait, weakness, and reluctance to stand or move. Upon arising, there is trembling and the animal stands with an arched back. When the animal is forced to move, its heart rate accelerates, and there is a tendency for it to regurgitate, so that frothy, green ruminal contents accumulate on the lips, nares, and mouth. There is slow development of the disease, with accompanying marked weight loss, sometimes inhalation pneumonia, and death after a week or more.

■ anorexia, weakness, stiff gait, arched back; sometimes frothy, green saliva on lips

Pathology—The lesions are not distinctive. There may be a few ecchymotic hemorrhages on the heart and diaphragm. Microscopically, there may be mild renal tubular epithelial flattening with proteinaceous casts and indications of myocardial and hepatocellular degeneration.

■ no specific lesions

Treatment—When *Baileya* intoxication occurs, the animal should be removed from the forage and given supportive nursing care. At present, range management to control plant growth or prevent exposure of animals to the plants is the only known method for reducing losses. Alteration of diet and administration of various drugs prior to the appearance of clinical signs may be beneficial. A complete discussion of treatment and prevention is presented in the treatment of *Hymenoxys*, in this chapter.

■ supportive, nursing care

Centaurea L.

■ *C. melitensis*
C. repens
C. solstitialis

knapweed
Maltese star thistle
Napa thistle
Russian knapweed
St. Barnaby's thistle
star thistle
tocalote
Turkestan thistle
yellow star thistle

Taxonomy and Morphology—Comprising 450–500 species, *Centaurea*, a member of the tribe Cardueae, exhibits greatest diversity in the Mediterranean region. A few species are native to the Americas and Australia. Some European botanists divide the genus into a number of segregate genera that are poorly differentiated, but most employ a traditional approach, as is taken here. Commonly known as star thistles or knapweeds, many of the species are cultivated as ornamentals, whereas others are noxious weeds. In North America, 29 species are present. Most have been inadvertently introduced, the consequences of which have been disastrous for much of our range, especially in the Northwest and northern Great Plains. Only 2 and perhaps a third species pose toxicologic risk (see plate 13.1):

C. melitensis L.	Maltese star thistle, Napa thistle, tocalote
C. repens L. (= *C. picris* L.) (= *Acroptilon repens* DC.)	Russian knapweed, Turkestan thistle, Russian star thistle
C. solstitialis L.	yellow star thistle, St. Barnaby's thistle

Plants herbs or subshrubs; annuals or biennials or perennials; from taproots or rhizomes or caudices. **Stems** erect or ascending; 30–175 cm tall; branched or not branched. **Leaves** simple; alternate; upper ones usually smaller; blades linear to oblanceolate; margins entire to pinnately dissected. **Heads** 1 to numerous; small to large; borne at branch ends; cylindrical or globose; phyllaries in several series, imbricate, shape variable, apices spine-tipped or with a scarious or hyaline, erose or fimbriate or pectinate appendage; receptacles flat to slightly convex; receptacular bracts bristly. **Ray Florets** absent. **Disk Florets** numerous; outer enlarged, sterile, somewhat resembling ray florets; inner perfect, fertile; corollas magenta or purple or blue or pink or yellow or white. **Pappus** of capillary bristles or scales or rarely absent. **Achenes** elliptic-oblong; flattened; attached obliquely.

■ herbs or subshrubs; leaves simple, alternate; disk florets only, yellow, pink, blue, or purple

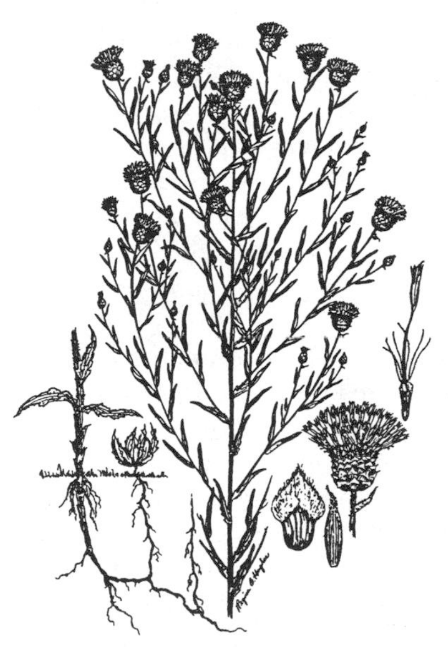

Centaurea repens

Centaurea solstitialis

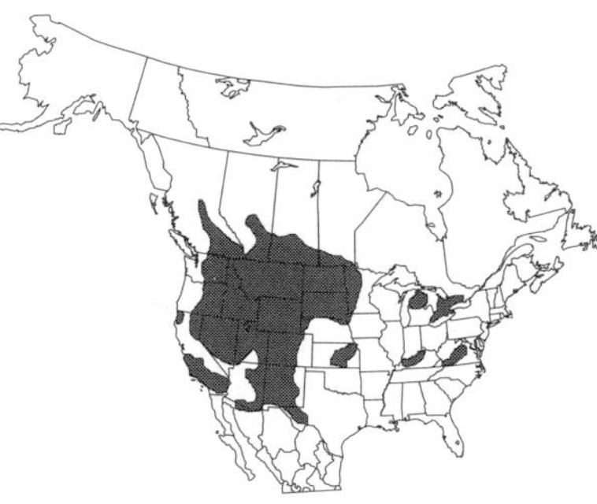

Centaurea repens

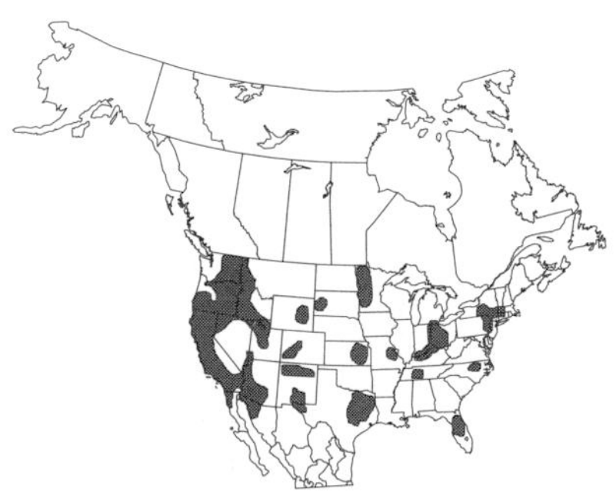

Centaurea solstitialis

1. Plants perennials. Rhizomes present. Florets pink-purple or blue or white. Phyllary apices not spine tipped . . . . *C. repens*
1. Plants annuals. Rhizomes absent. Florets yellow. Phyllary apices spine tipped.
   2. Central spines of phyllaries 10–25 mm long . . . . *C. solstitialis*
   2. Central spines of phyllaries 3–9 mm long . . . . *C. melitensis*

Distribution and Habitat—*Centaurea repens* is distributed mainly in the intermountain states. It is adventive in a few sites farther east. *Centaurea solstitialis* is most abundant from central California to Washington and Idaho. Scattered populations occur elsewhere, and it appears to be increasing in abundance. *Centaurea melitensis* appears sporadically across the continent.

■ aggressive pasture and waste area invaders

All three species have broad ecological tolerance and readily infest roadsides, waste areas, and other sites with disturbed soils. Once established, *C. solstitialis* spreads aggressively because it is a prolific seed producer. *Centaurea repens* spreads by deep rhizomes and forms extensive, dense colonies, especially in sandy washes where there is little competition.

■ aggressive, allelopathic, severely limit growth of desirable pasture and range forages

Disease Problems—Species of *Centaurea* are prized both for the beauty of their heads and despised for their ability to invade sites as noxious weeds. In addition to the neurotoxic effects of some species (Fowler 1965), their introduction has been an agronomic disaster in the western states and provinces with millions of acres of rangeland and pasture degraded. Introduced in the mid-1800s to early 1900s, probably as contaminants in alfalfa seed, they have spread vigorously and proven very difficult to control by conventional means (Maddox 1979; Strang et al. 1979). They are particularly noxious because they are allelopathic, thus further limiting the productivity of grasslands. It is difficult for grasses and other forages to compete with them (Sheley et al. 1993). The worst offenders have been *C. diffusa, C. maculosa, C. repens,* and *C. solstitialis*. From 1958 to 1985 the acreage in California alone infested by *C. solstitialis* increased by 6-fold to 7.5 million acres (Maddox & Mayfield 1985). By 1985, *C. solstitialis* was found in 209 counties in 23 states and was continuing to spread, while *C. repens* was present in 412 counties in 21 states (Maddox et al. 1985). Because *C. solstitialis* is considered to be the more serious problem, various biological control methods are being used to control its infestations (Maddox 1979).

■ horses; neurologic disease affecting prehension by lips and tongue; continuous, long-term ingestion of very large amounts of plant material

Thus far, only *C. repens* and *C. solstitialis* have proven to cause intoxications in North America, but this may simply reflect the large amount of plant material that must be eaten to cause disease and the low palatability of the genus in general, rather than restriction of toxicants to them. In addition, intoxications occur only in horses; thus it is difficult to evaluate experimentally other species of the genus for toxicity. Involvement of horses also has been reported as a problem in South America and Australia (Martin et al. 1971; Gard et al. 1973; Perdomo et al. 1978).

It is of interest that although *C. repens* is considered to have no adverse effects on ruminants and to be useful forage for sheep in the United States and Australia (Cordy 1954; Gard et al. 1973), in other areas of the world it has been described as toxic to them (Steyn 1933a,b; Dil'bazi 1974). In Azerbaijan, *C. repens,* commonly called choking mustard, is reported to have caused severe disease problems in buffalo (Dil'bazi 1974). The disease described for buffalo is similar to that appearing in horses; bizarre lip and tongue movements, ataxia, twitching, circling, and diarrhea are observed. However, no distinctive neuropathology was noted, as it is with horses. In South Africa, *C. repens* given in a dosage of 600 g to sheep for 1–2 days was lethal. It caused acute digestive disturbance and fluid accumulation in the lungs and body cavities (Steyn 1933a,b). Interestingly, when it was given in gradually increasing doses over a week or more, tolerance developed. Thus, in some circumstances *C. repens* appears to be nontoxic.

■ interruptions of several weeks or more in eating the plant may be protective

Disease caused by *C. repens* and *C. solstitialis* appears to be the same, albeit *C. repens* is the more toxic (Young et al. 1970b). With both, it appears mainly during the dry summer and fall, when animal subsistence is dependent upon grazing these plants. It is not a problem on improved or irrigated pastures. The toxic effects are apparently not cumulative, given that prolonged feeding periods interrupted at 1- or 2-week intervals do not result in disease (Young et al. 1970a). There seems to be cumulative storage of the toxins rather than cumulative effect or damage. Thus, amounts of the plant that might otherwise produce disease—60% or more of b.w. of *C. repens* and 100% of b.w. of *C. solstitialis*—may be eaten without deleterious effect if not consumed on a continuous basis.

Centaurea solstitialis was seemingly known to the ancients. Some medicinal value is implied by the discovery of heads of *C. solstitialis* among other plants in a Neanderthal grave in Iraq (Lietava 1992).

- equine parkinsonism, rapid development of basal ganglia lesions

Disease Genesis—Since its first description as a disease in 1954, studies on the pathogenesis of *Centaurea* intoxication have been limited by the lack of reproducibility of the disease in species other than the horse (Cordy 1954, 1978). However, the specificity of the lesions for the basal ganglia has prompted more investigations in concert with attempts to unravel the mysteries of human diseases associated with dopaminergic pathways, e.g., Parkinson's disease, Huntington's disease, tardive dyskinesia. This renewed interest has prompted the development of in vitro techniques where limitations of in vivo species susceptibility are not apparent (Riopelle & Stevens 1993). This is a rather unusual disease in that it occurs almost as an all-or-none event. The lesions develop quickly and completely. Progressive stages of degeneration are not apparent in the affected areas, except for some changes in the neurons adjacent to the necrotic foci in the globus pallidus and the pars reticularis of the substantia nigra (Cordy 1954). The disease, which is sometimes called equine parkinsonism, is typical of those involving the basal ganglia, with loss of inhibitory input producing involuntary movements such as tremor, writhing movements, chorea, and persistent postural effects such as dystonia (Cote & Crutcher 1991).

Repin

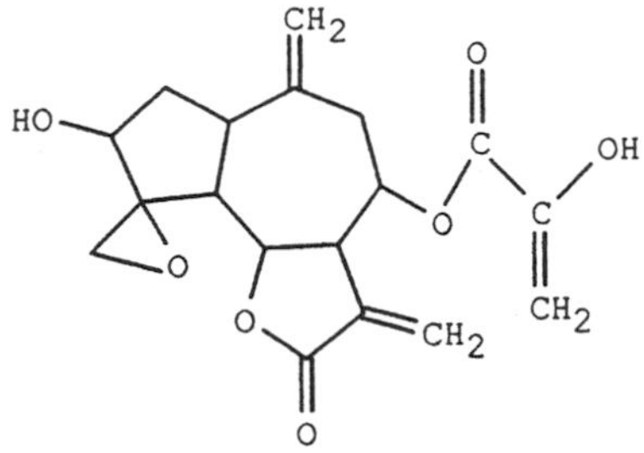

Janerin

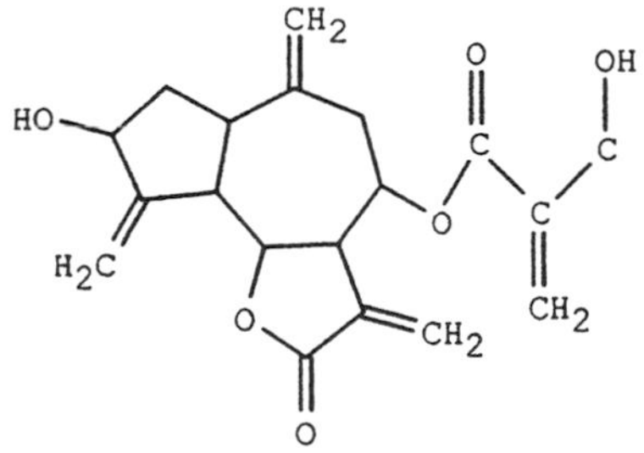

Cynaropicrin

Solstitialin

There are dozens of volatile compounds and scopoletin in the various species of *Centaurea,* but the components of most intense interest are the guaianolide sesquiterpene lactones, of which there are likewise dozens (Stevens 1982; Stevens & Merrill 1985; Binder et al. 1990a,b; Bubenchikova 1990; Kaij-a-Kamb et al. 1992). It is suggested that many of them may be artifacts of the isolation procedures, e.g., some of the chlorohydrins (Hamburger et al. 1993). In spite of the possible artificiality of some, they give an indication of the types of toxicants that may be present in the plants or formed during or after absorption in the digestive tract. Some of the lactones appear to be responsible for enhancing *Centaurea*'s growth and to be allelopathic. Lactones cytotoxic against neuronal cells in vitro include repin, subluteolide (an epimer of repin), janerin, cynaropicrin (0.04% in *C. repens*), 13-O-acetylsolstitialin, and acroptilin (Stevens 1982; Wang et al. 1991; Cheng et al. 1992; Riopelle et al. 1992). The epimers repin and subluteolide are unique in that they are highly reactive epoxides. Thus, it is tempting to consider them as the culpable toxins because they are found mainly in the two confirmed neurotoxic species *C. repens* and *C. solstitialis.* Their toxicity was ranked by an in vitro assay as repin > subluteolide > janerin > cynaropicrin > acroptilin > solstitialin (Riopelle & Stevens 1993). Both cynaropicrin and solstitialin A-13-acetate have been shown to be toxic in primary cultures of fetal rat substantia nigra cells (Wang et al. 1991; Cheng et al. 1992). These sesquiterpenes are also quite unstable, and it is suggested that they may be precursors of the ultimate neurotoxicants (Hamburger et al. 1993). Aspartic and glutamic acids, neuroexcitatory amino acids, also are present (Roy et al. 1995).

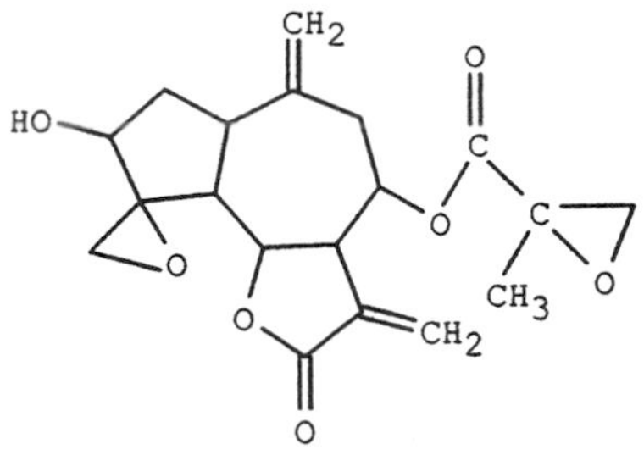

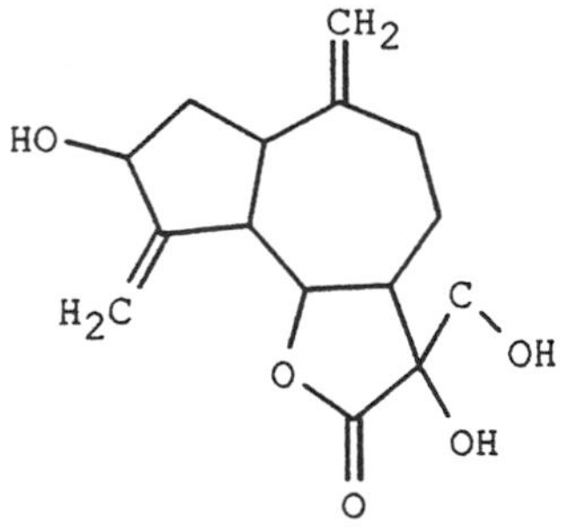

- numerous volatile compounds of unknown significance; several guaianolide sesquiterpene lactones of possible significance

Clinical Signs—Following ingestion of *Centaurea* for several months or more, there is abrupt onset of impaired eating and/or drinking. This may be accompanied by apprehension, excitement, and apparent confusion. Within a day or two, there is pronounced depression and, upon close examination, hypertonicity (not flaccid paralysis) of the lips and tongue. The increased tone may produce a fixed facial expression and/or constant chewing almost as if there was a throat obstruction—hence the use of the name *chewing disease.* The mouth may be held open or tightly closed, with the tongue protruding. The increased tone of the tongue causes it to be curled to form a groove or <bss>V<xss> shape. Rhythmic and/or writhing movements of the tongue

■ abrupt onset of impaired eating and drinking; apprehension, depression, hypertonicity, and rhythmic movements of lips and tongue

and lips also may be seen. Other neurologic signs such as locomotor difficulties, aimless walking, and drowsy inactivity with the head held very low also are seen. In most cases, the ensuing disease is permanent, but not progressive, after the first few days. In a few cases, there may be some apparent but not complete recovery, and animals may survive with limited ability to eat and drink. More often the disease leads to dehydration and starvation, with the animal exhibiting bizarre behavior such as submerging its head to allow water to flow into the esophagus.

■ gross pathology, softening and/or cavitation of globus pallidus and substantia nigra

Pathology—Although there may be traumatic lesions in the mouth, distinctive changes are limited to the brain. Involvement is primarily in the globus pallidus and the substantia nigra, where distinct pale yellowish to buff-colored foci of softening and/or cavitation up to 1 cm in diameter are seen. They are typically bilateral and symmetrical. Microscopically there is locally extensive necrosis of neurons, glia, and capillaries within sharply defined margins. Occasionally lesions may be seen in other sites of both gray and white matter.

■ general nursing care

Treatment—For the most part, treatment is limited to general nursing care, providing good feed and easy access to water, and giving supplemental vitamins. If the animal is promptly removed from the area containing *Centaurea,* there may be some apparent improvement, but permanent impairment should be expected. In Argentina, horses with obvious signs caused by *C. solstitialis* have been treated with reported success using glutamine synthetase and a bovine brain ganglioside extract given daily i.m. for a month (Selfero et al. 1989).

Erechtites Raf.

■ *E. glomerata*
E. hieracifolia
E. minima
burnweed
fireweed

Taxonomy and Morphology—Comprising 15–25 species, *Erechtites,* a member of the tribe Senecioneae, is native to the Americas and Australasia (Barkley and Cronquist 1978). In North America, 1 native and 2 introduced species are of toxicologic concern:

E. glomerata (Poir.) DC.	cut-leaved coast fireweed, cut-leaved coast burnweed
E. hieracifolia (L.) Raf.	eastern fireweed, American burnweed
E. minima (Poir.) DC.	toothed coast fireweed, toothed coast burnweed

■ herbs; leaves simple, alternate; heads numerous; disk florets only, yellow

Plants herbs; annuals; from taproots; often malodorous. **Stems** erect or ascending; 60–200 cm tall; often branched above. **Leaves** simple; alternate; lower larger, subpetiolate; upper smaller, sessile, bases clasping; blades lanceolate or oblanceolate; margins entire or coarsely and irregularly double serrate to pinnately lobed or dissected. **Heads** numerous; borne in elongate or flat-topped panicles or corymbs; cylindrical or urceolate; phyllaries in 2-series; outer few, short, setaceous; inner 13 or 14, in 2 rows, equal, linear; receptacles flat; receptacular bracts absent. **Ray Florets** absent. **Disk Florets** numerous; marginal ones pistillate, fertile; inner ones perfect; fertile; corollas yellowish white or light yellow. **Pappus** of capillary bristles; numerous; white; deciduous. **Achenes** cylindrical; 5-angled or 10-to-20-ribbed.

1. Heads urceolate or campanulate.
   Phyllaries 19–17 mm long . . . . *E. hieracifolia*
1. Heads cylindrical. Phyllaries 4–8 mm long.
   2. Leaves pinnately lobed. Inner phyllaries 12–15 . . . . *E. glomerata*
   2. Leaves serrate. Inner phyllaries 7–9 . . . . *E. minima*

Erechtites glomerata

Erechtites hieracifolia

■ open woodlands and grasslands

Erechtites glomerata

Erechtites hieracifolia

Distribution and Habitat—*Erechtites glomerata and E. minima* are Australasian species that have naturalized in the woodlands and grasslands of the Pacific Coast states. *Erechtites hieracifolia* is native and common in the open woods, especially muddy sites, of the eastern half of North America. It is adventive in the west.

■ hepatotoxicity, closed diester retronecine-type pyrrolizidine alkaloids

Disease Problems and Genesis—Closely related to *Senecio, E. hieracifolia* causes pyrrolizidine-type hepatotoxicity characteristic of that genus (Riet-Correa et al. 1998). It contains the closed diester retronecines; senecionine and senecicannabine, which represent a risk when sufficient plant material is eaten. The presence of these compounds is consistent with the occurrence of senecionine, seneciphylline, and retrorsine, mainly as *N*-oxides, in the Australian species *E. quadridentata* (Culvenor & Smith 1955). A complete discussion of pyrrolizidine compounds and aspects of their hepatotoxicity is presented in the treatment of *Senecio,* in this chapter.

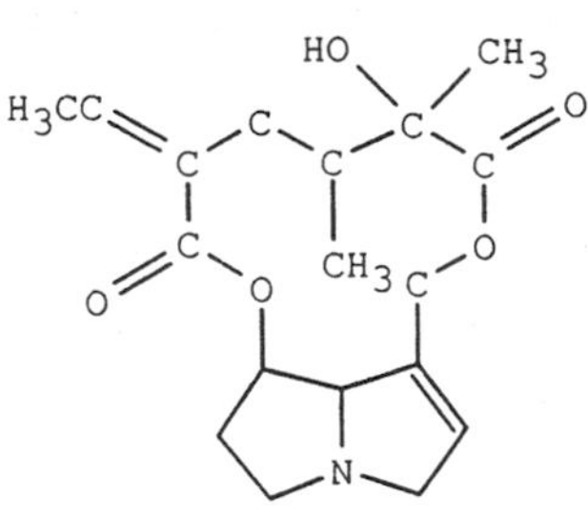

Senecionine

Senecicannabine

Eupatorium L.

Taxonomy and Morphology—Principally a genus of the New World, *Eupatorium,* a member of the tribe Eupatorieae, comprises 600–1,000 species. The generic name is taken from Eupator, who was a king of ancient Pontus, a part of modern northeastern

■ *E. rugosum*

boneset
fall poison
thoroughwort
poolroot
poolwort
richweed
squaw-weed
white sanicle
white snakeroot

Turkey; Eupator is reported by Pliny to have used one species of the genus as an antidote for poison. In recent years, some taxonomists have proposed dividing the genus into several smaller genera, perhaps better regarded as subgenera or sections. The traditional classification is used in this treatment. Of the approximately 75 species that occur in North America, only one is toxicologic importance (see plate 13.2):

E. rugosum Houtt. (= *E. ageratoides* L.f.) (= *E. urticaefolium* Reichard) (= *Ageratina altissima* [L.] R.M.King & H.Rob.)	white snakeroot, richweed, white sanicle, fall poison, thoroughwort, whitetop, boneset, squaw-weed, poolwort, poolroot

■ erect, perennial herbs; leaves simple, opposite, conspicuously 3-ribbed; heads numerous; disk florets only, white

Plants herbs; perennials; from fibrous rooted crowns or short rhizomes. **Stems** erect; 30–150 cm tall; often branched. **Leaves** simple; opposite; blades ovate to lanceolate; conspicuously 3-ribbed; margins serrate; petioles long. **Heads** numerous; small; borne in loose hemispheric or flat-topped clusters at ends of branches; obconic; phyllaries in 2-series, subequal; linear; receptacles flat; receptacular bracts absent. **Ray Florets** absent. **Disk Florets** 15–25; perfect; fertile; corollas white. **Pappus** of capillary bristles. **Achenes** cylindrical; 5-ribbed; blackish.

■ shaded or wooded sites

Distribution and Habitat—Distributed throughout the eastern half of North America, *Eupatorium rugosum* is usually found in low, moist, partially shaded sites such as wooded stream banks and terraces. Large dense stands may develop, and it may become the dominant understory species in the area. Unlike other species of the genus, *E. rugosum* is seldom found in open pastures or disturbed sites. Plants cannot withstand cultivation, and thus populations are most abundant in shaded areas that remain uncultivated. Growth begins in May or June, and flowering occurs in the fall.

■ milk sickness in humans and trembles in animals; historical interest, loss of many animals and humans in late 1700s and early 1800s

Eupatorium rugosum

Disease Problems—The diseases called milk sickness in humans and trembles in animals are of considerable historical interest because they caused great suffering and panic among the early nineteenth-century settlers in the Mississippi and Ohio River valleys. Sometimes decimating the populations of villages, these diseases reached a peak in the early 1800s and then declined as cultivation and settlement progressed (Christensen 1965; Duffy 1990). In some areas of Indiana and Ohio, one-fourth to one-half of the deaths in the early 1800s were attributed to milk sickness, and the ravages of the disease were especially severe in Illinois, Indiana, Kentucky, North Carolina, Ohio, and Tennessee (Wolf et al. 1918; Doyle & Walkey 1923; Couch 1933). An epidemic of milk sickness in 1818 in the area of Pigeon Creek, Indiana, is reputed to be the cause of death of Abraham Lincoln's mother, Nancy Hanks Lincoln, and several of her relatives, neighbors, and their livestock (Sandburg 1926; Duffy 1990). An example of the severity of the havoc wrought by this plant occurred in Indiana in the early 1800s (Hansen 1928). Several of the pioneer migration routes from the populous East to the middle West converged at a beautiful site on the White River in southern Indiana known as Hindustan Falls, and eventually the thriving community of Hindustan developed at the site. However, within a few years a mysterious malady of unknown cause had killed much of the livestock and eventually a substantial portion of the human population as well. The town was abandoned to the cause, at the time unknown, leaving the remnants of habitation and a luxuriant stand of *E. rugosum* along the riverbanks.

During the 1800s, numerous causes for milk sickness, including the bacterium

Bacillus lactimorbi, were proposed. Many astute observers, such as Anna Pierce and John Rowe in the 1830s, reported a clear association between *E. rugosum* and the disease, but their findings were either ignored or dismissed (Snively & Furbee 1966). Even by the beginning of the twentieth century, when there was reasonable experimental evidence to incriminate *E. rugosum,* the role of the plant and whether it was the sole cause continued to be disputed. It was not until after 1910 that *E. rugosum* was confirmed unequivocally as the culprit (Marsh & Clawson 1917; Wolf et al. 1917, 1918). Milk sickness continues to be a sporadic problem for rural people even as we enter the twenty-first century (Pammel 1911; Richardson 1931; Doyle 1947; Hartmann et al. 1963).

- cumulative effect, large amounts of plant for a few days or smaller amounts for several weeks; metabolic derangement; cardiotoxic in horses; hepatotoxic in goats

Eupatorium rugosum

Consumption of green plant material equivalent to 5–10% of the animal's b.w. is required to produce intoxication. Because the toxicant's effects are cumulative, small amounts ingested for several weeks may be almost as toxic as a large amount eaten in several days (Marsh & Clawson 1917). There is slow loss of toxicity with heat and/or drying of the plant; however, this is variable, and even frost-damaged plants are reported to remain toxic for some time (Smetzer et al. 1983). Although all livestock species are susceptible, the disease has been of greatest concern in dairy cattle, and occasional episodes are still reported (Stotts 1984). There are many similarities in clinical signs among the various animal species, but there also are some notable differences in horses and goats. As noted in early descriptions, swallowing difficulties, sweating, and cardiac pathology are distinctive in horses (Doyle & Walkey 1923; Graham & Boughton 1925; Smetzer et al. 1983; Olson et al. 1984; Thompson 1989). In goats, severe, acute hepatic necrosis accompanied by photosensitization in animals that lived for several days was associated with ingestion of large amounts of the foliage of *E. rugosum* in central Texas (Reagor et al. 1989). Experimentally, the single-dose LD_{50} for fresh *E. rugosum* was about 5 mg/kg b.w. for Angora and Spanish goats. There was severe centrilobular hepatic necrosis, with head pressing, paddling seizures, depression, and anorexia. Muscle tremors were rarely seen.

Extracts of *Eupatorium* species have been used for a variety of medicinal purposes, especially as tonics and vermifuges (Pammel 1911). *Eupatorium perfoliatum* and *E. purpureum* are among the most widely used plants for medicinal purposes; decoctions are especially popular as tonics. In large doses, they are reported to cause digestive disturbance with colic and also neurotoxic effects (Millspaugh 1974).

- crude toxin mixture; tremetol; composed of methyl ketone benzofuran derivatives; metabolic impairment; inhibition of TCA cycle

Tremetone

Dehydrotremetone

Hydroxytremetone

Disease Genesis—Tremetol, an extract of the viscous oily fraction of the plant, originally was associated with the toxic effects and named after the predominant clinical sign of muscular trembling (Couch 1927). An aromatic with an odor similar to that of clovers or nutmeg, tremetol is actually a complex mixture of sterols and derivatives of methyl ketone benzofuran such as tremetone, dehydrotremetone, hydroxytremetone, and desmethylencecalin (Bonner & Degraw 1962; DeGraw & Bonner 1962; Zalkow & Burke 1963; Zalkow et al. 1977). With very low aqueous solubility, these ketones are similar in chemical structure to the insecticide and piscicide rotenone, and most are likewise toxic to fish. Although tremetone is the predominant ketone, accounting for nearly half of the ketone content by weight, its toxicity has been questioned. Administration of the crude tremetol, but not tremetone, was reported to be toxic to chickens (Bowen et al. 1963). However, subsequent studies seem to substantiate the importance of tremetone as a toxic component (Beier & Norman 1990; Beier et al. 1993). Tremetol causes marked metabolic alterations in chickens, seemingly focused on impairment of the tricarboxylic acid (TCA) cycle and decreased utilization of glucose (Wu et al. 1973). There is an initial increase in blood glucose and free fatty acids, and then a decrease in concert with inhibition of citrate synthase activity. This selective enzyme inhibition results in acidosis and ketonemia characterized by sustained, marked elevation of serum β-hydroxybutyrate concentrations. A hypoglycemia develops without significant alteration of liver glycogen. Based on in vitro experiments, it appears that toxicity is due to hepatic biotransformation of tremetone and perhaps other constituents of the tremetol mixture to the ultimate toxicants and further to nontoxic metabolites such as dehydrotremetone (Beier et al. 1987, 1993). There are also indications that biliary elimination and intestinal reabsorption (enterohepatic recycling) may play a significant role in the cumulative action of the toxicant (Smetzer et al. 1983).

■ disease occurrence mainly in the central and southeastern portions of the plant's distribution

Tremetol concentration appears to decrease slowly when the plant is dried, but intoxications have occurred following consumption of hay. Tremetol content and toxicity potential vary considerably among plants from different areas. Probably for this reason, there is considerable difference among distributional ranges for the species and the disease. *Eupatorium rugosum* is widespread throughout the eastern half of North America and is often very abundant locally; yet the occurrence of disease is much more restricted, occurring mainly in the central and southeastern portions of its range. In many regions, where the plants are common and abundant, few cases of disease are encountered.

■ fat-soluble toxins excreted in milk; may cause illness in calves or humans without signs in the cow

One of the most important aspects of the disease in cattle is the excretion of tremetol in milk. Although this protects the lactating animal, it places calves and humans at risk. Both may develop clinical signs of intoxication prior to or without signs appearing in the cow. The toxin is not necessarily destroyed by pasteurization. Fortunately, however, the commercial practice of mixing milk from numerous sources reduces and almost eliminates the hazard for humans. Because the toxin is fat soluble, butter also is toxic. Although experimental evidence is lacking, there are anecdotal indications of toxicant transmission via consumption of muscle or internal organs of affected animals by carnivores (Wolf et al. 1918; Marsh 1929).

In addition to tremetol, an array of sesquiterpene lactones with cytotoxic or other biological activity are reported for various species of *Eupatorium* (Picman 1986). However, these do not appear to be of concern in *E. rugosum.*

■ many descriptive names for the disease: trembles, sick stomach, tires, slows, swamp sickness, river fever, puking fever, stiff joints, colica dementia

Clinical Signs—The disease syndrome in livestock is known by many descriptive names, although trembles is the most common. The colorful vernacular names, many dating from the 1700s and 1800s, exemplify many of the clinical signs, e.g., sick stomach, tires, slows, swamp sickness, river fever, puking fever, stiff joints, paralysis intestinalis, colica dementia, caeconemia lacemesis, mukosoma, morbeo lacteo, and ergodeleteria. In some instances, these names apply to humans as well where listlessness, leg pains and cramps, vomiting, loss of appetite, constipation, cold extremities, a white coating on the tongue, and an acetone odor to the breath are typical signs.

■ gradual onset, depression, weakness, stiffness, constipation, tremors, acetone odor to breath, collapse

The disease may vary from an acute or subacute to chronic form, depending on the amount and toxicity of the plants ingested. In cattle, the chronic form, characterized by marked weight loss, is most common. The initial signs, which develop after several weeks of plant ingestion, are listlessness and a reluctance to move. Weakness—evidenced by the animal's lying down most of the time—and muscular tremors, when the animal is forced to stand, develop later. Tremors are very prominent and are consistently elicited by forced movement. Later these signs may be accompanied by constipation, apparent joint stiffness, and occasional belligerence. An acetone odor to the breath may be apparent. Terminal collapse and coma occur several days to weeks after clinical signs appear. Signs may appear in nursing calves even in the absence of signs in the cow.

■ sheep and pigs, prominent weakness

Sheep develop similar signs but usually suffer a shorter disease course, with death occurring several days after the onset of signs. Grinding of the teeth and a peculiar jerky type of respiratory difficulty are early signs. Weakness is very prominent; the animal often lies with its head extended and jaws resting on the ground. The signs in pigs are similar.

Horses appear to be particularly susceptible to the effects of *E. rugosum*, often dying 1–2 days after onset of signs. Swallowing difficulties with slobbering and dyspnea are common. A weak unsteady gait, constipation, patchy sweating, and trembling also are seen. Weakness may be very noticeable, with the head held in a low drooping position. Cardiac effects are diagnostically important. There are marked irregularities of cardiac rhythm, jugular vein distension and pulsation, and ventral edema. Electrocardiographic examination may reveal atrioventricular block, premature ventricular contractions, amplitude changes, and other indications of myocardial damage. These effects may be accompanied by appropriate changes in serum muscle enzyme activities, i.e., a marked increase in LDH and CK (Smetzer et al. 1983; Olson et al. 1984).

- horses, disease more acute; sweating, labored respiration, trembling, weakness, cardiac irregularities

Upon ingestion of large amounts of plant material by goats, there may be loss of appetite with depression and, in some animals, head pressing, paddling seizures, and photosensitization.

- goats, large amounts of plant, depression, head pressing, seizures, photosensitization

Pathology—In most animal species, gross and microscopic changes are minimal. There may be nonspecific passive congestion of the brain, liver, and kidneys, with some degree of fatty liver and subepicardial hemorrhages. In contrast, in the horse, pale grayish myocardial streaks with ventricular dilation may be present as distinctive lesions. Microscopically, the pale areas represent myocardial degeneration and necrosis. In goats, severe centrolobular hepatic necrosis is the principal finding.

- horses, myocardial necrosis; goats, hepatic necrosis; otherwise few distinctive lesions

Treatment—Therapy is nonspecific and entails good nursing care, symptomatic treatment, and provision of high-quality forage to sustain the animal for the protracted recovery period. The most important consideration is alleviation of the severe ketosis and acidosis (Couch 1928; Hartmann et al. 1963). This may be tied to administration of parenteral carbohydrates for control of the low blood glucose levels. Activated charcoal should be given (2–3 g/kg b.w. orally) to reduce absorption of the toxin, including that eliminated in the bile (prevention of enterohepatic cycling). It is imperative to prevent access to the plants during the growing season.

- good nursing care, high-quality feed, parenteral glucose, activated charcoal

Flourensia DC.

Taxonomy and Morphology—Comprising about 24 species in North and South America, *Flourensia* is a member of the tribe Heliantheae. In North America, 2 species occur, but only 1 is of toxicologic importance:

F. cernua DC. tarbush, blackbrush, hojase, hojasen, varnish bush

- *F. cernua*

  blackbrush
  hojase
  hojasen
  tarbush
  varnish bush

Plants shrubs; perennials; herbage aromatic with an odor of tar, resinous-glutinous. **Stems** erect or ascending; 100–200 cm tall; highly branched; branchlets puberulent. **Leaves** simple; alternate; blades elliptic; apices acute; margins entire; bases acute. **Heads** small; borne in axils of leaves on short curved peduncles; campanulate; phyllaries in 3-series, imbricate, linear, resinous-glutinous, apices often spreading; receptacles flat or slightly convex; receptacular bracts present, partially enfolding florets and dropping with them. **Ray Florets** absent. **Disk Florets** numerous; perfect; fertile; corollas dull yellow. **Pappus** of 2 ciliate, unequal awns. **Achenes** oblong-cuneate; slightly flattened; silky-glabrous.

- aromatic shrubs; leaves simple, alternate, margins entire; heads small, axillary; disk florets only, yellow

■ dry, sandy or gravelly soils

Distribution and Habitat—*Flourensia cernua* inhabits dry, gravelly or sandy soils of the deserts and plains of the Southwest.

Flourensia cernua

Disease Problems—Causing extensive winter losses of sheep and goats, *F. cernua* became a serious problem in far West Texas in the 1930s and early 1940s (Mathews 1944). Cattle are at risk as well (Dollahite & Allen 1975). The risk of problems appears to be increasing as rangelands deteriorate and *Flourensia* increases in abundance. Fredrickson and co-workers (1994) described the ongoing increased dominance of *F. cernua* in the Chihuahuan Desert. Decline in the quality of grasslands likewise favors it. In the 1960s, there was an extensive die-off of pronghorn in West Texas, caused by severe range depreciation due to drought and overgrazing (Hailey et al. 1966). Typically, little *F. cernua* is browsed by these free-ranging herbivores, but during the drought, those pronghorn that had subsisted almost entirely on the species were readily detected because of the resulting limitations in their locomotion. Postmortem examination revealed tarbush to make up more than 94% of the ruminal contents. The animals also exhibited lesions of excess fluid accumulation, typical of disease caused by *F. cernua*.

■ flowers and fruits are most toxic; mainly eaten only by very hungry animals; digestive disturbance; metabolic derangement

The buds, heads, green achenes, and especially the mature achenes are the main problems. Dollahite and Allen (1975) speculated that the leaves were probably toxic as well, but that it would be difficult for animals to eat enough to cause intoxication. Experimentally, in sheep and goats, 1% b.w. of mature achenes produced severe digestive disturbance and other signs of illness and death within 1–2 days (Dollahite & Allen 1975). A lower dose was of little consequence even if given for several days. Thus the size of the first dose seems to be quite important. Under normal conditions, animals do not eat enough of the mature achenes, whereas hungry animals may feed extensively on them for the 12–24 hours needed to produce the severe effects. When ingested over a long period, the leaves may also cause disease, but in a somewhat different form from that caused by the fruit. In experimental studies, female lambs fed 15% *F. cernua* foliage in their diet for 120 days developed hepatic damage with elevation of serum GGT, AST, LDH, CK, and AP activity (Fredrickson et al. 1994). Larger amounts of plant material, up to 30%, in the diet for only 28 days did not cause problems (King et al. 1996). Consistent with the enzyme changes, detrimental effects on skeletal muscle were apparent, as well as increased deaths from other causes. Interestingly, the dried leaves and heads of *F. cernua* have been used as an herbal treatment for indigestion in humans (Pesman 1962).

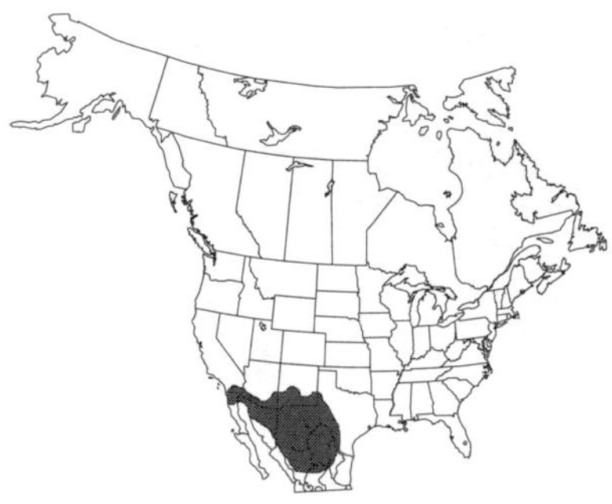

Flourensia cernua

■ specific toxin unknown; possibly polyphenolics

Disease Genesis—The specific cause of the disease produced by *F. cernua* is not known. Although the crude protein levels are high (>18%), there are substantial amounts of polyphenolics in the leaves: 54 mg/g total phenolics and 4 mg/g condensed tannins. These types of compounds may play a significant role when eaten in large amounts.

■ abrupt onset, anorexia, depression, labored respiration, reluctance to move

Clinical Signs—Following ingestion of the fruits, typical signs are loss of appetite, listlessness, depression, labored respiration, and groaning. The animal stands with an arched back and "tucked-up abdomen." It is reluctant to move but also avoids lying down. Death occurs within 24 hours, or the animal exhibits severe illness for a week and then makes a rapid recovery. Long-term ingestion of foliage may result in a milder response, mainly loss of appetite and lethargy. Serum enzymes suggestive of liver and muscle involvement (GGT, AST, LDH, CK, AP) may be elevated.

■ gross pathology, reddened and ulcerated gut mucosa

Pathology—With intoxication due to ingestion of the fruits, reddening and ulceration of the mucosa of the abomasum and small intestine are prominent and may be severe enough to result in perforation. There also may be hemorrhagic effusions in the lungs and hydrothorax. Microscopically, most noteworthy is the endothelial degeneration of pulmonary blood vessels. Long-term ingestion of leaves may cause liver and skeletal muscle damage.

Treatment—Aimed at relief of the signs, treatment is mainly nonspecific nursing care.

■ general nursing care

Gutierrezia Lag.

Taxonomy and Morphology—Comprising 9–30 species, *Gutierrezia,* a member of the tribe Astereae, is a New World genus native to the western United States, southern Canada, Mexico, and temperate South America. The generic name honors Pedro Gutierrez de Salceda, a Spanish botanist working in Mexico. Taxonomists differ markedly in their interpretation of the boundaries of the genus and its relationship to the genera *Xanthocephalum* and *Amphiachyris.* Thus most species have more than one scientific binomial. The traditional view—one large genus—is employed here. Species are primarily weedy, invading overgrazed prairies and pastures and often increasing enormously in abundance (Pieper 1993). The annuals are often referred to as broomweeds and the perennials as snakeweeds or matchweeds. Only the perennial species are of toxicologic significance. The following species are of interest (see plate 13.3):

■ *G. microcephala*
G. sarothrae

broomweed
coyaye
hierba de vibora
matchbrush
matchweed
slinkweed
snakeweed
turpentine weed

G. microcephala (DC.) A.Gray (= *X. microcephalum* [DC.] Shinners)	threadleaf snakeweed, small-headed matchweed, small-headed matchbrush, sticky snakeweed
G. sarothrae (Pursh) Britton & Rusby (= *X. sarothrae* [Pursh] Shinners) (= *G. longipappa* Blake) (= *G. tenuis* Greene)	snakeweed, broom snakeweed, matchweed, matchbrush, turpentine weed, coyaye, perennial broomweed, slinkweed, turpentine weed, hierba de vibora, hierba de San Nicolas

■ subshrubs and shrubs; leaves simple, alternate; heads small, numerous, terminal; ray and disk florets yellow; foliage bright green early in season

Plants subshrubs or shrubs; perennials; flat-topped or corymbiform; from woody rootstocks; herbage usually glutinous, glabrous or scabrous-puberulent. **Stems** tufted; erect or ascending; 10–60 cm tall; profusely branched above. **Leaves** simple; alternate; blades linear to oblanceolate; often punctate; margins entire. **Heads** numerous; small; borne singly or in small glomerules at ends of branches; cylindrical or campanulate or obconic; phyllaries in 3- or 4-series, imbricate, linear-lanceolate, glutinous; receptacles flat; receptacular bracts absent. **Ray Florets** 1–13; pistillate; usually fertile; corollas yellow. **Disk Florets** 1–13; perfect or functionally staminate; fertile or sterile; corollas yellow. **Pappus** of 8–10 white scales. **Achenes** small; obpyramidal; multinerved; glabrous or puberulent.

Early in the growing season, the foliage of both species has a distinctive bright green appearance and is quite aromatic. Late in the growing season, the foliage is a dull tan or gray-brown. The common name *matchweed* reflects the ease and quickness with which the resinous herbage burns.

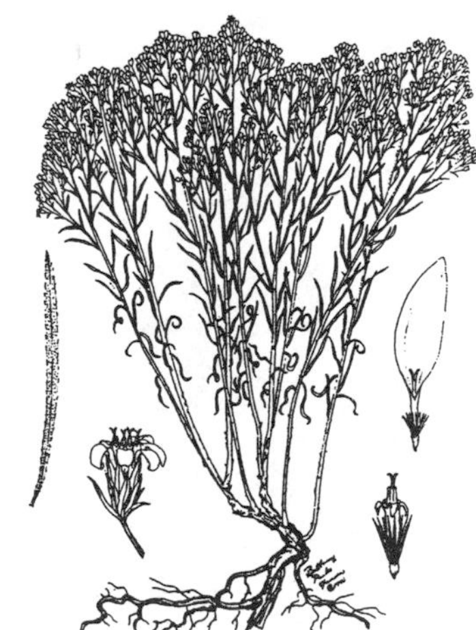

Gutierrezia sarothrae

1. Heads cylindrical. Phyllaries 4–6. Ray florets 1 or 2. Disk florets 1 or 2; sterile . . . . *G. microcephala*
1. Heads obconic or campanulate. Phyllaries 8–21. Ray florets 3–7. Disk florets 2–6; fertile . . . . *G. sarothrae*

■ open plains, dry ranges

Gutierrezia microcephala

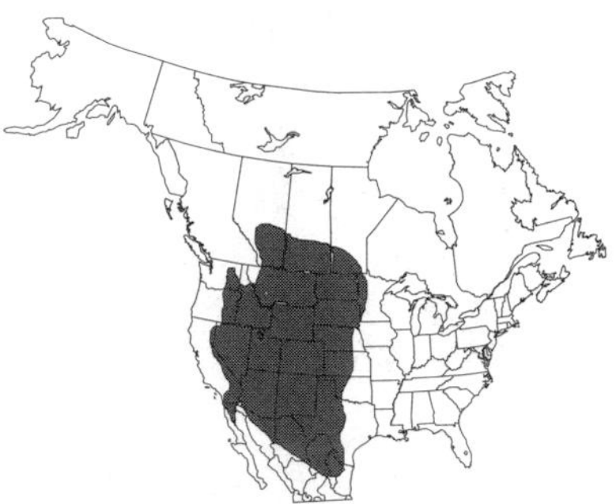

Gutierrezia sarothrae

■ reproductive disease, abortion, weak calves or lambs; impaired reproduction; generalized systemic disease, especially with plants growing on sandy soils

■ heavy plant infestations, up to 16% abortions; calf crop down to 60%

Distribution and Habitat—*Gutierrezia microcephala* occurs in the Southwest, whereas *G. sarothrae* is more widespread, throughout the western half of the continent.

Disease Problems—*Gutierrezia* has had a reputation for causing abortion and premature calves in West Texas since the early 1900s—hence the common name *slinkweed*. In addition, mature animals also may show signs of disease. However, there were many inconsistencies in disease occurrences. On many ranges, species of *Gutierrezia* are an increasingly dominant forb but are typically little eaten (Rosiere et al. 1975; Hakkila et al. 1987; Harrington & Pieper 1993). In addition, experimental reproduction of the disease was difficult, and early studies in which the plant was given at low dosage for 50–60 days were negative (Schmidt 1931; Boughton & Hardy 1933). Producers, however, continued to experience problems that they attributed to this genus, but only from specific locations; in other areas where the plant was plentiful, adverse effects were not common. There was also considerable variation in disease occurrence from season to season and year to year. The reality of the perceived difference in disease incidence between years was subsequently borne out in experimental studies in which results were negative in 1933 but distinctly positive in 1934 (Mathews 1936). In these studies, 3–10% b.w. of green plant given over 3–5 days was lethal in sheep and cattle. Follow-up studies in the 1950s not only confirmed the acute toxicity but also clearly established an association between *G. microcephala* and abortion and/or the birth of small, weak calves (Dollahite & Anthony 1955, 1956, 1957). At least some of the variability in occurrence of disease was accounted for by the apparent increased risk from plants growing in sandy soils as opposed to those from loamy soils, although this may be too broad a generalization. Pregnant cows given about 0.5 kg/day of the plants for several weeks delivered small, weak calves or aborted small, near-term fetuses. In many instances, plants were eaten for 2–3 months prior to evidence of reproductive effects. However, stage of pregnancy may be a factor as well as duration of dosage. The same or slightly larger dosage was associated with acute disease and sometimes death of the cow. Sheep and cattle seem to be at greatest risk and goats somewhat less so. Other species, including pigs and laboratory animals, are also susceptible (Matthews 1936; Dollahite & Allen 1959).

Although generally considered to be unpalatable, *Gutierrezia* is sometimes eaten by livestock without their being forced to do so (Pieper 1990; Harrington & Pieper 1993). The magnitude of the problem is illustrated by the results of a survey of 148 counties in central and West Texas in 1984 (McGinty & Welch 1987). About one-third of the rangeland was infested with *G. microcephala* and/or *G. sarothrae*, and nearly half of these infested rangelands were associated with disease losses. The High Plains and Trans-Pecos were areas of especially severe problems. Losses due to death and abortion were estimated at 21,120 and 24,550 for cattle and 12,270 and 13,900 for sheep, respectively. The economic significance is further illustrated by a study of the effects of *G. sarothrae* on cow-calf operations in the southern plains (Carpenter et al. 1991). With heavy plant infestations, abortion rates were as high as 16.6% and weaning weights decreased as much as 20.6%. In the typical cow-calf operation, the percentage of calf crop at birth was reduced from 85 to 71.9% with medium infestation and to 59.1% with heavy infestation; medium infestation was about half the number of plants of heavy infestation. Similarly, weaning weights were decreased from 198.5 kg to 176 kg for medium and to 157.5 kg for heavy infestations. Cow deaths increased only with heavy infestations, from 1 to 4.7%. No adverse effects were detectable when plant infestations were light (plant number 20% of heavy). The lack of effect of low infestations is consistent with experimental studies in which dried herbage of *Gutierrezia* at 20–30% of the diet did not impair reproduction in sheep or cattle (Martinez et al. 1993). That there may be economic considerations even in the absence of distinct reproductive and acute intoxication effects is borne out by the small but significant decrease in weight gain in breeding-age heifers with small amounts (0.58–1.16 kg/day) of *G. microcephala* in the diet (Williams et al. 1992). This seems to occur because of some factor other than simply a decrease in feed intake. Ruminal pH is decreased somewhat, but there is little apparent effect on ruminal volatile fatty acids (Edrington et al. 1993c). Studies evaluating factors affecting the occurrence of disease confirm variation in toxicity and importance of soil type, and increased risk with new growth (Edrington et al. 1991a, 1992; Williams et al. 1992). Control of the disease through biological control of plant populations continues to be explored and may yet become reality (Cordo 1985).

Disease Genesis—A number of different types of possible toxicants have been isolated from species of *Gutierrezia*. Early studies indicated that saponins might be responsible for the reproductive effects observed. Following parenteral administration of a crude saponin extract from *G. microcephala*, it was noted that periodic vulvar swelling and marked mammary swelling preceded abortions (Dollahite et al. 1962). The extract contained an unsaturated triterpenoid sapogenin with one or more hydroxyl and carboxyl groups (Shaver et al. 1963). This saponin extract was given orally to mice and rabbits and in vitro on uterine smooth muscle. It produced abortions in rabbits but failed to increase uterine weight in mice, as would be expected with an estrogenic toxicant. In addition, it induced or promoted uterine contractions, in contrast to the commercially available saponin, which produced abortion when given parenterally. Thus the investigators concluded that there was reasonable evidence to suspect saponins as the cause of *Gutierrezia*'s toxicity and that the saponins produced their effects by a mode other than estrogenic action.

- toxicants unknown, steroidal effect of saponins? monoterpenes? diterpene acids as found in *Pinus*?

The structural similarities of saponins to steroids has led to continued speculation about a role for the latter in causing abortion or premature parturition. This led to an extensive series of studies at New Mexico State University employing rats and sheep (Flores-Rodriguez et al. 1989; Edrington et al. 1990; 1991b, 1993a,c). In rats, which appear to be very sensitive to *Gutierrezia,* there are effects on both the male and female. In the male, when fed plant material for more than 20 days, the number of abnormal sperm increased and fertility decreased. In the female, progesterone at mating decreased, ovulation decreased, and the number of live births decreased, while numbers of dead embryos increased. Although decreased feed intake and malnutrition resulted from *Gutierrezia* intake, they were not factors in the embryonic mortality in rats. Uterine weight also decreased, a finding that supports earlier reports noting a lack of estrogenic activity with *Gutierrezia* extracts (Shaver et al. 1963; Edrington et al. 1993b). There seem to be no residual effects on rat pups born of dams fed *Gutierrezia,* as shown by subsequent reproductive trials (Staley et al. 1995). Suppression of steroid production may occur because of the hypocholesterolemic activity of the toxicant similar to that shown for saponins in general (Sidhu & Oakenfull 1986). Furthermore, marginally toxic doses of *Gutierrezia* for 30 days in ewes resulted in slightly increased estrous cycle length and in decreased estrus detection and breeding by rams (Oetting et al. 1990). In contrast to the effects in rats, levels of blood progesterone were not decreased in ewes.

In spite of the association of disease with saponins and considering the ubiquitous nature of these compounds in plants, the search has continued for alternative possibilities as the definitive toxins. An array of volatile monoterpenes have been identified in *Gutierrezia* (Molyneux et al. 1980; Epstein & Seidel 1989). There seem to be marked seasonal, yearly (1.24–2.7% dry weight), and site variations in monoterpene content and composition. In some instances, pinenes are quite low, whereas in others they are the predominant types. It is suggested that site variations may be due to the occurrence of ecotypes that could also be a factor in expression of disease. Labdane resin acids and furano-diterpene acids are present. Reminiscent of the bioactive compounds of *Pinus,* several diterpene acids, including polyalthic acid, have been isolated from the exudate from trichomes of *G. sarothrae* (Roitman et al. 1994). There is considerable intersite variation for these compounds as well. Numerous highly oxygenated flavonals are also present, some of which have biological activity (Roitman & James 1985). The role of these or other alternative toxicants remains to clarified. Interestingly, the sesquiterpene lactones so common among the genera of this family are not present in *Gutierrezia* (Seaman 1982).

- acute systemic disease, nasal discharge, weakness, anorexia; reproductive disease, vulvar swelling and mammary development long before expected parturition, abortion, stillborns, birth of small weak calves or lambs

Clinical Signs—The acute disease produced by *Gutierrezia* in the adult is manifested as a crusting of the nasal pad, mucopurulent nasal discharge, decreased appetite, weakness and listlessness (head low and back arched), and sometimes hematuria and icterus. A stringy mucoid secretion from the vulva may be seen, even in nonpregnant animals. Diarrhea and/or constipation with foul-smelling mucus containing feces is also common. Clinical chemistry changes include a decrease in serum cholesterol and iron and increased creatinine and bilirubin (both direct and indirect). There also may be a mild increase in serum hepatic enzymes (AST, GGT, AP). These signs may terminate in death within a few days.

The more typical reproductive effects of *Gutierrezia* are preceded by vulvar swelling and mammary development several weeks or months prior to expected parturition. Within a few days there follows abortion, stillbirth, or the premature birth of small weak calves, which often survive less than 48 hours. Typically calves of 12–20 kg b.w. are seen. Retention of placental membranes is common.

- gross pathology, icterus, mild reddening of gut mucosa

Pathology—In the cow, grossly there may be icterus, swollen, pale yellowish brown kidneys, a mottled yellowish liver, few scattered splotchy hemorrhages, and mild to moderate reddening of the mucosa of the stomach and small intestine. Microscopically, there will be mild renal tubular flattening and dilation, interstitial hemorrhage, hepatic necrosis, and myocardial degeneration.

- general nursing care

Treatment—There is no specific treatment. Good nursing care remains the appropriate approach. A high level of nutrition may provide protection or alleviation of many of the adverse effects associated with ingestion of *Gutierrezia* (Williams et al. 1992; Smith et al. 1994; Staley et al. 1996). Experiments in rats demonstrated a lack of protection for progesterone but fortuitously revealed a beneficial effect for the carrier, safflower seed oil, when administered parenterally (Chambers et al. 1993; Smith et al. 1993). The benefit may in part be due to the nutritional value of the oil. The possible use of this discovery via incorporation of the oil in livestock feeds remains to be elucidated. Activated charcoal (1–1.5g/kg b.w.) may be of value in preventing toxicosis when fed as a supplement (Poage et al. 2000). Administration of progesterone or vitamin A is not effective in alleviating the reproductive effects.

Helenium amarum

Helenium L.

Taxonomy and Morphology—Comprising about 40 species native to the New World, *Helenium,* a member of the tribe Helenieae, contains both weeds and popular ornamentals cultivated for their showy heads with a range of bright colors. The generic name is from the Greek *helenion* and is believed to honor either Helen of Troy or Helenus, son of Priam, the last Trojan king. Species are generally known as the sneezeweeds. The most commonly encountered species of toxicologic significance are the following (see plate 13.4):

H. amarum (Raf.) H.Rock	bitter sneezeweed, Spanish daisy
H. autumnale L.	common sneezeweed, autumn sneezeweed, sneezewort, staggerwort, staggerweed, yellow star, oxeye, false sunflower, swamp sunflower
H. flexuosum Raf. (= *H. nudiflorum* Nutt.)	purplehead sneezeweed, southern sneezeweed
H. hoopesii A.Gray (= *Dugaldia hoopesii* [A.Gray] Rydb.)	orange sneezeweed
H. microcephalum DC.	smallhead sneezeweed
H. quadridentatum Labill.	rosilla

■ *H. amarum*, *H. autumnale*, *H. flexuosum*, *H. hoopesii*, *H. microcephalum*, *H. quadridentatum*

bitterweed, false sunflower, oxeye, rosilla, sneezeweed, sneezewort, Spanish daisy, staggerweed, staggerwort, swamp sunflower, yellow star

Plants herbs; annuals or biennials or perennials; from fibrous roots or taproots or crowns; herbage often aromatic when crushed; glabrous or variously indumented. **Stems** erect or ascending; 10–150 cm tall; branched or not branched; winged or not winged. **Leaves** simple; alternate or appearing fascicled; sessile or subsessile; blades linear-filiform to lanceolate or spathulate; margins entire or pinnatifid or laciniate. **Heads** showy; numerous; borne solitary or in corymbs at ends of elongate peduncles; globose or hemispheric; phyllaries 16, in 2-series, lanceolate, glabrous to pubescent, glandular-punctate, reflexed in fruit; receptacles spherical; receptacular bracts absent. **Ray Florets** absent or present; 8–10; typically reflexed at maturity; pistillate or neutral; fertile or sterile; corollas 3-lobed, yellow or red to reddish brown or orange at base. **Disk Florets** numerous; perfect; fertile; corollas yellow or yellow-green or brown or purple. **Pappus** of 5–10 scales; scarious or hyaline; typically awn-tipped; rarely absent. **Achenes** obpyramidal; 4- or 5-ribbed; sericeous.

■ herbs, some aromatic; stems some winged; leaves simple, alternate; ray florets when present 3-lobed, yellow, orange, or reddish brown; disk florets yellow to brown or purple

Helenium autumnale

1. Plants annuals; from taproots.
   2. Stems not winged. Leaves linear-filiform; margins pinnately lobed . . . . *H. amarum*
   2. Stems winged. Leaves elliptic or linear-lanceolate; margins entire or serrate- or dentate-undulate.
      3. Heads borne among uppermost leaves. Leaves elliptic; margins serrate- or dentate-undulate . . . . *H. microcephalum*
      3. Heads borne above uppermost leaves. Leaves lanceolate; margins entire . . . . *H. quadridentatum*
1. Plants perennials; from fibrous roots or short rhizomes or crowns.
   4. Stems not winged. Pappus scales 2.5–4 mm long . . . . *H. hoopesii*
   4. Stems winged. Pappus scales 0.5–2 mm long.
      5. Disk florets purple or brownish purple. Ray florets neutral, sterile, or absent . . . . *H. flexuosum*
      5. Disk florets yellow. Ray florets pistillate, fertile . . . . . . *H. autumnale*

Distribution and Habitat—Species of *Helenium* are adapted to a variety of habitats and soil types. Occurring in the eastern half of the continent with greatest abundance in the southeast quarter, *H. amarum* is a dominant of the early stages of secondary succession and a classic indicator of disturbed soil and overgrazing. It often forms large populations. *Helenium flexuosum* is likewise distributed in the eastern half of the continent in moist and disturbed sites. In contrast, *H. hoopesii* is distributed in the west on slopes and in meadows at higher elevations. Typically abundant in

■ variety of habitats

overgrazed areas, *H. microcephalum* and *H. quadridentatum* are southwestern species. Occurring across the northern half of the continent in moist or low wet sites, *H. autumnale* has the broadest distributional range.

■ metabolic derangement; digestive disturbance; neurologic and cardiotoxic effects; large dosage, acute disease; low dosage, cumulative action with abrupt or gradual onset in up to several weeks; milk taint

Helenium hoopesii

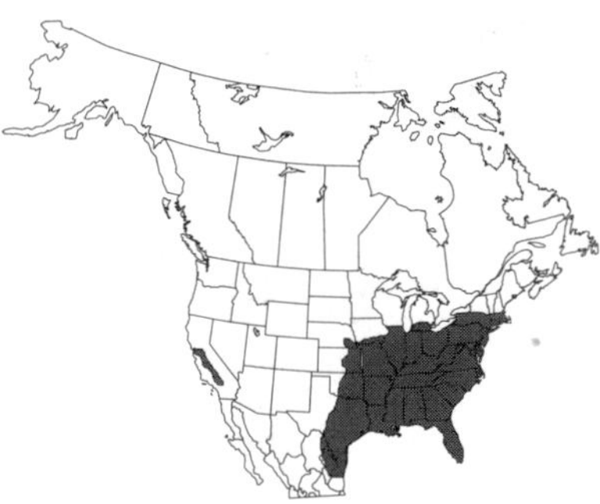

Helenium amarum

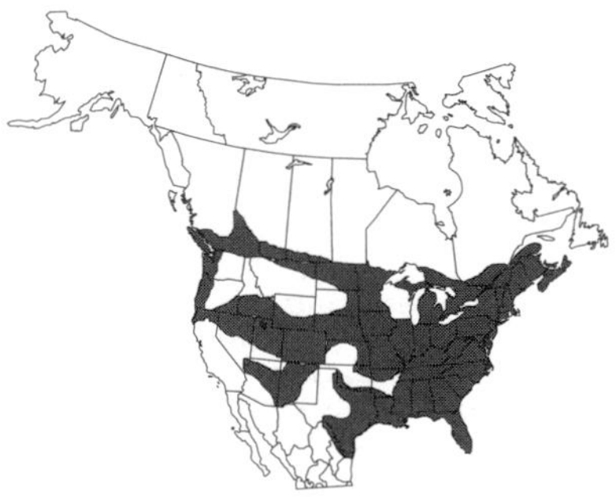

Helenium autumnale

Disease Problems—*Helenium autumnale* was the first species to be associated with biological effects. In the 1800s it was recommended for a variety of medicinal purposes, e.g., as a tonic, a cold remedy, and a control for fever (Millspaugh 1974). Its use in colonial times for decoration was associated with sneezing and irritation of the nose and mouth. Because it was rarely eaten, the species was seldom a toxicologic problem in livestock, although there are a few reports of neurologic problems in horses, mules, and sheep (Phares 1889; Lamson 1913). Experimentally, when given orally to dogs, it caused diarrhea (with straining), vomiting, and death (Lamson 1913). There were also effects on the heart—in vitro studies showed paralysis of heart muscle.

By the early 1900s, a disease of high-altitude summer ranges in the Rocky Mountains began to be recognized as a serious problem in sheep. It was called spewing disease because of the purported regurgitation and vomition (Marsh et al. 1921). Originally attributed to *Zigadenus,* in the lily family, the effects were later shown to be caused by *H. hoopesii,* and the regurgitation and vomition were discovered to be an atypical sign occasionally seen in the later stages of the subacute or chronic form of the disease. These findings were made in the first investigations conducted at the newly established USDA experiment station at Salina, Utah, in 1915. By the 1940s the disease was estimated to cause 8,000 deaths per year and to reduce significantly the productivity of sheep grazing the 1.5 million acres infested with *H. hoopesii* in Colorado (Doran & Cassady 1944). It was of similar seriousness in Utah.

Because of the similarity in the diseases caused by *H. autumnale* and *H. hoopesii,* other species of the genus were carefully evaluated for toxicity. *Helenium microcephalum* appears to be the most toxic; a dose of 0.125% b.w. of mature green stems and leaves causes illness, and a dose of 0.25% b.w. is lethal, whether given as a single dose or divided into two doses given 24 hours apart (Boughton & Hardy 1939, 1940; Dollahite et al. 1964). The leaves of the winter rosettes seem to be less toxic than the stems and leaves of mature plants. *Helenium autumnale* and *H. flexuosum* are intermediate in toxicity; *H. hoopesii* and *H. amarum* are lower (Phares 1889; Marsh et al. 1921; Dollahite et al. 1973). *Helenium amarum* is of concern mainly because it is a cause of bitter-tasting milk, butter, and meat. The bitterness of the meat has, on occasion, prompted charges of attempted food poisoning. Paradoxically, it had, at one time, a reputation as a tonic for promoting digestion when taken in small amounts (Phares 1899).

The toxic effects of the species of *Helenium* are cumulative; 1.5–2% b.w. or more of green *H. hoopesii* daily for several weeks results in illness and sometimes death (Marsh et al. 1921). Although dosage below 1.5% b.w. did not result in disease, it is not clear that there is a dose below which cumulative effects do not eventuate. A similar amount is required of *H. amarum,* given for several days to produce intoxication (Dollahite et al. 1973). *Helenium quadridentatum* was thought to be of low toxicity (Phares 1889), but plants in Cuba produced acute toxicity when 3–4% b.w. of plant were given experimentally to 45-day-old calves (Alfonso et al. 1986).

All species are unpalatable and are eaten only under special circumstances. Thus those species that grow in situations where alternative forage is limited or where animals become less selective pose a risk. This accounts for the extensive losses from *H. hoopesii,* which is of relatively low toxicity but is grazed under circumstances sometimes encouraging ingestion. Because of this relatively low toxicity potential, disease control is amenable to management. Prevention of the disease can be accomplished by open, quiet herding, slow grazing, uniform range use, careful selection of bed grounds, and careful use of salt to avoid indiscriminate feeding (Cassady 1940; Doran & Cassady 1944). It is also very important to limit range use to times of plentiful fresh forage.

Helenium flexuosum

■ pseudoguaianolide sesquiterpene lactones

Disease Genesis—Sesquiterpene lactones are implicated as a cause of disease for a number of genera in the Asteraceae, including *Helenium* (Picman 1986). Seaman (1982) lists 1,350 sesquiterpene structures from all but two tribes of the family. These compounds, accumulating in glandular trichromes, have considerable taxonomic value. The toxic types found in the tribe Helenieae (now considered to be part of the Heliantheae by some taxonomists) are pseudoguaianolides of complex structure, closely related to furanoeremophilanes (Herz et al. 1963; Herz 1968). Those present in *Helenium* are the *trans*-fused helenanolides and seco-helenanolides. Helenalin, one of the more common of the toxic types, is found in the following species: *H. amarum, H. aromaticum, H. autumnale, H. campestre, H. chihuahuensis, H. laciniatum, H. mexicanum, H. microcephalum, H. puberulum, H. quadridentatum, H. scorzoneraefolia,* and *H. vernale* (Romo & de Vivar 1967; Herz 1978; Seaman 1982). Additional helenanolides are found in most species of *Helenium,* and in a few species these may be present without helenalin, e.g., neohelenalin (mexicanin D) and flexuosins A and B in *H. flexuosum,* and carabone and linifolin in *H. integrifolium* (Vivar et al. 1987). *Helenium microcephalum* contains an array of the lactones, including a novel dimeric type, microlenin acetate (Imakura et al. 1980). Tenulin is also common to many species, but only in *H. amarum* is it the major helenanolide responsible for the toxic effects. In addition to being toxic, tenulin is very bitter and is excreted in very small amounts (>0.1%) in milk; it imparts a bitter taste for several hours after the milk is ingested (Ivie et al. 1975b). Interestingly, hymenoxon from *Hymenoxys* is also found in low concentrations in *H. hoopesii* (Ivie et al. 1976; Hill et al. 1977).

Helenium hoopesii

Helenanolide concentrations in most species are in the range of 0.5–3% d.w. of plant material. Typically concentrations are highest in the flowers, except for tenulin concentration, which is highest in the leaves (Kim et al. 1987). Based upon intraperitoneal LD_{50}'s in mice, mexicanin E (microhelenalin D) is most toxic (3.1 mg/kg b.w.), compared with helenalin (9.9 mg/kg b.w.) and tenulin (184.6 mg/kg b.w.) (Kim et al. 1987). In vitro assays of cytotoxicity show the helenanolides and mexicanins to be apparently of similar potency (Beekman et al. 1997). For livestock, the lethal doses for helenalin (100–125 mg/kg b.w.) and hymenoxon (75–100 mg/kg b.w.) are similar, in contrast to that for tenulin, which is 1 g/kg b.w. or more (Ivie et al. 1975a,b; Witzel et al. 1976; Terry et al. 1981). Thus it is not surprising to find a similar range for lethal doses of the plants containing these toxicants. In spite of the differences in toxicity, with sufficient dosage the signs produced by the various helenanolides are similar to one another, including those caused by hymenovin.

Helenium microcephalum

The highly reactive α-methylene and cyclopentanone groups of the helenanolides cause a variety of effects. Sesquiterpene lactones with the α-methylene group inhibit smooth muscle contractility (Hay et al. 1994). Some of the helenanolides are potent cytotoxins, and their potential antineoplastic effects have provided insights into mechanisms of action. Cytotoxicity appears to be directly related to alkylation of sulfhydryl groups by O=CC=CH—reactive sites provided by the exocyclic α-methylene-γ-lactone and/or cyclopentanone moieties of the helenanolides (Kupchan et al. 1970, 1971; Lee et al. 1971; Hall et al. 1977). Reaction with cysteine and other thiol compounds markedly reduces activity (Hall et al. 1989). The α-methylene-γ-lactone group also is involved in degranulation of mast cells and inhibition of adenylate cyclase

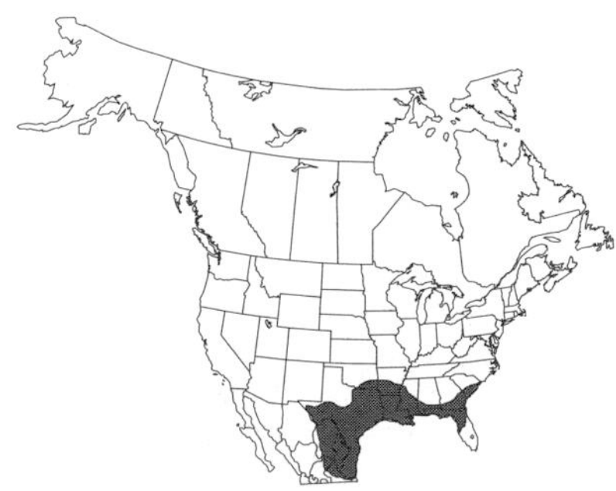

Helenium quadridentatum

- metabolic derangement, inhibition of numerous enzymes of aerobic and anaerobic metabolism; additional neurologic, cardiac, and other effects

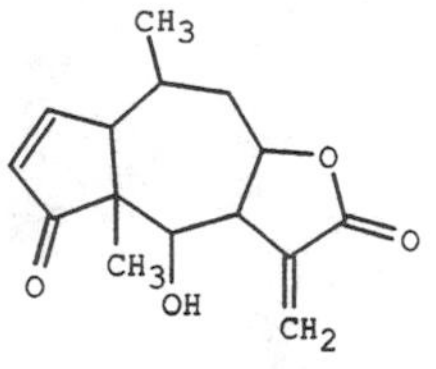

Helenalin

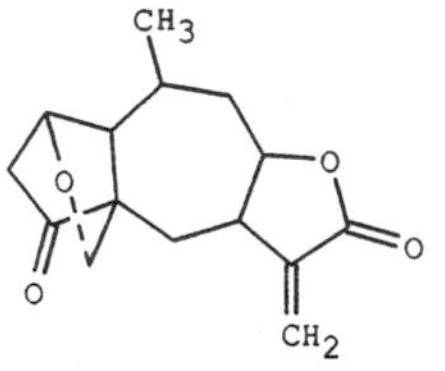

Microhelenalin

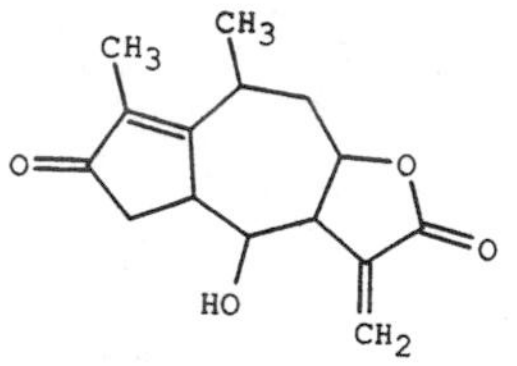

Neohelenalin or mexicanin D

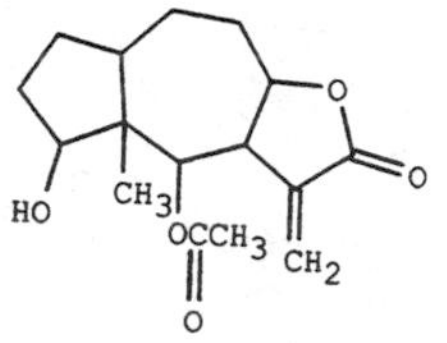

Flexuosin A

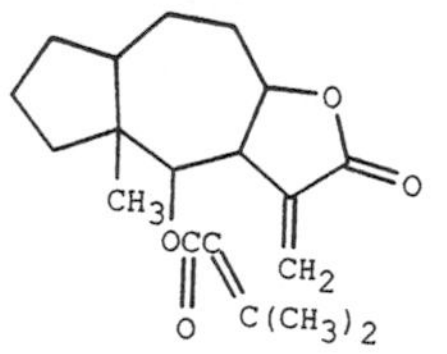

Flexuosin B

Tenulin

caused by these compounds. Tenulin, which lacks this moiety, has low activity relative to the activities of helenalin and hymenoxon (Elissalde et al. 1983; Elissalde & Ivie 1987). Variations in structure may account for some of these differences in activity. Helenalin contains both reactive sites, tenulin only the cyclopentanone, and hymenoxon the exocyclic methylene plus a reactive dihemiacetal in place of the cyclopentanone (Manners et al. 1978). In vivo systemic mechanisms are similar in that toxicity due to the helenanolides is accompanied by depletion of glutathione. An artificial elevation of glutathione protects against toxicity for a few hours (Merrill et al. 1985, 1988). Moreover, as discussed in the treatment section of *Hymenoxys* in this chapter, thiols are intimately involved with prevention. However, effects are not limited to thiol binding, because sulfhydryl inhibitors do not diminish hymenoxon binding in erythrocytes, and mutagenicity is apparent only for hymenoxon, presumably due to the reactive dihemiacetal (Manners et al. 1978; Hill et al. 1980). Furthermore, hymenoxon is a more potent degranulator of mast cells than helenalin, perhaps due to the influence of the dihemiacetal (Elissalde et al. 1983).

The overall action of the helenanolides seems to be directed mainly toward inactivation of essential metabolic enzymes. Cytotoxicity does not appear to be due to lipid peroxidation (Merrill et al. 1985). Key enzymes affected include phosphofructokinase, hexokinase, glycogen synthetase, aldolase, phosphorylase A, glucose-6-phosphatase, and succinic dehydrogenase (Lee et al. 1977; Hall et al. 1978; Gaspar et al. 1986). Thus, there are marked effects on both aerobic and anaerobic metabolism. There are additional effects on many other enzymes, including those involved with nucleotide synthesis and function (Lee et al. 1977; Page et al. 1987). There also are effects on mitochondrial ATPase activity and oxidative phosphorylation (Narasimhan et al. 1989).

Helenanolides containing the exocyclic methylene interfere with cyclic AMP function by inhibiting adenylate cyclase (Elissalde & Ivie 1987). Of interest is that cardiac effects with high dosage also are related to cyclic AMP; however, in this situation there is an elevation resulting in increased calcium influx into the myocardium and a positive inotropic effect (Willuhn & Rottger 1982; Takeya et al. 1983; Itoigawa et al. 1987). Although the underlying mechanism of intoxication seems to be related to enzyme inhibition, there are substantial contributions from other actions. Calcium entry in other cells is facilitated at lower dosage and inhibited at higher dosage, but not via an effect on protein kinase C (Powis et al. 1994). In acute intoxications, cholinergic and cardiovascular derangements are important (Lamson 1913; Szabuniewicz & Kim 1972). Treatment directed at the cholinergic stimulation, neurologic depression, and cardiac depression and arrhythmias may provide some immediate relief. These other effects are of lesser importance in the more chronic cases, in which the insidious effects of metabolic disturbances predominate.

The influence of mixed-function oxidases (MFOs) and other biotransformation pathways on toxicity potential is not clear. Most of the toxic effects are due directly to the parent helenanolides, including the mutagenic activity of hymenoxon (MacGregor 1977). Inducers and inhibitors of biotransformation reportedly have little influence on toxicity, although pretreatment with carbon tetrachloride, which would affect other pathways, does afford some protection (Jones & Kim 1981). The lack of effect on these pathways occurs in spite of the dependence on biotransformation in the liver and excretion in urine for clearance of these toxicants (Hill et al. 1980; Terry et al. 1983). However, even if toxicant disposition is not influenced by alterations in these processes, there is the potential for helenanolides to act on and alter biotransformation of other compounds. This is borne out by the decrease in P450 content or activity and in MFO pathways such as alkylresorufin *O*-dealkylation reported for *Helenium* and helenalin (Dalvi & McGowan 1982; Chapman et al. 1988; Eissa et al. 1995, 1996). Thus the helenanolide-containing plants may have profound effects on the disposition of other toxicants.

Clinical Signs, Pathology, and Treatment—The clinical signs, pathological changes, and nature of treatment are essentially the same as those for *Hymenoxys* and are described in that section in this chapter. *Helenium hoopesii* is less likely to produce an abrupt onset of the disease than are other species of *Helenium*. Changes reported for *H. hoopesii* include a 10-fold increase in AST activity in subacute cases (Buck et al. 1961).

■ disease as for *Hymenoxys*

Hymenoxys Cass.

Taxonomy and Morphology—Exclusively a New World genus, *Hymenoxys,* a member of the tribe Helenieae, comprises 27–30 species classified in 3 subgenera that are regarded by some taxonomists as distinct genera. The generic name is derived from the Greek roots *hymen* and *oxys,* meaning "membrane" and "sharp" and alluding to the pointed pappus scales. Most of the species in North America are of little toxicologic significance; some were used as chewing gum by Native Americans. Intoxication problems have been associated with 4 species (see plate 13.5):

■ *H. lemmonii*
H. odorata
H. richardsonii
H. subintegra

bitterweed
hymenoxys
pingue
rubberweed

H. lemmonii (Greene) Cockerell	alkali hymenoxys, Lemmon's hymenoxys
H. odorata DC.	bitterweed, bitter rubberweed
H. richardsonii (Hook.) Cockerell	Colorado rubberweed, pingue
H. subintegra Cockerell	

Plants herbs; annuals or biennials or perennials; from taproots or branched woody caudices or rarely rhizomes; herbage aromatic when crushed; dotted with microscopic globules of resinlike exudate. **Stems** erect or ascending; 10–60 cm tall; branched or not branched; sulcate. **Leaves** simple; alternate; primarily basal; blades of various shapes; punctate; margins entire to pinnatifid. **Heads** showy; numerous; borne solitary or in corymbs at ends of elongate peduncles; hemispheric or cylindrical; phyllaries in 2- or 3-series, similar or different, outer rigid, typically fused at base, inner coriaceous, free, elliptic; receptacles conical; receptacular bracts absent. **Ray Florets** 5–35; 3-lobed; typically reflexed at maturity; pistillate; fertile; corollas yellow, becoming white. **Disk Florets** numerous; perfect; fertile; corollas yellow. **Pappus** of 5–8 scales; hyaline; aristate or awned. **Achenes** obpyramidal; sericeous.

■ aromatic herbs; resinous exudate; leaves simple, alternate; heads showy, ray and disk florets yellow, rays 3-lobed

1. Plant bases conspicuously long woolly-villous. Branching woody caudices present . . . . *H. richardsonii*
1. Plant bases glabrous or canescent. Branching woody caudices absent.
    2. Leaves entire or with 3 small lobes. Leaves sericeous or strigose. Pappus scales 3 mm long . . . . *H. subintegra*
    2. Leaves pinnately lobed. Leaves glabrous or glandular-scurfy or canescent. Pappus scales 1–2.5 mm long.
        3. Plants biennials or short-lived perennials. Leaves succulent. Corollas of ray florets 10–20 mm long . . . . . *H. lemmonii*
        3. Plants annuals. Leaves not succulent. Corollas of ray florets 5–10 mm long . . . . *H. odorata*

Hymenoxys odorata

■ mainly open sites

Hymenoxys richardsonii

Hymenoxys lemmonii

■ generally unpalatable; serious losses mainly in sheep, due to *H. odorata*; disease the same as with species of *Helenium*

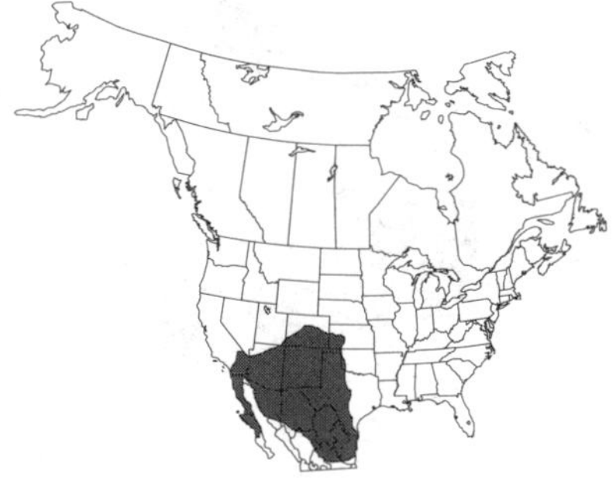

Hymenoxys odorata

Distribution and Habitat—The center of distribution of *Hymenoxys* is in the Rocky Mountains and plains, with species distributed from southwestern Canada southward to the state of Veracruz in Mexico. A second center is in Peru and Bolivia in South America. Typically occurring in soils derived from limestone or igneous formations, most species are widely distributed and often dominate extensive local areas. *H. lemmonii* is a species of moist or wet alkaline meadows up to 2,700 m elevation. *Hymenoxys richardsonii* is a forb of higher elevations, to 3,100 m, in dry desert sagebrush communities. *Hymenoxys subintegra* is limited in distribution to dry meadows and openings in the pine forests of southern Utah and northern Arizona. A weed of overgrazed plains, rangelands, and roadsides, *H. odorata* occurs throughout the Southwest. It does not compete well with grasses.

Disease Problems—In the early 1900s a serious problem of range deterioration developed in Texas that was related to the marked increase in the number of grazing livestock. Between 1913 and 1928, there was a 3-fold increase in sheep and a 2-fold increase in goats (Hardy et al. 1931). As the range deteriorated, *H. odorata* flourished. The first losses due to *H. odorata* were reported in the early 1920s, and thereafter the problem became very serious, with exceptional losses in 1925 and 1926 (Clawson 1931; Rowe et al. 1973). At about the same time, *H. richardsonii* emerged as a problem in the Southwest (Pammel 1919a; Parker 1936). Of lesser importance, *H. lemmonii* was recognized as a problem in the same region (Fleming et al. 1934a,b). Although containing the same toxicants, *H. subintegra* is less abundant and does not pose the same risk as the other three species. *Hymenoxys odorata* has since been recognized as the most important of the pseudoguaianolide-containing species because of the devastating losses it causes year after year in the Edwards Plateau area of Southwest Texas, where annual losses are estimated to be in the millions of dollars (Rowe et al. 1973). The magnitude of the problem is exemplified by the continued efforts directed toward developing treatment and prevention protocols to reduce ongoing losses.

Species of *Hymenoxys* are generally unpalatable, but animals may eat them in the absence of preferred forage and even may develop a taste for them under such conditions (Rowe et al. 1973). They are said to be most toxic at or near the time of flowering, which occurs in the spring for *H. odorata* and summer for *H. richardsonii* and *H. lemmonii*. Losses are primarily in the winter or early spring for *H. odorata* and spring or fall for the others. Sheep and goats are at greatest risk, but cattle and horses may also be affected, whether the plants are eaten fresh or dried in hay (Camp 1985). The acute LD_{50} of green plants for sheep is 1.3% of b.w. (or 0.3–0.8% d.w.), but during times of drought this may drop to 0.5% b.w. (Clawson 1931; Boughton & Hardy 1937; Rowe et al. 1973). This difference seems to be related to the physiological condition of the plant rather than the animal's nutrition. The level of nutrition, excluding protein levels, is considered not to be a factor, although ensuring that the animal has adequate calcium and phosphorus has been suggested as a nonspecific preventive of the disease (Parker 1936; Rowe et al. 1973). The estimated LD_{50} for subacute repeated ingestion of smaller amounts of the whole dried plant in sheep ranges from 2.9% to more than 8.5% b.w. (Rowe et al. 1973). Goats are less susceptible, requiring nearly 2 times more plant material. *Hymenoxys richardsonii* and *H. lemmonii* are of similar toxicity; 1.5% and 1% b.w. of green plant, respectively, cause toxicity in several days or weeks (Fleming et al. 1934a; Aanes 1961).

Depending upon the rapidity of plant consumption, three forms of the disease are observed. Relatively uncommon, the acute form causes severe clinical signs and death in 24–48 hours. In the more common subacute form, signs are less severe and death occurs in 4–15 days. In the third, more chronic form, there are few distinct clinical signs, but eventual death due to starvation and dehydration is seen when small amounts of *Hymenoxys* are consumed over a prolonged period of time.

- acute disease, rare, death in 1–2 days; subacute disease, most common, death in 4–15 days; chronic disease, gradual starvation

Disease Genesis—As in the closely related genus *Helenium*, the toxins in *Hymenoxys* are a series of pseudoguaianolide sesquiterpene lactones, including hymenoxon, hymenoxynin (a glucoside), hymenolide, vermeerin, paucin, and floribundin (Herz et al. 1970). Some of these lactones are quite novel. The glandular trichromes of some species such as *H. richardsonii* and *H. subintegra* contain nearly a dozen lactones (Ahmed et al. 1995). Similar lactones are found in many other genera closely related to *Hymenoxys* (Spring et al. 1994). The seco-helenanolide, hymenoxon, is the principal toxin. It is one of several dihemiacetals epimeric at C-3 and C-4, the mixture of which is referred to as hymenovin (Ivie et al. 1975a; Petterson & Kim 1976; Herz 1978; Hill et al. 1979). The oral toxic dose of hymenoxon is about 100 mg/kg, causing death in 2–4 days, with signs and pathology the same as those produced by the plants (Ivie et al. 1975a). Hymenoxon concentrations in *H. odorata* may vary considerably between locations regardless of climate. The concentrations may be high in both the leaves and flowers throughout the season, but in some instances are reported to be highest (up to 4% d.w.) early in the season, i.e., late winter or early spring, and then gradually declining, with a sharp drop after plants have gone to seed (Kim et al. 1987; Pfeiffer & Calhoun 1987).

- pseudoguaianolide sesquiterpene lactones, dihemiacetals

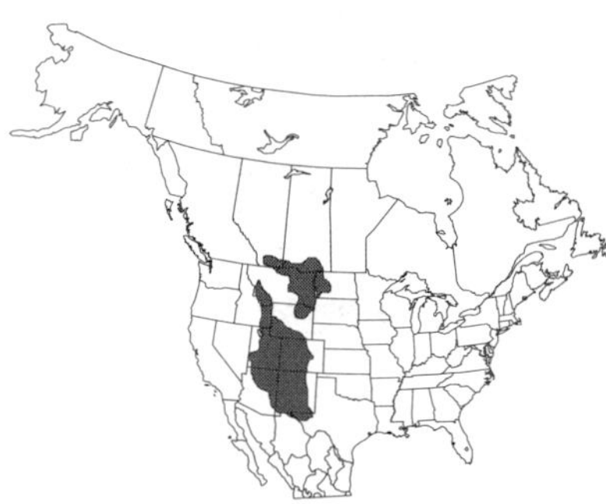

Hymenoxys richardsonii

Hymenoxys subintegra

The toxins of *Hymenoxys* appear to act by inhibition of essential metabolic enzymes (Witzel et al. 1974; Calhoun et al. 1981). The effects are apparently due to the exocyclic methylene, given that hymenoxon, the principal toxin, lacks the cyclopentanone moiety. The effects are cumulative; thus, animals may become intoxicated either acutely, by quickly eating large amounts of plant material, or chronically, by eating small amounts over a period of several weeks (Clawson 1931; Boughton & Hardy 1937; Witzel et al. 1977; Calhoun et al. 1981). There may be a threshold level below which toxicity does not develop, but that level is not known. Additional effects from alteration of ruminal microflora may also contribute, because some of the sesquiterpenes are potent antimicrobials (Ivie et al. 1975a). The presence of the dihemiacetal moiety in hymenoxon produces direct alkylating effects on DNA, including mutagenicity, that are not apparent for helenanolides (Manners et al. 1978; Sylvia et al. 1985).

- inhibition of essential metabolic enzymes

Clinical Signs—The various species of *Hymenoxys* produce signs that are similar to those produced by species of *Helenium, Baileya, Psilostrophe,* and *Sartwellia.* Signs of the uncommon acute disease may appear within 4 hours of consumption of large amounts of plant material for 1–4 days. Early signs include depression, decreased appetite, ruminal atony, and mild bloat. The animal may stand with an arched back, walk with difficulty, and grind its teeth. A serous nasal discharge, coughing, and wheezing are common, and there may be some evidence of slight regurgitation such as green stains of the muzzle and repeated swallowing movements. Gradually, fine muscular tremors develop, as well as dyspnea with an expiratory groan. Terminally, head pressing develops, then recumbency with clonic and tonic convulsions, and finally death in 24–48 hours.

- acute and subacute disease, depression, anorexia, decreased ruminal activity, cough, walk with difficulty; exercise intolerance; regurgitation of ruminal contents giving appearance of foamy green excess saliva

  chronic disease, weight loss, dehydration

In these acute cases, prominent changes in the animal's chemistry include hypoglycemia (blood sugar down to about one-half normal), metabolic acidosis (to pH 7.0), and elevation of blood lactic acid (15-fold) and pyruvic acid (2-fold) (Witzel et al. 1974). Hemoconcentration and an increase in blood urea nitrogen and creatinine may also occur. A clotting defect associated with a factor X deficiency and

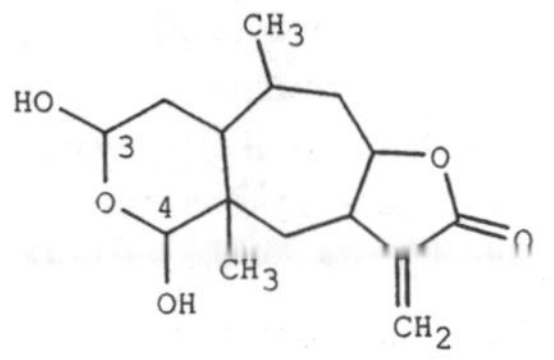

Hymenoxon

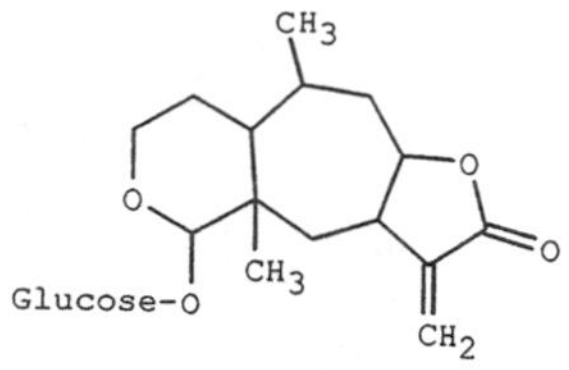

Hymenoxynin

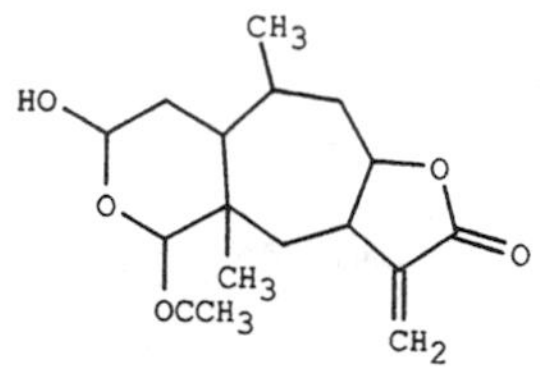

Hymenolide

an increase in prothrombin time to approximately 20 seconds and partial thromboplastin time to approximately 30 seconds, respectively, may also be seen (Steel et al. 1976).

If insufficient plant material is consumed to produce the acute disease, the more typical form may prevail following a few weeks' ingestion. Signs are milder than in the acute form, and death occurs in 4–15 days. Vomiting or regurgitation are more likely to occur with the slower onset of disease, and perhaps with the less toxic plant species, but are not seen in all instances.

Ingestion of small amounts of plant material over a prolonged time may produce the disease form in which there are few clinical signs other than gradual deterioration, weight loss, and dehydration. Death is due to dehydration and starvation.

- few if any distinctive lesions

Pathology—Neither the gross nor the microscopic lesions are particularly distinctive. The primary gross lesions are associated with the gastrointestinal tract and the lungs. Gaseous distension of the gut may be marked. Edema and congestion of the abomasal mucosa and various areas of the forestomach are common. These changes may include the proximal portion of the small intestine. Mucosal hemorrhage with free luminal blood, renal tubular degeneration with proteinaceous casts, and some hepatic degeneration may also be evident. Quite commonly there is congestion, edema, and hemorrhage in the lungs. Additional changes include subendocardial and epicardial petechial hemorrhages, as well as excess pericardial fluid (Rowe et al. 1973; Witzel et al. 1977; Camp 1985).

- general nursing care; parenteral nutritional supplementation
- prevention, supplementation with proteins? sulfur amino acids? antioxidants? or good feed

Treatment—Treatment protocols and recommendations for control of the disease due to *Hymenoxys* also are applicable to *Helenium, Baileya, Parthenium, Psilostrophe,* and *Sartwellia.* With acute intoxication, the animal should be removed from the forage and given supportive nursing care. This care may involve administration of anticholinergic, antidepressant, and antiarrhythmic drugs, although these remedies may provide only temporary relief (Szabuniewicz & Kim 1972). Protection is afforded by administration of parenteral L-cysteine but is not effective once clinical signs occur (Ivie et al. 1975a; Rowe et al. 1980; Merrill et al. 1988). Protein supplementation with or without methionine, soybean meal, and sodium sulfate, or other dietary changes to increase thiols in the animal, have proven beneficial in some studies, contrary to reported producer experiences (Bridges et al. 1980; Rowe et al. 1980; Calhoun et al. 1989; Hall et al. 1989; Post & Bailey 1992). Antioxidants such as ethoxyquin, preferably in combination with methionine, are of value as preventives, but they must be given prior to and concurrently with consumption of bitterweed (Kim et al. 1982, 1983; Calhoun et al. 1989). Of interest is that ethoxyquin has been of no preventive value against intoxication caused by a South African species of *Geigeria,* also a member of the tribe Inuleae, which contains similar sesquiterpene lactones (van Heerden et al. 1993).

At present, range management to control plant growth or to prevent exposure of animals to the plants is the only known method for reducing losses. This becomes particularly important as the plants approach flowering, with respect to both pasture use and hay production. Combining cattle, sheep, and goats in a grazing program with moderate stocking rates also is helpful in reducing losses. This can be further augmented using a system of pasture rotation. These approaches reduce losses, but they do not eliminate the problem entirely (Merrill & Schuster 1978). Providing sorghum silage or even quantities of spineless cactus pads has proven to be of benefit as alternative forage (Dameron & Carpenter 1938).

Hypochaeris L.

Taxonomy and Morphology—Comprising 50–100 species native to the Mediterranean region and South America, *Hypochaeris,* a member of the tribe Lactuceae, is commonly called cat's-ear. The genus name, sometimes incorrectly spelled *Hypochoeris,* was used by Theophrastus and Pliny. Its European species are sometimes grown in wildflower gardens and are attractive to bees and butterflies. Four introduced species are present in North America, only one of which is of possible toxicologic significance:

■ *H. radicata*
flatweed
cat's-ear
gosmore

H. radicata L.	spotted cat's-ear, hairy cat's-ear, gosmore, flatweed, rough cat's-ear

Plants herbs; perennials; acaulescent; from caudices and several large roots or occasionally fibrous roots; sap viscid, white. **Stems** absent. **Leaves** simple; alternate; in basal rosettes; blades oblanceolate; surfaces hispid; margins pinnately toothed or lobed. **Heads** 1–3; showy; borne on elongate scapes; scapes 1 to several per plant, 15–60 cm long, branched or not branched, bracteoles present or absent; heads cylindrical to obconical; phyllaries in 2-series, imbricate, linear-lanceolate, glabrous or midribs hispid; receptacles flat; receptacular bracts present, membranous scales. **Ligulate Florets** numerous; perfect; fertile; corollas 5-toothed, bright yellow. **Pappus** of 1–2 rows of plumose bristles. **Achenes** terete; 10-ribbed, barbellate, beaked.

■ herbs with milky, viscid sap; leaves simple, alternate, in basal rosettes; heads showy, borne on elongate scapes; ligulate florets bright yellow

Distribution and Habitat—A native of Eurasia, *H. radicata* is an invader of disturbed sites, roadsides, lawns, and fields in the Pacific states, mainly west of the Cascades. It is especially abundant as an invader of the eruption area on Mount St. Helens. It also is widely scattered in eastern North America.

■ weedy invader species

Disease Problems—*Hypochaeris* has been tenuously linked to a neurologic disease of horses known as stringhalt. This disease causes selective degeneration of the peripheral nerves (Cahill et al. 1986). Apparently recognized for centuries, this disease was mentioned in *Henry VIII* by Shakespeare, who referred to it as springhalt. In the 1860s, a type of stringhalt emerged as a serious problem in Australia (Pemberton & Caple 1980). It also appeared in New Zealand and much later and less commonly in the western United States. Characteristically occurring in the form of outbreaks without predisposing leg injuries, this type is now referred to as Australian stringhalt. It also differs from the ordinary form of stringhalt by being reversible (Pemberton & Caple 1980). It occurs in the late summer and autumn of drier years as a neurologic problem of exaggerated hock flexion with varying degrees of severity. It is characterized by hypermetria and hyperflexion in the pelvic limbs. Backward movement becomes very difficult. The disease may develop quickly, with severe signs prominent within 2–3 weeks.

■ Horses, a possible but unconfirmed cause of reversible stringhalt; mainly late summer or fall

Although the disease is described here, there is as yet no compelling evidence to incriminate *H. radicata* as the specific cause. In several of the cases in North America, the plant has been present and its involvement suggested (Galey et al. 1991; Gay et al. 1993). Likewise, the disease in Australia has often been associated with the species, but in many instances, plants were not present in the outbreak area and other plant species or factors seemed to be involved (Huntington et al. 1989). Furthermore, investigators have been unable to reproduce the disease experimentally with *H. radicata* (Cahill et al. 1985). In three outbreaks in Chile with signs typical of stringhalt, *H. radicata* was not uniformly present and available to all affected animals (Araya et al. 1998). The cases all occurred in the summer or fall, during particularly dry years.

Hypochaeris radicata

- toxin unknown; mainly degeneration of large myelinated nerves; jerky rear-leg action

Hypochaeris radicata

Disease Genesis—Stringhalt is a disease of unknown cause. Flavonoids and several sesquiterpene lactone guaianolides of unknown toxicologic significance have been isolated from *H. radicata* (Bohlmann & Bohlmann 1980). The lactones are hyporadiolide cinnamate and methylacrylate esters (Seaman 1982). A mycotoxin has also been suggested as the possible cause, although a specific fungus has not been incriminated (Pemberton & Caple 1980).

The effect of the unknown toxinis primarily peripheral nerve and axonal degeneration, resulting in neurogenic atrophy. The large myelinated nerves, especially their distal portions, are most severely affected (Cahill et al. 1986). Axonal degeneration precedes that of myelin; it occurs first in the nerves of the pelvic limbs and then in those of the thoracic limbs, although the left recurrent laryngeal nerve, the longest nerve in the body, is typically the most severely affected. Electromyographic activity of the long digital extensor is exaggerated, even at rest; peroneal nerve conduction velocity is decreased, as is response to repetitive stimuli (Huntington et al. 1989). The jerky pelvic limb action is very similar to that reported to occur in some intoxications in horses due to ingestion of *Lathyrus* and *Pisum* (Mayhew 1989). It is suggested that the problem ultimately may be due to an aberration in the reflex arc from muscle spindles in the distal muscles of the pelvic limb, perhaps mediated through the red nucleus (Pemberton & Caple 1980; Cahill & Goulden 1992).

- abrupt onset, exaggerated flexion and delayed extension of pelvic limbs; hopping gait; recovery in weeks or months

Clinical Signs—There is typically abrupt onset of sudden and involuntary flexion and delayed extension of the hocks, usually bilateral. These effects are especially noticeable when the animal is backing or turning. In severe cases, the legs strike against the abdomen during flexion. There may be a peculiar gait, typically described as a bunny hop. The thoracic limbs may be affected, with decreased flexion, stumbling, toe scuffing, and carpal knuckling. Sometimes accompanied by laryngeal hemiplegia, stringhalt usually exhibits an outbreak type of incidence. It is a disease of some duration but is not necessarily permanent. Recovery may occur spontaneously weeks, months, or even years later.

- microscopic, peripheral nerve degeneration, peroneal, sciatic, recurrent laryngeal; hypertrophy of extensor and flexor muscles

Pathology—Few gross changes are observed in animals affected by stringhalt. Microscopically, there is Wallerian degeneration, especially of the large-diameter myelinated fibers of the peripheral nerves, including the peroneal, sciatic, and recurrent laryngeal (Cahill et al. 1985; Robertson-Smith et al. 1985; Slocombe et al. 1992). There may also be lower spinal axonal degeneration. This leads apparently to atrophy of some muscle fibers, fatty replacement, and hypertrophy of the long and lateral digital extensors and lateral deep digital flexor muscles, among others.

- spontaneous recovery; limited success with drug treatment

Treatment—Because stringhalt is a disease of spontaneous recovery, it is somewhat difficult to evaluate therapeutic approaches. Surgery, involving tenotomy of the lateral digital extensors of the pelvic limb, has been used for many years with some success, but this technique really only modifies expression of the disease. Various other approaches, including the use of tranquilizers, sedatives, relaxants such as mephenesin, and thiamine, also have been used, with little or limited success. Phenytoin (15 mg/kg b.w.) given orally as a paste mixed with feed and given daily for 2 weeks seems to be of value in improving the gait and attenuating the electromyographic changes (Huntington et al. 1991). Additional benefit was noted with a second daily dosage of 10 mg/kg given each evening of the second week. Baclofen may also be of potential benefit (Cahill & Goulden 1992).

Isocoma Nutt.

■ *I. coronopifolia*
I. plurifolia
I. tenuisecta

burroweed
jimmyweed
rayless goldenrod

Taxonomy and Morphology—Comprising 16 species, *Isocoma*, a member of the tribe Astereae, is native to the southwestern United States and northern Mexico. Since its original description in 1840 by Thomas Nuttall, there has been considerable taxonomic debate as to whether it should be treated as a distinct genus or merely as a section of *Haplopappus*. The treatment of Nesom (1991) is followed here. Three species are of toxicologic importance (see plate 13.6):

I. coronopifolia (A.Gray) Greene
(= *Haplopappus fruticosus* [Rose & Standl.] Blake)

I. pluriflora (Torr. & A.Gray) Greene — jimmyweed, rayless goldenrod
(= *I. wrightii* [A.Gray] Rydb.)
(= *Haplopappus heterophyllus* [A.Gray] Blake)

I. tenuisecta Greene — burroweed
(= *I. fruticosa* Rose & Standl.)
(= *Haplopappus tenuisectus* [Greene] Blake)

■ perennial subshrubs; profusely branched; leaves simple; disk florets only, yellow

Plants subshrubs; perennials; from woody rootstocks; herbage glutinous, glabrous. **Stems** erect or decumbent-ascending; 30–120 cm tall; profusely branched. **Leaves** simple; alternate or fascicled; blades oblanceolate or linear-oblanceolate; glandular punctate to stipitate glandular; margins entire or pinnatifid or serrate. **Heads** numerous; small; borne in flat or hemispheric corymbs; turbinate or campanulate; phyllaries in 3- or 4-series, imbricate, linear-lanceolate, glutinous; receptacles convex, alveolate; receptacular bracts absent. **Ray Florets** absent. **Disk Florets** 8–25; perfect; fertile; corollas yellow. **Pappus** of bristles; coarse; unequal; brownish white; barbellate. **Achenes** obpyramidal; 4–10 ribbed.

1. Leaf margins pinnatifid.
   2. Upper stems glabrous . . . . *I. coronopifolia*
   2. Upper stems hispidulous or puberulous . . . . *I. tenuisecta*
1. Leaf margins entire or serrate.
   3. Heads 4–5.3 mm long. Corollas 4.1–6 mm long . . . . *I. pluriflora*
   3. Heads 6–7 mm long. Corollas 6–9.5 mm long . . . . *I. coronopifolia*

Isocoma coronopifolia

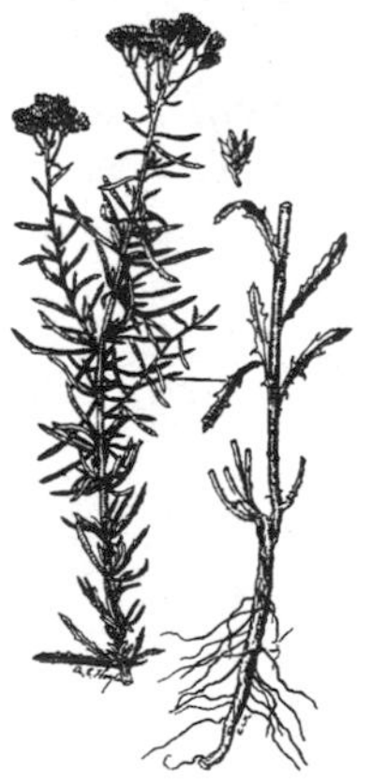
Isocoma pluriflora

Isocoma tenuisectus

Isocoma pluriflora

- dry, open sites, alkaline soils

Isocoma tenuisecta

Distribution and Habitat—*Isocoma coronopifolia* is found in dry, open, calcareous or igneous sites in southern Texas and Nuevo León. The other two species are of the Southwest. *Isocoma pluriflora* occupies dry, open sites and may be locally abundant in alkaline areas. *Isocoma tenuisecta* is present in sandy soils of grasslands and *Larrea* stands, plants typically occurring up to 1,700 m elevation.

- alkali disease, milk sickness; same disease as trembles, caused by *Eupatorium rugosum;* cumulative effect, large amounts of plant for a few days or smaller amounts for several weeks; metabolic derangement

Disease Problems—In the early 1900s a disease called alkali disease or milk sickness was noted in cattle and horses in the Pecos Valley of New Mexico. The disease, which seemed to be encountered more often after the first frosts, was characterized by violent trembling of the animals, induced by exercise. Problems in humans fortunately were minimal because of early recognition of the disease's similarity to the milk sickness of the eastern states, caused by *Eupatorium rugosum.* In at least one case, the disease was diagnosed by a physician based on a description of "milk sick" made by his father, also a physician, many years earlier in Indiana (Couch 1926). Subsequent experiments confirmed *I. pluriflora* (reported as *I. wrightii*) to be the cause of the disease in cattle, horses, and sheep when green plants were given in daily dosage of 1–2% b.w. for 1 to several weeks. The disease also appeared in nursing young. Its clinical signs were essentially identical to those caused by *E. rugosum,* with the exception that *E. rugosum* also produces heart and liver syndromes in horses and goats, respectively. Increased incidence of the disease following frosts appears to be related to increased palatability of *Isocoma* (Marsh et al. 1926). Both *I. coronopifolia* and *I. tenuisecta* have been suspected of causing the same disease syndrome (Buehrer et al. 1939).

- crude toxin tremetol, a mixture composed of methyl ketone benzofuran derivatives; metabolic impairment, inhibition of TCA cycle

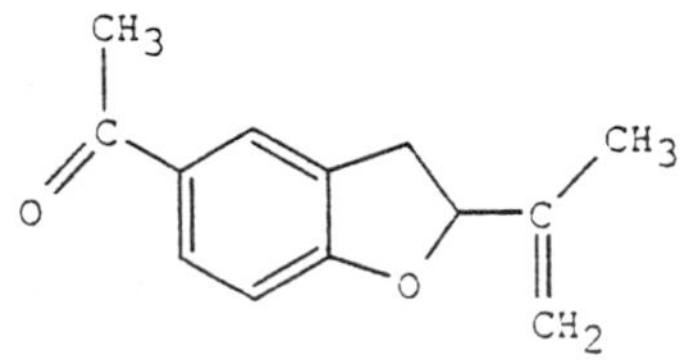

Tremetone

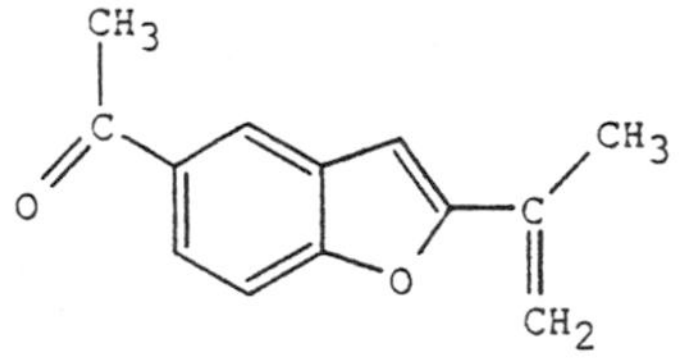

Dehydrotremetone

Disease Genesis—Shortly after its identification in *E. rugosum,* tremetol was confirmed to be present in *I. pluriflora* (Couch 1929, 1930). A number of monoterpenes and sesquiterpenes of unknown toxicologic significance also have been isolated from the species (Zalkow et al. 1979). The seriousness of the disease problem caused by *I. pluriflora* in the 1930s prompted several studies on tremetol and development of a useful chicken model, which was subsequently used for further studies on the disease (Butler 1945). With use of the chicken model, tremetol has been shown to cause metabolic alternations, including impairment of the tricarboxylic acid (TCA) cycle and glucose utilization (Wu et al. 1973). Isolatable even from dried plant material, the slightly volatile tremetol was shown to be a complex mixture of four or more components (Lathrop 1939; Dermer & Cleverdon 1943). The essential oil, composed of various terpenes, was nontoxic, whereas the higher-molecular-weight fractions of the crude tremetol extract were the most toxic portions; toxicity was induced by biotransformation products of the extracts (Beier et al. 1993). Extracts of *I. pluriflora,* which contained trematol, caused trembles (cf. *Eupatorium rugosum,* in *Eupatorium* treatment in this chapter). The extract mixture contained dehydrotremetone, subsequently shown to be an isolation artifact of tremetone (Burke 1966; Cabat 1968; Beier et al. 1993). Interestingly, although intoxication was readily induced using 1 g/kg b.w. of the extract, lesser amounts were associated with tolerance rather than toxicity (Butler 1945). This may be an important consideration in accounting for the infrequency of disease in many situations. Of unknown significance is the presence of pyridine (2% of dry plant material) reported to occur in other species of *Isocoma* (Buehrer et al. 1939).

- gradual onset, depression, weakness, stiffness, constipation, tremors, acetone odor to breath, collapse

Clinical Signs—*Isocoma* produces signs similar to those of *Eupatorium rugosum.* They include depression, a humped stance in sheep, foreleg stiffness, constipation with some blood, and labored respiration in the late stages. The most characteristic feature of the disease is trembling of variable intensity induced by exercise.

Pathology—Grossly, there are few distinctive changes. The liver will be pale and the gall bladder distended with thick, dark bile. Microscopically, mild to moderate renal tubular degeneration and hepatic fatty change are present.

- no distinctive lesions

Treatment—Similar to treatment for intoxication due to *E. rugosum,* therapy is nonspecific and entails good nursing care and provision of high-quality forage to sustain the animal during the protracted recovery period. The most important consideration is alleviation of the severe ketosis and acidosis (Couch 1928; Hartmann et al. 1963). This may be tied to administration of parenteral carbohydrates for control of the low glucose levels in the blood. To prevent enterohepatic recycling, activated charcoal may be given orally (2–3 g/kg b.w.) to reduce reabsorption of the toxin, including that eliminated in the bile. It is imperative to prevent access to the plants during the growing season.

- good nursing care, high-quality feed, parenteral glucose, activated charcoal

Iva L.

Taxonomy and Morphology—Native to North America, *Iva,* a member of the tribe Heliantheae, comprises 15–19 species, many of which are associated with wet sites—hence the common names *marsh elder* and *sumpweed.* Toxicologic problems have been associated with 2 species:

- *I. acerosa*
  I. angustifolia

  copperweed
  sumpweed

I. acerosa (Nutt.) R.C.Jacks. (= *Oxytenia acerosa* Nutt.)	copperweed
I. angustifolia Nutt. ex DC.	narrowleaf sumpweed

Plants herbs or subshrubs; annuals or perennials; from taproots or crowns. **Stems** erect or ascending; 50–200 cm tall; branched above. **Leaves** simple; alternate or opposite below and alternate above; blades linear to linear-lanceolate; margins entire or serrate or pinnately lobed. **Heads** numerous; small; sessile or subsessile; borne in axils of leaves or bracts; campanulate or turbinate; phyllaries in 1- or 2-series, broadly ovate or linear-lanceolate; receptacles flat; receptacular bracts as large as phyllaries. **Ray Florets** absent. **Disk Florets** 5–7; marginal one pistillate, fertile, corollas absent or minute; central ones functionally staminate, sterile, corollas translucent. **Pappus** absent. **Achenes** obovate; ribbed; rugose; black; glabrous or white villous-tomentose.

- herbs or subshrubs; leaves simple, linear; heads small, numerous, axillary; disk florets only, translucent

1. Plants herbs; annuals. Leaf margins entire or serrate.
   Marginal pistillate florets 1–2 . . . . *I. angustifolia*
1. Plants subshrubs; perennials. Leaf margins pinnately lobed.
   Marginal pistillate florets 5 . . . . *I. acerosa*

Iva acerosa

Distribution and Habitat—*Iva acerosa* is found in moist alkaline seeps, springs, and dry washes and on stream banks of the West. *Iva angustifolia* is typically found in well-drained, upland areas in the eastern part of its range and along drainage ditches, along creek beds, and in moist disturbed sites in the south-central United States.

Intoxication problems associated with the two species of *Iva* are quite different. *Iva acerosa* causes a metabolic disease affecting primarily sheep, whereas *I. angustifolia* causes reproductive problems in a number of species. Aspects of each are discussed separately.

Iva acerosa

Metabolic Disease

- metabolic derangement; digestive disturbance

Disease Problems—First recognized as causing a disease in the early 1920s, intoxication due to *I. acerosa* is a sporadic problem of southwestern ranges, affecting cattle and, to a lesser extent, sheep. The plant is eaten only when other forage is scarce. Cattle are most likely to eat it when they are trailed from summer range in the fall. The foliage is especially toxic, and the dry fallen leaves remain a hazard for sheep in the fall and winter (James et al. 1980). Approximately two-thirds of the animals affected die in 30 minutes to 10 hours after ingesting the plants. The remainder die in 7–10 days (Thorp et al. 1940a). Affected animals rarely recover. The lethal dose in cattle is about 0.5% b.w. In lambs, administration of 200 g or about 1% b.w. is lethal (Thorp et al. 1940b; James et al. 1980).

- specific toxins unknown; possibly sesquiterpene lactones?

Disease Genesis—The specific toxicants of *I. acerosa* have not been identified, but pseudoguaianolide sesquiterpene lactones are well represented in species of *Iva*, and the clinical signs are consistent with the effects of these types of compounds; thus their involvement in the disease cannot be excluded (Seaman 1982).

- anorexia, depression, weakness, nervousness, collapse; no distinctive lesions; no specific treatment

Clinical Signs, Pathology, and Treatment—Loss of appetite, depression, weakness, and coma are consistently seen. In some instances there may be signs of struggling, nervousness, or excitement. The lesions are not distinctive and include small hemorrhages of the mucosa of the abomasum and small intestine and congestion of the liver. Microscopically, there are mild degenerative changes in the liver, including cloudy swelling and fatty change. There is no specific treatment.

Reproductive Disease

Iva angustifolia

Disease Problems—*Iva angustifolia* in all stages of growth appears capable of causing reproductive effects, although the seedling stage is responsible for most field problems (Murphy et al. 1983). Intoxication requires several weeks of ingestion and is a problem during the last half of gestation. Reproductive effects produced in rabbits and cows include premature, small, weak, or stillborn offspring and probably abortions (Murphy et al. 1983). These effects are preceded by early mammary development and, in some instances, lactogenesis with dripping milk. Placental retention is not a prominent feature, but excess placental mineralization is common. The toxicity potential of other species in not known; however, the dried foliage of *I. annua* has been administered at 25% of the diet to pregnant rats throughout pregnancy without causing adverse effects on the fetuses (Burrows et al. 1998).

Disease Genesis—The specific toxicant has not been identified. Using rat embryo cultures, plant extracts were shown to inhibit DNA synthesis (Irvin et al. 1986). This system excludes the possibility of metabolic activation. Two pseudoguaianolide sesquiterpene lactones—ivangustin, a xantholide, and ivangulin, an ambrosanolide—are present in in *I. angustifolia* (Herz 1968; Connolly & Hill 1991). However, because these compounds are of low reactivity, they seem to represent minimal risk.

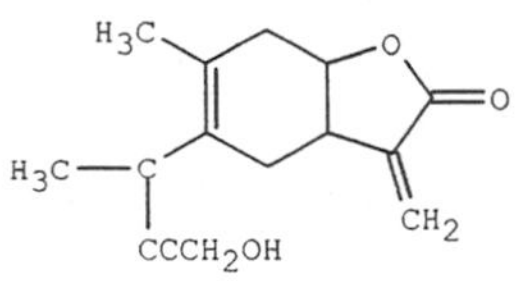

Ivangulin

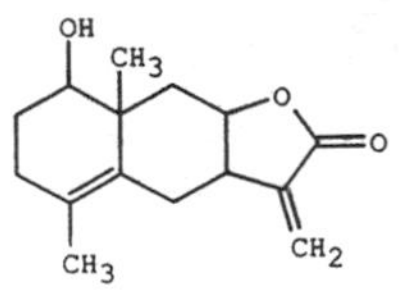

Ivangustin

Clinical Signs, Pathology, and Treatment—Premature mammary development, dripping milk, and excess vaginal mucus precede abortion or birth of premature and/or weak calves. Placental retention is not a prominent feature. Placental mineralization is common and may be useful in diagnosis, although it is not a specific diagnostic feature. Disease control has been accomplished by delaying breeding so that consumption of seedlings, if unavoidable, does not occur during the last half of gestation.

- late gestation, abortion, stillbirth, premature births of small weak calves
- specific toxin unknown; low reactive sesquiterpene lactones present
- time breeding to avoid plants in late gestation

Parthenium L.

Taxonomy and Morphology—Comprising 15–16 species, *Parthenium,* a member of the tribe Heliantheae, is a New World genus. The generic name is derived from the Greek *parthenos* meaning "virgin," reputedly alluding to the production of achenes only by the ray florets. One species, *P. integrifolium,* was used by some tribes of American Indians to treat burns, and *P. argentatum,* commonly known as guayule, has been grown as a source of rubber. Of the 9 species present in North America, only 1 is of toxicologic interest:

- *P. hysterophorus*

  false ragweed
  feverfew
  Santa Maria
  cicutilla

P. hysterophorus L.	Santa Maria, false ragweed, cicutilla, Santa Maria feverfew

Plants herbs; annuals or perennials; from taproots; herbage hirsute, aromatic. **Stems** erect; 30–100 cm tall; branched above; longitudinally striate. **Leaves** simple; alternate; both basal and cauline, basal forming a rosette; lower subpetiolate, upper sessile; blades oblong; margins entire or pinnatifid or bipinnatifid. **Heads** numerous; small; borne in open panicles or corymbs; hemispheric; phyllaries in 2-to-4-series, imbricate, papery; receptacles convex to flat; receptacular bracts present, scales. **Ray Florets** 5; inconspicuous; pistillate; fertile; corollas white or cream, persistent on achenes. **Disk Florets** numerous; functionally staminate; sterile; corollas white or cream-colored. **Pappus** of 2 petaloid scales. **Achenes** obovate; flattened; black; fused to adjacent receptacular bracts, disk florets, and phyllaries, the aggregation dropping as a unit at maturity.

- aromatic herbs; leaves simple, alternate, in basal rosettes; ray and disk florets white or cream-colored

Distribution and Habitat—The native species of *Parthenium* are most abundant in the southern United States and Mexico. Naturalized from tropical America, *P. hysterophorus* is locally abundant in waste areas and other disturbed sites throughout the southeastern one-fourth of the United States but is occasionally adventive as far north as Michigan and Massachusetts.

Parthenium hysteropherus

Disease Problems—Of little apparent problem in North America, *P. hysterophorus* has been associated with disease in India, where it has inadvertently been introduced and naturalized. Problems appear when it is eaten in substantial amounts for several weeks (Narasimhan et al. 1977a, b). In the field, it is reported to be readily eaten by goats but sparingly by cattle and water buffalo. In experimental studies, however, it is eaten without much reluctance by cattle and buffalo calves. It causes both systemic toxicity and contact dermatitis (presumably allergic) in some animals and in humans. The development of signs of intoxication has been variable. When constituting 5–10% of the diet for cattle, it resulted in deaths in a few weeks, whereas 50% in the diet caused deaths in a few days (Narasimhan et al. 1980). However, in another study in which buffalo and crossbred beef calves were fed 50% green *P. hysterophorus* in their diet for 4 weeks, there were no adverse effects in the beef calves and only dermatitis

- digestive disturbance; metabolic derangement

- sesquiterpene lactones; ambrosanolide; disease not totally typical of others caused by the lactones

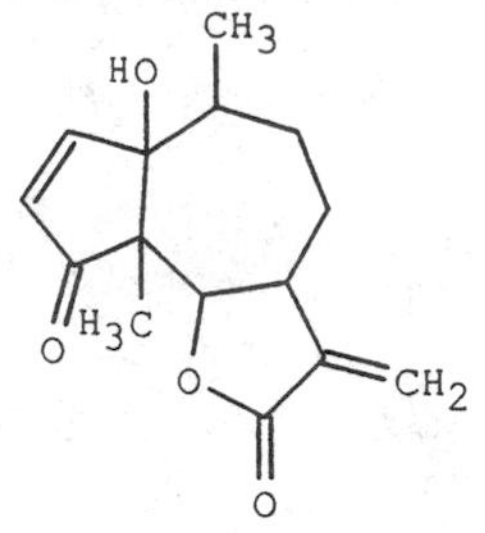

and digestive tract problems (diarrhea and poor weight gain) in the buffalo calves (More et al. 1982). A study carried out in India indicated that about 4% of humans exposed to *P. hysterophorus* as an occupational hazard developed a delayed type of hypersensitivity with contact dermatitis and intense itching (Subba Rao et al. 1977).

Parthenin

Disease Genesis—The disease appears to be associated with sesquiterpene lactones, principally the ambrosanolide parthenin, which is found in concentrations up to 3.5% d.w. in the plant. However, the signs are not entirely consistent with those caused by other taxa containing the lactones—there is considerably more digestive tract involvement with *Parthenium*. Interestingly, when ensiled for 5 weeks, *P. hysterophorus* no longer contained detectable parthenin and was not toxic when fed for as long as 12 weeks (Narasimhan et al. 1993). An array of sesquiterpene lactones other than parthenin are present in the other species of *Parthenium*. They are cytotoxic or otherwise bioactive but are not necessarily intoxication risks (Picman 1986).

- transient diarrhea; loss of condition; muscular twitching; increasingly excitable

Clinical Signs—Following consumption of plant material, a mild or moderate diarrhea develops and then subsides in 2–3 days. In the following 1–3 weeks, the animal's general condition progressively worsens. During the last 1–2 days before death, muscular twitching and excitability may be apparent. Skin changes in the neck, thoracic, and abdominal areas include loss of hair, intense itching, and reddening.

- gross pathology, few distinct lesions; possibly ulcerations along the digestive tract

Pathology—The most consistent lesions are those of apparent irritation: ulcerations of the digestive tract from the muzzle, dental pad, tongue, palate through the esophagus, abomasum, and small intestine. Microscopically, there is moderate renal tubular and hepatocellular degeneration and necrosis.

- general nursing care

Treatment—There is no specific treatment for the problems produced by *P. hysterophorus,* but possible approaches are discussed in the treatment of *Hymenoxys* in this chapter.

Psathyrotes A.Gray

- *P. annua*

  desert velvet
  mealy rosettes
  velvet rosettes
  velvet brittle-stem

Taxonomy and Morphology—Comprising 3–5 species native to the western United States, *Psathyrotes* has traditionally been placed in the tribe Senecioneae, but some taxonomists now position it in the tribe Heliantheae. Its generic name is derived from the Greek word for "brittleness" and is indicative of the brittle stems of the species. Only 1 species is of toxicologic interest:

P. annua (Nutt.) A.Gray	desert velvet, mealy rosettes, velvet rosettes, velvet brittle-stem

- scurfy herbs; leaves simple, alternate; heads axillary; disk florets only, yellow to purplish red

Plants herbs; annuals or short-lived perennials; forming loose mats or low mounds to 30 cm in diameter; herbage scurfy. **Stems** decumbent to ascending; 5–15 cm tall; freely branching; brittle. **Leaves** simple; alternate; gray-green; sparsely tomentose; nerves 3; blades suborbicular to rhomboidal; margins sparely serrate-crenate. **Heads** borne singly in axils of leaves or in forks of stems; turbinate; phyllaries 13 or 18; in 2-series, imbricate, outer erect or recurved; receptacles flat; receptacular bracts absent. **Ray Florets** absent. **Disk Florets** 13–16; perfect; fertile; corollas light yellow, sometimes purplish red, distally arachnoid-villous. **Pappus** of 35–50 bristles; in 1-series; coarse; red-brown; smooth to barbellate. **Achenes** cylindrical or obconic; 10-ribbed; covered with antrorse tawny or reddish brown hairs.

Psathyrotes annua

Psathyrotes annua

Distribution and Habitat—All species of *Psathyrotes* are restricted to the southwestern United States. *Psathyrotes annua* is found in dry sandy scrub and alkali flats.

- dry, open sites, some in alkaline flats

Disease Problems and Genesis—Although not palatable, *P. annua* under dire conditions is eaten and produces a serious, acute disease and death. Experimentally, acute disease can be reproduced with a dose of 0.2% b.w. or repeated daily doses of 0.05–0.1% b.w. of d.w. plant material. The liver seems to be the organ affected in either instance (Binns et al. 1962). Feeding dried plant material to ewes for 10 days in gradually increasing dosage (0.125–0.25% b.w.) resulted in a 20-fold increase in serum AST by day 3 and a 40-fold increase by day 8 (Buck et al. 1961). The specific cause of intoxication is not known, but sesquiterpene lactones may be a factor. Desacetylisotenulin has been isolated from *P. ramosissima* and shown to be molluscicidal (Kubo & Matsumoto 1984).

- acute hepatotoxicosis; toxin unknown

Clinical Signs, Pathology, and Treatment—Following ingestion of *P. annua,* there is abrupt onset of weakness, depression, incoordination, and icterus, terminating in death in a few days. The heart rate may be increased, and there may be arrhythmias. A marked increase in serum hepatic enzymes can be expected, and BSP clearance will be markedly prolonged. Grossly, the liver is swollen and friable and has a nutmeg appearance. Microscopically, there is hepatocellular degeneration and necrosis, and the bile ducts are distended with bile. Treatment is directed toward relief of the signs.

- abrupt onset, weakness, depression, ataxia, icterus; marked increase in serum hepatic enzymes

  gross pathology, liver friable

  microscopic, hepatic necrosis

Psilostrophe DC.

Taxonomy and Morphology—Comprising 7 species native to the western United States and northern Mexico, *Psilostrophe,* a member of the tribe Helenieae, is closely related to *Baileya* (Heiser 1944; Brown 1978). Commonly known as paper-daisy or paperflower because of the appearance of its dried ray florets, its generic name is derived from the Greek roots *psilos* and *strophe,* meaning "naked" and "twist," or "turn." The following species are of toxicologic interest:

- *P. bakeri*
  P. cooperi
  P. gnaphalodes
  P. mexicana
  P. sparsiflora
  P. tagetina
  P. villosa

  paper daisy
  paperflower

P. bakeri Greene	Colorado paperflower
P. cooperi (A.Gray) Greene	white-stem paperflower
P. gnaphalodes DC.	cudweed paperflower

P. mexicana R.C.Br.	Mexican paperflower
P. sparsiflora (A.Gray) A.Nelson	green-stem paperflower
P. tagetina (Nutt.) Greene	woolly paperflower, marigold paperflower
P. villosa Rydb.	hairy paperflower

■ herbs or subshrubs; leaves simple, alternate; heads showy, terminal; ray and disk florets yellow or yellow-orange

Plants herbs or subshrubs; perennials or biennials; from woody taproots; herbage sparsely pilose to densely woolly or floccose. **Stems** erect; 1 to many; 10–60 cm tall. **Leaves** simple; alternate; upper smaller, sessile; lower larger, petioled; blades linear to oblanceolate or spathulate; margins entire to pinnately lobed; **Heads** showy; borne solitary in corymbose clusters at ends of branches; cylindrical; phyllaries 5–10; in 1- or 2-series, outer 1–7 sometimes scarious and smaller; receptacles flat; receptacular bracts absent. **Ray Florets** 2–8; pistillate; fertile; corollas 3-lobed, broad, yellow or yellow-orange, reflexed and papery when mature; persistent on mature achenes. **Disk Florets** 5–25; perfect; fertile; corollas yellow or yellow-orange, lobes glandular-papillose or puberulent. **Pappus** 4–6 hyaline, lanceolate scales. **Achenes** linear; terete or slightly angular; conspicuously striate; glabrous to villous.

Psilostrophe tagetina

1. Plants subshrubs. Lower leaves linear . . . . *P. cooperi*
1. Plants herbs. Lower leaves oblanceolate to spathulate.
    2. Stems and leaves sparsely pilose; green. Corollas of ray florets reflexed against phyllaries when mature . . . . *P. sparsiflora*
    2. Stems and leaves floccose or villous-tomentose; gray or gray-green. Corollas of ray florets spreading horizontally when mature.
        3. Ray florets 5–6. Disk florets 10–20. Pappus scales one-fourth to one-half the length of corollas of disk florets . . . . *P. bakeri*
        3. Ray florets 2–5. Disk florets 5–12. Pappus scales one-half to equal length of corollas of disk florets.
            4. Pappus scales long villous; lacerate-dissected. Achenes villous.
                5. Heads subsessile. Peduncles 0.5–8 mm long. Corollas of disk florets 3.5–4 mm long . . . . *P. gnaphalodes*
                5. Heads peduncled. Peduncles 8–30 mm long. Corollas of disk florets 4.5–5 mm long . . . . *P. mexicana*
            4. Pappus scales glabrous; entire or erose. Pappus scales glabrous; entire or erose. Achenes glabrous or with a few scattered hairs.
                6. Heads subsessile. Peduncles 0.5–5 mm long. Corollas of ray florets 3–5 mm long; deeply 3-lobed . . . . *P. villosa*
                6. Heads peduncled. Peduncles 10–40 mm long. Corollas of ray florets 5–14 mm long; shallowly 3-lobed . . . . *P. tagetina*

■ dry, open sites in semiarid regions

Distribution and Habitat—Growing in the semiarid region of the southwestern United States and adjacent Mexico, species of *Psilostrophe* are often conspicuous roadside weeds in addition to occupying a variety of soil types in dry, open sites in desert, grasslands, woodlands, and shrublands. *Psilostrophe cooperi* is a species of the Mojave and northern Sonoran deserts. Occurring at elevations of 1,350–2,000 m, *P. bakeri* is restricted to rocky, often somewhat alkaline slopes in western Colorado. A species of the Chihuahuan Desert, *P. gnaphalodes* occurs in southwestern Texas and the northern third of Mexico. *Psilostrophe sparsiflora* is present in sagebrush and pinyon-juniper communities in the Great Basin Desert. *Psilostrophe mexicana* also is a species of the Chihuahuan Desert and is found in southern Chihuahua and northern Durango. The morphologically variable *P. tagetina* is the most widespread. *Psilostrophe villosa* is the easternmost species of the genus.

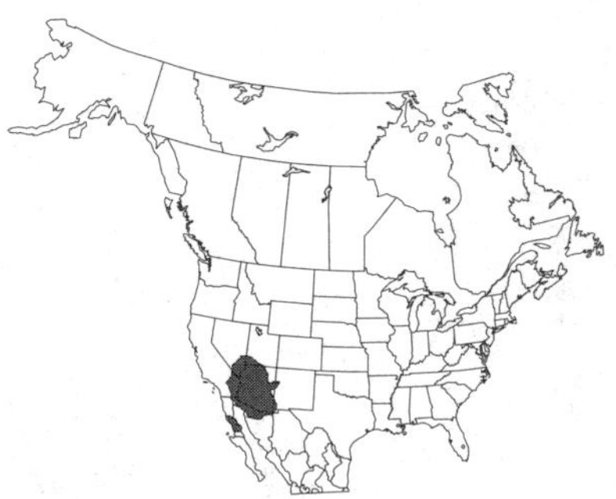
Psilostrophe cooperi

Disease Problems—Scattered losses of sheep due to *Psilostrophe* were reported in the early 1930s (Mathews 1934). In some instances, the morbidity rate of a flock was as high as 50%, with death losses of up to 20% of affected animals. The disease problem is primarily one of sheep, mainly in the spring and fall when other forage is scarce and plants of the genus are eaten (Huffman & Couch 1942). Cattle and goats are at lower risk because of lower susceptibility and/or because they are less likely to eat the plants. A very large amount of plant material is required for intoxication, and because of this, early feeding trials were generally negative or inconclusive. Eventually, *P. tagetina* at a dosage of nearly 100% of b.w. given over many days was shown to produce disease and in some instances death (Mathews 1934). Appearance of the disease with *P. gnaphalodes* required somewhat higher dosage. Although a large amount of plant material is required, that is not always a limiting factor. Even though the plants are apparently not relished, they may be eaten without great reluctance, especially the putatively more toxic younger forms.

- metabolic and digestive derangement, gradual onset, especially in sheep; large amounts of plant material required

Psilostrophe sparsiflora

Disease Genesis—All of the species contain helenanolides such as helenalin, vermeerin, floribundin (psilotrophin, a C-8 epimer of vermeerin), and derivatives of hymenoxon (Seaman 1982). Several other helenanolides, and a novel enol-type pseudoguaianolide, cooperin, have been isolated from *P. cooperi* (Stuppner et al. 1988). Thus toxicity of *Psilostrophe* is essentially that for *Helenium* and *Hymenoxys*. A complete discussion of aspects of the disease genesis is presented in the treatment of *Helenium*, in this chapter.

- sesquiterpene lactones, helenanolides

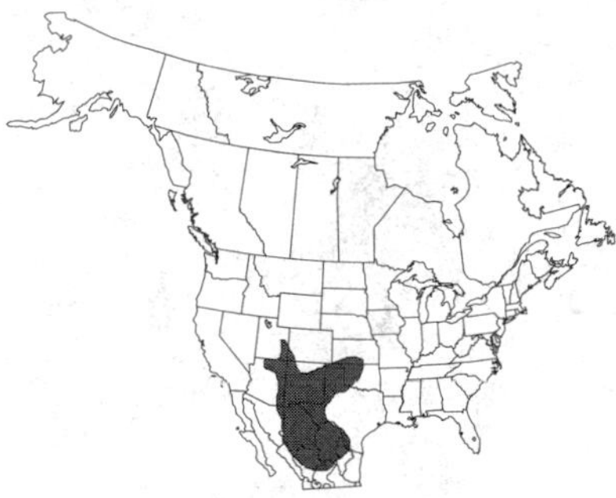

Psilostrophe tagetina

Vermeerin

Clinical Signs—Following ingestion of *Psilostrophe*, onset of the disease is gradual, with depression and weakness being observed. Affected animals become sluggish in their movements and lag behind the remainder of the flock. As the disease increases in severity, emaciation, coughing, and sometimes regurgitation of greenish ingesta that accumulates on the lips and nasal area are seen. Animals are not disposed to move about, and when forced to do so, they stumble and quickly become exhausted, with heart and respiration rates markedly elevated.

- gradual onset, depression, weakness, loss of body condition, coughing, sometimes frothy, green saliva on lips

Pathology and Treatment—Few obvious changes other than poor body condition are associated with *Psilostrophe* intoxication. Microscopic changes are limited to mild renal tubular and hepatocellular degeneration. There is no specific treatment, but approaches are described in the treatment of *Hymenoxys* in this chapter.

- no distinct lesions

general nursing care

Rudbeckia L.

Taxonomy and Morphology—Comprising 15–18 species native to North America, *Rudbeckia* is a member of the tribe Heliantheae. Its name honors Olaus Rudbeck, professor of botany and Linnaeus's mentor at the University of Uppsala in Sweden. Among the showiest of wildflowers, its species are cultivated widely as garden

■ R. grandiflora
R. hirta
R. laciniata
R. maxima
R. mollis
R. occidentalis
R. subtomentosa

blackhead
black-eyed Susan
brown-eyed Susan
coneflower
golden glow

ornamentals for their showy heads. The following species are representative of the genus and of toxicologic significance:

R. grandiflora (Sweet) DC.	rough coneflower
R. hirta L.	brown-eyed Susan, black-eyed Susan
R. laciniata L.	cutleaf coneflower, green-headed coneflower, golden glow
R. maxima Nutt.	great coneflower
R. mollis Elliot	
R. occidentalis Nutt.	blackhead, western coneflower, black coneflower
R. subtomentosa Pursh	sweet coneflower

■ herbs; leaves simple; alternate; heads terminal; ray florets yellow, orange, or orange-brown; disk florets purplish black or brown or yellowish or greenish

Plants herbs; annuals or biennials or perennials; from taproots or fibrous roots or rhizomes or woody caudices. **Stems** erect; 30–300 cm tall; branched or not branched; terete; ribbed. **Leaves** simple; alternate; basal and cauline; petiolate or sessile; blades of various shapes; glabrous or variously indumented; margins entire or toothed or lobed. **Heads** showy; few; large; borne singly at ends of peduncles terminating stems and branches; hemispheric to subglobose; phyllaries in 2- or 3-series, imbricate, linear to lanceolate, herbaceous, spreading or reflexed; receptacles enlarged, hemispheric to conical or columnar; receptacular bracts oblong, folded about achenes. **Ray Florets** 5–30 or rarely absent; neutral; sterile; corollas yellow or orange or dark orange-brown, apices entire or 3-toothed. **Disk Florets** numerous; perfect; fertile; corollas purplish black or brown, or yellowish or greenish. **Pappus** absent or coroniform. **Achenes** oblong; 4-angled; glabrous.

Rudbeckia hirta

1. Ray florets absent . . . . *R. occidentalis*
1. Ray florets present.
    2. Stems and leaves hirsute. Pappus absent . . . . *R. hirta*
    2. Stems and leaves glabrous or pubescent. Pappus coroniform.
        3. Leaf margins deeply lobed or laciniate.
            4. Stems glabrous. Corollas of disk florets greenish yellow. Receptacles columnar at head maturity . . . . *R. laciniata*
            4. Stems densely pubescent. Corollas of disk florets purplish brown. Receptacles hemispheric at head maturity . . . . . *R. subtomentosa*
        3. Leaf margins entire or toothed.
            5. Plants 125–300 cm tall. Leaf bases clasping . . . . *R. maxima*
            5. Plants 40–100 cm tall. Leaf bases not clasping.
                6. Plants annuals or biennials. Taproots present. Rhizomes or woody caudices absent . . . . *R. mollis*
                6. Plants perennials. Taproots absent. Rhizomes or woody caudices present . . . . *R. grandiflora*

■ wide variety of habitats

Distribution and Habitat—Species of *Rudbeckia* occupy a variety of soil types and habitats, both disturbed and undisturbed, across the continent. Forming large populations in overgrazed prairies, along roadsides, and in waste areas, *R. hirta* is found across the continent, with greatest abundance in the East. *Rudbeckia laciniata* occurs in moist areas and along stream banks throughout the eastern three-fourths of the continent. In contrast, *R. occidentalis* is a western species found in meadows and open woods from Washington to Wyoming and Colorado. A south-central species, *R. maxima* favors moist, open sites, as does *R. grandiflora*. *Rudbeckia subtomentosa* occupies prairies and low areas in the Midwest. Present only in the extreme Southeast, *R. mollis* is found in sandy, open woods.

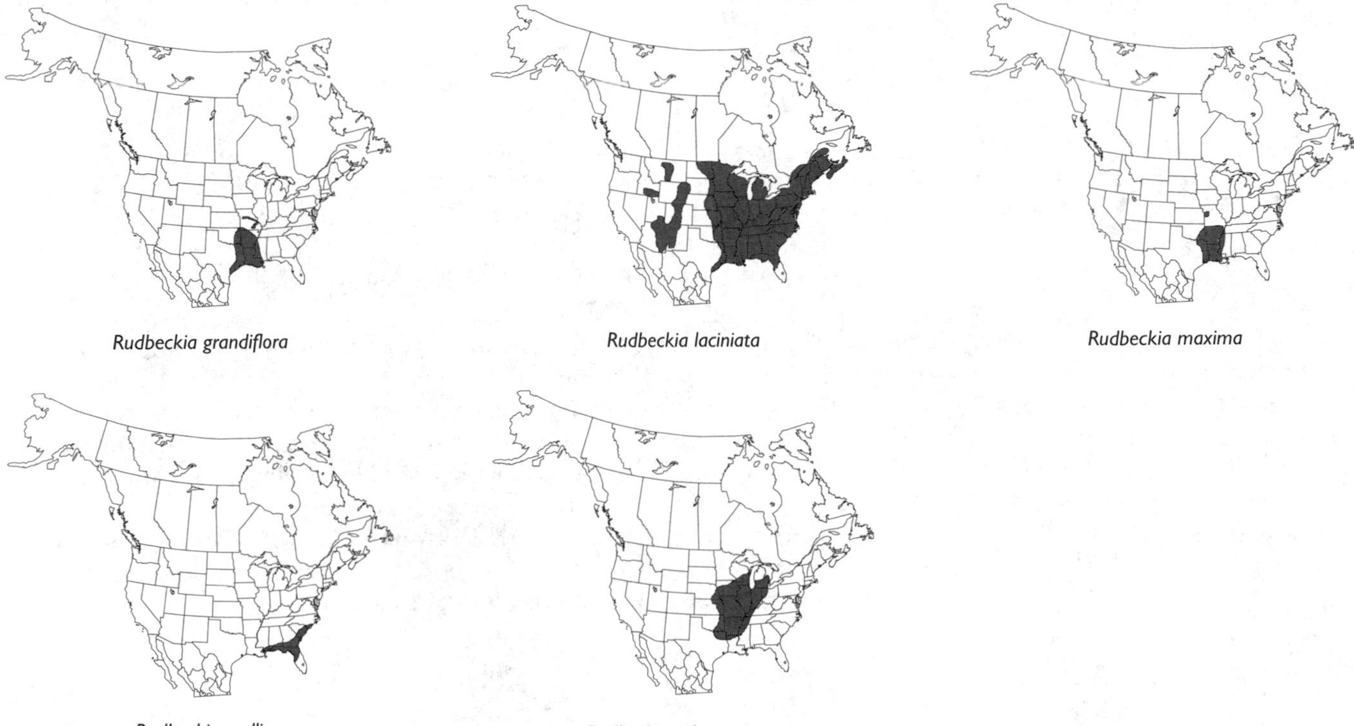

Rudbeckia grandiflora

Rudbeckia laciniata

Rudbeckia maxima

Rudbeckia mollis

Rudbeckia subtomentosa

Disease Problems— Anecdotal information indicating the toxicity of *Rudbeckia,* and especially *R. laciniata,* is abundant. In one report, 27 sheep that ate plants of the species for 2 days became ill, and 7 of them subsequently died (Chesnut & Wilcox 1901). In another case, pigs foraging in a field containing almost entirely *R. laciniata* became ill, with signs said to be suggestive of belladonna, including dry mucous membranes and convulsive seizures (Gates 1930a). A number of the pigs died. The disease was reproduced in a pig that died after being fed *R. laciniata. Rudbeckia* also is reported to be toxic to cattle (Chesnut 1898; Pammel 1911, 1921). *Rudbeckia maxima* has long been considered to be a toxic species and continues to have this reputation.

- intoxication potential not clear; possible neurologic effects; digestive disturbance

Experimental efforts to reproduce intoxication by *R. laciniata* revealed that although it caused acute adverse effects, these were only transient (Skidmore & Peterson 1932). Attempts to cause intoxications in pigs and sheep fed about 3% b.w of green plant material resulted in transient signs, but continued feeding was accompanied by resolution and no further problem. These observations suggest limited toxicity potential. However, rabbits and cavies fed plants for 4–5 days or more became paralyzed and died. Part of the difficulty in evaluating the species lies with the unpalatability of the plants; they have a very disagreeable odor and taste. Animals must be very hungry before they will eat even a limited amount.

Disease Genesis—Bioactive ambrosanolide sesquiterpene lactones are reported to occur in *R. grandiflora, R. mollis,* and *R. subtomentosa* but not in *R. hirta* or *R. laciniata* (Gutierrez & Herz 1990). Similar derivatives of alantolactone and ivangustin are present in *R. laciniata,* while rudmollin and rudmollitrin occur in *R. mollis* (Seaman 1982).

- toxins unknown; bioactive sesquiterpene lactones are consistently represented

■ depression anorexia, ataxia, no lesions

treatment, general nursing care

Clinical Signs, Pathology and Treatment—Signs of *Rudbeckia* intoxication include depression, loss of appetite, incoordination, and increased respiration rate. There are no distinctive pathologic lesions. Treatment is based upon relief of the presenting signs.

Sartwellia A.Gray

■ *S. flaveriae*

sartwellia

Taxonomy and Morphology—A small genus of only 4 species native to the southwestern United States and northern Mexico, *Sartwellia* is a member of the Heliantheae (Turner 1971). Only 1 species has been implicated in toxicologic problems:

S. flaveriae A.Gray threadleaf sartwellia, downy sartwellia

■ annual herbs; leaves simple, opposite, margins entire; heads woolly, terminal; ray and disk florets yellow

Plants herbs; annuals; from taproots; herbage glabrous. **Stems** ascending; several from the base; 10–40 cm tall; branched above. **Leaves** simple; opposite; sessile; blades linear-filiform; alternate; of various shapes; margins entire. **Heads** numerous; small; borne in corymbose cymes at top; campanulate to turbinate; phyllaries 5, in 2-series, broadly oval or oblong to elliptic; yellowish; keeled below; linear-lanceolate, densely tomentose; receptacles flat to convex; receptacular bracts absent. **Ray Florets** 3–5; pistillate; fertile; corollas yellow, ovate, apices truncate or 2- or 3-toothed. **Disk Florets** 6–15; perfect; fertile; corollas yellow. **Pappus** of 5 scales and 5 bristles, fused basally; stramineous. **Achenes** cylindrical; 10-ribbed; hispidulous; black.

Sartwellia flaveriae

Sartwellia flaveriae

■ gypsum soils, desert flats

Distribution and Habitat—Plants of *Sartwellia* are characteristic of gypsum soils of desert flats. Other species of the genus are encountered infrequently in the northern states of Mexico.

■ metabolic disturbance; generally requires long-term ingestion; toxins not known, sesquiterpene lactones suspected

Disease Problems and Genesis—*Sartwellia flaveriae* causes a systemic disease in goats, typically after they have consumed plants for 2–3 months (Mathews 1940). Observed effects varied, depending on the duration of ingestion. Consumption for a short time produced renal and hepatic effects, whereas longer-term consumption resulted in a chronic syndrome, with fluid accumulation in the chest cavity. In addition to producing the systemic toxicity, this species has also caused a photodermatitis in individuals collecting it. The toxins in *Sartwellia* have not been identified, but the signs are consistent with those caused by sesquiterpene lactones. A complete discussion of these compounds is presented in the treatments of *Helenium* and *Hymenoxys,* in this chapter.

Clinical Signs, Pathology, and Treatment—After a prolonged period of eating *Sartwellia,* animals exhibit a gradual loss of appetite, a decline in general body condition, listlessness, and a degree of exercise intolerance. With forced movement, there is an undue increase in respiratory rate. When ingestion is short-term, the principal changes are mild to moderate focal necrosis of the kidneys and liver. In the chronic form, the changes are those of fluid accumulation—excess fluid in the chest and extensive pulmonary edema. There may also be renal and hepatic degeneration without distinctive focal necrosis. There is no specific treatment, but approaches as described for *Helenium/Hymenoxys* may be appropriate.

- gradual loss of body condition, listlessness, exercise intolerance

  gross pathology, fluid accumulation in the chest

  treatment, general nursing care

Senecio L.

Taxonomy and Morphology—Cosmopolitan in distribution and comprising 1,000–3,000 species, *Senecio* is one of the largest genera of flowering plants, is a member of the tribe Senecioneae, and is extraordinarily diverse in growth form and habitat. Its name is derived from the Latin *senex,* meaning "old man," a presumed allusion to the white or gray pappus hairs. Species are commonly referred to as the groundsels, ragworts, or butterweeds. Biosystematic studies indicate that the genus is paraphyletic, and in recent years, taxonomists have proposed its division into as many as 19 genera (Barkley et al. 1996; Vincent 1996). Most notable of these segregates is *Packera,* a genus with about 60 species that encompasses many of the common prairie and woodland species of North America. Toxicologic reports use the name *Senecio,* and thus the traditionally broad circumscription of the genus is employed in this treatment.

Cultivated for both their foliage and showy heads, a number of species, including *S. confusus* (Mexican flame-thrower vine), *S.* x*hybridus* (florists' cineraria pot plant), and *S. mikanioides* (German ivy, or parlor ivy), are cultivated as ornamentals and are not of known toxicologic risk. Some species are members of stable plant communities, whereas others, such as *S. jacobaea* (tansy ragwort), are noxious weeds and readily invade disturbed sites. In North America, some 120 native and introduced species are present. Of toxicologic importance or known to contain toxic, unsaturated pyrrolizidine alkaloids are the following (see plates 13.7 and 13.8):

S. ampullaceus Hook.	Texas groundsel
S. aureus L.	golden ragwort, squaw-weed, heart-leaved groundsel
S. bicolor (Willd.) Tod. (= *S. cineraria* DC.)	dusty miller
S. candidissimus Greene	
S. congestus (R.Br.) DC.	swamp ragwort, northern swamp groundsel
S. cymbalarioides H.Buek	alpine meadow butterweed, few-leaved groundsel
S. dimorphophyllus Greene	different groundsel, Payson's groundsel
S. eremophilus Richardson	desert groundsel, dryland ragwort
S. flaccidus Less. (= *S. douglasii* DC.) (= *S. longilobus* Benth.)	threadleaf groundsel, woolly groundsel

- *S. ampullaceus*
  S. aureus
  S. bicolor
  S. candidissimus
  S. congestus
  S. cymbalarioides
  S. dimorphophyllus
  S. eremophilus
  S. flaccidus
  S. fremontii
  S. glabellus
  S. integerrimus
  S. jacobaea
  S. plattensis
  S. pseudoarnica
  S. riddellii
  S. sanguisorboides
  S. serra
  S. spartioides
  S. squalidus
  S. tomentosus
  S. triangularis
  S. viscosus
  S. vulgaris
  S. werneriifolius

  butterweed
  dusty miller
  gauge plant
  groundsel
  ragwort
  squaw-weed
  stinking willy
  yellowtop

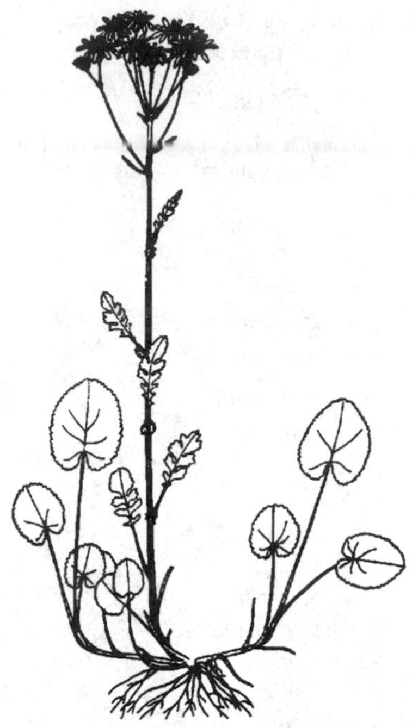

Senecio aureus

- herbs, shrubs, or vines; leaves simple, alternate; heads numerous, terminal; ray florets yellow or red or purple or white, rarely absent; disk florets yellow or rarely white or purple

S. fremontii Torr. & A.Gray	Fremont's groundsel, dwarf mountain butterweed
S. glabellus Poir.	butterweed, yellowtop
S. integerrimus Nutt.	western groundsel, gauge plant, wet-the-bed, one-stemmed butterweed
S. jacobaea L.	tansy ragwort, stinking willy
S. plattensis Nutt.	prairie ragwort, prairie groundsel, Platte groundsel
S. pseudoarnica Less.	seaside groundsel
S. riddellii Torr. & A.Gray	sand groundsel, Riddell's groundsel
S. sanguisorboides Rydb.	
S. serra Hook.	tall butterweed, saw groundsel, butterweed groundsel
S. spartioides Torr. & A.Gray	broom groundsel, narrow-leaved butterweed
S. squalidus L.	
S. tomentosus Michx.	southern woolly groundsel, woolly groundsel
S. triangularis Hook.	arrowleaf groundsel
S. viscosus L.	sticky groundsel
S. vulgaris L.	common groundsel, old-man-in-the-spring
S. werneriifolius (A.Gray) A.Gray	rock butterweed, montane groundsel

Plants herbs or shrubs or vines; annuals or perennials; from woody rootstocks or fibrous roots. **Stems** erect or ascending; 10–130 cm tall; typically branching above. **Leaves** simple; alternate; cauline and/or basal; blades of various shapes; margins entire or toothed or lobed or pinnately dissected. **Heads** numerous; borne in terminal corymbs or rarely singly; campanulate or obconic or urceolate; phyllaries 12–25, in 1- or 2-series, linear, outer when present short, curled, setaceous, inner erect, equal in length; receptacles flat; receptacular bracts absent. **Ray Florets** present or rarely absent; pistillate; fertile; corollas yellow or red or purple or white. **Disk Florets** numerous; perfect; fertile; corollas yellow or rarely white or purple. **Pappus** of numerous capillary bristles; soft; white or gray; entire or minutely barbellate. **Achenes** cylindrical; 5–10 ribbed; glabrous or hispidulous.

Hybridization is common in *Senecio,* and many species intergrade with others. Identification of taxa is based on both vegetative and head characters. Because of the large number of species in the genus and of toxicologic interest, a taxonomic key is not presented here; rather the reader is encouraged to use floristic treatments specific for his or her area.

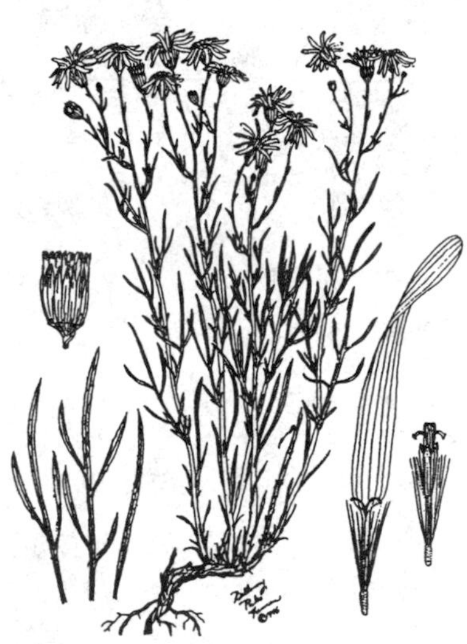

Senecio flaccidus

- exceptional diversity in habitats

Distribution and Habitat—*Senecio* exhibits exceptional environmental diversity, and its species are encountered in a variety of diverse plant communities, habitats, and soil conditions. Often for convenience, North American taxa are informally designated as either prairie or woodland species. Representative of the prairies and of particular interest because of their role in livestock losses are these:

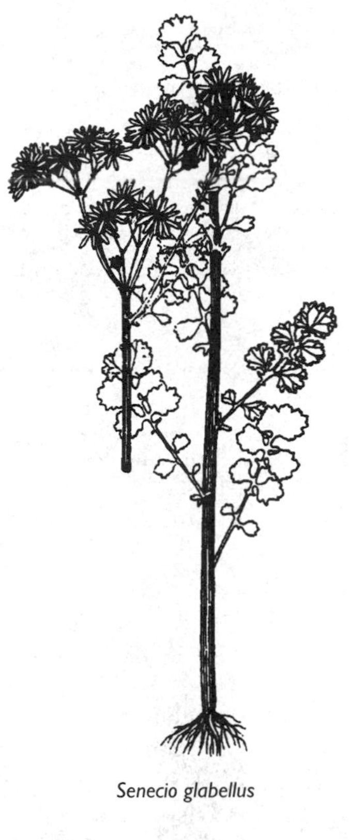
Senecio glabellus

Senecio integerrimus

Senecio jacobaea

Senecio plattensis

Senecio riddellii

Senecio vulgaris

S. flaccidus	widespread in dry plains and deserts of the Southwest; plants perennials
S. integerrimus	prairies and open woods of western half of continent; plants biennials or short-lived perennials
S. plattensis	prairies, roadsides, and other dry places from Great Plains; plants biennials or short-lived perennials
S. riddellii	widespread in sandy soils of western part of Great Plains; plants perennials
S. spartioides	dry, open, rocky sites to 3,700 m elevation from the Great Plains westward; plants perennials

Senecio aureus

The following species are representative of woodlands and also of particular interest:

S. aureus	moist woods and swampy sites in eastern half of continent, especially in the Midwest; plants perennials
S. glabellus	moist or wet soils, often weedy in low fields of the southeastern quarter of the continent; plants annuals

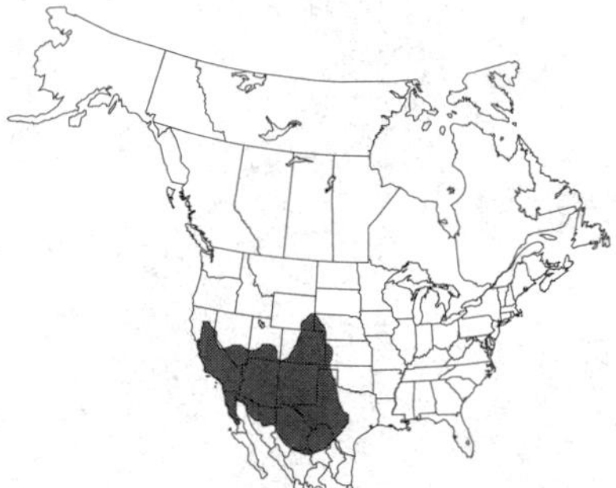

Senecio flaccidus

The other species of toxicologic interest have localized distributions. At times, they form dense populations, especially in mountain meadows. The two important introduced species, *S. jacobaea* and *S. vulgaris,* may be found in open pastures, waste areas, and open woodlands. *Senecio jacobaea* is especially common along highway and railway rights-of-way, power lines, fencerows, and unused cutover timberland left for grazing. Although both species may be found in various areas of North America, they are serious problems only in the inland valleys of the Pacific Coast—*S. vulgaris* in California and *S. jacobaea* from northern California to British Columbia. The latter also is present in eastern North America, especially in Nova Scotia.

■ liver disease risk from species with long growing seasons, abundant herbage, and/or locally abundant; of most concern, *S. jacobaea, S. flaccidus, S. riddellii, S. vulgaris*

mainly horses and cattle; more-resistant species such as sheep used to control plant populations

Disease Problems—The species of *Senecio* are of great economic importance as causes of liver disease in livestock. In addition, humans appear to be highly susceptible, and the plants have been responsible for severe intoxications, especially in infants and older adults (Huxtable 1980, 1989). Exposure has often been via the consumption of herbal teas and dietary supplements for medicinal purposes. Bah and co-workers (1994) have pointed out the serious risk of liver disease with use of the Mexican medicinal plant, *S. candidissimus,* which is widely used as an anticancer herbal product. A South American species employed as a cough remedy also has been associated with adverse effects—a veno-occlusive liver disease (Tomioka et al. 1995). There is also a suggestion that *Senecio* species are responsible for primary liver neoplasia in areas of South Africa (Huxtable 1989), but this association is questionable because of the high potential for exposure of the study population to aflatoxins and other mycotoxins.

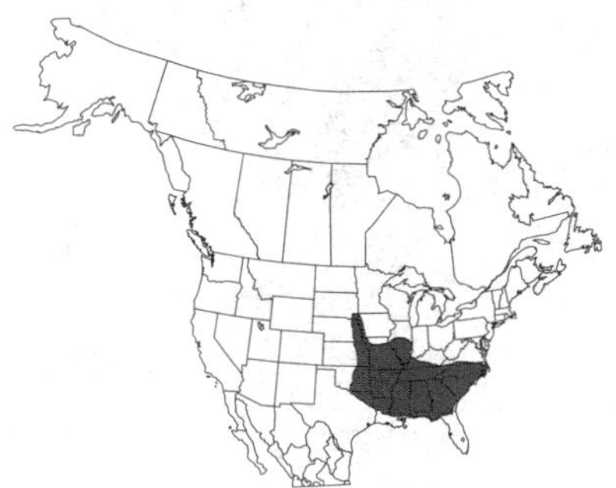

Senecio glabellus

Species of toxicologic interest for livestock are those in areas with long growing seasons, those producing abundant herbage, and/or those that are locally abundant. These factors become more important as one considers the large amount of plant material required to induce toxicosis. Many species are spring flowering and represent a hazard only during the brief period they flourish; they are often of low palatability. Thus many factors influence the likelihood of intoxication. Of the numerous species native to or naturalized in North America, four pose the most distinct threats to livestock: *S. jacobaea, S. flaccidus, S. riddellii,* and *S. vulgaris.* These species persist for longer periods, often blooming in summer or fall, and in some cases may readily be eaten in hay or cubes if not fresh. Invading alfalfa fields seriously afflicted by the alfalfa weevil, *S. vulgaris* has been particularly troublesome. The species inhabiting the drier western ranges, *S. riddellii* and especially *S. flaccidus,* may be eaten throughout much of the year, particularly when alternative forage is scarce or dry. *Senecio flaccidus* is the most hazardous because it is evergreen and foliage is present throughout the year, whereas *S. riddellii,* although a perennial, dies back and loses its leaves during the winter. The season of heaviest livestock loss for most species varies somewhat with the geographic area but typically is late spring or summer. Species of short season and/or sparse foliage cannot be entirely dismissed as disease threats; even the small *S. plattensis* may occasionally present problems for livestock (Lilley 1980; Johnson et al. 1985a). For additional details on the ecology and biology of the various *Senecio* species of toxicologic interest, the reader is referred to the treatises of Sharrow et al. (1988) and Coombs et al. (1991).

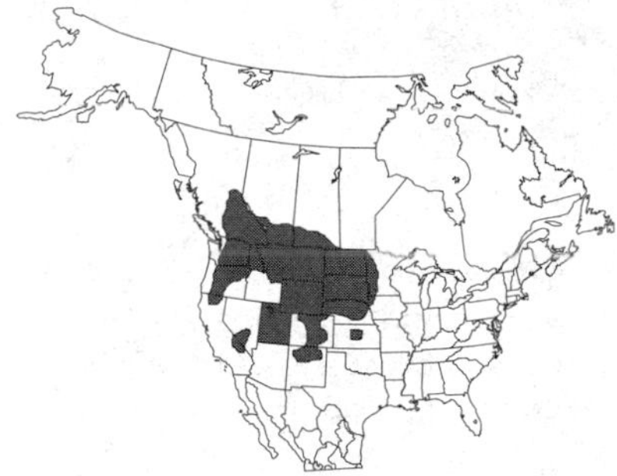

Senecio integerrimus

In rare instances, an animal might eat *Senecio* equivalent to 5–10% of b.w. in a few days or weeks and develop acute liver disease. However, most cases are due to ingestion of smaller doses or a total dosage of 25–50% of b.w. over several months, which results in a more chronic type of disease (Clawson 1933; Mathews 1933a; Dollahite 1972). Thus palatability, access to large amounts of plant material, and availability of other forage are of great importance. During the early years of awareness of *Senecio*-associated disease, it was recognized that sheep were useful in controlling plants of the genus while suffering little from overt illness (Pethick 1906; Pethick & Rutherford 1907; Rutherford 1909). The validity of this observation has been borne out subsequently, and it is now recognized that although all animal species are susceptible, the required toxic doses differ considerably among them. Horses and cattle are most often poisoned, whereas sheep and goats are much less sensitive, especially to chronic exposure (Hooper 1978). Intoxication in sheep requires about 10–15 times greater dosage of *S. jacobaea* than for cattle. Interestingly, although sheep and goats are resistant to intoxication from chronic exposure, they also are more predisposed to eat the plants than are cattle and horses; thus they are useful for controlling the growth of *Senecio* (Dollahite 1972; Cheeke 1989). Cattle eat *Senecio* when other feed is scarce or following storms (Fowler 1968; Dollahite 1972). Horses are similarly reluctant to eat the plants, although they will do so to some extent when plants contaminate hay, especially alfalfa hay cubes/wafers, as does *S. vulgaris.* Even in the latter instance, there is decreased feed intake.

- rarely 5–10% b.w. may cause acute liver necrosis and failure; most commonly 25–50% b.w. over several weeks or months, less acute liver fibrosis and failure

Senecio jacobaea

The most devastating problems associated with *Senecio* were a consequence of the inadvertent introduction of *S. jacobaea* into North America in the 1860s. It has a long flowering period and is a prolific seed producer. By the 1950s it had become well established from Northern California to British Columbia, occupying millions of acres and causing serious livestock losses. The magnitude of the problem has been reduced considerably through the use of biological controls and possibly because of the natural cyclic nature of plant populations (Tilt 1969; Cheeke 1989). An insect herbivore, the cinnabar moth *(Tyria jacobaeae),* specific for ragwort was introduced into northern California in 1959 and into other areas thereafter. This resulted in partial control of the weed, especially the flowering stages (Harris et al. 1976; Isaacson 1976). However, even better control was attained by using pests feeding on the sequential stages of plant growth; the cinnabar moth on the summer foliage and the flea beetle *(Longitarsus jacobaeae)* on the stems, roots, and petioles in winter (Hawkes & Johnson 1976; McEvoy 1985; RR James et al. 1992). The ragwort seed fly *(Pegohylemyia)* also has been used to reduce populations, along with cleaning of equipment prior to movement, to reduce the likelihood of transporting seeds from one area to another (Coombs et al. 1991). The species has been controlled but not eliminated. The presence of rosettes and vegetative reproduction allow persistence.

- biological controls have reduced *S. jacobaea* problems

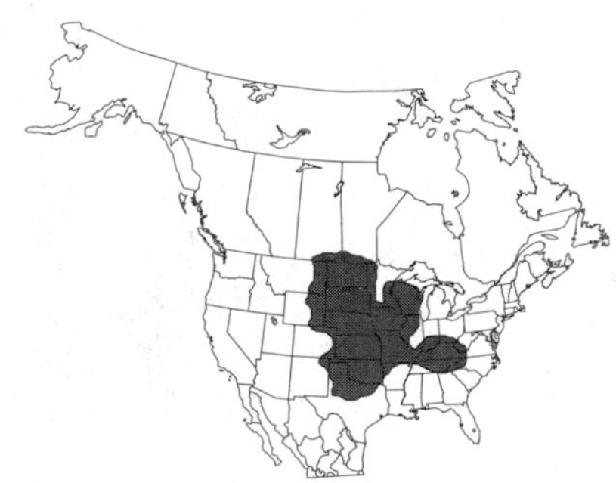

Senecio plattensis

Thousands of acres continue to be infested and to be a threat to livestock populations (Coombs et al. 1991). However, populations may be reduced sufficiently to increase perennial grasses markedly and to minimize the likelihood of animals eating sufficient *Senecio* to cause disease.

■ pyrrolizidine alkaloids, free and N-oxides; unsaturated and esterified; the most toxic are 12-membered retronecine macrocyclic diesters

Disease Genesis—Intoxications are due to a group of alkaloids collectively known as pyrrolizidine alkaloids (PAs) found as either the free base or the *N*-oxide, both of which are hepatotoxic (Molyneux et al. 1991). The disease pyrrolizidine alkalosis is known from most areas of North America but is not always caused by *Senecio*. Other genera and families, including *Crotalaria* in the Fabaceae and genera in the Boraginaceae, also contain toxic PAs. These hepatotoxic PAs, of which there are numerous examples, are unsaturated at the 1–2 position of the necine nucleus and esterified (usually with branching) at the CH_2OH side chain group at the C-9 position. Alpha unsubstitution and esterification further increase toxicity, and most of the more toxic members are 12-membered macrocyclic diesters such as jacobine, retrorsine, riddelliine, senecionine, and seneciphylline. There are nearly 100 of these types of compounds (Hartmann & Witte 1995). There are even more numerous non-toxic types lacking the requisite 1–2 unsaturation, including bulgarsenine, macrophylline, nemorensine, platyphylline, procerine, and rosmarinine (Robins 1982).

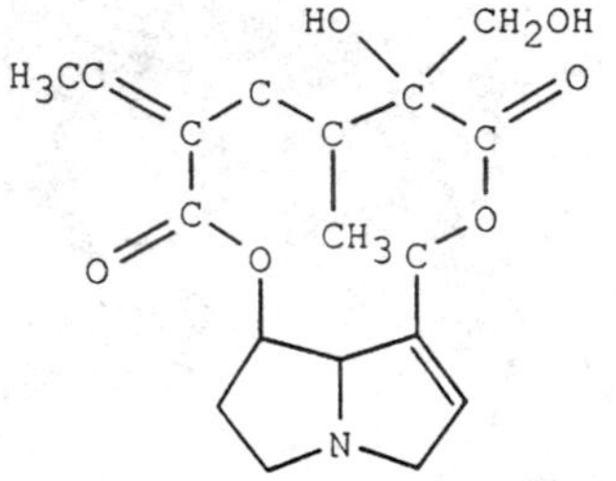

Retrorsine

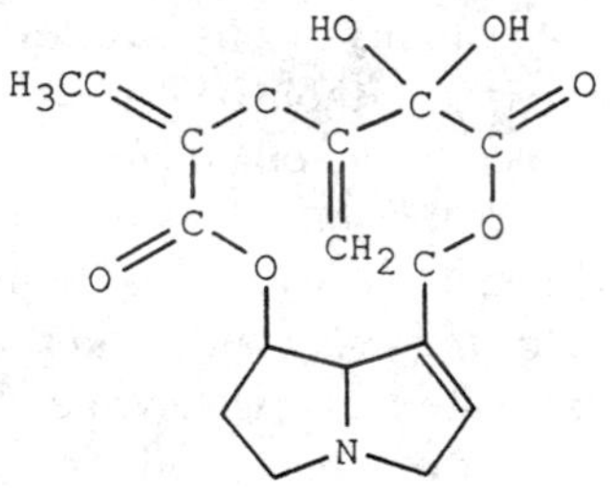

Riddelline

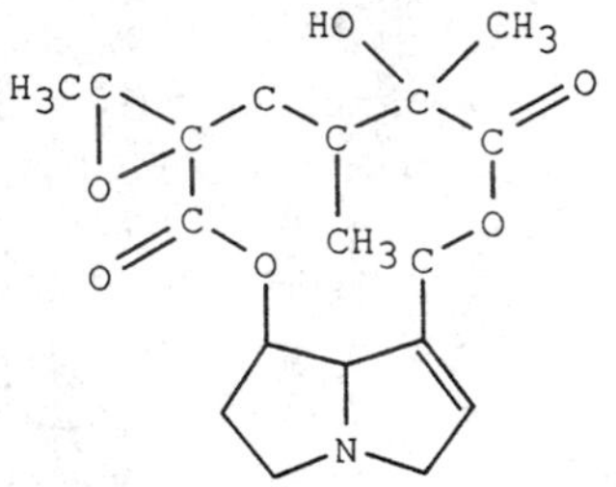

Jacobine

■ flower and upper leaves, high risk; variability in alkaloid content, some plant populations may have very low levels of toxic PAs

Considering *S. vulgaris* to be representative of the genus, the PAs are synthesized in the roots and translocated to the aerial portions in the form of *N*-oxides (Hartmann et al. 1989; Hartmann & Witte 1995). The inflorescences are the major sites of accumulation, followed by the upper stems, leaves, and lower stems. The same general order of PA distribution seems to be true for the perennials, although synthesis also seems to occur in the aerial portions of *S. flaccidus* and *S. riddellii* (Johnson et al. 1985b; Birecka et al. 1988). In general, aside from the inflorescences, the highest concentrations of PAs are found in the youngest tissues. Some variation among species may be expected because there are annuals, biennials, perennials that die back to their subterranean crowns in winter, and perennials that retain their foliage throughout the year. The PA concentrations of the perennials varies by 10-fold or more during the course of the growing season, with lowest concentrations during winter dormancy and highest just before flowering (Johnson et al. 1985a). The annual *S. vulgaris* and biennial *S. jacobaea* seem to have more constant PA levels through the course of the growing season, again with the highest concentrations during budding and flowering. Thus for all species, the onset of flowering heralds the period of greatest risk. However, considering that in many instances ingestion will occur over long periods, there is risk anytime the plants are available for consumption.

Senecio riddellii

There is considerable variation in total unsaturated PA content as well as variation in the alkaloids represented and in the proportion of alkaloids present as the free base or *N*-oxides (table 13.4.). Even though there is considerable variation in the proportions of free base to *N*-oxide forms, some of it may be an artifact of the extraction procedures. It is suggested that the *N*-oxide form should be viewed as the dominant form in the vegetative parts of plants containing pyrrolizidine alkaloids, including most species of *Senecio* (Hartmann & Witte 1995).

Senecio spartioides

Table 13.4. Alkaloid content and approximate toxic dosages for species of *Senecio*

Species	PA %dw	Free Base/ N-oxide ratio	Toxic Dose (mgPA/kg/day for cattle)
S. candidissimus	0.36%	1/5	?
S. dimorphophyllus	0.17	?	?
S. flaccidus	1-3 (8.7)*	20-50/50-80%	>10, 20+ days
S. jacobaea	0.1-0.9	73/27	2.5, 20 days; 1.3, 60+ days
S. plattensis	0.09-0.56	50/50	?
S. riddellii	2-9 (18)*	50-20/80-95%	15-20, 20d
S. serra	<0.01	?	?
S. spartioides	0.3-8	?	?
S. vulgaris	0.16-0.25	25-50/50-75	1-2, 3-17days 1-2, 100 days (horse)

Sources: Molyneux et al. 1979, 1991; Johnson 1982; Johnson & Smart 1983; Johnson & Molyneux 1984; Johnson et al. 1985a,b; Mendel et al. 1988; Craig et al. 1991; Stelljes et al. 1991.
*Extreme value.

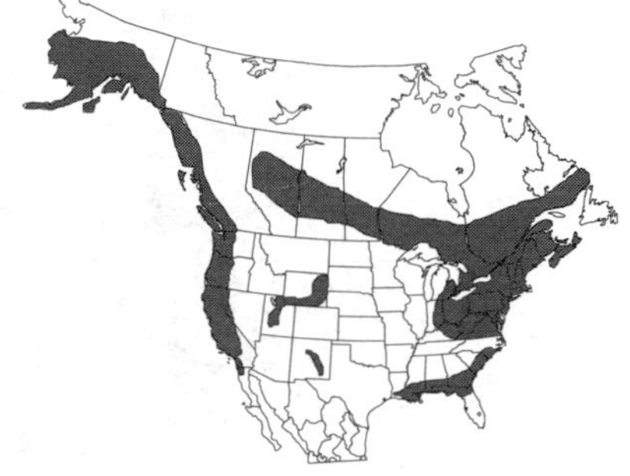

Senecio vulgaris

The unsaturated PAs in *S. plattensis* have not been specifically identified; all of the remaining species except *S. riddellii* and *S. serra* contain senecionine and other retronecine-type PAs. *Senecio riddellii* contains retrorsine and riddelline in common with other species.

In some instances the array of alkaloids present in a species may vary from one location to another. For example, senecionine is present in *S. triangularis* collected from California but not in plants collected in Canada (Rueger & Benn 1983). Some populations of *S. flaccidus,* one of the most if not the most hazardous species, have been shown to be devoid of toxic PAs (Ray et al. 1987). There remains the possibility of threshold dosage, but because of the variation in PA content and animal sensitivity, almost any daily dosage must be considered a risk if continued long enough (Johnson & Molyneux 1984). *Senecio dimorphophyllus* is an example of a toxic species of low hazard because plants are widely dispersed at high elevations and are unlikely to be eaten in quantity by livestock. Similarly, *S. glabellus* contains senecionine and is nearly equal in toxicity risk to *S. jacobaea* and *S. vulgaris,* but the plants are so bitter and unpalatable that they are almost never eaten (Adams & Van Duuren 1953; Goeger et al. 1983).

- PA metabolism, ester hydrolysis, and metabolism in liver to either nontoxic products or highly reactive pyrroles

Upon ingestion, the *N*-oxide PA is reduced to the more lipophilic free base in the digestive tract (Mattocks 1986; Molyneux et al. 1991). Any of the *N*-oxide PA present in the systemic circulation may also be converted in the liver to the parent PA (Muller et al. 1992; Couet et al. 1996). Both the free base and *N*-oxide PAs are toxic, but the toxicity of the latter is several times less (Culvenor et al. 1976). The free base PAs derived either from the *N*-oxide or directly from the plant may be further reduced to less toxic metabolites in the digestive tract or absorbed into the systemic circulation. The PAs are detoxified in the liver by ester hydrolysis and/or *N*-oxidation via cytochrome P450–mixed-function oxidase (MFO) and subsequent dehydrogenation (Mattocks & White 1971; Guengerich 1977; Hirono 1981; Mattocks 1986; Ramsdell et al. 1987). By an alternative independent MFO pathway, the 1–2 unsaturated PAs are dehydrogenated to form short-lived but toxic chemically reactive pyrroles and possibly other metabolites that act as biological alkylating agents (Jago et al. 1970; McLean 1970; Mattocks 1986; Winter & Segall 1989). Specific bioactivating cytochrome forms have been identified, such as P450 UT-A in rats and P4502B in

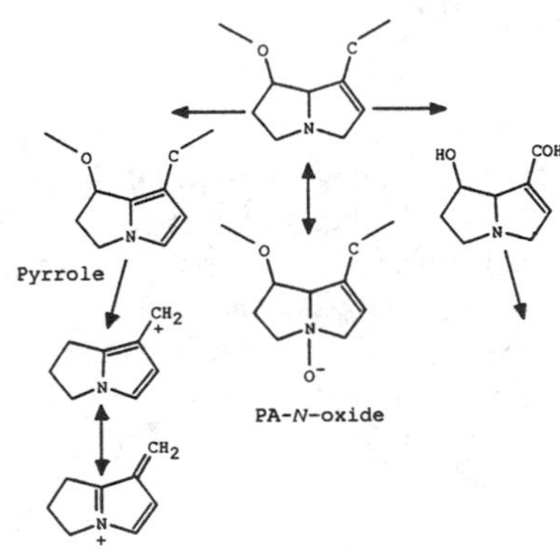

PA metabolism

guinea pigs (Williams et al. 1989; Chung et al. 1995). The parent PAs are essentially unreactive. PA pyrroles are hypothesized to form highly electrophilic C-9 carbonium intermediates capable of reacting with and alkylating proteins, DNA, and other cellular nucleophiles (Hsu et al. 1975; Wickramanayake et al. 1985; Buhler et al. 1986; Reed et al. 1988). Stable products of pyrrolic thioester alkylation may persist for long periods. The PA pyrroles also are subject to ester hydrolysis to form the less reactive but still toxic 7,9-hydroxy (7-hydroxy, 1-hydroxymethyl) product (DHP) either before or after conjugation at the 7 position with glutathione (White 1976; Buhler et al. 1991; Reed et al. 1992). Toxicity of the closed-ring diester PAs is enhanced by steric hindrance to ester hydrolysis. Pyrroles in the liver peak at about 2 hours and then gradually decline over the next 1–2 days. Attempts have been made to correlate toxicity with rate of pyrrole formation, and in some instances there is good correlation. However, this is fraught with difficulty because toxicity is determined by a balance of all the foregoing pathways rather than any individual one (Winter & Segall 1989). Thus, increased esterase activity may counteract the increased rate of pyrrole formation (Dueker et al. 1992b). Species differences in the activities of various pathways may well account for the variation in responses to the many PA combinations in plants (Lame et al. 1991; Miranda et al. 1991; Dueker et al. 1992a). In addition, there may be considerable difference in the rates of formation of toxic pyrroles from the various parent PAs (Yan et al. 1995).

Although the liver is central to both toxicant formation and detoxication, the role of induction of mixed-function oxidase activity is not clear. In guinea pigs, pyrrole formation and/or toxicity of the alkaloids is increased by phenobarbital pretreatment, whereas in sheep there is no effect at dosages that increase cytochrome P450 and activity of aminopyrine *N*-demethylase (White et al. 1973; Swick et al. 1982, 1983a; Chu et al. 1993). Changes in toxicity are correlated with changes in concentrations of pyrroles bound in the liver (White et al. 1973). Animal species sensitivity to the various alkaloids seems to correlate with rate of pyrrole formation, which is high in laboratory rodents and horses and low in chickens and sheep (Shull et al. 1976; Mattocks 1986; Winter et al. 1988). The order of sensitivity in males is hamster > rabbit > mouse > rat > steer > bull > lamb > chicken > quail, which may be compared with the order of acute toxicity for retrorsine of rat > mouse > hamster and chicken > quail > guinea pig (Mattocks 1986). However, the order of susceptibility to plant material may vary from that of individual alkaloids, as was shown for chickens and guinea pigs (Hooper 1978). In general, of the livestock species, pigs seem to be the most susceptible, and sheep and goats the least. As is so clearly pointed out by Blythe and Craig (1994), the apparent resistance of sheep and goats to the toxic effects is likely the result of the biotransformation of the cyclic diester PAs by ruminal bacteria and differences in rate of pyrrole production and/or rate of hydrolysis (Lanigan & Smith 1970; Swick et al. 1983b; Craig et al. 1986, 1992; Wachenheim et al. 1992a,b).

- pyrroles react to alkylate DNA, alter RNA and protein synthesis; impairment in cell division leads to slow-growing cells that cannot divide; cell and nucleus become very large; hepatocytomegaly, karyomegaly

The reactive metabolites interact with various tissue components, resulting in a severe decrease in DNA-mediated RNA and protein synthesis and an increase in nucleolar segregation and lysosomal enzyme activity (McLean 1970). Interaction with cell components may produce necrosis, hepatocyte degeneration and slow replacement, and/or impairment of cell division due to alkylation of DNA (Bull et al. 1968). Closed diesters are especially potent in cross-linking DNA strands to each other and to proteins and in exerting antimitotic effects, presumably as a result of greater pyrrole production (Petry et al. 1986; Hincks et al. 1991; Coulombe et al. 1992). With chronic exposure, there are increases in rough endoplastic reticulum, the number of centrioles (>2), and apparent cellular reproductive activity. Some of the cells continue to grow even though they are unable to divide. Both the nucleus and cytoplasm increase, sometimes as much as 10-to-20-fold, resulting in hepatocytomegaly and karyomegaly. Cell size is magnified most in horses and least in chickens (Hooper 1978). The impairment of cell division is superimposed on a stimulus for regeneration. The eventual disease manifestations are due to hepatic insufficiency resulting from parenchymal necrosis, atrophy, fibrosis, bile duct proliferation, and hepatocytomegaly. In spite of their cytotoxic effects, DNA interactions, genotoxic effects in mice, and apparent embryotoxic and teratogenic effects in laboratory rats, the PAs of *Senecio* have not been shown to be direct reproductive toxins in livestock, presumably because of the short-lived and localized effects of the pyrroles (Johnson & Smart 1983; Mirsalis et al. 1993). This does not preclude the rare case of hepatotoxic effects on the fetus or neonate, because PAs do cross the placenta (Small et al. 1993).

Under some conditions, dietary supplements affecting the liver, such as cysteine and other thiol agents and/or the antioxidants and glutathione-*S*-transferase inducers, butylated hydroxyanisole (BHA) and ethoxyquin have shown promise in protecting laboratory animals and chickens, but the results have not been consistent (Buckmaster et al. 1976; Kim & Jones 1982; Garrett & Cheeke 1984; Kim 1984). In addition, the results have not been as promising upon direct application to target species such as ponies (Garrett et al. 1984). Increased protein in the diet is somewhat protective, mainly because of the presence of sulfur amino acids (Cheeke & Garman 1974). Experiments with chickens indicate that protection provided by BHA is associated with changes other than a decrease in pyrrole formation (Dickinson & Braun 1987). Although there is little reason to favor use of antioxidants, the role of zinc as for hepatotoxic mycotoxins (see *Lolium,* in chapter 58) remains to be clarified (Cheeke 1989).

The pyrrolizidine alkaloids typically produce their effects over many weeks; continued exposure to the toxicant produces increasing injury to the liver. Eventually failure of the liver to clear metabolic products results in an excess of ammonia in the circulation and tissues. However, it is not entirely clear that the ensuing neurologic signs are due solely to ammonia, because other changes such as neurotransmitter alterations also occur (Maddison 1992; Mair 1997). Within a short period, there is development of signs typical of hepatoencephalopathy.

- gradual increase in insult to liver leads to hepatic insufficiency and neurologic manifestations in weeks, months, or years; stress such as pregnancy or malnutrition may be a factor in precipitating onset of signs

In addition, an important feature is the potential for progressive liver disease to result eventually in hepatic failure, months or a year or more after exposure to the toxins ceases (Molyneux et al. 1988). Thus, it may not be sufficient to look for plants in the animal's forage or environment at the time clinical signs are seen. Pregnancy, lactation, transport, nutrition, and other forms of stress may precipitate onset of disease-related signs produced by ingestion of plants several months previously (Johnson & Smart 1983; Synge & Stephen 1993). In addition, in those individuals recovering from the hepatotoxic effects, there may be continued and perhaps permanent debilitation, which may in some measure be due to the long-term presence of sulfur-bound pyrroles in the body (Lessard et al. 1986; Mattocks & Jukes 1990; Seawright et al. 1991a). Depression of vitamin A levels in plasma and liver secondary to chronic PA ingestion may also be a factor in animal health (Moghaddam & Cheeke 1989).

Because pyrrolizidine alkaloids from *Heliotropium,* in the Boraginaceae, are reported to predispose sheep to accumulate copper in the liver, there is concern that the same factors may be true for those alkaloids of *Senecio*. This is all the more important because of the use of sheep for biological control of *Senecio*. Although copper accumulation in laboratory animals given PAs has been reported, it has not been clearly demonstrated that *Senecio* predisposes accumulation in sheep (White et al. 1984b; Cheeke 1989). However, there is a report that copper accumulation resulted in hemolytic crisis in two horses that consumed *S. jacobaea* in New Zealand (Dewes & Lowe 1985).

■ public health concern due to mutagenic and carcinogenic potential of pyrroles

As DNA-reactive mutagens and carcinogens capable of causing chromosomal damage, the PAs are of considerable public health concern (Schoental 1968; Clark 1976; Yamanaka et al. 1979; Mattocks 1986; Huxtable 1989; Rubiolo et al. 1992; Bah et al. 1994). Of greatest concern is direct consumption of plant products. PAs of *Senecio* have produced severe intoxications in humans when consumed in herbal teas or when achenes contaminate cereal grains and are eaten in bread products and other foods. Some of the herbal products are especially popular in Mexican American communities. For example, *S. candidissimus* is used as an anticancer medication but contains significant amounts of 12-membered macrocyclic PAs (Pereda-Miranda 1995). The PAs are readily absorbed from the intestine and pass to the liver via the portal circulation. They pass the placenta and also are excreted in milk and urine. Thus they pose a hazard to the fetus and the neonate (Dickinson et al. 1976; Deinzer et al. 1982; Goeger et al. 1982; White et al. 1984a; Cheeke 1989; Candrian et al. 1991). Although there are clearly direct toxicosis risks, the role of PAs as carcinogens remains in some doubt (Chan et al. 1994). In a rare case, liver lesions typical of PAs were found in a 2-month-old foal from a mare that grazed *Senecio*-infested pasture during pregnancy (Small et al. 1993). Sulfur-bound pyrroles were identified in fixed liver tissue from the foal.

■ limited risk with PA excretion in milk

The hazard with milk is limited by the low dosage potential. Even with continuous high dosage of *S. jacobaea* (1% b.w./day), an amount toxic to the lactating cow or goat, concentrations in the milk are less than 1 ppm and probably in most instances are on the order of 1–10 ppb. In the rat, intoxication via milk requires continuous exposure for 6 months (Goeger et al. 1982). The PAs may also be found in concentrations up to 3.9 ppm in honey from bees feeding on *S. jacobaea* (Deinzer et al. 1977). The potential for problems in humans, however, is reduced by the bitter taste and atypical color of such honey.

Wildlife, especially free-ranging ruminants, seem to be at little risk from PAs when these plants are browsed. They are at least as resistant as sheep to the effects (Dean & Winward 1974; Basson 1987). This may be related to coevolution of wildlife with the plants in their environment (Laycock 1978). The same resistance does not necessarily hold for captive or relocated species, as illustrated by the appearance of disease in captive Père David's deer when *S. jacobaea* was consumed (Fowler 1981). The PAs also seem to play a role in influencing insect, especially butterfly, associations with flowering plants. In some instances the PAs promote, and in other cases they deter, nectar feeding (Pliske et al. 1976; Masters 1991). They may play a role in protecting the host insect and in ensuring reproductive success with involvement in sexual communication and development (Boppre 1986; 1990).

■ depression, anorexia, weight loss, increase in serum hepatic enzymes; rarely photosensitization

Clinical Signs—Although the intoxication mechanisms are essentially the same in all animals, there are important differences in signs and their onset and duration, within and among animal species. These differences depend on the rate of toxin ingestion. Depression and loss of appetite and weight are typical of early signs in most cases, regardless of animal species. Icterus is a less consistent early sign. As the disease develops, marked elevation of serum hepatic enzymes and abnormalities in tests of hepatic function will become apparent in all animal species. Photosensitization is not typically seen. In some instances intoxication may occur within days or weeks of ingestion, especially with *S. riddellii* or *S. longilobus*. The more typical onset is after months of eating the plant and perhaps several or more months after exposure ceases, especially with *S. jacobaea* and *S. vulgaris*. Death may occur in some instances without being preceded by distinctive signs.

In horses, the onset of signs is typically abrupt, regardless of the length of time of plant ingestion. The terminal signs are consistent with and suggestive of hepatoencephalopathy and include head pressing, incessant or aimless walking or pacing, persistent chewing of fences, yawning, drowsiness, rectal straining, incoordination, ascites, and diarrhea or constipation. Increased thirst and hemoglobin in the urine are also seen occasionally. The many vernacular names associated with the various plants containing pyrrolizidine alkaloids—e.g., *Walla Walla walking disease, sleepy staggers,* and *walkabout*—attest to the distinctiveness of the neurologic signs. Horses may recover after many months, but there may be residual, perhaps permanent effects such as exercise intolerance (Lessard et al. 1986; Seawright et al. 1991a). Laryngeal or pharyngeal paralysis has been reported as an uncommon sign with *S. jacobaea* hepatoencephalopathy in horses (Pearson 1991).

■ horses, abrupt onset; head pressing, aimless pacing, ataxia, chewing, yawning, intermittent drowsiness, rectal straining, and either constipation or diarrhea

Similar signs occur in cattle, but the disease is typically subacute, with less striking manifestations. A decrease in appetite and milk production is followed by marked weight loss and an emaciated appearance and finally by weakness leading to recumbency. This course of events may occur in several days or take a week or more and may terminate in neurologic signs and mania. Rectal straining is common and is often associated with prolapse and either diarrhea or constipation.

■ cattle, slower onset; weight loss, weakness, less pronounced neurologic signs, rectal straining with or without prolapse

Diagnosis is aided by determination of a composite of serum hepatic enzyme activities. Unfortunately these activities are not always sustained throughout the course of disease. This seems to be true especially for GGT, which may increase as much as 5-to-10-fold in the early stages but not necessarily later; thus it is of limited value for prognosis (Craig et al. 1978, 1991; Lessard et al. 1986). Evaluation based on a composite of GLDH, GGT, AP, and AST appears to represent the most reliable approach, keeping in mind that in some instances declines in activity may be apparent terminally. SDH and LDH appear to be less consistently markedly elevated. Because severe histopathologic changes precede the appearance of signs, hepatic function tests such as clearance rate of serum bromsulfophthalein will be markedly decreased, but this is often late in the course of the disease. The use of serum hepatic enzymes for both diagnosis and prognosis can be augmented by determination of plasma bile acids and amino acids. Elevation of levels of bile acids above 50 µmol/l is indicative of a poor prognosis (Mendel et al. 1988). Many amino acids also are increased in terminal liver disease, with the exception of the branched-chain amino acids leucine, isoleucine, and valine, which are little affected (Gulick et al. 1980). A ratio of the sum of these branched-chain amino acids divided by the sum of phenylalanine and tyrosine (BCR) of 3 or less is suggestive of but not specific for PA intoxication, and a value of less than 2 indicates a poor prognosis (Gulick et al. 1980; Garrett et al. 1984). Alternatively, plasma tyrosine greater than 120 nmol/ml and phenylalanine greater than 95 mol/ml can be used. To this may be added ornithine greater than 160 mol/ml to ensure reliability of the prognosis tendered (Gulick et al. 1980).

■ diagnosis, increase in serum hepatic enzymes; slow BSP clearance; liver biopsy; increase in serum bile acids; plasma amino acid ratio changes; test for PAs in plant or animal tissues; test for sulfur-bound pyrroles in animal tissues

Because the microscopic lesions of pyrrolizidine alkaloid intoxication are quite suggestive, hepatic biopsy is a very useful aid in diagnosis. Furthermore it may provide a basis for prognosis.

Diagnosis may be further aided by the use of various tests for the presence of PAs in the plant, the liver, or other animal tissues. A rapid colorimetric field test for plant tissues is available that employs nitroprusside to convert the parent PA to a pyrrole and Ehrlich's reagent for color development (Mattocks & Jukes 1987). The test is reasonably specific for unsaturated PAs. Enzyme-linked immunoabsorbent assay (ELISA) also can be used on plant tissues to detect specific PAs (Roseman et al. 1992; Langer et al. 1996). Whereas the unreacted pyrroles are short-lived, those that have reacted with sulfur-containing amino acids and become sulfur-bound are long-lived and may be extracted from liver, either fresh or formalin-fixed, and identified chromatographically (Mattocks & Jukes 1990, 1992; Winter et al. 1990, 1992). PA exposure also can be monitored by identification of sulfur-bound pyrrolic metabolites in other tissues such as hemoglobin in RBCs (Bryant & Osterman-Golkar 1991; Seawright et al. 1991a,b; Winter et al. 1993).

- gross pathology, liver small and firm, edema of the omentum, increased fluid in the abdomen

  microscopic, liver, centrilobular necrosis; bile duct proliferation and portal fibrosis; megalocytosis and karyomegaly; central vein fibrosis

Pathology—Although not always present, the most striking features seen grossly following chronic ingestion of the toxicant are a firm, fibrotic, shrunken, grayish blue to yellowish liver, with marked accentuation of the lobular pattern; a thick, bluish gray, edematous omentum; and/or an edematous appearance of the viscera. There may be abundant ascitic fluid; distension of the gallbladder, with thickened walls; and scattered hemorrhages over the intestinal serosa. Other changes of interest include patchy and mild gastroenteritis. In rare instances, where large amounts of the plants are eaten in a few weeks, the liver may be enlarged and congested due to hepatocellular necrosis and intralobular hemorrhage.

Microscopically, four types of hepatic changes are considered representative of exposure to PAs: centrilobular necrosis, bile duct proliferation and portal fibrosis, megalocytosis/karyomegaly up to 20 times normal size, and fibrosis around and within central veins. Early in the course of disease, there may be random small foci of necrosis, but this is usually minimal late in the disease in comparison with the severe disruption in lobular architecture due to bile duct proliferation and extensive portal fibrosis. In cattle, prominent and sometimes numerous intranuclear inclusions due to cytoplasmic invagination are seen as an early change (Craig et al. 1991). A ring appearance of the nuclei is seen with margination of chromatin, leaving a pale pink center. Megalocytosis, which is especially prominent in horses, is not an exclusive feature of PA toxicosis, given that other toxicants, including aflatoxins, also may be a cause (Jago 1969). Binucleated hepatocytes may also be seen. In instances of long-term but interrupted exposure, there may be regenerative nodules. Diagnostically noteworthy are the megalocytosis and the distinct fibrosis within central veins.

In humans, the disease is known as veno-occlusive disease because of fibrosis and occlusion of hepatic vein branches. This leads to congestion and extensive fibrosis around the central veins (Huxtable 1989). The restriction in hepatic blood flow results in fluid accumulation in tissues and in the abdomen.

- minimal extrahepatic lesions; mild renal and rarely pulmonary lesions

Extrahepatic lesions with *Senecio* are not common, although myocardial necrosis, colonic edema and hemorrhage, and adrenal cortical hypertrophy are reported (Mendel et al. 1988). Occasionally, mild to moderate nephrosis with megalocytosis may be apparent. Lung changes are more likely with other sources of pyrrolizidine alkaloids such as *Crotalaria* in the Fabaceae.

- general nursing care; sedation, fluids, high-quality and high-energy feeds with low but high-quality protein; supplementation with sulfur amino acids is of questionable value

Treatment—It is important to keep in mind that the effects of these toxic alkaloids are cumulative, at least until compensatory hepatic regeneration occurs. Because of this, prevention is important especially with respect to exposure via hay.

Treatment as outlined by Divers (1996) for liver failure is supportive and typically includes sedation, parenteral fluids, and dietary management to supply energy needs while keeping protein levels low to reduce the ammonia load to the liver. Energy needs may be met with molasses, corn, and sorghum, in conjunction with grass hay or beet pulp. However, treatment is generally of little value in most cases because, by the time signs appear, the hepatic changes are extensive and severe. Although decreasing protein in the diet may be used in treatment once liver disease is apparent, this may not be appropriate for prevention, given that increased toxicity is apparent with protein deficiency (Mattocks 1986). Sulfur-containing amino acids such as cysteine and methionine have been reported to be of some value in treatment, but this has not been confirmed experimentally (Shull et al. 1976). Although the branched-chain amino acid/phenylalanine-tyrosine ratio is altered, therapy with branched-chain amino acids has not proven of value (Mendel et al. 1988). Possibly any approaches that increase hepatic glutathione content may be helpful.

Alkaloid concentrations can be reduced appreciably by ensiling the forage rather than making hay, but the PAs are not entirely eliminated (Candrian et al. 1984). Silage with a particularly high percentage of *Senecio* should be fed with caution because intoxications have been reported (Donald & Shanks 1956). Supplementation with minerals, especially phosphorus, copper, and cobalt, which has been advocated for reducing *Senecio* intake, has not proven to be of value experimentally (Johnson 1982).

The public health hazard should be considered, especially for cattle, because the alkaloids are potential carcinogens and can be present in low levels in milk and possibly the liver.

Solidago L.

Taxonomy and Morphology—A mainly North American genus of about 100 species, *Solidago* is member of the tribe Astereae and is commonly encountered across the continent as a constituent in a variety of plant communities, as an ornamental, and as a weed. Its name comes from the Latin *solidus* and *ago,* meaning "to make whole," alluding to its reputed healing properties. Paradoxically, its pollen is often mistakenly thought to be the cause of hay fever. The pollen is not wind-borne, but because *Solidago*'s showy heads are in full bloom at the time individuals begin to suffer, it is blamed. The actual culprits are species of *Ambrosia,* the ragweeds, which have inconspicuous heads but produce copious amounts of wind-borne pollen. *Solidago* species of toxicologic interest include the following:

- *S. flexicaulis*
  S. hispida
  S. juncea
  S. missouriensis
  S. mollis
  S. rigida
  S. spectabilis

  goldenrod

S. flexicaulis L.	broad-leaved goldenrod
S. hispida Muhl. ex Willd.	hairy goldenrod
S. juncea Aiton	early goldenrod
S. missouriensis Nutt.	golden goldenrod, Missouri goldenrod
S. mollis Bartl.	ashy goldenrod, velvety goldenrod, soft goldenrod
S. rigida L.	rigid goldenrod, stiff goldenrod
S. spectabilis (Eaton) A.Gray	basin goldenrod, Nevada goldenrod, western goldenrod

Plants herbs; perennials; from rhizomes or caudices. **Stems** erect or ascending; 10–200 cm tall; branched above. **Leaves** simple; alternate; primarily cauline or both cauline and basal; blades of various shapes, commonly oblanceolate or obovate; margins entire or serrate or dentate. **Heads** numerous; small; borne in panicles or corymbs or axillary clusters; secund or not secund; subcylindric to narrowly campanulate; phyllaries in several-series, imbricate, linear to lanceolate, bases chartaceous; receptacles flat or slightly convex, alveolate; receptacular bracts absent. **Ray Florets** 3–25; pistillate; fertile; corollas yellow or rarely white. **Disk Florets** 3–60; perfect; fertile; corollas yellow. **Pappus** of capillary bristles; usually white. **Achenes** cylindrical; terete or anular; many-ribbed; glabrous or indumented.

- perennial herbs; leaves simple, alternate; heads small, numerous; ray florets yellow or rarely white; disk florets yellow

Species of *Solidago* are notorious for their hybridization and introgression of characters; thus identification of individual plants can be difficult. In general, they can be

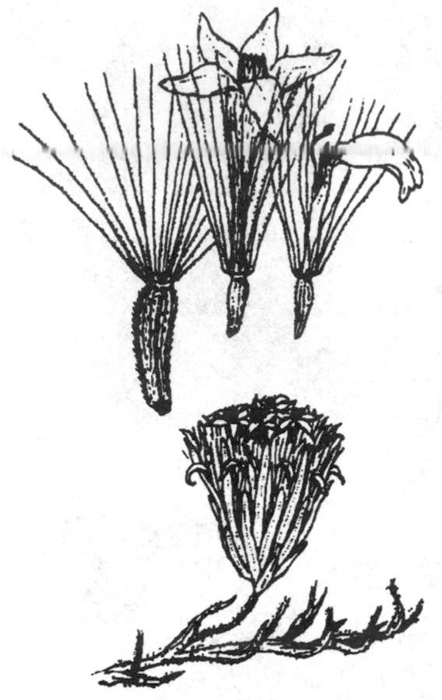

Solidago canadensis

Solidago canadensis

Solidago missouriensis

Solidago mollis

Solidago rigida

divided into three groups on the basis of their underground organs, the appearance and distribution of their leaves, and the appearance of their inflorescences.

1. Heads borne in terminal corymbs; large, 7–13 mm in diameter . . . *S. rigida*
1. Heads borne in terminal panicles or terminal racemes or axillary clusters; small, 2–5 mm in diameter.
    2. Heads borne in terminal and axillary racemes . . . . *S. spectabilis*
    2. Heads borne in axillary clusters or terminal panicles.
        3. Heads borne in axillary clusters; not secund.
            4. Caudices present; branched. Rhizomes absent. Upper stems hispid . . . . *S. hispida*

4. Caudices absent. Rhizomes present.
   Upper stems glabrous . . . . *S. flexicaulis*
3. Heads borne in terminal panicles; secund.
   5. Leaves and upper stems densely hirtellous . . . . *S. mollis*
   5. Leaves and upper stems glabrous or sparsely hirsute.
      6. Leaves conspicuously 3-nerved. Achenes glabrous or sparsely indumented. Basal leaves deciduous . . *S. missouriensis*
      6. Leaves 1-nerved. Achenes densely indumented. Basal leaves persistent . . . . *S. juncea*

Distribution and Habitat—*Solidago* exhibits greatest diversity in the eastern half of the continent, with numerous species occupying a variety of soil types in dry or moist, open sites in the eastern forests and grasslands from Canada south to the Gulf of Mexico. *Solidago spectabilis* is one of the relatively few species found in the western half. It occurs in saline marshes and seeps of the Great Basin area.

■ wide variety of habitats

Solidago flexicaulis

Solidago hispida

Solidago juncea

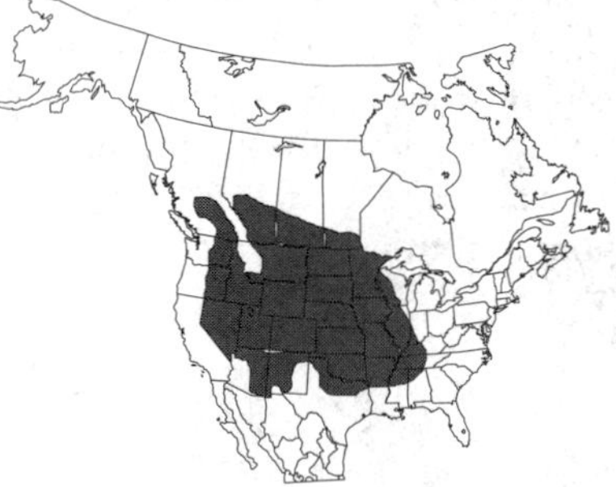
Solidago missouriensis

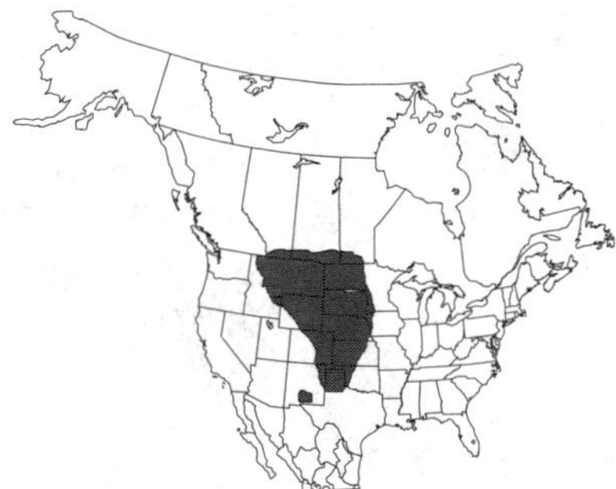
Solidago mollis

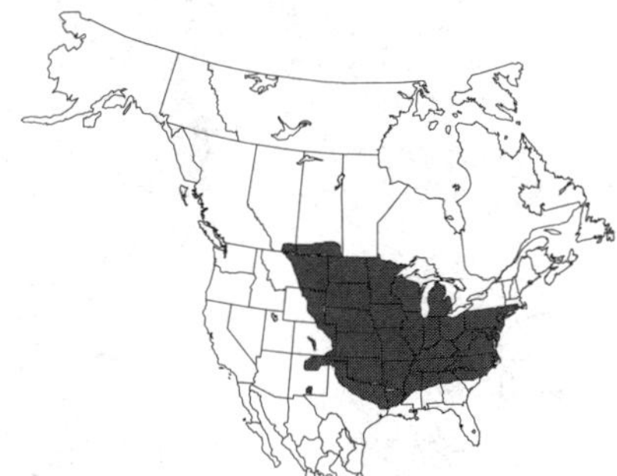
Solidago rigida

Disease Problems—In the late 1800s and early 1900s in Wisconsin, a chronic disease problem in horses was described (Pammel 1911). The animals exhibited depression, emaciation, incoordination, pale mucous membranes, and edema of the extremities. Death occurred in 2–8 weeks. Circumstantial evidence indicated the cause to be an unidentified species of *Solidago,* but other observations revealed the presence of fungi on the suspect plants in some instances (Scott 1895; Rusby 1896). Subsequent observations in other areas revealed that some species of the genus are indeed toxic. Species of *Solidago* are of limited palatability, but when eaten in early summer, they may provide useful forage. Reports indicate that they are mainly a problem in the fall and winter while they are still green, succulent, and attractive to hungry sheep (Fleming et al. 1931; Welch & Morris 1952). They are also toxic when dried in hay (Fleming

Solidago spectabilis

■ acute to subacute neurotoxicosis, variable in expression

1918). *Solidago spectabilis* is the species most often implicated. Lockett (1917) reported a number of instances of neurointoxication in sheep involving this species. Typically there was hyperesthesia, excitement, head shaking, excess salivation, seizures, and then death. Sedation was effective in relieving the signs and preventing lethality. The disease was reproduced experimentally; 500 g of plant material given to a 6-to-7-month-old lamb produced similar clinical signs and death about 15 hours after the plant was eaten. However, in another study, sheep and calves given 1.25 and 3.4 kg, respectively, of *S. spectabilis* showed little or no adverse effects (Fleming 1918, 1920). In contrast, *S. mollis* and *S. missouriensis* (reported as *S. concinna*) are reported to have caused death of sheep in 12 hours when given two 0.45 kg doses of partially air-dried plant material 4 hours apart (Beath et al. 1939). Smaller amounts of these species in the feed of sheep caused weight loss. These apparent differences in toxicity may be partially accounted for by changes in growth, given that Beath et al. (1953) suggested that dwarfed types of the various species were the most toxic.

■ toxin unknown; possibly related to block of GABA inhibitory activity

Disease Genesis—Specific toxicants have not been identified in *Solidago,* but the clinical signs are suggestive of a GABA inhibitory effect. Unlike so many other genera of the family, sesquiterpene lactones have not been identified as toxic compounds in *Solidago* (Seaman 1982). However, intraperitoneal administration of aqueous or alcoholic extracts of eight species of the genus to mice caused illness and, in some instances, death (Worthley et al. 1967). In this assay the more potent species were *S. flexicaulis, S. hispida,* and *S. juncea.* Species of *Solidago* continue to be suspected as neurotoxic, particularly for sheep.

■ abrupt onset, excess salivation, bizarre lip and jaw movements, muscle quivering, back arched, seizures

Clinical Signs, Pathology, and Treatment—Animals affected by *Solidago* exhibit excess salivation, constant lip and jaw movements, and quivering of their muscles that begins about the ears and head and progresses to other areas. Spasms cause shaking of the entire body. These seizures may be induced by noise or touch. The animals stand with the back arched and legs up under the body. They stand or wander about in a dazed manner. Just before death, they fall and have seizures. Upon necropsy, there are no distinctive or specific lesions. Treatment is primarily symptomatic in response to signs that appear. Sedation to prevent or reduce the severity of the seizures may prevent death in some instances. Animals should be allowed to rest with minimal stress during their recovery.

Tetradymia DC.

■ *T. canescens*
T. glabrata
T. spinosa
T. stenolepis

coal-oil bush
dog brush
horsebrush
lizard shade
rabbit brush
rat brush
skink brush

Taxonomy and Morphology—Commonly known as horsebrushes or cotton-thorns, the 10 species of *Tetradymia,* a member of the tribe Senecioneae, are all native to the western half of North America (Strother 1974). The generic name is derived from the Greek *tetradymos,* meaning "four together," a reference to heads with four florets, which are characteristic of some species of the genus. Of greatest toxicologic interest are the following species (see plate 13.9):

T. canescens DC.	spineless horsebrush, common horsebrush, little gray horsebrush
T. glabrata Torr. & A.Gray	lizard shade, rat brush, dog brush, skink brush, littleleaf horsebrush, spring rabbit brush, coal-oil brush

T. spinosa Hook. & Arnold	catclaw horsebrush, thorny horsebrush, cottonclaw horsebrush, cottonhorn horsebrush
T. stenolepis Greene	Mojave horsebrush

Plants shrubs; perennials; armed or not armed with spines. **Stems** erect or ascending; 10–150 cm tall; profusely branched; typically canescent-tomentose. **Leaves** of 1 or 2 forms; simple; alternate, small and all alike or small leaves fascicled in axils of spinose leaves; small leaves linear to oblanceolate, glabrous to tomentose or canescent; margins entire. **Heads** numerous; borne singly or in pairs on axillary peduncles, forming racemose or corymbose clusters; cylindrical to hemispheric; phyllaries 4–6, in 1- or 2-series, oval-elliptic to lanceolate, canescent-tomentose to glabrous; receptacles flat; receptacular bracts absent. **Ray Florets** absent. **Disk Florets** 4–8; perfect; fertile; corollas yellow to cream-colored. **Pappus** absent or of numerous bristles or scales. **Achenes** obconic or fusiform; terete or angled; obscurely 5-nerved; glabrous or long pilose.

■ profusely branched shrubs; leaves simple, alternate; heads numerous, single or in pairs; disk florets only, yellow to cream-colored

Tetradymia canescens

1. Spines present.
   2. Stems uniformly canescent-tomentose. Clustered leaves glabrous or sparsely pilose. Pappus of scales. Achenes densely villous; hairs long, white .... *T. spinosa*
   2. Stems not uniformly canescent-tomentose; glabrous streaks present. Clustered leaves silvery-sericeous. Pappus of bristles. Achenes sparsely villous; hairs short .... *T. stenolepis*
1. Spines absent.
   3. Primary leaves densely canescent-tomentose; oblanceolate; spreading. Clustered axillary leaves oblanceolate .... *T. canescens*
   3. Primary leaves glabrous or sparsely tomentose; subulate; appressed. Clustered axillary leaves linear .... *T. glabrata*

Distribution and Habitat—Flowering in the spring and summer, species of *Tetradymia* are common, sometimes conspicuous elements in several vegetation types in the intermountain region from British Columbia to Baja California. *T. canescens* is most widespread, occupying barren, rocky/sandy plains, brushy sites and open pine and aspen woods of mountain foothills. *T. glabrata* is almost as wide ranging. *Tetradymia stenolepis* is a species of the Mojave Desert of California. Scattered populations of *T. spinosa* occur in the Great Basin.

■ often open dry sites

Tetradymia glabrata

Disease Problems—At one time, species of *Tetradymia* were among the most devastating of the phototoxic plants, causing, for example, losses up to 1,000 sheep in a single episode (Fleming et al. 1922). The most toxic is *T. glabrata,* which becomes green early in spring and is eaten in the bud stage at that time. Thus for a few weeks in the early spring it poses a serious hazard. *Tetradymia canescens* becomes green during the summer when there is an abundance of other forages and is not so likely to be eaten. It also appears to be less toxic. Geographic distribution of the disease corresponds generally to the range of these two species, but other species of the genus in the area also should be considered potentially toxic (Clawson & Huffman 1936). *Tetradymia stenolepis* is suspected of causing intoxications of sheep in California (Fuller & McClintock 1986).

■ liver disease and hepatogenous photosensitivity, especially in sheep, bighead; bud stage of plant in spring or early summer; possibly due to ingestion of combination of *Artemisia* and *Tetradymia* species

Although the acute disease originally was attributed to *Tetradymia* alone, it now appears that there are complicating factors (Fleming et al. 1922; Jennings et al. 1978; Johnson 1978). It was difficult experimentally to consistently produce all features of

Tetradymia spinosa

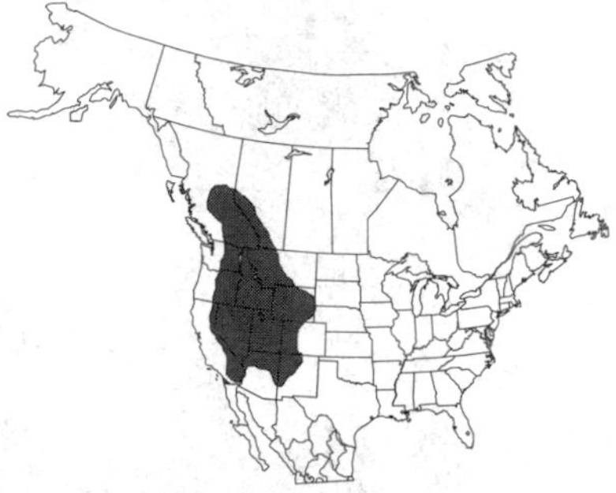

Tetradymia canescens

- toxin not clearly established but possibly furanoeremophilane derivatives such as tetradymol, which may be biotransformed to a toxin causing liver disease with accumulation of photodynamic phylloerythrin

Tetradymia glabrata

Tetradymia spinosa

the disease when only plants of the genus were fed. However, when preceded by ingestion of *Artemisia nova* or *A. tridentata, Tetradymia* was much more likely to cause severe disease, including photosensitization. Fleming et al. (1922) described an acute disease with liver involvement following ingestion of 1% b.w. or more in a single dose or 0.3–0.4% for 10 days or more. Later studies implicated large doses as causing severe acute disease and smaller doses (0.5%) causing illness accompanied by photosensitization (Clawson & Huffman 1937). Similarly, in the original studies it was thought that only fresh, green plants were toxic, but later work revealed that not only are all stages of growth toxic but also toxicity was retained to a large extent following drying (Clawson & Huffman 1936; Johnson 1974a).

Occurring in sheep, the disease is called bighead because of the swelling of the skin of the face, ears, and lips. This swelling may be so severe that tissues crack and serum exudes from the skin. In some instances, animals become permanently blind (Johnson 1974a). There is distinct and severe liver damage accompanying the onset of skin lesions; thus the disease is one of hepatogenous or secondary photosensitization. In addition to liver involvement, Welch and Morris (1952) described an apparent paralysis associated with abrupt onset of signs of intoxication. Severe liver disease and death are likely with the more toxic *T. glabrata,* whereas with *T. canescens* there is greater opportunity for development of photosensitization (Johnson 1974b). As a practical matter, this disease is restricted to sheep, although they are not necessarily the only species susceptible.

Disease Genesis—Although not definitely confirmed, toxicity seems to be due to furanoeremophilane derivatives, which are structurally closely related to the sesquiterpene lactones. Tetradymol, which is hepatotoxic in laboratory animals and sheep, is the most abundant of these derivatives (Jennings et al. 1974, 1978). Tetradymol is a lipid-soluble, type 1 mixed-function oxidase binder that weakly uncouples oxidative phosphorylation. The toxicity of tetradymol is enhanced by biotransformation inducers and inhibited by blockers. The supposition is that toxicity is therefore due to the products of hepatic biotransformation (Jennings et al. 1978). This is consistent with *Tetradymia*'s apparent interaction with *Artemisia,* because species of the latter genus seem to influence activity of the hepatic mixed-function oxidase pathway (Eissa et al. 1995, 1996). The overall effect is a debilitating liver disease; the cause of the disease, and presumably like that occurring with *Tribulus* (a member of the Zygophyllaceae, in chapter 76), is an inability of the liver to clear the photodynamic chlorophyll derivative phylloerythrin. This photodynamic derivative is activated by UV light in those body areas most susceptible to exposure.

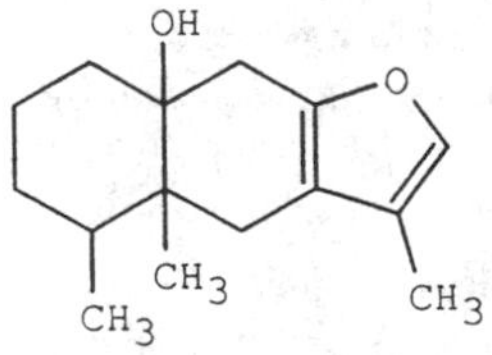

Tetradymol

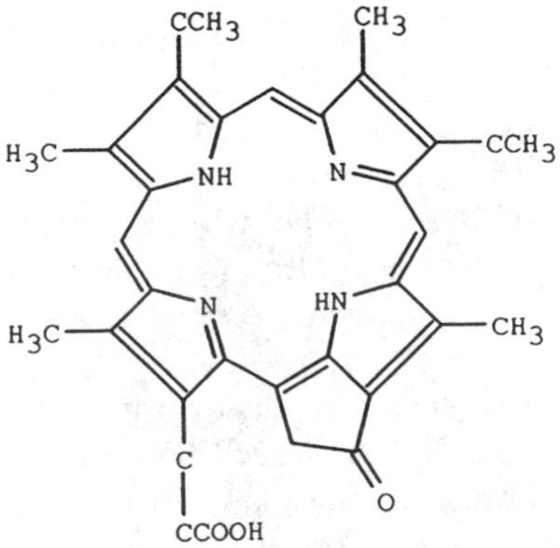

Phylloerythrin

Tetradymia stenolepis

Clinical Signs—When large amounts of *Tetradymia* are eaten, there is abrupt onset of loss of appetite, lethargy/depression coupled with uneasiness, tremors, incoordination, and labored respiration. In some instances the signs—e.g., head pressing and icterus—are more directly associated with liver disease. A few animals may die in 24–48 hours, some may do poorly for several weeks before recovery, and others may only develop swelling and erythema of the nose, ears, lips, and skin around the eyes. This leads to peeling and sloughing of the skin in the affected areas and in some cases loss of ears or eyes. The signs in the latter instance are those typical of photosensitization. The disease seems to be clearly the hepatogenous type, because serum hepatic enzymes (AST, SDH, GGT) are elevated 10–20 times and BSP clearance time is increased 10-fold.

- abrupt onset, anorexia, depression, uneasiness, swelling and erythema of skin of head especially

Pathology—Grossly, the liver is tan and friable and the gallbladder is distended with bile. There also may be reddening of the mucosa of the abomasum and small intestine, with isolated small hemorrhages. Emphysema of the lungs also may be prominent. Microscopically, there is marked centrilobular hepatic necrosis.

- gross pathology, liver tan and friable

  microscopic, hepatic necrosis

Treatment—In instances of intoxication due to *Tetradymia,* there is little to be done other than good nursing care to enable the animal to survive long enough for regeneration of hepatic function. It is not clear whether limiting intake of green forage to reduce the chlorophyll load is directly helpful.

- general nursing care

Verbesina L.

Taxonomy and Morphology—Comprising approximately 60 species native to the New World, *Verbesina,* a member of the tribe Heliantheae, is a morphologically diverse genus of herbs, shrubs, and trees. Of the 17 species in North America, only one is of toxicologic significance:

- *V. encelioides*

  American dogweed
  butter daisy
  cowpen daisy
  golden crownbeard

V. encelioides (Cav.) Benth. & Hook.	golden crownbeard, butter daisy, cowpen daisy, American dogweed

Plants herbs; annuals; from taproots; herbage strigose-canescent; gray-green. **Stems** erect; 20–100 cm tall. **Leaves** simple; alternate above and opposite below; blades ovate to deltoid; margins coarsely dentate; petioles present or absent and bases clasping stems. **Heads** large; borne singly or in clusters of 2 or 3 at ends of peduncles; hemispheric; phyllaries in 2-series, imbricate, oblong to lanceolate, strigose; receptacles convex; receptacular bracts present, folded. **Ray Florets** 10–15; pistillate; fertile; corollas yellow, apices 3-toothed. **Disk Florets** numerous; perfect; fertile; corollas yellow. **Pappus** absent on ray achenes and of 2 awns on disk achenes. **Achenes** of 2 forms; ray achenes small, obconic, rugose; disk achenes obovate, flattened; winged; hirsutellous.

- annual herbs; leaves simple, alternate below, opposite above; heads terminal; ray and disk florets yellow; ray apices 3-toothed

Verbesina encelioides is readily distinguished from other members of the genus by its annual habit and taproot system. The other species are perennials with a fibrous root system. Two varieties are generally recognized. Occurring primarily in the eastern half of the continent, especially in the Southeast, var. *encelioides* has leaves with clasping bases and longer phyllaries. Occupying drier sites in the western part of the continent, var. *exauriculata* has petioled leaves and shorter phyllaries.

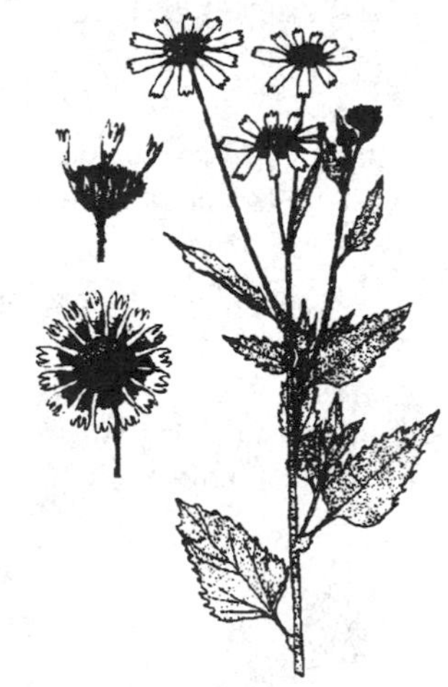
Verbesina encelioides

Distribution and Habitat—Believed to be native to Mexico and the southwestern United States, *V. encelioides* is found across the southern half of the continent.

■ disturbed-site invader

Adapted to a variety of soil types, it often forms large populations in disturbed sites. It is adventive and naturalized in other parts of the world.

■ rare problem in North America; marked fluid accumulation indicative of cardiac insufficiency

Disease Problems—Although *V. encelioides* is native to western North America, it has proven to be a much more serious problem for sheep and cattle in areas where it has naturalized in Australia (Oelrichs et al. 1985). Because it is not readily eaten, it is mainly a problem during times of drought or when hungry animals being moved from one area to another are allowed to browse along the way. A dose of 0.5% b.w. of green plant material is lethal in 1–2 days; sheep are at greatest risk. The signs are suggestive of a cardiovascular problem, but this has not been confirmed. Intoxication also has been experimentally reproduced in sheep in Argentina with a dose of 5 g/kg b.w. (Lopez et al. 1996). Dosage of 6.3 g dried plant material/kg b.w. produced a disease with signs and pathology similar to those seen in Australia. Force feeding of 4.5% b.w. of another species of the genus, *V. nana,* over a period of 3 days to a sheep failed to produce adverse effects (Boughton & Hardy 1941).

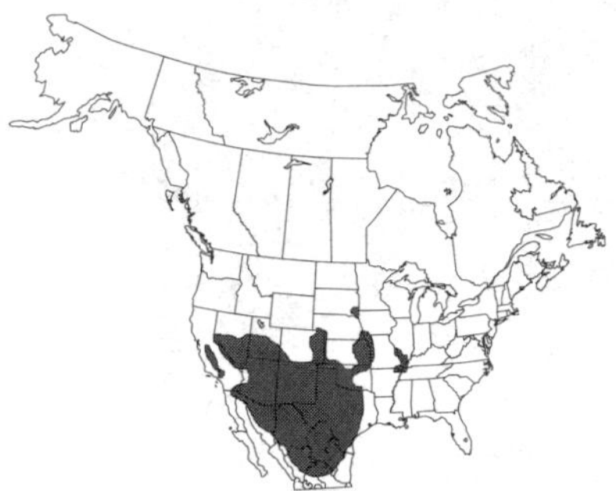

Verbesina encelioides

■ toxin is galegine; also found in *Galega* of Fabaceae

$$(H_3C)_2C{=}C{-}C{-}N{-}C({=}NH){-}NH_2$$

Galegine

Disease Genesis—In *V. encelioides,* the toxicant is galegine, 3-methyl-2-butenyl-guanidine, the same compound found in *Galega officinalis,* of the Fabaceae (see chapter 37), and in *Schoenus asperocarpus* (poison sedge), a member of the Cyperaceae found in western Australia (Eichholzer et al. 1982; Oelrichs et al. 1985; Huxtable et al. 1993). There also may be nitrate accumulation. The toxin is not cumulative—small amounts ingested daily do not produce adverse effects. A cardiovascular effect for galegine is supported by the observed effects of a structurally similar but more complex compound, caracasanamide, from *V. caracasana,* a shrubby South American species (Delle Monarche et al. 1993). Caracasanamide, a dimethoxyphenyl compound linked via an ethylene amide side chain to a galegine moiety, causes positive inotropic effects, vasodilation, and respiratory stimulation at low dosage. However, overdosage results in respiratory depression and cardiac failure.

Toxin concentrations are quite variable, but in general they increase with plant maturity up to 0.25% d.w. Concentrations of galegine are low in *Verbesina,* averaging about 0.08%, relative to 0.46% for *Galega,* but the broader distribution of *V. encelioides* in North America makes it a greater potential risk for livestock (Keeler et al. 1992).

■ abrupt onset, depression, listlessness, frothy nasal exudate, exercise intolerance, death in 1–2 days

Clinical Signs—Animals ingesting *V. encelioides* typically die in 1–2 days. In some instances signs such as depression, listlessness, frothy nasal exudate, and markedly elevated heart and respiratory rates may be observed.

■ gross pathology, fluid accumulation throughout body, pulmonary edema and froth in airways

microscopic, pulmonary edema and hepatic necrosis

Pathology—The gross changes are very distinctive and are suggestive of cardiac failure, with accumulation of fluid in the thoracic cavity and around the heart, and considerable froth in the airways coupled with congestion and edema of the lungs. There also may be congestion of the liver. Microscopically, there is hepatic fatty change, hepatocellular degeneration, and hemorrhage and edema of the lungs.

■ diuretics, general nursing care

Treatment—In instances in which a diagnosis can be made based upon clinical signs, diuretics and treatment to relieve the respiratory distress may be of value.

Viguiera Kunth

Taxonomy and Morphology—Comprising 150–180 species distributed from the western United States to central Argentina, *Viguiera* is a member of the tribe

Heliantheae. It is morphologically quite similar to *Helianthus* (sunflowers) but can be distinguished by the appearance of its receptacle, involucre, and pappus. The generic name honors the French botanist and physician Alexandre Viguier, and species are commonly called goldeneyes. In North America, 13 or 14 species are present. Some taxonomists, however, segregate some of them into the genus *Heliomeris*. Two species are of toxicologic interest:

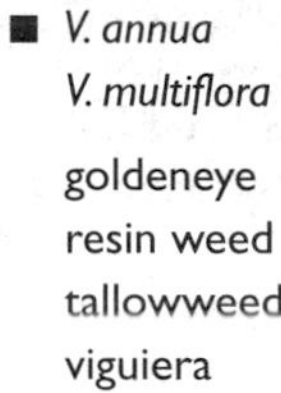

■ *V. annua*
V. multiflora

goldeneye
resin weed
tallowweed
viguiera

V. annua (M.E.Jones) S.F.Blake	annual goldeneye, resinweed, tallowweed, southern goldeneye
V. multiflora (Nutt.) S.F.Blake (= *Heliomeris multiflora* Nutt.)	showy goldeneye, viguiera, perennial goldeneye

Plants herbs or shrubs; annuals or perennials; from compact caudices or taproots. **Stems** erect or ascending; 15–130 cm tall; branched; surfaces strigose. **Leaves** simple; opposite below and alternate or opposite above; blades linear to lanceolate-linear; margins entire or sparsely toothed. **Heads** showy; borne on elongate peduncles; hemispheric; phyllaries in 2- to several-series, imbricate, linear to lanceolate; receptacles flat to convex; receptacular bracts present, clasping achenes, 3-lobed or not lobed. **Ray Florets** 8–16; neutral; sterile; corollas yellow, apices 3-toothed. **Disk Florets** numerous; perfect; fertile; corollas yellow. **Pappus** absent or of 2 awns and several short scales. **Achenes** turbinate or oblong; somewhat flattened; glabrous; black.

■ herbs or shrubs; leaves simple, opposite below, variable above; heads terminal; ray and disk florets yellow; ray 3-toothed

Viguiera multiflora

■ seasonally dry, arid uplands

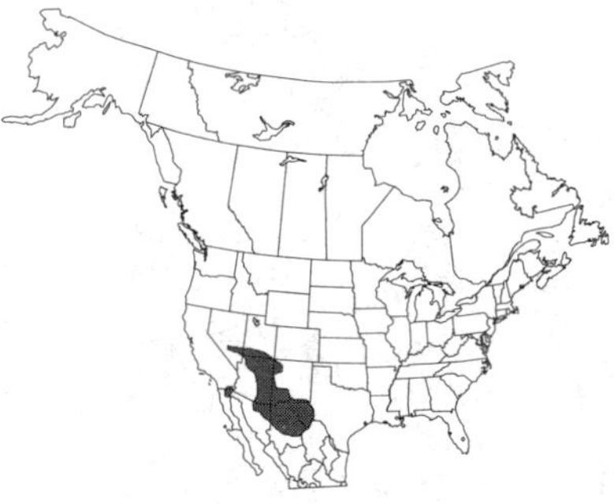

Viguiera annua

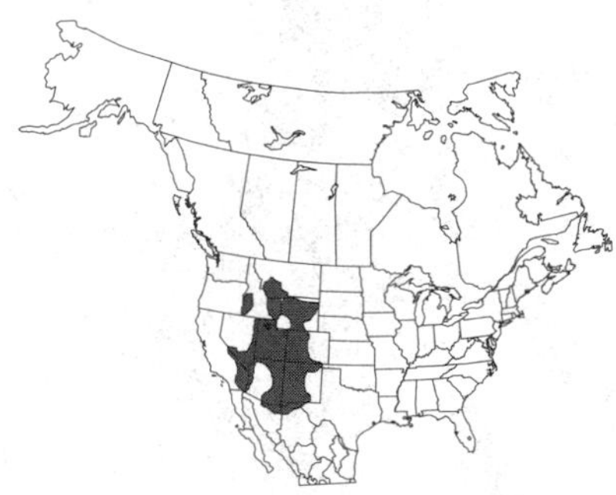

Viguiera multiflora

1. Plants annuals. Taproots present . . . . *V. annua*
1. Plants perennials. Fibrous roots present . . . . *V. multiflora*

Distribution and Habitat—*Viguiera* occurs primarily in seasonally dry, arid, elevated areas, with some of its members extending into tropical lowlands. Populations of *V. annua* occupy gravelly or sandy soils up to 2,300 m elevation. Plants are sometimes found along roadsides and in other disturbed sites and are especially locally abundant in southern Arizona. *Viguiera multiflora* occurs at higher elevations, with sagebrush and in open pine-juniper woodlands.

■ abrupt onset, cellular oxygen deficiency due to presence of MHb secondary to formation of nitrite from nitrate accumulation in plants

Disease Problems and Genesis—Sporadic but significant cattle losses are reported for animals eating *V. annua* when it forms large, dense stands following abundant germination of the achenes after autumn rains (Schmutz et al. 1968; Williams 1989). It and *V. multiflora* accumulate nitrate when growing in areas with excess nitrogen due to fertilizers or animal excreta (Williams 1989). Concentrations range as high as 4% KNO_3 or more, especially in the stems.

■ abrupt onset, labored breathing, weakness, trembling, collapse, quick death

Clinical Signs, Pathology, and Treatment—Within a few hours of plant ingestion, signs are abruptly apparent: labored respiration, trembling, weakness, and eventual collapse. If not treated immediately, most animals showing these signs die within minutes or an hour. A complete discussion of nitrate/nitrite intoxication, including the use of methylene blue for treatment is presented in the treatment of *Sorghum,* in chapter 58.

Xanthium L.

■ *X. ambrosioides*
X. spinosum
X. strumarium

abrojo
Bathurst bur
bur thistle
burweed
buttonbur
clotbur
cocklebur
ditchbur
hedgehog bur
sheep bur

Taxonomy and Morphology—A member of the tribe Heliantheae, *Xanthium* is apparently native to both the Old World and the New World. Its generic name is from the Greek *xanthos,* meaning "yellow," and possibly refers to the yellow dye extracted from its burs. Although more than 50 species have been recognized in the past 100 years, there is now general agreement among taxonomists that the genus comprises only 3:

X. ambrosioides Hook. & Arn.	
X. spinosum L.	spiny cocklebur, clotbur
X. strumarium L.	common cocklebur, burweed, abrojo, sheep bur, Bathurst bur, bur thistle, buttonbur, hedgehog bur, ditchbur

■ annual herbs; leaves simple, alternate; heads staminate or pistillate; florets inconspicuous, borne in prominent prickly burs

Xanthium strumarium

Plants herbs; annuals; from taproots; herbage scabrous; armed or not armed with spines; monoecious. **Stems** erect or ascending or decumbent; 20–200 cm tall. **Leaves** simple; alternate; blades lanceolate or obovate or ovate to deltoid; margins entire or toothed or lobed or pinnatifid; petioles long. **Heads** of 2 types, staminate and pistillate different. **Staminate Heads** borne in clusters, terminal or in upper leaf axils; hemispheric; phyllaries absent or in 1–3 series, free; receptacles conical; receptacular bracts present. **Pistillate Heads** borne singly or in clusters in middle and lower leaf axils; phyllaries fused and completely enclosing florets to form a bur with hooked prickles. **Ray Florets** absent. **Disk Florets** numerous in staminate heads, 2 in pistillate heads; corollas absent or white to cream-colored; stamens 3–6, filaments united, anthers free. **Pappus** absent. **Achenes** remaining within burs; fusiform; black. **Burs** ellipsoidal; brown or yellowish brown; prickle bases pilose or glandular.

1. Stems not bearing spines in leaf axils. Leaves deltoid to suborbicular; bases cordate. Burs with 2 beaks . . . . *X. strumarium*
1. Stems bearing spines in leaf axils. Leaves lanceolate or obovate; bases acute or cuneate. Burs with 1 beak.
   2. Stems erect or ascending. Leaves lanceolate. Burs puberulent . . . . *X. spinosum*
   2. Stems decumbent. Leaves obovate. Burs densely villous . . . . *X. ambrosioides*

The plethora of species names that have been published apply for the most part to X. *strumarium.* Now recognized to be an extremely variable species both morphologically and ecologically, it was historically divided into numerous species primarily on the basis of differences in bur morphology. Variation, however, has been determined to be continuous among the described forms, and intergradation so complete that even recognition of varieties is tenuous.

Distribution and Habitat—Adapted to a variety of clay, loam, and sandy soils, species of *Xanthium* are now distributed throughout the warmer parts of the world as weeds of disturbed sites. They are classic indicators of damp soils and often form large, dense populations in floodplains, along the shorelines of ponds and reservoirs, and in the highly disturbed soils of feedlots. *Xanthium strumarium* is now believed to be an Old World species, whereas X. *spinosum* and X. *ambrosioides* are of the New World. It has been hypothesized that X. *spinosum* originated in South America. *Xanthium ambrosioides* is native to the central Argentine plains of Patagonia and is rarely encountered in North America, in contrast to the other two species.

■ weedy invader species, disturbed sites; especially abundant in sandy floodplains and shorelines

Disease Problems—Ubiquitous in distribution, *Xanthium* also is cosmopolitan in consistently causing intoxication problems. At one time, mechanical injury from the spines of the burs was considered of prime importance (Hansen 1920). Creation of sores about the face and mouth, irritation and blockage of the throat and digestive system, tangles in the throat and digestive system, and tangles in wool or hair were said to be the most likely consequences of animal contact with the burs. Subsequently it became apparent that the plants were toxic.

■ rapid-acting; seasonal metabolic derangement due to ingestion of seeds or new sprouts or cotyledons; all animal species, especially pigs and calves; about 1% b.w. of sprouts causes disease, almost always fatal; lower dosage typically causes no apparent effects

In the early 1900s, cocklebur toxicity became a particularly serious problem in the Great Plains and Mississippi Valley; losses were especially severe in New Mexico, Oklahoma, and Texas (Marsh et al. 1923). The disease continues to be a significant seasonal problem. It is encountered mainly in the spring and early summer when both moisture and warm temperatures are conducive to massive sprouting of the burs. Pigs and calves are most commonly involved but other species, including sheep, horses, and chickens, although rarely affected, are also at risk (Reynard & Norton 1942; Watt & Breyer-Brandwijk 1962; Martin et al. 1986).

The toxic dose in both pigs and calves is about 1% b.w. of the fresh green sprouts (Marsh et al. 1923; Stuart et al. 1981; Martin et al. 1992). Sheep appear to require a higher dosage, 1.5–2% b.w. (Mendez et al. 1998). The seeds or burs also may be of concern but are less often associated with losses. A dose of 0.14–0.2% b.w. of the seeds is reported to be toxic in calves, but feed concentrations of 5% seeds were without effect in poultry (Bierer & Rhodes 1960; Witte et al. 1990). The burs themselves are seldom a problem but are reported to cause mechanical injury in chicks and may cause systemic intoxications in pigs and calves on the rare occasions when the diet consists of 20% ground burs or 30% whole burs either in grain or hay (Bierer & Rhodes 1960; Stuart et al. 1981; Witte et al. 1990). It is of interest that in toxicity due to burs in hay, no masses or clusters of burs were seen in the digestive tract of calves (Witte et al. 1990). The three species of *Xanthium* are much alike and probably all cause similar adverse effects (Sampson & Malmsten 1942).

The seeds have been suspected to be a cause of spiking mortality in chickens, but this has not been confirmed experimentally with a dosage of 25% of feed concentration given for 21 days (Goodwin et al. 1992). Decreased weight gain but not liver necrosis or hypoglycemia was observed. These results are not unexpected, because studies in pigs indicate that intoxications occur from large single doses but not from repeated administration of smaller ones.

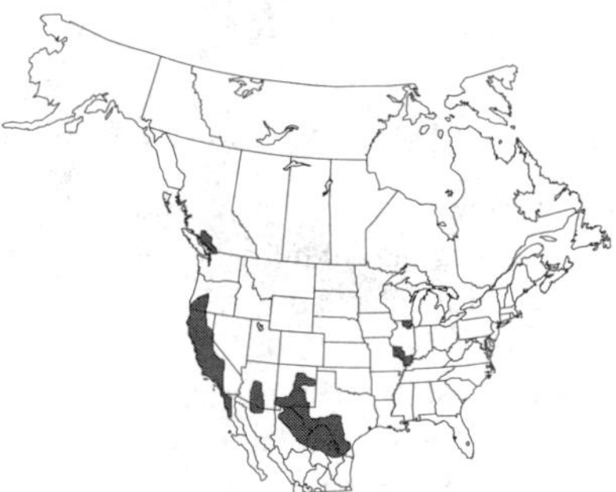

Xanthium spinosum

Xanthium strumarium

- disease due to a series of novel diterpene glycosides; CAT and others

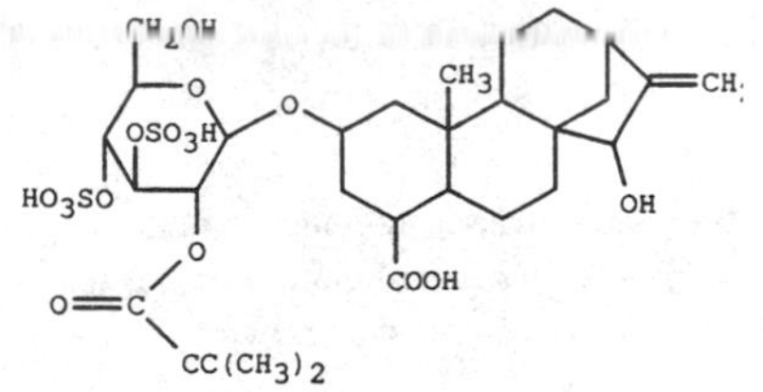

Atractyloside

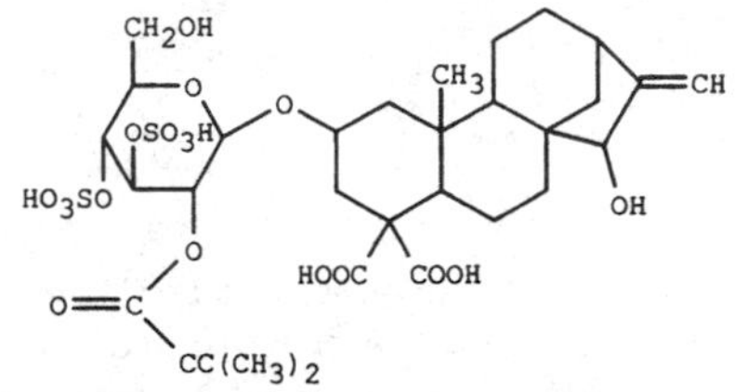

Carboxyatractyloside

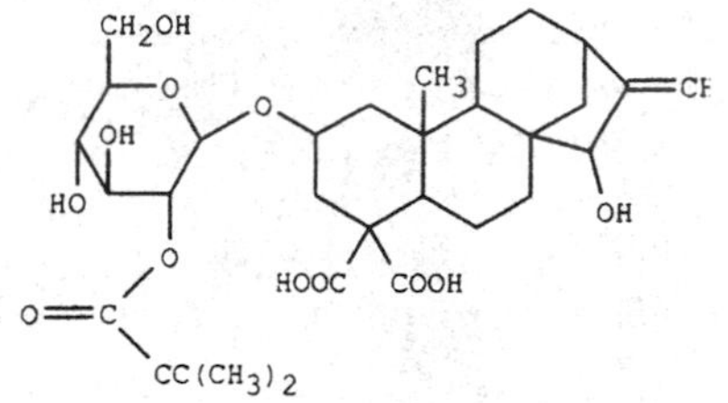

Kaurene glycoside

- inhibition of ADP translocase on cytosolic side of mitochondrial membrane; inhibition of oxidative phosphorylation and ATP formation; variety of tissues affected, brain, liver; signs not all due to liver disease but develop concurrently with it

Disease Genesis—The toxin in *Xanthium* was originally thought to be hydroquinone (Kuzel & Miller 1950), but later studies failed either to demonstrate its presence or to replicate signs of the disease when it was administered. In addition, it produced hyperglycemia instead of hypoglycemia (Cole et al. 1980; Stuart et al. 1981). In the meantime, the novel diterpene glycosides, atractyloside and carboxyatractyloside (CAT, or gummiferin) were isolated from the Old World composite *Atractylis gummifera,* and CAT was subsequently shown to be responsible for the hypoglycemic effects of *X. strumarium* (Danieli et al. 1972; Kupiecki et al. 1974; Craig et al. 1978). The full range of toxic effects was eventually related to CAT, which is concentrated in the embryonic and cotyledonary tissues. It is thus found only in the seeds within the bur (0.46%) and in the two-leaf seedling stage (0.12%) (Cole et al. 1980). By the time the first foliage leaves appear (four-leaf stage), CAT concentrations in those portions are negligible. It is of interest that species of the composite genus *Wedelia*—*W. glauca* and *W. asperrima*—in South America and Australia (also devoid of hydroquinone) cause clinical signs and pathologic changes very similar to those of *Xanthium* (Klingenberg et al. 1985; Seawright 1989; Oelrichs et al. 1992; Collazo & Riet-Correa 1996; Tapia et al. 1996). These species contain atractyloside, wedeloside, and other toxins very similar to CAT, but in the tissues of the mature plants rather than in the seeds or seedlings (Schteingart & Pomilio 1984; Oelrichs et al. 1992). Wedeloside and its congeners differ from CAT in that the aglycone portion is hydroxylated at C-13, and a complex amino sugar is present rather than a sulfated sugar. The complexity of the situation is further underscored by the finding of two additional analogs of wedeloside that are rhamnosyl biosides (MacLeod et al. 1990a,b). Further studies showed additional diterpene kaurene glycosides in *X. pungens* and *X. spinosum,* which share the carboxyatractylogenin but differ in that the glycoside is nonsulfated (MacLeod et al. 1990a). Thus it appears that the toxins should be considered a series of diterpene glycosides of which CAT is a prototype. The respective roles of the aglycone and glycoside in determining toxicity, and the effect of the glycoside and influence of the substituent groups on configuration of the carboxyl groups of the aglycone on translocase binding, are yet to be clarified (Vignais et al. 1966; Allmann et al. 1967a; Klingenberg 1984).

Intoxication is the result of a unique interaction in cellular bioenergetics. Much of our understanding of the actions of CAT date back to the 1800s and to recognition of atractyloside as a powerful tool in the biochemistry laboratory. Atractyloside was eventually shown to inhibit oxidative phosphorylation by blocking transport of phosphoryl groups across the mitochondrial membrane, effects antagonized by ADP (Bruni et al. 1964; Allmann et al. 1967a,b). The target translocase is a membrane-bound protein that binds ADP on the cytosolic side and undergoes a conformation change to facilitate passage of the charged ADP through the inner mitochondrial membrane and to promote an exchange of the outward diffusion process for ATP (Vignais et al. 1973; Brandolin et al. 1974; Vignais et al. 1985). Transmembranal exchange of ADP/ATP is driven by membrane potential to neutralize the excess charge to the cytosol. CAT binds to the translocase from the cytosolic side to block ADP binding and transport. CAT is much more potent than atractyloside, probably because it is little affected by ADP concentrations (Luciani et al. 1971). Because the ADP translocase step is rate limiting in phosphorylation and serves as the exclusive link between energy production inside mitochondria and energy utilization in the cytosol, CAT produces profound effects on cell function and viability (Akerboom et al. 1977; Klingenberg 1984). Other enzyme systems, such as fatty acid oxidation, also are affected and may contribute to the overall effects. Thiols antagonize some of these additional actions of CAT; however, cysteine and glutathione are not protective against the lethal inhibition of ADP/ATP translocase (Allmann et al. 1967a; Hatch et al. 1982). These toxins are not cumulative, and the effects seem to be almost all-or-none, with some animals developing an apparent tolerance (Marsh et al. 1923; Stuart et al. 1981; Mendez et al. 1998).

Since the 1800s, extracts from *Atractylis* have been used for both medicinal and nefarious purposes. Experimentally, in laboratory animals, atractyloside was shown to cause dyspnea, regurgitation, and seizures (Tocco 1922). The metabolic effects were similar to those ascribed to inhibition of the TCA cycle and included an initial increase in blood glucose, lactate, and pyruvate, leading to severe hypoglycemia, decrease in glycogen synthesis, and depletion of glycogen in muscle and liver (Santi 1958, 1964). Concurrently, intracellular ADP, lactate, and pyruvate were greatly increased, and ATP decreased (Pocchiari & Silano 1968). An example of the effects of these toxicants can be seen in the case of a 7-year-old boy given an extract of the root of *Atractylis gummifera* (Georgiou et al. 1988). Within 2 days after ingesting the decoction, he became severely ill and died. There was marked elevation of serum AST, ALT, and bilirubin; diffuse hepatocellular necrosis; and disruption of the hepatic architecture. Although hepatotoxicity is a common sequela to intoxication, it is not necessarily the direct cause of death. Animals may die without these lesions, and a few may develop liver lesions without dying (Stuart et al. 1981; Martin et al. 1992). In addition, experimental studies revealed that protection against the liver effects did not directly influence lethality (Hatch et al. 1982).

Species of *Xanthium* also contain sesquiterpene lactones, some of which are biologically active against insects but are of unknown consequence as toxicants (Deuel & Geissman 1957; Picman 1986). Xanthatin and xantholides are cytotoxic in vivo but are of low toxicity, with an LD_{50} of 800 mg/kg b.w. in mice (Roussakis et al. 1994).

Clinical Signs—Within a few hours of sprout consumption, signs in pigs appear and include depression, weakness, loss of appetite, vomiting, a tucked-up appearance, incoordination, spasmodic muscular activity, and recumbency. The disease may progress to convulsive paddling, coma, and death within a few hours or may take several days. Hypothermia has been noted in some cases. A peculiar gait or posture, with the head held in an erect manner, seems to be typical in pigs but not calves. Except for a lack of vomiting, the signs in calves are otherwise similar to those in pigs, although onset of signs is slightly delayed in older calves with a functional rumen, in some cases for a day or more.

■ abrupt onset, depression, weakness, ataxia, vomiting, seizures, and sometimes hypothermia

sharp decrease in blood glucose, increase in hepatic enzymes

Clinical pathologic changes are very distinctive. These include a 10-fold or greater increase in serum hepatic enzymes (AST, SDH). Blood glucose levels decline markedly (to as low as 10 mg/dl), especially in calves. Unfortunately, these changes are most discernible when the animal is close to death, because the disease often proceeds quite rapidly. The initial rise in blood glucose is short-lived and seldom seen in clinical situations.

■ gross pathology, fluid accumulations, edema of the gallbladder wall and surfaces of liver and gut; liver swollen

microscopic, hepatic necrosis

Pathology—The most prominent gross changes associated with *Xanthium* intoxication are extensive accumulation of protein-rich fluid in the abdominal cavity and, to a lesser extent, in the pericardial sac and thoracic cavity. There is also gelatinous edema of the gallbladder wall, hepatic capsule, and subserosa of the small intestine, with fibrin clots/tags on visceral surfaces. The liver is typically swollen and has accentuated markings, especially on the cut surfaces. Scattered splotchy hemorrhages may be present on the surfaces of the heart and intestines.

Microscopically, the most distinctive lesion is a severe, diffuse centrilobular hepatocellular necrosis involving almost entire lobules. In some instances, few cells or plates remain. Neuronal degeneration and cerebral edema are inconsistent findings. Similar changes are seen in all animal species. Because of the acuteness of the disease, it is likely that plant fragments or intact sprouts will be discernible in the stomach contents. As with all such diseases in pigs, dehydration due to reduced water intake may complicate the diagnosis because of similarities to sodium intoxication and water deprivation.

■ diagnosis aided by plant fragments in stomach; assay of urine or stomach contents for CAT

Diagnosis may be aided by assay for carboxyatractyloside by using thin-layer chromatography. Concentrations of 100–200 ppm in the rumen and 0.1 ppm in urine are reported for intoxicated calves (Witte et al. 1990). There is also potential for the use of an ELISA based on a diterpene glycoside-protein conjugate for diagnostic purposes (Bye et al. 1990).

■ activated charcoal, parenteral glucose, but poor prognosis; neostigmine or physostigmine helpful but not curative

Treatment—It should be stressed that these toxicants are very potent disrupters of cellular function and are poorly amenable to therapeutic intervention. At present there is no specific treatment to counteract the ADP/ATP mitochondrial imbalance. Parenteral glucose and bicarbonate are reported to be helpful but not curative (Santi 1958). The value of these and other parenteral sugars such as fructose remains to be fully evaluated. However, supportive measures can be taken to reduce continued absorption of the toxin. Because fatty diets—i.e., milk or lard—were noted to be protective in pigs eating the sprouts, they have been recommended for treatment, with some measure of success (Marsh et al. 1923). Activated charcoal may also be useful in decreasing toxin absorption, but it should not be used concurrently with the fatty substances. Neostigmine or physostigmine given intramuscularly may provide temporary alleviation of muscular problems and improvement of overall attitude, but these drugs are not curative and are probably mainly of diagnostic benefit.it.

Xylorhiza Nutt.

■ *X. glabriuscula*
parry aster
woody aster

Taxonomy and Morphology—Comprising 8 species of small shrubs and herbs of the xeric regions of the western United States and Mexico, *Xylorhiza* is a member of the tribe Astereae. Its generic name is sometimes erroneously spelled as *Xylorrhiza.* Taxonomists differ in their opinions as to whether it should be recognized as a distinct genus or as a section of *Machaeranthera* or *Aster,* but Watson (1977) presents convincing evidence for its recognition as a distinct genus. Only 1 species is of toxicologic concern:

X. glabriuscula Nutt. — parry aster, woody aster
(= *Machaeranthera glabriuscula* [Nutt.] Cronquist & D.D.Keck)
(= *Aster parryi* A.Gray)
(= *Aster xylorhiza* Torr. & A.Gray)

■ perennial herbs; leaves simple, alternate; heads terminal; ray florets white or blue; disk florets yellow

Plants herbs; perennials; from branched caudices and woody taproots; herbage densely puberulent-villous to glabrous. **Stems** erect; several from caudices; 10–40 cm tall; usually not branched. **Leaves** simple; alternate; blades linear to lanceolate or oblanceolate; margins entire or rarely sparsely toothed. **Heads** borne singly at ends of branches or stems; broadly campanulate or hemispheric; phyllaries in 2- or 3-series, imbricate, linear or lanceolate, glabrous to densely villous; receptacles convex; receptacular bracts absent. **Ray Florets** 12–30; pistillate; fertile; corollas white or pale blue. **Disk Florets** 35–125; perfect; fertile; corollas yellow. **Pappus** of capillary bristles; barbellate; often brown. **Achenes** linear or clavate; slightly compressed; velutinous-sericeous.

Distribution and Habitat—*Xylorhiza glabriuscula* occupies barren, open, alkaline sites of the northern Great Plains and Rocky Mountains principally in Wyoming. Populations also occur in Montana, South Dakota, Colorado, and Utah.

■ barren, open, alkaline sites

Xylorhiza glabriuscula

Disease Problems—Plants of *X. glabriuscula* have a disagreeable odor and bitter taste. They are eaten only when animals are very hungry, for example, by sheep immediately following shearing when they have been penned for some time without feed (Beath 1920; Marsh 1924). Plants also seem to be more readily eaten at senescence when they begin to dry; they are also less toxic at this time (Prien & Raiford 1911). A high proportion of acutely affected animals die, especially lambs and older pregnant ewes. Consumption of 0.5 kg b.w. of green plant material by sheep is reported to cause illness (Huffman & Couch 1942), and dosage of 0.3% b.w. of green plants/day for 3–4 days or a single dose of 0.875% b.w. is lethal in sheep (Beath et al. 1939, 1953).

■ abrupt onset, apparent cardiac failure, excess selenium in the plants, risk especially in sheep eating young plants

Disease Genesis—Disease is due to the high levels of selenium accumulated in the plants. *Xylorhiza glabriuscula* is considered to be an indicator species for the metal (Beath et al. 1939; Trelease & Beath 1949). Concentrations are highest in young plants and decline with the onset of flowering and advancing maturity (Beath 1920). A more complete discussion of aspects of selenium intoxication is presented in the treatment of *Astragalus,* in chapter 37.

Clinical Signs—Because *Xylorhiza* is eaten mainly by sheep, it is most likely to produce an acute disease. There is severe muscular weakness, with the animal's head held down and the ears drooping. Onset seems to occur within a few hours of ingestion. Colic, bloat, frothy and bloody discharge from the nose, stiffness and staggering when walking, and marked elevation of heart and respiration rates are other features of the disease. Prostration and coma follow, with death in a few hours or in 3–4 days.

■ abrupt onset, weakness, depression, frothy nasal discharge, exercise intolerance

Pathology and Treatment—Grossly, there is mild reddening of the digestive tract, marked visceral congestion, excess fluid in the thoracic cavity and pericardial sac, and pulmonary edema. Microscopically, degeneration and necrosis of the renal tubular epithelium and liver parenchymal cells are likely to be present. There is no specific treatment other than good nursing care. The abrupt onset in sheep and the peracute disease course make treatment in general of little benefit.

■ gross pathology, excess fluid accumulation throughout body

microscopic, renal and hepatic necrosis

Genera of Questionable Toxicity

Achillea L.

Comprising 75–85 species native primarily to the Northern Hemisphere of the Old World, *Achillea* is a member of the tribe Anthemideae. In North America, *A. millefolium* L., popularly known as yarrow or milfoil, is most commonly encountered. As presently circumscribed, it includes taxa formerly recognized as *A. lanulosa* Nutt. and *A. borealis* Bong. Plants of this species are aromatic, herbaceous perennials with erect stems; pinnately dissected, fernlike leaves; and numerous, small heads with white or pinkish white ray florets borne in flat-topped or hemispheric corymbs. The indumentum of the herbage ranges from glabrous to woolly-villous. A highly variable, circumboreal species, it occurs in a variety of habitats and vegetation types across the continent, with greatest abundance in grasslands, open woods, and pastures of the Great Plains.

■ *A. millefolium*

milfoil
yarrow

■ aromatic, perennial herbs; stems erect; leaves fernlike; heads numerous; ray and disk florets white or pinkish white

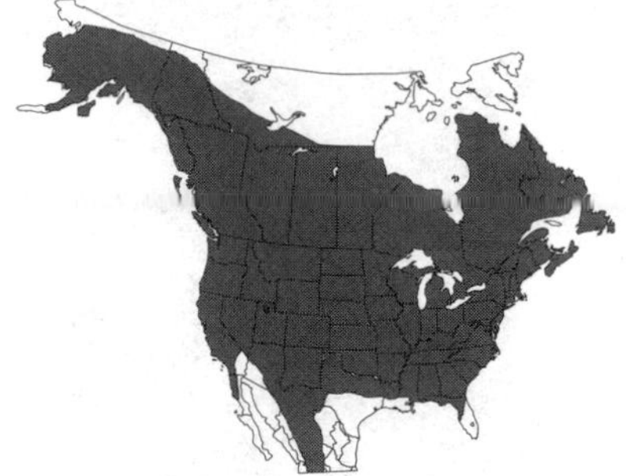

Achillea millefolium

Achillea millefolium

- folk remedy to control bleeding; achillein
- achillein or monoterpenes, possible cause of digestive disturbance; diarrhea and colic; milk taint

Achillein

One of the most widely used folk remedies, *A. millefolium* is believed to be good for a variety of ailments and especially bleeding disorders and other problems involving blood vessels. Its leaves contain achillein, an *N*-methyl glycoalkaloid that has hemostatic activity in rabbits (Miller & Chow 1954). The foliage also contains a bitter astringent that causes irritation of the digestive tract. Large amounts when eaten by sheep produce digestive disturbance (Pammel 1911; Millspaugh 1974). These effects may be caused by the array of monoterpenes present, such as camphor, cineole, and thujenes (Falk et al. 1974; Seaman 1982; Hanlidou et al. 1992). Choline is present but in very low concentrations, 0.025% d.w. (Bauer et al. 1957). When administered subcutaneously in mice, an unknown alkaloid that has been isolated from an Old World species, *A. santolina,* causes neurologic manifestations, including pelvic limb paralysis and rigidity of the tail (Etman et al. 1987).

Achillea millefolium also contains several sesquiterpene lactones, achillin and its hydroxy and acetoxy derivatives, and desacetyl matricarin (Seaman 1982). Although not specifically associated with disease, they may represent a risk if sufficient plant material is eaten. Some sesquiterpene guaianolides, such as α-peroxyachifolid, cause allergic contact dermatitis in humans (Hausen et al. 1991). The foliage of some cultivars yields a very strong positive picrate reaction because of the presence of the cyanogenic prunasin (Fikenscher et al. 1980). The astringent-related digestive disturbances may result in diarrhea and colic and perhaps milk tainting. Treatment should be directed toward relief of the signs and in most instances will not be needed.

Ageratina Spach

- *A. adenophora*

  sticky eupatorium
  crofton weed

- shrubby perennials; stems erect; leaves opposite, serrate; heads terminal; disk florets only, white

Native to the New World and comprising approximately 200 species, *Ageratina,* a member of the tribe Eupatorieae, is considered by some taxonomists to be only a section or subgenus of *Eupatorium.* If the genus is recognized as distinct, some 15 species are present in North America. Only 1 is of toxicologic interest, *A. adenophora* (Spreng.) R.M.King & H.Rob. (= *Eupatorium adenophorum*) (sticky eupatorium, or crofton weed, as it is called in Australia). Plants of this species are shrubby perennials with erect, glandular stems; opposite, serrate leaves; and heads with only white disk florets borne in flat-topped clusters at the ends of the branches. Native to southern Mexico, this species is found in disturbed sites as far north as central California, especially in coastal areas. It also is adventive and naturalized elsewhere in the world.

Although native to Central America, *A. adenophora* causes problems primarily in the areas in which it has been introduced. Plants of the species apparently are eaten only by horses, and the disease produced is known as blowing disease. It was reported in the 1920s in Hawaii, in the 1940s in Australia, and in the 1970s in New Zealand (O'Sullivan et al. 1985). Following sustained long-term ingestion of primarily flowering plants, there is gradual onset of apparently permanent, severe respiratory difficulty and exercise intolerance (Everist 1981; Seawright 1989).

- Australia, long-term ingestion by horses, respiratory disease; of questionable concern in North America

- toxin, derivative of agerophorone that is converted to the ultimate pulmonary toxicant; pulmonary edema and emphysema

Experimentally, 10–15 g green, flowering plants/kg b.w. daily (2–3 kg) fed to horses for 44 days, or often much longer, caused signs of severe, persistent respiratory disease. In some instances, many months of feeding were required for the disease to appear. The closely related *A. riparia,* also introduced from Central America, produces a similar problem. In North America, there is little likelihood of intoxications occurring, because of the large amounts of flowering plant material required. In most areas, insufficient numbers of plants will be present.

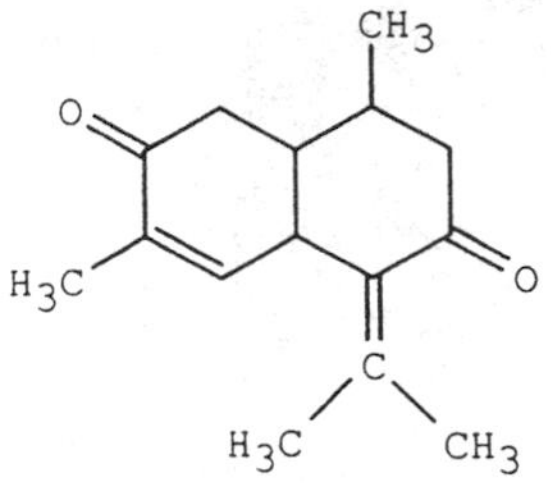

9-Oxy-10,11-dehydroagerophorone

Ageratina adenophora

Ageratina adenophora

The toxin is 9-oxy-10,11-dehydroagerophorone, which is biotransformed to the ultimate toxicant (Oelrichs et al. 1995; 1998). The site of action varies with the animal species. Experiments with mice administered single large doses and repeated smaller amounts in their feed produced effects on the liver rather than the lungs (Sani et al. 1992; Oelrichs et al. 1998). The effects included hepatic necrosis and bile duct hyperplasia. This difference in tissue affected suggests reactions similar to those occurring in *Myoporum* in the Myoporaceae and *Perilla* in the Lamiaceae (see chapters 52 and 45). As happens to the perilla ketone, biotransformation to form the ultimate toxicant may produce lesions in the lungs of some animal species and in the livers of others.

The initial clinical sign is coughing, especially with exercise. This is followed by the gradual onset of other signs such as exercise intolerance, difficult respiration (heaving when breathing), colic, and a decrease in animal condition. Upon auscultation, respiratory rales and even cardiac arrhythmias may be noted. A "heave line" along the lateral abdominal musculature may slowly develop.

At necropsy, the lungs do not collapse normally, because of pulmonary edema, emphysema, and interstitial and subpleural fibrosis. There is excess fluid in the thorax and pericardial sac, and the heart is dilated. In some instances abscesses have been present in the lungs. Microscopically, interstitial fibrosis and proliferation of alveolar lining cells are prominent.

Because the lung lesions are persistent, there is little likelihood of correcting the problem with treatment. Reduction of exercise stress and general supportive care are indicated. The basic changes are in the terminal air exchange units; therefore, improving air flow elsewhere will have only limited beneficial effects.

Ageratum L.

■ A. conyzoides
A. corymbosum
A. houstonianum
A. littorale
billy-goat weed
cielitos
mota morada

An American genus of some 40 species, *Ageratum,* a member of the tribe Eupatorieae, is morphologically very similar to *Eupatorium* and is perhaps the most widely known genus of the tribe (Johnson 1971). This familiarity is due to extensive cultivation of *A. conyzoides* L., commonly known as billy-goat weed, and *A. houstonianum* Mill. Both species are adventive throughout the world. Two other species, *A. corymbosum* Zuccagni (cielitos, mota morada, or flattop ageratum), and *A. littorale* A.Gray occur in North America. All 4 are found in Mexico and the extreme southern portion of the U.S.

■ herbs or subshrubs; leaves simple, opposite; disk florets only, white, blue, or lavender

Plants of the genus are annual or perennial herbs or subshrubs with branched stems; opposite, simple, leaves; and clusters of small heads with white, blue, or lavender disk florets borne at ends of branches. Ray florets are absent. Morphologically very similar to *Eupatorium, Ageratum* is distinguished primarily by having a pappus of five awns rather than capillary bristles.

Ageratum corymbosum

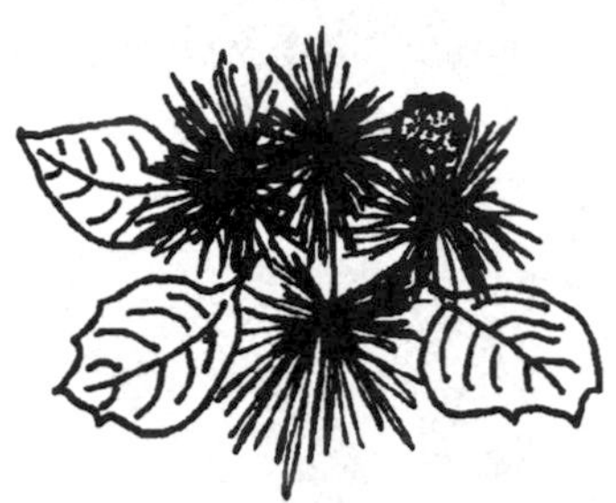

Ageratum houstonianum

■ not known to cause disease in North America; liver disease with hemorrhage and photosensitization in Cuba

Forage contaminated with *A. houstonianum* is reported to cause an acute hemorrhagic disease and photosensitization in cattle in Cuba (Alfonso et al. 1989). Aqueous extracts of *A. conyzoides* given orally to rats and mice produced sedative-type effects (Yamamoto et al. 1991). Various species of the genus are reported to cause livestock intoxication problems in India (Sharma 1994). There also were problems for people working in infested fields. *Ageratum,* however, has not yet been implicated in disease problems in North America.

Ageratum houstonianum

Ageratum houstonianum contains compounds—precocenes I and II—that act as hormones regulating insect growth and cause extensive hepatocellular necrosis in rats in experimental studies (Hsia et al. 1981; Ravindranath et al. 1987) It is thought that they are bioactivated in the liver to reactive metabolites. Because the dosage required for toxicity is in the range of 200–300 mg/kg b.w. and the compounds are of limited concentration in the plants, a role for them as toxicants for livestock is unlikely.

Ambrosia L.

■ A. artemisiifolia
A. psilostachya
ragweed

A predominantly American genus consisting of some 40 species, *Ambrosia,* a member of the tribe Heliantheae, is one of the most notorious genera of the Asteraceae. Even though its generic name means "nectar of the gods," its allergenic, wind-borne pollen causes misery and even danger for millions of hay fever sufferers in the late summer and early fall. Of the approximately 25 species in North America, the more common species such as *A. artemisiifolia* L. (common or short ragweed) and *A. psilostachya* DC. (western ragweed) may be eaten in some circumstances and possibly pose a toxicologic risk.

Species of the genus are herbaceous or woody annuals or perennials from taproots or woody rhizomelike roots. Their leaves are opposite or alternate or opposite below and alternate above, simple, and with entire to variously toothed or lobed or dissected margins. The plants are monoecious, and the disk florets inconspicuous. Ray florets are absent. The staminate heads are borne in terminal racemes on short recurved peduncles, and the pistillate heads form hardened burs in the axils of the upper leaves.

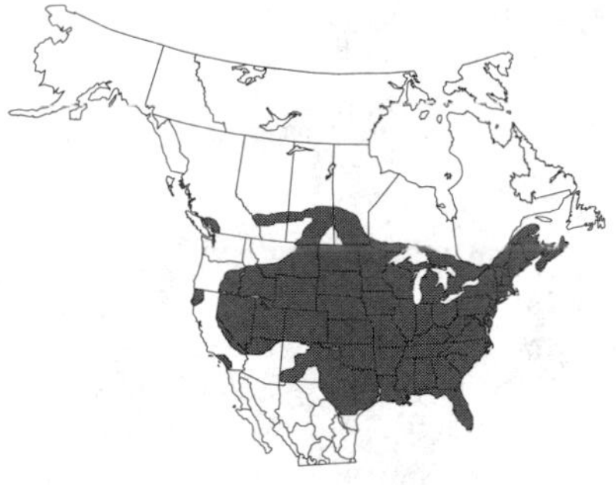

Ambrosia artemisiifolia

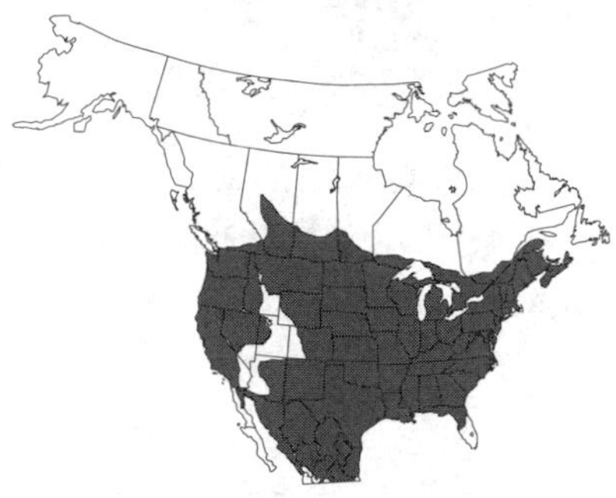

Ambrosia psilostachya

Ambrosia artemisiifolia

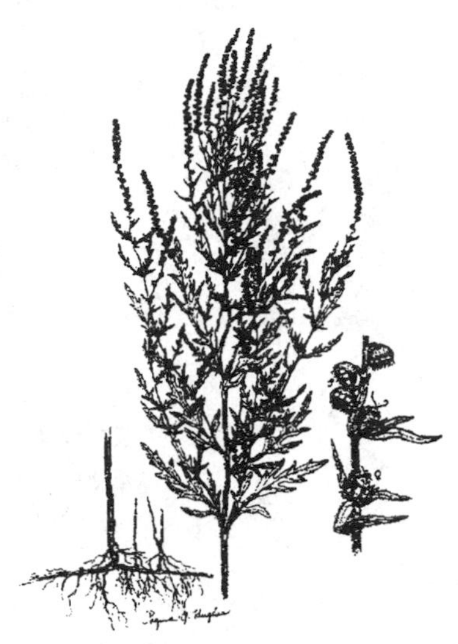

Ambrosia psilostachya

■ classic weedy waste-area invaders

■ possible cause of digestive disturbance and primary photosensitization due to sesquiterpene lactones, polyacetylenes, thiophenes

A thiophenic polyine

Occupying a variety of soil types in a variety of vegetation types, species of the genus are classic indicators of land abuse. Present across most of North America, *A. artemisiifolia* is the most common. *Ambrosia psilostachya* is characteristic of prairies and dry pastures across the continent except in the high mountains and the mid-Atlantic coastal states.

Although toxicity of these species of *Ambrosia* has not been confirmed, there is the possibility of photosensitization and/or digestive disturbances. They contain polyacetylenes and thiophenes that are phototoxic for a variety of bacteria, in some instances even more potent than 8-methoxypsoralen, which is one of the phototoxic compounds of the Apiaceae (Marchant & Cooper 1987; Downum et al. 1989). In this instance the interaction is with cell membranes rather than with DNA, as occurs with furanocoumarins. Species of the genus also contain pseudoguaianolide sesquiterpene lactones such as coronopilin, parthenin, psilostachyins, ambrosin, and parthenolide (Romo & deVivar 1967; Seaman 1982). These compounds may account for toxicity in the rare event that large amounts of the plants are eaten. Some species are very bitter and have strong irritant properties (Millspaugh 1974). Thus the possible signs of intoxication due to species of *Ambrosia* range from skin necrosis to nausea, vomiting, and abdominal pain.

Anthemis L.

■ *A. arvensis*
A. cotula

chamomile
dog daisy
dog fennel
mayweed

An Old World genus with 60–100 species, *Anthemis,* a member of the tribe Anthemideae, is commonly known as chamomile. In North America, 5 morphologically variable species are naturalized. Most common of the 5 species and widespread in fields and waste areas across the continent are *A. arvensis* L., commonly known as scentless chamomile or corn chamomile, and *A. cotula* L., known variously as

- annual aromatic herbs; leaves finely dissected
- possible contact irritation and milk taint

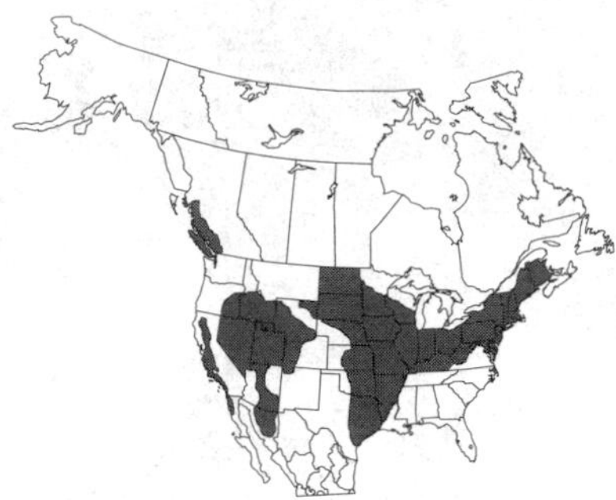

Anthemis cotula

mayweed, dog fennel, dog daisy, dog's chamomile, or stinking chamomile. Both are aromatic, herbaceous annuals with erect stems and alternate, incised-dentate to pinnately dissected leaves. Borne singly at the ends of branches or peduncles, the heads have yellow or white ray florets and yellow disk florets subtended by subequal, imbricate phyllaries.

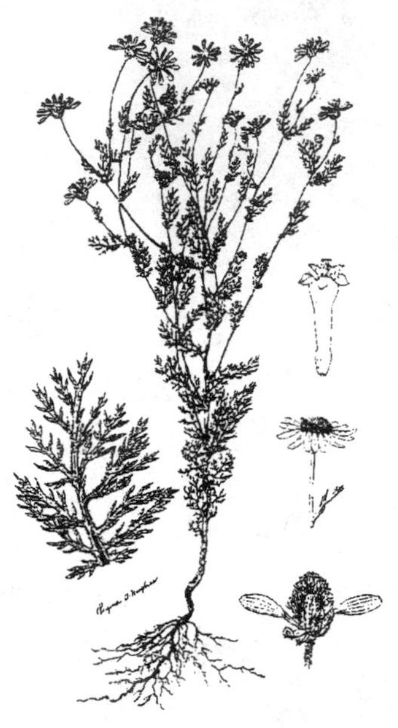

Anthemis cotula

The species of *Anthemis* contain irritant substances that affect the skin and mucous membranes and taint milk and possibly meat (Watt & Breyer-Brandwijk 1962; Whitson et al. 1991). *Anthemis altissima* has tested strongly positive for cyanogenesis in some studies, but it has not been a clinical problem (Fikenscher et al. 1980). Clinical signs, if they occur, reflect contact irritation and include skin rash, and soreness and blistering of the mouth and lips. Treatment involves topical preparations for relief of the irritation and pain.

- *C. nauseosus*
  C. parryi
  C. viscidiflorus

chamisa
horsebrush
rabbit brush

Chrysothamnus nauseosus

Chrysothamnus Nutt.

Commonly known as rabbit brush, *Chrysothamnus,* a member of the tribe Astereae, comprises 16 species of shrubs and subshrubs native to the western half of the continent, primarily in the United States. The generic name is derived from the Greek roots *chryso* and *thamnos,* meaning "golden" and "shrub," which reflect the showy yellow heads of the genus. Most commonly encountered are *C. nauseosus* (Pall.) Britton, generally known as bitter rabbit brush, rubber rabbit brush, chamisa, gray horsebrush, or common horsebrush; *C. parryi* (A.Gray) Greene; and *C. viscidiflorus* (Hook.) Nutt., or yellow rabbit brush. All have several morphological and ecological varieties and are profusely branched shrubs with erect or upward-arching stems. The branchlets are distinctively white tomentose in *C. nauseosus* and *C. parryi.* The simple, alternate leaves are sessile and linear to oblanceolate with entire margins. Borne in terminal cymose clusters, the cylindrical heads have only yellow disk florets subtended by imbricated phyllaries. The pappus is of numerous capillary bristles. Occupying a variety of soil types and tolerating alkaline conditions, all species of the genus are found in open, dry habitats of deserts, plains, canyons, and mountains in the western half of the continent. Increasing in abundance with overgrazing, plants sometimes form dense stands of several acres to a few square miles.

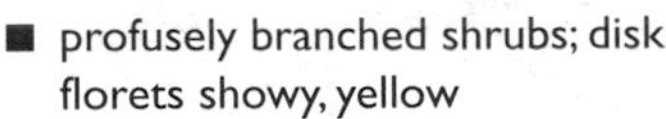

■ profusely branched shrubs; disk florets showy, yellow

Chrysothamnus nauseosus

The species name of one member of the genus is indicative of the strong odor of the foliage. However, in spite of this odor, the plants are a valuable winter forage for deer and less so for sheep and other livestock, with good palatability and nutritive value (Bhat et al. 1990). Very little plant material is eaten in the summer. This is perhaps a reflection of the high concentrations of secondary volatile compounds at this time, in contrast to low levels in the winter (Halls et al. 1994). Thus the risk of intoxication may be greater in summer. Some investigators believe the species to be toxic only under some circumstances, as when, for example, animals are forced to subsist almost entirely on them without a prior acclimation period (Sampson & Malmsten 1942). *Chrysothamnus* also has been suggested as a cause of abortion (LF James et al. 1992).

■ eaten as winter forage; under some conditions when large amounts are eaten without prior adaptation, may cause digestive disorder

Cichorium L.

A member of the tribe Lactuceae and native to the Mediterranean region, *Cichorium* comprises 7–9 species of weeds, ornamentals, and salad greens. In North America, only 2 introduced species are present, *C. endivia* L. (endive) and *C. intybus* L., commonly known as chicory, succory, witloof, wild chicory, blue chicory, or blue-sailors. Now a cosmopolitan weed of roadsides, abused fields, and waste areas, *C. intybus* is a deep-rooted perennial 30–170 cm tall, with erect, deeply grooved stems, the upper branches of which are essentially leafless. Its sap is a milky latex, and the alternate leaves are generally pinnatifid below and entire or toothed above. The showy, sessile, primarily axillary heads consist of only ligulate florets with azure blue or rarely white corollas.

■ *C. endivia*
C. intybus

chicory
endive
succory
witloof

■ erect perennial; milky sap; heads axillary; ligulate florets blue or white

Various parts of the plant have been used widely for food and drink. The leaves are eaten in salads, and the roots are used as a substitute for coffee or an adulterant to enhance its flavor (Millspaugh 1974). Because of its tolerance to poor soils, the species also has been promoted as an alternative forage for livestock, with use of the aerial portions either fresh or dried for all livestock species and the roots for pigs (Kains 1898). However, consumption of the leaves or roots has been occasionally associated with acute disease and death in livestock (Bubien et al. 1962; Wachnik 1962; Cooper & Johnson 1984). The effects seem to be primarily on the digestive tract. Cattle fed almost exclusively on the roots for 5 days developed severe digestive disturbances. A few animals died, but the remainder recovered quickly after feeding

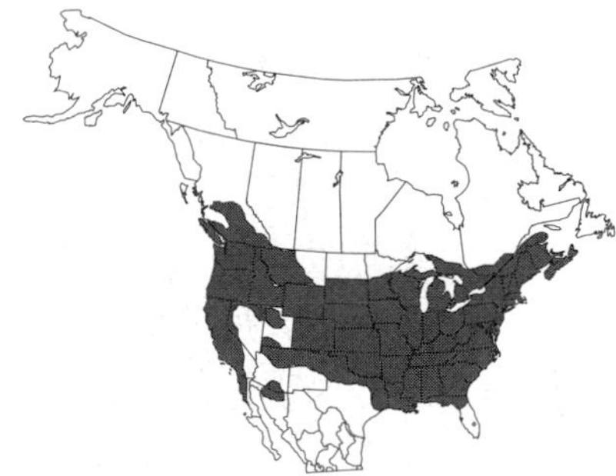

Chichorium intybus

Chichorium intybus

- large amounts may cause digestive disorder until animals adapt; milk taint? effects may be due to sesquiterpene lactones

of the roots was discontinued. Large quantities fed to dairy cattle or goats may result in bitter milk (Kains 1898).

Cichorium intybus contains lactucin and lactucopicrin, the same bitter sesquiterpene lactones found in *Lactuca* (Seaman 1982; Picman 1986). At the very least, these lactones should be considered milk tainters, and when eaten in large amounts, they may cause loss of appetite, abdominal pain, and diarrhea. Pathologic changes are limited to irritative reddening of the digestive tract. In most cases, the signs will be of short duration and will not require treatment.

Lactucin

Lactucopicrin

Conyza Less.

- *C.canadensis*
  C. coulteri

  fleabane
  horseweed
  mare's tail

- erect annuals; heads small, inconspicuous; ray florets inconspicuous, white; disk florets yellow

- rare intoxication; not confirmed experimentally

Primarily tropical and subtropical in distribution, *Conyza,* a member of the tribe Astereae, comprises more than 50 species. Of some toxicologic interest are *C. canadensis* (L.) Cronquist (= *Erigeron canadensis*), commonly known as horseweed, mare's tail, and Canada fleabane, and *C. coulteri* A. Gray, or Coulter's conyza. Both species are annuals with taproots and erect, branched, stems 30–175 cm tall. Alternate, simple, and numerous, the leaves are linear to oblanceolate, with entire to toothed margins and clasping or nonclasping bases. The numerous, small, campanulate heads, with inconspicuous or absent white ray florets and yellow disk florets subtended by phyllaries in 2- or 3-series, are borne in terminal panicles. The obovate achenes bear a pappus of capillary bristles at their apices. A classic indicator of barren, recently disturbed soils, *C. canadensis* is found in waste areas and abused pastures, prairies, and woodlands throughout most of North America. Widespread in northern Mexico, *C. coulteri* occurs on ditch banks, dry stream beds, and disturbed sites.

Conyza canadensis

Conyza canadensis

Conyza canadensis force-fed to a lamb at 3% b.w. was lethal in 3 days (Boughton & Hardy 1941). In a case involving cattle, the plant was suspected to be the cause of blindness, increased salivation, depression, nervous twitching, circling, and a reluctance to move 7–8 days after the animals were moved to new pastures containing plants (Dollahite 1978). Morbidity reputedly reached 90%, and mortality was 50%. However, attempts to reproduce the disease experimentally were negative. The foliage has generally been negative for cyanogenesis in our experience.

Conyza coulteri

Florestina Cass.

A Mexican and Central American genus of 8 species, *Florestina,* a member of the tribe Heliantheae, is represented in North America by a single species *F. tripteris* DC. (= *Palafoxia tripteris*). Commonly known as sticky florestina or sticky palafoxia, it is an erect, single stemmed, annual herb 10–60 cm tall with opposite leaves below and alternate leaves above. Its turbinate heads are borne in a dichotomously branched cyme and bear only white disk florets subtended by purple-tinged phyllaries in 1-series. Comprising 8–10 scales, the pappus is present at the apex of obpyramidal, 4-sided achenes with either smooth or 3-ribbed faces. The species is distributed in the plains of northern Mexico. Boughton and Hardy (1939) conducted feeding trials and found that foliage equivalent to 0.5% b.w. caused death in sheep in 33 minutes. The foliage was positive for cyanide when tested via the picrate test, and the clinical signs appearing were consistent with cyanide as the cause of death.

- *F. tripteris*

  sticky florestina
  sticky palafoxia
- erect annual herbs; disk florets only, white
- possibly cyanogenic under some circumstances

Florestina tripteris

Helianthus L.

An American genus of some 65 species and numerous hybrids, *Helianthus,* a member of the tribe Heliantheae, is one of the most commonly encountered and readily recognized members of the family. Representative species include *H. annuus* L. (annual or common sunflower), *H. maximiliani* Schrad. (Maximilian's sunflower), *H. mollis* Lam. (ashy sunflower), and *H. tuberosus* L. (Jerusalem artichoke). Many of the species, especially *H. annuus,* are cultivated for their oil, achenes, rhizomes,

- *H. annuus*
  H. maximiliani
  H. mollis
  H. tuberosus

  sunflower
  Jerusalem artichoke
- erect herbs; heads large, showy; ray florets yellow; disk florets brown to purple-black

Helianthus annuus

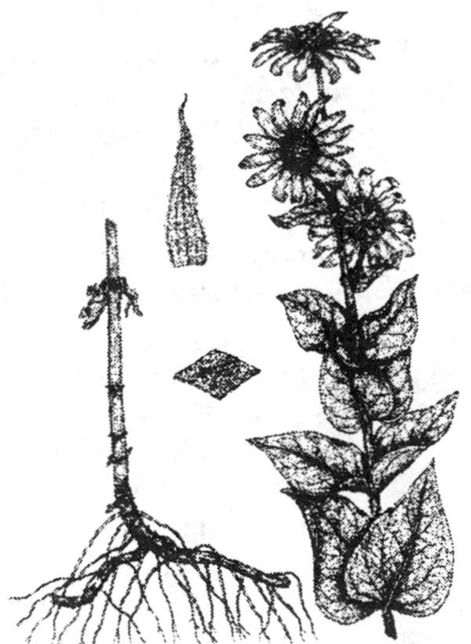

Helianthus mollis

- commonly eaten without adverse effect; under some circumstances may present a risk due to nitrate accumulation or bioactive sesquiterpene lactones

Helianthus annuus

and beauty. Distributed across the continent from southern Canada to Mexico, they also are an important resource for wildlife.

Species of the genus are herbaceous annuals and perennials, from either taproots or thickened rootstocks. The erect stems, 60–300 cm tall, bear simple leaves that are opposite or opposite below and alternate above, of various shapes, and variously indumented. The large showy heads are borne on long or short peduncles at the ends of the branches or in the axils of the leaves. They have yellow, pistillate but sterile ray florets and dark brown to perfect, fertile, purple-black disk florets subtended by phyllaries in 2- to 4-series and receptacular bracts. Two typically caducous awns and often 2 small scales compose the pappus. The achenes are obovoid and flattened.

Although palatable and readily sought out by livestock, species of *Helianthus* are of low nutritional value. Young plants have been considered by some to be toxic. Ataxia, excitement, circulatory problems, and eventually collapse occur 3–4 hours after young plants with immature achenes are eaten (Cooper & Johnson 1984). Because the disease is accompanied by dark-colored blood and the signs are acute, nitrate is suspected to be the cause of intoxication. *Helianthus annuus* is quite saline tolerant, and under appropriate conditions where soil sulfates are high, the plants may accumulate up to 800 ppm sulfur (Mayland & Robbins 1994). Species of *Helianthus* also contain sesquiterpene lactones such as orizabin and pinnatifidin derivatives, which are of unknown toxicologic significance (Seaman 1982). Experimental studies on insects suggest that they produce neurotoxic effects via picrotoxin-like GABA/chloride channel antagonism (Mullin et al. 1991). Such effects also could account for the observed clinical signs; however, they have not been confirmed in higher animals.

Lactuca L.

- *L. canadensis*
  L. serriola

  prickly lettuce
  wild lettuce
  tall lettuce
  trumpetweed

- erect herbs; milky sap; heads terminal; ligulate florets yellow, blue, white, or lavender

Lactuca canadensis

The many cultivated varieties of the cultivated lettuce, *L. sativa,* are probably the best-known members of the Old World genus *Lactuca,* a member of the Lactuceae that comprises 50–100 species and hybrids. The generic name is derived from the Latin *lac,* for "milk," which reflects the milky sap present in all species. In addition to the cultivated taxa, several species are common weeds in disturbed soils of waste areas, roadsides, rights-of-way, and abused pastures. In North America, some 13 introduced species are present. Representative are *L. canadensis* L. (wild lettuce, trumpetweed, tall lettuce) and *L. serriola* L. (prickly lettuce).

Plants of the genus are herbaceous annuals, biennials, and perennials with a milky sap. The erect stems, 20–250 cm tall, bear both basal and cauline, alternate, simple leaves with blades of various shapes and marginal dissection. Borne in terminal panicles, the cylindrical or urceolate heads contain 5–35 yellow, blue, white, or lavender ligulate florets that are subtended by imbricate phyllaries in 2- to several-series. The pappus is of capillary bristles, and the achenes are flattened with a long or short beak.

Reports of intoxications due to *Lactuca* are largely anecdotal with few authenticated incidents. Since the 1700s the milky sap has had a reputation as a substitute for opium (Coxe 1799). Large doses of sap extracts were reported to produce narcotic effects in frogs. However, this reputation has been questioned for more than 100 years, and it now appears that it has no chemical basis; it is now viewed as a psychotropic drug fraud (Tyler 1980; Der Marderosian et al. 1988). The action is psychological rather than physiological, and Tyler (1980) has termed modern promotion of the use of *Lactuca* sap a hoax. Mullins and Horowitz (1998) report of severe adverse effects associated with i.v. administration of a lettuce powder "tea" but no narcotic activity. In one of the few instances of field observations in animals, Gregory (1997) associated consumption of a butterhead variety of lettuce with transient narcosis in a duckling. Beath and co-workers (1953) reported the occurrence of what appears to be tryptophan-type pulmonary emphysema (atypical interstitial pneumonia) in cattle moved from dry summer range to irrigated fall pasture containing abundant *L. serriola.*

Lactucin

Lactucopicrin

Lactuca does contain several toxicants: lactucoral, lactucrin, lacturin, lactucol, and a small amount of hyoscyamine (Pammel 1911; Millspaugh 1974). Two cytotoxic sesquiterpene lactones, lactucin and lactucopicrin, also are present in the milky sap in concentrations of approximately 2% and 3.5%, respectively (Schmitt 1940; Seaman 1982; Picman 1986). Some of these compounds are intensely bitter, and they may react with oxidants to cause unusually colored and bitter milk from cows eating the plants.

Except for the labored respiration associated with the pulmonary emphysema, there have been no distinct diseases with which to associate signs, pathology, or therapy. A complete discussion of the type of pulmonary emphysema characteristic of *Lactuca* is presented in the treatment of *Brassica,* in chapter 16.

Silybum Adans.

Comprising only 2 species and native to the Mediterranean region, *Silybum,* a member of the tribe Cardueae, is represented in North America by *S. marianum* (L.) Gaertn. Commonly known as milk thistle, blessed milk thistle, or St. Mary's thistle, it is an herbaceous annual or biennial, 100–200 cm tall, with glabrous or slightly tomentose herbage. The erect stems bear large, simple, alternate, spiny-dentate, often pinnately lobed leaves that are dark green, blotched with white. The leaves typically are petiolate below, becoming sessile and auriculate above. Borne singly at the branch ends, the large heads contain numerous purple to pink disk florets that are subtended by imbricate phyllaries in several series, the outer of which are spinose. Ray florets are absent. The pappus is of numerous flat, barbellate, capillary bristles. The achenes are glabrous and slightly flattened. Plants are encountered infrequently across the continent as weeds along roadsides, in abused pastures, and in waste areas. They are sometimes cultivated as ornamentals, and the achenes have been used as a coffee substitute.

Both nitrate and cyanogenic glycosides may be accumulated by *S. marianum;* however, they have only rarely been shown to cause disease problems (Kendrick et al. 1955). Because of the propensity to accumulate nitrates, care should be taken to avoid (for 1 week) grazing pastures infested with *S. marianum* following application of herbicides such as 2,4-D, which may cause a temporary increase nitrate levels (Harradine 1990). The plant also contains silybin, silychristin, silydianin, and silymarin, which

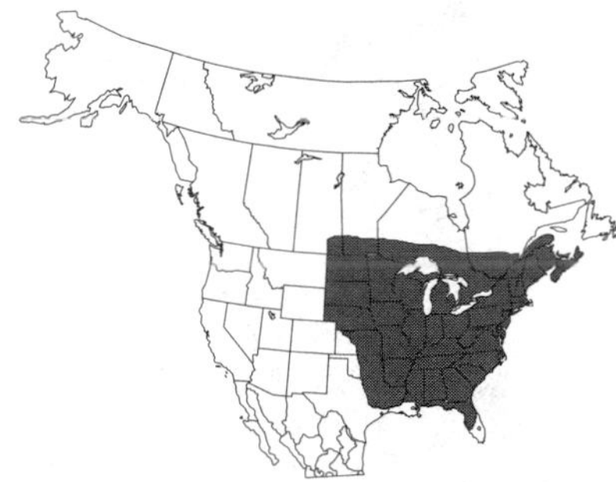

Lactuca canadensis

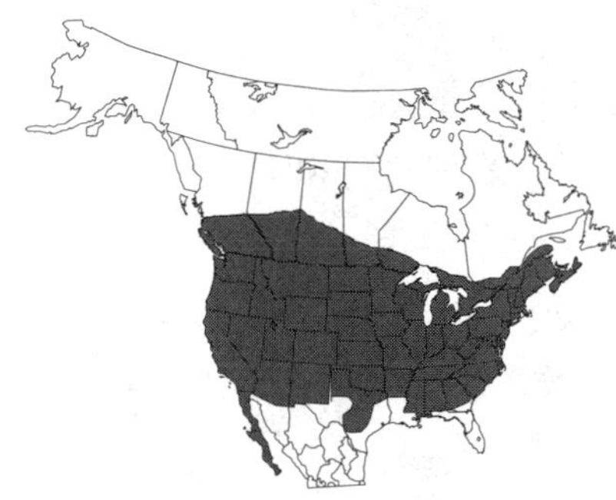

Lactuca serriola

- unsubstantiated narcotic effect attributed to the milky sap; sesquiterpene lactones, lactucin, and lactucopicrin; digestive disorder or milk taint

- *S. marianum*

  milk thistle
  St. Mary's thistle

- erect herbs; leaves spiny; heads terminal; disk florets only, purple to pink; phyllaries spiny, giving thistle appearance

Silybum marianum

- toxicity not reported, but plants may contain nitrate and cyanogenic glycosides

are referred to as the silymarin-flavonoid complex. These compounds, especially silybin, act to stabilize membranes by preventing or inhibiting peroxidation processes and also inhibiting prostaglandin synthesis (Rui 1991). Presumably these effects are factors in the protection afforded by silymarin against liver injury such as that induced by the algal toxin microcystin-LR (Mereish et al. 1991). The achenes have been fed with success to cattle as a preventive against ketosis (Vojtisek et al. 1991).

Tanacetum L.

- *T. vulgare*

  tansy
  common tansy

- coarse, aromatic, perennial herbs; leaves pinnately dissected; heads showy, numerous; disk florets only, yellow

Comprising perhaps 50 species of the Northern Hemisphere, *Tanacetum,* a member of the tribe Anthemideae, is closely related to *Chrysanthemum,* and the boundary between them is questioned. Of the 4 introduced species in North America, *T. vulgare* L. (tansy or common tansy) is most commonly encountered throughout much of the United States and Canada. It is a coarse, aromatic, herbaceous, glabrous perennial, 40–150 cm tall, from a stout rhizome. The simple, alternate leaves are glandular-punctate and pinnately dissected with a broadly winged rachis. The numerous heads are borne in flat-topped or hemispheric corymbs and have numerous yellow disk florets. Ray florets are absent. The pappus is absent or present as a short crown, and the achenes are 4- or 5-ribbed.

Tanacetum vulgare

Tanacetum vulgare

- rarely eaten; large dosage possibly neurotoxic or cardiotoxic

Although *T. vulgare* is used extensively as a folk remedy tonic, there are several instances of fatal overdoses of the tea or oil. Neurointoxication and cardiac/respiratory depression are the predominant effects (Millspaugh 1974). The toxicant has not been identified. Sesquiterpene lactones are poorly represented and appear not to be important factors in the toxicity of *T. vulgare* (Seaman 1982).

Vernonia Schreb.

- *V. baldwinii*
  V. gigantea
  V. noveboracensis

  ironweed

- perennial herbs; heads terminal; disk florets only, purple, pink, or white

Comprising about 1,000 species in both the Old World and the New World, *Vernonia,* a member of the tribe Vernonieae, is commonly known as ironweed. In North America the genus is characterized by extensive hybridization, and local hybrid swarms are common. Approximately 29 species and named hybrids are present; among the most commonly encountered are *V. baldwinii* Torr. (western ironweed), *V. gigantea* (Walter) Trel. (= *V. altissima* Nutt.; tall ironweed), and *V. noveboracensis* (L.) Michx. (New York ironweed). Plants of the genus are herbaceous perennials with tough, fibrous stems; simple, alternate leaves with entire or toothed margins; and numerous heads borne in flat-topped or hemispheric clusters at ends of the branches. The heads contain only purple, pink, or white disk florets, which are subtended by imbricate phyllaries in several series. The pappus is in 2-series, with long bristles and short bristles or scales. The cylindrical achenes are ribbed. Flowering in the late summer and fall, species of the genus are found in prairies and open woods. They are often weedy in overgrazed pastures and disturbed habitats.

Although *Vernonia* has not been incriminated as toxic in North America, *V. mollisima* in South America has been shown to cause disease in cattle, sheep, goats, and rabbits (Dobereiner et al. 1976; Tokarnia et al. 1986). Experimental administration of 1–2% b.w. of the species produced muscular tremors, labored respiration, and paddling seizures accompanied by severe hepatocellular necrosis and finally death over the course of 48 hours in 4 of 7 cows. Studies with other species in other areas of the world failed to demonstrate toxic effects (Watt & Breyer-Brandwijk 1962). North American species contain cytotoxic sesquiterpene lactones similar to those of the South American species (Catalan et al. 1986; Jakupovic et al. 1986; Picman 1986). Vernolepin from an African species, *V. hymenolepis,* is a cytotoxic α-methylene-γ-lactone that reacts with thiol groups to inhibit tumors. It also inactivates various enzymes by thiol interaction (e.g., glycogen synthase), has antihistaminic activity, and inhibits platelet aggregation induced by arachidonic acid (Smith et al. 1972; Laekeman et al. 1983). Although the plants are not known specifically to cause disease, the possibility of the occurrence of intoxications should not be entirely dismissed, given intake of sufficient plant material.

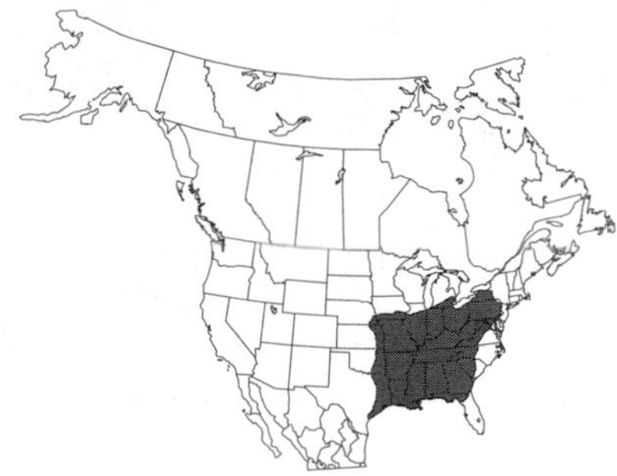

Vernonia gigantea

- no toxicity confirmed but sesquiterpene lactones present; South American species causes disease very similar to that caused by *Xanthium*

References

Aanes WA: Pingue *(Hymenoxys richardsonii)* poisoning in sheep. *Am J Vet Res* 22;47–52, 1961

Adams R, Van Duuren BL: Usaramoensine, the alkaloid in *Crotalaria usaramoensis* E.G.Baker, integerrimine from *Crotalaria incana* Linn, and senecionine from *Senecio glabellus* DC.: stereochemical relationships. *J Am Chem Soc* 19;4631–4635, 1953

Ahmed AA, Spring O, Abd El-Razek MH, Hussein NS, Mabry TJ: Sesquiterpene lactones and other constituents from *Hymenoxys richardsonii* and *H. subintegra. Phytochemistry* 39;1127–1131, 1995

Akerboom TPM, Bookelman H, Tager JM: Control of ATP transport across the mitochondrial membrane in isolated rat-liver cells. *FEBS Lett* 74;50–54, 1977

Alfonso HA, Figueredo M de los A, Rodriguez J, Fuentes O, Gomez BC, Aguilera JM, Perez M, Perez N, Sanchez LM: Toxicity of *Helenium quadridentatum* in calves and male white mice. *Rev Salud Anim* 8;295–296, 1986 (*Vet Bull* 57;7291, 1987)

Alfonso HA, Rivera M, Aparicio JM, Ancisar J, Marrero E, Cabrera JM: Natural and experimental poisoning by *Ageratum houstonianum* Mill. in cattle. *Rev Cubana Cienc Vet* 20;113–120, 1989

Allmann DW, Harris RA, Green DE: Site of action of atractyloside in mitochondria. 1. Inhibition of outer membrane functions by atractyloside. *Arch Biochem Biophys* 120;693–702, 1967a

Allmann DW, Harris RA, Green DE: Site of action of atractyloside in mitochondria. 2. Inhibition of oxidative phosphorylation. *Arch Biochem Biophys* 122;766–782, 1967b

Araya O, Krause A, Solis de Ovando M: Outbreaks of stringhalt in southern Chile. *Vet Rec* 142;462–463, 1998

Bah M, Bye R, Pereda-Miranda R: Hepatotoxic pyrrolizidine alkaloids in the Mexican medicinal plant *Packera candidissima* (Asteraceae: Senecionine). *J Ethnopharmacol* 43;19–30, 1994

Barkley TM, Cronquist A: *Erechites. North Am Flora,* ser 2, 10;139–142, 1978

Barkley TM, Clark BL, Funston AM: The segregate genera of *Senecio* sensu lato and *Cacalia* sensu lato (Asteraceae: Senecioneae) in Mexico and Central America, Compositae: Systematics. In *Proceedings of the International Compositae Conference,*

Kew, 1994, Hind DJN, Beentje HJ eds, Royal Botanic Gardens, Kew, vol 1, pp613–620, 1996

Basson PA: Poisoning of wildlife in southern Africa. *J S Afr Vet Assoc* 58;219–228, 1987

Bauer S, Masler L, Orszagh S: Origin of the hypotensive effect of *Achillea millefolium*. *Chem Zvesti* 10;528–532, 1956 (*Chem Abstr* 51;6952, 1957)

Beath OA: *Chemical and Pharmacological Examination of the Woody Aster*. Wyo Agric Exp Stn Bull 123;39–66, 1920

Beath OA, Eppson HF, Gilbert CS, Bradley WB: *Poisonous Plants and Livestock Poisoning*. Wyo Agric Exp Stn Bull 231, 1939

Beath OA, Gilbert CS, Eppson HF, Rosenfeld I: *Poisonous Plants and Livestock Poisoning*. Wyo Agric Exp Stn Bull 324, 1953

Beekman AC, Woerdenbag HJ, Uden W van, Pras N, Konings AWT, Wikstrom HV, Schmidt TJ: Structure-cytotoxicity relationships of some helenanolide-type sesquiterpene lactones. *J Nat Prod* 60;252–257, 1997

Beier RC, Norman JO: The toxic factor in white snakeroot: identity, analysis, and prevention. *Vet Hum Toxicol* 32 (Suppl);81–88, 1990

Beier RC, Norman JO, Irwin TR, Witzel DA: Microsomal activation of constituents of white snakeroot (*Eupatorium rugosum* Houtt) to form toxic products. *Am J Vet Res* 48;583–585, 1987

Beier RC, Norman JO, Reagor JC, Rees MS, Mundy BP: Isolation of the major component in white snakeroot that is toxic after microsomal activation: possible explanation of sporadic toxicity of white snakeroot plants and extracts. *Nat Toxins* 1;286–293, 1993

Berg RT, McElroy LW: Effect of 2,4-D on the nitrate content of forage crops and weeds. *Can J Agric Sci* 33;354–358, 1953

Bhat RB, Welch BL, Weber DJ, McArther ED: Mid-winter protein, phosphorus, and digestibility of *Chrysothamnus nauseosus* subspecies. *J Range Manage* 43;177–179, 1990

Bierer BW, Rhodes WH: Corn cockle, cockleburs, and coffee beans as poultry ration contaminants. *J Am Vet Med Assoc* 137;352–353, 1960

Binder RG, Turner CE, Flath RA: Comparison of yellow starthistle volatiles from different plant parts. *J Agric Food Chem* 38;764–767, 1990a

Binder RG, Turner CE, Flath RA: Volatile components of purple starthistle. *J Agric Food Chem* 38;1053–1055, 1990b

Binns W, James LF, Shupe JL: *Psathyrotes annua*, poisonous plant for sheep. *Vet Med* 57;509–511, 1962

Birecka H, Birecki M, Cohen EJ, Bitonti AJ, McCann PD: Ornithine decarboxylase, polyamines, and pyrrolizidine alkaloids in *Senecio* and *Crotalaria*. *Plant Physiol* 86;224–230, 1988

Blythe LL, Craig AM: Role of the liver in detoxification of poisonous plants [letter]. *Vet Hum Toxicol* 36;564–566, 1994

Bohlmann F, Bohlmann R: Three guaianolides from *Hypochoeris radicata*. *Phytochemistry* 19;2045–2046, 1980

Bonner WA, DeGraw JI Jr: Ketones from "white snakeroot" *Eupatorium urticaefolium*. *Tetrahedron* 18;1295–1309, 1962

Boppre M: Insects pharmacophagously utilizing defensive plant chemicals (pyrrolizidine alkaloids). *Naturwissenschaften* 73;17–26, 1986

Boppre M: Lepidoptera and pyrrolizidine alkaloids: exemplification of complexity in chemical ecology. *J Chem Ecol* 16;165–185, 1990

Boughton IB, Hardy WT: *Miscellaneous Suspected Plants*. Tex Agric Exp Stn Annu Rep 46;211, 1933

Boughton IB, Hardy WT: *Toxicity of Bitterweed for Sheep*. Tex Agric Exp Stn Bull 552, 1937

Boughton IB, Hardy WT: *Feeding Trials of Suspected Plants*. Tex Agric Exp Stn Annu Rep 52;239–240, 1939

Boughton IB, Hardy WT: *Feeding Trials of Suspected Plants*. Tex Agric Exp Stn Annu Rep 53;236–237, 1940

Boughton IB, Hardy WT: *Feeding Trials of Suspected Plants.* Tex Agric Exp Stn Annu Rep 54;159, 1941

Bowen DM, DeGraw JI Jr, Shah VR, Bonner WA: The synthesis and pharmacologic action of tremetone. *J Med Chem* 6;315–319, 1963

Brandolin G, Meyer C, Defaye G, Vignais PM, Vignais PV: Partial purification of an atractyloside-binding protein from mitochondria. *FEBS Lett* 46;149–153, 1974

Bremer K: *Asteraceae Cladistics and Classification.* Timber Press, Portland, Oreg, 1994

Bridges GW, Bailey EM, Camp BJ: Prevention of bitterweed intoxication of sheep. *Vet Hum Toxicol* 22;87–90, 1980

Brown RC: Biosystematics of *Psilostrophe* (Compositae: Helenieae). 2. Artificial hybridization and systematic treatment. *Madrono* 25;187–201, 1978

Bruni A, Luciani S, Contessa AR: Inhibition by atractyloside of the binding of adenine-nucleotides to rat-liver mitochondria. *Nature* 201;1219–1220, 1964

Bryant MS, Osterman-Golkar SM: Hemoglobin adducts as dosimeters of exposure to DNA-reactive chemicals. *Chem Ind Inst Toxicol* 11(10);1–9, 1991

Bubenchikova VN: Coumarins of plants of the genus *Centaurea. Chem Nat Compd* 26;709, 1990 (from *Khim Prir Soedin* 26;829–830, 1990)

Bubien Z, Wachnik Z, Zuchowski A: Chicory root poisoning in heifers. *Med Wet* 18;603–605, 1962 (*Vet Bull* 33;2137, 1963)

Buck WB, James LF, Binns W: Changes in serum transaminase activities associated with plant and mineral toxicity in sheep and cattle. *Cornell Vet* 51;568–585, 1961

Buckmaster GW, Cheeke PR, Shull LR: Pyrrolizidine alkaloid poisoning in rats: protective effects of dietary cysteine. *J Anim Sci* 43;464–473, 1976

Buehrer TF, Mason CM, Crowder JA: The chemical composition of rayless goldenrod *(Aplopappus hartwegi). Am J Pharm* 111;105–112, 1939

Buhler DR, Kedzierski B: Biological reactive intermediates of pyrrolizidine alkaloids. In *Advances in Experimental Medicine and Biology*, Plenum Press, New York, vol 197, pp611–620, 1986

Buhler DR, Miranda CL, Kedzierski B, Reed RL: Mechanisms for pyrrolizidine alkaloid activation and detoxification. In *Advances in Experimental Medicine and Biology*, Plenum Press, New York, pp597–603, 1991

Bull LB, Culvenor CCJ, Dick AT: *Frontiers of Biology.* Vol 9, *The Pyrrolizidine Alkaloids,* North-Holland Publ, Amsterdam, 1968

Burke NI: An investigation of the toxic plant—rayless goldenrod *(Aplopappus heterophyllus).* PhD thesis, Oklahoma State Univ, 1966 (*Diss Abstr B* 27[12];4302, 1967)

Burrows GE, Schwab RP, Stein LE, Kirkpatrick JG, Qualls C, Tyrl RJ: Comparison of the reproductive effects in rats of the plants *Baptisia australis*, *Iva annua*, and *Sophora nuttalliana.* In *Toxic Plants and Other Toxicants*, Garland T, Barr AC eds, CAB International, New York, pp297–302, 1998

Busam L, Habermehl GG: Accumulation of mycotoxins by *Baccharis cordifolia:* a reason for livestock poisoning. *Naturwissenschaften* 69;392–393, 1982

Butler SO: Fractions of trematol and their toxicities. Master's thesis, Oklahoma A&M Coll, 1945

Bye SN, Coetzer THT, Dutton MF: An enzyme immunoassay for atractyloside, the nephrotoxin of *Callilepis laureola* (impila). *Toxicon* 28;997–1000, 1990

Cabat GA: The isolation and identification of several constituents of rayless goldenrod, the reactions of azides with bicyclo(2.2.1)-2-heptene and the reactions of benzenesulfonyl azide with aromatic compounds. PhD thesis, Oklahoma State Univ, 1968 (*Diss Abstr B* 28[12];4926–4927, 1968)

Cahill JI, Goulden BE: Stringhalt—current thoughts on aetiology and pathogenesis. *Equine Vet J* 24;161–162, 1992

Cahill JI, Goulden BE, Pearce HG: A review and some observations on stringhalt. *N Z Vet J* 33;101–104, 1985

Cahill JI, Goulden BE, Jolly RD: Stringhalt in horses: a distal axonopathy. *Neuropathol Appl Neurobiol* 12;459–475, 1986

Calhoun MC, Ueckert DN, Livingston CW, Baldwin BC Jr: Effects of bitterweed

(Hymenoxys odorata) on voluntary feed intake and serum constituents of sheep. *Am J Vet Res* 42;1713–1717, 1981

Calhoun MC, Baldwin BC, Kuhlmann SW, Kim HL: Experimental prevention of bitterweed *(Hymenoxys odorata)* poisoning in sheep. *Am J Vet Res* 50;1642–1646, 1989

Camp BJ: The toxic principle of bitterweed *(Hymenoxon odorata)*. In *Plant Toxicology—Proceedings of the Australia-USA Poisonous Plants Symposiumm*, Seawright AA, Hegarty MP, James LF, Keeler RF eds, Queensland Poisonous Plants Committee, Animal Research Institute, Yeerongpilly, Australia, pp473–478, 1985

Candrian U, Luthy J, Schmid P, Schlatter C, Gallasz E: Stability of pyrrolizidine alkaloids in hay and silage. *J Agric Food Chem* 32;935–937, 1984

Candrian U, Zweifel U, Luthy J, Schlatter C: Transfer of orally administered [3H] seneciphylline into cow's milk. *J Agric Food Chem* 39;930–933, 1991

Carpenter BD, Ethridge DE, Sosebee RE: Economics of broom snakeweed control on the southern plains. *J Range Manage* 44;232–237, 1991

Cassady JT: Management for prevention of sneezeweed poisoning. *Nat Wool Grow* 30;12–14, 1940

Catalan CAN, de Iglesias DIA, Kavka J, Sosa VE, Herz W: Sesquiterpene lactones and other constituents of *Vernonia mollissima* and *Vernonia squamulosa*. *J Nat Prod* 49;351–353, 1986

Chambers SW, Smith GS, Stavanja MS, Staley EC, Thilsted JP, Hallford DM: Safflower oil protects rats from embryotoxins of ingested snakeweed foliage [abstract]. *J Anim Sci* 71 (Suppl 1);446, 1993

Chan PC, Mahler J, Bucker JR, Travlos GS, Reid JB: Toxicity and carcinogenicity of riddelline following 13 weeks of treatment to rats and mice. *Toxicon* 32;891–908, 1994

Chapman DE, Roberts GB, Reynolds DJ, Grippo AA, Holbrook DJ, Hall IH, Chaney SG, Chang J, Lee KH: Acute toxicity of helenalin in BDF1 mice. *Fund Appl Toxicol* 10;302–312, 1988

Cheeke PR: Pyrrolizidine alkaloid toxicity and metabolism in laboratory animals and livestock. In *Toxicants of Plant Origin*, vol 1, *Alkaloids,* Cheeke PR ed, CRC Press, Boca Raton, Fla, pp1–22, 1989

Cheeke PR, Garman GR: Influence of dietary protein and sulfur amino acid levels on the toxicity of *Senecio jacobaea* (tansy ragwort) to rats. *Nutr Rep Int* 9;197–207, 1974

Cheng CH, Costall B, Hamburger M, Hostettmann K, Naylor RJ, Wang Y, Jenner P: Toxic effects of solstitialin A 13-acetate and cynaropicrin from *Centaurea solstitialis* L. (Asteraceae) in cell cultures of foetal rat brain. *Neuropharmacology* 31;271–277, 1992

Chesnut VK: *Principal Poisonous Plants of the United States.* USDA Div Bot Bull 20, 1898

Chesnut VK, Wilcox EV: *The Stock-Poisoning Plants of Montana.* USDA Div Bot Bull 26, 1901

Christensen WI: Milk sickness: a review of the literature. *Econ Bot* 19;293–300, 1965

Chu PS, Lame MW, Segall HJ: In vivo metabolism of retrorsine and retrorsine-*N*-oxide. *Arch Toxicol* 67;39–43, 1993

Chung W-G, Miranda CL, Buhler DR: A cytochrome P4502B form is the major bioactivation enzyme for the pyrrolizidine alkaloid senecionine in guinea pig. *Xenobiotica* 25;929–939, 1995

Clark AM: Naturally occurring mutagens. *Mutat Res* 32;361–374, 1976

Clawson AB: A preliminary report on the poisonous effects of bitter rubberweed *(Actinea odorata)* on sheep. *J Agric Res* 43;693–701, 1931

Clawson AB: The American groundsel species of *Senecio* as stock poisoning plants. *Vet Med* 28;105–110, 1933

Clawson AB, Huffman WT: Further notes on the study of bighead. *Nat Wool Grow* 26;18–21, 1936

Clawson AB, Huffman WT: Bighead in sheep caused by plant poisoning. *Nat Wool Grow* 27;13–16, 1937

Clawson AG, Bunyea H, Couch JF: *Treatment for Livestock Poisoned by Hydrocyanic*

Acid. USDA Yearb Agric, 1935

Cole RJ, Stuart BP, Lansden JA, Cox RH: Isolation and redefinition of the toxic agent from cocklebur *(Xanthium strumarium). J Agric Food Chem* 28;1330–1332, 1980

Collazo L, Riet-Correa F: Experimental intoxication of sheep and cattle with *Wedelia glauca. Vet Hum Toxicol* 38;200–203, 1996

Connolly JD, Hill RA: *Dictionary of Terpenoids.* Vol 1, *Mono Terpenoids*, Chapman & Hall, London, 1991

Cook CW, Stoddart LA, Harris LE: *The Nutritive Value of Winter Range Plants in the Great Basin.* Utah Agric Exp Stn Bull 372, 1954

Coombs EM, Bedell TE, McEvoy PB: Tansy-ragwort *(Senecio jacobaea):* importance, distribution, and control in Oregon. In *Noxious Range Weeds*, James LF, Evans JO, Ralphs MH, Child RD eds, Westview Press, Boulder, Colo, pp419–428, 1991

Cooper MR, Johnson AW: *Poisonous Plants of Britain and Their Effects on Animals and Man.* Her Majesty's Stationery Office, London, pp83–90, 1984

Cordo HA: Host-specificity studies of the Argentine weevil, *Heilodus ventralis*, for the biological control of snakeweeds (*Gutierrezia* spp). In *Proceedings of VI International Symposium on Biological Control of Weeds*, Delfosse ES ed, Agriculture Canada, Canadian Government Publ Centre, Ottawa, pp709–720, 1985

Cordy DR: Nigropallidal encephalomalacia in horses associated with ingestion of yellow star thistle. *J Neuropathol Exp Neurol* 13;330–342, 1954

Cordy DR: *Centaurea* species and equine nigropallidal encephalomalacia. In *Effects of Poisonous Plants on Livestock*, Keeler RF, Van Kampen KR, James LF eds, Academic Press, New York, pp327–336, 1978

Cote L, Crutcher MD: Basal ganglia. In *Principles of Neural Science*, 3d ed, Kandel ER, Schwartz JH, Jessell TM eds, Elsevier, New York, pp647–659, 1991

Couch JF: Acidosis, trembles, and milksickness. *Science* 64;456–457, 1926

Couch JF: The toxic constituents of richweed or white snakeroot *(Eupatorium urticaefolium). J Agric Res (USDA)* 35;547–576, 1927

Couch JF: The toxicity of tremetol. *J Am Vet Med Assoc* 73;603–607, 1928

Couch JF: Tremetol, the compound that produces "trembles" (milksickness). *J Am Chem Soc* 51;3617–3619, 1929

Couch JF: The toxic constituents of rayless goldenrod. *J Agric Res (USDA)* 40;649–658, 1930

Couch JF: *Trembles (or Milk Sickness).* USDA Circ 306, 1933

Couet CE, Hopley J, Hanley AB: Metabolic activation of pyrrolizidine alkaloids by human, rat, and avocado microsomes. *Toxicon* 34;1058–1061, 1996

Coulombe RA, Hincks JR, Kim HY, Segall HJ, Stermitz FR, Molyneux RJ: DNA crosslinking by pyrrolizidine alkaloids. In *Poisonous Plants: Proceedings of the Third International Symposium*, James LF, Keeler RF, Bailey EM Jr, Cheeke PR, Hegarty MP eds, Iowa State Press, Ames, pp181–185, 1992

Coxe JR: An inquiry into the comparative effects of the *Opium officinarum*, extracted from *Papaver somniferum,* or white poppy of Linnaeus; and of that procured from the *Lactuca sativa*, or common cultivated lettuce of the same author. *Trans Am Philos Soc* 4;387–414,1799

Craig AM, Meyer C, Koller LD, Schmitz JA: Serum enzyme tests for pyrrolizidine alkaloid toxicosis. *Proc Am Assoc Vet Lab Diagn* 21;161–178, 1978

Craig AM, Blythe LL, Lassen ED, Slizeski ML: Resistance of sheep to *p*-pyrrolizidine alkaloids. *Isr J Vet Med* 42;376–384, 1986

Craig AM, Pearson EG, Meyer C, Schmitz JA: Serum liver enzyme and histopathologic changes in calves with chronic and chronic-delayed *Senecio jacobaea* toxicosis. *Am J Vet Res* 52;1969–1978, 1991

Craig AM, Latham CJ, Blythe LL, Schmotzer WB, O'Connor OA: Metabolism of toxic pyrrolizidine alkaloids from tansy ragwort *(Senecio jacobaea)* in ovine ruminal fluid under anaerobic conditions. *Appl Environ Microbiol* 58;2730–2736, 1992

Craig JC Jr, Mole ML, Billets S, El-Feraly F: Isolation and identification of the hypoglycemic agent, carboxyatractyloside from *Xanthium strumarium*. *Phytochemistry* 15;1178, 1978

Cronquist A: *An Integrated System of Classification of Flowering Plants*. Columbia Univ Press, New York, 1981

Culvenor CCJ, Smith LW: The alkaloids of *Erichtites quadridentata* DC. *Aust J Chem* 8;556–561, 1955

Culvenor CCJ, Edgar JA, Jago MV, Outteridge A, Peterson JE, Smith LW: Hepato- and pneumotoxicity of pyrrolizidine alkaloids and derivatives in relation to molecular structure. *Chem Biol Interact* 12;299–324, 1976

Dalvi RR, McGowan C: Helenin inhibition of liver microsomal enzymes. *J Agric Food Chem* 30;988–989, 1982

Dameron WH, Carpenter OL: *Bitterweed Toxicity Prevention*. Tex Agric Exp Stn Annu Rep 51;214–215, 1938

Danieli B, Bombardelli E, Bonati A, Gabetta B: Structure of the diterpenoid carboxyatractyloside. *Phytochemistry* 11;3501–3504, 1972

Dean RE, Winward AH: An investigation into the possibility of tansy ragwort poisoning of black-tailed deer. *J Wildl Dis* 10;166–169, 1974

Deem AW, Thorp F Jr, Durrell LW: Range plant newly found to be poisonous [letter]. *Science* 89;435, 1939

DeGraw JI Jr, Bonner WA: The synthesis of dihydrotremetone. *Tetrahedron* 18;1311–1314, 1962

Deinzer ML, Thomson PA, Burgett DM, Isaacson DL: Pyrrolizidine alkaloids: their occurrence in honey from tansy ragwort (*Senecio jacobaea* L). *Science* 195;497–499, 1977

Deinzer ML, Arbogast BL, Buhler DR: Gas chromatographic determination of pyrrolizidine alkaloids in goat's milk. *Anal Chem* 54;1811–1814, 1982

de Lima TCM, Morato GS, Takahashi RN: Evaluation of the central properties of *Artemisia verlotorum*. *Planta Med* 59;326–329, 1993

Delle Monarche G, Botta B, Delle Monarche F, Espinal R, De Bonnevaux, De Luca C, Botta M, Corelli F, Carmignani M: Novel hypotensive agents from *Verbesina caracasana*. 2. Synthesis and pharmacology of caracasanamide. *J Med Chem* 36;2956–2963, 1993

Der Marderosian A, Liberli LE: *Natural Product Medicine*. GF Stuckley, Philadelphia, 1988

Dermer OC, Cleverdon R: Studies on the chemical constituents of rayless goldenrod *(Aplopappus heterophyllus)*. *Proc Okla Acad Sci* 23;63–66, 1943

Deuel PG, Geissman TA: Xanthanin. 2. Structures of xanthanin and xanthatin. *Am Chem Soc J* 79;3778–3783, 1957

Dewes HF, Lowe MD: Haemolytic crisis associated with ragwort poisoning and rail chewing in two thoroughbred fillies. *N Z Vet J* 33;159–160, 1985

Dickinson JO, Braun RC: Effect of 2(3)-tert-butyl-4-hydroxyanisole (BHA) and 2-chloroethanol against pyrrole production and chronic toxicity of monocrotaline in chickens. *Vet Hum Toxicol* 29;11–15, 1987

Dickinson JO, Cooke MP, King RR, Mohamed PA: Milk transfer of pyrrolizidine alkaloids in cattle. *J Am Vet Med Assoc* 169;1192–1196, 1976

Dil'bazi GI: Poisoning of buffalos eating hay containing choking mustard. *Veterinariya (Moscow)* 2;106–107, 1974

Divers TJ: Therapy of liver failure. In *Large Animal Internal Medicine*, 2d ed, Smith BP ed, Mosby, St Louis, pp948–950, 1996

Dobereiner J, Tokarnia CH, Purisco E: *Vernonia mollissima*, a toxic plant responsible for mortality of cattle in south of Mato Grosso. *Pesq Agropecu Bras, Ser Vet* 11;49–58, 1976

Dollahite JW: *Desert Baileya Poisoning in Sheep, Goats, and Rabbits*. Tex Agric Exp Stn Prog Rep 2149, 1960

Dollahite JW: The use of sheep and goats to control *Senecio* poisoning in cattle. *Southwest Vet* 25;223–226, 1972

Dollahite JW: *Research and Observations on Toxic Plants in Texas 1932–1972*. Tex Agric Exp Stn Tech Rep 78–1, 1978

Dollahite JW, Allen TJ: *Feeding Perennial Broomweed to Cattle, Swine, Sheep, Goats, Rabbits, Guinea Pigs, and Chickens.* Tex Agric Exp Stn Prog Rep 2105, 1959

Dollahite JW, Allen TJ: The toxicity of the fruit of *Flourensia cernua* (tarbush)(blackbrush). *Southwest Vet* 28;113–117, 1975

Dollahite JW, Anthony WV: *Experimental Production of Premature Calves and Retained Placentas by Feeding a Species of Perennial Broomweed.* Tex Agric Exp Stn Prog Rep 1825, 1955

Dollahite JW, Anthony WV: *Experimental Production of Abortion, Premature Calves, and Retained Placentas by Feeding a Species of Perennial Broomweed.* Tex Agric Exp Stn Prog Rep 1884, 1956

Dollahite JW, Anthony WV: Poisoning of cattle with *Gutierrezia microcephala*, a perennial broomweed. *J Am Vet Med Assoc* 130;525–530, 1957

Dollahite JW, Shaver T, Camp BJ: Injected saponins as abortifacients. *Am J Vet Res* 23;1261–1263, 1962

Dollahite JW, Hardy WT, Henson JB: Toxicity of *Helenium microcephalum* (smallhead sneezeweed). *J Am Vet Med Assoc* 145;694–696, 1964

Dollahite JW, Rowe LD, Kim HL, Camp BJ: Toxicity of *Helenium amarum* (bitter sneezeweed) to sheep. *Southwest Vet* 26;135–137, 1973

Donald LG, Shanks PL: Ragwort poisoning from silage. *Br Vet J* 112;307–311, 1956

Doran CW, Cassady JT: *Management of Sheep on Range Infested with Orange Sneezeweed.* USDA Circ 691, 1944

Downum KR, Villegas S, Rodriguez E, Keil DJ: Plant photosensitizers: a survey of their occurrence in arid and semiarid plants from North America. *J Chem Ecol* 15;345–355, 1989

Doyle JT: Milk sickness. *N C Med* J 8;404, 1947

Doyle LP, Walkey FL: *White Snakeroot* (Eupatorium urticaefolium) *Poisoning in Livestock.* Purdue Univ Agric Exp Stn Bull 270, 1923

Dueker SR, Lame MW, Morin D, Wilson DW, Segall HJ: Guinea pig and rat hepatic microsomal metabolism of monocrotaline. *Drug Metab Dispos* 20;275–280, 1992a

Dueker SR, Lame MW, Segall HJ: Hydrolysis of pyrrolizidine alkaloids by guinea pig hepatic carboxylesterases. *Toxicol Appl Pharmacol* 117;116–121, 1992b

Duffy DC: Land of milk and poison. *Nat Hist* 7;4–8, 1990

Duncan WH, Piercy PL, Feurt SD, Starling R: Toxicological studies of southeastern plants. 2. Compositae. *Econ Bot* 11;75–85, 1957

Edrington TS, Flores-Rodriguez GJ, Smith GS: Effects of ingested snakeweed *(Gutierrezia microcephala)* herbage on size of testes and seminal vesicles and on semen quality in rats. *Proc West Sect Am Soc Anim Sci* 41;88–90, 1990

Edrington TS, Smith GS, Reynolds PR, Samford MD, Ross TT: Comparative toxicity of snakeweed *(Gutierrezia microcephala)* collected from two sites. *Proc West Sec Am Soc Anim Sci* 42;173–176, 1991a

Edrington TS, Smith GS, Samford MD, Medrano J, Ross TT, Thilsted JP: Ingested snakeweed *(Gutierrezia microcephala)* foliage related to embryonic and fetal mortality of albino rats. *Proc West Sec Am Soc Anim Sci* 42;12–14, 1991b

Edrington TS, Smith GS, Ross TT, Velastegui W, Campos D: Toxicity of new growth and whole plant foliage of snakeweed *(Gutierrezia microcephala)* as changed by drying. *Proc West Sec Am Soc Anim Sci* 43;414–417, 1992

Edrington TS, Flores-Rodriguez GI, Smith GS, Hallford DM: Effect of ingested snakeweed *(Gutierrezia microcephala)* foliage on reproduction, semen quality, and serum clinical profiles of male rats. *J Anim Sci* 71;1520–1525, 1993a

Edrington TS, Ross TT, Smith GS, Hall L, Oetting BC, Williams JL: Hepato-renal toxicosis characterized in sheep dosed intraruminally with snakeweed foliage (*Gutierrezia* spp). *Vet Hum Toxicol* 35;506–508, 1993b

Edrington TS, Smith GS, Ross TT, Hallford DM, Samford MD, Thilsted JP: Embryonic mortality in Sprague-Dawley rats induced by snakeweed *(Gutierrezia microcephala)*. *J Anim Sci* 71;2193–2198, 1993c

Eichholzer JV, Lewis IAS, MacLeod JK, Oelrichs PB, Vallely PJ: Galegine and a new dihydroxyalkylacetamide from *Verbesina encelioides. Phytochemistry* 21;97–99, 1982

Eissa FZ, Burrows GE, Clarke CR: The effects of *Artemisia filifolia* and *Helenium flexuosum* on disposition of antipyrine in rabbits. *Vet Hum Toxicol* 37;47–49, 1995

Eissa FZ, Qualls CW, Burrows GE, Lish JW: Effects of feeding *Artemisia filifolia* and *Helenium flexuosum* on rabbit cytochrome P450 isozymes. *Vet Hum Toxicol* 38;19–23, 1996

El Bahri L, Belguith J, Djegham M, Bellil H: *Artemisia campestris* L.: a poisonous plant of North Africa. *Vet Hum Toxicol* 39;311–312, 1997

Elissalde MH, Ivie GW: Inhibition of macrophage adenylate cyclase by the α-methylene-γ-lactone moiety of sesquiterpene lactones from forage plants. *Am J Vet Res* 48;148–152, 1987

Elissalde MH, Ivie GW, Rowe LD, Elissalde GS: Considerations of the structure of sesquiterpene lactones on biological activity: influence of the α-methylene-γ-lactone moiety on mast cell degranulation. *Am J Vet Res* 44;1894–1897, 1983

Epstein WW, Seidel JL: Monoterpenes of *Gutierrezia sarothrae. J Agric Food Chem* 37;651–654, 1989

Etman MA, Hassan AB, Atta AH, Fahmy GE, Youssef AK, Nadia AEB: Phytochemical and pharmacological studies on *Achillea santolina* L. plant. *Egypt J Vet Sci* 24;181–189, 1987

Everist SL: *Poisonous Plants of Australia*, 2d ed. Angus & Robertson, Sydney, Australia, 1981

Falk AJ, Bauer L, Bell CL, Smolenski SJ: The constituents of the essential oil from *Achillea millefolium* L. *Lloydia* 37;598–602, 1974

Fikenscher LH, Hegnauer R, Ruijgrok HWL: Die verbreitung der blausaure bei den cormophyten. 14. Mitteilung. Neue beobachtungen zur cyanogenese bei den compositen. *Planta Med* 40;202–211, 1980

Fleming CE: *Range Plants Poisonous to Sheep and Cattle in Nevada.* Nev Agric Exp Stn Bull 95;25–27, 1918

Fleming CE: *Poisonous Range Plants.* Nev Agric Exp Stn Annu Rep 1919;39–43, 1920

Fleming CE, Wright LH, Miller MR: *Poisonous Plants.* Nev Agric Exp Stn Annu Rep 1919;39–43, 1920

Fleming CE, Miller MR, Vawter LR: *The Spring Rabbit-brush* (Tetradymia glabrata). Nev Agric Exp Stn Bull 104, 1922

Fleming CE, Miller MR, Vawter LR, Young A: *Poisonous Range Plants.* Nev Agric Exp Stn Annu Rep 1930;12–13, 1931

Fleming CE, Miller MR, Vawter LR, Young A: *Poisonous Plants.* Nev Agric Exp Stn Annu Rep 1933;10, 1934a

Fleming CE, Miller MR, Vawter LR, Young A: *Poisonous Range Plants.* Nev Agric Exp Stn Annu Rep 1934;25–26, 1934b

Flores-Rodriguez GI, Smith GS, McDaniel KC: Effects of ingested snakeweed *(Gutierrezia microcephala)* herbage on reproduction, serum progesterone, and blood constituents of female albino rats. *Proc West Sect Am Soc Anim Sci* 40;217–221, 1989

Fowler ME: Nigropallidal encephalomalacia in the horse. *J Am Vet Med Assoc* 147;607–616, 1965

Fowler ME: Pyrrolizidine alkaloid poisoning in calves. *J Am Vet Med Assoc* 152;1131–1137, 1968

Fowler ME: Plant poisoning in captive nondomestic animals. *J Zoo Anim Med* 12;134–137, 1981

Fredrickson E, Thilsted J, Estell R, Havstad K: Effect of chronic ingestion of tarbush *(Florensia cernua)* on ewe lambs. *Vet Hum Toxicol* 36;409–415, 1994

Fuller TC, McClintock E: *Poisonous Plants of California.* Univ California Press, Berkeley, 1986

Galey FD, Hullinger PJ, McCaskill J: Outbreaks of stringhalt in northern California [letter]. *Vet Hum Toxicol* 33;176–177, 1991

Gard GP, de Sarem WG, Ahrens PJ: Nigropallidal encephalomalacia in horses in New South Wales. *Aust Vet J* 49;107–108, 1973 (Ga Agric Exp Stn, 1980)

Garrett BJ, Cheeke PR: Evaluation of amino acids, B vitamins, and butylated hydroxyanisole as protective agents against pyrrolizidine alkaloid toxicity in rats. *J Anim Sci* 58;138–144, 1984

Garrett BJ, Holtan DW, Cheeke PR, Schmitz JA, Rogers QR: Effects of dietary supplementation with butylated hydroxyanisole, cysteine, and vitamins B on tansy ragwort *(Senecio jacobaea)* toxicosis in ponies. *Am J Vet Res* 45;459–464, 1984

Gaspar AR, Potgieter DJ, Vermeulen NM: The effect of sesquiterpene lactones from *Geigeria* on glycolytic enzymes. *Biochem Pharmacol* 35;493–497, 1986

Gates FC: Botanical notes, notes in 1929. *Trans Kans Acad Sci* 33;27–28, 1930a

Gates FC: *Principal Poisonous Plants in Kansas.* Kans Agric Exp Stn Tech Bull 25, 1930b

Gay CC, Fransen S, Richards J, Holler S: *Hypochoeris*-associated stringhalt in North America. *Equine Vet J* 25;456–457, 1993

Georgiou M, Sianidou L, Hatzis T, Papadatos J, Koutselinis A: Hepatotoxicity due to *Atractylis gummifera* L. *Clin Toxicol* 26;487–493, 1988

Gill S, Dembinska-Migas W, Sliwinska E, Daniewski M, Bohlmann F: Pseudoguaianolides from *Gaillardia grandiflora. Phytochemistry* 19;2049–2051, 1980

Goeger DE, Cheeke PR, Schmitz JA, Buhler DR: Effect of feeding milk from goats fed tansy ragwort *(Senecio jacobaea)* to rats and calves. *Am J Vet Res* 43;1631–1633, 1982

Goeger DE, Cheeke PR, Ramsdell HS, Nicholson SS, Buhler DR: Comparison of the toxicities of *Senecio jacobaea*, *Senecio vulgaris*, and *Senecio glabellus* in rats. *Toxicol Lett* 15;19–23, 1983

Goodwin MA, Mallinson ET, Brown J, Player EC, Latimer KS, Dale N, Shaff WV, Dickson TG: Toxicological pathology of cockleburs (*Xanthium* spp.) for broiler chickens. *Avian Dis* 36;444–446, 1992

Graham R, Boughton IB: *White Snakeroot Poisoning.* Univ Ill Agric Exp Stn Circ 295, 1925

Gregory M: Lettuce as a suspected cause of narcosis in a duckling [letter]. *Vet Rec* 141;316, 1997

Guengerich FP: Separation and purification of multiple forms of microsomal cytochrome P-450: activities of different forms of cytochrome P-450 towards several compounds of environmental interest. *J Biol Chem* 252;3970–3979, 1977

Gulick BA, Liu IKM, Qualls CWjr, Gribble DH, Rogers QH: Effect of pyrrolizidine alkaloid-induced hepatic disease on plasma amino acid patterns in the horse. *Am J Vet Res* 41;1894–1898, 1980

Gutierrez AB, Herz W: Ambrosanolides and other constituents of two *Rudbeckia* species. *Planta Med* 56;295–297, 1990

Habermehl GG: Poisonous plants of Brazil. *Toxicon* 32;143–156, 1994

Hailey TL, Thomas JW, Robinson RM: Pronghorn die-off in trans-Pecos Texas. *J Wildl Manage* 30;488–496, 1966

Hakkila MD, Holechek JL, Wallace JD, Anderson DM, Cardenas M: Diet and forage intake of cattle on desert grassland range. *J Range Manage* 40;339–342, 1987

Hall IH, Lee KH, Mar EC, Starnes CO, Waddell TG: Antitumor agents. 21. A proposed mechanism for inhibition of cancer growth by tenulin and helenaline and related cyclopentanones. *J Med Chem* 20;333–337, 1977

Hall IH, Lee KH, Eigebaly SA: Antitumor agents. 27. Effects of helenalin on anaerobic and aerobic metabolism of Ehrlich ascites cells. *J Pharm Sci* 67;552–554, 1978

Hall IH, Grippo AA, Holbrook DJ, Roberts G, Lin H-C, Kim HL: Role of thiol agents in protecting against the toxicity of helenalin in tumor-bearing mice. *Planta Med* 55;513–517, 1989

Hall MC: Uses and limitations of santonin. *Vet Med* 16;50, 1921

Halls SC, Gang DR, Weber DJ: Seasonal variation in volatile secondary compounds of *Chrysothamnus nauseosus* (Pallas) Britt,: Asteraceae ssp. *hololeucus* (Gray) Hall. and Clem. influences herbivory. *J Chem Ecol* 20;2055–2063, 1994

Hamburger M, Wolfender J-L, Hostettmann K: Search for chlorinated sesquiterpene lactones in the neurotoxic thistle *Centaurea solstitialis* by liquid chromatography-mass spectrometry, and model studies on their possible artifactual formation. *Nat Toxins* 1;315–327, 1993

Hanlidou E, Kokkalou E, Kokkini S: Volatile constituents of *Achillea grandifolia. Planta Med* 58;105–107, 1992

Hansen AA: *Cocklebur.* USDA Circ 109, 1920

Hansen AA: The tragedy of Hindustan. *Nature* 11;385–386, 1928

Hardy WT, Cory VL, Schmidt H, Dameron WH: *Bitterweed Poisoning in Sheep.* Tex Agric Exp Stn Bull 433, 1931

Harradine AR: Effect of 2,4-D ester on the nitrate concentration of variegated thistle (*Silybum marianum* [L.] gaertn.). *Aust J Exp Agric* 30;221–226, 1990

Harrington B, Pieper RD: Snakeweed on ranges in central New Mexico. In *Snakeweed Research Update and Highlights,* Sterling TM, Thompson DC eds, N M State Univ Res Rep 674;15, 1993

Harris P, Thompson LS, Wilkinson ATS, Neary ME: Reproductive biology of tansy ragwort, climate, and biological control by the cinnabar moth in Canada. *Proceedings of IV International Symposium on Biological Control of Weeds,* Freeman TE ed, Univ of Florida, Gainesville, pp163–173, 1976

Hartmann AF Sr, Hartmann AF Jr, Purkerson ML, Wesley ME: Tremetol poisoning—not yet extinct. *J Am Med Assoc* 185;706–709, 1963

Hartmann T, Witte L: Chemistry, biology, and chemoecology of the pyrrolizidine alkaloids. In *Alkaloids: Chemical and Biological Perspectives*, Pelletier SW ed, Pergamon Press, Tarrytown, NY, pp155–233, 1995

Hartmann T, Ehmke A, Eilert U, von Borstel K, Theuring C: Sites of synthesis, translocation, and accumulation of pyrrolizidine alkaloid N-oxides in *Senecio vulgaris* L. *Planta* 177;98–107, 1989

Hatch RC, Jain AV, Weiss R, Clark JD: Toxicologic study of carboxyatractyloside (active principle in cocklebur—*Xanthium strumarium*) in rats treated with enzyme inducers and inhibitors and glutathione precursor and depletory. *Am J Vet Res* 43;111–116, 1982

Hausen BM, Breuer J, Weglewski J, Rucker G: Alpha-peroxychifolid and other new sensitizing sesquiterpene lactones from yarrow (*Achillea millefolium* L., Compositae). *Contact Dermatitis* 24;274–280, 1991

Hawkes RB, Johnson GR: *Longitarsus jacobaea* aids moth in the biological control of tansy ragwort. *Proceedings of IV International Symposium on Biological Control of Weeds,* Freeman TE ed, Univ of Florida, Gainesville, pp193–196, 1976

Hay AJ, Hamburger M, Hostettmann K, Hoult JR: Toxic inhibition of smooth muscle contractility by plant-derived sesquiterpenes caused by their chemically reactive alphamethylene butyrolactone functions. *Br J Pharmacol* 112;9–12, 1994

Hegnauer R: The chemistry of the Compositae. In *The Biology and Chemistry of the Compositae*, vol 1, Heywood JB, Turner BL, eds, Academic Press, New York, pp283–336, 1977

Heizer CB: Monograph of *Psilostrophe. Ann Mo Bot Gard* 31:279–300, 1944

Herz W: Pseudoguaianolides in Compositae. In *Recent Advances in Phytochemistry*, vol 1, Mabry TJ ed, Appleton-Century-Crofts, New York, pp229–269, 1968

Herz W: Sesquiterpene lactones from livestock poisons. In *Effects of Poisonous Plants on Livestock*, Keeler RF, Van Kampen KR, James LF eds, Academic Press, New York, pp487–497, 1978

Herz W, deVivar AR, Romo J, Viswanathan N: Constituents of *Helenium* species. 13. The structures of helenolin and mexicanin A. *Am Chem Soc J* 85;19–26, 1963

Herz W, Aota K, Holub M, Samek Z: Sesquiterpene lactones and lactone glucosides from *Hymenoxys* species. *J Org Chem* 35;2611–2624, 1970

Heywood VH, Harborne JB, Turner BL: An overture to the Compositae. In *The Biology and Chemistry of the Compositae*, vol 1, Heywood JB, Turner BL eds, Academic Press, New York, pp1–20, 1977

Hill DW, Kim HL, Martin CL, Camp BJ: Hymenoxon found in both *Baileya multiradiata* and *Helenium hoopesii. J Agric Food Chem* 25;1304–1307, 1977

Hill DW, Kim HL, Camp BJ: Quantitative analysis of hymenoxon in plant tissue. *J Agric Food Chem* 27;885–887, 1979

Hill DW, Bailey EM, Camp BJ: Tissue distribution and disposition of hymenoxon. *J Agric Food Chem* 28;1269–1273, 1980

Hincks JR, Kim HY, Segall HJ, Molyneux RJ, Stermitz FR, Coulombe RAjr: DNA cross-linking in mammalian cells by pyrrolizidine alkaloids: structure-activity relationships. *Toxicol Appl Pharmacol* 111;90–98, 1991

Hirono I: Natural carcinogenic products of plant origin. *CRC Crit Rev Toxicol* 8;235–277, 1981

Hooper PT: Pyrrolizidine alkaloid poisoning—pathology with particular reference to differences in animal and plant species. In *Effects of Poisonous Plants in Livestock*, Keeler RF, Van Kampen KR, James LF eds, Academic Press, New York, pp161–176, 1978

Hsia MTS, Grossman S, Schrankel KR: Hepatotoxicity of the anti-juvenile hormone precocene II and the generation of dihydrodiol metabolites. *Chem Biol Interact* 37;265–277, 1981

Hsu IC, Robertson KA, Shumaker RC, Allen JR: Binding of tritiated dehydronecine to macromolecules. *Res Commun Chem Pathol Pharmacol* 11;99–106, 1975

Huffman WT, Couch JF: Plants poisonous to livestock. *USDA Yearb Agric* 1942;354–373, 1942

Huntington PJ, Jeffcott LB, Friend SCE, Luff AR, Finkelstein DI, Flynn RJ: Australian stringhalt—epidemiological, clinical, and neurological investigations. *Equine Vet J* 21;266–273, 1989

Huntington PJ, Seneque S, Slocombe RF, Jeffcott LB, Mclean A, Luff AR: Use of phenytoin to treat horses with Australian stringhalt. *Aust Vet J* 68;221–224, 1991

Huxtable CR, Dorling PR, Colegate SM: Identification of galegine, an isoprenyl guanidine, as the toxic principle of *Schoenus asperocarpus* (poison sedge). *Aust Vet J* 70;169–171, 1993

Huxtable RJ: Herbal teas and toxins: novel aspects of pyrrolizidine poisoning in the United States. *Perspect Biol Med* 24;1–14, 1980

Huxtable RJ: Human health implications of pyrrolizidine alkaloids and herbs containing them. In *Toxicants of Plant Origin*, vol 1, *Alkaloids*, Cheeke PR ed, CRC Press, Boca Raton, Fla, pp41–86, 1989

Imakura Y, Lee K-H, Sims D, Wu R-Y, Hall IH, Furukawa H, Itoigawa M, Yonaha K: Antitumor agents. 36. Structural elucidation of sesquiterpene lactones microhelanins-A, B, and C, microlenin acetate, and plenolin from *Helenium microcephalum. J Pharm Sci* 69;1044–1049, 1980

Irvin TR, Murphy M, Ray A, Bratton GR, Reagor J: Identification of the toxic constituents underlying the prenatal toxicity of *Iva angustifolia* employing postimplantation rodent embryo culture systems. *Proc Am Assoc Vet Lab Diagn* 29;199–208, 1986

Isaacson DL: The role of biological agents in integrated control of tansy ragwort. *Proceedings of IV International Symposium on Biological Control of Weeds*, Freeman TE ed, Univ of Florida, Gainesville, pp189–192, 1976

Itoigawa M, Takeya K, Furukawa H, Ito K: Mode of cardiotonic action of helenalin, a sesquiterpene lactone, on guinea pig ventricular myocardium. *J Cardiovasc Pharmacol* 9;193–201, 1987

Ivie GW, Witzel DA, Herz W, Kannan R, Norman JO, Rushing DD, Johnson JH, Rowe LD, Veech JA: Hymenovin. Major toxic constituent of western bitterweed (*Hymenoxys odorata* DC.). *J Agric Food Chem* 23;841–845, 1975a

Ivie GW, Witzel DA, Rushing DD: Toxicity and milk bittering properties of tenulin, the major sesquiterpene lactone constituent of *Helenium amarum* (bitter sneezeweed). *J Agric Food Chem* 23;845–849, 1975b

Ivie GW, Witzel DA, Herz W, Sharma RP, Johnson AE: Isolation of hymenovin from

Hymenoxys richardsonii (pingue) and *Dugaldia hoopesii* (orange sneezeweed). *J Agric Food Chem* 24;681–682, 1976

Jago MV: The development of the hepatic megalocytosis of chronic pyrrolizidine alkaloid poisoning. *Am J Pathol* 56;405–422, 1969

Jago MV, Edgar JA, Smith LW, Culvenor CCJ. Metabolic conversion of heliotridine based pyrrolizidine alkaloids to dihydroheliotridine. *Mol Pharmacol* 6;402–406, 1970

Jakupovic J, Gage DA, Bohlmann F, Mabry TJ: Sesquiterpene lactones from *Vernonia marginata*. *Phytochemistry* 25;1179–1183, 1986

James LF, Keeler RF, Johnson AE, Williams MC, Cronin EH, Olsen JD: *Plants Poisonous to Livestock in the Western States.* USDA Agric Inf Bull 415, 1980

James LF, Panter KE, Nielsen DB, Molyneux RJ: The effect of natural toxins on reproduction in livestock. *J Anim Sci* 70;1573–1579, 1992

James RR, McEvoy PB, Cox CS: Combining the cinnabar moth *(Tyria jacobaea)* and the ragwort flea beetle *(Longitarsus jacobaea)* for control of ragwort *(Senecio jacobaea)*: an experimental analysis. *J Appl Ecol* 29;589–596, 1992

Jarvis BB, Wang S, Cox C, Madhusudane Rao M, Philip V, Varaschin MS, Barros CS: Brazilian *Baccharis* toxins: livestock poisoning and the isolation of macrocyclic trichothecene glucosides. *Nat Toxins* 4;58–71, 1996

Jennings PW, Reeder SK, Hurley JC, Caughlan CN, Smith GD: Isolation and structure determination of one of the toxic constituents from *Tetradymia glabrata*. *J Org Chem* 39;3392–3398, 1974

Jennings PW, Reeder SK, Hurley JC, Robbins JE, Holian SK, Holian A, Lee P, Pribanic JAS, Hull M: Toxic constituents and hepatotoxicity of the plant *Tetradymia glabrata* (Asteraceae). In *Effects of Poisonous Plants on Livestock*, Keeler RF, Kampen KR, James LF eds, Academic Press, New York, pp217–228, 1978

Johnson AE: Experimental photosensitization and toxicity in sheep produced by *Tetradymia glabrata*. *Can J Comp Med* 38;406–410, 1974a

Johnson AE: Predisposing influence of range plants on *Tetradymia*-related photosensitization in sheep: work of Drs. A. B. Clawson and W. T. Huffman. *Am J Vet Res* 35;1583–1585, 1974b

Johnson AE: *Tetradymia* toxicity—a new look at an old problem. In *Effects of Poisonous Plants on Livestock*, Keeler RF, Van Kampen KR, James LF eds, Academic Press, New York, pp209–216, 1978

Johnson AE: Failure of mineral-vitamin supplements to prevent tansy ragwort *(Senecio jacobaea)* toxicosis in cattle. *Am J Vet Res* 43;718–723, 1982

Johnson AE, Molyneux RJ: Toxicity of threadleaf groundsel *(Senecio douglasii* var. *longilobus)* to cattle. *Am J Vet Res* 45;26–31, 1984

Johnson AE, Smart RA: Effects on cattle and their calves of tansy ragwort *(Senecio jacobaea)* fed in early gestation. *Am J Vet Res* 44;1215–1219, 1983

Johnson AE, James LF, Spillett J: The abortifacient and toxic effects of big sagebrush *(Artemisia tridentata)* and juniper *(Juniperus osteosperma)* on domestic sheep. *J Range Manage* 29;278–280, 1976

Johnson AE, Molyneux RJ, Merrill JC: Chemistry of toxic range plants: variation in pyrrolizidine alkaloid content of *Senecio*, *Amsinckia*, and *Crotalaria* species. *J Agric Food Chem* 33;50–55, 1985a

Johnson AE, Molyneux RJ, Stuart LD: Toxicity of Riddell's groundsel *(Senecio riddellii)* to cattle. *Am J Vet Res* 46;577–582, 1985b

Johnson MF: A monograph of the genus *Ageratum* L. (Compositae—Eupatorieae). *Ann Mo Bot Gard* 58;6–88, 1971

Jones DH, Kim HL: Toxicity of hymenoxon in Swiss white mice following treatment with microsomal enzyme inducers, inhibitors, and carbon tetrachloride. *Res Commun Chem Pathol Pharmacol* 33;361–366, 1981

Kaij-a-Kamb M, Amoros M, Girre L: Chimie et activites biologiques du genre *Centaurea*. *Pharm Acta Helv* 67;178–188, 1992

Kains MG: *Chicory Growing.* USDA Div Bot Bull 19, 1898

Keeler RF, Baker DC, Panter KE: Concentration of galegine in *Verbesina encelioides* and *Galega officinalis* and the toxic and pathologic effects induced by the plants. *J Environ Pathol Toxicol Oncol* 11;75–81, 1992

Kendrick JW, Tucker J, Peoples SA: Nitrate poisoning in cattle due to ingestion of variegated thistle, *Silybum marianum*. *J Am Vet Med Assoc* 126;53–56, 1955

Kim HL: Tissue thiol induction in mice and protective effect against pyrrolizidine alkaloids by dietary ethoxyquin and methionine hydroxy analog. *Vet Hum Toxicol* 26;314–316, 1984

Kim HL, Jones LP: Protective effects of butylated hydroxyanisole, ethoxyquin, and disulfiram on acute pyrrolizidine alkaloids poisoning in mice. *Res Commun Chem Path Pharmacol* 36;341–344, 1982

Kim HL, Anderson AC, Herrig BW, Jones LP, Calhoun MC: Protective effects of antioxidants on bitterweed (*Hymenoxys odorata* DC.) toxicity in sheep. *Am J Vet Res* 43;1945–1950, 1982

Kim HL, Herrig BW, Anderson AC, Jones LP, Calhoun MC: Elimination of adverse effects of ethoxyquin (EQ) by methionine hydroxy analog (MHA). *Toxicol Lett* 16;23–29, 1983

Kim HL, Safe SH, Calhoun MC: Quantitation of hymenoxon and related sesquiterpene lactones. *J Agric Food Chem* 35;891–895, 1987

King DW, Estell RE, Fredrickson EL, Havstad KM, Wallace JD, Murray LW: Effects of *Flourensia cernua* ingestion on intake, digestion kinetics, and ruminal fermentation of sheep consuming tabosa. *J Range Manage* 49;325–330, 1996

Klingenberg M: The ADP-ATP carrier in mitochondrial membranes. In *The Enzymes of Biological Membranes*, vol 3, *Membrane Transport*, Martonosi A ed, Plenum Press, New York, pp383–438, 1984

Klingenberg M, Appel M, Oelrichs: Wedeloside, a powerful inhibitor and ligand of the mitochondrial ADP/ATP carrier. *FEBS Lett* 189;245–249, 1985

Krisken J, Willuhn G: Weitere sesquiterpenlactone aus den bluten von *Arnica chamissonis* [abstract]. *Planta Med* 45;132, 1982

Kubo I, Matsumoto T: Desacetylisotenulin, a molluscicide from the desert plant *Psathyrotes ramosissima*. *Agric Biol Chem* 48;3147–3149, 1984

Kupchan SM, Fessler DC, Eakin MA, Giacobbe TJ: Reactions of alpha methylene lactone tumor inhibitors with model biological nucleophiles. *Science* 168;376–378, 1970

Kupchan SM, Eakin MA, Thomas AM: Tumor inhibitors. 69. Structure-cytotoxicity relationships among the sesquiterpene lactones. *J Med Chem* 14;1147–1152, 1971

Kupiecki FP, Ogzewalla CD, Schell FM: Isolation and characterization of a hypoglycemic agent from *Xanthium strumarium*. *J Pharm Sci* 63;1166–1169, 1974

Kuzel NR, Miller CE: A phytochemical study of *Xanthium canadense*. *J Am Pharm Assoc* 39;202–204, 1950

Laekeman GM, Mertens J, Totte J, Bult H, Vlietinck AJ, Herman AG: Isolation and pharmacological characterization of vernolepin. *J Nat Prod* 46;161–169, 1983

Lame MW, Jones AD, Morin D, Segall HJ,: Metabolism of [14C]monocrotaline by isolated perfused rat liver. *Drug Metab Dispos* 19;516–524, 1991

Lamson PD: On the pharmacological action of helenin, the active principle of *Helenium autumnale*. *J Pharmacol Exp Toxicol* 4;471–489, 1913

Langer T, Mostl E, Chizzola R, Gutleb R: A competitive enzyme immunoassay for the pyrrolizidine alkaloids of the senecionine type. *Planta Med* 62;267–271, 1996

Lanigan GW, Smith LW: Metabolism of pyrrolizidine alkaloids in the ovine rumen. 1. Formation of 7α-hydroxy-1α-methyl-8α-pyrrolizidine from heliotrine and lasiocarpine. *Aust J Agric Res* 21;493–500, 1970

Lathrop CA: Isolation and fractionation of tremetol from rayless goldenrod. Master's thesis, Oklahoma A&M Coll, 1939

Laycock WA: Coevolution of poisonous plants and large herbivores on rangelands. *J Range Manage* 31;335–342, 1978

Lee KH, Huang ES, Piantadosi C, Pagano JS, Geissman TA: Cytotoxicity of sesquiterpene lactones. *Cancer Res* 31;1649–1654, 1971

Lee KH, Hall IH, Mar EC, Starnes CO, ElGebaly SA, Waddell TG, Ruffner CG, Weidner I: Sesquiterpene antitumor agents: inhibitors of cellular metabolism. *Science* 196;533–536, 1977

Lessard P, Wilson WD, Olander HJ, Rogers QR, Mendel VE: Clinicopathological study of horses surviving pyrrolizidine alkaloid *(Senecio vulgaris)* toxicosis. *Am J Vet Res* 47;1776–1780, 1986

Lietava J: Medicinal plants in a middle paleolithic grave Shanidar IV? *J Ethnopharmacol* 35;263–266, 1992

Lilley CW: Pyrrolizidine alkaloid toxicosis. *Equine Pract* 2 [letter];6, 1980

Lockett S: Sheep poisoned by western goldenrod. *J Am Vet Med Assoc* 51;214–221, 1917

Looman J: *111 Range and Forage Plants of the Canadian Prairies.* Agric Can Publ 1751, 1983

Lopez TA, Campero CM, Chayer R, Cosentino B, Caracino M: Experimental toxicity of *Verbesina encelioides* in sheep and isolation of galegine. *Vet Hum Toxicol* 38;417–419, 1996

Luciani S, Martini N, Santi R: Effects of carboxyatractyloside a structural analogue of atractyloside on mitochondrial oxidative phosphorylation. *Life Sci* 10;961–968, 1971

MacGregor JT: Mutagenic activity of hymenovin, a sesquiterpene lactone from western bitterweed. *Food Cosmet Toxicol* 15;225–227, 1977

MacLachlan D, Arnason T, Lam J: Structure-function relationships in the phototoxicity of acetylenes from the Asteraceae. *Biochem Syst Ecol* 14;17–23, 1986

MacLeod JK, Lewis IAS, Moeller PDR, Oelrichs PB: The toxic extractives from *Wedelia asperrima.* 3. Structures of two naturally occurring rhamnosyl analogues of wedeloside. *J Nat Prod* 53;1256–1261, 1990a

MacLeod JK, Moeller PDR, Franke FP: Two toxic kaurene glycosides from the burrs of *Xanthium pungens. J Nat Prod* 53;451–455, 1990b

Maddison JE: Hepatic encephalopathy. *J Vet Intern Med* 6;341–353, 1992

Maddox DM: The knapweeds: their economics and biological control in the western states, USA. *Rangelands* 1;139–141, 1979

Maddox DM, Mayfield A: Yellow starthistle infestations are on the increase. *Calif Agric* 39;10–12, 1985

Maddox DM, Mayfield A, Poritz NH: Distribution of yellow starthistle *(Centaurea solstitialis)* and Russian thistle *(Centaurea repens). Weed Sci* 33;315–327, 1985

Mair TS: Ammonia and encephalopathy in the horse. *Equine Vet J* 29;1–2, 1997

Manley GD, Edds GT, Sundlof SF: Cattle deaths from poisonous plants. *Fla Vet J* 11;20, 1982

Manners GD, Ivie GW, MacGregor JT: Mutagenic activity of hymenovin in *Salmonella typhimurium:* association with the bishemiacetal functional group. *Toxicol Appl Pharmacol* 45;629–633, 1978

Marchant YY, Cooper GK: Structure and function relationships in polyacetylene photoactivity. In *Light-Activated Pesticides*, Heitz JR, Downum KR eds, American Chemical Society, Washington DC, pp241–254, 1987

Marsh CD: *Stock Poisoning Plants of the Range.* USDA Bull 1245, 1924

Marsh CD: *Trembles.* USDA Farmers Bull 1593, 1929

Marsh CD, Clawson AB: *Eupatorium urticaefolium* as a poisonous plant. *J Agric Res* 11;699–715, 1917

Marsh CD, Clawson AB, Eggleston WW: *Baccharis pteronioides* as a poisonous plant of the Southwest. *J Am Vet Med Assoc* 57;430–434, 1920

Marsh CD, Clawson AB, Couch JF, Marsh H: *Western Sneezeweed* (Helenium hoopesii) *as a Poisonous Plant.* USDA Bull 947, 1921

Marsh CD, Roe GC, Clawson AB: *Livestock Poisoning by Cocklebur.* USDA Circ 283, 1923

Marsh CD, Roe GC, Clawson AB: *Rayless Goldenrod* (Aplopappus heterophyllus) *as a Poisonous Plant.* USDA Bull 1391, 1926

Martin AA, Yanarella FG, Maurel RA, Ruager J: Nigropallidal encephalomalacia in horses due to chronic intoxication with yellow star thistle *(Centaurea solstitialis). An Vet* 3;43–53, 1971

Martin T, Stair EL, Dawson L: Cocklebur poisoning in cattle. *J Am Vet Med Assoc* 189;562–563, 1986

Martin T, Johnson BJ, Sangiah S, Burrows GE: Experimental cocklebur *(Xanthium strumarium)* intoxication in calves. In *Poisonous Plants: Proceedings of the Third International Symposium*, James LF, Keeler RF, Bailey EM Jr, Cheeke PR, Hegarty MP eds, Iowa State Univ Press, Ames, pp489–494, 1992

Martinez JH, Ross TT, Becker KA, Smith GS: Ingested dry snakeweed foliage did not impair reproduction in ewes and heifers during late gestation. *Proc West Sect Am Soc Anim Sci* 44;32–35, 1993

Masters AR: Dual role of pyrrolizidine alkaloids in nectar. *J Chem Ecol* 17;195–205, 1991

Mathews FP: *Poisoning of Cattle by Species of Groundsel (*Senecio longilobus *Benth. and* Senecio riddellii *Torr. & Gray).* Tex Agric Exp Stn Bull 481, 1933a

Mathews FP: The toxicity of *Baileya multiradiata* for sheep and goats. *J Am Vet Med Assoc* 83;673–679, 1933b

Mathews FP: Psilostrophe tagetinae *and* Psilostrophe gnaphalodes, *Two Plants Poisonous to Sheep and Cattle on the Ranges of the Southwest.* Tex Agric Exp Stn Bull 500, 1934

Mathews FP: The toxicity of broomweed *(Gutierrezia microcephala)* for sheep, cattle, and goats. *J Am Vet Med Assoc* 88;54–61, 1936

Mathews FP: The toxicity of *Sartwellia flaveriae* to goats. *J Agric Res* 61;287–293, 1940

Mathews FP: *The Toxicity of the Ripe Fruit of Blackbrush or Tarbush* (Flourensia cernua) *for Sheep and Goats.* Tex Agric Exp Stn Bull 664, 1944

Mattocks AR: *Chemistry and Toxicology of Pyrrolizidine Alkaloids.* Academic Press, London, 1986

Mattocks AR, Jukes R: Improved field tests for toxic pyrrolizidine alkaloids. *J Nat Prod* 50;161–166, 1987

Mattocks AR, Jukes R: Recovery of the pyrrolic nucleus of pyrrolizidine alkaloid metabolites from sulphur conjugates in tissues and body fluids. *Chem Biol Interact* 75;225–239, 1990

Mattocks AR, Jukes R: Detection of sulphur-conjugated pyrrolic metabolites in blood and fresh or fixed liver tissue from rats given a variety of toxic pyrrolizidine alkaloids. *Toxicol Lett* 63;47–55, 1992

Mattocks AR, White INH: The conversion of pyrrolizidine alkaloids to *N*-oxides and to dehydropyrrolizidine derivatives by rat-liver microsomes in vitro. *Chem Biol Interact* 3;383–396, 1971

Mayhew IG: Neurologic patients, 4: 1, Stringhalt, lathyrism, and shivering. In *Large Animal Neurology*, Lea & Febiger, Philadelphia, pp219–224, 1989

Mayland HF, Robbins CW: Sulfate uptake by salinity-tolerant plant species. *Commun Soil Sci Plant Anal* 25;2523–2541, 1994

McEvoy PB: Depression in ragwort *(Senecio jacobaea)* abundance following introduction of *Tyria jacobaea* and *Longitarsus jacobaea* on the central coast of Oregon. *Proceedings of VI International Symposium on Biological Control of Weeds,* Delfosse ES ed, Agriculture Canada, Canadian Government Publ Centre, Ottawa, pp57–64, 1985

McGinty A, Welch TG: Perennial broomweed and Texas ranching. *Rangelands* 9;246–249, 1987

McLean EK: The toxic actions of pyrrolizidine *(Senecio)* alkaloids. *Pharm Rev* 22;429–483, 1970

Mendel VE, Witt MR, Gitchell BS, Gribble DN, Rogers QR, Segall HJ, Knight HD: Pyrrolizidine alkaloid-induced liver disease in horses: an early diagnosis. *Am J Vet Res* 49;572–578, 1988

Mendez MC, Santos RC, Riet-Correa F: Intoxication by *Xanthium cavanillesii* in cattle and sheep in southern Brazil. *Vet Hum Toxicol* 40;144–147, 1998

Mereish KA, Bunner DL, Ragland DR, Creasia DA: Protection against microcystin-LR/Ninduced hepatotoxicity by silymarin: biochemistry, histopathology, and lethality. *Pharm Res* 8;273–277, 1991

Merrill J, Kim HL, Safe S: Helenalin: mechanism of toxic action. *Adv Exp Med Biol* 197;891–896, 1985

Merrill JC, Schuster JL: Grazing management practices affect livestock losses from poisonous plants. *J Range Manage* 31;351–354, 1978

Merrill JC, Kim HL, Safe S, Murray CA, Hayes MA: Role of glutathione in the toxicity of the sesquiterpene lactones hymenoxon and helenalin. *J Toxicol Environ Health* 23,159–169, 1988

Miller FM, Chow LM: Alkaloids of *Achillea millefolium* L. 1. Isolation and characterization of achilleine. *J Am Chem Soc* 76;1353–1354, 1954

Millspaugh CF: *American Medicinal Plants.* Dover, pp305–377, 1974 (original 1892)

Miranda CL, Chung W, Reed RE, Zhao X, Henderson MC, Wang J, Williams DE, Buhler DR: Flavin-containing monooxygenase: a major detoxifying enzyme for the pyrrolizidine alkaloid senecionine in guinea pig tissues. *Biochem Biophys Res Commun* 178;546–552, 1991

Mirsalis JC, Steinmetz KL, Blazak WF, Spalding JW: Evaluation of the potential of riddelliine to induce unscheduled DNA synthesis, S-phase synthesis, or micronuclei following in vivo treatment with multiple doses. *Environ Mol Mutagen* 21;265–271, 1993

Moghaddam MF, Cheeke PR: Effects of dietary pyrrolizidine *(Senecio)* alkaloids on vitamin A metabolism in rats. *Toxicol Lett* 45;149–156, 1989

Molyneux RJ, Johnson AE, Roitman JN, Benson ME: Chemistry of toxic range plants: determination of pyrrolizidine alkaloid content and composition in *Senecio* species by nuclear magnetic resonance spectroscopy. *J Agric Food Chem* 27;494–499, 1979

Molyneux RJ, Stevens KL, James LF: Chemistry of toxic range plants. Volatile constituents of broomweed *(Gutierrezia sarothrae)*. *J Agric Food Chem* 28;1332–1333, 1980

Molyneux RJ, Johnson AE, Stuart LD: Delayed manifestation of *Senecio*-induced pyrrolizidine alkaloidosis in cattle: case reports. *Vet Hum Toxicol* 30;201–205, 1988

Molyneux RJ, Johnson AE, Olsen JD, Baker DC: Toxicity of pyrrolizidine alkaloids from Riddell groundsel *(Senecio riddellii)* to cattle. *Am J Vet Res* 52;146–151, 1991

More PR, Vadlamudi VP, Qureshi MI: Note on the toxicity of *Parthenium hysterophorus* in livestock. *Indian J Anim Sci* 52;456–457, 1982

Muller L, Kasper P, Kaufmann G: The clastogenic potential in vitro of pyrrolizidine alkaloids employing hepatocyte metabolism. *Mutat Res* 282;169–176, 1992

Mullin CA, Alfatafta AA, Harman JL, Everett SL, Serino AA: Feeding and toxic effects of floral sesquiterpene lactones, diterpenes, and phenolics from sunflower (*Helianthus annuus* L.) on western corn rootworm. *J Agric Food Chem* 39;2293–2299, 1991

Mullins ME, Horowitz BZ: The case of the salad shooters: intravenous injection of wild lettuce extract. *Vet Hum Toxicol* 40;290–291, 1998

Murphy MJ, Reagor JC, Ray AC, Rowe LD: Bovine abortion associated with ingestion of *Iva angustifolia* (narrowleaf sumpweed). *Proc Am Assoc Vet Lab Diagn* 26;161–166, 1983

Nagy JG, Tengerdy RP: Antibacterial action of essential oils of *Artemisia* as an ecological factor. 2. Antibacterial action of the volatile oils of *Artemisia tridentata* (big sagebrush) on bacteria from the rumen of mule deer. *Appl Microbiol* 16;441–444, 1968

Narasimhan TR, Ananth M, Narayana Swamy M, Rajendra Babu M, Mangala A, Subba Rao PV: Toxicity of *Parthenium hysterophorus* L. [letter]. *Current Sci* 46;15–16, 1977a

Narasimhan TR, Ananth M, Narayana Swamy M, Rajendra Babu M, Mangala A, Subba Rao PV: Toxicity of *Parthenium hysterophorus* L. to cattle and buffaloes. *Experientia* 33;1358–1359, 1977b

Narasimhan TR, Ananth M, Narayana Swamy M, Rajendra Babu M, Mangala A, Subba Rao PV: Toxicity of *Parthenium hysterophorus* L.: partheniosis in cattle and buffaloes. *Indian J Anim Sci* 50;173–178, 1980

Narasimhan TR, Kim HL, Safe SH: Effects of sesquiterpene lactones on mitochondrial oxidative phosphorylation. *Gen Pharmacol* 20;681–687, 1989

Narasimhan TR, Keshava Murthy BS, Subba Rao PV: Nutritional evaluation of silage made from the toxic weed *Parthenium hysterophorus* in animals. *Food Chem Toxicol* 31;509–515, 1993

Nesom GL: Taxonomy of *Isocoma* (Compositae: Astereae). *Phytologia* 70;69–114, 1991

Oelrichs PB, Vallely PJ, MacLeod JK, Lewis IAS: Chemistry and toxic effects of *Verbesina enceloiodes*. In *Plant Toxicology—Proceedings of the Australia-USA Poisonous Plants Symposiumm*, Seawright AA, Hegarty MP, James LF, Keeler RF eds, Queensland Poisonous Plants Committee, Animal Research Institute, Yeerongpilly, Australia, pp479–483, 1985

Oelrichs PB, MacLeod JK, Lewis IAS, Klingenberg M: Chemistry and toxic effects of the active compounds from *Wedelia asperrima*. In *Poisonous Plants: Proceedings of the Third International Symposium*, James LF, Keeler RF, Bailey EM Jr, Cheeke PR, Hegarty MP eds, Iowa State Univ Press, Ames, pp282–287, 1992

Oelrichs PB, Calanasan CA, MacLeod JK, Seawright AA, Ng JC: Isolation of a compound from *Eupatorium adenophorum* (Spreng.) [*Ageratina adenophora* (Spreng.)] causing hepatotoxicity in mice. *Nat Toxins* 3;350–354, 1995

Oelrichs PB, Seawright AA, MacLeod JK, Ng JC: The isolation of a hepatotoxic compound from *Eupatorium adenophorum* (Spreng.). *Toxic Plants and Other Natural Toxicants*, Garland T, Barr AC eds, New York, pp271–275, 1998

Oetting BC, Ross TT, Walraven K, Kloppenburg P, Smith GS, Hallford DM: Effects of ingested snakeweed herbage on estrual activity, blood progesterone, and serum clinical profiles of fine-wool ewes. *Proc West Sec Am Soc An Sci* 41;23–26, 1990

Olson CT, Keller WC, Gerken DF, Reed SM: Suspected tremetol poisoning in horses. *J Am Vet Med Assoc* 185;1001–1003, 1984

O'Sullivan BM, Gibson JA, McKenzie RA: Intoxication of horses by *Eupatorium adenophorum* and *E. riparium* in Australia. In *Plant Toxicology—Proceedings of the Australia-USA Poisonous Plant Symposiumm*, Seawright AA, Hegarty MP, James LF, Keeler RF eds, Animal Research Institute, Yeerongpilly, Australia, pp423–426, 1985

Page JD, Chaney SG, Hall IH, Lee KH, Holbrook DJ: Inhibition of inosine monophosphate dehydrogenase by sesquiterpene lactones. *Biochim Biophys Acta* 926;186–194, 1987

Pammel LH: *A Manual of Poisonous Plants*. Torch Press, Cedar Rapids, Iowa, pp753–802, 1911

Pammel LH: Pingue poisoning in Arizona. *Am J Vet Med* 14;32, 1919a

Pammel LH: White snakeroot poisonous in Illinois. *Am J Vet Med* 14;31, 1919b

Pammel LH: Golden glow poisons. *Vet Med* 16;50, 1921

Parker KW: *Prevention of Death Losses in Sheep on Areas Infested with Pingue* (Actinea richardsoni): *Appendix, Observation of Feeding Habits of Sheep on the Range*. N M Agric Exp Stn Bull 241, 1936

Pearson EG: Liver failure attributable to pyrrolizidine alkaloid toxicosis and associated with inspiratory dyspnea in ponies: three cases (1982–1988). *J Am Vet Med Assoc* 198;1651–1654, 1991

Pemberton DH, Caple IW: Australian stringhalt in horses. *Vet Annu* 20;167–171, 1980

Perdomo E, deFreitas A, deFreitas A: Poisoning in horses by *Centaurea solstitialis* L. *Veterinaria (Montevideo)* 14;137–140, 1978

Pereda-Miranda R: Bioactive natural products from traditionally used Mexican plants. In *Phytochemistry of Medicinal Plants: Recent Advances in Phytochemistry*, vol 29, Arnason JT, Mata R, Romea JT eds, pp83–112, 1995

Perkins KD, Payne WW: *Guide to the Poisonous and Irritant Plants of Florida*. Fla Coop Ext Serv Circ 441, 1978

Pesman MW: *Meet Flora Mexicana*. DS King, Globe, Ariz, 1962

Pethick WH: *Report of the Director General*. Can Dept Agric 1905, pp90–100, 1906

Pethick WH, Rutherford JG: *Special Report on Pictou Cattle Disease, 1906*. Can Dept Agric, 1907

Petry TW, Bowden GT, Buhler DR, Sipes IG: Genotoxicity of the pyrrolizidine alkaloid jacobine in rats. *Toxicol Lett* 32;275–281, 1986

Petterson RD, Kim HL: X-ray structures of hymenoxon and hymenolane: pseudoguaianolides isolated from *Hymenoxys odorata* DC. (bitterweed). *J Chem Soc Perkin Trans* 2;1399–1403, 1976

Pfeiffer FA, Calhoun MC: Effects of environmental, site, and phenological factors on hymenoxon content of bitterweed *(Hymenoxys odorata). J Anim Sci* 65;1553–1562, 1987

Phares DL: *Diseases of Sheep and Calves—Bitterweed.* Miss Agric Exp Stn Bull 9, 1889

Picman AK: Biological activities of sesquiterpene lactones. *Biochem Syst Ecol* 14;255–281, 1986

Pieper RD: Brown snakeweed content of herbivore diets. In *Snakeweeds: Problems and Perspectives,* N M Agric Exp Stn Bull 751, pp203–210, 1990

Pieper RD: Livestock grazing influences snakeweed density. In *Snakeweed Research Update and Highlights,* Sterling TM, Thompson DC eds, N M State Univ Res Rep 674;14, 1993

Pliske TE, Edgar JA, Culvenor CCJ: The chemical basis of attraction of ithomiine butterflies to plants containing pyrrolizidine alkaloids. *J Chem Ecol* 2;255–262, 1976

Poage GW III, Scott CB, Bisson MG, Hartmann FS: Activated charcoal attenuates bitterweed toxicosis in sheep. *J Range Manage* 53;73–78, 2000

Pocchiari F, Silano V: Effect of atractyloside on glucose and pyruvate metabolism in rat diaphragm muscle. *Biochem J* 107;305–309, 1968

Poisonous Plants of the Southern United States. Ga Coop Ext Serv, 1980

Post LO, Bailey EM: The effect of dietary supplements on chronic bitterweed *(Hymenoxys odorata)* poisoning in sheep. *Vet Hum Toxicol* 34;209–213, 1992

Powis G, Gallegos A, Abraham RT, Ashendel CL, Zalkow LH, Grindey GB, Bonjouklian R: Increased intracellular Ca2+ signaling caused by the antitumor agent helenalin and its analogues. *Cancer Chemother Pharmacol* 34;344–350, 1994

Prien OL, Raiford LC: *Woody Aster,* Xylorrhiza parryi *Gray.* Wyo Agric Exp Stn Bull 88, 1911

Ramsdell HS, Kedzierski B, Buhler DR: Microsomal metabolism of pyrrolizidine alkaloids from *Senecio jacobaea:* isolation and quantitation 6,7-dihydro-7-hydroxy-1-hydroxymethyl-5H-pyrrolizidine and *N*-oxides by high performance liquid chromatography. *Drug Metab Dispos* 15;32–36, 1987

Ravindranath V, Boyd MR, Jerina DM: Hepatotoxicity of precocene I in rats: role of metabolic activation in vivo. *Biochem Pharmacol* 36;441–446, 1987

Ray AC, Williams HJ, Reagor JC: Pyrrolizidine alkaloids from *Senecio longilobus* and *Senecio glabellus. Phytochemistry* 26;2431–2433, 1987

Reagor J, Jones LP, Ray AC, Bailey EM Jr: *Eupatorium rugosum,* white snakeroot poisoning in goats. Presented at Third International Symposium on Poisonous Plants, Logan, Utah, 1989

Reed RL, Ahern KG, Pearson GD, Buhler DR: Crosslinking of DNA by dehydroretronecine, a metabolite of pyrrolizidine alkaloids. *Carcinogenesis* 9;1355–1361, 1988

Reed RL, Miranda CL, Kedzierski B, Henderson MC, Buhler DR: Microsomal formation of a pyrrolic alcohol glutathione conjugate of the pyrrolizidine alkaloid senecionine. *Xenobiotica* 22;1321–1327, 1992

Reynard GB, Norton JBS: *Poisonous Plants of Maryland in Relation to Livestock.* Md Agric Exp Stn Bull A10;249–312, 1942

Richardson WP: Milk sickness or trembles: report of five cases. *South Med Surg* 93;731–732, 1931

Riet-Correa F, Rivero R, Dutra F, Timm CD, Mendez MC: Recently encountered poisonous plants of Rio Grande do Sul and Uruguay. In *Toxic Plants and Other Toxicants,* Garland T, Barr AC eds, CAB International, New York, pp1–5, 1998

Riopelle RJ, Stevens KL: In vitro neurotoxicity bioassay: neurotoxicity of sesquiterpene lactones. In *Bioactive Natural Products—Detection, Isolation, and Structural Determinationn,* Colegate SM, Molyneux RJ eds, CRC Press, Boca Raton, Fla, pp457–463, 1993

Riopelle RJ, Boegman RJ, Little PB, Stevens KL: Neurotoxicity of sesquiterpene lactones. In *Poisonous Plants, Proceedings of the Third International Symposium,* James LF, Keeler RF, Bailey EM Jr, Cheeke PR, Hegarty MP eds, Iowa State Press, Ames, pp298–303, 1992

Robertson-Smith RG, Jeffcott LB, Friend SCE, Badcoe LM: An unusual incidence of neurological disease affecting horses during a drought. *Aust Vet J* 62;6–12, 1985
Robins DJ: The pyrrolizidine alkaloids. In *Progress in the Chemistry of Organic Natural Products*, no. 41, Herz W, Grisebach H, Kirby GW eds, Springer-Verlag, Vienna, pp115–203, 1982
Rodriguez E, Towers GHN, Mitchell JC: Biological activities of sesquiterpene lactones. *Phytochemistry* 15;1573–1580, 1976
Roitman JN, James LF: Chemistry of toxic range plants: highly oxygenated flavonol methyl ethers from *Gutierrezia microcephala*. *Phytochemistry* 24;835–848, 1985
Roitman JN, James LF, Panter KE: Constituents of broom snakeweed *(Gutierrezia sarothrae)*, an abortifacient rangeland plant. In *Plant-Associated Toxins: Agricultural, Phytochemical, and Ecological Aspects*, Colegate SM, Dorling PR eds, CAB International, Wallingford, UK, pp345–350, 1994
Romo J, deVivar AR: The pseudoguaianolides. *Prog Chem Org Nat Prod* 25;90, 1967
Roseman DM, Wu X, Milco LA, Bober M, Miller RB, Kurth MJ: Development of a class-specific comparative enzyme-linked immunosorbent assay for the detection of pyrrolizidine alkaloids in vitro. *J Agric Food Chem* 40;1008–1014, 1992
Rosiere RE, Beck RF, Wallace JD: Cattle diets on semidesert grassland: botanical composition. *J Range Manage* 28;89–93, 1975
Roussakis Ch, Chinou I, Vayas C, Harvala C, Verbist JF: Cytotoxic activity of xanthatin and the crude extracts of *Xanthium strumarium* [letter]. *Planta Med* 60;473–474, 1994
Rowe LD, Dollahite JW, Kim HL, Camp BJ: *Hymenoxys odorata* (bitterweed) poisoning in sheep. *Southwest Vet* 26;287–293, 1973
Rowe LD, Kim HL, Camp BJ: The antagonistic effect of L-cysteine in experimental hymenoxon intoxication in sheep. *Am J Vet Res* 41;484–486, 1980
Roy DN, Peyton DH, Spencer PS: Isolation and identification of two potent neurotoxins, aspartic acid and glutamic acid, from yellow star thistle *(Cetaurea solstitialis)*. *Nat Toxins* 3;174–180, 1995
Rubiolo P, Pieters L, Calomme M, Bicchi C, Vlietinck A, Vanden Berghe D: Mutagenicity of pyrrolizidine alkaloids in the *Salmonella typhimurium*/mammalian microsome system. *Mutat Res* 281;143–147, 1992
Rueger H, Benn MH: The alkaloids of *Senecio triangularis* Hook. *Can J Chem* 61;2526–2529, 1983
Rui YC: Advances in pharmacological studies of silymarin. *Mem Inst Oswaldo Cruz* 86 (Suppl 2);79–85, 1991
Rusby HH: How golden-rods may be poisonous. *Am Vet Rev* 20;5–7, 1896
Rutherford JG: Pictou cattle disease. In *Rep Vet Dir Gen & Livestock Comm for Two Years Ending 31 March 1908*, Can Dept Agric, pp21–23, 1909
Sampson AW, Malmsten HE: *Stock-Poisoning Plants of California*. Calif Agric Exp Stn Bull 593, 1935 (reprint 1942)
Sandburg C: *Abraham Lincoln—The Prairie Yearss*, vol 1. Harcourt, Brace & Co, New York, pp39–41, 1926
Sani Y, Harper PAW, Cook RL, Seawright AA, Ng JC: The toxicity of *Eupatorium adenophorum* for the liver of the mouse. In *Poisonous Plants, Proceedings of the Third International Symposium*, James LF, Keeler RF, Bailey EM Jr, Cheeke PR, Hegarty MP eds, Iowa State Press, Ames, pp626–629, 1992
Santi R: Potassium atractylate, a new inhibitor of the tricarboxylic acid cycle. *Nature* 182;257, 1958
Santi R: Pharmacological properties and mechanism of action of atractyloside. *J Pharm Pharmacol* 16;437–438, 1964
Schmidt H: *Poisonous Plant Investigations*. Tex Agric Exp Stn Annu Rep 44;14, 1931
Schmitt A: Bitter principles of the milk juice of Compositae, e.g. *Lactuca virosa*. *Bot Arch* 40;516–559, 1940 (*Br Abstr* BII;954, 1949)
Schmutz EM, Freeman BN, Reed RE: *Livestock-Poisoning Plants of Arizona*. Univ Arizona Press, 1968

Schoental R: Toxicology and carcinogenic action of pyrrolizidine alkaloids. *Cancer Res* 28;2237–2246, 1968

Schteingart CD, Pomilio AB: Atractyloside, toxic compound from *Wedelia glauca*. *J Nat Prod* 47;1046–1047, 1984

Scott JL. Golden-rod killing horses. *Gard Forest* 8;477–478, 1895

Seaman FC: Sesquiterpene lactones as taxonomic characters in the Asteraceae. *Bot Rev* 48;121–595, 1982

Seawright AA: *Animal Health in Australia*. Vol 2, *Chemical and Plant Poisons*. Australian Government Publ Serv, Canberra, 1989

Seawright AA, Hrdlicka J, Wright JD, Kerr DR, Mattocks AR, Jukes R: The identification of hepatotoxic pyrrolizidine alkaloid exposure in horses by the demonstration of sulphur-bound pyrrolic metabolites on their hemoglobin [letter]. *Vet Hum Toxicol* 33;286–287, 1991a

Seawright AA, Kelly WR, Hrdlicka J, McMahon P, Mattocks AR, Jukes R: Pyrrolizidine alkalosis in cattle due to *Senecio* spp. in Australia. *Vet Rec* 129;198–199, 1991b

Selfero NA, Merlassino JL, Audisio S: Intoxication by *Centaurea solstitialis:* therapeutic evaluation. *Therios* 13;42–44, 1989

Sharma OP: Plant toxicoses in north-western India. In *Plant-Associated Toxins: Agricultural, Phytochemical, and Ecological Aspects,* Colegate SM, Dorling PR eds, CAB International, Wallingford, UK, pp19–24, 1994

Sharrow SH, Ueckert DN, Johnson AE: Ecology and toxicology of *Senecio* species with special reference to *Senecio jacobaea* and *Senecio longilobus*. In *Ecology and Economic Impact of Poisonous Plants on Livestock Production*, James LF, Ralphs MH, Nielsen DB eds, Westview Press, Boulder, Colo, pp181–196, 1988

Shaver TN, Camp BJ, Dollahite JW: The chemistry of a toxic constituent of *Xanthocephalum* species. *Ann N Y Acad Sci* 111;737–743, 1963

Sheley RL, Larson LL, Johnson DE: Germination and root dynamics of range weeds and forage species. *Weed Technol* 7;234–237, 1993

Shull LR, Buckmaster GW, Cheeke PR: Factors influencing pyrrolizidine *(Senecio)* alkaloid metabolism: species, liver sulfhydryls, and rumen fermentation. *J Anim Sci* 43;1247–1253, 1976

Sidhu GS, Oakenfull DG: A mechanism for the hypocholesterolaemic activity of saponins. *Br J Nutr* 55;643–649, 1986

Skidmore LV, Peterson NF: Observations on the toxicity of golden glow *(Rudbeckia laciniata)* to swine and other animals. *J Am Vet Med Assoc* 81;655–662, 1932

Slocombe RF, Huntington PJ, Friend SCE, Jeffcott LB, Luff AR, Finkelstein DK: Pathological aspects of Australian stringhalt. *Equine Vet J* 24;174–183, 1992

Small AC, Kelly WR, Seawright AA, Mattocks AR, Jukes R: Pyrrolizidine alkaloidosis in a two month old foal. *Zentralbl Veterinaermed* A 40;213–218, 1993

Smetzer DL, Coppock RW, Ely RW, Duckett WM, Buck WB: Cardiac effects of white snakeroot intoxication in horses. *Equine Pract* 5;26–32, 1983

Smith CH, Larner J, Thomas AM, Kupchan SM: Inactivation of glycogen synthase by the tumor inhibitor vernolepin. *Biochim Biophys Acta* 276;94–104, 1972

Smith GS, Hallford DM, Stavanja S, Staley EC, Thilsted JP, Greenberg JA: Safflower seed oil improved embryo-fetal tolerance of toxins in snakeweed foliage ingested by rats [abstract]. *J Anim Sci* 71;445, 1993

Smith GS, Ross TT, Hallford DM, Thilsted JP, Staley EC: Efforts to improve animal tolerance of ingested snakeweed foliage. In *Plant-Associated Toxins: Agricultural, Phytochemical, and Ecological Aspects*, Colegate SM, Dorling PR eds, CAB International, Wallingford, UK, pp495–500, 1994

Snively WD, Furbee L: Discoverer of the cause of milk sickness. *J Am Med Assoc* 196;1055–1060, 1966

Sorensen NA: The taxonomic significance of acetylenic compounds. In *Recent Advances in Phytochemistry,* vol 1, pp187–227, 1968

Spring O, Zitterell-Haid B, Bierner MW, Mabry TJ: Chemistry of glandular trichromes in *Hymenoxys* and related genera. *Biochem Syst Ecol* 22;171–195, 1994

Staley EC, Smith GS, Greenberg JA: Effects on reproduction in female offspring from Sprague-Dawley rats fed 10% snakeweed *(Gutierrezia microcephala)* throughout pregnancy and concurrent treatment with safflower oil. *Vet Hum Toxicol* 37;440–442, 1995

Staley EC, Smith GS, Greenberg JA: Decreased reproductive effects from snakeweed *(Gutierrezia microcephala)* in Sprague-Dawley rats with increased dietary snakeweed consumption. *Vet Hum Toxicol* 38;259–264, 1996

Steel EG, Witzel DA, Blanks A: Acquired coagulation factor X activity connected with *Hymenoxys odorata* DC. (Compositae), bitterweed poisoning in sheep. *Am J Vet Res* 37;1383–1386, 1976

Stelljes ME, Kelley RB, Molyneux RJ, Seiber JN: GC-MS determination of pyrrolizidine alkaloids in four *Senecio* species. *J Nat Prod* 54;759–773, 1991

Stevens KL: Sesquiterpene lactones from *Centaurea repens*. *Phytochemistry* 21;1093–1098, 1982

Stevens KL, Merrill GB: Sesquiterpene lactones and allelochemicals from *Centaurea* species. In *The Chemistry of Allelopathy: Biochemical Interactions among Plants*, Thompson AC ed, American Chemical Society, Washington DC, pp83–98, 1985

Steyn DG: Plant poisoning in stock and the development of tolerance. *Onderstepoort J Vet Sci Anim Ind* 1;149–156, 1933a

Steyn DG: Poisoning of human beings by weeds contained in cereals (bread poisoning). *Onderstepoort J Vet Sci Anim Ind* 1;219–265, 1933b

Stotts R: White snakeroot toxicity in dairy cattle. *Vet Med* 79;118–120, 1984

Strang RM, Lindsay KM, Price RS: Knapweeds: British Columbia's undesirable aliens. *Rangelands* 1;141–143, 1979

Strother JL: Taxonomy of *Tetradymia* (Compositae: Senecioneae), *Brittonia* 26;177–202, 1974

Stuart BP, Cole RJ, Gosser HS: Cocklebur *(Xanthium strumarium* L. var. *strumarium)* intoxication in swine: review and redefinition of the toxic principle. *Vet Pathol* 18;368–383, 1981

Stuppner H, Stuppner H, Rodriguez E: A novel enol-pseudoguaianolide from *Psilostrophe cooperi*. *Phytochemistry* 27;2681–2684, 1988

Subba Rao PV, Mangala A, Subba Rao BS, Prakish KM: Clinical and immunological studies on persons exposed to *Parthenium hysterophorus* L. *Experientia* 33;1387–1388, 1977

Swick RA, Cheeke PR, Goeger DE, Buhler DR: Effect of dietary *Senecio jacobaea* and injected *Senecio* alkaloids and monocrotaline on guinea pigs. *J Anim Sci* 55;1411–1416, 1982

Swick RA, Miranda CL, Cheeke PR, Buhler DR: Effect of phenobarbital on toxicity of pyrrolizidine *(Senecio)* alkaloids in sheep. *J Anim Sci* 56;887–894, 1983a

Swick RA, Cheeke PR, Ramsdell HS, Buhler DR: Effect of sheep rumen fermentation and methane inhibition on the toxicity of *Senecio jacobaea*. *J Anim Sci* 56;645–651, 1983b

Sylvia VL, Joe CO, Stipanovic RD, Kim HL, Busbee DL: Alkylation of deoxyguanosine by the sesquiterpene lactone hymenoxon. *Toxicol Lett* 29;69–76, 1985

Synge BA, Stephen FB: Delayed ragwort poisoning associated with lactation stress in cows. *Vet Rec* 132;327, 1993

Szabuniewicz M, Kim HL: The pharmacodynamic and toxic action of *Helenium microcephalum* extract and helenaline. *Southwest Vet* 25;305–311, 1972

Takeya K, Itoigawa M, Furukawa H: Positive inotropic effect of helenalin, a sesquiterpene lactone, on guinea-pig myocardium. *Chem Pharm Bull* 31;1719–1725, 1983

Tapia MO, Cravero A, Cadenazzi GE, Koch O: Liver necrosis induced by *Wedelia glauca* in rats: a light and electron microscopy study. *Vet Hum Toxicol* 38;10–14, 1996

Terry MK, Kim HL, Corrier DE, Bailey EM Jr: The acute oral toxicity of hymenoxon in sheep. *Res Commun Chem Pathol Pharmacol* 31;181–84, 1981

Terry MK, Williams HG, Kim HL, Post LO, Bailey EM Jr: Ovine urinary metabolites of hymenoxon, a toxic sesquiterpene lactone isolated from *Hymenoxys odorata* DC. (bitterweed). *J Agric Food Chem* 31;1208–1210, 1983

Thomas MD, Hendricks RH, Hill GR: Sulfur content of vegetation. *Soil Sci* 70;9–18, 1950

Thompson LJ: Depression and choke in a horse: probable white snakeroot toxicosis [letter]. *Vet Hum Toxicol* 31;321–322, 1989

Thorp F, Durrell LW, Harshfield GS, Barr CG: Oxytenia *Found to Be Poisonous to Livestock*. Colo Agric Exp Stn Farm Bull 2;18–19, 1940a

Thorp F, Harshfield GS, Durrell LW, Barr CG: *Oxytenia acerosa*—a plant poisonous for livestock [letter]. *J Am Vet Med Assoc* 96;97, 1940b

Tilt SE: Ragwort toxicosis in a heifer. *Can Vet J* 10;302–306, 1969

Tocco L: Sull' avvelenamento per *Carlina gummifera* Nota II Ricerche farmacologiche sul principio attivo della *Carlina gummifera* (atractylato di K). *Arch Int Pharmacodyn Ther* 26;171–186, 1922

Tokarnia CH, Dobereiner J, Amorim PR, Gava A, Consorte LB: Experimental poisoning by *Vernonia mollissima* (Compositae) in rabbits. *Pesq Vet Bras* 6;5–10, 1986

Tokarnia CH, Peixoto PV, Gava A, Barros CSL: Experimental poisoning of bovines by *Baccharis megapotamica* var. *megapotamica* and var. *weirii* (Compositae). *Pesq Vet Bras* 12;19–31, 1992

Tomioka M, Calvo F, Siguas A, Sanchez L: Hepatic veno-occlusive disease associated with ingestion of *Senecio tephrosiodes*. *Rev Gastroenterol Peru* 15;299–302, 1995

Trelease SF, Beath OA: *Selenium*. Trelease & Beath, New York, 1949

Turner BL: Taxonomy of *Sartwellia* (Compositae Helenieae). *SIDA* 4;265–273, 1971

Tyler VE: Hallucinogenic drug hoaxes of the American Hippies. In *Natural Products as Medicinal Agents*, Hippokrates Verlag, Stuttgart, pp339–350, 1980

van Heerden J, van der Lugt JJ, Durante E: Experimental vermersiekte (*Geigeria ornativa* O. Hoffm.) poisoning in sheep. 1. An evaluation of diagnostic aids and an assessment of the preventive effect of ethoxyquin. *J S Afr Vet Assoc* 64;76–81, 1993

Vignais PV, Duee ED, Vignais PM, Huet J: Effects of atractyligenin and its structural analogues on oxidative phosphorylation and on the translocation of adenine nucleotides in mitochondria. *Biochim Biophys Acta* 118;465–483, 1966

Vignais PV, Vignais PM, Lauquin G, Morel F: Binding of adenine diphosphate and of antagonist ligands to the mitochondrial ADP carrier. *Biochimie* 55;763–778, 1973

Vignais PV, Block MR, Boulay F, Brandolin G, Lauquin GJ-M: Molecular aspects of structure-function relationships in mitochondrial adenine nucleotide carrier. In *Structure and Properties of Cell Membranes*, vol 2, Benga G ed, CRC Press, Boca Raton, Fla, pp139–179, 1985

Vincent PLD: Progress on clarifying the generic concept of *Senecio* based on an extensive world-wide sample of taxa, Compositae: systematics. In *Proceedings of the International Compositae Conference, Kew, 1994*, Hind DJN, Beentje HJ eds, Royal Botanic Gardens, Kew, vol 1, pp597–611, 1996

Vivar ARde, Delgado G, Soriano-Garcia M, Toscano RA, Huerta E, Reza-Garduno RG, De Vivar AR: Sesquiterpene lactones from a population of *Helenium integrifolium*. *J Nat Prod* 50;284–286, 1987

Vojtisek B, Hronova B, Hamrik J, Jankova B: Milk thistle (*Silybum marianum* L. Gaertn.) in the feed of ketotic cows. *Vet Med Praha* 36;321–330, 1991

Wachenheim DE, Blythe LL, Craig AM: Characterization of rumen bacterial pyrrolizidine alkaloid biotransformation in ruminants of various species. *Vet Hum Toxicol* 34;513–517, 1992a

Wachenheim DE, Blythe LL, Craig AM: Effects of antibacterial agents on in vitro ovine ruminal biotransformation of the hepatotoxic pyrrolizidine alkaloid jacobine. *Appl Environ Microbiol* 58;2559–2564, 1992b

Wachnik Z: Poisoning of pigs with leaves of chicory, *Cichorium intybus*. *Med Wet* 18;493–495, 1962 (*Vet Bull* 33;980, 1963)

Waddell TG, Geissman TA: Paucin, a sesquiterpene lactone glucoside. *Tetrahedron Lett* 7;515–517, 1969

Wang Y, Hamburger M, Cheng CHK, Costall B, Naylor RJ, Jenner P, Hostettmann K: Neurotoxic sesquiterpenoids from the yellow star thistle *Centaurea solstitialis* L. (Asteraceae). *Helv Chim Acta* 74;117–123, 1991

Watson TJ: The taxonomy of Xylorhiza (Asteraceae-Astereae). *Brittonia* 29;199–216, 1977

Watt JM, Breyer-Brandwijk MG: *The Medicinal and Poisonous Plants of Southern Africa.* E & S Livingston, Edinburgh, pp197–305, 1962

Welch BL, Pederson JC: In vitro digestibility among accessions of big sagebrush by wild mule deer and its relationship to monoterpenoid content. *J Range Manage* 34;497–500, 1981

Welch H, Morris HE: *Range Plants Poisonous to Livestock in Montana.* Mont Agric Exp Stn Circ 197, 1952

White INH: The role of liver glutathione in the acute toxicity of retrorsine to rats. *Chem Biol Interact* 13;333–342, 1976

White INH, Mattocks AR, Butler WH: The conversion of the pyrrolizidine alkaloid retrorsine to pyrrolic derivatives in vivo and in vitro and its acute toxicity to various animal species. *Chem Biol Interact* 6;207–218, 1973

White RD, Krumperman PH, Cheeke PR, Deinzer ML, Buhler DR: Mutagenic responses of tansy ragwort *(Senecio jacobaea)* plant, pyrrolizidine alkaloids, and metabolites in goat milk with the *Salmonella*/mammalian-microsome mutagenicity test. *J Anim Sci* 58;1245–1254, 1984a

White RD, Swick RA, Cheeke PR: Effects of dietary copper and molybdenum on tansy ragwort *(Senecio jacobaea)* toxicity in sheep. *Am J Vet* 45;159–161, 1984b

Whitson TD, Burrill LC, Dewey SA, Cudney DW, Nelson BE, Lee RD, Parker R: *Weeds of the West.* Western Society of Weed Science, 1991

Wickramanayake PP, Arbogast BL, Buhler DR, Deinzer ML, Burlingame AL: Alkylation of nucleosides and nucleotides by dehydroretronecine: characterization of adducts by liquid secondary ion mass spectrometry. *J Am Chem Soc* 107;2485–2488, 1985

Williams DE, Reed RL, Kedzierski B, Dannan GA, Guengerich FP, Buhler DR: Bioactivation and detoxication of the pyrrolizidine alkaloid senecionine by cytochrome P-450 enzymes in rat liver. *Drug Metab Dispos* 17;387–392, 1989

Williams JL, Campos D, Ross TT, Becker KA, Martinez JM, Oetting BC, Smith GS: Snakeweed (*Gutierrezia* spp.) toxicosis in beef heifers. *Proc West Sect Am Soc Anim Sci* 43;67–69, 1992

Williams MC: Accumulation of nitrate by annual goldeneye and showy goldeneye. *J Range Manage* 42;196–198, 1989

Willuhn G, Rottger PM: Helenalin und seine derivate, die herzwirksamen verbindungen der arnikabluten [abstract]. *Planta Med* 45;131, 1982

Winter CK, Segall HJ: Metabolism of pyrrolizidine alkaloids. In *Toxicants of Plant Origin*, vol 1, Cheeke PR ed, CRC Press, Boca Raton, Fla, pp23–40, 1989

Winter CK, Segall HJ, Jones AD: Species differences in the hepatic microsomal metabolism of the pyrrolizidine alkaloid senecionine. *Comp Biochem Physiol* 90C;429–433, 1988

Winter H, Seawright AA, Mattocks AR, Jukes R, Tshewang U, Gurung BJ: Pyrrolizidine alkaloid poisoning in yaks: first report and confirmation by identification of sulphur-bound pyrrolic metabolites of the alkaloids in preserved liver tissue. *Aust Vet J* 67;411–412, 1990

Winter H, Seawright AA, Hrdlicka J, Tshewang U, Gurung BJ: Pyrrolizidine alkaloid poisoning of yaks *(Bos grunniens)* and confirmation by recovery of pyrrolic metabolites from formalin-fixed liver tissue. *Res Vet Sci* 52;187–194, 1992

Winter H, Seawright AA, Hrdlicka J, Mattocks AR, Jukes R, Wangdi K, Gurung KB: Pyrrolizidine poisoning of yaks: diagnosis of pyrrolizidine alkaloid exposure by the demonstration of sulphur-conjugated pyrrolic metabolites of the alkaloid in circulating haemoglobin. *Aust Vet J* 70;312–313, 1993

Witte ST, Osweiler GD, Stahr HM, Mobley G: Cocklebur toxicosis in cattle associated with the consumption of mature *Xanthium strumarium*. *J Vet Diagn Invest* 2;263–267, 1990

Witzel DA, Rowe LD, Clark DE: Physiopathologic studies on acute *Hymenoxys odorata* (bitterweed) poisoning in sheep. *Am J Vet Res* 35;931–934, 1974

Witzel DA, Ivie GW, Dollahite JW: Mammalian toxicity of helenalin, the toxic principle of *Helenium microcephalum* DC. (smallhead sneezeweed). *Am J Vet Res* 37;859–861, 1976

Witzel DA, Jones LP, Ivie GW: Pathology of subacute bitterweed *(Hymenoxys odorata)* poisoning in sheep. *Vet Pathol* 14;73–78, 1977

Wolf FA, Curtis RS, Kaupp BF: Studies on trembles or milksickness and white snakeroot. *J Am Vet Med Assoc* 52;820–827, 1917

Wolf FA, Curtis RS, Kaupp BF: *A Monograph on Trembles or Sickness and White Snakeroot.* N C Agric Exp Stn Tech Bull 15, 1918

Wood MK, Buchanan BA, Skeet W: Shrub preference and utilization by big game on New Mexico reclaimed mine land. *J Range Manage* 48;431–437, 1995

Worthley EG, Schott CD, Hauptmann GA: Toxicity of some goldenrods. *Econ Bot* 21;238–242, 1967

Wu CH, Lampe KF, Mende TJ: Metabolic changes induced in chickens by the administration of tremetol. *Biochem Pharmacol* 22;2835–2841, 1973

Yamamoto LA, Soldera JC, Emim JA, Godhino RO, Souccar C, Lapa AJ: Pharmacological screening of *Ageratum conyzoides* L. (Mentrasto). *Med Inst Oswaldo Cruz* 86;145–147, 1991

Yamanaka H, Nagao M, Sugimura T, Furuya T, Shirai A, Matsushima T: Mutagenicity of pyrrolizidine alkaloids in the *Salmonella*/mammalian-microsome system. *Mutat Res* 68;211–216, 1979

Yan CC, Cooper RA, Huxtable RJ: The comparative metabolism of the four pyrrolizidine alkaloids, seneciphylline, retrorsine, monocrotaline, and trichodesmine in the isolated, perfused rat liver. *Toxicol Appl Pharmacol* 133;277–284, 1995

Young S, Brown WW, Klinger B: Nigropallidal encephalomalacia in horses caused by ingestion of weeds of the genus *Centaurea. J Am Vet Med Assoc* 157;1602–1605, 1970a

Young S, Brown WW, Klinger B: Nigropallidal encephalomalacia in horses fed Russian knapweed (*Centaurea repens* L.). *Am J Vet Res* 31;1393–1404, 1970b

Zalkow LH, Burke N: Constitution of toxol: a toxic constituent of *Aplopappus heterophyllus. Chem Ind* 1963;292–293, 1963

Zalkow LH, Gelbaum L, Ghosal M, Fleischmann TJ: The co-occurrence of desmethylencecalin and hydroxytrematone in *Eupatorium rugosum. Phytochemistry* 16;1313, 1977

Zalkow LH, Harris RN III, Burke NI: The lower terpenoids of *Isocoma wrightii. J Nat Prod* 42;96–102, 1979

Berberidaceae Juss. 14

Widespread in the temperate regions of the Northern Hemisphere, the Berberidaceae comprises 13–15 genera and 570–650 species in both the Old World and the New World. Commonly known as the barberry family, it is prized economically for its many ornamental shrubs, with some 100 species and 9 genera offered in the horticultural trade. Closely related to the Ranunculaceae, or buttercup family, and sharing the common alkaloid berberine, the family is characterized by having groups of genera so closely related that generic boundaries are tenuous, as illustrated by the relationship of *Berberis* and *Mahonia,* described below. In North America, 8 genera and 33 native species are present (Whetstone et al. 1997). Intoxication problems are associated with 3 genera.

- Barberry Family

  Berberis
  Mahonia
  Nandina
  Podophyllum

Plants herbs or shrubs; perennials; evergreen or deciduous; rhizomes present or absent. **Stems** armed or not armed; glabrous. **Leaves** simple or 1-to-3-pinnately compound; alternate or opposite or whorled; venation palmate or pinnipalmate or pinnate; glabrous; stipules absent. **Inflorescences** solitary flowers or panicles or racemes or umbels; axillary or terminal. **Flowers** small; perfect; perianths in 2-series. **Sepals** 6; caducous or persistent; in 2 whorls. **Corollas** radially symmetrical; bowl- or saucer-shaped. **Petals** 6 or 9; in 2 or 3 whorls; free. **Stamens** 6 or 12 or 18. **Pistils** 1; simple, carpels 1; stigmas 1; styles 1 or 0; ovaries superior, ovoid in longitudinal section; locules 1; placentation parietal or basal. **Fruits** berries, may rupture and release drupe-like seeds. **Seeds** numerous or 1 or 2.

- perennial herbs or shrubs; flowers small; petals 6 or 9, in 2 or 3 whorls; fruits berries

Berberis L.

Taxonomy and Morphology—Commonly known as barberry, *Berberis* comprises some 500 species. Its name is derived from *berberys,* from the Arabic name for its fruits. Prized for its many ornamental shrubs, it also is of economic importance because some species, in particular *B. vulgaris,* serve as hosts for *Puccinia graminis,* the black stem-rust of wheat. Taxonomists and horticulturalists have strong opinions about division of the genus into two genera—*Berberis* and *Mahonia* (Whetstone et al. 1997). Species with simple leaves and spiny stems are positioned by some individuals in *Berberis,* and those with compound leaves and nonspiny stems are placed in *Mahonia.* There also are differences between the two taxa in terms of morphology of the stems, appearance of the inflorescences, and susceptibility to *Puccinia.* However, because numerous species in other parts of the world exhibit intermediate combinations of characters, most taxonomists typically recognize one large genus, as is done in this treatment. Species of the genus also are quite similar toxicologically. In contrast, horticulturalists continue to recognize two genera. In North America, more than 25 native and introduced species are present. The following are representatives:

- *B. aquifolium*
  B. canadensis
  B. fremontii
  B. julianae
  B. koreana
  B. repens
  B. xstenophylla
  B. thunbergii
  B. vulgaris

  barberry
  berberis
  mahonia
  Oregon grape

Berberis aquifolium

B. aquifolium Pursh (= *Mahonia aquifolium* [Pursh] Nutt.)	Oregon grape
B. canadensis Mill.	American barberry
B. fremontii Torr. (= *M. fremontii* [Torr.] Fedde)	desert mahonia
B. julianae C.K.Schneid.	wintergreen barberry
B. koreana Palib.	Korean barberry
B. repens Lindl. (= *M. repens* [Lindl.] G.Don)	creeping barberry, trailing mahonia
B. x*stenophylla* Lindl.	rosemary barberry, coral barberry
B. thunbergii DC.	Japanese barberry
B. vulgaris L.	common barberry, berberry, European berberis

■ shrubs; leaves simple or 1-pinnately compound, alternate; flowers small; petals 6, yellow, dark orange-red or white; berries yellow-red or black-blue

Plants shrubs; evergreen or deciduous. **Stems** 10–450 cm tall; armed or not armed with spines; spur branches present or absent. **Leaves** simple or 1-pinnately compound; alternate; blades elliptic or lanceolate or orbicular or oblanceolate or obovate. **Inflorescences** terminal racemes or umbels or rarely solitary flowers. **Flowers** small. **Sepals** 6; yellow. **Petals** 6; yellow to dark orange-red or rarely white. **Stamens** 6. **Berries** globose to ovoid or cylindrical; yellow-red or black-blue; typically glaucous. **Seeds** 1–10; tan to red-brown or black.

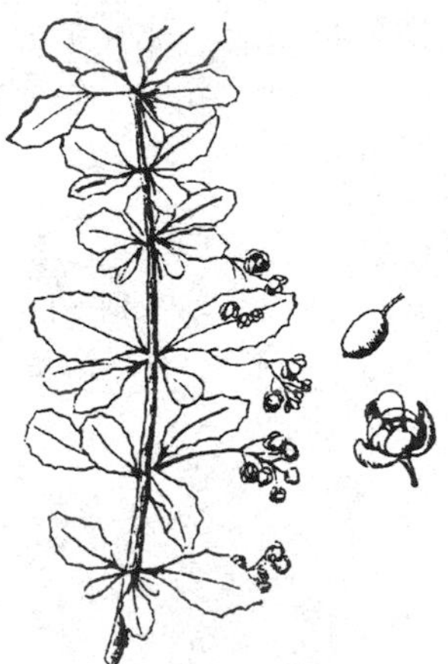
Berberis canadensis

1. Stems not spiny. Leaves compound.
    2. Racemes 1-to-11-flowered. Bracts acuminate . . . . *B. fremontii*
    2. Racemes 25-to-70-flowered. Bracts obtuse or acute.
        3. Abaxial surfaces of leaflets glabrous; shiny . . . . *B. aquifolium*
        3. Abaxial surfaces of leaflets papillate; dull . . . . *B. repens*
1. Stems spiny. Leaves simple.
    4. Plants evergreen. Berries blue-black.
        5. Leaf margins entire . . . . *B.* x*stenophylla*
        5. Leaf margins dentate . . . . *B. julianae*
    4. Plants deciduous. Berries red to purple.
        6. Leaf margins entire. Inflorescences umbels or solitary flowers . . . . *B. thunbergii*
        6. Leaf margins dentate. Inflorescences racemes.
            7. Branchlets reddish . . . . *B. koreana*
            7. Branchlets grayish.
                8. Blade teeth 16–30. Racemes 10-to-20-flowered . . . *B. vulgaris*
                8. Blade teeth 3–12. Racemes 3-to-12-flowered . . . *B. canadensis*

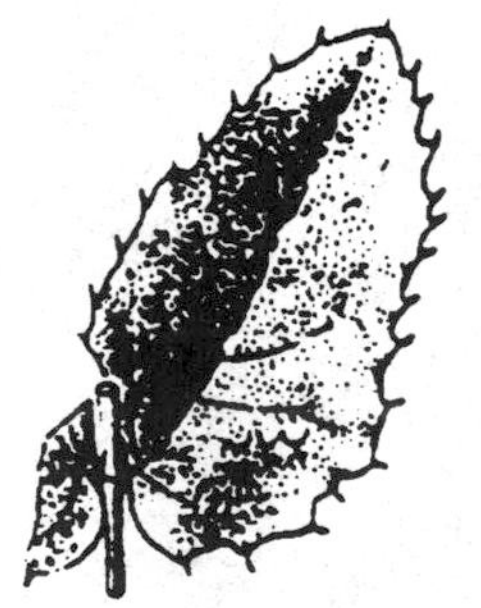
Berberis repens

■ shallow, dry soils, also ornamentals

Distribution and Habitat—Native to the woodlands of northern Asia, Europe, and America, species of *Berberis* are adapted to a variety of soil types, often those that are shallow and dry. Most of our native species have relatively restricted distributions; diversity is greatest in the Pacific Coast states. Introduced ornamentals include *B. julianae* (China), *B. koreana* (Korea), and *B. thunbergii* (Japan). Also naturalized and widespread in the same region is *B. vulgaris* from Eurasia; it is susceptible to *Puccinia graminis* and has been subjected to eradication efforts in wheat-growing areas in an attempt to interrupt the rust's life cycle.

Berberis aquifolium

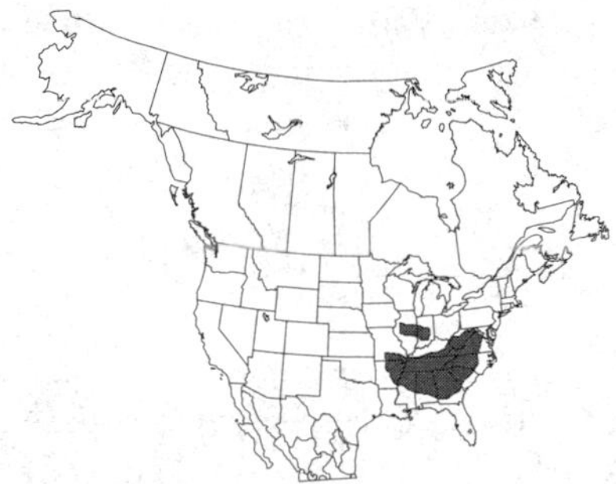
Berberis canadensis

Berberis fremontii

Disease Problems—*Berberis* is perhaps best known for its numerous ornamental shrubs, but it also provides food and, in the past, medicines. Its berries are sour but edible and used for jams and jellies (Hart 1992). Roots were used by American Indians and settlers to make medicinal tinctures for stomach trouble, hemorrhage, tuberculosis, and eye trouble. Today *B. aquifolium* is used to treat psoriasis (Hansel 1992). As early as 1911, Pammel reported that *B. aquifolium* contains irritant principles that cause inflammation of the digestive tract. Irritation of the kidneys also was reported (Millspaugh 1974). Plants of the genus, however, are generally bitter and unpalatable and thus typically represent little hazard.

- foliage or berries, bitter and unpalatable; rarely irritation, digestive disturbance, possibly urinary system

Disease Genesis—Several species of *Berberis* have been shown to contain protoberberine alkaloids (Bhakuni & Jain 1986). These alkaloids include berberine, canadine, columbamine, corypalmine, jatrorrhizine, and palmatine. Berberine is identical with umbellatine and chemically related to hydrastine. Hydrastinine is a decomposition product of either berberine or hydrastine. Although these alkaloids are similar in structure, they differ in the effects they produce. In general, protoberberine alkaloids relax smooth muscle, causing a transient decrease in blood pressure and inhibition of acetylcholinesterase (Southon & Buckingham 1989). Berberine may cause seizures in higher dosage and also inhibits enzymes such as tyrosine decarboxylase and tryptophanase. The specific agent responsible for the therapeutic action of *B. aquifolium* on psoriasis is not known (Hansel 1992).

- numerous protoberberine alkaloids, berberine, canadine, low risk, high-dosage neurotoxic

Berberine

Canadine

Columbamine

Corypalmine

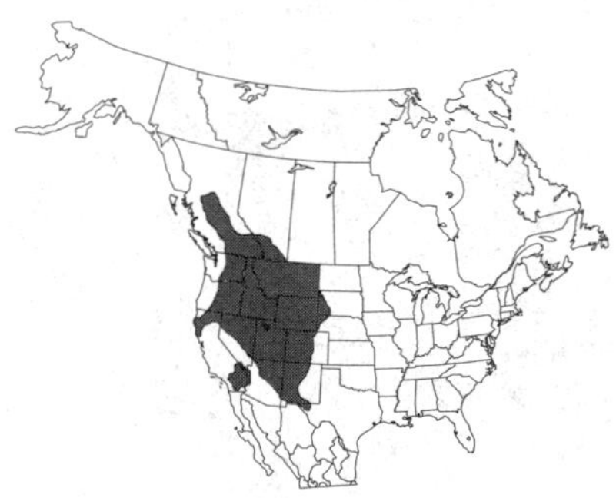
Berberis repens

Berberis vulgaris

- nausea, vomiting, diarrhea

gross pathology, mild reddening of digestive tract

treatment not needed

Clinical Signs, Pathology, and Treatment—Signs of *Berberis* intoxication reflect mild to moderately severe irritation of the digestive tract and include nausea, vomiting, and diarrhea. The results of ingestion are generally mild, not lethal; thus treatment, if needed at all, is limited to temporary relief of the diarrhea.

Nandina Thunb.

- *N. domestica*

heavenly bamboo
nandina
sacred bamboo

Taxonomy and Morphology—*Nandina,* a monotypic genus native to China and Japan, gets its name from *nandin,* the Japanese common name for the species. With more than 60 highly esteemed cultivars, the single species is as follows:

N. domestica Thunb. heavenly bamboo, nandina, Chinese sacred bamboo

- erect shrubs, reddish branchlets; leaves 3-pinnately compound, alternate, red-purple in winter; flowers numerous, tiny, cream to white; berries red to purple-red in winter

Plants shrubs; evergreen or semideciduous. **Stems** erect; clumped; typically not branched; 100–200 cm tall; branchlets reddish. **Leaves** 2- or 3-pinnately compound; alternate; leaflet blades elliptic to lanceolate to ovate, turning red-purple in winter; margins entire; petiole bases clasping. **Inflorescences** terminal or axillary panicles; erect or arching. **Flowers** small; numerous. **Sepals and Petals** intergrading; cream to white. **Stamens** 6. **Berries** globose; red to purple-red. **Seeds** 1–3; grayish or brownish.

- ornamental

Distribution and Habitat—Very common and widely cultivated throughout North America, *N. domestica* is quite tolerant of a variety of environmental conditions and is prized for its winter foliage of reddish purple leaves and red berries.

- cyanogenic glycosides, peracute intoxication, ruminants, pets, rare; protoberberine alkaloids; serotonin antagonist, nantenine

Disease Problems and Genesis—Many but not all cultivars of *Nandina* are strongly cyanogenic and represent a hazard to ruminants that gain access to them. Occasionally other animals, such as puppies, may be at risk (Bradley et al. 1988). Cyanogenesis is due to the glucosides *p*-hydroxymandelonitrile and *p*-glucosyl-oxymandelonitrile, which are derived from phenylalanine and tyrosine, respectively. Under the appropriate conditions in the plant or in the rumen, they subsequently are hydrolyzed to the aglycone or sugar-free glycone and then to free HCN (Seigler 1977).

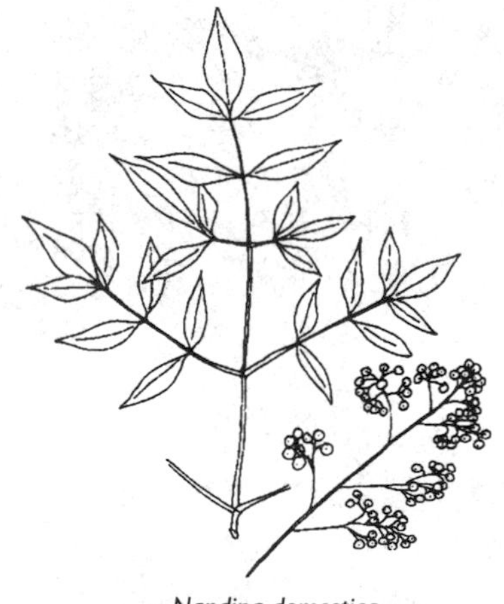

Nandina domestica

p-Hydroxymandelonitrile

Nantenine

Several protoberberine alkaloids are present in various tissues of *N. domestica,* including berberine, domesticine, dometine, nandinine, and protopine (Manske & Ashford 1954). A serotonin receptor antagonist, nantenine (*O*-methyldomesticine), is also present in the fruits (Shoji et al. 1984). The role of these compounds as toxicants following ingestion of a few berries is not known. For most cultivars, cyanogenesis is the more important intoxication factor.

Clinical Signs—Typical of cyanide intoxication, abrupt onset of apprehension and distress occurs within a few minutes of the animal's browsing the foliage of *Nandina*. This is quickly followed by weakness, incoordination, and labored respiration. If intoxication is severe, the animal may be unable to stand and may lay in lateral recumbency, with periodic paddling and tetanic seizures. The entire sequence of signs may occur in a 5-to-15-minute period, with either death or recovery at the end of this time. Complete recovery may require several hours to overnight.

- abrupt onset, apprehension, distress, weakness, ataxia, labored respiration, collapse, seizures, rapid sequence

Pathology and Treatment—There are few distinctive changes. For the most part they are limited to congestion of the abdominal viscera and scattered, small splotchy hemorrhages. The primary antidote for ruminants is sodium thiosulfate given i.v. at a dose of 0.25–0.5 g/kg b.w. Sodium nitrite, 10 mg/kg b.w. i.v., can be used to augment the effects of the sodium thiosulfate, but it is not required for effective treatment except in dogs. A more complete discussion is presented in the treatment of the Rosaceae, in chapter 64.

- no lesions

  treatment, ruminants, sodium thiosulfate; dogs, sodium nitrite with sodium thiosulfate

Podophyllum L.

Taxonomy and Morphology—One of the classic harbingers of spring in the deciduous forests of eastern North America, *Podophyllum* comprises 2–7 species native to both the Old World and the New World. Its name is derived from the Greek roots *podos* and *phyllon,* meaning "foot" and "leaf." Taxonomists differ in their opinions of the taxon's circumscription, some preferring to position the Asian species in separate genera. In North America, 1 species is present:

P. peltatum L. mayapple, American mandrake, Indian apple, wild lemon, duck's-foot, ground lemon, vegetable calomel, podophyllum, podophyllin, devil's-apple, vegetable mercury, umbrella leaf

- *P. peltatum*

  American mandrake
  devil's-apple
  duck's-foot
  ground lemon
  Indian apple
  mayapple
  podophyllin
  podophyllum
  umbrella leaf
  vegetable calomel
  vegetable mercury
  wild lemon

Plants herbs; rhizomes present, segmented, producing 1 leaf or flowering shoot per year. **Stems** erect; 20–60 cm tall. **Leaves** simple; solitary and basal or paired and cauline; opposite; venation palmate; glossy green; blades reniform-orbicular; peltate; margins palmately 5-to-7-lobed or parted; margins of lobes or parts entire or dentate. **Inflorescences** solitary flowers; arising in angle between petioles. **Flowers** fragrant or malodorous; nodding. **Sepals** 6; caducous; white or pale green. **Petals** 6 or 9; white or rarely pink. **Stamens** 6 or 12 or 18. **Berries** ovoid to ellipsoid; yellow or orange or red or maroon. **Seeds** 20–50; yellow or orange or red or maroon.

- erect herbs; leaves simple, solitary or paired, opposite; petals 6 or 9, white or pink; berries yellow, orange, red or maroon

Distribution and Habitat—Species of *Podophyllum* occur in the forests of eastern North America and eastern Asia. Adapted to the shaded, rich, moist soils of the forest floors, *P. peltatum* is distributed throughout the eastern half of the continent. One of the first species to begin growth in the spring, it typically forms dense stands and thus is conspicuous in the otherwise leafless woods. Flowering occurs before the canopy closes.

- rich, moist woods

Disease Problems—Although called American mandrake, *P. peltatum* is not related to the legendary mandrake of Eurasia, *Mandragora officinarum* in the Solanaceae, which so captivated Shakespeare. Recounting some of the ancient tales surrounding the use of the European mandrake, Mack (1992) pointed out that *Podophyllum* has

- roots well-known emetic, used commercially as podophyllin; all plant parts, irritation, digestive disturbance, uncommon

Podophyllum peltatum

Podophyllum peltatum

■ livestock, toxic lignans, podophyllotoxin, irritant, digestive tract

humans, podophyllotoxin binds tubulin of microtubules, arrests cell mitosis and cell division, disruption of long axon structure, neurotoxic with long-term use

Podophyllotoxin

quite different effects. The plant was used as an emetic in the 1700s. The roots were well known to Indian tribes and colonists alike as effective purgatives (Harshberger 1920). In small dosage they caused bile-stained, liquid feces. With larger dosage, intestinal muscular activity and pain increased considerably. The purgative effect is transmitted in milk to neonates (Harshberger 1920). The root has been officially recognized as a cathartic and cholagogue in the U.S. Pharmacopoeia. The resin extract called podophyllin has been similarly used, and in the 1900s it was incorporated in proprietary medicines such as Carter's Little Liver Pills (Graham & Chandler 1990). It is now used topically for wart removal. Even the edible fruits, with their attractive strawberry-like flavor, can be a problem. This risk is illustrated by Medsger (1940), who related that, as a boy, he experienced a severe but short-lasting colic when he ate excessive numbers of the fruits.

Podophyllum peltatum is not a common cause of intoxication, but occasionally it causes irritation of the digestive tract in livestock when rhizomes and young shoots are eaten in the early spring (McIntosh 1928; Hansen 1930). Intoxications are more likely to be seen in humans from accidental or suicidal ingestions than from the plant's use as an herbal remedy or from prolonged topical application (West et al. 1982; Holdright & Jahangiri 1990; Kao et al. 1992). However, although the species is of limited importance toxicologically, it is of considerable interest experimentally because of its unique effects on cells.

Disease Genesis—Use of podophyllin or its derivatives without proper medical supervision has led to serious adverse effects in humans and even death in a few instances. Most of these effects of *P. peltatum* result from the activity of a series of toxic lignan genins and glycosides, of which the genin podophyllotoxin and its stereoisomer picropodophyllin, an artifact formed during extraction, are the most well known. Podophylloresin is the crude product containing these compounds. Concentrations of podophylloresin and podophyllotoxin are highest when the plant is in flower and decline as the fruit matures. These lignans bind with the protein tubulin of the microtubules, preventing polymerization and arresting cell division in metaphase (Jardine 1980). Axonal flow of nutrients also is disrupted. These effects are almost identical to those of colchicine, and their respective binding sites on tubulin are in proximity. Inhibition of RNA, DNA, and protein synthesis occur almost independently and may be responsible for effects on the liver and gastrointestinal epithelium (Yang et al. 1994). Additional effects include inhibition of axoplasmic transport and, at higher dosages, inhibition of intracellular nucleoside transport (Jardine 1980). Because of the variety of effects and tissues affected, a wide range of clinical manifestations may be anticipated during episodes of intoxication.

One of the most prominent effects, severe purgation, is due mainly to podophyllin or podophylloresin rather than to any type of polyphenolic effect on the digestive tract (Chatterjee 1952).

The resin extract has been used topically for wart removal. It is especially effective against anogenital warts or condyloma acuminata and is officially approved for prescription use for this purpose under the name *podofilox* (Condylox®) (Hartwell & Schrecker 1958; Mack 1992; Bonnez et al. 1994; Ferenczy 1995). Because of its high lipid solubility, podophyllin is readily absorbed from topical application, and thus it is not unexpected that, when used repeatedly for removal of genital warts, toxic systemic effects may occur. Such effects include bone marrow depression, liver dysfunction, peripheral neuropathy, depression of the central nervous system, and pancytopenia, including thrombocytopenia and leukopenia (Chamberlain et al. 1972; Stoehr et al. 1978; Moher & Maurer 1979; Rate et al. 1979; Filley et al. 1982).

Intrauterine fetal death or mild teratogenic effects also have been reported (Chamberlain et al. 1972; Karol et al. 1980). Less commonly, exposures to large amounts of podophyllin or the purified products are accompanied, after a delay of 8–10 hours, by neurologic problems, including loss of reflexes, seizures, coma, and gradual onset of a delayed peripheral neuropathy (McFarland & McFarland 1981; Cassidy et al. 1982; Filley et al. 1982).

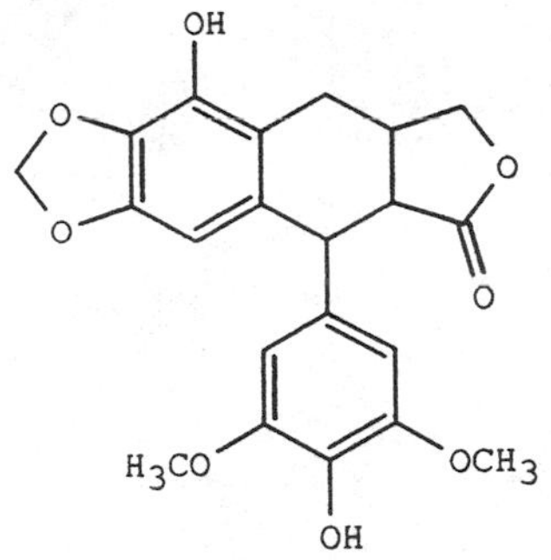

Experimentally, podophyllotoxin given daily orally to rats for 1 week, caused typical irritation of the digestive tract. It also caused severe degenerative changes in the liver, kidneys, pancreas, and testes (Chang et al. 1992b). A single i.p. injection of podophyllotoxin to rats caused neurotoxic effects that included degenerative changes in the dorsal root ganglion. This is possibly indicative of a cause of the clinical sensory aberrations noted in humans (Chang et al. 1992a).

Podophyllin also has been shown to be effective in treating rheumatoid arthritis, although toxicity limits its use for this purpose (Larsen et al. 1989).

OH

O O O O

H_3CO OCH_3

OH

α-Peltatin

Clinical Signs and Treatment—The effects of ingestion of *P. peltatum* by animals are limited to the irritant effects: loss of appetite, profuse salivation, colic, and diarrhea. There may be edema of the skin of the muzzle, face, submaxillary area, and around the eyes. Rarely there may be nervous signs such as excitement, incoordination, and peripheral numbness and/or tingling. Recovery is generally uneventful.

In humans the most common signs are nausea, vomiting, abdominal pain, and diarrhea, which may be severe and protracted. Pancytopenia with notable reduction in granulocytes and platelets and with peripheral neuropathy may be of concern. Less commonly, cerebral neurologic manifestations such as confusion, disorientation, difficulty walking, and visual or auditory hallucinations may develop. There may be a marked increase in cerebrospinal fluid protein concentrations during the first week or so. The signs generally resolve over several weeks. However, a delayed peripheral neuropathy may develop and persist for several months or more.

■ livestock, anorexia, profuse salivation, colic, diarrhea, rarely neurologic signs

humans, nausea, vomiting, colic, diarrhea; long-term use, pancytopenia, especially decreased platelets; difficulty standing or walking

Pathology—In animals, intoxication due to *P. peltatum* is normally a self-limiting disease of short duration following ingestion of plant parts. In humans, fatalities are not common but have nevertheless occurred from either medicinal use or accidental or intentional ingestion of large amounts of podophyllin. A wide variety of pathologic changes may be apparent, especially reddening and edema of the mucosa of the stomach and small intestine. Microscopically, there may be hepatocellular necrosis, and a high number of cells with mitotic figures may be apparent in various tissues such as those of the liver and gut.

■ livestock, gross pathology, reddened mucosa of stomach and intestines

microscopic, liver necrosis, numerous cells with mitotic figures

Treatment—Treatment should be directed to maintain fluid and electrolyte balance. Unfortunately the lipid solubility and rapid gastrointestinal absorption limit the value of emetics and lavage, but charcoal may provide some benefit. In humans, hemoperfusion may be helpful in eliminating the toxin.

■ activated charcoal, fluids, electrolytes

References

Bhakuni DS, Jain S: Protoberberine alkaloids. In *The Alkaloids: Chemistry and Pharmacology*, vol 28, Brossi A ed, Academic Press, Orlando, Fla, pp95–181, 1986

Bonnez W, Elswick RK, Bailey-Farchione A, Hallahan D, Bell R, Isenberg R, Stoler MH, Reichman RC: Efficacy and safety of 0.5% podofilox solution in the treatment and suppression of anogenital warts. *Am J Med* 96;420–425, 1994

Bradley M, Neiman LJ, Burrows GE: Seizures in a puppy [letter]. *Vet Hum Toxicol* 30;121, 1988

Cassidy DE, Drewry J, Fanning JP: *Podophyllum* toxicity: a report of a fatal case and review of the literature. *J Toxicol Clin Toxicol* 19;35–44, 1982

Chamberlain MJ, Reynolds AL, Yeoman WB: Toxic effect of *Podophyllum* application in pregnancy. *Br Med J* 3;391–392, 1972

Chang LW, Yang CM, Chen CF, Deng JF: Experimental podophyllotoxin (bajiaolian) poisoning. 1. Effects on the nervous system. *Biomed Environ Sci* 5;283–292, 1992a

Chang LW, Yang CM, Chen CF, Deng JF: Experimental podophyllotoxin (bajiaolian) poisoning. 2. Effects on the liver, intestine, kidney, pancreas, and testis. *Biomed Environ Sci* 5;293–302, 1992b

Chatterjee R: Indian *Podophyllum. Econ Bot* 6;342–354, 1952

Ferenczy A: Epidemiology and clinical pathophysiology of condylomata acuminata. *Am J Obstet Gynecol* 172;1331–1339, 1995

Filley CM, Graff-Radford NR, Lacey JR, Heitner MA, Earnest MP: Neurologic manifestations of podophyllin toxicity. *Neurology* 32;308–311, 1982

Graham NA, Chandler RF: *Podophyllum. Can Pharm J* 123;330–331, 1990

Hansel R: *Mahonia aquifolium:* plant antipsoriatic agent. *Dtsch Apoth Ztg* 132;2095–2097, 1992

Hansen AA: *Indiana Plants Injurious to Livestock*. Indiana Agric Exp Stn Circ 175, 1930

Harshberger JW: *Text-Book of Pastoral and Agricultural Botany.* Blakiston's Son & Co, Philadelphia, 1920

Hart J: *Montana Native Plants and Early Peoples.* Montana Historical Society Press, 1992

Hartwell JL, Schrecker AW: The chemistry of podophyllum. In *Progress in the Chemistry of Organic Natural Products,* Zechmeister L ed, Springer-Verlag, Vienna, pp83–166, 1958

Holdright DR, Jahangiri M: Accidental poisoning with podophyllin. *Hum Exp Toxicol* 9;55–56, 1990

Jardine I: Podophyllotoxins. In *Anticancer Agents Based on Natural Product Models,* Cassady JM, Douros JD eds, Academic Press, New York, pp319–351, 1980

Kao WF, Hung DZ, Tsai WJ, Lin KP, Deng JF: Podophyllotoxin intoxication: toxic effect of bajiaolian in herbal therapeutics. *Hum Exp Toxicol* 11;480–487, 1992

Karol MD, Conner CS, Watanabe AS, Murphrey KJ: *Podophyllum:* suspected teratogenicity from topical application. *Clin Toxicol* 16;283–286, 1980

Larsen A, Petersson I, Svensson B: *Podophyllum* derivatives (CPH 82) compared with placebo in the treatment of rheumatoid arthritis. *Br J Rheumatol* 28;124–127, 1989

Mack RB: Living mortals run mad, mandrake *(Podophyllum)* poisoning. *N C Med J* 53;98–99, 1992

Manske RHF, Ashford WR: The protoberberine alkaloids. In *The Alkaloids: Chemistry and Physiology,* vol 4, Manske RHF, Holmes HL eds, Academic Press, New York, pp77–118, 1954

McFarland MF, McFarland J: Accidental ingestion of *Podophyllum. Clin Toxicol* 18;973–977, 1981

McIntosh RA: Mandrake (may apple) poisoning. *North Am Vet* 9;31–32, 1928

Medsger OP: *Edible Wild Plants.* Macmillan, New York, 1940

Millspaugh CF: *American Medicinal Plants.* Dover, New York, 1974 (reprint from 1892)

Moher LM, Maurer SA: *Podophyllum* toxicity: case report and literature review. *J Fam Pract* 9;237–240, 1979

Pammel LH: *Manual of Poisonous Plants.* Torch Press, Cedar Rapids, Iowa, 1911

Rate RG, Leche J, Chervenak C: Podophyllin toxicity. *Ann Intern Med* 90;723, 1979

Seigler DS: The naturally occurring cyanogenic glycosides. In *Progress in Phytochemistry,* vol 4, Reinhold L, Harborne JB, Swain T eds, Pergamon Press, Oxford, pp83–120, 1977

Shoji N, Umeyama A, Takemoto T, Ohizumi Y: Serotonergic receptor antagonist from *Nandina domestica* Thunberg. *J Pharm Sci* 73;568–570, 1984

Southon IW, Buckingham J: *Dictionary of Alkaloids.* Chapman & Hall, London, 1989

Stoehr GP, Peterson AL, Taylor WJ: Systemic complications of local podophyllin therapy. *Ann Intern Med* 89;362–363, 1978

West WM, Ridgeway NA, Morris AJ, Sides PJ: Fatal podophyllin ingestion. *South Med J* 75;1269–1270, 1982

Whetstone RD, Atkinson TA, Spaulding DD: Berberidaceae Jussieu—Barberry family. In *Flora of North America North of Mexico,* vol 3, Flora of North America Editorial Comm, Oxford Univ Press, New York, 1997

Yang CM, Deng J-F, Chen CF, Chang LW: Experimental podophyllotoxin (bajiaolian) poisoning. 3. Biochemical bases for toxic effects. *Biomed Environ Sci* 7;259–265, 1994

15 Boraginaceae Juss.

- Forget-me-not Family

  Amsinckia
  Cynoglossum
  Echium
  Heliotropium
  Symphytum

  QUESTIONABLE
  Borago
  Hackelia
  Mertensia
  Myosotis

Cosmopolitan in distribution but especially diverse in western North America and the Mediterranean region, the Boraginaceae comprises approximately 100 genera and 2,000 species. Known as the borage or forget-me-not family, the Boraginaceae is of little economic importance. Some 30 genera are cultivated as ornamentals, e.g., heliotropes, forget-me-nots, and hound's-tongue. Several species are sources of dyes—e.g., alkanet to color cosmetics, medicines, wood, marble, and wines—or are eaten as a potherb (comfrey) or used as medicines by American Indians and settlers. Decoctions of the western *Lithospermum ruderale* have contraceptive action, and knowledge of its effect is said to have inspired the development of modern oral contraceptives (Cronquist 1981). The traditional garden herb *Borago officinalis,* or borago, has been used since the Middle Ages for its reputed culinary and medicinal value.

In North America, the family is represented by 38 native and introduced genera and about 400 species. Of these, the species of 11 genera may pose a toxicologic risk.

- herbs, shrubs, or trees; leaves mostly alternate; petals 5, fused; fruits nutlets

Plants herbs or shrubs or trees; annuals or biennials or perennials; herbage typically scabrous or hispid. **Leaves** simple; alternate or occasionally alternate above and opposite below; venation pinnate or a single vein; stipules absent. **Inflorescences** solitary flowers or simple cymes or helicoid (scorpioid) cymes; terminal or axillary. **Flowers** perfect; perianths in 2-series. **Sepals** 5; fused or free. **Corollas** radially or bilaterally symmetrical; tubular or salverform or funnelform. **Petals** 5; fused; of various colors. **Stamens** 5; exserted beyond or included within perianths; epipetalous. **Pistils** 1; compound, carpels 2; stigmas 1, not lobed or 2-lobed; styles 1; ovaries superior, 4-lobed or terete in cross section; placentation axile. **Fruits** nutlets 4 or rarely 1–3; attached laterally or vertically; smooth or rugose or tuberculate or echinate or prickly. **Seeds** 1.

- toxins, pyrrolizidine alkaloids; plant species of highest risk are those containing macrocyclic closed diesters; of lesser risk are those with open diesters; plants with only monoesters are generally of low risk

Many genera of the Boraginaceae contain pyrrolizidine alkaloids (PAs); however, only a few have been associated with clinical disease. This lack of disease in large measure is due to the relatively low toxicity potential for many of the PAs. The structural influences on PA toxicity include unsaturation of the 3-pyrroline ring; the presence of one or two hydroxy groups attached to the rings at C-7 and C-9; esterification of the hydroxy groups; and branching of the ester chain. The potency of the PA structures, arranged in general order of descending potency, is macrocyclic "closed" diesters > open diesters > monoesters with a 7-OH group > monoesters (Peterson et al. 1983; Mattocks 1986; Kim et al. 1993). A somewhat confusing variety of PAs is represented in the Boraginaceae, although in general the PAs are of lesser toxicity than the closed-diester macrocyclic PAs of *Senecio* in the Asteraceae and *Crotalaria* in the Fabaceae. They are confusing in part because of the many stereoisomers that are present. For example, echiumine and symlandine, lycopsamine and intermedine, and symphytine and myoscorpine are 3′ epimer pairs, and echimidine and heliosupine are 2′ epimers. Some genera contain primarily the less potent structural types, and therefore large amounts of plant material would be required for intoxication—in some instances much larger amounts than are likely to be eaten, considering plant availability and palatability. In addition to ingestion of foliage, excretion of the toxins in milk also represents a risk. A complete discussion of PA toxicity is presented in the treatment of *Senecio* (see chapter 13) and in a review by Peterson and Culvenor 1983. In addition to the genera described below, *Lappula, Lithospermum,* and *Plagiobothrys* may be similarly toxic.

Amsinckia Lehm.

Taxonomy and Morphology—Commonly known as fiddleneck because of its distinctive cymes resembling the abdomen of a scorpion, *Amsinckia* comprises 15–50 species native to western North America, South America, and Europe. Some taxonomists restrict the circumscription of the genus and consider it to be strictly New World. Hybridization among species is common, and identification is typically difficult. Well known as toxic are the following species:

■ *A. intermedia*
A. lycopsoides
A. menziesii
A. tesselata

burweed
devil's lettuce
fiddleneck
fingerweed
forget-me-not
tarweed

A. intermedia Fisch. & C.A.Mey. (= *A. arizonica* Suksd.) (= *A. echinata* A.Gray) (= *A. intactilis* J.F.Macbr.) (= *A. microphylla* Suksd.) (= *A. nana* Suksd.) (= *A. rigida* Suksd.)	tarweed, yellow tarweed, coast fiddleneck, yellow burweed, medium fiddleneck, fingerweed, rancher's fiddleneck, fireweed fiddleneck, yellow forget-me-not, yellow burweed
A. lycopsoides Lehm. (= *A. barbata* Greene) (= *A. hispida* [Ruiz & Pav.] I.M.Johnst.) (= *A. idahoensis* M.E.Jones) (= *A. parviflora* A.Heller) (= *Benthamia lycopsoides* [Lehm.] Lindl. ex Druce)	tarweed fiddleneck, bugloss fiddleneck
A. menziesii (Lehm.) A.Nelson & J.F.Macbr. (= *A. micrantha* Suksd.) (= *Echium menziesii* Lehm.)	Menzies fiddleneck, small-flowered fiddleneck
A. tesselata A.Gray (= *A. gloriosa* Suksd.)	devil's lettuce, rough fiddleneck

Amsinckia intermedia

■ annual herbs; inflorescence in shape of neck of fiddle; petals yellow or orange; nutlets trigonous

Plants herbs; annuals; from taproots. **Stems** erect or rarely decumbent; 15–120 cm tall; branched or not branched. **Leaves** both basal and cauline; sessile or short petiolate; surfaces hispid; blades linear to narrowly ovate; margins entire or occasionally dentate. **Inflorescences** helicoid cymes; terminal or axillary; bracts absent. **Flowers** tubular or salverform. **Sepals** 5, but sometimes appearing 2–4 due to fusion; hispid or pilose. **Petals** yellow or orange. **Nutlets** 4, trigonous; tubercles round or sharp.

1. Sepal lobes 2–4. Corolla tubes 20-veined at bases . . . . *A. tesselata*
1. Sepal lobes 5. Corolla tubes 10-veined at bases.
   2. Corollas salverform; tubes closed by hairy bulges. Styles included in corolla tubes . . . . *A. lycopsoides*
   2. Corollas funnelform; tubes open. Styles exserted beyond corolla tubes.
      3. Corollas orange or orange-yellow; 7–10 mm long; longer than calyces . . . . *A. intermedia*

Amsinckia intermedia

3. Corollas light yellow; 4–7 mm long; shorter than or equal to calyces . . . . *A. menziesii*

■ dry, open sites

Amsinckia lycopsoides

Distribution and Habitat—In North America, greatest diversity in the genus occurs in the southwestern United States and Mexico. Eastward, 3 species reach the Great Plains. Typically occupying dry, open sites, *Amsinckia* is adapted to a variety of soil types. Also adapted to disturbance, some species are weedy and encountered in rights-of-way and along roadsides. *Amsinckia lycopsoides* is sometimes cultivated as an ornamental in the East.

■ hepatic disease, caused mainly by eating seeds; may be accompanied by neurologic manifestations

Disease Problems—The cause of aimless wandering in horses, a disease eventually known as walking disease was first noted in Washington (Kalkus et al. 1925). Produced by the consumption of the nutlets of *A. intermedia,* it was given other common names that also typified its features, e.g., hard liver disease, Walla Walla hard liver disease, protein poisoning, or winter-wheat poisoning. The disease also was seen in pigs and cattle. Intoxication problems are mainly associated with hepatic fibrosis, the result of contamination of feed grain or screenings with *A. intermedia* seeds (10–25%) (Woolsey et al. 1952; Kennedy 1957). Based on experiments with a limited number of subjects, chickens and sheep appear to be somewhat resistant to the effects, given that the 10–25% contamination rate produced only a decrease in weight gain (McCullock 1940; Muth 1941).

■ moderate intoxication risk due to open-diester PAs such as echiumine

Disease Genesis—Species of *Amsinckia* contain a mixture of open diesters and monoesters of retronecine; the potency is low to moderate, and the toxic dosage requirements high. *Amsinckia intermedia* contains the open diester echiumine and the monoesters intermedine, lycopsamine, and sincamidine. Total PA concentrations in the seeds of *A. intermedia* are erratic, in some instances being undetectable. In the leaves they may be as much as 0.25% d.w. (Johnson & Molyneux 1985). *Amsinckia lycopsoides* also contains echiumine, intermedine, and lycopsamine, whereas *A. menziesii* contains intermedine, lycopsamine, and several open-diester lycopsamine derivatives (Roitman 1983). There are at least 6 PAs and their acetylated derivatives present in *A. tesselata,* including lycopsamine and tessellatine, which account for 90% of the total (Roitman 1983; Kelley & Seiber 1992).

Amsinckia menziesii

Echiumine

Intermedine

■ signs most distinctive in horses; abrupt onset, head pressing, aimless pacing, ataxia, chewing, yawning, intermittent drowsiness, rectal straining, either constipation or diarrhea

Clinical Signs—Signs of *Amsinckia* intoxication are most distinctive in horses and begin with the animals' refusal of grain and the appearance of icterus and oral ulcers. Thereafter, they may be observed to be unthrifty, sluggish/sleepy, and emaciated and perhaps have abdominal distension due to fluid accumulation. They may yawn and lean on solid objects. Often there are episodes of furious delirium or restlessness with incessant and aimless walking, even through fences and other obstructions. They eventually become comatose and die. Other species are less likely to show the nervous signs. The disease in cattle may escape notice until the livers are examined after slaughter. Anemia and hemoglobinuria are additional signs noted in pigs.

Pathology and Treatment—A complete description of the disease's pathology and treatment is presented in the following treatment of *Cynoglossum.*

Amsinckia tesselata

Cynoglossum L.

Taxonomy and Morphology—Comprising 80–90 species distributed widely in the temperate regions of the world, *Cynoglossum* is characterized by its prickly nutlets, which readily become entangled in hair, fur, wool, or clothing; hence the colloquial names *beggar's-lice, sheep-lice,* and *sticktight.* Its generic name is derived from the Greek roots *kyon,* for "dog," and *glossa,* for "tongue," which is indicative of the similarity of the basal leaves to a dog's tongue and the basis for the common name *hound's-tongue.* A few species are cultivated as ornamentals for their intense blue petals, which often exhibit a range of colors as they mature (see plate 15.1).

■ *C. amabile*
C. grande
C. occidentale
C. officinale

beggar's-lice
Chinese forget-me-not
dog bur
dog's-tongue
glovewort
hound's-tongue
sheep-lice
sticktight
woolmat

C. amabile Stapf & J.R.Drumm.	Chinese forget-me-not
C. grande Douglas ex Lehm.	Pacific hound's-tongue
C. occidentale A.Gray	western hound's-tongue
C. officinale L.	common hound's-tongue, beggar's-lice, dog bur, woolmat, dog's-tongue, sheep-lice, sticktight, glovewort

Plants herbs; biennials or perennials; from taproots; herbage typically hispid. **Stems** erect; 30–120 cm tall. **Leaves** basal and/or cauline; petiolate or sessile; blades oblanceolate to elliptic; margins entire. **Inflorescences** terminal helicoid cymes; bracts absent. **Flowers** funnelform or salverform; with conspicuous crests or scales in throats. **Sepals** enlarged in fruit. **Petals** blue or purple or white or pink. **Nutlets** 4; attached laterally; prickly, forming burs.

■ erect herbs; leaves basal and/or cauline; inflorescence in shape of neck of fiddle; petals blue or white or pink; nutlets prickly

1. Cymes subtended by leaves. Plants biennials.
   2. Petioles of basal leaves not winged.
      Petals reddish purple . . . . *C. officinale*
   2. Petioles of basal leaves winged.
      Petals blue or pink or white . . . . *C. amabile*
1. Cymes borne above leaves. Plants perennials.
   3. Stems glabrous. Leaves mostly basal; long-petioled;
      wings absent . . . . *C. grande*
   3. Stems hispid. Leaves mostly cauline; sessile or subsessile;
      wings present . . . . *C. occidentale*

Distribution and Habitat—Introduced from Eurasia, *C. officinale* has now spread throughout much of North America, including all provinces of Canada except those of the extreme east. It forms dense stands and is a particularly troublesome weed in the intermountain area and the western Great Plains. As an introduced plant, it seems to be continuing to expand its range of infestation. It is shade tolerant and tends to favor more moist sites, even in rocky areas. *Cynoglossum amabile* from eastern Asia also is naturalized in many areas of North America. The native *C. grande* is found on the dry, wooded slopes of the Pacific Coast states and British Columbia. *Cynoglossum occidentale* favors pine forests of California and Oregon.

Cynoglossum officinale

- dense stands in moist sites

As a biennial, C. *officinale* typically develops only a rosette in the first year after seed germination and then flowers the second year. In the southern parts of its range, it may behave as an annual and flower in one season.

- mainly hay contaminated by *Cynoglossum;* acute digestive disturbance or, with long-term ingestion, liver fibrosis and insufficiency, accompanying neurologic manifestations

Disease Problems—The species of primary toxicologic interest, *C. officinale,* is a serious weed problem in some areas of North America. Dispersal is facilitated by the propensity of the prickly nutlets to become entangled in hair or wool, permitting grazing animals to become transporters of the species on rangelands (Upadhyaya & Cranston 1991; DeClerck-Floate 1997). The nutlets occasionally cause eye irritation. Despite having been introduced into North America in the early 1900s, the toxicity problems caused by the species were not really recognized until C. *officinale* was associated with a liver disease in Colorado (Knight et al. 1984). The disease occurs naturally in calves and also has been experimentally reproduced (Baker et al. 1989, 1991). Reports of Old World occurrences describe an acute ruminal/abomasal irritation and severe diarrhea, as well as a subacute/chronic wasting type of disease (Greatorex 1966; Mandryka 1979). Acute digestive tract disease is not inconsistent with the findings of Stegelmeier and co-workers (1996) in horses. Several species are thought to be toxic, but only C. *officinale* has been associated with clinical disease—primarily due to contamination of hay, because fresh plants have a noxious odor and are seldom eaten.

Cynoglossum grande

- moderate to high risk due to high levels of open diesters, heliosupine, and monoesters
- PA levels are high in seeds and young leaves; serious risk to horses, ingestion of a few plants each day for a few weeks may result in liver disease

Disease Genesis—Species of *Cynoglossum* contain a mixture of open diesters and monoesters of heliotridine and supinidine. *Cynoglossum amabile* contains amabiline (supinidine) and echinatine (heliotridine), which are both monoesters and therefore of quite low hazard. *Cynoglossum officinale* contains the more toxic open diester heliosupine (heliotridine), acetylheliosupine, and several monoesters, including echinatine; it is of considerably greater hazard than are other species of the genus. Concentrations of PAs are present in both seeds and leaves; those in leaves are highest early in the growing season of the plant's first or second year (Baker et al. 1989; Pfister et al. 1992). Developing young leaves may have 10–50 times greater total PA concentrations than older leaves, in large measure due to translocation of the alkaloids to them, presumably as protection against herbivorous insects (van Dam et al. 1994). Total alkaloid concentrations are typically 0.5–2% d.w., with about two-thirds being heliosupine. The *N*-oxide form predominates over the free base form by a ratio of 3:1 or more (Knight et al. 1984; Baker et al. 1989, 1991; Pfister et al. 1992).

Amabiline

Heliosupine

The typical effects attributed to PAs, hepatic fibrosis and biliary hyperplasia, are produced by C. *officinale* when the dosage is low and protracted. These plants represent a very serious risk, especially for horses. Dosage of 15 mg total PAs/kg of b.w. to horses daily for 1 week resulted in severe acute hepatic necrosis with high activity of serum hepatic enzymes, but with few of the typical lesions (Stegelmeier et al. 1994, 1996). Lower dosage (5 mg PA/kg) produced little effect on enzyme activity, and the detrimental effects of weight loss and alteration of behavior were resolved in 2 months. It was suggested that ingestion of as few as 1–2 plants/day may represent a serious risk for horses (Stegelmeier et al. 1996).

Although chemical evidence is lacking, *C. grande* and *C. occidentale* may also contain PAs because they serve as alternative host plants for *Gnophaela latipennis,* a moth known to feed on plants containing PAs and to sequester the alkaloids as a defensive mechanism (Godfrey & Crabtree 1986; L'Empereur et al. 1989).

Clinical Signs—Signs are typical of those associated with ingestion of PAs. Occasionally the disease is acute, but most commonly it is chronic. Poor body condition and weight loss are the most consistent signs among the various animal species. Additional signs in horses include icterus, red (hemoglobin-containing) urine, and neurologic manifestations such as yawning, pacing, and licking of objects. Less commonly, abrupt onset may be associated with liver necrosis and fulminant failure, a hemogram indicating a degenerative left shift, and occasionally the necrotizing dermatitis of photosensitization (Stegelmeier et al. 1996).

Serum hepatic enzymes, e.g., AST, GGT, AP, and other tests of hepatic function and integrity, may be altered, but these changes are not striking in all stages of the disease. They will be most severely changed with abrupt onset of hepatic necrosis. As with toxicity due to PAs from other genera, diagnosis may be facilitated by determination of the activity of the serum hepatic enzymes and by the isolation and identification of sulfur-conjugated pyrroles in blood (bound to hemoglobin) or the detection of pyrroles in fresh or fixed liver tissue (Mattocks & Jukes 1992; Stegelmeier et al. 1996).

- poor body condition, weight loss, and especially in horses, abrupt onset; head pressing, aimless pacing, ataxia, chewing, yawning, intermittent drowsiness, rectal straining, and either constipation or diarrhea

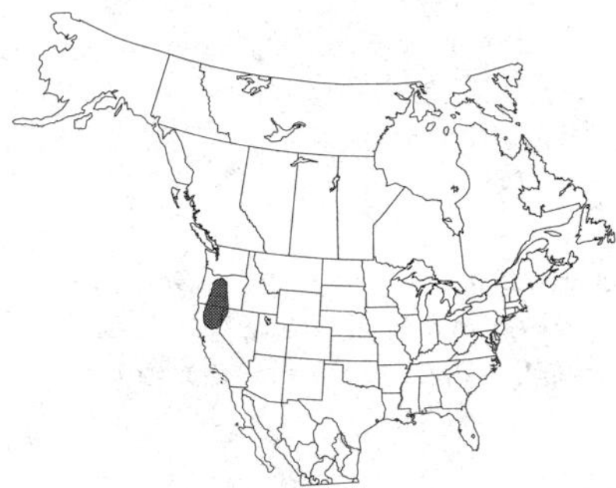

Cynoglossum occidentale

Pathology—Although not always present, the most striking features seen grossly following chronic ingestion of the toxicant are a firm, fibrotic, shrunken, grayish blue to yellowish liver with a marked accentuation of lobular pattern; a thick, bluish gray, edematous omentum; and/or an edematous appearance of the abdominal viscera. There may be abundant ascitic fluid; distension of the gallbladder, with thickened walls; and scattered hemorrhages over the intestinal serosa. In rare instances when large amounts of the plants are eaten in a short period, the liver may be enlarged and congested because of hepatocellular necrosis and intralobular hemorrhage, and there may be scattered small, splotchy hemorrhages of the serosal surfaces of the gut mucosa and viscera. Cecal and ileal infarctions also may be present in horses having the acute disease.

Microscopically, there will typically be mild, diffuse necrosis, hyperplasia of the portal bile duct, severe disruption of cord architecture, and an increase in cell and nuclear size of hepatocytes and biliary epithelial cells. In cattle, enlarged renal tubular epithelial cells may be noted.

- gross pathology, liver, shrunken and firm, edema of the omentum, increased fluid in the abdomen

  microscopic, liver, centrilobular necrosis; bile duct proliferation, portal fibrosis; megalocytosis, karyomegaly; central vein fibrosis

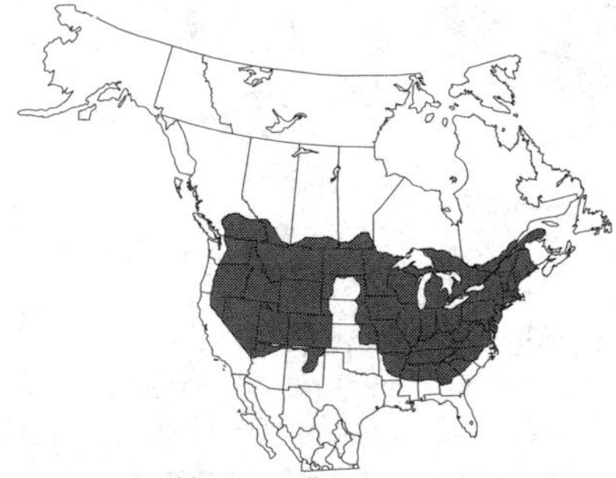

Cynoglossum officinale

Treatment—It is important to keep in mind that the effects of the toxic PAs are cumulative, at least until compensatory hepatic regeneration occurs. Because of this, prevention of further animal access to the plants is important, especially with respect to exposure via hay. Sulfur-containing amino acids such as cysteine and methionine have been reported to be of some value in treatment, but this has not been confirmed experimentally (Shull et al. 1976). Although there is a disturbance of the branched-chain amino acid/phenylalanine-tyrosine ratio, therapy with branched-chain amino acids has not proven to be of value (Mendel et al. 1988). It is possible that any approaches to increase hepatic glutathione content may be helpful.

Treatment is supportive and typically includes parenteral fluids and a low-protein diet to reduce the ammonia load to the liver. However, these measures are generally of little value in most cases because, by the time distinctive signs appear, the hepatic changes are extensive and severe. Although decreasing the protein in the diet may be used in treatment, this may not be appropriate for prevention, because increased toxicity is apparent with protein deficiency (Mattocks 1986).

The public health hazard should be considered, especially for cattle, because these alkaloids are potential carcinogens and can be present in low levels in milk and possibly the liver.

- general nursing care; sedation, fluids, high-quality and high-energy feeds with low but high-quality protein; supplementation with sulfur amino acids is of questionable value

Echium L.

- *E. plantagineum*
  E. vulgare
  blue devil
  blue thistle
  blueweed
  Paterson's curse
  salvation jane
  viper's bugloss

Taxonomy and Morphology—Important as a cause of toxicity in other parts of the world, *Echium* comprises 40–50 species native to the Old World. Commonly known as viper's bugloss, its name was used by Dioscorides and comes from the Greek root *echis*, for "viper." Two introduced species have naturalized in North America and are encountered in roadsides, fields, and other disturbed sites.

E. plantagineum L. (= *E. lycopsis* L.)	Paterson's curse, salvation jane
E. vulgare L.	blueweed, blue thistle, blue devil, viper's bugloss

- erect herbs or shrubs; leaves in basal rosettes and on stems; petals blue, pink or white; nutlets tuberculate

Plants herbs or shrubs; annuals or biennials or perennials; herbage hispid to strigose or canescent. **Stems** erect; 30–80 cm tall; not branched or rarely so. **Leaves** both basal and cauline; basal leaves forming rosette, oblanceolate, petiolate; cauline leaves linear or elliptic or oblong, sessile or subsessile. **Inflorescences** terminal panicles of helicoid cymes; bracts small or foliaceous. **Flowers** radially to bilaterally symmetrical; funnelform. **Sepals** fused. **Petals** blue or pink or white. **Stamens** exserted beyond corollas. **Nutlets** 1–4; ovoid; trigonous; attached vertically; tuberculate.

Echium vulgare

1. Exserted stamens 4 .... *E. vulgare*
1. Exserted stamens 2 or 3 .... *E. plantagineum*

Distribution and Habitat—*Echium* is native to the Canary Islands and Mediterranean region. *Echium vulgare* is widespread across the continent, while *E. plantagineum* is found in the coastal mountains of California. Both species are adapted to a variety of soil types and favor disturbed areas.

- little problem as yet in North America; if plant populations become large, liver disease is a risk with long-term ingestion, especially in horses

Disease Problems—Although species of *Echium* are important as causes of disease in other parts of the world, as yet they have not proven to be other than troublesome weed pests here in North America. The lack of field cases may be the a result of too little plant material available and/or conditions not favoring its consumption.

- toxins are open-diester PAs, heliosupine

Disease Genesis—Species of *Echium* contain mainly open diesters of retronecine and heliotridine; thus the hazard is moderate to high. *Echium italicum* contains echimidine, an open diester of retronecine. For *E. plantagineum*, echimidine, acetyllycopsamine, and acetylintermedine account for 75% of the PAs, with lesser amounts of echiumine, lycopsamine, and uplandicine present. Total PA concentrations are typically about 0.1% d.w. but may range as high as 0.9% (Peterson 1985). *Echium vulgare* contains asperumine, heliosupine, and acetylheliosupine, all open diesters of heliotridine. Disease problems in horses have been described in Australia, another area where the genus is naturalized (Sharrock 1969; Seaman 1978). In general, monogastrics appear to be at greatest risk because detoxication of PAs is favored in ruminants (Peterson 1985). As with other PA-containing plants, such as *Senecio* in the Asteraceae, there is marginal risk in sheep, as shown in experimental studies where extensive consumption affected only a few animals (Culvenor et al. 1984). In sheep the primary risk seems to be increased susceptibility to chronic copper intoxication and subsequent hemolytic crisis predisposed by the effects of the PAs (St. George-Grambauer & Rac 1962; Seaman & Dixon 1989). In general, the disease is similar to that produced by other members of the family containing toxic PAs (Seaman & Walker 1985; Seaman 1987).

Echium plantagineum

Heliosupine

Echium vulgare

■ in general, weight loss, depression; horses, abrupt onset; head pressing, aimless pacing, ataxia, chewing, yawning, intermittent drowsiness, rectal straining, and either constipation or diarrhea

Clinical Signs—Decreased appetite, depression, and icterus are common in both the acute and the more typical chronic form. In horses neurologic signs include incoordination, head pressing, and incessant walking or pacing. These signs are accompanied by the elevations in serum hepatic enzymes and other chemistry changes typical of the pyrrolizidine alkaloids (Giesecke 1986).

Pathology and Treatment—A complete description of the disease's pathology and treatment is presented in the preceding treatment of *Cynoglossum*.

Heliotropium L.

■ *H. amplexicaule*
H. arborescens
H. convolvulaceum
H. curassavicum
H. europaeum
H. indicum

Chinese pulsey
cola de mico
heliotrope
quail plant
turnsole

■ herbs or shrubs; inflorescence often in shape of neck of fiddle; petals white, purple, blue or rarely yellow; nutlets separate or paired

Taxonomy and Morphology—Comprising 200–250 species distributed primarily in tropical and warm-temperate regions of the world, *Heliotropium* is noted for its ornamentals widely grown as both border and specimen plants. Its generic name is derived from the Greek roots *helio* and *tropos,* meaning "sun" and "turn," which reflects an older misconception that the inflorescences turned with the sun. Numerous species are recognized as toxic. In North America, 27 native and introduced species are present. The following are of known toxicity:

H. amplexicaule Vahl	violet heliotrope, blue heliotrope
H. arborescens L.	common heliotrope
H. convolvulaceum (Nutt.) A.Gray	bindweed heliotrope
H. curassavicum L.	Chinese pulsey, quail plant, seaside heliotrope, salt heliotrope, cola de mico
H. europaeum L.	European heliotrope
H. indicum L.	India heliotrope, turnsole

Plants herbs or shrubs; annuals or perennials; herbage typically villous or scabrous or hispid. **Stems** erect or prostrate to ascending; 20–200 cm tall. **Leaves** alternate or rarely subopposite; generally cauline; sessile or petiolate; blades linear to ovate-deltoid or obovate. **Inflorescences** helicoid (scorpioid) cymes or solitary flowers; terminal or axillary; bracts present or absent. **Flowers** small; funnelform to salverform. **Sepals** free or nearly so; persistent or deciduous. **Petals** white or blue or purple or occasionally yellow. **Stamens** included within corolla tubes. **Nutlets** 4; separate or united in pairs; attached vertically; ovoid; smooth or rugose or tuberculate.

1. Plants perennials.
   2. Plants shrubs . . . . *H. arborescens*
   2. Plants herbs.
      3. Stems and leaves succulent; glabrous. Petals white to blue . . . . *H. curassavicum*
      3. Stems and leaves not succulent; villous. Petals purple . . . . *H. amplexicaule*

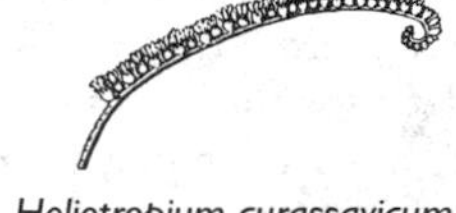
Heliotropium curassavicum

■ most in dry, open sites, except for *H. curassavicum* in moist soils

1. Plants annuals.
   4. Corollas 8–20 mm in diameter . . . . *H. convolvulaceum*
   4. Corollas 2–5 mm in diameter.
      5. Nutlets paired at maturity. Ovaries 2-lobed . . . . *H. indicum*
      5. Nutlets separate at maturity. Ovaries 4-lobed . . . . *H. europaeum*

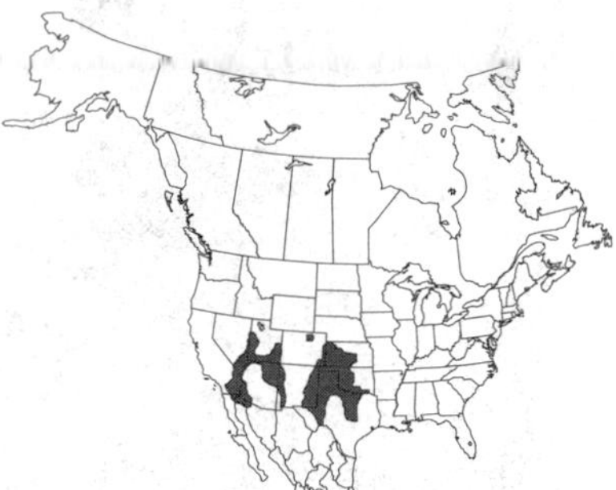

Heliotropium convolvulaceum

Distribution and Habitat—With the exception of *H. convolvulaceum,* which is native to the southwestern United States and northern Mexico, the toxic species of *Heliotropium* are naturalized introductions: *H. amplexicaule* from Argentina, *H. arborescens* from Peru, *H. indicum* from Brazil, *H. curassavicum* from Central America, and *H. europaeum* from Europe and central Asia. All characteristically occupy dry, open sites and are especially common in sandy soils. Succulent *H. curassavicum* occurs in damp saline soils in dry marshes or along the seashore—hence its common name. All are somewhat weedy but do not occur in large populations and thus do not appear to represent a significant toxicologic hazard.

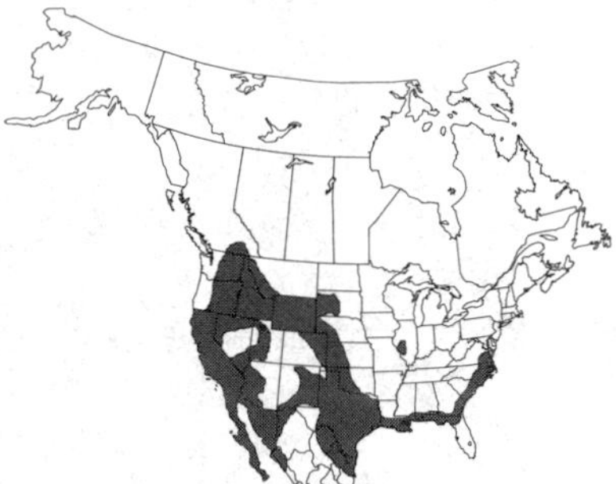

Heliotropium curassavicum

Disease Problems—The genus is of considerable importance as a naturalized taxon in Australia but has yet to be incriminated in intoxications in North America. As with *Echium,* there simply may be insufficient plant material available in any given area, and/or conditions may not favor its consumption. If very large populations eventually become established, clinical disease may become a more real threat. Plants pose a risk either fresh or dried. Even contamination of straw bedding with 20% *H. europaeum* caused death in preruminant calves after several weeks of illness (Harper et al. 1985). In addition to the foliage, the seeds may pose a risk as contaminants in grains or their screenings. Contamination of 4,980 seeds/kg of wheat caused death of cattle in a feed yard (Hill et al. 1997).

Severe veno-occlusive liver disease occurs occasionally in humans following inadvertent consumption of the seeds or from teas and herbal products made from species of *Heliotropium* (Mohabbat et al. 1976; Culvenor et al. 1986). In an incident in Tadjikistan, 4,000 people became intoxicated and a few died when wheat contaminated with the seeds of *H. lasocarpum* was used for making bread (Chauvin et al. 1994). The hepatic lesions are primarily vascular oriented and are termed heliotrope hepatoangiopathy (Mansurov 1995).

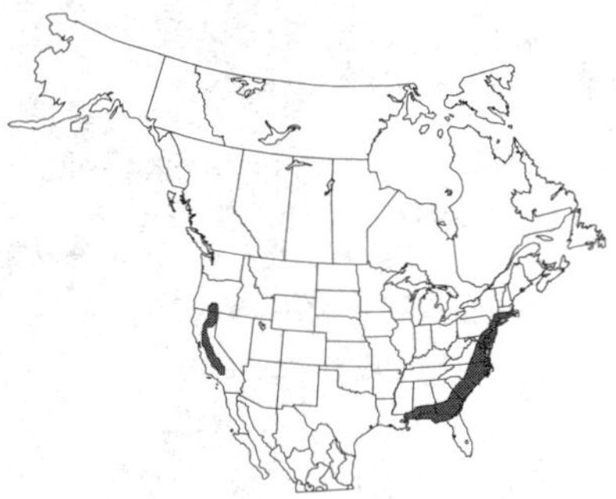

Heliotropium europaeum

Disease Genesis—Species of *Heliotropium* contain esters of retronecine and mainly heliotridine (O'Dowd & Edgar 1989). *Heliotropium amplexicaule* contains only indicine, a monoester of retronecine and thus is of little hazard. *Heliotropium arborescens, H. curassavica,* and *H. indicum* contain lasiocarpine, an open diester of heliotridine, and monoesters of heliotridine. Greater risks are posed by *H. europaeum,* which contains lasiocarpine and acetyllasiocarpine, as well as monoesters such as heliotrine and europine, and by the Old World species *H. supinum,* which contains lasiocarpine, other diesters (including heliosupine), and several monoesters. These alkaloids occur mainly as *N*-oxides and typically decrease with plant maturity. The relative proportions of diesters and monoesters may vary with environmental conditions; increasing rainfall seems to increase the proportion of the diesters (O'Dowd & Edgar 1989).

■ little problem as yet in North America; if plant populations become large, liver disease is a risk with long-term ingestion, especially in horses; in other areas of the world, seeds and herbal products are a risk to humans

In Australia, intoxications occur occasionally in cattle, but primarily sheep are affected; the plants are not palatable to most livestock species (Bull et al. 1961). Seeds are also a hazard, especially for chickens and ducks (Pass et al. 1979). In Australia, sheep eating *Heliotropium* for sustained periods are predisposed to accumulate copper in the liver, and the signs of disease are a composite of those associated with the

PAs and chronic copper intoxication (Bull et al. 1956). Symptoms include megalocytosis, elevated liver copper, and large amounts of free hemoglobin in blood and urine. It is of interest to note that most of the PAs in *Heliotropium* are degraded in the sheep rumen and that manipulation of the ruminal flora, such as with chronic iodoform administration, to decrease methane production attenuates the toxicity risk (Lanigan et al. 1978; Peterson et al. 1992). Furthermore, there seems to be a genetic predisposition either for or against PA susceptibility.

■ toxins are open-diester PAs, lasiocarpine

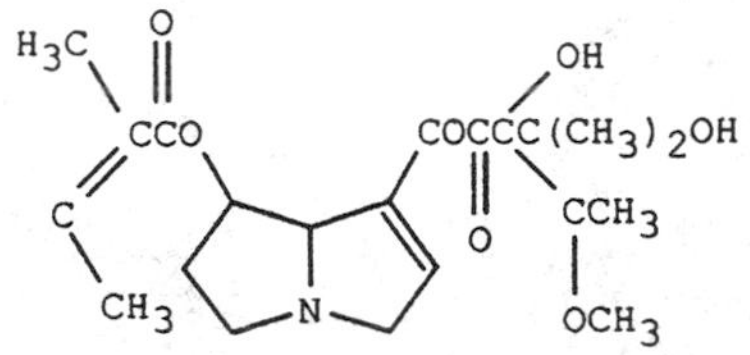

H_3C O CCO C CH_3 N OH COCCC$(CH_3)_2$OH O CCH_3 OCH_3

Lasiocarpine

The slower action of repeated ingestion of small doses of plant material tends to produce the typical PA response of fibrosis and biliary proliferation. In contrast, consumption of large amounts of plant material may pose an acute toxicity risk, as shown by the rapid onset of neurologic signs, respiratory difficulties, and death in mice given i.p. doses of aqueous extracts from the African species *H. scottae* (Wahome et al. 1994).

Other pharmacologic effects are attributed to the PAs of *Heliotropium,* such as lasiocarpine and heliotropine (Bull et al. 1968). There may be blockage of oxidative phosphorylation, as shown in magnesium-deficient systems, but this appears to be of little consequence toxicologically (Gallagher 1968). Developmental skeletal anomalies have been induced in rats given a large i.p. dosage of heliotrine in the second week or dehydroheliotridine on the 14th day of pregnancy (Green & Christie 1961; Peterson & Jago 1980).

The alkaloids may be present in honey at concentrations not exceeding 1 mg/kg. The risk is not clear, but honey has not been associated with disease or any adverse effects in people.

Clinical Signs, Pathology, and Treatment—In the event that livestock intoxications occur in North America, the clinical signs will be similar to those associated with *Cynoglossum*—weight loss, depression, ataxia, weakness leading to recumbency, and in some instances episodic aggression. Gross lesions include a hard, yellow-brown, fibrotic liver, ascites, and edema of the mesenteries, while microscopically in the liver megalocytosis, bile ductile hyperplasia in portal triads, periportal fibrosis, and a few regenerative nodules are seen.

■ in general, weight loss, depression; horses, abrupt onset; head pressing, aimless pacing, ataxia, chewing, yawning, intermittent drowsiness, rectal straining, and either constipation or diarrhea

The signs of hepatic veno-occlusive disease in humans are divided into four stages (Chauvin et al. 1994): stage 1—signs indicative of effects on the digestive tract, such as nausea, colic, vomiting, and weakness, predominate; stage 2—digestive tract signs, plus the appearance of a swollen, tender liver; stage 3—the previous signs, plus accumulation of fluid in the abdomen; stage 4—sedated appearance or coma.

■ veno-occlusive disease in humans, digestive disturbance and liver disease

Symphytum L.

Taxonomy and Morphology—Commonly known as comfrey, *Symphytum* comprises 25–35 species native to Europe and especially the Mediterranean region. Its name is derived from the Greek roots *syn* and *phyton,* meaning "together" and "plant," presumably referring to the healing properties of the genus. Rich in essential minerals, *S. officinale* is grown for green manure in Europe, as an ornamental, and as a medicinal plant. In North America, 4 species are naturalized or adventive:

■ *S. asperum*
S. officinale
S. tuberosum
S. xuplandicum

comfrey

S. asperum Lepech.	rough comfrey, prickly comfrey
S. officinale L.	common comfrey
S. tuberosum L.	tuberous comfrey
S. xuplandicum Nyman (= *S. asperum* x *S. officinale*)	Russian comfrey

■ erect, perennial herbs; petals blue, pink, white, or rarely yellow-white; nutlets tuberculate

Symphytum officinale

Plants herbs; perennials; from taproots or rhizomes; herbage typically hispid. **Stems** erect; 30–200 cm tall; branched or not branched; winged or not winged. **Leaves** alternate or uppermost crowded and opposite; basal or cauline; basal leaves petiolate; cauline leaves sessile or subsessile; bases decurrent or not decurrent. **Inflorescences** short 1-sided cymes; terminal or axillary; often branched; bracts absent. **Flowers** urceolate or campanulate; throats with 5 scales. **Sepals** fused. **Petals** blue or pink or white or rarely yellow-white. **Stamens** inserted within corolla tubes. **Nutlets** 4; attached vertically; smooth to rugose; tuberculate; bases with thickened collarlike rings.

1. Stems not branched. Plants rhizomatous . . . . *S. tuberosum*
1. Stems branched. Plants not rhizomatous.
   2. Stems winged. Nutlets smooth. Corollas urceolate . . . . *S. officinale*
   2. Stems not winged. Nutlets rugose. Corollas campanulate.
      3 Sepals 3 mm long. Flower buds red. Petals blue . . . . *S. asperum*
      3. Sepals 5 mm long. Flower buds purple or pink. Petals pink or purple . . . . *S.* x*uplandicum*

■ all species adventive or naturalized; waste area invaders

Distribution and Habitat—All species of *Symphytum* are adventive or naturalized from Europe primarily in the northern half of the continent. *Symphytum asperum* and *S.* x*uplandicum* are found in waste areas from British Columbia to Quebec as occasional weeds. *Symphytum officinale* is a garden escape found in the eastern half of North America and on the Pacific Coast. *Symphytum tuberosum* is reported to be adventive in New England.

■ not a livestock problem

in humans, possibility of liver disease with long-term excessive herbal use

Disease Problems—*Symphytum officinale* and *S.* x*uplandicum* are widely revered for medicinal use, and *S. officinale* is sometimes cultivated as a garden herb. These species are described as having nearly universal healing powers and are considered by some individuals to be "wonder medicaments." They are commonly used in the form of teas, poultices, and extracts (Tyler 1993). *Symphytum asperum* and especially *S.* x*uplandicum* are reported to be useful forages in their countries of origin, but clearly, as is reflected in their chemical constituents, they are hazardous in some situations.

■ the risk is moderate; extensive array of open-diester PAs; variable PA concentrations in herbal products

Symphytum asperum

Disease Genesis—Species of *Symphytum* contain a variety of open diesters of heliotridine and retronecine and are of moderate to high risk. In addition to the monoester echinatine, *S. asperum* contains asperumine, acetylechinatine, heliosupine, and acetylheliosupine, all open diesters of heliotridine. *Symphytum officinale* also contains a series of open diesters, including echimidine (the most toxic), heliosupine, acetylheliosupine, lasiocarpine, and symphytine (Peterson 1985). Alkaloid concentrations are highest in the roots, at 1,000–8,000 ppm; concentrations are less than 100 ppm in the leaves under some conditions (Couet et al. 1996). *Symphytum* x*uplandicum* contains a similarly impressive array of open diesters of retronecine, including echimidine, acetylintermedine, lycopsamine, acetyllycopsamine, uplandicine, and symlandine and its isomer symphytine, as well as a few monoesters.

Heliosupine

Symphytine

Although these species are toxic, the risk is low for livestock because of the low likelihood of animal access to significant amounts of the plants. However, the risk for humans due accidental ingestion is substantial (Ridker et al. 1985). In a survey of commercial comfrey products from health food stores, Betz et al. (1994) found up to 400 ppm pyrrolizidine alkaloids. Couet et al. (1996) reported PA concentrations up to 5,200 ppm for tablets of comfrey, clearly a health risk.

- PAs cause hepatic veno-occlusive disease in humans; 4 stages, not necessarily progressive—
  (1) digestive disturbance, nausea, colic, vomiting
  (2) enlarged tender liver
  (3) abdominal fluid accumulation
  (4) sedation or coma

The alkaloids *S. officinale* are genotoxic when present as a group but are not necessarily genotoxic individually (Couet et al. 1996). Other pharmacologic effects are reported for PAs of *Symphytum*, but they appear to be of little consequence toxicologically (cf. *Heliotropium*).

Clinical Signs, Pathology, and Treatment—Intoxication due to species of *Symphytum* is not likely to be a clinical problem, but if it occurs, the clinical signs and pathologic changes will be the same as those produced by *Cynoglossum* in livestock and *Heliotropium* in humans. Likewise, treatment will be the same.

Genera of Questionable Toxicity

Borago L.

Comprising 3 species native to the Old World, *Borago* is represented in North America by *B. officinalis* L., commonly known as borage or tailwort. Long used in herbal medicines as a demulcent and diuretic, plants are rich in potassium and calcium and are cooked as a potherb or sometimes put in salads. It is an herbaceous, taprooted, hispid-setose annual 20–60 cm tall, with petiolate, broadly elliptic or ovate to oblanceolate lower leaves and smaller, sessile, clasping upper leaves. Its blue flowers are borne in loose, terminal, few-flowered cymes and produce 4 obovoid, rugose-tuberculate nutlets.

- *B. officinalis*

  borage
  tailwort

- annual herbs; upper leaves clasping; petals blue; nutlets rugose-tuberculate

Borago officinalis

Introduced by the Spanish in the 1500s, *B. officinalis* has naturalized sparingly in the West and more abundantly in the northeastern part of the continent (Nuez & Hernandez 1994). Although it has not yet proven to be a toxicologic problem, it contains a mixture of the monoesters intermedine, lycopsamine, and their diester acetylated derivatives, in addition to amabiline, a monoester lacking the C-7 hydroxyl

- naturalized sparingly; potentially a low risk due to monoester PAs

(Dodson & Stermitz 1986). The hazard is low, and *B. officinalis* is not likely to cause clinical problems, but if so, the clinical signs and pathologic changes will be the same as those produced by *Cynoglossum* in livestock and *Heliotropium* in humans. Likewise, treatment will be the same.

Hackelia Opiz

- *H. californica*
  H. floribunda
  forget-me-not
  stickseed
- higher elevations
- not a problem as yet in livestock; possible risk from open-diester PAs

A genus of about 45 species widespread in both the New World and the Old World, *Hackelia* exhibits greatest diversity in the western United States. Plants are taprooted perennials or biennials with stems typically arising from much-branched caudices and bearing both basal and cauline leaves. The small blue or white flowers are borne in sympodial cymes (false racemes) and bear glochidiate-prickly nutlets. Species of toxicologic interest are *H. californica* (A.Gray) I.M.Johnst., commonly known as California stickseed, and *H. floribunda* (Lehm.) I.M.Johnst., commonly known as many-flowered stickseed, showy stickseed, or biennial forget-me-not. Both species occur in thickets, meadows, and stream banks at higher elevations in western mountains.

Hackelia californica

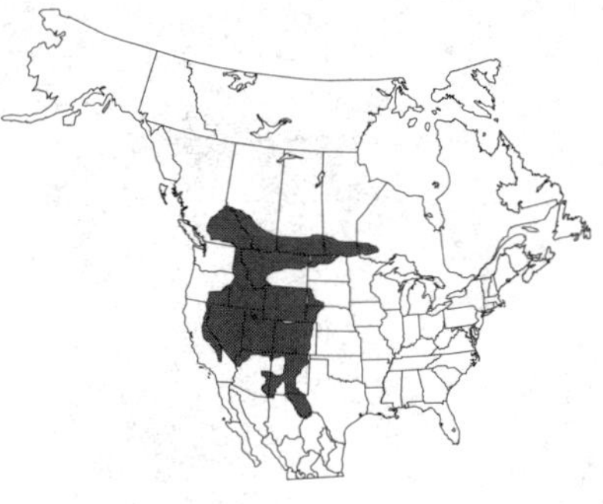

Hackelia floribunda

Species of *Hackelia* have a reputation for toxicity among the Navajo (Hagglund et al. 1985) but have not been associated with adverse effects in livestock. However, because of the presence of toxic alkaloids in them, they should be considered to be toxic under appropriate circumstances. The potentially toxic PAs in these species are open diesters and monoesters of retronecine, present primarily (90%) as *N*-oxides. *Hackelia floribunda* contains the open diester latifoline (0.4% d.w. in buds and flowers); *H. californica* contains two diester analogs of latifoline and serves as a host plant for a moth that sequesters PAs as a protective mechanism (Hagglund et al. 1985; Godfrey & Crabtree 1986; L'Empereur et al. 1989). If intoxications occur, the clinical signs and pathologic changes would be similar to those produced by *Cynoglossum*.

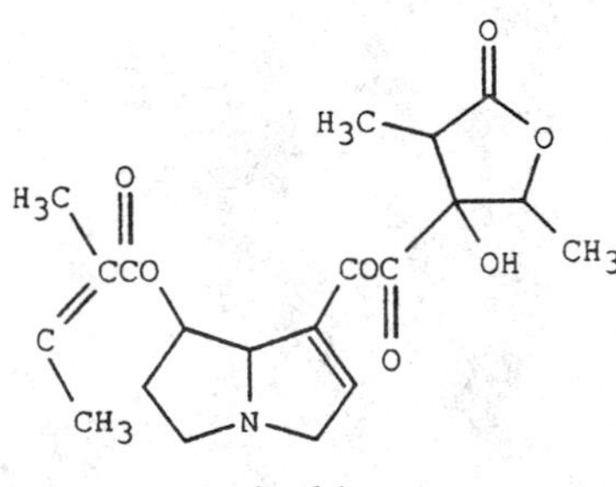

Latifoline

Mertensia Roth

- *M. ciliata*
  M. lanceolata
  bluebell
- erect perennial herbs; petals blue, white, or pink

Commonly known as bluebells or lungworts, species of *Mertensia* are native to both North America and temperate Eurasia. There are 35–40 species in the genus, of which 20–24 are present in North America, mostly in the West. All are perennial herbs arising from woody, branched caudices. The glabrous to strigose or hirsute leaves, both petiolate basal and sessile cauline, arise from erect stems 40–150 cm tall that terminate in small bractless cymes, which may also arise from the axils of the uppermost leaves. The blue or rarely white or pink flowers are cylindric or funnelform and produce 4 rugose nutlets.

Of possible toxicologic interest are *M. ciliata* (James ex Torr.) G.Don, commonly known as streamside bluebell, and *M. lanceolata* (Pursh) DC., or mountain bluebell,

Mertensia ciliata

Mertensia ciliata

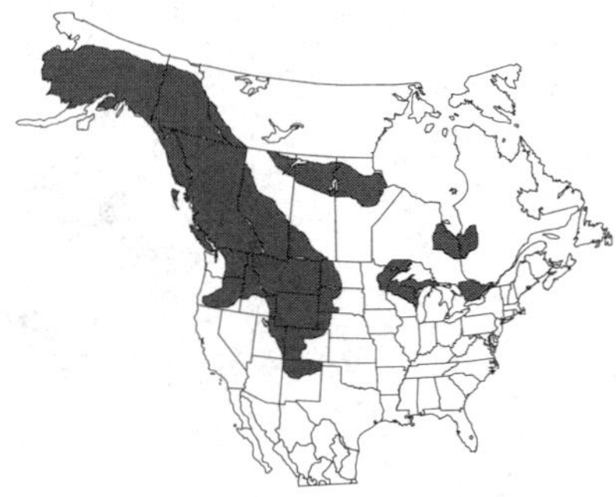

Mertensia lanceolata

which now includes 5 previously recognized species. These 2 taxa contain only monoesters of retronecine, intermedine, and/or lycopsamine as *N*-oxides and thus are probably of little hazard (Li & Stermitz 1988); concentrations of 0.3% for *M. ciliata* and 1.7% for *M. lanceolata* (cited as *M. bakeri*) are reported. Species of *Mertensia* serve as host plants for moths that sequester PAs as a defensive mechanism (Godfrey & Crabtree 1986).

■ low risk due to monoesters PAs

Myosotis L.

Comprising about 50 species in temperate and boreal regions, *Myosotis* (forget-me-not, scorpion grass) is represented in North America by approximately 12 native and naturalized species. Plants of the genus are annual or perennial, glabrous to hirsute herbs with alternate, entire, mostly cauline leaves borne on slender stems 10–60 cm tall and terminating in bractless helicoid cymes. With blue or less often pink or white petals, the flowers produce 4 smooth, shiny nutlets, each with a distinctive raised encircling margin.

■ *M. scorpioides*
M. sylvatica

forget-me-not

■ erect herbs; petals blue, pink, or white

Myosotis scorpioides

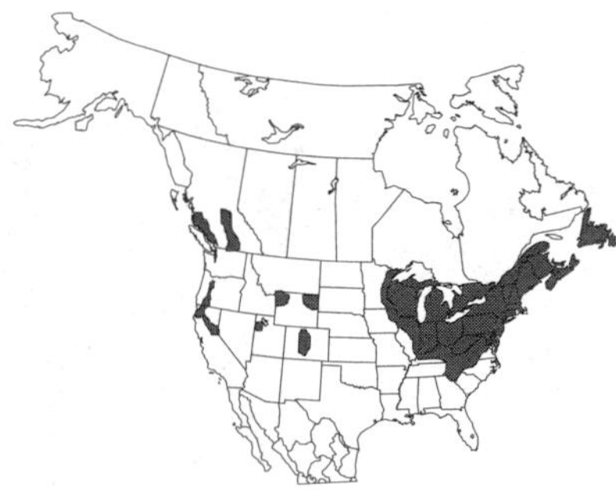

Myosotis scorpioides

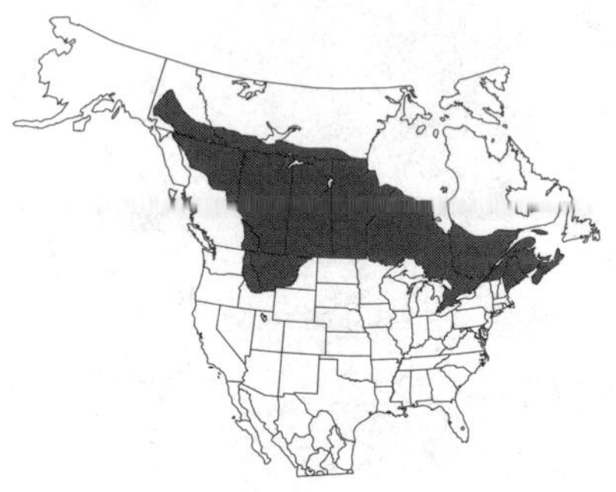

Myosotis sylvatica

- potentially moderate risk due to open-diester PAs

Introduced from Eurasia as garden ornamentals and now naturalized across the continent, *M. scorpioides* L. (common forget-me-not) and *M. sylvatica* Ehrh. ex Hoffm. (garden forget-me-not, wood forget-me-not) have not been implicated in disease problems, but the potential exists because of the presence of mono- and open diesters of retronecine. *Myosotis scorpioides* contains the open diesters symphytine, acetylscorpioidine, and myoscorpine and the monoester scorpioidine. *Myosotis sylvatica* contains the two diesters heliosupine and acetylheliosupine and a retronecine monoester. Species of *Myosotis* serve as host plants for moths that sequester PAs as a defensive mechanism (Godfrey & Crabtree 1986). In the event that intoxications occur, the clinical signs and pathology would be similar to those produced by *Cynoglossum*.

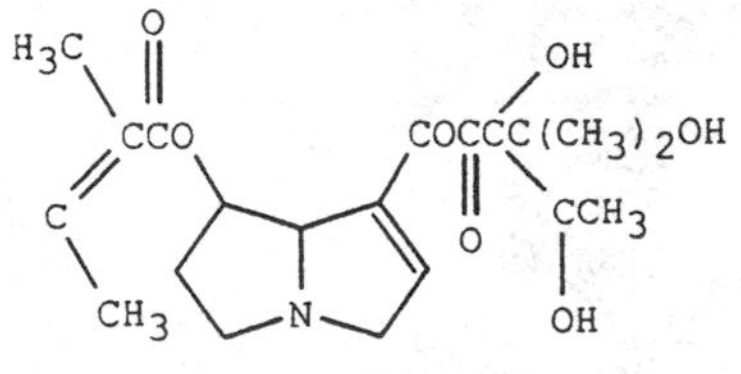

Heliosupine

References

Baker DC, Smart RA, Ralphs M, Molyneux RJ: Hound's-tongue *(Cynoglossum officinale)* poisoning in a calf. *J Am Vet Med Assoc* 194;929–930, 1989

Baker DC, Pfister JA, Molyneux RJ, Kechele P: *Cynoglossum officinale* toxicity in calves. *J Comp Pathol* 104;403–410, 1991

Betz JM, Eppley RM, Taylor WC, Andrzejewski D: Determination of pyrrolizidine alkaloids in commercial comfrey products (*Symphytum* sp.). *J Pharm Sci* 83;649–653, 1994

Bull LB, Dick AT, Keast JC, Edgar G: An experimental investigation of the hepatotoxic and other effects on sheep of consumption of *Heliotropium europaeum* L.: heliotrope poisoning of sheep. *Aust J Agric Res* 7;281–332, 1956

Bull LB, Rogers ES, Keast JC, Dick AT: *Heliotropium* poisoning in cattle. *Aust Vet J* 37;37–40, 1961

Bull LB, Culvenor CCJ, Dick AT: *The Pyrrolizidine Alkaloids: Their Chemistry, Pathogenicity, and Other Biological Properties.* North-Holland Publ, Amsterdam, 1968

Chauvin P, Dillon JC, Moren A: An outbreak of heliotrope food poisoning, Tadjikistan, November 1992/NMarch 1993. *Sante* 4;263–268, 1994

Couet CE, Crews C, Hanley AB: Analysis, separation, and bioassay of pyrrolizidine alkaloids from comfrey *(Symphytum officinale). Nat Toxins* 4;163–167, 1996

Cronquist A: *An Integrated System of Classification of Flowering Plants.* Columbia University Press, New York, 1981

Culvenor CCJ, Jago MV, Peterson JE, Smith LW, Payne AL, Campbell DG, Edgar JA, Frahn JL: Toxicity of *Echium plantagineum* (Paterson's curse). 1. Marginal toxic effects in Merino wethers from long-term feeding. *Aust J Agric Res* 35;293–304, 1984

Culvenor CCJ, Edgar JA, Smith LW, Kumana CR, Lin HJ: *Heliotropium lasiocarpum* Fisch. and Mey. identified as cause of veno-occlusive disease due to a herbal tea [letter]. *Lancet* 1;978, 1986

DeClerck-Floate R: Cattle as dispersers of hound's-tongue on rangelands in southeastern British Columbia. *J Range Manage* 50;239–243, 1997

Dodson GD, Stermitz FR: Pyrrolizidine alkaloids from borage *(Borago officinalis)* seeds and flowers. *J Nat Prod* 49;727– 728, 1986

Gallagher CH: The effects of neuromuscular blocking agents on mitochondria. 4. Effects of *d*-tubocurarine, pyrrolizidine alkaloids, and magnesium on oxidative phosphorylation. *Biochem Pharmacol* 17;533–538, 1968
Giesecke PR: Serum biochemistry in horses with *Echium* poisoning. *Austr Vet J* 63;90–91, 1986
Godfrey GL, Crabtree L: Natural history of *Gnophaela latipennis* (Boisduval) (Arctiidae: Pericopinae) in northern California. *J Lepidopterists' Soc* 40;206–213, 1986
Greatorex JC: Some unusual cases of plant poisoning in animals. *Vet Rec* 78;725–727, 1966
Green CR, Christie GS: Malformations in foetal rats induced by the pyrrolizidine alkaloid, heliotridine. *Br J Exp Pathol* 42;369–378, 1961
Hagglund KM, L'Empereur KM, Roby MR, Stermitz FR: Latifoline and latifoline-*N*-oxide from *Hackelia floribunda,* the western false forget-me-not. *J Nat Prod* 48;638–639, 1985
Harper PAW, Walker KH, Krahenbuhl RE, Christie BM: Pyrrolizidine alkaloid poisoning in calves due to contamination of straw by *Heliotropium europaeum. Aust Vet J* 62;382–383, 1985
Hill BD, Gaul KL, Noble JW: Poisoning of feedlot cattle by seeds of *Heliotropium europaeum. Aust Vet J* 75;360–361, 1997
Johnson AE, Molyneux RJ: Variation in toxic pyrrolizidine alkaloid content of plants, associated with site, stage of growth, and environmental conditions. In *Plant Toxicology—Proceedings of the Australia-USA Poisonous Plants Symposium,* Seawright AA, Hegarty MP, James LF, Keeler RF eds, Animal Research Institute, Yeerongpilly, pp209–218, 1985
Kalkus JW, Tripper HA, Fuller JR: Enzootic hepatic cirrhosis of horses (walking disease) in the Pacific Northwest. *J Am Vet Med Assoc* 68;285–298, 1925
Kelley RB, Seiber JN: Pyrrolizidine alkaloids from *Amsinckia gloriosa.* In *Poisonous Plants: Proceedings of the Third International Symposium,* James LF, Keeler RF, Bailey EM Jr, Cheeke PR, Hegarty MP eds, Iowa State Press, Ames, Iowa, pp534–538, 1992
Kennedy PC: Tarweed poisoning in swine: Symposium on Poisoning. *J Am Vet Med Assoc* 130;305–306, 1957
Kim HY, Stermitz FR, Molyneux RJ, Wilson DW, Taylor D, Coulombe RA Jr: Structural influences on pyrrolizidine alkaloid-induced cytopathology. *Toxicol Appl Pharmacol* 122;61–69, 1993
Knight AP, Kimberling CV, Stermitz FR, Roby MR: *Cynoglossum officinale* (hound's-tongue)—a cause of pyrrolizidine alkaloid poisoning in horses. *J Am Vet Med Assoc* 185;647–650, 1984
L'Empereur KM, Li Y, Stermitz FR: Pyrrolizidine alkaloids from *Hackelia californica* and *Gnophaela latipennis,* an *H. californica*/Nhosted arctiid moth. *J Nat Prod* 52;360–366, 1989
Lanigan GW, Payne AL, Peterson JE: Antimethanogenic drugs and *Heliotropium europaeum* poisoning in penned sheep. *Aust J Agric Res* 29;1281–1292, 1978
Li Y, Stermitz FR: Pyrrolizidine alkaloids from *Mertensia* species of Colorado. *J Nat Prod* 51;1289–1290, 1988
Mandryka II: *Cynoglossum officinale* (hounds tongue) as a poisonous plant. *Veterinaria (Moscow)* 9;69–70, 1979 (*Vet Bull* 50;2343, 1980)
Mansurov KK: New data on heliotropic dystrophy of the liver. *Klin Med (Moscow)* 73;15–18 1995
Mattocks AR: *Chemistry and Toxicity of Pyrrolizidine Alkaloids.* Academic Press, London, 1986
Mattocks AR, Jukes R: Detection of sulphur-conjugated pyrrolic metabolites in blood and fresh or fixed liver tissue from rats given a variety of toxic pyrrolizidine alkaloids. *Toxicol Lett* 63;47–55, 1992
McCulloch EC: Hepatic cirrhosis of horses, swine, and cattle due to the seeds of the tarweed, *Amsinckia intermedia. J Am Vet Med Assoc* 96;5–18, 1940

Mendel VE, Witt MR, Gitchell BS, Gribble DN, Rogers QR, Segall HJ, Knight HD: Pyrrolizidine alkaloid-induced liver disease in horses: an early diagnosis. *Am J Vet Res* 49;572–578, 1988

Mohabbat O, Younos MS, Merzad AA, Srivastava RN, Sediq GG, Aram GN: An outbreak of hepatic veno occlusive disease in north western Afghanistan. *Lancet* 2;269–270, 1976

Muth OH: An attempt to determine the toxicity of *Amsinckia intermedia* (tarweed) for fattening lambs. *J Am Vet Med Assoc* 99;145–146, 1941

Nuez F, Hernandez Bermejo JE: Neglected horticultural crops. In *Neglected Crops: 1492 from a Different Perspective,* Hernandez Bermejo JE ed, Food and Agriculture Organization, United Nations, Rome, pp303–332, 1994

O'Dowd DJ, Edgar JA: Seasonal dynamics in the pyrrolizidine alkaloids of *Heliotropium europaeum. Aust J Ecol* 14;95–105, 1989

Pass DA, Hogg GG, Russell RG, Edgar JA, Tence IM, Rikard-Bell L: Poisoning of chickens and ducks by pyrrolizidine alkaloids of *Heliotropium europaeum. Aust Vet J* 55;284–288, 1979

Peterson JE: The toxicity of *Echium plantagineum* (Paterson's curse). In *Plant Toxicology—Proceedings of the Australia–USA Poisonous Plants Symposium,* Seawright AA, Hegarty MP, James LF, Keeler RF eds, Animal Research Institute, Yeerongpilly, pp191–199, 1985

Peterson JE, Culvenor CC: Hepatotoxic pyrrolizidine alkaloids. In *Handbook of Natural Toxins,* vol 1, Keeler R, Tu AT eds, Marcel Dekker, New York, pp637–671, 1983

Peterson JE, Jago MV: Comparison of the toxic effects of dehydroheliotridine and heliotrine in pregnant rats and their embryos. *J Pathol* 131;339–355, 1980

Peterson JE, Payne AL, Culvenor CCJ: *Heliotropium europaeum* poisoning of sheep with low liver copper concentrations and the preventive efficacy of cobalt and antimethanogen. *Aust Vet J* 69;51–56, 1992

Pfister JA, Molyneux RJ, Baker DC: Pyrrolizidine alkaloid content of houndstongue (*Cynoglossum officinale* L.). *J Range Manage* 45;254–256, 1992

Ridker PM, Ohkuma S, McDermott WV, Trey C, Huxtable RJ: Hepatic venoocclusive disease associated with the consumption of pyrrolizidine-containing dietary supplements. *Gastroenterology* 88;1050–1054, 1985

Roitman JN: The pyrrolizidine alkaloids of *Amsinckia menziesii. Aust J Chem* 36;769–768, 1983

St George-Grambauer TD, Rac R: Hepatogenous chronic copper poisoning in sheep in south Australia due to the consumption of *Echium plantagineum* L. (salvation jane). *Aust Vet J* 38;288–293, 1962

Seaman JT: Pyrrolizidine alkaloid poisoning of horses. *Aust Vet J* 54;150, 1978

Seaman JT: Pyrrolizidine alkaloid poisoning of sheep in New South Wales. *Aust Vet J* 64;164–167, 1987

Seaman JT, Dixon RJ: Investigations into the toxicity of *Echium plantagineum* in sheep. 2. Pen feeding experiments. *Aust Vet J* 66;286–292, 1989

Seaman JT, Walker KH: Pyrrolizidine alkaloid poisoning of cattle and horses in New South Wales. In *Plant Toxicology—Proceedings of the Australia-USA Poisonous Plants Symposium,,* Seawright AA, Hegarty MP, James LF, Keeler RF eds, Animal Research Institute, Yeerongpilly, pp235–246, 1985

Seaman JT, Turvey WS, Ottaway SJ, Dixon RJ, Gilmour AR: Investigations into the toxicity of *Echium plantagineum* in sheep. 1. Field grazing experiments. *Aust Vet J* 66;279–285, 1989

Sharrock AG: Pyrrolizidine alkaloid poisoning in a horse in New South Wales. *Aust Vet J* 45;388, 1969

Shull LR, Buckmaster GW, Cheeke PR: Factors influencing pyrrolizidine *(Senecio)* alkaloid metabolism: species, liver sulfhydryls, and rumen fermentation. *J Anim Sci* 43;1247–1253, 1976

Stegelmeier BL, Gardner DR, Molyneux RJ, Pfister JA, James LF: The clinicopathologic changes of *Cynoglossum officinale* (houndstongue) intoxication in horses. In *Plant-Associated Toxins: Agricultural, Phytochemical, and Ecological Aspects,* Colegate SM, Dorling PR eds, CAB International, Wallingford, UK, pp297–302, 1994

Stegelmeier BL, Gardner DR, James LF, Molyneux RJ: Pyrrole detection and the pathological progression of *Cynoglossum officinale* (houndstongue) poisoning in horses. *J Vet Diagn Invest* 8;81–90, 1996

Tyler VE: *Honest Herbal: A Sensible Guide to the Use of Herbs and Related Remedies,* 3d ed. Pharmaceutical Products Press, New York, 1993

Upadhyaya MK, Cranston RS: Distribution, biology, and control of hound's-tongue in British Columbia. *Rangelands* 13;103–106, 1991

van Dam NM, Verpoorte R, van der Meijden E: Extreme differences in pyrrolizidine alkaloid levels between leaves of *Cynoglossum officinale. Phytochemistry* 37;1013–1016, 1994

Wahome WM, Muchiri DJ, Mugera GM: An acute toxicity study of *Heliotropium scottae* Rendle in mice. *Vet Hum Toxicol* 36;295–297, 1994

Woolsey JH, Jasper DE, Cordy DR, Christensen JF: Two outbreaks of hepatic cirrhosis in swine in California, with evidence incriminating the tarweed, *Amsinckia intermedia. Vet Med* 47;55–58, 1952

16 Brassicaceae Burnett

- Mustard Family
  Crucifer Family

  Berteroa
  Brassica
  Descurainia
  Erysimum
  Sinapis
  Thlaspi

  QUESTIONABLE
  Crambe
  Raphanus
  Stanleya

A large, well-defined family of major economic importance, the Brassicaceae comprises 350–400 genera and 3,000–3,350 species. An equally acceptable, older, alternative name is Cruciferae. Members of the family are commonly called mustards or crucifers and are distributed primarily in the temperate regions of the Northern and Southern Hemispheres. Very few taxa occur in the tropics. The greatest concentration of genera and species is in central Asia and the Mediterranean region. In North America, 99 genera, 778 species, and 248 subspecies and varieties are present (Rollins 1993). Of these, 39 genera are endemic and 677 species are native. Greatest abundance of the native taxa is in the western third of the continent.

Many species are cultivated for food, for animal feed, and as ornamentals. Others are noxious weeds. For humans, members of the Brassicaceae are more important as vegetables, condiments, and garnishes than as dietary staples, as is characteristic of members of the Poaceae (the grasses) and the Fabaceae (the legumes). Cabbage, broccoli, cauliflower, turnips, mustard, horseradish, watercress, and radishes are cruciferous crops. For livestock, crucifers provide forage in the field, silage, seed meal, and root/stem fodder. In some European countries up to 30% of the vegetable acreage is dedicated to growing members of the family. With about 50 genera in horticultural trade, familiar ornamentals include sweet alyssum, wallflower, candytuft, dame's rocket, and honesty. Unfortunately, weedy species are present in the family as well. Especially troublesome are the mustards, tansy mustards, peppergrasses, false flaxes, and shepherd's purses. Toxicologic problems are associated with several genera, which are discussed collectively and individually below.

- herbs; flowers crosslike; petals 4 or 0, yellow, white, lavender, or purple; fruits siliques or silicles

Plants herbs; annuals or biennials or perennials; caulescent or acaulescent; strongly aromatic or not aromatic. **Leaves** simple or 1-pinnately compound or rarely palmately compound; alternate or rarely opposite; basal or forming a basal rosette or cauline; petiolate or sessile; clasping or not clasping; venation pinnate; margins entire or dentate or incised or pinnately lobed or pinnatifid; stipules absent. **Inflorescences** racemes; terminal or axillary. **Flowers** perfect; perianths in 2- or 1-series. **Sepals** 4; free. **Corollas** radially or rarely bilaterally symmetrical. **Petals** 4 or 0; persistent or caducous; free; clawed or not clawed; of various colors, but commonly yellow or white or lavender or purple. **Stamens** 6 or 4 or 2; tetradynamous or of equal length. **Pistils** 1; compound, carpels 2; stigmas 1, often 2-lobed; styles 1 or 0; ovaries superior; locules 2; placentation parietal. **Fruits** siliques or silicles; dehiscent or indehiscent. **Seeds** 2 to numerous.

Quick identification of members of the Brassicaceae is facilitated by the cruciform (cross) arrangement of the four petals—hence the older family name; the six stamens, four long and two short; and distinctive silique or silicle, which normally dehisce, leaving a central, thin partition.

Disease Problems—A wide array of disease syndromes has been associated with the Brassicaceae. Some are of a general nature and not unique to the family or to the toxicants present, such as acute respiratory distress syndrome (ARDS), nitrate accumulation, photosensitization, digestive disturbance and bloat, and polioencephalomalacia. Other diseases somewhat specific for the family are thyroid enlargement/goiter, antinutritional effects, liver hemorrhage, blindness, hemolytic anemia, reproductive problems, and tainting of eggs, milk, and meat. Some of these diseases are almost exclusively caused by the genus *Brassica* and its many varieties because large amounts are eaten and because of the exclusiveness of the toxicants present. Because they are mainly cool-season crops, the brassicas often compose nearly the entire diet during periods when most other forages are dormant. Thus the toxicants are likely to be ingested in large quantities. Many of the diseases characteristic of the family are discussed in the treatment of *Brassica.*

- valued for food

several disease problems associated with family—
ARDS
digestive
nitrate
photosensitization
PEM
thyroid
antinutritional
liver
blindness
anemia
reproductive
milk taint

Other problems caused by members of the family include tainting of eggs, milk, or meat and photosensitization, both of which may occur alone or together with the irritant effects. The occurrence of peculiar flavors or taint is mostly correlated with feeding the seed products but also has been reported for various forages, including those of *Coronopus, Lepidium,* and *Thlaspi* (Park 1969; Park et al. 1969; Fenwick et al. 1989). The mustards are well known as anticarcinogens, but remarkably they also exert some neoplastic promoter activity (Stoewsand 1995).

- uncommon problems, some mainly *Brassica* others nonspecific

Common to many genera in the family is irritation of the digestive tract. It is most often seen with inadvertent consumption of wild species. Such situations sometimes occur because the wild species often predominate when stressful growing conditions, such as lack of moisture, prevail. In most instances where death is the outcome of intoxication, seeds are involved. Severe problems have occurred in cattle when seeds of various species were dumped, intentionally or otherwise, in pastures (Mason & Lucas 1983; Kernaleguen et al. 1989; Semalulu & Rousseaux 1989). Most members of the Brassicaceae are probably capable of causing severe digestive disturbances, but *Sinapis arvense* is one of the more common causes (Hughes 1924; Eaton 1941; Holmes 1965). Others that are of concern as possible digestive tract irritants are listed in table 16.1. Digestive tract problems may be accompanied by signs of neurologic involvement, as has been described in cases involving *Barbarea vulgaris* (Hansen 1930).

Table 16.1. Genera and species of the Brassicaceae that may cause digestive tract irritation

Armoracia rusticana	horseradish
Barbarea vulgaris	yellow rocket, winter cress
Camelina sativa	false flax
Capsella bursa-pastoris	shepherd's purse
Cardaria draba	pepperwort, hoary cress
Lepidium virginicum	large peppergrass
Lesquerella spp.	bladderpod
Sisymbrium altissimum	tumbling mustard
S. officinale	common hedge mustard

Disease Genesis—Two types of sulfur-containing toxicants are found in the Brassicaceae—glucosinolates and *S*-methylcysteine sulfoxide (SMCO) (Stoewsand 1995). The glucosinolates, formerly known as thioglucosides or mustard oil glucosides, are found in all members (Fenwick et al. 1989). Three of the parent compounds are sinigrin or prop-2-enyl glucosinolate; gluconapin or but-3-enyl glucosinolate; and glucosinalbin or 3-indolylmethyl glucosinolate. The trivial names are a derivation often employing the prefix *gluco-* in combination with the genus name of the species from which the compound was originally isolated (Fenwick et al. 1983). There are numerous parent glucosinolates of the same basic structure. They are hydrolyzed by myrosinase, an enzyme in the plant tissue or in the ruminal microenvironment (Majak

- 2 general toxin types—sulfur-containing (SMCO), mustard oil glucoside glucosinolates

of the glucosinolates, 3 types of parent compounds—sinigrin, gluconapin, glucosinalbin

these yield 3 classes of hydrolysis products—isothiocyanate, oxazolidinethione, thiocyanate et al. 1991). In the short term, ruminal pH may favor increased hydrolysis to produce some types of irritant glucosinolates (Shires et al. 1982; Smith & Crowe 1987). The three main classes of hydrolysis products are isothiocyanates (mustard oils), oxazolidinethiones, and thiocyanates (Virtanen 1965; Fenwick et al. 1989). The side chain on the parent compound determines the potency of the product of enzyme hydrolysis; these products are largely responsible for the pungency and flavor of many condiments. Typically, the glucosinolate is hydrolyzed to an allylisothiocyanate that imparts the mustardlike odor. In contrast, in a few species, the hydrolysis product is an allylthiocyanate that imparts a garliclike odor to the plant (Gmelin & Virtanen 1959; Majak et al. 1991).

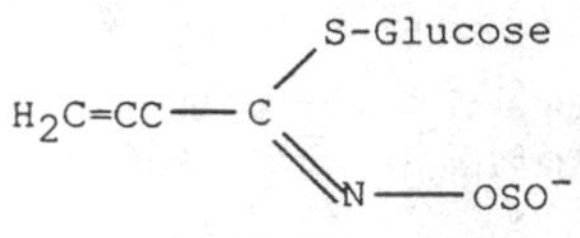

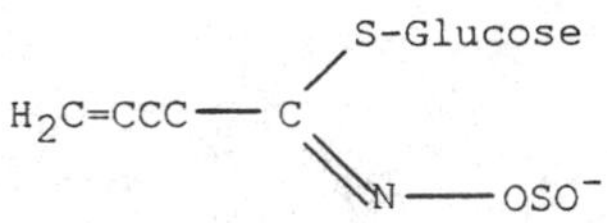

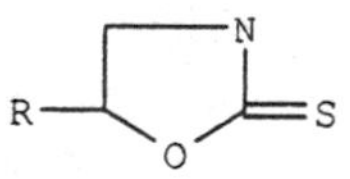

The indoles of the glucosinolates are mainly nitriles rather than carbinols and are thought by some to play a major role in modifying biotransformation enzyme activity and in other beneficial biological effects (Pantuck 1976; Wall et al. 1988).

The glucosinolates in low concentrations are appetite stimulants, but at higher levels are potent irritants, which, when applied to foods, make them very hot to the taste, as illustrated by the effects of the condiments white and black pepper. Because of their potent effects, the mustards have long been used as topical irritants in humans or to make blisters for treatment of various locomotor ailments in horses. Digestive disturbances due to glucosinolates may be caused by either cultivated species or wild species present in the pasture or range.

Concentrations of total glucosinolates are usually low to moderate in foliage, high in some roots, and quite high in seeds (Fenwick et al. 1989). Concentrations are typically about 1,000 ppm for foliage, 30,000+ ppm for horseradish root, 40,000 ppm for rapeseed, and 20,000–60,000 ppm for mustard and pepper seeds. Brussels sprouts have unusually high levels of glucosinolates, sometimes 2,000–3,000 ppm. Concentrations are markedly reduced by ensiling, e.g., decreasing 10-fold in rape, from 0.3% to 0.03%, in 30 days (Fales et al. 1987). Although glucosinolates are present in all members of the family, some genera and species seem to be more likely to be involved in intoxications. They are treated individually here.

Neither the glucosinolates nor their products are passed in milk in appreciable amounts (VanEtten & Tookey 1978; Papas et al. 1979). Although not passed in milk in quantities sufficient to cause thyroid effects, they may nonetheless be important in causing "off" tastes in milk. Some of the glucosinolate products have antibacterial and antifungal activity (Virtanen 1965).

The glucosinolates are probably responsible for several disease problems, both experimental and clinical, in addition to irritation of the digestive tract. Stoewsand (1995) listed thyroid enlargement, embryonal death, decreased fetal weights, adrenal enlargement, and liver, pancreatic, and renal lesions as consequences of experimental administration of glucosinolates to rats. Of the clinically significant problems, growth retardation, liver hemorrhage, and egg taint are seen in poultry given feeds containing *Brassica* seeds. Digestive disturbances also occur with inadvertent ingestion of wild plants or seeds with high levels of glucosinolates. Thyroid enlargement is of importance only in those instances in which livestock graze cruciferous crops or eat various seed products/meals for prolonged periods, because a long period of ingestion is required to produce the adverse effect. Ruminants appear to be less susceptible to the long-term detrimental effects of glucosinolates than are pigs and other monogastrics, based upon results of feeding rapeseed and crambe meals (Lambert et al. 1970; Bell 1984). In most instances thyroid problems are of little significance. As an example, *Rapistrum rugosum* (turnipweed), fed in the preflowering stage to sheep for 11 weeks, failed to cause any increase in thyroid size (Stocks et al. 1984).

Effects on the thyroid range from non-iodine responsive block of tyrosine iodina-

S-Glucose
$H_2C{=}CC$—C
N—OSO^-
Sinigrin

S-Glucose
$H_2C{=}CCC$—C
N—OSO^-
Gluconapin

S-Glucose
C—C
N
N—OSO^-
Glucobrassicin

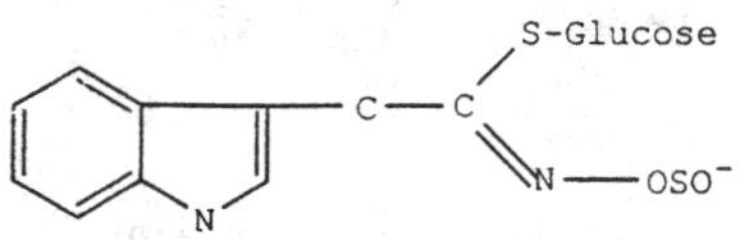

■ glucosinolates, potent, pepperlike irritants, highest in seeds, mustard and pepper seeds; also high in horseradish root and leaves of brussels sprouts; low levels in milk, taint

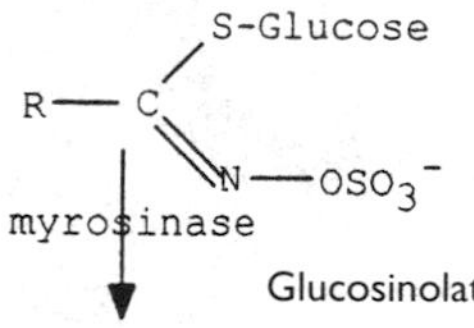

S-Glucose
R—C
N—OSO_3^-
myrosinase
Glucosinolate

R—N=C=S
Isothiocyanate

or

R, N, O, S
Oxazolidinethione

or

R—S—C≡N
Thiocyanate

■ glucosinolates, digestive tract irritant; poultry, decreased growth, liver problems; livestock, thyroid inhibition, rare in North America because long-term ingestion is required

■ high sulfur in plants may limit copper absorption

tion and inhibition of T4 formation—due to goitrin, the breakdown product of thiooxazolidine—to simple iodine responsive inhibition of thyroid iodine uptake due to thiocyanate formed from the indole glucobrassicin (Virtanen 1965; Paxman & Hill 1974b). Goitrogenesis in kale varieties is mainly due to thiocyanate (Paxman & Hill 1974a). In many instances the effects on the thyroid are not an impediment to weight gains, but under some circumstances neonate survivability may markedly decrease (Sinclair & Andrews 1961; Russel 1967). Under some dietary conditions, especially with marginal iodine levels in the diet, there may be an augmented effect on the thyroid (David 1976). Although impairment of iodine availability may not reduce growth, iodine supplementation may improve wool growth (Barry et al. 1983).

An additional effect of considerable importance with the high content of sulfur-containing compounds in plants of the family is the potential for long-term ingestion to limit copper absorption and to predispose to deficiency of the mineral. This factor is discussed specifically in the treatment of polioencephalomalacia (PEM) due to *Brassica*.

Perhaps the most important products associated with the glucosinolates and their adverse effects are rapeseed seed, oil, and meal derived from *Brassica napus, B. campestris, B. juncea, B. sarson,* and other species (Hill 1979; Campbell 1984). The discovery of cultivars of rapeseed low in glucosinolates and their subsequent genetic manipulation to reduce erucic acid levels led to the development of the now widely acclaimed canola oil and meal products (Bell 1982, 1984). The oil has less than 5% erucic acid and less than 3 mg/g of common glucosinolates, and the meal about 38% protein (Campbell 1984). Similar low glucosinolate varieties are available in other countries (Lajolo et al. 1991). It is desirable to have minimal amounts of erucic acid because experimental evidence links it with cardiomyopathy in rats (McCutcheon et al. 1976). Soaking mustard oil cake in water may be useful in reducing the potential for adverse effects; when it was soaked for 2–6 hours, signs of glucosinolate activity were reduced by 36% (Tyagi et al. 1997).

- SMCO, metabolic products, oxidant effects, hemolysis

$H_3C-S(=O)-C(NH_2)COOH$ (SMCO) —lyase→ $H_3C-S-OH$ or $H_3C-S(=O)-S-CH_3$ or others

The other sulfur-containing toxicant, *S*-methylcysteine sulfoxide (SMCO) is metabolized by lyase or in the rumen to dimethyl sulfide, which produces oxidant effects serving primarily to hemolyze RBCs in a manner similar to the oxidants of *Allium* (onions) in the Liliaceae and *Acer rubrum* (red maple) in the Aceraceae (Benevenga et al. 1989). Thus, ruminants are the only species at risk. Hemolysis is seen after consumption of the plants for several weeks and is typically not a life-threatening disease. This problem is caused by various species of *Brassica* and *Raphanus*. In contrast to the glucosinolates, ensiling results in only minor reduction of SMCO concentrations (Fales et al. 1987).

Thyroid Enlargement

Clinical Signs—The thyroids are noticeably enlarged. In severe cases there is loss of weight or decreased weight gain, and in cases of grazing throughout pregnancy, neonate mortality may increase. Changes in serum chemistries of lambs include a transient rise in serum thyroxin (T4), an increase in serum glucose, and a decrease in serum cholesterol, triglycerides, and urea.

- weight loss, poor neonate survival

Pathology and Treatment—Thyroid enlargement is associated with hypertrophy of the epithelia lining the follicles from low cuboidal follicular to tall columnar cells. The amount of colloid is reduced, and in some instances lumens of the follicles are quite small. In most instances, thyroid effects are readily attenuated or prevented by administration of or supplementation with iodine during the grazing period.

- microscopic, hypertrophy of thyroid follicular epithelium
- supplement with iodine

Digestive Disturbance

■ anorexia, salivation, colic, diarrhea

Clinical Signs—Signs indicative of irritation of the digestive tract are usually abrupt in onset and reflect abdominal pain. Horses may be up and down, kicking at their abdomens, and may lack a desire to eat. In cattle, the irritant effects may be indicated by colic, mouth kept open with the tongue out, and a marked increase in salivation. The occurrence of diarrhea, with or without blood, is variable. There may also be neurologic signs such as circling, staggering backward, recumbency, and opisthotonus.

■ gross pathology, sloughing of ruminal epithelium, reddened mucosa, stomach, small intestine

Pathology—The most obvious effects of irritation are on the stomachs and intestines: ruminal epithelial degeneration and sloughing, and reddening of the mucosae of the abomasum, small intestine, and sometimes the caecum and large intestine. Edema and hemorrhage may also be present, as well as mild renal tubular necrosis.

■ nursing care, pain relief

Treatment—Symptomatic relief of the pain and inflammation is the primary aim of treatment. In most instances the disease can be prevented or reduced by a gradual introduction to the feed or by preventing animal access to the seed products.

Berteroa DC.

■ *B. incana*
hoary alyssum
hoary false alyssum

Taxonomy and Morphology—Native to temperate Eurasia, *Berteroa* comprises 5 species. Its name honors an Italian botanist of the early 1800s. In North America, 2 species occur, albeit no collections of *B. mutabilis* have been made for decades, and it may not be present (Rollins 1993). Of toxicologic importance is a single species:

B. incana (L.) DC. hoary alyssum, hoary false alyssum

■ stiffly erect annuals; leaves simple; flowers small; petals white

Plants annuals or biennials; caulescent. **Stems** stiffly erect; 10–70 cm tall; branched above. **Leaves** simple; cauline; sessile; blades elliptic to oblanceolate; apices acute to obtuse; margins entire. **Racemes** terminal; elongating in fruit. **Flowers** small. **Sepals** equal; ascending. **Petals** white; apices deeply bifid. **Stamens** 6; tetradynamous. **Fruits** silicles; elliptic; slightly flattened, sides convex. **Seeds** 3–6 per locule; round; brown; narrowly winged.

■ waste areas, hay meadows

Distribution and Habitat—*Berteroa incana* occupies disturbed sites, waste areas, and hay meadows across northern North America.

Berteroa incana

Disease Problems and Genesis—*Berteroa incana* has been associated with severe edema of the distal limbs and less often with laminitis in horses in the upper Midwest (Becker et al. 1991; Geor et al. 1992). As yet problems have been seen only in horses eating the seeds of plants in heavily infested pastures or hay contaminated with 20–30%, or perhaps as little as 10–15%, of the species (Ellison 1992; Geor et al. 1992). Within 36 hours of ingestion, there may be increased body and foot temperature, and edema, which may disappear in 2–4 days (Murphy 1997). The morbidity rate may be higher than 50%, but most animals recover readily if exposure to the plants is stopped immediately (Hovda & Rose 1993). If animals are allowed to continue to eat the contaminated hay or pasture after signs develop, the disease may result in debilitating laminitis and or severe digestive disturbances. The foliage is typically of low palatability and is eaten reluctantly (Becker et al. 1991). In one instance mares foaled either weak or premature foals (Ellison 1992). Abortion at 9–10 months of gestation also has been reported (Hovda & Rose 1993).

The cause of the edema and other systemic signs is not known. They are not typical of those caused by glucosinolates, which are so prevalent in the family. Horses intoxicated by other taxa, e.g., seeds of *Raphanus raphanistrum,* exhibit predominantly colic and other evidences of digestive tract irritation (Eliore 1922).

- horses, rapid onset, limb edema, laminitis, transient most cases; cause unknown

Berteroa incana

Clinical Signs—The earliest sign of intoxication due to *B. incana* is increased foot temperature in 24 hours and edema a few hours later. In most instances the temperature increase will not be noticed, and the most prominent effect will be the abrupt onset of pitting edema of the distal portions of one or more limbs. In most cases this is the extent of the disease. In a few animals, when large amounts of plant material are eaten or when the hay contains 50% or more *B. incana,* other signs develop such as colic, bloody diarrhea, slow capillary refill time, fever to 40°C, increased heart and respiration rate, dehydration, and red discoloration of the urine. Within a day or two, obvious mild to severe laminitis may become apparent, the horse standing with its forelegs extended forward and its weight borne on the rear legs. In a few instances the disease may be of such severity that euthanasia may be necessary. The onset of signs may correlate with a change in hay source or pasture in the previous 24 hours.

Hematologic changes include evidence of hemolysis and dehydration; the packed cell volume may increase substantially to greater than 50% in spite of the lysis of RBCs. Serum chemistry changes are also indicative of dehydration, with elevation of creatinine, urea nitrogen, and total protein. Serum hepatic enzymes and serum phosphorus may be mildly elevated.

- foot temperature increased, abrupt onset, pitting edema of limbs, rarely severe; colic, diarrhea, increased capillary refill time and heart and respiration rates, dark urine, laminitis

Pathology—Grossly, there is subcutaneous edema, pulmonary edema, and edema of the kidney interstitium. The mucosa of the stomach and small intestine may be ulcerated, and small splotchy hemorrhages may be apparent on the mucosal and serosal surfaces. Unusually large amounts of calcium carbonate crystals are seen in the bladder.

- gross pathology, s.c. and pulmonary edema; ulceration, mucosa of stomach, small intestine

Treatment—No specific agents are known to reverse the disease process. Typically therapy is directed toward relief of pain, inflammation, and edema and includes the use of phenylbutazone, flunixin meglumine, and furosimide or other diuretics.

- diuretics, pain relief, anti-inflammatory drugs

Brassica L.

Taxonomy and Morphology—On Old World genus of 30–35 species and numerous cultivars, *Brassica* is commonly known as mustard or cole. The genus name was used by Pliny, and species of the genus have been cultivated since antiquity for both food and animal fodder. As illustrated in the list given below, many of our common vegetables are members of this taxon. Some of them have escaped cultivation, naturalized, and lost the traits relished by humans. Some have become troublesome weeds in other crops. In North America, 7 introduced species are commonly encountered:

B. elongata Ehrh.	
B. juncea (L.) Czern.	India mustard, brown mustard
B. napus L.	rape, colza
var. *napobrassica*	rutabaga, swede
B. nigra (L.) W.D.J.Koch	black mustard

- *B. elongata*
  B. juncea
  B. napus
  B. nigra
  B. oleracea
  B. rapa
  B. tournefortii

broccoli	mustard
brussels sprouts	pak-choi
cabbage	pe-tsai
cauliflower	rape
collards	rutabaga
colza	swede
kale	tendergreen
kohlrabi	turnip

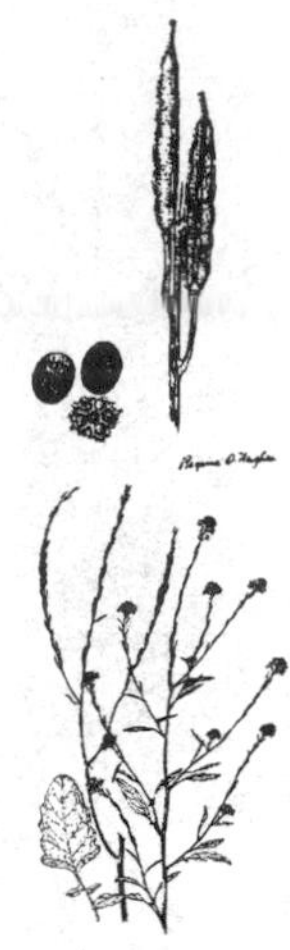

Brassica nigra

■ erect, branched, smooth herbs; leaves simple; petals conspicuous yellow or yellow-white

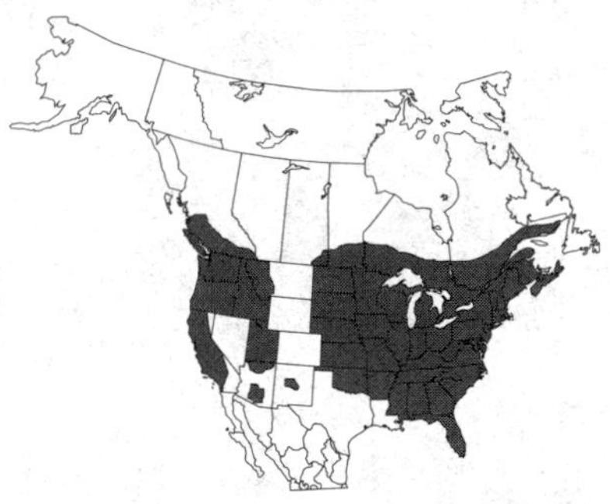

Brassica nigra

■ naturalized weeds in waste areas

■ 6 serious disease problems—
anemia
ARDS
blindness
bloat
PEM
reproductive

others of very low risk—
photosensitization
nitrate
oxalates

B. oleracea L.	wild cabbage
var. *acephala*	kale, collards
var. *alboglabra*	Chinese kale, Chinese broccoli
var. *botrytis*	broccoli, cauliflower
var. *capitata*	cabbage
var. *gemmifera*	brussels sprouts
var. *gongylodes*	kohlrabi
var. *sabauda*	savoy cabbage
B. rapa L.	field mustard, bird's rape
var. *chinensis*	pak-choi, Chinese cabbage
var. *olifera*	turnip rape
var. *pekinensis*	pe-tsai
var. *perviridis*	tendergreen, spinach mustard
var. *rapifera*	turnip
B. tournefortii Gouan	wild turnip rape

Plants herbs; annuals or biennials or rarely perennials; caulescent. **Stems** erect; 30–150 cm tall; branched; glabrous. **Leaves** simple; basal and cauline; petiolate or sessile; blades of various shapes; margins entire or margins of lower leaves pinnatifid or dentate. **Racemes** terminal; elongating in fruit. **Flowers** usually conspicuous. **Sepals** erect or spreading; saccate. **Petals** obovate; clawed; yellow or rarely yellow-white. **Stamens** 6; tetradynamous. **Fruits** siliques; subterete or angled; beaked; often torulose. **Seeds** numerous; in 1 row; subglobose.

Taxonomists differ in their opinions as to whether white mustard and charlock should be placed in *Brassica* or segregated in the genus *Sinapis*. Following the treatments of Rollins (1993) and Tsunoda and co-workers (1980), the two are positioned in *Sinapis* here.

Extensive hybridization in the genus often makes positive identification of individual plants difficult. A key to species is not presented here; rather the reader is encouraged to use a taxonomic key covering the plants in his or her geographical area.

Distribution and Habitat—All species of the genus are native to Eurasia. Most have naturalized as weeds in fields, waste grounds, roadsides, and other disturbed sites; some are troublesome in cultivated fields.

Disease Problems—Species of *Brassica* produce a variety of adverse effects. Cote (1944) initially recognized four types—respiratory, digestive, neurologic, and urinary—and noted that an animal could exhibit signs of two or more at the same time. Other types have been described since then. In the following subsections, five very different disease problems—anemia, acute respiratory distress, blindness, polioencephalomalacia, and adverse effects on reproduction are discussed separately. Problems of lesser importance—taint, photosensitization, nitrate and oxalate accumulation, and other syndromes also are reviewed.

Increased rainfall and early frosts seem to be conducive to a special risk (Cote 1944; Evans 1951). In South Africa, Jersey cattle grazing for 5–6 months on *B. oleracea* with high sulfur content developed poor body condition, Heinz body anemia, and poor reproductive performance, mainly of poor conception rate and a few abortions (Taljaard 1993). In another report, liver dysfunction, shown by an increase in GGT and glutamate dehydrogenase, was prominent in dairy cattle that were fed fresh rape (Barnouin & Paccard 1988). In general, rape and kale are the most common causes of adverse effects, in part because they are the most often used for feeding livestock.

Anemia

Disease Problems—The most consistent problem in ruminants grazing species of *Brassica* for prolonged periods is the development of hemolytic anemia (Clegg & Evans 1962; Penny et al. 1964). It may occur even when grazing is only for a restricted time (Stamp & Stewart 1953). The anemia is most severe in cattle but is seldom life threatening. Onset is typically heralded by the occurrence of Heinz bodies in RBCs 2–3 weeks after consumption of rape, kale, turnips, or other brassica forage begins. In some instances in cattle, the packed cell volume (PCV) may decline 30% in 6 weeks and then stabilize or perhaps return to nearly normal. In most instances, however, even with consumption of as much as 40–70 kg of the offending forage per day, the hemolysis is less severe. Most commonly, PCV and HB decrease mildly (Grant et al. 1968a; Backgren & Jonsson 1969; Tucker 1969; Young et al. 1982). The response in sheep is less severe, and in goats it may be severe but not sustained as long as in cattle (Penny et al. 1964; Grant et al. 1968b; Greenhalgh et al. 1969). The onset of hemolysis is often manifested as a dark discoloration of the urine due to the presence of free hemoglobin, but a decline in PCV is usually hardly noticeable.

- hemolytic anemia, Heinz bodies, typically transient, seldom serious, economic loss, cattle on rape, kale, turnips, sheep less severe

The hemolysis is generally of older RBCs, accompanied by a decreasing myeloid/erythroid ratio in the bone marrow as destruction becomes balanced by erythropoiesis (Tucker 1969; Young et al. 1982). Recovery occurs within a few weeks of the animal's ceasing to eat *Brassica*; there may be partial recovery even with continued ingestion. Although in some instances economic losses may be substantial, the disease is often rather mild. In a 4-year experiment, lambs grazed kale, turnips, or Chinese cabbage for prolonged periods during the fall and winter. All experienced good gains, as good as or better than those experienced by animals eating grass or clover forages (Reid et al. 1994). Heinz body formation increased markedly, but RBC packed cell volumes decreased only slightly (Cox-Ganser et al. 1994). Serum thyroxin (T4) decreased initially but only temporarily, and thyroid weights increased modestly. Rape with purple-tinged foliage is thought to be more likely to cause toxicity problems (Cote 1944; Perrett 1947).

- hemolysis, effects of oxidant sulfides derived from SMCO; SMCO itself inactive, increasing with plant maturity, high in flowers

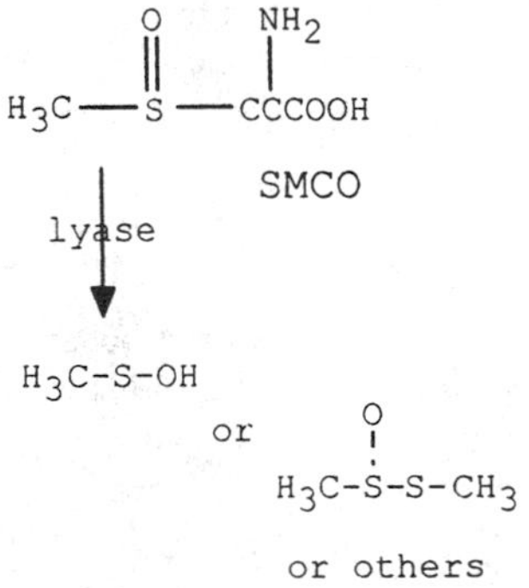

Disease Genesis—The toxicants that are responsible for hemolysis are derived from *S*-methyl-L-cysteine sulfoxide (SMCO) in the plant (Smith et al. 1978). When the plant is macerated, a lyase enzyme much like alliinase in onions is liberated. This results in breakdown of SMCO and the formation of methyl methane thiosulfinate and then various sulfides, disulfides, trisulfides, and thiosulfonates (Smith 1977; Marks et al. 1992). These reactions are pH dependent, inhibited in an acidic medium, and favored by an alkaline environment (Whittle et al. 1976). SMCO is relatively inactive, and most adverse effects are associated with dimethyl disulfide or its oxide (Smith 1980). Blood concentrations of dimethyl disulfide peak at the time of high Heinz body presence and maximal hemolysis, while there is no detectable SMCO (Earl & Smith 1982). The formation of the toxic end products in ruminants is due to action of the enzyme produced by the plant tissues or by ruminal microorganisms (Smith 1978, 1980). Formation does not occur in the acidic environment of the stomach of monogastrics. Toxic effects have been observed in sheep, goats, cattle, and roe deer (Schoon et al. 1989). Whether from fresh or dried kale or from silage, SMCO is equally toxic (Pelletier & Martin 1973).

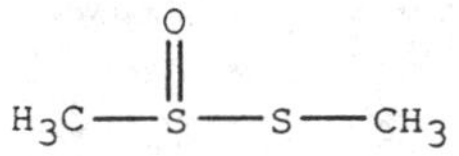

$H_3C-S(=O)-C(NH_2)COOH$

SMCO

↓ lyase

$H_3C-S-OH$ or $H_3C-S(=O)-S-CH_3$ or others

$H_3C-S(=O)-S-CH_3$

Dimethyl disulfide oxide

The various reaction products, when absorbed, exert oxidant effects on RBCs in much the same manner as *n*-propyl disulfide from onions; they react with SH groups on hemoglobin to cause lysis directly or an agglutinating or denaturing action manifested as Heinz bodies, which are then subject to splenic removal. Reduced glutathione (GSH) is protective against these effects. The hemolytic effect and the decrease in weight gain depend upon the dose of SMCO. Dosage of 100 mg/kg b.w. results in low-grade anemia, and 150–200 mg/kg causes acute hemolysis (Smith et al. 1978). Smith (1980) suggested that the dosage of SMCO be limited to 100 mg/kg live weight to avoid depression of weight gain.

Concentrations of SMCO are variable in *Brassica*. They increase with plant maturity. For example, in *B. oleracea,* they increase from 6,000 ppm in October to 9,000–12,000 ppm in winter, and the concentrations are highest in flowers (Whittle et al. 1976; Giovanni et al. 1989). Concentrations of SMCO also vary with the species and variety of *Brassica* and in general are about 10 times greater than glucosinolate levels (Young et al. 1982; Marks et al. 1992). Concentrations of SMCO can be manipulated by nitrogen and sulfur fertilization practices, because it appears to function as a sulfur storage form (McDonald et al. 1981). Low levels of soil sulfur result in as much as a 50% reduction of SMCO in the plant (Barry et al. 1984).

The effects of *Brassica* are not necessarily devastating. Even in the presence of substantial amounts of SMCO (4,000–7,000 ppm dry matter) and thiocyanate (400–500 ppm dry matter), lambs may experience good growth rates (Young et al. 1982). Anemia also occurs in free-roaming roe deer (Onderscheka et al. 1987; Schoon et al. 1989). In general, contamination of feed with up to 20% *Brassica* seeds does not appear to represent a serious risk (Shires et al. 1982).

- no signs in most instances; with moderate severity, decreased appetite, weakness, dark urine

Clinical Signs—Signs related to RBC lysis include loss of appetite, weakness, dark-colored urine, decrease in milk production, increase in heart and respiratory rate, and sometimes diarrhea. The RBC numbers will be reduced as reflected in decreased packed cell volume (PCV), blood hemoglobin, and RBC count.

- gross pathology, spleen enlarged, kidneys dark metallic color

  microscopic, hemosiderin, liver, kidneys, mild renal tubular necrosis

Pathology—Typically, with hemolytic anemia the spleen is enlarged, the kidneys are dark metallic in color, and the liver is slightly enlarged. Microscopically, there is iron and hemosiderin accumulation in the liver and kidney and mild renal tubular degeneration, with epithelial necrosis and pigment in the tubules.

- recovery usually without treatment; if severe, change forage

Treatment—Seldom is the anemia severe enough to require therapy. Removal from the *Brassica* forage will usually result in a return to normal RBC numbers.

Acute Respiratory Distress Syndrome (ARDS)

- ARDS, ruminants, presumably from sudden increase in tryptophan in rumen, metabolized to 3-MI, absorbed and further metabolized in lung to the ultimate pulmonary toxin

Disease Problems and Genesis—*Brassica* produces the same disease as do members of the Fabaceae and Poaceae, presumably a result of a sudden increase in ruminal tryptophan levels, predisposing formation of excess 3-methylindole (3-MI) (for a more detailed description of the disease, see chapters 37 and 58). The original descriptions of the disease were prompted mainly by the continuing problems occurring in cattle when they were first allowed to graze rape or kale in the fall of the year (Schofield 1948).

- 7–10 days after begin grazing rape, kale, or turnips; abrupt onset, respiratory distress, open-mouthed breathing, head extended, increased heart and respiratory rates

Clinical Signs—Typically, signs occur 7–10 days after animals begin to graze rape, kale, or turnips. Rarely cases may be seen as early as 5 days or as late as 4 weeks. The animals are in great distress, with labored, open-mouthed breathing with heads extended and a noticeable expiratory grunt. The heart and respiratory rates are increased, and there may be either constipation or fetid diarrhea. Evidences of emphysema are often accompanied by subcutaneous accumulation of air along the back and over the ribs.

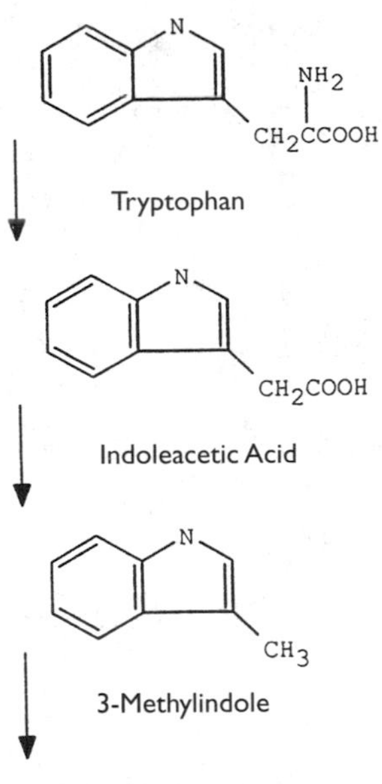

Pathology—The lungs are wet, heavy, rubbery, and distended, with pockets of air in the interlobular septa. On cut surfaces, the lung is dark red and firm, with numerous irregular sections divided by interlobular emphysema. There is edema and congestion throughout the lungs, and frothy exudate in the trachea and bronchi. The liver may be pale, friable, and mottled; the gallbladder, distended with bile; and the kidneys, pale. Microscopically, rupture, collapse, and filling of the alveoli with serous exudate are evident, along with edema and air pockets in the interlobular connective tissue. There may be extensive areas of degeneration and necrosis around the central veins of the liver.

- gross pathology, lungs wet, rubbery, distended; froth in airways

  microscopic, alveolar distension and rupture, edema and air pockets in septa

Treatment—It is very important to minimize stress during the period of severe respiratory distress. Numerous drugs have been used, including corticosteroids, antihistamines, nonsteroidal anti-inflammatories, atropine, diuretics, and antibiotics, but therapy appears to be of limited value. Often, handling the animal for treatment causes stress and death due to acute respiratory embarrassment.

The disease can be prevented by management techniques aimed at more-gradual changes in pasture quality. Such recommendations include cutting the forage and allowing it to wilt before turning the animals onto a new lush pasture; limiting grazing time on the new pasture; and gradually introducing a new pasture by rotating the animals every few days between a lush pasture and a dry pasture or drylot with hay (Potchoiba et al. 1992). When these approaches are not possible, diet supplementation with 200 mg/animal of an ionophore such as monensin or lasalocid may be used to prevent or at least attenuate the severity of the disease. The ionophore should be fed beginning a few days before and continuing for 10 days following the change in pasture (Hammond et al 1980; Nocerini et al. 1985; Potchoiba et al. 1992). This is a sustained protective effect, but it is not invariably preventive for all animals (Honeyfield et al. 1985).

- avoid stress; prevent by gradual pasture change, limit grazing time, ionophores

Blindness

Disease Problems and Genesis—Temporary blindness has been reported in cattle several weeks after they begin grazing rape. Recovery was gradual and complete within 6 weeks after cessation of grazing (Dalton 1953). The pupils were dilated and not responsive to light, and no changes were apparent upon ophthalmoscopic examination (Perrett 1947; Crawshaw 1953). In other instances, corneal opacity was a problem (Michael 1953). The blindness, which is of unknown cause, does not appear to be due to polioencephalomalacia, because it is transient and not associated with lethality.

- cattle, temporary blindness several weeks after beginning grazing, nonlethal, cause unknown

Clinical Signs, Pathology, and Treatment—The blindness is typically temporary, lasting a few weeks. It may be accompanied by aimless wandering and sometimes belligerent or violent actions in which inanimate objects are battered. In roe deer the effects are more extensive, with distinct pathologic changes, which include myocardial degeneration, cerebral hyperemia and edema, and necrosis of the cerebrum. Treatment is for relief of the symptoms.

- aimless walking, aggression

Bloat

■ usually free-gas, nonfrothy bloat in first few days of grazing

Disease Problems and Genesis—This disease may occur at any time while grazing the *Brassica* forages, but often it occurs within a short time after the initiation of grazing (Morton & Campbell 1997). Bloat associated with grazing turnips is the free-gas type rather than frothy bloat (Wikse et al. 1987). This may be due to a rapid rate of fermentation and production of gas. It is readily treated if observed early but can progress rapidly to respiratory compromise and death. In contrast, bloat as seen in roe deer grazing rape is of the frothy type (Onderscheka et al. 1987).

■ abdominal distension, labored respiration

Clinical Signs—Regardless of whether the bloat is free or frothy, there is noticeable distension of the abdominal area, especially in the left flank area. Respiration is labored, and the rate is increased markedly.

■ head and neck tissues congested with blood

Pathology—Grossly, congestion and hemorrhage of the tissues of the head and neck are evident. The thoracic esophagus may be pale, while the cervical portion is congested, producing a characteristic "bloat line" at the thoracic inlet.

■ release free gas, tube or trocar

Treatment—The free gas is readily discharged by passage of a stomach tube or by cannulation of the rumen through the left flank just posterior to the last rib. In the latter case, a cannula can be allowed to remain in place to correct the problem in chronic bloaters.

Frothy bloat requires administration of a chemical to reduce surface tension of the liquid and break up the bubbles. The primary aim is to relieve the distension. This is accomplished by mechanical release of the gas and destabilization of the ruminal foam in which the air bubbles are trapped. The surface activity can be reduced effectively by the use of oils, fats, detergents, and other surface active agents given by drench. The use of oils and fats is limited because they are subject to rapid degradation. In common use are the pluronic surfactants, of which the polyoxypropylene-polyoxyethylene types are the best examples. All of these antifoaming agents can also be used prophylactically to lower surface tension, but they require continuous daily administration. Poloxalene (Bloatguard®) is water soluble and can be given in a variety of ways, including in liquid molasses or in mineral and other types of blocks that are licked. Monensin and lasalocid are partially protective against both frothy pasture bloat and bloat due to a high-grain diet (Bartley et al. 1983).

Pasture management and grazing control are other ways to prevent the occurrence of bloat. Pastures with a mixture of productive high-bloat forages and low-bloat types such as sainfoin or bird's-foot trefoil are especially useful if they can be maintained as true mixtures.

Polioencephalomalacia (PEM)

■ rare, affecting few individuals in herd soon after feed change, often fatal

Disease Problems—In rare instances cattle, sheep, or other ruminants grazing species of *Brassica* will develop a neurotoxicosis or die suddenly with little warning (Wikse et al. 1987; Schoon et al. 1989; Hill & Ebbett 1997). Sometimes termed polio (PEM) or cerebrocortical necrosis (CCN), the disease often follows a change in feed to lusher forage or larger amounts, e.g., turnips for cattle and rape for roe deer (Wikse et al. 1987; Schoon et al. 1989). It is very similar to PEM caused by high dietary sulfate and is quite often fatal (Raisbeck 1982; Gooneratne et al. 1991). Fortunately the disease is of low and sporadic occurrence. A peculiar form of neurologic disease, perhaps comparable to polio in cattle, occurs in free-roaming roe deer, in which they undergo a marked behavior change, lose their fear of humans, and become oblivious to their environment, with no reaction to sound or sight (Onderscheka et al. 1987; Schoon et al. 1989). Some recover after a few weeks.

Disease Genesis—The specific cause of these neurotoxic effects has not been definitely identified in all instances, but a likely factor is the high sulfur content of *Brassica* plants. Species of the genus with sulfur concentrations as high as 8,500 ppm (0.85%) have been associated with blindness, ataxia, and sudden death in ruminants and with pathologic changes typical of PEM from other causes (Gooneratne et al. 1991, 1992; Hill & Ebbett 1997). These levels are consistent with levels of sulfate (0.63%) that cause experimental dietary-induced PEM (Gould et al. 1991; Rousseaux et al. 1991; Olkowski et al. 1992). The neurotoxic effects of sulfur are shown by the PEM that occurred in a flock of 2,200 ewes grazed on an alfalfa pasture shortly after it had been sprayed with a 35% suspension of elemental sulfur (Bulgin et al. 1996). Two percent of the animals developed signs or had specific brain lesions consistent with PEM. In the ruminal bacterial environment, sulfate is reduced to sulfide (Lewis 1954; Olkowski et al. 1992; Cummings et al. 1995). Presumably the organic sulfur compounds in *Brassica* serve as similar substrates for reduction. Once reduced, the sulfide may interact with sulfite and may serve as a thiaminase (Leichter & Joslyn 1969), but the neurotoxicosis is more likely due to additional causes, possibly direct effects on the brain via lipid peroxidation. Not only are the pathologic changes produced by both *Brassica* and high dietary levels of sulfur quite similar, but with both there are minimal changes in tissue thiamine levels. In neither instance is the disease clearly thiamine responsive, as with other causes of PEM (Thomas 1986; Gooneratne et al. 1992; Olkowski et al. 1992; Hill & Ebbett 1997). The sulfur from plants may be acting in concert with that in water or other sources to increase the risk of development of PEM. In areas with minimal sulfate in the water, PEM due to *Brassica* may be a rare disease. An additional factor of concern with long-term ingestion of high-sulfur forages is the potential for copper deficiency (Taljaard 1993). The detrimental effects of decreasing bioavailability of copper in ruminants due to dietary sulfur is well documented (Gooneratne et al. 1989a). Sulfur from various sources may be reduced in the rumen to sulfide, which reacts with copper, or sulfur may interact with molybdenum to form poorly absorbed copper thiomolybdate complexes (Suttle 1974, 1991; Qi et al. 1993).

- probably related to high sulfur in plants and water; sulfur reduced to sulfide in rumen; direct effects or via thiaminase activity in rumen; not clearly thiamine responsive; sulfur also interferes with copper bioavailability

Clinical Signs—PEM is a disease of abrupt onset. Initial signs are twitching of the ears and muscles of the face and champing of the jaws, which are accompanied by profuse salivation. Animals stand motionless in an apparent stupor, circle, or wander aimlessly forward or backward. Blindness and/or lack of awareness of surroundings and inability to eat properly are consistent features of the disease. Once animals become recumbent, especially if there also are seizures, the prognosis is unfavorable for recovery, even with treatment.

- abrupt onset, ear twitching, champing jaws, salivation, circling or aimless wandering, blindness, recumbency, seizures

Pathology—Grossly, severe cerebral edema causes flattening of the gyri and compression of the sulci. The accompanying malacia of gray matter is obvious in the cerebral cortex as well as other areas, including the basal ganglia, the thalamus, and the cerebellum. The parietal cortex may be yellowish gray in color. The areas of cerebral cortical malacia often fluoresce under ultraviolet light, especially at white and gray matter junctions. Microscopically, there is prominent laminar necrosis and a lack of inflammatory response. Shrunken and eosinophilic neurons with pyknotic nuclei are

- gross pathology, brain, edema, flattened gyri, compressed sulci, parietal cortex yellowish gray, fluorescence with UV light

  microscopic, neuronal necrosis

scattered in multiple foci with perivascular hemorrhage and edema and with necrosis of some small arterioles.

In roe deer the lesions include myocardial degeneration, cerebral hyperemia and edema, and necrosis of the cerebrum.

■ parenteral thiamine, may not always give complete relief

Treatment—Although parenteral administration of 10 mg/kg b.w. of thiamine HCl may provide some minor beneficial effect or be preventive, it does not eliminate the signs. Symptomatic treatment and control of the cerebral edema may also be of benefit.

Reproductive Effects

Disease Problems and Genesis—Decreases in fertility in both cattle and sheep have been marginal and sometimes difficult to detect (Boyd & Reed 1961). The effects, which are accompanied by evidence of hematologic changes, are seen after animals graze kale for 2 or more months. Conception rate decreases, so an increase in the number of services is required (Melrose & Brown 1962; Williams et al. 1965). Some of the reproductive effects appear to be secondary to thyroid enlargement; increased gestation period and decreased birth weight are reported for ewes (Sinclair & Andrews 1958). Mallard (1981) reported the birth of weak and stillborn kids and gestation prolonged 1–2 weeks in goats wintered on a daily diet of 4–6 kg kale. In an epidemiologic study in France, an increase in perinatal calf mortality was noted when rape or kale were fed to dams during the last 90 days of pregnancy (Barnouin et al. 1992). Lambing difficulties with weak labor, uterine inertia, and stillborns also were reported for a flock of sheep fed for 5–6 weeks on forage composed mostly of *Sinapis arvensis* and smaller amounts of *Capsella bursa-pastoris* and *Raphanus raphanistrum* (Mihalka 1982). Further studies suggested the potential for mustard glucosinolates' predisposing sheep to *Chlamydia* infection, with subsequent reproductive losses (Mihalka 1983).

■ marginally decreased fertility, even without thyroid effects; possibly effects of glucosinolate

The cause of the reproductive effects is not clear. Experimental administration of acetone or alcohol extracts of kale orally or subcutaneously to rats revealed no evidence of estrogenic activity (Pickard & Crighton 1967). The occurrence of difficulties produced by genera other than *Brassica* may be an indication that this is indeed a glucosinolate-related effect. That this is the case is substantiated further by studies demonstrating delayed sexual maturity in gilts, attributable to glucosinolate goitrogenic activity (Fenwick et al. 1989).

Other Disease Problems and Genesis

■ taint of milk, eggs, meat; rarely primary photosensitization

Undesirable odors and flavors of meat have been noted in animals fed rape (Wheeler et al. 1974). Similarly a fishy odor of eggs, especially brown eggs, due to accumulation of trimethylamine, and liver hemorrhage have been problems in poultry fed rapeseed meal (Fenwick & Curtis 1980; Butler et al. 1982). Odor and flavor problems seem to accrue from the glucosinolates and have been markedly reduced or eliminated by selection programs to reduce glucosinolate concentrations in the plants.

Photosensitization is a rare adverse effect associated with rape in North America. In Australia, however, the primary type apparently has been a common problem in dairy cattle grazing various brassicas (Morton & Campbell 1997). Case (1957) reported a type with little pigment correlation. A mild form with typical head distribution in sheep also has been described (Vermunt et al. 1993).

Species of *Brassica* are prone to accumulate toxic concentrations of nitrate and oxalate under appropriate conditions. Mejia et al. (1985) reported concentrations of oxalate of 0.6–7.9% d.w. for *B. napus*.

Descurainia Webb & Berthold

- *D. pinnata*
  D. sophia

  flixweed
  tansy mustard

Taxonomy and Morphology—Comprising approximately 40 species in both the Old World and the New World, *Descurainia* is commonly known as tansy mustard. In North America, 14 native and introduced species are present, only 2 of which are of toxicologic concern:

- erect annual herbs; leaves simple, in basal rosettes that whither, cauline pinnatifid; flowers small; petals yellow or whitish

D. pinnata (Walter) Britton	tansy mustard
D. sophia (L.) Webb ex Prantl	flixweed

Plants annuals or biennials; caulescent; herbage bearing branched hairs. **Stems** erect; 20–80 cm tall; canescent; branched above. **Leaves** simple; cauline and forming basal rosettes that wither as plants mature; upper leaves gradually smaller; blades ovate or oblong to oblanceolate; margins 1-to-3-pinnatifid into linear segments. **Racemes** terminal; elongating in fruit. **Flowers** small. **Sepals** ovate; green or yellow, sometimes rose tinged. **Petals** obovate or spathulate; clawed; yellow or whitish. **Stamens** 6; tetradynamous. **Fruits** siliques; linear or clavate; terete or subterete. **Seeds** numerous; elliptic; yellowish to reddish brown.

1. Upper leaves 1-pinnatifid. Rachises glandular. Siliques clavate . . . . *D. pinnata*
1. Upper leaves 2- or 3-pinnatifid. Rachises not glandular. Siliques linear . . . . *D. sophia*

Descurainia pinnata

Distribution and Habitat—Occupying a diversity of sites, including dry open prairies, roadsides, fields, and waste areas, *D. pinnata* is native to much of North America, especially the western half. *Descurainia sophia* from Eurasia is also widespread in fields and waste areas except for the southeastern states.

- dry, open sites

Disease Problems—*Descurainia pinnata* was first reported as toxic in the early 1940s, when it was associated with a disease of the tongue called wooden tongue or tongue paralysis (Hershey 1945). This disease has been described as a field problem that appears in years with an overabundance of winter moisture; it has not been experimentally reproduced (Hershey 1945; Rosiere et al. 1975; Staley 1976). The plants appear to be toxic prior to and when in bloom in the late winter or early spring but not thereafter. This time of toxicity also coincides with the period in which they are likely to be eaten. In two studies evaluating forage utilization by cattle on New Mexico ranges, 25–31% of the diet was composed of *D. pinnata* (Rosiere et al. 1975; Hakkila et al. 1987). Thus, because of its moderate palatability and likelihood of being eaten, *D. pinnata* is a substantial risk, although few animals develop signs of intoxication. *Descurainia sophia* is a reduced risk because it appears to be less likely eaten (Pfister et al. 1990).

- livestock, neurologic like the originally described blind staggers; tongue paralysis, especially plants in bloom; toxicity not confirmed experimentally; photosensitization, hepatogenous?

Like other genera of the Brassicaceae, *D. pinnata* in Montana is suspected to cause hepatogenous photosensitization, but this has not been confirmed (Pfister et al. 1989, 1990). Photosensitization also has been observed in eastern Colorado with prolific growth of *D. sophia* (Knight 1996).

Descurainia sophia

Disease Genesis—Because the disease caused by *Descurainia* is of such sporadic occurrence, only limited efforts have been expended to determine its cause. It has not been experimentally reproduced, and the specific toxin is unknown. Although many of the

- neurologic effects, cause unknown; possibly sulfur in plant; not highly lethal as with PEM; selenium levels not high

Descurainia pinnata

- anorexia, blindness, circling, difficult to drive, tongue paralysis, weight loss, weakness, tremors, head bobbing
- good feed, provide water by stomach tube

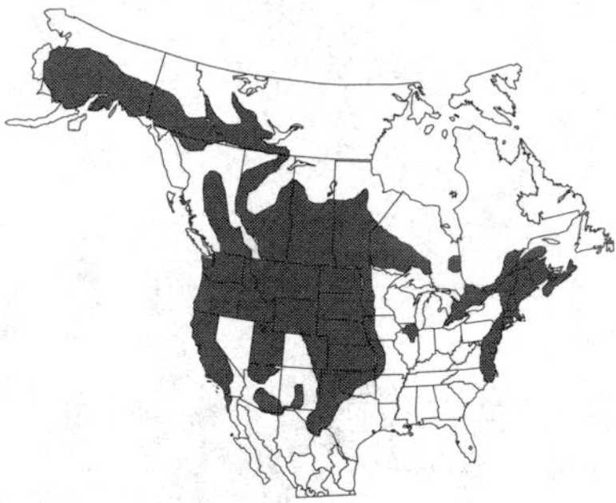

Descurainia sophia

- *E. asperum*
  E. capitatum
  E. cheiranthoides
  E. cheiri
  E. hieraciifolium
  E. repandum

  prairie rocket
  treacle mustard
  wallflower
  wormseed mustard

signs are suggestive of blind staggers, tested selenium levels were only in the 1–3 ppm range, and thus selenium appears not to be a cause (Hershey 1945). It may be a serious disease, but typically the signs are slowly reversible when the animal stops eating the plants. However, in some instances animals are said to become permanent "space-gazers" (Staley 1981).

Sulfur may play a role in the cause of neurologic disease due to *D. pinnata*. Under some conditions, plants may accumulate high concentrations of sulfur, up to 1% d.w. of plant, mainly as nonprotein organic forms (Mayland & Robbins 1994). This circumstance may be complicated when drinking water contains appreciable sulfur. However, the neurologic signs attributable to *D. pinnata* are not entirely consistent with typical PEM.

Clinical Signs—Initially there are varying degrees of blindness, and the animals walk in circles until they are exhausted or they encounter an obstruction; their behavior is much the same as with hepatoencephalopathy. The animals become difficult to drive. Later, tongue paralysis develops, and the animals are unable to eat or drink properly, the rumen becomes atonic, and they lose weight and are weak and dull. There may also be spasms of the muscles of the head, ears, and neck, leading to tremors and head bobbing, suggestive of lead poisoning in cattle; however, these signs are most common in the goat and horse. Some cattle develop a persistent attitude of appearing to gaze into space.

Treatment—Water should be given by stomach tube, 2–3 gallons twice daily in cattle. There is usually a favorable response to maintaining water balance and providing good feed. The ruminal atony may be treated with 100 ml of ethanol/450 kg b.w. diluted in 500 ml of Ringer's solution, a dose that should provide mild incoordination (Staley 1976).

Erysimum L.

Taxonomy and Morphology—Most numerous in eastern Europe and southwestern Asia, *Erysimum* comprises 180–200 species and is now circumscribed to include species placed in the genera *Cheiranthus* and *Cheirinia*. Its name was used by Hippocrates and comes from the Greek root *eryo,* meaning "to draw out"; some species produce skin blisters. Diverse in habit and occurring in a variety of habitats, species of the genus are grown as ornamentals primarily for the beauty of their flowers. Some, however, are grown for their scent or their attractive evergreen foliage. In North America, 19 native and introduced species occur. Of toxicologic interest are the following:

E. asperum (Nutt.) DC.	western wallflower
E. capitatum (Douglas *ex* Hook.) Greene	western wallflower, prairie rocket
E. cheiranthoides L.	wormseed mustard, treacle mustard wormseed wallflower
E. cheiri (L.) Crantz (= *Cheiranthus cheiri* L.)	English wallflower
E. hieraciifolium L.	Siberian wallflower
E. repandum L.	spreading wallflower, bushy wallflower

Plants annuals or biennials or perennials; caulescent; herbage with appressed branched hairs. **Stems** erect; 10–100 cm tall; branched or not branched above. **Leaves** simple; cauline and forming basal rosettes; petiolate or sessile; blades linear to oblanceolate; margins entire to dentate or pinnatifid. **Racemes** terminal; corymbose; elongating in fruit. **Flowers** small or large; narrow. **Sepals** erect; linear or oblong. **Petals** obovate or spathulate; clawed; orange to bright yellow, or white or rarely pink, purple, red, brown, or lavender. **Stamens** 6; strongly tetradynamous. **Fruits** siliques; linear; terete or 4-angled; smooth or weakly torulose. **Seeds** numerous; oblong; terete or compressed.

■ erect herbs; leaves simple, basal rosettes; petals orange, bright yellow, white, rarely other colors

1. Petals 3–10 mm long; 2–3.5 mm wide.
    2. Sepals 2–3.5 mm long. Petals 3.5–5 mm long .... *E. cheiranthoides*
    2. Sepals 4.5–6 mm long. Petals 6–10 mm long.
        3. Plants annuals. Pedicel diameters equal to silique diameters. Siliques diverging at right angles from rachises .... *E. repandum*
        3. Plants perennials or biennials. Pedicel diameters less than silique diameters. Siliques erect or ascending ...... *E. hieraciifolium*
1. Petals 13–35 mm long; 5–15 mm wide.
    4. Siliques flattened. Seeds in 2 rows. Nectaries absent .... *E. cheiri*
    4. Siliques terete or 4-angled. Seeds in 1 row. Nectaries present.
        5. Pedicel diameters equal to silique diameters. Siliques diverging at right angles from rachises .... *E. asperum*
        5. Pedicel diameters less than silique diameters. Siliques erect or ascending .... *E. capitatum*

Erysimum asperum

Erysimum cheiranthoides

■ open sites, plains, prairies

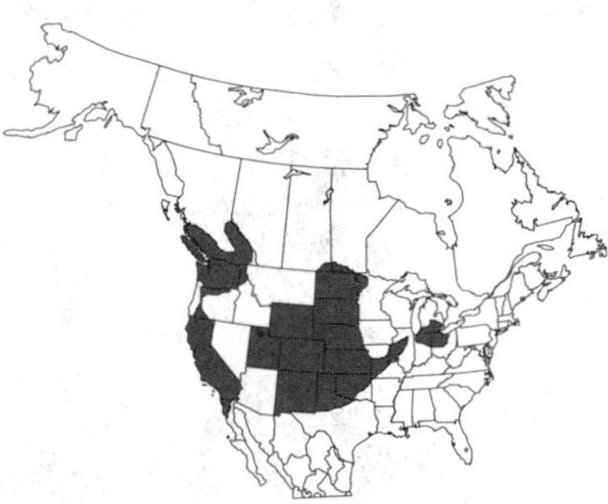

Erysimum asperum

Erysimum cheiranthoides

Distribution and Habitat—The native species of *Erysimum* are adapted to a variety of soil types and open habitats. *Erysimum capitatum,* with 5 varieties, is more commonly encountered in the western half of the continent, while *E. asperum* is more common in the Great Plains. *Erysimum cheiranthoides* occurs across the continent; unresolved is whether it is native or introduced (Rollins 1993). Of the definitely introduced species, *E. repandum* is a weed in disturbed soils across the continent; *E. hieraciifolium* is weedy primarily in the Northeast; and *E. cheiri* is a prized showy ornamental popularly planted in rock gardens and terraces for its fragrant spring display of colorful flowers.

Disease Problems and Genesis—Species of *Erysimum* are not well known as toxic plants, and few problems have been attributed to them. A few anecdotal accounts,

- mainly digestive disturbance; cardenolides, glycosides of uzarigenin, strophanthidin, especially in seeds, flowers

O
O
CH_3
OCH
OH
O
OH
H or Sugars

Strophanthidin

such as a loss of pigs in Canada, suggest their potential as toxicants, but there is otherwise little information about their toxicity (Lewis and Elvin-Lewis 1977). For the most part it is their toxicant content that portends potential for problems.

Species of *Erysimum* contain an array of potent cardenolides. Cheirosides A, C, D, E, F, and G are glycosides of uzarigenin, whereas cheirotoxin, deglucocheirotoxin, alliside, erysimoside, and helveticoside are strophanthidin derivatives, some of which are quite potent (Shoppee 1964; Coffey 1970; Singh and Rastogi 1970; Belokon & Makarevich 1980; Makarevich et al. 1980; Joubert 1989; Park et al. 1989). Most species contain several to as many as 13 cardenolides (Kowalewski 1960; Fursa et al. 1984). Cardenolide concentrations are very high in seeds, intermediate in flowers, low in leaves, and very low in stems and roots. The concentrations range, e.g., from 1.4% in seeds to 0.03% in roots for the Old World species *E. wahlenbergii* (Kortus & Kowalewski 1977). Their distribution in the plant probably accounts for the lack of clinical problems, but should the plants be eaten in large quantities, the signs would be similar to those of other cardenolide-containing plants, e.g., *Nerium oleander* in the Apocynaceae (see chapter 8).

Sinapis L.

- *S. alba*
  S. arvensis

  charlock
  crunchweed
  white mustard
  wild mustard

Taxonomy and Morphology—The genus *Sinapis* comprises 7 species native to the Mediterranean region. Two species are introduced weeds in North America:

S. alba L. (= *Brassica alba* [L.] Rabenh.) (= *Brassica hirta* Moench)	white mustard
S. arvensis L. (= *Brassica kaber* [DC.] L.C.Wheeler)	charlock, wild mustard, crunchweed

Sinapis alba

Taxonomists differ in their opinions as to whether these two species should be placed in *Sinapis* or *Brassica*. Cytogenetic data indicate that they are different from species of *Brassica* (Tsunoda et al. 1980). Following the treatment of Rollins (1993) and Tsunoda and his co-workers (1980), the two are positioned here in *Sinapis*.

- erect annuals; leaves simple; petals yellow

Plants annuals; herbage glabrous or indumented with unbranched hairs. **Stems** erect; 20–60 cm tall; branched or not branched above. **Leaves** simple; cauline only; lower petiolate, upper sessile; blades elliptic or obovate; margins entire to pinnatifid or lyrate. **Racemes** terminal; corymbose; elongating in fruit. **Flowers** small or large. **Sepals** linear or oblong; spreading; not saccate; yellowish. **Petals** obovate; yellow. **Stamens** 6; tetradynamous. **Fruits** siliques; linear or oblong; terete or angled; conspicuously beaked. **Seeds** 1–12; globose; yellow or brown or black.

1. Pedicels spreading at right angles to rachises. Siliques hirsute. Beaks of siliques linear; flattened; 15–30 cm long . . . . *S. alba*
1. Pedicels erect or ascending. Siliques glabrous. Beaks of siliques conical; terete or angled; 5–10 mm long . . . . *S. arvensis*

- weedy invaders

Distribution and Habitat—Both species of *Sinapis* occur in disturbed soils of grain fields, orchards, roadsides, and waste areas. *Sinapis alba* is cultivated but escapes. *Sinapis arvensis* is well known as a seed contaminant.

Disease Problems and Genesis, Clinical Signs, Pathology, and Treatment—*Sinapis alba* is quite toxic, and animals eating its seeds or grazing plants in fruit may suffer severe distress and die in a few days (Hughes 1924; Eaton 1941; Gallie & Paterson 1945; Holmes 1965; Kernaleguen et al. 1989). They may give off a conspicuous mustard odor. Other clinical signs and the pathologic changes are typical of the disease caused by the glucosinolates. Treatment is likewise the same.

In some circumstances, *S. arvensis* will occasionally accumulate toxic concentrations of nitrate. Troxler (1981) reported a case in which 19 of 48 cows died about 2 hours after being fed the species. The plants had been harvested in the preflowering stage and fed to the animals because of the onset of cold weather and snowfall. Nitrate concentration was 6.2% d.w.

- severe digestive tract irritation from seeds; glucosinolates, profuse salivation, colic, diarrhea, acute nitrate intoxication

Sinapis alba

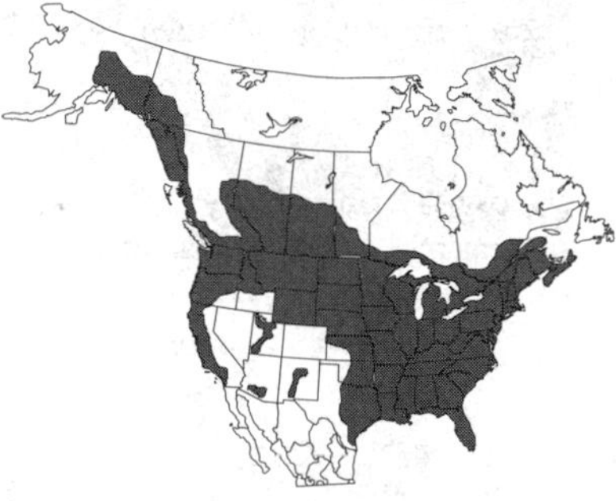

Sinapis arvensis

Thlaspi L.

Taxonomy and Morphology—A genus of about 75 species with greatest diversity in central Europe and western Asia, *Thlaspi,* commonly known as pennycress, is represented in North America by 9 native and introduced species. Its name is from the Greek roots *thalo* and *aspis,* meaning "to flatten" and "shield," alluding to the appearance of its silicles. A few species are cultivated as ornamentals; others are weeds of cultivated fields and waste areas. Only 1 species is of toxicologic interest:

T. arvense L. fanweed, field pennycress, Frenchweed, stinkweed

- *T. arvense*

  fanweed
  field pennycress
  Frenchweed
  stinkweed

Plants annuals; herbage glabrous, malodorous. **Stems** erect; 10–50 cm tall; branched or not branched. **Leaves** simple; primarily cauline; sessile; clasping; blades lanceolate or oblanceolate; margins dentate or sinuately lobed. **Racemes** elongate. **Flowers** small. **Sepals** erect; not saccate; green, often purple tinged. **Petals** spathulate; clawed; white. **Stamens** 6; tetradynamous. **Fruits** silicles; orbicular to broadly elliptic; flattened; winged; apices notched. **Seeds** 2–8 per locule.

- erect annuals; leaves simple; flowers small; petals white

Thlaspi arvense

■ weedy invader

Distribution and Habitat—Naturalized from Europe, *T. arvense* is a widely distributed, troublesome weed of grain fields, pastures, gardens, roadsides, and open waste areas.

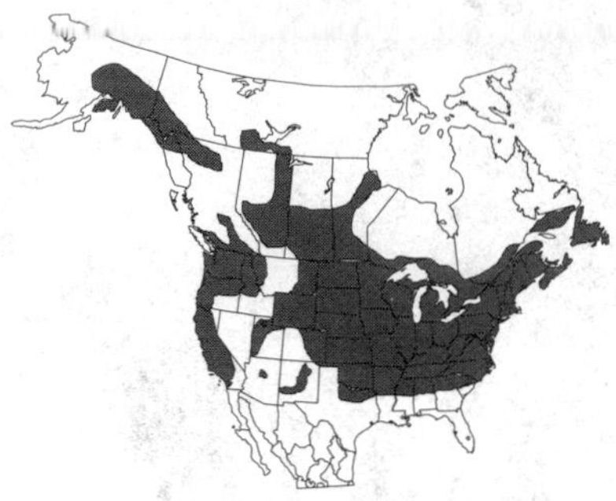

Thlaspi arvense

■ irritant effects, digestive disturbance; photosensitization, unknown type

Disease Problems—Consumption of substantial quantities of fanweed by cattle causes serious digestive system problems, which are mainly due to irritation (Smith & Crowe 1987). Clinical signs occur within a few hours after ingestion. In severe cases where fanweed has composed nearly 100% of the diet, the signs may not disappear entirely when the animals discontinue eating the plants. Ingestion of smaller quantities by lactating cattle is associated with bitter or noxious-flavored milk.

Typical of the family, *Thlaspi* also is reported as a cause of photosensitization in cattle and pigs (Case 1957; Martin & Morgan 1987). Case (1957) considered it to be a common cause of the disease, with seemingly increased potential for several days after a summer rain. Photosensitization was produced experimentally in guinea pigs fed the seeds of *T. arvense* as 20% or more of their diet (Baksi & Case 1971). In these instances it appeared to be secondary to liver disease. However, in another instance of photosensitization of cattle on a pasture of Bermuda grass with heavy infestation of *T. arvense,* only serum LDH was elevated and not GGT, AP, or bilirubin (Martin & Morgan 1987).

■ glucosinolate, allylthiocyanate, from sinigrin, high in seeds, flowers

$$N{\equiv}C{-}S{-}CC{=}CH_2$$

Allylthiocyanate

Disease Genesis—Irritation of the digestive tract is presumably due to release of the irritant glucosinolate allylthiocyanate, enzymatically from its precursor, sinigrin. In this instance, in contrast to most other members of the Brassicaceae, the allylthio compound is almost the only glucosinolate present (Gmelin & Virtanan 1959; Shires et al. 1982; Majak et al. 1991). Hydrolysis to form the allyl compound is favored at the pH of the rumen and is markedly reduced as the pH drops to 5 or below (Smith & Crowe 1987). Thus there is limited risk of irritant release at the pH of the monogastric stomach, as shown experimentally when seeds were fed to mice (Shires et al. 1982). The greatest risk lies with seeds because glucosinolate concentrations are highest in seeds, intermediate in flowers, and lowest in vegetative parts or when plants are in the rosette stage (Majak et al. 1991). There may be considerable variation from year to year in both thiocyanate content and enzymatic release.

Photosensitization due to allylthiocyanate is consistent with the effects of glucosinolates as described for other genera of the family.

■ diarrhea, colic, blood in urine

Clinical Signs—Depending on the amount and toxicity of the plants eaten, the signs will range from mild digestive irritation and colic and bloody urine to severe debilitating colic and perhaps even abortion or other manifestations of systemic disease.

■ gross pathology, abomasal reddening

Pathology—In rare instances of *T. arvense* ingestion, the disease may be fatal. Grossly, there are indications of digestive irritation, with marked reddening and submucosal edema of the abomasum. In the case of photosensitization, extensive skin necrosis is found, especially in the less pigmented areas. In addition, in some instances, there will be hepatocellular degeneration and necrosis.

■ nursing care, rumen acidifiers, vinegar

Treatment—For the digestive disturbance, the most common approach is directed toward relief of the pain and irritation. Early in the course of disease, lowering of the rumen pH to 5 or below will reduce the enzymatic release of additional irritant. The use of piperazine to react with and inactivate the allylisothiocyanate has also been suggested (Smith & Crowe 1987). Treatment of the skin necrosis involves providing shade or protection from sunlight and perhaps application of some type of skin ointment such as carron oil or charcoal in a liquid medium (Case 1957).

Genera of Questionable Toxicity

Crambe L.

Native to the Abyssinian highlands and plains of northern Africa, *C. abyssinica* L., commonly called Abyssinian kale or crambe, is occasionally cultivated in North America for its seed. It is a glaucous annual with large pinnately lobed or pinnatisect leaves arising from an erect stem that terminates in an elongate, corymbose raceme. The white flowers bear hard, globose, indehiscent silicles, each with only 1 large seed.

Crambe seeds and the pericarps of the silicles are processed into oil and a high-protein meal. Similar to that of rapeseed, the oil has a high erucic acid content and has been of considerable interest as a lubricant (Mustakas et al. 1968). Meal made from the seeds typically has greater than 40% crude protein and is considered a useful substitute for soybean meal (Anderson et al. 1993). Its palatability is markedly improved by treatment with heat and sodium carbonate (Mustakas et al. 1968; Lambert et al. 1970). Containing up to 4% glucosinolates, the meal can be fed to feedlot cattle at up to 4.2% of their rations (Price et al. 1993). The thioglucosides present in it can be removed by various processes to improve its usefulness as animal feed (Tookey et al. 1965). Toxicity is not a problem as long as the correct proportion of the meal in the supplement is maintained.

- *C. abyssinica*

  Abyssinian kale
  crambe

- seed meal good in small amounts, especially if glucosinolates removed

Raphanus L.

Believed to be native to the Mediterranean region, *Raphanus* comprises 3 species, 2 of which are present in North America. Weedy *R. raphanistrum* L.—commonly known as wild radish, jointed charlock, or white charlock—has naturalized across the United States and southern Canada. *Raphanus sativus* L., the cultivated radish, occasionally escapes but does not persist. Both species are annuals or biennials from stout or thickened taproots typically with basal rosettes of leaves. Their erect stems 30–80 cm tall bear obovate to oblong lower leaves with coarsely pinnatifid margins. Their upper cauline leaves are smaller, lanceolate to oblong, with typically dentate or entire margins. Bearing white flowers and then beaked, linear, indehiscent siliques, the racemes are elongate.

- *R. raphanistrum*
  R. sativus

  jointed charlock
  radish
  white charlock

- seeds, digestive tract irritation, colic, diarrhea

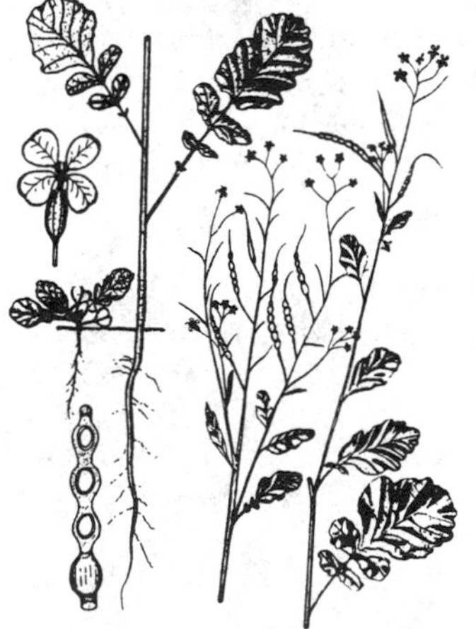

Raphanus raphanistrum

Raphanus sativus

Raphanus raphanistrum

Although these species have been of little problem, the seeds of *R. raphanistrum* are capable of irritating the digestive tract in the same fashion as other members of the family. Reddening of the mucosa of the stomach and intestine and death followed experimental administration of the seeds to rabbits (Steyn 1933). Horses developed colic after eating feed contaminated with 25–30% *Raphanus* seeds for several days (Eloire 1922).

Stanleya Nutt.

- *S. albescens*
  S. pinnata
  S. viridiflora

  prince's-plume

- selenium accumulators, unpalatable but possibly increases selenium availability to other nearby plants

Commonly known as prince's-plume, *Stanleya* comprises 6 species, all native to the western half of North America. Three species are of some toxicologic interest: *S. albescens* M.E.Jones, *S. pinnata* (Pursh) Britton, and *S. viridiflora* Nutt. ex Torr. & A.Gray. *Stanleya albescens* is an herbaceous biennial; the other 2 are subshrubs. The herbage of all is glabrous or indumented with unbranched hairs. Their cauline leaves are petiolate or sessile and clasping with entire to highly dissected margins. The dense, plumose racemes bear white or yellow flowers, which give rise to linear siliques.

Because they favor selenium-rich soils, species of *Stanleya* are generally considered to be indicators of readily available soil selenium (Trelease & Beath 1949). Selenocystathionine and Se-methyl-selenocysteine are synthesized and present in appreciable concentrations in *S. pinnata* (Rosenthal 1982). However, plants of the genus are of little toxicity risk because the dimethylselenides in their foliage typically render them unpalatable. The plants may pose a problem by increasing the availability of selenium in the surface soil for passive accumulators, such as corn, wheat, and other grains, that may be growing nearby and are likely to be eaten (Trelease & Beath 1949).

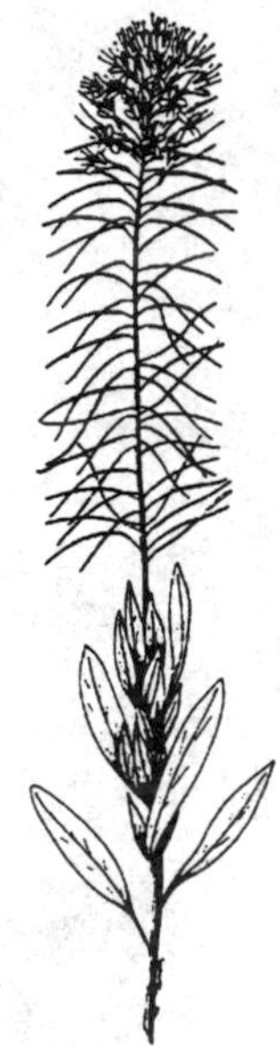

Stanleya pinnata

Stanleya albescens

References

Anderson VL, Slanger WD, Boyles SL, Berg PT: Crambe meal is equivalent to soybean meal for backgrounding and finishing beef steers. *J Anim Sci* 71;2608–2613, 1993

Backgren AW, Jonsson G: Blood and bone marrow studies in cattle feeding on *Brassica* species. *Acta Vet Scand* 10;309–318, 1969

Baksi SN, Case AA: Photosensitization in guinea-pigs due to ingestion of *Thlaspi arvense* (penny-cress mustard) seed. *Indian Vet J* 48;1001–1006, 1971

Barnouin J, Paccard P: Facteurs de risque nutritionnels de la pathologie hepatique dans les troupeaux bovins laitiers en France. *Can Vet J* 29;915–920, 1988

Barnouin J, Chalus T, Lescourret F: Increased perinatal French dairy calf mortality associated with fresh rape in the prepartum diet. *Prev Vet Med* 12;111–120, 1992

Barry TN, Duncan SJ, Sadler WA, Millar KR, Sheppard AD: Iodine metabolism and thyroid hormone relationships in growing sheep fed on kale *(Brassica oleracea)* and ryegrass *(Lolium perenne)*–clover *(Trifolium repens)* fresh-forage diets. *Br J Nutr* 49;241–253, 1983

Barry TN, Manley TR, Millar KR, Smith RH: The relative feeding value of kale *(Brassica oleracea)* containing normal and low concentrations of *S*-methyl-L-cysteine sulphoxide (SMCO). *J Agric Sci (Camb)* 102;635–643, 1984

Bartley EE, Nagaraja TG, Pressman ES, Dayton AD, Katz MP, Fina LR: Effects of lasalocid or monensin on legume or grain (feedlot) bloat. *J Anim Sci* 56;1400–1406, 1983

Becker RL, Martin NP, Murphy MJ: *Hoary Alyssum: Toxicity to Horses, Forage Quality, and Control.* Minn Agric Ext AG-FS-5567-A, 1991

Bell JM: From rapeseed to canola: a brief history of research for superior meal and edible oil. *Poult Sci* 61;613–622, 1982

Bell JM: Nutrients and toxicants in rapeseed meal: a review. *J Anim Sci* 58;996–1010, 1984

Belokon VF, Makarevich IF: Cardiac glycosides from *Cheiranthus cheiri*. 2. *Khim Prir Soedin* 1980(3);424–425 (*Chem Abstr* 93:128778x, 1980)

Benevenga NJ, Case GL, Steele RD: Occurrence and metabolism of *S*-methyl-L-cysteine and *S*-methyl-L-cysteine sulfoxide in plants and their toxicity and metabolism in animals. In *Toxicants of Plant Origin,* vol 3, *Proteins and Amino Acids,* Cheeke PR ed, CRC Press, Boca Raton, Fla, pp203–228, 1989

Boyd H, Reed HCB: Investigation into the incidence and causes of infertility in dairy cattle—influence of kale feeding, milk production, and management factors associated with "farming intensity." *Br Vet J* 117;192–200, 1961

Bulgin MS, Lincoln SD, Mather G: Elemental sulfur toxicosis in a flock of sheep. *J Am Vet Med Assoc* 208;1063–1065, 1996

Butler EJ, Pearson AW, Fenwick GR: Problems which limit the use of rapeseed meal as a protein source in poultry diets. *J Sci Food Agric* 33;866–875, 1982

Campbell SJ: Quality control in a canola crushing plant. *J Am Oil Chem Soc* 61;1097–1101, 1984

Case AA: Photosensitization syndrome in cattle, sheep, and swine. *North Am Vet* 38;161–165, 1957

Clegg FG, Evans RK: Haemoglobinaemia of cattle associated with the feeding of Brassicae species. *Vet Rec* 74;1169–1176, 1962

Coffey S: *Rodd's Chemistry of Carbon Compounds,* 2d ed. Vol 2, part D, *Alicyclic Compounds,* Elsevier, Amsterdam, pp360–421, 1970

Cote FT: Rape poisoning in cattle. *Can J Comp Med* 8;38–41, 1944

Cox-Ganser JM, Jung GA, Pushkin RT, Reid RL: Evaluation of brassicas in grazing systems for sheep. 2. Blood composition and nutrient status. *J Anim Sci* 72;1832–1841, 1994

Crawshaw HA: Rape blindness [letter]. *Vet Rec* 65;254, 1953

Cummings BA, Gould DH, Caldwell DR, Hamar DW: Ruminal microbial alterations associated with sulfide generation in steers with dietary sulfate-induced polioencephalomalacia. *Am J Vet Res* 56;1390–1395, 1995

Dalton PJ: Rape blindness [letter]. *Vet Rec* 65;298, 1953

David JSE: The effect of prolonged kale feeding on the thyroid glands of sheep. *J Comp Pathol* 86;235–241, 1976

Earl CRA, Smith RH: Dimethyl disulphide in the blood of cattle fed on brassicas. *J Sci Food Agric* 34;23–28, 1982

Eaton G: Suspected poisoning of bullocks by white mustard. *Vet Rec* 53;146, 1941

Ellison SP: Possible toxicity caused by hoary alyssum *(Berteroa incana)*. *Vet Med* 87;473–475, 1992

Eloire A: Poisoning of cart horses by field wallflower (wild radish). *Vet J* 78;209, 1922

Evans ETR: Kale and rape poisoning in cattle [letter]. *Vet Rec* 63;348–349, 1951

Fales SL, Gustine DL, Bosworth SC, Hoover RJ: Concentrations of glucosinolates and *S*-methylcysteine sulfoxide in ensiled rap (*Brassica napus* L.). *J Dairy Sci* 70;2402–2405, 1987

Fenwick GR, Curtis RF: Rapeseed meal and its use in poultry diets: a review. *Anim Feed Sci Tech* 5;255–298, 1980

Fenwick GR, Heaney RK, Mullin WJ: Glucosinolates and their breakdown products in food and food plants. *CRC Crit Rev Food Sci Nutr* 18;123–201, 1983

Fenwick GR, Heaney RK, Mawson R: Glucosinolates. In *Toxicants of Plant Origin,* vol 2, *Glycosides,* Cheeke PR ed, CRC Press, Boca Raton, Fla, pp1–41, 1989

Fursa NS, Dolya VS, Litvinenko VI: Chemical composition of seeds of some species of the genus *Erysimum*. *Rastit Resur* 20;244–248, 1984 (*Chem Abstr* 101:3977p, 1984)

Gallie JGE, Paterson JD: Charlock poisoning of lambs. *Vet Rec* 57;198, 1945

Geor RJ, Becker RL, Kanara EW, Hovda LR, Sweeney WH, Winter TF, Rorick JK, Ruth GR, Hope E, Murphy MJ: Toxicosis in horses after ingestion of hoary alyssum. *J Am Vet Med Assoc* 201;63–67, 1992

Giovanni R, Barbedette D, Allez M, Viroben G: The anemia factor in cruciferous forages. 1. *S*-Methylcysteinesulphoxide content. *Fourrages* 117;49–63, 1989

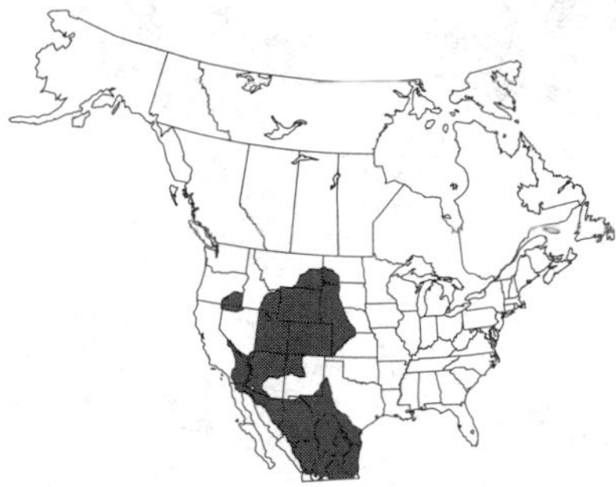

Stanleya pinnata

Stanleya viridiflora

Gmelin R, Virtanen AI: A new type of enzymatic cleavage of mustard oil glucosides: formation of allylthiocyanate in *Thlaspi arvense* L. and benzylthiocyanate in *Lepidium ruderale* L. and *Lepidium sativum* L. *Acta Chem Scand* 13;1474–1475, 1959

Gooneratne SR, Buckley WT, Christensen DA: Review of copper deficiency and metabolism in ruminants. *Can J Anim Sci* 69;819–845, 1989a

Gooneratne SR, Olkowski AA, Klemmer RG, Kessler GA, Christensen DA: High sulphur related thiamine deficiency in cattle: a field study. *Can Vet J* 30;139–146, 1989b

Gooneratne SR, Olkowski AA, Christensen DA, Rousseaux CG, Chauvet A: Ovine polioencephalomalacia associated with dietary sulphur intake. *J Vet Med* 38;229–239, 1991

Gooneratne SR, Olkowski AA, Rousseaux CG, Christensen DA: Role of thiamine status in sulfur induced polioencephalomalacia in sheep. *Res Vet Sci* 52;78–85, 1992

Gould DH, McAllister MM, Savage JC, Hamar DW: High sulfide concentrations in rumen fluid associated with nutritionally induced polioencephalomalacia in calves. *Am J Vet Res* 52;1164–1169, 1991

Grant CA, Holtenius P, Jonsson G, Thorell CB: Kale anemia in ruminants. 1. Survey of the literature and experimental induction of kale anaemia in lactating cows. *Acta Vet Scand* 9;126–140, 1968a

Grant CA, Holtenius P, Jonsson G, Luthman J, Thorell CB: Kale anemia in ruminants. 2. Observations on kale-fed sheep. *Acta Vet Scand* 9;141–150, 1968b

Greenhalgh JFD, Sharman GAM, Aitken JN: Kale anemia. 1. The toxicity to various species of animal of three types of kale. *Res Vet Sci* 10;64–72, 1969

Hakkila MD, Holechek JL, Wallace JD, Anderson DM, Cardenas M: Diet and forage intake of cattle on desert grassland range. *J Range Manage* 40;339–342, 1987

Hammond AC, Carlson JR, Breeze RG: Prevention of tryptophan- induced acute bovine pulmonary oedema and emphysema (fog fever). *Vet Rec* 107;322–325, 1980

Hansen AA: *Indiana Plants Injurious to Livestock.* Indiana Agric Exp Stn Circ 175, 1930

Hershey AL: Some poisonous plant problems of New Mexico. *N M Agric Exp Stn Bull* 322;20–21, 1945

Hill FI, Ebbett PC: Polioencephalomalacia in cattle in New Zealand fed chou moellier *(Brassica oleracea). N Z Vet J* 45;37–39, 1997

Hill R: A review of the "toxic" effects of rapeseed meals with observations on meal from improved varieties. *Br Vet J* 135;3–16, 1979

Holmes RG: A case of suspected poisoning of dairy cows by white mustard seeds *(Sinapis alba). Vet Rec* 77;480–481, 1965

Honeyfield DC, Carlson JR, Nocerini MR, Breeze RG: Duration of inhibition of 3-methylindole production by monensin. *J Anim Sci* 60;226–231, 1985

Hovda LR, Rose ML: Hoary alyssum *(Berteroa incana)* toxicity in a herd of broodmare horses. *Vet Hum Toxicol* 35;39–40, 1993

Hughes HT: Mustard poisoning. *Vet J* 80;43–45, 1924

Joubert JPJ: Cardiac glycosides. In *Toxicants of Plant Origin,* vol 2, *Glycosides,* Cheeke PR ed, CRC Press, Boca Raton, Fla, pp61–96, 1989

Kernaleguen A, Smith RA, Yong CW: Acute mustard seed toxicosis in beef cattle: Saskatchewan. *Can Vet J* 30;524, 1989

Knight AP: An update on plant poisoning. In *Proceedings, 14th ACVIM Forum,* pp509–511, 1996

Kortus M, Kowalewski Z: Cardenolide compounds in *Erysimum wahlenbergii* (Asch. et Eng.) *Borb. Ann Pharm (Poznan)* 12;103–111, 1977 (*Chem Abstr* 87:130532g 1977)

Kowalewski Z: Papierchromatographische untersuchung der cardenolide von 8 *Erysimum-arten* und zwei vertretern verwandter gattungen. *Helv Chim Acta* 43;1314–1321, 1960

Lajolo FM, Marquez UML, Filisetti TMCC, McGregor DI: Chemical composition and toxic compounds in rapeseed (*Brassica napus* L.) cultivars grown in Brazil. *J Agric Food Chem* 39;1933–1937, 1991

Lambert JL, Clanton DC, Wolff IA, Mustakas GC: Crambe meal protein and hulls in beef cattle rations. *J Anim Sci* 31;601–607, 1970

Leichter J, Joslyn MA: Kinetics of thiamine cleavage by sulphite. *Biochem J* 113;611–615, 1969

Lewis D: The reduction of sulphate in the rumen of sheep. *Biochem J* 56;391–399, 1954

Lewis WH, Elvin-Lewis MPF: *Medical Botany—Plants Affecting Man's Health.* Wiley, New York, p36, 1977

Majak W, McDiarmid RE, Benn MH, Willms WD: Autolysis of *Thlaspi arvense* in bovine rumen fluid. *Phytochemistry* 30;127–129, 1991

Makarevich IF, Zolot'ko ZS, Kolesnikov DD: Extraction of strophanthidin and accompanying cardenolides from *Erysimum canescens* Roth. *Farm Zh (Kiev)* 1980(6);33–35, 1980

Mallard C: Toxicity of Cruciferae and pregnancy disorders in a herd of goats. *Bull Group Tech Vet* 3;23–25, 1981

Marks HS, Hilson JA, Leichtweis HC, Stoewsand GS: *S*-Methylcysteine sulfoxide in *Brassica* vegetables and formation of methyl methanethiosulfinate from brussels sprouts. *J Agric Food Chem* 40;2098–2101, 1992

Martin T, Morgan S: What caused the photosensitivity in these dairy heifers? *Vet Med* 82;848–850, 1987

Mason RW, Lucas P: Acute poisoning in cattle after eating old non-viable seed of chou moellier *(Brassica oleracea* convar. *acephala). Aust Vet J* 6;272–273, 1983

Mayland HF, Robbins CW: Sulfate uptake by salinity-tolerant plant species. *Commun Soil Sci Plant Anal* 25;2523–2541, 1994

McCutcheon JS, Umermura T, Bhatnagar MK, Walker BL: Cardiopathogenicity of rapeseed oils and oil blends differing in erucic, linoleic, and linolenic acid content. *Lipids* 11;545–552, 1976

McDonald RC, Manley TR, Barry TN, Forss DA, Sinclair AG: Nutritional evaluation of kale *(Brassica oleracea)* diets. 3. Changes in plant composition induced by fertility practices, with special reference to SMCO and glucosinolate concentrations. *J Agric Sci (Camb)* 97;13–23, 1981

Mejia MR, Montalvo VR, Martinez RR: Levels of oxalates in wild forages coming from the states of Hidalgo, Guanajuato, Mexico, Tlaxcala, and the Federal District. *Vet Mex* 16;21–25, 1985

Melrose DR, Brown BB: Some observations on the possible effect of kale feeding on fertility in dairy cattle. *J Reprod Fert* 4;232, 1962

Michael DT: Rape or cole: some observations on its management in relation to the health of sheep grazing on it. *Vet Rec* 65;231–232, 1953

Mihalka S: Puerperal losses caused by plant poisoning in a sheep flock. *Magy Allatorv Lapja* 37;456–458, 1982

Mihalka S: Reproductive losses caused by chlamydiosis and concurrent plant poisoning *(Sinapis arvensis)* on a sheep farm. *Magy Allatorv Lapja* 38;625–627, 1983

Morton JM, Campbell PH: Disease signs reported in south-eastern Australian dairy cattle while grazing *Brassica* species. *Aust Vet J* 75;109–113, 1997

Murphy MJ: Hoary alyssum *(Berteroa incana)* toxicosis. Presented at Fifth International Symposium on Poisonous Plants, San Angelo, Tex, 1997

Mustakas GC, Kirk LD, Griffin ELjr: Crambe seed processing: improved feed meal by soda ash. *J Am Oil Chem Soc* 45;53–57, 1968

Nocerini MR, Honeyfield DC, Carlson JR, Breeze RG: Reduction of 3-methylindole production and prevention of acute bovine pulmonary edema and emphysema with lasalocid. *J Anim Sci* 60;232–238, 1985

Olkowski AA, Gooneratne SR, Rousseaux CG, Christensen DA: Role of thiamine status in sulphur induced polioencephalomalacia in sheep. *Res Vet Sci* 52;78–85, 1992

Onderscheka K, Tataruch F, Steineck T, Klansek E, Vodnansky M, Wagner J, Echsel H: Studies on losses in wild roe deer caused by eating rape. *Wien Tierarztl Monatsschr* 74;369–379, 1987

Pantuck EJ, Hsiao K-C, Loub WD, Wattenberg LW, Kuntzman R, Conney AH: Stimulatory effect of vegetables on intestinal drug metabolism in the rat. *J Pharmacol Exp Ther* 198;278–283, 1976

Park JBS, Kubelka W, Jentzsch K: Cardenolides of the seeds of wallflower, *Erysimum cheiri. Sci Pharm* 57;445–449, 1989 (*Chem Abstr* 112:136033b, 1980)

Papas A, Ingalls JR, Campbell LD: Studies on the effects of rapeseed meal on thyroid status of cattle, glucosinolate and iodine content of milk, and other parameters. *J Nutr* 109;1129–1139, 1979

Park JBS, Kubelka W, Jentzsch K: Cardenolides of the seeds of wallflower, *Erysimum cheiri. Sci Pharm* 57;445–449, 1989 (*Chem Abstr* 112:136033b, 1990)

Park RJ: Weed taints in dairy produce. 1. *Lepidium* taint. *J Dairy Res* 36;31–35, 1969

Park RJ, Armitt JD, Stark W: Weed taints in dairy produce. 2. *Coronopus* or land cress taint in milk. *J Dairy Res* 36;37–46, 1969

Paxman PJ, Hill R: The goitrogenicity of kale and its relation to thiocyanate content. *J Sci Food Agric* 25;329–337, 1974a

Paxman PJ, Hill R: Thiocyanate content of kale. *J Sci Food Agric* 25;323–328, 1974b

Pelletier G, Martin LJ: The blood picture of sheep fed on fresh and dried marrow-stem kale. *Can J Anim Sci* 53;229–236, 1973

Penny RHC, David JSE, Wright AI: Observations on the blood picture of cattle, sheep, and rabbits fed on kale. *Vet Rec* 76;1053–1059, 1964

Perrett DR: Suspected rape poisoning in cattle. *Vet Rec* 59;674, 1947

Pfister JA, Baker DC, Lacey JR, Brownson R: Photosensitization of cattle in Montana: is *Descurainia pinnata* the culprit? *Vet Hum Toxicol* 31;225–227, 1989

Pfister JA, Lacey JR, Baker DC, James LF, Brownson R: Is tansymustard causing photosensitization of cattle in Montana? *Rangelands* 12;170–172, 1990

Pickard DW, Crighton DB: An investigation into the possible oestrogenic effect of kale. *Br Vet J* 123;64–69, 1967

Potchoiba MJ, Carlson JR, Nocerini MR, Breeze RG: Effect of monensin and supplemental hay on ruminal 3-methylindole formation in adult cows after abrupt change to lush pasture. *Am J Vet Res* 53;129–133, 1992

Price WD, Lovell RA, McChesney DG: Naturally occurring toxins in feedstuffs: Center for Veterinary Medicine perspective. *J Anim Sci* 71;2556–2562, 1993

Qi K, Lu CD, Owens FN: Sulfate supplementation of Angora goats: sulfur metabolism and interactions with zinc, copper, and molybdenum. *Small Ruminant Res* 11;209–225, 1993

Raisbeck MF: Is polioencephalomalacia associated with high-sulfate diets. *J Am Vet Med Assoc* 180;1303–1305, 1982

Reid RL, Puoli JR, Jung GA, Cox-Ganser JM, McCoy A: Evaluation of brassicas in grazing systems for sheep, 1. Quality of forage and animal performance. *J Anim Sci* 72;1823–1831, 1994

Rollins RC: *The Cruciferae of North America.* Stanford Univ Press, Stanford, Calif, 1993

Rosenthal GA: *Plant Nonprotein Amino and Imino Acids.* Academic Press, New York, 1982

Rosiere RE, Beck RF, Wallace JD: Cattle diets on semidesert grassland: botanical composition. *J Range Manage* 28;89–93;1975

Rousseaux CG, Olkowski AA, Chauvet A, Gooneratne SR, Christenson DA: Ovine polioencephalomalacia associated with dietary sulphur intake. *J Vet Med* A 38;229–239, 1991

Russel AJF: A note on goitre in lambs grazing rape *(Brassica napus)*. *Anim Prod* 9;131–133, 1967

Schofield FW: Acute pulmonary emphysema of cattle. *J Am Vet Med Assoc* 112;254–259, 1948

Schoon HA, Brunckhorst D, Fehlberg U: Rape intoxication in roe deer. *Prakt Tierarzt* 70;50–52, 1989

Semalulu SS, Rousseaux CG: Suspected oriental mustard seed *(Brassica juncea)* poisoning in cattle: Saskatchewan. *Can Vet J* 30;595–596, 1989

Shires A, Bell JM, Keith MO, McGregor DI: Rapeseed dockage: effects of feeding raw and processed wild mustard and stinkweed seed on growth and feed utilization of mice. *Can J Anim Sci* 62;275–285, 1982

Shoppee CW: *Chemistry of the Steroids,* 2d ed. Butterworths, Washington DC, pp326–397, 1964
Sinclair DP, Andrews ED: Prevention of goitre in new-born lambs from kale-fed ewes. *N Z Vet J* 6;87–95, 1958
Sinclair DP, Andrews ED: Deaths due to goitre in new-born lambs prevented by iodized poppy-seed oil. *N Z Vet J* 9;96–100, 1961
Singh B, Rastogi RP: Cardenolides—glycosides and genins. *Phytochemistry* 9;315–331, 1970
Smith RA, Crowe SP: Fanweed toxicosis in cattle: case history, analytical method, suggested treatment, and fanweed detoxification. *Vet Hum Toxicol* 29;155–156, 1987
Smith RH: Kale and rape poisoning. *Vet Ann* 17;28–33, 1977
Smith RH: Kale poisoning: the brassica anaemia factor. *Vet Rec* 107;12–15, 1980
Smith RH, Kay M, Matheson NA, Lawson W: *S*-Methylcysteine sulphoxide, the ruminant kale anemia factor. *J Sci Food Agric* 29;414–416, 1978
Staley E: A treatment for tansy mustard poisoning. *Okla Vet* 28;74, 1976
Stamp JT, Stewart J: Haemolytic anaemia with jaundice in sheep. *J Comp Pathol* 63;48–52, 1953
Steyn DG: Poisoning of human beings by weeds contained in cereals (bread poisoning). *Onderstepoort J Vet Sci Anim Ind* 1;219–269, 1933
Stocks DC, Dunster PJ, Gibson JA: Observations on the effects of *Rapistrum rugosum* on thyroid function. *Aust Vet J* 61;264–265, 1984
Stoewsand GS: Bioactive organosulfur phytochemicals in *Brassica oleracea* vegetables—a review. *Food Chem Toxicol* 33;537–543, 1995
Suttle NF: Effects of organic and inorganic sulphur on the availability of dietary copper to sheep. *Br J Nutr* 32;559–568, 1974
Suttle NF: The interactions between copper, molybdenum, and sulphur in ruminant nutrition. *Ann Rev Nutr* 11:121–140, 1991
Taljaard TL: Cabbage poisoning in ruminants. *J S Afr Vet Assoc* 64;96–100, 1993
Thomas KW: Oral treatment of polioencephalomalacia and subclinical thiamine deficiency with thiamine propyl disulphide and thiamine hydrochloride. *J Vet Pharmacol Ther* 9;402–411, 1986
Tookey HL, VanEtten CH, Peters JE, Wolff IA: Evaluation of enzyme-modified, solvent-extracted crambe seed meal by chemical analysis and rat feeding. *Cereal Chem* 42;507–514, 1965
Trelease SF, Beath OA: *Selenium.* By the Authors, New York, 1949
Troxler J: Fatal poisoning of 19 heifers fed fresh white mustard (*Sinapis alba* L.). *Schweiz Arch Tierheilkd* 123;495–497, 1981
Tsunoda S, Hinata K, Gomez-Campo C: *Brassica Crops and Wild Allies.* Japan Scientific Societies Press, Tokyo, 1980
Tucker EM: The onset of anaemia and the production of haemoglobin C in sheep fed on kale. *Br Vet J* 125;472–479, 1969
Tyagi AK, Tripathi MK, Karim SA: Effect of soaking on glucosinolate of mustard-cake and its utilization in sheep feeding. *Indian J Anim* Sci 67;76–77, 1997
VanEtten CH, Tookey HL: Glucosinolates from cruciferous plants. In *Effects of Poisonous Plants on Livestock,* Keeler RF, Van Kampen KR, James LF eds, Academic Press, New York, pp507–520, 1978
Vermunt JJ, West DM, Cooke MM: Rape poisoning in sheep. *N Z Vet J* 41;151–152, 1993
Virtanen AI: Studies on organic sulphur compounds and other labile substances in plants. *Phytochemistry* 4;207–228, 1965
Wall ME, Taylor H, Perera P, Wani MC: Indoles in edible members of the Cruciferae. *J Nat Prod* 51;129–135, 1988
Wheeler JL, Park RJ, Spurway RA, Ford AL: Variation in the effects of rape forage on meat flavor in sheep. *J Agric Sci* 83;569–571, 1974
Whittle PJ, Smith RH, McIntosh A: Estimation of *S*-methylcysteine sulphoxide (kale

anaemia factor) and its distribution among *Brassica* forage and root crops. *J Sci Food Agric* 27;633–642, 1976

Wikse SE, Leathers CW, Parish SM: Diseases of cattle that graze turnips. *Compend Contin Educ Food Anim* 9;F112–F121, 1987

Williams HL, Hill R, Alderman G: The effects of feeding kale to breeding ewes. *Br Vet J* 121;2–17, 1965

Young NE, Austin AR, Orr RJ, Newton JE, Taylor RJ: A comparison of hybrid stubble turnip (cv. Appin) with other cruciferous catch crops for lamb fattening. 2. Animal performance and toxicological evaluation. *Grass Forage Sci* 37;39–46, 1982

Calycanthaceae Lindl. 17

Comprising 3 genera and 6–9 species native to temperate North America and eastern Asia, the Calycanthaceae, or strawberry-shrub family, is of little economic importance other than for the ornamental value of its species, which are grown for the aromatic fragrance of their showy flowers (Nicely 1965). The family resembles the Magnoliaceae and Annonaceae in its spirally arranged perianth parts and its numerous stamens and pistils. In North America, the family is represented by the native genus *Calycanthus*, which contains a strychninelike alkaloid. Occasionally cultivated is the Asiatic shrub *Chimonanthus*.

■ Strawberry-shrub Family
Calycanthus

Plants shrubs; deciduous; rhizomatous; aromatic. **Leaves** simple; opposite; petiolate; stipules absent. **Inflorescences** solitary flowers; borne at ends of short, leafy branches. **Flowers** perfect; perianths in 1-series; radially symmetrical. **Perianth Parts** 15–30; petaloid; fleshy; spirally arranged. **Stamens** 10–20; spirally arranged; staminodia 10–25, inside stamens. **Pistils** 5–35; simple; spirally arranged; stigmas 1, decurrent; styles 1, filiform; ovaries superior; locules 1; ovules 2; placentation parietal. **Hypanthia** urceolate or campanulate. **Fruits** achenes; 1–35; enclosed within fleshy hypanthia. **Seeds** 1.

■ aromatic shrubs; leaves simple, opposite

Calycanthus L.

Taxonomy and Morphology—Commonly known as strawberry shrub, sweet shrub, or spicebush, *Calycanthus* contains only 2 species. Its name is derived from the Greek roots *kalyx* and *anthos,* meaning "covering" or "cup" and "flower," and is believed to reflect the uniseriate perianth. In the nineteenth and early twentieth centuries, plants of the genus were frequently grown in gardens for their pleasant scent. Both species are native to North America and of toxicologic interest:

■ *C. floridus*
C. occidentalis

allspice
strawberry shrub
sweet shrub

C. floridus L.	Carolina allspice, sweet shrub, strawberry shrub
C. occidentalis Hook. & Arn.	California allspice, California sweet shrub

Plants aromatic. **Stems** erect or ascending; 100–400 cm tall; branchlets terete or 4-angled; pubescent or glabrous; lateral buds exposed or hidden by petioles; bud scales absent. **Leaves** 2-ranked; apices acute or acuminate or obtuse; margins entire or rarely serrate; bases acute or truncate or obtuse. **Flowers** aromatic, with strawberry or pineapple scent; 4–7 cm in diameter. **Perianth Parts** linear to oblanceolate or spathulate; maroon or reddish brown or green tipped. **Stamens** 10–20; linear to oblong, ribbon-shaped. **Achenes** cylindrical; dark reddish brown.

■ aromatic shrubs; flowers aromatic, strawberry scent; perianth parts maroon, reddish brown, or green tipped

Calycanthus occidentalis

- wooded sites
- neurologic disease, tetanic seizures, strychninelike, mainly from eating fruits
- alkaloids, calycanthine; indole alkaloids, folicanthine, neurotoxic, CNS stimulant, exaggerated reflexes

Calycanthus floridus

Calycanthus occidentalis

1. Plants of Appalachian Mountains region. Lateral buds hidden by petiole bases. Perianth parts oblong-elliptic to obovate-lanceolate; apices acute . . . . *C. floridus*
1. Plants of Sierra-Cascade and Coastal Mountains region. Lateral buds exposed. Perianth parts linear to linear spathulate or ovate-elliptic; apices rounded . . . . *C. occidentalis*

Distribution and Habitat—Both species of *Calycanthus* occupy ecologically similar habitats along streams and at the edges of woods. *Calycanthus floridus* occurs in the eastern third of the continent, while *C. occidentalis* is found on the coast and in the Sierra-Cascade Mountains of the Pacific Coast states.

Disease Problems—Various tribes of American Indians used the 2 species of *Calycanthus* to treat urinary problems and severe colds. In the late 1800s, a disease problem involving *C. floridus* was described (Chesnut 1898). Limited mainly to the southern Appalachian Mountain region of Georgia and Alabama and adjacent Florida, it was reputed to have caused the death of hundreds of cattle and sheep when the animals ate 5–10 fruits (Sterns 1888). Anecdotal information indicated that only ruminants were affected, mainly after drinking water. Although the plants were not normally eaten, animals were believed to acquire a taste for plants of the species (Sterns 1888; Dayton 1931). Animals were said to die suddenly or to be ill for several weeks before dying. It is now known that intoxication is related primarily to ingestion of the fruits, which cause a neurologic disease characterized by tetanic seizures, protrusion of the third eyelid, hyperesthesia, and, in general, signs reminiscent of strychnine intoxication (Bradley & Jones 1963).

Disease Genesis—The toxicants of *Calycanthus* have long been known to be alkaloids, but it was not until the 1950s and 1960s that their structures were resolved. The principal alkaloid is calycanthine, a very complex nonindole structure (Manske 1965). The remaining three toxins are the closely related indole alkaloids chimonanthine, calycanthidine, and folicanthine. They differ only in the presence of an increasing number of methyl groups—two, three, and four, respectively.

Calycanthine in many ways acts in a manner similar to strychnine. It is a stimulant of the central nervous system and causes exaggerated reflexes; thus it was considered to have strychninelike action (Chen et al. 1942). In the cockroach's nervous system, although not changing pre- or postsynaptic transmission directly, it does depress excess acetylcholine depolarization and thereby decreases the efficiency of synaptic transmission (Adjibade et al. 1991). Unlike strychnine, it lacks potassium channel effects on the axonal membrane. In vitro experiments show inhibition of contraction of tracheal strips from hypersensitized guinea pigs, suggesting inhibition of histamine release (Bertrand et al. 1989).

McGuigan and Hess (1912) evaluated the effects of isocalycanthine, a compound named because of its slight differences in melting point and other points from a previously isolated fraction of calycanthine. At low dosage in frogs it caused tetanic spasms, and at high dosage the result was paralysis. In cats it caused increased response to stimuli, markedly increased muscle tone, spasticity and tremors, and, in some instances, clonic-type seizures. It also caused cardiac ventricular depression. Epinephrine provided some protection against lethality.

Plants also contain small amounts of the β-carboline harmine (Allen & Holmstedt 1980).

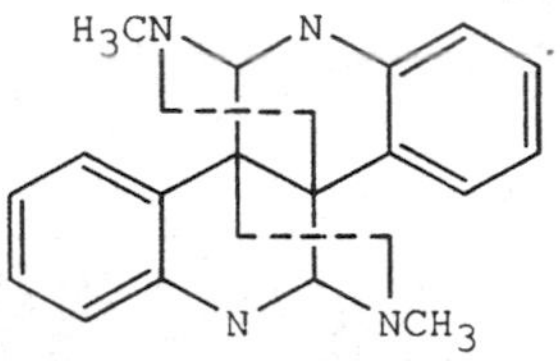

Calycanthine

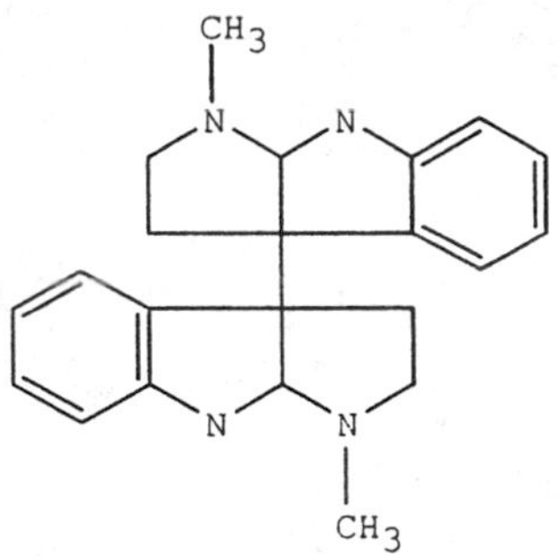

Chimonanthine

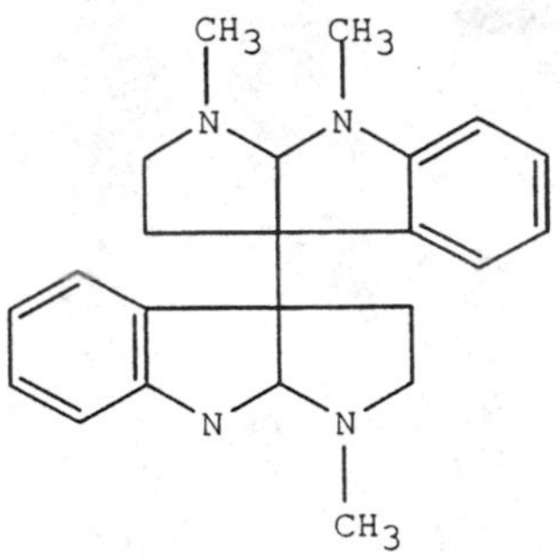

Calycanthidine

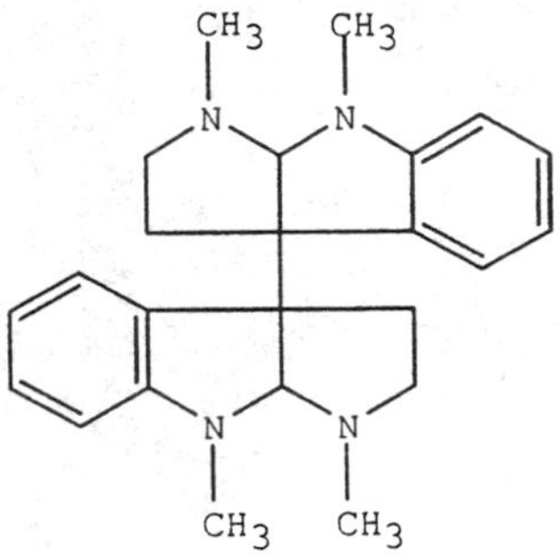

Folicanthine

Clinical Signs—The signs of *Calycanthus* intoxication are abrupt in onset and include muscle fasciculations, increased sensitivity to touch or other stimuli, incoordination that is especially noticeable when running, and seizures. The seizures are tetanic or rigid in nature and quickly render the animal unable to stand, after which the seizures worsen and the animal lies in lateral recumbency with legs rigid and the head twisted back. The third eyelid may protrude somewhat. Serum chemistries and hematology are both normal.

Pathology—Gross lesions are few, and those present are limited to scattered, small splotchy hemorrhages, suggestive of an agonal death. Examination of the animal's ruminal contents may reveal portions of the plant's fleshy hypanthia and cylindrical fruits.

Treatment—The seizures require administration of sedatives in a manner similar to that of the general treatment of strychnine intoxication. It may also be helpful to keep the animal in a quiet, dark area to reduce external stimuli and their inducement of seizures.

- abrubt onset, muscle tremors, hypersensitivity, ataxia, rigidity, tetanic seizures, protrusion of third eyelid
- no lesions, may have visible plant material in rumen
- sedation, quiet environment

References

Adjibade Y, Hue B, Pelhate M, Anton R: Action of calycanthine on nervous transmission in cockroach central nervous system. *Planta Med* 57;99–101, 1991

Allen JRF, Holmstedt BR: The simple -carboline alkaloids. *Phytochemistry* 19;1573–1582, 1980

Bertrand C, Adjibade Y, Gies JP, Kuballa B, Landry Y, Anton R: Pharmacological effects of calycanthine hydrochloride in guinea-pig airways. *Planta Med* 55;648, 1989

Bradley RE, Jones TJ: Strychnine-like toxicity of *Calycanthus*. *Southwest Vet* 14;40,71,73, 1963

Chen AL, Powell CE, Chen KK: The action of calycanthine. *J Am Pharm Assoc* 31;513–516, 1942

Chesnut VK: *Preliminary Catalogue of Plants Poisonous to Stock*. USDA Bur Anim Ind 15th Annu Rep, pp387–420, 1898

Dayton WA: *Important Western Browse Plants*. USDA Misc Publ 101, 1931

Manske RHF: The alkaloids of Calycanthaceae. In *The Alkaloids: Chemistry and Physiology*, vol 8, Manske RHF ed, Academic Press, New York, pp581–589, 1965

McGuigan H, von Hess CL: Isocalycanthine and its quaternary base. *J Pharmacol Exp Ther* 3;441–454, 1912

Nicely KA: A monographic study of the Calycanthaceae. *Castanea* 30;38–81, 1965

Sterns EE: The fruit of *Calycanthus* L. *Bull Torrey Bot Club* 15;205–209, 1888

18 Campanulaceae Juss.

- Bluebell Family
  Bellflower Family
  Lobelia

Comprising 70–85 genera and 1,950–2,000 species, the Campanulaceae, or bellflower family, is cosmopolitan in distribution. It is divided into 2 distinct subfamilies, the Campanuloideae and Lobelioideae, which differ in the symmetry and orientation of their flowers and the fusion of their anthers. Because of these differences, taxa of the Lobelioideae are segregated by some taxonomists into their own family; however, numerous intermediates exist and thus one large family is used in this treatise. Producing generally large, showy flowers, more than 140 species are cultivated as easy-to-grow ornamentals, especially in rock gardens. They are particularly prized for their many shades of blue. Representative are Canterbury bells, harebells, lobelias, Scottish bluebells, and cardinal flowers. In North America, approximately 25 native and introduced genera and 240 species are present. Toxicologic problems are associated with only one genus.

- herbs, shrubs, small trees; leaves simple, alternate; petals 5 or 0, fused, blue, red, white, or purple; fruits capsules

Plants herbs or rarely shrubs or small trees; annuals or biennials or perennials. **Leaves** simple; alternate; primarily cauline; venation pinnate; stipules absent. **Inflorescences** spikes or racemes or panicles or solitary flowers; terminal or axillary. **Flowers** perfect; perianths in 2- or 1-series. **Sepals** 5 or 3 or 4; fused. **Corollas** radially or bilaterally symmetrical; bowl-shaped or bilabiate, with 2 petals above and 3 below. **Petals** 5 or 0; all alike or of 2 forms; fused; blue or red or white or purple. **Androecia** radially or bilaterally symmetrical. **Stamens** 5; of equal length or of 2 lengths; free or fused by filaments or by anthers. **Pistils** 1; compound, carpels 2–5; stigmas 2–5, capitate; styles 1; ovaries inferior; locules 2–5; placentation axile or parietal. **Nectaries** present; staminal or borne on pistils at bases of styles. **Fruits** capsules; poricidal or loculicidal. **Seeds** numerous.

- some *Campanula* contain toxic calystegins

Toxicologically, the family is known mainly for the toxic pyridine alkaloids in *Lobelia,* but some species of *Campanula* contain the toxic glycosidase-inhibiting calystegins. However, an association with disease has not been shown for *Campanula,* in contrast to the effects of the calystegins in *Solanum* (Nash et al. 1998).

Lobelia L.

- *L. berlandieri*
  L. cardinalis
  L. inflata
  L. siphilitica
  cardinal flower
  Indian pink
  Indian tobacco
  lobelia

Taxonomy and Morphology—Primarily a genus of the Southern Hemisphere, *Lobelia* was introduced to European botanists in the early 1600s. Comprising 250–365 species, its name honors Matthias de l'Obel, a sixteenth-century Flemish botanist. Quite diverse in form and habitat, species vary from montane trees to small herbaceous weeds. Many are cultivated as edging plants, in window boxes, or as trailing vines in hanging baskets. Approximately 45 native and introduced species and named hybrids are present in North America. Of toxicologic interest are the following:

L. berlandieri A.DC.	Berlandier lobelia
L. cardinalis L.	cardinal flower, Indian pink
L. inflata L.	Indian tobacco
L. siphilitica L.	great lobelia, blue lobelia, Louisiana lobelia, blue cardinal flower

■ erect herbs, sap milky in some; flowers bilabiate; petals blue, lavender, purple, red, or white

Plants herbs; annuals or perennials from short basal shoots; roots fibrous; sap often white and acrid. **Stems** erect; 60–150 cm tall; branched or not branched. **Leaves** sessile or subsessile or petiolate; blades lanceolate or elliptic to oblanceolate; margins entire or serrate or dentate. **Inflorescences** terminal racemes or rarely solitary flowers in axils of uppermost leaves; bracts present; pedicels twisted before anthesis. **Flowers** in 2-series; bilabiate. **Sepals** 5. **Petals** 5; fused, but tubes split to bases; lower 3 lobes large, spreading; upper 2 lobes small, recurved; fenestrate or not fenestrate; blue or lavender or purple or red or white. **Stamens** of 2 lengths; fused by upper filaments and anthers. **Pistils** 1; stigmas 2-lobed; locules 2; placentation axile. **Capsules** loculicidal, valves recurved. **Seeds** minute.

1. Plants perennials. Flowers 20–50 mm long.
   2. Petals red . . . . *L. cardinalis*
   2. Petals blue to white . . . . *L. siphilitica*
1. Plants annuals. Flowers 6–15 mm long.
   3. Plants of northeastern and eastern portions of continent. Flowers 6–8 mm long . . . . *L. inflata*
   3. Plants of southern Texas and adjacent Mexico. Flowers 10–15 mm long . . . . *L. berlandieri*

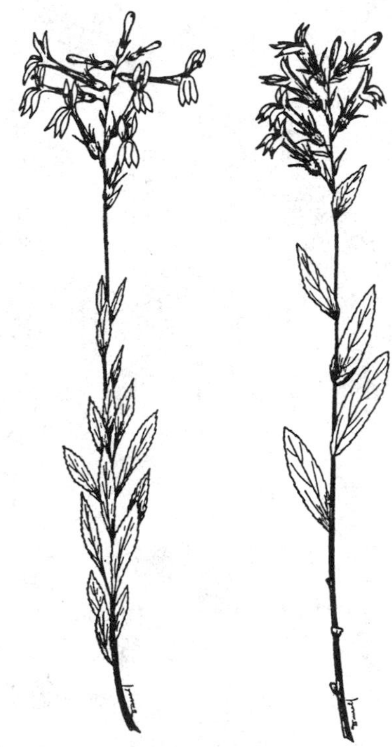

Lobelia cardinalis *Lobelia syphilitica*

Distribution and Habitat—Species of *Lobelia* are most abundant in the eastern half of the continent and are typically encountered in the wet soils of stream banks, gravel bars, bogs, and swamps. *Lobelia cardinalis* occurs across the continent. *Lobelia siphilitica* extends westward to the eastern slope of the Rocky Mountains. A western species, *L. berlandieri* is present in south Texas and adjacent Mexico, while *L. inflata* is encountered primarily in the northeastern and extreme eastern part of the continent.

■ wet soils, shaded sites

Disease Problems—Species of *Lobelia* are noteworthy mainly because of their use for medicinal and recreational purposes. Used in a manner similar to tobacco as well as for homicidal intent, *L. inflata* is best known for these purposes (Millspaugh 1974). In part, its popularity is based upon its mood-lifting or antidepressant effects (Subarnas et al. 1992). Its use for medicinal purposes is much reduced since its classification as poisonous by the U.S. Food and Drug Administration (Ballentine et al. 1985), which has led to seizure of herbal products by FDA officials. As its name implies, *L. syphilitica* was once used in the treatment of syphilis.

■ humans, neurotoxic, well-known antidepressant, mood-lifting effects, esp. *L. inflata*

livestock, *L. berlandieri*, neurotoxic, digestive disturbance

Species of *Lobelia* are seldom causes of problems in livestock, with the exception of *L. berlandieri*. It intermittently causes extensive cattle losses in southern Texas and Nuevo León and Tamaulipas in northern Mexico (Dollahite & Allen 1962; Dollahite 1978; Williams et al. 1987; Lopez et al. 1994). Intoxications occur mainly in late winter to early spring. In some cases losses due to death are extensive, while in others the disease is mainly one of incapacitation rather than death. Experimentally in sheep, a dosage of 0.6–2.2% of b.w. causes the development of signs of intoxication in 1–2 days and death in 3–9 days (Lopez et al. 1994).

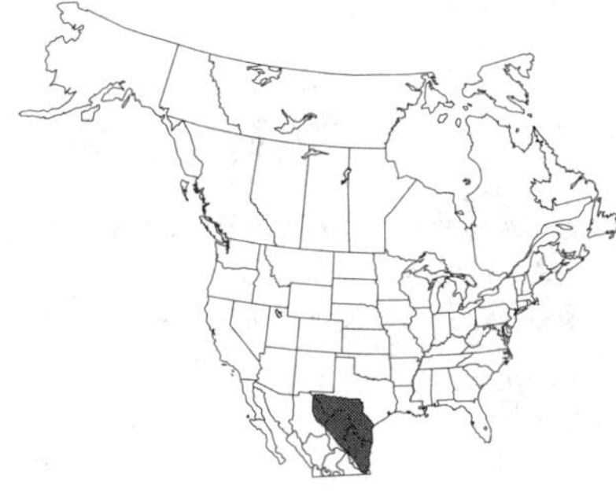

Lobelia berlandieri

- pyridine alkaloids, three groups—lobelines, lelobines, lobinines; nicotinic ganglionic stimulation, blockade, less potent than nicotine; neurotoxic, decreased blood pressure

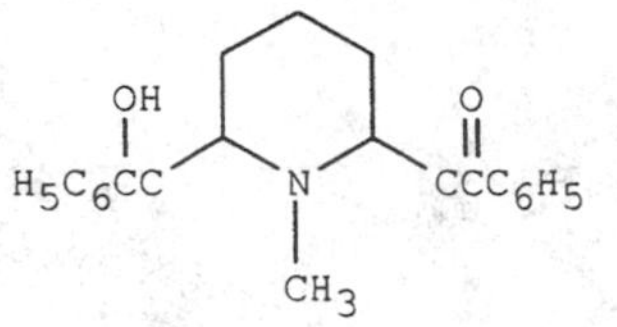

Lobeline

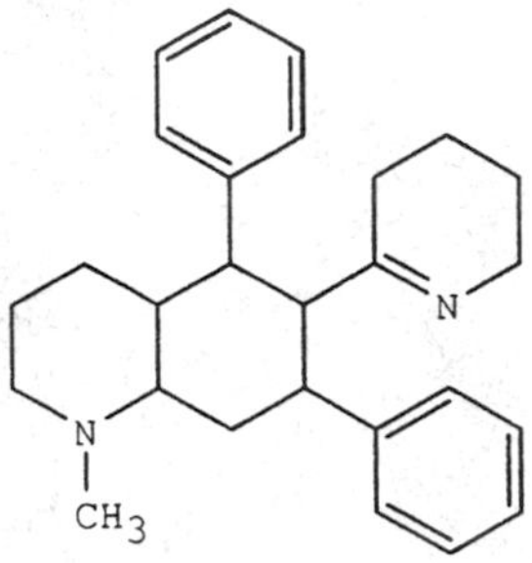

Lobinaline

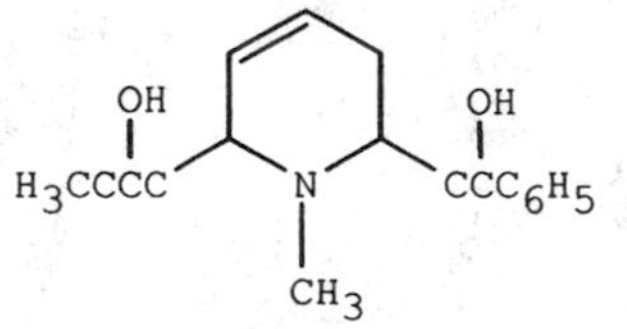

Lobinanidine

Disease Genesis—An array of pyridine alkaloids somewhat akin to nicotine is responsible for the medicinal and toxic effects produced by *Lobelia*. They are divided into three groups: the lobelines, lelobines, and lobinines. Lobeline is the major alkaloid in *L. inflata;* maximum concentrations are attained before and during flowering, at which time they may reach 0.4% in the leaves and 1% in the flowers and fruits (Marion 1950). Total alkaloid concentrations reported for leaves are 0.58% in *L. inflata*, 0.44% in *L. cardinalis*, 0.54% in *L. syphilitica*, and 0.08% in *L. berlandieri* (Marion 1950; Dollahite and Allen 1962; Krochmal et al. 1972). The number of alkaloids present in a species seems to increase with plant maturity (Lopez et al. 1994). In addition, their relative proportions differ among the species. In contrast to *L. inflata*, *L. syphilitica* and probably *L. berlandieri* contain pyridines other than lobeline (Marion 1960; Williams et al. 1987). They are typically of lesser potency than lobeline (Curtis 1926).

In the early 1800s the Thompsonian movement in herbal medicine, which advocated the extensive use of natural products for numerous maladies, promulgated the use of lobeline as a general cure-all. In several instances, overdosing caused disastrous effects, including death (Locock 1992). It later attained some reputable use as a general stimulant, a respiratory stimulant, and an antiasthmatic, but its use for these purposes has now been abandoned. It has been used as a smoking deterrent, with little if any demonstrable benefit (Kozlowski 1984).

Lobeline is a potent nicotinic ganglionic stimulant but also has atropinic activity (Barlow & Franks 1971). As with nicotine, the overall effects on nicotinic ganglionic receptors are stimulation at low dosage and blockade or paralysis at high dosage (Mansuri et al. 1973). Lobeline is a carotid body stimulant, decreasing heart rate and sometimes causing irregularities in beat (Korczyn et al. 1969). The effects are similar to but less marked than those of nicotine in reducing heart rate and blood pressure experimentally in rats (Sloan et al. 1988). Lobeline effects on the central nervous system also are much diminished relative to those of nicotine. In ponies, lobeline administration causes a transient increase in respiratory rate and tidal volume (Art et al. 1991).

Lobelia inflata also contains β-amyrin palmitate, which may in part be the basis for its antidepressant action (Subarnas et al. 1993a,b).

- livestock, *L. berlandieri,* diarrhea, weakness, depression, anorexia, ataxia, labored respiration, coma, death, cardiopulmonary collapse

Clinical Signs—The following clinical signs, pathology, and treatment relate primarily to intoxication due to *L. berlandieri*. After an animal has eaten plants for several days, there is diarrhea, excess nasal discharge, extension of the neck, and drooping of the ears. These signs are followed in a few hours by depression, loss of appetite, pupillary dilation, labored respiration, and incoordination. During this period there may be marked elevation of serum hepatic enzyme activity (AP, AST). The signs worsen over the next few days and terminate in coma and death, probably due to cardiopulmonary collapse. In instances when sublethal amounts are eaten, there may be prolonged narcosis.

- gross pathology, hyperinflated lungs, excess peritoneal fluid; rumen and abomasum mucosal ulcers

  microscopic, mild necrosis, liver, kidney tubules

Pathology—Grossly, there is hyperinflation of the lungs, congestion of the brain and lungs, scattered splotchy hemorrhages, distension of the gallbladder with bile, and excess reddish peritoneal fluid. In some animals ulceration of the mucosa of the rumen and abomasum has been prominent. Microscopically, the ruminal, reticular, and abomasal ulcerations are distinctive, with severe parakeratotic hyperkeratosis of the mucosa. There may also be necrosis of the mucosa of the small intestine and ileum. Degeneration and necrosis may also be apparent in the liver and kidney tubules. Neuronal degeneration and spongiosis may be present in some instances.

Treatment—Good nursing care, as well as hand feeding and watering, may promote survival in animals exhibiting prolonged narcosis.

■ nursing care, hand feed and water

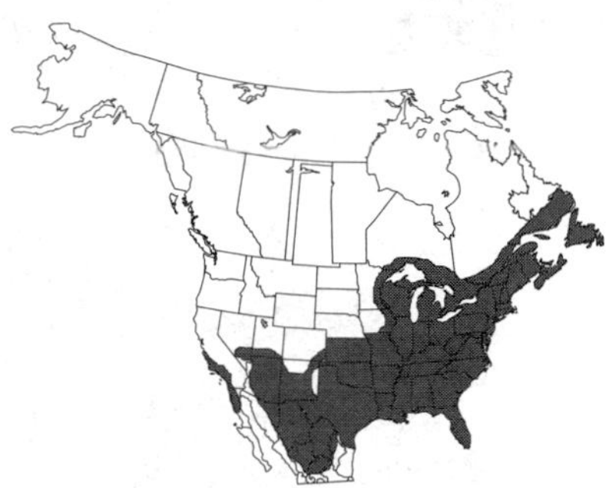

Lobelia cardinalis

Lobelia inflata

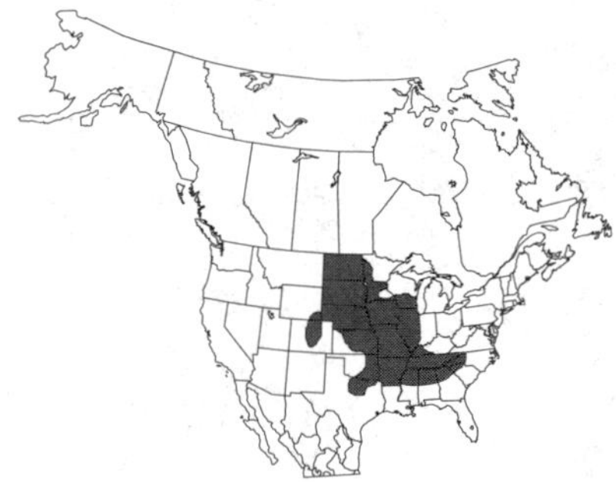

Lobelia syphilitica

References

Art T, Desmecht D, Amory H, Lekeux P: Lobeline-induced hyperpnea in equids: comparison with rebreathing bag and exercise. *Zentralbl Veterinarmed* A 38;148–152, 1991

Ballentine C, Hecht A, Janiger H, Maifarth S: Pretty but poisonous. *FDA Consum* 19;41, 1985

Barlow RB, Franks F: Specificity of some ganglionic stimulants. *Br J Pharmacol* 42;137–142, 1971

Curtis FR: Action of lobeline. *Lancet* 211;1255–1258, 1926

Dollahite JW: *Research and Observations on Toxic Plants in Texas 1932–1972*. Tex Agric Exp Stn Tech Rep 78–1, 1978

Dollahite JW: Allen TJ: Poisoning of cattle, sheep, and goats with *Lobelia* and *Centaurium* species. *Southwest Vet* 15;126–130, 1962

Korczyn AD, Bruderman I, Braun K: Cardiovascular effects of lobeline. *Arch Int Pharmacodyn Ther* 182;370–375, 1969

Kozlowski LT: Pharmacological smoking deterrents. *Can Pharm J* 117;152–155, 1984

Krochmal A, Wilken L, Chien M: Lobeline content of four Appalachian lobelias. *Lloydia* 35;303–304, 1972

Locock RA: Lobelia. *Can Pharm J* 125;33–34, 1992

Lopez R, Martinez-Burnes J, Vargas G, Loredo J, Medellin J, Rosiles R: Taxonomical, clinical, and pathological findings in moradilla (*Lobelia*-like) poisoning in sheep. *Vet Hum Toxicol* 36;195–198, 1994

Mansuri SM, Kelkar VV, Jindal MN: Some pharmacological characteristics of ganglionic activity of lobeline. *Arzneim Forsch* 23;1721–1725, 1973

Marion L: The pyridine alkaloids. In *The Alkaloids,* vol 1, Manske RHF, Holmes HL eds, Academic Press, New York, pp165–269, 1950

Marion L: The pyridine alkaloids. In *The Alkaloids,* vol 6, Manske RHF ed, Academic Press, New York, pp123–144, 1960

Millspaugh CF: *American Medicinal Plants*. Dover, New York, pp378–389, 1974 (reprinted from 1892)

Nash RJ, Watson AA, Winters AL, Fleet GWJ, Wormald MR, Dealler S, Lees E, Asano N, Molyneux RJ: Glycosidase inhibitors in British plants as causes of livestock disorders. *Toxic Plants and Other Natural Toxicants,* Garland T, Barr AC eds, CAB International, New York, pp276–284, 1998

Sloan JW, Martin WR, Bostwick M, Hook R, Wala E: The comparative binding characteristics of nicotine ligands and their pharmacology. *Pharmacol Biochem Behav* 30;255–267, 1988

Subarnas A, Oshima Y, Sidik, Ohizumi Y: An antidepressant principle of *Lobelia inflata* L. (Campanulaceae). *J Pharm Sci* 81;620–621, 1992

Subarnas A, Tadano T, Nakahata N, Arai Y, Kinemuchi H, Oshima Y, Kisara K, Ohizumi Y: A possible mechanism of antidepressant activity of beta-amyrin palmitate isolated from *Lobelia inflata* leaves in the forced swimming test. *Life Sci* 52;289–296, 1993a

Subarnas A, Tadano T, Oshima Y, Kisara K, Ohizuma Y: Pharmacological properties of beta-amyrin palmitate, a novel centrally acting compound, isolated from *Lobelia inflata* leaves. *J Pharm Pharmacol* 45;545–550, 1993b

Williams HJ, Ray AC, Kim HL: Δ^3-Piperideine alkaloids from the toxic plant *Lobelia berlandieri*. *J Agric Food Chem* 35;19–22, 1987

19 Cannabaceae Endl.

■ Hemp Family

Cannabis

Commonly known as the hemp family, the Cannabaceae comprises 2 small, sharply delimited genera of substantial economic importance. *Humulus,* with 2 species native to northern temperate regions, is the source of hops used in brewing beer. *Cannabis,* with 1 polymorphic species, is the well-known source of hemp and psychotropic drugs. The family is closely related to the Moraceae, or mulberry family, and is included in it by some taxonomists. Cannabinaceae, an illegitimate spelling of the name, appears in some older books. *Cannabis* is of primary toxicologic interest.

■ annual herbs, vines; petals absent

Plants herbaceous vines or herbs; annuals; strongly aromatic; dioecious; wind-pollinated. **Leaves** simple or palmately compound; alternate above and opposite below; venation pinnate; stipules present. **Inflorescences** of 2 types; staminate and pistillate different; axillary; bracts present. **Staminate Inflorescences** panicles. **Pistillate Inflorescences** spikes or clusters; bracts foliaceous, enclosing flowers and fruits. **Flowers** imperfect, staminate and pistillate different; perianths in 1-series. **Calyces** radially symmetrical. **Sepals** 5; free or fused. **Petals** absent. **Stamens** 5. **Pistils** 1; compound, carpels 2; stigmas 2, linear; styles 1, short; ovaries superior; locules 1; placentation apical. **Fruits** achenes; enclosed by subtending bracts. **Seeds** 1.

Although the family is known mainly for the intoxicating effects of *C. sativa,* spent hops *(H. lupulus)* have been reported to cause a potentially lethal malignant hyperthermia-like reaction in dogs, especially greyhounds (Duncan et al. 1997). Several hours after ingestion, signs such as panting, elevated heart and respiratory rates, brown urine, and high body temperature may be apparent and require immediate medical attention. Treatment is nonspecific and directed toward alleviation of the clinical signs.

Cannabis L.

■ *C. sativa*

cannabis
dagga
ganja
hashish
hemp
kif
marijuana
pot

Taxonomy and Morphology—Although several species were originally described, the current taxonomic consensus is that *Cannabis* consists of a single, highly variable species that has diversified under cultivation into a more northern subspecies cultivated primarily for fiber and a more tropical subspecies cultivated primarily for its drug content. The generic name is derived from the Greek *kannabis,* which is believed to come from the Arabic *kinnab,* for "hemp." The following is the single species:

C. sativa L. — hemp, Indian hemp, marijuana, cannabis, kif, hashish, dagga, ganja, pot

■ odoriferous, erect herbs; leaves palmately compound, leaflets 5–11, alternate above, opposite below

Plants herbs; from taproots; odoriferous; dioecious or rarely monoecious. **Stems** erect; 100–550 cm tall; branched; surfaces antrorsely strigose. **Leaves** palmately compound; alternate above and opposite below; leaflets 5–11; blades linear to lanceolate to elliptic; adaxial surfaces dark green, with pustulate hairs; abaxially surfaces whitish green with yellowish brown resin glands; apices acuminate, margins serrate; stipules lanceolate. **Inflorescences** borne on short, leafy branches from axils of uppermost leaves; staminate flowers pedicelled; pistillate flowers sessile. **Flowers** imperfect, staminate and pistillate different. **Sepals** 5; free or fused; puberulent or pilose. **Petals** absent. **Stamens** 5. **Pistils** 1. **Achenes** ovoid; white or greenish, purple mottled; subtended by bract and bracteole.

Cannabis sativa

■ crop plant, naturalized

Distribution and Habitat—Believed to be native to central Asia, *C. sativa* is now naturalized throughout the world. Adapted to a variety of soil types, populations are encountered in disturbed sites or in illegal, clandestine cultivation. In the upper Midwest, it is a common roadside weed in some areas.

■ neurotoxic, plant fresh, dried or resin products, all animal species; impairment of reaction time, motor coordination, and visual perception, plus effects on other systems

Disease Problems—*Cannabis sativa* has been used medicinally for millennia. Believed to have been first grown in China about 6,000 years ago, it is one of the oldest cultivated plants. Historically, it was used in the treatment of insomnia, asthma, migraine headaches, and depression; today it is employed to treat glaucoma and relieve the side effects of modern cancer therapies (Simpson & Conner-Ogorzaly 1995). It was introduced into Europe in the early 1500s and was cultivated in colonial North America for its tough and durable fibers (hemp), which were used to make ropes, cordage, sails, mats, and sacks. Escape and establishment in disturbed sites likely account for its present naturalized distribution on the continent. In the twentieth century, primary interest has been in its psychoactive effects, and its use is regulated by federal and state acts. A plethora of colloquial names are used for *C. sativa*. Typically *hemp* is used for the plant and the fibers extracted from it, while *marijuana, hashish, ganja,* and others are related mainly to its illicit use (Small 1997).

Intoxications involving *Cannabis* are primarily related to its use for psychoactive experiences by humans and either inadvertent or intentional exposure of pets. Most animal species are probably susceptible. Intoxications have been reported for cattle, horses, dogs, and ferrets (Cardassis 1951; Frye 1968; Merriwether 1969; Clarke et al. 1971; Silverman 1974; Jones 1978; Godbold et al. 1979; Crow & Sokolow 1980; Jain & Arora 1988; Valentine 1992). Pets may eat fresh plants, dried plant products, or the more refined resins and related products. The overall effects of *Cannabis* include impairment of reaction time, motor coordination, and visual perception. In humans these effects cause emotional disturbances. It also affects respiration, endocrine function, and the activity of other systems. A diagnosis may be made more difficult by the understandable reluctance of some animal owners to admit to exposure of pets to an illicit drug. Surprisingly, this reluctance is not always apparent (Schwartz & Riddile 1985; Smith 1988). Schwartz and Riddile (1985) commented on the common practice of teenagers blowing smoke into the nostrils of pets for amusement. Recovery of the pets was prompt, but a few cases were serious enough to warrant medical assistance.

■ cannabinoids Δ-9-THC, Δ-8-THC, cannabinol; flowers and leaves; neurologic effects, depression, rarely severe, nonlethal; plant or crude product, "cannabis"

Disease Genesis—Numerous chemical compounds are present in *Cannabis*, but the most important toxicologically are the cannabinoids. Of the more than 60 present, the most important is Δ-9-tetrahydrocannabinol (THC). Of lesser importance are Δ-8-THC, cannabinol (a metabolic product of THC), and cannabidiol (Mason & McBay 1985). THC concentrations, highest in the flowers and leaves, vary considerably among the various cultivars.

The cannabinoids are absorbed rapidly from smoke and slowly and erratically following ingestion (Mason & McBay 1985). They are metabolized rapidly by the

Δ-9-THC

Cannabinol

Cannabidiol

- depression, drowsiness; periodic hyperactivity tremors; may appear dizzy, disoriented
- no treatment, or activated charcoal

mixed-function oxidase system in the liver, and Δ-9-THC is subject to first-pass effect via the liver, resulting in the formation of 11-hydroxy-THC. The form and dosage of THC have a marked influence on the severity and duration of the intoxication. Inhalation of smoke by pets from that exhaled by people generally results in an intoxication problem of only a few hours' duration. Problems following ingestion of leaves may last a little longer but are still of short duration. Consumption of more purified products such as resins will result in more severe signs and much longer duration because of the higher dosage of THC.

In spite of the presence of moderate signs at low dosage, the overall toxicity of the cannabinoids is low. Experimentally, lethal effects could be shown in rats at around 1 g/kg orally with THC or *Cannabis* extracts, but lethality was not observed in beagle dogs or rhesus monkeys at 3–9 g/kg b.w. (Thompson et al. 1973). Massive doses produced depression and prostration and other neurologic signs in dogs, and huddled posture and anesthesia in monkeys, but not lethality. With nominal dosage, recovery typically occurred within 24 hours, but with very large doses, it was prolonged up to 120 hours.

Clinical Signs—The signs of *Cannabis* intoxication are quite variable, but in general the dominant effect is depression and drowsiness interspersed with periods of arousal, sometimes with bizarre actions, including hyperactivity, tremors, and hyperresponsiveness to touch and sound. Animals may appear to be glassy-eyed, dizzy, or disoriented. Vomiting and to a lesser extent diarrhea are seen but are not usually of a serious nature. Other effects that have been observed include compulsive eating, bumping into furniture, biting objects, shivering, decreased body temperature, and a rigid sawhorse stance. Recovery usually occurs in a few to 24 hours.

Pathology and Treatment—*Cannabis* does not generally produce fatal disease, and pathologic changes are not anticipated. Mild intoxication requires only careful observation or perhaps some sedation of the affected animal. In more severe cases, induced emesis, activated charcoal given orally, bowel evacuation, and/or purgatives may be helpful in limiting the severity and duration of the effects. It is important to maintain body temperature, and stimulants may also be appropriate.

References

Cardassis J: Poisoning of horses by *Cannabis indica*. *Rec Med Vet (France)* 127;971, 1951

Clarke EGC, Greatorex JC, Potter R: *Cannabis* poisoning in the dog [letter]. *Vet Rec* 88;694, 1971

Crow SE, Sokolow V: Marijuana intoxication [letter]. *J Am Vet Med Assoc* 176;388, 1980

Duncan KL, Hare WR, Buck WB: Malignant hyperthermia-like reaction secondary to ingestion of hops in five dogs. *J Am Vet Med Assoc* 201;51–54, 1997

Frye FL: Acute *Cannabis* intoxication in a pup. *J Am Vet Med Assoc* 152;472, 1968

Godbold JC, Hawkins BJ, Woodward MG: Acute oral marijuana poisoning in the dog. *J Am Vet Med Assoc* 175;1101–1102, 1979

Jain MC, Arora N: Ganja *(Cannabis sativa)* refuse as cattle feed. *Indian J Anim Sci* 58;865–867, 1988

Jones DL: A case of canine *Cannabis* ingestion. *N Z Vet J* 26;135–136, 1978

Mason AP, McBay AJ: *Cannabis*: pharmacology and interpretation of effects. *J Forensic Sci* 30;615–631, 1985

Meriwether WF: Acute marijuana toxicity in a dog (a case report). *Vet Med/Sm Anim Clin* 64;577–578, 1969

Schwartz RH, Riddile M: Marijuana intoxication in pets [letter]. *J Am Vet Med Assoc* 187;206, 1985

Silverman J: Possible hashish intoxication in a dog. *Am Anim Hosp Assoc J* 10;517–519, 1974

Simpson BB, Conner-Ogorzaly M: *Economic Botany: Plants in Our World,* 2d ed. McGraw-Hill, New York, 1995

Small E: Cannabaceae Endlicher: Hemp family. In *Flora of North America,* vol 3, *Magnoliophyta: Magnoliidae and Hamamelidae.* Flora of North America Editorial Comm, Oxford Univ Press, New York, pp381–387, 1997

Smith RA: Coma in a ferret after ingestion of *Cannabis* [letter]. *Vet Hum Toxicol* 30;486, 1988

Thompson GR, Rosenkrantz H, Schaeppi UH, Braude MC: Comparison of acute oral toxicity of cannabinoids in rats, dogs, and monkeys. *Toxicol Appl Pharmacol* 25;363–372, 1973

Valentine J: Unusual poisoning in a dog [letter]. *Vet Rec* 130;307, 1992

20 Caprifoliaceae Juss.

■ Honeysuckle Family

Lonicera
Sambucus

QUESTIONABLE
Symphoricarpos

Exhibiting greatest diversity in eastern North America and eastern Asia, the Caprifoliaceae, commonly known as the honeysuckle family, comprises 15–18 genera and 365–450 species. It is economically important because of the numerous species of ornamental shrubs; more than 100 species and innumerable varieties of 13 genera are in the horticultural trade of North America. A few species, e.g., *S. canadensis* (elderberry), are used to make wine and jellies. Toxicologic problems are more likely to involve humans than livestock. Three genera are involved.

■ herbs, shrubs, vines, trees; leaves opposite; petals 5, fused; fruits berries or berrylike drupes

Plants shrubs or woody vines or small trees or herbs; perennials. **Stems** often twining or climbing or trailing. **Leaves** simple or 1-pinnately compound; opposite; petiolate or sessile; venation pinnate; blades elliptic to lanceolate or obovate; margins entire or serrate; stipules absent or rarely present. **Inflorescences** simple or compound cymes; terminal or axillary; bracts present or absent; bracteoles present or absent. **Flowers** perfect; perianths in 2-series. **Calyces** radially or bilaterally symmetrical; rotate or tubular. **Sepals** 5; fused. **Corollas** radially or bilaterally symmetrical; rotate or tubular or bilabiate. **Petals** 5; fused; frequently gibbous; of various colors. **Stamens** 5; epipetalous. **Pistils** 1; compound, carpels 2–5; stigmas 1–3, not lobed or 2-lobed, capitate; styles 0 or 1; ovaries inferior; locules 1–5; placentation axile. **Fruits** berries or berrylike drupes. **Seeds** 1 to numerous.

With the exception of the Rubiaceae, the Caprifoliaceae is readily distinguished from other families by its opposite leaves, inferior ovaries, and multicarpellate pistils. The general absence of stipules in the family is useful is separating it from the Rubiaceae.

Lonicera L.

■ *L. involucrata*
L. japonica
L. periclymenum
L. sempervirens
L. tatarica

gold-and-silver flower
honeysuckle
twinberry
woodbine

Taxonomy and Morphology—Comprising 150–180 species distributed primarily in the temperate and subtropical regions of the Northern Hemisphere, *Lonicera*, commonly known as honeysuckle, was named for Adam Lonitzer, a sixteenth-century German naturalist and author of a popular herbal in the mid-1500s. The genus is prized for both its attractive foliage and its showy, sweet-scented flowers that are attractive to birds and insects. Numerous cultivars have been developed. In North America, approximately 35 native and introduced species and named hybrids are present. Commonly encountered and representative of the genus are the following:

L. involucrata (Richardson) Banks ex Spreng.	black twinberry
L. japonica Thunb.	Japanese honeysuckle, gold-and-silver flower
L. periclymenum L.	woodbine, common honeysuckle

L. sempervirens L.	trumpet honeysuckle, coral honeysuckle, evergreen honeysuckle
L. tatarica L.	Tartarian honeysuckle

Plants shrubs or woody vines. **Stems** erect or twining or trailing. **Leaves** simple; often coriaceous. **Inflorescences** verticils of 6 flowers on terminal or axillary rachises or pairs of flowers borne on axillary peduncles; bracts present. **Flowers** radially or bilaterally symmetrical; tubular or bilabiate with a 4-lobed upper lip and an unlobed lower lip. **Petals** yellow or red or pink or orange or white. **Pistils** 1; stigmas capitate; styles 1, elongate. **Fruits** berries; globose; red or blue or black. **Seeds** few.

■ shrubs, vines, erect or twining; leaves simple; flowers with upper lip 4-lobed, lower unlobed; petals yellow, red, pink, orange, or white; berries red, blue, or black

A key for identification is not presented here. The reader is encouraged to use those appearing in floras specific for his or her region.

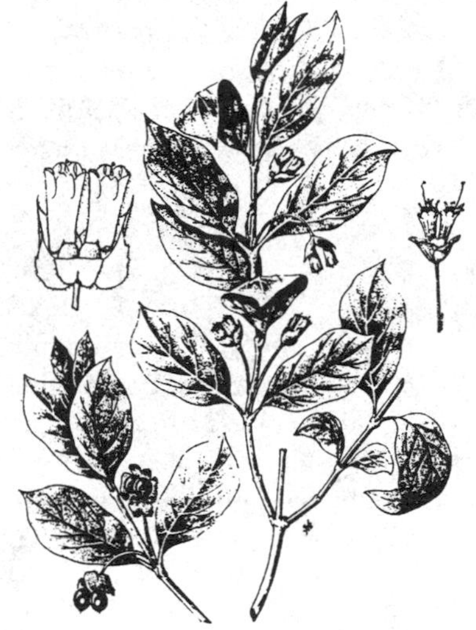

Lonicera involucrata

Lonicera japonica

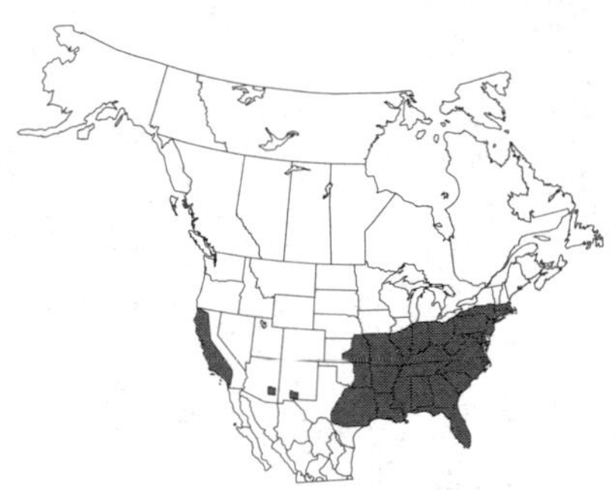

Lonicera japonica

Distribution and Habitat—*Lonicera involucrata* and *L. sempervirens* are natives. The former is found in moist sites across Canada and throughout the western United States, while the latter is the common species of open woods of the eastern states, especially the Southeast. Introductions from Europe and Asia, *L. japonica, L. tatarica,* and *L. periclymenum* are ornamentals that have escaped at various sites throughout eastern North America. *Lonicera japonica* is well established in the Southeast and is a troublesome weed in some areas.

■ woodlands

Lonicera sempervirens

Disease Problems—Although the berries of some species, e.g., *L. involucrata,* are known to be edible, those of others have been associated with digestive tract disturbances when eaten by children in Europe (O'Leary 1964; Frohn & Pfander 1984). In the United States, their ingestion by children prompts numerous telephone calls to Poison Control Centers, but clinical signs are seldom seen (Lawrence 1989). Frohn and Pfander (1984) considered the European species to be slight toxicity risks, but only when large numbers of berries are consumed.

■ humans, berries, digestive disturbance; livestock, little risk

Toxicity in livestock does not appear to be a problem. Ferguson (1929) reported three instances of digestive disturbance in cows due to ingestion of wild honeysuckle, presumably *L. sempervirens,* in North Carolina. However, Hansen in the same reference, disputed that *Lonicera* was the cause. He further stated that although honeysuckle may occasionally cause problems in horses, it was not a problem in cattle.

The twigs and bark of *L. tatarica* are eaten and provide sustenance for deer, without causing problems (Gill & Healy 1973).

- irritant terpenoid saponins, lonicerosides

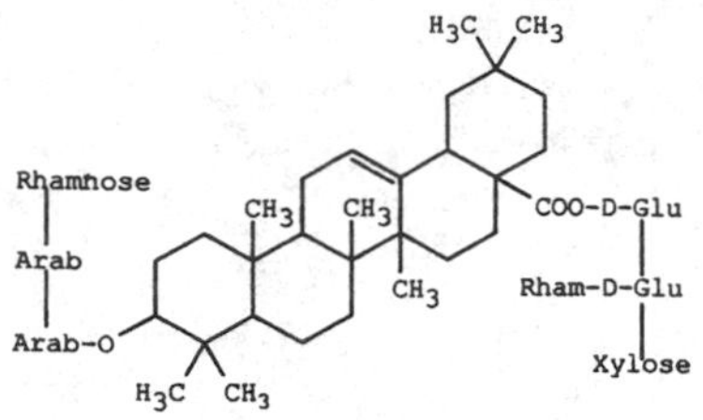

Lonicerocide B

Disease Genesis—Although the specific toxicants of *Lonicera* have not been definitely identified, *L. japonica* has been shown to contain carotenoid tetraterpenoids such as webbiaxanthin and loniceraxanthin, and triterpenoid saponins of hederagenin called lonicerosides A and B (Son et al. 1994). Similar to the terpenoids of *Hedera* in the Araliaceae (see chapter 11), these compounds may have the potential to irritate the mucosa of the digestive tract, but this has not been confirmed. There also are one or more iridoid glucosides of limited potency as mild purgatives, which may contribute to the irregularly occurring digestive tract disturbances (Inouye et al. 1974). A role for saponins is given further credibility by reports of a correlation between the foam index of an extract and its toxicity with high dosage in mice (Frohn & Pfander 1984).

- nausea, vomiting, diarrhea

treat with activated charcoal, fluids

Clinical Signs and Treatment—Seen mainly in the children, the signs are related to irritation of the digestive tract and include nausea, vomiting, and diarrhea. Treatment is aimed at prevention of toxicant absorption via activated charcoal and toward relief of any persistent vomiting.

Sambucus L.

- *S. caerulea*
  S. canadensis
  S. ebulus
  S. mexicana
  S. nigra
  S. racemosa

bourtree
elder
elderberry
sauco
tapiro

Taxonomy and Morphology—Comprising 20–25 species, *Sambucus,* commonly known as elderberry or elder, is a cosmopolitan genus. Species are sources of wine and jelly, several are cultivated ornamentals, and the wood of some is used to make musical instruments. Native Americans and settlers used the plants medicinally for a variety of ailments. In North America, 6 native and introduced species are present:

S. caerulea Raf. (= *S. glauca* Nutt.) (= *S. neomexicana* Woot.)	blue elderberry
S. canadensis L.	common elderberry, common elder, sweet elder, American elder
S. ebulus L.	
S. mexicana C.Presl ex DC.	Mexican elder, tapiro, sauco, blueberry elder
S. nigra L.	black elder, bourtree, European elder
S. racemosa L. (= *S. callicarpa* Greene) (= *S. microbotrys* Rydb.) (= *S. melanocarpa* [A.Gray] McMinn) (= *S. pubens* Michx.)	red elderberry, European red elder, stinking elderberry

- erect herbs, shrubs, small trees; stems ill-scented when bruised; leaves pinnately compound, terminal leaflets present; petals white or cream-colored; fruits berrylike drupes, purple-black, dark blue, red, yellow, or white

Plants shrubs or small trees or herbs; deciduous; often spreading by root suckers. **Stems** erect; or ascending; 1–8 m tall; lenticels present; malodorous when bruised; pith spongy; branchlets stout, sometimes hollow. **Leaves** 1- or rarely 2-pinnately compound; terminal leaflet present; leaflets 5–11, blades lanceolate to ovate or elliptic, apices acuminate, margins serrate. **Inflorescences** terminal corymbose or paniculate cymes. **Flowers** numerous; radially symmetrical; rotate. **Sepals** inconspicuous. **Petals** white or cream. **Pistils** 1; stigmas 3-to-5-lobed; styles 1, short. **Fruits** berrylike drupes; with 3–5 stones; purple-black or dark blue or red or yellow or white.

Sambucus canadensis

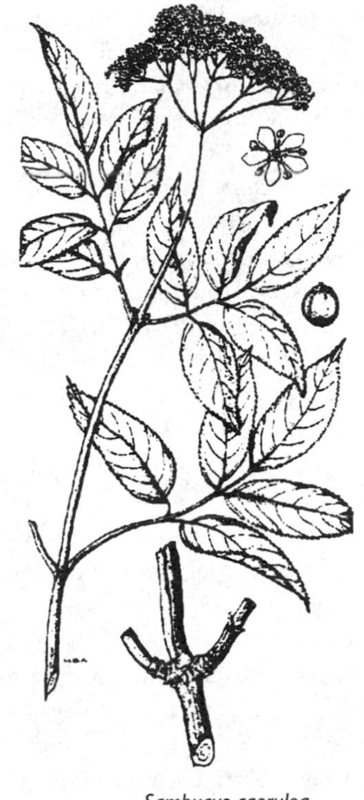

Sambucus caerulea

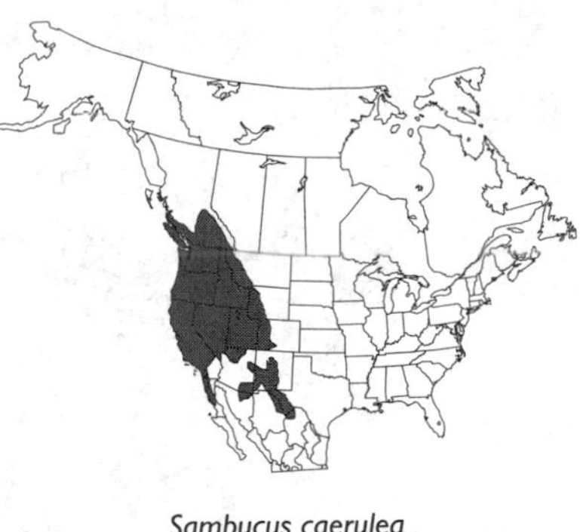

Sambucus caerulea

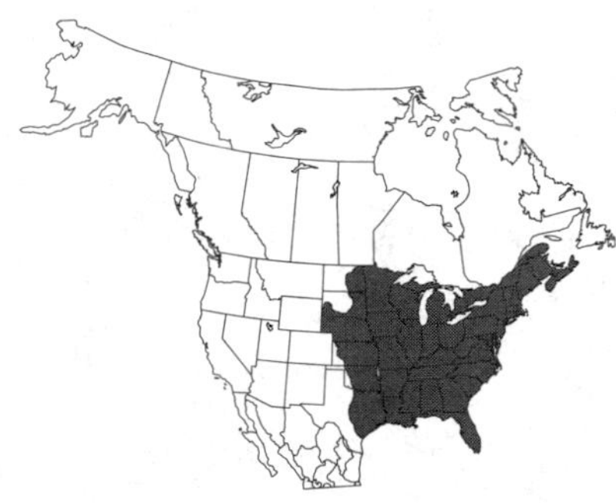

Sambucus canadensis

Sambucus mexicana

1. Plants herbs. Stipules well developed; ovate .... *S. ebulus*
1. Plants shrubs or trees. Stipules minute or absent.
    2. Plants trees.
        3. Plants of southwestern portion of continent. Drupes glaucous .... *S. mexicana*
        3. Plants of northeastern portion of continent. Drupes not glaucous .... *S. nigra*
    2. Plants shrubs.
        4. Inflorescences paniculate; pyramidal. Pith brown. Drupes red or yellow or white .... *S. racemosa*
        4. Inflorescences corymbose; flat-topped. Pith white. Drupes purple black or dark blue.
            5. Plants of the western half of the continent. Drupes glaucous .... *S. caerulea*
            5. Plants of the eastern half of the continent. Drupes not glaucous .... *S. canadensis*

Distribution and Habitat—Ornamental introductions from Europe, *S. nigra* and *S. ebulus* occasionally escape from cultivation. *Sambucus caerulea* is found in valleys and on slopes in the open woodlands of western North America. *Sambucus canadensis* is the common elderberry of moist fields and woods of eastern North America. *Sambucus racemosa* is found in open areas, streamsides, and woodlands throughout the northern portion of the continent. As its specific epithet implies, *S. mexicana* is distributed in the Southwest. All species are sometimes cultivated as ornamentals.

■ open woodlands, cultivated ornamentals

Disease Problems—Species of *Sambucus* are used for food and medicines. The edible drupes are used to make pies, wine, and jelly, and the numerous medicinal uses of

■ all plant parts, low risk, digestive disturbance; low risk of abrupt neurologic cyanide effects

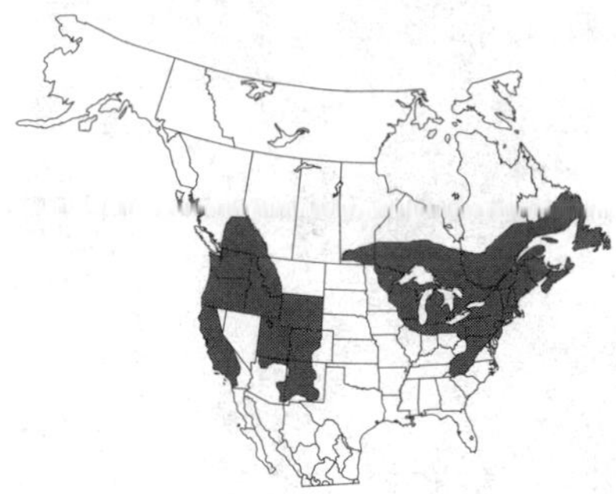
Sambucus racemosa

the genus have caused some individuals to consider it a complete pharmacy in itself (Millspaugh 1974). The leaf buds were considered to be potent cathartics, and the sap a laxative. Tinctures made from the leaves and flowers caused diuresis and circulatory problems, terminating in exhaustion and profuse sweating (Millspaugh 1974).

Although it is clear that species contain bioactive constituents, they are uncommon causes of disease. Ingestion of the roots and/or stems has been associated with digestive tract problems (Cooper & Johnson 1984). The roots and stems produce purgative effects, and the drupes, when eaten raw, may produce similar results, including nausea and vomiting (Pammel 1911). In a case in the 1800s, a boy in Scotland developed severe vomiting and bloody diarrhea after eating leaves of *Sambucus*. A second child exhibited mild neurologic signs after eating the flowers (Pammel 1911). In another episode, 11 of 25 people who drank elderberry juice made 2 days previously developed nausea and vomiting (Kunitz et al. 1984). Other signs seen in some individuals included abdominal pain, weakness, dizziness, and numbness. One individual became stuporous and required hospitalization. The severity of signs was directly correlated with the amount of juice consumed. Cyanide was not detected in the blood of those affected.

In some circumstances the leaves of *Sambucus* are cyanogenic, and the stems have been associated with accumulation of nitrate, but these conditions have not been demonstrated to pose a substantial risk to livestock.

- irritant terpenoids; cyanogenesis, prunasin, sambunigrin, low risk

C-Glucose
C≡N
Prunasin

CC≡N
O-Glucose
Sambunigrin

Disease Genesis—The toxicants responsible for the digestive tract problems have not been identified, although triterpenoids, such as oleanolic acid, are present in the leaves of *S. nigra* and *S. canadensis* (Inoue & Sato 1975). The stones/seeds contain a heat-labile resinous substance (Frohn & Pfander 1984). Lectins or hemagglutinins are present in both the bark and fruit of *S. nigra* (Kaku et al. 1990; Mach et al. 1991). Any of these types of toxicants could be responsible for irritation of the digestive tract.

Species of *Sambucus* have traditionally been thought to be toxic, because of the presence of cyanogenic glucosides in the foliage and fruit. *Sambucus nigra* contains several phenylalanine-derived glucosides, including holocalin, prunasin, sambunigrin, and zierin (Jensen & Nielsen 1973; Seigler 1977). Presumably other species contain a similar array of these compounds. The risk of intoxication, however, is quite low but cannot be entirely ignored. For the most part, the propensity to produce adverse effects of any type is lost when the berries are cooked or fermented to make jellies or wine (Pogorzelski 1982; Cooper & Johnson 1984). The digestive tract problems are not consistent with cyanide intoxication.

- humans, rarely abrupt onset, vomiting, colic, profuse salivation, diarrhea

livestock rarely, abrupt onset, weakness, apprehension, ataxia, labored respiration, collapse, seizures

Clinical Signs—In cases involving irritation of the digestive tract, there is abrupt onset of vomiting, colic, excess salivation, and diarrhea. These problems may be accompanied by increased heart and respiratory rates, tremors, and paralysis.

When cyanide intoxication occurs in livestock, the clinical signs typically appear soon after consumption of plant material and include weakness, apprehension, incoordination, labored respiration, collapse, and tetanic seizures. A more detailed discussion of the signs and diagnosis is presented in the treatment of the Rosaceae (see chapter 64).

- no lesions

activated charcoal; sodium thiosulfate

Pathology and Treatment—There are few if any distinctive pathologic changes. A few scattered, small hemorrhages on the heart and visceral surfaces may be present. Prevention of toxicant absorption via activated charcoal and relief of any persistent vomiting are important considerations in treatment. For cyanogenesis, the standard, well-established response employing sodium thiosulfate with or without sodium nitrite is appropriate. A complete discussion of this approach is presented in the treatment of the Rosaceae (see chapter 64).

Genera of Questionable Toxicity

Symphoricarpos Duhamel

Comprising 8–11 species in North America and 1 in China, *Symphoricarpos*, commonly known as snowberry or coralberry, is a familiar sight across the continent in a variety of vegetation types and in cultivation as an ornamental. Commonly encountered species are *S. albus* (L.) S.F.Blake (snowberry); *S. orbiculatus* Moench (coralberry or buckbrush); *S. occidentalis* Hook. (wolfberry), and *S. oreophilus* A.Gray (mountain snowberry). All species of the genus are low, bushy shrubs, colonial by root sprouts. The stems typically have exfoliating bark and bear opposite, short, petiolate, simple leaves. In the leaf axils are borne verticils of small, 5- or 4-merous, white to pink, campanulate to subsalverform flowers, which in turn give rise to white or purple-red or red berrylike drupes with 2 stones.

■ *S. albus* *S. orbiculatus* *S. occidentalis* *S. oreophilus*
buckbrush coralberry snowberry wolfberry

Symphoricarpos albus

Symphoricarpos orbiculatus

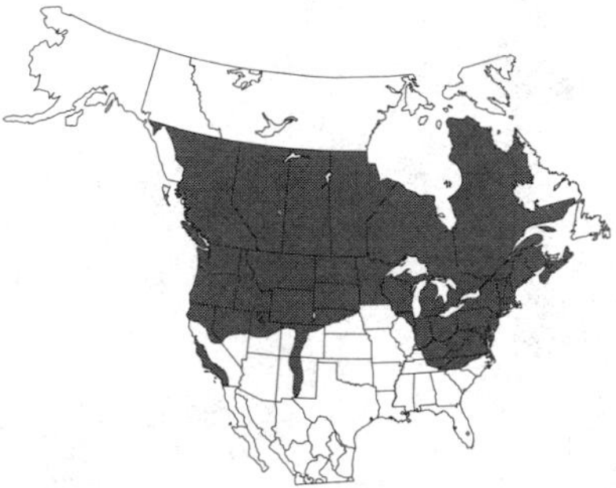
Symphoricarpos albus

Symphoricarpos orbiculatus

■ humans, digestive disturbance, low risk

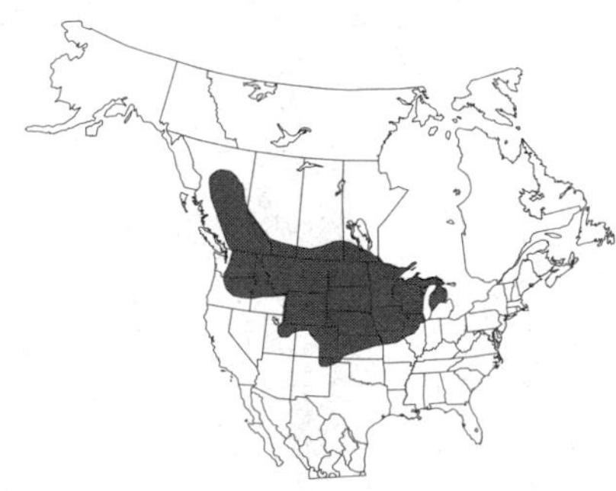
Symphoricarpos occidentalis

The foliage and fruit of various species of *Symphoricarpos* have long been suspected by some individuals to be mildly toxic to children and livestock, but Marsh and co-workers (1928) were unable to cause adverse effects in sheep given the fruits of *S. oreophilus* (reported as *S. vaccinoides*) up to 4% b.w. of the animal per day in two feedings. Likewise, European reports suggest that the fruits have purgative effects, but the risk is questionable (Lewis 1979; Frohn & Pfander 1984). Most ingestions are asymptomatic; in only one case was vomiting of short duration observed (Lewis 1979; Cooper & Johnson 1984). Rarely are signs other than those related to irritation of the digestive tract, e.g., dizziness and sedation, noted. Experimentally, extracts of the drupes caused only slight sedative effects (Szaufer et al. 1978).

The foliage of *S. albus* cultivated in Poland has been shown to contain small amounts of the alkaloid chelidonine, which appears to be of little consequence toxicologically (Szaufer et al. 1978). This species and others are used as a food source by wildlife, the drupes by birds and the foliage by deer and elk (Kingery et al. 1996).

In summary, intoxications are rare and probably of minimal severity. Clinical signs are likely to be limited to those of mild digestive tract upset such as vomiting and possibly mild sedation.

References

Cooper MR, Johnson AW: *Poisonous Plants in Britain and Their Effects on Animals and Man.* Ministry of Agriculture, Fisheries and Food, Her Majesty's Stationery Office, London, pp75–77, 1984

Ferguson GA: Three cases of honeysuckle poisoning. *North Am Vet* 10;18–19, 1929

Frohn D, Pfander HJ: *A Colour Atlas of Poisonous Plants.* Wolfe Science, London, 1984

Gill JD, Healy WM: *Shrubs and Vines for Northeastern Wildlife.* USDA Forest Service, Gen Tech Rep NE-9, 1973

Inoue T, Sato K: Triterpenoids of *Sambucus nigra* and *S. canadensis. Phytochemistry* 14;1871–1872, 1975

Inouye H, Takeda Y, Uobe K, Yamauchi K, Yabuuchi N, Kuwano S: Purgative activities of iridoid glucosides. *Planta Med* 25;285–288, 1974

Jensen SR, Nielsen BJ: Cyanogenic glucosides in *Sambucus nigra* L. *Acta Chem Scand* 27;2661–2685, 1973

Kaku H, Peumans WJ, Goldstein IJ: Isolation and characterization of a second lectin (SNA-II) present in elderberry (*Sambucus nigra* L.) bark. *Arch Biochem Biophys* 277;255–262, 1990

Kingery JL, Mosley JC, Bordwell KC: Dietary overlap among cattle and cervids in northern Idaho forests. *J Range Manage* 49;8–15, 1996

Kunitz S, Melton RJ, Updyke T, Breedlove D, Werner SB: Poisoning from elderberry juice—California. *Morb Mortal Wkly Rep* 33;173–174, 1984

Lawrence RA: Common "red berry" calls to the poison control center [abstract]. *Vet Hum Toxicol* 31;361, 1989

Lewis WH: Snowberry *(Symphoricarpos)* poisoning in children. *J Am Med Assoc* 242;2663, 1979

Mach L, Scherf W, Ammann M, Poetsch J, Bertsch W, Marz L, Glossl J: Purification and partial characterization of a novel lectin from elder (*Sambucus nigra* L.) fruit. *Biochem J* 278;667–671, 1991

Marsh CD, Clawson AB, Roe GC: *Four Species of Range Plants Not Poisonous to Livestock.* USDA Tech Bull 93, 1928

Millspaugh CF: *American Medicinal Plants.* Dover, New York, 1974 (reprint from 1892)

O'Leary SB: Poisoning in man from eating poisonous plants. *Arch Environ Health* 9;216–242, 1964

Pammel LH: *A Manual of Poisonous Plants.* Torch Press, Cedar Rapids, Iowa, pp744–748, 1911

Pogorzelski E: Formation of cyanide as a product of decomposition of cyanogenic glucosides in the treatment of elderberry fruit *(Sambucus nigra). J Sci Food Agric* 33;496–498, 1982

Seigler DS: The naturally occurring cyanogenic glycosides. In *Progress in Phytochemistry,* vol 4, Reinhold L, Harborne JB, Swain T eds, Pergamon Press, Oxford, pp83–120, 1977

Son KH, Jung KY, Chang HW, Kim HP, Kang SS: Triterpenoid saponins from aerial parts of *Lonicera japonica. Phytochemistry* 35;1005–1008, 1994

Szaufer M, Kowalewski Z, Phillipson JD: Chelidonine from *Symphoricarpos albus. Phytochemistry* 17;1446–1447, 1978

Caryophyllaceae Juss.

21

Comprising 75–90 genera and 2,000–2,100 species, the Caryophyllaceae, or pink family, is widespread in distribution, but with greatest diversity in the temperate and warm-temperate regions of the Northern Hemisphere. The Mediterranean region is a principal center of distribution. In North America, approximately 40 native and introduced genera and 350 species are present.

The family is divided into 3 distinct subfamilies—Alsinoideae, Silenoideae, Paronychioideae—which differ in the presence of stipules and fusion of sepals. Because of these differences, taxa of the Alsinoideae and Paronychioideae are segregated by a few taxonomists into their own families, the Alsinaceae and the Illecebraceae. Because their unifying characters are more significant, a single large family is recognized in this book, as is the case in most of the recent taxonomic treatments.

The Caryophyllaceae is noted for its large number of ornamentals, both for the garden and as cut flowers. Most familiar are *Dianthus* (carnation), *Gypsophila* (baby's breath), *Lychnis* (Maltese cross), and *Silene* (catchfly or campion). Also familiar are the numerous annual weeds, such as *Cerastium, Arenaria,* and *Stellaria,* commonly known as the chickweeds, which are found in lawns, flower beds, and other disturbed sites in the early spring. Species of a few genera are used as fodder in the Old World.

- Pink Family

  Agrostemma
  Drymaria
  Saponaria
  Vaccaria

- common ornamentals, dianthus, carnation, baby's breath

Plants herbs or rarely subshrubs; annuals or perennials; nodes often swollen. **Stems** not branched or dichotomously branched. **Leaves** simple; opposite or whorled or fascicled; herbaceous or indurate; cauline or basal or forming a basal rosette; sessile or petiolate; connate or not connate or connected by a transverse line; blades linear to lanceolate to elliptic; venation pinnate or appearing parallel-veined or with a single vein; margins entire; stipules absent or present. **Inflorescences** solitary flowers or cymes or umbellate clusters; terminal or axillary; bracts present or absent. **Flowers** perfect or rarely imperfect; perianths in 2- or 1-series. **Sepals** 4 or 5; free or fused. **Corollas** radially symmetrical. **Petals** 5 or 0; free; clawed or not clawed; typically white or pink to red. **Stamens** 1–10. **Pistils** 1; compound, carpels 2–5; stigmas 2–5; styles 2–5; ovaries superior; locules 1; placentation free-central or basal. **Hypanthia** absent or present; cup-shaped. **Fruits** capsules or utricles; dehiscent or indehiscent; valvate. **Seeds** numerous or 1.

- herbs or subshrubs; leaves simple; petals 5 or 0, white, pink, or red

The Caryophyllaceae is distinguished by the combination of opposite leaves with bases connate and sheathing or connected by a transverse line; the 2–5 carpellate, 1-locular ovaries with free-central or basal placentation; and the capsules dehiscing by valves or teeth at their apices.

Four genera in this family are known to cause intoxication problems. They contain steroidal saponins that are glycosides of pentacyclic oleananes. Githagenin is the genin typical of the family (Kon & Soper 1940; Abubakirov & Amanmuradov 1964). Saponins similar to those present in the known toxic genera also are found in a few other genera, e.g., *Dianthus* and *Gypsophila* (Connolly & Hill 1991). Although species of these taxa have not been reported to be toxic, they may in unusual circumstances cause adverse effects when large amounts of them are ingested.

Agrostemma L.

■ *A. githago*
corn campion
corn cockle

Taxonomy and Morphology—Native to Eurasia and comprising 2 species, *Agrostemma* is naturalized throughout much of North America. Reflecting the showy appearance of the flowers, its name is derived from the Greek roots *agros* and *stemma*, meaning "field" and "wreath" or "garland." Of toxicologic interest is the following species:

A. githago L. corn cockle, common corn cockle, corn campion

■ erect annual herbs; flowers solitary; petals 5, red, reddish purple, or white

Plants annuals; from taproots; herbage with whitish, appressed hairs. **Stems** erect; 30–100 cm tall; not branched or with a few branches; often 4-angled. **Leaves** cauline; sessile; veins typically 1; blades linear to linear-lanceolate; apices acute or long acuminate. **Inflorescences** solitary flowers; borne on elongate peduncles; bracts absent. **Flowers** large; perianths in 2-series. **Sepals** 5; fused; tubes 10-ribbed; lobes long, linear, green. **Petals** 5; clawed; red or reddish purple or rarely white; lobes obovate to oblanceolate; apices emarginate. **Stamens** 10. **Styles** 5. **Capsules** ovoid; dehiscent; apices 5-toothed. **Seeds** black; tuberculate.

Agrostemma githago

Distribution and Habitat—Now naturalized, *A. githago* is widely distributed across the continent in grain fields and waste areas. Also cultivated as a garden ornamental, it is especially abundant in the northern half. Once a common weed, its abundance in fields has declined with increased use of herbicides and improved seed cleaning.

■ weedy ornamental, naturalized

■ livestock, seeds in feed, irritant digestive disturbance, rarely also neurologic effects

■ seeds, irritant steroidal saponins, glycosides of githagenin

Disease Problems—The primary problem occurs when the seeds of *A. githago* contaminate stocks of wheat, other cereal grains, or soybeans. Animals eating large amounts of contaminated feed are likely to develop an irritated digestive tract and occasionally die. The species is particularly a problem when contaminated feeds are ground and fed to poultry. Birds typically avoid eating whole seeds, but if force-fed, they can be toxic (Quigley & Waite 1931; Heuser & Schumacher 1942). As little as 0.25% ground seed was toxic to newborn chicks, whereas 6-week-old chicks tolerated up to 5% in their feed, adjusting over several weeks to any slight adverse effects (Quigley & Waite 1931; Heuser & Schumacher 1942; Bierer & Rhodes 1960). Larger amounts in the feed result in marked reduction of feed intake. Intoxications also are suspected in cattle consuming contaminated forage and having *Agrostemma* seeds in their rumens (Ballarini 1985). Hay containing the capsular fruits and seeds also has been associated with disease in cattle (Smith et al. 1997). The roots are suspected to cause death in pigs (Hansen 1924). All animal species, including poultry, ducks, geese, and even foxes, are at risk mainly from seeds (Kotz 1965).

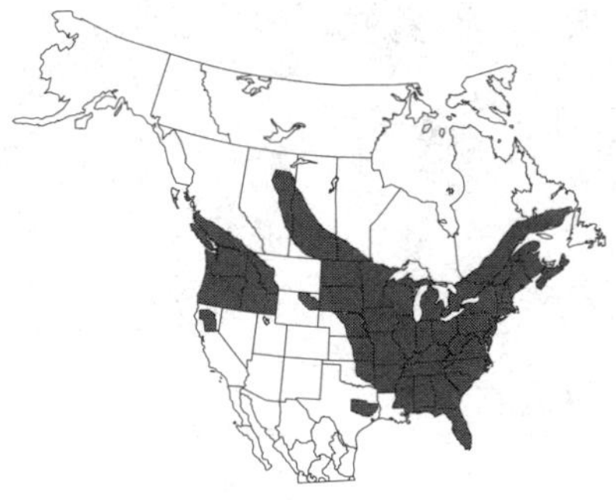
Agrostemma githago

Disease Genesis—The cause of disease are steroidal saponins, referred to by various names such as agrostemmin, githagoside, githagin, sapotoxin, or smilacin. These saponins are glycosides of the pentacyclic oleanane terpenoid githagenin, the genin typical of the family Caryophyllaceae (Wedekind & Schicke 1929, 1930; Kon & Soper 1940; Abubakirov & Amanmuradov 1964). Saponin concentrations of up to 6% are reported for the seeds (Harshberger 1920; Wedekind & Krecke 1926).

Clinical Signs—In chicks, the effects produced by *A. githago* are primarily on the upper digestive tract and include depression and a very pronounced drop in appetite.

In other species signs are a pronounced drop in appetite, increased salivation, bloat, colic, and diarrhea. A decrease in rumen motility accompanies the drop in appetite in cattle. Systemic signs such as muscle spasms and tremors and paralysis may also be present. In severe cases the animals may die in 1 or 2 days.

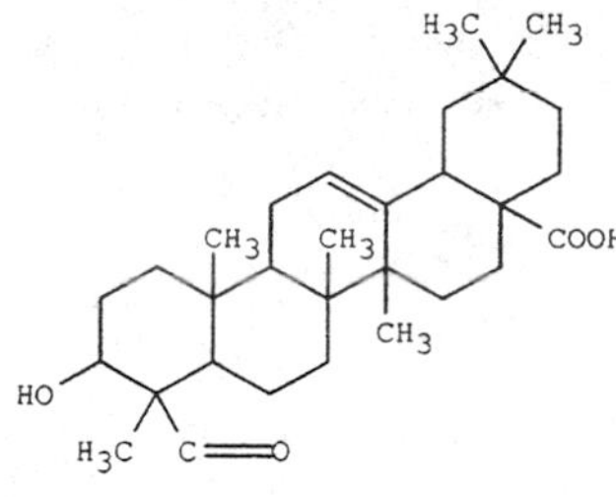

Githagenin

- depression, anorexia, increased salivation, bloat, colic, diarrhea; rarely tremors, paralysis

Pathology—Ingestion for a few days by chicks results in the development of severe reddening of the mucous membranes of the mouth, tongue, and pharynx, erosions of the gizzard, and fluid accumulation around the heart. In other species there is reddening of the mucosa of the lower digestive tract, with hemorrhage possibly occurring.

- gross pathology, reddened mucosa digestive tract

Treatment—In cases of *A. githago* intoxication, the main consideration in treatment is maintenance of hydration with fluids and electrolytes. Control of the diarrhea may not be necessary in all instances.

- fluids, electrolytes

Drymaria Willd. ex Roem. & Schult.

- *D. arenarioides*
  D. pachyphylla

  alfombrillo
  drymary
  inkweed

Taxonomy and Morphology—Comprising 48–50 species, all but 2 of which are American, *Drymaria,* commonly known as drymary, is primarily a subtropical genus and thus somewhat different from most other members of the Caryophyllaceae. Occurring from the southwestern United States to Argentina and Chile, its species are often conspicuous elements of the herbaceous vegetation in some areas (Duke 1961). Some species are planted as ground covers, and others are used in folk remedies to cure a variety of ailments. Of toxicologic importance are the following:

D. arenarioides Willd. ex Roem. & Schult.	alfombrillo, drymary
D. pachyphylla Wooton & Standl.	thickleaf drymary, inkweed

- spreading herbs; leaves opposite or appearing whorled; petals 5, white

Plants annuals or perennials; from taproots. **Stems** prostrate or spreading to erect; 10–25 cm long or tall; branched or not branched. **Leaves** opposite or appearing whorled; cauline; sessile or long petiolate; somewhat fleshy; blades linear to elliptic or suborbicular; stipules present or absent. **Inflorescences** solitary flowers or umbellate cymes; terminal or axillary. **Flowers** small; perianths in 2-series. **Sepals** 5; free; fused; green; margins scarious. **Petals** 5; white; obovate or oblanceolate; bifid or 1- or 2-cleft. **Stamens** 5. **Styles** 3. **Capsules** ovoid or globose; dehiscent; 3-toothed. **Seeds** 1 to many; of various shapes; tuberculate.

1. Stems spreading to erect. Plants glaucous. Leaves broadly elliptic to suborbicular. Seeds dull-black . . . . *D. pachyphylla*
1. Stems prostrate. Plants not glaucous. Leaves linear to narrowly elliptic. Seeds dull brown . . . . *D. arenarioides*

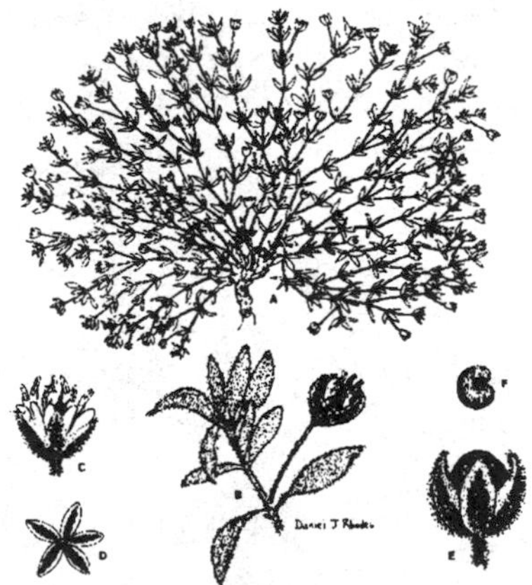

Drymaria arenarioides

- desert stream beds, gravel bars

Distribution and Habitat—*Drymaria pachyphylla* is found in barren alkaline sites where water collects seasonally, i.e., gravel bars and silty stream beds across the Southwest from Texas to Arizona and into Mexico. There is a strong tendency for plants to proliferate after summer rains (Little 1937). *Drymaria arenarioides* favors acid, sandy soils in the same types of habitats in the Chihuahuan and Sonoran Deserts in the Mexican states of Chihuahua, Sonora, Durango, Zacatecas, and San Luis Potosí (Dollahite 1959). It tends to be a toxicologic problem mainly in heavily overgrazed sites.

■ livestock, seldom eaten, digestive disturbance, from foliage green or dry, neurologic effects also

Drymaria arenarioides

Drymaria pachyphylla

Disease Problems—Both species of *Drymaria* have been recognized as causing digestive and neurologic disease since the early 1900s in Mexico and southern New Mexico (Lantow 1929). The foliage is quite unpalatable and is eaten only under adverse conditions, which, however, are not unusual in the hot, dry areas where these taxa grow. In some episodes involving hungry and thirsty cattle, *D. arenarioides* caused losses of 25–40% of affected herds (Sperry & Walker 1957). Livestock losses due to *D. pachyphylla* are related in part to an increase in the abundance of the species in the early 1920s (Little 1937). Although relatively unpalatable, plants of this species, which wilt somewhat in the heat of the day, are more turgid and succulent in the early morning hours, when they are more likely to be eaten. They are toxic whether green or dry, even after frosts. *Drymaria pachyphylla* is more likely to cause problems during the summer, whereas *D. arenarioides* is a problem mainly in the fall and early winter.

The toxicity of fresh foliage of *D. arenarioides* is 0.1–0.5% b.w. in sheep and cattle and slightly more (0.3–1%) in goats (Dollahite & Anthony 1956; Jacoby & Morton 1974; Larios & Jabalera 1976, 1983; Williams 1978). Animals quite often die in less than 24 hours. Toxicity is similar (0.5–0.8%) for either the foliage or flowers and seeds of *D. pachyphylla* (Lantow 1929; Mathews 1933). In chicks, 2–3% in the diet is toxic (Williams 1978).

■ irritant saponins, typical of the family

Disease Genesis—The disease appears to be due to saponins found in all parts of the plant of both species. Williams (1978) discovered at least six saponins to be present, with a total concentration of 3% in *D. arenarioides*. This is consistent with the toxicity of aqueous extracts and the lethality of saponin extracts given to sheep in amounts equivalent to the 3% level in plants (Lantow 1929; Williams 1978). The exact structures of the saponins are not known, but they are likely to be glycosides of githagenin, which is the characteristic genin of the family (Kon & Soper 1940; Abubakirov & Amanmuradov 1964).

Experimentally, the plant is toxic to mice at 15% in the diet and causes hepatic disease and possibly abortions. Commercially obtained saponin was similarly toxic but not clearly abortifacient (Ocampo 1972).

■ abrupt onset, restlessness, diarrhea, colic, trembling, salivation, hypothermia; later labored respiration, violent seizures

Clinical Signs—Following consumption of *Drymaria,* there is abrupt onset of restlessness, diarrhea, colic, trembling, increased salivation, and hypothermia. These signs are followed by depression, labored respiration, increased heart and respiration rates, prostration, and coma. Late in the course of signs, the head may be twisted backward and violent muscular contraction seizures occur; however, some animals may die with little evidence of struggle. Death may occur within a few hours of the onset of signs or 12–24 hours after ingestion of the plants.

■ gross pathology, reddening, edema, mucosa of abomasum, small intestine

microscopic, mild necrosis renal tubules, liver

Pathology—Grossly, there is congestion of the lungs, liver, and spleen and splotchy hemorrhages on the surfaces of the heart, lungs, and diaphragm. Edema and possibly small hemorrhages of the wall of the gallbladder may also be apparent and may extend some distance up the bile ducts. The mucosa of the abomasum and small intestine may be reddened and edematous, with blood in the lumens. Free fluid may also accumulate in the abdomen. Microscopically, there is necrosis of the mucosa of the digestive tract and epithelium of renal tubules, and hemorrhage and necrosis around the central veins of the liver. In some animals a yellow pigment is apparent in the myocardium. Intestinal mucosal infarcts are sometimes apparent in goats, possibly because of the animals' longer survival time.

Treatment—Specific treatment is lacking, but attempts should be made toward relief of the predominant signs. However, in general, response to treatment of any type is poor. The low palatability of the plants is a major factor in control of the disease. In times of summer drought and scarcity of other forage, losses can be minimized by providing water and food. There is generally little problem with well-fed and well-watered animals.

- nursing care

Saponaria L.

Taxonomy and Morphology—Native to Eurasia and Africa, *Saponaria,* or soapwort, comprises 20–30 perennial species. Its name is from the medieval Latin *sapo,* for "soap," because the sap of some species makes a lather with water. Some members are cultivated as ornamentals. In North America, 3 introduced species are present. Only 1 is of toxicologic significance:

- *S. officinalis*

  bouncing bet
  fuller's herb
  soapwort

S. officinalis L. bouncing bet, soapwort, fuller's herb

Plants perennials; annuals; colonial from rhizomes; herbage typically glabrous. **Stems** erect; 30–90 cm tall; not branched or sparsely branched above. **Leaves** cauline; short-petiolate or sessile; connected by a transverse ridge; veins 3, prominent, parallel; blades ovate to lanceolate-elliptic; apices acute or acuminate. **Inflorescences** capitate or condensed cymes; terminal or axillary; bracts present. **Flowers** large; perianths in 2-series; cylindrical. **Sepals** 5; fused; tubes not ribbed; lobes triangular; green or purple tinged. **Petals** 5, sometimes 10; long-clawed; with 2 small appendages; white or pink; lobes oblong to obovate; emarginate to retuse. **Stamens** 10; exserted. **Styles** 2. **Capsules** ovoid; dehiscent; apices 4-toothed. **Seeds** numerous; rotund-reniform; dark brown; reticulate-papillose.

- perennial erect herbs, little branched; leaves prominently 3-veined; flowers large; petals 5, white or pink; capsules ovoid

Saponaria officinalis

Saponaria is now circumscribed to include only perennial species. *Saponaria vaccaria,* an annual and the subject of several toxicologic studies, is now positioned in the genus *Vaccaria,* which is described below. Toxicants and toxicologic problems produced are the same for both genera.

Distribution and Habitat—Cultivated for its showy flowers, *S. officinalis* is naturalized throughout most of temperate North America but is less abundant in the drier Southwest. It is weedy in grain fields, along roadsides, and in waste areas. Plants are especially abundant in the northwestern United States and the Canadian prairie provinces.

- naturalized ornamental

Disease Problems—The seeds of *S. officinalis* cause a disease of the digestive tract when eaten as a contaminant in grain feeds. Although rarely eaten, feeding studies conducted by Huffman indicated that sheep may be severely intoxicated when fed amounts equivalent to 3% b.w. (Kingsbury 1964).

- livestock, seeds, rarely digestive disturbance

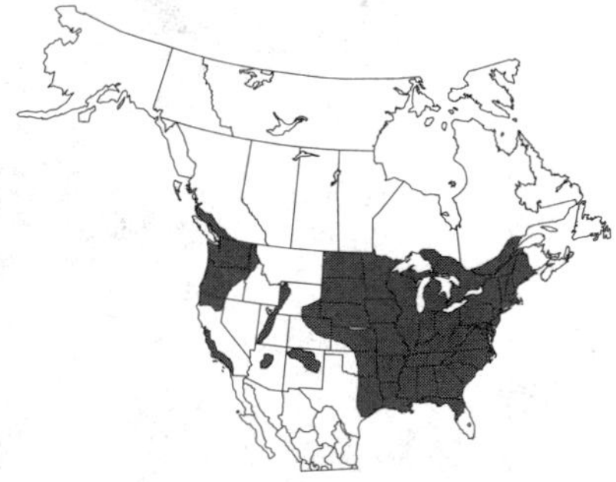

Saponaria officinalis

Disease Genesis—The seeds of *S. officinalis* are rich in saponins, such as the sapindosides. They are similar to those in *Vaccaria* and include glycosides of the pentacyclic oleanane diterpenoid githagenin (or gypsogenin), the genin typical of the family Caryophyllaceae (Kon & Soper 1940; Connolly & Hill 1991). In addition, there is saporin, a ribosomal inactivating protein that acts by removing an adenine residue from the 28s RNA. It acts in a fashion similar to that of ricin and abrin extracted from *Ricinus* in the Euphorbiaceae and *Abrus* in the Fabaceae. It is doubtful,

- irritant saponins, glycosides of githagenin

however, whether saporin plays any role in the toxicity of *Saponaria,* because it lacks a facilitator or cell-uptake peptide portion (Ippoliti et al. 1992).

- depression, anorexia, excess salivation, bloat, colic, diarrhea

reddening, mucosa, digestive tract

fluids, electrolytes

Clinical Signs, Pathology, and Treatment—Ingestion of the seeds of *S. officinalis* produces depression, a pronounced drop in appetite and rumen motility, increased salivation, bloat, colic, and diarrhea. These signs are accompanied by severe reddening of the mucosa of the digestive tract and possibly hemorrhage. The main consideration in treatment is maintenance of fluid balance with fluids and electrolytes. Control of the diarrhea may not be necessary in all instances.

Vaccaria Wolf

- *V. hispanica*

  China cockle
  cow cockle
  cowherb
  dairy pink
  pink cockle
  spring cockle

Taxonomy and Morphology—Native to the Old World, *Vaccaria* comprises 3 or 4 species. Its generic name is derived from the Latin *vacca,* for "cow," presumably in reference to its use as fodder for cattle or its common occurrence in pastures. Originally positioned in the genus *Saponaria,* 1 species is present in North America:

V. hispanica (Mill.) Rauschert (= *Saponaria vaccaria* L.) (= *V. pyramidata* Medik.) (= *V. segetalis* [Neck.] Asch.) (= *V. vulgaris* Host)	cow cockle, spring cockle, cowherb, pink cockle, China cockle, dairy pink

- erect annuals, sparsely branched; petals 5, pink to reddish; capsules ovoid

Plants annuals; from taproots; herbage glabrous, glaucous. **Stems** erect; 20–80 cm tall; not branched or sparsely branched above. **Leaves** cauline; lowest short-petiolate, upper sessile, clasping or lower connate; blades lanceolate to ovate; apices acute. **Inflorescences** open cymes; terminal or axillary; bracts absent. **Flowers** small; perianths in 2-series; urceolate to cylindrical. **Sepals** 5; fused; tube 5-angled and winged; lobes triangular, acuminate; green. **Petals** 5; long-clawed; appendages absent; pink to reddish; lobes obovate to oblanceolate; retuse. **Stamens** 10; exserted. **Styles** 2. **Capsules** ovoid; dehiscent; apices 4-toothed. **Seeds** numerous; globose; reddish brown to black; tuberculate.

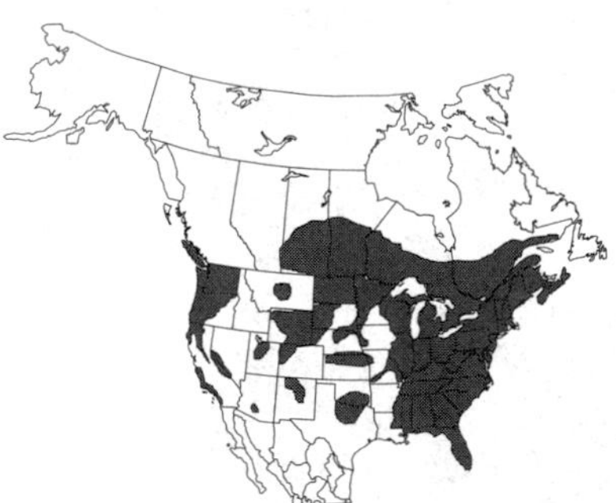

Vaccaria hispanica

- weedy, naturalized

Distribution and Habitat—Native to Europe, *V. hispanica* is a common weedy species throughout most of temperate North America. Plants occur in grain fields, along roadsides, and in waste areas. Plants are especially abundant in the northwestern United States and the Canadian prairie provinces.

- livestock, seeds in feed, digestive disturbance

Disease Problems—Because of its favorable growth and other agronomic characteristics, *Vaccaria* has been evaluated as a possible feed seed (Goering et al. 1966). However, the seeds of *V. hispanica* (reported as *Saponaria vaccaria)* cause a disease of the digestive tract when eaten as a contaminant in grain feeds.

- irritant diterpenoid saponins, glycosides of githagenin, vaccaroside, vacsegoside

Disease Genesis—The seeds of *V. hispanica* are rich in saponins, 0.64% in the whole seeds and 0.96% in the seed coats (Mazza et al. 1992). These saponins, such as vaccaroside and vacsegoside, are glycosides of the pentacyclic oleanane diterpenoid githagenin (or gypsogenin), the genin typical of the family Caryophyllaceae (Kon & Soper 1940; Abubakirov & Amanmuradov 1964; Connolly & Hill 1991).

Clinical Signs, Pathology, and Treatment—Ingestion of the seeds of *V. hispanica* produces the same signs as caused by *Saponaria officinalis:* depression, a pronounced drop in appetite and rumen motility, increased salivation, bloat, colic, and diarrhea. Severe reddening of the digestive tract mucosa and possibly hemorrhage develop. The main consideration in treatment is maintenance of hydration with fluids and electrolytes. Control of the diarrhea may not be necessary in all instances.

- depression, anorexia, excess salivation, bloat, colic, diarrhea

reddening, gut mucosa

fluids, electrolytes

Vaccaria hispanica

Vaccaroside

References

Abubakirov NK, Amanmuradov K: Vaccaroside, a triterpene glycoside from *Saponaria vaccaria. Zh Obshchei Khim* 34;1661–1665, 1964 (*Chem Abstr* 61;8564, 1964)

Ballarini G: *Agrostemma githago* poisoning in cattle. *Atti Soc Ital Buiatria* 17;187–190, 1985 (*Vet Bull* 56;2439, 1986)

Bierer BW, Rhodes WH: Poultry ration contaminants. *J Am Vet Med Assoc* 137;352–353, 1960

Connolly JD, Hill RA: *Dictionary of Terpenoids.* Vol 2, *Di- and Higher Terpenoids.* Chapman & Hall, London, 1991

Dollahite JW: Toxicity of *Drymaria arenarioides* for cattle, sheep, and goats. *J Am Vet Med Assoc* 135;125–127, 1959

Dollahite JW, Anthony WV: *Toxicity of* Drymaria arenarioides *for Cattle, Sheep, and Goats.* Tex Agric Exp Stn Prog Rep 1911, pp1–4, 1956

Duke JA: Preliminary revision of the genus *Drymaria. Ann Mo Bot Gard* 68;173–268, 1961

Goering KJ, Eslick RF, Watson CA, Keng J: Utilization and agronomic studies of cow cockle *(Saponaria vaccaria). Econ Bot* 20;429–433, 1966

Hansen AA: The poison plant situation in Indiana. 3. *J Am Vet Med Assoc* 66;351–362, 1924

Harshberger JW: *Pastoral and Agricultural Botany.* Blakiston's Son & Co, Philadelphia, pp57–58, 1920

Heuser GF, Schumacher AE: The feeding of corn cockle to chickens. *Poult Sci* 21;86–93, 1942

Ippoliti R, Lendaro E, Bellelli A, Brunori M: A ribosomal protein is specifically recognized by saporin, a plant toxin which inhibits protein synthesis. *FEBS Lett* 298;145–148, 1992

Jacoby PW Jr, Morton HL: *Alfombrilla.* Ariz Coop Ext Serv Agric File Q-352, 1974

Kingsbury JM: *Poisonous Plants of the United States and Canada.* Prentice-Hall, Englewood Cliffs, NJ, 1964

Kon GAR, Soper HR: Sapogenins. 8. The sapogenins of fuller's herb. *J Chem Soc* 1940;617–620, 1940

Kotz J: Morphology and pathogenesis of corn cockle, *Lychnis (Agrostemma githago)*, poisoning in poultry. 3 and 4. *Med Wet* 21;520–524,730–734, 1965 (*Vet Bull* 36;2793, 1966)

Lantow JL: *The Poisoning of Livestock by* Drymaria pachyphylla. N M Agric Exp Stn Bull 173, 1929

Larios F, Jabalera J: Gross lesions, haemogram, and blood chemistry in experimental acute *Drymaria arenarioides* poisoning in cattle. *Tech Pecu Mex* 30;111, 1976

Larios F, Jabalera J: Toxicity of *Drymaria arenarioides* (Caryophyllaceae) when fed to cattle. *Tech Pecu Mex* 44;86–91, 1983

Little EL: A study of poisonous drymaria on southern New Mexico ranges. *Ecology* 18;416–426, 1937

Mathews FP: The toxicity of *Drymaria pachyphylla* for cattle, sheep, and goats. *J Am Vet Med Assoc* 83;255–260, 1933

Mazza G, Biliaderis CG, Przybylski R, Oomah BD: Compositional and morphological characteristics of cow cockle *(Saponaria vaccaria)* seed, a potential alternative crop. *J Agric Food Chem* 40;1520–1523, 1992

Ocampo CL: Biomedical investigations of the toxic and abortifacient effects of saponins from *Drymaria arenarioides*, in pregnant mice. *Vet Mex* 3;94–97, 1972

Quigley CD, Waite RH: Miscellaneous feeding trials with poultry. 1. Effects of corn cockle on poultry. *Md Bull* 325;343–354, 1931

Smith RA, Miller RE, Lang DE: Presumptive intoxication of cattle by corn cockle, *Agrostemma githago* (L.) Scop. [letter]. *Vet Hum Toxicol* 39;250, 1997

Sperry OE, Walker AH: Alfombrillo causing heavy losses to livestock in Chihuahua, Mexico. *Cattleman* 43;36,60, 1957

Wedekind E, Krecke R: Uber das githagenin, das endsapogenin aus *Agrostemma githago*. 1. Mitteilung uber die bestandteile des kornradesamens. *Hoppe-Seylers Z Physiol Chem* 155;122–136, 1926

Wedekind E, Schicke W: Uber githagenin und githaginsaure. 2. Mitteilung uber die bestandteile des kornradesamens. *Hoppe- Seylers Z Physiol Chem* 182;72–81, 1929

Wedekind E, Schicke W: Uber githagosaure und githagonolsaure. 3. Mitteilung uber die bestandteile des kornradesamens. *Hoppe-Seylers Z Physiol Chem* 190;1–14, 1930

Williams MC: Toxicity of saponins in alfombrilla *(Drymaria arenarioides)*. *J Range Manage* 31;182–184, 1978

Celastraceae R.Br.

22

Comprising 50–90 genera and 800–1,300 species, the Celastraceae, commonly known as the bittersweet, staff-tree, or spindle-tree family, is primarily pantropical in distribution, with relatively few taxa in temperate regions. It is somewhat closely related to the Aquifoliaceae but is distinguished by the presence of a glandular hypanthial disk and seeds with a brightly colored aril (Hou 1955). With the exception of 8 genera widely grown as ornamental shrubs, the family is of slight economic importance as a source of carving wood, yellow dye, oil, and herbal teas. In the Philippines, species of the genus *Lophopetalum* are used as arrow poisons (Wagner et al. 1984). In the Middle East and North Africa the leaves of *Catha edulis* (qat or khat) are chewed for their stimulatory effects, which are due to pseudoephedrine and similar compounds that are not present in other genera of the family. Of the approximately 13 native and introduced genera and 35 species present in North America, 2 genera are of minor toxicologic importance because of the presence of cardenolides.

■ Bittersweet Family
Spindle-tree Family
Staff-tree Family

Celastrus
Euonymus

Plants shrubs or small trees or woody vines; bearing perfect flowers or dioecious or polygamodioecious. **Stems** erect or trailing or climbing. **Leaves** simple; opposite or alternate; herbaceous or coriaceous; venation pinnate; blades lanceolate to ovate; surfaces glabrous or puberulent; apices acute to acuminate; margins crenulate or serrate or entire; stipules absent or present, minute, caducous. **Inflorescences** compound cymes or panicles or solitary flowers; terminal or axillary; peduncles jointed. **Flowers** perfect or functionally unisexual; perianths in 2-series. **Sepals** 4 or 5; free or fused. **Corollas** radially symmetrical. **Petals** 4 or 5; free; short clawed; ovate to orbicular; spreading or reflexed; white to greenish white or greenish purple to dark purple. **Stamens** 4 or 5; attached at edge of prominent nectar disk. **Pistils** 1; compound, carpels 2–5; stigmas 1, not lobed or 2-to-5-lobed, capitate; styles 1 or 0; ovaries superior, often appearing partly inferior because of thickness of nectary disk; locules 2–5; placentation axile; ovules 2–6 per locule. **Fruits** septicidal capsules or berries or drupes or achenes. **Seeds** 2–10; with showy arils.

■ shrubs, vines, small trees; petals 4 or 5, white, greenish white, greenish purple, or dark purple

Celastrus L.

Taxonomy and Morphology—Cosmopolitan in distribution, albeit mostly tropical, *Celastrus* comprises approximately 30 species. Its name is from the ancient Greek vernacular name *kelastros*. Some species are occasionally grown as ornamental climbers. In North America, 2 species are of some toxicologic concern:

C. orbiculatus Thunb.	oriental bittersweet
C. scandens L.	American bittersweet, false bittersweet, waxwork, staff vine

■ *C. orbiculatus*
C. scandens

bittersweet
staff vine
waxwork

- climbing, woody vines; leaves alternate; petals 5, greenish; capsules globose, yellow or orange

Celastrus scandens

Plants woody vines; spreading via root suckers; dioecious or polygamodioecious. **Stems** climbing; to 15 m long; branches slender, terete or angular, surfaces smooth or lenticels present, pith solid or chambered. **Leaves** alternate; blades elliptic to ovate or obovate or suborbicular. **Inflorescences** terminal panicles or axillary cymes. **Flowers** functionally unisexual; small. **Sepals** 5; fused. **Petals** 5; oblong-ovate; greenish. **Stamens** 5. **Pistils** 1; carpels 3; stigmas 3-lobed, styles 1, short; locules 3; ovules 2 per locule. **Fruits** capsules; globose; yellow or orange. **Seeds** 3–6; ellipsoidal; reddish brown; arils orange to red.

1. Leaves suborbicular to broadly oblong-obovate. Flowers 2 or 3 in axillary cymes . . . . *C. orbiculatus*
1. Leaves ovate to ovate-lanceolate. Flowers numerous in terminal panicles . . . . *C. scandens*

- woodlands

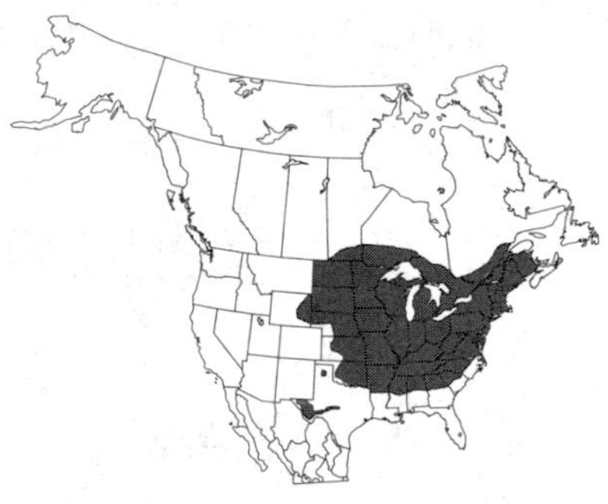

Celastrus scandens

Distribution and Habitat—As their common names indicate, *C. scandens* is native, whereas *C. orbiculatus* is introduced from eastern Asia. *Celastrus scandens* is found in woods and along streams throughout the eastern half of the continent. Introduced from Japan and China and widely used as an ornamental, *C. orbiculatus* has escaped and occurs sporadically throughout the eastern half of the continent.

- foliage and fruit, digestive disturbance, steroidal cardenolides; alkaloids, celastrine, toxicity unknown

Disease Problems and Genesis—There is little information regarding the toxicity potential of *Celastrus*. Although *C. scandens* is reported to contain cardenolides, specific information on toxicity is lacking (Cook et al. 1944; Bruning & Wagner 1978). Whether *C. orbiculatus* similarly contains cardenolides is unknown, but other Old World species are reported to have cardiac effects (Neogi & Baliga 1964–1965). Plants of these species should be considered capable of at least causing the digestive disturbances typical of cardenolides, especially with consumption of the fruits and seeds. Although the fruits contain these toxic compounds, some animals such as rabbits and squirrels are able to eat them without problem (Gill & Healy 1973).

The alkaloids celastrine and paniculatine also have been determined to occur in *Celastrus*, but there are no indications that they play a role in intoxications (Basei & Pabrai 1946).

- nausea, vomiting, diarrhea; mild

no treatment needed

Clinical Signs, Pathology, and Treatment—Depending upon the animal species involved, expected clinical signs would include nausea, vomiting, and diarrhea. The effects are likely to be mild and not likely to elicit pathologic changes or require treatment.

Euonymus L.

- *E. alatus*, *E. americanus*, *E. atropurpureus*, *E. bungeanus*, *E. europaeus*, *E. fortunei*, *E. hamiltonianus*, *E. japonicus*, *E. kiautschovicus*, *E. obovatus*

arrowwood, burning bush, bursting heart, euonymus, pegwood, spindle tree, strawberry bush, wahoo, wintergreen

Taxonomy and Morphology—Comprising 175–200 species and cosmopolitan in distribution, *Euonymus* is a genus of popular ornamentals grown for their attractive foliage and showy fruits. Some species are used as ground covers. The genus name is taken from an ancient Greek vernacular name for one of the species. In some publications, the spelling *Evonymus* is used, but this is incorrect. Taxonomists disagree as to whether the endings of the specific epithets should be *-a* or *-us*. For example, wahoo may be *E. atropurpurea* in one reference and *E. atropurpureus* in another. Because *Euonymus* is a masculine Latin noun, the *-us* ending is used here. In North America, approximately 15 species are present. The native species include the following:

E. americanus L.	strawberry bush, bursting heart, American burning bush
E. atropurpureus Jacq.	wahoo, burning bush, arrowwood
E. obovatus L.	running strawberry bush

The ornamental species include the following:

E. alatus Siebold	winged euonymus, burning bush, winged spindle tree
E. bungeanus Maxim.	winterberry euonymus
E. europaeus L.	European spindle tree, pegwood
E. fortunei (Turcz.) Hand.-Mazz.	evergreen wintergreen
E. hamiltonianus Wall.	
E. japonicus Thunb.	evergreen euonymus
E. kiautschovicus Loes.	spreading evergreen euonymus

Plants shrubs or rarely small trees; deciduous or evergreen. **Stems** erect or prostrate or climbing; branches terete or angular. **Leaves** opposite; petiolate; blades elliptic to ovate or obovate; margins serrate or serrulate. **Inflorescences** solitary flowers or cymes in leaf axils. **Flowers** perfect or functionally imperfect; small. **Sepals** 4 or 5; fused. **Petals** 4 or 5; oblong-ovate; greenish to purplish. **Stamens** 5. **Pistils** 1; carpels 3–5; stigmas 3-to-5-lobed, styles 1, short; ovaries embedded in nectary disks; locules 3–5; ovules 1–6 per locule. **Fruits** capsules; typically 3-to-5-lobed; yellow or red to brown. **Seeds** 1–6; yellowish brown or red or white or black; arils orange to red.

■ shrubs to small trees; leaves opposite; flowers small; petals 4 or 5, greenish to purplish; capsules yellow, red, or brown

1. Branches corky winged . . . . *E. alatus*
1. Branches not corky winged.
    2. Leaves evergreen.
        3. Leaf apices long acuminate . . . . *E. bungeanus*
        3. Leaf apices short acuminate or acute or obtuse.
            4. Cymes loose. Leaves thin; herbaceous . . . . *E. kiautschovicus*
            4. Cymes dense. Leaves thick; coriaceous.
                5. Stems prostrate or climbing; rooting at nodes. Branches warty . . . . *E. fortunei*
                5. Stems erect; not rooting at nodes. Branches smooth . . . . *E. japonicus*
    2. Leaves deciduous.
        6. Abaxial leaf surfaces puberulent. Petals purplish . . *E. atropurpureus*
        6. Abaxial leaf surfaces glabrous. Petals greenish or yellowish.
            7. Capsules tuberculate.
                8. Stems erect. Uppermost leaves lanceolate to ovate . . . . *E. americanus*
                8. Stems prostrate. Uppermost leaves obovate . . . . . . *E. obovatus*
            7. Capsules smooth.
                9. Capsules not lobed . . . . *E. kiautschovicus*
                9. Capsules 4-or-5-lobed.
                    10. Leaf apices long-acuminate. Petioles 12–25 mm long . . . . *E. bungeanus*
                    10. Leaf apices acute or short-acuminate. Petioles 6–10 mm long.
                        11. Branches angular. Seeds white . . . . *E. europaeus*
                        11. Branches terete. Seeds pink to red . . *E. hamiltonianus*

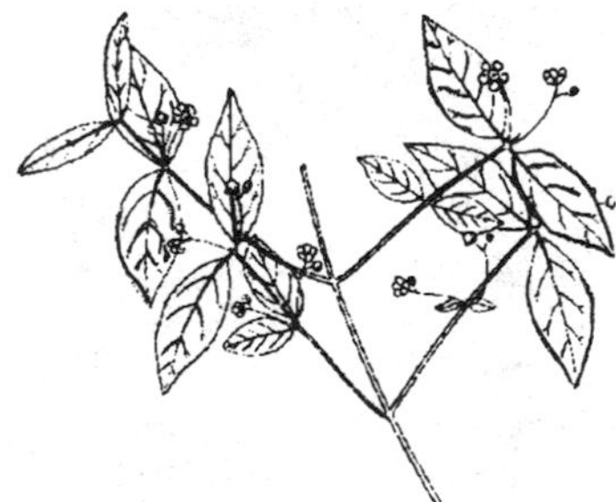

Euonymus americanus

Euonymus atropurpureus

■ moist, deciduous woodlands

Distribution and Habitat—The three native species are characteristically encountered in moist soils of the deciduous forests of the eastern half of the continent. Typically the plants occur in shade. The introduced Eurasian species have escaped cultivation and occur sporadically throughout the forests as well.

■ foliage, fruits cardiotoxic, digestive disturbance

Disease Problems—In spite of the popularity of this cardiotoxic genus as an ornamental, reports of intoxications have been surprisingly few. *Euonymus europaeus*, for example, has long been considered toxic, but few incidences have been recorded (Cooper & Johnson 1984). The clinical signs reported are consistent with those produced by cardiotoxins in general, although perhaps not as severe as those of other cardiotoxic genera. Indeed *E. europaeus* is reported to have been used as an emetic, a purgative, and a cardiotonic folk medicine (Smith 1977). A few instances of intoxication have occurred in horses as well as ruminant species, with signs related mainly to the digestive tract (Perkins & Payne 1978; Bartik & Piskac 1981; Cooper & Johnson 1984). Gastrointestinal involvement has varied from constipation in a case involving two horses to diarrhea in other animals.

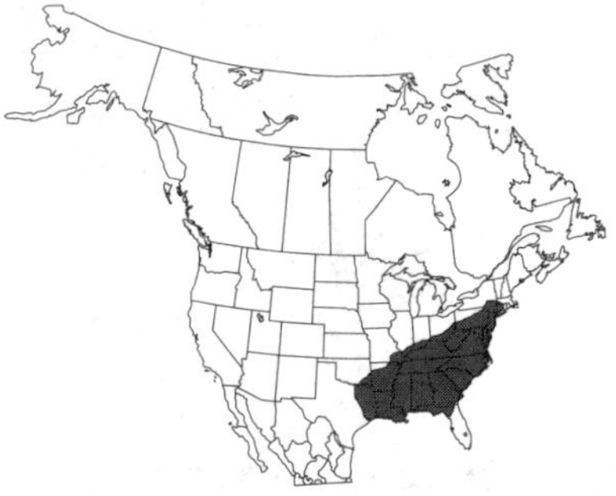

Euonymus americanus

Euonymus atropurpureus

Disease Genesis—Several different types of potential toxicants, including both cardenolides and alkaloids, are present in the *Euonymus*. The similarity in their nomenclature, as illustrated in table 22.1, has been a cause of considerable confusion.

Table 22.1. Representative cardenolides and alkaloids in species of *Euonymus*

Species	Cardenolides	Alkaloids
E. alatus	two types	evonine, euonymine, wilfordine
E. atropurpureus	evatroside evatromonoside	three types
E. europaeus	evonoside evomonoside evobioside	evorine, isoevorine evonoline, franganine frangulanine, frangufoline, isoevonine, evozine
E. hamiltonianus	euonymoside A	euonine, euonymine, evonine, isoevorine, neoevorine, neoeuonymine

Sources: Bishay et al. 1973; Smith 1977; Fung 1986; Baek et al. 1994.

■ cardenolides, evonoside, euonymoside A, mainly digestive disturbance

alkaloids, complex derivatives of nicotinic acid, evonine, toxicity unknown

methylxanthines, caffeine

Most of the cardenolides, sometimes spelled with the prefix *eu-* rather than *ev-*, are glycosides of digitoxigenin (= evonogenin) and differ in the number of attached sugars (Shoppee 1964; Bruning & Wagner 1978; Joubert 1989). Euonymoside A is a glycoside of acovenosigenin or 1β-hydroxydigitoxigenin (Baek et al. 1994). Typically, these compounds have been isolated from seeds (ranging in concentration from 8 to occasionally 100 mg%), but more-recent work indicates that the glycosides are present in all plant parts and that most species contain them (Fung 1986). Thus, although leaves may have a lower content, they nonetheless represent a hazard. In addition, many species contain several cardenolides; as many as a dozen are present in *E. europaeus*. Because of the large number of genera containing cardenolides, it seems reasonable to assume that these types of compounds are typical of the genus. All of the cardenolides have not yet been identified, and it has been suggested that those in the leaves may not contain the digitoxigenin nucleus (Fung 1986). However, it seems unlikely that the seeds and leaves would contain totally different genins.

The alkaloids are esters and diesters of nicotinic acid, e.g., evonine, often as complex, closed diesters or sesquiterpenes. They are found in the roots, bark, leaves, fruits, and seeds of many species (Smith 1977; Ishiwata et al. 1983). The role of these com-

Rhamnose-gluc-glucose

Evonoside

Rhamnose-rham-glucose

Euonymoside A

Evonine

pounds in intoxications is not clear; the alkaloids are reported to be devoid of biological activity by themselves in rats, although they are insecticidal (Smith 1977). In addition, theobromine is present in seeds, and theophylline and caffeine in the leaves, of *E. europaeus* (Bohinc et al. 1976). Triterpenoids such as α-amyrin also are present in *E. europaeus* (Pasich et al. 1980).

Clinical Signs, Pathology, and Treatment—In monogastric as well as in ruminant species, the signs include vomiting/regurgitation, colic, chills, and weakness. Gastrointestinal involvement ranges from constipation to diarrhea. Changes at necropsy will likely be limited primarily to the digestive tract. Treatment is symptomatic for alleviation of the digestive disturbance and in most instances will not be required. In severe cases control of cardiac arrhythmias may be required.

- vomiting, colic, chills, weakness, constipation or diarrhea

gross pathology, reddened gut mucosa

treatment seldom needed

References

Baek NI, Lee YH, Park JD, Kim SI, Ahn BZ: Euonymoside A: a new cytotoxic cardenolide glycoside from the bark of *Euonymus sieboldianus*. *Planta Med* 60;26–28, 1994

Bartik M, Piskac A: *Veterinary Toxicology*. Elsevier, Amsterdam, 1981

Basei NK, Pabrai PR: Chemical investigation of *Celastrus paniculata* Willd. *J Am Pharm Assoc* 35;272–273, 1946

Bishay DW, Kowalewski Z, Phillipson JD: Peptide and tetrahydroisoquinoline alkaloids from *Euonymus europaeus*. *Phytochemistry* 12;693–698, 1973

Bohinc P, Colnaric M, Hidajat S, Primozic J: Xanthic alkaloids of the pegwood—*Euonymus europaeus* L. *Farm Vestn (Ljubljana)* 27;149–156, 1976 (*Chem Abstr* 86:103077k, 1977)

Bruning R, Wagner H: Ubersicht uber die Celastraceen-inhaltsstoffe: chemie, chemotaxonomie, biosynthese, pharmakologie. *Phytochemistry* 17;1821–1858, 1978

Cook DL, Parks LM, Dunker MFW, Uhl AH: Phytochemical study of the leaves of *Celastrus scandens* Linne. 1. A preliminary study. *J Am Pharm Assoc* 33;15–17, 1944

Cooper MR, Johnson AW: *Poisonous Plants in Britain and Their Effects on Animals and Man*. Ref Book 161. Ministry of Agriculture, Fisheries and Food, Her Majesty's Stationery Office, London, pp79–80, 1984

Fung SY: Alkaloids and cardenolides in sixteen *Euonymus* taxa. *Biochem Syst Ecol* 14;371–373, 1986

Gill JD, Healy WM: *Shrubs and Vines for Northeastern Wildlife*. USDA Forest Service, Gen Tech Rep NE-9, 1973

Hou D: A revision of the genus *Celastrus*. *Ann Mo Bot Gard* 42;215–302, 1955

Ishiwata H, Shizuri Y, Yamada K: Three sesquiterpene alkaloids from *Euonymus alatus* forma striatus. *Phytochemistry* 22;2839–2841, 1983

Euonymus europaeus

Euonymus obovatus

Joubert JPJ: Cardiac glycosides. In *Toxicants of Plant Origin,* vol 2, *Glycosides,* Cheeke PR ed, CRC Press, Boca Raton, Fla, pp61–96, 1989

Neogi NC, Baliga AK: Studies of *Celastrus paniculata. J Sci Res Banaras Hindu Univ* 15;135–141, 1964–1965 (*Chem Abstr* 64:5649f, 1966)

Pasich B, Bishay DW, Kowalewski Z, Rompel H: Chemical investigation of *Euonymus europeus. Planta Med* 39;391–395, 1980

Perkins KD, Payne WW: *Guide to the Poisonous and Irritant Plants of Florida.* Fla Coop Ext Serv Circ 441, 1978

Shoppee CW: *Chemistry of the Steroids,* 2d ed. Butterworths, Washington, DC, pp326–397, 1964

Smith RM: The Celastraceae alkaloids. In *The Alkaloids,* Manske RHF ed, Academic Press, New York, pp215–248, 1977

Wagner H, Habermeier H, Schulten HR: The cardiac glycosides of the arrow poison of *Lophopetalum toxicum* Loher. *Helv Chim Acta* 67;54–64, 1984

Chenopodiaceae Vent.

23

Comprising about 100 genera and 1,400–1,500 species, the Chenopodiaceae, commonly known as the goosefoot family, is cosmopolitan, but especially abundant in saline or alkaline habitats. Principal centers of distribution are the plains and deserts of North America, the pampas of South America, the shores of the Red, Caspian, and Mediterranean Seas, the central Asiatic basin, the Karroo of South Africa, and the salt plains of Australia. Economically important genera include the sugar and forage beet *(Beta),* spinach *(Spinacia),* and Swiss chard *(Beta).* As noxious weeds of cultivated fields, likewise important are Russian thistle *(Salsola),* goosefoot *(Chenopodium),* and kochia *(Kochia).* Several genera—e.g., shad scale *(Atriplex),* greasewood *(Sarcobatus),* and winter fat *(Eurotia)* are important elements of the continent's desert flora. In North America, approximately 27 genera and 200 native and introduced species are present. Intoxication problems are associated with more than 10 genera.

■ Goosefoot Family

Bassia
Beta
Chenopodium
Halogeton
Kochia
Salsola
Sarcobatus
Suckleya

QUESTIONABLE
Monolepis

Plants herbs or subshrubs or shrubs; annuals or biennials or perennials; strongly aromatic or not aromatic; bearing perfect flowers or gynomonoecious or monoecious or dioecious. **Stems** fleshy or not fleshy; erect or decumbent or prostrate; not jointed or rarely jointed. **Leaves** simple; opposite or alternate; cauline or basal and forming a rosette; petiolate or sessile; blades flat or terete; fleshy or herbaceous; venation pinnate or pinnipalmate or not apparent; surfaces glabrous or farinose or indumented; margins entire or dentate or lobed or incised; stipules absent. **Inflorescences** of solitary flowers or cymes or glomerules; spicate or paniculate; axillary or terminal; bracts absent or present; bracteoles present or absent. **Flowers** inconspicuous; perfect or imperfect, perfect and staminate and pistillate similar; perianths in 1-series or absent; typically pentagonal. **Calyces** radially symmetrical; funnelform or campanulate or urceolate. **Sepals** 1–5 or 0; free or fused; appressed; usually green. **Petals** absent. **Stamens** 1–5; opposite the sepals. **Pistils** 1; compound, carpels 2–5; stigmas 2 or 3 or rarely 5; styles 2 or 3 or rarely 5; ovaries superior; locules 1; placentation basal. **Fruits** utricles or achenes; usually enclosed within persistent sepals; pericarps free or fused to seeds; indehiscent or dehiscent. **Seeds** 1.

■ herbs, shrubs, subshrubs; leaves simple; flowers green; petals absent

Many genera of the Chenopodiaceae are considered facultative halophytes and grow in highly saline or alkaline soils (Chapman 1975). Growth in such soils may not be detrimental to forage quality and in some instances—as with *Halogeton* and *Salsola,* for example—may even improve it (Williams 1960; Fowler et al. 1992). Some species of *Salicornia,* grown in seawater, have been used with success as feed for livestock (Jefferies & Rudmik 1984). This tolerance involves uptake of sodium and magnesium salts, mainly as carbonates, chlorides, and sulfates. Because of this, some taxa have a propensity to accumulate one or more of the compounds selenium, oxalate, sulfur, and nitrate.

■ many are useful feed when only part of diet; problems mainly in sheep subsisting exclusively on these plants

grow in saline, alkaline soils; may accumulate selenium, oxalate, sulfur, or nitrate

Selenium accumulation is dependent upon the levels and form of the element in the soil (Trelease & Beath 1949). Aspects of these factors are described more fully in

■ selenium, *Atriplex, Grayia*

the treatment of *Astragalus* in the Fabaceae (see chapter 37). Genera of the Chenopodiaceae noted for their accumulation of selenium are *Grayia* and *Atriplex*. Among the taxa that accumulate selenium, the amounts taken up vary considerably. For example, *G. spinosa* generally has low levels, whereas *G. brandegei* occasionally has high concentrations. *Atriplex canescens* has a tendency to accumulate selenium, primarily as selenate, in the range of 100–200 ppm. In contrast, *A. argentea, A. confertifolia,* and *A. pabularis* generally have less than 50 ppm even on soils high in selenium (Trelease & Beath 1949; Davis 1972). In general, selenium intoxication is not a significant risk with any of these plants.

■ oxalate is most consistent risk; *Atriplex, Bassia, Chenopodium, Salsola* plus others, soluble oxalates, mainly sodium and acid sodium; calcium oxalate also present

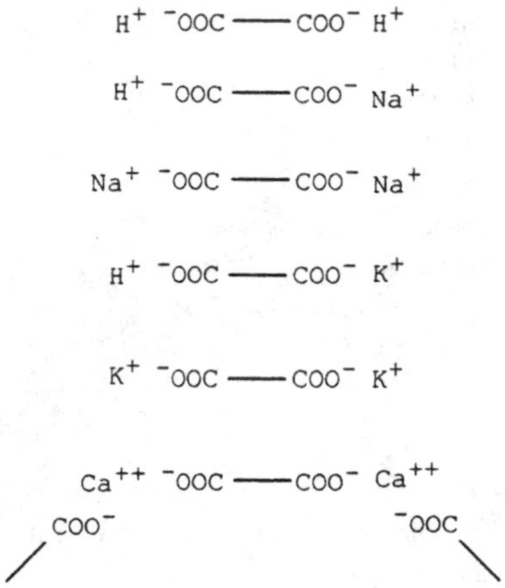

$H^+ \; ^-OOC — COO^- \; H^+$

$H^+ \; ^-OOC — COO^- \; Na^+$

$Na^+ \; ^-OOC — COO^- \; Na^+$

$H^+ \; ^-OOC — COO^- \; K^+$

$K^+ \; ^-OOC — COO^- \; K^+$

$Ca^{++} \; ^-OOC — COO^- \; Ca^{++}$

COO^- / ^-OOC

Oxalic acid and its salts

Oxalate accumulation is the most consistent feature of the family and of greatest concern toxicologically. The disease produced is essentially the same regardless of the taxon. Species of *Atriplex, Bassia, Chenopodium,* and *Salsola* have been shown to be accumulators (Mathams & Sutherland 1952). In addition, *Monolepis* (povertyweed), *Salicornia* (glasswort), *Spinacia* (cultivated spinach), and *Suaeda* (seepweed) have the potential to be oxalate accumulators but normally do not represent threats to livestock, because of other factors. For example, the leaves of *Spinacia oleracea* may have 1–8% soluble salts and up to 12% total oxalates d.w. (Libert & Franceschi 1987). However, care must be taken in interpreting values of oxalate concentration because they may vary considerably, depending upon the method of extraction used (Mathams & Sutherland 1952). In the Chenopodiaceae, soluble oxalates are present primarily as disodium and acid/sodium oxalate. The total of oxalates includes the insoluble salts such as calcium oxalate. Plants growing where there are high levels of nitrogenous runoff may pose a special risk because in conditions of high nitrogen, high potassium, and low phosphorus, oxalate accumulation may increase (Libert & Franceschi 1987).

■ sulfur accumulation, *Atriplex, Kochia, Salicornia, Sarcobatus*

nitrate accumulation, *Beta, Chenopodium, Salsola*

Taxa of interest as sulfur accumulators, up to 3.5% d.w., include *Atriplex, Kochia, Salicornia,* and *Sarcobatus* (Thomas & Hendricks 1950; Mayland & Robbins 1994). These plants, however, are not necessarily a risk for disease unless large amounts of them containing high levels of sulfur are eaten for a week or more. Of these genera, only *Kochia* has proven to be a field problem, possibly contributing to the occurrence of polioencephalomalacia.

Genera noted for their accumulation of nitrate include *Beta, Chenopodium,* and *Salsola* (Gilbert et al. 1946; Case 1957).

Bassia All.

■ *B. hirsuta*
B. hyssopifolia

bassia
smotherweed

Taxonomy and Morphology—Comprising 5–20 species all native to the Old World, *Bassia* is named for Ferdinando Bassi, an eighteenth-century Italian botanist. Morphologically very similar to and easily confused with *Kochia,* European taxonomists generally combine the two genera. They can be distinguished, however, by the presence of a hooked spine or tubercle on each sepal of *Bassia* and their absence in *Kochia.* In North America, 2 introduced species are present:

B. hirsuta (L.) Asch.	three-spined bassia
B. hyssopifolia (Pall.) Kuntze	five-hooked bassia, smotherweed

■ annual herbs, branched, villous, red at maturity; leaves alternate; flowers in spikes

Plants herbs; annuals; from taproots; gynomonoecious. **Stems** erect; 20–100 cm tall; profusely branched at bases; surfaces tomentose and villous; branches terete or angled; often turning red with maturity. **Leaves** alternate; cauline; sessile; herbaceous or fleshy; blades flat or subterete, linear to oblanceolate; surfaces silky villous to appressed pilose; margins entire. **Inflorescences** solitary flowers or short spikes in axils of uppermost leaves or bracts. **Flowers** perfect and pistillate, intermixed in spikes; perianths in 1-series. **Sepals** 5; fused; membranous; villous-lanate; bearing abaxial spine or tubercule; lobes incurved; enlarging in fruit. **Stamens** 5. **Pistils** 1; stigmas 2, exserted. **Fruits** utricles; enclosed in sepals; obovate; plano-convex; brown to black.

1. Sepals 5; bearing abaxial hooked spines. Leaves flat and herbaceous . . . . *B. hyssopifolia*
1. Sepals 3; bearing abaxial tubercules. Leaves subterete and fleshy . . . . *B. hirsuta*

Bassia hyssopifolia

■ weedy, naturalized

Distribution and Habitat—Both species are naturalized in North America. *Bassia hyssopifolia* occurs across the continent but is most common in the intermountain region and is considered a noxious weed in irrigated fields. It also is present in the saline soils of beaches and salt marshes. *Bassia hirsuta* is ecologically similar but occurs only along the beaches and marshes of the Atlantic coast.

Bassia hirsuta

Bassia hyssopifolia

Disease Problems—*Bassia hyssopifolia* is a reasonable forage for sheep in areas where it does not constitute the sole food for a sustained period of time. It is readily eaten by sheep and is a good source of protein and phosphorus, but it is erratic in productivity, depending upon the summer rains in the dry, alkaline areas where it is most common (Cook et al. 1956). One drawback to its use is the occurrence of diarrhea in sheep that are sustained solely on it for more than 2 weeks. In addition, under some circumstances *Bassia* also may cause an acute neurologic disease that is often fatal (James et al. 1975).

■ sheep, if exclusive forage, digestive disturbance; neurologic effects

Disease Genesis—The cause of the diarrhea is not known, but a similar effect is observed with the closely related *Salsola tragus (S. kali)*. The acute neurologic disease is most likely due to the presence of high concentrations of soluble (probably potassium) oxalates in the plant tissues. James et al. (1976) fed 7 sheep *B. hyssopifolia* with an oxalate content of 6.1% and caused death of all animals, 5 within 24 hours. The remaining 2 died 17 and 22 days later. Oxalates appeared to be the cause of death, at least in the 5 that died within 24 hours, but possibly not in those with the chronic disease. If oxalates are involved in the chronic cases, it indicates an obvious lack of microbial adaptation in the rumen, in contrast to animal response to other oxalate accumulators such as *Halogeton*.

■ digestive disturbance cause unknown, perhaps due to excessive salts, sulfates; neurologic effects, soluble oxalates

Clinical Signs—Most likely to be seen as a disease problem in sheep, ingestion of *Bassia* typically results in abrupt onset of weakness, incoordination, tetany, and coma. In chronic cases, weakness and incoordination are accompanied by a hopping gait, sometimes photosensitization, and/or a bloody nasal discharge.

In acute cases, serum calcium levels are markedly decreased to less than 4 mg/dl, serum urea nitrogen is little changed, and serum AST is increased about 2-fold. In the chronic cases, serum calcium is little affected, but AST is increased as in the acute disease.

■ diarrhea and/or weakness, ataxia, tetany, coma; decreased blood calcium

Pathology— Scattered splotchy hemorrhages are evident over the serosal surfaces of the rumen and kidneys. Most of the distinctive changes involve the kidneys, which are pale and slightly swollen. There may also be edema of various tissues, including the lungs. Microscopically, nephrosis with tubular epithelial desquamation and necrosis is seen, along with dilated tubules.

■ microscopic, kidney, tubular epithelial necrosis, dilation

■ parenteral calcium

Treatment—Prompt administration of parenteral calcium solutions will generally result in immediate relief of many of the signs, and animals will become ambulatory within a few hours. Relapses may occur, requiring additional treatment, and some animals may eventually die of renal failure. Dilute lime water can also be given to bind the oxalates remaining in the digestive tract.

Beta L.

■ *B. vulgaris*
beet
beetroot
fodder beet
mangel
mangold
sugar beet
remolacha

Taxonomy and Morphology—Comprising 6–12 species, *Beta* is native to the Mediterranean region and western Asia. Its name is derived from the Celtic word *bett,* meaning "red," an obvious reference to the intense purple-red color of the taproots. Cultivated for both its large taproots and its profusion of large leaves, 1 species is of economic importance as a source of food and sugar for humans and fodder for livestock:

B. vulgaris L.	beet, beetroot, mangel, sugar beet, fodder beet, mangold, remolacha

■ erect herbs from large taproots; leaves alternate

Plants herbs; annuals or biennials or perennials; from large taproots. **Stems** erect; 60–120 cm tall; branched above; surface usually glabrous; green or turning red with maturity. **Leaves** alternate; both basal forming large rosettes and cauline; petiolate; herbaceous to fleshy; blades flat, oval or ovate-oblong; surfaces glabrous; margins typically undulate. **Inflorescences** terminal spikes or glomerules in axils of leaves; bracts generally absent. **Flowers** perfect; perianths in 1-series; urceolate. **Sepals** 5; fused; becoming indurate in fruit. **Stamens** 5. **Pistils** 1; stigmas 2–5. **Fruits** achenes; enclosed in indurate shell formed by enlarging receptacle and sepal bases.

Beta vulgaris

Distribution and Habitat—Native to southern Europe, *B. vulgaris* is widely cultivated. Plants occasionally escape but apparently do not become established.

■ crop plant

■ intoxications uncommon, leaves, diarrhea, and/or decreased blood calcium, coagulation problems; roots, digestive disturbance, ruminal acidosis

Disease Problems—Normally beet roots and tops are safe and valuable feeds. However, a variety of adverse effects may occur, although they are rare. They include digestive upset and decreased blood calcium. In some instances more than one type of intoxication may occur in the same herd (Chambers 1944). One of the earliest problems associated with beets was the development of renal calculi composed of uric acid, phosphoric acid, and lime (Pammel 1911). In an experimental study, calculi were found in the kidneys or bladders of sheep fed corn, hay, and either mangels or sugar beets. Steyn (1934) reported the occurrence of decreased coagulability of blood and excess bleeding due to a decrease in blood calcium when beet tops were eaten in large quantities. Pigs fed the tops became incoordinated and had tremors and spasms in addition to the inhibition of coagulation (Steyn 1934). Cows became depressed and developed diarrhea. Similar responses were noted for sheep, in which there was diarrhea (with or without blood), decreased coagulability of blood, bloody urine, and seizures that occurred over several days (Pammel 1919). There are even reports of horses dying several days after eating the frozen tops of sugar beets (Pammel 1919).

One of the most consistent problems, a digestive disturbance creating ruminal acidosis, arises when large amounts of beets are fed or there is a change from one type to another, especially a change to sugar beets. Beet pulp, which has proven toxic to pigs, may also be toxic when ensiled (Novy 1960). Beet pulp fed to bulls for 4 months was associated with but not necessarily proven to be a cause of optic nerve degeneration (Alibasoglu et al. 1973). In this instance a deficiency of vitamin A was considered a possible factor.

Disease Genesis—Beet tops serve as sources of both nitrate, up to 88,000 ppm KNO_3, and oxalate, up to 12% (Everist 1981). Although not common, intoxications due to nitrate accumulation occur, primarily in cattle, sometimes under unusual circumstances such as after the tops are frozen and the NO_3 concentration is 10,000–60,000 ppm (Gilbert et al. 1946; Savage 1949). Inadvertent exposure of the beet tops to 2,4-D herbicide produced an increase in nitrate from 2,200 ppm up to as high as 88,000 ppm (Stahler & Whitehead 1950; Whitehead & Moxon 1952). In addition, nitrites, formed from nitrates when beets are cooked at low temperatures or for a short time, may be a cause of intoxication in pigs (Cooper & Johnson 1984).

■ leaves, nitrate or oxalate, acute disease with methemoglobinemia or decreased blood calcium; rarely chronic, possible renal calculi; roots, acute disease, highly fermentable carbohydrates

Some of the more important problems posed by beets stem from the large amounts of soluble oxalates in the leaves and the presence of highly fermentable carbohydrates in the roots. In both instances the diseases are of abrupt onset and are acute in nature. A calcium-responsive disease may occur with consumption of either the leaves or the roots. The cause is presumably oxalates that cause a hypocalcemia-type disease much like milk fever. There is a marked drop in blood calcium, which is readily responsive to parenteral calcium preparations (Chambers 1944; Worden et al. 1954). Although the upper limits of oxalate concentrations in beet leaves are generally about 10% soluble and 14% total, extremes of 30–33% total oxalates have been reported in Mexico (Mejia et al. 1985; Libert & Franceschi 1987). However, it is important to keep in mind that differences in assay methodology may cause large differences in reported values. Beet leaves may produce either an acute or a chronic type of disease (Gorisek 1962; 1963). In the latter there are bony changes with lameness, a decrease in serum phosphorus, and an increase in serum alkaline phosphatase. A role for saponins in addition to oxalates has been proposed. There is also the possibility of coagulation problems due to the lowered calcium levels (Steyn 1934; Gorisek 1960).

There may be obvious digestive problems along with the hypocalcemia, or the digestive problems may occur without any abnormalities in mineral metabolism (Penny 1954; Scarisbrick 1954; Simesen & Konggaard 1970). The digestive disturbances related to carbohydrates often occur when there has been an abrupt increase in the amount of beets in the diet or a change to varieties with high dry matter content. Experimentally, in cows dosed with beet juice, rumen pH quickly dropped to less than 4.5 and blood lactate levels rose markedly (Williams & Coup 1959).

The presence of a highly fermentable carbohydrate is important for other animal species as well. Pigs (35+ kg b.w.) fed 3+ kg/day each of pulped beets sans tops, lost condition, developed diarrhea, and became quite nervous after about 1 month. These changes seemed to coincide with a change to beets with higher dry matter content, i.e., sugar beets (Baxter 1956).

Clinical Signs—Digestive disturbances are manifested by diarrhea, laminitis, a marked drop in ruminal pH, and an increase in lactate and volatile fatty acids. In some animals there are accompanying seizures. These signs may be seen in any animal species. Acute oxalate toxicity is most likely to be seen as a disease problem in sheep, typically resulting in abrupt onset of weakness, incoordination, tetany, and coma. Nitrate intoxication occurs mainly in cattle and is indicated by dark mucous membranes, dark or brown blood, depression, rapid respiration, incoordination, and death in a short time.

■ diarrhea; and/or oxalate intoxication with weakness, ataxia, tetany, coma, decrease in blood calcium

nitrate intoxication with depression, weakness, rapid respiration, ataxia, dark blood and mucous membranes

- gross pathology, digestive disturbance, reddened and sloughing digestive tract mucosa; kidney lesions with oxalate; dark or brownish tissues sometimes with nitrate

Pathology—Digestive disturbances may result in degeneration and sloughing of the epithelium of the rumen and sometimes the mucosa of the abomasum and small intestine. There will be few changes due to nitrate intoxication other than perhaps dark-colored or brownish tissues in some animals.

In the event of acute oxalate intoxication, grossly the kidneys will be pale and swollen, and body cavities will contain excess fluid. There may also be edema of the abdominal viscera and pulmonary congestion and edema. Microscopically, renal tubular dilation and flattened, necrotic tubular epithelium will be noted.

- ruminal alkalinizers, parenteral calcium or methylene blue i.v.

Treatment—Oxalate intoxication can be treated with parenteral calcium preparations, and nitrate intoxication with methylene blue i.v., 2–4 mg/kg b.w. Digestive disturbances may be responsive to ruminal alkalinizers.

It has been recommended that beets be stored for several months or the tops allowed to wilt for several days to reduce the likelihood of occurrence of diarrheic-type digestive disturbances (Cooper & Johnson 1984).

Chenopodium L.

- *C. album*
  C. ambrosioides
  C. carinatum
  C. glaucum

  goosefoot
  lamb's-quarters
  Mexican-tea
  nabo blanco
  pata de ganso
  pigweed
  Spanish tea
  wormseed

Taxonomy and Morphology—Cosmopolitan in distribution, *Chenopodium* comprises 100–150 species. The characteristic shape of the leaves of most species is reflected both in its generic name—which is derived from the Greek roots *chen,* for "goose," and *pous,* for "foot"—and in its common name *goosefoot.* The name *pigweed* also is used but is perhaps more properly applied to *Amaranthus.* Many species are common but not especially noxious weeds. Several are gathered from the wild and cooked as potherbs, and a few are cultivated as culinary herbs or for their seeds, which have a low gluten content and are cooked like rice. Native Americans employed one taxon as a vermifuge. Of the approximately 50 species in North America, most are introduced from Eurasia. Representative species include the following:

C. album L.	lamb's-quarters, white goosefoot, pata de ganso, quelite cenizo, nabo blanco, pigweed
C. ambrosioides L.	wormseed, Mexican-tea, Spanish tea, American wormseed
C. carinatum R.Br.	keeled goosefoot, ridged goosefoot
C. glaucum L.	oak-leaved goosefoot

- herbs; leaves alternate; flowers in numerous terminal or axillary glomerules

Plants herbs; annuals or perennials; from taproots; herbage farinose or glandular. **Stems** erect or prostrate or decumbent; 10–300 cm tall; branched at bases or above or not branched; often turning red- or maroon-striped with maturity. **Leaves** alternate; primarily cauline; usually petiolate; herbaceous to fleshy; blades flat, linear to rhombic-hastate or deltoid; surfaces glabrous to farinose or puberulent or gland-dotted; margins entire to sinuate-dentate or sinuate-lobed. **Inflorescences** numerous glomerules borne in spikes or panicles; terminal or axillary; bracts present or absent. **Flowers** perfect or rarely some pistillate; perianths in 1-series; pentagonal. **Sepals** 5 or rarely 2–4; fused; typically conspicuously farinose or gland-dotted. **Stamens** 5. **Pistils** 1; styles 2 or 3 or rarely 4 or 5. **Fruits** utricles; enclosed by persistent sepals; pericarps free or attached to seeds.

Some individuals of the various species of *Chenopodium* are difficult to identify, because many of their characters exhibit phenotypic plasticity and there is considerable intergradation due to hybridization. Mature fruits and lower cauline leaves are

Chenopodium album

Chenopodium album

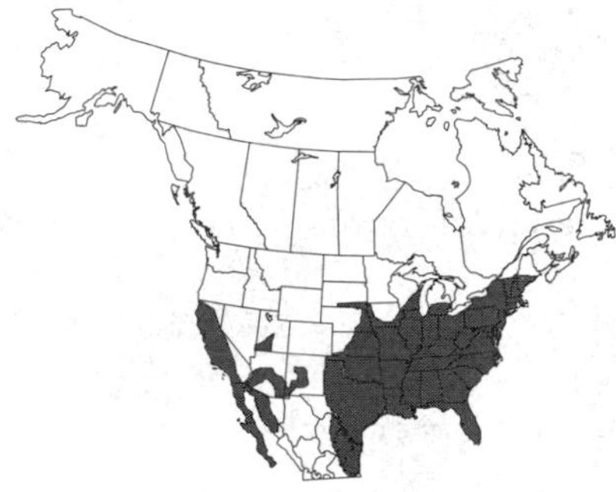

Chenopodium ambrosioides

almost essential for precise identification. A taxonomic key to species is not presented here, rather the reader is encouraged to use floristic treatments specific for his or her area.

Distribution and Habitat—Adapted to a variety of soil types and climatic conditions, species of *Chenopodium* are found throughout the world. Most are characteristically encountered in highly disturbed sites.

■ waste areas, disturbed sites

Disease Problems—Species of *Chenopodium* are considered by many to be edible, although *C. ambrosioides* may have a noxious odor. The leaves of *C. album* are especially valued and have been used in many cultures as potherbs. The grain produced by some species has a high protein content (15–19%) and has been exploited as a traditional food by people in the Andes (Mujica 1994). Very rarely are members of this genus a cause of intoxications, but under unusual circumstances they may cause or contribute to an abrupt onset of weakness, incoordination and tremors, hypocalcemia-type signs, within a few hours after large amounts are eaten (Herweijer & Den Houter 1970). Problems are most likely to occur in sheep but may also be seen in cattle and other ruminants. There is also potential for milk taint with fresh goosefoot (Everist 1981).

■ leaves edible, rarely with large amounts eaten, sheep and other ruminants, neurologic effects, tremors, ataxia, decrease in blood calcium

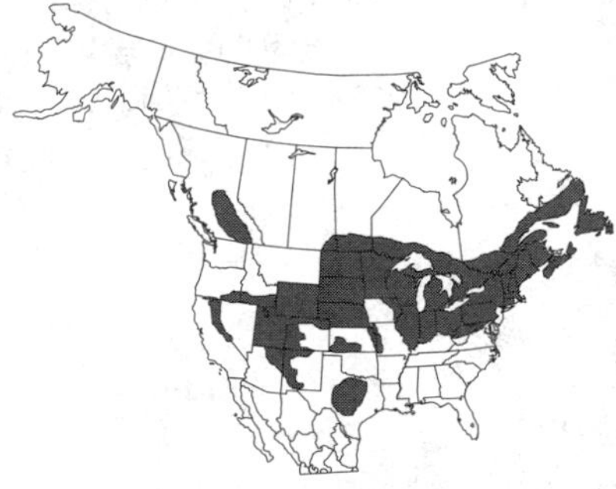

Chenopodium glaucum

Disease Genesis—Several potential toxicants are present in *Chenopodium,* including oxalates, nitrates, and cyanogenic glycosides. Oxalates are typically 1–2%. Concentrations, however, may be up to 7–10% soluble and 18% total oxalates, and extremes in the range of 30–33% total oxalates have been reported in Mexico (Herweijer & Den Houter 1970; Mejia et al. 1985; Libert & Franceschi 1987; Guil et al. 1996). The toxic role of oxalates is illustrated by the hypocalcemia-type disease produced in cows dosed with sodium oxalate. They developed tremors and seizures typical of the disease produced by *C. album* (Watt & Breyer-Brandwijk 1962). Nitrates may be 10,000–30,000 ppm or even as high as 70,000 ppm NO_3 (Olson & Whithead

■ acute oxalate intoxication; nitrate problems unlikely

1940; Gilbert et al. 1946). Cyanogenic glycosides, although present, appear to be of little significance toxicologically, at least in *C. album and C. ambrosioides* (Everist 1981).

Chenopodium ambrosioides is quite aromatic because of the presence of oil of wormseed, or oil of chenopodium, which contains an unsaturated terpene peroxide called ascaridole. This oil is a neurotoxin that causes vomiting, weakness, sleepiness, seizures, and cardiac irregularities (Pammel 1911; Watt & Breyer-Brandwijk 1962).

■ oxalate; depression, weakness, tremors, labored respiration

Clinical Signs—Signs of oxalate intoxication occur abruptly a few hours after consumption of plants, and animals may die within 12 hours. There is a progression of signs from depression, weakness, tremors, weak pulse, and labored respiration to prostration, coma, and death in a few hours. Nitrate intoxication occurs mainly in cattle and is indicated by dark mucous membranes, dark or brown blood, depression, rapid respiration, incoordination, and death in a short time.

■ microscopic, kidney, tubular epithelial necrosis, dilation

Pathology—Grossly, the kidneys are pale and swollen, and excess fluid is present in the body cavities. There may also be edema of various tissues, including the lungs. Microscopically, there is nephrosis with tubular epithelial desquamation and tubule dilation and atrophy.

■ parenteral calcium

Treatment—Prompt administration of parenteral calcium solutions will generally result in immediate relief of many of the signs of oxalate intoxication, and the animals will become ambulatory within a few hours. Relapses may occur, and some animals may eventually die of renal failure. Nitrate intoxication should be treated with methylene blue i.v., 2–4 mg/kg b.w.

Halogeton C.A.Mey.

■ *H. glomeratus*
halogeton

Taxonomy and Morphology—Native to the Mediterranean region and central Asia, *Halogeton* comprises 3–5 species. Reflecting its habitat, its generic name is derived from the Greek roots *halos,* or "salt," and *geiton,* or "neighbor." The genus is represented in North America by 1 species (see plate 23.1):

H. glomeratus (M.Bieb.) C.A.Mey. halogeton

■ annual herb, blue-green when young; leaves alternate, fleshy, bristle-tipped; flowers inconspicuous axillary glomerules

Plants herbs; annuals; from taproots; polygamous; herbage dull green to blue-green when young and stramineous after frosts. **Stems** prostrate or decumbent to ascending; 10–50 cm long; profusely branched at bases; glaucous; glabrous or with tomentum in leaf axils; often red or purple. **Leaves** alternate; primarily cauline; sessile; fleshy; blades terete, linear, surfaces glabrous; apices spinose or bristle-tipped. **Inflorescences** sessile glomerules borne in axils of leaves and 2–5, fleshy, sepaloid bracteoles. **Flowers** perfect or pistillate; perianths in 1-series. **Sepals** 5; fused; membranous-winged or apiculate in some flowers. **Stamens** 3–5. **Pistils** 1; styles 2. **Fruits** utricles; orbicular; flattened; enclosed in sepals and bracteoles; pericarps attached to seeds.

■ aggressive invader, spreading in range

Distribution and Habitat—Introduced from central Asia in the early 1930s, *H. glomeratus* is a prolific seed producer and moisture absorber. It aggressively invades and forms dense stands, in some instances excluding all other species at a site. Found initially in Elko County, Nevada, in 1934, it spread into the adjacent deserts of Utah and Idaho by 1940 and into northern California by 1946 and has since expanded its range into other western states (Bellue 1949; Vawter 1950; Stoddart et al. 1951;

Pemberton 1986). Continuing to spread, it occurs in desert saltbush and sagebrush communities and pinyon-juniper woodlands up to more than 2,000 m elevation and where annual precipitation is 7–50 cm.

Dispersal appears to be via sheep droppings along trails (Cook & Stoddart 1953; Blackwell et al. 1979). *Halogeton glomeratus* is considered an aggressive invader of depleted rangeland, where it can easily establish itself (Cronin & Williams 1966). It is particularly a problem along trails, railroad rights-of-way, on bare knolls with loose soil, and in sheep bedding areas. Because the seeds may remain viable for many years, it is difficult to eradicate (Cronin & Williams 1966).

Halogeton glomeratus

- reasonable feed if eaten in moderation with good rumen fill, large amounts eaten in short period with empty rumen, abrupt onset, lethal neurologic effects, mainly sheep, less severe in cattle
- soluble oxalates very high, mainly sodium salts, deadly with rapid ingestion on an empty rumen, more severe neurologic effects than with typical oxalate intoxication, fewer muscular problems

Disease Problems—Losses caused by *Halogeton* were first noted in the early 1940s; in some instances devastating losses of 500 or more sheep occurred in a matter of hours (Cook & Stoddart 1953). The typical scenario of intoxication involved moving sheep to winter grazing and allowing them to rest and feed in a large stand of *Halogeton*. This is particularly a problem after frosts when plants are more palatable and when the animals are hungry. Availability of water is a major factor in the occurrence of intoxication (James et al. 1970). Thirst decreases feed consumption, and when water becomes available, not only will the animals eat avidly but the empty rumen will also render them more susceptible to intoxication (Cook & Stoddart 1953). The disease in sheep is acute and highly lethal. *Halogeton* is less of a problem in other animal species, although it may cause a problem in cattle when they are moved from noninfested to infested range or when hungry animals are moved through infested areas (Bellue 1949). In cattle, the disease may be of abrupt onset, as in sheep, or may present as subacute locomotor difficulties unlikely to be lethal (James 1970; Lincoln & Black 1980). Again the problem is amplified when *Halogeton* is particularly abundant around watering areas. Thus, it is clear that feeding and watering management is very important in control of the disease.

Disease Genesis—*Halogeton* presents a major problem because it contains high concentrations of soluble oxalates, mainly as the sodium salt. These levels are generally quite high in early growth in early to midsummer, decrease slightly in the fall and in rainy conditions, and decline to low levels in late winter (Morton et al. 1959a,b; Williams & Cronin 1966). In a study in northern Idaho, soluble oxalate concentrations ranged from 20% in late summer to 5% in weathered winter plants (Morton et al. 1959b). When plants are snow-covered or the winter remains dry, there is little leaching of the oxalates and their concentrations remain high through the winter. In addition, when the soil has high levels of sodium chloride, growth of *Halogeton* may be very vigorous, and its tissues may accumulate high levels of soluble oxalates (Williams 1960). Thus, under some conditions during the winter grazing season, the soluble oxalate levels may be up to 3.6% in stems, 28.1% in leaves, and 8.1% in seeds. That soluble oxalates are closely involved as a cause of the disease is demonstrated by the close similarity of the disease to that caused by i.v. infusion of sodium oxalate (James & Johnson 1970).

The problem in sheep is almost an all-or-none phenomenon. It is difficult to get sheep to eat a diet composed entirely of *Halogeton*, and even a diet of up to 36.1% *Halogeton* (16.6% soluble oxalate in the plants) for 100 days, if introduced gradually, causes few adverse effects other than decreased feed intake and weight gain (James & Butcher 1972). Experimentally in sheep, when the rumen is poorly filled following fasting, 12 oz of *Halogeton* (8.7% oxalate content) or 1 oz of soluble oxalate may be lethal. In contrast, with normal ruminal fill, the lethal dose increases to 18 oz of *Halogeton* or 1.6 oz of soluble oxalate (Cook & Stoddart 1953). If the plant is

Halogeton glomeratus

$H^+ \ ^-OOC—COO^- \ H^+$

$H^+ \ ^-OOC—COO^- \ Na^+$

$Na^+ \ ^-OOC—COO^- \ Na^+$

$H^+ \ ^-OOC—COO^- \ K^+$

$K^+ \ ^-OOC—COO^- \ K^+$

$Ca^{++} \ ^-OOC—COO^- \ Ca^{++}$

COO^- ^-OOC

Oxalic acid and its salts

eaten during the course of a full day, tolerance increases by a factor of 4. Sheep can adjust to increasing amounts of oxalates in their diet because microbial degradation of oxalates to carbonates increases as the ruminal microflora adapts, so the major risk with *Halogeton* is that of abrupt exposure (Morris & Garcia-Rivera 1955; James et al. 1967; Allison et al. 1977; Duncan et al. 1997).

The effects of the soluble oxalates are several and rather complex: digestive disturbances due to the irritation effects with large dosage; calcium interactions affecting blood calcium, blood coagulation, nerve and muscle function (tetany), and precipitates forming in the kidney tubules; and formation of urinary calculi with long-term low dosage (Rimington & Steyn 1933; Willson 1934). There are subtle differences among the diseases caused by various oxalate-containing plants. These differences in a large measure are due to variations in the form of the oxalate with its two carboxy groups. Thus oxalate may be present with various cations such as H^+, K^+, or Na^+ or a combination of them. *Halogeton* with an acid sap of pH 6 and oxalates mainly in the disodium and acid/sodium forms appears to produce quite different effects than most of the other plants with acid sap of pH 2 or with other forms of oxalate (Dye 1956; Williams 1960; James 1972). Some contrast is provided by a comparison of the effects of ammonium oxalate used to model intoxication from *Oxalis* and sodium oxalate as a model for *Halogeton* toxicosis in which tetany is not a prominent feature (James & Johnson 1970; James et al. 1971). Clearly there are effects on calcium homeostasis. However, despite the precipitous drop in blood calcium typical of oxalate intoxication, there is, unique to *Halogeton*, little or no tendency for tetany, in contrast to disease induced by *Rumex* (James 1968). Furthermore, parenteral calcium preparations lack efficacy against *Halogeton*-induced disease. Although there are effects on the kidney, e.g., calcium oxalate precipitates are formed, these do not appear to be primary factors in the development of the acute lethal effects (Shupe & James 1969). Metabolic disturbances secondary to inactivation of enzymes such as lactate dehydrogenase or succinic dehydrogenase may be more important factors (Emerson et al. 1964; Van Kampen & James 1969).

The cause of the differences in effects between various plants may be due to factors in addition to oxalate. Most of the oxalate in the plant is formed via a glycolate to glyoxylate to oxalate pathway (Franceschi & Horner 1980). If, with extremely high levels of oxalate, an appreciable amount of glycolate and glyoxylate are present in the plant, perhaps influenced by feedback inhibition of glycolate oxidase, then these potent neurotoxins may contribute to disease development (Bove 1966; Bachmann & Goldberg 1971).

- sheep, within a few hours, dullness, head held low, anorexia, white froth from mouth; shortly, weakness, stiffness, rapid respiration, ataxia, comatose, extensor seizures

  cattle, similar, belligerence, subacute; with forced movement, stiffness, difficult movement evident; sharp drop in blood calcium

Clinical Signs—About 4 hours after ingestion of *Halogeton* by sheep, the initial signs appear: dullness, lowering of the head, loss of appetite, and reluctance to follow the flock. The animal may also cough and appear to drool, with white froth around the mouth. By 5–6 hours after ingestion, the signs increase in severity, with weakness and stiffness, incoordination of the pelvic limbs, and rapid/shallow breathing. With an obvious lack of coordination, the animal may repeatedly lie down and stand up. Eventually it becomes comatose, with jerky extensor rigidity and irregular respiration. Death may occur within a few hours or, less commonly, up to 24 or more hours after onset of signs.

In cattle, onset of signs may occur within a few hours after eating the plants and are similar to those seen in sheep; incoordination, apprehension, belligerence, excess salivation, recumbency, coma, bloat, cyanosis, and death. The subacute locomotor difficulties will be observed mainly when the animals are forced to move. Stiffness is apparent first in the forelegs and then progresses to all limbs, making movement difficult. They may become recumbent but will recover within a few days if given feed and water. Calves may exhibit more-severe signs, including hypersensitivity to stimulation and seizures.

Typically blood calcium sharply declines, to as low as 1.4 mg/dl, during which time blood magnesium and phosphorus may double. Other changes in blood chemistry include moderate increases in serum lactate dehydrogenase (up to 2–3 times) and slight increases in blood urea nitrogen (up to 2 times).

Pathology—Grossly, excess free fluid is present in the abdominal and chest cavities and the heart sac. The lungs may be filled with blood-tinged froth, appearing red or purple rather than pink, and splotchy hemorrhages may be scattered on the surfaces of the heart, rumen, and other viscera. The kidneys will be pale and swollen/edematous and may yield birefringent crystals from scrapings of cut surfaces. Microscopically, the renal tubules will be filled with proteinaceous casts and calcium oxalate crystals, and the epithelium flattened and necrotic. Even in animals dying within an hour or two of onset of signs, tubular degeneration and oxalate crystals may be apparent. There may also be edema and hemorrhage of the ruminal wall, and oxalate crystals present in the submucosa and in the walls of vessels.

■ gross pathology, excess free fluid in chest and abdomen; kidneys pale and edematous

microscopic, kidneys, tubules dilated with proteinaceous casts, oxalate crystals, epithelium flattened; oxalate crystals in ruminal wall microvessels

Treatment—For most oxalate-plant intoxications, parenteral calcium (calcium borogluconate i.v.) is the standard treatment. With *Halogeton,* this approach seems to be somewhat beneficial for cattle, whereas in sheep it may delay deaths but not necessarily result in survival. Prevention of consumption by proper feed and water management is the most effective control. This approach may be aided by the use of calcium supplements, the most effective of which is a pellet of 83% alfalfa, 15% calcium carbonate, and 2% molasses (Cook & Stoddart 1953). Calcium chloride or dicalcium phosphate also are beneficial, but regardless of which is used, they are effective only when given shortly before the oxalate challenge; the protection is short-lived.

Halogeton is easily controlled by 2,4-D herbicide in the first few months of growth, but the extensive distributional range of the plant now makes this an untenable approach to control its growth except in restricted areas.

■ parenteral calcium, effectiveness limited in sheep, better in cattle; increase rumen fill to reduce sudden ingestion of large amounts of plant

Kochia Roth

Taxonomy and Morphology—Cosmopolitan in distribution, *Kochia* comprises 20–30 species. Its name honors Wilhelm Koch, an eighteenth-century German botanist. Although *Kochia* is typically thought of as a weedy genus in North America, some of its species are planted as ornamentals and for forage. Morphologically very similar to and easily confused with *Bassia,* European taxonomists generally combine the two genera. They can be distinguished by the presence of a hooked spine or tubercle on each sepal of *Bassia* and their absence in *Kochia.* Four species of *Kochia,* 2 native and 2 introduced, occur in North America (see plate 23.2):

K. americana S.Watson	red sage, red molly
K. californica S.Watson	Mojave red sage
K. prostrata (L.) Schrad.	forage kochia
K. scoparia (L.) Roth	burning bush, summer cypress, fireball, fireweed, Mexican fireweed, belvedere, kochia, morenita, poor man's alfalfa

■ *K. americana*
K. californica
K. prostrata
K. scoparia

belvedere
burning bush
fireball
fireweed
kochia
morenita
poor man's alfalfa
red molly
red sage
summer cypress

- herbs or subshrubs, profusely branched, becoming red with maturity; leaves alternate; flowers small, axillary in leaflike bracts at branch ends

Plants herbs or subshrubs; annuals or perennials; from large taproots; polygamous. **Stems** prostrate or decumbent to erect; 8–200 cm tall; profusely branched at bases; branches spreading to ascending; at first yellowish green, then streaked with red, and finally red-purple at maturity. **Leaves** alternate; cauline; petiolate to sessile; herbaceous or fleshy; blades flat or subterete, linear to narrowly lanceolate, surfaces typically pilose or villous to hirsute. **Inflorescences** spicate; of solitary or clustered flowers borne in axils of leaflike bracts at ends of branches. **Flowers** perfect or pistillate; perianths in 1-series; pentagonal. **Sepals** 5; fused; membranous-winged. **Stamens** 5. **Pistils** 1; styles 2 or 3, villous. **Fruits** utricles; orbicular; flattened; enclosed in sepals; pericarps not attached to seeds.

1. Plants annuals. Leaf veins 3–5 at blade bases . . . . *K. scoparia*
1. Plants perennials. Leaf veins 1 at blade bases.
    2. Stems prostrate . . . . *K. prostrata*
    2. Stems erect or ascending.
        3. Stems branched only at bases. Leaves terete or subterete. Stems glabrous or white-tomentose . . . . *K. americana*
        3. Stems branched above bases. Leaves flattened. Stems gray-brown puberulent . . . . *K. californica*

Kochia scoparia

- weedy invaders, disturbed sites

Distribution and Habitat—A native of Europe, *K. scoparia* is now common at the edges of cultivated fields and in waste areas, disturbed roadsides, and railroad rights-of-way throughout North America. It is particularly common in the Great Plains. It also is cultivated in drier areas of Colorado, Oklahoma, and Texas, the immature plants providing valuable livestock forage with a high protein content. *Kochia prostrata* is a rare introduction from Eurasia, whereas both *K. americana* and *K. californica* are native to western North America.

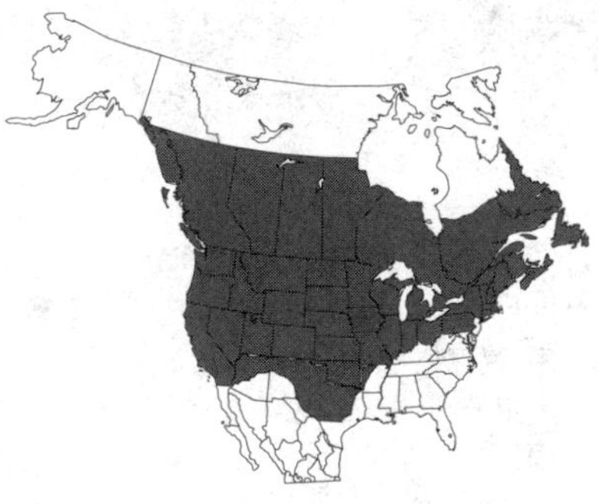

Kochia scoparia

- good feed in moderation; especially useful in areas with saline waters; if more than 50% of diet, poor weight gain, mild liver disease with photosensitization, PEM

oxalate and nitrate accumulation are little problem

Disease Problems—It is primarily the introduced *K. scoparia* that is of toxicologic interest. The native species are not sufficiently abundant to represent a risk. Since its introduction in the late 1800s, *K. scoparia* has generally been regarded as a crop weed to be controlled. In the 1940s its use as a potential forage in times of drought began to be appreciated. Following research in South Dakota (Erickson 1947), it was promoted as an alternative to the more traditional forages, especially when adverse conditions prevail (Burchard 1947; Montgomery 1947; Streeter 1947). Since that time its overall value has been somewhat ambiguous because it has been shown to have useful forage value yet under some circumstances it can be quite toxic. It is very water efficient with exceptional salt tolerance. It has such tolerance to salinity that it is considered to be a prime indicator plant for identifying areas of exceptionally high salinity in the Great Plains, such as saline seeps where groundwater may have 30,000 ppm sulfate (Brown et al. 1983). Experimentally, there was no decrease in growth of *Kochia* when irrigated with saline water of 14,000 ppm, an amount equivalent to 40% percent seawater. Yield was highest with water of 10,500 ppm salinity. Under such conditions, crude protein (20%) and total digestible nutrients (TDN) were not decreased (Fuehring et al. 1985). *Kochia* has been favorably compared with alfalfa as a forage, with similar protein content and digestibility and slightly lower gross energy (Sherrod 1971, 1973). Because of its obvious advantages it has been promoted as an arid-land forage, and efforts are under way to improve its nutritive value by selective breeding (Fuehring et al. 1986). This is necessary because, in spite of its useful characteristics, poor weight gains have been continually reported as a disadvantage (Rankins 1989; Thilsted et al. 1989). There is considerable variability in its effects on weight (Cohen et al. 1989). Dry-matter intake by beef cattle decreased markedly when *Kochia* exceeded 60% of the diet; this change was accompanied by a marked decrease in nitrogen retention.

In addition to weight loss, liver disease with photosensitization and, less commonly, polioencephalomalacia (PEM) sporadically accompany the use of *K. scoparia* as a forage (Galitzer & Oehme 1978; Dickie & Berryman 1979; Dickie & James 1983; Thilsted et al. 1989). Aspects of these two most important disease problems will be discussed separately. There also is a propensity for accumulation of oxalate and/or nitrate, but this has not been a significant problem.

In a series of experiments of feeding sheep and cattle for several weeks with *K. scoparia* hay, a number of changes in clinical chemistry were noted, including a decrease in serum prolactin from 12.4 to 1.5 ng/ml, a decrease in insulin from 0.53 to 0.23 ng/ml, and moderate increases in blood glucose, serum bilirubin, and hepatic enzymes (Rankins & Smith 1991; Rankins et al. 1991a,b,c). However, the changes were not statistically significant. The decrease in serum prolactin is similar to that seen with endophyte-infected fescue (see chapter 58). However, in contrast to what might be expected with fungal infection, metoclopramide was not protective against the toxic effects. Other treatments such as zinc sulfate and *N*-acetyl-L-cysteine also were ineffective in preventing the chemical changes. The changes indicative of mild hepatotoxicosis are consistent with the moderate but transient increases in serum GGT noted for sheep grazing *K. scoparia* pasture for about 10 weeks (Kirkpatrick et al. 1998).

Early reports indicated that intoxication problems were associated with the flowering stage (Galitzer & Oehme 1978), but in subsequent studies rapidly growing, irrigated, vegetative *Kochia* also proved to be toxic (Thilsted et al. 1989). It should be emphasized that generally *Kochia* is not to be considered a toxic plant except in situations where it constitutes almost the entire diet. Farmers surveyed in Alberta typically considered *Kochia* to be a problem only when it constituted more than 50% of the diet (Cohen et al. 1989). Interest in the introduced *K. prostrata* for its forage value may prompt its use as a primary feed (Clements et al. 1997). If it is fed in situations where it is the main dietary constituent, it may have adverse effects similar to those of *K. scoparia*.

Seeds have some detrimental effect but not for all animal species. They were toxic to rats and turkeys but not chukar partridges in a 14-day trial (Coxworth & Salmon 1972; Smith et al. 1993). Seeds fed as a grain supplement may depress feed intake (Coxworth et al. 1969).

Photosensitization and Weight Loss

Disease Problems and Genesis—*Kochia* has been shown to contain several potential toxicants, including nitrates, oxalates, saponins, and alkaloids, and possibly to have thiaminase activity. Of these, nitrates and oxalates appear to be of little risk because the problems are related to consumption of plants for several or more weeks. Moreover, toxicity often does not correlate with the levels of these two toxicants in the forage (Thilsted et al. 1989). Oxalate concentrations are usually 2–6%, although occasionally may be as high as 11% (Sprowls 1981; Fuehring et al. 1986; Thilsted et al. 1989). However, most are present as insoluble calcium oxalate, with the soluble oxalates seldom exceeding 3% (Dickie et al. 1989). Kidney changes are not accompanied by oxalate crystals in the tubules. Although oxalates do not appear to be directly related to toxicity, they may play a role. In the diet of monogastrics they decrease calcium availability, and in ruminants the oxalates are apparently metabolized in the rumen to yield carbonates and bicarbonates in sufficient quantities to produce alkalosis and diuresis, with chronic alkalosis possibly affecting growth

■ most common adverse effect, weight loss, possibly due to oleanolic acid genin saponins; liver disease and photosensitivity cause unknown, possibly saponins

(Talapatra et al. 1948). Nitrate contents of 1.5% and 3.7% as NO_3 are reported for *K. americana* and *K. scoparia,* respectively (Gilbert et al. 1946).

That saponins are of importance is shown by the decrease in toxicity of the seeds to turkey poults when they are washed in sodium hydroxide (Coxworth & Salmon 1972). Furthermore, the liver lesions are somewhat similar to those produced by *Agave* (see chapter 3) and other saponin-containing toxic plants. The saponins may also play a role in depression of feed intake and/or weight gain (Coxworth et al. 1969; Rankins 1989; Thilsted et al. 1989). However, their concentrations appear to vary considerably, inasmuch as the hemolysis potential of mature fruits may range from nil to marked. The saponins are glycosides of oleanolic acid, a triterpene sapogenin (Souto & Milano 1966; Kernan et al. 1973). The photosensitization has not been shown to be clearly either primary or secondary.

Harman and harmine, toxic β-carboline alkaloids, have been isolated from *Kochia* but in such small amounts as to make them unlikely prospective toxicants (Drost-Karbowska & Kowalewski 1978; Allen & Holmstedt 1980). In general, alkaloids, which are poorly represented in the family (Raffauf 1996), are of questionable significance. However, they cannot be excluded as possible complicating factors in causing disease. In some studies, crude measures, such as the Dragendorff test showed an increase in alkaloid content with maturity (Fuehring et al. 1986).

- decreased appetite; skin erythema, edema, and sloughing, nonpigmented areas; increased serum hepatic enzymes, bilirubin

Clinical Signs—As indicated above, grazing of *Kochia* for 2 or more weeks may result in several adverse consequences. In many animals the only consequence will be diminished weight gain. More-severe involvement will be reflected by greater impairment of appetite, depression, swollen eyelids, crusty areas of the nasal pad, erythema and epithelial sloughing of nonpigmented areas of skin, and necrosis of teat ends. These changes will be accompanied by changes in serum chemistry, including a mild to moderate increase in hepatic enzymes (GGT, AST, LDH, SDH, creatine kinase, AP) and a marked increase in both direct and indirect bilirubin. As the disease continues its course, icterus, depression, weakness, diarrhea, and dehydration may be noted. Additional serum chemistry changes seen are a mild to moderate increase in blood urea nitrogen, creatinine, and calcium and a decrease in insulin and prolactin (Rankins 1989; Thilsted et al. 1989). These signs may occur in ruminants and horses, with a morbidity of up to 40% and an appreciable death loss (Sprowls 1981).

- gross pathology, skin lesions

  microscopic, kidney tubule dilation, flattened epithelium; liver, mild necrosis and fibrosis

Pathology—The most noteworthy changes will be of the skin, liver, and kidney. Changes typical of photosensitization are skin thickening and necrosis of the less pigmented areas. The kidneys may be swollen and pale. Microscopically, diffuse tubular dilation will be seen, with foamy, flattened or necrotic epithelial cells, tubular lumina with proteinaceous fluid, hyaline casts, cell debris, and hemoglobin casts. There may also be interstitial fibrosis and mononuclear cell infiltration, but no oxalate crystals (Thilsted et al. 1989). The liver changes are generally mild to moderate, with hepatocellular swelling (foamy/granular cytoplasm), bile stasis, and mild necrosis and/or fibrosis (Sprowls 1981). Of interest are the round golden brown concretions observed in bile ducts, which are birefringent with polarized light (Thilsted et al. 1989).

- general nursing care, provide shade, supplement kochia with other feeds

Treatment—Disease affecting kidney and liver function and accompanying photosensitization are treated with general nursing care. For the most part the disease will resolve with prevention of animal access to *Kochia*. Although the skin changes are related to problems of liver function, the kidney disease is typically an equally serious threat to survival of the animal (Thilsted et al. 1989). Prevention or treatment employing zinc sulfate as described for *Nolina* (chapter 3) has not been effective. Similarly *N*-acetyl-L-cysteine and *trans*-stilbene oxide to stimulate glutathione levels also have been ineffective (Rankins 1989). In most cases supplemental feeding with an additional forage to reduce the consumption of *Kochia* has been effective in minimizing disease problems.

Neurologic Disease

Disease Problems and Genesis—The sporadic occurrence of polioencephalomalacia (PEM) with ingestion of *Kochia* appears to be the result of a complex interaction between the plant and sulfate. Like other members of the family, *K. scoparia* grows well on saline soils and under such conditions may accumulate high concentrations of sulfur (up to 0.9% total) as sulfates and organic compounds (Mayland & Robbins 1994). Experimentally, high sulfur or sulfate levels in the diet and/or water have neurologic effects and appear to be important factors in the development of PEM, although the specific mechanism is not known (Olkowski et al. 1990, 1992; Rousseaux et al. 1991; McAllister et al. 1997). The neurotoxic effects of sulfur are demonstrated by the occurrence of PEM in a flock of 2,200 ewes grazed on an alfalfa pasture that had been sprayed with a 35% suspension of elemental sulfur (Bulgin et al. 1996). About 2% of the animals developed signs of PEM or had the specific pathologic lesions.

- PEM, few animals, high plant and water sulfates, possible thiaminase activity but disease not very thiamine responsive

In the ruminal bacterial environment, sulfate is reduced to sulfide (Lewis 1954; Gould et al. 1991) and presumably the organic sulfur compounds in plants serve as similar substrates for reduction. Once reduced, the sulfide may interact with sulfite and act as a thiaminase (Leichter & Joslyn 1969; Chick et al. 1989). There is, however, little decrease in blood thiamine, and the neurotoxicosis is more likely due to additional causes, such as direct effects on the brain via lipid peroxidation (Gooneratne et al. 1989b, 1992; Gould 1998). Thus it appears that in the saline conditions in which *Kochia* is often grown for forage, the presence of high levels of sulfur in the plants and probably in the animals' drinking water have the potential in combination to cause PEM. In addition, *Kochia* may contain a thiaminase factor, although this has not been confirmed (Dickie & James 1983). Of considerable importance is the comparative toxicity of fresh forage, hay, and silage. Hepatic effects and weight loss are associated with hay as well as green forage (Rankins & Smith 1991), but as yet neurologic effects are known only with green forage.

High sulfur or sulfate levels in the diet also may have an effect on limiting the availability of copper in ruminants. Sulfate and/or its reduction product sulfide interact with molybdenum and copper to form a copper-thiomolybdate complex (Suttle 1974, 1991; Gooneratne et al. 1989a; Qi et al. 1993). Thus the potential exists for copper deficiency with long-term ingestion of high-sulfate diets.

- high sulfate levels may reduce Cu availability, deficiency

Clinical Signs—A small number of animals may show signs of neurologic involvement, including incoordination, muscle tremors, head pressing, blindness, circling, aggressive/excitable behavior, neck twisting, and convulsive seizures. These signs are reported only for ruminants and usually end in death. Elevation of serum pyruvic acid and lactic acid are observed even in the absence of neurologic signs (Dickie & James 1983).

- ataxia, tremors, circling, blindness, excitability, seizures

Pathology—Changes associated with the neurologic effects of *Kochia* include herniation of the cerebellum into the foramen magnum; swollen, flattened, and edematous gyri; and softness and liquefaction of the cerebral cortex gray matter but not white

- gross pathology, brain, soft gray cortex, flattened gyri

  microscopic, edema, neuronal necrosis cerebrum, granular layer degeneration of cerebellum matter. Microscopically, there will be edema, neuronal necrosis and gliosis in the cerebrum, and degeneration of the granular layers and Purkinje's cells of the cerebellum (Dickie & Berryman 1979; Raisbeck 1982). Diagnosis may be aided grossly by examining the brain for fluorescence under ultraviolet light, a nonspecific test for cerebrocortical necrosis (Anderson 1984).

■ parenteral thiamine

Treatment—Neurologic disease will respond to some extent to parenteral thiamine HCl, 5–10 mg/kg b.w. In less severely affected animals, 1 g of oral thiamine per sheep or more in cattle may produce longer-duration attenuation of the problem by inhibiting thiaminase I activity in the gut (Thomas 1986). Because of the apparent interrelationship between high sulfur and low copper levels in depletion of blood thiamine, it may be of value to supplement with copper as a preventive against polioencephalomalacia when feeding *Kochia* (Gooneratne et al. 1989a).

Salsola L.

■ *S. collina*
S. kali
S. paulsenii
S. soda
S. tragus
S. vermiculata

katune
Russian thistle
saltwort
tumbleweed

Taxonomy and Morphology—Cosmopolitan in distribution but with greatest diversity in Eurasia, *Salsola* comprises approximately 50 species. Its name is derived from the Latin *salus*, for "salty," and refers either to the habitats characteristically occupied or to the taste of most species. Morphologically quite similar and possibly able to hybridize, all of the species present in North America are introductions from Eurasia. There has been considerable confusion among taxonomists as to the identities and origins of the species. The identity of the weedy Russian thistle has been particularly troublesome; 6 species names have been used for it at one time or another. *Salsola kali* is the name most commonly used in the taxonomic and toxicologic literature. Unfortunately, it appears that this name was misapplied to plants of *S. tragus* and *S. collina*, 2 other Eurasian introductions. At present, the taxonomic consensus is that *S. tragus* is the correct name for the widespread Russian thistle, and *S. kali* should be used only for plants occurring on the beaches of the Atlantic and Gulf coasts. The almost indistinguishable *S. collina* is found primarily in the northern Great Plains. Six species occur in North America:

S. collina Pall. (= *S. kali* L. misapplied)	katune, tumbleweed, Russian thistle
S. kali L. sensu stricto	saltwort
S. paulsenii Litv.	
S. soda L.	
S. tragus L. (= *S. australis* R.Br.) (= *S. iberica* Sennen & Pau) (= *S. kali* L. misapplied) (= *S. pestifer* A.Nelson) (= *S. ruthenica* Iljin)	Russian thistle, tumbleweed
S. vermiculata L.	

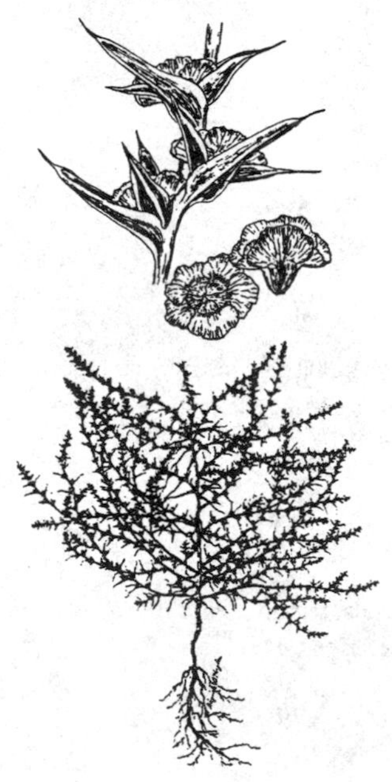

Salsola tragus

■ globose, erect, annual herbs, red with maturity; leaves alternate, stiff, apices spinulose; flowers axillary

Plants globose herbs; annuals; from taproots; herbage glabrous or villous. **Stems** erect; 30–100 cm tall; profusely branched at bases; branches arching upward to form globose shape; surfaces striate, turning purplish red with maturity. **Leaves** alternate; cauline; sessile; fleshy; blades linear to filiform; subterete; stiff; apices spinulose. **Inflorescences** solitary flowers borne in axils of leaflike bracts; bracteoles 2, spreading. **Flowers** perfect; perianths in 1-series; pentagonal. **Sepals** 4 or 5; fused; membranous-winged. **Stamens** 5; exserted. **Pistils** 1; styles 2 or 3, exserted. **Fruits** utricles; enclosed in sepals; suborbicular; flattened.

1. Plants shrubs. Leaves 3–9 mm long; not reduced above .... *S. vermiculata*
1. Plants herbs. Leaves 10–60 mm long; reduced above.
   2. Plants of coastal beaches, mudflats, and marshes; fleshy in fruit.
      3. Plants of Atlantic or Gulf Coast. Sepal apices spinulose ...... *S. kali*
      3. Plants of Pacific Coast. Sepal apices not spinulose .... *S. soda*
   2. Plants of inland habitats; not fleshy in fruit.
      4. Leaves yellow-green. Branch tips papillate.
Sepal wings in fruit 2.5–4.5 mm wide .... *S. paulsenii*
      4. Leaves green or purple-green. Branch tips smooth.
Sepal wings in fruit 0.5–2.5 mm wide or absent.
         5. Flower-bearing portions of branches 15–40 cm long.
Bracts appressed or only tips recurved; scabrous.
Sepal wings absent or 0.1–0.5 mm wide .... *S. collina*
         5. Flower-bearing portions of branches 2–10 cm long.
Bracts spreading; glabrous.
Sepal wings 0.7–2.5 mm wide .... *S. tragus*

Distribution and Habitat—Adapted to a variety of soil types and tolerant of drought and saline conditions, species of *Salsola* occupy coastal mudflats, beaches, and open areas in salt marshes. Others are quite weedy and classic indicators of highly disturbed soils such as cultivated fields, construction sites, feedlots, rights-of-way, and waste areas. A noxious and perhaps the most abundant weed of semidesert range and wheatland in western North America, *S. tragus* is a prolific seed producer. The globose growth form and abscission layer at the base of the primary stem facilitate seed dispersal as the entire plant breaks loose and blows before the wind as a tumbleweed.

■ weedy invaders, disturbed sites

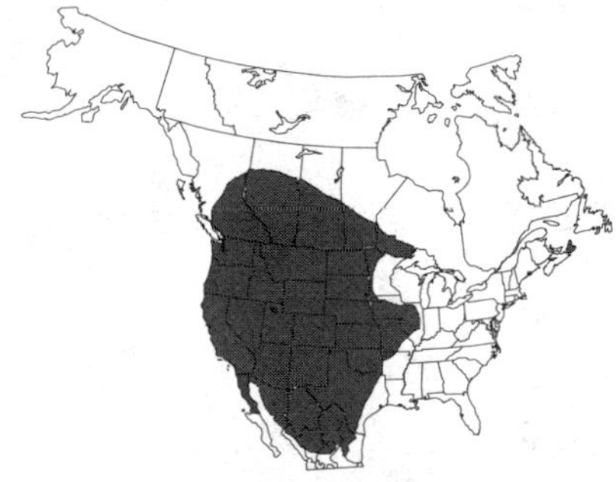

Salsola tragus

Disease Problems—As noted above, *Salsola* is the name generally used in the toxicologic literature of North America rather than the binomials *S. tragus* or *S. collina*. In the following paragraphs, only the genus name *Salsola* is used to avoid confusion. Saline tolerant, *Salsola* flourishes in conditions that would cause problems for most other plants. Because of this, it is an arid-lands forage of considerable value (Fowler et al. 1992). When young and green, it is palatable and nutritious for sheep and cattle but not necessarily for horses (Cook et al. 1956; Hageman et al. 1988). In the later, more mature stages of growth, its stiff branches and spiny leaves discourage consumption. The spines may cause mechanical injury to horses and dogs. Although diarrhea may be a consequence when sheep subsist on *Salsola* for several weeks, the most important disease problem is a rare, acute neurologic/metabolic problem following ingestion of large amounts in only a few hours.

■ reasonable forage when young and green; eaten in excess, diarrhea, rarely neurologic effects

Although *Salsola* has not yet proven to be a problem in North America, it is worthy of note that the genus (at least 6 species) in South Africa is associated with reproductive problems in sheep. When the leaves or stems are eaten in large quantities, especially in the last trimester of pregnancy, gestation is prolonged; lambing is sometimes delayed for a month or more, with severe difficulties at birth (Basson et al. 1969; Joubert et al. 1972; Kellerman et al. 1988). There is an increase in serum cortisol and decreased mammary development, but the effects are apparently not related to estrogenic activity or lack thereof. The problem is more common when dormant plants are eaten during drought.

- rarely acute halogeton-like oxalate-induced neurologic effects

Disease Genesis—*Salsola* in some conditions may accumulate both nitrates and oxalates. Levels of nitrate may be about 5,000 ppm as NO_3 and rise with increased salinity (Fowler et al. 1992). Even higher levels may be produced experimentally; however, soil nitrates in areas where the plant grows are seldom sufficient for this to occur (Gilbert et al. 1946; Berg & McElroy 1953). Concentrations of oxalate up to 15% d.w. total and 10% or more of soluble salts have been reported (Mathams & Sutherland 1952; Hageman et al. 1988), but levels of 2–3% are more likely (Fowler et al. 1992). Oxalates are reported to be highest before or early in flowering and decline thereafter. The neurologic signs seen are consistent with oxalate-induced toxicosis and are quite similar to those caused by *Halogeton glomeratus.* They also are exhibited by animals after ingestion of a species closely related to *Salsola, Threlkeldia proceriflora* (soda bush), which causes extensive sheep losses in Australia, with most deaths occurring a few hours after the species is eaten (Everist 1981).

- diarrhea, abrupt onset, depression, weakness, labored respiration, prostration, seizures, coma

Clinical Signs—In sheep, diarrhea may occur as a consequence of subsistence on *Salsola* for several weeks. Signs of oxalate intoxication occur abruptly a few hours after consumption, and animals may die within 12 hours. There is a progression of signs in a few hours from depression, weakness, weak pulse, and labored respiration to prostration, coma, and death. Typically, levels of serum calcium will be markedly decreased.

- microscopic, kidney, tubular epithelial necrosis, dilation

Pathology—Grossly, the kidneys may be pale and swollen, and there may be excess fluid in the body cavities. There may also be edema of the abdominal viscera and pulmonary congestion and edema. Microscopically, there will be some renal tubular dilation and flattened necrotic tubular epithelium.

- parenteral calcium, response variable

Treatment—Prompt administration of parenteral calcium solutions may result in immediate relief of many of the signs, and animals may become ambulatory within a few hours. Relapses occur, and animals may eventually die of the neurotoxic effects as well as renal failure.

Sarcobatus Nees

- *S. vermiculatus* greasewood
- erect deciduous shrubs; branches spine-tipped; staminate flowers in terminal spikes; pistillate flowers in leaf axils

Taxonomy and Morphology—A monotypic genus, *Sarcobatus* is native to western North America. Its name is derived from the Greek roots *sarko,* for "flesh," and *batus,* for "bramble," in reference to its stiff, spiny leaves and branches. The single species is the following:

S. vermiculatus (Hook.) Torr. greasewood, black greasewood

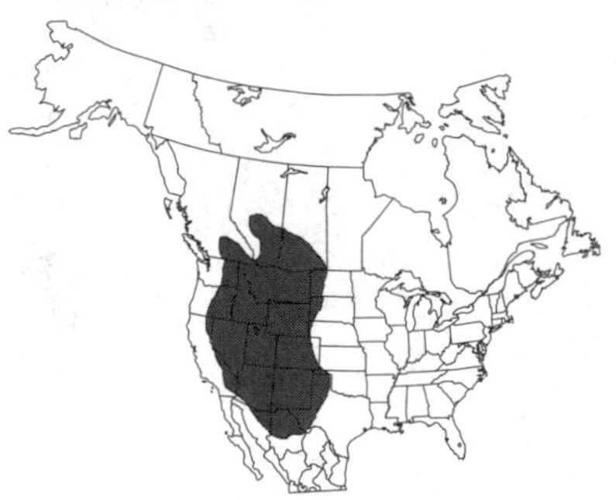

Sarcobatus vermiculatus

Plants shrubs, often globose; perennials; deciduous; monoecious. **Stems** erect; 30–300 cm tall; profusely branched; branches spreading, stout, rigid, spine-tipped; yellowish to light gray. **Leaves** opposite below and alternate above; cauline; sessile; succulent; blades semiterete, flat above, linear to linear-filiform; glabrous or sparsely stellate; apices of some spinose; margins entire. **Inflorescences** of 2 types; bracts present or absent. **Flowers** imperfect, staminate and pistillate different; perianths in 1-series or absent. **Staminate Flowers** numerous; borne in terminal spikes; spirally arranged; covered by peltate, rhombic scales; perianths absent. **Pistillate Flowers** 1 or 2 in leaf axils; perianths cuplike, winged, fused with basal portion of ovaries; styles 2. **Fruits** achenes; conical or turbinate; coriaceous.

Distribution and Habitat—*Sarcobatus vermiculatus* is quite tolerant of highly alkaline soils and high water tables. It occurs in the semiarid valleys of the intermountain West.

Sarcobatus vermiculatus

Disease Problems—Browsed extensively by sheep, the foliage of *Sarcobatus* has a salty, alkaline taste but without acidity (Couch 1922). It is typically eaten without problem. In the early 1900s it was the cause of extensive losses of sheep and sometimes cattle; e.g., one episode involved the overnight deaths of 1,000 of 1,700 hungry sheep feeding exclusively on it (Marsh et al. 1923; Sampson & Malmsten 1942; Hershey 1945). Losses typically occur when animals that are very hungry and thirsty eat large amounts of *Sarcobatus* quickly, conditions similar to those that favor *Halogeton* intoxication (Fleming et al. 1928). About 1.5% b.w. is toxic if eaten quickly (Marsh et al. 1923). Fleming and co-workers (1928) found that when *Sarcobatus* was eaten exclusively, it was a problem, but when mixed with other forages, it was not. Even amounts up to 1.8 kg/45 kg b.w. were not a problem if given with other forage, but 2.5 kg/45 kg was lethal.

- alkaline soils
- reasonable forage unless eaten in excessive amounts, mainly sheep, cattle less risk

Disease Genesis—*Sarcobatus* contains a mixture of neutral sodium and potassium oxalates—about 10–15% d.w. in leaves and smaller amounts in the stems and fruits (Couch 1922; Fleming et al. 1928). In exceptional cases oxalate concentrations may exceed 20% d.w. (Willson 1934). Oxalate concentrations reach a peak in early fall (Fleming et al. 1928). Aqueous extracts of the plant containing 40% oxalates, when given to a sheep, produced the same signs seen when the plant is eaten. Neither saponins nor cyanide were found in the species (Couch 1922).

- soluble oxalates, neutral sodium and potassium salts

Clinical Signs—The signs appear abruptly a few hours after consumption, and animals may die within 12 hours. Signs progress from depression, weakness, weak pulse, and labored respiration to prostration, coma, and death in a few hours.

- abrupt onset, depression, weakness, labored respiration, prostration, seizures, coma

Pathology—Grossly, the kidneys are pale and swollen, and there is excess fluid in the body cavities. There may also be edema of the abdominal viscera, as well as pulmonary congestion and edema. Microscopically, renal tubular dilation and flattened necrotic tubular epithelium are seen. In addition to numerous oxalate crystals in the tubules, crystals may also be seen in the walls of small arteries in the rumen and less often in the brain.

- microscopic, kidney, tubular epithelial necrosis, dilation

Treatment—Prompt administration of parenteral calcium solutions may result in relief of many of the signs, and animals may become ambulatory within a few hours. Relapses occur, and animals may eventually die of neurologic effects and renal failure.

- parenteral calcium, response variable

Suckleya A.Gray

Taxonomy and Morphology—A monotypic genus, *Suckleya* is native to North America. The following is the single species:

- *S. suckleyana*

  poison suckleya

S. suckleyana (Torr.) Rydb. poison suckleya

Plants fleshy herbs; annuals; monoecious. **Stems** prostrate or ascending; 10–50 cm long; freely branched; terete; glabrous or farinose; often reddish. **Leaves** alternate; cauline; petiolate; blades orbicular to rhombic-ovate; surfaces farinose when young,

- annual fleshy herbs, often reddish; leaves alternate; imperfect flowers in axillary glomerules

then glabrous; apices rounded; margins repand-dentate. **Inflorescences** glomerules in axils of leaves. **Flowers** imperfect, staminate and pistillate different; perianths in 1-series or absent. **Staminate Flowers** borne in upper leaf axils; bracts absent; calyces subglobose, 3-or-4-parted, 2 sepals larger and spathulate; stamens 3 or 4. **Pistillate Flowers** borne in lower leaf axils; bracts 2, conduplicate, ovate-rhombic, fused below; sepals absent or reduced; styles 2. **Fruits** utricles; enclosed by bracts; compressed; pericarps free from seeds.

Suckleya suckleyana

- water edges, around shallow ponds and lakes

Distribution and Habitat—*Suckleya* is distributed primarily on the eastern front of the Rocky Mountains (Chu et al. 1991). Although not common, it may form dense populations in low areas, such as around shallow ponds, dry lake beds, edges of lakes, and water tanks.

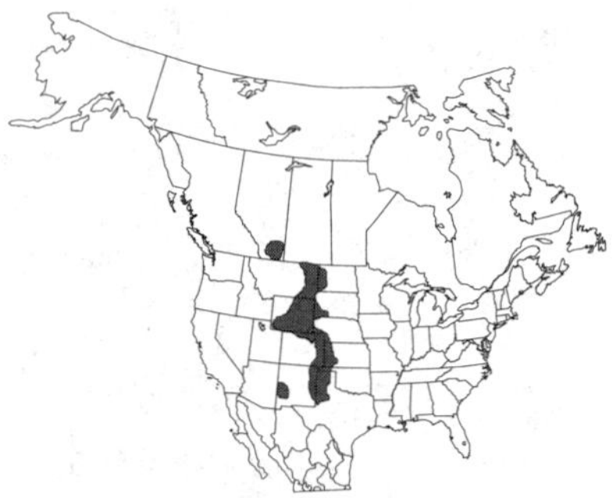

Suckleya suckleyana

- abrupt onset, acute illness, death quickly; risk is generally low most years

Disease Problems—*Suckleya suckleyana* is of only minor toxicologic importance, only sporadically does it cause deaths in sheep and cattle. Episodes typically occur in years when dense stands of plants are established around the edges of shallow stock ponds as the water slowly recedes. If livestock are allowed access to these areas, they may consume sufficient plants to cause intoxication, especially in years of limited forage due to drought or grasshopper invasion and predation (Berry & Gonzales 1986). However, in most instances there is little risk that significant problems will occur.

- cyanogenic glycosides, dhurrin, highest risk in September, October

N
C
O-β-D-Glucose
HO

Dhurrin

Disease Genesis—The acute onset of signs is due to the presence of the cyanogenic diglucoside dhurrin in the foliage (Nahrstedt et al. 1993). The rapidity of generation of hydrogen cyanide from the foliage indicates the presence of a β-glucosidase as well. The cyanogenic potential is usually less than 100 ppm but in some situations may be in excess of 400 ppm, especially in September and October. Such levels are clearly hazardous, because greater than 200 ppm is considered to be the "risk break" for toxicity due to cyanogenic glycosides (Hindmarsh 1930; Thorp et al. 1937; Barr et al. 1939; Hershey 1945). Experimentally, intoxications occurred when sheep or cattle were dosed orally with chopped plants containing 364 ppm HCN potential but not with plants containing 110 ppm (Thorp & Deem 1939). The risk of intoxications seems to be increased if animals drink shortly before or after ingestion of the plant (Stout 1939).

- abrupt onset, distress, rapid respiration, ataxia, collapse, seizures; death quickly

Clinical Signs—There is peracute onset of distress, muscular twitching, incoordination, licking of the lips, urine dribbling, staggering, and then collapse. Initially the respiratory rate increases, but within a few minutes it becomes slow and labored. The heart rate becomes rapid and irregular. When the animal is down, there are tonic seizures with leg paddling, the head held back, and death typically in a few minutes.

- no lesions, plant in rumen

Pathology—There are no distinctive changes. For the most part the changes are limited to congestion of the abdominal viscera and scattered, small splotchy hemorrhages. The stems and leaves are readily identifiable when the rumen contents are examined postmortem.

- sodium thiosulfate, with or without sodium nitrite

Treatment—The primary antidote for ruminants is sodium thiosulfate given i.v. at a dosage of 0.25–0.5 g/kg b.w. Sodium nitrite i.v., at a dosage of 10 mg/kg b.w., can be used to augment the effects of the sodium thiosulfate, but it is not required for effective treatment. A complete discussion of cyanogenesis and its treatment is presented in the treatment of Rosaceae (see chapter 64).

Genera of Questionable Toxicity

Monolepis Schrad.

Comprising 6 species in both the Old and New World, *Monolepis* is represented in western North America by 3 species. All are desert taxa of the intermountain region. Commonly known as povertyweed or patata, *M. nuttalliana* (Schult.) Greene also occurs at lower elevations in the mountains and as a weed in disturbed soils in the Midwest. It is a fleshy, herbaceous, prostrate to ascending, much branched annual from a taproot. Its alternate leaves are short petiolate or subsessile and have lanceolate to triangular blades with a pair of divergent lobes toward the base. Clustered in the axils of the uppermost leaves, the perfect or pistillate flowers have 1 sepal, 1 stamen, and a 2-styled ovary. The utricles are ovoid and compressed, and their pericarps are pitted at maturity.

- *M. nuttalliana*

  patata
  povertyweed

- possible nitrate accumulator

Sometimes forming rather dense populations in disturbed soils, *M. nuttalliana* may contain up to 4.2% nitrate salts. Heretofore, it has not been a problem for livestock (Schmutz et al. 1968).

References

Alibasoglu M, Erturk E, Meric I: Encephalo-ocular syndrome in young bulls fed with beet pulp. *Vet Fak Derg* 20;239–255, 1973 (*Vet Bull* 44;2406, 1974

Allen JRF, Holmstedt BR: The simple -carboline alkaloids. *Phytochemistry* 19;1573–1582, 1980

Allison MJ, Littledike ET, James LF: Changes in ruminal degradation rates associated with adaptation to oxalate ingestion. *J Anim Sci* 45;1173–1179, 1977

Anderson BC: Ultraviolet light: a tool for detecting cerebrocortical necrosis. *Vet Med* 79;1309–1310, 1984

Bachmann E, Goldberg L: Reappraisal of the toxicology of ethylene glycol. 3. Mitochondrial effects. *Food Cosmet Toxicol* 9;39–55, 1971

Barr CG, Reuszer HW, Thorp F: The chemical composition of *Suckleya suckleyana*. *Science* 90;497–498, 1939

Basson PA, Morgenthal JC, Bilbrough RB, Marais JL, Kruger SP, Van Der Merwe JLdeB: "Grootlamsiekte," a specific syndrome of prolonged gestation in sheep caused by a shrub, *Salsola tuberculata* (Fenzl ex Moq.) Schinz var. *tomentosa* C.A.Smith ex Aellen. *Onderstepoort J Vet Res* 36;59–104, 1969

Baxter JT: Suspected fodder-beet poisoning in pigs. *Vet Rec* 68;236–237, 1956

Bellue MK: Halogeton invades California. *Calif Dept Agric Bull* 38;33–36, 1949

Berg RT, McElroy LW: Effect of 2,4-D on the nitrate content of forage crops and weeds. *Can J Agric Sci* 33;354–358, 1953

Berry TJ, Gonzales P: Do your cattle-owning clients know about this poisonous range plant? *Vet Med* 81;1055–1056, 1986

Blackwell WH, Haacke JD, Hopkins CO: *Halogeton* (Chenopodiaceae) in North America. *SIDA* 8;157–169, 1979

Bove KE: Ethylene glycol toxicity. *Am J Clin Pathol* 45;46–50, 1966

Brown PL, Halvorsen AD, Siddoway FH, Mayland HF, Miller MR: *Saline Seeps—Diagnosis, Control, and Reclamation*. USDA Conserv Res Rep 30, 1983

Bulgin MS, Lincoln SD, Mather G: Elemental sulfur toxicosis in a flock of sheep. *J Am Vet Med Assoc* 208;1063–1065, 1996

Burchard GA: Kochia new forage crop. *Aberdeen Angus J* 28;50,116, 118, 1947

Case AA: Some aspects of nitrate intoxication in livestock. *J Am Vet Med Assoc* 130;323–329, 1957

Chambers F: Sugar beet poisoning in cattle. *Vet Rec* 56;128, 1944
Chapman VJ: The salinity problem in general, its importance, and distribution with special reference to natural halophytes. In *Plants in Saline Environments,* Poljakoff-Mayber A, Gale J eds, Springer-Verlag, New York, pp7–24, 1975
Chick BF, McCleary BV, Beckett RJ: Thiaminases. In *Toxicants of Plant Origin,* vol 3, *Proteins and Amino Acids,* Cheeke PR ed, CRC Press, Boca Raton, Fla, pp73–91, 1989
Chu GL, Stutz HC, Sanderson SC: Morphology and taxonomic position of *Suckleya suckleyana* (Chenopodiaceae). *Am J Bot* 78;63–68, 1991
Clements CD, Gray KJ, Young JA: Forage kochia: to seed or not to seed. *Rangelands* 19;29–31, 1997
Cohen RDH, Iwaasa AD, Mann ME, Coxworth E, Kernan JA: Studies on the feeding value of *Kochia scoparia* (L.) Schrad. hay for beef cattle. *Can J Anim Sci* 69;735–743, 1989
Connolly JD, Hill RA: *Dictionary of Terpenoids.* Vol 2, *Di- and Higher Terpenoids.* Chapman & Hall, London, 1991
Cook CW, Stoddart LA: *The Halogeton Problem in Utah.* Utah Agric Exp Stn Bull 364;1953
Cook CW, Stoddart LA, Harris LE: *Comparative Nutritive Value and Palatability of Some Introduced and Native Forage Plants for Spring and Summer Grazing.* Utah Agric Exp Stn Bull 385, 1956
Cooper MR, Johnson AW: *Poisonous Plants of Britain and Their Effects on Animals and Man.* Her Majesty's Stationery Office, London, pp80–83, 1984
Couch JF: The toxic constituents of greasewood *(Sarcobatus vermiculatus). Am J Pharm* 94;631–641, 1922
Coxworth ECM, Salmon RE: Kochia seed as a component of the diet of turkey poults: effects of different methods of saponin removal or inactivation. *Can J Anim Sci* 52;721–729, 1972
Coxworth ECM, Bell JM, Ashford R: Preliminary evaluation of Russian thistle, *Kochia,* and garden atriplex as potential high protein content seed crops for semiarid areas. *Can J Plant Sci* 49;427–434, 1969
Cronin EH, Williams MC: Principles for managing ranges infested with halogeton. *J Range Manage* 19;226–227, 1966
Davis AM: Selenium accumulation in a collection of *Atriplex* species. *Agron J* 64;823–824, 1972
Dickie CW, Berryman JR: Polioencephalomalacia and photosensitization associated with *Kochia scoparia* consumption in range cattle. *J Am Vet Med Assoc* 175;463–465, 1979
Dickie CW, James LF: *Kochia scoparia* poisoning in cattle. *J Am Vet Med Assoc* 183;765–768, 1983
Dickie CW, Gerlach ML, Hamar DW: *Kochia scoparia* oxalate content. *Vet Hum Toxicol* 31;240–242, 1989
Drost-Karbowska K, Kowalewski Z: Isolation of harmane and harmine from *Kochia scoparia. Lloydia* 41;289–290, 1978
Duncan AJ, Frutos P, Young SA: Rates of oxalic acid degradation in the rumen of sheep and goats in response to different levels of oxalic acid administration. *Anim Sci* 65;451–455, 1997
Dye WB: Chemical studies on *Halogeton glomeratus. Weeds* 4;55–60, 1956
Emerson PM, Wilkinson JH, Withycombe WA: Effect of oxalate on the activity of lactate dehydrogenase isoenzymes. *Nature* 202;1337–1338, 1964
Erickson EL: *Forage from Kochia.* S D Agric Exp Stn Bull 384, 1947
Everist SL: *Poisonous Plants of Australia,* 2d ed. Angus & Robertson, Sydney, Australia, 1981
Fleming CE, Miller MR, Vawter LR: *The Greasewood* (Sarcobatus vermiculatus): *A Range Plant Poisonous to Sheep.* Nev Agric Exp Stn Bull 115, 1928
Fowler JL, Hageman JH, Moore KJ, Suzukida M, Assadian H, Valenzuela M: Salinity effects on forage quality of Russian thistle. *J Range Manage* 45;559–563, 1992

Franceschi VR, Horner HT: Calcium oxalate crystals in plants. *Bot Rev* 46;361–427, 1980

Fuehring HD, Finkner RE, Oty GW: *Yield and Composition of Kochia Forage as Affected by Salinity of Water and Percent Leaching.* N M Water Resour Res Inst Tech Completion Rep 199, 1985

Fuehring HD, Finkner RE, French C, Pierce W, Stallings W: *Selective Breeding Program to Improve the Water-Use Efficiency and Nutritive Acceptability of Kochia as a Forage Grazing Crop.* N M Water Resour Res Inst Tech Completion Rep 213, 1986

Galitzer SJ, Oehme FW: *Kochia scoparia* (L.) Schrad. toxicity in cattle: a literature review. *Vet Hum Toxicol* 20;421–423, 1978

Gilbert CS, Eppson HF, Bradley WB, Beath OA: *Nitrate Accumulation in Cultivated Plants and Weeds.* Wyo Agric Exp Stn Bull 277, 1946

Gooneratne SR, Buckley WT, Christensen DA: Review of copper deficiency and metabolism in ruminants. *Can J Anim Sci* 69;819–845, 1989a

Gooneratne SR, Olkowski AA, Klemmer RG, Kessler GA, Christensen DA: High sulfur related thiamine deficiency in cattle: a field study. *Can Vet J* 30;139–146, 1989b

Gooneratne SR, Olkowski AA, Rousseaux CG, Christensen DA: Role of thiamine status in sulphur induced polioencephalomalacia. *Res Vet Sci* 52;78–85, 1992

Gorisek J: Blood coagulation and calcium content of blood and urine in cows fed sugarbeet leaves. *Vet Arh* 30;300–306, 1960 (*Vet Bull* 31;1890, 1961)

Gorisek J: Metabolic disorders in dairy cows after feeding sugar beet tops and dosing with saponin. *Dtsch Tierarztl Wochenschr* 69;325–327, 1962 (*Vet Bull* 32;4300, 1962)

Gorisek J: Stoffwechselstorungen bei milchkuhen im zusammenhang mit der verfutterung von zuckerrubenblatt. *17th World Vet Congr* 2;1343–1344, 1963

Gould DH: Polioencephalomalacia. *J Anim Sci* 76;309–314, 1998

Gould DH, McAllister MM, Savage JC, Hamar DW: High sulfide concentrations in rumen fluid associated with nutritionally induced polioencephalomalacia in calves. *Am J Vet Res* 52;1164–1169, 1991

Guil JL, Torija ME, Gimenez JJ, Rodriguez-Garcia I, Gimenez A: Oxalic acid and calcium determination in wild edible plants. *J Agric Food Chem* 44;1821–1823, 1996

Hageman JH, Fowler JL, Suzukida M, Salas V, Lecaptain R: Analysis of Russian thistle (*Salsola* species) selections for factors affecting nutritional value. *J Range Manage* 41;155–158, 1988

Hershey AL: *Some Poisonous Plant Problems of New Mexico.* N M Agric Exp Stn 322, 1945

Herweijer CH, Den Houter LF: Poisoning in sheep caused by *Chenopodium album. Tijdschr Diergeneeskd* 95;1134–1136, 1970

Hindmarsh WL: The lethal dose of hydrocyanic acid for ruminants. *J Counc Sci Ind Res Aust* 3;12–13, 1930

James LF: Serum electrolyte, acid-base balance, and enzyme changes in acute *Halogeton glomeratus* poisoning in sheep. *Can J Comp Med* 32;539–543, 1968

James LF: Locomotor disturbance of cattle grazing *Halogeton glomeratus. J Am Vet Med Assoc* 156;1310–1312, 1970

James LF: Oxalate toxicosis. *Clin Toxicol* 5;231–243, 1972

James LF, Butcher JE: Halogeton poisoning of sheep: effect of high level of oxalate intake. *J Anim Sci* 35:1233–1238, 1972

James LF, Johnson AE: Prevention of fatal *Halogeton glomeratus* poisoning in sheep. *J Am Vet Med Assoc* 157;437–442, 1970

James LF, Butcher JE, Van Kampen KR: Relationship between *Halogeton glomeratus* consumption and water intake by sheep. *J Range Manage* 23;123–127, 1970

James LF, Street JC, Butcher JE: In vitro degradation of oxalate and cellulose by rumen ingesta from sheep fed *Halogeton glomeratus. J Anim Sci* 26;1438–1444, 1967

James LF, Williams MC, Bleak AT: Toxicity of *Bassia hyssopifolia* to sheep. *J Range Manage* 29;284–285, 1976

James MP, Seawright AA, Steele DP: Experimental acute ammonium oxalate poisoning of sheep. *Aust Vet J* 47;9–17, 1971

Jefferies RL, Rudmik T: The response of halophytes to salinity: an ecological perspective. In *Salinity Tolerance in Plants,* Staples RC, Toenniessen GH eds, Wiley & Sons, New York, pp213–227, 1984

Joubert JPJ, Basson PA, Lucks HJ, Burger JHS: "Grootlamsiekte," a specific syndrome of prolonged gestation in sheep: further investigations. *Understepoort J Vet Res* 39;59–70, 1972

Kellerman TS, Coetzer JAW, Naude TW: *Plant Poisonings and Mycotoxicoses of Livestock in Southern Africa.* Oxford Univ Press, Cape Town, South Africa, 1988

Kernan JA, Coxworth E, Fleming S: Microdetermination of triterpene sapogenin content of *Kochia scoparia* seed using gas-liquid chromatography. *J Agric Food Chem* 21;232–234, 1973

Kirkpatrick JG, Helman RG, Burrows GE, Tyrl RJ, VonTungeln D: Transient hepatotoxicity in sheep grazing *Kochia scoparia.* In *Toxic Plants and Other Natural Toxicants,* Garland T, Barr AC eds, CAB International, New York, pp504–508, 1998

Leichter J, Joslyn MA: Kinetics of thiamine cleavage by sulphite. *Biochem J* 113;611–615, 1969

Lewis D: The reduction of sulphate in the rumen of sheep. *Biochem J* 56;391–399, 1954

Libert B, Franceschi VR: Oxalate in crop plants. *J Agric Food Chem* 35;926–938, 1987

Lincoln SD, Black B: Halogeton poisoning in range cattle. *J Am Vet Med Assoc* 176;717–718, 1980

Marsh CD, Clawson AB, Couch JF: *Greasewood as a Poisonous Plant.* USDA Dept Circ 279, 1923

Mathams RH, Sutherland AK: The oxalate content of some Queensland pasture plants. *Queensl J Agric Sci* 9;317–334, 1952

Mayland HF, Robbins CW: Sulfate uptake by salinity-tolerant plant species. *Commun Soil Sci Plant Anal* 25;2523–2541, 1994

McAllister MM, Gould DH, Raisbeck MF, Cummings BA, Loneragan GH: Evaluation of ruminal sulfide concentrations and seasonal outbreaks of polioencephalomalacia in beef cattle in a feedlot. *J Am Vet Med Assoc* 211;1275–1279, 1997

Mejia MR, Montalvo VR, Martinez RR: Niveles de oxalatos en forrajes silvestres procedentes de los estados de Hidalgo, Guanajuato, Mexico, Tlaxcala y Distrito Federal. *Vet Mex* 16;21–25, 1985

Montgomery GA: New high-protein forage crop. *Capper's Farmer* 58;9,48, 1947

Morris MP, Garcia-Rivera J: The destruction of oxalates by the rumen contents of cows. *J Dairy Sci* 38;1169, 1955

Morton HL, Hass RH, Erickson LC: *Halogeton and Its Control.* Idaho Agric Exp Stn Bull 307, 1959a

Morton HL, Hass RH, Erickson LC: Oxalate and mineral contents of *Halogeton glomeratus. Weeds* 7;255–264, 1959b

Mujica A: Andean grains and legumes. In *Neglected Crops: 1492 from a Different Perspective,* Hernandez Bermejo JE ed, Food and Agriculture Organization, United Nations, Rome, pp131–148, 1994

Nahrstedt A, Lechtenberg M, Brinker A, Seigler DS, Hegnauer R: 4-Hydroxymandelonitrile glucosides, dhurrin in *Suckleya suckleyana* and taxiphyllin in *Girgensohnia oppositiflora* (Chenopodiaceae). *Phytochemistry* 33;847–850, 1993

Novy J: Ensiled sugar beet pulp poisoning in cattle. *Vet Cas* 9;439–446, 1960 (*Vet Bull* 31;1251, 1961)

Olkowski AA, Gooneratne SR, Crichlow EC, Rousseaux CG, Christensen DA: Effects of high dietary sulfur on brain functions using evoked potentials technique. *Can J Vet Res* 54;113–118, 1990

Olkowski AA, Gooneratne SR, Rousseaux CG, Christensen DA: Role of thiamine status in sulphur induced polioencephalomalacia in sheep. *Res Vet Sci* 52;78–85, 1992

Olson OE, Whithead E: Nitrate content of some South Dakota plants. *S D Acad Sci Proc* 20;95–101, 1940

Pammel LH: *A Manual of Poisonous Plants.* Torch Press, Cedar Rapids, Iowa, 1911

Pammel LH: Range plants poisonous to sheep and cattle in Nevada. *Am J Vet Med* 14;243–246, 1919

Pemberton RW: The distribution of halogeton in North America. *J Range Manage* 39;281–282, 1986

Penny RHC: Suspected poisoning by fodder-beet in the bovine. *Vet Rec* 66;134, 1954

Qi K, Lu CD, Owens FN: Sulfate supplementation of Angora goats: sulfur metabolism and interactions with zinc, copper, and molybdenum. *Small Ruminant Res* 11;209–225, 1993

Raffauf RF: *Plant Alkaloids: A Guide to Their Discovery and Distribution.* Haworth Press, New York, 1996

Raisbeck MF: Is polioencephalomalacia associated with high- sulfate diets? *J Am Vet Med Assoc* 180;1303–1305, 1982

Rankins DL Jr: Evaluation of *Kochia scoparia* (L.) Schrad. toxicosis in sheep and cattle. PhD thesis, New Mexico State Univ, 1989 (*Diss Abstr Int* 50:3228B)

Rankins DL Jr, Smith GS: Nutritional and toxicological evaluations of kochia hay *(Kochia scoparia)* fed to lambs. *J Anim Sci* 69;2925–2931, 1991

Rankins DL Jr, Smith GS, Hallford DM: Altered metabolic hormones, impaired nitrogen retention, and hepatotoxicosis in lambs fed *Kochia scoparia. J Anim Sci* 69;2932–2940, 1991a

Rankins DL Jr, Smith GS, Hallford DM: Effects of metoclopramide on steers fed *Kochia scoparia* hay. *J Anim Sci* 69;3699–3706, 1991b

Rankins DL Jr, Smith GS, Hallford DM: Serum constituents and metabolic hormones in sheep and cattle fed *Kochia scoparia* hay. *J Anim Sci* 69;2941–2946, 1991c

Rimington C, Steyn DG: *Psilocaulon absimile* N.E.Br. as a stock poison: determination of oxalic, malic, tartaric acids, etc. *Onderstepoort J Vet Sci Anim Ind* 1;439–455, 1933

Rousseaux CG, Olkowski AA, Chauvet A, Gooneratne SR, Christenson DA: Ovine polioencephalomalacia associated with dietary sulphur intake. *J Vet Med* A 38;229–239, 1991

Sampson AW, Malmsten HE: *Stock-Poisoning Plants of California.* Calif Agric Exp Stn Bull 593, 1942

Savage A: Nitrate poisoning from sugar beet tops. *Can J Comp Med* 13;9–10, 1949

Scarisbrick R: Acid indigestion in a sheep fed on mangolds. *Vet Rec* 66;131–132, 1954

Schmutz EM, Freeman BN, Reed RE: *Livestock-Poisoning Plants of Arizona,* Univ Arizona Press, Tucson, 1968

Sherrod LB: Nutritive value of *Kochia scoparia.* 1. Yield and chemical composition at three stages of maturity. *Agron J* 63;343–345, 1971

Sherrod LB: Nutritive value of *Kochia scopari.* 3. Digestibility of kochia hay compared with alfalfa. *J Dairy Sci* 56:923–926, 1973

Shupe JL, James LF: Additional physiopathologic changes in *Halogeton glomeratus* (oxalate) poisoning in sheep. *Cornell Vet* 59;41–55, 1969

Simesen MG, Konggaard SP: Experimental investigations on fodder-beet poisoning in cattle. *Nord Vet Med* 22;174–185, 1970

Smith GS, Zornes ML, Schemnitz SD: Nutritional and toxicological evaluation of seeds from *Kochia scoparia:* chukar bioassay [abstract]. *J Anim Sci* 71;444, 1993

Souto J, Milano VA: Triterpenic saponin in ripe fruits of *Kochia scoparia* (morenita). *Rev Invest Agropecu,* ser 2 *(Biol Prod Veg),* 3;367–383, 1966

Sprowls RW: Problems observed in horses, cattle, and sheep grazing kochia. *Proc Am Assoc Vet Lab Diagn* 24;397–406, 1981

Stahler LM, Whitehead EI: The effect of 2,4-D on potassium nitrate levels in leaves of sugar beets. *Science* 112;749–751, 1950

Steyn DG: *The Toxicology of Plants of South Africa.* Central News Agency, South Africa, pp85–92, 1934

Stoddart LA, Clegg BS, Markham BS, Stewart G: The halogeton problem on Utah ranges. *J Range Manage* 4;223–225, 1951

Stout EN: Suckleya suckleyana. Colo Agric Exp Stn Bull 359-A, 1939

Streeter CP: Kochia—the weed that's feed. *Farm J* 71;33, 1947
Suttle NF: Effects of organic and inorganic sulphur on availability of dietary copper to sheep. *Br J Nutr* 32;559–568, 1974
Suttle NF: The interactions between copper, molybdenum, and sulphur in ruminant nutrition. *Annu Rev Nutr* 11;121–140, 1991
Talapatra SK, Ray SC, Sen KC: Calcium assimilation in ruminants on oxalate-rich diet. *J Agric Sci* 38;163–173, 1948
Thilsted J, Hibbs C, Kiesling H, Hallford D, Kirksey R, Meininger A, Tompkins J: Kochia *(Kochia scoparia)* toxicosis in cattle: results of four experimental grazing trials. *Vet Hum Toxicol* 31;34–41, 1989
Thomas KW: Oral treatment of polioencephalomalacia and subclinical thiamine deficiency with thiamine propyl disulphide and thiamine hydrochloride. *J Vet Pharmacol Ther* 9;402–411, 1986
Thomas MD, Hendricks RH, Hill GR: Sulfur content of vegetation. *Soil Sci* 70;9–18, 1950
Thorp F, Deem AW: *Suckleya suckleyana,* a poisonous plant. *J Am Vet Med Assoc* 94;192–197, 1939
Thorp F, Deem AW, Harrington HD, Tobiska JW: Suckleya suckleyana: *a poisonous plant.* Colo Agric Exp Stn Tech Bull 22, 1937
Trelease SF, Beath OA: *Selenium.* By the Authors, New York, 1949
Van Kampen KR, James LF: Acute halogeton poisoning of sheep: pathogenesis of lesions. *Am J Vet Res* 30;1779–1783, 1969
Vawter LR: *Halogeton* glomeratus—a range plant poisonous to sheep and cattle. *Calif Vet* 3;12–15, 1950
Watt JM, Breyer-Brandwijk MG: *The Medicinal and Poisonous Plants of Southern and Eastern Africa.* E & S Livingston, Edinburgh, pp536–552, 1962
Whitehead EI, Moxon AL: *Nitrate Poisoning.* S D Agric Exp Stn Bull 424, 1952
Williams MC: Effect of sodium and potassium salts on growth and oxalate content of halogeton. *Plant Physiol* 35;500–505, 1960
Williams MC, Cronin EH: Five poisonous range weeds—when and why they are dangerous. *J Range Manage* 19;274–279, 1966
Williams VJ, Coup MR: Preliminary studies on the toxicity of fodder beet to sheep. *N Z Vet J* 7;8–14, 1959
Willson VA: Toxic properties of greasewood. *J Am Vet Med Assoc* 85;76–81, 1934
Worden AN, Bunyan J, Pickup J: A fatal hypocalcemia-like syndrome in dairy cows following the excess consumption of fodder-beet. *Vet Rec* 66;133–134, 1954

Clusiaceae Lindl. 24

Almost cosmopolitan in distribution, the Clusiaceae, commonly known as the St.-John's-wort or mangosteen family, comprises 40–50 genera and 1,000–1,200 species. Greatest diversity is in moist tropical regions and the northern temperate zone. An equally acceptable, older, alternative name is Guttiferae. Two well-marked subfamilies, the Clusioideae and the Hypericoideae, are typically recognized. In general, taxa of the former are woody, are tropical in distribution, have nonpunctate glandular leaves, bear imperfect flowers, and have arillate seeds. In contrast, taxa of the latter subfamily are typically herbaceous or woody, are temperate in distribution, have punctate glandular leaves, bear perfect flowers, and have nonarillate seeds. Taxonomists differ in their opinion as to whether these subfamilies should be recognized as distinct families. The conservative interpretation of one family is employed in this treatise. Economically, the family is a minor source of ornamentals, timber, drugs, gums, oils, and food. The edible mangosteen and mammee apple are widely eaten in the tropics. Present in North America are 6 native and introduced genera and some 70 species. Photosensitization problems are associated with 1 genus.

- Mangosteen Family
  St.-John's-wort Family
  Hypericum

Plants herbs or shrubs; perennials or annuals; sap viscous, often brightly colored. **Leaves** simple; opposite or rarely whorled; sessile or subsessile; venation pinnate; surfaces with glandular punctations present or absent; margins entire; stipules absent. **Inflorescences** simple or compound cymes or solitary flowers; terminal; bracteoles present. **Flowers** perfect or imperfect; perianths in 2-series. **Sepals** 2 or 4 or 5; 5 in 1 whorl and all equal, or 2 or 4 in 2 whorls and of 2 sizes; free. **Corollas** radially symmetrical. **Petals** 4 or 5; free; usually yellow or pink. **Stamens** 9 to numerous; not fascicled or fascicled in 3–5 groups; free or fused by filaments. **Pistils** 1; compound, carpels 2–5; stigmas 2–5; styles 2–5; fused at bases; ovaries superior; locules 1 or 3–5; placentation axile or parietal. **Fruits** capsules or berries or drupes. **Seeds** numerous.

- herbs, shrubs; leaves simple, opposite or whorled; petals 4 or 5, yellow or pink

Hypericum L.

Taxonomy and Morphology—A morphologically diverse genus of 200–400 species, *Hypericum* is commonly known as St.-John's-wort. Used by Dioscorides, its name is Greek but of obscure meaning. In North America, approximately 60 native and introduced species are present in all types of vegetation. Some species are planted as ground covers and border plants, whereas others are noxious weeds. Representative of the few North American species of toxicologic importance are the following species (see plate 24.1):

H. concinnum Benth.	goldwire
H. formosum Kunth	western St.-John's-wort

- *H. concinnum*
  H. formosum
  H. perforatum
  H. pseudomaculatum
  H. punctatum

amber, cammock, eola weed, goatsbeard, goatweed, goldwire, grace of God, herb john, Klamath weed, penny john, rosin rose, St.-John's-wort, tipton weed

- erect herbs, shrubs; leaves often glandular-punctate, glands black or translucent; flowers showy; petals 5, bright yellow, some glandular-punctate; fruits capsules

H. perforatum L.	common St.-John's-wort, Klamath weed, amber, goatweed, goatsbeard, tipton weed, eola weed, penny john, herb john, rosin rose, cammock, grace of God
H. pseudomaculatum Bush	large spotted St.-John's-wort
H. punctatum Lam.	spotted St.-John's-wort

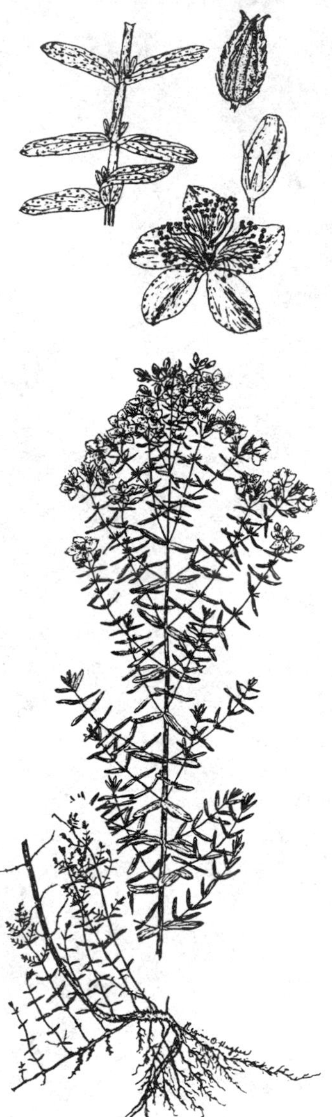

Hypericum perforatum

Plants herbs or shrubs; perennials or annuals; from taproots or rhizomes or woody caudices. **Stems** erect; 15–120 cm tall. **Leaves** typically glandular-punctate; glands translucent or amber or black. **Inflorescences** solitary flowers or cymes. **Flowers** showy; perfect. **Sepals** 5; glandular-punctate or not glandular-punctate. **Petals** 5; bright yellow; glandular-punctate or not glandular-punctate; convolute in bud. **Stamens** numerous; fascicled in 3–5 groups opposite petals. **Pistils** 1; carpels 3–5; stigmas minute or capitate; styles 3–5; locules 1 or 3–5; placentation axile or parietal. **Fruits** capsules; globose or ovoid or conical. **Seeds** short-cylindrical; pitted or alveolate.

1. Stems many from woody caudices. Leaves folded; linear to lanceolate; apices acute . . . . *H. concinnum*
1. Stems few from taproots or rhizomes. Leaves flat; elliptic or oblong; apices obtuse.
    2. Rhizomes and stolons present . . . . *H. formosum*
    2. Rhizomes and stolons absent.
        3. Branches angular or ridged immediately below nodes. Seeds 1–1.3 mm long . . . . *H. perforatum*
        3. Branches terete immediately below nodes. Seeds 0.5–0.8 mm long.
            4. Sepals 2.5–4 mm long. Petals 4–7 mm long. Styles 2–4 mm long . . . . *H. punctatum*
            4. Sepals 5–7 mm long. Petals 8–12 mm long. Styles 6–10 mm long . . . . *H. pseudomaculatum*

- open meadows and woods

Distribution and Habitat—Species of *Hypericum* occur in a diversity of generally open habitats in different vegetation types across the continent. *Hypericum concinnum* is strictly Californian. Also western, *H. formosum* occurs in meadows, brushy areas, and open pine or aspen woods at higher elevations as far east as the Rocky Mountains. Populations of *H. punctatum* and *H. pseudomaculatum* occur in moist or dry soils of fields and open woods in the eastern half of the continent.

Native to Europe but now widely distributed around the world, *H. perforatum* is abundant as a weed across North America. Introduced in the mid- to late 1700s, it has become especially abundant in the Pacific Coast states and infests millions of acres. It is aggressive and quite competitive with native species; however, biological control via the Klamath weed flea beetles (*Chrysolina quadrigemina* and *C. hyperici*) has reduced its abundance (Huffaker & Kennett 1959; Crompton et al. 1988). Not all areas are amenable to such control, because their climatic conditions may not favor survival of the beetles (Williams 1985).

- humans, early medicinal and herbal uses; treatment for anxiety, depression

  livestock toxicity, primary photosensitization, some systemic effects

Disease Problems—Although causing photosensitization when eaten by livestock in large amounts, species of *Hypericum*, and in particular *H. perforatum*, have a long history of medicinal use in the Old World—as an astringent, a diuretic, and a sedative (Mitich 1994). Used extensively in Germany, where herbal medications are popular, it is approved for the treatment of anxiety, depression, and sleep disorders; 66

million daily dosage prescriptions were written in 1 year (De Smet & Nolen 1996; Linde et al. 1996). In recent years, *Hypericum* has gained considerable popularity as an antidepressant, and this use is extensively reviewed by Linde and associates (1996).

Considerable mysticism is associated with *Hypericum*, beginning with references to it in Greek and Roman mythology. It was recognized as a potential problem for livestock, and in some regions of the world, black sheep were preferred because of the susceptibility of white sheep to photosensitization after eating these plants. Arabs are said to have darkened the white spots on horses with henna to protect them from the sun on occasions of exposure to the plant (Rogers 1914; Giese 1980). Most animal species may be affected, with severe blistering effects on the skin and, in some cases, accompanying systemic signs. Although the disease is seldom lethal, it markedly affects production of meat, milk, and wool and often causes the affected animal considerable distress. In contrast to other animals, deer and goats seem to eat the plants readily and with considerable impunity (Sampson & Parker 1930).

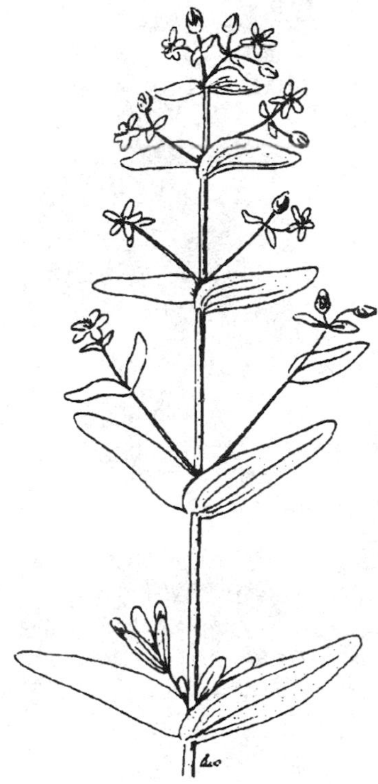

Hypericum punctatum

Although it seems clear that the effects on the skin are not due to a decreased ability of the liver to clear a photodynamic substance, systemic effects have been reported. Early medicinal uses of the genus may be related to the systemic actions. Sheep given 4–16 g/kg b.w. of *H. perforatum* exhibited a 9-fold increase in blood urea nitrogen, a rise in serum enzymes (AST, ALT, CK, LDH), and a decrease in blood glucose, hemoglobin, packed cell volume, and total protein (Kako et al. 1993). In contrast, calves suffering from photosensitization after being given 3 g/kg b.w. of dried plant material did not exhibit changes in levels of serum hepatic enzymes and other parameters of hepatic function (Araya & Ford 1981). This difference may simply reflect the larger dosage given to the sheep.

Dosage of 1% b.w. or 0.23 kg of plant material in 230–318 kg calves for several days produced prominent signs when the animals were exposed to sunlight (Marsh & Clawson 1930). Sheep are less susceptible, requiring 0.7–0.9 kg for several days.

Disease Genesis—The leaves, stems, and floral organs of most species of *Hypericum* have minute glands embedded in their surfaces. These glands contain either red fluorescent pigments such as hypericin and pseudohypericin or their precursor protohypericin. Glands containing hypericin and the other pigments appear black, whereas those containing the precursor are translucent or clear. The abundance of these glands varies from species to species. The most conspicuous attribute of the 5 species of toxicologic concern are the numerous black glands on the sepals and petals. In general, most other North American species of the genus lack these black glands and have relatively small amounts of hypericin. In contrast, more than 50% of the Old World species contain the pigment (Mathis & Ourisson 1963).

- highly condensed quinone, hypericin, photodynamic, activated by UV light; in glands in leaves and flowers, increasing with plant maturity, variable with geographic location; increased serotonin systemic activity

Hypericin

Hypericin is the toxicant responsible for photosensitization. In *H. perforatum*, the most extensively studied species of the genus, concentrations are low in young shoots (54 mg/kg); they increase to 1.2 g/kg as the plants mature and temperatures rise, to peak in midsummer (Horsley 1934; Jensen et al. 1995). However, this increase in toxin level may be offset by the apparent decrease in palatability of the plants to livestock as they mature (Horsley 1934; Gilkey & Dennis 1969). There is considerable geographical variation in hypericin concentrations across North America, apparently related to the original Old World sources of these plant populations. Toxin concentrations averaged 195 ppm d.w. in Nova Scotia but were 2–3 times higher in plants from British Columbia (Jensen et al. 1995). As is so often the case, there also is marked intraspecific variation in toxin concentration. In Australia, narrow-leaved varieties are reported to have 2–3 times more hypericin than those with broader leaves (1,000 versus 400 ppm) (Southwell & Campbell 1991).

Hypericum concinnum

Hypericin, a naphthodianthrone, is a highly condensed quinone with many hydroxyl groups (Pace & Mackinney 1941; Brockmann 1957; Brockmann & Sanne 1957). It is absorbed into the systemic circulation and is activated in the peripheral circulation in areas most readily penetrated by ultraviolet light of 540–610 nm wavelength. It undergoes a type II reaction and becomes a potent generator of singlet oxygen capable of reacting with amino acids, nucleic acids, and membrane lipids of the capillary endothelium and the adjacent perfused tissues (Pace 1942; Dodge & Knox 1986). Generally the reaction effects are limited to the skin, as opposed to type I reactions, from which more-extensive changes may be noted. Although hypericin concentrations are reported to be much reduced when cut plants dry in the sun, intoxication may sometimes occur when hay containing them is fed (Marsh & Clawson 1930; Araya & Ford 1981; Kumper 1988).

The antidepressant activity of *H. perforatum* extracts appears to be related to serotonin activity. Experimentally, the extracts inhibited serotonin uptake by rat synaptosomes, and it was suggested that inhibition of serotonin uptake by postsynaptic receptors is a key factor in the neurologic activity (Perovic & Muller 1995).

- skin erythema, edema, itching, sloughing, lightly haired or less pigmented areas; mild diarrhea, restlessness, anorexia

Clinical Signs—In cattle and most other species, the most striking effects of *Hypericum*-induced photosensitization are severe pruritis, reddening, swelling, edema, and sloughing of the skin of the teats, udder, escutcheon, and the less pigmented and lightly haired areas. There is often a striking demarcation in distribution of the lesions, coinciding exactly with the lighter-colored areas of the skin and not extending into the darker areas. In addition to the obvious skin changes, there may be accompanying signs of diarrhea, restlessness, agitation, loss of appetite, and increased respiratory rate. Occasionally violent behavior during the onset of the disease may be a serious problem. In sheep, because of the protective effects of wool, the skin lesions may be limited to ears, eyelids, face, lips, and coronets. The signs may last for 2 weeks, and subsequent healing take 2 or more months.

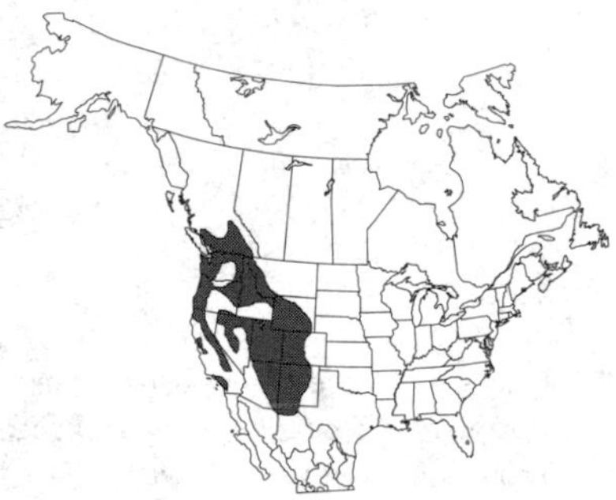
Hypericum formosum

- gross pathology, skin crusty, ulcerative, proliferative areas

Pathology—Skin changes typically are limited to the lesser pigmented areas or to the head in sheep. Depending upon the amount of plant material eaten, the changes range from minimal reddening and blisterlike effects to extensive areas of sloughing. In more-severe cases, the skin will be thickened with a gelatinous appearance extending down into the deeper coriumm. Crusty and ulcerative or proliferative areas may be present, especially around the face. Microscopically the edema of the skin will be accompanied by necrosis and a polymorphonuclear infiltrate in the deeper corium.

- treatment not needed, provide shade

Treatment—Prevention of animal access to the plants is usually sufficient for recovery to begin within a few days, although signs may regress slowly with reepithelialization of the denuded areas. In some cases relief of pain may be helpful, in addition to soothing skin medications.

Hypericum perforatum

References

Araya OS, Ford EJH: An investigation of the type of photosensitization caused by the ingestion of St. John's wort *(Hypericum perforatum)* by calves. *J Comp Pathol* 91;135–141, 1981

Brockmann H: Photodynamically active plant pigments. *Proc Chem Soc Lond* 1957;304–312, 1957

Brockmann H, Sanne W: Zur kenntnis des hypericins und pseudo- hypericins. *Chem Ber* 90;2480–2491, 1957

Crompton CW, Hall IV, Jensen KIN, Hildebrand PD: The biology of Canadian weeds. 83. *Hypericum perforatum* L. *Can J Plant Sci* 68;149–162, 1988

De Smet PAGM, Nolen WA: St. John's wort as an antidepressant. *Br Med J* 313;241–242, 1996

Dodge AD, Knox JP: Photosensitizers from plants. *Pestic Sci* 17;579–586, 1986

Giese AC: Hypericism. *Photochem Photobiol Rev* 5;229–255, 1980

Gilkey HM, Dennis LRJ: *Livestock-Poisoning Plants of Oregon.* Oreg Agric Exp Stn Man 1, pp75–76, 1969

Horsley CH: Investigation into the action of St. John's wort. *J Pharmacol Exp Ther* 50;310–322, 1934

Huffaker CB, Kennett CE: A ten-year study of vegetational changes associated with biological control of Klamath weed. *J Range Manage* 12;69–82, 1959

Jensen KIN, Gaul SO, Specht EG, Doohan DJ: Hypericin content of Nova Scotia biotypes of *Hypericum perforatum* L. *Can J Plant Sci* 75;923–926, 1995

Kako MDN, Al-Sultan II, Saleem AN: Studies of sheep experimentally poisoned with *Hypericum perforatum. Vet Hum Toxicol* 35;298–300, 1993

Kumper H: *Hypericum perforatum* poisoning in sheep [Hypericismus bei Schafen]. In *Tagung der Fachgruppe "Krankheiten der kleinen Wiederkauer," Giessen,* 10 Juni 1988, pp86–89, 1988

Linde K, Ramirez G, Muldrow CD, Pauls A, Weidenhammer W, Melchart D: St. John's wort for depression—an overview and meta- analysis of randomized clinical trials. *Br Med J* 313;241–242, 1996

Marsh CD, Clawson AB: *Toxic Effect of St. Johnswort (Hypericum perforatum) on Cattle and Sheep.* USDA Tech Bull 202, 1930

Mathis C, Ourisson G: Etude chimio-taxonomique du genre *Hypericum* I. Repartition de l'hypéricine. *Phytochemistry* 2;157–171, 1963

Mitich LW: Common St. Johnswort. *Weed Technol* 8;658–661, 1994

Pace N: The etiology of hypericism, a photosensitivity produced by St. Johnswort. *Am J Physiol* 136;650–656, 1942

Pace N, Mackinney G: Hypericin, the photodynamic pigment from St. John's wort. *J Am Chem Soc* 63;2570–2574, 1941

Perovic S, Muller WE: Pharmacological profile of hypericum extract: effect on serotonin uptake by postsynaptic receptors. *Arzneimittelforschung* 45;1145–1148, 1995

Rogers TB: On the action of St. John's wort as a sensitizing agent for non-pigmented skin. *Am Vet Rev* 46;145–162, 1914

Sampson AW, Parker KW: *St. Johnswort on Range Lands of California.* Calif Agric Exp Stn Bull 503, 1930

Southwell IA, Campbell MH: Hypericin content variation in *Hypericum perforatum* in Australia. *Phytochemistry* 30;475–478, 1991

Williams KS: Climatic influences on weeds and their herbivores: biological control of St. John's wort in British Columbia. In *Proceedings of VI International Symposium on Biological Control of Weeds,* Delfosse ES ed, Agriculture Canada, Canadian Government Publ Centre, Ottawa, pp127–132, 1985

Hypericum pseudomaculatum

Hypericum punctatum

25 Convolvulaceae Juss.

- Morning-glory Family
  Convolvulus
  Ipomoea

Nearly cosmopolitan in distribution, but most abundant in tropical and subtropical regions, the Convolvulaceae, or morning-glory family, comprises 50 genera and 1,500–1,800 species. It is economically quite important because sweet potato *(Ipomoea batatas)*, with hundreds of cultivars and an important staple in the tropics, is a member. The family also is noted for ornamentals, such as morning glory, wood rose, moonflower, and cypress vine, and for noxious weeds, in particular the bindweeds. In North America, approximately 17 genera and 140 species, both native and introduced, are present. Intoxication problems are associated with 2 genera.

- herbs or vines; petals 5, fused, white, shades of blue, purple, pink, or red; fruits capsules

Plants herbaceous vines climbing by twining or herbs; annuals or perennials. **Root Systems** fibrous or with a central taproot; often fleshy. **Stems** typically climbing or trailing or occasionally ascending to erect. **Leaves** simple or palmately compound; alternate; petiolate or sessile; blades of various shapes, but typically cordate; venation pinnate or palmate; margins entire or sinuate or pinnately cleft or palmately lobed; leaflets of compound leaves 3–7; stipules absent. **Inflorescences** solitary flowers or simple or compound cymes; terminal or axillary; bracts present, typically paired, sometimes forming an involucre. **Flowers** perfect; perianths in 2-series. **Sepals** 5; free or fused. **Corollas** radially symmetrical; funnelform or campanulate or salverform or rotate; convolute in bud. **Petals** 5; fused; white to shades of blue or purple or pink or red, colors often darker at bases or apices. **Stamens** 5; didynamous; exserted beyond or included within corollas. **Pistils** 1; compound, carpels 2 or rarely 3–5; stigmas 1 or 2 or rarely 3–5, not lobed or 2-lobed, capitate or linear; styles 1 or 2 or rarely 3–5; ovaries superior; locules 2; placentation axile or basal. **Fruits** capsules; dehiscent or occasionally indehiscent; valvate. **Seeds** 1–4 or rarely 6 or 10.

- seeds well known for hallucinogenic activity; emerging interest as source of calystegins

Heretofore, the Convolvulaceae has been known primarily for the hallucinogenic effects of the seeds of some species of *Ipomoea*. However, *Convolvulus,* containing different toxicants, is toxic as well. In addition, there is increasing interest in a new class of alkaloids known as the calystegins. These compounds, initially isolated from species of *Calystegia,* such as *C. sepium* (hedge bindweed), are present in other genera of the family (Tepfer et al. 1988). Also present in taxa of the Solanaceae and believed to be the probable cause of crazy cow syndrome (Molyneux et al. 1994), they have not yet been shown to cause disease problems in the Convolvulaceae.

Convolvulus L.

- *C. arvensis*
  bindweed
  obsession vine

Taxonomy and Morphology—Commonly known as bindweed because of the tendency of its twining stems to form tangles around other plants or vertical supports, *Convolvulus* comprises 200–250 species. Its name comes from the Latin *convolvere,* meaning "to twine." Taxonomists differ in their opinion as to whether it should be broadly or narrowly circumscribed. Some include the genera *Calystegia* and

Jacquemontia, while others keep the 3 taxa separate, an interpretation followed in this treatise. Because of this difference in opinion, some commonly encountered species have two technical names. For example, hedge bindweed may be *Convolvulus sepium* in one reference and *Calystegia sepium* in another. Many species of *Convolvulus* are showy ornamentals and others noxious weeds. Of the 6 species in North America, only 1 is of toxicologic concern:

C. arvensis L.	bindweed, European bindweed, obsession vine, field bindweed

Plants herbaceous vines; perennials; from deep taproots and spreading rhizomes. **Stems** trailing or climbing; to 150 cm long. **Leaves** simple; petiolate; blades sagittate or hastate; margins entire or undulate; bases truncate or cordate. **Inflorescences** of 1–3 flowers borne at ends of elongate axillary peduncles. **Flowers** showy; funnelform or campanulate. **Sepals** fused. **Petals** white or pink. **Pistils** 1; stigmas 2-lobed, linear; styles 1; locules 2. **Capsules** ovoid to globose; valves 2–4. **Seeds** usually 4; ovoid; glabrous.

- perennial climbing vines; leaves simple; flowers showy; petals white or pink

Convolvulus arvensis

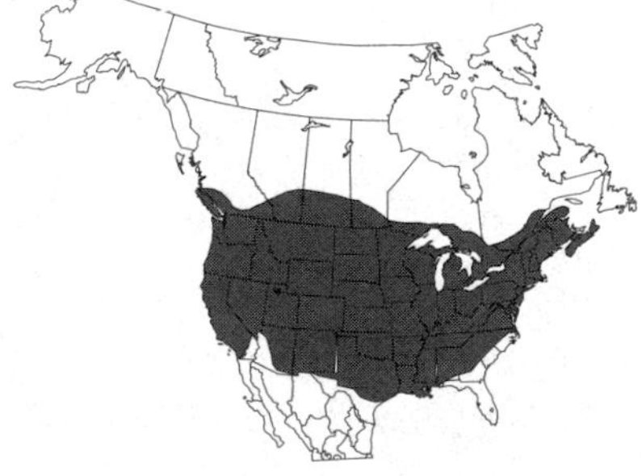

Convolvulus arvensis

Distribution and Habitat—A native of Europe, *C. arvensis* is now an aggressive weed in cultivated fields, along roadsides, and in waste places throughout much of North America. Often forming tangled mats in grain fields, it can clog harvesting equipment and easily fulfill the expectations suggested by its common names.

- waste-area weeds

Disease Problems—Species of *Convolvulus* are not generally a problem, although the tangled mats of the stems of the weedy species such as *C. arvensis* have been considered capable of causing a tangled obstructive mass in the digestive tract. Occasionally ingestion of the whole plant causes problems. The roots are irritant laxatives and also have diuretic action (Millspaugh 1974). The seeds and foliage also have purgative activity (Watt & Breyer-Brandwijk 1962). Whole plants of *C. arvensis,* fed to mice as 100% of their diet for 4 days, produced excitement initially and then depression a few hours or a day later (Schultheiss et al. 1995). Residual effects persisted for 1 day after the animals were returned to the control diet.

- seeds and foliage, mild digestive disturbance; possible neurologic effects

- digestive disturbance toxins; also enzyme-inhibiting alkaloids, calystegins, significance unknown

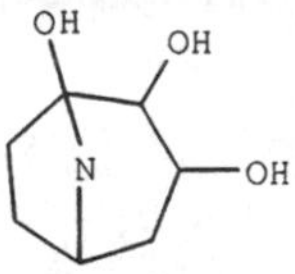

Calystegins

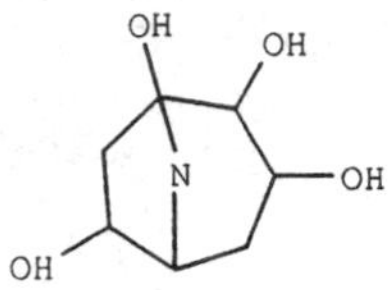

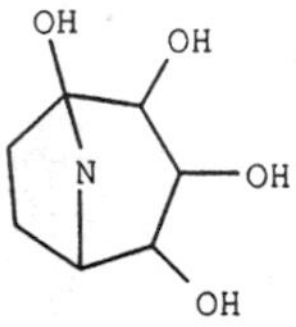

Disease Genesis—The purgative action of *C. arvensis* is reported to be due to a resinoid, convolvulin (Millspaugh 1974). However, some of the ulcerative effects on the digestive tract seen experimentally in mice are suggested to be due to the anticholinergic effects of tropane alkaloids (Schultheiss et al. 1995). It is of interest that although the species has not been associated with cerebellar disease, potent inhibitors of lysosomal carbohydrate metabolism have been isolated from the roots (Tepfer et al. 1988; Molyneux et al. 1993). These compounds, calystegins A and B, also isolated from *Solanum dimidiatum* as the probable cause of crazy cow syndrome, are potent competitive inhibitors of β-glucosidase and α-galactosidase. They act in a manner somewhat analogous to swainsonine in locoweed species of *Astragalus* (Molyneux et al. 1994). As a cause of disease, plant ingestion for a prolonged period would be necessary.

- diarrhea, colic

Clinical Signs—Livestock grazing extensively on the foliage of *C. arvensis* may develop manifestations indicative of changes in the digestive tract. With small to moderate amounts, diarrhea is the likely sign, whereas with large amounts stomach and intestinal motility may decrease, especially in horses. Secondary ulcerative lesions may be present, as indicated by experiments in mice (Schutheiss et al. 1995).

- mild reddening of gut mucosa; treatment not needed

Pathology and Treatment—Experimentally in mice the lesions included dehydration; reddening of the stomach mucosa, with erosions and ulcerations; and some accumulation of fluid and gas in the small intestine. There were also multiple foci of liver necrosis. Because intoxication with this genus has not been a problem, specific treatments have not been reported. Relief of the main symptoms is indicated, whether diarrhea or decreased digestive tract motility.

Ipomoea L.

- *I. arborescens*
  I. batatas
  I. carnea
  I. hederacea
  I. pes-caprae
  I. tricolor

  baros
  bay hops
  hierba de la India
  hierba de la raya
  morning glory
  palo blanco
  palo santo de castilla
  pata de vaca
  railroad vine
  sweet potato

Taxonomy and Morphology—A large and variable genus most abundant in tropical and subtropical regions, *Ipomoea,* commonly called morning glory, comprises 600–700 species (Austin & Huaman 1996). Its name comes from the Greek roots *ips* and *homoios,* meaning "worm" and "resembling," an obvious allusion to the appearance of the stems. Some taxonomists have attempted to divide it into 6–10 genera, but the taxa are essentially indistinguishable, and such a classification is impractical. The genus is valued for food, as ornamentals, and as soil binders. The large tubers of *I. batatas* (sweet potato) and several other species are an important carbohydrate source in the tropics. Prized for their showy flowers, several species with numerous cultivars are popular ornamentals and frequently used to hide or disguise the appearance of walls, fences, or dilapidated buildings. In North America, approximately 60 native and introduced species are present. Of toxicologic concern are the following:

I. arborescens (Humb. & Bonpl.) G.Don	palo blanco
I. batatas (L.) Lam.	sweet potato
I. carnea Jacq. (= *I. fistulosa* Mart.) (= *I. crassicaulis* Benth.)	tree morning glory, hierba de la India, shrubby morning glory, baros, palo santo de castilla
I. hederacea Jacq.	ivy-leaved morning glory
I. pes-caprae (L.) R.Br. (= *I. brasiliensis* [L.] Sweet)	beach morning glory, railroad vine, pata de vaca, goatsfoot morning glory, bay hops, soil morning glory, hierba de la raya

I. tricolor Cav. (= *I. violacea* L. in part)	numerous named varieties

■ herbs, vines, shrubs, small trees, sap sometimes milky; flowers showy; capsules ovoid to globose

The binomial *I. violacea* appears in some toxicological reports; however, the plants actually belong to *I. tricolor*. *Ipomoea violacea* is a name of disputed application and has been applied to plants of both *I. tricolor*, native to Mexico, and *I. tuba*, a species of the West Indies and the coastal United States.

Plants herbaceous vines or herbs or shrubs or small trees; annuals or perennials; sap sometimes milky. **Stems** climbing or prostrate to erect. **Leaves** simple or palmately compound; sessile or long petiolate; blades of various shapes; margins entire or toothed or lobed. **Inflorescences** solitary flowers or cymes; terminal or axillary. **Flowers** showy; usually large; salverform to funnelform or campanulate. **Sepals** imbricate, often unequal. **Petals** of various colors. **Pistils** 1; stigmas 2-or-3-lobed or capitate; styles 1; locules 1–4. **Capsules** ovoid to globose; valves 2–4. **Seeds** 1–4 or rarely 6; tan or brown to black; glabrous or tomentose.

Ipomoea hederacea

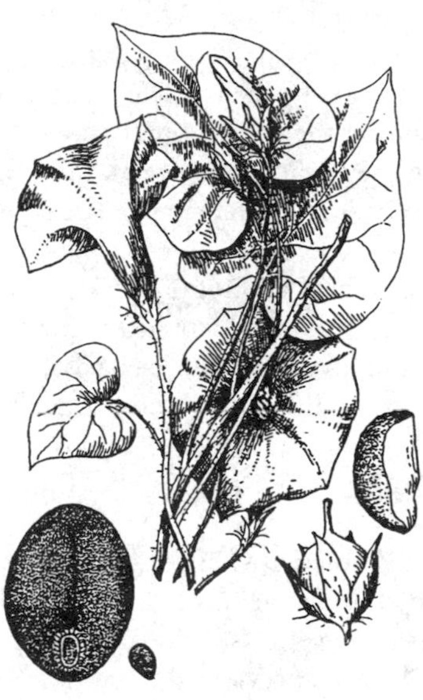

Ipomoea purpurea

Ipomoea arborescens

Ipomoea carnea

Ipomoea hederacea

Ipomoea tricolor

1. Plants shrubs or small trees. Stems erect; stout.
    2. Seeds with black hairs on all surfaces .... *I. carnea*
    2. Seeds with white hairs on dorsal angles .... *I. arborescens*
1. Plants herbs. Stems twining or trailing; slender.
    3. Peduncles and pedicels with reflexed hairs .... *I. hederacea*
    3. Peduncles and pedicels glabrous or with spreading hairs.
        4. Plants of beaches and sand dunes of Gulf Coast. Stems rooting at nodes. Sepal apices obtuse .... *I. pes-caprae*
        4. Plants in cultivation. Stems not rooting at nodes. Sepal apices acute or acuminate.
            5. Sepal margins scarious; white or pale .... *I. tricolor*
            5. Sepal margins herbaceous; green .... *I. batatas*

Distribution and Habitat—With greatest diversity in the Americas (Austin & Huaman 1996), species of *Ipomoea* occupy a variety of habitats and vegetation types. *Ipomoea batatas, I. carnea,* and *I. pes-caprae* are pantropical in distribution. *Ipomoea tricolor* and *I. arborescens* are native to Mexico and Central America. *Ipomoea hederacea* occurs from the southern United States to Argentina.

- 3 disease problems—
  digestive disturbance
  neurologic effects
  respiratory distress

Disease Problems—Although *Ipomoea* is perhaps best known toxicologically for the hallucinogenic effects associated with consumption of the seeds of some species, it is associated with three different disease syndromes—digestive, neurologic, and respiratory. Aspects of each of these problems are discussed separately in the following subsections.

Digestive Tract Effects

- some plant species, root and foliage, mild digestive disturbance, cause unknown

Disease Problems and Genesis—Both the root and the foliage of *I. pes-caprae* are reported to be purgatives (Watt & Breyer-Brandwijk 1962). Also considered toxic is *I. arborescens* (Standley 1926). However, many *Ipomoea* species are not toxic. For example, the large tuberous root of *I. leptophylla* (bush morning glory), the state flower of Montana, is reported to be very sweet and crisp but tender whether eaten raw, boiled, or roasted (Ebeling 1986). The fruit of *I. arborescens* and the leaves of *I. purpurea* also are eaten without harm.

In contrast to the relatively low toxicity of the leaves of North American species of *Ipomoea*, the foliage of *I. carnea* was the cause of severe anemia and degeneration of multiple organs in livestock in North Africa. Necroses of livers and kidneys were especially notable after repeated dosage of leaves to sheep and goats (Damir et al. 1987).

- diarrhea

Clinical Signs, Pathology, and Treatment—In the rare instance of intoxication due to ingestion of foliage, diarrhea is the most likely adverse effect. Ingestion of seeds by animals in quantities sufficient to cause adverse effects is unlikely. Such an intoxication would not be fatal, and treatment is not generally required.

Neurologic Effects

- humans, neurotoxic, seeds long used in rituals, hallucinogenic

  laboratory animals, large amounts, excitement; smaller amounts long-term, diarrhea, weight loss, death

Disease Problems—The seeds of *Ipomoea* are of primary toxicologic concern whether consumed in single large or multiple lower dosages. They produce both neurologic and purgative effects (O'Leary 1964) and have been used ritualistically for centuries by the Indians of Mexico (Schultes 1969; Diaz 1977). Ololiuqui, which is obtained from the seeds of *Rivea corymbosa*, another member of the family, was the primary hallucinogen, but the seeds of *I. tricolor* (including *I. violacea*) were used as well. They continue to be acclaimed and used as a hallucinogen in some communities.

With respect to the potency of the toxic varieties in humans, 20–50 seeds are considered a low dose, 100–150 seeds medium, and 200–500 seeds high (Der Marderosian 1967). The effects can be devastating in some instances, as exemplified by a case in which a young man suffered repeated bizarre experiences after chewing 300 seeds of the variety 'Heavenly Blue' and eventually committed suicide by intentionally crashing his car (Cohen 1964). Acute intoxication in mice is clearly neurologic and causes central excitation and motor dysfunction (Rice & Genest 1965). In contrast, long-term administration produces digestive rather than neurologic effects. A mixture of *I. hederacea* and *I. lacunosa* seeds fed to rats at 8% of the diet for 90 days caused a marked increase in mortality, diarrhea, lethargy, rough hair coat, and weight loss, but no distinctive changes in serum chemistry. The effects of 0.8% in the diet were minimal and those of 2.53% were perhaps marginal (Dugan & Gumbmann 1990). The large dosage required indicates that feeding grain contaminated with *Ipomoea* seeds is unlikely to cause intoxication in livestock.

Ipomoea muelleri, an Australian species commonly known as weir vine, has caused problems of a different nature. The foliage caused extensive losses of sheep. The problem was reproduced experimentally by feeding the relatively unpalatable plants for several weeks (Gardiner et al. 1965). The clinical signs produced were mainly neurologic, with ataxia, a jerky gait with knuckling of the hind limbs, exercise intolerance, and respiratory difficulties with excess frothy salivation. There were no distinctive lesions at necropsy.

Disease Genesis—The effects of ingestion of the seeds are due to the presence of ergot-type indole alkaloids similar to those produced by members of the fungal family Clavicipitaceae (see chapter 58). In the seeds of *I. tricolor,* the compounds are primarily lysergic acid amide (ergine), isolysergic acid amide (isoergine), and chanoclavine (Hofmann & Tscherter 1960). Additional alkaloids found in the seeds of other toxic species of *Ipomoea* include tryptophan, elymoclavine, penniclavine, ergometrine, and ergometrinine (Taber et al. 1963). The array of alkaloids is often greater in those species or varieties with higher alkaloid content (Witters 1975). Neurotoxic varieties of *I. tricolor,* such as 'Pearly Gates' not only have higher concentrations of alkaloids but also the toxic dosage is lower than those with smaller amounts of alkaloids (Rice & Genest 1965). The effects are quite similar to those caused by pure LSD. The alkaloids are heat labile, and when they are baked for an hour at 121°C, losses may be 25–40% when mixed with wheat flour or up to 70% when alone (Friedman & Dao 1990). A complete discussion of the structural differences and effects of these various alkaloids is presented in the treatment of *Festuca* in the Poaceae (see chapter 58).

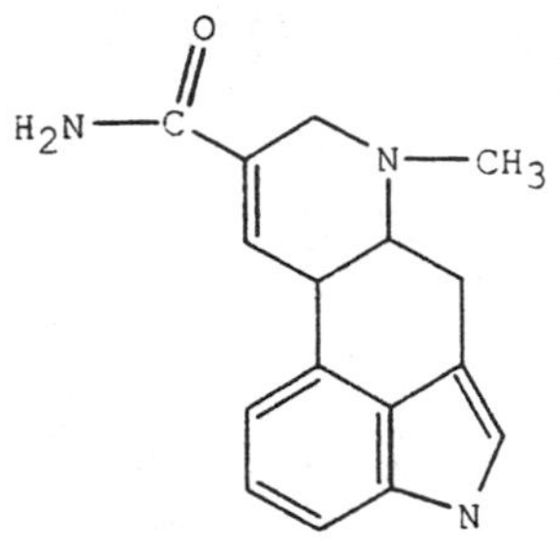

■ ergot-type neurotoxic indole alkaloids, lysergamide, elymoclavine, chanoclavine

Lysergamide
(Lysergic acid amide)

Elymoclavine

Chanoclavine I

Some species of *Ipomoea* have relatively low alkaloid content (table 25.1.), albeit there can be considerable variation among varieties of a species. Even though alkaloid content is low, this does not mean a lack of risk. Although both *I. hederacea* and *I. lacunosa* have low alkaloid content, they are capable of causing intoxication when consumed long-term as demonstrated in mice (Dugan & Gumbmann 1990). In addition, the array of compounds in a low-alkaloid species may be such that some of the more toxic alkaloids may be present in concentrations similar to those in the toxic varieties. Such is the case with ergonovine, which in some instances is in concentrations in wild species equivalent to those in the 'Heavenly Blue' variety of *I. tricolor* (Wilkinson et al. 1986). This may be significant because of the uterine stimulant effects of ergonovine.

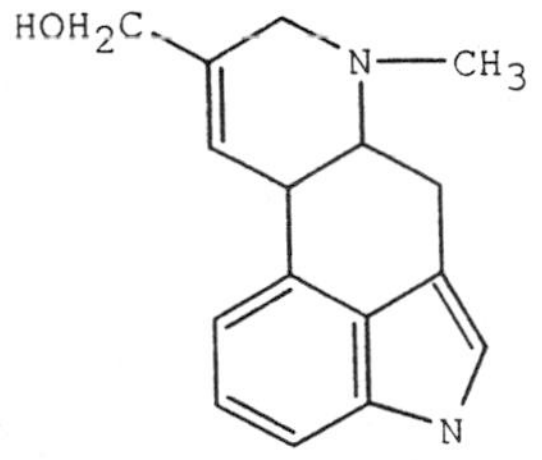

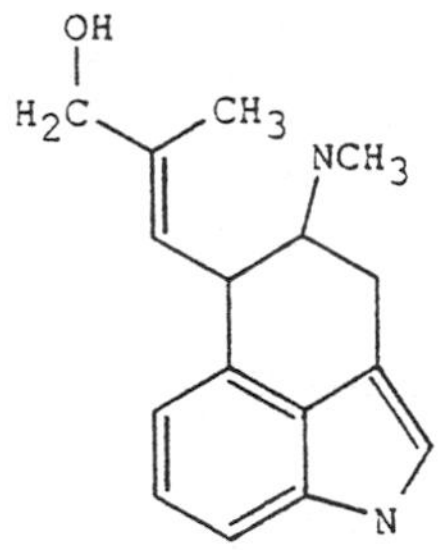

Table 25.1. Species of *Ipomoea* with low total or hallucinatory alkaloid content

I. alba	moonvine
I. coccinea	red morning glory
I. hederacea	ivy-leaved morning glory
I. hederifolia	Texas red morning glory
I. lacunosa	pitted morning glory
I. nil	Scarlett O'Hara
I. purpurea	tall morning glory
I. quamoclit	cypress vine morning glory
I. x*sloteri*	cardinal climber
I. trichocarpa	cotton morning glory
I. wrightii	palm-leaf morning glory

Sources: Genest & Sahasrabudhe 1966; Wilkinson et al. 1986, 1987.

Some of the more toxic horticultural varieties of *I. tricolor* are 'Heavenly Blue,' 'Pearly Gates,' 'Summer Skies,' 'Blue Star,' 'Flying Saucers,' and 'Wedding Bells' (Der Marderosian 1967). These varieties differ from one another mainly in flower color.

Conversely, varieties of other species low in alkaloid content include 'Candy Pink,' 'Crimson Rambler,' 'Darling,' 'Early Call,' 'Giant,' 'Imperialis,' 'Imperialis Mixed,' 'O'Hara,' 'Royal Marine,' and 'Scarlett O'Hara' (Der Marderosian 1967; Friedman et al. 1989).

■ presence of enzyme inhibitors in some plant species, causing metabolic effects

Some Old World species of *Ipomoea* contain calystegins A and B, similar to those described for *Convolvulus arvensis* and those isolated from *Solanum dimidiatum* as the probable cause of crazy cow syndrome. These potent alkaloidal competitive inhibitors of β-glucosidase and α-galactosidase act in a manner somewhat analogous to swainsonine, the α-mannosidase inhibitor present in locoweed species of *Astragalus* (Molyneux et al. 1994; De Balogh et al. 1998).

Although North American species of *Ipomoea* do not appear to present a significant risk to livestock, species from Australia and Africa, *I. muelleri* and *I. carnea,* are a risk. These species contain the enzyme inhibitors swainsonine and calystegin and cause neurotoxic effects with signs such as ataxia, tremors, nystagmus, aimless walking, awkward stiff posture, and perhaps paralysis. The main pathologic lesions—degeneration and loss of Purkinje cells in the cerebellum and less prominent cytoplasmic vacuolation-type degenerative lesions in the cerebrum, spinal cord, liver, and kidneys—are consistent with a lysosomal storage disease as may be expected with calystegins from *Solanum* or swainsonine from *Astragalus* (Molyneux et al. 1995; Srilatha et al. 1997; De Balogh et al. 1998).

■ humans, seeds, auditory and visual distortion, mood elevation, nausea, sluggishness

Clinical Signs—Because animal consumption of the seeds in high dosage is unlikely and the alkaloid levels are typically too low in foliage to cause neurotoxicity, clinical signs are not seen. Human consumption of the seeds is generally intentional. Ingestion of low dosage of seeds of the hallucinogenic species causes slight increases in awareness and other sensory effects without visual distortion. Medium dosage results in auditory and visual hallucinations with intense and/or distorted imagery and mood elevation. High dosage causes similar but more sustained effects and supposedly increased philosophical insight. Side effects include nausea, sluggishness, and coldness of extremities.

■ effects transient; nursing care

Pathology and Treatment—The neurologic effects produced by *Ipomoea* are not lethal, except for accidental death resulting from the individual's behavior while hallucinating. Treatment is directed toward relief of the neurotoxic signs.

Respiratory Effects

■ livestock, especially cattle, acute respiratory distress, mold-damaged sweet potatoes

Disease Problems—A respiratory disease associated with rotten sweet potatoes has been recognized in the southeastern United States since the early 1900s (Hansen 1928), when it was common practice, especially in the spring, to feed the spoiled tubers to livestock. The disease also has been observed in other areas of the world (Hill & Wright 1992). When ingested by cattle, the mold-damaged sweet potatoes cause an acute respiratory distress syndrome (ARDS) identical to that produced by *Perilla frutescens* in the Lamiaceae and by other forages such as rape, kale, grasses, and legumes (Hansen 1928; Monlux et al. 1953; Peckham et al. 1972). The disease is abrupt in onset and often fatal. Cattle are most commonly affected, but other livestock are also susceptible. A complete discussion of this disease is presented in the treatment of the Poaceae (see chapter 58) and the treatment of *Perilla frutescens* (see chapter 45).

Disease Genesis—The respiratory syndrome is associated with a lung edema factor produced when the sweet potatoes are infected with a mold such as *Fusarium solani*

(Wilson et al. 1971; Peckham et al. 1972). The cause is a series of furanoterpenoids, mainly 4-ipomeonol (Wilson et al. 1970; Boyd & Wilson 1972; Boyd et al. 1974, 1975; Wilson & Burka 1983). In the infected tuber, ipomeamarone, a stress metabolite or phytoallexin, is formed initially by the tuber in the presence of the fungus. Ipomeamarone, which is hepatotoxic, is then biotransformed by the fungus to the pneumotoxic ipomeanol. Bioactivation of 4-ipomeanol, primarily in the Clara cells and macrophages of the lung rather than in cells in the liver, results in the formation of reactive metabolite intermediates (Boyd 1977). These reactive metabolites bind to proteins in the lungs and cause tissue destruction, especially of type I pneumocytes and epithelial cells (Boyd & Burka 1978). Under the stimulus of factors released by macrophages, type II pneumocytes proliferate to replace the lost type I cells and in the process cause formation of foci with massive increase in cellularity. The disease and its pathophysiology are similar to those caused by *Perilla* ketone and 3-methylindole (Slaughter et al. 1983). An apparently identical disease is seen with scarlet runner beans infected with *Fusarium semitectum* (Linnabary & Tarrier 1988).

- furanoterpenoids, 4-ipomeonol, metabolized in lungs to destructive toxicants

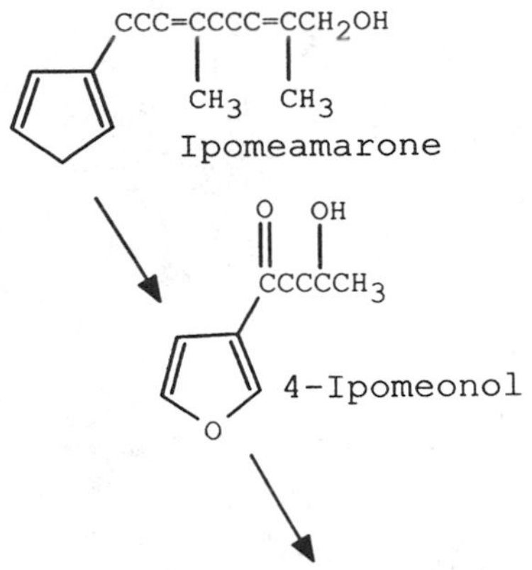

CCC=CCCC=CCH$_2$OH
CH$_3$ CH$_3$
Ipomeamarone

O OH
CCCCCH$_3$
4-Ipomeonol

Toxic metabolite

Clinical Signs—Ingestion of mold-damaged sweet potatoes results in abrupt onset of labored, open-mouthed, head-extended respiration, with foamy saliva stringing from the mouth. Distress is obvious, and animals may become difficult to handle and sometimes are belligerent. Death often occurs within a few days after onset of signs.

- severe respiratory distress, open-mouthed breathing, head extended, often fatal

Pathology—Death due to respiratory distress is associated with distended lungs that fail to collapse on opening of the chest and exhibit areas of firmness and froth-filled airways. The interlobular septa and subpleura may be distended with fluid or gelatinous edema and air bubbless. Cut surfaces will ooze fluid and reveal patchy marbled areas of firm and reddened lobules with prominent edema fluid in connective tissue and septa. Microscopically, alveolar epithelialization and macrophage accumulation, and hyaline membrane formation, are typical. These conditions will be shown by alveoli lined by basophilic type II pneumocytes with numerous mitotic figures and thickened septal walls containing infiltrative mononuclear cells. The alveolar lumens will be filled with fibrinous exudate.

- gross pathology, lungs heavy, wet, rubbery, distended; edema, frothy exudate in airways; liver pale, friable, mottled

  microscopic, alveoli filled with serous exudate, edema, and air pockets in interlobular septa; alveolar epithelialization

Treatment—The severity of the respiratory distress may suggest therapeutic intervention with atropine or bronchodilators. It is important to avoid stressing the animals when handling them.

- avoid stress

References

Austin DF, Huaman Z: A synopsis of *Ipomoea* (Convolvulaceae) in the Americas. *Taxon* 45;3–38, 1996

Boyd MR: Evidence for the Clara cell as a site of cytochrome P450-dependent mixed-function oxidase activity in lung. *Nature* 269;713–715, 1977

Boyd MR, Burka LT: In vivo studies on the relationship between target organ alkylation and the pulmonary toxicity of a chemically reactive metabolite of 4-ipomeanol. *J Pharmacol Exp Ther* 207;687–697, 1978

Boyd MR, Wilson BJ: Isolation and characterization of 4-ipomoenol, a lung-toxic furanoterpenoid produced by sweet potatoes *(Ipomoea batatas)*. *J Agric Food Chem* 20;428–430, 1972

Boyd MR, Burka LT, Harris TM, Wilson BJ: Lung-toxic furanoterpenoids produced by sweet potatoes *(Ipomoea batatas)* following microbial infection. *Biochim Biophys Acta* 337;184–195, 1974

Boyd MR, Burka LT, Wilson BJ: Distribution, excretion, and binding of radioactivity in the rat after intraperitoneal administration of the lung-toxic furan [14C] 4-ipomeanol. *Toxicol Appl Pharmacol* 32;147–157, 1975

Cohen S: Suicide following morning glory seed ingestion. *Am J Psychiatry* 120;1024–1025, 1964

Damir HA, Adam SEI, Tartour G: The effects of *Ipomoea carnea* on goats and sheep. *Vet Hum Toxicol* 29;316–319, 1987

De Balogh KKIM, Dimande AP, Van Der Lugt JJ, Molyneux RJ, Naude TW, Welman WG: *Ipomoea carnea:* the cause of a lysosomal storage disease in goats in Mozambique. In *Toxic Plants and Other Natural Toxicants,* Garland T, Barr AC eds, New York, pp428–434, 1998

Der Marderosian A: Psychotomimetic indoles in the Convolvulaceae. *Am J Pharm* 139;19–26, 1967

Diaz JL: Ethnopharmacology of sacred psychoactive plants used by Indians of Mexico. *Annu Rev Pharmacol Toxicol* 17;647–675, 1977

Dugan GM, Gumbmann MR: Toxicological evaluation of morning glory seed: subchronic 90-day feeding study. *Food Chem Toxicol* 28;553–559, 1990

Ebeling W: *Handbook of Indian Foods and Fibers of Arid America.* Univ California Press, Berkeley, p757, 1986

Friedman M, Dao L: Effect of autoclaving and conventional and microwave baking on the ergot alkaloid and chlorogenic acid contents of morning glory (*Ipomoea tricolor* Cav.cv.) heavenly blue seeds. *J Agric Food Chem* 38;805–808, 1990

Friedman M, Dao L, Gumbmann MR: Ergot alkaloid and chlorogenic acid content in different varieties of morning-glory (*Ipomoea* spp.) seeds. *J Agric Food Chem* 37;708–712, 1989

Gardiner MR, Royce R, Oldroyd B: *Ipomoea muelleri* intoxication of sheep in western *Australia. Br Vet J* 121;272–277, 1965

Genest K, Sahasrabudhe MRS: Alkaloids and lipids of *Ipomoea, Rivea,* and *Convolvulus* and their application to chemotaxonomy. *Econ Bot* 20;416–428, 1966

Hansen AA: Potato poisoning. *North Am Vet* 9;31–34, 1928

Hill BD, Wright HF: Acute interstitial pneumonia in cattle associated with consumption of mould-damaged sweet potatoes *(Ipomoea batatas). Aust Vet J* 69;36–37, 1992

Hofmann A, Tscherter H: Isolation of the lysergic acid-alkaloid from the ancient Aztec drug ololiuqui (*Rivea corymbosa* [L.] Hall.f.). *Experientia* 16;414, 1960

Linnabary RD, Tarrier MP: Acute bovine pulmonary emphysema caused by the fungus *Fusarium semitectum. Vet Hum Toxicol* 30;255– 256, 1988

Millspaugh CF: *American Medicinal Plants.* Dover, New York, pp479–481, 1974 (reprint from 1892)

Molyneux RJ, Pan YT, Goldmann A, Tepfer DA, Elbein AD: Calystegins, a novel class of alkaloid glycosidase inhibitors. *Arch Biochem Biophys* 304;81–88, 1993

Molyneux RJ, James LF, Ralphs MH, Pfister JA, Panter KE, Nash RJ: Polyhydroxy alkaloid glycosidase inhibitors from poisonous plants of global distribution: analysis and identification. In *Plant-Associated Toxins: Agricultural, Phytochemical, and Ecological Aspects,* Colegate SM, Dorling PR eds, CAB International, Wallingford, UK, pp107–112, 1994

Molyneux RJ, McKenzie RA, O'Sullivan BM, Elbein AD: Identification of the glycosidase inhibitors swainsonine and calystegin B2 in weir vine (*Ipomoea* sp. Q6 [aff. Calobra]) and correlation with toxicity. *J Nat Prod* 58;878–886, 1995

Monlux W, Fitte J, Kendrick G, Dubuisson H: Progressive pulmonary adenomatosis in cattle. *Southwest Vet* 6;267–269, 1953

O'Leary SB: Poisoning in man from eating poisonous plants. *Arch Environ Health* 9;216–242, 1964

Peckham JC, Mitchell FE, Jones OH, Doupnik B Jr: Atypical interstitial pneumonia in cattle fed mouldy sweet potatoes. *J Am Vet Med Assoc* 160;169–172, 1972

Rice WB, Genest K: Acute toxicity of extracts of morning-glory seeds in mice. *Nature* 207;302–303, 1965

Schultes RE: Hallucinogens of plant origin. *Science* 163;245–254, 1969

Schultheiss PC, Knight AP, Traub-Dargatz JL, Todd FG, Stermitz FR: Toxicity of field bindweed *(Convolvulus arvensis)* to mice. *Vet Hum Toxicol* 37;452–454, 1995

Slaughter SR, Statham CN, Philpot RM, Boyd MR: Covalent binding of metabolites of 4-ipomoenol to rabbit pulmonary and hepatic microsomal proteins and to the enzymes of the pulmonary cytochrome P-450-dependent monooxygenase system. *J Pharmacol Exp Ther* 224;252–257, 1983

Srilatha CH, Gopal Naidu NR, Rama Rao P: Pathology of *Ipomoea carnea* toxicity in goats. *Indian J Anim Sci* 67;253–254, 1997

Taber WA, Vining LC, Heacock RA: Clavine and lysergic acid alkaloids in varieties of morning glory. *Phytochemistry* 2;65–70, 1963

Tepfer D, Goldmann A, Pamboukdjian N, Maille M, Lepingle A, Chevalier D, Denarie J, Rosenberg C: A plasmid of *Rhizobium meliloti* 41 encodes catabolism of two compounds from root exudate of *Calystegium sepium. J Bacteriol* 170;1153–1161, 1988

Watt JM, Breyer-Brandwijk MG: *The Medicinal and Poisonous Plants of Southern and Eastern Africa,* 2d ed. Livingston, Edinburgh, pp306–310, 1962

Wilkinson RE, Hardcastle WS, McCormick CS: Ergot alkaloid contents of *Ipomoea lacunosa, I. hederacea, I. trichocarpa,* and *I. purpurea* seed. *Can J Plant Sci* 66;339–343, 1986

Wilkinson RE, Hardcastle WS, McCormick CS: Seed ergot alkaloid contents of *Ipomoea hederifolia, I. quamoclit, I. coccinea,* and *I. wrightii. J Agric Food Sci* 39;335–339, 1987

Wilson BJ, Burka LT: Sweet potato toxins and related toxic furans. In *Handbook of Natural Toxins,* vol 1, *Plant and Fungal Toxins,* Keeler RF, Tu AT eds, Marcel Dekker, New York, pp3–41, 1983

Wilson BJ, Yang DTC, Boyd MR: Toxicity of mold-damaged sweet potatoes *(Ipomoea batatas). Nature* 227;521–522, 1970

Wilson BJ, Boyd MR, Harris TM, Yang DTC: A lung oedema factor from moldy sweet potatoes *(Ipomoea batatas). Nature* 231;52–53, 1971

Witters WL: Extraction and identification of clavine and lysergic acid alkaloids from morning glories. *Ohio J Sci* 75;198–201, 1975

26 Coriariaceae DC.

■ Coriaria Family
Coriaria

A monogeneric family of warm-temperate regions, the Coriariaceae, or coriaria family, is a taxonomic enigma. It is not closely related to other families, and its taxonomic position thus is uncertain. In stem anatomy and seed coat morphology it somewhat resembles the Ranunculaceae, while chemically it is like the Menispermaceae. It also has an interesting, discontinuous geographical distribution; species are present in Central and South America, the western Mediterranean region, New Zealand, the Himalayan Mountains, eastern Asia, New Guinea, and Japan. The family is of little economic importance. Plants are rich in tannins and are used for curing leather and making indelible inks, and when crushed in water, the fruits of one species produce a fly poison. Plants of several species are planted as ornamentals.

Plants shrubs or small trees or subshrubs or herbs; perennials; bearing perfect flowers or monoecious or polygamomonoecious. **Stems** erect or ascending; 60–400 m tall; branches angular, often arranged so as to form flat frondlike sprays; lenticels numerous, corky. **Leaves** simple; opposite or rarely whorled; venation palmate or convergent-parallel; margins entire; stipules minute, caducous. **Inflorescences** racemes; terminal or axillary. **Flowers** perfect or imperfect; perianths in 2-series. **Sepals** 5; fused. **Corollas** radially symmetrical. **Petals** 5; free; shorter than sepals; enlarging and becoming fleshy in fruit; green at anthesis; red or dark brown or black or purple in fruit. **Stamens** 10; in 2 whorls; free or fused to petals. **Pistils** 5–10; simple; carpels 1; stigmas 1, diffuse; styles 1, elongate; ovaries superior; locules 1; placentation parietal; ovules 1. **Fruits** achenes; flattened; enclosed by fleshy petals.

Coriaria L.

■ *C. japonica*
C. microphylla
C. ruscifolia
C. terminalis

■ herbs, shrubs, trees; leaves simple, opposite or whorled; petals 5, red, dark brown, black, purple

Taxonomy and Morphology—The only genus of the family, *Coriaria* comprises 5–8 species. From the Latin *corium,* meaning "hide" or "leather," its name alludes to the use of some species in tanning. Plants are grown as ornamentals for their autumn foliage and pendant or arching racemes of brightly colored, fleshy flowers when in fruit. Species cultivated as ornamentals in North America include the following:

C. *japonica* A.Gray

C. *microphylla* Poir.

C. *ruscifolia* L.

C. *terminalis* Hemsl.

The morphological characters of the genus are those of the family and thus not repeated here.

1. Plants herbs or subshrubs. Rhizomes present . . . . *C. terminalis*
1. Plants shrubs or small trees. Rhizomes absent.
   2. Leaves 1–2 cm long. Fruiting corollas purplish or dark brown . . . . *C. microphylla*
   2. Leaves 2.5–9 cm long. Fruiting corollas red or purple-black or black.
      3. Leaf apices acuminate. Racemes 4–6.5 cm long. Flowers imperfect . . . . *C. japonica*
      3. Leaf apices acute. Racemes 10–30 cm long. Flowers perfect . . . . *C. ruscifolia*

Coriaria terminalis

Distribution and Habitat—As noted in the family comments, species of *Coriaria* collectively have an unusual interrupted distributional pattern. Their use as ornamentals in North America is limited. Because it is the most frost hardy, *C. terminalis*, native to western China, is the species most likely to be encountered.

Coriaria myrtifolia

- uncommon ornamentals
- neurotoxic, fruits, foliage, mainly humans and pets
- picrotoxinin-like compounds, coriamyrtin, tutin, GABA inhibitory

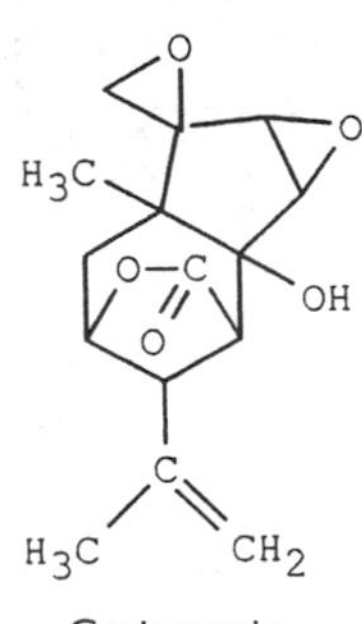

Coriamyrtin

Disease Problems—Species of *Coriaria* are recognized to cause deaths in humans, primarily from ingestion of decoctions (Swanson & Chen 1936). The fruits have been used in Mexico to poison dogs but also have caused accidental deaths in children (Standley 1926). Chopra and co-workers (1940) stated that an Indian species is toxic and that the seeds are always avoided when the fruit is eaten. *Coriaria myrtifolia* is reported to be a continuing toxicity risk to goats and to a lesser extent other livestock in southern France, especially the Mediterranean region (Faliu et al. 1985). The unripe fruit and foliage of a New Zealand species have been a continuing risk for both humans and livestock (Graham & Cartridge 1961). Wherever species of *Coriaria* are present, there is a serious neurotoxicosis risk (Jean-Blain 1981); however, in North America they are generally of little risk to animals because of their sparse use as ornamentals.

Bees feeding on these plants produce toxic honey. Interestingly, the toxicity is due not to the nectar but rather to the honeydew produced by a leafhopper parasite, *Scolypopa australis* (Patwardhan & White 1973; McBarron 1976).

Disease Genesis—Although the leaves of *Coriaria* are strongly tanniferous and have been used extensively for tanning, there seems to be little risk of intoxication due to these polyphenolic complexes. Toxicity is due to coriamyrtin present in the foliage. It is a convulsant very similar to picrotoxin but of greater potency (Marshall 1913; Bleckwenn et al. 1940). Experimentally, it is several times more toxic than picrotoxin when given either i.v. or s.c. to mice or rabbits; a lethal dose is less than 1 mg/kg b.w. i.v. (Swanson & Chen 1936). Although coriamyrtin is more potent, its activity is short, only a few minutes in duration, and it is therefore of less value as an analeptic than is picrotoxin (Maloney 1936a; 1936b). Tutin, a similar convulsant, is found in the foliage but not in the fruit pulp or juice of *C. ruscifolia*, a New Zealand species (Fitchett & Malcolm 1909). Coriamyrtin and tutin are similar structurally to picrotoxinin obtained from *Anamirta cocculus* in the Menispermaceae, and presumably all inhibit GABA. The effects are, thus, unchecked expression of excitatory muscular activity. Coriamyrtin, corianin, coriatin (8,9-dihydro-8-hydroxy coriamyrtin), and tutin have been isolated from *C. microphylla* (Aguirre-Galviz & Templeton 1990). These convulsants are a risk for all animal species, as illustrated by the appearance of neurotoxic signs in two dogs in New Zealand that consumed leaves of *Coriaria* from the ruminal contents of a lamb dying from the disease. The presence of tutin in the vomitus of the dogs was used to confirm the diagnosis (Graham & Cartridge 1961).

Humans are at risk via honey contamination. The principal toxicants in honey produced by bees feeding on the honeydew on leaves of *C. ruscifolia* are an insect

- humans, possible honey contaminate, low risk

metabolite of tutin called hyenanchin (hydroxylated tutin, or mellitoxin), and lesser amounts of tutin (Patwardhan & White 1973). Although hyenanchin is 5–10 times less toxic than tutin, severe intoxications in humans were said to result from ingestion of as little as 1 teaspoon of toxic honey. There seems to be little risk from toxic honey in North America because of the limited number of plants present in any one area.

- twitching, tremors, jerky body movements; higher dosage, severe tonic seizures

Clinical Signs—In the rare event of toxicity due to *Coriaria,* the signs are distinctly neurologic, with twitching, tremors, and jerky body movements at low dosage. There also will be a decrease in appetite, increased salivation, and sometimes bloat. At higher dosage, there are severe tonic seizures of several minutes duration, with the animal's head twisted backward and its legs extended.

- no lesions

Pathology—Because this is a peracute toxicity, the only changes to be expected are scattered splotchy hemorrhages on the surfaces of the abdominal viscera and heart, and congestion and edema of various abdominal organs and the meninges.

- treat with sedation, diazepam, pentobarbital

Treatment—Control of the seizures is paramount and need only be maintained for a short period, probably an hour or less. Diazepam is a specific antagonist of the GABA inhibitory effects of these convulsants. A single administration of 20 mg/kg b.w. should suffice. Alternatively, pentobarbital (20–25 mg/kg i.v.); xylazine (0.05–0.1 mg/kg i.v. or 0.1–0.2 mg/kg i.m.); or acepromazine (0.1–0.2 mg/kg i.v. or 0.2–0.4 mg/kg i.m.) may be used.

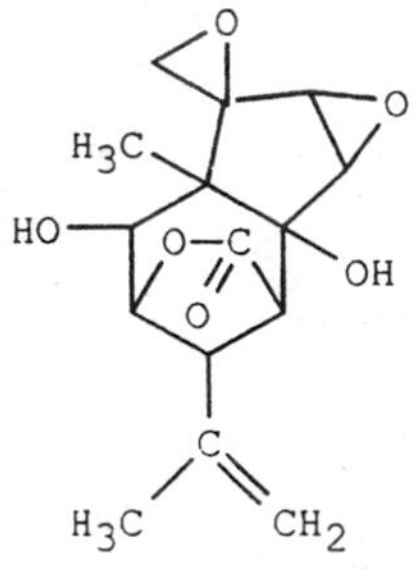

Tutin

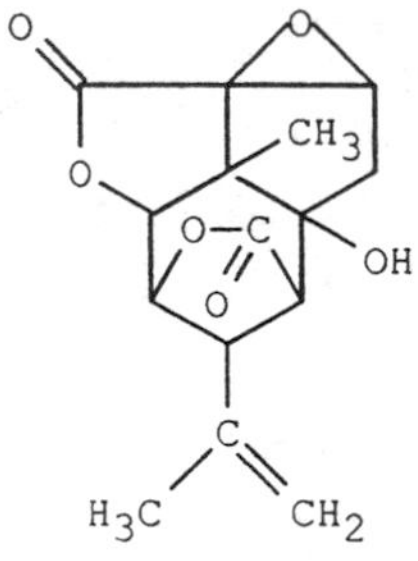

Picrotoxinin

References

Aguirre-Galviz LE, Templeton W: Toxic sesquiterpenoid lactones from the leaves of *Coriaria microphylla. Planta Med* 56;244, 1990

Bleckwenn WJ, Hodgson ER, Herwick RP: A clinical comparison of picrotoxin, metrazol, and coriamyrtin used as analeptics and as convulsants. *J Pharmacol Exp Ther* 69;81–88, 1940

Chopra RN, Badhwar RL, Ghosh S: *Poisonous Plants of India,* vol 1. Academic Publ, Jaipur, India, 1940 (reprint 1984)

Faliu L, Puyt JD, Jean-Blain C: Poisonous plants: *Coriaria myrtifolia. Point Vet* 17;531–532, 1985

Fitchett F, Malcolm J: On the physiological action of tutin. *Q J Exp Physiol* 2,335–362, 1909

Graham JM, Cartridge MEA: Tutu poisoning in dogs. *N Z Vet J* 9;45, 1961

Jean-Blain C: Tableau des baies et fruits bacciformes les plus couramment rencontrés avec indication le leur toxicite. *Notes Toxicol Vet* 4;67–78, 1981

Maloney AH: Barbiturate-coriamyrtin antagonism. *Proc Soc Exp Biol Med* 34;59–593, 1936a

Maloney AH: Studies on the pharmacological action of coriamyrtin. *J Pharmacol Exp Ther* 57;361–375, 1936b

Marshall CR: The pharmacological action of coriamyrtin. *J Pharmacol Exp Ther* 4;135–150, 1913

McBarron EJ: *Medical and Veterinary Aspects of Plant Poisons in New South Wales.* Department of Agriculture, New South Wales, Glenfield, Australia, 1976

Patwardhan VN, White JW: Problems associated with particular foods. In *Toxicants Occurring Naturally in Foods,* 2d ed, National Research Council, Washington, DC, pp477–507, 1973

Standley PC: *Trees and Shrubs of Mexico,* parts 1–5. U.S. National Museum, Washington, DC, 1926

Swanson EE, Chen KK: The pharmacological action of coriamyrtin. *J Pharmacol Exp Ther* 57;410–418, 1936

Crassulaceae DC.

27

Comprising 25–35 genera and 900–1,500 species, the Crassulaceae, or stonecrop family, is of limited toxicologic importance except for several genera in South Africa, the center of species diversity. Also known as the orpine family, it is cosmopolitan with the exception of Australia and Polynesia and is most common in arid, temperate or warm-temperate regions. The succulent habit and crassulacean acid metabolism are adaptations to the dry habitat.

■ Stonecrop Family

Cotyledon
Kalanchoe

QUESTIONABLE
Crassula
Sedum

Members of the family are of considerable interest as ornamentals because of their succulence and minimal water requirements. *Sedum,* the stonecrops, and *Crassula,* the jade plants, have long been popular; *Kalanchoe,* the kalanchoes, is now becoming so as well. Of considerable toxicologic importance in South Africa, *Cotyledon* and *Tylecodon* are seldom encountered as ornamentals in North America. In North America, approximately 15 genera and 117 species, both native and introduced, are present. Toxicologic problems are associated primarily with 2 genera.

Plants herbs or subshrubs or small shrubs; annuals or perennials; succulent; glabrous. **Leaves** simple; alternate or opposite or rarely whorled; sessile or petiolate; blades of various shapes; venation not apparent; stipules absent. **Inflorescences** cymes or solitary flowers; terminal or axillary. **Flowers** perfect or rarely imperfect; perianths in 2-series. **Sepals** 4 or 5, or rarely 3 or 6; fused slightly at bases. **Corollas** radially symmetrical. **Petals** 4 or 5, or rarely 3 or 6; free or rarely fused; white or green or pink or purple or yellow. **Stamens** 4 or 5 or 8 or 10; in 1 or 2 whorls. **Pistils** 4 or 5, equal in number to number of petals; simple, carpels 1; stigmas 1; styles 1, slender or short, subulate; ovaries superior; locules 1; placentation parietal; nectaries 4 or 5; small; scalelike, subtending each pistil. **Fruits** follicles or aggregates of follicles; loculicidal. **Seeds** 1 to numerous; small.

■ herbs, shrubs, subshrubs; leaves simple; petals white, green, pink, purple, or yellow; fruits follicles

Cotyledon L.

Taxonomy and Morphology—Of considerable importance as a toxic genus in South Africa, *Cotyledon* comprises 9–40 species native to Africa and southern Arabia. Its name is derived from the Greek *kotyle,* meaning "cavity," which alludes to the deeply cupped leaves. In North America, only 1 species is of toxicologic interest:

C. orbiculata L. pig's ears

■ *C. orbiculata*

pig's ears

■ erect shrubs; leaves opposite, gray-green; petals 4 or 5, fused, orange, deep red, or purple

Plants shrubs; herbage glaucous; gray-green. **Stems** erect; 50–150 cm tall; branched below. **Leaves** opposite; sessile; terete or plano-convex; blades obovate to linear; margins typically red or purple. **Inflorescences** contracted, paniculate cymes. **Flowers** numerous; nodding; campanulate. **Sepals** 5; fused. **Petals** 5; fused; 4; fused, with reflexed lobes; orange to deep red or purple. **Stamens** 10; in 2 whorls; filaments hairy. **Pistils** 5. **Fruits** aggregates of follicles.

Cotyledon orbiculata

■ ornamental

Distribution and Habitat—Introduced from South Africa, the single species represented in North America is limited to warmer areas in the southern United States and in Mexico, where it is not subject to killing frosts. *Cotyledon orbiculata* has escaped and become established along the coasts of California.

■ cardiotoxic, all plant parts, bufadienolides, digestive and cardiac disturbance

Disease Problems and Genesis—Two cases of toxicity in California due to *C. orbiculata* have been reported (Fuller & McClintock 1986). In each incident, plants were trimmed after flowering, and the trimmings discarded in a pasture where they were consumed by sheep. In one case 3 sheep died; in the other 21 sheep died. Toxicity is due to bufadienolide cardiotoxins (cotyledoside), which have structures and effects similar to the toxicants present in *Kalanchoe* (Naude & Anitra Shultz 1982; Kellerman et al. 1988). Ingestion of large amounts may produce an acute disease, whereas smaller amounts eaten several times result in subacute or chronic problems.

■ acute disease, abrupt onset, diarrhea, weakness, respiratory difficulty, cardiac arrhythmias

subacute/chronic, slower onset, neurologic effects, postural changes, seizures

Clinical Signs, Pathology, and Treatment—As described in South African cases, the signs of acute intoxication due to *C. orbiculata* are typical of cardiac glucosides—ruminal stasis, diarrhea, weakness, respiratory difficulty, cardiac arrhythmias, and electrocardiographic changes. When the intoxication is subacute, the cardiac changes are less evident and neurologic signs more pronounced. Postural changes are obvious, with the animal's back arched and its neck extended. In addition, the animal may be dehydrated, unable to swallow, and very susceptible to stress such as forced movement. In the chronic form, the postural effects are more marked, especially paresis, and may persist for weeks or months. Other neurologic manifestations such as spasms or seizures may also be noted (Kellerman et al. 1988). Interestingly, putative closely related species in the Western Hemisphere, e.g., those of *Echeveria,* have not been implicated in toxicity problems. Pathology and treatment are similar to those for *Kalanchoe,* mainly correction of the cardiac effects.

Kalanchoe Adans.

■ *K. beharensis*
K. blossfeldiana
K. daigremontiana
K. delagonensis
K. fedtschenkoi
K. pinnata
K. prolifera

air plant
chandelier plant
devil's backbone
feltbush
floppers
kalanchoe
lavender scallops
life plant
love plant
Mexican hat plant
mother-of-millions
Palm Beach bells
velvetleaf

Taxonomy and Morphology—Comprising 125–200 species, *Kalanchoe* is primarily a genus of Madagascar and southern Africa. Several species have become naturalized in Australia and the Americas. In North America, introduced species are grown as houseplants in the north and out of doors in the south. The genus name is derived from a Chinese vernacular name for one of the species. It is divided into three sections, which were treated in the past as three separate genera—*Kalanchoe, Bryophyllum,* and *Kitchingia.* Currently, one broadly circumscribed genus is recognized. In North America, several species with numerous cultivars have gained popularity. Some are grown for their succulent foliage, others for their showy flowers, and others as curiosities because they produce numerous plantlets. Commonly encountered, representative species include the following:

K. beharensis Drake	feltbush, velvetleaf
K. blossfeldiana Poelln.	florist's kalanchoe, Palm Beach bells
K. daigremontiana Hamet & Perrier	mother-of-millions, devil's backbone, Mexican hat plant
K. delagonensis Eckl. & Zeyh. (= *K. tubiflora* [Harv.] Hamet) (= *K. verticillata* Elliot)	mother-of-millions, chandelier plant
K. fedtschenkoi Hamet & Perrier	South American air plant, lavender scallops

K. pinnata (Lam.) Pers.	air plant, sinalo todo, floppers, life plant, Mexican love plant
K. prolifera (Bowie) Hamet	

A hybrid of *K. daigremontiana* and *K. delagonensis* also is cultivated. Most toxicologic reports use the binomial *K. tubiflora,* which is now regarded as a taxonomic synonym of *K. delagonensis.* Other common names also applied to species of the genus include cathedral bells, live-forever, and panda plant.

Plants herbs or subshrubs; producing or not producing plantlets on leaf margins and in inflorescences. **Stems** erect; 20–100 cm tall; often woody at bases. **Leaves** opposite or whorled; sessile or petiolate; flat or terete; blades linear to broadly ovate; margins entire to pinnatifid; bases often clasping stems. **Inflorescences** paniculate cymes; few to many flowered. **Flowers** showy; salverform or urceolate. **Sepals** 4; fleshy. **Petals** 4; fused; white or green or violet or yellow or red; lobes spreading or recurved. **Stamens** 8; in 2 whorls. **Pistils** 4. **Fruits** follicles.

- succulent, erect herbs or subshrubs; leaves opposite or whorled; flowers showy; petals 4, fused, white, green, violet, red, or yellow

Because the species cited are simply representative of the numerous cultivated introductions of the genus, a key for their identification is not presented. Rather, the reader is encouraged to use standard horticultural references if positive identification is needed.

Distribution and Habitat—All species of the genus in North America are introductions from Madagascar and Africa. In the southern United States and in Mexico, where they are not subject to killing frosts, they may naturalize. *Kalanchoe delagonensis* (reported as *K. verticillata*) is established in the Rio Grande Valley (Correll & Johnston 1979).

- cultivated ornamentals

Disease Problems—Species of *Kalanchoe* are well known as toxic to livestock in South Africa and Australia, where they grow as natives or are naturalized. They are eaten when other foods are scarce, and animals, primarily cattle, die suddenly or after a period of persistent diarrhea. Studies in Australia indicate that *K. delagonensis* (reported as *K. tubiflora*) and the hybrid between *K. daigremontiana* and *K. delagonensis* are the most toxic, while *K. daigremontiana, K. prolifera, K. pinnata,* and *K. fedtschenkoi* are less of a risk. The lethal dosage for *K. delagonensis* in calves was estimated to be 7 g flowers/kg b.w. and 40 g leaves wet weight/kg b.w. (McKenzie & Dunster 1986). In Africa, additional toxic species include *K. lanceolata, K. paniculata, K. rotundifolia,* and *K. thyrsiflora* (Anderson et al. 1983). Based on North American studies using chickens, K. delagonensis, *K. daigremontiana,* and *K. fedtschenkoi* were the most toxic; the toxic dose was 8–12 g/kg b.w. (Williams & Smith 1984). *Kalanchoe tomentosa,* a hybrid of *K. tomentosa* x *K. beharensis,* and several cultivars of *K. blossfeldiana* were not toxic at up to 20 g/kg b.w. per day for 2 days. However, in the study by Williams & Smith (1984), leaves were used rather than the more toxic flowers. In addition, results in chickens may be somewhat misleading because birds, when compared with other livestock, do not appear to be a good model system for examining effects of cardiotoxins, as was discovered in studies of *Asclepias* in the Asclepiadaceae (Ogden et al. 1992). *Kalanchoe blossfeldiana* was suspected to cause the death of an iguana that consumed an entire plant (Clay 1977).

- severe digestive disturbance and cardiotoxic effects, especially when flowers are eaten; dogs at high risk

Kalanchoe delagonensis

In North America, species of *Kalanchoe* represent a minimal hazard to livestock because of their relative inaccessibility in most instances. Household pets are at greater risk, and dogs are reported to be particularly susceptible (Kellerman et al. 1988).

However, it is doubtful that pets would normally eat enough plant material to become intoxicated except possibly when the plants are in flower during the winter. Because of their unique asexual reproduction involving the formation of plantlets, some species of the genus are popular as houseplants or as classroom curiosities; thus the likelihood of pet or human exposures is increased. If sufficient plant material is eaten, an acute cardiotoxicosis may develop.

- cardiotoxic bufadienolides, in many plant species; little risk with the most common species, *K. blossfeldiana*, florist's kalanchoe

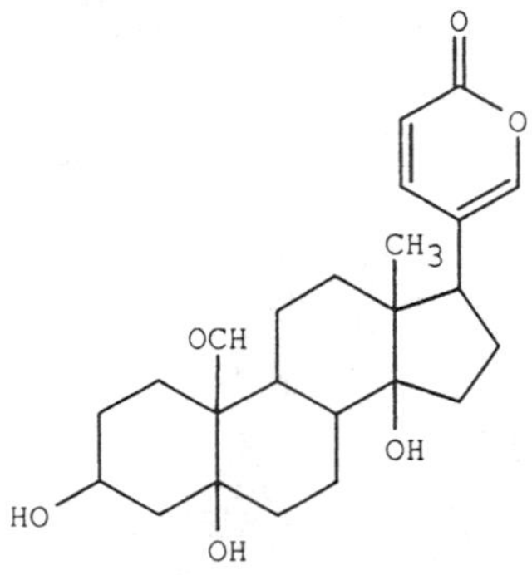

Disease Genesis—The toxic effects of *Kalanchoe* are due mainly to a family of cardiotoxic bufadienolides that are present in all plant organs but in highest concentration in the flowers (Anderson et al. 1984; Capon et al. 1986; Yamagishi et al. 1989; Oelrichs et al. 1992). With similar structures, these potent toxins include the bryotoxins, bryophyllins, and bersalgenins, which are found in many species. There are three bryotoxins, designated A, B, and C; B and C are orthoacetates of A. In *K. lanceolata,* a Madagascan species, lanceotoxins A and the glycosidic B, based upon acetylhellebrigenin, are present. Unique in the extent of their acetylation, these compounds are very similar in structure to cotyledoside.

Disease seems to be the direct consequence of these compounds, because dosage with bryotoxins produces the same disease as do the plants (McKenzie et al. 1989). Studies in Australia indicate that *K. delagonensis* and the hybrid, which are the most toxic, contain the highest concentrations of the bufadienolides—366–472 mg bryotoxins/kg fresh flowers (McKenzie et al. 1987). *Kalanchoe pinnata, K. prolifera,* and *K. fedtschenkoi* produce similar effects, but toxin levels are much lower and the threat of serious disease is reduced, although an extract from *K. pinnata* caused depression of the central nervous system and analgesia (McKenzie et al. 1987; Pal & Nag Chaudhuri 1989; Reppas 1995). All of these species, as well as *K. lanceolata* of South Africa, produce cardiac effects consistent with cardenolides (Anderson et al. 1983; Masvingwe & Mavenyengwa 1997). *Kalanchoe integra* of India, although not known to contain bufadienolides, nevertheless produces signs compatible with such toxins (Varma et al. 1981; Varma 1986). Although there appears to be considerable variation in toxin content among species, all should be considered toxic. Because the bufadienolides have an extended sojourn in the body, ingestion several times within a few days may produce subacute or chronic disease.

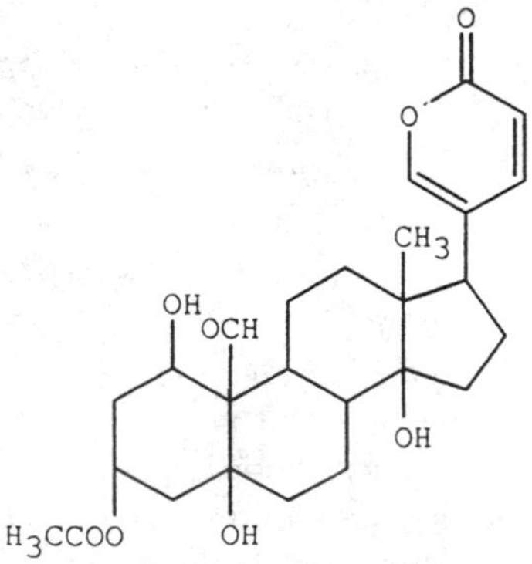

O
O
CH_3
OCH
OH
HO
OH

Hellebrigenin

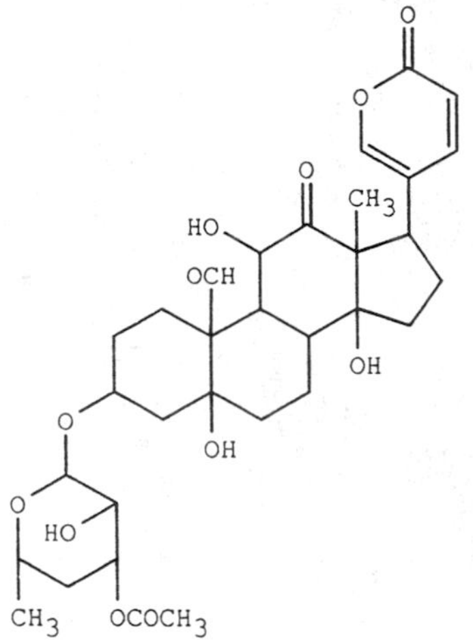

O
O
O
CH_3
HO
OCH
OH
O
OH
O
HO
CH_3 $OCOCH_3$

Bryotoxin A

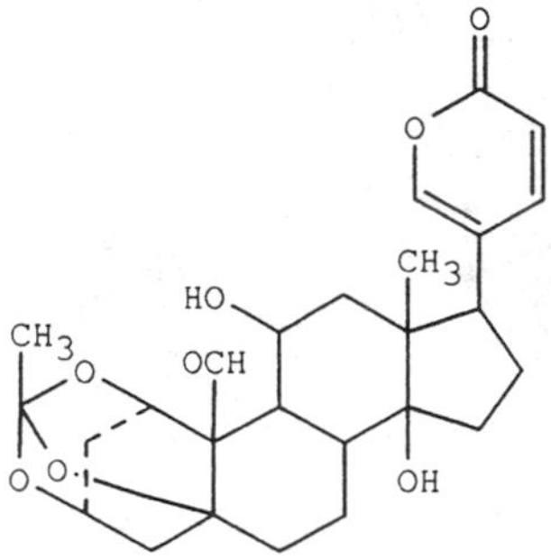

O
O
CH_3
HO
CH_3 O
OCH
O O
OH

Bryotoxin C

O
O
CH_3
OH OCH
OH
H_3CCOO OH

Bersalgenin-3-acetate

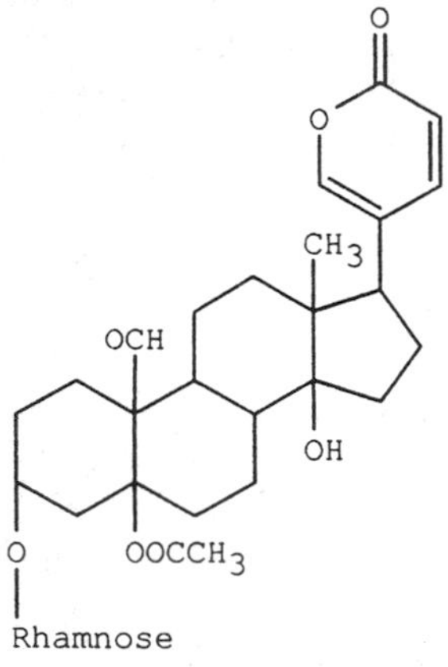

O
O
CH_3
OCH
OH
O $OOCCH_3$
Rhamnose

Lanceotoxin B

Clinical Signs—Beginning a few hours after ingestion, early signs include depression, loss of interest in food or water, excess salivation, and diarrhea. The last is probably the most consistent sign and after a few additional hours may begin to be mucoid and bloody. The diarrhea may persist for several days without further signs developing. If sufficient plant material is eaten or the more toxic species are consumed, there may be an increase in heart rate, intermittent drooling, and labored respiration with expiratory grunting. During the second day of illness, cardiac abnormalities may be seen: arrhythmias, slow heart rate, and heart block, with dropped beats and jugular pulse (McKenzie & Dunster 1986; McKenzie et al. 1989). By this time severe weakness is typical. The animal may collapse and die after 4 or 5 days of illness. When the more toxic *K. delagonensis* is involved, the animal may die within hours (McKenzie & Dunster 1986). The signs may vary somewhat, depending on the animal species and plant parts consumed, but in most cases persistent diarrhea is observed. During the course of the disease, blood urea nitrogen and creatinine may increase 2-to-20-fold as a consequence of prerenal effects, because only mild nephrosis is observed (McKenzie & Dunster 1986). As much as a 2-fold increase in blood glucose may also be noted.

■ abrupt onset, depression, excess salivation, severe and persistent diarrhea; severe cases, by next day, labored respiration, alterations in cardiac rate and rhythm, weakness

Pathology—There are few distinctive lesions; scattered small hemorrhages of the ruminal wall and the abomasal mucosa and perhaps several ulcers may be seen. Ulceration and necrosis of the folds of the omasum and constriction of the bronchiolar lumens with mild pulmonary emphysema also may be evident. Cardiac lesions will be present in many instances, such as scattered ecchymotic hemorrhages on the heart surfaces and foci of myocardial degeneration. Microscopically, these foci will contain degenerating fibers, necrosis, edema, and mononuclear cell infiltrates (with a few neutrophils).

■ gross pathology, reddening, ulceration of epithelium and mucosa, rumen, stomach and small intestine

microscopic, myocardial necrosis

Treatment—Cattle intoxicated by *K. delagonensis* respond favorably to activated charcoal (5 g/kg) and electrolyte solution (150 ml/kg daily) given orally and symptomatic treatment for the cardiac abnormalities. The latter includes atropine, 0.5 mg/kg, divided i.v. and s.c., to increase heart rate and reverse heart block and arrhythmias, or the more specific propranolol, 5 mg slowly i.v., to slow the heart. Their administration can be repeated, atropine every 6–14 hours as needed or propranolol every 2–4 hours. Therapy instituted the first day of illness is very effective (McKenzie & Dunster 1987). Intoxication by the less toxic species of the genus may require treatment only for the diarrhea.

■ activated charcoal, atropine

Genera of Questionable Toxicity

Crassula L.

A large and diverse genus of 250–300 species, *Crassula,* or the jade plants, contains numerous ornamentals widely grown for their foliage and interesting succulent growth forms, which range from tiny cushionlike herbs to robust shrubs. Often woody, the stems bear opposite, sessile or subsessile leaves with flat or terete, entire or finely toothed blades and bases that are often fused. Borne singly or in small cymes, the small tubular or saucer-shaped flowers bear 5 or occasionally 4 sepals and petals, the latter being pink or red or white or yellow.

■ *Crassula*

jade plant

■ common succulents, little reason to consider toxic

Although little is known of its toxicity potential, *Crassula* is often the subject of inquiries to poison control centers because of ingestions by children. Effects are limited to transient nausea and digestive upset (Spoerke & Smolinske 1990). Prompted by inquiries regarding suspected intoxications, we have attempted experimentally without success to produce intoxication in cats by administration of a homogenized slurry of the leaves.

Sedum L.

- *Sedum*

  golden carpet
  orpine
  stonecrop

- irritant sap, perhaps mild digestive disturbance

Sedum stenopetalum

The largest genus of the Crassulaceae, containing 300–600 species, *Sedum,* commonly known as stonecrop or orpine, is almost exclusively a taxon of the Northern Hemisphere. Its species occupy a diversity of habitats ranging from marshlands to deserts. Many are popular ornamentals, as both houseplants and garden plants. Some are low succulents cultivated for their prostrate habit especially in rock gardens. Plants of the genus are herbs or subshrubs with erect or decumbent stems bearing alternate, opposite, or whorled, terete or plano-convex leaves. The sap is generally acrid to the taste. The usually terminal, compound cymes bear typically 5-merous flowers with white or yellow, rarely red or purple petals.

As with *Crassula,* information about the toxicologic potential of *Sedum* species is limited. Native to the Old World, *S. acre* L., commonly known as golden carpet or mossy stonecrop, is naturalized throughout the eastern third of the continent. Its acrid sap is reported to be very irritating (Pammel 1911; Bartik & Piskac 1981). It is said to have been used to remove warts and corns. Similar properties are described for *S. purpureum* Tausch, live-forever, and *S. stenopetalum* Pursh, worm-leaf sedum or narrow-petaled stonecrop (Pammel 1911). *Sedum acre* contains the piperidine alkaloids sediene and sediendione, which are of unknown importance but are potential teratogens based on their chemical structure (Keeler & Crowe 1985; Maksimovic et al. 1990). Other species contain serotonin and norepinephrine (Kimbrough et al. 1987).

When discarded trimmings of *S. spectabile* H.Ohba, a tall ornamental from China and Korea, were eaten by several gilts, depression, loss of appetite, fever, and incoordination were observed. Treated with an emetic, a laxative, and an antibiotic, the animals exhibited complete recovery by the following day (Vegh 1985).

References

Anderson LAP, Anitra Shultz R, Joubert JPJ, Prozesky L, Kellerman TS, Erasmus GL, Procos J: Krimpsiekte and acute cardiac glycoside poisoning in sheep caused by bufadienolides from the plant *Kalanchoe lanceolata* Forssk. *Onderstepoort J Vet Res* 50;295–300, 1983

Anderson LAP, Steyn PS, van Heerden FR: The characterization of two novel bufadienolides, lanceotoxins A and B from *Kalanchoe lanceolata* [Forssk.] Pers. *J Chem Soc Perkin Trans* 1;1573–1575, 1984

Bartik M, Piskac A: *Veterinary Toxicology.* Elsevier, Amsterdam, pp244–245, 1981

Capon RJ, MacLeod JK, Oelrichs PB: Bryotoxins B and C, toxic bufadienolide orthoacetates from the flowers of *Bryophyllum tubiflorum* (Crassulaceae). *Aust J Chem* 39;1711–1715, 1986

Clay BR: Poisoning and injury by plants. In *Current Veterinary Therapy,* vol 6, *Small Animal Practice,* Kirk RW ed, Saunders, Philadelphia, pp179–184, 1977

Correll DS, Johnston MC: *Manual of the Vascular Plants of Texas,* 2d printing. Univ Texas at Dallas, Richardson, Tex, 1979

Davis DE, Kates AH, McCormack J: *Poisonous Plants of the Southern United States.* Georgia Coop Extension Service, 1980

Fuller TC, McClintock E: *Poisonous Plants of California.* Univ Calif Press, Berkeley, pp124–126, 1986

Keeler RF, Crowe MW: Anabasine, a teratogen from the *Nicotiana* genus. In *Plant Toxicology—Proceedings of the Australia-USA Poisonous Plants Symposium,*

Seawright AA, Hegerty MP, James LF, Keeler RF eds, Animal Research Institute, Yeerongpilly, Australia, pp324–333, 1985

Kellerman TS, Coetzer JAW, Naude TW: *Plant Poisonings and Mycotoxicoses of Livestock in Southern Africa.* Oxford Univ Press, Cape Town, South Africa, 1988

Kimbrough TD, Reynolds JD, Humphreys KJ, Weekley LB: Diurnal changes in tissue leaf levels of tryptophan, tyrosine, and amine metabolites in *Sedum morganianum* and *Sedum podophyllum. Biochem Physiol Pflanz* 182;67–72, 1987

Maksimovic M, Sober M, Nikolin B: Gas chromatography/mass spectrometry in the elucidation of the structure of piperidine alkaloids. *Rapid Comm Mass Spectrom* 4;503–504, 1990

Masvingwe C, Mavenyengwa M: *Kalanchoe lanceolata* poisoning in Brahman cattle in Zimbabwe: the first field outbreak. *J S Afr Vet Assoc* 68;18–20, 1997

McKenzie RA, Dunster PJ: Hearts and flowers: *Bryophyllum* poisoning of cattle. *Aust Vet J* 63;222–227, 1986

McKenzie RA, Dunster PJ: Curing experimental *Bryophyllum tubiflorum* poisoning of cattle with activated carbon, electrolyte replacement solution, and antiarrhythmic drugs. *Aust Vet J* 64;211–214, 1987

McKenzie RA, Franke FP, Dunster PJ: The toxicity to cattle and bufadienolide content of six *Bryophyllum* species. *Aust Vet J* 64;298–301, 1987

McKenzie RA, Franke FP, Dunster PJ: The toxicity for cattle of bufadienolide cardiac glycosides from *Bryophyllum tubiflorum* flowers. *Aust Vet J* 66;374–376, 1989

Naude TW, Anitra Shultz R: Studies on South African cardiac glycosides II observations on the clinical and haemodynamic effects of cotyledoside. *Onderstepoort J Vet Res* 49;247–254, 1982

Oelrichs PB, MacLeod JK, Capon RJ: Isolation and identification of the toxic principles in *Bryophyllum tubiflorum (Kalanchoe).* In *Poisonous Plants: Proceedings of the Third International Symposium,* James LF, Keeler RF, Bailey EM Jr, Cheeke PR, Hegarty MP eds, Iowa State Press, Ames, pp288–292, 1992

Ogden L, Burrows GE, Tyrl RJ: Comparison of *Asclepias* species based on their toxic effects on chickens. In *Poisonous Plants: Proceedings of the Third International Symposium,* James LF, Keeler RF, Bailey EM Jr, Cheeke PR, Hegarty MP, eds, Iowa State Press, Ames, pp500–505, 1992

Pal S, Nag Chaudhuri AK: Some pharmacological actions of *Bryophyllum pinnatum. Planta Med* 55;647–648, 1989

Pammel LH: *Manual of Poisonous Plants.* Torch Press, Cedar Rapids, Iowa, pp501–503, 1911

Reppas GP: *Bryophyllum pinnatum* poisoning of cattle. *Aust Vet J* 72;425–427, 1995

Spoerke DG, Smolinske SC: *Toxicity of Houseplants.* CRC Press, Boca Raton, Fla, pp104–105, 1990

Varma RK: Antidotal treatment of experimental *Kalanchoe integra* poisoning. *Indian J Anim Sci* 56;411, 1986

Varma RK, Garg BD, Kharole MU, Ahmad A: Chronic toxicity studies on *Kalanchoe integra* in sheep. *Indian J Anim Sci* 51;522–526, 1981

Vegh E: Ornamental plant *(Sedum spectabile)* poisoning in pigs. *Magy Allatorv Lapja* 40;505–506, 1985 (*Vet Bull* 56;1142, 1986)

Williams MC, Smith MC: Toxicity of *Kalanchoe* spp., to chicks. *Am J Vet Res* 45;543–546, 1984

Yamagishi T, Haruna M, Yan X-Z, Chang J-J, Lee K-H: Antitumor agents, 110: bryophyllin B, a novel potent cytotoxic bufadienolide from *Bryophyllum pinnatum. J Nat Prod* 52;1071–1079, 1989

28 Cucurbitaceae Juss.

■ Gourd Family

Bryonia
Cucurbita
Momordica

QUESTIONABLE
Cucumis

Widespread in tropical and subtropical regions of both the Old World and the New World, the Cucurbitaceae, or gourd family, comprises 118 genera and 825 species, a large number of which are known only in cultivation (Jeffrey 1980, 1990). Archeological evidence indicates a long history of use of the family for food, fodder, ornaments, medicines, and containers, especially by the Indian tribes of the southwestern United States and Mexico (Jeffrey 1980; Ebeling 1986). Quite familiar are the fruits of *Cucurbita,* the pumpkins and squashes; *Citrullus,* the watermelons and citrons; *Cucumis,* the cucumbers and muskmelons; *Lagenaria,* the gourds; and *Luffa,* the vegetable sponge. In unusual circumstances these commonly cultivated species may accumulate noxious compounds or toxins, be extremely bitter-tasting, and cause problems of the digestive tract (Allen 1943; Rymal et al. 1984; Fuller & McClintock 1986; Kirschman & Suber 1989). In North America, approximately 28 genera and 75 species, both native and introduced, are present. Toxicologic problems are associated primarily with 3 genera.

■ herbaceous trailing vines; flowers staminate and pistillate, similar; petals 5 or 6, fused, white, greenish white, or dark yellow; fruits pepos or berries, often fleshy

Plants herbaceous vines; perennials or annuals; from fleshy taproots; strongly aromatic or not aromatic; monoecious or dioecious. **Stems** trailing or climbing; bearing coiling, often branched tendrils at nodes. **Leaves** simple or palmately compound; alternate; venation palmate; blades cordate or oval; surfaces glabrous or pubescent or hispid; apices acuminate or obtuse; margins dentate or palmately lobed or palmately cleft; bases usually cordate; stipules absent. **Inflorescences** of 1 or 2 types; staminate and pistillate similar or different; solitary flowers or racemes or corymbs or clusters; axillary; bracts absent or present. **Flowers** imperfect, staminate and pistillate similar; perianths in 2-series. **Sepals** 5 or 6; fused. **Corollas** radially symmetrical; campanulate or funnelform or rotate. **Petals** 5 or 6; fused; white or greenish white to dark yellow. **Stamens** 5 or 3; fused by filaments or anthers or free. **Pistils** 1; compound, carpels 3; stigmas 3–5, not lobed or 2- or 3-lobed, capitate or linear; styles 1–3; ovaries inferior, terete or flattened in cross section, lobes 5 or 6; locules 1; placentation parietal. **Fruits** pepos or berries; pericarps dry or fleshy; smooth or armed with prickles. **Seeds** 1 or 3 or numerous; typically flattened; oval; smooth to granular.

The Cucurbitaceae is readily recognized by its prostrate or trailing habit; tendril-bearing, herbaceous stems; and imperfect flowers with inferior ovaries and three parietal placentae.

Bryonia L.

■ *B. alba*
B. dioica

bryony
wild hop

Taxonomy and Morphology—An Old World genus, *Bryonia* comprises 10–12 species. Its name is taken from the Greek root *bruein,* meaning “to sprout,” a reference to the aerial growth arising from the roots each spring. Its quick-growing foliage is used to cover arbors and fences. The genus is represented in North America by only 2 species:

B. alba L.	white bryony
B. dioica Jacq.	red bryony, wild hop

In some publications, the binomial *B. cretica* L. is used for red bryony, with *B. dioica* cited as a taxonomic synonym. Jeffrey (1969) treated them as distinct species, an interpretation with which we concur.

Bryonia alba

Plants perennials; roots fleshy; dioecious or less often monoecious. **Stems** climbing; tendrils not branched; surfaces glabrous. **Leaves** simple; petiolate; blades triangular; 3- or 5-lobed. **Staminate Inflorescences** peduncled racemes or cymes. **Pistillate Inflorescences** solitary or axillary clusters of flowers. **Flowers** rotate. **Petals** white or yellow. **Stamens** with anthers free or fused; staminodia 3 or 5. **Pistils** with 3 stigmas; styles slender. **Fruits** small berries; malodorous; smooth; black or red or orange. **Seeds** 3–6; ovoid to oblong.

1. Stigmas glabrous. Mature berries black . . . . *B. alba*
1. Stigmas papillose-pubescent. Mature berries red or orange . . . . . . *B. dioica*

■ climbing perennial vines with unbranched tendrils; leaves simple, 3- or 5-lobed; petals white or yellow

Distribution and Habitat—Both species are native to Europe. *Bryonia alba* also occurs in western Asia, and *B. dioica* occurs in North Africa. Introduced as an ornamental in North America, *B. dioica* is naturalized in disturbed sites along the central California coast just south of the San Francisco area, while *B. alba* has escaped and become established in Utah.

■ ornamentals

Disease Problems—Used by the Greeks, Romans, and medieval herbalists, the dried roots of both species of *Bryonia* were prepared as purgatives. Today, intoxications are rare but produce a similar effect—severe disturbance of the digestive tract. In one instance of suspected intoxication in a dog, the main presenting signs were abrupt onset of vomiting and diarrhea (Whur 1986). Berries of *B. alba* were present in the vomitus and seeds subsequently found in the intestinal contents. The animal died about 24 hours after ingestion of the berries. Upon necropsy, lesions of extensive fluid accumulation in the abdomen and thorax, and scattered ecchymotic hemorrhages, were discovered.

■ digestive disturbance, all plant parts, especially bitter fruit

Disease Genesis—Species of *Bryonia* contain a variety of biologically active constituents. Of toxicological interest are the various tetracyclic cucurbitane terpenoid saponins from genins such as cucurbitacin J in *B. alba* (Connolly & Hill 1991). Present in *B. dioica* are several genins, such as bryogenin, bryosigenin, bryodulcosigenin, and cucurbitacin L, which differ only in subtle changes at C-24 or C-25 (Tunmann et al. 1966; Ripperger 1976; Oobayashi et al. 1992). Responsible for the extreme bitterness of the fruits, they are found in the plant as glycosides. Bryonoside, bryoside, and bryodulcoside are glycosides of bryodulcosigenin, and bryoamaride and the

■ irritant tetracyclic terpenoid saponins, glycosides of bryogenin, cucurbitacin J; ribosomal inhibiting glycoprotein, single chain, active only in cell-free systems

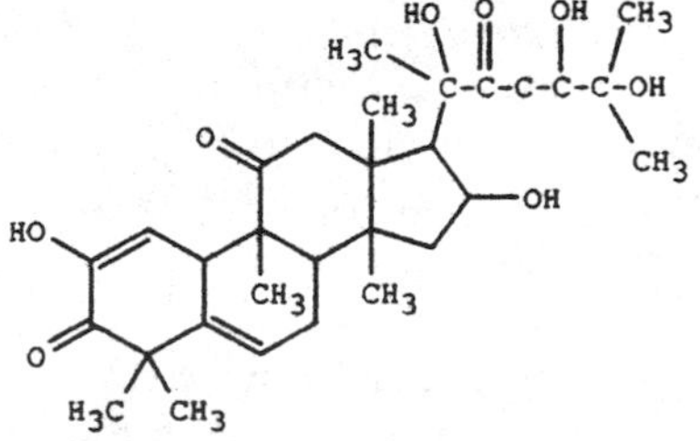

Cucurbitacin J

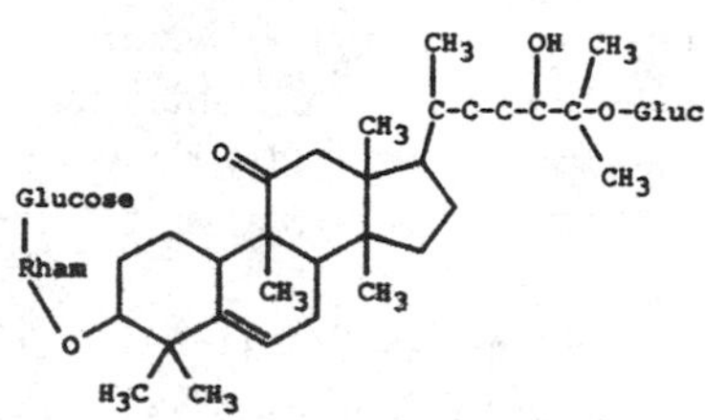

Bryoside

Bryoamaride

brydiosides are glycosides of cucurbitacin L. They are unique glycosides in that one or more sugars are attached at two or more different sites on the genin. This structure increases the likelihood of extraction artifacts and problems in identification.

Other toxicants in *B. dioica* include brydiofin, a protein of 66,000 MW present in the fruits but not the roots and foliage (Munoz et al. 1992). When administered i.p., extracts containing brydiofin killed mice. There also are ribosome- inhibiting glycoproteins called bryodin 1 and 2. They have a molecular weight of 27,000–29,000 daltons and a single A chain similar in size and action to the effectomer chain of ricin (Bolognesi et al. 1990; Wachinger et al. 1993; Siegall et al. 1994). However, unlike ricin, there is no accompanying facilitator chain present to promote cell entry and absorption from the digestive tract. Thus these proteins are primarily active in cell-free lysates; very high concentrations are required in vitro to affect whole cells (Gasperi-Campani et al. 1980; Bolognesi et al. 1990). It is suggested that these glycoproteins from the roots and leaves are isoforms and should be designated bryodin-R and bryodin-L, respectively (Bolognesi et al. 1990).

- irritant effects, nausea, vomiting, diarrhea; reddened mucosa stomach, intestines

  treatment not needed

Clinical Signs, Pathology, and Treatment—The toxins in *Bryonia* have the potential to cause digestive tract disturbances such as nausea, vomiting, and diarrhea. These are usually not lethal toxicants but do cause reddening of the mucosa of the stomach and intestines. Typically the effects will not require treatment. In instances of more severe disease, relief of the general symptoms may be needed on an individual basis.

Cucurbita L.

- *C. foetidissima*

  buffalo gourd
  calabacilla loco
  stinking cucumber
  stinking gourd
  wild pumpkin

Taxonomy and Morphology—Native to the New World, *Cucurbita* comprises 15–30 species. Greatest diversity is in tropical and subtropical regions. Its name is a Latin word for a type of gourd. Representative of the 8 species in North America is the following:

C. foetidissima Kunth	buffalo gourd, calabacilla loco, stinking gourd, stinking cucumber, wild pumpkin

- prostrate or climbing perennial; herbage malodorous; root large, tuberous; leaves simple, grayish green; flowers large, showy; petals 5, yellow; fruits pepos, green with light stripes when young, lemon yellow at maturity

Cucurbita foetidissima

Plants perennials; from massive, fusiform roots; herbage malodorous when bruised, scabrous, hispid; monoecious. **Stems** numerous; prostrate or occasionally climbing; 3–6 m long; tendrils branched. **Leaves** simple; blades triangular-ovate; thick; grayish green; apices acute or acuminate; margins entire or angular-lobed at bases; bases cordate. **Inflorescences** solitary flowers in leaf axils; staminate flowers long-peduncled; pistillate flowers sessile or short-peduncled. **Flowers** large; showy; staminate and pistillate similar; campanulate. **Sepals** 5. **Petals** 5; yellow; 5–12 cm long. **Stamens** connivent. **Pistils** with 3 two-lobed stigmas. **Fruits** pepos; globose; green with light stripes when young; lemon yellow at maturity. **Seeds** numerous; oval to ovate; white to tawny or black.

Distribution and Habitat—*Cucurbita foetidissima* is a common vine of the semiarid regions of the Southwest. Plants occur in clay, sandy, and gravelly soils of disturbed sites—waste areas, roadsides, rights-of-way, and edges of cultivated fields—from west Texas, Missouri and Nebraska to Arizona, southern California, and Mexico.

Disease Problems—Various species of *Cucurbita* have been used or cultivated for centuries in the New World, especially in Central America, for food, livestock fodder, ornaments, and medicines (Allen 1943; Saade & Hernandez 1994). Human selection produced the numerous cultivars of squashes and pumpkins now prized as food. Except for the flowers and seeds, the wild species of the genus are generally inedible because of their bitterness (Ebeling 1986). If eaten, they are capable of causing severe disturbance of the digestive tract. However, the seeds of *C. maxima*, a species used in folk medicine in Brazil, failed to cause any changes in serum chemistry, urinalysis, hematology, or pathology when given orally to rats and pigs in a single large dose or daily for 30 days (Queiroz-Neto et al. 1994).

- sandy gravelly waste areas

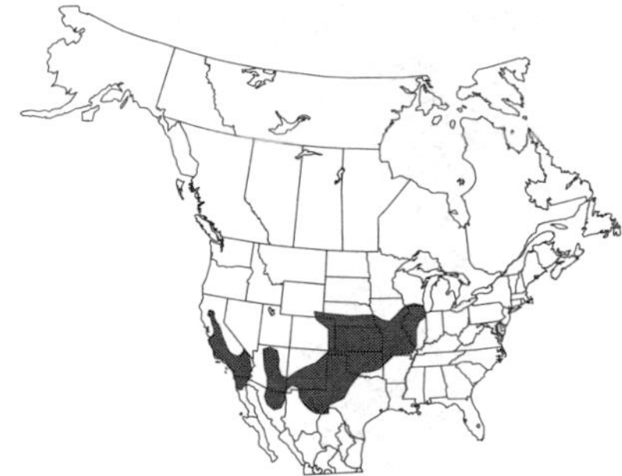

Cucurbita foetidissima

Because of its widespread distribution and availability, *C. foetidissima* has been used for millennia by American Indians of the southwestern United States and of Mexico (Bemis et al. 1979). The seeds were used for food and the fruit as a source of soapy detergent. However, as is typical of the family, the extreme bitterness of the fruit limited its use. There is now renewed interest in the seeds as a source of oil rich in linoleic acid, the large roots as a source of starch, and the foliage, with 10–13% crude protein, as livestock feed. This interest may foster genetic improvement via selection and/or hybridization to reduce the extreme bitterness. There is also considerable interest in continued development of other species for domestic use (Saade & Hernandez 1994). The dried pepos, with the pulp removed, are still used as vessels for water storage and drinking in some regions.

- humans, some species used for food, squashes, pumpkins; wild species, very bitter fruits, seeds used for food, fruits used as detergent

Disease Genesis—Typically the domesticated species are not toxicologic risks because genetic manipulation has reduced their content of the cucurbitacins to negligible levels. However, the extreme bitterness sometimes encountered in an individual squash is a reminder of the potential toxicity (Enslin 1954; Rymal et al. 1984; Kirschman & Suber 1989). This bitterness is due to the presence of cucurbitacins, most likely E. Ingestion of squash with high concentrations of these bitter compounds may cause severe abdominal pain.

- bitterness of fruit of cultivated species or *C. foetidissima* due to irritant tetracyclic terpenoid saponins, cucurbitacins

Cucurbitacin E

Clinical Signs, Pathology, and Treatment—Ingestion of squashes with high concentrations of cucurbitacins causes abrupt onset of severe abdominal pain from cramps, vomiting, diarrhea, and prostration. Because of the extreme bitterness associated with these compounds, it is unlikely that sufficient plant material will be eaten to cause death. Although not fatal, the toxicants may cause reddening of the mucosa of the stomach and intestines. Typically the effects will not require treatment; however, in a few individuals relief of the general symptoms may be needed.

- abrupt onset, colic, retching, vomiting, diarrhea, prostration

  irritant effect, reddened mucosa stomach, intestines

  treatment generally not needed

Momordica L.

Taxonomy and Morphology—Comprising 35–45 species, *Momordica* is native to the moist tropics of the Old World, particularly Africa. Its name is from the Latin *morder*, meaning "to bite," an allusion to the jagged edges of the seeds, which appear to have been bitten. Only 2 naturalized species are present in North America:

M. balsamina L.	balsam apple
M. charantia L.	balsam pear, bitter gourd, cundeamor, bitter cucumber, wild balsam apple, balsamina

- *M. balsamina*
  M. charantia

  balsam apple, balsam pear, balsamina, bitter cucumber, bitter gourd, cundeamor

Plants annuals; monoecious or occasionally dioecious. **Stems** prostrate or climbing; 1–5 m long; tendrils not branched. **Leaves** simple; blades triangular-ovate to cordate-suborbicular; margins palmately 3-to-7-lobed or dissected. **Inflorescences** solitary flowers or staminate flowers sometimes clustered; axillary; peduncles bearing 1 bract.

- prostrate or climbing annuals with unbranched tendrils; leaves simple; flowers showy; petals 5, yellow or yellow-orange, centers often darker; fruits pendulous, tuberculate, yellow-orange pepos

Flowers showy; staminate and pistillate similar; rotate or campanulate. **Sepals** 5. **Petals** 5; yellow or yellow-orange; centers often darker. **Stamens** 3; free. **Pistils** with 3 stigmas. **Fruits** pendulous pepos; ellipsoid or ovoid; tuberculate or rugose-tuberculate; green when young, orange or orange-yellow at maturity; dehiscing at maturity to expose seeds. **Seeds** numerous; flattened; elliptic to ovate; arils bright red or purple; fleshy.

1. Stems pubescent. Bracts attached at middles or bases of peduncles; margins entire .... *M. charantia*
1. Stems glabrous. Bracts attached at apices of peduncles; margins serrate .... *M. balsamina*

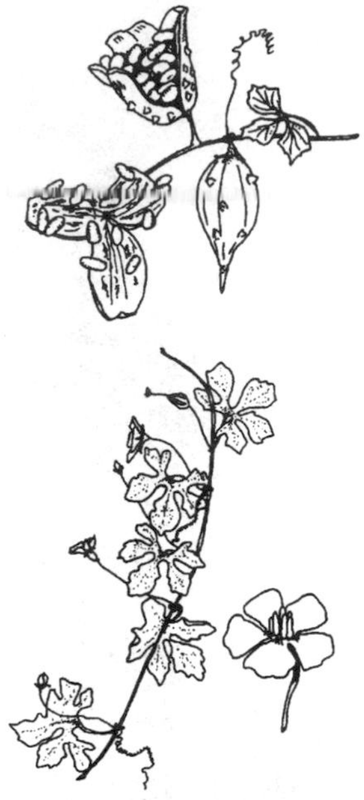

Momordica charantia

- mature fruits and seeds, irritant especially in dogs; some neurologic effects

Momordica balsamina

- irritant tetracyclic cucurbitane terpenoids, glycosides of momordicin I, momordicin II; ribosomal-inhibiting glycoprotein, momordin, single chain, active only in cell-free systems

Momordica charantia

Distribution and Habitat—*Momordica charantia,* native to the Old World tropics, has naturalized in thickets and waste areas along the Gulf Coast. *Momordica balsamina* is more limited in distribution along the coast.

Disease Problems—These species of *Momordica* have been a source of medicants, especially decoctions of the roots, leaves, and fruit for topical use (Diaz 1936; Lal et al. 1990). They also were reputed to be effective in reducing the high blood sugar associated with diabetes mellitus (Diaz 1936; Rivera 1941). Both the ripe fruit and infusions of the vegetative tissues have been used to induce abortion (Morton 1982). In addition to being a medicinal herb in tropical America, *M. charantia* is especially appreciated as a food source in the Orient (Morton 1967). When immature, the cooked pepos are edible and the seeds are used in curries. The mature seeds and fruits, however, are believed to cause severe digestive tract disturbance. They appear to be a particular problem in dogs, which also may exhibit prominent neurologic signs (Foley 1976; Morton 1982). Juice from the fruit and extracts from the seeds, when administered orally and daily for 30 days, caused a significant increase in serum GGT and AP but produced no microscopic pathologic changes in the liver (Tennekoon et al. 1994).

Disease Genesis—The biologically active constituents were originally thought to be alkaloids or alkaloid-like compounds in the foliage (Diaz 1936; Rivera 1941). A saponin-like toxin also was isolated that caused liver damage as well as digestive disturbance (Rivera 1942). The toxins are now thought to be tetracyclic cucurbitane terpenoids; the genin is momordicin I and its glycoside is momordicin II (Connolly & Hill 1991). *Momordica charantia* is also the source of at least two single-peptide chain lectins. One is momordin, a potent inhibitor of protein synthesis in cell-free systems, and the other is momordica hemagglutinin (Lin et al. 1978; Barbieri et al. 1979). Momordin is toxic, with an LD_{50} of 5 mg/kg b.w. i.p. in mice (Lin et al. 1978).

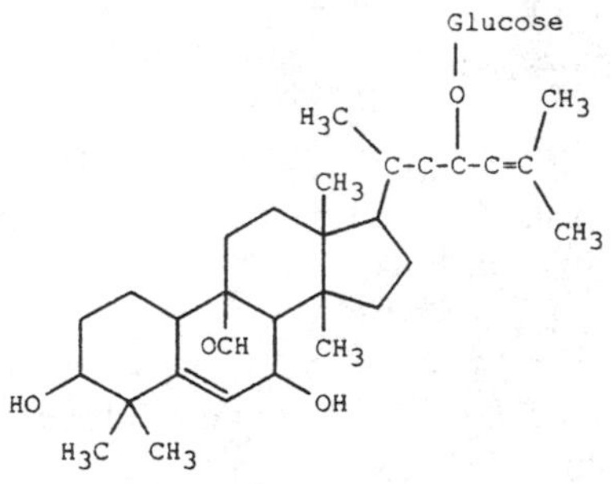

Momordicin II

Clinical Signs—Following ingestion of *Momordica,* there is abrupt onset of obvious digestive tract problems, with the animal exhibiting a painful or tender abdomen, vomiting, and diarrhea. In more severe cases, there may be neurologic signs such as depression, muscle spasms, and seizures ranging from paddling to extensor rigidity with the head held in a "looking-back" position.

- tender abdomen, vomiting, diarrhea; if severe, depression, muscle spasms, seizures

Pathology and Treatment—The changes observed are generally limited to reddening of the mucosa of the stomach and small intestine. Irritation is generally of limited duration, but in severe intoxications, control of neurologic signs by administration of sedatives, such as diazepam, and/or an emetic may be indicated.

- reddened mucosa, stomach, intestines

treatment if severe, sedation

Genera of Questionable Toxicity

Cucumis L.

Native to the tropics of Africa, Asia, and Australia, *Cucumis* comprises 32 species (Kirkbride 1993). Eaten and used medicinally since before the time of the Egyptian dynasties, the genus has two commercial crops—cucumbers and muskmelons—both of which are commonly grown in home gardens. Of the 6 introduced species in North America, , *C. myriocarpus* Naudin, commonly known as prickly paddy melon, and *C. sativus* L., or the common cucumber, may pose a potential risk because of the occasional presence of cucurbitacins. Plants of the genus are annuals or perennials that are monoecious or rarely dioecious. The trailing or climbing, typically indumented stems bear unbranched tendrils and simple, petiolate, 3- or 5-lobed, indumented leaves with blades of various shapes and entire or dissected margins. Borne singly or in small clusters in the leaf axils, thc campanulate to rotate flowers are 5-merous or rarely 4-merous. The yellow petals are fused only near the bases. Globose to elongate, the fruits are pepos with glabrous, indumented or echinate pericarps.

- *C. myriocarpus*
  C. sativus

cucumber
prickly paddy melon

Native to Africa, *C. myriocarpus* has naturalized in fields and disturbed sites in California, where it is considered a noxious weed. *Cucumis sativus* is native to Southeast Asia and is now cultivated worldwide. It occasionally escapes but does not naturalize in North America.

Cucumis myriocarpus

Fuller and McClintock (1986) reported that in California, *C. myriocarpus* is readily eaten by cattle and deer without apparent adverse effects. However, in South Africa and Australia, it is a cause of serious digestive tract disease with irritation and a bloody diarrhea (Watt & Breyer-Brandwijk 1962; McKenzie et al. 1988; Carter 1990). There may also be effects on the liver and kidney and hemolysis, which results in dark urine containing hemoglobin. The foliage of these two species is quite bitter and generally not eaten. However, even when it becomes more palatable several weeks after a frost, it remains toxic (Carter 1990).

- foliage and fruits when bitter, glycosides of cucurbitacins A and C; irritant, diarrhea ± blood, reddened digestive tract, mild liver and kidney necrosis

Both species contain tetracyclic cucurbitane terpenoids; *C. myriocarpus* has cucurbitacin A, and *C. sativus* has cucurbitacin C (Connolly & Hill 1991). These genins are accompanied by their respective sugars to form glycosides in the plants, the cause of the extreme bitter taste of the fruits. Some plants have a beta-glucosidase present that allows hydrolysis of any cucurbitacins present in the fruit pulp (Watt & Breyer-Brandwijk 1962).

Cucurbitacin A

After ingestion, there is abrupt onset of severe diarrhea, which may be bloody. Grossly, there is severe reddening of the mucosa of the abomasum or stomach and the small intestine. Microscopically, there may be degeneration of the tubular epithelium in the kidney and liver cell degeneration and necrosis. Myocardial necrosis is

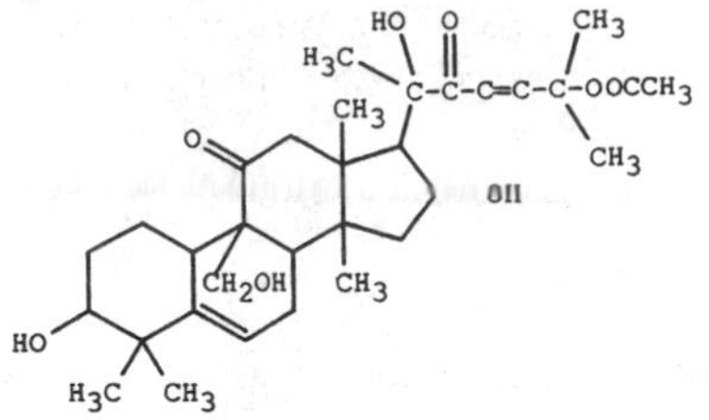

Cucurbitacin C

also seen occasionally. In some individuals it may be necessary to provide relief of the diarrhea and the irritation of the digestive tract mucosa.

References

Allen PH: Poisonous and injurious plants of Panama. *Am J Trop Med* 23 (Suppl);1–76, 1943

Barbieri L, Lorenzoni E, Stirpe F: Inhibition of protein synthesis in vitro by a lectin from *Momordica charantia* and by other hemagglutinins. *Biochem J* 182;633–635, 1979

Bemis WP, Berry JW, Weber CW: The buffalo gourd, a potential arid land crop. In *New Agricultural Crops,* Ritchie GA ed, Westview Press, Boulder, Colo, pp65–87, 1979

Bolognesi A, Barbieri L, Abbondanza A, Falasca AI, Carnicelli D, Battelli MG, Stirpe F: Purification and properties of new ribosome-inactivating proteins with RNA *N*-glycosidase activity. *Biochim Biophys Acta* 1087;293–302, 1990

Carter GI: Prickly paddy melon *(Cucumis myriocarpus)* poisoning of cattle [letter]. *Aust Vet J* 67;276, 1990

Connolly JD, Hill RA: *Dictionary of Terpenoids.* Vol 2, *Di- and Higher Terpenoids.* Chapman & Hall, London, pp1173–1180, 1991

Diaz LT: A preliminary study of an alkaloid-like material obtained from cundeamor or *Momordica charantia* L. *P R J Public Health Trop Med* 11;812–816, 1936

Ebeling W: *Handbook of Indian Foods and Fibers of Arid America.* Univ California Press, Berkeley, pp613–620,705–707, 1986

Enslin PR: Bitter principles of the Cucurbitaceae. 1. Observations on the chemistry of cucurbitacin A. *J Sci Food Agric* 5;410–416, 1954

Foley RH: Acute poisoning in a puppy caused by the balsam pear *(Momordica charantia). Vet Med/Small Anim Clin* 71;761–762, 1976

Fuller TC, McClintock E: *Poisonous Plants of California.* Univ California Press, Berkeley, pp126–127, 1986

Gasperi-Campani A, Barbieri L, Morelli P, Stirpe F: Seed extracts inhibiting protein synthesis in vitro. *Biochem J* 186;439– 441, 1980

Jeffrey C: A review of the genus *Bryonia* L. (Cucurbitaceae). *Kew Bull* 23:441–461, 1969

Jeffrey C: A review of the Cucurbitaceae. *Bot J Linn* Soc 81;233–247, 1980

Jeffrey C: Systematics of Cucurbitaceae: an overview. In *Biology and Utilization of the Cucurbitaceae,* Bates DM, Robinson RW, and Jeffrey C eds, Cornell Univ Press, Ithaca, NY, pp3–9,449–463, 1990

Kirkbride JH: *Biosystematic Monograph of the Genus Cucumis (Cucurbitaceae).* Parkway Publishers, Boone, NC, 1993

Kirschman JC, Suber RL: Recent food poisonings from cucurbitacin in traditionally bred squash [letter]. *Food Chem Toxicol* 27;555–556, 1989

Lal J, Chandra S, Gupta S, Tandan SK: Studies on anticonvulsant and antiinflammatory actions of extracts of *Momordica charantia. Indian Vet J* 67;82–83, 1990

Lin J-Y, Hou M-J, Chen Y-C: Isolation of toxic and non-toxic lectins from the bitter pear melon *Momordica charantia* Linn. *Toxicon* 16;653–660, 1978

McKenzie RA, Newman RD, Rayner AC, Dunster PJ: Prickly paddy melon *(Cucumis myriocarpus)* poisoning of cattle. *Aust Vet J* 65;167–170, 1988

Morton JF: The balsam pear—an edible, medicinal, and toxic plant. *Econ Bot* 21;57–68, 1967

Morton JF: *Plants Poisonous to People in Florida and Other Warm Areas.* Morton, Miami, 1982

Munoz SM, Salvarelli SM, Saiz MI, Conde FP: A toxic protein from *Bryonia dioica* Jacq. fruits: the brydiofin. *Biochem Biophys Res Commun* 183;1011–1018, 1992

Oobayashi K, Yoshikawa K, Arihara S: Structural revision of bryonoside and structure elucidation of minor saponins from *Bryonia dioica. Phytochemistry* 31;943–946, 1992

Queiroz-Neto A de, Mataqueiro MI, Santana AE, Alessi AC: Toxicologic evaluation of acute and subacute oral administration of *Cucurbita maxima* seed extracts to rats and swine. *J Ethnopharmacol* 43;45–51, 1994

Ripperger H: Isolierung und struktur der cucurbitacine aus *Bryonia dioica. Tetrahedron* 32;1567–1569, 1976

Rivera G: Preliminary chemical and pharmacological studies on "cundeamor," *Momordica charantia* L., part 1. *Am J Pharm* 113;281–297, 1941

Rivera G: Preliminary chemical and pharmacological studies on "cundeamor," *Momordica charantia* L., part 2. *Am J Pharm* 114;72–87, 1942

Rymal KS, Chambliss OL, Bond MD, Smith DA: Squash containing toxic cucurbitacin compounds occurring in California and Alabama. *J Food Prot* 47;270–271, 1984

Saade RL, Hernandez SM: Cucurbits (*Cucurbita* spp.). In *Neglected Crops: 1492 from a Different Perspective,* Bermejo JEH, Leon J eds, Food and Agriculture Organization, United Nations, Rome, pp63–77, 1994

Siegall CB, Gawlak SL, Chace D, Wolff EA, Mixan B, Marquardt H: Characterization of ribosome-inactivating proteins isolated from *Bryonia dioica* and their utility as carcinoma-reactive immune conjugates. *Bioconjug Chem* 5;423–429, 1994

Tennekoon KH, Jeevathayaparan S, Angunawala P, Karunanayake EH, Jayasinghe KSA: Effect of *Momordica charantia* on key hepatic enzymes. *J Ethnopharmacol* 44;93–97, 1994

Tunmann P, Gerner W, Stapel G: Konstitution des bryodulcosigenins. *Liebigs Ann Chem* 694;162–168, 1966

Wachinger M, Samtleben R, Gerhauser C, Wagner H, Erfle V: Bryodin, a single-chain ribosome-inactivating protein, selectively inhibits the growth of HIV-1–infected cells and reduces HIV-1 production. *Res Exp Med (Berl)* 193;1–12, 1993

Watt JM, Breyer-Brandwijk MG: *The Medicinal and Poisonous Plants of Southern and Eastern Africa.* E & S Livingston, Edinburgh, pp336–368, 1962

Whur P: White bryony poisoning in a dog. *Vet Rec* 119;411, 1986

29 Cupressaceae Bartl.

■ Cypress Family

Cupressus
Juniperus

Widespread in temperate regions of both the Old World and the New World, the Cupressaceae, commonly known as the cypress family, comprises 25–30 genera and 110–140 species (Watson & Eckenwalder 1997). Taxonomists differ in their opinions of its circumscription; some individuals segregate the alternate-leaved taxa into their own family, the Taxodiaceae. Similarities in cone anatomy, however, support union of alternate- and opposite-leaved taxa into one broad family, the interpretation employed here. With a fossil record extending to the Jurassic period of the Mesozoic, the Cupressaceae is an ancient family (Miller 1988). The majority of genera are monotypic and exhibit disjunct or relictual distributions substantially smaller than during the Tertiary. Economically, the family is a source of timber, cabinet wood, cedar shingles, pencils, oils, and numerous evergreen ornamentals. Toxicologic problems are associated with 2 of the 9 genera present in North America.

■ shrubs or trees; leaves simple, 2- or 4-ranked, linear or scalelike; pollen and seed cones

Plants trees or shrubs; evergreen or deciduous; strongly aromatic or not aromatic; resinous; producing pollen cones and seed cones; monoecious or dioecious; pneumatophores absent or present. **Leaves** simple; opposite or alternate; 2- or 4-ranked; persistent for 1 or 2–12 years; petiolate or sessile; linear or scalelike or awl-like; blades all alike or of 2 forms, juvenile and adult; stipules absent. **Pollen Cones** solitary or rarely clusters of 2–5; terminal or axillary; globose to oblong; maturing in 1 season; microsporophylls overlapping; microsporangia 2–10 per microsporophyll; pollen spherical; not winged. **Seed Cones** solitary or in clusters of 2–5; terminal; globose; woody or fleshy; maturing in 1–2 years; ovuliferous scales overlapping or abutting, fused to bracts; scale-bract complexes peltate, oblong or cuneate, fleshy or woody at maturity; megasporangia 2–9 per scale. **Seeds** 1–20 per scale; wings present or absent.

Cupressus L.

■ *C. macrocarpa*

Monterey cypress

Taxonomy and Morphology—Comprising 10–26 species in warm temperate regions, *Cupressus* is commonly known as cypress. In North America, 7 species are native in the far west (Watson & Eckenwalder 1997). The bald cypress of the southeastern United States is *Taxodium distichum,* not a species of *Cupressus* as its vernacular name suggests. Toxicologic problems have been associated with only 1 *Cupressus* species:

C. macrocarpa Hartw. Monterey cypress

■ evergreen trees, open crown, monoecious; leaves opposite, scalelike, 4-ranked; pollen cones yellow; seed cones woody

Plants trees; evergreen; monoecious; crowns open, broadly spreading, typically with a few major branches near ground. **Stems** 20–25 m tall; bark rough, fibrous, dark brown to ash gray; branchlets in decussate arrays, terete. **Leaves** opposite; 4-ranked; sessile; scalelike; appressed; overlapping; blades of 2 forms, juvenile and adult; abaxial gland absent or inconspicuous. **Pollen Cones** yellow; microsporophylls 8–20; microsporangia 3–10 per microsporophyll. **Seed Cones** woody; oblong; maturing in 1–2 years, generally persisting on tree unopened for many years. **Seeds** dark brown.

Cupressus macrocarpa

Distribution and Habitat—As its common name implies, *C. macrocarpa* is native to the Monterey Peninsula. With respect to natural populations, it is the geographically most restricted species in the genus, although the fossil record indicates that it was once much more widespread. It has been carried elsewhere as an ornamental and has naturalized in some areas.

■ ornamental

Disease Problems—Problems caused by *C. macrocarpa* seem to be very similar to the reproductive problems caused by *Pinus ponderosa* (see chapter 57). Pregnant cattle in the last trimester of gestation may undergo premature parturition a few days to weeks after eating foliage of the species (MacDonald 1956; Mason 1974; Sloss & Brady 1983). Calves may be stillborn or may be weak and die shortly after birth. MacDonald (1956) described a case in which 9 cows aborted 2–4 weeks prior to term with retained placentae and enlarged cotyledons. Two cows died. The fetal placental membranes were retained for 5–7 days. In another instance, trimmings discarded in a pasture were suspected to be the cause of abrupt death of a cow that exhibited few pathologic changes (O'Scanaill 1986). The disease like that produced by *P. ponderosa* is not reproducible experimentally in sheep (Mason 1984). There may be a risk with other *Cupressus* species, given that death of cattle has been reported due to *C. sempervirens* (Gould 1962). It seems appropriate to avoid allowing pregnant cattle access to the foliage of any species of *Cupressus*.

■ cattle, reproductive effects, last third of pregnancy, a few days or weeks after ingestion; effects on fetus and dam, premature parturition; toxicity of other *Cupressus* species or in other animal species unknown

Disease Genesis—The cause of the reproductive effects produced by *C. macrocarpa* is not known at present, but a role for diterpene acids, such as isocupressic acid, is likely because of their involvement in reproductive problems caused by *Pinus* and *Juniperus*. In particular, there are many similarities to *Pinus*: premature parturition limited to the last trimester, enlarged cotyledons, effects in cattle but not sheep, and severe illness and death in some dams. In one instance, 2 of 3 fetuses had malacia of the white matter of the brain, consistent with hypoxia of placental origin such as might be expected with the type of disease induced by *Pinus* (Munday et al. 1973).

■ cause unknown; probably diterpene acids as in *Pinus*, many similarities in effects

Clinical Signs, Pathology, and Treatment—The effects of *C. macrocarpa* range from stillborn calves to those so weak that they are unable to suckle or to follow their mothers or that they exhibit increased susceptibility to other diseases and death. Although not necessarily specific for this disease, cerebral leukomalacia of 2 fetuses has been noted in one case. In some instances the cows may be weak and in poor condition and die shortly after birth of the affected calf. Prevention of the reproductive effects requires that cattle be pastured where there is no or limited access to trees of the species, at least during pregnancy.

■ stillborn, small or weak newborn, poor survival; dam may be ill

Juniperus L.

Taxonomy and Morphology—A circumboreal genus comprising 45–60 species, *Juniperus*, or juniper, is a familiar sight both in cultivation and in the wild. A plethora of cultivars of different growth forms, textures, colors, and odors are used extensively in landscaping. The wood is valued for its aromatic, insect-repelling properties and is used to make cedar chests and to line clothes closets. The distinctive flavor of gin is due to the volatile oil extracted from one species. In older publications, the genus

■ *J. communis*, *J. deppeana*, *J. horizontalis*, *J. occidentalis*, *J. osteosperma*, *J. sabina*, *J. scopulorum*, *J. virginiana*

juniper, red cedar, savin

Juniperus

- strongly aromatic evergreen shrubs or trees; most dioecious; leaves simple, opposite or whorled; pollen cones yellow; seed cones fleshy, berrylike

name *Sabina* may occasionally be encountered for this taxon. Some species have the common name *cedar,* which also is applied to species of *Thuja, Calocedrus, Cedrus,* and *Chamaecyparis.* In North America, 13 native and numerous introduced species are present. The following are representative taxa:

J. communis L. (= *J. canadensis* Lodd. ex Burgsd.)	common juniper, little juniper, dwarf juniper
J. deppeana Steud.	alligator juniper
J. horizontalis Moench	creeping juniper, trailing juniper
J. occidentalis Hook.	western juniper, Sierra juniper
J. osteosperma (Torr.) Little (= *J. utahensis* [Engelm.] Lemmon)	Utah juniper, little Utah juniper
J. sabina L.	savin
J. scopulorum Sarg.	Rocky Mountain juniper, red cedar
J. virginiana L.	eastern red cedar, Virginia juniper

Plants trees or shrubs; evergreen; strongly aromatic; dioecious or rarely monoecious; crowns dense, of various shapes. **Stems** 3–30 m tall; bark rough, fibrous, brown or reddish brown, shredding and exfoliating in thin strips; branchlets 3-to-6-angled, not oriented in flattened sprays. **Leaves** simple; opposite in 4 ranks or whorled; persistent for 2–12 years; sessile; scalelike or awl-like; appressed; overlapping; blades of 2 forms, juvenile and adult, abaxial gland present. **Pollen Cones** yellow; microsporophylls 6–21; microsporangia 2–8 per microsporophyll. **Seed Cones** globose to ovoid; fleshy, berrylike; maturing in 1–2 years; usually glaucous. **Seeds** 1–3 per scale; round or angled; wings absent.

Because *Juniperus* is a gymnosperm and not a flowering plant, it is technically incorrect to call its mature seed cones "berries" or "fruits," which by definition are the mature ovaries of flowers. The terms, however, have been used for decades in both the popular and the scientific literature and most likely will continue to be employed.

The species listed above are merely representative of the genus, and intoxication problems may be caused by any member of the genus. Therefore, a taxonomic key for their identification is not presented here. The reader is directed to use floristic treatments specific for his or her area.

Juniperus communis

Juniperus occidentalis

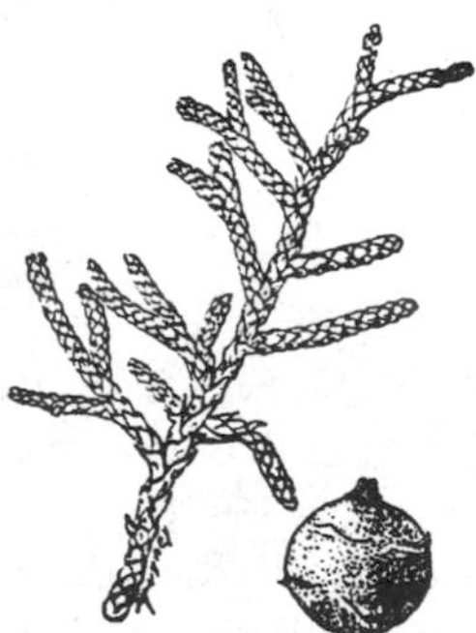

Juniperus osteosperma

Juniperus scopulorum

Distribution and Habitat—Occupying a variety of soil types and climatic regimes, species of *Juniperus* are present in all vegetation types and a diversity of habitats across the continent. Some aggressively invade abused grasslands and are considered noxious weeds. Dispersal is via birds eating the seed cones and voiding the seeds in their feces. Providing both winter cover and food, they are important wildlife species.

- most types of habitats, spread by seed-eating birds

Disease Problems—The fleshy seed cones of *Juniperus* are an important winter food for birds, rodents, and deer (Ebeling 1986). The winter and spring foliage is eaten to some extent by deer and other wildlife, but they seem to eat only small amounts except in dire circumstances (Krausman et al. 1997). The sweet seed cones were also an important food for members of various American Indian tribes, especially those in the West, and in some instances were consumed in large amounts. Eaten raw, dried and made into a mush or cake, or boiled, they provided much needed fall and winter sustenance (Ebeling 1986). The cones also were used as flavoring and seasoning for tea and meats; e.g., they were used in the making of pemmican, a jerky-type product. The cones of *J. osteosperma* and *J. deppeana* were considered to be particularly desirable (Harrington 1967). Although not relished, the inner bark was used as an emergency food.

- seed cones, foliage, food for wildlife; small amounts, food and medicine for humans; large amounts, reproductive effects in cattle, abortions

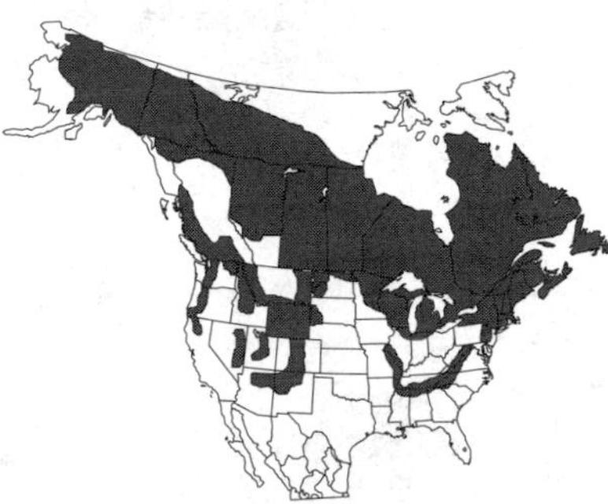

Juniperus communis

In spite of its widespread use as food, the foliage and to a lesser extent the seed cones of some species are purported to have diuretic effects (Janku et al. 1957; Koedam & Looman 1980). Tea made from the leaves of *J. scopulorum* has a reputation as a uterine stimulant among American Indians (Hart 1992). The foliage also is suspected of causing abortion in cattle (Norris & Valentine 1954; James et al. 1992). Experimentally, the foliage of *J. osteosperma* fed at a daily rate of 0.45 kg/kg b.w. for a sustained period caused abortion in 2 of 6 sheep at 87 and 125 days of gestation (Johnson et al. 1976). Attempts to repeat this effect using lesser amounts of foliage were not successful. Gardner and co-workers (1998a) have produced abortions in cattle given 7.6–9.8 g/kg b.w. of the foliage of *J. communis.*

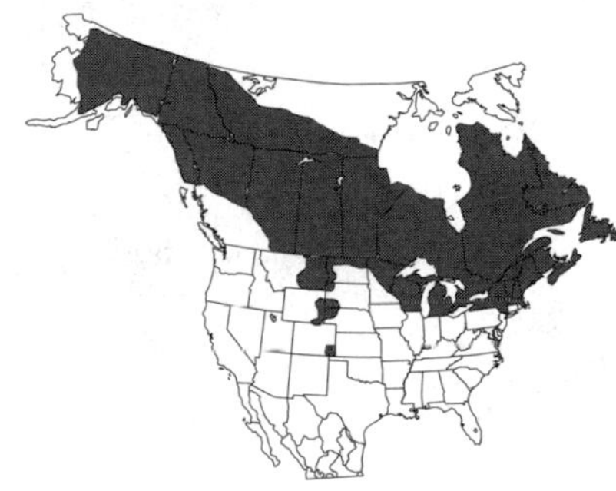

Juniperus horizontalis

Disease Genesis—Many compounds are present in *Juniperus;* the seed cones, shoots, and foliage contain various mono- and diterpenes, mainly labdane and pimarane types (San Feliciano et al. 1991a,b). Species of the genus appear to be divisible into two groups—those with oils composed almost entirely of pinane derivatives, α-pinene, e.g., *J. communis,* and those with mainly sabinane derivatives such as sabinyl acetate. Reproductive effects are most likely due to diterpene acids, such as isocupressic acid, the toxin from *P. ponderosa,* which is present in *J. communis* at high concentrations (2% d.w.) and also in *J. scopulorum* (Gardner et al. 1998a,b). Other bioactive compounds, however, are present in various species of *Juniperus.* Juniper berry oil from *J. communis* contains terpine-4-ol, which appears to be responsible for the diuretic effects (Janku et al. 1957). When extracted, the foliage yields a volatile/essential oil that is readily absorbed through the skin or mucous membranes and is thought to cause diuresis and abortion (Macht 1938). Koedam and Looman (1980) identified 27 compounds in the volatile oil of the European species *J. sabina*; almost all were monoterpenes or their derivatives. This volatile oil has been associated with increased fetal resorption in mice (Pages et al. 1989). The reproductive effects may also be due to sabinyl acetate, because the leaf oil of *Plectranthus fruticosus* in the family Lamiaceae, which is 60% sabinyl acetate and a mixture of small amounts of many other monoterpenes, is both embryotoxic and teratogenic in rats (Fournier et al. 1986; Pages et al. 1988). Malformations are mainly those of the brain and eyes—hydrocephalus, anophthalmia, and microphthalmia. In a study of the essential oil of *Salvia lavandulifolia,* also a member of the Lamiaceae, a fraction rich in sabinyl acetate

- numerous compounds, pinanes, terpineol, sabinyl acetate, diterpene acids; reproductive effect likely due to isocupressic acid, as with *Pinus*; possibly also a role for sabinyl acetate

Juniperus occidentalis

Juniperus osteosperma

Juniperus scopulorum

■ premature births, stillborns, small or weak newborns; dam may be ill

Juniperus virginiana

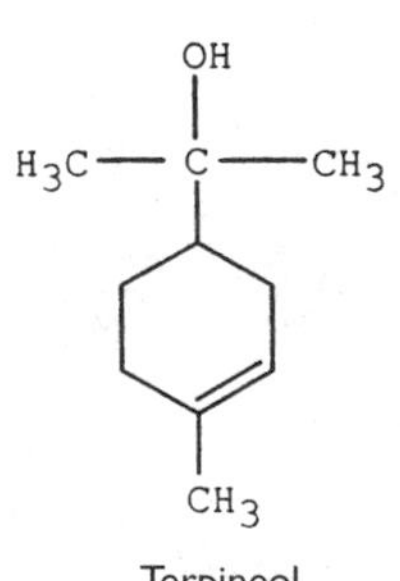

Terpineol

decreased weight gain and liver weight and caused abortions in mice (Pages et al. 1992). The volatile oils from the foliage of *J. sabina* and *J. virginiana* seem to be somewhat unique in the genus because they contain a large amount of sabinyl acetate, 37–38% and 30%, respectively (Banthorpe et al. 1973; Koedam & Looman 1980). The oil composition of *J. osteosperma* is not known, but based upon the experimental inducement of abortion in sheep (Johnson et al. 1976), it appears likely that its oil is similar to that of *J. sabina* and *J. virginiana.* The effects on sheep in contrast to the specific reproductive effects of *Pinus* on cattle may be the result of the very high concentrations of isocupressic acid in *Juniperus* or may further indicate a role for additional toxicants. Thus there seems to be a risk of abortion and other reproductive effects with high dosage of some species but not necessarily all junipers. Because the oils cause an increase in uterine tone, they should not be consumed by pregnant women.

It is of interest that antimitotic activity is among the many actions attributed to this genus. *Juniperus occidentalis, J. scopulorum, J. silicicola, J. virginiana,* and ornamentals *J. lucayana* and *J. sabina* are reported to contain podophyllotoxin, which binds with tubulin in the cell, preventing its polymerization and thus arresting cell division in metaphase (see *Podophyllum,* in chapter 14) (Hartwell & Schrecker 1958; Kupchan et al. 1965; Jardine 1980).

Clinical Signs, Pathology, and Treatment—In the event that large amounts of *Juniperus* are eaten, there may be premature parturition and the birth of small, weak, or stillborn calves or lambs. There are no pathologic changes specific for these reproductive effects. The only way to prevent the effects is to ensure that pregnant animals do not have access to junipers or have cause to eat significant amounts of the foliage.

References

Banthorpe DV, Davies HS, Gatford C, Williams SR: Monoterpene patterns in *Juniperus* and *Thuja* species. *Planta Med* 23;64–69, 1973

Ebeling W: *Handbook of Indian Foods and Fibers of Arid America.* Univ California Press, Berkeley, 1986

Fournier G, Paris M, Dumitresco SM, Pages N, Boudene C: Contribution to the study of *Plectranthus fruticosus* leaf essential oil. *Planta Med* 52;486–488, 1986

Gardner DR, Panter KE, James LF, Stegelmeier BL: Abortifacient effects of lodgepole pine *(Pinus contorta)* and common juniper *(Juniperus communis)* on cattle. *Vet Hum Toxicol* 40;260–263, 1998a

Gardner DR, Panter KE, James LF, Stegelmeier BL, Pfister JA: Diterpene acid chemistry of ponderosa pine and implications for late-term induced abortions in cattle. In *Toxic Plants and Other Natural Toxicants,* Garland T, Barr AC eds, New York, pp339–344, 1998b

Gould CM: Cypress poisoning. *Vet Rec* 74;743, 1962

Harrington HD: *Edible Native Plants of the Rocky Mountains.* Univ New Mexico Press, Albuquerque, pp242–244, 1967

Hart J: *Montana Native Plants and Early Peoples.* Montana Historical Society Press, Helena, 1992

Hartwell JL, Schrecker AW: The chemistry of *Podophyllum.* In *Progress in the Chemistry of Organic Natural Products,* Zechmeister L ed, Springer-Verlag, Vienna, pp83–166, 1958

James LF, Panter KE, Nielsen DB, Molyneux RJ: The effect of natural toxins on reproduction in livestock. *J Anim Sci* 70;1573–1579, 1992

Janku I, Hava M, Motl O: Ein diuretisch wirksamer Stoff aus wacholder (*Juniperus communis* L.). *Experientia* 13;255–256, 1957

Jardine I: Podophyllotoxins. In *Anticancer Agents Based on Natural Product Models,* Cassady JM, Douros JD eds, Academic Press, New York, pp319–351, 1980

Johnson AE, James LF, Spillett J: The abortifacient and toxic effects of big sagebrush *(Artemisia tridentata)* and juniper *(Juniperus osteosperma)* on domestic sheep. *J Range Manage* 29;278–280, 1976

Koedam A, Looman A: Effect of pH during distillation on the composition of the volatile oil from *Juniperus sabina. Planta Med* 40;22–28, 1980

Krausman PR, Kuenzi AJ, Etchberger RC, Rautenstrauch KR, Ordway LL, Hervert JJ: Diets of desert mule deer. *J Range Manage* 50;513–522, 1997

Kupchan SM, Hemingway JC, Knox JR: Tumor inhibitors. 7. Podophyllotoxin, the active principle of *Juniperus virginiana. J Pharm Sci* 54;659–660, 1965

MacDonald J: Macrocarpa poisoning. *N Z Vet J* 4;40, 1956

Macht DI: The absorption of drugs and poisons through the skin and mucous membranes. *J Am Med Assoc* 110;409–414, 1938

Mason RW: Foetal cerebral leucomalacia associated with *Cupressus macrocarpa* abortion in cattle. *Aust Vet J* 50;419, 1974

Mason RW: Attempted induction of premature birth in sheep fed *Cupressus macrocarpa. Aust Vet J* 61;192–193, 1984

Miller CN Jr: The origin of modern conifer families. In *Origin and Evolution of Gymnosperms,* Beck CB ed, Columbia Univ Press, New York, pp448–486, 1988

Munday BL, Mason RW, Cumming R: Observations on diseases of the central nervous system of cattle in Tasmania. *Aust Vet J* 49;451–455, 1973

Norris JJ, Valentine KA: *Principal Livestock Poisoning Plants of New Mexico Ranges.* N M Agric Exp Stn Bull 390, 1954

O'Scanaill T: Suspected cypress poisoning in a cow [letter]. *Irish Vet J* 40;15, 1986

Pages N, Salazar M, Chamorro G, Fournier G, Paris M, Dumitresco SM, Boudene C: Teratological evaluation of *Plectranthus fruticosus* leaf essential oil. *Planta Med* 54;296–298, 1988

Pages N, Fournier G, Chamorro G, Salazar M, Paris M, Boudene C: Teratological evaluation of *Juniperus sabina* essential oil in mice. *Planta Med* 55;144–146, 1989

Pages N, Fournier G, Velut V, Imbert C: Potential teratogenicity in mice of the essential oil of *Salvia lavandulifolia* Vahl.: study of a fraction rich in sabinyl acetate. *Phytother Res* 6;80–83, 1992

San Feliciano A, Caballero E, del Ray B, Sancho I: Diterpene acids from *Juniperus communis* subsp. *hemisphaerica. Phytochemistry* 30;3134–3136, 1991a

San Feliciano A, Miguel del Corral JM, Gordaliza M, Castro MA: Two diterpenes from leaves of *Juniperus sabina. Phytochemistry* 30;695–697, 1991b

Sloss V, Brady JW: Abnormal births in cattle following ingestion of *Cupressus macrocarpa* foliage. *Aust Vet J* 60;223, 1983

Watson FD, Eckenwalder JE: Cupressaceae. In *Flora of North America,* vol 2, *Pteridophytes and Gymnosperms,* Flora of North America Editorial Committee eds, Oxford Univ Press, New York, pp399–422, 1993

Sabinyl acetate

Isocupressic acid

30 Cycadaceae L.

■ Cycad Family
Cycas

Considered to be one of the most evolutionarily primitive families of seed-bearing plants, the Cycadaceae, or cycad family, is an ancient taxon with a well-documented fossil history extending to the Permian period of the Paleozoic (Norstog & Nicholls 1997). In the Jurassic and early Cretaceous periods its members and related taxa dominated the world's vegetation. The Mesozoic is sometimes referred to as the Age of the Cycads, albeit zoologists would likely prefer to say the Age of the Dinosaurs. Described by Linnaeus in 1753, the family was initially large and heterogeneous and contained all extant genera of cycads. However, on the basis of detailed comparative studies of the 11 genera, it was subsequently divided into 4 families and is now considered to contain only 1 genus (Johnson 1959; Stevenson 1992).

Plants treelike or herbs; perennials; evergreen; caulescent or acaulescent; producing sporangia borne in pollen and seed cones; dioecious. **Stems** subterranean or aerial; erect; columnar; 2–10 m tall; not branched; covered with persistent bases of old leaves. **Leaves** of 2 forms, foliage and scale (cataphylls) different. **Foliage Leaves** large, 2–3 m long; 1-pinnately compound, spirally clustered at stem apices, petiolate; blades oblong; leaflets circinate in bud, linear-lanceolate to falcate, 1-nerved, coriaceous to papery; petiole bases sheathing. **Cataphylls** prominent; alternating with foliage leaves; stiff; apices spinose. **Pollen Cones** terminal; erect; cylindrical; rusty yellow; microsporophylls numerous, spirally arranged; microsporangia numerous, in 2–5 groups; pollen with motile sperm. **Seed Cones** open; megasporophylls grouped loosely at stem apices, consisting of basal linear stalk and expanded apical lobe, deflexed at maturity; ovules 2 per megasporophyll, attached to stalk. **Seeds** drupelike; ovoid; flattened; seed coat differentiated into indurate inner and fleshy outer layers.

Cycas L.

■ *C. circinalis*
C. revoluta

cycad
fern palm
nut palm
sago cycas
sago palm

Taxonomy and Morphology—Among the most striking ornamentals in tropical and subtropical gardens, *Cycas* comprises about 45 species. Its name is derived from the Greek word *koikos,* a name apparently used by Theophrastus for the doum palm. The word is believed to have been transliterated to *kykas* and subsequently latinized to *Cycas* (Jones 1993). Our understanding of species relationships in the genus is incomplete, and systematic work is definitely needed (Jones 1993; Norstog & Nicholls 1997). Numerous names have been published, and there is considerable confusion about their application. It is also likely that additional species have yet to be discovered and described. Two introduced species are encountered in North America:

C. circinalis L.	crozier cycad, fern palm, false sago palm
C. revoluta Thunb.	sago cycas, sago palm, Australian nut palm

As their common names imply, these cycads, and other species of the genus, are superficially similar to palms in the family Arecaceae, and they are often mistaken for them. However, cycads and palms have very few features in common. The leaves of *Cycas* are frequently used in decoration of church altars and in funerals and other religious ceremonies (Jones 1993).

■ perennial, evergreen, herbs, treelike; leaves of 2 forms—foliage 1-pinnately compound, scale spinose; dioecious; pollen and seed cones

One of the most primitive of the seed-bearing plants, *Cycas* has motile sperm, which move to fertilize the ovule in a manner analogous to fertilization in animals. Morphological features of the genus are the same as those of the family and thus are not repeated here.

1. Margins of leaflets flat. Pollen cones 10–14 cm in diameter. Mature seeds pale reddish yellow . . . . *C. circinalis*
1. Margins of leaflets revolute. Pollen cones 4–6 cm in diameter. Mature seeds bright orange to red . . . . *C. revoluta*

Distribution and Habitat—*Cycas* is the most widely distributed cycad genus and occurs from Madagascar, and perhaps the east coast of Africa, to northern Australia and southern Japan (Norstog & Nicholls 1997). In North America, *C. circinalis* and *C. revoluta* are grown as ornamentals only in southern Florida, along the Gulf Coast, and less frequently in California. They are cultivated elsewhere as tub plants.

■ ornamentals

Disease Problems—Records of toxicity due to *Cycas* and other cycad genera begin with the report of Captain James Cook's exploration of what is now Australia (Whiting 1963; Hall 1987). His men ate roasted seeds found in an aboriginal campfire; the empty husks were apparently thought to be remnants of portions eaten by the aborigines. However, much to the sailors displeasure, the seeds were noxious and caused severe diarrhea and vomiting. Likewise, when fed to pigs, they produced severe signs several weeks later and 2 died. What was unknown to Cook's men was that the aborigines thoroughly washed and dried the seeds to remove the toxins prior to eating them (Hall 1987). On numerous occasions other Australian explorers experienced similar episodes (Low 1987). Interestingly it appears that native peoples in Australia, the Pacific Islands, and South Africa all learned to make the seeds of *Cycas* and flour from them edible by repeated washing/soaking and drying/ageing (Whiting 1963; Botha et al. 1991). The seeds of similar genera are reputedly eaten by peccaries and bears in Mexico.

■ food source, seeds eaten by humans around the world with special preparation

hepatotoxic in North America, foliage, main risk, dogs; neurotoxic also, leaves, seeds

Cycas revoluta

Eventually the seeds, leaves, and pulp of *Cycas* and other cycads were recognized to be toxic to livestock, typically producing a neurologic disease in cattle and hepatotoxicosis in sheep (Anderson & Hall 1964; Hall & McGavin 1968; Gabbedy et al. 1975; Hooper 1978). The seeds also have been suspected to be the cause of amyotrophic lateral sclerosis (ALS) in the Chamorro people of the Mariana Islands of the Pacific (Whiting 1963). A number of cases of hepatotoxicity in dogs consuming the leaves, stems, or seeds of *Cycas* have occurred (Senior et al. 1985; Botha et al. 1991; Albretsen et al. 1998). Dogs chewing the stems developed marked effects within 1 hour. There is typically evidence of liver involvement, and animals often recover. Ingestion of seeds by dogs often seems to be associated with neurologic as well hepatic effects (Albretsen et al. 1998). Cattle seem to be predisposed to eat the leaves; and sheep, both seeds and leaves. In Okinawa, cattle eating the leaves of *C. revoluta* developed ataxia and paralysis of the pelvic limbs and lesions in the white matter of the cervical and lumbar spinal cord (Kobayashi et al. 1984). In some animals, both hepatotoxic and neurotoxic effects occurred. The type of disease manifested may be a function of the plant part eaten, although this has not been confirmed experimentally.

- hepatotoxin, glycosides cycasin, macrozamin, leaves, seeds, hydrolysis in plant or animal to MAM

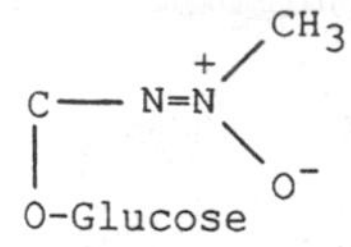

Cycasin

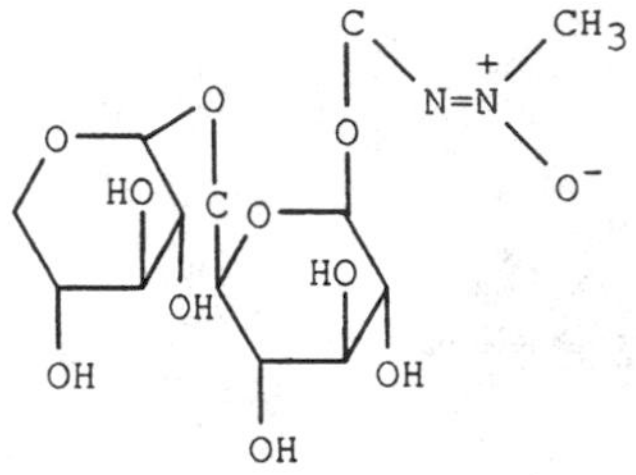

Macrozamin

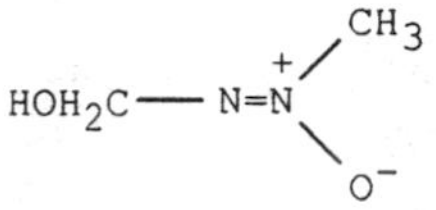

Methylazoxymethanol (MAM)

Disease Genesis—*Cycas* contains two types of toxins—glycosides of methylazoxymethanol (MAM) and a neurotoxic amino acid, β-*N*-methylamino-L-alanine (L-BMAA). The principal glycosides present are cycasin, which is a β-glucoside, and macrozamin, which is a β-primeveroside. Several other minor glycosides, neocycasins, also are present (Yagi & Tadera 1987). All of these glycosides are subject to hydrolysis by β-glycosidases in the plant or, more important, by bacterial enzymes in the animal's digestive tract to yield the aglycone and sugars (Spatz et al. 1967). The aglycone MAM is the toxic portion, alkylating DNA and RNA and causing hepatotoxicosis (Laqueur & Spatz 1968). The type of disease produced varies somewhat from species to species, both cycad and animal, but in general, hepatotoxicosis seems to be the major concern. Seawright and co-workers (1998) suggested that the early work demonstrating neurotoxic rather than hepatotoxic effects in cattle caused by *Bowenia*, another cycad, might have been the result of storage of the plants at cold temperatures, which caused the loss of MAM. When fed immediately, fresh plants, containing MAM, regularly cause hepatotoxic rather than neurotoxic effects.

As might be expected, MAM with its capacity to alkylate DNA and RNA is carcinogenic, mutagenic, neurotoxic, and teratogenic, in addition to being hepatotoxic. Whether this capability causes significant problems in animals is unknown (Smith 1966; Spatz et al. 1967; Tustin 1983). The teratogenic potential may be of particular concern (Keeler 1972). Glycosides are found in all parts of *Cycas*. Cycasin concentrations are highest in the leaves, lower in the seeds, and lowest in the pith of the trunk. In other cycad genera, the highest concentrations of macrozamin are found in the seeds (Yagi & Tadera 1987), and the same might be expected in *Cycas*.

- neurotoxin, L-BMAA, excitatory activity on NMDA glutamate receptors, resulting in neuronal degeneration; MAM may also be neurotoxic

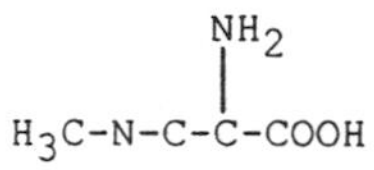

L-BMAA

The second toxicant present in *Cycas*, L-BMAA, is an unusual neurotoxic amino acid found in the seeds. Other unusual amino acids such as cycasindene and cycasthioamide also are present in the seeds of *C. revoluta* (Pan et al. 1997). L-BMAA is present at levels of 0.5–1 mg/g wet weight in *C. circinalis* and about 0.3 mg/g in *C. revoluta* (Duncan et al. 1989; Charlton et al. 1992). Levels in fresh leaves of *C. circinalis* are 1.3–1.7 mg/g. L-BMAA is similar in structure and action to the neurolathyrogens, except that the glutamate receptor of major concern is the *N*-methyl-D-aspartate (NMDA) type (Ross et al. 1987). However, L-BMAA, which acts as a weak excitotoxin in some circumstances, also affects the kainate- and quisqualate-type glutamate receptors (Weiss et al. 1989). It has been suggested that L-BMAA itself may not be the culpable toxin but rather that, in the presence of bicarbonate, the α-carbamate form of the L-α–amino acid forms a more specific structural analog of NMDA and interactant at the receptor (Nunn et al. 1991). It also has been suggested that the problem in diseases such as ALS may be a defect in L-α–amino acid–cysteine metabolism and that the other changes occur as secondary problems (Olney et al. 1991). A role for these types of compounds as inducers of apoptosis in the brain is suggested by studies with a cycad from another family, *Lepidozamia peroffskyana* (Gobe 1994).

L-BMAA is subject to active uptake and transport across the blood-brain barrier via the neutral amino acid carrier (Smith et al. 1992). High dosage of L-BMAA fed to macaques produced motor neuron dysfunction in 2–12 weeks (Spencer et al. 1987). This is consistent with a supportive role for cycads as etiologic agents of ALS in humans (Spencer et al. 1993). There remains doubt, however, about the role of these plants in the disease (Duncan et al. 1988; Garruto et al. 1988). Only very low concentrations remain in the seed flour after processing. In addition, the effects in other experiments appear to be more closely associated with other factors such as decreased calcium and increased aluminum and manganese in the diet. These observations do not preclude the possibility of L-BMAA acting as an acute neurotoxicant in animals.

Although it is less potent than the neurolathyrogen L-BOAA, it is nevertheless an excitatory neurotoxicant; however, it takes hours rather than minutes to produce the presumably indirect actions (Nunn et al. 1987).

The respective roles of MAM and L-BMAA in the neurotoxic effects of cycad toxicity are not clear. From the work of Hirono et al. (1969) in the 1960s, it became apparent that both MAM and its hydrolysis product cycasin were neurotoxic. However, in these studies the effect occurred only with oral or parenteral administration in the first 24 hours after birth. Lesions were confined to the cerebellum, especially the granular layers (Hirono et al. 1969; Jones et al. 1972; Sanger et al. 1972). Similar effects have been produced in several animal species, although susceptibility and manifestations vary (Haddad et al. 1972). There is necrosis and cell loss in the granular layers and long-term disorganization of Purkinje cells. Several weeks after administration of MAM to newborn animals, signs such as tremors, swaying, goose-stepping, and falling may be apparent. The role of MAM in the adult is not clear, but it appears, based on experiments with newborns, that the compound is prevented access to the central nervous system by the blood-brain barrier, a barrier that is incomplete at birth. There is an active transport mechanism for cycasin across the intestinal mucosa via the Na^+/glucose transporter (Hirayama et al. 1994), but whether there are factors or conditions that promote access to the central nervous system in adults is not clear. It is of interest that L-BMAA administered to young rats also causes selective cerebellar damage, although not to granule cells (Seawright et al. 1990).

The exclusion of MAM from the brain may be less of a factor if it or its hydrolysis products are further biotransformed in the rumen to specific neurotoxins, as suggested by Seawright and co-workers (1998). Further evidence of a neurotoxic role for cycasin is epidemiological data about the incidence of ALS in the Chamorro people of Guam. Concentrations of cycasin and L-BMAA in the flour made from cycas seeds are quite variable, depending on the preparation. In areas with a high incidence of ALS, the preparation methods do not adequately reduce the toxin levels (Kisby et al. 1992). Sometimes cycasin concentrations are reduced to equal those of L-BMAA, but in other instances there is 10 times more cycasin remaining. Furthermore, daily ingestion of cycasin for several months produced spinal lesions in goats—axonal swelling and demyelination in the lateral and dorsal funiculi (Shimizu et al. 1986). Interestingly, 2 of 4 goats displaying spinal lesions also developed liver necrosis.

Clinical Signs—Because *Cycas* is so limited in distribution and abundance in North America, intoxications are most likely to occur in pets eating the seeds or leaves. In dogs, the initial sign is vomiting, which may begin within minutes of ingestion and may persist, with periodic episodes, for hours. There may also be increased salivation and increased water consumption. During the next few days anorexia, diarrhea or constipation, and icterus will be observed, in conjunction with a marked decrease in blood platelets and an increase in serum bilirubin (especially conjugated), hepatic enzymes, urea nitrogen, and creatinine. At this point the prognosis is guarded because liver involvement may be life threatening. Ingestion of the seeds may be accompanied by neurotoxic as well as hepatotoxic effects.

■ dogs, hepatotoxic, persistent vomiting, diarrhea or constipation, colic, anorexia, icterus, depression; neurologic signs in some cases

neurologic effects, weakness, ataxia, limb placement problems

A similar disease occurs in sheep but may, on occasion, be subacute to almost chronic. Emaciation, pale to icteric mucous membranes, and other evidence of liver disease are observed. The animals will be lethargic, but there is seldom diarrhea. In cattle, the neurologic disease occurs after consumption of the leaves for several weeks or more. The signs are those related to a proprioceptive deficit and progressive weakness in the pelvic limbs: stumbling, staggering, incoordination, and sometimes goose-stepping. This disease has been observed in the Caribbean and other tropical areas but not in continental North America (Hall & McGavin 1968; Mason & Whiting 1968). The disease in primates and presumably humans is a motor neuron dysfunction.

■ gross pathology, enlarged, nutmeg liver

microscopic, hepatic necrosis or fibrosis and bile duct proliferation; neurologic, microscopic, axonal and myelin degeneration, spinal

Pathology—In dogs, sheep, and cattle displaying both liver and brain involvement, the characteristic gross change is an enlarged, nutmeg liver. Microscopically, there is hepatic necrosis, which typically is centrilobular. In more-chronic cases fibrosis, bile duct proliferation, bile stasis, fatty change, and moderately enlarged hepatocytes may be noted. Additionally, there may be scattered small splotchy hemorrhages on viscera, as well as mild tubular nephrosis.

Cattle exhibit no distinctive gross changes in instances in which the disease is exclusively neurologic, but there are numerous microscopic lesions. Bilaterally symmetrical degeneration in the ventral, lateral, and dorsal columns is present through the length of the cord but is especially evident in the cervical portion (Reams et al. 1993). Axonal and myelin degeneration is particularly noticeable in the fasciculus gracilis and the dorsal spinocerebellar tracts. This implies severe effects on proprioception and sensory input to the cerebellum related to the pelvic limbs (Hall & McGavin 1968; Mason & Whiting 1968). The lesions in primates are more closely related to motor neuron dysfunction, as there is degeneration of giant Betz cells and pyramidal neurons in the motor cortex and of the large anterior horn neurons of the cord (Spencer et al. 1987).

■ no specific treatment, activated charcoal, nursing care

Treatment—Supportive care as for hepatic disease in general is indicated. No specific treatments are known for either the hepatic or the neurologic syndromes.

References

Albretsen JC, Khan SA, Richardson JA: Cycad palm toxicosis in dogs: 60 cases (1987–1997). *J Am Vet Med Assoc* 213;99–101, 1998

Anderson JL, Hall WT: Neurotoxic effects from cycad leaves. *Fed Proc* 23;1349, 1964

Botha CJ, Naude TW, Swan GE, Ashton MM, Van Der Wateren JF: Suspected cycad *(Cycas revoluta)* intoxication in dogs. *J S Afr Vet Assoc* 62;189–190, 1991

Charlton TS, Marini AM, Markey SP, Norstog K, Duncan MW: Quantification of the neurotoxin 2-amino-3-(methylamino)-propanoic acid (BMAA) in Cycadales. *Phytochemistry* 31;3429–3432, 1992

Duncan MW, Kopin IJ, Garruto RM, Lavine L, Markey SP: 2-Amino-3 (methylamino)-propionic acid in cycad-derived foods is an unlikely cause of amyotrophic lateral sclerosis/parkinsonism. *Lancet* 1;631–632, 1988

Duncan MW, Kopin IJ, Crowley JS, Jones SM, Markey SP: Quantification of the putative neurotoxin 2-amino-3-(methylamino)-propanoic acid (BMAA) in Cycadales: analysis of the seeds of some members of the family Cycadaceae [corrected]. *J Anal Toxicol* 13;Suppl A/NG, 1989

Gabbedy BJ, Meyer EP, Dickson J: Zamia palm *(Macrozamia reidlei)* poisoning of sheep. *Aust Vet J* 51;303–305, 1975

Garruto RM, Yanagihara R, Gajdusek DC: Cycads and amyotrophic lateral sclerosis/parkinsonism dementia. *Lancet* 2;1079, 1988

Gobe GC: Apoptosis in brain and gut tissue of mice fed a seed preparation of the cycad *Lepidozamia peroffskyana. Biochem Biophys Res Commun* 205;327–333, 1994

Haddad RK, Rabe A, Dumas R: Comparison of effects of methylazoxymethanol acetate on brain development in different species. *Fed Proc* 31;1520–1523, 1972

Hall WTK: Cycad *(Zamia)* poisoning in Australia. *Aust Vet J* 64;149–151, 1987

Hall WTK, McGavin MD: Clinical and neuropathological changes in cattle eating the leaves of *Macrozamia lucida* or *Bowenia serrulata* (family Zamiaceae). *Pathol Vet* 5;26–34, 1968

Hirayama B, Hazama A, Loo DF, Wright EM, Kisby GE: Transport of cycasin by the intestinal Na+/glucose cotransporter. *Biochem Biophys Acta* 1193;151–154, 1994

Hirono I, Shibuya C, Hayashi K: Induction of a cerebellar disorder with cycasin in newborn mice and hamsters. *Proc Soc Exp Med Biol* 131;593–599, 1969

Hooper PT: Cycad poisoning in Australia—etiology and pathology. In *Effects of Poisonous Plants on Livestock,* Keeler RF, Van Kampen KR, James LF eds, Academic Press, New York, pp337–347, 1978

Johnson LAS: The families of cycads and Zamiaceae of Australia. *Proc Linn Soc N S W* 84:64–117, 1959

Jones DL: *Cycads of the World.* Smithsonian Institution Press, Washington, DC, 1993

Jones M, Yang M, Mickelsen O: Effects of methylazoxymethanol glucoside and methylazoxymethanol acetate on the cerebellum of the prenatal Swiss albino mouse. *Fed Proc* 31;1508–1511, 1972

Keeler RF: Known and suspected teratogenic hazards in range plants. *Clin Toxicol* 5;529–565, 1972

Kisby GE, Ellison M, Spencer PS: Content of the neurotoxins cycasin (methylazoxymethanol β-D-glucoside) and BMAA (β-*N*-methylamino-L-alanine) in cycad flour prepared by Guam Chamorros. *Neurology* 42;1336–1340, 1992

Kobayashi A, Tadera K, Yagi F, Kono I, Sakamoto T, Yasuda N: Studies on poisoning of grazing cattle due to ingestion of *Cycas revoluta* Thunb. *Bull Fac Agric, Kagoshima Univ* 34;119–129, 1984

Laqueur GL, Spatz M: Toxicology of cycasin. *Cancer Res* 28;2262–2267, 1968

Low T: Explorers and poisonous plants. In *Toxic Plants and Animals: A Guide for Australia,* Covacevick J, Davie P, Pearn J eds, Queensland Museum, South Brisbane, Australia, pp65–70, 1987

Mason MM, Whiting MG: Caudal motor weakness and ataxia in cattle in the Caribbean area following ingestion of cycads. *Cornell Vet* 58;541–554, 1968

Norstog KJ, Nicholls TJ: *The Biology of the Cycads.* Cornell Univ Press, Ithaca, NY, 1997

Nunn PB, Davis AJ, O'Brien P: Carbamate formation and the neurotoxicity of L-α–amino acids. *Science* 251;1619, 1991

Nunn PB, Seelig M, Zagoren JC, Spencer PS: Stereospecific acute neuronotoxicity of "uncommon" plant amino acids linked to human motor-system diseases. *Brain Res* 410;375–379, 1987

Olney JW, Zorumski C, Price MT, Labruyere J: Response to carbamate formation and the neurotoxicity of L-α–amino acids. *Science* 251;1619–1620, 1991

Pan M, Mabry TJ, Beale JM, Mamiya BM: Nonprotein amino acids from *Cycas revoluta. Phytochemistry* 45;517–519, 1997

Reams RY, Janovitz EB, Robinson FR, Sullivan JM, Casanova CR, Mas E: Cycad *(Zamia puertoriquensis)* toxicosis in a group of dairy heifers in Puerto Rico. *J Vet Diagn Invest* 5;488–494, 1993

Ross SM, Seelig M, Spencer PS: Specific antagonism of excitotoxic action of "uncommon" amino acids assayed in organotypic mouse cultures. *Brain Res* 425;120–127, 1987

Sanger VL, Yang M, Mickelsen O: Cycasin-induced central nervous system lesions in postnatal mice. *Fed Proc* 31;1524–1528, 1972

Seawright AA, Brown AW, Nolan CC, Cavanagh JB: Selective degeneration of cerebellar cortical neurons caused by cycad neurotoxin, L-β–methylaminoalanine (L-BMAA), in rats. *Neurol Appl Neurobiol* 16;153–169, 1990

Seawright AA, Oelrichs PB, Ng JC, Sani Y: The toxicity of the Australian cycad *Bowenia serrulata* to cattle. In *Toxic Plants and Other Natural Toxicants,* Garland T, Barr AC eds, CAB International, New York, pp447–452, 1998

Senior DF, Sundlof SF, Buergelt CD, Hines SA, O'Neil-Foil CS, Meyer DJ: Cycad poisoning in the dog. *J Am Anim Hosp Assoc* 21;103–109, 1985

Shimizu T, Yasuda N, Kono I, Yagi F, Tadera K, Kobayashi A: Hepatic and spinal lesions in goats chronically intoxicated with cycasin. *Jpn J Vet Sci* 48;1291–1295, 1986

Smith DWE: Mutagenicity of cycasin aglycone (methylazoxy-methanol), a naturally occurring carcinogen. *Science* 152;1273–1274, 1966

Smith QR, Nagura H, Takada Y, Duncan MW: Facilitated transport of the neurotoxin, β-*N*-methylamino-L-alanine, across the blood-brain barrier. *J Neurochem* 58;1330–1337, 1992

Spatz M, Smith DWE, McDaniel EG, Laqueur GL: Role of intestinal microorganisms in determining cycasin toxicity. *Proc Soc Exp Biol Med* 124;691–697, 1967

Spencer PS, Nunn PB, Hugon J, Ludolph AC, Ross SM, Roy DN, Robertson RC: Guam amyotrophic lateral sclerosis-parkinsonism-dementia linked to a plant excitant neurotoxin. *Science* 237;517–522, 1987

Spencer PS, Kisby GE, Ross SM, Roy DN, Hugon J, Ludolph AC, Nunn PB: Guam ALS-PDC: possible causes [letter]. *Science* 262;825–826, 1993

Stevenson D: A formal classification of the extant cycads. *Brittonia* 44;220–223, 1992

Tustin RC: Notes on the toxicity and carcinogenicity of some South African cycad species with special reference to that of *Encephalartos lanatus*. *J S Afr Vet Assoc* 54;33–42, 1983

Weiss JH, Koh J-Y, Choi DW: Neurotoxicity of β-*N*-methylamino-L-alanine (BMAA) and β-*N*-oxalylamino-L-alanine (BOAA) on cultured cortical neurons. *Brain Res* 497;64–71, 1989

Whiting MG: Toxicity of cycads. *Econ Bot* 17;271–302, 1963

Yagi F, Tadera K: Azoxyglycoside contents in seeds of several cycad species and various parts of Japanese cycad. *Agric Biol Chem* 51;1719–1721, 1987

Datiscaceae Lindl. 31

A small family of 3 genera and 4 species, the Datiscaceae is of little economic value. It is generally considered related to Begoniaceae. Two monotypic genera are arborescent and occur in India and Southeast Asia. The third genus is herbaceous and occurs in southwestern North America. The morphological features of the family are given in the genus treatment.

■ Datisca Family
Datisca

Datisca L.

Taxonomy and Morphology—*Datisca* comprises only 2 species, 1 in Asia and 1 in North America, where it is sometimes grown as novelty.

■ *D. glomerata*
durango root

D. glomerata (C.Presl) Baill. durango root

Plants herbs; perennials; herbage glabrous; dioecious or polygamous. **Stems** erect; 100–200 cm tall; clustered; branched above. **Leaves** 1-pinnately compound; alternate above and appearing opposite or whorled below; petiolate; margins serrate to dentate; stipules absent. **Inflorescences** axillary clusters of flowers; bracts absent. **Flowers** imperfect or perfect; staminate and pistillate flowers similar; radially symmetrical; perianths in 1-series. **Sepals** 3–8. **Petals** absent. **Stamens** 8–12; opposite sepals. **Pistils** 1; compound; carpels 3; stigmas 3; styles 3, bifid; ovaries inferior; locules 1; placentation parietal. **Fruits** capsules; ovoid; 3-angled. **Seeds** numerous; small; ellipsoidal; pitted in rows; light brown.

■ erect, perennial herbs; leaves 1-pinnately compound, upper alternate; flowers in axillary clusters; petals absent; fruits ovoid, 3-angled capsules

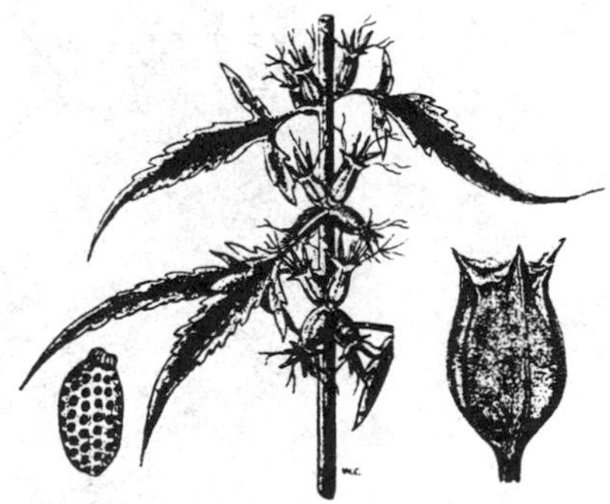

Datisca glomerata

Distribution and Habitat—*Datisca glomerata* favors dry stream beds and washes in the foothills and mountains up to 2,000 m.

■ dry stream beds and washes

Disease Problems—Essentially limited to California as a cause of disease, *D. glomerata* is an occasional cause of digestive tract problems in cattle and possibly sheep and other ruminants (Wagnon & Hart 1945; Fuller & McClintock 1986; Galey et al. 1990). Like so many other taxa causing intoxications, it seems to be a problem only when all other forage has been depleted. Fuller and McClintock (1986) described a number of cases in which small numbers of cattle died in such circumstances. Association of the species with disease was confirmed experimentally by administration of dry leaves and capsules to cattle (Wagnon & Hart 1945). Sheep were very reluctant to eat the plants and appear to be at less risk. In most instances only a small amount of plant material is eaten, and the response is likely to be transient diarrhea of only a day's duration. However, *Datisca* may cause a more severe systemic response not entirely due to the apparent digestive tract effects, as shown by animals that die quickly and quietly after displaying extreme lethargy and weakness (Galey et al. 1990).

■ ruminants digestive disturbance, transient diarrhea, sometimes more severe with extreme weakness, may be fatal

Datisca glomerata

- terpenoids, datiscosides, glycosides of cucurbitacins; severe irritation

Datiscosides

- abrupt onset, lethargy, diarrhea, weakness, dehydration

- gross pathology, severe reddening, abomasum, small intestine, leaves in rumen

  treat with activated charcoal, fluids, electrolytes

Disease Genesis—*Datisca glomerata* contains a series of terpenoids. Datiscosides A-H are glycosides of cucurbitacin D except for B and H, which are glycosides of cucurbitacin F. The latter differs in having a hydroxy rather than an oxy group at C-3 (Connolly & Hill 1991). Plants also contain a nonglycosidic triterpenoid, datiscacin. These compounds, similar to those of the Cucurbitaceae, appear to be likely causes of the digestive problems and perhaps the lethal effects, because they are known to cause other plant-associated intoxications.

Clinical Signs—Following ingestion of *D. glomerata,* there is abrupt onset of lethargy, diarrhea, and subsequent dehydration. This leads to recumbency and death within a few hours in severe cases. The weakness may be severe enough that animals are unable to raise their heads off the ground, even when lying down.

Pathology and Treatment—There are few significant lesions other than the severe irritant effects and reddening of the mucosa of the abomasum and small intestine. The leaves of *D. glomerata* are likely to be identifiable in the rumen. Supportive care, including fluids and electrolytes, will in most instances be adequate.

References

Connolly JD, Hill RA: *Dictionary of Terpenoids.* Vol 2, *Di- and Higher Terpenoids,* Chapman & Hall, London, pp1174–1256, 1991

Fuller TC, McClintock E: *Poisonous Plants of California.* Univ California Press, Berkeley, pp127–128, 1986

Galey FD, Johnson BJ, Bruce R, McCaskill J: A case of *Datisca glomerata* poisoning in beef cattle. *Vet Hum Toxicol* 32;575– 576, 1990

Wagnon KA, Hart CH: Durango root *(Datisca glomerata)* poisoning of range stock. *J Am Vet Med Assoc* 107;3–5, 1945

Dennstaedtiaceae Ching

32

Cosmopolitan in distribution and comprising about 20 genera and 400 species, the Dennstaedtiaceae is commonly known as the bracken fern family. With greatest abundance in the tropics, the family is circumscribed either broadly or narrowly, with the number of included genera ranging from 8 to more than 60 (Holttum 1947; Page 1976). In North America, 4 genera and 6 species are present; only one is of toxicologic interest.

■ Bracken Family
Pteridium

Plants herbs; perennials; producing sporangia in sori on abaxial surfaces of fronds. **Fronds** all alike; vernation circinate, 1-to-4-pinnately compound; stipitate; erect to spreading; veins free or anastomosing; pinnae stalked, opposite or subopposite; stipes jointed, grooved or not grooved adaxially. **Sori** present; borne on revolute margins; linear; false indusia formed by revolute margins, partially concealing sporangia. **Sporangia** all alike. **Spores** all alike; tetrahedral to globose. **Gametophytes** all alike; green; cordate.

■ ferns; sporangia in sori borne on revolute margins

Pteridium Gled. ex Scop.

Taxonomy and Morphology—*Pteridium* is now recognized to comprise only a single, polymorphic species (Tryon 1941; Page 1976). Its name is derived from the Greek *pteridion,* meaning "small fern," somewhat of a misnomer because the fronds may be as long as 4.5 m. Two subspecies and 12 varieties are distinguished. In older references, the varieties were typically treated as species and binomials such as *P. caudatum, P. latiusculum,* and *P. pseudocaudatum* appear. In other areas of the world, other species names sometimes used are *P. esculentum* and *P. revolutum.* Four varieties are present in North America:

■ *P. aquilinum*
bracken
bracken fern
brake fern
braken

P. aquilinum (L.) Kuhn (= *Pteris aquilina* L.)	bracken fern, brake fern, bracken, braken
var. *caudatum* (L.) Sadeb.	lacy bracken
var. *latiusculum* (Desv.) Underw. ex A.Heller	eastern bracken
var. *pseudocaudatum* (Clute) A.Heller	tailed bracken
var. *pubescens* Underw.	western bracken

Plants from rhizomes, forming large, dense colonies. **Roots** dark brown or black; branching; thick. **Rhizomes** black; woody; branching; scales absent. **Fronds** erect; 0.5–4.5 m long; deciduous; croziers silver, clawlike; stipitate; erect to spreading; blades deltoid; veins not anastomosing; pinnae stalked, opposite or subopposite, proximally 2-pinnately compound and distally 1-pinnately compound; rachises yellow-brown, curved; stipes jointed, grooved adaxially. **Sori** present; rare; contiguous; borne on

■ deciduous fern from black, woody rhizomes; fronds erect, pinnately compound; sporangia partially concealed by revolute margins of fronds

revolute margins; linear; false indusia formed by revolute margins, partially concealing sporangia.

■ extensive stands in well drained soils of open woodlands

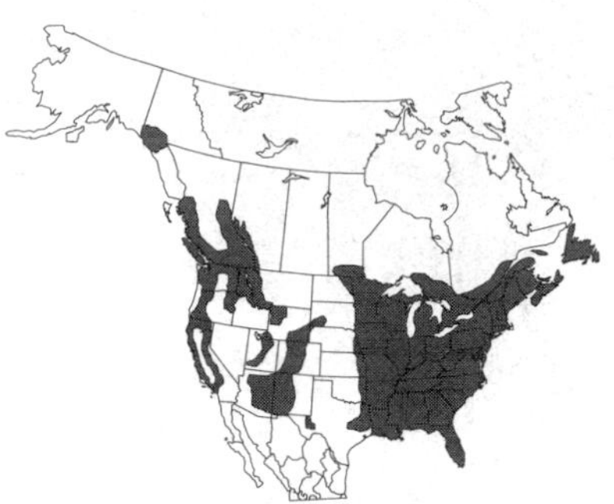

Pteridium aquilinum

Distribution and Habitat—*Pteridium aquilinum* is nearly cosmopolitan in distribution. Page (1976) contended that it is one of the most widespread species of vascular plants and is exceeded in geographical range only by a few annual weeds. It is found throughout North America. As their common names indicate, var. *pubescens* occurs in the western half and var. *latiusculum* in the eastern half. Variety *pseudocaudatum* is present in the southeastern quarter of the continent, and var. *caudatum* is restricted to southern Florida. Occupying a diversity of sites, plants will typically be found in semishaded habitats such as dry, open woodlands. Occasionally there may be extensive stands along pasture edges, in logged areas, and along roadsides. Moisture requirements are modest in comparison with those of other ferns, and good drainage is a necessity. Plants may grow well on rocky or gravelly areas with low soil fertility, but poorly drained areas will not sustain growth.

■ horses, neurologic

ruminants, bone marrow depression, neoplasia, retinal degeneration

Disease Problems—Widespread throughout the world, *P. aquilinum* is not only one of the most abundant plants but also one of the most extensively studied in terms of its toxicity (Page 1976; Taylor 1990). It is common in many parts of North America and has been associated with several different disease syndromes of both monogastric animals (mainly horses) and ruminants. The primary problem in horses is a deficiency of thiamine, whereas in ruminants it is bone marrow depression and/or neoplasia.

■ horses, neurologic, plants fresh or dry, ingestion for 2–3 weeks, slow onset of incapacitation

In the 1800s a disease in horses different from the typical syndrome in cattle was recognized in the United Kingdom and the Pacific Coast states of North America. Typically seen in the winter months, it caused in some instances a number of deaths. Eventually *Pteridium* contamination of hay was incriminated as the cause, and experimental studies confirmed the association (Hadwen & Bruce 1920; Roberts et al. 1949). A diet composed of 20–25% bracken for 3 or more weeks resulted in the development of a subacute neurotoxicosis. Following onset, the signs progressively worsened, leading to death in about 1 week. Although the most common source of plants is contaminated hay, intoxication occasionally follows consumption of green bracken in poor pastures or when it contaminates bedding. Higher levels of the fern in the forage may result in quicker onset or more rapidly developing signs. Although the neurologic syndrome is the problem typically associated with horses, *P. aquilinum* was suspected as a cause of a very low hematocrit and dark-colored urine in 2 horses eating plants growing in a pasture in the northeastern United States (Kelleway & Geovjian 1978).

Pigs are less sensitive to intoxication, and they readily consume the rhizomes, the most toxic portion of the plant (Harvey et al. 1944). It is only in rare instances where plants are plowed up, exposing the rhizomes, that pigs subsisting primarily on them for a prolonged time may develop signs of thiamine deficiency. Polioencephalomalacia (PEM) has been reported in sheep feeding heavily on the fronds or the rhizomes for a prolonged period, but this is a rare occurrence (Bakker et al. 1980; Chick et al. 1981).

Pteridium aquilinum

The first distinct although perhaps not conclusive linking of *P. aquilinum* with disease in cattle was in the United Kingdom in a series of experiments in the early 1900s, which were a follow-up to field observations reported in 1894 (Almond 1894; Stockman 1917; 1922). The clinical description so lucidly detailed by Stockman was thereafter amplified in long-term studies of the disease by investigators in Wales (WC Evans et al. 1954a,b,c, 1958). The disease has since been recognized as one of the most ubiquitous of all the plant intoxication problems. Research is conducted throughout the world, generating numerous publications and international meetings.

The most common disease problem in ruminants is severe bone marrow depression or the so-called radiomimetic effect (Evans 1968). Appearance of the problem requires ingestion of *P. aquilinum* over several weeks or months in amounts similar to those producing thiamine deficiency in horses. It is not, however, a thiamine-responsive disease. The disease is characterized by severe hypoplasia of the marrow, with all cellular elements affected. Because of the rapid turnover and short lifespan of certain cell types, the disease is manifested primarily as a deficiency of blood platelets and neutrophils. There may be an early mild decline in red blood cells due to hemorrhage, but anemia due to decreased cell production occurs only terminally, because of the long life span of the circulating mature red blood cell. Usually the deficiency of platelets and neutrophils is of such severity that the animal does not survive long enough to develop an anemia. In addition there are marked increases in heparin-like activity in blood and an increase in fibrinogen and capillary fragility, as well as platelet changes affecting coagulation and clot retraction (Evans et al. 1954b; Evans 1968; Yamane et al. 1975a,b,c).

■ ruminants, large amounts ingested for several weeks or more, bone marrow depression, platelet deficiency, coagulation problems, hemorrhage

Because of the coagulation changes there is concern about possible effects on pregnancy when hemorrhage could precipitate embryonic or fetal death. Although there appear to be no direct lethal effects on the fetus, it has been difficult to confirm or rule out possible teratogenic effects of *Pteridium* (Yasuda et al. 1974). Growth suppression may be apparent with retardation of growth and ossification, effects most likely related to thiaminase activity. Further studies have shown a decrease in fertility of the dam and developmental retardation and neurobehavioral disturbances in the offspring when female rats, beginning at 35 days of age, were fed a diet composed of 30% bracken for 7.5 months (Gerenutti et al. 1992). Epidermal-derived tissues seem to be especially vulnerable.

During the early years of bracken studies in the United Kingdom, a disease of the bladder in ruminants was being recorded in various areas of the world but was not associated with a specific cause (Kalkus 1913). This disease, commonly called enzootic hematuria, cystic hematuria, or cystic hemangiomatosis was also given the more colorful name *red water*. It was eventually shown to be due to consumption of small amounts of bracken for prolonged periods and to involve bladder lesions, some of which were neoplastic (Rosenberger 1971). In areas where *P. aquilinum* is abundant, there continue to be occurrences of bladder neoplasia in cattle (McKenzie 1978; Karcher et al. 1988). Experimentally, high dosages of bracken in cattle for 5–6 months caused acute death, while lower dosages for 9–36 months produced the bladder lesions, some of which were neoplastic (Pamukcu et al. 1967a). These chronic effects on the bladder, which may persist in a cyclic manner for a long period, eventually lead to inflammation, erosions, and papillomatous growths and may predispose the development, in a few years, of a variety of neoplasms such as hemangiomas and more-invasive carcinomas (Price & Pamukcu 1968; Pamukcu et al. 1976). The frequency and malignancy of the tumors increase with larger dosage of the fern (Yoshikawa et al. 1981). Hematuria and bladder neoplasms predominate in cattle, but the tissues involved in sheep are more varied, including papillomas, sarcomas, and carcinomas of the jaw and digestive tract (McCrea & Head 1978; 1981). Tumor distribution varies among animal species, possibly related to influences of intestinal pH differences. For example, neoplasms also occur in the digestive tract of the rat (Schacham et al. 1970). The range of species susceptibility to bracken carcinogenicity is exemplified by the gut, liver, and kidney tumors occurring in toads that were force-fed bracken for 2–5 months (El-Mofty et al. 1980).

■ cattle, small amounts, long-term ingestion, "enzootic hematuria," cyclic inflammation of urinary bladder with papillomas and other neoplasms

sheep and other animals, neoplasms in intestine and other tissues

The etiology of the bladder neoplasms has been a continuing concern. They are rather uncommon in most areas of North America. In one study of slaughter cattle in Wisconsin, only 25 bladders (11 neoplasms) with proliferative lesions were noted in 83,000 cattle (Brobst & Olson 1963). Later studies revealed the presence of virus particles (papilloma-like or Herpes-type) in bovine bladder neoplasms (Olson et al. 1965; Kaminjolo et al. 1972). Studies in the Scottish Highlands on the various neoplasms associated with bracken, sometimes several in the same animals, led to further speculation on the role of viruses (Jarrett 1978; Jarrett et al. 1978). Although bracken fern certainly is a factor in the development of enzootic hematuria, its relationship with viruses in the development of neoplasia is not clear. Initial studies failed to demonstrate any protective effect against enzootic hematuria when formalinized wart (papilloma) vaccine was administered repeatedly (Pamukcu et al. 1967b), but immunosuppression allowing expression of tumors induced by bovine papilloma virus (BPV) tumors remains a possibility (Rao et al. 1987). In one study, type-2 BPV DNA could be found in about half of the natural and experimental cases (Campo et al. 1992). Other studies have shown an interaction between viruses and the plant flavonoid quercetin (Pennie & Campo 1992). Should viruses be directly involved with papilloma development, there is the opportunity for the use of vaccines (Hopkins 1986).

■ humans, possibly neoplastic, esophagus, stomach, intestine; from eating fern croziers or drinking milk from cows eating bracken

Pteridium aquilinum has a long history of use by humans. The rhizomes were widely used in Europe by poorer classes as flour for making bread (Rymer 1976; Veitch 1990). Maoris in New Zealand and other peoples in Australasia used the rhizomes in a similar manner. Interesting folklore developed around the species, including a belief that burning it caused rain. In 1636 a request in England was made for all to abstain from burning bracken while the king was passing Staffordshire, to ensure fair weather for his journey (Rymer 1976). Portions of the plant continue to be used as food in several areas of the world; however, in this instance it is the young developing fronds or croziers rather than the rhizomes that are eaten. The croziers are held in high esteem, especially in Brazil and Japan, where they are eaten fresh, lightly cooked, or preserved (Hirono 1990; Santos et al. 1990; Veitch 1990). They are in such demand in Japan that they are imported. Because they are well-recognized carcinogens (Evans & Mason 1965; Evans 1979), there are important health concerns relating both to direct consumption of bracken fern itself and to indirect consumption via meat and dairy products from animals eating the plants (Smith & Seawright 1995). As emphasized in several review articles, the ferns are eaten by people in various parts of the world and may be the cause of neoplasms of the esophagus, stomach, and white blood cells (Fenwick 1988; Evans & Galpin 1990; Santos et al. 1990; Veitch 1990). It is also clear that toxicant concentrations are quite high in the newly developing fronds, i.e., the fiddleheads or croziers ((Alonso-Amelot et al. 1992). Furthermore, although preparation of the fern for human use reduces the concentrations of toxicants, it is still capable of inducing neoplasia when fed to rats. In certain areas of Japan where the young fronds of bracken are commonly eaten, there is an increased incidence of esophageal neoplasia (Hirono 1990).

The health risks due to indirect consumption are not completely understood (Hodgson 1991). Milk from cows eating bracken presents a potential problem because it may contain the toxicant ptaquiloside (Smith 1990). Milk from bracken-fed cows has been shown to be capable of inducing neoplasms when fed for a prolonged time to mice, rats, and a calf (Evans et al. 1972; Pamucku et al. 1978; Villalobos-Salazar et al. 1988). Interestingly, a broader range of tumors is seen with milk than with the plant itself, possibly indicative of the role of metabolites (Villalobos-Salazar et al. 1990). Public health concerns are further highlighted by epidemiologic studies showing greater incidence of gastric neoplasia in regions of Central and South America where bracken is abundant than in areas where it is absent (Villalobos-Salazar 1987; Villalobos-Salazar et al. 1989, 1990).

An additional disease problem that may be related to the foregoing effects in ruminants is the development of progressive retinal degeneration in sheep and other animal species (Hirono 1990; Xu 1992). This nontransmissible disease, known as bright blindness, is characterized by degeneration of the retinal neuroepithelium (Watson et al. 1965; Watson & Barnett 1970). There is severe atrophy of the layer of rods and cones and of the outer nuclear layer, especially in the tapetal region. Experimentally the lesions can be reproduced in sheep by feeding 1 kg or more of bracken daily for 6–12 months or the dried, powdered plant for 4 months (Barnett & Watson 1970; Watson et al. 1972; Hirono et al. 1993). Because of the long latent period, the disease is seen in animals more than 1 year old, and recovery is possible when the sheep are removed from the offending pasture early in the disease course. During the course of disease, hematologic changes suggestive of the radiomimetic effects may also be seen (Barnett & Watson 1970; Watson et al. 1972).

■ sheep, very long-term ingestion, retinal degeneration, or bright blindness

Because of the length of feeding time required to produce some of these effects, and the scattered availability and relative scarcity of extensive bracken growth in many parts of North America, the most likely problem to be encountered in ruminants is abrupt onset of acute hemorrhagic disease. It is more likely to occur in drier years when alternative forage is lacking. This bladder-related disease occurs with continuous exposure to *P. aquilinum* in pastures or range throughout the year. Hence, it is not common in most areas of North America, where exposure, although high, is sustained for only several months at a time. The exception to this exposure is in the coastal areas of Oregon, Washington, and British Columbia (Kalkus 1913; Plummer 1944). In spite of the obvious risk associated with bracken, it is used successfully as a livestock feed (Garcia et al. 1991, 1992). When used as such, it should make up about 25% of the roughage (no more than 50%) and/or be alternated with other forages every few weeks. Use at these levels obviates some of the palatability and nutritional shortcomings of bracken as a livestock feed.

■ although toxic, when given in small amounts for brief periods it has been used as feed without problems

Disease Genesis—Following the establishment of *Pteridium* as the cause of the neurologic disease, research efforts to determine the mechanism of action culminated in experiments on rats, which showed a rapid reversal of signs when thiamine was administered (Weswig et al. 1946). This was subsequently confirmed in further studies with rats and then in horses (Evans & Evans 1949; Roberts et al. 1949; Evans et al. 1951). The results of the studies in rats suggested the possibility of thiaminase activity rather than a simple deficiency, and this was subsequently borne out in further studies.

Thiamine deficiency results from the action of a type I thiaminase found in the protein fraction of the plant tissues. The enzyme catalyzes a base exchange reaction in which there is substitution of the active thiazole moiety by a nitrogen-containing ring, such as proline or pyridine, onto the methylene of the pyrimidine ring (Brent & Bartley 1984; Chick et al. 1985, 1989). Thus, rather than simply destroying thiamine, as does a type II thiaminase, there is an additional effect in which an analog is created to inhibit thiamine-requiring metabolic reactions. The overall antithiamine effect is a competitive-type inhibition of thiamine cofactor activity. Substantial and continuous ingestion of the inhibitor or plant is required to produce disease. The possibility that a more complex set of events could occur is raised by the discovery of activity of both type I and II thiaminases in a South African variety of *P. aquilinum* (Meyer 1989).

■ horses, thiaminase type I, destructive of thiamine and inhibitory to its reactions

Many causes for the radiomimetic and carcinogenic effects have been suggested

- ruminants, bone marrow and neoplastic effects, ptaquiloside, and other illudanes, hydrolysis to reactive dienone, binds and cleaves DNA

(Hopkins 1986; Fenwick 1988). The most plausible, if not the sole toxin, is ptaquiloside, an unstable norsesquiterpene glycoside with an illudane skeleton (Niwa et al. 1983; Van der Hoeven et al. 1983; Agnew & Lauren 1991; Oelrichs et al. 1995). Other illudanes and pterosins are also present (Castillo et al. 1997). The variation in expression of carcinogenesis and bladder effects seems to be closely related to differences in fern ptaquiloside content (Smith et al. 1988). Prakash and co-workers (1996) fed calves dried, powdered bracken for 28 days. In addition to hematologic changes such as a decrease in platelets and neutrophils, DNA adducts with ptaquiloside were detected in the ileum and bladder but not in other tissues. The effects of wrong base insertion were suggested to be due to activation of the H-*ras* oncogene (Prakash et al. 1996).

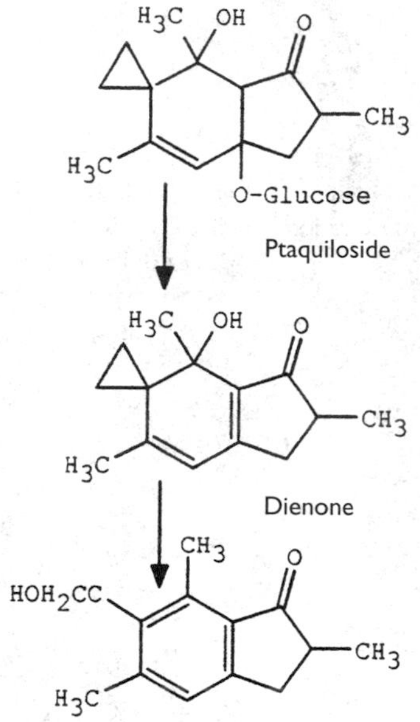

Ptaquiloside is a direct-acting carcinogen, forming a dienone with a highly reactive cyclopropane ring following hydrolysis of the glucose. This very reactive electrophile is then capable of alkylating amines, nucleotides, and other nucleophilic substrates (Ojika et al. 1987). The dienone is capable of covalently binding and cleaving DNA (Ojika et al. 1989), possibly as a result of the formation of opening of the cyclopropyl ring to yield a highly reactive center (Smith et al. 1994b). The activated dienone is much more toxic than the parent ptaquiloside; when given i.v., only 1/10 the dose is required to cause carcinogenesis (Shahin et al. 1998). The unstable dienone may also react with water, degrading to pterosins, of which there are two dozen or more in bracken and 100 or more in other genera (Fukuoka et al. 1978; Murakami & Tanaka 1988). Given in increasing dosage for 6 weeks, ptaquiloside caused pancytopenia in a calf (Hirono et al. 1984). Progressive retinal degeneration has been reproduced experimentally in sheep by feeding ptaquiloside (Hirono et al. 1993).

H_3C OH O CH_3 H_3C O-Glucose

Ptaquiloside

H_3C OH O CH_3 H_3C

Dienone

CH_3 O HOH_2CC CH_3 H_3C

Pterosin B

The illudane norsesquiterpene glycosides are not unique to *P. aquilinum*. Ptaquiloside or its analogs have been isolated from *Histiopteris incisa, Pteris cretica, Hypolepis punctata, Dennstaedtia,* and other related ferns (Saito et al. 1990). Thus there is potential for disease to be caused by other fern species, although they are probably unlikely to be eaten by livestock in sufficient quantities to pose a problem.

Ptaquiloside concentrations are highest in the young fronds of *P. aquilinum,* ranging from 0.7 to as much as 3.9 mg/g d.w. (Smith et al. 1990; Alonso-Amelot et al. 1992, 1995). Although concentrations are quite variable, they typically are about 1 mg/g in plant material collected from different areas of the world (Smith et al. 1994b). There appears to be considerable variation in ptaquiloside content among the different varieties of *P. aquilinum*—high in var. *caudatum* and low in var. *arachnoideum* (Alonzo-Amelot et al. 1995). In some instances concentrations in the young fronds may exceed 5 mg/g of plant material (Smith et al. 1994a). These levels further underscore the public health concerns about consumption of young fronds by humans. The potential for problems due to milk contamination is illustrated by the excretion of about 10% of ingested ptaquiloside by cows in their milk. The incidence of neoplasia is much higher in rats fed bracken with high ptaquiloside concentrations as compared with those fed plants with low levels (Smith et al. 1988). Other factors may also play a role, because even the spores are possible carcinogens, having caused leukemias in experiments with mice (Evans & Galpin 1990; Trotter 1990). Interestingly, early observations by stockmen suggested that the problem of enzootic hematuria was greater when spores were present (Kalkus 1913).

Other compounds in bracken that have been considered to be of possible toxicologic importance include quercetin and shikimic acid. Although some experiments have shown a carcinogenic effect for quercetin (Pamukcu et al. 1980), other results indicate that these compounds are not the carcinogens of bracken (Hirono 1992; Hirono et al. 1977). There is also a lack of interaction between the two compounds to augment their effects (Ngomuo & Jones 1996). A lack of effects is not unexpected, because these compounds are very common constituents of plants. Although not of direct concern, it is possible that there may be an indirect action on ptaquiloside analogous to the role of quercetin in increasing estradiol-induced tumorigenesis in hamsters (Zhu & Liehr 1994).

Clinical Signs—In the horse, signs develop over several days and begin with depression, constipation, and a slightly unsteady gait. The incoordination becomes noticeably worse over the next few days; eventually there may be difficulty standing. In addition there is a marked loss of body condition, weakness, cold extremities, a tendency for watery stools, and muscular twitching, which causes jerky movements of the head, neck, limbs, and tail. Raising the head may cause the horse to assume a crouch position. It becomes increasingly difficult for the animal to stand. It goes down, in some instances after much nervousness and possibly violence. Once recumbent, there may be seizures and bizarre posture of the neck. At this point, death is close and treatment may be too late. The course of these signs may last from 2 days to more than a week. Initially the appetite may remain good but fades as the signs worsen.

■ horse, neurologic, onset over several days, depression, unsteady gait, ataxia, loss of condition, weakness, twitching, collapse

Because of the inhibition of the pyruvate dehydrogenase reaction, blood pyruvate and lactate increase several-fold. There will also be a decrease in blood thiamine. Transketolase activity, another inhibited thiamine cofactor reaction, will decrease. It is most readily measured in blood.

The radiomimetic or acute hemorrhagic syndrome in cattle presents with bloody nasal discharge, bloody feces, a decrease in appetite, labored respiration, listlessness, and an increase in body temperature. These signs are accompanied by a marked decrease in circulating neutrophils and blood platelets. Terminally, there may also be an increase in plasma heparin, histamine, and fibrinogen concurrent with a large increase in mast cells and tissue infiltration of eosinophils, monocytes, lymphocytes, and plasma cells. The bone marrow changes are accompanied by increased capillary fragility, increased bleeding time, and defective clot retraction.

■ ruminants, bone marrow depression, bloody diarrhea, bloody discharges, anorexia, increased capillary fragility, marked decrease in platelets

Enzootic hematuria, which is primarily a disease of cattle 3–4 years of age, is indicated by an intermittent problem of pinkish to brownish urine. Weeks or months may occur between episodes, and signs are sometimes precipitated by parturition, transport, or other stress and last for several weeks. Initially, the early urine stream may be clear, and the last, red. There is increased frequency of urination. Urinalysis will show increased protein and white blood cells in addition to the numerous red blood cells. The appetite may not be affected unless there is marked inflammation of the bladder, but eventually body condition declines, and the number of circulating red blood cells may decrease (decreased PCV and hemoglobin). The disease generally progresses slowly and lasts up to several years.

■ cattle, repeated episodes of bloody urine, evidences of bladder inflammation

Sheep affected by bright blindness will have a glassy-eyed appearance and increased alertness. The pupils will be dilated, and their pupillary reaction to light poor to absent; the pupils will constrict only in bright light. Upon ophthalmoscopic examination, there is increased reflection from the tapetum lucidum, the optic disc is pale, and the vessels are narrowed. These signs may occur concurrently with signs of acute hemorrhagic disease, including neutropenia and thrombocytopenia.

■ sheep, pupillary dilation, glassy-eyed, increased alertness, changes in the optic disc

Pathology—In horses with thiamine deficiency, there is congestion of the brain and viscera, with few distinct lesions. The cerebrum may be swollen and edematous. Microscopically, there is necrosis of some neurones.

■ horses, no distinct lesions, scattered neuronal necrosis

The acute hemorrhagic disease in cattle is typified by areas of extensive hemorrhage in various tissues, including those of the digestive tract, splotchy hemorrhages of the mucous membranes, and subcutaneous edema and hemorrhage.

With enzootic hematuria, the bladder epithelium is desquamated and proliferative.

- cattle, areas of hemorrhage in body cavities, digestive tract, and other sites; bladder, areas of desquamation, papillomatous growths, foci of hemorrhage, rarely invasive neoplasms

  sheep, atrophy of outer layers of the retina

Areas of vascularized epithelial proliferation appear as polypoid, papillary, or fungoid reddened foci (often multiple) with a thickened submucosa in the areas between. Microscopically, columns of transitional epithelium infiltrate into the lamina propria along with mononuclear cells. The blood vessels in affected areas are dilated, forming angiomatous cavities. Neoplasia, which may eventually occur, is most commonly of the noninvasive papillomatous type, but other papillary or sessile types of transitional cell carcinoma, squamous cell carcinoma, adenocarcinoma, or hemangioma may also develop.

Retinal degeneration in sheep with bright blindness involves mainly atrophy of the outer layers—the outer nuclear layer and the layer of rods and cones. They are affected mainly in the central areas, the periphery being nearly normal in some instances. There is little or no inflammatory reaction. Of some interest diagnostically is the decrease in activity of retinal lactate dehydrogenase (LDH) (Patterson et al. 1970; Watson et al. 1972). Type 1 LDH activity is increased, but types 2, 3, and 4 are decreased sufficiently that overall there is a decrease.

- horses, thiamine i.v., i.m.; nursing care

Treatment—For horses, treatment is best accomplished by administration of 0.5–1 g of thiamine parenterally, followed by decreasing amounts over the next few days. The animals should be kept quiet. The remaining therapy is good nursing care, highly palatable feeds, and, if needed, a laxative other than mineral oil.

- nursing care, systemic antibiotics, transfusion to supply platelets

In ruminants, the bone marrow suppression and deficiency of blood platelets and neutrophils is best treated with nursing care and antimicrobials to counteract any bacterial infection that may occur as a result of diminished host defenses. Batyl alcohol—glyceryl ether linked with stearyl alcohol—is of dubious value to control the coagulation deficit. Initially it appeared to be of some value both experimentally and in the field, but further studies failed to demonstrate any efficacy, although platelet adhesiveness improved (IA Evans et al. 1958; Tunkl et al. 1963; Dalton 1964; Philp & Gowdey 1967).

References

Agnew MP, Lauren DR: Determination of ptaquiloside in bracken fern *(Pteridium esculentum). J Chromatogr* 538;462–468, 1991

Almond N: Fern poisoning. *J Comp Pathol Ther* 7;165–167, 1894

Alonso-Amelot ME, Perez-Mena M, Calcagno MP, Jaimes-Espinoza R, Castillo U: Ontogenic variation of biologically active metabolites of *Pteridium aquilinum* (L. Kuhn), pterosins A and B, and ptaquiloside in a bracken population of the tropical Andes. *J Chem Ecol* 18;1405–1420, 1992

Alonso-Amelot ME, Rodulfo-Baechler S, Jaimes-Espinosa R: Comparative dynamics of ptaquiloside and pterosin B in the two varieties *(caudatum* and *arachnoideum)* of neotropical bracken fern (*Pteridium aquilinum* L. Kuhn). *Biochem Syst Ecol* 23;709–716, 1995

Bakker HJ, Dickson J, Steele P, Nottle MC: Experimental induction of ovine polioencephalomalacia. *Vet Rec* 107;464–466, 1980

Barnett KC, Watson WA: Bright blindness in sheep: a primary retinopathy due to feeding bracken *(Pteris aquilina). Res Vet Sci* 11;289–290, 1970

Brent BE, Bartley EE: Thiamin and niacin in the rumen. *J Anim Sci* 59;813–822, 1984

Brobst DF, Olson C: Neoplastic and proliferative lesions of the bovine urinary bladder. *Am J Vet Res* 24;105–111, 1963

Campo MS, Jarrett WFH, Barron R, O'Neil BW, Smith KT: Association of bovine papillomavirus type 2 and bracken fern with bladder cancer in cattle. *Cancer Res* 52;6898–6904, 1992

Castillo UF, Wilkins AL, Lauren DR, Smith BL, Towers NR, Alonso-Amelot ME, Jaimes-Espinosa R: Isoptaquiloside and caudatoside, illudane-type sesquiterpene glucosides from *Pteridium aquilinum* var. *caudatum*. *Phytochemistry* 44;901–906, 1997

Chick BF, Carroll SN, Kennedy C, McCleary BV: Some biochemical features of an outbreak of polioencephalomalacia in sheep. *Aust Vet J* 57;251–252, 1981

Chick BP, Quinn C, McCleary BV: Pteridophyte intoxication of livestock in Australia. In *Plant Toxicology—Proceedings of the Australia-USA Poisonous Plants Symposium*, Seawright AA, Hegarty MP, James LF, Keeler RF eds, Animal Research Institute, Yeerongpilly, Australia, pp453–464, 1985

Chick BF, McCleary BV, Beckett RJ: Thiaminases. In *Toxicants of Plant Origin*, vol 3, *Proteins and Amino Acids*, Cheeke PR ed, CRC Press, Boca Raton, Fla, pp73–91, 1989

Dalton RG: The effects of batyl alcohol on the haemetology of cattle poisoned by bracken. *Vet Rec* 76;411–416, 1964

El-Mofty MM, Sadeh IA, Bayoumi S: Improvement in detecting the carcinogenicity of bracken fern using an Egyptian toad. *Oncology* 37;424–425, 1980

Evans ETR, Evans WC, Roberts HE: Studies on bracken poisoning in the horse. *Br Vet J* 107;364–371, 1951

Evans IA: The radiomimetic nature of bracken toxin. *Cancer Res* 28;2252–2261, 1968

Evans IA: Bracken carcinogenicity. *Res Vet Sci* 26;339–348, 1979

Evans IA, Galpin OP: Bracken and leukaemia. *Lancet* 335;231, 1990

Evans IA, Mason J: Carcinogenic activity of bracken. *Nature* 208;913–914, 1965

Evans IA, Thomas AJ, Evans WC, Edwards CM: Studies on bracken poisoning in cattle, part 5. *Br Vet J* 114;253–267, 1958

Evans IA, Jones RS, Mainwaring-Burton R: Passage of bracken fern toxicity into milk. *Nature* 237;107–108, 1972

Evans WC, Evans ETR: The effects of the inclusion of bracken *(Pteris aquilina)* in the diet of rats, and the problem of bracken poisoning in farm animals. *Br Vet J* 105;175–186, 1949

Evans WC, Evans ETR, Hughes LE: Studies on bracken poisoning in cattle, part 1. *Br Vet J* 110;295–306, 1954a

Evans WC, Evans ETR, Hughes LE: Studies on bracken poisoning in cattle, part 2: 1950 bracken poisoning experiments (Lluest farm). *Br Vet J* 110;365–380, 1954b

Evans WC, Evans ETR, Hughes LE: Studies on bracken poisoning in cattle, part 3: field outbreaks of bovine bracken poisoning. *Br Vet J* 110;426–442, 1954c

Evans WC, Evans IA, Thomas AJ, Watkins JE, Chamberlain AG: Studies on bracken poisoning in cattle, part 4. *Br Vet J* 114;180–198, 1958

Fenwick GR: Bracken *(Pteridium aquilinum)*—toxic effects and toxic constituents. *J Sci Food Agric* 46;147–173, 1988

Fukuoka M, Kuroyanagi M, Yoshihira K, Natori S: Chemical and toxicological studies on bracken fern, *Pteridium aquilinum* var. *latiusculum*. 2. Structures of pterosins, sesquiterpenes having 1-indanone skeleton. *Chem Pharm Bull* 26;2365–2385, 1978

Garcia R, Obeid JA, Almeida SA, Tafuri ML: Feno de samambaia *(Pteridium aquilinum)* como volumoso para bovinos. *Rev Soc Bras Zootec* 20;365–372, 1991

Garcia R, Obeid JA, Daza LA, Rodrigues MT, Tafuri ML: Substituicao parcial do feno de campim-jaragua (*Hyparrhenia rufa* [Ness] Stapf) pelo feno de samambaia (*Pteridium aquilinum* [L.] Kuhn) como alimento alternativo para caprinos. *Rev Soc Bras Zootec* 21;158–165, 1992

Gerenutti M, Spinosa H de S, Bernardi MM: Effects of bracken fern (*Pteridium aquilinum* L. Kuhn) feeding during the development of female rats and their offspring. *Vet Hum Toxicol* 34;307–310, 1992

Hadwen S, Bruce EA: The poisoning of horses by the common bracken (*Pteris aquilina* L.). *Vet J* 76;98–109, 1920

Harvey RB, Larson AH, Landon RH, Boyd WL, Erickson LC: *Weeds Poisonous to Livestock*. Minn Agric Exp Stn Bull 388, pp3–5, 1944

Hirono I: Carcinogenicity of bracken fern and its causal principle. In *Bracken Biology and*

Management, Thomson JA, Smith RT eds, Aust Inst Agric Sci Occas Publ 40, Sydney, Australia, pp233–241, 1990

Hirono I: Is quercetin carcinogenic? *Jpn J Cancer Res* 83;313–314, 1992

Hirono I, Fushimi K, Matsubara N: Carcinogenicity test of shikimic acid in rats. *Toxicol Lett* 1;9–10, 1977

Hirono I, Kono Y, Takahashi K, Yamada K, Niwa H, Ojika M, Kigoshi H, Niiyama K, Uosaki Y: Reproduction of acute bracken poisoning in a calf with ptaquiloside, a bracken constituent. *Vet Rec* 115;375–378, 1984

Hirono I, Yagyu S, Haga M: Reproduction of progressive retinal degeneration (bright blindness) in sheep by administration of ptaquiloside contained in bracken. *J Vet Med Sci* 55;979–983, 1993

Hodgson ES: Is bracken a health hazard? [letter]. *Lancet* 337;493, 1991

Holttum RE: A revised classification of the leptosporangiate ferns. *J Linn Soc Bot* 53;123–158, 1947

Hopkins NCG: Aetiology of enzootic haematuria. *Vet Rec* 118;715– 717, 1986

Jarrett WFH: Transformation of warts to malignancy in alimentary carcinoma in cattle. *Bull Cancer* 65;191–194, 1978

Jarrett WHF, McNeil PE, Grimshaw WTR, Selman IE, McIntyre WIM: High incidence area of cattle cancer with a possible interaction between an environmental carcinogen and a papilloma virus. *Nature* 274;215–217, 1978

Kalkus JW: *A Preliminary Report on the Investigations of Bovine red Water (Cystic Hematuria) in Washington.* Wash Agric Exp Stn Bull 112, 1913

Kaminjolo JS, Mugera JM, Rostel AF: Isolation of a Herpes-type virus from some tumours of bovine origin. *Zentrbl Vet Med* B 19;626–632, 1972

Karcher LF, Anderson WI, Dietze AE: Urinary bladder carcinoma suggestive of enzootic hematuria with secondary hydronephrosis in a holstein cow. *Bovine Pract* 23;94–96, 1988

Kelleway RA, Geovjian L: Acute bracken fern poisoning in a 14-month-old horse. *Vet Med/Small Anim Clin* 73;295–296, 1978

Lott DG: The use of thiamine in mare's tail poisoning of horses. *Can J Comp Med* 15;274–276, 1951

McCrea CT, Head KW: Sheep tumours in North East Yorkshire. 1. Prevalence on seven moorland farms. *Br Vet J* 134;454–461, 1978

McCrea CT, Head KW: Sheep tumours in North East Yorkshire. 2. Experimental production of tumours. *Br Vet J* 137;21–30, 1981

McKenzie RA: Bovine enzootic haematuria in Queensland. *Aust Vet J* 54;61–64, 1978

Meyer P: Thiaminase activities and thiamine content of *Pteridium aquilinum, Equisetum ramosissimum, Malva parviflora, Pennisetum clandestinum,* and *Medicago sativa. Onderstepoort J Vet Res* 56;145–146, 1989

Murakami T, Tanaka N: Occurrence, structure, and taxonomic implications of fern constituents. In *Progress in the Chemistry of Organic Natural Products 54,* Herz W, Grisebach H, Kirby GW, Tamm C eds, Springer-Verlag, Vienna, pp1–353, 1988

Ngomuo AJ, Jones RS: Genotoxicity studies of quercetin and shikimate in vivo in the bone marrow of mice and gastric mucosal cells of rats. *Vet Hum Toxicol* 38;176–180, 1996

Niwa H, Ojika M, Wakamatsu K, Yamada K, Hirono I, Matsushita K: Ptaquiloside, a novel norsesquiterpene glucoside from bracken, *Pteridium aquilinum* var. *latiusculum. Tetrahedron Lett* 24;4117–4120, 1983

Oelrichs PB, Ng JC, Bartley J: Purification of ptaquiloside, a carcinogen from *Pteridium aquilinum. Phytochemistry* 40;53–56, 1995

Ojika M, Wakamatsu K, Niwa H, Yamada K: Ptaquiloside, a potent carcinogen isolated from bracken fern *Pteridium aquilinum* var. *latiusculum*: structure elucidation based on chemical and spectral evidence, and reactions with amino acids, nucleosides, and nucleotides. *Tetrahedron* 43;5261–5274, 1987

Ojika M, Sugimoto K, Okazaki T, Yamada K: Modification and cleavage of DNA by ptaquiloside: a new potent carcinogen isolated from bracken fern. *Chem Soc Chem Commun* 1989;1775–1777, 1989

Olson C, Pamukcu AM, Brobst DF: Papilloma-like virus from bovine urinary bladder tumors. *Cancer Res* 25;840–849, 1965

Page CN: The taxonomy and phytography of bracken—a review. *Bot J Linn Soc* 73;1–34, 1976

Pamukcu AM, Goksay SK, Price JM: Urinary bladder neoplasms induced by feeding bracken fern *(Pteris aquilinum)* to cows. *Cancer Res* 27;917–924, 1967a

Pamukcu AM, Olson C, Goksay SK: Influence of a papilloma vaccine on chronic bovine enzootic hematuria. *Cancer Res* 27;2197–2200, 1967b

Pamukcu AM, Price JM, Bryan GT: Naturally occurring and bracken fern- induced bovine urinary bladder tumors. *Vet Pathol* 13;110–122, 1976

Pamukcu AM, Erturk E, Yalciner S, Milli U, Bryan GT: Carcinogenic and mutagenic activities of milk from cows fed bracken fern *(Pteridium aquilinum). Cancer Res* 38;1556–1560, 1978

Pamukcu AM, Yalciner S, Hatcher JF, Bryan GT: Quercetin, a rat intestinal and bladder carcinogen present in bracken fern *(Pteridium aquilinum). Cancer Res* 40;3468–3472, 1980

Patterson DSP, Sweasey D, Terlecki S: Changes in lactate dehydrogenase (LDH) isoenzymes associated with retinal degeneration of ewes affected with bright blindness. *Vet Rec* 87;760–761, 1970

Pennie WD, Campo MS: Synergism between bovine papillomavirus type 4 and the flavonoid quercetin in cell transformation in vitro. *Virology* 190;861–865, 1992

Philp RB, Gowdey CW: Anaemia, thrombocytopenia, and reduced platelet adhesiveness in rats fed bracken fern, and protective effects of batyl alcohol. *Nature* 216;498–499, 1967

Plummer PJG: Histopathology of enzootic haematuria. *Can J Comp Med* 7;153–158, 1944

Prakash AS, Pereira TN, Smith BL, Shaw G, Seawright AA: Mechanism of bracken fern carcinogenesis: evidence for H-*ras* activation via initial adenine alkylation by ptaquiloside. *Nat Toxins* 4;221–227, 1996

Price JM, Pamukcu AM: The induction of neoplasms of the urinary bladder of the cow and the small intestine of the rat by feeding bracken fern *(Pteris aquilinum). Cancer Res* 28;2247– 2251, 1968

Rao DST, Joshi HC, Kumar M: Immune status of calves suffering from bracken-fern–induced haematuria. *Indian J Anim Sci* 57;1207–1208, 1987

Roberts HE, Evans ETR, Evans WC: The production of "bracken staggers" in the horse and its treatment by vitamin B1 therapy. *Vet Rec* 61;549–550, 1949

Rosenberger G: Nature, manifestations, cause and control of chronic haematuria in cattle. *Vet Rev* 1971;189–206, 1971

Rymer L: The history and ethnobotany of bracken. *Bot J Linn Soc* 73;151–176, 1976

Saito K, Nagao T, Takatsuki S, Koyama K, Natori S: The sesquiterpene carcinogen of bracken fern, and some analogues, from the Pteridaceae. *Phytochemistry* 29;1475–1479, 1990

Santos RC, Brasileiro-Filho G, Silva ME: Oncogenicity of bracken fern *(Pteridium aquilinum)* from Preto, Brazil, for rats fed low animal protein diets. In *Bracken Biology and Management,* Aust Inst Agric Sci Occas Publ 40, Thomson JA, Smith BL eds, Sydney, Australia, pp253–257, 1990

Schacham P, Philip RB, Gowdey CW: Antihematopoietic and carcinogenic effects of bracken fern *(Pteridium aquilinum)* in rats. *Am J Vet Res* 31;191–197, 1970

Shahin M, Smith BL, Oelrichs PB, Moore MR, Worral S, Seawright AA, Prakash AS: Induction of mammary carcinoma, monocytosis, and type II pneumonocyte proliferation by activated ptaquiloside, a bracken fern carcinogen. In *Toxic Plants and Other Natural Toxicants,* Garland T, Barr AC eds, CAB International, New York, pp329–333, 1998

Smith BL: Bracken fern and animal health in Australia and New Zealand. In *Bracken Biology and Management,* Aust Inst Agric Sci Occas Publ 40, Thomson JA, Smith BL eds, Sydney, Australia, pp227–232, 1990

Smith BL, Seawright AA: Bracken fern (*Pteridium* spp.) carcinogenicity and human health—a brief review.. *Nat Toxins* 3;1–5, 1995

Smith BL, Embling PP, Agnew MP, Lauren DR, Holland PT: Carcinogenicity of bracken fern *(Pteridium aquilinum)* in New Zealand. *N Z Vet J* 36;56–58, 1988

Smith BL, Lauren DR, Embling PP, Agnew MP: Ptaquiloside in Australian and New Zealand ferns as a cause of neoplasia. In *Bracken Biology and Management,* Aust Inst Agric Sci Occas Publ 40, Thomson JA, Smith BL eds, Sydney, Australia, pp241–246, 1990

Smith BL, Seawright AA, Ng JC, Hertle AT, Thomson JA, Bostock PD: Concentration of ptaquiloside, a major carcinogen in bracken fern (*Pteridium* spp.), from eastern Australia and from a cultivated worldwide collection held in Sydney, Australia. *Nat Toxins* 2;347–353, 1994a

Smith BL, Shaw G, Prakash A, Seawright AA: Studies on DNA adduct formation by ptaquiloside, the carcinogen of bracken fern (*Pteridium* spp.). In *Plant-Associated Toxins: Agricultural, Phytochemical and Ecological Aspects,* Colegate SM, Dorling PR eds, CAB International, Wallingford, UK, pp167–172, 1994b

Stockman S: Bracken poisoning in cattle in Great Britain. *J Comp Pathol Ther* 30;311–316, 1917

Stockman S: Bracken poisoning in cattle in Great Britain. *J Comp Pathol Ther* 35;273–275, 1922

Taylor JA: The bracken problem: a global perspective. In *Bracken Biology and Management,* Aust Inst Agric Sci Occas Publ 40, Thomson JA, Smith BL eds, Sydney, Australia, pp3–19, 1990

Thomas KW: Oral treatment of polioencephalomalacia and subclinical thiamine deficiency with thiamine propyl disulphide and thiamine hydrochloride. *J Vet Pharmacol Ther* 9;402–411, 1986

Trotter WR: Is bracken a health hazard? *Lancet* 336;1563–1565, 1990

Tryon RM: A revision of the genus *Pteridium. Rhodora* 43;1–31, 1941

Tunkl B, Aleraj Z, Orlic N: Therapy of bracken poisoning in cattle: batyl alcohol. *Vet Glas* 17;945–947, 1963 (*Vet Bull* 34;2309, 1964)

Van der Hoeven JCM, Lagerweij WJ, Posthumus MA, van Veldhuizen A, Holterman HAJ: A new mutagenic compound isolated from bracken fern (*Pteridium aquilinum* [L.] Kuhn). *Carcinogenesis* 4;1587–1590, 1983

Veitch B: Aspects of aboriginal use and manipulation of bracken fern. In *Bracken Biology and Management,* Aust Inst Agric Sci Occas Publ 40, Thomson JA, Smith BL eds, Sydney, Australia, pp215–226, 1990

Villalobos-Salazar J: *Pteridium aquilinum* and gastric cancer [abstract]. *XXIII World Vet Congress,* Montreal, Canada, p90, 1987

Villalobos-Salazar J, Hasbun C, Meneses A, Rojas JL, Sigaram M: Carcinogenicity of *Pteridium aquilinum* var. *caudatum* in mice [abstract]. In *9th World Congress on Animal, Plant, and Microbial Toxins,* International Society on Toxinology, Stillwater, Okla, p92, 1988

Villalobos-Salazar J, Meneses A, Rojas JL, Mora J, Porras RE, Herrero MV: Bracken derived carcinogens as affecting animal health and human health in Costa Rica. In *Bracken Toxicity and Carcinogenicity as Related to Animal and Human Health,* Taylor JA ed, Inst Earth Stud Publ 44, Univ Coll Wales, Aberystwyth, Dyfed, UK, pp40–51, 1989

Villalobos-Salazar J, Meneses A, Salas J: Carcinogenic effects in mice of milk from cows fed on bracken fern *(Pteridium aquilinum)*. In *Bracken Biology and Management,* Aust Inst Agric Sci Occas Publ 40, Thomson JA, Smith BL eds, Sydney, Australia, pp247–251, 1990

Watson WA, Barnett KC: Bright blindness in sheep—the relationship to environment. *Br Vet J* 126;482–486, 1970

Watson WA, Barlow RM, Barnett KC: Bright blindness—a condition prevalent in Yorkshire Hill Sheep. *Vet Rec* 77;1060–1069, 1965

Watson WA, Terlecki S, Patterson DSP, Sweasey D, Hebert CN, Done JT: Experimentally produced progressive retinal degeneration (bright blindness) in sheep. *Br Vet J* 128;457–468, 1972

Weswig PH, Freed AM, Haag JR: Antithiamine activity of plant materials. *J Biol Chem* 165;737–738, 1946

Xu LR: Bracken poisoning and enzootic haematuria in cattle in China. *Res Vet Sci* 53;116–121, 1992

Yamane O, Hayashi T, Sako S: Studies on blood coagulation disorders in domestic animals thromboelastograms of normal cattle and cattle affected with bracken poisoning. *Jpn J Vet Sci* 37;577–583, 1975a

Yamane O, Hayashi T, Sako S, Kihara T, Koyama M: Studies on hemorrhagic diathesis of experimental bovine bracken poisoning. 1. Detection of circulating anticoagulants. *Jpn J Vet Sci* 37;335–340, 1975b

Yamane O, Hayashi T, Sako S, Tatematsu S, Takeda K, Fukushima H: Studies on hemorrhagic diathesis of experimental bovine bracken poisoning. 2. Heparin-like substance level in blood. *Jpn J Vet Sci* 37;341–347, 1975c

Yasuda Y, Kihara T, Nishimura H: Embryotoxic effects of feeding bracken fern *(Pteridium aquilinum)* to pregnant mice. *Toxicol Appl Pharmacol* 28;264–268, 1974

Yoshikawa T, Oyamada T, Yoshikawa H, Sakaguchi M: Histopathogenesis of bracken-fern induced experimental tumor of urinary bladder. *Jpn J Vet Sci* 43;875–885, 1981

Zhu BT, Liehr JG: Quercetin increases the severity of estradiol-induced tumorigenesis in hamster kidney. *Toxicol Appl Pharmacol* 125;149–158, 1994

33 Ebenaceae Gürke

■ Ebony Family
Diospyros

Widespread in the tropics and subtropics of both the Old World and the New World, the Ebenaceae, or ebony family, comprises 5 genera and 450–550 species. The majority of species belong to *Diospyros;* the other 4 genera are small and restricted to Africa and Madagascar. The family is important for its hard woods, most notably ebony, and its juicy, edible fruits—persimmons and date plums. Intoxication problems have been described only for *Diospyros*.

Plants trees; solitary or colonial; dioecious or rarely polygamodioecious. **Stems** pubescent or glabrous; terminal buds absent. **Leaves** simple; alternate; petiolate or subsessile; blades thick or thin, coriaceous, dorsal surfaces with dark glands or necrotic spots; venation pinnate; surfaces glabrous or sparsely pubescent; apices acuminate or rounded; margins entire; bases cuneate or rounded; stipules absent. **Inflorescences** of 2 types, staminate and pistillate different; axillary. **Staminate Inflorescences** simple cymes, 2- or 3-flowered, bracts present, caducous; bracteoles present, caducous. **Pistillate Inflorescences** solitary; on short recurved pedicels; bracts present. **Flowers** imperfect or rarely perfect, staminate and pistillate similar, pistillate larger; perianths in 2-series. **Sepals** 4, fused, deeply lobed. **Corollas** radially symmetrical; urceolate; convolute. **Petals** 4; fused; yellowish green to creamy white or orange. **Stamens** 16 or 24; in 2 or 3 whorls; equaling perianths; epipetalous; androecial rudiments present in pistillate flowers, 8–16. **Pistils** 1; compound, carpels 4; stigmas 4; styles 4, fused one-half length; ovaries superior; locules 8; placentation apical-axile; gynoecial rudiments present in staminate flowers. **Fruits** berries; subglobose; yellow or pale orange to red-orange or black; glabrous or pubescent; glaucous. **Seeds** 8; oblong; thick; flat; hard; brown; rugose.

Diospyros L.

■ *D. kaki*
D. texana
D. virginiana
date palm
kaki
persimmon
possumwood

Taxonomy and Morphology—Comprising 400–500 species, *Diospyros* is the only genus of the family in North America. Its name is derived from the Greek roots *dios* and *pyros,* meaning "divine" and "grain" or "grain of Zeus," alluding to edibility of its fruits. Greatest diversity is in the lowland rain forests of Malaysia, tropical Africa, and Central America. Relatively few species occur in the north temperate zone. Two species are native to the United States, and several species are present in Mexico. The most important species of toxicologic interest are these:

D. kaki L.f.	Chinese persimmon, Japanese persimmon, kaki, Japanese date plum
D. texana Scheele	Texas persimmon, black persimmon, Mexican persimmon
D. virginiana L.	common persimmon, eastern persimmon, possumwood

Because *Diospyros* is the only representative of the family in North America, the morphological features of the genus are those of the family and thus are not repeated here.

- trees; leaves simple, alternate; petals 4, yellowish green, creamy white, or orange; berries yellow, pale orange, red-orange or black

1. Leaves subsessile. Mature berries black; pubescent. Plants of western Texas and northeastern Mexico . . . . *D. texana*
1. Leaves petiolate. Mature berries yellow-orange to orange-red; glabrous. Plants of the eastern half of United States or cultivated.
   2. Abaxial surfaces of leaves pubescent. Petioles winged. Plants cultivated in orchards . . . . *D. kaki*
   2. Abaxial surfaces of leaves glabrous. Petioles not winged. Plants growing in wild or grown as ornamentals . . . . *D. virginiana*

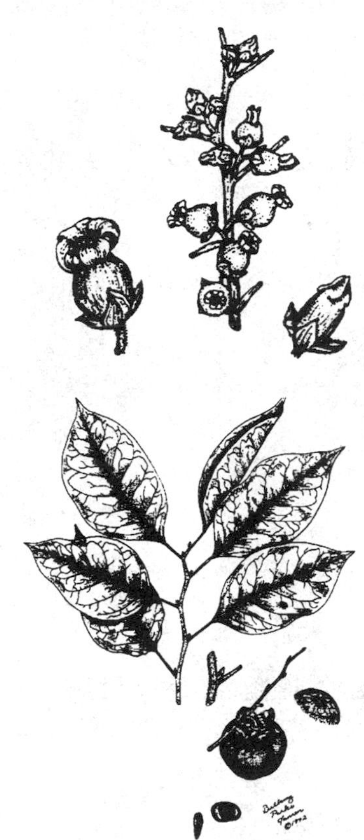

Diospyros virginiana

Distribution and Habitat—Native to Asia, *D. kaki,* with several varieties, is cultivated as a commercial crop in California and Florida. A slow-growing, short-lived, native species, *D. virginiana* occupies a variety of soils and habitats, including rocky, dry uplands and swampy bottomlands. Weedy, it is capable of infesting overgrazed pastures, old fields, and abandoned cropland. It often forms, via root sprouts, dense thickets of tall, straight, spindly trees. It is distributed in the eastern half of the United States. When it is grown as an ornamental, its well-shaped crown provides good shade and beautiful autumn color. Trees of *D. texana* are encountered in rocky soils of arroyos and slopes in open woodlands of western Texas and northern Mexico.

- ornamentals, or variety of open woodlands

Disease Problems—*Diospyros kali* has been grown for centuries in the Orient for its large, edible fruits, and extracts of its leaves have been used for their hemostatic, diuretic, laxative, and hypotensive effects. *Diospyros virginiana* and *D. texana* are likewise eaten; the former is especially relished after fall frosts in rural regions of the southeastern United States. The most consistent problem associated with their consumption by both animals and humans is the propensity of the fruit tissues to congeal into obstructing masses, termed phytobezoars, in the stomach and small intestine (Haslam & Lilley 1988). In regions where wild persimmons are plentiful, the development of gastric impactions in horses and cattle that eat large numbers of them is a continuing problem in the fall of the year (Honnas & Schumacher 1985; Morgan & Bellamy 1994; Cummings et al. 1997). The phytobezoars that form from the seed masses may cause ulcerations of the stomach mucosa. The fruits are most hazardous when eaten prior to ripening or before the first frosts. Episodes may occur when the fruits are eaten in large numbers immediately following the first frosts, at which time they lose some of their astringency and develop a slightly sweet taste. After the fruits ripen completely, they are less of a problem, unless very large numbers are eaten. The risk of impactions may be complicated by seeds or fragments of other plants consumed at the same time (Honnas & Schumacher 1985).

- livestock, especially horses; fruits, large amounts, form mass in stomach, impaction, ulceration

  humans, phytobezoars, especially when there has been previous stomach surgery

Persimmons are the most common cause of phytobezoars in humans; thus it is not surprising that the majority of cases of bezoar formation occur within the range of the 2 native species, the persimmon belt of the southeastern United States—eastward from Kansas, Oklahoma, and Texas (De Bakey & Ochsner 1939; Chont 1942; Lou et al. 1990). Previous stomach surgery is also a factor contributing to the risk of bezoar formation (Morey et al. 1955; Moriel et al. 1983; Krausz et al. 1986; Verstrandig et al. 1989; Chisholm et al. 1992). Unpeeled and immature berries are the biggest risk, but ripe fruit also may play a role (Choi & Kang 1988; Benharroch et al. 1993). The bezoars may form in the stomach, but most will be in the small intestine from the duodenum to the proximal ileum.

- condensed tannins, soluble cause astringency in mouth; whole fruit, tannins cross-link in stomach to form congealed mass of fruit and seeds; astringent action on stomach mucosa, ulcerations

Example
condensed tannin proanthocyanidin made up of leukoanthocyanidin units

- other toxins, terpenoids, 7-methyljugalone diospyrol; some may complement tannin effects

Diospyros texana

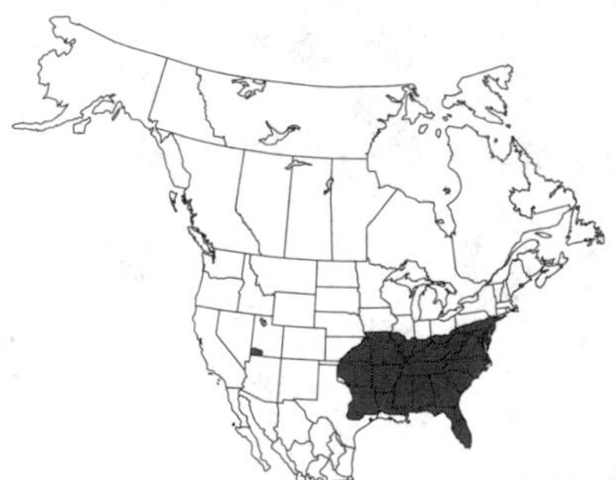

Diospyros virginiana

Disease Genesis—Species of *Diospyros* are rich in bioactive compounds. The most well known in the fruits are proanthocyanidins (condensed tannins) consisting of catechins and gallocatechins (Matsuo & Ito 1978; Balza et al. 1989). Some of these tannins seem to have medicinal value due to their capacity to scavenge oxygen-free radicals, thereby reducing lipid peroxidation in cells and tissues (Uchida et al. 1990). A beneficial effect of persimmon tannins is illustrated by the lengthening of the life span of stroke-prone, spontaneously hypertensive rats (Piretti et al. 1985).

The proanthocyanidins exist as both soluble, short-chain types and insoluble, longer, complex oligomers. The more-soluble polyphenolic proanthocyanidins, called shibuol or phlorbutannin, are responsible for the astringency of persimmons, because of their high affinity for proteins (Izumi et al. 1933; Matsuo & Ito 1978). These moderately large tannins cross-link and interact with proline-rich protein residues in the mouth to produce the astringency sensation (Goldstein & Swain 1963; Lea 1991). With ripening or after frosts, there is increasing cross-linking of proanthocyanidin oligomers due to the presence of acetaldehyde groups and possibly other compounds, which provide the necessary reactant linkages (Izumi et al. 1933; Haslam & Lilley 1988). Polymerization results in a decrease in proanthocyanidin solubility and reactivity with proteins, thus decreasing astringency. There also may be some binding of proanthocyanidins to soluble polysaccharides, such as pectin, which greatly increase during ripening (Haslam & Lilley 1988).

Although the immature fruit is quite astringent and seemingly unpalatable, it is sometimes eaten in large numbers by horses and cattle and thus produces severe digestive dysfunction. When large numbers of the immature berries are eaten, there is the potential for extensive cross-linking or polymerization, to form a matrix of tannin, cellulose, hemicellulose, and protein, especially in the acidic environment of the stomach (Izumi et al. 1933; Chont 1942; Holloway et al. 1980). The resulting mass of fruit tissues and seeds may cause partial or complete blockage of the digestive tract. The problem in the stomach is exacerbated by astringent effects on the mucosa. The mass may eventually result in erosion or ulceration of the mucosa and/or facilitate absorption of other toxicants that might not otherwise be a problem. The ripe fruits also may cause problems when large numbers are eaten, but they are less likely to do so than the immature fruits. In humans the problems may be increased when the whole fruit is eaten, because the proanthocyanidin content is highest just below the exocarp or skin of the berry.

The hypotensive effects of *D. kaki* are attributed to kaempferol 3-glucoside (astragalin) and quercetin 3-β-glucoside (Funayama & Hikino 1979). These compounds also have laxative and diuretic actions. Other compounds present in *Diospyros* include an array of terpenoids or saponins; glycosides of genins such as the oleananes allobetulin and oxyallobetulin; a host of taraxeranes such as sawamilletin; and several ursanes such as epiuvaol and marsformosanone (Connolly & Hill 1991). Bioactive naphthoquinones are also abundantly represented in the genus (Thomson 1971; Gafner et al. 1987). The wood and fruit of *D. virginiana* contain 7-methyljugalone, a dimeric derivative of naphthoquinone, which is a termiticidal agent (Carter et al. 1978). The Asian and African species *D. chloroxylon, D. mespiliformis,* and *D. montana* contain diospyrin, a binaphthyl derivative bisquinone (Thomson 1971). The effects on the stomach may be due to the combination of the astringent proanthocyanidins and the irritant terpenoids or reactive naphthoquinones.

Some Old World species, such as *D. mollis,* are employed as anthelmintics in humans, the active principles presumed to be diospyrol, its monomer, and their glycosides (Borsub et al. 1976). Diospyrol is similar to stypandrol and dianellidin, isolated from *Stypandra imbricata* in the monocot family Phormiaceae, except for the absence of an acetyl functional group ortho to the phenolic hydroxyl and aryl methyl groups (Colegate et al. 1992). Because blindness has been reported in Thailand that is apparently due to optic neuritis associated with the use of aqueous decoctions from *Diospyros* species, there is considerable interest in the chemical constituents of these plants. The binaphthalene-tetrol stypandrol causes a unique disease in Australia (Huxtable et al. 1992). The acute paretic effects produced include brain edema, wallerian degeneration of optic nerves, and central myelin disruption. In chronic cases, there is resolution of the myelin problem, but degeneration of optic nerves and loss of retinal photoreceptors continues. However, diospyrol, though acutely lethal at high dosage, has not been shown to produce blindness (Dorling et al. 1992; Huxtable et al. 1992). Furthermore, two species of *Diospyros* have been examined for the presence of stypandrol and dianellidin, with negative results (Colegate et al. 1992). Thus there is doubt that these compounds play a significant role in the toxicity of *Diospyros*.

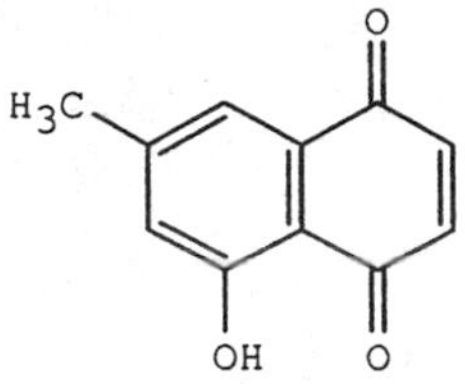

7-Methyljugalone

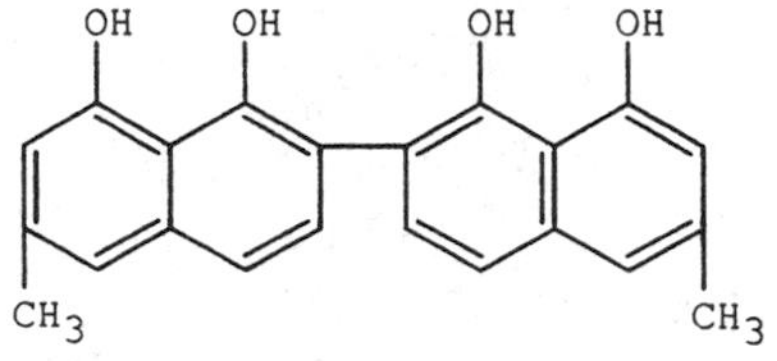

Diospyrol

Clinical Signs—The signs, resulting from impaction and irritation of the digestive tract, are most pronounced in horses. Colic of some severity is likely; in some animals the pain may be severe, as shown by a reluctance to stand or walk and by rolling when down. Gut motility will be nil, the heart rate will be elevated markedly, and there may be a variety of changes in hematology and serum chemistry, including leukocytosis. Perforation of ulcerations leads to peritonitis, followed by circulatory collapse and death.

- horses, colic, rolling, gut motility absent, increased heart rate, weakness; may be lethal if ulcers perforate

humans, persistent vomiting

In humans, there also may be persistent vomiting and gas distention of the small intestine. The disease may develop over several weeks in individuals who have had previous stomach surgery.

Pathology—There is typically a large mass of the pulp, skins, and especially seeds of *Diospyros* in the stomach, with reddening and perhaps ulceration of the mucosa. In humans the masses may be in the stomach or small intestine.

- gross pathology, mass of pulp, skins, seeds in stomach, mucosal ulcerations

Treatment—The general approach is directed toward alleviation of the digestive problems. Oral administration of mineral oil will facilitate passage of a large number of persimmon seeds in the feces, but this may not solve the problem. In some individuals the problem may be resolved by mechanical breakup of the bezoars using an endoscope or by dissolving them with cellulose and papain (Holloway et al. 1980; Krausz et al. 1986). Otherwise, surgical intervention will be required to remove the mass. The possible use of polyethylene glycol or polyvinylpyrrolidone to bind with and reduce the reactivity of the polyphenolics is a possible approach in the early stages. This approach has been used with some success to reduce the toxicity of the hydrolyzable tannins (Kumar & Vaithiyanathan 1990).

- mineral oil, or low-molecular-weight polyethylene glycol orally; endoscopic-facilitated mechanical breakup of bezoars

References

Balza F, Abramowski Z, Towers GHN, Wiriyachirita P: Identification of proanthocyanidin polymers as the piscicidal constituents of *Mammea siamensis, Polygonum stagninum,* and *Diospyros diepenhorstii. Phytochemistry* 28;1827–1830, 1989

Benharroch D, Krugliak P, Porath A, Zurgil E, Niv Y: Pathogenetic aspects of persimmon bezoars, a case-control retrospective study. *J Clin Gastroenterol* 17;149–152, 1993

Borsub L, Thebtaraonth Y, Ruchirawat S, Sadavongvivad C: A new diglucoside from the anthelmintic berries of *Diospyros mollis. Tetrahedron Lett* 1976;105–108, 1976

Carter FL, Garlo AM, Stanley JB: Termiticidal components of wood extracts: 7-methyljugalone from *Diospyros virginiana. J Agric Food Chem* 26;869–873, 1978

Chisholm EM, Leong HT, Chung SC, Li AK: Phytobezoars: an uncommon cause of small bowel obstruction. *Ann Roy Coll Surg Engl* 74;342–344, 1992

Choi S-O, Kang J-S: Gastrointestinal phytobezoars in childhood. *J Pediatr Surg* 23;338–341, 1988

Chont LK: Phytobezoar and its formation in vitro. *Radiology* 38;14–21, 1942

Colegate SM, Dorling PR, Huxtable CR: *Stypandra* species (blindgrass) toxicity: chemistry, biosynthesis, and chemotaxonomy. In *Poisonous Plants: Proceedings of the Third International Symposium,* James LF, Keeler Rf, Bailey EM Jr, Cheeke PR, Hegarty MP eds, Iowa State Univ Press, Ames, pp459–463, 1992

Connolly JD, Hill RA: *Dictionary of Terpenoids.* Vol 2, *Di- and Higher Terpenoids,* Chapman & Hall, London, pp655–1460, 1991

Cummings CA, Copedge KJ, Confer AW: Equine gastric impaction, ulceration, and perforation due to persimmon *(Diospyros virginiana)* ingestion. *J Vet Diagn Invest* 9;311–313, 1997

De Bakey M, Ochsner A: Bezoars and concretions: a comprehensive review of the literature with an analysis of 303 collected cases and a presentation of 8 additional cases. *Surgery* 4;132–160, 1939

Dorling PR, Colegate SM, Huxtable CR: The molecular mechanism of stypandrol poisoning. In *Poisonous Plants: Proceedings of the Third International Symposium,* James LF, Keeler Rf, Bailey EM Jr, Cheeke PR, Hegarty MP eds, Iowa State Univ Press, Ames, pp469–473, 1992

Funayama S, Hikino H: Hypotensive principles of *Diospyros kaki* leaves. *Chem Pharm Bull* 27;2865–2868, 1979

Gafner F, Chapuis J-C, Msonthi JD, Hostettmann K: Cytotoxic naphthoquinones, molluscicidal saponins, and flavonols from *Diospyros zombensis. Phytochemistry* 26;2501–2503, 1987

Goldstein JL, Swain T: Changes in tannins in ripening fruits. *Phytochemistry* 2;371–383, 1963

Haslam E, Lilley TH: Natural astringency in foodstuffs—a molecular interpretation. *CRC Crit Rev Food Sci Nutr* 27;1–40, 1988

Holloway WD, Lee SP, Nicholson GI: The composition and dissolution of phytobezoars. *Arch Pathol Lab Med* 104;159–161, 1980

Honnas CM, Schumacher J: Primary gastric impaction in a pony. *J Am Vet Med Assoc* 187;501–502, 1985

Huxtable CR, Dorling PR, Colegate SM: *Stypandra* poisoning: the role of stypandrol and proposed pathomechanisms. In *Poisonous Plants: Proceedings of the Third International Symposium,* James LF, Keeler RF, Bailey EM Jr, Cheeke PR, Hegarty MP eds, Iowa State Univ Press, Ames, pp464–468, 1992

Izumi S, Isida K, Iwamoto M: The mechanism of the formation of phytobezoars, with special reference to the persimmon ball. *Jpn J Med Sci Trans* 2 (*Biochem* 2);21–35, 1933

Krausz MM, Moriel EZ, Ayalon A, Pode D, Durst AL: Surgical aspects of gastrointestinal persimmon phytobezoar treatment. *Am J Surg* 152;526–529, 1986

Kumar R, Vaithiyanathan S: Occurrence, nutritional significance, and effect on animal productivity of tannins in tree leaves. *Anim Feed Sci Technol* 30;21–38, 1990

Lea AGH: Flavor, color, and stability in fruit products: the effect of polyphenols. In *Plant Polyphenols,* Hemingway RW, Laks PE eds, Plenum Press, New York, pp827–847, 1991

Lou CC, Lin JN, Wang KL: Gastrointestinal phytobezoar in children. *Chang Keng I Hsueh* 13;208–213, 1990

Matsuo T, Ito S: The chemical structure of kaki-tannin from immature fruit of the persimmon (*Diospyros kaki* L.). *Agric Biol Chem* 42;1637–1643, 1978

Morey DAJ, Means RL, Hirsley EL: Diospyrobezoar in the postgastrectomy stomach. *Arch Surgery* 71;946–948, 1955

Morgan SE, Bellamy J: Persimmon colic in a mare. *Equine Pract* 16;8–10, 1994

Moriel EZ, Ayalon A, Eid A, Rachmilewitz D, Krausz MM, Durst AL: An unusually high incidence of gastrointestinal obstruction by persimmon bezoars in Israeli patients after ulcer surgery. *Gastroenterology* 84;752–755, 1983

Piretti MV, Pistore R, Razzoboni C: On the chemical constitution of kaki tannin. *Ann Chim (Rome)* 75;137–144, 1985 (*Chem Abstr* 103;211101s, 1985)

Thomson RH: *Naturally Occurring Quinones,* 2d ed. Academic Press, London, pp227–239, 1971

Uchida S, Ohta H, Niwa M, Mori A, Nonaka G, Nishioka I, Ozaki M: Prolongation of life span of stroke-prone spontaneously hypertensive rats (SHRSP) ingesting persimmon tannin. *Chem Pharm Bull* 38;1049–1052, 1990

Verstandig AG, Klin B, Bloom RA, Hadas I, Libson E: Small bowel phytobezoars: detection with radiography. *Radiology* 172;705–707, 1989

34 Equisetaceae Michx. ex DC.

■ Horsetail Family
Equisetum

A monogeneric family, the Equisetaceae, or horsetail family, is nearly cosmopolitan in distribution, absent only in Australia and New Zealand. Greatest abundance is in the Northern Hemisphere. Classified in its own division, the Sphenophyta, the family has been present since the Devonian period of the Paleozoic.

Plants herbs; perennials; from creeping and branching rhizomes; terrestrial to emergent aquatics; producing sporangia on whorled, peltate sporophylls borne in terminal strobili. **Stems** all alike or of 2 forms, fertile and sterile; persisting for 1 or more growing seasons; jointed; longitudinally ribbed; branched or not branched; internodes hollow. **Leaves** small; scalelike; chlorophyll absent; whorled; fused to form toothed sheaths at nodes; sheaths with 1 or 2 dark bands. **Strobili** terminal; sessile or peduncled; erect; elliptic to cylindrical; apices rounded or acute or apiculate. **Sporophylls** all alike; peltate; polygonal; with 5–10 sporangia. **Sporangia** borne on inner surfaces; pendulous. **Spores** all alike; greenish or white; with 4 elaters. **Gametophytes** all alike; green.

Equisetum L.

■ *E. arvense*
E. fluviatile
E. hyemale
E. laevigatum
E. palustre
E. pratense
E. sylvaticum
E. telmateia
horsetail
scouring rush

Taxonomy and Morphology—Essentially unchanged since the Devonian and sometimes called a living fossil, *Equisetum* comprises 2 distinct subgenera and 15 species (Hauke 1993). Its name is derived from the Latin roots *equis* and *seta,* for "horse" and "bristle," which allude to the bushy, tail-like appearance of the shoots of some species. It also is known as scouring rush because of the abrasive texture of the stems, which were used by settlers to clean their cooking pots and kettles—the precursor of the modern Brillo™ or SOS™ pad. Used by various Indian tribes for food and crafts, plants of the genus are now occasionally cultivated as botanical curiosities. However, they are aggressively weedy and, once established in an area, are almost impossible to eradicate. Although of little wildlife significance, they are sometimes browsed by cattle, deer, and muskrats. In North America, 11 species are present. Of toxicologic importance are the following:

E. arvense L.	field horsetail, common horsetail
E. fluviatile L.	water horsetail, swamp horsetail
E. hyemale L.	common scouring rush, rough horsetail
E. laevigatum A.Braun	smooth scouring rush
E. palustre L.	marsh horsetail
E. pratense Ehrh.	meadow horsetail, shady horsetail
E. sylvaticum L.	wood horsetail, woodland horsetail
E. telmateia Ehrh.	giant horsetail

The morphological features of the genus are those of the family and thus are not repeated here. The taxonomic key presented below is modified from that of Hauke (1993).

- perennial herbs from rhizomes; stems jointed, ribbed, some branched, some with terminal strobili; leaves scalelike

1. Stems perennial. Teeth of sheaths articulated.
   2. Sheaths cylindrical; tightly appressed; ash gray with 2 black bands. Strobilus apices apiculate .... *E. hyemale*
   2. Sheaths obconical; not appressed; green with 1 black band. Strobilus apices rounded .... *E. laevigatum*
1. Stems annual. Teeth of sheaths not articulated.
   3. Stems branched; whorls of branches present.
      4. First internodes of branches shorter than subtending stem sheaths.
         5. Branches solid .... *E. telmateia*
         5. Branches hollow.
            6. Stem furrows 9–32; shallow. Sheath teeth 12–24. Central cavities more than one-half diameter of stems .... *E. fluviatile*
            6. Stem furrows 5–10; deep. Sheath teeth 5–10. Central cavities less than one-third diameter of stems .... *E. palustre*
      4. First internodes of branches longer than or equal to subtending stem sheaths.
         7. Secondary branches present. Sheath teeth reddish; papery; coherent in 3 or 4 groups. Stem ridges with recurved spicules .... *E. sylvaticum*
         7. Secondary branches absent. Sheath teeth dark brown-black; firm, not papery; coherent in 6–12 groups. Stem ridges with tubercules or transverse ridges.
            8. Branches spreading or drooping. Branch sheath teeth acute .... *E. pratense*
            8. Branches ascending. Branch sheath teeth attenuate .... *E. arvense*
   3. Stems not branched; whorls of branches absent.
      9. Fertile stems green.
         10. Stem furrows 9–32; shallow. Sheath teeth 12–24. Central cavities more than one-half diameter of stems .... *E. fluviatile*
         10. Stem furrows 5–10; deep. Sheath teeth 5–10. Central cavities less than one-third diameter of stems .... *E. palustre*
      9. Fertile stems brown or white.
         11. Sheath teeth reddish; papery; coherent in 3 or 4 groups. Stem ridges with recurved spicules ...... *E. sylvaticum*
         11. Sheath teeth dark brown-black; firm, not papery; coherent in 6–12 groups. Stem ridges with tubercules or transverse ridges.
             12. Fertile stems turning green and persisting after spores shed .... *E. pratense*
             12. Fertile stems remaining white or brown and dying after spores shed.
                 13. Sterile stems 5–20 mm in diameter. Sheath teeth 14–36 .... *E. telmateia*
                 13. Sterile stems 1.5–5 mm in diameter. Sheath teeth 4–14 .... *E. arvense*

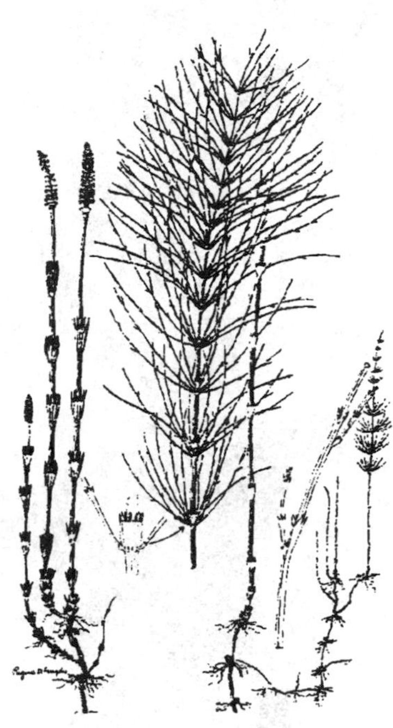

Equisetum arvense

Equisetum hyemale

Equisetum laevigatum

■ moist sites; sometimes large, dense stands

Distribution and Habitat—Plants of *Equisetum* are characteristically encountered in damp or wet areas—sandbars, seeps, borrow ditches, stream terraces, and, occasionally, standing water. The habitat for some species is similar to that occupied by *Typha*, the cattails. They may form large, dense populations via vigorous rhizomatous growth. Distributed across the continent, *E. hyemale* and *E. arvense* have the greatest distributional ranges, as their common names imply. *Equisetum laevigatum* ranges from the extreme southwest to Quebec. The other species are distributed more or less across the northern half of the continent.

■ neurotoxic, mainly horses, large amount in hay for 2–3 weeks, slow onset of incapacitation

large amounts fresh, cattle, diarrhea, rarely neurologic

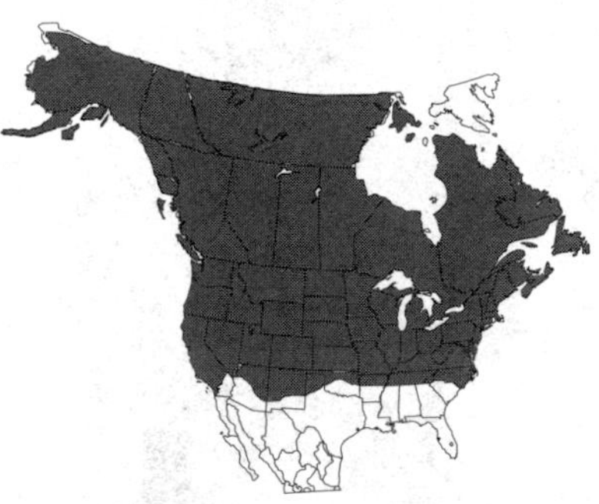

Equisetum arvense

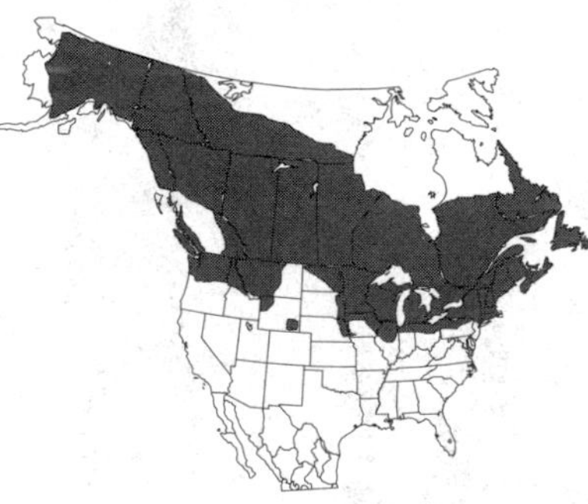

Equisetum fluviatile

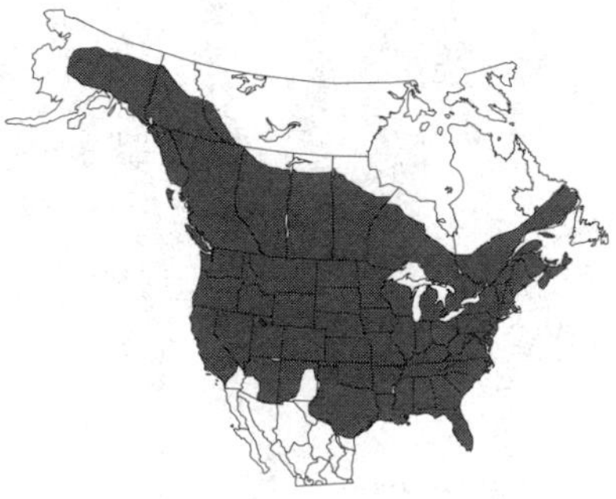

Equisetum hyemale

Disease Problems—Intoxications due to consumption of *Equisetum* are unusual, although European species have a reputation for toxicity. In the 1800s the genus acquired some recognition as toxic in North America, and the disease produced was termed equisetosis (Pammel 1911, 1921; Rapp 1954). Probably all species are neurotoxic, causing muscle weakness, trembling, and collapse (Harvey et al. 1944). Only rarely are plants of the genus eaten in quantities sufficient to cause disease; they are a very coarse forage because of the high concentrations of silicates in the stems. The texture of the stems and the absence of typical foliage leaves likely limit the hazard of intoxication. However, because they remain green in winter, they are eaten, especially the tops, by pastured cattle, sheep, and probably horses in some circumstances (Cary et al. 1924). In his 1819 travels through the Mississippi River basin, Thomas Nuttall (1980) observed that vast fields of *E. hyemale* intermixed with cane provided winter grazing for cattle in some areas and that, when their diet comprised excessive amounts of *Equisetum*, cattle developed scours and became chronically debilitated. Early spring growth seems especially attractive (Edwards 1949). However, because of the quantities required for intoxication, contamination of hay is the more common route of ingestion, especially for horses, because the plants seem to be more readily eaten when dried.

Although other animal species can be intoxicated, for the most part, the disease is one of horses. On numerous occasions both horses and cattle have had access to the same contaminated hay, but only the horses developed the disease (Rich & Jones 1902; Hudson 1924; Edwards 1949). Young horses seem to be particularly at risk. Typically, intoxication follows ingestion for 3–4 weeks of hay containing 20% or more of *Equisetum*. Occasionally, disease may occur following only a week or so of ingestion when larger amounts contaminate the hay. Most cases in North America have involved *E. arvense*, probably as a result of both its broad distribution and the diversity of habitats occupied (Chesnut & Wilcox 1901; Rich & Jones 1902). *Equisetum arvense*, *E. palustre*, and *E. pratense* are the species most likely to cause intoxications. However, the other species should be considered to be toxic as well, because deaths in cattle are reported to be caused even by *E. hyemale* (Pammel 1911).

Disease Genesis—*Equisetum* contains small amounts of the pyridine alkaloids 3-methoxypyridine and l-nicotine, but these seem to be of little significance (Marion 1950). There is little doubt that toxicity is due to the action of a thiaminase that is destroyed by high temperature but not by drying, as in hay (Henderson et al. 1952). This is clearly a thiamine-responsive toxicosis, but it is not clear whether the thiaminase is a type I or II. It has been assumed that the genus has a type I thiaminase similar to that in *Pteridium*, because the clinical signs are so similar. Type I thiaminases require cosubstrates for the methyltransferase not only to destroy the thiamine activity but also to create a thiamine analog to antagonize the actions of thiamine (Brent & Bartley 1984; Chick et al. 1989). Type II thiaminases hydrolyze and inactivate thiamine. Studies on a South African species of *Equisetum* in which both type I and II thiaminases are present suggest that the same may be true for other species as well (Meyer 1989).

- neurologic, horses, thiaminase activity, unknown whether type I or II; thiaminase activity less likely in cattle

  diarrhea due to coarse nature of stems

  neurotoxic alkaloids of unknown significance

Clinical Signs—The disease produced is usually of slow onset—developing over several days—and begins with a poor and scurfy appearance and slight incoordination. As the incoordination worsens, there is a loss of muscular control, a staggering gait, and great difficulty in negotiating turns without falling. With the loss of equilibrium, animals become uneasy, nervous, and occasionally belligerent. After a few days they go down, may periodically struggle and/or have tonic/clonic seizures, and then die from exhaustion. This progression of signs may take several days to a week. During the first few days they will continue to eat and drink. Serum chemistries will show a marked increase in blood lactate and pyruvate and a decrease in blood transketolase. Cattle may show similar signs, except that diarrhea seems more common (Hudson 1924; Edwards 1949).

- horses, slow onset, poor condition, ataxia, poor muscle control, nervous, weakness, recumbent

Pathology—Generally, few distinctive changes are associated with *Equisetum* intoxication. There may be passive congestion of the lungs, liver, brain, and spinal cord, with scattered petechia on the epicardium.

- no distinctive lesions

Treatment—If intervention occurs early in the course of the disease, good nursing care and laxatives will result in recovery in many instances. However, the most appropriate treatment is 500 mg to 1 g/day of thiamine given parenterally for several days. This will typically result in alleviation of signs, beginning within 24 hours; recovery will be nearly complete within a few days. Feeding grain seems to be somewhat protective against the development of thiamine deficiency (Pammel 1911).

- thiamine parenterally, quick response

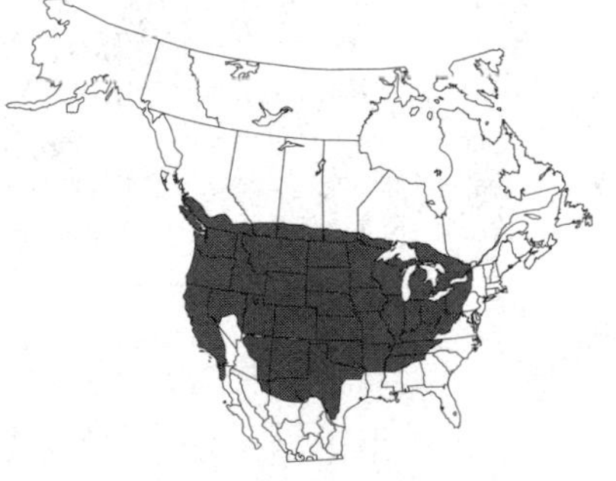

Equisetum laevigatum

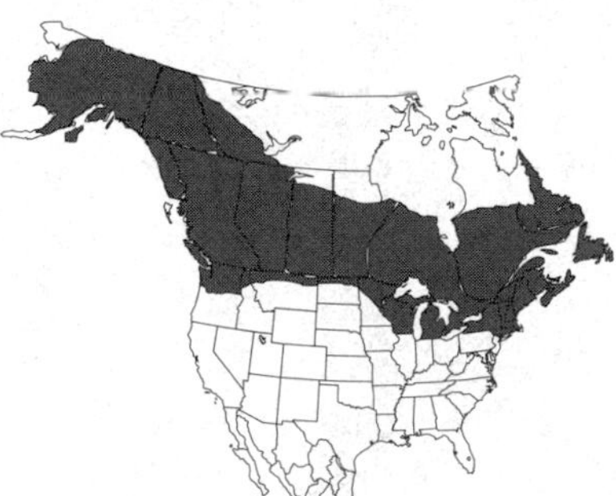

Equisetum palustre

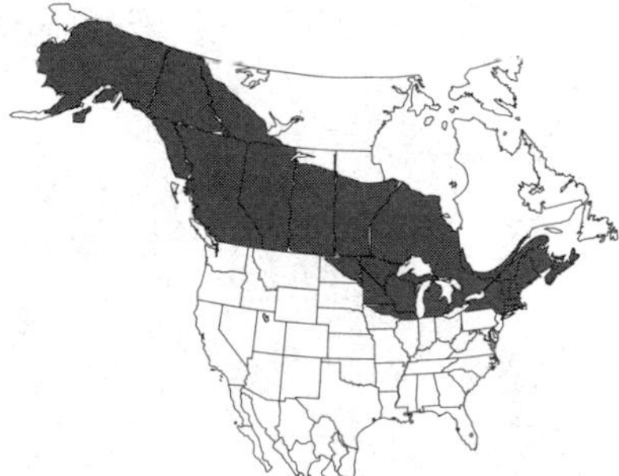

Equisetum pratense

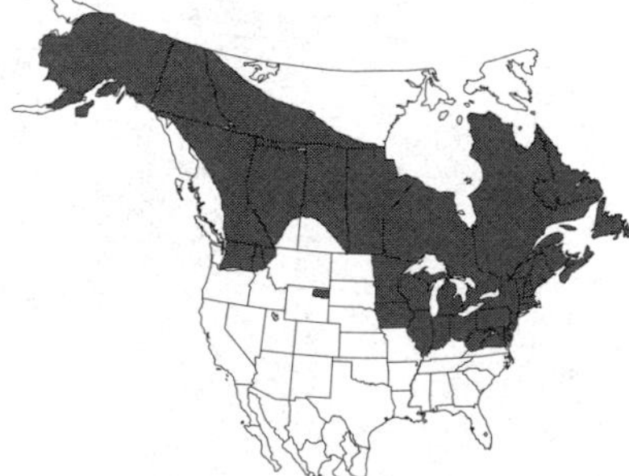

Equisetum sylvaticum

Equisetum telmateia

References

Brent BE, Bartley EE: Thiamin and niacin in the rumen. *J Anim Sci* 59;813–822, 1984

Cary CA, Miller ER, Johnstone GR: *Poisonous Plants of Alabama.* Ala Polytech Inst Ext Serv Circ 71, 1924

Chesnut VK, Wilcox EV: *The Stock-Poisoning Plants of Montana.* USDA Div Bot Bull 26, 1901

Chick BF, McCleary BV, Beckett RJ: Thiaminases. In *Toxicants of Plant Origin,* vol 3, Cheeke PR ed, CRC Press, Boca Raton, Fla, pp73–91, 1989

Edwards CM: Some observations on plant poisoning in grazing animals. *Vet Rec* 61;864–865, 1949

Harvey RB, Larson AH, Landon RH, Boyd WL, Erickson LC: *Weeds Poisonous to Livestock.* Minn Agric Exp Stn Bull 388, pp3–5, 1944

Hauke RL: Equisetaceae. In *Flora of North America,* vol 2, *Pteridophytes and Gymnosperms,* Flora of North America Editorial Comm, Oxford Univ Press, New York, pp76–84, 1993

Henderson JA, Evans EV, McIntosh RA: The antithiamine action of *Equisetum. J Am Vet Med Assoc* 120;375–378, 1952

Hudson R: Poisoning by horsetail *(Equisetum arvense). Vet J* 80;40, 1924

Marion L: The pyridine alkaloids. In *The Alkaloids,* vol 1, Manske RHF, Holmes HL eds, Academic Press, New York, pp165–269, 1950

Meyer P: Thiaminase activities and thiamine content of *Pteridium aquilinum, Equisetum ramosissimum, Malva parviflora, Pennisetum clandestinum,* and *Medicago sativa. Onderstepoort J Vet Res* 56;145–146, 1989

Nuttall T: *A Journal of Travels into the Arkansas Territory during the Year 1819,* Lottinville S ed, Univ Oklahoma Press, Norman, pp68,89, 1980

Pammel LH: *Manual of Poisonous Plants.* Torch Press, Cedar Rapids, Iowa, pp323–325, 1911

Pammel LH: Equisetosis or horsetail poisoning. *Vet Med* 16(4);43, 1921

Rapp WF Jr: The toxicity of *Equisetum. Am Fern J* 44;148–154, 1954

Rich FA, Jones LR: *A Poisonous Plant, the Common Horsetail* (Equisetum arvense). Vt Agric Exp Stn Bull 95;185–192, 1902

Ericaceae Juss. 35

Prized for its showy ornamentals and the edible fruits of some of its members, the Ericaceae, commonly known as the heath family, comprises 100–125 genera and 3,300–3,500 species. In North America, approximately 33 genera and 215 species and named hybrids are present. Distributed throughout the world, species are characteristically encountered in acidic soils, with greatest abundance in the cool temperate regions of both the Northern and the Southern Hemispheres. In the tropics, taxa of the family are present at higher elevations. Species of *Erica* are an important element of the moors in Europe, and species of *Arctostaphylos* are important in the Californian chaparral. Many members of the family are ornamental shrubs and are widely cultivated in gardens. Familiar taxa include species of *Rhododendron* (azaleas and rhododendrons), *Erica* (heaths), *Kalmia* (mountain laurels), *Calluna* (heathers), and *Arbutus* (madrones). The fruits of *Vaccinium*—cranberries, bilberries, and blueberries—are relished, as are those of *Gaylussacia,* the huckleberry genus. *Gaultheria* was the original source of wintergreen oil.

■ Heath Family

Andromeda
Kalmia
Ledum
Leucothoe
Lyonia
Menziesia
Pieris
Rhododendron

Plants mainly shrubs or small trees or subshrubs or perennial herbs; evergreen or deciduous; generally mycotrophic. **Leaves** simple; alternate or rarely opposite; mainly coriaceous; venation pinnate; margins entire to serrate; stipules absent. **Inflorescences** panicles or racemes or umbellate clusters or corymbs or flowers solitary; terminal or axillary; bracts present; bracteoles 2. **Flowers** perfect; perianths in 2-series; radially or slightly bilaterally symmetrical; funnelform or campanulate or urceolate. **Sepals** 3–7, typically 5; free or fused. **Petals** 3–7, typically 5; fused or rarely free; of various colors. **Stamens** twice the number of petals or rarely 4 or 5 or 20; anthers appendaged or with terminal awn or tubules; dorsifixed; dehiscing poricidally or rarely longitudinally. **Pistils** 1; compound, carpels 4 or 5; stigmas 1, capitate or peltate; styles 1; ovaries superior or inferior; locules 4 or 5 or 8 or 10 (doubling by false partitions); placentation axile. **Nectaries** borne on intrastaminal disks. **Fruits** berries or capsules. **Seeds** numerous, small.

■ herbs, shrubs, subshrubs, small trees; leaves simple; petals 3–7, mainly fused; fruits berries or capsules

Disease Problems—Members of the Ericaceae produce an array of secondary compounds, including iridoids, phenol heterosides, triterpenoids, proanthocyanins, 5-methoxy-flavanols, and the definitely toxic diterpenoids (Cronquist 1981). Some genera—e.g., *Vaccinium,* with its edible fruits—are not toxic, and others are. Likewise within a genus, some species may be toxic and others not. Even among the so-called toxic species, the hazard varies. Because their toxicity is qualitatively, if not quantitatively, the same (Marsh 1924; Sampson & Malmsten 1942), an overview of the toxicants and the disease problems they produce are presented in the following paragraphs. When appropriate, additional information about the compounds and/or problems is presented in the generic treatments that follow. For the most part, the plants of the family are a problem only in special circumstances. Although the toxic species

■ toxic diterpenoids, numerous other compounds

are not generally palatable, most of them are evergreen and therefore pose a greater risk in winter. Unfortunately, some of the more toxic genera, especially *Kalmia* and *Rhododendron,* are prized for their beauty and are widely cultivated, thus increasing the risk of accidental ingestion by pets and children.

- toxic nature of family known from ancient times; especially related to toxic honey; digestive disturbance

Our knowledge of the toxicity of members of the Ericaceae begins in antiquity. Since the time of the Greeks, species of *Rhododendron* have been known for their propensity to render toxic the honey made from the nectar of their flowers. Soldiers of Xenophon's army in 401 B.C. near Trebizond on the Black Sea coast of Turkey reputedly were intoxicated by honey produced from *R. luteum* or *R. ponticum* (Howes 1950; Carey et al. 1959; Leach 1966). Those who consumed large quantities developed severe digestive distress. There were no deaths, only incapacitation for 1–2 days. In 67 B.C., during Pompey's foray against Mithridates, intoxication of soldiers again occurred. In the Black Sea region, this annual scourge, characterized by digestive problems, dizziness, heart problems, weakness, and visual derangement became known as a response to "mad honey" (Leach 1966). The region continues to be a source of problem honey (Sutilupinar 1993). Since those early reports, in other areas of the world, numerous incidences have been reported of intoxication in humans from honey and in livestock from foliage. Illustrative of this are episodes of toxic honey in the state of Washington in 1942, 1952, and 1968 ("Toxic honey," 1969). *Kalmia* and *Rhododendron* seem to be particularly troublesome sources (Howes 1950; Patwardhan & White 1973).

Toxic members of the Ericaceae also are purported to cause secondary adverse effects in animals or humans who eat meat from animals dying from intoxication (Leach 1967; Nanobashvili & Shavgulidze 1987). The viscera of chickens fed laurels is said to be toxic to cats (Chesnut, 1898). In contrast to the toxic effects, the leaves of some species of *Rhododendron* have been used as a folk remedy for hypertension (Hikino et al. 1979). Unfortunately, they also were used at one time as ill-advised cures for arthritis and other ailments, with disastrous results (Leach 1967).

- most animal species susceptible; foliage unpalatable, but evergreen and eaten in winter when other feed is scarce

  toxic dosage as low as 0.12% b.w. for some plant species

  intoxications in humans due to honey contamination, persistent vomiting, cardiac effects

Since the mid-1700s, attention has been drawn to the toxic effects of the various species of eastern North America, especially those of *Kalmia.* It was well recognized that most plants were eaten only when other forage was lacking. Thus they were seldom a problem except in unique circumstances, such as when used as ornamentals in zoos or when trimmings of the foliage were discarded where they could be eaten (Halsted 1895). Most animal species are susceptible, including horses, goats, sheep, cattle, laboratory rats, llamas, and even kangaroos (Miller 1981; Fowler 1985; Thiemann 1991; Hough 1997). Many intoxications, as the numerous reports attest, involve sheep or goats, often but not invariably in winter when there is snow cover and/or the animals are experiencing cold stress (Pritchard 1956; Lodge et al. 1968; Smith 1978; Higgins et al. 1985; Shannon 1985; Hollands & Hughes 1986; Hosie et al. 1986; Casteel & Wagstaff 1989; Gregory 1990; Black 1991; Power et al. 1991; Plumlee et al. 1992). Species of western North America also were especially important causes of sheep losses in California and the Northwest (Marsh 1924). Occasionally, ornamental species are suspected in intoxications of household pets (Rose et al. 1988; Frape & Ward 1993). Deer have sometimes been considered to be resistant; they can be intoxicated but generally are very reluctant to eat enough foliage of *Kalmia* or *Rhododendron* to be affected (1.75% b.w.), except in the coldest weather and only when available alternative feed is limited (Chesnut 1898; Forbes & Bechdel 1931). Occasionally the foliage is a risk for children. A 2-year-old child ate a cluster of buds of *Pieris japonica* and 1 hour later experienced sneezing, a burning sensation of the mouth, drooling, and vomiting (Tschekunow et al. 1989). The heart rate was slowed to 65/min and blood pressure was reduced to 78/30; these signs were corrected to 105/min and 90/70 with atropine treatment.

The problem with toxic honey is probably somewhat exaggerated, but nevertheless there is some risk, especially with that involving the more toxic species such as *Kalmia latifolia* (White & Riethof 1959). The adverse effects are typically rather mild, with the exception of persistent vomiting in a few patients (Klein-Schwartz & Litovitz 1985; Geroulanos et al. 1992). In the rare more serious intoxications, there are cardiac arrhythmias, decreased blood pressure, and rarely seizures (Schindler 1962; Gossinger et al. 1983). In addition to the toxicity imparted to honey, there are direct toxic effects on bees. The bees become weak and sluggish and fly only in spurts, during which they may fall prostrate on the ground (Carey et al. 1959). In some instances the bees appear to be forcibly expelled from the hive, landing on their sides, spinning and twitching, and dying in 8–10 minutes (Olszowy 1977). The entire colony may be decimated. In such a situation, bees are found with crumpled wings and incomplete pupal development.

Species of *Leucothoe* appear to be the most toxic, but they are less commonly the cause of intoxications than the more abundant *Kalmia* and *Rhododendron*. Toxicity may vary considerably among collections of the same species and also among many of the ornamental cultivars (Tallent et al. 1957; Verlangieri et al. 1976). Comparatively, the toxic dosages for sheep range from 0.12% b.w. for *L. davisiae*, to 0.15–0.2% b.w. for *K. angustifolia*, to 0.3–0.4% b.w. for *K. latifolia* and *K. microphylla* (Crawford 1908; Fleming 1920; Marsh 1924; Marsh & Clawson 1930; Clawson 1933). The lethal dosages are generally 2-to-4-fold greater. Generally the deciduous-leaved azaleas are of lesser risk than the evergreen rhododendrons. Experimentally, the ornamental azalea *R. indicum* caused moderate signs of intoxication at 1% b.w., severe signs at 2% b.w., and death at approximately 3% b.w. in cattle (Tokarnia et al. 1996).

■ extensive series of diterpenoid grayanotoxins

Disease Genesis—Genera of the Ericaceae contain large concentrations of phenolics, including tannins (Frohne & Pfander 1984), but the toxic principles are an extensive array of diterpenic polyalcohols or diterpenoids (table 35.1.). The first isolated was called andromedotoxin by Plugge in the 1880s, at about the same time that it was isolated and named asebotoxin by Eykman, after the Japanese name for the plant of origin, *Andromeda japonica* (Hardikar 1922). The nomenclature of these compounds is quite confusing, because numerous names are applied to the same compounds. Andromedotoxin and asebotoxin are also called rhodotoxin and acetylandromedol, but all these names have now been superseded by the name grayanotoxin I. Similarly anhydroandromedol, or Δ-10(18)-andromedenol, is now referred to as grayanotoxin II (Tallent et al. 1957; Schindler 1962). The specific stereochemistry of the diterpenoids has been determined (Kakisawa et al. 1965; Kumazawa & Iriye 1970). There also are variations in the basic ring structure represented such as hexahydroxy-B-homo-C-nor-grayanoid, which likewise are toxic (Burke & Doskotch 1989). In some instances, the diterpenoids occur as glycosides: grayanoside B and pierosides A and B. The aglycone of grayanoside B is grayanotoxin XVIII (Sakakibara et al. 1979). The nomenclature employing the grayanotoxin designations is confusing because the term *grayanotoxin* is used to designate specific toxins as well as being used as a general term. The term *ericaceous diterpenoids* is preferred when referring to the toxins in general. In addition to the diterpenoids, a sesquiterpene, germacrone, has been identified in *Ledum;* it acts as a potent antifeedant to rabbits (Reichardt et al. 1990).

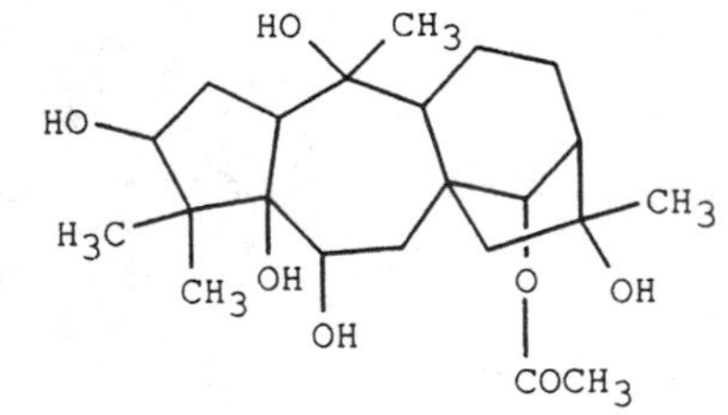

Grayanotoxin I

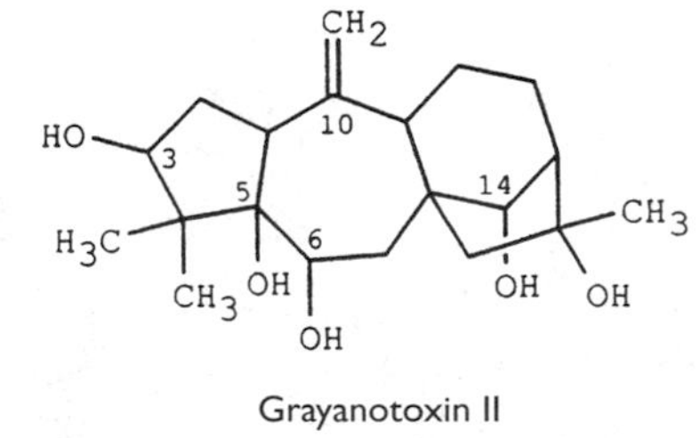

Grayanotoxin II

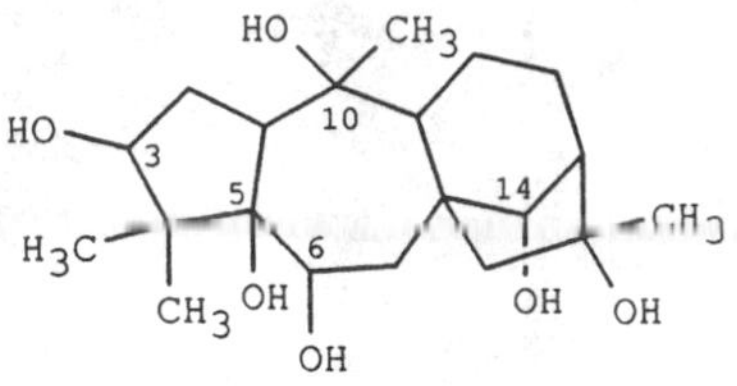

Grayanotoxin III

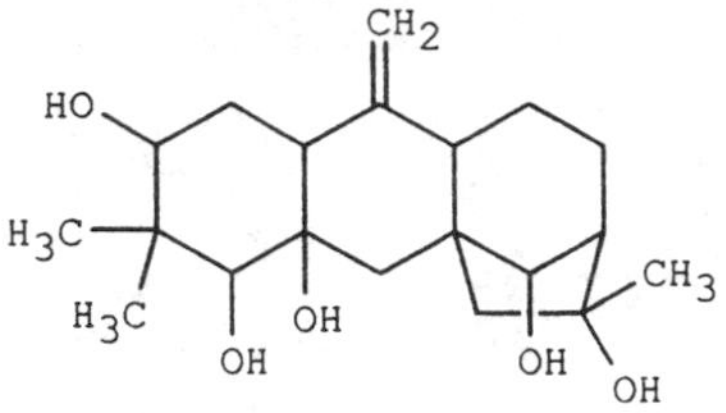

Leucothol B

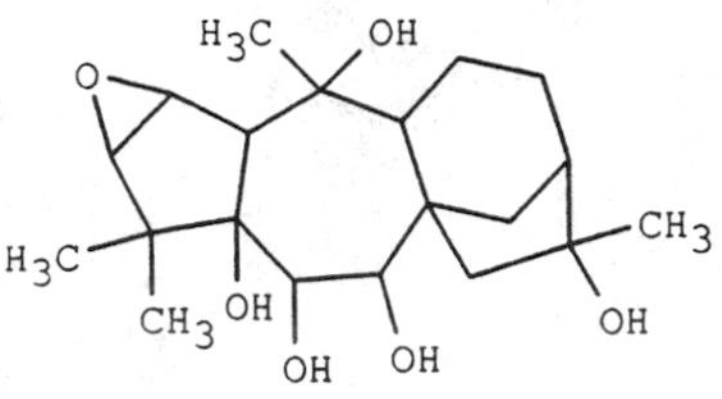

Lyoniatoxin

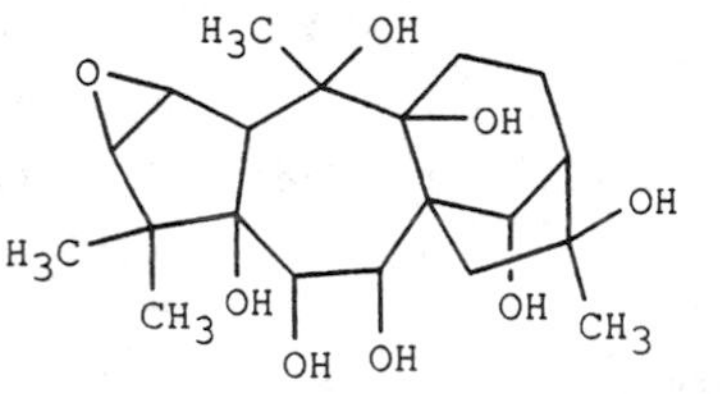

Pieristoxin

■ toxins in all plant parts, including nectar

Table 35.1. Diterpenoid toxins of genera of the Ericaceae

Genus	Toxins	References
Kalmia	GTX I, II, III	Tallent et al. 1957
Leucothoe	GTX I, II, III, VIII, IX, X, XI, XII, XIII, XVIII grayanosides A, B leucothol B, D	Tallent et al. 1957; Hikino et al. 1976; Sakakibara et al. 1979, 1981
Lyonia	GTX I lyoniatoxin	Tallent et al. 1957; Hikino et al. 1976
Pieris	pieristoxin A, B, C asebotoxins I, II, III, IV, V, VII GTX XVIII grayanoside B pieroside B	Tallent et al. 1957; Schindler 1962; Hikino et al. 1969a,b, 1976; Sakakibara et al. 1981
Rhododendron	GTX I, III rhodojaponin I, II, III, IV, V, VII	Hardikar 1922; Tallent et al. 1957; Hikino et al. 1969b, 1976

Note: GTX = grayanotoxin.

Toxicity varies considerably among the various ericaceous diterpenoids. When tested by i.p. administration in mice, the most toxic were asebotoxin I and III, rhodojaponin III and V, deacetyllyoniatoxin, and grayanotoxin III and VI, with LD_{50} of 1 mg/kg b.w. or less (Scott et al. 1971; Hikino et al. 1976). The LD_{50} of grayanotoxin I was slightly more than 1 mg/kg, while the LD_{50}s of grayanotoxin II, VII, and VIII and rhodojaponin II were greater than 10 mg/kg, and those of grayanotoxin V, XI, and XII and rhodojaponin IV were greater than 100 mg/kg.

The presence of hydroxyl groups on the andromedane skeleton, especially in the beta configuration, is essential for activity, and five appears to be the optimal number of groups (Masutani et al. 1981; Terai et al. 1996). The most important positions are C-3, C-5, and C-6. A beta methyl group at C-10 and a 14-β-hydroxyl, or preferably acylation with propionic acid or lactic acid, are also important for maximal activity (Hikino et al. 1976; Mager et al. 1981; Masutani et al. 1981). The structural requirements of reactive oxygen groups are similar to those for veratridine and aconitine (Masutani et al. 1981).

The toxins are found in all parts of the plants, including nectar from the flowers. It is suggested that, at least for *Kalmia,* toxicity is lower in the summer and fall of the year (Verlangieri et al. 1976), but for the most part it is not clear what variations occur because of stage of growth or season of the year. The presence of the toxins is variable in both a genus and among the species. In some instances high concentrations have been noted, whereas other species may lack detectable levels (Tallent et al. 1957). Thus, even some species of *Kalmia* and *Rhododendron* may be of little risk. Grayanotoxins have been isolated from honey in concentrations sufficient to cause clinical problems at least of short duration and mild nature (Scott et al. 1971; Onat et al. 1991; Ozesmi et al. 1994). Interestingly, uncapped honey is reported to be toxic, but once the comb cells in the hive are capped by the bees and the honey ripened, it is apparently safe to eat (Carey et al. 1959; Olszowy 1977).

In one of the early reports on effects of the toxins, the action was surmised to be centered on the vagus nerve, initially stimulating and eventually paralyzing it (Hardikar 1922). The result was a slowing of respiration, as well as muscle fatigue and paralysis of the diaphragm. Dyspnea due to bronchiole spasm and increased bronchiolar secretions was prominent. Heart rate and blood pressure decreased, leading to a reflex increase in both and possibly arrhythmias. The effects on decreasing blood pressure, heart rate, and respiration have been confirmed for extracts of *Kalmia angustifolia* and *Rhododendron maximum* (Waud 1940; Moran et al. 1954; Wood et al. 1954). Blood pressure effects appear to be at least partly due to reflex vasodilation. With large doses, cardiac effects are prominent. These include EKG changes, conduction impairment, and AV nodal and ventricular tachycardia (Moran et al. 1954). Neurologic effects are also produced with large doses: stimulation initially and then depression.

In an extensive series of experiments using squid axons, grayanotoxin was shown to increase sodium permeability by nearly 100-fold, an action that was antagonized effectively by the well-known sodium channel blocker tetrodotoxin (Seyama & Narahashi 1973; Narahashi & Seyama 1974). Receptor binding specificity for grayanotoxin, veratridine, and aconitine appears to involve a common receptor site (Catterall 1980; Catterall et al. 1982), which is slightly removed from that of the pyrethroids (Takeda & Narahashi 1988). While tetrodotoxin and saxitoxin block the channel opening, grayanotoxin and the other plant toxins bind at site 2 on the outward side to modify sodium channel characteristics (Catterall 1986). Closed channels open more slowly and open channels close slowly, resulting in persistent activation of a portion of the sodium channels. Selectivity of the channels is markedly decreased, such that they become increasingly permeable to potassium and other cations (Seyama & Narahashi 1981). At high dosage there may be effects on calcium and other channels (Romey & Lazdunski 1982). The overall result is that the sodium channels open at more-negative potentials, and the slow closure results in persistent depolarization. Similar actions are evident in frog skeletal muscle, where there is persistent depolarization (Masutani et al. 1981), and at the neuromuscular junction, where there is increased transmitter release, depolarization, and increased frequency and amplitude of spontaneous miniature end point potentials (MEPPs) (Zushi et al. 1983). There have been suggestions that other mechanisms may also be involved, because depolarization still occurs after inactivation of sodium channel gating proteins (Seyama et al. 1985). Additional neurotoxic actions may accrue by these other mechanisms. Locomotor depression appears to be associated with antagonism of central glutaminergic receptors, given that these effects are reversed by quisqualic acid in mice (Ohgaki et al. 1988). Moreover, other sodium channel activators do not cause such neurologic depression. There also seem to be cholinergic responses, because the increased salivation as well as the decreased heart rate are responsive to atropine (Moran et al. 1954; Onat et al. 1991).

Cardiac effects include negative chronotropic actions with decreased sinoatrial node activity, leading to sinus arrest due to persistence of open channels (as shown in rabbits) (Masakazu & Seyama 1984). There is also depolarization of the Purkinje fibers in either the resting or the stimulated state, and long-lasting depolarization of the ventricles in the stimulated state (Akera et al. 1976; Nakao & Seyama 1984; Ito et al. 1985). Vagal-mediated decreased heart rate and a variety of cardiac arrhythmias ranging from AV blocks to intermittent sinus node arrest are typical of intoxication due to diterpenoids in honey (Gossinger et al. 1983; Onat et al. 1991).

The toxic effects are primarily acute; there seems to be little or no cumulative action, except perhaps on body weight and condition (Hikino et al. 1979). However, sometimes there is persistence of the signs for up to a week, delayed onset of signs, or secondary effects, such as effects on the fetus, which may give the impression of subacute to chronic effects (Smith 1979; Gregory 1990).

Experimentally, in mice and chickens, grayanotoxin I, although lethal in adults, was not associated with embryotoxicity or teratogenicity (Kobayashi et al. 1990). Long-term use as a folk remedy is also sometimes accompanied by numbness of the hands (Hikino et al. 1979). Also, in vitro cancer screening has shown the expressed sap of *Kalmia latifolia* or extracts from it to be cytotoxic (Mancini & Edwards 1979).

- toxins bind sodium channels, slowing their opening and closing and decreasing their ion selectivity

effects on many cells, especially excitable cells; neurologic, cardiac, muscular effects; persistent action, depolarization easier to trigger, leads to increased reflex activity, increased vagal influence, decreased blood pressure and heart rate, inhibition of cardiac conduction, arrhythmias; large dosage, neurologic signs

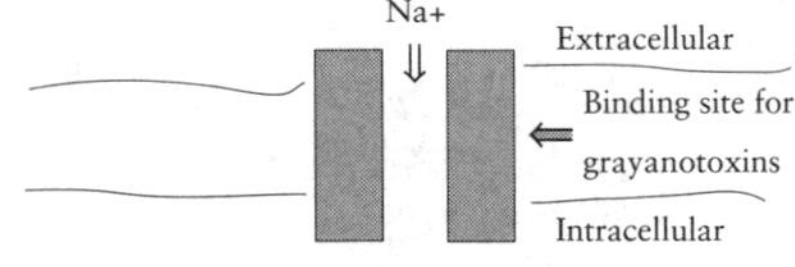

Sodium channel
(after Catterall 1986)

- anorexia, profuse salivation, repeated swallowing, persistent vomiting, retching, colic, irregular respiration, bellowing; effects last up to 24 hours

Clinical Signs—The signs develop several hours after ingestion, whether due to ingestion of foliage or honey, and include anorexia, profuse salivation and nasal secretions (with the appearance of foam around the mouth and nose), repeated swallowing, persistent retching/vomiting, bloat, severe abdominal pain, and irregular respiration. In some cases regurgitation or vomiting may be forceful (projectile). Diarrhea is not typical and in fact seems to be protective, as it has been noted that sheep are seldom fatally intoxicated when diarrhea develops (Crawford 1908). There is often bellowing or other indications of pain and odontoprisis. Partial blindness and seizures have also been reported and eventually, in fatal intoxications, animals become weak, reluctant or unable to stand, and almost paralyzed. The toxins also produce hypotension and respiratory depression.

The acute effects typically last for several hours up to 24 hours, but weakness or neurologic effects may persist for 2–3 days. Marked loss of body weight has also been reported (Casteel & Wagstaff 1989).

Several procedures have been described for detecting the ericaceous diterpenoids. The first employs a saturated solution of antimony trichloride in chloroform for detection on thin-layer chromatography with observation under UV light (Schindler 1962; Frohne & Pfander 1984). The limit of detection is 2 µg. Another more recent method involves extraction of the toxins from rumen/stomach contents with 2–3 vol of methanol and then chloroform, chromatographing on silica gel developed in 9:1 vol/vol chloroform/methanol and detection using Godin's reagent (Humphreys & Stodulski 1986). Godin's reagent is a 1% wt/vol ethanolic vanillin solution mixed with an equal volume of aqueous 3% wt/vol perchloric acid. The presence of andromedotoxin is detected by development of a purple color after spraying and warming in an oven at 85°C for 5 minutes.

- no distinctive lesions, leaves in stomach

Pathology—The changes, which in most cases are rather mild and not distinctive, include edema of the lungs and renal tubular and hepatocellular necrosis. Typically leaves will be present in the stomach, especially the tough leaves of *Rhododendron*.

- activated charcoal, atropine, control of pain, stimulants have been used

Treatment—There are no specific antidotes; treatment is primarily directed toward relief of the symptoms. Analgesics and various stimulants, including ephedrine and the methylxanthines (caffeine, theobromine), have been suggested, but their efficacy has not been confirmed experimentally; drenches of coffee and tea have been used. Atropine has been used with some success. Yi and co-workers (1985) reported success in cattle with a combination of 10–20 mg atropine sulfate and 15–20 ml of 10% sodium camphorsulfonate given subcutaneously. Because of the acute nature of the disease, oral administration of charcoal may be of some benefit in limiting absorption of the toxins. The increased salivation and some of the cardiac effects, including the decreased heart rate, are responsive to atropine. Although not confirmed experimentally, sodium and possibly calcium channel blockers may also be of some benefit.

Andromeda L.

- *A. polifolia*
  andromeda
  bog rosemary
  marsh andromeda

Taxonomy and Morphology—A monotypic genus named for the daughter of Cepheus and Cassiope in Greek mythology, *Andromeda* is circumboreal in the Arctic and cool-temperate regions. Some taxonomists regard the North American populations as distinct from the European ones and use the name *A. glaucophylla*. Their morphological and ecological similarity, however, favors recognition of just 1 species:

A. polifolia L.
(= *A. glaucophylla* Link) bog rosemary, marsh andromeda, andromeda

Plants shrubs; evergreen; compact or spreading. **Stems** erect to decumbent; 10–70 cm tall; wiry or not wiry; glabrous or glaucous. **Leaves** alternate; coriaceous; blades linear to narrow elliptic; abaxial surfaces sparsely white tomentose; margins entire, revolute. **Inflorescences** terminal umbellate corymbs; pedicels recurved. **Flowers** urceolate. **Sepals** 5; fused. **Petals** 5; fused; white or pink; lobes short, recurved or spreading. **Stamens** 10; anthers with divergent awns. **Pistils** 1; ovaries superior; locules 5. **Fruits** subglobose capsules.

- evergreen shrubs; leaves alternate, margins entire and revolute; flowers urceolate; petals 5, fused, white or pink

Andromeda polifolia

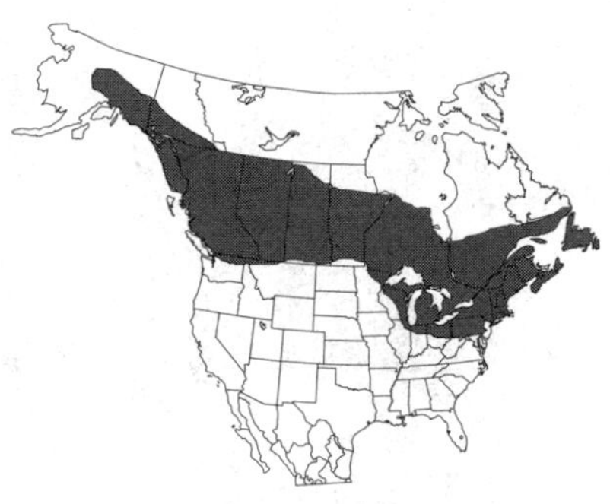

Andromeda polifolia

Distribution and Habitat—Populations of *A. polifolia* are typically encountered in sphagnum peat bogs and wet, acidic sites across Canada.

- wet, acidic sites

Kalmia L.

Taxonomy and Morphology—Named for Peter Kalm, a Swedish botanist and student of Linnaeus who collected in eastern North America in the 1740s, *Kalmia* is a North American genus of 10 species and more than 80 cultivars (Southall & Hardin 1990; Jaynes 1997). Species of toxicologic interest include the following:

K. angustifolia L.	sheep laurel, lambkill, wicky, sheep poison, lamb laurel, low laurel, dwarf laurel
K. latifolia L.	mountain laurel, laurel, calico bush, ivybush, spoonwood, sheep laurel, poison laurel, small laurel, rose laurel, high laurel, roundleaf laurel, American laurel
K. microphylla (Hook.) A.Heller (= *K. occidentalis* Small)	small-leaved laurel, alpine laurel, bog laurel, western swamp laurel, western swamp kalmia, pale laurel, western laurel
K. polifolia Wangenh.	swamp laurel, bog laurel

- *K. angustifolia*
  K. latifolia
  K. microphylla
  K. polifolia

  alpine laurel
  American laurel
  bog laurel
  calico bush
  dwarf laurel
  high laurel
  ivybush
  kalmia
  lamb laurel
  lambkill
  low laurel
  mountain laurel
  poison laurel
  rose laurel
  spoonwood
  swamp laurel
  wicky

Plants shrubs or rarely small trees; evergreen or deciduous; foliage open or dense. **Stems** erect or ascending; 15–300 tall; bark cinnamon red or light brown or gray; exfoliating with age; twigs terete or 2- or 3-angled. **Leaves** alternate or opposite or

- shrubs or small trees; leaf margins entire; flowers rotate to campanulate; petals 5, fused, white, pink, light rose, or rose purple

whorled; coriaceous; blades of various shapes; abaxial surfaces variously indumented; margins entire, revolute or not revolute. **Inflorescences** corymbs or solitary flowers; terminal or axillary. **Flowers** rotate to campanulate. **Sepals** 5; fused. **Petals** 5; fused; white or pink or light rose or rose purple; with a red ring at base of lobes; keeled; with 10 pouches. **Stamens** 10; anthers positioned in petal pouches, without awns. **Pistils** 1; ovaries superior; locules 5. **Fruits** ovoid to globose capsules.

Kalmia latifolia

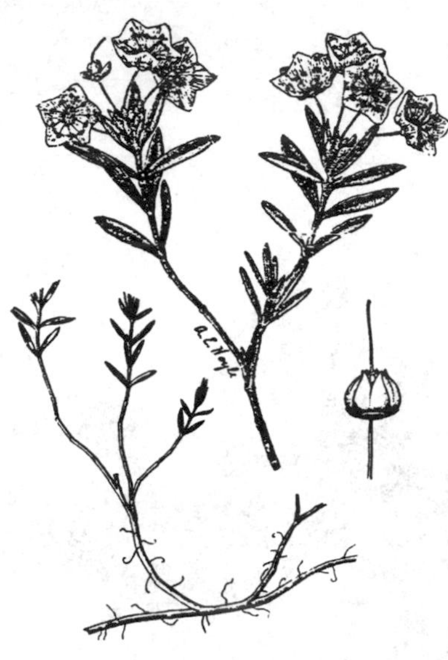

Kalmia microphylla

Kalmia polifolia

- moist acidic soils, edge of bogs and swamps

Kalmia angustifolia

1. Leaves alternate . . . . *K. latifolia*
1. Leaves opposite or whorled.
   2. Plants 10–20 cm tall . . . . *K. microphylla*
   2. Plants 25–120 cm tall.
      3. Inflorescences terminal corymbs or solitary flowers. Leaves subsessile. Capsule valves 2-cleft . . . . *K. polifolia*
      3. Inflorescences axillary racemes. Leaves petiolate. Capsule valves not cleft . . . . *K. angustifolia*

Distribution and Habitat—Of the 10 species in *Kalmia,* 7 are present on the continent and 3 are native to Cuba. In the east, *K. angustifolia* is found along the edges of bogs and swamps and in moist, acid soils of the coastal plains and mountains. The species also is naturalized in Germany and northwestern England. *Kalmia latifolia*

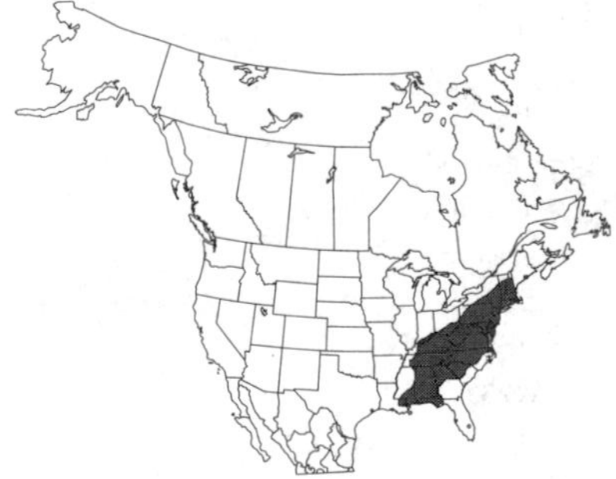

Kalmia latifolia

Kalmia microphylla

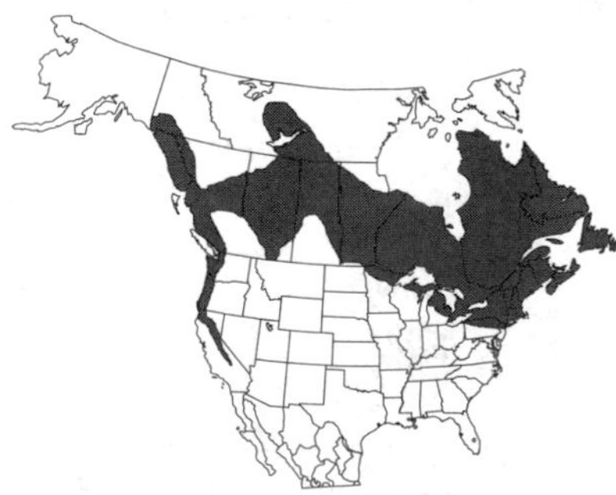

Kalmia polifolia

forms dense thickets in sandy, rocky acid soils of lowland hills, stream banks, slopes, and mountainsides from Maine to Florida. Populations are most abundant in the Appalachian Mountains.

Kalmia polifolia has an exceptionally wide range, occupying bogs and moist meadows up to 3,500 m elevation from Alaska to Labrador. It has naturalized in Scotland and southeastern England. Strictly western, *K. microphylla* is present in subalpine and alpine mountain meadows and bogs up to 3,750 m elevation.

Ledum L.

- *L. glandulosum*
  L. groenlandicum
  L. xcolumbianum
  L. palustre

  Labrador tea
  trapper's tea

- evergreen shrubs; leaves alternate, margins entire and revolute; flowers rotate; petals 5, white

- bogs, swamps in coniferous forests

Taxonomy and Morphology—Comprising 3–5 species in both North America and eastern Asia, *Ledum* is closely related to *Rhododendron.* Although grown as an ornamental, it is less showy than other members of the family. The common name *Labrador tea* reflects its use. During the Revolutionary War, a widely consumed beverage was made from the leaves of *L. groenlandicum.*

L. glandulosum Nutt.	trapper's tea, smooth Labrador tea, glandular-leaved Labrador tea
L. groenlandicum Oeder	bog Labrador tea, common Labrador tea
L. x*columbianum* Piper	hybrid of preceding 2 species
L. palustre L.	northern Labrador tea

Plants shrubs; evergreen; compact or spreading. **Stems** erect or decumbent; 50–200 cm tall. **Leaves** alternate; coriaceous; sessile or petiolate; blades linear to oblong or ovate; narrowly elliptic; abaxial surfaces lanate-tomentose or glandular-scaly; margins entire, revolute. **Inflorescences** terminal umbellate corymbs; pedicels erect in anthesis, recurved in fruit. **Flowers** rotate. **Sepals** 5; fused only at base. **Petals** 5; free; white; lobes oblong or obovate. **Stamens** 5–10; anthers without awns; filaments filiform. **Pistils** 1; ovaries superior; locules 5. **Fruits** ovoid capsules.

Ledum glandulosum

Ledum glandulosum

Ledum groenlandicum

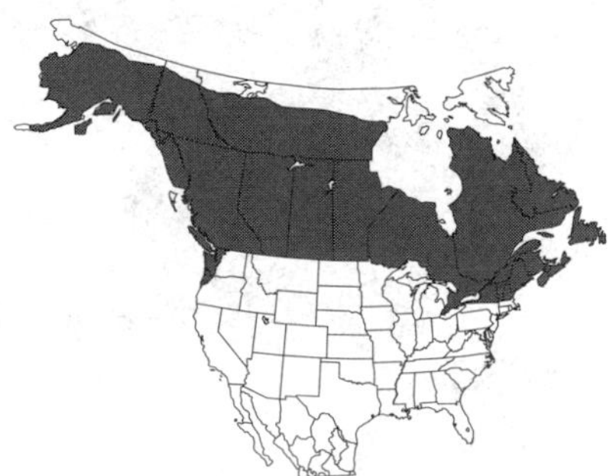

Ledum groenlandicum

Ledum xcolumbianum

1. Abaxial surfaces of leaves glandular-scaly. Stamens shorter than styles.
   2. Capsules subglobose; 1–3.5 mm long. Leaves plane or slightly revolute . . . . *L. glandulosum*
   2. Capsules ovoid; 4–4.5 mm long. Leaves strongly revolute . . . . *L. xcolumbianum*
1. Abaxial surfaces of leaves rusty lanate-tomentose. Stamens longer than styles.
   3. Leaves elliptic to oblong; stamens 5–7. Capsules 5–7 mm long . . . . *L. groenlandicum*
   3. Leaves linear. Stamens 8–11. Capsules 3–5 mm long . . . . . . *L. palustre*

Distribution and Habitat—Like other members of the family, *Ledum* typically occupies bogs, swamps, and wet sites in coniferous forests. *Ledum groenlandicum* is found across the northern part of the continent. *Ledum palustre* is circumpolar in the northern portions of the Canadian provinces. Two species are strictly western: *L. glandulosum* is abundant in moist mountain meadows and boggy areas of open woods up to 3,300 m elevation. The hybrid species *L.* x*columbianum* is confined to the Pacific Coast. It also is cultivated as an ornamental in the West.

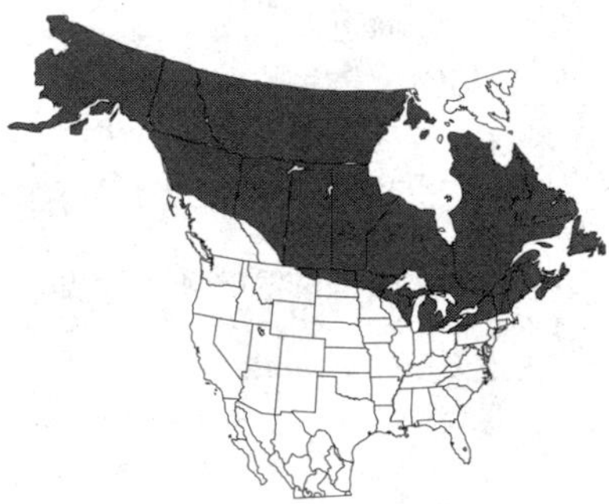
Ledum palustre

Leucothoe D.Don

■ *L. axillaris*, *L. davisiae*, *L. fontanesiana*, *L. populifolia*, *L. racemosa*, *L. recurva*

black laurel, calfkill, dog laurel, fetterbush, kill-calf, leucothoe, Sierra laurel

■ shrubs or small trees; leaves alternate; petals 5, fused, white

Taxonomy and Morphology—Present in both the Old World and the New World, *Leucothoe* comprises 40–45 species, many of which are shade tolerant and grown for their showy flowers and glossy foliage. The genus is named for the daughter of Orchamus, a king of Babylon. Some taxonomists segregate the deciduous species into the genus *Eubotrys;* the traditional circumscription, however, is maintained here. In North America, 8–12 species are present. Of toxicologic interest are the following:

L. axillaris (Lam.) D.Don	dog laurel, hemlock, calfkill, leucothoe, swamp dog-laurel
L. davisiae Torr. ex A.Gray	black laurel, Sierra laurel
L. fontanesiana (Steud.) Sleumer	drooping leucothoe, mountain dog-laurel
L. populifolia	swamp leucothoe
L. racemosa (L.) A.Gray (= *Eubotrys racemosa* [L.] Nutt.)	fetterbush, calfkill, kill-calf, deciduous swamp fetterbush
L. recurva (Buckley) A.Gray (= *Eubotrys recurva* [Buckley] Britton)	deciduous mountain fetterbush

Plants shrubs or small trees; evergreen or deciduous. **Stems** erect; 100–400 cm tall. **Leaves** alternate; coriaceous; petiolate; blades elliptic to oblong-lanceolate; margins serrate or dentate, not revolute. **Inflorescences** axillary racemes at uppermost nodes; bracteoles present or absent. **Flowers** urceolate or tubular or ovoid or campanulate. **Sepals** 5; fused. **Petals** 5; fused; white; lobes short. **Stamens** 10; anthers with or without awns. **Pistils** 1; ovaries superior; locules 5. **Fruits** depressed globose capsules.

Leucothoe davisiae

1. Leaves deciduous.
   2. Racemes straight. Sepals lanceolate. Anther awns 4 . . . . . . *L. racemosa*
   2. Racemes curved. Sepals ovate. Anther awns 2 . . . . *L. recurva*
1. Leaves evergreen.
   3. Plants of the Sierra-Cascade Mountains . . . . *L. davisiae*

3. Plants of the Appalachian Mountains and eastern coastal plain.
   4. Corollas 8–10 mm long. Calyx lobes triangular; 1 mm long. Anther awns absent . . . . *L. populifolia*
   4. Corollas 5–7 mm long. Calyx lobes ovate or lanceolate; 1.5–2.5 mm long. Anther awns present; 2.
      5. Racemes 2–4 cm long. Calyx lobes broadly ovate; apices obtuse or rounded . . . . *L. axillaris*
      5. Racemes 4–10 cm long. Calyx lobes lanceolate; apices acute or subacute . . . . *L. fontanesiana*

■ bogs, swamps, moist woods

Distribution and Habitat—All but *L. davisiae* are species of the eastern United States. *Leucothoe racemosa* is found in the moist acidic soils of woods and barrens of the eastern coastal plain. The morphologically similar *L. recurva* occurs in moist woods at higher elevations. *Leucothoe axillaris* forms dense thickets in moist woods of the coastal plain. The similar *L. fontanesiana* is found along mountain streams. Relatively rare and restricted in distribution, *L. populifolia* occurs at the edges of swamps.

Leucothoe davisiae is a western species, found in bogs and other wet areas with acid soils in the Sierra Nevada Mountains at elevations of 1,000–2,600 m.

Leucothoe axillaris

Leucothoe davisiae

Leucothoe fontanesiana

Leucothoe populifolia

Leucothoe racemosa

Leucothoe recurva

Lyonia Nutt.

■ *L. ligustrina*
L. mariana

big-boy
he-huckleberry
maleberry
male blueberry
staggerbush

Taxonomy and Morphology—Named for John Lyon, an enthusiastic collector of American plants in the late 1700s and early 1800s, *Lyonia* comprises 30–35 species native to the eastern United States, the West Indies, and eastern Asia. Several species are cultivated for their attractive, shiny, leathery foliage. In North America, 7 species are present, 2 of which are of toxicologic interest:

L. ligustrina (L.) DC.	maleberry, male blueberry, he-huckleberry, big-boy
L. mariana (L.) D.Don	staggerbush

■ shrubs or small trees; leaves alternate; petals 5, fused, white, pink, or rose

Plants shrubs or rarely small trees; evergreen or deciduous. **Stems** erect; 100–400 cm tall; branches angular or terete; surfaces glabrous or pilose. **Leaves** alternate; coriaceous; short petiolate; blades elliptic to oblong or obovate; margins entire or serrulate, not revolute. **Inflorescences** axillary; umbellate or racemose clusters or panicles of racemes; peduncles short or long; pedicels long. **Flowers** campanulate or urceolate or cylindrical. **Sepals** 5; fused. **Petals** 5; fused; white or pink or rose; lobes short. **Stamens** 10; anthers without awns. **Pistils** 1; ovaries superior; locules 5. **Fruits** globose or depressed globose capsules; weakly 5-ribbed.

Lyonia ligustrina

Lyonia mariana

Lyonia mariana

1. Leaf margins serrulate. Inflorescences panicles of racemes. Corollas globose-urceolate; 3–5 mm long .... *L. ligustrina*
1. Leaf margins entire. Inflorescences umbellate or racemose clusters. Corollas cylindrical; 7–13 mm long .... *L. mariana*

■ moist sites

Distribution and Habitat—Distribution of both species is the eastern half of the United States. *Lyonia ligustrina* occurs in swamps and other wet areas, and *L. mariana* in moist sandy sites.

■ *M. ferruginea*
M. pilosa

fool's huckleberry
menziesia
minniebush
mock azalea
tree lyonia

Menziesia Sm.

Taxonomy and Morphology—Comprising 7 species in North America and eastern Asia, *Menziesia,* commonly known as mock azalea, honors Archibald Menzies, a naturalist on the Vancouver Expedition of 1790–1795. Species of the genus are cultivated in rock and woodland gardens for their pendent clusters of flowers. Two species are of toxicologic interest:

M. ferruginea Sm.	rusty menziesia, rusty-leaved menziesia, mock azalea, fool's huckleberry, tree lyonia, smooth Pacific menziesia
M. pilosa (Michx.) Juss.	minniebush

Plants shrubs; deciduous. **Stems** erect or spreading; 50–400 cm tall; bark finely shredding; branchlets pilose or glandular-pilose. **Leaves** alternate; thin; blades elliptic to oblong-obovate; abaxial surfaces typically glandular; margins entire, not revolute. **Inflorescences** terminal umbellate or corymbose clusters; pedicels recurved. **Flowers** urceolate or campanulate or cylindrical. **Sepals** 4 or 5; fused; glandular-ciliate. **Petals** 4 or 5; fused; yellowish green or greenish white to pinkish or reddish yellow; lobes short. **Stamens** 8; anthers without awns. **Pistils** 1; ovaries superior; locules 4 or 5. **Fruits** oblong or globose capsules; coriaceous.

■ deciduous shrubs; leaves alternate, thin; petals 4 or 5, fused, yellowish green, greenish white, pinkish to reddish yellow

■ open woods

Menziesia ferruginea

Menziesia ferruginea

Menziesia pilosa

1. Plants of the Rocky and Sierra-Cascade Mountains. Leaf margins glandular-ciliate . . . . *M. ferruginea*
1. Plants of the Appalachian Mountains. Leaf margins ciliate . . . . . *M. pilosa*

Distribution and Habitat—As is apparent in the key, one species is western and the other is eastern. *Menziesia ferruginea* is found in woods and along stream banks, especially on the north slopes of hillsides up to 300 m elevation in the West. This species also is cultivated as an ornamental. *Menziesia pilosa* is an eastern species of mountain woods.

Pieris D.Don

Taxonomy and Morphology—Comprising 7–10 species and numerous cultivars, *Pieris* is native to North America and eastern Asia. Its species, especially *P. japonica,* are cultivated for their showy foliage and inflorescences. Of the 3 species—2 native and 1 introduced—present in North America, 2 are of toxicologic interest:

■ *P. floribunda*
P. japonica

lily-of-the-valley bush
mountain fetterbush
mountain pieris

P. floribunda (Pursh) Benth. & Hook.	mountain fetterbush, mountain pieris
P. japonica (Thunb.) D.Don ex G.Don	lily-of-the-valley bush

Plants shrubs or small trees; evergreen. **Stems** erect; freely branched; 100–400 cm tall; branchlets terete or angular; strigose or glabrous. **Leaves** alternate; coriaceous; short

■ shrubs or small trees, freely branched; leaves alternate; flowers fragrant; petals 5, fused, white or pink

petiolate; blades ovate to elliptic to obovate; margins entire or minutely serrulate, not revolute. **Inflorescences** terminal panicles or racemes, typically 1-sided; pedicels recurved. **Flowers** fragrant; ovoid to urceolate or cylindrical; 5-saccate at base. **Sepals** 5; fused; coriaceous. **Petals** 5; fused; white or pink; lobes short, recurved. **Stamens** 10; anthers with 2 divergent awns near bases. **Pistils** 1; ovaries superior; locules 5. **Fruits** globose capsules.

Pieris floribunda

Pieris floribunda

1. Panicles erect. Leaves ovate to elliptic or oblong. Branches and petioles strigose . . . . *P. floribunda*
1. Panicles drooping. Leaves obovate to lanceolate. Branches and petioles glabrous . . . . *P. japonica*

■ mountain woods or ornamental

Distribution and Habitat—Native to Japan, Taiwan, and eastern China, *P. japonica* is widely planted as an ornamental in eastern North America. *Pieris floribunda* is native to and found in mountain woods of the eastern United States.

Rhododendron L.

■ *R. albiflorum*
R. catawbiense
R. macrophyllum
R. maximum
R. occidentale

azalea
great laurel
red laurel
rhododendron
rosebay
rosebay laurel
white laurel

Taxonomy and Morphology—One of the most popular genera of ornamental shrubs in regions where the environment permits its growth, *Rhododendron* comprises 700–1,000 species and innumerable cultivars. Its name is derived from the Greek roots *rhodos* and *dendron,* meaning "rose" and "tree." Distributed primarily in the cool-temperate regions of the Northern Hemisphere, it extends into the Southern Hemisphere in Southeast Asia. Its species are divided taxonomically into 8 subgenera and numerous sections, and a number of common names—rhododendron, azalea, rosebay, laurel—are employed. Traditionally, the deciduous species are collectively referred to as the azaleas, and the evergreen taxa as the rhododendrons, albeit there are exceptions. In North America, some 30 native and introduced species and named hybrids are present. Representative and of toxicologic interest are the following (see plate 35.1):

R. albiflorum Hook.	white rhododendron, Cascade rhododendron, white-flowered rhododendron
R. catawbiense Michx.	mountain rosebay, red laurel, purple laurel, rosebay laurel, catawba rhododendron

R. macrophyllum D.Don ex G.Don (= *R. californicum* Hook.)	Pacific rhododendron, western rhododendron, California rosebay
R. maximum L.	rosebay, great laurel, white laurel
R. occidentale (Torr. & A.Gray) A.Gray	western azalea

Plants shrubs or rarely small trees; evergreen or deciduous. **Stems** erect or ascending or prostrate; bark smooth or rough. **Leaves** alternate, but often crowded into false whorls at branch ends; coriaceous or herbaceous; thick or thin; blades of various shapes; surfaces glabrous or variously indumented; margins entire or minutely crenulate, revolute. **Inflorescences** umbellate or racemose or corymbose clusters; terminal or axillary. **Flowers** radially or bilaterally symmetrical; rotate or campanulate or funnelform. **Sepals** 5; fused. **Petals** 5; fused; of various colors except blue. **Stamens** 5 or 10; anthers without awns. **Pistils** 1; ovaries superior; locules 5. **Fruits** capsules; hard or soft.

■ shrubs or small trees; leaves alternate, margins revolute; petals 5, fused, various colors

Rhododendron albiflorum

Rhododendron maximum

Rhododendron occidentale

1. Plants of the Appalachian Mountains.
   2. Leaf apices acuminate or acute. Pedicels stipitate-glandular. Calyx lobes 2–6 mm long . . . . *R. maximum*
   2. Leaf apices obtuse or rounded. Pedicels hirsutellous. Calyx lobes 0.5–1 mm long . . . . *R. catawbiense*
1. Plants of the Rocky and Sierra-Cascade Mountains.
   3. Leaves evergreen; coriaceous; thick . . . . *R. macrophyllum*
   3. Leaves deciduous; herbaceous; thin.
      4. Flowers 5–20 in terminal corymbose clusters. Corollas narrowly funnelform. Stamens 5 . . . . *R. occidentale*
      4. Flowers 1–4 in axillary clusters. Corollas campanulate. Stamens 10 . . . . *R. albiflorum*

■ moist woods and forests, ornamentals

Rhododendron albiflorum

Rhododendron catawbiense

Rhododendron macrophyllum

Distribution and Habitat—The 2 eastern species are readily recognized by their striking floral displays in the spring. Both *R. catawbiense* and *R. maximum* are found in mountain woods, the latter sometimes forming dense stands. The 3 western species are summer flowering. *Rhododendron albiflorum* is found along stream banks and on moist, montane to subalpine slopes of the higher mountains. *Rhododendron macrophyllum* is found in moist forests to 1100 m elevation in the coastal mountains and on the western slope of the Cascade Mountains. *Rhododendron occidentale* is found along stream banks and in moist woods to 2200 m elevation. Both *R. macrophyllum,* and especially *R. occidentale,* are cultivated as ornamentals.

References

Akera T, Ku DD, Frank M, Brody TM, Iwasa J: Effects of grayanotoxin I on cardiac membrane potential and on myocardial contractile force. *J Pharmacol Exp Ther* 199;247–254, 1976

Black DH: Rhododendron poisoning in sheep [letter]. *Vet Rec* 128;363–364, 1991

Burke JW, Doskotch RW: Kalmanol: a cardiotoxic diterpene with a new ring skeleton from the plant *Kalmia angustifolia* [abstract]. *Am Chem Soc Natl Meet* 198;39, 1989

Carey FM, Lewis JJ, MacGregor JL, Martin-Smith M: Pharmacological and chemical observations on some toxic nectars. *J Pharm Pharmacol* 11;269T–274T, 1959

Casteel S, Wagstaff J: *Rhododendron macrophyllum* poisoning in a group of goats and sheep [letter]. *Vet Hum Toxicol* 31;176–177, 1989

Catterall WA: Neurotoxins that act on voltage-sensitive sodium channels in excitable membranes. *Annu Rev Pharmacol Toxicol* 20;15–43, 1980

Catterall WA: Molecular properties of voltage-sensitive sodium channels. *Annu Rev Biochem* 55;953–985, 1986

Catterall WA, Hartshorne RP, Beneski DA: Molecular properties of neurotoxin receptor sites associated with sodium channels from mammalian brain. *Toxicon* 20;27–40, 1982

Chesnut VK: *Principle Poisonous Plants of the United States.* USDA Div Bot Bull 20, pp44–48, 1898

Clawson AB: *Alpine kalmia* (Kalmia microphylla) *as a stock- poisoning plant.* USDA Tech Bull 391, 1933

Crawford AC: *Mountain Laurel, a Poisonous Plant.* USDA Bur Plant Ind Bull 121, Misc Pap, pp21–35, 1908

Cronquist A: *An Integrated System of Classification of Flowering Plants.* Columbia Univ Press, New York, 1981

Fleming CE: *Poisonous Range Plants.* Nev Agric Exp Stn Annu Rep 1919, p42, 1920

Forbes EB, Bechdel SI: Mountain laurel and rhododendron as foods for white tailed deer. *Ecology* 12:323–333, 1931

Fowler ME: Plant poisoning in two pack llamas. *Calif Vet* 39;17–20, 1985

Frape D, Ward A: Suspected rhododendron poisoning in dogs [letter]. *Vet Rec* 132;515–516, 1993

Frohne D, Pfander HJ: *A Colour Atlas of Poisonous Plants.* Wolfe Publ, London, pp105–112, 1984

Geroulanos S, Attinger B, Cakmakci M: Honey induced poisoning. *Schweiz Rundsch Med Prax* 81;535–540, 1992

Gossinger H, Hruby K, Haubenstock A, Pohl A, Davogg S: Cardiac arrhythmias in a patient with grayanotoxin-honey poisoning. *Vet Hum Toxicol* 25;328–329, 1983

Gregory NF: "Delayed" rhododendron poisoning. *Goat Vet Soc J* 11;41, 1990

Halsted BD: *Poisonous Plants of New Jersey.* N J Agric Exp Stn Annu Rep 1895;401–419,1895

Hardikar SW: On rhododendron poisoning. *J Pharmacol Exp Ther* 20;17–44, 1922

Higgins RJ, Hannam DAR, Humphreys DJ, Stodulski JBJ: Rhododendron poisoning in sheep. *Vet Rec* 116;294–295, 1985

Rhododendron maximum

Rhododendron occidentale

Hikino H, Ito K, Takemoto T: Stereostructure of asebotoxin I and II, toxins of *Pieris japonica. Chem Pharm Bull* 17;854–855, 1969a

Hikino H, Ito K, Ohta T, Takemoto T: Stereostructure of rhodojaponin I, II, and III, toxins of *Rhododendron japonicum,* and of asebotoxin III, toxin of *Pieris japonica. Chem Pharm Bull* 17;1078–1079, 1969b

Hikino H, Ohta T, Ogura M, Ohizumi Y, Konno C, Takemoto T: Structure-activity relationship of ericaceous toxins on acute toxicity in mice. *Toxicol Appl Pharmacol* 35;303–310, 1976

Hikino H, Ohizumi Y, Konno C, Hashimoto K, Wakasa H: Subchronic toxicity of ericaceous toxins and *Rhododendron* leaves. *Chem Pharm Bull* 27;874–879, 1979

Hollands RD, Hughes M-C: *Pieris formosanum* poisoning in the goat [letter]. *Vet Rec* 118;407–408, 1986

Hosie BD, Mullen CM, Gillespie ID, Cochrane GW: Rhododendron poisoning in lambs [letter]. *Vet Rec* 118;110, 1986

Hough I: Rhododendron poisoning in a western grey kangaroo. *Aust Vet J* 75;175–177, 1997

Howes FN: Sources of poisonous honey. *Kew Bull* 1949;167–171, 1950

Humphreys DJ, Stodulski JBJ: Detection of andromedotoxins for the diagnosis of rhododendron poisoning in animals. *J Appl Toxicol* 6;121–122, 1986

Ito K, Saruwatari N, Mitani K, Enomoto Y: Characterization of depolarization induced by palytoxin and grayanotoxin-I in isolated cardiac tissues from dogs and guinea pigs. *Naunyn- Schmiedberg's Arch Pharmacol* 330;67–73, 1985

Jaynes RA: *Kalmia Mountain Laurel and Related Species.* Timber Press, Portland, Oreg, 1997

Kakisawa H, Kozima T. Yanai M, Nakanishi K: Stereochemistry of grayanotoxins. *Tetrahedron* 21;3091–3104, 1965

Klein-Schwartz W, Litovitz T: Azalea toxicity: an overrated problem. *Clin Toxicol* 23;91–101, 1985

Kobayashi T, Yasuda M, Seyama I: Developmental toxicity potential of grayanotoxin I in mice and chicks. *J Toxicol Sci* 15;227–234, 1990

Kumazawa Z, Iriye R: Stereochemistry of grayanotoxin-II. *Tetrahedron Lett* 1970;927–930, 1970

Leach DG: The history of rhododendron poisoning. *Gard J* 1966;215–217,239, 1966

Leach DG: The history of rhododendron poisoning, part 2. *Gard J* 1967;15–18,33, 1967

Lodge RW, McLean A, Johnston A: *Stock-Poisoning Plants of Western Canada.* Can Dep Agric Publ 1361, 1968

Mager PP, Seese A, Takeya K: Structure-toxicity relationships applied to grayanotoxins. *Pharmazie* 36;381–382, 1981

Mancini SD, Edwards JM: Cytotoxic principles from the sap of *Kalmia latifolia. J Nat Prod* 42;483–488, 1979

Marsh CD: *Stock-Poisoning Plants of the Range.* USDA Bull 1245, 1924

Marsh CD, Clawson AB: *Mountain-Laurel* (Kalmia latifolia) *and Sheep Laurel* (Kalmia angustifolia) *as Stock-Poisoning Plants.* USDA Tech Bull 219, 1930

Masakazu N, Seyama I: Effect of α-dihydro-grayanotoxin-II on the electrical activity of the rabbit sino-atrial node. *J Physiol* 357;79–91, 1984

Masutani T, Seyama I, Narahashi T, Iwasa J: Structure-activity relationship for grayanotoxin derivatives in frog skeletal muscle. *J Pharmacol Exp Ther* 217;812–819, 1981

Miller RM: Azalea poisoning in a llama: a case report. *Vet Med/Small Anim Clin* 76;104, 1981

Moran NC, Dresel PE, Perkins ME, Richardson AP: The pharmacological actions of andromedotoxin, an active principle from *Rhododendron maximum. J Pharmacol Exp Ther* 110;415–432, 1954

Nakao M, Seyama I: Effect of α-dihydro-grayanotoxin-II on the electrical activity of the rabbit sino-atrial node. *J Physiol* 357;79–91, 1984

Nanobashvili VI, Shavgulidze MM: Toxicity of the shrub *Azalea pontica* for sheep, and its effect on meat quality. *Veterinariya (Moscow)* 1;62–63, 1987 (*Vet Bull* 57;4529, 1987)

Narahashi T, Seyama I: Mechanism of nerve membrane depolarization caused by grayanotoxin I. *J Physiol* 242;471–487, 1974

Ohgaki T, Uchida S, Meguri H, Ogita K, Yoneda Y: Preventive action of quisqualic acid against grayanotoxin-induced suppression of locomotor activity in mice. *Neuropharmacology* 27;1045–1053, 1988

Olszowy DR: Of bees, rhododendrons, and honey. *Am Bee J* 117;498–500, 1977

Onat F, Yegen BC, Lawrence R, Oktay A, Oktay S: Site of action of grayanotoxins in mad honey in rats. *J Appl Toxicol* 11;199–201, 1991

Ozesmi C, Aydogan S, Ascioglu M: Effects of honey containing grayanotoxin I on frog gastrocnemius muscle. *Vet Hum Toxicol* 36;117–119, 1994

Patwardhan VN, White JW: Problems associated with particular foods. In *Toxicants Occurring Naturally in Foods,* 2d ed, Comm Food Protection, National Research Council, National Academy of Sciences, Washington, DC, pp477–507, 1973

Plumlee KH, VanAlstine WG, Sullivan JM: Japanese pieris toxicosis of goats. *J Vet Diagn Invest* 4;363–364, 1992

Power SB, O'Donnell PG, Quirk EG: Pieris poisoning in sheep [letter]. *Vet Rec* 126;599–600, 1991

Pritchard WR: Laurel *(Kalmia angustifolia)* poisoning of sheep. *North Am Vet* 37;461–462, 1956

Reichardt PB, Bryant JP, Anderson BJ, Phillips D, Clausen TP, Meyer M, Frisby K: Germacrone defends Labrador tea from browsing by snowshoe hares. *J Chem Ecol* 16;1961–1970, 1990

Romey G, Lazdunski M: Lipid soluble toxins thought to be specific for Na^+ channels block Ca^{++} channels of neuronal cells. *Nature* 297;79–80, 1982

Rose A, Pitchford W, Monin T, Burrows GE: Acute weakness and death in a cat [letter]. *Vet Hum Toxicol* 30;334–335, 1988

Sakakibara J, Shirai N, Kaiya T, Nakata H: Grayanotoxin-XVIII and its grayanoside B, a new A-Nor-B-Homo-Ent-Kaurine and its glucoside from *Leucothoe grayana. Phytochemistry* 18;135–137, 1979

Sakakibara J, Shirai N, Kaiya T: Diterpene glycosides from *Pieris japonica. Phytochemistry* 20;1744–1745, 1981

Sampson AW, Malmsten HE: *Stock-Poisoning Plants of California.* Calif Agric Exp Stn Bull 593, rev ed, pp41–45, 1942

Schindler H: Uber acetylandromedol (andromedotoxin) in verschiedenen Ericaceen, insbesondere in rhododendron, und seine annahernde bestimmung. *Planta Med* 10;232–237, 1962

Scott PM, Coldwell BB, Wiberg GS: Grayanotoxins: occurrence and analysis in honey and a comparison of toxicities in mice. *Food Cosmet Toxicol* 9;179–184, 1971

Seyama I, Narahashi T: Increase in sodium permeability of squid axon membranes by α-dihydrograyanotoxin II. *J Pharmacol Exp Ther* 184;299–307, 1973

Seyama I, Narahashi T: Modulation of sodium channels of squid nerve membranes by grayanotoxin I. *J Pharmacol Exp Ther* 219;614–624, 1981

Seyama I, Yamaoka K, Yakehiro M, Yoshioka Y, Morihara K: Is the site of action of grayanotoxin the sodium channel gating of squid axon? *Jpn J Physiol* 35;401–410, 1985

Shannon D: Rhododendron poisoning in sheep [letter]. *Vet Rec* 116;451, 1985

Smith MC: Japanese pieris poisoning in the goat. *J Am Vet Med Assoc* 173;78–79, 1978

Smith MC: Fetal mummification in a goat due to Japanese pieris *(Pieris japonica)* poisoning. *Cornell Vet* 69;85–87, 1979

Southall RM, Hardin JW: A taxonomic revision of *Kalmia* (Ericaceae). *J Elisha Mitchell Soc* 90;1–23, 1990

Sutilupinar N, Mat A, Satganoglu Y: Poisoning by toxic honey in Turkey [letter]. *Arch Toxicol* 67;148–150, 1993

Takeda K, Narahashi T: Chemical modification of sodium channel inactivation: separate sites for the action of grayanotoxin and tetramethrin. *Brain Res* 448;308–312, 1988

Tallent WH, Riethof ML, Horning EC: Studies on the occurrence and structure of acetylandromedol (andromedotoxin). *J Am Chem Soc* 79;4548–4552, 1957

Terai T, Sato M, Narama I, Matuura T, Katakawa J, Tetsumi M: Transformation of grayanotoxin III to 10-epi-grayanotoxin III: its X-ray crystallographic analysis and acute toxicity in mice. *Chem Pharm Bull* 44;1245–1247, 1996

Thiemann AK: Rhododendron poisoning [letter]. *Vet Rec* 128;411, 1991

Tokarnia CH, Armien AG, Peixoto PV, Barbosa JD, Brito M de F, Dobereiner J: Experiments on the toxicity of some ornamental plants in cattle. *Pesq Vet Bras* 16;5–20, 1996

Toxic honey. *New Scientist* 41;218, 1969

Tschekunow H, Klug S, Marcus S: "Systematic poisoning" from ingestion of *Pieris japonica:* a case report [abstract]. *Vet Hum Toxicol* 31;360, 1989

Verlangieri AJ, Gawlikowski JN, Shapiro R: Acute toxicity of *Kalmia angustifolia* (sheep laurel) extracts in the rat. *Vet Toxicol* 18;122–124, 1976

Waud RA: The action of *Kalmia angustifolia* (lambkill). *J Pharmacol Exp Ther* 69;103–111, 1940

White JW, Riethof ML: The composition of honey. 3. Detection of acetylandromedol in toxic honeys. *Arch Biochem Biophys* 79;165–167, 1959

Wood HB, Stromberg VL, Keresztesy JC, Horning EC: Andromedotoxin: a potent hypotensive agent from *Rhododendron maximum. J Am Chem Soc* 76;5689–5692, 1954

Yi HS, Yuan H, Wan YR, Fang YC, Zhou BR: Treatment of *Rhododendron molle* poisoning in cattle. *Chin J Vet Sci Technol* 8;32–34, 1985 (*Vet Bull* 56;5577, 1986)

Zushi S, Miyagawa J-I, Yamamoto M, Kataoka K, Seyama I: Effect of grayanotoxin on the frog neuromuscular junction. *J Pharmacol Exp Ther* 226;269–275, 1983

36 Euphorbiaceae Juss.

■ Spurge Family

Acalypha
Aleurites
Cnidoscolus
Croton
Eremocarpus
Euphorbia
Hippomane
Hura
Jatropha
Manihot
Mercurialis
Pedilanthus
Phyllanthus
Reverchonia
Ricinus
Sapium
Stillingia
Synadenium
Tragia

QUESTIONABLE
Codiaeum

■ herbs, shrubs, or trees; flowers imperfect; perianths radially symmetrical

■ large family, economic importance; some greatly feared species, toxicity due mainly to 3 groups of diterpene esters—phorbols, ingenols, and daphnanes—all potent irritants

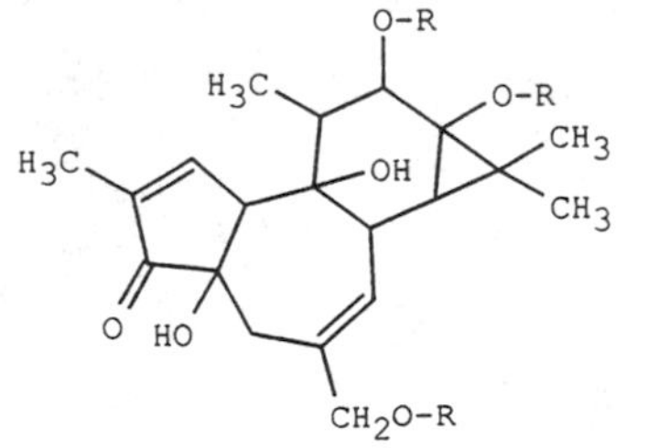

Phorbols

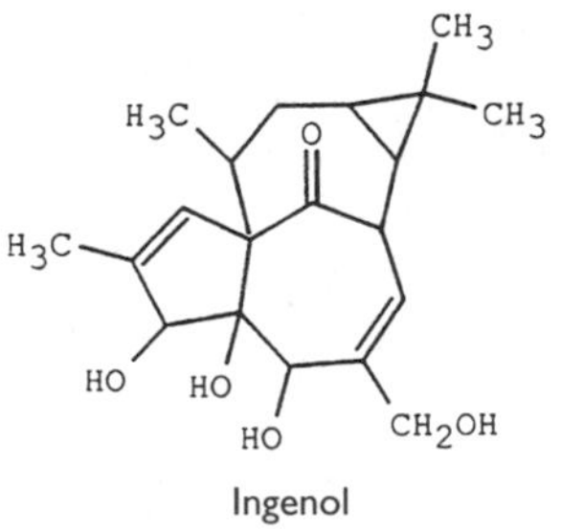

Ingenol

Highly diversified morphologically, chemically, and ecologically, the Euphorbiaceae, or spurge family, is cosmopolitan in distribution but exhibits greatest diversity in the tropics and subtropics, especially tropical America and the Indomalaysian region. It is one of the 6 largest families of flowering plants and comprises 300–321 genera and 7,500–8,000 species. The family is economically quite important, being the source of timber, dyes, natural rubber, food for both humans and wildlife, oils, and medicines. In North America, approximately 45 genera and 375 species are prized as ornamentals or despised as noxious weeds and toxic plants.

Plants herbs or shrubs or trees; annuals or perennials or biennials; armed or not armed with stinging hairs; monoecious or dioecious; sap colorless or of various colors, thin or viscid. **Stems** erect or prostrate or decumbent or trailing; occasionally succulent. **Leaves** simple or rarely compound; alternate or opposite, or opposite above and alternate below, or rarely whorled; petiolate or sessile; venation pinnate or palmate; margins entire or variously toothed or palmately lobed; stipules present or absent. **Inflorescences** of 1 or 2 types, staminate and pistillate similar or different; simple cymes or compound cymes (spicate, racemose, paniculate, or clustered) or cyathia; terminal or axillary; bracts present or absent, herbaceous or petaloid. **Flowers** imperfect, staminate and pistillate similar or different; perianths absent or in 1- or 2-series; radially symmetrical. **Sepals** typically 5 or 0 or 3–8; free or fused; herbaceous or petaloid. **Petals** 0 or 5 or rarely 6; free; white. **Stamens** 1 to many; free or fused by filaments. **Pistils** 1; compound, carpels 3 or 2 or 5–20; stigmas 3, 2- or 3-lobed or not lobed; styles 3 or 2–20; ovaries superior, lobes 3 or 0 or 2–20; locules 3 or 1–20; placentation axile; ovules 1 or 2 per locule. **Nectaries** absent or present; 1–5; sepaliferous or receptacular or glands on cyathia, with or without petaloid appendages. **Fruits** capsular schizocarps or achenes or drupes; dehiscent or indehiscent. **Seeds** typically 3 or 1–20.

The Euphorbiaceae exhibits considerable diversity. Although most of its taxa are thought to be at least mildly toxic because of their irritating milky sap, several are infamous as toxicants and others have been used for medicinal purposes. *Hippomane* and *Ricinus* have been greatly feared as toxic plants, whereas *Euphorbia antisyphilitica* was once purported to have the value its name implies. Species of *Phyllanthus* are well known for their antiviral and anticarcinogenic effects. Although alkaloids are reported to be present in many species, the irritant effects are for the most part due to diterpene esters (Willaman & Schubert 1961; Kinghorn 1985). These esters can be grouped into three categories: the tiglianes (phorbols), ingenanes (ingenols), and daphnanes (huratoxins) (Evans & Soper 1978; Spoerke & Smolinske 1990). All produce similar effects, but their irritant potency may vary considerably; the effects of some on the skin may persist for several days or more (Hickey et al. 1981). These species are of considerable importance as skin irritants in humans, and there is interest in evaluating their potency. However, the mouse ear and human skin tests commonly used to measure irritant potency may not be representative of the irritation they cause to the mucosa of the digestive tract. Genera of the family not discussed below may be expected to produce irritant effects occasionally.

Additional toxic effects of considerable interest are the inhibitory actions of proteins or proteinaceous toxins on ribosomes and on protein synthesis in several species. The best-known of these toxins is ricin, from *Ricinus,* because of its great potency. The leaves and/or seeds of several other species contain similar toxins, but these toxins appear to lack a component facilitating their entry into the cell. They thus lack cytotoxicity and are toxic only in cell-free systems.

Daphnanes

Acalypha L.

■ *A. virginica*

three-seeded mercury
Virginia copperleaf

Taxonomy and Morphology—Native to both the Old World and the New World, *Acalypha,* commonly known as copperleaf or three-seeded mercury, comprises 390–450 species. Its name was used by Hippocrates for a nettle. A few species, most notably the shrubs *A. hispida* (chenille plant) and *A. wilkesiana* (Joseph's coat), are grown for their showy foliage and inflorescences in the southern portions of the continent. A few other species are weeds, but most are simply minor components of different vegetation types. In North America, 20 species both native and introduced are present. Toxicologic problems have been reported for only 1:

A. virginica L. three-seeded mercury, Virginia copperleaf

■ erect, sparsely branched, annual herbs; leaves simple, alternate; flowers axillary

Plants herbs; annuals; not armed with stinging hairs; monoecious; sap colorless, thin. **Stems** erect; 10–60 cm tall; sparsely branched; surfaces hispid; hairs straight or recurved. **Leaves** simple; alternate; long petiolate; venation pinnate; blades narrowly rhombic to broadly lanceolate; margins crenate; stipules small, lanceolate. **Inflorescences** spicate; pistillate flowers below staminate; axillary; bracts 1–3, herbaceous, 10–16 lobed or not lobed, hirsute to glabrous. **Flowers** with perianths in 1-series. **Sepals** 4 in staminate flowers; 3–5 in pistillate flowers; fused, herbaceous. **Petals** absent. **Stamens** 4–10. **Pistils** with carpels 3; styles 3, slender, branched; lobes 3 or 0; locules 3. **Fruits** capsular schizocarps; indumented. **Seeds** 3; ovoid; brown; shallowly pitted in rows.

Acalypha virginica

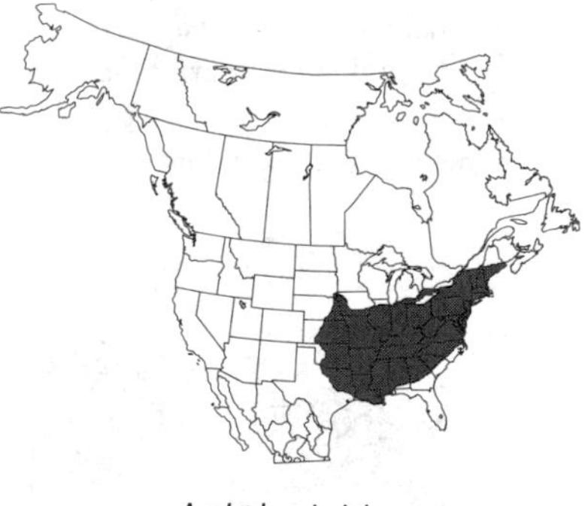

Acalypha virginica

■ sandy soils

Distribution and Habitat—*Acalypha virginica* is found in moist or dry, sandy soils of open woodlands, fields, and roadsides in the eastern half of North America.

■ irritant action on digestive tract due to diterpene esters

Disease Problems and Genesis—Probably all species of *Acalypha* cause irritation, but only *A. virginica* is of sufficient potency to cause clinical problems; the others are of low potential at best (Mitchell & Rook 1979; Spoerke & Smolinske 1990). As in other members of the family, the problem is due to the presence of diterpene esters, which cause irritation of the mucosa of the digestive tract.

■ diarrhea, colic

treat to relieve signs

Clinical Signs, Pathology, and Treatment—The irritation produced by *A. virginica* on the digestive system results in purgation and colic. Relief of the inflammation and pain associated with the irritation may be necessary in some instances.

Aleurites J.R.Forst. & G.Forst.

■ *A. cordata*
A. fordii
A. moluccana
A. montana
A. trisperma

banucalad nut
candleberry
candlenut
China wood oil tree
Japan oil tree
kukui
lumbang nut
mu-oil tree
tung tree
tung oil tree
varnish tree
wood oil tree

Taxonomy and Morphology—Comprising 5 or 6 species, *Aleurites* is commercially important for its oils—tung, candlenut, and Japanese wood—used in paints and varnish. Soot from burning tung oil is the basic ingredient of India ink. Dyes also are made from several species, and the stones or seeds, once their oil is removed, are used as fertilizer. The fresh stones, resembling walnuts, are said to taste like chestnuts, and stored stones, like almonds but with a rancid aftertaste (DeBusk 1978). The genus name is derived from the Greek root *aleuron,* meaning "floury," because all parts of the plants typically are covered with an indumentum that suggests a dusting of flour. Some taxonomists divide the genus into 2 genera and place all species except *A. moluccana* in the genus *Vernicia.* However, the distinctive, unusual multistoned drupes are common to all species, and thus a broadly circumscribed genus is recognized here. Introduced species of toxicologic interest are as follows:

A. cordata (Thunb.) R.Br. ex Steud.	wood oil tree, Japan oil tree
A. fordii Hemsl.	tung tree, tung oil tree, China wood oil tree
A. moluccana (L.) Willd.	candlenut, candleberry, varnish tree, kukui
A. montana (Lour.) E.H.Wilson	mu-oil tree
A. trisperma Blanco	lumbang nut, banucalad nut

■ trees; sap white, viscid; leaves simple, alternate, palmate venation; flowers small, terminal; petals 5, white, greenish, or reddish white

Aleurites fordii

Plants trees; perennials; evergreen or deciduous; crowns rounded; foliage typically farinose; monoecious; sap white, viscid. **Stems** erect; 7–60 m tall. **Leaves** simple; alternate; long petiolate; venation palmate, veins 5–7; blades suborbicular to ovate; apices acute to acuminate; margins entire or crenate or 3-to-5-lobed; petioles bearing 2 nectaries at apices. **Inflorescences** paniculate; terminal; lax; bracts absent. **Flowers** small; appearing before or simultaneously with leaves; perianths in 2-series. **Sepals** 2–5; fused; herbaceous; splitting into 2 or 3 valvate lobes at anthesis. **Petals** 5; white or greenish or reddish white. **Stamens** 8–20; borne in 1–4 whorls on conical receptacles. **Pistils** with carpels 2–5; styles 1, divided into linear branches. **Fruits** drupes; subglobose or ovoid; stones 2–5. **Seeds** 1 per stone.

1. Sepal and ovary surfaces stellate .... *A. moluccana*
1. Sepal and ovary surfaces glabrous or puberulent or tomentose.
    2. Leaves deciduous .... *A. trisperma*
    2. Leaves evergreen.
        3. Drupe surfaces tuberculate .... *A. cordata*
        3. Drupe surfaces smooth or ridged.

4. Flowers appearing before leaves. Ovary surfaces glabrous or puberulent . . . . *A. fordii*
4. Flowers appearing with leaves. Ovary surfaces tomentose . . . . *A. montana*

Distribution and Habitat—All species of *Aleurites* are native to Southeast Asia and the Pacific Islands. They have been cultivated in North America as shade trees or crops for their oil. All are generally restricted to the warmer, southern parts of the continent, especially the coastal areas of the southeastern states (Balthrop et al. 1953). The most common is *A. fordii*, which was formerly grown in plantations on the Gulf coastal plain. *Aleurites moluccana* is the state tree of Hawaii.

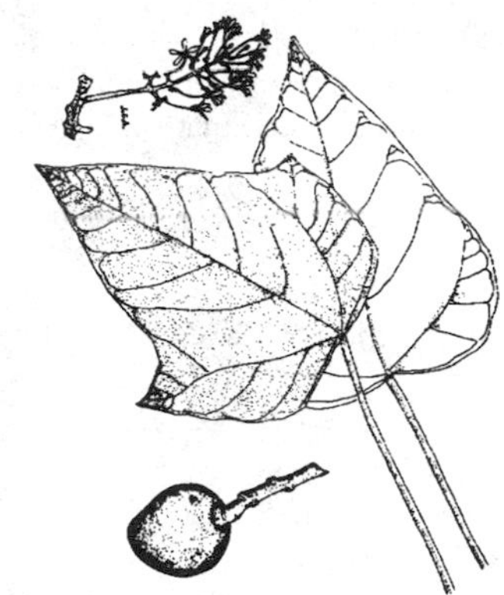

Aleurites moluccana

Aleurites fordii

- originally crop plants

Disease Problems—Upon ingestion of as few as 1–3 drupes of *Aleurites*, a very painful digestive disorder with vomiting and profuse diarrhea occurs in humans (Palmer 1933; Erickson & Brown 1942). For example, consumption of the drupes/stones by 64 schoolchildren in two separate episodes resulted in the development of nausea and vomiting within 1–2 hours; recovery took 2–3 days (Lin et al. 1994). When stomachs were empty, onset of the disease was more rapid. Typical signs included abdominal pain and diarrhea; dizziness, fatigue, and tremor were seen less often (Lin et al. 1996). Although the digestive system signs are marked and rarely include systemic effects, complete recovery may be expected within 24 hours (Lampe & Fagerstrom 1968). Toxicity from the oil differs from that of the stones/seeds in that it is less potent and in experimental animals causes little or no irritation but may result in neurotoxic and other systemic actions. The pulp residue following oil extraction has 22–25% protein but also contains heat-labile toxin(s) (Holmes & Rayner 1958). The foliage is also toxic and causes similar effects. Cattle eating the apparently palatable leaves develop profuse watery diarrhea in several days (Emmel et al. 1942). *Aleurites fordii* is the most toxic, the others being considerably less potent (Emmel 1947).

- humans, cattle; severe irritation, digestive disturbance, from fruits, seeds or leaves

Disease Genesis—Toxicity was originally thought to be due to saponins and other toxins of a similar nature (Emmel 1947). Eventually, however, the toxins were shown to be a series of hydroxyphorbol tigliane diterpenoid esters very similar to those present in other genera of the Euphorbiaceae (Okuda et al. 1974, 1975; Connolly & Hill 1991). In addition to being potent irritants, some of the diterpenes are piscicides comparable in activity to rotenone (Hiroto et al. 1979). The rather innocuous oil of *Aleurites* is octadecatrienoic acid (Balthrop et al. 1953). Of unknown importance is a ribosomal-inhibiting protein (RIP) that is extracted from the stone/seed of *A. fordii* (Reynolds & Irvin 1980). Inhibiting protein synthesis in cell-free systems, it is a class 1 RIP—it lacks cytotoxicity because of the absence of a B chain (which is present in ricin and abrin) to facilitate absorption and entry into the cell (Stirpe et al. 1976).

- tigliane phorbol esters

13-Acetyl-16-hydroxyphorbol

Clinical Signs—Regardless of whether the foliage, drupes, or stones/seeds of *Aleurites* are eaten, a profuse watery diarrhea, sometimes with blood, is the main attribute of the disease. This is usually accompanied by other manifestations of the irritating nature of the toxins—excess salivation, anorexia, tenesmus, listlessness, emaciation, and in severe cases death. This irritation also is indicated by a reddening and cracking of the mucous membranes around the mouth. In cattle, several days elapse after the animals eat the plants before the signs reach their peak of severity. The signs may occur more quickly in other animal species and include vomiting and severe abdominal pain. In children, nausea and vomiting are the main signs. Appearing less frequently in children are lethargy/tiredness, diarrhea, headache, tremor, dry mouth, sweating, and chest tightness.

- profuse diarrhea, excess salivation, anorexia, listlessness, tenesmus, emaciation; several days' duration

■ gross pathology, severe reddening and hemorrhage of mucosa of stomach and duodenum

Pathology—As the clinical signs indicate, the main lesions are related to the irritant effects, producing severe reddening and hemorrhage of the mucosa of the stomach/abomasum and duodenum. The remainder of the intestine is only slightly affected. There may also be reddening of the serosal surfaces of the viscera, distension of the gallbladder with thick bile, and congestion of the abdominal viscera.

■ activated charcoal, fluids, electrolytes

Treatment—It is important to alleviate the irritation and pain, to decrease intestinal motility, and to restore fluid and electrolyte balance. The use of oral activated charcoal also may be beneficial.

Cnidoscolus Pohl

■ *C. aconitifolius*
C. angustidens
C. stimulosus
C. texanus

bull nettle
chaya
copapayo
fingerroot
mala mujer
spurge nettle
tread-softly

Taxonomy and Morphology—A New World genus, *Cnidoscolus,* known as bull nettle or spurge nettle, comprises 50–75 species. Its name is derived from the Greek *knide,* for "nettle," and *skolos,* for "point," reflecting its abundant stinging hairs. Primarily tropical in distribution, the genus is represented in North America by 4 species:

C. aconitifolius (Mill.) I.M.Johnst.	chaya, copapayo
C. angustidens Torr.	mala mujer
C. stimulosus (Michx.) Engelm. & A.Gray	bull nettle, fingerroot, tread-softly
C. texanus (Meull.Arg.) Small	bull nettle, mala mujer, spurge nettle

■ herbs, shrubs, or trees, often from massive rootstocks; sap white, viscid; stout stinging hairs; leaves simple, alternate, palmate venation; flowers terminal, white; petals absent

Plants herbs or shrubs or trees; perennials, often from massive rootstocks; herbage armed with stout pustulate-based stinging hairs; monoecious or rarely dioecious; sap white, viscid. **Stems** erect or decumbent; herbs 30–100 cm tall; branched. **Leaves** simple; alternate; long petiolate; venation palmate; blades suborbicular; margins palmately 3-to-5-lobed or divided or entire; stipules persistent. **Inflorescences** terminal cymes; bracts present. **Flowers** with perianths in 1-series. **Sepals** 5 or rarely 4; fused; petaloid; white; calyces salverform in staminate flowers, bowl-shaped in pistillate flowers. **Petals** absent. **Stamens** 10–30. **Pistils** with carpels 3; styles 3, branched, each with 3–5 lobes. **Fruits** capsular schizocarps; ovoid; 3-lobed; armed with stinging hairs. **Seeds** 3; oblong; brownish white; caruncles conspicuous.

Cnidoscolus texanus

1. Filaments of outer whorl of stamens fused.
   2. Plants shrubs or trees. Leaf blades lobed more than halfway to midribs . . . . *C. aconitifolius*
   2. Plants herbs. Leaf blades lobed less than halfway to midribs . . . . *C. angustidens*
1. Filaments of outer whorl of stamens free.
   3. Staminate calyces glabrous or puberulent. Calyx tubes 8–10 mm long. Seeds 6–8 mm long. Plants of southeastern United States . . . . *C. stimulosus*
   3. Staminate calyces hirsute or hispid. Calyx tubes 15–25 mm long. Seeds 14–18 mm long. Plants of south-central United States and eastern Mexico . . . . *C. texanus*

■ deep sandy soils

Distribution and Habitat—Primarily tropical in distribution, *Cnidoscolus* is present only in the deep sandy soils of the southern states. *Cnidoscolus stimulosus* occurs on the Atlantic Coast and the Gulf coastal plain. *Cnidoscolus texanus* is present from southern Arkansas to Tamaulipas. *Cnidoscolus angustidens* occurs in the Southwest, and *C. aconitifolius* is distributed from northern Mexico to Central America.

Disease Problems and Genesis—Species of *Cnidoscolus* bear stinging hairs (trichomes) that consist of a multicellular pedestal and a single elongated, tapered, hollow cell with a slightly swollen tip (Thurston & Lersten 1969; Lookadoo & Pollard 1991). Upon contact, the end shears off at an oblique angle as the hair penetrates the skin. The cellular contents are released into the wound, producing a burning sensation and swelling. These hairs are similar to those present in some members of the Urticaceae, or nettle family (see chapter 72). They are suggested to be a defensive adaptation against mammalian herbivores (Pollard 1986). Disease produced by contact with the stinging hairs is essentially the same as for *Urtica.* Although evidence on the precise identity of the chemical toxicants is conflicting, the hairs contain compounds that appear to be very similar in activity to serotonin (Lookadoo & Pollard 1991). The latex sap may also be irritating. The plants also are slightly cyanogenic; they contain low concentrations of linamarin, but this is of little consequence, because they are unlikely to be eaten (Seigler 1977).

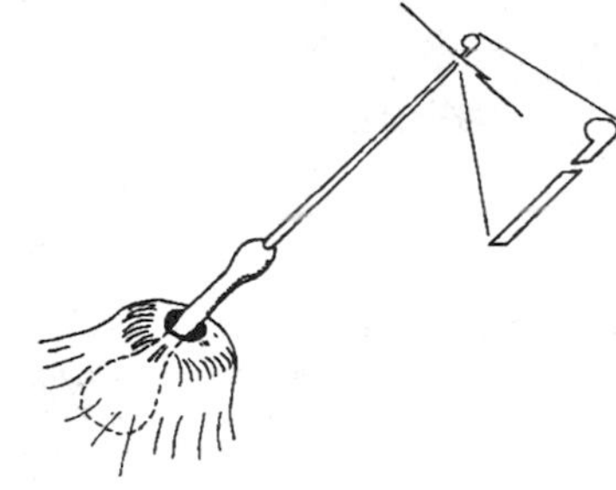
Stinging hair

Cnidoscolus angustidens

Cnidoscolus stimulosus

Cnidoscolus texanus

Clinical Signs—The signs produced by contact with species of *Cnidoscolus* are the same as caused by *Urtica,* except that massive exposure is very rare because of the relatively low density of plants in a population. Instead, transient erythema, swelling, and pruritis lasting for 20–45 minutes are typical following contact with any part of the plant.

- stinging hairs, local reaction due to contact irritation and chemical mediators
- transient erythema, swelling, itching; 20–45 minutes' duration

Treatment—Typically the localized reaction does not require treatment, but some relief may be obtained by application of a local anesthetic.

- treatment generally not needed

Croton L.

Taxonomy and Morphology—Widely distributed in warm regions of the world, *Croton* is a genus of more than 600 species. Its common name is *croton,* which is also applied to *Codiaeum variegatum,* a popular ornamental species of this family. *Croton* is an important wildlife genus, the lipid-rich seeds being consumed by doves and other birds—hence the common name *doveweed.* The foliage is normally not eaten by livestock even when other forage is scarce, and thus species of the genus tend to increase greatly in abundance when pastures or rangelands are overgrazed. Some species have been used in folk medicines. The potent purgative croton oil is obtained from *C. tiglium,* an Old World species. In North America, about 45 species and numerous varieties occur. Commonly encountered species representative of the genus are as follows:

- *C. capitatus*
  C. glandulosus
  C. monanthogynus
  C. texensis

croton
doveweed
hogwort
prairie tea
skunkweed

C. *capitatus* Michx.	woolly croton, hogwort, doveweed
C. *glandulosus* L.	tropic croton, tooth-leaved croton
C. *monanthogynus* Michx.	one-seed croton, prairie tea
C. *texensis* (Klotsch) Meull.Arg.	Texas croton, doveweed, skunkweed

■ branched, erect herbs or shrubs; leaves simple; fruits often 3-angled

Plants herbs or shrubs; annuals or perennials; monoecious or dioecious; herbage aromatic or not aromatic, with stellate hairs or peltate scales; sap colorless, thin or viscid. **Stems** erect; typically branched above. **Leaves** simple; alternate or appearing opposite or whorled immediately below inflorescence; petiolate; blades of various shapes; margins entire or serrate; stipules caducous. **Inflorescences** spicate or racemose; terminal or axillary; bracts absent. **Flowers** small; perianths in 1- or 2-series. **Sepals** 5, or rarely 4 or 6; fused; herbaceous. **Petals** present or absent; when present, as many as sepal lobes; white. **Stamens** 3–20. **Pistils** with carpels 3 or 2; styles 3 or 2, each divided 1–4 times; locules 3 or 1. **Fruits** capsular schizocarps; 3-angular or slightly flattened. **Seeds** 3 or 1.

Because the species listed are only representative of the genus, a key for their identification is not presented. The reader is directed to taxonomic manuals specific for his or her area.

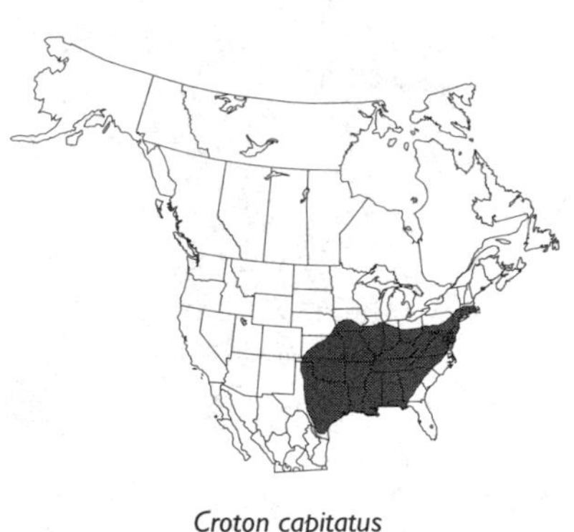

Croton capitatus

Croton capitatus

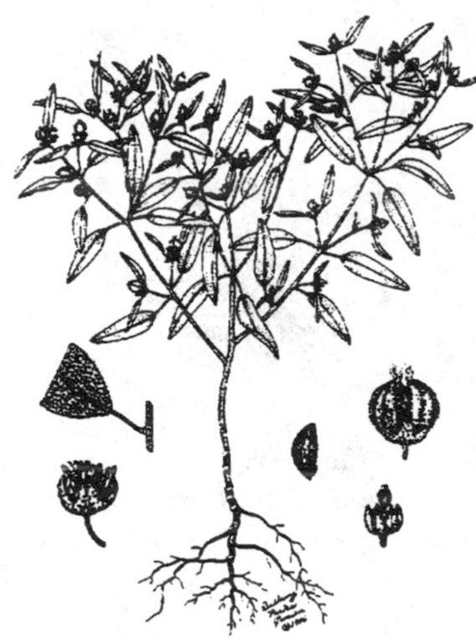

Croton texensis

■ diverse habitats, often dryland invaders

Distribution and Habitat—*Croton* is most diverse in warm regions of the world. In North America, most species are distributed in Mexico and the southern tier of states, with some as far north as South Dakota, Ohio, and Indiana. Plants occupy a variety of habitats in different vegetation types.

■ irritant tigliane phorbol esters; digestive disturbance

Crotofolin A

Disease Problems—The crotons are potent irritants, but because they are very unpalatable, most problems are limited to skin irritation. Plants are more likely to be eaten when they contaminate hay. When this occurs, they may cause irritation of the digestive tract (Tehon et al. 1946). Species of *Croton* contain irritating phorbol esters, especially the strongly purgative croton oil (Kinghorn 1985). These tigliane diterpenoids include numerous derivatives of phorbol and a series of crotofolanes designated crotofolin A, C, and E (Connolly & Hill 1991). In addition there are potent class 1 ribosomal-inhibiting proteins (RIPs) present, at least in the Old World *C. tiglium*, that are capable of inhibiting protein synthesis in cell-free systems but are of little concern as a cause of intoxication (Sperti et al. 1976).

Clinical Signs and Treatment—Irritation of the digestive tract results in purgation and colic. Relief of the inflammation and pain of the digestive tract by administration of demulcents or emollients may be necessary in some instances.

■ diarrhea, colic; nursing care

Eremocarpus Benth.

Taxonomy and Morphology—Occurring only in western North America, *Eremocarpus* is a monotypic genus. It was initially described as a species of *Croton* but differs in having only one carpel. This trait is reflected in its name, which is derived from the Greek roots *eremos* and *karpos,* meaning "lonely fruit." The single species is the following:

■ *E. setigerus*
doveweed
turkey mullein

E. setigerus (Hook.) Benth. turkey mullein, doveweed

Plants herbs; annuals; monoecious; herbage malodorous, grayish green, stellate-hispid; sap colorless, thin. **Stems** prostrate to ascending; 2–10 cm tall; 5–80 cm long; dichotomously branched; matted. **Leaves** simple; alternate; lower long petiolate; upper forming rosettes at branchlet ends; blades ovate to deltoid-ovate; 3-nerved; margins entire; stipules minute. **Inflorescences** small; subtended by rosette leaves of 2 types; staminate terminal, racemose; pistillate axillary, spicate. **Flowers** tiny; perianths in 1-series or absent. **Sepals** 5 or 6 or absent in pistillate flowers; herbaceous. **Petals** absent. **Stamens** 5–9. **Pistils** conspicuously stellate; carpels 1; styles 1, slender; locules 1. **Fruits** capsulelike. **Seeds** 1; ellipsoidal; gray-mottled; smooth; shiny.

■ malodorous, annual herbs; leaves simple, alternate, forming rosettes at branchlet ends; flowers tiny; fruits capsulelike; seeds I

Distribution and Habitat—Occurring primarily in semidesert conditions, *E. setigerus* is distributed from southern Washington to Baja California.

■ semidesert areas

Disease Problems and Genesis—As its common names imply, the seeds of *E. setigerus* are eaten by birds. The foliage, however, may be toxic to livestock, especially when present in hay. The principal problem is impediment and blockage of the movement of materials through the digestive tract. Because of the fibrous nature of the plant tissues and the dense stiff hairs, there is a tendency for tangled plant masses or phytobezoars to form in the stomachs of sheep, cattle, and other species (Sampson & Malmsten 1942; Kingsbury 1964). Like other members of the Euphorbiaceae, *E. setigerus* contains diterpenes. Among them is eremone, which is reputed to stun fish (Jolad et al. 1982; Ebeling 1986). Plants were crushed and thrown into streams by some Indian tribes. The irritating nature of the diterpene may be a problem for livestock as well. The plant also contains tiliroside, a bioactive glycoside (antineoplastic and insect feeding deterrent) of unknown toxicologic significance (Bajaj et al. 1986).

■ forms fibrous bezoars in stomach or rumen; irritant diterpenes, eremone

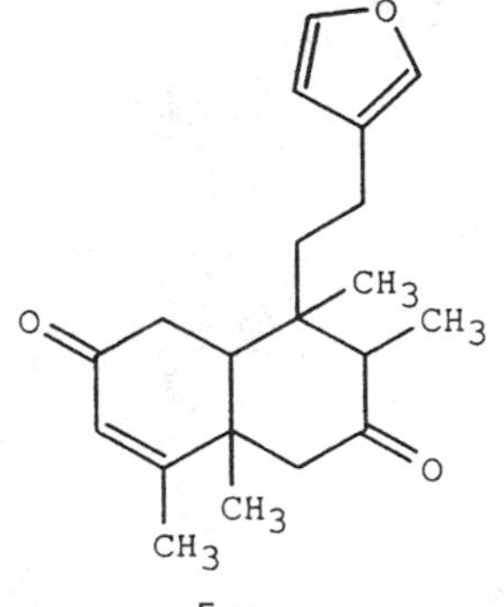

Eremone

Eremocarpus setigerus

Eremocarpus setigerus

■ erratic appetite, diarrhea

medical or surgical relief of the bezoar

Clinical Signs and Treatment—Following consumption of *E. setigerus,* erratic appetite, diarrhea, and poor productivity, including decreased milk yield, are observed. In animals with moderate signs, treatment to promote passage of the plant materials through the digestive tract may be employed; otherwise, mechanical removal of the bezoars via a rumenotomy may be required.

Euphorbia L.

■ very large and diverse genus; sap milky, viscid

Taxonomy and Morphology—Commonly known as spurge or euphorbia, *Euphorbia,* with 1,500–2,000 species, is one of the 6 largest genera of flowering plants. Its name honors Euphorbus, a Greek physician to King Juba II of Mauritania in the first century A.D. Juba used the milky sap medicinally and wrote a treatise on its value (Turner 1995). Virtually cosmopolitan in distribution, the genus is morphologically and ecologically quite diverse—small to large, herbaceous or woody or cactuslike, leafless or leafy, and inhabiting deserts, forests, or grasslands. Common to all species, however, is the milky, latexlike sap and the cyathium, a highly modified inflorescence.

■ cyathium

Characteristic only of *Euphorbia* and closely related genera, the cyathium, which resembles a single flower, is actually an inflorescence comprising one pedicelled pistillate flower surrounded by numerous staminate flowers, all enclosed in a cup- or urn-shaped involucre. The flowers lack sepals and petals, and the staminate flowers are reduced to a single stamen. The involucre bears one or more glands on its rim. In some species, such as *E. pulcherrima* (poinsettia), the leaves immediately below the cyathia are brightly colored and often mistaken for petals of a large complex flower.

■ *E. canariensis*
E. candelabrum
E. corollata
E. cyparissias
E. dentata
E. esula
E. fulgens
E. globosa
E. heterophylla
E. ipecacuanhae
E. lactea
E. lathyris
E. lucida
E. maculata
E. marginata
E. milii
E. myrsinites
E. prostrata
E. pulcherrima

candelabra cactus
crown of thorns
dragon bones
euphorb
ghost weed
gopher plant
hat-rack cactus
lepadena
Mexican fire plant
milk purslane
mole plant
mole weed
poinsettia
snow-on-the-mountain
springwort
spurge

The genus is traditionally divided into 5–10 subgenera based on the succulence and woodiness of the stems, the presence or absence of spines, and the appearance of the cyathia (Turner 1995). Some taxonomists circumscribe the genus more narrowly than others and treat some of the subgenera as distinct genera. Thus the generic names *Chamaesyce, Poinsettia, Tithymalopsis, Agaloma,* and *Tithymalus* are encountered in some publications. Because the cyathium is such a unifying feature and boundaries among these smaller genera are tenuous, the traditional, broad circumscription of *Euphorbia* is maintained in this treatment. Thus in North America, some 170 species, both native and introduced are present. Representative of the genus are the following:

Shrubby Ornamentals

E. fulgens Karw.	scarlet plume
E. milii Des Moul.	crown of thorns
E. myrsinites L.	creeping spurge
E. pulcherrima Willd. ex Klotzsch	poinsettia

Succulent, Cactuslike Ornamentals

E. canariensis L.	
E. candelabrum Kotschy	candelabra cactus
E. globosa (Haw.) Sims	
E. lactea Haw.	hat-rack cactus, dragon bones

Noncultivated Perennials

E. corollata L.	flowering spurge
E. cyparissias L.	cypress spurge

E. esula L.	leafy spurge, esula spurge
E. ipecacuanhae L.	ipecac spurge
E. lucida Waldst. & Kit.	broad-leaved spurge
Noncultivated Annuals	
E. dentata Michx.	toothed spurge
E. heterophylla L.	green euphorb, pointed spurge, Mexican fire plant
E. lathyris L.	caper spurge, mole plant, mole weed, gopher plant, springwort, myrtle spurge
E. maculata L. (= *E. supina* Raf.)	prostrate spurge, spotted spurge, milk purslane
E. marginata Pursh	snow-on-the-mountain, ghost weed, lepadena
E. prostrata Aiton	prostrate spurge

Plants herbs or shrubs; annuals or perennials; not armed with stinging hairs; monoecious; sap white, viscid. **Stems** erect or prostrate; succulent or not succulent. **Leaves** simple; alternate or opposite; petiolate or sessile; venation pinnate; margins entire or dentate. **Inflorescences** cyathia; borne in terminal or axillary cymes; bracts and bracteoles present or absent, opposite or whorled, herbaceous or petaloid; involucres cup-shaped or urceolate, lobes 4 or 5, entire or laciniate; nectar glands 1–5, of various shapes and colors, petaloid appendages present or absent. **Flowers** with perianths absent; staminate flowers 2–15 per cyathia, comprising 1 stamen, borne on inner surface of involucral cup; pistillate flowers 1 per cyathium, long pedicelled in fruit. **Sepals** absent. **Petals** absent. **Stamens** 1. **Pistils** with carpels 3; styles 3, divided. **Fruits** capsular schizocarps; 3-lobed; glabrous or indumented. **Seeds** 3.

- herbs or shrubs; sap white, viscid; leaves simple; inflorescences cyathia; fruits 3-lobed; seeds 3

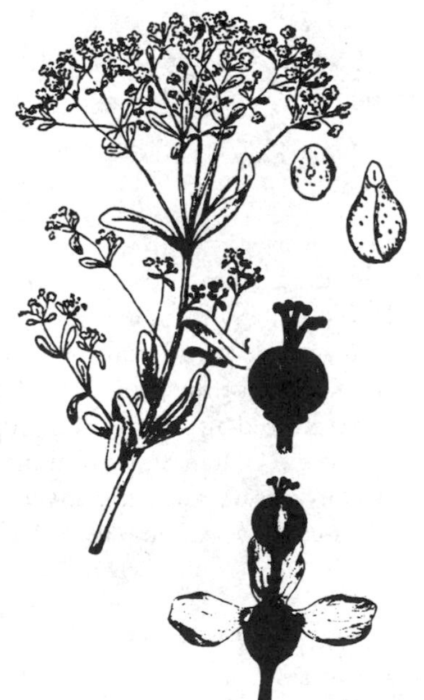

Euphorbia corollata

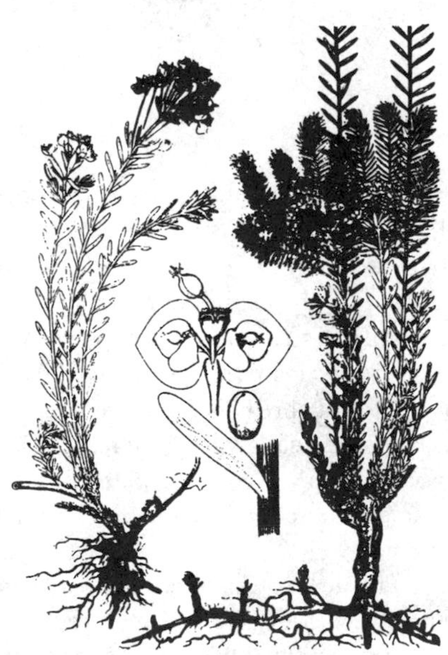

Euphorbia cyparissias

Euphorbia dentata

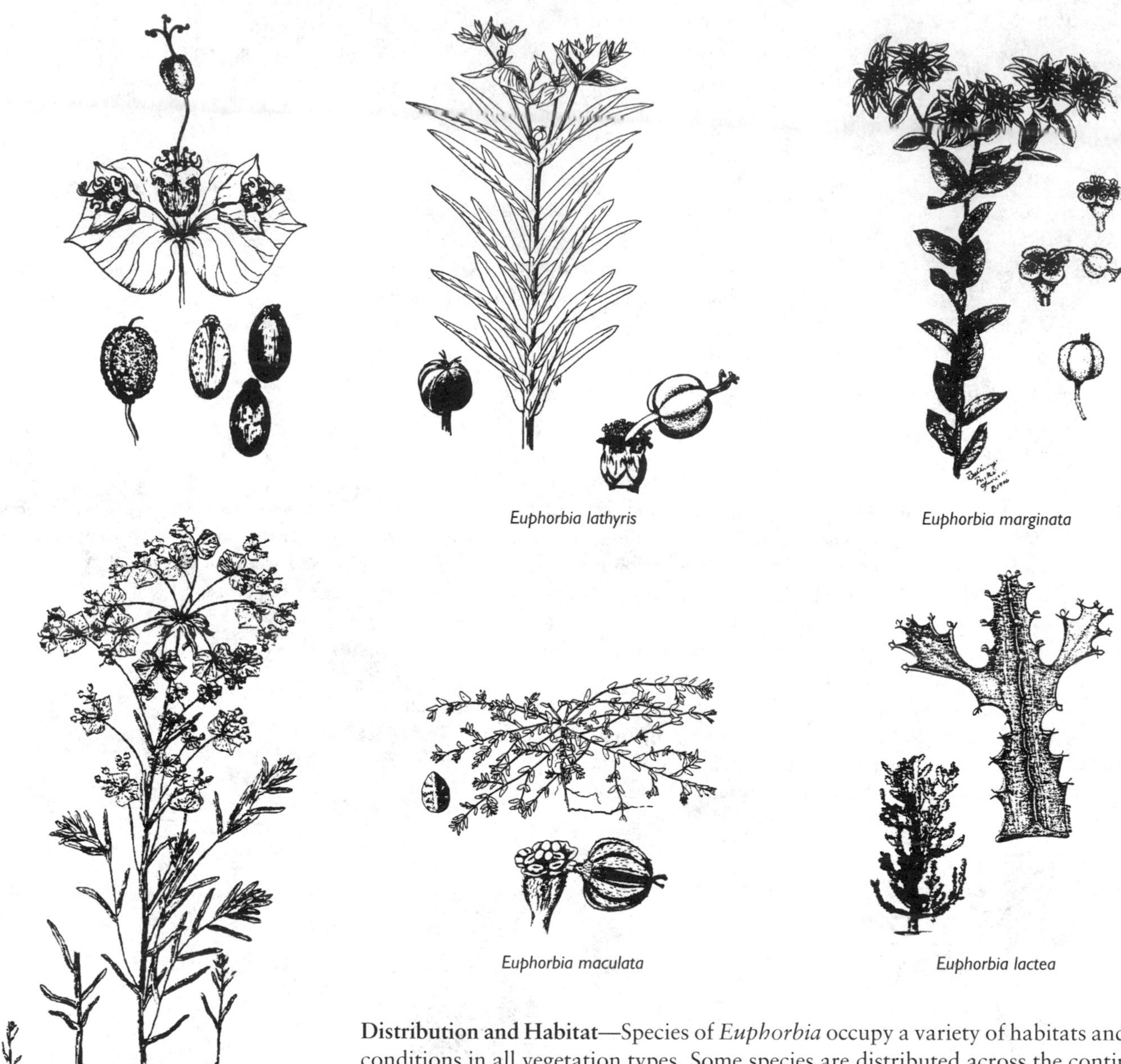

Euphorbia lathyris

Euphorbia marginata

Euphorbia esula

Euphorbia maculata

Euphorbia lactea

Distribution and Habitat—Species of *Euphorbia* occupy a variety of habitats and soil conditions in all vegetation types. Some species are distributed across the continent, whereas others are fairly restricted. Many are weedy and quickly form dense populations in disturbed soils; some are quite aggressive and considered noxious pests. A good example is *E. esula,* perhaps the most important species of the genus because of the economic losses it causes. Native to Eurasia and introduced into North America in the 1820s, it spread widely across the northern half of the United States and into Canada. It occupies a variety of habitats, except coniferous forests, and often invades ranges and pastures, forming extensive stands and reducing their productivity as well as causing intoxication problems (Dunn 1979).

- irritant milky sap of most species is not a severe risk to humans or animals; economic problem limiting productivity of pastures

Disease Problems—The irritant properties of the milky sap of *Euphorbia* and their association with disease are well recognized. The sap of some species was used to enhance the effects of arrow poisons (Verdcourt & Trump 1969). In a few instances a decoction of the roots when chewed for medicinal purposes has caused very serious illness and death. Adverse reactions, however, are typically limited to less severe irritation. The range of such effects is illustrated by the following examples. Species such as *E. lathyris, E. marginata,* and *E. prostrata* are reported to cause severe purgation if sufficient fruits or foliage are eaten. *Euphorbia lathyris* is reported to be quite irritating to the hands and face and in the absence of other forage may be eaten by livestock, which subsequently exhibit irritation of the digestive tract (Pullar 1939). It is sometimes cultivated for its purported value as a deterrent for moles and gophers; the deterrent property is presumably due to the effects of the sap exuding from broken or bruised roots. Although perhaps difficult to accept as true, *E. marginata* is reputed to be of such potency that it was once used as a branding agent for cattle (Chesnut 1898). The toxic effect may also be imparted in the milk of cows that eat the plants (Sampson & Malmsten 1942). Consumption of *E. maculata* has been associated with a dermatitis akin to photosensitization, but with significant mortality (Case 1957). However, a similar-appearing disease in humans has been ascribed to a combination of irritation and sunburn (Mitchell & Rook 1979). Sperry et al. (1977) reported that 3 kg of either *E. prostrata* or *E. marginata* fed to cattle caused severe diarrhea and emaciation that lasted for weeks. In other studies, *E. tirucalli* fed at dosages of 21.5–30 g fresh plant/kg b.w. caused only slight irritation in cattle (Tokarnia et al. 1996). *Euphorbia myrsinites* is reported to be a cause of skin swelling (edema) and blistering of several days' duration in children (Spoerke & Temple 1979). In South Africa, cattle moved through a thick stand of a euphorb appeared as if they had been exposed to a severe causticant, with marked effects on facial skin, eyes, tongue, and mouth (Kellerman et al. 1988).

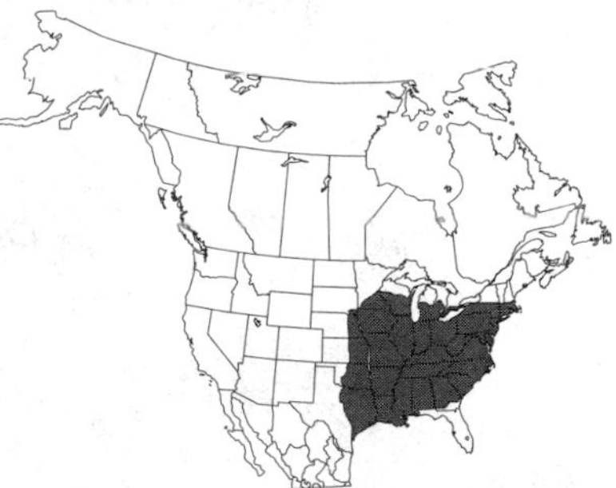

Euphorbia corollata

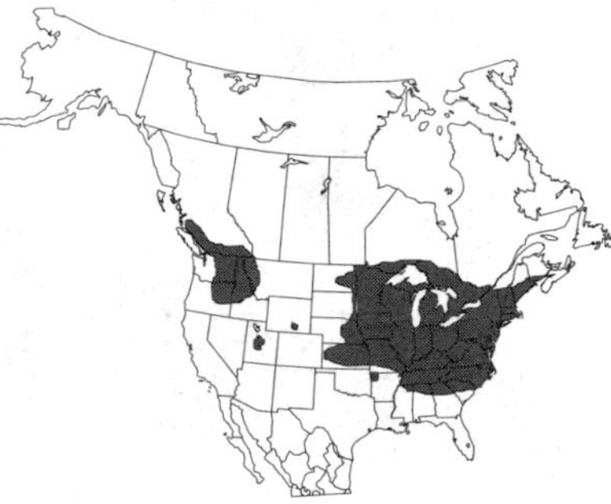

Euphorbia cyparissias

There seems little doubt that the various species of *Euphorbia* are capable of causing irritation of the skin, mucous membranes, conjunctiva, and digestive tract, especially with prolonged exposure, but the hazard is much exaggerated. In most cases of exposure, signs of intoxication are absent. When present, the effects are generally limited to excess salivation, mild vomiting, and diarrhea, and only after consumption of large quantities of plant (Borys et al. 1987; Hornfeldt 1989). The irritant potency of the latex sap seems to vary with season (Frohne & Pfander 1984).

Perhaps the species of *Euphorbia* of greatest interest is the much beloved *E. pulcherrima,* or poinsettia. An extensive folklore about the dangers of this species has developed. It includes the death of a child and severe digestive disturbance in a dog (Frohne & Pfander 1984). In reality, most accounts of poinsettia poisoning are unconfirmed and not consistent with experimental evidence and carefully evaluated clinical experiences. Analysis of 22,793 cases of reported human exposures to poinsettia revealed that 93.3% were in children, that there were no fatalities, and that in 92.4% there were no signs of adverse effects (Krenzelok et al. 1996). Poinsettia is probably one of the less toxic species of *Euphorbia* (Stone & Collins 1971; Winek 1978; Mitchell & Rook 1979; Runyon 1980; Spoerke & Smolinske 1990). In large dosages, i.e., 15.8–30 g fresh plant/kg b.w., it can cause severe irritation of several days' duration in cattle (Tokarnia et al. 1979). In contrast, up to 20 g/kg has little effect on sheep (Armien & Tokarnia 1994). As a houseplant, the amount of plant material available for consumption is so small that the species is definitely overrated as a hazard.

Euphorbia dentata

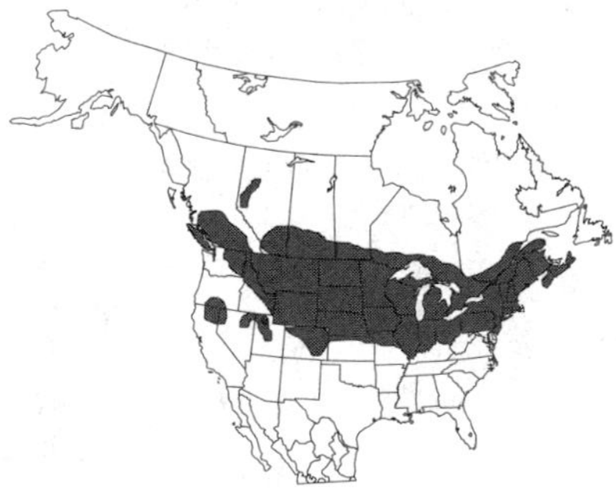

Euphorbia esula

That problems with the euphorbs are not limited to direct toxic activity is illustrated by the difficulties posed by *E. esula* (leafy spurge). Introduced from Eurasia, this noxious perennial has devastated millions of acres of pasture and rangeland in the north-central United States and adjacent Canada (Bakke 1936). It spreads via prolific production of long-lived seeds and thereafter forms extensive root systems to 5 m deep, which permit it to infest a pasture and eventually reduce its grazing value to nothing (Muenscher 1930; Selleck et al. 1962; Lym & Messersmith 1985). In some areas it may represent 50% of the plant coverage. *Euphorbia esula* is very difficult to control with herbicides and is very competitive with grasses. It is a good example of the difficulty in typifying the negative attributes of the genus. When mature, the plant is clearly toxic to cattle and is eaten only reluctantly. However, when it is young or animals are gradually introduced to it as a part of their diet, it is more readily eaten. Grazing by sheep and goats has been shown to be an effective way to control pasture infestations without producing ill effects in the animals (Johnston & Peake 1960; Hanson et al. 1992; Walker et al. 1992; Walker & Kronberg 1992). Its nutritional value compares favorably with that of alfalfa in chemical composition. Available crude protein is 26%, digestibility 80%, and phosphorus 0.53% (Fox et al. 1991). It may require several weeks for sheep or goats to adapt to grazing *E. esula*, but the weight gains and overall performance of sheep will not be impaired (Landgraf et al. 1984; Olson et al. 1996). It continues to be the subject of intense efforts to control its spread, and the results of biological control with species of *Aphthona* (flea beetle) are promising (Carlson & Mundal 1990; Messersmith & Lym 1990). A combination of grazing and herbicides also has shown promising results (Lym et al. 1997). Some care must be exercised with grazing, because sheep may aid spread of plants via seed in their fleece and feces (Olson et al. 1997).

Euphorbia lathyris

Euphorbia maculata

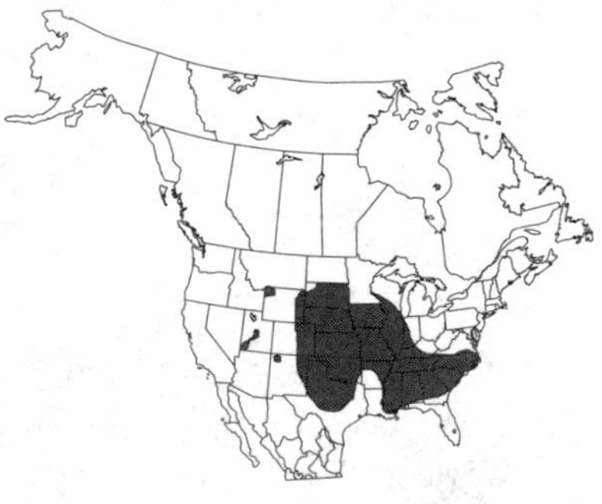

Euphorbia marginata

- irritant complex diterpene euphorbol esters, esulones, miliamines
- potent capsaicin-like diterpene resiniferatoxin

Disease Genesis—The irritant effects of *Euphorbia* are due to an extensive series of complex diterpenoid euphorbol esters similar to those found in other genera of the family (Evans & Soper 1978; Natori et al. 1981; Kinghorn 1985). Some species contain only a few esters; e.g., *E. esula* has only the esulones A–C and the similar but more highly esterified esulatins A–C (Connolly & Hill 1991; Hohmann et al. 1997). Other species such as *E. maculata* (reported as *E. supina*) may have a dozen or more esters, and *E. milii* contains a series of unique ingenols, the miliamines. The marked variation in the array of diterpenoids may account for the differences in irritant potency among species of the genus.

Honey produced by bees feeding extensively on *Euphorbia* may be very irritating because of the diterpenoids (Sosath et al. 1988). In addition to the diterpenoids, several steroids of unknown significance are present in the sap (Nielsen 1979). Because many of the steroids have saponin-like activity, they may contribute to the irritation. The diterpenoids are also toxic to fish and have been used in a manner similar to the use of rotenone (Kellerman et al. 1988).

An analog of capsaicin, the toxin of peppers (*Capsicum* spp.), has been isolated from the sap of several tropical, succulent, cactuslike species of *Euphorbia*, e.g., *E.*

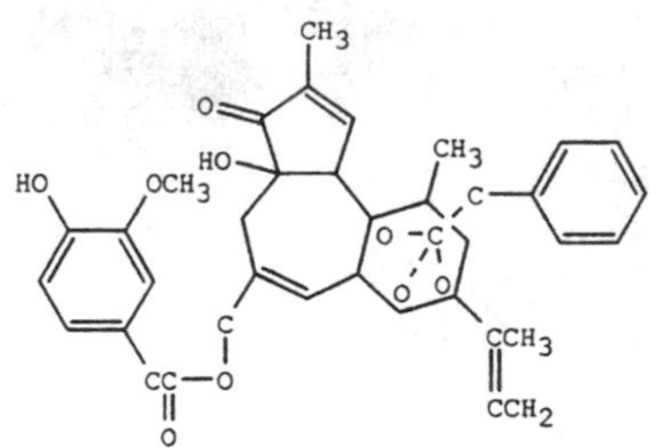

Resiniferatoxin

Esulone A

Miliamine A

resinifera and *E. poissonii* (Szallasi & Blumberg 1989) but is of unknown clinical significance. This compound, resiniferatoxin, is a phorbol-related diterpene of the daphnane type, which, unlike other diterpenes of the genus, contains a homovanillic acid ester functional group that is responsible for the activation and desensitization actions on the small primary sensory afferent neurons (Szolcsanyi et al. 1991). The very potent resiniferatoxin acts in an identical manner on the same neurons as capsaicin, but at about 100 times lower concentration (Winter et al. 1990).

Clinical Signs and Pathology—The signs produced by *Euphorbia* reflect the localized irritation of the skin, mucous membranes, and digestive tract and include excess salivation, vomiting, conjunctivitis, and more rarely diarrhea. Pathology generally is not of concern because the disease is rarely fatal. The changes are limited to those areas of direct contact.

- excess salivation, vomiting, reddened eyes

Treatment—This is symptomatic and aimed to control the irritation and perhaps, in rare cases, to restore fluid and electrolyte balance. Treatment of the topical effects includes washing with alcohol to remove the sap, followed by ointments to relieve the inflammation.

- wash affected skin with alcohol, relief of pain

Hippomane L.

Taxonomy and Morphology—Comprising 2 or 3 species, *Hippomane* is native to tropical America and the Caribbean Islands. One species is of toxicologic interest; it is widespread and said to be one of the most infamous poisonous plants in tropical America (Little & Wadsworth 1964). Translated literally, the name of this species means "the little apple, which makes horses mad" (Tomlinson 1980):

H. mancinella L. manchineel, manzanillo, little apple, beach apple, poison guava

- *H. mancinella*

  beach apple
  little apple
  manchineel
  manzanillo
  poison guava

Plants trees; perennials; evergreen or deciduous; monoecious; crowns broad, spreading; sap white, viscid. **Stems** to 15 m tall; trunks to 0.6 m in diameter; bark dark brown or gray, scaly or fissured or warty; branches widely forking; branchlets green when young. **Leaves** simple; alternate; long petiolate; venation pinnate; glossy dark green; blades ovate to cordate; apices acute or acuminate; margins crenate to serrate; petioles bearing gland at apices; stipules linear. **Inflorescences** spicate; terminal; staminate flowers above pistillate in clusters of 5–15; pistillate flowers solitary or paired; bracts present; glands present. **Flowers** inconspicuous; perianths in 1-series; greenish. **Sepals** 2 or 3; fused. **Petals** absent. **Stamens** 2 or 3. **Pistils** with carpels 6–8; styles 6–8, brown, curved; locules 6–8. **Fruits** drupes; applelike; yellow-green or reddish green; sweet-scented. **Seeds** 6–8; dark brown; enclosed in 1 stone.

- trees; sap white, viscid; leaves simple alternate; flowers small greenish; fruits applelike, sweet-scented, 6–8 seeds per stone

Hippomane mancinella

Distribution and Habitat—*Hippomane mancinella* is distributed from southern Florida through the Caribbean to Central America and northern South America. In Florida it is found in low ground and hummocks along the coast, in the Everglades, and on the Keys. Once forming beach-front thickets, it is now almost extirpated in many areas (Morton 1974).

Disease Problems—*Hippomane mancinella* has long had a reputation as a toxic plant. Early European explorers and sailors who ate its fruits were reported to experience severe irritation and death (Little & Wadsworth 1964). It was greatly feared by early colonists who tried to eradicate its threat by cutting it down, albeit with caution

- long reputed as deadly, contact with the sap evoked great fear; severe irritation of skin, eyes and the digestive tract

Hippomane mancinella

- diterpenes, huratoxin, mancinellin

Huratoxin

Mancinellin

because the sap and wood chips were considered dangerous (Lauter et al. 1952). Smoke from the burning wood was considered dangerous to the eyes (Little & Wadsworth 1964). Even the shade of the trees was to be avoided, especially if there was the possibility of dew or raindrops falling from the leaves. Warnings were given to avoid use of the trees for shelter during rainstorms (Pitts et al. 1993), and it was claimed that individuals sleeping under them would be injured, blinded, or even killed (Little & Wadsworth 1964).

Few plants have invoked such fear. So much folklore has evolved about the species that it is difficult to distinguish fact from fiction. Deaths may have occurred, but in most cases the effects are those of transient severe irritation of the skin, eyes, or digestive tract (Lauter et al. 1952; Blohm 1962). Severe corneal damage is of particular concern (Pitts et al. 1993). Although both the fruit and the foliage are capable of causing irritation of the digestive tract, the seriousness of ingestion has been considerably exaggerated, because irritation of the mucous membranes usually limits intake.

Disease Genesis—Several toxins appear to be present in *H. mancinella.* An aqueous extract of the wood yields a somewhat thermolabile toxin, which, when administered parenterally, causes a drop in blood pressure and an increase in salivation; these effects are antagonized by epinephrine and atropine, respectively (Schaeffer et al. 1954; Carroll et al. 1957). Extraction of the leaves and twigs produces a series of toxins, the major, stable one of which is a polyhydroxyl tannin glycoside, methyl tri-O-methyl gallate (Rao 1974, 1977). The fruit contains an acetylcholinesterase-inhibiting, physostigmine-like alkaloid (Lauter & Foote 1955).

In spite of the apparent diversity of toxins in the plant, the more important irritating or caustic effects are caused in a large measure by diterpene esters of the tigliane (mancinellin) and daphnane (huratoxin) types, as is characteristic of other members of the family (Adolph & Hecker 1984).

- burning sensation, excess salivation and tears, vomiting, diarrhea

Clinical Signs and Pathology—The signs, which develop within minutes of contact with *H. mancinella,* reflect the localized irritation of the skin, mucous membranes, and digestive tract. They include a burning sensation, excess salivation, vomiting, conjunctivitis, excess lacrimation, and diarrhea. Contact with the cornea results in edema, epithelial defects, pain, and blepharospasm of several days' duration. Pathology is generally not of concern, because the disease is rarely fatal. The changes are limited to areas of direct contact.

- wash affected skin with alcohol, relief of pain and inflammation; ointments; nursing care

Treatment—Treatment is for symptoms and is aimed to control the irritation and perhaps, in rare cases, to restore fluid and electrolyte balance. Treatment of the topical effects includes washing with alcohol to remove the sap, followed by application of ointments to relieve the irritative inflammation. Ocular problems will typically respond to conservative treatment with atropine and antimicrobials.

Hura L.

- *H. crepitans*
  H. polyandra

hura
jabillo
molinillo
monkey-pistol
possumwood
sandbox tree

Taxonomy and Morphology—Comprising only 2 species, *Hura,* commonly known as sandbox tree, is native to tropical America and the West Indies. The common name reflects the use of the mature, hollowed-out capsules as boxes for sand to blot ink in earlier times, before the advent of absorbent writing paper. It North America, it is planted as an ornamental and for shade in southern Florida and California. Elsewhere it is used for fence posts, firewood, framing, veneer, and crates. These are the 2 species:

H. crepitans L.	sandbox tree, molinillo, hura, monkey-pistol, possumwood
H. polyandra Baill.	jabillo, sandbox tree

■ trees, trunks massive; leaves simple, alternate, glossy green, white veins; staminate flowers dark red or white; fruits with explosive ejection of seeds

Plants trees; perennials; evergreen or deciduous; armed with blackish spines; monoecious; crowns globose, dense; sap colorless or white, viscid or thin. **Stems** to 30 m tall; trunks massive, to 2 m diameter; bark pale gray; branchlets glabrous to pubescent. **Leaves** simple; alternate; long petiolate; venation pinnate, veins white; glossy green; blades cordate to ovate or orbicular; apices acuminate; margins entire or denticulate; stipules linear-lanceolate, 10–20 mm long. **Inflorescences** of 2 types; staminate flowers borne in terminal spicate cluster; pistillate flowers solitary in distal leaf axils; bracts absent. **Flowers** cupulate; perianths in 1-series; dark red or white. **Sepals** 5; fused; coriaceous. **Petals** absent. **Stamens** 8–20; borne in 2 or 3 whorls. **Pistils** with carpels 5–20; stigmas 5–20 lobed; styles 1; locules 5–20. **Fruits** capsular schizocarps; 5–20 lobed; explosively dehiscent. **Seeds** 5–20; flattened; round; brown.

Hura crepitans

1. Staminate flowers dark red . . . . *H. crepitans*
1. Staminate flowers white . . . . *H. polyandra*

Distribution and Habitat—*Hura crepitans* is native to the West Indies and tropical America and is distributed from Costa Rica to Brazil, Peru, and Bolivia. It is planted in southern Florida and California and has naturalized in the Miami area. *Hura polyandra* extends from the southern edge of our range in Sinaloa, Mexico, south to Costa Rica.

Hura crepitans

■ irritation of skin, eyes, and digestive tract; diterpene huratoxin

Disease Problems and Genesis—The seeds of *H. crepitans,* which are apparently not distasteful, are used as a purgative in humans; only 2 or 3 are needed. In spite of their apparent toxicity, some animals, such as macaws, eat them as food (Blohm 1962). The seeds, the sap, and the bark all have emetic and cathartic activity. In addition, the sap causes inflammation and blisters upon contact with the skin of some individuals, and it is very irritating to the eyes, reportedly causing temporary blindness (Little & Wadsworth 1964). Foresters discourage planting of the trees as ornamentals because of the danger of the sap.

Toxicity of the genus is due to a diterpene derivative known as huratoxin, or hippomane A, which is also a potent piscicide (Sakata et al. 1971). Crepitin, a protein much like ricin, is also reported to be present (Blohm 1962). It is most likely a class 1 RIP, similar in nature to those in *Jatropha, Acalypha,* and *Aleurites* in that it lacks cytotoxicity because of the absence of a B chain that facilitates absorption into the cell.

Huratoxin

■ nausea, vomiting, diarrhea; reddened, blistered skin

Clinical Signs, Pathology, and Treatment—As is typical of the family, the irritating effects of *Hura* on the digestive tract are manifested as nausea, vomiting, and diarrhea. Contact of the sap with the skin produces reddening and blistering. The pathology and treatment are the same as those for other taxa of the family.

Jatropha L.

Taxonomy and Morphology—Native to the tropics and subtropics, with greatest diversity in South America, *Jatropha* comprises 170–200 species (McVaugh 1945). Its name is derived from the Greek roots *iatros* and *trophe,* meaning "physician" and "food," alluding to its medicinal properties, which also are reflected in some of its

■ *J. cathartica*
J. curcas
J. dioica
J. macrorhiza
J. multifida

Barbados nut	purge nut
coral plant	rubber plant
emetico	rubber stem
jicamilla	sangre de drago
leather stem	tartago
physic nut	tuba curcas bean

■ erect herbs, shrubs, small trees; sap viscid; leaves simple; flowers small; petals 5, yellow, purple to red

Jatropha curcas

Jatropha cathartica

Jatropha curcas

common names, e.g., gout plant, purge nut, and emetico. Some species are grown as ornamentals or hedging for their foliage. Although toxic when raw, the seeds of *J. curcas* and *J. multifida* are baked and eaten in some regions. Oil is obtained from *J. curcas*. In North America, about 12 species are present. The following are representative of the genus:

J. cathartica Terán & Berl.	jicamilla
J. curcas L.	physic nut, purge nut, tartago, tuba curcas bean, Barbados nut
J. dioica Sessé ex Cerv.	leather stem, rubber stem, rubber plant, sangre de drago
J. macrorhiza Benth.	jicamilla
J. multifida (L.) Pohl	tartago, emetico, coral plant

Plants herbs or shrubs or small trees; perennials; monoecious or dioecious; sap pale to colored, viscid. **Stems** erect; to 8 m tall; from thickened, woody or fleshy roots. **Leaves** simple; alternate or fascicled; petiolate; venation pinnate or palmate; blades cordate to ovate or reniform; margins entire or variously toothed or palmately lobed; stipules present. **Inflorescences** compound cymes or small clusters; terminal or axillary; peduncled; flat-topped; dichotomously branched; bracts absent. **Flowers** small; perianths in 2-series. **Sepals** 5; free or fused; each with an adjacent gland. **Petals** 5; free or fused; yellow or purple to red. **Stamens** 7–12; in 2 whorls. **Pistils** with carpels 3 or rarely 2–5; styles 3 or rarely 1–5, free or fused. **Fruits** capsular schizocarps. **Seeds** 1–3.

1. Plants herbs.
   2. Leaves entire or 2- or 3-lobed. Plants dioecious. Cymes lateral . . . . *J. dioica*
   2. Leaves 5-to-9-lobed. Plants monoecious. Cymes terminal.
      3. Petals red . . . . *J. cathartica*
      3. Petals whitish . . . . *J. macrorhiza*
1. Plants shrubs or trees.
   4. Leaves deeply 7-to-11-lobed. Petals red to purple . . . . *J. multifida*
   4. Leaves entire or 3-to-5-lobed. Petals yellow-green or whitish.
      5. Plants dioecious. Cymes lateral. Plants of Texas and Mexico . . . . *J. dioica*
      5. Plants monoecious. Cymes terminal. Plants of southern Florida . . . . *J. curcas*

Distribution and Habitat—A genus of warm regions, *Jatropha* occurs only as far north as southern Florida, Texas, and Arizona. The woody species *J. curcas* and *J. multifida* are native to Central America and have naturalized in disturbed sites in southern Florida. *Jatropha curcas* also is occasionally encountered in Sinaloa, Mexico. *Jatropha cathartica* occupies brushy areas in southwestern Texas and adjacent Mexico. *Jatropha dioica* is a Chihuahuan Desert species. *Jatropha macrorhiza* is encountered in the grasslands of mesas and plains of central Mexico north to Texas, New Mexico, and Arizona.

Disease Problems—As the common names indicate, the primary effects of these fruits or seeds, especially *J. curcas* and *J. multifida*, when ingested, are on the digestive tract. The seeds of *J. curcas* fed for 4 weeks to chicks as 0.5% of their diet produced weight loss and degeneration of the liver, kidneys, and digestive tract (Samia et al. 1992). They are also toxic to cattle, sheep, and goats, causing severe purgation following a single large dose or repeated lower doses. The effects are mainly due to irritation of the digestive tract and, in the case of cattle, marked liver and kidney degeneration as well (Ahmed & Adam 1979a,b). Seeds of *J. dioica* were lethal in sheep at 1% b.w. when given 3 times in 12 days (Boughton & Hardy 1939). Leaves are also toxic but less so; 2% b.w. was toxic but not fatal in a sheep. Ingestion of both immature and ripe *Jatropha* seeds by children caused transient digestive dysfunction with signs such as persistent vomiting, abdominal pain, nausea, and less commonly other systemic signs (Ho 1960; Joubert et al. 1984; Makalinao 1993).

Jatropha dioica

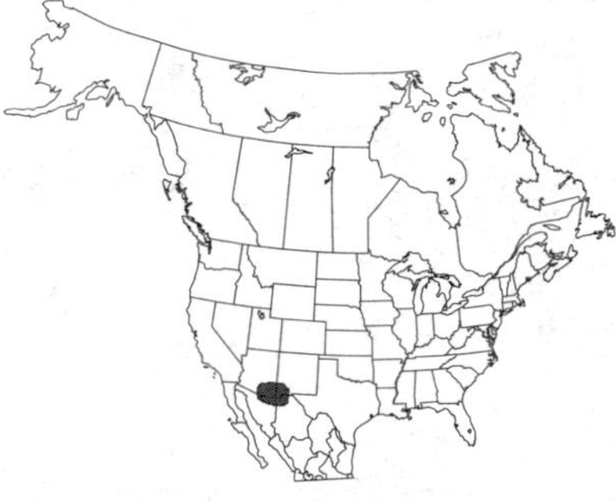
Jatropha macrorhiza

Disease Genesis—The seeds of *Jatropha* contain a wide variety of diterpenoid esters, especially the tigliane-type typical of the euphorbs (Adolph et al. 1984). These esters include citlalitrione, curculathyrane, curcusones A–D, and riolozatrione (Connolly & Hill 1991). Although systemic effects such as cardiac, respiratory, and intestinal depression have been shown in in vitro studies, intoxication nevertheless results primarily from irritation of the digestive tract, from the mouth through the intestines (Blohm 1962). Toxic effects are well illustrated by the common names *bellyache bush, purge nut,* and *physic nut.* The seeds and/or leaves of several species also contain proteins that inhibit protein synthesis, similar to ricin's action; curcin, for example, is present. However, they are class 1 RIPs, lack B chain–type activity, and thus have limited if any effects in the animal (Sperti et al. 1976; Stirpe et al. 1976; Gasperi-Campani et al. 1980; Reynolds & Irvin 1980). Some species may also have cyanogenic potential, but the clinical relevance of this is not clear (Blohm 1962).

- tigliane-type diterpenoid esters, citlalitrione, curcusones, riolozatrione

CHAIN-TYPE ACTIVITY

Citlalitrione

Curcusone A

Riolozatrione

Clinical Signs, Pathology, and Treatment—As is typical of the family, the irritation of the digestive tract is manifested as nausea, vomiting, and diarrhea. The pathology and treatment are the same as for other genera and species in the family.

- irritation of the digestive tract; minor effects on the liver and kidney
- nausea, vomiting, diarrhea

Manihot Mill.

Taxonomy and Morphology—Comprising 150–170 species, *Manihot* is a New World genus with greatest diversity in Brazil. Its name is derived from the native Brazilian name *manioc.* Some species yield rubber, but more importantly, *M. esculenta* is one of the most valued food crop plants in tropical countries. Originally dispersed by Portuguese colonists and now cultivated around the world, its fleshy roots yield carbohydrates from which tapioca, Brazilian arrowroot flour, and other food products are made. In North America, 6 species are present. Only 1 is of toxicologic importance:

- *M. esculenta*

cassava
gari
mandioca
manioc
tapioca plant
yuca

M. esculenta Crantz (= *M. utilissima* Pohl)	cassava, manioc, tapioca plant, yuca, gari, mandioca, bitter cassava

■ erect, perennial shrubs, sap white, viscid; tuberous roots; leaves simple, alternate, lobed; flowers small; petals absent

Plants shrubs; perennials; monoecious; sap white, viscid. **Stems** erect; 100–200 cm tall; soft; brittle; from fleshy, elongate, tuberous roots. **Leaves** simple; alternate; long petiolate; venation palmate; blades deeply parted into 3–7 spathulate or linear-lanceolate lobes; surfaces glabrous or veins minutely puberulent. **Inflorescences** paniculate; terminal. **Flowers** small; perianths in 1-series; campanulate. **Sepals** 5; fused; green or colored. **Petals** absent. **Stamens** 10; in 2 whorls. **Pistils** with carpels 3; stigmas dilated; styles 3; locules 3. **Fruits** capsular schizocarps; globose; 6-angular, winged. **Seeds** 3.

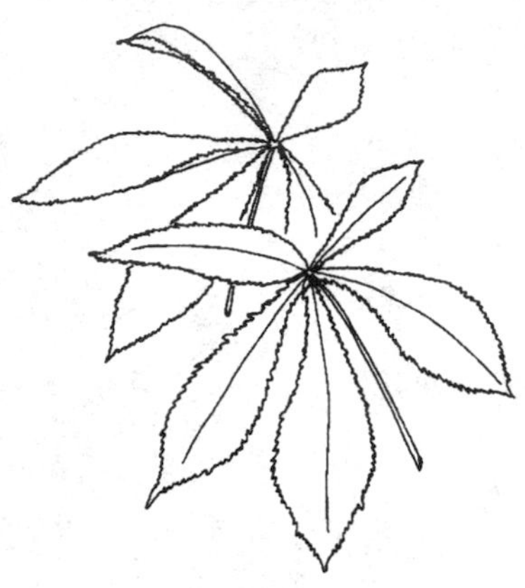

Manihot esculenta

■ uncommon, southern warmer areas

Distribution and Habitat—*Manihot esculenta* is seldom encountered in North America because there are few areas where it is sufficiently hardy to survive. It is reported to occur in disturbed sites in southern Florida and the Keys. The other species of the genus present in North America occur in Mexico, Arizona, and southern Texas.

■ tuberous root, tropical food staple; all plant parts cyanogenic

Disease Problems—Very starchy and high in caloric content, the large, tuberous roots of *M. esculenta* are used as food. They offer a higher yield of calories per unit of land than any of the cereal grains and are relatively easy to grow (Balagopalan et al. 1988). The tubers are low in protein, but the leaves are high in protein, being similar to alfalfa in this regard, and low only in methionine. The drawback to the species is that all parts of the plants are cyanogenic, especially the roots of some varieties. Although widely used as a food staple in the tropics, cyanide intoxication, including deaths, is a continuing problem (Akintonwa 1996). HCN potential also is a major limiting factor in the use of the tubers or leaves for animal feed. In many instances, they can be safely used only when dried or made into silage. Indeed, consumption of the roots in the tropics is associated with several chronic disease problems, such as tropical ataxic neuropathy, as discussed in the treatment of the Rosaceae (see chapter 64) (Balagopalan et al. 1988).

■ cyanogenic glycosides, linamarin and lotaustralin; especially in outer peel of root; leaves also a risk when wilted but not when sun-dried

```
           C≡N
          /
Glucose-O—C—CH3
          \
           CH3
```

Linamarin

```
           C≡N
          /
Glucose-O—C—CH3
          \
           CH2CH3
```

Lotaustralin

■ peracute intoxication and possibly chronic disease in humans

Disease Genesis—All plant parts contain the cyanogenic glucosides linamarin and lotaustralin, and the glucosidase linamarinase, which is highly compartmentalized in mesophyll tissue (Seigler 1977). HCN potential may be as high as 2,000 ppm d.w. in the root cortex of some varieties, especially the so-called bitter types. The leaves are of less hazard but nevertheless are a problem, with up to 180–240 ppm HCN potential and a mortality rate as high as 40% in goats fed fresh leaves (Devendra 1977). The HCN potential is typically maintained throughout the growth cycle. Fresh root peels are often unsafe for feeding; they require processing to reduce their HCN potential. They can be dried or, even better, made into silage to render them less toxic (Tewe & Egbunike 1992). It is also advisable to dry the leaves. Sun drying is preferred, because oven drying results in loss of the glucosidase activity before the HCN potential is fully eliminated (Omole 1992). Because of the glucosidase activity, HCN potential actually increases during the first few hours of drying but eventually drops to less than half after 24 hours (Tewe & Egbunike 1992). Sun drying eventually may result in the loss of 80–90% of cyanogenic potential (Gomez & Valdivieso 1985). HCN potential also increases the first 3–4 hours after wilting, then declines during the next few days (Hahn et al. 1992).

Intoxication with HCN is usually peracute, but there are exceptions. In some instances in humans there is apparently slow hydrolysis and absorption, so that the signs occur over a more protracted time. Also in humans, chronic intoxication due to long-term ingestion of cyanogenic tropical foods, such as cassava and to a lesser extent tropical lima beans, is a serious problem, causing several among a series of similar chronic neurologic syndromes. They include tropical ataxic neuropathy, tropical amblyopia, West Indian amblyopia, ataxic neuropathy of Nigeria, and tobacco amblyopia (MacKenzie & Phillips 1968; Osuntokun 1968; Montgomery 1965). Bilateral optic atrophy and nerve deafness, posterior column myelopathy (spinal), and sometimes polyneuropathy occur. There may be severe sensory deficit in the legs. It is not clear whether the neurologic effects are due solely to the direct effects of cyanide—depletion of cysteine and other sulfur compounds in the tissues—or are complications of nutritional deficiencies. The role of cyanide as a cause of chronic intoxication in livestock is not clear. Such effects have not heretofore been recognized in the field as associated with cyanide, with the possible exception of the ataxia-urinary incontinence syndrome in horses on sorghum pastures.

Possible teratogenic effects related to cassava ingestion have been suggested. Rats fed 80% of their diet as cassava for the first 15 days of pregnancy had an increased number of embryonic deaths and resorptions, as well as some effects on the fetuses, such as decreased growth, limb defects, and microcephaly (Singh 1981). However, other studies in hamsters indicated that maternal toxicity and fetal vertebral and rib abnormalities occurred only with very high dosages (Frakes et al. 1985).

Clinical Signs and Pathology—Within a few minutes of eating *M. esculenta,* there is abrupt onset of apprehension and distress, quickly followed by weakness, incoordination, and labored respiration. If intoxication is severe, the individual may be unable to stand and may lay in lateral recumbency, with periodic paddling and tetanic seizures. The entire sequence of signs may occur in a 5-to-15-minute period, with either death or recovery at the end of this time. The chronic, neurologic manifestations reported for humans in tropical areas are not as well recognized in livestock. There are few distinctive pathologic changes.

- abrupt onset, apprehension, distress, weakness, ataxia, labored respiration, recumbency, seizures; rapid development of all signs

Treatment—The primary antidote for ruminants is sodium thiosulfate given i.v. at a dose of 0.25–0.5 g/kg b.w. Sodium nitrite, 10 mg/kg body weight i.v., can be used to augment the effects of the sodium thiosulfate, but it is not required for effective treatment of ruminants. Both compounds are required in humans. A more complete discussion is offered in the treatment of the Rosaceae (see chapter 64). Crushing, parboiling, and cooking in stews for human consumption effectively eliminates the cyanide (Bassir & Fafunso 1976).

- ruminants, sodium thiosulfate with or without sodium nitrite; humans, both compounds

Mercurialis L.

Taxonomy and Morphology—Comprising 7 or 8 species, *Mercurialis* is native to the Old World. It is commonly known as herb mercury, a name derived from the Roman *herba mercurialis* and believed to reflect the Romans' belief that the medicinal virtues of the plants had been discovered by the god Mercury. Although plants of the genus have been employed as an emollient, the genus is not now a source of medicinal drugs. The common name *mercury* also is used for species of *Croton.* In North America, 1 naturalized species of the genus is present:

M. annua L. annual mercury, herb mercury

- *M. annua*

annual mercury
herb mercury

Plants herbs; annuals; herbage glabrous; dioecious; sap colorless, thin. **Stems** spreading to erect; 10–50 cm tall; branched. **Leaves** simple; opposite; short petiolate; venation pinnate; blades lanceolate to ovate; margins serrate; stipules lanceolate.

- branched annual herbs; leaves simple, opposite; flowers small, green; petals absent

Mercurialis annua

Mercurialis annua

- waste area invader
- irritation of digestive tract and moderate systemic effects but toxicants unknown
- weakness lethargy, diarrhea, anemia, icterus, weight loss, blood or hemoglobin in urine
- gross pathology, reddened mucosa of stomach and intestines, excess fluid

  microscopic, lung edema, liver and kidney necrosis
- activated charcoal, fluids, electrolytes

Inflorescences of 2 types, staminate and pistillate different; axillary; staminate flowers borne in long-peduncled spikes; pistillate flowers borne in small, subsessile clusters. **Flowers** small; green; perianths in 1-series. **Sepals** 3; fused. **Petals** absent. **Stamens** 8–20. **Pistils** with carpels 2, styles 2, not branched, locules 2. **Fruits** capsular schizocarps; subglobose; slightly 2-lobed; hirsute. **Seeds** 2.

Distribution and Habitat—Native to Europe, western Asia, and northern Africa, *M. annua* has naturalized in disturbed soils of roadsides, fields, and waste areas in the northeastern United States and adjacent Canada and in coastal central California.

Disease Problems and Genesis—Lacking the milky latex sap so typical of the family, *M. annua* nevertheless has an unpleasant odor and is generally considered to be unpalatable. However, in some instances it is eaten by sheep with little reluctance (Sendil 1978). It appears to be more toxic to calves, but the greatest risk is typically to sheep, and when grazed for several days or weeks, the plant may cause serious digestive tract problems. In addition there may be lysis of red blood cells and other systemic effects (Senf & Seffner 1965; Landau et al. 1973; Bizzeti et al. 1987; Welchman et al. 1995). Milk production typically declines or ceases and may not resume even if the other signs subside (Bizzeti et al. 1987). The plant is toxic mainly when in flower or fruit, and although unlikely to be eaten fresh, it may be a hazard when present in hay. Despite its reputed use as a potherb, plants are toxic even when boiled (Frohne & Pfander 1984). In flocks of sheep eating plants of *M. annua,* 25% or more of the animals may be affected, but the death rate is usually much lower, perhaps 2–5%. Not present in North America, *M. perennis* (perennial mercury, or dog's mercury) causes similar intoxication effects (Watson 1998).

The toxicants of *M. annua* are unknown. It lacks the latex sap and the diterpenes so typical of other members of the Euphorbiaceae.

Clinical Signs—The principal effects of *M. annua* are indicative of irritation of the digestive tract and include weakness, lethargy, and watery diarrhea. Anemia is a common complication. There also may be icterus, hemoglobinuria, hematuria, and in some animals incoordination, wool loss, itching, and substantial weight loss. The severe decline in milk production may not be reversible. These effects may be produced by consumption of a large amount of plant material in a short period or smaller amounts eaten repeatedly for a month or more.

Pathology—The most typical changes are reddening and hemorrhage of the mucosa of the stomach and small intestine and excess fluid in the chest, abdominal cavity, and heart sac. In addition the lungs may be red and edematous, the heart muscle pale and mottled, the liver congested and nutmeg-colored, and the kidneys dark and swollen. Microscopically, there may be edema of the lung and destruction of some alveolar walls. The liver may be congested with foci of necrosis and megalocytosis. There may be tubular degeneration of the kidneys with proteinaceous casts and some interstitial fibrosis. There may also be focal cardiac myositis.

Treatment—It is important to alleviate the irritation and pain, to decrease intestinal motility, and to restore fluid and electrolyte balance. The use of oral activated charcoal also may be beneficial.

Pedilanthus Neck. ex Poit.

Taxonomy and Morphology—Comprising about 14 species distributed from southern North America to northern South America, *Pedilanthus* is commonly known as slipper flower. Both its common and generic names, derived from the Greek roots *pedilon* and *anthos,* meaning "sandal" and "flower," reflect the distinctive appearance of the highly modified inflorescence. Closely related to the previously described *Euphorbia, Pedilanthus* has its flowers borne in cyathia. Resembling a single flower, the cyathium is actually one pedicelled pistillate flower surrounded by numerous staminate flowers, all enclosed in a campanulate or tubular involucre. The flowers lack sepals and petals, and the staminate flowers are reduced to a single stamen. The involucre bears 2–6 glands on its inner surface. Unlike the radially symmetrical cyathia in *Euphorbia,* the cyathia of *Pedilanthus* are bilaterally symmetrical and bear spurs. They fancifully resemble tiny red birds—hence the common name *redbird cactus.* In North America, 2 species are present:

■ *P. macrocarpus*
P. tithymaloides

candelillo
Christmas candle
devil's backbone
fiddle flower
jewbush
pinipini
redbird cactus
ribbon cactus
slipper flower
tuturutu

P. macrocarpus Benth.	candelillo
P. tithymaloides (L.) Poit.	Christmas candle, ribbon cactus, redbird cactus, slipper flower, fiddle flower, jewbush, pinipini, tuturutu, devil's backbone

Plants succulent shrubs; perennials; evergreen or deciduous; monoecious; sap white, viscid. **Stems** erect; 1–2.5 m tall; terete; jointed; branches dark green, zigzag. **Leaves** simple; alternate; often caducous; petiolate or sessile; midribs of blades keeled or not keeled. **Inflorescences** cyathia; borne in dense terminal cymes; bracts caducous or absent; involucres bilaterally symmetrical; campanulate or tubular; red or pink or green; spurs short or long; nectar glands 2–6, borne inside. **Flowers** with perianths absent or in 1-series; staminate flowers 8–15 per cyathia, comprising 1 stamen, borne on inner surface of involucral cup; pistillate flowers 1 per cyathium, long pedicelled in fruit. **Sepals** 0 or 3, scalelike. **Petals** absent. **Stamens** 1. **Pistils** with carpels 3; styles 3, divided. **Fruits** capsular schizocarps; 3-lobed. **Seeds** 3.

■ succulent, erect, perennial shrubs; sap white, viscid; leaves simple, alternate; inflorescences cyathia; fruits 3-lobed

Pedilanthus tithymaloides

Pedilanthus macrocarpus

Pedilanthus tithymaloides

1. Leaves ovate or ovate-oblong; midribs keeled.
   Plants of southern Florida and eastern Mexico . . . . *P. tithymaloides*
1. Leaves linear; midribs not keeled.
   Plants of western Mexico and Baja California . . . . *P. macrocarpus*

Distribution and Habitat—*Pedilanthus macrocarpus* is a Sonoran Desert species. *Pedilanthus tithymaloides* is indigenous to tropical areas from Florida to Venezuela.

It is also cultivated as an ornamental in California, as a hedge plant in Florida, and as a houseplant throughout North America.

■ caustic irritant sap

Disease Problems and Genesis—Both the seeds and the caustic milky sap are toxic when consumed in large amounts, this is presumably due to saponins (Blohm 1962; Oakes & Butcher 1962). Terpenoids have not been identified in the sap. The clinical signs, pathology, and treatment are consistent with those of other genera in the family.

Phyllanthus L.

■ *P. abnormis*
leaf-flower

■ annual herbs; sap white, viscid; leaves simple, alternate, 2-ranked; flowers small, staminate and pistillate

Taxonomy and Morphology—Comprising approximately 750 species in tropical and warm-temperate regions, *Phyllanthus,* commonly known as leaf-flower, is native to both the Old World and the New World, with greatest diversity in the former (Webster 1970). Quite diverse in habit, its species provide edible fruits, timber, and ornamentals. Some are used medicinally. The generic name is derived from the Greek roots *phyllon* and *anthos,* for "leaf" and "flower," alluding to the characteristic that flowers of some species appear to be borne on the margins of the leaves; in reality the leaves are absent, and the stems are flattened, leaflike structures termed cladophylls. About 24 species, both native and introduced, occur in North America. Toxic effects have been associated with only 1 species:

P. abnormis Baill. (=*P. garberi* Small)	leaf-flower

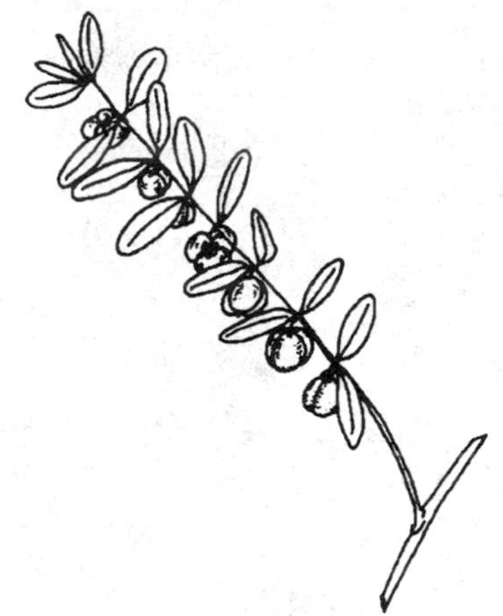

Phyllanthus abnormis

Plants herbs; annuals; monoecious; sap white, viscid. **Stems** erect or prostrate; 10–50 cm tall. **Leaves** simple; alternate; 2-ranked; subsessile; blades elliptic to oblong; apices obtuse or emarginate; bases cuneate to subcordate; stipules ovate to linear-lanceolate. **Inflorescences** axillary; staminate flowers paired at proximal nodes; 1 staminate and 1 pistillate flower at distal nodes. **Flowers** small; staminate and pistillate different; perianths in 1-series. **Sepals** of staminate flowers 4 or rarely 5 or 6, fused, yellowish or reddish tinged; sepals of pistillate flowers 5 or 6, ovate or elliptic or obovate, greenish with pale margins. **Petals** absent. **Stamens** 2; filaments fused; anthers connivent. **Pistils** with carpels 3; styles 3, divided. **Fruits** capsular schizocarps; elastically dehiscent; often reddish and glaucous. **Seeds** 6; light brown; longitudinally ribbed.

■ dry sandy soils

Distribution and Habitat—*Phyllanthus abnormis* is a species of dry, sandy soils of western Oklahoma and Texas south to Tamaulipas. There also are numerous populations in the coastal sand dunes of peninsular Florida.

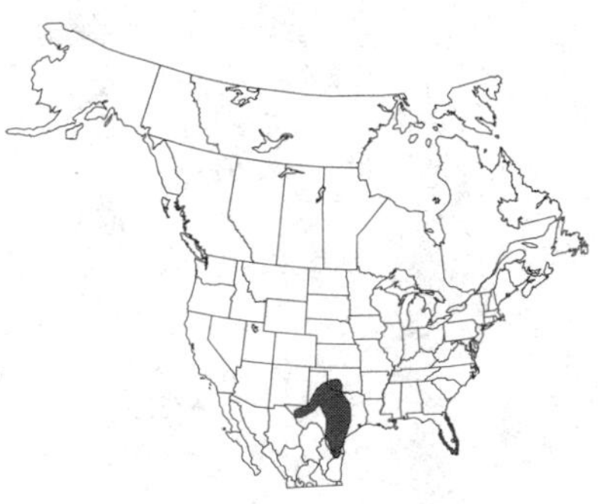

Phyllanthus abnormis

■ liver disease, mainly cattle, fresh forage

Disease Problems—Extensively grazed by livestock, *P. abnormis* causes acute to chronic hepatic degeneration in cattle but apparently has less effect in sheep and goats. It may not be specific for cattle, inasmuch as a suspect case has been encountered involving a horse with grazing restricted to a site with abundant *P. abnormis.* Drying is reported to decrease toxicity, so that disease is likely to appear only with consumption of fresh forage (Mathews 1945). Experimentally, 1–2% b.w. of fresh plant material fed daily for 4–15 days caused severe liver disease (Mathews 1945). Some African species are well respected for their toxicity potential; the leaves, bark, and roots pose a risk to both livestock and humans (Verdcourt & Trump 1969). Even the smoke from burning bark was considered to be highly toxic. Other species of

Phyllanthus are known for a host of medically useful biological effects, including hepatic antiviral and antineoplastic activity, but it is not known if these effects are related to the toxicity of this species (Blumberg et al. 1989).

Disease Genesis—*Phyllanthus abnormis* lacks diterpenes, and the specific toxicant is unknown. There is some similarity of the disease it produces to that caused by *Senecio* spp., but pyrrolizidine alkaloids are not present. Other alkaloids have been reported to be present in other species (Mensah et al. 1988), but *P. abnormis* appears to contain little if any. Terpenoids have not been identified in species of *Phyllanthus*.

■ toxin unknown

The toxicant is apparently passed in milk, because nursing calves are also affected. In some cases, secondary photosensitization may be a primary manifestation of the disease (Dollahite 1978). Antiviral and antineoplastic effects are attributable to a series of 6-phyllanthostatin glycosides and the related phyllanthoside from the roots of *P. acuminatus*, a Costa Rican tree (Pettit et al. 1984, 1985, 1990). These compounds have now been synthesized in the laboratory.

Clinical Signs—Following consumption of *P. abnormis*, cattle initially are listless, have a poor appetite, and lose weight and condition rapidly. These signs are followed by onset of a clay-colored diarrhea with tenesmus, drop in milk production, and sometimes continuous walking. There may be icterus and/or photosensitization.

■ listless, anorexia, weight loss, diarrhea, straining, icterus, photosensitization

Pathology—Typically the liver will appear yellow and may vary from slightly swollen to small and firm, depending on the length of the disease course. The hepatic lymph nodes are usually enlarged, and the mesentery and gallbladder wall edematous. Microscopically, there is hepatocellular necrosis and regeneration, plus fibroblastic and bile duct proliferation without megalocytosis. Yellow lipoid pigment may be observed in reticuloendothelial cells of the enlarged hepatic lymph nodes. There may also be mild renal tubular degeneration.

■ gross pathology, liver yellow, gallbladder wall edematous

microscopic, liver necrosis, fibrosis and bile duct proliferation

Treatment—Nursing care to allow hepatocellular regeneration is the primary concern. In many cases this will not be feasible, because recovery requires several months.

■ nursing care

Reverchonia A.Gray

Taxonomy and Morphology—Closely related to *Phyllanthus*, *Reverchonia* is a monotypic genus indigenous to the Southwest (Webster & Miller 1963). The following is the single species:

■ *R. arenaria*

sand reverchonia

R. arenaria A.Gray sand reverchonia

Plants herbs; annuals; from sparsely branched taproots; monoecious; herbage glabrous; sap colorless, thin. **Stems** erect; 20–50 cm tall; branched; glaucous-white. **Leaves** simple; alternate; petiolate; thickened; venation not apparent; blades elliptic to oblong-linear; apices apiculate; bases cuneate; stipules lanceolate, papery, reddish. **Inflorescences** cymose clusters; axillary; borne only on branches; each cluster with 4–6 staminate flowers and 1 pistillate flower in center. **Flowers** small; staminate and pistillate different; perianths in 1-series; reddish. **Sepals** of staminate flowers 4, fused, ovate-oblong, purplish or pinkish with medial green strip; sepals of pistillate flowers 6 or rarely 5, fused, oblong, reddish. **Petals** absent. **Stamens** 2; filaments free. **Pistils** with carpels 3; styles 1, deeply divided. **Fruits** capsular schizocarps; ovoid-globose; smooth; yellowish. **Seeds** trigonous; dark or reddish brown; tangential surfaces smooth; radial surfaces papillate.

■ erect, annual herbs; leaves simple, alternate; flowers small, pistillate and staminate

Reverchonia arenaria

- probable cause of liver disease
- signs, pathology, and treatment, likely as for *Phyllanthus*

Distribution and Habitat—A highly specialized desert annual, *R. arenaria* is found in sandy soils of dunes, prairies, and stream banks.

Disease Problems and Genesis—Little is known about the toxicity of *R. arenaria* other than experimental evidence indicating a risk when fresh plants are fed to sheep or cattle at about 1% animal b.w. (Kingsbury 1964). The primary effects are on the liver and kidney, and the disease is apparently quite similar to that produced by *Phyllanthus abnormis*. Diterpenes are not present, and the cause of the disease is not known.

Clinical Signs, Pathology, and Treatment—The clinical signs and pathological changes exhibited by animals consuming *R. arenaria* are essentially the same as those observed in animals eating *Phyllanthus abnormis* and are discussed in the treatment of that species.

Ricinus L.

- *R. communis*

castor, castor bean, castor-oil plant, higuerilla, palma Christi, steadfast, wonder tree

- erect annual herbs, almost treelike in some areas; leaves simple, alternate, large, palmately lobed, odor pungent; flowers terminal; petals absent; pistils red or purplish red; fruits globose, spiny; seeds mottled, ticklike

Taxonomy and Morphology—A monotypic genus believed to be native to Africa and possibly southwest Asia, *Ricinus* is now distributed throughout the world. Its name comes from the classical Latin name for the common European wood tick, *Ixodes ricinus*, because the mottled seeds resemble the arachnid when it is engorged with blood. Cultivation of the genus coincides with the domestication of other Old World crop plants at about 7,000 B.C., given that its seeds have been found at early agricultural sites in western Asia (Moshkin 1986b). Seeds also were discovered in the Egyptian pyramids (Frohne & Pfander 1984). Today, numerous cultivars are planted as ornamentals and as a crop for the oil contained in the seeds. In addition to its familiar medicinal use as a purgative, castor oil is used as an industrial lubricant and coolant. It is used in the manufacture of rubber, textiles, plastics, paints, soaps, printing inks, and cosmetics (Moshkin 1986a). The single polymorphic species is the following:

R. communis L. — castor bean, castor-oil plant, castor, higuerilla, palma Christi, steadfast, wonder tree

Plants herbs; annuals (perennial in tropics); monoecious; herbage glabrous; sap colorless; thin. **Stems** erect; 1–5 m tall (to 13 m tall in tropics); branched above, branches ascending; internodes hollow, often glaucous, red at maturity. **Leaves** simple; alternate; long petiolate; peltate; dark glossy green or red; odor pungent; blades large, 30–90 cm wide; palmately 5-to-11-lobed, lobes lanceolate or ovate-oblong, apices acuminate; petioles with conspicuous glands. **Inflorescences** paniculate; terminal; staminate flowers below pistillate. **Flowers** with perianths in 1-series. **Sepals** 3–5; fused; caducous in pistillate flowers. **Petals** absent. **Stamens** numerous; filaments branched. **Pistils** with carpels 3; styles 3, divided, red or purplish red. **Fruits** capsular schizocarps; globose; spiny. **Seeds** 3; ellipsoidal; mottled.

Ricinus communis

Distribution and Habitat—*Ricinus communis* is cultivated as an ornamental for its striking foliage. It may become a weed, occasionally growing wild in vacant city lots and annually reseeding itself. Typically, it overwinters only in the warmer, southernmost extent of North America. Plants are grown commercially in some areas for the production of castor oil from the seeds.

Disease Problems—The seeds of *R. communis* have long been appreciated for their ornamental value. However, there is little question about their toxicity, because occasional cases of severe digestive tract disturbances with diarrhea have been reported (Wedin et al. 1986). All animals are susceptible, and cases of intoxication are documented in all types of livestock, including poultry (Clarke 1947; Geary 1950; Balint 1974). In the 1970s, several thousand wild ducks in the Texas Panhandle died apparently as a result of feeding on remnants of cultivated castor beans (Jensen & Allen 1981). Estimates have been made of the toxicity of the seeds in different animals, and it seems clear that the horse is most susceptible (Steyn 1934; Tokarnia et al. 1979). The lethal dose ranges from 0.1 g/kg b.w. in the horse, 0.4 g/kg b.w. for geese, and 1–1.4 g/kg b.w. for rabbits, sheep, and pigs to 5.5 g/kg b.w. in goats and 14 g/kg b.w. in chickens (Steyn 1934; Tokarnia et al. 1979). However, others have suggested these estimates may not be correct (Geary 1950). Indeed the danger may be overrated; consumption of the intact seeds is unlikely to be as serious as ingestion of ground seeds, because they are very hard and difficult to chew. This is illustrated by a case in which a man partly chewed and swallowed 12 seeds. Although he experienced profuse diarrhea, he recovered following conservative treatment (Spyker et al. 1982).

■ ornamental

Ricinus communis

■ risk of severe digestive disturbance; a few seeds well chewed may be lethal but risk is generally limited

■ numerous cases of intoxication in humans but few fatalities; most recover with conservative treatment

The seeds seem to be a particular hazard for children; intoxications are sometimes ascribed to as few as 1 or 2 seeds (Morton 1982). Numerous cases are reported in Europe; Frohne and Pfander (1984) described an incident involving schoolboys in which the severity of the disease produced correlated with the number of seeds eaten, from part of 1 seed up to 6 whole seeds. There also is considerable variation in the individual's response to ingestion of the seeds. Three cases reported by Calloner and McCarron (1990) illustrate the response that may develop. In the first case, a 3-year-old boy vomited and experienced abdominal pain about 1 hour after eating some seeds. Diarrhea occurred at about 8 hours, and fluids were given at 20 hours to correct dehydration. Nausea and vomiting recurred several times for the first 2 days, and the diarrhea continued for 5 days. In the second case, a 28-year-old man developed abdominal pain and explosive diarrhea 4 hours after intentionally eating 4 seeds as a cathartic. A 4-day sojourn in the hospital was required. In the third case, a 39-year-old man who had often eaten roasted castor beans, ate 4 thoroughly chewed, raw seeds. In 15 minutes he began to vomit blood-flecked fluid. Later he experienced abdominal pain, a burning pain in the throat, and watery diarrhea. The signs resolved in 12 hours. Kinamore and co-workers (1980) reported two ingestions by children in which vigorous, early emesis was carried out and the patients did not develop signs. Rauber and Heard (1985), in reviewing the older literature, suggested that the dangers of the seeds are often overstated. In support of their view, they described two cases. One involved 9 children, and ingestion of even 2 well-chewed beans produced only minor diarrhea. The second case was an attempted suicide in which an adult ate 24 chopped beans but did not exhibit signs when treated conservatively. Challoner and McCarron (1990), in their literature review, noted that few deaths have been reported since the 1940s. In most cases, the vomiting and diarrhea have been treated with vigorous fluid therapy with good success. Occasionally hemolysis and liver and/or kidney problems have been observed.

Ingestion of the leaves and/or pericarps of the fruits produces a different array of signs—neurologic rather than digestive tract disturbance (Dobereiner et al. 1981). The lethal doses of leaves are reported to be 20 g/kg b.w. and of pericarps 5 g/kg b.w. The signs of muscular weakness commence several hours after ingestion.

■ 3 types of toxicants—
an agglutinin
ricin
ricinin, an alkaloid in leaves, neuromuscular effects

Disease Genesis—The castor oil of commercial use is 90% ricinoleic acid. The residue, or castor meal, after oil extraction has a high protein content and potential use as livestock feed in tropical countries (Fuller et al. 1971). Its use is limited by the presence of several noxious compounds, including the alkaloid ricinin, an agglutinin

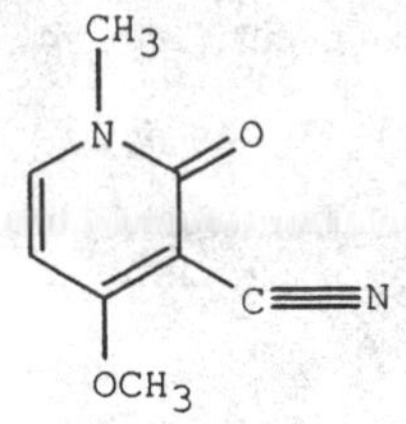

CH_3–N, O, C≡N, OCH_3

Ricinin

■ agglutinin protein, risk?

■ ricin in seeds, glycoprotein, poorly absorbed orally, low toxicity for ruminants, greater risk for monogastrics such as humans and pets

■ ricin, composed of a facilitator B chain to promote cell entry and an effector A chain that inhibits ribosomes and prevents protein synthesis

protein, and ricin (Jenkins 1963). Ricinin is found in the pericarp of the fruits and in the leaves (Marion 1950; Tokarnia et al. 1979; Dobereiner et al. 1981). It appears to be the cause of the convulsive seizures and muscular weakness. Ricinin may cause the seizures by binding to a GABAa receptor site (Ribeiro et al. 1997). In addition its pyridine structure suggests a neuromuscular effect, possibly due to antagonism of acetylcholine at nicotinic receptors.

The agglutinin is a tetramere composed of two heterodimers similar to but with a slightly higher mass than ricin (Lord et al. 1994). This protein has strong red blood cell–agglutinating activity but appears otherwise to be of limited toxicologic significance, except perhaps as a contributing factor in the allergic response.

Ricin is a heterodimeric protein found in the protein body (storage unit) of the seed. It is a glycoprotein composed of an enzymatic A chain of 267 amino acid residues and a lectin B chain of 262 residues with a slightly higher mass. The chains are linked by a single disulfide bond (Olsnes & Pihl 1978, 1982). The molecular mass of the protein, depending on the isoform, is about 65,000 daltons, and the respective structures have been described in detail (Katzin et al. 1991; Rutenber et al. 1991). Typical of proteins, ricin is poorly absorbed, and unless the seed is well pulverized, little toxin passes into the animal's body. When ricin is given orally, most of it passes through the digestive tract intact or is degraded to peptides (Ishiguro et al. 1983, 1992b). The amount absorbed is probably less than 1%. This accounts for the 60-to-100-fold greater dosage required for toxicity via oral versus parenteral administration (Steyn 1934; Sekine et al. 1986).

Ruminants appear to be less susceptible than other animals. Even the well-ground seed when mixed with feed has little effect on cattle. Experimentally, supplementation of feed with ricin-rich castor bean meal (37.7% crude protein) for 113 days produced only decreased performance (Albin et al. 1968). In another experiment, a 227 kg heifer fed 2.3 kg of a mixture of green leaves, stems, and partly ripened seeds developed only a transient purgative effect the following day (Sanders et al. 1942). In humans, severe allergic reactions occur in a few individuals following exposure to the pulverized seed. Experimentally, repeated exposure to ricin produces an immunologic response, and protection is readily accomplished using either ricin or its subunits (Godal et al. 1983; Foxwell et al. 1985; Lemley et al. 1994). However, ricin may not be the exclusive cause of the allergic reaction, because other similar proteins are present in the seed.

Following absorption in the digestive tract, most of the ricin passes via the blood to the liver, less to the spleen, and less still to other tissues (Ishiguro et al. 1992b). It enters a cell by a unique process in which the lectin B chain binds, via at least two and possibly three or more binding domains, to galactoses in the cell membrane (Rutenber & Robertus 1991; Frankel et al. 1996). This facilitates internalization of the ricin, after which it is translocated via endosomes to the Golgi network and the endoplasmic reticulum. As this occurs, the facilitator B chain is removed by hydrolysis prior to reaching the cytosol. It has been suggested that this pathway is an integral step toward the ultimate toxic effect (Lord et al. 1994; Sandvig & Van Deurs 1996). However, the binding of the B chain, other than facilitating entry of the A chain, is otherwise not essential for expression of toxicity (Morino et al. 1995). The actual expression of toxicity involves A chain enzymatic hydrolyses of an adenine from an *N*-glycosidic site on 28S rRNA. This prevents peptide elongation, inactivates the eukaryotic 60S ribosome, inhibits protein synthesis, and secondarily inhibits DNA and RNA synthesis (Endo & Tsurugi 1987; Endo et al. 1987). A detailed review of the structure of ricin and its mechanism of action is presented by Lord and coworkers (1994).

Because of the presence of both facilitator and effector chains, ricin is considered a prototypical class 2 ribosomal-inhibiting protein (RIP). In all respects it is very similar to abrin, another class 2 RIP present in *Abrus precatorius* in the Fabaceae (see chapter 37) (Olsnes 1977). Ribosomal-inactivating proteins have been identified in numerous plants and in some animal tissues as well (Youle 1995). The importance of the role of the B chain in ricin is underscored by the numerous class 1 RIPs in other plants; these RIPs are not cytotoxic, because the toxic moiety (enzymatic A chain) lacks an accompanying lectin B chain to facilitate cell entry (Gasperi-Campani et al. 1980). The importance of the B chain is further illustrated by the observation that ricin activity is blocked by interaction of the B chain with halogenated glycopeptides (Lambert et al. 1991). Because of these features, ricin has potential as an adjunct anticarcinogenic drug, and considerable effort is being expended to elucidate the precise molecular mechanisms of its action (Lord et al. 1994). This activity was the initial basis for the development of immunotoxins for antitumor activity (Vitetta et al. 1987).

The toxic effects of ricin appear a few hours after administration. Experimentally in rats, severe diarrhea and loss of intestinal villi occurred 5 hours after administration (Ishiguro et al. 1983). The effects appear to be due to a combination of actions. Binding of the facilitator B chain to epithelial cells results in impairment of absorption of sugars and a loss of cell viability (Ishiguro etal. 1984, 1992a). This, in addition to the lectin irritant action on the mucosal surface and inhibition of protein synthesis in the epithelial cells, produces the lesions of the digestive tract (Ishiguro et al. 1992c). There also are systemic effects, as illustrated by the progressive paralysis observed in ducks experimentally intoxicated with *R. communis* (Jensen & Allen 1981). Depending upon dosage, fatal liver and kidney effects occur in laboratory animals (Dirheimer et al. 1966, 1968). There is also the potential for cardiovascular effects; they are primarily indirect, causing peripheral decrease in vascular tone, hypotension, and vasospasms (Christiansen et al. 1994; Zhang et al. 1994). However, cardiovascular effects, though a possible problem in long-term immunotherapy, are unlikely to occur in typical oral intoxications. There are no changes in electrocardiographic parameters, and it is suggested that the effects are due primarily to increased vascular permeability (Ma et al. 1996). The primary action appears to be on the digestive tract, with effects on other systems observed only in those animals receiving less than a rapidly fatal dose or when the toxin is used therapeutically.

Ricin is so potent that its nefarious use would seem to be an appropriate plot for a James Bond movie. Reality, however, has rivaled fantasy. In the late 1970s a Bulgarian broadcaster in London was assassinated in a bizarre fashion. It is hypothesized that a pin-sized metal sphere was embedded in the victim's thigh during an "accidental" collision with the assassin and that an umbrella served as the delivery weapon. The broadcaster became ill within hours and died 2 days later. An empty compartment in the sphere, discovered after an autopsy, indicated the possibility of a toxin; ricin was the prime suspect because of its remarkable potency (Knight 1979). Subsequently the role of the Russian KGB in the killing was confirmed. It had provided technical assistance to the Bulgarian Secret Service, which had used ricin as suspected (Rich 1992; Reuters 1993). The KGB was very much involved in developing devices to deliver compounds such as ricin clandestinely, and ricin appears to have been used in assassination attempts on other famous individuals (Remnick 1992). Because of its great potency, the possession and/or threatened use of ricin in crude or purified form in the United States may result in prosecution under the Biological Weapons Anti-Terrorism Act of 1989 ("Man held for poison possession," 1995). Its possible use by militant groups opposed to governmental authority has been of concern (de Fiebre 1995).

- chewed seeds, after several hours; severe watery diarrhea, nausea, colic, vomiting, sweating, excess salivation, weakness, collapse

leaves, weakness, tremors, ataxia

Clinical Signs—In most cases there will be few if any signs of intoxication because the hard seeds are difficult to chew. On the rare occasions when the seeds are well chewed, the most consistent effect, following a delay of 6 or more hours, is severe watery diarrhea, which may be bloody, and abdominal straining. These signs will be accompanied by nausea, severe abdominal pain, decreased blood pressure, weakness with trembling, anorexia, sweating, vomiting, and sudden collapse. There also may be copious thick, ropy salivation and hypothermia. On occasion, neurologic signs such as seizures and coma in the terminal stages have been reported. In animals exhibiting marked clinical signs, there is likely to be elevation of serum hepatic enzymes such as AST, ALT, LDH, and others that reflect liver damage. In humans, there is an increase in blood urea nitrogen (BUN), calcium, and phosphorus and a decrease in glucose in blood.

When leaves are ingested, the clinical signs of muscular weakness, tremors, ataxia, exercise intolerance, and copious salivation are likely to be of a transient nature, followed by rapid recovery after a few hours. Rarely there may be more serious intoxication with seizures and possibly death.

- gross pathology, reddening, edema, necrosis, mucosa of stomach and small intestine

Pathology—The most characteristic lesion following seed ingestion is severe reddening, edema, and necrosis of the mucosa of the digestive tract, especially the stomach and the small intestine. There may be considerable blood in the lumen of the intestine. The adjacent regional lymph nodes are edematous and necrotic, and there also may be splenic lymphoid necrosis. Scattered small, splotchy hemorrhages may be seen in the viscera and heart. Under some conditions and in some species, there may by degeneration of other tissues, including those of the liver and kidney. Pathologic changes when there is intoxication due to ingestion of the leaves are minimal.

- nursing care, activated charcoal, fluids, electrolytes

Treatment—In cases of *Ricinus* intoxication, the basic approach is supportive and includes good nursing care, administration of fluids with electrolytes and glucose, and diuresis. Activated charcoal may be of some value, but the primary goal of treatment is to allow the animal to survive long enough for regeneration of the affected tissues. Unfortunately the effects of ricin are very debilitating, and once severe signs are apparent, the prognosis is poor because of the persistence of the lesions.

Based on experimental in vitro studies, glycosidase inhibitors, such as *N*-bromocetyl-alpha-D-galactopyranosylamine, show promise in preventing ricin cytotoxicity either by inhibiting cellular uptake or by blocking glycosidase activity (Hassoun et al. 1992).

Experimentally, survival time in mice is increased with administration of dexamethasone and/or difluoromethylornithine and to a lesser extent with the antioxidants vitamin E succinate and hydroxyanisole (Muldoon & Stohs 1994).

Sapium P.Browne

- *S. sebiferum*

popcorn tree
tallow tree
vegetable tallow

Taxonomy and Morphology—Comprising 100–120 species in the tropics and subtropics of both the Old World and the New World, *Sapium* is commonly known as tallow tree. Named by Pliny, a Roman natural historian, it is of minor economic importance as a source of rubber, black dye, and wax for candles, soap, and cloth dressing. Some species are planted as shade and specimen trees. In North America, 5 introduced species are present. Only 1 is of known toxicologic significance:

S. sebiferum (L.) Roxb.	Chinese tallow tree, popcorn tree, vegetable tallow tree, vegetable tallow

Plants trees; perennials; deciduous; foliage glabrous; monoecious; sap white, viscid. **Stems** erect; to 15 m tall; branches brittle, drooping, slender. **Leaves** simple; alternate; petiolate; turning bright red or yellow-red in autumn; blades rhombic ovate; apices acuminate; margins entire; bases cuneate or truncate; petioles with 2 glands at apices. **Inflorescences** spicate; terminal or axillary; staminate flowers in clusters of 3s above solitary pistillate flowers; bracteoles present, with 2 glands. **Flowers** with perianths in 1-series. **Sepals** 3; fused; small. **Petals** absent. **Stamens** 2 or 3; filaments free. **Pistils** with carpels 3; styles 3, spreading, not divided. **Fruits** capsular schizocarps; fleshy; 3-lobed. **Seeds** 3; large; black; covered with thick layer of white wax; persistent on central column after capsule dehiscence.

- deciduous trees; sap white, viscid; leaves simple, alternate, red in autumn; fruits 3-lobed; seeds black

Sapium sebiferum

Distribution and Habitat—Native to China and Japan, *S. sebiferum* is planted as an ornamental along the Atlantic and Gulf coasts from South Carolina to Texas, as well as in California. It has naturalized in these southern areas, and in some it has become a noxious weed that degrades the native vegetation. Distribution is limited to regions where the minimum temperature exceeds 20°F.

- ornamental

Disease Problems and Genesis—The green foliage and fruits of *S. sebiferum* cause severe irritation of the digestive tract and a purgative effect in cattle but apparently not in sheep or goats (Russell et al. 1969). The toxicity is only moderate, because it takes approximately 1% b.w. of fresh plants fed for 5 or more days to produce the irritation. In a suspected case in a monkey, the signs were subacute and less distinctive. There was anemia, lethargy, and a loss of appetite but no pronounced diarrhea; these effects are perhaps due to a lower, more protracted dosage (Soifer 1973).

The foliage contains several biologically active compounds, including gallic acid and astragalin; however, the more important toxicants are tigliane-type, diterpene-ester irritants similar to those in other members of the Euphorbiaceae (Ohigashi et al. 1983; Liu et al. 1988). The oil in the seeds and the tallow covering the seed are nontoxic; the latter is used somewhat as an edible fat.

- irritation of digestive tract; tigliane diterpene phorbol

Phorbol

Clinical Signs, Pathology, and Treatment—The effects of *S. sebiferum* are those of a potent purgative; severe diarrhea may be accompanied by blood in some cases. These effects in most animals are self-limiting. When death occurs, the inflammatory effects on the digestive tract may be accompanied by mild renal and hepatic degeneration. Treatment is mainly symptomatic to control diarrhea and alleviate dehydration.

- severe diarrhea; reddened gut mucosa; treat with activated charcoal, fluids, electrolytes

Stillingia Garden ex L.

Taxonomy and Morphology—Named in honor of eighteenth-century English botanist Benjamin Stillingfeet, *Stillingia* comprises approximately 30 species in the warmer regions of both the Old World and the New World (Rogers 1951). Of essentially no economic importance other than their toxicity, 7 species are present in North America. Of them, 3 are worthy of mention (see plate 36.1):

S. sylvatica Garden ex L.	queen's delight, queen's root, yau root
S. texana I.M.Johnst.	Texas stillingia
S. treculiana (Meull.Arg.) I.M.Johnst.	trecul stillingia

- *S. sylvatica*
  S. texana
  S. treculiana

  queen's delight
  queen's root
  stillingia
  yau root

Plants herbs; perennials; from thick, erect, woody caudices; herbage glabrous; monoecious; sap white, viscid. **Stems** erect; 10–75 cm tall. **Leaves** simple; alternate; ascend-

- erect perennial herbs; sap white, viscid; leaves simple, alternate; flowers terminal; fruits 3-lobed

- dry soils

ing; sessile or subsessile; blades linear to lanceolate or elliptic or oblanceolate; apices acute; margins crenulate or serrulate, a small gland in each notch; stipules glandular. **Inflorescences** spicate; terminal; staminate flowers solitary or clustered at upper nodes; pistillate flowers solitary or clustered at bases; bracteoles present, with 2 glands. **Flowers** of 2 types, staminate and pistillate different; perianths in 1-series. **Sepals** 2 in staminate flowers, 3 in pistillate flowers; fused. **Petals** absent. **Stamens** 1–3. **Pistils** with carpels 3; styles 3, not divided. **Fruits** capsular schizocarps; subglobose; 3-lobed. **Seeds** 3; ovate-oblong; whitish; smooth.

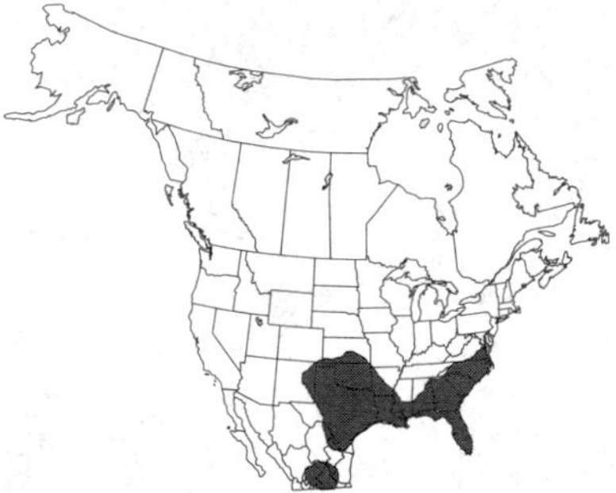

Stillingia sylvatica

Stillingia sylvatica

Stillingia texana

1. Staminate flowers 1 at upper nodes of spikes. Styles 2 mm long . . . . *S. treculiana*
1. Staminate flowers 5–15 at upper nodes. Styles 4–5 mm long.
   2. Leaves linear or linear-lanceolate. Schizocarps 5–7 mm in diameter . . . . *S. texana*
   2. Leaves elliptic to lanceolate or oblanceolate. Schizocarps 13–15 mm in diameter . . . . *S. sylvatica*

Stillingia treculiana

Distribution and Habitat—*Stillingia sylvatica* is a species of dry, sandy soils distributed from Virginia to eastern New Mexico. More limited in distribution, *S. texana* occurs in calcareous soils from Oklahoma south to Coahuila. *Stillingia treculiana* is present on calcareous soils of west Texas and northern Mexico.

- irritant effects; daphnane diterpenes, gnidilatidin

  peracute disease from cyanogenic glycosides; rapidly fatal

Disease Problems and Genesis—Species of *Stillingia* are cyanogenic and possibly cause irritation of the digestive tract. In feeding trials of suspected toxic plants, Boughton and Hardy (1939, 1940) fed *S. treculiana* to a sheep at a dose of 0.5% b.w. The animal died in 70 minutes. In an attempt to cause liver disease in sheep, they fed animals 0.25% b.w. daily for 5 days. All animals died from what appeared to be cyanide intoxication. The rapid death in 70 minutes is consistent with cyanogenesis, but the effects of the repeated doses are not. The milky sap of *Stillingia* is irritating to the skin and mucous membranes (Mitchell & Rook 1979), and *S. treculiana* contains the daphnane diterpenoid gnidilatidin, which may represent a risk to the digestive tract (Connolly & Hill 1991).

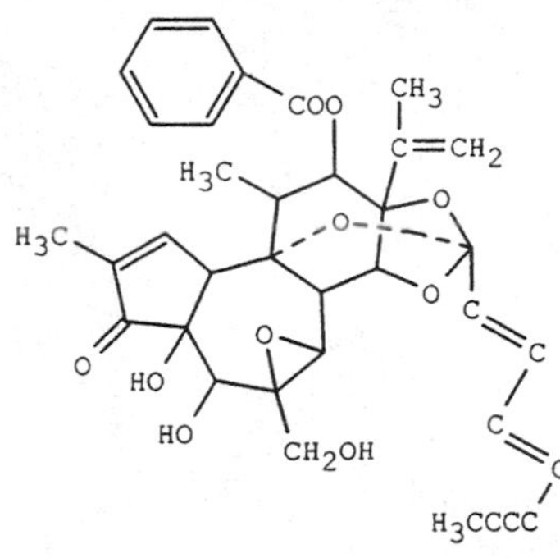

Gnidilatidin

Clinical Signs and Pathology—Within a few minutes of grazing the foliage of *Stillingia*, there is abrupt onset of apprehension and distress, which is quickly followed by weakness, incoordination, and labored respiration. If intoxication is severe, the animal may be unable to stand and may lay in lateral recumbency, with periodic paddling and tetanic seizures. The entire sequence of signs may occur in a 5-to-15-minute period, with either death or recovery at the end of this time. Because the disease progresses so rapidly, there are few distinctive pathologic changes. For the most part they are limited to congestion of the viscera and scattered, small splotchy hemorrhages.

- abrupt onset, apprehension, distress, weakness, ataxia, labored respiration, recumbency, seizures; rapid development of all signs

Treatment—The primary antidote for cyanogenesis in ruminants is sodium thiosulfate given i.v. at a dose of 0.25–0.5 g/kg b.w. Sodium nitrite, 10 mg/kg b.w. i.v., can be used to augment the effects of the sodium thiosulfate, but it is not required for effective treatment. A complete discussion of care is presented in the treatment of the Rosaceae (see chapter 64).

- sodium thiosulfate, with or without sodium nitrite

Synadenium Boiss.

Taxonomy and Morphology—Comprising 15–19 species native to Africa, Madagascar, and the Mascarene Islands, *Synadenium* is closely related to the previously described *Euphorbia* and *Pedilanthus*. Like them, it has flowers that are borne in cyathia. Resembling a single flower, the cyathium is actually one pedicelled pistillate flower surrounded by numerous staminate flowers, all enclosed in a cup-shaped involucre. The flowers lack sepals and petals, and the staminate flowers are reduced to a single stamen. Unlike *Euphorbia*, which has separate glands, *Synadenium* has two glands that are united into a ring on the rim of the involucre. The generic name reflects this fusion, being derived from the Greek roots *syn* for "united" and *aden* for "gland." One species is commonly cultivated in North America:

- *S. grantii*

  African milkbush

S. grantii Hook.f. African milkbush

- erect, profusely branched, succulent shrubs; sap white, viscid; leaves simple, alternate; inflorescences cyathia, axillary, dark red; fruits 3-lobed

Plants succulent shrubs; perennials; monoecious; sap white, viscid. **Stems** erect; profusely branched; branches erect, thick, dark green. **Leaves** simple; alternate; succulent; short petiolate or subsessile; blades oblanceolate to obovate or spathulate; adaxial surfaces dark glossy green; abaxial surfaces pale green; apices rounded to apiculate; margins entire; bases attenuate; midribs reddish. **Inflorescences** cyathia; borne in dense axillary clusters; involucres cup-shaped; shallow; margins entire or lobed; lobes ciliate or dentate; nectar glands 2, fused into a ring, dark red. **Flowers** with perianths absent or in 1-series; staminate flowers numerous, borne on inner surface of involucral cup and separated by membranous partitions; pistillate flowers 1 per cyathium, occasionally absent; pedicelled in fruit. **Sepals** absent in staminate flowers; 3 and fused in pistillate flowers. **Petals** absent. **Stamens** 1. **Pistils** with carpels 3; styles 3. **Fruits** capsular schizocarps; 3-lobed. **Seeds** 3.

Synadenium grantii

Distribution and Habitat—Native to East Africa and restricted to hot, sunny, frost-free regions, *S. grantii* is cultivated as a houseplant.

- ornamental

Disease Problems and Genesis—*Synadenium* has a very irritating milky sap, which on occasion causes oral irritation and severe dermatitis with erythema and blistering, especially in young children (Spoerke et al. 1985). As is typical of most euphorbs, the

- severe irritant action of sap, diterpene phorbol

toxicant is a diterpene phorbol and, in this genus, is of considerable potency (Kinghorn 1980).

■ erythema skin, diarrhea

Clinical Signs, Pathology, and Treatment—The clinical signs of irritation are typical of the family. Treatment to alleviate them are the same as employed for other members of the family.

Tragia L.

■ *T. betonicifolia*
T. ramosa
T. saxicola

noseburn

■ perennial herbs with stinging hairs; leaves simple, alternate; flowers small, inconspicuous

Taxonomy and Morphology—Armed with stinging hairs and causing a rather mild dermatitis when touched, *Tragia,* commonly known as noseburn, comprises approximately 150 species in tropical and warm-temperate regions of both the Old World and the New World (Miller & Webster 1967). The center of species diversity is Brazil, and in North America, 14–20 species are present. The following are representative and commonly encountered:

T. betonicifolia Nutt.	betony noseburn
T. ramosa Torr.	bushy noseburn
T. saxicola Small	

Tragia betonicifolia

Plants herbs, occasionally becoming suffrutescent; perennials; from caudices of woody taproots; armed with stinging hairs; monoecious; sap colorless, thin. **Stems** erect or decumbent or trailing; solitary or numerous from caudices; branched. **Leaves** simple; alternate; petiolate or sessile; venation pinnate; blades suborbicular or ovate to lanceolate or oblong; margins entire or serrate or toothed or divided; stipules lanceolate. **Inflorescences** racemose; opposite the leaves at upper nodes; staminate flowers above pistillate; bracteoles present. **Flowers** small; inconspicuous; staminate and pistillate similar; perianths in 1-series. **Sepals** 3–6; fused. **Petals** absent. **Stamens** 2–6; filaments fused at bases. **Pistils** with carpels 3; styles 3, divided. **Fruits** capsular schizocarps; subglobose; surfaces hispid, with stiff stinging hairs. **Seeds** 3; brownish black or mottled.

1. Plants of extreme southern Florida and Keys. Leaf blades suborbicular . . . . *T. saxicola*
1. Plants of south-central states. Leaf blades ovate or triangular or narrowly lanceolate or oblong.
    2. Leaf blades ovate or triangular. Calyx lobes of pistillate flowers longer than pistils at anthesis . . . . *T. betonicifolia*
    2. Leaf blades narrowly lanceolate to oblong. Calyx lobes of pistillate flowers shorter than pistils at anthesis . . . . *T. ramosa*

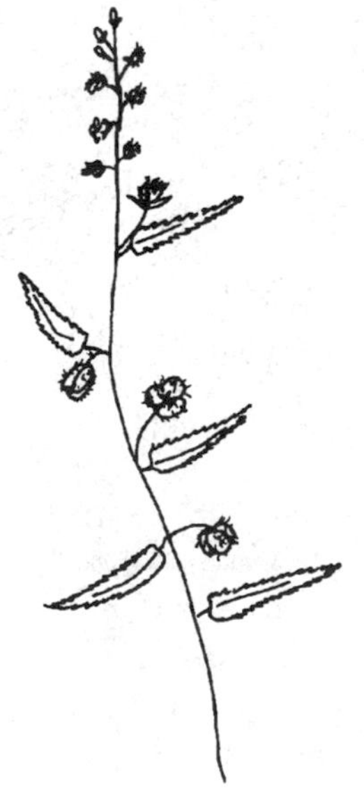

Tragia ramosa

Distribution and Habitat—The genus is most abundant in the southern half of the continent. *Tragia betonicifolia* is found in sandy, open woods. Also in open sites, especially in grasslands, *T. ramosa* occupies the southwestern corner of the continent. *Tragia saxicola* is endemic to the pinewoods, typically on limestone, of extreme southern Florida and the Keys (Miller & Webster 1967).

Disease Problems—Frequently overlooked until touched and a burning sensation announces their presence, species of *Tragia* are inconspicuous in appearance. Their stinging hairs are unique in structure, differing from those of other genera such as

Cnidoscolus and *Urtica*. The stinging hair lacks a pedestal and has no bulbous tip. Instead there is a complex structure composed of a primary stinging cell up to 0.4 mm long, with a central vacuole containing a proteinaceous substance and surrounded by three shorter lateral accessory cells. These complex structures are found on the stems, leaves, and fruits. On contact with the skin, the wall of the primary cell is pushed back, exposing a slender, pointed calcium oxalate crystal in the tip. This crystal is extruded on contact, and it penetrates the skin and allows the cellular contents to enter the wound (Thurston & Lersten 1969; Thurston 1976).

■ sandy soils

Tragia betonicifolia

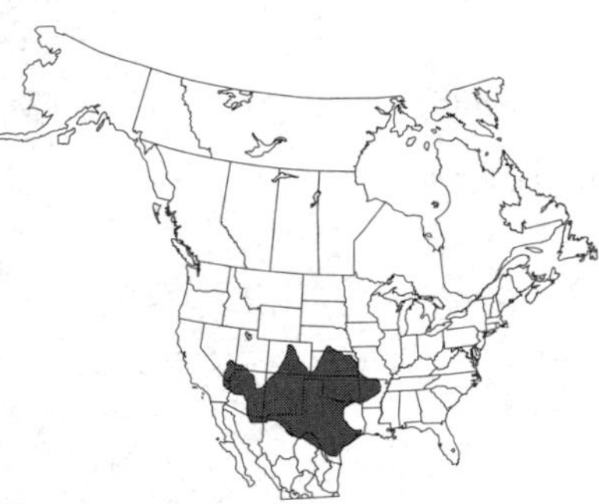

Tragia ramosa

Tragia saxicola

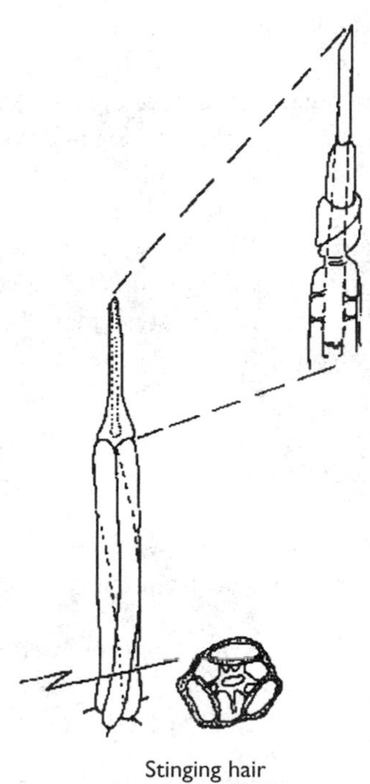

Stinging hair
(after Thurston 1976)

Disease Genesis—The identity of the toxicant is unknown. The proteinaceous substance of the central vacuole appears to act as or to contain the toxicant (Thurston 1976). Although the morphology of *Tragia*'s stinging hair is different from the morphologies of *Cnidoscolus* and *Urtica,* the transient burning sensation is similar, which suggests that a serotonin-like compound is the likely agent, as has been proposed for the other two genera (Lookadoo & Pollard 1991). There also may be quantitative differences in the compounds present, because the reaction elicited by *Tragia* is typically less pronounced than that of *Urtica.*

■ contact with stinging hairs, mild skin irritation; unique hair structure

■ chemical from hairs, serotonin-like; transient burning sensation

Clinical Signs—Contact with the stinging hairs of *Tragia* characteristically, and particularly in humans, produces erythema, swelling, pain, and pruritis at the contact site. These signs typically persist for a few minutes, although in especially sensitive individuals they may last an hour or more. Systemic signs are very unusual.

■ skin, erythema, swelling, pain, itching; few minutes' duration

Treatment—If administered early, atropine and antihistamines may be of value. Otherwise, treatment is symptomatic and generally not needed.

■ treatment not needed

Genera of Questionable Toxicity

Codiaeum A.Juss.

- *C. variegatum*
  croton

Comprising 6 species native to Southeast Asia and the Pacific Islands, *Codiaeum* is commonly known as croton. This common name also is applied to species of the genus *Croton,* another member of the family. Only 1 introduced *Codiaeum* species is of toxicologic interest in North America:

C. variegatum (L.) A.Juss. croton
var. *pictum* (Lodd.) Meull.Arg.

- erect, evergreen shrubs or trees; leaves simple, alternate, leathery, variegated white and yellow, typically turning red with age
- ornamentals
- mildly irritant effects

Codiaeum variegatum is a perennial, evergreen shrub or tree with thin, colorless sap and simple, alternate, thick, leathery leaves. Variegated white and yellow, the leaves typically turn red with age. Grown for its brightly colored foliage, *C. variegatum* is planted as screening hedges or borders in the South and as tub plants elsewhere.

Lacking the white latex sap typical of most members of the family, species of *Codiaeum* are only mildly irritating, although allergic reactions may occur (Mitchell & Rook 1979). Experimentally, up to 20 g of fresh plant/kg b.w. given in single doses to sheep and cattle caused no adverse effects (Armien & Tokarnia 1994; Tokarnia et al. 1996).

References

Adolph W, Hecker E: On the active principles of the spurge family. 10. skin irritants, cocarcinogens, and cryptic cocarcinogens from the latex of the manchineel tree. *J Nat Prod* 47;482–496, 1984

Adolph W, Opferkuch HJ, Hecker E: Irritant phorbol derivatives from four *Jatropha* species. *Phytochemistry* 23;129–132, 1984

Ahmed OMM, Adam SEI: Effects of *Jatropha curcas* on calves. *Vet Path* 16;476–482, 1979a

Ahmed OMM, Adam SEI: Toxicity of *Jatropha curcas* in sheep and goats. *Res Vet Sci* 27;89–96, 1979b

Akintonwa A: Acute cyanide poisoning attributed to *Manihot esculenta* [abstract]. *J Toxicol Clin Toxicol* 34;599, 1996

Albin RC, Harbaugh FG, Zinn DW: Castorbean meal of three ricin levels for cattle [abstract]. *J Anim Sci* 27;288, 1968

Armien AG, Tokarnia CH: Experiments on the toxicity of some ornamental plants in sheep. *Pesq Vet Bras* 14;69–73, 1994

Bajaj R, Chang CJ, McLaughlin JL, Powell RG, Smith CR Jr: Tiliroside from the seeds of *Eremocarpus setigerus. J Nat Prod* 49;1174–1175, 1986

Bakke AL: Leafy spurge, *Euphorbia esula* L. *Iowa Agric Exp Stn Res Bull* 198;207–246, 1936

Balagopalan C, Padmaja G, Nanda SK, Moorthy SN: *Cassava in Food, Feed, and Industry.* CRC Press, Boca Raton, Fla, 1988

Balint GA: Ricin: the toxic protein of castor oil seeds. *Toxicology* 2;77–102, 1974

Balthrop E, Gallagher WB, McDonald TF, Camariotes S: Tung nut poisoning. *J Fla Med Assoc* 40;813–820, 1953

Bassir O, Fafunso M: Effect of pre-cooking processing on the cyanide contents of the leaves of eight cultivars of the cassava plant *Manihot esculenta. Plant Foods Man* 2;91–94, 1976

Bizzeti M, Lucia PG de, Corazza M, Santini PP: Experimental *Mercurialis annua* poisoning in sheep. *Ann Fac Med Vet Pisa* 60;165–175, 1987

Blohm H: *Poisonous Plants of Venezuela.* Harvard Univ Press, Cambridge, pp47–61, 1962

Blumberg BS, Millman I, Venkateswaran PS, Thyagarajan SP: Hepatitis B virus and hepatocellular carcinoma—treatment of HBV carriers with *Phyllanthus amarus. Cancer Detect Prev* 14;195–201, 1989

Borys DJ, Setzer SC, Hornfeldt CS: A retrospective study of plant exposures reported to the Hennepin Regional Poison Control Center in 1985. *Vet Hum Toxicol* 29;83–85, 1987

Boughton IB, Hardy WT: Feeding trials of suspected plants. *Texas Agric Exp Stn Annu Rep* 52;239–240, 1939

Boughton IB, Hardy WT: Feeding trials of suspected plants. *Texas Agric Exp Stn Annu Rep* 53;236–237, 1940

Carlson RB, Mundal D: Introduction of insects for the biological control of leafy spurge in North Dakota. *N D Farm Res Bull* 47(6);7–9, 1990

Carroll MN Jr, Fox LE, Ariail WT: Investigation of the toxic principles of *Hippomane mancinella* L. 3. Toxic actions of extracts of *Hippomane mancinella* L. *J Am Pharm Assoc* 46;93–97, 1957

Case AA: Photosensitization syndrome in cattle, sheep, and swine. *North Am Vet* 38; 161–165, 1957

Challoner KR, McCarron MM: Castor bean intoxication. *Ann Emerg Med* 19;1177–1183, 1990

Chesnut VK: *Principal Poisonous Plants of the United States.* USDA Div Bot Bull 20, 1898

Christiansen VJ, Hsu C-H, Dormer KJ, Robinson CP: The cardiovascular effects of ricin in rabbits. *Pharmacol Toxicol* 74;148–152, 1994

Clarke ECG: Poisoning by castor seed. *Vet J* 103;273–278, 1947

Connolly JD, Hill RA: *Dictionary of Terpenoids.* Vol 2, *Di- and Higher Terpenoids.* Chapman & Hall, London, pp1277–1279, 1991

DeBusk F: Plants of importance in north and west Florida. *J Fla Med Assoc* 65;205–206, 1978

de Fiebre C: Two convicted of possessing deadly poison. *Minneap Star-Trib,* 1 March, p1B, 1995

Devendra C: Cassava as a feed source for ruminants. In *Cassava as Animal Feed,* Cassava as Animal Feed Workshop Proceedings, Nestel B, Graham M eds, International Development Research Centre, Ontario, Canada, pp107–119, 1977

Dirheimer G, Haas F, Metais P: Modifications biochimiques du sang au cours de l'intoxication du rat par la ricine. *C R Soc Biol (Paris)* 160;2458, 1966

Dirheimer G, Haas F, Metais P: Hepatonephrite experimentale provoquee par la ricine. In *Hepatonephrites Toxiques,* Toxicol Med Ser B, Roche L ed, Masson et Cie, Paris, pp45–50, 1968

Dobereiner J, Tokarnia CH, Canella CFC: Experimental poisoning of cattle by the pericarp of the fruit of *Ricinus communis. Pesq Vet Bras* 1;95–97, 1981

Dollahite JW: *Research and Observations on Toxic Plants in Texas—1932–1972.* Tex Agric Exp Stn Tech Rep 78–1, 1978

Dunn PH: The distribution of leafy spurge *(Euphorbia esula)* and other weedy *Euphorbia* spp. in the United States. *Weed Sci* 27;509–516, 1979

Ebeling W: *Handbook of Indian Foods and Fibers of Arid America.* Univ California Press, Berkeley, p242, 1986

Emmel MW: The toxic principle of the species *Aleurites. J Am Vet Med Assoc* 111;386–387, 1947

Emmel MW, Sanders DA, Swanson LE: The toxicity of *Aleurites fordi* for cattle. *J Am Vet Med Assoc* 101;136–137, 1942

Endo Y, Tsurugi K: RNA *N*-glycosidase activity of ricin A-chain. *J Biol Chem* 262;8128–8130, 1987

Endo Y, Matsui K, Motizuki M, Tsurugi K: The mechanism of action of ricin and related toxic lectins on eukaryotic ribosomes. *J Biol Chem* 262;5908–5912, 1987

Erickson JLE, Brown JH: A study of the toxic properties of tung-nuts. *J Pharmacol Exp Ther* 74;114–117, 1942

Evans FJ, Soper CJ: The tigliane, daphnane, and ingenane diterpenes: their chemistry, distribution, and biological activities. *Lloydia* 41;193–237, 1978

Fox D, Kirby D, Lym RG, Caton J, Rabbenhoft K: Chemical composition of leafy spurge and alfalfa. *N D Farm Res Bull* 48(6);7–9, 1991

Foxwell B, Detre S, Donova T, Thorpe P: The use of anti-ricin antibodies to protect mice intoxicated with ricin. *Toxicology* 95;79–88, 1985

Frakes RA, Sharma RP, Willhite CC: Developmental toxicity of the cyanogenic glycoside linamarin in the golden hamster. *Teratology* 31;241–246, 1985

Frankel AE, Burbage C, Fu T, Tagge E, Chandler J, Willingham MC: Ricin toxin contains at least three galactose-binding sites located in B chain subdomains 1 alpha, 1 beta, and 2 gamma. *Biochemistry* 35;14749–14756, 1996

Frohne D, Pfander HJ: *A Colour Atlas of Poisonous Plants.* Wolfe Publ, London, 1984

Fuller G, Walker HG Jr, Mottolo AC, Kuzinicki DD, Kohler GO, Vohra P: Potential for detoxified castor meal. *J Am Oil Chem Soc* 48;616–618, 1971

Gasperi-Campani A, Barbieri L, Morelli P, Stirpe F: Seed extracts inhibiting protein synthesis in vitro. *Biochem J* 186;439–441, 1980

Geary T: Castor bean poisoning. *Vet Rec* 32;472–473, 1950

Godal A, Fodstad O, Pihl A: Antibody formation against the cytotoxic proteins abrin and ricin in humans and mice. *Int J Cancer* 32;515–521, 1983

Gomez G, Valdivieso M: Cassava foliage: chemical composition, cyanide content, and effect of drying on cyanide elimination. *J Sci Food Agric* 36;433–441, 1985

Hahn SK, Reynolds L, Egbunike GN: *Cassava as Livestock Feed in Africa.* International Institute of Tropical Agriculture, Ibadan, Nigeria, pp149–154, 1992

Hanson T, Kirby D, Hull CS, Potts L: Biological control of leafy spurge using Angora goats. Leafy Spurge Symposium, Lincoln, Nebraska, July 1992, Great Plains Agric Counc Publ 144, pp2–53/2–54, 1992

Hassoun EA, Bagchi D, Roche VF, Stohs SJ: An assessment of potential chemoprotectant activity against ricin toxicity by mechanism based glycosidase inhibitors in macrophage J744A.1 cell cultures. *Toxicon* 30;1545–1554, 1992

Hickey TA, Worobec SM, West DP, Kinghorn AD: Irritant contact dermatitis in humans from phorbol and related esters. *Toxicon* 19;841–850, 1981

Hiroto M, Ohigashi H, Koshimizu K: Piscicidal constituents and related diterpene esters from *Aleurites fordii. Agric Biol Chem* 43;2523–2529, 1979

Ho RKB: Acute poisoning from the ingestion of seeds of *Jatropha curcas. Hawaii Med J* 19;421–423, 1960

Hohmann J, Vasas A, Gunther G, Mathe I, Evanics F, Dombi G, Jerkovich G: Macrocyclic diterpene polyesters of the jatrophane type from *Euphorbia esula. J Nat Prod* 60;331–335, 1997

Holmes RL, Rayner ET: Isolation of two nitrogen-free toxins from tung kernels. *J Am Oil Chem Soc* 35;586–589, 1958

Hornfeldt CS: Confusion over toxicity of poinsettia [letter]. *J Am Vet Med Assoc* 194;1004, 1989

Ishiguro M, Mitarai M, Harada H, Sekine I, Nishimori I, Kikutani M: Biochemical studies on oral toxicity of ricin. 1. Ricin administered orally can impair sugar absorption by rat small intestine. *Chem Pharm Bull* 31;3222–3227, 1983

Ishiguro M, Harada H, Ichiki O, Sekine I, Nishimori I, Kikutani M: Biochemical studies on oral toxicity of ricin. 2. Effects of ricin, a protein toxin, on glucose absorption by rat small intestine. *Chem Pharm Bull* 32;3141–3147, 1984

Ishiguro M, Nakashima H, Tanabe S, Sakakibara R: Interaction of toxic lectin ricin with epithelial cells of rat small intestine in vitro. *Chem Pharm Bull* 40;441–445, 1992a

Ishiguro M, Tanabe S, Matori Y, Sakakibara R: Biochemical studies on oral toxicity of ricin. 4. A fate of orally administered ricin in rats. *J Pharmacobio-Dyn* 15;147–156, 1992b

Ishiguro M, Matori Y, Tanabe S, Kawase Y, Sekine I, Sakakibara R: Biochemical studies on oral toxicity of ricin. 5. The role of lectin activity in the intestinal absorption of ricin. *Chem Pharm Bull* 40;1216–1220, 1992c

Jenkins FP: Allergenic and toxic components of castor bean meal: review of the literature and studies of the inactivation of these components. *J Sci Food Agric* 14;773–780, 1963

Jensen WI, Allen JP: Naturally occurring and experimentally induced castor bean *(Ricinus communis)* poisoning in ducks. *Avian Dis* 25;184–194, 1981

Johnston A, Peake RW: Effect of selective grazing by sheep on the control of leafy spurge (*Euphorbia esula* L.). *J Range Manage* 12;192–195, 1960

Jolad SD, Hoffman JJ, Schram KH, Cole JR, Tempesta MS, Bates RB: Constituents of *Eremcarpus setigerus* (Euphorbiaceae): a new diterpene, eremone, and hautriwaic acid. *J Org Chem* 47;1356–1358, 1982

Joubert PH, Brown JMM, Hay IT, Sebata PDB: Acute poisoning with *Jatropha curcas* (purging nut) in children. *S Afr Med J* 65;729–730, 1984

Katzin BJ, Collins EJ, Robertus JD: Structure of ricin A-chain at 2.5 A. *Proteins* 10;251–259, 1991

Kellerman TS, Coetzer JAW, Naude TW: *Plant Poisonings and Mycotoxicoses of Livestock in Southern Africa.* Oxford Univ Press, Cape Town, South Africa, pp220–221, 1988

Kinamore PA, Jaeger RW, de Castro FJ: *Abrus* and *Ricinus* ingestion: management of three cases. *Clin Toxicol* 17;401–405, 1980

Kinghorn AD: Major skin-irritant principle from *Synadenium grantii. J Pharm Sci* 69;1446–1447, 1980

Kinghorn AD: Skin-irritant and tumor promoting compounds of plants of the Euphorbiaceae. In *Plant Toxicology—Proceedings of the Australia–USA Poisonous Plant Symposium,* Seawright AA, Hegarty MP, James LF, Keeler RF eds, Animal Research Institute, Yeerongpilly, Australia, pp357–366, 1985

Kingsbury JM: *Poisonous Plants of the United States and Canada.* Prentice-Hall, Englewood Cliffs, NJ, 1964

Knight B: Ricin—a potent homicidal poison. *Br Med J* 1;350–351, 1979

Krenzelok EP, Jacobsen TD, Aronis JM: Poinsettia exposures have good outcomes . . . just as we thought. *Am J Emerg Med* 14;671–674, 1996

Lambert JM, McIntyre G, Ganthier MN, Zullo D, Rao V, Steeves RM, Goldmacher VS, Blattler WA: The galactose-binding sites of the cytotoxic lectin ricin can be chemically blocked in high yield with reactive ligands prepared by chemical modification of glycopeptides containing triantennary *N*-linked oligosaccharides. *Biochemistry* 30;3234–3247, 1991

Lampe KF, Fagerstrom R: *Plant Toxicity and Dermatitis.* Williams & Wilkins, Baltimore, 1968

Landau M, Egyed MN, Flesh D: *Mercurialis annua* poisoning in housed sheep. *Refu Vet* 30;131–135, 1973

Landgraf BK, Fay PK, Havstad KM: Utilization of leafy spurge *(Euphorbia esula)* by sheep. *Weed Sci* 32;348–352, 1984

Lauter WM, Foote PA: Investigation of the toxic principles of Hippomane mancinella l. 2. Preliminary isolation of a toxic principle of the fruit. *J Am Pharm Assoc* 44;361–363, 1955

Lauter WM, Fox LE, Ariail WT: Investigation of the toxic principles of *Hippomane mancinella* L. 1. Historical review. *J Am Pharm Assoc* 41;199–201, 1952

Lemley PV, Creasia DA, Plesha M, Walters D, O'Neill A: Ricin sub-unit vaccination in mice and protection from challenge [abstract]. *Toxicon* 32;545, 1994

Lin TJ, Liao HJ, Lin WL, Sa JH, Hung DZ, Deng JF: An outbreak of *Aleurites fordii* poisoning [abstract]. *Toxicon* 32;545, 1994

Lin TJ, Hsu CI, Lee KH, Shui LL, Deng JF: Two outbreaks of acute tung nut *(Aleurites fordii)* poisoning. *Clin Toxicol* 34;87–92, 1996

Little EL Jr, Wadsworth FH: *Common Trees of Puerto Rico and the Virgin Islands.* USDA Forest Service, Agric Handb 240, 1964

Liu S-Q, Pezzuto JM, Kinghorn AD: Additional biologically active constituents of the Chinese tallow tree *(Sapium sebiferum). J Nat Prod* 51;619–620, 1988

Lookadoo SE, Pollard AJ: Chemical constituents of stinging trichromes of *Cnidoscolus texanus. J Chem Ecol* 17;1909–1916, 1991

Lord JM, Roberts LM, Robertus JD: Ricin: structure, mode of action, and some current applications. *FASEB J* 8;201–208, 1994

Lym RG, Messersmith CG: A summary of leafy spurge control with herbicides in North Dakota since 1963. *N D Farm Res Bull* 43;3–6, 1985

Lym RG, Sedivec KK, Kirby DR: Leafy spurge control with angora goats and herbicides. *J Range Manage* 50;123–128, 1997

Ma L, Hsu C-H, Patterson E, Thadani U, Robinson CP: Ricin depresses cardiac function in the rabbit. *Toxicol Appl Pharmacol* 138;72–76, 1996

MacKenzie AD, Phillips CI: West Indian amblyopia. *Brain* 91;249–260, 1968

Makalinao IR: A descriptive study on the clinical profile of *Jatropha* seed poisoning [abstract 55]. *Vet Hum Toxicol* 35;330, 1993

Man held for poison possession kills self. *Stillwater News Press,* 25 December 1995

Marion L: The pyridine alkaloids. In *The Alkaloids,* vol 1, Manske RHF, Holmes HL eds, Academic Press, New York, pp165–269, 1950

Mathews FP: The toxicity of a spurge *(Phyllanthus abnormis). Cornell Vet* 35;336–346, 1945

McVaugh R: The jatrophas of Cervantes and of the Sesse and Mocino Herbarium. *Bull Torrey Bot Club* 72;31–41, 1945

Mensah JL, Gleye J, Moulis C, Fouraste I: Alkaloids from the leaves of *Phyllanthus discoideus. J Nat Prod* 51;1113–1115, 1988

Messersmith CG, Lym RG: Leafy spurge control: 10 years of research enhancement. *N D Farm Res Bull* 47(6);3–6, 1990

Miller KI, Webster GL: A preliminary revision of *Tragia* (Euphorbiaceae) in the United States. *Rhodora* 69;241–305, 1967

Mitchell J, Rook A: *Botanical Dermatology.* Greengrass, Vancouver, BC, pp261–302, 1979

Montgomery RD: The medical significance of cyanogen in plant foodstuffs. *Am J Clin Nutr* 17;103–113, 1965

Morino H, Sakakibara R, Ishiguro M: The binding of ricin to its receptor is not required for the expression of its toxicity. *Biol Pharm Bull* 18;1770–1772, 1995

Morton JF: *500 Plants of South Florida.* Seeman Publ, Miami, 1974

Morton JF: *Plants Poisonous to People in Florida and Other Warm Areas.* JF Morton, Miami, 1982

Moshkin VA: Economic importance and regions of cultivation of castor. In *Castor,* Moshkin VA ed, Russ Transl Ser 43, AA Balkema Publ, Rotterdam, pp1–5, 1986a

Moshkin VA: History and origin of castor. In *Castor,* Moshkin VA ed, Russ Transl Ser 43, AA Balkema Publ, Rotterdam, pp6–10, 1986b

Muenscher WC: *Leafy Spurge and Related Weeds.* Cornell Ext Bull 192, 1930

Muldoon DF, Stohs SJ: Modulation of ricin toxicity in mice by biologically active substances. *J Appl Toxicol* 14;81–86, 1994

Natori S, Ikekawa N, Suzuki M: *Advances in Natural Product Chemistry.* Halsted Press, New York, pp327–340, 1981

Nielsen PE: Steroids from *Euphorbia* and other latex-bearing plants. *Phytochemistry* 18;103–104, 1979

Oakes AJ, Butcher JO: *Poisonous and Injurious Plants of the Virgin Islands.* USDA Misc Publ 882, 1962

Ohigashi H, Ohtsuka T, Hirota M, Koshimizu K, Tokuda H, Ito Y: Tigliane type diterpene-esters with Epstein-Barr virus-inducing activity from *Sapium sebiferum. Agric Biol Chem* 47;1617–1622, 1983

Okuda T, Yoshida T, Koike S, Toh N: The toxic constituent of the fruits of *Aleurites fordi. Chem Pharm Bull* 22;971–972, 1974

Okuda T, Yoshida T, Koike S, Toh N: New diterpene esters from *Aleurites fordii* fruits. *Phytochemistry* 14;509–515, 1975

Olsnes S: Abrin and ricin: two toxic lectins inactivating eukaryotic ribosomes. In *Perspectives in Toxinology,* Bernheimer AW ed, J Wiley & Sons, New York, pp123–147, 1977

Olsnes S, Pihl A: Abrin and ricin—two toxic lectins. *Trends Biochem Sci* 3;7–10, 1978
Olsnes S, Pihl A: Toxic lectins and related proteins. In *Molecular Action of Toxins and Viruses,* Cohen P, van Heyningen S eds, Elsevier Biomed, Amsterdam, pp51–105, 1982
Olson BE, Wallander RT, Thomas VM, Kott RW: Effect of previous experience on sheep grazing leafy spurge. *Appl Anim Behav Sci* 50;161–176, 1996
Olson BE, Wallander RT, Kott RW: Recovery of leafy spurge seed from sheep. *J Range Manage* 50;10–15, 1997
Omole TA: The use of cassava for feeding rabbits. In *Cassava as Livestock Feed in Africa,* Hahn SK, Reynolds L, Egbunike GN eds, International Institute of Tropical Agriculture, Ibadan, Nigeria, pp58–71, 1992
Osuntokun BO: An ataxic neuropathy in Nigeria: a clinical, biochemical, and electrophysiological study. *Brain* 91;215–248, 1968
Palmer HE: Are the seeds of the tung oil tree poisonous when eaten by man or animal? *J Fla Med Assoc* 20;13–15, 1933
Pettit GR, Cragg GM, Suffness MI, Gust D, Boettner FE, Williams M, Saenz-Renauld JA, Brown P, Schmidt JM, Ellis PD: Antineoplastic agents. 104. Isolation and structure of the *Phyllanthus acuminatus* Vahl (Euphorbiaceae) glycosides. *J Org Chem* 49;4258–4266, 1984
Pettit GR, Cragg GM, Suffness M: Phyllanthostatin 1-phyllanthoside orthoacid rearrangement. *J Org Chem* 50;5060–5063, 1985
Pettit GR, Schaufelberger DE, Nieman RA, Dufresne C, Saenz-Renauld JA: Antineoplastic agents. 177. Isolation and structure of phyllanthostatin 6. *J Nat Prod* 53;1406–1413, 1990
Pitts JF, Barker NH, Gibbons DC, Jay JL: Manchineel keratoconjunctivitis. *Br J Ophthalmol* 77;284–288, 1993
Pollard AJ: Variation in *Cnidoscolus texanus* in relation to herbivory. *Oecologia* 70;411–413, 1986
Pullar EM: Studies on five suspected poisonous plants. *Aust Vet J* 15;19–23, 1939
Rao KV: Toxic principles of *Hippomane mancinella. Planta Med* 25;166–171, 1974
Rao KV: Toxic principles of *Hippomane mancinella.* 2. Structure of hippomanin A. *Lloydia* 40;169–172, 1977
Rauber A, Heard J: Castor bean toxicity re-examined: a new perspective. *Vet Hum Toxicol* 27;498–502, 1985
Remnick D: KGB plot to assassinate Solzhenitsyn reported. *Wash Post,* 21 April, pD1, 1992
Reuters News Service: Ex-KGB aide is questioned in poison umbrella killing. *N Y Times,* 1 November, pA3, 1993
Reynolds R, Irvin JD: A survey of selected plants for the presence of eukaryotic protein biosynthesis inhibitors. *Tex J Sci* 32;55–57, 1980
Ribeiro RL, De Mello N, Ferraz AC, Oliveira BH, Da Cunha C: Effect of the alkaloid ricin upon the binding of the [3H]-flunitrazepam to the central benzodiazepine receptor [abstract]. *J Venom Anim Toxins* 3;200, 1997
Rich V: Murderous experiments of Stalin's police chief. *New Sci* 135;8, 1992
Robertus JD, Yan X, Ernst S, Monzingo A, Worley S, Day P, Hollis T, Svinth M: Structural analysis of ricin and implications for inhibitor design. *Toxicon* 34;1325–1334, 1996
Rogers DJ: A revision of *Stillingia* in the New World. *Ann Mo Bot Gard* 38;207–259, 1951
Runyon R: Toxicity of fresh poinsettia *(Euphorbia pulcherrima)* to Sprague-Dawley rats. *Clin Toxicol* 16;167–173, 1980
Russell LH, Schwartz WL, Dollahite JW: Toxicity of Chinese tallow tree *(Sapium sebiferum)* for ruminants. *Am J Vet Res* 30;1233–1238, 1969
Rutenber E, Robertus JD: Structure of ricin B-chain at 2.5 A resolution. *Proteins* 10;260–269, 1991
Rutenber E, Katzlin BJ, Ernst S, Collins EJ, Mlsna D, Ready MP, Robertus JD: Crystallographic refinement of ricin to 2.5 A. *Proteins* 10;240–250, 1991

Sakata K, Kawazu K, Mitsui T: Studies on a piscicidal constituent of *Hura crepitans*. 2. Chemical structure of huratoxin. *Agric Biol Chem* 35;2113–2126, 1971

Samia MA, El Badwi, Adam SEI, Hapke HJ: Toxic effects of low levels of dietary *Jatropha curcas* seed on Brown Hisex chicks. *Vet Hum Toxicol* 34;112–115, 1992

Sampson AW, Malmsten HE: *Stock-Poisoning Plants of California*. Calif Agric Exp Bull 593, 1942 (reprint)

Sanders DA, Emmel MW, Swanson LE: *Tung Tree* Aleurites fordi *Hemsl. Foliage Poisoning of Cattle*. Fla Agric Exp Stn Bull 376, 1942

Sandvig K, Van Deurs B: Endocytosis, intracellular transport, and cytotoxic action of Shiga toxin and ricin. *Physiol Rev* 76;949–966, 1996

Schaeffer HJ, Lauter WM, Foote PA: A preliminary phytochemical study of *Hippomane mancinella* L. *J Am Pharm Assoc* 43;43–45, 1954

Seigler DS: The naturally occurring cyanogenic glycosides. In *Progress in Phytochemistry*, Reinhold L, Harbone JB, Swain T eds, Pergamon Press, Oxford, pp83–120, 1977

Sekine I, Kawase Y, Nishimori I, Mitarai M, Harada H, Ishiguro M, Kikutani M: Pathological study on mucosal changes in small intestine of rat by oral administration of ricin. 1. Microscopic observation. *Acta Pathol Japan* 36;1205–1212, 1986

Selleck GW, Coupland RT, Frankton C: Leafy spurge in Saskatchewan. *Ecol Monogr* 32;1–29, 1962

Sendil C: Studies on the clinical findings and therapy of experimentally induced *Mercurialis annua* poisoning in sheep and calves. *Vet Fak Derg* 25;480–499, 1978 (*Vet Bull* 49;6875, 1979)

Senf W, Seffner W: *Mercurialis annua* poisoning in sheep. *Monatsh Veterinarmed* 20;622–625, 1965 (*Vet Bull* 36;320, 1966)

Singh JD: The teratogenic effects of dietary cassava on the pregnant albino rat: a preliminary report. *Teratology* 24;289–291, 1981

Soifer FK: Tallow tree poisoning in a monkey. *J Zoo Anim Med* 4;13, 1973

Sosath S, Ott HH, Hecker E: Irritant principles of the spurge family (Euphorbiaceae). 13. Oligocyclic and macrocyclic diterpene esters from latices of some *Euphorbia* species utilized as source plants of honey. *J Nat Prod* 51;1062–1074, 1988

Sperry OE, Dollahite JW, Hoffman GO, Camp BJ: *Texas Plants Poisonous to Livestock*. Tex Agric Exp Stn Bull B-1028, 1977

Sperti S, Montanaro L, Mattioli A, Testoni G, Stirpe F: Inhibition of protein synthesis in vitro by crotins and ricin. *Biochem J* 156;7–13, 1976

Spoerke DG, Smolinske SC: *Toxicity of Houseplants*. CRC Press, Boca Raton, Fla, 1990

Spoerke DG, Temple AR: Creeping spurge—*Euphorbia myrsinites*. *Am J Dis Child* 132;28, 1979

Spoerke DG, Montanio CD, Rumack BH: Pediatric exposure to the houseplant *Synadenium grantii*. *Vet Hum Toxicol* 28;283–284, 1985

Spyker DA, Sauer K, Kell SQ, Guerrant RL: A castor bean poisoning and a widely available bioassay for ricin [abstract]. *Vet Hum Toxicol* 24;293, 1982

Steyn DG: *The Toxicology of Plants in South Africa*. Central News Agency, South Africa, pp297–303, 1934

Stirpe F, Pession-Brizzi A, Lorenzoni E, Strocchi P, Montanaro L, Sperti S: Studies on the proteins from the seeds of *Croton tiglium* and of *Jatropha curcas*. *Biochem J* 156;1–6, 1976

Stone RP, Collins WJ: *Euphorbia pulcherrima*: toxicity to rats. *Toxicon* 9;301–302, 1971

Szallasi A, Blumberg PM: Resiniferatoxin, a phorbol-related diterpene, acts as an ultrapotent analog of capsaicin, the irritant constituent in red pepper. *Neuroscience* 30;515–520, 1989

Szolcsanyi J, Szallasi A, Szallasi Z, Joo F, Blumberg PM: An ultrapotent neurotoxin of capsaicin-sensitive primary afferent neurons. *Ann N Y Acad Sci* 632;473–475, 1991

Tehon LR, Morrill CC, Graham R: *Illinois Plants Poisonous to livestock*. Ill Agric Exp Stn Bull 599, 1946

Tewe OO, Egbunike GN: Utilization of cassava in nonruminant livestock feeds. In *Cassava as Livestock Feed in Africa,* Hahn SK, Reynolds L, Egbunike GN eds, International Institute of Tropical Agriculture, Ibadan, Nigeria, pp28–38, 1992

Thurston EL: Morphology, fine structure, and ontogeny of the stinging emergence of *Tragia ramosa* and *T. saxicola* (Euphorbiaceae). *Am J Bot* 63;710–718, 1976

Thurston EL, Lersten NR: The morphology and toxicology of plant stinging hairs. *Bot Rev* 35;393–412, 1969

Tokarnia CH, Dobereiner J, da Silva MF: *Plantas Toxicas da Amazonia.* Instituto Nacional de Pesquisas da Amazonia, Manaus, Brazil, pp58–62, 1979

Tokarnia CH, Armien AG, Peixoto PV, Barbosa JD, Brito M de F, Dobereiner J: Experiments on the toxicity of some ornamental plants in cattle. *Pesq Vet Bras* 16;15–20, 1996

Tomlinson PB: *The Biology of Trees Native to Tropical Florida.* Harvard Univ Printing Office, Allston, Mass, 1980

Turner R: *Euphorbias.* BT Batsford Publ, London, 1995

Verdcourt B, Trump EC: *Common Poisonous Plants of East Africa.* Collins, London, 1969

Vitetta ES, Fulton RL, May RD, Till M, Uhr JW: Redesigning nature's poisons to create anti-tumor reagents. *Science* 238;1098–1104, 1987

Walker J, Kronberg S: Managing leafy spurge with livestock. Leafy Spurge Symposium, Lincoln, Nebraska, July 1992, Great Plains Agric Council Publ 144, pp4–23/4–28, 1992

Walker JW, Hemenway KG, Hatfield PG, Glimp HA: Training lambs to be weed eaters: studies with leafy spurge. *J Range Manage* 45;245–249, 1992

Watson PJ: Suspected dog's mercury *(Mercurialis perennis)* poisoning in cattle. *Vet Rec* 142;116–117, 1998

Webster GL: A revision of *Phyllanthus* (Euphorbiaceae) in the continental United States. *Brittonia* 22;44–76, 1970

Webster GL, Miller KI: The genus *Reverchonia* (Euphorbiaceae). *Rhodora* 65;193–207, 1963

Wedin GP, Neal JS, Everson GW, Krenzelok EP: Castor bean poisoning. *Am J Emerg Med* 4;259–261, 1986

Welchman DdeB, Gibbens JC, Giles N, Piercy DWT, Skinner PH: Suspected annual mercury *(Mercurialis annua)* poisoning of lambs grazing fallow arable land. *Vet Rec* 137;592–593, 1995

Willaman JJ, Schubert BG: *Alkaloid-Bearing Plants and Their Contained Alkaloids.* USDA Tech Bull 1234, 1961

Winek CL, Butala J, Shanor SP, Fochtman FW: Toxicology of poinsettia. *Clin Toxicol* 13;27–45, 1978

Winter J, Dray A, Wood JN, Yeats JC, Bevan S: Cellular mechanism of action of resiniferatoxin: a potent sensory neuron excitotoxin. *Brain Res* 520;131–140, 1990

Youle RJ: Ribosome inactivating enzymes in castor bean and man. In 5th Pan-American Symposium, International Society on Toxinology, Fredrick, Md, August 1995

Zhang L, Hsu C-H, Robinson CP: Effects of ricin administration to rabbits on the ability of their coronary arteries to contract and relax in vitro. *Toxicol Appl Pharmacol* 129;16– 22, 1994

37 Fabaceae Lindl.

■ Legume Family
Pea Family
Bean Family

Abrus
Acacia
Albizia
Astragalus
Baptisia
Caesalpinia
Canavalia
Cassia
Chamaecrista
Coronilla
Crotalaria
Cytisus
Erythrina
Galega
Genista
Glottidium
Gymnocladus
Indigofera
Laburnum
Lathyrus
Leucaena
Lotus
Lupinus
Medicago
Melilotus
Oxytropis
Pisum
Prosopis
Robinia
Senna
Sesbania
Sophora
Spartium
Thermopsis
Trifolium
Vicia
Wisteria

QUESTIONABLE
Desmanthus
Mimosa
Mucuna
Pachyrhizus
Phaseolus
Psoralea
Tephrosia

One of the largest families of flowering plants, the Fabaceae comprises some 700 genera and 17,000 species of trees, shrubs, and herbs occupying a variety of habitats. Cosmopolitan in distribution, the family also is economically one of the most important, second only to the grass family. The fruits and seeds of many taxa are sources of protein-rich foods both for humans and domesticated animals, e.g., the numerous species and varieties of peas, beans, soybeans, and peanuts. The clovers, lespedezas, and alfalfas provide forage. Woody species, especially in the tropics, provide lumber, and many have been taken into cultivation as ornamentals, including redbud, honey locust, black locust, and silk tree. Aggressive weeds such as kudzu and some of the clovers also are members of the family. Symbiotic nitrogen fixation by almost all taxa is especially important ecologically, often determining the rate of succession in prairies.

Genera of the Fabaceae are traditionally grouped by taxonomists in 3 subfamilies on the basis of differences in leaf division, inflorescence type, and floral morphology as illustrated in the following key.

1. Corollas radially symmetrical. Filaments of stamens exserted beyond petals .... subfamily Mimosoideae (Mimosaceae)
1. Corollas bilaterally symmetrical. Filaments of stamens included within petals.
    2. Flowers strongly bilaterally symmetrical. Corollas papilionaceous. Upper (adaxial) petal enclosing other petals in bud .... subfamily Papilionoideae (Fabaceae)
    2. Flowers weakly bilaterally symmetrical. Corollas not papilionaceous. Upper (adaxial) petal enclosed by the petals in bud .... subfamily Caesalpinioideae (Caesalpiniaceae)

In recent years, however, there has been a tendency to recognize the 3 subfamilies as separate families in one order. Regardless of the classification scheme, the legumes as a whole form a morphologically well-defined group. The traditional approach of 1 family and 3 subfamilies is employed in this treatise. Typically encountered in the older literature, an alternative family name is Leguminosae. It reflects the characteristic fruit of the family.

■ herbs, vines, shrubs, or trees; leaves compound or rarely simple, stipulate; flowers typically papilionaceous; fruits legumes

Plants herbs or vines or shrubs or trees; annuals or perennials; caulescent or nearly acaulescent; armed or not armed with spines or prickles; with or without tendrils. **Stems** erect or decumbent or trailing or twining. **Leaves** simple or palmately or 1- or 2-pinnately compound; alternate; venation pinnate; leaflets 3 to numerous, terminal leaflet absent or present; margins entire or serrate or pinnately lobed; stipules present, fused to petioles or free. **Inflorescences** spikes or racemes or heads or umbels or solitary flowers; terminal or axillary; bracts present or absent; bracteoles present or absent. **Flowers** perfect; perianths in 2-series. **Calyces** bilaterally or radially symmetrical. **Sepals** 5 or appearing 4 because of fusion of lobes. **Corollas** bilaterally or radially symmetrical; papilionaceous or not papilionaceous. **Petals** 5 or 1; all alike or of 3 forms—upper (adaxial, termed the standard or banner), 2 lateral (termed the wings) free or fused to keel, and 2 lower (termed the keel) fused distally and enclosing stamens and pistil. **Androecia** radially or bilaterally symmetrical. **Stamens** 10 or 5 or numerous; exserted or included within keels; fused by filaments or free; diadelphous or monadelphous; anthers all alike or of 2 forms. **Pistils** 1; simple; carpels 1; sessile or stipitate; stigmas 1, capitate or linear; styles 1; ovaries superior; locules 1; placentation parietal. **Hypanthia** present; small; cup-shaped. **Nectaries** absent or present, staminal or receptacular. **Fruits** legumes. **Seeds** 1 to numerous.

Intoxication problems produced by members of the Fabaceae are as varied in their nature as the species that compose the family. They can, however, be grouped into six general categories: gastrointestinal irritation, cyanogenic, neurotoxic, teratogenic, photosensitization, and hemorrhagic.

■ problems—
digestive irritation
cyanogenesis
neurotoxicosis,
teratogenesis
photosensitization
hemorrhage

Abrus Adans.

Taxonomy and Morphology—A member of the subfamily Papilionoideae and comprising 4 or 5 species widely distributed in the tropics, the genus is represented in North America by this introduced species:

A. precatorius L	rosary pea, prayer bean, crab's-eye, crab's stone, gidee-gidee, guinea pea, Indian licorice, love bean, lucky bean, jequirity bean, jumble beads, Seminole bead, ojo de pájaro, eye-of-the-bird

■ *A. precatorius*

crab's-eye	jumble beads
crab's stone	love bean
eye-of-the-bird	lucky bean
gidee-gidee	ojo de pájaro
guinea pea	prayer bean
Indian licorice	rosary pea
jequirity bean	Seminole bead

The variety of colloquial names of the species reflects the appearance and uses of its seeds, as does its scientific name. The specific epithet *precatorius* comes from the Latin *precator,* which means "one who prays" and reflects the use of the beans in rosaries.

Plants woody vines; without tendrils. **Stems** twining; slender. **Leaves** 1-pinnately compound; pairs of leaflets 8–15; terminal leaflet absent; leaflets oblong, apices apiculate, margins entire. **Inflorescences** racemes; 1–3 cm long; dense; borne on axillary peduncles. **Corollas** papilionaceous. **Petals** lavender-pink to pale reddish purple or rarely white; the standards enclosing the wings and keels. **Stamens** 9; included within keels; filaments fused; monadelphous. **Legumes** flat; broad; beaked; pubescent; more or less septate between seeds. **Seeds** 4–8; ovoid; shiny; bright red with black spot at one end.

■ twining vines; leaves 1-pinnately compound, lacking tendrils; flowers pealike; petals lavender-pink, reddish purple, or white; legumes flat, broad; seeds distinctive scarlet with black spot

Abrus precatorius

Distribution and Habitat—*Abrus precatorius* is native to India but has been introduced widely throughout the warmer regions of the world because of its attractiveness. In North America, it is an ornamental and is also an escape in Florida and other more tropical regions.

Disease Problems—Contrary to one of its common names, *A. precatorius* does not bring good luck if it is eaten. It has a reputation as one of the most toxic plants in the world. Its brilliantly scarlet and black seeds contain the toxicant and pose the

■ seeds commonly encountered, many uses; jewelry, rosaries, rattles, charms

hazard because they are used in variety of ways, e.g., in rattles, rosaries, jewelry, clothing ornaments, and good luck charms (Sullivan & Chavez 1981). The seeds formerly were imported from tropical America, Africa, and India but are now banned because of their toxicity. This ban has not eliminated intoxication problems. The greatest hazard is for children and pets, which may encounter the seeds in the home.

■ ingestion of well-chewed seeds may cause severe digestive disturbance; risk is limited by hard seed covering; seeds difficult to chew and more likely to be swallowed intact; generally risk is for humans and pets

Ingestion of the seeds typically results in adverse effects on the gastrointestinal tract when they are chewed enough to break the seed coats. All animals are susceptible, with the lethal dose ranging from about 100 mg/kg b.w. in the horse to 600 mg/kg in cattle and more than 2 g/kg in goats (Chopra et al. 1984; Tokarnia et al. 1979; Barri et al. 1990). Death of children eating 1 or 2 seeds has been reported (Morton 1982). Although digestive problems are most likely, tetany, flushed appearance to the skin, and pupillary dilation were reported in 2 children who chewed the seeds (Gunsolus 1955). In addition, repeated ingestion of sublethal doses may produce detrimental effects on the fetus; malformations and deaths have been demonstrated experimentally in mice (El-Shabrawy et al. 1987). Because of its reputation as an abortifacient plant in India, aqueous extracts of the seeds were evaluated in rats and found to cause resorption of half of the fetuses from 7 rats and several developmental anomalies (cleft palate, limb clubbing, and joint defects) in nonresorbed fetuses (Sethi et al. 1990; Nath et al. 1992). Toxicologic interactions between *Abrus* and other plants that produce digestive disturbances, e.g., *Cassia,* also represent a potential problem in other parts of the world when their seeds contaminate livestock feed (Omer et al. 1992).

■ toxicity reduced by heating seeds

Abrus precatorius's nefarious reputation as one of the most toxic plants, however, does not appear to be deserved. The seed coat is very hard and difficult to rupture when chewed. An intact seed appears to produce no toxic effects. When more than 20 seeds were ingested by a 15-month-old child and passed intact in his stools for several days, there were no adverse effects (Swanson-Biearman et al. 1992). In fact, in societies such as that of the Andaman Islands of India, the seeds are even boiled and eaten as a food source (Rajaram & Janardhanan 1992a). Other parts of the plant pose little or no problem. The root is used as a licorice substitute, and the foliage as a sweetener because of the presence of several triterpene glycosides termed abrusides (Choi et al. 1989).

In addition, it has been observed that substantial doses of *Abrus* in ruminants may cause little effect other than a transient increase in body temperature (Simpson & Banerjee 1932). Likewise, particularly large doses, several grams or many seeds, are required to produce adverse effects in dogs (Simpson & Banerjee 1932). As with castor bean *(Ricinus),* the risk by ingestion may be overstated because very few exposures have resulted in clinical problems (Kinamore et al. 1980).

■ toxins proteins called abrins, nearly identical with ricins from castor bean; highly toxic i.v., i.m., and i.p., much less orally

Disease Genesis—Early studies (Martin 1887; Ghatak & Kaul 1932) indicated the toxicity of *Abrus* to be due to one or more proteins, although other compounds, including the alkaloid-like abrine or precatorine, are present. Later workers extracted several proteins from the cotyledons, some of which were discovered to be potent hemagglutinins (Misra et al. 1968; Humphreys 1969). It is now recognized that the toxins of *Abrus* are type-2 ribosomal-inhibiting proteins very similar to ricin (Olsnes 1977; Olsnes & Pihl 1978, 1982; Hegde et al. 1991; Citores et al. 1993). These compounds and their isoforms are called abrins. They are of similar size (63,000–64,000 daltons) and consist of two polypeptide chains, designated A and B, joined by a single disulfide bond. Abrins range in potency from an LD_{50} of 0.06 μg/kg for abrin II to 0.55 μg/kg for abrin I given i.p. in mice (Hegde et al. 1991). The overall LD_{50} is in the range of 0.1–0.2 μg/kg b.w. There is approximately 0.5 mg abrins per gram of seed.

- abrin made up of A and B chains; A chain inactivates ribosomes, inhibiting protein synthesis; B chain facilitates A chain entry into cells; effects mainly seen as digestive disturbance

The A chain, with a molecular weight of 30,000, inhibits initiation and elongation of peptide chains by inactivating the 60S eukaryotic ribosomal subunit. This is accomplished by hydrolysis of an adenine from the 28S rRNA. The B chain, with a molecular weight of 35,000, binds to complex carbohydrates on the cell surface, facilitating entry of the A chain into the cell (Olsnes & Pihl 1982). Its mode of action is essentially the same as that of ricin, and the molecules are so similar that the A and B chains may substitute for one another, the B chain from ricin serving to facilitate cell entry of the A chain from abrin (Olsnes 1977). *Abrus precatorius* also contains larger 4-chained agglutinins that have limited RIP activity, but they are much less toxic because the B chains are ineffective in facilitating cellular entry of the A chains (Hegde et al. 1991; Citores et al. 1993).

Although the effects of abrin are not cell specific, rapidly growing and dividing cells are most likely to be affected. This includes the cells of the mucosal surface of the digestive tract. Signs of intoxication are delayed because of slow absorption of abrin and its indirect action, which destroys cells through inhibition of synthesis of vital proteins.

Tolerance to the adverse effects of *Abrus* has been noted in ponies given increasing daily doses (Simpson & Banerjee 1932). This observation is in agreement with reports (Chopra et al. 1940; Tokarnia et al. 1979) that farmers in India, and later in Brazil, prevented intoxications by feeding their animals increasing amounts of the plant prior to allowing them to forage in areas where it was abundant.

Because the toxin is a protein, its hazard is much reduced when ingested, because of poor absorption. As with ricin, the oral toxic dose is 100 times or more greater than the parenteral dose. When given parenterally, it is extremely potent and may perhaps be exploited, like ricin, in schemes of commercial intrigue. Hide merchants in India are said to have secretly impaled abrin-coated wooden spikes in the necks of cows to cause death indistinguishable from that caused by clostridial disease (Rahman & Salam Mia 1972). The supply of hides was thereby increased in a society not given to butchering cattle.

Clinical Signs—Six to 8 hours or more after ingestion of *Abrus,* signs related primarily to irritation and dysfunction of the digestive tract develop. They include excess salivation, vomiting, abdominal pain, bloat, watery diarrhea (often bloody), and dehydration. In addition, depression and respiratory distress may be observed. There may be an elevation of body temperature to 104°F and, just before death, hyperexcitability and convulsive seizures.

- abrupt onset of excess salivation, colic, vomiting, severe diarrhea, dehydration

Pathology—Grossly, the mucosa of the digestive tract from the stomach to the rectum may show reddening, with splotchy hemorrhages and ulceration. The mucosal surface of the urinary bladder may present a similar appearance. Pulmonary edema and emphysema and congestion of various organs also are common. Retinal hemorrhage is less often noted. Microscopic lesions will confirm the gross changes; they consist of vascular hyperemia, surface lesions with ulcerations, and hemorrhage. Mild to moderate hepatic and renal tubular degeneration are typical.

- gross pathology, severe reddening, hemorrhage, ulceration, mucosa of stomach and intestines

Treatment—Symptomatic alleviation of dehydration with fluid and electrolytes is the main treatment. This may be augmented by orally administered activated charcoal. Laxatives or purgatives may be useful as a preventive in instances of known ingestion, prior to development of clinical signs.

- activated charcoal, fluids and electrolytes

Acacia Mill.

- *A. berlandieri*
  A. constricta
  A. farnesiana
  A. greggii
  A. rigidula
  A. roemeriana
  A. schaffneri
  A. tortuosa

  acacia
  blackbrush
  catclaw
  guajillo
  huisache
  huisachillo
  uña de gato

Taxonomy and Morphology—*Acacia,* a large genus of some 700–1,200 species in the subfamily Mimosoideae and commonly known as mimosa or wattle, is most abundant in the tropical and subtropical regions of the world. Occupying a diversity of habitats from coastal beaches to arid deserts, it comprises trees, shrubs, herbs, and vines. Of the approximately 46 species present in North America, only a few, most notably *A. berlandieri,* are of clear toxicologic concern. Species of interest include the following:

A. berlandieri Benth.	guajillo
A. constricta Benth.	mescat acacia, whitethorn acacia
A. farnesiana (L.) Willd.	sweet acacia, huisache
A. greggii A.Gray	catclaw acacia, devil's catclaw, Gregg's catclaw, uña de gato
A. rigidula Benth.	blackbrush, chaparro prieto
A. roemeriana Scheele	catclaw, Roemer's acacia
A. schaffneri (S.Watson) F.J.Herm.	twisted acacia
A. tortuosa (L.) Willd.	huisachillo, twisted acacia

- shrubs and trees; leaves 2-pinnately compound, terminal leaflet absent; petals yellow, white, or orange; legumes irregularly constricted

Plants trees or shrubs or rarely herbs; deciduous; perennials; sometimes rhizomatous and forming clones; armed or not armed with thorns or spines. **Leaves** 2-pinnately compound; pairs of leaflets many per pinna; terminal leaflet absent; leaflets linear or oblong, folding when touched; stipules herbaceous or forming woody spines.

Acacia constricta

Acacia farnesiana

Acacia greggii

Inflorescences spikes; globose or cylindrical; borne in terminal or axillary racemes or panicles. **Flowers** small; symmetry radial. **Sepals** 4 or 5; fused. **Petals** 4 or 5; fused at bases; yellow or white or orange. **Stamens** many; conspicuous; filaments free or fused at bases, elongated, exserted beyond petals. **Legumes** flat or terete; irregularly constricted; straight or curved; glabrous to pubescent. **Seeds** flat; orbicular.

Species of *Acacia* are, in general, fast-growing, deciduous, loosely branched, armed shrubs or trees, with bipinnately compound leaves. They are distinguished primarily by differences in the appearance of their leaves.

- ▪ wide range of habitats; deserts to woodlands

1. Inflorescences elongate spikes; 2–8 cm long; 0.5–1 cm wide.
    2. Spines present; stipular; straight. Prickles absent. Legumes terete or subterete; 5–7 mm in diameter . . . . *A. rigidula*
    2. Spines absent. Prickles recurved; internodal. Legumes flat; 15–20 mm wide . . . . *A. greggii*
1. Inflorescences globose or subglobose heads; 1 cm in diameter.
    3. Stipules not spinose.
        4. Leaflets 20–50 pairs per pinna. Stems bearing straight prickles . . . . *A. berlandieri*
        4. Leaflets 4–15 pairs per pinna. Stems bearing curved prickles or prickles absent.
            5. Pinnae 4–7. Petals yellow. Stems not bearing curved prickles . . . . *A. constricta*
            5. Pinnae 1–3. Petals cream white. Stems bearing curved prickles . . . . *A. roemeriana*
    3. Stipules spinose.
        6. Bracts borne near middle of peduncles . . . . *A. constricta*
        6. Bracts borne at apices of peduncles.
            7. Plants large shrubs or small trees; 200–400 cm tall. Gland borne near middle of petiole or absent. Legumes glabrous . . . . *A. farnesiana*
            7. Plants small shrubs; 50–150 cm tall. Gland borne between lowest pinnae. Legumes velvety-pubescent.
                8. Pinnae 5–8 pairs. Peduncles 10–40 mm long . . . . . *A. tortuosa*
                8. Pinnae 2–3 pairs. Peduncles 2–3 mm long . . . . . . *A. schaffneri*

Acacia berlandieri

Acacia constricta

Distribution and Habitat—Species of *Acacia* are most commonly encountered in the tropical and subtropical regions of the world, and in Australia it is the largest genus of flowering plants (Isely 1969). Many produce a profusion of fragrant yellow inflorescences and are cultivated as ornamentals. In North America, the native acacias are considered to be northern outliers of tropical American species (Isely 1969). Locally very abundant on limestone ridges and plains, *A. berlandieri* is distributed throughout southern Texas and adjacent Mexico. *Acacia constricta* is found on the plains of the Southwest. From either the Old World or Central America, *A. farnesiana* now ranges through much of Mexico. *Acacia greggii* is a species of the deserts and forests of the Southwest. *Acacia roemeriana* and *A. schaffneri* (originally called *A. tortuosa*) are found in brushy sites in southwestern Texas and adjacent Mexico. A West Indian species, the morphologically very similar *A. tortuosa* is now believed to occur in only a few localities in southern Florida (Isely 1969).

Acacia farnesiana

Disease Problems—Species of *Acacia* are valued for food for both humans and livestock (Felger 1977; Coppock 1993). The seeds are made into a mush or meal, and during times of drought, the leaves and pods are used extensively for livestock forage. Most species are an important source of protein; e.g., *A. berlandieri* is a valued

Acacia greggii

- pods and leaves source of food for humans and livestock; prolonged subsistence of livestock on *Acacia* may result in locomotor and reproductive problems; foliage may be cyanogenic

forage with up to 17% crude protein in the pods and 16% in the leaves. However, despite its value, extensive reliance on *Acacia* for forage may lead to reproductive problems. As an example, cattle maintained on semiarid rangeland in Texas dominated by *A. berlandieri* had poor reproductive performance relative to matched herd mates on grass pastures (Stec et al. 1992).

Of the many species in the genus, only a few cause overt intoxication problems, of which there are two types: a neurotoxic condition known as guajillo wobbles, and cyanogenesis.

Acacia rigidula

Acacia roemeriana

Acacia schaffneri

- ruminants subsisting on leaves and pods of *A. berlandieri* for 6–9 months or more may develop peculiar gait, guajillo wobbles or limberleg; reversible

Guajillo Wobbles

Disease Problems—Characteristically producing a peculiar movement of the pelvic limbs in a sideways direction, the disease guajillo wobbles, or limberleg, is associated primarily with ingestion of *A. berlandieri.* As noted above, this species is a valued forage because of its high protein content. Intoxication primarily occurs during dry periods when grass and other forages are not available and animals are required to subsist almost exclusively on its leaves and pods for periods of 6–9 months (Price & Hardy 1953). Losses are typically low but increase as the severity of the drought increases. If animals are moved to good forages before clinical signs become severe, the disease is generally reversible. Goats are at high risk because they are browsers and, even in good years, may forage extensively on these shrubs or trees.

The disease was produced experimentally when sheep or goats were fed *A. berlandieri* for 8–10 months, given only alfalfa for several months, and then fed *Acacia* again for several months (Boughton & Hardy 1938, 1941). Treatment of this type of intoxication with vitamins A, D, and B complex for 25 days was without apparent value.

Disease Genesis—In North American species of *Acacia,* an extensive array of amines and alkaloids have been identified, e.g., 44 in *A. rigidula* and 50 in *A. berlandieri* (Clement & Forbes 1998; Clement et al. 1998). This array of compounds included amines of known physiologic activity (Smith 1977). These same compounds also are found in many other plant genera, especially those of the Cactaceae. In addition, alkaloids such as nicotine, nornicotine, and mescaline have been identified in the foliage (Clement & Forbes 1998; Clement et al. 1998), and willardine, a heterocyclic excitatory amino acid, has been isolated from the seeds (Patneau et al. 1992).

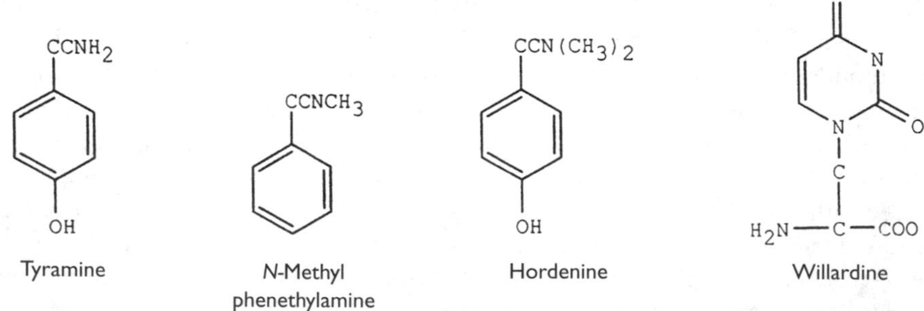

Tyramine N-Methyl phenethylamine Hordenine Willardine

The sympathomimetic amines *N*-methyltyramine, tyramine, *N*-methyl-phenethylamine (NMP), phenethylamine, and hordenine are abundant in *A. berlandieri* and *A. rigidula.* Concentrations in the leaves of *A. berlandieri* range from 0.66% d.w. in May to 0.28–0.46% in September and October (Camp & Lyman 1956; Camp & Moore 1960; Camp et al. 1964; Adams & Camp 1966; Pemberton et al. 1993; Clement

& Forbes 1998). Other species contain these amines, but in such low concentrations (<0.1% d.w.) that they are unlikely to cause problems (Camp & Norvell 1966). The amines cause stimulation of smooth muscle and cardiovascular pressor effects, resulting in increased blood pressure, pupillary dilation, and decreased urinary bladder and uterine (cat) muscle tone (Barger & Dale 1910; Hjort 1934). In rats, toxicosis resulted in piloerection, increased respiration and heart rates, apprehension, and, at higher dosage, bulging of the eyes, rigidity of the tail, and a tendency to move backward (Camp et al. 1964). Stimulation of the release of norepinephrine is believed to be at least partly responsible for the pressor effects (Evans et al. 1979). The toxicological significance of these amines is not clear, because they have not been used to reproduce the disease. Their effects are transient, and although effects may be pronounced immediately after administration, their long-term importance is questionable (Camp 1970).

- toxins appear to be extensive array of sympathetic amines, tyramine, phenethylamine, and hordenine; amino acid willardine also present in seeds
- amines, long-term effects unclear, role in disease unconfirmed, may be responsible for decreased reproductive efficiency

These sympathetic amines also may cause reproductive problems. Vera-Avila and co-workers (1996) have reviewed the complex interactions associated with increasing consumption of plants containing high amounts of phenolic amines during times of nutritional stress, and although clear adverse effects are not always evident, there is good reason to anticipate problems under some circumstances. These effects are shown in experiments with the most abundant of the amines, NMP. Female Angora goats given NMP i.v. for 45 days, in dosage simulating ingestion in the field, experienced a consistent decline in serum progesterone and decreased pregnancy rates (Forbes et al. 1993). Similarly, male Angora goats fed *A. berlandieri* and *A. rigidula* had a reduction in testicular and adrenal size and function (Vera-Avila et al. 1996).

Willardine is a non-NMDA glutamate receptor agonist acting on kainate and to a lesser extent on AMPA receptor subtypes in the cerebrum (Evans et al. 1980; Patneau et al. 1992; Hawkins et al. 1995). If willardine causes desensitization of kainate receptors, it could be of toxicologic significance with long-term ingestion. Its role, however, has not been explored.

Clinical Signs, Pathology, and Treatment—After consumption of *Acacia* for many months, there is loss of weight, but otherwise the animal appears alert and continues to have a reasonable appetite. The affected animal develops a peculiar and unique ataxia, as if its pelvic limbs and joints were made of rubber. This results in a wobbly gait in which it moves laterally (sideways)—hence the disease name *guajillo wobbles*. The problem is accentuated with forced movement, and the animal is reluctant to move and may fall if forced to do so (Price & Hardy 1953). There is also a proprioceptive deficit in the thoracic limbs. In spite of the distinctiveness of the neurologic effects, pathologic changes are minimal, and they are generally reversible if the animal is immediately fed a high-quality, acacia-free diet.

- gradual onset of weight loss, wobbly ataxia, difficulty in placement of thoracic limbs; reversible

Cyanogenesis

Disease Problems—Surveys of the cyanogenic potential of species of *Acacia* have revealed more than 90 species in both the Old World and the New World that produce cyanogenic compounds (Watt & Breyer-Bradwijk 1962; Gupta & Nauriyal 1966; Vihan & Panwar 1987; Seigler et al. 1989; Burrows et al. 1992). New foliage is especially prone to have high cyanogenic potential. In North America, even though all 8 species listed in the taxonomy section are considered cyanogenic (Seigler et al. 1989), few clinical problems have been associated with them, despite their use as forage in times of drought. The risk of intoxication seems to increase after the first frosts in the fall (Schmutz et al. 1968).

- occasional occurrences, abrupt cyanide intoxication, especially with new foliage or in autumn after first frosts

- cyanogenic glycosides, linamarin, and others

```
                  C≡N
                 /
Glucose-O—C—CH3
                 \
                  CH3
```

Linamarin

Disease Genesis—At least nine cyanogenic compounds are produced by species of *Acacia,* including four that are unique to the genus—proacacipetalin, epiproacacipetalin, proacaiberin, and 3-hydroxyheterodendrin (Secor et al. 1987; Seigler et al. 1989). In addition, prunasin, linamarin, lotaustralin, and sambunigrin are present. Unfortunately, little is known about the hazards they pose in North American species, other than the observation that instances of intoxication are rare. A possible explanation of this may be the absence of β-glucosidases in the plant tissues. These enzymes, which play an important role in hydrolysis of the glycosides and release of cyanide, are apparently lacking in many of the Australian species of the genus (Finnemore & Gledhill 1928a,b; Conn & Maslin 1982; Conn et al. 1985).

- abrupt onset of weakness, apprehension, rapid respiration, recumbency, seizures, and death, all within a few minutes

  treat with sodium thiosulfate with or without sodium nitrite

Clinical Signs, Pathology, and Treatment—Again, little is known about the species of *Acacia* in North America because of the small number of reported poisonings. Signs of intoxication are those characteristic of cyanide poisoning and include a sequence of signs progressing rapidly from polypnea and apprehension to severe weakness, recumbency, convulsive seizures, and dyspnea to coma and death. Likewise, treatment is the same as for intoxications due to other cyanogenic plants, including use of the classical antidotes, sodium nitrite and sodium thiosulfate. Interestingly, there is poor therapeutic response to treatment in cases of intoxication by an Asiatic species, *A. leucophloea* (Vihan & Panwar 1987).

Albizia Durazz.

- *A. julibrissin*

  mimosa tree
  pink siris
  silk tree

Taxonomy and Morphology—Comprising approximately 150 species of trees, shrubs, or woody vines in the subfamily Mimosoideae, *Albizia* is a genus of tropical and subtropical regions and occurs in savannas and along sandy riverbeds. In North America, 8 species have been introduced as ornamentals. Most commonly encountered is the following:

A. julibrissin Durazz. silk tree, mimosa tree, pink siris

The common name *mimosa* also is often used for *A. julibrissin,* which unfortunately suggests a relationship to species of the genus *Mimosa.* The generic name is often misspelled as *Albizzia.*

- small spreading trees or shrubs; leaves 2-pinnately compound, no terminal leaflet; petals 4 or 5, pink or white; legumes thin-walled, flat

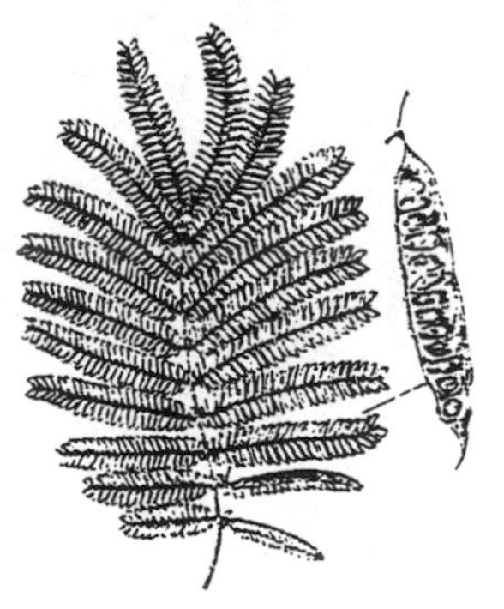

Albizia julibrissin

Plants small trees to 9 m tall or shrubs; deciduous; bark smooth, pale gray-brown; branches spreading, prone to splitting with age; crown broad, flattened. **Leaves** 2-pinnately compound; large, appearing lacy or fernlike; pairs of pinnae 5–18; pairs of leaflets 20–30 per pinna; terminal leaflet absent; leaflets falcate oblong, strongly asymmetrical, folding when touched, apices mucronate; stipules herbaceous. **Inflorescences** spikes; globose; borne in terminal racemes. **Flowers** small; symmetry radial. **Sepals** 4 or 5; fused. **Petals** 4 or 5; fused at bases; pink or white. **Stamens** 20–40; conspicuous; filaments fused at bases; elongate; exserted well beyond petals. **Legumes** flat; 1 cm wide; 10–20 cm long; green and then brown at maturity; thin-walled, the seeds causing bulges; glabrous. **Seeds** flattened; ovate-oblong; brown.

Distribution and Habitat—A native of southern Asia and occurring from Iran to Japan, *A. julibrissin* was introduced into North America as an ornamental yard tree. A prolific seed producer, it readily escapes cultivation and is now naturalized in the southeastern quarter of the continent. The plants typically are short-lived. Flowering occurs from June until August.

Disease Problems—The actual hazard posed by *A. julibrissin* is not fully known, because reports of intoxications have been relatively few. However, neurotoxic effects have been produced experimentally in sheep fed the legumes (Dollahite 1978; Robinson et al. 1998). Acute neurotoxicosis occurred when 1–1.5% b.w. of the dried ground legumes were administered to sheep. The lethal dose is 1.5–2% b.w. of the dry, mature legumes. The intoxication risk due to ingestion of leaves or of repeated ingestion of small amounts of the legumes is not known.

■ naturalized ornamental

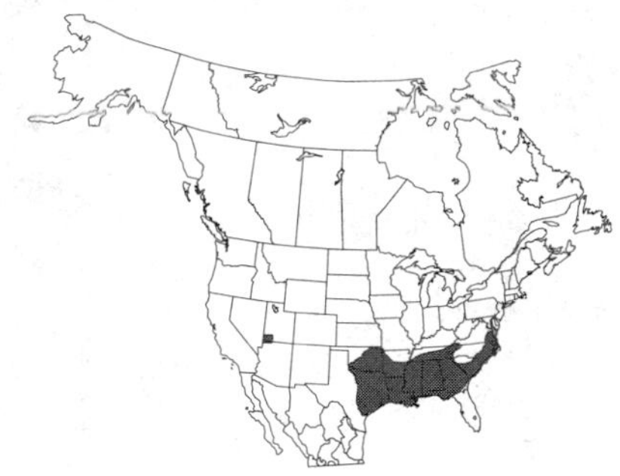

Albizia julibrissin

Although experimental work on *A. julibrissin* in North America is limited, there is considerable information on *A. tanganyicensis* and *A. versicolor,* two species of importance in southern Africa (Needham & Lawrence 1966; Basson et al. 1970; Soldan et al. 1996). Ingestion of the legumes of these species by cattle, sheep, and goats causes hyperthermia, hyperesthesia, and clonic/tetanic seizures at a dosage of approximately 0.5–0.6 kg legumes/100 kg b.w. There is a degenerative cardiac and skeletal myopathy with accompanying elevation of serum creatinine phosphokinase. Interestingly, in contrast to these three species with toxic effects, some tropical species of *Albizia* are useful as livestock feed (Lowry 1989; Dwatmadgi 1992).

■ livestock, legumes, neurotoxicosis with 1–2% b.w. in sheep; other plant species well known as toxic in South Africa

Disease Genesis—The toxins of the African species are alkaloids, such as 4-methoxypyridoxine, which are structural analogs of pyridoxine, or vitamin B_6. Their adverse effects can be antagonized readily with two administrations of 20–25 mg pyridoxine HCl given parenterally 8 hours apart (Steyn et al. 1987; Gummow et al. 1992). *Albizia julibrissin* contains similar toxins, and its toxic effects, likewise, are ameliorated by pyridoxine (Robinson et al. 1998). Sheep administered up to 2% b.w. of the dried legumes, normally a lethal dose, were protected from adverse affects with simultaneous parenteral administration of pyridoxine HCl.

■ toxic alkaloids, structural analogs of pyridoxine; mechanism of action unknown but seizures may result from interference with GABA synthesis

4-Methoxypyridoxine

$H_2NC—C—C—COOH$

γ-Aminobutyric acid (GABA)

The specific mode of action of the toxins is not fully understood, but pyridoxine antagonists such as 4-deoxypyridoxol and 4-methoxymethylpyridoxol HCl are well-recognized causes of seizures in laboratory animals (Holtz & Palm 1964; Kosower & Rock 1968; Sawaya et al. 1978). Pyridoxine serves as a cofactor with glutamic acid decarboxylase in the formation of GABA. It also serves as a cofactor with GABA transaminase to catalyze breakdown of GABA to succinic acid (Dakshinamurti 1977). This has led to the suggestion that the seizures produced are due to the impairment of the synthesis of this inhibitory pathway neurotransmitter (Horton & Meldrum 1973). In mice, administration of 4-deoxypyridoxine results in a decrease in glutamic acid decarboxylase and GABA in the brain (Sawaya et al. 1978). However, there also is a decrease in L-dopa decarboxylase associated with a decrease in dopa formation. Furthermore, 4-deoxypyridoxine causes a decrease in GABA transport, which is antagonized by pyridoxal phosphate (Snodgrass & Iversen 1973). The antagonists also impair activity of GABA transaminase and the degradation of GABA. Thus the toxins may cause effects at several points. The synthesis of biogenic amines appears not to play a role in the acute toxicity of these pyridoxal antagonists.

Clinical Signs—Following ingestion of large amounts of legumes in a short time, the onset of signs occurs in 12–24 hours. Initially, there is an exaggerated response to tactile, auditory, and visual stimuli. This is closely followed by the onset of muscular twitching and tremors and then seizures characterized by deep, labored respiration, apparent excessive salivation or foaminess due to jaw movements, backing up or turning, bizarre neck postures, including opisthotonus, and collapse with extended thoracic limbs and paddling pelvic limbs. The seizures may last 1–2 minutes and may occur again in several minutes.

■ abrupt onset of hyperesthesia, tremors, excess salivation, repeated seizures

■ treat with pyridoxine HCl

Pathology and Treatment—No distinctive gross or microscopic lesions are associated with ingestion of *A. julibrissin.* In experimental studies (Robinson et al. 1998), the disease responds readily to parenteral administration of 10–20 mg/kg b.w. of pyridoxine HCl. Several days may be required for complete recovery, and a second dose of pyridoxine 8–12 hours after the first may be helpful.

Astragalus L.

■ *A. bisulcatus*
A. canadensis
A. emoryanus
A. lentiginosus
A. miser
A. mollissimus
A. pattersonii
A. praelongus
A. wootonii
garbancillo
loco
locoweed
milk vetch
peavine

Taxonomy and Morphology—Comprising approximately 2,000 species, *Astragalus,* a member of the subfamily Papilionoideae, is the largest genus of the Fabaceae. Its members are commonly referred to as milk vetches or locoweeds. The genus name was used by Dioscorides and is derived from the classical Greek word *astragalos,* meaning "anklebone." This bone was used in ancient times as a form of dice. One conjecture is that the name was applied because the rattle of the seeds in the dry legumes may have resembled the sound of dice in a thrower (Barneby 1964a). It also has been supposed that the name reflects the similarity of the seed or rootstock shape to the anklebone. With greatest abundance in the northern temperate zone, *Astragalus* species are of minor economic importance, with the exception of the intoxication problems they cause. Some are a source of tragacanth gum, which is used in cosmetics and pharmaceuticals, and a few are cultivated as ornamentals. In North America, the genus is represented by 340–375 species and about 250 varieties (Barneby 1964a,b). Of these numerous taxa, several are particularly important toxicologically. Species representative of the various disease problems produced by *Astragalus* include the following (see plate 37.1):

■ annual or perennial herbs; leaves 1-pinnately or rarely palmately compound, terminal leaflet present; flowers pealike; petals 5, white, yellowish, pink, or violet; legumes sometimes inflated

A. bisulcatus (Hook.) A.Gray	two-grooved milk vetch
A. canadensis L.	Canada milk vetch
A. emoryanus (Rydb.) Cory	Emory's milk vetch, red-stemmed peavine
A. lentiginosus Douglas ex Hook.	speckled loco, freckled milk vetch
A. miser Douglas	timber milk vetch, weedy milk vetch
A. mollissimus Torr.	woolly loco, purple loco, woolly locoweed
A. pattersonii A.Gray	Patterson's milk vetch
A. praelongus E.Sheld.	stinking milk vetch
A. wootonii E.Sheld.	Wooton's locoweed, garbancillo

Astragalus bisulcatus

Plants annuals or perennials; herbaceous; caulescent or acaulescent. **Stems** decumbent or ascending to erect. **Leaves** 1-pinnately or rarely palmately compound; pairs of leaflets 1–20; terminal leaflet present; petiolules present; stipules herbaceous, well developed or forming sheaths. **Inflorescences** elongate or capitate racemes borne at ends of peduncles arising in axils of leaves; bracts present. **Sepals** 5; fused. **Corollas** papilionaceous. **Petals** 5; white or yellowish or pink or violet; banners oblanceolate, reflexed; wings clawed, auriculate; keels rounded, not abruptly beaked. **Stamens** 10; diadelphous; enclosed in keel. **Legumes** of various shapes and sizes; 1- or 2-loculed. **Seeds** 1 to many.

Astragalus is morphologically quite similar to the closely related *Oxytropis,* also known as locoweed or crazyweed. Several species of the latter were originally described as species of *Astragalus.* The two genera differ in the nature of their keels. In *Astragalus* the blades of the keel petals are incurved to form a rounded, deltoid or triangular apex, whereas in *Oxytropis* the blades of the keel petals are abruptly contracted into an erect or recurved subulate beak or conical point.

The species of *Astragalus* are morphologically quite similar and difficult to identify, especially in the absence of mature fruits, which are the primary basis for differentiation (Barneby 1964a,b). The nature of the indumentum, the appearance of the stipules, and the presence of a stipe or gynophore below the mature legume are also used.

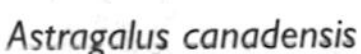

Astragalus canadensis

Astragalus miser

Astragalus mollissimus

1. Plants annuals.
    2. Legumes inflated, bladderlike; obliquely ovoid or semiovoid. Seeds not filling locule.
        3. Racemes with 2–10 flowers. Banners 4.5–7.5 mm long. Keels 4–6.5 mm long. Bracteoles 2 . . . . *A. wootonii*
        3. Racemes with 10–40 flowers. Banners 7.5–22 mm long. Keels 6.5–16.5 mm long. Bracteoles 0 . . . . *A. lentiginosus*
    2. Legumes not inflated, not bladderlike; linear-oblong or narrowly ellipsoidal. Seeds filling locule . . . . *A. emoryanus*
1. Plants perennials.
    4. Plants acaulescent. Stems shorter than leaves or inflorescences.
        5. Stipules of lowest leaves free. Calyces 8–14 mm long. Banners 13–25 mm long. Keels shorter than wings . . . *A. mollissimus*
        5. Stipules of lowest leaves fused and forming a sheath. Calyces 2–6 mm long. Banners 6–12 mm long. Keels longer than wings . . . . *A. miser*
    4. Plants conspicuously caulescent. Stems longer than leaves or inflorescences.
        6. Leaflet hairs dolabriform, 2 ends free.
            7. Rhizomes present. Flowers descending at anthesis. Legumes 2-locular . . . . *A. canadensis*
            7. Rhizomes absent. Flowers ascending at anthesis. Legumes 1-locular . . . . *A. miser*
        6. Leaflet hairs basifixed, 1 end free.

- great range of habitats in western half of North America

Astragalus bisulcatus

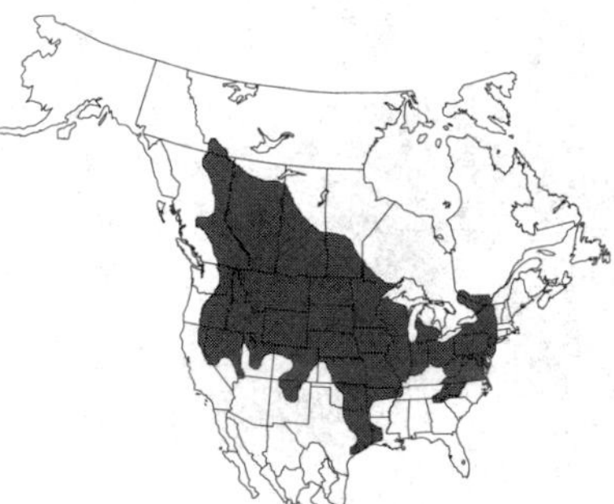

Astragalus canadensis

Astragalus emoryanus

- 5 types of problems—
  locoism
  cracker-heels disease
  selenium accumulation
  photosensitization
  thiamine deficiency

8. Stipules of lowest leaves fused and forming a papery-membranous sheath .... *A. bisulcatus*
8. Stipules of lowest leaves free.
   9. Legumes inflated, bladderlike; obliquely ovoid or semiovoid; 2-locular or almost so .... *A. lentiginosus*
   9. Legumes not inflated, not bladderlike; elliptic to oblong or obovate; 1-locular.
      10. Petals white. Calyx tubes 4.5–6.5 mm long. Ovules 40–84 per legume .... *A. praelongus*
      10. Petals yellowish white. Calyx tubes 6–9 mm long. Ovules 22–28 per legume .... *A. pattersonii*

Distribution and Habitat—Species of *Astragalus* occur in a wide range of vegetation types of the northern temperate region: grasslands, deserts, alpine tundra, coastal dunes, and open woods. The toxic species, however, are almost exclusively found in the western half of the continent, i.e., from the Great Plains westward. They are seldom dominant, instead more often occurring only as minor components except on unusual soil types such as selenium shales. Typically germinating in the autumn, plants overwinter and flourish the following spring or summer. Many of the perennial species are relatively short-lived because of the erratic rainfall patterns of the semiarid West. However, their long-lived seeds persist until the moisture cycle again favors plant establishment and growth (Ogden et al. 1988; Welsh 1989). The cyclic nature of *Astragalus* populations is a major factor in the occurrence of disease outbreaks (Welsh 1989; Ralphs & James 1999). Three species—*A. lentiginosus, A. miser, A. mollissimus*—are of particular interest because of their role in livestock losses and their wide distribution and abundance.

A polymorphic species exhibiting a tremendous range of variation and comprising 34 or more named varieties, *A. lentiginosus* occurs from below sea level in the Salton Sea basin to more than 3,600 m elevation in the Sierra Nevada. It is especially ubiquitous in the creosote scrub, juniper mountain brushlands of the Great Basin, and plants grow in a variety of soils.

As one of its common names implies, *A. miser* is found in aspen or conifer woodlands, mountain brush, or alpine meadows to 3,570 m elevation. Like *A. lentiginosus*, it is morphologically variable, with 8 poorly differentiated varieties recognized.

Astragalus mollissimus is a short-lived perennial primarily of the Great Plains and Rocky Mountain foothills to about 2,300 m elevation. Occurring as either scattered plants or small localized populations, it occupies a variety of clay, loam, and sandy soils.

Disease Problems—*Astragalus* represents probably one of the most important toxic plant genera in North America (James & Nielsen 1988). Many of its species are of significant concern to ranchers. At least five disease problems are produced by species of the genus. Two—locoism and cracker-heels disease—are of major economic importance, and aspects of each are discussed separately in the following subsections. Of lesser importance are the problems associated with selenium accumulation by certain species of *Astragalus* and the photosensitization that appears sporadically after cattle and sheep graze *A. cicer* (chickpea milk vetch) for 3–30 days (Walter-Hansen et al. 1986; Martin et al. 1987). The specific type of photosensitization is not yet determined, but initial studies indicated that despite substantial elevations of phylloerythrin in serum, there were no apparent liver lesions.

A fifth problem, a neurotoxicity quite different from locoism or cracker-heels disease, has been associated with an Old World species of *Astragalus* in Spain. Sheep

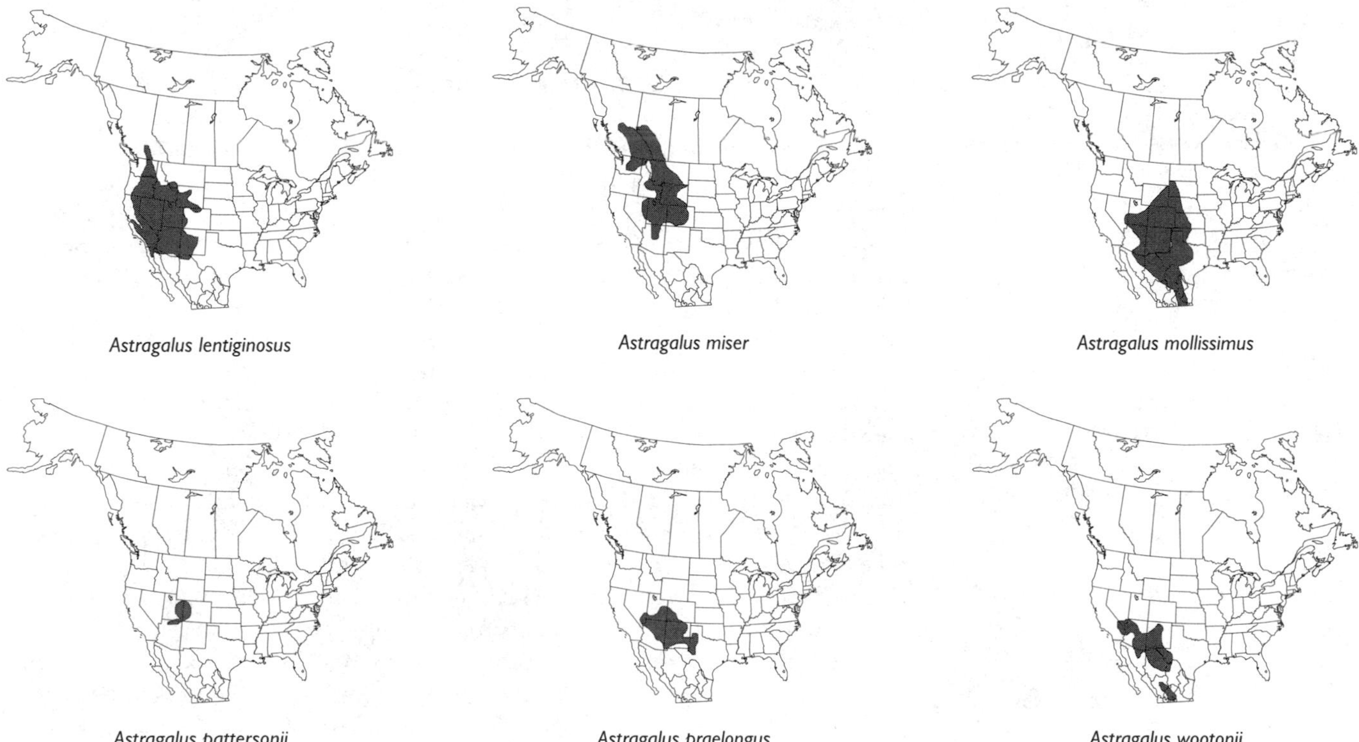
Astragalus lentiginosus

Astragalus miser

Astragalus mollissimus

Astragalus pattersonii

Astragalus praelongus

Astragalus wootonii

eating plants of *A. lusitanicus* for 1 or more days exhibited abrupt onset of a fatal neurologic disease involving decreased blood thiamine levels (Roman et al. 1987). Blood pyruvate was increased, and the disease was responsive to parenteral thiamine administration. With lower dosage of plant material over several weeks, there were no alterations in blood pyruvate or thiamine levels and no clinical signs of disease (Rodriguez et al. 1990). However, there were distinct cerebrocortical and cerebellar degenerative changes, including changes in Purkinje cells. The lesions were consistent with cerebrocortical necrosis/polioencephalomalacia rather than with locoism or nitrotoxicosis. Although this disease has not been reported to occur in North America, it seems unlikely that it is restricted to a single species of the genus.

It must be emphasized that not all species of *Astragalus* are toxic. Some, such as *A. cicer, A. nuttallianus,* and possibly *A. riparius* Barneby (Piper's milk vetch) are useful forage plants, although there may be problems with digestibility (James et al. 1981; White & Wight 1981; Marten et al. 1990; Townsend 1993; Williams & Molyneux 1988; Weimer et al. 1993). Toxicity potential may also differ considerably among varieties of the same species, as is shown for the toxic *A. miser,* some varieties of which may be useful forage (Barneby 1964a). In addition species such as *A. crassicarpus,* considered toxic by some, are apparently readily eaten without problems in the Canadian prairies (Looman 1983). Because of the differences among varieties, some early information on toxicity may need to be reassessed. Furthermore, there is interest in introducing Old World species of the genus as soil builders and forages (Davis 1973; Williams 1981a). Their use, however, should not be construed to indicate a total lack of adverse effects, as illustrated by the sporadic photosensitization problem occurring with ingestion of *A. cicer* (Marten et al. 1987, 1990).

■ not all species toxic; some exceptions—*A. cicer, A. nuttallianus, A. riparius*

Locoism

- importance of locoism is shown by long history of attempts to deal with it, including laboratories established specifically to investigate it

Disease Problems—In the 1870s and 1880s, species of *Astragalus* and *Oxytropis,* most notably *A. mollissimus* and *O. lambertii,* emerged as a serious toxicologic problem for ranchers maintaining livestock on the Great Plains and eastern Front Range of the Rocky Mountains (Marsh 1909). Because of the pronounced behavioral changes induced in animals eating them, they were called locoweeds, and the disease, locoism; these terms are derived from the Spanish word *loco,* for "crazy." The term *locoism* originally was applied rather loosely to a variety of neurologic diseases in which animals exhibited stupor, but eventually it came to be associated specifically with changes produced by ingestion of *Astragalus* and the closely related *Oxytropis.*

- early perceptions: mainly in late winter or early spring; requires prolonged ingestion; animals become accustomed to eating the plants and a preference persists in a few individuals; affects a variety of animal species

The livestock industry of Colorado was particularly devastated by locoism, and as a consequence, a state law was passed offering a bounty for each pound of locoweed dug below the root crown. Eventually costing the state $50,000 per year in payments and resulting in numerous cases of fraud, the law was repealed in 1885 (Marsh 1909; Cook & Redente 1993). The disease was of widespread concern throughout the West, with cases in Arizona and California being described in 1873 by the commissioner of the U.S. Department of Agriculture (Marsh 1909). In the 1880s and 1890s, numerous anecdotal reports were published that detailed many aspects of locoism. By 1886 it was described, in a USDA report, as a disease of the West from Kansas and Nebraska westward that had become more prevalent during the 1880s (Stalker 1887). It was described as appearing primarily in late winter and early spring and as posing a problem for individual animals that acquired a taste for locoweeds and subsequently sought them out. It was noted that most animals ordinarily did not eat the plants. Once in the habit of eating locoweeds, the animals were said to be "loco" or "locoed," even though they might not be severely affected for several years unless they were able to eat large amounts of plant material in a short time. Those not severely affected showed signs only when handled; they became intensely excited or frantic. It was reputed by many ranchers that horses, once "locoed," never fully recovered (Stalker 1887).

The great interest in locoism culminated in establishment of a USDA investigative laboratory at Hugo, Colorado, in 1905 under the direction of C. D. Marsh. Its specific purpose was to investigate this devastating disease of livestock. The work conducted there resulted in a landmark publication (Marsh 1909) on the nature of locoism and the role of *A. mollissimus* and *O. lambertii* (or what may have been *O. sericea,* according to Ralphs and James [1999]) as its cause in horses, cattle, and sheep. We owe much of our understanding of the signs and aspects of the disease to the careful observations of Marsh and later F. P. Mathews (1932), working in Texas.

Marsh and his co-workers reported the effects of the locoweeds to be cumulative, requiring 180 kg or more of plant material eaten for 4–6 weeks to cause the disease in mature horses or cattle. *Astragalus mollissimus* was most toxic but seldom eaten, whereas *O. lambertii* was much more readily eaten by sheep and cattle and was therefore a more serious hazard to these animals. It was observed that some animals began to eat locoweed early in the growing season when the grass forage was poor. Marsh's observations confirmed previous perceptions that the habit persisted even after the grass flourished, and in some instances the animals seemed to seek out the plants over considerable distances (Marsh 1909; Fleming 1918). Additional research revealed that a variety of species, including pigs, poultry, and antelope were susceptible to the effects of locoweeds. Young animals seemed to be the most affected.

The economic significance of locoweed toxicosis was so great that a second research facility, the Locoweed Research Laboratory, was established in Alpine, Texas, in 1930. Its principal research objective was the study of ways to combat the livestock

Table 37.1. Species and varieties of *Astragalus* known or suspected to produce locoism in livestock

Species	Common name
A. adsurgens Pall.	standing milk vetch (? tox.)
A. agnicidus Barneby	lambkill milk vetch, Humboldt's milk vetch
A. allochrous A.Gray	half-moon locoweed, rattleweed
A. arizonicus A.Gray	Arizona milk vetch
A. arrectus A.Gray (= *A. palousensis* Piper)	Palouse milk vetch, hanging-pod milk vetch
A. asymmetricus E.Sheld.	horse locoweed
A. bisulcatus (Hook.) A.Gray (= *A. haydenianus* A.Gray ex Brand)	two-grooved milk vetch
A. canadensis L.	Canada milk vetch
var. *mortonii* (= *A. mortonii* Nutt.)	Morton's locoweed, short-tooth Canada milk vetch
A. crassicarpus Nutt. (= *A. caryocarpus* Ker Gawl.)	ground plum, buffalo plum, Indian pea
A. douglasii (Torr. & A.Gray) A.Gray	Douglas's locoweed
A. emoryanus (Rydb.) Cory	Emory's milk vetch, red-stemmed peavine
A. flexuosus (Hook.) Douglas ex G.Don	prairie milk vetch, pliant milk vetch
A. hornii A.Gray	Horn's milk vetch
A. humistratus A.Gray	ground-cover milk vetch
A. lentiginosus Douglas ex Hook.	speckled loco, specklepod milk vetch, freckled milk vetch
var. *diphysus* (A.Gray) M.E.Jones (= *A. diphysus* A.Gray)	blue locoweed, rattleweed
var. *lentiginosus*	spotted locoweed
var. *wahweapensis* S.L.Welsh	Wahweap milk vetch
A. lonchocarpus Torr.	great rushy milk vetch
A. lotiflorus Hook.	lotus milk vetch, low milk vetch
A. missouriensis Nutt.	Missouri milk vetch, Nuttall's milk vetch
A. mollissimus Torr.	woolly loco, purple loco, woolly locoweed
var. *coryi* Tidestr. (= *A. argillophilus* Cory)	half-moon locoweed
var. *earlei* (Greene ex Rydb.) Tidestr. (= *A. earlei* Greene ex Rydb.)	Earle's locoweed
var. *mollissimus*	woolly locoweed
var. *thompsonae* (S.Watson) Barneby (= *A. thompsonae* S.Watson)	Thompson's woolly loco
A. nothoxys A.Gray	sheep locoweed, beaked milk vetch
A. nuttallii (Torr. & A.Gray) J.T.Howell (= *A. menziesii* A.Gray)	Nuttall's milk vetch
A. oocarpus A.Gray	San Diego milk vetch
A. oxyphysus A.Gray	diablo locoweed
A. pattersonii A.Gray	Patterson's milk vetch, Stanislaus milk vetch
A. pectinatus (Hook.) Douglas ex G.Don	tine-leaved milk vetch (? tox.)
A. praelongus E.Sheld.	stinking milk vetch, straight-stem locoweed
A. pubentissimus Torr. & A.Gray	Green River milk vetch
A. spaldingii A.Gray	Spalding milk vetch
A. succumbens Douglas ex Hook. (= *A. doryycnoides* [Douglas] ex G.Don)	crouching milk vetch, Columbia milk vetch
A. tenellus Pursh	pulse milk vetch (? tox.)
A. tephrodes A.Gray	ashen milk vetch
A. thurberi A.Gray	Thurber's locoweed
A. tridactylicus A.Gray	three-fingered milk vetch, foothill orophaca
A. vexilliflexus E.Sheld.	bent-flowered milk vetch (? tox.)
A. wootonii E.Sheld.	Wooton's locoweed, Wooton's milk vetch, garbancillo

Sources: Buck et al. 1961; Cory 1930; Crawford 1908; Fox et al. 1998; Fuller & McClintock 1986; James et al. 1981; Marsh 1909; Marsh & Clawson 1919; Mathews 1932; McCracken et al. 1970; Molyneux et al. 1985, 1989, 1991; Ralphs 1992; Ralphs et al. 1988; Schmutz et al. 1968; Smith et al. 1992.

Note: Suspected species are designated by "? tox." Nomenclature follows that of Barneby (1964a,b). Species names in parentheses were used in initial reports of toxicity.

losses due to this disease (Mathews 1932). Within a few years of the publication of the initial studies confirming the association between locoweeds and locoism, it became apparent that the various species of *Astragalus* were the cause of additional disease syndromes (Marsh & Clawson 1920a).

■ not all species referred to as locoweeds cause locoism

It must be emphasized that the common name *locoweed* has been applied loosely to a variety of taxa, including many species of *Astragalus* and other genera and thus cannot be used as a reliable indicator of whether a particular species is toxic or which disease syndrome will be produced. The species of *Astragalus* confirmed or suspected of causing locoism are listed in table 37.1. Although many species are indicated, most intoxications are due to *A. lentiginosus, A. mollissimus,* and *A. wootonii* as well as species of *Oxytropis.*

■ locoism, variety of effects—
neurologic
cardiac
reproductive

■ neurologic problems of behavior, coordination, difficulty eating, body weight loss

Locoism is a systemic disease that produces a variety of effects, most notably on the nervous system, the heart, and the reproductive system. Neurologic problems are manifested by alterations in behavior, coordination, eating and drinking, vision, weight, and body condition (James et al. 1969; James 1983; James & Panter 1989).

■ cardiac, right-heart failure at higher elevations

Locoism may have severe effects on the heart. When cattle at high elevations of 2,100 m or more graze extensively on certain locoweeds such as *A. lentiginosus,* the risk of development of right-heart failure as an early manifestation of locoism is significant. Right-heart failure, or brisket disease, is primarily a problem in cattle moved from lower elevations up to summer range in mountain areas where there is an abundance of locoweeds (James et al. 1983b, 1985). The disease is more common and serious in calves; morbidity rates range up to almost 100% in calves, with case fatality rates of 10–15%. Although the signs are not typical of locoism, many of the pathologic changes are. Among mature cows, a few may show signs of heart failure, whereas others exhibit typical locoism signs and/or abortions (James et al. 1985; Panter et al. 1988). Prompt transfer to lower elevations lessens myocardial stress and provides relief from the cardiac signs.

■ reproductive, fetal deformities, abortions, small/weak neonates, impaired fertility of males and females

Changes in reproduction also may accompany locoism and may involve sires, dams, fetuses, and neonates (Panter et al. 1999b). The assortment and severity of these reproductive effects are exemplified by the loss of lambs, sometimes numbering in the hundreds or thousands, under certain range conditions (James et al. 1967). Males exhibit a decrease in libido, spermatogenesis, sperm viability, and sperm motility after eating locoweed for 30 days (James & Van Kampen 1971b; Panter et al. 1989). These effects are the result of degenerative changes in Sertoli cells, spermatogonia, spermatocytes, and epithelia of the accessory sex glands (James & Van Kampen 1971b). Similar changes, including ovarian and placental lesions, occur in females, but these changes develop more quickly than those in males, with the earliest effects appearing in about 1 week (Van Kampen & James 1971b). Cytoplasmic vacuolation of ovarian luteal cells is prominent (Van Kampen & James 1971b). Serum progesterone levels decline substantially, in some instances to as little as 20–25% of basal levels. Possibly, in combination with the reported estrogenic effects (James & Foote 1972), a vicious cycle is set up whereby the imbalance in estrogen/progesterone leads to an increase in cotyledonary PGF_{1a} and PGF_{2a} (Ellis et al. 1985). The levels of the latter are particularly high in ewes carrying dead lambs. The effects are reversible within several weeks, but by then the pregnancies may have terminated.

The tissue changes in fetuses due to locoweed parallel those occurring in the dams and are manifested as abortion, small/weak neonates, and deformed and/or developmentally impaired neonates (James et al. 1967; Balls & James 1973; Hartley & James 1973, 1975; James 1976; Astorga 1992; Ralphs et al. 1994b). Abortions appear to follow in utero death of the fetuses, as supported by prostaglandin changes and ultrasonic observations (Ellis et al. 1985; Panter et al. 1987). Fetuses seem to be especially susceptible to the effects of locoweed at 30–60 days of gestation (James 1972). If the feeding of locoweed is stopped, the tissue changes in the dam and fetus resolve in 30–40 days (Hartley & James 1975). In horses, similar changes are reported for foals, albeit less commonly: limb deformities, contracture-like problems of flexor tendons, and arthrogryposis (McIlwraith & James 1982).

It is of interest to note that in contrast to the distinctive reproductive effects caused by locoweeds in livestock, fetal effects are not readily expressed in mice and Syrian golden hamsters, species commonly used in tests of teratogenicity (Berry et al. 1989). Even in rats, the effects seem to be limited to changes in fetal viability and weight change rather than deformities (Nelson et al. 1980).

In addition to its effects on the nervous system, heart, and reproductive system, early observations of locoism suggested an influence on immune function because the incidence of other diseases associated with it seemed to increase (Stegelmeier et al. 1999b). This has not been confirmed experimentally, but there is suppression of mitogen-induced responses, suggestive of decreased T-cell function, and marginal decreases in lymphocyte numbers (Sharma et al. 1984).

- all animal species affected, horses at highest risk; disease typically in late winter to early summer, less commonly in autumn after plentiful summer rains

Horses are at greatest risk, but significant losses occur among all livestock species, including pigs and poultry (Marsh 1909; Fleming 1918). In most instances, only a few animals are affected, but these cases are serious. Death or injury also may result from misadventure associated with behavior changes caused by locoism, e.g., drowning (Oehme et al. 1968). Electroencephalographic changes (James et al. 1977) and renal disease (Van Kampen & James 1970) may begin within days after animals commence eating locoweeds. Signs, however, are more often noted after 2–9 weeks when the animals have eaten 30–50% of their b.w. in d.w. of plant material (Marsh 1909; Mathews 1932; Beath et al. 1935). Although weight loss and poor body condition are important manifestations of locoweed intoxication, the opposite may be true initially when animals begin to graze plants (Marsh 1909; James et al. 1968b, 1970). Because of the high protein content (up to 20% crude protein) of these succulent cool-season forbs, the effects on weight gain begin only after several weeks of grazing, when the toxic effects supervene, and they may persist for a month or more after grazing ceases (Ralphs et al., 1993, 2000).

In the southern plains and across the deserts of the southwestern United States and adjacent Mexico, locoism is typically a problem in the winter, spring, and early summer, following an autumn with good rains. Less commonly it appears in the autumn, except in years with plentiful rainfall in the summer or early autumn that produces a late-season flush of plant growth, especially of *A. mollissimus*. Elsewhere, at higher elevations, locoism is a problem during the summer (Marsh 1909; Mathews 1932).

- extensive investigations on habituation to plants

apparently animals begin eating locoweeds early in spring, a few continue through growing season, even with abundant grass available; once preference is acquired, increased future risk with any true locoweed

The numerous anecdotal reports and suggestions of livestock habitually or addictively eating locoweeds prompted an extensive series of studies on the factors involved in intoxication by both *Astragalus* and *Oxytropis*. The important question is, why do some animals eat locoweeds? Observations of cattle on ranges containing *A. mollissimus* revealed that, early in the season, animals ate the plant only when grasses and other forages had not yet begun to grow (Ralphs et al. 1993). Once they started to eat the locoweed, a few animals continued to do so even when good forage became available. They ceased eating locoweeds only when the plants had matured (Ralphs et al. 1991). If initial exposure occurred later, during the growing season, cattle ate the plant only as grazing pressure and lack of alternative forage increased. In some instances, i.e., as with *A. lentiginosus,* animals may readily eat the dry stems long after the growing season (Ralphs et al. 1988). It was eventually confirmed that, as had long been suspected, cattle also began to eat locoweed ("learned") when sharing range with animals that had already acquired a taste for the plants early in the season (Marsh 1909; Patterson 1982; Ralphs et al. 1994a). However, maternal influence does not seem to predispose lambs to eat locoweeds (Pfister & Price 1996). It was noted that sheep fed locoweed for more than 1 year were more easily and seriously intoxicated in subsequent years (Balls & James 1973). Signs may occur under some circumstances long after feeding on locoweeds has stopped (Ralphs et al. 1991). Thus, it appears that under certain circumstances, animals begin to eat or acquire a taste for locoweeds (Ralphs et al. 1990). Once so habituated or adapted, they are at greater risk of developing the disease at least during the same season. The acquired taste also may extend from one locoweed species to another (Mathews 1932). Once livestock have become accustomed to eating locoweed, care must be taken in moving them from one range to another. Environmental factors such as weather, e.g., unseasonable frosts, may contribute to shifting appetites to locoweed (Ralphs & Molyneux 1989). Even though the animals may not be predisposed to eat plants again, they tend to exhibit signs of the disease more readily in the future.

- toxic indolizidine alkaloids, especially swainsonine and its *N*-oxide; found in all plant parts and highest in the legumes

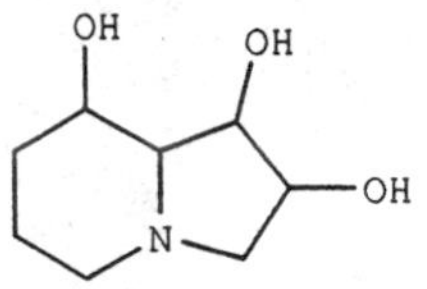

OH OH OH N

Swainsonine

Disease Genesis—For many years, the toxicants responsible for locoism eluded characterization. There was even a suggestion that barium carbonate was a cause. Eventually, the similarity of the clinical signs of locoism to those caused by the Australian genus *Swainsona,* also a member of the Fabaceae, led to the identification of swainsonine and related compounds as the toxicants in *Astragalus* and *Oxytropis* (Molyneux & James 1982; Dorling et al. 1989). These toxicants are the indolizidine alkaloids, dihydroxy lentiginosine and 2-epilentiginosine, and the trihydroxy swainsonine (Molyneux 1990). Swainsonine and its *N*-oxide are the toxicants of greatest interest because their administration in a pure state produces the same biochemical and pathological changes in the animal as ingested plant material (Molyneux & James 1982; Tulsiani et al. 1985; James et al. 1991).

The concentration of swainsonine is highest in seeds and legumes, lower in flowers, and lowest in foliage. Levels in the leaves are half or less than those in the legumes (Molyneux et al. 1989). Even the pollen contains the compounds and is toxic (James et al. 1978). Dried stems retain enough swainsonine to pose a serious risk even after a year or more (Mathews 1932; Molyneux et al. 1985, 1989; Ralphs et al. 1988). Concentration among varieties of a species also varies, as is demonstrated by comparison of *A. flavus* var. *flavus,* which has very little swainsonine in contrast to var. *argillosus,* which has levels similar to other toxic species. These differences can be of such magnitude as to mitigate against extrapolating from one variety to another and ascribing toxicity potential to all varieties of a single species (Molyneux et al. 1989).

Concentrations of swainsonine are quite low even in toxic species of *Astragalus,* typically in the range of 0.05–0.1% (Molyneux et al. 1994). For example, swainsonine concentrations may be up to 0.36% in seeds, 0.037% in the green stems, 0.023% in the senescent stems, and 0.005% in the dry stems of *A. lentiginosus;* 0.035% in the vegetative parts of *A. flavus;* and 0.28% in the flowers of *A. mollissimus* (Molyneux et al. 1985, 1989, 1994). Molyneux and his co-workers (1994) suggested that concentrations of swainsonine greater than 0.001% are of concern. The threshold dosage for the appearance of signs of toxicity is estimated to be about 0.3 mg swainsonine/kg b.w. (Stegelmeier et al. 1995a). Lesser doses may cause a decrease in productivity but not produce overt signs of toxicity. Species listed in table 37.1 contain significant concentrations of swainsonine. However, those designated with "? tox" in the table are of questionable toxicity, even though the alkaloids are present. For example, var. *robustior* of *A. adsurgens* was reported to be nontoxic (Marsh 1919) but contains significant concentrations to be a risk. The role of swainsonine for the plant is not understood, but the concentrations are so low that it does not appear to be a protective factor limiting palatability to livestock (Ralphs & Molyneux 1989; Ralphs et al. 1986).

Swainsonine is water soluble and is rapidly absorbed and excreted from the animal's body. With a half-life in the body of 16–20 hours in cattle and sheep, it appears to be eliminated by first-order kinetics (Stegelmeier et al. 1995a). Thus very little remains after about 5 days, accounting for the rapid resolution of many of the cellular effects. It is excreted primarily in the urine. Although cattle and sheep are less susceptible than horses, this does not seem to be the result of ruminal biotransformation (Miller et al. 1992).

- swainsonine, water soluble, readily excreted; half-life 16–20 hours, little remains in body after 5 days; ingestion must be consistent to produce cumulative toxic effects; also excreted in milk, risk for nursing young

Once absorbed, swainsonine is distributed widely throughout the body, including the brain, gonads, and fetuses if present. Swainsonine concentrations, determined by measuring α-mannosidase activity, were highest in the liver (3–4 μg/ml) and kidney (1–2 μg/ml) in sheep given 10% *A. lentiginosus* in their diet (Stegelmeier et al. 1994, 1995b). Appreciable α-D-mannosidase (0.2–0.5 μg/ml) activity also was apparent in the blood. Swainsonine is present in the dam's milk at such concentrations as to pose an intoxication risk to offspring or other animals given it experimentally (James & Van Kampen 1976; James & Hartley 1977; Molyneux et al. 1985). Its presence in the milk may, in large measure, account for the reported occurrence of clinical signs in nursing calves prior to their appearance in the dams (James et al. 1967).

First isolated from *Swainsona canescens* in Australia, the protonated swainsonine cation (all three -OH groups are required) mimics the half-chair conformation of the mannosyl cation and thereby serves as a reversible competitive inhibitor of α-D-mannosidase (Dorling et al. 1980, 1983; Colegate et al. 1989). Because it concentrates in an acid medium, its primary effects are on lysosomal acid mannosidase. The catabolism of mannose-containing oligosaccharides, mainly derived from breakdown of cell components, is therefore inhibited. This results in the accumulation of a variety of water-soluble/osmotically active oligosaccharides, with as many as nine mannose residues, in lysosomes of cells throughout the animal's body, including those of fetal tissues (Dorling et al. 1978, 1989; Warren et al. 1989).

- swainsonine closely resembles mannosyl cation, competitive inhibitor of α-D-mannosidase, inability to break down mannose-containing oligosaccharides; foamy cytoplasmic vacuolations; cellular distension compromises function and integrity

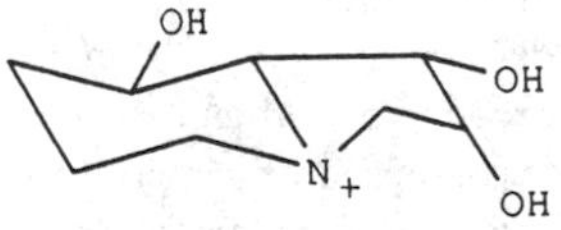

Swainsonine cation

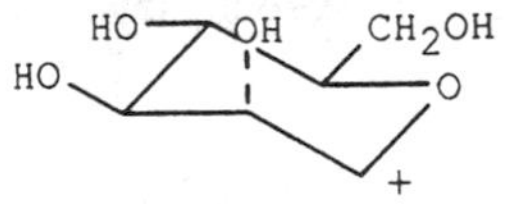

Mannosyl cation

The accumulation of these compounds results in proliferation and hypertrophy of lysosomes to the extent that the cytoplasm of many cells is completely filled. This phenomenon appears with conventional staining as foamy vacuolation. Accumulation of mannose/acetylglucosamine (5:2) complexes is particularly noteworthy in neurons throughout the brain (Tulsiani et al. 1989). In this respect, locoism is very similar if not identical to inherited bovine mannosidosis (Jolly et al. 1989).

In addition to effects in the lysosomes, there also is noncompetitive inhibition of Golgi mannosidase II, which results in interference with mannosyl trimming of complex glycoproteins (Tulsiani et al. 1982). Although this is a contributing factor, it seems to play a rather minor role in the overall manifestations of toxicity (Abraham et al. 1983). The functional capacity of cells heavily laden with the space-occupying cellular accumulations is not clearly understood, but it seems reasonable to assume that it is decreased. Swainsonine-induced neuronal mannoside storage is closely associated with degenerative changes in axons, which regress after exposure to the toxin is stopped (Huxtable et al. 1982). Weight loss, which in rats appears to be due to appetite suppression (Pritchard et al. 1989), may be associated with problems in cell function. The accumulation of oligosaccharides and synthesis of abnormal *N*-linked glycoproteins in neurones undoubtedly affects function, but how and what aspect are not known. In addition, the presence or absence of typical neuronal vacuolation does not correlate well with the effects exhibited by the animal, because even with their apparently complete regression, there are residual neurologic effects (Hartley & James 1975).

- all cells affected, but neurologic manifestations largely due to effects on cerebellum and Purkinje cells

Characteristic of neuronal storage diseases, ectopic dendritogenesis is apparent with ingestion of swainsonine (Walkley 1988). These unusual structural changes in neurons are apparent in horses and cattle. Focal axonal swellings and aspiny meganeurites are seen in both thalamic type 1 cells and cortical pyramidal cells (Walkley & Siegel 1989). The neuronal change—neurite growth from axon hillocks—induced by swainsonine experimentally in cats persists long after the animals have otherwise recovered from oligosaccharide accumulation (Walkley et al. 1987). Purkinje cells seem to be particularly sensitive, perhaps in part because of their exceptionally expansive dendritic trees, which are predisposed to enlargement (Walkley 1988), and possible aberrant synapse formation. Even though only limited functional deficits are demonstrable, i.e., accumulation of obsolete organelles due to difficulties in retrograde axonal transport (Pritchard et al. 1989), it seems reasonable to expect that accumulations of oligosaccharides are responsible for more than all-or-nothing neuronal cell death as a cause of cerebellar dysfunction and proprioceptive deficits. It might be anticipated that, in some measure, the decrease in appetite is an expression of a functional change in the central nervous system. Functional changes of a serious nature may occur in other tissues, as shown by the problem of cardiac failure despite the relatively little tendency for either skeletal or cardiac muscle to be influenced directly by α-D-mannosidase inhibition and accumulation of oligosaccharides (Van Kampen & James 1970; Warren et al. 1989).

Swainsonine has other unique actions such as augmentation of natural killer-cell activity when administered experimentally in mice (Humphries et al. 1988). There also are effects on other immune-modulated responses (Stegelmeier et al. 1999b).

- depression, weight loss, ataxia, stiff gait, exaggerated limb movement, interference with eating

  horses, dramatic response to handling or confinement; animals become violent when restrained; extreme measures required to control animal; violent response not controllable with tranquilizers; reserpine may be helpful but not curative; affected horses permanently unsound for riding

- cattle, similar signs, less violent than horses, more likely to wander; sheep, more gradual onset of signs, loss of flocking instinct

Clinical Signs—Not only are horses more susceptible to locoism than are other species, but they also display more distinctive signs. Depression is prominent as affected horses stand very still, with head and neck extended down as if asleep. At other times, horses may stare as if blind. Gaits are awkwardly stiff-legged, with exaggerated movement of the limbs. Eating is sometimes slow and difficult, with chewing movements so ineffective that the animal appears to be sucking on the plant stems while trying to eat. There may be nibbling with the lips quivering, the jaws moving spastically, and the head nodding.

When affected horses are handled, especially about the head, the most striking signs of the disease are displayed. Otherwise docile animals may become quite violent, almost maniacal, especially when restraint is applied or confinement attempted. Some horses may shy violently, be very difficult to back up or lead, or they may rear up and fall backward when startled. Death may occur during such episodes or 1–2 weeks later from the disease itself or to misadventure, such as drowning or fatal falls.

Less severely affected animals may have a temporary increase in appetite, especially for salt (James et al. 1970). However, weight loss and rough coat appearance may occur in spite of this good appetite. Horses are uncoordinated and unreliable for riding. Those that seem to recover, especially from severe cases of locoism, may exhibit impaired vision or hearing, incoordination, and a high-stepping gait for years.

In cattle, the signs are similar, but although the animals may become belligerent, they are less prone to violent behavior. Weakness, depression, dull eyes, rough hair coat, and weight loss are consistent features. They may have a gait similar to that in horses, as well as show incoordination, loss of balance, and a tendency to isolate themselves from herd mates. They may have difficulty eating, exhibiting abnormal tongue movements, exaggerated chewing motions, and head shaking. There may be sudden jumps or movements, as if startled without apparent cause.

Reproductive signs of locoism include an increase in the number of nonpregnant cows with irregular estrous cycles and difficulties in labor. Cows also exhibit a loss of mothering behavior. Diarrhea may accompany the other signs in nursing calves. If cattle are prevented from further access to locoweed plants, the neurologic signs of the disease may resolve and disappear in 6–7 weeks, and the weight regained 2–3 weeks thereafter.

Signs in sheep are usually less conspicuous than those observed in other species. They include a gradual onset of nervousness and abnormalities of movement, such as stiffness, pacing gait with the head held high, and incoordination. Their sense of direction and flocking instinct are lost, and they tend to separate from other individuals. Less commonly, they are excitable and/or belligerent. Like other species, affected sheep are predisposed to accidents around water or on rough terrain. Late in the course of the disease, weakness, weight loss, and patchy or generalized loss of wool may be prominent. Just before death, paddling seizures are likely.

The clinical signs of locoism in goats are similar to those in sheep (Mathews 1932). Transitory paresis or paralysis of the pelvic limbs is evident when the animals are startled. This may become complete in later stages of the disease.

In all species, 4-to-5-fold elevations of serum AST and alkaline phosphatase are typical, especially prior to the onset of severe signs. Diagnosis may be aided by the presence of cytoplasmic vacuolation in peripheral lymphocytes, but they may vanish 10–12 days after the animal stops eating locoweed. Otherwise, distinctive hematologic changes are lacking (James et al. 1970). Significant depression of serum α-mannosidase activity has been reported for range cattle ingesting *A. mollissimus* (Stegelmeier et al. 1994). As might be expected with a toxin that causes nonspecific oligosaccharide accumulation in a variety of cells, many changes in the physiology of the animal may be seen—as illustrated by a 50% increase in serum thyroxine and triiodothyronine—as probably a direct effect on the thyroid (Pulsipher et al. 1994).

- cattle eating plants at higher elevations may have labored respiration, jugular pulse, brisket edema, indicative of cardiac insufficiency; alleviated by moving animals to lower altitude

As noted earlier, locoism, in addition to producing effects on the nervous system, also affects the heart and reproductive system. Clinical signs of right-heart failure typically occur in calves at high elevations on summer range and include depression, dry and rough hair coat, dullness of the eyes, subcutaneous edema of the jaw and brisket, labored respiration, and jugular pulse. Elevation of serum AST activity can also be expected. These signs may be seen in conjunction with those more typical of locoism in adult animals. They generally are quickly alleviated by moving the animals to a lower altitude.

Clinical signs of the reproductive effects may occur alone or with the other manifestations of locoism in all species. In males, there is a decrease in libido after about 30 days of locoweed ingestion. In addition to the pathologic changes mentioned earlier, there is cytoplasmic vacuolation of Sertoli cells, spermatogonia, spermatocytes, and epithelia of the accessory sex glands (James & Van Kampen 1971a). The result is a decrease in spermatogenesis and an increased number of dead sperm, many of which exhibit head or tail loss. Sperm motility gradually declines to poor levels by the 45th day of ingestion (Panter et al. 1989).

In females, as noted earlier, similar changes, including ovarian and placental lesions, occur, but they develop more quickly than those in males, with the earliest effects detectable in about 1 week. Serum progesterone levels decline substantially, in some instances to as little as 20–25% of basal level.

- reproductive effects, decrease in male libido, lowered reproductive performance in females; deformed, small or weak newborns with poor survival rates; deformities, limb rotation, tendon contraction

Deformed, small, or weak neonates with poor survival rates are typical. Deformities include lateral rotation of the thoracic limbs, contracted tendons, carpal and tarsal flexures, and increased mobility (looseness) of the hock joints. If the animal survives, it may at least partially recover from some of these deformities. Some apparently fully formed neonates are weak, have difficulty nursing, or lack the desire to nurse and thus fail to thrive after birth (James et al. 1967; James 1972; Balls & James 1973; Astorga 1992; Ralphs et al. 1994b). Abortions and weak neonates with poor survival chances are more common than deformities, especially in sheep.

- gross pathology, few lesions, occasional stomach mucosal erosions

  microscopic, widespread cytoplasmic vacuolation of cells in most tissues, adrenal, kidney, pancreas, liver, Purkinje cells

Pathology—Specific gross changes are limited, but there may be occasional localized erosions of the mucosa of the stomach or abomasum near the pylorus, and the liver and kidneys may be pale or tan. If the animal has continued to eat locoweeds up to the time of death, there will be prominent microscopic changes, including widespread cytoplasmic vacuolation of cells throughout the viscera and nervous system, a condition typically designated as visceral cytopathic vacuolation. These changes are prominent in the thyroid follicular epithelium, parathyroid chief cells, pancreatic acinar and beta cells, adrenal cortical and medullary cells, pituitary cells, renal tubular epithelium, liver cells, lymph nodes, reticuloendothelial cells of the spleen and thymus, urinary bladder epithelium, and glandular epithelium of the stomach/abomasum. The vacuolar lesions in most organs disappear or diminish within a few weeks after ingestion of locoweed ceases. However, the extensive neurologic changes are much more persistent, and any cellular losses are permanent. Degenerated, vacuolated neurons are seen throughout the central and peripheral nervous systems. Vacuolation of Purkinje cells of the cerebellum persists for a year or more, and many are lost, replaced by glial cells. This is especially noticeable in horses and may be responsible for the permanency of signs in some instances. There also are lesions in the eyes and cytoplasmic vacuolation of the ganglion cells and bipolar neurons of the inner retina, epithelial covering of the ciliary body, and secretory epithelium of the lacrimal glands.

- feed well; allow animals to recover body condition gradually; salvage for slaughter; horses may be used for reproductive purposes; prevent future access to plants

Treatment—The violent behavior in horses is not controllable by administration of tranquilizers or other traditional approaches. However, 3–4 mg of reserpine administered i.m. to an adult horse may allow it to be more safely handled (Staley 1978; Kirkpatrick & Burrows 1990). This is not a cure and should be relied upon only for a temporary change in behavior. Strychnine (9–12 mg s.c.) for adult cattle and Fowler's solution (15–20 ml orally) for adult horses were recommended by Marsh (1909), but they are palliative at best. Prevention has been attempted through the use of protein supplements, but this seems to worsen the condition (Mathews 1932; James & Van Kampen 1974). Similarly, a variety of vitamin and mineral supplements, even those purportedly formulated specifically for locoism, including clinoptilolite (a natural zeolite clay with high levels of aluminum and iron) and several bentonite clay preparations, are without apparent benefit (Keeler & James 1971; James & Van Kampen 1974; Bachman et al. 1992; Pulsipher et al. 1994). Determination of tissue concentrations indicates that little benefit from supplementation with copper or other minerals may be expected (Galyean et al. 1996).

In order to reduce the likelihood of intoxication due to locoism, it has been recommended that access to plants be limited to the early pod stage and that locoweed-infested range be heavily stocked so that the legumes are eaten quickly by a larger number of animals to reduce the amount and time they are eaten by any single animal (Ralphs 1987). Animals also may be allowed to graze locoweed-infested range at intervals (cyclic grazing), but care must be taken to allow resolution of the toxic effects before returning them to the infested range (Pfister et al. 1996). This may require a rest period twice as long as the exposure period (for example, 2–3 weeks grazing locoweed range followed by 4–6 weeks on clean range before returning them to the infested range).

Another preventive approach involves a single oral dose of 200 mg/kg b.w. of lithium chloride while feeding locoweed forage to avert the cows from eating of the plants (Ralphs & Olsen 1998). This may prevent ingestion of locoweed for a year or more, provided the averted animals graze independently of nonaverted cattle. Prevention of intoxication also may be accomplished by carefully observing grazing animals to identify and remove any animals eating locoweeds. Those animals with a taste for locoweeds can then be conditioned, using multiple lithium chloride dosage of 200 mg/kg b.w., to avoid eating the plants in the future (Ralphs et al. 1997). It should be anticipated that conditioned aversion will require repeated dosage, given that experimentally it is difficult to avert animals already familiar with or accustomed to eating the plants.

Because of the lack of appropriate therapeutic measures, animals, in most instances, can be salvaged for slaughter after being fed well for 1–2 months. Horses should be considered unsafe for riding, but with appropriate caution they may be used for reproductive purposes.

Nitrotoxicosis

- acute to subacute neurologic disease of livestock on mountain pastures without signs of locoism; cracker-heels disease, associated with *Astragalus* timber milk vetches rather than locoweeds

Disease Problems—During the late 1800s and early 1900s most of the intoxication problems associated with *Astragalus* were simply referred to as locoism. However, between 1915 and 1930, it became apparent that other disease problems were caused by species of the genus. In 1915–1916 an apparent neurologic illness of cattle, attributed to ingestion of *Astragalus tetrapterus,* occurred in southwestern Utah. The clinical signs were quite different from those of locoism (Marsh & Clawson 1920a). Subsequently, similar disease signs were noted in cattle in other areas of western North America in the 1920s and 1930s. The most serious problems perhaps were in the Pacific Northwest, most notably British Columbia, where repeated losses occurred on forest ranges due to ingestion of *Astragalus* plants (Bruce 1927).

The plant species involved in these instances were generally different from those causing locoism and occupied different habitats, typically forested areas—hence the application of the common name *timber milk vetch* to some of them. The principal culprit subsequently was identified as *A. miser* var. *hylophilus.* The disease, which lacks the behavioral changes of locoism, is known by a variety of common names in cattle or sheep—timber trouble, timber grass poisoning, timber paralysis, jack-pine fever, husky knocking disease, roaring disease, lungers, wheezers, and Kamloops cattle disease. In horses, similar names for the disease are employed—mountain fever, roaring disease, timber paralysis, and Clinton horse disease. The name perhaps most typically associated with the disease in cattle is cracker-heels, which is indicative of the peculiar noise produced by the pelvic limbs rubbing against each other when the animal walks (Newsome et al. 1936). Several descriptive reviews of the disease have appeared (Bruce 1927; Newsome et al. 1936; Mathews 1940; McDonald 1952; Williams et al. 1979; James et al. 1980).

Seen mainly in midsummer, the disease may occur in any animal species, but cattle, especially lactating cows, are most likely to be affected (Bruce 1927; Newsome et al. 1936; McDonald 1952). In sheep and horses, the disease is more acute but does not necessarily cause higher mortality. Ingestion of about 6 kg/day of plant for 5 days is sufficient to cause illness; 1% of b.w. of *A. emoryanus* for 6 days was toxic to sheep and goats (Mathews 1940).

In most instances, the species of *Astragalus* causing cracker-heels disease are not those causing locoism. There are, however, a few that are known for their potential to cause either disease, e.g., *A. emoryanus* (Sperry et al. 1952; Molyneux et al. 1989). However, the amount of plant material eaten is clearly an important factor in determining which disease syndrome is produced.

Because plants of *Astragalus* often have high protein content and grow well in adverse environmental conditions, there is the temptation to promote their use as forages. For example, *A. nuttallianus,* closely related to the toxic *A. emoryanus,* is now recognized as a useful forage well accepted by livestock. Sheep fed 335% of b.w. of plants of var. *nuttallianus* for 31 days and 150% of b.w. of var. *macilentus* for 20 days did not develop disease (Boughton & Hardy 1940). There also is the temptation to introduce species from the Old World; however, with this there is the risk of introducing toxic forages, as in the case of *A. falcatus.* A native of southeastern Europe, it has been introduced repeatedly but has been found to contain toxic compounds (Barneby 1964b; Davis 1973; Williams 1981a).

- neurologic problem, glycosides of 3-NPA and 3-NPOH, miserotoxin; 3-NPOH more completely absorbed and thus more toxic

$O_2N—C—C—COOH$

3-Nitropropionic acid (3-NPA)

$O_2N—C—C—COH$

3-Nitropropanol (3-NPOH)

Disease Genesis—Cracker-heels disease is caused by accumulation of toxic glycosides of 3-nitropropionic acid (3-NPA) and 3-nitropropanol (3-NPOH), which are commonly referred to as nitrotoxins (Williams & Barneby 1977a; Williams 1982; Majak & Benn 1992). Surveys of the distribution of these compounds in *Astragalus* reveal that they occur in both New World and Old World species (Williams & Barneby 1977a,b; Williams 1982). Of the more than 500 species and varieties in North America, about half contain them (Williams & James 1978). In most species, only one of the two types is present, and usually all species within a taxonomic section of the genus have the same type of glycoside (Williams 1982). For example, species of 10 North American sections contained glycosides of 3-NPOH and 21 contained those of 3-NPA. In a few species both types are present in equal amounts (Williams et al. 1975). Species commonly causing problems or possessing the potential to do so because of unusually high concentrations of 3-NPA glycosides are presented in table 37.2.

The glycosides of 3-NPOH are of greater toxicological importance in livestock than those of 3-NPA. However, when 3-NPA glycosides are present in unusually high concentrations, the species may be dangerous. Toxic *A. canadensis,* for example, has unusually high concentrations of 3-NPA types. *Astragalus falcatus,* which has even higher concentrations of 3-NPA, should be considered as similarly hazardous (Williams & Barneby 1977a). The toxicity potential of *A. canadensis* and *A. falcatus* also may be enhanced by the presence of small amounts of 3-NPOH types as well as those of 3-NPA. A ranking system has been proposed wherein highly hazardous species of *Astragalus* are classified as class I and those of lower risk are class II, with the dividing points dependent upon animal species because cattle are more susceptible than sheep (Williams & James 1975; Stermitz & Yost 1978).

Both 3-NPA and 3-NPOH occur in plants of *Astragalus* primarily as glycosides. Miserotoxin, which is the β-D-glucoside of 3-nitro-1-propanol is the most important 3-NPOH type (Stermitz et al. 1969). Other 3-NPA types are cibarian, corollin, coronarian, coronillin, and karakin (Majak & Bose 1976; Moyer et al. 1977). These parent glycosides are hydrolyzed to the free aglycones 3-NPA and/or 3-NPOH by esterases and β-glucosidase in the rumen or gut of herbivores (Williams et al. 1970). They are both readily absorbed, 3-NPOH more rapidly than 3-NPA (Majak et al. 1984; Pass et al. 1984). Even though both 3-NPA and 3-NPOH are toxic when given parenterally, 3-NPA is of lesser toxicologic importance because, unlike 3-NPOH, much of it is biodegraded in the digestive tract before absorption (Williams et al. 1969, 1970; Williams & James 1976; Gustine et al. 1977; Majak & Clark 1980). In vitro evidence suggests that both 3-NPA and 3-NPOH are subject to nitroreduction in the ruminal environment; 3-NPA to β-alanine and 3-NPOH to 3-amino-1-propanol (Anderson et al. 1993a,b). Although only a small amount of 3-NPOH is typically biodegraded to aminopropanol by ruminal microorganisms, the reaction can be enhanced by various factors, such as increased availability of ferrous and sulfide ions. This enhance-

Table 37.2. Representative species of *Astragalus* containing nitrotoxins

3-Nitropropanol (3-NPOH)	
A. atropubescens Coult. & Fisher	Kelsey's milk vetch
A. clevelandii Greene	Cleveland's milk vetch
A. convallarius Greene	lesser rushy milk vetch
A. diversifolius A.Gray	mesic milk vetch
A. emoryanus (Rydb.) Cory	Emory's milk vetch, Emory's locoweed
A leptocarpus Torr. & A.Gray	slim-pod milk vetch
A. michauxii (Kuntze) F.J.Herm.	Michaux's milk vetch
A. miser Douglas	
var. *hylophilus* (Rydb.) Barneby	forest milk vetch
var. *oblongifolius* (Rydb.) Cronquist	Rydberg's weedy milk vetch
var. *serotinus* (A.Gray) Barneby	Cooper's milk vetch
A. pachypus Greene	bush milk vetch
A. pterocarpus S.Watson	winged milk vetch
A. reventus A.Gray	Blue Mountains milk vetch
A. serenoi (Kuntze) E.Sheld.	naked milk vetch
A. sheldonii (Kuntze) Barneby	Sheldon's milk vetch
A. terminalis S.Watson	railhead milk vetch
A. tetrapterus A.Gray	four-wing poison vetch
A toanus M.E.Jones	Toano milk vetch
A. whitneyi A.Gray	balloon milk vetch
3-Nitropropionic Acid (3-NPA) (High Concentrations)	
A. camptopus Barneby	Bruneau milk vetch
A. canadensis L.	Canada milk vetch
A. cibarius E.Sheld.	browse milk vetch
A. curvicarpus (A.Heller) J.F.Macbr.	sickle milk vetch
A. falcatus Lam.	Russian sickle milk vetch
A. flexuosus (Hook.) Douglas ex G.Don	pliant milk vetch
A. greggii S.Watson	Gregg's milk vetch
A. hallii A.Gray	Hall's milk vetch
A. quinqueflorus S.Watson	five-flowered milk vetch
A. tweedyi Canby	Tweedy's milk vetch

Sources: Williams & Barneby 1977a; Williams 1982.

ment of degradation may represent a means of protecting animals ingesting *Astragalus* if it can be artificially induced (Anderson et al. 1993a,b). NPA metabolism is not similarly enhanced. Interestingly, some animal species susceptible to 3-NPOH are little affected by its parent miserotoxin because they lack the capacity to hydrolyze the glucoside (Majak et al. 1983).

Upon absorption, 3-NPOH is biodegraded to 3-NPA in the liver but not in blood (Majak et al. 1981; Muir et al. 1984), a process that can be blocked by administration of alcohol dehydrogenase inhibitors (Pass et al. 1985). Biodegradation of 3-NPOH to 3-NPA is accompanied by the formation of a small amount of NO_2, which accounts for the observed formation of methemoglobin.

Despite being less of a threat than 3-NPOH when given orally, 3-NPA is ultimately responsible for cracker-heels disease (McDiarmid et al. 1992). This seeming contradiction is due to the already described differences in degradation in the digestive tract.

■ once absorbed, 3-NPOH is metabolized in liver to 3-NPA, which causes neurologic effects

The neurotoxic effects of 3-NPA appear to be due to irreversible inhibition of mitochondrial succinate dehydrogenase by the 3-nitroproprionate carbanion, which interferes with the production of cellular energy (Alston et al. 1977; Pass et al. 1994). This inhibition and possibly other effects result in activation of ATP-sensitive potassium channels and depolarization, as observed in pyramidal cells of the rat hippocampus (Riepe et al. 1992). Increased activity of superoxide dismutase and glutathione peroxidase and lipid peroxidation in liver and brain also occurs with 3-NPA intoxication in rats (Fu et al. 1995). Although both 3-NPA and 3-NPOH given to rats produce similar effects, higher doses of 3-NPOH are required because it must be converted to 3-NPA to provide the reactive carbanion (Gould et al. 1985). Brain lesions produced by 3-NPA in rats and mice include neuronal degeneration, axonal and dendritic swelling, and myelin degeneration of the input nuclei (caudate nucleus and putamen) and output nuclei (globus pallidus and substantia nigra) of the basal ganglia. This degeneration also occurs in the hippocampus and thalamus (Gould & Gustine 1982; Hamilton & Gould 1987; Majak & Pass 1989). There seems to be a selective inhibition of the TCA cycle in GABA neurons, which may account for the action on the caudoputamen, an area rich in these types of neurons (Hassel & Sonnewald 1995). Because these basal areas are important for motor control, motor dysfunction would be expected following these alterations caused by 3-NPA.

In humans, 3-NPA derived from stored, moldy sugarcane acts as a striatal toxin (Wenjuan et al. 1986; Ludolph et al. 1992). A few hours after ingestion of the infected sugarcane, individuals exhibited digestive disturbances, and then neurologic signs such as dizziness, lethargy, seizures, and coma (Ming 1995). Putaminal lesions developed, producing dystonia and chorea (Ludolph et al. 1991). These changes are similar to those described for cattle eating various nitrotoxin-containing species of *Astragalus* (James et al. 1980). Experimentally, subcutaneous administration of 3-NPA for 30 days to rats resulted in striatal necrosis with a reduction of expression of enkephalin and substance P in specific neurons (Wullner et al. 1994). In addition, results with an *N*-methyl-D-aspartate (NMDA) receptor antagonist suggested that excitotoxicity due to activation of NMDA receptors also might be involved. It is hypothesized that the decrease in energy availability in brain, secondary to inhibition of succinate dehydrogenase, renders the tissues more susceptible to glutamate-type toxicity, possibly mediated through the NMDA receptor (Novelli et al. 1988; Beal et al. 1991; Ludolph et al. 1992; Brouillet et al. 1993).

In addition to neurotoxicosis, 3-NPA also produces hypotension via a mixture of vasodilatory and cardiodepressor activities (Castillo et al. 1993). These may contribute to the neurotoxic effects.

- disease control difficult to accomplish by animal or pasture management without risk of overgrazing

Cracker-heels is a difficult disease to control by animal management. Plants of *Astragalus* are not preferred forages and are eaten only as the abundance of other forbs declines. Thus, early-season grazing, which reduces miserotoxin concentrations later in the season, may not be practical or may result in overgrazing of and a decline in preferred forbs (Majak et al. 1988; Quinton & Majak 1992). In the Northwest, plants of *A. miser* var. *serotinus* growing in open grasslands tend to have higher miserotoxin levels than populations growing in forested sites; concentrations peak in the bud stage and decline after pod formation (Majak et al. 1974; Majak & McLean 1975). It also is noted that there is less grazing of *A. miser* by cattle in these grasslands than in forested areas, perhaps related to the higher miserotoxin concentrations as well as to a lack of other desirable forages in the forested sites (Majak et al. 1996). Although miserotoxin concentrations are lowest late in the growing season, *Astragalus* may be so abundant and alternative forage so limited that the increasing amounts eaten compensate for the lower toxin levels in the plants. In more southern areas, rainfall patterns are of increasing importance because they dictate relative abundance of *Astragalus* plants. Toxicity potential also increases with rainfall after a prolonged dry spell (Majak & Pass 1989).

- typically gradual onset, ataxia, excess salivation, tremors, pelvic-limb placement abnormality, clicking sound from pelvic limbs when walking; wheezing respiration; posterior paresis; when startled, animal may become uncoordinated or collapse

Clinical Signs—The severity and rapidity of appearance of signs of cracker-heels disease vary considerably. Morbidity is usually 10–15%. Typically, the disease commences after the animals have been eating the forage for a week or more. In acute cases, signs may be limited to constipation and slight ataxia developing into a staggering walk, and thence to recumbency and death in some animals. Death is probably due to cardiac and respiratory failure. Animals also may appear to be temporarily blind.

Most commonly, the signs will appear gradually: weakness, ataxia, increased salivation, muscle spasms and tremors, and a dazed and docile appearance. There will be a proprioceptive deficit, with placement problems especially noticeable in the pelvic limbs. Other particularly noteworthy changes include difficult breathing accompanied by wheezing, roaring, or coughing; a diminished and huskier voice; continual flexion of the rear fetlocks; and the distinctive clicking sound made when the dewclaws or hooves strike the opposite pelvic limbs during movement. Animals also may urinate in spurts while moving. There may be posterior paresis or paralysis that is especially evident with stress or excitement. When startled, the animal may collapse on its pelvic limbs or be uncoordinated when running and eventually fall due to apparent paresis. The case fatality rate may be up to 30%. In more-chronic cases and/or in some of the animals recovering from the more acute signs, weight loss and a rough-coat appearance may be long-lasting and perhaps permanent. There are no hematologic or serum chemistry changes of diagnostic value, but methemoglobin concentrations may be 20% or higher (Maricle et al. 1996). Horses may exhibit sweating and paroxysmal episodes of distress, staggering, and falling.

Pathology—Cracker-heels produces few distinctive gross changes. There may be nonspecific hepatic congestion, pulmonary emphysema and edema, excess pericardial and/or pleural fluid, and abomasal ulceration. Microscopic findings are similarly nonspecific, with the exception of changes in the brains of cattle with chronic cases of the disease. These changes include the presence of foci of necrosis in the thalamus, moderate necrosis of Purkinje cells in a few cerebellar folia, spongiosis in the white matter of the globus pallidus, distension of the lateral ventricles, and wallerian degeneration of the ventral and lateral columns of the spinal cord and the sciatic nerve.

- gross pathology, no lesions

  microscopic, in more-chronic cases, foci of necrosis in various regions of brain and degenerative lesions of spinal cord and sciatic nerve

Treatment—Providing alternative forage to animals exhibiting signs is the only treatment. Administration of thiamine has been suggested, but it has not proven of value. Induction of alcohol dehydrogenase is only preventive and is of questionable practical value. Artificial induction of transferable 3-NPOH degradation by ruminal microorganisms (enhanced by a 3-NPOH analog, nitroethane) is a promising approach, but it has yet to be fully developed and tested (Majak & Cheng 1983; Majak & Pass 1989; Majak 1991; Majak et al. 1998).

- nursing care, supplemental feeding

Selenosis

Disease Problems—Like locoism, selenium toxicosis appeared as a problem for ranchers in western North America in the second half of the nineteenth century as cattle and sheep ranching developed. There has been general agreement that the first recorded incidence of what appears to be the disease occurred in 1856 at Fort Randall, just west of the Missouri River, in what is now Gregory County, South Dakota. As recorded by an army surgeon, T. C. Madison, after only a 10-day encampment around the fort, horses suffered the loss of mane and tail hair and developed such severe hoof problems that they were prevented from effective grazing (Moxon 1937). Subsequently, the disease, albeit of unknown cause, became well recognized by stockmen in the area as a cause of chronic debilitation. In other areas of South Dakota, early settlers, encountering similar problems with grazing animals, associated the disease with mineral salts or alkaline water and began to call it alkali disease. As recorded in numerous annual reports of the Wyoming Agricultural Experiment Station in the

- originally, disease with hair loss and hoof deformities seen in areas with alkali flats and called alkali disease, thought to be similar to "blind staggers"

late 1800s and early 1900s, a similar disease was presenting problems for stockmen in Wyoming. Problems also appeared in Utah and Montana.

■ concept developed of selenium-induced disease complex composed of 3 syndromes—acute toxicity, blind staggers, chronic alkali disease

Studies on these disease problems commenced in earnest in the early 1920s. In South Dakota, they were conducted primarily by K. W. Franke from 1929–1936 with help from others, including personnel from the U.S. Department of Agriculture. In Wyoming, O. E. Beath was instrumental in leading the effort to elucidate the factors involved in the disease. Eventually it was shown to be associated with the presence of selenium in feed grains (Robinson 1933; Franke 1934; Byers 1935). At about the same time, several genera of plants, including *Astragalus,* were shown to be accumulators of high levels of selenium (Beath et al. 1932). The disease syndromes became known as alkali disease in South Dakota and blind staggers in Wyoming. Initially, stockmen in Wyoming used the terms *alkali disease* and *blind staggers* interchangeably because the diseases were quite similar, but eventually the names came to represent distinctly different combinations of signs (Beath et al. 1934). It also was noted that blind staggers resembled locoism and that under some experimental conditions, locoism was produced rather than blind staggers (Beath et al. 1932).

■ acute disease, rare, primarily due to excess selenium added to feed

subacute disease, blind staggers, a neurologic problem

chronic alkali disease, slow and reversible loss of mane and tail hair and deformed hooves

Eventually selenium toxicosis came to be regarded as a disease complex comprising three distinct syndromes that were due to an excess of selenium—acute selenosis, subacute/chronic blind staggers, and chronic alkali disease (Draize & Beath 1934; 1935; Beath et al. 1935). Typically appearing in the summer or fall, acute selenosis had a high case-fatality rate. Seen from early spring to late fall, blind staggers was described to be of sudden onset following prolonged ingestion and led to blindness, paresis/paralysis, and a moderate case-fatality rate. It was originally described as a chronic syndrome, but its classification was questioned because it seemed to represent a more acute condition than the chronic alkali disease. Based on observations made after its initial description, blind staggers was later designated a subacute form (Hershey 1945). Chronic alkali disease was considered to appear slowly after prolonged ingestion of plants and to have a low case-fatality rate. Affected animals exhibited reversible loss of hair and abnormalities of hoof growth. Although all three syndromes were initially reported to have been reproduced experimentally with plant sources of selenium (Trelease & Beath 1935), only acute selenosis and alkali disease subsequently have been consistently reproduced by administration of selenium. Furthermore, the neurologic signs of blind staggers have been difficult to reproduce with selenium compounds (Rhian & Moxon 1943).

■ blind staggers no longer considered syndrome of selenium excess

It is now recognized that whereas alkali disease is a manifestation of selenium excess, additional factors are involved with the disease called blind staggers, as is discussed below. It is important to note that perhaps all of this concern about the number of forms of selenium toxicosis is of little consequence, because the economic importance of selenium intoxication has diminished significantly. The disease seems to have assumed a rather minor status even in the so-called seleniferous areas of western North America, and the few cases that are seen are, for the most part, readily reversible (Raisbeck et al. 1993; Witte et al. 1993).

■ the toxins, forms of selenium found in plant tissues; mainly selenates

Disease Genesis—As its common name implies, selenosis is caused by high levels of selenium in the soil and its subsequent accumulation in the tissues of certain plant species growing in it. Selenium-rich soils occur throughout much of North America, especially in the eastern half. However, there, because of low soil pH and high moisture, most of it is tied up in ferric complexes and unavailable to plants (Berrow & Ure 1989). In contrast, in the western half of the continent, selenium occurs in a form (selenates) available for uptake by plants in shale-derived soils of arid/semiarid, mainly alkaline areas where leaching is minimal (Byers 1935; Williams et al. 1940, 1941;

Telease & Beath 1949). It must noted that selenosis may occur in the eastern half of the continent, albeit rarely, when soil and climatic factors interact in such a way as to permit uptake of selenium (Witte & Will 1993).

Concentrations of selenium vary with soil depth (Beath et al. 1935). Maximum levels usually occur 90–120 cm below the surface. High levels may occur as far down as 240 cm. Soils presenting a risk typically have selenium concentrations of 1–2 ppm; some may have them as high as 100 ppm (Rosenfeld & Beath 1964). In general, soils with 0.5 ppm or greater selenium should be considered hazardous. Selenium in the soil may occur in one of several forms: elemental, selenide (Se^{2+}), selenite (Se^{4+}), selenate (Se^{6+}), and organic compounds. Selenate is the most soluble form and selenide the least, in part because of the formation of metallic selenides. Whereas the uptake of selenium is restricted in nonaccumulating plant species, accumulating taxa may even take up selenites, some of which are bound by soil iron (Beath 1937; Moxon 1937; Williams & Mayland 1992). The selenates are readily absorbed and represent the most available source, yielding higher plant concentrations than even water-soluble organic seleniums (Johnson et al. 1967). Selenium accumulation by a plant often imparts a noxious sulfurous odor to the foliage; this is due mainly to dimethylselenide.

- plant species vary in proclivity to accumulate selenium; 4 groups—
  (1) primary, require Se for growth
  (2) secondary accumulators
  (3) passive accumulators, low Se levels
  (4) nonaccumulators

Differences among the three syndromes of selenosis were initially attributed to variation in selenium concentration or chemical form in the plants. It was suggested that water-soluble, organic selenium in plants produced blind staggers and that water-insoluble, bound selenium caused alkali disease (Moxon 1937). It was also noted that some species of *Astragalus* were very tolerant of high soil selenium and in fact grew preferentially in its presence (Trelease & Trelease 1938). This observation substantiated those made by Beath and led to a division of plants into three groups—high, moderate, and low accumulators—based on their proclivity to accumulate selenium (Trelease & Beath 1949). This classification can be further modified, and plants classified into at least four categories: primary accumulators, indicators, and converters; secondary accumulators; low or passive accumulators; and nonaccumulators. Species of *Astragalus* accumulating or indicating selenium are presented in table 37.3. Other species noted for selenium uptake include those of the genera *Mentzelia, Oonopsis* (= *Haplopappus*), *Stanleya,* and *Xylorhiza* (= *Aster*) (Trelease & Beath 1949; Rosenfield & Beath 1964; Ihnat 1989).

A primary accumulator, indicator, or converter species was described as having up to several thousand ppm of selenium, e.g., 15,000 ppm or 1.5% d.w. in *A. racemosus* (Wilber 1980). Indicator species are found typically in association with elevated levels of readily available soil selenium but do not necessarily have a high selenium content; for example, some populations of *A. bisulcatus,* traditionally described as an indicator species, lack the sulfurous odor characteristic of selenium plants (Welsh et al. 1987). Furthermore, some of the so-called indicator species may grow but not persist in nonseleniferous soils. Selenium concentrations may be low enough that other toxicants such as swainsonine may represent a greater hazard. Converter species, also known as translocator species, such as some of the deep-rooted species of *Astragalus,* appear to bring selenium up from the deeper soil layers and thus increase its availability to nearby plants such as corn, wheat, and other grains, which are more likely to be eaten (Trelease & Beath 1949).

In contrast, secondary accumulators have levels up to several hundred ppm. They include species of a number of genera such as *Aster, Atriplex, Grindelia, Gutierrezia, Iva, Machaeranthera, Penstemon,* and *Solidago* (Trelease & Beath 1949). Species that are low or passive accumulators typically have concentrations in the range of 5–10 ppm or up to 50 ppm. They are termed low or passive because their uptake of selenium is restricted relative to that of accumulators, and they include the cereal grains

Table 37.3. Species of *Astragalus* accumulating or indicating presence of selenium

A. albulus Wooton & Standl.	Cibola milk vetch
A. beathii Ced.Porter	Beath milk vetch
A. bisulcatus (Hook.) A.Gray	two-grooved milk vetch
var. *bisulcatus*	
(= *A. diholcos* Tidestr.)	
var. *haydenianus* (A.Gray ex Brandegee) Barneby	Hayden milk vetch
(= *A. haydenianus* A.Gray ex Brandegee)	
A. crotalariae (Benth.) A.Gray	Salton milk vetch
A. eastwoodiae M.E.Jones	Eastwood milk vetch
A. flavus Nutt.	yellow milk vetch
(= *A. argillosus* M.E.Jones)	
(= *A. confertiflorus* A.Gray)	
A. grayi Parry ex S.Watson	Gray's milk vetch
A. moencoppensis M.E.Jones	Osterhout's milk vetch
A. oocalycis M.E.Jones	Arboles milk vetch
A. osterhoutii M.E.Jones	Moenkopi milk vetch
A. pattersonii A.Gray	Patterson's milk vetch
A. pectinatus (Hook.) Douglas ex G.Don	tine-leaved milk vetch
A. praelongus E.Sheld.	stinking milk vetch, straight-stem locoweed
(= *A. recedens* Ced.Porter)	
A. preussii A.Gray	Preuss milk vetch
A. racemosus Pursh	creamy locoweed, alkali milk vetch
A. sabulosus M.E. Jones	straight-stem poison vetch
A. saurinus Barneby	dinosaur milk vetch
A. tenellus Pursh	pulse milk vetch
A. toanus M.E.Jones	Toano milk vetch

Sources: Trelease & Beath 1949; Rosenfeld & Beath 1964).
Note: Nomenclature follows that of Barneby (1964a,b).

and various forage plants such as alfalfa. The cereal grains, especially wheat, can represent a hazard, with selenium concentrations sometimes 5–10 ppm or more (Robinson 1933). The fourth category is needed to include those species with selenium levels at less than 0.1 ppm, inasmuch as some species of *Astragalus,* the closely related *Oxytropis,* and *Lupinus* do not seem to accumulate selenium under any soil conditions (Davis 1986).

■ animals most likely to eat passive accumulators; plants with higher levels have sulfurous odor and are unpalatable

Most native grasses accumulate little selenium, while some native forbs may be found with concentrations higher than 3,000–4,000 ppm. In spite of the remarkably high concentrations of selenium in indicator species, it appears that selenosis occurs most often when animals consume the passive accumulators that are more likely to be eaten, such as the native grasses and grain forages. Even though species of *Astragalus* are traditionally associated with selenosis, there is some question as to whether they have a direct and major role in causing the disease, because most are unpalatable. Their role as converters or translocators is thought to be more significant (Trelease & Beath 1949). The greater feed bulk provided by the passive accumulator grasses and grain forages also is an important factor.

Once taken up by the plant, regardless of what type of accumulator it is, selenium is found in all organs, with high concentrations in flowers and seeds (Rosenfeld & Beath 1964; Cowgill 1981). In general, plant tissues with 5 ppm or greater selenium should be considered hazardous. Selenium content and chemical form in the foliage varies with stage of growth. In many of the passive accumulators, concentrations are high early in growth, decline slowly until maturity, and decline rapidly thereafter (Olson et al. 1942). In contrast, peak concentrations in other plants, including the indicator species of *Astragalus,* occur just prior to and during flowering (Trelease & Beath 1949; Rosenfeld & Beath 1964). After maturity, selenium levels decline markedly, but the plant crown and roots retain sufficient levels so that succulent new growth tends to have high concentrations.

Once absorbed by plants, selenium regardless of its form in the soil, moves through the metabolic pathways for sulfur and is incorporated into various sulfur-amino acids and proteins (Olson et al. 1970; Gunther 1974). The forms in which it occurs in plants vary with the plant species. Some effects detrimental to the plant are produced, e.g., impairment of protein synthesis (Eustice et al. 1981). The accumulator species of *Astragalus* are more tolerant of selenium because incorporation into proteins is limited (Brown & Shrift 1981). The organic forms of selenium, such as selenomethionine, are the most toxic. Sodium selenite is only about half as toxic as an equivalent amount of the selenium in wheat (Fitzhugh et al. 1944; O'Toole & Raisbeck 1995).

Selenium absorbed by the animal is widely distributed throughout the body. It is transported in blood, as selenite bound to red blood cells and as selenide in plasma proteins. Some biotransformation of selenomethionine and selenocysteine occurs in the rumen. Selenium is eliminated as selenide, mainly in the urine as trimethylselenide, and to a lesser extent in milk and bile as metalloselenides and in expired air as sulfurous- or garlicky-smelling dimethylselenide (Dudley 1936; Moxon 1937; McConnell 1948; McConnell & Cooper 1950; Ewan 1989). Biliary excretion is augmented by arsenic and various metals (Levander & Baumann 1966; Hill 1975). Under some conditions, monogastric animals may be at greater risk because they lack the capacity for the selenite reduction that occurs in the rumen (Crinion & O'Connor 1978; Goyer 1991). In general, dosage of plant material supplying 2–3 mg selenium/kg b.w. per day is toxic if given for a sufficient duration (Rosenfeld & Beath 1964).

■ found in organic forms, selenomethionine, selenocysteine; incorporated as amino acids into protein; may render cells more susceptible to destructive oxidative processes

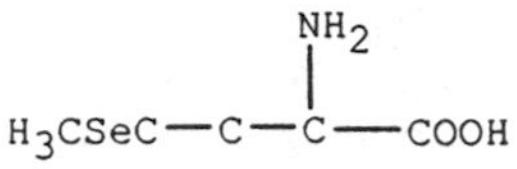

$$H_3CSeC-C-\underset{}{\overset{NH_2}{\overset{|}{C}}}-COOH$$

Selenomethionine

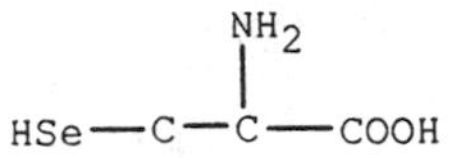

$$HSe-C-\overset{NH_2}{\overset{|}{C}}-COOH$$

Selenocysteine

Selenium is toxic to the cell in some forms; free cysteine forms are particularly so (Martin & Hurlbut 1976). As a protective response, selenocysteine is incorporated into glutathione peroxidase or converted to other organoselenium forms, including selenomethionine (Burk 1991). As in plants, selenium in animals follows sulfur pathways, including its incorporation into proteins as analogs of sulfur amino acids where there is risk of disrupting disulfide linkages (McConnell & Wabnitz 1957; Brown & Shrift 1980). Initially, selenium is found as selenotrisulfides where it is linked between two sulfurs. It is eventually converted to more-permanent amino acid/protein forms (Gunther 1974). Sulfhydryl-dependent glycolytic and cellular respiratory enzymes seem to be particularly vulnerable to selenium (Wright 1940; Rosenfeld & Beath 1964). It is tempting to ascribe all toxic effects of selenium to its incorporation into proteins because of its association with sulfur-amino acids and its subsequent interference with the structure and function of proteins/enzymes, but this has not been conclusively shown. Certainly, selenium is found in appreciable concentrations in those tissues most affected in disease, but the precise role it plays in the degenerative effects is not clear. It has been hypothesized that selenium compounds are toxic because of their propensity to form superoxide, hydrogen peroxide, and other oxyradicals, with formation countered in tissues by various defense mechanisms, including methylation reactions (Spallholz 1994).

Although consideration of selenium as a toxicant usually brings to mind the acute and chronic syndromes, additional ones need to be considered. The effects on hooves, which are distinct from chronic laminitis, may cause interference with effective grazing, thereby resulting in weight loss; fortunately this is usually reversible (Moxon et al. 1944; O'Toole & Raisbeck 1995). In chickens, the effects of excess selenium include decreased production, weight, and hatchability of eggs; deformities of beaks and eyes; edema; and wiry down (Moxon 1937; Ort & Latshaw 1978). The potential for selenium-induced reproductive effects in birds is further underscored by the bizarre anomalies ascribed to the "Kesterson effect or syndrome," e.g., multiple deformities, including missing eyes, beaks, and extremities (Ohlendorf et al. 1986; Harris 1989; Presser 1994). Although teratogenic effects due to excess selenium have been

■ reproductive effects of some concern, based upon effects reported in wild birds subjected to environmental selenium; these effects unconfirmed for livestock grazing seleniferous forages

suspected in pigs, sheep, and cattle, they have not been confirmed experimentally (Trelease & Beath 1949; Wahlstrom & Olson 1959; Dinkel et al. 1963; Panter et al. 1995; Yaeger et al. 1998). Studies in hamsters revealed that malformations (mainly encephaloceles) occurred only with dosage that produced significant maternal intoxication (Ferm et al. 1990; Willhite et al. 1990).

■ blind staggers as seen at present is more likely a manifestation of bovine polio—PEM

There is little question of the broad range of disease conditions caused by selenium—from acute to chronic. However, because of the difficulty in reproducing the syndrome known as blind staggers, whether it really is a disease caused by excess selenium is still a question (Olson 1978; Van Kampen & James 1978; James et al. 1992). As described in the preceding sections in this treatment of *Astragalus,* it is well known that other toxicants in addition to selenium are present in some species and, under appropriate conditions, may play a primary role in disease expression. In particular, the similarity between locoism and blind staggers had been noted in early investigations (Beath et al. 1934; Hershey 1945). The possibility of multiple toxins seems especially true of *A. bisulcatus,* one of the most notorious selenium accumulators but quite variable in its toxicity. In an early description of the disease caused by this species, cattle were said to "act crazy," especially when startled (Beath & Lehnert 1917). Beath and his co-workers (1935) considered *A. bisulcatus* to be the most important toxic plant in Wyoming and noted that it was toxic in all soil types, although more so in selenium-rich soils. Furthermore, it was not invariably malodorous and at times was readily eaten, even sought out and eaten in preference to good forage (Beath et al. 1932). It eventually was suggested that the loco toxin and selenium coexisted in some species of *Astragalus* (Moxon 1937; Maag & Glenn 1967). A number of studies have been carried out to elucidate the role of the various toxins in the diseases caused by selenium-accumulating species of *Astragalus.* Experimentally, pigs, cattle, and sheep fed *A. bisulcatus* or *A. praelongus* invariably developed signs of locoism, while those given selenium showed signs typical of selenism (Maag & Glenn 1967; Van Kampen & James 1978; James et al. 1983b; Baker et al. 1989; James et al. 1994). These results are supported by field studies and are in contrast to earlier studies in which indicator species such as *A. bisulcatus* produced lesions typical of selenosis in pigs (Hartley et al. 1985). In a 3-year survey in Wyoming, 6 cases of neurologic disease in cattle were diagnosed in the field as blind staggers, but none were confirmed with necropsies. Five of the 6 cases were confirmed as PEM, with only 1 animal exhibiting elevation of tissue selenium (Raisbeck et al. 1993). Four of the cases were associated with high sulfate levels (2,800–6,000 ppm) in the animals' drinking water. These findings are not totally unexpected, because the signs of blind staggers and PEM are quite similar, albeit the latter has a higher case-fatality rate (Jensen et al. 1956). The concentrations of sulfur rather than selenium, and those of swainsonine in the plants, may be the determinants of which disease prevails. Moderation of toxicity or development of tolerance due to ruminal degradation of selenium such that the development of locoism would be favored is also a possibility (Rasmussen & James 1993, 1994).

The pathophysiology of selenosis in the animal may be helpful in deciphering the disease complex. Selenium accumulates in and causes degeneration of selected tissues, especially of the heart, liver, and kidney. Thus, in sheep dying of acute selenosis after prolonged ingestion, the lesions are those of myocardial degeneration (flabby "dishrag" heart) and also of the liver and kidney, either from the selenium itself or secondary to cardiac failure (Glenn et al. 1964b; James et al. 1982). The same types of lesions are present in chronic alkali disease, but in most instances they are not severe enough to cause death. Instead, hair and hoof changes are more noteworthy. The opposite situation also occurs, such as in pigs exhibiting acute intoxication due to excess selenium in the feed. In this situation, the lesions are mainly associated with paralytic spinal poliomalacia, but they also include, of lesser significance, coronary band changes, hoof separation, and hair loss (Miller & Schoening 1938; Harrison et al. 1983; Mahan & Moxon 1984; Casteel et al. 1985; Hartley et al. 1985). In horses, even in a case of acute selenosis caused by the inadvertent administration of sodium selenite for only 5 days, hair and hoof changes predominated (Dewes & Lowe 1987). This supports the historical view that selenium toxicosis, varying from acute to chronic, is one disease with lesions shifting from mild to severe (Draize & Beath 1935). This same continuum has been apparent in studies in dogs, in which signs of both blind staggers and alkali disease were reported (Rhian & Moxon 1943).

Based on our present understanding of selenium toxicosis, a revised classification of its disease syndromes in livestock is proposed here. Two syndromes are recognized (Raisbeck et al. 1998a). One is acute selenosis, which is characteristic of sheep and rarely occurs in other species. Regardless of the time period of selenium ingestion, sheep die very quickly with few signs (Beath et al. 1932; Glenn et al. 1964a,b; James et al. 1982).

- animal species vary in responses; 2 syndromes—
  (1) acute deaths, cardiac failure in sheep
  (2) chronic disease, hoof and hair changes in horses, cattle, and pigs; subacute blind staggers excluded as disease of selenium excess

The second syndrome typically seen in other animal species is chronic selenosis, or alkali disease. In contrast to the disease in sheep, the disease in horses, whether exposure is months, weeks, or a few days, seems always to eventuate in the hoof and hair changes of alkali disease. This has consistently been seen in North America, as well as in Australia and Ireland (Knott & McCray 1959; Crinion & O'Connor 1978; Hultine et al. 1979; McLaughlin & Cullen 1986; Dewes & Lowe 1987). Neurologic effects also may occur as an expression of selenium toxicity. They are clearly part of the syndrome in pigs, whether the ingestion period is long or short, and can be related to lesions in the spinal cord and brainstem. Perhaps a third syndrome or effect, associated with compromised immune function, eventually may be recognized (Raisbeck et al. 1998b).

In this treatment, blind staggers is excluded as a syndrome of the selenium disease complex. As discussed in depth by O'Toole and co-workers (1996), there appears to be little justification for considering it to be a consequence of selenium ingestion. Raisbeck and co-workers (1993) contended that it is probably misdiagnosed much of the time and in reality is more likely to be PEM or locoism, even though early investigators were quite explicit in their descriptions of the disease and were sure that it was different from locoism (Draize & Beath 1935; Hershey 1945).

Clinical Signs—The acute form of selenosis is quite rare and is more likely to be caused by ingestion of feed additives rather than plants. Sheep may exhibit signs of cardiac failure, such as labored respiration and exercise intolerance, but they die quickly and signs are generally not observed. The acute disease is otherwise mainly a toxicosis of the digestive system except in pigs, in which there is a paralytic effect due to spinal poliomalacia. In cattle, clinical signs include watery diarrhea, colic, bloody froth from the nose, bloat, prostration, and eventually death due to respiratory failure.

- acute disease, sheep, labored respiration, exercise intolerance, sudden death; pigs, paralysis; others, diarrhea

Chronic selenosis, or alkali disease, is the more commonly encountered form of selenium intoxication and is especially characteristic of horses. It is not necessarily a syndrome of long-term ingestion; signs in some instances occur after only a few weeks. Affected animals become stiff in their movements, dull in behavior, and lack overall vitality. Hair coats are rough and unkempt in appearance, the mucous membranes are pale, and there is marked weight loss. The most characteristic changes are those involving the hair and hooves. Hair loss is primarily of the long hairs of the mane and tail, and animals exhibiting severe cases are sometimes said to have "bobtail disease." The effects on the hooves are initially soreness and cracking, followed by

- chronic, stiffness, rough hair, loss of mane and tail hair, bobtail; hoof cracking, soreness, and deformed growth

deformities in growth and eventual sloughing of the hooves. Hoof or nail deformities and hair loss (body hair in pigs) are typical of the disease in all species except sheep.

■ diagnosis, tissue concentrations of selenium, blood, liver, heart, hair

Diagnosis can be aided by determining concentrations of tissue selenium, which should be less than 1 ppm in all tissues. Normal concentrations are 0.1–0.5 ppm in fresh tissues and 4 times higher in dried tissues (Gunther 1974). Concentrations with intoxication will be 5–10 ppm or more in the liver and kidney and 1–5 ppm in other tissues such as blood, heart, lung, and spleen. Selenium levels in the hair are quite variable, but values greater than 5 ppm are considered elevated. In some instances, more than 1 ppm may be significant. Because most of the selenium in blood is bound to the red blood cells, whole blood is preferred to serum for analysis (Miller & Williams 1940; Rosenfeld & Beath 1945; Glenn et al. 1964c; Maag & Glenn 1967; Witte et al. 1993; Raisbeck et al. 1998a). Selenium determination can be made from formalin-fixed as well as fresh liver (Sullivan et al. 1993). Values in fixed pig liver were shown to be 86% of those in fresh frozen liver. Selenium concentrations in the urine should be less than 100 µg/l.

The original descriptions of the clinical signs of blind staggers were probably based exclusively on observations of the disease in cattle. Onset of the disease varied from abrupt to somewhat gradual. Signs began with the animal wandering away from herd mates, stumbling, and exhibiting apparent impairment of vision. Secondarily, blindness became pronounced; an appetite craving bone, metal, or wood developed; and the gait became unsteady because of weakness of the thoracic limbs. In a few days, there was aimless wandering, circling, impaired swallowing, tongue paralysis, colic, labored respiration, and the appearance of swollen eyelids and cloudy corneas. As is apparent, the clinical signs of blind staggers are quite different from those of acute and chronic selenosis, thus supporting the contention that it is not due to selenium.

■ gross pathology, sheep, flabby heart, pulmonary edema

pigs, spinal poliomalacia

others, reddening of intestinal mucosa; chronic hair and hoof changes

microscopic, moderate degenerative-type lesions of heart, liver, kidneys, spleen, lungs

Pathology—In most instances, the pathologic changes caused by selenium will be mainly in the skin, hooves, mane, and tail. Microscopically, as observed in cattle, the tubules of the stratum medium are replaced by parakeratotic cellular debris mixed among the normal matrix, and there are various other abnormalities of the hoof epithelium as well (Raisbeck et al. 1993; O'Toole & Raisbeck 1995; Raisbeck & O'Toole 1998). Atrophic hair follicles, areas of dyskeratosis, and superficial follicular keratosis may be present in the tails of affected animals. There also may be multifocal myocardial necrosis and spongiform change in the area of the basal ganglia and brainstem.

In the acute deaths of sheep the pathologic changes are suggestive of cardiac involvement and include a flabby "dishrag" heart and pulmonary congestion and edema, with blood-tinged froth in the upper airways.

In addition to the hair and hoof lesions described for cattle and horses, pigs develop degenerative lesions of the spinal cord (Baker et al. 1989; Stowe et al. 1992). The changes, which have been reproduced with both organic and inorganic forms of selenium, are characterized as poliomalacia, which is especially prominent in the cervical and lumbar intumescences. There also may be polioencephalomalacia. In addition, selenosis in pigs is associated with skeletal and focal myocardial degeneration, leading to fibrosis, and degenerative hepatic changes.

■ reversible with change in diet; arsenic added to diet has protective effect

Treatment—Chronic selenosis, or alkali disease, is typically reversible after elimination of the animal's source of selenium. Increased protein in the diet may be beneficial because low protein is associated with increased susceptibility (Rosenfeld & Beath 1964). Metals such as copper form complexes with selenium, possibly aiding excretion (Stowe 1980). Arsenic, provided in some form, as in the salt or mineral mix, may help to increase biliary excretion and act as a preventive (Moxon et al. 1944; Minyard et al. 1960; Hill 1975). Hershey (1945) has suggested a mix of 4.2 g of sodium arsenite with 100 kg of salt, or 8.4 g of sodium arsenite with 100 kg of salt and 100 kg of mineral mix. The originally recommended treatment consisted of an oral drench of 4–10 l of water and 5 mg of strychnine given s.c. every 2 hours for 3 or 4 doses for an adult cow. This treatment was continued to the point of causing slight hyperesthesia.

Baptisia Vent.

- *B. alba*
  B. australis
  B. xbicolor
  B. bracteata
  B. tinctoria

false indigo
wild indigo

Taxonomy and Morphology—Comprising approximately 17 species and 6 named hybrids, *Baptisia,* a member of the subfamily Papilionoideae, is a North American genus of perennial herbs. The genus name is derived from the Greek root *bapto,* meaning "to dye," which is indicative of the use of some species as a substitute for indigo (blue dye). This use also is reflected in common names of the species. Species of toxicological interest include the following:

B. alba (L.) R.Br.	white wild indigo, Atlantic wild indigo, prairie false indigo
B. australis (L.) R.Br.	blue wild indigo, blue false indigo
B. x*bicolor* Greenm. & Larisey	
B. bracteata Muhl. ex Elliott	plains wild indigo
B. tinctoria (L.) R.Br.	yellow wild indigo

Baptisia australis

Several important changes in classification and nomenclature within this genus have occurred. Originally described as a species, *B. bicolor* is now recognized to be a naturally occurring hybrid between *B. australis* and *B. bracteata.* The binomials *B. leucophaea* and *B. leucantha,* used in early reports of intoxication problems with the genus, are now synonyms of *B. bracteata* and *B. alba,* respectively.

- perennial herbs; leaves palmately compound, stipules large in some; flowers pealike; petals white, cream, yellow or blue-violet; legumes with conspicuous beak

Plants perennials; herbaceous; caulescent; from large woody caudices. **Stems** usually solitary; branched above. **Leaves** palmately compound; sessile or short-petioled; leaflets 3, oblanceolate to obovate, margins entire; stipules large or small, persistent or caducous. **Inflorescences** terminal or axillary racemes; few to many flowered; bracts persistent or caducous. **Sepals** 5; fused. **Corollas** papilionaceous. **Petals** 5; white or cream or yellow or blue-violet; banners reniform to suborbicular, reflexed; wings clawed, straight; keels straight; petals nearly straight. **Stamens** 10; separate; enclosed in keels. **Legumes** stipitate; globose to oblong; conspicuously beaked; membranous to indurate. **Seeds** 1 to many.

Species of the genus are readily recognized by the palmately compound leaves and plump legumes with curved beak. In addition, the plants characteristically turn a black-gray or silvery gray as they dry in the field or in hay. Identification of individuals is sometimes difficult because of the hybridization and introgression among the species.

1. Banners blue-violet or purple.
   2. Racemes erect. Wings and keels blue-violet or purple . . . . . . *B. australis*
   2. Racemes ascending. Wings and keels cream or yellow . . . . . *B.* x*bicolor*
1. Banners yellow or white or cream.
   3. Petals yellow . . . . *B. tinctoria*
   3. Petals white or cream.

Baptisia bracteata

4. Leaves and legumes glabrous. Flowers secund.
   Pedicels at maturity 5–10 mm long . . . . *B. alba*
4. Leaves and legumes villous. Flowers not secund.
   Pedicels at maturity 15–35 mm long . . . . *B. bracteata*

■ open woods and prairies

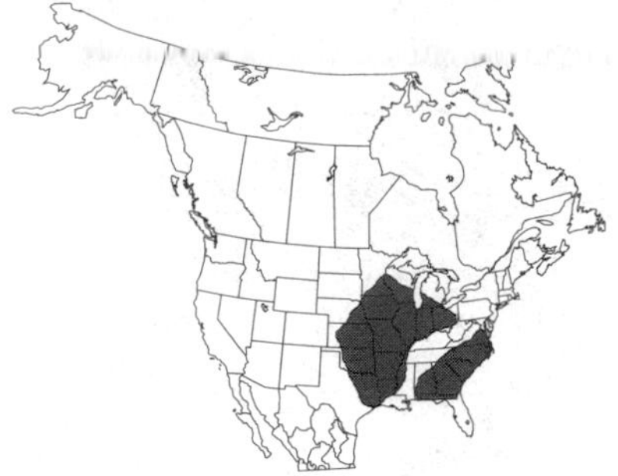

Baptisia alba

Distribution and Habitat—*Baptisia* is a genus of the eastern half of North America. With the exception of *B. tinctoria,* which is strictly an open-woods species, all species are encountered in both prairies and open woods. In general, *B. australis* prefers clay and loam soils, whereas *B. alba* typically is found on sand. *Baptisia bracteata* is found in clay, loam, and sand.

■ plants toxic but rarely eaten, especially the more toxic, bitter legumes; may cause digestive disturbance or neurologic problems

Disease Problems—There is uncertainty as to whether species of *Baptisia* pose a significant disease problem. Little experimental work has been conducted. The foliage of *Baptisia* is commonly present in hay but is not normally eaten by horses or cattle. It may be eaten by goats, but with no reported adverse effects. Feeding studies of 6 species from the southeastern United States produced no adverse effects in chicks at doses up to 3% b.w. (Duncan et al. 1955). However, as early as 1917, *B. alba* (as *B. leucantha*) had been reported to contain an emetic or cathartic toxin (Pammel 1917b; Gates 1930), and in 1930 it was reported to cause severe diarrhea and death in cattle consuming it fresh or in hay (Hansen 1930). Pregnant rats experimentally fed the seeds and legumes of *B. australis* daily at levels of 15% of the animals' diet developed tremors and a rough unkempt appearance and either died or aborted all fetuses after 10–12 days (Burrows et al. 1998).

■ quinolizidine alkaloids, cytisine types, act upon cholinergic receptors, especially nicotinic; also present are teratogenic types anagyrine and baptifoline

Disease Genesis—Studies of the chemical constituents of *Baptisia* revealed the presence of quinolizidine alkaloids (Marion & Cockburn 1948; Marion & Ouellet 1948; Marion & Turcotte 1948; Turcotte et al. 1953; Cranmer & Mabry 1966). Cytisine subsequently was confirmed as common to all species and, more often than not, the major alkaloid present. In addition to cytisine, *N*-methylcytisine is common to most species. Anagyrine also is abundant, being present in more than half the species and, in some, representing 25–50% of the total alkaloid concentrations. Baptifoline also is present in a few species. A number of other alkaloids have been isolated, but most are present only in small concentrations. In *B. australis,* total alkaloid concentrations range from 100 to 5,000 ppm (Oung-Boran et al. 1969; Wink et al. 1983). The main alkaloids are *N*-methylcytisine 74%, cytisine 12.9%, and anagyrine 6.6%. Several others, including sparteine, tinctorine, rhombifoline, and lupanine, are present at less than 2%.

The primary effect of these quinolizidine alkaloids is upon nicotinic, and to a lesser extent on muscarinic, acetylcholine receptors (Schmeller et al. 1994). Extracts of *B. australis* in vitro, caused a decrease in the rate and force of contractions of the right atria of guinea pigs, and cardiac arrest at higher dosage (Oung-Boran et al. 1969). In rabbits, in vivo, there was a decrease in blood pressure. Based upon these results, the aerial portions, whether fresh or dried in hay, should be considered toxic if eaten in sufficient quantities.

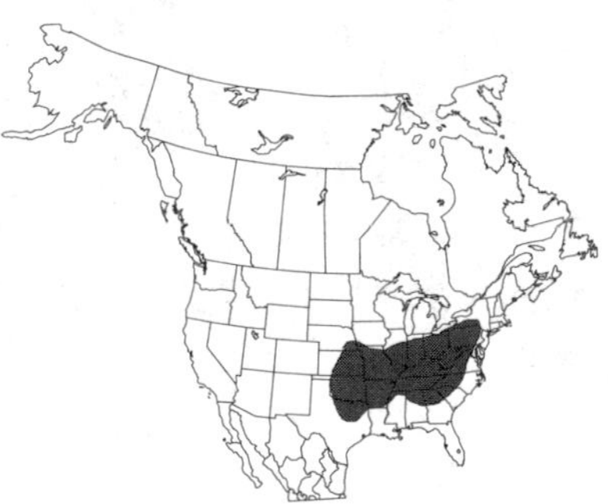

Baptisia australis

Baptisia xbicolor

Cytisine

Anagyrine

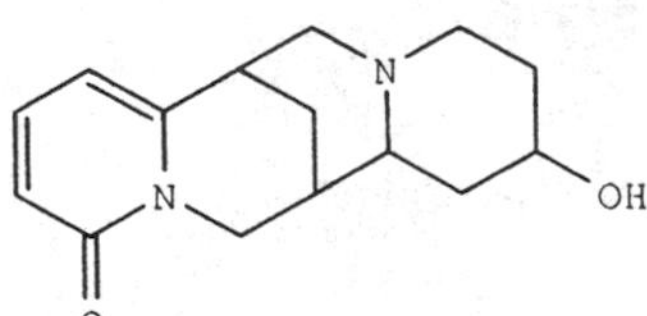

Baptifoline

The presence of pyridone alkaloids such as anagyrine and baptifoline in high concentrations in many species suggests a teratogenic potential (Leonard 1953; Cranmer & Mabry 1966; Wink et al. 1981). Isoflavone glycosides—formononetin and luteolin—also have been reported to occur in *B. australis,* implying a potential for detrimental reproductive effects (Oung-Boran et al. 1969). A more complete discussion of these alkaloids is presented in the treatment of *Laburnum,* in this chapter.

■ abrupt onset, diarrhea, ataxia, tremors; possibly fetal deformities

treatment is general nursing care

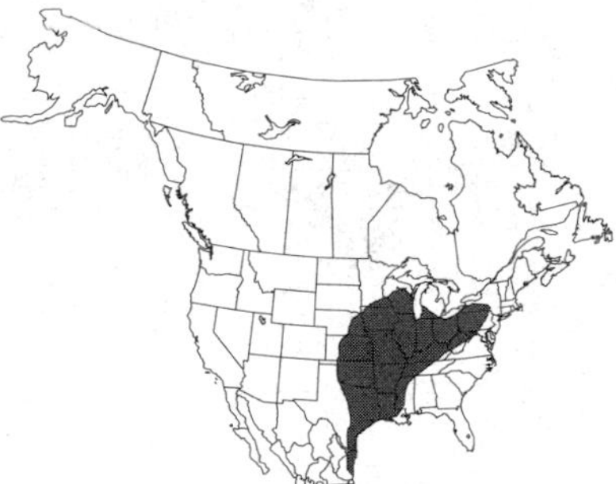

Baptisia bracteata

Clinical Signs, Pathology, and Treatment—Except for severe diarrhea in cattle, specific signs have not been described for animals eating *Baptisia.* However, based upon the types of toxins present in the genus, it may be anticipated that the signs would be similar to those caused by *Laburnum.* Those include decreased appetite, excess salivation, incoordination, and tremors. Fetal abnormalities also may be expected.

Because the effects of these quinolizidine alkaloids are physiologic, primarily on nicotinic receptors, there is unlikely to be any distinctive gross or microscopic pathology.

Conservative supportive care is the only treatment indicated.

Baptisia tinctoria

Caesalpinia L.

Taxonomy and Morphology—Comprising between 70–150 species, *Caesalpinia,* a member of the subfamily Caesalpinioideae, is a large heterogenous genus distributed throughout the world. The generic name honors Andrea Cesalpini, an Italian taxonomist. In North America, it is represented by some 25 native and introduced species, of which 3 are of possible toxicologic significance:

C. coriaria (Jacq.) Willd.	divi-divi
C. gilliesii (Wallich ex Hook.) Benth.	bird-of-paradise, yellow bird-of-paradise, paradise poinciana
C. pulcherrima (L.) Swartz	Barbados pride, dwarf poinciana, maravilla, tabachin, flor de Camaron, peacock flower, Barbados flower fence, chinchemalinche

■ *C. coriaria*
C. gilliesii
C. pulcherrima

Barbados pride	flower fence
bird-of-paradise	maravilla
chinchemalinche	peacock flower
divi-divi	poinciana
flor de Camaron	tabachin

In older treatments, the generic name *Poinciana* was used for several species of *Caesalpinia.*

Plants perennials; trees or shrubs or herbs; armed or not armed with thorns or prickles. **Stems** erect or climbing. **Leaves** 2-pinnately compound; pinnae 3–30; pairs of leaflets 2–25 per pinna; terminal leaflet absent; petiolules present; stipules herbaceous, caducous or persistent; typically producing a disagreeable odor when crushed. **Inflorescences** terminal racemes or panicles; bracts caducous. **Flowers** showy; symmetry slightly bilateral. **Sepals** 5; fused. **Petals** 5; subequal; yellow to orange or red; clawed. **Stamens** 10; filaments free; long exserted beyond petals. **Legumes** ovate to lanceolate; flattened; straight or twisted; pericarps thick-walled or membranous. **Seeds** 2 to many.

■ erect herbs, shrubs, trees; leaves 2-pinnately compound, ill-scented when crushed; flowers showy; petals yellow, orange, or red; stamens 10, prominent, long

1. Pairs of leaflets 15–25 per pinna. Stamen filaments green. Legumes curved . . . . *C. coriaria*
1. Pairs of leaflets 6–10 per pinna. Stamen filaments red. Legumes straight.
    2. Young branches glandular. Leaflets 2 mm wide; 8 mm long; oblong . . . . *C. gilliesii*
    2. Young branches not glandular. Leaflets 15 mm wide; 25 mm long; elliptic to obovate . . . . *C. pulcherrima*

Caesalpinia gilliesii

Caesalpinia pulcherrima

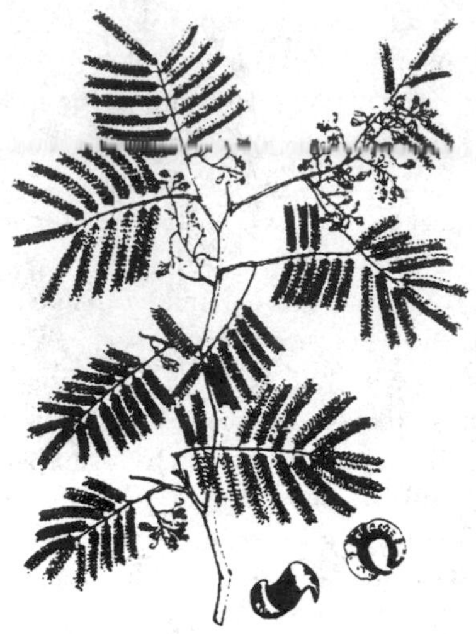

Caesalpinia coriaria

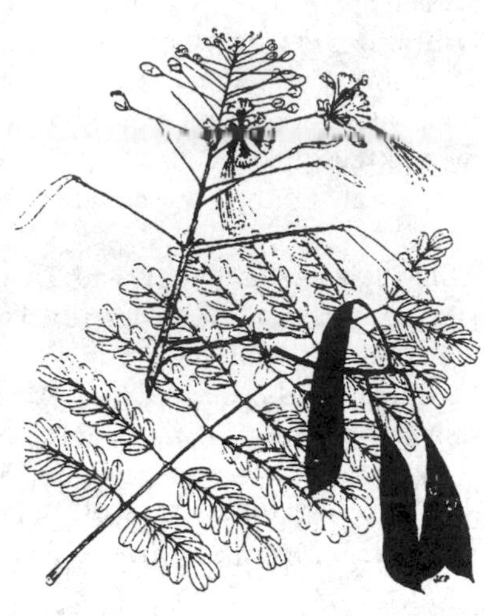

Caesalpinia pulcherrima

- cultivated ornamentals in warmer areas; naturalized

Distribution and Habitat—Encountered in both the Old World and the New World, *Caesalpinia* is primarily a tropical genus of dry, open scrub or lowland rain forests. In North America, native species are found in the Southwest. Introduced ornamentals are cultivated only in the warmer, less frost-prone areas.

- transient digestive disturbance
- toxins acting on digestive tract unknown; may be tannins, irritant diterpenoids, or proteins

O, CCOOH, CH_3, CH_3, OH, H_3C, OH, CH_3, OH

Pulcherralpin

Disease Problems—Although the seeds of species of *Caesalpinia* were eaten by members of western American Indian tribes after appropriate preparation (Ebeling 1986), ingestion of the legumes can cause severe gastrointestinal problems. Two boys who ate 5 legumes each of *C. gilliesii,* which are very similar in appearance to garden peapods, suffered severe digestive problems lasting for 24 hours (Cann & Verhulst 1958). The disease, which was experimentally reproduced in animals, included profuse diarrhea. In addition, dry leaves of *C. pulcherrima* fed to rabbits have been reported to cause pulmonary edema and hydrothorax (Watt & Breyer-Brandwijk 1962).

Disease Genesis—The legumes of *C. coriaria* contain high concentrations (30–50%) of gallotannins, which produce potent astringent effects (Watt & Breyer-Brandwijk 1962). Leroux (1986) attributed vomiting and diarrhea, sometimes with blood, to phytohemagglutinins in the flowers and fruits of mimosa du Japon, another species of *Caesalpinia.* There also are diterpenoids, pulcherralpin, and caesalpin, which may play a role in causing digestive tract problems (Connolly & Hill 1991). Little experimental work, however, has been conducted to elucidate the toxicity of this genus.

- nausea, vomiting, diarrhea

  treatment, control vomiting and diarrhea

Clinical Signs and Treatment—Nausea and persistent vomiting may occur a few hours after ingestion of the seeds, followed later by diarrhea. Treatment, if needed, is directed toward relief of the vomiting and diarrhea.

Canavalia DC.

- *C. ensiformis*

  horse bean
  jack bean
  passionflower
  stock bean
  wonder bean

Taxonomy and Morphology—Native primarily to the tropics and subtropics of the New World, *Canavalia,* a member of the subfamily Papilionoideae, comprises some 50 species. Of the 14 species reported for North America, only 1 appears to be of toxicological importance:

C. ensiformis (L.) DC. jack bean, horse bean, wonder bean, giant stock bean, passionflower

- annual viny climbers; leaves 1-pinnately compound; flowers pealike; petals lavender to purple-pink; legumes reddish purple to lavender

Plants annuals; viny climbers. **Stems** erect or suberect; to 2 m long. **Leaves** 1-pinnately compound; leaflets 3; ovate-oblong to elliptic; apices acute or acuminate or mucronate; margins entire; stipules and stipels small, caducous. **Inflorescences** pendulous racemes; borne on axillary peduncles. **Flowers** large. **Sepals** 5; fused; calyces campanulate, 2-lipped. **Corollas** papilionaceous. **Petals** 5; lavender to purple-pink; banners orbicular, sides reflexed; wings clawed, free; keels curved. **Stamens** 10; diadelphous; enclosed in keels. **Legumes** linear; flattened; 3 cm wide; 20–40 cm long; reddish purple to lavender. **Seeds** 4–15; flattened, but plump; white with a brown line half the length.

Canavalia ensiformis

Distribution and Habitat—Native to South America, *C. ensiformis* is cultivated in the warmer regions of North America as an ornamental, a cover crop, and a food crop for its beans.

- ornamental or crop plant

Disease Problems—Although capable of causing irritation of the digestive tract, *C. ensiformis* is of little toxicological importance in North America because of its limited use and the large number of seeds required for intoxication. In India, the seeds, with proper preparation, are eaten by some native peoples (Rajaran & Janardhanan 1992b). Digestive tract disease, possibly with death, has been observed in rats, chicks, and cattle fed large amounts of the seeds or beans (Orru & Cesaris Demel 1948; Shone & Drummond 1961; Jayne-Williams 1973). In Zimbabwe, 3 oz of seeds/kg b.w. fed to cattle over 3 days caused severe diarrhea and dehydration (Shone 1961). After heating, the seeds were no longer toxic when fed to cattle. In field cases from the same area, involving cattle eating *C. ensiformis*, lameness, prostration, centrilobular liver necrosis, and gallbladder ulcerations were observed (Affleck 1961). In contrast, Watt and Breyer-Brandjwick (1962) reported a lack of toxicity for *C. ensiformis*.

- irritation of the digestive tract and diarrhea

Disease Genesis—Three classes of toxins are present in the seeds of *C. ensiformis*: the proteins canatoxin and concanavalin A (Con A), the amino acids canaline and canavanine, and the cyanogenic glycosides linamarin and lotaustralin. Canatoxin is a metalloprotein with a molecular weight of 115,000 daltons that appears to interfere with Ca^{2+} movement, possibly by interaction with Ca^{2+}-ATPase (Carlini & Guimaraes 1981, 1991; Rosenthal 1982; Alves et al. 1992). Con A is a globular protein of 2, 4, or more subunits, each with a molecular weight of 25,500. It forms saccharide-binding lectins that are widely used as biochemical markers (Nicolson 1976).

- present are proteins, amino acids, cyanogenic glycosides; proteins canatoxin and concanavalin A are the most likely toxins, given that heated beans are much less toxic; roles for canaline, canavanine, and linamarin are unclear

It is not uncommon for toxic proteins and lectins to occur together, especially in leguminous seeds (Carlini et al. 1988). There is considerable homology between canatoxin and Con A, but they differ in binding sites; the monovalent canatoxin binds to gangliosides and the multivalent Con A binds to mannose/glucose residues (Carlini & Guimaraes 1991). Canatoxin is neurotoxic, and Con A, as a lectin, agglutinates red blood cells. Canatoxin in vitro inhibits DNA synthesis and is cytolytic (Campos et al. 1991). It also causes a sustained decrease in blood glucose and an increase in insulin in rats, responses that can be abolished by a variety of opioid and parasympathetic antagonists (Ribeiro-DaSilva & Prado 1993). Con A also is capable of causing adverse effects. Experimentally in rats, it binds to glycoconjugates of enterocytes lining the intestines, causing reduction of appetite and possibly interference with intestinal absorption (Larue-Achagiotis et al. 1992). Experimentally, Con A administered

$$H_2N{-}O{-}C{-}C{-}\overset{NH_2}{\overset{|}{C}}{-}COOH$$

Canaline

$$H_2N{-}\overset{NH}{\overset{\|}{C}}{-}N{-}O{-}C{-}C{-}\overset{NH_2}{\overset{|}{C}}{-}COOH$$

Canavanine

in large parenteral doses causes a variety of developmental defects in rats. These apparently result from interference with neural crest cells during embryonic development (Nishida et al. 1996).

Both canatoxin and Con A are poorly absorbed, and adverse systemic effects are unlikely to occur with oral ingestion; thus *C. ensiformis* is of limited hazard. This is borne out by the large doses required for intoxication. For example, in cows, greater than 3% b.w. of the bean meal is needed for signs to be produced (Shone & Drummond 1961). In addition, the toxic effects of the seeds are attenuated by heat (Shone 1961).

Composing up to 5% of the dry mass of the seeds of *C. ensiformis,* the amino acid canavanine is an antagonist of arginine metabolism and is weakly hepatotoxic upon long-term administration. However, it does not appear to be of direct importance as a toxicant (Hegarty 1978). Although canaline, a pyridoxal antagonist, is present, the disease problem caused by *C. ensiformis* does not resemble that caused by other pyridoxal antagonists such as those found in species of *Albizia.* The role of this amino acid is not clear. The cyanogenic glycosides linamarin and lotaustralin, though present, appear not to be factors of toxicologic risk (Seigler et al. 1989).

■ diarrhea

treatment, fluids and electrolytes

Clinical Signs, Pathology, and Treatment—The predominant effect of ingestion of large numbers of seeds is a severe and persistent diarrhea accompanied by lack of appetite. There are few distinctive lesions, and reddening of the mucosa of the stomach and small intestine is the main pathologic change. Treatment is directed toward relief of the signs of the digestive tract dysfunction.

Cassia L.

■ confusing nomenclature; taxonomic revisions of *Cassia, Chamaecrista,* and *Senna* have created some problems

■ *C. fistula*

golden rain
golden shower
Indian laburnum
Midas tree
pudding-pipe tree
purging cassia
purging fistula

Taxonomy and Morphology—A member of the subfamily Caesalpinioideae, *Cassia* was originally interpreted by taxonomists to be a relatively large genus with 3 well-defined subgenera and more than 500 species. Subsequent systematic studies, however, resulted in recognition of its 3 subgenera as different genera—*Cassia, Chamaecrista,* and *Senna* (Irwin & Barneby 1982). *Cassia* is now interpreted to include only about 30 species distributed primarily in the tropics. *Chamaecrista* comprises some 250 species distributed principally in the neotropics, and *Senna* has approximately 240 taxa in both the tropics and temperate regions.

Because reports of intoxication problems use all three generic names, a summary of the nomenclature of the species most commonly involved is given in table 37.4. Information about the toxicity of species now recognized as members of *Chamaecrista* and *Senna* is presented in the treatments of those two genera.

Table 37.4. Summary of present nomenclature of species formerly classified as *Cassia*

Early Classification	Current Classification
Cassia alata	*Senna alata*
Cassia bicapsularis	*Senna bicapsularis*
Cassia fasciculata	*Chamaecrista fasciculata*
Cassia fistula	*Cassia fistula*
Cassia lindheimeriana	*Senna lindheimeriana*
Cassia marilandica	*Senna marilandica*
Cassia nictitans	*Chamaecrista nictitans*
Cassia obtusifolia	*Senna obtusifolia*
Cassia occidentalis	*Senna occidentalis*
Cassia roemeriana	*Senna roemeriana*
Cassia siamea	*Senna siamea*
Cassia tora	*Senna obtusifolia*

With respect to toxicity problems, species of *Cassia* in the strict sense, although capable of causing gastrointestinal dysfunction, pose relatively little risk. In North America, only 1 introduced species is worthy of mention:

C. fistula L.	golden shower, Midas tree, purging cassia, purging fistula, Indian laburnum, golden rain, pudding-pipe tree

Plants perennials; trees; 10–20 m tall; deciduous or semievergreen; bark smooth, gray. **Stems** erect; branchlets ridged with corky spots. **Leaves** 1-pinnately compound; 15–60 cm long; pairs of leaflets 3–8; terminal leaflet absent; leaflets ovate to lanceolate, apices acuminate, bases cuneate; stipules subulate, 1–2 mm long; petioles and rachises without glands. **Inflorescences** pendulous racemes; 20–40 cm long; bracts and bracteoles subulate-cordate. **Flowers** showy; symmetry slightly bilateral. **Sepals** 5; free; overlapping; unequal in size; brown or pink; villous. **Petals** 5; pale to bright yellow; ovate-elliptic. to orange or red. **Stamens** 10; upper 3 sterile or greatly reduced; anthers with soft hairs. **Legumes** cylindrical; 1.5–2.5 cm in diameter; 30–60 cm long; black; hirsute or subglabrous; divided into compartments. **Seeds** numerous.

■ trees; leaves 1-pinnately compound; flowers showy; petals pale to bright yellow, orange, or red; legumes cylindrical

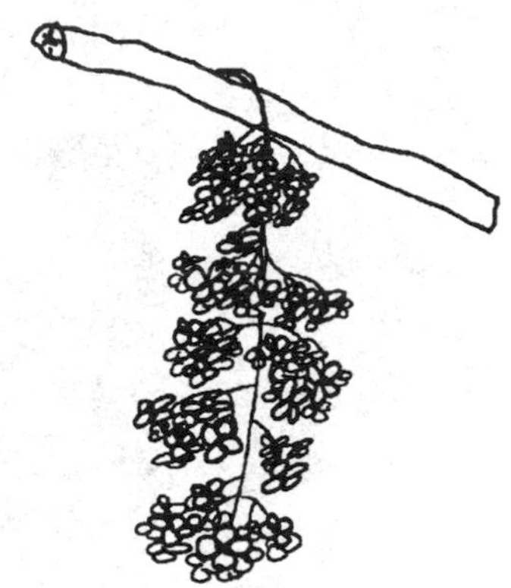

Cassia fistula

Distribution and Habitat—Reputed to be native to India, *C. fistula* is distributed in tropical Central and South America, Southeast Asia, northern Australia, and the Pacific Islands. In North America, it is strictly an ornamental, cultivated in Florida and southern California for its long-lasting displays of flowers borne in long, pendulous racemes.

■ cultivated ornamental

Disease Problems and Genesis—Typical of the cassias and sennas, the fruits and seeds of *C. fistula* contain a number of anthraquinones that may cause irritation of the digestive tract. Considerable quantities of plant material are required, and the problem is generally self-limiting and transient. A more complete discussion of the problem and its origins is presented in the treatment of *Senna,* in this chapter.

■ digestive tract irritation by anthraquinones, generally not severe

Clinical Signs, Pathology, and Treatment—Clinical signs include nausea, vomiting, diarrhea, and colic. Unlike some of the species of *Senna, C. fistula* causes only digestive tract dysfunction and does not cause muscular weakness; thus the pathology typical of *Senna* ingestion is not seen. Treatment is directed toward alleviating the digestive tract effects.

■ mild to moderate vomiting, diarrhea, perhaps colic

Chamaecrista Moench

Taxonomy and Morphology—Comprising approximately 250 species, *Chamaecrista,* a member of the subfamily Caesalpinioideae, was long interpreted by taxonomists to be a subgenus of *Cassia.* Systematic studies have resulted in its recognition once again as a distinct genus (Irwin & Barneby 1982). Reports of intoxication problems have employed both generic names. In North America, approximately 16 native and introduced species are present; 2 are worthy of discussion:

■ *C. fasciculata*
C. nictitans

locust weed
partridge pea
sensitive pea
sensitive plant

C. fasciculata (Michx.) Greene (= *Cassia fasciculata* Michx.)	showy partridge pea, locust weed
C. nictitans (L.) Moench (= *Cassia nictitans* L.)	sensitive partridge pea, wild sensitive plant, sensitive pea

- annual herbs; leaves 1-pinnately compound, terminal leaflet absent, conspicuous stipules; flowers showy; petals yellow, often red-spotted basally

Plants perennials or annuals; herbs or shrubs or small trees. **Stems** erect or ascending or spreading; glabrous or indumented. **Leaves** 1-pinnately compound; pairs of leaflets 5–20; terminal leaflet absent; leaflets oblong, apices obtuse to acute or aristate, acuminate, glabrous or indumented, typically folding when touched; stipules conspicuous; petioles and rachises with or without glands. **Inflorescences** axillary racemes; bracteoles 2, borne near or above middle of pedicels. **Flowers** showy; symmetry slightly bilateral. **Sepals** 5; free; lanceolate. **Petals** 5; yellow; often red-spotted basally; suborbicular to ovate. **Stamens** 5 or 10; anthers basifixed. **Legumes** flat; oblong; walls papery or coriaceous. **Seeds** numerous.

Chamaecrista fasciculata and *C. nictitans* are both annual herbs from taproots with reddish brown legumes. They differ in several features.

1. Pedicels 10–25 mm long. Petals subequal in length; 10–20 mm long. Stamens 10 . . . . *C. fasciculata*
1. Pedicels 2–4 mm long. Petals very unequal in length; 3–8 mm long. Stamens 5 . . . . *C. nictitans*

Chamaecrista fasciculata

- sandy soils of prairies, open woods

Distribution and Habitat—Distribution of *Chamaecrista* is primarily neotropical, with greatest abundance in the Americas. In both the Old World and the New World, a few species such as *C. fasciculata* and *C. nictitans* occur in temperate regions. The former is distributed throughout the eastern half of the continent. A classic indicator of sandy soils in prairies and open woods, it also is found along roadsides and in fallow fields. It is somewhat weedy and increases in abundance with disturbance. *Chamaecrista nictitans* likewise occurs in the eastern half of the continent but in general is more southern in distribution. It occupies the same habitats as *C. fasciculata*.

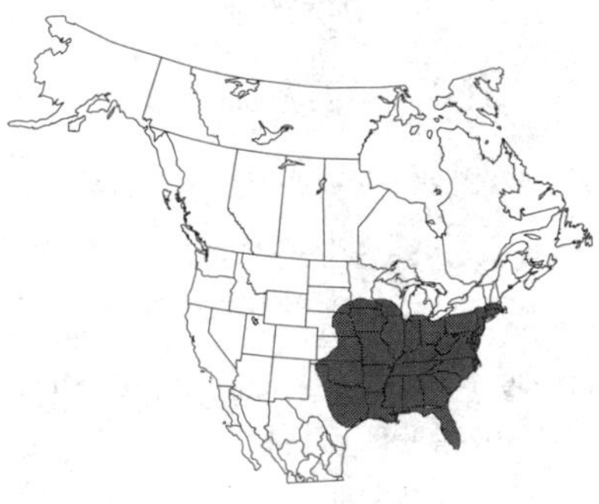

Chamaecrista fasciculata

- digestive tract irritation by anthraquinones, not severe; unlikely because toxins are mainly in legumes

Disease Problems and Genesis—Typical of the cassias and sennas, the fruits and seeds of *C. fasciculata* and *C. nictitans* contain anthraquinones that may cause irritation of the digestive tract. It is very unlikely that problems will occur, because large quantities of the legumes must be consumed; problems are generally self-limiting and of short duration. Otherwise the foliage is nutritious and readily eaten by cattle and sheep (Phillips 1959b). Interestingly, the seeds of *C. fascicularis* are especially useful as feed for quail (Ball et al. 1991). A more complete discussion of the problem and its origins is presented in the treatment of *Senna*, in this chapter.

O

O

Anthraquinone

- diarrhea of mild degree and short duration

Clinical Signs, Pathology, and Treatment—In the rare event of intoxication, clinical signs may include diarrhea and perhaps some indication of colic. Unlike some of the species of *Senna*, *C. fasciculata* and *C. nictitans* cause only digestive dysfunction and do not cause muscular weakness; thus the pathology characteristic of *Senna* ingestion is not seen (see *Senna*, in this chapter). Treatment is directed toward alleviating the diarrhea but is seldom needed.

Coronilla L.

■ *C. scorpioides*
C. varia
coronilla
crown vetch

Taxonomy and Morphology—Comprising approximately 20 species, *Coronilla*, a member of the subfamily Papilionoideae, is an Old World genus. Its name is derived from the diminutive of the Latin word *corona*, which reflects the crownlike clusters of flowers. In North America, 5 species have been introduced as ornamentals and ground covers. Most commonly encountered are the following 2:

C. scorpioides (L.) Koch	coronilla
C. varia L.	crown vetch

■ herbs; flowers pealike; petals 5, yellow, white, purple or pink; legumes segmented

Plants annuals or perennials; herbaceous; caulescent. **Stems** prostrate to ascending or erect. **Leaves** simple or 1-pinnately compound; sessile; pairs of leaflets 3–14; terminal leaflet present; petiolules absent; stipules free. **Inflorescences** capitate umbels borne at ends of peduncles arising in axils of leaves; 2–15 flowers per umbel. **Sepals** 5; fused. **Corollas** papilionaceous. **Petals** 5; yellow or white or purple or pink; banners suborbicular; wings oblong; keels beaked. **Stamens** 10; diadelphous; enclosed in keels. **Legumes** segmented; cylindrical or tetragonal. **Seeds** numerous.

1. Plants annuals. Leaves simple or trifoliate.
   Flowers yellow; 2–5 per umbel . . . . *C. scorpioides*
1. Plants perennials. Leaves 1-pinnately compound.
   Flowers white or purple or pink; 10–20 per umbel . . . . *C. varia*

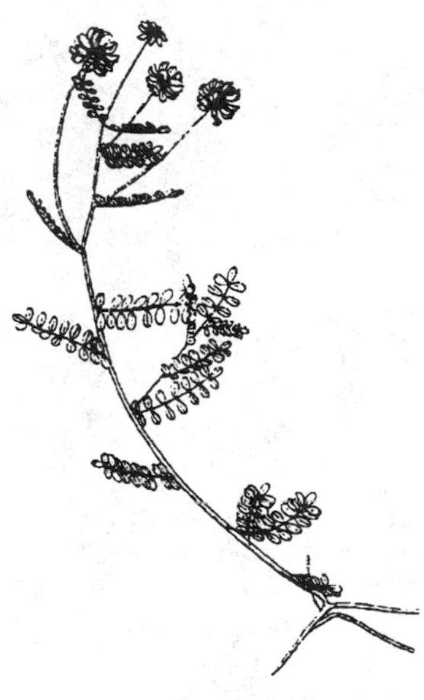

Coronilla varia

Distribution and Habitat—A native of the Mediterranean region, *A. varia* is cold-hardy and is naturalized across the northern half of the United States, extending well into Canada. A vigorous perennial, it has been widely used as a ground cover for erosion control along highways. It also is used as a forage crop. Native to southern Europe, the rare *C. scorpioides* is adventive in extreme eastern Canada and the adjacent northeastern corner of the United States.

■ cultivated for forage, ground cover, especially along highways

■ unlikely cause of toxicity in livestock

Disease Problems—*Coronilla varia* is a palatable and useful forage of somewhat lesser value than alfalfa, but its commercial use in North America has been curtailed in part because of concerns about its toxicologic potential from the presence of cardiotoxins and nitrotoxins (Reynolds et al. 1967, 1969). Disease problems have been reproduced experimentally in voles, chicks, and pigs and have been observed as a field problem in coypu (Shenk et al. 1976; Salyi et al. 1988). In spite of the long concern about toxicity, Reynolds and co-workers (1967, 1969) were unable to produce adverse effects with prolonged feeding of *C. varia* fresh or dried and pelleted. Furthermore, the plants were readily eaten, even when in flower and with seed. Although the possibility of toxicologic effects in ruminants is unlikely, horses may be at greater risk, but this has not been demonstrated.

■ 2 types of toxicants—
(1) weakly active cardenolides, hyrcanoside, probably of little consequence
(2) esters of 3-NPA, risk to monogastric animals but not ruminants

Disease Genesis—Two types of toxicants are present in *Coronilla*, cardenolides and nitro compounds. The first comprises an assortment of weakly active glycosides of coroglaucigenin and corotoxigenin, including hyrcanoside, and deglucohyrcanoside (Feiser & Feiser 1959; Williams & Cassady 1976; Hembree et al. 1979; Sovova 1979; Joubert 1989). Typical of cardenolides, the mechanism of action is inhibition of cell membrane Na^+,K^+-ATPase (Mraz et al. 1992). Because the cardenolides are found primarily in the seeds, there is little risk to livestock (Sovova et al. 1984). These compounds also are present in low concentrations in other tissues of the plant, such as the roots. Unless harvested and fed directly, the seeds are of little hazard because the dose to foraging animals would be quite low.

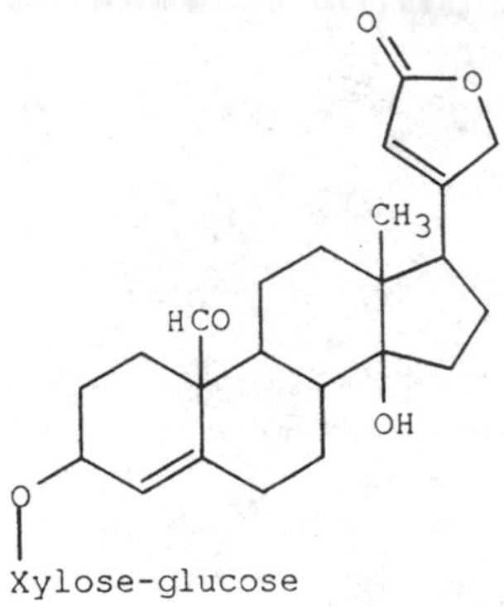

Of greater importance is the presence of nitrotoxins, specifically esters of 3-nitropropionic acid (3-NPA) in the herbage of *Coronilla.* These are mono-, di-, and triesters such as cibarian, corollin, coronarian, coronillin, and karakin (Majak & Bose 1976, Moyer et al. 1977). Although 3-NPA concentrations may vary considerably depending on environmental conditions, concentrations up to 3–4% of dry matter may occur, especially in stem tips and leaflets (Faix et al. 1978; Salyi et al. 1988). This represents a hazard to monogastric herbivores such as meadow voles and nutria (Gustine et al. 1974, 1976; Shenk et al. 1976; Bustine & Moyer 1983; Salyi et al. 1988) but not to ruminants, because 3-NPA is subject to ruminal degradation (Gustine et al. 1977; Majak & Clark 1980). This does not mean that 3-NPA is nontoxic for ruminants, but that the concentrations in *C. varia* are too low to cause problems. Levels of blood methemoglobin may be in the range of 20–30%. These levels are not clinically significant themselves, but they may be a complicating factor in some situations. For a more complete discussion of nitrotoxins, see the detailed discussion in the treatment of *Astragalus,* in this chapter.

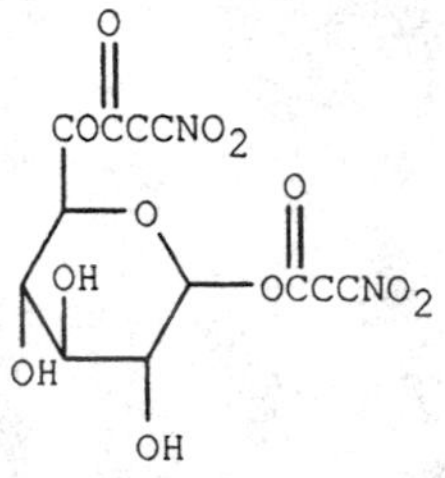

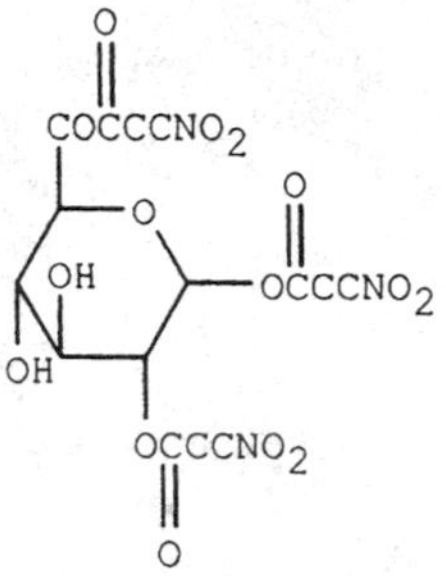

Hyrcanoside

Cibarian

Coronillin

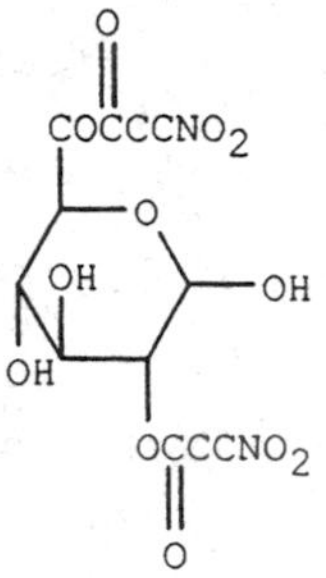

Coronarian

- anorexia, ataxia, recumbency, seizures with rhythmic thoracic limb extension, paralysis

Clinical Signs—In small mammals, there is depressed appetite and ataxia in mild cases. In more severe cases, circling, rolling, recumbency with seizures and rhythmic thoracic limb extension, paralysis of the pelvic limbs, and labored respiration will be seen. Levels of blood methemoglobin may exceed 25%. Adverse effects due to ingestion of *Coronilla* have not been seen in ruminants.

- few lesions in most cases

Pathology—In most instances there are few distinctive lesions. However, in field cases involving coypu, there were gross lesions consistent with cardiac insufficiency, such as ascites, hydrothorax, rounded heart, and pulmonary edema. Microscopically there was cerebellar edema and degeneration of the Purkinje cell layer.

- treatment not needed

Treatment—Nonspecific treatment of symptoms is indicated. The methemoglobinemia produced in severe cases is not sufficient to warrant treatment with methylene blue.

Crotalaria L.

- *C. incana*, *C. pallida*, *C. pumila*, *C. retusa*, *C. rotundifolia*, *C. sagittalis*, *C. spectabilis*

chipilin, crotalaria, rabbit-bells, rattlebox, rattlepod

Taxonomy and Morphology—Comprising approximately 600 species, *Crotalaria,* a member of the subfamily Papilionoideae, is the largest genus in tropical Africa and widespread in tropical regions elsewhere (Polhill 1961). Its members are commonly referred to as rattleboxes. The genus name is derived from the classical Greek word *kratalon,* meaning "castanet," which probably reflects the sound produced by the loose seeds in the inflated legumes when shaken. Many species of the genus are recognized to cause serious livestock intoxications. In North America, *Crotalaria* is represented by some 25 native and introduced species (Senn 1939; Windler 1974); only a few of these are of toxicological importance (see plate 37.2):

C. incana L.	chipilin, shackshack crotalaria
C. pallida Aiton (= *C. mucronata* Desv.) (= *C. striata* DC.)	smooth crotalaria
C. pumila Ortega	
C. retusa L.	wedge-leaf rattlebox
C. rotundifolia J.F.Gmel. (= *C. angulata* Mill.)	low rattlebox, rabbit-bells
C. sagittalis L.	rattlebox, rattlepod, arrow crotalaria, weedy rattlebox
C. spectabilis Roth	showy crotalaria

Crotalaria sagittalis

Other species of unknown toxicologic significance have been introduced as ornamentals or used as cover crops or soil builders. They are not consistently part of the flora of North America. They include the following:

C. assamica Benth.	
C. brevidens Benth. (= *C. intermedia* Kotschy)	
C. juncea L.	sunn hemp
C. saltiana Andr.	
C. virgulata Klotzsch (= *C. grantiana* Harv.)	
C. zanzibarica Benth. (= *C. usaramoensis* E.G.Baker)	

Crotalaria spectabilis

■ erect annuals or perennials; flowers pealike; petals yellow, olive green, blue, or white; legumes inflated

Plants annuals or short-lived perennials; herbaceous. **Stems** erect or decumbent. **Leaves** simple or palmately compound; leaflets 3; stipules absent or present, herbaceous or filiform. **Inflorescences** terminal or axillary racemes; bracts present, leaflike; bracteoles paired or absent. **Sepals** 5; free or fused lobes unequal; calyces campanulate, lobes unequal. **Corollas** papilionaceous. **Petals** 5; yellow or olive green or sometimes blue or white; banners orbicular to ovate, reflexed, clawed; wings oblong, clawed; keels curved, beaked. **Stamens** 10; monadelphous; anthers of 2 forms, linear and globose; enclosed in keels. **Legumes** globose to oblong; swollen to inflated. **Seeds** 1 to many; usually cordate to oblong-reniform.

Most commonly encountered species of toxicological significance are differentiated by the appearance of their leaves, stipules, indumentum, and banners.

1. Leaves palmately compound.
   2. Petioles with spreading hairs. Calyces 3/4 to 5/4 length of keels; densely tomentose, hairs spreading. Legumes fusiform-clavate . . . . *C. incana*
   2. Petioles glabrous or with appressed hairs. Calyces 1/4 to 1/2 length of keels; pubescent, hairs appressed. Legumes cylindrical.
      3. Racemes 1-to-5-flowered. Keels 6–11 mm long. Legumes 12–15 mm long . . . . *C. pumila*
      3. Racemes 25-to-40-flowered. Keels 12–15 mm long. Legumes 6–8 mm long . . . . *C. pallida*

■ mainly open areas

Crotalaria incana

Crotalaria pumila

Crotalaria retusa

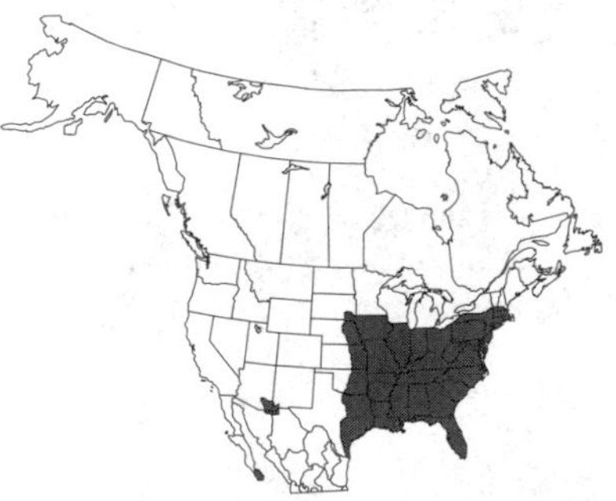
Crotalaria sagittalis

1. Leaves simple.
    4. Leaf blades lanceolate or linear to elliptic. Stipules triangular; fused to stem. Plants not cultivated. Native species.
        5. Plants annual. Stems erect; winged. Stipules well developed . . . . *C. sagittalis*
        5. Plants perennial. Stems decumbent or spreading; not winged. Stipules small or absent . . . . *C. rotundifolia*
    4. Leaf blades obovate to oblanceolate. Stipules ovate or setaceous or absent. Plants cultivated. Introduced species.
        6. Stipules ovate; conspicuous. Bracts of racemes present; large; ovate . . . . *C. spectabilis*
        6. Stipules absent or inconspicuous; setaceous. Bracts of racemes absent or small; subulate . . . . *C. retusa*

Distribution and Habitat—Native *C. sagittalis* is found throughout the eastern half of the United States, extending southward into tropical America. Somewhat weedy, it is especially common in the southeastern United States and may form extensive populations in rocky prairies, sandy open areas, or open woodlands. Pantropic and so readily naturalizing that its origin is uncertain, *C. incana* is probably native to the Americas and is found in localized populations in the sandy soils of the Texas Gulf Coast. Present in South and Central America, *C. pumila* is found in the southwestern part of the continent. Also present in tropical America, *C. rotundifolia* occurs in sandy soils of dry, open woods of the extreme Southeast. *Crotalaria spectabilis, C. retusa,* and *C. pallida,* which were introduced as soil builders in the 1930s, infrequently escape from cultivation in the southeastern United States. They probably do not persist for long at a site. *Crotalaria juncea,* which was introduced from Asia where it has been cultivated as a source of fiber for thousands of years, occurs under similar circumstances.

Disease Problems—Species of *Crotalaria* have long been known as causes of hepatotoxicity, especially of a chronic nature. An example of this is the early accounts of a devastating disease affecting mainly horses, which became apparent in the 1880s in the Missouri River valley of South Dakota, Iowa, and Nebraska (Stalker 1894). Known as river-bottom disease, it was a slowly progressive but fatal disease of scattered occurrence. In some instances, all horses on a farm were affected. Signs developed over weeks or months, with the animals exhibiting a loss of vigor, emaciation, stupor or sleepy appearance, and, in some instances, a loss of awareness of their surroundings. The disease was eventually related to liver disease caused by consumption of *C. sagittalis,* which was common in the bottomlands and apparently eaten fresh or dried in hay. The disease continued to be a problem in the area for many years but has since become rare (Williams 1893). Occasional occurrences of this severe liver disease due to various species of *Crotalaria* continue to be seen in livestock in the southeastern states (Gibbons et al. 1953; Peckham et al. 1974).

In the early 1900s, there was interest in introducing legumes into the southeastern states as soil builders (Ritchey & McKee 1941). As a result of this interest, two dozen species of *Crotalaria* were evaluated for possible introduction (McKee & Enlow 1931). Most species in this and subsequent studies proved to be of little or no problem (Becker et al. 1935; Duncan et al. 1955). Unfortunately, two species, *C. pallida* and *C. spectabilis,* had already been introduced prior to evaluation of their toxicity potential (McKee & Enlow 1931; Thomas 1934). By the time studies revealed its toxicity, the robust and productive *C. spectabilis* was in use as a cover crop/soil builder (Becker et al. 1935). Serious losses in livestock occurred, for example, when it was used as a cover crop in tung tree plantations (Piercy & Rusoff 1946). Because of subsequent attempts to eliminate its use, it has become much less of a threat, although there are occasional problems in horses and other animals (Piercy & Rusoff 1945; Gibbons et al. 1950, 1953; Cox et al. 1958). The introduced *C. retusa,* when eaten fresh, has also been associated with the disease in horses in Australia (Bull et al. 1968), but it has not been a problem in North America.

- *C. sagittalis* of historical interest as a cause of river-bottom disease, liver disease mainly of horses
- numerous species of *Crotalaria* introduced or tested as soil-building legumes; some were shown to be toxic

All plant parts of *Crotalaria,* fresh or dried, are toxic. Fortunately, palatability appears to be limited for the more toxic species, especially *C. spectabilis* (Becker et al. 1935). In most instances, liver fibrosis and failure develop after consumption of the fresh plants, contaminated hay, and possibly contaminated silage for several months. Seeds pose a particular hazard in poultry and swine; they are quite small and easily overlooked in screenings of grains (Smith & Osborne 1962). Contamination of feed with as little as 0.05% of the seeds of *C. retusa,* 0.1% of *C. spectabilis,* or 1% of *C. pallida* may cause intoxication in chicks if fed for several weeks, whereas 0.01% of *C. spectabilis* seeds did not cause adverse effects (Schmittle et al. 1959; Bierer et al. 1960; Kelly et al. 1961; Hooper & Scanlan 1977). Five seeds per day or 10 seeds every other day of *C. retusa* or *C. spectabilis* to poultry were lethal in 18 or more days (Thomas 1934; Emmel 1937). Large, single doses of 80–100 seeds may be lethal in several weeks. Pigs are at similar risk with concentrations of as little as 0.5–1% of *C. spectabilis* seeds in feed or 0.02% of *C. retusa* seeds, causing problems when fed for several months (Emmel et al. 1935a,b; Bierer et al. 1960; Peckham et al. 1974; Hooper & Scanlan 1977). Again, larger amounts are toxic in a shorter period of time. Nutritional factors, such as increased intake of sulfur-containing amino acids, seem to play some role in attenuating the severity of the adverse effects (Yan & Huxtable 1995c).

- large amounts of plant, fresh or in hay, eaten for several weeks or months, liver fibrosis, failure; seeds may be special risk for pigs, poultry; rarely large amounts eaten in short period, acute liver necrosis, failure

Disease Genesis—The toxicants of *Crotalaria* are mainly 11-membered, macrocyclic "closed" diester pyrrolizidine alkaloids (PAs) such as monocrotaline and anacrotine. Also present are a few 12-membered PAs typical of those found in *Senecio,* another important hepatotoxic genus. Most of these PAs are derivatives of retronecine (table 37.5) (Neal et al. 1935; Adams et al. 1953; Willette & Cammarato 1972; Pilbeam et al. 1979; Peterson & Culvenor 1983; Mattocks 1986).

Table 37.5. Pyrrolizidine alkaloids of *Crotalaria*

Species	Alkaloids
C. incana	anacrotine (= crotanecine), integerrimine, usaramine
C. pallida	integerrimine, nilgirine, acetyl nilgirine, usaramine
C. retusa	monocrotaline, retusamine (= otonecine)
C. sagittalis	monocrotaline
C. spectabilis	monocrotaline, spectabiline, retusine

- toxicants, mainly 11-membered macrocyclic closed diester pyrrolizidine alkaloids

The most toxic of the PAs in *Crotalaria,* monocrotaline is representative of the toxins of the genus. It differs somewhat from the PAs of *Senecio,* in that it is less cytotoxic and its metabolites seem to be less susceptible to hydrolysis, perhaps because it has one less member of the closed diester chain—11 rather than 12 (Kim et al. 1993). Like other PAs, monocrotaline is subject to various biotransformation processes in the liver, where its pronounced effects are similar to those of other PAs (see complete discussion of PAs in the treatment of *Senecio,* in chapter 13).

- PAs, metabolized in liver to nontoxic products or highly reactive pyrroles, which alkylate DNA and alter RNA and protein synthesis; cells grow slowly but are unable to divide; cells and nuclei greatly enlarge; hepatocytomegaly, karyomegaly, eventually hepatic insufficiency; may take weeks, months, or years

Although the dominant effects of *Crotalaria,* like other PA-containing plants, are on the liver, there are significant extrahepatic effects as well. The genus produces the broadest range of lesions in an animal of all PA-containing plants (Hooper 1978). In some animal species, significant amounts of pyrrole metabolite(s) are excreted into

- dominant effects of PAs are on liver; lung may be affected in some animal species such as pigs; experimentally in dogs monocrotaline may cause pulmonary hypertension

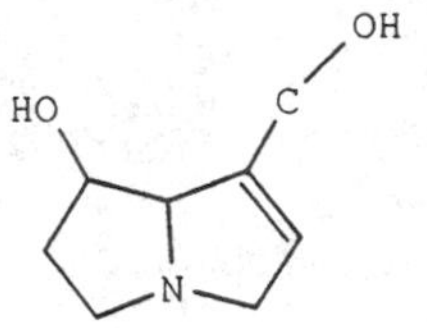

Retronecine

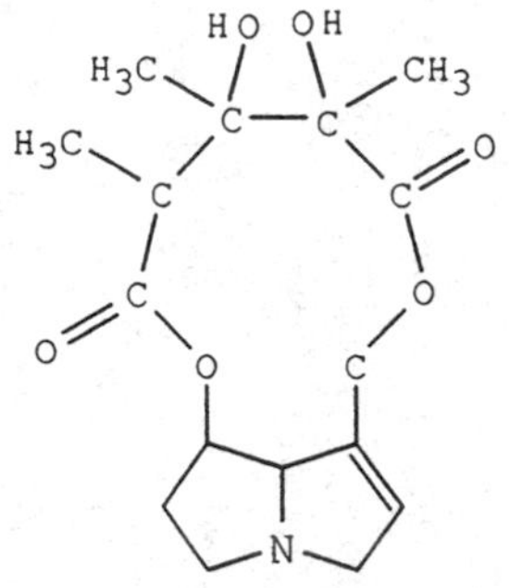

Monocrotaline

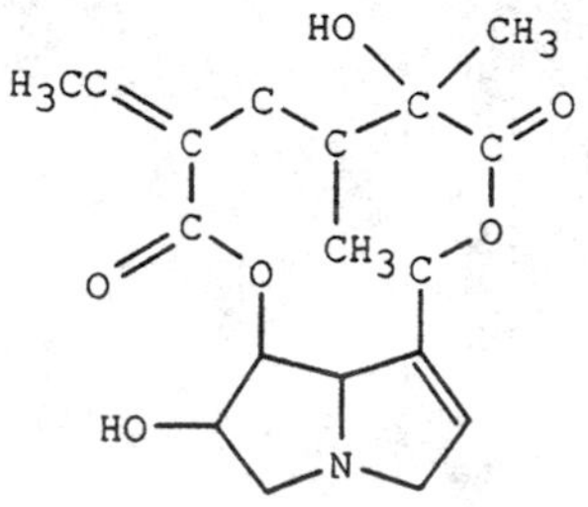

Anacrotine

bile and blood, where they are transported, bound to erythrocytes, to the lung (Pan et al. 1991; Schultze et al. 1991). Thus there is a propensity for these longer-lived, electrophilic, alkylation reactants to produce extrahepatic changes—pneumotoxic and nephrotoxic—in addition to hepatotoxic effects typical of PAs. Endothelial cells are not responsible for the biotransformation of monocrotaline to the toxic metabolite (Taylor et al. 1997). The effects on the lung appear to be attributable to dehydromonocrotaline, which previously had been shown to produce pneumotoxic effects but was thought to be too short-lived (Bruner et al. 1983, 1986; Hilliker & Roth 1985; Mattocks et al. 1991; Glowaz et al. 1992; Mattocks 1992a; Yan & Huxtable 1994, 1995a,b). Monocrotaline also reacts with glutathione in the liver to form glutathionyl-dehydroretronecine, which, unlike conjugates of other PAs, circulates to cause lung damage (Nigra & Huxtable 1992). The reactive toxicants act mainly on the pulmonary vasculature (Wilson et al. 1992; Thomas et al. 1996). Ultimately, the pyrrole binds to epithelial and vascular endothelial cells in the lung and elsewhere to cause toxic effects (Hoorn et al. 1993). In contrast to the toxicants of *Perilla* (see chapter 45), which react where they are formed in the lung to stimulate development of type II pneumocytes and cause an acute respiratory distress syndrome, the PAs of *Crotalaria* inhibit mitosis and cause a decrease in numbers of these cells (Thomas et al. 1996). A discussion of the role of antioxidants in preventing the pulmonary and other effects of the PAs is presented in the treatment of *Senecio,* in chapter 13.

In rats, monocrotaline produces pulmonary structural and functional changes, including megalocytosis of type II pneumocytes. This effect is similar to the increased size but not to the number of hepatocytes so characteristic of PAs (Molteni et al. 1984; Wilson & Segall 1990). The increased size of type II cells is consistent with DNA cross-linking as seen in vitro in endothelial cells (Wagner et al. 1993; Periera et al. 1998). Pulmonary arterial walls are thickened because of hypertrophy and hyperplasia of smooth muscle, edema, and increased connective tissue. This condition is accompanied by a decrease in plasminogen activator and an increase in prostacyclin production. The endothelial lining of blood vessels proliferates. Doses of monocrotaline insufficient to cause liver disease may yet cause injury to the lung's vascular system, resulting in increased pulmonary arterial pressure, leakage of fluid into the lung tissue, and increased lung weight. The leakage of fluid into lung tissue appears not to be secondary to inflammatory changes such as may be mediated by thromboxane A2 or serotonin or via platelet/fibrin thrombi that accompany the pulmonary microvascular changes (Roth & Ganey 1988; White & Roth 1988; Taylor et al. 1997). A role for decreased fibrinolytic activity is not clear (Schultze et al. 1991; Schultze & Roth 1993a,b). The effects on type II pneumocytes in rats are accompanied by derangement of polyamine regulatory mechanisms (Baybutt et al. 1994). The overall effects of monocrotaline are progressive lung degeneration with decreased pulmonary arterial contractile activity (Shubat et al. 1987; Reindel et al. 1990). There are indications that alterations of Na^+,K^+-ATPase may be involved in the changes in vascular contractility (Shubat et al. 1990).

As clinical entities, the extrahepatic effects are mainly of importance in pigs, in which significant renal and minor pulmonary effects are observed (Sanders et al. 1936; Peckham et al. 1974; Hooper & Scanlan 1977). Renal effects in pigs include glomerular as well as tubular degeneration (McGrath et al. 1975). An example of the extrahepatic effects in other animal species is the lung edema, congestion, and consolidation in association with diffuse fibrosing alveolitis that occurred in horses fed *C. juncea* for 30–60 days, and in which hepatic changes were mild (Nobre et al. 1994). Extrahepatic effects are mainly of experimental interest as a model for primary pulmonary hypertension in humans (Wilson et al. 1992). However, it should not be inferred that monocrotaline or other PAs of *Crotalaria* are causes of this disease in humans. Significant pulmonary lesions are seen in dogs, monkeys, and rats given monocrotaline experimentally but not in humans inadvertently intoxicated by *Crotalaria* (Allen et al. 1967; Allen & Chesney 1972; Miller et al. 1978; Mattocks 1986).

Synthesis of the PAs takes place in the leaves, and high concentrations are found in leaves, flowers, and fruits (Birecka et al. 1988). A few species seem prone to have particularly high concentrations of PAs, and they represent a high risk. The seeds of *C. spectabilis* have up to 5% d.w. and the leaves 1.5% (two-thirds or more free base). The seeds of *C. retusa* have 2.7% d.w., whereas those of *C. alata, C. argyrolobioides,* and *C. quinquefolia* contain about 1% d.w. PAs. Most other species evaluated, including *C. sagittalis* and *C. saltiana,* contain concentrations in the range of 0.1–0.2% (Willette & Cammarato 1972; Johnson & Molyneux 1985; Williams & Molyneux 1987). Probably all species of *Crotalaria* contain some PAs, but only low amounts and little if any monocrotaline (Pilbeam et al. 1983). This of course does not mean that all of these species are harmless, as illustrated by the previously described intoxication problem caused by *C. sagittalis.* Likewise, *C. saltiana* also produces signs of intoxication consistent with those due to PAs (Barri & Adam 1981; Barri et al. 1984).

As discussed in the treatment of *Senecio,* all PA-containing plants, including species of *Crotalaria,* are of public health concern. Because they can easily occur as contaminants in cereal grains, seeds of *Crotalaria* are especially a problem, and their long-term ingestion may lead to veno-occlusive liver disease in humans (Tandon et al. 1976). Other effects of *Crotalaria* also are of concern. Extracts from the legumes of *C. retusa* produce chromosomal aberrations in bone marrow cells of mice (Ribeiro et al. 1993). Although there appears to be a threshold dose for toxic effects in rats of 10–13% of the LD_{50} (Shubat & Huxtable 1992), repeated or continual exposure of humans to any dosage of PAs should be considered a potential cause of veno-occlusive liver disease (Huxtable 1989). These alkaloids should be considered potential carcinogens and can be passed in low levels in the milk of dairy cows and possibly in the liver of slaughter animals (Shumaker et al. 1976).

Clinical Signs—The first evidence of intoxication is poor appetite and weight loss. This may be followed in a few days or weeks by other signs. Loss of vigor, incoordination, icterus, and eventually stupor with severe weakness are typical of horses. The animals may appear sleepy, drag their pelvic limbs when walking, and be so weak as to rest their head on any handy solid object such as a fence or tree limb. There may be episodes of unmanageable behavior, restlessness, self-mutilation, persistent walking or circling, or leaning/pushing against solid objects such as fences or buildings. These signs may develop over a period of several days, weeks, or months and are sometimes seen several months after consumption of *Crotalaria* for a few weeks.

- anorexia, weight loss; horses, weakness, head pressing, aimless pacing, ataxia, chewing, yawning, intermittent drowsiness, rectal straining, and either constipation or diarrhea

The signs in cattle are similar to those in horses and include some of the following: depression, salivation, ataxia, bloat, icterus, straining or tenesmus with partial eversion of the rectum, and diarrhea that is sometimes tarlike because of the presence of blood.

- cattle, weight loss, anorexia, depression, excess salivation, ataxia, icterus, tenesmus

Loss of hair, especially pigmented hair, is an early sign in pigs. This is followed by a stunted and unthrifty appearance, loss of appetite, failure to gain weight, lethargy, and labored respiration.

- pigs, additionally, hair loss, labored respiration

Intoxication in poultry due to seeds may occur within a few days or weeks following ingestion. In acute cases, the comb is congested and dark red, and there may be a watery discharge from the mouth and nostrils. In chronic cases, the comb becomes pale purplish and scaly, and the animal has a very unthrifty appearance.

As with toxicity due to PAs of other species (see *Senecio,* in chapter 13), diagnosis is facilitated by determination of profiles of serum hepatic enzymes and by isolation and identification of sulfur-conjugated metabolites bound to hemoglobin in blood or from fresh or fixed liver tissue (Mattocks 1992b; Mattocks & Jukes 1992).

- gross pathology, liver, small, firm, edema of omentum, increased fluid in abdomen

  microscopic, liver, mild centrilobular necrosis, bile duct proliferation and portal fibrosis, megalocytosis and karyomegaly, central vein fibrosis

  some animals, mild to moderate pulmonary emphysema, edema, minor renal tubular necrosis

Pathology—Typically there are scattered splotchy hemorrhages on the heart surfaces, mesentery, fat, gut mucosa, and viscera; congestion of the lungs; and excess free fluid (or gel) in the chest, heart, and/or abdominal cavities. The spleen may be enlarged, in combination with an obvious anemia. The liver is firm, slightly shrunken, yellow or tan (purple in more acute cases), and mottled with irregular reddish brown foci, a condition commonly referred to as a hobnail, or nutmeg, liver. In cattle, the gallbladder is often distended with bile, and the walls are edematous. Microscopic lesions reveal the more distinctive changes associated with *Crotalaria* intoxication. There may be mixed pulmonary emphysema and atelectasis with interstitial edema. There is often renal tubular degeneration with interstitial fibrosis, but it is the appearance of the liver that typifies the disease, as with other PA-caused intoxications. Liver lesions include mild centrilobular necrosis, inter- and intralobular fibrosis, bile duct proliferation, portal and central vein fibrosis, hepatocellular hypertrophy and karyomegaly, and hyperplastic nodules.

The pig differs from other species in the presence of gastric mucosal ulceration and submucosal edema, pulmonary congestion and edema, and chronic nephritis. Microscopically, the kidney is congested and edematous, with tubular degeneration and interstitial fibrous connective tissue proliferation. In some cases, tubular epithelial megalocytosis may be apparent. In the lungs, the alveolar walls are thickened because of epithelial proliferation, septal cell hypertrophy, and fibrosis. Microscopic lesions in the liver, including megalocytosis, are typical of those in other species. In general the cellular hypertrophy in the liver and kidneys are among the most useful diagnostic lesions (Peckham et al. 1974). In poultry, there are numerous, small splotchy hemorrhages on the heart surfaces, and gelatinous fluid in the abdomen, especially around the liver. In chronic cases, the liver is small and dull gray, with rounded edges and a thickened capsule.

- general nursing care; sedation, fluids, high-quality and high-energy feeds with low but high-quality protein; sulfur amino acids are of questionable value

Treatment—Treatment is supportive and typically includes parenteral fluids and a low-protein diet to reduce the nitrogenous load to the liver. However, treatment is generally of little value in most cases because, by the time signs appear, the hepatic changes are extensive and severe. Although decreasing protein in the diet may be used in treatment, it should be selective, because increased toxicity is apparent with protein deficiency (Mattocks 1986). Sulfur-containing amino acids such as cysteine and methionine have been reported to be of some value in treatment of the liver disease, but this has not been confirmed experimentally (Shull et al. 1976). However, monocrotaline-induced pulmonary effects in rats appear to be attenuated by the sulfonic amino acid taurine (Yan & Huxtable 1996). Although the ratio of branched chain amino acid to phenylalanine-tyrosine is altered, therapy with branched-chain amino acids has not proven of value (Mendel et al. 1988). Any approaches to increase hepatic glutathione content may be helpful.

It is important to keep in mind that the effects of these toxic alkaloids are cumulative, at least until compensatory hepatic regeneration occurs. Because of this, prevention of access to plant material of *Crotalaria* is especially important when exposure is via contaminated hay. Alkaloid concentrations can be reduced appreciably by ensiling the forage rather than making hay, but the PAs are not entirely eliminated (Candrian et al. 1984). Antioxidants such as butylated hydroxyanisole have afforded some protection against the acute toxicity of monocrotaline in mice and chickens, but care must be taken in extrapolating these results, because similar effects have not been consistently demonstrated for livestock with either the acute or the more typical chronic intoxication (Miranda et al. 1981; Dickinson & Braun 1987).

Cytisus L.

■ *C. scoparius*

broom
Scotch broom

Taxonomy and Morphology—Comprising approximately 33 species, *Cytisus,* commonly known as broom and a member of the subfamily Papilionoideae, is an Old World genus. Because of their wide ecological tolerance, many of its members are grown as ground covers and ornamentals, whereas others are weedy. It forms a taxonomic complex with the genera *Genista* (broom or woadwaxen) and *Spartium* (Spanish broom), and taxonomists have struggled for more than 100 years in attempting to clarify their relationships and boundaries (Bisby 1981). Because of this uncertainty, many species have been moved from one genus to another several times. As presently interpreted, *Cytisus* in North America consists of 6 introduced species and hybrid taxa, *Genista* of 5, and *Spartium* of 1. Reports of intoxication problems may use one generic name or another; thus the three are described together in this treatment. *Cytisus* also is closely related to *Laburnum,* which is a member of the complex as well; species of the 2 genera hybridize, and the intoxication problems posed by both are similar.

■ shrubs or small trees; leaves palmately compound, leaflets 3; flowers pealike; petals yellow or white

The most commonly encountered species in North America is the following:

C. scoparius (L.) Link   Scotch broom
(= *G. scoparia* [L.] Lam.)
(= *S. scoparium* L.)

Cytisus scoparius

Plants perennials; evergreen or deciduous shrubs or rarely small trees; branches frequently ribbed. **Leaves** palmately compound or rarely simple; leaflets 3; stipules minute or absent. **Inflorescences** terminal racemes or clustered in axils of leaves. **Sepals** 5; fused; calyces 2-lipped. **Corollas** papilionaceous. **Petals** 5; yellow or white; banners broadly ovate, reflexed; wings elliptic, clawed, auriculate; keels oblong. **Stamens** 10; fused; monadelphous; enclosed in keels. **Legumes** oblong; flat; papery or leathery. **Seeds** 2 to many.

Cytisus scoparius is an erect, freely branching, deciduous shrub to 3 m tall. Its leaflets are elliptic to obovate, and its bright yellow flowers are solitary or clustered in leaf axils at the ends of the branches. The legumes, to 5 cm long, are villous along the sutures.

■ ornamental; naturalized noxious weed

Cytisus scoparius

Distribution and Habitat—Introduced from Europe as an ornamental, *C. scoparius* is now naturalized in dry, sandy soils along the East Coast from Nova Scotia to Virginia and along the Pacific Coast. It has become particularly troublesome as a noxious weed from north-central California to British Columbia. Other species of the genus are ornamentals.

■ rarely a cause of intoxications; possible neurologic, reproductive, and digestive disturbances; low risk, in part due to bitterness

Disease Problems—*Cytisus* has somewhat of a reputation, probably not entirely deserved, as a psychedelic, as do *Genista* and *Spartium.* Their dried flowers apparently do produce some psychedelic effects, especially those of *G. canariensis* (Canary Island broom), a native of the Canary Islands, which has been proposed as a tobacco substitute (Fadiman 1965). Ingestion of *C. scoparius* is reported to cause neurologic effects such as excitement and ataxia in horses (Miller et al. 1980). In addition to neurologic dysfunction, ingestion of *Cytisus* may cause digestive tract and reproductive/teratogenic effects. These are illustrated by a California case in which 4 boys who ate the flowers of *C. scoparius* subsequently developed nausea and vomiting (Fuller & McClintock 1986) and by experimental studies, in which *C. scoparia* produced malformed lambs (Keeler 1972). However, the actual hazard due to consumption of the species is questionable. Reports of intoxications are few, and in New Zealand, studies of the forage potential of the plant revealed that it is a nutritious browse, albeit of low to medium acceptability to sheep and goats (Lambert et al. 1989a,b).

■ quinolizidine alkaloids, cytisine types, which act upon cholinergic receptors, especially nicotinic; teratogenic alkaloids anagyrine and ammodendrine

Disease Genesis—Leaves of *C. scoparius* contain an array of quinolizidine alkaloids such as lupanine, anagyrine, and sparteine, which predominates. Concentrations as high as 0.1% d.w. occur (Wink et al. 1983). Other species may contain a different array of alkaloids in addition to cytisine. The closely related *Genista pilosa* also contains sparteine and others. The pyridine ammodendrine also is present.

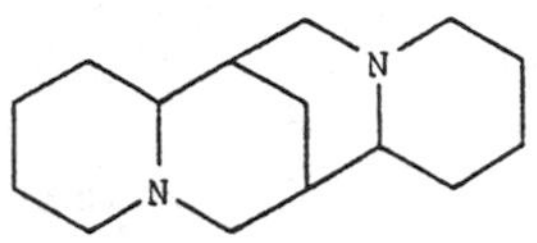

Sparteine

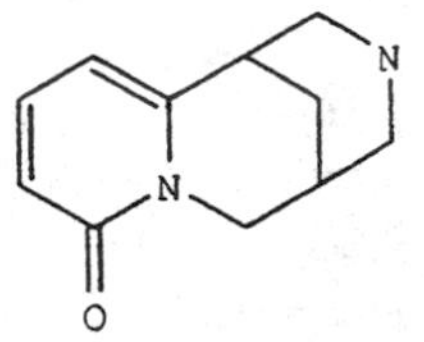

Cytisine

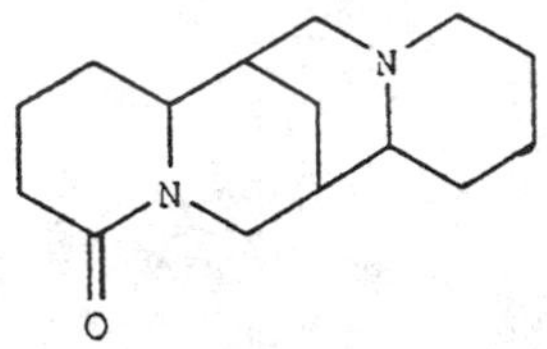

Lupanine

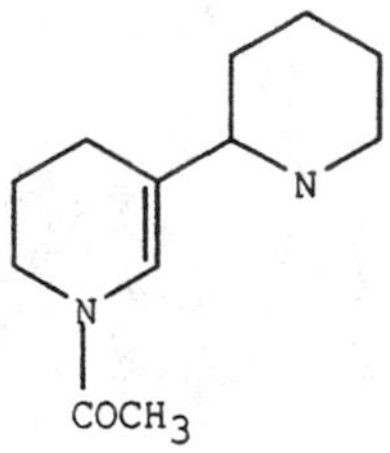

$COCH_3$

Ammodendrine

The primary effect of these alkaloids is to bind with nicotinic and muscarinic acetylcholine receptors to produce the neurologic and digestive tract effects. Sparteine has relatively low to moderate binding affinity for cholinergic receptors, somewhat favoring muscarinic over nicotinic types (Schmeller et al. 1994). At the dosage likely from ingestion of *Cytisus,* sparteine and lupanine probably have only a weak sedative effect (Pothier et al. 1998). These alkaloids also impart a bitter taste to the plant, as perceived by humans, and thus may render the species unpalatable to grazing animals (Wink 1994). The occurrence of malformed lambs following ingestion of *C. scoparius* (Keeler 1972) is not surprising in view of its content of the teratogenic anagyrine and ammodendrine.

A more complete discussion of these quinolizidine alkaloids and their effects is presented in the treatment of *Laburnum,* in this chapter.

■ most likely, abrupt onset, diarrhea, ataxia, tremors; possibly fetal deformities

treatment, general nursing care

Clinical Signs, Pathology, and Treatment—Because the alkaloids present in *Cytisus* are similar to those of *Laburnum,* it generally has been assumed that the signs, pathology, and treatment of disease are likewise similar. However, this assumption has not been confirmed. They may be quite different because of differences in the types and proportions of the alkaloids.

Erythrina L.

■ *E. americana* — *E. flabelliformis*
E. coralloides — *E. herbacea*
E. crista-galli — *E. poeppigiana*

cardinal spear — coral bean
Cherokee bean — coral tree
colorin — corcho
mountain immortelle — crybaby tree

Taxonomy and Morphology—Comprising 108–112 species, 70 of which occur in the New World, *Erythrina,* is a pantropical genus and a member of the subfamily Papilionoideae (Krukoff & Barneby 1974; Neill 1988). There also are numerous stabilized hybrids, both natural and manmade. Commonly known as coral tree, the genus name is derived from the Greek root *erythros,* which reflects the numerous shades of luminous red exhibited by the flowers. Most of the New World species are pollinated by hummingbirds. There are only 2, or possibly 3, species native to North America, the others present having been introduced as ornamentals from Central and

South America and Africa for their spectacular racemes of flowers. Representative species include the following:

E. americana Miller	coral tree, coral bean, colorin
E. coralloides A.DC.	flame coral tree, naked coral tree
E. crista-galli L.	cockspur coral tree, crybaby tree
E. flabelliformis Kearney	colorin, corcho
E. herbacea L.	cardinal spear, Cherokee bean, eastern coral bean
E. poeppigiana (Walp.) Cook	mountain immortelle

Erythrina crista-galli

Plants perennials; evergreen or deciduous shrubs or tree or rarely perennial herbs; typically bearing blunt conical thorns or recurved prickles. **Leaves** pinnately compound; leaflets 3, ovate to rhomboid, margins entire; stipules small, ovate or linear. **Inflorescences** large, showy, terminal or axillary racemes; bracts present, caducous. **Sepals** 5; fused; calyces 2-lipped. **Corollas** papilionaceous. **Petals** 5; of various shades of red or orange or yellow; banners of various shapes, erect or reflexed; wings ovate or oblong, short-clawed; keel petals free or fused only at end. **Stamens** 10; fused; monadelphous or diadelphous; protruding at maturity. **Legumes** linear; flat or cylindrical; coriaceous or ligneous. **Seeds** 1 to many; ovoid; red or carmine or brown, sometimes black-spotted.

■ shrubs or trees; leaves pinnately compound; flowers pealike; petals red, orange, or yellow; seeds red, carmine, or brown, sometimes black-spotted

Distribution and Habitat—Native to the United States are *E. herbacea* and *E. flabelliformis.* The former is found in sandy, wooded coastal sites from the southeastern United States along the Gulf Coast into eastern Mexico. It also occurs in scattered areas of western Mexico. *Erythrina flabelliformis* is found in the desert grasslands and woodlands of the Southwest. *Erythrina americana* and *E. coralloides* are species of central and southern Mexico commonly cultivated as ornamentals in Mexico and southern California and rarely elsewhere in North America. Some taxonomists contend that *E. coralloides* occurs naturally in southern Arizona. *Erythrina crista-galli* and *E. poeppigiana* are South American species that are cultivated as ornamentals in the warmer parts of North America.

■ cultivated ornamentals; coastal areas, desert grasslands, woodlands

Erythrina coralloides

Disease Problems—The seeds of *Erythrina* have a potent paralytic action, and there are references to their purported use to poison rats and dogs in southern and central Mexico, where *E. americana* is indigenous and also used as an ornamental (Blohm 1962; Perkins & Payne 1978). Even though the seeds have had a longstanding reputation for toxicity, it was not until the 1930s that Lehman (1935) confirmed the curare-like effects of alcoholic extracts of *E. americana.* In spite of the potency of the seeds as toxicants, there are very few reports of problems associated with their ingestion. In Mexico, despite their reputation as a weak hypnotic, the flowers of *E. americana* and other species are boiled and eaten as a vegetable (Aguilar et al. 1981; Sotelo et al. 1993; Morton 1994). In a few instances, species of *Erythrina* were suspected to cause intoxication, but their involvement was not confirmed (Perkins & Payne 1978; Everist 1981).

■ seeds well known as toxic; curare-like paralytic and hypnotic action, but flowers eaten as vegetables

Disease Genesis—The toxicity of the seeds is due to the presence of an extensive series of tetracyclic isoquinoline erythrinane alkaloids, unique to *Erythrina* (Folkers & Major 1937; Folkers et al. 1942). All of these tertiary bases are active when given orally. They are divided into three groups, primarily on the basis of differences in the

■ seeds; extensive series of isoquinoline alkaloids, absorbed orally—dienoids, lactonics, and alkenoids

Erythrina flabelliformis

Erythrina herbacea

substituents of the D ring, and the degree of saturation of the A and B rings. The eryso-group comprises phenolics and dienoid alkaloids; the erythroi-group, lactonic alkaloids; and the erythra-group, alkenoid-parent alkaloids (Cordell 1981; Chawla & Kapoor 1995). A few of these compounds occur as glycosides.

As illustrated in table 37.6, most species contain a formidable array of these alkaloids, all of which probably have toxicologic potential (Hargreaves et al. 1974; Abdullah et al. 1979; Southon & Buckingham 1989; Sotelo et al. 1993; Chawla & Kapoor 1995). *Erythrina americana* contains the highest concentrations of alkaloids; the content of seeds is about 2% and 10-fold less in the flowers. Interestingly only α- and β-erythroidines are present in the flowers (Abdullah et al. 1979; Aguilar et al. 1981). It is important to note that the alkaloids are not heat-labile and may be found in the cooked flowers. These alkaloids may have sedative effects, but they are eaten without serious problems, as noted above (Morton 1994). The seeds and foliage also have been considered to have cyanogenic potential, but this has not been confirmed (Perkins & Payne 1978; Sotelo et al. 1993).

Table 37.6. Alkaloids of *Erythrina*

Species	Alkaloids
E. americana	α- and β-erythroidines, erysovine, erythravine
E. coralloides	erysotrine, erythraline, erysodine, erysovine
E. crista-galli	cristadine, erysodine, erysopine, erysovine, erythraline, erythrinine
E. flabelliformis	erysotrine, erysodine, erysovine
E. herbacea	erybidine, erysodine, erysopine, erysotrine

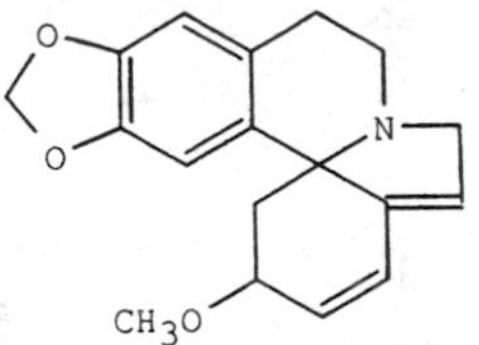

Erythraline (dienoid)

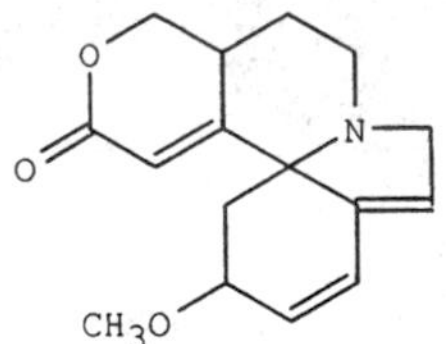

α-Erythroidine (lactonic)

Erythramine (alkenoid)

The alkaloids of *Erythrina* have a curare-like effect, causing motor paralysis in mice (Folkers & Major 1937). Administered intraperitoneally or intravenously, alcoholic extracts of the seeds of *E. americana* caused decreased blood pressure and respiratory paralysis (Lehman 1935). Exposure in vitro resulted in nerve paralysis, and the extract was considered to have curare-like activity, albeit somewhat weaker and of shorter duration (Blohm 1962). β-Erythroidine traditionally has been considered to be the most potent of the alkaloids, but depending on the parameter, both dihydro-β-erythroidine and erysodine have been shown to be quite potent. Probably all of the alkaloids have activity as antagonists at nicotinic acetylcholine receptors (Decker et al. 1995; Tsuda & Sano 1996). Both peripheral and central neural receptors are affected, and one result is the release of nicotinic-mediated neuronal dopamine. This is important because at least some of these alkaloids cross the blood-brain barrier. The lethal dose for mice and rats is 600–675 mg crude extract/kg b.w. when given i.p. and 100 mg/kg when given i.v. (Lehman 1935). The oral LD_{50} in mice for erythroidine HCl, as isolated originally, was 120 mg/kg b.w. (Folkers & Major 1937).

The alkaloids of *Erythrina* represent a public health hazard either directly via ingestion of the seeds or indirectly via consumption of milk containing the compounds (Payne & Foley 1991; Soto-Hernandez & Jackson 1993). Home-produced goat's milk may be particularly hazardous because of the likelihood of goats ingesting seeds during grazing/foraging. Because the seeds of *E. americana* are high in protein and fat, they have been evaluated for their food potential (Sotelo et al. 1993). Flour made from the seeds may be rendered nontoxic by extracting the alkaloids with methanol prior to milling.

- alkaloids, curare-like motor paralysis due to effects on nicotinic cholinergic receptors; alkaloids excreted in milk

Clinical Signs, Pathology, and Treatment—Because the toxicologic risk of *Erythrina* is low, detailed descriptions of clinical signs and pathology are not available. Signs indicative of curare-induced paralysis of the muscles, e.g, weakness, could be expected.

Because of the alkaloids' mode of action, there would be few if any distinctive lesions in the unlikely instance of a fatal intoxication. Treatment, likewise because of the nature of the disease problem, should involve supportive care, including administration of cholinomimetics and cholinesterase inhibitors.

- weakness, paralysis

treatment, cholinesterase inhibitors

Galega L.

Taxonomy and Morphology—An Old World genus of 6 species, *Galega* is a member of the subfamily Papilionoideae and is related to *Astragalus* and *Oxytropis*. The 3 genera are classified in the same tribe, albeit different subtribes. They differ in the venation of the legumes and leaflets, fusion of the stamen filaments, and appearance of the keel petals. In North America, *Galega* is represented by 1 introduced species:

- *G. officinalis*

French honeysuckle
French lilac
goat's rue

G. officinalis L. goat's rue, French lilac, French honeysuckle

Plants perennials; herbaceous. **Stems** erect; several arising from short caudices; 40–140 cm tall. **Leaves** 1-pinnately compound; pairs of leaflets 4–10; terminal leaflet present; blades narrowly ovate to lanceolate or linear-elliptic; apices obtuse or emarginate, mucronate; petiolules absent; stipules herbaceous, toothed or lobed, sagittate. **Inflorescences** elongate racemes borne at ends of peduncles arising in axils of leaves; bracts present; pedicels filiform. **Sepals** 5; fused. **Corollas** papilionaceous. **Petals** 5; bluish lilac, reddish purple, or white; banners oblanceolate, reflexed; wings clawed; keels broad, rounded; not auriculate. **Stamens** 10; monadelphous; enclosed in keels. **Legumes** terete; cylindrical; somewhat constricted between seeds; **Seeds** 8.

- erect perennial herbs; leaves 1-pinnately compound, terminal leaflet present; stipules lobed or toothed; flowers pealike; petals bluish lilac, reddish purple, or white

Galega resembles *Astragalus* but can be distinguished by its stipules, which are toothed or lobed; leaflets with veins that extend to the margins; and legumes with oblique veins. In contrast, *Astragalus* has unlobed stipules, leaflets with veins that do not reach the margins, and legumes with transverse veins.

Galega officinalis

Distribution and Habitat—Originally brought to North America from the Middle East in the 1890s as a possible livestock forage, *G. officinalis* escaped and spread quickly as a weed via waterways in several counties of northern Utah. It is sometimes called professor plant by farmers in the area, who attribute its introduction to faculty at Utah State University. Subsequently the subject of a considerable eradication effort, the species is now limited in distribution to irrigated areas of Cache County, Utah.

- weed of irrigated areas of Cache County, Utah

Disease Problems—Although closely related to *Astragalus* and *Oxytropis*, *G. officinalis* causes a different disease syndrome, a massive pulmonary congestion. However, the species is of little toxicologic significance in North America because of its limited distribution. Even in its natural range in Europe, intoxications are rare (Puyt et al. 1981; Puyt & Faliu 1984). It is seldom eaten by cattle, and the few cases of intoxication have involved sheep. Five hundred grams of fresh, green plants or 0.1–0.2% b.w. of dry plant material is sufficient to kill a sheep (Puyt et al. 1981; Keeler et al. 1986b). Although intoxication due to *Galega* is a rare problem, it is very serious when it does occur, because it is rapidly fatal and the mortality rate may range from 10% to 50% (Faliu et al. 1981). Plants become less palatable and more toxic with increasing maturity; there is little problem with ingestion of young, leafy plants (Tingey 1971). The disease seems to be almost an all-or-none phenomenon, the animals either becoming ill and dying within 24 hours of eating the plants or surviving and tolerating a 5-to-10-fold increase in dosage so that even continued consumption does not result in death (Keeler et al. 1986b). The tolerance appears to be due to extreme

- rapidly fatal massive pulmonary edema; fluid accumulation in other tissues

biological variation in response to the toxin, given that experimentally toxicity ranged from death at 10 mg/kg b.w. of dried plant to no adverse effects in other sheep at 24 mg/kg (Keeler et al. 1988).

- toxic guanidine derivatives, galegin, hydroxygalegin

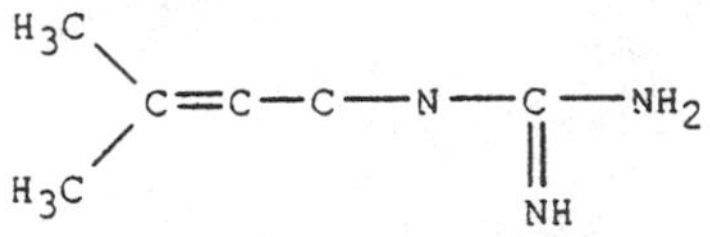

$(H_3C)_2C{=}C{-}C{-}N{-}C({=}NH){-}NH_2$

Galegin

Disease Genesis—Toxicity of *G. officinalis* appears to be due to the guanidine derivatives galegin and hydroxygalegin, although the mechanism by which these compounds cause the severe fluid accumulations that characterize the disease is not known (Puyt et al. 1981). Galegine also has been shown to be hypotensive and neurotoxic (Keeler et al. 1986b). Further evidence of a role for galegin in the disease is its presence in two other plants that cause an identical disease syndrome: *Schoenus asperocarpus* (poison sedge, family Cyperaceae) and *Verbesina encelioides* (cowpen daisy, family Asteraceae) in Australia (Eichholzer et al. 1982; Huxtable et al. 1993).

- abrupt onset, respiratory distress, protruding tongue, foamy nasal discharge, abdominal distension

Clinical Signs—*Galega officinalis* produces an acute disease that principally affects the lungs. It is manifested as severe respiratory distress, with the tongue protruding, a foamy white nasal discharge, gray mucous membranes, abdominal distension, and signs of struggling before death. Animals literally drown in their own body fluids. There are no consistent changes in serum chemistry.

- gross pathology, pulmonary edema, froth in airways; fluid in chest and around heart; hemorrhages on heart surfaces

Pathology—Grossly, there is massive pulmonary congestion and edema with a white foamy fluid exuding from the lung tissue and airways. The chest cavity is filled with 1–1.5 l of clear amber-colored, rapidly clotting thoracic fluid (Puyt et al. 1981; Keeler et al. 1986b; Gresham & Booth 1991). There is excess fluid in the pericardial sac and small splotchy hemorrhages on the heart surfaces. Microscopically, there is widespread passive congestion of tissues and little inflammatory reaction.

- general nursing care; diuretics

Treatment—Treatment of *Galega* intoxication focuses upon elimination of the fluid from the lungs with diuretics such as furosemide, in addition to nonspecific general nursing care. However, treatment has typically been of limited effectiveness.

Genista

See treatment of *Cytisus,* in this chapter.

Glottidium Desv.

- *G. vesicarium*

  bagpod
  bladderpod
  coffee-bean
  coffeeweed

Taxonomy and Morphology—A monotypic genus, *Glottidium* is a member of the subfamily Papilionoideae. Because it is superficially similar to *Sesbania* (treated in this chapter), the two genera have been merged by some taxonomists. Careful study of their morphology, however, revealed numerous significant differences, and the two genera are now recognized as distinct (Polhill and Sousa 1981). The single species of the genus is the following (see plate 37.3):

G. vesicarium (Jacq.) R.M.Harper (= *Sesbania vesicaria* [Jacq.] Elliott)	bagpod, bladderpod, coffee-bean, coffeeweed, bagpod sesbania

- annual herbs; leaves 1-pinnately compound, terminal leaflet absent; flowers pealike; petals 5, yellow; legumes inflated, persistent baglike endocarp around seeds

Plants annuals; herbaceous. **Stems** erect; 150–300 cm tall; slender; bright green; branches few, prominent. **Leaves** 1-pinnately compound; leaflets 20–52; terminal leaflet absent; blades elliptic; margins entire; stipules caducous. **Inflorescences** axillary, drooping racemes bearing 3–12 flowers; bracts and bracteoles present, caducous. **Sepals** 5; fused. **Corollas** papilionaceous. **Petals** 5; yellow; banners broadly obovate to reniform, apices emarginate, short-clawed, reflexed; wings short-clawed, auriculate; keels rounded, long-clawed, auriculate. **Stamens** 10; diadelphous; enclosed in keels. **Legumes** oblong to elliptic; inflated, but slightly compressed; stipitate, beaked; yellow; pericarp separating at maturity into exocarp, which falls, and papery endocarp, which persists around seeds. **Seeds** 1 or 2; reniform.

The intact, yellow legumes and the papery endocarps of dehisced legumes persist on the tall, naked stems of the plants long after the leaves have fallen. Their appearance has been poetically likened to watchmen with lanterns in the bleakness of the winterscape.

Because of the differences in taxonomic opinion as to the relationship of *Glottidium* and *Sesbania,* the binomial *Sesbania vesicaria* appears in some literature reports of the toxicological problems caused by bagpod.

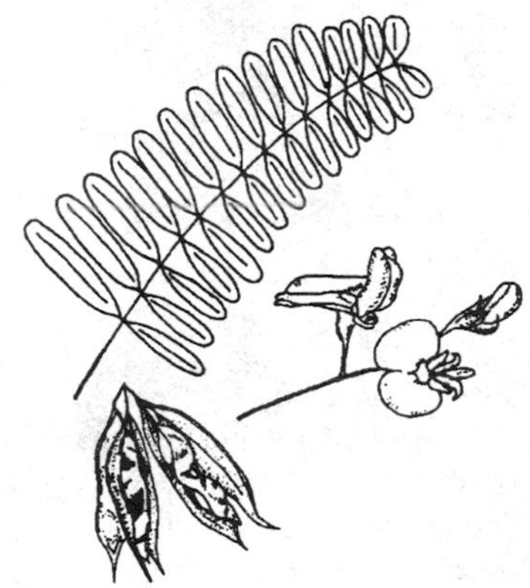

Glottidium vesicarium

Distribution and Habitat—Often forming large, dense populations in low areas subject to flooding and along roadsides and sandy stream banks, *G. vesicarium* occurs in the coastal plains of the southeastern United States. Some authors propose that it was introduced from the West Indies.

- dense populations in sandy areas subject to flooding

Glottidium vesicarium

Disease Problems—*Glottidium vesicarium* is well known as a cause of digestive tract irritation and diarrhea, and the disease is essentially the same as that produced by species of *Sesbania.* Both green and ripe seeds (beans) are toxic, the lethal dose being 1.1–2% b.w. or about 150 seeds for an adult chicken (Emmel 1935; Featherly et al. 1943; West & Emmel 1950; Duncan et al. 1955). However, Emmel (1944) later reported that immature seeds were 2–3 times more toxic than mature seeds. The lethal dose for cattle is 1–2% b.w. or less of mature seeds (Boughton & Hardy 1939a,b; Featherly et al. 1943; West & Emmel 1950). About a bushel of the pods are sufficient to kill a calf weighing 250–300 kg. The more thoroughly the animal chews the seeds, the greater their toxicity. In sheep 0.125% b.w. of seeds was rapidly fatal and 0.062% b.w. was fatal in 53 hours (Boughton & Hardy 1939b).

In late fall or early winter, animals newly introduced into an area containing large populations of *G. vesicarium,* may readily and perhaps avidly consume large numbers of seeds in a short time (Specht et al. 1973; Reagor & Jones 1980). As a consequence, they may develop clinical signs within a few days of being placed in the infested pasture. If the beans are consumed slowly, they produce sufficient irritation of the animal's digestive tract to limit subsequent intake to less than that required for intoxication. After the first 3–4 days following introduction to infested pastures, the hazard declines rapidly because the animals usually will no longer eat amounts sufficient to produce intoxication. Intoxications occasionally occur when animals are denied access to their normal forage for several days because of heavy snowfall.

- severe digestive disturbance, often fatal, from seeds, not foliage; eaten mainly by newly introduced animals in late fall or winter when foliage is absent and legumes with seeds hang on plant; if exploring calves eat enough, they become ill

The green leaves and stems are of low palatability, and their toxicity is 1% that of the seeds. They represent little hazard other than causing slight digestive discomfort. It is only during the late fall and winter, when other forage is scarce and the fruits hang conspicuously and readily available from the starkly naked branches of the persistent stems, that the hazard of *G. vesicarium* becomes significant. Even under these conditions, cattle seldom eat the legumes and seeds.

Disease Genesis—Identification of the toxicants in *G. vesicarium* has been difficult, but apparently the same compounds are present as those in *Sesbania.* They include saponins, sesbanine, and sesbanimides. The sesbanimides, especially A, appear to be the toxicants responsible for the severe digestive tract disease and death (Powell et al. 1990; Kim et al. 1992). Concentrations of sesbanimides are in the range of 18–30

- various compounds, but toxic effects seem to be due to sesbanimides; saponins and others may be contributing factors

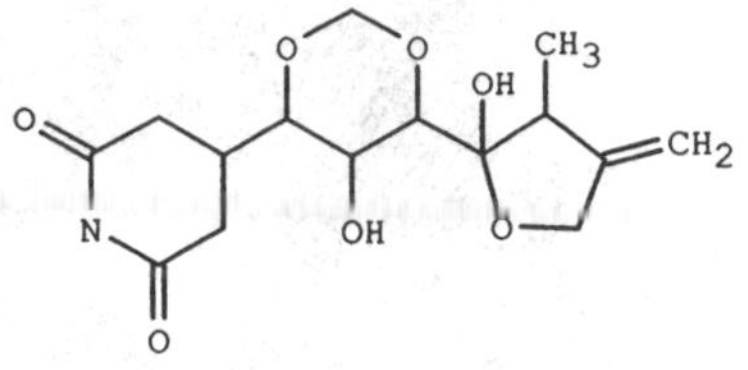

Sesbanimide A

μg/g of seed. Heating the seeds reduces the toxicity potential considerably (Foote & Gramling 1940). A complete discussion of the nature and effects of the various compounds is presented in the treatment of *Sesbania,* in this chapter.

- abrupt onset, severe watery diarrhea ± blood, anorexia, absence of ruminal activity, weakness, depression, colic; diarrhea may become mucoid

Clinical Signs—Most cases involve cattle deaths in December, January, or February with few premonitory signs other than loss of appetite and watery diarrhea. With close attention, after an initial delay of 12–24 hours, the earliest signs—depression, ruminal inactivity, abdominal pain, weakness, and mucoid feces—may be noted. The heart rate is increased and sometimes irregular. The animal stands with back arched, head down, and neck extended. Less commonly there is dyspnea and tremors. Neutrophilia with many immature forms may accompany the other signs. There may be a several-fold increase in blood urea nitrogen and hepatic enzymes, but these changes are generally not sufficient or even consistent enough to be confirmatory. In chickens there is depression, purple comb, muscular twitching, and profuse diarrhea.

- severe reddening of mucosa of abomasum and small intestine; necrotic enteritis in poultry

Pathology—Gross changes are limited to moderate to severe abomasal and intestinal reddening, congestion, and edema. Less commonly the reticulum, rumen, and omasum are also affected. A few scattered ecchymotic hemorrhages on the heart surfaces and pulmonary congestion may also be present. Microscopically, there is mild to moderate degeneration of liver cells and renal tubular degeneration and necrosis.

In chickens, the dominant changes are a necrotic enteritis involving the gizzard and duodenum and fatty degeneration of the liver. The crop is filled with fluid, the intestines are filled with gas, and the feathers pull out easily. In most cases there is little to indicate a specific cause of death, and the pathology often belies the severity of the disease. Intact or sprouted beans may readily be identified in ruminal contents.

- activated charcoal, fluids, electrolytes

Treatment—Because the specific cause of the intoxication is unknown, treatment is symptomatic. Activated charcoal may be given orally to absorb the toxins, accompanied by fluids and electrolytes for dehydration.

Gymnocladus Lam.

- *G. dioica*

  American coffee berry
  chicot
  coffee-bean
  Kentucky coffee tree
  Kentucky mahogany
  nicker tree
  stump tree

Taxonomy and Morphology—Comprising 4 species in eastern Asia, *Gymnocladus,* a member of the subfamily Caesalpinioideae, is represented in eastern North America by a single species:

G. dioica (L.) K.Koch	Kentucky coffee tree, chicot, coffee-bean, Kentucky mahogany, stump tree, nicker tree, American coffee berry

- medium trees; leaves 2-pinnately compound, terminal leaflet absent; flowers inconspicuous; petals greenish white; legumes hard, thick-walled, dark brown; seeds in glutinous pulp

Plants perennials; medium-sized to large trees; polygamodioecious. **Leaves** large, 30–60 cm wide, 30–100 cm long; deciduous; 2-pinnately compound; pinnae 3–6; pairs of leaflets 4–7 per pinna; terminal leaflet absent; petiolules present; stipules small, deciduous. **Inflorescences** terminal racemes or panicles. **Flowers** perfect or imperfect; symmetry radial; hypanthia present, tubular. **Sepals** 5; free. **Petals** 5; equal; oblong or oblanceolate; greenish white. **Stamens** 10; filaments free; in 2-series, long and short. **Legumes** oblong; flattened, but thick; straight or curved; pericarps thick-walled; dark brown; not dehiscing for 1 or 2 years. **Seeds** 5–8; dark olive brown; thick-walled; hard; embedded in glutinous pulp.

- moist forests

Distribution and Habitat—*Gymnocladus dioica* is distributed throughout the eastern half of the continent. Seldom abundant, it occurs as scattered trees in rich, moist forests.

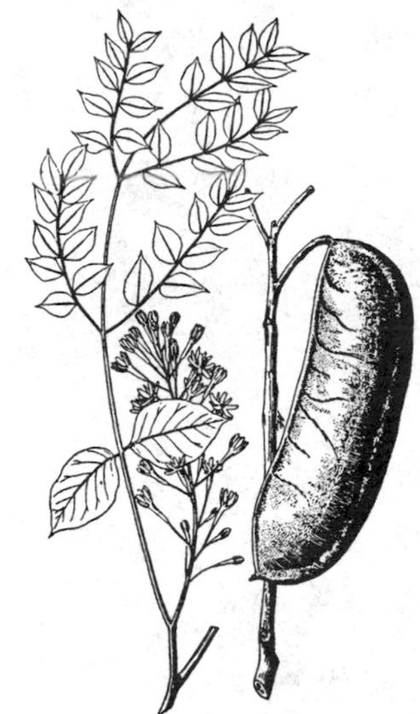

Gymnocladus dioica

Disease Problems—As its common name implies, the beans or seeds of G. *dioica* were used as a coffee substitute by early settlers. Compared with chicory, they were a poor substitute and thus never attained equal popularity (Nuttall 1980). They are not toxic when parched. Toxicity in humans arises from consumption of the seeds, whereas in livestock it is associated with ingestion of new foliage in the late spring, especially that arising from stumps. Trimmings also may pose a hazard, as shown by the death of sheep in 24 hours following their ingestion (Reynard & Norton 1942). Hansen (1924) reported a case in which the leaves from a wind-toppled tree were eaten, resulting in the deaths of several sheep. In an unusual instance, intoxication occurred when seeds were eaten by a woman who had mistaken the legumes for those of *Gleditsia triacanthos* (honey locust). Gastrointestinal irritation and narcotic effects are produced.

- ingestion of seeds or foliage; digestive disturbance; trimmings or sprouts from stumps may present special risk

Disease Genesis—The toxicity of this species has been attributed to an alkaloid, possibly cytisine, which is found in all parts of the plant, including the glutinous pulp surrounding the seeds in the legumes (Chesnut 1898a,b). However, the presence of cytisine has not been confirmed. Instead, unusual amino acids such as 4-methylglutamic acid derivatives and an alkaloid, 5-hydroxypipecolic acid are reported to be present in seeds (Evans & Bell 1978; Southon 1994). However, these compounds are not likely to be of toxicologic significance, because they also are present in similar concentrations in the legumes of *Gleditsia triacanthos* (honey locust), the pods of which are relished by children and livestock (Pammel 1911). The seeds also contain a series of complex glycosides of hydroxylated oleanolic acid terpenoids or gymnocladdosaponins, such as gymnocladdosides, but the toxicity of these compounds is not known (Connolly & Hill 1991). The toxin is apparently heat-labile—hence the lack of toxicity following the roasting of the beans. The toxin also seems to be water soluble, because intoxications have been reported to follow after animals drink from water tanks into which the pods have fallen (Gates 1930). Crude extracts of the toxin are reported to cause narcotic effects and have been employed to poison flies (Chesnut 1898a).

Gymnocladus dioica

- toxicant unknown; possibly glycosides of oleanolic acid, gymnocladdosides

Gymnocladdoside genin

Clinical Signs—Signs indicative of gastrointestinal dysfunction are characteristic of animals that ingest *G. dioica.* Most commonly seen are excess salivation, diarrhea, colic, weakness, and depression.

- excess salivation; diarrhea, colic, weakness

Pathology and Treatment—Gross lesions are usually restricted to reddening of the mucosal surfaces of the stomach and small intestine, although degenerative lesions may be found in other organs. Treatment is symptomatic, directed toward alleviating

- gross pathology, reddened mucosa of stomach and intestine

  treatment, activated charcoal, fluids, electrolytes

the pain and inflammation accompanying the gastroenteritis. When given early, activated charcoal may be beneficial.

Indigofera L.

■ *I. spicata*
I. suffruticosa
indigo

Taxonomy and Morphology—Comprising approximately 700 species in the tropical and subtropical regions of the world, *Indigofera,* a member of the subfamily Papilionoideae, is one of the largest genera of the family. Its name is derived from the Latin words *indigus,* meaning "indigo," and *ferus,* or "bearing." As the name implies, plants yield a vivid blue dye. The cloth of Egyptian mummies dating from 2300 B.C. were dyed with *I. tinctoria,* a species of Southeast Asia; and in the early colonial period, the dye was more valuable than cloves, with plantations of *I. tinctoria* being established by the Dutch East Indies Company in India and Sumatra. Development of a synthetic indigo by Baeyer in 1878 led to a decline in importance of the natural dye. Some species are grown as ornamentals for their foliage and dense racemes of flowers.

In North America the genus is represented by some 15 native and introduced species. Only 2 are of toxicologic importance:

I. spicata Forssk. (= *I. endecaphylla* Jacq. ex Poit.)	creeping indigo, trailing indigo
I. suffruticosa Mill.	indigo, anil indigo

Indigofera spicata

■ herbs, shrubs or trees; leaves compound; flowers pealike; petals purple, rose, or white

Plants annuals or perennials; shrubs or small trees or herbs; herbage typically hirsute. **Leaves** palmately or 1-pinnately compound or rarely simple; leaflets 3 to many, terminal leaflet present, margins entire, sessile or petiolules present, stipels present or absent; stipules bristlelike, fused to petioles at bases. **Inflorescences** dense or loose racemes or spikes, rarely solitary; peduncles long; bracts small, caducous. **Flowers** small. **Sepals** 5; fused; calyces 2-lipped. **Corollas** papilionaceous. **Petals** 5; purple to rose to white; banners orbicular to ovate, reflexed, often short-clawed; wings obliquely oblanceolate to obovate or oblong, clawed, adhering to keels; keels rounded, clawed, apices blunt or acute, auriculate. **Stamens** 10; diadelphous; enclosed in keels. **Legumes** oblong or linear; cylindrical or sometimes 4-angled; straight or curved. **Seeds** 1 to many; spherical or cylindrical; separated by thin partitions.

Indigofera spicata is a perennial from a large taproot, with prostrate, yellowish brown or yellowish green stems to 100 cm long. It has 4–9 leaflets and red to purple petals. In contrast, *I. suffruticosa* is a shrub 1–2 m tall with sparsely strigose branches, 9–17 leaflets, and orange petals.

Indigofera suffruticosa

■ cultivated, ground covers or as source of indigo; naturalized

Distribution and Habitat—Native to tropical America and introduced as a source of commercial indigo, *I. suffruticosa* has become naturalized in the southeastern coastal plains. Of little value as a source of indigo, *I. spicata* is a weedy species originally from Africa. Because of its potential as a ground cover and forage, it has been introduced into a number of areas of the world, including Florida. Its abundance in Florida and adjacent Gulf states is undoubtedly due to its escape from cultivation and subsequent establishment along roadsides and other disturbed sites. It is an invader in pastures and sometimes is unnoticed among the taller grasses. Its continued spread in southeastern states is likely (Lievens 1992).

■ foliage, ruminants, liver disease; horses, neurotoxic

Disease Problems—Many species of *Indigofera* are of considerable interest to agronomists because they are palatable, high-protein forages with promising potential (Aylward et al. 1987). Fortunately, only a few species are toxic. They include *I. spi-*

cata, which has been recognized as a disease problem in Hawaii and Florida (Nordfeldt et al. 1952; Morton 1989), *I. linnaei* in Australia (Seawright 1989), *I. teysmanni* in India (Krishna et al. 1986), and *I. hochstetteri* in Africa (Suliman et al. 1983). These species cause liver disease, although renal disease and even hematuria may be produced in some instances. *Indigofera suffruticosa* has been shown experimentally to be toxic in mice (Ribeiro et al. 1991), and *I. spicata* is toxic to most animal species, causing abortions and/or severe liver disease after it has been consumed in large amounts for several weeks (Nordfeldt et al. 1952; Aylward et al. 1987). In horses, however, it is neurotoxic rather than hepatotoxic and produces effects similar to those of Birdsville disease caused by *I. linnaei* in Australia. *Indigofera spicata* may have feed value if its toxicity is attenuated by ruminal bacteria in a transferable manner similar to that for the genus *Leucaena* (lead tree), a member of the subfamily Mimosoideae and treated in this chapter (Jones 1985).

Disease Genesis—*Indigofera spicata* contains 3-nitropropionic acid (3-NPA) and various mono-, di-, and triester derivatives of NPA-endecaphyllins, some of which are identical with karakin of *Corynocarpus laevigatus* in the family Corynocarpaceae (see chapter 10) and hiptagin from taxa in other genera (Finnegan & Stephani 1968). Also present are the nonprotein α-amino acids indospicine and canavanine (Finnegan & Mueller 1965; Aylward et al. 1987). Concentrations of 3-NPA are low and, in most instances, are unlikely to be a hazard. Although 3-NPA concentrations are low in most species, it cannot be assumed that such is the case for all members of the genus, because some taxa contain in excess of 10 mg NO_2/g of plant material (Williams 1981b). A large amount of 3-NPA is required for intoxication because much of it is biodegraded in the digestive tract before absorption (Williams et al. 1970; Gustine et al. 1977; Majak & Clark 1980). Unusually high dosage may cause neurotoxic effects via inhibition of mitochondrial succinate dehydrogenase (Alston et al. 1977). For additional discussion of 3-NPA toxic effects, see "Nitrotoxicosis" in the treatment of *Astragalus* in this chapter.

- low levels of 3-NPA and derivatives, little risk; nonprotein α-amino acids, indospicine, canavanine, arginine analogs, appear to cause the hepatotoxic effects; neurotoxic cause unknown

$O_2N—C—C—COOH$

3-Nitropropionic acid (3-NPA)

$H_2N—C(=NH)—C—C—C—C—C(NH_2)—COOH$

Indospicine

Indospicine, which seems to be consistently present in species of *Indigofera* (Aylward et al. 1987), is a hepatotoxic analog of arginine whose toxic effects are antagonized by arginine (Hegarty & Pound 1968 1970). The reverse is also true. Indospicine acts as an in vitro arginine antagonist and inhibits incorporation of arginine and other amino acids in liver cells (Christie et al. 1971; Hegarty 1978). Canavanine, also an arginine analog and its antagonist in some instances, is only weakly hepatotoxic and does not seem to influence the effects of indospicine (Hegarty & Pound 1970; Rosenthal & Dahlman 1991).

Both indospicine and, to a lesser extent, canavanine inhibit synthesis of nitric oxide and decrease glutathione levels. This may result in an increase in superoxides in hepatocytes, in contrast to the effect of arginine, which increases nitric oxide in cells (Pass et al. 1996). Although indospicine appears to be the culpable toxicant in most animals, it is not clear that it is the cause of the neurotoxicosis in horses. In this respect, it is of interest to note that widespread cloudy swelling of the liver was observed in 3 horses with experimentally induced Birdsville disease (Hooper et al. 1971). Also, meat from horses dying of this disease contains sufficient indospicine to cause liver disease in dogs (Hegarty et al. 1988; Kelly et al. 1992). Indospicine also is a potential reproductive risk for livestock. Experimentally it has produced cleft palate and increased fetal resorptions in rats (Keeler 1972).

- cattle, weight loss, weakness, icterus, head pressing, increased serum hepatic enzymes, perhaps abortion

  horses, ataxia, seizures

Clinical Signs—In most species, abortion and/or signs of liver insufficiency predominate. Signs of the latter include weight loss, weakness, icterus, and head pressing.

The neurotoxic signs in horses include ataxia, staggering (especially when turning), periodic seizures, and, less commonly, corneal opacities.

■ gross pathology, liver small, firm, nodular

microscopic, liver fibrosis, cholestasis

Pathology—Hepatic effects are typically chronic; thus a small, firm, yellow, cirrhotic, and perhaps nodular liver can be expected. Microscopically, the changes include moderate fibrosis and cholestasis. Increases in serum hepatic enzymes (AST, ALT, GGT, and SDH) also will be present; otherwise serum chemistries and hematology are little affected.

■ cattle, general nursing care; horses, sedation; supplements, molasses or peanut meal

Treatment—In horses, control of seizures and other detrimental changes in behavior is of immediate concern. Although, the specific basis of the neurologic disease caused by *Indigofera* has not been identified, supplementation with gelatin, molasses, or peanut meal seems to be beneficial in reversing the disease process. This is presumably due to the high levels of arginine in these products (Hooper et al. 1971; Morton 1989). Treatment of the disease in other animal species is based upon nursing care to allow regeneration of hepatic function.

Laburnum Medik.

■ *L. alpinum*
L. anagyroides
L. xwatereri

bean tree
golden-chain tree
laburnum
golden-rain tree

■ shrubs or small trees; leaves palmately compound; flowers pealike; petals lemon yellow

Taxonomy and Morphology—Comprising 2 species and 1 hybrid species, *Laburnum,* a member of the subfamily Papilionoideae, is an Old World genus closely related to *Cytisus.* Species of the 2 genera hybridize, and the intoxication problems posed by all are similar.

L. alpinum (Mill.) Bercht. & J.Presl	Scotch laburnum, alpine golden-chain
L. anagyroides Medik. (= *L. vulgare* L.)	golden-chain tree, golden-rain tree, bean tree, laburnum
L. x*watereri* (Kirchn.) Dippel (= *L. anagyroides* x *L. alpinum*)	golden-chain tree, Voss's laburnum

Laburnum anagyroides

Plants perennials; deciduous shrubs or small trees; bark smooth, gray tinged with green. **Leaves** palmately compound; leaflets 3; stipules absent. **Inflorescences** axillary racemes; long; pendulous. **Sepals** 5; fused; calyces 2-lipped. **Corollas** papilionaceous. **Petals** 5; pale to lemon yellow; banners rounded to obovate, reflexed; wings obovate, clawed, auriculate; keels convex, shorter than wings. **Stamens** 10; fused; monadelphous; enclosed in keels. **Legumes** linear; flat. **Seeds** several.

1. Twig ends glabrous. Legumes glabrous. Seeds brown . . . . *L. alpinum*
1. Twig ends pubescent. Legumes pubescent. Seeds black.
   2. Upper sutures of legumes winged . . . . *L.* x*watereri*
   2. Upper sutures of legumes thickened and keeled, but not winged . . . . *L. anagyroides*

■ ornamentals

Distribution and Habitat—Native to south-central and southeastern Europe and to western Asia, all 3 taxa of *Laburnum* are cultivated as ornamentals primarily in the eastern portion of North America for their showy racemes.

■ neurologic disease, all animal species and humans; leaves, legumes, and seeds

Disease Problems—Most reports of intoxications due to *Laburnum* are from Europe, where trees are commonly planted as ornamentals. Cases have been described for pigs, horses, cows, dogs, and humans following consumption of its leaves or its legumes and seeds. Horses are most susceptible, with the lethal dosage of seeds being

0.5 g/kg b.w., whereas in dogs it is 6 g/kg (Clarke & Clarke 1967). Typically, the genus produces signs of mild to moderate neurologic dysfunction. Death is due to respiratory failure (Allen 1948; Auchterlonie 1948; Clarke et al. 1971; Leyland 1981; Cooper & Johnson 1984; Habermehl 1992). There also are instances in which plant material has been eaten in large quantities without development of intoxication (Whitty 1948). In England, species of *Laburnum* have been responsible for numerous instances of concern in children, but only rarely were signs of intoxication observed, and, for the most part, they were mild and disappeared in a few hours (Mitchell 1951; Forrester 1979; Bramley & Goulding 1981). Although serious intoxications in humans are rare, fatalities have been reported. In one instance, an adult male with a mental disorder was found dead. Subsequent examination revealed 23 legumes of *L. anagyroides* in his stomach. In this instance, the onset of signs was apparently very abrupt and quickly fatal (Richards & Stephens 1970).

Disease Genesis—The toxicants of *Laburnum* are an array of quinolizidine alkaloids of the cytisine type. Composing 50% or more of the total alkaloid content of the plant, cytisine and/or *N*-methylcytisine predominate. A dozen others, including the teratogenic anagyrine (1% of the total) and other pyridone-nucleus quinolizidines, may be present. In *L. anagyroides,* alkaloid concentrations are relatively low in the leaves (0.36 mg/g) and green legumes (3.16 mg/g), but very high in mature seeds (12+ mg/g) (Szentesi & Wink 1991). Concentrations also are moderately high in buds. Overall, they are lowest in midsummer (Greinwald et al. 1990). *Laburnum alpinum* and *L.* x*watereri* are similarly toxic (Wink et al. 1983). In *L.* x*watereri,* epibaptifoline, a pyridone-nucleus quinolizidine believed to be a teratogen, is the main alkaloid present in the leaves (2 mg/g) during the summer and fall when other alkaloid concentrations are minimal (Greinwald et al. 1990). The toxic dosage of seeds is estimated at 0.5 g/kg b.w. in horses, which, based upon alkaloid concentrations of 1%, is equivalent to 5 mg/kg b.w. of cytisine (Clarke et al. 1971). The toxic dosage in other animal species is estimated to be several-fold greater. In the human fatality reported by Richards and Stephens (1970), the estimated dosage of cytisine, based upon tissue analysis, was 35–50 mg, equivalent to 0.5–1 mg/kg b.w. In calves, gavage with a total dosage of cytisine alkaloidal extract of *L. anagyroides* seeds and pods of 30.5 mg/kg b.w. given daily over 3 days caused moderate signs of intoxication (Keeler & Baker 1990).

■ quinolizidine alkaloids of cytisine type; activate central and peripheral nicotinic cholinergic receptors; cause stimulation-type effect; possibly teratogenic because of presence of anagyrine

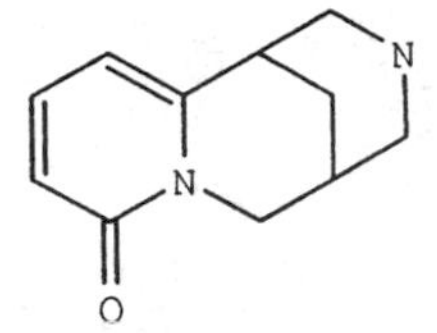

Cytisine

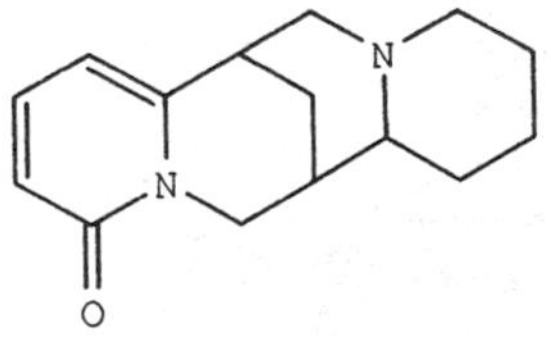

Anagyrine

The alkaloids represented in *Laburnum* bind with considerably greater affinity to nicotinic acetylcholine receptors than to muscarinic types (Schmeller et al. 1994). Cytisine is a nicotinic receptor agonist that is capable of causing curare-like, depolarizing ganglionic blockade at high dosage. However, the more common signs are of stimulation at lower dosage. It is more readily absorbed, and of similar or greater potency than nicotine, depending on the parameter of comparison. Effects on the central nervous system are much reduced in comparison with the effects of nicotine (Sloan et al. 1988). Although it seems to have greater affinity for the central nicotinic receptor, as measured in vitro in porcine brain, *N*-methylcytisine is 10 times less potent in vivo than either cytisine or nicotine (Dale & Laidlaw 1912; Scott & Chen 1943; Barlow & McLeod 1969; Beani et al. 1985; Schmeller et al. 1994). Cytisine, like nicotine, is rapidly absorbed, metabolized, and eliminated; hence there is rapid onset of signs, but they are of short duration. In addition to the expected neurologic effects, myodegeneration is reported and suggested to be an effect caused by the α-pyridone quinolizidine alkaloids such as cytisine and anagyrine (Keeler & Baker 1990).

Because of the presence of substantial concentrations of anagyrine in the seeds, legumes, and possibly other tissues in *L. anagyroides* and of epibaptifoline in *L.*

xwatereri, teratogenesis is a risk with repeated ingestion of what may otherwise be nontoxic amounts of plant material.

- increased salivation, anorexia, ataxia, tremors, colic, walk with knuckling of fetlocks

  humans and dogs, colic, vomiting

Clinical Signs—The signs observed depend on whether the leaves or the legumes and seeds are ingested and time in the growing season. Typically, they include increased salivation, anorexia, incoordination, tremors, and colic. Abnormal flexion causing knuckling of the fetlocks has been seen in cattle eating the leaves. Consumption of the leaves, especially late in the season, may lead to a myopathy and predispose to reproductive and/or teratogenic effects because of the increased presence of the pyridone alkaloid types. In children and dogs, abdominal pain and vomiting without diarrhea are the most typical clinical signs, accompanied by a variety of other signs such as drowsiness, weakness, ataxia, and a rapid heart rate. There is early respiratory stimulation, but later respiration is depressed.

- no lesions

Pathology—Few if any gross lesions have been reported with *Laburnum* intoxication. Microscopically, there may be mild degenerative skeletal myopathy.

- activated charcoal, nursing care

Treatment—*Laburnum* produces a rather acute disease of short duration with little likelihood of death. Thus, a conservative approach employing the oral administration of charcoal and supportive care is appropriate. In species in which it is appropriate, induction of vomiting may be of value unless that induced by the disease itself is deemed sufficient.

Lathyrus L.

- *L. cicera*
  L. hirsutus
  L. latifolius
  L. odoratus
  L. palustris
  L. pusillus
  L. sativus
  L. splendens
  L. sylvestris

  annual sweet pea
  chickpea
  Caley pea
  Campo pea
  chickling pea
  everlasting pea
  flat pea
  flat-podded vetch
  forest vetchling
  marsh pea
  peavine
  perennial pea
  pride of California
  riga pea
  singletary pea
  wild winter pea
  wood pea

Taxonomy and Morphology—Comprising approximately 150–160 species and numerous varieties, *Lathyrus,* a member of the subfamily Papilionoideae, is closely related to *Vicia* (vetch), *Pisum* (pea), and *Lens* (lentil). Commonly known as sweet pea, vetchling, or peavine, the genus name was used in the third century B.C. by Theophrastus in his treatise on plants and is derived from the Greek roots *la,* meaning "very," and *thoures,* meaning "stimulant," in reference to the reputed stimulating or irritating effect of the seeds. Its members are cultivated for food and forage, and in some parts of the world they are an important part of the diet of humans. Other species are cultivated as ornamentals, such as *L. odoratus* (sweet pea), which is familiar to many.

In North America, some 42 species occur; 27 are native, and the others are cultivated or naturalized (Hitchcock 1952). The most commonly encountered are the following (see plate 37.4):

L. cicera L.	chickpea, flat-podded vetch
L. hirsutus L.	Caley pea, wild winter pea
L. latifolius L.	everlasting pea, perennial pea
L. odoratus L.	annual sweet pea
L. palustris L.	marsh pea
L. pusillus Elliot	singletary pea
L. sativus L.	chickling pea, riga pea
L. splendens Kellogg	pride of California, Campo pea
L. sylvestris L.	flat pea, wood pea, forest vetchling, narrow-leaved everlasting pea, flat peavine

Plants annuals or perennials; herbaceous; typically climbing or trailing vines by means of tendrils; taprooted or rhizomatous. **Stems** trailing or sometimes erect; often winged and angular. **Leaves** subsessile; 1-pinnately compound; terminal leaflet absent; leaflets 2–14, blades linear or oblong, margins entire, petiolules absent; rachises ending in tendrils; stipules leaflike. **Inflorescences** axillary racemes or solitary flowers; bracts absent. **Sepals** 5; fused; calyces radially symmetrical or 2-lipped; 10-nerved. **Corollas** papilionaceous. **Petals** 5; white or pink or blue or purple or red or yellow; banners orbicular, reflexed; wings clawed, auriculate, free from keels or nearly so; keels rounded, not abruptly beaked. **Stamens** 10; diadelphous; enclosed in keels. **Legumes** linear to oblong; sessile or stipitate; usually flattened or somewhat terete; valves coiling upon dehiscence. **Seeds** 2 to many.

- climbing or trailing vines; stems sometimes winged; leaves 1-pinnately compound, leaflets 2–14, with terminal tendrils, stipules leaflike; flowers pealike; petals white, pink, blue, purple, red, or yellow

Lathyrus is morphologically quite similar to *Vicia*. The two genera differ in the appearance of their styles and the winging of their stems. In *Lathyrus* the style is flattened and pubescent only on its adaxial or inner surface, whereas in *Vicia* the style typically is terete and either pubescent all the way around or only on the abaxial or outer surface. The stems of *Lathyrus* are typically winged, and those of *Vicia* are never so.

Species of *Lathyrus* are readily distinguished by vegetative characters such as presence or absence of tendrils, appearance of leaflets, and appearance of stipules.

Lathyrus hirsutus

1. Plants annuals.
   2. Corollas 20–30 mm long . . . . *L. odoratus*
   2. Corollas 8–15 mm long.
      3. Legumes papillose-hirsute. Calyx tubes as long as calyx lobes . . . . *L. hirsutus*
      3. Legumes glabrous. Calyx tubes much shorter than calyx lobes.
         4. Corollas white and blue. Stipules linear . . . . *L. sativus*
         4. Corollas purple or red-purple. Stipules lanceolate to ovate.
            5. Racemes 2-flowered . . . . *L. pusillus*
            5. Racemes 1-flowered . . . . *L. cicera*
1. Plants perennials.
   6. Leaflets 2.
      7. Corollas 20–30 mm long . . . . *L. latifolius*
      7. Corollas 10–20 mm long . . . . *L. sylvestris*
   6. Leaflets 4–10.
      8. Corollas 15–22 mm long. Banners reflexed 90 degrees . . . . *L. palustris*
      8. Corollas 25–40 mm long. Banners reflexed 120–180 degrees . . . . *L. splendens*

Lathyrus latifolius

Distribution and Habitat—Primarily circumboreal in distribution, *Lathyrus* has three apparent centers of diversity: northern North America, southern South America, and southern Eurasia at the eastern end of the Mediterranean and adjacent northern Africa. The distribution and habitat of the species commonly encountered are as follows:

L. cicera—from Eurasia; uncommon in disturbed areas; from California to southeastern states except in deserts

L. hirsutus—from Eurasia; along roadsides and in disturbed sites, wet meadows, and creek beds; southern third of continent

L. latifolius—from Eurasia; cultivated as ornamental; occasionally escaping and naturalizing in the southeastern portion of the continent

Lathyrus palustris

Lathyrus pusillus

Lathyrus sylvestris

L. *odoratus*—from Italy and Sicily; cultivated as ornamental, cosmopolitan in horticultural trade; occasionally escaping but not persisting
L. *palustris*—native; in wet meadows, swamps, and wet woods across the continent; circumboreal
L. *pusillus*—native; in moist prairies and along roadsides
L. *sativus*—from Eurasia; cultivated as ornamental in North America and for food and forage in other parts of world
L. *splendens*—native; uncommon in chaparral of southern California and Baja California
L. *sylvestris*—from Eurasia; cultivated as ornamental; occasionally escaping and persisting in eastern and northwestern portions of the continent

Disease Problems—The disease in humans associated with this genus, termed lathyrism, is known from early Hindu and Greek writings, including those of Hippocrates (Roy & Spencer 1989). Thus it is one of the oldest neurotoxic diseases known. It was a continuing and serious problem in Europe and Asia during the 1700s and 1800s, with most cases attributable to the use of pea flour in breadmaking (Gardner & Sakiewicz 1963). For centuries, it has been a problem on the Indian subcontinent and today remains a threat in Bangladesh, India, and Ethiopia. It is a problem for those least able to deal with drought and famine, the poor. *Lathyrus* may be lifesaving during such difficult times because plants produce fruits even in the harshest years. However, when individuals are compelled to subsist on peas for sustained periods, the threat of lathyrism is significantly greater. Such conditions are a prime environment for epidemics of the disease.

■ neurolathyrism, an ancient disease in humans from prolonged subsistence on a diet of peas; especially prevalent in times of famine

■ humans, insidious or abrupt onset, spastic paresis of legs, mainly in young males; involuntary spasms of calf muscles during sleep; also occurs in most animal species, especially horses

Lathyrism is a disease of profound social impact because it produces more or less permanent physical incapacitation in individuals already living under demanding and wretched conditions. There may be an insidious or, more often, a sudden onset of muscle spasms, stiffness, and loss of muscular control. The main problem is spastic paraparesis, with increased tone of the quadriceps and gastrocnemius muscles. The degree of spasticity is correlated with slowing of peripheral nerve conduction, especially a slowing of central motor conduction to the tibialis anterior muscle (Misra & Sharma 1994). In some cases prior to the onset of spasticity and rigidity, there are prodromal signs, including numbness, cramps, general weakness, and reflex irritability of muscles. Especially noteworthy are the very painful cramps or involuntary contractions of the calf muscles while the individual is asleep (Paissios & Demopoulos 1962; Gardner & Sakiewicz 1963). Recovery from early signs is possible, but often the disease is permanent, with eventual deformities preventing normal limb support and then causing muscle atrophy. Typically, there are no sensory deficits or loss of bladder or bowel control, but problems with urinary frequency and impotency have been reported (Hugon et al. 1988; Haimanot & Kidane 1990). Teenage males are at particularly high risk. In an epidemiologic study in Bangladesh, where the disease remains a risk, the prevalence rate in the population was 14 per 10,000 (Haque et al. 1996). The gender proportions were 87.1% in males and 12.9% in females; 90% of both sexes were less than 30 years of age at the time of onset, with some families having several affected members. The disease is reported to be reproducible in guinea pigs and primates subclinically deficient in ascorbic acid before feeding the seeds of *L. sativus* (Jahan & Ahmad 1993).

In addition to humans, lathyrism occurs in a variety of animal species, including horses, cattle, sheep, pigs, monkeys, and elephants (Stockman 1929; Chopra et al. 1940; Coulson et al. 1969). It has been the subject of several reviews (Stockman 1929; Selye 1957; Barrow et al. 1974; Roy & Spencer 1989; Haimanot & Kidane 1990).

■ osteolathyrism is mainly an experimental disease

Although lathyrism in humans is quite consistent in its signs and progression, there are differences in its appearance among animal species. In addition, its appearance in experimental studies involving laboratory animals is quite different. The terms *osteolathyrism* and *neurolathyrism* are generally used to reflect these differences between nature and the laboratory (Selye 1957). Osteolathyrism is mainly an experimental anomaly, with few clinical corollaries except for reproductive effects, although findings suggestive of osteolathyric bone changes in a few individuals from Bangladesh exhibiting neurolathyrism have been reported (Haque et al. 1994). The most prominent osteolathyric findings in 2 individuals were failure of fusion of vertebral and iliac epiphyses (Haque et al. 1997a). Most clinical manifestations are representative of neurolathyrism. There are apparent differences in susceptibility between monogastric animals and ruminants, possibly because of ruminal degradation of the toxins. Horses are very susceptible and develop a disease resembling that in humans. It begins with weakness and incoordination, followed by paraplegia, which in this species is more flaccid than rigid. Disease in horses, either observed or suspected, is reported following the feeding of the peas or forage of *L. incanus, L. nissolia, L. latifolius,* and *L. sativus* (Leather 1885; Hughes 1895; Steyn 1933, 1934; Beath et al. 1953; Greatorex 1966; Lippegaus et al. 1992). Recovery may be rapid and complete in early cases. A unique and consistent finding in horses is degeneration of the vagus and recurrent laryngeal nerves, causing both inspiratory and expiratory dyspnea (Tel'tsov 1940).

Lathyrus hirsutus

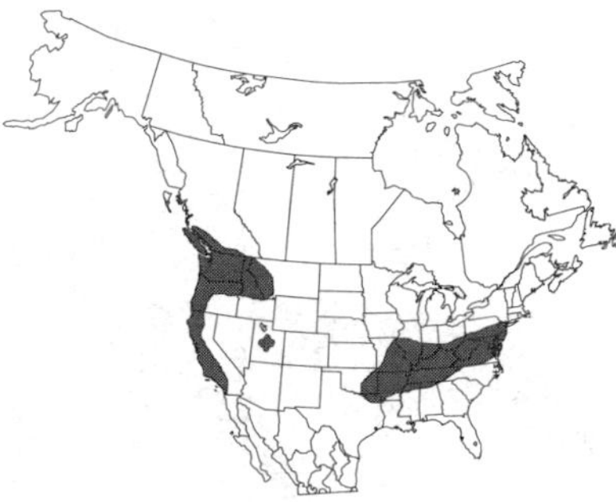

Lathyrus latifolius

■ avoid feeding toxic *Lathyrus* species when maturing legumes and seeds are present

Disease in cattle and sheep has been associated with forage of *L. sylvestris, L. hirsutus,* and *L. latifolius,* especially when the plants are eaten as the seeds are developing in the legumes (Sugg et al. 1944; Foster 1990; Rasmussen et al. 1992, 1993; Burrows et al. 1993; Rowe et al. 1993). Intoxication may require a large amount of forage eaten for a prolonged period, particularly in sheep. When *L. sylvestris* is fed as less than half of the diet, is given for only for 1–2 weeks, or is fed in the vegetative–early flower bud stage, there is little problem. In fact, it is a palatable, highly nutritious forage with high protein in the prebloom or early bloom stages (Daniel et al. 1946; Forster et al. 1991). The ruminal flora of sheep also is able to adapt somewhat when the animal is fed appreciable quantities of plant material for several months (Foster 1990; Rasmussen et al. 1992, 1993). However, this protection may not be dependable, because other factors—e.g., administration of the ionophore monensin—may supervene to prevent or reverse the adaptation (Rasmussen et al. 1992, 1993; Rowe et al. 1993). It is of interest that although the disease in ruminants due to *L. sylvestris* may be lethal, few or no pathological changes are present. This is in contrast to disease caused by *L. latifolius,* in which calves, as well as horses, exhibit a distinct neuropathology (Burrows et al. 1993). As a general rule, forage of any of these species of *Lathyrus* should not be fed when the maturing legumes and seeds are present on the plants. Other species such as *L. venosus* are a reliable source of good, palatable forage (Looman 1983).

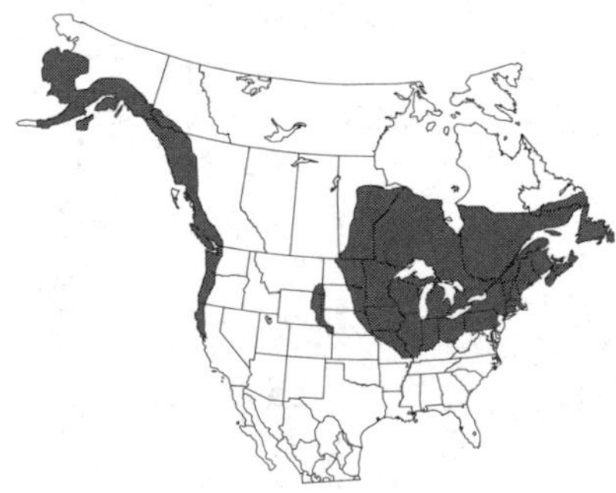

Lathyrus palustris

■ 3 syndromes—
(1) acute, often lethal with distinctive neuropathology
(2) acute, often lethal without distinctive neuropathology
(3) acute, generally nonlethal transient neurologic effects

The division of lathyrism into two types of syndromes is complicated by a variety of experimental findings. Baby chicks fed large amounts of the seeds or seed extracts of *L. sativus* develop distinct neurologic signs. Adults are much more resistant (Moslehuddin et al. 1987; Chowdhury 1988). Chicks fed large amounts of *L. odoratus* develop skeletal and neurologic changes, with considerable mortality (Raharjo et al. 1988). Similarly, rats given *L. odoratus* or *L. pusillus* develop skeletal changes (including osteoporosis), paralysis, and aortic aneurysms (Lee 1950). Those given other species of *Lathyrus* show distinct neurologic signs (Lewis et al. 1948; Lewis & Schulert 1949; Ponseti & Baird 1952; Schulert & Lewis 1952). In the case of *L. sylvestris,* both the seeds and the forage produce neurologic disease.

As the information presented above indicates, animals exhibit at least three, some-

Lathyrus pusillus

Lathyrus splendens

- neurolathyrism, related to novel amino acids mainly in seeds; ODAP (= BOAA), DABA, OxDABA

HOOC—C(=O)—N—C—C(NH$_2$)—COOH

ODAP (= BOAA)

H$_2$N—C—C—C(NH$_2$)—COOH

DABA

HOOC—C(=O)—N—C—C—C(NH$_2$)—COOH

OxDABA

what distinct neurologic syndromes associated with ingestion of *Lathyrus*. The first is an acute, often lethal disease with distinctive neuropathologic changes, as illustrated by the effects of *L. latifolius* in horses. The second is an acute, often lethal disease without distinct neuropathologic changes, as seen in sheep that eat *L. sylvestris*. The distinction between the two syndromes may depend upon whether the pathologic effects occur primary or secondary to the development of signs. The third syndrome is an acute, generally nonlethal disease with a set of clinical and pathological changes different from either of the first two. This last is best exemplified by the disease appearing in cattle after they eat *L. hirsutus*. Whether these three syndromes represent different stages of the same disease or are related to a common mechanism activated to different extents by various compounds or dosage is unknown. It may be that with some animal or plant species, not enough plant material can be eaten to induce severe signs or for pathologic changes to develop.

Disease Genesis—Disease problems caused by *Lathyrus* are due to a variety of novel amino acids that occur in the genus and also to the presence of nitriles (Bell & Foster 1962; Bell 1963, 1973; Rao et al. 1963; Roy & Spencer 1989; Riepe et al. 1995). Neurolathyrism in its various forms seems to be closely associated with β-*N*-oxalylamino-α,β-diaminoproprionic acid (ODAP) and possibly with α,γ-diaminobutyric acid (DABA) and its oxalylated derivative, γ-*N*-oxalyl-L-α,γ-diaminobutyric acid (OxDABA)(Jacob et al. 1967; Olney et al. 1976). Using alternative nomenclature, ODAP is synonymous with β-*N*-oxalylamino-L-alanine (BOAA) and L-3-oxalylamino-2-aminoproprionic acid (OAP or OxDAPA) (Roy & Spencer 1989). Concentrations of ODAP in different species of *Lathyrus* may range up to several milligrams per gram of seed (Siddique et al. 1996). Distribution of these amino acids among species of *Lathyrus* is summarized in table 37.7. They are apparently somewhat heat labile, as shown by experiments with autoclaved *L. sativus* fed to chicks (Rotter et al. 1990).

Table 37.7. Distribution of novel amino acids in various species of *Lathyrus*

ODAP	DABA/OxDABA	DABA/OxDABA + ODAP	BAPN
L. alatus	*L. aurantius*	*L. cirrhosus*	*L. hirsutus*
L. articulatus	*L. laevigatus* var. *aureus*	*L. gorgoni*	*L. odoratus*
L. arvense	*L. luteus*	*L. grandiflorus*	*L. pusillus*
L. cicera		*L. heterophyllus*	*L. roseus*
L. clymenum		*L. latifolius*	
L. megallanicus		*L. multiflora*	
L. ochrus		*L. rotundifolius*	
L. pannonicus		*L. sylvestris*	
L. pseudocicera		*L. tuberosus*	
L. quadrimarginatus		*L. undulatus*	
L. sativus			
L. setifolius			
L. szowitzii			
L. tingitanus			
L. tremolsianus			

Source: Bell 1973.

- presumed mechanism, novel amino acids act as agonists at central glutamate kainate and quisqualate excitatory receptors, and excessive activity of receptors leads to neuronal degeneration

Understanding the presumed mechanism of action requires an appreciation of excitatory transmission in the brain. Depending on the classification used, there are four or more classes of glutamate excitatory receptors, three of which directly gate ion channels and are of concern in lathyrism (Monaghan et al. 1989; Kandel & Schwartz 1991). All employ glutamate as the compartmentalized transmitter, but each is unique in response to alternative agonists and the conditions for gating effects; these factors then are the basis for separating the receptors.

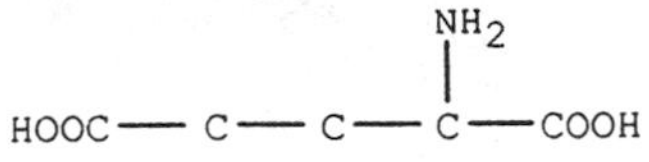

Glutamic acid

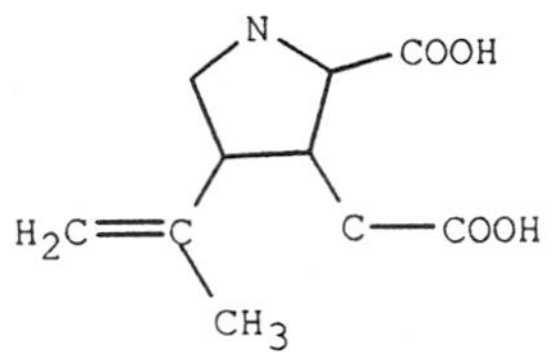

Kainic acid

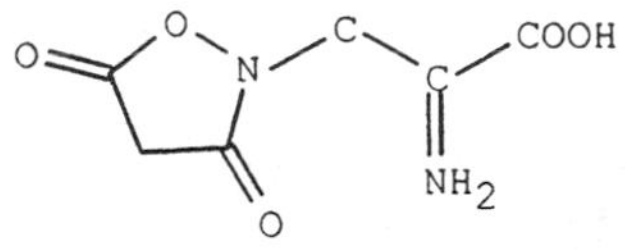

Quisqualic acid

The two glutamate receptors of primary importance as sites of lathyrogen binding and activation are the kainate and quisqualate A receptors. Both are gated to Na^+ and K^+ channels and are sometimes termed AMPA receptors because they are also bound by α-amino-3-hydroxy-5-methyl-4-isoxazole-proprionate (AMPA) (Kandel & Schwartz 1991). These are the major excitatory receptors of the motor neurons, and it is these two receptors that appear to be affected by ODAP and possibly by DABA and OxDABA (Olney et al. 1976; MacDonald & Morris 1984; Zagoren et al. 1985). The neurotoxic effects of ODAP are blocked specifically by kainate and quisqualate glutamate receptor antagonists (Ross et al. 1987; Weiss et al. 1989). The NMDA receptor, which is both ion channel–gated and voltage-gated does not appear to be a primary site of action of the lathyrogenic neurotoxins, inasmuch as NMDA receptor antagonists are not protective against damage wrought by ODAP (Weiss et al. 1989). Selective accumulation of ODAP in the cerebellum of monkeys may increase their susceptibility (Mehta et al. 1980). This is especially significant because ODAP stimulates glutamate receptors in the cerebellum, elevating cGMP levels—actions that are prevented by excitatory amino acid antagonists (La Bella et al. 1993). Other plants also are known that contain kainate receptor agonist toxins, e.g., willardiine in *Acacia* and *Mimosa,* and stizolobic acid in *Stizolobium* (Patneau et al. 1992; Maruyama & Takeda 1993).

Presumably, excessive agonist activity on the excitatory receptors leads to a type of glutamate toxicity with degeneration of motor neurons. Continuous depolarization of the voltage-independent kainate and quisqualate receptors, which causes distinct postsynaptic vacuolation and degeneration, may result in indirect Ca^{2+} activation (Olney et al. 1976; MacDonald & Morris 1984; Ross et al. 1987). Excitotoxins, including glutamate, also have been implicated as possible factors in several degenerative diseases of humans, especially in the elderly, as reviewed in detail by Olney (1994) and Dawson and co-workers (1995). There is a close similarity between the disease produced in macaques given ODAP for many months and the naturally occurring neurolathyrism in humans. The clinical effects are apparently secondary to pathologic changes (Spencer et al. 1986). In addition, ODAP under appropriate conditions causes acute disease directly, without underlying lesions (Parker et al. 1979; Mehta et al. 1983). This also is apparent for DABA. Lambs given large doses of DABA orally have abrupt onset of nervous signs such as weakness, rigidity, and paddling seizures as well as hepatic degeneration (Keeler et al. 1967). Similarly, intramuscular administration of DABA causes weakness, muscle tremors, limb flexure, and death in 36 hours. If repeated doses are given for several days, joint flexure occurs, persisting for several weeks or more. These are important considerations because, clinically, both acute reversible disease and a subacute to chronic irreversible disease are noted in animals. Numerous other actions of ODAP are now becoming apparent, and some may be contributing factors. These include other changes in glutamate disposition and direct toxic effects on astrocytes (Lindstrom et al. 1990; Ormandy & Jope 1990; Bridges et al. 1991). The latter effects may be particularly relevant because it is now recognized that astrocyte degeneration precedes a marked increase in vulnerability of neurons to glutamate toxicity (Rosenberg & Aizenman 1989). Even in low concentrations, ODAP increases activity of glutamate synthetase, causing a derangement of glutamate regulation in astrocytes (Miller et al. 1993).

The marked susceptibility of neurons with long axons to toxin-induced degenerative processes is detailed in a lucid discussion by Cavanagh (1984). He noted that these cells were rendered open to axonal transport failures because of their lengths and the high levels of metabolic activity required for their maintenance.

There is considerable similarity between the disease signs when either *L. sylvestris* is fed or DABA is administered to the animal (Keeler et al. 1967; Rowe et al. 1993). Because DABA interferes with the urea cycle by inhibiting hepatic ornithine transcarbamoylase, it was thought that hyperammonemia might account for the neurotoxic effects (O'Neal et al. 1968). Subsequent studies revealed that DABA was neurotoxic when administered intracisternally as well intraperitoneally (Chen et al. 1972). This observation, together with other results indicating that protection afforded by thyroxine against DABA intoxication is accompanied by markedly decreased concentrations of DABA in the brain, increases the likelihood that DABA acts mainly or exclusively as a direct neurotoxicant (Ressler et al. 1961; Vivanco et al. 1966). There also may be a role for nitric oxide, because it is involved with NMDA-induced free amino acid accumulation and GABA/Cl^- receptor destruction in the cerebral cortex (Strosznajder et al. 1994). It should be pointed out that DABA is found normally in very small amounts in bovine brain, possibly as a precursor to the more toxic diaminopropane, thus accounting for the delay in onset of signs of toxicity (Nakajima et al. 1967; Chen et al. 1972).

The likelihood that intoxication will occur may be influenced by environmental conditions. ODAP is suggested to have a role in the plant as a zinc carrier, and under conditions of soil deficiency it may increase in compensation (Lambein et al. 1994). Despite the vast amount of information substantiating the neurotoxic effects of ODAP and DABA, it is not clear that they are the direct cause in all instances of disease involving *Lathyrus*.

- osteolathyrism, inhibition of lysyl oxidase by BAPN; prevents cross-linking of collagen and elastin in developing cartilage and bone; fetal abnormalities; mainly an experimental disease

$N{\equiv}C{-}C{-}NH_2$

BAPN

Osteolathyrism is associated with the presence of β-aminoproprionitrile (BAPN) and related nitriles in *L. odoratus* and a few other species. These nitriles act by irreversibly inhibiting lysyl oxidase, a cupric metalloenzyme that catalyzes conversion of lysine residues to amino residues in the cross-linking of collagen and elastin (Barrow et al. 1974). Thus, changes are mainly manifested as perturbations in developing cartilage and bone, such as in the developing fetus. As confirmation of this potential, when aminoacetonitrile, a synthetic osteolathyrogen, was given to pregnant ewes, it caused a variety of reproductive effects (Keeler & James 1971). These effects included abortions, stillbirths, and various fetal deformities such as carpal flexure, contracted pasterns, twisted or bowed limbs, spinal abnormalities, and increased bone fragility. In poultry, BAPN decreases ovarian activity, egg production, and shell quality, the latter apparently due to inhibition of lysyl oxidase (Chowdhury 1988; Chowdhury & Davis 1989).

It is not clear if osteolathyrogens are capable of causing neurotoxic effects, as has been suggested (Lee et al. 1956). It was originally assumed that the paralysis that accompanies skeletal changes was secondary to spinal changes. That this is not the case is shown by the effects of dietary supplements of casein, which attenuate the skeletal changes but worsen the reversible paralysis (Lee et al. 1956). This is an indication that the paralysis is due to some other action of BAPN. Osteolathyrogenic effects have been produced by a synthetic neurolathyrogens, but osteolathyrogens do not produce neurologic effects (Mennin & Thomas 1970). Thus, the osteolathyrogen BAPN may not be responsible for the acute neurotoxicosis associated with *L. hirsutus* in cattle. Perhaps other uncommon amino acids are accountable for these effects (Roy 1981).

- horses, stiff gait, stringhalt-like, laryngeal paresis, ataxia, hyperexcitability, weakness; eventually flaccid paraplegia

  sheep and cattle, abrupt onset, lethargy, anorexia, ataxia, pacing, circling, tremors, increased salivation, seizures, collapse

  mild form in cattle from *L. hirsutus*, transient lameness and ataxia

Clinical Signs—Signs of lathyrism vary somewhat among the different species of *Lathyrus*, which also exhibit considerable variation in their toxicity. The neurotoxic signs of the disease are most prominent in the horse, and death or severe incapacitation is not an unlikely sequela in this species. Signs such as a stiff or stringhalt-like gait, incoordination, hyperexcitability, weakness, and struggling to stand represent progressive stages of the disease. When these signs are present, the disease may still be reversible, but more permanent, obvious neuromuscular incapacitation or flaccid paraplegia may supervene with continued plant ingestion. Concomitantly, and unique to the horse, paresis or paralysis of the larynx, with loud inspiratory and expiratory noises, may become quite noticeable.

In sheep, the disease maybe mistaken for enterotoxemia with acute onset of lethargy, anorexia, incoordination, pacing, circling, head pressing, tremors, increased salivation, grinding of the teeth, and seizures. The seizures occur as a sudden extension of the thoracic limbs, head, and neck, with the spasms lasting several minutes and leading to collapse. There may be recovery but also recurrence. Deaths may occur, but typically few animals at risk will become affected or die.

In cattle, the disease may occur as in sheep, possibly with paraplegia developing. A milder form has been reported in animals eating *L. hirsutus* when legumes are present. In this milder form, the animals appear to be lame, stand hunched up with the legs placed well under the body, and are reluctant to walk. They prefer to lie down, and when they are forced to stand, muscular tremors may be quite noticeable. This is a transient disease, and recovery is usually complete in a few days. It is of interest to note that this form has been associated with *L. hirsutus* even though this species has been reported to be osteolathyrogenic rather than neurotoxic.

Chickens and ducks develop prominent neurologic signs such as distorted head and neck postures, stiffening, and seizures. Pigeons seem to be somewhat resistant.

In humans, the disease begins with weakness of the legs, and when the affected individual attempts to walk, the muscles become tremulous. There is impaired ability to walk, and further degrees of lower spastic paraparesis may develop, resulting in limb and gait abnormalities and eventually incapacitation.

- humans, weakness of legs, quivering muscles, spasticity of leg muscles, impaired ability to walk

Given the lack of field observations, the relevance of *Lathyrus* as a cause of clinically significant reproductive abnormalities is not clear. Experimentally, fetal skeletal malformations can be produced with osteolathyrogens; changes include pastern and carpal joint contractures, lateral limb rotation, and spinal deformities. These defects occur with exposure to the toxins for as little as 10 days during a surprisingly long time span of susceptibility, e.g., in the first two-thirds of gestation for lambs (Keeler 1975).

In humans, the signs of neurolathyrism may be mistaken or confused with those of tropical spastic paresis (TSP) in some instances. However, out of 444 cases of lathyrism in Bangladesh, only 4 were found seropositive to the viral cause of TSP (Haque et al. 1995).

Pathology—Other than the physical deformities produced by the neuromuscular alterations, there are no gross lesions. However, extensive microscopic neurologic changes may be present. In the horse, degenerated motor neurons in the central brainstem, the medulla oblongata (especially the vagal and accessory nuclei), and the spinal cord are numerous. Degeneration of spinal horn neurons (especially lateral and ventral) and spinal tract axons with multifocal demyelination and sclerosis also may be prominent. Few neutrophils are present, but there is evidence of macrophage and astrocyte activity. There also may be atrophy and demyelination of the recurrent laryngeal nerve, especially the left one (Leather 1885; Lippegaus et al. 1992). In some instances in cattle and sheep, there are no pathologic changes.

- gross pathology, no lesions

  microscopic, horses, degeneration of motor neurons in brainstem, medulla, spinal tracts, with focal demyelination; recurrent laryngeal nerve demyelination

  humans, degeneration of Betz cells in motor cortex and pyramidal tract of spinal columns

  experimental osteolathyrism, osteoporosis, cartilage and bone defects

In humans, there are few detailed studies of the histopathology; many of the brain effects were inferred from the clinical observations or described many years after disease onset (Gardner & Sakiewicz 1963; Ludolph et al. 1987). Except for the lack of laryngeal nerve degeneration, the lesions are similar to those in the horse. They include degeneration of the Betz cells (large pyramidal neurons of the motor cortex) and degeneration and sclerosis of the myelinated fibers of the dorsal and lateral (especially the pyramidal tract) spinal columns. The ventral horns appear to be less affected (Gardner & Saklewicz 1963; Hirano et al. 1976). In monkeys, neural degeneration is more widespread, scattered in all columns of the spinal cord and in the brain as well (Selye 1957). Spondylolysis is also seen in vertebrae L_2–L_5 (Paissios & Demopoulos 1962).

In experimentally produced osteolathyrism in rats, there is osteoporosis (not preventable with calcium), plus various bone and cartilage defects and degenerative arthritis (Gardner 1959).

■ sedation for seizures, otherwise no treatment, although long-term use of centrally acting muscle relaxants is of value in humans; livestock disease with some species of *Lathyrus* is readily reversible

Treatment—Neurolathyrism in animals is generally not responsive to treatment except to control seizures. The milder forms in cattle and sheep typically are readily reversible upon removal from the causative forages.

In most instances of neurolathyrism in humans, treatment also has been of little value. Surgical intervention involving severance of the adductor muscles has been attempted to relieve the contracture (Haque et al. 1994). Of greater interest is the daily administration of a centrally acting muscle relaxant, tolperisone HCl. Administration of 150 mg/kg b.w. daily for several months resulted in marked improvement and prevention of the progressive nature of the disease (Haque et al. 1994).

Supplemental calcium, corticoids, and possibly reserpine appear to attenuate experimental osteolathyrism in chickens, but this is a disease of unknown significance clinically (Selye & Bois 1957; Naber & Blackwood 1962; Waibel et al. 1962).

Leucaena Benth.

■ *L. leucocephala*
L. pulverulenta
L. retusa

guacis
jumby bean
lead tree
tepeguaje
white popinac

Taxonomy and Morphology—Comprising approximately 40 species of acacia-like trees and shrubs, *Leucaena*, a member of the subfamily Mimosoideae, is a New World genus. Its generic name is derived from the Greek root *leukos,* which means "white" and reflects the flower color of many of the species. In North America, 3 species are generally encountered:

L. leucocephala (Lam.) de Wit	white popinac, lead tree, guacis, jumby bean
L. pulverulenta (Schltdl.) Benth.	great lead tree, tepeguaje
L. retusa A.Gray	littleleaf lead tree

■ evergreen shrubs or trees; leaves 2-pinnately compound, terminal leaflet absent; flowers small; petals 5, cream white or yellow

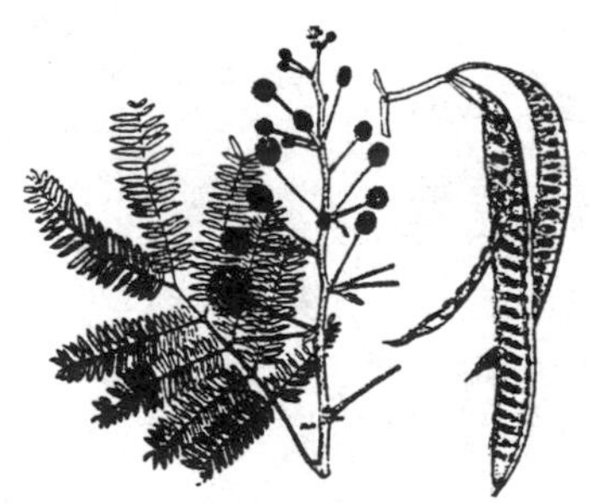
Leucaena leucocephala

Plants trees or shrubs; evergreen; perennials; not armed with thorns or spines. **Leaves** 2-pinnately compound; pairs of pinnae 3–25; leaflets 8–80 per pinna; terminal leaflet absent; petioles elongate; with 1 gland on adaxial side; stipules inconspicuous, deciduous. **Inflorescences** spikes; globose; borne in terminal racemes or panicles. **Flowers** small; symmetry radial. **Sepals** 5; fused. **Petals** 5; free; cream white or yellow. **Stamens** 10; conspicuous; free; exserted well beyond petals. **Legumes** flat; linear; short stipitate. **Seeds** 10 or more; obovate; brown.

1. Pinnae 4–8 per leaf. Leaflets 8–18 per pinna.
   Inflorescences bright-yellow . . . . *L. retusa*
1. Pinnae 8–50 per leaf. Leaflets 20–50 per pinna.
   Inflorescences cream-white.
   2. Pinnae 8–18 per leaf. Leaflets 22–34 per pinna;
      1.5–4 mm wide . . . . *L. leucocephala*

2. Pinnae 20–50 per leaf. Leaflets 30–80 per pinna; 0.7–1 mm wide . . . . *L. pulverulenta*

For more than a century, the binomial *L. glauca* was used erroneously for white popinac in North America, and thus many older literature references use that name rather than the correct *L. leucocephala*. When planted together, *L. leucocephala* and *L. pulverulenta* may hybridize.

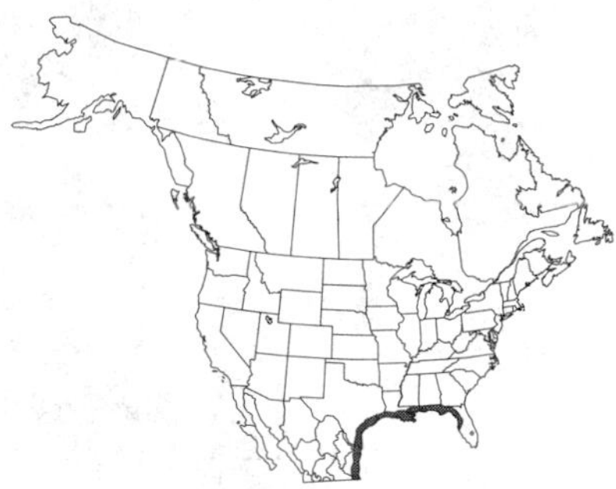

Leucaena leucocephala

Distribution and Habitat—Species of *Leucaena* are distributed throughout tropical and subtropical America and in Polynesia. Used in plantings, *L. leucocephala* is now found around the world. In North America, it occurs along the Gulf Coast. *Leucaena pulverulenta* and *L. retusa* are present in southern Texas and adjacent Mexico.

Disease Problems—The only species of toxicologic importance is *L. leucocephala*, which in some tropical areas is used for timber, for fuel, and as a livestock forage. It is a palatable source of protein and minerals and is comparable in quality to alfalfa. In addition, it is high in tannins, minimizing problems with bloat (Jones 1979). Toxicologically, the primary drawbacks to its use are growth depression and temporary hair loss in horses, mules, pigs, and rabbits (Owen 1958).

- good protein source forage; problem when large amounts are fed, especially in monogastric animals; growth depression and hair loss

Ruminants occasionally experience hair loss and/or hemorrhagic cystitis, but in some areas of the world where *L. leucocephala* has gained popularity as a forage, the disease is almost unknown in this group of animals. Resistance is due to ruminal degradation of the toxicant (Jones 1981). However, in other areas, there is a problem when animals are fed almost an exclusive diet of the species. When *Leucaena* makes up 30% or less of the diet of most animal species or no more than 50% of the diet of goats, there is little problem, and it may be a very useful supplemental feed (Jones 1979; Haque et al. 1997b).

Leucaena pulverulenta

Disease Genesis—The principal toxicant in *Leucaena* is mimosine {β-[*N*-(3-hydroxy-4-pyridone)]-α-aminopropionic acid}, which is degraded to the toxic 3-hydroxy-4(1H)-pyridone (3,4-DHP) by enzymes in the plant when its tissues are macerated or by bacteria in the rumen (Hegarty et al. 1964; Lowry et al. 1983). Highest concentrations of mimosine are found in young shoots (8–12%), moderate levels in young leaves (4–6%) and pods and seeds (4–5%), and lowest levels in older leaves (1%) (Jones 1985).

Leucaena retusa

Mimosine is a structural analog and minor antagonist of enzymatic reactions requiring pyridoxine. Both mimosine and 3,4-DHP act as inhibitors of DNA synthesis and, to a lesser extent, RNA synthesis, probably by virtue of their propensity for chelation of metal cofactors such as iron, copper, and zinc via their keto groups (Tsai & Ling 1971; Hegarty et al. 1978; Stuenzi et al. 1979). Their effects on DNA appear to be by Fe^{2+} chelation, stopping the cell cycle at G_1/S transition (Lalonde 1990; Kulp & Vulliet 1996). Mimosine is the cause of the characteristic hair loss, and even though the mechanism of this loss is unknown, the compound has been studied for use as a parenterally administered defleecing agent (Hegarty 1978). Both free and conjugated 3,4-DHPs inhibit iodine binding. This prevents tyrosine iodination and synthesis of thyroxine, with a resultant decrease in its levels and development of hyperplastic goiter that is not responsive to supplemental iodine (Hegarty 1978; Hegarty et al. 1979; Jones 1979; Holmes 1981). Edema of the corium also may be a factor contributing to hair loss. Both mimosine and 3,4-DHP are excreted in urine, much of the latter as the glucuronide conjugate (Hegarty et al. 1964, 1979).

- toxin, mimosine, in all plant parts, structural analog and minor antagonist of pyridoxine, degraded to 3,4-DHP in plant or by ruminal microorganisms; mimosine and 3,4-DHP inhibit DNA and RNA synthesis

Some of the effects of the disease may be due to inhibition of various biochemical

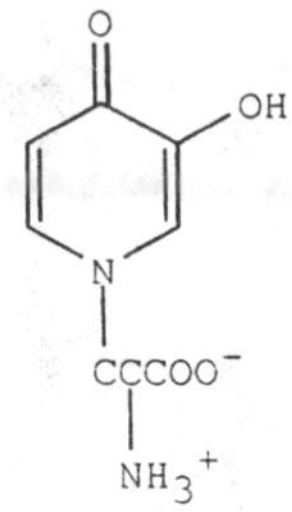

Mimosine

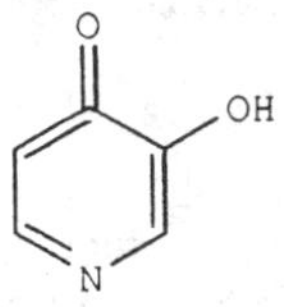

3,4-DHP

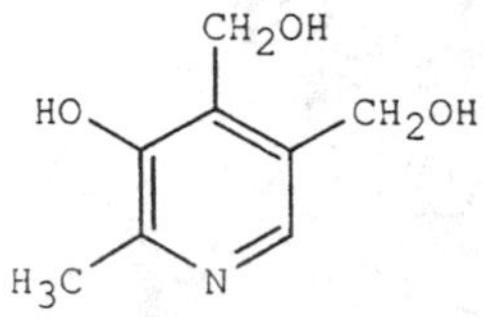

Pyridoxine

processes and enzymes (Hegarty 1978). Fortunately, all of these effects are readily reversible. Additional changes of unknown clinical significance are the mild reproductive effects, including fetal resorptions and terata, seen in experimental studies of rats and pigs fed either the plant or mimosine (Deweede & Wayman 1970; Wayman et al. 1970; Keeler 1972).

Degradation of both mimosine and 3,4-DHP by ruminal bacteria seems to account for the lack of disease in ruminants in some countries (Jones 1981). This resistance can be transferred from one animal to another by ruminal inoculation (Jones & Lowry 1984; Jones & Megarrity 1986; Hammond et al. 1989). Such a transfer represents a means of preventing the occurrence of intoxications and allowing the use of this tropical forage by livestock (Romeo 1989). Experimentally, signs of intoxication, such as hair loss, decreased appetite, and uncoordinated chewing, could be produced only in goats that had been previously exposed to *Leucaena* (Martinez & Seifert 1991). Previous exposure resulted in increased consumption of the plant and uptake of mimosine but with minimized effects on serum thyroxine (T4).

- ruminants, disease resistance via ruminal degradation of 3,4-DHP, transferable

Clinical Signs—The principal effect after the plant is eaten for a week or more is depilatory—loss of fleece in sheep; of hair on the tail, ears, head, and sheath in cattle; and of hair of the mane and tail in horses and mules. Additional effects include weakness, depressed appetite, enlarged thyroid, and ridges on the hooves of horses. Hemorrhagic cystitis is occasionally seen in ruminants. Lambs in experimental studies also had mild anemia, as indicated by decreased packed cell volume and hemoglobin concentration (Prasad 1988).

- wool and hair loss; horses, mane and tail; cattle, tail, ears, face, sheath

Pathology—Grossly, there is emaciation; hair loss; gum, tongue and esophageal erosions; and enlarged thyroids. Microscopically, hyperplasia and focal proliferation of the lining epithelium account for the thyroid enlargement. There also may be some renal tubular and hepatic degeneration and necrosis, as observed experimentally in lambs (Prasad & Paliwal 1989).

- gross pathology, hair loss, emaciation, oral mucosal erosions

  microscopic, thyroid, focal hyperplasia of lining epithelium

Treatment—Specific treatment other than supportive care is not necessary, because this is a readily reversible disease upon elimination of *Leucaena* from the animal's diet. Experimentally, supplementation with ferric sulfate and polyethylene glycol has been shown to improve growth in chicks and to prevent reproductive effects in pigs fed *Leucaena* meal (Wayman et al. 1970; D'Mello 1982).

- disease generally reversible; no treatment needed; use ferric sulfate supplementation for prevention

Lotus L.

Taxonomy and Morphology—Primarily a warm-temperate genus of the Northern Hemisphere, *Lotus*, a member of the subfamily Papilionoideae, comprises approximately 100 species, commonly known as bird's-foot trefoils, deer vetches, and lotuses. The generic name was used by Theophrastus and Dioscorides. It is closely related to *Coronilla*, and the 2 genera differ morphologically only in the appearance of the fruit. Species are used as ornamentals in gardens and as border plants; a few are edible; some are used for hay and forage; and others are good nectar sources for bees and butterflies. In North America, the genus is represented by 44 native and introduced species. Only 1 is of toxicologic significance:

- *L. corniculatus*

  bird's-foot trefoil

L. corniculatus L. bird's-foot trefoil

Plants annuals or perennials; herbaceous, sometimes slightly woody at base; caulescent. Stems decumbent or ascending or erect. **Leaves** 1-pinnately compound; leaflets

- perennial herb; leaves 1-pinnately compound, leaflets 5; flowers pealike; petals 5, golden yellow fading to orange suffused with red

3–15; terminal leaflet present; stipules well developed or minute and glandular. **Inflorescences** solitary or paired flowers or umbels; axillary; peduncled or subsessile; bracts 1, typically present. **Sepals** 5; fused. **Corollas** papilionaceous. **Petals** 5; yellow or white, with red, purple, or pink areas; banners broadly ovate, clawed, reflexed; wings obovate or oblique-oblong, clawed; keels obtuse to acute, beaked. **Stamens** 10; diadelphous; enclosed in keels. **Legumes** linear to oblong; terete or flattened; indehiscent or dehiscent with 2 valves coiling; 1- or multiloculed. **Seeds** 2 to many.

Lotus corniculatus is a perennial from a woody taproot, with many decumbent or ascending stems 20–50 cm tall. It has 5 leaflets, 3 terminal and 2 proximal; the proximal leaflets are sometimes interpreted as stipules. The stipules, however, are inconspicuous and glandlike. The leaflets are obovate to oblanceolate or rhombic-obovate. Borne in umbels, the flowers are initially bright golden yellow but later appear orange suffused with red.

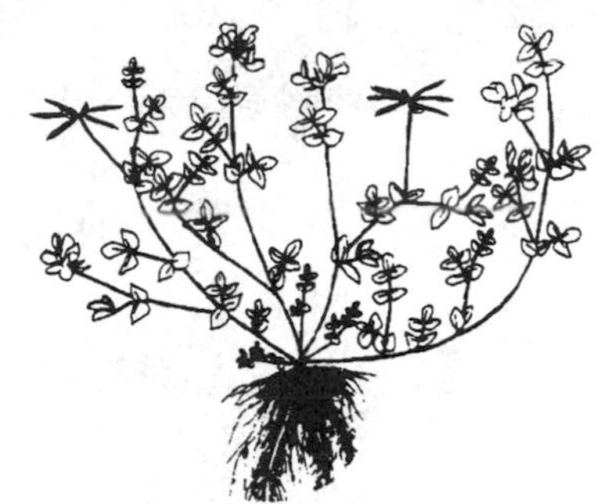

Lotus corniculatus

Distribution and Habitat—The best-known species of the genus, *L. corniculatus* is widely used as a hay and forage crop. Native to Eurasia, it has escaped and is naturalized in meadows and disturbed sites throughout much of eastern North America.

- crop plant

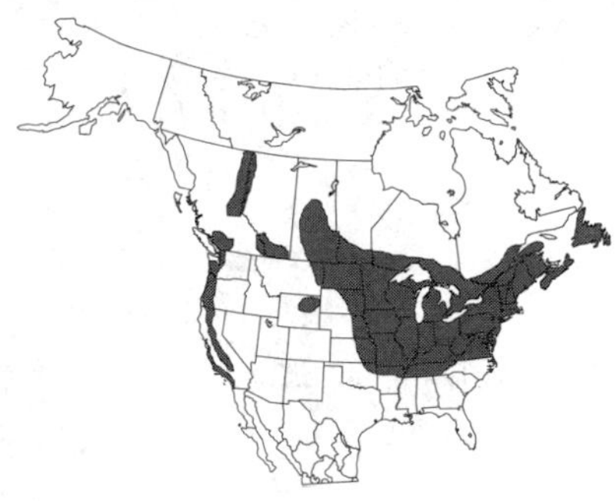

Lotus corniculatus

Disease Problems—Species of *Lotus* are widely respected forages. *L. corniculatus* is highly palatable and nutritious and has a reputation as a bloat-safe legume (Martin et al. 1987; Howarth et al. 1991). Although the plant is not generally thought of as toxic, cases of cyanide poisoning have been reported from other areas of the world (Everist 1981). It was early noted that an Egyptian species, *L. arabicus,* was cyanogenic in its early stages of growth and just before ripening of the legumes (Dunstan & Henry 1901). In addition, photosensitization of unknown etiology occasionally occurs in animals feeding almost exclusively on the luxuriant foliage of this species.

- bloat-safe legume; rarely photosensitization or cyanide toxicosis

Disease Genesis—Species of the genus contain two classes of compounds—cyanogenic and nitropropionic—that may be toxic under unusual circumstances and thus represent a potential hazard. Experimentally, an extract of the foliage of of *L. corniculatus* was lethal to sheep when given in large amounts via ruminal cannulae (Dougherty & Christensen 1953). The extract contained 55.2–66.5 mg cyanide /100 ml. Its removal by distillation eliminated or substantially reduced the toxicity of the plant extract. Some of the species that have been identified as cyanogenic are listed in table 37.8. However, undue concern is not warranted because in most species the HCN potential in general is low.

- very low risk of cyanogenesis with *L. corniculatus*; linamarin and lotaustralin

other species, 3-NPA, low risk only for nonruminants

Table 37.8. Species of *Lotus* with cyanogenic potential

L. angustissimus	
L. corniculatus	bird's-foot trefoil
L. grandiflorus	
L. incanus	
L. krylovii	
L. mearnsii	
L. nevadensis	Nevada deer vetch
L. parviflorus	
L. tenuis	narrow bird's-foot trefoil, slender trefoil

Cyanogenesis is due to the presence of linamarin and lotaustralin, in ratios ranging from 1:1 to 1:9 (Butler 1965). These glucosides are derived from valine, leucine, and isoleucine (Seigler 1977; Saupe 1981). Although cyanogenesis is under genetic control, phenotypic expression is variable and may increase with increased moisture and temperature (Abbott 1977; Jones et al. 1978). However, even under the most favorable conditions, the risk to livestock is minimal.

$C{\equiv}N$
/
Glucose-O—C—CH_3
\
CH_3

Linamarin

$C{\equiv}N$
/
Glucose-O—C—CH_3
\
CH_2CH_3

Lotaustralin

$O_2N—C—C—COOH$

3-Nitropropionic acid
3-NPA

In addition to cyanogenetic compounds, some species of *Lotus* contain nitrotoxins that produce neurotoxic effects. Introduced *L. pedunculatus* and native *L. argyraeus, L. greenei,* and *L. strigosus* var. *tomentellus* have been found to contain concentrations of 3-nitropropionic acid (3-NPA) in excess of 25 mg NO_2/g plant, which represents a hazard at least to nonruminants (Williams 1983). The 3-NPA is present as glucoside esters hiptagin, karakin, cibarian, and coronarian, as shown for *L. pedunculatus* (Gnanasunderam & Sutherland 1986). As with cyanogenesis, this potential risk has not given rise to actual disease problems. For a more complete discussion of nitrotoxins, see the detailed discussion of them in the treatment of *Astragalus,* in this chapter.

Although *L. corniculatus* sometimes appears to produce a slight reproductive effect, i.e., increasing the number of services required for sheep, its measurable estrogenic activity is little or none and is one-half or less that of alfalfa *(Medicago sativa)* (Pieterse & Andrews 1956; Engle et al. 1957).

■ cyanide, abrupt onset, distress, weakness, ataxia, labored respiration, seizures

Clinical Signs—In the rare event of cyanide intoxication caused by *Lotus,* there is abrupt onset of apprehension and distress, which occurs within a few minutes of the animal's grazing forage of the species. This is followed by weakness, incoordination, and labored respiration. If the intoxication is severe, the animal may be unable to stand and may lay in lateral recumbency with periodic paddling and tetanic seizures. The entire sequence of signs may occur over a 5-to-15-minute period, with either death at this time or most likely recovery over the next few hours.

■ no pathological lesions

Pathology—There are few distinctive changes. For the most part, the changes are limited to acute passive congestion of the liver, lungs, and other tissues, with scattered, small splotchy hemorrhages.

■ cyanide, sodium thiosulfate

Treatment—The primary antidote for ruminants is sodium thiosulfate given i.v. at a dose of 0.25–0.5 g/kg b.w. Sodium nitrite i.v., at 10 mg/kg b.w., can be used to augment the effects of the sodium thiosulfate, but it is not required for effective treatment. A more complete discussion of this treatment is given in the discussion of the family Rosaceae (see chapter 64).

Lupinus L.

■ *L. arbustus*
L. argenteus
L. bingenensis
L. formosus
L. latifolius
L. leucophyllus
L. onustus
L. sericeus
L. sulphureus

lupine

Taxonomy and Morphology—Comprising more than 200 species, *Lupinus,* a member of the subfamily Papilionoideae, is a genus with a complex pattern of variation involving hybridization and polyploidy. More than 600 names at the specific, subspecific, and varietal levels have been published, and there have been numerous classificatory revisions as taxonomists have attempted to develop consensus on how to interpret the pattern of variation. The genus name is derived from the Latin word *lupus,* meaning "wolf," a fanciful reference to the misconception that lupines take over an area and deplete the fertility of the soil. In reality, members of the genus are soil enhancers; some species are grown as winter cover crops to add nitrogen and phosphorus to soils of low fertility in the southern United States.

In North America, more than 150 species and a plethora of subspecies and varieties are recognized (Barneby 1989a). Most of these taxa have not been evaluated for toxicity, and many are probably of little or no risk. In this treatise, only those few species that have been clearly associated with toxicity are discussed (Marsh et al. 1916; Kingsbury 1964) (see plate 37.5):

L. arbustus Douglas ex Lindl. (= *L. laxiflorus* Douglas ex Lindl. in part)	spur lupine
L. argenteus Pursh (= *L.* ×*alpestris* A.Nelson) (= *L. caudatus* Kellogg) (= *L. laxiflorus* Douglas ex Lindl. in part)	silvery lupine, tailcup lupine, grassland lupine
L. bingenensis Suksd. (= *L. leucopsis* J.Agardh in part)	
L. formosus Greene (= *L. greenei* A.Heller)	grassland lupine
L. latifolius Lindl. *ex* J.Agardh	broad-leaved lupine
L. leucophyllus Douglas ex Lindl.	woolly-leaved lupine, western lupine, velvet lupine, white-leaved lupine, poison lupine
L. onustus S.Watson	woodland lupine, Plumas lupine
L. sericeus Pursh (= *L. bakeri* Greene) (= *L. comatus* Rydb.) (= *L. leucopsis* J.Agardh in part)	silky lupine, sink lupine
L. sulphureus Douglas ex Hook.	sulfur lupine

Lupinus argenteus

As indicated by their parenthetical placement, several of the species cited in early reports of toxicity are now recognized to be conspecific with other taxa. In addition, 2 species, *L. laxiflorus* and *L. leucopsis,* have been divided and their varieties positioned in other species.

Lupinus leucophyllus

Plants annuals or perennials; herbs or subshrubs; caulescent or acaulescent. **Stems** erect; typically indumented. **Leaves** palmately compound; leaflets 3–17; petioles elongate; stipules fused to bases of petioles. **Inflorescences** terminal racemes or spikes; bracts present, caducous in some species; bracteoles 2. **Flowers** showy; large. **Sepals** 5; fused; calyces 2-lipped. **Corollas** papilionaceous. **Petals** 5; blue or violet or lilac or crimson or white or yellow; banners orbicular or elliptic-ovate, erect or reflexed, margins reflexed; wings falcate-oblong or obovate, fused at apices; keels curved, beaked. **Stamens** 10; monadelphous; anthers of 2 forms, long and short; enclosed in keels. **Legumes** oblong-elliptic; flattened; sessile; sericeous or hirsute; somewhat torulose. **Seeds** 2–12.

- erect herbs or subshrubs; leaves palmately compound, leaflets 3–17 but mainly 5 or more; flowers terminal, pealike; petals 5, blue, violet, lilac, crimson, white, or yellow
- wide range of habitats, open to woodlands

Distribution and Habitat—*Lupinus* has two principal areas of distribution in the world. One is the Mediterranean region, and the second is the cordilleran axis of western North and South America. In North America, the genus is most abundant in the western half. Only a few species occur in the eastern half. *Lupinus perennis* is distributed widely, whereas *L. diffusus, L. westianus,* and a few other species are present only in the extreme Southeast.

These species occupy a variety of habitats, including meadows, brushlands, and open pine, fir, spruce, and aspen forests and woodlands. Plants typically are encountered in stony and low-fertility soils and from low to high elevations, e.g., populations of *L. argenteus* and *L. onustus* are present at 3,000 m or more.

- many species provide high-protein forage; cultivated for food and forage in other areas of world; seeds from sweet lupines and/or debittered seeds used for flour

Lupinus arbustus

Lupinus argenteus

Lupinus bingenensis

- lupinosis, an Old World hepatotoxicosis, not a problem as yet in North America

- acute disease of North America, neurologic disease mainly of sheep; ingestion of mature plants with legumes

Disease Problems—Lupines have been recognized for their useful as well as their toxic characteristics since ancient times. They have been cultivated for centuries as seed crops, soil builders, and fodder crops (Johnson et al. 1986). Unfortunately, their high concentrations of crude protein (30–40%) are offset by high concentrations of bitter alkaloids (>3%) in many species. Thus, low-alkaloid, or "sweet," species and cultivars have long been sought. Two such species, *L. albus* and *L. angustifolius* have alkaloid concentrations well below 0.1% and are used as protein supplements similar to soybean meal. Other cultivated species are *L. luteus* and *L. mutabilis.*

The Greeks and others were aware of the potential adverse effects of lupines, and eventually debittering processes, involving soaking or boiling of the seeds in water, were devised to allow their use in flour for breads. There continues to be strong interest in the use of lupine seeds for food in some areas of the world. In Australia, where *L. angustifolius* seeds are widely used, alkaloid content of 0.015% is typical, and there have been few problems with their use (Petterson et al. 1987). Indeed, evaluation of the toxicity of extracts from seeds of *L. albus* and *L. angustifolius* indicates remarkably low toxicity (<2 g/kg b.w. orally) in spite of alkaloid concentrations up to 10% in the seeds (Petterson et al. 1987; Stobiecki et al. 1993). However, in studies with different cultivars, some risk is apparent even with "sweet" species such as *L. albus.* When more than 10% of its seeds were present in feed fed to swine, there was feed refusal, decreased weight gain, and other problems associated with intestinal dysfunction (Casper et al. 1991). Higher concentrations of seeds (20–30%) could be fed to poultry especially when the ration was supplemented with DL-methionine and L-lysine (Perez-Escamilla et al. 1988; Watkins et al. 1988). Extrusion of the seeds in the preparation of lupine meal did not decrease toxicity, but soaking in water markedly reduced the occurrence of detrimental effects. However, in spite of these drawbacks, lupines remain a remarkable potential food source. Intoxications in humans are rare (Smith 1987; Marquez & Gutierrez-Rave 1991).

Lupinosis is a chronic hepatotoxic disease of the Old World, common in Europe, Australasia, and South Africa. It has not proven to be a problem in North America, but its occurrence here cannot be dismissed entirely. Rare instances of apparent hepatic disease with icterus in horses have been reported (Marsh et al. 1916). Lupinosis is primarily a disease of sheep grazing lupine stubble in the summer and often appears after summer rains.

The increasing use of rangeland in western North America for livestock production was accompanied by devastating losses due to toxic plants. Foremost among these plants were a host of lupine species. In the late 1800s, losses of sheep due to lupines were sometimes catastrophic, e.g., 1,900 animals in a flock of 3,000, and 1,150 of 2,500 animals in another (Chesnut & Wilcox 1901). Losses occurred mainly the first day after animals began to eat the plants, but the losses continued for 2 more days. Case fatality rates were about 50%. In most instances of losses, the sheep had been allowed to forage on mature plants in flower and fruit. Hay with large amounts of lupine legumes presented a similarly serious problem, albeit the resulting losses were not quite as severe.

Acute lupine toxicity, a neurologic disease, is mainly a disease of sheep, although horses and cattle also are at risk. This may be a reflection of differences in the eating preferences of the animals. *Lupinus leucophyllus* and *L. sericeus* are reported to be eaten readily by horses and cattle when plants are immature but not when mature. In contrast, sheep do not readily eat lupines until the plants are mature, with legumes present. Likewise, cattle in the Californian foothills appear to eat mature *L. argenteus* (reported as *L. laxiflorus*) readily (Wagnon 1960). Lupines often form a large portion of the diets of mule deer, up to 36% of the summer diet and 16% in winter in various areas of Arizona (Krausman et al. 1997). The palatability of various species of *Lupinus* also seems to vary.

Of considerable interest to livestock producers are the various reproductive effects that result in abortions, stillbirths, or congenital defects. A seemingly specific problem of this nature, called crooked calf disease, vexed cattle producers in western range states for many years (Palotay 1959). These effects were ascribed to an assortment of causes, as illustrated in one instance by the name *acorn calves* for the disease (Hart et al. 1947). Lupines were suspected as one of the possible causes (Wagnon 1960), but their role was not confirmed until the studies of Binns and James (1961). These studies revealed that the critical time period in this disease generally was days 40–70 of gestation (Shupe et al. 1968). In some instances, this period extended to day 100. Because this period is later than the period of organogenesis in fetal development, the disease was interpreted to be one of deformation rather than malformation (Shupe et al. 1967a,b). These findings led to studies using ultrasound techniques in which deformities were related to prolonged reduction of fetal movement following daily ingestion of the foliage of *Lupinus* (Bunch et al. 1986, 1992; Panter et al. 1990a; Panter 1993; Panter & Keeler 1993). Fetuses between days 38 and 50 were especially vulnerable. Typically, 10–15% of such calves have limb, vertebral, and palate defects of such severity that many die due to difficult birth or inability to follow their dams and to nurse. Deformed neonates have serious limb defects that become more limiting as they increase in size and weight (Keeler et al. 1976). Experimentally, those species of *Lupinus* that did not decrease fetal movement did not cause malformations and had only minimal effects on the dams (Panter et al. 1990a). Genera in other families—e.g., *Conium* in the Apiaceae and *Nicotiana* in the Solanaceae—produce the same reduction in fetal movement and subsequent deformities.

Lupinus formosus

Lupinus latifolius

■ reproductive, cattle, crooked calf disease, stillbirths, weak or deformed calves

■ chronic hepatotoxic lupinosis, due to fungal infestation; formation of toxic peptides, phomopsins; lesions and disease similar to that caused by pyrrolizidine alkaloids

Disease Genesis—Initial studies on lupine-induced disease in Europe were complicated by the diversity of clinical signs that ingestion of the plants produced (Gardiner 1967a). Eventually, this variation in signs was shown to be the result of two dissimilar types of toxicants present in different species. The chronic hepatotoxic disease known as lupinosis in the Old World is the result of infestation of lupine tissues by a fungus and the subsequent formation of toxic peptides (Gardiner 1966). The more commonly encountered disease in North America is an acute neurologic disease produced by quinolizidine alkaloids. In addition, one of these alkaloids, anagyrine, causes the reproductive effects described above.

As noted above, lupinosis is primarily a disease of sheep grazing lupine stubble. It is seen to a lesser extent in cattle and horses (Gardiner & Seddon 1966; Arnold et al. 1976). The causative factor is infestation of the stubble by the fungus *Diaporthe toxica* (formerly *Phomopsis leptostromiformis*) and its production of linear hexapeptide toxins known as phomopsins A, B, and C (Gardiner 1967a; van Warmelo et al. 1970; Wood et al. 1973; Jago et al. 1982; Peterson et al. 1987; Allen 1992; Williamson et al. 1998). Phomopsins bind to tubulin, inhibiting microtubule polymerization/assembly and stability (Tonsing et al. 1984). As a result, there is interference with mitosis and arrest of the process in metaphase. The decrease in cell division and subsequent continued cell growth results in the formation of multinucleated cells, an effect somewhat akin to that produced by pyrrolizidine alkaloids (see discussion in *Senecio*, in chapter 13) (Petterson & Coackley 1973).

Lupinus leucophyllus

Like the pyrrolizidine alkaloids, the phomopsins also interact with copper to predispose its accumulation in the liver of sheep (Gardiner 1967b). Interference with vitamin E and selenium metabolism also may be a factor in the changes observed, especially those involving muscle (Allen 1992). The main adverse effects involve

■ acute, neurologic disease, quinolizidine alkaloids

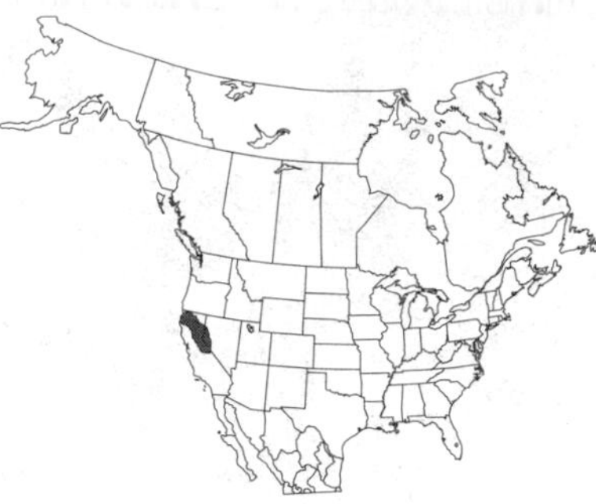

Lupinus onustus

Lupinus plattensis

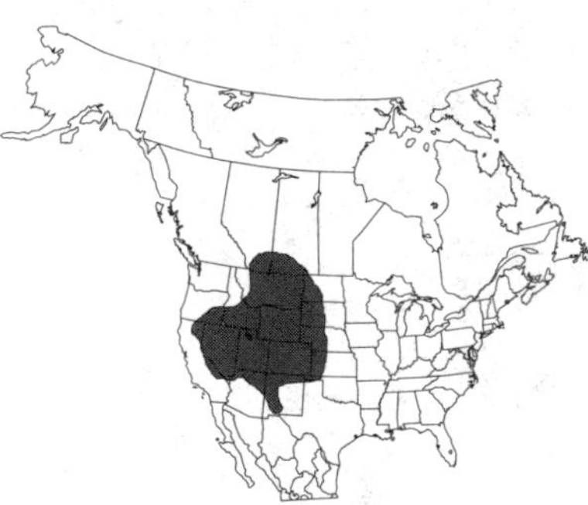

Lupinus pusillus

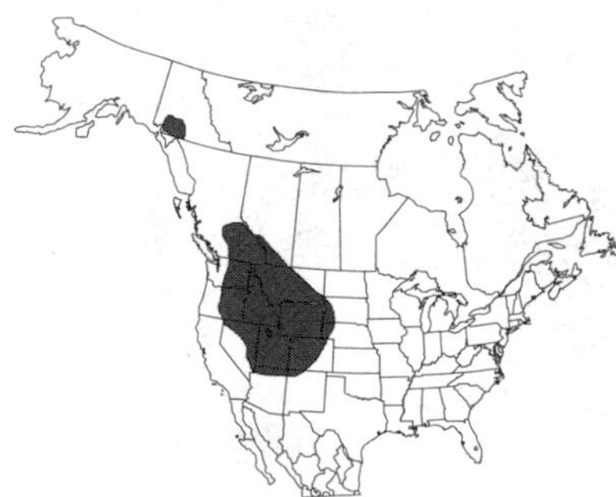

Lupinus sericeus

Table 37.9. Species of *Lupinus* reported to be toxic or to contain high concentrations of toxic alkaloids

Species	Common Name	Lupanine	Anagyrine
L. albicaulis	pine lupine, sickle-keeled lupine		++
L. xalpestris	mountain silvery lupine		++
L. andersonii	Anderson's lupine		++, ammod
L. arbustus	spur lupine		ammod
(= varieties of *L. laxiflorus*)			
subsp. *calcaratus*			+, ammod, gramod
subsp. *silvicola* (= *L. laxiflorus* var. *silvicola*)			+
L. arcticus		+	+, ammod
L. argenteus	silvery lupine	+	++
var. *argenteus* (= *L. parviflorus*			+
var. *myrianthus*) (= *L. myrianthus*)			
var. *holosericeus*	spurred lupine, tailcup lupine	+	++
(= *L. caudatus*)	little flowered lupine, silky lupine	+	++
(= *L. laxiflorus* var. *argophyllus*)	grassland lupine		
var. *montigenus*			++
(= *L. montigenus*)	Mr. Rose Lupine		++
var. *utahensis*		?	
(= *L. spathulatus*)			
L. bingenensis (= several varieties of *L. leucopsis*)		+	
L. diffusus (= *L. cumulicola*)	sandhill lupine	+	
L. evermannii	Evermann's lupine		++
L. formosus	grassland lupine		ammod
(= *L. greenei*)			ammod
L. latifolius	broad-leaved lupine	+	++
L. lepidus	stemless lupine, prairie lupine	+	
L. leucophyllus	woolly-leaved lupine, western lupine velvet lupine, white-leaved lupine, poison lupine	+	
(= *L. cyaneus*)		+	++
(= *L. erectus*)			++
L. littoralis	seashore lupine		+
L. longifolius		+	+
L. mollis (= *L. leucopsis* var. *mollis*)			+
L. nootkatensis	Nootka lupine	+	?
L. onustus	Plumas lupine, woodland lupine	+	
L. ornatus		+	
L. polyphyllus	showy lupine, big lupine, many-leaved lupine, Washington lupine	+	++, ammod
(= *L. burkei*)	Burke's lupine	+	++
(= *L. wyethii*)	lowland lupine, Wyeth lupine	+	
(= *L. humicola*)			
L. pusillus	low lupine, dwarf lupine, rusty lupine	+	
L. sericeus	silky lupine, sink lupine	+	++, ammod
(= *L. bakeri*)			
(= *L. bakeri* subsp. *amplus*)		+	
(= *L. comatus*)			
(varieties of *L. leucopsis*)			
L. sulphureus	sulfur lupine		++, ammod

Note: Synonyms in parentheses were used in original publications. +, alkaloid type present; ++, concentrations of 1.44g/kg present; ?, alkaloids possibly present but confirmation needed; ammod, ammodendrine present; gramod, gramodendrine present.

Sources: Marsh et al. 1916; Beath 1925; Clawson 1931; Beath et al. 1934; Kingsbury 1964; Leipold et al. 1977; Hatfield et al. 1980; Kinghorn et al. 1980; Davis 1982; Davis & Stout 1986; Meeker & Kilgore 1987; Majak et al. 1994; Panter et al. 1997, Wink et al. 1995.

Table 37.10. Species of *Lupinus* reported to contain toxic alkaloids but not confirmed to contain sufficient quantities to cause intoxication.

Species	Common Name	Alkaloid Type	
		Lupanine	Anagyrine
L. agardhianus A.Heller		+	
L. albifrons Benth. ex Lindl.	white-leaved lupine	+	
L. arboreus Sims	tree lupine, yellow bush lupine	+	
L. arizonicus (S.Watson) S.Watson	Arizona lupine	+	ammod
L. bicolor.	miniature lupine	+	
L. chamissonis Eschsch.		+	ammod
L. concinnus J.Agardh	Bajada lupine, annual lupine	+	
L. densiflorus Benth.		+	+
L. elegans.	hierba loca	+	
L. excubitus	grape soda lupine		
L. hartwegii	garbancillo	+	
L. hirsutissimus Benth.	stinging lupine	+	ammod
L. nanus Douglas ex Benth.		+	+
L. perennis L.	eastern lupine, sundial lupine	+	
L. plattensis S.Watson	Platte lupine	+	
L. polycarpus Greene		+	+
L. pubescens	hairy lupine	+	
L. rivularis Douglas ex Lindl.	stream lupine, streambank lupine	+	
L. ruber A.Heller	red lupine	+	
L. sparsiflorus Benth.	Mojave lupine, Coulter's lupine	+	+, ammod
L. subcarnosus	Texas bluebonnet	+	
L. succulentus	arroyo lupine	+	
L. texensis	Texas bluebonnet	+	
L. truncatus Nutt. ex Hook. & Arnold		+	ammod

Note: +, alkaloid type present; ammod, ammodendrine present.

Lupinus sulphureus

Lupinine

Sparteine

Lupanine

Anagyrine

Baptifoline

necrosis and fibrosis of the liver, tubular degeneration of the kidneys, and degeneration of skeletal and cardiac muscles (Allen et al. 1992). Just as nutritive values differ among the different species of *Lupinus,* there are differences in susceptibility to *Diaporthe* infestation and tendency to cause lupinosis (Arnold et al. 1976). Thus, a search for plants that pose a lower risk has continued.

The acute neurologic disease is due to the presence of quinolizidine, and possibly indole- and piperidine-derived, alkaloids. It is the disease of primary concern in North America. Numerous species of *Lupinus* contain these toxic alkaloids, but only a few have actually been implicated in cases of intoxication. Typically, western species are of most concern because of the relative scarcity of alternative forages under some conditions. However, some southeastern species, such as *L. diffusus* and *L. westianus* also have been shown to contain these toxicants and may cause intoxication if sufficient amounts of the plants are eaten (Goldberg & Moates 1967; Goldberg & Sahli 1967). Experimentally, seeds of *L. diffusus* fed to mice were toxic at about 0.8% b.w. and lethal at 1.3+% b.w. (Duncan et al. 1955). Species that have caused acute disease or deformities or that contain toxic alkaloids in concentrations sufficient to cause disease are listed in tables 37.9. and 37.10.

Like other genera of the Fabaceae, such as *Baptisia, Cytisus, Laburnum,* and *Sophora,* species of *Lupinus* contain a diverse array of quinolizidine alkaloids. Some are bicyclic (lupinine and epilupinine), and others are tricyclic (cytisine and *N*-methylcytisine) or tetracyclic (sparteine, lupanine, aphylline, and anagyrine) (Leonard 1953; Bohlmann & Schumann 1967; Aslanov et al. 1987; Keeler 1989; Wink et al. 1995). Based on differences in their chemical structure, they are classified into five groups: lupinines, sparteine-lupanine, anagyrine-cytisine, aphylline-multiflorine, and matrine

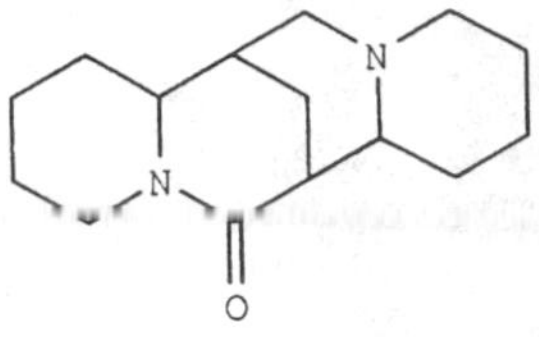

Aphylline

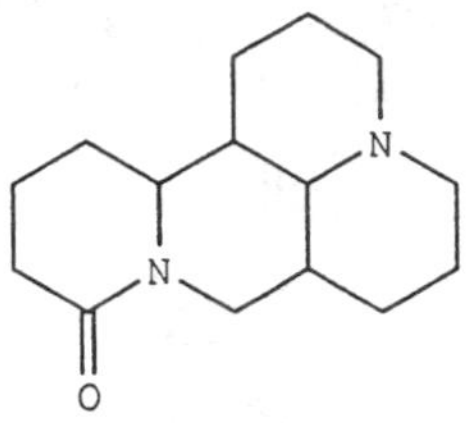

Matrine

(Kinghorn & Balandrin 1984; Ohmiya et al. 1995). A number of unusual types also are present in the genus.

In addition to variation among species with respect to the specific alkaloids present, there also is considerable variation in alkaloid content from one locality to another. This is perhaps illustrated by the variability in risk with *Lupinus argenteus*. There is considerable risk when this species is eaten in the Rocky Mountains, in contrast to its being eaten in the Canadian prairies with little or no problem, especially in hay (Looman 1983). Specific alkaloids may be absent in plants from one location and represent a serious risk in those from other sites. Total alkaloid concentrations reported for *L. argenteus* (reported as *L. caudatus*) range from 0.5–2.5% d.w. for vegetative parts and 1.2–9.5% for seeds (Keeler et al. 1976; Davis 1982). The range of concentrations for *L. sericeus* is similar. In *L. argenteus,* there also is conspicuous variation in the presence of anagyrine, with 0–0.6% in the vegetative parts and 0–1.3% in seeds. This type of variation is also shown with *L. latifolius*; some collections have high levels, 1.14% d.w. in the foliage (Meeker & Kilgore 1987), whereas in others, anagyrine is not present (Panter et al. 1997).

The leaves are the site of synthesis of these alkaloids, which are then transported to other tissues and organs of the plant (Wink et al. 1984). Concentrations of total alkaloids and anagyrine in the foliage are high in the early stages of growth and decline as the plant matures, especially in the later half of the growing season. By the time of flowering and fruit and seed development, foliage concentrations may be only 20–25% of their early-season values. In contrast, concentrations in the seeds are 2–10 times greater than in the foliage (Keeler et al. 1976; Davis 1982). There also is considerable diurnal variation in alkaloid concentrations in all plant parts (Wink & Witte 1984). Concentrations in late afternoon may be twice those at night.

Quinolizidine alkaloids may exert an effect on plant palatability because some, such as sparteine and, to a lesser extent, lupanine, are quite bitter when tasted by humans (Wink 1994). It has been suggested that they represent a chemical defense against herbivory (Wink 1985). Some ester alkaloids excreted via roots are allelopathic, preventing growth of other seedlings (Wink 1994).

- neurologic effects, alkaloids interact with nicotinic cholinergic receptors, central and peripheral, also impart bitterness

Quinolizidine alkaloids bind to and exert agonist effects on acetylcholine receptors. The most prominent actions are on the nicotinic (cytisine and lupanine) and muscarinic types (sparteine) (Sloan et al. 1988; Schmeller et al. 1994). Extracts of the seeds of *L. leucophyllus* and *L. sericeus* or the pure quinolizidine alkaloids from them, given i.p. to guinea pigs, caused depression, incoordination, tremors, increased response to stimuli, exaggerated movements, prolonged decrease in systemic blood pressure, and eventually death, presumably due to depression of the respiratory centers (Marsh et al. 1916; Couch 1926). The principal alkaloids of the lupanine-type given to rats in the same way caused tremors and paddling seizures (Petterson et al. 1987). At low dosage, sparteine and lupanine are weak sedatives (Pothier et al. 1998).

Susceptibility differs among animal species. The lethal dose for guinea pigs is 30–100 mg/kg, in contrast to 100–200 mg/kg for rats. The alkaloids are rapidly eliminated from the body, minimizing the possibility of cumulative effects. This lack of alkaloid accumulation is illustrated by experiments in which as much as 125 kg of *L. argenteus* or 200 kg of *L. plattensis* given over a 2-month period to a 450 kg horse failed to produce adverse effects. A total dose of 295 kg (0.45–3.6 kg/day) over 3 months was similarly nontoxic (Van Es et al. 1929). There even may be development of a degree of tolerance with continued ingestion of the plant material (James et al. 1968).

Sparteine, which has low acute toxicity, causes diuresis and hypoglycemia, increases uterine tone, and controls cardiac rhythm irregularities. It is particularly active in prolonging the action potential by slowing K^+ conductance; lupanine is less potent (Ohta & Narahashi 1973; Kleinhaus 1980). Sparteine forms complexes or chelates metals such as calcium, copper, cobalt, and zinc, which may play a role in its effects on muscle (Mason & Peacock 1973; Boschmann et al. 1974). Sparteine also interacts with prostaglandins (F) in causing uterine contractions (Abtahi et al. 1978). Lupanine causes a decrease in blood pressure.

Toxic effects caused by *Lupinus* are not limited to quinolizidine alkaloids. Indole and piperidine derivatives also are present in some species and may play a role in both acute toxicity and fetal deformities (Fitch et al. 1974; Keeler & Panter 1989). Ammodendrine, a piperidine derivative, has been found in a number of species, as indicated in tables 37.9. and 37.10. It and gramodendrine, a bipiperidyl-indole, are the principal alkaloids present in *L. arbustus* subsp. *calcaratus* (Hatfield et al. 1985). In mice, these compounds reduce motor activity and cause depression of the central nervous system. They have about half the potency of the lupanine-type alkaloids (Hatfield et al. 1985).

Investigations of episodes of crooked calf disease revealed that although several quinolizidine alkaloids were usually present in the species of *Lupinus* responsible, anagyrine was common to all (Keeler 1973a,b). The hypothesis that anagyrine is the teratogen was tested by experiments using extracts rich in specific alkaloids (Keeler 1976). It appears that anagyrine (a tetracyclic α-pyridone) and possibly others such as thermopsine and baptifoline, which have a similar chemical structure (specifically an unsaturated piperidine ring), have the capability to induce teratogenic effects, possibly by depression of fetal activity, when exposure to them occurs in early pregnancy. Keeler (1973) has suggested that anagyrine must be present at a concentration of 1.44 g/kg plant material, and the plants eaten for 30 days or more, in order for teratogenic effects to be produced.

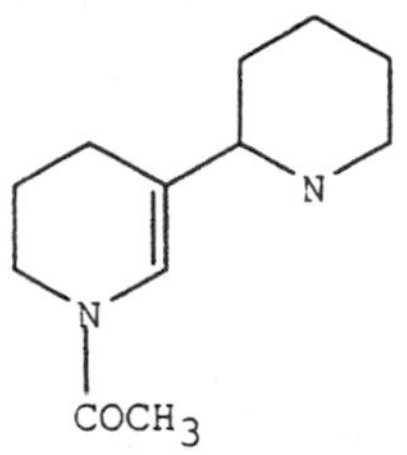

■ reproductive, quinolizidine alkaloids such as anagyrine (1.44 g/kg plant material) and baptifoline; also piperidine-derivative alkaloids, ammodendrine, *N*-acetylhystrine; all have depressant actions on fetus

N
N
$COCH_3$

Ammodendrine

N
N
$COCH_3$

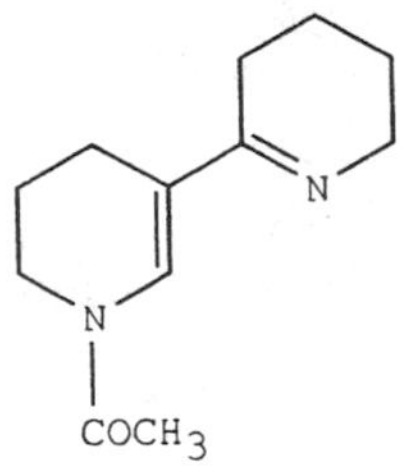

N-Acetylhystrine

Lupinus formosus and *L. arbustus,* which experimentally cause deformed calves, have very little or no anagyrine but rather have as their major alkaloid ammodendrine, which is structurally related to anabasine, the teratogen from *Nicotiana* (Keeler & Panter 1989; Panter et al. 1998). Ammodendrine, which is a major alkaloid in many lupine species, and gramodendrine present in *L. arbustus* caused a marked decrease in locomotor activity in mice in experimental studies (Hatfield et al. 1985). If present during days 40–70 of gestation at concentrations of approximately 0.5 g/kg plant material or higher (Keeler & Panter 1989; Panter et al. 1998), they could possibly produce fetal depression sufficient to cause deformities as was proposed by Panter and co-workers (1990a,b). It has been suggested that anagyrine may in reality function as a teratogen only after its biotransformation, via A ring cleavage in the bovine rumen, to an α-substituted piperidine. If this conversion is specific for the bovine rumen, it would explain the lack of teratogenic effects in sheep and goats for some lupine species rich in anagyrine. Biotransformation is not required for teratogenicity with ammodendrine.

There is an apparent difference in susceptibility of goats to these types of teratogens. Experimentally, *L. formosus,* which contains *N*-acetylhystrine in addition to ammodendrine, is a potent cause of deformities, whereas *L. argenteus* (reported as *L. caudatus*), which contains anagyrine and is a potent teratogen in cattle, is without effect in goats even at doses intoxicating the dam (Panter et al. 1990b, 1998). It may be a matter of dosage, given that *L. latifolius,* exceptionally rich in anagyrine, has been linked with terata in goats (Kilgore et al. 1981). However, the hypothesis of ruminal biotransformation as the basis for differences among species is difficult to reconcile with plasma alkaloid data in sheep, goats, and cattle fed foliage of *L. argenteus* (= *caudatus*). In sheep and goats given 7.8 g/kg b.w. and cows given 3 g/kg b.w. (2 animals for each), peak blood levels of anagyrine and lupanine were high in the sheep (0.85 μg/ml and 0.58 μg/ml, respectively), intermediate in cows (0.47 μg/ml and 3 µg/ml), and low in goats (0.3 µg/ml and 0.23 µg/ml). These levels peaked at 3 hours and remained high for 8–10 hours (Gardner & Panter 1993). These data are perhaps an indication of high absorption and/or low biotransformation of anagyrine in sheep. The alkaloids were virtually undetectable 18 hours after a single dose. Signs of intoxication were noted only in the cows. In spite of the apparent role for teratogenic chemicals, fetal effects seem as much due to inhibition of fetal movement as to specific actions on tissue development.

- public health risk, teratogenic alkaloids are excreted in milk

The presence of these quinolizidine alkaloids in the milk of animals eating *Lupinus* may represent a public health hazard in unique situations. In an isolated incident in northern California, an infant with deformities of the hands and arms and aplasia of red blood cells was born to a woman in the habit of drinking the milk of goats foraging heavily on *L. latifolius* (Kilgore et al. 1981; Ortega & Lazerson 1987). Deformities also occurred in the kids of the goats and in puppies whose dams were given milk from them.

- neurologic; abrupt onset, depression, nervousness, ataxia, muscle twitching, labored respiration;

  sheep, more-severe disease, excitable, head pressing, head butting, frenzy, rigid seizures

  horses, colic, exaggerated gait

Clinical Signs—The acute neurologic disease produced by *Lupinus* is seen mainly in sheep, although other species, including cattle and horses, also are at risk. Early signs, and those seen in mild cases, include depression (the ears drooping in sheep), incoordination, muscle twitching, nervousness, and some breathing difficulty. As the disease progresses, depression and muscular weakness become more prominent and respiration more labored. Sheep are prone to become quite excitable with little cause, exhibiting head pressing, head butting, and frenzy. There also may be seizures of the extensor rigidity type. In severe cases, the animal collapses within hours of the onset of signs. Death may occur in the first few hours or not for 2–3 days. Constipation is typical. Horses show signs of colic and move with a peculiar walk in which they lift their legs very high. There are no distinctive changes in serum chemistry or hematology.

- chronic hepatic disease of Old World; signs typical of liver problems

Lupinosis due to fungal infestation of the stubble produces signs typical of chronic liver disease. There is a loss of appetite, progressive weakness, listlessness and reluctance to move, icterus, constipation, occasional photosensitization, and darkening of the urine due to hemoglobinuria. These signs develop over several days to a week or more and may be accompanied by indicators of hepatoencephalopathy such as head pressing and wandering into fences.

- terata; skeletal, thoracic limb joints fixed in flexion, lateral rotations, vertebral curvatures, cleft palate

For the most part, terata are a problem only in calves. Rarely are they observed in goats. The fetal deformities are mainly skeletal, most commonly arthrogryposis affecting the thoracic limbs. The elbow joint is immobile, flexed, and unable to be extended because of misalignment and malpositioning of the ulnar-humeral articulation. There is usually lateral rotation distal to the elbow. Additional deformities may also be present, including vertebral curvatures involving the neck and back (lordosis, scoliosis, and torticollis), and cleft palate. These deformities occur with or without the presence of obvious signs of disease in the dam.

- no lesions except in chronic liver disease

Pathology—There are no distinctive lesions with acute intoxication due to alkaloids. In fungal lupinosis the liver in chronic cases is usually shrunken and firm from fibrosis and may be discolored grayish orange. Microscopically, distinctive changes in the liver include variable fatty change and progressive fibrosis with biliary hyperplasia.

- neurologic; sedation, avoid access to lupines with legumes present

Treatment—For the acute disease, treatment is nonspecific, directed at providing relief for the primary neurologic signs. To reduce the probability of intoxication, it is recommended that if it is necessary to move sheep through a stand of lupines in fruit, the animals be allowed to move leisurely so they can graze selectively. If hurried, they may take bites from plants they pass (Welch & Morris 1952). In addition, it is recommended that animals not be bedded overnight near stands of *Lupinus* nor allowed to graze near them in the early morning when they are hungry and less selective in their grazing.

To reduce the incidence of terata, various mineral supplements—including cobalt, copper, iron, manganese, and zinc, as well as vitamins A and E—have been tried, but without apparent success (Keeler et al. 1977).

Medicago L.

Taxonomy and Morphology—An Old World genus and a member of the subfamily Papilionoideae, *Medicago* comprises some 56 species. In North America, 18 introduced species are present, most of which are naturalized. The genus name is derived from the Latin word *Medica,* meaning "plant of Media," the area in which alfalfa is reputed to have been brought into cultivation as fodder for horses. Some species are cultivated for forage, and others are weeds. Only 1 species is of toxicologic significance in North America:

■ *M. sativa*
alfalfa
lucerne

M. sativa L. alfalfa, lucerne

Plants perennials; herbaceous; from deep taproots. **Stems** erect; 50–100 cm tall; glabrous or indumented. **Leaves** 1-pinnately compound; leaflets 3; blades obovate to linear; apices rounded; margins dentate distally; bases cuneate; stipules fused to bases of petioles, lanceolate to linear. **Inflorescences** axillary racemes; 5-to-40-flowered; bracts present, setaceous. **Flowers** small. **Sepals** 5; fused; lobes subequal. **Corollas** papilionaceous. **Petals** 5; blue to purple; banners oblong or obovate; reflexed; wings oblong; keels curved, obtuse. **Stamens** 10; diadelphous; enclosed in keels, explosively exserted when corollas probed. **Legumes** spirally coiled 1–3 turns; puberulent or glabrous; not prickly. **Seeds** 1 to several.

Other species of the genus are distinguished by yellow flowers, annual habit, prickly legumes, and noncoiled legumes.

■ deep-rooted perennial herbs; leaves 1-pinnately compound, leaflets 3; flowers in axillary racemes, small, pealike; petals blue to purple

Medicago sativa

Distribution and Habitat—A Eurasian genus, *Medicago* has its center of diversity in the Mediterranean basin. Many of the species introduced deliberately or accidentally into North America have naturalized. *Medicago sativa* escapes cultivation and is found as solitary plants or small populations along roadsides, in fencerows, and in fallow fields, especially in the northern half of the continent.

■ crop plant, occasionally escapes

Disease Problems—Although one of the most important forage crops because of its productivity and nitrogen content, *M. sativa* is associated with four very different disease problems. Two—bloat and acute respiratory distress syndrome (ARDS)—are of primary importance. Of lesser importance are the disease problems of photosensitization and estrogenic effects. Aspects of each problem are discussed separately in the following subsections.

■ disease problems: bloat, ARDS, photosensitization, reproductive

Bloat

Disease Problems—Recognized for centuries as a hazard in livestock production, pasture bloat continues to be a serious problem associated with the use of alfalfa, certain other legumes, and some grasses for forage (Clarke & Reid 1974; Howarth 1975).

■ bloat, foamy distension of rumen; pasture, less commonly with hay

A number of names have been used interchangeably with bloat for the disease, including *hoven, tympany, hoove, blown,* and *meteorisme* (Howarth et al. 1991). The name *pasture bloat* implies association of the problem with fresh, green plants, but the disease also occasionally appears when animals eat alfalfa hay. It is a problem primarily in cattle and is observed to a lesser extent in sheep and other ruminants, including wildlife species.

Bloat is caused by the animal's inability to expel gases produced during ruminal fermentation. As the gases accumulate, the rumen is distended, and eructation is further inhibited. If the gases are allowed to continue accumulating, eventually the pressure produced in the rumen will interfere with movement of the diaphragm and lungs, and the animal will suffocate. This disease remains a serious impediment to animal productivity.

■ complex causes; lack of tannins allows rapid cellular breakdown and release of proteins, which increase surface tension to stabilize the foam produced

Disease Genesis—Pasture bloat is a complex disease problem involving a combination of factors. Because numerous reviews of bloat have been published (Howarth 1975; Howarth et al. 1991), only an overview is presented here. The primary factor in the appearance of bloat is the rapid release of soluble plant proteins via mechanical and digestive degradation of the ingested plant cells (Howarth et al. 1977; Sant & Wilson 1982). Forages that undergo rapid breakdown are often bloat risks. The presence of these proteins increases the surface tension of the ruminal solution and creates a stabilized foam as small bubbles of gas are trapped. The foam prevents effective eructation of the gas by inhibiting proper function of the cardia to facilitate movement of the gas into the esophagus. The foam may further block the movement of the gas in the esophagus.

A number of factors influence the release and action of the surface-acting compounds liberated during cellular degradation. In spite of this role for surface-acting agents, saponins do not seem to be of particular importance in the development of bloat (Majak et al. 1995). The rates of cell disruption and digestion are major factors, and condensed tannins seem to play a dominant role in slowing protein release and preventing the abrupt accumulation of ruminal gases (Fay et al. 1980; Lees 1992). Their presence may be a factor in the decrease in bloat potential observed as plants mature.

■ ruminal distension, left flank, respiratory distress, open-mouth breathing, foamy salivation, colic

Clinical Signs—The most obvious sign of bloat caused by *M. sativa* is the rapid onset of ruminal distension, initially seen in the area of the left flank but quickly seeming to involve the entire abdomen. This typically leads to respiratory embarrassment and distress and, if uncorrected, will shortly cause death. There is discomfort and colic, and the animal lies down, rolls, and kicks at its abdomen. Respiration is labored, and there is obvious distress, with open-mouth breathing, protruding tongue, extension of the head, and excessive foamy salivation.

■ head and neck tissues congested with blood

Pathology—Grossly, there is congestion and hemorrhage of the tissues of the head and neck. The thoracic esophagus may be pale while the cervical portion is congested, producing a characteristic "bloat line" at the thoracic inlet.

■ mechanical relief of distension; drench of agents to reduce surface tension, oils, detergents

prevent with surfactants, poloxalene

Treatment—The primary aim of treatment is to relieve the distension. This is accomplished by mechanical release of the gas and/or destabilization of the ruminal foam in which the air bubbles are trapped. Mechanical release is accomplished via passage of a stomach tube or ruminal trocar. The surface activity and the amount of foam present can be effectively reduced by the use of oils, fats, detergents, and other surface-active agents given by drench. Use of oils and fats is limited because they are subject to rapid degradation. Commonly employed are the pluronic surfactants, of which polyoxypropylene-polyoxyethylene types are the best examples. All of these antifoaming agents also can be used prophylactically to lower surface tension but require continuous daily administration. Poloxalene (Bloatguard®) is water soluble and can be given in a variety of ways, e.g., in liquid molasses or in mineral blocks or other type of blocks that are licked. Attempts to prevent bloat using various mineral mixtures have not been successful but confirm the effectiveness of poloxalene (Hall et al. 1994; Majak et al. 1995). Ionophores such as monensin and lasalocid also may be used to reduce the incidence of bloat, but they are less effective than poloxalene (Bartley et al. 1983; Lowe et al. 1991).

Pasture management and grazing control are used to prevent the occurrence of bloat. Pastures with a mixture of productive, high-bloat forages and highly palatable low-bloat types such as *Onobrychis viciifolia* (sainfoin); and *Lotus corniculatus* (bird's-foot trefoil) are especially useful if they can be maintained as true mixtures (Marten et al. 1987; Howarth et al. 1991).

Acute Respiratory Distress Syndrome

Disease Problems—Sometimes referred to as acute bovine pulmonary emphysema, the respiratory problems caused by *M. sativa* are categorized here, as recommended by Pierson and Kainer (1980), as a type of acute respiratory distress syndrome (ARDS). The disease, originally recognized in the 1920s, causes considerable loss of livestock in North America (Schofield 1948). The important factor in its appearance is the abrupt change in forage quality that occurs when animals are moved from a poor pasture to a lush green one. An acute, pneumonia-like disease appears in 5–10 days. The disease most often affects mature cattle. The time of year varies with geographic location. The most common scenario involves movement of cattle in late summer and early fall from dry pastures or ranges to irrigated pastures of alfalfa or grass (Schofield 1948; Mackey 1952; Hyslop 1969; Blake & Thomas 1971; Urrutia et al. 1997). The disease has been readily reproduced experimentally by modeling these conditions (Tucker & Maki 1962). However, it can occur at other times, as illustrated by its common appearance in April, May, and June in cattle moved to lush *Cynodon dactylon* (coastal Bermuda grass) in eastern Texas (Williams et al. 1977). The morbidity rate is usually low but occasionally may exceed 50%. The case fatality rate is typically 25–50%.

- ARDS, abrupt change in forage quality, from poor to good, precipitates severe respiratory distress in 5–10 days; often in the fall when cattle are moved to lush pasture

Various grasses and other forages such as members of the Brassicaceae (rape, kale, turnip tops), *Perilla frutescens* (purple mint), and *Fusarium,* when causing mold of sweet potatoes or runner beans, also can cause this disease syndrome.

- abrupt increase in tryptophan in diet of ruminants, converted to compounds toxic to pulmonary tissues

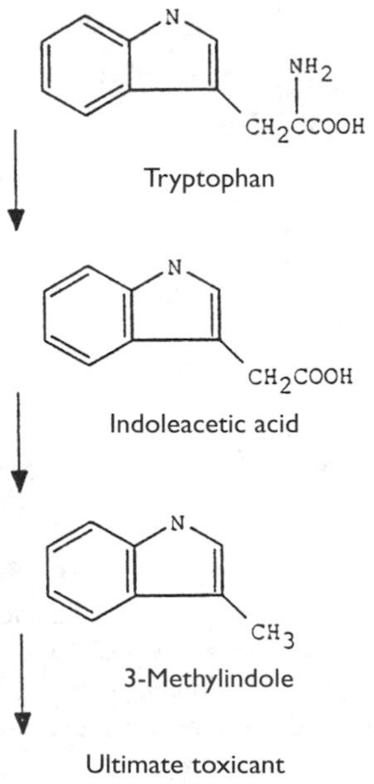

NH_2

CH_2CCOOH

Tryptophan

↓

CH_2COOH

Indoleacetic acid

↓

CH_3

3-Methylindole

↓

Ultimate toxicant

Disease Genesis—ARDS associated with plants is the result of a sudden change in forage quality that produces an abrupt increase of L-tryptophan in the rumen, where it is converted by bacteria to indolepyruvate, indoleacetate, and then 3-methylindole (3-MI) (Carlson & Dickinson 1978). The 3-MI is absorbed into the circulatory system and subsequently is biotransformed by the mixed-function oxidase system in the tissues of the lung. This biotransformation produces a short-lived, reactive compound that in turn produces toxic effects at its formation site in the bronchiolar epithelium. In animals accustomed to lush forage, tryptophan is converted mainly to indole, which is not bioactivated in the lung to the toxicant (Kirkland & Bray 1984). Experimentally, ARDS can be reproduced by oral administration of tryptophan or by oral or intravenous administration of 3-MI (Carlson et al. 1968, 1972, 1975). Not all closely related indoles, e.g., tryptamine, are subject to degradation to 3-MI or similar compounds

(Yokoyama & Carlson 1974). The disease can be prevented by reducing ruminal conversion of tryptophan to 3-MI via the use of ionophore/polyether antibiotics such as lasalocid, monensin, narasin, nigericin, and salinomycin (Hammond & Carlson 1980; Hammond et al. 1980). A more complete discussion of the specific details of the mechanism of intoxication is presented in the treatment of the Poaceae (chapter 58).

■ abrupt onset, labored respiration, frothy salivation, s.c. emphysema along neck and shoulders

Clinical Signs—The disease is characterized by rapid onset of dyspnea with loud expiratory grunts and open-mouth respiration accompanied by frothy, excessive salivation. Harsh respiratory sounds are evident on auscultation. The animal subsequently may develop subcutaneous emphysema along the neck, shoulders, and back. Death often occurs 1–2 days after onset of signs or, in a few cases of pastured cattle, without signs being observed.

■ gross pathology, heavy firm lungs, distended with fluid and air bubbles

microscopic, alveolar epithelial proliferation

Pathology—At necropsy, there may be a variety of minor changes, including a pale and friable liver and distension of the gallbladder, but these are of little significance in comparison to the effects on the respiratory system. The lungs are often markedly distended. Upon opening of the chest, the lungs fail to collapse completely and are heavy and firm, with froth-filled airways. The lymphatics are usually dilated and the interlobular septa are distended with edema fluid, fibrin, and air bubbles. Cut surfaces reveal an almost marbled appearance of patchy areas of firm, reddened to purple lobules and prominent edema fluid in connective tissue and septa. Air bubbles and/or large bullae may be apparent. Microscopically, alveolar epithelialization, accumulations of alveolar macrophages, and formation of hyaline membranes are commonly observed. Alveolar epithelial changes are due to proliferation of type II pneumocytes, which produce an almost glandular-like appearance. There may be a few eosinophils; otherwise granulocytic inflammatory cells are rare.

■ avoid stress, diuretics, atropine and antiinflammatory drugs may be used but are seldom effective

prevention, pasture management, ionophores

Treatment—It is very important to minimize stress during the period of severe respiratory distress. Numerous drugs have been used, including corticosteroids, antihistamines, nonsteroidal antiinflammatories, atropine, diuretics, and antibiotics, but therapy appears to be of limited value. Often, handling the animal for treatment causes stress and death due to acute respiratory embarrassment.

The disease can be prevented by management techniques aimed at gradual changes in pasture quality. Recommendations to accomplish this include cutting the forage of a lush pasture and allowing it to wilt before permitting animals to graze; limiting grazing time on the new pasture; and rotating the animals every few days between the lush and dry pastures or between the lush pasture and a drylot with hay provided (Potchoiba et al. 1992). When these approaches are not possible, diet supplementation with 200 mg/animal per day of an ionophore such as monensin or lasalocid may be used to prevent or at least attenuate the severity of the disease. Feeding of the ionophore should begin a few days before the pasture is changed and continue for 10 days after the change (Hammond et al. 1980; Nocerini et al. 1985; Potchoiba et al. 1992). This approach has a sustained protective effect but does not prevent appearance of the disease in all animals (Honeyfield et al. 1985). The efficacy of the ionophore can be enhanced by combining its use with the use of hay to reduce the intake of the lush forage (Potchoiba et al. 1992).

■ hepatogenous photosensitization; unknown cause but possibly a mycotoxicosis, mainly with hay, often damaged; phylloerythrin presumed the ultimate cause

Photosensitization

Disease Problems and Genesis—Species of *Medicago* in other areas of the world, e.g., *M. denticulata* in Australia, have been associated with primary photosensitization (type I) when fresh plants are eaten (Dodd 1916; Bull & Macindoe 1926). In North America, photosensitization due to *M. sativa* has been a problem mainly with hay and silage rather than with fresh forage. Because most cases are accompanied by evidence of hepatic involvement, it is classified as type III—secondary or hepatogenous photosensitivity (Clare 1955; Monlux et al. 1963; BL Glenn et al. 1964, 1965; Putnam et al. 1986; Scruggs & Blue 1994).

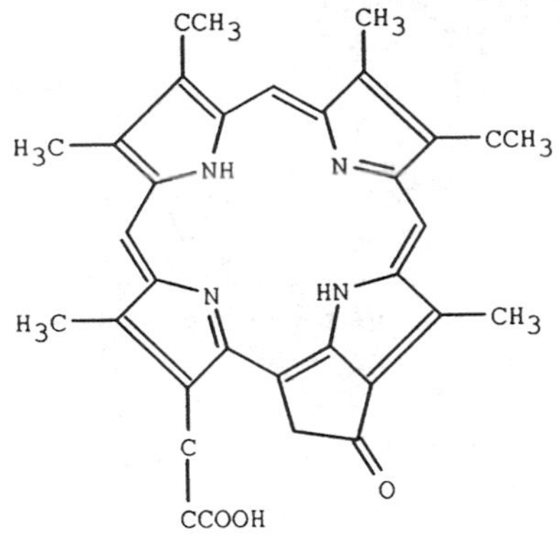

Phylloerythrin is considered to be the photodynamic agent. It is formed in the gut during microbial degradation of chlorophyll and normally is eliminated by the liver via the biliary system. With hepatic dysfunction, it accumulates in the circulation in sufficient quantities to function as a photodynamic agent activated by wavelengths of 400–600 nm to form reactive intermediates destructive to tissues.

The clinical signs appearing in particular environmental circumstances—e.g., when plants are water damaged or insect infested—often suggest mycotoxins as a cause, but these compounds are seldom isolated from the forage of *M. sativa* (Monlux et al. 1963; Mohamed et al. 1977; Putnam et al. 1986; Scruggs & Blue 1994).

In one instance, alfalfa silage was identified as a cause of primary photosensitization in cattle (House et al. 1996). There was no evidence of hepatic involvement, although the signs and skin pathology were identical to those of hepatogenous-type photosensitization. It was suggested that fermentation of the silage may have been responsible for the formation of photodynamic products during the breakdown of chlorophyll.

CCH_3 CH_3 H_3C CCH_3 NH N N HN H_3C CH_3 C O CCOOH

Phylloerythrin

Clinical Signs—In cases of photosensitization, skin changes are typically limited to the nonpigmented areas and include loss of hair and reddening. The skin over the affected areas is covered with serous exudate and is cracked. Necrosis, sloughing of large patches, and scar formation follow. In addition, there may be obvious loss of body weight, reddening of the sclera, and corneal edema resulting in temporary vision problems, including increased sensitivity to light. In sheep that have recovered, their ears may be absent and occasionally their eyes may be obliterated.

- less pigmented areas of skin; cracking with serous exudate, sloughing of patchy areas; sensitivity to light; icterus and depression

In the hepatogenous type, there may be icterus (yellowing) of the mucous membranes. Typically there is elevation of serum hepatic enzymes and/or indications of decrease in hepatic function, such as delayed clearance of BSP.

Pathology—In addition to the skin changes, in the case of hepatogenous photosensitization the tissues are often icteric. Depending on the cause and duration of the liver disease, gross and microscopic lesions may vary from acute to chronic hepatitis and may include fatty change, necrosis, fibrosis, and cirrhosis.

- gross pathology; obvious skin lesions; liver lesions

Treatment—Because of the nature of photosensitization, treatment is limited to supportive care. Animals should be protected from direct sunlight by moving them to shaded areas or indoors, although this is generally not feasible. Because the disease is due to a metabolic product of the breakdown of chlorophyll in the animal's rumen, reduction in the animal's intake of green plant material seems appropriate. There may be some cases in which even a small reduction in the amount of phylloerythrin is of some benefit, but it may not always be efficacious, because skin effects are reported to persist even when the animals are being fed concentrate rations (Clare 1955).

- supportive care, good feed, provide shade

Estrogenic Effects

Disease Problems—Even though widely used for forage, silage, and hay and commonly fed to cows, *M. sativa* rarely produces estrogenic effects. Nevertheless, irregular estrus,

- reproductive problems; rare, estrus irregularities

cystic ovaries, and precocious mammary and genital enlargement are occasionally reported in cattle (Adler & Trainin 1960; Lotan & Adler 1966; Romero et al. 1997). Estrogenic activity also has been confirmed experimentally (Bickoff et al. 1969).

■ low concentration of phytoestrogens coumestans, coumestrol; weak estrogen, pasture or hay; plant content varies with cutting, year, maturity, and variety

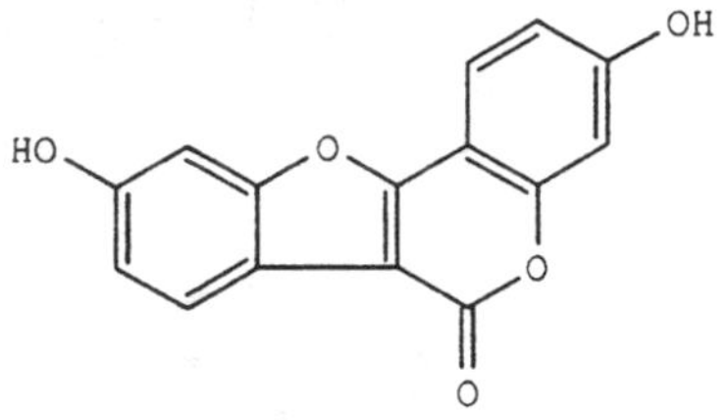

HO OH O O O

Coumestrol

Disease Genesis—Some species of *Medicago* contain low concentrations of coumestans, flavonoids that also are found in other legumes such as *Trifolium* and are sometimes referred to as phytoestrogens. Coumestrol is the principal one present. It is a weak estrogen that is absorbed, conjugated, and eliminated in the urine (Bickoff et al. 1964; 1969; Cox 1978 1985). In *M. sativa,* coumestrol concentrations are normally in the range of 1–20 ppm in the vegetative stages of growth and increase as the plants mature; they are highest at flowering. There may be variation in its estrogenic activity among the different cuttings; highest in the first cutting and especially low in the third and fourth. There also may be considerable variation in the levels from year to year (Cheng et al. 1953; Pieterse & Andrews 1956; Kitts et al. 1959). Some varieties may have higher coumestrol content than others. In plants from Finland, coumestrol concentrations of 25–65 ppm produced distinct estrogenic effects on rat uteri (Saloniemi et al. 1995).

In some instances, phytoestrogenic activity sufficient to cause disease seems to be a consequence of the plants having foliage with fungal disease (Saba et al. 1972; Adams 1989). Hay poses less risk because drying reduces the estrogenic activity as much as 75%. However, because of the numerous factors influencing coumestrol concentration in the plant tissues, hay cannot be dismissed entirely as a potential problem (Bickoff et al. 1960, 1969).

It has been estimated that coumestrol concentrations of 25–40 ppm may be sufficient to exert estrogenic effects such as altering production of gonadotropins, depressing ovarian activity, and delaying puberty of female lambs (Wong et al. 1971; Newsome & Kitts 1977, 1980; Adams 1989). Romero and co-workers (1997), investigating a reproductive problem in dairy cattle, reported that 608 cows inseminated 1,264 times had only 376 pregnancies, of which only 238 were normal. There were 36 abortions, and the remaining 102 exhibited other problems such as cervical mucous, repeat breeders, ovarian cysts, and metritis. These problems were attributed to coumestrol concentrations of about 66.8 ppm in the dry alfalfa. It thus appears that, although rare, estrogenic effects produced by *M. sativa* may be a cause of low-grade reproductive problems in livestock. Because of this rarity, however, other causes of reproductive problems must be eliminated before coumestrols are incriminated.

■ decreased reproductive efficiency; reversible, no treatment needed

Clinical Signs and Treatment—On some occasions, while plants of *M. sativa* are being consumed, there may be a short-term decrease in estrus, ovulation, and secretion of ovarian estrogen. A more complete discussion of the estrogenic effects of coumestans in presented in the treatment of *Trifolium,* in this chapter. Because the estrogenic effects are slight and readily reversed by decreasing or stopping consumption of the alfalfa, specific treatment is not needed.

■ *M. alba*
M. indica
M. officinalis
hubam
Indian melilot
sour clover
sweet clover

Melilotus Mill.

Taxonomy and Morphology—Comprising about 20 species, *Melilotus,* a member of the subfamily Papilionoideae, is an Old World genus. The genus name is derived from the Latin words *mel,* meaning "honey," and *lotus,* meaning "plant," and is believed to reflect the fragrance of the foliage. Used as forage and green manure, 7 species are present in North America. Three of them are naturalized and widespread as weeds.

M. alba Medik.	white sweet clover, hubam
M. indica All.	sour clover, Indian melilot
M. officinalis (L.) Lam.	yellow sweet clover

Plants annuals or biennials; herbaceous; from taproots; herbage coumarin-scented when dry. **Stems** erect; 100–200 cm tall; glabrous or sparsely indumented. **Leaves** 1-pinnately compound; leaflets 3; blades obovate to linear; apices rounded; margins dentate or serrulate; bases acute; stipules fused to bases of petioles. **Inflorescences** elongate, axillary racemes; many-flowered; bracts minute or absent. **Flowers** small. **Sepals** 5; fused; lobes subequal. **Corollas** papilionaceous. **Petals** 5; white or yellow; banners oblong to obovate; reflexed; wings oblong, coherent to keels; keels curved, obtuse. **Stamens** 10; diadelphous; enclosed in keels, not explosively exserted when corolla probed. **Legumes** indehiscent; ovate or suborbicular; flattened; stipitate or sessile; surface reticulate-veined or ribbed, glabrous or puberulent. **Seeds** 1.

- herbs with coumarin scent when dry; leaves 1-pinnately compound, leaflets 3, margins dentate; flowers pealike; petals white or yellow

1. Petals white. Legumes dark brown or black at maturity . . . . *M. alba*
1. Petals yellow. Legumes gray or light brown at maturity.
    2. Flowers 3–5 mm long. Pedicels recurved.
    Sepal lobes long, acute. Legumes cross-ribbed . . . . *M. officinalis*
    2. Flowers 1–2.5 mm long. Pedicels ascending.
    Sepal lobes short, blunt. Legumes reticulate-veined . . . . *M. indica*

Melilotus alba

The binomial *M. albus* Lam. appears in many publications but was published 10 years later than Medikus's *M. alba* and thus is illegitimate according to the rules of botanical nomenclature.

Distribution and Habitat—The center of diversity of *Melilotus* appears to be the Mediterranean basin and southwest Asia. *Melilotus alba* and *M. officinalis* apparently made it to North America with early colonists, because these species are cited as present in floristic works of the 1730s. The sweet clovers are used extensively as forage crops or as soil builders, especially in the northern Great Plains. Often establishing dense populations in disturbed soils, they also are common weeds along roadsides and in waste areas throughout much of North America. *Melilotus alba* and *M. officinalis* are common; *M. indica* is less so.

- crop plants, common waste area and roadside weeds

Cut for silage or hay, the sweet clovers compare favorably with alfalfa in quantity, utility, and nutrition. They have many attributes that favor their continued agronomic use—drought resistance, rapid growth, high yield, relative saline tolerance, and exceptional honey production. They are of moderate hay value because the succulent stems are difficult to dry even when conditioned. The yellow-flowered species are favored because of their finer stems. They make good silage and are exceptional deep-rooted soil builders (Goplen 1980).

Disease Problems—The sweet clovers became increasingly popular as forages in the early 1920s, about the same time that a seemingly new clostridial-type hemorrhagic or blackleg-type disease was being recognized. Eventually the feeding of sweet clover hay was linked to the disease (Schofield 1924; Roderick & Schalk 1931). It was noted that there was no problem when sweet clover pasture was grazed. The characteristic manifestation of the disease was intractable bleeding due to inhibition or retardation of blood coagulation (Roderick 1929, 1931). Typically, morbidity is low (10–15%) and the case-fatality rate high (>50%) (Blakley 1985). The most devastating episodes

- good hay and silage forages; succulent stems dry with difficulty; when not completely cured, interference with coagulation of blood, hemorrhage, especially after surgery, primarily in cattle

occur following surgery, dehorning, and castration. There may be no indication prior to surgery of the animal's predisposition for clotting problems, but following the surgery, it may be seemingly impossible to stop the bleeding. In other instances, large areas of subcutaneous hemorrhage—hematomas or blood blisters—may alert the producer to possible problems. Handling animals in squeeze chutes or similar types of facilities also predisposes trauma and subsequent development of subcutaneous hemorrhage. Of special concern is the potential for hemorrhage during parturition or in the newborn animal. Anemia and hemorrhage may be problems in the newborn in the absence of overt signs other than prolonged prothrombin time in the dam (Fraser & Nelson 1959). In a rather unusual case, "downer" cattle were encountered in which hemorrhage was associated with grub (*Hypoderma* spp.) migration in and around the lumbar spinal cord (Meads et al. 1964).

Disease problems are more likely to appear in the winter when animals are fed *Melilotus* hay put up in large round bales during a wet summer. These conditions of large bales and moist plants make curing of the tougher stems more difficult. Problems also may appear when more-mature, coarser hay is put up (Roderick & Schalk 1931). Because species of *Melilotus* often readily invade sites, they may present problems as weedy contaminants in grass hays (Yamini et al. 1995). Silage is less often a problem, but it is still a risk, especially in dairy cattle (Puschner et al. 1998).

Although all animals are at some risk, this hemorrhagic disease occurs primarily in cattle that have eaten the improperly cured plants for several weeks. It appears infrequently in horses and other livestock (McDonald 1980). Sheep are less susceptible than are cattle (Linton et al. 1963). Thus, there is little likelihood of problems at lambing and in newborns.

Melilotus officinalis

- high concentration of coumarin, vanilla-like smell to hay, converted to hydroxycoumarin and dicoumarol by fungi in partially cured stems; dicoumarol prevents vitamin K regeneration and depletes its availability

Coumarin

4-Hydroxycoumarin

Disease Genesis—Species of *Melilotus* contain coumarin, which imparts a pleasant vanilla-like fragrance and a slightly bitter taste to the plant. It is this compound that produces the pleasant odor we associate with recently mowed roadsides or meadows. Coumarin itself is of relatively low toxicity, but at high and prolonged dosage it is hepatotoxic in laboratory rodents, possibly as a result of limited hydroxylation capacity (Cohen 1979). In other species, such as baboons and humans, hepatotoxicity is not apparent. Under certain conditions when the plants are harvested and incompletely dried for hay or fermented in silage, a toxic compound that predisposes the animal to hemorrhage is formed (Campbell & Link 1941). This formation is associated with fungi growing within the plant stems. The relatively high moisture in the stems promotes the growth of various soil fungi. Because the fungi typically grow only within the stems, toxic hay may appear clean. In other instances, it may be blackened and have a musty tobacco-like odor (Roderick & Schalk 1931; Puschner et al. 1998). However, hay with an obvious spoiled appearance also may be of low toxicity. The fungi involved include species of *Penicillium, Aspergillus, Fusarium, Mucor, Humicolor,* and *Arthrobacter.*

Within the plant stems, coumarin accumulates secondary to the formation of *o*-hydroxycinnamic acid from L-phenylalanine. Fungal growth in the partially cured stems initiates the conversion of *o*-coumaric acid and melilotic acid glucosides to β-hydroxymelilotic acid, then to 4-hydroxycoumarin, and finally, through a condensation reaction, to the toxic bishydroxycoumarin or dicoumarol (Scheel 1978; Kosuge & Gilchrist 1977; Murray et al. 1982). Dicoumarin, dicumarol, and dicumarin are synonyms of dicoumarol.

Dicoumarol interferes with the proper functioning of vitamin K. Under normal circumstances, in the process of vitamin K–induced carboxylation of coagulation cofactors, vitamin K 2,3-epoxide is formed. To maintain adequate levels, regeneration of the active form of vitamin K_1 is accomplished by an epoxidase enzyme; it is this enzyme that is competitively inhibited by dicoumarol (Shearer et al. 1976; Ren et al. 1977). Thus, in the presence of dicoumarol, there is insufficient vitamin K available for activation of prothrombin and factors VII, IX, and X. Continued ingestion of hay with dicoumarol concentrations as low as 10 ppm may result in marginal depletion of coagulation factors (Linton et al. 1963). Toxic effects are clearly demonstrated at 20–30 ppm (Casper et al. 1982). Dicoumarol concentrations in the plants may be as high as 165 ppm or more, especially in hay with high protein levels (Benson et al. 1981). Typically the disease occurs after the animal has consumed the toxic hay for 4–5 weeks and the coagulation factors have been depressed to such an extent that hemorrhage occurs (Blakley 1985). Dicoumarol may remain at toxic concentrations in the hay for 3–4 years or more. Its concentrations typically are much lower in silage (Benson et al. 1981).

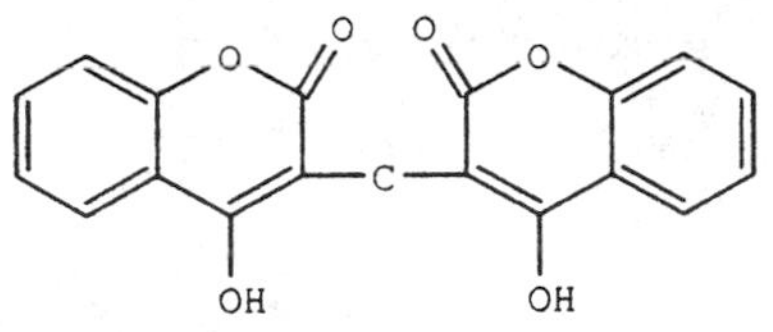

Dicoumarol

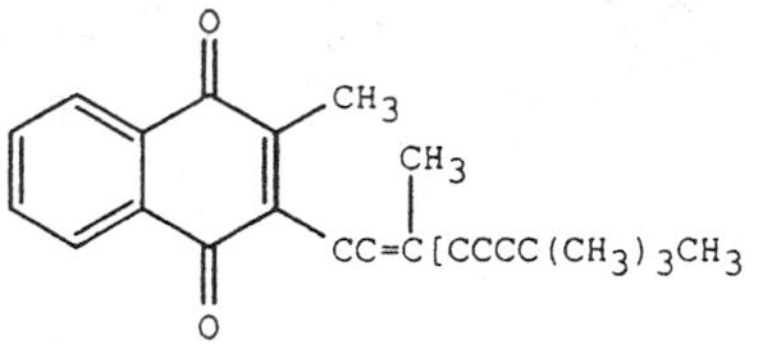

Vitamin K_1

Dicoumarol produces other effects. It inhibits or uncouples oxidative phosphorylation, transepithelial sugar transport, and activity of mixed-function oxidases in the liver. The clinical significance of these effects is not clear (Martius & Nitz-Litzow 1953; Kosuge & Gilchrist 1977; Murray et al. 1982).

Both high- and low-coumarin cultivars of *Melilotus* are recognized, there being about a 20-fold difference in dicoumarol content in hay made from them (Goplen 1971; Sanderson et al. 1986). Low-coumarin cultivars such as 'Polara,' 'Denta,' and 'Norgold' have dicoumarol concentrations of about 5 ppm. The disadvantage to the use of these cultivars, however, is that they produce less forage than the high-coumarin cultivars, almost 10% in some instances. It is possible to prevent fungal growth and dicoumarol formation during curing of the forage by the application of fungal growth inhibitors such as 2 ppm propionic acid (Sanderson et al. 1986).

Melilotus species are rather unusual in providing the appropriate substrate (coumarin) and environment for the formation of dicoumarol, but on rare occasions the problem may occur with other forages such as *Anthoxanthum odoratum,* vernalgrass (Kosuge & Gilchrist 1977; Pritchard et al. 1983).

■ hematomas, intractable bleeding, anemia

diagnosis, prothrombin or coagulation times

Clinical Signs—The clinical signs typically observed are indicative of severe hemorrhage in the animal's subcutaneous tissue or body cavities. They range from intractable spontaneous or surgically induced bleeding to weakness due to anemia, lameness, or dyspnea. The site of the hemorrhage influences the signs that appear. Often animals are found dead in the pasture or feedlot, and the hemorrhages are not detected until necropsy. Other signs noted include abortion and stillbirths, weakness, and neurologic abnormalities (Blakley 1985). As noted above, the disease occurs primarily in cattle; clinical signs are rarely seen in horses or sheep.

In addition to the clinical signs of excessive hemorrhage, the disease is readily diagnosed by laboratory determination of coagulation parameters. Prothrombin time is prolonged from approximately 15 seconds to 100–200 seconds and is a better diagnostic test than either bleeding time or whole-blood clotting time. The cause of the coagulation changes may not be readily determined, because most anticoagulants, including warfarin rodenticides, produce the same syndrome and because the forage may not be visibly moldy when growth of the fungus is within the plant stems.

Diagnosis may be aided by analysis of blood or plant tissues (especially stems) for dicoumarol. In most cases with distinct hemorrhagic problems, dicoumarol concentrations in plants will exceed 50 ppm of plant tissue (Casper et al. 1982; Puschner et al. 1998). Lower levels, 20–50 ppm, may still be toxic but may require a much longer time of ingestion to cause problems.

- accumulations of blood, large and/or scattered

Pathology—Other than large and/or scattered sites of hemorrhage, there are few distinct lesions. The tissues may be pale because of the anemia, and there may be some hypoxic degeneration of hepatocytes, but these are mild changes and are not of diagnostic significance.

- whole blood or plasma, vitamin K_1, supplement with good alfalfa hay

  prevent by alternating forage, using alfalfa or other good-quality hay

Treatment—In severe cases, treatment is best accomplished by transfusion of 20 ml/kg of whole blood or 10 ml/kg of plasma to supply coagulation factors and/or red blood cells. Vitamin K_1 is given at a dosage of 1–5 mg/kg b.w. either i.v., i.m., or s.c. depending on the severity of the disease and whether or not coagulation factors have been transfused. Vitamin K_1 therapy may be necessary on a daily basis for a week or more (Goplen & Bell 1967). Vitamin K_3 has been used with success in some instances at double the dose for vitamin K_1 and has the advantage of being considerably less expensive and perhaps longer lasting (Radostits et al. 1980; Alstad et al. 1985). It is especially useful as follow-up to initial K_1 therapy. Overall though, vitamin K_1 remains the more effective therapy (Goplen & Bell 1967). Prolongation of prothrombin times may persist for 1 or more weeks following removal of the fungus-infected hay. During this period, animals should be fed good-quality alfalfa. Suspect sweet clover hay may be fed if it is alternated with good alfalfa hay every 10–14 days. Pregnant cows should not be fed hay or silage containing *Melilotus* unless these forages have been evaluated for dicoumarol content. It would be prudent to forgo feeding sweet clover at least 1 month prior to calving.

It has been recommended that if sweet clover is to be used as a forage, it should be sown heavily to promote dense growth with thin stems, cut early while the stems are thin, dried well, and packed well as silage (Schofield 1924). This old advice remains timely.

Oxytropis DC.

- *O. campestris*
  O. deflexa
  O. lagopus
  O. lambertii
  O. sericea
  O. splendens

  crazyweed
  loco
  locoweed
  oxytrope

Taxonomy and Morphology—Comprising approximately 300 species, *Oxytropis,* a member of the subfamily Papilionoideae, is a circumboreal genus with greatest diversity in the interior plateaus of Asia. The genus name is derived from the Greek roots *oxys,* meaning "sharp," and *tropis,* meaning "keel," a reference to the beak or projection at the apex of the flower's keel petals. Its members are commonly referred to as locoweeds, crazyweeds, locos, or point vetches. The name *locoweed* also is applied to species of the closely related genus *Astragalus;* taxa of both genera cause locoism. In some taxonomic references, the common name *oxytrope* is used. With the exception of the intoxication problems they cause, species of the genus are of little economic importance. A few are grown as ornamentals, and some provide palatable, albeit sparse forage. In North America, the genus is represented by 20–25 species and numerous varieties (Barneby 1952). Of these, 6 species have been implicated in toxicological problems (see plate 37.6):

O. campestris (L.) DC. (= *O. monticola* A.Gray)	field crazyweed, yellow locoweed, slender crazyweed
O. deflexa (Pall.) DC.	stemmed oxytrope, nodding locoweed, pendent-pod crazyweed
O. lagopus Nutt.	rabbit-foot crazyweed, hare's locoweed
O. lambertii Pursh	Lambert's locoweed, Lambert's crazyweed, purple locoweed, purple loco, stemless locoweed

O. sericea Nutt. ex Torr. & A.Gray (= *O. saximontana* A.Nelson)	white loco, white point locoweed
O. splendens Douglas ex Hook.	showy crazyweed

■ perennial herbs; leaves 1-pinnately compound, leaflets 9–41, stipules present; flowers pea-like; petals 5, keels sharply beaked, white, yellowish, pink, violet, or purple

Plants perennials; herbaceous; acaulescent (1 species caulescent). **Leaves** 1-pinnately compound; pairs of leaflets 9–41; terminal leaflet present; petiolules absent; stipules present, herbaceous, free or fused to petioles. **Inflorescences** elongate or capitate racemes borne at ends of elongate peduncles; bracts present or absent. **Sepals** 5; fused. **Corollas** papilionaceous. **Petals** 5; white or yellowish or pink or violet or purple; banners oblanceolate, reflexed; wings clawed, auriculate; keels rounded, abruptly beaked at apex. **Stamens** 10; diadelphous; enclosed in keels. **Legumes** cylindrical to lanceolate or inflated; 1- or 2-loculed. **Seeds** several.

Oxytropis is morphologically quite similar to the closely related *Astragalus,* and several species of *Oxytropis* were originally described as species of *Astragalus.* The two genera differ in the nature of their keels. In *Oxytropis* the blades of the keel petals are abruptly contracted into an erect or recurved subulate beak or conical point—hence the common name *point vetch* for some species. In contrast, the blades of the keel petals are incurved to form a rounded, deltoid or triangular apex in *Astragalus.* Species of *Oxytropis* are morphologically quite variable, and minor discontinuities among them make consistent recognition tenuous. Interspecific hybridization further complicates their taxonomy.

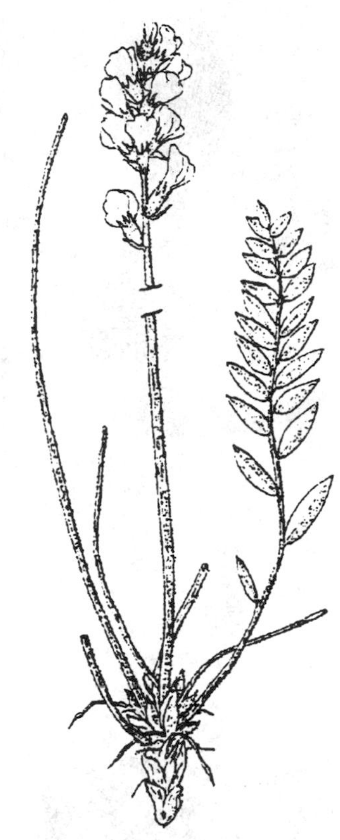
Oxytropis campestris
(after Welsh 1960)

1. Stipules fused one-fourth length or less to petioles. Legumes descending or pendulous at maturity; 1-locular ...... *O. deflexa*
1. Stipules fused one-half length or more to petioles. Legumes ascending or erect at maturity; 2-locular.
  2. Flowers white or yellowish white.
    3. Leaflet hairs dolabriform, 2 ends free .... *O. lambertii*
    3. Leaflet hairs basifixed, 1 end free.
      4. Legumes coriaceous; rigid at maturity .... *O. sericea*
      4. Legumes membranous-chartaceous; flexible at maturity .... *O. campestris*
  2. Flowers purple or pink purple or carmine.
    5. Leaflet hairs dolabriform, 2 ends free .... *O. lambertii*
    5. Leaflet hairs basifixed, 1 end free.
      6. Legumes coriaceous; rigid at maturity .... *O. sericea*
      6. Legumes membranous-chartaceous; flexible at maturity.
        7. Leaflets fascicled in verticils of 3–6. Foliage silky-villous .... *O. splendens*
        7. Leaflets alternate or subopposite. Foliage silky-pilose or hispid-hirsute.
          8. Calyces silky-hirsute. Flowers pink-purple or blue-purple .... *O. lagopus*
          8. Calyces pilose. Flowers purple .... *O. campestris*

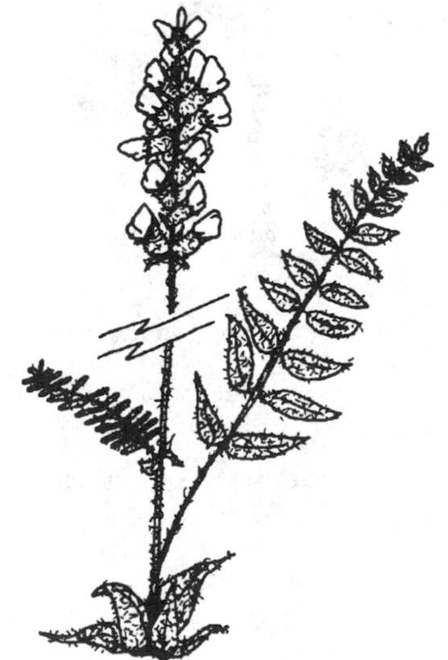
Oxytropis deflexa

Distribution and Habitat—Species of *Oxytropis* are found primarily in arctic, montane, and prairie habitats in North America. Greatest abundance is in the western half of the continent. Plants of the genus do not dominate their habitat but rather form localized populations. Three species—*O. campestris, O. lambertii, O. sericea*—are of particular interest because of their role in livestock losses, their wide distribution, and their relative abundance.

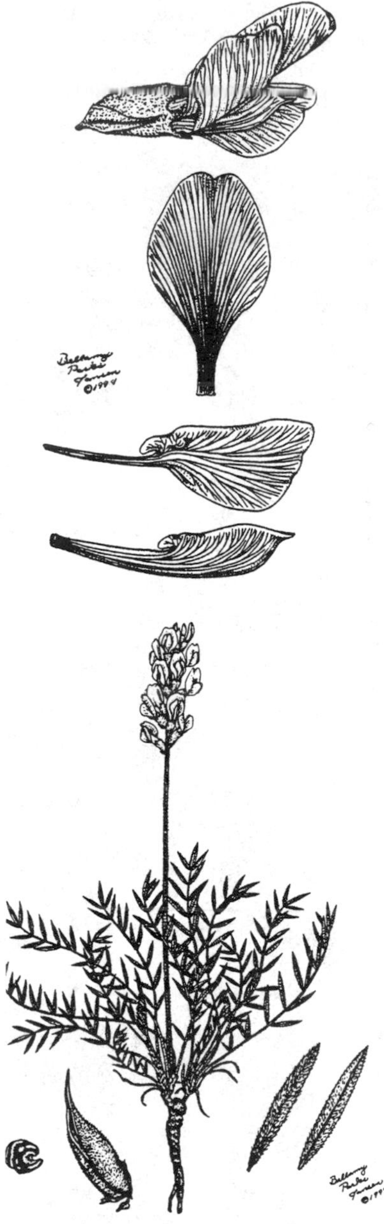

Oxytropis lambertii

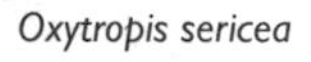

Oxytropis sericea

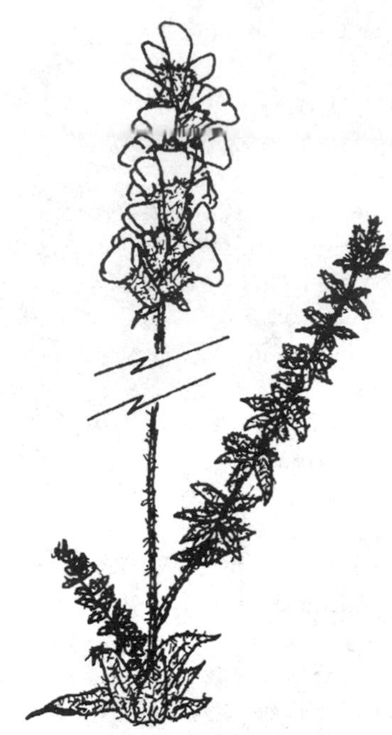

Oxytropis splendens

- arctic, montane, prairie habitats
- disease mainly due to *O. lambertii, O. sericea*

some species such as *O. riparia* are nontoxic

A circumboreal species, *O. campestris* occurs across the northern part of the continent. Plants are found in prairies, meadows, and open woodlands up to elevations of 2,600 m.

Described by some range managers as the most dangerous of the locoweeds, *O. lambertii* often forms extensive populations in a variety of habitats and vegetation types: shortgrass prairie, desert grassland, sagebrush, and desert shrub communities. Plants also are found in pine forests up to 3,150 m in elevation. Plants grow in a variety of soils derived from sandstone, limestone, and shale; they may be found on dunes, ridges, scarps, and badlands. Purple locoweed is one of the most widespread of the North America species.

Oxytropis sericea exhibits an ecology similar to that of *O. lambertii,* and the two species are sympatric through much of their ranges. Plants are encountered in rocky prairies, plains, sagebrush to juniper grasslands at elevations up to 3,300 m. Populations do not occur as far south as those of *O. lambertii* but are very widespread in the Rocky Mountains and Plains.

Disease Problems—*Oxytropis* has been described as one of the most important toxic plants in North America (James & Nielsen 1988). Causing locoism, it and the closely related *Astragalus* became a serious problem for ranchers maintaining livestock on the Great Plains and eastern front range of the Rocky Mountains in the 1870s and 1880s (Marsh 1909). All of the species listed are of toxicologic concern, in some instances originally referred to by the alternative names indicated under the taxonomy (Crawford 1908; Marsh 1909; James et al. 1970, 1981; Harries et al. 1972). However, *O. lambertii* and *O. sericea* are the species most commonly implicated in locoism. Marsh (1929) considered *O. lambertii* to be the most destructive of all the locoweeds because of its abundance and wide distribution. A complete discussion of the aspects of this disease is presented in the treatment of *Astragalus,* in this chapter. Aspects of locoism specific to *Oxytropis* are summarized below.

It must be emphasized that not all species of *Oxytropis* are toxic. *Oxytropis riparia* Litv. (oxus oxytrope, or Ruby Valley point vetch), an introduced Eurasian species in the Rocky Mountain region, provides palatable forage (Williams & Molyneux 1988; Barneby 1989a). Likewise, based upon extensive experimental evidence, *O. besseyi* (Rydb.) Blank. was reported by Marsh (1929) to be nontoxic.

Identification of species of *Oxytropis,* in particular *O. lambertii* and *O. sericea,* as one of the causes of locoism in horses, cattle, and sheep was the result of intensive studies by Marsh (1909). Subsequent work by Mathews (1932) contributed to our understanding of the signs and aspects of the disease. Ralphs and co-workers (1986) noted that consumption of *Oxytropis* by grazing animals is similar to that of *Astragalus.* They observed that early in the growing season little foliage of *O. sericea* was eaten by cattle; most consumption occurred in the last few weeks of the season. Free-ranging cattle readily ate large amounts of the immature legumes, but flowers were eaten only when the availability of grass and forbs was poor (Ralphs et al. 1986, 1987, 1989). In some instances, when the succulent immature legumes were available, they constituted up to 50% of the diet. Except for a few so-called loco eaters, cattle previously exposed to the plants and those encountering it for the first time exhibited little difference in locoweed consumption.

Observations of sheep indicated that there was no initial preference for *O. sericea,* but a few ewes acquired a taste for the species (Ralphs et al. 1991). Ewes exhibiting locoism the previous year did not eat more plants than other animals, but the neurologic effects of the disease were apparent when the animals lowered their heads to graze. This movement precipitated transient seizures manifested by trembling of the head, bowing or tucking of the head under the chest, and moving of the head in a bobbing motion.

In addition to the neurologic problems of locoism, *Oxytropis* also causes the associated right-heart failure. When cattle at high elevations of 2,100 m or more graze extensively on *O. sericea,* there is a significant risk (James et al. 1983a). Right-heart failure, or brisket disease, is primarily a problem in cattle moved from lower elevations up to summer range in mountain areas where *Oxytropis* and species of *Astragalus* are abundant. The disease is more common and more serious in calves; the morbidity rate ranges up to almost 100% in calves, with a case-fatality rate of 10–15%. Although the signs are not typical of locoism, many of the pathologic changes are. Among mature cows, a few may show signs of heart failure, while others exhibit typical locoism signs and/or abortions (James et al. 1986; Panter et al. 1988). Prompt transfer to lower elevations lessens myocardial stress and provides relief from the cardiac signs.

Although similar with respect to locoism, *Oxytropis* is significantly different from *Astragalus* in that it does not accumulate selenium and does not cause cracker-heels disease.

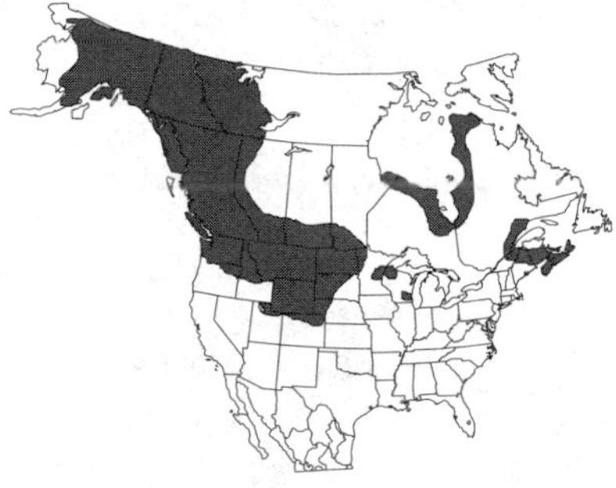

Oxytropis campestris

■ a few animals develop preference for locoweeds, eat sufficient to cause neurologic disease; on high-elevation ranges, may be right-heart failure

Oxytropis deflexa

Oxytropis lagopus

Oxytropis lambertii

Oxytropis sericea

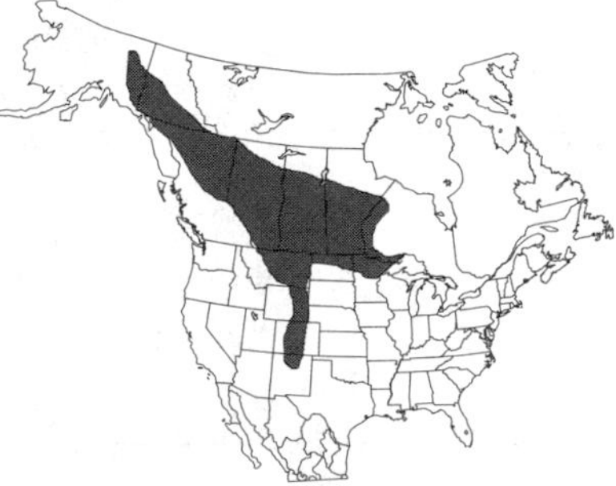

Oxytropis splendens

- toxins are indolizidine alkaloids, especially swainsonine and its *N*-oxide
- swainsonine closely resembles mannosyl cation, acts as competitive inhibitor of α-D-mannosidase; inability of breakdown of mannose-containing oligosaccharides in cells

  all cells affected, but neurologic manifestations largely due to effects on cerebellum and Purkinje cells

Disease Genesis—The similarity of the clinical signs of locoism, caused by *Oxytropis* and the closely related *Astragalus,* to those caused by the Australian genus *Swainsona,* also a member of the Fabaceae, led to the identification of swainsonine and related compounds as the toxicants in *Oxytropis* (Molyneux & James 1982). Swainsonine, dihydroxy lentiginosine, and 2-epilentiginosine are indolizidine alkaloids (Molyneux 1990). A complete discussion of the effects of these compounds is presented in the treatment of *Astragalus,* in this chapter. However, several aspects of disease genesis characteristic of *Oxytropis* are presented here.

Concentrations of swainsonine in the legumes of *O. sericea* are in the range of 0.007–0.1% or more, well above the less than 0.001% concentrations deemed hazardous (Molyneux et al. 1989; Molyneux et al. 1994). Concentrations also vary with the stage of plant maturity and from year to year. In *O. sericea,* swainsonine in leaves in one year increased from 0.025% in early growth to 0.04% by the flower and legume stages and decreased thereafter, and in the next year concentrations ranged from 0.064% to 0.058% and 0.018% at the same stages (Ralphs & Molyneux 1989). Concentrations in flowers were similar or somewhat higher. Experimentally, 20% *O. sericea* in the diet of sheep or cattle is sufficient to cause signs of intoxication in 3–4 weeks. In cows, recovery from the reproductive effects after ingestion of plants for 30 days required 2 or more weeks (Panter et al. 1999a). If given long enough, even a dosage of swainsonine as low as 0.2 mg/kg/day is sufficient to cause intoxication (Stegelmeier et al. 1999a). Some of the nontoxic species, such as *O. riparia,* have been shown to lack swainsonine (Williams & Molyneux 1988).

The pattern of swainsonine absorption and excretion in intoxications caused by *Oxytropis* is the same as that produced by *Astragalus.* Tissue concentrations, as measured by α-mannosidase activity, were highest in the liver and kidney (Stegelmeier et al. 1995a). Appreciable α-mannosidase concentrations of 0.14–0.24 μg/ml occur in the blood (Pulsipher et al. 1994).

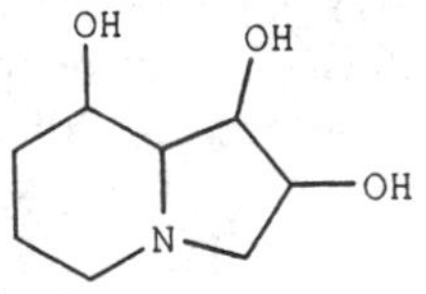

Swainsonine

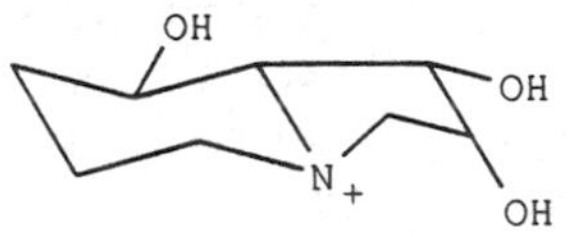

Swainsonine cation

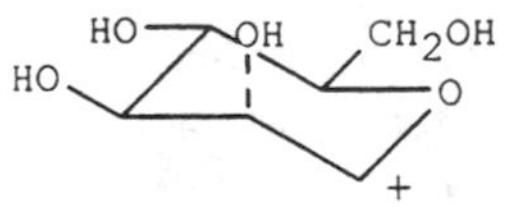

Mannosyl cation

- general signs, depression, weight loss, ataxia, stiff gait, exaggerated limb movement, interference with eating

  most noteworthy change in horses, dramatic response to handling or confinement; may become violent when restrained; violence not controlled by tranquilizers

  cattle, similar signs but less violent, more likely to wander; sheep, more gradual onset of signs, loss of flocking instinct
- animals eating plants at higher elevations, labored respiration, jugular pulse, brisket edema, indicative of cardiac insufficiency; alleviated by moving to lower altitude
- gross pathology, few lesions

  microscopic, widespread cytoplasmic vacuolation of cells in many tissues
- treatment, feed well, allow animals to recover body condition gradually; salvage for slaughter

Clinical Signs, Pathology, and Treatment—The clinical signs and pathology of locoism caused by ingestion of *Oxytropis* are identical to those produced by species of *Astragalus.* As with *Astragalus,* prevention and treatment have been attempted with the use of mineral mixtures (Silent Herder), bentonite clay, and banamine without success (Dugarte-Stavanja et al. 1997). In addition, evaluation of liver mineral concentrations following grazing of *O. sericea* does not indicate any likelihood for alteration of the disease with mineral supplements (Galyean et al. 1996). Much of the initial work was conducted on *Astragalus* and *Oxytropis* simultaneously, and this information is presented in depth in the treatment of *Astragalus,* in this chapter.

Vitamin E and selenium also have been tried but are not effective (Richards et al. 1999).

Prevention of intoxication from *Oxytropis* may be accomplished by carefully observing grazing animals to identify and remove any animals eating the plant. Those animals with a taste for *Oxytropis* can then be conditioned using multiple dosage of 200 mg/kg b.w. of lithium chloride with some plant material to prompt them to avoid further ingestion (Ralphs et al. 1997). It should be anticipated that conditioned aversion will require repeated dosage, given that experimentally it is difficult to avert animals already familiar with or accustomed to eating the plants.

Pisum L.

Taxonomy and Morphology—Comprising 5 species native to the Old World, *Pisum*, whose genus name was used by Virgil, is a member of the subfamily Papilionoideae and related to *Vicia* and *Lathyrus*. One species of the genus is of minor toxicologic interest in North America:

- *P. sativum*

  Austrian pea
  garden pea

P. sativum L. garden pea, Austrian pea

With several subspecies, varieties, and races, *P. sativum* is cultivated around the world for food and forage. Variety *arvense* (field pea) is grown for forage.

- climbing annual herbs; leaves 1-pinnately compound, leaflets 1–4, terminal tendrils; flowers pealike, axillary; petals white, lilac or pink

Plants annuals; herbaceous; caulescent; glabrous; glaucous. **Stems** climbing by means of branched tendrils; 100–200 cm long. **Leaves** 1-pinnately compound; pairs of leaflets 1–4; terminal leaflet absent; leaflets suborbicular to elliptic or oblong, margins entire or dentate; rachises ending in tendrils; petiolules absent; stipules conspicuous, typically larger than leaflets, obliquely ovate, margins dentate, bases amplexicaul. **Inflorescences** solitary flowers or 2- or 3-flowered racemes; axillary; peduncles elongate; bracts absent. **Flowers** large; showy. **Sepals** 5; fused. **Corollas** papilionaceous. **Petals** 5; white or lilac or pink; banners orbicular, reflexed; wings clawed, auriculate, fused to keels; keels rounded. **Stamens** 10; diadelphous; enclosed in keels. **Legumes** linear to oblong-linear; flattened or somewhat terete; inflated. **Seeds** 2–10; subglobose.

Pisum is distinguished from *Lathyrus* and *Vicia* by the appearance of its style, which is folded longitudinally in contrast to the round or oval styles of the other two genera.

Pisum sativum

- crop plant

Distribution and Habitat—Species of *Pisum* are believed to be native to the Mediterranean region and western Asia. Now widely planted for food and forage, *P. sativum* does not escape cultivation.

- rarely neurologic effects; locomotor difficulties

Disease Problems—The seeds (peas) of *P. sativum* are a popular food throughout the world. In addition, the foliage, typically after the legumes are harvested, is used to make silage. Although widely fed without apparent problems, there have been a few reports of neurointoxication problems. In one case, peavine silage was suspected to cause tremors and incoordination in cattle. The signs were very similar to those of the various staggers syndromes produced by *Paspalum dilatatum* (Dallis grass), and *Lolium perenne* (perennial rye). The affected animals subsequently recovered in a few days (Shaw & Muth 1949). In another case, newborn lambs from ewes fed peavine silage developed similar signs (Whiting et al. 1957). No adverse effects were noted in the ewes. The problem of incoordination and/or paralysis appeared in lambs for several lambing seasons. The signs appeared 2–3 days after birth. The lambs had seizures that terminated in collapse, and after resting for a few minutes, they were able to rise and appeared normal until another episode occurred. In most cases, the lambs gradually recovered over the course of a year. The disease was experimentally reproduced in lambs of several breeds and with silage from several farms. The toxin was apparently transmitted in the milk, because lambs given cow milk from birth did not develop the disease. The disease also has been seen in older lambs from ewes fed for several days on peavine silage that was black rather than dark green ("Research reports," 1963).

The disease problem also has appeared in pigs. Hogs pastured on *P. sativum* for 5–6 weeks developed a neuromuscular problem in which locomotion became very difficult (Smith 1943). The animals walked in a crouching posture and fell when they attempted to run.

The foliage of *P. sativum* is seldom dried and fed as hay. There is one report that it caused intestinal impactions in horses (Kermen 1939).

■ toxin unknown, mycotoxin?

Disease Genesis—Neurointoxication due to *P. sativum* has not been a significant or continuing toxicologic problem for producers. A specific toxicant and a physiological basis for the effects have not been identified. Appearance of the disease when animals consume black rather than green silage may be an indication of toxins formed during the ensilage process or from fungi.

Willardiine and isowillardiine have been isolated from the seeds, and even though willardiine is a potent non-NMDA glutamate receptor agonist, these toxins appear to be unlikely causes of problems with the foliage, because of their low concentrations (Lambein & Van Parijs 1968; Evans et al. 1980; Hawkins et al. 1995). β-Cyanoalanine, a neurotoxin present in *Vicia*, also occurs in *P. sativum*, but at low concentrations in the bound form (Ressler et al. 1969). Peas have been fed raw to rats, which exhibited reasonable growth especially when their diet was supplemented with methionine (Hove et al. 1978). In contrast, peas fed to mice impaired growth and caused immunologic disturbances such as increased titers of gamma-globulins and immunoglobulin G and increased T lymphocytes (Martinez et al. 1995).

■ ataxia, paresis; generally reversible; treatment not needed

Clinical Signs, Pathology, and Treatment—On the rare occasions of intoxication in nursing lambs, the clinical signs included incoordination, paresis, and perhaps paralysis. The signs were readily reversible in most individuals when the silage no longer was fed to the ewes. Degeneration of the Purkinje cells in the cerebellar cortex was the principal lesion observed. There also was degeneration of other cerebellar and spinal neurons but no demyelination. Diet supplementation with copper, cobalt, and vitamins A, D, or E was not effective in preventing the disease (Whiting et al. 1957).

Prosopis L.

■ *P. glandulosa*
algarroba
mesquite
mezquite

Taxonomy and Morphology—A member of the subfamily Mimosoideae, *Prosopis* comprises 44 species in the drier, warmer regions of the world, where they may dominate the vegetation. Their wood is used for timber, flooring, and firewood. Producing abundant nectar, the flowers are readily visited by bees, which produce a sweet honey. The legumes are fed to livestock in some regions. Although considered a noxious weed to be eradicated in the southwestern United States, trees are planted for shade and as ornamentals elsewhere. Of the 8 species occurring in North America, only 1 has been associated with toxicologic problems:

P. glandulosa Torr.	honey mesquite, mesquite, mezquite, Texas mesquite, algarroba

In many older publications, the binomials *P. juliflora* and *P. chilensis* have been used erroneously for honey mesquite. These two species occur in Central and South America, not in North America.

Plants perennials; irregularly shaped trees or large shrubs; foliage dull, yellow-green. **Stems** erect or ascending; bark dark brown, rough, furrowed; branches stiff, grayish brown, zigzag, with short spurs, armed with 1 or 2 stout spines at nodes, terminal buds absent. **Leaves** alternate, fascicled at ends of spurs; 2-pinnately compound, dichotomously forked; leaflets 12–36, blades linear to oblong, apices obtuse or mucronate, margins entire; stipules modified into spines. **Inflorescences** elongate racemes or spikes; borne on spurs; typically pendulous. **Flowers** fragrant; numerous; small; perfect; radially symmetrical. **Sepals** 5; fused. **Petals** 5; free; creamy white to pale yellow. **Stamens** 10. **Legumes** cylindrical; constricted slightly between seeds; 7–25 cm long; yellow-tan; indehiscent. **Seeds** 5–18; hard; brown; shiny.

Other species of the genus are distinguished on the basis of leaflet number and appearance of the inflorescence and legume.

■ shrubs or trees; stout spines at nodes; leaves alternate, 2-pinnately compound; flowers fragrant; petals creamy white to pale yellow; legumes long, cylindrical, constricted between seeds

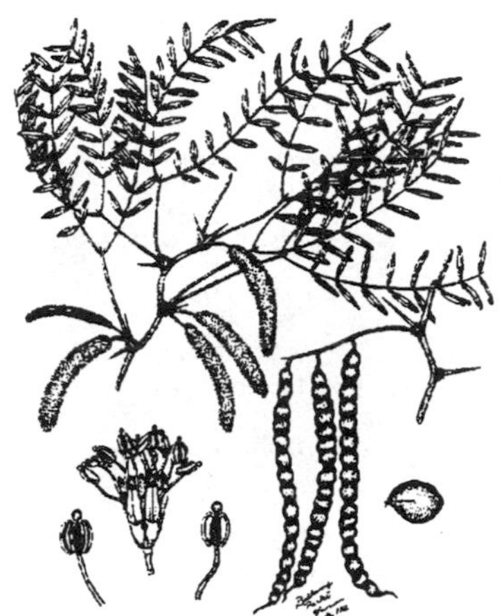

Prosopis glandulosa

Distribution and Habitat—Species of *Prosopis* occur in southwest Asia, Africa, and the Americas, where diversity is greatest. Plants are deep-rooted, drought and heat tolerant, and typically found in arid and semiarid areas. Habitats occupied include washes, rocky ridges, and grasslands. Drought, suppression of fire, and overgrazing of prairies, combined with the tendency of the legumes and seeds to pass undigested in the feces of grazing animals, especially horses and cattle, resulted in a great increase in the abundance of *P. glandulosa* in the late 1800s and early 1900s (Fisher 1977; Jacoby & Ansley 1991). It is distributed in the southwestern portion of the United States and in adjacent Mexico. It occurs as far north as southern Kansas, Colorado, and Utah and as far east as eastern Oklahoma and Texas. Isolated populations are the result of seed dispersal by cattle. Some ecologists now contend that, contrary to popular belief, its geographic range is much the same as it was before the advent of Europeans and that it has increased only in abundance within its range.

■ arid and semiarid areas

Prosopis glandulosa

Disease Problems—The *Prosopis* species are viewed as either noxious pests or as food and ecosystem resources. Closely linked with various American Indian cultures of the Southwest, these trees and shrubs were used for fuel, shelter, weapons, tools, fiber, and medicine (Felker 1979). The wood is very hard and useful for tools. The species are prolific pod producers, and many of these peoples were dependent upon the pods for subsistence in much the same way that coastal peoples depended upon acorns. As a staple of life, the pods were used for making beverages (fermented or otherwise), eaten directly, and roasted for grinding into a meal called pinole (Felker 1979). The pods, often sweet and aromatic, were gathered at various stages of ripeness. The pods of the various mesquites have 9–13% protein and 13–36% sucrose. The seeds are of even higher protein concentration. Thus it appears that members of this genus have positive nutritional qualities (Walton 1923). Because of their composition, the pods are considered to be constipating and, when used for livestock, should be fed in combination with some other forage such as alfalfa (Felker 1979). Unfortunately, when livestock are forced to subsist on *Prosopis* for several months or more they display a wasting, malnutrition-like syndrome with muscle atrophy, weight loss, and anemia. There is a marked decrease in blood glucose (down to 20 mg% or less) and liver glycogen and an increase in blood ketones such as β-hydroxybutyric acid (Hendershot 1946; Adler 1949). Goats as browsers are especially susceptible, and in some areas of Mexico up to 2% of flocks may be affected, with nearly a 100% case-fatality rate

■ pods generally accepted as food for humans and livestock; when animals subsist on plants for several months, wasting malnutrition syndrome may develop, generally reversible if corrected early; goats at increased risk of serious disease

(Gomez 1977). Goats may survive for 2–3 weeks after onset of signs. The disease is reversible in cattle and sheep, but there may be residual effects (Dollahite 1964).

■ toxin unknown; possibly tyramines

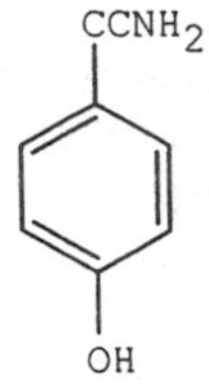

CCNH$_2$

OH

Tyramine

CCNCH$_3$

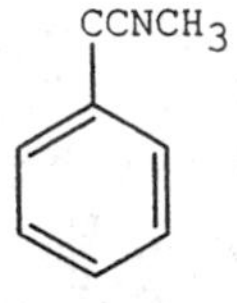

N-Methyl phenethylamine

Disease Genesis—The toxicant responsible for the deleterious effects is unknown. The high concentrations of sucrose are suggested as a cause of decreased cellulose digestion, leading to chronic indigestion (Adler 1949). It has also been suggested that the disease might be a form of starvation from ruminal impaction and stasis due to the indigestible fibrous pods (Cheeke & Shull 1985). Physiologically active tyramines, similar to those of the *Acacia,* are present in some species, including *P. glandulosa* and *P. velutina* (Camp & Norvell 1966; Cates & Rhoads 1977). Concentrations are highest in young leaves (up to 1.5% d.w.) and decline gradually with increasing maturity. As with the *Acacia,* the role of these amines as toxicants is not clear. Pipecolic acid is present in the leaves of *P. velutina* (Cates & Rhoads 1977). Cyanogenic glycosides have been isolated from the foliage of *P. glandulosa* (Seigler et al. 1989).

■ gradual onset; weight loss, hair loss, anemia, anorexia; awkward jaw and tongue movements, continuous chewing, profuse salivation

Clinical Signs—There is gradual onset of weight loss, hair loss, anemia, and loss of appetite. Animals become nervous and have difficulty coordinating jaw and tongue movements, with the tongue often protruding several centimeters from the mouth. They chew continuously and may regurgitate beans, giving the impression of indigestion. The rumen is inactive, and the saliva is profuse and stringy. Animals may recover but are subject to relapse under stress. In severe cases the muscular damage is of such severity that recovery is difficult and the case-fatality rate may be high.

■ gross pathology, wasting of muscles, rumen may be impacted

Pathology—The gross pathologic changes are not distinctive; they are those of a wasting, starvation type disease in spite of a continued appetite in some instances. There is muscle wasting, particularly of the masseters. The rumen may be impacted with pods and seeds, especially the fibrous bean and seed coat, and the omasum and abomasum are nearly empty.

■ good forage, inoculate rumen with ingesta from healthy animals; parenteral glucose and electrolytes

Treatment—Parenteral glucose solutions, inoculation with ruminal microbial contents from another animal, and good carbohydrate rations may be effective in reversing the malnutrition-like signs. However, in some instances, especially with goats, treatment may only delay death. Vitamin A does not seem to be of any additional value.

Robinia L.

■ *R. pseudoacacia*
black locust
false acacia
locust tree
post locust
yellow locust

Taxonomy and Morphology—Comprising some 10 species and hybrids, *Robinia,* a member of the subfamily Papilionoideae, is a North American genus of deciduous trees and shrubs. The pattern of variation within the genus is complex because of hybridization, and estimates of the actual number of species range from 4 to 20. Four species are widely distributed and commonly encountered. Although other species may be risks, only 1 has clearly been associated with toxicologic problems:

R. pseudoacacia L. — black locust, yellow locust, post locust, locust tree, false acacia

■ trees; prominent pairs of spines at nodes; leaves 1-pinnately compound, leaflets 7–19; flowers borne in pendulous racemes, large, pealike; petals white; legumes papery

Plants perennials; medium-sized trees; often forming groves via root sprouts; crowns irregular, spreading. **Stems** erect; bark gray, deeply furrowed; branches stout, angled; zigzag, puberulent, becoming glabrous, bearing pairs of spines at nodes; terminal buds absent; lateral buds without bud scales. **Leaves** 1-pinnately compound; leaflets 7–19;

terminal leaflet present; blades elliptic or oval; lower surfaces puberulent, becoming glabrous; apices rounded; margins entire; stipules modified into stout spines. **Inflorescences** pendulous racemes borne on short peduncles in axils of leaves. **Flowers** large. **Sepals** 5; fused; calyces slightly 2-lipped. **Corollas** papilionaceous. **Petals** 5; clawed; banners obcordate or suborbicular, reflexed; wings oblong to obovate, free; keels curved; white. **Stamens** 10; diadelphous; enclosed in keels. **Legumes** linear; straight or slightly curved; flattened; stipitate; brown; pericarp somewhat papery; persisting in early winter. **Seeds** 3–10; reniform; flattened; brown.

The specific epithet is generally hyphenated as *pseudo-acacia* in older publications. In his 1753 *Species Plantarum,* Linnaeus published the epithet as hyphenated. In his 1763 edition, however, the hyphen was absent and is believed to reflect his editorial preference. Modern rules of botanical nomenclature require that the hyphen not be used.

Robinia pseudoacacia

■ widely distributed because of extensive planting

Distribution and Habitat—*Robinia pseudoacacia* was distributed originally in the east-central part of the continent. Its range is now across the continent because it has been planted extensively in windbreaks, shelterbelts, and woodlots and as an ornamental. It has escaped and naturalized in some areas. Adapted to a variety of clay, loam, and sandy soils and rapid-growing, it may form dense groves of spindly trees by root sprouting. Somewhat weedy, it can dominate the landscape in some areas.

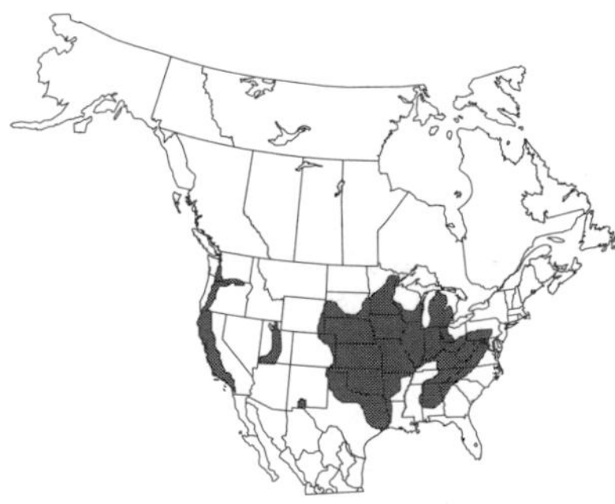

Robinia pseudoacacia

■ mainly from eating bark; leaves and seeds of low risk; digestive disturbance; rarely cardiac effects; horses seem to be highest risk, but all animal species affected

Disease Problems—Appreciation of *R. pseudoacacia* as a toxicologic problem stems from early observations of toxicity in horses when they were tied to locust trees for prolonged periods. Boredom as well as hunger probably caused them to strip and chew the bark of the trunk and branches within reach. The animals developed severe weakness, weak pulse, blue to rusty yellow-colored congestion of the mucous membranes (sclera of the eye), and pupillary dilation (Waldron & Tecumseh 1908). Deaths appeared to be due to cardiac failure. The risk posed by *R. pseudoacacia* was confirmed experimentally by the toxicity of i.v. infusions of bark extracts, which caused diarrhea, labored respiration, and paralysis of the pelvic limbs (Tasaki & Tanaka 1918).

Although most cases of intoxication involve horses eating the bark, all animal species are probably susceptible (Waldron & Tecumseh 1908; Cary et al. 1924; Hansen 1924, 1928a; Landolt et al. 1997). For example, in an experimental study, 2 oz of a bark decoction killed a cat in a few minutes. Humans also are at risk. Thirty-two boys in an orphanage suffered severe illness after chewing the inner bark of black locust from tree posts (Emery 1887). The signs in this episode ranged from vomiting to a feeble and irregular heart rate. In another case, children eating only threads of bark exhibited abdominal pain, nausea, weakness, and bile-stained vomitus (Mejia et al. 1991). As indicated by the signs given above, the toxic effects of *Robinia* are directed toward the digestive system and the heart. Laminitis is an additional complication in horses (Keller & Dewitz 1969).

In addition to the bark, the leaves or root sprouts, which are readily eaten, may also be toxic, but not consistently so. They are sometimes eaten with apparent impunity, especially when mature (Chesnut & Wilcox 1901; Barnes 1921; Hansen 1928a). Rare episodes of intoxication have occurred in cattle, sheep, and chickens following ingestion of the leaves or pods (Barnes 1921; Cary et al. 1924; Hansen 1924, 1928a). Consumption of leaves by chickens resulted in diarrhea, a heavy thumping respiration, and paralysis (Barnes 1921). Interestingly, budgerigars given leaves of *R. pseudoacacia* were intoxicated (vomiting, labored respiration) only by fall growth and not by spring growth (Shropshire et al. 1992). However, in spite of these episodes, the leaves are generally considered to pose a low toxicity risk. Some authors have even suggested that the leaves are potentially useful forage for livestock (Wood et al. 1926). The leaves are of interest because of their high nitrogen content due to the species' ability to fix nitrogen symbiotically. Meal of dried and pelleted locust leaves with 20–24% crude protein is used as feed in China. Unfortunately, research indicates that feeding meal composed of 20% locust leaves causes growth retardation in chickens and rabbits and decreased protein digestibility in lambs (Cheeke & Shull 1985).

The seeds also may pose a risk. Two children, 7 and 9 years old, had multiple vomiting episodes after eating 2 and 30 seeds, respectively (Filandrinos & Sioris 1992). The 7-year-old exhibited a slight increase in LDH and AP. Both children responded well to conservative treatment with charcoal, fluids, and sodium bicarbonate. Flowers, which have been thought by some to be nontoxic, are reported to be slightly narcotic (Pammel 1911; Wood et al. 1926). However, bark is the part of *Robinia* that produces the most consistent toxic effects.

- specific toxin unknown; glycoprotein lectins and triterpene glycosides in bark may cause irritant effects on digestive tract; other compounds, such as proteins, glucosides, and a camphorol, robinin

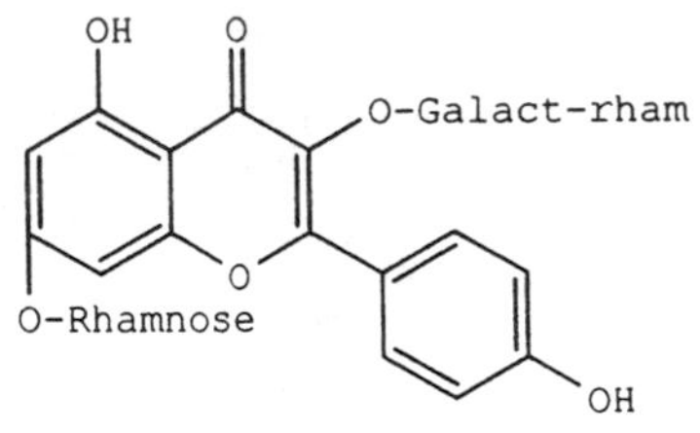

Robinin

Disease Genesis—The specific toxic principle of *R. pseudoacacia* has not been definitely identified, although a variety of toxins have been reported. The nomenclature of these compounds is confusing. Power (1913) isolated a protein from bark, which he termed robin, that caused severe vomiting and purgation in dogs. Later a nonalkaloid/nonprotein glucoside called robitin was isolated from the inner bark. It caused labored respiration, diarrhea, and paralysis of the pelvic limbs in horses, rabbits, and cattle (Tasaki & Tanaka 1918). Cary and co-workers (1924) produced the following signs in horses by giving an aqueous suspension of 1,250 g of bark: diaphragm spasms, diarrhea, weakness, nervous excitement, and irregular and violent heart action. A series of triterpene glycosides, termed robinisides, also has been isolated from bark (Cui et al. 1992; 1993). Robinin, a glycoside of camphorol, is reported from flowers (Pammel 1911; Wood et al. 1926).

This confusion in names and chemistry is further complicated by the isolation of different glycoprotein lectins from the bark and seeds of *Robinia.* The lectin from bark appears to be a tetramer composed of two subunits of 29,000–31,000 daltons (Markowitz 1971; Horejsi et al. 1978; McPherson & Hoover 1979; Tazaki & Yoshida 1992). It lacks disulfide bridges and has galactose activity. Although it does not cross-react immunologically with lectins from the leaves and seeds, it does with those from the bark of *Sophora japonica,* also a member of the subfamily Papilionoideae. Lectin concentrations in bark vary markedly during the year. They are low in June and increase markedly, by about 50-fold or more, to reach maximal values by November (Nsimba-Lubaki & Peumans 1986). The glycoprotein hemagglutinin extracted from seeds has many sugars but no sulfur-containing amino acids (Bourrillon & Font 1965). The relationship of these lectins to the toxicity of *Robinia* is not clear. Further complicating identification of the principal toxicant is the observation that the seeds of most species of *Robinia* contain canavanine (Lavin 1986), although this is unlikely to be of toxicologic significance.

There is some question of the toxicity of other species of *Robinia* because feeding tests with leaves and flowers in chicks and mice failed to detect a hazard for *R. boyntonii* Ashe and *R. pedunculata* Ashe at doses up to 2.7% b.w. (Duncan et al. 1955).

- rapid onset, colic, diarrhea, weakness; if severe, labored respiration, weak pulse, arrhythmias
- in humans, retching, vomiting

Clinical Signs—The signs produced by *R. pseudoacacia* are primarily indicative of cardiac and gastrointestinal dysfunction. Signs most commonly seen include colic, diarrhea, weakness, mydriasis, cold extremities, severe congestion of the oral mucous membranes, weak pulse, and arrhythmias. In some cases, dyspnea, accentuated heartbeat, and constipation may be present. Cattle may become quite belligerent. The disease is often not fatal, but recovery may require several weeks. In horses, recovery may be complicated by laminitis.

The signs described in children who ate the bark ranged from nausea, weakness, vomiting, flushed face, dry throat, and pupillary dilation to more serious retching, vomiting (bile stained or not), epigastric pain, feeble and irregular heartbeat, and stupor (Emery 1887; Mejia et al. 1991).

Pathology and Treatment—Necropsy lesions are usually restricted to reddening, edema, and perhaps small hemorrhages of the mucosa of the stomach and small intestine, although degenerative lesions may be found in other organs. Treatment is symptomatic, directed toward alleviating the pain and inflammation accompanying the digestive tract irritation. When given early, activated charcoal may be beneficial.

- gross pathology, reddening and edema of mucosa of stomach and small intestine

  treatment, activated charcoal, antidiarrheal drugs

Senna Mill.

Taxonomy and Morphology—Comprising approximately 240 species in both the tropical and the warm-temperate regions of the world, *Senna,* a member of the subfamily Caesalpinioideae, was long interpreted by taxonomists to be a subgenus of *Cassia*. Systematic studies have resulted in its recognition once again as a distinct genus (Irwin & Barneby 1982). Reports of intoxication problems have employed both generic names. The name *Ditremexia* also has been used for some species. In North America, approximately 45 native and introduced species are present. Commonly known as the sennas, they can be divided into two groups on the basis of their toxicity. Species of the first group pose little risk and include the following:

S. alata (L.) Roxb. (= *Cassia alata* L.)	candle bush, seven golden candles, ringworm cassia
S. bicapsularis (L.) Roxb. (= *Cassia bicapsularis* L.)	Christmas senna
S. marilandica (L.) Link (= *Cassia marilandica* L.) (= *Ditremexia marilandica* [L.] Britton & Rose)	American senna, locust plant
S. siamea (Lam.) H.S.Irwin & Barneby (= *Cassia siamea* L.)	Siamese shower, kassod tree

More-serious intoxication problems are associated with the second group, which includes the following (see plate 37.7):

S. lindheimeriana (Scheele) H.S.Irwin & Barneby (= *Cassia lindheimeriana* Scheele)	Lindheimer's senna
S. obtusifolia (L.) H.S.Irwin & Barneby (= *Cassia tora* auct. non L.) (= *Cassia obtusifolia* L.)	sicklepod, coffee-pod
S. occidentalis (L.) Link (= *Cassia occidentalis* L.)	coffeeweed, coffee senna, wild coffee, styptic weed
S. roemeriana (Scheele) H.S.Irwin & Barneby (= *Cassia roemeriana* Scheele)	two-leaved senna, twin-leaved senna

- *S. alata*
  S. bicapsularis
  S. lindheimeriana
  S. marilandica
  S. obtusifolia
  S. occidentalis
  S. roemeriana
  S. siamea

  candle bush
  cassia
  coffee-pod
  coffee senna
  coffeeweed
  golden candles
  Kassod tree
  locust plant
  senna
  Siamese shower
  sicklepod
  styptic weed
  wild coffee

- erect herbs; leaves 1-pinnately compound, leaflets 1-many, ill-scented; flowers in axillary racemes; petals 5, yellow or white
- distinctive curved legumes of *S. obtusifolia, S. occidentalis*

Senna obtusifolia

Senna occidentalis

- waste-area weedy species, often sandy soils

Plants annuals or perennials; all herbs in North America. **Stems** erect. **Leaves** 1-pinnately compound; pairs of leaflets 1-many; terminal leaflet absent; leaflets of various shapes; stipules caducous; petioles and rachises with glands. **Inflorescences** axillary racemes; bracts and bracteoles absent. **Flowers** showy; symmetry slightly bilateral. **Sepals** 5; free; overlapping; unequal in size; brown or pink; villous. **Petals** 5; yellow or white. **Stamens** 9 or 10; upper 3 sterile or greatly reduced; anthers basifixed, opening by an apical pore or a short slit, with soft hairs. **Legumes** cylindrical or flat or 4-sided; glabrous or indumented; divided or not divided into compartments. **Seeds** numerous.

Recognition of the species causing serious intoxication problems is based on leaflet shape and size, appearance of the petiolar gland, and features of the stamens and legumes. Characteristically, the foliage of most species is malodorous.

1. Plants annuals.
   2. Gland at bases of petioles. Leaflets ovate to elliptic. Pairs of leaflets 3–6. Legumes flattened; straight or curved . . . *S. occidentalis*
   2. Gland between 2 lowest leaflets. Leaflets obovate. Pairs of leaflets 2 or 3. Legumes tetragonal; arcuate . . . . *S. obtusifolia*
1. Plants perennials.
   3. Pairs of leaflets 1. Plants flowering in spring . . . . *S. roemeriana*
   3. Pairs of leaflets 4–8. Plants flowering in fall . . . . *S. lindheimeriana*

Distribution and Habitat—*Senna occidentalis,* which has worldwide distribution, occurs throughout southeastern North America. It may be found along roadsides, in waste areas, and in old pastures, especially in partially shady sites. *Senna obtusifolia* also is found in the Southeast. Weedy, it prefers sandy soils and often becomes a serious pest of cultivated crops. These two taxa are the most widespread species. In contrast, *S. roemeriana* and *S. lindheimeriana* are more restricted in their distribution.

Disease Problems—Most of the sennas found in North America are capable of producing laxative effects if mature leaves or legumes are eaten in sufficient quantity. However, the problem is generally self-limiting and transient. This attribute has commercial application. Proprietary laxative medicaments are prepared from extracts of the tropical sennas, with *S. acutifolia* and *S. angustifolia* being most important, although *S. alata* is apparently of equal potency. The native *S. occidentalis* is reported to have numerous medicinal uses, and its seeds have been used as a coffee substitute after roasting (Irwin & Barneby 1982). Indeed, the sennas contain appreciable crude protein, 16.5% in *S. occidentalis* and 21.6% in *S. obtusifolia,* and probably have some value when eaten in small amounts (Hoveland et al. 1986).

Disease problems of digestive disturbance and skeletal and cardiomyopathy associated with all four species of the coffee and two-leaved sennas are sporadic and typically limited to only a few animals at a time (Dollahite & Henson 1965; Henson & Dollahite 1966). Rarely are large numbers affected; such instances generally occur when plants of *Senna* are cut and inadvertently fed with green chopped forage (Nicholson et al. 1977; McCormack & Neisler 1980). Most intoxications are caused by the two eastern species, *S. obtusifolia* and *S. occidentalis,* when cattle eat wilted plants after the first frosts in the fall. In contrast, the western *S. roemeriana* is a problem when it flourishes in the spring. Although *S. lindheimeriana* is one of the commonest sennas of central and west Texas, the season when it is most troublesome has not been reported, but it is likely to be a problem in the fall, when it flowers (Boughton & Hardy 1939b).

Under most conditions, the plants are not eaten by livestock. The conditions that predispose plant consumption are not known. Starvation does not seem to be an initiating factor, because animals in good nutritional condition and grazing good pasture may become intoxicated (Henson et al. 1965). Seeds are more toxic than foliage, and contamination of feeds with as little as 1–2% seeds of some *Senna* species can pose a serious problem in pigs and chickens (Simpson et al. 1971; Charles & Muller 1975; Page et al. 1977; Suliman et al. 1982a; Colvin et al. 1986; Flory et al. 1992). For example, grain sorghum contaminated with 3.7% *S. occidentalis* and 1.6% *S. obtusifolia* seeds by weight is suspected of causing death in pigs. Experimentally, the same feed given for 16 days to young chicks caused severe weight loss, debilitation, and skeletal and cardiac muscle degeneration (Flory et al. 1992). Similarly, 10–20% feed contamination caused the death of pigs 7–8 days after the start of an experiment intended to reproduce conditions in which 420 pigs died out of a group of 1,200 in one episode and 38 out of 800 in another (Martins et al. 1986). However, such serious effects are not always produced. Feed containing 5% ground seeds of *S. obtusifolia* failed to produce adverse effects in 5-week-old chicks (Bierer et al. 1960). In other studies, rats fed seeds of *S. obtusifolia* for 13 weeks developed slight hematologic abnormalities at concentrations greater than 0.15% by weight and exhibited diarrhea at 5% levels (Voss & Brennecke 1991). The "no-observed effect" level was less than 0.15%. In horses, as little a dose as 0.15% b.w. of the seeds of *C. occidentalis* or lesser amounts given daily may be lethal (Irigoyen et al. 1991). Similarly 0.2% b.w. was lethal to ponies (Martin et al. 1981).

Cattle seem to be less susceptible but may nevertheless show severe effects with feeds containing approximately 10% *S. occidentalis* seeds, 0.25% b.w. (Rogers et al. 1979; Barros et al. 1990). Feed contaminated with 21.5% *S. occidentalis* seeds by weight, fed to cattle, caused progressive muscle weakness, incoordination, paresis of the pelvic limbs, reluctance to move, and eventual sternal and then lateral recumbency within 10 days (Barth et al. 1994). There was no diarrhea or myoglobinuria, but myofiber degeneration and necrosis was found throughout the muscles of the limbs and back. Toxicity of *S. obtusifolia* seeds for calves is reported to be lower than that for *S. occidentalis,* i.e., requiring about 40% contamination of the feed; however, the difference may be due to other factors, because direct comparisons between the two species were not made (Putnam et al. 1988). Similar severe effects on skeletal and cardiac muscle as well as liver necrosis with or without diarrhea also occur in cattle and goats eating the foliage of the sennas (Mercer et al. 1967; O'Hara et al. 1969; Schmitz & Denton 1977; Rogers et al. 1979; Rowe et al. 1987).

Intoxication of goats by the leaves *S. occidentalis* required several months at a daily dosage of 5 g/kg b.w., but the same amount of seeds was toxic in 9 days (Suliman et al. 1982a). The hazard posed by seeds of *S. occidentalis* appears to be only slightly diminished by heating (Suliman & Shommein 1986).

As has been shown for *Abrus* and *Senna,* toxicologic interactions between plants associated with digestive disturbances are a potential problem, especially with seeds that may contaminate feed (Omer et al. 1992).

Disease Genesis—The species of *Senna* contain a series of anthraquinones that exert toxic effects on numerous tissues, including those of the digestive tract and skeletal muscle. The sennas of commerce, *C. acutifolia* and *C. angustifolia,* contain bianthrone (dimers) anthraquinones of emodin and rhein. There are at least four bianthrones: sennosides A and B composed of rhein diglucoside isomers, and sennosides C and D composed of aloe-emodin/rhein diglucoside isomers. These compounds are most abundant in the seeds (Thomson 1971; Franz 1993). The anthraquinones of *S. obtusifolia*

Senna roemeriana

- some species, sources of commercial laxatives; ingestion of seeds or foliage by livestock, sporadic causes of digestive disturbance

  if sufficient plant is eaten, skeletal and perhaps cardiac muscle degeneration; may occur in a few days

Senna lindheimeriana

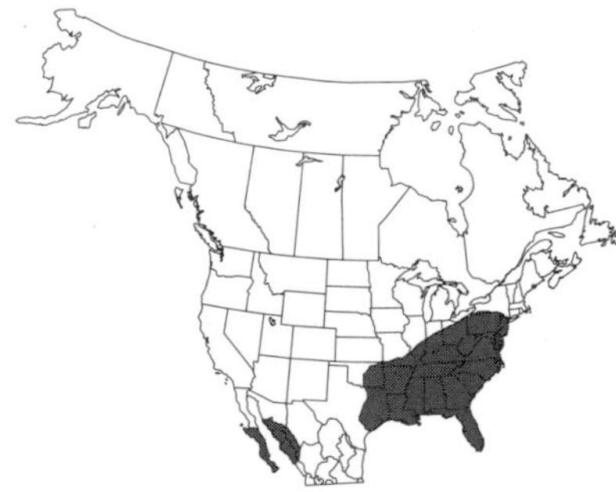

Senna obtusifolia

- digestive disturbance; anthraquinones; obtusin, obtusifolin, questin, and bianthrone sennosides; toxins causing muscle effects unknown, perhaps anthraquinones

Senna occidentalis

Senna roemeriana

include chrysophanic acid, emodin, obtusifolin, obtusin, auriantoobtusin, chrysoobtusin, physicion, and questin. In most instances, they probably occur as glucosidic dimers (Takido 1958, 1960; Lewis & Shibamoto 1989; Crawford et al. 1990).

Anaerobic gut bacteria seem to play an important role in the activity of these compounds. In humans, in a reductive process in the intestine, individual gluco-anthrones are formed from the glucosidic bianthrones. This is followed by hydrolysis of the glucosidic bond and oxidation to bianthrone genins or aglycones (Kobashi et al. 1980; Robinson 1991). The purgative potency of these three types of toxicants is bianthrone > anthrone > anthraquinone.

The myotoxins of *Senna* are water soluble, but they have not been completely identified (Graziano et al. 1983; Hebert et al. 1983). Some of the anthraquinones, such as emodin, inhibit electron transport in muscle mitochondria (Graziano et al. 1983; Lewis & Shibamoto 1989), possibly through oxygen radical formation catalyzed by the semiquinone in a manner akin to that proposed for adriamycin (Davies et al. 1983; Nohl & Jordan 1983). This is consistent with the observations of O'Hara and Pierce (1974a,b) of muscle mitochondrial damage and possible uncoupling of oxidative phosphorylation in experimental intoxication with *S. occidentalis.* Similarly, chronic intoxication in chickens due to *S. occidentalis* results in degenerative myopathy (Calore et al. 1997). In 1-day-old chicks given a diet of 4% *S. occidentalis* seeds for 11 days, many mitochondria were enlarged, with disrupted cristae, and contained lamellar structures (Cavaliere et al. 1997). The degenerative process was considered to be secondary to energy deficits. In addition to mitochondrial myopathy, extensive axonal damage and secondary myelin degeneration of peripheral nerves also have been shown in chicks fed seeds of *S. occidentalis* (Calore et al. 1998).

Obtusin

Obtusifolin

Questin

Sennoside A

Alternative reactants may be produced by hepatic hydroxylation of emodin at various ring sites to yield highly reactive mutagens, e.g., 2-hydroxy-emodin (Tanaka et al. 1987). Extracts of *S. obtusifolia* are highly mutagenic, suggestive of carcinogenic potential (Friedman & Henika 1991).

Clinical Signs—The disease produced by sennas is characterized by early digestive dysfunction and later by progressive muscular weakness. By the second day of plant consumption, mild to moderate diarrhea may seen. Although this may be accompanied by tenesmus and evidence of abdominal pain, it often goes unnoticed unless animals are being observed regularly and carefully. In many cases in cattle, the disease is not detected until weakness and/or recumbency are evident. Appetite is decreased and depression develops, and by the fourth or fifth day, the animals are weak and reluctant to move. At this time, they exhibit muscle fasciculations, stumble as they walk, and are unable to support themselves when turned sharply. They become recumbent and may not be able to stand even with assistance. Because they are bright and alert and continue to eat somewhat and drink, they may be considered to be "downer cows" due to metabolic dysfunction. They may die within a few hours or live for several days. Cardiac and respiratory distress and myoglobinuria also may be apparent.

■ day after eating plant, diarrhea, straining; if keep eating, anorexia, depression, muscle weakness, recumbent; signs of cardiac insufficiency may occur, dark-colored urine due to myoglobin

Death seems almost inevitable when the disease progresses to the stage of recumbency. In some cases, sufficient quantities of plant may be consumed to produce death more rapidly, and occasionally animals may be found dead without premonitory signs. Alternatively, ingestion of smaller amounts of plant may result in a delay for a week or more of the onset of milder signs. During the period of muscle involvement, marked elevations of serum AST and CK are common.

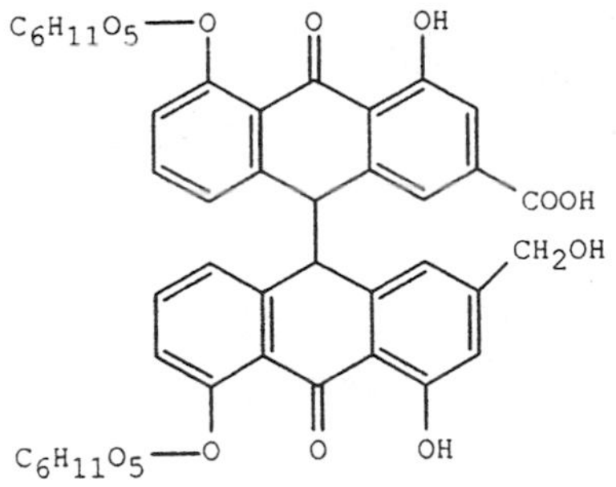

$C_6H_{11}O_5$—O O OH COOH CH_2OH $C_6H_{11}O_5$—O O OH

Sennoside C

Pathology—Grossly, there will be reddening of the mucosa of the stomach and small intestines. The skeletal muscles will be pale, especially those of the pelvic limbs. The presence of pale streaks and transversely banded stippling of various muscles in the pelvic limbs intermixed with normal-appearing muscles is common. Cardiac muscle may be flabby and have pale streaks in addition to hemorrhages on the heart surfaces and excess fluid in the pericardial sac. Pulmonary congestion and edema with excess pleural fluid are also common findings. There is often passive congestion of the liver. Microscopically, in muscle, there is swelling and vacuolation of the sarcoplasm with degeneration and rupture of sarcolemmal membranes; degenerative and apparently normal myofibers are intermixed. The extent of myocardial involvement varies, but it may be similar to that described for skeletal muscle. Mild centrilobular hepatocytic vacuolation with necrosis, mild renal tubular epithelial degeneration, and pulmonary congestion and edema also are seen frequently. Hepatic and renal lesions are probably secondary to myocardial insufficiency.

- gross pathology, reddening of gut mucosa; skeletal muscle, pale streaks, stippling; cardiac muscle, flabby, pale streaks; mild pulmonary edema

  microscopic, mild renal and hepatic lesions, muscle lesions

Treatment—Treatment is primarily symptomatic to alleviate the cathartic effects. Animals removed from the plants at the earliest indications of problems may not develop the subsequent and more serious muscular effects. Although the muscular degeneration is similar to that seen with white muscle disease, vitamin E/selenium therapy is contraindicated because in experimental cases it resulted in increased muscular damage and higher death rates (O'Hara et al. 1970).

- general nursing care

Sesbania Scop.

Taxonomy and Morphology—Present in both the Old and New World, *Sesbania,* a member of the subfamily Papilionoideae, comprises about 50 species. In North America, 10 native and introduced species are present. The taxonomic boundaries of the genus have been considerably revised as taxonomists have examined its relationship to other genera; for example, it has been combined with other genera such as *Glottidium,* and several of its species have been placed in the genera *Daubentonia* and *Agati.* Several species have well-recognized toxicity potential:

S. drummondii (Rydb.) Cory	poison bean, coffee-bean, rattlebush, false poinciana, Drummond rattlebush, rattlebox
S. punicea (Cav.) Benth. (= *Daubentonia punicea* [Cav.] DC.)	purple sesban, purple rattlebush, rattlebush, purple rattlebox, daubentonia, false poinciana

Other species are of little potential for intoxication:

S. emerus (Aubl.) Urb.	coffeeweed
S. exaltata (Raf.) Rydb. ex A.W.Hill (= *S. macrocarpa* Muhl. ex Raf.)	coffeeweed, bequilla, sesban

- *S. drummondii*
  S. emerus
  S. exaltata
  S. grandiflora
  S. punicea
  S. sericea
  S. sesban

  bequilla
  coffee-bean
  coffeeweed
  corkwood tree
  daubentonia
  false poinciana
  parrot flower
  poison bean
  rattlebox
  rattlebush
  sesban

- herbs, subshrubs, shrubs, trees; leaves 1-pinnately compound, terminal leaflet absent; flowers pealike; petals yellow, dull scarlet, purple, white, or multicolored; legumes sometimes 4-winged or 4-angled

S. grandiflora (L.) Poir. (= *Agati grandiflora* [L.] Desv.)	corkwood tree, parrot flower; scarlet wistaria tree, vegetable hummingbird
S. sericea (Willd.) Link	
S. sesban (L.) Merr.	sesban

The binomial *S. vesicaria* also appears in the literature in reports on the toxicity of bagpod or bladderpod. This species, however, is now classified as *Glottidium vesicarium*. It is toxicologically similar to the species of *Sesbania* and is described in the *Glottidium* treatment in this chapter.

Sesbania punicea

Plants annuals or perennials; herbs or subshrubs or shrubs or small trees to 10 m tall; evergreen or deciduous. **Leaves** 1-pinnately compound; leaflets 20–70; terminal leaflet absent; blades oblong; margins entire; bases typically asymmetrical; stipules subulate, caducous or persistent. **Inflorescences** short, axillary racemes; bracts and bracteoles caducous. **Sepals** 5; fused. **Corollas** papilionaceous. **Petals** 5; yellow or dull scarlet or purple or multicolored or white; long-clawed; banners orbicular or ovate, spreading or reflexed, auriculate; wings oblong or linear, falcate, auriculate; keels curved, rounded, apex obtuse or occasionally acuminate. **Stamens** 10; diadelphous; enclosed in keels. **Legumes** linear or rarely oblong; cylindrical or compressed; sometimes 4-angled or 4-winged; transversely septate between seeds; often persisting on plant in winter. **Seeds** 5 to numerous; oblong or subquadrate.

The species most commonly encountered in North America are distinguished by their habit, racemes, and legumes.

1. Plants herbs; annuals. Racemes 2-to-6-flowered. Legumes 10–20 cm long; not winged . . . . *S. exaltata*
1. Plants shrubs; perennials. Racemes 10-to-30-flowered. Legumes 5–8 cm long; 4-winged.
   2. Banners yellow; 12–15 mm long. Legumes 10 mm in diameter . . . . *S. drummondii*
   2. Banners orange-red or rose or purple; 15–20 mm long. Legumes 12–15 mm in diameter . . . . *S. punicea*

Sesbania drummondii

- toxic species mainly of the Gulf coastal plains

Distribution and Habitat—Species of *Sesbania* are widespread in tropical and subtropical regions. Plants grow both wild and cultivated, being used for ground cover, windbreaks, shade in coffee and tea plantations, ornamentals, fiber sources, forage, fodder, food, and medicine. They generally occur in seasonal or permanently wet habitats—e.g., borrow ditches, sandbars, and gravel bars—and around springs. *Sesbania exaltata* occurs in the southern fourth of the continent from the Carolinas west to southern California. Adventive populations appear farther north. *Sesbania drummondii* is a species of the Gulf coastal plains. It may die back to the ground in the northernmost parts of its range. Native to South America, *S. punicea* is an ornamental that sometimes escapes cultivation and naturalizes along the Gulf Coast. *Sesbania sesban* is cultivated as an ornamental.

- *S. drummondii, S. punicea* most toxic; severe irritation, digestive disturbance, from green or ripe seeds, often lethal; leaves of lower risk; other species much less toxic

Disease Problems—Species of *Sesbania,* some of which are introduced from the West Indies or South America, are well known as causes of digestive tract irritation and diarrhea. Both green and ripe seeds are toxic, in contrast to the nontoxic foliage (Duncan et al. 1955). Although toxicity may vary somewhat, intoxications for all plant species are quite similar. *Sesbania punicea* is the most toxic species, with 1–2 dozen seeds being sufficient to kill a chicken in 24–72 hours (Marsh & Clawson

1920b; Shealy & Thomas 1928; West & Emmel 1950; Terblanche et al. 1966). The lethal dose in sheep and cattle is less than 0.1% b.w., even when spread over several days. The lethal dose of *S. drummondii* for chickens is 1% b.w. and is even less for sheep, as low as 0.22% b.w. (Marsh & Clawson 1920b; Marsh 1929; Flory & Hebert 1984). In contrast the seeds of *S. macrocarpa* have been fed at up to 10% of the diet in quail with little adverse effect (Flunker et al. 1991). *Sesbania exaltata* is not only one of the least toxic of the indigenous *Sesbania,* but it also has crude protein levels up to 31% (Hoveland et al. 1986). Dosage of up to 7% seeds in the diet of hens causes only slight mortality. Lesser amounts (2–3%) depress food intake and egg production (Damron et al. 1988). However, the seeds are little risk in the field because they readily fall from the mature pods and do not persist on the plant as in the case of *Glottidium vesicarium.*

Except for *S. punicea,* the green leaves and stems are of poor palatability and are about 100-fold less toxic than the seeds. They represent little hazard other than causing slight digestive discomfort. In fact, the leaves of an Old World species, *S. sesban,* contain 24–30% crude protein and 2%+ calcium and are reportedly used as livestock forage in Africa (Brown et al. 1987). Foliage from the same species has been successfully used in India as part of the diet for lactating cattle (Maity et al. 1997). However, experimentally, significant mortality occurred when chickens were fed 20–30% *S. sesban* leaf meal (Brown et al. 1987; Shqueir et al. 1989a). With the more toxic *S. punicea,* the leaves are more hazardous; the lethal dose for sheep is 10 g/kg b.w. per day for 4–8 days (Terblanche et al. 1966).

■ several possible toxicants; saponins, sesbanine, sesbanimides

Occasionally the seeds of species of *Sesbania* are a problem in humans. Morton (1982) cited 2 boys, ages 2.5 and 6 years, who, after eating the seeds of *S. punicea,* developed a severe digestive disturbance with diarrhea that required hospitalization.

sesbanimides, smooth muscle inhibition, vascular pooling, decreased gut motility and transport

Disease Genesis—Identification of the toxicants has been difficult. Snyders and Pieterse (1968) were unable to isolate a toxin from *Sesbania punicea* in methanol, and it appeared to be unstable in ethanol. Although the chemical nature of the toxicants is not completely characterized, the main compounds of interest have been saponins, which are present primarily in the seeds of the various species (Foote & Gramling 1940; Nuessle & Lauter 1958; Hsu 1968; Shqueir et al. 1989a). Three glucuronide saponins of oleanolic acid have been identified in *Sesbania sesban* (Dorsaz et al. 1988). However, although the abundance of saponins in the plants is unquestioned, the quest to associate toxicity with these compounds has been beset with difficulties, and often in the process of their isolation, toxicity was lost en route. Considerable taxonomic and chemotaxonomic confusion continues regarding the various species and their chemical constituents. Other toxicants that have been isolated include the cytotoxic sesbanine (or sesbananine) from *S. drummondii* and sesbanimides from *S. drummondii* and *S. punicea* and canavanine from *S. sesban* (Powell et al. 1979, 1983; Gorst-Allman et al. 1984; Shqueir et al. 1989a). The toxicant is water soluble, but unfortunately it appears to be unstable and has been difficult to isolate (Snyders & Pieterse 1968). Among the compounds of potential concern, including the cytotoxic sesbanine (or sesbananine) from *S. drummondii* and sesbanimides (A, B, and C) from *S. drummondii* and *S. punicea,* and canavanine from *S. sesban* (Powell et al. 1983; Gorst-Allman et al. 1984; Shqueir et al. 1989a); sesbanimide A appears to be most likely responsible for lethality. Experimentally, sesbanimide A concentrations correlate well with the lethality of various species of *Sesbania* in mice (Powell et al. 1990). Concentrations of sesbanimide were highest in the highly toxic *S. punicea* (90–100 μg/g of seed), moderate (18–52 μg/g of seed) in the less toxic *S. drummondii,* and nil in species of low toxicity, *S. exaltata* and *S. sesban* (Powell et

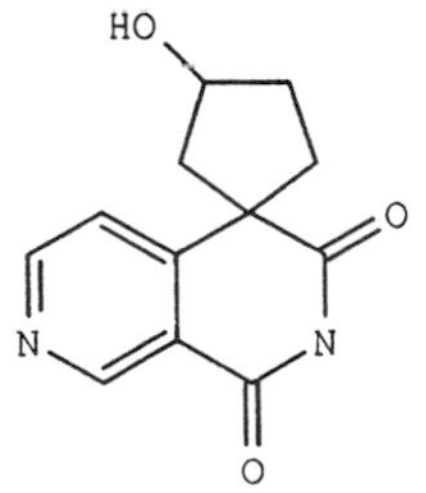

Sesbanine

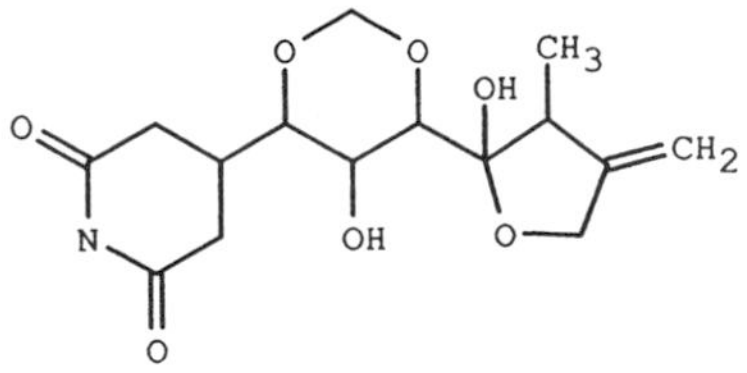

Sesbanimide A

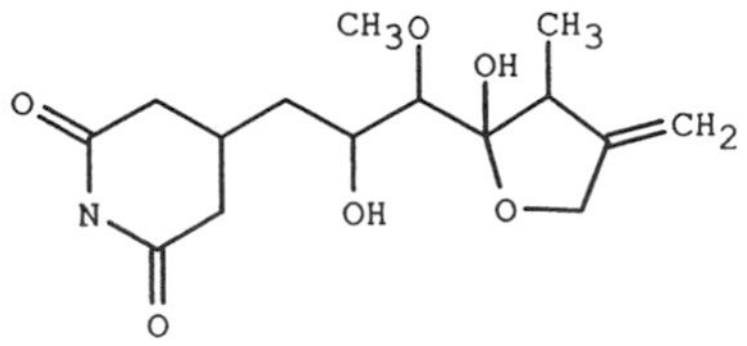

Sesbanimide C

al. 1990). Because of the interest in its anticancer activity, the sesbanimides have been synthesized, providing further confirmation of their structures (Matsuda & Terashima 1988).

It is quite likely that although the sesbanimides are the main toxicants, saponins and sesbanine may be contributing factors. Marceau-Day (1988) compared the effects of plant extracts with those caused by cyclohexamide and found that the effects associated with protein synthesis inhibition were similar but not identical. Furthermore, even those species without detectable sesbanimide have been associated with toxicity and occasional lethality. Heating the seeds, autoclaving with ammonia, or adding 1–2% cholesterol minimizes any adverse effects of *S. sesban* (Shqueir et al. 1989b).

Extracts of the seeds of the various species produce several effects. They augment or inhibit liver mixed-function oxidases, depending on the animal species, and they decrease plasma acetylcholinesterase activity (Flory & Hebert 1984; Banton et al. 1989). Perhaps the most consistent effect is the decreased motility and transport of contents of the digestive system. There is also mild to moderate liver and kidney degeneration and congestion and edema of the lungs, possibly secondary to heart failure (Terblanche et al. 1966). The predominant mechanism underlying these effects is inhibition of smooth muscle—vascular, intestinal, and pulmonary (Venugopalan et al. 1984, 1987). The vascular effects are seen as vasodilation, leading to decreased peripheral circulation, pooling of venous blood, and weakness, as well as the moderation of intestinal motility. The saponins are molluscicidal, spermicidal, and hemolytic (Dorsaz et al. 1988).

■ depression, ruminal atony, colic, diarrhea, weakness

Clinical Signs—After an initial delay of 12–24 hours in onset, depression, ruminal inactivity, abdominal pain, weakness, and mucoid feces may be noted (with close attention) as the earliest signs. Within a short time, diarrhea becomes severe, and the heart rate is increased and sometimes irregular. The animal stands with back arched, head down, and neck extended. Less commonly there is dyspnea and tremors. Neutrophilia, with many immature forms, may accompany the other signs. There may be a several-fold increase in blood urea nitrogen and hepatic enzymes, but these changes are generally not sufficient or even consistent enough to be confirmatory. In chickens there is depression, purple comb, muscular twitching, and profuse diarrhea.

■ gross pathology, reddening of mucosa of abomasum and intestines, pulmonary congestion, seeds in stomach

microscopic, gut lesions, mild renal tubular necrosis

Pathology—Gross changes are limited to moderate to severe abomasal and intestinal reddening. Less commonly the reticulum, rumen, and omasum are also affected. There may also be a few ecchymotic hemorrhages on the heart surfaces and pulmonary congestion. Microscopically, there is mild to moderate renal tubular degeneration and necrosis. In chickens, the dominant changes are a necrotic enteritis involving the gizzard and duodenum and fatty degeneration of the liver. The crop is fluid filled, the intestines are gas filled, and the feathers pull out easily. In most cases there is little to indicate a specific cause of death, and the pathology often belies the severity of the disease. Intact or sprouted beans may readily be identified in ruminal contents.

■ activated charcoal, fluids, electrolytes

Treatment—Because the specific cause is unknown, treatment is symptomatic. Activated charcoal may be given orally to absorb the toxins, accompanied by fluids and electrolytes for dehydration.

Sophora L.

Taxonomy and Morphology—Comprising some 40–70 species, *Sophora*, a member of the subfamily Papilionoideae, is a morphologically and ecologically heterogeneous genus distributed primarily in the Northern Hemisphere in both the Old World and the New World. Taxonomists have differed markedly in their delineations of its boundaries. Some individuals, for example, place the widely cultivated Japanese pagoda tree *(S. japonica),* in the genus *Styphnolobium.* In North America, the genus is represented by 12 native and introduced species, including these commonly encountered representatives (see plates 37.8 and 37.9):

■ *S. affinis*
S. arizonica
S. davidii
S. formosa
S. japonica
S. nuttalliana
S. secundiflora
S. stenophylla
S. tetraptera
S. tomentosa

blue laurel
Chinese scholar tree
coral bean
Eve's necklace
frijolito
Japanese pagoda tree
kowhai
mescal bean
sophora
Texas mountain laurel
white loco

S. affinis Torr. & A.Gray	Eve's necklace, Texas sophora
S. arizonica S.Watson	
S. davidii (Franch.) Skeels (= *S. vicifolia* Hance)	
S. formosa Kearney & Peebles	
S. japonica L.	Japanese pagoda tree, Chinese scholar tree
S. nuttalliana B.L.Turner (= *S. sericea* Nutt.)	silky sophora, white sophora, white loco
S. secundiflora (Ortega) Lag. ex DC.	mescal bean, Texas mountain laurel, coral bean, frijolito
S. stenophylla A.Gray	silvery sophora, blue sophora
S. tetraptera J.F.Mill.	kowhai, yellow kowhai, four-wing sophora
S. tomentosa L.	yellow sophora

Plants perennials; trees or shrubs or herbs; deciduous or evergreen. **Stems** erect or ascending; 0.2–20 m tall. **Leaves** 1-pinnately compound; leaflets 5–61; terminal leaflet present; stipules minute, caducous. **Inflorescences terminal** racemes or panicles; bracts present or absent. **Sepals** 5; fused; lobes short. **Corollas** papilionaceous. **Petals** 5; white or yellow or blue-violet; clawed; banners orbicular or broadly ovate or obovate, erect or reflexed; wings oblique-oblong; keels rounded or acuminate, petals overlapping or fused. **Stamens** 10; free; enclosed in keels. **Legumes** linear; terete; torose or torulose; 4-winged or not winged; fleshy or coriaceous or woody; indehiscent or dehiscent. **Seeds** few to many; oblong-reniform or subglobose; often yellow.

■ herbs, shrubs, or trees; leaves 1-pinnately compound, terminal leaflets; flowers pealike; petals 5, white, yellow, or blue-violet; legumes markedly constricted between seeds

1. Plants herbs.
    2. Leaflets linear or narrowly oblanceolate; flat. Petals blue-violet . . . . *S. stenophylla*
    2. Leaflets broadly oblanceolate or obovate; conduplicate. Petals white or pale cream . . . . *S. nuttalliana*
1. Plants trees or shrubs.
    3. Plants trees.
        4. Branches bearing spines . . . . *S. davidii*
        4. Branches not bearing spines.
            5. Leaves deciduous. Petals cream white or rose-pink.
                6. Inflorescences panicles. Legumes stramineous-brown or tan. Plants introduced ornamentals . . . . *S. japonica*
                6. Inflorescences racemes. Legumes black. Plants native . . . . *S. affinis*
            5. Leaves evergreen. Petals golden yellow or blue-violet.
                7. Leaflets 6–10. Petals blue-violet. Legumes not winged. Plants native . . . . *S. secundiflora*

Sophora nutalliana

Sophora secundiflora

7. Leaflets 10–40. Petals golden yellow. Legumes 4-winged. Plants introduced ornamentals ...... *S. tetraptera*

3. Plants shrubs.
   8. Leaves deciduous. Petals yellow or white to rosy pink. Seeds not red.
      9. Leaves densely pubescent. Petals yellow. Legumes tan or brown .... *S. tomentosa*
      9. Leaves glabrous or sparsely pubescent. Petals white or rose-pink. Legumes black .... *S. affinis*
   8. Leaves evergreen. Petals blue-violet. Seeds red.
      10. Leaflets 5–10 mm wide. Banners obovate. Claws of keel petals one-fourth to one-half length of blades .... *S. arizonica*
      10. Leaflets 10–25 wide. Banners oval. Claws of keel petals one-half to two-thirds length of blades.
         11. Leaflets 17–25 mm wide; upper surfaces glabrous .... *S. secundiflora*
         11. Leaflets 10–15 mm wide; upper surfaces pubescent .... *S. formosa*

In flower, the widespread *S. nuttalliana* is often mistaken for a member of the genus *Astragalus*. However, its terminal inflorescences, free stamens, acuminate keels, and coherent keel petals distinguish it from *Astragalus,* which has axillary racemes, diadelphous stamens, rounded keels, fused keel petals, and constricted legumes.

■ cultivated ornamentals, or in open areas

Sophora arizonica

Distribution and Habitat—*Sophora nuttalliana* is found in the open grasslands of the Great Plains and foothills of the Rocky Mountains. Populations also occur in eastern Utah. *Sophora stenophylla* favors open sandy areas of the intermountain deserts. As its name implies, *S. arizonica* is found only in Arizona. Widely distributed in warm coastal areas of the world, *S. tomentosa* is restricted to the coastal counties of Texas in North America. Taxonomists differ in their opinions as to whether it is native in North America. *Sophora affinis* occurs in Oklahoma, Arkansas, Texas, and Louisiana. Found on hillsides, rocky ledges, and in canyons, *S. secundiflora* occurs across southern Texas into New Mexico and Mexico. *Sophora formosa* is found from west Texas to southeastern Arizona. Some taxonomists believe that it might be only a form of *S. secundiflora* because of their morphological similarity. Both *S. formosa* and *S. arizonica* are rather uncommon, only occasionally becoming locally abundant.

Three species are introduced and cultivated as ornamentals for their showy inflorescences and attractive foliage. Native to New Zealand, *S. tetraptera* is the country's national flower. Both *S. davidii* and *S. japonica* are native to China and Korea, the latter at one time being important in the oriental batik and silk industries because of a yellow dye extracted from its legumes.

■ *S. secundiflora,* humans, seeds, neurologic effects, dizziness, stupor, visions

livestock, pets, foliage or seeds, neurologic and cardiac effects, exercise-induced seizures

Disease Problems—The physiological effects, both medicinal and toxic, of species of *Sophora* have been known and exploited for millennia. In the New World, *S. secundiflora* is among the most notorious of the supposed hallucinogenic plants. Its seeds are believed to have been used for thousands of years by American Indians living in what is now Texas and northern Mexico (Schultes 1969; Diaz 1977). They were used in tribal ceremonies and as articles of trade. Their ingestion clearly produces dizziness, numbness, and weakness and also is reported to cause stupor and visions. However, the hallucinogenic potential is questionable, and the seeds are now little used for this purpose (Hatfield et al. 1977).

The seeds, however, remain a toxicologic concern because of their continued use as ornamentation. Known as colorines, coral beans, mescal beans, red beans, red hots, or, in Spanish, frijolitos ("small beans"), the seeds are an attractive bright scarlet color. Their use in charms or musical instruments poses a risk to children and pets (Sullivan & Chavez 1981). A dog suspected of eating the seeds showed signs of exercise intolerance manifested by severe labored respiration and prostration (Knauer et al. 1995). Periodic seizures induced by exercise followed, and the dog would stiffen, tremble, and then collapse.

There also is toxicologic concern because of recreational drug abuse. Life-threatening interactions may occur when the seeds are mixed with neurotoxic drugs. A man intoxicated by a combination of 30–50 mescal beans, alcohol, and marijuana became wild and combative and subsequently developed myoclonic jerking and paralysis (Wells 1993).

The foliage, especially mature leaves, as well as the seeds, of *S. secundiflora* poses a problem for livestock (Boughton & Hardy 1935; Sperry et al. 1977). Cattle are at greatest risk, with ingestion of less than 1% b.w. causing severe neurotoxic effects. The disease is typically acute, appearing after the animal has eaten plant material for a short time. Calves may die in a few hours or days. Small amounts ingested for a week or more are not cumulative in effect. Sheep and goats are less susceptible and seldom show other than transient signs.

Other species of the genus are equally toxic. *Sophora alopecuroides,* an Old World species, was considered by the U.S. Department of Agriculture for use as a soil builder until it was discovered to be toxic at 1% b.w. and lethal at 2% b.w. (Radeleff 1970). However, not all species are hazardous; the foliage, flowers, and bark of *S. chrysophylla* in Hawaii are preferred browse for feral sheep and goats and Mouflon sheep even though they contain the toxicants characteristic of the genus (Scowcroft & Sakai 1983).

Sophora nuttalliana has long been suspected of being neurotoxic, but this suspicion has never been confirmed experimentally (Pammel 1911; Durrell et al. 1950; Sperry et al. 1977). This lack of confirmation may be a matter of dosage. However, teratogenic effects may be a more important concern. Studies with rats reveal that the species may cause a delay of ossification and developmental problems of the skull when the fetus is exposed during early pregnancy (Burrows et al. 1998). Based on limited experimental results, *S. arizonica* and *S. stenophylla* also appear to be teratogenic (Keeler 1972).

Disease Genesis—Species of *Sophora* contain an extensive array of quinolizidine alkaloids, which, although sometimes termed lupine alkaloids, actually differ considerably from those typical of *Lupinus* (Murakoshi et al. 1981, 1986). Because of the purported hallucinogenic properties of *S. secundiflora* and the highly valued medicinal role of some of the oriental species, the occurrence of these alkaloids has been determined for many species of the genus, and the mode of action of a few has been studied extensively. Unfortunately, limited toxicological information is available for most of the species in North America. These alkaloids are classified into two types and three series based on their chemical similarity to cytisine, sparteine, and matrine (table 37.11).

The diversity of alkaloids present in the germs is illustrated dramatically by those in the medicinally acclaimed, oriental species, *S. flavescens* (Ohmiya et al. 1974; Reyes et al. 1988). It contains methylcytisine, dehydrolupanine, anagyrine, and baptifoline of the cytisine series in combination with more than a dozen compounds of the matrine series, including sophocarpine, sophoridine, matrine, and oxymatrine (Bohlmann &

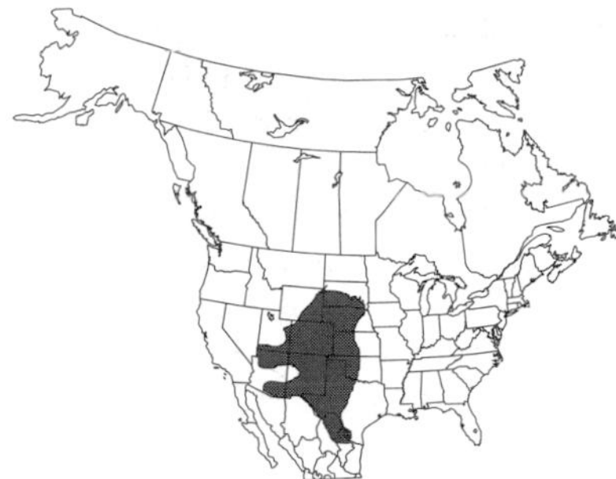

Sophora nuttalliana

Sophora secundiflora

■ other plant species, some nontoxic, others toxic but unknown risk; teratogenic?

Sophora stenophylla

Sophora tomentosa

- quinolizidine alkaloids, 2 types—
  (1) tricyclic cytisine, mainly neurotoxic, also cardiotoxic
  (2) tetracyclic lupanine, neurotoxic; matrine, sophocarpine, cardiotoxic, neurotoxic

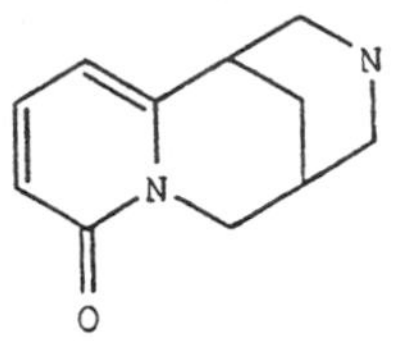

Cytisine

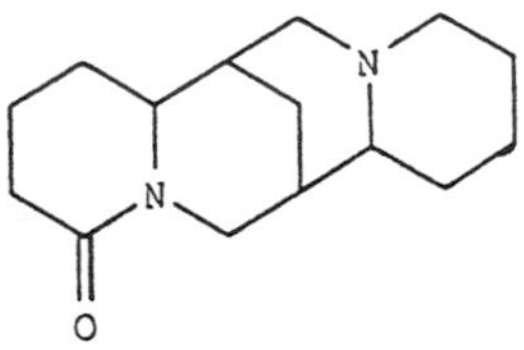

Lupanine

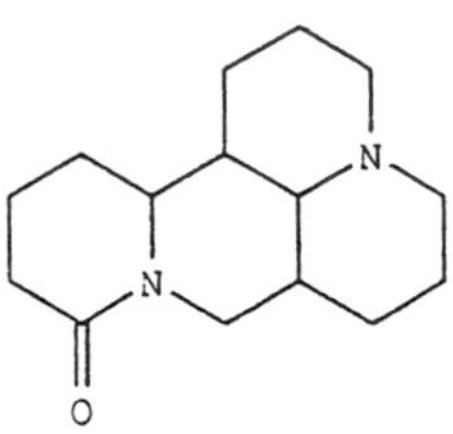

Matrine

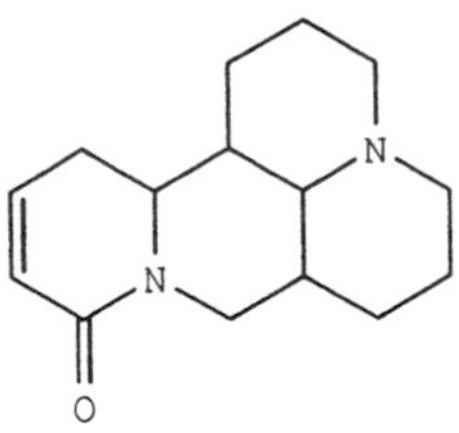

Sophocarpine

- teratogenic quinolizidines anagyrine, baptifoline in some species

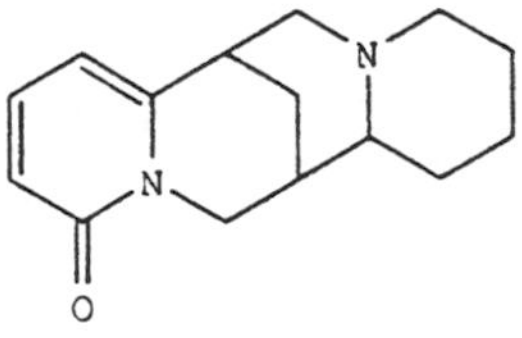

Anagyrine

Table 37.11. Quinolizidine alkaloids present in native and introduced species of *Sophora* in North America

Species	Tricyclic Type Cytisine Series	Tetracyclic Type Sparteine Series	Matrine Series
S. affinis	–	–	–
S. arizonica	+	–	–
S. formosa	+	–	–
S. japonica	cytisine *N*-methylcytisine	lupanine	matrine
S. nuttalliana	+	?	sophocarpine
S. secundiflora	cytisine (+)11-oxocytisine allylcytisine methylcytisine	epilupanine dehydrolupanine anagyrine sparteine	–
S. stenophylla	+	?	+
S. tetraptera	cytisine *N*-methylcytisine	baptifoline	matrine
S. tomentosa	cytisine *N*-acetylcytisine *N*-methylcytisine	epilamprolobine anagyrine baptifoline	matrine sophocarpine

Note: +, alkaloid type present but not identified; ?, alkaloid type possibly present but confirmation needed; –, alkaloid type not present.

Schumann 1967; Ueno et al. 1978; Aslanov et al. 1987; Saito et al. 1989; Jin et al. 1993). Even though a broad range of alkaloids may be present, one or two may predominate—for example, cytisine in *S. secundiflora* (Keller 1975; Hatfield et al. 1977; Izaddoost 1975, 1979; Izaddoost et al. 1976; Chavez & Sullivan 1984), lupanine in *S. japonica* (Abdusalamov et al. 1973; Wink et al. 1983), and sophocarpine in *S. nuttalliana* (Khan et al. 1992). In addition, the array of alkaloids and, thus, signs of intoxication may change with maturation of the plants (Keller & Hatfield 1979).

In addition to quinolizidine alkaloids, the bark of *S. japonica* contains a series of tetrameric lectins similar to the bark lectin of *Robinia pseudoacacia* (Hankins et al. 1988; Ueno 1991; Tazaki & Yoshida 1992).

Cytisine is the most toxic of the tricyclic alkaloids. It produces nicotinic-like effects stimulating and then blocking ganglia (Barlow & McLeod 1969), in addition to producing oxytocic and positive-cardiac inotropic effects (Reynolds 1955; Li & Chen 1986). Like nicotine, it causes rapid onset of ataxia, excess salivation, and short-lived seizures (Keeler & Baker 1990). In large doses, cytisine also may cause cardiac arrhythmias. It and other alkaloids produce neurologic effects in rats similar to those caused by known hallucinogenic drugs such as mescaline (Bourn et al. 1979).

The tetracyclic quinolizidines all seem to have inherent cardiac activity. Sparteine and the lupanines decrease heart rate by increasing the refractory period, and they increase the strength of contraction (McCawley 1955; Engelmann et al. 1974; Raschack 1974). In excess, they may cause arrhythmias. They, and sparteine in particular, also are neurotoxic, causing incoordination and muscular paralysis with tremors of the shoulder and neck, and a head bob (Couch 1926). Sparteine also has oxytocic activity possibly related to prostaglandins, given that it is blocked by indomethacin (Abtahi et al. 1978). The α-pyridone-A-ring tetracyclics like anagyrine, baptifoline, and thermopsine have neurologic and cardiac actions, but they are most well known as teratogens (McCawley 1955; Keeler 1976; Keeler & Baker 1990).

Quinolizidines of the matrine series, particularly matrine, allomatrine, and sophoridine, have strong cardiac effects, slowing the rate, prolonging the PR, QT, and other electrocardiographic intervals, and increasing contraction strength in a dose-dependent manner (Yin 1928; Yamazaki & Arai 1985; Li et al. 1986; Dai et al. 1987; Xin & Liu 1987; Cui & Zhang 1988; Kimura et al. 1989; B Zhang et al. 1990;

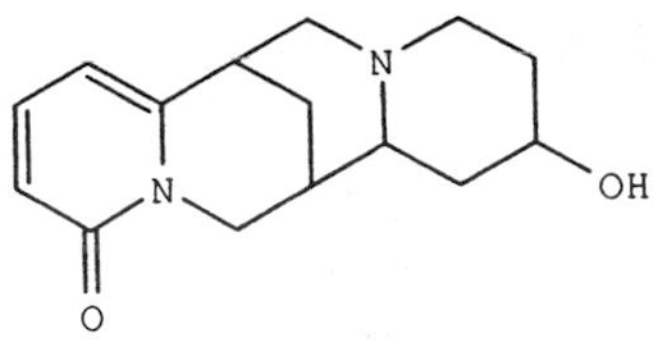

SS Zhang 1990; Wang et al. 1992). The mechanism seems to be related to calcium channels, because there is increased myocardial calcium uptake that can be blocked by the calcium channel blocker verapamil. The increased force of contraction following matrine administration in the rat also is blocked by verapamil (Wei et al. 1985). Matrine and, to a lesser extent, oxymatrine inhibit excitatory synapses mediated by glutamate and quisqualate in crayfish, which is perhaps indicative of their general reactivity (Ishida & Shinozaki 1984). However, there is a distinct lack of effects on GABA-induced responses in crayfish. Matrine is also antipyretic (Cho et al. 1986). Bitter tea made partly of *S. flavescens* and containing large amounts of D-matrine, was associated with neurologic impairment—dizziness, weakness, headache—in a family of four, one of whom developed progressive paralysis of one side of the body that persisted for several months (Tsai et al. 1993). Oxymatrine and sophocarpine also have cardiac effects but are more noteworthy as potent neurodepressants and hypothermics (Li & Chen 1986; C Yuan et al. 1987; H Yuan et al. 1986, 1987).

The quinolizidine alkaloids undergo hepatic biotransformation, especially *N*-oxidation, and are eliminated mainly in urine with half-lives of one to several hours (Barlow & McCleod 1969; Liu et al. 1986; Song et al. 1986). Both sophocarpine and its *N*-oxide are eliminated rapidly, with half-lives of elimination in rabbits of approximately 60 and 30 minutes, respectively (Wang et al. 1992). Interestingly, about 5% of the human population cannot metabolize sparteine by *N*-oxidation, and thus the compound accumulates in the body, causing adverse effects such as dizziness, blurred or double vision, and headache (Eichelbaum et al. 1979).

N
N
OH
O

Baptifoline

Clinical Signs—Because of the diversity in the alkaloids present and their relative abundance, the different species of *Sophora* are likely to produce different clinical signs in the animal. However, neurologic signs generally predominate. *Sophora secundiflora* is the only species for which signs are clearly known (Boughton & Hardy 1935; Sperry et al. 1977). Cattle exhibit the most severe signs, which begin with stiffness of the pelvic limbs and trembling of the rump and shoulder muscles. The animals may fall and be unable to rise. Sheep and goats also may be affected, but they are more likely to be able to walk after a brief period of somnolence. Forced exercise may precipitate occurrence of signs. *Sophora nuttalliana* is reportedly neurotoxic, with signs similar to those of locoism, but this has not been experimentally confirmed.

■ *S. secundiflora*, stiffness, trembling, collapse, exercise induced seizures

Pathology and Treatment—No specific pathologic changes for this acute disease problem have been documented, although serum phosphorus concentrations are reportedly doubled. Treatment is nonspecific and is directed toward relief of the symptoms that appear.

■ pathology, no lesions

treatment, sedation, general nursing care

Spartium L.

See treatment of *Cytisus,* in this chapter.

Thermopsis R.Br. ex W.T.Aiton

Taxonomy and Morphology—Most closely related to *Baptisia* and perhaps giving rise to it, *Thermopsis,* a member of the subfamily Papilionoideae, comprises 12–23 species, with greatest diversity in central and eastern Asia. Its genus name is derived from the Greek roots *thermos,* meaning "lupine," and *opsis,* meaning "appearance," which is indicative of the resemblance of its foliage and inflorescences to those of *Lupinus*. Plants of *Thermopsis* are cultivated to a minor extent for their silvery foliage

■ *T. mollis*
T. rhombifolia
T. villosa

buckbean
false lupine
golden banner
mountain peas
thermopsis
yellow pea

Thermopsis rhombifolia

and lupinelike racemes of showy flowers. In North America, 3 species occur—2 in the Southeast, which are of little toxicologic interest, and 1 in the West, which is important (see plate 37.10):

T. mollis (Michx.) M.A.Curtis ex A.Gray (= *T. fraxinifolia* M.A.Curtis)	Piedmont buckbean
T. rhombifolia (Nutt. ex Pursh) Nutt. ex Richardson	round-leaved thermopsis, prairie buckbean, yellow pea, false lupine, golden banner, mountain thermopsis, mountain peas
T. villosa (Walter) Fernald & Schub. (= *T. caroliniana* M.A.Curtis)	Blue Ridge buckbean

Thermopsis rhombifolia is now recognized to be a highly polymorphic, widely distributed species that encompasses taxa previously recognized as distinct, including in part *T. argentata, T. arenosa, T. californica, T. divaricarpa, T. gracilis, T. macrophylla, T. montana, T. ovata,* and *T. pinetorum* (Barneby 1989b). Reports of intoxication typically use the binomial *T. montana.*

■ erect perennial herbs; leaves palmately compound, leaflets 3, stipules; flowers showy, pealike; petals 5, yellow or purple

Plants perennials; herbaceous; rootstocks woody, rhizomatous; herbage typically pubescent. **Stems** erect; 40–100 cm tall; clustered; typically leafless at bases. **Leaves** palmately compound; leaflets 3, blades linear to elliptic to oblanceolate or obovate; stipules of 2 forms, distal leaflike, basal forming sheaths. **Inflorescences** terminal, elongate racemes; bracts present. **Flowers** showy. **Sepals** 5; fused; calyces 4-lobed, 2-lipped. **Corollas** papilionaceous. **Petals** 5; pale to vivid yellow or purple, sometimes brown or pinkish brown when dry; short-clawed; banners suborbicular, margins reflexed; wings auriculate; keels rounded, straight or curved; petals adherent. **Stamens** 10; free; enclosed in keels. **Legumes** linear to oblong; straight or curved; flattened; slightly constricted between seeds. **Seeds** 1–10; oblong-reniform.

1. Plants of the western United States .... *T. rhombifolia*
1. Plants of the southeastern United States.
   2. Stipules 3.5–6.5 cm long; auriculate. Legumes appressed against rachises; surfaces villous-tomentose ...... *T. villosa*
   2. Stipules 1.2–3.2 cm long; not auriculate. Legumes diverging from rachises; surfaces appressed pubescent .... *T. mollis*

■ western grasslands, southeastern woodlands

Distribution and Habitat—Adapted to a variety of soil types and habitats, including mountain meadows, desert shrub communities, and grasslands, *T. rhombifolia* is distributed from the eastern side of the Rocky Mountains to the Pacific Coast. *Thermopsis mollis* and *T. villosa* are found in forest openings in the piedmont and mountains of the Southeast.

■ livestock, foliage, skeletal muscle degeneration; plants most toxic when very young and when mature with legumes

humans, seeds, neurotoxic, digestive disturbance

Disease Problems—Ingestion of dry foliage of *T. rhombifolia* (reported as *T. montana*) by cattle at a dosage of 0.6–2.8 g/kg b.w. for several days or more produces severe muscle degeneration typified by persistent recumbency (Keeler et al. 1986a). Depending upon stage of growth, 300–400 g of dry plant material for 3–4 days will severely intoxicate cattle, and slightly more may be lethal (Chase & Keeler 1983). Toxicity potential is high in young plants, declines as they grow, and then reaches a peak at the time of seed formation (Keeler et al. 1986a). This pattern of toxicity is similar to that exhibited by species of *Lupinus.* If animals are fed and watered during the period of recumbency, they may recover.

Humans also are at risk. Children who ate a few seeds or flowers suffered digestive tract problems, weakness, and neurologic effects (Spoerke et al. 1988). The risk for humans is further illustrated by a report of 23 cases involving ingestion of 1 or 2 seeds or legumes or a handful of flowers (McGrath-Hill & Vicas 1996). However, the plants were not positively identified in all cases. In most cases there was vomiting, and dizziness and drowsiness in a few individuals.

Thermopsis mollis

Disease Genesis—The myopathic effects of *Thermopsis* are caused by a series of α-pyridone quinolizidine alkaloids, including, in decreasing order of relative abundance in plant tissues, anagyrine, thermopsine, 5,6-dehydrolupanine, cytisine, *N*-methylcytisine, lupanine, and 17-oxosparteine (Keller & Cole 1969; Cho & Martin 1971; Keeler et al. 1986a; Keeler & Baker 1990). Purified alkaloids cause the same signs of intoxication as the whole plant (Keeler & Baker 1990). Their concentrations mirror the toxicity potential of the plant: highest in young plants, declining, and then increasing again in the flowers and legumes. Alkaloid concentrations are about equal among leaves, stems, flowers, and fruit (Keller & Cole 1969). Alkaloid dosage in the range of 1.1–11.3 mg/kg per day for 2–3 days will cause the disease (Keeler et al. 1986a; Baker & Keeler 1989).

- quinolizidine alkaloids, anagyrine, thermopsine, neurotoxic, myotoxic

  high concentrations of anagyrine, teratogenic effects likely

Thermopsine

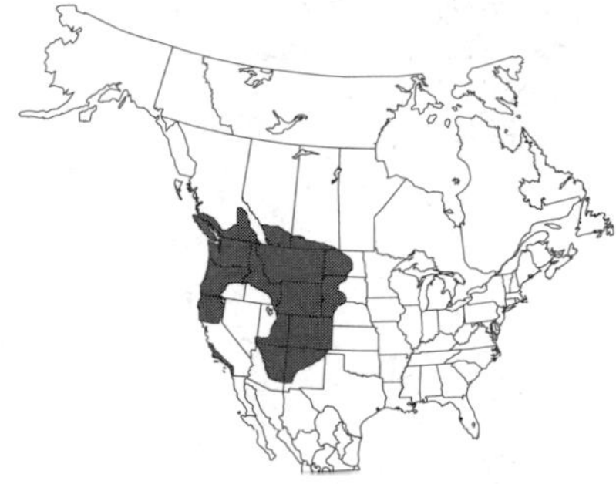

Thermopsis rhombifolia

The presence of high concentrations of anagyrine in *Thermopsis* suggests that it may cause teratogenic effects such as arthrogryposis, defects of limb rotation, and other terata similar to those caused by *Lupinus.* These effects have not been observed in clinical cases, but preliminary experiments in cattle fed *Thermopsis* indicated a risk (Keeler 1972).

Thermopsis villosa

Clinical Signs—After the animal has ingested plant material for 3 or more days, depression, anorexia, arched back, tremors, rough hair coat, drawn-in flanks, and swollen eyelids may be seen. Shortly thereafter, the animal will become recumbent and may remain so for several days to a week or more. During this period, it will eat and drink if food and water are made available. Except for GGT, serum enzymes such as CK, LDH, and AST may be markedly elevated.

Intoxications in humans have involved mainly children eating a few seeds or flowers. Protracted, intermittent vomiting for several hours after ingestion is the main problem. There also may be colic, weakness, headache, and, less commonly, dizziness or ataxia and drowsiness. The time from ingestion to signs is usually 1 to several hours, with uneventful recovery in 6–12 hours.

- livestock, onset 1-2 days, depression, anorexia, tremors, recumbency, elevation of serum CK, LDH

  humans, shortly after eating seeds, vomiting, colic, headache, dizziness, ataxia

Pathology—Other than pale streaks in muscles, few gross lesions are associated with ingestion of *Thermopsis.* Microscopically there is hyaline degeneration and necrosis of skeletal muscle in acute cases that result in death. In animals less severely affected, degeneration of muscle fibers is followed by regeneration.

- gross pathology, pale muscles

  microscopic, muscle necrosis, fibrosis

■ activated charcoal, nursing care; disease generally reversible

Treatment—No specific treatment is recommended, although in acute cases activated charcoal may be of benefit. Supportive nursing care to sustain the recumbent animal will provide good results. In humans, gastric lavage and oral administration of charcoal typically have been employed, but in most cases recovery will readily occur without treatment.

Trifolium L.

■ *T. hybridum*
T. pratense
T. repens
T. subterraneum

alsike clover
Dutch clover
ladino clover
purple clover
red clover
subterranean clover
subterranean trefoil
trefoil
white clover

Taxonomy and Morphology—Comprising 250–300 species, *Trifolium,* a member of the subfamily Papilionoideae, is almost a cosmopolitan genus, being found throughout the world with the exception of Australasia. Indicative of its palmately compound leaves, its generic name is derived from the Latin roots *tres,* meaning "three," and *folius,* for "leaf." Its members are commonly referred to as the clovers. It must be kept in mind that the name *clover* is applied to species of other legume genera as well—e.g., *Melilotus, Lespedeza,* and *Petalostemon*—albeit a qualifying adjective is usually appended. The genus is important throughout the world for forage, green manure, and ground cover. Land reclamation often involves seeding with grasses and species of *Trifolium.* In addition, the genus is an excellent nectar source for bees. A few species are weedy.

Of the more than 100 species, varieties, and cultivars in North America, approximately 85 are native. Most intoxication problems are associated with the perennials, many of which are introduced and naturalized. The numerous annuals that are widely cultivated for forage and as soil builders are seldom associated with disease problems. Of toxicologic interest are the following:

T. hybridum L.	alsike clover
T. pratense L.	red clover, purple clover
T. repens L.	white clover, Dutch clover, ladino clover
T. subterraneum L.	subterranean clover, subterranean trefoil

■ herbs; leaves palmately compound, leaflets 3; flowers pealike; petals white, red, pink, yellow, or purplish

Trifolium pratense

Plants annuals or perennials; herbaceous; sometimes stoloniferous. **Stems** erect or ascending or prostrate or creeping. **Leaves** palmately compound; leaflets 3, rarely 5 or 7; blades obcordate or obovate, margins entire or dentate; stipules fused to petioles. **Inflorescences** racemes or spikes; capitate or umbellate; axillary or terminal; sessile or peduncled; bracts present or absent. **Sepals** 5; fused; teeth equal or subequal; calyces 2-lipped, sometimes inflated in fruit. **Corollas** papilionaceous. **Petals** 5; white or red or pink or yellow or purplish; 4 or 5 fused to staminal tube; withering and turning brown rather than falling; banners oblong to ovate, straight or reflexed; not clawed, free or fused to staminal column; wings narrowly oblong, clawed, auriculate; keels oblong, rounded, clawed. **Stamens** 10; diadelphous; enclosed in keels; 5 or 10 filaments dilated distally. **Legumes** oblong or ovate; indehiscent; membranous or coriaceous; often protruding from calyces at maturity. **Seeds** 1 or 2, rarely 4–8; globose or ovoid-oblong or reniform.

1. Plants annuals. Stems prostrate, radiating from central taproot, forming circular clumps . . . . *T. subterraneum*
1. Plants perennials. Stems erect or ascending or creeping from rootstock, not forming circular clumps.
   2. Flowers sessile in capitate spikes. Orifices of calyces pilose. Basal margin of banners fused into tube . . . . *T. pratense*
   2. Flowers pedicelled in capitate racemes. Orifices of calyces glabrous. Basal margins of banners free, not fused into tube.

3. Stems erect, not rooting at nodes. Stipules not forming sheath above level of fusion with petioles . . . . *T. hybridum*
3. Stems creeping, rooting at nodes. Stipules forming membranous sheath above level of fusion with petioles . . . *T. repens*

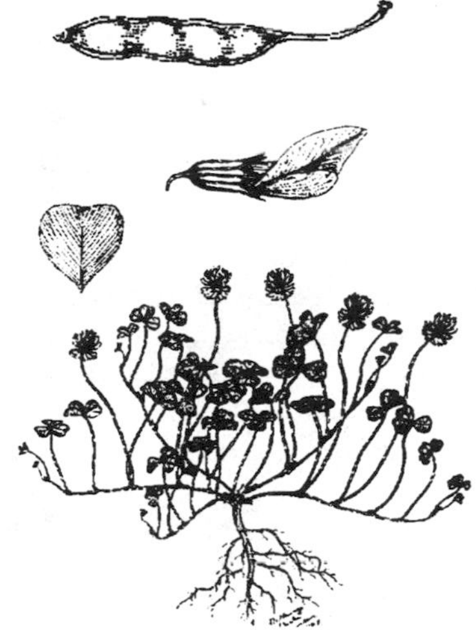

Trifolium repens

Distribution and Habitat—Especially abundant in the Northern Hemisphere, *Trifolium* is a genus of temperate and subtropical regions. Centers of diversity appear to be western Asia, the eastern end of the Mediterranean basin, and northwestern North America. Occurring predominantly in grassy habitats, the genus is adapted to a variety of soil types. The 4 species of toxicologic interest are native to Eurasia but have naturalized throughout North America. *Trifolium hybridum* is cultivated and naturalized primarily in the northern states and Canada. *Trifolium pratense* is widely cultivated and naturalized in much of the midwestern and eastern areas of North America. Especially common in lawns, *T. repens* occurs throughout North America in fields and moist roadsides. Introduced from the Old World, *T. subterraneum* is cultivated and rarely escapes.

- crop plants
- 5 disease problems—
  estrogenic
  slobbers,
  bloat
  photosensitization,
  cyanogenesis

Disease Problems—Species of *Trifolium* are commonly planted in combination with cool-season grasses and cereal grains to provide forage for livestock. If they make up 25% or perhaps even as much as 50% of the available forage, they normally pose little toxicological risk. However, in certain circumstances, they are capable of producing five very different disease problems—estrogenic effects, slobbers, bloat, photosensitization, and cyanogenesis. Bloat, a disease caused by the animal's inability to expel gases produced during ruminal fermentation, is most commonly associated with *T. repens.* Unlike the other diseases produced by the genus, it is due more to mechanically caused physical changes in the animal than to the presence of specific plant toxicants. A complete discussion of bloat is presented in the treatment of *Medicago,* in this chapter. Aspects of the remaining four problems are discussed separately in the present treatment.

Of interest is a report from New Zealand that in red deer grazing *T. repens,* urine turned scarlet red upon standing for about an hour (Niezen et al. 1992). However, there was no apparent affect on animal health.

Estrogenic Effects

- reproductive, "clover disease," decreased conception in sheep, less risk in cattle

Disease Problems—In countries where species of *Trifolium* are used extensively as livestock forages, estrogenic effects are the most significant of the problems caused by the genus. Serious limitations in reproductive performance are produced. Most commonly associated with these estrogenic effects is *T. subterraneum* in Australia and *T. repens* in Scandinavia; less frequently, other species are involved. Species of *Medicago* also cause these same effects. In Australia, where both *Trifolium* and *Medicago* forages present serious problems in sheep, the disease is called clover disease (Moule 1961; Thain 1968; Cox 1978, 1985; Adams 1989). Manifestations include decreased conception rates, maternal dystocias, increased postnatal mortality, uterine prolapse and lactation in nonpregnant ewes, and bulbourethral gland metaplasia in wethers. These signs occur in various combinations. Exposure of the dam also may predispose neuroendocrine problems in the developing female neonate (Bindon et al. 1982). Although suspected as a cause of abortions, estrogenic effects of clovers rich in phytoestrogens seem to be less important in cattle, even with long-term ingestion whether on pasture, hay, or silage (Rankin 1963; Lundh 1995). However, transient effects such as irregular estrus, cystic ovaries, and precocious mammary and genital development are reported for cattle grazing alfalfa or clovers (Adler & Trainin 1960; Thain 1965).

In North America, decreased conception rates have been reported. Cattle grazing cultivar 'Mt. Barker' of *T. subterraneum* were affected (Donaldson 1983), and sheep eating *T. repens* exhibited delayed onset of estrus and an increase in the number of breedings needed in yearling ewes (Engle et al. 1957).

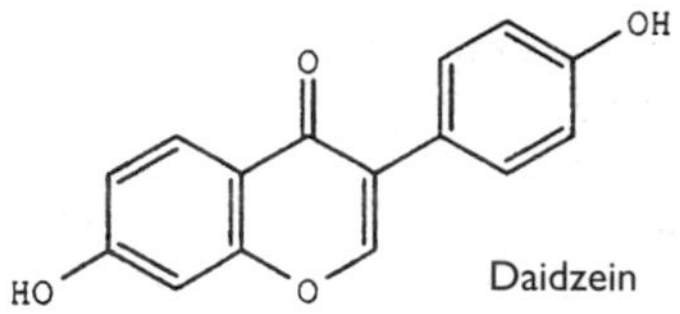

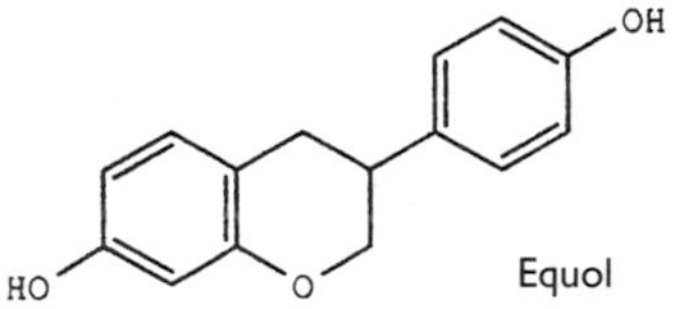

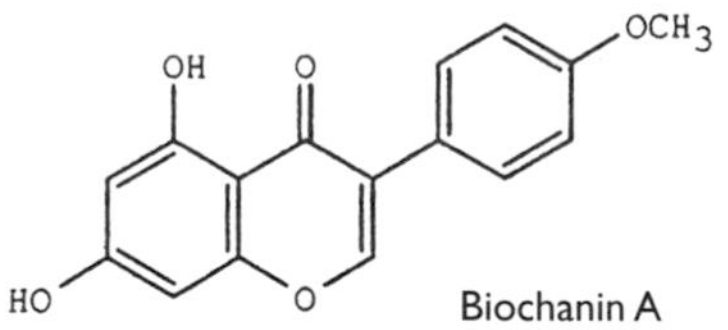

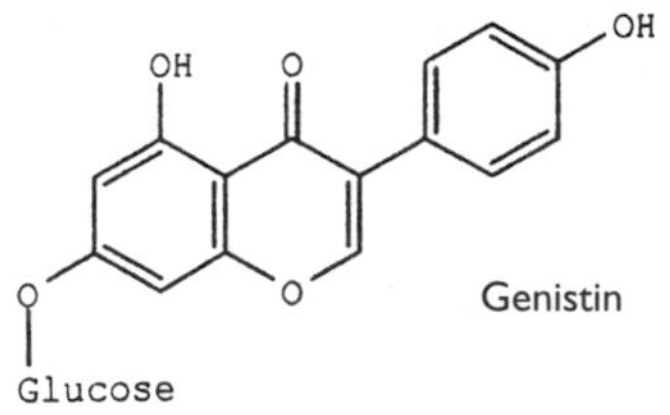

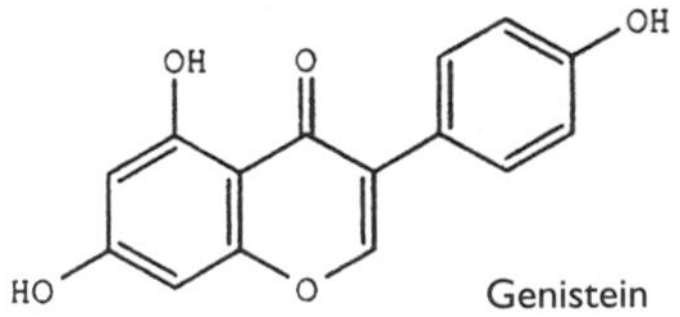

■ phytoestrogens, weak estrogens, isoflavone glycosides, formononetin, biochanin, genistin, metabolized to other weak estrogens; coumestans, coumestrol

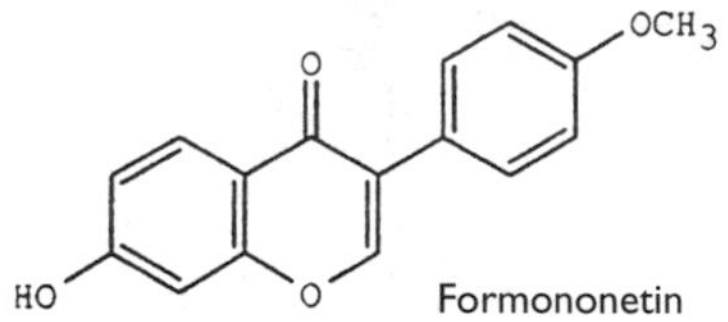

Disease Genesis—The causes of the estrogenic effects are phytoestrogens; isoflavone glycosides in *T. pratense, T. hybridum,* and *T. subterraneum*; and coumestans in *T. repens* (Bickoff et al. 1969; Wong et al. 1971; Adams 1989). Phytoestrogens are not uncommon in plants in general, but in most instances they are of little toxicologic importance because they are in such low concentrations (Bergeron & Goulet 1980). All of the phytoestrogens are weak estrogens, as measured by techniques such as the uterine response assay or the uterine estrogen receptor binding assay. Depending on the procedure used, coumestrol, the most potent, has only about one-thousandth the potency of diethylstilbestrol or 17β-estradiol (Cheng et al. 1954; Bickoff et al. 1962; Wong & Flux 1962; Braden et al. 1967; Shemesh et al. 1972; Shutt & Cox 1972; Cox 1978; Livingston 1978; Welshons et al. 1990). The relative ranking of these compounds using other systems such as estrogenic receptors in human breast-cancer cells also shows coumestrol (activity equal to that of zearalenone) to be more potent than genistein, and formononetin to be the least potent (Martin et al. 1978). Although they are not the most potent, the isoflavones seem to be the most important because they are found in the highest concentrations, up to 2–4% d.w. (Rossiter 1970). The coumestans are of lesser importance even though they are of the highest potency (Cox 1978). The parent isoflavones in *Trifolium* are the glycosides, formononetin, biochanin A, and genistin. The first is biotransformed to equol and the latter two to genistein. Coumestrol, the most important of the coumestans, is 2–10 times as potent as genistein and equol, depending on the assay system used. The glycosidic parent compounds are virtually inactive.

Found in concentrations up to 1–2% d.w. in some cultivars of *T. subterraneum* and 0.5–1% in *T. repens,* formononetin is bioactivated by demethylation to daidzein in the rumen and, to a lesser extent, in the liver and then further reduced to the estrogenic equol, which is absorbed, conjugated (90%+), and excreted in the urine (Shutt & Braden 1968; Lindsay & Kelly 1970; Adams 1989; Lundh et al. 1990; Saloniemi et al. 1995). This sequence of changes occurs in sheep and probably cattle as well (Braden et al. 1967). Plasma concentrations of conjugated isoflavones several hours after ingestion may be in excess of 200–300 μg/dl and for total equol are in the range of 100–150 μg/dl at 24 hours. Coumestrol also is found mainly in the conjugated form in plasma (Bickoff et al. 1969). In most situations, formononetin is of greatest concern, and concentrations greater than 0.5% in the plant are considered to be potentially estrogenic in sheep.

Biochanin A and genistin are metabolized in the rumen by demethylation and/or glycoside hydrolysis to genistein and thence to the nonestrogenic paraethylphenol. The sheep's ruminal microflora are able to adapt within hours to cause ruminal reduction and inactivation of biochanin A and genistein (Nillson et al. 1967; Morley et al. 1969). However, in unusual circumstances when ruminal reduction is limited—e.g., restricted food intake—estrogenic effects even from biochanin A and genistein may become apparent (Davies 1989). This is illustrated by the appearance of the disease in cattle eating cultivar 'Mt. Barker' of *T. subterraneum,* which is low in formononetin but high in genistein (Donaldson 1983; Davies 1989).

As noted in the discussion of disease problems, the effects of these phytoestrogens are most pronounced in sheep and less so in cattle. Decreased susceptibility of cattle was thought to result from differences in metabolism of the phytoestrogens (Braden et al. 1967; Shutt et al. 1967; Shutt & Braden 1968). However, other studies indicate that there is little difference between sheep and cattle in ruminal degradation of formononetin. In fact, in spite of their increased susceptibility, conjugation activity is much greater in sheep and concentrations of unbound equol are much lower than in cattle (Lundh 1990; Lundh et al. 1990). Thus, the difference in activity seems more likely due to decreased sensitivity of the estrogen receptors in cattle to these compounds (Lundh et al. 1990).

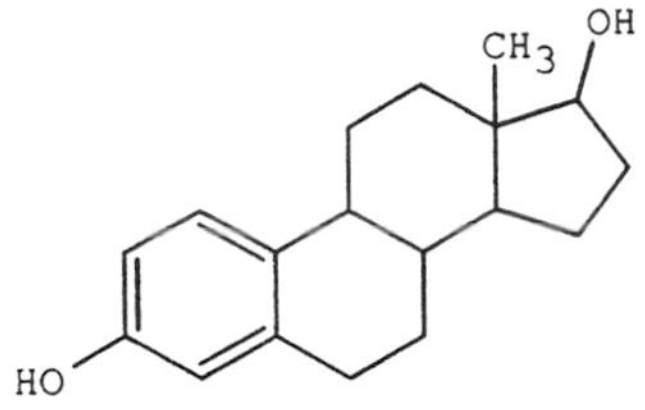

CH3 OH

HO

Estradiol

Estrogenic activity varies with the cutting, stage of growth, disease, plant health, environment (drought), phosphate level, and genetic constitution of the forage. Of these variables, foliar diseases caused by fungi and viruses are especially noteworthy causes of increased phytoestrogen concentrations (Bickoff et al. 1969; Saba et al. 1972; Cox 1978). This is especially true for *T. repens,* which is usually of little risk (Wong et al. 1971). It is suggested that drought-induced increases in phytoestrogen concentrations may serve a regulatory or protective function in maintaining appropriate-sized populations of wildlife (Leopold et al. 1976). As an example, phytoestrogen levels may have an influence in reducing clutch size in California quail in dry years when food is limited. Phytoestrogen concentrations also are increased as much as 2-fold with phosphate deficiency (Rossiter & Beck 1966). Genetic makeup also is important, given that cultivars low in formononetin—e.g., 'Daliak,' 'Northam A,' 'Seaton Park,' and 'Mt. Barker'—are available. These taxa, however, are usually high in biochanin A or genistein (Davies et al. 1970; Rossiter 1970). The effects of plant maturity on phytoestrogen level are unclear because experimental data are conflicting. The ambiguity is due in part to the superimposed effects of repeated cutting at various times. In some studies, the highest concentrations of the compounds are reported to be when the plants are in the vegetative stages, whereas in other studies, phytoestrogen concentrations seem to vary little with age (Pieterse & Andrews 1956; Kitts et al. 1959). Grazed *T. repens* is reported to have the highest estrogenic activity in late fall and early winter (Saba et al. 1974).

- estrogenic activity of forages, variability due to many factors—cultivar, cutting, plant disease, soil phosphate

risk with pasture, hay, or silage

Because of all these variables, it is difficult to predict the toxicity potential of either hay or silage made from toxic pastures. Generally, drying reduces the estrogenic activity of coumestrol forages but has less effect on those with isoflavones (Bickoff et al. 1960). In some instances, ensiling has appeared to reduce the problem for both sheep and cattle (Wright 1960; Austin et al. 1982). However, it is clear that under the appropriate circumstances both hay and silage still represent some degree of risk (Cheng et al. 1953; Adler & Trainin 1960; Livingston 1978; Kallela et al. 1984; Adams 1989).

Effects of varying severity may be noted. In the most serious cases in sheep that are grazed on pastures in which the isoflavones predominate, there may be permanent infertility due to long-term or persistent endometrial cystic glandular hyperplasia. Increased thickness of mucous secretions greatly limits sperm penetration of the cervix, and physiologic changes of the cervix make it more like the uterus. The number of estrogen receptor binding sites is increased in the cervix but not elsewhere, contributing to impaired sperm transport (Tang & Adams 1981). These signs persist long after the animals have been removed from the *Trifolium* pastures (Lightfoot & Wroth 1974; Adams 1990). The effects are mediated by a complex series of events involving the neuroendocrine and reproductive tissues. Luteinizing hormone (LH) secretion is disturbed, and there appears to be interference with feedback regulation of estrogen by the hypothalamic-pituitary axis (Rodgers et al. 1980). There is a decrease in pulsatile release of LH and a lack of fluctuation in its secretion between the breeding and nonbreeding seasons (Chamley et al. 1981). Binding of estrogen receptors results in variable effects on progesterone; however, duration of the corpus luteum and its secretion of progesterone are generally decreased (Shemesh et al. 1972; Adams et al. 1981; Kaplanski et al. 1981). A discussion of the possible mechanisms of dietary phytoestrogens has been detailed by Hughes (Chapin et al. 1996).

Eventually there is decreased responsiveness or desensitization of the uterus to estrogens, which may lead to increased susceptibility of the uterus to infection and may predispose to low-grade endometritis (Tang & Adams 1982). The phytoestrogens have also been suggested as a possible factor in the etiology of ringwomb in sheep (Ward 1975). In contrast, short-term grazing of pastures with either isoflavones or coumestans in the forage, may result in temporary infertility that is evident only while the animals graze. In this situation, decreased fertilization results from irregularities in ova transport and/or implantation and possible neuroendocrine changes (Livingston 1978). Temporary infertility in ewes may be manifested as an increase in the number of animals requiring repeat breeding and of maiden ewes displaying mammary development before breeding (Maxwell 1979). In some instances, only marginal, difficult-to-confirm effects occur (Engle et al. 1957). There also may be developmental effects on the female reproductive tract of newborns. The milder, short-term grazing form of the disease problem is seen in cattle as well as sheep and appears to be the form likely to be seen in North America. The isoflavones also are suggested as a contributing factor in the incidence of buller cattle in feed yards (Edwards 1995). Pigs may be at considerable risk of adverse effects when grazing phytoestrogenic forages, because although formononetin is rapidly absorbed and biotransformed, there seems to be a decreased capacity to conjugate equol (Lundh 1995). An increased risk also is suggested by the high sensitivity of pigs to the effects of zearalenone.

Other effects of phytoestrogens include uncoupling of oxidative phosphorylation by biochanin A and formononetin, about one-tenth that caused by 2,4-dinitrophenol. These isoflavones also inhibit the respiratory chain and electron transport at the same site as rotenone (Lundh & Lundgren 1991). Genistein, which is also found in bacteria, is a tyrosine kinase inhibitor (Akiyama et al. 1987) and is also cytotoxic by virtue of its inhibition of DNA-topoisomerase (Markovits et al. 1989). Metabolic and carcinogenic effects also are reported (Adams 1989).

- effect, mainly temporary increased number of services for conception; especially sheep, cattle less affected

Clinical Signs and Pathology—In sheep, a wide variety of reproductive disorders caused by the phytoestrogens of *Trifolium* occur. They include abnormal development of genitalia, infertility, abnormal estrous cycles, follicular cysts, endometrial hypoplasia, uterine prolapse, and dystocia of maternal origin. Manifestations of the disease in North America, where forages are likely to contain only low concentrations of the phytoestrogens, will mainly be restricted to low-grade infertility problems, perhaps with abnormal estrous cycles. The lambing season may be abnormally prolonged, with an overall decrease in the birth rate as well. Effects in cattle are only marginal. Because of the rarity of this disease, it should not be considered the cause of infertility problems until exhaustive efforts have been made to eliminate more common infectious or nutritional causes.

Diagnosis can be aided by chemical assay of the suspect forage for phytoestrogens. However, because of various constraints with the chemical assay, bioassays measuring the change in DNA induced in breast cancer–derived MCF-7 cells or the increase in weight of the mouse uterus generally may be more useful (Welshons et al. 1990; Galey et al. 1993). However, some care must be taken, because the results of the mouse uterine assay may not be representative of effects in livestock (Saba et al. 1974).

- reversible when forage changed

Treatment—As indicated above, many of the effects caused by the phytoestrogens of *Trifolium* are readily reversible with removal of the animal from the offending forage. Another promising experimental approach is to control expression of the disease through immunization to provide antibodies against the phytoestrogens (Livingston 1978; Cox 1985). Prevention of the disease may be accomplished through the use of newer improved cultivars low in phytoestrogens (Davis et al. 1970).

Slobbers

Disease Problems—Among the diseases caused by *Trifolium* is one that is associated with infestation of *T. pratense* by the fungus *Rhizoctonia leguminicola*, which causes black patch disease of the plant. The disease name is derived from the bronze to black spots or rings, 1–10 mm diameter, that appear on the leaves and stems. Consumption of infected plants in hay or in the pasture produces an intoxication characterized by excessive salivation and generally called slobbers (Aust 1974). The fungus persists from growing season to growing season in both vegetative tissues and seed. Thus, once it occurs in an area, it is likely to be a continuing problem, primarily on second cuttings of hay when the weather is cool and moist. Mowing and allowing regrowth of pastures may reduce the amount of infection if moisture is not excessive. Other forages, including alsike and white clovers, lespedeza, and medics, also may be parasitized by the fungus, but usually only when they are growing adjacent to red clover. These other forages, however, are seldom a cause of the disease (Broquist 1985).

- *T. pratense* infected with fungus causing black patch disease of clover, pasture or hay, cholinergic effects in all livestock; not lethal, decreased productivity

Slobbers may be seen in cattle, sheep, and horses that ingest large amounts of infected forage, either fresh or as hay or silage, for several days or more. The disease is not usually of life-threatening severity but rather is primarily of economic importance because of animal weight loss and decreased milk production.

Disease Genesis—The disease slobbers, resulting from ingestion of *T. pratense* infected with black patch disease, is caused by the alkaloid (1S,6S,8aS)-acetoxy-6-aminooctahydroindolizine. It is produced by the fungus and is commonly referred to as slaframine, rhizotoxin, or simply the slobber factor (Broquist 1985). After absorption in the digestive tract, slaframine is biotransformed by flavoprotein oxidase in the liver to an acetylcholine-like quaternary amine, or ketoimine, capable of cholinergic stimulation of exocrine glands (Aust 1974; Guengerich & Aust 1977). Concentrations of slaframine may range as high as 50–100 ppm.

- fungus produces alkaloid, slaframine, metabolized in liver to cholinergic quaternary ketoimine

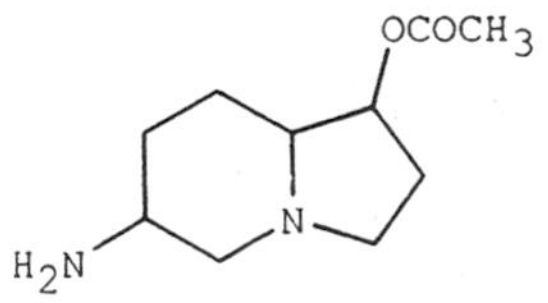

$OCOCH_3$ N H_2N

Slaframine

In addition, swainsonine, a competitive inhibitor of α-D-mannosidase and a major toxicant in *Astragalus* (treated in this chapter), has been detected. Concentrations up to 2.5 ppm occur, but their toxicologic significance in *T. pratense* is unknown (Broquist 1985).

Clinical Signs and Pathology—After the animal has eaten the fungus-infested stems and leaves for several days, excessive salivation with stringy, clear, viscid saliva hanging from the mouth may be seen. Although excess salivation is the predominant sign, other changes primarily referable to the parasympathetic nervous system also may be an important part of the disease syndrome. These additional signs include hypothermia, excess lacrimation, diarrhea, bloat, increased frequency of urination, stiffness, and inappetence. The most consistent clinical signs in horses include excessive salivation and lacrimation, colic, and diarrhea. Abortion in mares has been attributed to slaframine and has been experimentally reproduced in laboratory rodents.

- profuse salivation, lacrimation, diarrhea, bloat, stiffness, anorexia

Because slobbers is not a lethal disease, but rather one of economic impact due to physiologic changes in the animal, pathological changes are not distinctive.

Treatment—Slaframine intoxication may be at least partially responsive to atropine. Removal of the contaminated forage is usually followed by prompt alleviation of

- no treatment needed; signs readily reversible with change of feed signs of toxicosis, although a return to full milk production may be delayed. No effective means to detoxify contaminated feeds are known.

Photosensitization

- photosensitization, "trifoliosis," both primary and hepatogenous, the latter possibly due to mycotoxin

Disease Problems and Genesis—In some instances with *T. hybridum* and *T. repens,* photosensitization may be seen that appears to be a primary or direct type because it is readily reversible (Pammel 1920; Hansen 1928b; Fincher & Fuller 1942; Morrill 1943). However, some caution is warranted because it has not been specifically confirmed to be this type. Commonly referred to as trifoliosis or dew poisoning, it may be seen in any animal species but seems to be most common in horses and in young animals of other species. It may appear to be a contact dermatitis, but its expression is clearly correlated with the pigmentation pattern of the skin (Byrne 1937). It has been seen primarily in the wet season on warm, bright, sunny days. It is an acute disease with rapid recovery after animals are removed from the pasture. The cause of this sporadic photosensitization is not known, but it is quite similar in nature to that caused by other primary photosensitizers such as *Hypericum* and *Brassica.*

Another more chronic type of cutaneous disease also been reported to occur in horses grazing pastures of *T. hybridum.* This requires long-term ingestion—a year or more—and the skin changes appear to be due to hepatogenous photosensitization secondary to bile duct proliferation and periportal fibrosis (Schofield 1933; Traub et al. 1982; Nation 1989). An association between *T. hybridum* and liver disease was hypothesized, but the relationship has not been confirmed. The disease has been proposed to be a mycotic disease rather than a disease caused directly by the plants (Nation 1989). Infection of clover with the fungus *Cymodothea trifolii* has been associated with hepatogenous photosensitization in horses in Minnesota and Wisconsin, and mice that were fed infected foliage for 3 weeks had an increase in serum SDH (Murphy 1997).

- skin, erythema, swelling, edema, sloughing, pigmented areas; with or without elevation of serum hepatic enzymes

  in severe cases, head pressing, ataxia, other signs of neurologic disturbance

Clinical Signs—Signs include erythema, swelling, edema, and sloughing of the skin in the lightly pigmented areas. Often the extent of these signs is clearly demarcated at the edge of the darker areas. There may be accompanying pruritis, restlessness, loss of appetite, and diarrhea.

In the case of the hepatogenous type, marked elevation of serum hepatic enzymes (SDH, AST, AP) and blood ammonia may be expected, and in some cases icterus of the sclera and other mucous membranes is seen. There also may be signs of hepatoencephalopathy, such as incessant walking, head pressing, incoordination, depression alternating with excitement, coma, and death. In other situations a more chronic form may prevail, with weakness and weight loss.

- gross pathology, skin lesions, hepatogenous form, liver firm, nodular

  microscopic, liver necrosis, fibrosis, bile duct proliferation

Pathology—In addition to the crusty, ulcerative, and proliferative skin changes, there may be obvious liver involvement. In those instances the liver will be enlarged, firm, and nodular. Microscopically, some hepatocellular degeneration, vacuolation, and necrosis will be seen, but the most distinctive and dominating lesions are bile duct proliferation and periportal fibrosis. There also may be mild renal tubular nephrosis.

- nursing care, provide shade

Treatment—Preventing the animals from having access to the forage usually is sufficient to initiate recovery. The clinical signs of the disease regress slowly during healing of the skin. Liver involvement may result in slower resolution of the skin lesions and may require special dietary considerations.

Cyanogenesis

Disease Problems and Genesis—Although several species of *Trifolium* are reported to be cyanogenic (table 37.12), cyanogenesis is a rare consequence of ruminants grazing *T. repens* late in a growing season when rainfall has been poor or erratic (Coop 1949; Vickery et al. 1987).

- cyanogenic glycosides, linamarin, lotaustralin

Linamarin

Table 37.12. Cyanogenic species of *Trifolium*

T. hybridum	alsike clover
T. nigrescens	
T. pratense	red clover, purple clover
T. repens	white clover, Dutch clover, ladino clover
T. uniflorum	

Source: Seigler 1989.

The cyanogenic glucosides lotaustralin and linamarin are present in ladino clover in ratios ranging from 2:1 to 4:1, in addition to linamarase, a β-glucosidase, which, under certain environmental conditions or when the plant is mechanically damaged, is free to hydrolyze the glucosides (Conn 1979). Hydrolysis of the two glucosides results in formation of a cyanohydrin, which then can dissociate to form hydrogen cyanide. The reaction is catalyzed by hydroxynitrile lyase. Levels of these cyanogenic glycosides up to 1% fresh weight may be present in high-HCN genotypes of *T. repens* (Collinge & Hughes 1982). Expression of cyanogenesis is quite variable and is markedly influenced by environmental factors such as temperature and moisture stress. A complete discussion of these factors and cyanide toxicity is presented in the treatment of *Sorghum* (see chapter 58).

- low cyanogenic glycosides concentrations; very low risk

Trifolium repens exhibits considerable polymorphism for cyanogenesis. Populations are designated negative or positive with respect to their cyanogenetic potential. The negative types, or "sweet" types, seem to be more abundant, but positive populations, or "bitter" types, may be more productive and persistent (Rogers & Frykolm 1937). However, even in the positive populations, concentrations of the cyanogenic glucosides are seldom high enough to represent a risk. Unique circumstances are required for cyanide intoxication to occur (Gurnsey et al. 1977).

The polymorphic nature of *T. repens* is a function of the genes responsible for the synthesis of the two cyanogenic glucosides, as well as those for the hydrolytic glucosidase linamarase. Cyanogenic risk generally requires the presence of both of the dominant alleles of both genes. Possession of the dominant allele *Ac* produces moderate cyanogenesis; homozygous individuals lacking it are noncyanogenic (Corkill 1942). It is, therefore, a relatively straightforward task to breed selectively for low cyanogenesis. However, if, as hypothesized, cyanogenesis is a defensive strategy against herbivory, then *Trifolium* populations lacking cyanogenesis protection may not be as durable (Jones et al. 1978).

Even populations with glucoside concentrations too low to be a cyanogenic hazard may present a risk. In rare instances of prolonged ingestion of *T. repens,* formation of thiocyanate may present a goitrogenic risk. It also has been noted that lambs born of ewes grazing high-cyanide *T. repens* had markedly reduced erythrocyte glutathione peroxidase levels, suggesting a possible increased risk of nutritional myopathy (Lehmann et al. 1991).

Clinical Signs—In the rare event of toxicosis, an abrupt onset of apprehension and distress will occur within a few minutes of the animal's grazing the offending forage.

- abrupt onset of distress, apprehension, weakness, ataxia, labored respiration, collapse,

This is quickly followed by weakness, incoordination, and labored respiration. If the intoxication is severe, the animal may be unable to stand and may lay in lateral recumbency, with periodic paddling and tetanic seizures. The entire sequence of signs may occur over a 5-to-15-minute period, with either death or recovery at the end of this time.

■ no lesions

Pathology—Few distinctive changes are seen. For the most part, the changes are limited to congestion of the viscera and scattered, small splotchy hemorrhages.

■ sodium thiosulfate with or without sodium nitrate

Treatment—The primary antidote for ruminants is sodium thiosulfate given i.v. at a dose of 0.25–0.5 g/kg b.w. Sodium nitrite, given at 10 mg/kg b.w. i.v., can be used to augment the effects of the sodium thiosulfate, but it is not required for effective treatment. A more complete discussion of treatment for cyanogenesis is presented in the treatment of the Rosaceae (see chapter 64).

Vicia L.

■ *V. angustifolia*
V. faba
V. sativa
V. villosa

broad bean
common vetch
fava bean
hairy vetch
horse bean
narrow-leaved vetch
spring vetch
vetch
Windsor bean

Taxonomy and Morphology—Commonly known as vetch or tare, *Vicia,* a member of the subfamily Papilionoideae, comprises 140–150 species, occurring primarily in the northern temperate zone. Its generic name is derived from the Latin *vincire,* meaning "to bind," which reflects the clasping tendrils of most species. *Vicia* also is the classical Latin name for *tare*. Species of the genus are economically important as cover crops in orchards, erosion control along roadsides, green manure, livestock forage and silage, and wildlife food. One of the oldest cultivated crops, *V. faba* (broad bean) is used for food. In North America, some 30–35 native and introduced species are present. Intoxication problems have been associated with 4 species:

V. angustifolia L.	narrow-leaved vetch
V. faba L.	broad bean, horse bean, fava bean, Windsor bean
V. sativa L.	common vetch, spring vetch
V. villosa Roth	hairy vetch

■ herbs, vines; leaves 1-pinnately compound, tendrils rather than terminal leaflets; flowers pealike; petals blue, violet, purple, white, or yellow

Plants annuals or perennials; herbaceous; typically vines by means of tendrils. **Stems** climbing or trailing, or erect and short in a few species; terete or angular. **Leaves** subsessile; 1-pinnately compound; terminal leaflet absent; leaflets 4–20; blades linear or oblong, petiolules absent; rachises ending in tendrils; stipules small, leaflike, semisagittate. **Inflorescences** axillary racemes or 1 or 2 flowers in axils of leaves; bracts minute, caducous. **Sepals** 5; fused; calyces radially symmetrical or 2-lipped. **Corollas** papilionaceous. **Petals** 5; blue or violet or purple or white or yellow; clawed; banners obovate or oblong, emarginate, reflexed; wings oblique oblong, auriculate, adherent to keels; keels oblong, curved, obtuse. **Stamens** 10; diadelphous; enclosed in keels. **Legumes** linear to oblong; sessile or short stipitate; flattened or somewhat terete. **Seeds** 2 to many.

Classified in the same tribe, *Vicia* and *Lathyrus* are closely related and morphologically quite similar. The two genera differ in the appearance of their styles and winging of their stems. In *Vicia,* the style typically is terete and either pubescent all the way around or only on the abaxial or outer surface, whereas in *Lathyrus,* the style is flattened and pubescent only on its adaxial or inner surface. The stems of *Vicia* are never winged, whereas those of *Lathyrus* are typically so.

1. Inflorescences peduncled. Flowers 10–40; borne in elongate racemes. Stems and leaves densely villous . . . . *V. villosa*

1. Inflorescences sessile or subsessile. Flowers 1–5; borne solitary or in pairs or in short racemes. Stems and leaves glabrous or sparsely villous.
   2. Tendrils absent or reduced to setae. Legumes 8–20 cm long . . . *V. faba*
   2. Tendrils present. Legumes 1.5–7 cm long.
      3. Flowers 18–30 mm long. Mature legumes pale brown . . . . *V. sativa*
      3. Flowers 10–18 mm long. Mature legumes blackish . . *V. angustifolia*

Some taxonomists suggest that *V. angustifolia* should be treated as a variety of the highly variable *V. sativa.*

Vicia americana

Distribution and Habitat—Circumboreal in grassy and woodland habitats, *Vicia* has centers of diversity in the Mediterranean basin and eastern Asia. The 4 species of toxicologic interest in North America are native to Europe. *Vicia villosa, V. sativa,* and *V. angustifolia* are widely cultivated alone or with small grains such as rye and wheat. They have escaped and naturalized in abandoned fields, along roadsides, and in waste areas, often forming dense stands and readily reseeding. *Vicia faba* is occasionally cultivated in vegetable gardens and rarely escapes.

Disease Problems—Although the foliage of all species of *Vicia* is edible and palatable to livestock, three disparate types of disease problems have been associated with the genus; the toxins in the seeds apparently differ from those in the foliage. Two types—a neurologic disease and a dermatopathy—are described separately in the following subsections. A third, not reported to occur in North America, is favism, or fabism, an acute hemolytic anemia produced by the seeds of *V. faba,* especially when undercooked, in persons with the genetic trait of low levels of glucose-6-phosphate dehydrogenase (Watt & Breyer-Brandwijk 1962; Kosower & Kosower 1967). Interestingly, these genetically unusual individuals also develop the disease upon inhalation of the species' pollen. The seeds of *V. faba,* when fed uncooked, may cause small intestinal inflammatory lesions similar to those caused by the lectins in species of *Phaseolus,* treated below in this chapter (Rubio et al. 1989).

Neurologic Disease

Disease Problems—Ingestion of the seeds of *Vicia* produces a neurologic disease. In cattle, small amounts of *V. villosa* seeds caused signs similar to those of rabies, including bellowing and seizures and in some instances sudden death (Claughton & Claughton 1954; Panciera 1978). Incorporation of seeds of *V. sativa* or *V. villosa* in poultry rations produces growth depression and death in the birds. However, the effects are markedly attenuated by heating, similar to the heat-labile actions produced by lathyrogens (Thayer & Heller 1946; Kienholz et al. 1962; Harper & Arscott 1962; Arscott & Harper 1964). Seeds of *V. angustifolia* are also neurotoxic in ducks and monkeys (Anderson et al. 1924). In a reported ingestion of 50 or more seeds of *V. tetrasperma* by a 3-year-old child, no adverse effects were noted other than protracted vomiting from administration of syrup of ipecac (Wine & Johnson 1993).

Vicia villosa

- crop plants; naturalized
- 3 types of disease—neuropathy, dermatopathy, favism; favism not generally a problem in North America

Disease Genesis—Two types of toxicants have been identified in the seeds of *Vicia.* Cyanogenic glycosides—vicianine, a benzaldehyde-type, and l-mandelonitrile-vicianoside, an unusual diglucoside—have been isolated from *V. angustifolia* and *V. sativa* (Bertrand 1906; Roseveare 1948). However, experimental studies with rats failed to verify that these compounds are hazardous (Ruby et al. 1955).

Seeds of these two species also contain β-cyano-L-alanine and its bound form,

- livestock, rare disease from eating large amounts of seeds, neurologic; poultry growth depression
- possible causes, neurotoxic, β-cyanoalanine and its derivative

```
          NH2
          |
N≡C—C—C—COOH
```

β-Cyanoalanine

```
          NH-Glut
          |
N≡C—C—C—COOH
```

γ-Glutamyl-β-cyanoalanine

Vicia angustifolia

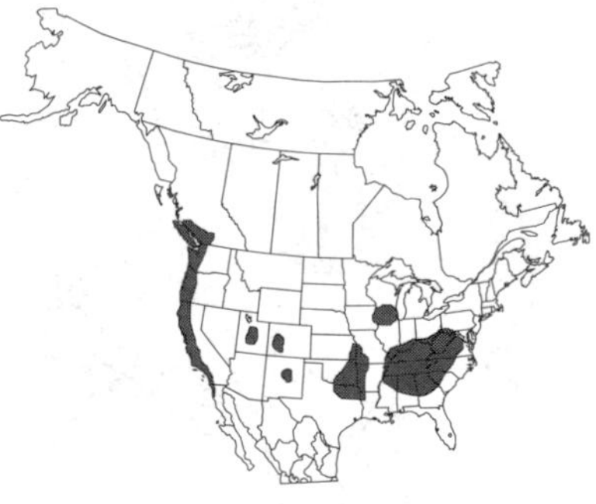

Vicia sativa

Vicia villosa

γ-L-glutamyl-β-cyano-L-alanine (Ressler et al. 1963). Given orally, they are most toxic in poultry. Typically, a diet of 50% seeds is lethal in chicks but produces little effect in rats (Ressler et al. 1969). These two toxins cause excitement, incoordination, respiratory distress, seizures characterized by head retraction and arched back, and death when fed for several days to poultry (Harper & Arscott 1962; Ressler 1962; Arscott & Harper 1963). The bound form is only about half as toxic as the free amino acid (Ressler et al. 1963). The levels of the toxins vary considerably, and in some cases they are low enough that seeds can be fed safely to poultry (Darre et al. 1998). However, in these instances, low concentrations of toxins may accumulate in the animal tissues. The seeds can be detoxified by cooking and washing (Ressler et al. 1997). β-Cyanoalanine given parenterally to rats is highly lethal but, when administered orally, causes only increased irritability, some overall increase in mortality, and an elevation of excretion of cystathionine in the urine. β-Cyanoalanine inhibits cystathionase and interferes with transsulfuration, thereby increasing the dietary requirements for sulfur amino acids (Pfeffer & Ressler 1967; Darre et al. 1998). Although these effects are similar to those of a pyridoxine deficiency, pyridoxine HCl is not clearly protective (Ressler et al. 1964, 1967). Pyridoxal HCl, but not pyridoxine HCl, is protective against parenterally administered β-cyanoalanine in rats, but not against the effects induced by oral doses in either rats or chickens (Ressler et al. 1967). Although the seeds of *V. villosa* can cause an acute, rapidly fatal disease in cattle, they have been shown to contain only small amounts of bound and no free β-cyanoalanine (Claughton & Claughton 1954; Panciera 1978).

Dermatopathy

Disease Problems—Although the seeds of *Vicia* produce neurologic effects, its foliage is associated with one or more subacute cutaneovisceral diseases when consumed in pastures (Panciera et al. 1966; Panciera 1978). The most consistently produced is a syndrome of dermatitis and visceral granulomatous disease appearing primarily during periods of vigorous plant growth (Panciera et al. 1992). Most cases have been from Oklahoma and Texas and involve *V. villosa,* but the disease has been seen in other states and countries and is associated with other species and hybrids, e.g., *V. dasycarpa* and *V. benghalensis* (Burroughs et al. 1984; Peet & Gardner 1986; Odriozola et al. 1991; Johnson et al. 1992; Harper et al. 1993). The occurrence of this disease is very sporadic, often appearing in pastures where *Vicia* has been grazed for many years without incident. Furthermore, the incidence of appearance is quite low relative to the vast acreage planted with vetch for forage. Most cases involve mature black cattle, such as Angus or Holstein, although horses also are apparently affected (Anderson & Divers 1983; Woods et al. 1992). Clinical signs appear 3–4 or more weeks after pasturing. The morbidity is usually low, approximately 10%, but may be as high as 40–50%. The case-fatality rate typically exceeds 50%.

Disease Genesis—The toxicant producing this dermatopathy is unknown, as are the factors that contribute to the disease's sporadic appearance. Experimental information likewise is limited. The disease has been experimentally produced in an Angus cow afflicted by the disease 1 year earlier, but not in her 1-year-old calf. It was suggested that the disease may be due to a type IV hypersensitivity reaction, with a granulomatous response and multinucleated giant cell formation (Panciera et al. 1992). T-lymphocyte activation by *Vicia* lectin is a possible alternative mechanism.

Clinical Signs—Manifestations of this vetch pasture disease begin as a rapidly progressive dermatitis with papular swellings, plaques, and thickening of the skin. This leads to an exudative dermatitis, with matting of the hair by a yellowish fluid and development of crusts covering areas of epithelial sloughing. Accompanying signs are pruritis and eventual loss of hair. A thickened and fissured or folded, lichenified appearance of the skin results. The changes occur initially on the neck, tailhead, escutcheon, udder, and teats and subsequently progress to the face, trunk, and limbs, independent of the animal's pigmentation pattern. As the skin changes progress, keratoconjunctivitis and/or diarrhea with or without blood may develop. The temperature is often elevated, and other signs such as cough and rapid breathing also may be seen. The animals usually have a diminished appetite and rapidly lose considerable weight, so that they are in poor condition by the time skin lesions are prominent. There may be a transient period of apparent recovery, with a renewed interest in eating after several days, but the usual outcome is death in 10–14 days. The disease may present a similar appearance to subacute to chronic streptothricosis caused by *Dermatophilus congolensis* or to severe primary photosensitization; however, the morbidity and case-fatality rates, lack of therapeutic response, and accompanying pathology readily differentiate these diseases.

- cattle, horses; mainly *V. villosa* in black cattle; dermatitis and visceral granulomatous disease, grazing vetch in pasture for several weeks; often lethal
- unknown cause, difficult to reproduce, possibly immunologic mediated
- abrupt onset, progressive, papular swellings, plaques, thickening, exudation, crusting of skin, independent of pigmentation; weight loss, diarrhea, anorexia, febrile; may appear to improve but relapse and die

Pathology—At necropsy, in addition to the obvious skin lesions, a yellowish gray infiltrate is commonly seen in the dermis, adrenals, myocardium, and renal cortex extending from the surface of the kidney. Less commonly, this infiltrate may be seen in the liver, spleen, lymph nodes, and thyroid. Microscopically, the infiltrate is composed of macrophages, lymphocytes, plasma cells, eosinophils, and multinucleated giant cells (Panciera 1978).

- gross pathology, skin lesions, gray streaks in many tissues

  microscopic, macrophage, lymphocyte, plasma cell infiltrate in liver, kidneys, spleen, lymph nodes, thyroid

Treatment—Treatment is nonspecific and similar to that for primary photosensitization. The prognosis must be considered poor. Recommendations for prevention of the disease are difficult to formulate because it is so sporadic in appearance and may not be seen again in the same pasture. However, all animals should be removed immediately from the pasture when skin abnormalities are noted.

- nursing care, no specific treatment

Wisteria Nutt.

Taxonomy and Morphology—A member of the subfamily Papilionoideae, *Wisteria* comprises 6–10 species in eastern North America and eastern Asia. Its generic name honors Caspar Wistar, a professor of anatomy at the University of Pennsylvania in the late 1700s and early 1800s. Vigorous climbers, the Asiatic species are commonly grown for their attractive foliage and quite showy racemes. In North America, 2 native and 2 introduced species are typically encountered:

- *W. floribunda*
  W. frutescens
  W. macrostachya
  W. sinensis

  wisteria

W. floribunda (Willd.) DC.	Japanese wisteria
W. frutescens (L.) Poir.	American wisteria, Atlantic wisteria
W. macrostachya (Torr. & A.Gray) Nutt.	Kentucky wisteria, Mississippi wisteria
W. sinensis (Sims) Sweet	Chinese wisteria

Some taxonomists consider the native species *W. frutescens* and *W. macrostachya* to be 2 varieties of a single species.

Plants perennials; woody vines; without tendrils. **Stems** twining; high climbing; 3–15 m long; older stems becoming trunklike, gnarled; bark gray-brown, fissured. **Leaves** 1-pinnately compound; leaflets 9–19; terminal leaflet present; petiolules present; stipels present; stipules small, caducous, herbaceous, well developed or forming

- perennial woody vines; leaves 1-pinnately compound, terminal leaflets, no tendrils; flowers pealike; petals blue, violet-blue, purple, pink, or white; legumes flattened

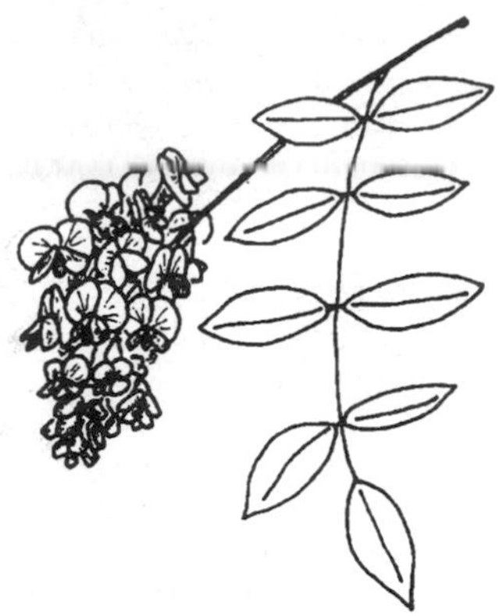

Wisteria floribunda

sheaths. **Inflorescences** pendulous racemes; borne at ends of short branches; elongate; bracts caducous. **Sepals** 5; fused; calyces 2-lipped. **Corollas** papilionaceous. **Petals** 5; blue or violet-blue or purple or pink or white; clawed; banners suborbicular, reflexed, with 2 hornlike lobes or appendages above claw; wings obovate, falcate, auriculate, apices coherent; keel petals free, curved, apices obtuse. **Stamens** 10; diadelphous; enclosed in keels. **Legumes** linear-oblong to clavate; flattened; slightly torulose; tardily dehiscent; pericarp coriaceous. **Seeds** 2–12; large; reniform.

1. Legumes glabrous. Wings with auricles parallel to claws. Plants native.
   2. Leaflets 9–15. Racemes 4–15 cm long. Upper calyx lobes shorter than tubes . . . . *W. frutescens*
   2. Leaflets 9. Racemes 15–30 cm long. Upper calyx lobes as long as tubes . . . . *W. macrostachya*
1. Legumes velvety-pubescent. Wings with auricles divergent from claws. Plants introduced.
   3. Leaflets 15–19. Stems twining clockwise. Flowers 1.5–2 cm in diameter . . . . *W. floribunda*
   3. Leaflets 7–11. Stems twining counterclockwise. Flowers 2.5–3 cm in diameter . . . . *W. sinensis*

■ cultivated ornamentals; woodlands

Distribution and Habitat—Members of the genus are native to the mild temperate parts of China, Japan, and the United States. They are usually found in low, moist alluvial soils in woods and along stream banks. The native species are distributed in the southeastern quarter of the continent. Extensively cultivated, the introduced ornamentals *W. floribunda* and *W. sinensis* are encountered occasionally as escapes in the same area, especially southward.

Wisteria frutescens

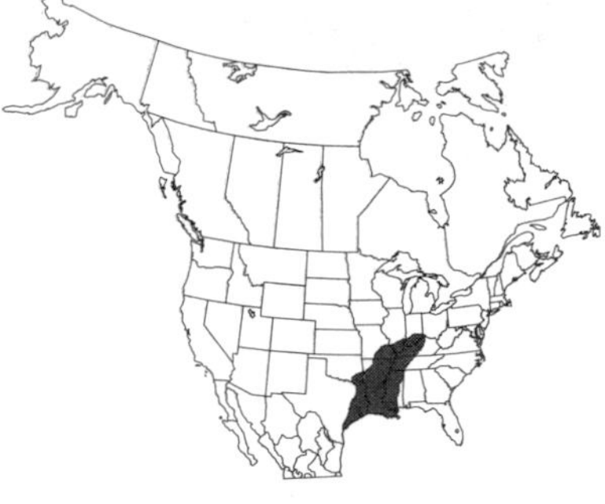

Wisteria macrostachya

Disease Problems—Prized for their beautiful blossoms in large pendulous racemes, plants of *Wisteria* are commonly cultivated. Mixed in batter to make fritters, the flowers also are eaten (Fernald & Kinsey 1943). Although there have been few reports of toxicologic problems, seeds of the genus have a reputation for toxicity. A 4-year-old and 5-year-old brother and sister developed nausea and bloodstained vomiting within 30 minutes of eating several seeds (Jacobziner & Raybin 1961). In an Italian case, 2 youths each ate 5 or 6 seeds and developed digestive problems (Piola et al. 1983). In both cases, recovery was complete 2–3 days after ingestion. In another instance, a 50-year-old woman developed nausea and a severe headache 3.5–4.5 hours after eating 10 seeds (Rondeau 1993). At 6 hours, the initial signs were followed by those of severe distress of the digestive system. She exhibited vomiting (with blood), diarrhea, colic, sweating, dizziness, and fainting. Her blood glucose was mildly elevated at 137 mg/dl, as was the number of polymorphonuclear white blood cells. The woman was treated in an emergency room and discharged, but signs of the problem persisted for 7 days.

■ animals or humans, digestive disturbance

Disease Genesis—Like the seeds of many members of this family, the seeds of *W. floribunda* contain a glycoprotein lectin, which in this case binds to *N*-acetyl-D-galactosamine (Kurokawa et al. 1976). This binding prevents replacement of the cells of the mucosal layer, especially in the small intestine, and is reflected in the severe effects on the digestive system.

■ proteinaceous lectin affecting mucosal cells of intestine; effects may be severe

Clinical Signs, Pathology, and Treatment—As illustrated in the case studies cited, the signs of *Wisteria* intoxication reflect gastrointestinal dysfunction. Most commonly seen are vomiting, diarrhea, colic, weakness, and perhaps mild neurologic changes.

Recovery may require several days. Necropsy lesions are usually restricted to reddening of the mucosa of the stomach and small intestine. Treatment is symptomatic, directed toward alleviating the pain and inflammation accompanying the gastroenteritis. When given early, activated charcoal may be beneficial.

- vomiting, diarrhea, colic, weakness, gross pathology, reddening of mucosa of intestine

  treat with activated charcoal, fluids, electrolytes, antidiarrheal drugs

Genera of Questionable Toxicity

Desmanthus Willd.

A relatively small genus of 24 species in the predominantly woody subfamily Mimosoideae, *Desmanthus* is unusual because most of its species are perennial herbs. Native to the New World and generally regarded as nutritious and readily eaten by livestock (Phillips 1959a), plants of this genus are not associated with intoxication problems with the exception of one report. A prairie and plains species from Texas to Kansas, *D. leptolobus* Torr. & A.Gray, commonly known as prairie or slender-lobed bundleflower, was suspected as the cause of severe colic in horses in Texas. However, subsequent experimental studies in 2 horses fed 4 g/kg b.w. (0.4%) of *D. leptolobus* for 9 days failed to confirm the species as a cause of the observed effects (Harvey et al. 1986). An indole alkaloid, 2-hydroxy-*N*-methyltryptamine, has been isolated from the roots of *D. illinoensis* (Miehx.) MacMill. ex Rob. & Fernald (Illinois bundleflower), but its toxicologic significance is not known (Southon & Buckingham 1989).

- *D. leptolobus*

  bundleflower

- perennial herbs, curved legumes clustered

  nontoxic

Mimosa L.

A member of the subfamily Mimosoideae, *Mimosa* is primarily a New World genus comprising 400–450 species, which are commonly referred to as the mimosas. This common name also is used for several species of the genus *Acacia* and for the rather common ornamental tree *Albizia julibrissin.* The plant known to florists as mimosa is a species of *Acacia.* In North America, there are 19 native and introduced species of the genus *Mimosa,* including the ornamental *M. pudica* L. (sensitive plant, touch-me-not, shame plant, live-and-die plant, humble plant, or mimosa). *Mimosa pudica* is often cultivated as a novelty because it exhibits "sleep" and "shame" when its leaflets fold and its petioles droop at night, when they are touched, or when temperature or air pressure drops abruptly. This movement is due to rapid loss of turgor in the pulvinus cells or to the action of contractile proteins.

- *M. pudica*

  humble plant
  mimosa
  sensitive plant
  shame plant
  touch-me-not

Although specific reports of intoxication problems caused by species of *Mimosa* are lacking, the plants have been found to contain several potential toxicants. The genus is included here because there is the possibility that it could cause problems similar to those caused by *Leucaena* because of the presence of mimosine, and because of the application of its common name to *Albizia* and *Acacia,* which do cause significant problems. As with *Leucaena,* problems would occur only if the plant is a major component of the diet, an event unlikely in most instances because of the spiny or thorny herbage. The principal toxicant is mimosine (β-*N*-[3-hydroxy-4-pyridone]-α-amino-propionic acid), which is degraded to the toxic 3-hydroxy-4-(1H)-pyridone (3,4-DHP) by enzymes in the plant when its tissues are macerated or by bacteria in the rumen (Hegarty et al. 1964; Lowry et al. 1983). Mimosine is a structural analogue and minor antagonist of enzymatic reactions requiring pyridoxine. Both mimosine and 3,4-DHP act as inhibitors of DNA and, to a lesser extent, of RNA synthesis, possibly by metal cofactor chelation (Tsai & Ling 1971).

- no reports of intoxication, contains mimosine, willardiine, saponins

In addition to mimosine, triterpenoid/steroidal saponins, known as mimonosides

A, B, and C, are present (Anton et al. 1993). Also present is willardiine, a heterocyclic excitatory amino acid acting on kainate receptors in the cerebrum and also present in the seeds of species of *Acacia* (Patneau et al. 1992). Bark extracts of the Central American species, *M. tenuifolia* Blanco, also contain alkaloids that have been shown experimentally to have an effect on smooth muscle in rats and guinea pigs, thus decreasing intestinal peristalsis and increasing gastric and uterine contractions (Meckes-Lozoya et al. 1990).

Mucuna Adans.

■ *M. pruriens*
velvet bean

■ seeds as charms
possibly neurotoxic?

A tropical genus of approximately 100 species in both the Old World and the New World, *Mucuna* is a member of the subfamily Papilionoideae. It is represented in North America by 4 introduced species limited in distribution to Florida, California, and parts of Mexico. Most commonly encountered is *M. pruriens* (L.) DC. (velvet bean), a twining vine with trifoliate leaves and hispid legumes grown as an ornamental and a forage. Used as good-luck charms, the large, brown seeds of *M. sloanei* Fawc. & Rendle (ojo de venado, or eye of the deer), are purported to contain high concentrations (350 mg) of L-dopa, which may pose a hazard to children and pets (Sullivan & Chavez 1981). However, this risk has not been confirmed experimentally or clinically. The foliage of some species is used for livestock forage, but in some instances digestive disturbances occur (Blohm 1962).

Pachyrhizus Rich. ex DC.

■ *P. erosus*
chopsui potato
jicama
yam bean

■ uncooked legumes, digestive disturbance, mild diarrhea

Comprising 6 species in the New World tropics, *Pachyrhizus* is a member of the subfamily Papilionoideae. It is placed in the same subtribe as the previously described *Canavalia* (jack bean). Its generic name is derived from the Greek roots *pachys* and *rhizon,* meaning "thick" and "root," which reflect its large, tuberous roots. In North America, the genus is represented by a single species, *P. erosus* (L.) Urb. (= *P. angulatus* Rich. ex DC., = *P. palmatilobus* Benth. ex Hook.f., yam bean, chopsui potato, or jicama. This perennial, herbaceous vine has naturalized in Florida and along the Gulf Coast, where it has been cultivated for its massive tuberous roots, which are eaten raw or cooked. The legumes also are eaten as a vegetable when thoroughly cooked beforehand. Ingestion of the raw legumes or seeds causes irritation of the digestive tract and mild to moderate diarrhea of short duration, because of the presence of terpenoids. Typically, treatment is unnecessary, except in instances when dehydration is severe enough to require administration of fluids. The seeds also are reported to contain rotenone and pachyrrhizid and have been used as fish poisons and insecticides (Blohm 1962).

Phaseolus L.

■ *P. lunatus*
P. vulgaris
black bean
Burma bean
Java bean
kidney bean
lima bean
navy bean
pinto bean

A large genus of the subfamily Papilionoideae, *Phaseolus* includes many of the cultivated bean plants. Problems are generally limited to *P. lunatus* L. (lima bean, Java bean, Burma bean), and *P. vulgaris* L. (kidney bean, black bean, pinto bean, navy bean). Two types of problems are observed. The first 2 species are cyanogenic under some conditions, and the third contains factors that alter animal growth and cause mortality. Rarely encountered in North America, the seeds and foliage of *P. lunatus* contain up to 20 mg/100 gm of linamarin, formerly known as phaseolunatin (Dunstan & Henry 1903; Pammel 1917a; Gibbs 1974). The seeds also contain lotaustralin. Of more importance is the occasional problem of cyanogenesis in the foliage of *P. limensis* Macfad. Although probably of rare occurrence, the foliage of cultivated lima beans has either caused death of livestock or been noted to have high cyanogenic activity (Pammel 1921; Reynard & Norton 1942; Fuller & McClintock 1986). This activity was noted a few hours after a frost or when the plants were harvested for the beans while still green. However, cyanogenesis seems to be an unlikely problem with *Phaseolus*. Although in one instance deaths of pigs was reported, the specific cause was not reported, and cyanide intoxication from forage is not typically a problem in pigs.

■ *P. lunatus*, foliage and legumes, weakly cyanogenic

P. vulgaris, lectins in seeds, beans toxic when eaten raw, digestive disturbance

Problems with *P. vulgaris* are related to lectins in the seeds. The raw beans, when eaten by rats, pigs, and other species produce antinutritional effects, limiting growth and increasing mortality (Jaffe & Vega Lette 1968; Banerjee & Hogue 1975; Hove et al. 1978; Bigbie & King 1981; Myer 1981). When the seeds are heated, the toxicity potential is eliminated. In humans, consumption of kidney beans that have been allowed to soften for a prolonged period in water but not cooked also may cause acute digestive disturbances (Epidemiological Research Laboratory 1976; Noah et al. 1980). The adverse effects are caused by heat-labile hemagglutinins that bind to and destroy the cells of the intestinal tract lining. Secretions from the goblet cells increase, and absorption and digestion are impaired (Pusztai 1989).

When fed in moderate amounts to rats and pigs, there was mainly weight loss, with some deaths, among the rats. When humans eat large amounts, they may experience vomiting and diarrhea. Treatment is aimed at relief of the signs.

Psoralea L.

■ *P. macrostachya*
P. subacaulis
P. tenuiflora

Indian pea
scurf pea

A member of the subfamily Papilionoideae, *Psoralea*, was long interpreted to consist of some 130 species present in both the Old World and the New World. Legume taxonomists, however, have divided the genus into 6 or 9 genera and restricted its circumscription to 20 species endemic to the Mediterranean area of Africa (Stirton 1981). The 35–50 North American species are thus classified in 4 genera: *Hoita*, *Orbexilum*, *Pediomelum*, and *Psoralidium*. Because not all taxonomists accept this classification and because most of the toxicological literature uses the name *Psoralea*, it is used in this treatise. Representative species of these herbaceous perennials include *P. macrostachya* DC. (= *Hoita macrostachya* [DC.] Rydb.); *P. subacaulis* Torr. & A.Gray (= *Pediomelum subacaule* [Torr. & A.Gray] Rydb.); and *P. tenuiflora* Pursh

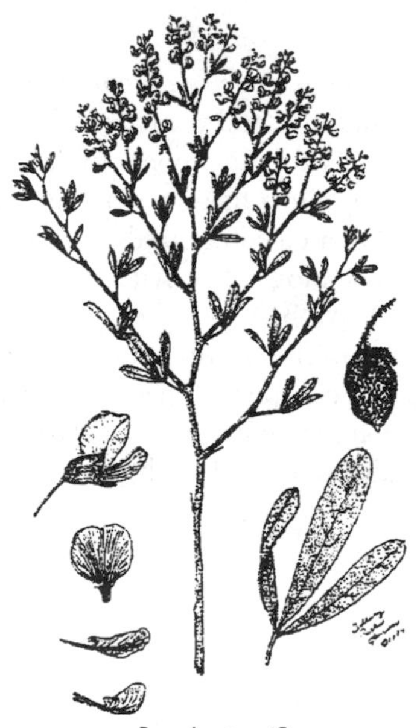

Psoralea tenuiflora

Psoralea macrostachya

Psoralea subacaulis

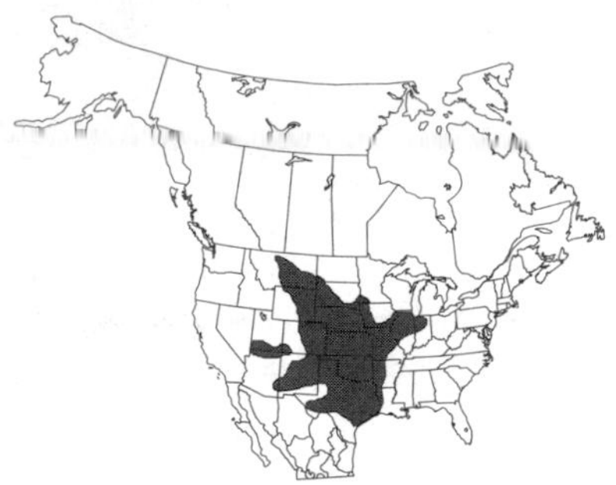

Psoralea tenuiflora

- furanocoumarins, psoralen, pseudopsoralen, possible photosensitization

Pseudopsoralen

(Indian turnip). The major groups of species in the genus are differentiated on the basis of leaf and leaflet appearance, calyx appearance, and nature of legume dehiscence. Species occupy a variety of habitats in open forests, grasslands, and desert shrub communities throughout North America.

The toxicologic potential of members of this genus is not clear. Marsh (1929) considered *P. tenuiflora* to be toxic to livestock but so distasteful that it was never eaten. Leaflets and young fruit of *P. canescens* and *P. subacaulis* fed to mice and chicks up to 2% b.w. failed to cause adverse effects (Duncan et al. 1955). However, *P. subacaulis* and several other species, including *P. acaulis* Stev. ex M.Bieb., *P. bituminosa* L., *P. corylifolia* L., *P. drupacea* Bunge, *P. macrostachya,* and *P. plicata* Delile contain phototoxic furanocoumarins (Murray et al. 1982). The risk for livestock is generally low because the furanocoumarins are present mainly in the fruits and seeds except in *P. subacaulis,* in which the furanocoumarin psoralen is present in all plant tissues, and *P. glandulosa* L. and *P. macrostachya,* which contain the linear psoralen and the toxic angular angelicin, respectively, in the leaves (Scott et al. 1976; Innocenti et al. 1991; Cappalletti et al. 1992). Transfer of these furanocoumarins to animals also may occur by topical contact with glands on the surfaces of the plants.

Furanocoumarins are activated by long-wavelength UV light (320–380 nm). The interaction between them and DNA in the cells of the animal results in tissue destruction due to alterations in the lipid portion of the membranes. This leads to signs such as erythema and edema of the skin of the muzzle, ears, udder, teats, and vulva. In some cases, the lesions may be limited to areas of direct contact with the plants; in others, they may be more widely distributed in less protected and minimally pigmented skin areas. Blisters may form in the skin, with exudation and subsequent ulceration. Involvement of the eyes may also be seen, with sensitivity to light, cloudy corneas, and inflammation of the cornea and conjunctiva observed. The disease is self-limiting, and treatment is not generally required. Prevention of animal access to the plants is sufficient to allow recovery. However, recovery in severe cases may require several weeks. In the meantime, nursing offspring may require supplemental feeding. A compete discussion of the role of furanocoumarins is given in the treatment of genera in the family Apiaceae that cause photosensitization (see chapter 7).

Tephrosia Pers.

- *T. virginiana*

  rabbit pea
  devil's shoestring

- toxicity potential unknown

Species of *Tephrosia* are of unknown toxicity potential in North America. Some are reported to be toxic to insects and fish, and the roots of *T. virginiana* (L.) Pers. (rabbit pea, devil's shoestring) were used by members of American Indian tribes in the southeastern states as a vermifuge and a fish poison (Sievers et al. 1938). Extracts of the roots contain rotenone and related compounds, but the levels appear to be quite variable, and in some locations biological activity is nil (Clark 1933; Sievers et al. 1938). When fed to chicks at 1.6% b.w., *T. virginiana* was nontoxic, which is not surprising, considering the low toxicity of rotenone and the variability in its concentrations (Duncan et al. 1955). In contrast, a species in Africa causes adverse effects on the brain, liver, kidney, and digestive system when large amounts (1–5% b.w.) are fed to goats for 1–2 weeks (Suliman et al. 1982b).

References

Abbott RJ: A quantitative association between soil moisture content and the frequency of the cyanogenic form of *Lotus corniculatus* L. at Birsay, Orkney. *Heredity* 38;397–400, 1977

Abdullah MI, Barakat IE, Games DE, Ludgate P, Mavraganis VG, Ratnayake VU, Jackson AH: Studies of *Erythrina* alkaloids, part 3: G.C./M.S. investigations of alkaloids in the seeds of a further fourteen species. *Ann Mo Bot Gard* 66;533–540, 1979

Abdusalamov BA, Aslanov KhA, Sadykov AS, Khoroshkova OA: Level of alkaloids in *Sophora japonica. Khim Prir Soedin* 1972;658, 1972 (*Chem Abstr* 78;108230Z, 1973)

Abraham DJ, Sidebotham R, Winchester BG, Dorling PR, Dell A: Swainsonine affects the processing of glycoproteins in vivo. *FEBS Lett* 163;110–113, 1983

Abtahi FS, Auletta FJ, Sadeghi D, Djahanguire B, Scommegna A: Effect of sparteine sulfate on uterine prostaglandin F in the rat. *Prostaglandins* 16;473–482, 1978

Adams HR, Camp BJ: The isolation and identification of three alkaloids from *Acacia berlandieri. Toxicon* 4;85–87, 1966

Adams NR: Phytoestrogens. In *Toxicants of Plant Origin,* vol 4, *Phenolics,* Cheeke PR ed, CRC Press, Boca Raton, Fla, pp23–51, 1989

Adams NR: Permanent infertility in ewes exposed to plant oestrogens. *Aust Vet J* 67;197–201, 1990

Adams NR, Hearnshaw H, Oldham CM: Abnormal function of the corpus luteum in some ewes with phyto-estrogenic infertility. *Aust J Biol Sci* 34;61–65, 1981

Adams R, Van Duuren BL: Usaramoensine, the alkaloid in *Crotalaria usaramoensis* E.G.Baker. Integerrimine from *Crotalaria incana* Linn and senecionine from *Senecio glabellus* D.C. Stereochemical relationships. *J Am Chem Soc* 75;4631–4635, 1953

Adler HE: Indigestion from an unbalanced kiawe (mesquite) bean diet. *J Am Vet Med Assoc* 115;263, 1949

Adler JH, Trainin D: A hyperoestrogenic syndrome in cattle. *Refu Vet* 17;108–115, 1960

Affleck H: Jack bean poisoning in cattle. *Rhod Agric J* 58;21, 1961

Aguilar MI, Giral F, Espejo O: Alkaloids from the flowers of *Erythrina americana. Phytochemistry* 20;2061–2062, 1981

Akiyama T, Ishida J, Nakagawa S, Watanabe S, Itoh N, Shibuya M, Fukami Y: Genistein, a specific inhibitor of tyrosine-specific protein kinases. *J Biol Chem* 262;5592–5595, 1987

Allen DR: *Laburnum* poisoning in cattle. *Vet Rec* 60;586, 1948

Allen JG: Recent advances with cultivated lupines with emphasis on toxicological aspects. In *Poisonous Plants: Proceedings of the Third International Symposium,* James LF, Keeler RF, Bailey EM Jr, Cheeke PR, Hegarty MP eds, Iowa State Univ Press, Ames, pp229–233, 1992

Allen JG, Steele P, Masters HG, Lambe WJ: A lupinosis-associated myopathy in sheep and the effectiveness of treatments to prevent it. *Aust Vet J* 69;75–81, 1992

Allen JR, Chesney CF: Effect of age on development of cor pulmonale in nonhuman primates following pyrrolizidine alkaloid intoxication. *Exp Mol Pathol* 17;220–232, 1972

Allen JR, Carstens LA, Olson BE: Veno-occlusive disease in *Macaca speciosa* monkeys. *Am J Pathol* 50;653–667, 1967

Alstad AD, Casper HH, Johnson LJ: Vitamin K treatment of sweet clover poisoning in calves. *J Am Vet Med Assoc* 187;729–731, 1985

Alston TA, Mela L, Bright HJ: 3-Nitroproprionate, the toxic substance of *Indigofera,* is a suicide inactivator of succinate dehydrogenase. *Proc Natl Acad Sci* 74;3767–3771, 1977

Alves EW, Teixeira Da Silva AT, Teixeira Da Silva C, Carlini CR: Effects of canatoxin on the Ca^{2+}-ATPase of sarcoplasmic reticulum membranes. *Toxicon* 30;1411–1418, 1992

Anderson CA, Divers TJ: Systemic granulomatous inflammation in a horse grazing hairy vetch. *J Am Vet Med Assoc* 183;569–570, 1983

Anderson LAP, Howard A, Simonsen JL: Studies on lathyrism. *Indian J Med Res* 12;613–643, 1924

Anderson RC, Rasmussen MA, Allison MJ: Metabolism of the plant toxins nitropropionic acid and nitropropanol by ruminal microorganisms. *Appl Environ Microbiol* 59;3056–3061, 1993a

Anderson RC, Rasmussen MA, Allison MJ. Reduction of nitropropanol by ruminal microorganism [abstract]. *J Anim Sci* 71 (Suppl 1);204, 1993b

Anton R, Jiang Y, Weniger B, Beck JP, Rivier L: Pharmacognosy of *Mimosa tenuiflora* (Willd.) Poiret. *J Ethnopharmacol* 38;153–157, 1993

Arnold GW, Hill JL, Maller RA, Wallace SR, Carbon BA, Nairn M, Wood MR, Weeldenburg J: Comparison of lupine varieties for nutritive value as dry standing feed for weaner sheep and for incidence of lupinosis. *Aust J Agric Res* 27;423–435, 1976

Arscott GH, Harper JA: Relationship of 2,5-diamino-4,6-diketopyrimidine, 2,4-diaminobutyric acid, and a crude preparation of β-cyano-L-alanine to the toxicity of common and hairy vetch seed fed to chicks. *J Nutr* 80;251–254, 1963

Arscott GH, Harper JA: Evidence for a difference in toxicity between common and hairy vetch seed for chicks. *Poult Sci* 43;271–273, 1964

Aslanov KA, Kushmuradov YK, Sadykov AS: Lupine alkaloids. In *The Alkaloids,* vol 31, Brossi A ed, Academic Press, San Diego, Calif, pp117–192, 1987

Astorga JB: Maternal ingestion of locoweed: effects on lamb bonding and behavior. PhD dissertation, Utah State Univ, 1992

Auchterlonie L: *Laburnum* poisoning [letter]. *Vet Rec* 48;633, 1948

Aust SD: *Rhizoctonia leguminicola*—slaframine. In *Mycotoxins,* Purchase IFH ed, Elsevier Scientific, Amsterdam, pp97–109, 1974

Austin AR, Aston K, Drane HM, Saba N: The fertility of heifers consuming red clover silage. *Grass Forage Sci* 37;101–106, 1982

Aylward JH, Court RD, Haydock KP, Strickland RW, Hegarty MP: *Indigofera* species with agronomic potential in the tropics: rat toxicity studies. *Aust J Agric Res* 38;177–186, 1987

Bachman SE, Galyean ML, Smith GS, Hallford DM, Graham JD: Early aspects of locoweed toxicosis and evaluation of a mineral supplement or clinoptilolite as dietary treatments. *J Anim Sci* 70;3125–3132, 1992

Baker DC, Keeler RF: *Thermopsis montana*–induced myopathy in calves. *J Am Vet Med Assoc* 194;1269–1272, 1989

Baker DC, James LF, Hartley WJ, Panter KE, Maynard HF, Pfister J: Toxicosis in pigs fed selenium-accumulating *Astragalus* plant species or sodium selenate. *Am J Vet Res* 50;1396–1399, 1989

Ball DM, Hoveland CS, Lacefield GD: *Southern Forages,* Potash & Phosphate Institute, Norcross, Georgia, 1991

Balls LD, James LF: Effect of locoweed (*Astragalus* spp.) on reproductive performance of ewes. *J Am Vet Med Assoc* 162;291–292, 1973

Banerjee GC, Hogue DE: Nature of the anti–vitamin E factor of kidney-bean *(Phaseolus vulgaris)* fed to sheep. 1. Effect of feeding on plasma tocopherol. *Indian J Anim Sci* 45;551–553, 1975

Banton MI, Flory W, Jowett PLH, Winston GW: Comparison of the effects of *Sesbania drummondii* on the hepatic microsomal monooxygenase systems of chickens and rats. *Am J Vet Res* 50;1795–1983, 1989

Barger G, Dale HH: Chemical structure and sympathomimetic action of amines. *J Pharmacol Exp Ther* 41;19–59, 1910

Barlow RB, McCleod LJ: Some studies on cytisine and its methylated derivatives. *Br J Pharmacol* 35;161–174, 1969

Barneby RC: A revision of the North American species of *Oxytropis* DC. *Proc Calif Acad Sci,* ser 4, 27;177–309, 1952

Barneby RC: Atlas of North American *Astragalus,* part 1: the Phacoid and Homaloboid Astragali. *Mem N Y Bot Gard* 13;1–596, 1964a

Barneby RC: Atlas of North American *Astragalus,* part 2: the Cercidothrix, Hypoglottis, Piptoloboid, Trimeniaeus, and Orophaca Astragali. *Mem N Y Bot Gard* 13;597–1188, 1964b

Barneby RC: Fabales. In *Intermountain Flora,* vol 3, part B, Cronquist A, Holmgren AH, Holmgren NH, Reveal JL, Holmgren PK eds, New York Botanical Garden, New York, 1989a

Barneby RC: *Thermopsis.* In *Intermountain Flora,* vol 3, part B (Fabales), Cronquist A, Holmgren AH, Holmgren NH, Reveal JL, Holmgren PK eds, New York Botanical Garden, New York, pp235–236, 1989b

Barnes MF: Black locust poisoning of chickens. *J Am Vet Med Assoc* 59;370–372, 1921

Barri MES, Adam SEI: The toxicity of *Crotalaria saltiana* to calves. *J Comp Pathol* 91;621–627, 1981

Barri MES, Adam SEI, Omer OH: Effects of *Crotalaria saltiana* on Nubian goats. *Vet Hum Toxicol* 26;476–480, 1984

Barri MES, El Dirdiri NI, Abu Damir H, Idris OF: Toxicity of *Abrus precatorius* in Nubian goats. *Vet Hum Toxicol* 32;541–545, 1990

Barros CSL, Pilati C, Andujar MB, Graca DL, Irigoyen LF, Lopes ST, Santos CF: *Cassia occidentalis* (Leg. Caes.) poisoning in cattle. *Pesq Vet Bras* 10;47–58, 1990

Barrow MV, Simpson CF, Miller EJ: Lathyrism: a review. *Q Rev Biol* 49;101–128, 1974

Barth AT, Kommers GD, Salles MS, Wouters F, Lombardo de Barros CS: Coffee senna *(Senna occidentalis)* poisoning in cattle in Brazil. *Vet Hum Toxicol* 36;541–545, 1994

Bartley EE, Nagaraja TG, Pressman ES, Dayton AD, Katz MP, Fina LR: Effects of lasalocid or monensin on legume or grain (feedlot) bloat. *J Anim Sci* 56;1400–1406, 1983

Basson PA, Adelaar TF, Naude TW, Minnie JA: *Albizia* poisoning: report of the first outbreak and some experimental work in South Africa. *J S Afr Vet Med Assoc* 41;117–130, 1970

Baybutt RC, Aziz SM, Fagerland JA, Olson JW, Gillespie MN: Monocrotaline alters type II pneumocyte morphology. *Toxicol Appl Pharmacol* 129;188–195, 1994

Beal MF, Swartz KJ, Hyman BT, Storey E, Finn SF, Koroshetz W: Aminoacetic acid results in excitotoxin lesions by a novel indirect mechanism. *J Neurochem* 57;1068–1073, 1991

Beani L, Bianchi C, Nilsson L, Nordberg A, Romanelli L, Sivilotti L: The effect of nicotine and cytisine on 3H-acetylcholine release from cortical slices of guinea-pig brain. *Naunyn-Schmiedeberg's Arch Pharmacol* 331;293–296, 1985

Beath OA: *Lupine Studies. 2. The Silvery Lupine.* Wyo Agric Exp Stn Bull 144, 1925

Beath OA: *The Occurrence of Selenium and Seleniferous Vegetation in Wyoming. 2. Seleniferous Vegetation.* Wyo Agric Exp Stn Bull 221, pp29–64, 1937

Beath OA, Lehnert EH: *The Poisonous Properties of the Two-grooved Milkvetch (Astragalus bisulcatus).* Wyo Agric Exp Stn Bull 112, pp57–67, 1917

Beath OA, Draize JH, Eppson HF: *Three Poisonous Vetches.* Wyo Agric Exp Stn Bull 189, 1932

Beath OA, Draize JH, Gilbert CS: *Plants Poisonous to Livestock.* Wyo Agric Exp Stn Bull 200, 1934

Beath OA, Eppson HF, Gilbert CS: *Selenium and Other Toxic Minerals in Soils and Vegetation.* Wyo Agric Exp Stn Bull 206, 1935

Beath OA, Gilbert CS, Eppson HF, Rosenfeld I: *Poisonous Plants and Livestock Poisoning.* Wyo Agric Exp Stn Bull 324, p78, 1953

Becker RB, Neal WM, Dix Arnold PT, Shealy AL: A study of the palatability and possible toxicity of 11 species of *Crotalaria,* especially of *C. spectabilis* Roth. *J Agric Res* 50;911–922, 1935

Bell EA: A new amino-acid, γ-hydroxyhomoargenine in *Lathyrus. Nature* 199;70–71, 1963

Bell EA: Aminonitriles and amino acids not derived from proteins. In *Toxicants Occurring Naturally in Foods,* 2d ed, Comm Food Protection, National Research Council, National Academy of Sciences, Washington, DC, pp153–169, 1973

Bell EA, Foster RG: Structure of lathyrine. *Nature* 194;91–92, 1962

Benson ME, Casper HH, Johnson LJ: Occurrence and range of dicumarol concentrations in sweet clover. *Am J Vet Res* 42;2014–2015, 1981

Bergeron J-M, Goulet M: Etude des effets phytoestrogeniques et phytotoxiques de la vegetation de milieux ouverts sur la souris de laboratoire. *Can J Zool* 58;1575–1581, 1980

Berrow ML, Ure AM: Geological materials and soils. In *Occurrence and Distribution of Selenium,* Ihnat M ed, CRC Press, Boca Raton, Fla, pp213–242, 1989

Berry DL, Molyneux RJ, James LF, Willhite CC: Developmental toxicology of the indolizidine alkaloids castanospermine and swainsonine in rodents. In *Swainsonine and Related Glycosidase Inhibitors,* James LF, Elbein AD, Molyneux RJ, Warren CD eds, Iowa State Univ Press, Ames, pp417–424, 1989

Bertrand G: La vicianine, nouveau glycoside cyanhydrique contenu dans les graines de vesce. *C R* 143;832–834, 1906

Bickoff EM, Livingston AL, Booth AN, Hendrickson AP, Kohler GO: Estrogenic activity in dehydrated and sun cured forages. *J Anim Sci* 19;189–197, 1960

Bickoff EM, Livingston AL, Hendrickson AP, Booth AN: Relative potencies of several estrogen-like compounds found in forages. *Agric Food Chem* 10;410–412, 1962

Bickoff EM, Livingston AL, Witt SC, Knuckles BE, Guggolz J, Spencer RR: Isolation of coumestrol and other phenolics from alfalfa by countercurrent distribution. *J Pharm Sci* 53;1496–1499, 1964

Bickoff EM, Spencer RR, Witt SC, Knuckles BE: Studies on the Chemical and Biological Properties of Coumestrol and Related Compounds. USDA Tech Bull 1408, 1969

Bierer BW, Vickers CL, Rhodes WH, Thomas JB: Comparison of the toxic effects of *Crotalaria spectabilis* and *Crotalaria giant striata* as complete feed contaminants. *J Am Vet Med Assoc* 136;318–322, 1960

Bigbie R, King TP: The nutritional toxicity of the seed lectins of *Phaseolus vulgaris* in the pig. *Biochem Soc Trans* 9;145P, 1981

Bindon BM, Adams NR, Piper LR: Effects of oestrogenic pasture on luteinizing hormone levels and the response to GnRH in female lambs. *Anim Prod Sci* 5;7–13, 1982

Binns W, James LF: A congenital deformity in calves, similar to "crooked calf disease," has been experimentally produced by feeding heifers lupine and lead. *Proc West Sect Am Soc Anim Prod* 12;66–69, 1961

Birecka H, Birecka M, Cohen EJ, Bitonti AJ, McCann PP: Ornithine decarboxylase, polyamines, and pyrrolizidine alkaloids in *Senecio* and *Crotalaria. Plant Physiol* 86;224–230, 1988

Bisby FA: Genisteae. In *Advances in Legume Systematics,* part 1, Polhill RM, Raven PH eds, Royal Botanic Gardens, Kew, pp409–425, 1981

Blake JT, Thomas DW: Acute bovine pulmonary emphysema in Utah. *J Am Vet Med Assoc* 158;2047–2052, 1971

Blakley BR: Moldy sweet clover (dicoumarol) poisoning in Saskatchewan cattle. *Can Vet J* 26;357–360, 1985

Blohm H: *Poisonous Plants of Venezuela.* Harvard Univ Press, Cambridge, 1962

Bohlmann F, Schumann D: Lupine alkaloids. In *The Alkaloids,* vol 9, Manske RHF ed, Academic Press, New York, pp175–221, 1967

Boschmann E, Weinstock LM, Carmack M: Metal complexes of the three sparteine diastereoisomers: properties and reactivities of the copper (II) derivatives. *Inorg Chem* 13;1297–1300, 1974

Boughton IB, Hardy WT: *Mescalbean* (Sophora secundiflora) *Poisonous for Livestock.* Tex Agric Exp Stn Bull 519, 1935

Boughton IB, Hardy WT: *Suspected Plants.* Tex Agric Exp Stn Annu Rep 51;220–221, 1938

Boughton IB, Hardy WT: *Feeding Trials of Suspected Plants.* Tex Agric Exp Stn Annu Rep 52, pp239–240, 1939a

Boughton IB, Hardy WT: Toxicity of the coffee bean *(Sesbania vesicaria)* for sheep. *J Am Vet Med Assoc* 95;239–241, 1939b

Boughton IB, Hardy WT: *Feeding Trials of Suspected Plants.* Tex Agric Exp Stn Annu Rep 53, pp236–237, 1940

Boughton IB, Hardy WT: *Feeding Trials of Suspected Plants.* Tex Agric Exp Stn Annu Rep 54, p159, 1941

Bourn WM, Keller WJ, Bonfiglio JF: Comparisons of mescal bean alkaloids with mescaline, Δ9-THC and other psychotogens. *Life Sci* 25;1043–1054, 1979

Bourrillon R, Font J: Purification and physicochemical properties of a phytohemagglutinin from *Robinia pseudoacacia*—isolation of an active glycopeptide. *Biochim Biophys Acta* 154;28–39, 1965

Braden AWH, Hart NK, Lamberton JA: The oestrogenic activity and metabolism of certain isoflavones in sheep. *Aust J Agric Res* 18;335–348, 1967

Bramley A, Goulding R: *Laburnum* poisoning. *Br Med J* 283;1220–1221, 1981

Bridges RJ, Hatalski C, Shim SN, Nunn PB: Gliotoxic properties of the *Lathyrus* excitotoxin β-*N*-oxalyl-L-α,β-diaminoproprionic acid (β-L-ODAP). *Brain Res* 561;262–268, 1991

Broquist HP: The indolizidine alkaloids, slaframine and swainsonine: contaminants in animal forages. *Annu Rev Nutr* 5;391–409, 1985

Brouillet E, Jenkins BG, Hyman BT, Ferrante RJ, Kowall NW, Srivastava R, Roy DS, Rosen BR, Beal MF: Age-dependent vulnerability of the striatum to the mitochondrial toxin 3-nitroproprionic acid. *J Neurochem* 60;356–359, 1993

Brown DL, Barnes DA, Rezende SA, Klasing KC: Yield, composition, and feeding value of irrigated *Sesbania sesban* var. *nubia* leaves harvested at latitude 38°N during a Mediterranean summer. *Anim Feed Sci Technol* 18;247–255, 1987

Brown TA, Shrift A: Identification of selenocysteine in the protein of selenate-grown *Vigna radiata. Plant Physiol* 66;758–761, 1980

Brown TA, Shrift A: Exclusion of selenium from proteins of selenium-tolerant *Astragalus species. Plant Physiol* 67;1051–1053, 1981

Bruce EA: Astragalus campestris *and Other Stock Poisoning Plants of British Columbia.* Dom Can Dep Agric Bull 88, 1927

Bruner LH, Hilliker KS, Roth RA: Pulmonary hypertension and ECG changes from monocrotaline pyrrole in the rat. *Am J Physiol* 245;H300–H306, 1983

Bruner LH, Carpenter LJ, Hamlow P, Roth RA: Effect of a mixed function oxidase inducer and inhibitor on monocrotaline pyrrole pneumotoxicity. *Toxicol Appl Pharmacol* 85;416–427, 1986

Bull LB, Macindoe RHF: Photosensitization in sheep: trefoil dermatitis. *J Aust Vet Assoc* 2;85–91, 1926

Bull LB, Culvenor CCJ, Dick AT: *The Pyrrolizidine Alkaloids.* North-Holland Publ, Amsterdam, 1968

Bunch TD, Panter KE, James LF: Real-time ultrasound: a new way to study the effects of poisonous plants. *Utah Sci* (Winter);110–113, 1986

Bunch TD, Panter KE, James LF: Ultrasound studies of the effects of certain poisonous plants on uterine function and fetal development in livestock. *J Anim Sci* 70;1639–1643, 1992

Burk RF: Molecular biology of selenium with implications for its metabolism. *FASEB J* 5;2274–2279, 1991

Burroughs GW, Neser JA, Kellerman TS, van Niekerk FA: Suspected hybrid vetch (*Vicia villosa* crossed with *Vicia dasycarpa*) poisoning of cattle in the Republic of South Africa. In *Proceedings of the 13th World Congress on Diseases of Cattle,* vol 2, World Buiatrics Assoc, pp681–685, 1984

Burrows GE, Drummond RB, Basudde CDK: Cyanogenic plants of Zimbabwe. In *Poisonous Plants: Proceedings of the Third International Symposium,* James LF, Keeler RF, Bailey EM Jr, Cheeke PR, Hegarty MP eds, Iowa State Univ Press, Ames, pp561–566, 1992

Burrows GE, Tate LH, Tripp ML, Whitenack D, Edwards WC: Suspected intoxications due to *Lathyrus* [letter]. *Vet Hum Toxicol* 35;262–263, 1993

Burrows GE, Schwab RP, Stein LE, Kirkpatrick JG, Qualls C, Tyrl RJ: Comparison of the reproductive effects in rats of the plants *Baptisia australis, Iva annua,* and *Sophora nuttalliana.* In *Toxic Plants and Other Natural Toxicants,* Garland T, Barr AC eds, CAB International, New York, pp297–302, 1998

Bustine DL, Moyer BG: Review and mechanisms of toxicity of 3-nitropropanoic acid in nonruminant animals. *Proceedings of the 14th International Grassland Congress,* Smith JA, Hays VW eds, Westview Press, Boulder, Colo, pp736–738, 1983 [Note: The first author's surname, Gustine, is misspelled in the publication.]

Butler GW: The distribution of the cyanoglucosides linamarin and lotaustralin in higher plants. *Phytochemistry* 4;127–131, 1965

Byers HG: *Selenium Occurrence in Certain Soils in the United States with a Discussion of Related Topics.* USDA Tech Bull 482, 1935

Byrne KV: Dermatitis in white pigs due to photosensitization. *Aust Vet J* 13;74–75, 1937

Calore EE, Cavaliere MJ, Haraguchi M, Gorniak SL, Dagli MLZ, Raspantini PC, Calore NMP: Experimental mitochondrial myopathy induced by chronic intoxication by *Senna occidentalis* seeds. *J Neurol Sci* 146;1–6, 1997

Calore EE, Cavaliere MJ, Haraguchi M, Gorniak SL, Dagli ML, Raspantini PC, Calore NM, Weg R: Toxic peripheral neuropathy of chicks fed *Senna occidentalis* seeds. *Ecotoxicol Environ Saf* 39;27–30, 1998

Camp BJ: Action of *N*-methyltyramine and *N*-methyl-β-phenylethylamine on certain biological systems. *Am J Vet Res* 31;755–762, 1970

Camp BJ, Lyman CM: The isolation of *N*-methyl-β-phenethylamine from *Acacia berlandieri. J Am Pharm Assoc* 11;719–721, 1956

Camp BJ, Moore JA: A quantitative method for the alkaloid of *Acacia berlandieri. J Pharm Sci* 49;158–160, 1960

Camp BJ, Norvell MJ: The phenethylamine alkaloids of native range plants. *Econ Bot* 20;274–278, 1966

Camp BJ, Adams R, Dollahite JW: The chemistry of the toxic constituents of *Acacia berlandieri. N Y Acad Sci* 111;744–750, 1964

Campbell HA, Link KP: Studies on the hemorrhagic sweet clover disease. 4. The isolation and crystallization of the hemorrhagic agent. *J Biol Chem* 138;21–33, 1941

Campos MM, Carlini CR, Guimaraes JA, Marques-Silva VM, Rumjanek VM: Effect of canatoxin on cell cultures. *Cell Biol Int Rep* 15;581–594, 1991

Candrian U, Luthy J, Schmid P, Schlatter C, Gallasz E: Stability of pyrrolizidine alkaloids in hay and silage. *J Agric Food Chem* 32;935–937, 1984

Cann HM, Verhulst HL: *Poisonous Plants.* National Clearinghouse Poison Control Centers, 2 September 1958

Cappalletti EM, Innocenti G, Caporale G: Possible ecological significance of within-fruit and seed furocoumarin distribution in two *Psoralea* species. *J Chem Ecol* 18;155–164, 1992

Carlini CR, Guimaraes JA: Isolation and characterization of a toxic protein from *Canavalia ensiformis* (jack bean) seeds, distinct from concanavalin A. *Toxicon* 19;667–675, 1981

Carlini CR, Guimaraes JA: Plant and microbial toxic proteins as hemilectins: emphasis on canatoxin. *Toxicon* 29;791–806, 1991

Carlini CR, Barcellos GBS, Baeta-Neves ADV, Guimaraes JA: Immunoreactivity for canatoxin and concanavalin A among proteins of leguminous seeds. *Phytochemistry* 27;25–30, 1988

Carlson JR, Dickinson EO: Tryptophan-induced pulmonary edema and emphysema in ruminants. In *Effects of Poisonous Plants on Livestock,* Keeler RF, Van Kampen KR, James LF eds, Academic Press, New York, pp261–272, 1978

Carlson JR, Dyer IA, Johnson RJ: Tryptophan-induced interstitial pulmonary emphysema in cattle. *Am J Vet Res* 29;1983–1989, 1968

Carlson JR, Yokoyama MT, Dickinson EO: Induction of pulmonary edema and emphysema in cattle and goats with 3-methylindole. *Science* 176;298–299, 1972

Carlson JR, Dickinson EO, Yokoyama MT, Bradley BJ: Pulmonary edema and emphysema in cattle after intraruminal and intravenous administration of 3-methylindole. *Am J Vet Res* 36;1341–1347, 1975

Cary CA, Miller ER, Johnstone GR: *Poisonous Plants of Alabama.* Ala Agric Exp Stn Circ 71, 1924

Casper HH, Alstad AD, Monson SB: Dicumarol levels in sweet clover toxic to cattle. *Proc Am Assoc Vet Lab Diagn* 25;41–48, 1982

Casper HH, Berg IE, Crenshaw JD, Colville JL, Wass WM: Lupin bean meal toxicosis in swine. *J Vet Diagn Invest* 3;172–173, 1991

Casteel SW, Osweiler GD, Cook WO, Daniels G, Kadlec R: Selenium toxicosis in swine. *J Am Vet Med Assoc* 186;1084–1085, 1985

Castillo C, Valencia I, Reyes G, Hong E: An analysis of the antihypertensive properties of 3-nitropropionic acid, a compound from plants of the genus *Astragalus*. *Arch Inst Cardiol Mex* 63;11–16, 1993

Cates RG, Rhoads DF: *Prosopis* leaves as a resource for insects. In *Mesquite—Its Biology in Two Desert Scrub Ecosystems,*, Simpson BB ed, Dowder, Hutchinson & Ross, Stroudsburg, Pa, pp61–83, 1977

Cavaliere MJ, Calore EE, Haraguchi M, Gorniak SL, Dagli MLZ, Raspantini PC, Calore NMP, Weg R: Mitochondrial myopathy in *Senna occidentalis*–seed-fed chicken. *Ecotoxicol Environ Safe* 37;181–185, 1997

Cavanagh JB: The problems with neurons with long axons. *Lancet* 2;1284–1287, 1984

Chamley WA, Adams NR, Hooley RD, Carson R: Hypothalamic-pituitary function in normal ewes and ewes which grazed oestrogenic subterranean clover for several years. *Aust J Biol Sci* 34;239–244, 1981

Chapin RE, Srevens JT, Hughes CL, Kelce WR, Hess RA, Daston GP: Endocrine modulation of reproduction, a symposium overview. *Fundam Appl Toxicol* 29;1–17, 1996

Charles OW, Muller HD: Coffee weed *(Cassia obtusifolia)* toxicity response of laying hens [abstract]. *Poult Sci* 54;1745, 1975

Chase RL, Keeler RF: Mountain thermopsis toxicity in cattle. *Utah Sci* 44;28–31, 1983

Chavez PI, Sullivan G: Qualitative and quantitative comparison of the quinolizidine alkaloids of the fasciated and normal stems of *Sophora secundiflora*. *J Nat Prod* 47;735–736, 1984

Chawla AS, Kapoor VK: *Erythrina* alkaloids. In *Alkaloids: Chemical and Biological Perspectives,* vol 9, Pelletier SW ed, Elsevier, New York, pp85–153, 1995

Cheeke PR, Shull LR*: Natural Toxicants in Feeds and Poisonous Plants.* AVI Publishing, Westport, Conn, 1985

Chen C-H, Flory W, Koeppe RE: Variation of neurotoxicity of L-and D-2,4-diaminobutyric acid with route of administration. *Toxicol Appl Pharmacol* 23;334–338, 1972

Cheng E, Story CD, Payne LC, Yoder L, Burroughs W: Detection of estrogenic substances in alfalfa and clover hays fed to fattening lambs. *J Anim Sci* 12;507–513, 1953

Cheng E, Yoder L, Sotry CD, Burroughs W: Estrogenic activity of some isoflavone derivatives. *Science* 120;575–576, 1954

Chesnut VK: *Principal Poisonous Plants of the United States.* USDA Div Bot Bull 20, 1898a

Chesnut VK: Report on poisonous plants. *15th Annual Report, Bureau Animal Industries,* pp387–420, 1898b

Chesnut VK, Wilcox EV: *The Stock-Poisoning Plants of Montana.* USDA Div Bot Bull 26, 1901

Cho CH, Chuang CY, Chen CF: Study of the antipyretic activity of matrine: a lupine alkaloid isolated from *Sophora subprostrata. Planta Med* 52;343–345, 1986

Cho YD, Martin RO: Biosynthesis of *Thermopsis* alkaloids from carbon-14 dioxide: evidence for the formation of the pyridone bases from lupanine via 5,6-dehydrolupanine. *Can J Biochem* 49;971–977, 1971

Choi Y-H, Hussain RA, Pezzuto JM, Kinghorn AD, Morton JF: Abrusides A–D, four novel sweet-tasting triterpene glycosides from the leaves of *Abrus precatorius. J Nat Prod* 52;1118–1127, 1989

Chopra RN, Badhwar RL, Ghosh S: *Poisonous Plants of India.* Academic Publishers, Jaipur, India, pp345–350,381–388, 1984

Chowdhury SD: Lathyrism in poultry. *World's Poult Sci J* 44;7–16, 1988

Chowdhury SD, Davis RH: Comparison of the effects of two lathyrogens on the reproductive system of the laying hen. *Vet Rec* 124;240–242, 1989

Christie GS, DeMunk FG, Madsen NP, Hegarty MP: The effects of an arginine antagonist on stimulated human lymphocytes in culture. *Pathology* 3;139–144, 1971

Citores L, Ferreras JM, Iglesias R, Carbajales ML, Arias FJ, Jimenez P, Rojo MA, Girbes

T: Molecular mechanism of inhibition of mammalian protein synthesis by some four-chain agglutinins. *FEBS Lett* 329;59–62, 1993

Clare NT: Photosensitization in animals. *Adv Vet Sci* 2;182–211, 1955

Clark EP: The occurrence of rotenone and related compounds in the roots of *Cracca virginiana. Science* 77;311–312, 1933

Clarke EEC, Clarke ML: *Garner's Veterinary Toxicology,* 3d ed. Williams & Wilkins, Baltimore, 1967

Clarke ML, Clarke EGC, King T: Fatal *Laburnum* poisoning in a dog. *Vet Rec* 88;199–200, 1971

Clarke RTJ, Reid CSW: Foamy bloat of cattle: a review. *J Dairy Sci* 57;753–785, 1974

Claughton WP, Claughton HD: Vetch seed poisoning. *Auburn Vet* 10;125, 1954

Clawson AB: Two lupines shown to be poisonous to livestock. *Off Rec (USDA)* 10;71, 1931

Clement BA, Forbes TDA: Toxic amines and alkaloids from Texas acacias. In *Toxic Plants and Other Natural Toxicants,* Garland T, Barr AC eds, CAB International, New York, pp351–355, 1998

Clement BA, Goff CM, Forbes TDA: Toxic amines and alkaloids from *Acacia rigidula. Phytochemistry* 49;1377–1380, 1998

Cohen AJ: Critical review of the toxicology of coumarin with special reference to interspecies differences in metabolism and hepatotoxic response and their significance. *Food Cosmet Toxicol* 17;277–289, 1979

Colegate SM, Dorling PR, Huxtable CR: Structural elucidation of swainsonine: relationship of structure to activity. In *Swainsonine and Related Glycosidase Inhibitors,* James LF, Elbein AD, Molyneux RJ, Warren CD eds, Iowa State Univ Press, Ames, pp91–99, 1989

Collinge DB, Hughes MA: Developmental and physiological studies on the cyanogenic glucosides of white clover, *Trifolium repens* L. *J Exp Bot* 33;154–161, 1982

Colvin BM, Harrison LR, Sangster LT, Gosser HS: *Cassia occidentalis* toxicosis in growing pigs. *J Am Vet Med Assoc* 189;423–426, 1986

Conn EE: Cyanogenic glycosides. *Int Rev Biochem, Biochem Nutr IA* 27;21–43, 1979

Conn EE, Maslin BR: A preliminary report on cyanogenesis in Australian *Acacia* species. *Int Group Study Mimosoideae Bull* 10;26–31, 1982

Conn EE, Currey S, Conn ME: Cyanogenesis in Australian species of *Acacia:* surveys of herbarium specimens and living plants. *West Aust Herb Res Notes* 10;1–60, 1985

Connolly JD, Hill RA: *Dictionary of Terpenoids.* Vol 2, *Di- and Higher Terpenoids.* Chapman & Hall, London, pp657–1460, 1991

Cook CW, Redente EF: Development of the ranching industry in Colorado. *Rangelands* 15;204–207, 1993

Coop IE: The occurrence, metabolism, and toxicity of cyanogenic glycosides with special reference to white clover. *Proc Specialist Conf Agric Plant Anim Nutr Relat Soil Clim Factors* 1951;335–342, 1949

Cooper MR, Johnson AW: *Poisonous Plants of Britain.* Her Majesty's Stationery Office, London, pp144–148, 1984

Coppock DL: Grass hay and *Acacia* fruits: a local feeding system for improved calf performance in semi-arid Ethiopia. *Trop Anim Health Prod* 25;41–49, 1993

Cordell GA: *Introduction to Alkaloids.* Wiley & Sons, New York, 1981

Corkill L: Cyanogenesis in white clover (*Trifolium repens* L.). 5. The inheritance of cyanogenesis. *N Z J Sci Technol,* ser B, 23;178–193, 1942

Cory VL: A new loco from the Edwards Plateau of Texas. *Rhodora* 32;4–7, 1930

Couch JF: Relative toxicity of the lupine alkaloids. *J Agric Res* 32;51–67, 1926

Coulson W, Linker A, Bottcher E: Lathyrism in swine. *Arch Pathol* 87;411–417, 1969

Cowgill VM: Arsenic and selenium in Se-accumulator and non-accumulator plants. In *Selenium in Biology and Medicine,* Spallholz JE, Martin JL, Ganther HE eds, AVI Publishing, Westport, Conn, pp336–373, 1981

Cox DH, Harris DL, Richard TA: Chemical identification of crotalaria poisoning in horses. *J Am Vet Med Assoc* 133;425–426, 1958

Cox RI: Plant estrogens affecting livestock in Australia. In *Effects of Poisonous Plants on Livestock,* Keeler RF, Van Kampen KR, James LF eds, Academic Press, New York, pp451–464, 1978

Cox RI: Immunophysiological control of phyto-oestrogen toxicity. In *Plant Toxicology, Proceedings of the Australian-USA Poisonous Plants Symposium,* Seawright AA, Hegarty MP, James LF, Keeler RF eds, Animal Research Institute, Yeerongpilly, Brisbane, Australia, pp98–108, 1985

Cranmer MF, Mabry TJ: The lupine alkaloids of the genus *Baptisia* (Leguminosae). *Phytochemistry* 5;1133–1138, 1966

Crawford AC: *Barium, a Cause of the Locoweed Disease.* USDA Bur Plant Ind Bull 129, 1908

Crawford L, McDonald GM, Friedman M: Composition of sicklepod *(Cassia obtusifolia)* toxic weed seeds. *J Agric Food Chem* 38;2169–2175, 1990

Crinion RA, O'Connor JP: Selenium intoxication in horses. *Ir Vet J* 32;81–86, 1978

Cui B, Kinjo J, Nohara T: Triterpene glycosides from the bark of *Robinia pseudo-acacia* L., 1. *Chem Pharm Bull* 40;2995–2999, 1992

Cui B, Kinjo J, Nohara T: Triterpene glycosides from the bark of *Robinia pseudo-acacia* L., 2. *Chem Pharm Bull* 41;553–556, 1993

Cui L, Zhang B: The antiarrhythmic effects and mechanisms of sophoridine. *Zhongguo Yaolixue Yu Dulixue Zazhi* 1;3–10, 1986 (*Chem Abstr* 108;143128g, 1988)

Dai S, Chan M-Y, Lee S-S, Ogle CW: Effects of *Sophora flavescens* Ait. on hemodynamics and ventricular fibrillation threshold in anaesthetized dogs. *Am J Chin Med* 15;53–57, 1987

Dakshinamurti K: Neurobiology of pyridoxine. *Adv Nutr Res* 4;143–179, 1977

Dale HH, Laidlaw PP: The physiological action of cytisine, the active alkaloid of *Laburnum (Cytisus-Laburnum). J Pharmacol Exp Ther* 3;205–221, 1912

Damron BL, Flunker LK, Sundlof SF: *Sesbania* seed's effect on broiler chick and leghorn hen performance [abstract]. *Poult Sci* 67;11, 1988

Daniel TW, Wolberg FB, Miller VL, Alswager JH, Ensminger ME, Spielman AA: Chemical composition and digestibility of flat pea forage in three stages of maturity. *J Anim Sci* 5;80–86, 1946

Darre MJ, Minior DN, Tatake JG, Ressler C: Nutritional evaluation of detoxified and raw common vetch seed (*Vicia sativa* L.) using diets of broilers. *J Agric Food Chem* 46;4675–4679, 1998

Davies HL: Studies in uterotrophic activities in strains of subterranean clover. *Aust J Agric Res* 40;871–877, 1989

Davies HL, Rossiter RC, Maller R: The effects of different cultivars of subterranean clover *(T. subterraneum)* on sheep reproduction in the south-west of western Australia. *Aust J Agric Res* 21;359–369, 1970

Davies KJA, Doroshow JH, Hochstein P: Mitochondrial NADH dehydrogenase-catalysed oxygen radical production by adriamycin and the relative inactivity of 5-iminodaunorubicin. *FEBS Lett* 153;227–230, 1983

Davis AM: Protein, crude fiber, tannin, and oxalate concentrations of some introduced *Astragalus* species. *Agron J* 65;613–615, 1973

Davis AM: The occurrence of anagyrine in a collection of western American lupines. *J Range Manage* 35;81–84, 1982

Davis AM: Selenium uptake in *Astragalus* and *Lupinus* species. *Agron J* 78;727–729, 1986

Davis AM, Stout DM: Anagyrine in western American lupines. *J Range Manage* 39;29–30, 1986

Dawson R Jr, Beal MF, Bondy SC, Di Monte DA, Isom GE: Excitotoxins, aging, and environmental neurotoxins: implications for understanding human neurodegenerative diseases. *Toxicol Appl Pharmacol* 134;1–17, 1995

Decker MW, Anderson DJ, Brioni JD, Donnelly-Roberts DL, Kang CH, O'Neill AB, Piattoni-Kaplan M, Swanson S, Sullivan JP: Erysodine, a competitive antagonist at neuronal nicotinic acetylcholine receptors. *Eur J Pharmacol* 280;79–89, 1995

Deweede D, Wayman O: Effect of mimosine on the rat fetus. *Teratology* 3;21–28, 1970
Dewes HF, Lowe MD: Suspected selenium poisoning in a horse. *N Z Vet J* 35;53–54, 1987
Diaz JL: Ethnopharmacology of sacred psychoactive plants used by the Indians of Mexico. *Annu Rev Pharmacol Toxicol* 17;647–675, 1977
Dickinson JO, Braun RC: Effect of 2(3)-tert-butyl-4-hydroxyanisole (BHA) and 2-chloroethanol against pyrrole production and chronic toxicity of monocrotaline in chickens. *Vet Hum Toxicol* 29;11–15, 1987
Dinkel CA, Minyard JA, Ray DE: Effects of season of breeding on reproductive and weaning performance of beef cattle grazing seleniferous range. *J Anim Sci* 22;1043–1045, 1963
D'Mello JPF: Toxic factors in tropical legumes. *World Rev Anim Prod* 17;41–46, 1982
Dodd S: Trefoil dermatitis. *J Comp Pathol Ther* 29;47–62, 1916
Dollahite JW: Management of the disease produced in cattle on an unbalanced diet of mesquite beans. *Southwest Vet* 17;293–296, 1964
Dollahite JW: *Research and Observations on Toxic Plants in Texas 1932–1972*. Tex Agric Exp Stn Tech Rep 78–1, 1978
Dollahite JW, Henson JB: Toxic plants as the etiologic agent of myopathies in animals. *Am J Vet Res* 26;749–752, 1965
Donaldson LE: Clover disease in two Mississippi cattle herds. *J Am Vet Med Assoc* 182;412–413, 1983
Dorling PR, Huxtable CR, Vogel P: Lysosomal storage in *Swainsona* spp. toxicosis: an induced mannosidosis. *Neuropathol Appl Neurobiol* 4;285–295, 1978
Dorling PR, Huxtable CR, Colegate SM: Inhibition of lysosomal α-mannosidase by swainsonine, an indolizidine alkaloid isolated from *Swainsona canescens*. *Biochem J* 191;649–651, 1980
Dorling PR, Colegate SM, Huxtable CR: The biochemical activity of swainsonine: an indolizidine alkaloid isolated from *Swainsona canescens*. *Toxicon* (Suppl 3);93–96, 1983
Dorling PR, Colegate SM, Huxtable CR: Swainsonine: a toxic indolizidine alkaloid. In *Toxicants of Plant Origin,* vol 1, Cheeke PR ed, CRC Press, Boca Raton, Fla, pp237–256, 1989
Dorsaz A-C, Hostettmann M, Hostettmann K: Molluscicidal saponins from *Sesbania sesban*. *Planta Med* 54;225–227, 1988
Dougherty RW, Christensen RB: In vivo absorption studies of hydrocyanic acid of plant juice origin. *Cornell Vet* 43;481–486, 1953
Draize JH, Beath OA: *Plants Poisonous to Livestock*. Wyo Agric Exp Stn Bull 200, 1934
Draize JH, Beath OA: Observations on the pathology of blind staggers and alkali disease. *J Am Vet Med Assoc* 86;753–763, 1935
Dudley HC: Toxicology of selenium. 1. A study of the distribution of selenium in acute and chronic cases of selenium poisoning. *Am J Hyg* 23;169–180, 1936
Dugarte-Stavanja M, Smith GS, Edrington TS, Hallford DM: Failure of dietary bentonite clay, Silent Herder mineral supplement, or parenteral banamine to alleviate locoweed toxicosis in rats. *J Anim Sci* 75;1867–1875, 1997
Duncan WH, Piercy PL, Starling RJ: Toxicological studies of southeastern plants. 1. Leguminosae. *Econ Bot* 9;243–255, 1955
Dunstan WR, Henry TA: 11. The nature and origin of the poison of *Lotus arabicus*. *Roy Soc Lond Phil Trans B* 194;515–533, 1901
Dunstan WR, Henry TA: Cyanbildung in pflanzen. 3. Phaseolunatin, das cyanbildende glucosid von *Phaseolus lunatus*. *Chem Centralbl,* ser 2, 74;1333–1334, 1903
Dupuy HP, Lee JG: The isolation of a material capable of producing experimental lathyrism. *J Am Pharm Assoc* 43;61–62, 1954
Durrell LW, Jensen R, Klinger B: *Poisonous and Injurious Plants in Colorado*. Colo Agric Exp Stn Bull 412A, 1950
Dwatmadgi, Teleni E, Bird AR, Lowry JB: Nutritive value of *Albizia lebbeck* supplements for growing sheep. *Aust J Exp Agric* 32;273–278, 1992

Ebeling W: *Handbook of Indian Foods and Fibers of Arid America.* Univ California Press, Berkeley, 1986

Edwards TA: Buller syndrome. *Large Anim Vet* (July/August);6–8, 1995

Eichelbaum M, Spannbrucker N, Dengler HJ: Influence of defective metabolism of sparteine on its pharmacokinetics. *Eur J Clin Pharmacol* 16;189–194, 1979

Eichholzer JV, Lewis IAS, MacLeod JK, Oelrichs PB, Vallely PJ: Galegin and a new dihydroalkylacetamide from *Verbesina encelioides. Phytochemistry* 21;97–99, 1982

Ellis LC, James LF, McMullen RW, Panter KE: Reduced progesterone and altered cotyledonary prostaglandin values induced by locoweed *(Astragalus lentiginosus)* in sheep. *Am J Vet Res* 46;1903–1907, 1985

El-Shabrawy OA, El-Gengaihi S, Ali Ibrahim N: Toxicity and teratogenicity of abrin. *Egypt J Vet Sci* 24;135–142, 1987

Emery ZT: Report of thirty-two cases of poisoning by locust bark. *N Y Med J* 45;92–95, 1887

Emmel MW: The toxicity of *Glottidium vesicarium* (Jacq.) Harper seeds for the fowl. *J Am Vet Med Assoc* 87;13–21, 1935

Emmel MW: The toxicity of *Crotalaria retusa* L. seeds for the domestic fowl. *J Am Vet Med Assoc* 91;205–206, 1937

Emmel MW: The toxicity of *Glottidium vesicarium* (Jacq.) Harper for cattle. *J Am Vet Med Assoc* 104;222–223, 1944

Emmel MW, Sanders DA, Henley WW: Additional observations on the toxicity of *Crotalaria spectabilis* (Roth) for swine. *J Am Vet Med Assoc* 87;175–176, 1935a

Emmel MW, Sanders DA, Henley WW: *Crotalaria spectabilis* Roth seed poisoning in swine. *J Am Vet Med Assoc* 86;43–54, 1935b

Engelmann K, Raake W, Petter A: Importance of hydrophobic binding forces on the antiarrhythmic activity of alkyl substituted sparteines. *Arzneim-Forsch* 24;759–761, 1974

Engle PH, Bell DS, Davis RR: The effect of ladino clover, birdsfoot trefoil, and bluegrass pasture on the rate of conception among ewes. *J Anim Sci* 16;703–710, 1957

Epidemiological Research Laboratory: Unusual outbreak of food poisoning. *Br Med J* 2;1268, 1976

Eustice DC, Kull FJ, Shrift A: Selenium toxicity: aminoacylation and peptide bond formation with selenomethionine. *Plant Physiol* 67;1054–1058, 1981

Evans CS, Bell EA: Uncommon amino acids in the seeds of 64 species of Caesalpinieae. *Phytochemistry* 17;1127–1129, 1978

Evans CS, Bell EA, Johnson ES: *N*-Methyltyramine, a biologically active amine in *Acacia* seeds. *Phytochemistry* 18;2022–2023, 1979

Evans RH, Jones AW, Watkins JC: Willardiine: a potent quisqualate-like excitant. *J Physiol* 308;71–72, 1980

Everist SL: *Poisonous Plants of Australia,* 2d ed. Angus & Robertson, London, 1981

Ewan RC: Animal tissues. In *Occurrence and Distribution of Selenium,* Ihnat M ed, CRC Press, Boca Raton, Fla, pp121–167, 1989

Fadiman J: *Genista canariensis:* a minor psychedelic. *Econ Bot* 19;383, 1965

Faix JJ, Gustine DL, Wright MJ: Beta-nitropropionic acid concentration of crownvetch plant parts as affected by growth-temperature-maturation variables. *Agron J* 70;689–691, 1978

Faliu L, Puyt JD, Tainturier D: Le galega (*Galega officinalis* L.) une legumineuse tres dangereuse pour les ovins. *Rec Med Vet (France)* 157;419–426, 1981

Fay JP, Cheng K-J, Hanna MR, Howarth RE, Costerton JW: In vitro digestion of bloat-safe and bloat-causing legumes by rumen microorganisms: gas and foam production. *J Dairy Sci* 63;1273–1281, 1980

Featherly HI, Harnden EE, Dermer OC, Smith HC: *Glottidium vesicarium,* a poisonous plant in the Southwest. *Vet Med* 38;478–479, 1943

Feiser LF, Feiser M: *Steroids.* Reinhold Publ, New York, 1959

Felger RS: Mesquite in Indian cultures of Southwestern North America. In *Mesquite—Its*

Biology in Two Desert Scrub Ecosystems,, Simpson SS ed, Dowder, Hutchinson & Ross, Stroudsburg, Pa, pp150–176, 1977

Felker P: Mesquite: an all-purpose leguminous arid land tree. In *New Agricultural Crops,* Ritchie GA ed, Westview Press, Boulder, Colo, pp89–132, 1979

Ferm VH, Hanlon DP, Willhite CC, Choy WN, Book SA: Embryotoxicity and dose-response relationships of selenium in hamsters. *Reprod Toxicol* 4;183–190, 1990

Fernald ML, Kinsey AC: *Edible Wild Plants of North America.* Idlewild Press, Cornwall-on-Hudson, New York, 1943

Filandrinos DT, Sioris LJ: Transient elevation of liver function tests following ingestion of black locust seeds [abstract]. *Vet Hum Toxicol* 34;351, 1992

Fincher MG, Fuller HK: Photosensitization—trifoliosis—light sensitization. . *Cornell Vet* 32;95–98, 1942

Finnegan RA, Mueller WH: Chemical examination of a toxic extract of *Indigofera endecaphylla. J Pharm Sci* 54;1136–1144, 1965

Finnegan RA, Stephani RA: Structure of hiptagen as 1, 2, 4, 6, tetra-O-(3-nitropropanoyl)—D-glucopyranoside, its identity with endecaphyllin X, and the synthesis of its methyl ether. *J Pharm Sci* 57;353–354, 1968

Finnemore H, Gledhill WC: Poison plants: the presence of cyanogenetic glycosides in certain species of *Acacia. J Counc Sci Res Aust* 1;254, 1928a

Finnemore H, Gledhill WC: The presence of cyanogenetic glucosides in certain species of *Acacia. Aust J Pharm* 9;174–178, 1928b

Fisher CE: Mesquite and modern man in Southwestern North America. In *Mesquite—Its Biology in Two Desert Scrub Ecosystems,*, Simpson BB ed, Dowder, Hutchinson & Ross, Stroudsburg, Pa, pp177–188, 1977

Fitch WL, Dolinger PM, Djerassi C: Alkaloid studies. 68. Novel piperidyl alkaloids from *Lupinus formosus. J Org Chem* 39;2974–2979, 1974

Fitzhugh OG, Nelson AA, Bliss CI: The chronic oral toxicity of selenium. *J Pharmacol Exp Ther* 80;289–299, 1944

Fleming CE: *Range Plants Poisonous to Sheep and Cattle in Nevada.* Nev Agric Exp Stn Bull 95, 1918

Flory W, Hebert CD: Determination of the oral toxicity of *Sesbania drummondii* seeds in chickens. *Am J Vet Res* 45;955–958, 1984

Flory W, Spainhour CB, Colvin B, Herbert CD: The toxicologic investigation of a feed grain contaminated with seeds of the plant species *Cassia. J Vet Diagn Invest* 4;65–69, 1992

Flunker LK, Damron BL, Wilson HR: Research note: feeding various levels of ground *Sesbania macrocarpa* Muhl. seed to bobwhite quail. *Poult Sci* 70;658–659, 1991

Folkers K, Major RT: Isolation of erythroidine, an alkaloid of curare action, from *Erythrina americana* Mill. *J Am Chem Soc* 59;1580–1581, 1937

Folkers K, Koniuszy F, Shavel J Jr: Erythrina alkaloids. 13. Studies on the constitution of erythraline, erythramine, and erythratine. *J Am Chem Soc* 64;2146–2151, 1942

Foote PA, Gramling LG: A chemical investigation of the seeds of *Glottidium vesicarium* (Jacq.) Harper. *J Am Pharm Assoc* 29;311–312, 1940

Forbes TDA, Tolleson DR, Hensarling CM, Randel RD: Effects of exogenous amines on reproduction in female Angora goats. *S Afr J Anim Sci* 23;196–200, 1993

Forrester RM: Have you eaten *Laburnum*? *Lancet* 1;1073, 1979

Forster LA, Fontenot JP, Perry HD, Foster JG, Allen VG: Apparent digestibility and nutrient balance in lambs fed different levels of flatpea hay. *J Anim Sci* 69;1719–1725, 1991

Foster JG: Flatpea (*Lathyrus sylvestris* L.): a new forage species? A comprehensive review. *Adv Agric* 43;241–313, 1990

Fox WE, Allred KW, Roalson EH: *A Guide to the Common Locoweeds and Milkvetches of New Mexico.* N M Agric Exp Stn Circ 557, 1998

Franke KW: A new toxicant occurring naturally in certain samples of plant foodstuffs. *J Nutr* 8;597–608, 1934

Franz G: The senna drug and its chemistry. *Pharmacol* 47 (Suppl 1);2–6, 1993

Fraser CM, Nelson J: Sweet clover poisoning in newborn calves. *J Am Vet Med Assoc* 135;283–286, 1959

Friedman M, Henika PR: Mutagenicity of toxic weed seeds in the Ames test: jimson weed *(Datura stramonium)*, velvetleaf *(Abutilon theophrasti)*, morning glory (*Ipomoea* spp.), and sicklepod *(Cassia obtusifolia)*. *J Agric Food Chem* 39;494–501, 1991

Fu Y-T, He F-S, Zhang S-L, Zhang J-S: Lipid peroxidation in rats intoxicated with 3-nitropropionic acid. *Toxicon* 33;327–331, 1995

Fuller TC, McClintock E: *Poisonous Plants of California*. Univ of California Press, Berkeley, p158, 1986

Galey FD, Mendez LE, Whitehead WE, Holstege DM, Plumlee KH, Johnson B: Estrogenic activity in forages: diagnostic use of the classical mouse uterine bioassay. *J Vet Diagn Invest* 5;603–608, 1993

Galyean ML, Ralphs MH, Reif MN, Graham JD, Braselton WE Jr: Effects of previous grazing treatment and consumption of locoweed on liver mineral concentrations in beef steers. *J Anim Sci* 74;827–833, 1996

Gardiner MR: Fungus-induced toxicity in lupinosis. *Br Vet J* 122;508–516, 1966

Gardiner MR: Lupinosis. *Adv Vet Sci* 11;85–138, 1967a

Gardiner MR: The role of copper in the pathogenesis of subacute and chronic lupinosis of sheep. *Aust Vet J* 43;243–248, 1967b

Gardiner MR, Seddon HD: Equine lupinosis. *Aust Vet J* 42;242–244, 1966

Gardner AF: Experimental lathyrism—review of the literature. *Am J Clin Nutr* 7;213–223, 1959

Gardner AF, Sakiewicz N: A review of neurolathyrism including the Russian and Polish literature. *Exp Med Surg* 21;164–191, 1963

Gardner DR, Panter KE: Comparison of blood plasma alkaloid levels in cattle, sheep, and goats fed *Lupinus caudatus*. *J Nat Toxins* 2;1–11, 1993

Gates FC: *Poisonous Plants in Kansas*. Kans Agric Exp Stn Tech Bull 25, 1930

Ghatak N, Kaul R: Chemical examination of the seeds of *Abrus precatorius*, Linn, part 1. *J Indian Chem Soc* 9;383–387, 1932

Gibbons WJ, Hokanson JF, Wiggins AM, Schmitz MB: Cirrhosis of the liver in horses. *North Am Vet* 31;229–232, 1950

Gibbons WJ, Durr EH, Cox SA: An outbreak of cirrhosis of the liver in horses. *North Am Vet* 34;556–558, 1953

Gibbs RD: *Chemotaxonomy of Flowering Plants*, vol 1. McGill-Queens Univ Press, Montreal, 1974

Glenn BL, Monlux AW, Panciera RJ: A hepatogenous photosensitivity disease of cattle. 1. Experimental production and clinical aspects of the disease. *Pathol Vet* 1;469–484, 1964

Glenn BL, Panciera RJ, Monlux AW: A hepatogenous photosensitivity disease of cattle. 1. Histopathology and pathogenesis of the hepatic lesions. *Pathol Vet* 2;49–67, 1965

Glenn MW, Jensen R, Griner LA: Sodium selenate toxicosis: the effects of extended oral administration of sodium selenate on mortality, clinical signs, fertility, and early embryonic development in sheep. *Am J Vet Res* 25;1479–1485, 1964a

Glenn MW, Jensen R, Griner LA: Sodium selenate toxicosis: pathology and pathogenesis of sodium selenate toxicosis in sheep. *Am J Vet Res* 25;1486–1494, 1964b

Glenn MW, Martin JL, Cummins LM: Sodium selenate toxicosis: the distribution of selenium within the body after prolonged feeding of toxic quantities of sodium selenate to sheep. *Am J Vet Res* 25;1495–1499, 1964c

Glowaz SL, Michnika M, Huxtable RJ: Detection of a reactive pyrrole in the hepatic metabolism of the pyrrolizidine alkaloid, monocrotaline. *Toxicol Appl Pharmacol* 115;168–173, 1992

Gnanasunderam C, Sutherland ORW: Hiptagin and other aliphatic nitro esters in *Lotus pedunculatus*. *Phytochemistry* 25;409–410, 1986

Goldberg SI, Moates RF: Alkaloids of *Lupinus diffusus* Nutt. *Phytochemistry* 6;137–140, 1967

Goldberg SI, Sahli MS: Alkaloids of *Lupinus westianus* Small. *J Med Chem* 10;124–125, 1967

Gomez MVZF: Intoxication of goats by mesquite *(Prosopis juliflora). Int Symp Vet Diagn Lab* 1;581–585, 1977

Goplen BP: Polara, a low coumarin cultivar of sweetclover. *Can J Plant Sci* 51;249–251, 1971

Goplen BP: Sweetclover production and agronomy. *Can Vet J* 21;149–151, 1980

Goplen BP, Bell JM: Dicoumarol studies. 4. Antidotal and antagonistic properties of vitamins K1 and K3 in cattle. *Can J Anim Sci* 47;91–100, 1967

Gorst-Allman CP, Steyn PS, Vleggaar R: Structural elucidation of sesbanimide using high-field NMR spectroscopy. *Chem Soc Perkin Trans* 1;1311–1314, 1984

Gould DH, Gustine DL: Basal ganglia degeneration, myelin alterations, and enzyme inhibition induced in mice by the plant toxin 3-nitroprpanoic acid. *Neuropathol Appl Neurobiol* 8;377–393, 1982

Gould DH, Wilson MP, Hamar DW: Brain enzyme and clinical alterations induced in rats and mice by nitroaliphatic toxicants. *Toxicol Lett* 27;83–89, 1985

Goyer RA: Toxic effects of metals. In *Casarett and Doull's Toxicology,* 4th ed, Amdur MO, Doull J, Klaasen CD eds, Pergamon Press, New York, pp623–680, 1991

Graziano MJ, Flory W, Seger CL, Hebert CD: Effects of a *Cassia occidentalis* extract in the domestic chicken *(Gallus domesticus). Am J Vet Res* 44;1238–1244, 1983

Greatorex JC: Some unusual cases of plant poisoning in animals. *Vet Rec* 78;725–727, 1966

Greinwald R, Schultze W, Czygan FC: Alkaloid patterns of the above-ground parts of *Laburnum watereri* (Kirchn.) Dipp. *Biochem Physiol Pflanz* 186;1–10, 1990

Gresham ACJ, Booth K: Poisoning of sheep by goat's rue. *Vet Rec* 129;197–198, 1991

Guengerich FP, Aust SD: Activation of the parasympathomimetic alkaloid slaframine by microsomal and photochemical oxidation. *Mol Pharmacol* 13;185–195, 1977

Gummow B, Bastianello SS, Labuschagne L, Erasmus GL: Experimental *Albizia versicolor* poisoning in sheep and its successful treatment with pyridoxine hydrochloride. *Onderstepoort J Vet Res* 59;111–118, 1992

Gunsolus JM: Toxicity of jequirity beans [letter]. *J Am Med Assoc* 157;779, 1955

Gunther HE: Biochemistry of selenium. In *Selenium,* Zingaro RA, Cooper WC eds, Van Nostrand Reinhold, New York, pp546–614, 1974

Gupta I, Nauriyal MM: *Acacia leucophloea* Willd. (Raunja) poisoning in livestock. *Indian Vet J* 43;538–543, 1966

Gurnsey MP, Jones WT, Merrall M, Reid CSW: Cyanide poisoning in cattle: two unusual cases. *N Z Vet J* 25;128–130, 1977

Gustine DJ: Aliphatic nitro compounds in crownvetch: a review. *Crop Sci* 19;197–203, 1979

Gustine DJ, Shenk JS, Moyer BG, Barnes RF: Isolation of β-nitropropionic acid from crownvetch. *Agron J* 66;636–639, 1974

Gustine J, Gebble L, Barnes RF: Relationship between β-propionic acid content of crownvetch and toxicity in non-ruminant animals. *J Anim Sci* 42;616–621, 1976

Gustine DJ, Moyer BG, Wangsness PJ, Shenk JS: Ruminal metabolism of 3-nitropropanol-D-glucopyranoses from crownvetch. *J Anim Sci* 44;1107–1111, 1977

Habermehl GG: Poisonous plants of Europe. In *Poisonous Plants: Proceedings of the Third International Symposium,* James LF, Keeler RF, Bailey EM Jr, Cheeke PR, Hegarty MP eds, Iowa State Univ Press, Ames, pp74–83, 1992

Haimanot RT, Kidane Y: Lathyrism in rural northwestern Ethiopia: a highly prevalent neurotoxic disorder. *Int J Epidemiol* 19;664–672, 1990

Hall JW, Walker I, Majak W: Evaluation of two supplements for the prevention of bloat. *Can Vet J* 35;702–705, 1994

Hamilton BF, Gould DH: Nature and distribution of brain lesions in rats intoxicated with 3-nitropropionic acid: a type of hypoxic (energy deficient) brain damage. *Acta Neuropathol (Berl)* 72;286–297, 1987

Hammond AC, Carlson JR: Inhibition of ruminal degradation of L-tryptophan to 3-methylindole in vitro. *J Anim Sci* 51;207–214, 1980

Hammond AC, Carlson JR, Breeze RG: Prevention of tryptophan-induced acute bovine pulmonary oedema and emphysema (fog fever). *Vet Rec* 107;322–325, 1980

Hammond AC, Allison MJ, Williams MJ, Prine GM, Bates DB: Prevention of *Leucaena* toxicosis of cattle in Florida by ruminal inoculation with 3-hydroxy-4-(1H)-pyridone-degrading bacteria. *Am J Vet Res* 50;2176–2180, 1989

Hankins CN, Kindinger J, Shannon LM: The lectin of *Sophora japonica:* purification, properties, and *N*-terminal amino acid sequences of five lectins from bark. *Plant Physiol* 86;67–70, 1988

Hansen AA: The poisonous plant situation in Indiana, 3. *J Am Vet Med Assoc* 66;351–362, 1924

Hansen AA: Toxic trees. *North Am Vet* 9;49–53, 1928a

Hansen AA: Trifoliosis and similar livestock diseases. *North Am Vet* 9;34–36, 1928b

Hansen AA: Indiana plants injurious to livestock. *Indiana Agric Exp Stn Circ* 175, 1930

Haque A, Hossain M, Khan JK, Kuo YH, Lambein F, De Reuck J: New findings and symptomatic treatment for neurolathyrism, a motor neuron disease occurring in North West Bangladesh. *Paraplegia* 32;193–195, 1994

Haque A, Khan JK, Wouters G, Hossain M, Lambein F: Study of HTLV-I antibodies in CSF and serum of neurolathyrism patients in Bangladesh. *Ann Soc Belg Med Trop* 75;131–134, 1995

Haque A, Hossain M, Wouters G, Lambein F: Epidemiological study of lathyrism in north-western districts of Bangladesh. *Neuroepidemiology* 15;83–91, 1996

Haque A, Hossain M, Lambein F, Bell EA: Evidence of osteolathyrism among patients suffering from neurolathyrism in Bangladesh. *Nat Toxins* 5;43–46, 1997a

Haque N, Khan MY, Lal M, Parihar NS: Nutritional and histopathological studies on Jamunapari goats fed on concentrate diet supplemented with various levels of leucaena leaves. *Indian J Anim Sci* 67;524–527, 1997b

Hargreaves RT, Johnson RD, Millington DS, Mondal MH, Beavers W, Becker L, Young C, Rinehart KL Jr: Alkaloids of American species of *Erythrina. Lloydia* 37;569–580, 1974

Harper JA, Arscott GH: Toxicity of common and hairy vetch seed for poults and chicks. *Poult Sci* 41;1968–1974, 1962

Harper PA, Cook RW, Gill PA, Fraser GC, Badcoe LM, Power JM: Vetch toxicosis in cattle grazing *Vicia villosa* ssp. *dasycarpa* and *V. benghalensis. Aust Vet J* 70;140–144, 1993

Harries WN, Baker FP, Johnston A: An outbreak of locoweed poisoning in horses in south-western Alberta. *Can Vet J* 13;141–145, 1972

Harris T: The Kesterson syndrome. *Amicus* 11;4–9, 1989

Harrison LH, Colvin BM, Stuart BP, Sangster LT, Gorgacz EJ, Gosser HS: Paralysis in swine due to focal symmetrical poliomalacia: possible selenium toxicosis. *Vet Pathol* 20;265–273, 1983

Hart GH, Guilbert HR, Wagnon KA, Goss H: *Acorn Calves.* Calif Agric Exp Stn Bull 699, 1947

Hartley WJ, James LF: Microscopic lesions in fetuses of ewes ingesting locoweed *(Astragalus lentiginosus). Am J Vet Res* 34;209–211, 1973

Hartley WJ, James LF: Fetal and maternal lesions in pregnant ewes ingesting locoweed *(Astragalus lentiginosus). Am J Vet Res* 36;825–826, 1975

Hartley WJ, James LF, Broquist H, Panter KE: Pathology of experimental locoweed and selenium poisoning in pigs. In *Plant Toxicology, Proceedings of the Australia-USA Poisonous Plants Symposium,* Seawright AA, Hegarty MP, James LF, Keeler RF eds, Animal Research Institute, Yeerongpilly, Brisbane, Australia, pp141–149, 1985

Harvey RB, Rowe LD, Reagor JC: Suspected toxicity of *Desmanthus leptolobus* (prairie bundleflower, prairie mimosa) to horses. *Southwest Vet* 37;143–144, 1986

Hassel B, Sonnewald U: Selective inhibition of the tricarboxylic acid cycle of GABAergic neurons with 3-nitropropionic acid in vivo. *J Neurochem* 65;1184–1191, 1995

Hatfield GM, Valdes LJJ, Keller WJ, Merrill WL, Jones VH: An investigation of *Sophora secundiflora* seeds (mescalbeans). *Lloydia* 40;374–383, 1977

Hatfield GM, Yang DJ, Ferguson PW, Keller WJ: Identification of toxic alkaloids from the *calcaratus* subspecies of *Lupinus arbustus. J Agric Food Chem* 33;909–912, 1985

Hawkins LM, Beaver KM, Jane DE, Taylor PM, Sunter DC, Roberts PJ: Binding of the new radioligand *(S)*-(3H)AMPA to rat brain synaptic membranes: effects of a series of structural analogues of the non-NMDA receptor agonist willardiine. *Neuropharmacology* 34;405–410, 1995

Hebert CD, Flory N, Seger C, Blanchard RE: Preliminary isolation of a myodegenerative toxic principle from *Cassia occidentalis. Am J Vet Res* 44;1370–1374, 1983

Hegarty MP: Toxic amino acids of plant origin. In *Effects of Poisonous Plants in Livestock,* Keeler RF, Van Kampen KR, James LF eds, Academic Press, New York, pp575–585, 1978

Hegarty MP, Pound AW: Indospicine, a new hepatotoxic amino-acid from *Indigofera spicata. Nature* 217;354–355, 1968

Hegarty MP, Pound AW: Indospicine, a hepatotoxic amino acid from *Indigofera spicata:* isolation, structure, and biological studies. *Aust J Biol Sci* 23;831–842, 1970

Hegarty MP, Schinckel PG, Court RD: Reaction of sheep to the consumption of *Leucaena glauca* Benth. and its toxic principle mimosine. *Aust J Agric Res* 15;153–167, 1964

Hegarty MP, Lee CP, Christie GS, DeMunk FG, Court RD: Comparative toxicities of mimosine and some chemically related compounds to mouse bone marrow cells in liquid culture. *Aust J Biol Sci* 31;115–121, 1978

Hegarty MP, Lee CP, Christie GS, Court RD, Haydock KP: The goitrogen 3-hydroxy-4(1H)-pyridone, a ruminal metabolite from *Luecaena leucocephala:* effects in mice and rats. *Aust J Biol Sci* 32;27–40, 1979

Hegarty MP, Kelly WR, McEwan D, Williams OJ, Cameron R: Hepatotoxicity to dogs of horse meat contaminated with indospicine. *Aust Vet J* 65;337–340, 1988

Hegde R, Maiti TK, Podder SK: Purification and characterization of three toxins and two agglutinins from *Abrus precatorius* seed by using lactamyl-sepharose affinity chromatography. *Anal Biochem* 194;101–109, 1991

Hembree JA, Chang CJ, McLaughlin JL, Peck G, Cassady JM: Potential antitumor agents: a cytotoxic cardenolide from *Coronilla varia. J Nat Prod* 42;293–298, 1979

Hendershot JM: Ketosis in the Hawaiian Islands. *J Am Vet Med Assoc* 108;74–75, 1946

Henson JB, Dollahite JW: Toxic myodegeneration in calves produced by experimental *Cassia occidentalis* intoxication. *Am J Vet Res* 27;947–949, 1966

Henson JB, Dollahite JW, Bridges CH, Rao RR: Myodegeneration in cattle grazing *Cassia* species. *J Am Vet Med Assoc* 147;142–145, 1965

Hershey AL: *Some Poisonous Plant Problems of New Mexico.* N M Agric Exp Stn Bull 322, 1945

Hill CH: Interrelationships of selenium with other trace elements. *Fed Proc* 34;2096–2100, 1975

Hilliker KS, Roth RA: Injury to the isolated, perfused lung by exposure in vitro to monocrotaline pyrrole. *Exp Lung Res* 8;201–212, 1985

Hirano A, Llena JF, Streifler M, Cohn DF: Anterior horn cell changes in a case of neurolathyrism. *Acta Neuropathol (Berl)* 35;277–283, 1976

Hironaka R: Peavine silage paralysis. *Feedstuffs* 35;24, 1963

Hitchcock CL: A revision of the North American species of *Lathyrus. Univ Wash Publ Biol* 15:1–104, 1952

Hjort AM: Some physiological properties of certain *N*-methylated-β-phenylethylamines. *J Pharmacol Exp Ther* 52;101–112, 1934

Holmes JHJ: Toxicity of *Leucaena leucocephala* for steers in the wet tropics. *Trop Anim Health Prod* 13;94–100, 1981

Holtz P, Palm D: Pharmacological aspects of vitamin B6. *Pharmacol Rev* 16;113–178, 1964

Honeyfield DC, Carlson JR, Nocerini MR, Breeze RG: Duration of inhibition of 3-methylindole production by monensin. *J Anim Sci* 60;226–231, 1985

Hooper PT: Pyrrolizidine alkaloid poisoning–pathology with particular reference to differences in animal and plant species. In *Effects of Poisonous Plants in Livestock,* Keeler RF, Van Kampen KR, James LF eds, Academic Press, New York, pp161–176, 1978

Hooper PT, Scanlan WA: *Crotalaria retusa* poisoning in pigs and poultry. *Aust Vet J* 53;109–114, 1977

Hooper PT, Hart B, Smith GW: The prevention and treatment of Birdsville disease of horses. *Aust Vet J* 47;326–329, 1971

Hoorn CM, Wagner JG, Roth RA: Effects of monocrotaline pyrrole on cultured rat pulmonary endothelium. *Toxicol Appl Pharmacol* 120;281–287, 1993

Horejsi V, Haskovec C, Kocourek J: Studies on lectins. 38. Isolation and characterization of the lectin from black locust bark (*Robinia pseudoacacia* L.). *Biochem Biophys Acta* 532;98–104, 1978

Horton RW, Meldrum BS: Seizures induced by allylglycine, 3-mercaptopropionic acid and 4-deoxypyridoxine in mice and photosensitive baboons, and different modes of inhibition of cerebral glutamic acid decarboxylase. *Br J Pharmacol* 49;52–63, 1973

House JK, George LW, Oslund KL, Galey FD, Stannard AW, Koch LM: Primary photosensitization related to ingestion of alfalfa silage by cattle. *J Am Vet Med Assoc* 209;1604–1607, 1996

Hove EL, King S, Hill GD: Composition, protein quality, and toxins of seeds of the grain legumes *Gjycine max, Lupinus* spp., *Phaseolus* spp., *Pisum sativum,* and *Vicia faba. N Z J Agric Res* 21;457–462, 1978

Hoveland CS, Buchanan GA, Bosworth SC, Bailey IJ: *Forage Nutritive Quality of Weeds in Alabama.* Ala Agric Exp Stn Bull 577, 1986

Howarth RE: A review of bloat in cattle. *Can Vet J* 16;281–294, 1975

Howarth RE, Majak W, Waldern DE, Brandt SA, Fesser AC, Goplen BP, Spurr DT: Relationships between ruminant bloat and the chemical composition of alfalfa herbage. 1. Nitrogen and protein fractions. *Can J Anim Sci* 57;345–357, 1977

Howarth RE, Chaplin RK, Cheng K-J, Goplen BP, Hall JW, Hironaka R, Majak W, Radostits OM: *Bloat in Cattle.* Agriculture Canada Publ 1858/E, 1991

Hsu C-C: Isolation, purification, and structural characterization of the saponins from *Glottidium vesicarium.* PhD Thesis, Oklahoma State Univ, 1968

Hughes J: Feeding experiments with Indian peas containing *Lathyrus sativus. Analyst* 20;169–173, 1895

Hugon J, Ludolph A, Roy DN, Schaumberg HH, Spencer PS: Studies on the etiology and pathogenesis of motor neuron diseases. 2. Clinical and electrophysiologic features of pyramidal dysfunction in macaques fed *Lathyrus sativus* and IDPN. *Neurology* 38;435–442, 1988

Hultine JD, Mount ME, Easley KJ, Oehme FW: Selenium toxicosis in the horse. *Equine Pract* 1;57–60, 1979

Humphreys DJ: The toxic factors of the seeds of *Abrus precatorius. Phytochemistry* 8;2397–2400, 1969

Humphries MJ, Matsumoto K, White SL, Molyneux RJ, Olden K: Augmentation of murine natural killer cell activity by swainsonine, a new antimetastatic immunomodulator. *Cancer Res* 48;1410–1415, 1988

Huxtable CR, Dorling PR, Walkley SU: Onset and regression of neuroaxonal lesions in sheep with mannosidase induced experimentally with swainsonine. *Acta Neuropathol* 58;27–33, 1982

Huxtable CR, Dorling PR, Colegate SM: Identification of galegine, an isoprenyl guanidine, as the toxic principle of *Schoenus asperocarpus* (poison sedge). *Aust Vet J* 70;169–171, 1993

Huxtable RJ: Human health implications of pyrrolizidine alkaloids and herbs containing them. In *Toxicants of Plant Origin,* vol 1, Cheeke PR ed, CRC Press, Boca Raton, Fla, pp41–86, 1989

Hyslop NStG: Bovine pulmonary emphysema. *Can Vet J* 10;251–257, 1969

Ihnat M: Plants and agricultural materials. In *Occurrence and Distribution of Selenium,* Ihnat M ed, CRC Press, Boca Raton, Fla, pp33–105, 1989

Innocenti G, Cappelletti EM, Caporale G: Furocoumarin contents in the vegetative organs of cultivated *Psoralea* species. *Int J Pharmacol* 29;311–316, 1991

Irigoyen LF, Graca DL, Barros CSL: Experimental poisoning by *Cassia occidentalis* (Leg.Caes.) in horses. *Pesq Vet Bras* 11;35–44, 1991

Irwin HS, Barneby RC: The American Cassiinae: a synoptical revision of Leguminosae, tribe Cassieae, subtribe Cassiinae in the New World. *Mem N Y Bot Gard* 35, 1982

Isely D: Legumes of the United States. 1. Native *Acacia*. *SIDA* 3;365–386, 1969

Ishida M, Shinozaki H: Glutamate inhibitory action of matrine at the crayfish neuromuscular junction. *Br J Pharmacol* 82;523–531, 1984

Izaddoost M: Alkaloid chemotaxonomy of the genus *Sophora*. *Phytochemistry* 14;203–204, 1975

Izaddoost M: Synergistic effects of alkaloids and pipecolic acid on the toxicity of the seeds of *Sophora secundiflora*. *Tex J Sci* 31;319–323, 1979

Izaddoost M, Harris BG, Gracy RW: Structure and toxicity of alkaloids and amino acids of *Sophora secundiflora*. *J Pharm Sci* 65;352–354, 1976

Jacob E, Patel AJ, Ramakrishnan CV: Effect of neurotoxin from the seeds of *Lathyrus sativus* on glutamate metabolism in chick brain. *J Neurochem* 14;1091–1094, 1967

Jacoby PW Jr, Ansley RJ: Mesquite: classification, distribution, ecology, and control. In *Noxious Range Weeds,* James LF, Evans JO, Ralphs MH, Child RD eds, Westview Press, Boulder, Colo, pp364–376, 1991

Jacobziner H, Raybin HW: Plant and insecticide poisonings. *N Y State Med J* 61;2463–2466, 1961

Jaffe WG, Vega Lette CL: Heat-labile growth-inhibiting factors in beans *(Phaseolus vulgaris)*. *J Nutr* 94;103–110, 1968

Jago VM, Peterson JE, Payne AL: Lupinosis: response of sheep to different doses of phomopsin. *Aust J Exp Biol Med Sci* 60;239–251, 1982

Jahan K, Ahmad K: Studies on neurolathyrism. *Environ Res* 60;259–266, 1993

James LF: Effect of locoweed on fetal development: preliminary study in sheep. *Am J Vet Res* 33;835–842, 1972

James LF: Effect of locoweed *(Astragalus lentiginosus)* feeding on fetal lamb development. *Can J Comp Med* 40;380–384, 1976

James LF: Neurotoxins and other toxins from *Astragalus* and related genera. In *Handbook of Natural Toxins,* vol 1, *Plant and Fungal Toxins,* Keeler RF, Tu AT eds, Marcel Dekker, New York, pp445–462, 1983

James LF, Foote W: Estrogenic properties of locoweed *(Astragalus lentiginosus)*. *Can J Comp Med* 36;360–365, 1972

James LF, Hartley WJ: Effects of milk from animals fed locoweed in kittens, calves, and lambs. *Am J Vet Res* 38;1263–1265, 1977

James LF, Nielsen DB: Locoweeds: assessment of the problem on western U.S. rangelands. In *The Ecology and Economic Impact of Poisonous Plants on Livestock Production,* James LF, Ralphs MH, Nielsen DB eds, Westview Press, Boulder, Colo, pp171–180, 1988

James LF, Panter KE: Locoweed poisoning in livestock. In *Swainsonine and Related Glycosidase Inhibitors,* James LF, Elbein AD, Molyneux RJ, Warren CD, eds, Iowa State Univ Press, Ames, pp23–38, 1989

James LF, Van Kampen KR: Acute and residual lesions of locoweed poisoning in cattle and horses. *J Am Vet Med Assoc* 158;614–618, 1971a

James LF, Van Kampen KR: Effects of locoweed intoxication on the genital tract of the ram. *Am J Vet Res* 32;1253–1256, 1971b

James LF, Van Kampen KR: Effect of protein and mineral supplementation on potential locoweed (*Astragalus* spp.) poisoning in sheep. *J Am Vet Med Assoc* 164;1042–1043, 1974

James LF, Van Kampen KR: Effects of locoweed toxin on rats. *Am J Vet Res* 37;845–846, 1976

James LF, Shupe JL, Binns W, Keeler RF: Abortive and teratogenic effects of locoweed on sheep and cattle. *Am J Vet Res* 28;1379–1388, 1967

James LF, Binns W, Shupe JL: Blood changes in cattle and sheep fed lupine. *Am J Vet Res* 29;557–560, 1968

James LF, Van Kampen KR, Staker GR: Locoweed *(Astragalus lentiginosus)* poisoning in cattle and horses. *J Am Vet Med Assoc* 155;525–530, 1969

James LF, Van Kampen KR, Johnson AE: Physiopathologic changes in locoweed poisoning of livestock. *Am J Vet Res* 31;663–672, 1970

James LF, Olsen JD, Sharma RP: Locoweed poisoning in sheep: electroencephalographic and brain amine changes. *Clin Toxicol* 11;53–60, 1977

James LF, Foote W, Nye W, Hartley WJ: Effects of feeding *Oxytropis* and *Astragalus* pollen to mice and *Astragalus* seeds to rats. *Am J Vet Res* 39;711–712, 1978

James LF, Hartley WJ, Williams MC, Van Kampen KR: Field and experimental studies in cattle and sheep poisoned by nitro-bearing *Astragalus* or their toxins. *Am J Vet Res* 41;377–382, 1980

James LF, Hartley WJ, Van Kampen KR: Syndromes of *Astragalus* poisoning in livestock. *J Am Vet Med Assoc* 178;146–150, 1981

James LF, Smart RA, Shupe JL, Bowns J, Schoenfeld J: Suspected phytogenic selenium poisoning in sheep. *J Am Vet Med Assoc* 180;1478–1481, 1982

James LF, Hartley WF, Van Kampen KR, Nielsen D: Relationship between ingestion of the locoweed *Oxytropis sericea* and congestive right-sided heart failure in cattle. *Am J Vet Res* 44;254–259, 1983a

James LF, Van Kampen KR, Hartley WJ: *Astragalus bisulcatus*—a cause of selenium or locoweed poisoning. *Vet Hum Toxicol* 25;86–89, 1983b

James LF, Hartley WJ, Nielsen D, Allen S, Panter KE: Locoweed *(Oxytropis sericea)* poisoning and congestive heart failure in cattle. *J Am Vet Med Assoc* 189;1549–1556, 1985

James LF, Panter KE, Broquist HP, Hartley WJ: Swainsonine-induced high mountain disease in calves. *Vet Hum Toxicol* 33;217–219, 1991

James LF, Panter KE, Molyneux RJ: Selenium poisoning in livestock. In *Poisonous Plants: Proceedings of the Third International Symposium,* James LF, Keeler RF, Bailey EM Jr, Cheeke PR, Hegarty MP eds, Iowa State Univ Press, Ames, pp153–158, 1992

James LF, Hartley WF, Panter KE, Stegelmeier BL, Gould D, Mayland HF: Selenium poisoning in cattle. In *Plant-Associated Toxins: Agricultural, Phytochemical, and Ecological Aspects,* Colegate SM, Dorling PR eds, CAB International, Wallingford, UK, pp416–420, 1994

Jayne-Williams DJ: Influence of dietary jack beans *(Canavalia ensiformis)* and of concanavalin A on the growth of conventional and gnotobiotic Japanese quail *(Coturnix coturnix japonica)*. *Nat New Biol* 243;150–151, 1973

Jensen R, Griner LA, Adams OR: Polioencephalomalacia of cattle and sheep. *J Am Vet Med Assoc* 129;311–321, 1956

Jin LX, Cui YY, Zhang GD: HPLC analysis of alkaloids in *Sophora flavescens* Ait. *Yao Hsueh Hsueh Pao* 28;136–139, 1993

Johnson AE, Molyneux RJ: Variation in toxic pyrrolizidine alkaloid content of plants, associated with site, stage of growth, and environmental conditions. In *Plant Toxicology, Proceedings of the Australia-USA Poisonous Plants Symposium,* Seawright AA, Hegarty MP, James LF, Keeler RF eds, Animal Research Institute, Yeerongpilly, Brisbane, Australia, pp209–218, 1985

Johnson B, Moore J, Woods LW, Galey FD: Systemic granulomatous disease in cattle in California associated with grazing hairy vetch *(Vicia villosa)*. *J Vet Diagn Invest* 4;360–362, 1992

Johnson CM, Asher CJ, Broyer TC: Distribution of selenium in plants. In *Selenium in Biomedicine, First International Symposium, 1966, Muth, Ohio,* Oldfield JE, Weswig PH eds, AVI Publishing, Westport, Conn, pp57–75, 1967

Johnson JC, Miller JD, Bedell DM: Tifwhite-78 lupine seed as a feedstuff for cattle. *J Dairy Sci* 69;142–147, 1986

Jolly RD, Thompson KG, Van De Water NS, Winchester BG, Walkley SU, Daniel PF: Bovine mannosidosis. In *Swainsonine and Related Glycosidase Inhibitors,* James LF,

Elbein AD, Molyneux RJ, Warren CD eds, Iowa State Univ Press, Ames, pp291–303, 1989

Jones DA, Keymer RJ, Ellis WM: Cyanogenesis in plants and animal feeding. In *Biochemical Aspects of Plant and Animal Coevolution*, Harborne JB ed, Academic Press, London, pp21–34, 1978

Jones RJ: The value of *Leucaena leucocephala* as a feed for ruminants in the tropics. *World Anim Rev* 31;13–23, 1979

Jones RJ: Does ruminal metabolism of mimosine explain the absence of *Leucaena* toxicity in Hawaii? *Aust Vet J* 57;55, 1981

Jones RJ: *Leucaena* toxicity and the ruminal degradation of mimosine In *Plant Toxicology, Proceedings of the Australia-USA Poisonous Plants Symposium,* Seawright AA, Hegarty MP, James LF, Keeler RF eds, Animal Research Institute, Yeerongpilly, Brisbane, Australia, pp111–119, 1985

Jones RJ, Lowry JB: Australian goats detoxify the goitrogen 3-hydroxy-4(1H)pyridone (DHP) after rumen infusion from an Indonesian goat. *Experientia* 40;1435–1436, 1984

Jones RJ, Megarrity RG: Successful transfer of DHP-degrading bacteria from Hawaiian goats to Australian ruminants to overcome the toxicity of *Leucaena. Aust Vet J* 63;259–262, 1986

Joubert JPJ: Cardiac glycosides. In *Toxicants of Plant Origin,* vol 2, *Glycosides,* Cheeke PR ed, CRC Press, Boca Raton, Fla, pp61–96, 1989

Kallela K, Heinonen K, Saloniemi H: Plant oestrogens: the cause of decreased fertility in cows. *Nord Vet Med* 36;124–129, 1984

Kandel ER, Schwartz JH: Directly gated transmission at central synapses. In *Principles of Neural Science,* 3d ed, Kandel ER, Schwartz JH, Jessel TM eds, Elsevier, New York, pp153–173, 1991

Kaplanski O, Shemesh M, Berman A: Effects of phyto-oestrogens on progesterone synthesis by isolated bovine granulosa cells. *J Endocrinol* 89;343–348, 1981

Keeler RF: Known and suspected teratogenic hazards in range plants. *Clin Toxicol* 5;529–565, 1972

Keeler RF: Lupin alkaloids from teratogenic and nonteratogenic lupins. 1. Correlation of crooked calf disease incidence with alkaloid distribution determined by gas chromatography. *Teratology* 7;23–30, 1973a

Keeler RF: Lupin alkaloids from teratogenic and nonteratogenic lupins. 2. Identification of the major alkaloids by tandem gas chromatography–mass spectrometry in plants producing crooked calf disease. *Teratology* 7;31–36, 1973b

Keeler RF: Toxins and teratogens of higher plants. *Lloydia* 38;56–86, 1975

Keeler RF: Lupin alkaloids from teratogenic and nonteratogenic lupins. 3. Identification of anagyrine as the probable teratogen by feeding trials. *J Toxicol Environ Health* 1;887–898, 1976

Keeler RF: Quinolizidine alkaloids in range and grain lupines. In *Toxicants of Plant Origin,* vol 1, *Alkaloids,* Cheeke PR ed, CRC Press, Boca Raton, Fla, pp133–167, 1989

Keeler RF, Baker DC: Myopathy in cattle induced by alkaloid extracts from *Thermopsis montana, Laburnum anagyroides,* and a *Lupinus* sp. *J Comp Pathol* 103;169–182, 1990

Keeler RF, James LF: Experimental teratogenic lathyrism in sheep and further comparative aspects with teratogenic locoism. *Can J Comp Med* 35;332–341, 1971

Keeler RF, Panter KE: Piperidine alkaloid composition and relation to crooked calf disease–inducing potential of *Lupinus formosis. Teratology* 40;423–432, 1989

Keeler RF, James LF, Binns W, Shupe JL: An apparent relationship between locoism and lathyrism. *Can J Comp Med Vet Sci* 31;334–341, 1967

Keeler RF, Cronin EH, Shupe JL: Lupin alkaloids from teratogenic and nonteratogenic lupins. 4. Concentration of total alkaloids, individual major alkaloids, and the teratogen anagyrine as a function of plant part and stage of growth and their relationship to crooked calf disease. *J Toxicol Environ Health* 1;899–908, 1976

Keeler RF, James LF, Shupe JL, Van Kampen KR: Lupine-induced crooked calf disease and a management method to reduce incidence. *J Range Manage* 30;97–102, 1977

Keeler RF, Johnson AE, Chase RL: Toxicity of *Thermopsis montana* in cattle. *Cornell Vet* 76;115–127, 1986a

Keeler RF, Johnson AE, Stuart LD, Evans JO: Toxicosis from and possible adaptation to *Galega officinalis* in sheep and the relationship to *Verbesina encelioides* toxicosis. *Vet Hum Toxicol* 28;309–315, 1986b

Keeler RF, Baker DC, Evans JO: Individual animal susceptibilty and its relationship to induced adaptation or tolerance in sheep to *Galega officinalis* L. *Vet Hum Toxicol* 30;420–423, 1988

Keller H, Dewitz W: Poisoning of nine horses by the bark of false acacia *Robinia pseudoacacia. Dtsch Tierarztl Wochenschr* 76;115–117, 1969

Keller WJ: Alkaloids from *Sophora secundiflora. Phytochemistry* 14;2305–2306, 1975

Keller WJ, Cole FR: The alkaloids of *Thermopsis montana. Lloydia* 32;498–502, 1969

Keller WJ, Hatfield M: 11-Allylcytisine and other minor alkaloids from unripe *Sophora secundiflora* fruits. *Phytochemistry* 18;2068–2069, 1979

Kelly WR, Young MP, Hegarty MP, Simpson GD: The hepatotoxicity of indospicine in dogs. In *Poisonous Plants: Proceedings of the Third International Symposium,* James LF, Keeler RF, Bailey EM Jr, Cheeke PR, Hegarty MP eds, Iowa State Univ Press, Ames, pp126–130, 1992

Kelly WW, Barber CW, Pate DD, Hill CH: Effect of feeding crotalaria seed to young chickens. *J Am Vet Med Assoc* 139;1215–1217, 1961

Kermen WR: Impaction in horses due to ingestion of pea straw. *J Am Vet Med Assoc* 95;237–238, 1939

Khan MA, Burrows GE, Holt EM: (5a,6a,7a,11B)—13,14-sophocarpine monohydrate. *Acta Cryst* C48;2051–2053, 1992

Kienholz EW, Jensen LS, McGinnis: Evidence for chick growth inhibitors in several legume seeds. *Poult Sci* 41;367–371, 1962

Kilgore WW, Crosby DG, Craigmill AL, Poppen NK: Toxic plants as possible human teratogens. *Calif Agric* 35;6, 1981

Kim HL, Krakoff IH, Newman RA: Isolation of sesbanimide from the seed of *Sesbania vesicaria. Gen Pharmacol* 23;701–703, 1992

Kim HY, Stermitz FR, Molyneux RJ, Wilson DW, Taylor D, Coulombe RA Jr: Structural influences on pyrrolizidine alkaloid-induced cytopathology. *Toxicol Appl Pharmacol* 122;61–69, 1993

Kimura M, Kimura I, Li X, Kong X, Cai Y: Positive inotropic action and conformation of lupine alkaloids in isolated cardiac muscle of guinea pig and bullfrog. *Phytother Res* 3;101–105, 1989

Kinamore PA, Jaeger RW, de Castro FJ, Peck KO: *Abrus* and *Ricinus* ingestion: management of three cases. *Clin Toxicol* 17;401–405, 1980

Kinghorn AD, Balandrin MF: Quinolizidine alkaloids of the Leguminosae: structural types, analysis, chemotaxonomy, and biological activities. In *Alkaloids: Chemical and Biological Perspectives,* vol 2, Pelletier SW ed, John Wiley & Sons, New York, 105–148, 1984

Kinghorn AD, Selim MA, Smolenski SJ: Alkaloid distribution in some New World *Lupinus* species. *Phytochemistry* 19;1705–1710, 1980

Kingsbury JM: *Poisonous Plants of the United States and Canada,* Prentice-Hall, Englewood Cliffs, NJ, 1964

Kirkland JB, Bray TM: The effect of 3-methylindole on phospholipid synthesis in goat lung tissue slices. *Proc Soc Exp Biol Med* 175;30–34, 1984

Kirkpatrick JG, Burrows GE: Locoism in horses [letter]. *Vet Hum Toxicol* 32;168–169, 1990

Kitts WD, Swierstra E, Brink VC, Wood AJ: The estrogen-like substances in certain legumes and grasses. 2. The effect of stage of maturity and frequency of cutting on the estrogenic activity of some forages. *Can J Anim Sci* 39;158–163, 1959

Kleinhaus AL: Segregation of leech neurones by the effect of sparteine on action potential duration. *J Physiol* 299;309–321, 1980

Knauer KW, Reagor JC, Bailey EM Jr, Carriker L: Mescalbean *(Sophora secundiflora)* toxicity in a dog. *Vet Hum Toxicol* 37;237–239, 1995

Knott SG, McCray CWR: Two naturally occurring outbreaks of selenosis in Queensland. *Aust Vet J* 35;161–165, 1959

Kobashi K, Nishimura T, Kusaka M, Hattori M, Namba T: Metabolism of sennosides by human intestinal bacteria. *Planta Med* 40;226–236, 1980

Kosower SN, Kosower EM: Does 3,4-dihydroxyphenylalanine play a part in favism? *Nature* 215:285–286, 1967

Kosower NS, Rock RA: Seizures in experimental porphyria. *Nature* 217;565–567, 1968

Kosuge T, Gilchrist D: Chemistry of coumarin and related compounds. In *Mycotoxic Fungi, Mycotoxins, Mycotoxicosis: An Encyclopedic Handbook,* Wyllie TD, Morehouse LG eds, Marcel Dekker, New York, pp239–282, 1977

Krausman PR, Kuenzi AJ, Etchberger RC, Rautenstrauch KR, Ordway LL, Hervert JJ: Diets of mule deer. *J Range Manage* 50;513–522, 1997

Krishna L, Vaid J, Singh B: Pathological study on *Indigofera teysmanni* toxicity in sheep. *Indian J Comp Microbiol Immunol Inf Dis* 7;14–17, 1986

Krukoff BA, Barneby RC: Conspectus of species of the genus *Erythrina. Lloydia* 37;332–459, 1974

Kulp KS, Vulliet PR: Mimosine blocks cell cycle progression by chelating iron in asynchronus human breast cancer cells. *Toxicol Appl Pharmacol* 139;356–364, 1996

Kurokawa T, Tsuda M, Sugino Y: Purification and characterization of a lectin from *Wistaria floribunda* seeds. *J Biol Chem* 251;5686–5693, 1976

La Bella V, Brighina F, Piccoli F, Guarneri R: Effect of β-*N*-oxalylamino-L-alanine on cerebellar cGMP level in vivo. *Neurochem Res* 18;171–175, 1993

Lalonde M: A reversible arrest point in the late G1 phase of the mammalian cell cycle. *Exp Cell Res* 186;332–339, 1990

Lambein F, Van Parijs R: Isolation and characterization of 1-alanyl-uracil (willardiine) and 3-alanyl-uracil (iso-willardiine) from *Pisum sativum. Biochem Biophys Res Commun* 32;474–479, 1968

Lambein F, Haque R, Khan JK, Kebede N, Kuo Y-H: From soil to brain: zinc deficiency increases the neurotoxicity of *Lathyrus sativus* and may affect the susceptibility for the motorneurone disease neurolathyrism. *Toxicon* 32;461–466, 1994

Lambert MG, Jung GA, Fletcher RH, Budding PJ, Costall DA: Forage shrubs in North Island hill country. 2. Sheep and goat preferences. *N Z J Agric Res* 32;485–490, 1989a

Lambert MG, Jung GA, Harpster HW, Lee J: Forage shrubs in North Island hill country. 4. Chemical composition and conclusions. *N Z J Agric Res* 32;499–506, 1989b

Landolt G, Feige K, Schoberl M: Poisoning of horses by the bark of the false acacia *(Robinia pseudoacacia). Schweiz Arch Tierheilkd* 139;363–366, 1997

Larue-Achagiotis C, Picard M, Louis-Sylvestre J: Feeding behavior in rats on a complete diet containing concanavalin A. *Reprod Nutr Dev* 32;343–350, 1992

Lavin M: The occurrence of canavanine in seeds of the tribe Robinieae. *Biochem Syst Ecol* 14;71–73, 1986

Leather J: *Lathyrus*-poisoning in horses. *Vet J (London)* 20;233–239, 1885

Lee JG: Experimental lathyrism produced by feeding singletary pea *(Lathyrus pusillus)* seed. *J Nutr* 40;587–594, 1950

Lee JG, Dupuy HP, Rolfs HE: Dietary protein and the development of rat lathyrism. *J Nutr* 58;433–442, 1956

Lees GL: Condensed tannins in some forage legumes: their role in the prevention of ruminant pasture bloat. *Basic Life Sci* 59;915–934, 1992

Lehman AJ: Curare-like actions of *Erythrina americana. Proc Soc Exp Biol Med* 33;501–503, 1935

Leipold HW, Husby F, Brundage AL, Shupe JL: Congenital defects of calves on Kodiac Island. *J Am Vet Med Assoc* 170;1408–1410, 1977

Leonard NJ: Lupin alkaloids. In *The Alkaloids,* vol 3, Manske RHF, Holmes HL eds, Academic Press, New York, pp119–199, 1953

Leopold AS, Erwin M, Oh J, Browning B: Phytoestrogens: adverse effects on reproduction in California quail. *Science* 191;98–100, 1976

Leroux V: Intoxications des animaux de compagnie par les plantes d'appartement. *Le Point Vet* 18;45–55, 1986

Levander OA, Baumann A: Selenium metabolism. 6. Effect of arsenic on the excretion of selenium in bile. *Toxicol Appl Pharmacol* 9;106–115, 1966

Lewis DC, Shibamoto T: Effects of *Cassia obtusifolia* (sicklepod) extracts and their anthraquinones on muscle mitochondrial function. *Toxicon* 27;519–529, 1989

Lewis HB, Schulert AR: Experimental lathyrism in the white rat and mouse. *Proc Soc Exp Biol Med* 71;440–441, 1949

Lewis HB, Fajans RS, Esterer MB, Shen C-W, Oliphant M: The nutritive value of some legumes. Lathyrism in the rat. The sweet pea (*Lathyrus odoratus*), *Lathyrus sativus, Lathyrus cicera,* and some other species of *Lathyrus. J Nutr* 36;537–559, 1948

Leyland A: Laburnum *(Cytisus laburnum)* poisoning in two dogs. *Vet Rec* 109:287, 1981

Li H, Hong S, Ping Q, Qian D: Antiarrhythmic action of sophoridine. *Yaouri Fenxi Zazhi* 6;96–97, 1986 (*Chem Abstr* 105;433e, 1986)

Li R, Chen S: Effects of seven alkaloids of *Sophora alopecuroides* on contractility of papillary muscles of guinea pigs. *Zhongguo Yaoli Xuebao* 7;219–221, 1986 (*Chem Abstr* 105;18094e, 1986)

Lievens AW: Taxonomic treatment of *Indigofera* L. (Fabaceae: Faboideae) in the New World. PhD dissertation, Louisiana State Univ, Baton Rouge, 1992

Lightfoot RJ, Wroth RH: The mechanism of temporary infertility in ewes grazing oestrogenic subterranean clover prior to and during joining. *Proc Aust Soc Anim Prod* 10;130–134, 1974

Lindsay DR, Kelly RW: The metabolism of phyto-oestrogens in sheep. *Aust Vet J* 46;219–222, 1970

Lindstrom H, Luthman J, Mouton P, Spencer P, Olson L: Plant-derived neurotoxic amino acids (β,*N*-oxalylamino-L-alanine and β-*N*-methylamino-L-alanine): effects on central monoamine neurons. *J Neurochem* 55;941–949, 1990

Linton JH, Goplen BP, Bell JM, Jaques LB: Dicoumarol studies. 1. Oral administration of synthetic dicoumarol to various classes of sheep and cattle. *Can J Anim Sci* 43;344–352, 1963

Lippegaus K, Kahn B, Schoon H-A: Two cases of neurolathyrism in horses after consumption of seeds of *Lathyrus latifolius. Pferdeheilkunde* 8;181–186, 1992

Liu X, Yuan H, Huang S: Pharmacokinetics of matrine in rabbits. *Nanjing Yaoxueyuan Xuebao* 17;309–311, 1986

Livingston AL: Forage plant estrogens. *J Toxicol Environ Health* 4;301–324, 1978

Looman J: *111 Range and Forage Plants of the Canadian Prairies.* Agric Can Publ 1751, 1983

Lotan E, Adler JK: Early effects of excessive alfalfa feeding on bovine fertility. *Refu Vet* 23;102–110, 1966

Lowe LB, Ball GJ, Carruthers VJ, Dobos RC, Lynch GA, Moate PJ, Poole PR, Valentine SC: Monensin controlled-release intraruminal capsule for control of bloat in pastured dairy cattle. *Aust Vet J* 68;17–20, 1991

Lowry JB: Agronomy and forage quality of *Albizia lebbeck* in the semi-arid tropics. *Trop Grassl* 23;84–91, 1989

Lowry JB, Maryanto, Tangendjaja B: Autolysis of mimosine to 3-hydroxy-4-1(H)pyridone in green tissues of *Leucaena leucocephala. J Sci Food Agric* 34;529–533, 1983

Ludolph AC, Hugon J, Dwivedi MP, Schaumburg HH, Spencer PS: Studies on the aetiology and pathogenesis of motor neuron diseases. 1. Lathyrism: clinical findings in established cases. *Brain* 110;149–165, 1987

Ludolph AC, He F, Spencer PS, Hammerstad J, Sabri M: 3-Nitroproprionic acid—exogenous animal neurotoxin and possible human striatal toxin. *Can J Neurol Sci* 18;492–498, 1991

Ludolph AC, Seelig MO, Ludolph A, Novitt P, Allen CN, Spencer PS, Sabri MI:

3-Nitroproprionic acid decreases cellular energy levels and causes neuronal degeneration in cortical explants. *Neurodegeneration* 1;21–28, 1992
Lundh T: Metabolism of estrogenic isoflavones in domestic animals. *Proc Soc Exp Biol Med* 208,33–39, 1995
Lundh TJ-O: Conjugation of the plant estrogens formononetin and daidzein and their metabolite equol by gastrointestinal epithelium from cattle and sheep. *J Agric Food Chem* 38;1012–1016, 1990
Lundh TJ-O, Lundgren BO: Uncoupling and inhibition of the respiratory chain in rat liver mitochondria by some naturally occurring estrogens and their metabolites. *J Agric Food Chem* 39;736–739, 1991
Lundh TJ-O, Pettersson HI, Martinsson KA: Comparative levels of free and conjugated plant estrogens in blood plasma of sheep and cattle fed estrogenic silage. *J Agric Food Chem* 38;1530–1534, 1990
Maag DD, Glenn MW: Toxicity of selenium: farm animals. In *Selenium in Biomedicine, First International Symposium, 1966, Muth, Ohio,* Oldfield JE, Weswig PH eds, AVI Publishing, Westport, Conn, pp127–140, 1967
MacDonald JF, Morris ME: Lathyrus excitation: mechanism of neuronal excitation by L-2-oxalylamino-3-amino and L-3-oxalylamino-2-amino-proprionic acid. *Exp Brain Res* 57;158–166, 1984
MacDonald MA: Timber milk vetch poisoning on British Columbia ranges. *J Range Manage* 5;16–21, 1952
Mackey DR: Acute pulmonary emphysema in cattle. *Southwestern Vet* 5;342–344, 1952
Mahan DC, Moxon AL: Effect of inorganic selenium supplementation on selenosis in post-weaning swine. *J Anim Sci* 58;1216–1221, 1984
Maity SB, Rekib A, Shukla NP: *Sesbania sesban* as replacement for conventional concentrate mixture in silage-based ration for crossbred lactating cows. *Indian J Anim Sci* 67;158–159, 1997
Majak W: Metabolism and absorption of toxic glycosides by ruminants. *J Range Manage* 45;67–71, 1991
Majak W, Benn MH: New glycosides of 3-nitropropanol from *Astragalus miser* var. *serotinus.* In *Poisonous Plants: Proceedings of the Third International Symposium,* James LF, Keeler RF, Bailey EM Jr, Cheeke PR, Hegarty MP eds, Iowa State Univ Press, Ames, pp523–527, 1992
Majak W, Bose RJ: Nitropropanylglucopyranoses in *Coronilla varia. Phytochemistry* 15;415–417, 1976
Majak W, Cheng KJ: Recent studies on ruminal metabolism of 3-nitropropanol in cattle. *Toxicon* (Suppl 3), pp265–268, 1983
Majak W, Clark LJ: Metabolism of aliphatic nitro compounds in bovine rumen fluid. *Can J Anim Sci* 60;319–325, 1980
Majak W, McLean A: Variability of miserotoxin concentration in timber milkvetch. *J Range Manage* 28;362–364, 1975
Majak W, Pass MA: Aliphatic nitrocompounds. In *Toxicants of Plant Origin,* vol 3, Cheeke PR ed, CRC Press, Boca Raton, Fla, pp143–159, 1989
Majak W, McLean A, Pringle TP, Van Ryswyk AL: Fluctuations in miserotoxin concentration of timber milkvetch on rangelands in British Columbia. *J Range Manage* 27;363–366, 1974
Majak W, Udenberg T, McDiarmid RE, Douwes H: Toxicity and metabolic effects of intravenously administered 3-nitropropanol in cattle. *Can J Anim Sci* 61;639–647, 1981
Majak W, Pass MA, Madryga FJ: Toxicity of miserotoxin and its aglycone (3-nitropropanol) to rats. *Toxicol Lett* 19;171–178, 1983
Majak W, Pass MA, Muir AD, Rode LM: Absorption of 3-nitropropanol (miserotoxin aglycone) from the compound stomach of cattle. *Toxicol Lett* 23;9–15, 1984
Majak W, Quinton DA, Douwes HE, Hall JW, Muir AD: The effect of clipping on the growth and miserotoxin content of Columbia milkvetch. *J Range Manage* 41;26–29, 1988

Majak W, Keller WJ, Duan Z, Munro D, Smith RA, Davis AM, Ogilvie RT: Alkaloid distribution in two species of *Lupinus* in central British Columbia. *Phytochemistry* 36;883–885, 1994

Majak W, Hall JW, McCaughey WP: Pasture management strategies for reducing the risk of legume bloat in cattle. *J Anim Sci* 73;1493–1498, 1995

Majak W, Stroesser L, Hall JW, Quinton DA, Douwes HE: Seasonal grazing of Columbia milkvetch by cattle on rangelands in British Columbia. *J Range Manage* 49;223–227, 1996

Majak W, Hunter C, Stroesser L: Tolerance in cattle to timber milkvetch *(Astragalus miser* var. *serotinus)* due to changes in rumen microbial populations. In *Toxic Plants and Other Natural Toxicants,* Garland T, Barr AC eds, CAB International, New York, pp239–242, 1998

Marceau-Day M-L: A study on the toxicity of *Sesbania drummondii* in chickens. PhD dissertation, 1988 (*Diss Abstr Int* 49B;3045, 1989)

Maricle B, Tobey J, Majak W, Hall JW: Evaluation of clinicopathological parameters in cattle grazing timber milkvetch. *Can Vet J* 37;153–156, 1996

Marion L, Cockburn WF: The papilionaceous alkaloids. 5. *Baptisia minor,* Lehm. *J Am Chem Soc* 70;3472–3474, 1948

Marion L, Ouellet J: The papilionaceous alkaloids. 2. *Baptisia australis* (L.) R. Br. *J Am Chem Soc* 70;691–692, 1948

Marion L, Turcotte F: The papilionaceous alkaloids. 4. *Baptisia perfoliata* (L.) R. Br. *J Am Chem Soc* 70;3253–3254, 1948

Markovits J, Linassier C, Fosse P, Couprie J, Pierre J, Jacquemin-Sablon A, Saucier J-M, Le Pecq J-B, Larsen AK: Inhibitory effects of the tyrosine kinase inhibitor genistein on mammalian DNA topoisomerase, 2. *Cancer Res* 49;5111–5117, 1989

Markowitz H: Isolation and properties of robin, a phytohemagglutinin from *Robinia pseudoacacia. Fed Proc* 30;692 (abstract 2772), 1971

Marquez RL, Gutierrez-Rave M, Miranda FI: Acute poisoning by lupine seed debittering water. *Vet Hum Toxicol* 33;265–267, 1991

Marsh CD: *The Loco-weed Disease of the Plains.* USDA Bur Anim Ind Bull 112, 1909

Marsh CD: *The Loco-weed Disease.* USDA Farmers Bull 1054, 1919 (rev 1936)

Marsh CD: *Stock-Poisoning Plants of the Range,* rev ed. USDA Bull 1245, 1929

Marsh CD, Clawson AB: *The Locoweed Disease.* USDA Farmers Bull 1054, 1919

Marsh CD, Clawson AB: Astragalus tetrapterus, *a New Poisonous Plant of Utah and Nevada.* USDA Circ 81, 1920a

Marsh CD, Clawson AB: *Daubentonia longifolia* (coffee bean), a poisonous plant. *J Agric Res* 20;507–514, 1920b

Marsh CD, Clawson AB, Marsh H: *Lupines as Poisonous Plants.* USDA Bull 405, 1916

Marten GC, Ehle FR, Ristau EA: Performance and photosensitization of cattle related to forage quality of four legumes. *Crop Sci* 27;138–145, 1987

Marten GC, Jordan RM, Ristau EA: Performance and adverse response of sheep during grazing of four legumes. *Crop Sci* 30;860–866, 1990

Martin BW, Terry MK, Bridges CH, Bailey EM Jr: Toxicity of *Cassia occidentalis* in the horse. *Vet Hum Toxicol* 23;416–417, 1981

Martin JL, Hurlbut JA: Tissue selenium levels and growth responses of mice fed selenomethionine, Se-methylselenocysteine, or sodium selenite. *Phosphorus Sulfur* 1;295–300, 1976

Martin PM, Horwitz KB, Ryan DS, McGuire WL: Phytoestrogen interaction with estrogen receptors in human breast cancer cells. *Endocrinology* 103;1860–1867, 1978

Martin SHC: The proteids of the seeds of jequirity [abstract]. *J Chem Soc* 52;990, 1887

Martinez JA, Esparza ML, Larralde J: Immunological changes in growing mice fed on diets containing casein or peas *(Pisum sativum* var. *Belinda)* as the source of protein. *Br J Nutr* 73;87–97, 1995

Martinez MA, Seifert HS: The virulence of *L. leucocephala* for goats in northeastern Mexico. *Berl Muench Tierarztl Wochenschr* 104;257–262, 1991

Martins E, Martins VMV, Riet-Correa F, Sonconi RA, Paraboni SV: *Cassia occidentalis* poisoning in swine. *Pesq Vet Bras* 6;35–38, 1986

Martius C, Nitz-Litzow D: Uber den wirkungsmechanismus des dicumarols und verwandter verbindungen. *Biochim Biophys Acta* 12;134–140, 1953

Maruyama M, Takeda K: Stizolobic acid on frog spinal cord: possible species dependent activation of excitatory amino acid receptors. *Comp Biochem Physiol* 104C;439–444, 1993

Mason SF, Peacock RD: Complexes of some first-row transition elements with (–)-sparteine. *J Chem Soc Dalton Trans* 1973;226–228, 1973

Mathews FP: *Locoism in Domestic Animals.* Tex Agric Exp Stn Bull 456, 1932

Mathews FP: The toxicity of red-stemmed peavine (= *Astragalus emoryanus*) for cattle, sheep, and goats. *J Am Vet Med Assoc* 97;125–134, 1940

Matsuda F, Terashima S: Total synthesis of natural (+)-sesbanimide A and (–)-sebanimide B. *Tetrahedron* 44;4721–4736, 1988

Mattocks AR: *Chemistry and Toxicology of Pyrrolizidine Alkaloids.* Academic Press, London, 1986

Mattocks AR: Pyrrolizidine alkaloids: what metabolites are responsible for extrahepatic tissue damage in animals? In *Poisonous Plants: Proceedings of the Third International Symposium,* James LF, Keeler RF, Bailey EM Jr, Cheeke PR, Hegarty MP eds, Iowa State Univ Press, Ames, pp192–197, 1992a

Mattocks AR: Recent advances in pyrrolizidine research. In *Poisonous Plants: Proceedings of the Third International Symposium,* James LF, Keeler RF, Bailey EM Jr, Cheeke PR, Hegarty MP eds, Iowa State Univ Press, Ames, pp161–168, 1992b

Mattocks AR, Jukes R: Detection of sulphur-conjugated pyrrolic metabolites in blood and fresh or fixed liver tissue from rats given a variety of toxic pyrrolizidine alkaloids. *Toxicol Lett* 63;47–55, 1992

Mattocks AR, Croswell S, Jukes R, Huxtable RJ: Identity of a biliary metabolite formed from monocrotaline in isolated, perfused rat liver. *Toxicon* 29;409–415, 1991

Maxwell JAL: Temporary infertility in Merino ewes grazing improved pastures in Western Australia. *Aust Vet J* 55;381–384, 1979

McCawley EL: Cardioactive alkaloids. In *The Alkaloids,* vol 5, Manske RHF ed, Academic Press, New York, pp79–107, 1955

McConnell KP: Passage of selenium through the mammary glands of the white rat and the distribution of selenium in milk proteins after subcutaneous injection of sodium selenite. *J Biol Chem* 173;653–657, 1948

McConnell KP, Cooper BJ: Distribution of selenium in serum proteins and red blood cells after subcutaneous injection of sodium selenate containing radioselenium. *J Biol Chem* 183;459–466, 1950

McConnell KP, Wabnitz CH: Studies on the fixation of radioselenium in proteins. *J Biol Chem* 226;765–776, 1957

McCormack JE, Neisler WE: *Cassia obtusifolia* (sicklepod) toxicity in a dairy herd. *Vet Med/Small Anim Clin* 75, 1849–1851, 1980

McCracken DS, Schermeister LJ: Phytochemical and cytological evaluation of several *Astragalus* species of North Dakota. *Lloydia* 33;19–24, 1970

McDiarmid RE, Pass MA, Benn MH, Majak W: Biotransformation of 3-nitropropanol by alcohol dehydrogenase. In *Poisonous Plants: Proceedings of the Third International Symposium,* Iowa State Univ Press, Ames, pp131–136, 1992

McDonald GK: Moldy sweetclover poisoning in a horse. *Can Vet J* 21;250–251, 1980

McGrath JPM, Duncan JR, Munnell JF: *Crotalaria spectabilis* toxicity in swine: characterization of the renal glomerular lesion. *J Comp Pathol* 85;185–194, 1975

McGrath-Hill CA, Vicas IMO: Case series of *Thermopsis* exposures. J *Toxicol Clin Toxicol* 34;599, 1996

McIlwraith CW, James LF: Limb deformities in foals associated with ingestion of locoweed by mares. *J Am Vet Med Assoc* 181;255–258, 1982

McKee R, Enlow CR: *Crotalaria, a New Legume for the South.* USDA Circ 137, 1931

McLaughlin JG, Cullen J: Clinical cases of chronic selenosis in horses. *Ir Vet J* 40;136–138, 1986

McPherson A, Hoover S: Purification of mitogenic proteins from *Hura crepitans* and *Robinia pseudoacacia*. *Biochim Biophys Res Comm* 89;713–720, 1979

Meads EB, Taylor PA, Pallister WA: An unusual outbreak of sweet clover poisoning in cattle. *Can Vet J* 5;65–71, 1964

Meckes-Lozoya M, Lozoya X, Gonzales JL: In vitro pharmacological properties of some extracts of *Mimosa tenuiflora* (tepescohuite). *Arch Invest Med (Mex)* 21;163–169, 1990

Meeker JE, Kilgore WW: Identification and quantitation of the alkaloids of *Lupinus latifolius*. *J Agric Food Chem* 35;431–433, 1987

Mehta T, Parker AJ, Cusick PK, Zarghami NS, Haskell BE: The *Lathyrus sativus* neurotoxin: evidence of selective retention in monkey tissue. *Toxicol Appl Pharmacol* 52;54–61, 1980

Mehta T, Parker AJ, Cusick PK, Zarghami NS, Haskell BE: The *Lathyrus sativus* neurotoxin: resistance of the squirrel monkey to prolonged oral high doses. *Toxicol Appl Pharmacol* 69;480–484, 1983

Mejia MJ, Morales MM, Llopis A, Martinez I: Intoxication of school children by ornamental trees. *Aten Primaria* 8;88,90–91, 1991

Mendel VE, Witt MR, Gitchell BS, Gribble DN, Rogers QR, Segall HJ, Knight HD: Pyrrolizidine alkaloid-induced liver disease in horses: an early diagnosis. *Am J Vet Res* 49;572–578, 1988

Mennin S, Thomas DW: Comparative effects of an osteolathyrogen and a neurolathyrogen on brain and connective tissues. *Proc Soc Exp Biol Med* 134;489–491, 1970

Mercer HD, Neal FC, Himes JA, Edds GT: *Cassia occidentalis* toxicosis in cattle. *J Am Vet Med Assoc* 151;735–741, 1967

Miller JF, Davis DE, Kates AH, McCormack J: *Poisonous Plants of the Southern United States*. Ga Coop Ext Serv, 1980

Miller PR, Kiehl DE, Smith GS: Alpha-mannosidase inhibitory activity of locoweed extracts after incubation in rumen fluid. In *Poisonous Plants: Proceedings of the Third International Symposium*, James LF, Keeler RF, Bailey EM Jr, Cheeke PR, Hegarty MP eds, Iowa State Univ Press, Ames, pp137–142, 1992

Miller S, Nunn PB, Bridges RJ: Induction of astrocyte glutamine synthetase activity by the *Lathyrus* toxin β-*N*-oxalyl-L-α,β-diaminoproprionic acid (β-L-ODAP). *Glia* 7;329–336, 1993

Miller WC, Rice DL, Kreusel RG, Bedrossian CWM: Monocrotaline model of noncardiogenic pulmonary edema in dogs. *J Appl Physiol* 45;962–965, 1978

Miller WT, Schoening HW: Toxicity of selenium fed to swine in the form of sodium selenite. *J Agric Res* 56;831–842, 1938

Miller WT, Williams KT: Effect of feeding repeated small doses of selenium as sodium selenite to equines. *J Agric Res* 61;353–368, 1940

Ming L: Moldy sugarcane poisoning—a case report with a brief review. *Clin Toxicol* 33;363–367, 1995

Minyard JA, Dinkel CA, Olson OE: Effects of arsenilic acid in counteracting selenium poisoning in cattle. *J Anim Sci* 19;260–264, 1960

Miranda CL, Reed RL, Cheeke PR, Buhler DR: Protective effects of butylated hydroxyanisole against the acute toxicity of monocrotaline in mice. *Toxicol Appl Pharmacol* 59;424–430, 1981

Misra DS, Soni BK, Sharma D: Fractionation and characterization of haemagglutinating principle of *Abrus precatorius* L. *Indian J Exp Biol* 6;108–109, 1968

Misra UK, Sharma VP: Peripheral and central conduction studies in neurolathyrism. *J Neurol Neurosurg Psychiatry* 57;572–577, 1994

Mitchell RG: *Laburnum* poisoning in children: report on ten cases. *Lancet* 261;57–58, 1951

Mohamed FHA, Imbabi SE, Adam SEI: Hepatogenous photosensitization in horses due to *Aphis craccivora* on lucerne. *Bull Anim Health Prod Afr* 28;184–187, 1977

Molteni A, Ward WF, Ts'ao CH, Port CD, Solliday NH: Monocrotaline-induced pulmonary endothelial dysfunction in rats. *Proc Soc Exp Biol Med* 176;88–94, 1984

Molyneux RJ: Polyhydroxypyrrolizidine and indolizidine alkaloids of Leguminosae. *Planta Med* 56;520, 1990

Molyneux RJ, James LF: Loco intoxication: indolizidine alkaloids of spotted locoweed (*Astragalus lentiginosus*). *Science* 216;190-191, 1982

Molyneux RJ, James LF, Panter KE: Chemistry of toxic constituents of locoweed (*Astragalus* and *Oxytropis*) species. In *Plant Toxicology, Proceedings of the Australia-USA Poisonous Plants Symposium,* Seawright AA, Hegarty MP, James LF, Keeler RF eds, Animal Research Institute, Yeerongpilly, Brisbane, Australia, pp266–278, 1985

Molyneux RJ, James LF, Panter KE, Ralphs MH: The occurrence and detection of swainsonine in locoweeds. In *Swainsonine and Related Glycosidase Inhibitors,* James LF, Elbein AD, Molyneux RJ, Warren CD eds, Iowa State Univ Press, Ames, pp100–117, 1989

Molyneux RJ, James LF, Panter KE, Ralphs MH: Analysis and distribution of swainsonine and related polyhydroxyindolizidine alkaloids by thin layer chromatography. *Phytochem Anal* 2;125–129, 1991

Molyneux RJ, James LF, Ralphs MH, Pfister JA, Panter KP, Nash RJ: Polyhydroxy alkaloid glycosidase inhibitors from poisonous plants of global distribution: analysis and identification. In *Plant-Associated Toxins: Agricultural, Phytochemical, and Ecological Aspects,* Colegate SM, Dorling PR eds, CAB International, Wallingford, UK, pp107–112, 1994

Monaghan DT, Bridges RJ, Cotman CW: The excitatory amino acid receptors: their classes, pharmacology, and distinct properties in the function of the central nervous system. *Ann Rev Pharmacol Toxicol* 29;365–402, 1989

Monlux AW, Glenn BL, Panciera RJ, Corcoran JB: Bovine hepatogenous photosensitivity associated with the feeding of alfalfa hay. *J Am Vet Med Assoc* 142;989–994, 1963

Morley FHW, Bennett D, Axelsen A: Adaptation by sheep to isoflavones in clover. *Aust J Exp Agric Anim Husb* 9;569–573, 1969

Morrill CC: Clover sickness, or trifoliosis. *North Am Vet* 24;731–732, 1943

Morton JF: *Plants Poisonous to People in Florida and Other Warm Areas.* JF Morton, Miami, 1982

Morton JF: Creeping indigo (*Indigofera spicata* Forssk.) (Fabaceae)—a hazard to herbivores in Florida. *Econ Bot* 43;314–327, 1989

Morton JF: Pito (*Erythrina berteroana*) and chipilin (*Crotalaria longirostrata*) (Fabaceae), two soporific vegetables of Central America. *Econ Bot* 48;130–138, 1994

Moslehuddin ABM, Hang YD, Stoewsand GS: Evaluation of the toxicity of processed *Lathyrus sativus* seeds in chicks. *Nutr Rep Int* 36;851–855, 1987

Moule GR: The fertility of sheep grazing oestogenic pastures. *Aust Vet J* 37;109–114, 1961

Moxon AL: *Alkali Disease or Selenium Poisoning.* S D Agric Exp Stn Bull 311, 1937

Moxon AL, Rhian MA, Anderson HD, Olson OE: Growth of steers on seleniferous range. *J Anim Sci* 3;299–309, 1944

Moyer BG, Pfeffer PE, Moniot JL, Shamma M, Gustine DL: Corollin, coronillin, and coronarian: three new 3-nitropropanol-D-glucopyranoses from *Coronilla varia. Phytochemistry* 16;375–377, 1977

Mraz M, Opletal L, Sovova M, Drasar P, Havel M: Inhibition of Na^+,K^+-ATPase by glycosides from *Coronilla varia* [letter]. *Planta Med* 58;467–468, 1992

Muir AD, Majak W, Pass MA, Yost GS: Conversion of 3-nitropropanol (miserotoxin aglycone) to 3-nitropropionic acid in cattle and sheep. *Toxicol Lett* 20;137–141, 1984

Murakoshi I, Kidoguchi E, Nakamura M, Haginiwa J, Ohmiya S, Higashiyama K, Otomasu H: (–)-Epilamprolobine and its *N*-oxide, lupin alkaloids from *Sophora tomentosa. Phytochemistry* 20;1725–1730, 1981

Murakoshi I, Kubo H, Ikram M, Israr M, Shafi N, Ohmiya S, Otomasu H: (+)-11-Oxocytisine, a lupin alkaloid from leaves of *Sophora secundiflora. Phytochemistry* 25;2000–2002, 1986

Murphy MJ: Secondary photosensitization in horses ingesting *Cymodothea trifolii* infested clover. Fifth International Symposium on Poisonous Plants, San Angelo, Texas, May 1997

Murray RDH, Mendez J, Brown SA: *The Natural Coumarins—Occurrence, Chemistry, and Biochemistry*.. J Wiley & Sons, Chichester, pp504–505, 1982

Myer RO: Protein utilization and toxic effects of raw beans *(Phaseolus vulgaris)* in diets for young pigs and evaluation of heat processed cull beans including extruded cull bean-soybean mixtures in diets for young pigs and chicks. PhD thesis, Texas A&M Univ, Kingsville (*Diss Abstr Int* 42B:1826, 1981)

Naber EC, Blackwood UB: Partial reversal of β-aminoproprio-nitrile toxicity by calcium in the chick. *J Nutr* 78;207–214, 1962

Nakajima T, Wolfgram F, Clark WG: Identification of 1,4-methylhistamine, 1,3-diamino-propane, and 2,4-diaminobutyric acid in bovine brain. *J Neurochem* 14;1113–1118, 1967

Nath D, Sethi N, Singh RK, Jain AK: Commonly used Indian abortifacient plants with special reference to their teratologic effects in rats. *J Ethnopharmacol* 36;147–154, 1992

Nation PN: Alsike clover poisoning: a review. *Can Vet J* 30;410–415, 1989

Neal WM, Rusoff LL, Ahmann CF: The isolation and some properties of an alkaloid from *Crotalaria spectabilis* Roth. *J Am Chem Soc* 57;2560–2561, 1935

Needham AJE, Lawrence JA: The toxicity of *Albizia versicolor. Rhod Agric J* 63;137–140, 1966

Neill DA: Experimental studies on species relationships in *Erythrina* (Leguminosae: Papilionoideae). *Ann Mo Bot Gard* 75;886–969, 1988

Nelson BK, James LF, Sharma RB, Cheney CD: Locoweed embryotoxicity in rats. *Clin Toxicol* 16;149–166, 1980

Newsome FE, Kitts WD: Effects of alfalfa consumption on estrogen levels in ewes. *Can J Anim Sci* 57;531–535, 1977

Newsome FE, Kitts WD: The effects of feeding coumestrol on the reproductive organs of prepubertal lambs. *Can J Anim Sci* 60;53–58, 1980

Newsome IE, Cross F, McCrory BR, Groth AH, Tobiska JW, Balis E, Durrell LW, Smith EC, Stout EN: *Timber Milk Vetch as a Poisonous Plant.* Colo Agric Exp Stn Bull 425, 1936

Nicholson SS, Thornton JT, Rimes JT Sr: Toxic myopathy in dairy cattle caused by *Cassia obtusifolia* in greenchop. *Bovine Pract* 12;120, 1977

Nicolson GL, Concanavalin A: The tool, the techniques, and the problems. In *Concanavalin A as a Tool,* Bittiger H, Schebli HP eds, J Wiley & Sons, London, pp3–15, 1976

Niezen JH, Barry TN, Wilson PR, Lanr G: Red urine from deer grazed on pure red clover swards. *N Z Vet J* 40;164–167, 1992

Nigra L, Huxtable RJ: Hepatic glutathione concentrations and the release of pyrrolic metabolites of the pyrrolizidine alkaloid, monocrotaline, from the isolated perfused liver. *Toxicon* 30, 1195–1202, 1992

Nillson A, Hill JL, Davies HL: An in vitro study of formononetin and biochanin A metabolism in rumen fluid from sheep. *Biochim Biophys Acta* 148;92–98, 1967

Nishida A, Kobayashi T, Ariyuki F: Developmental toxicity of concanavalin A in rats: association with restricted migration of neural crest cells. *Food Chem Toxicol* 34;701–708, 1996

Noah ND, Bender AE, Reaidi GB, Gilbert RJ: Food poisoning from raw red kidney beans. *Br Med J* 281;236, 1980

Nobre D, Dagli MLZ, Haraguchi M: *Crotalaria juncea* intoxication in horses. *Vet Hum Toxicol* 36;445–448, 1994

Nocerini MR, Honeyfield DC, Carlson JR, Breeze RG: Reduction of 3-methylindole production and prevention of acute bovine pulmonary edema and emphysema with lasalocid. *J Anim Sci* 60;232–238, 1985

Nohl H, Jordan W: OH-generation by adriamycin semiquinone and H_2O_2: an explanation for the cardiotoxicity of anthracycline antibiotics. *Biochem Biophys Res Commun* 114;197–205, 1983

Nordfeldt S, Henke LA, Morita K, Matsumoto H, Takahashi M, Younge OR, Willers EH, Cross RF: *Feeding Tests with* Indigofera endecaphylla *Jacq. (Creeping Indigo) and Some Observations on Its Poisonous Effects on Domestic Animals*. Hawaii Agric Exp Stn Tech Bull 15, 1952

Novelli A, Reilly JA, Lysko PG, Henneberry RC: Glutamate becomes neurotoxic via the *N*-methyl-D-aspartate receptor when intracellular energy levels are reduced. *Brain Res* 451;205–212, 1988

Nsimba-Lubaki M, Peumans WJ: Seasonal fluctuations of lectins in barks of elderberry *(Sambucus nigra)* and black locust *(Robinia pseudoacacia)*. *Plant Physiol* 80;7470751, 1986

Nuessle NO, Lauter WM: Isolation of constituents of *Glottidium vesicarium*. *Econ Bot* 12;307–311, 1958

Nuttall T: *A Journal of Travels into the Arkansas Territory during the Year 1819*. Lottinville S ed, Univ Oklahoma Press, Norman, p49, 1980

Odriozola E, Paloma E, Lopez T, Campero C: An outbreak of *Vicia villosa* (hairy vetch) poisoning in grazing Aberdeen Angus bulls in Argentina. *Vet Hum Toxicol* 33;278–280, 1991

Oehme FW, Bailie WE, Hulbert LC: *Astragalus mollissimus* (locoweed) toxicosis of horses in western Kansas. *J Am Vet Med Assoc* 152;271–278, 1968

Ogden PR, Welsh SL, Williams MC, Ralphs MH: *Astragalus* and related genera-ecological considerations. In *The Ecological and Economic Impact of Poisonous Plants on Livestock Production*, James LF, Ralphs MH, Nielsen DB eds, Westview Press, Boulder, Colo, pp153–169, 1988

O'Hara PJ, Pierce KR: A toxic cardiomyopathy caused by *Cassia occidentalis*. 1. Morphologic studies in poisoned rabbits. *Vet Pathol* 11;97–109, 1974a

O'Hara PJ, Pierce KR: A toxic cardiomyopathy caused by *Cassia occidentalis*. 2. Biochemical studies in poisoned rabbits. *Vet Pathol* 11;110–124, 1974b

O'Hara PJ, Pierce KR, Read WK: Degenerative myopathy associated with ingestion of *Cassia occidentalis* L.: clinical and pathologic features of the experimentally induced disease. *Am J Vet Res* 30;2173–2180, 1969

O'Hara PJ, Pierce KR, Read WK: Effects of vitamin E and selenium on *Cassia occidentalis* intoxication in cattle. *Am J Vet Res* 31;2151–2156, 1970

Ohlendorf HM, Hoffman DJ, Saiki MK, Aldrich TW: Embryonic mortality and abnormalities of aquatic birds: apparent impacts of selenium from irrigation drainwater. *Sci Total Environ* 52;49–63, 1986

Ohmiya S, Otomasu H, Murakoshi I, Haginiwa J: *N*-Acetylcytisine from *Sophora tomentosa*. *Phytochemistry* 13;1016, 1974

Ohmiya S, Saito K, Murakoshi I: Lupine alkaloids. In *The Alkaloids: Chemistry and Pharmacology*, vol 47, Cordell GA ed, Academic Press, San Diego, Calif, pp1–114, 1995

Ohta M, Narahashi T: Sparteine interaction with nerve membrane potassium conductance. *J Pharmacol Exp Ther* 187;47–55, 1973

Olney JW: Excitotoxins in foods. *Neurotoxicology* 15;535–544, 1994

Olney JW, Misra CH, Rhee V: Brain and retinal damage from *Lathyrus* excitotoxin, β-*N*-oxalyl-L-α,β-diaminoproprionic acid. *Nature* 264;659–661, 1976

Olsnes S: Abrin and ricin: two toxic lectins inactivating eukaryotic ribosomes. In *Perspectives in Toxinology*, Bernheimer AW ed, J Wiley & Sons, New York, pp121–147, 1977

Olsnes S, Pihl A: Abrin and ricin—two toxic lectins. *Trends Biochem Sci* 3;7–10, 1978

Olsnes S, Pihl A: Toxic lectins and related proteins. In *Molecular Action of Toxins and Viruses*, Cohen P, Van Heyningen S eds, Elsevier Biomed, Amsterdam, pp51–105, 1982

Olson AE, Jornlin DF, Moxon AL: The selenium content of vegetation and the mapping of seleniferous soils. *J Am Soc Agron* 34;607–615, 1942

Olson OE: Selenium in plants as a cause of livestock poisoning. In *Effects of Poisonous Plants On Livestock,* Keeler RF, Van Kampen KR, James LF eds, Academic Press, New York, pp121–133, 1978

Olson OE, Novacek EJ, Whitehead EI, Palmer IS: Investigations on selenium in wheat. *Phytochemistry* 9;1181–1188, 1970

Omer SA, Ibrahim FH, Khalid SA, Adam SE: Toxicological interactions of *Abrus precatorius* and *Cassia senna* in the diet of Lohmann broiler chicks. *Vet Hum Toxicol* 34;310–313, 1992

O'Neal RM, Chen C-H, Reynolds CS, Meghal SK, Koeppe RE: The "neurotoxicity" of L-2,4-diaminobutyric acid. *Biochem J* 106;699–706, 1968

Ormandy GC, Jope RS: Inhibition of phosphoinositide hydrolysis by the novel neurotoxin β-*N*-oxalylamino-L-α,β-diaminoproprionic acid (L-BOAA). *Brain Res* 510;53–57, 1990

Orru A, Cesaris Demel V: Physiological and anatomo-pathological observations on rats fed on the seeds of sabre-bean or jack-bean. *Quad Nutr* 7;273–293, 1940 (*Nutr Abstr Rev* 17;4859, 1948)

Ort JF, Latshaw JD: The toxic level of sodium selenite in the diet of laying chickens. *J Nutr* 108;1114–1120, 1978

Ortega JA, Lazerson J: Anagyrine-induced red cell aplasia, vascular anomaly, and skeletal dysplasia. *J Pediatr* 111;87–89, 1987

O'Toole D, Raisbeck MF: Pathology of experimentally induced chronic selenosis (alkali disease) in yearling cattle. *J Vet Diagn Invest* 7;364–373, 1995

O'Toole D, Raisbeck M, Case JC, Whitson TD: Selenium-induced "blind staggers" and related myths: a commentary on the extent of historical losses attributed to selenosis on western US rangelands. *Vet Pathol* 33;104–116, 1996

Oung-Boran P, Lebtreton P, Netien G: Biochemistry and pharmacology of *Baptisia australis. Planta Med* 17;301–318, 1969

Owen LN: Hair loss and other toxic effects of *Leucaena glauca* ("Jumbey"). *Vet Rec* 70;454–456, 1958

Page RK, Vezey S, Charles OW, Hollifield T: Effects on feed consumption and egg production of coffee bean seed *(Cassia obtusifolia)* fed to white Leghorn hens. *Avian Dis* 21;90–96, 1977

Paissios CS, Demopoulos T: Human lathyrism: a clinical and skeletal study. *Clin Orthop* 23;236–249, 1962

Palotay JL: Crooked calves. *West Vet* 6;16–20, 1959

Pammel LH: *A Manual of Poisonous Plants.* Torch Press, Cedar Rapids, Iowa, 1911

Pammel LH: Poisoning from lima bean. *Am J Vet Med* 12;787–790, 1917a

Pammel LH: Poisonous plants in a pasture. *Am J Vet Med* 12;235–236, 1917b

Pammel LH: Alsike clover poisoning. *Am J Vet Med* 15;437–438, 1920

Pammel LH: Some forms of lima beans poisonous. *Vet Med* 16;45, 1921

Pan LC, Lame MW, Morin D, Wilson DW, Segall HJ: Red blood cells augment transport of reactive metabolites of monocrotaline from liver to lung in isolated and tandem liver and lung preparations. *Toxicol Appl Pharmacol* 110;336–346, 1991

Panciera RJ: Hairy vetch (*Vicia villosa* Roth) poisoning in cattle. In *Effects of Poisonous Plants on Livestock,* Keeler RF, Van Kampen KR, James LF eds, Academic Press, New York, pp555–563, 1978

Panciera RJ, Johnson L, Osburn BI: A disease of cattle grazing hairy vetch pasture. *J Am Vet Med Assoc* 148;804–808, 1966

Panciera RJ, Mosier DA, Ritchey JW: Hairy vetch (*Vicia villosa* Roth) poisoning in cattle: update and experimental induction of disease. *J Vet Diagn Invest* 4;318–325, 1992

Panter KE: Ultrasound imaging: a bioassay technique to monitor fetotoxicity of natural toxicants and teratogens. In *Bioactive Natural Products—Detection, Isolation, and Structural Determination,,* Colegate SM, Molyneux RJ eds, CRC Press, Boca Raton, Fla, pp465–480, 1993

Panter KE, Keeler RF: Quinolizidine and piperidine alkaloid teratogens from poisonous plants and their mechanism of action in animals. *Vet Clin North Am Food Anim* 9;33–40, 1993

Panter KE, Bunch TD, James LF, Sisson DV: Ultrasonographic imaging to monitor fetal and placental developments in ewes fed locoweed *(Astragalus lentiginosus)*. *Am J Vet Res* 48;686–690, 1987

Panter KE, James LF, Nielsen D, Molyneux RJ, Ralphs MH, Olsen JD: The relationship of *Oxytropis sericea* (green and dry) and *Astragalus lentiginosus* with high mountain disease in cattle. *Vet Hum Toxicol* 30;318–323, 1988

Panter KE, James LF, Hartley WJ: Transient testicular degeneration in rams fed locoweed *(Astragalus lentiginosus)*. *Vet Hum Toxicol* 31;42–46, 1989

Panter KE, Bunch TD, Keeler RF, Sisson DV, Callan RJ: Multiple congenital contractures (MCC) and cleft palate induced in goats by ingestion of piperidine alkaloid-containing plants: reduction in fetal movement as the probable cause. *Clin Toxicol* 28;69–83, 1990a

Panter KE, Keeler RF, Bunch TD, Callan RJ: Congenital skeletal malformations and cleft palate induced in goats by ingestion of *Lupinus, Conium,* and *Nicotiana* species. *Toxicon* 28;1377–1385, 1990b

Panter KE, James LF, Mayland HF: Reproductive response of ewes fed alfalfa pellets containing sodium selenate or *Astragalus bisulcatus* as a selenium source. *Vet Hum Toxicol* 37;30–32, 1995

Panter KE, Gardner DR, Gay CC, James LF, Mills R, Gay JM, Baldwin TJ: Observations of *Lupinus sulphureus*–induced "crooked calf disease." *J Range Manage* 50;587–592, 1997

Panter KE, Gardner DR, Molyneux RJ: Teratogenic and fetotoxic effects of the two piperidine alkaloid-containing lupines *(L. formosus* and *L. arbustus)* in cows. *J Nat Toxins* 7;131–140, 1998

Panter KE, Ralphs MH, James LF, Stegelmeier BL, Molyneux RJ: Effects of locoweed *(Oxytropis sericea)* on reproduction in cows with a history of locoweed consumption. *Vet Hum Toxicol* 41;282–286, 1999a

Panter KE, James LF, Stegelmeier BL, Ralphs MH, Pfister JA: Locoweeds: effects on reproduction in livestock. *J Nat Toxins* 8;53–62, 1999b

Parker AJ, Mehta T, Zarghami NS, Cusick PK, Haskell BE: Acute neurotoxicity of the *Lathyrus sativus* neurotoxin, L-3-oxalylamino-2-aminoproprionic acid, in the squirrel monkey. *Toxicol Appl Pharmacol* 47;135–143, 1979

Pass MA, Majak W, Muir AD, Yost GS: Absorption of 3-nitropropanol and 3-nitroproprionic acid from the digestive system of sheep. *Toxicol Lett* 23;1–7, 1984

Pass MA, Muir AD, Majak W, Yost GS: Effect of alcohol and aldehyde dehydrogenase inhibitors on the toxicity of 3-nitropropanol in rats. *Toxicol Appl Pharmacol* 78;310–315, 1985

Pass MA, Carlisle CH, Reuhl KR: 3-Nitropropionic acid toxicity in cultured murine embryonal carcinoma cells. *Nat Toxins* 2;386–394, 1994

Pass MA, Arab H, Pollitt S, Hegarty MP: Effects of the naturally occurring arginine analogues indospicine and canavanine on nitric oxide mediated functions in aortic endothelium and peritoneal macrophages. *Nat Toxins* 4;135–140, 1996

Patneau DK, Mayer ML, Jane DE, Watkins JC: Activation and desensitization of AMPA/kainate receptors by novel derivatives of willardiine. *J Neurosci* 12;595–606, 1992

Patterson PE: Loco, la yerba mala. *Rangelands* 4;147–148, 1982

Payne LD, Foley JP: The presence of dihydroerythroidines in the milk of goats fed *Erythrina poeppigiana* and *Erythrina berteroana* foliage. *Abstr Pap Am Chem Soc* 202;AGFD 129, 1991

Peckham JC, Sangster LT, Jones OH: *Crotalaria spectabilis* poisoning in swine. *J Am Vet Med Assoc* 165;633–638, 1974

Peet RL, Gardner JJ: Poisoning of cattle by hairy or woolly-pod vetch, *Vicia villosa* subspecies *dasycarpa*. *Aust Vet J* 63;381–382, 1986

Pemberton IJ, Smith GR, Forbes TDA, Hensarling CM: Technical note: an improved method for extraction and quantification of toxic phenethylamines from *Acacia berlandieri. J Anim Sci* 71;467–470, 1993

Perez-Escamilla R, Vohra P, Klasing K: Lupins *(Lupinus albus* var. *ultra)* as a replacement for soybean meal in diets for growing chickens and turkey poults. *Nutr Rep Int* 38;583–593, 1988

Periera TN, Seawright AA, Reilly PB, Cribb B, Webb R, Arungundrum SP: DNA alkylation properties of dehydromonocrotaline: sequence selective N7 guanine alkylation and heat and alkali-resistant multiple crosslinks. In *Toxic Plants and Other Natural Toxicants,* Garland T, Barr AC eds, CAB International, New York, pp249–254, 1998

Perkins KD, Payne WW: *Guide to Poisonous and Irritant Plants of Florida.* Fla Agric Ext Serv Circ 441, 1978

Peterson JE, Culvenor CC: Hepatotoxic pyrrolizidine alkaloids. In *Handbook of Natural Toxins,* vol 1, Keeler R, Tu AT eds, Marcel Dekker, New York, pp639–671, 1983

Peterson JE, Jago MV, Payne AL, Stewart PL: The toxicity of phomopsin for sheep. *Aust Vet J* 64;293–298, 1987

Petterson DS, Coackley W: Changes in cell cultures produced by toxic lupin extracts. *Aust J Exp Biol Med Sci* 51;513–520, 1973

Petterson DS, Ellis ZL, Harris DJ, Spadek ZE: Acute toxicity of the major alkaloids of cultivated *Lupinus angustifolius* seed to rats. *J Appl Toxicol* 7;51–53, 1987

Pfeffer M, Ressler C: β-Cyanoalanine, an inhibitor of rat liver cystathionase. *Biochem Pharmacol* 16:2299–2308, 1967

Pfister JA, Price KW: Lack of maternal influence on lamb consumption of locoweed *(Oxytropis sericea). J Anim Sci* 74;340–344, 1996

Pfister JA, Stegelmeier BL, Cheney CD, James LF, Molyneux RJ: Operant analysis of chronic locoweed intoxication in sheep. *J Anim Sci* 74;2622–2632, 1996

Phillips Petroleum Co: *Pasture and Range Plants,* section 1, 1959a

Phillips Petroleum Co: *Pasture and Range Plants,* section 2, 1959b

Piercy PL, Rusoff LL: *Livestock Poisoning by* Crotalaria spectabilis. La Agric Exp Stn Circ 36, 1945

Piercy PL, Rusoff LL: *Crotalaria spectabilis* poisoning in Louisiana livestock. *J Am Vet Med Assoc* 108;69–73, 1946

Pierson RE, Kainer RA: Clinical classification of pneumonias in cattle. *Bovine Pract* 15;73–79, 1980

Pieterse PJS, Andrews FN: Estrogenic activity of alfalfa and other feedstuffs. *J Anim Sci* 15;25–36, 1956

Pilbeam DJ, Polhill RM, Bell EA: Free amino acids and alkaloids of South American, Asian, and Australian *Crotalaria* species. *Bot J Linn Soc* 79;259–266, 1979

Pilbeam DJ, Lyon-Joyce AJ, Bell EA: Occurrence of the pyrrolizidine alkaloid monocrotaline in *Crotalaria* seeds. *J Nat Prod* 46;601–605, 1983

Piola C, Ravaglia M, Zoli MP: Intoxication from seeds of *Wisteria sinensis. Boll Soc Ital Farm Osp* 29;333–337, 1983

Polhill RM: Miscellaneous notes on African species of *Crotalaria* L., 2. *Kew Bull* 22;169–348, 1961

Polhill RM, Sousa M: Robinieae. In *Advances in Legume Systematics,* part I, Pohill RM, Raven PH eds, Royal Botanic Gardens, Kew, pp283–288, 1981

Ponseti IV, Baird WA: Scoliosis and dissecting aneurysm of the aorta in rats fed with *Lathyrus odoratus* seeds. *Am J Pathol* 28;1059–1077, 1952

Potchoiba MJ, Carlson JR, Nocerini MR, Breeze RG: Effect of monensin and supplemental hay on ruminal 3-methylindole formation in adult cows after abrupt change to lush pasture. *Am J Vet Res* 53;129–133, 1992

Pothier J, Cheav SL, Garland N, Dormeau C, Viel C: A comparative study of the effects of sparteine, lupanine, and lupin extract on the central nervous system of the mouse. *J Pharm Pharmacol* 50:949–954, 1998

Powell RG, Smith CR Jr, Weisleder D, Muthard DA, Clardy J: Sesbanine, a novel cytotoxic alkaloid from *Sesbania drummondii. J Am Chem Soc* 101;2784–2785, 1979

Powell RG, Smith CR Jr, Weisleder D: Sesbanimide, a potent antitumor substance from *Sesbania drummondii* seed. *J Am Chem Soc* 105;3739–3741, 1983

Powell RG, Plattner RD, Suffness M: Occurrence of sesbanimide in seeds of toxic *Sesbania* species. *Weed Sci* 38;148–152, 1990

Power FB: The poisonous constituent of the bark of *Robinia pseudoacacia. Am J Pharm* 85;339–344, 1913

Prasad J: Clinico-pathological effects of experimental *Leucaena* toxicity in lambs. *Indian J Anim Sci* 58;1181–1182, 1988

Prasad J, Paliwal OP: Pathological changes in experimentally induced *Leucaena* toxicity in lambs. *Indian Vet J* 66;711–714, 1989

Presser TS: "The Kesterson effect." *Environ Manage* 18;437–454, 1994

Price DA, Hardy WT: Guajillo poisoning of sheep. *J Am Vet Med Assoc* 122;223–225, 1953

Pritchard DG, Markson LM, Brush PJ, Sawtell JAA, Bloxham PA: Haemorrhagic syndrome of cattle associated with the feeding of sweet vernal *(Anthoxanthum odoratum)* hay containing dicoumarol. *Vet Rec* 113;78–84, 1983

Pritchard DH, Huxtable CR, Dorling PR, Colgate SM: Some characteristics of axonal dystrophy in swainsonine toxicosis. In *Swainsonine and Related Glycosidase Inhibitors,* James LF, Elbein AD, Molyneux RJ, Warren CD eds, Iowa State Univ Press, Ames, pp76–82, 1989

Pulsipher GD, Galyean ML, Hallford DM, Smith GS, Kiehl DE: Effects of graded levels of bentonite on serum clinical profiles, metabolic hormones, and serum swainsonine concentrations in lambs fed locoweed *(Oxytropis sericea). J Anim Sci* 72;1561–1569, 1994

Puschner B, Galey FD, Holstege DM, Palazoglu M: Sweet clover poisoning in dairy cattle in California. *J Am Vet Med Assoc* 212;857–859, 1998

Pusztai A: Lectins. In *Toxicants of Plant Origin,* vol 3, Cheeke PR ed, CRC Press, Boca Raton, Fla, pp29–71, 1989

Putnam MR, Qualls CW, Rice LE, Dawson LJ, Edwards WC: Hepatic enzyme changes in bovine hepatogenous photosensitivity caused by water-damaged alfalfa hay. *J Am Vet Med Assoc* 189;77–82, 1986

Putnam MR, Boosinger T, Spano J, Wright J, Wiggins A, D'Andrea G: Evaluation of *Cassia obtusifolia* (sicklepod) seed consumption in Holstein calves. *Vet Hum Toxicol* 30;316–318, 1988

Puyt JD, Faliu L: L'intoxication par *Galega officinalis* L. (French honeysuckle) chez le mouton: une intoxication en recrudescence en France. *Proceedings of the 13th World Congress on Diseases of Cattle,* vol 2, World Buiatrics Assoc, pp670–674, 1984

Puyt JD, Faliu L, Keck G, Gedfrain JC, Pinault L, Tainturier D: Fatal poisoning of sheep by *Galega officinalis* (French honeysuckle). *Vet Hum Toxicol* 23;410–412, 1981

Quinton DA, Majak W: The growth and miserotoxin content of Columbia milkvetch following early spring grazing or clipping. In *Poisonous Plants: Proceedings of the Third International Symposium,* James LF, Keeler RF, Bailey EM Jr, Cheeke PR, Hegarty MP eds, Iowa State Univ Press, Ames, pp148–152, 1992

Radeleff RD: *Veterinary Toxicology,* 2d ed. Lea & Febiger, Philadelphia, p109, 1970

Radostits OM, Searcy GP, Mitchall KG: Moldy sweetclover poisoning in cattle. *Can Vet J* 21;155–158, 1980

Raharjo YC, Cheeke PR, Arscott GH: Research note: effects of dietary butylated hydroxyanisole and cysteine on toxicity of *Lathyrus odoratus* to broiler and Japanese quail chicks. *Poult Sci* 67;153–155, 1988

Rahman A, Salam Mia A: *Abrus precatorius* poisoning in cattle. *Indian Vet J* 49;1045–1049, 1972

Raisbeck MF, O'Toole D: Morphologic studies of selenosis in herbivores. In *Toxic Plants and Other Natural Toxicants,* Garland T, Barr AC eds, CAB International, New York, pp380–387, 1998

Raisbeck MF, Dahl ER, Sanchez DA, Belden EL, O'Toole D: Naturally occurring selenosis in Wyoming. *J Vet Diagn Invest* 5;84–87, 1993

Raisbeck MF, O'Toole D, Belden EL, Waggoner JW: Chronic selenosis in ruminants. In *Toxic Plants and Other Natural Toxicants,* Garland T, Barr AC eds, CAB International, New York, pp389–396, 1998a

Raisbeck MF, Schamber RA, Belden EL: Immunotoxic effects of selenium in mammals. In *Toxic Plants and Other Natural Toxicants,* Garland T, Barr AC eds, CAB International, New York, pp260–266, 1998b

Rajaram N, Janardhanan K: The chemical composition and nutritional potential of the tribal pulse, *Abrus precatorius* L. *Plant Foods Hum Nutr* 42;285–290, 1992a

Rajaram N, Janardhanan K: Nutritional and chemical evaluation of raw seeds of *Canavalia gladiata* (Jacq.) DC. and *C. ensiformis* DC.: the under utilized food and fodder crops in India. *Plant Foods Hum Nutr* 42;329–336, 1992b

Ralphs MH: Cattle grazing white locoweed: influence of grazing pressure and palatability associated with phenological growth stage. *J Range Manage* 40;330–332, 1987

Ralphs MH: Ecology, control, and grazing management of locoweeds in the western U.S. In *Poisonous Plants: Proceedings of the Third International Symposium,* James LF, Keeler RF, Bailey EM Jr, Cheeke PR, Hegarty MP eds, Iowa State Univ Press, Ames, pp528–533, 1992

Ralphs MH, James LF: Locoweed grazing. *J Nat Toxins* 8;47–51, 1999

Ralphs MH, Molyneux RJ: Livestock grazing locoweed and the influence of swainsonine on locoweed palatability and habituation. In *Swainsonine and Related Glycosidase Inhibitors,* James LF, Elbein AD, Molyneux RJ, Warren CD eds, Iowa State Univ Press, Ames, pp39–49, 1989

Ralphs MH, Olsen JD: Conditioned food aversion: a management tool to prevent livestock poisoning, In *Toxic Plants and Other Natural Toxicants,* Garland T, Barr AC eds, CAB International, New York, pp227–231, 1998

Ralphs MH, James LF, Pfister JA: Utilization of white locoweed (*Oxytropis sericea* Nutt.) by range cattle. *J Range Manage* 39;344–347, 1986

Ralphs MH, Mickelsen LV, Turner DL: Cattle grazing white locoweed: diet selection patterns of native and introduced cattle. *J Range Manage* 40;333–335, 1987

Ralphs MH, James LF, Nielsen DB, Baker DC, Molyneux RJ: Cattle grazing Wahweap milkvetch in southeastern Utah. *J Anim Sci* 66;3124–3130, 1988

Ralphs MH, Panter KE, James LF: Feed preferences and habituation of sheep poisoned by locoweed. *J Anim Sci* 68;1354–1362, 1990

Ralphs MH, Panter KE, James LF: Grazing behavior and forage preference of sheep with chronic locoweed toxicosis suggest no addiction. *J Range Manage* 44;208–209, 1991

Ralphs MH, Graham D, Molyneux RJ, James LF: Seasonal grazing of locoweeds by cattle in northeastern New Mexico. *J Range Manage* 46;416–420, 1993

Ralphs MH, Graham D, James LF: Social facilitation influences cattle to graze locoweed. *J Range Manage* 47;123–126, 1994a

Ralphs MH, Graham D, James LF, Panter KE: Locoweed effects on a calf crop. *Rangelands* 16;35–37, 1994b

Ralphs MH, Graham D, Galyean ML, James LF: Creating aversions to locoweed in naive and familiar cattle. *J Range Manage* 50;361–366, 1997

Rankin JEF: Natural oestrogens associated with abortion in cattle in northern Ireland. *Br Vet J* 119;30–34, 1963

Rao SLN, Ramachandran LK, Adiga PR: The isolation and characterization of L-homoargenine from seeds of *Lathyrus sativus. Biochemistry* 2;298–300, 1963

Raschack M: Cardiovascular effects of sparteine and sparteine derivatives: *Arzneim-Forsch* 24;753–758, 1974

Rasmussen MA, James LF: Ruminal metabolism as a mechanism for selenium tolerance in ruminants. *J Anim Sci* 71 (Suppl 1);204, 1993

Rasmussen MA, James LF: Selenium metabolism in the rumen. In *Plant-Associated Toxins: Agricultural, Phytochemical and Ecological Aspects,* Colegate SM, Dorling PR eds, CAB International, Wallingford, UK, pp512–516, 1994

Rasmussen MA, Foster JG, Allison MJ: *Lathyrus sylvestris* (flatpea) toxicity in sheep and ruminal metabolism of flatpea neurolathyrogens. In *Poisonous Plants: Proceedings of the Third International Symposium,* James LF, Keeler RF, Bailey EM Jr, Cheeke PR, Hegarty MP eds, Iowa State Univ Press, Ames, pp377–381, 1992

Rasmussen MA, Allison MJ, Foster JG: Flatpea intoxication in sheep and indications of ruminal adaptation. *Vet Hum Toxicol* 35;123–127, 1993

Reagor JC, Jones LP: *Sesbania vesicaria* poisoning. *Southwest Vet* 33;112, 1980

Reindel JF, Ganey PE, Wagner JG, Slocombe RF, Roth RA: Development of morphologic, hemodynamic, and biochemical changes in lungs of rats given monocrotaline pyrrole. *Toxicol Appl Pharmacol* 106;179–200, 1990

Ren P, Stark PY, Johnson RL, Bell RG: Mechanism of action of anticoagulants: correlation between the inhibition of prothrombin synthesis and the regeneration of vitamin K_1 from vitamin K_1 epoxide. *J Pharmacol Exp Ther* 201;541–546, 1977

Research reports: sheep: peavine silage paralysis. *Feedstuffs* 35;24, 1963

Ressler C: Isolation and identification from common vetch of the neurotoxin β-cyano-L-alanine, a possible factor in neurolathyrism. *J Biol Chem* 237;733–735, 1962

Ressler C, Redstone PA, Erenberg RH: Isolation and identification of a neurotoxic factor from *Lathyrus latifolius. Science* 134;188–190, 1961

Ressler C, Nigam SN, Giza Y-H, Nelson J: Isolation and identification from common vetch of γ-L-glutamyl-β-cyano-L-alanine, a bound form of the neurotoxin β-cyano-L-alanine. *J Am Chem Soc* 85;3311–3312, 1963

Ressler C, Nelson J, Pfeffer M: A pyridoxal-β-cyanoalanine relation in the rat [letter]. *Nature* 203;1286–1287, 1964

Ressler C, Nelson J, Pfeffer M: Metabolism of β-cyanoalanine. *Biochem Pharmacol* 16;2309–2319, 1967

Ressler C, Nigam SM, Giza Y-H: Toxic principle in vetch. Isolation and identification of γ-L-glutamyl-L-β-cyanoalanine from common vetch seeds: distribution in some legumes. *J Am Chem Soc* 91;2758–2765, 1969

Ressler C, Tatake JG, Kaizer E, Putman BH: Neurotoxins in a vetch food: stability to cooking and removal of γ-glutamyl-β-cyanoalanine and β-cyanoalanine and acute toxicity from common vetch (*Vicia sativa* L.) legumes. *J Agric Food Chem* 45;189–194, 1997

Reyes A, Miranda N, Martinez R: Constituents of *Sophora tetraptera* sensu Reiche. *Rev Latinoam Quim* 19:32–35, 1988

Reynard GB, Norton JBS: *Poisonous Plants of Maryland in Relation to Livestock.* Md Agric Exp Stn Tech Bull A-10, 1942

Reynolds AK: Uterine stimulants. In *The Alkaloids,* vol 5, Manske RHF ed, Academic Press, New York, pp163–209, 1955

Reynolds PJ, Jackson CJr, Lindahl IL, Henson PR: Consumption and digestibility of crownvetch (*Coronilla varia* L.) forage by sheep. *Agron J* 59;589–591, 1967

Reynolds PJ, Jackson CJr, Henson PR: Comparison of the effects of crownvetch (*Coronilla varia* L.) and alfalfa hays on the liveweight gain of sheep. *Agron J* 61;187–190, 1969

Rhian M, Moxon AL: Chronic selenium poisoning in dogs and its prevention by arsenic. *J Pharmacol Exp Ther* 78;249–264, 1943

Ribeiro LR, Bautista AR, Silva AR, Sales LA, Salvadori DM, Maia PC: Toxicological and toxicogenetic effects of plants used in popular medicine and in cattle feed. *Mem Inst Oswaldo Cruz* 86 (Suppl 2);89–91, 1991

Ribeiro LR, Silva AR, Bautista AR, Costa SL, Sales LA, Rios AC, Salvadori DM: Clastogenic effect of extracts obtained from *Crotalaria retusa* L. and *Crotalaria mucronata* Desv. on mouse bone marrow cells. *Mutat Res* 300;253–258, 1993

Ribeiro-DaSilva G, Prado JF: Increased insulin circulating levels induced by canatoxin in rats. *Toxicon* 31;1131–1136, 1993

Richards HGH, Stephens A: A fatal case of laburnum seed poisoning. *Med Sci Law* 10;260–265, 1970

Richards JB, Hallford DN, Duff GC: Serum luteinizing hormone, testosterone, and thyroxine and growth responses of ram lambs fed locoweed *(Oxytropis sericea)* and treated with vitamin E/selenium. *Theriogenology* 52;1055–1066, 1999

Riepe M, Hori N, Ludolph AC, Carpenter DO, Spencer PS, Allen CN: Inhibition of energy metabolism by 3-nitropropionic acid activates ATP-sensitive potassium channels. *Brain Res* 586;61–66, 1992

Riepe M, Spencer PS, Lambein F, Ludolph AC, Allen CN: In vitro toxicological investigations of isoxazolinone amino acids of *Lathyrus sativus*. *Nat Toxins* 3;58–64, 1995

Ritchey GE, McKee R: Production of crotalaria for forage. In Crotalaria *for Forage,* Fla Agric Exp Stn Bull 361, pp1–33, 1941

Robinson GH, Burrows GE, Holt EM, Tyrl RJ, Schwab RP: Evaluation of the toxic effects of the legumes of mimosa *(Albizia julibrissin)* and identification of the toxicant. In *Toxic Plants and Other Natural Toxicants,* Garland T, Barr AC eds, CAB International, New York, pp453–458, 1998

Robinson T: *The Organic Constituents of Higher Plants,* 6th ed. Cordus Press, North Amherst, Mass, 1991

Robinson WO: Determination of selenium in wheat and soils. *J Assoc Off Agric Chem* 16;423–424, 1933

Roderick LM: The pathology of sweet clover disease in cattle. *J Am Vet Med Assoc* 70;314–326, 1929

Roderick LM: A problem in the coagulation of the blood "sweet clover disease of cattle." *Am J Physiol* 96;413–425, 1931

Roderick LM, Schalk AF: *Studies on Sweet Clover Disease.* N D Agric Exp Stn Tech Bull 250, 1931

Rodgers RJ, Clarke IJ, Findlay JK, Brown A, Cumming IA, Muller BD, Walker SK: Plasma LH and FSH in ewes that were either fertile or infertile after long-term grazing of oestrogenic pasture. *Aust J Biol Sci* 33;213–220, 1980

Rodriguez FS, Roman AG, Salvago MRM, Miranda FI: Effect of *Astragalus lusitanicus* Lam. in experimental subchronic neurotoxicity of lambs. *Vet Hum Toxicol* 32;551–554, 1990

Rogers CF, Frykolm OC: Observations on the variations in cyanogenetic power of white clover plants. *J Agric Res* 55;533–537, 1937

Rogers RJ, Gibson J, Reichmann KG: The toxicity of *Cassia occidentalis* for cattle. *Aust Vet J* 55;408–412, 1979

Roman AG, Salvago MR, Rodriguez FS, Miranda FI: Physiopathologic changes in lambs fed with *Astragalus lusitanicus* Lam. *Vet Hum Toxicol* 29;387–389, 1987

Romeo JT: Heterocyclic amino and imino acids (mimosine, azetidine-2-carboxylic acid, and pipecolic acid). In *Toxicants of Plant Origin,* vol 3, *Proteins and Amino Acids,* Cheeke PR ed, CRC Press, Boca Raton, Fla, pp126–139, 1989

Romero R, Castillanos M del RT, Mendoza RM, Reyes RA, Garcia AR: Sindrome estrogenico en vacas lecheras por consumo de alfalfas con grandes cantidades de coumestrol. *Vet Mex* 28;25030, 1997

Rondeau ES: Wisteria toxicity. *Clin Toxicol* 31;107–112, 1993

Rosenberg PA, Aizenman E: Hundred-fold increase in neuronal vulnerability to glutamate toxicity in astrocyte-poor cultures of rat cerebral cortex. *Neurosci Lett* 103;162–168, 1989

Rosenfeld I, Beath OA: The elimination and distribution of selenium in the tissues in experimental selenium poisoning. *J Nutr* 30;443–449, 1945

Rosenfeld I, Beath OA: *Selenium—Geobotany, Biochemistry, Toxicity, and Nutrition..* Academic Press, New York, 1964

Rosenthal GA: *Plant Nonprotein Amino and Imino Acids: Biological, Biochemical, and Toxicologic Properties.* Academic Press, New York, 1982

Rosenthal GA, Dahlman DL: Incorporation of L-canavanine into proteins and the expression of its antimetabolic effects. *J Agric Food Chem* 39;987–990, 1991

Roseveare GM: *The Grasslands of Latin America.* Imp Bur Pastures Field Crops Bull 36, Aberysturth, England, p291, 1948

Ross SM, Seelig M, Spencer PS: Specific antagonism of excitotoxic action of "uncommon" amino acids assayed in organotypic mouse cortical cultures. *Brain Res* 425;120–127, 1987

Rossiter RC: Factors affecting the oestrogenic content of subterranean clover pastures. *Aust Vet J* 46;141–148, 1970

Rossiter RC, Beck AB: Physiological and ecological studies on the oestrogenic isoflavones in subterranean clover (*T. subterraneum* L.). *Aust J Agric Res* 17;447–456, 1966

Roth RA, Ganey PE: Platelets and the puzzles of pulmonary pyrrolizidine poisoning. *Toxicol Appl Pharmacol* 93;463–471, 1988

Rotter RG, Marquardt RR, Low RK-C, Briggs CJ: Influence of autoclaving on the effects of *Lathyrus sativus* fed to chicks. *Can J Anim Sci* 70;739–741, 1990

Rowe LD, Corrier DE, Reagor JC, Jones LP: Experimentally induced *Cassia roemeriana* poisoning in cattle and goats. *Am J Vet Res* 48;992–997, 1987

Rowe LD, Ivie GW, DeLoach JR, Foster JG: The toxic effects of mature flatpea (*Lathyrus sylvestris* L. cv. Lathco) on sheep. *Vet Hum Toxicol* 35;127–133, 1993

Roy DN: Toxic amino acids and proteins from *Lathyrus* plants and other leguminous species: a literature review. *Nutr Abstr Rev A* 51;691–707, 1981

Roy DN, Spencer PS: Lathyrogens. In *Toxicants of Plant Origin*, vol 3, *Proteins and Amino Acids*, Cheeke PR ed, CRC Press, Boca Raton, Fla, pp169–201, 1989

Rubio LA, Brenes A, Castano M: Histological alterations to the pancreas and the intestinal tract produced by raw faba bean (*Vicia faba* L. minor) diets in growing chicks. *Br Poult Sci* 30;101–114, 1989

Ruby ES, Beasley J, Stephenson EL: Prussic acid poisoning in common vetch *(Vicia sativa)* seed. *Proc Arkansas Acad Sci* 7;18–20, 1955

Saba N, Drane HM, Hebert CN, Newton JE, Betts JE: Effect of disease on the oestrogenic activity and coumestrol content of white clover and lucerne. *J Agric Sci* 78;471–475, 1972

Saba N, Drane HM, Hebert CN, Holdsworth RJ: Seasonal variation in oestrogenic activity, coumestrol and formononetin content of white clover. *J Agric Sci* 83;505–510, 1974

Saito K, Yamazaki M, Yamakawa K, Fujisawa S, Takamatsu S, Kawaguchi A, Murakoshi I: Lupin alkaloids in tissue culture of *Sophora flavescens* var. *angustifolia:* greening induced production of matrine. *Chem Pharm Bull* 37;3001–3004, 1989

Saloniemi H, Wahala K, Nykanen-Kurki P, Kallela K, Saaskamoinen I: Phytoestrogen content and estrogenic effect of legume fodder. *Proc Soc Exp Biol Med* 208;13–17, 1995

Salyi G, Sztojkov V, Hilbertne Miklovics M: Crownvetch (*Coronilla varia* L.) poisoning of the coypu. *Magyar Allatorv Lapja* 43;313–316, 1988

Sanders DA, Shealy AL, Emmel MW: The pathology of *Crotalaria spectabilis* Roth poisoning in cattle. *J Am Vet Med Assoc* 89;150–155, 1936

Sanderson MA, Meyer DW, Casper HH: Dicoumarol concentrations in sweetclover hay treated with preservatives and in spoiled hay of high-and low-coumarin cultivars of sweetclover. *Anim Feed Sci Tech* 14;221–230, 1986

Sant FI, Wilson D: Use of a cellulolytic enzyme digestion technique to distinguish bloat-causing from non-bloat causing legumes and to select for speed of mesophyll cell wall disintegration in red clover (*Trifolium pratense* L.). *J Agric Sci* 98;99–102, 1982

Saupe SG: Cyanogenic compounds and angiosperm phylogeny. In *Phytochemistry and Angiosperm Phylogeny,* Young DA, Seigler DS eds, Praeger, New York, pp80–116, 1981

Sawaya C, Horton R, Meldrum B: Transmitter synthesis and convulsant drugs: effects of pyridoxal phosphate antagonists and allylglycine. *Biochem Pharmacol* 27;475–481, 1978

Scheel LD: The toxicology of sweet clover and coumarin anticoagulants. In *Mycotoxic Fungi, Mycotoxins, Mycotoxicoses: An Encyclopedic Handbook,* Wyllie TD, Morehouse LG eds, Marcel Dekker, New York, pp121–142, 1978

Schmeller T, Sauerwein M, Sporer F, Wink M, Muller WE: Binding of quinolizidine alkaloids to nicotinic and muscarinic acetylcholine receptors. *J Nat Prod* 57;1316–1319, 1994

Schmittle SC, Richey DJ, Tumlin JT: Toxicity of *Crotalaria spectabilis* seed in poultry. *Poult Sci* 38;1244–1245, 1959

Schmitz DG, Denton JH: Senna bean toxicity in cattle. *Southwest Vet* 30;165–170, 1977
Schmutz EM, Freeman BN, Reed RE: *Livestock Poisoning Plants of Arizona.* Univ Arizona Press, Tucson, 1968
Schofield FW: Damaged sweet clover: the cause of a new disease in cattle simulating hemorrhagic septicemia and blackleg. *J Am Vet Med Assoc* 64;553–575, 1924
Schofield FW: *Liver Disease of Horses (Big Liver) Caused by the Feeding of Alsike Clover.* Ont Vet Coll Circ 52, Ontario Department of Agriculture, 1933
Schofield FW: Acute pulmonary emphysema of cattle. *J Am Vet Med Assoc* 112;254–259, 1948
Schulert AR, Lewis HB: Experimental lathyrism. *Proc Soc Exp Biol Med* 81;86–89, 1952
Schultes RE: Hallucinogens of plant origin. *Science* 163;245–254, 1969
Schultze AE, Roth RA: Fibrinolytic activity in blood and lungs of rats treated with monocrotaline pyrrole. *Toxicol Appl Pharmacol* 121;129–137, 1993a
Schultze AE, Roth RA: Procoagulant and fibrinolytic properties of bovine endothelial cells treated with monocrotaline pyrrole. *Toxicol Appl Pharmacol* 122;7–15, 1993b
Schultze AE, Wagner JG, White SM, Roth RA: Early indications of monocrotaline pyrrole-induced lung injury in rats. *Toxicol Appl Pharmacol* 109;41–50, 1991
Scott BR, Pathak MA, Mohn GR: Molecular and genetic basis of furocoumarin reactions. *Mutat Res* 39;29–74, 1976
Scott CC, Chen KK: The pharmacological action of *N*-methylcytisine. *J Pharmacol Exp Ther* 7;334–339, 1943
Scowcroft PG, Sakai HF: Impact of feral herbivores on mamane forests of Mauna Kea, Hawaii: bark stripping and diameter class structure. *J Range Manage* 36;495–498, 1983
Scruggs DW, Blue GK: Toxic hepatopathy and photosensitization in cattle moldy alfalfa hay. *J Am Vet Med Assoc* 204;264–266, 1994
Seawright AA: *Animal Health in Australia,* vol 2, *Chemical and Plant Poisons.* Department of Primary Industries and Energy, Australian Government Public Service, Canberra, pp62–63, 1989
Secor JB, Conn EE, Dunn JE, Seigler DS: Detection and identification of cyanogenic glucosides in six species of *Acacia. Phytochemistry* 15;1703–1706, 1976
Seigler DS: The naturally occurring cyanogenic glycosides. In *Progress in Phytochemistry,* Reinhold L, Harbone JB, Swain T eds, Pergamon Press, Oxford, pp83–120, 1977
Seigler DS, Maslin BR, Conn EE: Cyanogenesis in the Leguminosae. *Adv Legume Biol Monogr Syst Bot Mo Bot Gard* 29;645–672, 1989
Selye H: Lathyrism. *Rev Can Biol* 16;1–82, 1957
Selye H, Bois P: Effect of corticoids upon experimental lathyrism. *Endocrinology* 60;507–513, 1957
Senn HA: The North American species of *Crotalaria. Rhodora* 41;317–370, 1939
Sethi N, Nath D, Singh RK: Teratological aspects of *Abrus precatorius* seeds in rats. *Fitoterapia* 61;61–63, 1990
Sharma RP, James LF, Molyneux RJ: Effect of repeated locoweed feeding on peripheral lymphocytic function and plasma proteins in sheep. *Am J Vet Res* 45;2090–2093, 1984
Shaw JN, Muth OH: Some types of forage poisoning in Oregon cattle and sheep. *J Am Vet Med Assoc* 114;315–317, 1949
Shealy AL, Thomas EF: *Daubentonia Seed Poisoning of Poultry.* Fla Agric Exp Stn Bull 196, 336–342, 1928
Shearer MJ, McBurney A, Breckenridge A, Barkhan P: Dose-response relationship between warfarin and the metabolism of phylloquinone (vitamin K_1) in man. *Int J Vitam Nutr Res* 46;215–219, 1976
Shemesh M, Lindner HR, Ayalon N: Affinity of rabbit uterine oestradiol receptor for phyto-oestrogens and its use in a competitive protein-binding radioassay for plasma coumestrol. *J Reprod Fertil* 29;1–9, 1972
Shenk JS, Wangsness PJ, Leach RM, Gustine DL, Gobble JL, Barnes RF: Relationship between β-nitropropionic acid content of crownvetch and toxicity in nonruminant animals. *J Anim Sci* 42;616–621, 1976

Shone DK: Toxicity of jack bean. *Rhod Agric J* 58;18–20, 1961

Shone DK, Drummond RB: *Poisonous Plants of Rhodesia.* Government Printing Office, Washington, DC, pp30–32, 1961

Shqueir AA, Brown DL, Klasing KC. Canavanine content and toxicity of *Sesbania* leaf meal for growing chicks. *Anim Feed Sci Technol* 25;137–147, 1989a

Shqueir AA, Brown DL, Taylor SJ, Rivkin I, Klasing KC: Effects of solvent extractions, heat treatments, and added cholesterol on *Sesbania sesban* toxicity in growing chicks. *Anim Feed Sci Technol* 27;127–135, 1989b

Shropshire CM, Stauber E, Arai M: Evaluation of selected plants for acute toxicosis in budgerigars. *J Am Vet Med Assoc* 200;936–939, 1992

Shubat PJ, Huxtable RJ: The threshold for toxicity of the pyrrolizidine alkaloid, monocrotaline. In *Poisonous Plants: Proceedings of the Third International Symposium,* James LF, Keeler RF, Bailey EM Jr, Cheeke PR, Hegarty MP eds, Iowa State Univ Press, Ames, pp186–191, 1992

Shubat PJ, Banner W Jr, Huxtable RJ: Pulmonary vascular responses induced by the pyrrolizidine alkaloid, monocrotaline, in rats. *Toxicon* 25;995–1002, 1987

Shubat PJ, Bowers RJ, Huxtable RJ: Na^+/K^+-adenosine triphosphate activity of pulmonary arteries after intoxication with the pyrrolizidine alkaloid, monocrotaline. *J Pharmacol Exp Ther* 252;70–76, 1990

Shull LR, Buckmaster GW, Cheeke PR: Factors influencing pyrrolizidine *(Senecio)* alkaloid metabolism: species, liver sulfhydryls, and rumen fermentation. *J Anim Sci* 43:1247–1253, 1976

Shumaker RC, Robertson KA, Hsu IC, Allen JR: Neoplastic transformation in tissues of rats exposed to monocrotaline or dehydroretronecine. *J Natl Cancer Inst* 56;787–789, 1976

Shupe JL, Binns W, James LF, Keeler RF: Lupine, a cause of crooked calf disease. *J Am Vet Med Assoc* 151;198–203, 1967a

Shupe JL, James LF, Binns W: Observations on crooked calf disease. *J Am Vet Med Assoc* 151;191–197, 1967b

Shupe JL, Binns W, James LF, Keeler RF: A congenital deformity in calves induced by the maternal consumption of lupin. *Aust J Agric Res* 19;335–340, 1968

Shutt DA, Braden AW: The significance of equol in relation to the oestrogenic responses in sheep ingesting clover with a high formononetin content. *Aust J Agric Res* 19;545–553, 1968

Shutt DA, Cox RI: Steroid and phyto-oestrogen binding to sheep uterine receptors in vitro. *J Endocrinol* 52;299–310, 1972

Shutt DA, Axelsen A, Lindner HR: Free and conjugated isoflavones in the plasma of sheep following ingestion of oestrogenic clover. *Aust J Agric Res* 18;647–655, 1967

Siddique KHM, Loss SP, Herwig SP, Wilson JM: Growth, yield, and neurotoxin (ODAP) concentration of three *Lathyrus* species in Mediterranean-type environments of western Australia. *Aust J Exp Agric* 36;209–218, 1996

Sievers AF, Russell GA, Lowman MS, Fowler ED, Erlanson CO, Little VA: *Studies on the Possibilities of Devil's Shoestring* (Tephrosia virginiana) *and Other Native Species of* Tephrosia *as Commercial Sources of Insecticides.* USDA Tech Bull 595, 1938

Simpson CF, Damron BL, Harms RH: Toxic myopathy of chicks fed *Cassia occidentalis* seeds. *Avian Dis* 15;284–290, 1971

Simpson KS, Banerjee PC: Cases of poisoning in the horse with ratti seeds *(Abrus precatorius),* by oral administration. *Indian J Vet Sci Anim Husb* 2;59–65, 1932

Sloan JW, Martin WR, Bostwick M, Hook R, Wala E: The comparative binding characteristics of nicotinic ligands and their pharmacology. *Pharmacol Biochem Behav* 30;255–267, 1988

Smith EV: The poisonous plant problem in the southeastern states. *North Am Vet* 24;345–353, 1943

Smith FH, Osborne JC: Toxic effects of crotalaria seed. *Vet Med* 57;234–237, 1962

Smith GS, Allred KW, Kiehl DE: Swainsonine content of New Mexican locoweeds. *Proc West Sect Am Soc Anim Sci* 43;405–407, 1992

Smith RA: Potential edible lupine poisonings in humans. *Vet Hum Toxicol* 29;444–445, 1987

Smith TA: Phenethylamine and related compounds in plants. *Phytochemistry* 16;9–18, 1977

Snodgrass SR, Iversen LL: Effects of amino-oxyacetic acid on [3H] GABA uptake by rat brain slices. *J Neurochem* 20;431–439, 1973

Snyders JH, Pieterse MJ: Isolation and identification of the active principle(s) in poisonous plants: *Sesbania punicea. Agric Res (Pretoria, S Afr)* 1968;83, 1968

Soldan AW, van Inzen C, Edelsten RM: *Albizia versicolor* poisoning of sheep and goats in Malawi. *J S Afr Vet Assoc* 67;217–221, 1996

Song Y, Dong Q, Chen H, Jin Y: The metabolism of 3H-oxymatrine in mice and rats. *Zhongguo Yixue Kexueyuan Xuebao* 8;261–265, 1986 (*Chem Abstr* 107;190310z, 1987)

Sotelo A, Soto M, Lucas B, Giral F: Comparative studies of the alkaloidal composition of two Mexican *Erythrina* species and nutritive value of the detoxified seeds. *J Agric Food Chem* 41;2340–2343, 1993

Soto-Hernandez M, Jackson AH: Studies of alkaloids in foliage of *Erythrina berteroana* and *Erythrina poeppigiana*: detection of β-erythroidine in goat's milk. *Phytochem Anal* 4;97–99, 1993

Southon IW: *Phytochemical Dictionary of the Leguminosae.* Vol 1, *Plants and Their Constituents.* Chapman & Hall, London, 1994

Southon IW, Buckingham J: *Dictionary of Alkaloids.* Chapman & Hall, London, 1989

Sovova M: Kardenolidy rodu *Coronilla.* 1. *Cesk Farm* 28,248–250, 1979

Sovova M, Opletal L, Zackova P, Stindlova L: *Coronilla varia* L. IV. Substances contained in the root. *Cesk Farm* 33;376–380, 1984

Spallholz JE: On the nature of selenium toxicity and carcinostatic activity. *Free Radic Biol Med* 17;45–64, 1994

Specht TE, Clay BR, Ward BC: Bagpod *(Glottidium vesicarium)* toxicity in southeastern Oklahoma cattle. *Okla Vet* 25;13–15, 1973

Spencer PS, Ludolph A, Dwivedi MP, Roy DN, Hugon J, Schaumburg HH: Lathyrism: evidence for role of the neuroexcitatory amino acid BOAA. *Lancet* 2;1066–1067, 1986

Sperry OE, Vardiman PH, Gray RG: *Peavine—A Poisonous Range Plant in Texas..* Tex Agric Exp Stn Prog Rep 1474, 1952

Sperry OE, Dollahite JW, Hoffman GO, Camp BJ: *Texas Poisonous Plants.* Tex Agric Exp Stn Ser B-1028, 1977

Spoerke DG, Murphy MM, Wruk KM, Rumack BH: Five cases of *Thermopsis* poisoning. *Clin Toxicol* 26;397–406, 1988

Staley EE: An approach to treatment of locoism in horses. *Vet Med/Small Anim Clin* 73;1205–1206, 1978

Stalker M: The "loco" plant and its effect on animals. In USDA Bur Anim Ind Annu Rep 1886, pp271–276, 1887

Stalker M: Crotalism—a new disease among horsess. In Rep Iowa Agric Coll Dept Bot, pp114–115, 1894

Stec J, Busbee DL, Holloway JW, Rouquette FM: Inhibition of DNA synthesis and decreased reproductive capacity in cattle maintained on semi-arid rangeland with *Acacia* species as a major component of available browse. *Vet Hum Toxicol* 34;43–48, 1992

Stegelmeier BL, Ralphs MH, Gardner DR, Molyneux RJ, James LF: Serum α-mannosidase activity and the clinicopathologic alterations of locoweed *(Astragalus mollissimus)* intoxication in range cattle. *J Vet Diagn Invest* 6;473–479, 1994

Stegelmeier BL, James LF, Panter KE, Molyneux RJ: Serum swainsonine concentration and α-mannosidase activity in cattle and sheep ingesting *Oxytropis sericea* and *Astragalus lentiginosus* (locoweeds). *Am J Vet Res* 56;149–154, 1995a

Stegelmeier BL, James LF, Panter KE, Molyneux RJ: Tissue and serum swainsonine concentrations in sheep ingesting *Astragalus lentiginosus* (locoweed). *Vet Hum Toxicol* 37;336–339, 1995b

Stegelmeier BL, James LF, Panter KE, Gardner DR, Pfister JA, Ralphs MH, Molyneux RJ: Dose response of sheep poisoned with locoweed *(Oxytropis sericea)*. *J Vet Diagn Invest* 11;448–456, 1999a

Stegelmeier BL, James LF, Panter KE, Ralphs MH, Gardner DR, Molyneux RJ, Pfister JA: The pathogenesis and toxicokinetics of locoweed (*Astragalus* and *Oxytropis* spp.) poisoning in livestock. *J Nat Toxins* 8;35–45, 1999b

Stermitz FR, Yost GS: Analysis and characterization of nitro compounds from *Astragalus* species. In *Effects of Poisonous Plants on Livestock*, Keeler RF, Van Kampen KR, James LF eds, Academic Press, New York, pp371–378, 1978

Stermitz FR, Norris FA, Williams MC: Miserotoxin, a new naturally occurring nitro compound. *J Am Chem Soc* 91;4599–4600, 1969

Steyn DG: *Lathyrus sativus* L. (chickling vetch: khesari: Indian pea) as a stock food. *Onderstepoort J Vet Sci Anim Husb* 1;163–171, 1933

Steyn DG: *The Toxicology of Plants of South Africa*. Central News Agency, South Africa, pp85–92, 1934

Steyn PS, Vleggaar R, Anderson LAP: Structure elucidation of two neurotoxins from *Albizia tanganyicensis*. *S Afr J Chem* 40;191–192, 1987

Stirton CH: Psoraleeae. In *Advances in Legume Systematics,* part 1, Polhill RM, Raven PH eds, Royal Botanic Gardens, Kew, pp337–343, 1981

Stobiecki M, Blaszczyk B, Kowalczyk-Bronisz SH, Gulewicz K: The toxicity of seed extracts and their fractions from *Lupinus angustifolius* L. and *Lupinus albus* L. *J Appl Toxicol* 13;347–352, 1993

Stockman R: Lathyrism. *J Pharmacol Exp Toxicol* 37;43–53, 1929

Stowe HD: Effects of copper pretreatment upon the toxicity of selenium in ponies. *Am J Vet Res* 41;1925–1928, 1980

Stowe HD, Eavey AJ, Granger L, Halstead S, Yamini B: Selenium toxicosis in feeder pigs. *J Am Vet Med Assoc* 201;292–295, 1992

Strosznajder J, Chalimoniuk M, Samochocki M, Gadamski R: Nitric oxide: a potent mediator of glutaminergic neurotoxicity in brain ischemia. *Ann N Y Acad Sci* 723,429–432, 1994

Stuenzi H, Perrin D, Teitei T, Harris R: Stability complexes for metal complexation by isomers of mimosine and related compounds. *Aust J Chem* 32;21–30, 1979

Sugg RS, Simms BT, Baker KG: Studies of toxicity of wild winter peas *(Lathyrus hirsutus)* for cattle. *Vet Med* 39;308–311, 1944

Suliman HB, Shommein AM: Toxic effect of the roasted and unroasted beans of *Cassia occidentalis* in goats. *Vet Hum Toxicol* 28;6–11, 1986

Suliman HB, Wasfi IA, Adam SEI: The toxicity of *Cassia occidentalis* to goats. *Vet Hum Toxicol* 24;326–330, 1982a

Suliman HB, Wasfi IA, Adam SEI: The toxic effects of *Tephrosia apollinea* on goats. *J Comp Pathol* 92;309–315, 1982b

Suliman HB, Wasfi IA, Tartour G, Adam SEI: The effects of *Indigofera hochstetteri* on goats. *Rev Elev Med Vet Pays Trop* 36;393–402, 1983

Sullivan G, Chavez PI: Mexican good-luck charm potentially dangerous. *Vet Hum Toxicol* 23;259–260, 1981

Sullivan JM, Pando KM, Everson RJ, Robinson FR: The effect of formalin fixation on the concentration of selenium in porcine liver. *J Vet Diagn Invest* 5;131–133, 1993

Swanson-Biearman B, Dean BS, Krenzelok EP: Failure of whole bowel irrigation to decontaminate the GI tract following massive jequirity bean ingestion [abstract]. *Vet Hum Toxicol* 34;352, 1992

Szentesi A, Wink M: Fate of quinolizidine alkaloids through three trophic levels: *Laburnum anagyroides* (Leguminosae) and associated organisms. *J Chem Ecol* 17;1557–1573, 1991

Takido M: Constituents of seeds of *Cassia obtusifolia*. 1. The structure of obtusifolin. *Chem Pharm Bull* 6;397–400, 1958

Takido M: Constituents of seeds of *Cassia obtusifolia*. 3. The structure of obtusin, chryso-obtusin, and aurantio-obtusin. *Chem Pharm Bull* 8;246–251, 1960

Tanaka H, Morooka N, Haraikawa K, Ueno Y: Metabolic activation of emodin in the reconstituted cytochrome P-450 system of the hepatic microsomes of rats. *Mutat Res* 176;165–170, 1987

Tandon BM, Tandon RK, Tandon HD, Narndranathan M, Joshi YK: An epidemic of veno-occlusive disease of liver in central India. *Lancet* 2;271–272, 1976

Tang BY, Adams NR: Oestrogen receptors and metabolic activity in the genital tract after ovariectomy of ewes with permanent infertility caused by exposure to phyto-oestrogens. *J Endocrinol* 89;365–370, 1981

Tang BY, Adams NR: Proportions of nucleic acids in uteri of ewes with clover disease and the effect of oestrogen after ovariectomy. *Aust J Biol Sci* 35;527–531, 1982

Tasaki B, Tanaka U: On the toxic constituents in the bark of *Robinia pseudoacacia* L. *J Coll Agric, Imp Univ Tokyo* 3;337–356, 1918

Taylor DW, Wilson DW, Lame MW, Dunston SD, Jones AD, Segall HJ: Comparative cytotoxicity of monocrotaline and its metabolites in cultured pulmonary artery endothelial cells. *Toxicol Appl Pharmacol* 143;196–204, 1997

Tazaki K, Yoshida K: The bark lectin of *Robinia pseudoacacia:* purification and partial characterization. *Plant Cell Physiol* 33;125–129, 1992

Tel'tsov AI: Lesions of the recurrent and vagus nerves at the larynx in *Lathyrus sativus* poisoning in horses. *Sov Vet* 11/12;83–86, 1940

Terblanche M, de Klerk WA, Smit JD, Adelaar TF: A toxicological study of the plant *Sesbania punicea* Benth. *J S Afr Vet Med Assoc* 37;191–197, 1966

Thain RI: Bovine infertility possibly caused by subterranean clover: a preliminary report. *Aust Vet J* 41;277–281, 1965

Thain RI: Residual herd infertility in cattle. *Aust Vet J* 44;218–222, 1968

Thayer R, Heller VG: *Vicia villosa* (vetch): chemical composition and use in poultry rations. *Proc Okla Acad Sci* 26;40–43, 1946

Thomas EF: The toxicity of certain species of *Crotalaria* seed for the chicken, quail, turkey, and dove. *J Am Vet Med Assoc* 85;617–622, 1934

Thomas HC, Lame MW, Wilson DW, Segall HJ: Cell cycle alterations associated with covalent binding of monocrotaline pyrrole to pulmonary artery endothelial cell DNA. *Toxicol Appl Pharmacol* 141;319–329, 1996

Thomson RH: *Naturally Occurring Quinones,* 2d ed. Academic Press, London, 1971

Tingey DC: Goatsrue, a potential forage crop, turned out to be a weed. *Utah Sci* 32;25–28, 1971

Tokarnia CH, Dobereiner J, da Silva MF: *Plantas toxicas da Amazonia a bovinos e outras herbivoros.* Instituto Nacional Pesquisas da Amazonia, Manaus, pp82–83, 1979

Townsend CE: Breeding, physiology, culture, and utilization of cicer milkvetch (*Astragalus cicer* L.). *Adv Agron* 49;253–308, 1993

Traub JL, Potter KA, Bayly WM, Reed SM: Alsike clover poisoning. *Mod Vet Pract* 63;307–309, 1982

Trelease SF, Beath OA: *Selenium.* By the Authors, New York, 1949

Trelease SF, Trelease HM: Selenium as a stimulating and possibly essential element for indicator plants. *Am J Bot* 25;372–380, 1938

Tsai KJ, Lin TJ, Lin WL, Deng JF: The potential neurotoxicity of bitter tea drinking—a case report.. *Vet Hum Toxicol* 35:330, 1993(abstract)

Tsai W-C, Ling K-H: Toxic action of mimosine. 1. Inhibition of mitosis and DNA synthesis of H.Ep-2 Cell by mimosine and 3,4-dihydroxypyridine. *Toxicon* 9;241–247, 1971

Tsuda Y, Sano T: *Erythrina* and related alkaloids. In *The Alkaloids,* vol 48, Cordell GA ed, Academic Press, San Diego, Calif, pp249–337, 1966

Tucker JO, Maki LR: Acute pulmonary emphysema of cattle. 1. Experimental production. *Am J Vet Res* 23;821–823, 1962

Tulsiani DRP, Harris TM, Touster O: Swainsonine inhibits the biosynthesis of complex glycoproteins by inhibition of Golgi mannosidase II. *J Biol Chem* 257;7936–7939, 1982

Tulsiani DRP, Broquist HP, James LF, Touster O: Biochemical and pathological effects of swainsonine and locoweed in the pig. In *Plant Toxicology, Proceedings of the Australia-USA Poisonous Plants Symposium,* Seawright AA, Hegarty MP, James LF, Keeler RF eds, Animal Research Institute, Yeerongpilly, Brisbane, Australia, pp279–289, 1985

Tulsiani DRP, Broquist HP, James LF, Touster O: Locoweed- and swainsonine-induced production of hybrid glycoproteins and oligosaccharides in brain. In *Swainsonine and Related Glycosidase Inhibitors,* James LF, Elbein AD, Molyneux RJ, Warren CD eds, Iowa State Univ Press, Ames, pp188–201, 1989

Turcotte F, Leduc R, Marion L: The papilionaceous alkaloids. 20. *Lupinus wyethii* S.Wats. and *Baptisia versicolor* Lodd. *Can J Chem* 31;387–390, 1953

Ueno A, Morinaga K, Fukushima S, Okuda S: Studies on the lupin alkaloids. 7. Isolation and structure of (–)-Δ7-dehydrosophoramine. *Chem Pharm Bull* 26;1832–1836, 1978

Ueno M, Ogawa H, Matsumoto I, Seno N: A novel mannose-specific and sugar specifically aggregatable lectin from the bark of the Japanese pagoda tree *(Sophora japonica)*. *J Biol Chem* 266;3146–3153, 1991

Urrutia H, Brevis C, Quezada M, Donoso S: An outbreak of acute bovine oedema and emphysema (ABPE) in Parral, Chile. *Arch Med Vet* 29;161–166, 1997

Van Es L, Cantwell LR, Martin HM, Kramer J: *On the Nature and Cause of the "Walking Disease" of Northwestern Nebraska.* Nebr Agric Exp Stn Res Bull 43, 1929

Van Kampen KR, James LF: Pathology of locoweed *(Astragalus lentiginosus)* poisoning in sheep. *Pathol Vet* 7;503–508, 1970

Van Kampen KR, James LF: Ophthalmic lesions in locoweed poisoning of cattle, sheep, and horses. *Am J Vet Res* 32;1293–1295, 1971a

Van Kampen KR, James LF: Ovarian and placental lesions in sheep from ingesting locoweed *(Astragalus lentiginosus)*. *Vet Pathol* 8;193–199, 1971b

Van Kampen KR, James LF: Manifestations of intoxication by selenium-accumulating plants. In *Effects of Poisonous Plants on Livestock,* Keeler RF, Van Kampen KR, James LF eds, Academic Press, New York, pp135–138, 1978

van Warmelo KT, Marasas WFO, Adelaar TF, Kellerman TS, van Rensburg IBJ, Minne JA: Experimental evidence that lupinosis of sheep is a mycotoxicosis caused by the fungus, *Phomopsis leptostromiformis* (Kuhn) Bubak. *J S Afr Vet Med Assoc* 41;235–247, 1970

Venugopalan CS, Flory W, Tucker TA, Hebert CD, Strain GM: Assessment of smooth muscle function in *Sesbania drummondii* toxicosis in *Gallus domesticus*. *Am J Vet Res* 45;764–768, 1984

Venugopalan CS, O'Malley NA, Marceau-Day ML, Flory W: Smooth muscle inhibition by an extract of *Sesbania drummondii*. *Am J Vet Res* 48;293–296, 1987

Vera-Avila HR, Forbes TDA, Randel RD: Plant phenolic amines: potential effects on sympathoadrenal medullary, hypothalamic-pituitary-adrenal, and hypothalamic-pituitary-gonadal function in ruminants. *Domest Anim Endocrinol* 13;285–296, 1996

Vickery PJ, Wheeler JL, Mulcahy C: Factors affecting the hydrogen cyanide potential of white clover (*Trifolium repens* L.). *Aust J Agric Res* 38;1053–1059, 1987

Vihan VS, Panwar HS: A note on toxicity of *Acacia leucophloea* (Rajma) in sheep. *Indian J Anim Res* 21;53–55, 1987

Vivanco F, Ramos F, Jimenez-Diaz C: Determination of γ-aminobutyric acid and other free amino acids in whole brains of rats poisoned with β,β-iminodiproprionitrile and α,γ-diaminobutyric acid with, or without, administration of thyroxine. *J Neurochem* 13;1461–1467, 1966

Voss KA, Brennecke LH: Toxicological and hematological effects of sicklepod *(Cassia obtusifolia)* seeds in Sprague-Dawley rats: a subchronic feeding study. *Toxicon* 29;1329–1336, 1991

Wagner JG, Petry TW, Roth RA: Characterization of monocrotaline pyrrole-induced DNA cross-linking in pulmonary artery epithelium. *Am J Physiol* 264;L517–L522, 1993

Wagnon KA: Lupine poisoning as a possible factor in congenital deformities in cattle. *J Range Manage* 13;89–91, 1960

Wahlstrom RC, Olson OE: The effect of selenium on reproduction in swine. *J Anim Sci* 18;141–145, 1959

Waibel PE, Burger RE, Krista LM: Influence of reserpine and antibiotics on incidence of dissecting aneurysm in turkeys as induced by β-aminoproprionitrile. *Poult Sci* 41;1554–1559, 1962

Waldron CA, Tecumseh VS: Poisoning from locust bark. *Am Vet Rev* 33;456–459, 1908

Walkley SU: Pathobiology of neuronal storage disease. *Int Rev Neurobiol* 29;191–244, 1988

Walkley SU, Siegel DA: Comparative studies of the CNS in swainsonine-induced and inherited feline α-mannosidosis. In *Swainsonine and Related Glycosidase Inhibitors,* James LF, Elbein AD, Molyneux RJ, Warren CD eds, Iowa State Univ Press, Ames, pp57–75, 1989

Walkley SU, Wurzelmann S, Siegel DA: Ectopic axon hillock-associated neurite-growth is maintained in metabolically reversed swainsonine-induced neuronal storage disease. *Brain Res* 410;89–96, 1987

Walter-Hansen K, Ruth G, Schwartz S, Ellefson M, Ehle F, Marten GC: Photosensitivity in calves and sheep fed cicer milkvetch *(Astragalus cicer). Proc Am Assoc Vet Lab Diagn* 29;209–220, 1986

Walton GP: *A Chemical and Structural Study of Mesquite, Carob, and Honey Locust Beans.* USDA Bull 1194, 1923

Wang X, Huang S, Liu R: Pharmacokinetics and pharmacodynamics of sophocarpine and oxysophocarpine. *J China Pharm Univ* 23;161–164, 1992

Ward WR: The aetiology of "ringwomb" or partial dilation of the cervix. *Vet Annu* 15;75–78, 1975

Warren CD, Bugge B, Daniel PF, Linsley KB, Daniels DK, James LF, Jeanloz RW: Locoweed toxicosis in sheep: oligosaccharides accumulated in fetal and maternal tissues. In *Swainsonine and Related Glycosidase Inhibitors,* James LF, Elbein AD, Molyneux RJ, Warren CD eds, Iowa State Univ Press, Ames, pp345–359, 1989

Watkins BA, Manning B, Al-Athari AK: The effects of *Lupinus albus* cultivar ultra on broiler performance. *Nutr Rep Int* 38;173–181, 1988

Watt JM, Breyer-Brandwijk MG: *The Medicinal and Poisonous Plants of Southern and Eastern Africa,* E & S Livingston, Edinburgh, pp536–552, 1962

Wayman O, Iwanga II, Hugh WI: Fetal resorption in swine caused by *Leucaena leucocephala* (Lam.) De Wit. in the rat. *J Anim Sci* 30;583–589, 1970

Wei JW, Liao JF, Chuang CY, Chen CF, Han PW: Cardiovascular effects of matrine isolated from the Chinese herb shan-dou-gen. *Proc Nat Sci Counc Repub China* 9;215–219, 1985

Weimer PJ, Hatfield RD, Buxton DR: Inhibition of ruminal cellulose fermentation by extracts of the perennial legume cicer milkvetch *(Astragalus cicer). Appl Environ Microbiol* 59;405–409, 1993

Weiss JH, Koh J-y, Choi DW: Neurotoxicity of β-*N*-methylamino-L-alanine (BMAA) and β-*N*-oxalylamino-L-alanine (BOAA) on cultured cortical neurons. *Brain Res* 497;64–71, 1989

Welch H, Morris HE: *Range Plants Poisonous to Livestock in Montana.* Mont Agric Exp Stn Circ 197, 1952

Wells SR: Intentional ingestion of the mescal bean *(Sophora secundiflora)* [abstract]. *Vet Hum Toxicol* 35;330, 1993

Welsh SL: Legumes of the north-central states: Galegeae. *Iowa State J Sci* 35;111–250, 1960

Welsh SL: *Astragalus* and *Oxytropis:* definitions, distributions, and ecological parameters. In *Swainsonine and Related Glycosidase Inhibitors,* James LF, Elbein AD, Molyneux RJ, Warren CD eds, Iowa State Univ Press, Ames, pp3–13, 1989

Welsh SL, Atwood ND, Goodrich S, Higgins LC: *Great Basin Naturalist Memoirs—A Utah Flora..* Brigham Young Univ, Provo, Utah, p356, 1987

Welshons WV, Rottinghaus GE, Nonneman DJ, Dolan-Timpe M, Ross PF: A sensitive bioassay for detection of dietary estrogens in animal feeds. *J Vet Diagn Invest* 2;268–273, 1990

Wenjuan H, Xiaotian L, Xiaoming C: Isolation and identification of 3-nitropropionic acid from *Arthrinium* as the causative agent of sugarcane poisoning. *Chin J Prev Med* 2013;321–323, 1986

West E, Emmel MW: *Some poisonous plants of Florida.* Fla Agric Exp Stn Bull 468, 1950

White LM, Wight JR: Seasonal dry matter yield and digestibility of seven grass species, alfalfa, and cicer milkvetch in eastern Montana. *Agron J* 73;457–462, 1981

White SM, Roth RA: Pulmonary platelet sequestration is increased following monocrotaline pyrrole treatment of rats. *Toxicol Appl Pharmacol* 96;465–475, 1988

Whiting F, Connell R, Plummer PJG, Clark RD: Incoordination (cerebellar ataxia) among lambs from ewes fed peavine silage. *Can J Comp Med* 21;77–84, 1957

Whitty JD: *Laburnum. Vet Rec* 48;634, 1948

Wilber CG: Toxicology of selenium: a review. *Clin Toxicol* 17;171–230, 1980

Willette RE, Cammarato LV: Phytochemical survey of Connecticut. 1. Isolation of monocrotaline from *Crotalaria sagittalis* L. fruit. *J Pharm Sci* 61:122–125, 1972

Willhite CC, Ferm VH, Zeise L: Route-dependent pharmacokinetics, distribution, and placental permeability of organic and inorganic selenium in hamsters. *Teratology* 42;359–371, 1990

Williams GD, Hatkin J, Jones LP: Acute respiratory distress syndrome occurring in Texas pastured cattle. *Proc Am Assoc Vet Lab Diagn* 20;327–338, 1977

Williams KT, Lakin HW, Byers HG: *Selenium Occurrence in Certain Soils in the United States, with a Discussion of Related Topics: Fourth Report.* USDA Tech Bull 702, 1940

Williams KT, Lakin HW, Byers HG: *Selenium Occurrence in Certain Soils in the United States, with a Discussion of Related Topics: Fifth Report.* USDA Tech Bull 758, 1941

Williams M, Cassady JM: Potential antitumor agents: a cytotoxic cardenolide from *Coronilla varia* L. *J Pharm Sci* 65;912–914, 1976

Williams MC: Nitro compounds in foreign species of *Astragalus. Weed Sci* 29;261–269, 1981a

Williams MC: Nitro compounds in *Indigofera* species. *Agron J* 73;434–436, 1981b

Williams MC: 3-Nitroproprionic acid and 3-nitro-1-propanol in species of *Astragalus. Can J Bot* 60;1956–1963, 1982

Williams MC: Toxic nitro compounds in *Lotus. Agron J* 75;520–522, 1983

Williams MC, Barneby RC: The occurrence of nitro-toxins in North American *Astragalus* (Fabaceae). *Brittonia* 29;310–326, 1977a

Williams MC, Barneby RC: The occurrence of nitro-toxins in Old World and South American *Astragalus Brittonia* 29;327–331, 1977b

Williams MC, James LF: Toxicity of nitro-containing *Astragalus* to sheep and chicks. *J Range Manage* 28;260–263, 1975

Williams MC, James LF: Poisoning in sheep from Emory milkvetch and nitro compounds. *J Range Manage* 29;165–167, 1976

Williams MC, James LF: Livestock poisoning from nitro-bearing *Astragalus.* In *Effects of Poisonous Plants on Livestock,* Keeler RF, Van Kampen KR, James LF eds, Academic Press, New York, pp379–389, 1978

Williams MC, Mayland HF: Selenium absorption by two-grooved milkvetch and western wheatgrass from selenomethionine, selenocystine, and selenite. *J Range Manage* 45;374–378, 1992

Williams MC, Molyneux RJ: Occurrence, concentration, and toxicity of pyrrolizidine alkaloids in *Crotalaria* seeds. *Weed Sci* 35;476–481, 1987

Williams MC, Molyneux RJ: Toxicological investigations on Ruby Valley pointvetch. *J Range Manage* 41;399–400, 1988

Williams MC, Van Kampen KR, Norris FA: Timber milkvetch poisoning in chickens, rabbits, and cattle. *Am J Vet Res* 30;2185–2190, 1969

Williams MC, Norris FA, Van Kampen KR: Metabolism of miserotoxin to 3-nitro-1-propanol in bovine and ovine ruminal fluids. *Am J Vet Res* 31;259–262, 1970

Williams MC, Stermitz FR, Thomas RD: Nitro compounds in *Astragalus* species. *Phytochemistry* 14;2306–2308, 1975

Williams MC, James LF, Bond BO: Emory milkvetch *(Astragalus emoryanus* var. *emoryanus)* poisoning in chicks, sheep, and cattle. *Am J Vet Res* 40;403–406, 1979

Williams TA: *Some Plants Injurious to Stock.* S D Agric Exp Stn Bull 33, 1893

Williamson PM, Highlet AS, Gams W, Sivasithamparam K, Cowling WA: *Diaporthe toxica* sp. nov., the cause of lupinosis in sheep. *Mycol Res* 98;1364–1368, 1994

Wilson DW, Segall HJ: Changes in type II cell populations in monocrotaline pneumotoxicity. *Am J Pathol* 136;1293–1299, 1990

Wilson DW, Segall HJ, Pan LC, Lame MW, Estep JE, Morin D: Mechanisms and pathology of monocrotaline pulmonary toxicity. *Crit Rev Toxicol* 22;307–325, 1992

Windler DR: A systematic treatment of the native crotalarias of North America (Leguminosae). *Rhodora* 76;151–204, 1974

Wine HA, Johnson PN: *Vicia tetrasperma* (four-seeded vetch) ingestion by a three-year-old child. *Vet Hum Toxicol* 35;436–437, 1993

Wink M: Chemical defense of lupines: biological function of quinolizidine alkaloids. *Plant Syst Evol* 150;65–81, 1985

Wink M: Allelochemical properties or the reason d'etre of alkaloids. In *The Alkaloids,* vol 43, Cordell GA ed, Academic Press, Boca Raton, Fla, pp1–118, 1994

Wink M, Witte L: Turnover and transport of quinolizidine alkaloids: diurnal fluctuations of lupanine in the phloem sap, leaves, and fruits of *Lupinus albus* L. *Planta Med* 161;519–524, 1984

Wink M, Meissner C, Witte L: Patterns of quinolizidine alkaloids in 56 species of the genus *Lupinus. Phytochem* 38; 139–153, 1995

Wink M, Hartmann T, Witte L, Schiebel HM: The alkaloid patterns of cell suspension cultures and differentiated plants of *Baptisia australis* and their biogenic implications. *J Nat Prod* 44;14–20, 1981

Wink M, Witte L, Hartmann T, Theuring C, Volz V: Accumulation of quinolizidine alkaloids in plants and cell suspension cultures: genera *Lupinus, Cytisus, Baptisia, Genista, Laburnum,* and *Sophora. Planta Med* 48;253–257, 1983

Wink M, Heinen HJ, Vogt H, Schiebel HM: Cellular localization of quinolizidine alkaloids by laser desorption mass spectrometry (LAMMA 1000). *Plant Cell Rep* 3;230–233, 1984

Witte ST, Will LA: Investigation of selenium sources associated with chronic selenosis in horses of western Iowa. *J Vet Diagn Invest* 5;128–131, 1993

Witte ST, Will LA, Olsen CR, Kinker JA, Miller-Graber P: Chronic selenosis in horses fed locally produced alfalfa hay. *J Am Vet Med Assoc* 202;406–409, 1993

Wong E, Flux DS: The oestrogenic activity of red clover isoflavones and some of their degradation products. *J Endocrinol* 24;341–348, 1962

Wong E, Flux DS, Latch GCM: The oestrogenic activity of white clover (*Trifolium repens* L.). *N Z J Agric Res* 14;639–645, 1971

Wood HC, LaWall CH, Youngken HW, Anderson JF, Griffith I: *The Dispensatory of the United States of America,* 21st ed. Lippincott, Philadelphia, 1926

Wood PMcR, Brown AGP, Petterson DS: Production of the lupinosis mycotoxin by *Phomopsis rossiana. Aust J Exp Biol Med Sci* 51;557–558, 1973

Woods LW, Johnson B, Hietala SK, Galey FD, Gillen D: Systemic granulomatous disease in a horse grazing pasture containing vetch (*Vicia* sp.). *J Vet Diagn Invest* 4;356–360, 1992

Wright CI: Effect of selenium on urease and arginase. *J Pharmacol Exp Ther* 68;220–230, 1940

Wright PA: Infertility in rabbits induced by feeding ladino clover. *Proc Soc Exp Biol Med* 105;428–430, 1960

Wullner U, Young AB, Penney JB, Beal MF: 3-Nitropropionic acid toxicity in the striatum. *J Neurochem* 63;1772–1781, 1994

Xin H, Liu S: Effects of matrine on myocardial contraction and arrhythmia in isolated heart atria. *Zhongguo Yaoli Xuebao* 8;501–505, 1987 (*Chem Abstr* 107;228850r, 1987)

Yaeger MJ, Neiger RD, Holler L, Fraser TL, Hurley DJ, Palmer IS: The effect of subclinical selenium toxicosis on pregnant beef cattle. *J Vet Diagn Invest* 10;268–273, 1998
Yamazaki M, Arai A: On the contractile response of fundus strip from rats to matrine, an alkaloidal component of *Sophora flavescens*. *J Pharmacobio Dyn* 8,513–517, 1985
Yamini B, Poppenga RH, Braselton WE Jr, Judge LJ: Dicoumarol (moldy sweet clover) toxicosis in a group of Holstein calves. *J Vet Diagn Invest* 7;420–422, 1995
Yan CC, Huxtable RJ: Quantitation of the hepatic release of metabolites of the pyrrolizidine alkaloid, monocrotaline. *Toxicol Appl Pharmacol* 127;58–63, 1994
Yan CC, Huxtable RJ: The effect of the pyrrolizidine alkaloids, monocrotaline and trichodesmine, on tissue pyrrole binding and glutathione metabolism in the rat. *Toxicon* 33;627–634, 1995a
Yan CC, Huxtable RJ: The relationship between the concentration of the pyrrolizidine alkaloid monocrotaline and the pattern of metabolites released from the isolated liver. *Toxicol Appl Pharmacol* 130;1–8, 1995b
Yan CC, Huxtable RJ: Relationship between glutathione concentration and metabolism of the pyrrolizidine alkaloid, monocrotaline, in the isolated, perfused liver. *Toxicol Appl Pharmacol* 130;132–139, 1995c
Yan CC, Huxtable RJ: Effects of taurine and guanidinoethane sulfonate on toxicity of the pyrrolizidine alkaloid monocrotaline. *Biochem Pharmacol* 51;321–329, 1996
Yin C: Pharmacological effect of matrine, main constituent of kujin. *Yakugaku-Zasshi* 8;42–52, 1928
Yokoyama MT, Carlson JR: Dissimilation of tryptophan and related indolic compounds by ruminal microorganisms, in vitro. *Appl Microbiol* 27;540–548, 1974
Yuan C, Lu S, Yao X: Oxymatrine induction of the hepatic microsomal drug-metabolizing enzymes in mice. *Shoudu Yixueyuan Xuebao* 8;23–26, 1987
Yuan H, Yin Y, He H, Zhao Y: Pharmacological studies on *Sophora alopecuroides (*II) neuropharmacological effects of oxymatrine. *Yaouri Fenxi Zazhi* 6;349–352, 1986 (*Chem Abstr* 106;96090k, 1987)
Yuan H, He H, Zhao Y, Wang Z: Inhibitory action of sophocarpine on the central nervous system. *Zhongyao Tongbao* 12;237–239, 1987 (*Chem Abstr* 107;17829c, 1987)
Zagoren JC, Seelig M, Roy DN, Bornstein MB, Spencer PS: Pathological changes in spinal cord induced by *Lathyrus* neurotoxin (β-*N*-oxalylamino-L-alanine) in vitro. *J Neuropathol Exp Neurol* 44;124, 1985
Zhang B, Wang N, Li X, Kong X, Cai Y: Antiarrhythmic effects of matrine. *Zhongguo Yaoli Xuebao* 11;253–257, 1990 (*Chem Abstr* 113;363x, 1990)
Zhang SS, Li SH, Liu SF: Relationship between the positive inotropic action of matrine and extracellular calcium. *Yaoxue Xuebao* 25;637–639, 1990

Fagaceae Dumort. 38

Comprising 8 or 9 genera and about 600–1,000 species distributed primarily in the Northern Hemisphere, the Fagaceae, commonly known as the beech or oak family, is represented in North America by 5 genera, 97 species, and innumerable hybrids of oak. Species of *Fagus* (beech) and *Quercus* (oak) dominate most of the eastern deciduous forests. Before its almost complete extinction due to chestnut blight, *Castanea* (American chestnut) also was one of the most important trees in eastern North America. In the West, evergreen species of *Quercus, Lithocarpus* (tan oak), and *Castanopsis* (giant chinquapin) are encountered. Prized for its wood—for flooring, furniture, barrels, and ship decking—the family also is economically important as food for both wildlife and humans. The fruits of *Fagus* (beechnuts) and *Quercus* (acorns) are a major component of the diets of many forest animals, and sweet chestnuts are roasted or included in purees, stuffings, and stews. Oil also is extracted from beechnuts.

■ Beech Family
Oak Family
Fagus
Quercus

Plants trees or shrubs; deciduous or evergreen; monoecious. **Leaves** simple; alternate; blades entire to variously toothed or lobed; awns present or absent; venation pinnate; stipules present, caducous. **Inflorescences** of 2 types, staminate and pistillate different; axillary. **Staminate Inflorescences** catkins; pendulous; elongate or globose; bracts present or absent, caducous. **Pistillate Inflorescences** solitary flowers or clusters of 2–4; bracts present, fused to form involucral caps or husks that completely or partially enclose flowers. **Flowers** produced before or simultaneously with or after leaves; imperfect, staminate and pistillate similar; perianths in 1-series; radially symmetrical. **Sepals** 4–8; fused. **Petals** absent. **Stamens** 4–20. **Pistils** 1; compound, carpels 3–7; stigmas 3; styles 3; ovaries inferior; locules 3–7; placentation axile; ovules 2 per locule; gynoecial rudiments occasionally present. **Fruits** nuts; subtended by involucral caps in *Quercus* (acorn) or enclosed in spiny or prickly husks that are dehiscent at maturity in *Castanea* and *Fagus*. **Seeds** 1.

■ shrubs and trees; leaves simple, alternate; flowers imperfect, staminate in catkins, pistillate solitary or clustered; fruits nuts

Fagus L.

Taxonomy and Morphology—Comprising 8–10 species in the Northern Hemisphere of both the Old World and the New World, *Fagus* is one of the dominant trees of deciduous forests. Its name is probably derived from the Greek *fagein,* meaning "to eat." In North America, the genus is represented by 1 native and 1 introduced species:

F. grandifolia Ehrh.	American beech, beechnut
F. sylvatica L.	European beech

■ *F. grandifolia*
F. sylvatica
beech
beechnut

Plants massive trees; trunks stout; crowns dense, rounded. **Stems** 18–25 m tall; bark gray, smooth at tree maturity; twigs slender, slightly zigzag, bearing short lateral or

- massive trees, dense crowns; leaves in 2 rows, serrated appearance due to protrusion of secondary veins; nuts 3-angled

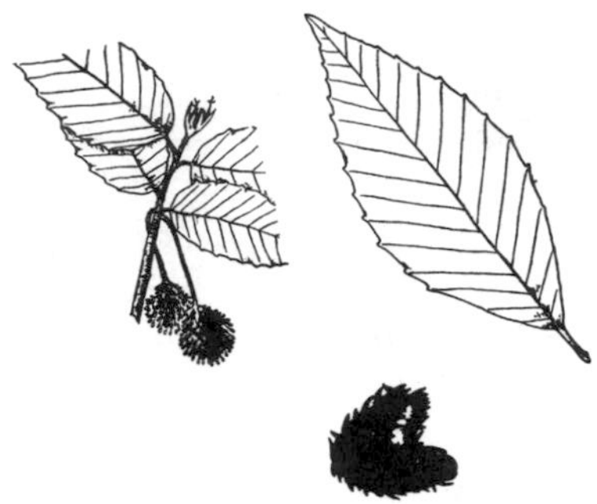

Fagus grandifolia

- eastern woodlands, ornamental

Fagus grandifolia

- low risk of digestive disturbance or neurologic effects
- toxicant unknown; possibly saponin-like irritant?
- probably signs related to irritation of digestive tract

spur branches; terminal buds elongated, narrow, golden brown, with numerous overlapping scales. **Leaves** borne in 2 rows or clustered on the spurs; blades ovate to lanceolate; secondary veins form conspicuous parallel ribs and protrude as teeth from margins, giving serrated appearance. **Staminate Flowers** borne in globose catkins on pendulous peduncles; appearing as leaves emerge. **Pistillate Flowers** typically borne in pairs at ends of short stalks; enclosed by numerous bracts that form a reddish-brown, spiny bur. **Nuts** 3-angled; brown; smooth; dropping from involucral burs at maturity.

Distribution and Habitat—The native *F. grandifolia* is distributed throughout the woodlands of the eastern half of North America. *Fagus sylvatica,* introduced from Europe, is cultivated as an ornamental primarily in the East and to a lesser extent in the Midwest and the West.

Disease Problems—Neither species of *Fagus* has proven to be a problem in North America. Garner (1961), however, reported that there have been intermittent poisonings in Europe of humans and livestock following consumption of beechnuts or the cake after the oil had been extracted. The disease involves irritation of the digestive tract and, less commonly, tetanic-type seizures. Wilkins and Cranwell (1990) described in detail an English case in 2 ponies that gorged on the nuts of *F. sylvatica* following a windstorm. They reported that the disease was characterized by signs of a neurotoxicosis, including ataxia, muscle fasciculation, hyperexcitability, extensor spasms of the pelvic limbs, and self-mutilation. The severity was such that euthanasia was necessary. Postmortem findings were few, other than petechial hemorrhages of the serosa and mucosa of the stomach and small intestine, and the mass of beechnuts in the stomach.

Disease Genesis—Because toxicity problems caused by *Fagus* are very rare, little is known about its toxicants. The effects in part appear to be due to an irritant/corrosive saponin-like substance that acts on the digestive tract. The foliage of beech has high levels of polyphenolic compounds similar to those of *Quercus,* and if sufficient plant material were eaten, signs of oak-type intoxication might be anticipated. *Fagus* also shares with *Quercus* a correlation with a high incidence of adenocarcinoma of the nose in people repeatedly exposed to their wood dust (Kleinsasser et al. 1991; Maciejewska et al. 1993). Such a predisposition, if true, suggests some similarity in disease potential between the two genera, even though this carcinogenic effect is due to an alcohol extractable, non-tannin fraction (Nelson et al. 1993). The neurotoxic effects—i.e., the tetanic type seizures—however, are not consistent with the action of polyphenolics (cf. next section). This suggests that *Fagus* may have a different array of toxicants from that of *Quercus.*

Clinical Signs, Pathology, and Treatment—Because species of *Fagus* have not been clearly linked with a specific disease problem, clinical signs and treatment cannot reliably be described. It appears that digestive tract derangement is most likely and should be treated to relieve the signs observed.

Quercus L.

Taxonomy and Morphology—Comprising approximately 85 native and introduced species and some 100 named hybrids, *Quercus* is one of the most important hardwood trees and is a familiar sight throughout North America. These woody perennials range in size from knee-high shrubs to massive trees. Most are deciduous and easily recognized by the presence of multiple buds at the twig ends and the characteristic acorn. Species of *Quercus* implicated specifically in toxicoses include the following; all species, however, should be considered potentially toxic.

Q. agrifolia Née	coast live oak, California live oak
Q. coccinea Muenchh.	scarlet oak
Q. douglasii Hook. & Arn.	blue oak, mountain white oak, iron oak
Q. durandii Buckley	Durand oak (see *Q. sinuata* var. *breviloba*)
Q. gambelii Nutt.	Gambel's oak, Utah white oak, Rocky Mountain white oak
Q. garryana Douglas ex Hook.	Oregon white oak, Garry oak, Oregon oak
Q. havardii Rydb.	shinnery oak, shin oak
Q. incana Bartram	bluejack oak, sand jack oak, upland willow oak, turkey oak
Q. marilandica Muenchh.	blackjack oak, jack oak, blackjack
Q. mohriana Buckley	Mohr oak
Q. prinus L.	chestnut oak, basket oak, rock chestnut oak, rock oak
Q. robur L.	English oak, European oak, common oak, pedunculate oak
Q. rubra L.	red oak, northern red oak, gray oak
Q. sinuata Walter var. *breviloba* (Torr.) C.H.Muell. (= *Q. durandii* Buckley)	bastard oak, bastard white oak, white shin oak, Durand oak, bluff oak
Q. stellata Wangenh.	post oak, iron oak
Q. turbinella Greene	shrub live oak, Sonoran scrub oak
Q. velutina Lam.	black oak, yellow-barked oak, yellow oak

Plants trees or shrubs; deciduous or evergreen; trunks slender or massive; crowns of various shapes. **Stems** 1–40 m tall; bark light or dark, smooth or rough; twigs slender or stout, terminal buds multiple, of various colors and shapes. **Leaves** diverse; blades of various shapes, texture, and color; margins entire to serrate or deeply

■ *Q. agrifolia*
Q. coccinea
Q. douglasii
Q. durandii
Q. gambelii
Q. garryana
Q. havardii
Q. incana
Q. marilandica
Q. mohriana
Q. prinus
Q. robur
Q. rubra
Q. sinuata
Q. stellata
Q. turbinella
Q. velutina

oak

■ shrubs or trees; leaves diverse in shape; staminate flowers in long catkins that appear as leaves emerge; nuts with scaly cap (acorns)

Quercus garryana

Quercus havardii

Quercus gambelii

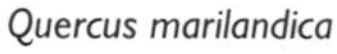

Quercus marilandica

Quercus mohriana

Quercus stellata

Quercus turbinella

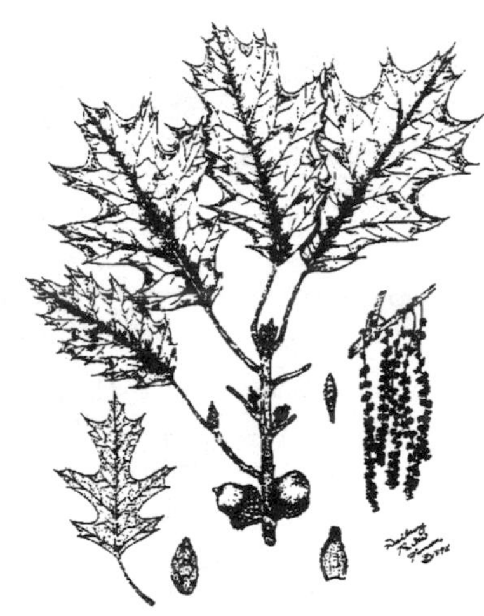

Quercus velutina

Quercus gambelii

Quercus garryana

pinnately lobed. **Staminate Flowers** borne in elongate, pendulous catkins; appearing as leaves emerge. **Pistillate Flowers** borne singly or in clusters of 2 or 3; inconspicuous before fruit development. **Nuts** partially enclosed by prominent scaly caps at bases (acorns); subspherical to oblong; dropping in late autumn or early winter.

Two sections, commonly known as the black/red oaks and the white oaks, are generally recognized. The black/red oaks (*Quercus* sect. *Lobatae*) have darker bark and acute, awned leaf lobes. The mature acorns occur on the previous year's wood, because it takes 2 years for them to develop. In contrast, the white oaks (*Quercus* sect. *Quercus*) typically have lighter bark; rounded, awnless leaf lobes; and acorns that mature in 1 year on the current season's wood. Identification of species is based on leaf and acorn features. Because of the large number of species in the genus and of toxicologic interest, a taxonomic key is not presented here; rather the reader is encouraged to use floristic treatments specific for his or her area.

Distribution and Habitat—Species of *Quercus* occur across North America from Canada to Mexico. They occupy a variety of climatic and edaphic regimes.

Disease Problems—Oak foliage and acorns provide valuable forage and mast for both livestock and wildlife during much of the year (Marsh et al. 1919; Ruyle et al. 1986).

Oak toxicosis, a urinary and digestive tract disease, occurs, however, when animals are forced to subsist almost entirely on plants of the genus for several days or more because of a lack of grasses and other forage (Marsh 1924). For example, this disease usually occurs in the spring in the Southwest and intermountain region within the distributional ranges of *Q. havardii* and *Q. gambelii*. Although they are among the most important browse species for livestock in their ranges, these two species also cause the most problems (Dayton 1931). *Quercus garryana* also is a very important browse plant.

- large amounts of foliage or acorns; eaten for 4–5 or more days; digestive disturbance and renal disease

otherwise animals may eat considerable amounts of oak forage without adverse effects

all oak species should be considered toxic

Cornevin's lucid descriptions in the 1890s revealed that oak buds and young leaves are the greatest risks and serve as the typical toxicant sources (Marsh et al. 1919). Occasionally in the late summer or fall when poor weather and/or forage conditions prevail, the acorns, whether green or mature, also create a serious problem (Sandusky et al. 1977; Kasari et al. 1986). The most important factor is the dosage of oak required whether as leaves or acorns. In early studies, disease was produced only erratically, because of the difficulty in giving a large enough dosage of the leaves (Marsh et al. 1919). In the United Kingdom, long, hot summers and high winds have been major factors in decreasing alternative forage and assuring plentiful available acorns as a cause of intoxication in cattle (Wiseman & Thompson 1984). An excellent example of conditions forcing consumption of lethal quantities of oak foliage occurred in northern California in the spring of 1985 when an April snowstorm and cold weather forced cattle to consume oak buds rather than other forages (Spier et al. 1987). As a result, 2,700 calves from 60 ranches in 3 counties died. At one time, some counties in west Texas experienced regular losses of more than 1,000 head of cattle per year due to consumption of oaks (Dollahite et al. 1963). The more typical situation involves occasional, scattered losses of a few animals in an area.

Quercus havardii

All species of *Quercus* should be considered potentially toxic (Kingsbury 1964; Harper et al. 1988). Some species are particularly hazardous because of their growth form—e.g., the rhizomatous low-growing *Q. havardii*—whereas others become so when brush-hogging or mowing results in large numbers of sprouts or bushy forms.

Quercus incana

Disease Genesis—Intoxications caused by *Quercus* have been associated with the actions of tannins—polyphenolic complexes that interact with and denature proteins (Dollahite et al. 1962). The oaks are renowned for their tannin content, and historically these compounds have long been used throughout the world to "tan" animal skins into leather (Gustavson 1954). Acorns, despite their high tannin content, were a dietary staple of many tribes of American Indians and early settlers. They were ground and the bitter tannins leached from the meal prior to cooking (Ebeling 1986).

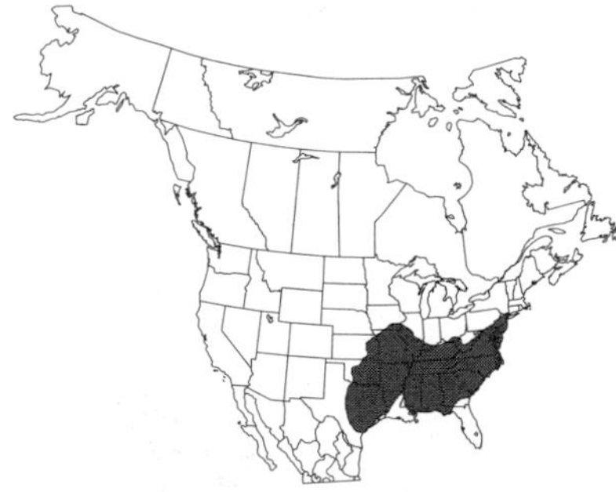

Quercus marilandica

Tannins are divided into two major groups, condensed and hydrolyzable, based on their structure and their reactions with acids or enzymes (Haslam 1966, 1989). This may be an oversimplistic representation of a very complex array of compounds, and sometimes a third category representing more-complex types is included (Kashiwada et al. 1992; Okuda et al. 1992). Hundreds of compounds are conveniently divided among the two or three groups. Vegetable tannins (500–3,000 molecular weight), with numerous free O-dihydroxy and O-trihydroxy phenolic nuclei that project toward the outer surface of the molecule, interact with carbohydrates and precipitate proteins (Adzet & Camarasa 1988; Haslam 1989). Polyphenolics of less than 500 or greater than 3,000 molecular weight are ineffective in protein precipitation and tanning (Salunkhe et al. 1990). The protein precipitation, or tanning effect, is due to the 1 or 2 hydroxyl groups per 100 molecular weight that are free to cross-link proteins with other macromolecules. The tannins or their polyphenolic products have an astringent effect on the mucosa of the small intestine (Adzet & Camarasa 1988).

- tannins, hydrolyzable; condensed tannins are not major factors; hydrolyzable are broken down to various polyphenolics—gallic, digallic, and ellagic acids

Quercus mohriana

Quercus stellata

Quercus turbinella

Quercus velutina

■ numerous phenolic, metabolic products; phenols, pyrocatechol, pyrogallol, phloroglucinol

The condensed tannins contain phenolic nuclei that may have polysaccharides or proteins linked to them. These tannins are resistant to hydrolysis and tend to polymerize to yield amorphous compounds known as phlobaphens. Condensed tannins, or proanthocyanidins, as they are referred to, for the most part pass unchanged through the digestive tract and, with few exceptions, appear to contribute little to disease problems (Haslam 1989; Oelrichs 1994). They bind to and limit availability of some proteins, however, producing broad nutritional effects, including decreased dry matter digestion (Robbins et al. 1987a,b; Salunkhe et al. 1990; Butler 1992; Kumar 1992; Reed 1995). In contrast, the hydrolyzable tannins, which are mixtures of phenolic amino acids esterified with sugars, are readily hydrolyzed by gut esterases to their constituent phenolics and sugars—e.g., gallic acid from gallotannins and ellagic acid from ellagotannins (Singleton & Kratzer 1973; McLeod 1974; Mehansho et al. 1987). Gallic acid may be further degraded to pyrogallol and other phenolics, which may then be conjugated in the liver with glucuronide (Murdiati et al. 1992).

Although the gallotannins from varying sources have much in common, they differ considerably in the degree of galloylation of the constituent units. The numerous hydroxyl groups offer a great variety to linkages in forming the array of different tannins with glucose cores, as shown by the illustrated galloylglucose unit. Even though the basic building blocks are the same, hydrolysis products vary greatly because of the differing ester linkages (Haslam 1966; Okuda et al. 1992). Tannins in the Fagaceae are typically composed of three gallic acid units joined by C—C bonds, all esterified with a unit derived from glucose. In general, tannin absorption is poor because the opportunities for protein binding are so great in the digestive tract (Robinson & Graessle 1943; Oler et al. 1976). The smaller-constituent phenolics are more favorably absorbed than the large-molecular-weight tannins. However, absorption probably increases for both when the intestinal epithelial barrier is somehow damaged. In addition to the high-molecular-weight tannins, low-molecular-weight unsaturated phenolics with numerous hydroxyl and methoxy groups have also been identified in oak leaves (Ishimaru et al. 1987).

Once absorbed, gallic acid is biotransformed to phenolic methoxy and polyhydroxy (catecholic) phenolic acids and conjugated for elimination (Booth et al. 1959; Tompsett 1959; Potter & Fuller 1968). Any tannic acid absorbed is probably biotransformed in the liver to phenolic subunits (Gupta & Dani 1987). Typically, phenolics may be hydroxylated, O-methylated, and/or conjugated with glucuronide or ethereal sulfate and subsequently eliminated by active tubular secretion in urine and bile (Williams 1959). Phenolics found in the urine of cattle exhibiting oak intoxication include *p*-cresol, hydroquinone, phenol, phloroglucinol, pyrocatechol, pyrogallol, and quinol; of these, only phenol is normally present (Lohan et al. 1983; Shi 1988). Experimentally, pyrogallol was found to be present only for the first few days after cattle began ingesting oak leaves (Plumlee 1998).

There is little question of the toxicity of tannic acid. However, it should be emphasized that the tannic acids of commerce are not uniform compounds and that they differ depending on source. In most studies the source and type of tannic acid are not clearly specified; thus ambiguity is considerable when comparing results. Given orally, tannic acids produce marked dose-dependent effects on the gut mucosa (Glick & Joslyn 1970; Mitjavila et al. 1977; Welsch et al. 1989). Tannins also markedly decrease ruminal urease and proteolytic enzyme activity (Lohan et al. 1981, 1983). When tannic acids are given parenterally, renal and hepatic alterations ensue (Arhelger et al. 1965; Boler et al. 1966; Konstantinov & Ivanov 1973). Tannic acids were formerly used for treatment of burns and in barium enemas, but these uses were discontinued because of the risk of hepatotoxicosis (McLure et al. 1944; Rambo et al. 1966). In rats, both tannic and gallic acids are reportedly absorbed from the colon, with the tannin producing hepatic effects and the gallic acid producing renal tubular necrosis when given in high dosage (Harris et al. 1966). Tannins produce similar liver lesions whether given orally or parenterally, but a much higher dosage (3–20 times) is required orally (Cameron et al. 1943; Boyd et al. 1965; Oler et al. 1976; Filippich et al. 1991). Experimental results in sheep are variable; in some instances there may be liver and/or kidney effects or neither, depending on the protocol (Tripathy et al. 1984; Zhu et al. 1992; Zhu & Filippich 1995). Zhu and co-workers (1995) administered tannic acid and various metabolites to sheep and concluded that liver and kidney lesions were caused by tannin rather than by the various metabolites, because the latter did not produce the lesions when given independently. A note of caution regarding experiments with commercially available tannins must be offered: in most studies the tannin or tannic acid used is not necessarily representative of those in oaks.

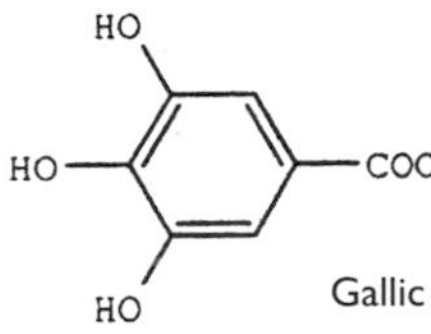

Many of the tannins are potent cytotoxins, as shown by their effects on various tumor cell lines (Kashiwada et al. 1992). Biotransformation products of tannin hydrolysates and other low molecular phenolics are also possible causes of disease (Shi & Hong 1992). These types of metabolites are considerably more toxic than their parent compounds, and some have proven to be toxic to renal tubules (Rayudu et al. 1970; Calder et al. 1971, 1975). Moreover, many are quite susceptible to autoxidation, forming semiquinone and superoxide radicals (Marklund & Marklund 1979; Ames 1983; Hodnick et al. 1988). Although these metabolites and their reactive radicals are quite toxic, a specific role in either the renal or hepatic effects caused by oaks has yet to be identified. In contrast to the foregoing, there are indications of antioxidant activity for polyphenolics in low dosages as scavengers of superoxide radicals and inhibitors of xanthine oxidase (Hatano et al. 1990; Costantino et al. 1992).

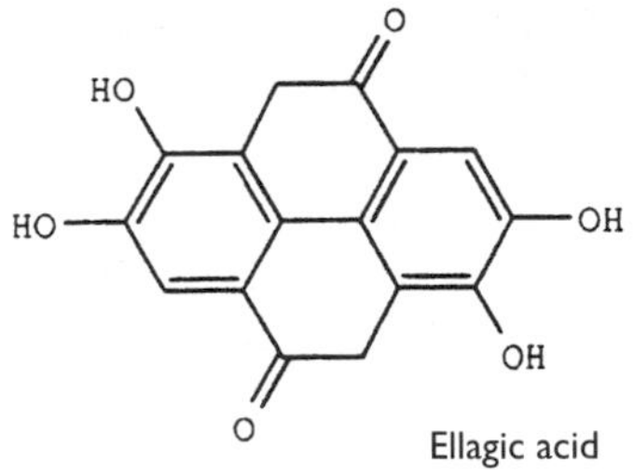

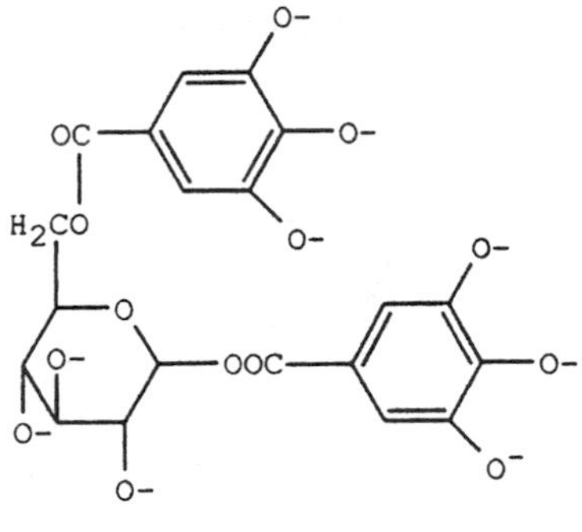

Although the specifics for the pathophysiology of oak intoxication may be lacking, nevertheless a plausible sequence of events can be formulated. Ingestion of oak forage in moderate quantities has little effect other than minor changes in the gastrointestinal epithelium. The absorbed phenolics are apparently eliminated without obvious detrimental effects on the host. Ingestion of large amounts of oak results in not only a considerable increase in the amount of phenolics present but also a greater likelihood of astringent effects on the gut mucosa, thereby increasing the potential for absorption. With a marked increase in the absorption of phenolics comes a concomitant rapid increase in biotransformation products. Tissue destruction occurs at the sites of highest toxin concentration—liver or kidney tubule—especially when conjugation capacity is exceeded. Variation among animal species as to the tissues affected seems to be considerable; however, the kidney is clearly the most susceptible tissue in ruminants. In many animals, there are marked effects on the gut, irrespective of the degree of renal involvement. From the observations of colonic ulceration when tannic acid was used in barium enemas, and the consistency of effects on the gut in animal species exhibiting liver rather than renal disease, it appears that gut changes are likely caused directly by the oak polyphenolics. This does not preclude additive effects from uremia. The disease is not strictly dose dependent, given that a threshold dose seems to be required and that there is considerable variation in susceptibility among individuals animals, and possibly between young and old as well (Housholder and Dollahite 1963; Govindwar & Dalvi 1990).

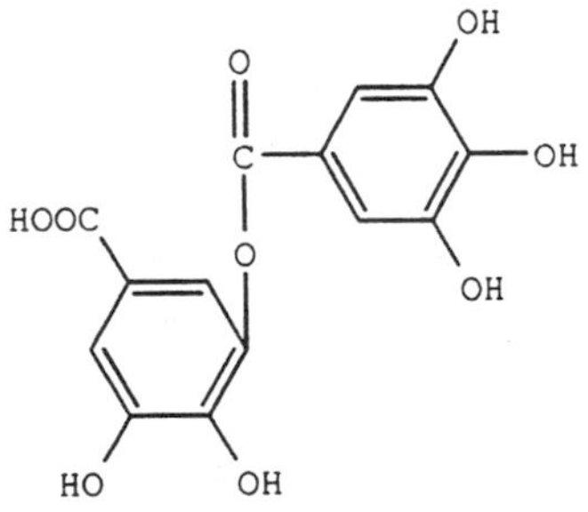

Other tannin effects of minor importance that have been reported include enhancement or depression in vitro of atrial contractility (Calixto et al. 1986); depression of the central nervous system (Takahashi et al. 1986); and competitive inhibition of catechol-O-methyltransferase by gallic acid, pyrogallol, and probably other related phenolics (Axelrod & Laroche 1959; Archer et al. 1960). There are variable vascular

Gallic acid

Ellagic acid

Galloylglucose tannin subunit

■ plausible sequence of events leading to disease; high levels of phenolic products, irritant action on digestive tract; once they are absorbed, increasing levels in renal tubule as urine is concentrated, destruction of tubular cells

Digallic acid

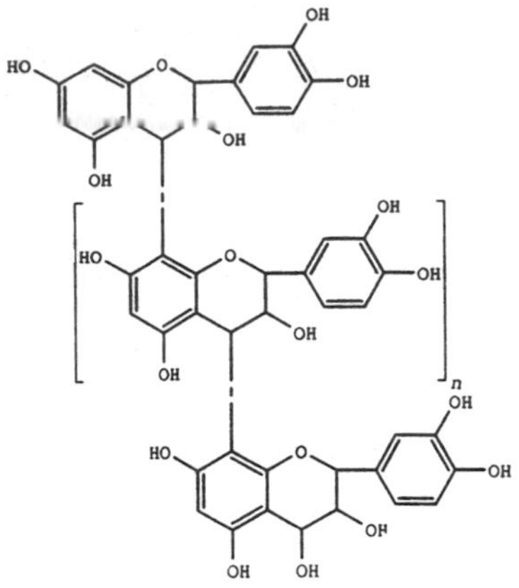

Example condensed tannin: proanthocyanidin, made up of leucoanthocyanidin units

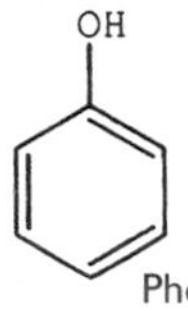

Phenol

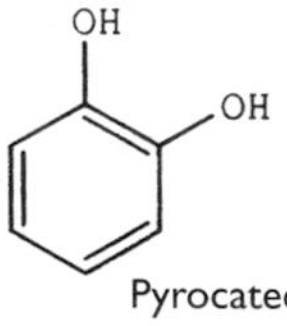

Pyrocatechol

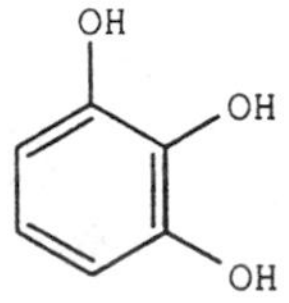

Pyrogallol

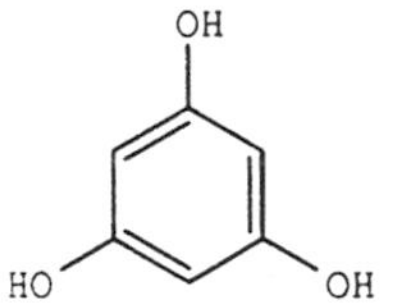

Phloroglucinol

effects due to inhibition of nitrogen oxide synthase; depending on the circumstances, contraction or relaxation is produced (Chiesi & Schwaller 1995). Interestingly, in contrast to their toxic effects, some tannin phenolics have an apparent beneficial effect in chronic renal failure (Yokozawa et al. 1991).

A variety of other effects has also been shown for tannins, including inhibition of mitochondrial respiration, (Nishizawa et al. 1983), gastric H^+,K^+-ATPase (Murakami et al. 1992), and lipid peroxidation–induced lipolysis (Kimura et al. 1983; Okuda et al. 1983). The above generalizations are of questionable validity because commercial tannins often used experimentally may be quite different from the higher-molecular-weight compounds found in many higher plants (Hagerman et al. 1992).

The oak species exhibit marked differences in both their total tannin content and the proportion that are of the hydrolyzable or condensed types (Basden & Dalvi 1987; Kleiner 1991). However, these differences in tannin level or type have not been specifically correlated with toxicity potential. Age of the leaves is also of apparent importance. With increasing leaf maturity, toughness and condensed tannins increase, while moisture, hydrolyzable tannins, protein precipitation capacity, and proteins decrease (Feeny 1970; Makkar et al. 1991). In immature leaves, the concentrations of tannins are high, and most are of the hydrolyzable type (Dawra et al. 1988; Faeth 1992). Members of the black/red oak group are purported to have higher levels of tannins than the white oaks. In some species, immature acorns have more tannins than mature acorns (Fleck & Layne 1990). The acorns of the white oaks are said to be sweeter and more edible than those of the black/red oaks (Medsger 1939).

Oaks may not be as unique in their chemical composition or phenolic content as they are in their likelihood of being eaten in large, potentially toxic amounts. Thus it is not surprising that tannins of some type are suspected as the toxicants in other plant families. In early studies, Marsh and co-workers (1919) found that new leaves of a number of trees were capable of causing problems similar to those caused by oaks, but the leaves were not eaten in sufficient quantities to cause disease. A hydrolyzable tannin, punicalagin, has been identified as one of the toxic principles in *Terminalia oblongata,* in the Combretaceae; it causes acute liver disease in cattle and sheep in Australia (Filippich et al. 1991). The effects of *T. oblongata* are an exception in that toxicity is apparently due to a combination of hydrolyzable and condensed tannins, the former causing liver disease, and the latter, kidney disease (Oelrichs et al. 1994). *Clidemia hirta* (Melastomataceae) in Indonesia also causes liver and kidney disease (Murdiati et al. 1990). In Brazil, tannins similar to those in oak are found in *Thiloa glaucocarpa* (also in the Combretaceae) and are suspected as the cause of intoxication of cattle (Itakura et al. 1987).

Although intoxications have been described for many animal species, risk differs considerably among them. Goats are able to use oak browse effectively, often subsisting on ranges where it composes the major part of their diet (Nastis & Malechek 1981; Villena & Pfister 1990; Dick & Urness 1991). Because of their apparent resistance, goats are often used to control brushy, oak-infested areas or are raised in place of cattle in high-risk areas. The apparent resistance is related to the presence of tannin-binding proteins in saliva and the lower digestive system. It has been estimated that up to 50% of dietary tannins may be bound by salivary and other proteins in goats (Provenza & Malechek 1984). Inducible tannase enzymatic activity in the rumen and ruminal mucosa of goats enables them to tolerate twice the tannin levels tolerated by cattle (Begovic et al. 1978). Nevertheless, goats may be affected, especially under heavy stocking rates or under conditions in which they are less able to be selective in diet (Merrill & Schuster 1978). In contrast, deer, also quite able to browse oak, have protective tannin-binding salivary glycoproteins (Robbins et al. 1987b;

Austin et al. 1989). Such proteins are not found in goats (Distel & Provenza 1991). The comparative tannin-neutralizing ability or protein-binding capacity is high in mule deer, and low to nil in sheep and cattle, seemingly related to eating habits of these species (Robbins et al. 1987b; Austin et al. 1989). Interestingly, mice and rats have protective and inducible proline-rich, tannin-binding, salivary proteins, but hamsters do not (Asquith et al. 1987).

Clinical Signs—Oak intoxication is typically a disease of cattle, and rarely are sheep, goats, or horses affected. The initial signs begin after the animals have been consuming large amounts of oak forage for 2–3 days or, more commonly, a week or longer. Anorexia, listlessness, rumen stasis, and constipation appear first but are often overlooked, and concern by the producer may be evoked only upon the onset of diarrhea (which is sometimes bloody), dehydration, colic, and subcutaneous edema of the neck, brisket, abdomen, and perineum (Panciera 1978). Frequency of urination may increase. At this time, there is usually evidence of altered renal function and metabolic acidosis; marked increase in blood urea nitrogen and creatinine; and alteration of serum electrolytes (Panciera 1978; Spier et al. 1987). The most consistent electrolyte changes are increased serum potassium and phosphate, and decreased sodium, chloride, and calcium. Typically, urinalysis will show low specific gravity (1.010), proteinuria, glucosuria, and hematuria. The urine may be clear and colorless or, in some early cases, dark colored because of polyphenolic metabolites. Severe elevations of blood urea nitrogen and creatinine are indications of a very poor prognosis. During the course of the disease, animals may become recumbent because of weakness. Many animals either die within a few days from subacute disease or become chronically debilitated from varying degrees of renal failure (Fowler & Richards 1965; Cedarvall et al. 1973; Stober et al. 1976; Cockrell & Beasley 1979; Holliman 1985). However, protracted disease is not inevitable, because regeneration of renal function may occur and animals that recover may experience a period of compensatory increased feed efficiency to offset the severe weight loss (Ostrowski et al. 1989). Those animals not experiencing a marked increase in blood urea nitrogen are likely to recover from the disease (Dixon et al. 1979).

- cattle, abrupt onset—
  1, anorexia, rumen stasis, constipation
  2, diarrhea, increased urination, s.c. edema neck, brisket, abdomen, perineum
  3, weakness, recumbency

  serum chemistry, increased blood urea nitrogen, creatinine, potassium;

  urine chemistry, low specific gravity

The clinical signs are similar for most animal species, the exceptions being sheep and goats, in which edema is apparently minimal (Dollahite 1961), and in horses, in which diarrhea (with or without blood), colic, and tenesmus are more consistent and severe (Duncan 1961; Broughton 1976; Daniels 1976). Renal effects also appear to be less severe in horses than in cattle (Anderson et al. 1983). Liver involvement may be apparent, possibly dependent on the animal species and/or the oak species. Involvement of both liver and kidney has been reported for rabbits (Camp et al. 1967), a cassowary (Kinde 1988), and horses (Broughton 1976; Warren & Vaughan 1985). Liver disease may also be seen on occasion in cattle, although of much less severity than the nephrosis (Garg et al. 1992).

- sheep and goats, less edema; horses more digestive disturbances

Pathology—The disease in cattle is characterized specifically by renal alterations, which are often accompanied by fluid accumulations and gastrointestinal tract pathology. Typically, the most striking gross changes are the fluid accumulations—ascites, hydrothorax, hydroperitoneum, subcutaneous edema, and edema of the mesentery (especially perirenal) and gut mucosa. In the gut, there is congestion and erosions of varying severity and location from the mouth to the colon; deep ulcerations and hemorrhage are possible. The kidneys are slightly swollen and pale, with petechiae on the surface extending into the cortex. The adjacent lymph nodes may be swollen and edematous. In chronic cases, the kidneys present a more roughened appearance.

- gross pathology, fluid accumulations, ascites; minor to major lesions of digestive tract, erosions, hemorrhage; kidneys swollen, pale

  microscopic, renal tubule dilation, epithelial cell necrosis

Microscopically, there will consistently be diffuse renal tubular degeneration; necrosis of cortical tubular epithelial cells; and dilated tubules with flattened cells or with only the basement membrane sans cells. The dilated tubules often contain hyaline, granular, or cellular casts of epithelial cells, neutrophils, and cellular debris. The medullary tubules are not as consistently or severely affected. More-chronic cases show nephron atrophy and interstitial fibrosis (Panciera 1978; Neser et al. 1982; Spier et al. 1987). In horses, the pathology is quite similar, but the gut lesions are often more severe, especially the edema of the cecal and colonic mucosa. Conversely, the renal disease may be less severe, and there may be some degree of hepatocellular degeneration (Broughton 1976; Warren & Vaughan 1985). In a cassowary, typical renal lesions were accompanied by less severe but nonetheless distinct hepatic degeneration (Kinde 1988).

Acorns, leaves (especially the tougher, more mature ones), or leaf fragments with their stellate hairs may be identifiable in the rumen contents, but these have little significance in most instances, because many animals consume oak mast in the late spring or early summer (Ruyle et al. 1986). Only when very large amounts are eaten does disease occur. The distinctive lesions, combined with knowledge of animal access to oak, are sufficient to make a diagnosis.

■ nursing care, supplemental feeding

Treatment—The major factor in survival of the animal is the extent of renal damage, and accompanying fluid and electrolyte imbalances, and acidosis. However, gastrointestinal changes cannot be discounted as important contributors to the illness, especially in horses. The basic goal is to assist the animal by symptomatic treatment to allow recovery of renal tubular function. This may require replenishment of fluid and electrolytes, although care must be taken because fluid leakage into body compartments may complicate organ function, e.g., respiratory distress. Recovery will require several weeks or more, during which time there may be considerable weight loss and a few animals may die. Alternatively, the renal disease may become chronic, with little chance for complete recovery. The likelihood of recovery seems to be much better if the animals are young, and they may even exhibit compensatory weight recovery during recuperation (Ostrowski et al. 1989).

The disease may be prevented to some extent by limiting continued consumption of oak buds and leaves, feeding supplements, or feeding some alfalfa hay. Supplements containing 5–10% calcium hydroxide appear to reduce the incidence of disease (Dollahite et al. 1963, 1966). Higher percentages may depress supplement intake. Studies with other animal species and plant genera support the value of this approach (Batu et al. 1978; Murdiati et al. 1990). The protective mechanism as shown for *Terminalia* in the Combretaceae and *Clidemia* in the Melastomataceae appears to be due to binding of calcium with formation of insoluble tannin complexes (Murdiati et al. 1990, 1991). Complex formation may be accomplished most effectively by calcium hydroxide, but activated charcoal, polyvinylpyrrolidone, and 4,000-molecular-weight polyethylene glycol (PEG) appear also to be of benefit against dietary tannins (D'Mello 1982; Pritchard et al. 1988; Kumar & Vaithiyanathan 1990; Murdiati et al. 1991). A dosage of 10 g or more per day of PEG added to the diet protected against the detrimental effects of oak forage (Silanikove et al. 1996). The value of alternative approaches based upon the use of more easily handled compounds such as magnesium oxide or calcium carbonate has not been evaluated. Keeping animals out of oak-infested areas or reducing grazing pressure to allow greater forage selectivity during the period of budding and early leaf growth is of course the most reliable method of preventing oak toxicosis.

References

Adzet T, Camarasa J: Pharmacokinetics of polyphenolic compounds. *Recent Adv Bot Hortic Pharmacol* 3;25–47, 1988

Ames BN: Dietary carcinogens and anticarcinogens. *Science* 221;1256–1264, 1983

Anderson GA, Mount ME, Vrins AA, Ziemer EL: Fatal acorn poisoning in the horse: pathologic findings and diagnostic considerations. *J Am Vet Med Assoc* 182;1105–1110, 1983

Archer S, Arnold A, Kullnig RK, Wylie DW: The enzymatic methylation of pyrogallol. *Arch Biochem* 87;153–154, 1960

Arhelger RB, Broom JS, Boler RK: Ultrastructural hepatic alterations following tannic acid administration to rabbits. *Am J Pathol* 46;409–434, 1965

Asquith TN, Uhlig J, Mehansho H, Putman L, Carlson DM, Butler L: Binding of condensed tannins to salivary proline-rich glycoproteins: the role of carbohydrate. *J Agric Food Chem* 35;331–334, 1987

Austin PJ, Suchar LA, Robbins CT, Hagerman AE: Tannin-binding proteins in saliva of deer and their absence in saliva of sheep and cattle. *J Chem Ecol* 15;1335–1347, 1989

Axelrod J, Laroche MJ: Inhibition of O-methylation of epinephrine and norepinephrine in vitro and in vivo. *Science* 130;800, 1959

Basden KW, Dalvi RR: Determination of total phenolics in acorns from different species of oak trees in conjunction with acorn poisoning of cattle. *Vet Hum Toxicol* 29;305–306, 1987

Batu A, Nadas UG, Gurel A: Experiments on the use of calcium hydroxide for treatment and prevention of oak leaf poisoning in rabbits. *Pendik Vet Bakteriyoloji Seroloji Enst Derg* 10;93–100, 1978 (*Vet Bull* 49:04119, 1979)

Begovic S, Duzic E, Sacirbegovic A, Tafro A: Examination of variation of tannase activity in ruminal content and mucosae of goats fed oak leaves and during ruminal application of 3 to 10% tannic acid. *Veterinaria (Yugoslavia)* 1978;445–457, 1978 (*Vet Bull* 50:04364, 1980)

Boler RK, Broom JS, Arhelger RB: Ultrastructural renal alterations following tannic acid administration to rabbits. *Am J Pathol* 49;15–32, 1966

Booth AN, Masri MS, Robbins DJ, Emerson OH, Jones FT, DeEds F: The metabolic fate of gallic acid and related compounds. *J Biol Chem* 234;3014–3016, 1959

Boyd EM, Bereczky K, Godi I: The acute toxicity of tannic acid administered intragastrically. *Can Med Assoc J* 92;1292–1297, 1965

Broughton JE: Acorn poisoning [letter]. *Vet Rec* 99;403–404, 1976

Butler LG: Antinutritional effects of condensed and hydrolysable tannins. In *Plant Polyphenols: Synthesis, Properties, and Significance,* Hemingway RW, Laks PE eds, Plenum Press, New York, pp693–698, 1992

Calder IC, Funder CC, Green CR, Ham KN, Tange JD: Comparative nephrotoxicity of aspirin and phenacetin derivatives. *Br Med J* 4;518–521, 1971

Calder IC, Williams PJ, Woods RA, Funder CC, Green CR, Ham KM, Tange JD: Nephrotoxicity and molecular structure. *Xenobiotica* 5;303–307, 1975

Calixto JB, Nicolau M, Rae GA: Pharmacological actions of tannic acid. 1. Effect on isolated smooth and cardiac muscle and on blood pressure. *Planta Med* 52;32–35, 1986

Cameron GR, Milton RF, Allenn JW: Toxicity of tannic acid. *Lancet* 2;179–186, 1943

Camp BJ, Steel E, Dollahite JW: Certain biochemical changes in blood and livers of rabbits fed oak tannin. *Am J Vet Res* 28;290–292, 1967

Cedarvall A, Johansson HE, Jonsson L: Acorn poisoning in cattle. *Nord Vet Med* 25;639–644, 1973

Chiesi M, Schwaller R: Inhibition of constitutive endothelial NO-synthase activity by tannin and quercetin. *Biochem Pharmacol* 49;495–501, 1995

Cockrell JM, Beasley JN: Renal damage to cattle during acorn poisoning. *Vet Med/Small Anim Clin* 74;82–85, 1979

Costantino L, Albasini A, Rastelli G, Benvenuti S: Activity of polyphenolic crude extracts as scavengers of superoxide radicals and inhibitors of xanthine oxidase. *Planta Med* 58;342–344, 1992

Daniels MG: Acorn poisoning [letter]. *Vet Rec* 99;465–466, 1976

Dawra RK, Makkar HPS, Singh B: Total phenolics, condensed tannins, and protein-precipitable phenolics in young and mature leaves of oak species. *J Agric Food Chem* 36;951–953, 1988

Dayton WA: *Important Western Browse Plants.* USDA Misc Publ 101, 1931

Dick BL, Urness PJ: Nutritional value of fresh Gambel oak browse for Spanish goats. *J Range Manage* 44;361–364, 1991

Distel RA, Provenza FD: Experience early in life affects voluntary intake of blackbrush by goats. *J Chem Ecol* 17;431–450, 1991

Dixon PM, McPherson EA, Rowland AC, Maclennan W: Acorn poisoning in cattle. *Vet Rec* 104;284–285, 1979

D'Mello JPF: Toxic factors in some tropical legumes. *World Rev Anim Prod* 18;41–46, 1982

Dollahite JW: Shin oak *(Quercus havardii)* poisoning in cattle. *Southwest Vet* 14;198–201, 1961

Dollahite JW, Pigeon RF, Camp BJ: The toxicity of gallic acid, pyrogallol, tannic acid, and *Quercus havardii* in the rabbit. *Am J Vet Res* 23;1264–1266, 1962

Dollahite JW, Housholder GT, Camp BJ: Calcium hydroxide, a possible antidote for shin oak *(Quercus havardii)* poisoning in cattle. *Southwest Vet* 16;115–117, 1963

Dollahite JW, Housholder GT, Camp BJ: Effect of calcium hydroxide on the toxicity of post oak *(Quercus stellata)* in calves. *J Am Vet Med Assoc* 148;908–912, 1966

Duncan CS: Oak leaf poisoning in two horses. *Cornell Vet* 51;159–162, 1961

Ebeling W: *Handbook of Indian Foods and Fibers of Arid America.* Univ California Press, Berkeley, pp210–219, 1986

Faeth SH: Do defoliation and subsequent phytochemical responses reduce future herbivory on oak trees? *J Chem Ecol* 18;915–925, 1992

Feeny P: Seasonal changes in oak leaf tannins and nutrients as a cause of spring feeding by moth caterpillars. *Ecology* 51;565–581, 1970

Filippich LJ, Zhu J, Oelrichs P, Alsalami MT, Doig AJ, Cao GR, English PB: Hepatotoxic and nephrotoxic principles in *Terminalia oblongata. Res Vet Sci* 50;170–177, 1991

Fleck DC, Layne JN: Variation in tannin activity of acorns of seven species of central Florida oaks. *J Chem Ecol* 16;2925–2934, 1990

Fowler ME, Richards WPC: Acorn poisoning in a cow and a sheep. *J Am Vet Med Assoc* 147;1215–1220, 1965

Garg SK, Makkar HPS, Nagal KB, Sharma SK, Wadhwa DR, Singh B: Oak *(Quercus incana)* leaf poisoning in cattle. *Vet Hum Toxicol* 34;161–164, 1992

Garner RJ: *Veterinary Toxicology,* 2d ed. Williams & Wilkins, Baltimore, 1961

Glick Z, Joslyn MA: Effect of tannic acid and related compounds on the absorption and utilization of proteins in the rat. *J Nutr* 100;516–520, 1970

Govindwar SP, Dalvi RR: Age-dependent toxicity of acorn extract in young and old male rats. *Vet Hum Toxicol* 32;23–26, 1990

Gupta MM, Dani HM: Characterization of tannic acid metabolites formed in vitro by rat liver microsomes and assay of their carcinogenicity by microsomal degranulation technique. *Chem Biol Interact* 63;39–45, 1987

Gustavson KH: Interaction of vegetable tannins with polyamides as proof of the dominant function of the peptide bond of collagen for its binding of tannins. *J Polym Sci* 7;317–324, 1954

Hagerman AE, Robbins CT, Weerasuriya Y, Wilson TC, Mcarthur C: Tannin chemistry in relation to digestion. *J Range Manage* 45;57–62, 1992

Harper KT, Ruyle GB, Rittenhouse LR: Toxicity problems associated with the grazing of oak in intermountain and Southwest USA. In *The Ecology and Economic Impact of Poisonous Plants on Livestock Production,* James LF, Ralphs MH, Nielsen DB eds, Westview Press, Boulder, Colo, pp197–206, 1988

Harris PA, Zboralske FF, Rambo ON, Margulis AR, Riegelman S: Toxicity studies on tannic acid administered by enema. 2. The colonic absorption and intraperitoneal toxicity of tannic acid on the colon and liver of rats. *Am J Roentgenol* 96;498–504, 1966

Haslam E: *Chemistry of the Vegetable Tannins.* Academic Press, London, pp10–12, 1966

Haslam E: *Plant Polyphenols: Vegetable Tannins Revisited.* Cambridge Univ Press, Cambridge, pp10–12, 1989

Hatano T, Yasuhara T, Yoshihara R, Agata I, Noro T, Okuda T: Effects of interaction of tannins with co-existing substances. 7. Inhibitory effects of tannins and related polyphenols on xanthine oxidase. *Chem Pharm Bull* 38;1224–1229, 1990

Hodnick WF, Kalyanaraman B, Pristos CA, Pardini RS: The production of hydroxyl and semiquinone free radicals during the autoxidation of redox active flavonoids. In *Oxygen Radicals in Biology and Medicine,* Simic MG, Taylor KA, Ward JF, von SonntAgric C eds, Plenum Press, New York, pp149–152, 1988

Holliman A: Acorn poisoning in ruminants. *Vet Rec* 116;546, 1985

Housholder GT, Dollahite JW: Some clinical biochemical changes in the blood serum of calves fed *Quercus havardii. Southwest Vet* 16;107–113, 1963

Ishimaru K, Nonaka G, Nishioka I: Phenolic glucoside gallates from *Quercus mongolica* and *Quercus acutissima. Phytochemistry* 26;1147–1152, 1987

Itakura Y, Habermehl G, Mebs D: Tannins occurring in the toxic Brazilian plant *Thiloa glaucocarpa. Toxicon* 25;1291–1300, 1987

Kasari TR, Pearson EG, Hultgren BD: Oak *(Quercus garryana)* poisoning of range cattle in southern Oregon. *Compend Contin Educ Food Anim* 8;F17–F24, 1986

Kashiwada Y, Nonaka G, Nishoka I, Chang JJ, Lee KH: Antitumor agents, 129: tannins and related compounds as selective cytotoxic agents. *J Nat Prod* 55;1033–1043, 1992

Kimura Y, Okuda H, Okuda T, Yoshida T, Hatano T, Arichi S: Studies on the activities of tannins and related compounds of medicinal plants and drugs. 2. Effects of various tannins and related compounds on adrenaline-induced lipolysis in fat cells. *Chem Pharm Bull* 31;2497–2500, 1983

Kinde H: A fatal case of oak poisoning in a double-wattled cassowary *(Casuarius casuarius). Avian Dis* 32;849–851, 1988

Kingsbury JM: *Poisonous Plants of the United States and Canada.* Prentice-Hall, Englewood Cliffs, NJ, 1964

Kleiner KW: Stability of phenolic and protein measures in excised oak foliage. *J Chem Ecol* 17;1243–1251, 1991

Kleinsasser O, Schroeder H-G, Mayer-Brix J: Preinvasive stages of adenocarcinoma of the nose after exposure to wood dust. *Eur Arch Otorhinolaryngol* 248;222–229, 1991

Konstantinov A, Ivanov A: Electron-microscopic changes of the liver and kidney in chickens treated with tannic acid. *Zentralbl Veterinaermed A* 20;426–433, 1973

Kumar R: *Prosopis cineraria* leaf tannins: their inhibitory effect upon ruminal cellulase and the recovery of inhibition by polyethylene glycol–4000. In *Plant Polyphenols: Synthesis, Properties, and Significance,* Hemingway RW, Laks PE eds, Plenum Press, New York, pp699–704, 1992

Kumar R, Vaithiyanathan S: Occurrence, nutritional significance, and effect on animal productivity of tannins in tree leaves. *Anim Feed Sci Technol* 30;21–38, 1990

Lohan OP, Lall D, Makkar HPS, Negi SS: Inhibition of rumen urease activity by tannins in oak leaves. *Indian J Anim Sci* 51;279–281, 1981

Lohan OP, Lall D, Vaid J, Negi SS: Utilization of oak tree *(Quercus incana)* fodder in cattle rations and fate of oak-leaf tannins in the ruminant system. *Indian J Anim Sci* 53;1057–1063, 1983

Maciejewska A, Wojtczak J, Bielichowska CG, Domanska A, Dutkiewicz J, Molocznik A: Biological effect of wood dust. *Med Pr* 44;277–288, 1993

Makkar HPS, Dawra RK, Singh B: Tannin levels in leaves of some oak species at different stages of maturity. *J Sci Food Agric* 54;513–519, 1991

Marklund S, Marklund G: Involvement of the superoxide anion radical in the autoxidation of pyrogallol and a convenient assay for superoxide dismutase. *Eur J Biochem* 47;469–474, 1979

Marsh CD: *Stock-Poisoning Plants of the Range.* USDA Bull 1245, p1245, 1924
Marsh CD, Clawson AB, Marsh H: *Oak-Leaf Poisoning of Domestic Animals.* USDA Bull 767, 1919
McLeod MN: Plant tannins—their role in forage quality. *Nutr Abstr Rev* 44;803–815, 1974
McLure RD, Lam CR, Romence H: Tannic acid treatment of burns: an obsequy. *Ann Surg* 120;387–398, 1944
Medsger OP: *Edible Wild Plants.* Macmillan, New York, 1939
Mehansho H, Butler LG, Carlson DM: Dietary tannins and salivary proline-rich proteins: interactions, induction, and defense mechanisms. *Annu Rev Nutr* 7;423–440, 1987
Merrill LB, Schuster JL: Grazing management practices affect livestock losses from poisonous plants. *J Range Manage* 31;351–354, 1978
Mitjavila S, Lacombe C, Carrera G, Derache R: Tannic acid and oxidized tannic acid on the functional state of rat intestinal epithelium. *J Nutr* 107;2113–2121, 1977
Murakami S, Muramatsu M, Otomo S: Inhibitory effect of tannic acid on gastric H^+,K^+-ATPase. *J Nat Prod* 55;513–516, 1992
Murdiati TB, McSweeny CS, Campbell RSF, Stoltz DS: Prevention of hydrolysable tannin toxicity in goats fed *Clidemia hirta* by calcium hydroxide supplementation. *J Appl Toxicol* 10;325–330, 1990
Murdiati TB, McSweeny CS, Lowry JB: Complexing of toxic hydrolysable tannins of yellow-wood *(Terminalia oblongata)* and harendong *(Clidemia hirta)* with reactive substances: an approach to preventing toxicity. *J Appl Toxicol* 11;333–338, 1991
Murdiati TB, McSweeny CS, Lowry JB: Metabolism in sheep of gallic acid, tannic acid, and hydolysable tannin from *Terminalia oblongata. Aust J Agric Res* 43;1307–1315, 1992
Nastis AS, Malechek JC: Digestion and utilization of nutrients in oak browse by goats. *J Anim Sci* 53;283–290, 1981
Nelson E, Zhou Z, Carmichael PL, Norpoth K, Fu J: Genotoxic effects of subacute treatments with wood dust extracts on the nasal epithelium of rats: assessment by the micronucleus and 32P-postlabelling. *Arch Toxicol* 67;586–589, 1993
Neser JA, Coetzer JAW, Boomker J, Cable H: Oak *(Quercus rubor)* poisoning in cattle. *J S Afr Vet Assoc* 53;151–155, 1982
Nishizawa M, Yamagishi T, Ohyama T: Inhibitory effect of gallotannins on the respiration of rat liver mitochondria. *Chem Pharm Bull* 31;2150–2152, 1983
Oelrichs PB, Pearce CM, Zhu J, Filippich LJ: Isolation and structure of terminalin, a toxic condensed tannin from *Terminalia oblongata. Nat Toxins* 2;144–150, 1994
Okuda T, Kimura Y, Yoshida T, Hatano T, Okuda H, Arichi S: Studies on the activities of tannins and related compounds from medicinal plants and drugs. 1. Inhibitory effects on lipid peroxidation in mitochondria and microsomes of liver. *Chem Pharm Bull* 31;1625–1631, 1983
Okuda T, Yoshida T, Hatano T: Pharmacologically active tannins isolated from medicinal plants. In *Plant Polyphenols: Synthesis, Properties, and Significance,* Hemingway RW, Laks PE eds, Plenum Press, New York, pp539–569, 1992
Oler A, Neal MW, Mitchell EK: Tannic acid: acute hepatotoxicity following administration by feeding tube. *Food Cosmet Toxicol* 14;565–569, 1976
Ostrowski SR, Smith BP, Spier SJ, Norman BB, Oliver MN: Compensatory weight gain in steers recovered from oak bud toxicosis. *J Am Vet Med Assoc* 195;481–484, 1989
Panciera RJ: Oak poisoning in cattle. In *Effects of Poisonous Plants on Livestock,* Keeler RF, van Kampen KR, James LF eds, Academic Press, New York, pp499–506, 1978
Plumlee K: Comparison of disease in calves dosed orally with oak or commercial tannic acid. In *Toxic Plants and Other Natural Toxicants,* Garland T, Barr AC eds, CAB International, New York, pp549–553, 1998
Potter DK, Fuller HL: Metabolic fate of dietary tannins in chickens. *J Nutr* 96;187–191, 1968
Pritchard DA, Stocks DC, O'Sullivan BM, Martin PR, Hurwood IS, O'Rourke PK: The

effect of polyethylene glycol (PEG) on wool growth and live weight of sheep consuming a mulga *(Acacia aneura)* diet. *Proc Aust Soc Anim Prod* 17;290–293, 1988

Provenza FD, Malechek JC: Diet selection by domestic goats in relation to blackbrush twig chemistry. *J Appl Ecol* 21;831–841, 1984

Rambo ON, Zboralske FF, Harris PA, Riegelman S, Margulis AR: Toxicity studies on tannic acid administered by enema. 1. Effect of enema-administered tannic acid on the colon and liver of rats. *Am J Roentgenol* 96;488–497, 1966

Rayudu GVN, Kadirvel R, Vohra P, Kratzer FH: Toxicity of tannic acid and its metabolites for chickens. *Poult Sci* 49;957–960, 1970

Reed JD: Nutritional toxicology of tannins and related polyphenols in forage legumes. *J Anim Sci* 73;1516–1528, 1995

Robbins CT, Hanley TA, Hagerman AE, Hjeljord O, Baker DL, Schwartz CC, Mautz WW: Role of tannins in defending plants against ruminants: reduction in protein availability. *Ecology* 68;98–107, 1987a

Robbins CT, Mole S, Hagerman AE, Hanley TA: Role of tannins in defending plants against ruminants: reduction in dry matter digestion. *Ecology* 68;1606–1615, 1987b

Robinson HJ, Graessle OE: Toxicity of tannic acid. *J Pharmacol Exp Ther* 77;63–69, 1943

Ruyle GB, Grumbles MJ, Murphy RC, Cline RC: Oak consumption by cattle in Arizona. *Rangelands* 8;124–126, 1986

Salunkhe DK, Chavan JK, Kadam SS: *Dietary Tannins: Consequences and Remedies.* CRC Press, Boca Raton, Fla, 1990

Sandusky GE, Fosnaugh CJ, Smith JB, Mohan R: Oak poisoning of cattle in Ohio. *J Am Vet Med Assoc* 171;627–629, 1977

Shi ZC: Identification of the phenolic substances in bovine urine associated with oak leaf poisoning. *Res Vet Sci* 45;152–155, 1988

Shi ZC, Hong ZL: Theory of tannin bioactivation and its application. 2. Confirmation of tannin bioactivation hypothesis by experiments [abstract]. 10th World Congress on Animal, Plant and Microbial Toxins, Singapore, 1991. *Toxicon* 30;550, 1992

Silanikove N, Gilboa N, Nir I, Perevolotsky A, Nitsan Z: Effect of a daily supplementation of polyethylene glycol on intake and digestion of tannin-containing leaves *(Quercus calliprinos, Pistacia lentiscus,* and *Ceratonia siliqua)* by goats. *J Agric Food Chem* 44;199–205, 1996

Singleton VL, Kratzer FH: Plant phenolics. In *Toxicants Occurring Naturally in Foods,* 2d ed, Comm Food Protection, National Research Council, National Academy of Sciences, Washington, DC, pp309–345, 1973

Spier SJ, Smith BP, Seawright AA, Norman BB, Ostrowski SR, Oliver MN: Oak toxicosis in cattle in northern California: clinical and pathologic findings. *J Am Vet Med Assoc* 191;958–964, 1987

Stober M, Ziegler HP, von Benten K: Acorn poisoning in cattle. *Bovine Pract* 11;36–41, 1976

Takahashi RN, deLima TCM, Morato GS: Pharmacological actions of tannic acid. 2. Evaluation of CNS activity in animals. *Planta Med* 52;272–275, 1986

Tompsett SL: Polyhydroxy (catecholic) phenolic acids—the formation of *m*-hydroxy-and methoxy-derivatives in man. *J Pharm Pharmacol* 11;32–35, 1959

Tripathy KC, Sahu BK, Panda NC, Nayak BC: Toxicity of tannic acid in goats. *Indian J Anim Sci* 54;1091–1093, 1984

Villena F, Pfister JA: Sand shinnery oak as forage for Angora and Spanish goats. *J Range Manage* 43;116–122, 1990

Warren CGB, Vaughan SM: Acorn poisoning [letter]. *Vet Rec* 116;82, 1985

Welsch CA, Lachance PA, Wasserman BP: Effects of native and oxidized phenolic compounds on sucrose activity in rat brush border membrane vesicles. *J Nutr* 119;1737–1740, 1989

Wilkins WM, Cranwell MP: Beechmast poisoning in ponies [letter]. *Vet Rec* 127;435, 1990

Williams RT: The fate of phenolic compounds in the body. In *The Pharmacology of Plant Phenolics,* Fairbairn JW ed, Academic Press, London, pp13–26, 1959

Wiseman A, Thompson H: Acorn poisoning [letter]. *Vet Rec* 115;605, 1984

Yokozawa T, Fujioka K, Oura H, Nonaka G I, Nishioka I: Effects of rhubarb tannins on uremic toxins. *Nephron* 58;155–160, 1991

Zhu J, Filippich LJ: Acute intra-abomasal toxicity of tannic acid in sheep. *Vet Hum Toxicol* 37;50–54, 1995

Zhu J, Filippich LJ, Alsalami MT: Tannic acid intoxication in sheep and mice. *Res Vet Sci* 53;280–292, 1992

Zhu J, Filippich LJ, Ng J: Rumen involvement in sheep tannic acid metabolism. *Vet Hum Toxicol* 37;436–440, 1995

Fumariaceae L. 39

With greatest diversity in temperate Eurasia, the Fumariaceae, or fumitory family, comprises 19 genera and 400–450 species. Taxonomists differ in their opinions as to whether it should be maintained as a distinct taxon or combined with the Papaveraceae and treated as a subfamily (Ownbey 1947). The two families are closely related. They both, for example, are particularly rich in similar isoquinoline alkaloids, compounds of considerable importance because of their medicinal promise. In general, European taxonomists combine the two families, whereas Americans, with a few exceptions, keep them separate on the basis of their differences in floral symmetry, sap chemistry, and stamen number and fusion. The family is of little economic importance except for a few ornamentals and a few agricultural weeds. In North America, 4 genera and 23 species are present (Stern 1997). Two genera are of toxicologic interest.

- Fumitory Family

  Corydalis
  Dicentra
  Fumaria

Plants herbs; annuals or biennials or perennials; from rhizomes or tubers or bulblets or taproots; caulescent or acaulescent; strongly aromatic; sap watery. **Stems** brittle; erect or ascending or decumbent. **Leaves** 1- or 2-pinnately compound or rarely simple; alternate; cauline and/or basal; petiolate; terminal leaflet present; venation pinnate; surfaces glabrous or with crystalline vesicles; margins pinnately cleft; stipules absent. **Inflorescences** racemes or panicles or corymbs; terminal or axillary. **Flowers** perfect; chasmogamous or rarely cleistogamous; perianths in 2-series. **Sepals** 2; caducous or persistent; free. **Corollas** bilaterally symmetrical in 1- or 2-planes. **Petals** 4; in 2 whorls, of 2 forms; the outer 2 free, 1 or both gibbous or spurred, keeled; the inner 2 fused at apices and covering stigmas; yellow or white or pinkish or purple-red. **Stamens** 6; free or fused by filaments; in fascicles; anthers of 2 forms, bilocular and unilocular, filaments spurred. **Pistils** 1; compound, carpels 2; stigmas 1 or 2, 2- or 3-lobed, horned; styles 1; ovaries superior; locules 1; placentation parietal. **Fruits** capsules; dehiscent or indehiscent. **Seeds** 1 to numerous; small.

- herbs; leaves alternate; inflorescences terminal or axillary; petals 4, in 2 whorls; fruits capsules

Genera of the Fumariaceae are rich in isoquinoline alkaloids and have long been used in folk medicine; e.g., drug fumitory (European *Fumaria officinalis*) is well known for its medicinal properties. A single species may contain a dozen or more alkaloids belonging to one or more of six classes or types. Playing a role in the medicinal and toxic effects, these six classes are known as the aporphines, cularines, phthalides (phthalideisoquinolines), benzophenanthridines, protoberberines, and protopines. The aporphines—bulbocapnine, corydine, dicentrine, glaucine, isoboldine, isocorydine, magnaflorine—bind stereospecifically to dopamine receptors and for the most part are inhibitory (Seeman 1981; Neumeyer 1985). They are most common in *Dicentra,* although they may be found in all of the genera. The effects of the individual alkaloids vary from cataleptic rigidity (especially at low doses) to strychnine-like seizures with tremors and twitching (Waud 1934, 1935, 1936). Bulbocapnine is especially noteworthy as a cause of the cataleptoid state (Katzenelbogen & Meehan 1933; Seeman 1981). Because of the variety of dopaminergic receptors, it is difficult

Fumaria officinalis

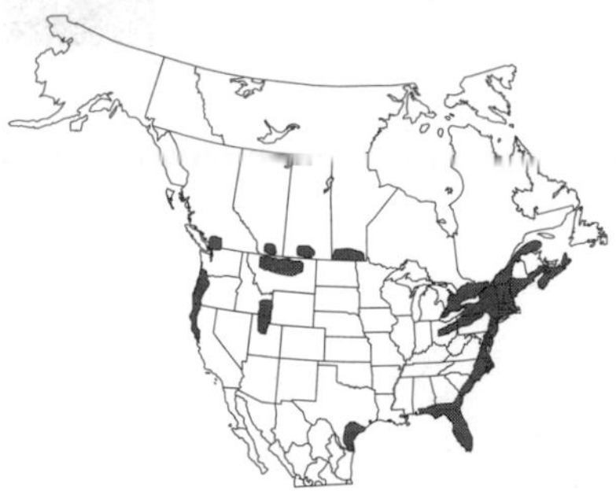

Fumaria officinalis

- isoquinoline alkaloids of 6 classes:
  aporphines
  cularines
  phthalides
  benzophenanthridines
  protoberberines
  protopines
- aporphines, dopamine receptor, mainly inhibitory; cularines of low potency

to predict the specific effects of the various alkaloids and their differing actions. Cularines, which are common in *Dicentra,* are of low potency but with high dosage may cause lethargy and sometimes clonic seizures (Reynolds 1940).

The phthalides (phthalideisoquinolines) have been of particular interest because some of the isomers of bicuculline, corlumine, hydrastine, and isocoryne are competitive inhibitors of GABAa binding (Curtis et al. 1970; Johnston 1978; Chernevskaja et al. 1990; Huang & Johnston 1990). The 1*S*, 9*R* configuration provides the appropriate intramolecular binding sites to prevent GABA activation of chloride channels, especially at the low-affinity sites. Bicuculline has been particularly widely used for molecular studies, but it is not necessarily the most potent (Huang & Johnston 1990). All act as convulsants, causing strychnine-like effects but not invariably with increased sensitivity to touch and sound (Welch & Hendersen 1934). They are similar in action but bind to a site different from that for noncompetitive GABA inhibitors such as picrotoxin, chlorinated hydrocarbon insecticides, and type II pyrethroids (Olsen 1982; Eldefrawi & Eldefrawi 1987). These alkaloids occur most frequently in *Corydalis.* Because of the stereospecific requirements and the presence of two asymmetric carbons, the spectrum of biological effects may vary considerably among some of the members of this group. This is illustrated by the contrast between adlumine and its stereoisomer corlumine. Adlumine has high cardiovascular and low convulsant potency; corlumine has the reverse (Rice 1938).

Isoboldine (an aporphine)

Corydine

Dopamine

- phthalides, potent inhibitors of GABAa binding; convulsants

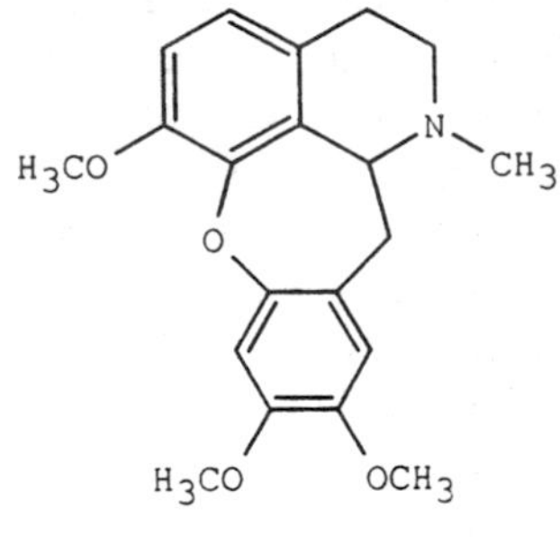

Cularine

Bicuculline (a phthalideisoquinilone)

- benzophenanthridines, low potency
  protoberberines, cardiovascular, and neurologic actions
  protopines similar in effect

Benzophenanthridines are of little toxicologic importance, except for sanguinarine, which causes an increase in intraocular pressure/glaucoma and is reportedly passed in the milk (Hakim et al. 1961; Lindner 1985). The mechanism of action is unknown, although there are effects on cardiac Na^+,K^+-ATPase, resulting in positive inotropic actions.

The protoberberines—capaurine, capauridine, scoulerine, palmatine, and tetrahydropalmatine—produce cardiovascular effects, increasing strength of contraction and decreasing heart rate and peripheral resistance (Lindner 1985). These alkaloids also have analgesic and hypnotic effects, especially palmatine and tetrahydropalmatine

(Hsieh et al. 1989). Severe depression of the central nervous and cardiovascular systems of life-threatening magnitude is reported for tetrahydropalmatine (Feldhaus et al. 1993). The decreased heart rate and blood pressure appear to be due to dopamine D_2 receptor antagonism (Chueh et al. 1995).

Protopines are similar in effect, although they have additional acetylcholine inhibitory actions similar to but much less potent than those of atropine (Shamma 1972; Zong et al. 1986).

Many of the effects of these plants and alkaloids—i.e., rigidity, difficulty in initiating movement, and postural abnormalities—are suggestive of a basal ganglia disease such as Parkinson's disease (Cote & Crutcher 1991). Involvement of the extrapyramidal motor system would be expected with dopaminergic antagonists and may also be accentuated by GABA inhibitors at low dosage. At higher dosage the poorly absorbed but potent convulsant GABA inhibitors may overwhelm the system.

Other effects, such as antithrombotic actions and lipid peroxidation inhibition, have been described, but these are not clearly produced by a specific alkaloid type (Matsuda et al. 1988; Martinez et al. 1992).

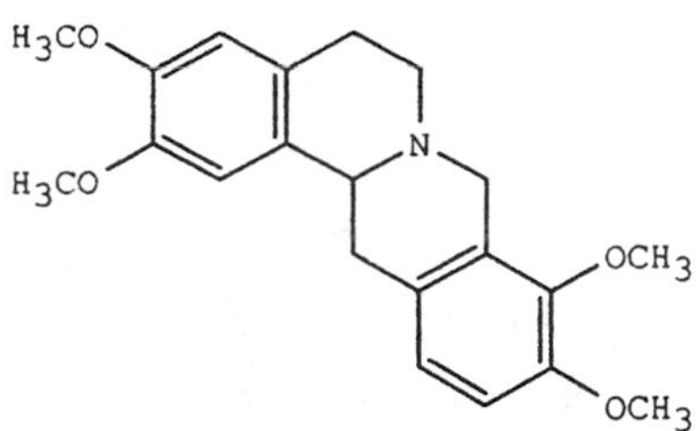

Tetrahydropalmatine

Protopine

Corydalis DC.

Taxonomy and Morphology—Comprising 100–300 species worldwide, *Corydalis* is represented in North America by only 10 species. Because of the resemblance of its flowers to the head of a crested lark, its name is derived from *korydallis,* the Greek name for the bird. The genus name *Capnoides* was used for the taxon in earlier publications. A few species are grown as ornamentals, primarily in rock and wall gardens, for their lacy, fernlike foliage and tubular, spurred flowers. The following are representative species of toxicologic interest:

■ *C. aurea*
C. caseana
C. curvisiliqua
C. flavula
C. lutea
C. micrantha
C. sempervirens

corydalis
fitweed
fumewort
golden fumeroot
golden smoke
scrambled eggs
smallflower
yellow fumeroot
yellow harlequin

C. aurea Willd.	golden corydalis, scrambled eggs, golden smoke, golden fumeroot
C. caseana A.Gray	fitweed
C. curvisiliqua Engelm.	curvepod corydalis
C. flavula (Raf.) DC.	yellow corydalis, yellow harlequin, yellow fumeroot, pale corydalis
C. lutea (L.) DC.	yellow corydalis
C. micrantha (Engelm. ex A.Gray) A.Gray	smallflower, southern corydalis, slender fumewort, slender corydalis
C. sempervirens (L.) Pers.	pink corydalis, rock fumewort

■ caulescent herbs; leaves pinnately compound; petals yellow, white, or pinkish, with median keel

Plants caulescent. **Stems** erect or ascending or decumbent; 5–150 cm tall; solitary or clustered; branched or not branched. **Leaves** 1- or 2-pinnately compound or rarely simple; cauline and/or basal; glabrous or with crystalline vesicles. **Inflorescences** racemes or panicles; terminal or axillary. **Flowers** bilaterally symmetrical in 1-plane. **Sepals** caducous or persistent. **Petals** yellow or white or pinkish; outer dissimilar, each with median adaxial crest or keel, sometimes with marginal distal wing, 1 basally spurred, the other not spurred or gibbous. **Capsules** dehiscent; 2-valved. **Seeds** 3 to numerous; reniform to subglobose.

1. Plants perennials.
   2. Petals yellow . . . . *C. lutea*
   2. Petals light pink to white with reddish purple apices . . . . *C. caseana*

Corydalis aurea

Corydalis caseana

1. Plants annuals or biennials.
   3. Petals pink or pink purple with yellowish apices . . . . . . *C. sempervirens*
   3. Petals yellow.
      4. Spurred petals 7–9 mm long. Spurs curved; 1–2 mm long. Fruiting pedicels 7–15 mm long . . . . *C. flavula*
      4. Spurred petals 10–22 mm long. Spurs straight; 4–8 mm long. Fruiting pedicels 1–5 mm long.
         5. Seeds 1.5 mm in diameter. Spurred petals 11–15 mm long . . . . *C. micrantha*
         5. Seeds 2 mm in diameter. Spurred petals 13–18 mm long.
            6. Seed coats roughened under magnification . . . . *C. curvisiliqua*
            6. Seed coats smooth under magnification . . . . *C. aurea*

Corydalis flavula

Corydalis micrantha

■ variety of habitats

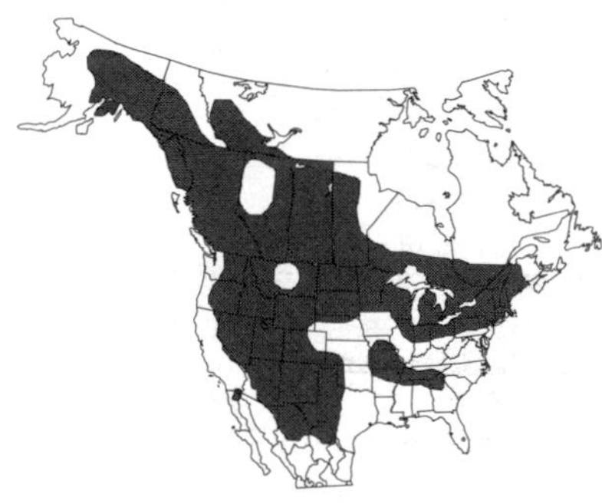

Corydalis aurea

Corydalis caseana

Distribution and Habitat—Most species of *Corydalis* in North America have fairly restricted distributions; *C. aurea* and *C. sempervirens* have the broadest. Occurring only sporadically, *C. caseana* is a western species with populations as far east as the Rocky Mountains. *Corydalis curvisiliqua* occurs in the southern Great Plains, whereas *C. micrantha* is a species of the Midwest and southeastern coastal states. *Corydalis flavula* is present throughout the eastern half of the continent. Cultivated as an ornamental, Eurasian *C. lutea* is reported to be adventive in Oregon and New York. A variety of soil types and habitats are occupied by species of the genus. Populations are typically small, albeit sometimes large ones are encountered in disturbed sites.

Disease Problems—Species of *Corydalis* have long been used for medicinal and recreational uses. Extracts from *C. cava* are employed as herbal sedatives, and those from *C. lutea* have antispasmodic activity on intestinal smooth muscle (Schafer et al. 1995; Boegge et al. 1996). Species indigenous to North America were not perceived as hazardous until the 1900s. The neurologic problem appeared mainly in conjunction with grazing of the mountain rangelands of the West. Losses were never extensive; rather there were sporadic, sudden deaths of only a few cattle or sheep at a time. The sparse, scattered populations of plants both limited the losses and facilitated confirmation of the association between animal losses and species occurrence (Fleming et al. 1931). The most toxicologically troublesome, *C. caseana* is quite succulent, with a moisture content in excess of 95%. It is apparently relished by livestock and grazed extensively when populations are encountered. Plants are hazardous throughout the grazing season, with approximately 3–4% b.w. wet weight causing severe disease. *Corydalis aureus* is reported to be a particular problem in some areas in years with heavy spring rains or snows.

- neurotoxic, cattle and sheep
- numerous isoquinoline alkaloids; effects mainly due to GABA antagonism by bicuculline and other phthalides

Corydalis curvisiliqua

Disease Genesis—As is apparent from table 39.1, the protoberberines exhibit the greatest diversity of types present in most species. Their role as toxicants, however, is not fully understood, because neurologic effects are not generally regarded as characteristic of them. Phthalideisoquinolines, or phthalides, are present in most species and, as GABA antagonists, could account for the signs produced, perhaps acting in concert with other types, including the aporphines. That the alkaloids are responsible for the disease is shown by the similarities among signs produced by the whole plant or crude alkaloid extracts of *C. caseana* (Miller 1931).

Table 39.1. Isoquinoline alkaloids present in species of *Corydalis*

Species	Aporphines	Phthalides	Protoberberines	Others
C. aurea	isoboldine	bicuculline cordrastine	capauridine capaurine corydaline dehydrocorydaline tetrahydropalmatine	protopine allocryptopine
C. caseana		bicuculline	caseadine caseamine caseanadine corypalmine isocorypalmine scoulerine tetrahydropalmatine	protopine allocryptopine
C. micrantha			capauridine capaurine scoularine tetrahydropalmatine	protopine
C. sempervirens		adlumine bicuculline capnoidine	berberine coptisine	adlumiceine adlumidiceine cryptopine protopine oxysanguinarine
C. ochroleuca	glaucine isocorydine	adlumine bicuculline	corypalmine isocorypalmine scoulerine sinactine tetrahydropalmatine	fumaramine ochrobirin protopine

Source: Preininger 1986.

Corydalis flavula

Corydalis micrantha

Clinical Signs—Within several hours of ingestion of *Corydalis,* signs of intoxication appear and within 1 hour become severe. The animals initially appear depressed or dull, and then the respiratory rate becomes very high (greater than 100/min). Heart rate also increases but not as markedly. These changes are quickly followed by the onset of neurologic signs: apparent disorientation; muscular twitching, especially of the lips, face, and eyelids; staggering; and collapse with seizures. While standing or lying down, the animal may bite or chew at objects such as sticks or the soil. Seizures, which may be precipitated by environmental stimuli, cause the animal to fall and

- depression, increased respiratory and heart rates; twitching of lips, face, eyelids; staggering, collapse, seizures

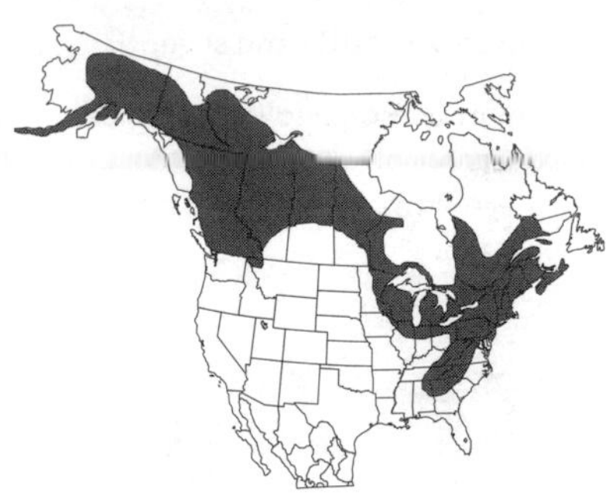
Corydalis sempervirens

display periodic episodes of paddling movements, which terminate in rigidity. As the seizures worsen, the respiration and heart rates become quite low. With *C. aureus,* a peculiar extreme stiffness, especially of the thoracic limbs, has been reported (Smith & Lewis 1990).

- no lesions

Pathology—Because of the rapid progress of the disease, there are few distinctive pathologic findings, especially if the animal dies within a few hours. There may be a few scattered splotchy hemorrhages.

- sedation, activated charcoal

Treatment—Treatment involves the use of activated charcoal to limit absorption of the alkaloids, and sedation to counteract the stiffness or seizures.

Dicentra Bernh.

- *D. canadensis*
  D. cucullaria
  D. eximia
  D. formosa
  D. ochroleuca
  D. spectabilis
  D. uniflora

bleeding-heart
butterfly banners
colicweed
Dutchman's-breeches
kitten-breeches
pearl harlequin
soldier's-cap
squirrel corn
stagger
staggerweed
steer's-head
turkey corn
turkey pea
white eardrops
white hearts

Taxonomy and Morphology—Comprising 15–20 species native to both North America and Asia, *Dicentra* is a popular ornamental, with plants grown for their showy flowers and lacy, fernlike, often glaucous foliage. Its name is derived from the Greek roots *dis* and *kentron,* meaning "twice" and "spur," an obvious allusion to the distinctive spurred, heart-shaped flowers. The genus name *Bikukulla* was used for the taxon in older publications. In North America, 9 species are present (Stern 1993). Of toxicologic interest are the following (see plate 39.1):

D. canadensis (Goldie) Walp.	squirrel corn, turkey corn, turkey pea, staggerweed, little blue stagger, trembling stagger, colicweed
D. cucullaria (L.) Bernh.	Dutchman's-breeches, soldier's-cap, white hearts, butterfly banners, little staggerweed, pearl harlequin, kitten-breeches, wild bleeding-heart
D. eximia (Ker Gawl.) Torr.	fringed bleeding-heart, staggerweed, wild bleeding-heart, turkey corn
D. formosa (Haw.) Walp.	western bleeding-heart, Pacific bleeding-heart
D. ochroleuca Engelm.	white eardrops
D. spectabilis (L.) Lem.	bleeding heart
D. uniflora Kellogg	steer's-head

- erect herbs; leaves 1- or 2-pinnately compound; flowers yellow, white, or purple-red

Plants acaulescent or caulescent. **Stems** erect; 10–40 cm tall; solitary or clustered; branched or not branched. **Leaves** 1- or 2-pinnately compound; cauline and/or basal; glabrous; often glaucous. **Inflorescences** of solitary flowers or racemes or panicles or corymbs; terminal or axillary or extra-axillary. **Flowers** bilaterally symmetrical in 2-planes; cordate or oblong in outline. **Sepals** caducous. **Petals** yellow or white or purple-red; outer similar, spurred or saccate. **Capsules** indehiscent or dehiscent; 2-valved. **Seeds** 3 to numerous; reniform; tuberculate or reticulate.

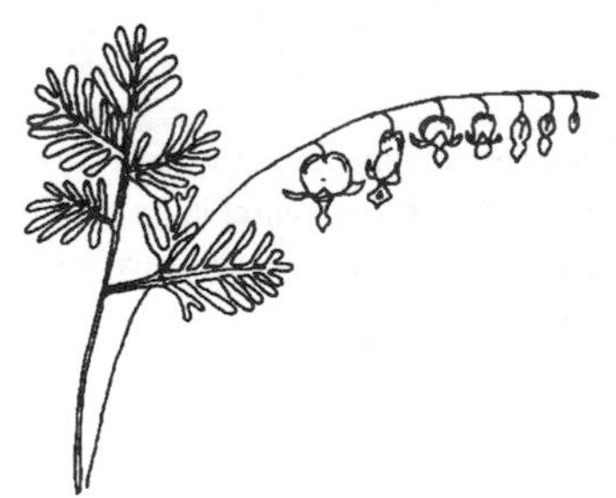
Dicentra canadensis

1. Plants caulescent.
   2. Flowers erect. Petals gray- or yellow-white.
      Seed coats tuberculate . . . . *D. ochroleuca*
   2. Flowers pendulous. Petals rose-purple to pink.
      Seed coats reticulate . . . . *D. spectabilis*
1. Plants acaulescent.

3. Rhizomes present. Bulblets and tubers absent.
   4. Reflexed portions of outer petals 4–8 mm long. Plants of eastern United States . . . . *D. eximia*
   4. Reflexed portions of outer petals 2–5 mm long. Plants of Pacific Coast . . . . *D. formosa*
3. Rhizomes absent. Bulblets and tubers present.
   5. Inflorescences solitary flowers. Scapes equal to or shorter than leaves . . . . *D. uniflora*
   5. Inflorescences racemes. Scapes longer than leaves.
      6. Flowers cordate-triangular. Spurs divergent; broadly acute. Bulblets white or pink; ovoid . . . . . . *D. cucullaria*
      6. Flowers cordate-ovate. Spurs not divergent; rounded. Bulblets yellow; globose . . . . *D. canadensis*

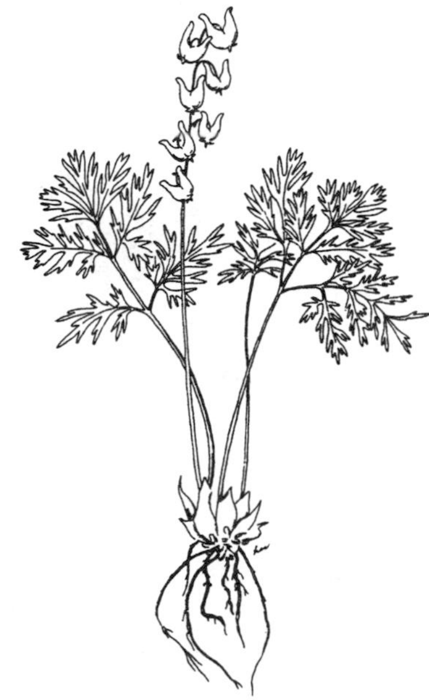

Dicentra cucullaria

Distribution and Habitat—*Dicentra* is mainly a woodland genus, preferring moist, rich soils of stream banks and floodplains. Species are present across the continent. *Dicentra formosa, D. ochroleuca,* and *D. uniflora* occur in the far West, whereas *D. canadensis, D. eximia,* and *D. cucullaria* are species of the East. Disjunct populations of the last taxon occur in eastern Washington and Oregon and adjacent Idaho. Old World *D. spectabilis* is widely cultivated and occasionally escapes but does not appear to establish itself (Stern 1961, 1997).

■ rich, moist woodlands

■ rare cause of neurologic effects

Disease Problems—Species of *Dicentra* were suspected to be toxic in the late 1800s and eventually shown to be the cause of livestock losses in Virginia and Indiana (Hansen 1924; Massey 1963). Although the plants are succulent and readily eaten, populations occur only sporadically, and the plants seldom are sufficiently abundant to cause disease. As a result, disease is rare and deaths most unusual. As some of its common names imply, *Dicentra* causes a short-term disease problem of trembling, staggering, and seizures (Harvey et al. 1944). *Dicentra cucullaria* seems to be the most toxic, requiring approximately 1.6 kg wet weight of plant per 100 kg (1.6%) b.w. to produce distinct signs of intoxication (Black et al. 1923). The other species seem to be less toxic (Eggleston et al. 1929; Black et al. 1930). All animal species are probably at risk, but cattle and sheep are more likely to be affected.

Dicentra canadensis

Disease Genesis—Like *Corydalis,* the species of *Dicentra* contain an array of isoquinoline alkaloids (table 39.2). The variety of toxic effects presumably depend upon the array of alkaloids present in the different species. Although the cularines are common to most species, they are generally considered not to be of toxicologic importance. The aporphines, which are dopamine antagonists, are more important, but they are not common to all the toxic species. The relative paucity of potent isoquinoline types is probably reflected in the lesser toxicity of this genus as compared with *Corydalis.* Only the Asiatic species *D. peregrina* contains phthalideisoquinolines.

■ numerous isoquinoline alkaloids; mainly aporphines, dopamine antagonism

Clinical Signs, Pathology, and Treatment—Signs produced by *Dicentra* are mainly referable to the nervous system, the most prominent being trembling and staggering. The animal may be quite active, staggering around with its head held high and "spitting" frothy saliva and regurgitated ingesta. The trembling may quickly become more violent and lead to lateral recumbency and seizures, with the head held back and the legs rigidly extended. The entire episode may last less than 30 minutes in some cases; return to apparent normalcy is relatively rapid. There may be several of these episodes before the animal recovers completely or in very rare cases dies. Generally the animal is not able to eat enough of the succulent plants to cause severe signs.

■ trembling staggering, seizures

sedation

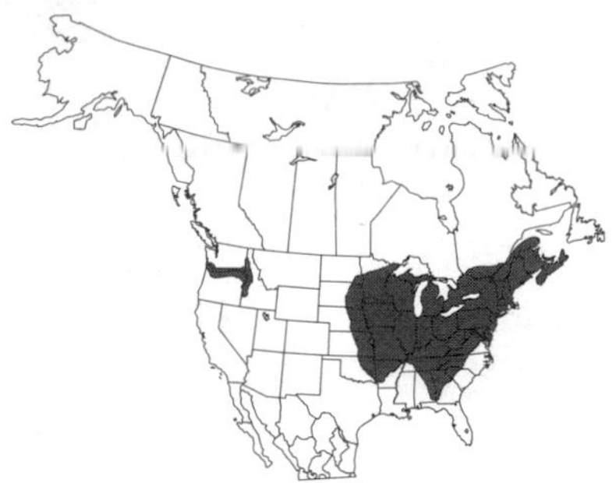

Dicentra cucullaria

Dicentra eximia

Dicentra formosa

Table 39.2. Isoquinoline alkaloids present in species of *Dicentra*

Species	Aporphines	Cularines	Protoberberines	Others
D. canadensis		cancentrine dehydrocancentrine		
D. cucullaria		cularidine cularine		ochotensine
D. eximia	corydine dicentrine glaucine	cularimine cularine		norprotosinomenine
D. formosa		cularine		
D. spectabilis	corydine		cheilanthifoline coptisine scoulerine	protopine chelilutine chelirubine chelerythrine sanguinarine dehydrosanguinarine

Source: Preininger 1986.

References

Black OF, Eggleston WW, Kelly JW, Turner HC: Poisonous properties of *Bikukulla cucullaria* (Dutchman's-breeches) and *B. canadensis* (squirrel-corn) *J Agric Res* 23;69–80, 1923

Black OF, Eggleston WW, Kelly JW: Toxicity of *Bikukulla formosa* (western bleeding-heart). *J Agric Res* 40;917–920, 1930

Boegge SC, Kesper S, Verspohl EJ, Nahrstedt A: Reduction of ACh-induced contraction of rat isolated ileum by coptisine, (+)-caffeoylmalic acid, *Chelidonium majus,* and *Corydalis lutea* extracts. *Planta Med* 62;173–174, 1996

Chernevskaja NI, Krishtal OA, Valeyev AY: Inhibitions of the GABA-induced currents of rat neurons by the alkaloid isocoryne from the plant *Corydalis pseudoadunca. Toxicon* 28;727–730, 1990

Chueh FY, Hsieh MT, Chen CF, Lin MT: DL-Tetrahydropalmitine-produced hypotension and bradycardia in rats through the inhibition of central nervous dopaminergic mechanisms. *Pharmacology* 51;237–244, 1995

Cote L, Crutcher MD: Basal ganglia. In *Principles of Neural Science,* 3d ed, Kandel ER, Schwartz JH, Jessell TM eds, Elsevier, New York, pp647–659, 1991

Curtis DR, Duggan AW, Felix D, Johnston GAR: GABA, bicuculline, and central inhibition. *Nature* 226;1222–1224, 1970

Eggleston WW, Black OF, Kelly JW: A botanical and chemical study of *Bikukulla exima,* with a key to North American species of *Bikukulla. J Agric Res* 39;477–481, 1929

Eldefrawi AT, Eldefrawi ME: Receptors for γ-aminobutyric acid and voltage-dependent chloride channels as targets for drugs and toxicants. *FASEB J* 1:262–271, 1987

Feldhaus KM, Horowitz RS, Dart RC, Brent J, Gomez H, Moore L, Fulton B, Stermitz FR, Beck JJ: Life-threatening toxicity from tetrahydropalmatine (THP) in an herbal medicine product. *Vet Hum Toxicol* 35;329 (abstr 51), 1993

Fleming CE, Miller MR, Vawter LR: *The Fitweed* (Capnoides caseana). Nev Agric Res Stn Bull 121, 1931

Hakim SAE, Mijovic V, Walker J: Distribution of certain poppy-fumaria alkaloids and a possible link with the incidence of glaucoma. *Nature* 189;198–201, 1961

Hansen AA: The poisonous plant situation in Indiana—3. *J Am Vet Med Assoc* 66;351–362, 1924

Harvey RB, Larson AH, Landon RH, Boyd WL, Erickson LC: *Weeds Poisonous to Livestock.* Minn Agric Exp Stn Bull 388, 1944

Hsieh M-T, Hung H-J, Ysai H-Y, Chiu N-Y, Hwang K-F, Lai J-S, Chen H-C, Wu T-S: The effects of the bioactive constituents of *Corydalis* tuber on the central nevous system. *Chin Pharm J* 41;227–237, 1989

Huang J-H, Johnston GAR: (+)-Hydrastine, a potent competitive antagonist at mammalian GABAa receptors. *Br J Pharmacol* 99;727–730, 1990

Johnston GAR: Neuropharmacology of amino acid inhibitory transmitters. *Annu Rev Pharmacol Toxicol* 18;269–289, 1978

Katzenelbogen S, Meehan MC: The chemistry of the blood and the cerebrospinal fluid, with special reference to calcium, in the cataleptoid state induced by bulbocapnine. *J Pharmacol Exp Ther* 47;131–139, 1933

Lindner E: Structure activities and pharmacological properties of the opium alkaloids. In *Chemistry and Biology of Isoquinoline Alkaloids,* Phillipson JD, Roberts MF, Zenk MH eds, Springer-Verlag, Berlin, pp38–46, 1985

Martinez LA, Rios JL, Paya M, Alcaraz MJ: Inhibition of nonenzymic lipid peroxidation by benzylisoquinoline alkaloids. *Free Radic Biol Med* 12;287–292, 1992

Massey AB: *Poisonous Plants.* Va Polytech Inst Agric Ext Serv Bull 222;9–10, 1963 (4th rev)

Matsuda H, Shiomoto H, Naruto S, Namba K, Kubo M: Anti-thrombic action of methanol extract and alkaloidal components from *Corydalis* tuber. *Planta Med* 54;27–33, 1988

Miller MR: The toxicity of *Corydalis caseana. J Agric Res* 42;239–243, 1931

Neumeyer JL: Synthesis and structure-activity relationships of aporphines as dopamine receptor agonists and antagonists. In *Chemistry and Biology of Isoquinoline Alkaloids,* Phillipson JD, Roberts MF, Zenk MH eds, Springer-Verlag, Berlin, pp146–170, 1985

Olsen RW: Drug interactions at the GABA receptor-ionophore complex. *Annu Rev Pharmacol Toxicol* 22;245–277, 1982

Ownbey GB: Monograph of the North American species of *Corydalis. Ann Mo Bot Gard* 34;187–258, 1947

Preininger V: Chemotaxonomy of Papaveraceae and Fumariaceae. In *The Alkaloids,* vol 29, Brossi A ed, Academic Press, Orlando, Fla, pp1–98, 1986

Reynolds AK, The pharmacological actions of cularine. *J Pharmacol Exp Ther* 69;112–116, 1940

Rice HV: Pharmacology of corlumine. *J Pharmacol Exp Ther* 63;329–334, 1938

Schafer HL, Schafer H, Schneider W, Elstner EF: Sedative action of extract combinations of *Eschscholtzia californica* and *Corydalis cava. Arzneim-Forsch* 45;124–126, 1995

Seeman P: Brain dopamine receptors. *Pharmacol Rev* 32;229–313, 1981

Shamma M: *The Isoquinoline Alkaloids.* Vol 25, *Organic Chemistry.* Academic Press, New York, 1972

Smith RA, Lewis D: Apparent *Corydalis aurea* intoxication in cattle [letter]. *Vet Hum Toxicol* 32;63–64, 1990

Stern KR: Revision of *Dicentra* (Fumariaceae). *Brittonia* 13:1–57, 1961

Stern KR: Fumariaceae. In *Flora of North America,* vol 3, *Magnoliophyta: Magnoliidae and Hamamelidae,* Flora of North America Editorial Committee ed, Oxford Univ Press, New York, pp340–357, 1997

Waud RA: The pharmacological action of the alkaloids of fumaraceous plants. 1. Isocorydine. *J Pharmacol Exp Ther* 50;100–107, 1934

Waud RA: The pharmacological action of the alkaloids of fumaraceous plants. 2. Corydine. *J Pharmacol Exp Ther* 55;40–45, 1935

Waud RA: The pharmacological action of the alkaloids of fumaraceous plants. 3. Dicentrine methine hydrochloride. *J Pharmacol Exp Ther* 58;332–336, 1936

Welch AD, Hendersen VE: A comparative study of hydrastine, bicuculline and adlumine. *J Pharmacol Exp Ther* 51;482–491, 1934

Zong R, Shi R, Huang L, Liu H, Tao J, Zhu X, Jiang M, Yang X, Zhao H: Spasmolytic effect of protopine and xiatianwu *(Corydalis decumbens)* in isolated cat ciliary muscle and guinea pig ileum. *Zhongcaoyao* 17;303–306, 1986 (*Chem Abstr* 105;164856w, 1986)

40 Ginkgoaceae Engl.

■ Ginkgo Family
Maidenhair Tree Family
Ginkgo

A monogeneric family, the Ginkgoaceae is commonly known as the ginkgo or maidenhair tree family. It is classified in its own division, the Ginkgophyta, and has been present since the late Triassic or early Jurassic period of the Mesozoic era (Major 1967).

Plants trees; perennials; deciduous; dioecious; resinous; trunks stout; crowns ovoid or obovoid, typically asymmetrical. **Stems** to 30 m tall; bark gray, furrowed, ridges flattened; primary branches ascending; spur branches present, thick, knoblike; buds brown, globose. **Leaves** simple; alternate or in fascicles of 3–5; venation dichotomous; blades flabellate, striate; surfaces glabrous; apices cleft to truncate; margins undulate. **Pollen Cones** borne on spur branches; cylindrical; microsporophylls numerous; microsporangia paired. **Seed Cones** absent; ovules 2, borne at broadened apices of branched peduncles arising from spur branches, subtended by collar of 2 modified megasporophylls. **Seeds** 1 or 2 per peduncle; drupelike; obovoid to ellipsoid; yellow to orange; glaucous; rugose; outer coat fleshy, malodorous; inner coat hard; 2-angled; creamy white.

Ginkgo L.

■ *G. biloba*
gingko
kew tree
maidenhair tree
silver apricot

■ trees deciduous, dioecious; leaves simple, fan-shaped; pollen cones; seeds not enclosed, borne on stalks, drupelike, outer coat fleshy, malodorous, yellow to orange; inner coat hard, creamy white

Taxonomy and Morphology—Called a living fossil by Darwin, *Ginkgo* is the sole survivor of a group of plants that originated in the late Paleozoic, became widespread and moderately abundant in the Mesozoic, and then declined in the Cenozoic (Major 1967). The fossil record indicates that the single remaining species has been essentially unchanged for 200 million years. The plant is valued as an ornamental because of the graceful arching of the branches, striking butter-yellow foliage in the fall, and resistance to disease pests and city pollution, and its wood is used to make chess pieces and oriental lacquer ware. Its seeds are prized as food and medicine. The generic name is a Latinization of the Chinese name *yin-kuo* and Japanese pronunciation *ginkoh*. The single species is the following:

G. biloba L. ginkgo, maidenhair tree, kew tree, silver apricot, ginko

Morphological features of the genus are the same as those of the family and thus are not repeated here.

■ ornamental

Distribution and Habitat—Native to China, *G. biloba* is now almost cosmopolitan in distribution. In North America it is cultivated as an ornamental throughout most of the United States and Mexico except for the harsh climates of the north. Botanists disagree as to whether it still exists in nature or only in cultivation. It is questioned whether isolated trees growing in the mountains of southeastern China are actually wild or merely escapes (Major 1967).

Disease Problems—As a gymnosperm, *G. biloba* produces seeds that are not enclosed in a ripened ovary, or fruit, at maturity, as is characteristic of flowering plants. Its "fruit" is actually a seed with a fleshy outer coat and a hard, angled, creamy white inner coat containing the gametophyte and embryo, which are edible. The "seeds" sold in markets as "silver almonds" or "white nuts" have had the outer coat removed (Whetstone 1993). Although these seeds are edible, they can cause seizures and death when consumed in large numbers. The intoxication is called gin-nan food poisoning (Yagi et al. 1993). Known since the 1600s in Japan, it was of particular concern during and after World War II when food was scarce. A number of young children died after eating the seeds. It continues to be a sporadic problem. Because *G. biloba* is dioecious, the relative scarcity of pistillate trees bearing seeds limits the occurrence of intoxications.

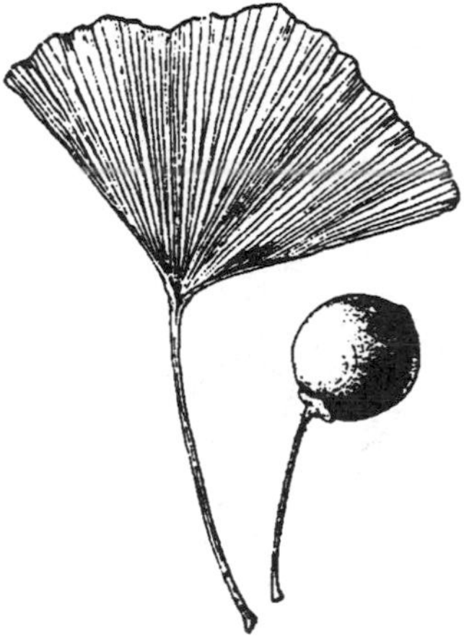

Ginkgo biloba

■ leaf extracts, popular herbal medicaments, improving blood flow

seeds, outer coat, allergic dermatitis, cross-reactive with allergens from *Toxicodendron*; inner portion neurotoxic

The outer, fleshy layer of the seed has a foul odor and is astringent for the skin, the digestive tract, and, when its absorbed breakdown products are eliminated, the urinary tract. It is also hemolytic (Watt & Breyer-Brandwijk 1962) and produces a contact dermatitis in some individuals (Mitchell & Rook 1979). The principal problem associated with *G. biloba,* this dermatitis is essentially the same as that caused by species of *Toxicodendron* (poison ivy, poison oak), in the Anacardiaceae (see chapter 6), and is characterized by severe itching and a weeping skin rash (Bolis 1939; Sowers et al. 1965; Becker & Skipworth 1975). Large doses of leaf extracts of *G. biloba* cause mild digestive upset with nausea, vomiting, and diarrhea (Tyler 1993).

In spite of ginkgo's potential for toxicity, preparations from its various parts are widely used in herbal medicines. With an extensive array of benefits attributed to them, medicaments are popular in many areas of the world (Oberpichler-Schwenk & Krieglstein 1992). Standardized extracts of the leaves of *G. biloba* are among the most widely prescribed medications, especially in Germany, where they are used to promote blood flow (Tyler 1993). The extract is particularly favored for relief of cerebral insufficiency and seems to produce positive effects in moderating the severity of Alzheimer's disease (LeBars et al. 1997). Although the mechanism of action is not known, it is suggested that reductions in free-radical activity and lipid peroxidation may be factors, as well as general enhancement of blood flow (Kleijnen & Knipschild 1992; Warburton 1993; Oyama et al. 1996). Many of these beneficial actions are consistent with experimental evidence that *Ginkgo* extracts are particularly active on the cardiovascular system, with effects on permeability, tone, blood viscosity, and cellular elements (Koch & Chatterjee 1993). Some of the cardiovascular actions of leaf extracts may be the result of inhibition of monoamine oxidase, which is in part responsible for limiting the action of neurohumoral biogenic amines (White et al. 1996). Chronic ingestion of *Ginkgo* has been associated with subdural hematoma (Rowin & Lewis 1996).

Disease Genesis—The neurotoxin in the inner portion of the seeds that causes gin-nan food poisoning is of considerable importance (Yagi et al. 1993). This toxin, 4-*O*-methylpyridoxine, is found only in the seeds, not in the foliage (Wada et al. 1993). It has been identified as the toxin also responsible for antagonism of pyridoxine (vitamin B_6) in *Albizia,* in the Fabaceae (see chapter 37) (Steyn et al. 1987). The toxin is thought to inhibit cerebral glutamate decarboxylase, which requires pyridoxal phosphate for the decarboxylation of glutamate to form γ-aminobutyric acid (GABA). This is consistent with the competitive inhibitory effects of the convulsant 4-deoxypyridoxine hydrochloride (Horton & Meldrum 1973). Even though pyridoxine is involved with the synthesis and metabolism of many of the neurotransmitters, seizure activity is thought to involve GABA (Dakshinamurti 1982). However, in the

■ neurotoxin, 4-*O*-methylpyridoxine, antagonist of pyridoxine

allergen, bilobol

herbal use, bilobalide, ginkgolide

CH_2OCH_3 HO CH_2OH H_3C N

4-*O*-Methylpyridoxine

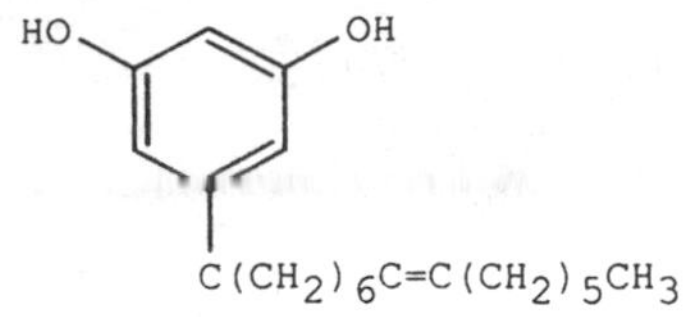

Bilobol

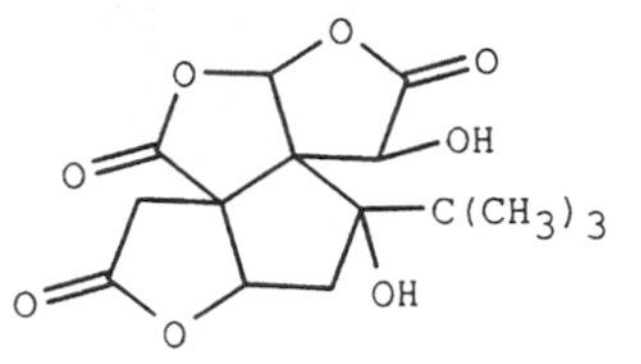

Bilobalide A

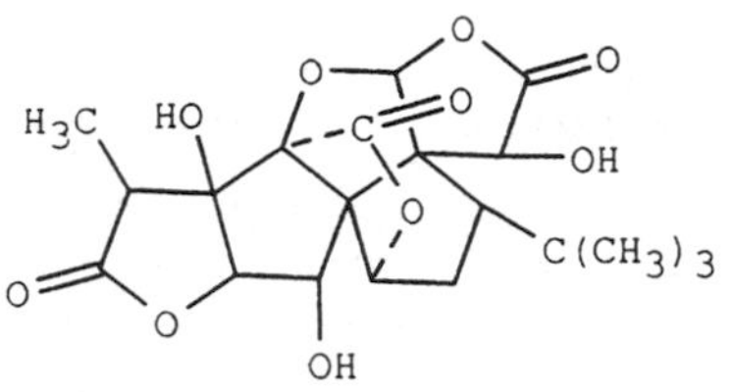

Ginkgolide B

- seeds, abrupt onset, seizures; no lesions
- seizures, sedation, pyridoxine HCl, activated charcoal

case of South African species of *Albizia,* pyridoxine is effective therapeutically but not pyridoxal or GABA (Gummow et al. 1992).

The presence of ginkgolic acid and bilobol in the fleshy, outer layer of the seeds may account for the apparent contact allergic response of the skin and mucous membranes (Watt & Breyer-Brandwijk 1962; Becker & Skipworth 1975). There is cross-reactivity between the allergens of *Ginkgo* and those of *Toxicodendron* (Sowers et al. 1965).

Because of the history of *Ginkgo*'s medicinal uses, numerous studies have been undertaken to confirm its neuroprotective actions. Bilobalide and ginkgolide A, the terpenoids found in the leaves, have been shown to act as neuroprotectants in mice and rats (Krieglstein et al. 1995). They also shorten the sleep time associated with various injectable anesthetics, possibly by altering their biotransformation rates (Wada et al. 1993). Because of the intense medical interest, some of the compounds, such as ginkgolides, have been synthesized (Corey et al. 1988).

Clinical Signs and Pathology—In humans, the signs of gin-nan food poisoning are abrupt onset of seizures, both clonic and tonic, and loss of consciousness. In acute disease, few if any distinctive changes are seen either grossly or microscopically.

Treatment—The periodic seizures may be controlled by diazepam, and toxin absorption may be reduced by oral administration of activated charcoal. Parenteral pyridoxine HCl, 20–25 mg/kg b.w., should be administered, as is used in cases of *Albizia* intoxication.

References

Becker LE, Skipworth GB: Ginkgo-tree dermatitis, stomatitis, and proctitis. *J Am Med Assoc* 231;1162–1163, 1975

Bolis M: Dermatitis venenata due to ginkgo berries. *Arch Derm Syphilol* 39;530, 1939

Corey EJ, Kang M-C, Desai MC, Ghosh AK, Houpis IN: Total synthesis of (±)-ginkgolide B. *J Am Chem Soc* 110;649–651, 1988

Dakshinamurti K: *Neurobiology of pyridoxine.* In *Advances in Nutritional Research,* vol 4, Draper HH ed, Plenum Press, New York, pp143–179, 1982

Gummow B, Bastianello SS, Labuschagne L, Erasmus GL: Experimental *Albizia versicolor* poisoning in sheep and its successful treatment with pyridoxine hydrochloride. *Onderstepoort J Vet Res* 59;111–118, 1992

Horton RW, Meldrum BS: Seizures induced by allylglycine, 3-mercaptopropionic acid, and 4-deoxypyridoxine in mice and photosensitive baboons, and different modes of inhibition of cerebral glutamic acid decarboxylase. *Br J Pharmacol* 49;52–63, 1973

Kleijnen J, Knipschild P: *Ginkgo biloba* for cerebral insufficiency. *Br J Clin Pharmacol* 34;352–358, 1992

Koch E, Chatterjee SS: Experimental evidence for the therapeutic use of ginkgo extract. *Haemostaseologie* 13;1–6, 1993

Krieglstein J, Ausmeier F, El-Abhar H, Lippert K, Welsch M, Rupalla K, Henrich-Noack P: Neuroprotective effects of *Ginkgo biloba* constituents. *Eur J Pharm Sci* 3;39–48, 1995

LeBars PL, Katz MM, Berman N, Itil TM, Freedman AM, Schatzberg AF: A placebo-controlled, double-blind, randomized trial of an extract of *Ginkgo biloba* for dementia. *J Am Med Assoc* 278;1327–1332, 1997

Major RT: The ginkgo, the most ancient living tree. *Science* 157;1270–1273, 1967

Mitchell J, Rook A: *Botanical Dermatology—Plants and Plant Products Injurious to the Skin.* Greengrass, Vancouver, BC, 1979

Oberpichler-Schwenk H, Krieglstein J: Pharmakologische wirkungen von *Ginkgo biloba*–extrakt und -inhaltsstoffen. *Pharm Unserer Zeit* 21;224–235, 1992

Oyama Y, Chikahisa L, Ueha T, Kanemaru K, Noda K: *Ginkgo biloba* extract protects brain neurons against oxidative stress induced by hydrogen peroxide. *Brain Res* 712;349–352, 1996

Rowin J, Lewis SL: Spontaneous bilateral subdural hematomas associated with chronic *Ginkgo biloba* ingestion. *Neurology* 46;1775–1776, 1996

Sowers EF, Weary PE, Collins OD: Ginkgo tree dermatitis. *Arch Dermatol* 91;452–456, 1965

Steyn PS, Vleggaar R, Anderson LAP: Structure elucidation of two neurotoxins from *Albizia tanganyicensis. S Afr J Chem* 40;191–192, 1987

Tyler VE: *The Honest Herbal,* 3d ed. Pharmaceutical Products Press, New York, 1993

Wada K, Sasaki K, Miura K, Yagi M, Kubota Y, Matsumoto T, Haga M: Bilobalide and ginkgolide A, isolated from *Ginkgo biloba* L., shortened the sleeping time induced in mice by anesthetics. *Biol Pharm Bull* 16;210–212, 1993

Warburton DM: *Ginkgo biloba* extract and cognitive decline [letter]. *Br J Clin Pharmacol* 36;137, 1993

Watt JM, Breyer-Brandwijk MG: *The Medicinal and Poisonous Plants of Southern and Eastern Africa.* E & S Livingston, Edinburgh, pp456–457, 1962

Whetstone RD: Ginkgoaceae. In *Flora of North America.* Vol 2, *Pteridophytes and Gymnosperms,* Flora of North America Editorial Committee ed, Oxford Univ Press, New York, pp350–351, 1993

White HL, Scates PW, Cooper BR: Extracts of *Ginkgo biloba* leaves inhibit monoamine oxidase. *Life Sci* 58;1315–1321, 1996

Yagi M, Wada K, Sakata M, Kokubo M, Haga M: Studies on the constituents of edible and medicinal plants. 4. Determination of 4-O-methylpyridoxine in serum of the patient with gin-nan food poisoning. *Yakugaku-Zasshi* 113;596–599, 1993

41 Hippocastanaceae DC.

■ Horse Chestnut Family
Aesculus

Comprising 2 genera and 15 species, the Hippocastanaceae, commonly known as the horse chestnut or buckeye family, is distributed primarily in the Northern Hemisphere of both the Old World and the New World. *Billia,* with 2 evergreen species, occurs only from southern Mexico to northern Colombia (Hardin 1957). *Aesculus* is widespread and of toxicologic interest.

■ shrubs or trees; deciduous; leaves palmately compound, opposite, leaflets 5–11; flowers in panicles; petals 4 or 5, free, clawed, white, greenish yellow, or red; fruits capsules; seeds 1–3, dark brown or light orange

Plants trees or shrubs; deciduous; andromonoecious. **Stems** 2–30 m tall; bark gray or dark brown, smooth or rough, scaly; terminal buds large, glutinous, ovoid; lateral buds conspicuous, mixed vegetative and floral, ovoid, solitary; bud scales conspicuous, 7 or more; lenticels large, pale brown to orange. **Leaves** palmately compound; opposite; leaflets 5–11; blades elliptic to obovate; venation pinnate; surfaces rugose, glabrous or tomentose or lanate; apices acute to acuminate; margins irregularly serrate; bases cuneate; stipules absent. **Inflorescences** panicles. **Flowers** produced simultaneously with or after leaves; perfect or imperfect, perfect and staminate similar; perianths in 2-series. **Calyces** bilaterally symmetrical; campanulate. **Sepals** 5; fused; gibbous or oblique. **Corollas** bilaterally symmetrical. **Petals** 4 or 5; unequal; free; clawed; white or greenish yellow or red. **Stamens** 5–8; free; filaments long, slender, unequal in length, curved upward. **Pistils** 1; compound, carpels 3; stigmas 1; styles 1; ovaries superior; locules 3; placentation axile; ovules 2 per locule. **Fruits** capsules; loculicidal. **Seeds** 1–3; dark brown or light orange-brown.

Aesculus L.

■ *A. californica*
A. flava
A. glabra
A. hippocastanum
A. pavia
A. sylvatica
buckeye
horse chestnut

Taxonomy and Morphology—*Aesculus* comprises 13 species. Hybridization among them is extensive, and identification of individual plants can be difficult. Numerous hybrids have been formally named. A few species provide timber or wood for crates and charcoal, but the genus is most prized for its showy ornamentals. Best known is the Old World *A. hippocastanum* (horse chestnut). Grown for its massive stature, large leaves, and striking inflorescences, this species has large, shiny, brown seeds that are the "conkers" played with by European children. The pale brown seed scar resembles the half-open eye of a deer—hence the common name *buckeye* for most of the American species. Introduced ornamentals occasionally encountered are *A. indica* and *A. turbinata,* the Indian and Japanese horse chestnuts. Representative of the genus are the following (see plate 41.1):

A. californica (Spach) Nutt.	California buckeye
A. flava Aiton (= *A. octandra* Marsh)	sweet buckeye, yellow buckeye
A. glabra Willd. (= *A. arguta* Buckley)	western buckeye, white buckeye, Ohio buckeye, Texas buckeye, fetid buckeye

A. hippocastanum L.	European horse chestnut, horse chestnut
A. pavia L.	red buckeye
A. sylvatica Bartram	painted buckeye, Georgia buckeye

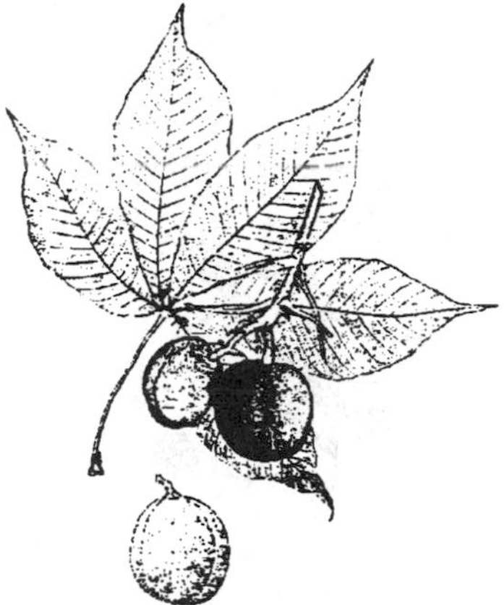
Aesculus flava

Morphological features of the genus are the same as those of the family and thus are not repeated.

1. Winter buds resinous. Claws of upper petals shorter than calyces.
   2. Capsules spiny or warty. Leaflets sessile. Calyces 5-lobed . . . . *A. hippocastanum*
   2. Capsules smooth. Leaflets petiolulate. Calyces 1- or 2-lobed . . . . *A. californica*
1. Winter buds not resinous. Claws of upper longer than calyces.
   3. Capsules spiny or warty. Petals equal or subequal. Stamens long exserted from corollas . . . . *A. glabra*
   3. Capsules smooth. Petals unequal. Stamens included within or as long as corollas.
      4. Petal margins stipitate-glandular . . . . *A. pavia*
      4. Petal margins villous.
         5. Pedicels and calyces stipitate-glandular . . . . *A. flava*
         5. Pedicels and calyces puberulent, not glandular . . . *A. sylvatica*

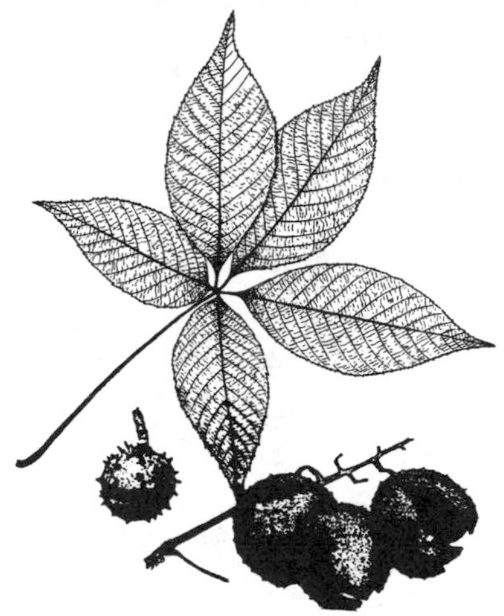
Aesculus glabra

Distribution and Habitat—The buckeyes are found in a variety of soils and habitats—in rocky and sandy soils, woods, and thickets and along fencerows and streams. *Aesculus flava* occurs in the rich moist woods of the Appalachian Mountains and Ohio River valley. *Aesculus glabra* is the most common species, occupying moist but well-drained soils of woods and thickets of the central United States. *Aesculus pavia* is an uncommon species of moist woods mainly of the southeastern United States. *Aesculus sylvatica* is an Appalachian species. Native to Greece, the Balkan Peninsula, and the Caucasus Mountains of northeastern Turkey, *A. hippocastanum* has long been cultivated as a garden and park tree in Europe and North America (Hardin 1957). The buckeyes may also be cultivated as ornamentals although they are not as widely used for this purpose as *A. hippocastanum.*

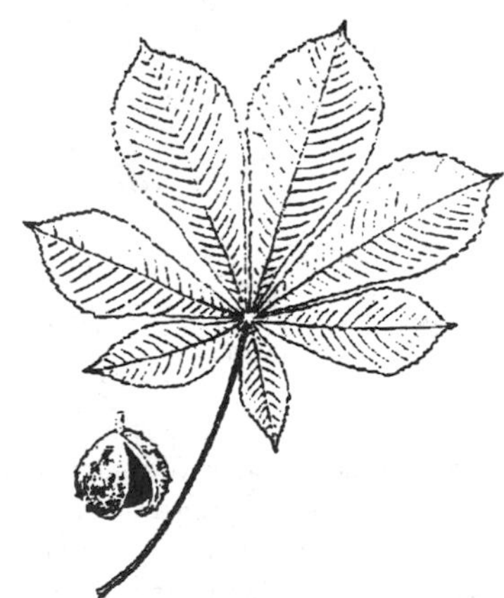
Aesculus hippocastanum

Disease Problems—In Europe where trees of *A. hippocastanum* are common, the seeds are well known as a cause of intoxication. Primarily a problem in children, ingestion of the seeds may cause mild to moderate neurologic signs. The relative toxicity of some of the species has been confirmed experimentally in chicks and hamsters: *A. hippocastanum* > *A. flava* > *A. glabra* > *A. pavia* (Williams & Olsen 1984). Although a lethal dose for some species is about 1% b.w. given for a day or more in calves, in most animals the onset of signs decreases intake, so that the disease is typically self-limiting. Serious intoxications, including occasional deaths, are most likely to be caused by *A. hippocastanum* because both its leaves and seeds are lethal (Reynard & Norton 1942; Williams & Olsen 1984). Seeds of *A. flava* at 0.5% b.w. caused serious but not lethal neurotoxic effects in calves, whereas 1% b.w. was lethal (Magnusson et al. 1983). *Aesculus glabra* appears to be less toxic, but it is nevertheless capable of causing distinct neurotoxic effects that are unlikely to be fatal except through misadventure (Casteel et al. 1992). In contrast, Cary (1922), using the leaves, roots, or seeds from *A. pavia,* found it difficult to give sufficient material to cause toxic manifestations in pigs, cattle, or horses.

Intoxications may be seen in the fall when fruits are consumed or in the spring following ingestion of leaves and buds. Consumption of mature leaves at other times is much less of a problem. Intoxication may be seen in a variety of animal species, especially in ruminants. In the latter, marked signs of neurointoxication occur, ranging from incoordination to seizures and rarely death (Cary 1922; Reynard & Norton 1942; Kornheiser 1983; Magnusson et al. 1983; Casteel et al. 1992). In chickens, the effects range from depression of growth to generalized depression and weakness (Creek & Vasaitis 1963; Williams & Olsen 1984). Similarly, weakness and depression are prominent in pigs and horses (Cary 1922). The pulverized seeds were reportedly used to immobilize fish, and this has subsequently been experimentally confirmed (Pammel 1911; Cary 1922).

Disease Genesis—Several compounds of interest are found in *Aesculus,* including various unsaturated sterols (sitosterol, stigmasterol), triterpenoid alcohols (taraxerol) and alkaloids (dicaffeoylspermidine, di-4-coumaroylspermidine) (Macbeth 1931; Mukherjee et al. 1983; Stankovic et al. 1984, 1985; Southon & Buckingham 1989). There is also a coumarin derivative (esculin or aesculin) and its glycoside (esculetin). However, the toxic effects appear to be due to a mixture of 30 or more pentacyclic oleanane triterpenoids, called saponins when complexed as glycosides with glucose and glucuronic acid. Found in both seeds and leaves, these saponins are acylated mono-, bi-, and tridesmosidic glycosides of several genins, including escigenin, barringtogenol C, hippocaesculin, and protoescigenin (Hoppe et al. 1968; Wulff & Tschesche 1969; Wagner et al. 1970; Aurada et al. 1984; Konoshima & Lee 1986; Singh et al. 1986, 1987; Sati & Rana 1987; Schrutka-Rechtenstamm et al. 1988; Profumo et al. 1991). The complexity of the array of saponins present is increased even more because the genins also may be esterified with tiglic acid, angelic acid, α-methylbutyric acid, or isobutyric acid. The nomenclature of these compounds can be quite confusing because of the similarity in names applied to the numerous

Escigenin

Barringtogenol C

Protoescigenin

β-Escin

Aesculus pavia

- ornamental, variety of habitats
- seeds and foliage, neurotoxic, mainly transient ataxia

Aesculus californica

Aesculus flava

- triterpenoid saponins of genins such as escigenin, barringtogenol C, protoescigenin; variously esterified; example β-escin; some medicinal uses glycosides and their respective free aglycones or genins, e.g., *escin, aescin, aesculin,* and *aesculuside.* This is due, in some measure, to the similarity in names applied to the glycosides and to the respective free aglycones or genins.

The best-known and best-studied escin is a mixture of saponins sometimes referred to as α-escin and the natural form β-escin. It may compose up to 13% d.w. of the seeds of *A. hippocastanum.* Depending upon the species, it is composed mainly of diesters as well as other aglycones, including barringtogenol C (= escinidin) and its diangeloyl diester derivative. Also present are escigenin and isoescigenin, which are derivatives of barringenol R_1 angeloyl diester (Connolly & Hill 1991).

When administered parenterally, these saponins are neurotoxic in low dosage and hemolytic in high dosage (Vogel et al. 1970; Wagner et al. 1970). The escins also have hypoglycemic activity (Yoshikawa et al. 1996). Some of the escin fractions, especially β-escin, have beneficial effects on the heart, kidneys, and other tissues and are used medicinally (Panigati 1992). Available in some countries under the trade names Venoplant, Reparil, Venostasin, and Vasotonin, escin is used to control inflammation, edema, and cellulitis, especially in peripheral vascular disease secondary to restoration of the plasma-lymph barrier (Vogel et al. 1970; Rothkopf & Vogel 1976; Annoni et al. 1979). It also is used as a antineoplastic cytotoxin (Konoshima & Lee 1986). It seems to have beneficial effects in decreasing capillary fragility and is sometimes referred to as capillaro-protective. Fortunately, the medicinal fractions are only rarely hepatotoxic (Takegoshi et al. 1986). Both escin and barringtogenol C have been used as fish poisons.

Escin has been the subject of numerous studies because of interest in its medicinal uses. As with other saponins, it is poorly absorbed when given orally but causes irritation of the intestinal mucosa. In rats much of what is absorbed is extensively biotransformed (Lang & Mennicke 1972). Plasma concentrations may persist for 24 or more hours following oral administration of extracts of β-escin (Kunz et al. 1991). There is an apparent increased sensitivity for ruminants to the toxic effects of the saponins. This may be due to the reducing environment of the rumen, which increases the likelihood of hydrolysis of the parent glycoside to allow for absorption of the free aglycone—presumably the neurotoxin, although not confirmed experimentally (Casteel et al. 1992).

Aesculus glabra

Aesculus pavia

Aesculus sylvatica

Clinical Signs—The disease usually begins with the animal exhibiting a sawhorse stance and a reluctance to stand or move. When the animal is required to do so, there is marked incoordination with an uneasy staggering gait, accompanied by hypermetria or an exaggerated motion and hyperesthesia. Seizures with muscle spasms, postural abnormalities, extensor rigidity, and collapse may be precipitated. Severe depression, colic, mydriasis, and trembling may also be seen. The disease is seldom fatal but may result in transient incapacitation, especially in the spring, when young foliage may be abundant and palatable. The signs are typically self-limiting, lasting perhaps 12–48 hours, but because of their nature they may nonetheless be quite alarming to the animal's owner. Clinical chemistry parameters are usually not remarkable except for elevated blood glucose as seen in calves experimentally (less than 150 g/dl) (Magnusson et al. 1983).

- sawhorse stance, ataxia, stagger, hyperesthesia, muscle spasms, trembling; typically lasting 24 hours

- no lesions, treatment rarely needed

Pathology and Treatment—There are no specific pathologic lesions, and necropsy is seldom performed because the disease is rarely fatal. Treatment is required only in exceptionally severe cases, and prevention of access to the toxic forage will be followed by recovery within several days.

References

Annoni F, Mauri A, Marincola F, Resele LF: Venotonic effect of escin on the human saphenous vein. *Arzneim-Forsch* 29;672–675, 1979

Aurada E, Jurenitsch J, Kubelka W: Structure of triterpene-sapogenins of *Aesculus glabra*. *Planta Med* 50;391–394, 1984

Cary CA: *Poisonous Action of Red Buckeye on Horses, Mules, Cattle, Hogs, and Fish*. Ala Agric Exp Stn Bull 218, 1922

Casteel SW, Johnson GC, Wagstaff DJ: *Aesculus glabra* intoxication in cattle [letter]. *Vet Hum Toxicol* 34;55, 1992

Connolly JD, Hill RA: *Dictionary of Terpenoids*. Vol 2, *Di- and Higher Terpenoids*, Chapman & Hall, London, 1991

Creek RD, Vasaitis V: The toxicity of the Ohio buckeye to the chick. *Poult Sci* 42;1473–1474, 1963

Hardin JW: A revision of the American Hippocastanaceae—2. *Brittonia* 9;173–195, 1957

Hoppe W, Gieren A, Brodherr N, Tschesche R, Wulff G: Structure of the principal aglycon of horse chestnut saponin. *Angew Chem Int Ed* 7;547–548, 1968

Konoshima T, Lee KH: Antitumor agents, 82. Cytotoxic sapogenols from *Aesculus hippocastanum*. *J Nat Prod* 49;650–656, 1986

Kornheiser KM: Buckeye poisoning in cattle. *Vet Med/Small Anim Clin* 78;769–770, 1983

Kunz C, Schaffler K, Biber A, Wauschkuhn CH: Bioverfugbarkeit von β-aescin nach oraler gabe zweier *Aesculus*-extract enthaltender darreichungsformen an gesunden prolanden. *Pharmazie* 46;145, 1991

Lang W, Mennicke WH: Pharmacokinetic study on tritiated escin in the mouse and rat. *Arzneim-Forsch* 22;1928–1932, 1972

Macbeth AK: Studies of the glucosides. Part 4. Aesculin. *J Chem Soc (London)* 1931; 1288–1290, 1931

Magnusson RA, Whittier WD, Veit HP, Easley KJ, Meldrum JB, Jortner BS, Chickering WR: Yellow buckeye (*Aesculus octandra* Marsh) toxicity in calves. *Bovine Pract* 18;195–199, 1983

Mukherjee KS, Bhattacharya MK: Phytochemical investigation of *Aesculus indica* Linn. and *Fragaria indica* Andr. *J Indian Chem Soc* 60;507–508, 1983

Pammel LH: *A Manual of Poisonous Plants*. Torch Press, Cedar Rapids, Iowa, pp616–618, 1911

Panigati D: Farmacologia dell' escina, saponina dell' *Aesculus hippocastanum* L. Parte 3. Farmacocinetica e tossicologia. *Boll Chim Farm* 131;320–321, 1992

Profumo P, Caviglia AM, Gastaldo P, Dameri RM: Aescin content in embryogenic callus and in embryoids from leaf explants of *Aesculus hippocastanum*. *Planta Med* 57;50–52, 1991

Reynard GB, Norton JBS: *Poisonous Plants of Maryland in Relation to Livestock*. Md Agric Exp Stn Bull A10, 1942

Rothkopf VM, Vogel G: Further results on efficacy and mode of action of the horse chestnut saponin escin. *Arzneim-Forsch* 26;225–235, 1976

Sati OP, Rana U: Triterpenoids of *Aesculus indica*. *Pharmazie* 42;141, 1987

Schrutka-Rechtenstamm R, Robien W, Jurenitsch J: Struktur der sapogenine aus samen von *Aesculus pavia* L. *Pharmazie* 43;208–210, 1988

Singh B, Agrawal PK, Thakur RS: Aesculuside-A, a new triterpene glycoside from *Aesculus indica*. *Planta Med* 52;409–410, 1986

Singh B, Agrawal PK, Thakur RS: Aesculuside-B, a new triterpene glycoside from *Aesculus indica*. *J Nat Prod* 50;781–783, 1987

Southon IW, Buckingham J: *Dictionary of Alkaloids*. Chapman & Hall, London, 1989

Stankovic SK, Bastic MB, Jovanovic JA: Composition of the sterol fraction in horse chestnut. *Phytochemistry* 23;2677–2679, 1984

Stankovic SK, Bastic MB, Jovanovic JA: Composition of the triterpene alcohol fraction of horse chestnut seed. *Phytochemistry* 24;119–121, 1985

Takegoshi K, Tohyama T, Okuda K, Suzuki K, Ohta G: A case of venoplant-induced hepatic injury. *Gastroenterol Jpn* 21;62–65, 1986

Vogel G, Marek ML, Oertner R: Studies on the mechanisms of therapeutic and toxic actions of the horse chestnut saponin escin. *Arzneim-Forsch* 20;699–703, 1970

Wagner J, Schlemmer W, Hoffmann H: On the compounds of horse chestnut seeds. 9. Structure and properties of triterpene-glycosides. *Arzneim-Forsch* 20;205–209, 1970

Williams MC, Olsen JD: Toxicity of seeds of three *Aesculus* spp. to chicks and hamsters. *Am J Vet Res* 45;539–542, 1984

Wulff G, Tschesche R: Uber Triterpene—26. Uber die struktur der rosskastaniensaponine (aescin) und die aglykone verwandter glykosidee. *Tetrahedron* 25;415–436, 1969

Yoshikawa M, Murakami T, Matsuda H, Yamahara J, Murakami N, Kitagawa I: Bioactive saponins and glycosides. 3. Horse chestnut. (1) The structures, inhibitory effects on ethanol absorption, and hypoglycemic activity of escins Ia, Ib, Iia, Iib, and IIIa from the seeds of *Aesculus hippocastanum* L. *Chem Pharm Bull* 44;1454–1464, 1996

42 Iridaceae Juss.

■ Iris Family
Gladiolus
Iris
QUESTIONABLE
Crocus

Cosmopolitan except for the far north, the Iridaceae, or iris family, comprises 80–88 genera and 1,500–1,800 species classified into 11 tribes. It is most abundant and diversified in southern Africa, the Mediterranean region, Central America, and South America. Horticulturally one of the most important families, it contains numerous garden and indoor ornamentals, such as crocus, freesia, gladiolus, and iris. It is also the source of the cooking condiment saffron and the cosmetic additive orris root. In North America, approximately 23 genera and 125 species, both native and introduced, are present. Four genera pose a threat. *Homeria* and *Moraea,* rarely used ornamentals from South Africa, contain potent cardiotoxic bufadienolides, and are capable of causing digestive and cardiac problems similar to those produced by *Bowiea* and *Urginea* in the Liliaceae (Kellerman et al. 1988). Intoxication problems, however, are caused by members of 2 commonly encountered genera.

■ herbs, from rhizomes, bulbs, or corms; leaves simple, alternate, margins entire, venation parallel; flowers terminal; fruits capsules

Plants herbs; perennials or annuals; from rhizomes or bulbs or corms; caulescent or acaulescent. **Root Systems** fibrous. **Leaves** simple; alternate; sessile; typically equitant; with open basal sheaths; blades linear to lanceolate or ensiform; venation parallel or parallel-convergent; margins entire; stipules absent. **Inflorescences** solitary flowers, or spicate or racemose or paniculate cymes; terminal; spathes present, consisting of 2 foliaceous bracts. **Flowers** often large and showy; perfect; perianths in 2-series; radially or bilaterally symmetrical. **Perianth Parts** 6; all alike; petaloid; free or fused; clawed or not clawed; spreading or reflexed. **Stamens** 3; free or fused. **Pistils** 1; compound, carpels 3; stigmas 3; styles 1, branched, sometimes petaloid; ovaries inferior; locules 3; placentation axile; hypanthia present or absent. **Fruits** capsules; loculicidal. **Seeds** numerous.

The Iridaceae is distinguished from the morphologically similar Liliaceae by its three stamens and equitant leaves typical of most taxa.

■ *G. cardinalis*
G. communis
G. xcolvillei
G. xgandavensis
G. italicus
G. tristis

Colville's gladiolus
field gladiolus
gladiolus
marsh Afrikaner
new year lily
waterfall gladiolus

Gladiolus L.

Taxonomy and Morphology—Comprising more than 250 species and innumerable cultivars, *Gladiolus* is one of the largest genera of the family and was named by the Romans (Goldblatt 1996). Its name is from the Latin *gladiolus,* which means "little sword, alluding to the flattened, sword-shaped leaves characteristic of the genus. Almost all garden gladioli or those sold as cut flowers are hybrids, derived for the most part from 7 or 8 South African species (Goldblatt 1996). Collection of these African taxa was extensive in the late 1700s, and hybrids appeared in the 1820s. By 1900, thousands of cultivars had been developed. The following species are representative of the genus:

G. cardinalis Curtis	waterfall gladiolus, new year lily
G. communis L.	
G. x*colvillei* Sweet	Colville's gladiolus
G. x*gandavensis* Van Houtte	
G. italicus Mill.	field gladiolus
G. tristis L.	marsh Afrikaner

Gladiolus corm

Plants perennials; from corms; caulescent. **Stems** erect; 50–100 cm tall; generally not branched; terete. **Leaves** with blades ensiform. **Inflorescences** spikes; branched or not branched; flowers secund. **Flowers** bilaterally symmetrical; funnelform. **Perianth Parts** fused; tubes curved; unequal, upper 3 larger than lower 3; of various colors. **Stamens** with filaments arching. **Pistils** with styles 1; stigmas 3. **Capsules** ellipsoid to oblong. **Seeds** brown; winged or not winged.

■ perennials, from corms, erect, unbranched; flowers funnelform

Because the variation exhibited by *Gladiolus* is extensive and the species and hybrids listed above merely representative, a taxonomic key is not presented. Should exact identification be required, horticultural treatments of the genus should be consulted.

Distribution and Habitat—An Old World genus, *Gladiolus* exhibits greatest diversity in southern Africa, with more 150 species believed to occur there (Goldblatt 1996). In North America, taxa rarely escape cultivation; *G. italicus* and *G. tristis,* for example, are occasionally encountered in orchards, old fields, and roadsides in California.

■ ornamentals

Disease Problems and Genesis—Intoxication problems are only occasionally caused by species of *Gladiolus.* In one group of cases in Italy, 3 herds with 166 cattle were affected; 44 became ill and 15 died (Meli et al. 1983). The signs reflected primarily severe digestive tract problems. Of the 15 fatalities, 2 animals died in the first 24 hours, 9 in 5–6 days, and 4 in 15–16 days. The toxic dosage of members of the genus is not well known, but a South African species caused acute congestive heart failure and death of rats in a dosage range of 3–6 g/kg of b.w., the corms being more toxic than the leaves (Van Dyk et al. 1994). The specific toxicants are not known, but digestive tract irritation is generally recognized as the primary feature of intoxication by species of this genus (Mendez et al. 1993).

■ digestive disturbance; specific toxins unknown

Clinical Signs—The signs, indicative of irritation of the digestive tract, include depression, loss of appetite, decreased motility of the stomach or rumen, increased salivation, hard black feces or diarrhea, and decreased body temperature. The animals also may be uncoordinated and anemic.

■ depression, anorexia, excessive salivation, diarrhea

Pathology—Gross pathologic changes are related primarily to the digestive tract—reddening, edema, ulcerations, and hemorrhages of the mucosa of the abomasum and small intestine. There may be excess blood-tinged peritoneal fluid and multifocal hemorrhages in the kidneys, the liver, and the perirenal tissues.

■ gross pathology, reddened gut mucosa

Treatment—Dehydration is a main threat to the well-being of all affected animals and requires judicious administration of fluids and electrolytes. In most instances other antidiarrheal treatment will not be required.

■ fluids, electrolytes

Iris L.

- *I. douglasiana*
  I. missouriensis
  I. pseudacorus
  I. versicolor
  I. virginica

  flag
  iris

Taxonomy and Morphology—Horticulturally quite important, with a plethora of cultivars, *Iris* comprises 225–300 species classified in 6 subgenera and 12 sections. Hybridization is common. The plants are commonly known as iris, flag, fleur-de-lis, or sword lily, and the generic name is indicative of the diversity of flower colors. It is derived from the Greek *iris,* meaning "rainbow." In North America, about 50 species and hybrids, both native and introduced, are present. The following are representative:

I. douglasiana Herb.	Douglas's flag
I. missouriensis Nutt.	western blue flag
I. pseudacorus L.	yellow flag, yellow water iris
I. versicolor L.	northern blue flag, wild iris, blue flag, poison flag
I. virginica L.	southern blue flag, Virginia iris

- erect, unbranched, perennials from bulbs or rhizomes; flowers showy, terminal

Plants perennials; from rhizomes or bulbs; rhizomes horizontal or vertical, short or long; bulb tunicas netted or leathery or papery. **Stems** erect; 15–100 cm tall; typically not branched. **Leaves** chiefly basal; blades linear to ensiform. **Inflorescences** terminal clusters; flowers 1 to several. **Flowers** radially symmetrical. **Perianth Parts** fused; tubes long or short; of various colors; in 2 whorls; outer 3 parts ("falls") reflexed or spreading, adaxial surfaces smooth or crested or ridged or bearded; inner 3 parts ("standards") erect, clawed. **Stamens** with filaments arching. **Pistils** with styles 1, 3-branched, expanded, petaloid, colored, spreading, lobes covering stamens, bifid or crested at apices. **Capsules** ellipsoid to oblong; coriaceous or chartaceous; 3- or 6-angled; ribbed or smooth. **Seeds** in 1 or 2 rows per locule.

Because the species listed above are cited only as representative of the numerous species present, a taxonomic key is not presented. If exact identification is need, the reader should use manuals specific for his or her area.

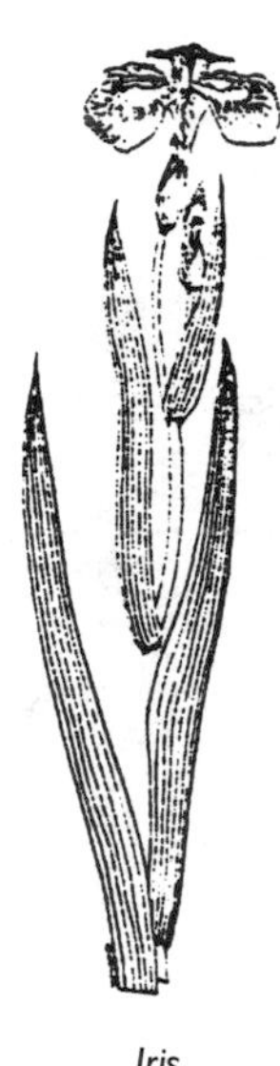

Iris

Distribution and Habitat—*Iris* is a genus primarily of the Northern Hemisphere and exhibits greatest diversity in Eurasia. In North America, species are present across the continent; some are widespread, and others, such as *I. douglasiana,* which is found near the coasts of northern California and southern Oregon, are restricted in distribution. The genus is especially abundant in the delta of the Mississippi River. Some Eurasian introductions, e.g., *I. pseudacorus,* have naturalized.

Iris douglasiana

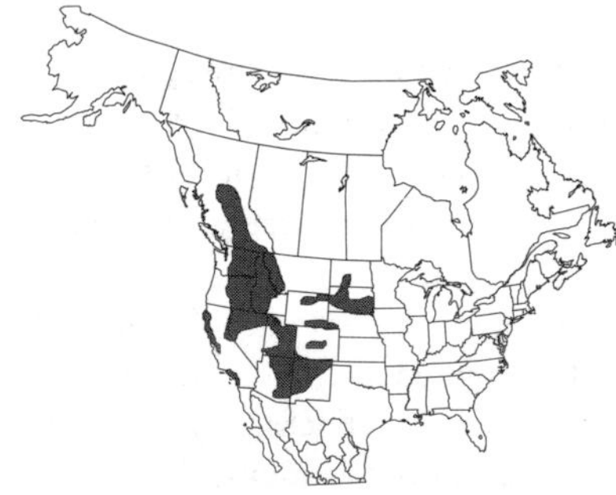

Iris missouriensis

Disease Problems—*Iris* is well known as a toxic genus in the Old World, and most of the rhizomatous species appear to be toxic to all animal species (Cooper & Johnson 1984). Small quantities of the rhizomes or large amounts of the leaves, even when dried, have a potent purgative effect and may even cause death (Bruce 1919; Boddie 1947). Harrington (1967) described the experience of an individual who tasted the peeled and cooked rhizome of *I. missouriensis* and said that it was sweet at first, but then terribly astringent and caused his throat to burn. In addition to the purgative effects, depression and respiratory problems have also been associated with ingestion of large quantities of iris leaves (Beath et al. 1953). Puppies and older dogs kept in a garage where the rhizomes might be stored temporarily are at special risk. The shallow rhizomes are easily dug up by the inquisitive animal.

■ digestive disturbance, leaves and especially rhizomes; puppies at special risk

Iris pseudacorus

Disease Genesis—Iris toxicity was thought to be due to a potent acrid resinous purgative called irisin, which also may be responsible for the relative unpalatability of plants, especially their rhizomes (Pammel 1911). However, a more likely cause of the toxic effects appears to be three closely related hopene-type pentacyclic terpenoids—missourin, zeorin, and missouriensin—plus quinines such as irisoquin, which have been isolated from the rhizomes of of *I. missouriensis* (Wong et al. 1985, 1986a,b; Connolly & Hill 1991). Although these compounds are cytotoxic, they have not been confirmed as the equivalent of irisin nor as the cause of the purgation. Similarly, a series of iridal-type terpenoids different from the hopene types, are present in *I. germanica, I. pseudacorus, I. versicolor,* and other species (Connolly & Hill 1991). Again, the role of these terpenoids as toxicants has not been shown.

■ hopene-type terpenoids, missourin, zeorin, missouriensin

Iris versicolor

Missourin

Zeorin

Missouriensin

Iris virginica

Clinical Signs—Shortly after ingestion of *Iris,* there is increased salivation and abdominal pain, leading to profuse diarrhea, often with blood. There may also be burnlike sores on the lips and muzzle. In most animals, however, the disease will be of only mild to moderate severity, lasting only a few hours to a day. When severe, dehydration occurs.

■ colic, excess salivation, diarrhea

Pathology—Intoxication caused by *Iris* results in pathologic changes similar to those produced by *Gladiolus,* i.e., gross changes that include reddening, edema, ulcerations, and hemorrhages of the mucosa of the abomasum and small intestine. There also may be excess blood-tinged peritoneal fluid and a few hemorrhages in the kidneys, the liver, and the perirenal tissues.

■ gross pathology, reddened gut mucosa

■ fluids, electrolytes, activated charcoal

Treatment—Dehydration is a serious threat to the well-being of all affected animals and will require judicious administration of fluids and electrolytes. In most instances, other antidiarrheal treatment will not be required.

Genera of Questionable Toxicity

Crocus L.

■ *C. sativus*
crocus
saffron

Crocus sativus

Comprising 75–80 species native to the Mediterranean region and southwestern Asia, *Crocus* is now widely cultivated and prized for the beauty of its late-winter and early-spring flowers. Its name is derived from the Greek name *krokas* for *C. sativus* L., commonly known as saffron. Plants of this species are acaulescent perennial herbs from corms, which produce linear, grasslike leaves and 1–4 radially symmetrical, lilac or reddish purple or white flowers arising from sheaths composed of 3 bracts. The 3 aromatic, cuneate, toothed style branches are blood red and much longer than the stamens. Interestingly, the species is sterile; the capsules do not develop, and reproduction is via offshoots of the corms. Cultivated since the time of the ancient Greeks, *C. sativus* is the source of saffron, at one time quite an economically important commodity, which is used for coloring and flavoring of food, liqueurs, and medicines. When consumed in large amounts, it may cause a digestive disturbance (Frohne & Pfander 1984). *Crocus sativus* and other species of the genus occasionally escape cultivation in North America. They pose minimal threat to grazing livestock, however, because they are seldom abundant.

■ large amounts, digestive disturbance

References

Beath OA, Gilbert CS, Eppson HF, Rosenfeld I: *Poisonous Plants and Livestock Poisoning.* Wyo Agric Exp Stn Bull 324, 1953

Boddie GF: Toxicological problems in veterinary practice. *Vet Rec* 59;471–486, 1947

Bruce EA: Iris poisoning of calves. *J Am Vet Med Assoc* 56;72–74, 1919

Connolly JD, Hill RA: *Dictionary of Terpenoids.* Vol 2, *Di- and Higher Terpenoids,* Chapman & Hall, London, pp1382–1501, 1991

Cooper MR, Johnson AW: *Poisonous Plants in Britain and Their Effects on Animals and Man.* Her Majesty's Stationery Office, London, pp134–135, 1984

Frohne D, Pfander HJ: *A Colour Atlas of Poisonous Plants,* 2d ed. Wolfe Science, London, 1984

Goldblatt P: Gladiolus *in Tropical Africa.* Timber Press, Portland, Oreg, 1996

Harrington HD: *Edible Native Plants of the Rocky Mountains.* Univ New Mexico Press, Albuquerque, p34, 1967

Kellerman TS, Coetzer JAW, Naude TW: *Plant Poisonings and Mycotoxicoses of Livestock in Southern Africa.* Oxford Univ Press, Cape Town, South Africa, pp87–90, 1988

Meli F, Braca G, Richetti A, Catarsini O: Outbreak of spontaneous poisoning by *Gladiolus segetum* Ker Gawl in cattle at pasture. *Atti Soc Ital Buiatria* 14;365–377, 1983

Mendez MDC, Delgago PE, Santos R, Sechin A, Reit-Correa F: Experimental intoxication by *Sisyrinchium platense* (Iridaceae) in cattle. *Pesq Vet Bras* 13;77–81, 1993

Pammel LH: *A Manual of Poisonous Plants.* Torch Press, Cedar Rapids, Iowa, pp388–390, 1911

Van Dyk S, Gerritsma-Van Der Vijver LM, Van Der Nest DG: The toxicity of *Gladiolus dalenii* van Geel. *S-Afr Tydskr Natuurwet Tegnolog* 13;125–134, 1994

Wong S-M, Pezzuto JM, Fong HHS, Farnsworth NR: Isolation, structural elucidation, and chemical synthesis of 2-hydroxy-octadecyl-5-methoxy-1,4-benzoquinone (irisoquin), a cytotoxic constituent of *Iris missouriensis. J Pharm Sci* 74;1114–1116, 1985

Wong S-M, Oshima Y, Pezzuto JM, Fong HHS, Farnsworth NR: Plant anticancer agents. 39. Triterpenes from *Iris missouriensis* (Iridaceae). *J Pharm Sci* 75;317–320, 1986a

Wong S-M, Pezzuto JM, Fong HHS, Farnsworth NR: Isolation and characterization of a new triterpene from *Iris missouriensis. J Nat Prod* 49;330–333, 1986b

Juglandaceae A.Rich. ex Kunth

43

Comprising 8 or 9 genera and about 60 species distributed primarily in the Northern Hemisphere, the Juglandaceae, commonly known as the walnut or hickory family, is an important component of the eastern deciduous forests. Its trees are major elements of the canopy, and their nuts are important to a variety of wildlife species. The family is also of considerable economic importance. With a beautiful grain and high polish, the wood is used for furniture, cabinets, paneling, gunstocks, and bowls. In North America, 2 genera—*Carya* (hickory) and *Juglans* (walnut)—are present. Intoxication problems are associated with the latter genus.

■ Walnut Family

Juglans

Plants shrubs or trees; monoecious. **Leaves** aromatic; 1-pinnately compound; alternate; leaflets abaxially gland-dotted, terminal leaflet present; venation pinnate; stipules absent. **Inflorescences** of 2 types, staminate and pistillate different; staminate inflorescences catkins, borne on previous year's twigs; pistillate inflorescences solitary flowers or clusters, borne on current year's twigs, bracts present, bracteoles 2 or 3. **Flowers** imperfect, staminate and pistillate different. **Staminate Flowers** with perianths in 1-series; radially symmetrical; perianth parts 2–6, fused, yellowish green to yellowish brown; stamens 3 to numerous. **Pistillate Flowers** with perianths in 1-series; bilaterally symmetrical; perianth parts 4, fused; pistils 1, compound, carpels 2 or 3, stigmas 2, styles 2, ovaries inferior, locules 1, placentation apical. **Fruits** nuts; surrounded by dehiscent or indehiscent, fibrous-fleshy husks derived from involucre and perianth parts. **Seeds** 1.

■ shrubs or trees; leaves aromatic, 1-pinnately compound, alternate; staminate flowers in catkins, pistillate flowers solitary or clustered; fruits nuts

Juglans L.

Taxonomy and Morphology—Comprising 20 or 21 species, *Juglans* includes many familiar and economically important taxa. Its name is derived from the ancient Latin name for the Persian walnut—*Jovis glans,* meaning "nut of Jove" or "nut of Jupiter," alluding to the delicious taste of the nutmeat. All species produce edible nuts, but quality is highest in the commercial varieties of the English walnut. Native Americans used black walnut medicinally in a variety of ways. In North America, 9 species, both native and introduced, are present. Named hybrids also occur. Most commonly encountered are the following:

■ *J. cineraria*
J. nigra
J. regia

butternut
walnut

J. cineraria L.	butternut, white walnut
J. nigra L.	black walnut
J. regia L.	English walnut, Persian walnut, Madeira walnut, Circassian walnut

Plants trees. **Stems** 3–50 m tall; bark light to dark gray or gray-brown, smooth or split into ridges or plates; twigs stout, terete, glandular puberulent; leaf scars large,

■ trees; leaflets 5–25

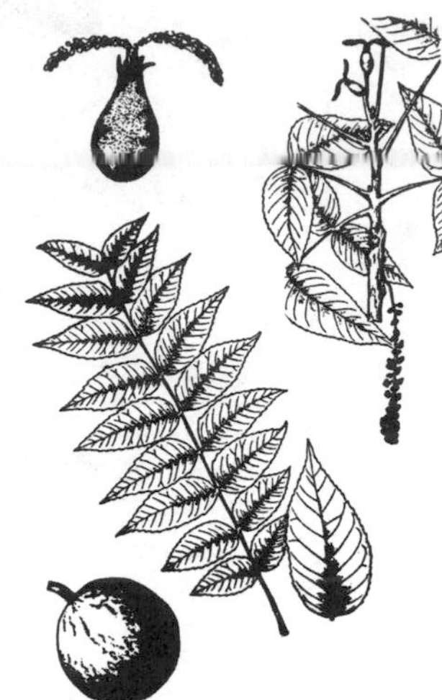
Juglans nigra

Juglans nigra

heart-shaped or 3-lobed; pith chambered. **Leaves** with leaflets 5–25, sessile or subsessile, uniform in size or medial slightly larger; terminal leaflet present or absent. **Stamens** 7–50. **Pistils** with plumose stigmas. **Nuts** with involucral husks indehiscent or dehiscent.

1. Leaflets 7–9; margins entire. Involucral husks dehiscent . . . . *J. regia*
1. Leaflets 11–23; margins serrate. Involucral husks indehiscent.
    2. Involucral husks pubescent but not sticky. Nuts globose. Upper edges of leaf scars glabrous . . . . *J. nigra*
    2. Involucral husks sticky-pubescent. Nuts ellipsoid to ovoid. Upper edges of leaf scars ciliate-fimbriate . . . . *J. cinerea*

Distribution and Habitat—*Juglans nigra and J. cinerea* are widely distributed in bottomlands of the eastern half of the continent. The other native species have restricted distributions in the West and Southwest. Native to southeastern Europe and adjacent Asia, *J. regia* is planted as an ornamental and a crop across the continent. It apparently has not naturalized.

■ horses, fresh shavings, new or old wood, stall bedding; laminitis in 12–24 hours

Disease Problems—Laminitis has been a problem in the management of horses probably for as long as horses have been domesticated. In most instances it involves only solitary individuals. However, in the 1970s, several incidents of the problem occurred in which large numbers of horses bedded on wood shavings in stalls were affected (True et al. 1979). The common factor among the farms was the presence of black walnut shavings. The resulting disease affected most of the exposed horses, causing limb edema and/or laminitis, especially of the thoracic limbs. The problem was limited to exposure to fresh shavings from new or old wood. Aged shavings, even those only 1 month old, presented little or no risk.

Experimentally, the disease could not be reproduced in ponies exposed to the shavings. Further evaluation revealed that the shavings themselves, and not contaminant varnishes or glues, were involved. Shavings of various soft woods, such as pine, are commonly used for bedding, and occasional problems with inadvertent contamination with small amounts of black walnut shavings continues to present problems (Ralston & Rich 1983; Uhlinger 1989). As little as 20% contamination has proven to be a problem (Ralston & Rich 1983). The signs typically develop within 12–24 hours and in most instances are readily reversible within 1–2 days after removal from the offending bedding. However, in some instances severe foot problems develop even with intensive therapy (Thomsen et al. 2000).

■ naphthoquinone, juglone of unknown significance; more likely an aqueous soluble toxin; enhanced vasoconstrictive activity of epinephrine and norepinephrine, worsened by hydrocortisone, reversed by α_1-adrenergic blockers, prazocin

Disease Genesis—Because the signs were associated only with fresh shavings, it was assumed that the agent in the shavings was either volatile or subject to decomposition upon exposure to air (True et al. 1979). The most likely culprit appeared to be juglone, a naphthoquinone long known to be a constituent of black walnut tissues. Juglone had been incriminated as a possible cause of the allelopathic effects of various species of *Juglans* whereby germination and growth of other plants within the root zone of the trees were inhibited (Massey 1925; Davis 1928). Juglone seemed to be particularly attractive as a factor in the development of laminitis, because it was known to be readily oxidized on exposure to air (Gries 1943). The role of juglone as the sole cause of the allelopathic effects continues to be questioned (MacDaniels & Pinnow 1976). It is also present in a number of species of *Carya*.

Experimentally, juglone administered to laboratory animals caused marked sedation but no changes suggestive of edema formation (Westfall et al. 1961). Attempts at reproducing the laminitis syndrome using either shavings or juglone have been rather unsuccessful with topical exposure and inconsistent with oral exposure (True & Lowe 1980). Aqueous extracts of black walnut heartwood, however, do contain an agent or agents capable of consistently causing the laminitis syndrome (Minnick et al. 1987; Uhlinger 1989; Galey et al. 1990b). It is still difficult to produce the disease with topical exposure; oral exposure remains the most consistent way to induce signs of laminitis. This finding is at odds with the original occurrences, which were of such magnitude and uniformity of onset as to be most likely due to topical exposure rather than ingestion. The aqueous extract appears to provide one of the most useful models for the study of equine laminitis. The laminitis that is induced by orally administered extracts of the heartwood results in increased capillary pressure and perfusion, leading to transvascular movement of fluid and subsequent increase in tissue edema and fluid pressure (Eaton et al. 1995). There is a 20% decrease in central venous pressure but no change in overall blood pressure or capillary fragility.

The specific toxicant in the aqueous extract remains to be identified, but in vitro experiments indicate that it produces its effects by enhancement of the vasoconstrictive actions of circulating neurohumoral agents such as epinephrine (Galey et al. 1990a). Furthermore, in vivo results of γ-scintigraphy show a decrease in perfusion of the foot, which is reversible by selective α_1-adrenergic blockade (Galey et al. 1990b). In vitro reversal of equine digital contraction induced by a combination of epinephrine, hydrocortisone, and the aqueous extract of black walnut was accomplished by prazocin, isoxsuprine, or nifedipine (Galey et al. 1989). This indicates a role for either α_1-adrenergic receptor blockade or calcium channel antagonism in treatment. In contrast to the vasoconstrictive effects of *J. nigra,* aqueous extracts of the leaves of *J. regia* have a vasorelaxant effect on rat aorta, inhibiting the contractile response induced by norepinephrine (Perusquia et al. 1995).

Clinical Signs—Over the course of a few hours, there is onset of depression, edema of the pelvic limbs, and signs of distress. Acute laminitis, most evident in the thoracic limbs, will be expressed as warm, sensitive hooves, and the appearance of a myositis-like stiff gait or stance, reluctance to move, and weight bearing mainly by the pelvic limbs. Distress may be indicated by flaring of the nostrils and signs of abdominal pain. There may be edema of the ventral aspects of the neck and chest. Body temperature and respiration and pulse rates typically increase.

- depression, distress, edema of limbs, warm hooves, stiff gait, reluctance to move, weight shifted to pelvic limbs

Pathology—*Juglans* produces a disabling rather than fatal disease. However, in the event of euthanasia of a suffering animal, rotation of the third digit within the hoof may be evident.

- rotation of digit within hoof

Treatment—Although newer drugs such as adrenergic blockers (prazosin) and calcium channel blockers (nifedipine) appear promising, traditional nonsteroidal anti-inflammatory drugs remain a mainstay of treatment. They include phenylbutazone and/or flunixin meglumine i.v. at least initially. Corticosteroids are generally avoided because they may enhance the existing vasoconstriction (Eyre et al. 1979; Galey et al. 1990a). Sedation may be necessary, and mineral oil may be helpful to relieve the abdominal discomfort and to aid in elimination of any foreign material from the digestive tract. In most instances removal of the offending shavings results in alleviation of signs within hours. It may be advisable in those instances where complete avoidance is not feasible to stockpile shavings so that they are allowed to age for several weeks or more prior to use.

- phenylbutazone, flunixin meglumine, sedation, mineral oil orally, remove from new shavings, possible use of adrenergic blockers

References

Davis EF: The toxic principle of *Juglans nigra* as identified with synthetic juglone, and its toxic effects on tomato and alfalfa plants. *Am J Bot* 15;620, 1928

Eaton SA, Allen D, Eades SC, Schneider DA: Digital Starling forces and hemodynamics during early laminitis induced by an aqueous extract of black walnut *(Juglans nigra)* in horses. *Am J Vet Res* 56;1338–1344, 1995

Eyre P, Elmes PJ, Strickland S: Corticosteroid-potentiated vascular responses of the equine digit: a possible pharmacologic basis for laminitis. *Am J Vet Res* 40;135–138, 1979

Galey FD, Beasley VR, Schaeffer DJ, Davis LE: Antagonism in isolated equine digital vessels of contraction induced by epinephrine in the presence of hydrocortisone and an aqueous extract of black walnut *(Juglans nigra)*. *J Vet Pharmacol Ther* 12;411–420, 1989

Galey FD, Beasley VR, Schaeffer DJ, Davis LE: Effect of aqueous extract of black walnut *(Juglans nigra)* on isolated equine digital vessels. *Am J Vet Res* 51;83–88, 1990a

Galey FD, Twardock AR, Goetz TE, Schaeffer DJ, Hall JO, Beasley VR: Gamma scintigraphic analysis of the distribution of perfusion of blood in the equine foot during black walnut *(Juglans nigra)*–induced laminitis. *Am J Vet Res* 51;688–695, 1990b

Gries GA: Juglone—the active agent in walnut toxicity. *Annu Rep North Nut Grow Assoc* 34;52–55, 1943

MacDaniels LH, Pinnow DL: Walnut toxicity, an unresolved problem. *Annu Rep North Nut Grow Assoc* 67;114–122, 1976

Massey AB: Antagonism of the walnuts (*Juglans nigra* L. and *Juglans cinerea* L.) in certain plant associations. *Phytopathology* 15;773–784, 1925

Minnick PD, Brown CM, Braselton WE, Meerdink GL, Slanker MR: The induction of equine laminitis with an aqueous extract of the heartwood of black walnut *(Juglans nigra)*. *Vet Hum Toxicol* 29;230–233, 1987

Perusquia M, Mendoza S, Bye R, Mata R: Vasoactive effects of aqueous extracts from five Mexican medicinal plants on isolated rat aorta. *J Ethnopharmacol* 46;63–69, 1995

Ralston SL, Rich VA: Black walnut toxicosis in horses. *J Am Vet Med Assoc* 183;1095, 1983

Thomsen ME, Davis EG, Rush BR: Black walnut induced laminitis. *Vet Hum Toxicol* 42;8–11, 2000

True RG, Lowe JE: Induced juglone toxicosis in ponies and horses. *Am J Vet Res* 41;944–945, 1980

True RG, Lowe JE, Heissen J, Bradley W: Black walnut shavings as a cause of acute laminitis. *Proc Annu Meet Am Assoc Equine Pract* 24;511–516, 1979

Uhlinger C: Black walnut toxicosis in ten horses. *J Am Vet Med Assoc* 195;343–344, 1989

Westfall BA, Russell RL, Auyong TK: Depressant agent from walnut hulls. *Science* 134;1617, 1961

Plate 7.1. *Cicuta maculata.*

Plate 8.1. *Nerium oleander.*

Plate 12.1. *Asclepias fascicularis.*

Plate 12.2. *Asclepias labriformis.*

Plate 12.3. *Asclepias latifolia.*

Plate 12.4. *Asclepias speciosa.*

Plate 12.5. *Asclepias subverticillata.*

Plate 13.1. *Centaurea solstitialis.*

Plate 13.2. *Eupatorium rugosum.*

Plate 13.3. *Gutierrezia sarothrae.*

Plate 13.4. *Helenium flexuosum.*

Plate 13.5. *Hymenoxys odorata.*

Plate 13.6. *Isocoma pluriflora.*

Plate 13.7. *Senecio flaccidus.*

Plate 13.8. *Senecio riddellii.*

Plate 13.9. *Tetradymia canescens.*

Plate 15.1. *Cynoglossum officinale.*

Plate 23.1. *Halogeton glomeratus.*

Plate 23.2. *Kochia scoparia.*

Plate 24.1. *Hypericum perforatum.*

Plate 35.1. *Rhododendron catawbiense.*

Plate 36.1. *Stillingia sylvatica.*

Plate 37.1. *Astragalus mollissimus.*

Plate 37.2. *Crotalaria spectabilis.*

Plate 37.3. *Glottidium vesicarium.*

Plate 37.4. *Lathyrus latifolius.*

Plate 37.5. *Lupinus sericeus.*

Plate 37.6. *Oxytropis lambertii.*

Plate 37.7. *Senna obtusifolia.*

Plate 37.8. *Sophora nuttalliana.*

Plate 37.9. *Sophora secundiflora.*

Plate 37.10. *Thermopsis rhombifolia.*

Plate 39.1. *Dicentra eximia.*

Plate 41.1. *Aesculus glabra.*

Plate 44.1. *Triglochin maritima.*

Plate 45.1. *Perilla frutescens.*

Plate 47.1. *Convallaria majalis.*

Plate 47.2. *Veratrum californicum.*

Plate 47.3. *Zigadenus nuttallii.*

Plate 49.1. *Gelsemium sempervirens.*

Plate 59.1. *Rumex crispus.*

Plate 62.1. *Aconitum columbianum.*

Plate 62.2. *Delphinium carolinianum.*

Plate 68.1. *Digitalis purpurea.*

Plate 69.1. *Cestrum diurnum.*

Plate 69.2. *Nicotiana glauca.*

Plate 69.3. *Solanum elaeagnifolium.*

Plate 73.1. *Lantana camara.*

Juncaginaceae Rich. 44

Comprising 4 genera and 14–17 species, the Juncaginaceae is a family of perennial and annual marsh plants found primarily in the temperate and cold regions of both hemispheres, especially along the coasts. The common name *arrowgrass family* is indicative of the superficial grasslike appearance of the plants, which are of no economic importance other than some being cyanogenic. In North America, only 1 genus occurs.

■ Arrowgrass Family
Triglochin

Plants cespitose herbs; perennials; acaulescent; from rhizomes; 20–80 cm tall. **Leaves** simple; alternate; basal; with basal membranous sheaths, margins overlapping; blades linear, elongate, fleshy, lenticular in cross section; ligules present. **Inflorescences** racemes or spikes; borne on elongate scapes; bracts absent. **Flowers** perfect; perianths in 1-series; radially symmetrical. **Perianth Parts** 3 or 6; in 1 or 2 whorls; free; concave; greenish. **Stamens** 3 or 6. **Pistils** 1; carpels 3 or 6; stigmas 3 or 6, plumose or papillose; styles absent; ovaries superior; locules 3 or 6; placentation basal or apical. **Fruits** follicular schizocarps, the mature follicles separating from the persistent central axis. **Seeds** 3 or 6; 1 per carpel.

■ perennial herbs from rhizomes; leaves simple, basal, fleshy, crescent-shaped in cross section; flowers in spikes or racemes

Triglochin L.

Taxonomy and Morphology—Almost cosmopolitan in distribution, *Triglochin* comprises about 12 species (Haynes & Hellquist 2000). Its name is derived from the Greek roots *treis* and *glochis*, for "three" and "point," which alludes to the three-pointed fruits of some species. The genus is represented in North America by 4 species (see plate 44.1):

■ *T. gaspensis*
T. maritima
T. palustris
T. striata
arrowgrass

T. gaspensis Lieth & D.Löve (= *T. gaspense*)	gaspe arrowgrass
T. maritima L. (= *T. maritimum*) (= *T. concinnum* Burtt Davy)	seaside arrowgrass, low arrowgrass, graceful arrowgrass
T. palustris L. (= *T. palustre*)	marsh arrowgrass, small arrowgrass
T. striata Ruiz & Pav. (= *T. striatum*)	three-ribbed arrowgrass, southern arrowgrass

In some taxonomic treatments, *Triglochin* is treated as a neuter noun—hence the *-um* ending for the specific epithets. Haynes and Hellquist (2000) treated the genus as feminine in accordance with the international rules of botanical nomenclature; an interpretation followed in this treatise.

Morphological features of the genus are the same as those of the family and are not repeated here.

Triglochin maritima

1. Stigmas 6. Ovary lobes 6. Central axes of schizocarps terete; not 3-winged. Ligule apices 2-lobed.
   2. Racemes 2–7 cm long. Leaves arching outward from sheaths . . . . *T. gaspensis*
   2. Racemes 6–45 cm long. Leaves erect from sheaths . . . . *T. maritima*
1. Stigmas 3. Ovary lobes 3. Central axes of schizocarps 3-angled; 3-winged. Ligule apices entire.
   3. Fruits linear-clavate; 7–8.3 mm long . . . . *T. palustris*
   3. Fruits globose; 1–2 mm long . . . . *T. striata*

■ marshy sites, edges of brackish waters, saline marshes

Distribution and Habitat—Species of the genus grow in tufts, sometimes forming rather dense and/or extensive patches in marshy sites, along the edges of brackish waters, and in depressions in areas of wet saline marshland soils from coast to coast across the northern half of North America. Along the coasts, plants can be found as far south as lower California and Delaware. They begin growth early in the spring and flower in midsummer.

■ ruminants, fresh or poorly dried hay; abrupt onset of cyanide toxicosis

Disease Problems—Species of *Triglochin* have long been recognized as toxic in northern Europe. They became of concern in North America with the advent of the livestock industry in the northern Great Plains and the intermountain region. The toxicity problem was accentuated as harvesting native grasses for livestock production increased in importance. While the grasses readily dried and cured, the more succulent *Triglochin* remained in a semigreen state. When the hay was subsequently consumed, intoxications occurred. Regrowth after cutting may be even more toxic whether eaten fresh or consumed in hay (Marsh et al. 1929; Fleming et al. 1933). Realization of these problems prompted considerable interest in the genus (Fleming et al. 1920; Marsh et al. 1929). Both sheep and cattle readily eat fresh plants, although sheep are more likely to do so. Consumption may be fostered by the high salt content of the plants, especially by animals newly introduced to arrowgrass-infested saline marshlands and pastures (Looman 1976).

■ cyanogenic glycoside, triglochinin, little risk with well-cured/dried hay

Disease Genesis—Toxicity is due primarily to triglochinin, a tyrosine-derived cyanogenic glucoside that has been isolated from both *T. maritima* and *T. palustris* (Eyjolfsson 1970; Ettlinger & Eyjolfsson 1972). Other glucosides—e.g., taxiphyllin in seedlings—are present, but relative to triglochinin, these are of little consequence (Ettlinger & Eyjolfsson 1972; Nahrstedt et al. 1979). Hydrolytic β-glucosidases specific for the glucosides are also present (Nahrstedt et al. 1979). Dry weight concentrations of triglochinin in *T. maritima* are reported to range from 1.3 to 3% or more in leaves and up to almost 7% in the flowering spikes (Majak et al. 1980). In studies by Fleming and co-workers (1920) and Beath and co-workers (1933), toxicity, expressed as HCN potential in the plant rather than as dosage, was up to 550 ppm wet weight or 2,640 ppm d.w. Toxicity of *T. palustris* appears to be lower, only 135 ppm d.w. (Moran et al. 1940). Depending upon plant HCN potential, the lethal dose of green *T. maritima* ranges from 0.5–2% b.w. in sheep and slightly less (to 1.5% b.w.) in cattle (Fleming et al. 1920; Marsh et al. 1929; Beath 1939; Moran 1954). When plants of low toxicity were initially dried, it took only 0.25% b.w. to cause death in sheep. This represented an apparent increase in toxicity, because the loss of 75–80% of its weight with drying would be expected to decrease the dose only to about 0.5% b.w. (Fleming et al. 1920; Beath et al. 1933).

Triglochin palustris

The effects of *Triglochin* are not cumulative; thus, small repeated doses are of little risk. Hay gradually loses its toxicity potential, but it may retain some HCN potential for a long period. Well-cured hay presents little problem, but when grow-

Triglochinin

Taxiphyllin

ing in swampy areas, the arrowgrass does not dry out well and may remain dangerous for an extended period of time (Fleming et al. 1928). As with other cyanogenic plants, new growth and that during moisture stress have increased triglochinin and toxicity (Clawson & Moran 1937; Beath 1939; Nahrstedt et al. 1979, 1984; Majak et al. 1980). Although increased salinity is associated with increased production of nitrogenous compounds, the toxicity potential seems to be greater in less alkaline environments (Majak et al. 1980; Jefferies & Rudmik 1984).

Clinical Signs—Ingestion of *Triglochin* produces the clinical signs characteristic of cyanide intoxication, including increased respiration rate and anxiety. These typically occur quickly, within 5–30 minutes after ingestion. If death does not occur during the following few hours, the chances of recovery are high; however, it may take up to 24 hours for complete resolution of clinical signs.

Pathology and Treatment—Because of the peracute nature of the disease, pathologic changes are limited to scattered small hemorrhages on various organ surfaces. There are no distinctive changes. Treatment entails the use of sodium thiosulfate alone or in combination with sodium nitrite both given i.v. A complete discussion of treatment is presented in the discussion of the Rosaceae (see chapter 64).

Triglochin gaspensis

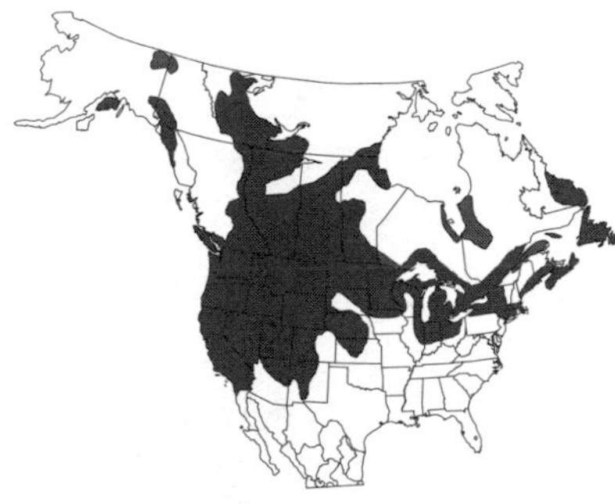

Triglochin maritima

- abrupt onset, apprehension, distress, weakness, ataxia, labored respiration, collapse, seizures, rapid sequence
- no lesions; treat with sodium thiosulfate

Triglochin palustris

Triglochin striata

References

Beath OA: *Poisonous Plants and Livestock Poisoning.* Wyo Agric Exp Stn Bull 231, 1939

Beath OA, Draize JH, Eppson HF: *Arrow Grass—Chemical and Physiological Considerations.* Wyo Agric Exp Stn Bull 193, 1933

Clawson AB, Moran EA: *Toxicity of Arrowgrass for Sheep and Remedial Treatment.* USDA Tech Bull 580, 1937

Ettlinger M, Eyjolfsson R: Revision of the structure of the cyanogenic glucoside triglochinin. *J Chem Soc Chem Commun* 1972;572–573, 1972

Eyjolfsson R: Isolation and structure determination of triglochinin, a new cyanogenic glucoside from *Triglochin maritima. Phytochemistry* 9;845–851, 1970

Fleming CE, Peterson NF, Miller MR, Wright LH, Louck RC: *Arrow-grass, A New Stock Poisoning Plant* (Triglochin maritima). Nev Agric Exp Stn Bull 98, 1920

Fleming CE, Miller MR, Vawter LR, Young A: *Poisonous Range Plants.* Nev Agric Exp Stn Annu Rep 1928;21–22, 1928

Fleming CE, Miller MR, Vawter LR, Young A: *Poisonous Range Plants.* Nev Agric Exp Stn Annu Rep 1933;10–11, 1933

Haynes RR, Hellquist CB: Juncaginaceae Richard: Arrow-grass family. In *Flora of North America,* vol 22, *Magnoliophyta: Alismatidae, Arecidae, Commelinidae (in part), and Zingiberidae,* Flora of North America Editorial Committee ed, Oxford Univ Press, New York, pp43–46, 2000

Jefferies RL, Rudmik T: The responses of halophytes to salinity: an ecological perspective. In *Salinity Tolerance in Plants,* Staples RC, Toenniessen GH eds, Wiley & Sons, New York, pp213–227, 1984

Looman J: Biological flora of the Canadian prairie provinces. 4. *Triglochin* L., the genus. *Can J Plant Sci* 56;725–732, 1976

Majak W, McDiarmid RE, Hall JW, Van Ryswyk AL: Seasonal variation in the cyanide potential of arrowgrass *(Triglochin maritima). Can J Plant Sci* 60;1235–1241, 1980

Marsh CD, Clawson AB, Roe GC: *Arrow Grass* (Triglochin maritima) *as a Stock-Poisoning Plant.* USDA Tech Bull 113, 1929

Moran EA: Cyanogenetic compounds in plants and their significance in animal industry. *Am J Vet Res* 15;171–176, 1954

Moran EA, Briese RR, Couch JF: Some new cyanogenetic plants. *J Wash Acad Sci* 30;237–239, 1940

Nahrstedt A, Hosel W, Walther A: Characterization of cyanogenic glycosides and β-glucosidases in *Triglochin maritima* seedlings. *Phytochemistry* 18;1137–1141, 1979

Nahrstedt A, Kant J-D, Hosel W: Aspects on the biosynthesis of the cyanogenic glucoside triglochinin in *Triglochin maritima. Planta Med* 50;394–398, 1984

Lamiaceae Lindl. 45

Cosmopolitan in distribution with greatest abundance in the Mediterranean region, the Lamiaceae, or mint family, comprises about 200 genera and 3,200 species. An alternative name is Labiatae, one of eight family names not ending in *-aceae*. Economically the family is of considerable importance as a source of aromatic oils (lavender, rosemary, mint), culinary herbs (oregano, thyme, savory, sage, basil), and ornamentals (salvia, coleus, bugloss, skullcap). In North America, about 70 genera and 500 species, both native and introduced, are present. Seven genera have been implicated in toxicologic problems.

- Mint Family

  Galeopsis
  Glechoma
  Lamium
  Perilla
  Salvia
  Stachys

  QUESTIONABLE
  Hedeoma

Plants herbs or shrubs; annuals or perennials; strongly aromatic or not aromatic. **Stems** 4-sided. **Leaves** simple or rarely compound; opposite or rarely whorled; typically decussate; cauline or forming a basal rosette; petiolate or sessile; venation pinnate or palmate; margins usually serrate; stipules absent. **Inflorescences** solitary flowers or simple cymes or spikes or racemes or verticils; terminal or axillary; bracts present. **Flowers** perfect; perianths in 2-series. **Calyces** bilaterally or radially symmetrical. **Sepals** 5 or 4; fused. **Corollas** bilaterally or rarely radially symmetrical; typically bilabiate with 2-lobed upper lips and 3-lobed lower lips. **Petals** 5 or rarely 4; fused; of various colors. **Stamens** 2 or 4; didynamous or of equal lengths; exserted beyond or included within corollas; epipetalous. **Pistils** compound, carpels 2, each divided in half; stigmas 2; styles 1; gynobasic or rarely apical; ovaries superior; locules 4; placentation axile, often appearing basal; nectaries present, receptacular, annular. **Fruits** nutlets; 4. **Seeds** 1 per nutlet.

- herbs or shrubs, stems 4-sided; corollas typically bilabiate; petals 5; fruits nutlets

The Lamiaceae is readily distinguished from all other flowering families, except the Verbenaceae, by its distinctive gynoecium of four nutlets with a gynobasic style. This attribute combined with the characters of square stems, opposite leaves, and bilabiate corollas is diagnostic.

Galeopsis tetrahit

Galeopsis L.

Taxonomy and Morphology—Native to temperate Eurasia, *Galeopsis* comprises 10 species. In North America, 1 species is of toxicologic interest:

- *G. tetrahit*

  hemp-nettle

G. tetrahit L. hemp-nettle

Plants herbs; annuals; from taproots. **Stems** erect; 15–70 cm tall; branched; surfaces retrorsely hispid and setose. **Leaves** petiolate; blades ovate to lanceolate or rhombic; surfaces hirsute or strigose; apices acuminate; margins coarsely crenate-serrate; bases cuneate; petioles hirsute. **Inflorescences** dense verticils; uppermost crowded, lower distant; bracteoles subulate. **Calyces** tubular-campanulate; 10-nerved. **Sepals** 5; lobes spinose, hirsute. **Corollas** bilabiate; upper lips hooded and pubescent; lower lips

- erect, branched, annual herbs; leaves petiolate; petals purple, pink, or white

Galeopsis tetrahit

3-lobed with 2 conical protrusions at base. **Petals** 5; purple or pink or white, with darker markings. **Stamens** 4; didynamous; included within corollas. **Nutlets** trigonous; mottled brown.

Some taxonomists use the binomial *G. bifida* Boenn. for this taxon.

Distribution and Habitat—A native of Europe, *G. tetrahit* has naturalized across the northern half of the continent. It occurs in both disturbed and undisturbed habitats.

- irritant digestive disturbance; toxin unknown, possibly galiridoside

Disease Problems and Genesis—Although it affects other animal species, *G. tetrahit* has primarily been a problem when its seeds contaminate feed for horses (Cooper & Johnson 1984). Feed containing 1.1–2.5% seeds, fed for about 1 week, was fatal to 4 of 18 horses and 79 of 150 pigs (Frolkin 1965). The signs were typical of digestive tract irritation. The toxin is unknown, but it appears to be an irritant type such as galiridoside.

HO
O
O
CH_3 O-Glucose

Galiridoside

- anorexia, sweating, tremors, constipation or diarrhea

  reddened gut mucosa

Clinical Signs, Pathology, and Treatment—With intoxications caused by *G. tetrahit*, there is initially loss of appetite, followed by abrupt onset of sweating and tremors. These signs are accompanied first by constipation and then diarrhea. The pathologic lesions are nonspecific and include reddening and hemorrhage of the mucosa of the stomach and small intestine. There also may be subcutaneous edema and icterus. Relief from the irritant effects may require the use of pain moderators as well symptomatic treatment of the digestive disturbance.

Glechoma L.

- *G. hederacea*

  alefoot
  alehoof
  creeping Charlie
  field balm
  gill-over-the-ground
  ground ivy
  runaway robin

Taxonomy and Morphology—Comprising 10–12 species, *Glechoma* is native to temperate Eurasia. Its name is a Latinization of *glechon,* the ancient Greek name for pennyroyal. Some taxonomists contend that *Glecoma* is the correct spelling according to the rules of botanical nomenclature. One species is present in North America:

G. hederacea L.	gill-over-the-ground, creeping Charlie, runaway robin, field balm, alefoot, alehoof, ground ivy

- mat-forming, branched, perennial herbs; leaves with long petioles; petals blue-violet, white, or pink

Plants herbs; perennials; mat-forming; from rhizomes or stolons; root systems fibrous. **Stems** lax; 10–40 cm tall; branched; surfaces glabrous to sparsely hirsute, pilose at nodes. **Leaves** long petiolate; blades suborbicular to reniform; surfaces glabrous or sparsely hirsute; margins coarsely crenate; bases cordate. **Inflorescences** axillary verticils; flowers 2–6, pedicelled. **Calyces** tubular to obconic; 15-nerved. **Sepals** 5; hirtellous-scabrous; adaxial lobes longer. **Corollas** bilabiate; upper lips concave; lower lips 3-lobed, spreading, middle lobe larger. **Petals** 5; blue-violet or occasionally white or pink. **Stamens** 4; didynamous; included within corollas. **Nutlets** brownish; smooth.

Distribution and Habitat—Introduced from Europe, *G. hederacea* has become a noxious lawn weed in North America, sometimes forming dense mats, especially in the northern half of the continent. A variegated leaf form is cultivated for its trailing stems and used in hanging baskets and window boxes.

Glechoma hederacea

Disease Problems and Genesis—A problem mainly in Europe, *G. hederacea* when ingested by horses in large amounts, either fresh or in hay, causes irritation of the digestive tract (Cooper & Johnson 1984). The disease has been experimentally reproduced in horses with both fresh and wilted plant material. It also affects cattle, causing respiratory distress. Although the species is common, it is rarely a cause of disease in North America (Kingsbury 1964). The toxin is unknown, but it appears to be an irritant type.

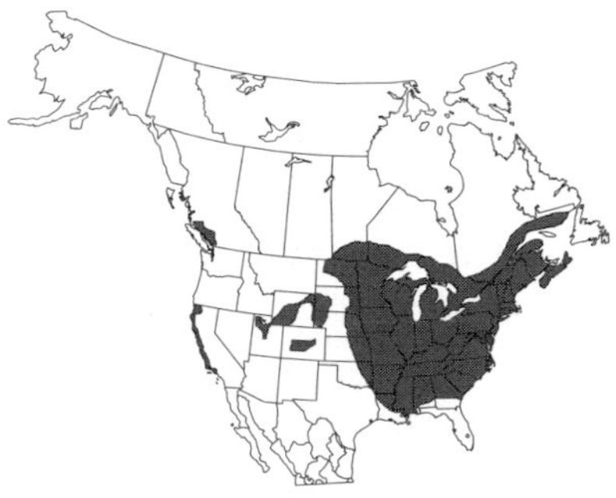

Glechoma hederacea

Glechomanolide

Sapogenin

- irritant, digestive disturbance; toxin unknown, possibly glechomanolide or sapogenin

Clinical Signs, Pathology, and Treatment—With intoxication due to *G. hederacea,* there is abrupt onset of colic with labored respiration, sweating, excess salivation, and apparent dizziness. The disease may be fatal. Grossly, there is reddening, with some hemorrhage of the mucosa of the stomach and small intestine. There may also be pulmonary edema and emphysema. Relief from the irritant effects may require the use of pain moderators as well as symptomatic treatment of the digestive disturbance.

- abrupt onset, colic, sweating, excess salivation

  reddened gut mucosa

Lamium L.

- *L. amplexicaule*

  henbit
  henbit dead nettle

Taxonomy and Morphology—Native to the Old World and especially the Mediterranean region, *Lamium,* commonly known as dead nettle, comprises 40–50 species. Used by the Roman natural historian Pliny, *Lamium* gets its name from the Greek *laimos,* or *lamos,* meaning "throat," which alludes to the tubular appearance of the corolla. Some species are cultivated as ground covers, and several cultivars have been developed. In North America, 5 species have been introduced. One is of toxicologic interest:

L. amplexicaule L. henbit, henbit dead nettle

- annual herbs; petals pinkish purple to red

Plants herbs; annuals; from short taproots; herbage glabrous to hirsute. **Stems** decumbent to ascending; 10–40 cm tall; branched at bases. **Leaves** cordate to suborbicular or reniform; lower leaves petiolate; upper leaves sessile; apices rounded; margins coarsely crenate; bases cordate or truncate. **Inflorescences** axillary verticils; flowers 6–12, sessile. **Calyces** tubular to campanulate; 10-nerved. **Sepals** 5; hirsute; lobes equal. **Corollas** bilabiate; upper lips hooded, abaxial surfaces pubescent; lower lips spreading, 3-lobed, middle lobe larger and constricted at base. **Petals** 5; pinkish purple to red. **Stamens** 4; didynamous; included within corollas. **Nutlets** trigonous; brownish; smooth.

Lamium amplexicaule

■ invasive weed

Distribution and Habitat—A Eurasian introduction that has become widespread over North America, *L. amplexicaule* is a noxious weed of disturbed sites and infests lawns, gardens, and croplands in the spring.

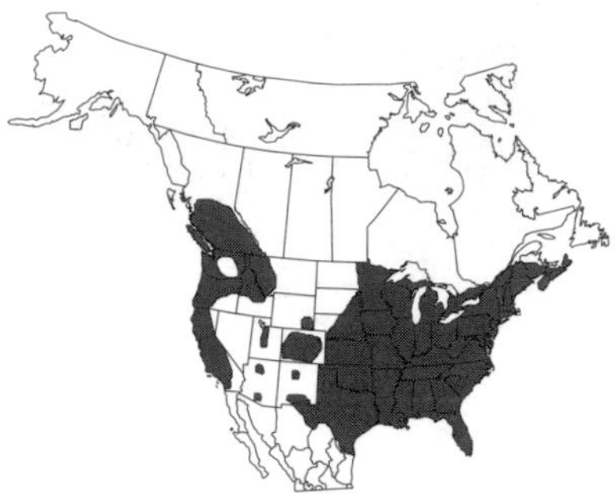

Lamium amplexicaule

■ fresh but not dried plant; neurologic actions, mainly in sheep, as yet not reported in North America

Disease Problems and Genesis—Neurologic problems caused by *L. amplexicaule* have been limited to Australia (Everist 1981). Signs of intoxication become evident after consumption of fresh plants for several days and animals are forced to move some distance, such as being driven from one pasture to another. Experimental reproduction of the disease has not been consistent. One attempt required 128 kg of plant material to be fed for 13 days to a sheep before signs were observed. However, predisposition to develop the neurologic response when stressed lasted for almost 3 weeks after feeding *Lamium* ceased (Dodd & Henry 1922). Nursing lambs developed the signs when ewes were fed the species. Sheep appear to be more prone to develop the disease, although it has occurred in horses and cattle as well. As yet the disease has not been reported in North America, although *L. amplexicaule* is a very common winter and spring annual and often develops immense populations in disturbed soils. Marsh and co-workers (1928) attempted to reproduce the disease identical to that caused by *Lamium* and *Malva parviflora* in Australia (Dodd & Henry 1922) but were unable to do so with 154% b.w. over 26 days, using the dried plant. Apparently the dried plants are not a risk. The specific cause of disease is not known. Lamiol is present but of unknown significance.

OH OH CH_3 O HO CH_3 O-Glucose

Lamiol

■ stilted gait, tremors, collapse; transient

Clinical Signs—After a period of stress, the animal exhibits a stilted gait with the head held high. Tremors begin in the shoulders and thighs, and eventually the animal loses control of the limbs. When the animal is down, there is flaccidity and little response to stimuli or pain. It may recover after as little as 5 minutes rest in some instances.

■ no lesions, treatment not needed

Pathology and Treatment—The disease produced by *L. amplexicaule* is rarely fatal, and there are no distinctive lesions. Typically, with rest, the signs are readily reversible without treatment.

Perilla L.

■ *P. frutescens*
beefsteak plant
blueweed
Joseph's coat
perilla
perilla mint
purple mint
rattlesnake weed
wild coleus

Taxonomy and Morphology—Comprising 4–6 species native to Southeast Asia, *Perilla* is morphologically similar to *Coleus* and is sometimes mistaken for it. Like coleus, it is cultivated for its summer foliage. Its name is a Latinization of a Hindu name. In North America, a single species has escaped cultivation and naturalized (see plate 45.1):

P. frutescens (L.) Britton	purple mint, perilla, perilla mint, beefsteak plant, wild coleus, Joseph's coat, rattlesnake weed, blueweed

■ herbs, annuals, strongly aromatic, erect; leaves often purplish especially undersides; corollas weakly bilabiate; petals purple, lavender, or white

Plants herbs; annuals; herbage strongly aromatic; often purplish. **Stems** erect; 20–175 cm tall; branched or unbranched above; surfaces glabrous or sparsely pubescent or pilose. **Leaves** petiolate; blades broadly ovate or elliptic; apices acute or acuminate; margins serrate; bases cuneate to attenuate. **Inflorescences** elongate racemes; terminal or axillary; flowers solitary in axils of ovate to elliptic bracts. **Calyces** weakly bilabiate; campanulate; 10-nerved. **Sepals** 5; villous; lobes subequal. **Corollas** nearly radially symmetrical; lobes subequal, rounded. **Petals** 5; purple to lavender or white. **Stamens** 4; of equal lengths; exserted beyond corollas. **Nutlets** globose; areolate; reddish brown.

Distribution and Habitat—Native to India, *P. frutescens* was later cultivated in China and Japan and subsequently brought by immigrants to the New World. It escaped cultivation and is now widely distributed throughout the southeastern United States. Forming dense stands, it is found in semishaded waste areas and damp woods, along open-wooded streams, and in seepage areas. In some areas it competes very well with other forbs and grasses and may dominate, especially during years of limited moisture.

Perilla frutescens

- dense stands in semishaded sites

Perilla frutescens

Disease Problems—*Perilla frutescens* is an attractive, aromatic herb that has a long history of use in Chinese traditional medicine and as a popular condiment in Japanese cooking. After its arrival in the New World, it found a home in the Appalachians and the Ozarks, where it was used for a variety of culinary purposes (Potts 1996). Studies by Petersen (1965) in the 1950s established it as a probable cause of extensive cattle losses in Arkansas and eastern Oklahoma in the 1940s and 1950s. It is now recognized as a cause of acute respiratory distress syndrome (ARDS), most commonly in cattle but also in other ruminants and horses (Linnabary et al. 1977; Lindley 1978; Kerr & Linnabary 1989). The hazard for acquiring this disease, also called panting disease, seems to be greatest in the late summer and early fall, and Petersen (1965) has used the phrase "fair time to frost" as the time of highest risk. Dried plants in hay are also toxic, but the risk is low compared with that of green plants in flower and/or fruit. Although this mint has often been regarded to be of low palatability, it is readily consumed at times. This may account for the higher frequency of the disease in some years, especially when other forage is limited. The specific amounts of plant material required to cause intoxication are not known, but an animal may eat sufficient amounts of fresh plants in a day to cause disease. Experimentally, as little as 2.3 kg of fresh plant caused in 24 hours the death of a calf weighing 79 kg (Kerr et al. 1986).

- mature plants, in flower or seed, fresh or dried; eaten for 1–3 days, ARDS; all livestock

Disease Genesis—Two cyanogenic glycosides, prunasin and an isomer of amygdalin, have been identified in *Perilla* (Aritomi et al. 1985). Large amounts of perillaldehyde also are present and are readily oxidized in air (Tada et al. 1996). However, the respiratory effects are due to several volatile, aromatic 3-substituted furan ketones, which are readily absorbed into the systemic circulation (Wilson et al. 1977; Garst & Wilson 1984). Large dosage of the ketones may cause diffuse alveolar injury and an immediate increase in fluid (edema) in the lungs (Breeze et al. 1984; Abernathy et al. 1994). It is not certain whether the ketones are directly toxic or must be biotransformed to reactive intermediates. In some studies it appears that the ketones are very quickly bioactivated at C-5 of the furan ring by the cytochrome P-450 oxidase system in the lungs (Boyd 1980; Garst & Wilson 1985). However, in an in vitro study, perilla ketones also appeared to exert a direct action on endothelial cells, causing a reversible increase in permeability by disruption of actin microfilaments (Waters et al. 1993). In the case of bioactivation, the resulting short-lived reactive intermediates are potent edematogenic agents, causing immediate damage to the pulmonary microvascular endothelial cells and to the type I pneumocytes that normally form 95% of the alveolar surface (Guerry-Force et al. 1988). This leads to significant increases in vascular permeability but not to an increase in pressure (Abernathy et al. 1992). There is subsequent stimulation of type II pneumocytes to replace the destroyed type I cells and to restore the epithelial barrier. Excessive proliferation of type II cells quickly leads to a glandular appearance of the tissue in 12–24 hours.

- volatile perilla furan ketones; form short-lived intermediates in lung; immediate damage to pneumocytes and endothelial cells; increased vascular permeability

O
C-C-C-C(CH3)2
O

Perilla ketone

Most animal species are susceptible, although in the dog and pig very high doses are required and the site of tissue destruction is the liver rather than the lung (Garst

Perillaldehyde

et al. 1985). The bioactivation rate, which is subject to steric and other influences, exerts some control over expression of the toxic effects (Garst et al. 1985; Wilson et al. 1990). The ketones also have a direct action on the small intestine, increasing peristaltic activity (Koezuka et al. 1985). The ketones are in highest concentration in the flowers and seeds. Thus, the mint is most toxic when it is flowering or fruiting in August and September. Prior to this time very large amounts are required for intoxication. There are also different chemotypes of *P. frutescens*, some containing the aldehyde forms of the toxin rather than the ketones, which may account for differences in toxicity (Koezuka et al. 1985).

- abrupt onset, labored open-mouth respiration, foamy salivation

Clinical Signs—The disease is characterized by abrupt onset of labored respiration manifested by a loud expiratory grunt, open-mouth respiration, and excess frothy salivation. Harsh respiratory sounds are evident on auscultation. Subcutaneous emphysema may subsequently develop along the neck, shoulders, and back. Death often occurs 1–2 days after onset of signs.

- gross pathology, lungs heavy, wet, rubbery, distended; edema, frothy exudate in airways; liver pale, friable, mottled

  microscopic, alveoli filled with serous exudate, edema, and air pockets in interlobular septa; alveolar epithelialization

Pathology—Grossly, when the chest is opened, the lungs fail to collapse completely and are heavy and firm, with froth-filled airways. Inflation of the lungs may be so severe that imprints of the ribs are apparent on the surfaces of the lung. The interlobular septa are distended with fluid, fibrin, and air bubbles. Cut surfaces reveal a granular or mottled (glandular) appearance of patchy areas of firm, reddened to purple lobules, and prominent edema fluid in connective tissue and septa. Air bubbles and/or large bullae may be apparent, and the lymphatics are usually dilated. Microscopically, alveolar epithelialization, accumulations of alveolar macrophages, and hyaline membrane formation are commonly observed. The pathologic changes are the same as those produced by high levels of tryptophan and moldy runner beans or sweet potatoes.

- avoid stressing animals; atropine

Treatment—Generally of very limited value, treatment with atropine may be beneficial, although it is important to avoid stressing the animals when handling them. This is particularly of concern when facilities are limited and/or the animals are intractable. Grazing in areas infested with *P. frutescens* should be limited during the flowering and fruiting period.

Salvia L.

- *S. coccinea*
  S. reflexa

  annual sage
  blue sage
  lance-leaf sage
  mirto
  red sage
  Rocky Mountain sage
  Texas sage
  tropical sage

Taxonomy and Morphology—Cosmopolitan in distribution but with greatest abundance in tropical America, *Salvia* is the largest genus of the family with about 900 species. Used by the Roman natural historian Pliny, its name is derived from the Latin *salveo,* meaning "to save," alluding to medicinal use of some species. It also is the source of the important cooking herb sage and is among the most popular of ornamentals, with long-lasting displays of both foliage and flowers. In North America, about 67 species are present. Of toxicologic interest are the following:

S. coccinea Juss. ex Murray	tropical sage, red salvia, mirto, mejorana, Texas sage
S. reflexa Hornem. (= *S. lanceaefolia* Poir.) (= *S. lanceolata* Willd.)	lance-leaf sage, blue sage, annual sage, Rocky Mountain sage

- erect herbs to shrubs; petals blue, purple, lavender, white, or red

Plants herbs or subshrubs or shrubs; annuals or biennials or perennials; herbage often aromatic, glabrous to glandular or variously indumented. **Stems** erect or ascending;

20–175 cm tall; branched or unbranched above. **Leaves** petiolate or sessile; blades of various shapes; margins entire or toothed or lobed. **Inflorescences** verticils; 2-to-40-flowered; borne in racemes or panicles; bracts and bracteoles typically present. **Calyces** bilabiate; tubular to campanulate; 10- or 15-nerved; upper lips entire or 3-lobed; lower lips deeply 1-lobed. **Sepals** 5; villous; lobes small, often spinose. **Corollas** bilabiate; tubular; upper lips 2-lobed or entire, hooded or straight or arched; lower lips 3-lobed, spreading, middle lobe large, often emarginate. **Petals** 5; blue to purple to lavender or white or red. **Stamens** 2; exserted beyond or included within corollas. **Nutlets** globose or ovoid; slightly trigonous; smooth or minutely tuberculate.

Salvia reflexa

1. Leaves linear to narrowly oblong-elliptic or lanceolate; bases cuneate. Corollas blue to white .... *S. reflexa*
1. Leaves deltoid-ovate to ovate-lanceolate; bases cordate or truncate or rounded. Corollas bright or deep red .... *S. coccinea*

Distribution and Habitat—Occupying a variety of habitats in different vegetation types, species of *Salvia* are distributed across the continent. *Salvia reflexa* is common in dry sandy or gravelly sites throughout western North America, while *S. coccinea* occupies sandy soils in brushy areas or open woods in the southeastern United States.

- sandy or gravelly soils

Salvia coccinea

Disease Problems—Some species of *Salvia*—e.g., *S. azurea,* the common blue sage of grasslands—are considered to be reasonable forages for livestock. Others are probably of value under most situations, but in unusual conditions species such as *S. coccinea* and *S. reflexa* appear to be capable of causing acute nitrate intoxication (Williams & Hines 1940; Everist 1981). In these situations, intoxication occurs when animals are forced to subsist almost entirely on *Salvia* for a day or more. However, nitrate may not be the only cause of intoxication; *S. reflexa* (reported as *S. lanceolata*) has been associated with an apparent problem of digestive tract irritation when present in alfalfa hay fed to cattle (Beath et al. 1953). The intoxication, also involving muscular weakness, was subsequently reproduced in sheep.

Disease Genesis—Nitrate concentrations may range as high as 25,000–30,000 ppm as NO_3 (Williams & Hines 1940; Everist 1981). On some occasions, exceptionally high levels have been reported. For example, in an intoxication incident of goats in India involving *S. coccinea,* nitrate levels in leaves were 94,000 ppm d.w. (Pal et al. 1997). Abortions have been observed, but at least in one instance the accompanying NO_3 was less than 5,000 ppm and may not have been a factor. Varshneya and coworkers (1995) reported an incident in goats of abortions, lethargy, and decreased appetite and ruminal activity. Nitrate concentration was 15,800 ppm, but the signs were not entirely consistent with this as a cause, especially considering the positive response to treatment with calcium borogluconate, metoclopramide, and magnesium sulfate. In most localities, it is unlikely that sufficient plant material will be available to allow the rapid consumption needed for nitrate intoxication to be a risk.

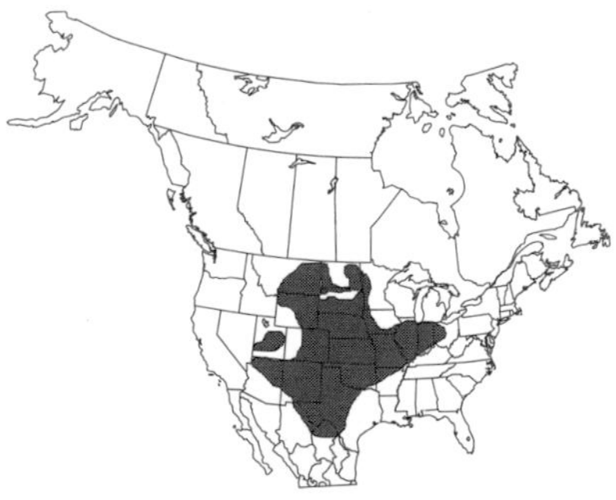

Salvia reflexa

- nitrate intoxication; large amounts of plant
- accumulation of nitrate, under conditions of elevated soil nitrogen
- abrupt onset, weakness, increased respiratory rate, twitching, dark mucous membranes

Clinical Signs—Of abrupt onset several hours after consumption of *Salvia,* the signs are quite often fatal within minutes or a few hours. The primary signs are weakness, increased respiratory rate, muscle twitching, and possibly dark-colored mucous membranes.

Pathology—Few distinctive lesions are seen at necropsy. The blood and tissues may be dark or chocolate colored, but this is not consistent. There also may be scattered,

- few lesions

small, splotchy hemorrhages and increased foamy fluid in the terminal airways of the lungs, but these changes are usually not extensive.

- methylene blue

Treatment—The primary antidote for treatment of nitrate intoxication is methylene blue, 2–4 mg/kg b.w., given as a 1–2% solution.

Stachys L.

- *S. arvensis*

  field hedge nettle
  staggerweed

- branched annual herbs; leaf margins coarsely serrate; flowers in terminal spikes; petals pale purple or white

Taxonomy and Morphology—Distributed primarily in temperate and warm-temperate regions, *Stachys,* commonly known as hedge nettle or betony, comprises about 300 species, some of which are cultivated as ornamentals. Its name is an ancient Greek one used by Dioscorides. In North America, approximately 30 species, both native and introduced, are present. Of toxicologic interest is the following:

S. arvensis L. staggerweed, field hedge nettle

Stachys arvensis

Plants herbs; annuals; from taproots; herbage sparsely hirsute. **Stems** decumbent to erect; 10–60 cm tall; branched. **Leaves** at lower nodes petiolate; sessile at upper nodes; blades broadly ovate; apices obtuse; margins coarsely serrate; bases truncate to cordate. **Inflorescences** verticils; 2–6 flowered; borne in terminal spikes; bracts present. **Calyces** radially symmetrical to slightly bilabiate; campanulate; 5- or 10-nerved. **Sepals** 5; lobes lanceolate, acute or aristate, purplish. **Corollas** bilabiate; tubular; upper lips entire or emarginate, straight or erect; lower lips 3-lobed, reflexed. **Petals** 5; pale purple or white. **Stamens** 4; exserted beyond or included within corollas. **Nutlets** oblong to ovoid; brown to black.

Distribution and Habitat—Of European origin, *S. arvensis* is now found worldwide. In North America it is encountered occasionally.

Stachys arvensis

- as with *Lamium,* fresh but not dried plant; neurologic actions, mainly in sheep, as yet not in North America

Disease Problems—Although *S. arvensis* has not yet been reported to cause problems in North America, it does so in Australia. It produces a disease very similar to that caused by *Lamium* (Everist 1981). It occurs primarily when sheep and horses are stressed or forced to move some distance after consuming plants of the species for several days or in some instances 40 or more days. If the animals are allowed to rest for a few days after appearance of the disease, recovery is rapid. There may also be abortion, diarrhea, ataxia, and meningitis (Watt & Breyer-Brandwijk 1962). Plants are probably not toxic when dried.

- stilted gait, tremors, collapse; transient

Clinical Signs, Pathology, and Treatment—With intoxication due to *S. arvensis,* there is abrupt onset of staggering and shivering when animals are stressed. The predisposition for such a response may persist for several days. There are no distinctive lesions. Typically, with rest, the signs are readily reversible without treatment.

Genera of Questionable Toxicity

Hedeoma Pers.

A New World genus, *Hedeoma* is represented in North America by 17 species. Present in the eastern half of the continent is *H. pulegioides* (L.) Pers., commonly known as American pennyroyal, squaw mint, or mosquito plant. A small annual herb 10–40 cm tall, it has branched stems that bear petiolate, lanceolate or elliptic to ovate or obovate leaves with entire or serrulate margins. The numerous, axillary, few-flowered verticils comprise small, blue, pedicelled flowers with flask-shaped, 13-nerved, 2-lipped calyces and slightly bilabiate corollas.

The leaves and flowers of *H. pulegioides* are a source of pennyroyal, a volatile, irritating oil. The plants are generally not a risk to livestock, although Standley (1926) reported that green plants or those dried in hay may cause slobbering, sweating, dilated pupils, and shortness of breath.

- *H. pulegioides*

  American pennyroyal
  mosquito plant
  squaw mint

- small annual herbs; flowers blue

Hedeoma pulegioides

- irritant oil, possibly digestive disturbance

References

Abernathy VJ, Roselli RJ, Parker RE, Pou NA: Effect of *Perilla* ketone on the in situ sheep lung. *J Appl Physiol* 72;505–514, 1992

Abernathy VJ, Pou NA, Parker RE, Roselli RJ: Evaluation of perilla ketone-induced unilateral lung injury using external gamma scanning. *J Appl Physiol* 76;138–145, 1994

Aritomi M, Kumori T, Kawasaki T: Cyanogenic glycosides in leaves of *Perilla frutescens* var. *acuta*. *Phytochemistry* 24;2438–2439, 1985

Beath OA, Gilbert CS, Eppson HF, Rosenfeld I: *Poisonous Plants and Livestock Poisoning*. Wyo Agric Exp Stn Bull 324, 1953

Boyd MR: Biochemical mechanisms in chemical-induced lung injury: role of metabolic activation. *CRC Crit Rev Toxicol* 7;103–188, 1980

Breeze RG, Laegreid WW, Bayly WM, Wilson BJ: Perilla ketone toxicity: a chemical model for the study of equine restrictive lung disease. *Equine Vet J* 16;180–184, 1984

Cooper MR, Johnson AW: *Poisonous Plants in Britain and Their Effects on Animals and Man*. Her Majesty's Stationery Office, London, 1984

Dodd S, Henry M: Staggers or shivers in livestock. *J Comp Pathol Ther* 35;41–61, 1922

Everist SL: *Poisonous Plants of Australia*, 2d ed. Angus & Robertson, London, pp380–390, 1981

Frolkin M: Poisoning of horses and pigs with seeds of hempnettle *(Galeopsis)*. *Veterinariya* 42;68, 1965 (*Vet Bull* 36;747, 1966)

Garst JE, Wilson BJ: Synthesis and analysis of various 3-furyl ketones from *Perilla frutescens*. *J Agric Food Chem* 32;1083–1087, 1984

Garst JE, Wilson WC: Alteration of the pulmonary toxicity of perilla ketone by ring methyl groups: evidence for the position of metabolism by lung cytochromes P-450 and a prospective inhibitor [abstract]. *J Anim Sci* 61 (Suppl 1);365, 1985

Garst JE, Wilson WC, Kristensen NC, Harrison PC, Corbin JE, Simon J, Philpot RM, Szabo RR: Species susceptibility to the pulmonary toxicity of 3-furyl isoamyl ketone (perilla ketone): in vivo support for involvement of the lung monooxygenase system. *J Anim Sci* 60;248–257, 1985

Guerry-Force ML, Coggeshall J, Snapper J, Meyrick B: Morphology of noncardiogenic pulmonary edema induced by perilla ketone in sheep. *Am J Pathol* 133;285–297, 1988

Kerr LA, Linnabary RD: A review of interstitial pneumonia in cattle. *Vet Hum Toxicol* 31;247–254, 1989

Kerr LA, Johnson BJ, Burrows GE: Intoxication of cattle by *Perilla frutescens* (purple mint). *Vet Hum Toxicol* 28;412–416, 1986

Kingsbury JM: *Poisonous Plants of the United States and Canada,* Prentice-Hall, Englewood Cliffs, NJ, 1964

Koezuka Y, Honda G, Tabata M: An intestinal propulsion promoting substance from *Perilla frutescens* and its mechanism of action. *Planta Med* 51;480–482, 1985

Lindley WH: Acute pulmonary edema. *Mod Vet Pract* 59;64–65, 1978

Linnabary RD, Wilson BJ, Garst JE, Holscher MA: Acute bovine pulmonary emphysema (ABPE): perilla ketone another cause. *Proc Am Assoc Vet Lab Diagn* 20;323–326, 1977 (*Vet Hum Toxicol* 20;325–326, 1978)

Marsh CD, Clawson AB, Roe GC: *Four Species of Range Plants Not Poisonous to Livestock*. USDA Tech Bull 93, 1928

Pal B, Varshneya C, Prasad B, Kishtwaria RS: Nitrate bearing *Salvia coccinea* poisoning in Gaddi goats in Kangra Valley. *Indian J Anim Sci* 67;37–38, 1997

Petersen DR: Bovine pulmonary emphysema caused by the plant *Perilla frutescens.* In *Proceedings of Symposium on Acute Bovine Pulmonary Emphysema, University of Wyoming,* R1–R13, 1965

Potts LK: Perilla. *Mo Conserv* 57;21, 1996

Standley PC: Trees and shrubs of Mexico. *Contrib U S Natl Herb* 23;1–1721, 1920–1926

Tada M, Matsumoto R, Chiba K: An oxidative metabolite of perillaldehyde from *Perilla frutescens. Phytochemistry* 43;803–804, 1996

Varshneya C, Kishtwaria RS, Prasad B: Kali phool *(Salvia coccinea)* poisoning in Gaddi goats. *Indian Vet J* 72;616–618, 1995

Waters CM, Alexander JS, Harris TR, Haselton FR: Perilla ketone increases endothelial cell monolayer permeability in vitro. *J Appl Physiol* 74;2493–2501, 1993

Watt JM, Breyer-Brandwijk MG: *The Medicinal and Poisonous Plants of Southern and Eastern Africa.* E & S Livingston, Edinburgh, pp514–529, 1962

Williams CH, Hines HJG: The toxic properties of *Salvia reflexa. Aust Vet J* 16;14–20, 1940

Wilson BJ, Garst JE, Linnabary RD, Channell RB: Perilla ketone: a potent lung toxin from the mint plant, *Perilla frutescens* Britton. *Science* 197;573–574, 1977

Wilson WC, Simon J, Garst JE: The effects of selected bulky substituents on the pulmonary toxicity of 3-furyl ketones in mice. *J Anim Sci* 68;1072–1076, 1990

Lauraceae Juss.

46

With centers of diversity in Southeast Asia and Brazil, the Lauraceae, or laurel family, is pantropical in distribution and comprises about 50 genera and 2,000–3,000 species. It is economically important for timber, aromatic oils, food, and a few ornamentals. Avocado, cinnamon, camphor, sassafras oil, and bay leaves are obtained from its members. Although the family is not generally thought of as alkaloid-producing, several alkaloids are present in low concentrations in various species. They are similar to some of those present in the Papaveraceae and include boldine, cassyfiline, cinnamolaurine, and laurotetanine (Southon & Buckingham 1989). Their toxicologic significance in unknown. In North America, 9 genera and 13 species, both native and introduced, are present. Toxicologic problems are associated with 2 genera.

■ Laurel Family

Cinnamomum
Persea

QUESTIONABLE
Cassytha
Sassafras
Umbellularia

Plants trees or shrubs; evergreen or rarely deciduous; strongly aromatic; dioecious or polygamodioecious or bearing perfect flowers. **Leaves** simple; alternate, or rarely opposite or whorled; petiolate; blades typically coriaceous; venation pinnate or pinnipalmate; margins entire or pinnately lobed; stipules absent. **Inflorescences** panicles or racemes or umbels or cymes; axillary; produced before or simultaneously with leaves; bracts present or absent. **Flowers** imperfect or perfect, staminate and pistillate similar; perianths in 1-series; radially symmetrical. **Sepals** 6 or rarely 4; free; in 2 whorls; petaloid; yellow or greenish yellow or white. **Petals** absent. **Stamens** 9 or 12, or rarely 3 or 6; in 3 or 4 whorls; 1–3 whorls of staminodia typically present. **Pistils** 1; simple, carpels 1; stigmas 1, capitate or discoid; styles 1, short; ovaries superior; placentation apical. **Hypanthia** well developed, resembling calyx tube. **Fruits** drupes or berries; red or blue-black; cupped or enclosed by hypanthium. **Seeds** 1; typically large.

■ aromatic shrubs or trees; leaves simple; staminate and pistillate flowers similar; petals absent; fruits drupes or berries

Cinnamomum Schaeff.

Taxonomy and Morphology—Comprising 200–300 species, *Cinnamomum* is the source of cooking cinnamon and camphor oil. Its name is derived from the Greek *kinnamomum,* a name used by Theophrastus. In North America, 1 introduced species is present:

■ *C. camphora*

alcanfor
camphor laurel
camphor tree

C. camphora (L.) J.Presl — camphor tree, camphor laurel, alcanfor

Plants trees; crowns dense, oval, spreading. **Stems** to 15 m tall; trunks stout; bark gray, rough, furrowed; twigs terete; bud scales orbicular, imbricated. **Leaves** alternate; blades ovate to elliptic or elliptic-lanceolate; primary veins 3; adaxial surfaces glabrous, lustrous; abaxial surfaces glaucous or pubescent in vein axils; apices acute; bases rounded to cuneate. **Inflorescences** panicles; bracts absent. **Flowers** perfect.

■ dense-crowned trees; leaves alternate, primary veins 3; fruits bluish black drupes

Cinnamomum camphora

- ornamental, naturalized southeastern United States
- irritant action, digestive disturbance; possible neurologic effects; camphor also alkaloid cinnamolaurine, of unknown significance

Sepals greenish white; adaxial surfaces pubescent; abaxial surfaces glabrous. **Stamens** 9; staminodia 3. **Fruits** drupes; subglobose; bluish black; seated in small cupules.

Distribution and Habitat—Native to eastern Asia, *C. camphora* is cultivated in states along the Gulf coast. It has naturalized in various areas from North Carolina to Texas.

Disease Problems and Genesis— Few toxicologic problems have been associated with the genus. However, when a camphor tree was planted in an aviary, 49 budgerigars died within 24 hours, apparently from its noxious fumes (Fuller & McClintock 1986). As implied by its common name, all parts of the tree contain camphor, approximately 1.5% in the leaf and 0.6% in the wood (Watt & Breyer-Brandwijk 1962). The toxicity of camphor has been known since the late 1800s and early 1900s, mainly from occasional nonfatal exposures generally associated with accidental ingestion of liniments containing the oil (Smith & Margolis 1954). Readily absorbed through the mucous membranes, camphor is a mild irritant, a cathartic, and a nervous system stimulant but is of low toxicity; the oral LD_{50} is 2 g/kg b.w. in rabbits and 1.8 g/kg in guinea pigs (Smith & Margolis 1954; Osol & Pratt 1973). The presence of camphor readily accounts for the deaths of the budgerigars, birds that are quite susceptible to noxious fumes. Also present are low concentrations of cinnamolaurine, an alkaloid very similar in structure to the alkaloids occurring in *Eschscholzia* in the Papaveraceae (see chapter 55).

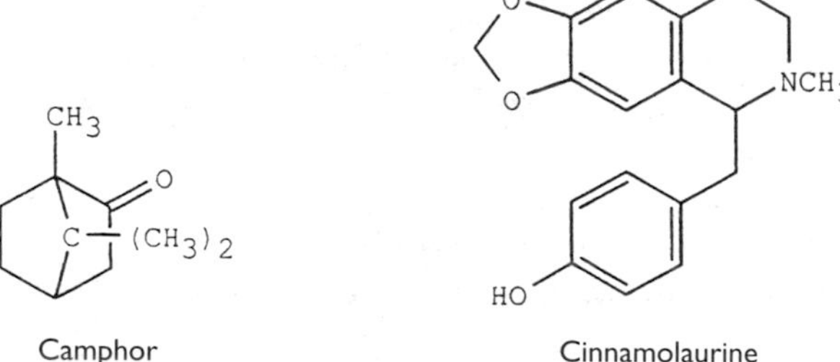

Camphor

Cinnamolaurine

- nausea, vomiting, depression
  seizures, spasticity

Clinical Signs—Signs will likely be limited to nausea, vomiting, and depression. In the most severe intoxications, systemic signs such as labored respiration, muscle twitching, tonic-clonic seizures, spasticity, and increased body temperature may be apparent.

- reddening of gut mucosa

Pathology—Most of the changes are nonspecific and include evidence of irritation of the digestive tract, visceral congestion, and edema of the brain. Microscopically, there may be widespread neuronal degeneration, sparing glia and blood vessels.

- nursing care

Treatment—Generalized care to alleviate the clinical symptoms is the typical treatment. When neurologic signs are prominent, pentobarbital may be used.

Persea Mill.

- *P. americana*
  aguacate
  alligator pear
  avocado

Taxonomy and Morphology—Comprising about 150 species in both the Old World and the New World, *Persea* is represented in North America by 3 native species and several introduced species. The best-known and of toxicologic importance is the following:

P. americana Mill. avocado, alligator pear, aguacate

Plants trees or shrubs; crowns dense, rounded. **Stems** to 30 m tall; bark brown or gray, slightly furrowed; twigs angular, green becoming brown. **Leaves** alternate; crowded near ends of twigs; blades ovate-elliptic; primary veins 1; apices abruptly acute to acuminate or aristate; bases cuneate or truncate. **Inflorescences** panicles; rufous; bracts absent. **Flowers** perfect. **Sepals** greenish yellow; surfaces indumented. **Stamens** 9; staminodia 3. **Fruits** berries; ovoid to pyriform; pericarps thick, coriaceous, glossy dark green, tuberculate. **Seeds** large.

■ dense-crowned shrubs or trees; leaves alternate, primary veins 1; fruits berries with 1 large seed

■ Guatemalan race, warty skin

Persea americana

The species comprises three ecological races, which are formally recognized as varieties by some taxonomists, and numerous cultivars (table 46.1.). These races differ in cold tolerance; the subtropical Mexican is the most cold-tolerant, and the tropical West Indian the least (Samson 1986). The skin of the fruits ranges from smooth and thin for Mexican to warty for Guatemalan and leathery for West Indian races. They also differ in oil content, flowering time, and appearance of the pericarp. Hybrids exist between the races. 'Fuerte,' a hybrid between Guatemalan and Mexican, is the most popular cultivar. 'Bacon' and 'Zutano' also are hybrids between Guatemalan and Mexican.

Table 46.1. Races and cultivars of *Persea americana*

Race	Cultivar
Guatemalan	'Anaheim,' 'Benik,' 'Gwen,' 'Hass,' 'Nabal,' 'Pinkerton,' 'Reed,' 'Rincon,' 'Taylor,' 'Whitesell,' 'Wurtz'
Mexican	'Duke,' 'Gottfried,' 'Jim,' 'Mexicola,' 'Pernod,' 'Santana'
West Indian	'Black Prince,' 'Pollock,' 'Simmonds,' 'Waldin'

Source: Samson 1986.

Distribution and Habitat—Believed to have its origin in southern Mexico and Central America, *P. americana* is now extensively cultivated in Mexico, California, and Florida. Plants are also grown as ornamentals in areas along the Gulf coast.

■ crop and ornamental plant

Disease Problems—From the early 1900s, avocados were suspected of being toxic, but it was not until the 1940s that reports of intoxications began to appear (Hurt 1943; Appelman 1944). Finally, in the 1980s, detailed observations revealed the nature of the disease caused by *P. americana.* As yet, only the Guatemalan type and its hybrids such as 'Fuerte' have been shown to be toxic. Leaves, fruits, and seeds are toxic, and the leaves especially so. The toxins have a unique predisposition to affect the lactating mammary gland, and in lactating animals this seems to serve as protection against other systemic effects. The goat seems to be particularly sensitive to these mammary effects. In lactating animals the primary effect of low dosage of leaves is a noninfectious mastitis with substantially diminished milk production (Craigmill et al. 1984, 1989). Mastitis is also apparent in mice fed avocado leaves and probably would occur in any lactating animal with the appropriate dosage (Sani et al. 1994). In the nonlactating animal or at higher dosage of leaves in the lactating animal, other effects such as cardiomyopathy become of increased importance (Sani et al. 1991; Stadler et al. 1991). In lactating goats, 20 g of fresh leaves/kg b.w. is likely to cause severe mastitis, and as the dosage is increased up to 30 g/kg or more, edema and cardiac effects become apparent. It appears that lower single doses are most likely to result in mastitis, whereas higher or repeated doses often lead to cardiac pathology (Burger et al. 1994b). Grant and co-workers (1991) fed sheep fresh leaves of the 'Fuerte' cultivar for 5 days and then those of other cultivars for differing periods of time. Dosage of 25 g/kg b.w. daily for 5 days caused acute cardiac failure, whereas dosage

■ Guatemalan type toxic; leaves and seeds, mastitis and/or cardiomyopathy

fruits toxic to birds

of 5.5 g/kg b.w. for 21 days caused chronic cardiac insufficiency and 2.5 g/kg for 32 days caused less pronounced cardiac effects.

Myocardial degeneration is probably the main destructive effect in all animal species with subcutaneous edema of the head and viscera sometimes present. Horses seem to be predisposed to edema of the head and chest (McKenzie & Brown 1991; Galey et al. 1993). Birds—i.e., canaries, budgerigars, cockatiels, and ostriches—suffer subcutaneous edema of the neck, chest, and pulmonary system (Hargis et al. 1989; Burger et al. 1994a). Experimentally, 1 g of avocado fruit caused agitation and feather pulling of short duration in budgerigars (Shropshire et al. 1992). Fresh or dried seed has been shown to be toxic and to cause death of mice in several days (Valeri & Gimeno 1953). Chickens and turkeys exhibit a lower order of sensitivity than most other birds. Many mammalian species are affected, including cattle, goats, horses, mice, rabbits, rats, and sheep.

■ persin, necrosis mammary gland, myocardium; cumulative effects

$H_3C-C(=O)-O-C-C(OH)-C-C(=O)(CH_2)_7C=C-C-C=C(CH_2)_4CH_3$

Persin

Disease Genesis—The necrosis of the mammary gland and myocardium is caused by the *R*-enantiomer of persin, a long-chain, unsaturated compound that has been isolated from the leaves (Oelrichs et al. 1995). Moderate oral dosage of persin to mice (60–100 mg/kg b.w.) causes mammary gland necrosis within a few hours. In higher dosage, myocardial necrosis results; the same range of effects is produced by the leaves (Oelrichs et al. 1995, 1998). The long chain is necessary but not unsaturation. Following oral administration, persin reaches peak blood levels in 1–2 hours but is not detectable 24 hours later. Concentrations reach a maximum in the mammary gland 6–7 hours after administration.

Unrefined oil from avocados when fed to rats causes fatty livers and a marked increase in serum alkaline phosphatase but not other enzymes indicative of hepatic problems (Werman et al. 1989). The oil makes up 15–30% of the fresh weight of the fruit and is mainly oleic acid. Its relationship to the other effects of avocado leaves and fruit is unknown. Avocado leaves also are a natural source of the insecticidal diterpene ryanodol (Gonzalez-Coloma et al. 1993).

■ livestock, abrupt onset, hard swollen udder, marked decrease in milk, subcutaneous edema, weakness, cardiac arrhythmias

Clinical Signs—Mastitis occurs within 24 hours after low dosage, as evidenced by a hard, swollen udder; an approximately 75% decrease in milk production; and watery, cheesy, curdled milk. As the dosage is increased, there is subcutaneous edema of the neck and brisket, an infrequent cough, weakness/depression, reluctance to move, respiratory distress, and cardiac arrhythmias. Serum hepatic enzymes such as LDH, CK, and AST will be elevated markedly.

■ birds, listless, feathers ruffled, labored respiration

Birds become listless, with feathers ruffled and respiration labored. The reaction in horses to large dosages is somewhat unique. There is severe edema of the head, neck, and ventral areas. The entire head is swollen, especially around the eyes, and the eyelids may be swollen shut, with the conjunctiva protruding. In one instance, the tongue was severely swollen and protruded rigidly from the mouth. In most animals the signs appear within a day of ingestion of avocado.

■ gross pathology, mammary glands, edematous; subcutaneous edema, excess fluid in body cavities

microscopic, heart, interstitial edema, myocardial necrosis; horses, myopathy, head and tongue

Pathology—Grossly, the mammary glands are edematous and reddened, with clots in the large ducts. In some individuals edema is prominent in many tissues, especially the subcutis, and in the lungs. There is general congestion of the lungs, liver, and spleen; free fluid in the heart sac, thorax, and abdomen; and edema of the gallbladder and perirenal tissues. The heart is pale and flabby.

Microscopically, there is edema of the mammary gland and degeneration and necrosis of the secretory acinar epithelium. In the heart there is interstitial edema and myocardial degeneration. Necrosis is especially noticeable in the ventricular walls and septum. In the horse, symmetrical ischemic myopathy of the head and tongue and ischemic myelomalacia of the lumbar spine have been reported (Galey et al. 1993).

Treatment—The approach to treatment is to relieve the signs using anti-inflammatories and diuretics. Antibiotics are sometimes added to control secondary infection in the mammary glands.

■ diuretics, nonsteroidal anti-inflammatory drugs

Genera of Questionable Toxicity

Cassytha L.

■ *C. filiformis*

laurel-vine
woe-weed

■ parasitic vine, threadlike stems; leaves scalelike

Comprising 17 species in both the Old World and the New World, *Cassytha* is quite different in vegetative features from other members of the Lauraceae and has been classified by some taxonomists in its own family, the Cassythaceae (Weber 1981). It is represented in North America by *C. filiformis* L. (laurel-vine, or woe-weed), which is a parasitic vine with yellowish or pale green threadlike stems and branches with a spicy odor. Its leaves are minute scales, and its spikes bear at irregular intervals, 1 to several perfect flowers subtended by minute bracts and bracteoles, which produce black, globose drupes. The morphology of the flowers unites the genus with other taxa of the Lauraceae. *Cassytha filiformis,* which can be confused with species of *Cuscuta* (dodder), parasitizes herbaceous and woody plants in coastal Texas and Florida.

■ alkaloids, such as cassyfiline, significance unknown

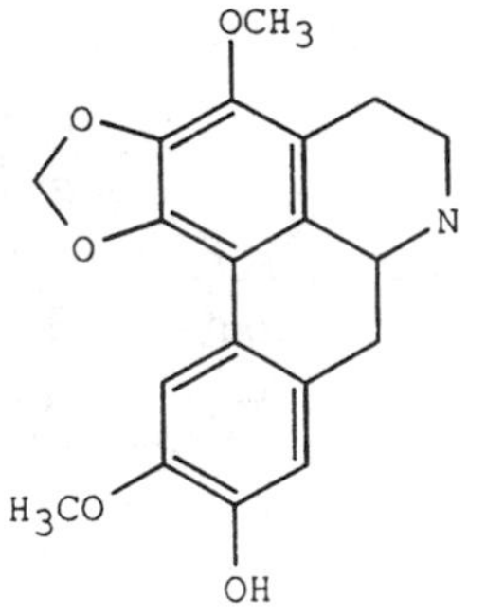

Cassytha is of interest because of its ancient Hindu medicinal use. Alkaloids such as cassyfiline and laurotetanine are present in low concentrations, i.e., only up to 0.1%. Laurotetanine can cause smooth muscle contractions and cramps. *Cassytha filiformis,* however, has not been incriminated as a problem, and there seems to be little likelihood of the occurrence of intoxications.

OCH_3 O O N H_3CO OH

Cassyfiline

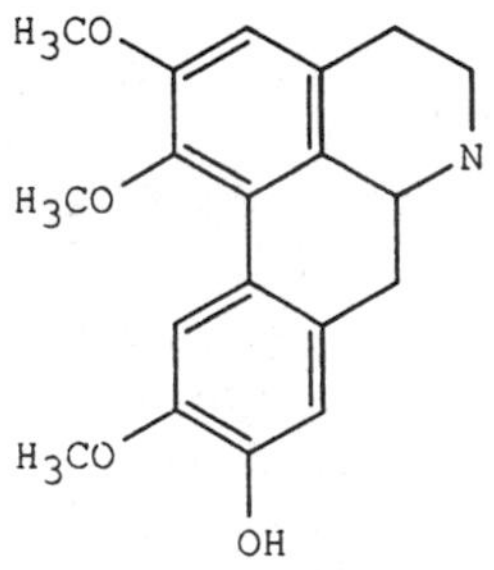

H_3CO H_3CO N H_3CO OH

Laurotetanine

Sassafras Nees

■ *S. albidum*

sassafras

■ deciduous trees, aromatic bark; leaves with palmate lobes and venation

Comprising 2 species, 1 in eastern North America and the other in eastern China, *Sassafras* is perhaps best known for "sassafras tea" made by steeping the bark or roots. The North American species is *S. albidum* (Nutt.) Nees. Flowering in the spring, it is a dioecious shrub or tree with spicy aromatic bark and mucilaginous, pale green twigs. Its alternate, deciduous leaves are ovate to elliptic, entire or more commonly palmately 1-, 2-, or 3-lobed, and have palmate or pinnipalmate venation. Borne in racemose or umbellate clusters, the imperfect, fragrant flowers are yellow or yellow-green and produce blue ovoid drupes borne on clavate, reddish pedicels. It is a woodland species.

Sassafras albidum

Sassafras albidum

■ oil has safrole, irritant, hepatotoxic

Safrole

Although used for food by wildlife (Gill & Healy 1973) and not a common cause of intoxication problems, *S. albidum* is the source of American oil, which is 80–90% safrole (Osol & Pratt 1973). The oil itself may cause minor problems such as vomiting, shakiness, and elevation of heart rate and blood pressure of a few hours' duration (Grande & Dannewitz 1987). Safrole is an irritant and in low dosage causes fatty change in the liver (Wade 1977). It is of low toxicity potential, with an oral LD_{50} of 2,350 mg/kg b.w. in mice and 1,950 mg/kg b.w. in rats (Hagan et al. 1965). With repeated dosage, liver lesions and adrenal enlargement become evident. With increasing severity dependent upon duration of dosage, there is hepatic fatty change, necrosis, fibrosis, bile duct proliferation, and nodular hepatocellular enlargement. With chronic administration, safrole is a hepatic carcinogen in mice and rats (Hagan et al. 1965; Miller et al. 1983).

Umbellularia (Nees) Nutt.

■ *U. californica*

California bay
California laurel
California olive
Oregon myrtle
pepperwood

■ evergreen shrubs or trees; leaves strongly aromatic; fruits dark purple drupes

■ irritant oil, umbellulone, possible digestive disturbance

Native to southern Oregon, California, and Baja California, *Umbellularia* is a monotypic genus. *Umbellularia californica* (Hook. & Arn.) Nutt. is commonly known as California bay, California laurel, California olive, Oregon myrtle, and pepperwood. An evergreen shrub or tree to 45 m tall, its thin bark is dark reddish brown. The very aromatic, oblong to elliptic leaves are a deep yellow-green. Borne in umbellate clusters that are enclosed in spirally arranged bracts, the perfect flowers with deciduous yellow sepals produce greenish, dark purple, globose drupes borne on small flat cupules. Plants of this species are encountered in canyon and valley habitats.

Like *Sassafras albidum, U. californica* is not known as a cause of intoxications, but it does contain an irritating oil known as umbellulone (Fuller & McClintock 1986). Strongly aromatic, the crushed leaves may cause sneezing. The oil causes hepatic fatty change but is of low toxicity. California Indians used the roasted drupes for food and the leaves to repel insects (Ebeling 1986).

Umbellulone

References

Umbellularia californica

Appelman D: Preliminary report on toxicity of avocado leaves. *Calif Avocado Soc Yearb* 1944;37, 1944

Burger WP, Naude TW, van Rensburg IBJ, Botha CJ, Pienaar ACE: Avocado *(Persea americana)* poisoning in ostriches. In *Plant- Associated Toxins: Agricultural, Phytochemical, and Ecological Aspects,* Colegate SM, Dorling PR eds, CAB International, Wallingford, UK, pp546–551, 1994a

Burger WP, Naude TW, Van Rensburg IB, Botha CJ, Pienaar AC: Cardiomyopathy in ostriches *(Struthio camelus)* due to avocado *(Persea americana* var. *guatemalensis)* intoxication. *J S Afr Vet Assoc* 65;113–118, 1994b

Craigmill AL, Eide RN, Shultz TA, Hedrick K: Toxicity of avocado (*Persea americana* [Guatamalan var.]) leaves: review and preliminary report. *Vet Hum Toxicol* 26; 381–383, 1984

Craigmill AL, Seawright AA, Mattila T, Frost AJ: Pathological changes in the mammary gland and biochemical changes in milk of the goat following oral dosing with leaf of the avocado *(Persea americana). Aust Vet J* 66;206–211, 1989

Ebeling W: *Handbook of Indian Foods and Fibers of Arid America.* Univ Calif Press, Berkeley, p227, 1986

Fuller TC, McClintock E: *Poisonous Plants of California.* Univ Calif Press, Berkeley, pp182–185, 1986

Galey FO, Read D, Hebbert B, Blumenshine KM: Facial edema with myopathy in horses exposed to avocado *(Persea americana)* plants. *Annu Meet Am Assoc Vet Lab Diagn* 37;33, 1993

Gill JD, Healy WM: *Shrubs and Vines for Northeastern Wildlife.* USDA Forest Service, Gen Tech Rep NE-9, 1973

Gonzalez-Coloma A, Cabrera R, Socorro Monzon AR, Fraga BM: *Persea indica* as a natural source of the insecticide ryanodol. *Phytochemistry* 34;397–400, 1993

Grande GA, Dannewitz SR: Symptomatic sassafras oil ingestion [abstract]. *Vet Hum Toxicol* 29;447, 1987

Grant R, Booker HH, Basson PA, Hofherr JB, Anthonnisen M: Cardiomyopathies caused by januariebos *(Gnidia polycephala)* and avocado *(Persea americana)* leaves. *J S Afr Vet Assoc* 59;101, 1988

Grant R, Basson PA, Booker HH, Hofherr JB, Anthonissen M: Cardiomyopathy caused by avocado (*Persea americana* Mill.) leaves. *J S Afr Vet Assoc* 62;21–22, 1991

Hagan EC, Jenner PM, Jones WI, Fitzhugh OG, Long EL, Brouwer JG, Webb WK: Toxic properties of compounds related to safrole. *Toxicol Appl Pharmacol* 7;18–24, 1965

Hargis AM, Stauber E, Casteel S, Eitner D: Avocado *(Persea americana)* intoxication in caged birds. *J Am Vet Med Assoc* 194;64–66, 1989

Hurt LM: *Avocado Poisoning.* Los Ang Cty Livest Dep Rep 43–44, 1942–1943

McKenzie RA, Brown OP: Avocado *(Persea americana)* poisoning of horses. *Aust Vet J* 68;77–78, 1991

Miller EC, Swanson AB, Phillips DH, Fletcher TL, Liem A, Miller JA: Structure-activity studies of the carcinogenicities in the mouse and rat of some naturally occurring and synthetic alkenylbenzene derivatives related to safrole and estragole. *Cancer Res* 43;1124–1134, 1983

Oelrichs PB, Ng JC, Seawright AA, Ward A, Schaffeler L, Macleod JK: Isolation and identification of a compound from avocado *(Persea americana)* leaves which causes necrosis of the acinar epithelium of the lactating mammary gland and the myocardium. *Nat Toxins* 3;344–349, 1995

Oelrichs PB, Kratzman S, Macleod JK, Ng JC, Seawright AA: A study of persin, the mammary cell necrosis agent from avocado *(Persea americana),* and its furan derivative in lactating mice. In *Toxic Plants and Other Natural Toxicants,* Garland T, Barr AC eds, CAB International, New York, pp215–219, 1998

Osol A, Pratt R: *The United States Dispensatory,* 27th ed. Lippincott, Philadelphia, p220, 1973

Samson JA: *Tropical Fruits,* 2d ed. Longman, New York, pp235–255, 1986

Sani Y, Atwell RB, Seawright AA: The cardiotoxicity of avocado leaves. *Aust Vet J* 68;150–151, 1991

Sani Y, Seawright AA, Ng JC, O'Brien G, Oelrichs PB: The toxicity of avocado leaves *(Persea americana)* for the heart and lactating mammary gland of the mouse. In *Plant-Associated Toxins: Agricultural, Phytochemical, and Ecological Aspects,* Colegate SM, Dorling PR eds, CAB International, Wallingford, UK, pp552–556, 1994

Shropshire CM, Stauber E, Arai M: Evaluation of selected plants for acute toxicosis in budgerigars. *J Am Vet Med Assoc* 200;936–939, 1992

Smith AG, Margolis G: Camphor poisoning—anatomical and pharmacologic study: report of a fatal case; experimental investigation of protective action of barbiturate. *Am J Pathol* 30;857–869, 1954

Southon IW, Buckingham J: *Dictionary of Alkaloids.* Chapman & Hall, New York, 1989

Stadler P, Van Rensburg IBJ, Naude TW: Suspected avocado *(Persea americana)* poisoning in goats. *J S Afr Vet Assoc* 62;186–188, 1991

Valeri H, Gimeno F: A phytochemical toxicological study of the fruit of the avocado *(Persea americana). Rev Med Vet Parasitol Caracas* 12;131–165, 1953

Wade A: *Martindale—The Extra Pharmacopoeia,* 27th ed. Pharmaceutical Press, London, pp294,1026, 1977

Watt JM, Breyer-Brandwijk MG: *The Medicinal and Poisonous Plants of Southern and Eastern Africa.* E & S Livingston, Edinburgh, pp529–531, 1962

Weber JL: A taxonomic revision of *Cassytha* (Lauraceae) in Australia. *J Adelaide Bot Gard* 3:187–262, 1981

Werman MJ, Mokady S, Neeman I, Auslaender L: The effect of avocado oils on some liver characteristics in growing rats. *Food Chem Toxicol* 27;279–282, 1989

Liliaceae Juss. 47

Exhibiting great diversity, the Liliaceae, or lily family, comprises 280–300 genera and 4,000–4,600 species distributed primarily in dry, temperate and subtropical habitats. Taxonomists differ in their opinions as to how to classify this diversity. Some recognize a broadly circumscribed taxon with 12–28 tribes, as is done here, while others prefer a narrow circumscription and divide the family into as many as 28 small families on the basis of such characters as inflorescence type, perianth fusion, and ovary position. Horticulturally, the Liliaceae is one of the most important families. There are more than 160 genera in the American horticultural trade alone; familiar ornamentals include lilies, tulips, hyacinths, daffodils, and amaryllises. Food plants include asparagus, garlic, onions, leeks, shallots, and chives. A variety of toxic species, some quite deadly, are also members of the family. In North America, approximately 90 genera and 525 species, both native and introduced, are present.

Plants herbs; perennials; from rhizomes or bulbs or tubers or corms or fleshy roots; caulescent or acaulescent. **Leaves** basal or forming a basal rosette or cauline; simple; alternate or whorled or rarely opposite; venation parallel or parallel-convergent or pinnipalmate; stipules absent. **Inflorescences** solitary flowers or spikes or racemes or panicles or umbels. **Flowers** typically large and showy; fragrant or not fragrant; perfect or rarely imperfect; perianths in 2-series or 1-series; radially symmetrical. **Perianth Parts** 6; all alike (petaloid) or of 2 forms (3 sepaloid and 3 petaloid); free or fused; coronas present or absent. **Stamens** 6 or rarely 3 or 12; fused to perianth parts or arising from receptacles. **Pistils** 1; compound; carpels 3, or rarely 2 or 4; stigmas 1 or 3; styles 1 or rarely 3; ovaries superior or inferior; locules 3; placentation axile or rarely parietal. **Hypanthia** absent or present; tubular. **Fruits** septicidal or loculicidal capsules or berries. **Seeds** 1 to numerous.

In addition to its horticultural importance, the Liliaceae is also of considerable toxicologic importance. Its genera produce several quite dissimilar types of disease problems—cardiotoxic, neurotoxic, hemolytic, teratogenic, cytotoxic, and irritative. The toxicants present are likewise dissimilar. Toxins with a steroid nucleus are common, especially steroidal saponins—glycosides that form a soapy lather (Heftmann 1970; Mahato et al. 1982). Members of the family are important sources of saponins, which in some instances appear to be responsible for the toxic effects. At least a dozen different sapogenins are represented in the family (Okanishi et al. 1975). Closely akin are the steroidal *Veratrum* alkaloids. Some genera produce several types of disease. This variation is the result of subtle chemical differences in the toxicants present, which produce marked differences in activity. In other genera producing more than one syndrome, the primary toxicants are not at all similar.

■ Lily Family

Allium
Amaryllis
Amianthium
Bowiea
Clivia
Colchicum
Convallaria
Cooperia
Crinum
Eucharis
Fritillaria
Galanthus
Gloriosa
Haemanthus
Hemerocallis
Hippeastrum
Hyacinthoides
Hyacinthus
Hymenocallis
Leucojum
Lilium
Narcissus
Nerine
Ornithogalum
Tulipa
Urginea
Veratrum
Zephyranthes
Zigadenus

QUESTIONABLE
Aletris
Asparagus
Erythronium
Melanthium
Pleea
Trillium

■ perennial herbs; leaves simple; flowers radially symmetrical, often large and showy; perianth parts 6

■ types of problems—
cardiotoxic
neurotoxic
hemolytic
teratogenic
cytotoxic
irritative

Allium L.

- *A. canadense*
- *A. cepa*
- *A. cernuum*
- *A. porrum*
- *A. sativum*
- *A. schoenoprasum*
- *A. validum*
- *A. vineale*

chive
cive
garlic
leek
onion

Allium canadense

Allium cernuum

Taxonomy and Morphology—Widely distributed in the Northern Hemisphere, *Allium* comprises 500–700 species. Derived from the Celtic word *all* meaning "hot," its name is an ancient Latin one for garlic. Several Old World species are cultivated for food or flavoring or as ornamentals. In North America, approximately 95 species are present. Of toxicologic interest are the following:

Cultivated Species

A. cepa L.	onion
A. porrum L.	leek
A. sativum L.	garlic
A. schoenoprasum L.	chive, cive

Wild Species

A. canadense L.	wild onion, Canada garlic, meadow leek
A. cernuum L.	nodding onion, lady's leek
A. validum S.Watson	wild onion, Pacific onion, swamp onion
A. vineale L.	wild garlic, field garlic, crow garlic, false garlic, stag's garlic

Plants from bulbs; acaulescent or caulescent; strongly aromatic, with odor of onions or garlic. **Bulbs** solitary or clustered; tunicas papery or fibrous. **Leaves** basal; blades linear to long lanceolate, flat or terete or grooved, solid or hollow; bases sheathing. **Scapes** erect; 10–150 cm tall; unbranched; solid or hollow. **Inflorescences** terminal umbels; flat-topped or hemispheric or globose; arising from scarious spathes of 1–3 bracts. **Flowers** few to many; perianths in 2-series; in some species replaced by bulblets. **Perianth Parts** 6; all alike; petaloid; free; white or pink or purple or greenish; withering and persisting. **Stamens** 6, fused to perianth parts; filaments wide, flat, fused or free. **Pistils** with stigmas 3-lobed or not lobed; styles 1; ovaries superior; placentation axile. **Fruits** loculicidal capsules; globose-ovoid; 3-lobed. **Seeds** numerous; angular or round or flat; black.

When crushed, the bulbs, foliage, and flowers emanate the characteristic onion or garlic odor, a feature that is useful in distinguishing *Allium* from morphologically similar genera, in particular *Zigadenus* (death camas). The two genera also are distinguished by their inflorescences; *Zigadenus* has a raceme rather than an umbel.

1. Leaves terete; hollow at bases.
    2. Scapes inflated below middles .... *A. cepa*
    2. Scapes not inflated below middles.
        3. Pedicels shorter than flower lengths.
           Bulblets never present .... *A. schoenoprasum*
        3. Pedicels longer than flower lengths.
           Bulblets usually present .... *A. vineale*
1. Leaves flat or keeled; solid at bases.
    4. Stamen filaments of 2 types, inner with a pair
       of lateral appendages, outer without .... *A. porrum*
    4. Stamen filaments all alike.

5. Bulb tunicas fibrous-reticulate .... *A. canadense*
5. Bulb tunicas not fibrous-reticulate; membranous or papery or fibers parallel.
   6. Bulbs compound, comprising 6–10 small bulblets (cloves) enclosed in common membrane ..... *A. sativum*
   6. Bulbs simple, not comprising small bulblets enclosed in membrane.
      7. Bulbs arising from stout rhizomes. Umbels erect . . *A. validum*
      7. Bulbs not arising from rhizomes. Umbels nodding .... *A. cernuum*

■ strongly aromatic herbs from bulbs; leaves basal, long, bases sheathing; fruits 3-lobed capsules

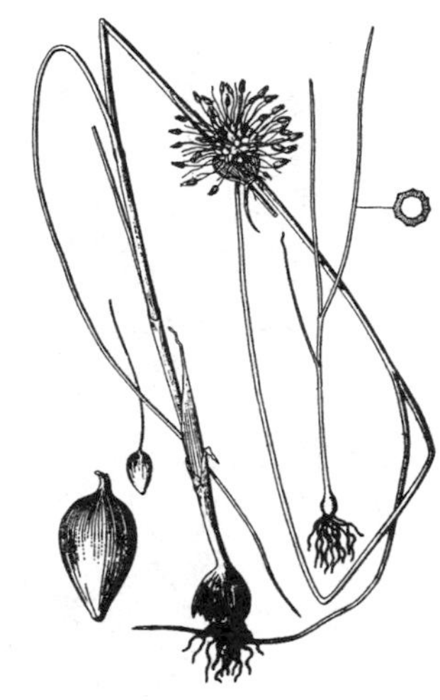

Allium vineale

Distribution and Habitat—Native species of *Allium* occur in a variety of habitats in different vegetation types across the continent. Some are widespread, others more restricted in their distributions. *Allium schoenoprasum,* for example, is circumboreal in North America. With numerous varieties, *A. canadense* is also one of the most widespread species, occupying prairies, meadows, open woods, and various disturbed sites throughout the eastern half of the continent. *Allium cernuum* also occurs in the eastern half but is not as abundant. *Allium validum* occurs mainly in moist sites, primarily in the Northwest. Native to Europe, *A. vineale* is naturalized as a noxious weed in lawns, playing fields, and meadows. Old World *A. cepa, A. porrum,* and *A. sativum* rarely escape cultivation, and although they may persist for a time, they do not naturalize.

■ variety of habitats

Disease Problems—Members of *Allium* have had important dietary and medicinal roles for millennia. Their use in early societies is probably best illustrated by the discovery of onions associated with mummified human remains (Fenwick & Hanley 1985b). It is now known that species of the genus contain chemical constituents with a wide range of beneficial medicinal effects. They have become the focus of intense interest as potential antineoplasia compounds, perhaps via their effects on biotransformation pathways (Sparnins et al. 1988; Dausch & Nixon 1990). Additionally, they exhibit marked effects on allergic reaction pathways (Dorsch et al. 1988).

■ large amounts, hemolytic anemia, Heinz bodies, especially cats, dogs, horses; cattle and sheep less severe; cumulative effects

Disease most commonly occurs when cultivated onions are used as animal feed. This is typically a problem in areas where onions are an important crop. Onion waste—culls or excess production—is often fed to cattle or sheep. Whether the harvested onions are fresh, spoiled, or cooked makes little apparent difference. Less commonly, wild plants may cause problems, especially in the spring when the leaves are most abundant. Problems may also develop when dehydrated onions, onion souffle, and onion soup are fed to pets (Fenwick & Hanley 1985d). The toxic effects are cumulative, and intoxication eventually ensues after repeated ingestion of large quantities of the plants.

Allium canadense

All animal species appear to be susceptible to the disease, albeit not equally so. It is difficult to compare animals, but it seems clear that sheep and goats are much less susceptible than most other animals (Koger 1956; Van Kampen et al. 1970; Kirk & Bulgin 1979). Susceptibility may correlate with the animal's relative propensity for Heinz body formation. Although cattle are said to be most susceptible (Fenwick & Hanley 1985d), it seems more likely that cats and then dogs are the most susceptible species. Sheep and goats, which are quite resistant, may consume very large amounts (10–20 lb/day) with only minimal signs developing (Kirk & Bulgin 1979). It is therefore not surprising that sheep relish onions, both the bulbs and the leaves, in spring pasture and only rarely develop signs of disease (Parker 1936; Stevens 1984). Cattle are also relatively resistant, as shown by the lack of adverse effects on performance (weight gains) when up to 25% cull onions (on a dry matter basis, or 46% of the

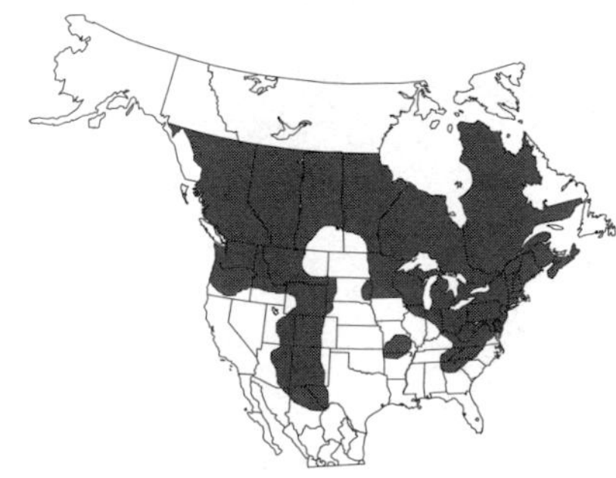

Allium cernuum

Allium validum

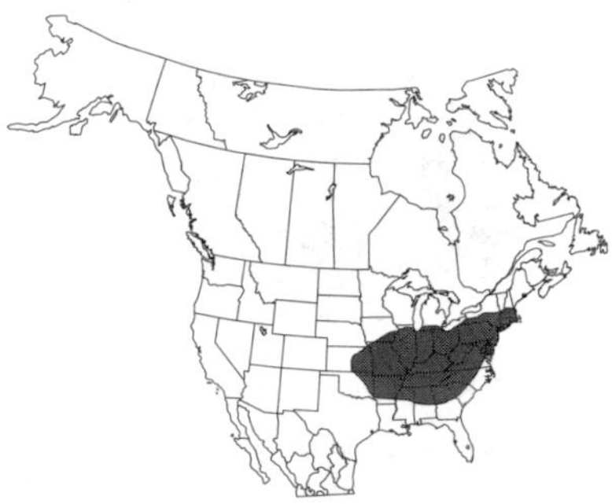
Allium vineale

feed) were fed in a balanced ration for approximately 4 months (Lincoln et al. 1992). During the first month, there was a decline in red blood cells and hematocrit but no obvious clinical anemia. Heinz bodies were also apparent in amounts proportional to dosage. These effects moderated after the first month. The important factor seems to be incorporation of chopped onions in a balanced ration so that the animals cannot selectively eat more of them. Problems are much more likely when cattle are fed onions free-choice, and some animals consume excessive amounts. For example, in an instance in which 85 calves and yearlings were given 1,000 kg onions per day, signs of intoxication occurred in 22 animals in 5 days (Verhoeff et al. 1985). Some animals apparently ate very large amounts. There was only 1 fatality, although the hematocrit dropped to as low as 12% in some animals. On pasture, the rarity of poisoning is in large measure due to the significant amount required over a sustained period, regardless of the animal species (Pierce et al. 1972; Hutchison 1977; Stevens 1984).

Both dogs and cats are very susceptible and have been intoxicated by cooked onions, whether boiled alone, included in soup, prepared as a souffle, or mixed with table scraps (Spice 1976; Kobayashi 1981; Stallbaumer 1981; Kay 1983; Yamato & Maede 1992; Houston & Myers 1993). Interestingly, onions were recognized for their toxic potential partly as a result of experiments in dogs to test their value against blacktongue disease. Dogs were given a daily ration that included onions, and it was noted that those animals that were given more than 15 g/kg b.w. of cooked or raw onions developed severe anemia (Sebrell 1930). In other experimental studies, dosages equivalent to 28–30 g/kg b.w. or more of raw onions fed once per day for 3 days have readily caused hematologic alterations in dogs and cats (Kobayashi 1981; Ogawa et al. 1986). In one instance, a male cat given onion extract orally equivalent to 3 g raw onion per milliliter developed mild hematologic changes after administration for 1–2 days (Kobayashi 1981). More-pronounced changes occurred after 4 ml (12 g) and 5 ml (15 g) of extract were given on days 3 and 6, respectively. The weight of the cat was not given, but assuming a weight of 3 kg, it appears that dosage of onions as low as 5 g/kg b.w. may cause marked hematologic alterations. Dogs appear to require somewhat more for intoxication to occur, but for either dogs or cats, when intoxication does occur, nearly all of the circulating erythrocytes may contain Heinz bodies (Kobayashi 1981; Ogawa et al. 1986). During the same period, methemoglobin levels in blood are typically low, in the range of 5% or less (Ogawa et al. 1986). The popularity of the use of baby food for sick cats has caused some problems because of the presence of onion powder in some instances. When fed baby food containing onion powder, cats may develop high numbers of Heinz bodies and reticulocytes, with decreased hematocrit (Robertson et al. 1998). Although perhaps not quite as susceptible as cats, some dogs may be profoundly affected by small doses of onions. A miniature poodle given a meal of discarded fried liver and onions developed dark red, coffee-colored urine 12 hours later and eventually a hematocrit of 20% and a moderate number Heinz bodies (Solter & Scott 1987). In another incident, 5.5 g/kg b.w. of minced dehydrated onions mixed with ground beef produced a marked reaction by the following day (Harvey & Rackear 1985). There were Heinz bodies, reticulocytes, and eccentrocytes, with a decrease in hematocrit to 20%.

Because of the strong pungent odor imparted by *Allium,* milk taint is an important concern with lactating animals. In addition, a serious problem may also occur with meat. Even the less potent members of the genus, such as leeks, may render to meat a strong, persistent, and unpleasant odor (Lund et al. 1991).

Disease Genesis—*Allium* contains a large number of sulfur-containing compounds, especially alk(en)ylcysteine sulfoxides. Trauma to the plant tissues allows enzymatic conversion of the sulfoxides to less stable compounds, followed by nonenzymatic conversion to an array of sulfides, disulfides, trisulfides, and thiosulfonates. The disulfides and trisulfides are chiefly responsible for odors and flavor, whereas thiopropanol-*S*-oxide is the lacrimation factor (Whitaker 1976; Fenwick & Hanley 1985c). Further biotransformation may occur in the liver; for example, the metabolism of allicin to diallyl disulfide, allyl mercaptan, vinyldithiins, and ajoenes in the isolated perfused rat liver (Egen-Schwind et al. 1992). Beneficial antimicrobial effects are due mainly to allyl thiosulfinates (Fenwick & Hanley 1985d; Weber et al. 1992). Similar to the cysteine derivatives from species of *Brassica,* some of the *Allium* sulfides such as the abundant *n*-dipropyl disulfide, and possibly dimethyl, di-1-propenyl, and di-2-propenyl disulfides, interact in the animal's body to form highly reactive oxidants (Gruhzit 1931a,b; Williams et al. 1941; Smith 1980; Fenwick & Hanley 1985a,c,d). The unsaturated 1-propylene appears to be a particularly potent hemolytic agent (Munday & Manns 1994). A variety of additional oxidative compounds capable of causing hemolysis may be present, such as sodium *n*-propylthiosulfate and either trans or cis sodium 1-propenylthiosulfate (Yuyama 1989; Miyata 1990; Yamato et al. 1994, 1998). All of these compounds are associated with effects on cell membranes and on hemoglobin. Depletion of the reductants of the glucose-6-phosphate–dependent reducing system and/or effects on G6P dehydrogenase reduces the protective capacity of the blood glutathione-reductase system and permits oxidation of hemoglobin and erythrocyte membranes (Fenwick & Hanley 1985d; Ogawa et al. 1985, 1986; Ahluwalia & Mohindroo 1989). The subsequent denaturation of hemoglobin with Heinz body formation and deformation of erythrocytes results in hemolysis and increased phagocytosis of these erythrocytes by the mononuclear phagocyte system. As a result, considerable hemosiderin may be seen in the spleen and Kupffer cells of the liver. The oxidative process results in the generation of small amounts of MHb. However, MHb is not a necessary precursor to the formation of Heinz bodies (Kiese 1974). In the case of *Allium*-induced effects, the processes of formation of MHb and Heinz bodies appear to be parallel (Ogawa et al. 1986).

■ highly reactive oxidants formed from sulfur compounds, various disulfides and thiosulfates

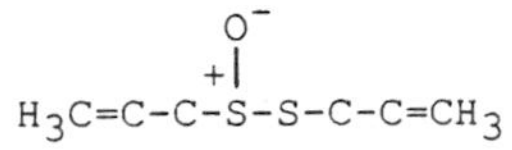

$H_3C{=}C{-}C{-}\overset{O^-}{\overset{+|}{S}}{-}S{-}C{-}C{=}CH_3$

Allicin

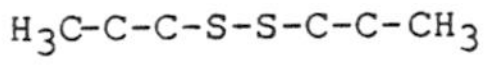

$H_3C{-}C{-}C{-}S{-}S{-}C{-}C{-}CH_3$

n-Dipropyl disulfide

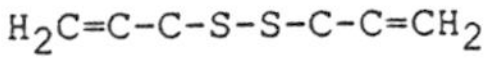

$H_2C{=}C{-}C{-}S{-}S{-}C{-}C{=}CH_2$

Di-2-propenyl disulfide

$H_3C{-}C{=}C{-}S{-}S{-}C{=}C{-}CH_3$

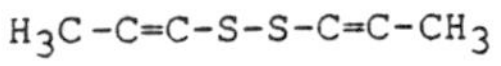

Di-1-propenyl disulfide

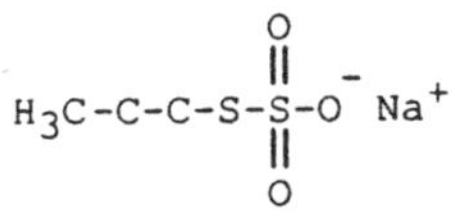

$H_3C{-}C{-}C{-}S{-}S(=O)_2{-}O^-\ Na^+$

Na *n*-propylthiosulfate

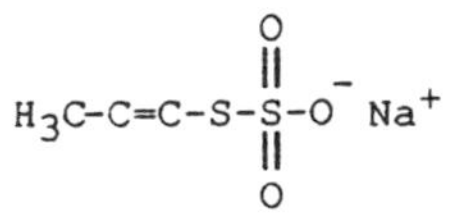

$H_3C{-}C{=}C{-}S{-}S(=O)_2{-}O^-\ Na^+$

Na *n*-propenylthiosulfate

The beneficial antimicrobial effects of this genus are apparently due mainly to allyl thiosulfinates (Fenwick & Hanley 1985b); Weber et al. 1992). Some of the reputed medicinal effects may be due to the ability of organosulfur compounds to modulate hepatic cytochrome P-450 pathways; diallyl sulfide, diallyl sulfoxide, and diallyl sulfone influence such activities (Brady et al. 1991).

It is of interest to note that some dogs in Japan seem to have a hereditary predisposition to the hemolytic effects of onions (Maede 1977; Yamoto & Maede 1992). This increase in susceptibility to oxidant effects is not necessarily associated with deficiencies in G6P dehydrogenase. Even high concentrations of glutathione in erythrocytes may not always be protective. Normally, it protects cells from toxic electrophiles formed by xenobiotics and participates with peroxidase to detoxify peroxides. These interactions are not always processes of detoxification, given that glutathione conjugation bioactivation of toxins has now been described (Monks et al. 1990). However, it is not clear that such a process is involved here.

Clinical Signs—In most cases the onset of intoxication will follow a prolonged period of onion ingestion, the length depending on the amount eaten and the animal species. The disease may appear as a peracute hemolytic episode but more commonly occurs with increasing severity of signs over 1–2 days. Hemoglobinuria, indicated by dark or coffee-colored urine is the sign that typically heralds onset. In some cases dogs may be observed to be depressed and lacking an appetite a day or more prior to the

■ abrupt onset, hemolysis, dark urine, depression, anorexia, weakness, Heinz bodies in RBCs hemolysis and hemoglobinuria, at which time numerous Heinz bodies may be present in a blood smear (Schouwenburg 1982). In some animals all erythrocytes are affected. Hemolysis will be accompanied by weakness, sometimes severe enough to result in tremors, and discoloration of the mucous membranes similar to but darker than the coloration produced with icterus. These signs may be noted in dogs, cats, cattle, and horses on occasion, but seldom will the full spectrum of signs be seen in sheep or goats (Sebrell 1930; Thorp & Harshfield 1939; Koger 1956; Van Kampen et al. 1970; Hothi et al. 1980). In most instances diagnosis is not difficult, because the animal's breath or body usually has a very strong onion odor. Recovery following removal of the source of onions is usually prompt and uneventful. In dogs, erythrocytes may be cleared of Heinz bodies within 7–10 days (Ogawa et al. 1986). However, in a few animals this has not been the case. Chronic debilitation follows months of feeding low to moderate amounts of onions to sheep, and sloughing of tails and feet occurs occasionally in cattle (Koger 1956; James & Binns 1966). Onions appear to have no teratogenic potential, and even though milk may have a distinctive odor, nursing young of severely affected dams are not similarly affected (James & Binns 1966; Pierce et al. 1972; Kirk & Bulgin 1979).

Laboratory findings are consistent with hemolysis and the accompanying response. The hematocrit will be decreased, marginally or in some cases to as low as 10–15%. However, even in the presence of what may appear to be severe hemoglobinuria, the hematocrit may be only slightly decreased. Within a few days a regenerative response will be evident, with polychromasia, reticulocytes, nucleated erythrocytes, increased cell size and hemoglobin content, and increased neutrophils with numerous bands. The response depends on the animal species affected (Van Kampen et al. 1970; Spice 1976). Prior to or during the hemolytic period, large numbers of Heinz bodies will be readily seen with appropriate staining, except possibly in sheep. Blood urea nitrogen is often decreased, possibly because of the low protein and high moisture content of onions (James & Binns 1966; Van Kampen et al. 1970; Kirk & Bulgin 1979). Within a few days after the onset of anemia there is usually a marked increase in erythrocyte G6P dehydrogenase and reduced glutathione (Ogawa et al. 1986).

- gross pathology, strong odor, plants in stomach

  microscopic, hemosiderin in liver, spleen, kidney; renal tubular necrosis and tubular pigment casts

Pathology—The most striking gross pathologic feature will be the strong pungent odor emanating from the body. Also onions will likely be present in abundance in the ingesta. The tissues may have a yellowish discoloration, and the spleen may be slightly swollen. Significant color changes will be seen in the liver, which becomes pale tan to golden brown and perhaps mottled, and the kidneys, which become brown to mottled black (Van Kampen et al. 1970; Pierce et al. 1972; Gill & Sergent 1981). Microscopically, hemosiderin in Kupffer cells of the liver, macrophages of the spleen, and possibly renal tubular cells may be striking. Mild to moderate hepatic necrosis, renal tubular nephrosis (pigment nephrosis) with tubular casts, and splenic hematopoietic foci may also be present, depending on the duration of the disease.

- nursing care; blood transfusion when severe

Treatment—Unless there is a very severe anemia requiring transfusion, treatment is not required. Once the onions are removed from the diet, the hemoglobinuria will subside in a day or two, and the erythrocyte mass will be replaced during the following weeks. During the period of anemia it is important not to stress the animals unduly, because they will have greatly diminished reserve. Moving livestock from pastures or pens to examine or treat them may precipitate severe problems (Koger 1956; Lazarus & Rajamani 1968).

Amianthium A.Gray

Taxonomy and Morphology—A monotypic genus indigenous to the eastern United States, *Amianthium* is placed in the segregate family Melanthiaceae by some taxonomists. Its name is derived from the Greek roots *amiantos* and *anthos,* meaning "without spots" and "flower," alluding to the glandless perianth. The single species is the following:

- *A. muscaetoxicum*
  - crow poison
  - gray fly poison
  - St.-Elmo's-feather
  - stagger grass

A. muscaetoxicum (Walter) A.Gray (= *Zigadenus muscaetoxicum* [Walter] Regel) (= *Chrosperma muscaetoxicum* [Walter] Kuntze)	gray fly poison, crow poison, stagger grass, St.-Elmo's-feather

- erect herbs from bulbs, not aromatic; leaves long, mainly basal; fruits 3-lobed capsules

Plants from bulbs; caulescent; not aromatic. **Bulbs** solitary; thick; tunicas papery. **Stems** erect; 30–120 cm tall. **Leaves** primarily basal; cauline leaves reduced, bract-like; blades of basal leaves linear, 2–3 cm wide, 30–45 cm long. **Inflorescences** terminal racemes; dense; conical to cylindrical; bracts present. **Flowers** numerous; perianths in 2-series. **Perianth Parts** 6; all alike; petaloid; free; white or pink; persisting and turning greenish; oblong-obovate; several-nerved. **Stamens** 6; filaments flat. **Pistils** with stigmas 3, small; styles 3; ovaries superior, deeply 3-lobed; placentation axile. **Fruits** septicidal capsules; 3-lobed. **Seeds** 3–6; oblong; purple-brown.

Amianthium muscaetoxicum

Distribution and Habitat—*Amianthium muscaetoxicum* is found in moist, open woodlands of the eastern United States. Only rarely is it planted as an ornamental.

- moist open woodlands

Disease Problems and Genesis—Early anecdotal information indicated *A. muscaetoxicum* to be a considerable hazard to cattle but much less so to sheep, horses, and pigs (Marsh et al. 1926a; Hoch 1969). Early spring seems to be the dangerous time, when the green leaves are most conspicuous and attractive in contrast to the surrounding dormant forage. Both leaves and bulbs are neurotoxic; the latter, when diced and mixed with honey, are reputed to be an effective poison for insects and birds—hence the common names. Toxicity is due to several cevanine-type, *Veratrum* ester alkaloids, such as amianthine, and a teratogenic jervanine type, jervine (Neuss 1953; Smith & Smith 1963; Smith et al. 1964). The role of these alkaloids in *Amianthium*'s toxicity is indicated by the similarity of the clinical signs it produces to those of *Zigadenus* and by observations of their effect on striated muscle—rapid fatigue and delayed relaxation (Alsberg 1912; Shaw et al. 1943). The toxic dose appears to be 0.1–0.2% b.w. of green plant for either sheep or cattle, and the lethal dose about 0.3% b.w.

- *Veratrum*-type alkaloids, neurotoxic and teratogenic as with *Zigadenus*; mainly cattle

Amianthium muscaetoxicum

Clinical Signs—A few hours after ingestion, profuse salivation and vomiting are seen as the early signs. As with *Zigadenus,* salivation is more prominent in sheep. During the following hours there is colic, weakness, incoordination, and rapid and labored respiration. The signs, including periodic vomiting, may persist for 1–2 days, during which time appetite will be markedly depressed. If enough plant material is eaten initially, the animal may die of respiratory failure the first day.

Although fetal deformities have yet to be reported in field cases, the presence of jervine in the plant indicates the potential for these effects (Neuss 1953). Malformations of the limbs would be expected (Keeler 1983, 1984).

- excess salivation, vomiting, weakness, colic, ataxia, rapid and labored respiration

  fetal deformities

■ no lesions

Pathology and Treatment—Distinctive pathologic changes are lacking, although small scattered hemorrhages may be expected. No effective treatment has been reported; an approach similar to that for *Zigadenus* intoxication may be appropriate.

Bowiea Harv. & Hook.f.

■ *B. volubilis*
climbing onion

Taxonomy and Morphology—Named for James Bowie, a dedicated nineteenth-century plant collector in South Africa, *Bowiea* comprises 2 or 3 species native to Africa. It is placed in the segregate family Hyacinthaceae by some taxonomists. It is quite distinctive because the leafless, twining, photosynthetic stems arise from bulbs that are mostly exposed aboveground. Of toxicologic interest is the following species:

B. volubilis Harv. & Hook.f. climbing onion

■ leafless vines from large bulbs; stems profusely branched, succulent, twining; flowers small, green-white

Plants from bulbs; caulescent; not aromatic. **Bulbs** large; solitary or clustered; partially exposed above soil surface; tunicas papery, scales large, silvery green. **Stems** twining; 2–3 m long; profusely branched; succulent; photosynthetic. **Leaves** small; caducous. **Inflorescences** twisting racemes; borne at ends of shoots; pedicels long; bracts absent. **Flowers** small; few; perianths in 2-series. **Perianth Parts** 6; all alike; petaloid; fused, cleft to base; reflexed; persistent; green-white. **Stamens** 6; filaments dilated. **Pistils** with stigmas 1, capitate; styles 1; ovaries superior. **Fruits** loculicidal capsules; short. **Seeds** 3–9.

Bowiea volubilis

■ ornamental

Distribution and Habitat—This South African species is encountered occasionally as a houseplant or cultivated as an ornamental in the warmest parts of North America.

■ cardiotoxic, bufadienolides, bovogenins, hellebrigenin

Disease Problems and Genesis—Because the bulbs are mainly aboveground, they are readily accessible for an animal to chew. *Bowiea volubilis* as a houseplant poses a special cardiotoxic risk for household pets because it contains monosidic cardiac glucosides. These cardiotoxins include bovosides A-E from bovogenins and other bufadienolides from genins such as hellebrigenin (Steyn 1934; Watt & Breyer-Brandwijk 1962; Kellerman et al. 1988). Raphides of calcium oxalate in the bulbs may also have a role in some of the adverse effects (Watt & Breyer-Brandwijk 1962). Although the bulb is said to be more potent than the foliage of *Digitalis,* the plants are generally of limited risk except for pets.

Bovogenin A

Hellebrigenin

■ vomiting, excess salivation, decreased heart rate, cardiac arrhythmias

Clinical Signs and Pathology—Following consumption of *B. volubilis,* the signs in dogs and cats are similar to those produced by other cardiotoxins. The initial signs indicate involvement of the digestive tract and include vomiting and salivation, but little diarrhea initially. More-severe signs include diuresis, decreased heart rate, arrhythmias, and seizures. There may be congestion of the viscera and scattered splotchy hemorrhages, but in general, distinctive changes are few.

Treatment—Orally administered charcoal may be helpful; otherwise treatment is directed toward relief of the effects on the digestive tract. Only very rarely is treatment for the cardiac effects needed. The cardiac signs are for the most part responsive to atropine.

■ activated charcoal, atropine

Colchicum L.

Taxonomy and Morphology—An Old World genus of temperate regions, *Colchicum,* commonly known as autumn crocus or naked lady, comprises 60–65 species. Its name is a Latinization of *Colchis,* the ancient name of the region east of the Black Sea. The common name *crocus* sometimes causes confusion because it is applied also to species of the genus *Crocus,* a member of the Iridaceae, or iris family. At first glance, members of the two genera indeed look alike, but close examination reveals that *Colchicum* has flowers with 6 stamens and a superior ovary, whereas those of *Crocus* have 3 stamens and an inferior ovary. Although there are exceptions, species of *Colchicum* typically flower in the autumn and those of *Crocus* in the spring. Of toxicologic interest is the following species:

■ *C. autumnale*
autumn crocus
meadow saffron

C. autumnale L. autumn crocus, meadow saffron

Plants from corms; acaulescent; not aromatic. **Corms** ovoid to subglobose; apices extended into tubular pseudostems; tunicas membranous to coriaceous, dark brown. **Leaves** basal; blades linear-lanceolate. **Inflorescences** of solitary flowers or clusters of 2–6; bracts 1, small. **Flowers** perianths in 2-series; campanulate. **Perianth Parts** 6; all alike; petaloid; fused, tubes long; purple or white or pink; narrowly elliptic to oblong-elliptic. **Stamens** 6; fused to perianth. **Pistils** with stigmas 3, decurrent; styles 3; ovaries superior; placentation axile. **Fruits** septicidal capsules; oblong-ovoid. **Seeds** numerous; subglobose.

■ herbs from corms; leaves basal; perianth parts purple, white, or pink

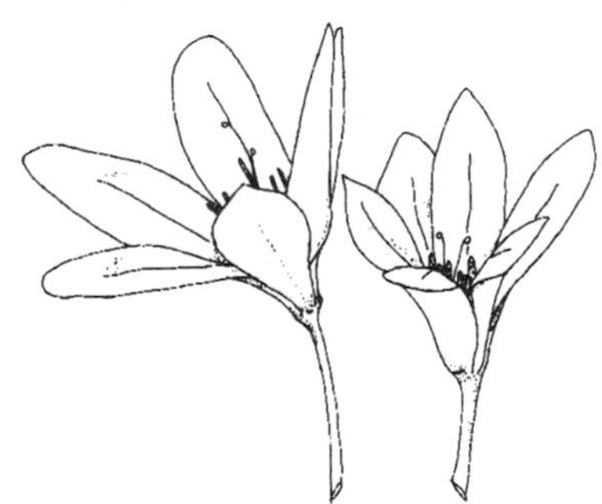

Colchicum autumnale

Distribution and Habitat—*Colchicum autumnale* is occasionally cultivated as an ornamental in North America, and even though it may escape, it represents little hazard for animals. It is more likely to occur in open sites.

■ ornamental

Disease Problems—*Colchicum* has been used for many centuries for the treatment of gout and other medical problems. Because of the relative scarcity of the plants in North America there has been little problem of toxicosis, except as a consequence of medical use. It is another matter in Europe and Eurasia, where *C. autumnale* and other indigenous species have caused numerous intoxication episodes, mainly digestive tract problems (Debarnot 1970; Tribunskii 1971; Chareyre et al. 1989; Yamada et al. 1998). In some instances, very high morbidity rates are observed, e.g., a case in which many animals were affected in a nomadic sheep flock that was allowed to graze an infested meadow while hungry (Panariti 1996). All animal species are susceptible, and intoxication episodes have been reported for cattle, dogs, goats, horses, pigs, and sheep. The lethal dose is approximately 6–10 g/kg b.w. of the fresh leaves or less if the capsular fruits are eaten (Tribunskii 1971; Chareyre et al. 1989). *Colchicum* is also toxic as a contaminant in hay (Kamphues & Meyer 1990). Occasionally, intox-

■ digestive disturbance, all plant parts

ications due to ingestion of the corms or flowers by humans occur (Spoerke & Smolinske 1990)

■ numerous tropolonic alkaloids, colchicine, colchiceine; interferes with cell division

Disease Genesis—The effects of *Colchicum* are ascribed mainly to colchicine, a tropolonic alkaloid. However, nearly three dozen other related alkaloids, including colchiceine and isocolchicine, are also present. These alkaloids have been used extensively in biochemical and biological investigations of cellular phenomena because of their very precise effects on mitosis. Colchicine interferes with microtubular-dependent cell functions by reversibly binding to tubulin protein and thus blocking polymerization. Chromosomal movement and transport of vesicles are prevented, arresting cell division in metaphase (Capraro & Brossi 1984). Another effect of this interference with microtubules is inhibition of migration and lysosomal degranulation of polymorphonuclear leukocytes. Microtubular effects in axons limit vesicle transport with consequent axonal degeneration and neuromuscular deficits (Anthony & Graham 1991). Other less well-known effects include conspicuous decreases in prolactin, insulin, catecholamine secretion, and glucose tolerance (Capraro & Brossi 1984). The alkaloids of *Colchicum* are found in all plant parts, especially flowers and seeds, in which the concentrations may exceed 0.2%. Although the concentration in the corms is much lower (0.05%), the mass of the organ is more than enough to provide sufficient amounts of the alkaloids to cause problems. Storage of large amounts occurs especially in the spring (Wildman 1960b).

Colchicine

Colchiceine

Colchicine is a lipid-soluble, rapidly absorbed toxin when given orally, reaching peak plasma concentrations in 1–2 hours (Murray et al. 1983). It is distributed widely throughout the body, especially to bone marrow, and the serum half-life in humans is 10–32 hours (Rochdi et al. 1992). Colchicine is biotransformed mainly in the liver, with about 30% eliminated unchanged in urine. High concentrations, in excess of 1 ppm, are present in milk from sheep, thus putting nursing young at risk (Cooper & Johnson 1984).

Because of the great interest in its many effects, colchicine is available as a synthetic mainly for experimental use. Its toxic effects greatly limit its medicinal use except for the treatment of gout. Unfortunately it is sometimes used with suicidal intent. As described by Stapczynski and co-workers (1981) and others, overdosage results in a three-phase sequence of effects. Initial digestive effects are followed by a 2-to-3-day period of life-threatening multiple organ involvement and a subsequent recovery period (Valenzuela et al. 1995). Effects on the organs, especially the liver and kidneys, may be manifested as cardiovascular, neuromuscular, hematologic, metabolic, and electrolyte alterations (Murray et al. 1983; Hood 1994; Valenzuela et al. 1995). Severe effects are noted with dosage of less than 0.5 mg/kg b.w. (Murray et al. 1983). It has been estimated that dosages of 0.8 mg/kg b.w. are almost 100% fatal (Rochdi et al. 1992). However, recovery even from dosage as high as 1 mg/kg b.w. may be possible with prompt vigorous treatment (Folpini & Furfori 1995). With the use of antibodies directed against colchicine, patients may survive even higher dosage (Terrien et al. 1990). The use of goat-derived colchicine-specific Fab fragments produces a dramatic reversal of the clinical signs and recovery without the sequential phases of intoxication (Scherrmann et al. 1992; Baud et al. 1995). As shown experimentally in mice, other treatments such as administration of aspartic acid and other amino acids are of questionable value (Wang et al. 1997).

■ depression, excess salivation, persistent vomiting, diarrhea, weakness, ataxia

Clinical Signs—Primarily indicative of the digestive disturbances, the initial signs are depression, hypersalivation, vomiting, milky diarrhea with straining, and indications of severe abdominal pain. Shortly thereafter, weakness, inappetence, incoordination, and paresis or collapse will be noted, especially if large amounts of *Colchicum* are eaten (Debarnot 1970; Tribunskii 1971; Chareyre 1989). In humans, multiple organ failure is a serious problem. A host of severe complications, including cardiovascular, pulmonary, renal, hematologic, metabolic, and neuromuscular disorders develop with acute intoxications (Murray et al. 1983). Marked elevation of serum enzymes such as LDH and CK, and evidence of coagulation problems such as an increase in prothrombin time, may be a consequence of the multiple organ involvement. Death may occur 1 to several days after ingestion. Although teratogenic effects have not been a field problem, they may be expected as cleft palate and craniofacial and other deformities have been noted in mice (Keeler 1972).

Pathology—Few distinctive gross changes are evident at postmortem except for reddening of the mucosa of the digestive tract from the tongue to the stomach and small intestine. There may be considerable fluid accumulation in the thoracic and abdominal cavities. Because death appears to be associated with a shocklike effect, there may be scattered small hemorrhages and congestion. Microscopically, experiments in calves indicate that necrosis may be present in the basal cell layer of the tongue, esophagus, and forestomachs, the gastric glands and intestinal crypts, and the epithelium of the renal pelvis and tubules and of the bladder (Yamada et al. 1998). Lymphocyte destruction in lymph nodes, spleen, and lymphoid follicles is also reported (Cooper & Johnson 1984).

■ gross pathology, reddened gut mucosa

Treatment—Activated charcoal should be given to reduce absorption of any alkaloid remaining in the digestive tract. Otherwise, general supportive measures should be instituted, such as administration of parenteral fluids containing dextrose and calcium. Aspartate may be added to the fluids if given early in the course of the disease. Atropine is sometimes used to control both digestive tract and cardiac problems. Although unlikely to be used in animals, colchicine-specific antibodies/Fab fragments have been used with success experimentally. Their use is similar to that of digoxin-specific antibodies (Terrien et al. 1990).

■ activated charcoal, fluids, electrolytes

Convallaria

Taxonomy and Morphology—A monotypic genus, *Convallaria* is a popular ornamental with several cultivars exhibiting variation in flower color and leaf variegation. Its name is derived from the Latin *convallis,* meaning "valley," which reflects the plant's typical habitat. Although 2 other species have been described, they are now considered to be localized variations. The single species is the following (see plate 47.1):

■ *C. majalis*
lily-of-the-valley

C. majalis L. — lily-of-the-valley
(= *C. majuscula* Greene)
(= *C. montana* Raf.)

Plants from rhizomes, typically forming dense colonies; acaulescent; not aromatic. **Stems** absent or short. **Leaves** 2 or 3; basal or subbasal, one above the other; blades broadly elliptic to oblong-oval; bases sheathing. **Scapes** erect or ascending; 10–20 cm long. **Inflorescences** terminal 1-sided racemes; pedicels recurved; bracts small, lanceolate. **Flowers** 5–13; perianths in 1-series; campanulate. **Perianth Parts** 6; all alike; petaloid; fused; lobes short, recurved; white. **Stamens** 6, fused to perianth parts. **Pistils** with stigmas 1, capitate, slightly lobed or not lobed; styles 1, elongate, straight; ovaries superior. **Fruits** berries; globose; red. **Seeds** numerous.

■ herbs from rhizomes; leaves basal or subbasal, sheathing; flowers in 1-sided racemes, urn-shaped; perianth parts white

Convallaria majalis

■ ornamental

Convallaria majalis

■ cardiotoxic but problems rare

■ extensive array of 23-C cardenolides; numerous glycosides of a few basic genins; cannogenin, periplogenin

Distribution and Habitat—Native to Eurasia and the southern Appalachian Mountains, *C. majalis* occurs in rich, moist soils of the forests. Some taxonomists contend that the North American populations are not native but rather early introductions from Europe. The species is now cultivated as an ornamental across the continent; it sometimes escapes or persists at abandoned homesites.

Disease Problems—In spite of the apparent cardiotoxic potency of *C. majalis,* surprisingly few intoxications have been reported. In a case involving a dog in which the plant species was identified postmortem in the intestine, there were few premonitory signs prior to death (Moxley et al. 1989). Inadvertent ingestion of *Convallaria* instead of *Allium* by a family produced a marked decrease in blood pressure and heart rate, with severe weakness (Edgerton et al. 1989). In spite of the apparent potent effects, serious intoxications are quite rare. Of 2,639 exposures to *C. majalis* reported to poison control centers in a 10-year period, only 3 individuals exhibited serious signs, and there were no fatalities (Krenzelok et al. 1996).

Disease Genesis—Possibly the most potent of all cardiotoxic plants, *Convallaria* also contains the greatest array of 23-C cardenolides. At least 38 have been identified from the aglycones bipindogenin, cannogenin, periplogenin, sarmentogenin, 19-hydroxysarmentogenin, sarmentologenin, nigrescigenin (= sarmentosigenin), strophanthidin, and strophanthidol (Kopp & Kubelka 1982a; 1982b). The array of potent cardenolides composed with differing sugars includes convallatoxin, convallatoxol, convallatoxoloside, convalloside, neoconvalloside, lokundjoside, locundeside, tholloside, and sarhamnoloside (Shoppee 1964; Singh & Rastogi 1970; Komissarenko & Stupakova 1986; Schrutka-Rechtenstamm et al. 1986; Joubert 1989). This diversity is in part due to interconversions among compounds in the plant as well as variation in the number of sugars; monosides, biosides, and triosides are present. The cardenolides are synthesized in the leaves and translocated to the rhizomes and roots, where they are stored (Schrutka-Rechtenstamm et al. 1985a). The array of cardenolides is most extensive at the end of the vegetative period. During fruiting, total concentration is lowest during this time (Schrutka-Rechtenstamm et al. 1985b). Cardenolide concentrations are highest in the roots, but all plant parts are of concern. Interestingly, not only is convallatoxin one of the more potent cardiotoxins, but it also is among the most cytotoxic cardenolides tested against nasopharyngeal carcinoma cells (Kelly et al. 1965).

Also present in *C. majalis* are spirostanol saponins glycosidic at the C-1, C-3, and/or C-5 positions (Hostettman & Marston 1995).

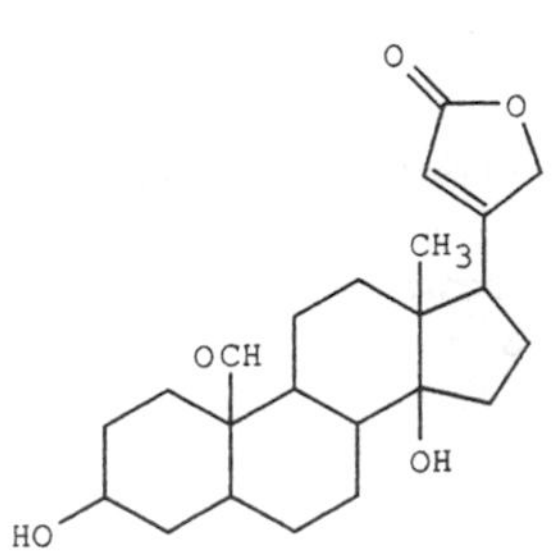

Cannogenin

Periplogenin

Sarmentogenin

Sarmentosigenin

Clinical Signs—In dogs and cats, the signs are similar to those produced by other cardiotoxins. The initial signs are indicative of the effects on the digestive tract and include vomiting and salivation but little diarrhea. More-severe signs include diuresis, decreased heart rate, arrhythmias, and seizures.

■ vomiting, excess salivation, decreased heart rate, cardiac arrhythmias

Pathology—Postmortem findings include scattered hemorrhages and reddening of the mucosa of the stomach and intestine. Plant parts, especially leaves, may be identifiable in the stomach or intestine.

■ reddened gut mucosa

Treatment—The cardiac signs are for the most part responsive to atropine. A more complete discussion of the signs, pathology, and treatment of cardenolide intoxication is presented in the treatment of *Digitalis,* in the Scrophulariaceae (see chapter 68).

■ activated charcoal, atropine

Cooperia Herb.

Taxonomy and Morphology—Comprising 6 or 7 species native to the New World, *Cooperia,* commonly known as rain lily or prairie lily, is closely related to *Zephyranthes,* and the 2 genera are merged by some taxonomists. Its name honors Joseph Cooper, an influential English gardener of the early 1800s. Two species are of toxicologic interest:

■ *C. drummondii*
C. pedunculata

cebolleta
prairie lily
rain lily

C. drummondii Herb. (= *Zephyranthes brazosensis* [Herb.] Taub)	Brazos rain lily, cebolleta
C. pedunculata Herb. (= *Zephyranthes drummondii* D.Don)	evening-star rain lily, prairie lily

Plants night-flowering; from bulbs; acaulescent. **Bulbs** large; globose or depressed globose; with short or long neck; tunicas papery, black. **Leaves** basal; blades linear, smooth, usually glaucous, erect or prostrate. **Scapes** erect; 10–35 cm long; hollow. **Inflorescences** solitary or paired flowers; spathes elongate, tubular. **Flowers** large and showy; fragrant; perianths in 1-series; salverform. **Perianth Parts** 6; all alike; petaloid; fused; white or yellow, abaxial surfaces typically red-tinged; oblong; apices obtuse. **Stamens** 6; fused to perianth parts; filaments short. **Pistils** with stigmas 1, 3-lobed; styles 1, elongate; ovaries inferior. **Fruits** capsules. **Seeds** numerous; flat; D-shaped; black.

■ night-flowering herbs from large black bulbs; leaves basal; flowers large, showy, fragrant, white or yellow

1. Leaves 6–12 mm wide. Perianth tubes 2.2–4 cm long. Styles 8–15 mm long . . . . *C. pedunculata*
1. Leaves 3–5 mm wide. Perianth tubes 8–18 cm long. Styles 40–100 mm long . . . . *C. drummondii*

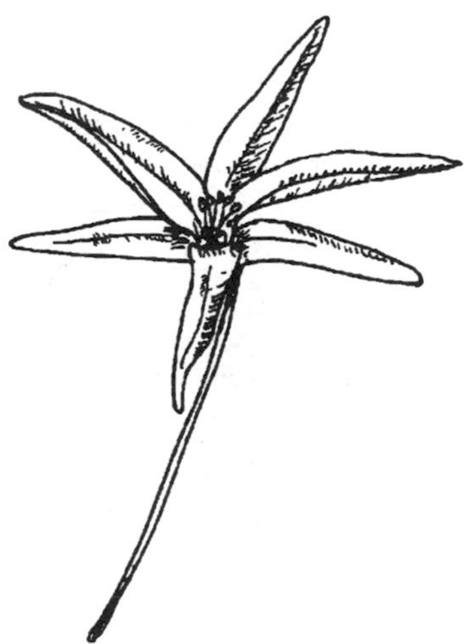
Cooperia drummondii

Distribution and Habitat—With the exception of one species in South America, *Cooperia* is a genus of northern Mexico and the south-central United States. The most widely distributed taxon of the genus, *C. drummondii* occurs with greatest abundance in Texas.

Disease Problems—Cattle in southern Texas have been observed to be affected by photosensitization from unknown causes since the early 1900s. Seldom fatal but responsible for severe decreases in productivity, most occurrences have been associated with rain and warm weather (Casteel et al. 1986). In vitro testing of various

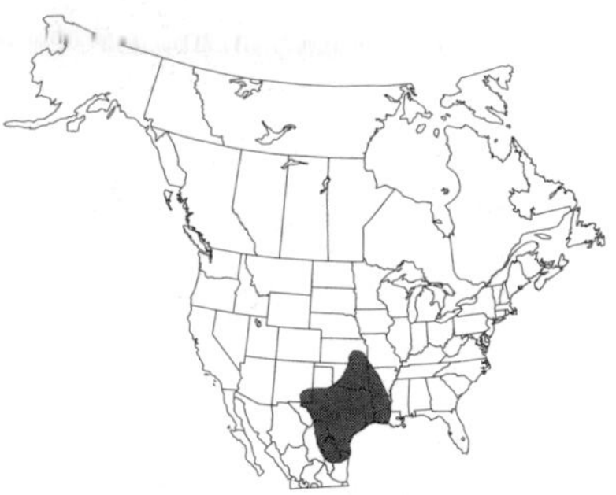

Cooperia drummondii

Cooperia pedunculata

plants, employing the fungus *Candida albicans* as the test organism, revealed dead leaves of *C. pedunculata* to be photoactive (Rowe et al. 1987). Further testing in mice demonstrated the species' potential for photosensitization. Fresh green leaves or dry dead leaves were not toxic; only wet, dead leaves after spring and fall rains were toxic. In an attempt to reproduce the disease using forage from a pasture in which the disease had occurred, cattle were given a mixture of 75% green and 25% dead leaves (Casteel et al. 1988). A heifer given 10 g/kg b.w. on day 1 and 5 g/kg on day 2 developed photosensitization. Continued treatment at the lower dosage for several more days increased the severity of the disease, and the animal died on day 23. Another heifer given 1.7 g/kg b.w. per day for 4 days developed mild skin problems. The syndrome appears to be one of primary or direct photosensitization, as indicated by profiles of serum chemistry and histopathologic evaluations (Casteel et al. 1986). *Cooperia* thus appears to be a cause of the photosensitization in cattle observed since the early 1900s in south Texas.

- dead leaves, photosensitization; bulbs, digestive disturbance
- unknown toxin, direct photosensitization

Disease Genesis—The specific photoactive toxin in *Cooperia* is unknown, but it appears to be direct-acting and not associated with liver disease and retention of phylloerythrin. *Cooperia drummondii* also is reported to contain irritating phenanthridine alkaloids such as lycorine and γ-lycorine (see table 47.1). Thus ingestion of the dead leaf residue may result in photosensitivity, whereas the bulbs are likely to be associated with digestive tract disturbance.

- skin lesions; erythema, edema, serum crusting, sloughing; light-colored areas of skin; itching; eyes reddened, conjunctiva and sclera

Clinical Signs—In cattle and most other species, the most striking effects are severe pruritis, reddening, swelling, edema, exudation and crusting of serum, and sloughing and peeling of the skin of the face, teats, udder, escutcheon, and the less pigmented and lightly haired areas. The skin around the nose and eyes is especially likely to be involved. There will be increased lacrimation, the muzzle may be reddened and cracked, and ulcerations may even be present on the tongue. There is often a very striking demarcation in distribution of the lesions, affecting lighter-colored areas right up to the edge of the darker areas of skin. The eyes may also be affected with edema of the cornea, reddening of the conjunctiva and sclera, and increased sensitivity to light. In addition to the obvious skin and eye involvement, there may be accompanying signs of diarrhea, restlessness, agitation, loss of appetite, and increased respiratory rate. Profiles of serum hepatic enzymes show little changes. The signs may last 2 weeks, while the healing process may take 2 or more months.

- gross pathology, crusty, ulcerative, proliferative skin lesions

Pathology—In more-severe cases, the skin will be thickened with a gelatinous appearance to the dermis. Crusty and ulcerative or proliferative areas may be present, especially around the face. The local lymph nodes may be enlarged. Microscopically, the edema of the skin will be accompanied by necrosis and a polymorphonuclear infiltrate and areas of hyperkeratosis and parakeratosis.

- remove animal from plant, gradual recovery

Treatment—Preventing animal access to the plants is usually sufficient for recovery to begin within a few days, albeit signs may regress slowly with reepithelialization of the denuded areas. In some cases, placing animals in shade and/or relief of pain may be helpful, in addition to soothing skin medications.

Fritillaria L.

Taxonomy and Morphology—A genus of the north temperate regions in both the Old World and the New World, *Fritillaria,* commonly known as fritillary, comprises approximately 100 species. Its name is derived from the Latin *fritillus,* meaning "checkerboard," and alludes to the shape and coloration of the mature capsule, which resembles a dicebox. With showy flowers, many species of the genus are prized ornamentals. In western North America, 17–21 species are present (Beetle 1944; Turrill & Sealy 1980). The following are representative taxa:

F. affinis (Schult.) Sealy (= *F. lanceolata* Pursh)	mission bells, rue-root, checker lily, rice-root fritillary
F. atropurpurea Nutt.	checker lily, chocolate lily, leopard lily, purple-spot fritillary
F. camschatcensis (L.) Ker Gawl.	black lily, Indian rice
F. pudica (Pursh) Spreng.	yellow bell, yellow fritillary

■ *F. affinis*
F. atropurpurea
F. camschatcensis
F. pudica

black lily
checker lily
chocolate lily
fritillary
Indian rice
leopard lily
mission bells
rue-root
yellow bell

Plants from bulbs; caulescent. **Bulbs** globose or ellipsoid; bulblets often numerous at bases; scales thick, fleshy; tunicas smooth. **Stems** erect; 10–150 cm tall; unbranched. **Leaves** alternate or whorled; sessile; blades linear to ovate. **Inflorescences** racemes or solitary flowers; bracts leaflike. **Flowers** typically large and showy; generally nodding; perianths in 2-series; campanulate to cup-shaped. **Perianth Parts** 6; all alike; petaloid; free; deciduous; of various colors. **Stamens** 6; fused to perianth parts; filaments slender. **Pistils** with stigmas 1 or 3; styles 1, 3-branched or not branched; ovaries superior. **Fruits** loculicidal capsules; membranous; 6-angled or 6-winged or rounded. **Seeds** numerous; flat; brownish.

■ erect, unbranched herbs from bulbs; leaves alternate or whorled; flowers large and showy

Because the species listed above are merely representative of the genus and not specifically implicated in toxicologic problems in North America, a key is not presented. If exact identification is needed, manuals specific for the reader's area should be consulted.

Fritillaria camschatcensis

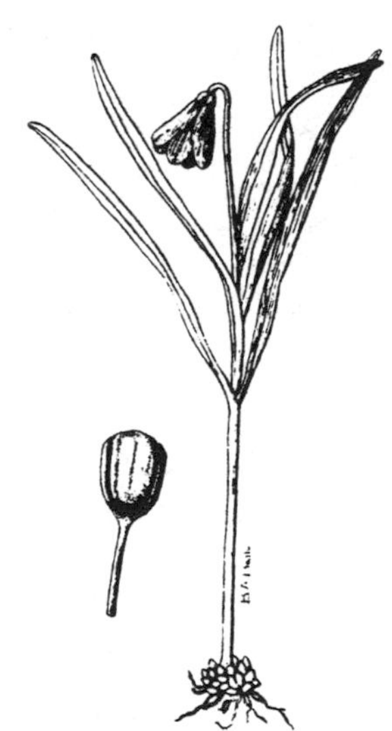
Fritillaria pudica

Fritillaria affinis

Fritillaria atropurpurea

Distribution and Habitat—In North America, species of *Fritillaria* occur primarily along the Pacific Coast from the Aleutian Islands to Northern Mexico (Beetle 1944). About half of the species are limited to California. A few occur as far east as western Nebraska and the Dakotas. They generally inhabit moist open to semiwooded

■ moist, open to semiwooded sites, ornamentals

sites. In addition, some species, both indigenous and introduced, are cultivated as garden ornamentals.

■ unknown risk, but similar to that of *Veratrum,* possibly neurotoxic and teratogenic

Disease Problems—This genus is known to contain toxic compounds, but the hazard it poses is not clear. Although the bulbs are apparently quite bitter, wild geese seek them out and avidly eat them (Beetle 1944). The bulbs and pods also are eaten by members of Indian tribes in Montana (Hart 1992).

■ *Veratrum*-type alkaloids, jervanine, imperialine

Disease Genesis—All plant parts of many species of *Fritillaria* contain *Veratrum* alkaloids, including the C-nor-D-homo steroidal cevanine and jervane types (Brown 1970; Kaneko et al. 1985; Hu et al. 1993). The cevanines of *Fritillaria,* unlike those of *Veratrum,* are glycosides with few hydroxyl groups and thus are not extensively esterified. *Fritillaria camschatcensis* contains jervanine, solanidine, solasodine, and solanidanine group alkaloids (Ripperger & Schreiber 1981; Mimaki & Sashida 1990). The jervanine kuroyurinidine is very similar to jervine, but its teratogenic potential has yet to be assessed. The cevanines, of which imperialine *(F. imperialis)* and verticine *(F. verticillata)* are the best known, appear similar in activity to those from *Veratrum.* They decrease blood pressure and heart rate and have a marked if not explicitly clear effect on excitable cells (Chen et al. 1935; Narumi 1935; Mir & Ghatak 1966; Etissami & Kachefiolasle 1971). Few of the *Fritillaria* species in North America have been studied for alkaloid content or physiologic activity. *Fritillaria camschatcensis* has been studied, and it appears to contain a potential teratogen but no hypotensive cevanine alkaloids (Mimaki & Sashida 1990).

Fritillaria camschatcensis

Fritillaria pudica

Imperialine

Verticine

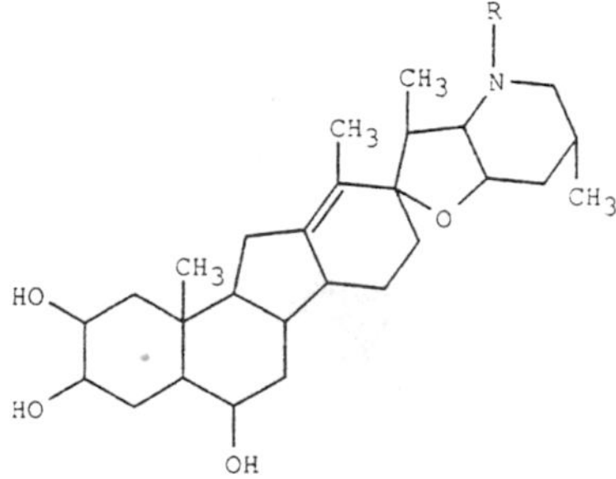

Kuroyurinidine

■ possibly as with *Veratrum, Zigadenus*

Clinical Signs, Pathology, and Treatment—Reports of actual cases of intoxication due to *Fritillaria* were not encountered in a survey of the literature. Thus, clinical signs and pathologic changes cannot be described, but they might be expected to be similar to the effects produced by extracts of the Asian species given parenterally in laboratory animals. Likewise, because of the similarity in their alkaloids, signs of acute intoxication might be anticipated to be similar to those produced by *Veratrum* and *Zigadenus* for some species of *Fritillaria,* whereas teratogenic effects may be caused by others. Complete discussions of clinical signs, pathology, and treatment are presented in the treatments of *Veratrum* and *Zigadenus* in this chapter.

Gloriosa L.

Taxonomy and Morphology—A monotypic genus native to tropical Africa and Asia, *Gloriosa* is grown in North America as an ornamental for its exotic, brilliantly colored flowers. Their beauty is reflected in the genus name, which is derived from the Latin *gloriosus* meaning "splendid" or "full of glory." The numerous species previously described are now treated taxonomically as cultivars of the single species:

G. superba L. glory lily, superb lily, climbing lily, creeping lily

- *G. superba*

  climbing lily
  creeping lily
  glory lily
  superb lily

Plants vines, climbing by means of tendrils at leaf apices; from tubers; caulescent; herbage glossy bright green. **Tubers** irregularly cylindrical; V- or L-shaped, producing aerial stems from ends; red-brown. **Stems** twining; to 2.5 m long; 1–4 per tuber; slender; sparsely branched or unbranched. **Leaves** alternate or opposite or whorled; blades ovate-lanceolate to oblong; apices tapering to a tendril 3–5 cm long. **Inflorescences** solitary flowers in axils of upper leaves; pedicels long, usually curved. **Flowers** showy; perianths in 2-series. **Perianth Parts** 6; all alike; petaloid; free; spreading; lanceolate to spathulate; yellow to red or purple or bicolored; apices reflexed; margins crisped. **Stamens** 6; fused to perianth parts or arising from receptacles; filaments long, curved. **Pistils** with stigmas 3, filiform; styles 1, 3-branched, bent near base at right angles to ovary; ovaries superior. **Fruits** loculicidal capsules. **Seeds** numerous.

- climbing vines from red-brown tubers, tendrils present; flowers solitary, axillary; perianth parts yellow, red, purple, or bicolored

Gloriosa superba

Distribution and Habitat—*Gloriosa superba* is cultivated as an outdoor ornamental in the warmest areas of North America and as a houseplant elsewhere.

- ornamental

Disease Problems and Genesis—Seldom producing toxicosis in North America, *G. superba* elsewhere is sometimes used in suicides. In lower dosage, it causes digestive tract problems (Aleem 1992). Like *Colchicum,* it contains colchicine and other tropolone alkaloids; 19 have been identified (Dvorackova et al. 1984). The tubers contain up to 0.36% colchicine and have proven lethal upon ingestion by humans (Nagaratnam et al. 1973); however, they seem to be of little toxicologic consequence in animals.

- digestive disturbance, colchicine, antimitotic activity

Colchicine

Clinical Signs, Pathology, and Treatment—The signs produced by *G. superba* are essentially the same as those for *Colchicum.* Profound and persistent vomiting is the most distinctive feature of the disease. There may also be diarrhea, sweating, and, in younger children, seizures. The problem is an acute one, with few if any distinctive pathologic changes. Aimed at relief of the adverse effects, conservative treatment is generally effective.

- persistent vomiting, diarrhea, sweating

Hemerocallis L.

Taxonomy and Morphology—Prized for its ornamentals, *Hemerocallis* comprises 22–24 species and more than 20,000 cultivars (Munson 1989). Its name is derived from the Greek roots *hemere* and *kallos,* meaning "day" and "beautiful," which allude to each flower's lasting only for 1 day before withering. Its common name—day lily—likewise reflects the brevity of each flower, albeit the plants produce flowers for quite a long time in the summer. Representative of the genus are the following species:

H. altissima Stout

H. esculenta Koidz.

- *H. altissima*
  H. esculenta
  H. fulva
  H. lilioasphodelus
  H. minor
  H. thunbergii

  day lily
  tawny lily

H. fulva (L.) L.	tawny lily, common orange day lily
H. lilioasphodelus L. (= *H. flava* L.)	lemon day lily, tall yellow day lily, custard day lily
H. minor Mill.	grass-leaved day lily
H. thunbergii Baker	Thunberg's day lily, late yellow day lily

■ herbs from rhizomes; leaves numerous, long-arching; flowers large, showy, funnelform; perianth parts yellow, red, orange, or magenta

Hemerocallis fulva

Plants from rhizomes; typically forming dense clumps; acaulescent or caulescent. **Roots** fibrous or tuberous, often enlarged at ends. **Stems** absent or present; short. **Leaves** numerous; basal or 2-ranked on short stems; sessile; blades linear, long, arching; bases flat or folded. **Scapes** erect; 1–2 m long; branched or unbranched. **Inflorescences** terminal racemes, bracts present. **Flowers** typically large and showy; each lasting for 1 day; perianths in 2-series; funnelform to campanulate. **Perianth Parts** 6; all alike; petaloid; fused at base; spreading or recurved; of various shades of yellow, orange, red, or magenta. **Stamens** 6, fused to perianth parts at tube summits; filaments elongate. **Pistils** with stigmas 1, capitate; styles 1; ovaries superior. **Fruits** loculicidal capsules; 3-angled. **Seeds** several; globose; black.

Because the species listed above are merely representative of the genus as a whole, a taxonomic key is not presented. If exact identification is needed, horticultural manuals should be consulted.

■ ornamentals, naturalized

Distribution and Habitat—*Hemerocallis* is native to eastern Asia, and many species have been introduced from China and Japan. Plants are hardy and tolerant of climatic extremes and thus are cultivated as ornamentals throughout most of North America. *Hemerocallis fulva,* the most common species, is naturalized in some areas of eastern North America and occurs as far west as Texas, Oklahoma, and Kansas.

■ roots possibly neurotoxic, blindness; as yet not known as toxic in North America

Disease Problems—Species of *Hemerocallis* are not recognized as toxic in much of the world, and as yet there have been no reports of toxicity in North America. In addition to being prized as ornamentals, they have been used for food. In China the flower buds are eaten raw in salads, cooked in soups, and dried and used as a flavoring (Everett 1981). The rhizomes and roots or concoctions made from them also are used to treat certain diseases, such as schistosomiasis, and in some instances adverse effects such as blindness, paresis, and even death have occurred (Shiao et al. 1962; Barlow et al. 1989). In addition, several species, including *H. altissima, H. esculenta, H. lilioasphodelus,* and *H. minor* are reported to have caused death of livestock (Wang et al. 1989). In mice the oral LD_{50} of the rootstocks of *H. thunbergii* is 4.1 and 2.45 mg/g b.w. for the whole rootstock and its cortex, respectively (Shiao et al. 1962). When given orally to mice, extracts of the leaves, flowers, and buds of *H. lilioasphodelus* were not toxic, whereas those made from the rootstocks were lethal in 2 of 6 animals (Der Marderosian & Roia 1979). Because of the great popularity of these plants as ornamentals, it is possible that in a unique circumstance livestock might gain access to rhizomes and roots of *Hemerocallis*. However, it is improbable that sufficient quantities would be available in most instances to cause intoxication.

■ binaphthalenic hemerocallin; brain edema, myelin disruption

Disease Genesis—The toxin in the rhizomes and roots appears to be a binaphthalenetetrol, hemerocallin, which is identical with stypandrol produced by *Stypandra imbricata,* an Australian member of the Liliaceae (Wang et al. 1989; Colegate et al. 1992). Hemerocallin has been identified specifically in the rootstocks of *H. lilioasphodelus* and *H. thunbergii.* The disease caused by stypandrol in Australia is unique (Dorling et al. 1992; Huxtable et al. 1992). The acute paretic effects of this disease involve brain edema, wallerian degeneration of optic nerves, and central myelin disruption. There appear to be direct effects on both the optic nerve and the retina, because animals surviving the acute effects exhibit resolution of the myelin damage but continued degeneration of optic nerves and loss of retinal photoreceptors (Huxtable et al. 1980; Main et al. 1981). Hemerocallin appears to cause the same disease in sheep (Barlow et al. 1989). Experimentally, when mice are fed hemerocallin for several days, it causes transient degeneration of astrocytic processes, intralamellar myelinic edema, and white matter vacuolation (Barlow & Middleton 1989). However, it is not yet known whether concentrations in the rhizomes, roots, and aerial parts of the species present in North America are sufficient to represent a hazard. Furthermore, the amount of plant material available may limit the risk. Animals must eat substantial amounts of *Stypandra* for several days before the disease develops, a situation quite unlikely to occur in North America with *Hemerocallis*.

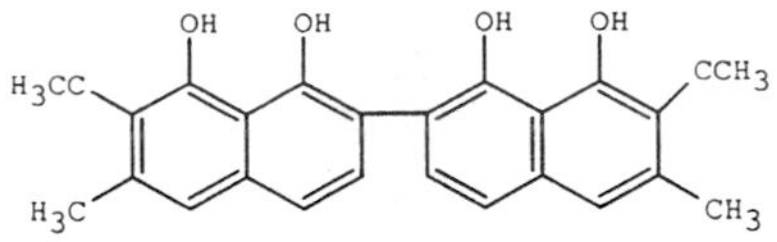

OH OH OH OH
H_3CC CCH_3
H_3C CH_3

Hemerocallin

Clinical Signs—Using the disease produced by *Stypandra* as a model, the following signs can be anticipated. After several days of plant consumption, the animal will exhibit incoordination, paresis, and apparent blindness. Some animals may die in a few days. Following prevention of access to the plants, there may be gradual recovery except for the blindness. With chronicity, pupillary light reflexes are lost. Upon ophthalmoscopic examination, hypertrophy of pigmented epithelium, especially of the posterior pole adjacent to the optic disk, is apparent. With retinal degeneration, the tapetum becomes very bright colored.

■ ataxia, paresis, possibly blindness

Pathology—Grossly, the most obvious acute lesion is edema of the brain. There are no gross lesions in the chronic form of the disease. Microscopically, three types of lesions are characteristic. Acutely, there is diffuse, spongy vacuolation of myelin in the brain and peripheral nerves. This change may resolve over several months, so various degrees may be observed, depending upon the stage of the disease. There will also be severe focal or diffuse degeneration, atrophy, and sclerosis of the optic nerve. Chronically, there is retinal degeneration and atrophy of the rods, cones, and outer nuclear layers. These changes become more apparent as the acute changes resolve.

■ gross pathology, brain edema
microscopic, myelin vacuolation, optic nerve degeneration, retinal degeneration

Treatment—General nursing and supportive care is appropriate for the acute disease, but there is no specific approach for reversing the effects on the eyes.

■ general nursing care

Hyacinthoides Heist. ex Fabr.

Taxonomy and Morphology—Native to western Europe and northwestern Africa, *Hyacinthoides,* commonly known as English bluebells or Spanish bluebells, comprises 4–6 species, which were placed in the closely related genus *Scilla* at one time. The generic name *Endymion* also has been used for the taxon. In North America, 1 introduced species is of toxicologic interest:

■ *H. non-scripta*
bluebells
English bluebell
harebell
wild hyacinth

H. non-scripta (L.) Chouard ex Rothm. (= *Scilla non-scripta* [L.] Hoffmanns. & Link) (= *Endymion non-scriptus* [L.] Garcke)	English bluebell, harebell, bluebells, wild hyacinth

Plants from bulbs; caulescent. **Bulbs** globose; tunicas membranous, white. **Leaves** basal; blades linear to linear-lanceolate, 25–45 cm long, fleshy; concave. **Scapes** erect; 20–50 cm long; fleshy. **Inflorescences** terminal racemes; somewhat 1-sided and drooping; bracts 2, linear-lanceolate, blue-tinged; bracteoles present. **Flowers** 6–12;

■ herbs from bulbs; leaves basal, fleshy, concave; flowers in drooping terminal racemes; perianth parts blue, violet-blue, or white

perianths in 2-series; tubular to campanulate. **Perianth Parts** 6; all alike; petaloid; free; erect or spreading, blue or violet blue or rarely white; 1-nerved. **Stamens** 6; fused to perianth parts. **Pistils** with stigmas 1, capitate; styles 1; ovaries superior. **Fruits** capsules; 3-angled. **Seeds** numerous; globose; black.

Hyacinthoides non-scripta

- ornamental
- digestive disturbance and rarely cardiotoxic activity from bulbs or seeds; cardiotoxins similar to those of *Urginea*
- vomiting, colic, diarrhea; rarely, depression, weakness, slow and irregular heart rate
- reddened gut mucosa

  activated charcoal

Distribution and Habitat—Native to western Europe and typically forming large populations in beech-oak forests, *H. non-scripta* is cultivated as an ornamental throughout North America. Plants occasionally escape.

Disease Problems and Genesis—Toxicity due to *Hyacinthoides* is an unusual problem occurring occasionally in livestock or humans when the bulbs are mistaken for onions (Cooper & Johnson 1984). Ingestion produces clinical signs indicative of effects on the digestive tract and rarely the heart. They are similar to those caused by other bulb-bearing taxa of the family. Intoxications may also occur with ingestion of the seeds and capsules.

The toxic effects appear to be due to cardiotoxins, compounds perhaps similar to the bufadienolides of *Urginea,* also a member of the family (see treatment in this chapter). Affecting most animal species, the few reports describe signs very similar to those produced by cardiotoxins (Thursby-Pelham 1967; Cooper & Johnson 1984).

Clinical Signs—With low to moderate dosage, digestive problems occur quickly and include retching, colic, and later diarrhea with or without blood, or constipation. Larger dosage may result in depression, weakness, exercise intolerance, decreased peripheral perfusion with cold extremities, and slow/irregular heart rate.

Pathology and Treatment—Unlikely to result in death, the changes are confined to reddening of the gastric and small intestinal mucosa. Orally administered charcoal may be helpful; otherwise treatment is directed toward relief of the digestive tract effects. Only rarely would treatment of the cardiac effects be indicated.

Hyacinthus L.

- *H. orientalis*

  Dutch hyacinth
  hyacinth

Taxonomy and Morphology—Now considered to be a monotypic genus, *Hyacinthus* is a familiar spring-flowering ornamental. Cited by Homer, its name is one of the oldest unchanged plant names in use today (Stearn 1990). In mythology, its flowers are said to have sprung up from the blood of Hyakinthos, a youth accidentally killed by Apollo. Its circumscription has been narrowed considerably since it was first described by Linnaeus; its species have been segregated in other genera such as *Hyacinthoides, Scilla,* and *Muscari.* It is placed in the segregate family Hyacinthaceae by some taxonomists. It must be noted that the common name *hyacinth* also is applied to plants of several other genera, e.g., *Muscari* and *Camassia,* also members of the Liliaceae, and to *Eichhornia* in the Pontederiaceae. The single species is the following:

H. orientalis L. hyacinth, Dutch hyacinth

- herbs from large bulbs; leaves basal, long, thick; flowers in terminal racemes, strongly scented; perianth parts blue, white, or pink

Plants from bulbs; acaulescent. **Bulbs** large; ovoid to depressed globose; tunicas papery-membranous, purple or pearly white. **Leaves** basal; blades linear to lanceolate or ovate; thick; apices hooded; margins involute, then spreading. **Scapes** erect until fruiting, then prostrate; 15–45 cm tall; thick; hollow. **Inflorescences** cylindrical terminal racemes; bracts small, 2-lobed. **Flowers** 2–40; typically waxy; strongly scented; perianths in 2-series; tubular-campanulate. **Perianth Parts** 6; all alike;

petaloid; fused; lobes oblong to spathulate, spreading or recurved; of various shades of blue, white, and pink. **Stamens** 6, fused to perianth tubes; included within perianths. **Pistils** with stigmas 1, capitate; styles 1, short, filiform; ovaries superior. **Fruits** loculicidal capsules; globose; 3-lobed or 3-angled. **Seeds** numerous; black; winged.

Hyacinthus orientalis

Distribution and Habitat—Native to the eastern Mediterranean region—Greece, Turkey, and Syria—*H. orientalis* is cultivated as a garden ornamental throughout North America. It is also a popular pot plant, often being forced to flower at Easter by florists.

- ornamental

Disease Problems and Genesis—The toxicity problems caused by *H. orientalis* are very similar to those produced by *Tulipa* (tulip) and include both dermatitis and digestive disturbances. Few instances of intoxication have been reported, the best known being the severe diarrhea that occurred when hyacinth bulbs were fed to cattle in the Netherlands during the difficult times of World War II (Cooper & Johnson 1984). Although the bulbs lack the onion odor, they may be mistaken for onions because of their similar appearance and inadvertently eaten. The causative factors for the skin and digestive problems have not been clearly identified, but raphides of calcium oxalate are present (Frohne & Pfander 1984). This may account, at least in part, for the irritant effects.

- irritant effects; dermatitis and digestive disturbance, similar to effects caused by tulips

Clinical Signs, Pathology, and Treatment—The effects of *Hyacinthus* ingestion are indicated by digestive tract problems, including retching, vomiting, and diarrhea. Death is unlikely to occur. Pathologic changes are confined to reddening of the mucosa of the stomach and small intestine. Orally administered charcoal may be helpful; otherwise treatment is directed toward relief of the irritation of the digestive tract.

- retching, vomiting, diarrhea
  reddened gut mucosa
  activated charcoal

Lilium L.

Taxonomy and Morphology—Comprising 80–100 species, numerous hybrids, and innumerable varieties, *Lilium* is a genus of the north temperate regions of both the Old World and the New World. Its name is a Latinization of *leirion,* the classical Greek name used by Theophrastus for the Madonna lily. Patterns of variation within the genus are complex because of hybridization and morphological similarity of the species. It must be kept in mind that the common name *lily* is applied to species of other genera as well, e.g., *Convallaria, Fritillaria, Hemerocallis,* and several others. In North America, approximately 30 species, both native and introduced, are present. The following are representative:

- *L. canadense*
  L. candidum
  L. catesbaei
  L. lancifolium
  L. longiflorum
  L. pardalinum
  L. superbum

  devil lily
  Easter lily
  leopard lily
  lily
  Madonna lily
  panther lily
  pine-lily
  tiger lily
  trumpet lily
  Turk's-cap lily
  white lily
  wild yellow lily

Introduced Species

L. candidum L.	Madonna lily, white lily
L. lancifolium Thunb.	tiger lily, devil lily
L. longiflorum Thunb.	Easter lily, trumpet lily
L. pardalinum Kellogg	leopard lily, panther lily

Native Species

L. canadense L.	wild yellow lily
L. catesbaei Walter	pine-lily
L. superbum L.	Turk's-cap lily

- erect herbs from white, yellow, or purple bulbs; flowers showy, perianth parts of various colors

Lilium longiflorum

Lilium pardalinum

Plants from bulbs; caulescent. **Bulbs** with scales; globose or elongate; large or small; scales imbricate, fleshy, white or yellow or purple. **Stems** erect; 30–250 cm tall; unbranched. **Leaves** alternate or whorled; sessile or subsessile; blades linear or lanceolate or elliptic. **Inflorescences** solitary flowers or racemes or umbels or verticils; flowers held erect or horizontal or inclined or pendulous; bracts present or absent. **Flowers** typically large and showy; perianths in 2-series; funnelform or cup-shaped or bowl-shaped. **Perianth Parts** 6; all alike; petaloid; erect to spreading or recurved; bases clawed or narrow; shades of white, yellow, orange, red, or maroon; adaxial surfaces often spotted. **Stamens** 6; fused to perianth parts or arising from receptacles. **Pistils** with stigmas 1, 3-lobed; styles 1, elongate; ovaries superior. **Fruits** loculicidal capsules; oblong. **Seeds** numerous; flat.

Because the species listed above are merely representative of the genus, a taxonomic key is not presented. If exact identification is needed, horticultural or floristic manuals should be consulted.

- ornamentals, some native across North America

Distribution and Habitat—Species of *Lilium* occupy a variety of habitats in the Northern Hemisphere. In North America, they are present across the continent, with an abundance in the deciduous forests of the East and the various vegetation types in California. They are cultivated in all parts of our range as garden plants or houseplants.

- cats, renal failure, ingestion of flowers and leaves

Disease Problems—Information about the toxicity potential of *Lilium* is limited. It is now known that *L. longiflorum, L. tigrinum,* and Asiatic hybrids cause acute renal failure in cats (Hall 1990; Gathings 1991; Carson et al. 1994). All parts of the plant represent a problem; the relative amounts needed are not yet fully known, but ingestion of a single bloom may cause intoxication (Carson et al. 1994). Most reported cases are severe, with signs appearing 2–4 days after ingestion of the flowers and/or other plant parts (Hall 1992). Animals quickly deteriorate, with death occurring within 5 days. Prompt aggressive treatment within a few hours of ingestion may markedly reduce the severity of the disease.

It appears that many of the species of *Lilium* are a risk to cats but not dogs (JO Hall, pers. comm., 1995). This may simply be due to a relatively higher dosage required for other animal species. Renal disease is not unknown in the Liliaceae. *Narthecium ossifragum,* commonly known as bog asphodel, in the United Kingdom and Norway and *N. asiaticum* in Japan cause digestive disturbances, diffuse renal tubular necrosis, and renal failure in cattle (Suzuki et al. 1985; Malone et al. 1992; Flaøyen et al. 1995). Disease caused by these species is similar to that caused by *Lilium.* In both instances the risk of renal damage is greatest with ingestion of the flowers (Malone et al. 1998).

- toxin unknown; possibly similar to furanone of *Narthecium*

Disease Genesis—Toxicants present in *Lilium* have not been identified. Like so many other genera of the family, the genus contains several steroidal saponins of unknown toxicologic significance (Shimomura et al. 1989). The toxin in *Narthecium* that causes kidney disease appears to be 3-methoxy-2-furanone, although tannin concentrations also are high (10%) (Suzuki et al. 1985; Flaøyen 1998; Langseth et al. 1999). Tubular damage due to oxalate crystals does not seem to be a problem, again similar to disease caused by *Lilium.*

- vomiting, depression, anorexia

elevated BUN, creatinine, potassium

Clinical Signs—Onset of signs occurs 12 or more hours after ingestion, with typical signs of vomiting, depression, and loss of appetite. Blood chemistries are markedly altered, with elevated levels of creatinine, blood urea nitrogen, phosphorus, and potassium (Hall 1990; Gathings 1991).

Pathology—Grossly, there is excess yellowish fluid in the chest and abdominal cavities, edema of the lungs, and slight swelling of the kidneys. Microscopically, the most distinctive lesions are in the kidneys. There is degeneration and necrosis of the proximal tubules with eosinophilic granular casts in many of them. Some tubules are dilated, and the cuboidal epithelial cells flattened. Renal interstitial edema may also be apparent. Less commonly there are minor hepatic changes such as congestion and biliary proliferation.

- gross pathology, excess fluid, throughout body

microscopic, kidney, necrosis, dilation proximal tubules, flattened epithelium

Treatment—The basic approach is for renal failure. If initiated within a few hours after ingestion, treatment consisting of administration of emetics, activated charcoal with a saline cathartic, and fluid diuresis have been effective in preventing renal failure (Hall 1992). Treatment initiated a day or more after onset of signs has been generally unsuccessful.

- immediate; emetic, activated charcoal, cathartic; otherwise, general nursing care

Narcissus L.

Taxonomy and Morphology—Native to Europe, northern Africa, and western Asia, *Narcissus* comprises 50–60 species and a plethora of hybrids and cultivars. Its name was used by Theophrastus and is believed to be derived from *narke,* meaning "numbness" or "torpor," alluding to its reputed narcotic properties. An alternative derivation is that its name comes from Narcissus in Greek mythology, who fell in love with his reflection in a pool and was turned into a lily by the gods.

- *N. jonquilla*
  N. poeticus
  N. pseudonarcissus
  N. tazetta

daffodil
jonquil
Lent lily
narcissus

One of the earliest plants to flower in the spring, the genus is horticulturally one of the most important and popular bulb plants. The common names *daffodil, narcissus,* and *jonquil* are used interchangeably for its species, and there is often confusion as to the difference. Traditional usage in the cut-flower trade is to call cultivars with long coronas, or trumpets, daffodils and to call those with short, cup-shaped coronas narcissi (or narcissuses). Although the name *jonquil* is used for cultivars with long coronas as well, it is more appropriately applied only to *N. jonquilla* and its hybrids. Simply treating the three vernacular names as synonyms is recommended. Horticulturalists classify members of *Narcissus* into 12 divisions based on appearance of the perianth (relative size, color, part number), number of flowers per scape, and origin. Representative of the genus are the following:

- erect herbs from ovoid brown bulbs; leaves basal; perianth parts of flowers yellow, white, or bicolored

N. jonquilla L.	jonquil
N. poeticus L.	poet's daffodil, poet's narcissus
N. pseudonarcissus L.	Lent lily, wild daffodil, trumpet narcissus
N. tazetta L.	polyanthus narcissus, bunchflower narcissus

Plants from bulbs; acaulescent. **Bulbs** large or small; solitary or in clusters; ovoid; tunicas papery-membranous, brown. **Leaves** basal; erect or spreading or prostrate; blades linear to narrowly lanceolate, flat. **Scapes** erect; 20–45 cm long. **Inflorescences** solitary flowers or umbels or clusters of 2–20 flowers; flowers held inclined or horizontal or declined, subtended by scarious spathe. **Flowers** typically large and showy; fragrant or not fragrant; perianths in 2-series; salverform. **Perianth Parts** 6; all alike; petaloid; fused; coronas conspicuous, long or short, trumpet- or cup-shaped, margins entire or erose or crisped; yellow or white or bicolor with coronas different from perianth parts. **Stamens** 6, in 2 whorls; fused to perianth parts. **Pistils** with stigmas 1, 3-lobed; styles 1; ovaries inferior. **Fruits** loculicidal capsules; globose or ellipsoidal. **Seeds** numerous.

Narcissus

A taxonomic key is not presented here. If exact identification is needed, horticultural manuals should be consulted.

■ ornamentals

Distribution and Habitat—In cool-temperate climates of the Old World, species of *Narcissus* occupy a variety of habitats from coastal beaches to alpine sites. They are cultivated as ornamentals throughout North America. *Narcissus poeticus* and *N. pseudonarcissus* occasionally escape, occur as waifs, or persist at abandoned homesteads.

■ digestive disturbance, eating bulbs; dermatitis from contact with bulbs, flowers, and stems of *Narcissus* and numerous other genera; rarely neurologic effects

Disease Problems—Although species of *Narcissus* are widely cultivated and highly prized, several adverse effects have been associated with this genus and other genera formerly grouped in the family Amaryllidaceae. Disease associated with *Narcissus* can be viewed as representative of other genera containing phenanthridine alkaloids. Both contact and allergic skin problems of the hands, arms, and face are problems for people who continually handle the bulbs and aerial portions (Walsh 1910; Mitchell & Rook 1979; Gude et al. 1988). Contact dermatitis has also been reported in a dog (Willemse & Vroom 1988). Consumption of the bulbs, flowers, or stems, even when cooked, results in digestive problems, typically of a transient nature (Wilson 1924; Litovitz & Fahey 1982; Vigneau et al. 1984; Goncalo et al. 1987).

■ phenanthridine alkaloids, digestive disturbance, neurologic effects; lycorine, crinin, lycorenine types

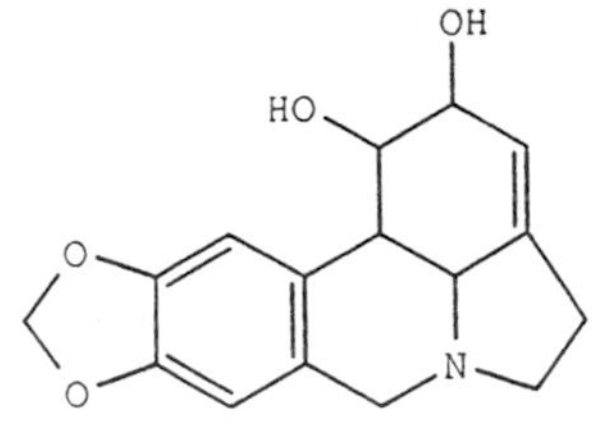

Disease Genesis—An extensive array of phenanthridine alkaloids (in excess of 100) of several types is found in *Narcissus* and other genera formerly grouped in the Amaryllidaceae (table 47.1). Lycorine (= narcissine) is a glucosidic alkaloid common to most species (Cook & Loudon 1952; Martin 1987). Other alkaloids of common occurrence include galanthine, galanthamine, haemanthamine, ambelline, clivonine, oduline, and pluviine (Cook & Loudon 1952; Wildman 1960a, 1968; Fuganti 1975; Martin 1987). These complex alkaloids and their isomers can be divided among several types. These types and examples for each include lycorine (lycorine), the most abundant; crinin (hippeastrine and tazettine); and lycorenine (galanthamine and hippeastrine). The same alkaloids also are found in the leaves and flowers in lesser amounts (Vigneau et al. 1982). Total alkaloid concentrations may be up to nearly 0.5% in the leaves and some portions of the bulbs (Moraes-Cerdeira et al. 1997). Levels are about half in the flowers and bulbils. Although the proportions of alkaloids may vary with the species and variety, in cultivar 'Ice Follies,' galanthamine, haemanthamine, and lycorine were the major alkaloids, in that order (Moraes-Cerdeira et al. 1997).

Lycorine

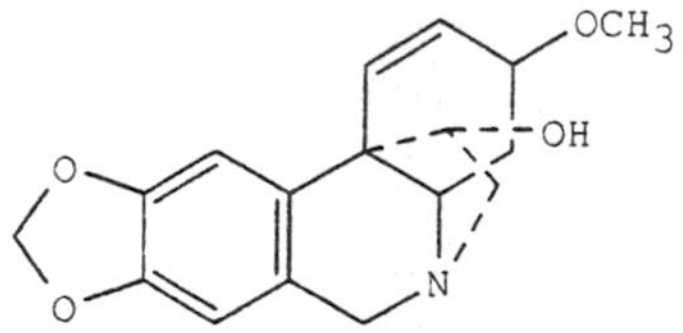

Haemanthamine

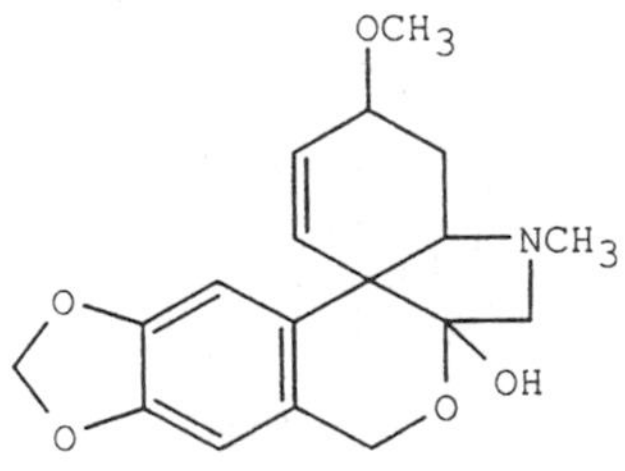

Tazettine

Because of the consistent presence of lycorine, problems associated with *Narcissus* can be viewed generally as representative of other genera containing phenanthridine alkaloids. In the original description of the action of the alkaloids in *Narcissus*, the flowering bulb was associated with anticholinergic-type effects and the nonflowering one with cholinergic actions (Ringer & Morshead 1879). However, in later studies by Laidlaw, the overall effects in the cat, attributed primarily to lycorine, were nausea, vomiting, excess salivation, and diarrhea (Ewins 1910). There was only a quantitative difference between the two growth stages when the flowering bulb was about half as toxic. Lycorine also is a centrally acting emetic and an inhibitor of protein synthesis in eukaryotic cells, by inhibition of peptide bond formation (Martin 1987; Ban et al. 1988). Other alkaloids may have different actions. Galanthamine is analgesic and cholinergic, with both central and peripheral reversible inhibition of acetylcholinesterase (Irwin & Smith 1960; Baraka & Harik 1977). It has been promoted as treatment for Alzheimer's disease, although it also causes a decrease in heart rate and some cardiac conduction abnormalities (Moraes-Cerdeira et al. 1997).

Table 47.1. Taxa of the Liliaceae containing phenanthridine alkaloids

Genus/Species	Common Names	Example Alkaloids Present
Amaryllis belladonna	belladonna lily, naked lady	lycorine, belarmine, ambelline, undulatine
Clivia		
C. miniata	kaffir lily	lycorine, clivonine, clivatine,
C. nobilis		miniatine, hippeastrine
Cooperia		
C. drummondii (= *Zephyranthes brazosensis*)	Brazos rain lily	lycorine, γ-lycorine
C. pedunculata (= *Zephyranthes drummondii*)	evening-star rain lily, prairie lily	
Crinum		
C. americanum	Florida crinum, southern swamp lily	lycorine, crinamine, haemanthamine
C. asiaticum		
C. zeylanicum		
Eucharis grandiflora	Amazon lily	lycorine
Galanthus nivalis	common snowdrop	lycorine, nivaline, tazettine, galanthamine, hippeastrine
Haemanthus		
H. coccineus	blood lily	lycorine, coccinine, montanine
H. multiflorus (= *Scadoxus multiflorus*)	powder puff lily	lycorine, haemultine, chlidanthine, haemanthidine, hippeastrine
Hippeastrum		
H. puniceum	amaryllis	lycorine, tazettine
H. vittatum	Barbados lily, common amaryllis	haemanthamine, hippeastrine, vittatine
Hymenocallis		
H. littoralis	spider lily, crown beauty, sea daffodil	lycorine, tazettine
H. occidentalis	inland hymenocallis	nivaline
Leucojum		
L. aestivum	meadow snowflake, summer snowflake	lycorine, lycorenine, galanthamine
L. autumnale	autumn snowflake	
L. roseum		
L. vernum	spring snowflake	
Narcissus spp.	daffodil, narcissus	lycorine, others
Nerine spp.	nerine, Guernsey lily	neronine, lycorine, tazettine
Zephyranthes		
Z. atamasco (= *Z. treatiae*)	atamasco lily, fairy lily, rain lily	lycorine, tazettine
Z. simpsonii	stagger grass, windflower, zephyr flower	galanthine

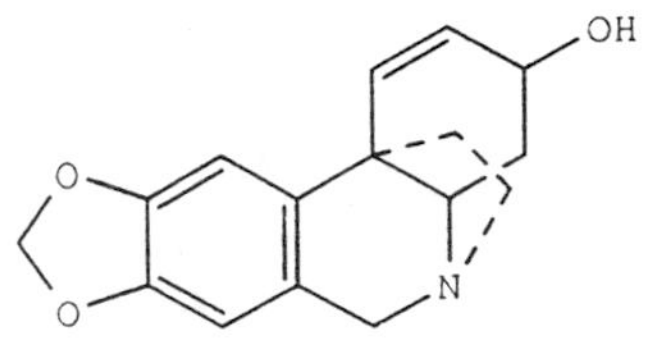

Crinamine and ambelline are hypotensive and respiratory depressants, whereas coccinine is a convulsant at high dosage (Southon & Buckingham 1989). These alkaloids may be quite toxic, with their LD_{50}s ranging from 5 mg/kg b.w. s.c. in mice for ambelline, 10 mg/kg i.v. in dogs for haemanthamine and crinin and 11 mg/kg s.c. in mice for galanthamine to 41–71 mg/kg i.v. in dogs for others (Southon & Buckingham 1989). Many of the phenanthridine alkaloids are also cytotoxic (Abd El Hafiz et al. 1991). However, in spite of the different and sometimes divergent actions attributed to the *Narcissus,* the overall effects of most plants containing these phenanthridine alkaloids are most typical of those due to lycorine.

Although the digestive tract problems observed are consistent with the results of experimental studies on the effects of the alkaloids present, a role for calcium oxalate crystals or raphides cannot be dismissed entirely (Ewins 1910; Gude et al. 1988). Calcium oxalate raphides may be significant factors in causing contact dermatitis in horticultural workers, as well as swelling and edema of the mouth and lips upon

Crinin

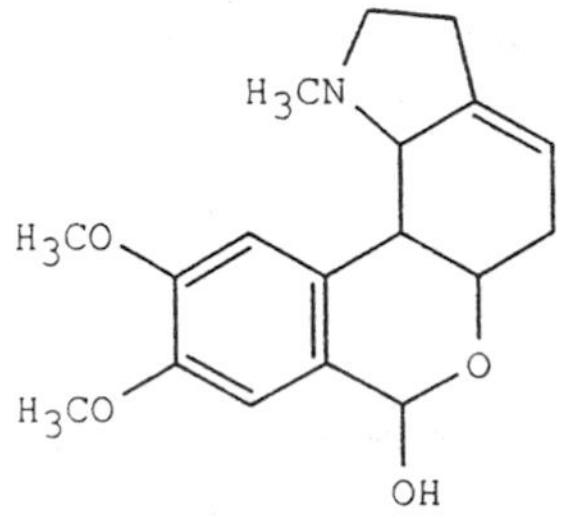

Lycorenine

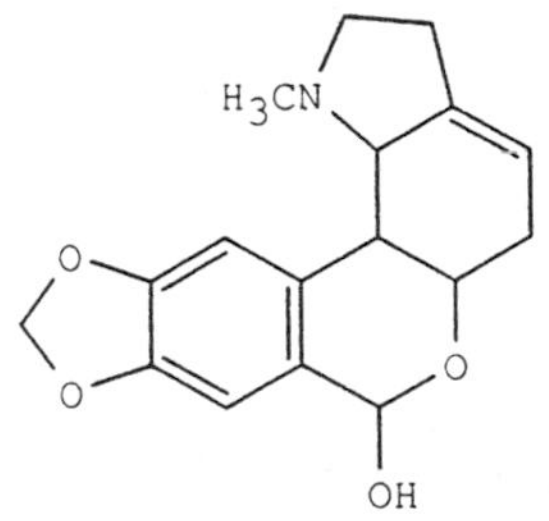

Oduline

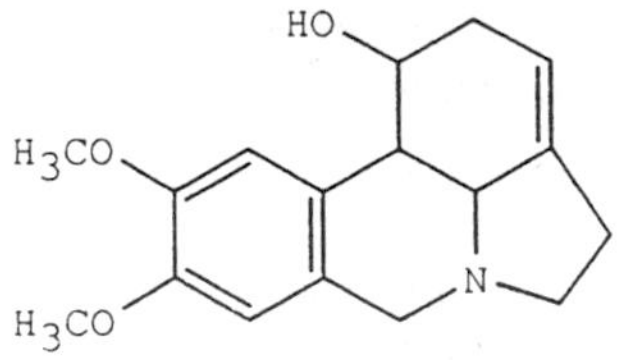

Pluviine

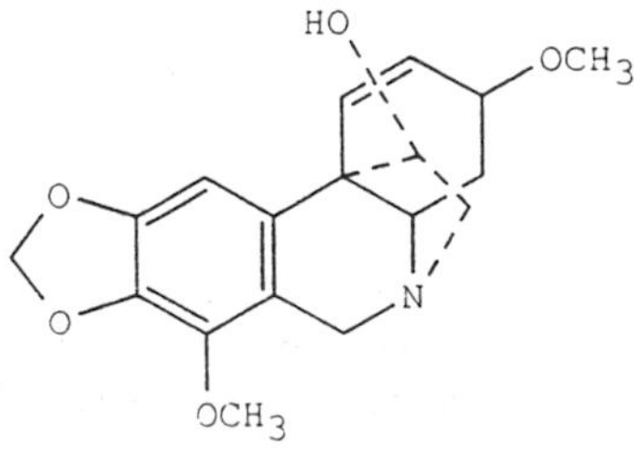

Ambelline

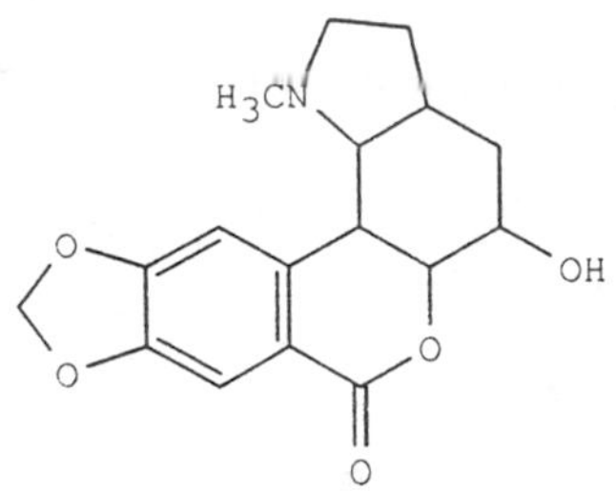

Clivonine

ingestion (Watt & Breyer-Brandwijk 1962; Hjorth & Wilkinson 1968; Frohne & Pfander 1984). In very unusual instances where very large amounts of plant material are eaten, especially the bulbs, additional systemic effects may be seen (Steyn 1934; Watt & Breyer-Brandwijk 1962). In this respect, the toxicity potential among some of the genera and species appears to vary considerably. Some are clearly hazardous; for example, species of *Clivia* are used as rat poison (Watt & Breyer-Brandwijk 1962). Similarly some species of *Haemanthus* are considered quite toxic, but the fruit of others is readily eaten by some animals (Watt & Breyer-Brandwijk 1962).

■ vomiting, excess salivation, diarrhea; dermatitis, erythema, loss of hair; ataxia, seizures

Clinical Signs—Irritation of the digestive tract, whether in humans or animals, is manifested by vomiting, excess salivation, restlessness, pain, diarrhea, and dyspnea. Not all the signs may be seen in one individual, and the outcome is rarely serious or even fatal. With ingestion of exceptionally large amounts of plant material, there may be depression, incoordination, seizures, paralysis, and various changes in cardiac function.

The dermatitis may appear as diffuse erythema, alopecia, and perhaps some papular and pustular lesions in various areas of the body.

■ reddened gut mucosa

Pathology—Lethal effects are rare, and even when they occur, few changes would be apparent other than reddening and edema of the mucosa of the stomach(s) and intestines, and perhaps ventricular dilation of the heart.

■ control of vomiting and diarrhea

Treatment—Relief of the irritation of the digestive tract is the primary aim of treatment. This includes antiemetics and antidiarrheals. Otherwise treatment is based upon relief of the signs. The dermatitis usually resolves promptly after exposure to the plant is eliminated.

Ornithogalum L.

■ *O. arabicum*
O. lacteum
O. nutans
O. saundersiae
O. thyrsoides
O. umbellatum

chinkerinchee
nap-at-noon
star-of-Bethlehem
summer snowflake
wonder flower

Taxonomy and Morphology—An Old World genus of 80–150 species, *Ornithogalum* is paradoxically prized for its striking ornamentals and despised for the invasive weedy nature of some of them. Used by Dioscorides, its name is derived from the Greek roots *ornis* for "bird," and *gala* for "milk," possibly alluding to the egglike color of the bulbs or flowers of some of its species. The following are representative of the species cultivated in North America:

O. arabicum L.	star-of-Bethlehem
O. lacteum Jacq.	
O. nutans L.	drooping star-of-Bethlehem
O. saundersiae Baker	giant chinkerinchee
O. thyrsoides Jacq.	chinkerinchee, wonder flower
O. umbellatum L.	star-of-Bethlehem, summer snowflake, nap-at-noon

■ erect herbs from bulbs; leaves basal, linear; flowers in racemes; perianth parts white, rarely yellow to red

Plants from bulbs; acaulescent. **Bulbs** ovoid or globose; solitary or clustered; producing numerous bulblets; tunicas white or brown, fleshy. **Leaves** basal; fleshy; blades linear. **Scapes** erect; 10–50 cm tall. **Inflorescences** racemes, elongate or corymbose; bracts present. **Flowers** 3–12; perianths in 2-series; campanulate. **Perianth Parts** 6; all alike; petaloid; free; equal or unequal; spreading or recurved; oblong to lanceolate; white or rarely yellow to red; keels present or absent. **Stamens** 6; arising from receptacles; filaments flat. **Pistils** with stigmas 1, 3-lobed; styles 1; ovaries superior. **Fruits**

loculicidal capsules; cylindrical to glabrous; 3-angled; yellow-green or purple-black or green-black. **Seeds** numerous.

1. Capsules purple-black or green-black.
   2. Bulbs globose. Racemes corymbose, pyramidal. Perianth parts 1–1.5 cm long . . . . *O. saundersiae*
   2. Bulbs ovoid. Racemes cylindrical or globose. Perianth parts 1.5–3.2 cm long . . . . *O. arabicum*
1. Capsules yellow-green.
   3. Perianth parts keeled; abaxial surfaces with green midstripe.
      4. Lower pedicels 2–6 cm long. Styles shorter than ovaries . . . . *O. umbellatum*
      4. Lower pedicels 0.5–1 cm long. Styles longer than ovaries . . . . *O. nutans*
   3. Perianth parts not keeled; abaxial surfaces without green midstripe.
      5. Leaf margins ciliate. Bases of stamen filaments not expanded . . . . *O. thyrsoides*
      5. Leaf margins entire. Bases of stamen filaments conspicuously expanded . . . . *O. lacteum*

Ornithogalum nutans

Ornithogalum umbellatum

Distribution and Habitat—Principal centers of species diversity for *Ornithogalum* are the Mediterranean region and South Africa. Most commonly planted in North America are *O. umbellatum* and *O. nutans,* which have escaped cultivation into roadsides, fields, and woods, especially in the Northeast and the Midwest (Hansen 1924). In some areas, *O. umbellatum* is considered a troublesome weed. Less commonly encountered, but considered by many individuals to be even more beautiful, are *O. arabicum, O. lacteum, O. saundersiae,* and *O. thyrsoides.*

Disease Problems—Intoxications are rare in North America, although a loss of 1,000 sheep is reported to have occurred during a rainy period in Maryland (Reynard & Norton 1942). It is mainly the bulbs that are a cause of digestive disturbance and cardiotoxicity. Plants have been more of a problem in South Africa, where many species are found in sufficient abundance to be eaten by grazing livestock. Ornamental varieties likewise pose a threat.

Disease Genesis—*Ornithogalum* is similar in toxicity to *Convallaria,* and the two genera have several cardenolides in common (Mrozik et al. 1959; Smith et al. 1967). At least eight cardenolides have been identified, with four aglycones (genins) represented: digitoxigenin, sarmentogenin, syriogenin, and uzarigenin (Smith et al. 1967; Singh & Rastogi 1970; Ghannamy et al. 1987). As is typical of cardenolides, much of the variation is due to the number of sugars—monosides and biosides—

- ornamentals, some naturalized in eastern North America
- digestive disturbance, rarely cardiotoxic effects
- extensive array of cardenolides, similar to those of *Convallaria*; mainly of 4 genins—digitoxigenin, sarmentogenin, syriogenin, uzarigenin

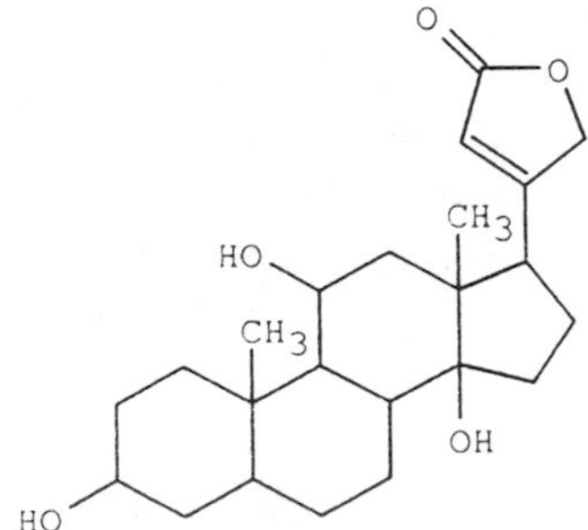

Sarmentogenin

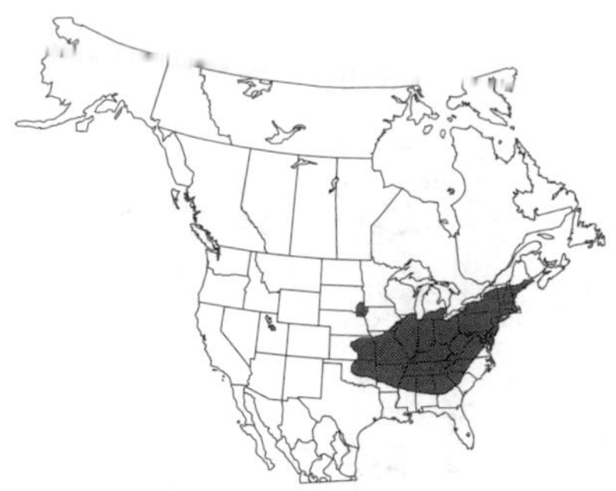

Ornithogalum umbellatum

represented. This potent array of cardiotoxins is found in all plant parts and especially in the bulbs. It is important to note that cardenolide isolation has been accomplished from both the European and the North American populations of *Ornithogalum.* The toxin or toxins responsible for disease associated with the African species have not been clearly identified, albeit many of the signs are consistent with those of the cardenolides (Kellerman et al. 1988). Also present in *Ornithogalum* are a group of spirostanol saponins, 18-norspirostanol glycosides of unknown toxicologic significance (Hostettmann & Marston 1995).

- diarrhea, anorexia, depression, colic

Clinical Signs—The most characteristic effect is severe diarrhea—fetid, watery, and sometimes bloody. Accompanying signs include inappetence, depression, colic, and increased respiration and heart rates. Surprisingly, blindness is a common sign for the disease in South Africa, which is known as chinkerinchee poisoning (Kellerman et al. 1988). This is a physiologic blindness, because it is not accompanied by microscopic changes. Blindness commences a week or more after ingestion of the plants and may last for several weeks or become permanent. Obvious cardiac problems have not been reported, although on some occasions deaths occur suddenly rather than in the more typical 2–5 days (Kellerman et al. 1988).

- reddened gut mucosa

Pathology—The most consistent finding is severe reddening and hemorrhage of the mucosa of the stomach and intestines. Other less distinctive changes include congestion and scattered hemorrhages on various other tissues (Breukink 1963; Kellerman et al. 1988).

- activated charcoal

Treatment—Directed toward relief of the cardiac problems, treatment may include the use of activated charcoal as with the other plants containing cardiac glycoside (Kellerman et al. 1988). Unfortunately, the onset of blindness, as seen in South Africa, seems to be unaffected by treatment.

Tulipa L.

- *T. fosteriana*
  T. gesneriana
  T. sylvestris
  tulip

Taxonomy and Morphology—Native to the north temperate Old World, especially central Asia, *Tulipa* comprises 100–150 species and numerous hybrids and cultivars. Its name is a Latinized derivative of the Arabic word *dulban,* meaning "turban" and alluding to the shape of the flowers. Apparently first cultivated extensively in Turkey, garden tulips were introduced into Europe in the 1550s. They became so popular in the 1620s and 1630s that they were treated as a precious commodity. Fortunes were made as individuals speculated and then lost when the market collapsed in 1637. The phenomenon is referred to as "tulipomania."

Standardization of tulip cultivar names began in 1913 under the auspices of the Royal Horticultural Society. The 1987 edition of the Dutch international register listed more than 2,300 names. Horticulturalists classify members of *Tulipa* into 15 divisions based on flowering time, number of flowers per stem, and perianth characters. It must be kept in mind that species other than those of the genus *Tulipa* also have the common name of *tulip,* e.g., species of *Calochortus, Homeria,* and *Papaver.* The following species are representative of the genus:

- fleshy, erect herbs from ovoid brown bulbs; leaves thick; flowers large and showy; perianth parts of various colors

T. fosteriana Hoog ex W.Irving	empress tulip
T. gesneriana L.	common tulip
T. sylvestris L.	Florentine tulip

Plants from bulbs; caulescent; herbage typically glaucous, fleshy. **Bulbs** ovoid; apices pointed; tunicas papery to coriaceous, brown, adaxial surfaces glabrous to coarsely ciliate or tomentose. **Stems** scapelike but bearing 2 or 3 smaller leaves; erect; 15–75 cm tall; unbranched; glabrous or pubescent. **Leaves** basal or cauline; thick; blades linear to oblong or broadly ovate. **Inflorescences** solitary flowers or clusters of 2–12 flowers. **Flowers** large and showy; perianths in 2-series; campanulate or cylindrical or cup-shaped. **Perianth Parts** 6; all alike; petaloid; free; erect or ascending; of various colors. **Stamens** 6; arising from receptacles; filaments flat, narrow. **Pistils** with stigmas 1, 3-lobed; styles absent or short; ovaries superior. **Fruits** loculicidal capsules; globose or ellipsoidal. **Seeds** numerous; flat.

Tulipa

Distribution and Habitat—Wild species of the genus in the Old World occupy a diversity of habitats. Garden tulips are cultivated as ornamentals throughout North America.

Disease Problems—The very popular garden tulips, of which there are numerous cultivars, are of limited importance toxicologically. The most common problem is that of a dermatitis in individuals who handle the bulbs commercially (Maretic et al. 1978). It is similar to the dermatitis produced by handling hyacinths, daffodils, and narcissi, but apparently tulips are the worst offenders (Overton 1926; Hjorth & Wilkinson 1968). Ingestion of the bulbs or to a lesser extent the foliage may cause digestive disturbances, as is true of these other genera. This is a problem in humans when the bulbs are mistaken for onions (Cooper & Johnson 1984).

- ornamentals
- dermatitis from handling bulbs; digestive disturbance from eating bulbs
- irritant and allergenic tulipins, formed from tuliposides

Disease Genesis—The cause of the effects on skin and the digestive tract are the irritant and possibly allergenic lactones tulipin A and B (Tschesche et al. 1968; Frohne & Pfander 1984; Spoerke & Smolinske 1990). The plant cells contain tuliposides A and B in their vacuoles. When the cells are ruptured, either chemical or enzymatic

```
            O
            ||
HO-C-C-C-C-O-Glucose
         ||
         CH2
```

Tuliposide A

↓

Tulipin A

```
      OH    O
      |     ||
HO-C-C-C-C-O-Glucose
         ||
         CH2
```

Tuliposide B

↓

Tulipin B

conversion of the precursors to the more toxic tulipin lactones occurs (Beijersbergen & Lommers 1972; Schonbeck & Schroeder 1972). *Tulipa* contains large amounts of the more toxic tuliposide A (Slob et al. 1975). Lectins are present and alkaloids have also been suggested as causes of problems, but their toxicologic significance is not clear (Leroux 1986; Oda & Minami 1986; Spoerke & Smolinske 1990).

■ nausea, vomiting, excess salivation; dermatitis, erythema, hair loss

Clinical Signs—The signs following ingestion of *Tulipa* are nausea, vomiting, increased salivation, labored respiration, and increased heart rate. These signs may persist for several days. Diarrhea is less commonly seen. The dermatitis may appear as diffuse erythema, alopecia, and perhaps some papular and pustular lesions in various areas of the body, including the hands.

■ reddened gut mucosa

activated charcoal

Pathology and Treatment—Unlikely to result in death, the changes are confined to reddening and inflammation of the mucosa of the stomach and small intestine. Orally administered charcoal may be helpful; otherwise treatment is directed toward relief of the digestive tract effects.

Urginea Steinh.

■ *U. maritima*

crusader's spears
mouse onion
sea onion
sea squill
squill
white squill

Taxonomy and Morphology—Comprising about 100 species native to the Mediterranean region and Africa, *Urginea* is positioned in the segregate family Hyacinthaceae by some taxonomists. Its name is derived from the name of an Afro-Arab tribe, the Ben Urgin. One species is cultivated in North America and is of toxicologic interest:

U. maritima (L.) Baker	squill, sea squill, sea onion, white squill, Mediterranean squill, mouse onion, crusader's spears

■ erect herbs from greenish or reddish bulbs; leaves basal; flowers in terminal racemes; perianth parts white with green or purple keel

Plants from bulbs; acaulescent. **Bulbs** massive, 10–20 cm in diameter; with scales; ovoid-globose; partially exposed above soil surface; scales free, greenish or reddish. **Leaves** basal; fleshy; glabrous; glaucous; blades linear to oblong, 30–50 cm long. **Scapes** erect; 50–350 cm tall; terete; often red tinged. **Inflorescences** dense, elongate terminal racemes; bracts subulate, caducous. **Flowers** with perianths in 2-series; campanulate or star-shaped. **Perianth Parts** 6; all alike; petaloid; free; spreading; oblong; white with green or purple keel. **Stamens** 6; filaments flat. **Pistils** with stigmas 1, capitate; styles 1; ovaries superior. **Fruits** capsules; oblong; 3-angled. **Seeds** numerous; flat; slightly winged; brown-black.

Urginea maritima

Distribution and Habitat—Native to Portugal, the Canary Islands, and the Mediterranean Basin, *U. maritima* is frost sensitive and occasionally cultivated as a late summer or early fall ornamental only in the warmest parts of North America or as a houseplant.

Disease Problems—Well known to the ancient Greeks and perhaps even to the Egyptians, squill was discussed specifically by followers of Hippocrates after Theophrastus (Stannard 1974). By the first century A.D. its medicinal uses began to be described. For example, the Roman medical writer Aulus Cornelius Celsus referred to the sucking of boiled squill as a treatment for dropsy, while Dioscorides in his *De Materia Medica,* devoted a chapter to its preparation and uses (Stannard 1974). These early writers also were aware of the danger of using *U. maritima* as a medicament. In subsequent centuries its medicinal use waxed and waned until a resurgence in its popularity as a treatment for dropsy and as an expectorant occurred in the 1700s and 1800s (Cowen 1974). For dropsy, its use was compared to that of digitalis, although its cardiac effects were not clearly recognized until the 1900s (Gemmill 1974). The 1900s brought a decline in interest for medicinal uses but exploitation of its potential as a rodenticide (Gentry et al. 1987). These uses led to the isolation and identification of the toxicants in the plant. Red squill, *U. indica,* exhibits similar activity and potency and has been widely used in India and exported to Europe for similar therapeutic purposes (Deb & Dasgupta 1976).

■ ornamental

■ medicinal and rodenticidal uses of bulbs; digestive disturbance; cardiotoxic effects from bulbs

There are occasional occurrences of intoxication in humans from accidental or deliberate ingestions of *U. maritima* (Tuncok et al. 1995). In such cases the clinical signs are typical of the effects of cardiac glycosides and include bradycardia and arrhythmias. Worldwide, species of *Urginea* are probably the most important cardiotoxic members of the Liliaceae; however, they are least important in North America, because of their limited presence here. They are a significant problem for livestock elsewhere—mainly South Africa—because of the widespread availability of numerous indigenous species.

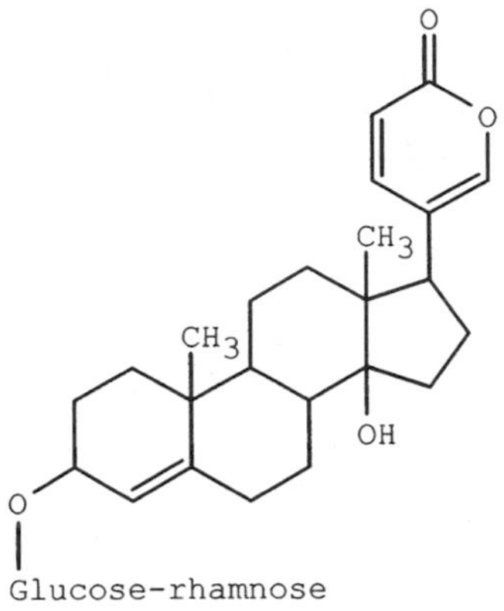

Scillaren A

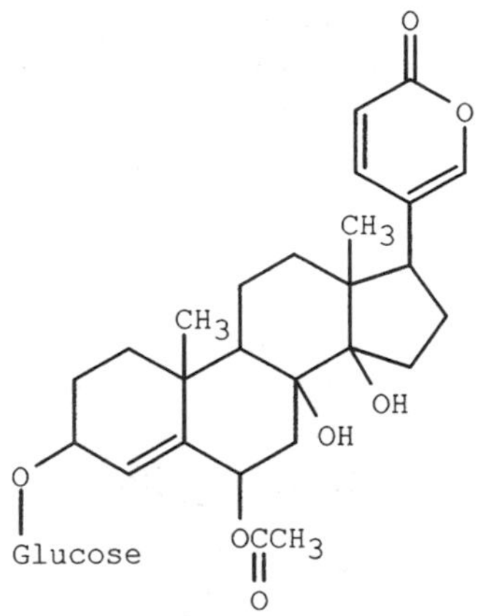

Scillaren B

Disease Genesis—The toxicants of interest in *Urginea* are a series of bufadienolide glycosides. They are sometimes referred to as scilla toxins or red squill toxins because these were the original generic and common names applied to the taxon. Although there are a number of bufadienolides in the plants, the most abundant and therefore the one of primary interest is scilliroside (Verbiscar et al. 1986). Scillarens A and B also are present. The cardiotoxins of squill are less potent, more rapid in onset of action, and shorter in duration of action than digoxin, which probably accounts for the paucity of problems except where plants are particularly abundant (Gemmill 1974; Hakim et al. 1976). There seems to be a renewed interest in squill as a commercial rodenticide, due in some measure to its very potent emetic effects, which help prevent toxicity in nontargeted animal species (Pahwa et al. 1982; Gentry et al. 1987). There is little doubt that all species of *Urginea* contain the same types of toxins and are capable of producing similar disease syndromes. In this respect it is of interest to note that some of the bufadienolides have a propensity to produce neurologic rather than cardiac effects. This is similar to the situation in the families Asclepiadaceae and Crassulaceae (see chapters 12 and 27) (Kellerman et al. 1988). The toxins are found in all plant parts, although as with other members of the Liliaceae, highest concentrations typically are in the bulb (Verbiscar et al. 1986). Flowering stems are also reported to be of particular hazard (Kellerman et al. 1988).

■ bufadienolide cardiotoxic glycosides; scillarens, all plant parts, especially bulbs

Clinical Signs and Pathology—Diarrhea, ruminal inactivity, and respiratory difficulty are the initial primary signs that appear 1–2 days after consumption of the plant (Nel et al. 1987; Kellerman et al. 1988). Subsequently, cardiac abnormalities such as arrhythmias (including AV block), increased heart rate, and fibrillation may occur. Signs indicative of effects on the nervous system—tremors and seizures—may be evident in the latter stages of the disease prior to death. There are few distinctive changes, although some scattered small splotchy hemorrhages may be present.

■ diarrhea, anorexia, ruminal inactivity; irregular cardiac rate

Treatment—Orally administered activated charcoal may be helpful; otherwise treatment is directed toward relief of the digestive tract effects. Only very rarely is treatment of the cardiac effects indicated. The cardiac signs are for the most part responsive to atropine.

■ activated charcoal

- *V. californicum*
  V. insolitum
  V. parviflorum
  V. viride
  V. woodii

  bear corn
  bugsbane
  corn lily
  devil's bite
  earth gall
  false hellebore
  hellebore
  Indian poke
  Indian uncus
  itchweed
  puppet root
  skunk cabbage
  varbell
  wolfsbane

- robust erect herbs from thick, black rhizomes; leaves numerous, alternate, some clasping; flowers in terminal panicles; perianth parts white, greenish, yellowish green, reddish brown, or purple

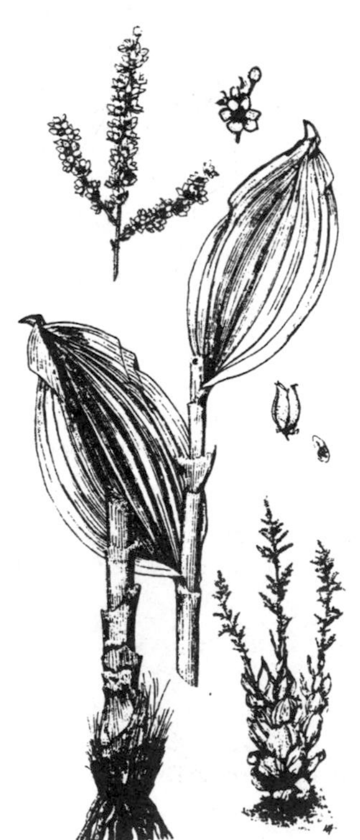

Veratrum californicum

Veratrum L.

Taxonomy and Morphology—Comprising 15–25 species in the temperate regions of both the Old World and the New World, *Veratrum* is commonly known as false or white hellebore. Its name is derived from the Latin roots *vere* and *aeter,* meaning "true" and "black," which reflect the dark black rhizomes. In North America, 5–8 species are present. Of toxicologic interest are the following (see plate 47.2):

V. californicum Durand	corn lily, skunk cabbage
V. insolitum Jeps.	Siskiyou false hellebore
V. parviflorum Michx. (= *Melanthium parviflorum* [Michx.] S.Watson)	hellebore
V. viride Aiton	American white hellebore, bear corn, swamp hellebore, devil's bite, green false hellebore, bugsbane, Indian uncus, Indian poke, meadow poke, puppet root, earth gall, itchweed, skunk cabbage, wolfsbane
V. woodii J.W.Robbins (= *Melanthium woodii* [J.W.Robbins] Bodkin)	varbell, Indian poke, Wood's false hellebore

To avoid confusion, care must be taken when using common names. The name *hellebore* also is used for the genus *Helleborus* in the Ranunculaceae, *poke* for *Phytolacca* in the Phytolaccaceae, and *skunk cabbage* for *Lysichiton* in the Araceae. Also toxic, these taxa produce different problems. *Veratrum parviflorum* and *V. woodii* are placed in the genus *Melanthium* by some taxonomists.

Plants robust; from rhizomes; caulescent; andromonoecious. **Rhizomes** thick; short; black. **Stems** erect; 100–200 cm tall; unbranched; hollow. **Leaves** numerous; alternate; cauline; sessile, often clasping; blades lanceolate to broadly ovate, typically plicate; multinerved. **Inflorescences** large terminal panicles; bracteoles present. **Flowers** perfect or imperfect, similar; perianths in 2-series; campanulate to saucer-shaped. **Perianth Parts** 6; all alike; petaloid; free; spreading; multinerved; oblong to obovate; white or greenish or yellowish green or reddish brown to purple. **Stamens** 6; arising from hypanthia; free or fused to perianths. **Pistils** with stigmas 3; styles 3; ovaries superior; hypanthia present, short. **Fruits** septicidal capsules; 3-lobed; membranous. **Seeds** numerous; flat; winged.

1. Ovaries and capsules tomentose. Margins of perianth parts fringed . . . . *V. insolitum*
1. Ovaries and capsules glabrous or sparsely pilose. Margins of perianth parts entire or denticulate.
   2. Panicles dense; branches ascending or spreading. Perianth parts white or greenish tinged . . . . *V. californicum*
   2. Panicles loose; branches drooping. Perianth parts yellow-green or green.
      3. Upper leaves oblong-elliptic. Pedicels pilose-villous . . . . *V. viride*
      3. Upper leaves linear. Pedicels glabrous.
         4. Perianth parts dark maroon. Apices of perianth parts obtuse . . . . *V. woodii*
         4. Perianth parts greenish or yellow green. Apices of perianth parts acute or acuminate . . . . . *V. parviflorum*

Distribution and Habitat—Although *Veratrum* is distributed across the continent, all species occupy similar moist, open sites in alpine meadows, open woodlands, swamps, bogs, and marshes and along waterways and lake edges in mountain ranges, often at higher elevations. The most widespread is *V. viride. Veratrum californicum* is a western species. The remaining species have more restricted distributions. *Veratrum woodii* is present from Ohio to Missouri, Oklahoma, and Arkansas.

Disease Problems—*Veratrum* species have long been recognized for both toxic and medicinal properties, especially the European species, *V. album* and *V. nigrum*. Extracts were used for hemorrhage, dropsy, intestinal parasites, and congestive heart failure. Unfortunately the potent preparations had very unfavorable therapeutic ratios and occasionally were lethal. Even standardized mixtures of alkaloids (veriloids) were likely to produce nausea and vomiting along with the desired antihypertensive action (Kauntze & Trounce 1951). Occasional incidents of toxicity in humans from *V. album* and *V. viride* continue to be reported, with nausea, vomiting, abdominal pain, hypotension, and slow or irregular heart rates as the predominant signs (Underhill 1959; Hruby et al. 1981; Garnier et al. 1982; Kulig & Rumack 1982; Mulligan & Munro 1987; Courtemanche et al. 1989; Jaffe et al. 1990). The plants may be eaten accidentally because they are mistaken for other plants such as *Phytolacca* or *Symplocarpus* (Courtemanche et al. 1989). The onset of vomiting may occur within 30 minutes of ingestion (Quatrehomme et al. 1993). A cupful of blanched stems and leaves of *A. viride* is sufficient to produce marked signs, including blurred vision lasting for 12 or more hours (Underhill 1959).

Initially it was noted that sheep feeding on *V. californicum* produced lambs with cyclopia and other types of cephalic malformations (Binns et al. 1963). This was subsequently shown to be associated specifically with consumption of the plant on the 14th day of gestation (Binns et al. 1965). The term *monkey face* was used to describe the cyclopic lambs. There was a single or double globe, with a reduced or nonexistent eye and a proboscis above or below the globe or entirely absent. Shortening of the maxilla/premaxilla was accompanied by an upward curvature of the mandible. Cebocephalus (short snout), various degrees of limited cerebral development, a single optic nerve, and prolonged gestation due to the absence of a pituitary gland were other features of the disease related to ingestion on the 14th day (Binns et al. 1965; Keeler 1988). A ewe carrying a deformed lamb may appear to be approaching parturition but then undergo involution of the udder and external genitalia and carry the lamb for a prolonged time, subsequently producing a very large stillborn lamb or one that dies soon after birth (James 1977). Although uncommon, similar teratogenic effects may occur in calves and other animal species (Bourque 1984).

Deformities are not limited to the head. Ingestion on days 19–21 results in embryonic death; on days 28–32, limb defects; and on days 31–33, tracheal stenosis that causes death at birth (Keeler et al. 1985, 1986; Keeler 1987). The limb defects involve bilateral shortening of the metacarpals, metatarsals, and tibias; arthrogryposis of fetlocks, carpals, hocks, and stifles; and deformities or fusing of the articulating surfaces (Keeler & Stuart 1987). Congenital tracheal stenosis is due to lateral flattening of the entire length, obliterating the lumen. The cartilaginous rings are decreased in thickness and number, not uniform in size and shape, irregularly spaced, and distorted in orientation (Keeler et al. 1985). The observed effects coincide with reported periods of embryonic development, as indicated in table 47.2.

The critical periods of limb development are the middle of week 4 through week 6 in the horse and cow, week 4 in sheep, and the middle of week 3 to the middle of week 4 in the pig (Noden & deLahunta 1985).

Veratrum viride

- moist meadows, swamps, marshes, and other waterway edges
- digestive disturbance; hypotension; fetal deformities, especially cephalic in sheep with ingestion in early pregnancy

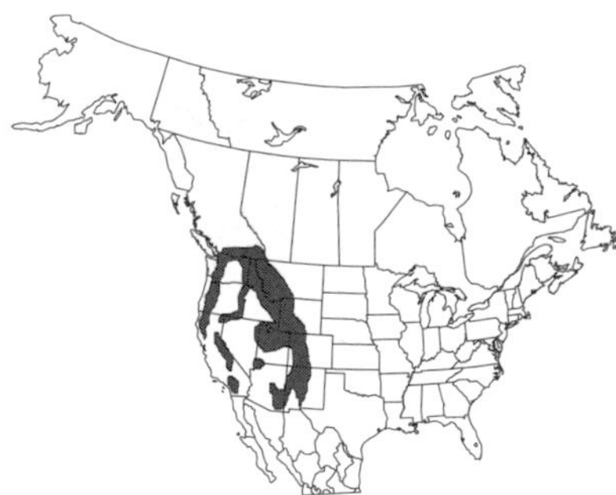

Veratrum californicum

Veratrum insolitum

Veratrum parviflorum

Veratrum viride

Veratramine (a veratranine)

- steroidal alkaloids; azasteroids; 5 groups—veratrines, cevanines, jervanines, solanidines, cholestanes and others

- veratrines and cevanines bind to sodium channels, delay closure, increase repetitive discharges and reflex responses

Table 47.2. Embryonic development by day of gestation

	Development (Days)			
Tissue	Horse	Cow	Sheep	Pig
limb buds	26	24/26	20/21	16/18
palate fusion	47	56	38	34.5
primitive streak	—	18	13/14	12
neural tube, open/close	—	19/23	15/16	13/16

Source: Evans & Sack 1973.

Table 47.3. Alkaloids of *Veratrum*

Group	Alkaloids	Derivatives
veratranine	verarine	
	veratramine	veratrosine (glycoside)
		Esters and C Position
cevanine (ceveratrum)	germine	germanitrine—C-3,-7,-15
		germitrine—C-3,-7,-15
		germitetrine
	protoverine	protoveratrine A, B—C-3,-6,-7,-15
	sabine	sabadine—C-3
	veracevine	cevacine—C-3
		veratridine—C-3
	veramarine	
	zygadenine	zygacine—C-3
		zygadenins—C-3
jervanine (jerveratrum)	cyclopamine (11-deoxojervine)	
	cycloposine (glycoside)	
	jervine	
	veratrobasine	
solanidanine	rubijervine	
	isorubijervine	
	solanidine	
	veralobine	
cholestane and others	etioline	
	veralozidine	
	veralozinine	
	veralozine	
	verazine	

Sources: Fried et al. 1949; Shoppee 1964; Kupchan & Flacke 1967; Heftmann 1970; Tomko & Voticky 1973; Greenhill & Grayshan 1992.

Disease Genesis—Species of *Veratrum* contain a series of nitrogenous steroidal alkaloids termed azasteroids (Fried & White 1949; Tomko & Voticky 1973; Greenhill & Grayshan 1992). Some of these are very potent and produce marked biological effects. Their nomenclature can be quite confusing, in part because different names are used for the base alkaloids and the mono- and polyester natural products. Also confusing is the application of the name *veratrine* to alkaloid mixtures rather than to a specific alkaloid. In this treatise the alkaloids are divided among five groups (table 47.3).

The first three groups are C-nor-D-homo steroids, so called because of the decrease in the size of ring C and the increase in ring D. They are of considerable toxicologic importance. Veratranines and especially cevanines, the most potent being the germine esters, are hypotensives, whereas jervanines are teratogens. The jervanines and veratranines are quite similar structurally and differ mainly in the ether bridge that forms ring E in jervanines and is apparently required for teratogenicity (Keeler 1983, 1984). The cevanines contain numerous hydroxyl groups—thus the array of mono- and polyesters, such as acyl, angeloyl, methoxybenzoyl, methyl butanoyl, and others. They are commonly referred to as the ester alkaloids and are the principal neurotoxicants and hypotensive agents. The jervanine alkaloids are not without neurologic effects. Jervinone and other jervine derivatives cause a transient decrease in blood pressure not associated with adrenoreceptors (Atta-ur-Rahman et al. 1993). Experimentally in rats, the decrease in blood pressure induced by jervinone, but not that of other derivatives, was prevented by atropine.

The neurotoxic *Veratrum* azasteroids act by binding open Na^+ channels, thus interfering with inactivation and delaying closure of the channels while allowing increased movement of Na^+, Ca^{2+}, and K^+ through the membrane. Because they act on selected voltage-sensitive rather than ligand-gated Na^+ channels, their effects are therefore modulated by action potentials and are blocked by tetrodotoxin (Catterall 1980). Brief but very potent binding and modification of Na^+ channels last several seconds, followed by slow inactivation thereafter (Leibowitz et al. 1987; Strichartz et al. 1987; Nanasi et al. 1990). This results in delay of repolarization and increased repetitive activity, with a single stimulus producing multiple discharges. All excitable cells are affected but most particularly afferent nerve endings, resulting in increased sensitivity, intensified activity, or even paralysis at high concentrations. The most sensitive cells appear to be vagal afferent fibers and endings, such as epimyocardial receptors. Increased activity of the reflex response via these receptors results in slowing of the heart rate, peripheral vasodilation, and hypotension (Bezold-Jarisch reflex). Additional effects are mediated through the pulmonary stretch and thoracic receptors (increased reflex activity producing respiratory depression), and the aortic arch and carotid sinus baroreceptors. Typically accompanying the cardiovascular effects is an initial transient cessation of respiration, followed by faster and deeper breathing. The area of the nodose, or distal, ganglion of the vagus is very sensitive to the changes in sodium conductance, resulting in increased afferent input to the vomiting center in the medulla. Thus, vomiting is characteristic of any route of alkaloid administration. High dosage of the alkaloids may produce digitalis-like effects on the heart, including arrhythmias, increased blood glucose, and marked EEG alterations (Buck et al. 1966). At least some of the effects may be mediated through serotonin activity (Nagata & Izumi 1991) Several excellent reviews detail the alkaloid mechanisms and effects (cf. Kupchan & Flacke 1967; Nickerson & Ruedy 1975; Honerjager 1983).

It is of interest that the protoveratrines appear to have marked beneficial effects in the treatment of hemorrhagic shock (Bertolini et al. 1990). Chemical modification of specific alkaloids has reduced the potential for some side effects and allowed their use against neuromuscular disorders such as myasthenia gravis (Flacke et al. 1966). Although these alkaloids are no longer of great interest medicinally, they are still in use as insecticidal sabadilla preparations (Ujvary 1991).

Perhaps the azasteroids of greatest toxicologic importance are the teratogenic jervanines, most notable in *V. californicum* but also present in *V. album* and probably other species as well (Atta-ur-Rahman et al. 1991). For *V. californicum*, cyclopamine appears to be the most important of the jervanine alkaloids, because it is the most abundant; it appears to be responsible for the cephalic if not the other deformities (Keeler & Stuart 1987). In this respect, *V. album* and others containing jervine should be considered capable of inducing fetal deformities other than cephalic (Keeler 1983; Atta-Ur-Rahman et al. 1991). The other alkaloid groups of *Veratrum* are not teratogenic (Keeler & Binns 1966; Keeler 1984). Affecting a broad range of animal species, albeit with varying types of deformities, the parent compounds are responsible for the effects of all the alkaloid groups, the jervanines acting unchanged upon the conceptus (Keeler 1983). The presence of unsaturation at C-5 and C-6 has been suggested to be a prime factor in the teratogenicity of the jervanines in addition to

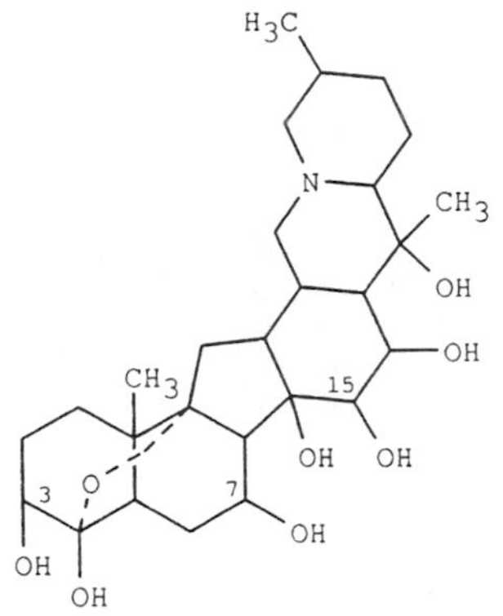

Germine (a cevanine)

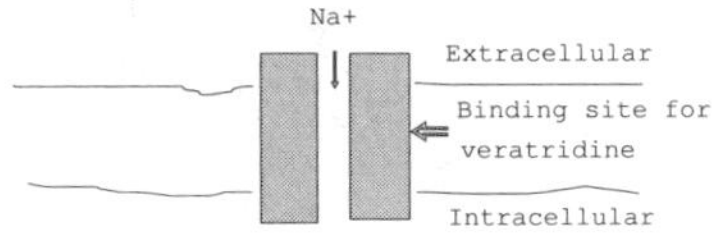

Sodium channel
(after Catterall 1980)

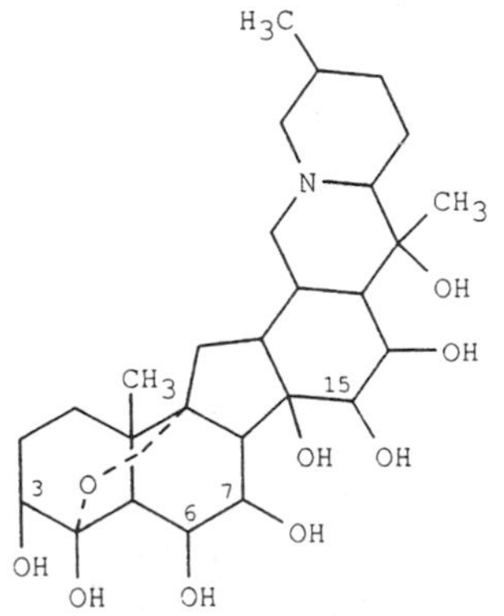

Protoverine (a cevanine)

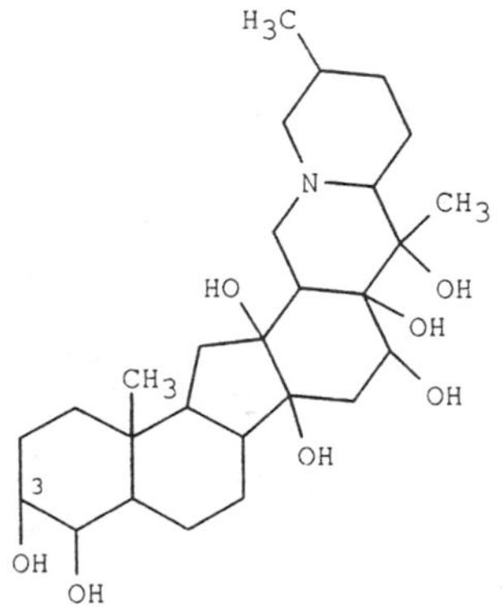

Sabine (a cevanine)

■ jervanines, teratogenic

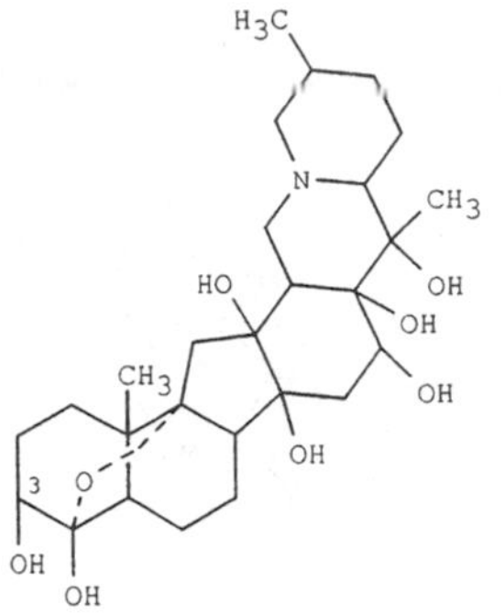

Veracevine (a cevanine)

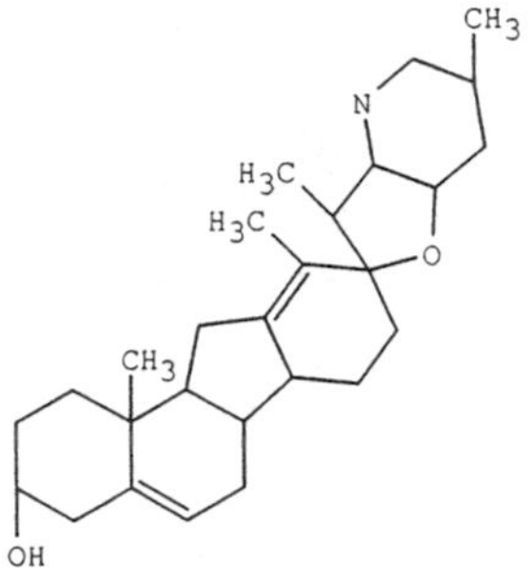

Cyclopamine (a jervanine)

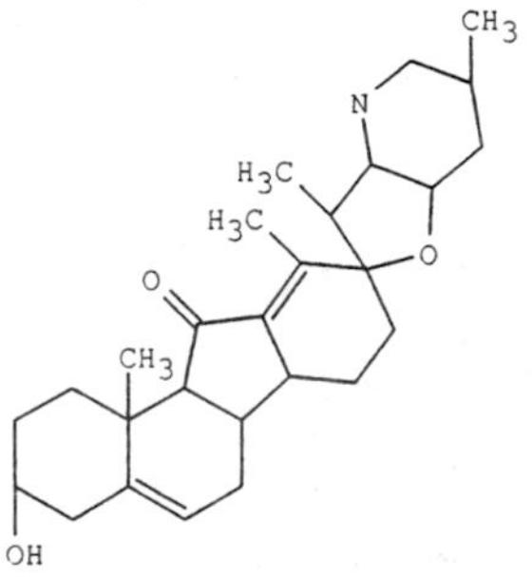

Jervine (a jervanine)

any role for the configuration of the nitrogen electrons (Gaffield & Keeler 1993). The mechanism by which teratogenic effects are produced is not known, although jervine appears to act on prechondrogenic stem cells during the early phase of mesenchymal differentiation into cartilage (Campbell et al. 1985). However, neurohumoral interactions involving inhibition of acetylcholine receptors have also been proposed as a mechanism of action (Sim et al. 1985). For an excellent in-depth review of teratogenesis due to toxic plants, see Keeler 1988.

All species should be considered capable of causing acute intoxication, because these effects are typical of a large number of the alkaloids present in many species. Both *V. viride* and *V. californicum* cause similar physiologic alterations (Buck et al. 1966).

Although the leaves and stems are reported to be acrid and burning, they are sometimes apparently willingly eaten and pose a risk for all classes of livestock: sheep, goats, cattle, horses, and chickens (Chesnut 1898a,b; Chesnut & Wilcox 1901; Gress 1935). Fortunately animals seldom eat sufficient plant material to cause toxic effects other than teratogenic effects, and these can be prevented by limiting animal access to the plants at the susceptible times. Alkaloid levels are highest in the leaves from June through early July and then decline to lower levels in August, when the roots attain their highest concentrations. The levels in stems seem to be intermediate between leaves and roots (Keeler & Binns 1971). Thus, like the cardenolides in other members of the Liliaceae, the alkaloids of *Veratrum* appear to be synthesized in the leaves and subsequently translocated to the underground portions. After frosts, the toxicosis risk is much reduced because the plants pose little attraction for animals (Keeler et al. 1986).

The habitat of *Veratrum* is of considerable importance, because it also seems to be a preferred habitat for deer, range livestock, and the host snail for *Fasciola hepatica* (Loft et al. 1986). Disease due to fasciolosis is capable of diminishing the capacity of the liver to eliminate toxicants (Galtier et al. 1986; Tufenkji et al. 1988; Burrows et al. 1992). If the *Veratrum* alkaloids undergo hepatic biotransformation, as seems likely from studies in insects, then animals infected with flukes and grazing in these areas for several months may be at greater toxicologic risk from the plants.

- abrupt onset, excess salivation, vomiting, weakness, trembling, ataxia, recumbency; irregular heart rate and rhythm

Clinical Signs—The acute signs of intoxication are due, for the most part, to the neurotoxic cevanine alkaloids and are probably similar for all species of *Veratrum*. Typically the signs are limited to excess salivation with froth around the mouth, frequent swallowing, and perhaps vomiting. In more severe intoxications there also may be indications of marked hypotension with apparent weakness, trembling, incoordination, limb paresis, and recumbency. These signs may be accompanied by slow respiration, irregular heart rate and rhythm, and cyanosis. Signs may begin 2–3 hours after ingestion and persist for several hours in the case of mild intoxications or for 1–4 days for more severe problems (Binns et al. 1965; Keeler 1990). Fetal deformities, embryonic deaths, and prolonged gestations may occur in the absence of signs of acute intoxication, at least in cases caused by *V. californicum* (Binns et al. 1965). The specific types of deformities will depend upon the time of gestation. There is little or no risk of teratogenic effects after the 6th week of gestation in sheep and pigs and after the 8th week in cattle and horses.

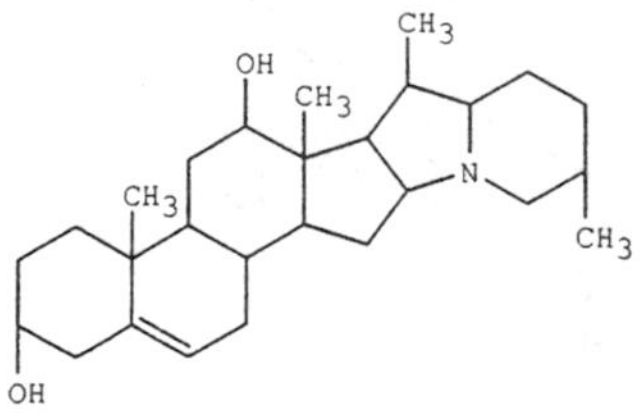

Rubijervine (a solanidanine)

- fetal deformities

Pathology—For acute intoxications the pathologic changes are limited to those characteristic of a shock syndrome, i.e., scattered small hemorrhages and congestion of some organs. The fetal deformities vary from severe cephalic malformations and cerebral and pituitary deficits to the tracheal and limb abnormalities described above.

Treatment—Atropine will provide relief of most of the more likely signs of intoxication. A more detailed discussion is presented in the treatment of *Zigadenus,* in this chapter.

- activated charcoal, atropine

Zephyranthes Herb.

Taxonomy and Morphology—Comprising 35–50 species, *Zephyranthes,* commonly known as zephyr lily, is a New World genus distributed in tropical, subtropical, and warm-temperate climatic regions. Its name is derived from the Greek roots *zephyros,* meaning "west wind," and *anthos,* meaning "flower," which reputedly reflect its origin in the Western Hemisphere. It was originally placed in the family Amaryllidaceae. Closely related to *Cooperia,* the two genera are merged by some taxonomists. In North America, 10–11 species are present. Of toxicologic significance is the following:

- *Z. atamasco*

  atamasco lily

- erect herbs from bulbs; leaves basal; flowers solitary; perianth parts white or purple-tinged

Z. atamasco (L.) Herb. atamasco lily
(= *Z. treatiae* S.Watson)
(= *Z. simpsonii* Chapm.)

Some taxonomists divide the taxon into 3 species primarily on the basis of geography, orientation of perianth parts, and appearance of leaf blades. Intergradation is extensive, however, and recognition of 1 species is preferable at the present time.

Zephyranthes atamasco

Plants from bulbs; acaulescent. **Bulbs** ovoid-oblong; with a short neck; tunicas papery. **Leaves** basal; light green; glossy; blades linear. **Scapes** erect; 10–30 cm long; hollow. **Inflorescences** of solitary flowers; spathes tubular, membranous, apices notched. **Flowers** held erect or slightly inclined; perianths in 1-series; funnelform. **Perianth Parts** 6; all alike; petaloid; fused; white or purple-tinged; lobes subequal, acute; scales in throats. **Stamens** 6; 3 long and 3 short; erect or spreading; fused to perianth parts at mouths of tubes. **Pistils** with stigmas 1, 3-lobed; styles 1, filiform; ovaries inferior. **Fruits** capsules; globose. **Seeds** numerous; flat; D-shaped; black.

Distribution and Habitat—Forming large populations, *Z. atamasco* is locally abundant in moist meadows and woods.

- moist woods and meadows

Zephyranthes atamasco

Disease Problems and Genesis—The bulb of *Z. atamasco* is toxic but not often eaten, because it is quite acrid. Ingestion produces severe irritation of the digestive tract and possibly neurologic effects (Cary et al. 1924; West & Emmel 1952). About 0.5% of b.w. is lethal for cattle, 0.9 kg of bulbs having caused death of steers weighing 136–182 kg. The specific toxicants are not known.

- digestive disturbance; unknown toxin

Clinical Signs, Pathology, and Treatment—A few hours after ingestion, there is diarrhea, which may be bloody. The animal may become ataxic and stagger, collapse, and die with little struggle. Pathologic changes have not been described but might be expected to be similar to those produced by other members of the family. Orally administered charcoal may be helpful; otherwise treatment is directed toward relief of the digestive tract effects.

- diarrhea

  reddened gut mucosa

  activated charcoal

Zigadenus Michx.

Taxonomy and Morphology—With the exception of 1 species in Asia, *Zigadenus,* commonly known as death camas, is indigenous to North and Central America, where

- *Z. brevibracteatus*
  Z. densus
  Z. elegans
  Z. exaltatus
  Z. fremontii
  Z. glaberrimus
  Z. leimanthoides
  Z. micranthus
  Z. nuttallii
  Z. paniculatus
  Z. vaginatus
  Z. venenosus
  Z. virescens

  alkali grass
  black snakeroot
  crow poison
  death camas
  poison camas
  sand-corn
  star lily
  St. Agnes' feather
  white camas

- erect herbs from dark brown or black bulbs or thick black rhizomes; leaves basal, long; flowers in terminal racemes or panicles; perianth parts white to greenish yellow, sometimes purple-tinged

14 species occur (Preece 1956). Although perhaps orthographically preferable, the name *Zygadenus*, used in some publications, is incorrect according to the rules of botanical nomenclature. The generic name is derived from the Greek roots *zygon* and *aden*, for "yoke" and "gland," alluding to the paired glands on the perianth parts. Placed in the segregate family Melanthiaceae by some taxonomists, *Zigadenus* has been subjected to considerable taxonomic revision (Preece 1956). As a result, several taxa for which toxicity information is available are no longer treated as species but rather as varieties; for example, *Z. gramineus*, *Z. intermedius*, and *Z. salinus* are now varieties of *Z. venenosus*. The generic names *Anticlea* and *Toxicoscordion* also appear in the literature. Present in North America are the following (see plate 47.3):

Z. brevibracteatus (M.E.Jones) H.M.Hall	
Z. densus (Desr.) Fernald	black snakeroot, crow poison, St. Agnes' feather, black death camas
Z. elegans Pursh (= *Z. glaucus* Nutt.)	elegant death camas, white camas, alkali grass, mountain death camas
Z. exaltatus Eastw.	
Z. fremontii Torr.	chaparral death camas, star lily
Z. glaberrimus Michx.	death camas, sand-bog death camas
Z. leimanthoides A.Gray	death camas, pine-barren death camas
Z. micranthus Eastw. (= *Z. fontanus* Eastw.)	
Z. nuttallii A.Gray	poison camas, Nuttall's death camas
Z. paniculatus (Nutt.) S.Watson	foothill death camas, sand-corn, panicled death camas
Z. vaginatus (Rydb.) J.F.Macbr.	alcove death camas
Z. venenosus S.Watson (= *Z. gramineus* Rydb.) (= *Z. intermedius* Rydb.) (= *Z. salinus* A.Nelson)	meadow death camas, grassy death camas
Z. virescens (Kunth) J.F.Macbr.	

Zigadenus elegans

Zigadenus nuttallii

Zigadenus paniculatus

Plants from rhizomes or bulbs; caulescent; bearing perfect flowers or andromonoecious. **Rhizomes** thick; short; black. **Bulbs** ovoid or globose; tunicas dark brown or black. **Stems** erect; scapelike; 20–120 cm tall; unbranched. **Leaves** basal and cauline; basal leaves long, linear to linear-lanceolate, keeled or conduplicate proximally; cauline leaves smaller, bractlike; glabrous. **Inflorescences** terminal panicles or racemes; pedicels recurved or ascending; bracts linear to ovate, scarious or foliaceous. **Flowers** perfect or staminate; perianths in 2-series; rotate or campanulate. **Perianth Parts** 6; all alike; petaloid; free; white to greenish yellow, sometimes purple-tinged; ovate to lanceolate; glands 1 or 2, yellow or yellow-green or yellow-brown; persistent after withering. **Stamens** 6, arising from receptacles; filaments dilated basally. **Pistils** with stigmas 3; styles 3; ovaries superior; hypanthia present or absent. **Fruits** septicidal capsules; ovoid to conical; 3-lobed. **Seeds** few to numerous; fusiform to oblong; smooth; angled; light brown.

Identification of species is based on features of the flowers and inflorescence. A key is not presented here. If exact identification is required, the monograph by Preece (1956) or manuals specific for the reader's area should be consulted.

Zigadenus venenosus

■ variety of habitats

Distribution and Habitat—*Zigadenus* typically begins growth early in the season. It flowers and fruits and the foliage withers by early summer, except for species in the high mountains or swamps. Species occupy a variety of habitats, including grasslands, alpine meadows, bogs, pine woods, beaches, and shrublands. Some have wide distributions, and others more restricted. In general, there is a marked correlation between sections of the genus and geographical distribution (Preece 1956). In the southeastern quarter of the continent *Z. glaberrimus, Z. densus, and Z. leimanthoides* occur. In the western half, especially California, are *Z. brevibracteatus, Z. exaltatus, Z. fremontii, Z. micranthus, Z. paniculatus, Z. nuttallii, and Z. venenosus.* Ranging across the northern parts of the continent and southward in the Rocky, Appalachian, and Ozarkian cordillera are *Z. elegans, Z. vaginatus,* and *Z. virescens.*

Zigadenus densus

■ digestive disturbance, hypotension, neurologic effects

Disease Problems—Early explorers, including Lewis and Clark, went to considerable effort to avoid eating *Zigadenus,* because its toxic effects were well known from the experiences of various Indian tribes who used bulbs and tubers of other genera for food. The common name *death camas* is used because the bulbs can be confused with the edible ones of *Camassia.* Occasional digestive disturbances and neurointoxication and, very rarely, deaths have been reported in humans who consumed the bulbs (Spoerke & Spoerke 1979; Wagstaff and Case 1987; Heilpern 1995). In children, 1 or 2 bulbs may cause severe illness, and 4 or 5 possibly death, depending on the species eaten. In the late 1800s and early 1900s incidences of intoxications in livestock, although uncommon, were quite severe, e.g., losses of up to 500 sheep in a single episode (Marsh et al. 1915). In 1900, more than 3,000 adult sheep were estimated to have been poisoned in Montana alone (Chesnut & Wilcox 1901). Even as recently as 1987, loss of 250 sheep in an episode over a 2-day period was reported (Panter et al. 1987).

Sheep eating the leaves are most commonly affected. There may be occasional losses of cattle that eat the leaves and flowering stalks, and even horses and domestic chickens may be intoxicated (Nieman 1928; Smith & Lewis 1991; Collett et al. 1996). Losses of cattle due to *Z. nuttallii* have been a problem in the tallgrass prairies of Oklahoma and Kansas (Marsh et al. 1926b; Gates 1930). In some instances, where the soil is soft, animals may pull up and ingest the more toxic bulbs.

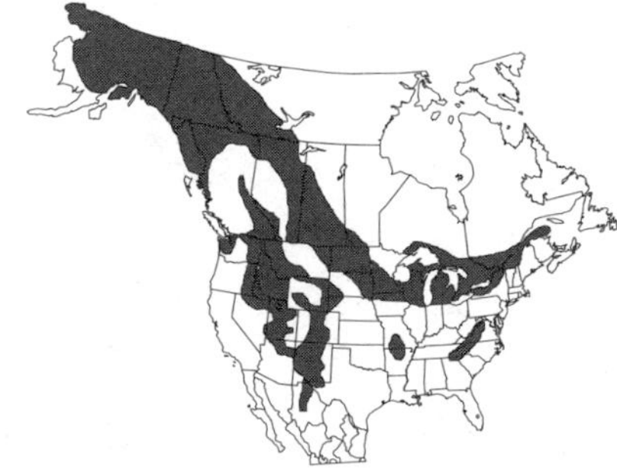

Zigadenus elegans

Disease Genesis—*Zigadenus* contains a series of cevanine-type *Veratrum* alkaloids. There are mono- and polyesters of germine, protoverine, and zygadenine, such as

- cevanine-type *Veratrum* azasteroid alkaloids; mono- and polyesters of germine, protoverine, and zygadenine, mixtures in various species; increased repetitive discharge and reflex activity; risk especially in sheep

Zigadenus exaltatus

Zigadenus fremontii

Zigadenus glaberrimus

germidine, neogermidine, neogermitrine, protoveratridine, protoveratrine, vanilloylzygadenine, veratroylzygadenine, and zygacine (Kupchan & Deliwala 1954; Kupchan et al. 1955a; Willaman & Schubert 1961). Although all of these alkaloids are not found in each species (*Z. venenosus* seemingly has the most extensive array), most species have an array sufficient to cause similar signs. The variety of possible combinations of alkaloids may in some measure be responsible for the considerable differences in potency among the species. Early experiments established the close relationship between the physiologic effects of *Zigadenus* and *Veratrum* alkaloids (Heyl et al. 1913; McLaughlin 1931; Kupchan & Deliwala 1953; Kupchan et al. 1955b). In dogs, there was marked hypotension, other effects on the heart and respiration, increased intestinal peristalsis, and, terminally, tetanic spasms, heart failure, and respiratory failure. Death was formerly attributed to heart failure but now is thought to be more closely associated with central respiratory depression (Mitchell & Smith 1911; Kupchan & Flacke 1967).

Zygadenine

The cevanine-type *Veratrum* alkaloids increase reflex activity and sensitize afferent pathway receptors seemingly in the absence of stimuli. They increase and extend the negative afterpotential and cause repetitive discharge in response to single stimuli in nerve and muscle. A more detailed account of the mechanism of action is given in the treatment of *Veratrum* in this chapter.

All species of *Zigadenus* should be considered toxic, albeit only a few represent a serious hazard for livestock. *Zigadenus nuttallii, Z. venenosus,* and *Z. paniculatus* are the most toxic, in that order. Because of the variability in total and individual alkaloid levels with growth stage and the variation among plants from different locations, it is difficult to give a precise estimate of toxicity (Beath et al. 1933, 1953). However, a conservative estimate of minimal lethal dose would be 1% b.w. of green plant in sheep for *Z. paniculatus* and *Z. venenosus* and slightly less for *Z. nuttallii.* The doses producing serious illness may be as low as 0.2–0.5% b.w. (Marsh & Clawson 1924; Beath et al. 1933, 1953; Panter & James 1989). Any difference in toxicity between *Z. paniculatus* and *Z. venenosus* is more than offset by the larger size of the former and, thus, more toxin available per plant (Beath et al. 1933). Range sheep that eat 4–5 kg of green forage per day would have little difficulty eating the 0.1–0.4 kg necessary to produce disease (Fleming et al. 1921). This amount represents approximately 30–100 plants, depending on the toxicity of the species (Chesnut & Wilcox 1901). Three to four times as much, or even more, of other less toxic species, such as *Z. elegans,* are required to produce disease. This greatly diminishes the hazard for species with low population densities, relative to that for the three most toxic species, which are often found in dense populations with sufficient plant material to represent a hazard. All plant parts, including the bulb, are apparently quite bitter and are generally avoided by grazing animals (Marsh & Clawson 1924; Spoerke & Spoerke 1979). Typically in an area, all of the grasses and other forbs are grazed down, while few, if any, of the death camas are eaten. However, because death camas is an early-season forb, little alternative forage may be available. Indeed, plants may be very conspicuous in the early spring in communities of dormant prairie grasses.

Zigadenus leimanthoides

Even the more toxic species may not always be such a risk. Nelson (1898) fed 0.45 kg of plant material to sheep of undisclosed body weight without causing adverse effects. However, the sheep had been fed smaller amounts the previous 3 days.

Although cattle appear to be of equal or greater susceptibility, sheep are at highest risk because they are most likely to eat the forbs, especially if they are particularly hungry (Marsh et al. 1915). The larger body size for cattle also is an important factor because it may be difficult for larger animals to find and consume sufficient plant to cause serious disease. The exception to this is *Z. nuttallii,* which seems to be toxic enough to pose an equally serious threat to cattle and sheep (Marsh et al. 1926b). Hay containing the offending plants seems to be the most likely source of intoxication for horses (Marsh & Clawson 1922; Fuller & McClintock 1986). All animal species are at risk, but those that effectively vomit (pigs) are difficult to intoxicate. Toxicity varies somewhat with stage of growth; the seeds and fruits are the most toxic. However, the leaves, which contain high concentrations of the alkaloids in early stages of growth, are the most serious hazard (Beath et al. 1933, 1953). Distribution of the alkaloids among the plant organs may vary during growth. Thus, under some conditions, especially in the fall after the aerial portions have withered, the bulbs, as storage organs, may become very toxic (Marsh et al. 1926b; Beath et al. 1933; Reid & Smith 1956). An additional factor of some importance is the response of plants to fire. Because *Z. nuttallii* is a forb of the prairies, it is subjected to range fires. After a fire, regrowth may have increased nitrogen levels in the leaves (Knapp 1986). Whether it is more toxic under such conditions is not known, but it may be more readily eaten. Interestingly, although the leaves and bulbs are quite bitter and reluctantly eaten by livestock, they are reported to be readily eaten—almost with impunity—by ground squirrels and other rodents (Fleming et al. 1921).

Zigadenus micranthus

Zigadenus nuttallii

Clinical Signs—Signs typically begin several hours or as long as a day after ingestion of the plants. Onset appears to be slowest and duration longest for *Z. nuttallii* (Marsh et al. 1926b). Onset is heralded by the development of profuse frothy salivation; ropy strings of saliva hang from the mouth. This is apparent in all animal species, although it is reported to be less severe in cattle (Fleming et al. 1921). Depression, nausea, vomiting, retching, colic, and grinding of the teeth accompany the excess salivation. In many cases, especially in cattle and horses, this will be the extent of the disease, and recovery in a day or two is typical (Fleming et al. 1921; Marsh & Clawson 1922; Gates 1930). As the disease develops further, there is loss of appetite, trembling, incoordination (especially in the pelvic limbs), and severe depression and weakness, with the head held low, the ears drooped, and the back arched. These signs are typical in sheep (Marsh et al. 1915; Fleming et al. 1921; Marsh & Clawson 1922; Marsh et al. 1926b; Panter et al. 1987). There may be hypothermia of several degrees, and increased intestinal peristalsis with frequent defecation and urination. Heart and respiration rates vary during the course of the disease, but in fatal cases respiration is usually increased and labored.

- abrupt onset, profuse salivation, depression, nausea, vomiting, retching, colic, grinding of teeth

  trembling, ataxia, severe weakness and depression

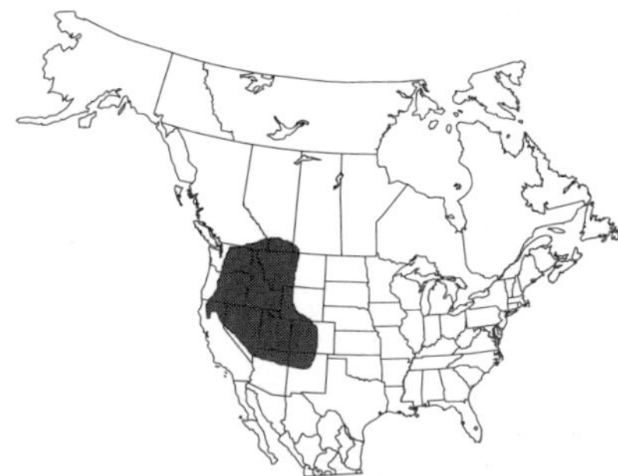

Zigadenus paniculatus

Zigadenus venenosus

Zigadenus virescens

In spite of the severe depression, sheep may be hyperexcitable. In one report, sheep after consuming *Z. venenosus* var. *gramineus* (reported as *Z. gramineus*) were laboriously moved about 1 km, which precipitated hyperexcitability. The animals had a stiff-legged walk or hop and would jump when touched, almost as if they had been touched with an electric prod (Avery et al. 1961). This case also illustrates the risk of forced movement in triggering the onset or worsening of signs. A few hours after onset, weakness may become so severe that animals prefer to lie down, and then they rise only with difficulty, if at all. Death is preceded by coma and spasmodic struggling for breath. The case fatality rate may be high in sheep but low in cattle and negligible in horses. Symptoms, especially those of the digestive disturbance, are reported to be more severe in lambs (Chesnut & Wilcox 1901).

- gross pathology, mild reddening gut mucosa

Pathology—Few distinctive lesions result from the disease caused by *Zigadenus*, even when animals are ill for several days before dying. Pulmonary congestion and edema with scattered small hemorrhages are most likely to be present, although there also may be reddening of the mucosa of the stomach and small intestine. Diagnosis may be aided by examination of ruminal contents for plant fragments and/or extraction of the contents for chemical identification of the alkaloids (Panter et al. 1987; Smith & Lewis 1991).

- activated charcoal, atropine

Treatment—A combination of 2 mg atropine and 8 mg of picrotoxin has been recommended for range sheep; preparations of this mixture were known in the past as death camas tablets (Beath et al. 1953). Atropine is administered to provide relief of the cardiovascular effects, although sympathomimetics are more effective for the hypotension. Picrotoxin to counteract the severe depression (including respiratory) may not be necessary. Activated charcoal may also provide considerable benefit.

Genera of Questionable Toxicity

Aletris L.

- *Aletris*

  colicroot
  star grass

- erect herbs from rhizomes; flowers in racemes; perianth parts white or yellow

Commonly known as colicroot or star grass, *Aletris* comprises 10–25 species native to eastern North America and Asia. All produce narrow linear-lanceolate leaves in basal rosettes that arise from short rhizomes. Their erect, scapelike stems bear bractlike leaves and terminate in a spicate raceme of tubular to campanulate, white or yellow flowers. The 6, fused perianth parts are wrinkled and roughened and tend to persist after withering. The superior ovary with its 3-branched style and minutely 2-lobed stigmas gives rise to an ovoid, beaked capsule containing numerous seeds. The genus is distributed throughout the eastern half of the continent.

- possibly digestive disturbance from steroidal saponins

The common name *colicroot* reflects the toxic properties of *Aletris*'s rhizomes. They appear to contain steroidal saponins, and an extract from *A. farinosa* L., also known as unicorn root, has been shown to depress uterine activity (Butler & Costello 1944). Otherwise, there is little evidence to implicate the genus in toxicologic problems.

Asparagus L.

- *A. densiflorus*
  A. officinalis
  A. setaceus

  asparagus
  asparagus fern
  emerald feather
  emerald fern

Comprising about 300 species, *Asparagus* is an Old World genus. Native to the seacoasts of Europe and Asia, *A. officinalis* L., with numerous cultivars, is the edible asparagus grown for food. It has sporadically escaped from cultivation and naturalized. A common houseplant, *A. setaceus* (Kunth.) Jessop (asparagus fern) is grown as an ornamental for its fine, feathery foliage, which is used in floral arrangements.

Asparagus densiflorus (Kunth.) Jessop (emerald fern or emerald feather) likewise is extensively cultivated as a potted houseplant. Plants of the genus are rhizomatous with erect or spreading or climbing stems that bear small, scalelike leaves and awl-like or filiform cladophylls. The perfect or functionally imperfect, green-yellow or white flowers are relatively inconspicuous and borne singly, paired, or clustered. Spreading to erect, the 6 perianth parts are fused about a superior ovary that gives rise to a bright red, 1-to-6-seeded berry.

■ herbs from rhizomes; leaves scale-like; fruits bright red berries

Because of their popularity as houseplants, *A. setaceus* and *A. densiflorus* when in fruit prompt numerous telephone calls to poison control centers when children and pets swallow the berries (Spoerke & Smolinske 1990). There is very little risk of intoxication, although various toxic properties, including irritant, cardiac, sedative, and diuretic effects, have been attributed to the genus. Some species contain significant concentrations of sapogenins (Pammel 1911; Watt & Breyer-Brandwijk 1962; Fuller & McClintock 1986; Kar & Sen 1986).

■ berries, sapogenins, possible mild irritant and other effects

Erythronium L.

■ *Erythronium*

adder's-tongue
dog's-tooth violet
fawn lily
glacier lily
trout lily

With the exception of 1 species in Eurasia, *Erythronium* is a North American genus comprising 15–25 species. It has several common names—dog's-tooth violet, trout lily, fawn lily, glacier lily, and adder's-tongue—which reflect morphological features of the corms and leaves. In plants of this genus, deep corms give rise to partially subterranean stems that bear 2 or 3 leaves at ground level, which thus appear basal or subbasal. Typically, the elliptic or lanceolate or oblanceolate blades are mottled, and the stem terminates in 1–5 large, showy, nodding flowers with white or yellow or pinkish, reflexed, free perianth parts. The fruits are loculicidal capsules, oblong to obovoid and 3-lobed. Species occupy a diversity of habitats across the continent, most frequently moist, shaded woodlands and rocky montane sites.

■ herbs from corms; flowers large and showy; perianth parts white, yellow, or pinkish

Erythronium oregonum

With the exception of one report (Kingsbury 1964) citing *E. oregonum* as a possible cause of intoxication in poultry, *Erythronium* has not been reported to cause toxicologic problems in North America. The flowers are commonly eaten by mountain hikers. However, the corms when eaten appear to cause effects similar to those produced by species of *Tulipa* (see treatment in this chapter). They include nausea, vomiting, increased salivation, labored respiration, and increased heart rate. Several species of the genus contain large amounts of tuliposides A and B (Slob et al. 1975). This indicates the possibility of the presence of the lactone tulipalin derivatives, which may cause mild contact irritation similar to that associated with tulips. Orally administered charcoal may be helpful; otherwise treatment is directed toward relief of the digestive tract effects.

■ possible digestive disturbance due to tuliposides

Melanthium L.

■ *M. hybridum*
M. virginicum

bunchflower

Native to North America, *Melanthium,* commonly known as bunchflower, comprises 5 species. Its name is derived from the Greek roots *melas* and *anthos,* meaning "black" and "flower," alluding to the persistent perianth parts, which become dark after flowering. It is closely related to *Veratrum* and *Zigadenus,* with *V. parviflorum* and *V. woodii* positioned within it by some taxonomists. Plants of the genus produce stout rhizomes that give rise to stems bearing primarily long basal leaves with linear to oblanceolate blades. Panicles with indumented branches and pedicels terminate the stems and bear greenish, perfect and staminate flowers with free, clawed, spreading perianth parts, each with a pair of glands at the base. The superior ovary, bearing 3 spreading styles, gives rise to an ovoid, septicidal capsule subtended by the withered

■ herbs from rhizomes; leaves basal; flowers in panicles; perianth parts greenish

Melanthium hybridum

Melanthium virginicum

perianth and enclosing several elliptic, flat, winged seeds. All species of the genus typically occur in wet soils of marshes and moist woods in the eastern half of the continent. Most commonly encountered are *M. hybridum* Walter and *M. virginicum* L.

- possible digestive disturbance; similar to effects of *Veratrum* alkaloids, profuse salivation, depression

The leaves and seeds of *Melanthium,* both fresh and in hay, are reported to pose a risk to sheep, cattle, and horses. Little is known, however, of the taxon's toxicity hazard or the specific toxicant. The clinical signs indicative of its effect on the digestive tract are suggestive of a *Veratrum*-type alkaloid (Pammel 1911; Hardin 1973). The genus also has a reputation for use in poisoning crows. Signs include rapid but weak heart rate, labored respiration, severe weakness, increased salivation, and retching. Most of these signs last only a few hours, but depression may persist for several days. There have been no deaths and, therefore, no pathology reported. As with *Veratrum* and *Zigadenus,* congestion and a few scattered splotchy hemorrhages may be all that can be expected. Generally, treatment would be based upon relief of the signs, but atropine may be of some value against the systemic effects, as with *Zigadenus.*

Pleea Michx.

- *P. tenuifolia*
  rush-featherling

- evergreen herbs from rhizomes; leaves linear; flowers in racemes

Endemic to the southeastern United States, *Pleea* comprises a single species, *P. tenuifolia* Michx., commonly known as rush-featherling. It is reputed to be toxic but with little evidence offered (Hardin 1973). Locally abundant in the Carolinas, Georgia, and Florida, plants of the genus occur in clumps arising from thick rhizomes. The equitant, evergreen, linear leaves are primarily basal, and the scapelike stems terminate in racemes with large, sheathing bracts subtending the flowers. The free, lanceolate, perianth parts persist in fruit about the ovoid to ellipsoid capsule derived from a superior ovary with 3 styles. The seeds are reddish black with wings.

Trillium L.

- *Trillium*
  birthroot
  trillium
  wake-robin

- erect herbs from rhizomes; leaves typically 3; flowers solitary, terminal; sepals reflexed; petals white, green-yellow, pink, or maroon

Known as trillium, birthroot, or wake-robin, *Trillium* comprises 30–50 species native to North America and Asia. Its name is derived from the Greek root *tris,* or "three," which reflects the arrangement of leaves and perianth parts, which are borne in threes. It is placed in the segregate family Trilliaceae by some taxonomists. Plants of the genus arise from short, stout rhizomes. The erect stems bear a single leaf or 3 sessile or petiolate, glossy leaves with blades that are elliptic to ovate, often mottled, and exhibiting net venation. The solitary flowers are terminal and, unlike most other members of the Liliaceae, have perianth parts of 2 forms. The 3 leaflike sepals are reflexed, and the 3 oblanceolate petals are green-yellow or white or pink or maroon. The superior ovary produces a many-seeded berry. Species typically occupy moist soils of forests across the continent. They also are grown as ornamentals.

The rhizomes of *Trillium* are potent emetics but have seldom been incriminated as causing problems. The causes of the toxic effects are not known, but most species contain steroidal saponins from sapogenins such as diosgenin and trillenogenin (trillin, trillarin) (Fieser & Fieser 1959; Nohara et al. 1975; Nakano et al. 1982). Some saponins are present as 18-norspirostanol glycosides of genins such as pennogenin. Dioscin is a representative trioside.

Clinical signs, apparently due to irritation of the digestive tract, may include violent vomiting and severe diarrhea (Pammel 1911; Perkins & Paine 1978). Distinctive pathologic changes are not likely to be present. Relief of the irritation is the primary aim of treatment, which includes antiemetics and antidiarrheals.

Trillium pusillum

Trillium cernuum

Trillium grandiflorum

References

Abd El Hafiz MA, Ramadan MA, Jung ML, Beck JP, Anton R: Cytotoxic activity of Amaryllidaceae alkaloids from *Crinum augustum* and *Crinum bulbispermum*. *Planta Med* 57;437–439, 1991

Ahluwalia P, Mohindroo A: Effect of oral ingestion of different fractions of *Allium cepa* on the blood and erythrocyte membrane lipids and certain membrane-bound enzymes in rats. *J Nutr Sci Vitaminol (Tokyo)* 35;155–161, 1989

Aleem HM: *Gloriosa superba* poisoning. *J Assoc Physic India* 40;541–542, 1992

Alsberg CL: The toxic action of *Amianthium muscaetoxicum* [abstract]. *J Pharm Exp Ther* 3;473, 1912

Anthony DC, Graham DG: Toxic responses of the nervous system. In *Casarett and Doull's Toxicology*, 4th ed, Amdur MO, Doull J, Klaassen CD eds, Pergamon Press, New York, pp407–429, 1991

Atta-ur-Rahman, Azhar Ali R, Parveen T, Choudhary MI, Sener B, Turkoz S: Alkaloids from *Veratrum album*. *Phytochemistry* 30;368–370, 1991

Atta-ur-Rahman, Azhar Ali R, Gilani A-ul-H, Choudhary MI, Aftab K, Sener B, Turkoz S: Isolation of antihypertensive alkaloids from the rhizomes of *Veratrum album*. *Planta Med* 59;569–571, 1993

Avery RJ, Niilo L, Kramer T, Johnston A, Darcel C leQ: Two cases of poisoning in livestock presenting difficulties in diagnosis. *Can Vet J* 2;250–252, 1961

Ban Y, Murakami Y, Iwasawa Y, Tsuchiya M, Takano N: Indole alkaloids in medicine. *Med Res Rev* 8;231–308, 1988

Baraka A, Harik S: Reversal of central anticholinergic syndrome by galanthamine. *J Am Med Assoc* 238;2293–2294, 1977

Barlow RM, Middleton DJ: Pathogenesis of *Hemerocallis* neurotoxicity in mice [abstract]. *Neuropathol Appl Neurobiol* 15;602, 1989

Barlow RM, Bedford PGC, Wang JH: Pathogenesis of "*Hemerocallis* blindness" in sheep [abstract]. *Neuropathol Appl Neurobiol* 15;592, 1989

Baud FJ, Sabouraud A, Vicaut E, Taboulet P, Lang J, Bismuth C, Rouzioux JM, Scherrmann J-M: Brief report: treatment of severe colchicine overdose with colchicine-specific Fab fragments. *N Engl J Med* 332;642–645, 1995

Beath OA, Eppson HF, Draize JH, Justice RS: *Three Species of* Zygadenus *(death camas)*. Wyo Agric Exp Stn Bull 194, 1933

Beath OA, Gilbert CS, Eppson HF, Rosenfeld I: *Poisonous Plants and Livestock Poisoning*. Wyo Agric Exp Stn Bull 324, p31–35, 1953

Beetle DE: A monograph of the North American species of *Fritillaria*. *Madrono* 7;133–159, 1944

Beijersbergen JCM, Lemmers CBG: Enzymic and non-enzymic liberation of tulipalin A (α-methylene butyrolactone) in extracts of tulip. *Physiol Plant Pathol* 2;265–270, 1972

Bertolini A, Ferrari W, Guarini S, Tagliavini S: Circulatory and respiratory consequences of massive hemorrhage are reversed by protoveratrines. *Experientia* 46;707–708, 1990

Binns W, James LF, Shupe JL, Everett G: A congenital cyclopian type malformation in

Trillium ovatum

Trillium pusillum

- digestive disturbance possibly due to irritant steroidal saponins from genins such as diosgenin, trillenogenin

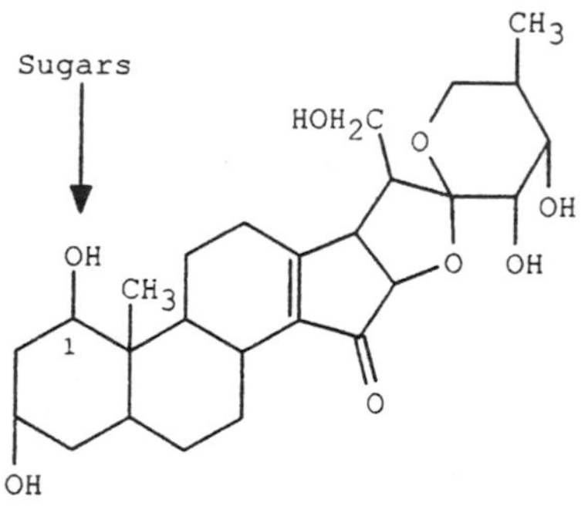

Trillenogenin

lambs induced by maternal ingestion of a range plant, *Veratrum californicum. Am J Vet Res* 24;1164–1175, 1963

Binns W, Shupe JL, Keeler RF, James LF: Chronologic evaluation of teratogenicity in sheep fed *Veratrum californicum. J Am Vet Med Assoc* 147;839–842, 1965

Bourque M: Congenital abnormality of the cyclops type in a calf. *Med Vet Que* 14;67–71, 1984

Brady JF, Wang MH, Hong JY, Xiao F, Li Y, Yoo JSH, Ning SM, Lee MJ, Fukuto JM, Gapac JM, Yang CS: Modulation of rat hepatic microsomal monooxygenase enzymes and cytotoxicity by diallyl sulfide. *Toxicol Appl Pharmacol* 108;342–354, 1991

Breukink HJ: *Ornithogalum thyrsoides* poisoning in sheep. *Tijdschr Diergeneeskd* 88;653–658, 1963

Brown KS: Abnormal steroidal alkaloids. In *Chemistry of the Alkaloids,* Pelletier SW ed, Van Nostrand Reinhold, New York, pp631–667, 1970

Buck WB, Keeler RF, Binns W: Some physiologic effects of *Veratrum* alkaloids in sheep and goats. *Am J Vet Res* 27;140–154, 1966

Burrows GE, Tasler J, Boray JC, Egerton J: Antipyrine, erythromycin, and oxytetracycline disposition in experimental fasciolosis. *Res Vet Sci* 53;191–197, 1992

Butler CL, Costello CH: Pharmacological studies. 1. *Aletris farinosa. J Am Pharm Assoc* 33;177–183, 1944

Campbell MA, Brown KS, Hassell JR, Horigan EA, Keeler RF: Inhibition of limb chondrogenesis by a *Veratrum* alkaloid: temporal specificity in vivo and in vitro. *Dev Biol* 111;464–470, 1985

Capraro HG, Brossi A: Tropolonic *Colchicum* alkaloids. In *The Alkaloids,* Brossi A ed, Academic Press, Orlando, Fla, pp1–70, 1984

Carson TL, Sanderson TP, Halbur PG: Acute nephrotoxicosis in cats following ingestion of lily (*Lilium* sp.). *Annu Meet Am Assoc Vet Lab Diagn* 37;43, 1994

Cary CA, Miller ER, Johnstone GR: *Poisonous Plants of Alabama.* Ala Polytech Ext Serv Circ 71, 1924

Casteel SW, Bailey EM, Reagor JC, Rowe LD: Photosensitization: an investigation and review of the problem in cattle of south Texas.·*Vet Hum Toxicol* 28;251–254, 1986

Casteel SW, Rowe LD, Bailey EM, Fiske RA, Schwartz WL, Reagor JC, Bridges CW: Experimentally induced photosensitization in cattle with *Cooperia pedunculata. Vet Hum Toxicol* 30;101–104, 1988

Catterall WA: Neurotoxins that act on voltage-sensitive sodium channels in excitable membranes. *Annu Rev Pharm Toxicol* 20;15–42, 1980

Chareyre S, Meram D, Pulce C, Descotes J: Acute poisoning of cows by autumn crocus. *Vet Hum Toxicol* 31;261–262, 1989

Chen KK, Rose CL, Anderson RC, Chou TQ: Pharmacological action of fritimine. *Chin J Physiol* 9;21–26, 1935

Chesnut VK: *Principal Poisonous Plants of the United States.* USDA Div Bot Bull 20, 1898a

Chesnut VK: *Thirty Poisonous Plants of the United States.* USDA Farmers Bull 86, 1898b

Chesnut VK, Wilcox EV: *The Stock-Poisoning Plants of Montana.* USDA Div Bot Bull 26, 1901

Colegate SM, Dorling PR, Huxtable CR: *Stypandra* species (blindgrass) toxicity: chemistry, biosynthesis, and chemotaxonomy. In *Poisonous Plants: Proceedings of the Third International Symposium,* James LF, Keeler Rf, Bailey EM Jr, Cheeke PR, Hegarty MP eds, Iowa State Univ Press, Ames, pp459–463, 1992

Collett S, Grotelueschen D, Smith R, Wilson R: Deaths of 23 adult cows attributed to intoxication by the alkaloids of *Zygadenus venenosus* (meadow death camas). *Agri-Practice* 17;5–9, 1996

Cook JW, Loudon JD: Alkaloids of the Amaryllidaceae. In *The Alkaloids: Chemistry and Physiology,* vol 2, Manske RHF, Holmes HL eds, Academic Press, New York, pp331–352, 1952

Cooper MR, Johnson AW: *Poisonous Plants in Britain and Their Effects on Animals and Man.* Her Majesty's Stationery Office, London, 1984

Courtemanche LA, Gephardt D, Jaffe AM: Poisoning due to *Veratrum viride* (false hellebore) ingestion [abstract]. *Vet Hum Toxicol* 31;360, 1989

Cowen DL: Squill in the seventeenth and eighteenth centuries. *Bull N Y Acad Med* 50;714–722, 1974

Dausch JG, Nixon DW: Garlic: a review of its relationship to malignant disease. *Prev Med* 19;346–361, 1990

Deb DB, Dasgupta S: Studies on Indian squill—*Urginea indica* (Roxb.) Kunth. *Q J Crude Drug Res* 14;49–60, 1976

Debarnot A: *Colchicum autumnale* (meadow saffron) poisoning. Thesis, Ecole Nationale Veterinaire, Alfort, Paris, 1968 (*Vet Bull* 40;2530, 1970)

Der Marderosian A, Roia FC Jr: Clinical management of household ornamental plants potentially toxic to humans. In *Toxic Plants,* Kinghorn AD ed, Columbia Univ Press, New York, pp103–135, 1979

Dorling PR, Colegate SM, Huxtable CR: The molecular mechanism of stypandrol poisoning. In *Poisonous Plants: Proceedings of the Third International Symposium,* James LF, Keeler Rf, Bailey EM Jr, Cheeke PR, Hegarty MP eds, Iowa State Univ Press, Ames, pp469–473, 1992

Dorsch W, Wagner H, Bayer T, Fessler B, Hein G, Ring J, Scheftner P, Sieber W, Strasser T, Weis E: Anti-asthmatic effects of onions. *Biochem Pharmacol* 37;4479–4486, 1988

Dvorackova S, Sedmera P, Potesilova H, Santavy F, Simanek V: Alkaloids of *Gloriosa superba* L. *Collect Czech Chem Commun* 49;1536–1542, 1984 (*Chem Abstr* 101;126895t, 1984)

Edgerton PH: Symptoms of digitalis-like toxicity in a family after accidental ingestion of lily of the valley plant. *J Emerg Nurs* 15;220–223, 1989

Egen-Schwind C, Eckard R, Kemper FH: Metabolism of garlic constituents in the isolated perfused rat liver. *Planta Med* 58;301–304, 1992

Etissami S, Kachefiolasle M: Effect of an alkaloid extract (SAF) of *Fritillaria imperialis* on lidocaine-induced conduction block. *Therapie* 26;1101–1105, 1971

Evans HE, Sack WO: Prenatal development of domestic and laboratory mammals: growth curves, external features, and selected references. *Zentralbl Vet Med C* 2;11–45, 1973

Everett TH: *Encyclopedia of Horticulture,* vol 5. Garland Publ, New York, pp1648–1651, 1981

Ewins AJ: Narcissine: an alkaloid from the bulb of the common daffodil *(Narcissus pseudonarcissus). J Chem Soc* 97;2406–2409, 1910

Fenwick GR, Hanley AB: *Allium* species poisoning [letter]. *Vet Rec* 116;28 1985a

Fenwick GR, Hanley AB: The genus *Allium*—part 1. *CRC Crit Rev Food Sci Nutr* 22;199–271, 1985b

Fenwick GR, Hanley AB: The genus *Allium*—part 2. *CRC Crit Rev Food Sci Nutr* 22;272–377, 1985c

Fenwick GR, Hanley AB: The genus *Allium*—part 3. *CRC Crit Rev Food Sci Nutr* 23;1–73, 1985d

Fieser LF, Fieser M: *Steroids.* Reinhold, New York, 1959

Flacke W, Caviness VS, Samaha FG: Treatment of myasthenia gravis with germine diacetate. *N Engl J Med* 275;1207–1214, 1966

Flaøyen A: *Narthecium ossifragum* associated nephrotoxicity in ruminants. In *Toxic Plants and Other Natural Toxicants,* Garland T, Barr AC eds, CAB International, New York, pp573–576, 1998

Flaøyen A, Binde M, Bratberg B, Djonne B, et al.: Nephrotoxicity of *Narthecium ossifragum* in cattle in Norway. *Vet Rec* 137;259–263, 1995

Fleming CE, Peterson NF, Miller MR, Wright LH: *Death Camas* (Zygadenus paniculatus *and* Zygadenus venenosus) *plants poisonous to sheep and cattle.* Univ Nev Agric Exp Stn Bull 101, 1921

Folpini A, Furfori P: Colchicine toxicity—clinical features and treatment: massive overdose case report. *Clin Toxicol* 33;71–77, 1995

Fried J, White HL, Wintersteiner O: Germidine and germitrine, two new ester alkaloids from *Veratrum viride. J Am Chem Soc* 71;3260–3261, 1949

Frohne D, Pfander HJ: *A Colour Atlas of Poisonous Plants,* 2d ed. Wolfe Science, London, pp149–151, 1984

Fuganti C: The Amaryllidaceae alkaloids. In *The Alkaloids. Chemistry and Physiology,* vol 15, Academic Press, New York, pp83–164, 1975

Fuller TC, McClintock E: *Poisonous Plants of California.* Univ Calif Press, Berkeley, pp289–292, 1986

Gaffield W, Keeler RF: Implication of C-5, C-6 unsaturation as a key structural factor in steroidal alkaloid-induced mammalian teratogenesis. *Experientia* 49;922–924, 1993

Galtier P, Larrieu G, Tufenkji AE, Franc M: Incidence of experimental fascioliasis on the activity of drug metabolizing enzymes in lamb liver. *Drug Metab Dispos* 14;137–141, 1986

Garnier R, Hoffelt J, Carlier P, Delmas-Riboulet G, Karchen J, Thimon D, Fournier E: Acute intoxication with home-made gentian wine—clinical and analytical findings [abstract]. *Vet Hum Toxicol* 24;36, 1982

Gates FC: *Principal Poisonous Plants in Kansas.* Kans Agric Exp Stn Tech Bull 25, 1930

Gathings AM: Acute renal failure in a cat: was it the tiger lily? *Napinet Rep* 3(2);2, 1 February 1991

Gemmill CL: The pharmacology of squill. *Bull N Y Acad Med* 50;747–750, 1974

Gentry HS, Verbiscar AJ, Banigan TF: Red squill (*Urginea maritima,* Liliaceae). *Econ Bot* 41;267–282, 1987

Ghannamy U, Kopp B, Robien W, Kubelka W: Cardenolides from *Ornithogalum boucheanum. Planta Med* 53;172–178, 1987

Gill PA, Sergent ESG: Onion poisoning in a bull [letter]. *Aust Vet J* 57;484, 1981

Goncalo S, Freitas JD, Sousa I: Contact dermatitis and respiratory symptoms from *Narcissus pseudonarcissus. Contact Dermatitis* 16;115–116, 1987

Greenhill JV, Grayshan P: The cevane group of *Veratrum* alkaloids. In *The Alkaloids: Chemistry and Pharmacology,* vol 41, Brossi A, Cordell GA eds, Academic Press, San Diego, Calif, pp177–237, 1992

Gress EM: *Poisonous Plants of Pennsylvania.* Pa Dept Agric Bull 531, 1935

Gruhzit OM: 1: anemia of dogs produced by feeding of the whole onions and of onion fractions. *Am J Med Sci* 181;812–814, 1931a

Gruhzit OM: 2: anemia of dogs produced by feeding disulfide compounds. *Am J Med Sci* 181;815–817, 1931b

Gude M, Hausen BM, Hietsch H, Konig WA: An investigation of the irritant and allergenic properties of daffodils (*Narcissus pseudonarcissus* L., Amaryllidaceae): a review of daffodil dermatitis. *Contact Dermatitis* 19;1–28, 1988

Hakim FS, Bowery NG, Evans FJ: Comparative potencies of European and Indian squill. *J Pharm Pharmacol* 28;81–82, 1976

Hall JO: Are Easter lilies toxic to cats? *Napinet Rep* 3(2);1, 16 July 1990

Hall JO: Nephrotoxicity of Easter lily *(Lilium longiflorum)* when ingested by the cat. *Proc Am Coll Vet Int Med Forum* 10;804, 1992

Hansen AA: The poisonous plant situation in Indiana—3. *J Am Vet Med Assoc* 66;351–362, 1924

Hardin JW: *Stock-Poisoning Plants of North Carolina.* N C Agric Exp Stn Bull 414, 1973

Hart J: *Montana Native Plants and Early Peoples.* Montana Historical Society Press, 1992

Harvey JW, Rackear D: Experimental onion-induced hemolytic anemia in dogs. *Vet Pathol* 22;387–392, 1985

Heftmann E: *Steroid Biochemistry.* Academic Press, New York, pp43–48, 1970

Heilpern KL: *Zigadenus* poisoning. *Ann Emerg Med* 25;259–262, 1995

Heyl FW, Hepner FE, Loy SK: Zygadenine, the crystallin alkaloid of *Zygadenus intermedius. J Am Chem Soc* 35;258–262, 1913

Hjorth N, Wilkinson DS: Contact dermatitis. 4. Tulip fingers, hyacinth itch, and lily rash. *Br J Dermatol* 80;696–698, 1968

Hoch JH: Poisonous plants of South Carolina. *Acta Phytother* 16;181–184, 1969

Honerjager P: Ceveratrum alkaloids: progress in understanding their membrane and inotropic actions. *TIPS* (June);258–262, 1983

Hood RL: Colchicine poisoning. *J Emerg Med* 12;171–177, 1994

Hostettmann K, Marston A: *Saponins*. Cambridge Univ Press, Cambridge, 1995

Hothi DS, Arneja JS, Chawla JS: Onion *(Allium cepa)* poisoning in bullocks. *Indian Vet J* 57;690–692, 1980

Houston DM, Myers SL: A review of Heinz-body anemia in the dog induced by toxins. *Vet Hum Toxicol* 35;158–161, 1993

Hruby K, Lenz K, Krausler J: *Veratrum album* poisoning. *Wiener Klin Wochenschr* 93; 517–519, 1981

Hu CH, Shang EN, Lin WH, Cai MS: Studies on the chemical constituents of *Fritillaria taipaiensis* L. *Yao Hsueh Hsueh Pao* 28;516–521, 1993

Hutchison TWS: Onions as a cause of Heinz body anemia and death in cattle. *Can Vet J* 18;358–360, 1977

Huxtable CR, Dorling PR, Slatter DH: Myelin oedema, optic neuropathy, and retinopathy in experimental *Stypandra imbricata* toxicosis. *Neuropathol Appl Neurobiol* 6;221–232, 1980

Huxtable CR, Dorling PR, Colegate SM: *Stypandra* poisoning: the role of stypandrol and proposed pathomechanisms. In *Poisonous Plants: Proceedings of the Third International Symposium,* James LF, Keeler Rf, Bailey EM Jr, Cheeke PR, Hegarty MP eds, Iowa State Univ Press, Ames, pp464–468, 1992

Irwin RL, Smith HJ III: Cholinesterase inhibition by galanthamine and lycoramine. *Biochem Pharmacol* 3;147–148, 1960

Jaffe AM, Gephardt D, Courtemanche L: Poisoning due to ingestion of *Veratrum viride* (false hellebore). *J Emerg Med* 8;161–167, 1990

James LF: Plant-induced congenital malformations in animals. *World Rev Nutr Diet* 26;208–224, 1977

James LF, Binns W: Effects of feeding wild onions *(Allium validum)* to bred ewes. *J Am Vet Med Assoc* 149;512–514, 1966

Joubert JPJ: Cardiac glycosides. In *Toxicants of Plant Origin,* vol 2, *Glycosides,* Cheeke PR ed, CRC Press, Boca Raton, Fla, pp61–96, 1989

Kamphues J, Meyer H: Contamination of hay by autumn crocus *(Colchicum autumnale)* and equine colic. *Tierarztl Praxis* 18;273–275, 1990 (*Vet Bull* 61;8010, 1991)

Kaneko K, Katsuhara T, Mitsuhashi H, Chen YP, Hsu HY, Shiro M: Isolation and structural elucidation of new alkaloids from *Fritillaria delavayi* Franch. *Chem Pharm Bull* 33;2614–2617, 1985

Kar OK, Sen S: Content of sapogenins in diploid, tetraploid, and hexaploid asparagus. *Int J Crude Drug Res* 24;131–133, 1986

Kauntze R, Trounce J: The hypotensive action of veriloid *(Veratrum viride),* a clinical investigation. *Lancet* 260;549–555, 1951

Kay JM: Onion toxicity in a dog. *Mod Vet Pract* 64;477–478, 1983

Keeler RF: Known and suspected teratogenic hazards in range plants. *Clin Toxicol* 5;529–565, 1972

Keeler RF: Plant metabolites that are teratogenic in offspring and toxic in the dam. *Toxicon* (Suppl 3);221–225, 1983

Keeler RF: Teratogens in plants. *J Anim Sci* 58;1029–1039, 1984

Keeler RF: *Prenatal and Neonatal Death of Lambs Due to Maternal Grazing of* Veratrum californicum. USDA Forest Serv Gen Tech Rep INT-222, pp118–122, 1987

Keeler RF: Livestock models of human birth defects, reviewed in relation to poisonous plants. *J Anim Sci* 66;2414–2427, 1988

Keeler RF: Early embryonic death in lambs induced by *Veratrum californicum. Cornell Vet* 80;203–207, 1990

Keeler RF, Binns W: Teratogenic compounds of *Veratrum californicum* (Durand). 2. Production of ovine fetal cyclopia by fractions and alkaloid preparations. *Can J Biochem* 44;829–838, 1966

Keeler RF, Binns W: Teratogenic compounds of *Veratrum californicum* as a function of plant part, stage, and site of growth. *Phytochemistry* 10;1765–1769, 1971

Keeler RF, Stuart LD: The nature of congenital limb defects induced in lambs by maternal ingestion of *Veratrum californicum. Clin Toxicol* 25;273–286, 1987

Keeler RF, Young S, Smart R: Congenital tracheal stenosis in lambs induced by maternal ingestion of *Veratrum californicum Teratology* 31;83–88, 1985

Keeler RF, Stuart LD, Young S: When ewes ingest poisonous plants: the teratogenic effects. *Vet Med* 81;449–454, 1986

Kellerman TS, Coetzer JAW, Naude TW: *Plant Poisonings and Mycotoxicoses of Livestock in Southern Africa.* Oxford Univ Press, Cape Town, South Africa, pp91–105,164, 1988

Kelly RB, Daniels EG, Spaulding LB: Cytotoxicity of cardiac principles. *J Med Chem* 8;547–548, 1965

Kiese M: *Methemoglobinemia: A Comprehensive Treatise.* CRC Press, Cleveland, 1974

Kingsbury JM: *Poisonous Plants of the United States and Canada.* Prentice-Hall, Englewood Cliffs, NJ, 1964

Kirk JH, Bulgin MS: Effects of feeding cull domestic onions *(Allium cepa)* to sheep. *Am J Vet Res* 40;397–399, 1979

Knapp AK: Ecophysiology of *Zigadenus nuttallii,* a toxic spring ephemeral in a warm season grassland: effect of defoliation and fire. *Oecologia (Berlin)* 71;69–74, 1986

Kobayashi K: Onion poisoning in the cat. *Feline Pract* 11;22–27, 1981

Koger LM: Onion poisoning in cattle. *J Am Vet Med Assoc* 129;75, 1956

Komissarenko NF, Stupakova EP: Neoconvalloside—a cardenolide glycoside from plants of the genus *Convallaria. Khim Prir Soedin* 1986;201–204, 1986 (*Chem Abstr* 105:197033d, 1986)

Kopp B, Kubelka W: New cardenolides of *Convallaria majalis. Planta Med* 45;87–94, 1982a

Kopp B, Kubelka W: New cardenolides of *Convallaria majalis. Planta Med* 45;195–202, 1982b

Krenzelok EP, Jacobsen TD, Aronis JM: Lily-of-the-valley *(Convallaria majalis)* exposures: are the outcomes consistent with the reputation? [abstract]. *J Toxicol Clin Toxicol* 34;601, 1996

Kulig K, Rumack BH: Severe *Veratrum* alkaloid poisoning [abstract]. *Vet Hum Toxicol* 24;36, 1982

Kupchan SM, Deliwala CV: *Zygadenus* alkaloids. 3. Active principles of *Zygadenus venenosus:* veratroylzygadenine and vanilloylzygadenine. *J Am Chem Soc* 75;1025–1029, 1953

Kupchan SM, Deliwala CV: Zygadenus alkaloids. 4. Active principles of *Zygadenus venenosus:* germine esters. *J Am Chem Soc* 76;5545–5547, 1954

Kupchan SM, Flacke WE: Hypotensive veratrum alkaloids. In *Antihypertensive Agents,* Schlittler E ed, Academic Press, New York, pp429–458, 1967

Kupchan SM, Deliwala CV, Zonis RD: Zygadenus alkaloids. 6. Active principles of *Zygadenus paniculatus. J Am Chem Soc* 77;755, 1955a

Kupchan SM, Lavie L, Zonis RD: *Zygadenus* alkaloids. 5. Active principles of *Zygadenus venenosus:* zygacine. *J Am Chem Soc* 77;689–691, 1955b

Langseth W, Torgersen T, Kolsaker P, Romming C, Jantsch TG, Mantle PG, Pearce J, Gibson SE, Goicochea MG, Flaoyen A: Isolation and characterization of 3-methoxy-2(5H)-furanone as the principal nephrotoxin from *Narthecium ossifragum* (L.) Huds. *Nat Toxins* 7;111–118, 1999

Lazarus AE, Rajamani S: Poisoning due to onion spoilage in cattle. *Indian Vet J* 45; 877–880, 1968

Leibowitz MD, Schwarz JR, Holan G, Hille B: Electrophysiological comparison of insecticide and alkaloid agonists of Na channels. *J Gen Physiol* 90;75–93, 1987

Leroux V: Poisoning of pets by house plants. *Point Vet* 18;45–55, 1986

Lincoln SD, Howell ME, Combs JJ, Hinman DD: Hematologic effects and feeding performance in cattle fed cull domestic onions *(Allium cepa). J Am Vet Med Assoc* 200;1090–1094, 1992

Litovitz TL, Fahey BA: Please don't eat the daffodils. *N Engl J Med* 306;547, 1982

Loft ER, Menke JW, Kie JG: Interaction of cattle and deer on mountain rangeland. *Calif Agric* 40;6–9, 1986

Lund LJ, Gibson GM, Turner RG: Leek taint in beef carcasses [letter]. *Vet Rec* 128; 263–264, 1991
Maede Y: High concentration of blood glutathione in dogs with acute hemolytic anemia. *Jpn J Vet Sci* 39;187–189, 1977
Mahato SB, Gangully AN, Sahu NP: Steroid saponins. *Phytochemistry* 21;959–978, 1982
Main DC, Slatter DH, Huxtable CR, Constable C, Dorling PR: *Stypandra imbricata* ("blindgrass") toxicosis in goats and sheep—clinical and pathologic findings in four field cases. *Aust Vet J* 57;132–135, 1981
Malone FE, Kennedy S, Reilly GAC, Woods FM: Bog asphodel *(Narthecium ossifragum)* poisoning in cattle. *Vet Rec* 131;100–103, 1992
Malone FE, Kennedy S, Reilly GAC, Woods FM: Bog asphodel *(Narthecium ossifragum)* poisoning in cattle. In *Toxic Plants and Other Natural Toxicants,* Garland T, Barr AC eds, CAB International, New York, pp564–568, 1998
Maretic Z, Russel FE, Ladavac J: Tulip bulb poisoning. *Period Biol* 80;141–143, 1978
Marsh CD, Clawson AB: *The Death Camas Species,* Zygadenus paniculatus *and* Zygadenus elegans, *as Poisonous Plants.* USDA Bull 1012, 1922
Marsh CD, Clawson AB: *The Meadow Death Camas* (Zygadenus venenosus) *as a Poisonous Plant.* USDA Bull 1240, 1924
Marsh CD, Clawson AB, Marsh H: Zygadenus, *or Death Camas.* USDA Bull 125, 1915
Marsh CD, Clawson AB, Marsh H: *Stagger Grass* (Chrosperma muscaetoxicum) *as a Poisonous Plant.* USDA Bull 710, 1918 and 1926a
Marsh CD, Clawson AB, Roe GC: *Nuttall's Death Camas* (Zygadenus nuttallii) *as a Poisonous Plant.* USDA Bull 1376, 1926b
Martin SF: The Amaryllidaceae alkaloids. In *The Alkaloids: Chemistry and Pharmacology,* vol 30, Brossi A ed, Academic Press, San Diego, Calif, pp251–376, 1987
McLaughlin AR: Responses of sheep to *Zygadenus gramineus,* "death camas." *Science* 73;135–136, 1931
Mimaki Y, Sashida Y: Studies on the chemical constituents of the bulbs of *Fritillaria camschatcensis. Chem Pharm Bull* 38;1090–1092, 1990
Mir GN, Ghatak BJR: Pharmacological studies on *Fritillaria imperialis. Indian J Exp Biol* 4;157–160, 1966
Mitchell J, Rook A: *Botanical Dermatology—Plants and Plant Products Injurious to the Skin,* Greengrass, Vancouver, BC, 1979
Mitchell PH, Smith G: The physiological effects of *Zygadenus intermedius. Am J Physiol* 28;318–329, 1911
Miyata D: Isolation of a new phenolic compound from the onion (*Allium cepa* L.) and its oxidative effect on erythrocytes [abstract]. *Jpn J Vet Res* 38;65, 1990
Monks TJ, Anders MW, Dekant W, Stevens JL, Lau SS, van Bladern PJ: Glutathione conjugate mediated toxicities. *Toxicol Appl Pharmacol* 106;1–19, 1990
Moraes-Cerdeira RM, Bastos JK, Burandt CL Jr, Nanayakkara NPD, Mikell J, McChesney JD: Alkaloid content of different bulb parts of *Narcissus* cv. Ice Follies [letter]. *Planta Med* 63;92–93, 1997
Moxley RA, Schneider NR, Steinegger DH, Carlson MS: Apparent toxicosis associated with lily-of-the-valley *(Convallaria majalis)* ingestion in a dog. *J Am Vet Med Assoc* 195;485–487, 1989
Mrozik H, Waud RA, Schindler O, Reichstein T: Die glykoside der zwiebeln von *Ornithogalum umbellatum* L sowie prufung der zwiebeln von *Ornithogalum prasinum* (Lindl). *Helv Chim Acta* 42;683–696, 1959
Mulligan GA, Munro DB: The biology of Canadian weeds. 77. *Veratrum viride* Ait. *Can J Plant Sci* 67;777–786, 1987
Munday R, Manns E: Comparative toxicity of prop(en)yl disulfides derived from Alliaceae: possible involvement of 1-propenyl disulfides in onion-induced hemolytic anemia. *J Agric Food Chem* 42;959–962, 1994
Munson RW Jr: *Hemerocallis, the Daylily.* Timber Press, Portland, Oreg, 1989

Murray SS, Kramlinger KG, McMichan JC, Mohr DN: Acute toxicity after excessive ingestion of colchicine. *Mayo Clin Proc* 58;528–532, 1983
Nagaratnam N, DeSilva DPKM, DeSilva N: Colchicine poisoning following ingestion of *Gloriosa superba* tubers. *Trop Geogr Med* 25;15–17, 1973
Nagata R, Izumi K: Veratramine-induced behavior associated with serotonergic hyperfunction in mice. *Jpn J Pharmacol* 55;129–137, 1991
Nakano K, Kashiwada Y, Nohara T: Steroidal saponins and sapogenins of underground parts of *Trillium kamtschaticum* Pall. *Yakugaku-Zasshi* 102;1031–1035, 1982 (*Chem Abstr* 98:104289r, 1983)
Nanasi PP, Kiss T, Danko M, Lathrop DA: Different actions of aconitine and veratrum alkaloids on frog skeletal muscle. *Gen Pharmacol* 21;863–868, 1990
Narumi Y: The effect on frogs of some alkaloids of *Fritillaria verticillata* Wild. *Tohoku J Exp Med* 26;365–380, 1935 (*Chem Abstr* 29;8132, 1935)
Nel PW, Anitra Schultz R, Jordaan P, Anderson LAP, Kellerman TS, Reid C: Cardiac glycoside poisoning in sheep caused by *Urginea physodes* (Jacq.) Bak. and the isolated physodine A. *Onderstepoort J Vet Res* 54;641–644, 1987
Nelson SB: Feeding wild plants to sheep. USDA Bur Anim Ind Ann Rep 15, pp421–425, 1898
Neuss N: A new alkaloid from *Amianthium muscaetoxicum* Gray. *J Am Chem Soc* 75;2772–2773, 1953
Nickerson M, Ruedy J: Antihypertensive agents and the drug therapy of hypertension. In *The Pharmacological Basis of Therapeutics,* 5th ed, Goodman LS, Gilman A eds, Macmillan, New York, pp705–726, 1975
Niemann KW: Report of an outbreak of poisoning in the domestic fowl, due to death camas. *J Am Vet Med Assoc* 73;627–630, 1928
Noden DM, deLahunta A: *The Embryology of Domestic Animals.* Williams & Wilkins, Baltimore, 1985
Nohara T, Miyahara K, Kawasaki T: Steroid saponins and sapogenins of underground parts of *Trillium kamtschaticum.* 2. Pennogenin and kryptogenin 3-O-glycosides, and related compounds. *Chem Pharm Bull* 23;872–885, 1975
Oda Y, Minami K: Isolation and characterization of a lectin from tulip bulbs, *Tulipa gesneriana. Eur J Biochem* 159;239–246, 1986
Ogawa E, Akahori F, Kobayashi K: In vitro studies on the breakdown of canine erythrocytes exposed to the onion extract. *Jpn J Vet Sci* 47;719–729, 1985
Ogawa E, Shinoki T, Akahori F, Masaoka T: Effect of onion ingestion on anti-oxidizing agents in dog erythrocytes. *Jpn J Vet Sci* 48;685–691, 1986
Okanishi T, Acahori A, Yasuda F, Takeuchi Y, Iwao T: Steroidal sapogenins of sixteen Liliaceae plants. *Chem Pharm Bull* 23;575–579, 1975
Overton SG: Dermatitis from handling flower bulbs. *Lancet* 2;1003, 1926
Pahwa R, Doharey RB, Varma BK: Laboratory evaluation of Silmurin (1% technical) powder against black rat, *Rattus rattus* (L). *Int Pest Control* 24;18–19, 1982
Pammel LH: *A Manual of Poisonous Plants.* Torch Press, Cedar Rapids, Iowa, 1911
Panariti E: Meadow saffron *(Colchicum autumnale)* intoxication in a nomadic Albanian sheep flock. *Vet Hum Toxicol* 38;227–228, 1996
Panter KE, James LF: Death camas—early grazing can be hazardous. *Rangelands* 11;147–149, 1989
Panter KE, Smart RA, Duelke B: Death camas poisoning in sheep: a case report. *Vet Hum Toxicol* 29;45–48, 1987
Parker KW: *Observation of Feeding Habits of Sheep on the Range.* N M Agric Exp Stn Bull 241, pp41–53, 1936
Perkins KD, Payne WW: *Guide to the Poisonous and Irritant Plants of Florida.* Fla Coop Ext Serv Circ 441, 1978
Pierce KR, Joyce JR, England RB, Jones LP: Acute hemolytic anemia caused by wild onion poisoning in horses. *J Am Vet Med Assoc* 160;323–327, 1972
Preece SH Jr: A cytotaxonomic study of the genus *Zigadenus* (Liliaceae). PhD dissertation, Washington State Univ, Pullman, 1956

Quatrehomme G, Bertrand F, Chauvet C, Ollier A: Intoxication from *Veratrum album. Hum Exp Toxicol* 12;111–115, 1993

Reid TL, Smith AG: The alkaloids of *Zygadenus gramineus. Proc S D Acad Sci* 35;124–130, 1956

Reynard GB, Norton JBS: *Poisonous Plants of Maryland in Relation to Livestock.* Md Agric Exp Stn Tech Bull A–10, 1942

Ringer S, Morshead EA: On the physiological action of narcissia, an alkaloid obtained from the bulb of the common daffodil *(Narcissus pseudo-narcissus). J Physiol* 1;437–451, 1879

Ripperger H, Schreiber K: *Solanum* steroid alkaloids. In *The Alkaloids,* vol 19, Rodrigo RGA ed, Academic Press, New York, pp81–192, 1981

Robertson JE, Christopher MM, Rogers QR: Heinz body formation in cats fed baby food containing onion powder. *J Am Vet Med Assoc* 212;1260–1266, 1998

Rochdi M, Sabouraud A, Baud FJ, Bismuth C, Scherrmann JM: Toxicokinetics of colchicine in humans: analysis of tissue, plasma, and urine data in ten cases. *Hum Exp Toxicol* 11;510–516, 1992

Rowe LD, Norman JO, Corrier DE, Casteel SW, Rector BS, Bailey EM, Schuster JL, Reagor JC: Photosensitization of cattle in southeast Texas: identification of phototoxic activity associated with *Cooperia pedunculata. Am J Vet Res* 48;1658–1661, 1987

Scherrmann JM, Sabouraud A, Urtizberea M, Rouzioux J, Lang J, Bismuth C: Clinical use of colchicine-specific Fab fragments in colchicine poisoning [abstract]. *Vet Hum Toxicol* 34;334, 1992

Schonbeck F, Schroeder C: Role of antimicrobial substances (tuliposides) in tulips attacked by *Botrytis* spp. *Physiol Plant Pathol* 2;91–99, 1972

Schouwenburg S van: Hemolytic anemia in a miniature dachshund caused by intake of large amounts of onions *(Allium cepa)* [letter]. *J S Afr Vet Med Assoc* 53;212, 1982

Schrutka-Rechtenstamm R, Kopp B, Loeffelhardt W: Origin of cardenolides in rhizomes and roots of *Convallaria majalis* L. iogenesis in situ or transport from leaves. *Z Naturforsch Reiche C* 40;627–630, 1985a

Schrutka-Rechtenstamm R, Kopp B, Loeffelhardt W: Turnover of cardenolides in *Convallaria majalis. Planta Med* 51;387–390, 1985b

Schrutka-Rechtenstamm R, Kopp B, Loeffelhardt W: Bioconversions leading to minor cardiac glycosides in *Convallaria majalis. Phytochemistry* 25;1107–1109, 1986

Sebrell WH: Public health reports. *U S Public Health Serv Rep* 45 (21);1175–1192, 1930

Shaw AO, Biswell HH, Foster JE, Collins RW: *Some Stock-Poisoning Plants of North Carolina.* N C Agric Exp Stn Bull 342, 1943

Shiao S-H, Shao B-R, Tseng S-Y, Chen C, Chang Y-C, Huang L-S: Studies on *Hemerocallis thunbergii* Baker. 1. The toxicity of *Hemerocallis thunbergii* Baker. *Acta Pharm Sin* 9;208–217, 1962

Shimomura H, Sashida Y, Mimaki Y: Steroidal saponins, pardarinoside A–G from the bulbs of *Lilium pardalinum. Phytochemistry* 28;3163–3170, 1989

Shoppee CW: *Chemistry of the Steroids,* 2d ed. Butterworths, Washington, DC, pp398–432, 1964

Sim FRP, Livett BG, Browne CA, Keeler RF: Studies on the mechanism of *Veratrum* teratogenicity. In *Plant Toxicology, Proceedings of the Australia-USA Poisonous Plants Symposium,* Seawright AA, Hegarty MP, James LF, Keeler RF eds, Animal Research Institute, Yeerongpilly, Brisbane, Australia, pp344–348, 1985

Singh B, Rastogi RP: Cardenolides—glycosides and genins. *Phytochemistry* 9;315–331, 1970

Slob A, Jekel B, de Jong B, Schlatmann E: On the occurrence of tuliposides in the Liliiflorae. *Phytochemistry* 14;1997–2005, 1975

Smith AG, Smith RL: Identification of isobutyric acid obtained from *Amianthium muscaetoxicum. Chem Ind (London)* 1963(40);1626, 1963

Smith AG, Miller LF, Smith RL: Alkamine G from *Amianthium muscaetoxicum. Chem Ind (London)* 1964(30);1362–1363, 1964

Smith JA, Paterson GR: Rhodexin A and rhodexoside in *Ornithogalum umbellatum. J Pharm Pharmacol* 19;221–225, 1967

Smith RA, Lewis D: Death camas poisoning in cattle. *Vet Hum Toxicol* 33;615–616, 1991
Smith RH: Kale poisoning: the brassica anemia factor. *Vet Rec* 107;12–15, 1980
Solter P, Scott R: Onion ingestion and subsequent Heinz body anemia in a dog: a case report. *J Am Anim Hosp Assoc* 23;544–546, 1987
Southon IW, Buckingham J: *Dictionary of Alkaloids.* Chapman & Hall, London, 1989
Sparnins VL, Barany G, Wattenberg LW: Effects of organosulfur compounds from garlic and onions on benzo(a)pyrene-induced neoplasia and glutathione *S*-transferase activity in the mouse. *Carcinogenesis* 9;131–134, 1988
Spice RN: Hemolytic anemia associated with ingestion of onions in a dog. *Can Vet J* 17;181–183, 1976
Spoerke DG, Smolinske SC: *Toxicity of Houseplants.* CRC Press, Boca Raton, Fla, p104–105, 1990
Spoerke DG, Spoerke SE: Three cases of *Zigadenus* (death camas) poisoning. *Vet Hum Toxicol* 21;346–347, 1979
Stallbaumer M: Onion poisoning in a dog [letter]. *Vet Rec* 108;523–524, 1981
Stannard J: Squill in ancient and medieval materia medica with special reference to its employment for dropsy. *Bull N Y Acad Med* 50;684–713, 1974
Stapczynski JS, Rothstein RJ, Gaye WA, Nieman JT: Colchicine overdosage: report of two cases and review of the literature. *Ann Emerg Med* 10;364–369, 1981
Stearn WT: The Linnean species of *Hyacinthus* (Liliaceae: Hyacinthaceae). *Ann Mus Goulandris* 8;181–122, 1990
Stevens H: Suspected wild garlic poisoning in sheep [letter]. *Vet Rec* 115;363, 1984
Steyn DG: *The Toxicology of Plants in South Africa.* Central News Agency, South Africa, 1934
Strichartz G, Rando T, Wang GK: An integrated view of the molecular toxinology of sodium channel gating in excitable cells. *Annu Rev Neurosci* 10;237–267, 1987
Suzuki K, Kobayashi M, Ito A, Nakgawa M: *Narthecium asiaticum* Maxim. poisoning of grazing cattle: observations on spontaneous and experimental cases. *Cornell Vet* 75;348–365, 1985
Terrien N, Urtizberea M, Scherrmann JM: Reversal of advanced colchicine toxicity in mice with goat colchicine-specific antibodies. *Toxicol Appl Pharmacol* 104;504–510, 1990
Thorp F Jr, Harshfield GS: Onion poisoning in horses. *J Am Vet Med Assoc* 94;52–53, 1939
Thursby-Pelham RHC: Suspected *Scilla non-scripta* (bluebell) poisoning in cattle. *Vet Rec* 80;709–710, 1967
Tomko J, Voticky Z: Steroidal alkaloids: the *Veratrum* and *Buxus* groups. In *The Alkaloids: Chemistry and Physiology,* vol 14, Manske RHF ed, Academic Press, New York, pp1–82, 1973
Tribunskii MP: *Colchicum autumnale* poisoning in lambs. *Veterinariya (Moscow)* 6;71–72, 1970 (*Vet Bull* 41;887, 1971)
Tschesche R, Kammerer F-J, Wulff G, Schonbeck F: Uber die antibiotisch wirksamen substanzen der tulpe *(Tulipa gesneriana). Tetrahedron Lett* 1968;701–706, 1968
Tufenkji AE, Alvinerie M, Pineau T, Boulard C, Galtier P: Incidence of a subclinical fascioliasis on antipyrine clearance and metabolite excretion in sheep. *Xenobiotica* 18;357–364, 1988
Tuncok Y, Kozan O, Cavdar C, Guven H, Fowler J: *Urginea maritima* (squill) toxicity. *Clin Toxicol* 33;83–86, 1995
Turrill WB, Sealy JR: Studies in the genus *Fritillaria* (Liliaceae). *Hooker's Icones Plant* 39;1–280, 1980
Ujvary I, Eya BK, Grendell RL, Toia RF, Casida JE: Insecticide activity of various 3-acyl and other derivatives of veracevine relative to the *Veratrum* alkaloids veratridine and cevadine. *J Agric Fd Chem* 39;1875–1881, 1991
Underhill JE: A case of hellebore poisoning [note]. *Can Field Nat* 73;128–129, 1959
Valenzuela P, Paris E, Oberpauer B, Rios JC, Concha F: Overdose of colchicine in a three-year-old child. *Vet Human Toxicol* 37;366–367, 1995

Van Kampen KR, James LF, Johnson AE: Hemolytic anemia in sheep fed wild onion *(Allium validum). J Am Vet Med Assoc* 156;328–332, 1970

Verbiscar AJ, Patel J, Banigan TF, Schatz RA: Scilliroside and other scilla compounds in red squill. *J Agric Food Chem* 34;973–979, 1986

Verhoeff J, Hajer R, van den Ingh TSGAM: Onion poisoning of young cattle. *Vet Rec* 117;497–498, 1985

Vigneau CH, Tsao J, Chamaillard C, Galzot J: Accidental absorption of daffodils *(Narcissus jonquilla):* two common intoxications [abstract]. *Vet Hum Toxicol* 24;35, 1982

Vigneau CH, Tsao J, Ducluzeau R, Galzot J: Deux cas cliniques d' intoxication par le jonquille. *J Toxicol Med* 4;21, 1984

Wagstaff DJ, Case AA: Human poisoning by *Zigadenus. Clin Toxicol* 25;361–367, 1987

Walsh D: Investigation of a dermatitis amongst flower-pickers in the Scilly Islands, the so-called "lily rash." *Br Med J* 2;854–856, 1910

Wang J-H, Humphreys DJ, Stodulski GBJ, Middleton DJ, Barlow RM, Lee JB: Structure and distribution of a neurotoxic principle, hemerocallin. *Phytochemistry* 28;1825–1826, 1989

Wang RY, Morasco R, Henry GC, Hoffman RS, Goldfrank LR: Antidotal efficacy of glutamate and aspartate for colchicine toxicity. *Vet Hum Toxicol* 39;207–210, 1997

Watt JM, Breyer-Brandwijk MG: *The Medicinal and Poisonous Plants of Southern and Eastern Africa,* 2d ed. E & S Livingston, Edinburgh, 1962

Weber ND, Andersen DO, North JA, Murray BK, Lawson LD, Hughes BG: In vitro viricidal effects of *Allium sativum* (garlic) extract and compounds. *Planta Med* 58;417–423, 1992

West E, Emmel MW: *Poisonous Plants in Florida.* Fla Agric Exp Stn Bull 510, 1952

Whitaker JR: Development of flavor, odor, and pungency in onion and garlic. *Adv Food Res* 22;73–133, 1976

Wildman WC: Alkaloids of the Amaryllidaceae. In *The Alkaloids: Chemistry and Physiology,* vol 6, Manske RHF ed, Academic Press, New York, pp289–413, 1960a

Wildman WC: Colchicine and related compounds. In *The Alkaloids: Chemistry and Physiology,* vol 6, Manske RHF ed, Academic Press, New York, p247–288, 1960b

Wildman WC: The Amaryllidaceae alkaloids. In *The Alkaloids: Chemistry and Physiology,* vol 11, Manske RHF ed, Academic Press, New York, pp307–405, 1968

Willaman JJ, Schubert BG: *Alkaloid-Bearing Plants and Their Contained Alkaloids.* USDA Tech Bull 1234, 1961

Willemse T, Vroom MA: Allergic dermatitis in a Great Dane due to contact with hippeastrum. *Vet Rec* 122;490491, 1988

Williams HH, Erickson BN, Beach EF, Macy IG: Biochemical studies of the blood of dogs with *N*-propyl disulfide anemia. *J Lab Clin Med* 26;996–1008, 1941

Wilson T: Poisoning caused by eating daffodil bulbs. *Bull Mo Bot Gard* 12;52, 1924

Yamada M, Nakagawa M, Haritani M, Kobayashi M, Furuoka H, Matsui T: Histopathological study of experimental acute poisoning of cattle by autumn crocus (*Colchicum autumnale* L.). *J Vet Med Sci* 60;949–952, 1998

Yamato O, Maede Y: Susceptibility to onion-induced hemolysis in dogs with hereditary high erythrocyte reduced glutathione and potassium concentrations. *Am J Vet Res* 53;134–137, 1992

Yamato O, Yoshihara T, Ichihara A, Maede Y: Novel Heinz body hemolysis factors in onion *(Allium cepa). Biosci Biotech Biochem* 58;221–222, 1994

Yamato O, Hayashi M, Yamasaki M, Maede Y: Induction of onion-induced haemolytic anaemia in dogs with sodium *n*-propylthiosulphate. *Vet Rec* 142;216–219, 1998

Yuyama M: Oxidative compounds in onion (*Allium cepa* L.): isolation and demonstration of their oxidative damage to erythrocytes. *Jpn J Vet Res* 37;144, 1989

48 Linaceae S.F.Gray

■ Flax Family
Linum

Widely distributed but most common in temperate and subtropical regions, the Linaceae, or flax family, comprises 6–13 genera and 200–300 species. Economically, it is important for its fiber and oil. A few taxa are cultivated as ornamentals. In North America, 4 genera and about 50 species, both native and introduced, are present. One genus is of toxicologic interest:

■ herbs or shrubs; leaves simple; petals 5 or 4, caducous

Plants herbs or subshrubs or shrubs; annuals or biennials or perennials. **Root Systems** with a central taproot. **Leaves** simple; alternate or alternate above and opposite below; sessile or subsessile; blades linear to ovate; stipules absent or present, glandular. **Inflorescences** solitary flowers, or simple or compound cymes; terminal and axillary. **Flowers** perfect; perianths in 2-series; radially symmetrical. **Sepals** 5 or 4; free or fused. **Petals** 5 or 4; caducous; free or fused. **Stamens** 5 or 4; fused by filaments; staminodia absent or present. **Pistils** 1; compound, carpels 2–5; stigmas 2–5, linear or capitate; styles 2–5; free or fused at bases; ovaries superior; locules 4–10; placentation axile. **Fruits** septicidal capsules. **Seeds** 8–10.

Linum L.

■ *L. catharticum*
L. grandiflorum
L. lewisii
L. neomexicanum
L. perenne
L. rigidum
L. usitatissimum
flax
linseed

Taxonomy and Morphology—Comprising 150–200 species in both the Old World and the New World, *Linum,* or flax, is best known for its stem fibers and seed oil. Used since prehistoric times, the durable fibers have high tensile strength and have been used to make linen, writing paper, and cigarette paper. The Egyptians wrapped their mummies in linen cloths. Used for poultices in folk medicine, flaxseed is also the source of linseed oil used in paints, varnishes, and printing inks. About 25 species of the genus are grown as ornamentals in rockeries or as border plants. In North America, approximately 36 species are present (Rogers 1963, 1968). Representative of the genus are the following:

L. catharticum L.	white flax
L. grandiflorum Desf.	flowering flax
L. lewisii Pursh	Lewis flax, blue flax, prairie flax
L. neomexicanum Greene	
L. perenne L.	wild blue flax, prairie flax
L. rigidum Pursh	stiffstem flax, stiffstem yellow flax
L. usitatissimum L.	common flax, linseed

■ erect herbs; leaves sessile, with a single vein; flowers saucer-shaped; petals 5, yellow with or without orange to red center, or red or blue

Plants herbs; annuals or biennials. **Stems** erect; 20–100 cm tall; branched or not branched. **Leaves** sessile; venation a single vein; margins entire or ciliate. **Inflorescences** solitary flowers or cymes. **Flowers** saucer-shaped. **Sepals** 5; free. **Petals** 5; caducous;

free or fused; yellow or yellow with an orange to red center or red or blue. **Stamens** 5. **Pistils** with carpels 5; stigmas 5; styles 5; locules 5 or 10 by development of additional septa. **Capsules** dehiscent into 10 or 5 parts. **Seeds** flat; oily.

Because the species cited above are merely representative of the genus, a taxonomic key is not presented. Should exact identification be required, appropriate manuals or the treatments of Rogers (1963, 1968) should be consulted.

Distribution and Habitat—*Linum* is distributed across the continent. Some species have fairly broad distributions; others are more restricted. Introductions from the Old World—e.g., *L. grandiflorum* and *L. rigidum*—have naturalized, as has *L. usitatissimum,* the commercial flax, which has been cultivated for years and is now widely distributed across Canada and the northern United States.

Disease Problems—Flax has been cultivated for millennia for its fibers and oil (Steyn 1934). Linseed oil is derived by pressing the seeds of *L. usitatissimum;* the residue remaining constitutes linseed meal, which is used to make cakes to feed livestock. Wild flax has also been widely used by members of Indian tribes as a source of fiber, and the seeds are considered to be pleasant-tasting edibles (Ebeling 1986). Feeding the seeds of *L. usitatissimum* to 7-day-old chicks at 2% or 10% of their diet for 6 weeks produced good growth rates but also some adverse effects, such as fatty livers and mild neutrophilic infiltration in the renal cortex and the intestinal lamina propria (Bakhiet & Adam 1995). Wild flaxes have long been suspected to be toxic because of the presence of cyanide, but this belief was based on anecdotal reports, some of which are difficult to interpret (Eggleston et al. 1930). For example, plants of *L. neomexicanum,* growing in great abundance on the site of previous livestock losses were negative when tested for HCN potential (Eggleston et al. 1930). However, aqueous and alcoholic extracts of the plants were highly lethal in rabbits and mice. The toxic effects were not consistent with HCN, because the onset of signs was delayed and animals died in about 18 hours.

Disease Genesis—Although the risk of intoxication is typically low, members of the genus do contain appreciable amounts of cyanogenic glucosides, such as linamarin and lotaustralin in the foliage, and the glucosidase linamarase needed for their hydrolysis to HCN (Butler 1965; Gibbs 1974; Seigler 1977). The predominant glucosides in the seeds are linustatin and neolinustatin, which have been found at concentrations of 2,000–3,500 ppm and 90–200 ppm, respectively, in Canadian cultivars (Smith et al. 1980; Oomah et al. 1992). Linamarin concentrations in plants were only 140–300 ppm. Linustatin and neolinustatin are bisglucosides of linamarin and lotaustralin, respectively (Smith et al. 1980; Fan & Conn 1985). Thus, the distribution of glycosides in *Linum* is similar to that in the Rosaceae, in which the bioside amygdalin occurs in the seeds and the monoside prunasin is in the leaves. Although cyanide is present, only in certain conditions does it become a significant hazard. As with the grasses, HCN potential of *Linum* is very high in the seedlings and the first few weeks of growth (1,000+ ppm) but then declines rapidly (Trione 1960). Flax straw is said to be a particular risk if it is second growth or contains green, immature stems (Stevens 1933). Chaff or green leaves at harvest and late-season volunteer flax pasture are also considered to be risks. Little risk is associated with the grazing of wild flax plants, as their densities are seldom sufficient to represent a significant problem.

Although rare, losses of both horses and calves has occurred when they have eaten linseed cake, especially when it was soaked in water at the time of feeding (Dunne 1924; Montgomerie 1924; Moore 1924). The clinical signs and their rapidity of onset

Linum lewisii

■ possible cyanide intoxication

Linum catharticum

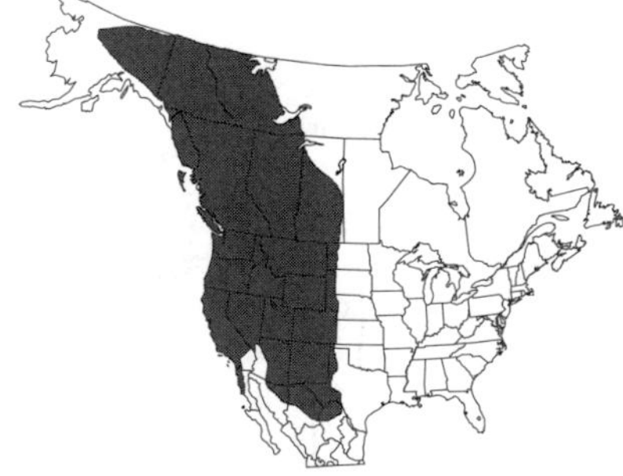

Linum lewisii

■ cyanogenic monoglycosides; linamarin and lotaustralin in foliage; bisglucosides linustatin and neolinustatin in seeds

$$\text{Glucose-O}-\underset{CH_3}{\overset{C\equiv N}{C}}-CH_3$$

Linamarin

$$\text{Glucose-O}-\underset{CH_2CH_3}{\overset{C\equiv N}{C}}-CH_3$$

Lotaustralin

were consistent with cyanide as the cause of death. The warm water apparently provided a medium for hydrolysis of the glucosides (Dunne 1924; Montgomerie 1924). In one instance, the meal contained 480 ppm HCN potential. It was suggested that heating to 100°C (boiling) would eliminate the risk.

The toxicity described for *L. neomexicana* appeared to be due to a saponin-like glycoside that was soluble in water or alcohol but not in typical organic solvents (Eggleston et al. 1930).

- lignans, metabolites of secoisolariciresinol possible estrogens

Flaxseed contains lignans that are biotransformed in the human intestinal tract to weakly estrogenic lignans (Phipps et al. 1993). Obermeyer and co-workers (1995) suggested the possibility that enterodiol and enterolactone—metabolites of secoisolariciresinol, the primary lignan in flaxseed—act as estrogens, such as diethylstilbesterol. This is of some concern for women, but there are no indications of any potential risk for grazing livestock.

Interestingly, the glucosides in linseed meal, especially linustatin and neolinustatin, seem to provide a protective effect against selenium toxicity (Halverson et al. 1955; Palmer et al. 1980; Smith et al. 1980).

- abrupt onset, distress, weakness, ataxia, collapse, seizures

Clinical Signs—Abrupt onset of apprehension and distress occurs within a few minutes of the animal's grazing the contaminated forage or the meal. This is quickly followed by weakness, incoordination, and labored respiration. If the intoxication is severe, the animal may be unable to stand and may lay in lateral recumbency with periodic paddling and tetanic seizures. The entire sequence of signs may occur over a 5-to-15-minute period, with either death or recovery at the end of this time.

- no lesions

Pathology—Few distinctive changes occur. For the most part the changes are limited to congestion of the viscera and scattered, small splotchy hemorrhages. The coarse stems may be identifiable when the rumen contents are examined during necropsy.

- sodium thiosulfate

Treatment—The primary antidote for ruminants is sodium thiosulfate given i.v. at a dose of 0.25–0.5 g/kg b.w. Sodium nitrite, 10 mg/kg b.w. i.v., can be used to augment the effects of the sodium thiosulfate, but it is not required for effective treatment of ruminants. A more complete discussion of aspects of treatment is presented in the treatments of the Rosaceae (see chapter 64).

OH, CH_2OH, OCH_3, C-C-C-C, OCH_3, CH_2OH, OH

Secoisolariciresinol

OH, CCH_3, C=C, CCH_3, OH

Diethylstilbesterol

References

Bakhiet AO, Adam SE: Response of Bovans chicks to low dietary levels of *Linum usitatissimum* seeds. *Vet Hum Toxicol* 37;534–536, 1995

Butler GW: The distribution of the cyanoglucosides linamarin and lotaustralin in higher plants. *Phytochemistry* 4;127–131, 1965

Dunne GT: Poisoning in calves by nascent hydrocyanic acid evolved by cake in solution. *Vet J* 80;40–42, 1924

Ebeling W: *Handbook of Indian Foods and Fibers of Arid America.* Univ California Press, Berkeley, 1986

Eggleston WW, Black OF, Kelly JW: *Linum neomexicanum* (yellow pine flax) and one of its poisonous constituents. *J Agric Res* 41;715–718, 1930

Fan TW-M, Conn EE: Isolation and characterization of two cyanogenic -glucosidases from flax seeds. *Arch Biochem Biophys* 243;361–373, 1985

Gibbs RD: *Chemotaxonomy of Flowering Plants,* vol 3. McGill-Queens Univ Press, Montreal, pp1621–1623, 1974

Halverson AW, Hendrick CM, Olson OE: Observations on the protective effect of linseed oil meal and some extracts against chronic selenium poisoning in rats. *J Nutr* 56;51–60, 1955

Montgomerie RF: Hydrocyanic acid generated from linseed cake meal: a case of poisoning in calves [letter]. *Vet J* 80;311–314, 1924

Moore J: Prussic acid poisoning from the use of linseed cake. *Vet J* 80;33–34, 1924

Obermeyer WR, Musser SM, Betz JM, Casey RE, Pohland AE, Page SW: Chemical studies of phytoestrogens and related compounds in dietary supplements: flax and chaparral. *Proc Soc Exp Biol Med* 208;6–12, 1995

Oomah BD, Mazza G, Kenaschuk EO: Cyanogenic compounds in flaxseed. *J Agric Food Chem* 40;1346–1348, 1992

Palmer IS, Olson OE, Halverson AW, Miller R, Smith C: Isolation of factors in linseed oil meal protective against chronic selenosis in rats. *J Nutr* 110;145–150, 1980

Phipps WR, Martini MC, Lampe JW, Slavin JL, Kurzer MS: Effect of flax seed ingestion on the menstrual cycle. *J Clin Endocrinol Metab* 77;1215–1219, 1993

Rogers CM: Yellow flowered species of *Linum* in eastern North America. *Brittonia* 15;97–122, 1963

Rogers CM: Yellow-flowered species of *Linum* in Central America and western North America. *Brittonia* 20;107–135, 1968

Seigler DS: The naturally occurring cyanogenic glycosides. In *Progress in Phytochemistry,* Reinhold L, Harbone JB, Swain T eds, Pergamon Press, Oxford, pp83–120, 1977

Smith CR Jr, Weisleder D, Miller RW, Palmer IS, Olson OE: Linustatin and neolinustatin: cyanogenic glycosides of linseed meal that protect animals against selenium toxicity. *J Org Chem* 45;507–510, 1980

Stevens OA: *Poisonous Plants and Plant Products.* N D Agric Exp Stn Bull 265, 1933

Steyn DG: *The Toxicology of Plants in South Africa.* Central News Agency, South Africa, 1934

Trione EJ: The HCN content of flax in relation to flax wilt resistance. *Phytopathology* 50;482–486, 1960

49 Loganiaceae Mart.

- Logania Family

  Gelsemium
  Spigelia
- well known for alkaloids in Old World and tropical taxa; strychnine, curare
- herbs, vines, or shrubs; leaves simple, opposite, with stipules; petals 4 or 5, fused

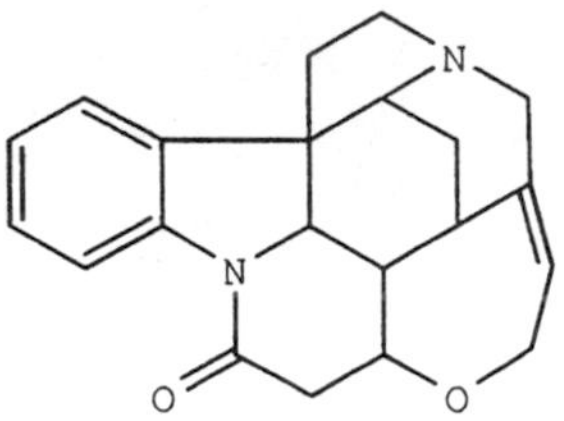

Most abundant in tropical and subtropical regions, the Loganiaceae, or logania family, comprises 20–32 genera and 500–800 species. Taxonomists differ as to whether it should be broadly circumscribed or divided into 2–6 smaller families. Many of its species are cultivated as ornamentals, most notably *Buddleja* (butterfly bush), *Gelsemium* (yellow jessamine), and *Spigelia* (pinkroot or Indian pink). In the tropics, some trees provide timber. The family, however, is most famous for the alkaloids derived from various species of *Strychnos*. The best known of these compounds are the strychnine alkaloids from Old World species and the curare alkaloids from tropical species of the New World (Bisset & Phillipson 1976). Several of the 4 or 5 genera present in North America also contain toxic alkaloids.

Strychnine

Plants herbs or woody vines or shrubs; perennials or annuals. **Leaves** simple; opposite; margins entire; stipules present, membranous or reduced to inconspicuous ridge between leaves. **Inflorescences** racemose or spicate cymes or solitary flowers; bracts and bracteoles present or absent. **Flowers** perfect; perianths in 2-series. **Sepals** 4 or 5; fused or free. **Corollas** radially symmetrical; funnelform or salverform or tubular. **Petals** 4 or 5; fused. **Stamens** 4 or 5; epipetalous. **Pistils** 1; compound, carpels 2 or 3; stigmas 1 or 2, capitate or lobed; styles 1; ovaries superior or partially inferior; lobed or flattened; locules 2 or 3; placentation axile or distally parietal. **Fruits** capsules; septicidal or loculicidal. **Seeds** numerous; winged or not winged.

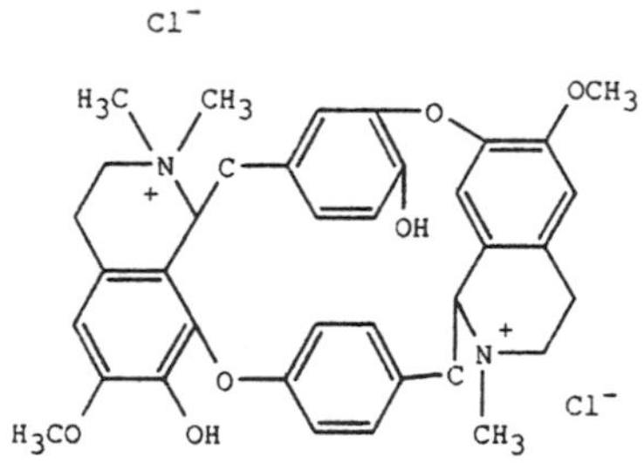

Tubocurarine chloride

Gelsemium Juss.

Taxonomy and Morphology—A small genus of only 3 species—2 in southeastern North America and 1 in eastern Asia and Indonesia—*Gelsemium* is commonly known as yellow jessamine. Its name is a Latinization of *gelsomino,* the Italian name for the true jasmine. The 2 North American species are the following (see plate 49.1):

G. rankinii Small	evening trumpet-flower
G. sempervirens (L.) W.T.Aiton	yellow jessamine, Carolina jessamine, wild jessamine, poor man's rope, evening trumpet vine, woodbine, false jasmine, Carolina wild woodbine

- *G. rankinii*
  G. sempervirens

  false jasmine
  jessamine
  poor man's rope
  trumpet flower
  trumpet vine
  wild woodbine
  woodbine

Plants woody vines; perennials; evergreen. **Stems** twining; climbing or trailing; to 20 m long; wiry; reddish brown. **Leaves** short petiolate; blades lanceolate to elliptic; surfaces glabrous; apices acute or acuminate; bases rounded to cuneate; stipules reduced to inconspicuous ridge between leaves. **Inflorescences** 2-to-5-flowered cymes or solitary flowers; bracts of pedicels scalelike. **Flowers** showy; fragrant or not fragrant; funnelform. **Sepals** 5; free. **Petals** 5; fused; yellow. **Stamens** 5; anthers sagittate. **Pistils**

with styles 4-lobed. **Capsules** ovoid or oblong; flattened; short-beaked. **Seeds** flattened; winged or not winged; dull brown.

1. Flowers fragrant. Sepal apices obtuse.
   Capsule beaks 1–2 mm long. Seeds winged . . . . *G. sempervirens*
1. Flowers not fragrant. Sepal apices acute.
   Capsule beaks 2–3 mm long. Seeds not winged . . . . *G. rankinii*

Distribution and Habitat—*Gelsemium sempervirens* is found in a variety of sites—open woods, thickets, swamps, low areas, and open fields—of the southeastern United States to Mexico and Guatemala. It also is cultivated as a cover for porches, fences, or banks in the southeastern states. *Gelsemium rankinii* is rare on the southeastern coastal plains.

Disease Problems—*Gelsemium* is well known as a neurotoxic genus in the southeastern United States, with all livestock species at risk (Shaw et al. 1943; West & Emmel 1952). Based upon an occurrence in the 1800s in which several children died, honey from its flowers was thought to be lethal; however, the accusation was never substantiated (Chesnut 1898; Patwardhan & White 1973). Dosage of 180 g of fresh leaves (approximately 5–10% of the diet) fed to chickens for 15 days caused death 5–10 days after the feeding ceased. The effects of the toxicant(s) were apparently cumulative (West & Emmel 1952). Conversely, leaves and stems fed to young chicks and poults for 40 days at up to 10% of the diet failed to cause any adverse effects other than depressed weight gain (Williamson et al. 1964). There seems to be little question of the toxicity of the genus because the Asian *G. elegans* was reportedly used in the 1800s to execute criminals in Hong Kong (Chesnut 1898). Because the plants are evergreen, they are especially hazardous in the winter when other forage is limited.

Disease Genesis—Alkaloids originally identified to be present in the genus were gelsemine and gelseminine (Pammel 1911; Schun & Cordell 1985). Gelsemine has been confirmed as the major alkaloid, whereas gelseminine has been discovered to be only an isolation artifact of gelsemine (Saxton 1965). Sempervirine, a complex indole, is also present and a significant component. Numerous minor alkaloids are also present, including gelsemicine (the most toxic) and gelsedine, both very complex indoles, and gelseverine (Schun & Cordell 1987).

Gelsemine

Gelsemicine

Sempervirine

- perennial, evergreen, twining, woody vines; flowers showy, funnelform; petals 5, fused, yellow

Gelsemium

- open woods, fields, swamps
- neurologic, seizures; all animal species
- potent neurotoxic alkaloids; gelsemine; also indole sempervirine

Gelsemium rankinii

Gelsemium sempervirens

These alkaloids are found in all organs—flowers, leaves, and especially the roots. Extracted from the roots, gelsemine was once used medicinally, but its misuse caused nausea, retching, laryngeal and pharyngeal spasms, and neurologic signs (Millspaugh 1974). Because the toxins cause paralysis of nerve endings, elixers made from the roots are used to treat neuralgia. Root extracts of *Gelsemium* also have been used in homeopathic medicine to induce catalepsy, a hypertonic trancelike state similar to but less severe than that produced by *Cannabis* (Sukul et al. 1986).

■ abrupt onset, weakness, clonic seizures, respiratory paralysis

Clinical Signs—Following continued ingestion of *Gelsemium*, there is abrupt onset of muscular weakness and clonic convulsive movements of the head, thoracic limbs, and pelvic limbs; a decrease in respiration rate and temperature; and eventually respiratory paralysis. Some animals may have considerable difficulty in swallowing. There also may be vision problems accompanied by pain. The appearance of clinical signs is usually indicative of a poor prognosis.

■ activated charcoal, sedation, respiratory assistance

Treatment—No clearly effective treatment is known. A number of different approaches to treatment have been suggested but without confirmation of their effectiveness. These include morphine, atropine, electroshock, administration of *Conium maculatum*, and administration of coffee (Chesnut 1898; Pammel 1911). Oral activated charcoal may be beneficial, and in severe intoxications respiratory assistance may be necessary.

Spigelia L.

■ *S. anthelmia*
S. gentianoides
S. lindheimeri
S. loganioides
S. marilandica
S. texana

Carolina pink
Indian pink
pinkroot
worm grass

Taxonomy and Morphology—An American genus of 35–50 species, *Spigelia*, commonly known as Indian pink or pinkroot, is classified by some taxonomists in its own family. Its name honors Adrian Van den Spieghel, a Dutch physician and botanist of the late sixteenth century. Six species are present in North America:

S. anthelmia L.	West Indian pinkroot
S. gentianoides Chapm. ex A.DC.	
S. lindheimeri A.Gray	
S. loganioides (Torr. & A.Gray ex Endl. & Fenzl) A.DC.	
S. marilandica (L.) L.	Carolina pink, pinkroot, Indian pink, worm grass
S. texana (Torr. & A.Gray) A.DC.	Texas pinkroot

■ herbs; petals 5, yellow, red, or purplish

Plants herbs; perennials or annuals. **Stems** erect or spreading; 10–90 cm tall; branched or not branched. **Leaves** sessile or short petiolate; blades ovate to elliptic or obovate; apices acute or obtuse; bases rounded; stipules small, membranous, caducous. **Inflorescences** 1-sided racemose or spicate cymes or rarely solitary flowers; bracts present or absent. **Flowers** funnelform or salverform or tubular. **Sepals** 5; fused. **Petals** 5; yellow or red or purplish. **Stamens** 5. **Pistils** with styles filiform; jointed near middles or at bases. **Capsules** globose; 2-lobed; valves dehiscent from cup-shaped bases. **Seeds** subglobose, angular.

Spigelia anthelmia

1. Styles jointed near middle.
   2. Corollas 8–10 mm long . . . . *S. lindheimeri*
   2. Corollas 12–14 mm long.
      3. Plants of Florida . . . . *S. loganioides*
      3. Plants of Texas and Mexico . . . . *S. texana*

1. Styles jointed near base.
   4. Plant annuals. Involucre of bracts present . . . . *S. anthelmia*
   4. Plants perennials. Involucre of bracts absent.
      5. Corollas red outside and yellow inside; 40–50 mm long. Stamens exserted beyond corollas . . . *S. marilandica*
      5. Corollas white or purplish outside and inside; 15–25 mm long. Stamens included within corollas . . . *S. gentianoides*

Spigelia anthelmia

Distribution and Habitat—*Spigelia anthelmia, S. gentianoides,* and *S. loganioides* are restricted in distribution to southern Florida and the Keys. *Spigelia lindheimeri* is a species of rocky breaks and prairies of southern Texas and adjacent northern Mexico, whereas *S. texana* is restricted to wooded slopes and riverbanks in southern Texas. *Spigelia marilandica* is the most widely distributed species and is encountered in moist, rich soils of woods from Maryland, Ohio, Missouri, and Oklahoma south to Florida and Texas.

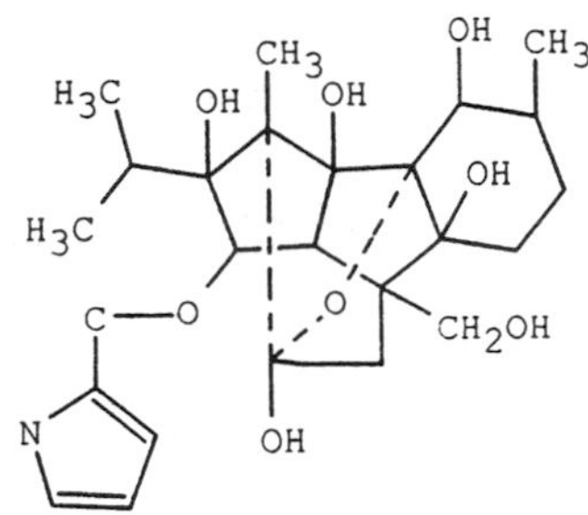

Spiganthine

Isochinolin

Ryanodine

Disease Problems and Genesis—Species of *Spigelia* have not presented problems in North America, but they can be considered to be toxic. Adverse reactions such as weakness and muscular rigidity in humans and livestock have occurred after administration of extracts of *S. anthelmia* to control intestinal worms. This vermifuge contains several bioactive compounds, primarily alkaloids such as spigeline, spiganthine, and isochinolin (Wagner et al. 1986; Achenbach et al. 1995). The latter two are unique cardiotoxic alkaloids. Spiganthine is of special interest because it is the main component of ethanolic extracts of the aerial portions of the plants that causes a delay in contraction development of ventricular papillary muscle and a slowed rate of development of myocardial contractions in guinea pigs (Achenbach et al. 1995). Spiganthine is similar chemically to ryanodine isolated from *Ryania speciosa* in the Flacourtiaceae, and it may act in a similar manner, possibly by maintaining calcium channels in the open position to slow accumulation of intracellular calcium and prevent muscle relaxation (Fleischer et al. 1985; Pessak et al. 1985; Vierling 1988). Induction of leakage of calcium from the sarcoplasmic reticulum of heart muscle appears to provide effects that are antagonistic to those of the cardiotonic *Digitalis* glycosides (Jenden & Fairhurst 1969). These mechanisms are consistent with the actions of spiganthine as well as ryanodine.

- cardiotoxic effects; alkaloids, spiganthine, isochinolin; decreased intracellular myocardial calcium

Clinical Signs, Pathology, and Treatment—Overdosage of the vermifuge drug produces dizziness, dilated pupils, diminished vision, eye and facial spasms, vomiting, seizures, and death (Blohm 1962; Perkins & Payne 1978). Pathologic changes have not been described. Treatment is based upon relief of the symptoms and may include administration of *Digitalis* glycosides.

- intoxications not reported; weakness, depression

References

Achenbach H, Hubner H, Vierling W, Brandt W, Reiter M: Spiganthine, the cardioactive principle of *Spigelia anthelmia. J Nat Prod* 58;1092–1096, 1995

Bisset NG, Phillipson JD: The Asian species of *Strychnos*. 4. The alkaloids. *Lloydia* 39;263–325, 1976

Blohm H: *Poisonous Plants of Venezuela*. Harvard Univ Press, Cambridge, 1962

Chesnut VK: *Principal Poisonous Plants of the United States*. USDA Div Bot Bull 20, 1898

Fleischer S, Ogunbunmi EM, Dixon MC, Fleer EAM: Localization of Ca^{2+} release channels with ryanodine in junctional terminal cisternae of sarcoplasmic reticulum of fast skeletal muscle. *Proc Nat Acad Sci* 82;7256–7259, 1985

Jenden DJ, Fairhurst AS: The pharmacology of ryanodine. *Pharmacol Rev* 21;1–25, 1969

Millspaugh CF: *American Medicinal Plants.* Dover, New York, pp517–521, 1974 (reprinted from 1892)

Pammel LH: *A Manual of Poisonous Plants.* Torch Press, Cedar Rapids, Iowa, pp688–689, 1911

Patwardhan VN, White JW Jr: Problems associated with particular foods. In *Toxicants Occurring Naturally in Foods,* 2d ed, National Academy of Sciences, Washington, DC, pp477–507, 1973

Perkins KD, Payne WW: *Guide to the Poisonous and Irritant Plants of Florida.* Fla Coop Ext Serv Circ 441, 1978

Pessak IN, Waterhouse AL, Casida JE: The calcium-ryanodine receptor complex of skeletal and cardiac muscle. *Biochem Biophys Res Comm* 128;449–456, 1985

Saxton JE: Alkaloids of *Gelsemium* species. In *The Alkaloids: Chemistry and Physiology,* vol 8, Manske RHF ed, Academic Press, New York, pp93–117, 1965

Schun Y, Cordell GA: Studies on the NMR spectroscopic properties of gelsemine—revisions and refinements. *J Nat Prod* 48;969–971, 1985

Schun Y, Cordell GA: Cytotoxic steroids of *Gelsemium sempervirens. J Nat Prod* 50;195–198, 1987

Shaw AO, Biswell HH, Foster JE, Collins RW: *Some Stock-Poisoning Plants of North Carolina.* N C Agric Exp Stn Bull 342, 1943

Sukul NC, Bala SK, Bhattacharyya B: Prolonged cataleptogenic effects of potentized homeopathic drugs. *Psychopharmacology* 89;338–339, 1986

Vierling W: Ryanodine in mammalian heart ventricular muscle: indication for the induction of calcium leakage from the sarcoplasmic reticulum. *Eur J Pharmacol* 145;329–334, 1988

Wagner H, Seegert K, Gupta MP, Esposito Avella M, Solis P: Cardiotonically active principles from *Spigelia anthelmia. Planta Med* 52;378–381, 1986

West E, Emmel MW: *Poisonous Plants in Florida.* Fla Agric Exp Stn Bull 510, 1952

Williamson JH, Craig FR, Barber CW, Cook FW: Some effects of feeding *Gelsemium sempervirens* (yellow jessamine) to young chickens and turkeys. *Avian Dis* 8;183–190, 1964

Malvaceae Juss.

50

Comprising 75–80 genera and 1,000–1,500 species, the Malvaceae, or mallow family, is cosmopolitan in distribution but particularly abundant in the American tropics. Of considerable economic importance, it includes a number of well-known ornamentals such as *Alcea* (hollyhock), *Malva* (mallow), *Hibiscus* (rose-of-Sharon), and numerous cultivars of *Hibiscus. Abelmoschus* (okra) is a cultivated member, whereas *Abutilon* (velvetleaf) and *Sida* (sida) are troublesome weeds. Most important is the genus *Gossypium,* whose seed fibers provide cotton. In North America, approximately 40 genera and 266 species, both native and introduced, are present.

■ Mallow Family

Abutilon
Gossypium
Malva
Modiola

Plants herbs or shrubs or rarely small trees; perennials or biennials or annuals; herbage typically stellate or lepidote; sap viscous, mucilaginous. **Leaves** simple; alternate; venation palmate; margins entire or crenulate or serrulate or palmately lobed or parted; stipules present. **Inflorescences** solitary flowers or simple cymes; axillary or terminal; bracts absent or present, often forming epicalyces. **Flowers** perfect or rarely imperfect; perianths in 2-series. **Sepals** 5; free or fused. **Corollas** radially symmetrical. **Petals** 5; free, but basally fused to the staminal tube; obovate; spreading to ascending; convolute; of various colors. **Stamens** numerous; fused by filaments into a tube. **Pistils** 1; compound, carpels 5 to numerous; stigmas 5 to numerous, capitate or decurrent; styles 5 to numerous; fused or free; ovaries superior, lobed or terete, lobes 5 to numerous; placentation axile. **Fruits** schizocarps or capsules or rarely berries; dehiscent or indehiscent; loculicidal or septicidal. **Seeds** 5 to numerous.

■ herbs, shrubs, or trees; leaves simple, alternate, palmate venation, stipules present; petals 5; stamens numerous, filaments fused into tube; styles 5 to numerous

The Malvaceae is of limited toxicologic importance, except for problems associated with the use of cottonseed meal as food for livestock. Three genera—*Abutilon, Malva, and Modiola*—produce disease problems of lesser importance, whereas other genera such as *Malvella* and *Sida* have only been suspected of causing disease. For example, *Malvella leprosa* (= *Sida leprosa*), commonly known as alkali mallow or whiteweed, was thought to be the cause of death in an episode involving sheep (Fuller & McClintock 1986). Steyn (1934) gave a sheep 200 g of dried leaves, stems, and seeds of the species, but disease signs did not appear. *Sida cordifolia* (heartleaf sida) and *S. rhombifolia* (arrowleaf sida or Queensland hemp) also have been suspected but not confirmed as toxic. Pammel (1911) indicated that chickens that ate capsules of *S. rhombifolia* died; however, later feeding tests by Steyn (1934) were negative. The discovery of ephedrine and gossypol in the aerial portions of the plants, especially the seeds, does suggest the possibility of adverse effects (Chopra et al. 1984). Other species of *Sida,* such as *S. spinosa,* however, do not have pigment glands, and gossypol is present in the seed tissues only at the same low levels (60 ppm) found in glandless cottonseed (Schmidt & Wells 1990).

■ *A. theophrasti*

abutilon hemp	cottonweed
American jute	elephant ears
butter print	Indian mallow
buttonweed	pie marker
China jute	Tientsin jute
chingma	velvetleaf

■ erect, velvety, annual herbs; leaves malodorous; petals yellow

Abutilon theophrasti

■ weed of waste areas

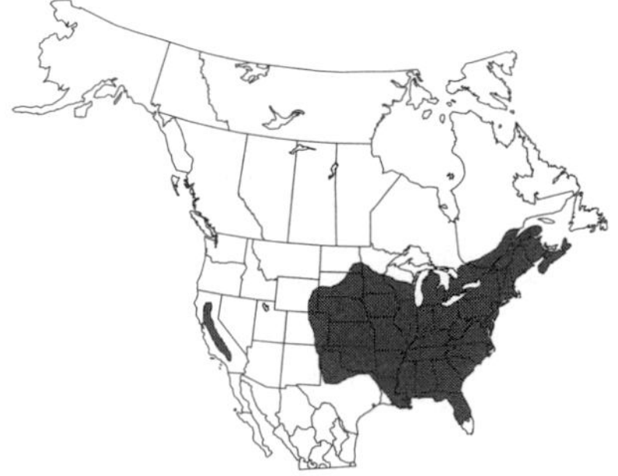

Abutilon theophrasti

Abutilon Mill.

Taxonomy and Morphology—Comprising 100–150 species, *Abutilon,* commonly known as Indian mallow, is widespread in tropical and subtropical regions of both the Old World and the New World. Some South American species and their hybrids are cultivated for their showy flowers, whereas other species are weedy. In North America, 26 native and introduced species are present. Only 1 is of possible toxicological interest:

A. theophrasti Medik. (= *Sida abutilon* L.)	velvetleaf, butter print, cottonweed, Indian mallow, pie marker, China jute, Tientsin jute, elephant ears, chingma, buttonweed, abutilon hemp, American jute

Plants herbs; annuals; from stout taproots; herbage velvety stellate. **Stems** erect; 100–200 cm tall; thick; sparsely branched. **Leaves** malodorous when crushed; long petiolate; blades ovate or cordate to orbicular; apices acuminate; margins entire or crenulate or dentate; stipules deciduous. **Inflorescences** small cymes in axils of upper leaves; bracts absent. **Sepals** fused; lobes acuminate to cuspidate. **Petals** yellow; apices truncate to retuse. **Pistils** with carpels 10–15, beaked, beaks divergent; stigmas capitate; styles clavate or filiform. **Fruits** schizocarps; mericarps villous to hirsute. **Seeds** 2–9 per mericarp; ovate; conspicuously notched; flat; black or gray-brown.

The common names of this taxon reflect its minor economic value as a source of fiber and its morphological features. The large leaves are quite soft to the touch, and the inflorescence in bud resembles the print mold used to stamp butter.

Distribution and Habitat—Native to tropical Asia, *A. theophrasti* was introduced from India and is now naturalized across North America. It is particularly common in waste areas, fields, and pastures and along roadsides. It commonly occurs with *Datura stramonium* as an invading annual of cultivated fields.

Disease Problems—Used in China as a fiber for making rope, bags, nets, and cordage, *A. theophrasti* was introduced into North America for the same purposes by colonists in the late 1700s or early 1800s (Spencer 1984). It soon became a troublesome weed, however, and is now considered a cause of extensive crop losses as an invader in corn, cotton, sorghum, and soybean plantings (Spencer 1984). Although Pammel (1911) thought *A. theophrasti* to be toxic, reported cases of intoxication are few, and even in its native India and China, it is not considered to be a problem (Chopra et al. 1984). In some countries the seeds are eaten (Warwick & Black 1988). However, limited clinical observations by others, and preliminary feeding tests with its foliage, suggested that toxicity was a risk (AA Case, pers. comm., 1979). Sows allowed to forage in an area dominated by mature plants became ill in 1–2 days, and some of them died over a 5-to-10-day period.

Disease Genesis—*Abutilon theophrasti* is a nitrate accumulator, a characteristic that is amplified when the plants grow in livestock areas with ample soil nitrogen. However, the signs and lesions described for pigs were not consistent with nitrate as the cause of death (AA Case, pers. comm., 1979). Rather, they were more suggestive of those produced by gossypol. However, *A. theophrasti* seeds are reported to lack pigment glands, and the levels of gossypol are quite low (Schmidt & Wells 1990). Furthermore, 10% seed in the diet of rats for 90 days caused only a small drop in feed consumption (Dugan & Gumbmann 1990). It thus seems unlikely that the foliage contains appreciable amounts of gossypol.

Clinical Signs—Rather indistinct, the signs of intoxication gradually appear over a period of 24–36 hours. There is depression, reluctance to move about, icterus, and the general appearance of unthriftiness. However, conspicuous abnormalities of any specific system are absent.

- possibly toxic but type not clear
- toxin unknown; nitrate accumulator
- depression, icterus

Pathology—Gross changes include excess fluid in the abdominal and thoracic cavities, icterus, and perhaps a friable liver. Microscopically, there is mild to moderate hepatocellular necrosis and fibrosis.

- gross pathology, fluid accumulation in body cavities

  microscopic, mild liver necrosis

Treatment—Because reports of toxicity are so few, a treatment protocol has not evolved. General supportive and nursing care is suggested as appropriate.

- nursing care

Gossypium L.

Taxonomy and Morphology—Distributed primarily in the tropics of both the Old World and the New World, *Gossypium* is a genus of 39 species, some of which have been cultivated for millennia (Fryxell 1979). Its name is an ancient one used for an Arabic species. The genus provides cotton, the most important of the plant fibers. Each seed has attached to it long fibers called lint or floss and collectively referred to as the staple. Nearly pure cellulose, these fibers are used to make the fabric of commerce. In addition, a dense covering of short hairs, or linters, may be present; their presence and appearance are used in distinguishing species of the genus. Historical records indicate that cotton was cultivated in the Indus Valley of Asia as early as 3,000 B.C. and introduced into Europe in the fifteenth century. Native Americans in Central and South America began cultivation of the New World species at about the same time. Representative of the genus are the following cultivated species:

- *G. arboreum*

  G. barbadense

  G. herbaceum

  G. hirsutum

  cotton
- erect subshrubs or shrubs; leaves petiolate; flowers large, showy; petals yellow, cream, or mauve

G. arboreum L.	Ceylon cotton, Chinese cotton, tree cotton
G. barbadense L.	sea island cotton
G. herbaceum L.	Levant cotton
G. hirsutum L.	upland cotton

Gossypium barbadense

Plants subshrubs or shrubs; perennials; herbage typically glandular-punctate; glabrous or puberulent or hirsute. **Stems** erect; 50–150 cm tall. **Leaves** petiolate; blades ovate to cordate; margins entire to palmately lobed; lobes 3–7; stipules filiform or subulate; deciduous or persistent. **Inflorescences** solitary flowers or small cymes in axils of upper leaves; bracts of epicalyces 3–7, foliaceous. **Flowers** large; showy. **Sepals** fused; lobes long or short; often conspicuously punctate. **Petals** yellow or cream or mauve. **Pistils** with carpels 3–5; stigmas decurrent, 3-to-5-lobed; styles clavate or filiform. **Fruits** capsules ("cotton boll"); locules 3–5. **Seeds** 2 to several per locule; comose, hairs white or brown.

Distribution and Habitat—Species of *Gossypium* occur in both hemispheres. As their common names imply, *G. arboreum* is native to Asia and northeastern Africa, and *G. herbaceum* is native to Asia Minor and Arabia. Indigenous to tropical America, *G. barbadense* and *G. hirsutum* are the two species widely cultivated in North America.

- crop plant

Disease Problems—The genus *Gossypium* is well known for its economic significance as a source of cotton. Less well known, however, is its importance as a food for both

- seed products, cumulative effects; congestive heart failure; pigs and young ruminants

humans and animals. Cottonseed oil is expressed from the seed and used as a cooking oil or made into margarine, and cottonseed meal, the residue from the pressing process, is used as an animal feed. More than 364 kg of cottonseed is produced concurrently with the production of a 227 kg bale of cotton fiber staple (Berardi & Goldblatt 1980). Because of its availability and the economics of its production, cottonseed meal is widely used as a protein feed supplement for livestock, especially in regions of high cotton production. In addition, whole cottonseed, which is considered a high-energy (91% TDN), high-protein (23%), high-fiber (17–20%) and high-fat (23%) feed is often fed to cattle (Lusas & Jividen 1987).

Unfortunately, because of the variability in quality of cottonseed meal and especially in gossypol content, toxicity problems occur, particularly in swine but also occasionally in cattle, especially young calves, and in lambs (Dinwiddie & Short 1911; Hollon et al. 1958). In an extensive series of experiments to isolate and/or eliminate the toxic effects of cottonseed, Dinwiddie & Short (1911) identified the lesions in swine as those of heart failure, but the combination of weakness and ocular lesions also suggested that vitamin A deficiency was a complicating factor. The toxic dosage of cottonseed meal was 6-to-7-fold greater for cattle than for swine. The toxic dose of cottonseed meal was later determined as 45% b.w. for swine (Withers & Carruth 1915).

Cottonseed meal may represent a serious risk for pets and other animal species when fed for a prolonged period of time. Sprinkled daily on the feed for several months, it was associated with the death of 4 dogs, with signs suggestive of congestive heart failure in 2 (Patton et al. 1985). The pathologic changes were consistent with experimentally induced cottonseed meal toxicosis in dogs and were composed of an enlarged and pale myocardium, thin left ventricular walls, interstitial edema, and granular myofiber degeneration (West 1940).

The root bark of cotton plants causes ergotlike uterine contracture and abortion (Pammel 1911). This feature, not present in other plant parts, has been exploited medicinally.

- binaphthalene pigments from glands in seeds, gossypol; free levels >100 ppm a risk in pigs and young ruminants; high levels a risk in mature ruminants

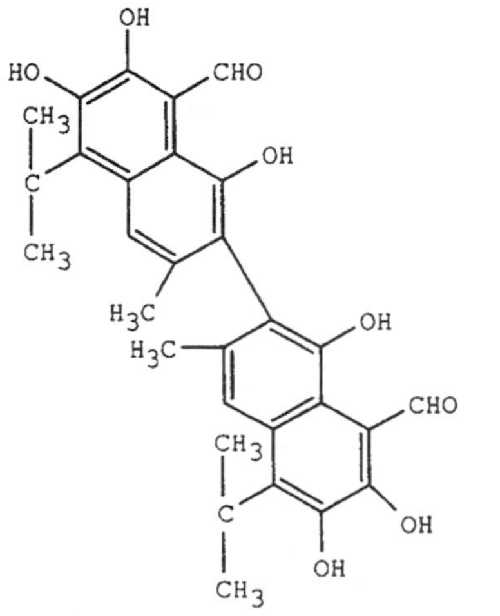

Disease Genesis—Problems with *Gossypium* are almost exclusively limited to ingestion of seed meal and other seed products containing substantial amounts of gossypol. The toxic effects of the meal were recognized as early as 1911, and the pigment gossypol was reported as a toxic component several years later (Dinwiddie & Short 1911; Withers & Carruth 1915; Schwartze & Alsberg 1924). Its correlation with disease, however, was not widely accepted for many years, because of problems in distinguishing its effects from those of vitamin A deficiency, particularly in ruminants. Cottonseed is now recognized to contain low levels of vitamin A.

Gossypol

Gossypol, unique to the cotton family and unknown elsewhere among the flowering plants, is a highly reactive polyphenolic binaphthalene derivative found primarily in the distinctive oil glands (generally termed gossypol glands) that are found throughout the aerial portions of the cotton plant and especially in the cotyledons of the seed (Lusas & Jividen 1987). The glands are spherical, usually black (varying from translucent yellow to dark red or purple), and appear to the unaided eye as small dark specks scattered throughout the seed tissue. They compose 2.4–4.8% of the seed weight and contain, in a water-soluble matrix, a mixture of toxic pigments, of which gossypol is the major constituent (Berardi & Goldblatt 1980; Yatsu et al. 1986). The seed content of gossypol may range from 0.4 to 1.7%, depending on a variety of factors such as the plant's genetic constitution, the soil type, and the climate. Plants of *G. barbadense* are believed to contain more glands and more gossypol than plants of *G. hirsutum*. Intensive breeding efforts have produced strains of the latter that lack glands and contain low levels of gossypol (Lusas & Jividen 1987; Fisher et al. 1988).

During the traditional extraction of cottonseed oil, the seeds are heated to high temperatures for a relatively long time before being mechanically pressed to expel the oil. During this process, the glands are ruptured and their contents released. Some of the released gossypol is rendered nontoxic during the milling process by reacting with protein amino groups and is termed bound gossypol (Clarke 1928). The remainder in the meal, termed the free gossypol, is the portion associated with intoxication problems. The type of processing has considerable influence on the gossypol content of the meal. Concentrations of free gossypol are highest with solvent extraction because of the water-soluble nature of the glandular matrix, and lowest with the commonly used screw pressing procedure. The final free gossypol content is dependent on both the seed content and the processing conditions. Purified cottonseed oil resulting from the processing is essentially free of gossypol. However, cottonseed hulls, if they contain large numbers of intact seeds or partial seeds, may contain toxicologically significant amounts of gossypol.

Low dietary iron increases the effects of gossypol. Conversely, iron can be added to attenuate its toxic effects to a limited extent (Eisele 1986; Haschek et al. 1989). Iron appears to interact with the gossypol in the digestive tract to reduce its absorption, and its accumulation in tissue.

The disposition of gossypol has been studied in considerable detail in swine (Abou-Donia & Dieckert 1975). Free gossypol is readily absorbed from the digestive tract and is found in highest concentrations in muscle, liver, adipose tissue, and red blood cells, respectively. It is metabolized in the liver to various compounds, conjugated as glucuronides and sulfates, and excreted in the feces (Abou-Donia 1989). Very little is eliminated in urine in any form. The half-life of elimination from the body is approximately 78 hours in pigs. In 3 lactating dairy cattle, the half-life ranged from 40.7 to 67.5 hours (Lin et al. 1991).

There is marked species variation in susceptibility, with rabbits, guinea pigs, and swine being much more susceptible than sheep and cattle. Experimentally, concentrations of 200–300 ppm free gossypol are highly lethal in swine. Mortality typically exceeds 50%, with signs that are consistent with cardiac insufficiency, including respiratory distress, elevation of serum CK and LDH, and depression of serum calcium, blood hemoglobin, and hematocrit (Haschek et al. 1989).

Cattle are able to tolerate relatively high concentrations of cottonseed products (Withers & Carruth 1915; Jones et al. 1941). Adult cattle tolerated up to 6.2 mg/kg b.w. free gossypol per day for 100 days, but 8.8 mg/kg caused problems (Lindsey et al. 1980). Calhoun and co-workers (1991) fed crossbred heifers a diet of 30% cottonseed for 56 days during breeding. Each animal received 3 kg/day, equal to 16.6 g, or 11.4 mg/kg b.w., of gossypol. The only evidence of adverse effects was an increase in red blood cell (RBC) osmotic fragility at 28 and 56 days and an increase in serum AP activity at 56 days. No reproductive effects were apparent, and the pregnancy rate was 95%. In other studies, moderate amounts of gossypol in adult cows caused increased RBC fragility and lysis, resulting in a decrease in hematocrit (Jimenez 1980; Lindsey et al. 1980). Respiratory distress may also increase with high environmental temperature and depressed feed intake. Because of these effects, it has been recommended that cottonseed in feed be limited to 2.7 kg/day for adult cows (Jimenez 1980; Smalley & Bicknell 1982). Although the adult ruminant is considered more resistant to the toxic effects of gossypol, typical clinical signs may occur when exceptionally high levels of cottonseed meal are fed for long periods of time (Jimenez 1979; Lindsey et al. 1980; Smalley & Bicknell 1982; Edwards et al. 1984; Hudson et al. 1988). The low susceptibility of the adult ruminant is not due to microbial or enzymatic action but instead is due to rapid and permanent chemical reaction of free gossypol with soluble protein components, especially lysine and possibly other amino acids, in the rumen liquor (Lyman et al. 1959; Reiser & Fu 1962).

The role of the rumen in attenuating the response to gossypol is further shown by the enhanced susceptibility of the preruminant calf or lamb to the toxic effects of high dietary cottonseed meal (Hollon et al. 1958; Holmberg et al. 1988; Morgan et al. 1988). The response of these younger animals (and of swine) is similar to that in older individuals; the only difference is the dosage of gossypol required. Sudden death may be a prominent feature, especially for preruminant calves or lambs, as demonstrated in young dairy calves fed a ration with 1,200 ppm free gossypol and steers weighing 182–320 kg on a diet with 2,200 ppm (Edwards et al. 1984). Lambs given a ration containing 900 ppm free gossypol for 3–4 weeks either died suddenly or died following signs of cardiac failure (Morgan et al. 1988). Similarly, dairy calves fed cottonseed meal with gossypol levels of 250–380 ppm for the first 10 weeks of life experienced high mortality (Holmberg et al. 1988). Little or no gossypol is eliminated in milk, so the newborn is not at risk from ingestion of cottonseed products by the dam (Lindsey et al. 1980).

The adult goat may not be as resistant to the effects of gossypol as cattle. Daily intake of 250–300 mg of free gossypol/day for several months produced typical signs of intoxication (East et al. 1994). Gossypol was present at 270 and 390 ppm free gossypol in the concentrate and protein supplements. The resulting signs and pathology were similar to those in other species.

■ reproductive effects; decreased fertility, male and female

Of additional concern are possible reproductive effects associated with gossypol. Transient, dose-related male infertility has been reported for some animal species and humans and is the subject of investigative efforts in China to determine its potential for contraceptive use in humans (Abou-Donia 1989). The effects are a decrease in testicular steroidogenesis and spermatogenesis and possibly direct action on spermatozoa (Lin et al. 1981; Saksena & Salmonsen 1982; Kramer et al. 1991). Although concentrations of gossypol are lower in testes than in most other tissues, some nonruminant animal species are apparently quite susceptible. In hamsters, a daily dosage of 10 mg/kg b.w. of gossypol causes degeneration of spermatocytes and marked decrease in fertility after 5–6 weeks (Hahn et al. 1981). Rats are only about half as sensitive. Recovery from the effects in both species occurs in a few weeks. Even a single dose of gossypol i.p. in rats caused rapid testicular degeneration with germ cell disorganization and fibrosis in hours (Thomas et al. 1991). Other species, such as mice and rabbits, are reported to have low susceptibility to the testicular effects (Hahn et al. 1981; Abou-Donia 1989). Overall, the effects on susceptible nonruminants include, in the female, altered estrous cycles, fertility, and embryo development (Randel et al. 1992). In the male there is germinal epithelial and sperm damage, with decreased sperm counts and motility and some degree of infertility.

Generally the reproductive effects of gossypol in pigs are not of concern because of the sensitivity of pigs to its other actions, albeit there may be decreased conception and litter size (Eisele 1986). Of greater concern are the effects on reproductive performance of rams or bulls. Detrimental effects have been shown in the spermatogenic tissues of young bulls fed cottonseed meal or whole cottonseed for 4 months (Arshami & Ruttle 1988). In other instances reduction in fertility with diets high in cottonseed meal was eliminated when the diet was changed (Jarrett 1983). Overall, ruminants may have some inhibition of ovarian steroidogenesis and embryo development, and germinal epithelial damage in the male, although there may be no obvious effect on semen quality (Randel et al. 1992). Even abortion has been suggested as a possible effect of consumption of cottonseed meal (Edwards et al. 1984). Seven abortions occurred among 120 dairy cows fed cottonseed meal containing 9,400 ppm gossypol. This effect was not confirmed experimentally in rats; no adverse effects were noted on pregnancy or the fetus (Beaudin 1985).

Clinical Signs—Swine are most commonly affected. After consumption of a ration high in cottonseed meal by swine for 1–2 months or more, there is abrupt onset of signs, including muscular weakness, frothy salivation, generalized edema, and hypothermia. The most prominent sign is pronounced labored breathing characterized as a gasping or thumping respiration. Most of the signs are readily referable to cardiac insufficiency, and animals often die within several days to a week following onset. There are terminal cyanosis and seizures related to cardiac failure. In many instances there is sudden death in apparently well animals.

- abrupt onset, weakness, labored gasping respiration, generalized edema

  in some instances, sudden death

The disease in calves and lambs is much the same as in swine. Prolonged ingestion of cottonseed meal produces abrupt onset of severe but vague clinical signs. Erratic appetite, muscular weakness, hypothermia, generalized edema, respiratory distress, and/or hemoglobinuria may be seen, but often the calves or lambs are found dead without prominent premonitory signs. Consumption of large quantities of cottonseed meal by adult cattle or functionally ruminant calves is less frequently associated with overt clinical disease; however, increased RBC fragility and occasional abrupt deaths without prominent clinical abnormalities may occur with diets of 3 kg or more per day of cottonseed meal. Disease manifestation has been reported during periods of heat stress after animals have been on highly supplemented rations for a prolonged period.

Poultry are of intermediate sensitivity to direct toxic effects of gossypol, but they are very prone to lay eggs with olive green yolks because of the interaction between yolk iron and gossypol (Schaible et al. 1933). Horses appear to be of low sensitivity to gossypol.

Pathology—Gross changes include widespread congestion and edema, and reddening and sloughing of the mucosal surface of the stomach and intestines. Discrete areas of ulceration are sometimes seen. There are usually large quantities of free fluid in the chest and abdominal cavities and pulmonary edema. The liver may have a mottled or nutmeg appearance, and the heart may appear dilated, flabby, and edematous. Microscopically in the heart, there is interstitial edema with separation and degeneration of the myofibers. Congestion and diffuse centrilobular necrosis of the liver and mild kidney degeneration are also typical.

- gross pathology, reddened mucosa, stomach, intestines; fluid in body cavities; heart flabby, dilated; liver, mottled

  microscopic, heart, edema of myocardium; liver necrosis

Treatment—Proper dietary management is the most effective method for reducing the probability of livestock losses due to gossypol intoxication. Cottonseed products remain excellent components in feed rations as long as they are not used in excess. For swine and preruminant calves and lambs, the level of free gossypol in the diet must be less than 100 ppm to ensure safety. A general rule, based on the average gossypol content of commercial cottonseed meal, is that the meal may be added to swine feed if limited to less than 9% of the total ration. Elevated gossypol concentrations in feed may be partially counteracted by additional iron, as ferrous sulfate, in the ration. The total intake of cottonseed products for adult cattle should be limited to 2.5–3 kg per head per day.

- nursing care; limit level of free gossypol in feed for pigs, lambs, and calves to 100 ppm

Because clinical signs usually do not appear for several weeks or months after cottonseed has been introduced into the diet, cattle feeders are advised to have feed samples analyzed for free gossypol content when rations high in cottonseed meal are utilized. Furthermore, it is important to note that because of the cumulative nature of the toxicant and the type of pathological changes, deaths often will continue for several days following cessation of feeding of the toxic ration.

Malva L.

- *M. neglecta*
  M. nicaeensis
  M. parviflora
  M. rotundifolia
  M. sylvestris
  cheeses
  cheeseweed
  mallow

Taxonomy and Morphology—A Eurasian genus of 25–30 species, *Malva* is represented in North America by 10 introduced taxa, most of which are naturalized weeds. Its name is derived from the Greek root *malachos,* which means "soothing," and is believed to reflect the medicinal properties of some of its species. Common names are mallow and cheeseweed; the latter is indicative of the shape of the fruit, which resembles an old-fashioned wheel of cheese. Representative and commonly encountered are the following species:

M. neglecta Wallr.	common mallow, cheeses, dwarf mallow
M. nicaeensis All.	bull mallow
M. parviflora L.	little mallow, cheeseweed, alkali mallow
M. rotundifolia L.	running mallow, roundleaf mallow
M. sylvestris L.	high mallow

- herbs; leaves long petiolate; flowers axillary; petals white, pink, lavender, or purple

Malva neglecta

Plants herbs; annuals or biennials or perennials; herbage glabrous or sparsely stellate. **Stems** erect or decumbent to ascending or trailing; 20–300 cm tall; branched or not branched. **Leaves** long petiolate; blades orbicular to reniform; shallow to deeply lobed or dissected; margins typically crenate; stipules persistent. **Inflorescences** solitary or fascicled flowers in axils of upper leaves; bracts of epicalyces 3, free, linear to ovate. **Flowers** small or large; inconspicuous or showy. **Sepals** fused. **Petals** white or pink or lavender to purple; obcordate; apices truncate to retuse. **Pistils** with carpels 6–20; not beaked; stigmas decurrent; styles filiform. **Fruits** schizocarps; depressed globose; mericarps smooth or corrugated; glabrous or indumented. **Seeds** 1 per mericarp.

1. Petals 15–30 mm long; 3–5 times longer than sepals .... *M. sylvestris*
1. Petals 4–12 mm long; 1–2 times longer than sepals.
   2. Petals 2 times longer than sepals. Mericarp margins rounded; backs smooth or faintly ridged .... *M. neglecta*
   2. Petals 1–1.3 times as long as sepals. Mericarp margins angular or winged; backs rugose-reticulate.
      3. Bracts of epicalyces lanceolate to ovate .... *M. nicaeensis*
      3. Bracts of epicalyces linear.
         4. Sepal lobes broadly ovate. Mericarp margins wavy-winged .... *M. parviflora*
         4. Sepal lobes triangular. Mericarp margins not wavy-winged .... *M. rotundifolia*

- weedy invaders

Malva neglecta

Distribution and Habitat—The North America species of *Malva* have been introduced, both deliberately and accidentally, from Eurasia. All are weedy, they readily escape cultivation, and occupy disturbed habitats across the continent. Some, such as *M. neglecta* and *M. sylvestris,* are widely distributed, whereas others, such as *M. nicaeensis,* have more restricted distributions.

Disease Problems—The green, immature fruits of *Malva* have a somewhat acceptable taste, especially if the subtending bracts and sepals are removed. They were used by members of various American Indian tribes as nibbles for children and as a base for salads, soups, relishes, and tea (Ebeling 1986).

Although the genus has been used as a food, there are indications of possible intoxication problems. An early report by Hester (1906) indicated a toxicity potential in cattle and horses that subsisted largely on *M. parviflora* and then were exercised heavily. Because of such reports, Marsh and co-workers (1928) administered 5.2% b.w. of *M. parviflora* to a sheep for 11 days, for a total dose of 57.2% b.w. without causing adverse effects. They repeated this in another sheep at 5.9% of b.w. for 26 days, for a total of 154% b.w., again without causing adverse effects. However, they used dried rather than fresh plants. Subsequent studies in Australia revealed that sheep, cattle, and horses did indeed develop a staggers type of syndrome when exercised extensively following ingestion of large amounts of fresh *M. parviflora* for several days or weeks (Dodd & Henry 1922). Dried or even wilted forage did not cause a problem. Systemic effects are typically lacking; there is no depression or appetite depression. However, *Malva* is generally of little consequence as a toxic plant because the special circumstances required are seldom met, i.e., ingestion of large amounts of plant material coupled with extensive exercise.

- rare neurologic problem with large amounts of fresh forage coupled with vigorous exercise; as yet not a problem in North America

When poultry eat the foliage or seeds of *M. parviflora,* there are changes in egg quality. The egg whites are sometimes pink, somewhat similar to the discolorations caused by gossypol in cottonseed meal (Sherwood 1928; Lorenz et al. 1933; Schaible et al. 1933).

Disease Genesis—The toxin present in *Malva* is not known but apparently is volatile, because it does not persist in dried plants. It is also passed in the milk. Horses, cattle, and sheep of all ages, but especially young animals, are affected. The elimination of the toxin in milk puts the nursing young at risk but also reduces the effects in the mother (Dodd & Henry 1922). In addition, both *M. parviflora* and *M. nicaeensis* accumulate nitrate in the range of 3–8% NO_3 d.w. (Shlosberg et al. 1975; Everist 1981), but this attribute has not proven to be a problem.

- toxin unknown; some species accumulate nitrate

The pink discoloration of egg whites appears to be the result of malvalic or sterculic acid (cyclopropene ring), which induces degeneration of vitelline membranes (Lorenz 1939; Shenstone & Vickery 1959). This allows diffusion of iron from the yolk to the white.

Clinical Signs—Sheep and cattle, especially lambs or calves, fed plants of the genus extensively for a week or more and forced to walk a considerable distance may become incapacitated. They tend to lag behind the other animals, move stiffly, and stand or move slowly, with the back arched and head extended. The muscles of the shoulders and rump quiver or tremble. Eventually the animals fall and struggle, sometimes regaining their feet but often are unable to stand. If allowed to rest for a while, they may be able to move away again without problem. The disease is much like the various staggers syndromes produced by *Paspalum* (dallis grass) and *Lolium* (perennial ryegrass) (see chapter 58).

- when exercised, animals move stiffly and slowly, back arched; trembling, collapse, struggle; stand after rest

In horses, sweating is marked and respiration is increased. Otherwise, the signs are similar to those of sheep and cattle.

Pathology and Treatment—There are few lesions other than scattered splotchy agonal hemorrhages and a small amount of fluid in the abdominal cavity. Alleviation of the signs can be readily accomplished with sedation, but animals should be allowed to rest for several days to a week to ensure that repeat episodes of distress do not occur.

- no lesions; no treatment needed

Modiola Moench

Taxonomy and Morphology—Native to tropical America and possibly southeastern North America, *Modiola* is a monotypic genus. The single species is the following:

- *M. caroliniana*
  - bristly mallow
  - Carolina modiola
  - ground ivy
  - red-flowered mallow

M. caroliniana (L.) G.Don (= *M. multifida* Moench) — Carolina modiola, ground ivy, wheel mallow, bristly mallow, red-flowered mallow

- prostrate herbs; leaves palmately cleft with 3–7 lobes; flowers small, axillary; petals salmon to purplish red

Plants herbs; perennials or rarely annuals; herbage hirsute. **Stems** prostrate or decumbent, 15–60 cm long; often rooting at nodes. **Leaves** petiolate; blades orbicular to reniform or triangular, margins palmately cleft, lobes 3–7, irregularly incised; stipules persistent. **Inflorescences** solitary flowers in axils of leaves; peduncles long; bracts of epicalyces 3, linear to obovate. **Flowers** small. **Sepals** fused; triangular-ovate. **Petals** salmon to purplish red; obcordate; apices truncate to retuse. **Pistils** with carpels 15–30, 2-beaked; stigmas capitate; styles filiform. **Fruits** schizocarps; depressed globose; mericarps reniform; hispid; black. **Seeds** 2 per mericarp.

- waste area weed

Modiola caroliniana

Distribution and Habitat—Introduced from tropical America, *Modiola* has spread over much of North America, especially in waste areas and lawns. Plants also occur at the edges of salt marshes. Some taxonomists suggest that the species may be native to the southeast portion of the continent and occurred as far north as South Carolina and possibly Virginia before the arrival of European colonists.

- neurologic effect; toxin unknown

Disease Problems and Genesis—In the few reported instances of intoxication, consumption of an appreciable amount of *Modiola* caused a neurologic disease in goats, cattle, and sheep (King 1920; Fuller & McClintock 1986). The cause of these effects is not known.

- ataxia, collapse, seizures
  - no lesions
  - sedation

Clinical Signs, Pathology, and Treatment—Signs begin with incoordination and then progress to prostration, seizures, and, in some animals, paralysis of the pelvic limbs. There are no distinctive lesions. The only known treatment is control of the neurologic effects by sedation.

References

Abou-Donia MB: Gossypol. In *Toxicants of Plant Origin,* vol 4, *Phenolics,* Cheeke PR ed, CRC Press, Boca Raton, Fla, pp1–22, 1989

Abou-Donia MB, Dieckert JW: Metabolic fate of gossypol: the metabolism of [14C]gossypol in swine. *Toxicol Appl Pharmacol* 31;32–46, 1975

Arshami J, Ruttle JL: Effects of diets containing gossypol on spermatogenic tissues of young bulls. *Proc West Sect Am Soc Anim Sci* 39;255–257, 1988

Beaudin AR: The embryotoxicity of gossypol. *Teratology* 32;251–257, 1985

Berardi LC, Goldblatt LA: Gossypol. In *Toxic Constituents of Plant Foodstuffs,* 2d ed, Liener IE ed, Academic Press, New York, pp183–237, 1980

Calhoun MC, Huston JE, Ueckert DN, Baldwin BC Jr, Kuhlmann SW, Engdahl BS: Performance of yearling heifers fed diets containing whole cottonseed. In *Beef Cattle Research in Texas,* Tex Agric Exp Stn Consol Rep PR–4839, 1991

Chopra RN, Badhwar RL, Ghosh S: *Poisonous Plants of India.* Academic Publishers, Jaipur, India, pp232–239, 1984

Clarke EP: Studies on gossypol. 2. Concerning the nature of Carruth's D gossypol. *J Biol Chem* 76;229–235, 1928

Dinwiddie RR, Short AK: *Cotton Seed Poisoning of Livestock.* Ark Agric Exp Stn Bull 108, pp393–410, 1911

Dodd S, Henry M: Staggers or shivers in livestock. *J Comp Pathol Ther* 35;41–61, 1922

Dugan GM, Gumbmann MR: Toxicological and nutritional evaluation of velvetleaf seed: subchronic 90-day feeding study and protein efficiency ratio assay. *Food Chem Toxicol* 28;95–99, 1990

East NE, Anderson M, Lowenstine LJ: Apparent gossypol-induced toxicosis in adult dairy goats. *J Am Vet Med Assoc* 204;642–643, 1994

Ebeling W: *Handbook of Indian Foods and Fibers of Arid America.* Univ California Press, Berkeley, 1986

Edwards WC, Burrows GE, Tyrl RJ: Toxic plants of Oklahoma. *Okla Vet* 36;14–16, 1984

Eisele GR: A perspective on gossypol ingestion in swine. *Vet Hum Toxicol* 28;118–122, 1986

Everist SL: *Poisonous Plants of Australia.* Angus & Robertson, London, pp515–521, 1981

Fisher GS, Frank AW, Cherry JP: Total gossypol content of glandless cottonseed. *J Agric Food Chem* 36;42–44, 1988

Fryxell PA: *The Natural History of the Cotton Tribe (Malvaceae, Tribe Gossypieae).* Texas A&M Univ Press, College Station, 1979

Fuller TC, McClintock E: *Poisonous Plants of California,* Univ California Press, Berkeley, pp188–190, 1986

Hahn DW, Rusticus C, Probst A, Homm R, Johnson AN: Antifertility and endocrine activities of gossypol in rodents. *Contraception* 24;97–105, 1981

Haschek WM, Beasley VR, Buck WB, Finnell JH: Cottonseed meal (gossypol) toxicosis in a swine herd. *J Am Vet Med Assoc* 195;613–615, 1989

Hester JH: The injurious effects of malva plant. *Am Vet Rev* 30;106–108, 1906

Hollon BF, Waugh RK, Wise GH, Smith FH: Cottonseed meals as the primary protein supplement in concentrate feeds for young calves. *J Dairy Sci* 41;286–294, 1958

Holmberg CA, Weaver LD, Guterbock WM, Genes J, Montgomery P: Pathological and toxicological studies of calves fed a high concentration cotton seed meal diet. *Vet Pathol* 25;147–153, 1988

Hudson LM, Kerr LA, Maslin WR: Gossypol toxicosis in a herd of beef calves. *J Am Vet Med Assoc* 192;1303–1305, 1988

Jarrett JA: Too much cottonseed led to poisoning. *Hoard's Dairyman* 128;568–569, 1983

Jimenez AA: Effects of gossypol in milking cow rations. *Feedstuffs* 52;28, 1979

Jimenez AA: Gossypol and the lactating cow. *Feedstuffs* 25;33, 1980

Jones JH, Jones JM, Bayles JJ: *Utilization of Home Grown Feeds in Fattening Steers in the Trans-Pecos Region.* Tex Agric Exp Stn Bull, 1941

King ED: Poisonous plants of the south. *J Am Vet Med Assoc* 57;302–313, 1920

Kramer RY, Garner DL, Ericsson SA, Wesen DA, Downing TW, Redelman D: The effects of cottonseed components on testicular development in pubescent rams. *Vet Hum Toxicol* 33;11–16, 1991

Lin T, Murono EP, Osterman J, Nankin HR, Coulsen PB: Gossypol inhibits testicular steroidogenesis. *Fertil Steril* 35;563–566, 1981

Lin YC, Nuber DC, Ju Y, Cutler G, Hinchcliff KW, Haibel G: Gossypol pharmacokinetics in mid-lactation Brown Swiss dairy cows. *Vet Res Commun* 15;379–385, 1991

Lindsey TO, Hawkins GE, Guthrie LD: Physiological responses of lactating cows to gossypol from cottonseed meal rations. *J Dairy Sci* 63;562–573, 1980

Lorenz FW: Egg deterioration due to ingestion by hens of malvaceous materials. *Poult Sci* 18;295–300, 1939

Lorenz FW, Almquist HJ, Hendry GW: Malvaceous plants as a cause of "pink white" in stored eggs. *Science* 77;606, 1933

Lusas EW, Jividen GM: Glandless cottonseed: a review of the first 25 years of processing and utilization research. *J Am Oil Chem Soc* 64;839–854, 1987

Lyman CM, Baliga BP, Slay MW: Reactions of proteins with gossypol. *Arch Biochem Biophys* 84;186–187, 1959

Marsh CD, Clawson AB, Roe GC: *Four Species of Range Plants Not Poisonous to Livestock.* USDA Tech Bull 93, 1928

Morgan S, Stair EL, Martin T, Edwards WC, Morgan GL: Clinical, clinicopathologic, pathologic, and toxicologic alterations associated with gossypol toxicosis in feeder lambs. *Am J Vet Res* 49;493–499, 1988

Pammel LH: *A Manual of Poisonous Plants.* Torch Press, Cedar Rapids, Iowa, 1911

Patton CS, Legendre AM, Gompf RE, Walker MA: Heart failure caused by gossypol poisoning in two dogs. *J Am Vet Med Assoc* 187;625–627, 1985

Randel RD, Chase CC Jr, Wyse SJ: Effects of gossypol and cottonseed products on reproduction of mammals. *J Anim Sci* 70;1628–1638, 1992

Reiser R, Fu HC: The mechanism of gossypol detoxification by ruminant animals. *J Nutr* 76;215–218, 1962

Saksena SK, Salmonsen RA: Antifertility effect of gossypol in male hamsters. *Fertil Steril* 37;686–690, 1982

Schaible PJ, Moore LA, Moore JM: Gossypol, a cause of discoloration in egg yolks from hens fed cottonseed meal. *Poult Sci* 12;334, 1933

Schmidt JH, Wells R: Evidence for the presence of gossypol in malvaceous plants other than those in the "cotton tribe." *J Agric Food* Chem 38;505–508, 1990

Schwartze EW, Alsberg CL: Pharmacology of gossypol. *J Agric Res* 28;191–201, 1924

Shenstone FS, Vickery JR: Substances in plants of the order Malvales causing pink whites in stored eggs. *Poult Sci* 38;1055–1070, 1959

Sherwood RM: *The Effect of Various Rations on Storage Quality of Eggs.* Tex Agric Exp Stn Bull 376, 1928

Shlosberg A, Egyed MN, Efron Y: *Malva* nicaeensis—a potentially poisonous plant for cattle. *Refu Vet* 32;155–156, 1975

Smalley SA, Bicknell EJ: Gossypol toxicity in dairy cattle. *Compend Contin Educ Pract Vet* 4;S378–S381, 1982

Spencer NR: Velvetleaf, *Abutilon theophrasti* (Malvaceae): history and economic impact in the United States. *Econ Bot* 38;407–416, 1984

Steyn DG: *The Toxicology of Plants in South Africa.* Central News Agency, South Africa, pp308–310, 1934

Thomas KD, Caxton-Martins AE, Elujoba AA, Oyelola OO: Effects of an aqueous extract of cotton seed (*Gossypium barbadense* Linn.) on adult male rats. *Adv Contraception* 7;353–362, 1991

Warwick SI, Black LD: The biology of Canadian weeds. 90. *Abutilon theophrasti. Can J Plant Sci* 68;1069–1085, 1988

West JL: Lesions of gossypol poisoning in the dog. *J Am Vet Med Assoc* 96;74–76, 1940

Withers WA, Carruth FE: Gossypol, the toxic substance in cottonseed meal. *J Agric Res* 5;261–288, 1915

Yatsu L, Jacks TJ, Kircher HW, Godshall MA: Chemical and microscopic studies of the matrix substance in pigment glands of cotton *(Gossypium hirsutum)* seeds. *J Am Oil Chem Soc* 63;534–537, 1986

Meliaceae Juss.

51

Comprising about 50 genera and 550 species, the Meliaceae, or mahogany family, is widespread in tropical and subtropical regions. Many species are common understory trees in the rain forests. Economically, the family is prized for the wood of *Swietenia* (mahogany) and *Cedrela* (West Indian cedar). In North America, 6 genera and 9 species, both native and introduced, are present. Intoxication problems are associated with 1 genus.

■ Mahogany Family

Melia

Plants trees or shrubs; deciduous or evergreen; typically aromatic; bearing perfect flowers or rarely polygamous. **Leaves** 1- or 2-pinnately compound; alternate; stipules absent. **Inflorescences** panicles; axillary or extra-axillary. **Flowers** perfect or rarely imperfect; perianths in 2-series. **Sepals** 4–6; fused; imbricate. **Corollas** radially symmetrical. **Petals** 4–6; free; valvate or imbricate. **Stamens** 8–12; filaments fused to form erect cylindrical tube with slender teeth. **Pistils** 1; compound, carpels 3–5; stigmas 1, capitate or discoid; styles 1 or absent; ovaries superior; locules 3–5; placentation axile. **Fruits** capsules or drupes. **Seeds** 1 to many.

■ shrubs or trees; leaves pinnately compound, alternate; flowers in panicles; petals 4–6

Melia L.

Taxonomy and Morphology—Native to Asia, *Melia* comprises 5–10 species. Its name is an ancient Greek one for ash *(Fraxinus)*, reputedly because of the similarity in leaves. One species is present in North America:

M. azedarach L.	Chinaberry, Persian lilac, white cedar, pride-of-India, Indian lilac, China tree, syringa berry tree, Texas umbrella tree, bead tree, paradise tree, paráiso, piocha, cavelon, Ceylon mahogany

■ *M. azedarach*

bead tree
cavelon
Ceylon mahogany
China tree
Chinaberry
Indian lilac
paradise tree
paráiso
Persian lilac
piocha
pride-of-India
syringa berry tree
Texas umbrella tree
white cedar

Plants trees; deciduous; trunks stout; crowns round, spreading. **Stems** to 15 m tall; bark dark or reddish brown, furrowed; twig surfaces stellate when young, glabrous when old; buds small, round, scales few. **Leaves** 2-pinnately compound; leaflets 50 or more; ovate to elliptic-lanceolate; venation pinnate; apices acuminate; margins crenate to serrate. **Inflorescences** axillary panicles; pedicels puberulent. **Flowers** small; fragrant. **Sepals** 5 or 6. **Petals** 5 or 6; oblanceolate or spathulate; white to pinkish lavender or purple. **Stamens** 10–12; anthers yellow; filament tubes purple. **Pistils** with stigmas 1, 3-to-6-lobed; styles absent; locules 3–6. **Fruits** drupes; globose or ovoid; pericarps smooth, thin, fleshy, yellow, or black upon drying. **Seeds** 1–6 per stone.

The common name Chinaberry also has been used for *Sapindus drummondii* in the Sapindaceae. Also toxic(see chapter 67), this species is more properly called soapberry.

■ deciduous trees; leaves 2-pinnately compound; flowers axillary; petals 5 or 6, white, pinkish lavender, or purple; fruits globose yellow drupes

Melia azedarach

■ ornamental, naturalized

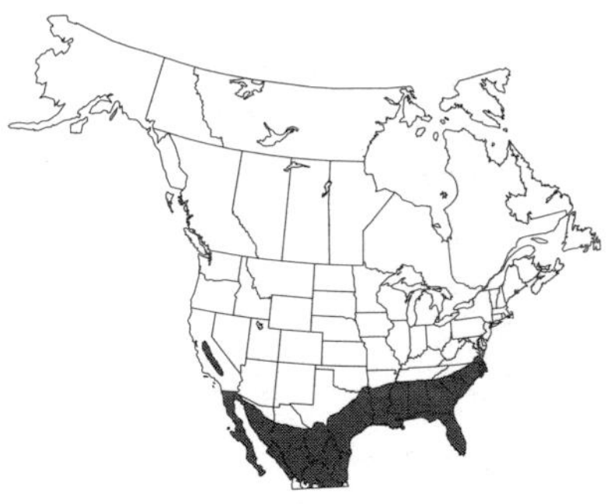

Melia azedarach

■ digestive disturbance; neurologic effects; fruits and flowers

Distribution and Habitat—Native to Asia, *M. azedarach* has been planted around the world as an ornamental. In North America it is limited to the southern states and Mexico, where it has escaped cultivation and naturalized.

Disease Problems—The medicinal properties of *Melia* are well known. Its drupes and bark or their extracts have been used extensively in horses as a tonic and to control intestinal parasites such as stomach bots (*Gastrophilus* spp.) (Standley 1926). Leaves placed in books are said to repel insects, and the drupes have been used to repel moths in closets and chests. However, toxicosis in humans and animals also has been reported. Fatalities are mainly in children who eat the drupes (Watt & Breyer-Brandwijk 1964). For example, a 3-year-old Venezuelan child reputedly died several days after eating the fruit of *M. azedarach* (Carratala 1939). That the species was definitely responsible was not confirmed, but aqueous extracts of the drupes subsequently given to rabbits caused digestive disturbance when given orally, and neurologic signs and death when administered subcutaneously.

Intoxications in animals may involve dogs, pigs, ruminants, rabbits, and other species. The disease may affect primarily the digestive system or there also may be a neurologic effect (Oelrichs et al. 1985). The potential of *Melia azedarach* to cause intoxications is quite variable, and in some areas where the species is common, disease problems may be rare or unknown. This is consistent with the reported variability in toxin content. The drupes and possibly the flowers are most commonly involved in appearance of the disease (Steyn 1934). The drupes are quite astringent when immature and green but become sweet when ripe and yellow. The minimal lethal dose is estimated to be 200 g for pigs and 800 g for sheep or goats.

Illness and death may follow consumption of the drupes in a few hours to a day or more (Kwatra et al. 1974). They are occasionally eaten by dogs; in a 10-year period (1987–1997), 24 cases were reported to the National Animal Poison Control Center (Hare et al. 1997). In most instances the dogs died 24–48 hours after ingestion. Animals are often found dead with their stomachs full of drupes. Smith (1940) recorded a case in which pigs, put in a pen with numerous chinaberries on the ground, ate all of them and were found dead the next day. However, pigs are said to be able, in some instances, to adapt to the toxicity of the drupes, eating a few and never developing the disease (Pammel 1921).

Rats fed dried leaves at 25% of their diet developed weakness and loss of appetite by the 14th day and then paralysis and death in another 2–10 days. There was prominent skeletal myopathy, including hyaline degeneration, fiber fragmentation, and necrosis (Bahri et al. 1992).

Because of the similarity in their names, *M. azedarach* has been confused with *M. azadirachta* in the toxicologic literature. Commonly known as margosa or neem, this evergreen tree is found in Indomalaysia. The taxon is now placed in its own genus and named *Azadirachta indica*. Used medicinally, it is also toxic when its leaves or drupes are eaten, but the toxicants are not necessarily the same (Ali & Salih 1982; Ali 1987). Both digestive and neurotoxic effects are apparent. Its fruits are the source of neem oil for topical and anthelmintic use, and the residue meal (neem cakes) was tested as a possible feed for lambs and chickens (Bhandari & Joshi 1974; Sadagopan et al. 1982; Chopra et al. 1984). The cakes produced detrimental effects, including severe degeneration of the digestive tract mucosa, growth depression, and death.

■ triterpene, epoxide meliatoxins; low dosage, diarrhea; high dosage, neurologic actions

Disease Genesis—Intoxications caused by *M. azedarach* appear to be due to several tetranortriterpenes called meliatoxins A_1, A_2, B_1, and B_2, which have been isolated from the fruits (Oelrichs et al. 1983). B_1 and B_2 are isomeric with A_1 and A_2, respectively, but lack the epoxide structure and are suggested to be merely isolation artifacts (MacLeod et al. 1990). The epoxide group may be a determinant of maximal activity, because A_2 is especially active in reducing insect herbivory. Given orally, these meliatoxins cause signs similar to those of the natural disease, i.e., muscular contractions, spasmodic quivering, collapse, and death. Sublethal dosage caused diarrhea, with or without blood, that lasted for 1–2 days. There is considerable variation in toxin content. In trees of some areas, the toxin is not present and the disease seems to be absent as well (Oelrichs et al. 1983; Bahri et al. 1992). There is considerable doubt about the toxicity potential of the foliage (Steyn 1934). This may account for its use as livestock forage despite the occurrence of severe toxicity in the same area for humans who eat the drupes (Chopra et al. 1984).

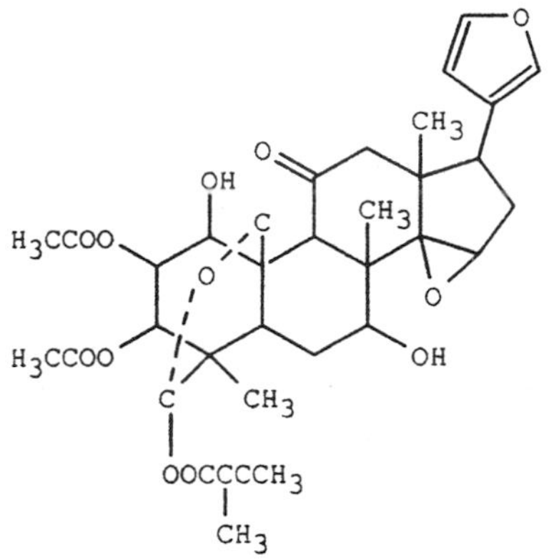

Meliatoxin A_1

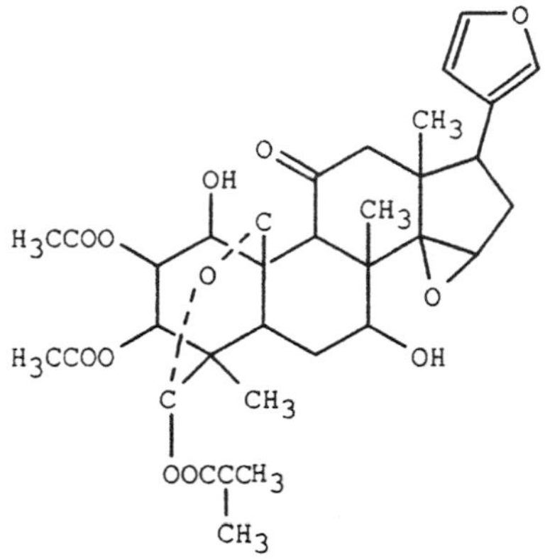

Meliatoxin A_2

Clinical Signs—The disease may present with increased salivation, retching, vomiting, anorexia, and diarrhea. There also may be early excitement and then paralysis along with the digestive disturbance. Respiration is labored and irregular, and weakness, muscle cramps, ataxia, and tetanic- or clonic-type seizures indicate more-serious signs. In some animals abrupt onset of neurologic signs and death may occur without prominent digestive disturbance. Gaseous distension of the stomach often occurs later in the disease.

- abrupt onset, retching, vomiting, excess salivation, diarrhea, weakness, trembling, ataxia, seizures

Pathology—The pathologic changes are mainly focal edema and reddening of the mucosa of the stomach and small intestine and gaseous distension of the stomach.

- gross pathology, edema and reddening of gut mucosa

Treatment—Induced emesis prior to onset of signs, gastric lavage, and/or administration of activated charcoal are important aspects of treatment, especially in dogs. Sedation also may be required in the case of marked neurologic effects. Because of the possibility of shock in the disease, fluids and immediate-acting steroids may be valuable adjuncts to treatment (Hare et al. 1997).

- activated charcoal, sedation

References

Ali BH: The toxicity of *Azadirachta indica* leaves in goats and guinea pigs. *Vet Hum Toxicol* 29;16–19, 1987

Ali BH, Salih AMM: Suspected *Azadirachta indica* toxicity in a sheep. *Vet Rec* 111;494, 1982

Bahri S, Sani Y, Hooper PT: Myodegeneration in rats fed *Melia azedarach*. *Aust Vet J* 69;33, 1992

Bhandari DS, Joshi MS: The effect of feeding deoiled neem cake on health of sheep. *Indian Vet J* 51;659–661, 1974

Carratala RE: Fatal intoxication by fruit from *Melia azedarach* L. *Rev Assoc Med Argent* 53;338–340, 1939

Chopra RN, Badhwar RL, Ghosh S: *Poisonous Plants of India*. Vol 1. Academic Publishers, Jaipur, India, pp284–289, 1984

Hare WR, Schutzman H, Lee BR, Knight MW: Chinaberry poisoning in two dogs. *J Am Vet Med Assoc* 210;1638–1640, 1997

Kwatra MS, Singh B, Hothi DS, Dhingra PN: Poisoning by *Melia azedarach* in pigs. *Vet Rec* 95;421, 1974

MacLeod JK, Moeller PDR, Molinski TF, Koul O: Antifeedant activity against *Spodoptera litura* larvae and [13C] NMR spectral assignments of the meliatoxins. *J Chem Ecol* 16;2511–2518, 1990

Oelrichs PB, Hill MW, Vallely PJ, MacLeod JK, Molinski TF: Toxic tetranortriterpenes of the fruit of *Melia azedarach*. *Phytochemistry* 22;531–534, 1983

Oelrichs PB, Hill MW, Vallely PJ, MacLeod JK, Molinski TF: The chemistry and pathology of meliatoxins A and B constituents from the fruit of *Melia azedarach* L. var. *australasica.* In *Plant Toxicology, Proceedings of the Australia-USA Poisonous Plant Symposium,* Seawright AA, Hegarty MP, James LF, Keeler RF eds, Animal Research Institute, Yeerongpilly, Queensland, pp387–394, 1985

Pammel LH: China tree poisonous. *Vet Med* 16;47, 1921

Sadagopan VR, Johri TS, Reddy VR, Panda BK: Feeding value of neem seed meal for starter chicks. *Indian Vet J* 59;462–465, 1982

Smith EV: Poisonous plants as a problem in southern livestock production. *Assoc South Agric Work Proc* 41;104–105, 1940

Standley PC: Trees and shrubs of Mexico. *Contrib U S Natl Herb* 23;1–1721, 1920–1926

Steyn DG: *The Toxicology of Plants in South Africa.* Central News Agency, South Africa, pp270–279, 1934

Watt JM, Breyer-Brandwijk MG: *The Medicinal and Poisonous Plants of Southern and Eastern Africa,* 2d ed. E & S Livingston, Edinburgh, pp744–751, 1964

Myoporaceae R.Br. 52

Comprising 3 or 4 genera and 110–125 species, the Myoporaceae, or myoporum family, is distributed primarily in Australasia, with 1 species in the New World and 2 in Africa. It is of minor economic importance, with a few species providing timber and a few grown as ornamentals. In North America, 2 introduced genera are present; intoxication problems are occasionally caused by species of 1 genus.

■ Myoporum Family

Myoporum

Plants shrubs or trees; evergreen; surfaces typically glandular or scaly or woolly. **Leaves** simple; alternate or opposite; stipules absent. **Inflorescences** cymes or solitary flowers; axillary. **Flowers** perfect; radially or bilaterally symmetrical; perianths in 2-series. **Sepals** 5; fused. **Petals** 5; fused. **Stamens** 4 or 5; staminodia present or absent. **Pistils** 1; compound; carpels 2; stigmas 1; styles 1; ovaries superior; locules 2–10; placentation axile. **Fruits** drupes. **Seeds** 2–10 per stone.

■ evergreen shrubs or trees; leaves simple; flowers in cymes or solitary; petals 5, fused; fruits drupes

Myoporum Banks & Sol. ex G.Forst.

Taxonomy and Morphology—The most widespread genus of the family, *Myoporum* comprises about 30 species. Its name is derived from the Greek root *myo,* meaning "to close," and the Latin root *porum,* meaning "pore," which reflect the translucent resinous glands of the leaves. Introduced species of toxicologic interest are the following:

M. insulare R.Br.	boobyalla
M. laetum G.Forst.	myoporum
M. montanum R.Br.	western boobyalla, waterbush
M. sandwicense (A.DC.) A.Gray	bastard sandalwood

■ *M. insulare*
M. laetum
M. montanum
M. sandwicense

bastard sandalwood
boobyalla
myoporum
waterbush

Plants shrubs or trees; evergreen; crowns spreading. **Stems** erect; 3–20 m tall; profusely branched; twigs bronze–green, glutinous. **Leaves** alternate or rarely opposite; petioled; blades linear to lanceolate or obovate; surfaces conspicuously glandular punctate; margins entire or toothed or sinuate. **Inflorescences** axillary cymes or rarely solitary flowers. **Flowers** radially or slightly bilaterally symmetrical; campanulate to funnelform. **Sepals** imbricate. **Petals** imbricate; white to pink, often spotted. **Stamens** 4; didynamous. **Drupes** globose to ovoid; smooth or rugose; white or yellow or purple.

■ profusely branched shrubs or trees; leaves conspicuously glandular punctate; petals white to pink, often spotted; drupes white, yellow, or purple

1. Leaf margins serrate or crenate above middles.
   2. Drupes reddish purple. Petals with purple spots . . . . *M. laetum*
   2. Drupes purplish black. Petals with pink spots . . . . *M. insulare*
1. Leaf margins entire or undulate above middles.
   3. Leaves 5–15 cm long. Drupes pale yellow to white . . . . *M. sandwicense*
   3. Leaves 2–5 cm long. Drupes purple . . . . *M. montanum*

Myoporum insulare

■ ornamentals

Distribution and Habitat—In North America these four species are not common. Frost-tender, they are grown as ornamentals primarily in southern Florida, Mexico, and southern California. *Myoporum insulare* and *M. montanum* are native to Australia, and *M. sandwicense* is Hawaiian. Native to New Zealand, *M. laetum* is cultivated near the coast and sometimes encountered in disturbed urban areas in southern California.

Myoporum laetum

■ liver disease, photosensitization, livestock

Disease Problems—Six Australian species of *Myoporum* are reported to cause severe liver disease in sheep and cattle and other livestock (Seawright 1989). In experiments carried out in conjunction with field intoxications, dosage of 30 g/kg b.w. of fresh leaves of *M. insulare* to a 160 kg heifer caused death in 70 hours (Jerrett & Chinnock 1983). In sheep, 20 g/kg b.w. of green leaves of *M. laetum* caused illness, including photosensitization, and 40 g/kg was lethal (Raposo et al. 1998). Low dosage was not cumulative in effect. The disease appears mainly in animals not adapted to the plant, because some tolerance occurs with consistent ingestion of small amounts. There also is considerable variation in toxicity among trees.

In North America the risk of intoxication due to species of *Myoporum* is minimal because of their limited distributional range. Cultivated as ornamentals, they are not typically accessible to livestock. However, a single tree may represent a threat. The evergreen nature of the foliage makes these trees a risk throughout the year, especially in winter. In Argentina, 10 heifers gained access to a tree of *M. laetum* and ate all of the leaves (Odriozola et al. 1987). Within 2 days, 2 of the heifers had died, and a third was ill but subsequently recovered. The main lesion was hepatic necrosis. In a few episodes, *M. laetum* has caused distress and death of cattle in California, where it is most widely cultivated (Fuller & McClintock 1986).

■ series of compounds, such as ngaione, metabolized in liver to toxicants

Ngaione

Disease Genesis—The cause of the liver disease produced by *Myoporum* is several furanosesquiterpenoids such as ngaione, which are biotransformed in hepatocytes to the ultimate in situ toxicant (Seawright 1989). Ngaione is the enantiomer of ipomeamarone, the phytoalexin found in fungal-infected, decaying sweet potatoes. Furanosesquiterpenoids are present in *M. insulare, M. laetum,* and *M. montanum* (Connolly & Hill 1991). The role of such toxicants in other species of the genus has not been confirmed. Myomontanone, which is found in some species and causes lung damage in laboratory animals, is not a problem in livestock (Hrdlicka et al. 1985; Seawright 1989).

■ depression, anorexia, icterus, photosensitization, increased serum liver enzymes

Clinical Signs—Occurring several days after ingestion of the leaves, the disease is manifested by depression, loss of appetite, icterus, constipation with or without blood, and increased serum liver enzymes such as GGT and AST. In some instances, photosensitization may be a prominent feature of the disease. Tremors also may occur.

■ gross pathology, liver enlarged, mottled; excess fluid in abdomen

microscopic, liver necrosis, bile duct proliferation

Pathology—The liver will be swollen and mottled yellow-tan. There will be considerable free fluid in the abdominal cavity and pulmonary edema in some animals. Hemorrhages of the pericardium, endocardium, and intestinal mucosa may be present. Microscopically, there is severe periportal hepatocellular necrosis, mild midzonal necrosis, and bile duct proliferation with mononuclear cell infiltrates. Degeneration of the tubular epithelium in the kidneys also may be apparent.

■ nursing care

Treatment—Relief of the signs and maintenance of the animal to allow regeneration of liver and kidney function are the treatment approach.

References

Connolly JD, Hill RA: *Dictionary of Terpenoids.* Vol 2, *Di-and Higher Terpenoids.* Chapman & Hall, London, pp655–1460, 1991

Fuller TC, McClintock E: *Poisonous Plants of California.* Univ California Press, Berkeley, pp194–195, 1986

Hrdlicka J, Seawright AA, Allen JG, Hastie J: The toxicity of myomontanone, a lung-damaging furanosesquiterpene from *Myoporum montanum.* In *Plant Toxicology, Proceedings of the Australia-USA Poisonous Plants Symposium,* Seawright AA, Hegarty MP, James LF, Keeler RF eds, Animal Research Institute, Yeerongpilly, Brisbane, Australia, pp465–472, 1985

Jerrett IV, Chinnock RJ: Outbreaks of photosensitization and deaths in cattle due to *Myoporum aff. insulare* R.Br. toxicity. *Aust Vet J* 60;183–186, 1983

Odriozola ER, Tapia MO, Lopez TA, Casaro AP, Calandra W: Natural poisoning of cattle with *Myoporum laetum. Rev Med Vet Argent* 68;230–232, 1987

Raposo JB, Mendez MC, Riet-Correa F, Andrade GB: Experimental intoxication by *Myoporum laetum* in sheep. *Vet Hum Toxicol* 40;132–135, 1998

Seawright AA: *Animal Health in Australia.* Vol 2, *Chemical and Plant Poisons,* 2d ed, Australian Government Publ Serv, Canberra, pp77–79, 1989

53 Oleaceae Hoffmanns. & Link

■ Olive Family
Ligustrum
QUESTIONABLE
Fraxinus
Jasminum

Nearly cosmopolitan in distribution, but with greatest diversity in southeastern Asia and Australasia, the Oleaceae, or olive family, comprises 22–30 genera and 500–600 species. Based primarily on differences in fruit type, 2 subfamilies and 7 tribes are typically recognized. The family is of considerable economic importance and is a source of food, high-quality oil, and wood for cabinetry, tools, and athletic equipment. Some members are popular flowering ornamentals—lilac, jasmine, and golden bells. Some, such as privet, are important hedging plants, and others, such as the ashes, are shade trees. In North America, approximately 12 genera and 70 species, both native and introduced, are present. Intoxication problems are caused by species of 1 genus and possibly by those of 2 other genera.

■ shrubs or trees; leaves opposite

Plants trees or shrubs; deciduous or evergreen; dioecious or polygamodioecious or bearing perfect flowers. **Leaves** simple or 1-pinnately compound; opposite or rarely subopposite; venation pinnate; stipules absent. **Inflorescences** paniculate or racemose cymes or glomerules; terminal or axillary. **Flowers** perfect or imperfect, staminate and pistillate similar; perianths absent or in 1-series or 2-series. **Calyces** radially symmetrical. **Sepals** 4 or 6–12 or absent; persistent or caducous; fused. **Corollas** radially symmetrical. **Petals** 4 or 6–12 or absent; fused or free. **Stamens** 2–5. **Pistils** 1; compound, carpels 2; stipitate or sessile; stigmas 1, 2-lobed; styles 1; ovaries superior; locules 1 or 2; placentation axile; ovules 1 or 2 per locule. **Fruits** drupes or samaras or berries or capsules. **Seeds** 1–4.

Ligustrum

■ *L. amurense*
L. japonicum
L. lucidum
L. obtusifolium
L. ovalifolium
L. quihoui
L. sinense
L. vulgare
privet
privet hedge

Taxonomy and Morphology—Native to eastern Asia and Australasia with the exception of 1 Mediterranean species, *Ligustrum,* commonly known as privet, comprises about 50 species, numerous cultivars, and several hybrids. Its species are important hedging or screening plants. In North America, 8 introduced species are present:

L. amurense Carr.	Amur privet, Amur River privet, north privet
L. japonicum Thunb.	Japanese privet, wax-leaf privet
L. lucidum W.T.Aiton	glossy privet, tree privet, large-leaf privet
L. obtusifolium Siebold & Zucc.	
L. ovalifolium Hassk.	California privet
L. quihoui Carr.	
L. sinense Lour.	Chinese privet
L. vulgare L.	privet, common privet, privet hedge

Plants shrubs or small trees; deciduous or evergreen. **Leaves** simple; blades elliptic to oblong or ovate; margins entire. **Inflorescences** terminal paniculate cymes. **Flowers** perfect; small; perianths in 2-series. **Sepals** 4; fused, lobes small; calyces campanulate. **Petals** 4; fused; funnelform; white. **Stamens** 2. **Fruits** drupelike berries; typically purple-black. **Seeds** 1–4.

■ shrubs or small trees; leaves simple; petals 4, fused, funnelform, white; fruits drupelike purple-black berries

Because of the extensive variation exhibited by the numerous cultivars and hybrids, a taxonomic key is not presented here. If exact identification is required, horticultural manuals should be consulted.

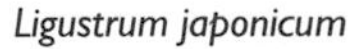

Ligustrum japonicum

Ligustrum sinense

■ ornamentals, naturalized in woodlands

Distribution and Habitat—Although typically grown as ornamental hedging, privet has occasionally escaped cultivation and become naturalized in open woods or along roadsides, especially in the eastern states. Plants also often persist at abandoned farm home sites.

Ligustrum sinense

Disease Problems—Reports of intoxication due to *Ligustrum* are not common, and there are few or no experimental data to confirm the risk it poses. The berries of various species have been considered a cause of purgation in children and animals since the late 1800s (Chesnut 1898; Harshberger 1920). In an anonymous report from New Zealand ("Yearling heifers," 1939), 5 heifers died after eating the leaves from a privet hedge. A lamb that was given clippings of *L. ovalifolium* developed severe digestive disease but recovered in 24 hours (Parkinson 1986). Horses are reported to have died within a day or two after eating privet leaves (Turner 1904; Cooper & Johnson 1984). Kerr and Kelch (1999) reported an instance in which several cows that ingested what was probably *L. amurense* (reported as *L. amurease*) died following signs suggestive of a metabolic disease and with cessation of rumination and a greenish nasal discharge. Intoxications are rare but can be severe. All livestock species are probably susceptible. Leaves, trimmings, and berries are all toxic, the berries perhaps more so. In contrast to the reports of intoxication in livestock, *L. sinense* seems to be an important browse for deer in the southeastern United States, with extensive consumption in the fall and winter (Stromayer et al. 1998). In addition, no adverse effects were apparent when the fruit of *L. vulgare* was fed at up to 40% of the diet for domestic turkey poults (DeBates & Lindzey 1982).

Ligustrum vulgare

In children, lethality has been reported, but this has been questioned as an actual risk (Frohn & Pfander 1984). Case reports of the European Toxicological Information

■ digestive disturbance; leaves or berries; possible cardiotoxic effects

Center indicate that vomiting and diarrhea may occur with ingestion of only a few berries; fatalities were not reported. However, Kozlov and Gylyava (1983) reported an incident in Russia involving 4 boys of ages 4–6 years who had apparently eaten fruits of *L. vulgare*. In 3 of the boys prominent signs of irritation in the digestive tract were successfully treated with charcoal and gastric lavage. However, 1 boy developed cardiac arrhythmia and circulatory deficit and died of cardiac arrest 3–4 hours after eating the berries. The pathologic changes were consistent with peracute death and included visceral congestion and scattered splotchy hemorrhages. Subsequently, aqueous extracts of the privet fruits were administered i.p. to mice and i.v. to rabbits (Kozlov & Gylyava 1983). The mice developed diarrhea and muscular twitching 30–40 minutes after administration but recovered within 24 hours. Two rabbits died of cardiopulmonary arrest within a few minutes after administration of the extract.

■ irritant terpenoid glycosides, ligustrins

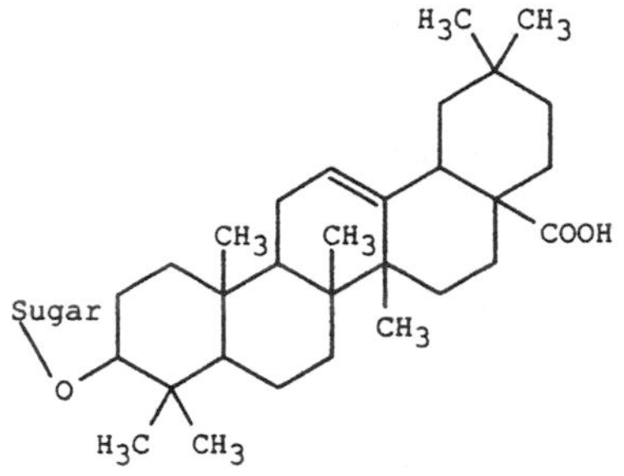

Ligustrins

Disease Genesis—The primary toxic effect of species of *Ligustrum* appears to be due to a digestive system irritant; however, the toxic principle is not definitely known. Pentacyclic terpenoid glycosides of oleanolic acid, such as ligustrins, part of a series of complex glycosides that have been isolated from the leaves of *L. japonicum* and *L. lucidum*, have been suggested as the likely toxins (Harshberger 1920; Cooper & Johnson 1984; Connolly & Hill 1991b). Other glucosides are present that appear to represent portions of more-complex structures (Inouye & Nishioka 1972). Derivatives of oleanolic acid also are found in the pericarp (Takemoto et al. 1955). Alkaloids, glycosides, and resins are present in the leaves (Ohta & Miyazaki 1953; Damirov 1960). Alcoholic extracts of the leaves have cardiac effects in vitro and in vivo that include augmentation of contractility and decreasing arterial pressure (Damirov 1960).

■ diarrhea, colic, vomiting, ataxia

reddened gut mucosa

fluids and electrolytes

Clinical Signs, Pathology, and Treatment—Signs of *Ligustrum* intoxication include diarrhea, colic, vomiting (depending on the animal species), and incoordination. Paralysis of the pelvic limbs is reported for horses. Death may occur in some cases. The pathologic changes are limited to intense reddening of the mucosa of the stomach/abomasum and proximal small intestine and/or mild alveolar edema and emphysema in the lungs. Treatment entails control of the purgation and fluid and electrolyte compensation for dehydration.

Genera of Questionable Toxicity

Fraxinus L.

■ *F. americana*
F. nigra
F. pennsylvanica
F. quadrangulata

ash

Fraxinus americana

Valued for its lumber, flowers, and foliage, *Fraxinus* comprises 60–65 species of deciduous trees mostly native to North America, eastern Asia, and the Mediterranean. In North America, approximately 18 native and introduced species are present. Commonly encountered are *F. americana* L. (white ash), *F. nigra* Marshall (black ash), *F. pennsylvanica* Marshall (green ash), and *F. quadrangulata* Michx.(blue ash).

Ashes grow to 40 m in height; bear opposite, 1-pinnately compound leaves; and produce small, perfect or imperfect flowers in racemose or paniculate cymes before the leaves appear. Petals may be present or absent, and the fruits are samaras with elongate wings.

No experimental studies have confirmed the toxicity of ash foliage, but there are indications of minor problems associated with its ingestion by livestock. Reeves (1966) described a problem in cattle that ate the leaves and fruit from a fallen branch of an ash tree, possibly *F. excelsior*, the introduced European ash. By the second day after consumption, the animals were unable to stand without assistance. They also exhibited abdominal pain and subcutaneous edema of the flanks and udder. The signs were almost resolved by the third day, and there appeared to be some response to administration of calcium borogluconate solution. The cause of the disease is not known, although glycosides such as aesculin and a monoterpenoid, oleuropein, are present (Cooper & Johnson 1984; Connolly & Hill 1991a). Ruminal impaction has also been considered as a problem with ingestion of the leaves by cattle (Cooper & Johnson 1984).

- trees; leaves opposite, pinnately compound; fruits samaras with elongate wings
- possible disease due to glycosides such as aesculin

HO O O Glucose

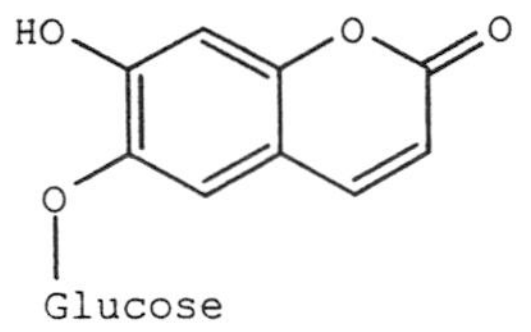

Aesculin

Jasminum L.

Commonly known as jasmine or jessamine, *Jasminum* comprises about 200 species in tropical and subtropical regions of the Old World. Its name is a Latinization of the ancient Arabic name *yasmin*. Now widely cultivated for their flower odor and color, plants of the genus are shrubs or woody climbers bearing opposite or rarely alternate, 1-pinnately compound leaves. Borne in terminal cymes or occasionally singly, their fragrant, salverform flowers have 4-to-9-lobed calyces and white, yellow, red, or pink 4-to-9-lobed corollas. The berries are 2-lobed and 2-to-4-seeded. In North America, 9 introduced species are present, including the familiar *J. nudiflorum* Lindl. (winter jasmine) and *J. officinale* L. (common white jasmine or poet's jasmine).

Although these species have not been incriminated in toxicity problems, they contain an extensive series of monoterpenoids and their glycosides—jasminin, jasminoside, jasmoside, and jasmolactones A–D—which may cause irritation of the digestive tract (Connolly & Hill 1991a).

- *J. nudiflorum*
  J. officinale

  common white jasmine
  jasmine
  jessamine
  poet's jasmine
  winter jasmine
- woody climbers; leaves pinnately compound; flowers terminal, fragrant; berries
- numerous possibly irritant monoterpenoids

References

Chesnut VK: *Preliminary Catalogue of Plants Poisonous to Stock.* USDA Bur Anim Ind 15th Ann Rep, pp387–420, 1898

Connolly JD, Hill RA: *Dictionary of Terpenoids.* Vol 1, *Mono- and Sesquiterpenoids.* Chapman & Hall, London, 1991a

Connolly JD, Hill RA: *Dictionary of Terpenoids.* Vol 2, *Di- and Higher Terpenoids.* Chapman & Hall, London, 1991b

Cooper MR, Johnson AW: *Poisonous Plants in Britain and Their Effects on Animals and Man.* Her Majesty's Stationery Office, London, pp176–177, 1984

Damirov IA: Chemical and pharmacological study of leaves of the Japanese privet grown in Apsheron. *Sb Tr Azerb Med Inst* 8;342, 1960 (*Chem Abstr* 56;9376b, 1962)

DeBates TK, Lindzey JS: food preferences of eastern wild turkeys and toxicity of common privet berries. *Res Briefs Pa Agric Exp Stn* 15(2);8–10, 1982

Frohn D, Pfander HJ: *A Colour Atlas of Poisonous Plants,* 2d ed. Wolfe Science, London, pp157–158, 1984

Harshberger JW: *Pastoral and Agricultural Botany.* Blakiston's Son & Co, Philadelphia, 1920

Inouye H, Nishioka T: Uber die monoterpenglucoside und verwandte naturstoffe—XIX. Uber die struktur des nüzhenids. Eines bitter schmeckenden glucosids aus *Ligustrum lucidum* sowie *Ligustrum japonicum. Tetrahedron* 28;4231–4237, 1972

Kerr LA, Kelch WJ: Fatal privet *(Ligustrum amurease)* toxicosis in Tennessee cows. *Vet Hum Toxicol* 41;391–392, 1999

Kozlov VA, Gylyava TN: Poisoning by the fruit of privet hedge. *Sud-Med Ekspertiza* 26;56–57, 1983

Ohta T, Miyazaki T: Components of the fruit of *Ligustrum japonicum. J Pharm Soc Jpn* 73;621–623, 1953 (*Chem Abstr* 47;9563d, 1953)

Parkinson SCJ: Suspected privet poisoning [letter]. *Vet Rec* 119;483–484, 1986

Reeves RJC: Cattle poisoning from ash leaves and fruits [letter]. *Vet Rec* 79;580, 1966
Stromayer KAK, Warren RJ, Johnson AS, Hale PE, Rogers CL, Tucker CL: Chinese privet and the feeding ecology of white-tailed deer: the role of an exotic plant. *J Wildl Manage* 62;1321–1329, 1998
Takemoto T, Yahagi N, Nishimoto N: Triterpenoids from the pericarp (with pulp) of *Ligustrum japonicum*. *J Pharm Soc Jpn* 75;737–739, 1955 (*Chem Abstr* 50;3492f, 1956)
Turner TW: Some interesting cases: poisoning by privet. *Vet Rec* 45;319–320, 1904
Yearling heifers poisoned by privet. *N Z J Agric* 59;430, 1939

Oxalidaceae R.Br.

54

Widespread in tropical and subtropical regions, the Oxalidaceae, commonly known as the wood sorrel family, comprises 7 or 8 genera and 900–1,000 species. There is considerable diversity of growth form—from tall trees to diminutive herbaceous annuals—and because of it, some taxonomists divide the family into 2 or 3 separate families. A broad interpretation is maintained in this treatise because of the similarity in all other characters.

- Wood Sorrel Family

  Oxalis

In North America, 2 genera are present. Represented by 2 species of evergreen trees introduced from tropical Southeast Asia, *Averrhoa* is grown for its edible fruits—carambolas—in southern Florida and southern California. *Oxalis* is the second genus present and has been implicated occasionally in intoxication problems.

Plants herbs or trees; annuals or perennials; from rhizomes or bulbs or stolons; caulescent or acaulescent. **Stems** erect or decumbent. **Leaves** palmately or 1-pinnately compound; alternate; cauline or basal; leaflets 3 to many; stipules present or absent. **Inflorescences** solitary flowers or simple cymes or umbels; axillary. **Flowers** perfect; chasmogamous or cleistogamous; perianths in 2-series. **Sepals** 5; free. **Corollas** radially symmetrical. **Petals** 5; free or coherent. **Stamens** 10; in 2 whorls; of 2 lengths; fused by filaments at bases; anthers dorsifixed. **Pistils** 1; compound, carpels 5; stigmas 5; styles 5, free; ovaries superior; locules 5; placentation axile. **Fruits** capsules or berries. **Seeds** numerous.

- herbs or trees; leaves compound; petals 5; stamens 10; styles 5

Related to the Geraniaceae and Linaceae, the Oxalidaceae is readily distinguished by its 10 stamens in 2 distinct whorls, 5 styles, 5-locular pistil, and compound leaves.

Oxalis L.

Taxonomy and Morphology—The largest genus in the family, *Oxalis* encompasses 800–850 species native to both the Old World and the New World. Its name is derived from the Greek root *oxys,* meaning "sour," which reflects the acrid, sour taste of the leaves. Economically, the genus is not of major importance. A few species are cultivated as garden or house ornamentals, the most famous being the Irish shamrock. A few are lawn and garden weeds, and a few are cultivated in South America for their edible tubers. In North America, approximately 30 native and introduced species are present. There has been considerable taxonomic confusion about the origins and relationships of these species, and numerous names have been published. Names have been misapplied as well causing considerable confusion. The treatment of Eiten (1963) is employed here. Of toxicologic interest are the following species:

- *O. corniculata*

  O. dillenii

  O. pes-caprae

  O. stricta

  O. violacea

  agrito

  Bermuda buttercup

  creeping lady's-sorrel

  jocoyote

  soursob

  wood sorrel

O. corniculata L.	creeping lady's-sorrel, creeping yellow wood sorrel, agrito, jocoyote
O. dillenii Jacq.	southern yellow wood sorrel

- sour-tasting herbs; leaves palmately compound, leaflets 3; petals yellow, yellowish orange, purplish pink, or white; fruits explosively dehiscent capsules

O. pes-caprae L. (= *O. cernua* Thunb.)	soursob, Bermuda buttercup, wood sorrel
O. stricta L. (= *O. europaea* Jord.)	common yellow wood sorrel, European wood sorrel
O. violacea L.	violet wood sorrel

Oxalis stricta

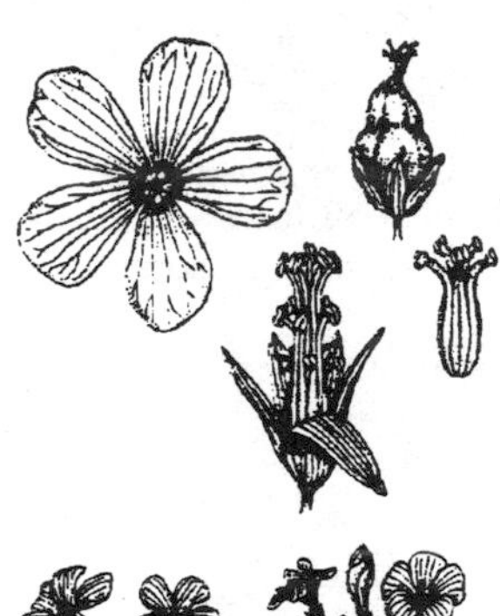

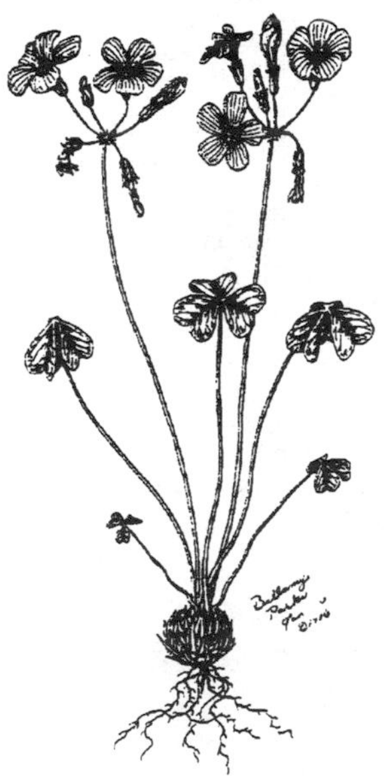

Oxalis violacea

Plants herbs; annuals or perennials; from rhizomes or scaly bulbs or taproots; caulescent or acaulescent; sap thin, sour. **Stems** erect or ascending or decumbent; 5–40 cm tall; surfaces glabrous or with septate hairs. **Leaves** palmately compound; cauline or basal; leaflets 3, obcordate, folding at night, venation pinnate. **Inflorescences** cymes or umbels or solitary flowers; bracteoles 2. **Flowers** campanulate. **Sepals** typically imbricate. **Petals** convolute; yellow to yellowish orange or purplish pink or white. **Fruits** loculicidal capsules; cylindrical to globose; explosively dehiscent. **Seeds** numerous; typically with basal arils and rugose surfaces.

1. Plants acaulescent. Leaves basal. Bulbs present.
   2. Petals yellow . . . . *O. pes-caprae*
   2. Petals violet . . . . *O. violacea*
1. Plants caulescent. Leaves cauline. Bulbs absent.
   3. Rhizomes present. Stipules absent. Hairs of stems, petioles, and pedicels septate; apices blunt . . . . *O. stricta*
   3. Rhizomes absent. Stipules present; conspicuous. Hairs of stems, petioles, and pedicels not septate; apices acute.
      4. Stems creeping; rooting at nodes. Seeds brown without white transverse ridges . . . . *O. corniculata*
      4. Stems decumbent or erect; not rooting at nodes. Seeds brown with white transverse ridges . . . . *O. dillenii*

- some species waste area weeds

Distribution and Habitat—Habitats occupied by species of *Oxalis* range from pristine natural ones in forests and grasslands to highly disturbed sites. Weedy species are encountered in lawns, gardens, waste areas, and greenhouses. The genus occurs across the continent, with some taxa exhibiting wide distributions and others narrow. Both *O. violacea* and *O. dillenii* are indigenous to the eastern half of the continent. *Oxalis stricta* is native to both eastern North America and eastern Asia, and *O. pes-caprae,* native to South Africa, is naturalized in California and Florida. Although described by Linnaeus from Italian specimens and generally cited as native to Europe, *O. corniculata* is almost cosmopolitan in distribution, and its origin best described as unknown (Eiten 1963).

- renal disease; metabolic tetany; ruminants, mainly sheep

Disease Problems—Species of *Oxalis* are typically rather innocuous and cause little problem in North America. In only rare instances of very large populations is there a risk of intoxication. The disease these species produce occurs worldwide in ruminants, mainly sheep. It may be acute, subacute, or chronic and occur after several days or months of grazing (Bull 1929; Smith 1951; McIntosh 1972; Rekhis & Amara 1990). The most typical effects, as reported in other areas of the world, are those of a subacute nephritis or chronic tetany, in some instances without a marked decrease in blood calcium. Experimentally, a number of changes were apparent in 6-month-old lambs fed exclusively *O. pes-caprae* for 6 weeks. There was a significant but not dangerous decrease in serum calcium, chloride, and albumen and an increase in phosphorus, magnesium, parathyroid hormone, and α- and γ-globulins (Manuta et al. 1985a,b). Ruminal pH decreased while urinary pH increased to 8, with increased calcium salts in urine and considerable oxalate deposition in the kidneys (Manuta &

Floris 1984; Manuta & Naitana 1984). These changes for the most part were readily reversed or prevented by providing an acidogenic feed. With long-term ingestion of plants containing high levels of oxalate, the disease is one of progressive kidney injury, and clinical signs develop when the changes culminate in renal failure.

Disease Genesis—Species of *Oxalis* are notorious as oxalate accumulators—hence the genus name. The levels of oxalate in *O. pes-caprae* are reported in ranges of 3.7–14.9% soluble and 5.9–16.6% total, and in *O. corniculata* 4.1% soluble and 7% total (Libert & Franceschi 1987). In the plants the sap pH is low, and the oxalate occurs as the free acid in the vacuoles and also as the acid/potassium salt (Michael 1959; Ranson 1965; Oke 1969). In this respect *Oxalis* is similar to *Rumex* in the Polygonaceae (see chapter 59). Blood calcium concentrations are not invariably decreased precipitously, although renal lesions of oxalate crystalluria are typical. Disease is a rarity with these plants in spite of the frequency with which they are encountered (Dodson 1959). As with other oxalate-accumulating plants, the rarity of intoxication is likely due in large measure to the activities of the ruminal microflora, which imparts tolerance when animals are gradually introduced to the plants rather than when they are suddenly required to subsist on them. Oxalate toxicity is much more likely to be caused by members of the Chenopodiaceae and Polygonaceae.

Although it may not have a role in field problems, an extract from *O. corniculata* has been shown to produce relaxant effects on smooth muscle and to cause hypotension in rats (Achola et al. 1995).

Oxalis stricta

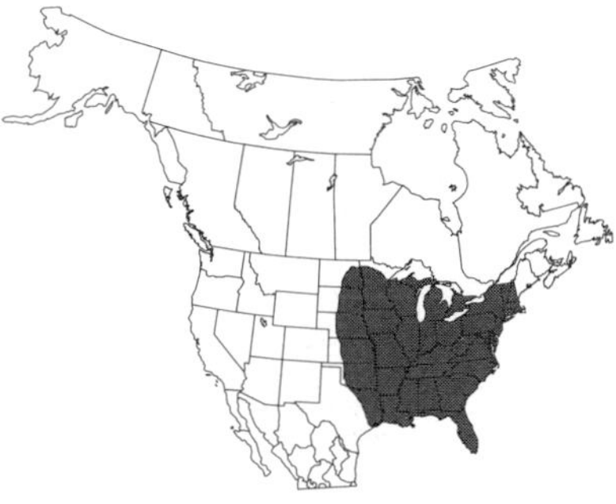

Oxalis violacea

■ accumulation of acidic oxalate salts; interaction with calcium; formation of insoluble calcium oxalate

Clinical Signs—Abrupt exposure of hungry animals to large amounts of *Oxalis* may result in onset of signs a few hours after ingestion, and the animals may die in 1–2 days. Signs progress from depression, weakness, weak pulse, and labored respiration to prostration, coma, and death.

With chronic intoxication the signs begin with a loss of weight accompanied by signs of renal failure: loss of appetite, rapid loss of condition, polyuria, edema, and elevation of blood urea nitrogen (BUN) and creatinine. Lack of urine-concentrating ability and the presence of oxalate crystals in urine sediment are also helpful aids to diagnosis.

■ abrupt onset, depression, weakness, labored respiration, collapse

slow onset, weight loss, increased urination, elevation of BUN creatinine

$H^+ \; {}^-OOC—COO^- \; H^+$

$H^+ \; {}^-OOC—COO^- \; Na^+$

$Na^+ \; {}^-OOC—COO^- \; Na^+$

$H^+ \; {}^-OOC—COO^- \; K^+$

$K^+ \; {}^-OOC—COO^- \; K^+$

$Ca^{++} \; {}^-OOC—COO^- \; Ca^{++}$

COO^- ... ${}^-OOC$

Oxalic acid and its salts

Pathology—In the more typical acute intoxications the kidneys are pale and swollen, and there is excess fluid in the body cavities. There may also be visceral edema and pulmonary congestion and edema. Microscopically, renal tubular epithelial degeneration and necrosis are seen. Other tubules are dilated and lined by a flattened epithelium.

Typically in the chronic disease the kidneys are small and fibrotic, exhibit a loss of tubules, and have scattered crystals in remaining tubules.

■ gross pathology, kidneys pale and swollen or small and fibrotic

microscopic, kidney tubules dilated, epithelium necrotic and flattened, loss of tubules

Treatment—Prompt administration of parenteral calcium solutions will generally result in immediate relief of many of the signs, and animals will become ambulatory within a few hours. Relapses may occur, and some animals may eventually die of acute renal failure. In the chronic form of the disease, general nursing care to assist the animal to survive long enough for tubular regeneration is appropriate.

■ calcium i.v., i.p.; nursing care

References

Achola KJ, Mwangi JW, Munenge RW: Pharmacological activity of *Oxalis corniculata*. *Int J Pharmacognosy* 33;247–249, 1995

Bull LB: Poisoning of sheep by soursobs *(Oxalis cernua)*. *Aust Vet J* 5;60–69, 1929
Dodson ME: Oxalate studies in the sheep. *Aust Vet J* 35;225–233, 1959
Eiten G: Taxonomy and regional variation of *Oxalis* section *Corniculatae*, 1: introduction, keys, and synopsis of the species. *Am Midl Nat* 69;257–309, 1963
Libert B, Franceschi VR: Oxalate in crop plants. *J Agric Food Chem* 35;926–938, 1987
Manuta G, Floris B: Experimental alimentary ovine oxalosis: behaviour of protein electrophoresis and alkali reserve of the blood. *Atti Soc Ital Sci Vet* 38;224–227, 1984
Manuta G, Naitana S: Experimental alimentary ovine oxalosis: hormonal and chemical changes in blood and urine. *Atti Soc Ital Sci Vet* 38;221–224, 1984
Manuta G, Naitana S, Floris B, Gallus M, Scala A: Parathyroid hormone, thyrocalcitonin, calcium, phosphorus, magnesium, zinc, copper, alkaline phosphatase, chloride, and osseous modifications in sheep fed with experimental oxalosis. *Clin Vet* 108;185–206, 1985a
Manuta G, Naitana S, Floris B, Nieddu A, Gioi F, Scala A: Haematic electrophoretic picture, alkalotic state, renal deposition of calcium salts, and their lysis in sheep sorrel or maize fed. *Clin Vet* 108;207–218, 1985b
McIntosh GH: Chronic oxalate poisoning in sheep. *Aust Vet J* 48;535, 1972
Michael PW: Oxalate ingestion studies in the sheep. *Aust Vet J* 35;431–432, 1959
Oke OL: Oxalic acid in plants and in nutrition. *World Rev Nutr Diet* 10;262–303, 1969
Ranson SL: The plant acids. In *Plant Biochemistry,* Bonner JF ed, Academic Press, New York, pp493–525, 1965
Rekhis J, Amara A: Two cases of food-poisoning by *Oxalis cernua* in goats. *Rev Med Vet* 141;8–9, 1990
Smith SW: Soursob poisoning in sheep. *J Agric South Aust* 54;377–378, 1951

Papaveraceae Juss. 55

Comprising 25–30 genera and 200–250 species, the Papaveraceae, commonly known as the poppy family, is cosmopolitan in distribution but is most abundant in the temperate and tropical regions of the Northern Hemisphere (Grey-Wilson 1993). Western North America and eastern Asia are centers of diversity. Taxonomists differ in their opinions as to whether the family should be combined with the closely related Fumariaceae (see chapter 39). They both, for example, possess similar isoquinoline alkaloids but differ in floral symmetry, sap chemistry, stamen number, and stamen fusion. In general, European taxonomists combine the 2 families, whereas Americans, with a few exceptions, keep them separate. The Papaveraceae is perhaps best known for *Papaver somniferum,* the source of opium. It also contains species that are used in herbal remedies, cultivated as ornamentals, or detested as noxious weeds. In North America, 17 genera and 63 species, native and introduced, are present (Kiger 1997). Intoxications are associated with 5 genera. Two are of questionable toxicity.

- Poppy Family

  Argemone
  Chelidonium
  Eschscholzia
  Glaucium
  Papaver

  QUESTIONABLE
  Sanguinaria
  Stylophorum

Plants herbs or subshrubs or shrubs or small trees; annuals or biennials or perennials; caulescent or acaulescent; from taproots or rhizomes; armed or not armed with prickles; sap typically viscous, yellow or orange or red or colorless. **Leaves** simple; alternate or opposite or whorled; basal and/or cauline; sessile or petiolate; venation pinnate or palmate; blades pinnately or palmately lobed or parted; margins often spinose; stipules absent. **Inflorescences** solitary flowers or cymes; terminal or axillary; bracts present or absent. **Flowers** perfect; perianths in 2-series. **Sepals** 2 or 3; caducous; free or fused; usually obovate. **Corollas** radially symmetrical; bowl-shaped; imbricate and crumpled in bud. **Petals** 4–16; in 2 or 3 whorls; free. **Stamens** numerous or rarely 4–15; in 3 or more whorls. **Pistils** 1; compound, carpels 2 to numerous; stigmas 2 to numerous; styles 1 or 0; ovaries superior; locules 1 to numerous; placentation parietal. **Fruits** capsules; poricidal or valvate or circumscissile. **Seeds** numerous; small.

- herbs, subshrubs, shrubs, or trees; sepals 2 or 3, quickly dropping; petals 4–16, in 2 or 3 whorls, free; fruits capsules
- extensive array of isoquinoline alkaloids, such as morphinans, protoberberines, benzophenanthridines, protopines

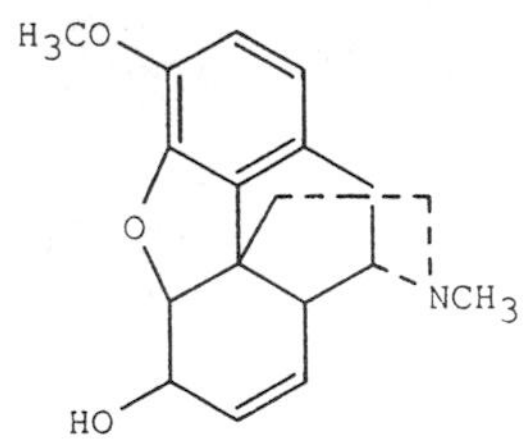

The Papaveraceae contains the greatest array of isoquinoline alkaloids of all plant families, and some of its species have been extensively exploited for social and medicinal purposes for millennia. The most important of these alkaloids are the morphinans (or morphinoids)—codeine, morphine, oripavine, and thebaine—but only one genus, *Papaver,* contains appreciable quantities of these compounds; thus, somnolence is often a feature of intoxication. The most common alkaloids of the family are the protoberberines (berberine, coptisine, palmatine, and scoulerine), the benzophenanthridines (chelirubine and sanguinarine), and the protopines (allocryptopine and protopine). Various effects reported to be produced by these common alkaloids include positive inotropic and negative chronotropic cardiac actions, weak acetylcholine or cholinesterase inhibition, and increased intraocular pressure (especially by sanguinarine) (Hakim et al. 1961; Shamma 1972; Lindner 1985; Bhakuni & Jain 1986). It should be pointed out that some of these alkaloids occur in the quaternary

H_3CO O NCH_3 HO

Codeine (a morphinan)

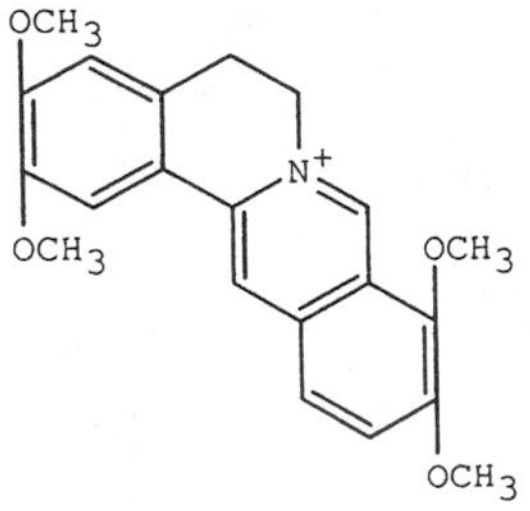

OCH_3 N^+ OCH_3 OCH_3 OCH_3

Palmatine (a protoberberine)

Chelirubine (a benzophenanthridine)

Protopine

form and thus are subject to exclusion by the blood brain barrier. However, some of the quaternary *N*-methoxy derivatives have neuromuscular blocking activity (Gozler 1987).

In contrast to the Fumariaceae, GABA-inhibitory phthalideisoquinolines are not present in the Papaveraceae, and dopaminergic inhibitory aporphines have only a limited presence (Blasko et al. 1982). A discussion of the isoquinoline alkaloid types is presented in the treatment of the Fumariaceae(see chapter 39).

Argemone L.

- *A. albiflora*
  A. mexicana
  A. munita
  A. polyanthemos

  cardo santo
  chicalote
  devil's fig
  prickly poppy
  yellow thistle

- herbs or subshrubs, prickles, white to orange viscid sap; leaves alternate; flowers conspicuous; petals 6, in 2 whorls, white, yellow, golden, bronze, or pale lavender

Taxonomy and Morphology—Native to the Americas, *Argemone,* commonly known as prickly poppy, comprises 20–32 species. Its name is derived from the Latin root *argema,* meaning "cataract of the eye." Species of the genus were reputedly once used as a treatment for the problem. Although a few species are grown as border plants for their showy flowers and prickly, thistlelike foliage, most are weedy and indicative of soil disturbance. In North America, 12 species are present. Of particular toxicologic interest are the following:

A. albiflora Hornem.	white prickly poppy
A. mexicana L.	Mexican prickly poppy, yellow thistle, yellow prickly poppy, devil's fig, chicalote, cardo santo
A. munita Durand & Hilg. (= *A. rotundata* Rydb.)	prickly poppy, armed prickly poppy, chicalote
A. polyanthemos (Fedde) G.B.Ownbey (= *A. intermedia* Sweet)	prickly poppy

Plants herbs or subshrubs; caulescent; from taproots; herbage glaucous, armed with prickles; sap viscous, white to orange. **Stems** erect; 40–150 cm tall; branched. **Leaves** alternate; basal and cauline; sessile; blades pinnately lobed or parted; margins dentate, teeth spinose. **Inflorescences** terminal cymes; bracts present, often foliaceous. **Flowers** conspicuous. **Sepals** 2 or 3; smooth or prickly; each with hollow, prickle-tipped horn. **Petals** 6; in 2 whorls; white or yellow or golden or bronze or pale lavender. **Stamens** 20–250 or more. **Pistils** with 3–7 carpels; stigmas 3-to-7-lobed; styles 1, short; locules 1. **Capsules** 3-to-7-valved; sutures grooved; prickly. **Seeds** subglobose; minutely pitted; arils present.

1. Petals yellow or golden or bronze. Stamens 25–50 .... *A. mexicana*
1. Petals white or pale lavender. Stamens 100–250.
   2. Flower buds obovoid or oblong. Sepal horns flattened ...... *A. munita*
   2. Flower buds subglobose. Sepal horns terete.
      3. Sepal horns 7–15 mm long .... *A. polyanthemos*
      3. Sepal horns 3–6 mm long .... *A. albiflora*

A key for the identification of all North American species of the genus is given by G. B. Ownbey in Kiger (1997).

Distribution and Habitat—All species of *Argemone* are native in the southern half of the continent. Most have relatively restricted distributions. *Argemone polyanthemos,* however, is widespread. Likewise, *A. mexicana* is widely distributed and is now considered a pantropical weed. Plants typically are found along roadsides, on sandy creek banks, in floodplains, and in disturbed or waste areas.

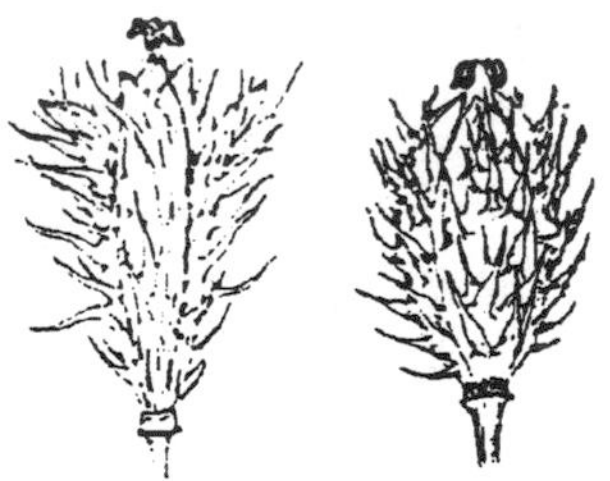

Argemone albiflora

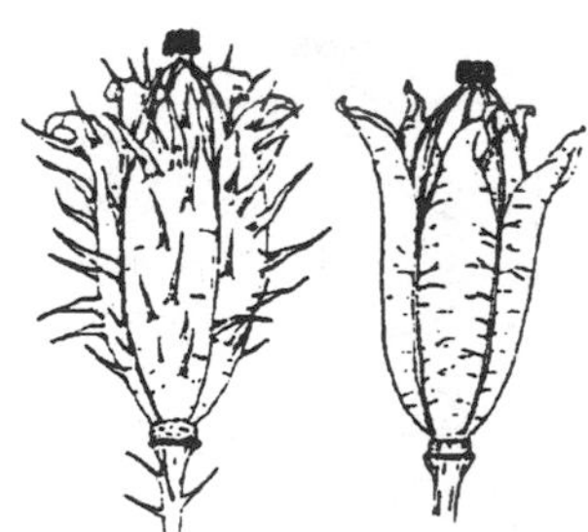

Argemone mexicana

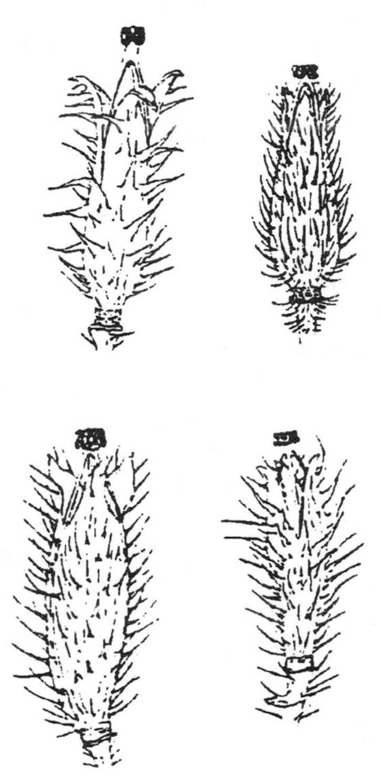

Argemone munita

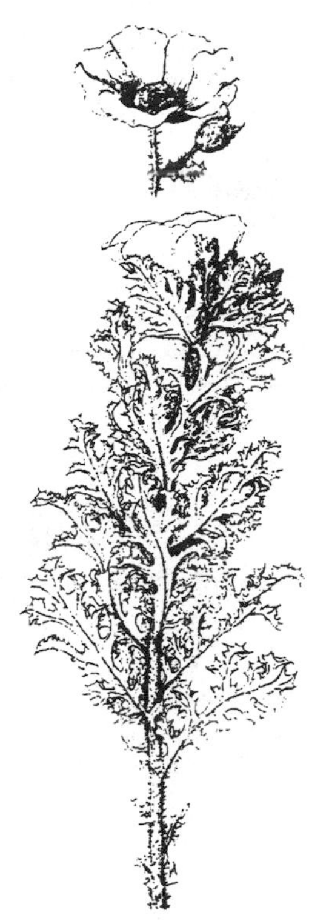

Argemone munita

Argemone polyanthemos

■ sandy soils, waste area invaders

Argemone albiflora

Argemone mexicana

Disease Problems—Highly unpalatable, the foliage of *Argemone* is rarely eaten by livestock, and the seeds seem to represent the greatest hazard when they contaminate grains fed to livestock, especially poultry. The seeds are similar in appearance to those of black mustard, and plants are common weeds in various cereal grains. As little as 1% seeds in the feed of day-old chicks caused growth depression and substantial death loss after 2 weeks of feeding (Norton & O'Rourke 1980). A diet of 100% *Argemone* seeds given to rats for several days caused sedation, depression, ventral edema, and death (Pahwa & Chatterjee 1989). Pathologic observations included mild to moderate hepatocellular necrosis and other degenerative changes in the liver and kidney, and gastric mucosal ulcerations. Similar effects were noted in cattle fed for several months with alfalfa hay contaminated with *Argemone* (Kellerman et al. 1988). There was ventral s.c. edema, ascites, hydrothorax, and centrilobular and portal fibrosis. Although *A. mexicana* is reputed to have antifertility effects in folk medicine of India, experimentally neither aqueous nor alcoholic extracts of the species had an effect on implantation in rats (Bodhanker et al. 1974).

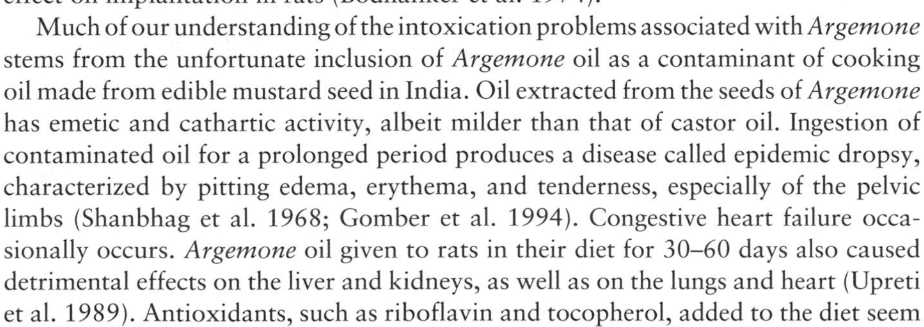

Much of our understanding of the intoxication problems associated with *Argemone* stems from the unfortunate inclusion of *Argemone* oil as a contaminant of cooking oil made from edible mustard seed in India. Oil extracted from the seeds of *Argemone* has emetic and cathartic activity, albeit milder than that of castor oil. Ingestion of contaminated oil for a prolonged period produces a disease called epidemic dropsy, characterized by pitting edema, erythema, and tenderness, especially of the pelvic limbs (Shanbhag et al. 1968; Gomber et al. 1994). Congestive heart failure occasionally occurs. *Argemone* oil given to rats in their diet for 30–60 days also caused detrimental effects on the liver and kidneys, as well as on the lungs and heart (Upreti et al. 1989). Antioxidants, such as riboflavin and tocopherol, added to the diet seem

- cardiac insufficiency; mainly from seeds and seed oil products

to protect against development of the adverse effects of *Argemone* oil in rats and to decrease severity of some of the signs of epidemic dropsy (Upreti et al. 1991; Kumar et al. 1992). Contamination of wheat with *Argemone* seeds also is of concern for humans, because it has been associated with constipation, generalized edema, glaucoma, and rarely death (Concon 1988).

Argemone munita

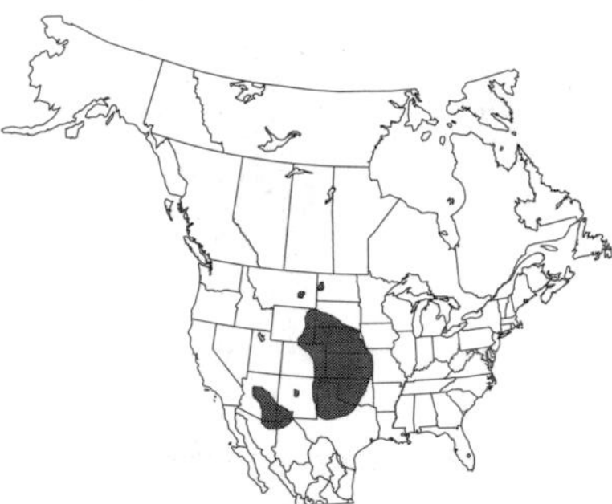

Argemone polyanthemos

- numerous alkaloids; reticuline, sanguinarine, dihydrosanguinarine

Disease Genesis—Alkaloids present in species of *Argemone* include protopines, protoberberines, and benzophenethridines. The protopine types are most common, although the berberines are also well represented (Stermitz 1968). In addition, *A. mexicana* contains reticuline, a simple isoquinoline, and *A. munita* has pavines (Preininger 1986). The seed oil contains mainly dihydrosanguinarine and lesser amounts of sanguinarine and berberine (Sarkar 1948; Fletcher et al. 1993). The toxic effects appear to be due to sanguinarine rather than to the much less toxic dihydrosanguinarine. These alkaloids act as enzyme inhibitors; they react with sulfhydryl groups and also probably cause oxygen radical lipid peroxidation (Sarkar 1948; Das et al. 1991; Upreti et al. 1991). In vitro studies indicate that sanguinarine inhibits Na^+,K^+-ATPase activity and glucose transport across the small intestinal wall (Tandon et al. 1993). Sanguinarine and other alkaloids cause an increase in intraocular pressure, and it has been suggested that extensive culinary use of seeds of *Argemone* and some of the other species of the Papaveraceae may increase the risk of glaucoma (Hakim et al. 1961).

Reticuline

Dihydrosanguinarine

- sluggishness, passiveness, weakness, sedation

Clinical Signs—The primary sign of *Argemone* intoxication in poultry is growth depression. Additional signs present in poultry and other species include sluggishness, passiveness, sedation, and either constipation or diarrhea.

- gross pathology, fluid accumulation

Pathology—The characteristic effect in all animals is fluid accumulation in the tissues. Thus, typical gross pathologic findings are edema of various tissues, including the subcutis; free fluid in the abdomen and chest; and congestion of the viscera. Hepatic fibrosis and erosions of the mucosa of the digestive tract are additional changes.

- nursing care

Treatment—Treatment is focused on good nursing care and alleviation of the specific symptoms.

Chelidonium L.

- *C. majus*

celandine
poppywort
rock poppy
swallowwort
wartwort

Taxonomy and Morphology—A monotypic genus, *Chelidonium* is native to Eurasia. Its name was used by Dioscorides and is derived from the Greek root *chelidon*, meaning "swallow" (the bird). Natural historians differ in their opinions as to what the name indicates. Some contend that it reflects the blooming of the flowers when the swallows arrive in the spring and their withering when the birds depart. Others suggest that the name reflects the folklore, as reported by Aristotle, that mother swallows bathe the eyes of their young with the taxon's sap (Kiger 1997). This seems unlikely, however, because the sap is quite irritating and can cause blistering of the skin. It was once used as a cure for warts. The single species is the following:

C. majus L. celandine, greater celandine, poppywort, wartwort, rock poppy, swallowwort

Plants herbs; biennials or perennials; caulescent; from stout rhizomes or taproots; herbage not armed with prickles; sap viscous, yellow to orange, typically turning red upon exposure to air. **Stems** erect; 30–100 cm tall; branched; ribbed; pilose-villous. **Leaves** alternate; basal and cauline; petioled; blades 1- or 2-pinnately parted; lobes 5–9; margins dentate or crenate. **Inflorescences** umbellate; terminal or axillary; few-flowered; bracts present. **Flowers** small. **Sepals** 2; free; glabrous or indumented. **Petals** 4; bright yellow; obovate to oblong. **Stamens** 12 to many. **Pistils** with 2 carpels; stigmas 2-lobed; styles 1, short; locules 1. **Capsules** linear to narrowly oblong; 2-valved; dehiscing from base; glabrous. **Seeds** subglobose; black; pitted; arils present.

- erect herbs, branched, not armed, yellow to orange viscous sap turns red upon exposure to air; leaves alternate, 5-to-9-lobed; flowers small; petals 4, bright yellow

In vegetative condition, *C. majus* is morphologically very similar to *Stylophorum diphyllum,* and the two species can be confused. Typically, *C. majus* has more leafy stems than *S. diphyllum,* which tends to have leafless stems except for 2 or 3 leaves subtending inflorescences. The flowers of the two taxa are quite different, and when they are present, identification is easy.

Chelidonium majus

Distribution and Habitat—Introduced from Eurasia, *C. majus* is naturalized in the northeastern quarter of North America. Scattered populations also occur in the Northwest. Rather weedy, it is found in woods, thickets, fields, fencerows, railroad rights-of-way, roadsides, and waste areas. Plants are occasionally cultivated in shady sites.

Disease Problems and Genesis—Plants of *C. majus* have a fetid odor and an acrid taste, which make them disagreeable. As a result they are not likely to be eaten in quantities sufficient to cause intoxications. However, a few cases have been reported, and the species also is used therapeutically to relieve bloat, nausea, and cramps due to intestinal spasms (Cooper & Johnson 1984). The effects are due to a wide array of aporphine, protoberberine, benzophenanthridine alkaloids, and sparteine (Preininger 1986). *Chelidonium majus* contains allocryptopine, berberine, chelerythrine, chelidonine, coptisine, protopine, sanguinarine, and stylopine (Haberlein et al. 1996). Spasmolytic activity is due to the effects of alkaloids, such as protopine, on the GABAa receptor complex, causing interference with the binding of agonists (Häberlein et al. 1996). The other alkaloids present seem to provide cooperative effects via modulation of the chloride channel.

Chelidonium majus

- digestive disturbance, neurologic effects; alkaloids such as berberine, sparteine

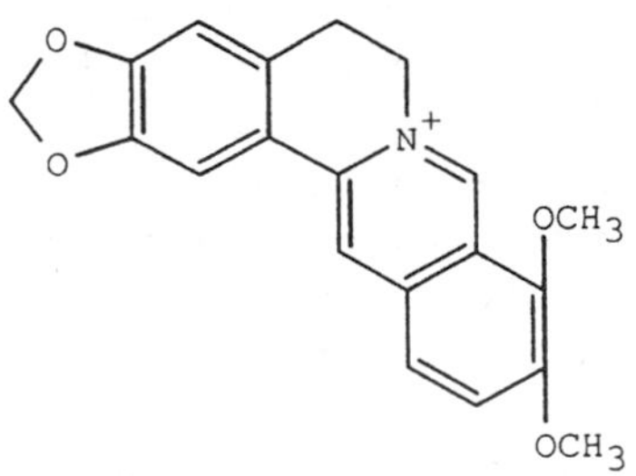

Berberine

Sparteine

- constipation, later diarrhea, excess salivation, drowsiness, ataxia

general nursing care

Clinical Signs, Pathology, and Treatment—There may be signs of purgation, but early in the course of the disease, constipation is likely. This may be accompanied by increased salivation and urination, drowsiness, and ataxia. When the animal is touched or aroused, mild seizures may be elicited. There are no distinctive gross or histopathologic changes. Treatment is focused on nursing care and relief of the specific symptoms.

Eschscholzia Cham.

- *E. californica*

California poppy

Taxonomy and Morphology—Comprising 12 species native to western North America, *Eschscholzia* is cultivated worldwide in warm-temperate regions as an ornamental. Its name honors Johann von Eschscholtz, an Estonian physician and naturalist, who botanized along the Pacific coast in 1816. Most familiar and representative of the genus is the following species:

E. californica Cham. California poppy

- herbs from large taproots, sap colorless; leaves alternate, 1-to-4-pinnately lobed; flowers large, showy; petals 4, yellow to orange, often with orange spot at base

Plants herbs; perennials or annuals; caulescent; from large taproots; herbage glaucous, blue-gray, not armed with prickles; sap colorless. **Stems** erect or spreading; 5–60 cm tall. **Leaves** alternate; basal and cauline; petioled; blades 1-to-4-pinnately lobed; lobes typically 3; margins deeply dissected. **Inflorescences** solitary flowers or cymes; terminal; bracts present. **Flowers** large; showy. **Sepals** 2; fused; forming a cap and deciduous as a unit; glabrous or glaucous. **Petals** 4; yellow to orange, typically with orange spot at base; obovate to oblanceolate. **Stamens** 12 to many. **Pistils** with 2 carpels; stigmas 4-to-8-lobed; styles absent; locules 1. **Hypanthia** present; receptacles expanded, forming flat or rimmed cup below calyx. **Capsules** cylindrical; 2-valved; dehiscing from base, often explosively. **Seeds** numerous; globose to ellipsoid; brown to black; reticulate; arils absent.

The taxon is highly variable, and numerous infraspecific taxa and cultivars have been described.

Eschscholzia californica

- dry, sandy, open sites

Distribution and Habitat—Distributed from the Columbia River south to northern Mexico and extreme southwestern Texas, *E. californica* occurs naturally in dry, grassy, open sites in upland areas. It is widely planted across the continent as an ornamental along roadsides and in wildflower plots. Sandy soils are preferred, and the species ranges into desert areas.

- toxicity unknown; known to be analgesic, soporific, sedative

Disease Problems—Species of *Eschscholzia,* although widespread in areas where livestock graze, have not been associated with intoxication problems. As with other genera of the family, plants of this genus probably are quite disagreeable to the taste and thus are seldom eaten. They were known to Indian tribes in California to be soporific and analgesic and were eaten after being leached and cooked (Ebeling 1986). Extracts are used as herbal medicament sedatives (Schafer et al. 1995). Experimentally, aqueous extracts in large dosage i.p. produced sedation in mice (Hakim et al. 1961). These properties are borne out by other studies in mice in which a dose of 100 mg/kg b.w. of dried plant caused a decrease in activity and increased the sleep time of a given dose of pentobarbital (Rolland et al. 1991). Interestingly, low doses produced an anxiolytic effect—increasing exploratory activity.

- numerous alkaloids; some specific for the genus; escholinine, eschscholtzidine

Disease Genesis—*Eschscholzia californica* contains aporphine, pavine, protopines (escholinine and eschscholtzidine), protoberberine, benzophenanthridine, and simple isoquinoline alkaloids. It lacks the opioid morphinans and papaverine (Preininger 1986). The alkaloidal content of plants of this taxon indicates that they should be considered potentially toxic and substantiates historical suspicions about their toxicity (Everist 1981).

Eschscholzia californica

Escholinine

Eschscholtzidine

Glaucium Mill.

Taxonomy and Morphology—Commonly known as horned poppy, because of the distinctive appearance of its fruit, *Glaucium* comprises 20–25 species native to the Old World. Its name is derived from the Greek *glaukos*, meaning "gray-green," which is indicative of the appearance of the herbage. Of importance in North America are the following species:

■ *G. corniculatum*
G. flavum
horned poppy
sea poppy
yellow horned poppy

G. corniculatum (L.) Rudolph	horned poppy
G. flavum Crantz	yellow horned poppy, sea poppy

Plants herbs; annuals or biennials or perennials; caulescent; from taproots; herbage glaucous, not armed with prickles; sap yellow. **Stems** erect or spreading; 30–90 cm tall; branched. **Leaves** alternate; basal and cauline; basal petiolate; cauline sessile; blades 1-to-2-pinnately lobed; lobes 7–9; margins serrate or incised; bases of cauline leaves clasping. **Inflorescences** solitary flowers; terminal or axillary; bracts present. **Flowers** showy. **Sepals** 2; free. **Petals** 4; yellow or orange to reddish orange, typically with blackish or reddish violet spot at base. **Stamens** numerous. **Pistils** with 2 carpels; stigmas 2-lobed; styles absent; locules 2. **Capsules** elongate; cylindrical; straight or curved; apices 2-lobed; 2-valved; dehiscing from base. **Seeds** numerous; dark brown; reticulate pitted; arils absent.

■ herbs with yellow sap; leaves alternate, 1-to-2-pinnately lobed; flowers showy; petals 4, yellow, orange, or reddish orange, often with blackish or reddish violet spot at base; capsules long, cylindrical, with 2-lobed apices

1. Capsules appressed pubescent. Petals orange to reddish orange; with blackish basal spot. Cauline leaf bases not clasping. Basal leaves few; glabrous or sparsely indumented . . . . *G. corniculatum*
1. Capsules glabrous or scabrous. Petals yellow or orange-yellow; with reddish violet basal spot. Cauline leaf bases clasping. Basal leaves numerous; densely indumented . . . . *G. flavum*

Glaucium flavum

Distribution and Habitat—Although sometimes planted as ornamentals in gardens, both species are more commonly encountered as weeds. Both have been introduced from Europe and become pests in pasture and cropland in the southern Great Plains. *Glaucium flavum* may be a particular problem in alfalfa fields.

■ pasture and cropland invaders

Glaucium flavum

■ toxicity unknown; numerous alkaloids with potential to cause adverse effects; glaucine

Disease Problems and Genesis—Little information is available about the toxicity of *G. flavum*, even though it has been used for medicinal purposes such as improvement of bile flow and alleviation of stomachache and ulcers (Grey-Wilson 1989). Plants contain protoberberines, protopines, aporphines (glaucine), and benzophenanthridines; the latter two types predominate (Preininger 1986; Daskalova et al. 1988). Reaching levels as high as 5.4% d.w., alkaloid concentrations are highest about the time of flowering and fruiting (Gasic & Popovic 1985; Gorunovic et al. 1985). Glaucine has dopaminergic inhibitory or adrenolytic activity and causes a variety of effects, including decreased blood pressure, depression of respiration, mild muscle relaxation, sedation, and seizures at high dosage (Southon & Buckingham 1989).

H_3CO, H_3CO, N, CH_3, H_3CO, OCH_3

Glaucine

Papaver L.

■ *P. nudicaule*
P. orientale
P. rhoeas
P. somniferum

Arctic poppy	Icelandic poppy
carnation poppy	opium poppy
corn poppy	oriental poppy
field poppy	poppy
Flanders poppy	white poppy

Taxonomy and Morphology—Primarily an Old World genus of 70–100 species, *Papaver*, commonly known as poppy, is prized as an ornamental for its showy flowers. It also is the source of opium. Its name is derived from the Greek root *pappa*, meaning "milk" or "food," an allusion to the thick, milky sap. In North America, 16 native and introduced species are present. Of particular toxicologic interest are the following:

P. nudicaule L.	Icelandic poppy, Arctic poppy
P. orientale L.	oriental poppy
P. rhoeas L.	field poppy, corn poppy, Flanders poppy
P. somniferum L.	opium poppy, carnation poppy, white poppy

■ herbs with white, yellow, orange, or red viscous sap; leaves alternate; flowers showy; petals 4–6, white, yellow, red, or purple

Plants herbs; annuals or biennials or perennials; caulescent or acaulescent; from taproots; not armed with prickles; sap viscous, white or yellow or orange or red. **Stems** erect or ascending; 5–150 cm tall; branched or unbranched; typically indumented. **Leaves** alternate; basal and/or cauline; sessile or petiolate; blades unlobed or 1-to-3-pinnately lobed or parted; margins entire or toothed or incised. **Inflorescences** solitary flowers or cymes; borne on long scapes or peduncles; bracts present. **Flowers** showy. **Sepals** 2 or rarely 3; persistent; free. **Petals** 4 or rarely 6; caducous or persistent; white or yellow to red or purple. **Stamens** numerous. **Pistils** with 3–22 carpels; stigmas 3-to-22-lobed, velvety; styles absent; locules 1. **Capsules** poricidal or valvate below stigma. **Seeds** numerous; minutely pitted; arils absent.

Papaver nudicaule

1. Plants acaulescent. Leaves all basal .... *P. nudicaule*
1. Plants caulescent. Leaves cauline and basal.
   2. Cauline leaf bases cordate clasping .... *P. somniferum*
   2. Cauline leaf bases not clasping.

3. Plants perennials. Petals 4–8 cm long . . . . *P. orientale*
3. Plants annuals. Petals 1.5–4 cm long . . . . *P. rhoeas*

A key for the identification of all North American species of the genus is given by R. W. Kiger and D. E. Murray in Kiger (1997).

Distribution and Habitat—Species of *Papaver* occur across the continent. Of the native species, some have broad distributions, whereas others are quite restricted. Introduced Old World species have naturalized as crop weeds and roadside waifs. It is illegal to cultivate *P. somniferum* in the United States.

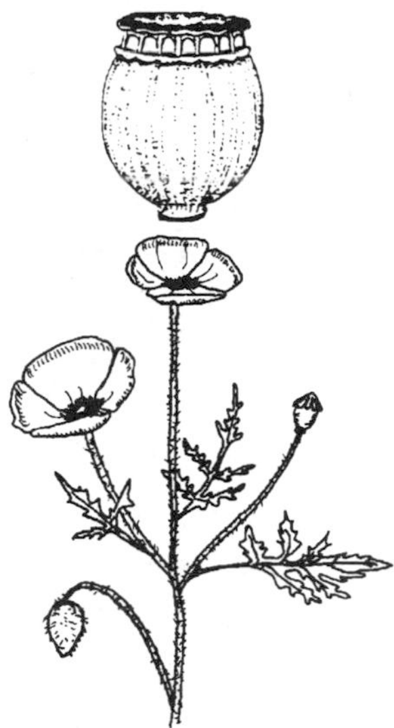

Papaver rhoeas

Disease Problems—Although various species of *Papaver* are cultivated and prized as ornamentals, the species of greatest economic importance is *P. somniferum*. Cultivated for millennia, it is considered either a blessing or a curse, depending upon whether the opium derived from its sap is used to combat insufferable pain or abused as a soporific. The Greek word *opium* means "juice" and reflects use of the sap, and indeed much of history is interwoven with this species. Records of ancient civilizations such as those of the Sumerians and Egyptians make reference to it, as do the writings of the Indians, Greeks, and Romans. Today, its cultivation and the physiological and mental addiction of thousands of individuals to its alkaloids is an apparently unresolvable societal problem. Another species, *P. rhoeas,* is immortalized as a symbol of World War I because so many soldiers died in European poppy fields.

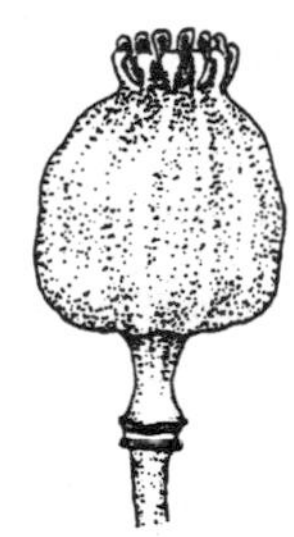

Papaver somniferum

- *P. somniferum* a major historical influence as source of opium
- digestive disturbance, neurologic effects

The alkaloids of *Papaver* produce marked effects on the digestive tract as well as the nervous system. Severe miosis is a prominent autonomic effect mediated through the oculomotor nerve nucleus. The poppy most likely to be a problem for animals in North America is *P. nudicaule.* Almost invariably problems involve accidental or intentional disposal of refuse or trimmings into an area where they can be eaten by sheep, cattle, or horses (McLennan 1929, 1930; Terblanche & Adelaar 1964; Malmanche 1970). Fresh green plants seem to be completely unacceptable, whereas wilted plants are eaten more readily. Because of the many alkaloids present, all species of *Papaver* should be considered toxic if eaten in quantity.

Although cultivation of the opium poppy and nonprescriptive use of its infamous alkaloids are forbidden in the United States, poppyseed toppings on cakes and breads contain sufficient alkaloids to cause positive urine tests. As little as 25 g of poppyseeds is sufficient to cause a positive urine screening test for opiates when a concentration of 300 ng/ml opioids is used as the threshold level, as recommended by the National Institute of Drug Abuse (Meneely 1992). However, this is not sufficient to cause opiate symptomatology in an adult person. In contrast, when the higher U.S. Department of Defense level of 4,000 ng/ml morphine or 2,000 ng/ml codeine was used, ingestion of poppyseed cake, bagels, muffins, or rolls was not sufficient to give a positive test, but Danish pastry or streusel with poppyseeds predisposed to a positive urinary test (Selavka 1991). These results should be taken as a warning against possible opioid intoxication in pets allowed to consume products with poppyseeds.

H_3CO O NCH_3 H_3CO

Thebaine

Disease Genesis—Species of *Papaver* are renowned for their morphinoid alkaloids, which are agonists of the opioid receptor. In addition, numerous other isoquinoline types are present in the genus (table 55.1).

Opium, the residue of the sap from *P. somniferum,* is a crude alkaloidal mixture of codeine, morphine, thebaine, papaverine, and other compounds of lesser toxicologic importance. These alkaloids produce effects distinctive to *P. somniferum* and

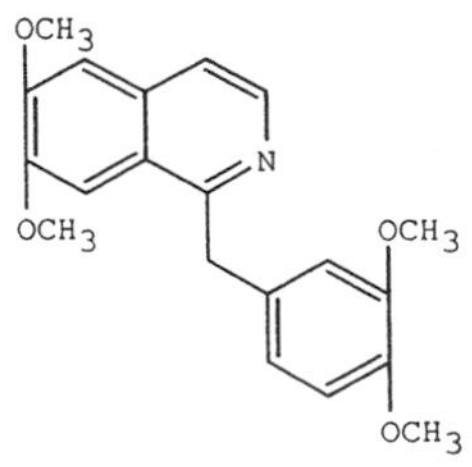

Papaverine

Papaver nudicaule

Papaver rhoeas

Table 55.1. Isoquinoline alkaloid types present in four species of *Papaver*

P. nudicaule	*P. orientale*	*P. rhoeas*	*P. somniferum*
proaporphines	aporphines	aporphines	aporphines
benzophenanthridines	proaporphines	benzophenanthridines	benzophenanthridines
isopavines	benzophenanthridines	morphinans (thebaine)	morphinans
protoberberines	morphinans (thebaine)	protoberberines	phthalideisoquinolines
protopines	promorphinans	protopines	promorphinans
promorphinans	protoberberines	rhoeadines	protoberberines
retroprotoberberines	retroprotoberberines	secophthalideisoquinolines	protopines
rhoeadines	benzylisoquinolines		benzylisoquinolines
			rhoeadines
			secophthalideisoquinolines

Sources: Guinaudeau & Shamma 1982; Preininger 1986.

perhaps a few other closely related species. Most of the other species of the genus contain a much different array of alkaloids; they lack the morphinoids, except for a small amount of thebaine. However, most contain sufficient alkaloids to produce some neurotoxic effects, mainly depressant in nature. The seeds may not present a risk from alkaloids, given that those of *P. somniferum* did not produce adverse effects when fed to rats for 10 days (Grant et al. 1995).

■ morphinoid alkaloids; depressant neurologic actions

Papaverine, a well-known smooth muscle spasmolytic found principally in *P. somniferum*, has been associated with chronic hepatotoxicity when used on a long-term basis (Lindner 1985). It is metabolized by the hepatic cytochrome P-450 system to conjugated phenolics, and glutathione depletion is suggested as a mechanism of intoxication (Davila et al. 1991).

■ *P. somniferum*, restlessness in some instances, analgesia, drowsiness, decreased respiration and gut activity

later coma, miosis

Clinical Signs—Signs associated with the morphinoids include analgesia, drowsiness, mood alteration, decreased respiration, decreased gastrointestinal motility, and in some cases nausea and vomiting. Other signs may prevail with large dosage, inconsistent with the extent of exposure that might be expected to occur in animals. The triad of signs suggestive of opioid exposure in humans is coma, miosis, and decreased respiration. With suspected exposure in dogs there was restlessness but with a preference for lying down, ataxia, a staring expression, and the appearance of being hypnotized (Odendaal 1986). Recovery was uneventful, although somewhat slow. In cattle, *P. somniferum* is reported to cause restlessness, loss of appetite, decreased temperature, and deep sleep. It was not fatal, but there was lingering debilitation (Cooper & Johnson 1984).

■ *P. nudicaule*, restlessness, increased activity, ataxia, stiffness, twitching, limb spasms

The signs with *P. nudicaule* are more suggestive of excitatory effects on the nervous system. Although in some cases animals are found dead, in others there is initial restlessness and increased activity followed by incoordination, stiffness, twitching, falling, and jerky limb spasms. The last sign may be of short duration but repeated frequently. Death is unlikely and rapid recovery is the rule, but in a few animals there may be lingering debilitation.

■ no lesions; sedation

Pathology and Treatment—Few gross or microscopic changes are evident other than visceral congestion and some scattered small hemorrhages. Alleviation of the signs is based mainly on symptomatic treatment, including the use of tranquilizers or sedatives. Acepromazine has been used with positive results (Malmanche 1970).

Genera of Questionable Toxicity

Sanguinaria L.

A monotypic genus, *Sanguinaria,* commonly known as bloodroot or red puccoon, is native to North America. Its name is derived from the Latin *sanguis,* meaning "blood," which reflects the deep orange to red color of the sap. The single species, *S. canadensis* L., lacks aerial stems and its leaves and flowering scapes arise from thick, branching rhizomes. The orbicular to reniform or cordate leaves are 5-to-7-palmately lobed with crenate margins. Arising at the ends of bractless scapes, the 1 or rarely 2 or 3 flowers are white to pinkish, with 2 free sepals; 6–12 unequal, oblong to oblanceolate petals; and numerous stamens. The 2-carpellate, 1-locular pistil has a distinct style and a 2-lobed stigma and produces an erect capsule that dehisces from the base. A perennial, the species is widespread in moist woods throughout eastern North America.

- *S. canadensis*

  bloodroot
  red puccoon

- herbs without aerial stems and with deep orange to red sap; leaves palmately lobed; flowers white to pinkish

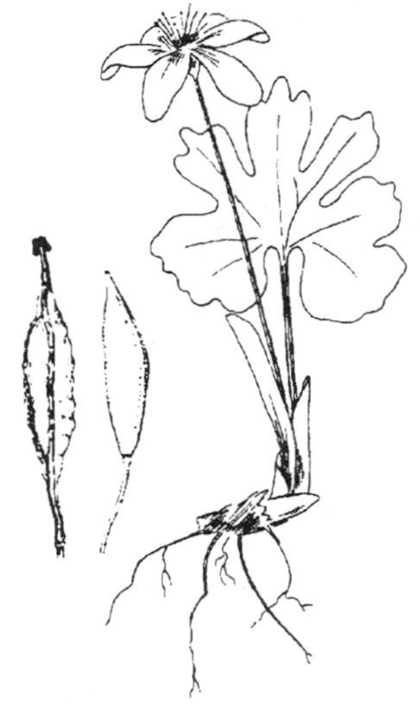

Sanguinaria canadensis

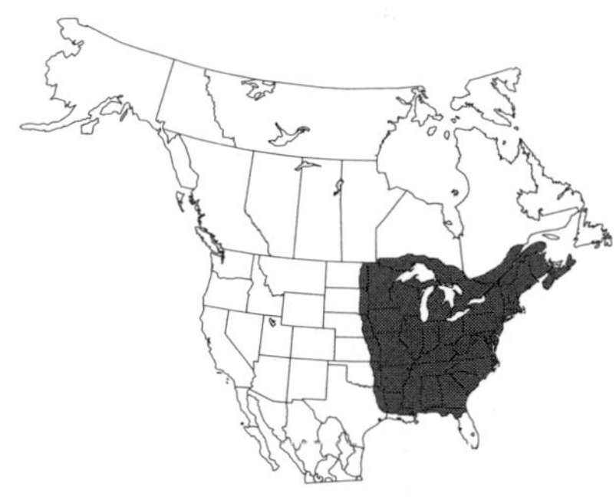

Sanguinaria canadensis

Although *Sanguinaria canadensis* is typically listed as a toxic plant, extracts from its rhizomes have long been used in folk medicines for a variety of ailments and as an ingredient in over-the-counter oral hygiene products (Mahady et al. 1993). Its use is somewhat controversial, and in the popular press, articles either extol its virtues or warn of its toxicity. *Sanguinaria canadensis* does contain protoberberines and benzophenanthridines (Preininger 1986), and sanguinarine is known to play a role in the toxicity of other members of the Papaveraceae, as described in other genus treatments above. It exhibits antibacterial, antitumor, and cytotoxic activity as well as suppression of collagenase, inhibition of Na^+,K^+-ATPase activity and inhibition of tubulin assembly (Hartwell 1960; Caolo & Stermitz 1979; Dzink & Socransky 1985; Tandon et al. 1993). It also increases intraocular pressure (Hakim et al. 1961).

- toxicity potential unclear; several alkaloids present known to cause adverse effects; sanguinarine

Stylophorum Nutt.

Comprising 3 species native to China and North America, *Stylophorum* is represented on this continent by a single species, *S. diphyllum* (Michx.) Nutt., which is commonly known as celandine poppy. Derived from the Greek roots *stylos* and *phoros,* which mean "style" and "bearing," its name reflects the long style, which is unusual in the

- *S. diphyllum*

  celandine poppy

- herbs; leaves 5-to-7-pinnately lobed; flowers showy; style elongate; petals yellow

Stylophorum diphyllum

- toxicity unknown; contains numerous alkaloids

Papaveraceae. At anthesis, the stems, which arise from rhizomes, are 30–50 cm tall. The leaves are both basal and cauline; the former are petiolate, and the latter sessile and subopposite. With irregularly dentate or crenate margins, the leaf blades are deeply 5-to-7-pinnately lobed. The inflorescences are umbellate, and leaflike bracts are present. The showy flowers have 2 free sepals; 4 yellow, obovate petals; many stamens; and a 2-to-4-carpellate pistil with 1 locule, an elongate style, and a 3-to-5-lobed stigma. The ellipsoid, pubescent capsules bear pale brown seeds.

A native perennial in rich moist woods of eastern North America, *Stylophorum* is vegetatively very similar to the Eurasian introduction *Chelidonium majus,* and the two species can be confused. Typically, *S. diphyllum* tends to have leafless stems except for the 2 or 3 leaves subtending the inflorescence, whereas *C. majus* has leafy stems. The flowers of the two taxa, however, are quite different, and when they are present, identification is easy.

Although *S. diphyllum* has not yet been associated with toxicosis, it should be considered a potential threat because it contains aporphines, protopines, protoberberines, and benzophenanthridines (Preininger 1986).

References

Bhakuni DS, Jain S: Protoberberine alkaloids. In *The Alkaloids,* vol 28, Brossi A ed, Academic Press, Orlando, Fla, pp95–181, 1986

Blasko G, Gula DJ, Shamma M: The phthalideisoquinoline alkaloids. *J Nat Prod* 45;105–122, 1982

Bodhankar SL, Garg SK, Mathur VS: Antifertility screening of plants. Part IX. Effect of five indigenous plants on early pregnancy in female albino rats. *Indian J Med Res* 62;831–837, 1974

Caola MA, Stermitz FR: Benzophenanthridinium salt equilibria. *Heterocycles* 12;11–14, 1979

Concon JM: *Food Toxicology.* Part B, *Contaminants and Additives.* Marcel Dekker, New York, pp1001–1031, 1988

Cooper MR, Johnson AW: *Poisonous Plants in Britain and Their on Animals and Man.* Ref Book 161. Her Majesty's Stationery Office, London, pp180–184, 1984

Das M, Upreti KK, Khanna SK: Biochemical toxicology of *Argemone* oil: role of reactive oxygen species in iron catalysed lipid peroxidation. *Bull Environ Contam Toxicol* 46;422–430, 1991

Daskalova E, Iskrenova E, Kiryakov HG, Evstatieva L: Minor alkaloids of *Glaucium flavum. Phytochemistry* 27;953–955, 1988

Davila JC, Davis PJ, Acosta D: Changes in glutathione and cellular energy as potential mechanisms of papaverine-induced hepatotoxicity in vitro. *Toxicol Appl Pharmacol* 108;28–36, 1991

Dzink JL, Socransky SS: Comparative in vitro activity of sanguinarine against oral microbial isolates. *Antimicrob Agents Chemother* 27;663–665, 1985

Ebeling W: *Handbook of Indian Foods and Fibers of Arid America.* Univ California Press, Berkeley, 1986

Everist SL: *Poisonous Plants of Australia,* 2d ed. Angus & Robertson, London, 1981

Fletcher MT, Takken G, Blaney BJ, Alberts V: Isoquinoline alkaloids and keto-fatty acids of *Argemone ochroleuca* and *Argemone mexicana* (Mexican poppy) seed. 1. An assay method and factors affecting their concentration. *Aust J Agric Res* 44;265–275, 1993

Gasic O, Popovic M: Dynamics of the glaucine content in *Glaucium flavum. Fitoterapia* 56;37–38, 1985

Gomber S, Daral TS, Sharma PP, Faridi MM: Epidemic dropsy in Trans Yamuma areas of Delhi and U.P. *Indian Pediatr* 31;671–674, 1994

Gorunovic M, Pekic N, Adamovic D, Lepojevic Z, Lukic P: New results of experimental cultivation of horned poppy in Siberia. *Pharmazie* 40;656–657, 1985

Gozler B: Pavine and isopavine alkaloids. In *The Alkaloids,* vol 31, Brossi A ed, Academic Press, San Diego, Calif, pp317–389, 1987

Grant G, More LJ, McKenzie NH, Dorward PM, Buchan WC, Telek L, Pusztai A: Nutritional and haemagglutination properties of several tropical seeds. *J Agric Sci* 124;437–445, 1995

Grey-Wilson C: *Glaucium flavum. Kew Mag* 6;147–151, 1989

Grey-Wilson C: *Poppies: A Guide to the Poppy Family in the Wild and in Cultivation.* Timber Press, Portland, Oreg, 1993

Guinaudeau H, Shamma M: The protopine alkaloids. *J Nat Prod* 45;237–246, 1982

Häberlein H, Tschiersch K-P, Boonen G, Hiller K-O: *Chelidonium majus* L.: components with in vitro affinity for the GABAa receptor. Positive cooperation of alkaloids. *Planta Med* 52;227–231, 1996

Hakim SAE, Mijovic V, Walker J: Distribution of certain poppy-fumaria alkaloids and a possible link with the incidence of glaucoma. *Nature* 189;198–201, 1961

Hartwell JL: Plant remedies for cancer. *Cancer Chemother Rep* 7;19–29, 1960

Kellerman TS, Coetzer JAW, Naude TW: *Plant Poisonings and Mycotoxicoses of Livestock in Southern Africa.* Oxford Univ Press, Cape Town, South Africa, 1988

Kiger RW: Papaveraceae. In *Flora of North America,* vol 3, *Magnoliophyta: Magnoliidae and Hamamelidae,* Flora of North America Editorial Committee ed, Oxford Univ Press, New York, pp300–339, 1997

Kumar A, Husain F, Das M, Khanna SK: An outbreak of epidemic dropsy in the Barabanki District of Uttar Pradesh, India: a limited trial for the scope of antioxidants in the management of symptoms. *Biomed Environ Sci* 5;251–256, 1992

Lindner E: Structure activities and pharmacological properties of opium alkaloids. In *The Chemistry and Biology of Isoquinoline Alkaloids,* Phillipson JD, Roberts MF, Zenk MH eds, Springer-Verlag, Berlin, pp38–46, 1985

Mahady GB, Schilling AB, Beecher CWW: *Sanguinaria canadensis* L. (Sanguinarius): in vitro culture and the production of benzophenanthridine alkaloids. In *Medicinal and Aromatic Plants 5, Biotechnology in Agriculture and Forestry,* vol 24, Bajaj YPS ed, Springer-Verlag, Berlin, pp313–328, 1993

Malmanche I de: Suspected *Papaver nudicaule* (Iceland poppy) poisoning in two ponies. *N Z Vet J* 18;96–97, 1970

McLennan GC: Poisoning of sheep by ingestion of Iceland poppies *(Papaver nudicaule). Aust Vet J* 5;117, 1929

McLennan GC: Iceland poppy poisoning [letter]. *Aust Vet J* 6;40, 1930

Meneely KD: Poppy seed ingestion: the Oregon perspective. *J Forensic Sci* 37;1158–1162, 1992

Norton JH, O'Rourke PK: Oedema disease in chickens caused by Mexican poppy *(Argemone mexicana)* seed. *Aust Vet J* 56;187–189, 1980

Odendaal JSJ: Suspected opium poisoning in two young dogs. *J S Afr Vet Assoc* 57;113–114, 1986

Pahwa R, Chatterjee VC: The toxicity of Mexican poppy (*Argemone mexicana* L.) seeds to rats. *Vet Hum Toxicol* 31;555–558, 1989

Preininger V: Chemotaxonomy of Papaveraceae and Fumariaceae. In *The Alkaloids,* vol 29, Brossi A ed, Academic Press, Orlando, Fla, pp1–98, 1986

Rolland A, Fleurentin J, Lanhers M-C, Younos C, Misslin R, Mortier F, Pelt JM: Behavioural effects of the American traditional plant *Eschscholzia californica*: sedative and anxiolytic properties. *Planta Med* 57;212–216, 1991

Sarkar SN: Isolation from Argemone oil of dihydrosanguinarine and sanguinarine: toxicity of sanguinarine. *Nature* 162;265–266 1948

Schafer HL, Schafer H, Schneider W, Elstner EF: Sedative action of extract combinations of *Eschscholtzia californica* and *Corydalis cava. Arzneim-Forsch* 45;124–126, 1995

Selavka CM: Poppy seed ingestion as a contributing factor to opiate-positive urinalysis results: the Pacific perspective. *J Forensic Sci* 36:685–696, 1991

Shamma M: The isoquinoline alkaloids. In *Organic Chemistry,* vol 25, Academic Press, New York, 1972

Shanbhag VV, Jha SS, Kekre MS, Rindani GJ: Epidemic dropsy in Bombay City and suburbs. *Indian J Med Sci* 22;226–231, 1968

Southon IW, Buckingham J: *Dictionary of Alkaloids*. Chapman & Hall, New York, 1989

Stermitz FR: Alkaloid chemistry and the systematics of *Papaver* and *Argemone*. *Recent Adv Phytochem* 1;161–183, 1968

Tandon S, Das M, Khanna SK: Effect of sanguinarine on the transport of essential nutrients in an everted gut sac model: role of Na^+,$K(^+)$-ATPase. *Nat Toxins* 1;235–240, 1993

Terblanche M, Adelaar TF: A note on the toxicity of *Papaver nudicaule* L. (Iceland poppy). *J S Afr Vet Med Assoc* 35;383–384, 1964

Upreti KK, Das M, Kumar A, Singh GB, Khanna SK: Biochemical toxicology of argemone oil. 4. Short-term oral feeding response in rats. *Toxicology* 58;285–298, 1989

Upreti KK, Das M, Khanna SK: Role of antioxidants and scavengers on *Argemone* oil–induced toxicity in rats. *Arch Environ Contam Toxicol* 20;531–537, 1991

Phytolaccaceae R.Br. 56

Widespread in tropical and subtropical regions, especially those of the New World, the Phytolaccaceae, or pokeweed family, comprises 18–22 genera and 100–125 species. Some taxonomists divide it into 4–6 families on the basis of differences in carpel fusion and fruit type. Of minor economic importance, the family is a source of dyes, folk medicines in South America, food, and ornamentals. In North America, 8 or 9 genera and 13 or 14 species are present. Only 1 genus is of toxicologic interest.

■ Pokeweed Family

Phytolacca

Plants herbs or shrubs or trees or vines; perennials; from fleshy or woody roots. **Leaves** simple; alternate; venation pinnate; margins entire to undulate; stipules absent or minute. **Inflorescences** racemes or spikes or panicles or cymes; axillary or terminal or extra-axillary. **Flowers** perfect or imperfect; perianths in 1-series; radially symmetrical. **Sepals** 5 or 4; free or fused. **Petals** absent. **Stamens** 4 to many. **Pistils** 1–16; simple or compound, carpels 1 to many; stigmas 1 to many; styles 1 to many; ovaries superior or inferior; locules 1 to many; placentation axile or basal. **Fruits** berries or drupes or achenes. **Seeds** 1 to many; winged or not winged.

■ herbs, shrubs, vines, or trees; leaves simple, alternate; petals absent

Phytolacca L.

Taxonomy and Morphology—Comprising about 35 species, *Phytolacca* is native to both the Old World and the New World and includes herbs, shrubs, and trees. Its name is derived from the Greek root *phyton,* meaning "plant," and the French root *lac,* for the color "crimson lake," alluding to the purple-red juice of the berries. Introduced herbaceous species of the genus are sometimes grown as garden ornamentals, while the arborescent evergreen *P. dioica,* commonly known as ombu, which is rarely cultivated in southern California and Florida. Only 1 native species is of significant toxicologic concern:

P. americana L. (= *P. decandra* L.) (= *P. rigida* Small)	poke, pokeweed, pokeberry, inkberry, scoke, poke sallet, pigeonberry, garget, cancerroot, crowberry, jalap, red ink plant

■ *P. americana*

cancerroot
crowberry
garget
inkberry
jalap
pigeonberry
poke
pokeberry
poke sallet
pokeweed
scoke

Plants herbs; perennials; from fleshy roots; herbage glabrous. **Stems** erect; 100–300 cm tall; fleshy; purplish red at maturity. **Leaves** with blades ovate to lanceolate. **Inflorescences** drooping or arching racemes; bracteoles present. **Flowers** perfect. **Sepals** 5; free; petaloid; greenish white or cream. **Stamens** 10. **Pistils** 1; compound; carpels 10; stigmas and styles 10; ovaries superior; locules 10; placentation axile. **Fruits** berries; purplish black; persistent after leaves have dropped. **Seeds** 10; shiny.

■ erect, perennial herbs from fleshy roots, stems purplish red at maturity; sepals greenish white or cream; fruits berries, purplish black

Distribution and Habitat—*Phytolacca americana* is native to the eastern half of the continent (Sauer 1952). It occurs sporadically farther west; populations are present in California, for example. Occupying a variety of clay, loam, and sand soils, it is

Phytolacca americana

Phytolacca americana

- foliage as potherb, berries other uses; digestive disturbance from leaves and roots; especially humans, rarely in livestock

typically associated with mounds of bulldozed soil and trees. It is a dominant of early serial stages of plant succession and is found in both prairies and open forests. A short-lived perennial, it is typically encountered as small localized populations.

Disease Problems—If properly cooked, the young shoots and leaves are edible and tasty; the famous native-plant enthusiast Euell Gibbons considered it a potherb par excellence. The new shoots are said to be asparagus-like in flavor. Hill-folk of the Appalachian and Ozark Mountains relish the plant and use it extensively (Thomas 1973). The ripe berries sans seeds are used to make jams and jellies and to provide coloring in cakes, candies, and beverages (Millspaugh 1974; Lewis & Smith 1979). The leaves also were esteemed by members of various American Indian tribes (Ebeling 1986). In spite of its longstanding use as a potherb, *Phytolacca* should be used with some caution because incidents of serious illness occur even with careful preparation according to methods reputed to ensure edibility. The risk is exemplified by a policy statement issued by the Herb Trade Association that, because of the plant's toxicity, all products containing pokeroot should be labeled clearly as to their toxicity (Herb Trade Association 1979). Several species in other areas of the world—e.g., *P. dioica, P. octandra, P. decandra,* and *P. dodecandra*—sometimes cause fatal digestive system effects (Storie et al. 1992; Spengel et al. 1995; Piexoto et al. 1997).

Consumption of either the leaves or the roots typically produces a transient digestive disease in both monogastric animals and ruminants. Dairy cattle fed pokeweed in green chop developed severe diarrhea by the next morning and exhibited a drastic drop in milk production and a decrease in body temperature (Kingsbury & Hillman 1965). The disease was subsequently reproduced by feeding about 2 kg of the plants to cattle. The roots are apparently the most toxic parts, and serious intoxications have occurred when pigs have uprooted plants and consumed them. The pigs exhibited neurologic signs in addition to those indicative of digestive tract involvement (Hansen 1924; Patterson 1929). Experimentally, the roots caused retching, tonic seizures, and coma in pigs (Patterson 1929). Although the berries are eaten by a variety of birds, experimental work with turkeys indicates some adverse effects in very young poults when given feed containing 5–10% pokeberries (Barnett 1975). There was decreased weight gain, ataxia, and 33–43% mortality in those fed 10%. Poults 9 weeks or older were not affected by 5% pokeberries in the feed (Cattley & Barnett 1977). These effects are comparable to those of sheep fed the more toxic *P. decandra* in Brazil (Peixoto et al. 1997). In the latter study, 20 g/kg b.w. of fresh leaves and sprouts caused death in about 5 hours whereas those given 10 g/kg b.w. died in 9 hours. A variety of signs were seen, including ataxia, head pressing, vision impairment, hyperesthesia, tremors, rapid and labored respiration, and ruminal mucosal reddening and necrosis.

Although animal disease is noted occasionally, in most instances consumption of *P. americana* is very limited and seldom associated with overt signs of intoxication. More-important considerations are the problems in humans eating the plant as a potherb. French (1900) recounted his experiences upon eating a small amount of the bitter root, which he had mistaken for horseradish. About 90 minutes after eating it, he experienced excess salivation, violent stomach cramps, retching, and vomiting that lasted for several hours and was accompanied at various times by tremors, prickly sensations, and prostration. There have been numerous reports from eastern North America of intoxications, mainly involving children (O'Leary 1964). In some instances the effects are only a minor nuisance, as, for example, in the case of Boy Scouts who ate about 1.5 cups of pokeberries mashed and cooked in pancakes (Edwards & Rodgers 1982). Several scouts developed a mild to moderate transient diarrhea in

5–12 hours. However, in other situations large numbers of people may be at risk. Callahan and co-workers (1981) reported an incident in which 61 people at a day camp were given poke as a potherb after it had been boiled, drained, and boiled again. Twenty-one became sick in 0.5–5.5 hours. Most had nausea, vomiting, and stomach cramps, and a few had diarrhea, dizziness, and headache. The signs lasted up to 48 hours, and 4 people exhibited protracted vomiting and dehydration. Rarely other more serious signs, such as those of cardiac problems, are evident. A woman who chewed a raw root instead of a boiled one for relief of a cough and sore throat, exhibited electrocardiographic changes that included abnormal waves and ST-T segment depression in addition to ventricular fibrillation and hypotension (Roberge et al. 1986). Heart block was noted in a man who had eaten a few uncooked leaves as well as some that were boiled (Hamilton et al. 1995). Fortunately, the block resolved when the vomiting subsided after treatment with promethazine. Even the liquid from the boiled leaves can be hazardous. About 100 ml of tea made from initial boiling of the leaves caused sweating, tremors, and fainting in a man 10–15 minutes after he drank it (Jaeckle & Freemon 1981). The signs were suggested to be those of cholinergic stimulation.

■ numerous irritant saponins, glycosides of phytolaccagenin, phytolaccinic acid, pokeberrygenin; oxalate salts also present

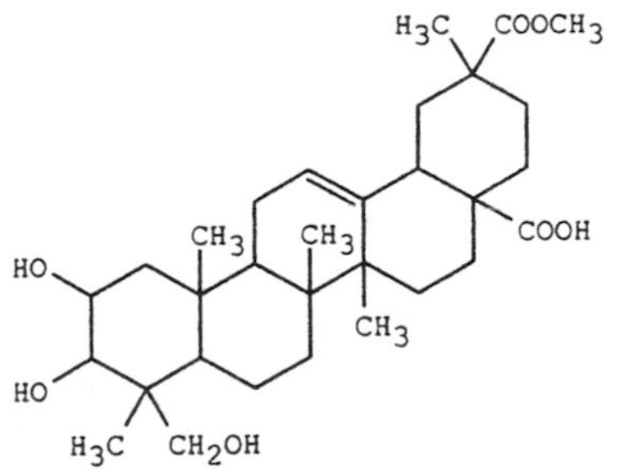

Disease Genesis—*Phytolacca* contains several toxicants, including saponins and oxalates. The problems have generally been attributed to 28,30-dicarboxylene-12-ene (triterpene) saponins, such as phytolaccagenin, phytolaccinic acid, and pokeberrygenin, and their respective glycosides or phytolaccosides (Stout et al. 1964; Johnson & Shimizu 1974; Kang & Woo 1980; Yi & Wang 1989). *Phytolacca dodecandra*, a more toxic African species, contains the apparently more toxic monocarboxyl saponin types, oleanolic acid, hederagenin, and bayogenin (Spengel et al. 1995). The saponins of *P. americana* irritate the mucosal surface of the digestive tract. Aqueous extracts, especially of the root, are very irritating and cause signs of severe digestive tract irritation and then paralysis (Macht 1937). When given parenterally to cats, there was impairment of liver function, respiration, and circulation, and, in large dosage, paralysis. The roots of *P. americana* also contain small amounts of GABA (0.05%) and substantial amounts of histamine (0.13–0.16% histamine) (Funayama & Hikino 1979). Oxalates also are present and may contribute to intoxication signs.

Phytolaccagenin

Possibly the best known toxicant in the berries is a mitogen that is noteworthy for its widespread use in immunologic studies (Farnes et al. 1964). It was noted that ingestion of the berries was followed within a few days by lymphocyte transformations to plasmacytoid lymphocytes, lymphocytes with mitotic figures, and plasmacytes (Barker et al. 1966). Hematologic aberrations included plasmacytosis, eosinophilia, platelet phagocytosis, and thrombocytopenia (Barker et al. 1967). This mitogen appeared to peak in activity at the time of *P. americana's* highest toxicity in August and October (Kell et al. 1982). It now appears that, rather than a single mitogen, several are present. Waxdal (1974) identified three major and two minor types from the roots of *P. americana*. Two, which are mitogenic for both B and T lymphocytes, contain a large number of disulfide bonds. One also acts as a hemagglutinin.

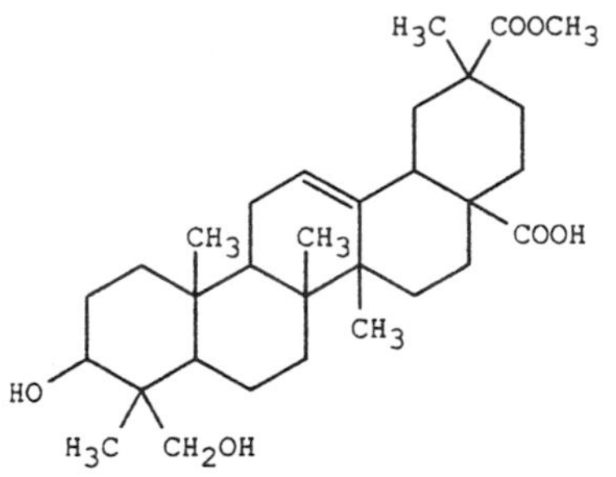

Phytolaccinic acid

■ pigs, diarrhea, rarely ataxia, sedation, seizures; poultry, decreased weight gain, swollen hock joints

Clinical Signs—The primary and sometimes sole sign of *Phytolacca* intoxication is diarrhea that occurs 3–4 hours after ingestion by monogastric animals and humans and 6–8 hours or more in ruminants. The diarrhea may be severe and may last 24 hours or longer. In cattle, it will be mild if the animals have not consumed large quantities of the plant. Some blood may be present, but in many cases what appears to be blood is merely the purple-red juice from the berries. Fecal tests for occult blood may be negative. Normal levels of milk production will resume in about 1 week.

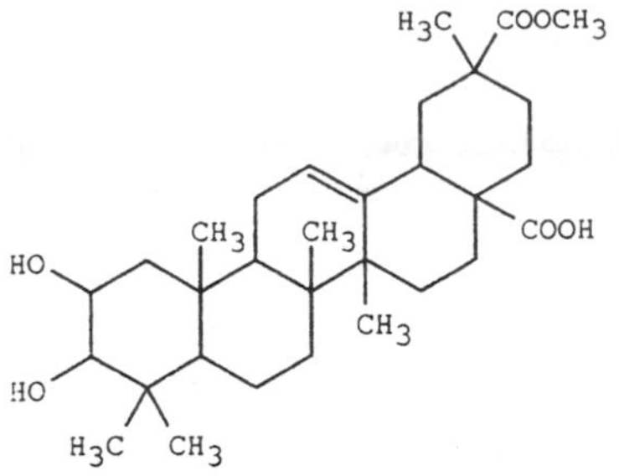

Pokeberrygenin

- gross pathology, reddened gut mucosa
- activated charcoal, fluids, electrolytes

The signs in pigs that ingest the roots are more severe and include incoordination, sedation, tonic seizures, paralysis, and death. Turkey poults 2–3 weeks old had decreased weight gain and swollen hock joints, and the birds were unsteady or unable to stand or walk.

Pathology—Gross lesions are limited to reddening of the mucosa of the stomach and small intestine with small focal erosions. In turkeys, there was ascites and enlarged gallbladders in addition to the swollen hock joints.

Treatment—There is no specific remedy for intoxication. Treatment is based mainly upon supportive care and relief of the symptoms. Fluids may be needed to correct dehydration. Activated charcoal may help to control absorption of the toxins. The patient should be monitored for hypotension and any other cardiac effects. Atropine has been of some value in alleviating the digestive as well as other effects.

References

Barker BE, Farnes P, LaMarche PH: Peripheral blood plasmacytosis following systemic exposure to *Phytolacca americana* (pokeweed). *Pediatrics* 38;490–493, 1966

Barker BE, Farnes P, LaMarche PH: Haematological effects of pokeweed. *Lancet* 1;437, 1967

Barnett BD: Toxicity of pokeberries (fruit of *Phytolacca americana* Large) for turkey poults. *Poult Sci* 54;1215–1217, 1975

Callahan R, Piccola F, Gensheimer K, Parkin WE, Prusakowski J, Scheiber G, Henry S: Plant poisoning—New Jersey. *Morb Mortal Wkly Rep* 30;65–66, 1981

Cattley RC, Barnett BD: The effect of pokeberry ingestion on immune response in turkeys. *Poult Sci* 56;246–248, 1977

Ebeling W: *Handbook of Indian Foods and Fibers of Arid America*. Univ California Press, Berkeley, p483, 1986

Edwards N, Rodgers GC Jr: Pokeberry pancake breakfast—or—it's gonna be a great day [abstract]. . *Vet Hum Toxicol* 24;293, 1982

Farnes P, Barker BE, Brownhill LE, Fanger H: Mitogenic activity in *Phytolacca americana* (pokeweed). *Lancet* 2;1100, 1964

French C: Pokeroot poisoning. *N Y Med J* 72;653–654, 1900

Funayama S, Hikino H: Hypotensive principles of *Phytolacca* roots. *J Nat Prod* 42;672–674, 1979

Hamilton RJ, Shih RD, Hoffman RS: Mobitz type I heart block after pokeweed ingestion. *Vet Hum Toxicol* 37;66–67, 1995

Hansen AA: The poison plant situation in Indiana, 3. *J Am Vet Med Assoc* 66;351–362, 1924

Herb Trade Association: *Policy Statement* 2. May 1979

Jaeckle KA, Freemon FR: Pokeweed poisoning. *South Med J* 74;639–640, 1981

Johnson A, Shimizu Y: Phytolaccinic acid, a new triterpene from *Phytolacca americana*. *Tetrahedron* 30;2033–2036, 1974

Kang SS, Woo WS: Triterpenes from the berries of *Phytolacca americana*. *J Nat Prod* 43;510–513, 1980

Kell SO, Rosenberg SA, Conlon TJ, Spyker DA: A peek at poke: mitogenicity and epidemiology [abstract]. *Vet Hum Toxicol* 24;294, 1982

Kingsbury JM, Hillman RB: Pokeweed *(Phytolacca)* poisoning in a dairy herd. *Cornell Vet* 55;534–538, 1965

Lewis WH, Smith PR: Poke root herbal tea poisoning. *J Am Med Assoc* 242:2759–2760, 1979

Macht DI: A pharmacological study of *Phytolacca*. *J Am Pharm Assoc* 26;594–599, 1937

Millspaugh CF: *American Medicinal Plants*. Dover Publ, New York, pp557–561, 1974 (reprinted from 1892)
O'Leary SB: Poisoning in man from eating poisonous plants. *Arch Environ Health* 9;216–242, 1964
Patterson FD: Pokeweed causes heavy losses in swine herd. *Vet Med* 24;114, 1929
Peixoto P, Wouters F, Lemos RA, Loretti AP: *Phytolacca decandra* poisoning in sheep in Southern Brazil. *Vet Hum Toxicol* 39;302–303, 1997
Roberge R, Brader E, Martin ML, Jehle D, Evans T, Harchelroad F, Magreni C, Gesualdi G, Belardi C, Sayre M, Hartmann A: The root of evil—pokeweed intoxication. *Ann Emerg Med* 15;470–473, 1986
Sauer JD: A geography of pokeweed. *Ann Mo Bot Gard* 39;113–125, 1952
Spengel SM, Luterbacher S, Schaffner W: New aspects on the chemotaxonomy of *Phytolacca dodecandra* with regard to the isolation of phytolaccagenin, phytolaccagenic acid, and their glycosides. *Planta Med* 61;385–386, 1995
Storie GJ, McKenzie RA, Fraser IR: Suspected packalacca *(Phytolacca dioica)* poisoning of cattle and chickens. *Aust Vet J* 69;21–22, 1992
Stout GH, Malofsky BM, Stout VF: Phytolaccagenin: a light atom X-ray structure proof using chemical information. *J Am Chem Soc* 86;957–958, 1964
Thomas F: The versatile poke plant. *Ozark Highw* 1973(2);22–23, 1973
Waxdal MJ: Isolation, characterization, and biological activities of five mitogens from pokeweed. *Biochemistry* 13;3671–3676, 1974
Yi Y-H, Wang C-L: A new active saponin from *Phytolacca esculenta*. *Planta Med* 55;551–552, 1989

57 Pinaceae Lindl.

- Pine Family
  Pinus

Comprising 10 genera and about 200 species, the Pinaceae, or pine family, dominates the temperate, coniferous forests of the Northern Hemisphere. Only a few species are present south of the equator, principally in the Malayan archipelago. A few taxonomists broadly circumscribe the family and include within it genera traditionally placed in the Cupressaceae and Taxaceae. Most individuals, however, recognize 3 families, as we have done in this treatise, because of major differences in leaf and seed cone morphology.

The family is of considerable economic importance. Its members are a source of timber, pulpwood, resins, and oils. They also are used extensively in landscape architectural designs. In North America, 6 genera and 66 species are present (Thieret 1993). Toxicologic problems are associated with 1 genus.

- resinous, aromatic trees, producing pollen and seed cones; leaves needlelike

Plants trees; evergreen or deciduous; resinous; aromatic; producing pollen cones and seed cones; monoecious. **Bark** smooth to scaly or furrowed. **Leaves** needlelike; persistent for 2–12 years; simple; alternate; 1- or 2-ranked or fascicled; stipules absent. **Pollen Cones** produced annually; axillary; solitary or clustered; ovoid to ellipsoid or cylindric; microsporophylls spirally arranged; microsporangia 2 per microsporophyll. **Seed Cones** produced annually, maturing in 1–3 years; subterminal or axillary; solitary or clustered; ovuliferous scales overlapping; megasporangia 2 per scale. **Seeds** 2 per scale; wings present or absent.

Pinus L.

- *P. contorta*
  P. jeffreyi
  P. ponderosa
  P. radiata

  Arizona pine
  Jeffrey pine
  limber pine
  lodgepole pine
  Monterey pine
  ponderosa pine
  western yellow pine

Taxonomy and Morphology—Comprising 90–110 species distributed from the Arctic Circle to Central America, northern Africa, and Sumatra in southeastern Asia, *Pinus*, commonly known as pine, exhibits the greatest diversity in habit and distribution of the needle-bearing conifers. Its species may be both dominants of climax forests—e.g., the oak-hickory-pine forests of the southeastern United States—or pioneers in early successional stages. Its name is derived from the Latin name for "pine." On the basis of needle anatomy, persistence of the scales of the fascicle sheaths, indumentum of the bud scales, and arming of the seed cones, 2 or 3 subgenera are generally recognized. In North America, 37 native and 1 widely naturalized species are present (Kral in Thieret 1993). Reproductive problems are associated with 4 species:

P. contorta Douglas ex Loudon	lodgepole pine, limber pine
P. jeffreyi Grev. & Balf.	Jeffrey pine
P. ponderosa Douglas ex Lawson & C.Lawson	ponderosa pine, western yellow pine
var. *arizonica* (Engelm.) Shaw	Arizona pine, Arizona ponderosa pine
var. *ponderosa*	Pacific ponderosa pine

var. *scopulorum* Engelm.	mountain ponderosa pine
P. *radiata* D.Don	Monterey pine

Plants trees or rarely shrubs; evergreen; crowns typically conical when young, rounded or flat-topped when mature. **Bark** furrowed or scaly. **Leaves** fascicled; in bundles of 2–5 or rarely 1 or 6; with surrounding basal sheaths; sessile; venation not apparent; terete or 2-to-3-angled; basal sheaths persistent or shed after first season. **Pollen Cones** small; borne in dense spicate clusters at bases of current year's growth; soft; tan to yellow or red, blue, or lavender. **Seed Cones** maturing in 2–3 years; sessile or stalked; erect or pendulous; quickly dropping or long persistent; opening to shed seeds soon after maturity or remaining closed for long periods; scales woody or pliable. **Seeds** with wings present or absent.

- evergreen trees; leaves in bundles of 2–5 or rarely 1 or 6; pollen cones small, soft, at base of current year's growth; seed cones mature in 2–3 years

1. Seed cones strongly asymmetrical; persistent on branches for many years after maturity; remaining closed. Apophyses conspicuously extended.
   2. Leaves 2 per mature fascicle .... *P. contorta*
   2. Leaves 3 per mature fascicle .... *P. radiata*
1. Seed cones symmetrical or slightly asymmetrical; deciduous immediately after maturity; opening quickly. Apophyses only slightly extended.
   3. Needles yellow-green. Lower surface of seed cone scales darker than upper surface .... *P. ponderosa*
   3. Needles grayish blue-green. Lower surface of seed cone scales not darker than upper surface .... *P. jeffreyi*

Pinus contorta

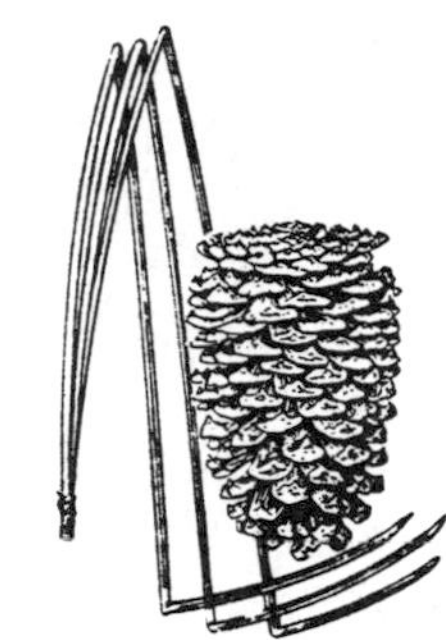

Pinus jeffreyi

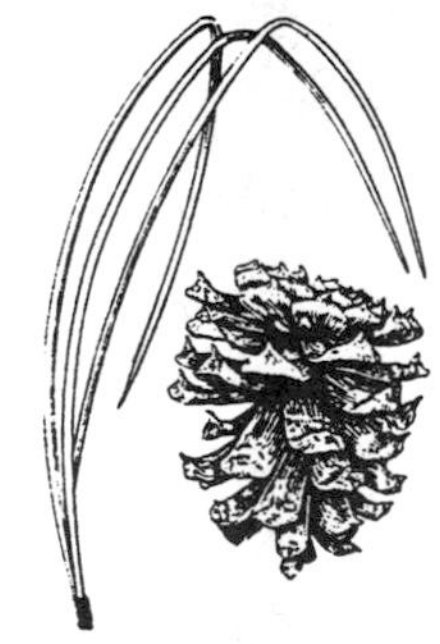

Pinus ponderosa

Distribution and Habitat—Typically occurring in drier mountainous sites, *P. ponderosa* is distributed throughout the western half of North America. Populations range as far east as the Great Plains. As its varietal name implies, var. *arizonica* is found in Arizona. It extends into New Mexico and the mountains of Chihuahua and Nuevo León. Variety *ponderosa* is found along the Pacific Coast from California to British Columbia, while var. *scopulorum* ranges through the Rocky Mountains. Although widely grown throughout the world as an ornamental and apparently capable of naturalizing, *P. radiata* has a rather restricted natural distribution. Populations occur in three coastal areas of California and on islands off the coast of Baja California in Mexico. *Pinus jeffreyi* is characteristic of high, dry montane forests of the Sierra Cascade Mountains at elevations of 2,000–2,500 m. With three varieties, *P. contorta* is found in the Rocky, Sierra-Cascade, and coastal mountains at a variety of elevations.

Disease Problems—Cattle producers in British Columbia had long suspected pine needles as a cause of reproductive problems in their animals (Bruce 1927), but it was not until the experimental observations of MacDonald (1952) that this association was confirmed. Cattle ate the needles of *P. ponderosa* even though other feed was available, and 2.3–2.9 kg/day of fresh buds and needles caused the birth of dead calves or small weak calves, many of which died shortly after birth. These effects are consistent with folklore that Indian women in Colorado, Oregon, and Washington used pine needle teas to induce abortions (Tucker 1961; Stevenson et al. 1972; James et al. 1989).

- reproductive disease, cattle, *P. ponderosa*, needles green or dry, illness in the dam, dead or small weak calves or abortions; risk increases with later stages of pregnancy, mainly in last 2 months; risk with 1 kg needles

sheep and deer little affected

P. contorta, P. radiata also risks

Ingestion of needles by pregnant cattle is now widely appreciated as a cause of serious losses estimated to be as high as $20 million per year (Miner et al. 1987). Lacey and co-workers (1988) estimated the value of the loss of 4,300 calves and 74

Pinus contorta

Pinus jeffreyi

Pinus ponderosa

Pinus radiata

cows per year to be $4.5 million. They suggested that such figures were probably underestimations of the problem. These losses are due in large measure to the extensive range of ponderosa pine and its dominance in western forests, an area estimated at 42 million acres (Burns 1983). Although toxicity may vary, all three varieties appear to pose a risk. The binomial name *P. arizonica* was used in early reports for individuals of var. *arizonica* (Gartner et al. 1988).

Needles of any age, green or dry, intact or partially decomposed, are toxic, although green ones may be more readily eaten (Jensen et al. 1989; Panter et al. 1992; Short et al. 1994). Budding branch tips and bark are even more toxic than needles (Panter et al. 1990). The needles are commonly eaten but often in insufficient amounts to cause problems. Increased ingestion appears to be somewhat weather related; it is favored by low winter temperatures (Pfister & Adams 1993; Pfister et al. 1998). Significant snow cover may be a major factor when it prevents animal access to normal forage and favors ingestion of fresh needles. Recent access to the needles after storms may also predispose their consumption (Stevenson et al. 1972). The role of stress in fostering consumption or effects is not clear, because early observations indicated that stress increased the effects, whereas later studies indicated that stress is actually protective (James et al. 1977; Short et al. 1994).

Initially sheep and deer were thought susceptible to the toxic effects of the needles, but this now appears not to be so (Call & James 1976; James et al. 1977). Experimentally, when needles made up 37% of the diet, sheep exhibited no indications of toxic effects other than a suggestion of bony malformations in the lambs (Call & James 1976). However, in other experiments, pine needles fed to ewes markedly increased the number of dead lambs at parturition (Short et al. 1995). Deer appear to be unaffected by pine needles, given that field observations have failed to reveal any effects even though needles may compose more than 50% of their diet at times (James et al. 1989). Reproductive effects are limited to cattle and bison, but the latter seem not to be predisposed to eat the needles (Short et al. 1992). The cause of the difference between cattle and sheep is not clear, but it appears to be related to response of the animals to the abortifacient rather than a difference in ruminal metabolism of the toxicants (Short et al. 1995). This is further shown by the lack of success in producing resistance to pine needle–induced abortion by transferring rumen microflora from elk (which are resistant) to cattle (Dehorty et al. 1999). Susceptibility in cattle increases with advancing stage of pregnancy; it begins about midgestation and reaches a zenith in the last month. Substantial amounts of the needles are required; small amounts (0.1–0.3 kg/day) are not toxic and in fact may induce some degree of tolerance (Short et al. 1992). Amounts exceeding 1 kg are generally required for intoxication, but as little as 0.68 kg has proven troublesome. A much higher frequency of adverse effects is seen with dosages in the range of 1.35–1.8 kg or more (Short et al. 1987, 1989, 1992). Ingestion of more than 2 kg/day for 2 or 3 days is sufficient to cause premature parturition in 24 hours to 3 weeks (Panter et al. 1992). Ingestion of the needles does not seem to affect overall fertility (Stevenson et al. 1972; Short et al. 1992). Calves are not directly affected by the toxins, and their survival depends on their maturity at birth (Short et al. 1992).

In addition to *P. ponderosa*, *P. contorta* and *P. radiata* are also considered to be reproductive risks for cattle. *Pinus radiata* is suspected as a cause of disease in New Zealand, where it has been cultivated (Knowles & Dewes 1980). The needles are readily eaten and are reported by producers to cause effects in the last 2 months of gestation. Dead newborn calves, small weak calves, and some ill cows have been observed. Although field reports of reproductive effects caused by *P. contorta* are limited, its toxicity potential is shown by the abortions produced in 2 cows after administration of 4.5–5.5.kg/day of needles for 8–10 days (Gardner et al. 1998a).

Systemic effects other than those on the fetus also are a serious consideration with consumption of pine needles. Ingested needles have marked effects on ruminal microorganisms. Dietary levels of 15–30% conspicuously alter fluid dynamics in the bovine rumen and also reduce volatile fatty acids and nitrogen retention and digestibility (Pfister et al. 1992). The needles have similar effects in lambs (Adams et al. 1992). The alterations in ruminal microflora permit the possibility of formation of metabolic products that might contribute to detrimental systemic effects.

Disease Genesis—The specific cause of the premature parturition in animals consuming *Pinus* appears to be related to disturbances in hormonal balance, but there are some inconsistencies in experimental results. Extracts of the needles show antiestrogenic activity, as measured by suppression of uterine weight increase and by competition with 17β-estradiol binding in the mouse uterus (Allen & Kitts 1961, 1964; Cook & Kitts 1964; Wagner & Jackson 1983). It seems apparent that although mice are susceptible to the embryotoxic effects of pine needles, these effects are not representative of those in cattle (Chow et al. 1972; Anderson & Lozano 1979; Cogswell & Kamstra 1980; Kubik & Jackson 1981). Nevertheless, these studies have proven useful in identifying the presence of antiestrogenic-type effects for the needles. Although it is clear that there are effects on estrogenic activity, the changes in circulating estrogens and progesterone are not consistent; alterations of varying degree and direction are reported. Elevations of cortisol, 17β-estradiol, and progesterone have been noted during the last weeks of the shortened gestation period (Short et al. 1989). New-growth needles at branch tips caused even greater serum cortisol increases (Panter et al. 1990). In other studies there was no apparent difference in serum estradiol and progesterone levels between control cows and those fed needles (Miner et al. 1987; Panter et al. 1990). However, in contrast, Jensen and co-workers (1989) fed needles to cows and observed increased necrosis of luteal cells, especially in the central zones of the corpus luteum, and a decrease in the number of progesterone-producing fetal binucleate placentome trophoblasts. Thus, there was a decrease in the number of placental and corpus luteal progesterone-producing cells, accompanied by a marked decrease in serum progesterone concentrations—changes incompatible with maintenance of pregnancy. Prostaglandins present in the needles themselves may also play some role in the reproductive effects (James et al. 1989).

- *P. ponderosa, P. contorta, P. jeffreyi,* all with diterpene acids, such as isocupressic acid and its esters, which decrease uterine blood flow, leading to decreased viability of fetus; with other abietane diterpenes such as isopimaric and neoabietic acids, may also cause illness in dam

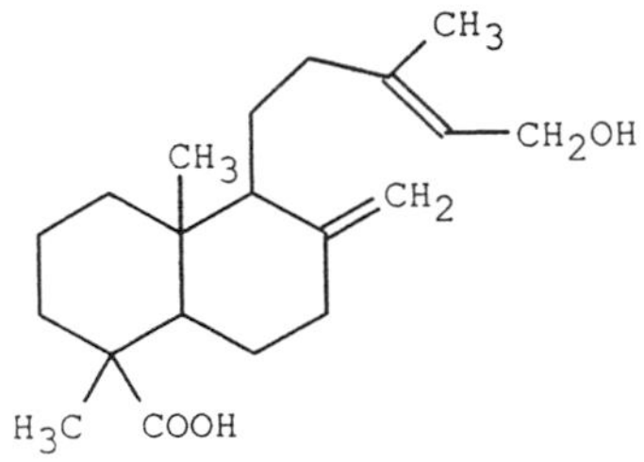

Isocupressic acid

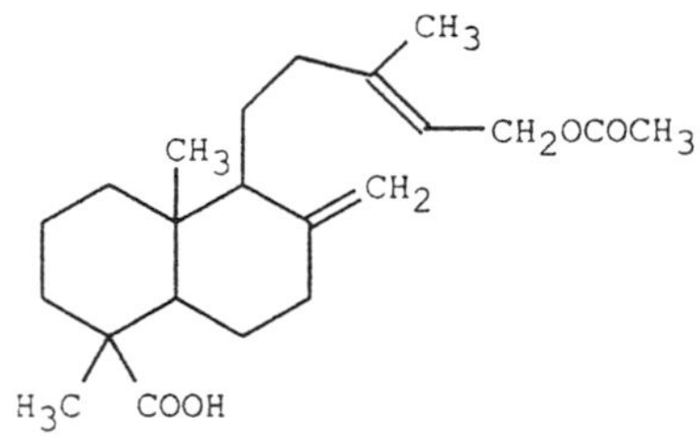

Acetylisocupressic acid

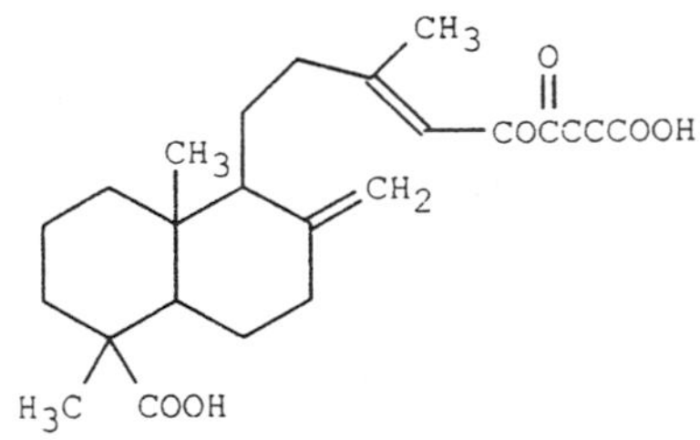

Succinylisocupressic acid

Vascular impairment of uterine blood flow appears to be a significant factor in the development of reproductive effects. Plasma from needle-fed cows causes increased caruncular arterial smooth muscle tone in vitro (Christenson et al. 1992a). In vivo, there is a decrease in uterine blood flow, which eventually leads to parturition (Christenson et al. 1992b; Ford et al. 1992). The toxin in needles appears to prevent the normal progressive catechol/estrogen–induced decrease in caruncular arterial tone or vasodilation, which is responsible for increasing blood flow with advancing pregnancy (Ford et al. 1997). It is suggested that the toxin causes failure of the block in potential sensitive calcium channel activity and subsequent calcium-mediated smooth muscle contraction, an effect normally moderated by estrogen metabolites (Christenson et al. 1993). These reproductive hormonal changes seem to be closely associated with the presence of isocupressic acid, a labdane diterpene, in pine needles (Zinkel & Magee 1991). Perhaps the diterpene acids or their metabolites alter estrogen control of uterine blood flow. It appears that the effects of pine needles are related more to modification of hormonal effects than to hormonal levels.

Gardner and co-workers (1994a) isolated a number of diterpene acids, including isocupressic acid and imbricataloic acid, from methylene chloride extracts of pine needles. When administered orally daily, isocupressic acid alone caused abortion in 4 of 5 cows dosed for 2–8 days, beginning on the 250th day of gestation. This is

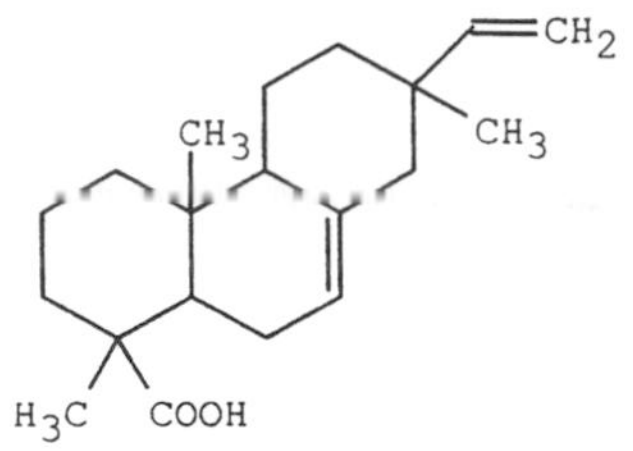

Isopimaric acid

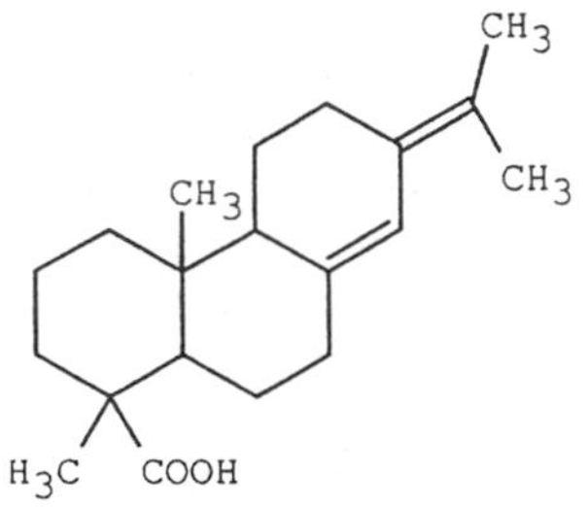

Neoabietic acid

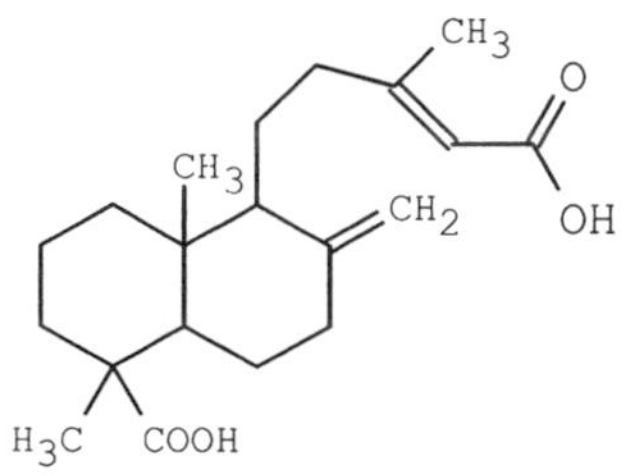

Dihydroagathic acid

consistent with the results of Kubik and Jackson (1981), who induced embryo resorption in the first week of gestation by using resin acids extracted from needles, and the results of James and co-workers (1994), who reported methylene chloride to be the most effective solvent for extraction of the abortifacient toxicant in needles. The related diterpene esters of isocupressic acid, acetylisocupressic and succinylisocupressic acids, which compose a major portion of the resin acids of pine needles, are also abortifacient in cattle (Zinkel & Magee 1991; Gardner et al. 1996). These esters are hydrolyzed in ruminal fluid to yield isocupressic acid. Isocupressic acid is itself metabolized in the rumen and liver to agathic and dihydroagathic acids, which may serve as the ultimate abortifacients (Lin et al. 1998; Gardner et al. 1999a,b). The marked variation in isocupressic acid concentrations, ranging from 0.5–2% to 13–15% of the total resin acids, suggests a considerable difference in toxicity of trees from different geographical locations (Zinkel & Magee 1991). Other abietane diterpenes—such as isopimaric, abietic, dehydroabietic, and neoabietic acids—that compose the resin gum residue of the needles produced systemic disease when administered orally, causing depression, anorexia, rumen stasis, tympany, respiratory distress, and peripheral neuropathy (Gardner et al. 1994b). These compounds may account for signs of illness in the dam.

Appreciable concentrations of isocupressic acid have also been shown for *P. contorta* and *P. jeffreyi* (Gardner et al. 1997, 1998b), and it would not be surprising to find other pine species, including *P. radiata,* with toxicity potential, because the resin acids are not unique to these species of pine. Intoxication may be a matter not only of species toxicity but also the likelihood of ingestion by cattle.

An additional class of vasoactive compounds, including alkanediols esterified with lauric and myristic acids, has been identified in pine needles, by means of an in vitro bioassay employing perfused late-term bovine placentomes (Al-Mahmoud et al. 1995). The most important of these vasoactive lipids is 1,12-dodecanedioyl-dimyristate (Ford et al. 1999). Vasoactive lipids may represent alternative or additional pathways for production of increased caruncular arterial tone. The abortifacient activity of fractions containing these compounds has been confirmed in guinea pigs; however, some difficulty has been encountered in producing abortions in cattle (Ford et al. 1994, 1999; Short et al. 1996).

Mycotoxins and bacteria have been suggested as possible contributing factors, but they do not appear to be significant factors (Chow et al. 1974; Adams et al. 1979; Panter et al. 1991).

- early vulvar swelling and mammary development, blood-tinged discharge from vulva, delivery of dead or weak calves, placental retention, metritis, flaccid uterus

  illness in the dam, anorexia, stasis of rumen, depression, weakness

Clinical Signs—Within a few days or up to 3 weeks after ingestion of pine needles during the last trimester of gestation, typical signs of impending parturition appear, i.e., swelling of the vulva, mammary development, and a blood-tinged mucoid discharge from the vulva. Delivery of a dead or weak calf follows within a few hours or days. In other animals there may be little mammary development or milk letdown and few indications of an incipient birth. Retention of the fetal membranes is common and is typically accompanied by infection and inflammation of the uterus. The uterus is flaccid, doughy, and fluid-filled. Calves often die shortly after birth from starvation, bronchopneumonia, or other secondary causes.

In the dam, signs of illness are vague and include loss of appetite, lack of ruminal activity, depression, and weakness.

- gross pathology, cow, copious uterine exudate; edema, ulceration, hemorrhage of endometrium; necrosis of caruncles

Pathology—In cows that die after abortion or the birth of a weak calf, there is copious uterine exudate that contains necrotic placental fragments. There is edema, focal ulceration, and splotchy hemorrhages of the endometrium, with necrotic areas in the caruncles. Otherwise, the lesions are rather nonspecific. Edema and splotchy hemorrhages of the abdominal serosa, the subcutis, and tissues around the heart may also be present (Stuart et al. 1989).

Treatment—If possible, the placenta should be removed and the cow treated for a uterine infection. It may also be necessary to treat the animal systemically to overcome dehydration, fight bacterial infection, and provide good nutrition. Although suggestions have been made promoting the value of vitamin A as a preventive, experimental studies have not substantiated its benefit (Short et al. 1992). Furthermore, protein, mineral, salt-sulfur, and bentonite supplements have no apparent protective benefit against the effects of pine needles (Short et al. 1994).

- remove retained placenta; treat metritis; good feed

References

Adams CJ, Neff TE, Jackson LL: Induction of *Listeria monocytogenes* infection by the consumption of ponderosa pine needles. *Infect Immun* 25;117–120, 1979

Adams DC, Pfister JA, Short RE, Cates RG, Knapp BW, Wiedmeier RD: Pine needle effects on in vivo and in vitro digestibility of crested wheatgrass. *J Range Manage* 45;249–253, 1992

Allen MR, Kitts WD: The effects of yellow pine (*Pinus ponderosa* Laws) needle on the reproductivity of the laboratory female mouse. *Can J Anim Sci* 41;1–8, 1961

Allison CA, Kitts WD: Further studies on the antiestrogenic activity of yellow pine needles. *J Anim Sci* 23;1155–1159, 1964

Al-Mahmoud MS, Ford SP, Short RE, Farley DB, Christenson L, Rosazza JPN: Isolation and characterization of vasoactive lipids from the needles of *Pinus ponderosa*. *J Agric Food Chem* 43;2154–2161, 1995

Anderson CK, Lozano EA: Embryotoxic effects of pine needles and pine needle extracts. *Cornell Vet* 69;169–175, 1979

Bruce EA: Astragalus serotinus *and other stock poisoning plants of British Columbia.* Dom Can Dep Agric Bull 88, 1927

Burns RM: *Silvicultural systems for the major forest types of the United States.* USDA Forest Serv Handb 445, 1983

Call JW, James LF: Effect of feeding pine needles on ovine reproduction. *J Am Vet Med Assoc* 169;1301–1302, 1976

Chow F-HC, Hanson KJ, Hamar DW, Udall RH: Reproductive failure of mice caused by pine needle ingestion. *J Reprod Fert* 30;169–172, 1972

Chow F-HC, Hamar DW, Udall RH: Mycotoxic effect on fetal development: pine needle abortion in mice. *J Reprod Fert* 40;203–204, 1974

Christenson LK, Short RE, Ford SP: Effects of ingestion of ponderosa pine needles by late-pregnant cows on uterine blood flow and steroid secretion. *J Anim Sci* 70;531–537, 1992a

Christenson LK, Short RE, Rosazza JP, Ford SP: Specific effects of blood plasma from beef cows fed pine needles during late pregnancy on increasing tone of caruncular arteries in vitro. *J Anim Sci* 70;525–530, 1992b

Christenson LK, Short RE, Farley DB, Ford SP: Effects of ingestion of pine needles *(Pinus ponderosa)* by late-pregnant beef cows on potential sensitive Ca^{2+} channel activity of caruncular arteries. *J Reprod Fert* 98;301–306, 1993

Cogswell C, Kamstra LD: Toxic extracts in ponderosa pine needles that produce abortion in mice. *J Range Manage* 33;46–48, 1980

Cook H, Kitts WD: Anti-oestrogenic activity in yellow pine needles *(Pinus ponderosa)*. *Acta Endocrinol* 45;33–39, 1964

Dehorty BA, Grings EE, Short RE: Effects of cross-inoculation from elk and feeding pine needles on the protozoan fauna of pregnant cows: occurrence of *Parentodinium africanum* in domestic U.S. cattle *(Bos taurus)*. *J Eukaryotic Microbiol* 46;632–636, 1999

Ford SP, Christenson LK, Rosazza JP, Short RE: Effects of ponderosa pine needle ingestion on uterine vascular function in late-gestation beef cows. *J Anim Sci* 70;1609–1614, 1992

Ford SP, Farley DB, Rosazza JPN: Use of the late-pregnant guinea pig as a bioassay for the abortifacient in ponderosa pine needles [abstract]. *J Anim Sci* 72 (Suppl 1);103, 1994

Ford SP, Rosazza JPN, Short RE: *Pinus ponderosa* needle-induced toxicity. In *Handbook of Plant and Fungal Toxicants,* D'Mello JPF ed, CRC Press, Boca Raton, Fla, pp219–229, 1997

Ford SP, Rosazza JPN, Al-Mahmoud MS, Lin S, Farley DB, Short RE: Abortifacient effects of a unique class of vasoactive lipids from *Pinus ponderosa* needles. *J Anim Sci* 77;2187–2193, 1999

Gardner DR, Molyneux RJ, James LF, Panter KE, Stegelmeier BL: Ponderosa pine needle-induced abortion in beef cattle: identification of isocupressic acid as the principal active compound. *J Agric Food Chem* 42;756–761, 1994a

Gardner DR, Molyneux RJ, James LF, Panter KE, Stegelmeier BL: Studies on the abortifacient principle of ponderosa pine needles (*Pinus ponderosa* Laws). In *Plant Associated Toxins: Agricultural, Phytochemical, and Ecological Aspects,* Colegate SM, Dorling PR eds, CAB International, Wallingford, UK, pp339–344, 1994b

Gardner DR, Panter KE, Molyneux RJ, James LF, Stegelmeier BL: Abortifacient activity in beef cattle of acetyl- and succinylisocupressic acid from ponderosa pine. *J Agric Food Chem* 44;3257–3261, 1996

Gardner DR, Panter KE, Molyneux RJ, James LF, Stegelmeier BL, Pfister JA: Isocupressic acid and related diterpene acids from *Pinus ponderosa* as abortifacient compounds in cattle. *J Nat Toxins* 6;1–10, 1997

Gardner DR, Panter KE, James LF, Stegelmeier BL: Abortifacient effects of lodgepole pine *(Pinus contorta)* and common juniper *(Juniperus communis)* on cattle. *Vet Hum Toxicol* 40;260–263, 1998a

Gardner DR, Panter KE, James LF, Stegelmeier BL, Pfister JA: Diterpene acid chemistry of ponderosa pine and implications for late-term induced abortions in cattle. In *Toxic Plants and Other Natural Toxicants,* Garland T, Barr AC eds, CAB International, New York, pp339–344, 1998b

Gardner DR, Panter KE, James LF: Pine needle abortion in cattle: metabolism of isocupressic acid. *J Agric Food Chem* 47;2891–2897, 1999a

Gardner DR, James LF, Panter KE, Pfister JA, Ralphs MH, Stegelmeier BL: Ponderosa pine and broom snakeweed: poisonous plants that affect livestock. *J Nat Toxins* 8;27–34, 1999b

Gartner FR, Johnson FD, Morgan P: Cattle abortion from ponderosa pine needles: ecological and range management considerations. In *The Ecology and Economic Impact of Poisonous Plants on Livestock Production,* James LF, Ralphs MH, Nielsen DB eds, Westview Press, Boulder, Colo, pp71–93, 1988

James LF, Call JW, Stevenson AH: Experimentally induced pine needle abortion in range cattle. *Cornell Vet* 67;294–299, 1977

James LF, Short RE, Panter KE, Molyneux RJ, Stuart LD, Bellows RA: Pine needle abortion in cattle: a review and report of 1973–1984 research. *Cornell Vet* 79;39–52, 1989

James LF, Molyneux RJ, Panter KE, Gardner DR, Stegelmeier BL: Effect of feeding ponderosa pine needle extracts and their residues to pregnant cattle. *Cornell Vet* 84;33–39, 1994

Jensen R, Pier AC, Kaltenbach CC, Murdoch WJ, Becerra VM, Mills KW, Robinson JL: Evaluation of histopathologic and physiologic changes in cows having premature births after consuming ponderosa pine needles. *Am J Vet Res* 50;285–289, 1989

Knowles RL, Dewes HF: *Pinus radiata* implicated in abortion. *N Z Vet J* 28;103, 1980

Kubik YM, Jackson LL: Embryo resorptions in mice induced by diterpene resin acids of *Pinus ponderosa* needles. *Cornell Vet* 71;34–42, 1981

Lacey JR, James LF, Short RE: Ponderosa pine: economic impact. In *The Ecology and Economic Impact of Poisonous Plants on Livestock Production,* James LF, Ralphs MH, Nielsen DB eds, Westview Press, Boulder, Colo, pp95–106, 1988

Lin S-J, Short RE, Ford SP, Grings EE, Rosazza JPN: In vitro biotransformations of isocupressic acid by cow rumen preparations: formation of agathic and dihydroagathic acids. *J Nat Prod* 61;51–56, 1998

MacDonald MA: Pine needle abortion in range beef cattle. *J Range Manage* 5;150–155, 1952

Miner JL, Staigmiller RB, Peterson MK, Short RE, James LF: Montana pine needles cause abortion in beef cattle. *Mont Agric Res* 4 (Winter);6–9, 1987

Panter KE, James LF, Molyneux RJ, Short RE, Sisson DV: Premature bovine parturition induced by ponderosa pine: effects of pine needles, bark, and branch tips. *Cornell Vet* 80;329–338, 1990

Panter KE, Galey FD, James LF, Molyneux RJ, Pausch RD, Staigmiller RB, Short RE: Ponderosa pine needle induced parturition in cattle: analysis for presence of mycotoxins. *J Agric Food Chem* 39;927–929, 1991

Panter KE, James LF, Molyneux RJ, Short RE: Ponderosa pine–induced premature bovine parturition. In *Poisonous Plants: Proceedings of the Third International Symposium,* James LF, Keeler RF, Bailey EM Jr, Cheeke PR, Hegarty MP eds, Iowa State Univ Press, Ames, pp387–391, 1992

Pfister JA, Adams DC: Factors influencing pine needle consumption by grazing cattle during winter. *J Range Manage* 46;394–398, 1993

Pfister JA, Adams DC, Wiedmeier RD, Cates RG: Adverse effects of pine needles on aspects of digestive performance in cattle. *J Range Manage* 45;528–533, 1992

Pfister JA, Panter KE, Gardner DR: Pine needle consumption by cattle during winter in South Dakota. *J Range Manage* 51;551–556, 1998

Short RE, Staigmiller RB, Bellows RA, Panter KE, James LF: Effects of various factors on abortions caused by pine needles. *Fort Keogh Livestock and Range Research Station, Miles City, Montana, 1987 Field Day,* USDA/Montana Agricultural Experiment Station, 1987

Short RE, James LF, Staigmiller RB, Panter KE: Pine needle abortion in cattle: associated changes in serum cortisol, estradiol and progesterone. *Cornell Vet* 79;53–60, 1989

Short RE, James LF, Panter KE, Staigmiller RB, Bellows RA, Malcolm J, Ford SP: Effects of feeding ponderosa pine needles during pregnancy: comparative studies with bison, cattle, goats, and sheep. *J Anim Sci* 70;3498–3504, 1992

Short RE, Bellows RA, Staigmiller RB, Ford SP: Pine needle abortion in cattle: effects of diet variables on consumption of pine needles and parturition response. *J Anim Sci* 72;805–810, 1994

Short RE, Ford SP, Grings EE, Kronberg SL: Abortifacient response and plasma vasoconstrictive activity after feeding needles from ponderosa pine trees to cattle and sheep. *J Anim Sci* 73;2102–2104, 1995

Short RE, Ford SP, Rosazza JPN, Farley DB, Klavons JA, Hall JB: Effects of feeding pine needle components to late pregnant cattle. *Proc West Sect Am Soc Anim Sci* 47;193–196, 1996

Stevenson AH, James LF, Call JW: Pine-needle *(Pinus ponderosa)*–induced abortion in range cattle. *Cornell Vet* 62;519–524, 1972

Stuart LD, James LF, Panter KE, Call JW, Short RE: Pine needle abortion in cattle: pathological observations. *Cornell Vet* 79;61–69, 1989

Thieret JW: Pinaceae. In *Flora of North America,* vol 3, *Pteridophytes and Gymnosperms,* Flora of North America Editorial Committee ed, Oxford Univ Press, New York, pp352–398, 1993

Tucker JS: Pine needle abortion in cattle. *Proc Am Coll Vet Toxicol* 1961;35–39, 1961

Wagner WD, Jackson LL: Phytoestrogen from *Pinus ponderosa* assayed by competitive binding with 17-beta-estradiol to mouse uterine cytosol. *Theriogenology* 19;507–515, 1983

Zinkel DF, Magee TV: Resin acids of *Pinus ponderosa* needles. *Phytochemistry* 30;845–848, 1991

58 Poaceae Barnhart

- Grass Family
  - *Anthoxanthum*
  - *Avena*
  - *Cynodon*
  - *Festuca*
  - *Glyceria*
  - *Hilaria*
  - *Lolium*
  - *Panicum*
  - *Paspalum*
  - *Pennisetum*
  - *Phalaris*
  - *Setaria*
  - *Sorghastrum*
  - *Sorghum*
  - *Stipa*
  - *Triticum*
  - *Zea*
  - QUESTIONABLE
  - *Brachiaria*
  - *Hordeum*

One of the 5 largest families of flowering plants, the Poaceae comprises some 650 genera and 10,000 species. Truly cosmopolitan in geographical and ecological distribution, its members, the grasses, occur from both polar regions to the equator, and from deserts and mountaintops into the oceans. In various parts of the world, especially the tropical and north-temperate semiarid regions with seasonal precipitation, they dominate the landscape and form grasslands.

Of all plants, the grasses are the most important to humans. Indeed, development of past civilizations and our present society is dependent upon them. They provide our staple foods—the cereals—and the bulk of our caloric intake. Wheat, rice, maize, sorghum, barley, oats, rye, and millet are all grasses. In addition, they are the principal forage for both wild and domesticated animals. Animal husbandry, the principal source of our protein, could not exist in its present form without grasses to be grazed or cereal grains to be fed. In addition, they are, as lawns and ornamentals, an integral part of our living environment.

Genera of this ubiquitous family are traditionally grouped by taxonomists into 5–8 subfamilies and 25 or more tribes on the basis of their differences in gross morphology, anatomy, cytology, physiology, and biochemistry (table 58.1). An equally acceptable and commonly used alternative family name is Gramineae.

Table 58.1. Major subfamilies and tribes of the Poaceae in North America

Subfamily	Tribe	Common Names
Chloridoideae	Cynodonteae	gramas
	Eragrostideae	lovegrasses
Oryzoideae	Oryzeae	rices
Panicoideae	Andropogoneae	bluestems, sorghums
	Paniceae	panicums, panicgrasses, millets
Pooideae	Aveneae	oats
	Bromeae	bromes
	Poeae	bluegrasses, fescues
	Triticeae	wheats, barleys, ryes

- specialized grass terminology—
  culms = stems
  spikelets = flowers and subtending bracts

A rather specialized terminology has evolved to describe the distinctive morphology of grasses. Most terms are those used for other families of flowering plants, but several important differences do exist. Most notable is the use of the word *culms* rather than *stems* for the erect, aerial axes of grass plants. *Spikelet* is another term unique to the Poaceae. This structural unit, comprising flowers, their subtending bracts, and an axis, is emphasized in the classification and identification of grasses.

Because of the extreme evolutionary reduction of the flowers, they are not used in identification as they are in other plant families.

- herbs or woody canes; root systems fibrous; leaves 2-ranked, with basal sheaths; flowers borne in spikelets; fruits caryopses

Plants herbs or woody canes; annuals or perennials; from rhizomes or stolons or caudices or crowns; terrestrial or emergent aquatics. **Root Systems** fibrous. **Culms (Stems)** branched or unbranched; jointed; internodes solid or hollow, terete or slightly to strongly flattened. **Leaves** basal or cauline; simple; alternate; with basal sheaths, margins overlapping or rarely fused; 2-ranked; blades linear to ensiform to ovate; venation parallel or parallel-convergent; ligules present or rarely absent; margins entire or scabrous. **Inflorescences** spikelets; borne in panicles or racemes or spikes or rames or enclosed in burs; terminal or axillary. **Spikelets** consisting of glumes and florets borne in 2 ranks on a shortened axis (rachilla); lowest bracts (glumes) 2, or rarely 1 or absent, subopposite, without flowers; florets 1 to many, borne above the glumes, consisting of paired bracts (lemmas and paleas), flowers typically present or rarely absent; lemmas enclosing paleas and flowers; paleas typically present or rarely absent. **Flowers** reduced; hidden by lemmas and paleas or glumes except at anthesis; perfect or imperfect; perianths in 1-series. **Sepals** (lodicules) 1–3; reduced to small appendages at bases of ovaries; free or fused. **Petals** absent. **Stamens** 3, or rarely 6 or 1. **Pistils** 1; compound, carpels 2 or 3; stigmas 2 or rarely 3, plumose; styles 2 or 1; ovaries superior; locules 1; placentation parietal. **Fruits** caryopses. **Seeds** 1.

A problem of nomenclatural ambiguity that has toxicologic implications exists in the Poaceae. Nine species in 7 genera in 3 tribes and 2 subfamilies are commonly referred to as millet (table 58.2). Because these grasses produce different disease problems, use of the common name alone when evaluating toxicologic risk or problem is ill-advised.

Table 58.2. Grasses commonly referred to as millet

Echinochloa	
E. crusgalli	barnyard millet, sawa millet
E. frumentacea	Japanese barnyard millet, sanwa millet
Eleusine coracana	finger millet, African millet
Panicum	
P. miliaceum	common millet, proso millet, hog millet, Russian millet, brown-corn millet
P. sumatrense	little millet
Paspalum scrobiculatum	kodo millet
Pennisetum glaucum	pearl millet, spiked millet, bulrush millet, cattail millet
Setaria italica	foxtail millet, Italian millet, German millet, Hungarian millet
Sorghum bicolor	great millet

- 6 major disease problems—
  photosensitization
  ARDS
  tetany
  nitrate toxicosis
  cyanide toxicosis
  fungal endophyte toxicosis

Disease Problems—Although an essential part of our existence by providing food for us or forage for our domesticated animals, some genera of the Poaceae cause significant intoxication problems. These problems include photosensitization, acute respiratory distress syndrome (ARDS), grass tetany, nitrate toxicosis, cyanide toxicosis, and disease due to fungal endophytes. Aspects of each of these diseases are described in the following subsections. When appropriate, additional information about the disease is presented in the generic treatments that follow. Other problems characteristic of certain genera, such as oxalate accumulation, are described in the treatments of those taxa.

- mechanical injury

In addition to intoxications, mechanical injuries to grazing animals sometimes are the result of contact with grasses that possess sharp and/or stiff organs or structures (table 58.3). Such injuries may result in obvious ulcerative to pyogranulomatous lesions of the mouth that may be reflected by a decreased appetite but usually without an effect on drinking.

Table 58.3. Examples of grasses causing mechanical injuries to grazing animals

Injury due to presence of awns	
Aristida spp.	threeawns
Avena fatua	wild oats
Bromus tectorum	downy brome
Hordeum	
H. jubatum	foxtail barley
H. murinum	wall barley
H. secalinum	little barley
Setaria spp.	bristlegrasses
Stipa spp.	needlegrasses, speargrasses
Injury due to presence of spines	
Cenchrus incertus	sandbur
Injury due to sharp margins of leaves	
Leersia spp.	cutgrasses
Spartina cynosuroides	big cordgrass

Sources: Pammel 1911; Bankowski et al. 1956; Rebhun et al. 1980; Peterson & Schultheiss 1984.

Photosensitization

■ primary photosensitization, cause unknown; rarely hepatogenous photosensitization due to infection of grasses with fungus *Pithomyces chartarum*

Disease Problems—A number of grass species are associated occasionally with photosensitization, a skin disease involving light-induced formation of molecules in the peripheral circulation and the subsequent damage they cause to surrounding tissues. This problem generally occurs when animals are consuming lush, green growth. Sperry and co-workers (1955) correlated Texas plants with the disease and observed that it typically occurred a week or two after a rainy period following a drought. It often involved annual grasses used for forage (table 58.4).

Table 58.4. Grass forages associated with direct, or primary, photosensitization

Avena sativa	oats
Cenchrus incertus	southern sandbur
Echinochloa crusgalli	barnyardgrass, barnyard millet, sawa millet
Eriochloa contracta	prairie cupgrass
Hordeum	
H. murinum	wall barley
H. vulgare	barley
Secale cereale	rye, annual rye, rye grain
Setaria italica	foxtail millet, Italian millet, German millet, Hungarian millet
Sorghum spp.	sorghum
Triticum aestivum	wheat, bread wheat, common wheat

Sources: Sperry et al. 1955; Rowe 1989; Horn & Burrows 1990.

Because icterus and other evidence of hepatic dysfunction do not accompany other clinical signs of photosensitization in most cases involving grasses, it has been assumed that the disease is the primary, or direct, type. The specific photodynamic agent is unknown, but the clinical signs indicate the presence of free-radical products that react with cellular membranes in a way much like that of hypericin from *Hypericum* (see chapter 24). However, secondary, or hepatic, photosensitization does occur, caused by species of *Panicum* (panicgrass) and other grasses, such as *Lolium* (ryegrass) infected with the fungus *Pithomyces chartarum*.

■ necrosis, sloughing of skin in areas with least protection from sunlight; areas of little wool or skin pigment

Clinical Signs—Appearance of the disease is manifested by the onset of necrosis and sloughing of the skin. Full expression of these signs may take several days. In most animal species, the lesions are scattered in the less pigmented areas. However, in sheep, the lesions are generally around the ears, eyes, mouth, and other areas of the body with less wool. Elsewhere the dense wool prevents light from reaching the skin to activate the circulating photodynamic agent in the capillaries. In addition to skin involvement, there may be reddening of the cornea of the eye and increased sensitivity to light. When an animal exhibits these skin changes, it is important to conduct a careful examination to ascertain the presence of icterus or to use serum chemistry to evaluate for possible involvement of the liver.

Treatment—There is little to be done for animals exhibiting photosensitization except to remove them from the offending grass forage and, if possible, to reduce the amount of exposure to direct sunlight by providing shade.

- remove from forage

Acute Respiratory Distress Syndrome

Disease Problems—Historically, pastured cattle experiencing acute respiratory distress were typically described to be suffering from acute bovine pulmonary emphysema. This disease and other respiratory problems, such as fog fever and atypical interstitial pneumonia, are categorized here as types of acute respiratory distress syndrome (ARDS), as recommended by Pierson and Kainer (1980). The disease, known as early as the 1920s, causes considerable loss of livestock in North America (Schofield 1948). It also is of importance in Europe and other areas (Begg & Whiteford 1948).

- ARDS, pneumonia-like but afebrile disease of ruminants, mainly cattle, 5–10 days after abrupt change from poor to lush forage; various grasses; commonly in autumn

The important feature in the appearance of the disease is an abrupt change in forage quality, which may occur when animals are moved from a poor pasture to a lush, green one, or when dry hay is replaced by lush, green forage. Less commonly, ARDS develops following a change from pasture to hay, such as occurred when a few steers in a group of 40 animals weighing 230 kg developed the disease about 1 week after being moved from low-quality pasture to hay of Old World bluestem (*Bothriochloa* spp.) (Phillips & Von Tungeln 1986).

An acute pneumonia-like but afebrile disease appears in 5–10 days. It most often affects mature cattle; the time of year varies with geographic location. The most common scenario involves movement of cattle in late summer and early fall from dry pastures or ranges to irrigated grass or alfalfa pastures (Schofield 1948; Mackey 1952; Hyslop 1969; Blake & Thomas 1971). The disease has been readily reproduced experimentally by modeling these conditions (Tucker & Maki 1962). However, it can occur at other times, as illustrated by its common appearance in April, May, and June in cattle grazing *Cynodon dactylon* (coastal Bermudagrass) in eastern Texas (Williams et al. 1977). The morbidity rate is usually low but occasionally may exceed 50%. The case-fatality rate is typically 25–50%. ARDS cases in Mexico have involved a variety of irrigated grasses such as *Pennisetum ciliare* (buffelgrass), *Cynodon* sp. (African stargrass), *Panicum coloratum* (Kleingrass), and *Cynodon dactylon* (Bermudagrass) (Romero et al. 1993).

No single grass genus is of special significance in the appearance of ARDS, and the disease is not discussed specifically in the following generic treatments. Other forages such as *Medicago sativa* (alfalfa), members of the Brassicaceae (rape, kale, turnip tops), and *Perilla frutescens* (purple mint) and the fungus *Fusarium,* when causing mold of sweet potatoes or runner beans, have been reported to cause ARDS (Peckham et al. 1972; Linnabary & Tarrier 1988; Kerr & Linnabary 1989).

Disease Genesis—ARDS associated with grasses is the result of a sudden change in forage quality that results in an abrupt increase of L-tryptophan in the rumen, where it is converted by bacteria to indolepyruvate, indoleacetate, and then to 3-methylindole

- disease appears to be due to tryptophan overload in rumen; metabolized to compounds toxic to lung, resulting in severe pulmonary emphysema and edema

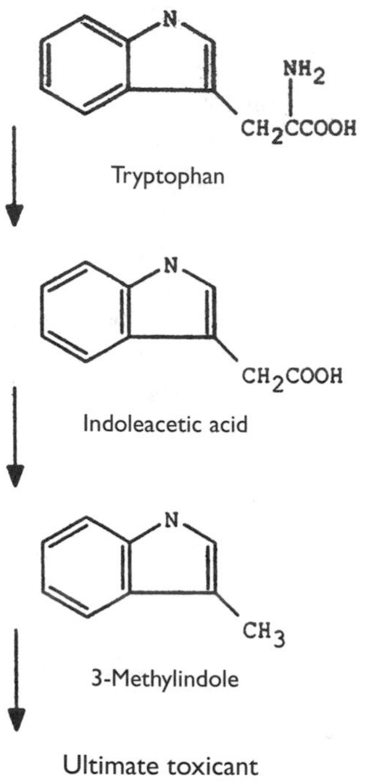

NH_2

CH_2CCOOH

Tryptophan

CH_2COOH

Indoleacetic acid

CH_3

3-Methylindole

Ultimate toxicant

(3-MI) (Carlson & Dickinson 1978; Hammond et al. 1979). The 3-MI is absorbed into the circulatory system and subsequently metabolized in the lungs to produce short-lived, reactive toxicants. In animals accustomed to lush grass forage, tryptophan is converted mainly to indole, which is not activated in the lung to the toxicant (Kirkland & Bray 1984). Experimentally, ARDS can be reproduced by oral administration of tryptophan or by oral or i.v. administration of 3-MI (Carlson et al. 1968, 1972, 1975). However, some closely related indoles, e.g., tryptamine, are not subject to degradation to 3-MI (Yokoyama & Carlson 1974). The disease can be prevented by minimizing ruminal conversion of tryptophan to 3-MI via the use of ionophores or polyether antibiotics such as lasalocid, monensin, narasin, nigericin, and salinomycin (Hammond & Carlson 1980; Hammond et al. 1980).

Once absorbed into the systemic circulation, 3-MI is biotransformed in the lungs by the cytochrome P-450 pathway in Clara cells and macrophages and to a lesser extent in type II pneumocytes in the lungs (Bray & Carlson 1979; Bray & Kirkland 1990; Nichols et al. 1990). This results in the formation of short-lived electrophilic intermediates. We are indebted to the extensive and elaborate studies of G. S. Yost and co-workers in the elucidation of the details of the formation and action of the reactive metabolites of 3-MI. Their studies indicate that the reactive intermediate is an imine methide electrophile (Nocerini et al. 1985c). Specific imines identified include 3-methyleneoxindole, 3-methyleneindolenine, or 3-hydroxy-3-methylindolenine (Appleton et al. 1991; Ruangyuttikarn et al. 1992; Skordos et al. 1998a,b). Prevention of their formation by inhibitors protects the animal against the disease (Bray & Carlson 1979; Hanafy & Bogan 1982; Nocerini et al. 1984, 1985a; Smith et al. 1993; Thornton-Manning et al. 1993). Although the specific free radical has not been identified, the formation of *C*-, *O*- and *N*-centered radicals has been shown in vitro and in vivo (Chen et al. 1994). Smith and co-workers (1993) have identified nearly a dozen metabolites in urine as sulfate, glucuronide, or mercapturate conjugates. The major metabolic pathways of 3-MI biotransformation involve formation of mono- and dihydroxy-3-methyloxindoles. The reactive intermediate or intermediates form adducts either with cellular proteins to produce the toxic effects or with glutathione as a protective conjugation mechanism.

3-MI appears to be selectively toxic to the lung rather than the liver because of more-rapid biotransformation rates in the lung (Nocerini et al. 1985a). Clara cells and type I pneumocytes are preferentially destroyed (Huang et al. 1977; Bradley & Carlson 1980). Type I cells are sloughed into the alveoli, and type II cells proliferate to restore the alveolar epithelium (Mason et al. 1977; Bradley & Carlson 1980). The proliferation of type II cells occurs quickly and accounts for the firmness and glandular appearance of the lungs when sectioned. During this period of proliferation, the ability of type II cells to synthesize and secrete functional surface-active phospholipids is decreased (Kirkland & Bray 1984, 1989). The surfactant produced no longer effectively lowers surface tension. Although sensitivity varies somewhat, experimental administration of 3-MI results in typical lesions in other species subject to biotransformational alterations (Turk & Thomas 1986). Experimental results in rats indicate the presence of inflammatory arachidonic acid mediators while alveolar septal injury is occurring (Woods et al. 1993). The role of prostaglandins in this inflammatory response is illustrated further by the partial protection provided by aspirin or indomethacin in goats exhibiting the disease after administration of 3-methylindole (Acton et al. 1992).

- abrupt onset; labored, open-mouth respiration; frothy salivation; severe distress

Clinical Signs—The disease is characterized by rapid onset of labored respiration with loud expiratory grunts and open-mouth breathing accompanied by frothy, excessive salivation. Harsh respiratory sounds are evident on auscultation. The animal subsequently may develop subcutaneous emphysema along the neck, shoulders, and back. Death often occurs 1–2 days after onset of signs. In a few cases, pastured cattle will be found dead without signs being observed.

Pathology—Grossly, a variety of minor changes may be evident, including a pale and friable liver and distension of the gallbladder, but these are of little significance in comparison with the effects on the respiratory system. The lungs are often remarkably distended. Upon opening the chest, the lungs fail to collapse completely and are heavy and firm, with froth-filled airways. The pulmonary lymphatics are usually dilated and the interlobular septa are distended with edema fluid, fibrin, and air bubbles. Cut surfaces reveal an almost marbled appearance of patchy areas of firm, reddened to purple lobules, and prominent edema in connective tissue and septa. Air bubbles and/or large bullae may be apparent. Microscopically, alveolar epithelialization, accumulations of alveolar macrophages, and hyaline membrane formation are commonly observed. Changes in the alveolar epithelium are due to proliferation of type II pneumocytes, producing an almost glandular-like appearance. A few eosinophils may be present; otherwise, granulocytic inflammatory cells are rare.

■ gross pathology, lungs distended, wet, heavy, froth in airways, marbled, with patchy firm, dark areas

microscopic, alveolar epithelial proliferation, glandular appearance

Treatment—Because of the pathologic changes that occur, ARDS is typically poorly responsive to treatment (Williams et al. 1977). It is very important to minimize stress during the period of severe respiratory distress. Numerous drugs have been used, including corticosteroids, antihistamines, nonsteroidal antiinflammatories, atropine, diuretics, and antibiotics. Therapy, however, appears to be of limited value, although antiprostaglandin drugs such as flunixin meglumine may alleviate the severity of the inflammatory response (Selman et al. 1985). Handling thc animal for treatment often causes stress and death due to acute respiratory embarrassment.

■ minimize stress; atropine, flunixin

prevent by gradual change to lush forage; ionophore pretreatment

ARDS can be prevented by management techniques aimed at ensuring gradual changes in pasture quality. Recommendations to accomplish this include (1) cutting the forage of a lush pasture and allowing it to wilt before permitting animals to graze; (2) limiting grazing time on the new pasture; and (3) rotating the animals every few days between the lush and dry pastures or between the lush pasture and a drylot with provided hay (Potchoiba et al. 1992). When these approaches are not possible, diet supplementation with 200 mg/day of an ionophore such as monensin or lasalocid may be used to prevent or at least attenuate the severity of the disease. Feeding of the ionophore should begin a week before the pasture is changed and continue for 10 days after the change (Hammond et al. 1980; Nocerini et al. 1985b; Potchoiba et al. 1992). This approach will provide a sustained protective effect, but it does not prevent the appearance of the disease in all animals (Honeyfield et al. 1985). The efficacy of the ionophore can be enhanced by combining it with the use of hay to reduce the intake of lush grass forage (Potchoiba et al. 1992).

Grass Tetany

Disease Problems—In the 1880s, a tetany syndrome involving tremors and recumbency in pastured cattle and less commonly sheep was recognized in Europe, especially the Netherlands (Sjollema 1932). The syndrome was seen with increasing frequency in the 1900s, most likely because of changes in management practices of the time (Stewart 1954). The clinical signs exhibited are somewhat similar to those of the hypocalcemia of "milk fever," and it has sometimes been called atypical milk fever. The two diseases differ in the presence of rigidity in the tetany syndrome. The tetany

■ tetany syndrome, cattle, less often in sheep; most commonly in lactating animals; rapidly growing cool-season grasses

syndrome occurs mainly in older cattle that are consuming rapidly growing, lush, cool-season forages and is known commonly as grass tetany, spring tetany, grass staggers, or hypomagnesemic tetany.

A distinction must be made between grass tetany, or spring tetany, and other types, such as winter tetany in animals overwintered inside or outside on hay or poor pasture (Stewart 1954). Common in some areas of North America, winter tetany is associated with consumption of dry winter rations rather than fresh plant material (Custer 1959; Horvath 1959; Leffel & Mason 1959; Marshak 1959). The two diseases are similar in appearance, but in this book, only aspects of grass tetany are discussed. Grass tetany must also be differentiated from the tremorgenic syndromes associated with species of *Lolium* and *Paspalum* (paspalum) and described below.

Typically but not exclusively appearing in the spring, grass tetany occurs in various areas of North America under different circumstances. Significant factors in the appearance of the disease are lactation; cool, wet, cloudy weather; and new forage growth. Signs are most commonly seen in newly lactating cows, both beef and dairy, and in some instances occur as quickly as 1–2 weeks after animals are moved onto a new pasture (Sjollema 1932).

A particularly devastating episode of grass tetany occurred in California during the winter of 1963–1964, when 4,000–6,000 cows died (Hjerpe 1964). Said to have been the most severe episode in the area for 75 years, the losses occurred in October-March in pregnant and lactating cows wintered on native-grass pasture. The disease was especially common among recently calved older cows. Except for those recumbent at the time of treatment, animals were generally responsive to calcium/magnesium therapy.

Tetany has been associated with many grasses, especially the cool-season species that use the C_3 photosynthetic pathway. No single genus is of special significance in the appearance of the disease, and it is not discussed specifically in the following generic treatments. Representative genera and species are listed in table 58.5, but it should be emphasized that the disease is not limited to these taxa. Of the cereal grain forages, wheat seems to be the most tetany prone and rye the least (Mayland et al. 1976).

Table 58.5. Grass forages associated with grass, or spring, tetany

Agropyron spp.	wheatgrasses
Avena sativa	oats
Bromus spp.	bromes
Dactylis glomerata	orchardgrass
Elymus spp.	wildryes
Festuca spp.	fescues
Hordeum spp.	barleys
Lolium perenne	perennial ryegrass
Phalaris spp.	canarygrasses
Phleum pratense	timothy
Secale cereale	rye, annual rye, rye grain
Triticum aestivum	wheat, bread wheat, common wheat

Sources: Grunes et al. 1970; Mayland & Grunes 1979; Sleper 1979.

Tetany is less often seen in pastures containing appreciable numbers of legumes or in poor pastures or when hay is fed (Muth & Haag 1945; Sleper 1979). The risk of disease seems to be greater for Holstein and Jersey cattle and less for Brahman (Green et al. 1989). A unique situation occurs in the southern Great Plains, where winter and spring grazing of wheat pasture is commonly practiced. Although wheat is highly nutritious, animal losses of 2–3% due to wheat-pasture tetany, or wheat-pasture poisoning as the disease also is known, are not uncommon. These losses typically occur from January to March or about 60–150 days after the animals begin grazing the pasture, and when they are in the last month of gestation or have calved within 2 months (Sims & Crookshank 1956).

Disease Genesis—The tremorgenic syndrome of grass tetany was eventually characterized as a hypomagnesemic tetany (Sjollema 1932). Decreased blood calcium, copper, potassium, phosphate, and other factors subsequently were shown also to be involved in some instances (Muth & Haag 1945; Singer et al. 1958). The disease came to be recognized as a complex metabolic disorder associated primarily with a magnesium deficit that occurred under a variety of circumstances. When magnesium alone was decreased, signs were often not severe, but when a magnesium deficit was accompanied by a decrease in calcium, tetany and hyperexcitability prevailed (Blakemore & Stewart 1933; Muth & Haag 1945). Barker (1939) described a range of syndromes of metabolic disturbances from the simple hypocalcemia of milk fever to the multiple effects of calcium and magnesium serum levels in grass tetany. Increased magnesium and decreased calcium resulted in paresis and narcosis; normal magnesium and decreased calcium produced paresis, paddling/tetanic spasms, and coma; and when both magnesium and calcium were decreased, there was abrupt onset of excitement, aggression, tetany, hyperesthesia, and recumbency. These same observations have been reaffirmed by others.

- complex metabolic disorder; mainly magnesium deficiency compounded by high potassium in plants and other factors, especially during rapid growth; danger zone, plasma magnesium less than 1.7 mg/dl

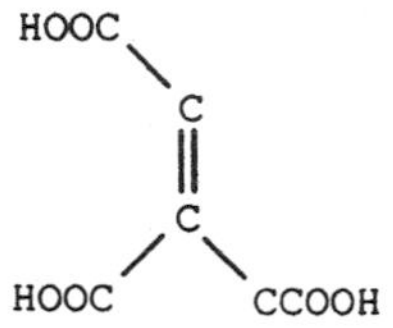

HOOC
C
C
HOOC CCOOH

Transaconitic acid

The complexity of the disease is exemplified by the variable effects on animals grazing wheat pasture. In some instances, a decrease in both magnesium and calcium seems to be primary, whereas in other situations a decrease in calcium seems foremost (Crookshank & Sims 1955; Bohman et al. 1983a). Decreased serum magnesium and calcium can occur without definite or with only marginal decreases of the elements in the forage. This does not mean that plant magnesium is not of concern. Tetany is unlikely to be caused by forages that have magnesium levels well above 0.25%, even though potassium and nitrogen levels also are high. When forage magnesium is below 0.25%, then factors such as low levels of digestible carbohydrates and vitamin D and high levels of potassium may limit its bioavailability and cause hypomagnesemia (Fontenot et al. 1989; Grunes & Welch 1989). In tetany-prone pastures, high levels of nitrogen and fatty acids and a decrease in total soluble carbohydrates result in decreased bioavailability of magnesium. Hypomagnesemic tetany is thus due to a conditioned or induced deficiency rather than reduced intake of magnesium (Metson et al. 1966). The disease is typically prevented by administration of magnesium salt supplements (Blakemore & Stewart 1935). Although previously suggested as a possible cause, aluminum in the diet seems to have little apparent effect (Allen & Robinson 1980; Kappel et al. 1983).

Continued examination of tetany-prone pastures eventually resulted in a proposed role for organic acids, such as citric acid or transaconitic acid, perhaps in combination with potassium (Burt & Thomas 1961; Bohman et al. 1969). Transaconitic acid also is present in many forages with concentrations in the range of 1–2.5% (Burau & Stout 1965). It is highest in young plants and in cold weather, especially in *Phalaris tuberosa* (canarygrass) and *Hordeum leporinum* (hare barley) (Burau & Stout 1965; Stout et al. 1967). In the animal, transaconitic acid is a competitive inhibitor of the aconitase enzyme, thereby blocking the TCA cycle and causing a buildup of aconitic and citric acid, especially in the liver and kidneys (Saffron & Prado 1949). Daily administration of transaconitic acid i.v. or orally to cattle produced little response except for a decrease in serum magnesium in one study and none in another (Camp et al. 1968; Kennedy 1968). However, when transaconitic acid was administered together with potassium chloride, a tetany syndrome was produced that was responsive to calcium/magnesium treatment (Bohman et al. 1969).

The occurrence of tetany was shown to be linked to growth spurts in the grass plant in which protein, potassium, digestibility, and transaconitic acid all increased suddenly and markedly (Bohman et al. 1983b). At these times, concentrations of calcium and magnesium may not be altered, but their absorption is somehow modified by formation of complexes so as to limit their availability to the animal. It was originally thought that increased concentrations of aconitic and citric acids might account for the binding of these divalent cations. However, this mechanism does not seem to be sufficient to cause such severe effects. An additional divalent cation binding factor is created by reduction of transaconitic acid to tricarballylate by an active bacterial reduction process in the rumen (Russell & Van Soest 1984; Russell 1985). This reaction can be attenuated by inoculation of the rumen with fermenting bacteria such as *Acidaminococcus,* which represents a potential preventive measure against the development of magnesium chelation and the resulting tetany (Cook et al. 1994).

Because of the complex nature and seriousness of the disease, various indices and procedures have been developed as measures of the tendency of pastures to produce tetany (Mayland et al. 1974). Ratios of the various elements involved in the disease have been used (Metson et al. 1966; Allcroft & Burns 1968; Grunes et al. 1970; Bohman et al. 1983a). For example, when the ratio of potassium equivalents to calcium and magnesium equivalents (K/[Ca + Mg]) exceeds 2.2, the forage is considered to be tetany-prone, a finding that has, for the most part, been consistent (Lewis & Sparrow 1991). Plasma magnesium less than 1.7 mg/dl indicates hypomagnesemia and less than 1 mg/dl warns that tetany is imminent (Monson 1963). Hartman's nomograph, devised for dairy cattle, can also be used for determining plasma magnesium and tetany danger based upon forage levels of magnesium, potassium, and crude protein (Mayland et al. 1976; Mayland & Grunes 1979).

Plant uptake of magnesium from the soil is controlled by numerous factors. It is thus difficult to increase, in an economical manner, the levels of the mineral in the plant herbage and its availability to the animal (Mayland & Wilkinson 1989).

- nervousness, twitching; progression to excitement, stiffness, ataxia, nervous apprehension, collapse, contraction of neck muscles twisting the head backward; paddling and/or tetanic seizures

Clinical Signs—In mild cases of grass tetany, the animal exhibits slight nervousness and some twitching. In more-severe cases, these signs progress to excitement, stiffness and incoordination, decreased appetite, muscular twitching of the face and ears, excess salivation, and grinding of the teeth. There is increasing nervous apprehension, with the head held high and the ears erect. The animal may be noticeably aggressive. Later it goes down, is unable to rise, and has contractions of the neck muscles that give the head a backward twist. The eyelids, including the third, may protrude and/or flicker. When the animal is prostrate, it exhibits thoracic limb paddling and general tetanic seizures. It is very responsive to stimuli, with noise and touch producing reflex seizures. Eventually respiration becomes labored, the heart pounds, and the animal becomes comatose. The case-fatality rate is high, and death may occur minutes to hours after onset of signs. Rarely, death may be delayed for a few days.

- few lesions

Pathology—There are few lesions other than scattered ecchymotic hemorrhages on the heart and other organs, distension of the gallbladder, and sometimes pulmonary emphysema.

- parenteral calcium, magnesium solutions; magnesium supplement in feed

Treatment—Grass tetany can be treated by i.v. administration of solutions of magnesium and calcium borogluconate. A favorable response may be expected in animals treated while still standing, but not necessarily in those that are recumbent.

Although treatment of cases of grass tetany is possible, prevention of the disease is more feasible, because of the problem in treating cattle in the pasture. Prevention may be accomplished by magnesium supplementation in the form of magnesium oxide (10–30 g/day, or 1–2 oz) combined with dry molasses, cottonseed meal, alfalfa meal, or another ingredient to increase palatability (Grunes et al. 1970; Robinson et al. 1989). A dosage of 3.7 g/day is needed for sheep. When magnesium carbonate is used, double the amount is required. Administration of magnesium supplements must be conducted while the animals are exposed to the tetany-prone pasture. Supplementation prior to, but not during, this time is not effective in preventing the occurrence of tetany (Ritter et al. 1984). Administration of magnesium in water, by boluses, or by foliar application is less effective, in part because of economic considerations. The value of administration of vitamin D is still open to question because of divergent effects on calcium and magnesium (Allcroft & Burns 1968; Moate et al. 1987; Robinson et al. 1989).

Other preventive approaches include improvements to pasture to reduce the likelihood of the problem (Robinson et al. 1989). Coarse soils can be limed and fertilized, and fine soils can be amended by improved drainage. It is important to avoid excessive fertilization, especially in the spring and summer; this is not so important in the fall. Specific cultivars of grasses are available that accumulate high concentrations of magnesium, such as the palatable endophyte-free tall fescue 'HiMag' *(Festuca arundinacea)* (Shewmaker et al. 1997). This experimental tall fescue cultivar is well accepted by cattle and has many favorable characteristics, such as increased levels of calcium and magnesium and a lower K/(Ca + Mg) ratio relative to those of other fescue cultivars (Crawford et al. 1998). Mixed-legume pastures may be of some value, and supplementation with good hay or grain to increase the digestible carbohydrates will be useful. For a discussion of the prevention of tetany, see Robinson et al. (1989).

Nitrate Accumulation

Disease Problems—Extensive use of nitrogen fertilizers, the use of forage species or varieties with a tendency to accumulate high concentrations of nitrate, and the use of large bales to reduce labor have exacerbated the problems associated with nitrate accumulation. Unusually dry growing conditions, such as during droughts, are also major factors in increasing the incidence of high nitrate concentrations (Pickrell et al. 1991). Nitrate intoxication constitutes a serious problem in part because of the large numbers of plant species with a propensity to accumulate nitrate. Foremost among genera causing this problem are *Sorghum* (in this chapter) and *Amaranthus* in the Amaranthaceae (see chapter 5). Both represent a potential problem, mainly in hay and to a lesser extent as fresh forage. Cereal grain hays and cornstalks likewise constitute a risk. Although numerous other species also accumulate nitrate under certain conditions—e.g., vines of *Ipomoea batatas* (sweet potato), with up to 25,000 ppm NO_3—most represent little hazard in most circumstances (Olson & Whitehead 1940; Brady et al. 1955; Hanway & Englehorn 1958). The use of hormonal herbicides may increase the accumulation of nitrate in many of the weedy plant species but does not seem to increase the problem with grasses and other cultivated forages (Berg & McElroy 1953). A number of excellent reviews on nitrate as a toxicant are available (Wright & Davison 1964; Bruning-Fann & Kaneene 1993), including many agricultural experiment station summaries and fact sheets (Whitehead & Moxon 1952; Hanway et al. 1963; Deeb & Sloan 1975; Strickland et al. 1995). A more complete discussion of all aspects of the problem is presented in the treatment of *Sorghum,* in this chapter.

- many plant species accumulate nitrate; climatic and management variables are important factors in nitrate accumulation

The number of known nitrate-accumulating grass genera and species is so extensive that it seems futile to present such a list, especially because so few have been incriminated in actual toxicologic problems. Species that have actually been implicated in disease are listed in table 58.6.

Table 58.6. Grass forages causing nitrate and/or nitrite intoxication

Avena sativa	oats
Cynodon spp.	Bermudagrasses, stargrasses
Echinochloa frumentacea	Japanese barnyard millet, sanwa millet
Hordeum	
H. jubatum	foxtail barley
H. vulgare	barley
Lolium spp.	ryegrasses
Pennisetum glaucum	pearl millet
Sorghum spp.	sorghums
Triticum aestivum	wheat, bread wheat, common wheat
Zea mays	corn, maize, Indian corn

Source: Montgomery & Hum 1995.

■ ruminants, nitrate reduced to nitrite by rumen microflora, which causes oxidation of hemoglobin to methemoglobin

Disease Genesis—The problem of nitrate intoxication is limited to ruminants, especially cattle, because of reduction of nitrate to nitrite by ruminal microorganisms and the potential for accumulation and absorption of nitrite from the rumen. Absorbed nitrite oxidizes iron in hemoglobin to the ferric form, resulting in the formation of methemoglobin (MHb). The compound is incapable of oxygen transport, and the animal literally starves for oxygen.

■ abrupt onset, rapid respiration, ataxia, weakness, dark mucous membranes

no specific lesions

treat with methylene blue

Clinical Signs, Pathology, and Treatment—Nitrate intoxication is characterized by abrupt onset of signs several hours after consumption of the forage. There is weakness, incoordination, rapid respiration, apprehension, and collapse. Recumbent animals remain so when approached or rise only with difficulty. Darkening of the mucous membranes may be seen upon close examination. Death may occur within minutes or a few hours.

At necropsy, few distinctive lesions are seen. The blood and tissues may be dark- or chocolate-colored, but this change is not consistently observed. There also may be scattered ecchymotic hemorrhages and increased foamy fluid in the terminal airways of the lungs, but these changes are not usually extensive.

The primary antidote for nitrate intoxication is i.v. administration of methylene blue, 2–4 mg/kg b.w. given as a 1–2% solution (Burrows 1980). Unfortunately, the rapid development of signs and the quickness of death provides little opportunity for treatment of animals that are not observed frequently. As described in the treatment of *Sorghum* (in this chapter), management strategies to minimize both accumulation of nitrate in the plant and animal consumption of high-nitrate forage are important.

Cyanogenesis

■ cyanogenic, mainly *Sorghum*

Disease Problems and Genesis—An appreciable number of grass genera are cyanogenic, but few are serious risks. Because the sorghums represent the most serious risk, cyanogenesis is described in depth in the treatment of that genus (in this chapter). Other genera of some importance as cyanide sources are indicated as such in their treatments. The grass taxa in which cyanide has been detected are listed in table 58.7. Most are not intoxication risks.

Table 58.7. Grasses determined to be cyanogenic and the principal glucosides present

Agrostis stolonifera	?	creeping bentgrass
Andropogon spp.	dhur	bluestems
Avena sativa	dhur	oats
Bambusa spp.	tax	bamboos
Bothriochloa		
B. bladhii (as *B. intermedia*)	?	Old World bluestem
B. ischaemum	?	yellow bluestem, plains bluestem
Bouteloua		
B. gracilis	dhur	blue grama
B. hirsuta	tri	hairy grama
Briza spp.	dhur	quakinggrasses
Catabrosa aquatica	dhur	brookgrass, water whorlgrass
Chloris		
C. truncata	tri	Australian fingergrass, Australian windmillgrass
C. ventricosa	dhur	Australian weepinggrass
Chrysopogon spp.	dhur	beardgrasses
Cortaderia spp.	tri	pampasgrasses
Cymbopogon spp.	dhur	lemongrasses
Cynodon		
C. aethiopicus	dhur	giant stargrass, Ethiopian dogstoothgrass
C. dactylon	dhur, tri	Bermudagrass
C. nlemfuensis	dhur	reuse kweekgrass, African Bermudagrass
C. plectostachyus	dhur	Naivasha stargrass, estrella, summer stargrass
C. transvaalensis	dhur	quickgrass, African dogstoothgrass
Dactyloctenium aegyptium	dhur	Egyptian crowfootgrass
Danthonia semiannularis	dhur	oatgrass
Dendrocalamus spp.	tax	bamboos
Digitaria eriantha	dhur	common finger grass
Eleusine		
E. coracana	tri	African millet
E. indica	dhur	goosegrass, yardgrass
Elymus spp.	dhur	wildryes
Festuca spp.	dhur	fescues
Glyceria		
G. canadensis	dhur	Canadian mannagrass
G. grandis	dhur	American mannagrass
G. septentrionalis	?	eastern mannagrass
Holcus lanatus	dhur	common velvetgrass
Hordeum vulgare	epi	barley
Lagurus ovatus	dhur	hare's-tail grass
Lamarckia aurea	dhur	goldentop
Leersia hexandra	dhur	clubhead cutgrass
Leptochloa dubia	dhur	green sprangletop
Lolium perenne	dhur	perennial ryegrass
Melica altissima	tri	oniongrass, melic
Molinia caerulea	tri	purple moorgrass
Oryza sativa	dhur	rice
Panicum		
P. maximum	dhur	guineagrass
P. muticum	dhur	paragrass
Poa pratensis	?	Kentucky bluegrass
Secale cereale	dhur	rye, rye grain, annual rye
Sorghastrum nutans	dhur	Indiangrass
Sorghum		
S. xalmum	dhur	Columbusgrass
S. bicolor	dhur	sorghum, Sudangrass, shattercane, milo
S. halepense	dhur	Johnsongrass
Stipa robusta	tri	sleepygrass
Tridens flavus	tri	purpletop
Triticum aestivum	dhur	wheat
Zea mays	dhur	corn, maize, Indian corn

Sources: Brunnich 1903; Pamell 1911; Leemann 1935; Moran 1940; Gibbs 1974a,b; Seigler 1976a,b, 1977, 1981; Haskins et al. 1979; Erb et al. 1981; Saupe 1981.

Note: dhur, dhurrin; *epi,* epiheterodendrin; *tax,* taxiphyllin; *tri,* triglochinin; *?,* specific glucoside not identified or not reported.

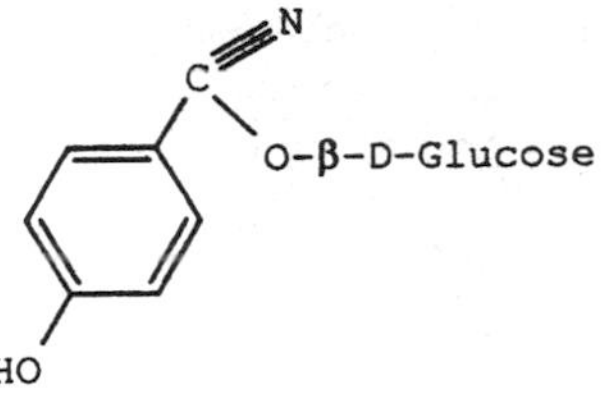

Dhurrin

■ abrupt onset, apprehension, distress, weakness, ataxia, labored respiration, collapse, seizures

Clinical Signs—In cases of cyanide intoxication, there is abrupt onset of apprehension and distress, which occur a few minutes after the animal consumes the offending grass. These signs are quickly followed by weakness, incoordination, and labored respiration. If the intoxication is severe, the animal may be unable to stand and may lay in lateral recumbency, with periodic paddling and tetanic seizures. The entire sequence of signs may occur over a 5-to-15-minute period, with either death or recovery at the end of this time.

■ no lesions

Pathology—There are few distinctive changes. For the most part the changes are limited to congestion of various tissues and scattered, small splotchy hemorrhages.

■ sodium thiosulfate, sodium nitrite

Treatment—The disease usually appears and progresses so rapidly that treatment is seldom available soon enough to be effective. This is not because the treatment itself is ineffective, but rather because cyanogenesis is so rapidly fatal. Toxicosis due to marginally lethal dosages may be responsive to general supportive therapy, but more-specific treatment is much preferred. The primary antidote for ruminants is sodium thiosulfate given i.v. at a dose of 0.25–0.5 g/kg b.w. Sodium nitrite, 10 mg/kg b.w. i.v., can be used to augment the effects of the sodium thiosulfate, but it is not required for effective treatment. Additional discussion of cyanogenesis is presented in the treatment of the Rosaceae (see chapter 64).

Fungal Infections

■ infection of the ovary or endophytes of stems and leaves

In addition to producing toxic compounds that cause the variety of disease problems outlined in this chapter, grasses are also hosts to disease-producing fungi. The types of fungal involvement described in this section are those that entail infection of the developing flower ovary (ergotism) and those in which other tissues of the grass plant are invaded (endophytic infection).

Ergotism

■ sporadic disease, of great historic importance in humans; at present mainly in livestock, rarely humans

Disease Problems—Ergotism occurs as a sporadic disease problem in livestock in various parts of the world, including the northern Great Plains and northwestern areas of North America (Burfening 1973). It is a disease of considerable discomfort and, in its severe gangrenous form, causes loss of the distal portions of the animal's extremities because of constriction of blood vessels. Humans are affected as well.

Ergotism is an ancient malady that has sometimes affected large populations of humans and even entire societies. References to the disease and the medicinal use of the involved alkaloids began with the Greeks, albeit these historical accounts did not directly associate the disease with the cereals being eaten (Bove 1970; van Rensburg & Altenkirk 1974). Epidemics of gangrenous necrosis of the limbs were prevalent in Europe from 800–1350 A.D., especially in France, where 40,000 people purportedly died in one epidemic. The disease, characterized by rotting of the flesh and screams of pain, became known as holy fire, St. Vitus's dance, or St. Anthony's fire. The connection with the legendary St. Anthony stems from the founding of the order of St. Anthony by Gaston in about 1090, after the recovery of his son from the disease, to care for victims of this gangrenous plague. However, these disease names were nonspecific and probably were applied to erysipelas as well (Carlton & Kunkel 1995). The gangrenous form was characterized by severe muscular pain in the limbs, followed in a few weeks by swelling, continued pain with unusual sensations, gradual numbing, and then sudden cessation of pain. The individual's skin became cold, wrinkled, black, and then mummified (Lemberger 1978).

In the 1500s a convulsive form of the disease became more common, appearing either alone or with the gangrenous form. Its neurologic signs began with tingling sensations of the skin, followed by itching, numbness, twitching, muscular spasms, and very painful seizures of alternating flexion and extension of the limbs caused by spinal lesions. Less severe signs occurred in some individuals, together with mild vomiting and diarrhea.

Linkage of the disease with the ergot from cereal grains was proposed by Dodart in 1673, but this was not clearly demonstrated until 1853 by Tulasne. Because of this knowledge and efforts to avoid contaminated grain, outbreaks became fewer and fewer. Ergotism now appears only sporadically. Unfortunately, these episodes occur among the poorest peoples, groups generally subsisting in poverty and/or marginal conditions. These difficult circumstances are strikingly similar to those of historical times when the poor, relying heavily on rye grains as a staple of their diet, were more commonly affected. One percent ergot contamination of flour produces the clinical signs of the disease, and 7% causes seizures and death (Bove 1970; van Rensburg & Altenkirk 1974). In 1977–1978 an episode of gangrenous ergot occurred in Ethiopia, with more than 100 individuals affected (King 1979). Most suffered the loss of distal portions of the feet and legs, and many died. An outbreak of ergotism in India in 1975 was attributed to the infestation of *Pennisetum glaucum*, or pearl millet (reported as *P. typhoides*), by the fungus *Claviceps fusiformis* (Krishnamachari & Bhat 1976). In contrast to the typical ergotism caused by *Claviceps purpurea*, the clinical signs were nausea, vomiting, and sleepiness, which appeared abruptly and lasted only 1–2 days. In general, human ergotism is becoming an increasingly rare disease.

■ parasitism by *Claviceps purpurea*, invasion of ovary, formation of sclerotia containing ergot alkaloids

Ergotism is caused by ingestion of alkaloid-containing sclerotia produced by the fungal genus *Claviceps*, an ascomycete (sac fungus) and member of the family Clavicepitaceae. The genus comprises some 32 species, all of which parasitize members of the Poaceae. *Claviceps purpurea* is the cause of typical ergotism (Bove 1970). It overwinters in the soil and releases ascospores at the time that grass flowers in the spring. When the airborne spores land on the stigmas of the flowers in the grass spikelets, they germinate and send hyphae into the flower ovary. A few days after infection, a copious sugary-sweet exudate termed *honeydew* appears. Produced by the fungal mycelium, it is the site of abundant conidia formation and may serve as a secondary source of infection via insects that visit it and disperse the conidia. The fungal tissue of *Claviceps* gradually replaces the ovary tissue of the grass, hardens, and evolves into a dark brown or black structure known as the sclerotium, or ergot, which resembles the grass grain. The term *ergot* derives from the French *l'ergot*, meaning "cockspur," because of the similarity in size and shape of the sclerotium to the avian appendage. It is the ergot that contains the toxic alkaloids, as much as 1.2% d.w. Proliferation of the fungal mycelium forces open the bracts of the floret and spikelet, thus exposing the sclerotia, which mature at the same time as the uninfected grains; both are harvested together. The fungal life cycle is completed when sclerotia fall to the ground, overwinter, germinate, and again produce asci and ascospores. The number of spikelets infected varies considerably. Growth of the fungus is promoted by warm, moist conditions; thus large numbers of spikelets typically are infected in wet or humid weather, and the likelihood of ergotism's appearance increases.

Traditionally ergot has been associated primarily with *Secale cereale* (rye), a member of the Triticeae, but numerous other grasses also may be infected (table 58.8). Although typically thought of as a problem of contaminated grain, cases in North America more often seem to be associated with animals that graze infected pasture grasses. In addition, problems may result from feed grain, silage, pasture, or even hay.

■ rare outbreaks, gangrenous effects in cattle

Ergotism has been recognized as a disease of livestock in North America since the late 1800s. In an extensive treatise, Salmon (1884) described outbreaks of gangrenous

Table 58.8. Grass forages implicated in field problems of ergot

Agrostis spp.	bentgrasses
Bromus spp.	bromes
Dactylis spp.	orchardgrasses
Elymus spp.	wildryes
Koeleria spp.	Junegrasses
Phalaris spp.	canarygrasses
Poa spp.	bluegrasses

Sources: Salmon 1884; Wilcox 1899; Coppock et al. 1989.

ergotism in Illinois, Kansas, and Missouri following the 1883 growing season. In these instances, forages rather than feed grain were involved. The disease was of such severity that there was concern over possible confusion with foot-and-mouth disease. There were many deaths and numerous maimed animals, mainly cattle, which exhibited loss of digits, hooves, and tail tips. Some animals had reddening and erosions of the oral mucosa. Similar outbreaks were reported in Iowa (Stalker 1892). Occasional outbreaks involving small numbers of animals may be expected to continue to occur sporadically in the central and northwestern United States and Canada.

■ several forms—
hyperthermia
reproductive
gangrenous
nervous

Ergotism affecting livestock occurs in several forms: (1) hyperthermia and production loss, (2) reproductive problems, (3) cutaneous and gangrenous disease, and (4) nervous or tremorgenic form. The form involving hyperthermia and production loss is associated with a body temperature of >40°C secondary to feeding grains such as *Hordeum vulgare* (barley) contaminated with 0.5% ergots (Jessup et al. 1987; Ross et al. 1989; Peet et al. 1991; Schneider et al. 1996). This form of ergotism has not been seen extensively in North America, possibly in part because of the limited use of small grains for fattening cattle. Reproductive problems such as lack of udder development and agalactia, small birth weights, and stillborns seem to occur most commonly in pigs (Lopez et al. 1997). Abortions have been alluded to but are not typical of ergot effects (Mantle & Gunner 1965). In cattle the gangrenous or cutaneous form is more common, although reproductive problems may also occur (Woods et al. 1966; Coppock et al. 1989; Hogg 1991). Ergotism in sheep is typically mild and not accompanied by gangrenous lesions at nominal dosages (Greatorex & Mantle 1973, 1974).

■ nervous form questionable

Neurologic signs similar to those described for human ergotism in the epidemics of the 1500s seem to be a rare manifestation of ergotism in livestock. Often cited reports of such occurrences leave considerable doubt as to the specific cause of the problems observed (Dillon 1955; Clegg 1959). Because of the lack of verifiable field cases, there is skepticism regarding the actual occurrence of neurotoxicosis in livestock (Bourke 1994). At least in some instances, the signs appear more akin to the tremorgenic syndromes associated with *Lolium perenne* (perennial rye) and *Paspalum dilatatum* (Dallisgrass). In some instances, the association between manifestations and ergot as a specific cause is difficult, and only a presumption of relationship can be made. Wilcox (1899) described a neurologic disease in horses in Montana that was presumed to be caused by severe ergot infection of several grasses. The disease was of abrupt onset, progressing from swallowing difficulties to incoordination and paralysis to death in 6–12 hours. In addition, the gangrenous form was noted in 2 other horses. The scarcity of neurologic episodes is perhaps a reflection of the large dosage required to cause these effects.

■ ergot alkaloids, clavines, peptides, carboxamides

Disease Genesis—Ergot alkaloids are a very complex group of compounds based on the ergoline ring system and derived from tryptophan (Rutschmann & Stadler 1978). They occur naturally in several fungal genera in addition to *Claviceps* and in a few flowering plants such as some members of the Convolvulaceae (morning glory family). All of the alkaloids of toxicologic importance are unsaturated at either the 8–9 position in the clavine types or the 9–10 position in the peptides and carboxamides (table 58.9). The dihydro derivatives generally do not occur naturally. Only the 8β-isomers are biologically active. The 8-α-isomers, identified by names ending in *-inine*, are not of toxicologic concern. A difference in the names used for one alkaloid exists; *ergonovine* is favored in North America, while *ergometrine* and *ergobasine* are preferred in Europe.

Table 58.9. Representative ergot alkaloids

Type	Alkaloid
Clavine	Elymoclavine
Carboxamide	
Lysergic acid group	Lysergic acid
Lysergic acid amide	Lysergic acid amide (= lysergamide)
	Ergonovine (= ergometrine, ergobasine)
Peptide	
Ergotamines	Ergovaline
	Ergosine
Ergoxine	Ergonine
	Ergostine
	Ergoptine
Ergotoxine	Ergocornine
	Ergocristine
	Ergocryptine

Sources: Berde & Sturmer 1978; Rutschmann & Stadler 1978.

The ergot alkaloids are complex not only in their chemical structure but also in their biological effects. As is apparent from their basic ergoline structure, these compounds are closely allied to the biogenic amines norepinephrine, serotonin, and dopamine. The alkaloids are potent α-agonists at low concentrations, but their maximal effects are considerably less than that of norepinephrine (Berde & Sturmer 1978). Ergotamine, ergonovine, and *N*-acetyl-loline are all vasoactive in horses but are less potent than either norepinephrine and phenylephrine (Abney et al. 1993). Lysergic acid amide is vasoconstrictive in cattle, possibly via action on serotonin-2 receptors (Oliver et al. 1993). Ergovaline, the predominant alkaloid is also a potent serotonin-2 receptor agonist (Dyer 1993). However, these direct actions on the vasculature are complicated by competition at the receptor, which causes partial antagonism and prevention of norepinephrine activity. α-Adrenergic interactions are also evident with other systems, such as blockade of hepatic glycolysis induced by epinephrine. The vasoconstrictor effects vary with different vascular beds. Other effects include increase in body temperature, emesis, neurologic/dopaminergic effects, and uterine stimulation. The rabbit is an especially sensitive model for the hyperthermic response; it is 10–20 times more sensitive than mice or rats (Griffith et al. 1978). Hyperthermic effects also have been noted in livestock fed large amounts of grain rations containing *C. purpurea* (Jessup et al. 1987; Ross et al. 1989; Peet et al. 1991). The alkaloids are acutely toxic in the range of 1–3 mg/kg b.w. i.v. in rabbits, but much less so when given orally.

The medicinal uses of the ergot alkaloids provide a view of some of their actions. Effects of the ergopeptines include uterine and cardiovascular pressor actions and hyperthermia (Clark et al. 1978; Loew et al. 1978). The uterine effects of ergonovine (= ergometrine), which has fewer other effects, have been of considerable medicinal value. The frequency and amplitude of uterine contractions are increased at low

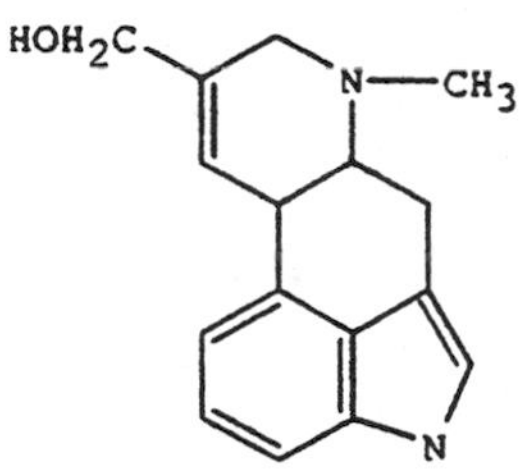

Elymoclavine

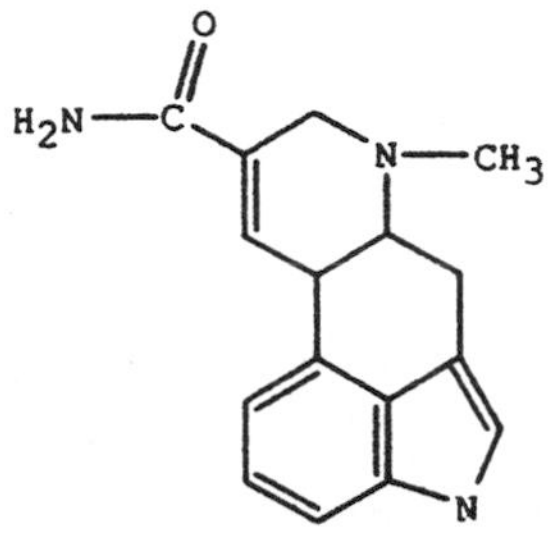

Lysergamide

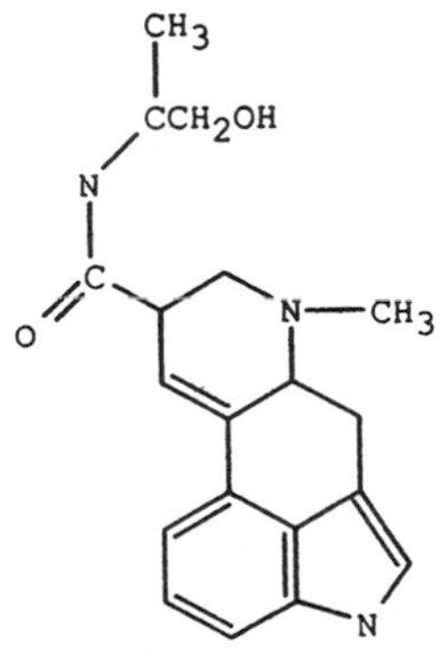

Ergonovine

■ alkaloids with adrenergic vasoactivity; serotonin activity; oxytocic-like activity on uterus

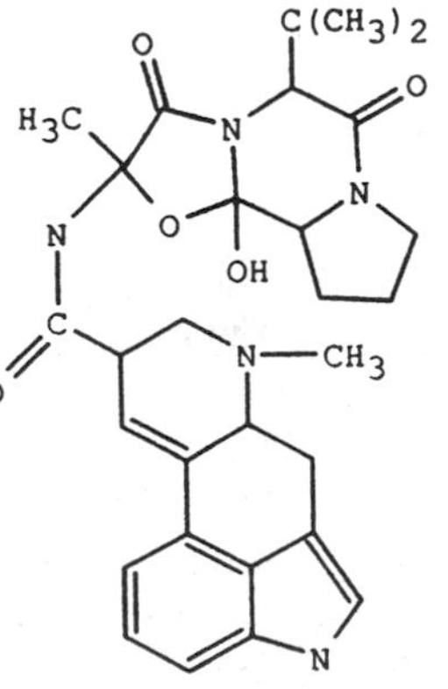

Ergovaline

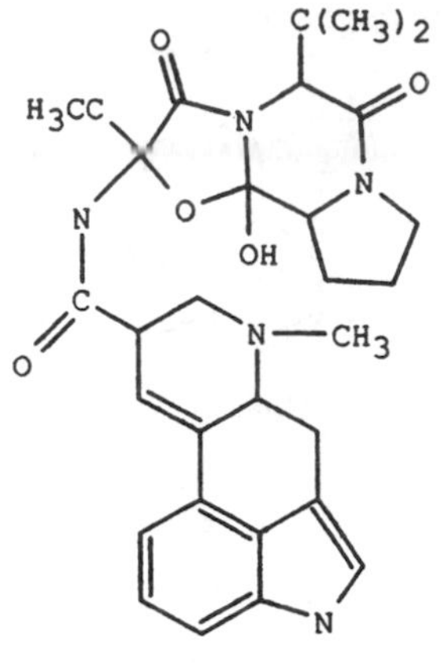

Ergonine

■ dopaminergic activity, inhibition of prolactin secretion

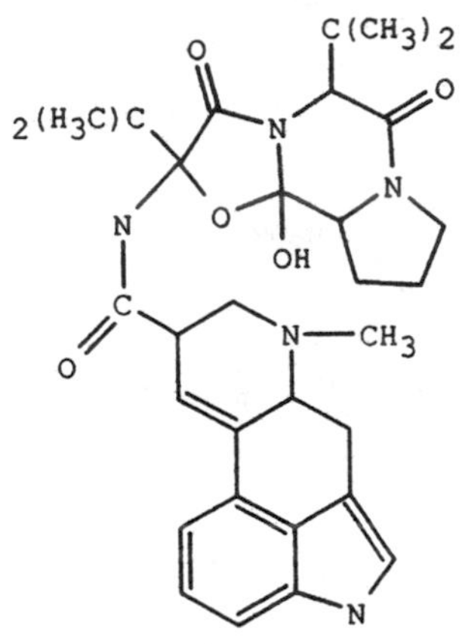

Ergocornine

■ reproductive action of ergopeptides, generally minor

dosage, and basal uterine tone is increased with higher dosage. This oxytocic-like effect is apparent with most of the 9–10 unsaturated peptides and a few of the others (Berde & Sturmer 1978). Vasoconstriction and ischemia are adverse effects, which in some instances may require therapeutic intervention with intra-arterial administration of vasodilators such as tolazoline, isoproterenol, or papaverine (Griffith et al. 1978). Other therapeutic uses include relief for migraine, antiserotonin activity, and especially control of postpartum uterine hemorrhage. The last use is probably where the alkaloids have found their most beneficial life-saving role.

The dopaminergic activity of the ergot alkaloids is an important factor in their effects (Strickland et al. 1993). Of the five or more dopamine receptors dispersed throughout the body, the alkaloids interact with D_1 and D_2 types but in an opposite manner. They are antagonistic to dopamine at the D_1 vasodilatory receptors and agonistic at the D_2 receptors in the anterior pituitary. The agonistic action at the D_2 receptors mimics the dopaminergic tonic inhibition of prolactin secretion, which occurs through dopamine secretion via the hypothalamic-hypophyseal portal system. Experimental studies in rats indicate the possibility of increased striatal D_2 receptor affinity (Mizinga et al. 1993). Rats fed 10–15% infected seed, with 325 ppb ergovaline, experienced a significant reduction of feed intake that was reversed when they were administered a D_2 dopamine antagonist (Larson et al. 1994). There was also a decrease in whole-brain D_2 receptor density.

The inhibition of prolactin secretion has far-ranging effects on the body, and it may be one of the most sensitive indicators of exposure to the ergopeptides (Jackson et al. 1988; Stamm et al. 1994). Secretion of melatonin also is reduced, but thyroid stimulating hormone, growth hormone, and ACTH are not affected (Berde & Sturmer 1978; Fluckiger & del Pozo 1978; Elsasser & Bolt 1987; Porter & Thompson 1992; Porter et al. 1993). Prolactin receptors are dispersed throughout the body in various tissues, including those of the liver, kidney, and brain. In addition to the well-known effects of prolactin on the mammary glands, it also has profound actions on lipid metabolism. The derangement of lipid metabolism secondary to the decrease in prolactin may be directly related to the subsequent cholesterol changes and fat necrosis, as well as indirectly to body temperature regulation.

The reproductive effects of ergot are clearly demonstrable in laboratory animals and pigs. The ergopeptides are primarily responsible, but the clavines are also active at high dosage. The alkaloids are mainly embryocidal, with only minor effects on skeletal formation (Grauwiler & Schon 1973; Witters et al. 1975; Griffith et al. 1978). Implantation may be prevented or early pregnancies terminated (Mantle 1969). Typically, there is a decrease in birth weights, an increased number of stillborns, and many piglets that die within 24 hours mainly due to the lack of mammary development and available milk (Nordskog & Clark 1945). However, even with obvious ergot ingestion there may be only minor reproductive effects (Campbell & Burfening 1972).

■ hyperthermia form, body temperature >40°C, salivation, increased respiration

Clinical Signs, Pathology, and Treatment—As described above, ergotism comprises four forms. The clinical signs of each form are different. In cattle having the mild hyperthermia form, body temperature is elevated to >40°C, salivation and respiration are increased, appetite and milk production are decreased, and death occurs in a few animals. These signs are most noticeable in high summer temperatures.

■ gangrenous form, lameness, sloughing of skin of distal portion of limbs, tail, ears

Clinical signs in the gangrenous or cutaneous form are more severe. Seen more often in winter, these signs include gradual onset of lameness, feet swollen and painful up to the coronet, decreased productivity, and in some instances diarrhea. Within 1–2 weeks, the feet become numb, the skin dead, and a crack appears in the skin over a joint of the foot or along the coronary band. No suppuration or break in the skin occurs otherwise. Eventually, the skin sloughs and the tendons rupture, or the entire distal portion of the limb may be lost. The pelvic limbs are most often or more severely affected. There also may be loss of several centimeters of the tail. Other features that have been observed in cattle include reddening and erosions of the oral mucosa, and necrosis and sloughing of the skin over various areas of the body. In sheep, necrosis of the tip of the tongue occurs with an otherwise less severe manifestation of the disease.

Reproductive effects include agalactia or decreased milk production, reduced fertility, decreased number of fetuses, and increased number of stillborns and perinatal deaths. As noted previously, signs of the neurologic form of ergotism are questionable and therefore are not elaborated upon here.

- reproductive form, agalactia, decreased fertility
- neurologic form, questionable

Ergotism appears only sporadically in the northern Great Plains and northwestern areas of North America (Burfening 1973). If the possibility of the disease is of concern, inspection of the grain and/or inflorescences of grass forages and appropriate management practices may be warranted. Ergots in grain are easily distinguished by their black or purplish color. They also are generally larger than the grass grains. When forage grasses are infected, close inspection of the spikelets with a magnifying lens is required to detect the ergotized flowers before sclerotia reach maturity.

Endophytic Fungi

Disease Problems and Genesis—In addition to *Claviceps purpurea,* which causes ergotism, the fungal family Clavicepitaceae includes several endophytic genera that also cause disease problems when associated with grasses. Knowledge of their biology and awareness of their importance in disease are increasing, and future work undoubtedly will reveal additional involvement in plant intoxications (Bacon et al. 1975; Bacon 1995). Grass endophytes are fungal taxa that complete their entire life cycle within the aerial portions of grasses by forming a systemic, usually intercellular association (Bacon & De Battista 1991). This relationship between the fungus and the grass plant is symbiotic. The fungus gains ready access to nutrients and a favorable environment for growth. In turn, the plant gains drought resistance, enhanced nitrogen uptake, and, because of the alkaloids produced by the fungus, increased insect resistance (Powell & Petroski 1992). Interestingly, pastures comprising endophyte-free grasses are more difficult to maintain than those infected with an endophyte (Bacon 1995).

- endophytes of stems and leaves, impart insect resistance to plants

Worldwide, at least 290 grass species have been confirmed to be hosts for clavicepitaceous endophytes, and it is estimated that the actual number of species infected is 25 times greater (Leuchtmann 1992). Species belonging to all grass subfamilies have been reported as hosts (Bacon & De Battista 1991). In North America, warm- and cool-season forages of a number of genera are well known as hosts (table 58.10).

Endophytic fungi of grasses are primarily species of the genera *Epichloe, Balansia, Atkinsonella,* and *Myriogenospora* and their anamorphic states *Acremonium, Ephelis,* and *Typhodium* (Bacon & De Battista 1991). Based on molecular data, Glenn and co-workers (1996) have proposed reclassifying the anamorphs of the polyphyletic *Epichloe* and closely related asexual grass endophytes in the new form genus *Neotyphodium.*

A variety of alkaloids are produced by these endophytic fungi and are sometimes grouped into three major categories (Bacon & De Battista 1991): (1) the ergot alkaloids, with both the clavine and peptide types such as ergopeptine; (2) the loline alkaloids, with pyrrolizidine bases and represented by loline; and (3) the lolitrems, which

- variety of alkaloids—
  clavine and peptides
  lolines
  lolitrems

Table 58.10. Principal grass forages in North America known to host endophytic fungi

Agropyron spp. (sensu lato)	wheatgrasses
Agrostis spp.	bentgrasses
Andropogon spp.	bluestems
Cynodon spp.	Bermudagrasses, stargrasses
Elymus spp.	wildryes
Festuca spp.	fescues
Glyceria spp.	mannagrasses
Holcus spp.	velvetgrasses
Hilaria spp.	hilarias, tobosagrasses, galletas
Hystrix spp.	bottlebrush grasses
Lolium spp.	ryegrasses
Paspalum spp.	paspalums, watergrasses
Poa spp.	bluegrasses
Sphenopholis spp.	wedgescales

Sources: White 1987; Leuchtmann 1992; TePaske et al. 1993.

are complex, isoprenoid-substituted indoles and are described in detail in the treatment of *Lolium* (in this chapter).

■ various syndromes—
tremorgenic
gangrenous
reproductive
physiologic

Disease syndromes associated with these endophytic genera of the Clavicepitaceae include those that are tremorgenic or gangrenous and those that cause alterations in reproductive or physiological attributes of the animal. The tremorgenic syndrome, or staggers, is one of the most common disease problems associated with endophytes. There are 20 or more fungal tremorgens (Gant et al. 1987). Several important grass forages are associated with tremorgens (table 58.11). Aspects of the disease are described more fully in the treatments of each grass genus. It is important to note that similar tremorgenic syndromes are caused by mycotoxins in grains or other substrates infected with *Penicillium, Fusarium,* and *Aspergillus. Zea mays* (maize or corn) is a good example of this situation (Cole & Dorner 1986).

Table 58.11. Grass forages associated with tremorgenic syndromes due to fungal endophytes

Cynodon dactylon	Bermudagrass
Hilaria	
H. jamesii	galleta, curlygrass
H. mutica	tobosa, tobosagrass
Lolium perenne	perennial ryegrass
Paspalum	
P. dilatatum	Dallisgrass, large watergrass
P. distichum	knotgrass, jointgrass
P. scrobiculatum	Indian paspalum, monagrass, kodo millet

Anthoxanthum L.

■ *A. aristatum*
A. odoratum

annual vernalgrass
sweet vernalgrass
vernalgrass

Taxonomy and Morphology—Comprising 15–20 cool-season species, *Anthoxanthum*, a member of the subfamily Pooideae and the tribe Aveneae, is an Old World genus prized for the distinctive, sweet, coumarin, or vanilla-like, fragrance of its herbage. The genus name is derived from the Greek roots *anthos,* meaning "flower," and *xanthos,* meaning "yellow," which reflect the color of the mature spikelets. In North America, 2 introduced species are present:

A. aristatum Boiss. (= *A. puelii* Lecoq & Lamotte)	annual vernalgrass
A. odoratum L.	sweet vernalgrass

Plants annuals or perennials; cespitose; herbage fragrant. **Culms** 10–75 cm tall; erect or decumbent; straight or geniculate; branched or not branched above; nodes glabrous or pubescent; internodes hollow. **Leaves** both basal and cauline. **Blades** flat; 2–15 mm wide. **Ligules** membranous. **Auricles** absent. **Sheaths** glabrous; margins free, overlapping. **Inflorescences** panicles; contracted or spicate. **Disarticulation** above the glumes; neuter florets dropping attached to perfect floret. **Spikelets** all alike; lanceolate; compression lateral; 5–10 mm long. **Florets** 3; of 2 forms; lower florets 2, neuter; upper florets 1, perfect. **Glumes** 2; unequal; different; longer than lemmas; awns absent; lower glumes ovate, nerves 1; upper glumes lanceolate, nerves 3. **Lemmas of Neuter Florets** longer than perfect lemmas; lanceolate; abaxial surfaces appressed-pilose; apices 2-lobed; awns 1, dorsal, straight or geniculate, twisted. **Paleas of Neuter Florets** absent. **Lemmas of Perfect Florets** broadly ovate to suborbicular; abaxial surfaces glabrous; shiny; awns absent. **Paleas of Perfect Florets** with nerves 1; backs rounded.

- fragrant; ligules membranous; inflorescences panicles; spikelets with 3 florets, 2 neuter below 1 perfect

1. Plants perennials. Awns of neuter florets straight . . . . *A. odoratum*
1. Plants annuals. Awns of neuter florets geniculate . . . . *A. aristatum*

Anthoxanthum odoratum

Distribution and Habitat—Species of *Anthoxanthum* are native to temperate Europe, Asia, and northern Africa. They are generally found in open habitats such as meadows and dry grasslands. In North America, both *A. odoratum* and *A. aristatum* are introduced and occur in waste areas, roadside ditches, lawns, and meadows. Albeit not particularly good forage, *A. odoratum* is sometimes planted in hay meadows to give a fragrance to the hay. It also is gaining popularity as a garden ornamental and a component of dried floral arrangements.

- invader species, waste areas

Anthoxanthum odoratum

Disease Problems—Although producing a pleasant, vanilla-like fragrance, *A. odoratum* is bitter-tasting and unpalatable. It is an unusual cause of hemorrhagic disease in cattle in Europe, where it is native, but it has only rarely been associated with toxicity in North America (Kosuge & Gilchrist 1977; Pritchard et al. 1983; Bartol et al. 2000). It is most likely to be a problem in preserved forages, mainly hay, following its consumption for several weeks. However, in England, carcasses of lambs grazing the taxon exhibited a high incidence of "blood splash," scattered areas of splotchy hemorrhages on the diaphragm, thoracic and abdominal walls, skeletal muscles, heart, and lungs (Restall 1980–1981). However, other plants cannot be excluded as causes or contributing factors in this problem of apparent increase in capillary fragility. Because of the scarcity, relative unpalatability, and low dicoumarol potential of *A. odoratum,* it is likely that the species will be associated with disease in North America, mainly as a contributing factor (Davies & Ashton 1964).

- hay, clotting deficiency, hemorrhagic disease, increased capillary fragility, cattle, less risk with sheep

Disease Genesis—The vanilla-like aroma is due to the presence of coumarins, which under unique circumstances might predispose to a hemorrhagic disease. When the forage is cut for hay and not sufficiently dried, 4-hydroxycoumarin may be formed from precursors by ring closure due to the action of fungi within the stems. A further condensation reaction leads to formation of bishydroxycoumarin (dicoumarol) (Davies & Ashton 1964). Dicoumarol, because of its similarity to vitamin K, competes for and inhibits the epoxidase enzyme responsible for regeneration of the active vitamin. As a result, there is a deficiency of coagulation factors, including prothrombin. For a more detailed discussion of this disease, see the treatment of *Melilotus* (sweet clover) in the Fabaceae (see chapter 37). *Hierochloe odorata* (holygrass) also contains appreciable coumarin and may be similarly toxic.

- coumarins, vanilla-like odor; formation of 4-hydroxycoumarin by fungi in stems, then to dicoumarol, inhibition of vitamin K regeneration, leads to coagulation factor deficiency

OH

4-Hydroxycoumarin

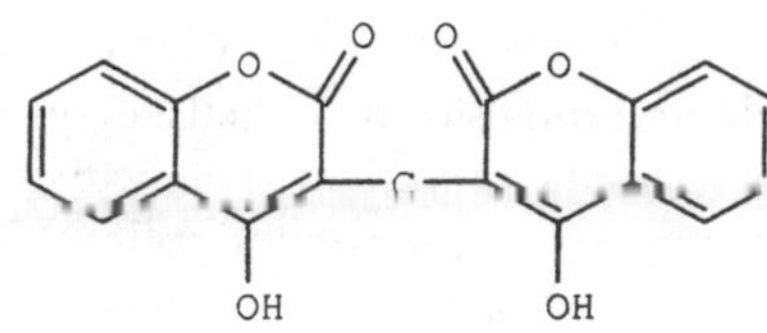

Dicoumarol

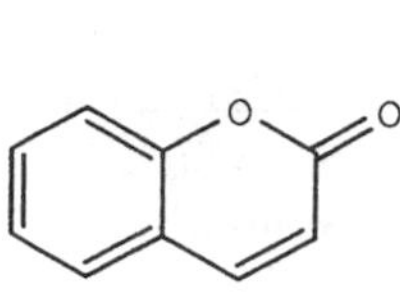

Coumarin

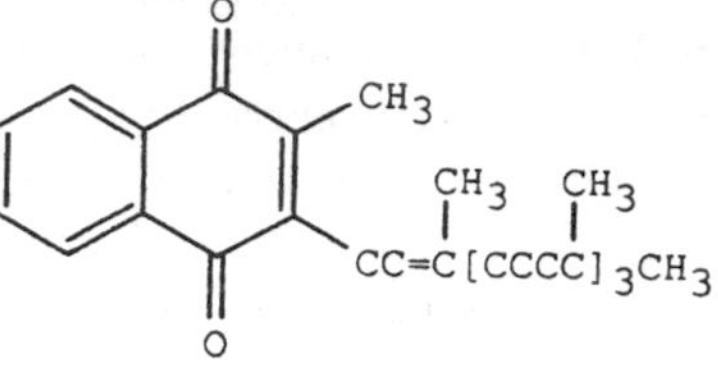

Vitamin K_1

- abrupt onset, hemorrhages, intractable bleeding

  accumulation of blood in body cavities

Clinical Signs and Pathology—Typically there is abrupt onset of hemorrhages at various sites under the skin, which produce large blisterlike swellings, bleeding into and accumulation in body cavities, and intractable bleeding from surgical or accidental wounds. The development of lesions is dependent upon the site of hemorrhage. The accumulations of blood in body cavities may be quite large and, in the case of abdominal masses, may cause signs of colic (Bartol et al. 2000).

- parenteral vitamin K_1, blood transfusion when severe

Treatment—Administration of vitamin K_1 parenterally is necessary, with the route depending upon the seriousness of the disease. In the most severe intoxications a blood transfusion may be needed (cf. treatment of *Melilotus*, in chapter 37).

Avena L.

- *A. sativa*

  oats

Taxonomy and Morphology—Comprising 10–27 cool-season species, *Avena*, a member of the subfamily Pooideae and the tribe Aveneae, is an Old World genus. There are 6 or 7 cultivated species, including the economically important *A. sativa* (oats), which was domesticated in northwestern Europe about 2000 B.C. Unfortunately, one of the world's most invasive weeds—*A. fatua* (wild oats)—also is a member of the genus. In North America, 6 species are present. Only 1 is of toxicologic significance:

A. sativa L. oats, cultivated oats

- ligules membranous; inflorescences open panicles; spikelets large; glumes longer than florets, membranous or papery

Plants annuals or perennials; cespitose. **Culms** 30–120 cm tall; erect; nodes glabrous; internodes hollow. **Leaves** primarily cauline. **Blades** flat; 4–8 mm wide; surfaces scabrous or sparsely pilose. **Ligules** membranous. **Auricles** absent. **Sheaths** glabrous or pilose; margins free, overlapping. **Inflorescences** panicles; open; pyramidal; branches ascending, spikelets borne on short bent pedicels. **Disarticulation** above the glumes or not occurring. **Spikelets** all alike; elliptic to lanceolate; compression lateral; 20–30 mm long. **Florets** 2–4; upper 1–2 occasionally rudimentary. **Glumes** 2; equal; alike; longer than lemmas; lanceolate; membranous or chartaceous; awns absent. **Lemmas** lanceolate; coriaceous to indurate; abaxial surfaces glabrous; nerves 7 or 9; awns absent or present, dorsal, straight. **Paleas** hyaline; nerves 2; keels 2, ciliate; backs rounded.

- crop plant

Distribution and Habitat—Species of *Avena* are most abundant in the Mediterranean region. They extend into northern Europe and are widely introduced in other temperate regions of the world. In North America, *A. sativa* is commonly cultivated. It may escape into nearby disturbed sites but does not persist.

Photosensitization

Disease Problems and Genesis—*Avena sativa* produces two disease problems of potential concern: (1) photosensitization from fresh forage following grazing for 1 to several weeks, and (2) acute nitrate intoxication, primarily from consumption of hay. The photosensitization caused by *A. sativa* is typical of that caused by most other grasses; it is the primary uncomplicated type not accompanied by icterus or other evidences of hepatic dysfunction (Schmidt 1931). Although there may be extensive areas of skin necrosis and sloughing, causing considerable discomfort, the disease is usually not a serious problem other than causing a temporary setback in productivity. The skin changes respond readily to elimination of the offending forage from the animals' diet. The specific photodynamic agent or the factors involved in its formation are unknown.

- 2 disease problems—
  photosensitization, direct, toxicant unknown
  nitrate intoxication

Clinical Signs—After consumption of *A. sativa,* there is onset of skin necrosis and sloughing, which may require several days for full expression. In most animal species the lesions are found scattered in the less pigmented areas of the skin. However, in sheep, because of the protective barrier of the wool, the lesions are generally on the head around the ears, eyes, mouth, and other areas of the body with less wool. In areas where light can penetrate the superficial layers of the skin, the circulating photodynamic agent is activated in the capillaries. In addition to skin involvement, there may be reddening of the cornea of the eyes and increased sensitivity to light. When such typical skin changes are observed, it is important to examine the animals carefully for the presence of icterus or to use serum chemistry to evaluate possible involvement of the liver.

- skin, necrosis, sloughing less pigmented areas; sheep, head around ears, eyes, and mouth; reddened corneas

Treatment—Remove animals from the pasture containing *A. sativa* and, if possible, reduce exposure to direct sunlight by providing shade.

- remove from *A. sativa* pasture, provide shade

Nitrate Accumulation

Disease Problems and Genesis—Like other cereal grain forages, *A. sativa* is capable of accumulating excess concentrations of nitrates under certain circumstances. Because of the use of heavy nitrogen fertilization to promote high grain yields, its hay or straw may have especially high levels. For example, up to 4.4%, or 44,000 ppm, NO_3 has been reported as associated with severe death losses (Newsom et al. 1937; Thorp 1938; Bradley et al. 1939). A more complete discussion of nitrate as a toxicant is presented in the treatment of *Sorghum* (in this chapter).

- nitrate accumulation, especially with heavy nitrogen fertilization

Clinical Signs—Onset of signs is abrupt and may appear several hours after consumption of *A. sativa.* Death may follow onset of signs within minutes to a few hours. The primary signs are weakness and possibly dark-colored mucous membranes.

- abrupt onset, weakness, dark-colored mucous membranes

Pathology—There are few distinctive lesions at necropsy. The blood and tissues may be dark- or chocolate-colored, but this is not consistently seen. There also may be scattered, small splotchy hemorrhages and increased foamy fluid in the terminal airways of the lungs, but these changes are not usually extensive.

- gross pathology, dark-colored blood and tissues

Treatment—The primary antidote for nitrate intoxication is methylene blue, 2–4 mg/kg b.w. given as a 1–2% solution.

- methylene blue

Cynodon Rich.

- *C. aethiopicus*
  C. dactylon
  C. nlemfuensis
  C. plectostachyus

  African Bermudagrass
  Bermudagrass
  estrella
  Ethiopian dogstoothgrass
  reuse kweekgrass
  stargrass

Taxonomy and Morphology—Comprising 8–10 warm season species, *Cynodon,* a member of the subfamily Chloridoideae and the tribe Cynodonteae, is a genus of the Old World tropics. The genus name is derived from the Greek roots *kuon,* meaning "dog," and *odous,* meaning "tooth," which reputedly refer to the sharp scale leaves of the rhizomes. In North America, 1 naturalized and 5 cultivated species are present. Of toxicologic importance are the following:

C. aethiopicus Clayton & J.R.Harlan	giant stargrass, Ethiopian dogstoothgrass
C. dactylon (L.) Pers.	Bermudagrass
C. nlemfuensis Vanderyst	reuse kweekgrass, African Bermudagrass
C. plectostachyus (K.Schum.) Pilg.	Naivasha stargrass, summer stargrass, estrella

- stoloniferous perennials; ligules ciliate; inflorescences 1-sided spikes, borne in 1 digitate whorl; spikelets laterally compressed with 1 floret

Plants perennials; stoloniferous, typically sod-forming. **Culms** 4–100 cm tall; prostrate or decumbent to erect; branched; stolons flattened; rhizomes present or absent; nodes glabrous; internodes hollow. **Leaves** cauline. **Blades** flat or folded; 1–7 mm wide. **Ligules** ciliate. **Auricles** absent. **Sheaths** margins free, overlapping. **Inflorescences** 1-sided spikes; 2–20; borne in a single digitate whorl at peduncle apices. **Disarticulation** above the glumes. **Spikelets** all alike; ovate to lanceolate; compression lateral; 1.7–3 mm long. **Florets** 1; perfect; rachillas prolonged, bearing rudimentary floret. **Glumes** 2; equal or unequal; alike; shorter than or as long as lemmas; lanceolate; nerves 1; awns absent. **Lemmas** lanceolate to ovate; nerves 1 or 3; cartilaginous; awns absent. **Paleas** hyaline; nerves 2; keels 2; backs rounded.

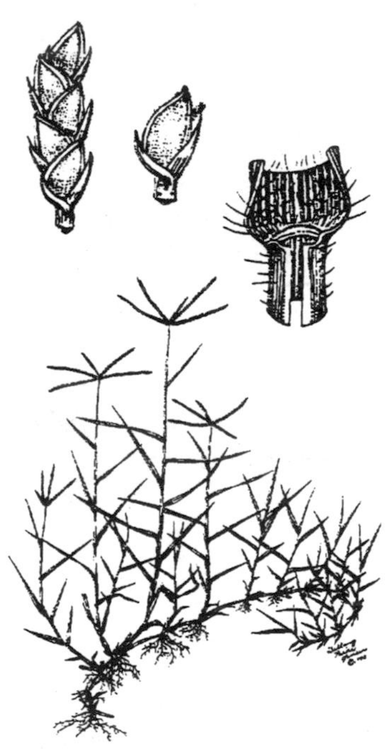

Cynodon dactylon

1. Rhizomes present. Spikes 4–6 per peduncle . . . . *C. dactylon*
1. Rhizomes absent. Spikes 5–20 per peduncle.
   2. Whorls of spikes 1 per peduncle.
      Lemma keels densely pubescent . . . . *C. nlemfuensis*
   2. Whorls of spikes 2–5 per peduncle.
      Lemma keels glabrous or sparsely pubescent.
      3. Stolons not arching. Glumes one-fourth to one-third length of spikelet . . . . *C. plectostachyus*
      3. Stolons arching. Glumes two-thirds to three-fourths length of spikelet . . . . *C. aethiopicus*

Distribution and Habitat—Species of *Cynodon* are native to the tropical regions of Africa and Asia. An important lawn and pasture grass in the southern half of the United States, *C. dactylon* has naturalized throughout the warm-temperate regions of the world. It can be a troublesome weed in cultivated fields. The other species are used for forage.

- cultivated plants

- disease problems—
  cyanogenesis
  nitrate intoxication
  photosensitization
  staggers syndrome
  ARDS

Disease Problems—Several problems are associated with species of *Cynodon,* but all are of minor importance. They are rare in occurrence, although in certain circumstances they may present serious problems. Cyanide intoxication, photosensitization, and the staggers syndrome are discussed here. Aspects of other problems such as nitrate intoxication and acute respiratory distress syndrome (ARDS) have been presented previously in the introductory remarks about disease problems in the family as a whole. It should be noted that nitrate accumulation is not a significant problem in *Cynodon*. Attempts to promote high levels in *C. aethiopicus* at a fertilization rate of 180 kg N/acre resulted in only 6,500 ppm nitrate in the stems (Rudert & Oliver 1978).

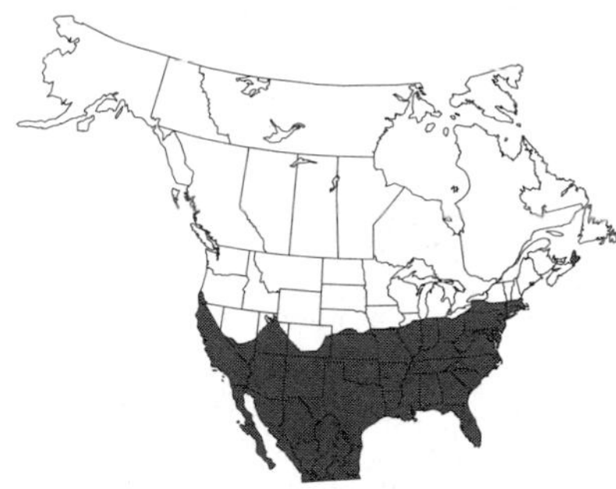

Cynodon dactylon

Cyanogenesis

Disease Problems—Species of this genus are considered to be particularly troublesome as cyanogenic risks in South Africa (Steyn 1934)). However, cyanogenesis has not proven to be a consistent problem in North America. In a large measure this is probably due to management practices of grazing more-mature stands of the grass.

- cyanogenesis, erratic problem, risk low in North America

Disease Genesis—All six species of *Cynodon* are probably cyanogenic to some degree. As in other grasses, cyanide is present as the glucoside dhurrin, which yields hydrocyanic acid (HCN) when hydrolyzed in the plant or rumen. Some strains of *C. plectostachyus* have HCN potential up to 1,000 ppm wet weight, a level exceeding that of the notoriously cyanogenic *Sorghum halepense* (Johnsongrass) grown under the same conditions (Van der Walt & Steyn 1941). As is true for the sorghums, HCN potential is high (550–750 ppm wet weight) in some varieties and strains of *Cynodon* in the early stages of growth and then declines markedly to 200 ppm or less during the later stages (Schroder 1977; Mislevy et al. 1983). In comparison, HCN potential of grain sorghums, under the same conditions, is reported to be 240–480 ppm wet weight and 3,000–6,000 ppm d.w. in the early stages of growth (Schroder 1977). However, such a precipitous decline is not observed in all varieties—e.g., *C. plectostachyus* (summer stargrass)—and is not apparent in fall regrowth of many of the varieties. HCN potential increases with warm dry weather, with wilting, and with heavy nitrogen fertilization (Van der Walt & Steyn 1941; Rudert & Oliver 1978). HCN potential declines quickly after a hard freeze (Mislevy et al. 1983). Drying at air temperature for 5 days results in an 8-fold reduction in HCN potential, whereas at refrigerator temperature the loss is about 2-fold for the same time period (Schroder 1977).

- cyanogenic glycoside, dhurrin; main risk early stages of plant growth, ruminants

N
C
O-β-D-glucose
HO

Dhurrin

Clinical Signs—Within a few minutes of grazing the toxic *Cynodon*, there is abrupt onset of apprehension and distress, quickly followed by weakness, incoordination, and labored respiration. If intoxication is severe, the animal may be unable to stand and may lay in lateral recumbency, with periodic paddling and tetanic seizures. The entire sequence of signs may occur over a 5-to-15-minute period, with either death or recovery at the end of this time.

- very abrupt onset; apprehension, distress, weakness, ataxia, labored respiration, collapse, seizures

Pathology—There are few distinctive changes. For the most part the changes are limited to congestion of the viscera and scattered, small splotchy hemorrhages.

- no lesions

Treatment—The primary antidote for ruminants is sodium thiosulfate given i.v. at a dose of 0.25–0.5 g/kg b.w. Sodium nitrite, 10 mg/kg b.w. i.v., can be used to augment the effects of the sodium thiosulfate, but it is not required for effective treatment. A more detailed discussion is presented in the treatment of the Rosaceae (see chapter 64).

- sodium thiosulfate, sodium nitrite

Photosensitization

Disease Problems—Both direct and hepatogenous types of photosensitization are reported for *C. dactylon*. The direct type is similar to that produced by other grasses;

it is seldom fatal. The hepatogenous type occurs in Florida and perhaps adjacent states, primarily several weeks after the upper leaves have been killed by frost (Kidder et al. 1951, 1961). The disease may occur at any time of the year but is most severe in winter and early spring. There is a greater risk of mortality than with the direct type of photosensitization. There may be serious weight loss of as much as 50 kg in cattle and up to 20% mortality.

A paralytic or so-called downer-type of disease of high mortality has been associated with *C. dactylon* under similar circumstances in the same area. Even though it is referred to as "Bermudagrass poisoning," it appears to be nonspecific and also produced with other forages (Gibbons 1953).

■ direct photosensitization, cause unknown; hepatogenous photosensitization, possibly fungal toxins

Disease Genesis—The cause of the hepatogenous skin effects appears to be related to fungal growth on the frost-killed leaves. When the plant is subsequently eaten after new growth occurs in a few weeks, the toxins are also consumed and signs are observed 1–2 weeks later. Several fungal taxa may cause the disease, one example being *Periconia minutissima,* a deuteromycete in the family Dematiaceae (Kidder et al. 1961). However, the disease has not been reproduced experimentally with the suspected fungi. The downer-type syndrome may be due to fungi that produce different toxins.

■ abrupt onset, scratching, rubbing; skin reddened, blistered; dark urine; may be icterus with hepatogenous photosensitization

Clinical Signs—After the animals graze *C. dactylon* for 1–2 weeks, there is depression, increased salivation and lacrimation, and diarrhea. The next day there may be licking and scratching of the skin. The skin irritation worsens, as indicated by incessant scratching, rubbing, and head shaking. White areas of the skin have a reddened, sunburned appearance. Other skin changes in the white or thin-skinned areas quickly become apparent, i.e., blistering of the ears, udder, and scrotum and around the eyes, mouth, muzzle, and anus. The edges of the ears are scabby and curled under. The urine may also appear dark-colored. Hoof wall distortions may become apparent several months after initial signs.

■ gross pathology, skin lesions

microscopic, liver connective tissue and bile duct proliferation

Pathology—Grossly, in addition to the skin effects the liver is granular in appearance and, depending upon the duration of effects, may be enlarged or small and firm. Microscopically, the liver changes range from cloudy swelling and increased hepatocyte size to fibrosis and bile duct proliferation.

■ remove from Bermudagrass; sodium thiosulfate i.v. or orally may be useful

Treatment—Preventing animal access to *C. dactylon* will aid in recovery. Sodium thiosulfate, 65g/100 kg b.w. i.v. or 130 g/100 kg orally has been recommended (Kidder et al. 1961). Mowing after a frost may help to prevent the disease.

Staggers Syndrome

■ staggers, rare problem in cattle, mature Bermudagrass, fresh or hay

Disease Problems—First reported during a particularly troublesome season in Oklahoma in 1950, the staggers syndrome is a rare problem associated with mature stands of common or coastal Bermudagrass. Occurring in the fall, cases have appeared mainly in Oklahoma and Louisiana and more rarely from Texas to Florida (Whitehair et al. 1951). Affecting 25,000 cattle, a particularly serious outbreak in Louisiana occurred in the fall of 1971 (Fichte 1972; Strain et al. 1982). Both fresh forage and dry hay are toxic, retaining this capacity for several years or more. The disease affects cattle of all ages and may in some instances involve up to 75% of a herd. Typically it follows consumption of the forage for 7–10 days or as little as 2.7% b.w. of the green plants (Nicholson 1989). In a goat, 600 g/day of toxic hay produced signs in 8 days, with the severity of the disease peaking in about 13 days (Strain et al. 1982). Mainly cattle are affected, with only mild effects observed in horses and goats. The disease seems to be associated with mature Bermudagrass with caryopses/seeds present. There may be marked loss of productivity in dairy cows and perhaps abortions (Fichte 1972). Deaths from this disease are unusual and typically due to misadventure, dehydration, or starvation.

Disease Genesis—The cause of the staggers syndrome has not been clearly identified; the syndrome is so erratic in occurrence that little work has been carried out to determine the cause. It is not a mineral imbalance or deficiency, however. It has been suggested that the effects are associated with destruction of B vitamins, and the clinical signs in some animals seem to be responsive to administration of nicotinamide and riboflavin (Blohm 1962). Much about its occurrence suggests that fungi are involved. Porter and co-workers (1974) have isolated a species of *Claviceps* producing ergot-type alkaloids from sclerotia obtained from common Bermudagrass that was responsible for a tremor problem in Mississippi in 1972. The major alkaloids isolated were ergonovine (30%) and ergonovinine (22%) in addition to penniclavine and chanoclavine. However, the isolation of such fungi in other outbreaks has not been as successful. Because of the similarity of the syndrome to those caused by other tremorgenic grasses, paspalitrem-type indole alkaloids continue to be prime suspects as the cause (Cole & Dorner 1986). This is consistent with field observations, which have resulted in the suggestion that *C. dactylon* in fruiting condition poses a greater risk and that mowing to remove the inflorescences may reduce the risk (Whitehair et al. 1951). The toxin is not transmitted in the milk. Further discussion of the syndrome is presented in the treatment of *Paspalum,* in this chapter, because the signs and treatment approach are identical.

- possibly alkaloids, such as ergonovine, from fungi; cattle, toxin not in milk

Clinical Signs and Pathology—The disease, like that caused by *Paspalum,* is characterized by tremors, head shaking, muscle fasciculations, incoordination, and stiff thoracic limbs. Most importantly, death typically does not occur. This disease is not accompanied by any distinctive pathologic changes, even in the unusual fatality.

- increased nervousness, seizures with ears erect, fine tremor of head and neck to shoulders; stiff, wobbly swaying movement of pelvic limbs; collapse, head twisted backward

Treatment—Prevention of animal access to the offending forage is usually sufficient to stop the syndrome, although in a few instances sedation may be helpful in moving the animals out of the pasture. In the event an animal falls and is unable to rise, food and water should be made available. Diazepam (0.8 mg/kg b.w. i.v.) is effective in providing sedation for several hours (Strain et al. 1982).

- remove from pasture; sedation

Festuca L.

Taxonomy and Morphology—Comprising 100–450 cool-season species, *Festuca,* a member of the subfamily Pooideae and tribe Poeae, is a large and variable genus for which a satisfactory classification is lacking. The generic name comes from an old Latin word meaning "weedy grass," which is ironic because of the taxon's substantial economic importance as a turf grass and a forage grass for both livestock and wildlife. Six of its species are listed among the world's 1,000 most important crop plants. In North America, some 44 native, introduced, and hybrid species occur. Numerous subspecies and varieties also have been described. Representative of the genus and of toxicologic significance are the following:

- *F. arundinacea*
  F. pratensis
  F. rubra

  Dovergrass
  English bluegrass
  fescue

F. arundinacea Schreb.	tall fescue, reed fescue, alta fescue
F. pratensis Huds. (= *F. elatior* L.)	meadow fescue, English bluegrass, Dovergrass

F. rubra L. var. *commutata* Gaudin	chewings fescue, red fescue

■ perennials; ligules membranous; inflorescences panicles; spikelets laterally compressed; glumes shorter than lemmas; lemma backs rounded

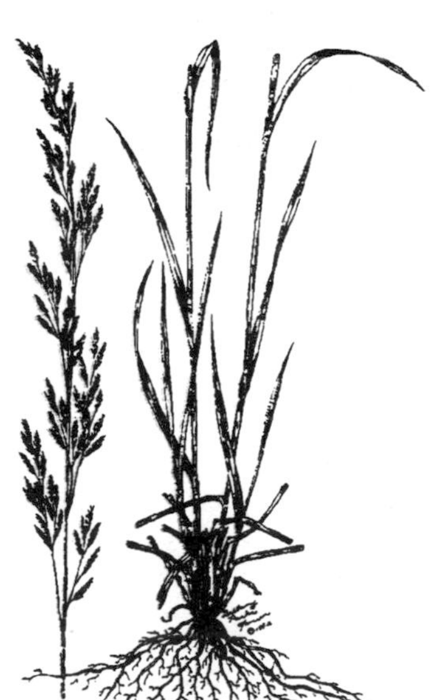

Festuca arundinacea

■ cultivated forages

Plants perennials; cespitose; bearing perfect flowers or rarely dioecious. **Culms** 50–150 cm tall; erect; not branched above; nodes glabrous; internodes hollow. **Leaves** both basal and cauline. **Blades** flat or folded or involute; 2–15 mm wide. **Ligules** membranous; lacerate; erose. **Auricles** present or absent. **Sheaths** glabrous; margins free or fused. **Inflorescences** panicles; open or contracted. **Disarticulation** above the glumes. **Spikelets** all alike or of 2 forms; staminate and pistillate similar; lanceolate; compression lateral; 3–25 mm long. **Florets** 2–25. **Glumes** 2; unequal; lanceolate; shorter than lemmas; awns absent; lower glumes with nerves 1 or 3; upper glumes with nerves 3 or 5. **Lemmas** lanceolate; backs rounded; abaxial surfaces glabrous; awns absent or present, 1, apical. **Paleas** hyaline; nerves 2; keels 2.

1. Blades folded or involute; 1–2.5 mm wide; stiff. Auricles absent . . *F. rubra*
1. Blades flat; 3–12 mm wide; lax. Auricles present.
   2. Old sheaths pale stramineous; intact. Auricles pilose. Short panicle branches with 3–10 spikelets. Lemma apices scabrous . . . . *F. arundinacea*
   2. Old sheaths brown; conspicuously shredding. Auricles glabrous. Short panicle branches with 1 or 2 spikelets. Lemma apices glabrous . . . . *F. pratensis*

Distribution and Habitat—Species of *Festuca* are abundant in the polar, temperate, and alpine regions of both hemispheres. Occurring in a diversity of floristic zones and habitats, they are adapted to different environmental conditions. Prized for their forage, both *F. arundinacea* and *F. pratensis* are introduced from Europe. *Festuca rubra* is used in mixtures of cool-season grasses for turf and sometimes in pasture plantings.

Disease Problems—*Festuca arundinacea,* or tall fescue, is a cool season perennial that is now grown on more than 35 million acres and said to be the most important cultivated pasture grass in the United States (Ball et al. 1991). It has many valuable attributes and is especially favored when planted in combination with cool-season legume forages. It and the other fescues pose little risk for the typical problems associated with grasses, i.e., nitrate accumulation, cyanogenesis, and photosensitization. Unfortunately, animals grazing fescue, especially tall fescue, are at risk of developing one or more of a group of rather unusual syndromes—summer slump, fat necrosis, fescue foot, or reproductive problems (Ball et al. 1995; Cross et al. 1995; Paterson et al. 1995). The syndrome produced is somewhat dependent on the environmental temperature and reproductive status of the animal (Jacobson et al. 1963). Diseases associated with fescues are probably the most important toxicologic problems of livestock in North America. Losses among all livestock species are estimated at nearly $1 billion per year (Hoveland 1993).

Occurring primarily in cattle in hot humid weather, the syndrome commonly known as summer slump, summer toxicosis, or summer fescue syndrome is the major problem and has the greatest economic impact. As the disease names imply, there is poor performance, mainly during the warm season when animals are likely to be heat stressed. This syndrome is reflected mainly as decreased weight gain and may occur to a lesser extent even without high temperatures (Beconi et al. 1995). In horses, systemic effects are less prominent than in cattle and are mainly seen as a slight diarrhea, increased sweating, and possibly a decrease in weight gain in young animals

(McCann et al. 1991; Aiken et al. 1993; Cross et al. 1995). Endophyte infection also seems to impart an apparent decrease in feed intake and digestibility for horses (Redmond et al. 1991). The fat necrosis syndrome, seen in cattle, may accompany summer slump and may persist long after the period of heat/humidity stress subsides. In contrast to the summer slump syndrome, which is not life threatening and is readily reversible, the fescue foot and reproductive syndromes can be quite devastating. Fescue foot, which occurs during the winter and early spring, is a gangrenous condition of the distal extremities. Its manifestation is very erratic in occurrence, and it is especially uncommon in the southern United States (Thompson & Porter 1990).

- disease syndromes—
  summer slump
  fat necrosis
  fescue foot
  reproductive problems

- summer slump, hot weather, decreased weight gain, cattle, less risk in other animals; may be accompanied by long-lasting necrosis of abdominal fat

  fescue foot, winter, cattle; gangrenous effect on feet, tail, and ears

Reproductive effects are particularly severe in horses (Taylor et al. 1985; Monroe et al. 1988). This is an important problem because of the large number of mares grazing fescue in the southeastern quadrant of the United States and the high percentage that will develop reproductive problems (Porter & Thompson 1992; Anas et al. 1998). In mild cases the primary effect is lack of udder development and poor lactation. There may be almost complete lack of milk production. As the disease's severity increases, effects may range from abortions or agalactia and weak foals to other serious gestational problems (Cruz et al. 1997). In the more severe cases, the delay in parturition, or the prolongation of gestation for several weeks, will result in the birth of exceptionally large foals. As a consequence, there may be dystocia, uterine rupture and other complications, and high foal mortality, just before, during, or shortly after birth (Poppenga et al. 1984; Taylor et al. 1985; Monroe et al. 1988; Putnam et al. 1991). In mares exposed to endophyte-infected fescue after gestation day 300, there is likely to be prolonged gestation and an enlarged thyroid with flattened cuboidal epithelia (Boosinger et al. 1995). Losses of both mares and foals may occur in the same herd, mainly due to delayed parturition and poor milk production. Reproductive effects in cattle and to a lesser extent in sheep are typically limited to decreased milk production, if there is exposure to the toxins just prior to parturition, and perhaps a decrease in conception rate (Petritz et al. 1980; Wallner et al. 1983; Siegel et al. 1985; Porter & Thompson 1992).

- reproductive effects, horses, common, poor udder development, agalactia

  more severe, delayed births, large foals, difficult birth, other complications

Festuca rubra

Disease Genesis—The deleterious effects due to ingestion of *Festuca* appear to be caused by ergot-type alkaloids produced by the endophytic fungus *Neotyphodium coenophialum* (formerly *Acremonium coenophialum*). Formerly classified in the genera *Epichloe* and *Balansia* (Bacon et al. 1975, 1977), the fungus grows entirely within the stems, leaf sheaths, and spikelets of the grass and does not alter the appearance of the plant. The life cycle of the fungus is completed entirely within the plant's tissues. There is no spore formation for external dissemination. Laboratory tests are required for detection of the fungus. Transmission of the fungus is strictly vertical via the grass caryopsis/seed. Thus, the level of infection in a pasture is relatively stable unless additional seed is used. Although endophyte infection poses problems for grazing livestock, it does impart an advantage to the fescue. Increasing productivity, it confers stress and pest tolerance to the grass plant (Bacon & Siegel 1988; Bacon 1995). A definite symbiotic relationship exists between the plant and the fungus (Hill et al. 1990). In the absence of the fungus, much closer management is needed to sustain a productive fescue pasture. The majority, and perhaps as much as 90% or more, of tall fescue plants in North America are infected with the fungus. In most instances the infection level in an individual pasture is greater than 50%. This is quite significant, because low infection levels—i.e., in the range of 10–20%—seem to produce little or no adverse effects in cattle (Bacon et al. 1977; Ghorbani et al. 1989). This may, however, not be true for the reproductive effects in horses. At levels of infection 20% or above, adverse effects such as increased water consumption begin to be

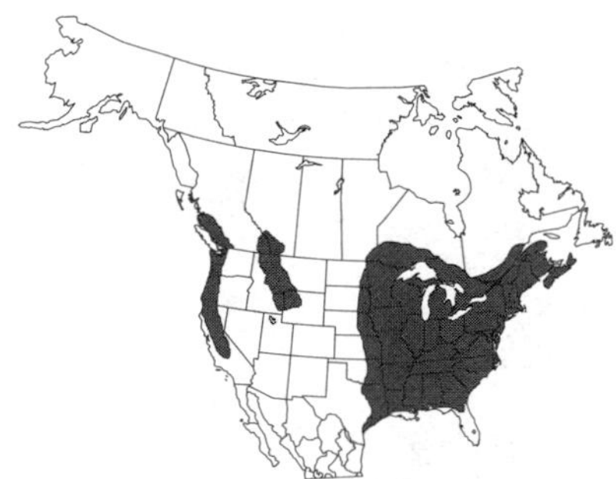

Festuca arundinacea

- fungal endophyte, in plant stems, leaves and spikelets, transmitted only via seed; most tall fescue pastures have level of infection 50% or more

noticeable in cattle (Aldrich et al. 1993a). There seems to be a strong relationship between level of endophyte infection and decreased daily weight gain (Aiken et al. 1993; Bransby 1997). It has been suggested that for each 10% increase in endophyte infection level there is a 45 g/day or 10% reduction in weight gain (Stuedemann et al. 1985; Crawford et al. 1989; Ball et al. 1991, 1995). This has been confirmed in analysis of data pooled from numerous studies over a 13-year period (Thompson et al. 1993). The effects on weight gain are not permanent; there may be compensatory weight gain in the feedlot following cessation of endophyte exposure (Ball et al. 1995; Beconi et al. 1995).

- alkaloids, ergopeptides, ergoclavines, and the amine halostachine

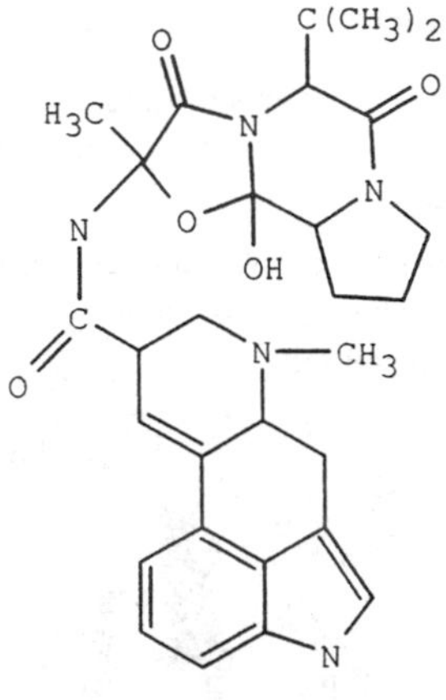

Ergovaline

Several toxic compounds have been isolated from species of *Festuca,* including perloline, beta-carbolines, and *N*-acetyl- and *N*-formyl-loline alkaloids (Bush et al. 1982; Hemken et al. 1984; Jackson et al. 1984). Some of these compounds may be associated with protection of plants against insect herbivory (Johnson et al. 1985; Siegel et al. 1985). Most of the signs associated with fescue toxicosis appear to be due mainly to various ergopeptides and, to a lesser extent, ergoclavine alkaloids typical of the Clavicepitaceae family. These findings are further strengthened by the observation that passive immunization of cattle by a monoclonal antibody developed in mice against the lysergic moiety of ergot alkaloids reversed the suppression of prolactin secretion (Hill et al. 1994). In addition to ergot-type alkaloids, halostachine, a sympathetic amine causing vasoconstrictive effects, may have additive actions in decreasing peripheral blood flow, such as occurs in fescue foot (Davis & Camp 1983).

- mainly ergovaline, highest concentrations and likely the cause of most effects, highest in seeds

Total alkaloid concentrations may be as high as 14 ppm, of which almost half are usually ergopeptides (Lyons et al. 1986). Compounds present include ergovaline, ergovalinine, ergonovine, ergosine, ergosinine, ergonine, ergocornine, ergopeptine, ergotamine, and ergotaminine (Lyons et al. 1986; Yates & Powell 1988). The ergotamine group of ergopeptides predominates in both the caryopses/seeds and other tissues (Yates et al. 1985). Ergovaline, levels of which may be as high as 5 ppm, and ergovalinine, the inactive α-isomer, typically account for the majority of the ergot alkaloids (Yates & Powell 1988; Hill et al. 1990). Ergovaline probably accounts for much of the toxic effects, because it causes adverse effects at concentrations as low as 0.15 ppm (Debessai et al. 1993). The clavine-type alkaloids are also represented—e.g., elymoclavine—but they are not as significant as the ergopeptides and cause mainly sedation (Fanchamps 1978; Porter et al. 1981).

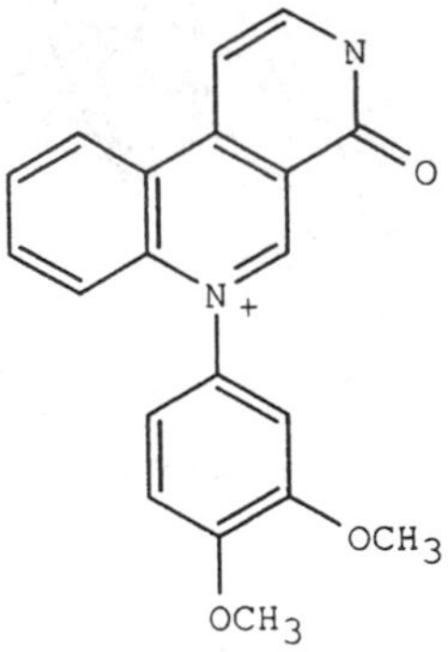

Perloline

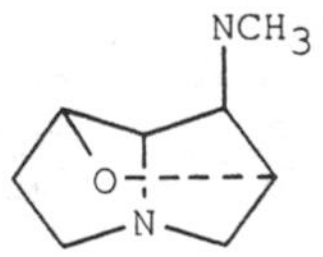

Concentrations of the endophyte and the ergot alkaloids are typically highest in the caryopses/seeds and leaf sheaths and considerably lower in the leaf blades (Siegel et al. 1984a; Lyons et al. 1986). Ergovaline concentrations in the stems and sheaths peak in late June and then decline as levels in the caryopses/seeds increase (Rottinghaus et al. 1991). Alkaloid concentrations are increased by nitrogen fertilization and seemingly decreased by heavy grazing pressure (Lyons et al. 1986; Bransby et al. 1988; Hill et al. 1990; Rottinghaus et al. 1991). The toxins are relatively stable, and fresh forage, hay, or caryopses/seeds are toxic, albeit effects are most pronounced with caryopses/seeds (Schmidt et al. 1982). Although the toxins are stable, the endophyte typically decreases in viability with time unless the plants are stored at low humidity and temperature (Siegel et al. 1985). There is essentially complete loss of viability after 2 years. The endophyte causes a slight decrease in digestibility of the forage for cattle and a consistent decrease in its palatability (van Santen 1992; Aldrich et al. 1993b).

Loline

The clinical expressions of fescue toxicity are well explained by the various actions of the alkaloids. Ergotamine causes an increase in plasma cortisol and triiodothyronine in cows and an accompanying increase in metabolic rate and heat production (Browning et al. 2000). Many of the alkaloids appear to be directly associated with the ability of some of the ergopeptides to reduce blood flow to the skin (Solomons et al. 1989; Rhodes et al. 1991). The summer syndrome is associated with impaired ability to dissipate body heat because of the vasoconstrictive effects of the toxin. A role also for decreased prolactin in this effect has been suggested (Thompson et al. 1987; Schillo et al. 1988). However, while a D_2-dopamine antagonist reversed the decrease in feed intake in endophyte-fed rats, similar treatment of cattle reversed the decrease in prolactin but did not protect against the decrease in feed intake or the increase in rectal temperature (Larson et al. 1994; Samford-Grigsby et al. 1997). An alteration in hepatic mixed-function oxidase (MFO) to increase heat production has also been proposed (Zanzalari et al. 1989). Cimetidine, an MFO inhibitor, lowers rectal temperature in affected animals. Heat stress is most noticeable at environmental temperatures above 32°C (Hemken et al. 1979, 1981). Stress effects appear to be less of a problem in the more heat-tolerant, Brahman-type cattle *(Bos indicus);* however, these breed differences are not clearly shown by plasma cortisol and prolactin changes (Browning 2000). Conversely, during periods of cold temperatures the vasoconstrictive effects are associated with dry gangrene of the extremities (fescue foot), which is similar to classical ergot intoxication. Consistent with heat and cold sensitivities, the effects of fescue foot seem to be more severe in Brahman-type cattle, which are considered to be less cold-hardy than the European breeds. There is also a suggestion that the vasoconstrictive effects are associated with increased risk of laminitis in horses (Rohrbach et al. 1995).

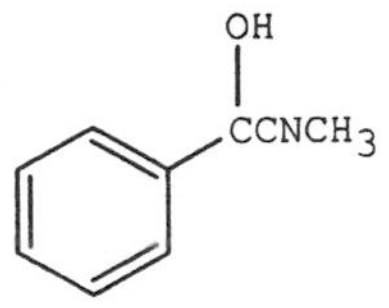

OH
CCNCH$_3$

Halostachine

■ summer slump, ergopeptides, decreased blood flow to skin, impaired body heat dissipation

The ergot alkaloids cause a decrease in prolactin secretion in cattle and a subsequent decrease in milk production (Wallner et al. 1983). These effects are not related to environmental temperature (Aldrich et al. 1993b). The horse is particularly sensitive to effects secondary to the decrease in prolactin secretion, probably the most sensitive indicator of intoxication (Jackson et al. 1988). Experimental administration of bromocriptine, a dopamine agonist, causes the same reproductive effects in the horse (Ireland et al. 1991). Some of the reproductive effects in cattle are suggested also to be related to the corpus luteum, with its degeneration leading to inability to maintain pregnancy (Porter & Thompson 1992).

■ reproductive effects, decreased prolactin secretion, decreased milk production

Endophyte infection with *Neotyphodium* spp. is common to many species of *Festuca,* including *F. arizonica, F. eastwoodae, F. pacifica, F. subulata,* and *F. versuta.* In addition, *N. coenophialum* will grow in *F. rubra* and *Lolium perenne.* Conversely, *N. lolii* will grow in *F. arundinacea* (Leuchtmann 1992).

Clinical Signs—The summer slump syndrome is the most consistent effect of the ergopeptide toxins and can be readily elicited by feeding infected hay or transferring animals to infected pastures. Signs are especially noticeable with temperatures exceeding 32°C and are indicative of a decreased ability to dissipate body heat. The signs include an increase in body temperature (0.5°C) and respiratory rate and a decrease in food intake, growth, milk production, and reproductive performance. Affected animals, particularly cattle, have a poor appearance because of weight loss and a rough hair coat that does not seem to shed. They tend to stand in the shade or in water instead of eating. The greater the heat and humidity stress, the more severe the syndrome. These signs are seen primarily in cattle; they are minor in sheep and are negligible in horses (Hemken et al. 1979; Putnam et al. 1991). Many minor alterations in hematologic and serum chemistry parameters can be shown experimentally, but these would be difficult to detect in the field (Oliver et al. 2000).

■ summer slump, increased body temperature and respiration rate, decreased growth and reproductive performance

Reproductive effects in horses include agalactia, decreased conception rate, prolonged gestation, weak foals, stillbirths, abortion, and firm thickened placentas. Foals have overgrown hooves, irregular incisor eruption, long hair coats, and poor muscling.

■ reproductive effects, horses, agalactia, decreased conception rates, prolonged gestation, weak foals, stillbirths, thickened placenta; large foals with overgrown hooves, long hair, poor muscling

Prolongation of gestation results in a large fetus, causing in some instances a difficult delivery with increased risks to both fetus and dam. There is a marked decrease in serum prolactin in all species 3–4 days after ingestion of the toxin. Reproductive effects in cattle and sheep are generally limited to decreased reproductive efficiency, conception delay, and possibly decreased birth weights (Bond et al. 1988; Bolt & Bond 1989).

■ fat necrosis, hard fat masses in abdominal cavity

Fat necrosis is often seen in animals that have had the summer slump syndrome, although it may be seen alone. Of long duration, it is characterized by the appearance of numerous hard fat masses in the adipose tissue of the abdominal cavity that are accompanied by a decrease in serum cholesterol and total lipids (Stuedemann et al. 1985). The effects are worse with increased nitrogen fertilization of the fescue. The firm masses may become large enough or occur in such numbers as to cause signs indicative of vascular or luminal obstruction in the intestine or urogenital system.

■ fescue foot, weight loss, rough hair, reluctance to move, soreness of limbs, most often pelvic limbs

Fescue foot is the most striking of the fescue toxicity syndromes. There is usually weight loss and development of a rough hair coat. Slight diarrhea and a mild temperature increase may also occur. The animal typically stands with an arched back and is reluctant to move, because of soreness in one or both pelvic limbs. Swelling and reddening of the coronary band(s), especially in the area between the hoof and dewclaws, is apparent upon close inspection. Eventually, a clear line of demarcation develops around the limb just above the fetlock. The swelling below this line progresses to dry gangrene and eventual loss of the foot. These changes rarely occur in the thoracic limbs. Similar lesions may occur in the tail, with eventual loss of the switch. The syndrome is indistinguishable from the dry gangrene characteristic of ergotism.

■ laboratory tests; detect endophyte in plant tissues, stain and identify; ELISA for alkaloids

As with *Lolium,* laboratory tests are available to detect endophyte presence in the tissues of *Festuca.* They include staining to permit observation and identification of the fungal elements in the plant tissues (Bacon et al. 1977; Smith 1987; Hignight et al. 1993). Most of these methods require overnight staining. However, a quick method employing rose bengal stain has been described for use with fresh forage (Saha et al. 1988). Fresh leaf sheath or stem pith is placed on a microscope slide, and 1–2 drops of a 0.5% solution of rose bengal in 5% aqueous ethanol applied. A cover slip is then applied, and the excess stain removed. The material is examined at 100–400x magnification, using a green filter; the fungal elements stain pink or red. If dried plant material is used, overnight soaking in an alkaline rose bengal solution is required. Other methods and stains are available, including culture of tissue from the seed and use of enzyme-linked immunosorbent assay (ELISA) (Johnson et al. 1982; Welty et al. 1986). Arrangements to evaluate caryopses/seeds for the presence of the endophytic fungus by staining or by an ELISA test should be made through local agricultural extension personnel or area agronomists or through agencies such as the Fescue Testing Laboratory at Auburn University, Auburn, Alabama.

Analysis of forage or seeds for the presence of ergovaline or other endophyte toxins also may be useful in confirming the cause of intoxication (Porter 1995).

■ fescue foot, gross pathology, necrosis, sloughing of skin and tissues, fetlocks and below

Pathology—Distinct pathologic changes are seen mainly with the fescue foot syndrome. There is sloughing and necrosis of the skin and underlying tissues distal to the line of demarcation above the fetlocks. Microscopically, the necrosis is accompanied by perivascular edema and thickened walls of blood vessels around the coronary band.

In horses exhibiting the reproductive effects of *Festuca,* the placenta is thick and tough. Microscopically, there is edema, colloid deposition, vascular degeneration, and fibrosis of the placenta.

Treatment—Treatment using vasodilator drugs has proven to be of limited value for both the summer slump and the fescue foot syndromes. Summer slump is reversible when alternative forages are provided. The summer-slump peripheral vasodilatory effects subside within about a week after exposure to the toxins ceases (Rhodes et al 1991). Some of the effects of fescue foot are also reversible, with the exception of well-developed lesions of the extremities, which cause almost permanent disability. Prophylactic and therapeutic selenium supplementation has been advocated, but experimental studies do not support its use (Taylor et al. 1985; Monroe et al. 1988). Copper oxide or progesterone-estradiol implants were also ineffective in preventing the effects of endophyte-infected fescue on animal performance and prolactin secretion (Coffey et al. 1992). In contrast, thiamine (1 g/day) has shown some potential as a protection against the increased temperature of summer slump (Dougherty et al. 1991). Of interest are the apparent protective effects of ivermectin in improving weight gains of cattle on high–endophyte-infected fescue, either alone or with concurrent use of implants of progesterone/estradiol benzoate (Bransby 1997). 17β-Estradiol implants alone may provide partial reversal of weight gain depression, but not equal to gains seen with endophyte-free forage (Beconi et al. 1995).

■ summer slump, generally reversible with change in forage; possible benefit from use of thiamine, estradiol, ivermectin, metoclopramide

fescue foot, dromperidone, perphenazine

Dopamine antagonists and other drugs have shown some promise as therapeutic agents. Metoclopramide, 15 mg/kg b.w. orally 3 times per week, increased prolactin secretion and weight gain in affected cattle and sheep (Lipham et al. 1989; Aldrich et al. 1993a). Simulation of the signs of fescue intoxication in mares with the dopamine agonist bromocriptine are reversible with perphenazine, 0.375 mg/kg b.w. orally twice per day for the last 30 days of gestation (Ireland et al. 1991). Perphenazine also increases prolactin secretion. Similarly, fluphenazine and dromperidone are also effective in preventing or reversing the reproductive effects of the ergopeptides in horses (Redmond et al. 1992; Bennett-Wimbush & Loch 1997). Uterine vasoconstriction in cattle is reversed by serotonin-2 inhibitors, including phenoxybenzamine (Dyer 1993). Passive or active immunization against ergot alkaloids may also be of some benefit in preventing the effects of these toxins (Hill et al. 1994; Filipov et al. 1998).

Avoidance or control of the fescue syndromes is difficult because the fungus is transmitted via the caryopsis/seed. Endophyte-free seed has been available since the early 1980s and may be used to develop new pastures. In addition, seed that has been stored for 1 or more years at room temperature, and in which the endophyte content has markedly decreased, may be used in establishing pastures. Unfortunately, fescue pastures developed from fungus-free seed are less hardy and more susceptible to insect damage. Reliance on endophyte-free fescue for livestock grazing requires careful management because it is more readily eaten and thus subject to overgrazing. If the choice is made to use endophyte-free seed, it would be advisable to consult with local agricultural extension personnel for information on the most appropriate practices for pasture development and management. Planting other grasses and legumes to establish mixed pastures is also a possibility. Treatment of seed with fungicides has been used but is costly (Siegel et al. 1984b). These approaches represent long-term measures, however, and do not provide a solution to the immediate toxicity problems.

■ control, endophyte-free seed, new pastures; difficult to maintain such pastures

At present, the best recommendation is to avoid relying entirely on fescue forage for any length of time. Supplementation of fescue forage with hay from other grasses or legumes may help to avoid problems. Management procedures can be implemented to minimize the risk to cattle. For example, Bransby and co-workers (1988) have shown that high stocking rates, which result in maintaining the forage at low heights, produce improved weight gains and minimal signs of summer slump. They suggest that endophyte and toxin levels are lower in the leafier short grass, and thus dosage to individual animals is lower. Therefore, producers may be able to develop ways to

■ control: supplement with other forages, increase stocking rates to keep fescue at low height, remove mares from fescue pasture the last 2 months of gestation diminish the problems and utilize the favorable characteristics of fescue. Proper management is particularly crucial with brood mares. They should be kept out of fescue pastures for at least the last 45 days of gestation. It is best to avoid extensive reliance on fescue to maintain brood mares. Otherwise, it is satisfactory forage for horses.

Infected forage harvested as hay may be rendered less toxic by treatment of the bales with anhydrous ammonia (Kerr et al. 1990). The hay may be fed to cattle if the animals are carefully monitored for endophyte effects, but it probably should not be relied upon for use in horses.

Glyceria R.Br.

- *G. canadensis*
  G. grandis
  G. septentrionalis
  G. striata

  floatgrass
  fowl meadowgrass
  mannagrass
  rattlesnake grass
  sweetgrass

Taxonomy and Morphology—Comprising 35–40 cool-season species, *Glyceria,* a member of the subfamily Pooideae and the tribe Meliceae, is most abundant in the Northern Hemisphere. The generic name comes from the Greek word *glykys,* meaning "sweet," which possibly reflects the taste of the caryopses of some species of the genus, as do the common names *sweetgrass* and *mannagrass.* Providing cover and food for birds and small mammals in wetland sites, species of the genus are highly palatable to livestock but are of little economic importance as forage because of their limited distribution in marshes or wetlands. In North America, 15–20 native and introduced species are present. Representative of the genus and of toxicologic significance are the following:

G. canadensis (Michx.) Trin.	Canadian mannagrass, rattlesnake grass
G. grandis S.Watson	American mannagrass, tall mannagrass
G. septentrionalis Hitchc.	eastern mannagrass, sweetgrass, floating mannagrass, floatgrass
G. striata (Lam.) Hitchc.	fowl mannagrass, fowl meadowgrass

- aquatic perennials, rhizomatous; leaves primarily basal; ligules membranous; inflorescences panicles; spikelets laterally compressed or terete; lemma nerves not converging at apices

Plants perennials; emergent aquatics; usually strongly rhizomatous. **Culms** 25–200 cm tall; erect or decumbent and rooting at nodes; not branched above; nodes glabrous; internodes hollow. **Leaves** primarily basal. **Blades** flat or folded; 2–20 mm wide. **Ligules** membranous. **Auricles** absent. **Sheaths** glabrous; margins fused. **Inflorescences** panicles; open or contracted; erect or nodding. **Disarticulation** above the glumes. **Spikelets** all alike; linear or oblong or ovate; terete or compression lateral; 2–35 mm long. **Florets** 3–16. **Glumes** 2; unequal; alike; shorter than lemmas; ovate; nerves 1; awns absent. **Lemmas** obovate or elliptic or oblong; backs rounded; nerves 7, or rarely 5 or 11, not converging at apex; apices rounded or obtuse; awns absent. **Paleas** hyaline; nerves 2; keels 2.

Glyceria canadensis

1. Spikelets linear; terete or nearly so; 10–40 mm long . . . . *G. septentrionalis*
1. Spikelets ovate or oblong; compression lateral; 2–8 mm long.
   2. Spikelets 3–4 mm wide. Nerves of lemmas not raised . . . . *G. canadensis*
   2. Spikelets 2–2.5 mm wide. Nerves of lemmas raised.
      3. Spikelets 2–4 mm long. Lower glumes acute; 0.5–1 mm long . . . . *G. striata*
      3. Spikelets 5–6 mm long. Lower glumes obtuse; 1.2–1.9 mm long . . . . *G. grandis*

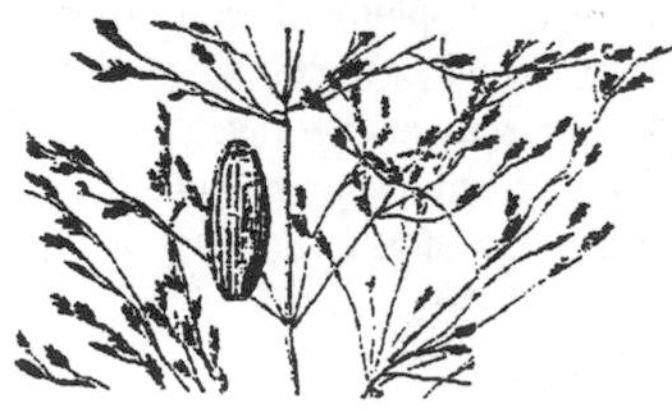

Glyceria grandis

Distribution and Habitat—Species of *Glyceria* are present in the temperate and subarctic regions of the Northern Hemisphere in both the Old World and the New World. Habitats include swamps, marshes, bogs, streamsides, and damp soils of meadows

Glyceria septentrionalis

Glyceria striata

and wet woods. *Glyceria striata* and *G. grandis* occur across the continent, the latter restricted to the northern half. *Glyceria septentrionalis* occurs in the eastern half of the continent, and *G. canadensis* in the northeastern quarter. Plants of various species are sometimes planted in bog gardens.

Disease Problems—Species of *Glyceria* are known to cause abrupt death in cattle when ingested rapidly in large amounts (Rainey 1946; Sharman 1967). Although cases are reported in North America, these grasses have not proven to be a major cyanogenic risk (Puls et al. 1978; Barton et al. 1983). There is the potential for problems when they are locally abundant, but more typically the small, scattered populations are a limited risk. Toxicity seems to be increased when the leaves are infected by the smut *Ustilago longissima,* which causes rusty brown lines on the foliage (Sikula 1981). As with other cyanogenic plants, the risk is greatest when livestock newly introduced into a pasture containing *Glyceria* consume the plants during periods of rapid plant growth.

Disease Genesis—Cyanogenic potential is appreciable in all species and in some instances may be in the range of 4,000–10,000 ppm (Reynard & Norton 1942; Rainey 1946; Sharman 1968). The HCN potential is associated with the glucoside dhurrin and appears to be subject to many of the same factors that influence cyanogenesis in the sorghums. HCN potential is highest in rapidly growing plant tissues, i.e., in spring growth and fall regrowth. In a European study covering 2 years, HCN potential increased steadily from May, reached a peak in July, and declined thereafter to low levels in October (Cotrau et al. 1977). The potential is much reduced in dry grass or hay and is especially low in the caryopses/seeds. The risk of intoxications is corre-

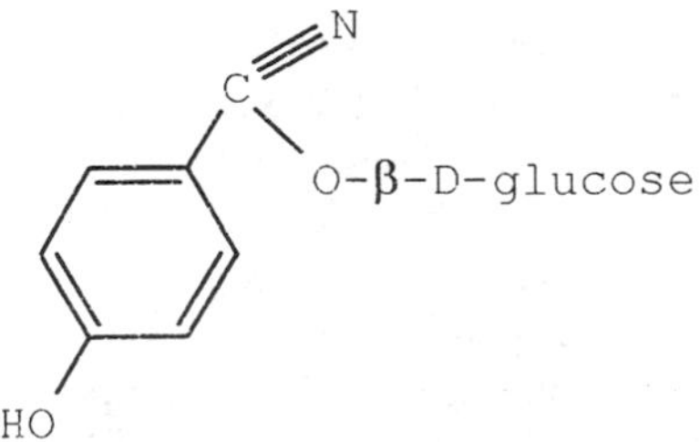

Dhurrin

■ wet areas, bogs, swamps, marshes, wet meadows

Glyceria canadensis

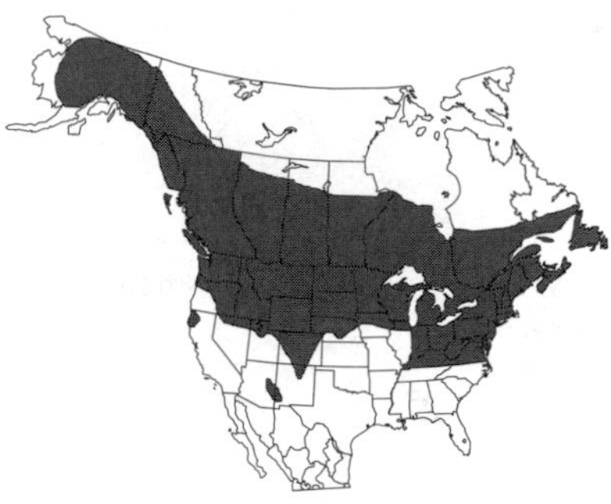

Glyceria grandis

■ cyanogenic, especially young plants, abrupt onset ruminants

■ glycoside dhurrin, especially in stages of rapid plant growth

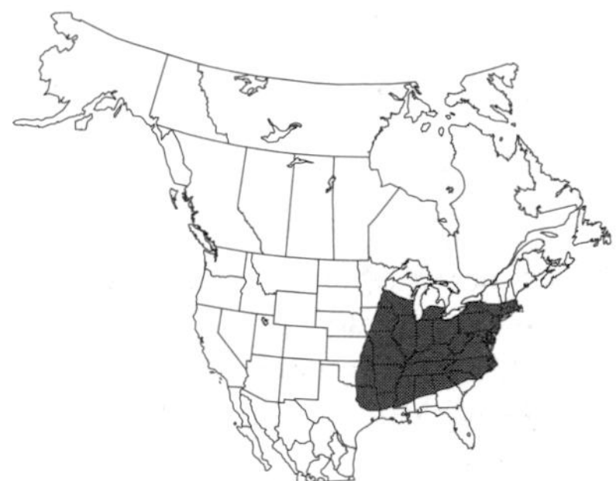

Glyceria septentrionalis

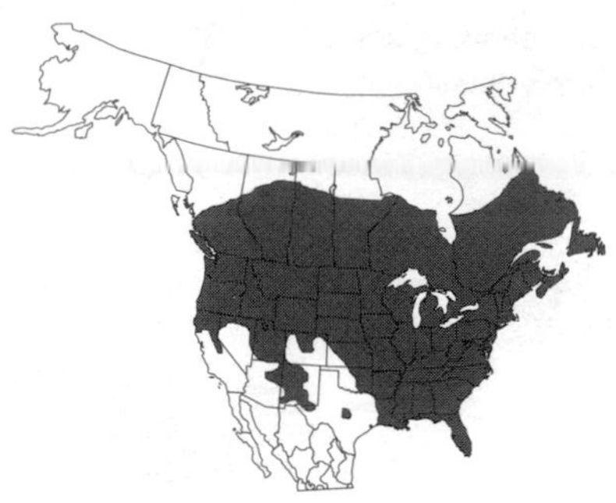
Glyceria striata

spondingly high during periods of active and productive growth (Barton et al. 1983). The high HCN potential in the plants results in high blood HCN, 10 ppm in some instances, and readily detectable levels of HCN in the rumen (Puls et al. 1978). Because plants of *Glyceria* are coarse grasses, the stems and leaves are typically evident when the rumen is examined postmortem (Barton et al 1983). An additional discussion of cyanogenesis is presented in the treatments of *Sorghum* (in this chapter) and the Rosaceae (see chapter 64).

- abrupt onset; apprehension, distress, weakness, ataxia, labored respiration, collapse, seizures

Clinical Signs—There is abrupt onset of apprehension and distress, which occur within a few minutes of grazing the offending grass. They are quickly followed by weakness, incoordination, and labored respiration. If intoxication is severe, the animal may be unable to stand and may lay in lateral recumbency with periodic paddling and tetanic seizures. The entire sequence of signs may occur over a 5-to-15-minute period, with either death or recovery at the end of this time.

- no lesions

Pathology—There are few distinctive changes. For the most part the changes are limited to congestion of the viscera and scattered, small splotchy hemorrhages. The stems and leaves are readily identifiable when the rumen contents are examined postmortem (Barton et al. 1983).

- sodium thiosulfate, sodium nitrite

Treatment—The primary antidote for ruminants is sodium thiosulfate given i.v. at a dose of 0.25–0.5 g/kg b.w. Sodium nitrite, 10 mg/kg b.w. i.v., can be used to augment the effects of the sodium thiosulfate, but it is not required for effective treatment. A more complete discussion is presented in the treatment of the Rosaceae (see chapter 64).

Hilaria Kunth

- *H. belangeri*
  H. jamesii
  H. mutica
  H. rigida
  H. swallenii

  big galleta
  curlygrass
  curly mesquite
  galleta
  tobosa
  tobosagrass

Taxonomy and Morphology—Strictly New World in distribution, *Hilaria*, a member of the subfamily Chloridoideae and the tribe Cynodonteae, comprises 9 warm-season species. Resistant to heavy grazing pressure, all are important range species in the Southwest. In North America, 5 species occur:

H. belangeri (Steud.) Nash	curly mesquite
H. jamesii (Torr.) Benth.	galleta, curlygrass
H. mutica (Buckley) Benth.	tobosa, tobosagrass
H. rigida (Thurb.) Benth ex Scribn.	big galleta
H. swallenii Cory	Swallen curly mesquite

- perennials; inflorescences spicate panicles; spikelets borne in clusters of 3; disarticulation below the glumes, clusters falling; spikelets of 2 forms

Plants perennials; rhizomatous or stoloniferous or cespitose. **Culms** 10–100 cm tall; erect or decumbent; stolons wiry; rhizomes short, woody, scaly; nodes glabrous or puberulent or villous; internodes solid. **Leaves** primarily basal or both basal and cauline. **Blades** flat or folded or involute; 1.5–5 mm wide. **Ligules** ciliate-membranous or ciliate. **Auricles** absent. **Sheaths** glabrous or villous; margins free, overlapping. **Inflorescences** spicate panicles; branches 2-ranked, short, appressed, deciduous, each bearing tight cluster of 3 spikelets; central axes zigzag. **Disarticulation** below the glumes; branches with attached spikelet clusters dropping. **Spikelets** of 2 forms; central 1 subsessile and almost hidden by lateral 2, compression lateral, florets 1, perfect or pistillate; lateral 2 sessile, compression dorsal, florets 1–5, staminate or neuter. **Florets** 1–5; perfect or pistillate or staminate or neuter. **Glumes** 2; unequal; shorter

Hilaria jamesii

Hilaria mutica

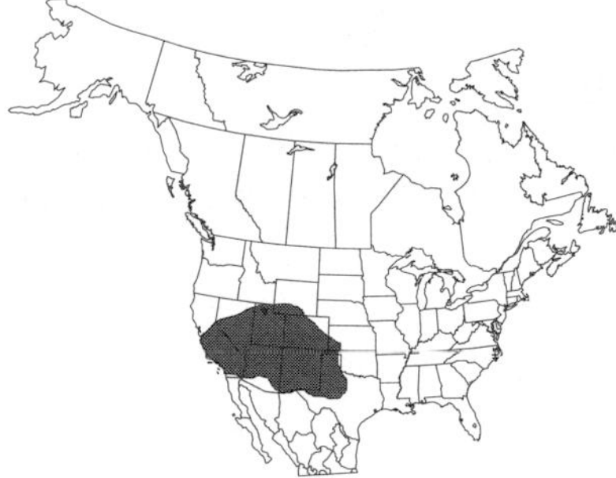

Hilaria jamesii

Hilaria mutica

than or as long as lemmas; lanceolate or flabellate; chartaceous or coriaceous; nerves 5 or 7; apices acute or lobed; awns present. **Lemmas** lanceolate; membranous; apices 2-lobed; awns absent or present. **Paleas** hyaline; nerves 2; keels 2.

1. Culms lanate . . . . *H. rigida*
1. Culms glabrous or indumented, but not lanate.
   2. Rhizomes present. Stolons absent. Blades involute, rarely flat.
      3. Glumes of lateral spikelets flabellate . . . . *H. mutica*
      3. Glumes of lateral spikelets elliptic . . . . *H. jamesii*
   2. Rhizomes absent. Stolons present. Blades flat.
      4. Clusters of spikelets 5–6 mm long. Glumes without numerous minute black glands; lower halves pale colored . . . . *H. belangeri*
      4. Clusters of spikelets 6.5–8 mm long. Glumes with numerous minute black glands; lower halves dark colored . . . . *H. swallenii*

Distribution and Habitat—Most abundant in the shortgrass prairies and desert grasslands of the southwestern United States and northern Mexico, species of *Hilaria* occur as far north as Wyoming and as far south as Venezuela. Plants are able to withstand prolonged droughts.

■ desert grasslands, shortgrass prairies

Disease Problems and Genesis—Species of *Hilaria* may cause an uncommon staggers syndrome identical to that caused by *Paspalum* or *Lolium*. Typically involving mature plants bearing fruiting spikelets, the incidence of this disease is promoted by increased

■ staggers syndrome, ergot alkaloids, similar to those of *Paspalum*

local rainfall. The fungus *Claviceps cinerea* invades the grass ovaries of several species of the genus and forms sclerotia that contain ergot alkaloids and are toxic when eaten by grazing livestock (Sprague 1950). A more complete discussion of the staggers syndrome is presented in the treatment of *Paspalum,* in this chapter.

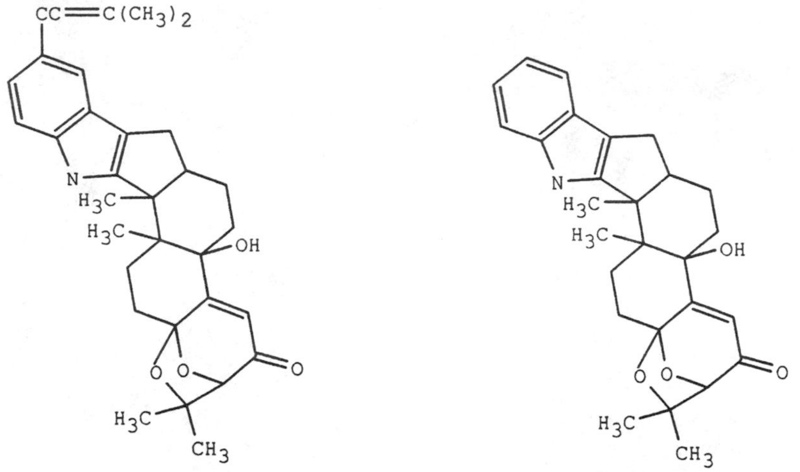

Paspalitrem A

Paspalinine

■ tremors, may not be able to walk or stand; recover after brief rest; rarely fatal

Clinical Signs—The primary manifestation is tremors, which may be severe enough to prevent walking or even standing in more severely affected animals. Forced movement intensifies the tremors and may cause the animal to fall. In most cases they are able to stand after a few minutes rest. The disease is rarely fatal, and in most instances the signs quickly resolve after the animals are removed from the toxic pasture.

■ no lesions

Pathology—There are no pathologic changes in the central or peripheral nervous systems. Observable effects are those related to the incapacitation caused by the toxin, i.e., dehydration and starvation.

■ remove from pasture, sedation

Treatment—If animals are moved out of the infected pasture when clinical signs are first noted, the disease usually is readily reversible within a few days. In cases in which the animals are temporarily incapacitated, recovery may be facilitated by placing feed and water within reach. In severe cases, sedation may be helpful.

Management practices that ensure that the animals eat young, growing grasses rather than mature plants with infected spikelets are effective in preventing appearance of the disease. However, once a severe infestation of *Claviceps* occurs in a pasture, subsequent infestations are likely whenever environmental conditions are appropriate, because the sclerotia continue to contaminate soil.

Lolium L.

■ *L. perenne*
L. persicum
L. rigidum
L. temulentum
darnel
ryegrass
tares

Taxonomy and Morphology—Comprising 8 cool-season species, *Lolium* is a Eurasian genus that has been introduced deliberately and accidentally throughout the rest of the world. Its generic name is an old Latin name for "darnel," reputed to be the tares sown by the enemy in the biblical parable. Species of the genus are economically important as pasture and turf species and as noxious weeds and toxic plants. A member of the subfamily Pooideae and the tribe Poeae, *Lolium* was originally classified in the tribe Hordeae, with wheat, barley, and rye, because of its 2-sided spike.

Subsequent taxonomic work, however, revealed a closer relationship to *Festuca* and other members of the Poeae—thus its present classification. In North America, 4 introduced species are commonly encountered:

L. perenne L. (= *L. multiflorum* Lam.)	perennial ryegrass, English ryegrass Italian ryegrass
L. persicum Boiss. & Hohen.	Persian ryegrass
L. rigidum Gaudin	
L. temulentum L.	darnel, tares

■ annuals or perennials; ligules membranous; inflorescences 2-sided spikes; spikelets with lemma midnerves against rachises; glumes 1

Plants annuals or perennials; cespitose. **Culms** 30–100 cm tall; erect; not branched above; nodes glabrous; internodes hollow. **Leaves** both basal and cauline. **Blades** flat or folded; 1–10 mm wide. **Ligules** membranous. **Auricles** present or rarely absent. **Sheaths** glabrous; margins free, overlapping. **Inflorescences** 2-sided spikes; spikelet attachment dorsal (midnerves of lemmas against rachises), somewhat sunken in rachises. **Disarticulation** above the glumes. **Spikelets** all alike; compression lateral; 7–26 mm long. **Florets** 2–25. **Glumes** 1; lower glumes absent except in terminal spikelet; upper glumes shorter or longer than lemmas; lanceolate; nerves 3 or 5. **Lemmas** lanceolate; backs rounded; glabrous; awns absent or present, 1. **Paleas** hyaline; nerves 2; keels 2.

Lolium perenne

1. Glumes shorter than florets . . . . *L. perenne*
1. Glumes longer than florets.
   2. Spikes subcylindric. Spikelets sunken in rachis cavities and concealed by upper glumes . . . . *L. rigidum*
   2. Spikelets flat. Spikelets not sunken in rachis cavities nor concealed by upper glumes.
      3. Spikes 8–12 cm long. Upper glumes three-fourths the length of spikelets. Lemmas 9–10 mm long . . . . *L. persicum*
      3. Spikes 15–25 cm long. Upper glumes as long as spikelets. Lemmas 5–8 mm long . . . . *L. temulentum*

Because of the annual habit of *L. multiflorum* (Italian ryegrass), some individuals recognize it as a distinct species. Most taxonomists, however, believe it to be nothing more than an agricultural selection by early Europeans from the otherwise morphologically indistinguishable perennial *L. perenne.*

Lolium perenne

Lolium temulentum

Distribution and Habitat—All species of *Lolium* are introduced from Eurasia. Believed to be the first species to be cultivated as a pasture grass in Europe and certainly one of the most important forage grasses there, *L. perenne* is cultivated as a pasture and lawn grass across the northern half of North America. It also is used to provide immediate cover on graded embankments. In the South it is planted as a winter forage. The other species of the genus are characteristically encountered as weeds in disturbed soils.

■ cultivated species or weedy invaders

Disease Problems—*Lolium* causes three significant intoxication problems: a disorientation syndrome known as tares, a staggers disease, and photosensitization. Aspects of each are presented in the following subsections. Although concentrations of up to 3.6% d.w. nitrate are reported (Everist 1981), nitrate accumulation is not a noteworthy problem in the genus.

■ 3 problems—
disorientation
staggers
photosensitization

Lolium temulentum

Tares

■ tares, drunken-like state, grain, all animals; foliage no risk

Disease Problems—*Lolium temulentum* is one of the most infamous of toxic plants because of the notoriety achieved from references to its noxious effects in various literary works, e.g., the Bible (Matthew 13:25–30) and Shakespeare's *Henry VI*. Many of these references emphasize the marked similarity of darnel to wheat, as well as its dreaded reputation for affecting the eyes or causing a type of drunkenness (Long 1927; Leemann 1933). Known since ancient times, toxicity in humans is the result of contamination of grains with the caryopses/seeds of darnel and subsequent ingestion of flour in baked products. The species' common names in French and German, Ivraie ("drunken") and Taumellolch ("giddiness"), are indicative of the type of disease produced (Leemann 1933). The green foliage is apparently acceptable forage; it is only the caryopses/seeds that are toxic (Long 1927). Cases are reported from South Africa and Europe, but in many instances are poorly authenticated (Watt & Breyer-Brandwijk 1962). Even in South Africa, where *L. temulentum* is eaten in large amounts by many groups of people, instances of intoxication are few (Steyn 1933a). Although the species is worldwide, it does not seem to have become a toxicity problem in Australia or North America (Pammel 1920; Everist 1981). Furthermore, although it is reported to affect most domestic animal species, actual cases of intoxication are rare (Watt & Breyer-Brandwijk 1962). Long (1927) reported the toxic doses to be 30 g in humans, 7 g/kg b.w. in horses, 15–18 g/kg b.w. in ruminants and poultry, and nil in pigs. Steyn (1933b) was unable to produce signs of intoxication with more than 3 kg fed over 38 days in rabbits, 1.25 kg over 10 days in rabbits, or 4 kg over 3 days in 7 month-old pigs. Deaths are rare, but Long (1927) reported that a horse was killed by 2 kg of seeds.

■ toxin unknown; possibly from a fungal endophyte

Disease Genesis—Early studies by Hofmeister indicated the presence of at least two types of toxicants—oily substituents/fatty acids affecting the digestive system and temulin (Guerin 1899). The alkaloids lolium and temulentum, and temulentic acid, have also been reported to be present in the caryopses/seeds (Steyn 1933b). Although the identity of the specific toxicant is not certain, the source may well be the fungus *Endoconidium temulentum*, which has been isolated from the caryopses/seeds (Long 1927). In one study this fungus was present consistently in samples from many areas of the world (Guerin 1899). It also was present in other species of *Lolium* but only rarely in *L. perenne*. Remarkably, it has been identified in seeds from the tomb of an Egyptian pharaoh constructed about 2400 B.C. (Leemann 1933). It has been suggested that temulin might be a fungal product, a mycotoxin perhaps. Thus, the disease appears to be an endophyte mycotoxicosis due to a heat-stable toxicant found mainly if not exclusively in the caryopsis/seed. The irregularity of effects is consistent with endophyte involvement.

■ generally mild disease, humans, dizziness, headache, mental confusion, somnolence; more rarely vomiting, colic, tremor, weakness

Clinical Signs and Pathology—In humans the signs include dizziness, headache, mental confusion, somnolence, and difficulties of vision, hearing, and speech. More-serious intoxications are accompanied by vomiting, abdominal pain, tremor, and pronounced weakness. If the disease does occur, it will probably only be mild. It is not fatal, and the effects are functional rather than morphologic.

■ general nursing care

Treatment—Elimination of the dietary source of the toxin will likely result in recovery without specific therapy. Otherwise, treatment is based upon the signs.

Ryegrass Staggers

■ staggers syndrome, uncommon, comparable to paspalum staggers

Disease Problems—*Lolium perenne* has been associated with a tremorgenic disease problem in Australia and New Zealand for many years. Occurring mainly in the summer and fall during hot, dry weather, the disease has become quite common in certain areas and affects all livestock species as well as deer (Mackintosh et al. 1982; Brooks & Cahill 1985; Smith 1987). In North America it has been a rather uncommon disease, but nevertheless significant episodes have been reported to be caused by consumption of fresh forage and possibly hay (Shaw & Muth 1949; Hunt et al. 1983; Galey et al. 1991). Periodic episodes of a staggers syndrome have occurred on the north coast of California, with serious problems in the 1980s (Humble 1996). The disease known as ryegrass staggers is essentially the same disease as paspalum staggers and Bermudagrass staggers, with few deaths reported. Ryegrass staggers is not the same syndrome as annual ryegrass toxicity, a devastating neurologic disease occurring mainly in Australia. This latter disease is caused by toxins produced by *Clavibacter,* a bacterium carried by the nematode *Anguina agrostis* to the spikelets of *Lolium* (McIntosh & Thomas 1967; Finnie & O'Shea 1988; Riley & Ophel 1992; Finnie 1994).

In addition to the tremorgenic syndrome, endophyte-infected *Lolium* has been associated with other effects similar to those caused by *Festuca.* Foote and co-workers (1994) have reviewed some of these other effects, such as increased deaths in newborn lambs from ewes grazing ryegrass, decreased milk production in cattle, decreased productivity in sheep, and a summer slump syndrome, with increased body temperature and respiration rate, in sheep. Ergotized inflorescences also are suspected of causing a fescue foot–like syndrome in Australia (Fraser & Dorling 1983). A summer-slump type of poor performance and reproductive problems also have been observed in horses (Powell 1996). The reproduction problems in horses appear to be very similar to those produced by *Festuca* forages and include poor udder development, agalactia, dystocia, weak foals, poor suck reflex, and retained placenta. These effects are responsive to dromperidone administration.

■ rarely reported effects, summer slump, gangrenous extremities, reproductive effects with decreased milk production

Disease Genesis—For many years fungal infection of *L. perenne* has been recognized. It is now known that the fungal endophyte *Neotyphodium lolii,* formerly *Acremonium lolii,* and its toxic alkaloids, lolitrem B and, to a lesser extent, lolitrems A, C, and D, are clearly associated with the disease (Neill 1940; Fletcher & Harvey 1981; Gallagher et al. 1981; 1984). Although lolitrem B is the major toxicant present, lolitrem A is similarly toxic (Munday-Finch et al. 1995). There may also be a role for other indole-diterpene mycotoxins, i.e., paxilline and lolitriol, which have been suggested as possible precursors to the lolitrems (Miles et al. 1992b). The effects

■ fungal endophyte; alkaloids; lolitrems, ergot-types ergovaline; decreased prolactin secretion, most common in cattle and sheep

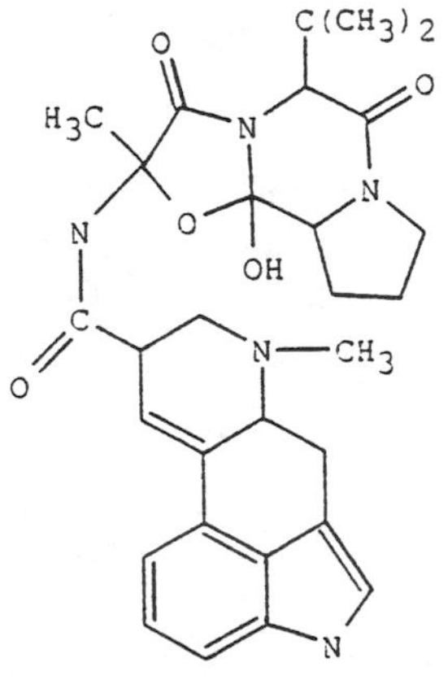

Ergovaline

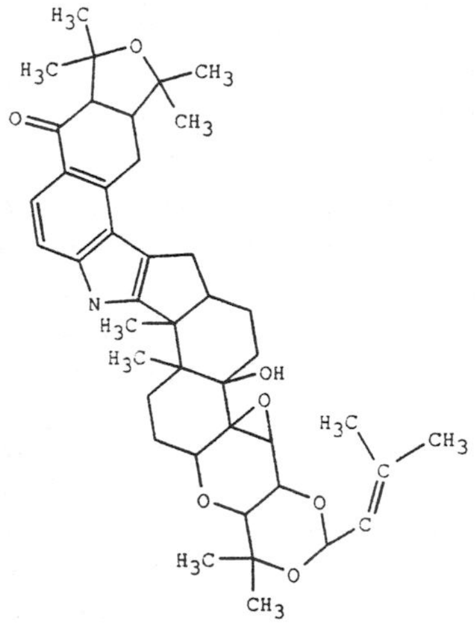

Lolitrem B

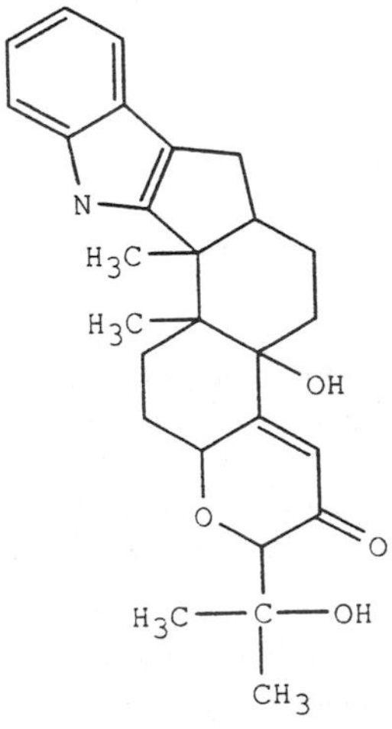

Paxilline

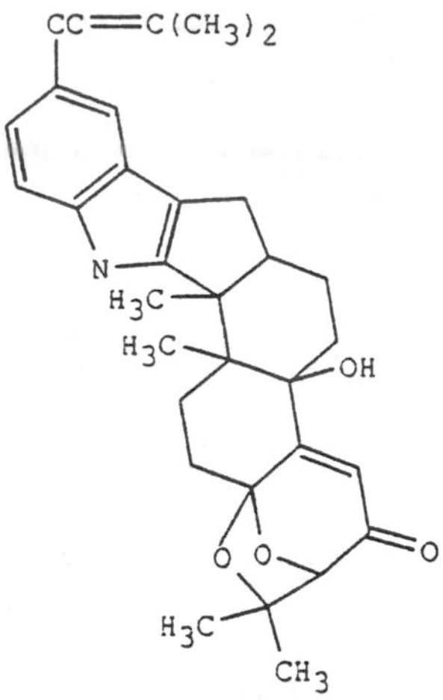

Paspalitrem A

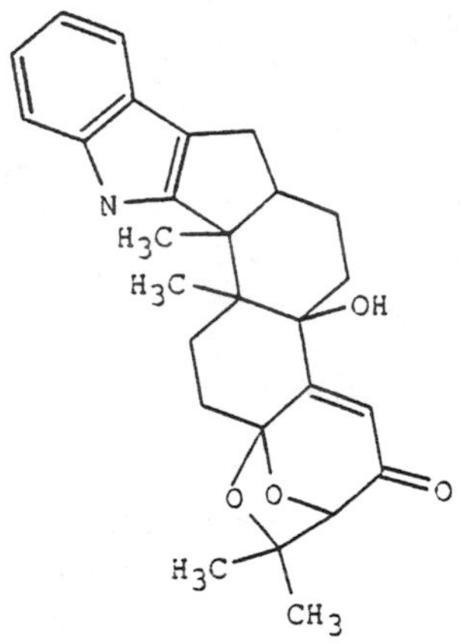

Paspalinine

of these alkaloids are clearly shown in sheep by the induction of tremors and electromyographic alterations for 24 hours following i.v. administration of 100 g/kg b.w. of lolitrem B, 2–4 hours following 2.2–7.5 g/kg b.w. of penitrem, or 1–2 hours following 0.66–1.5 mg/kg of paxilline (Smith et al. 1997; Smith & McLeay 1998). Ergot-type alkaloids such as ergovaline also are represented and are likely to play some role in the adverse effects (TePaske et al. 1993). Based on the type of signs observed, including tremors, it has been suggested that these tremorgens may impair inhibitory pathways in the nervous system, especially those mediated by GABA or glycine (Mantle & Penny 1981). This possibility has been examined in detail, and it appears that the fungal tremorgens inhibit GABA receptor function by binding close to the chloride channel (Gant et al. 1987). The indole alkaloids paspalitrems and paspalinine inhibit high-conductance Ca^{2+}-activated K^{+} channels, but this is apparently not associated with the tremors, because the nontremorgenic paspalicine is also an inhibitor (Knaus et al. 1994).

The neurotoxic effects are seen most commonly in cattle and sheep. Less commonly horses may be affected, as illustrated by a report of intoxication due to the feeding of hay containing lolitrem B (van Essen et al. 1995).

As with fescue intoxication, there is a decrease in serum prolactin and weight gain (Fletcher & Barrell 1984). Consumption of *Lolium* forage infected with *Neotyphodium lolii* may limit production because of heat stress at environmental temperatures in excess of 25°C, in much the same manner as does *Festuca* infected with *Neotyphodium coenophialum* (Fletcher et al. 1993). This is probably the cause of the reproductive problems in horses as well, because the effects are attenuated by dromperidone, a dopamine antagonist (Powell 1996). These alkaloids are subject to liver biotransformation and excretion in the bile (Mantle 1986).

■ endophyte and toxins, seeds, outer lower leaves

Neotyphodium lolii, much like *N. coenophialum,* is found in the caryopsis/seed and in the outer, lower leaf sheaths. It is transmitted only via the caryopsis/seed and imparts increased resistance against insect herbivory (Smith 1987). The presence of the endophyte is important, in some circumstances, for sustaining productivity of ryegrass pastures, because peramine (a pyrrolopyrazine alkaloid) and lolitrem are a deterrent and a toxicant, respectively, to the Argentine stem weevil, which limits production of the grass (Rowan 1993). Many cultivars of *L. perenne* are infected by the fungus. Its presence is particularly favored in turf types because the presence of the endophyte confers vigor and pest resistance. Even straw from stands used for grass seed production may present a risk when fed to livestock (Pearson et al. 1996). Because of the high fungal levels in the lower leaves, disease risk is also increased under conditions leading to grazing of the lower portions of the plants (Fletcher & Harvey 1981). In addition to leaves, caryopses/seeds from infected plants also may present a risk (Gallagher et al. 1982; Munday et al. 1985). Caryopsis/seed infection is lost with storage longer than 14 months.

Other fungal species infecting *L. perenne* and/or other species also may be factors in causing intoxications. Grass pastures containing *L. perenne* with low to moderate infection rates (16–30%) of a *Neotyphodium* species other than *N. lolii* and *N. typhinum* are suspected to have caused hyperthermia of cattle in the northwestern United States (Wilson et al. 1992).

■ increased nervousness, seizures with ears erect, fine tremor of head and neck to shoulders; move with stiff, wobbly swaying movement of pelvic limbs; collapse, head twisted backward

diagnosis confirmed by identifying fungal elements in plant tissue

Clinical Signs—The signs produced are the same as those of the other staggers syndromes, i.e., paspalum and Bermudagrass staggers. When animals are at rest, they may exhibit no obvious signs or only a fine tremor of the head and neck. When forced to move or disturbed, they move with a stiff, staggering, uncoordinated movement of the pelvic limbs. They may collapse but are usually able to rise again after a few minutes of rest. Often the head is twisted backward. If they rise again too quickly, the movements of the thoracic limbs are stiff and jerky, and the pelvic limbs weak and wobbly. Deaths are rare and due primarily to misadventure. Recovery is usually complete within a few days to a week. Occasionally, tremors are elicited in cattle several weeks later when they are handled.

As with fescue infected by *N. coenophialum,* diagnosis can be accomplished by identifying the fungal elements in the leaf sheath material, culturing for fungal growth, or using ELISA. Several methods have been described for staining and identifying the fungal elements (Smith 1987; Hignight et al. 1993). A quick method employing rose bengal is given in the treatment of *Festuca,* in this chapter (Saha et al. 1988). Analysis of foliage or caryopses/seeds for the presence of lolitrem B and endophyte toxins may be used to help establish a diagnosis (Porter 1995). It is suggested that 20 grass samples be pooled for analysis and that a concentration of approximately 2,000 ppb represents the threshold for risk (Tor-Agbidye et al. 1994).

Pathology—Grossly, there are no distinctive changes. In many animals there are no microscopic lesions; however, in a few instances in cattle, sheep, and deer, changes in the granular layer of the cerebellum are seen (Munday & Mason 1967; Mason 1968; Mackintosh et al. 1982). They include degeneration of Purkinje's cells, with rounded and elongated eosinophilic axonal swellings referred to as torpedoes.

■ rarely microscopic lesions in cerebellum

Treatment—If removed from the infected pasture, the animals experience prompt and complete recovery. On some occasions when animals are unable to walk without continued falling, food and water may be supplied until recovery allows them to be moved in a day or two. Prevention of the disease without compromising the productivity of the forage appears to be a possibility based upon the use of ryegrass cultivars selectively infected with an endophyte incapable of producing lolitrem (Clark 1992). For an extensive review of management prevention of ryegrass staggers, see Fletcher et al. (1993).

■ remove from pasture, provide food and water if unable to move animals

Photosensitization

Disease Problems—Undoubtedly a direct, primary type of photosensitization is produced by *L. perenne* (perennial rye), in much the same fashion as caused by grain forages, but the more important type is the hepatogenous type produced by infection of the species with the fungus *Pithomyces chartarum.* Known since the early 1900s as the cause of facial eczema in New Zealand, the fungus occurs in North America on *Lolium* and other grasses but has seldom been shown to be associated with disease (Mortimer & Ronaldson 1983). The disease is most commonly associated with, but not restricted to, species of *Lolium* (Alessi et al. 1994). There have been suspicious episodes with other forages (Monlux et al. 1963; Putnam et al. 1986), but they were not confirmed to be caused by the fungus and its toxin. A likely case involved lambs in western Oregon grazing perennial ryegrass with an exceptionally high concentration (600,000 spores/g) of *P. chartarum* in old growth (Hansen et al. 1994). There are apparently both toxin-producing and non-toxin-producing fungal strains; thus far mainly non-toxin-producing strains are known in North America (Ueno et al. 1974; Rowe 1989). The fungus seems to be a problem primarily in warm-temperate and tropical areas. Thus, its occurrence in North America is most likely to be in the South or the Southeast. The fungus is especially prone to grow in dead leaf litter, and warm rains and high humidity are important factors in disease occurrence. The risk of disease is increased with close grazing.

■ primary photosensitization; rarely hepatogenous photosensitization due to infection of plants by *Pithomyces chartarum*

- hepatogenous photosensitization, sporidesmins, necrosis, and bile duct injury; retention of photodynamic phylloerythrin, activated by ultraviolet light to cause tissue necrosis

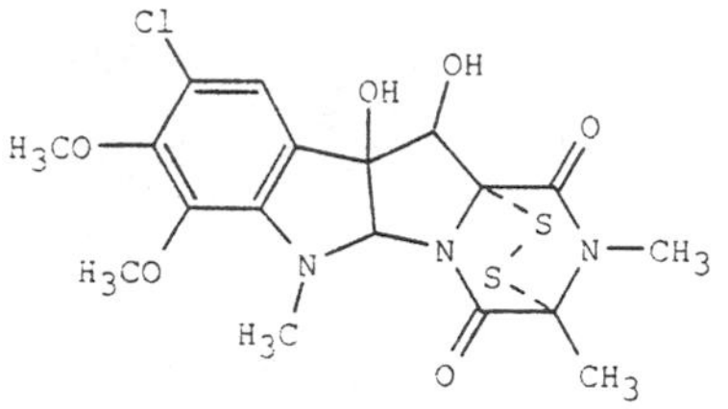

Sporidesmin A

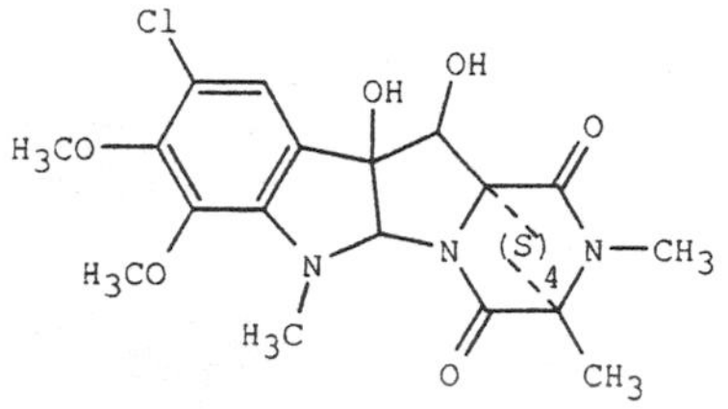

Sporidesmin G

Disease Genesis—The cause of the hepatogenous photosensitization is a series of 9 or more toxins, known as sporidesmins, which are produced by *P. chartarum.* Sporidesmin A accounts for about 90% or more and is among the most toxic. Sporidesmins E, G, and H are also quite toxic. These compounds are excreted in bile and in the process cause severe hepatocellular injury and necrotizing, obliterative cholangitis, and pericholangitis (Coetzer et al. 1983). Because of these effects on the bile ducts, the affected animal exhibits a relative inability to excrete bilirubin and presumably phylloerythrin, which is formed in the digestive tract during microbial degradation of chlorophyll and normally eliminated by the liver via the biliary system and in urine (Ford & Gopinath 1974, 1976). Phylloerythrin is a photodynamic compound that, when subjected to ultraviolet light penetrating the skin, causes tissue necrosis. The disease develops about 3 weeks after ingestion of the toxins. It is most severe in sheep, but other ruminants, rabbits, and guinea pigs are also susceptible. Some other animal species such as mice seem to be quite resistant to the effects of the toxin, suggesting perhaps differences in excretion pathways.

The fungus is not exclusive to *Lolium;* other grasses, such as *Brachiaria,* may be infected and toxic amounts of sporidesmin formed (Alessi et al. 1994).

- abrupt onset, scratching, rubbing; skin, reddened, blistered; dark urine; may be icterus and elevated serum hepatic enzyme activity with hepatogenous photosensitization
- "beaker test" for presence of mycotoxins

Clinical Signs—There is abrupt onset of edema and swelling of areas of skin exposed to light or to less pigmented areas where light may penetrate more readily to the capillary beds. In sheep these areas are the ears, eyelids, face, and lips. The skin will be reddened, warm, and inflamed. Irritation and accompanying pruritis are severe, with serum exuding from the sites and scab formation. Head and ear shaking may be almost violent. Rubbing of the sites produces excoriation and bleeding. There will often be distinct icterus of the mucous membranes of the mouth, vulva, and sclera and dark discoloration of urine. In other instances evaluation of serum chemistry will reveal marked elevation of the serum hepatic enzymes, especially GGT.

The disease is less severe in cattle and may be accompanied by intravascular hemolysis and hemoglobinuria in a few animals.

Diagnosis may be aided by the use of various tests to identify the presence of the fungus or its spores on grasses (Percival & Thornton 1958), e.g., the "beaker test" first described by Perrin (1959) and revised by Sandos and co-workers (1959). The test can be completed in 4 hours and involves extraction of dried, ground forage with acetone and eluted over an alumina column under slight pressure. When certain portions of the eluate, collected in beakers, are evaporated, a white deposit is positive for the nontoxic cyclic peptides pithomycolide and sporidesmolides, which indicate the presence of the fungus. It is a presumptive test that is strongly correlated with the presence of *Pithomyces.* Other tests, including bioassays in vitro employing cell culture, or in vivo using the rabbit cornea edema response or the whole animal response of guinea pigs, are also used.

- obvious skin lesions; hepatogenous photosensitization

  gross pathology, liver, large and dark or small and firm

  microscopic, liver necrosis, biliary proliferation, edema of walls of bile ducts

Pathology—Grossly, the liver may be dark and swollen in the early phase or pale, fibrous, and misshapen in long-term cases. There also is a mild to moderate cystitis and edematous thickening of the walls of the gallbladder and extrahepatic bile ducts. Some of the bile ducts may appear to be obliterated by scar tissue. Microscopically, there is hepatocellular necrosis, biliary hyperplasia and edema, and inflammatory cellular infiltrates around the interlobular bile ducts. The necrotizing cholangitis is associated with peribiliary fibrosis and necrotic debris occluding the intrahepatic bile ducts and impairing bile flow. There may also be adrenal cortical hyperplasia.

- general nursing care; prevention with zinc salts

Treatment—Recovery depends on general nursing care to allow the hepatic inflammation to subside and the hepatocytes to regenerate. The disease may be prevented or the severity reduced by prophylactic administration of zinc salts, but these are toxic, so they must be used with care (Smith & Towers 1985). Zinc oxide is given as a slurry drench, 1:2.5 wt/vol with water, at about 20 mg/kg b.w. per day in cattle. It can also be sprayed on the pasture. Zinc sulfate is administered in water at 30 g/cow per day. With either approach, administration can be daily to weekly. Zinc boluses for intraruminal use in cattle and sheep are available in New Zealand (Smith et al. 1998). Other studies have suggested a potential benefit for administration of molybdenum and sulfur, but the results were not conclusive (Gooneratne et al. 1994).

Panicum L.

Taxonomy and Morphology—Comprising 370–600 cool- and warm-season species, *Panicum,* a member of the subfamily Panicoideae and the tribe Paniceae, is the largest genus of the Poaceae. Its generic name is derived from the Latin *panus,* for a type of millet. It has relatively uniform spikelets but exhibits great variation in growth forms, with some species resembling woody shrubs and others diminutive pincushions. Leaf anatomy and photosynthetic pathways also vary considerably. Because of this variation, several classifications have been proposed. Some taxonomists recognize 1 genus and 3 subgenera, whereas others consider the 3 subgenera to be distinct genera. The broad circumscription of *Panicum* is employed in this treatise. Economically and ecologically, the genus is quite important. Its members are important components of the grasslands of North America. They provide considerable forage and hay for wildlife and livestock, and their caryopses also are an important part of the diet of many birds.

- *P. antidotale*
  P. coloratum
  P. dichotomiflorum
  P. maximum
  P. miliaceum
  P. virgatum

  blue panicum
  broomcorn millet
  broom millet
  brown corn
  common millet
  fall panicgrass
  fall panicum
  guineagrass
  hog millet
  Kleingrass
  proso millet
  Russian millet
  smooth witchgrass
  switchgrass

In North America, approximately 110 native and introduced species are present. Representative of the genus and associated with toxicological problems are the following:

P. antidotale Retz.	blue panicum
P. coloratum L.	Kleingrass
P. dichotomiflorum Michx.	fall panicum, smooth witchgrass, fall panicgrass
P. maximum Jacq.	guineagrass
P. miliaceum L.	common millet, broom millet, brown corn, hog millet, proso millet, Russian millet, broomcorn millet
P. virgatum L.	switchgrass

Plants annuals or perennials; cespitose or rhizomatous or stoloniferous. **Culms** 50–300 cm tall; erect or ascending or geniculate or decumbent; nodes glabrous or indumented; internodes hollow or solid. **Leaves** both basal and cauline or primarily cauline; forming or not forming a distinct basal rosette in late summer and fall. **Blades** flat or folded or involute; 1–30 mm wide. **Ligules** absent or present; membranous or ciliate-membranous or ciliate. **Auricles** absent. **Sheaths** glabrous or indumented; margins free, overlapping. **Inflorescences** panicles; open or contracted. **Disarticulation** below the glumes; at spikelet bases. **Spikelets** all alike; chasmogamous or cleistogamous; ovate to lanceolate or elliptic or obovate; compression dorsal; 1.4–6.5 mm long. **Florets** 2; of 2 forms; lower florets staminate or neuter; upper florets perfect. **Glumes** 2; unequal; different; lower glumes one-fourth to two-thirds the length of spikelets, broadly ovate to triangular or lanceolate, nerves 1 or 3 or 5; upper glumes similar to lower lemmas, ovate to lanceolate, nerves 5 or 7 or 9. **Lemmas of Lower Florets** ovate to lanceolate or obovate; membranous; abaxial surfaces glabrous or

- annuals or perennials; inflorescences panicles; spikelets dorsally compressed, with 2 florets; upper floret hard, shiny; glumes 2, unequal

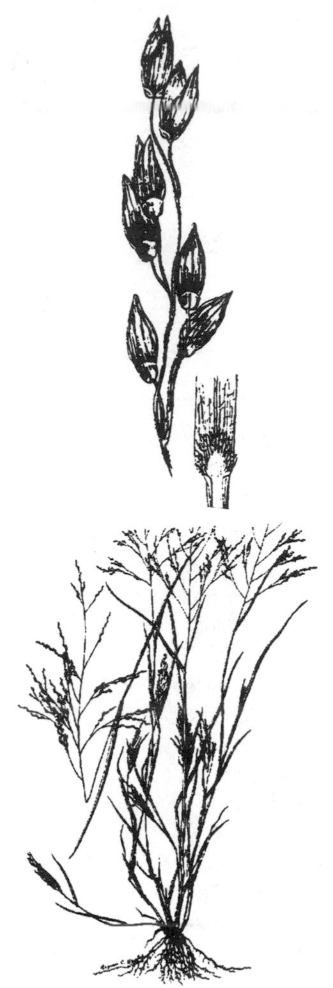

Panicum dichotomiflorum

- some are cultivated

indumented; awns absent. **Paleas of Lower Florets** absent or present; shorter than lemmas; hyaline. **Lemmas of Upper Florets** lanceolate or elliptic or suborbicular or obovate; indurate; abaxial surfaces smooth or rugose, glabrous, shiny. **Paleas of Upper Florets** present; as long as lemmas; tightly enclosed by lemma margins; indurate.

Identification of the numerous species of *Panicum* is based principally on vegetative features, e.g., presence or absence of a basal rosette of leaves formed in the autumn; branching of the culms; or the indumentum of sheaths. Because of the large number of species present in North America, a taxonomic key is not offered here. If exact identification is needed, floristic treatments specific for the reader's area should be used.

Distribution and Habitat—*Panicum* is pantropical in both hemispheres. Its species extend into the temperate regions of North America and occupy a variety of habitats, both natural and disturbed. Of the species listed above, *P. virgatum* and *P. dichotomiflorum* are native; the others are introduced from Africa or India.

- photosensitization, lambs, kids, 1 week or more of grazing mature grass, liver disease, crystalloid cholangiohepatopathy

Disease Problems—Species of *Panicum* are well known as causes of photosensitization in Australia and South Africa as well as North and South America. Now planted extensively as a livestock forage in south Texas, *P. coloratum* (Kleingrass) was introduced from South Africa into the United States in the 1950s, and cases of photosensitization began to be noted in the 1970s. They involved primarily lambs and kids and to a lesser extent mature sheep, Angora goats, and horses. The disease has not been reported to occur in cattle (Dollahite et al. 1977; Muchiri et al. 1980). The disease appears after animals have been grazing the pasture for a week or more and seems more likely to occur when the mature grass is in flower or seed; thus the highest incidence is from June to October. In horses the disease is reported to have occurred after animals have grazed Kleingrass for 1–3 years or experimentally been fed hay for 90 days (Cornick et al. 1988). The liver is clearly involved in the intoxication, but the lesions are not particularly severe or life threatening as in other types of liver disease.

The pathologic changes are consistent with those caused by *Agave lecheguilla* and *Nolina texana* (Agavaceae) in North America; *Tribulus terrestris* (Zygophyllaceae) in Australia, South Africa, and North America; and *Narthecium ossifragum* (Liliaceae) and *Brachiaria decumbens* (Poaceae) elsewhere in the world. Photosensitization caused by all of these species is associated with crystalloid cholangiohepatopathy. This implies liver involvement, but a specific and clear relationship between the lesions and the clinical signs remains to be shown.

Although plants of *Panicum* are sometimes cyanogenic, this is generally of little consequence.

- presence of calcium salts of glucuronidated saponins precipitated in bile ducts

Disease Genesis—Photosensitization is associated with but not necessarily caused by saponins such as dichotomin. Various species of the genus seem to be capable of causing the characteristic liver and skin changes: *P. coloratum, P. dichotomiflorum*, and *P. maximum* in North America and South Africa and *P. miliaceum* and *P. schinzii* in Australia and South America (Clare 1955; Bridges et al. 1987; Button et al. 1987; Bedotti et al. 1991). These species also have in common the presence of saponins of the furostanol and spirostanol types composed of diosgenin and/or yamogenin plus the requisite sugars (Patamalai et al. 1990; Miles et al. 1993; Munday et al. 1993). The characteristic crystals in the bile ducts have been shown to be calcium salts of saponins, but with a different structure from that of the parent saponins in the plants, and probably epismilagenin based (Holland et al. 1991; Lancaster et al. 1991; Miles et al. 1991, 1992a). It has been proposed that the parent genin, mainly diosgenin, is biotransformed in the body to smilagenin and then to epismilagenin, which is then glucuronidated to form the saponin. The saponin reacts with calcium and precipitates as the insoluble calcium salt in the bile ducts (Miles et al. 1992a). The precipitate is the calcium salt of epismilagenin β-D-glucuronide (Miles et al. 1992c).

The resulting disease is typified by characteristic needle-shaped biliary saponin crystals, which may wash out with alcohol fixation, leaving artifactual clefts. The liver pathology ranges from mild to moderate but is usually reversible if the animals are removed from the pasture containing *Panicum* at the first signs of disease.

Experimentally, diosgenin and smilagenin administered to rats resulted in icterus and crystalloid hepatopathy (Patamalai 1988). In sheep administered these compounds, 1 of 9 developed icterus, photosensitization, and crystals, while 3 of the 9 exhibited only crystals and a significant increase in serum GGT. Although a role for the saponins in the disease seems definite, additional factors may be required to cause liver damage. Thus, a role for sporidesmin or other hepatotoxins has been considered but not yet confirmed (Flaoyen et al. 1993). Additional discussion of this point has been presented in the treatment of *Tribulus* (see chapter 76). Even *P. virgatum* has been associated with photosensitization in lambs grazing it during an excessively hot, dry summer (Puoli et al. 1992). The disease was similar to that produced by other *Panicum* species. Skin and liver changes occurred, but crystalloid accumulations in the bile ducts were not observed.

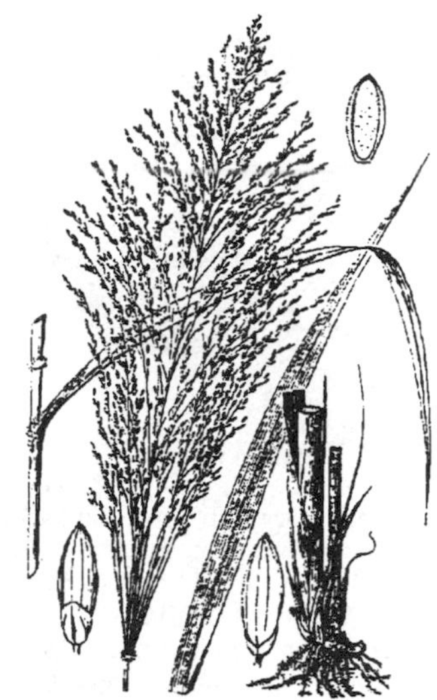

Panicum maximum

Dichotomin

Ca^{2+} salt of epismilagenin β-D-glucuronide

Clinical Signs—The signs are related primarily to necrosis of tissues around capillaries in the skin where UV light penetrates. The skin changes may develop quickly in 1–2 days or more gradually in a week or more. In sheep, most of the lesions are on the head and include swelling with marked loss of wool and sloughing of the skin on the ear tips, eyelids, around the eyes, near the lips and nose, and below the jaws. Serum oozes from these areas and dries, forming scabby crusts. There may also be reddening of the coronary bands, lameness, mild icterus, and depression. Serum hepatic enzymes such as GGT are elevated. If the animals continue grazing the pasture, an extensive number of deaths may occur. If moved out of the pastures, almost all affected animals will recover.

In horses there is chronic poor appetite and weight loss. Rarely there is hepatoencephalopathy with head pressing and periodic violent behavior. Serum GGT, blood ammonia, and total and direct bilirubin are markedly increased (Cornick et al. 1988). BSP clearance is slowed several-fold.

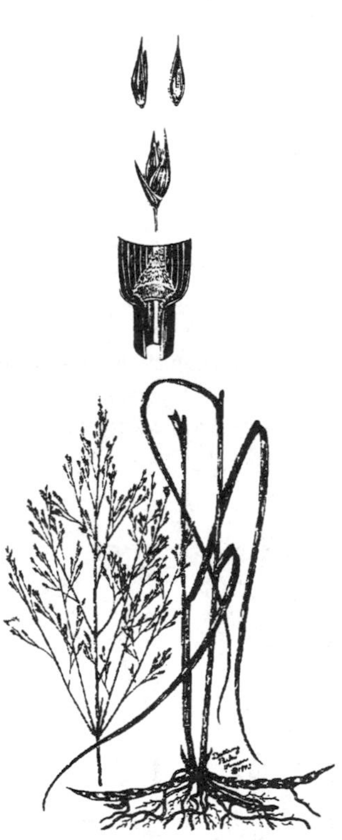

Panicum virgatum

- sheep, abrupt or gradual onset of skin lesions of ears, eyelids, face near lips and nose; loss of wool, serum oozing, sloughing, scabby crusts; depression, increased serum hepatic enzymes

Panicum virgatum

- gross pathology, obvious skin lesions

  microscopic, mild renal tubular necrosis; liver cell necrosis; numerous birefringent needlelike crystals or clefts in small bile ducts

- general nursing care, provide shade if possible

Pathology—Grossly, pathologic changes are limited to the skin as described above. Microscopically, there is mild necrosis of the proximal convoluted tubules of the kidneys, focal myocarditis and hemorrhages of the heart, and more pronounced degenerative changes in the liver. In the liver, there is swelling and necrosis of some hepatocytes, inflammation of the bile ducts, and numerous birefringent needlelike crystals or their artifactual clefts in the small bile ducts, bile canaliculi, and Kupffer cells. Clefts rather than the crystals may be observed, because the solubility of the crystals causes them to disappear from the slides in the mounting procedures. Some of the crystals or clefts are surrounded by phagocytes. Morphologically, the crystals resemble those of cholesterol (Bridges et al. 1987).

Treatment—The signs abate readily if the affected animals are removed from the offending forage. Placing the animals in shade, and symptomatic relief of the discomfort that accompanies sloughing of the skin as well as pruritis, are also helpful. There is little treatment for the liver disease other than providing general nursing care. Careful management to avoid grazing lambs or kids on *Panicum* has greatly minimized the problems associated with this forage (Bailey 1985).

Paspalum L.

- *P. dilatatum*
  P. notatum

  Bahiagrass
  Dallisgrass
  large watergrass

- annuals or perennials; inflorescences 1-sided racemes; spikelets planoconvex, with 2 florets; upper floret hard, shiny; glumes 1

Taxonomy and Morphology—Comprising 300–330 warm-season species, *Paspalum,* a member of the subfamily Panicoideae and the tribe Paniceae, is predominantly New World in distribution. Its generic name is derived from the Latin word *paspalos,* which means "millet." Several species provide excellent fodder and forage, others are weeds, and a few are cultivated as ornamentals. In North America, some 60 native and introduced species are present. With the exception of a few taxa, the native species are of little economic importance. Two species introduced from South America, however, provide considerable forage and are of toxicologic importance:

P. dilatatum Poir.	Dallisgrass, large watergrass
P. notatum Fluegge	Bahiagrass

Plants perennials or annuals; cespitose or rhizomatous or stoloniferous. **Culms** 10–200 cm tall; erect or ascending or geniculate or decumbent; not branched above; nodes glabrous or indumented; internodes hollow or solid. **Leaves** basal and cauline. **Blades** flat or folded or involute; 1.5–20 mm wide. **Ligules** membranous or ciliate. **Auricles** present or absent. **Sheaths** glabrous or indumented; margins free, overlapping. **Inflorescences** 1-sided racemes; racemes 1–20(60), borne at peduncle apices or attached laterally to an elongate axis. **Disarticulation** below the glumes; at spikelet bases. **Spikelets** all alike; planoconvex; ovate to suborbicular or obovate; compression dorsal; 1.1–5 mm long. **Florets** 2; of 2 forms; lower florets neuter; upper florets perfect. **Glumes** 1; lower glumes absent; upper glumes similar to lower lemmas, ovate to suborbicular or obovate, nerves 2 or 3 or 5. **Lemmas of Lower Florets** ovate to suborbicular or obovate; membranous; nerves 0 or 1 or 3 or 5; abaxial surfaces glabrous or indumented; awns absent. **Paleas of Lower Florets** absent. **Lemmas of Upper Florets** elliptic to suborbicular or obovate; indurate; abaxial surfaces smooth, glabrous; shiny. **Paleas of Upper Florets** present; as long as lemmas; tightly enclosed by lemma margins; indurate or coriaceous.

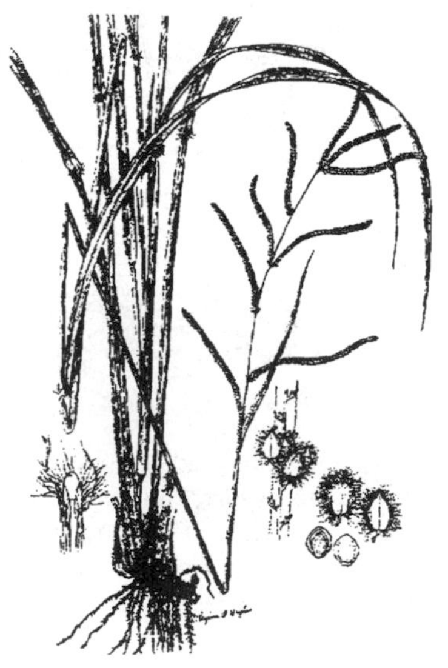
Paspalum dilatatum

1. Plants rhizomatous. Racemes 2 or 3 per peduncle, paired or whorled at apex. Spikelets borne in 2 rows, solitary . . . . *P. notatum*

1. Plants cespitose. Racemes 3–7 per peduncle, borne laterally along central axis. Spikelets borne in 4 rows, paired . . . . *P. dilatatum*

Distribution and Habitat—*Paspalum* is most abundant in the southeastern United States. Plants are characteristically encountered in damp or wet soils. Introduced from South America in the mid 1800s, *P. dilatatum* is now a valuable pasture grass throughout the south. It also may occur as a troublesome weed in lawns and moist disturbed sites. Also introduced into the southeastern states for forage and erosion control, *P. notatum* is native to Central and South America.

■ cultivated pasture plants

Paspalum dilatatum

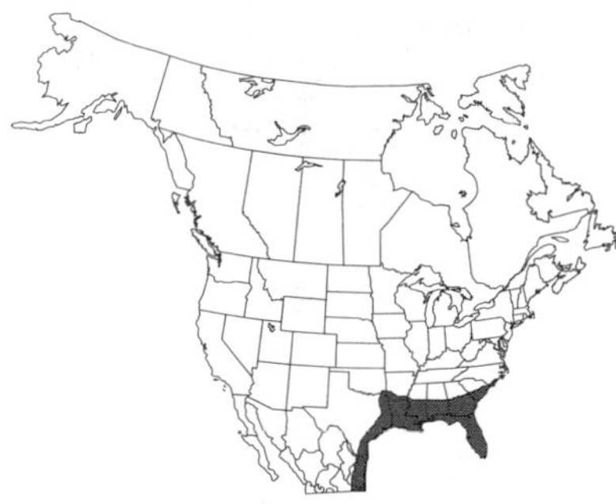

Paspalum notatum

Disease Problems—When infected with the endophytic fungus *Claviceps paspali*, species of *Paspalum* cause a tremorgenic syndrome known as staggers. This disease and the involvement of *C. paspali* have long been known to occur, from early reports in South Africa (Steyn 1933a). The disease is characterized by nervousness and tremors, increasing in severity within a few days to the point of incapacitation (Brown & Ranck 1915). Although sheep are susceptible, cattle are most likely to develop the disease (Mantle et al. 1977). This is due in large measure to their predisposition for eating the inflorescences, which are infected with the fungus and which are not ordinarily eaten by sheep and horses (Brown & Ranck 1915). The hazard increases with plant maturity and is greatest in the late summer or fall, especially after rainy or humid weather. It has been noted that some cows become prone to eating the diseased plants and appear to select for them (Brown & Ranck 1915). Toxicity potential is not decreased with drying of the plants, and thus hay is also a risk (Tyler et al. 1992). Onset of signs may occur within 1 to several days after cattle are introduced to a new pasture containing infected plants (Goodwin 1967; Noble 1985). Experimentally, 250 g of infected seed heads daily caused marked neurointoxication in cattle in 4 days (Lopez et al. 1986). Morbidity is low to moderate, with only occasional outbreaks exceeding 50%. With high dosage of the toxic forage, the effects may be lethal, but typically with onset of neurologic signs, ingestion of the infected forage ceases or is much reduced. Thus, case-fatality rates are negligible; most deaths are due to misadventures caused by the severe incoordination produced. The mere act of drinking seems to precipitate seizures, causing animals to sometimes fall into ponds (Brown & Ranck 1915). A case-fatality rate of greater than 10% should prompt a search for other causes of intoxication, e.g., lead poisoning.

■ neurologic problem, staggers syndrome, seizures, cattle mainly; for several days eat mature plants with caryopses/seeds present; increased incidence with rainy summers; moderate morbidity, low mortality

Disease Genesis—Species of *Paspalum* are parasitized by members of *Claviceps*, a fungal genus that also contains *C. purpurea*, the cause of ergot of rye and the disease known as ergotism. *Claviceps paspali* infects *Paspalum*. The fungus overwinters in the soil and releases spores at the time *Paspalum* is flowering. The airborne spores lodge among the flowers in the spikelets, germinate, and send hyphae into the ovaries. A few days after infection, a copious, sweet exudate termed honeydew appears; it is the result of abundant conidia formation. The fungal tissue gradually hardens and is transformed into a dark brown or black structure, known as the sclerotium or ergot, which resembles the caryopsis or grain of *Paspalum*. Proliferation of the fungal mycelium forces open the bracts surrounding the infected flowers, thus exposing the sclerotia. The sclerotia mature at the same time as the uninfected caryopses ripen, and both are harvested together. The fungal life cycle is completed when the sclerotia fall to the ground. The number of spikelets infected varies considerably; large numbers typically are affected in rainy or humid summers.

■ *Claviceps* fungi invade developing plant ovaries and form sclerotia containing neurotoxic indolic tremorgens, paspalinine, paspalitrems, inhibit GABA and glycine inhibitory pathways

Although *Claviceps* is infamous for the production of alkaloids responsible for ergotism in humans (St. Anthony's fire), it appears that paspalum staggers is caused

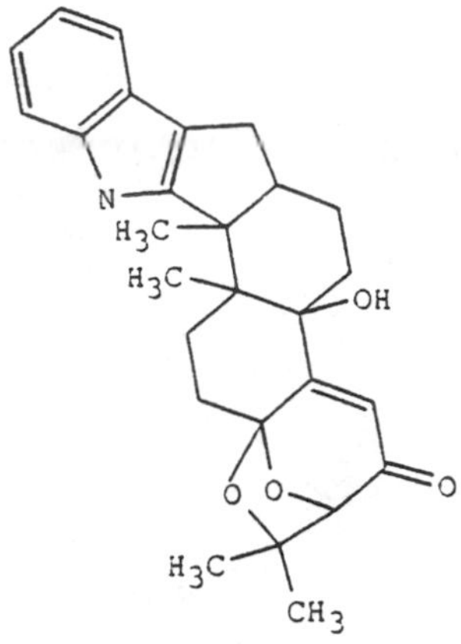

Paspalinine

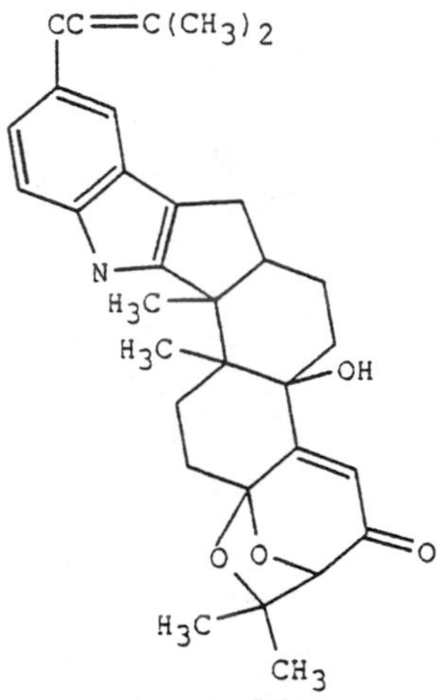

Paspalitrem A

by another group of neurotoxins (Groger et al. 1961). The sclerotia contain traces of the typical indole *Claviceps*-type toxins, but much larger amounts of indolic tremorgens, up to 1% d.w., are present (Mantle & Penney 1981). These indole derivatives include the nontremorgenic paspaline and paspalicine alkaloids (Gallagher et al. 1980b; Cole & Dorner 1986). Although not tremorgenic, paspalicine, like the other indole alkaloids, inhibits high conductance Ca^{2+}-activated K^{+} channels (Knaus et al. 1994). The tremorgenic paspalinine and its derivative paspalitrems A, B, and C are chemically very similar to the toxins causing ryegrass staggers (see treatment of *Lolium,* in this chapter). Based on the type of clinical signs observed, including tremors, it has been suggested that these tremorgens may impair inhibitory pathways in the nervous system, especially those mediated by GABA or glycine (Mantle & Penney 1981). This possibility has been examined in detail, and it appears that the fungal tremorgens inhibit GABA receptor function by binding close to the chloride channel (Gant et al. 1987). Peripheral nerve conduction seems to be unaffected by the alkaloids (Tyler et al. 1992).

The syndromes caused by *Lolium*, *Cynodon*, and overwintered, lodged *Zea* are essentially the same in appearance as the syndrome due to *Paspalum.* There may be some difference among these syndromes in the duration of signs because some of the alkaloids are eliminated by both liver metabolism and biliary secretion whereas paspalinines appear to be eliminated only by bile excretion (Mantle 1986). This leaves open the possibility of enterohepatic recycling's perpetuation of the activity of the alkaloids.

- increased nervousness, seizures with ears erect, fine tremor of head and neck to shoulders; move with stiff, wobbly swaying movement of pelvic limbs; collapse, head twisted backward

Clinical Signs—Signs usually appear several days after the animals have been introduced into a pasture containing infected plants. Initially there is increased nervousness, a "wild" facial expression with the ears held erect, and tremors of the lips, face, and ears, which rapidly progress posteriorly to tremors of the neck, shoulders, and pelvic limbs. The neck tremors may be accompanied by nodding head movements. Incoordination is prominent, with the animals exhibiting a swaying, stumbling, stiff-legged gait. When the animals are stationary, the pelvic limbs are typically spread and extended. Exercise markedly intensifies the head nodding and incoordination. Some animals may be easily excited and/or belligerent. The convulsive episodes may cause the animals to fall, but they are usually able to stand again in a few minutes. In more severe cases, the animals remain recumbent and may hold the head back in a twisted posture. Except for the most severely affected, they continue to eat and drink. Deaths are rare and usually through misadventure such as drowning or from dehydration and starvation.

- no lesions

Pathology—There are no distinctive pathologic changes in the central or peripheral nervous systems. In some animals there may be prominent congestion of the brain and meninges and a few small splotchy hemorrhages in the interstitium of the kidney. Observable effects are those related to the incapacitation caused by the toxin, i.e., dehydration and starvation.

- remove from pasture, provide water if recumbent

Treatment—If animals are moved from the infected pasture when signs are initially noted, the disease is usually readily reversible within a few days. Although recovery is usually uneventful, signs may remain inducible for several days thereafter, but recovery will be complete in 7–10 days (Brownie & Prassad 1987; Galey et al. 1991). In cases in which the animals are temporarily incapacitated, recovery may be facilitated by placing feed and water within reach. In severe cases, sedation may be helpful.

Management practices that ensure that the animals eat young, growing plants rather than mature ones with infected spikelets are effective in preventing appearance of the disease. However, once a severe infestation of *Claviceps* occurs in a paspalum pasture, subsequent infestations are likely whenever environmental conditions are appropriate, because the sclerotia continue to contaminate soil.

Pennisetum Rich.

Taxonomy and Morphology—A genus of about 80 warm-season species in the tropics and subtropics of both hemispheres, *Pennisetum* is a member of the subfamily Panicoideae and the tribe Paniceae. Its generic name is derived from the Latin words *penna* and *seta,* meaning "feather" and "bristle," which allude to the plumose bristles of some species. Morphologically quite similar to *Setaria* (bristlegrass) and *Cenchrus* (sandbur), its generic boundaries have been debated by taxonomists. Several species are used for forage and fodder. Most important is *P. glaucum* (pearl millet), which has been cultivated for human food as well as fodder for livestock. The most drought tolerant of the tropical cereals, it is believed to have been domesticated about 2000–3000 B.C. In addition, several species of the genus are popular garden ornamentals as well as being used in dried floral arrangements. Two disease problems—oxalate accumulation and nitrate intoxication—are produced by species of the genus. In North America, 5–13 introduced species are present. The following are those of toxicologic interest:

P. ciliare (L.) Link (= *Cenchrus ciliaris* L.)	buffelgrass
P. clandestinum Hochst. ex Chiov.	Kikuyugrass
P. glaucum (L.) R.Br. (= *P. americanum* [L.] Leeke)	pearl millet, horse millet, bulrush millet, spiked millet
P. purpureum Schumach.	napiergrass, elephantgrass

Two binomials—*P. glaucum* and *P. americanum*—continue to be used for pearl millet because of problems in Linnaeus's original description of the species and disagreement among taxonomists as to how the problems should be resolved.

Likewise, two binomials—*P. ciliare* and *Cenchrus ciliaris*—continue to be used for buffelgrass because of different taxonomic opinions as to how to divide the two genera. Numerous named cultivars of *P. ciliare*—e.g., 'Milgro,' 'Millex,' 'T-E Horsepower,' 'T-E Milgrazer'—have been developed.

Plants annuals or perennials; cespitose or rhizomatous or stoloniferous. **Culms** 30–250 cm tall; erect or ascending or geniculate; nodes glabrous or indumented; internodes hollow or solid. **Leaves** cauline or cauline and basal. **Blades** flat or folded or involute; 3–50 mm wide. **Ligules** ciliate-membranous or ciliate. **Auricles** absent. **Sheaths** glabrous or villous; margins free, overlapping. **Inflorescences** spicate panicles; spikelets borne in fascicles of 1–3 on short pedicels or branches subtended by 20–60 bristles (sterile panicle branches); bristles equal to or longer than spikelets, barbellate or plumose. **Disarticulation** below the glumes; spikelets and subtending bristles separating from rachis and falling as a unit. **Spikelets** all alike; lanceolate to oblong; compression dorsal; 3–11 mm long. **Florets** 2; of 2 forms; lower florets staminate or neuter; upper florets perfect. **Glumes** 1 or 2; unequal; different; hyaline; abaxial surfaces glabrous; lower glumes absent or to half the length of spikelets, broadly ovate, nerves 0 or 3 or 5; upper glumes one-fourth to equal the length of

■ *P. ciliare*
P. clandestinum
P. glaucum
P. purpureum

buffelgrass
bulrush millet
elephantgrass
horse millet
Kikuyugrass
napiergrass
pearl millet
spiked millet

■ annuals or perennials; inflorescences spicate panicles; spikelets borne singly or clustered, subtended by bristles, with 2 florets; upper floret leathery or papery

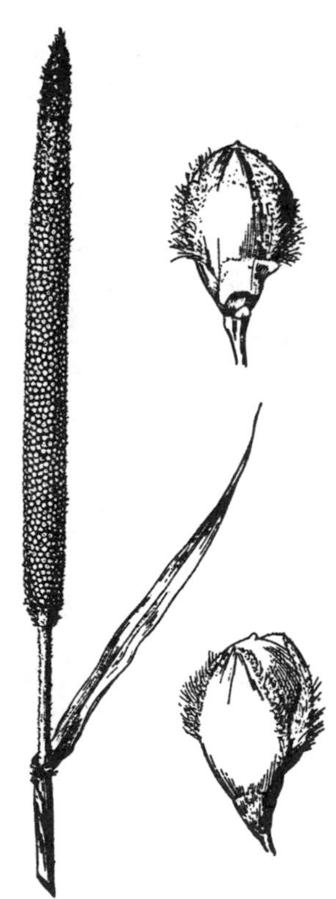

Pennisetum glaucum

Pennisetum ciliare

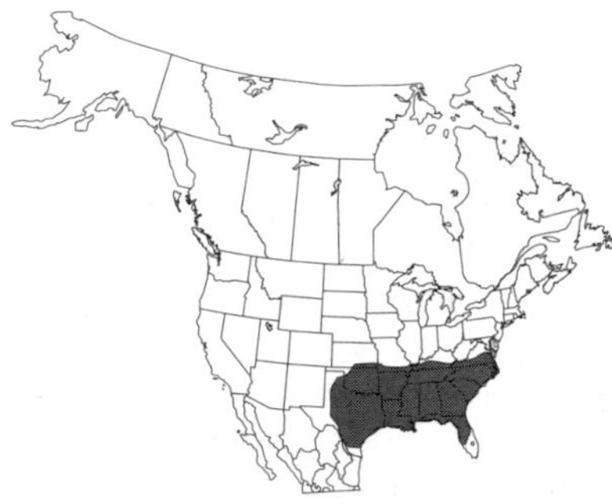

Pennisetum glaucum

lower lemmas, lanceolate to ovate, nerves 3 or 5. **Lemmas of Lower Florets** lanceolate; membranous; nerves 3 or 5 or 7; awns absent. **Paleas of Lower Florets** absent or present; as long as lemmas, hyaline. **Lemmas of Upper Florets** lanceolate or oblong to elliptic; chartaceous to coriaceous; nerves 5 or 7; abaxial surfaces smooth, glabrous or scabrous or villous. **Paleas of Upper Florets** present; as long as or longer than lemmas; chartaceous to coriaceous.

1. Bristles fused at bases to form inconspicuous disk . . . . *P. ciliare*
1. Bristles free at bases.
   2. Plants stoloniferous and rhizomatous. Panicles hidden; axillary; enclosed in uppermost sheaths . . . *P. clandestinum*
   2. Plants cespitose. Panicles visible; terminal; not enclosed in sheaths.
      3. Plants annuals. Involucral bristles as long as spikelets . . . . *P. glaucum*
      3. Plants perennials. Involucral bristles 2–5 times longer than spikelets . . . . *P. purpureum*

Distribution and Habitat—*Pennisetum* occupies a variety of habitats in savannas, woodlands, and disturbed areas. All North American species are introduced and typically encountered only in the southern states and Mexico because they are not conspicuously cold-hardy.

Oxalate Intoxication

- some cultivated for feed
- horses mainly, oxalate-induced Ca/P imbalance, bony changes, osteodystrophy

Disease Problems and Genesis—Reported mainly in Australia, oxalate accumulation is the cause of a chronic disease primarily of horses grazing lush pastures of *Pennisetum* for prolonged periods (Everist 1981; Seawright 1989). Associated with it are bone changes manifested as lameness and/or swellings of the face and jaw. Rarely is it a cause of acute oxalate intoxication in horses, sheep, and cattle. The disease has not yet been reported as a problem in North America. Concentrations of oxalates in the plants may vary up to 3.5% total and 2.2% soluble salts. The oxalate is bound to calcium in the plants and is relatively unavailable for absorption or as a cause of direct oxalate intoxication. This binding causes a reduction in calcium availability and thus the possibility of a calcium/phosphorus imbalance and fibrous osteodystrophy secondary to actions via the parathyroid.

- horses, lameness, weight loss, bony changes

  treat with calcium supplements

Clinical Signs, Pathology, and Treatment—The characteristic signs of oxalate intoxication in horses are lameness, weight loss, and prominent bony swellings of the mandibles and maxilla. Calcium supplementation in the diet will both prevent and alleviate the clinical signs.

Nitrate Accumulation

- high risk of dangerous nitrate accumulation, especially with dry weather

Disease Problems and Genesis—*Pennisetum glaucum* (pearl millet) is a widely cultivated annual forage of high nutritive value (Ball et al. 1991). It has gained favor in part because of its very low cyanogenic potential. However, this value is offset by its particularly hazardous propensity to accumulate nitrate (Selk et al. 1995). In an extensive field study, at three locations in Oklahoma and over a 2-year period, pearl millet was observed to consistently accumulate nitrate to a greater extent than sorghum forages grown under similar conditions (Selk et al. 1995). The comparison included 35 sorghum-Sudangrasses, 21 sorgo-Sudan grasses, 7 Sudan-Sudans, and 11 pearl millet varieties all grown under stress conditions. Thus, the use of pearl millet requires careful attention to growing conditions to avoid disastrous intoxications. Following a prolonged dry period, forage should be harvested about 2 weeks after a substantial rain if possible (Strickland et al. 1995). Additional discussion of nitrate intoxication is presented in the treatment of *Sorghum,* in this chapter.

Clinical Signs, Pathology, and Treatment—Signs occur with abrupt onset several hours after consumption of *Pennisetum* forage and quite often are fatal within minutes or a few hours. The primary effects are weakness, labored respiration, and possibly dark-colored mucous membranes. Few distinctive lesions are seen at necropsy. The blood and tissues may be dark or chocolate-colored, but this is not consistent. There also may be scattered, small splotchy hemorrhages and increased foamy fluid in the terminal airways of the lungs, but these changes are not usually extensive. The primary antidote is methylene blue, 2–4 mg/kg b.w. given as a 1–2% solution.

■ abrupt onset, weakness, labored respiration, dark mucous membranes

no lesions

treat with methylene blue

Phalaris L.

Taxonomy and Morphology—Comprising 15 or 16 cool-season species, *Phalaris,* a member of the subfamily Pooideae and the tribe Aveneae, occurs in the temperate regions of the both the Old World and the New World. Its species are cultivated for fodder, for birdseed, and as ornamentals. The inflorescences are often used in dried floral arrangements. In North America, 10 native and introduced species are present. Of toxicologic significance are the following:

P. angusta Nees ex Trin.	timothy canarygrass
P. aquatica L. (= *P. tuberosa* L.)	hardinggrass, toowoomba
P. arundinacea L.	reed canarygrass, ribbongrass, gardener's garters
P. canariensis L.	canarygrass, birdseedgrass
P. caroliniana Walter	Carolina canarygrass, southern canarygrass, Maygrass
P. minor Retz.	littleseed canarygrass

■ *P. angusta*
P. aquatica
P. arundinacea
P. canariensis
P. caroliniana
P. minor

birdseedgrass
canarygrass
gardener's garters
hardinggrass
Maygrass
ribbongrass
toowoomba

Plants annuals or perennials; cespitose or rhizomatous. **Culms** 10–200 cm tall; erect; not branched above; nodes glabrous; internodes hollow. **Leaves** cauline. **Blades** flat; 2–20 mm wide. **Ligules** membranous; entire or lacerate. **Auricles** absent. **Sheaths** glabrous; margins free, overlapping. **Inflorescences** spicate panicles. **Disarticulation** above the glumes; neuter florets dropping attached to perfect florets. **Spikelets** all alike; elliptic or oblanceolate; compression lateral; 3–10 mm long. **Florets** 2 or 3; of 2 forms; lower florets 1 or 2, neuter; upper floret 1, perfect. **Glumes** 2; equal or subequal; similar; longer than lemmas; nerves 3; backs keeled; wings present or absent; apices acuminate or acute; awns absent. **Lemmas of Neuter Florets** subulate or lanceolate; shorter than lemmas of perfect florets. **Paleas of Neuter Florets** absent. **Lemmas of Perfect Florets** lanceolate; coriaceous; nerves 5; abaxial surfaces sericeous or glabrous, shiny; awns absent. **Paleas of Perfect Florets** coriaceous; nerves 1 or 2; backs rounded.

■ annuals or perennials; ligules membranous; inflorescences spicate panicles; spikelets laterally compressed, with 1 fertile floret; glumes 2, longer than lemmas

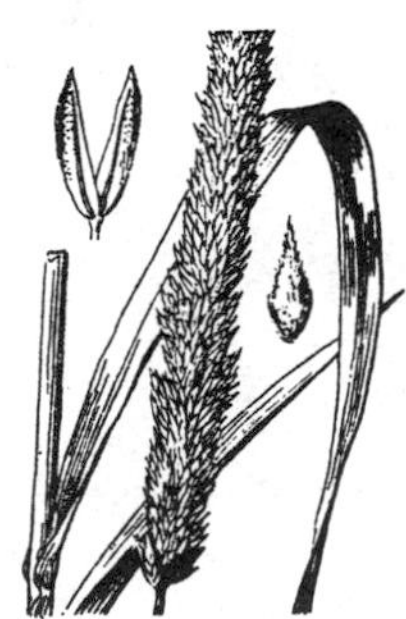

Phalaris angusta

1. Plants perennials. Rhizomes present.
   2. Glume keels winged. Neuter florets 1 . . . . *P. aquatica*
   2. Glume keels not winged. Neuter florets 2 . . . . *P. arundinacea*

Phalaris arundinacea

Phalaris canariensis

Phalaris caroliniana

Phalaris minor

- some used as pasture grasses
- 3 syndromes—
  sudden death
  staggers
  ARDS

1. Plants annuals. Rhizomes absent.
   3. Neuter florets 1 .... *P. minor*
   3. Neuter florets 2.
      4. Perfect florets 5–6 mm long. Neuter florets 2–3 mm long; lanceolate; glabrous to sparsely sericeous ...... *P. canariensis*
      4. Perfect florets 3–4 mm long. Neuter florets 1–2 mm long; subulate; densely sericeous.
         5. Panicles elliptic; 2–6 cm long. Lateral nerves of glumes glabrous or sparsely scaberulous, scaberules 0–5 .... *P. caroliniana*
         5. Panicles cylindrical; 7–15 cm long. Lateral nerves of glumes densely scaberulous, scaberules 9 to many .. *P. angusta*

Distribution and Habitat—The Mediterranean region and California appear to be centers of species diversity for *Phalaris.* Plants are encountered in dry weedy sites or in the moist soils of stream banks and low-lying areas of meadows. The caryopses of *P. canariensis* provide the canary seed of commerce.

Disease Problems—First reported as a problem in the early 1940s, species of *Phalaris* have proven to be important causes of disease in Australia (McDonald 1942; Gallagher et al. 1966, 1967a; Kerr 1972). They are much less of a problem in North America, although several of the species found here are clearly toxic (East & Higgins 1988; Nicholson 1989; Nicholson et al. 1989; Odriozola et al. 1991). Several different disease syndromes have been associated with livestock grazing *Phalaris* pastures in Australia. As suggested by Bourke and co-workers (1988), they seem to be resolvable into two types—a sudden death/cardiac syndrome and a neurologic/staggers syndrome. Both syndromes may occur in one group or even in one individual and in other regions of the world, including North America. The effects of *Phalaris* are seen mainly in sheep, although cattle also are affected. In addition to these syndromes, ARDS has also been associated with *Phalaris* as well as other forages.

Sudden Death/Cardiac Syndrome

Disease Problems—Signs of cardiac failure quickly terminate in death, or death with few premonitory signs occurs 1–2 days after hungry sheep are turned into a canarygrass or mixed pasture or, in some cases, within a few days after the pasture undergoes a flush of growth following rains. Young growing grass seems to be the common factor. Death occurs due to ventricular fibrillation and cardiac arrest and is not preventable by cobalt supplementation. This syndrome may also occur in other animal species and is suspected as a cause of death of a horse (Colegate et al. 1999).

Phalaris aquatica

Disease Genesis—*Phalaris* contains at least two types of potentially toxic alkaloids: indole tryptamine types, including β-carbolines and gramine, and tyramine types such as hordenine (Culvenor et al. 1964; Gallagher et al. 1966; Frahn & O'Keefe 1971). A full array of these compounds is not present in all plants. Sometimes the tryptamines, including carbolines, are found alone, and in other plants both tyramine (hordenine) and tryptamine types are present (Gander et al. 1976). The array seems to be consistent for all of the plants in a population (Marten et al. 1973). As is apparent from their structures, the tryptamine types are quite similar, and they may represent different steps in a synthesis sequence. The interrelationship is shown by the presence in the same plant of either dimethyltryptamine (DMT) or its 5-methoxy derivative (5-MDMT) and the respective carbolines or methoxycarboline (Simons & Marten 1971; Marten et al. 1973; Gander et al. 1976).

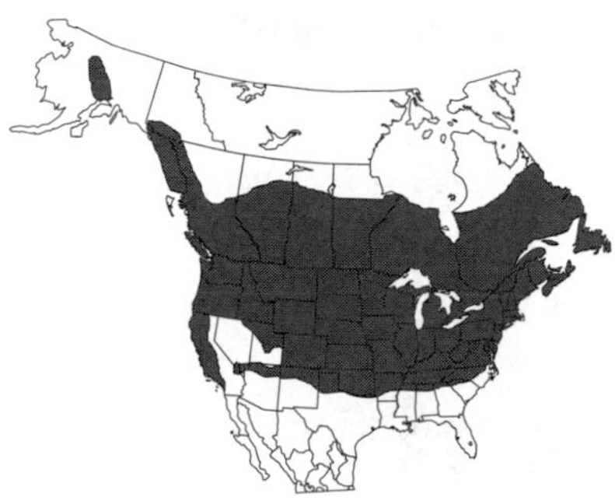

Phalaris arundinacea

Ranging from 100 to 27,500 ppm, total alkaloid concentrations are quite variable (Simons & Marten 1971). Gramine may range from a mean of 2,000 ppm up to 10,000 ppm for some populations (Coulman et al. 1977). Concentrations of 5-methoxy-*N*-methyltryptamine (5-MMT) may be as high as 4,000 ppm in some circumstances, yet sometimes under field conditions—e.g., wet meadows in British Columbia—total alkaloid levels may be very low, with a maximum of 1,000 ppm and a mean of 100 ppm (Majak et al. 1979). Tryptamine concentrations are high in young leaves, especially regrowth after harvest and after fertilization with ammonium salts. Levels typically decrease after frosts. The alkaloids in total seem to have a protective effect for the plants, given the strong negative correlation between alkaloid concentration and palatability (Simons & Marten 1971; Marten et al. 1973). Low or reduced alkaloid concentrations, especially in the range of less than 2,000 ppm, are associated with improved weight gains, whereas higher levels may cause diarrhea (Marten et al. 1981).

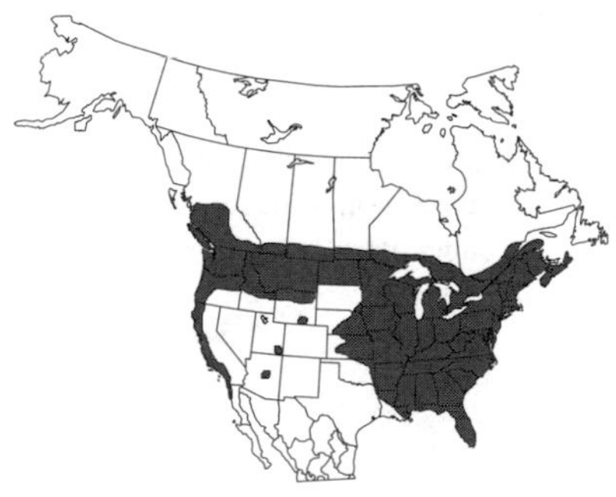

Phalaris canariensis

- sudden death due to cardiac failure, a few days after grazing a new pasture or after a flush of new grass growth
- alkaloids present—

  indole tryptamine types, tryptamine, β-carbolines, gramine

  tyramine types, hordenine, DMT, 5-MDMT

Early studies indicated that the tryptamine alkaloids could possibly account for the cardiorespiratory effects and sudden death (Gallagher et al. 1964, 1966, 1967b). This now seems unlikely, because the syndrome is seen with low-alkaloid varieties and is otherwise poorly correlated with alkaloid concentrations (Oram 1970; Kennedy et al. 1986). Although Gallagher and co-workers (1964) were able to show cardiorespiratory effects, these have subsequently proven difficult to reproduce experimentally with tryptamines (Bourke et al. 1988). Toxicants such as cyanide, nitrate, and other compounds may be contributing factors inasmuch as nitrate concentrations of 2,920 ppm and HCN potential of 360 ppm are reported (Bourke & Carrigan 1992). In some animals, polioencephalomalacia occurs as a complicating factor.

The tyramine alkaloids may be major factors in the cardiorespiratory effects. *N*-Methyltyramine, isolated from *P. aquatica* in pastures after an outbreak of sudden death, has been shown to cause serious cardiac effects in sheep with oral doses of 300–400 mg/kg b.w. (Anderton et al. 1994). Although deaths did not occur in the experimental animals, it was suggested, based on experimental findings, that inhibition of monoamine oxidase such as might occur in the pasture, would potentiate and markedly worsen the animals' response to tyramines. Tyramine and the monomethylated *N*-methyltyramine, but not the dimethylated hordenine, seem to be plausible factors in development of the cardiotoxicosis (Anderton et al. 1994). Thus the intoxication potential of forages may be assessed by using various analytical chromatographic procedures (Anderton et al. 1999).

- cardiac arrhythmias, respiratory distress, collapse, death, quickly

Clinical Signs—In the peracute sudden death syndrome, the primary signs are cardiac arrhythmias and insufficiency and respiratory distress, collapse, and death, all in a matter of a few hours. Death in many animals appears to be due to ventricular fibrillation and cardiac arrest. The signs are often precipitated by the stress of forced movement or during shearing.

- no lesions for cardiac failure

Pathology—Few changes are seen, other than congestion of the lungs and other viscera and perhaps small hemorrhages on the surfaces of the heart. Microscopically, when polioencephalomalacia occurs and complicates the pathology, there is perivascular and perineuronal vacuolation of the laminae of the cerebral cortex (especially the deep), with shrunken and acidophilic neurons. There is also cytoplasmic neuronal vacuolation of Purkinje's and other cells of that layer of the cerebellum (Bourke & Carrigan 1992).

- cobalt is not protective; treat based on signs

Treatment—Cobalt is not protective against the sudden death syndrome. Treatment is directed at symptomatic alleviation of signs. It may be of preventive benefit to avoid grazing pastures with rapidly growing plants or at least not allowing hungry animals access to rapidly growing forage.

Staggers/Nervous Syndrome

- acute or chronic disease; periodic seizures; in some instances seizures may begin after animals are removed from pasture

Disease Problems—The neurologic disease due to *Phalaris* may be acute or chronic; it also may begin as an acute disease and then persist as a chronic state (Gallagher et al. 1964). Intoxication may occur even when *Phalaris* forage makes up only half of the diet (Sousa & Irigoyen 1999). The syndrome is characterized by periodic seizures that may persist or, in some animals, begin weeks or months after removal from the pasture. The signs are variable and sometimes cyclic in intensity and may be more or less permanent. This form of *Phalaris* intoxication is preventable with cobalt administration. Cobalt appears to attenuate neurotoxin production or increase its destruction in the rumen (Lee & Kuchel 1953; Lee et al. 1956).

- tryptamine alkaloids, potentiate serotonin and catecholamine activity, cattle and sheep

C-N$(CH_3)_2$

N

Gramine

Disease Genesis—The tryptamine alkaloids DMT, 5-MDMT, and possibly bufotenine (5-OH-DMT) appear to be the most likely causes of this syndrome, although involvement of the β-carboline alkaloids cannot be excluded. Gramine and hordenine appear not to have sufficient potency to contribute appreciably to the overall toxic effects, although all of the alkaloid types are capable of producing neurologic signs similar to those caused by the plants (Bourke et al. 1988). Tryptamine alkaloids, especially 5-MDMT, are serotonin receptor agonists, while the β-carbolines are generally regarded as monoamine oxidase inhibitors that thereby potentiate serotonin and catecholamine activity (Montigny & Aghajanian 1977; McGeer et al. 1978). These mechanisms are consistent with those proposed for the neurologic effects of *Phalaris* (Gallagher et al. 1966). The signs produced in sheep are similar to the serotoninergic syndrome of tremors, rigidity, head weaving, and hyperreactivity produced in rats by the serotonin precursor L-tryptophan (Jacobs & Klemfuss 1975). Thus both the signs and the distribution of pathologic changes affecting specific nuclei are quite consistent with effects of the alkaloids on serotoninergic receptors in the brain and spinal cord (Gallagher et al. 1966; Bourke et al. 1990). The granular intracellular pigments appear to be a consequence of the accumulation of the toxins and not the cause of the clinical signs (Bourke et al. 1990). The tryptamines and/or carbolines also are associated with mild diarrhea in sheep (Marten et al. 1981).

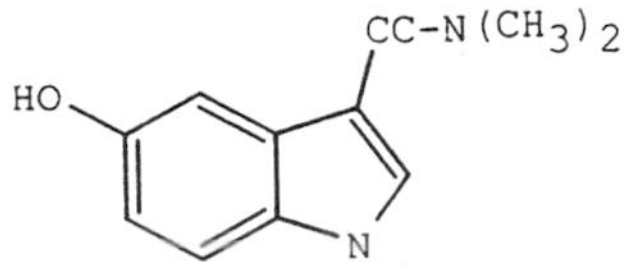

A disease with similar signs and pathology has been observed in feedlot sheep without exposure to *Phalaris* but with low levels of tryptamine alkaloids in other feed materials (Lean et al. 1989). This is not unexpected because tryptamine and related compounds are present in a large number of plants (Smith 1977).

The tryptamine alkaloids are apparently unstable under some conditions, because they are not a problem in hay (Gallagher et al. 1966). They also are a problem only in sheep and cattle, not horses. Significant degradation of these alkaloids apparently does not occur in the ruminal environment. The preventive effect of cobalt supplementation in ruminants appears to involve increasing ruminal destruction of the alkaloids (Lee & Kuchel 1953; Lee et al. 1956). Alkaloid concentrations in the plants may be influenced by soil type and increased by nitrogen fertilization and probably other environmental conditions (Gallagher et al. 1966, 1967a). Alkaloid concentrations are also subject to genetic manipulation, given that some strains and cultivars may have levels as little as one-tenth those in others (Kennedy et al. 1986). In addition, the array of alkaloid types may vary; some strains have only one type, such as gramine, present (Marten et al. 1981; Wittenberg et al. 1992). Because of the great interest in Australia in plant breeding to reduce toxicity of the *Phalaris*, ELISAs have been developed for identification of dimethyltryptamines and *N*-methyltyramines (Skerritt et al. 2000).

Bufotenin

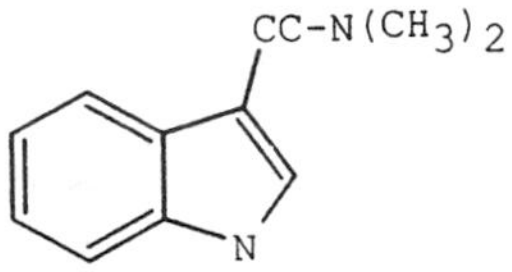

N-N-Dimethyltryptamine (DMT)

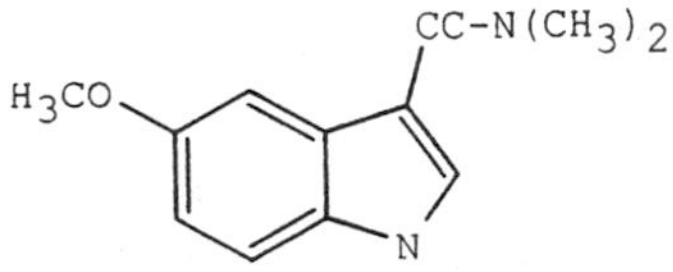

5-Methoxy DMT

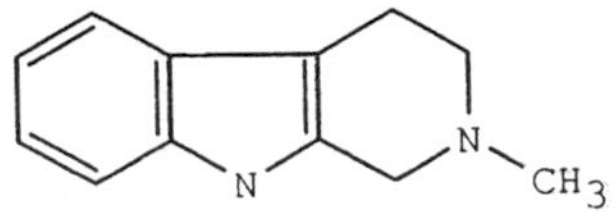

β-Carboline

Clinical Signs—The onset of signs, usually after the animals have been grazing the pasture for several weeks, may be abrupt or may develop slowly. With abrupt onset the signs are typical of other staggers syndromes in which the tremors are accompanied by excitement, a stiff hopping gait, shaking, and then collapse. When the animal is down, there may be seizures with extensor muscle rigidity and the head held in a backward flexed position. If allowed to remain undisturbed for a few minutes, it is usually able to stand. In addition, there may be exaggerated stepping, excitability, stiffness, continued chewing or abnormal lip and tongue movement, head nod, ear twitch, circling, incoordination, and excess salivation. Cardiac arrhythmias may also be noted. Forced movement or other forms of stress are factors in the onset of signs. In contrast to other staggers syndromes, the case-fatality rate for *Phalaris* staggers may be high, 30–40% or more.

- abrupt or slow onset, tremors, excitement, stiff gait, shaking, collapse; or recurrent episodes of head nod, ataxia, weakness; high case-fatality rate

In the chronic form, signs persist after removal from the *Phalaris* pasture and include recurrent episodes of head nodding, incoordination, and weakness of the thoracic limbs. In addition, in some animals the signs may not appear until several months after removal from the pasture.

Hematologic and serum chemistry changes are typically minimal except for mild changes in creatine kinase. Diagnosis may be aided by examination of cerebrospinal fluid for increased levels of protein and numbers of small mononuclear cells and for the presence of greenish intracytoplasmic granules in macrophages (East & Higgins 1988).

Pathology—Grossly, there are numerous sites in the brain with focal, bilateral, and symmetrical greenish gray-blue discolorationn. These include the cerebral cortex, thalamus, midbrain, pons, cerebellum, and medulla (Hartley 1978; East & Higgins 1988; Bourke et al. 1990). The pigment may also be present in the ventral gray matter horns throughout the spinal cord and in the renal cortex. The formalin tissue fixative

- gross pathology, focal, bilateral greenish gray-blue discoloration of various sites in brain

  microscopic, granular pigment in neurons of gray matter, spinal neuronal degeneration

may become pale blue after a few weeks (Hartley 1978). Microscopically, there is intracytoplasmic lysosomal accumulation of a greenish brown, granular pigment in neurons of gray matter nuclei, including the dorsal root ganglia, motor vagal, red, oculomotor, vestibular, pontine, facial, and olivary nuclei (East & Higgins 1988). This granular pigment is also present in macrophages. There is additional neuronal degeneration in the spinal cord with axonal necrosis, demyelination, and fibrillar astrocytosis in the medial longitudinal fasciculus and in the ventral, lateral, and ventromedial areas of white matter throughout the cord.

■ no specific treatment; prevent with oral cobalt

Treatment—In general, treatment is of little overall value, because losses may continue for several days after animal removal from the pasture, or the disease may become chronic. Cobalt (7 mg/sheep) given as a drench of cobalt chloride or sulfate may be protective in animals not yet showing signs. In Australia, cobalt is given as intraruminal pellets.

Acute Respiratory Distress Syndrome/ARDS

■ ARDS, cattle on pastures with lush regrowth

Disease Problems and Genesis—It has been suggested, but not confirmed, that tryptamine alkaloids may account for the occurrence of acute respiratory distress syndrome (ARDS) (Parmar & Brink 1976). ARDS has been a problem in cattle grazing meadows in British Columbia, often when the meadows have been harvested for hay and the regrowth subsequently grazed (Clapp 1962). In most instances *P. arundinacea* was the major forage species. However, tryptamine itself is not necessarily a factor, because it does not normally form 3-methylindole (Yokoyama & Carlson 1974). Aspects of ARDS are discussed under general syndromes of the Poaceae, earlier in this chapter.

Setaria P.Beauv.

■ *S. glauca*
S. italica

bristly foxtail
foxtail millet
German millet
Hungarian millet
Italian millet
Japanese millet
pigeongrass
Siberian millet
yellow bristlegrass
yellow foxtail

Taxonomy and Morphology—Comprising 110–125 warm-season species, *Setaria,* a member of the subfamily Panicoideae and tribe Paniceae, is a cosmopolitan genus that is most abundant in tropical Africa. Its generic name is derived from the Latin *seta,* which means "bristle," and is indicative of the bristlelike branches of the inflorescence that subtend the spikelets. Economically and ecologically, the genus is quite important. Some species have been cultivated as cereals since prehistoric times. In general, palatable and nutritious, other species provide forage for livestock and are important components of range forage in the Southwest. Others are noxious weeds in crops and lawns. Some are cultivated as ornamentals for their foliage and their bristly inflorescences. The caryopses are a most important component of the diet of many birds.

In North America, approximately 43 native and introduced species are present. Representative of the genus and associated with toxicological problems are the following:

S. glauca (L.) P.Beauv. (= *S. lutescens* [Weigel] F.T.Hubb.)	yellow foxtail, bristly foxtail, yellow bristlegrass, pigeongrass
S. italica (L.) P.Beauv.	foxtail millet, Italian millet, German millet, Siberian millet, Hungarian millet, Japanese millet

■ annuals or perennials; inflorescences spicate panicles; spikelets subtended by bristles, dorsally compressed, with 2 florets, upper perfect and hard or leathery

Plants annuals or perennials; cespitose or rhizomatous. **Culms** 25–120 cm tall; erect or geniculate; internodes hollow or solid. **Leaves** cauline. **Blades** flat or folded; 2–25

mm wide. **Ligules** ciliate-membranous or ciliate. **Auricles** absent. **Sheaths** with margins free, overlapping. **Inflorescences** spicate panicles; spikelets borne in fascicles on short pedicels or branches, subtended by bristles (sterile panicle branches). **Disarticulation** below the glumes; at spikelet bases. **Spikelets** all alike; lanceolate or ovate or elliptic; compression dorsal; 1.5–4 mm long. **Florets** 2; of 2 forms; lower florets staminate or neuter; upper florets perfect. **Glumes** 2; unequal; different; lower glumes one-third to one-half the length of spikelets, broadly ovate, nerves 1 or 3 or 5 or 7; upper glumes one-half to equal the length of lower lemmas, elliptic, nerves 5 or 7. **Lemmas of Lower Florets** elliptic; membranous or coriaceous; abaxial surfaces glabrous or indumented; awns absent. **Paleas of Lower Florets** absent or present; as long as lemmas; hyaline. **Lemmas of Upper Florets** elliptic; indurate or coriaceous-indurate; abaxial surfaces smooth or rugose, glabrous, shiny. **Paleas of Upper Florets** present; as long as lemmas; tightly enclosed by lemma margins; indurate.

1. Margins of upper sheaths not ciliate. Bristles subtending each spikelet 4–12. Spikelets solitary . . . . *S. glauca*
1. Margins of upper sheaths ciliate. Bristles subtending each fascicle 1–4. Spikelets 3–6, in fascicles . . . . *S. italica*

Setaria glauca

Distribution and Habitat—Species of *Setaria* are found in a variety of habitats in grasslands and woodlands. Many readily occupy disturbed sites. Introduced from Eurasia, *S. glauca* is widely distributed and is considered a pernicious weed in crops. Believed to be native to central China, *S. italica* is planted in the southern portion of the continent.

Disease Problems—Several minor toxicologic problems are associated with species of *Setaria*. They include primary photosensitization, myoarthropathy, and mechanical injury due to the bristles, which subtend the spikelets. Occurrence of the first two problems is rare.

Myoarthropathy

Disease Problems and Genesis—At one time, *S. italica* with mature inflorescences (headed out) was commonly cut for hay and used as feed for horses and other livestock. Unfortunately, even when mixed with equal amounts of another forage, it caused an acute syndrome of joint and muscle pain, and diuresis (Hinebauch 1892, 1896). The degenerative arthritis was sometimes severe. Recovery, even during the early stages of the disease, required a week or more following prompt elimination of the hay from the animal's diet. A glucoside from an aqueous extract of *Setaria* hay is reported to cause the characteristic signs, but this has not been confirmed (Pammel 1911).

- disease problems—
  photosensitization
  myoarthropathy
  mechanical injury
- rare, horses mainly; degenerative arthritis, hay from mature *S. italica*, cause unknown

Clinical Signs—There is abrupt onset of muscle and joint soreness and an increase in the frequency and volume of urination. The animal's temperature is increased moderately. Sweating and constipation are also seen. Soreness may be of such severity that the animal will not stand. In other instances, the onset is slower, with gradual onset of lameness and all joints painful to the touch.

- abrupt or slow onset, muscle and joint soreness, lameness, sweating

Pathology—If the initial period of *Setaria* consumption is of short duration, there may be few or minor pathologic changes. If prolonged, articular cartilage in joints of the limbs is degenerated and the synovial fluid is dark colored, necrosis of the renal

- cartilage degeneration, renal tubular necrosis

tubular epithelium may also be apparent. With repeated consumption there may be softening of the bones and rupture of tendonous attachments to the bones.

- mainly anti-inflammatory drugs

Treatment—Directed toward the signs, treatment includes anti-inflammatory drugs for relief of the joint pain and muscle degeneration.

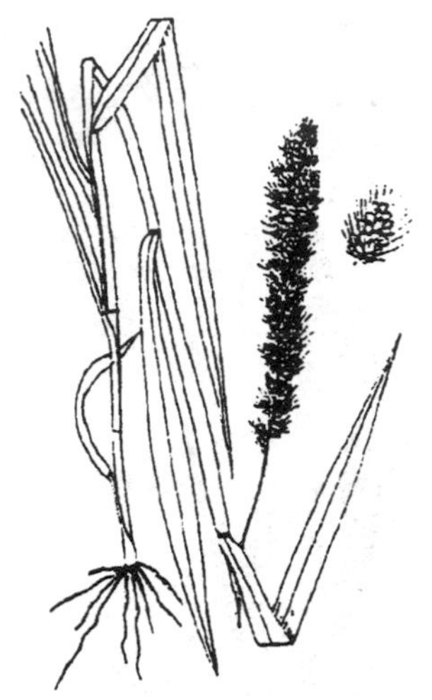

Setaria italica

Mechanical Injury

Disease Problems and Genesis—Most members of the genus have sharp and barbed bristles, which represent reduced, sterile panicle branches. They are particularly conspicuous in *S. glauca*. They are capable of penetrating the mucous membranes and causing serious erosions of the mouth (Bankowski et al. 1956; Rebhun et al. 1980; Peterson & Schultheiss 1984). When *Setaria is* harvested before inflorescence maturity, good hay may be produced. Problems occur when the grass is cut late and a substantial number of panicles are present. Mechanical injury is a particularly serious problem in horses, dogs, and cats. The barbed bristles may penetrate the skin and migrate through the tissues causing fistulous tracts and deep-seated infections. These effects are especially serious in the thoracic cavity.

- mechanical injury due to the sharp bristles in hay made from mature grass

Clinical Signs and Pathology—The signs produced by *Setaria* depend upon the site of penetration by the bristles. Lesions in the mouth will cause pain, resulting in a reluctance to eat and a loss of weight. There may also be an apparent increase in salivation and possibly fever if there is sufficient inflammatory response. The bristles cause deep ulcerations of the tongue, gums, dental pad, and cheeks, which may result in abscesses and caseation. The bristles may be present in the wounds.

- painful ulceration, abscesses, caseation, especially of mouth

Treatment—If possible, the offending barbed bristles (awns) should be removed and the site cleansed with a dilute hydrogen peroxide solution.

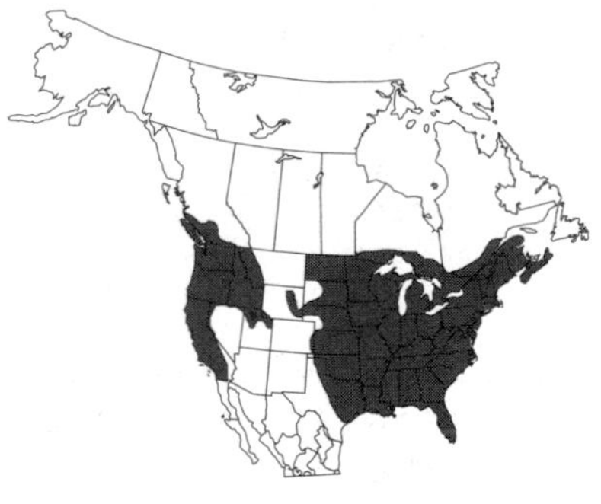

Setaria glauca

Sorghastrum Nash

Taxonomy and Morphology—Comprising 16–20 warm-season species in tropical Africa and America, *Sorghastrum* is a member of the subfamily Panicoideae and the tribe Andropogoneae. It is believed to be closely related to *Sorghum* and its generic name is derived from *Sorghum* and the Latin suffix *astrum,* which means "poor imitation of." It is represented in North America by 3 species, only 1 of which is commonly encountered and of toxicologic concern:

- *S. nutans*
  Indiangrass

S. nutans (L.) Nash (= *S. avenaceum* Nash)	Indiangrass

- perennials; inflorescences panicles; spikelets dorsally compressed, awned; glumes 2, longer than florets, leathery

Plants perennials; cespitose or rhizomatous. **Culms** 90–200 cm tall; erect; not branched above; nodes glabrous or indumented; internodes hollow. **Leaves** both basal and cauline. **Blades** flat; 3–15 mm wide; bases tapered. **Ligules** membranous or ciliate-membranous; frequently continuous with auricles. **Auricles** present. **Sheaths** glabrous or pilose; margins free, overlapping. **Inflorescences** panicles of rames; open or contracted; narrow to broadly rhombic; pedicelled spikelets absent; pedicels present; hirsute. **Disarticulation** below the glumes; rachises disarticulating; sessile spikelet, pedicel, and rachis joint falling as a unit. **Spikelets** all alike; lanceolate, compression dorsal, 5–8 mm long, awns present. **Florets** 1; perfect. **Glumes** 2; equal or subequal; different; longer than florets; golden or dark brown; coriaceous; nerves 5 or 7, obscure;

lower glumes hirsute; upper glumes glabrous or sparsely hirsute distally. **Lemmas** linear; hyaline; awns 1, apical, terete, twisted proximally, once or twice geniculate. **Paleas** absent or present; hyaline.

Distribution and Habitat—Species of *Sorghastrum* occupy habitats in grasslands, in savannas, and at the edges of woods. Occurring across North America except for the extreme west, *S. nutans* is a dominant of the continent's tallgrass prairies and thus a common constituent of wild or prairie hay.

Disease Problems and Genesis—*Sorghastrum* clearly represents a cyanogenic risk. The HCN potential varies in much the same manner as that of *Sorghum.* The potential is very high in the early stages of growth—up to 1,000 ppm or more in seedlings—and declines steadily to safe, low concentrations by the time the plants are typically grazed or cut for hay (Haskins et al. 1979; Vogel et al. 1987). In spite of the high dhurrin levels and hence high HCN potential, actual livestock losses are rare. This is probably due to management practices in which *Sorghastrum* is either fed when mature and minimally toxic or grazed in mixed pastures. The same factors determining toxicity in *Sorghum* appear to be important in this genus.

N
C
O-β-D-glucose
HO

Dhurrin

Clinical Signs—There is abrupt onset of apprehension and distress that occurs within a few minutes of grazing the plants. Weakness, incoordination, and labored respiration quickly follow. If intoxication is severe, the animal may be unable to stand and may lay in lateral recumbency with periodic paddling and tetanic seizures. The entire sequence of signs may occur over a 5-to-15-minute period, with either death or recovery at the end of this time.

Pathology—There are few distinctive changes. For the most part they are limited to congestion of the abdominal viscera and scattered, small splotchy hemorrhages.

Treatment—The primary antidote for ruminants is sodium thiosulfate given i.v. at a dose of 0.25–0.5 g/kg b.w. Sodium nitrite, 10 mg/kg b.w. i.v., can be used to augment the effects of the sodium thiosulfate, but it is not required for effective treatment. A more complete discussion is presented in the treatment of the Rosaceae (see chapter 64).

Sorghum Moench

Taxonomy and Morphology—Comprising 20–35 warm-season species, *Sorghum,* a member of the subfamily Panicoideae and the tribe Andropogoneae, is essentially an African genus. Its name is derived from *sorgo,* the Italian vernacular for plants of the

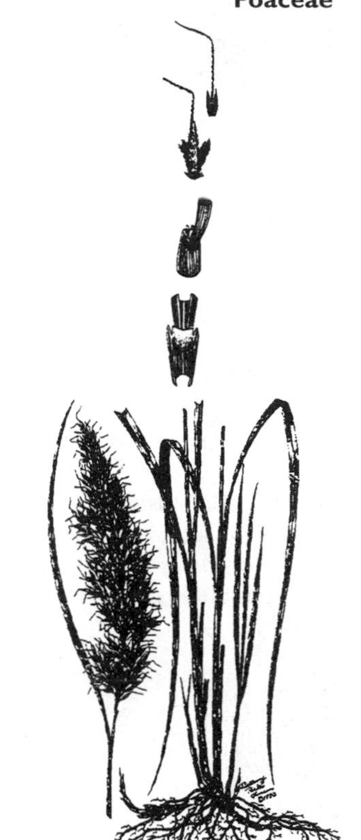

Sorghastrum nutans

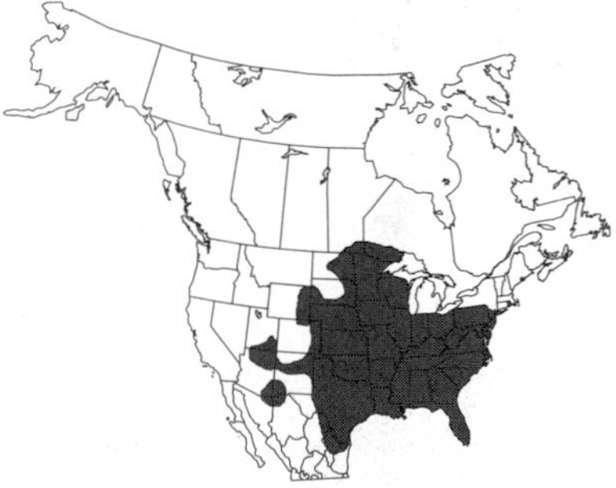

Sorghastrum nutans

- cyanogenic, especially early growth, low risk in mature forage
- abrupt onset; apprehension, distress, weakness, ataxia, labored respiration, collapse, seizures
- no lesions
- sodium thiosulfate, sodium nitrite

- *S. ×almum*
  S. bicolor
  S. halepense

  Columbusgrass
  great millet
  Johnsongrass
  milo
  shattercane
  sorghum
  sorghum almum
  Sudan
  Sudangrass

- annuals or perennials; inflorescences panicles of rames; spikelets dorsally compressed; glumes 2, equal, longer than florets, leathery

Sorghum bicolor

- *S. bicolor,* numerous races now grouped by usage

genus. Only 1 species, a Mexican endemic, is native to the New World. In North America, the genus is represented by 3 species of toxicologic importance:

S. ×almum Parodi	sorghum almum, Columbusgrass
S. bicolor (L.) Moench (= *S. arundinaceum* [Desv.] Stapf) (= *S. ×drummondii* Nees ex Steud.) (= *S. vulgare* Pers.)	sorghum, shattercane, Sudan, milo, great millet, Sudangrass
S. halepense (L.) Pers.	Johnsongrass

Plants annuals or perennials; cespitose or rhizomatous. **Culms** 60–300 cm tall; erect; not branched above; nodes glabrous or indumented; internodes solid or hollow below. **Leaves** cauline. **Blades** flat; 12–120 mm wide. **Ligules** membranous or ciliate-membranous or ciliate. **Auricles** absent. **Sheaths** glabrous; margins free, overlapping. **Inflorescences** panicles of rames; open or contracted; narrow to broadly rhombic; branches whorled. **Disarticulation** below the glumes; rachises disarticulating; sessile spikelet, pedicel, and rachis joint falling as a unit. **Spikelets** of 2 forms, sessile and pedicelled different. **Sessile Spikelets** elliptic to ovate; compression dorsal; 4–6 mm long; awns present. **Florets** 2; of 2 forms, lower neuter, upper perfect; glumes 2, equal, different, longer than florets, coriaceous, flat to convex, glabrous or sericeous; lemmas and paleas of lower florets hyaline; lemmas and paleas of upper florets linear, hyaline, awns 1, apical, terete, twisted proximally, once or twice geniculate. **Pedicelled Spikelets** well-developed or reduced to a glume; shorter or longer than sessile spikelets; staminate or neuter; lanceolate. **Lemmas** hyaline. **Paleas** hyaline.

1. Plants annuals. Rhizomes absent. Mature caryopses not protruding from glumes . . . . *S. bicolor*
1. Plants perennials. Rhizomes present. Mature caryopses protruding from glumes.
   2. Rhizomes long; forming an extensive system. Plants 1–2 m tall. Disarticulation of pedicelled spikelets at apex of pedicels . . . . *S. halepense*
   2. Rhizomes short; not forming an extensive system. Plants 3–4 m tall. Disarticulation of pedicelled spikelets in middle of pedicels . . . . *S. ×almum*

Species of *Sorghum* are economically quite important for forage, hay, silage, and grain. A major tropical cereal, *S. bicolor* is believed to have been first domesticated in the Ethiopia/Sudan region of northeastern Africa about 3,000 B.C. (Doggett 1988). Its wild progenitor was *S. arundinaceum,* which is now treated as one of its subspecies. As the use of *S. bicolor* spread through tropical Africa, into eastern Asia, and eventually throughout the world, scores of varieties, cultivars, races, and hybrids were developed (table 58.12). Differences among these taxa are sometimes slight and/or primarily physiological; identification of them is thus sometimes tenuous. To deal with this complexity, several classification schemes have been developed. In 1936, J. D. Snowden, an eminent *Sorghum* specialist, proposed recognizing 31 "species" to encompass the range of variation in the cultivated sorghums. His classification was revised by Murty and co-workers (Murty & Govil 1967; Murty et al. 1967), who proposed that the subgenus *Bicolor* (= *Eusorghum*) be divided into 8 "working groups" and 70 "groups." Further simplification by Harlan and de Wet (1972) led to the establishment of 5 basic races and 10 intermediate/hybrid races based on appearance of inflorescence, glumes, and caryopses. For convenience, these taxa can be grouped into four categories based on general usage (table 58.12). Of interest here is that the toxicologic potential differs considerably from taxon to taxon.

Table 58.12. Representative infraspecific taxa of *S. bicolor* grouped by general usage

Grass Sorghums—plants grown for forage use; tall, slender-stemmed, narrow-leaved	
Varieties	‘Piper,’ ‘Tift,’ ‘Wheeler’
Hybrids (sweet Sudans)	‘All Tex,’ ‘Grazer,’ ‘Sweet Sudan,’ ‘Greenleaf,’ ‘Haygrazer,’ ‘Hi Dan,’ ‘Planter’s Pride,’ ‘Rainbow Gold,’ ‘Sordan,’ ‘Sudax,’ ‘Sugar Dan,’ ‘Sugar Queen,’ ‘Sweet Sioux,’ ‘Trudan,’ ‘Zulu’
Grain Sorghums—plants grown for large, palatable grains and some forage	
Milo	dwarf, dwarf yellow, ‘Redlan,’ ‘Redbine,’ ‘Plainsman’
Kafir	‘Blackhull,’ ‘Dawn,’ ‘Early Red,’ ‘Pink,’ ‘Sharon,’ ‘Sunrise’
Hegari	dwarf hegari
Feterita	spur feterita
Hybrids	especially milo crosses
Others	durra and guinea corn (Africa), shallu (India), kaoliang (China)
Sorgos—plants grown for syrup, sugar, and forage; thick, juicy, and sweet culms, with up to 10% sucrose; grain of little value	
Varieties	‘Chinese Amber,’ ‘Gooseneck,’ ‘Honey,’ ‘Sumac,’ ‘Early Amber,’ ‘Black Amber,’ ‘Orange,’ ‘Sourless,’ ‘Sart,’ ‘Sugar Drip,’ ‘Planter,’ ‘White Africa’
Hybrids (sweet Sudans)	‘Atlas,’ ‘Early Sumac,’ ‘Norkan,’ ‘Leoti,’ ‘Brandes,’ ‘Rio’
Broomcorn Sorghums—grown for broom bristles; panicle branches and pedicels long and stiff	
	‘Acme,’ ‘Evergreen,’ ‘Scarborough,’ ‘Black Spanish’

Sources: Harlan & de Wet 1972; Doggett 1988; Purseglove 1988.

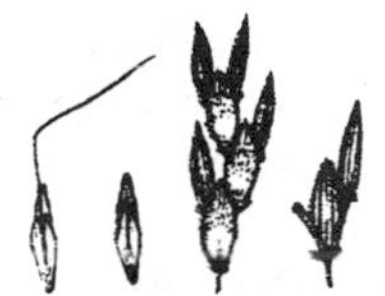

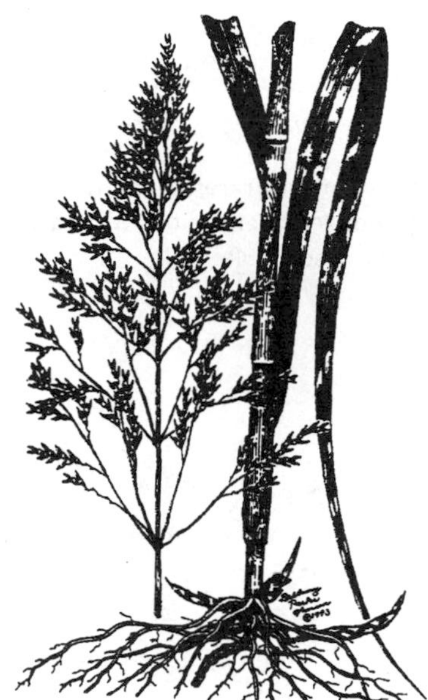

Sorghum halepense

Hybridization among members of the various categories has been extensive; e.g., grain x grass hybrids and sorgo x grass hybrids are common. In the older agronomic and taxonomic literature, the binomial *S. vulgare* Pers. was used for sorghum. It is illegitimate, however, because of the nomenclatural principle of priority and should not be used.

In contrast to annual *S. bicolor, S. halepense* is a vigorously rhizomatous perennial that is considered both a valuable forage and a pernicious weed. It is categorized as a “grass sorghum.” It has hybridized with *S. bicolor* to produce *S.* x*almum.*

Distribution and Habitat—Although *S. bicolor* is nominally a cultivated species, waifs occasionally occur along the roadsides. *Sorghum halepense* was originally thought to be deliberately introduced into North America about 1830 from the Mediterranean region as a forage and hay grass, but apparently it arrived earlier, possibly as a contaminant in other seed stocks. Its vigorous rhizomatous growth facilitates rapid establishment of the dense stands desired by cattle producers. However, this same trait makes it a troublesome weed of roadsides, ditches, and other disturbed sites. *Sorghum* x*almum* was originally described as a species in Argentina. It is generally treated as a distinct species rather than as a variety of either of its parents, because of its agronomic significance. It is used in the same way as *S. halepense* but is not a weedy pest.

■ crop plants

Sorghum halepense

Disease Problems—Valued for fresh forage, hay, silage, and grain, the sorghums historically have been associated with toxic effects in certain conditions. Extensive losses of cattle feeding on sorghum were recorded in India in the severe drought years of 1877, 1887, and 1895 (Vinall 1921). The toxic nature of *S. bicolor* was recognized in the late 1800s in Egypt, where it was grown as a protective shield around plantings of peanuts. In these circumstances it was observed that water deprivation

- 4 disease syndromes—
ataxia/cystitis/teratogenesis
photosensitization
nitrate intoxication
cyanide intoxication

markedly increased the toxicity of young plants (Dunstan & Henry 1902). Various cultivated plants of this genus were also serious problems in the Great Plains in the late 1800s and early 1900s. There were sporadic losses of livestock with otherwise useful forages or grain crops, especially kafir corn (Peters et al. 1903; Slade 1903; Francis 1915). Again it was noted that moisture stress markedly increased toxicity.

It has since become well accepted that *Sorghum* is toxic in certain environmental conditions. In many instances the risk of toxicity seems to be greater in times of drought, but this has not been a consistent finding. The genus produces four disease syndromes—ataxia, cystitis, and teratogenesis; photosensitization; nitrate intoxication; and cyanide intoxication. The first of these occurs primarily in horses, whereas nitrate and cyanide intoxication appear mainly if not exclusively in ruminants. Each syndrome is described below.

Ataxia/Cystitis/Teratogenesis

- horses mainly, rare; several weeks of grazing; fetal deformities very rare; cattle susceptible?

Disease Problems—The disease syndrome of ataxia/cystitis/teratogenesis, often called *Sorghum* cystitis, was first observed in the southern Great Plains in the early 1960s and has since been seen in horses of all ages and breeds (Romane et al. 1966). It occurs when animals graze any species of the genus or the hybrids. The offending plants may be in various stages of growth, ranging from young, rapidly growing individuals to those exhibiting mature second growth. The problem also has been seen in animals fed fresh-cut but not well-cured hay (Adams et al. 1969). It generally follows several weeks or months of grazing on sorghum pastures, although in rare instances it is reported after only a single week. In Australia the disease seems to be associated with drought, insect attack, or other stress on the plants (McKenzie & McMicking 1977).

The disease was originally described as occurring only in horses; cattle in the same pastures apparently were unaffected (Adams et al. 1969). Since that time however, several cases of what appears to be the same or a similar disease have been reported in cattle in Australia and possibly sheep in the United States (McKenzie & McMicking 1977; Bradley et al. 1995). Although the episodes in cattle and sheep have not been shown conclusively to be the same disease as in horses, *Sorghum* seems to be the common factor in all cases, and the fetal deformities and spinal pathology are similar in all species. If animals are promptly removed from the forage, there may be apparent improvement, but the pathologic changes persist. The risk of occurrence of this disease seems to be rather low, given that only intermittent episodes are reported (Morgan et al. 1990). Producers typically use sorghum varieties low in HCN potential, making the assumption that this will ensure avoidance of the ataxia and urinary tract effects, but the preventive value of this practice has not been confirmed.

Closely associated with the disease are fetal malformations and abortions, both sometimes occurring in the same animal (Knight 1968; Adams et al. 1969). In one instance both limb deformities and spinal pathology were seen in the same newborn calves (Seaman et al. 1981). These reproductive problems have been shown to occur after several weeks of grazing Sudan pastures either in horses in North America or cattle in Australia (Romane et al. 1966; Prichard & Voss 1967; Knight 1968; Adams et al. 1969; Seaman et al. 1981).

- cause unknown, cyanide unlikely; lathyrogenic nitriles?

Disease Genesis—The specific cause of this disease is unknown, but both cyanide and/or nitriles are suspected. The effects are related primarily to spinal cord pathology and are perhaps somewhat similar in nature to those caused by *Lathyrus* spp. in the Fabaceae. It seems unlikely that the effects are due to cyanide, because there is no link between toxicity and plant maturity. Experiments by one of us (GEB) to evaluate the effects of chronic cyanide exposure by administration of a substantial daily i.m. dose of cyanide to ponies throughout pregnancy resulted in no observable detrimental effects. Furthermore the production of developmental defects in hamsters requires oral dosage of another cyanogenic glycoside, linamarin, sufficient to cause obvious signs in the dams (Frakes et al. 1985). Neuronal changes—axonal spheroids—observed in sheep are considerably different from those of horses and cattle, and they have been suggested by Bradley and co-workers (1995) to be further evidence for the hypothesized lathyrogenic nature of *Sorghum* toxicity (Van Kampen 1970).

Clinical Signs—In horses, the signs develop gradually, beginning with incoordination and then urinary incontinence a day or two later, or both signs appear simultaneously. Occasionally, only the incoordination is seen. In particularly severe cases, flaccid paralysis of the tail and pelvic limbs may be seen within 24 hours after onset of signs. There may be considerable difficulty in walking. This is especially obvious when the animal is backed; it may sit down if it does not fall down. Abnormalities may not be apparent when the animal is standing, but the placing deficit of the rear feet becomes quite obvious with forced movement. In females, there is continual opening and closing of the vulva (as in estrus), with dribbling of urine. Scalds due to urine may be evident on the pelvic limbs. In the male, the penis is extended in a relaxed position, with continual dripping of urine. Body temperature and appetite are usually unaffected. Urinalysis may reveal that specific gravity and pH are in the normal ranges, but that occult blood, numerous cells, bacteria, and calcium carbonate crystals are present. This is not usually a fatal disease, but the subsequent limitations in performance often result in the animal's being euthanized.

- gradual onset of ataxia, urinary incontinence, dribbling of urine, paralysis of limbs and tail; teratogenic effects, arthrogryposis

The signs described for cattle are similar except for the inconsistency of the urinary incontinence. In sheep, as observed in feeder lambs, there was weakness, ataxia, head shaking, knuckling of the fetlocks, inability to rise, and opisthotonus (Bradley et al. 1995). Urinary tract effects were lacking.

The reproductive effects may present as abortion at any time in gestation, as a difficult delivery due to malformation of the fetus, or as a deformed and weak newborn. In the latter instance it may be difficult for the newborn to stand and suckle. The prominent deformity is arthrogryposis of both thoracic and pelvic limbs.

Pathology—Grossly, inflammation of the bladder is prominent. The bladder wall is reddened, thickened with edema, and sometimes ulcerated. The bladder may be filled with a sticky, granular-appearing, yellowish urine containing considerable calcium carbonate, which also tends to adhere to the bladder mucosa. There is often copious semicongealed pus in the bladder. In more chronic cases, the bladder mucosa is convoluted and may have large ulcerations. The vagina also may be reddened and coated with calcium carbonate. In the most serious cases there is evidence of an ascending urinary tract infection, sometimes with severe pyelonephritis. Microscopically, there are swollen degenerating axons, demyelination, and increased gitter cells in the lateral and especially ventral funiculi of all segments of the spinal cord. There may also be degenerative changes in the spinothalamic and pontocerebellar tracts.

- gross pathology, bladder wall reddened, edematous, ± pus

  microscopic, spinal demyelination, axonal degeneration

In affected newborn calves, there was extensive spinal involvement with Wallerian degeneration of the spinal axons in all but the dorsal columns at all levels and even extending up into the cerebellar white matter and brainstem (Seaman et al. 1981). In the single episode reported in sheep, the most distinctive changes were eosinophilic intranuclear and spheroidal axonal swellings in the medulla, cerebrum, midbrain, and ventral horn gray matter, with minimal Wallerian degeneration in the spinal cord.

Photosensitization

- primary photosensitization, rare, cause unknown

Disease Problems—A disease of very limited importance, photosensitization occurs only sporadically and is readily responsive to removal of the animals from the offending pasture. It is a primary photosensitization, with no icterus. The toxicant responsible has not been identified. Episodes are reported in sheep grazing regrowth sorghum with or without irrigation (Howarth 1931; Mathews 1938).

Nitrate Intoxication

- serious problem in livestock, especially with highly productive varieties and with extensive fertilization

Disease Problems—Nitrate intoxication is one of the most serious intoxication problems for the cattle industry in North America, especially in regions where the sorghums are used as a primary forage. High-producing sorghums are planted on approximately 18 million acres in the United States alone, and they represent a significant portion of the available livestock forage. They are used as silage, green chop, hay, bundle feed, and grazing forage. Although considerable research has been directed toward improvement of the sorghums as forage and reduction of their cyanogenic potential, the problem of nitrate accumulation has not been diminished and, in fact, may be enhanced by the development of higher-producing varieties (Clay et al. 1976). Death losses of 10–15% are not uncommon, especially in severe winters following dry summers, when conditions are favorable to both nitrate accumulation and expression of the disease (Dollahite & Rowe 1974; Haliburton & Edwards 1978). Losses may be devastating. Such was the case in an episode in Nebraska when, during severe December weather, 226 of 390 cattle died and an additional 42 cows subsequently aborted (Hibbs et al. 1978). This disaster was due to weedy *Amaranthus/Kochia* hay with a level of 4.9% KNO_3 and *Sorghum* hay with 8% KNO_3.

- nitrates accumulate under a variety of conditions

Disease Genesis—Most plants extract nitrogen from the soil in the form of nitrates. Normally, their tissues contain relatively small amounts of nitrates because they are rapidly reduced to nitrites, which in turn are reduced further to ammonia and subsequently incorporated in the formation of amino acids and proteins (Wright & Davison 1964). In the process, nitrogen undergoes a valence change from +5 in nitrate to –3 in ammonia. Under certain conditions, plants accumulate nitrate in their tissues, especially those of the stems. This accumulation may result from an increase in the rate of nitrate uptake or from interference in its reduction to nitrite. Uptake depends upon the amount of nitrogen, sulfur, and phosphorus in the soil; plant maturity; and plant genotype. Reduction depends, in part, upon light, moisture, temperature, concentrations of molybdenum and manganese, plant maturity, and plant genotype (Nason 1962; Wright & Davison 1964). Nitrate accumulation is markedly fostered by decreased soil moisture and light and attenuated by plant maturity (Whitehead & Moxon 1952). Factors that seem to predispose nitrate accumulation include plant diseases such as leaf rust, mildew or root rot; plant damage due to hail, wind, or other forces; and application of 2,4-D herbicides (Whitehead & Moxon 1952; Piening 1972; Hicks & Peterson 1976). In general, nitrate is increased by factors that inhibit photosynthesis and thus its utilization.

Nitrate concentrations in plants may be expressed in several ways. The following conversion factors may be used in evaluating some of the different ways to express nitrate:

$$KNO_3 \times 0.61 = NO_3$$
$$NO_3 \times 1.63 = KNO_3$$
$$NO_3\text{-N} \times 4.4 = NO_3$$
$$NO_3\text{-N} \times 7.2 = KNO_3$$

Typically nitrate concentrations, from highest to lowest, are distributed in the plant as follows: lower stems > upper stems > roots > leaves > flowers (Whitehead & Moxon 1952; Hanway et al. 1963). Nitrate concentrations in all plant parts generally decline appreciably after heading of the Sudangrasses (Mizukami et al. 1997). Because of all the factors involved in toxicity, it is difficult to establish a single critical level. Generally, a concentration of 10,000 ppm NO_3 in the stem is viewed as close to the danger level. In some instances, lower concentrations present a risk, in contrast to other situations in which feeds with 20,000 ppm are safely fed. The 1%, or 10,000 ppm, level should be viewed as a level at which the risk of complicating factors supervening to cause toxicosis is much increased. It has been proposed that concentrations in the total ration greater 1.5% cause overt disease, 1–1.5% cause abortions without other signs, and less than 1% may cause more insidious subacute to chronic effects (Case 1957; Johnson et al. 1994).

The multitude of factors and their interactions results in considerable variation in plant nitrate levels from year to year, from field to field, and even from area to area within a field (Selk et al. 1995). Thus, it is difficult to predict actual nitrate levels in hay from a field unless all bales are analyzed each year. Because of the possibility of concentrating in a few bales many of the plants with high levels of nitrates in a field, the use of large hay bales becomes a significant factor in intoxication. This is illustrated by a range of nitrate concentrations from 7,900 ppm to 43,600 ppm at various locations within a 5-acre pasture (Strickland et al. 1995). Intoxication is also a serious problem with rotation of crops when high levels of nitrogen fertilizer are applied to a crop such as wheat and then a nitrate accumulator such as sorghum or millet is planted the following year (Kralovec & Frycek 1974). Forages that are produced with irrigation from ponds or are subject to drainage from penned cattle or other livestock also may represent a considerable risk (Antoine et al. 1993).

■ ruminants, nitrate converted to nitrite by ruminal microflora, mainly from hay rather than fresh forage

Nitrate intoxication is a serious problem only in ruminants and particularly in cattle. The rate of feed intake in sheep is typically slower and is less likely to exceed the capacity of the ruminal microflora to utilize the nitrate and its reduction products. In addition, the half-life of nitrate is considerably longer in cattle than in sheep (Johnson et al. 1992). Although reported for horses, intoxication other than digestive dysfunction is unlikely in monogastrics except from nitrite itself. The important factors in development of toxicity have been grouped into four categories: (1) the amount of nitrate consumed, (2) the rate of intake, (3) the rate of nitrate release from the forage, and (4) the activity of the gastrointestinal microflora in reducing nitrate (Kemp et al. 1977a; Geurink et al. 1979). The role of nitrate availability is shown by the difference in the toxic dosage of nitrate for cattle of approximately 0.5g NO_3/kg b.w. when given as the salt and 1 g NO_3/kg b.w. or more when in forage (Crawford et al. 1966). Intoxication is also more likely to occur from ingestion of hay than from ingestion of fresh forage, because intake of nitrate on a dry-matter basis is more rapid, and release of nitrate from its intracellular sites is faster when the plant is dried (Geurink et al. 1979).

The susceptibility of ruminants is due to the activity of ruminal microorganisms, which, like higher plants, rapidly reduce nitrate to nitrite and then to ammonia, with possible reaction intermediates such as hydroxylamine, nitroxyl, nitramide, and/or hyponitrous acid (Lewis 1951a; Nason 1962; Winter 1962). The reactions occur in the reduced-oxygen atmosphere of the rumen as a form of nitrate respiration by the denitrifying facultative anaerobes, and the reaction products are readily absorbed from the rumen. Hydrogen gas and/or other readily available compounds such as succinate, formate, DL-lactate, citrate, and D-glucose serve as hydrogen donors in these reduction reactions (Lewis 1951b). Thus, readily fermentable carbohydrates in the diet promote reductive activity (Sapiro et al. 1949; Holtenius 1967). This is illustrated by the protective effect of feeding corn-based supplements to cattle, thus reducing nitrite accumulation (Nakamura et al. 1979; Burrows et al. 1987). Intoxications may not always be entirely due to nitrate intake, because high nitrite concentrations are occasionally reported in feeds, indicating possible problems from nitrite directly (Smith & Suleiman 1991).

- adaptation in several days to allow safe use of high-nitrate hay may be lost quickly with lack of feed

The balance between nitrate and nitrite is also influenced by acclimation of the ruminal microflora, because administration of even large amounts of nitrate to cattle without immediate previous exposure results in slow conversion to nitrite (Kemp et al. 1977b). The denitrifying microorganisms must be abundant enough to allow sufficiently rapid nitrite formation for toxicosis to develop. Many instances of nitrate intoxication occur during the winter when, because of a temporary lack of feed or adverse weather conditions, cattle become very hungry and voraciously consume nitrate-containing hay that otherwise would not cause problems when the animals are adapted to it. Acclimation occurs because, with increased amounts of nitrate in the diet, the ruminal microflora begin to adapt in 2–3 days, with an increased ability to fully utilize nitrate as a source of nitrogen (Allison & Reddy 1984). The adaptation can be lost just as quickly or more so. Maximal adaptation is present by 6 days. Typically, both nitrate and nitrite reduction are increased, with nitrite reduction being at least as fast as that of nitrate. However, with high dosage of nitrate, its reduction rate exceeds that of nitrite, allowing accumulation of nitrite in the rumen (Allison & Reddy 1984). A prime example of some of these factors involved in accumulation is an episode in Kansas (Brown et al. 1990). A mix of heifers and older cows were being fed a Sudan hay and, because of repairs being made on the feeder, were without feed overnight. The following morning a bale was placed in the feeder, and a day later some of the heifers were dead. It was surmised that the dominant older cows ate the leafy portions of the hay at a reasonable rate and that the very hungry heifers followed, voraciously consuming the less desirable stems, which contained the highest nitrate concentrations. The time without feed was a likely contributing factor, significantly increasing the appetites. Erratic exposure to increased nitrate concentrations may cause special problems because, although reductive capacity is maintained to some extent, utilization of the reaction products is reduced.

As in plants, many factors are involved in ruminal reduction of nitrate, including levels of copper, iron, and molybdenum; ruminal pH; and quality/energy of the rations (Lewis 1951b; Tillman et al. 1965; Holtenius 1967; Miyazaki & Kawashima 1976). Nitrate reductase is a molybdenum-containing enzyme, and either dietary deficiency or antagonism by administration of sodium tungstenate greatly reduces the risk of nitrite accumulation (Tillman et al. 1965; Korzeniowski et al. 1980, 1981). Cyanide, cyanate, and copper sulfate inhibit nitrate reductase activity in vivo as well, but only copper sulfate and tungstate inhibit nitrite reductase (Bicudo 1986). Various other additives also seem to influence nitrite levels. Penicillin has been shown in vitro to reduce nitrite accumulation, whereas monensin in vivo appears to increase it (Holtenius 1967; Malone 1978). Furthermore, methylene blue, given orally, enhances nitrite reduction to ammonia (Nason 1962). Yeast cultures have been promoted as a preventive, but other studies do not substantiate these claims (Stone 1974, 1976; Horn et al. 1984).

- many factors influence nitrate reduction to nitrite in rumen, mainly via changes in nitrate reductase activity

half-life of nitrite in blood is short

After ingestion of plants containing high levels of nitrates, nitrites begin to increase in the rumen, peaking in 3–4 hours. Concentrations in the blood similarly increase, but appreciable amounts of nitrite and other reaction intermediates such as hydroxylamine are of only fleeting existence in serum (Lewis 1951a; Wang et al. 1961; Winter 1962; Ishigami & Inoue 1976; Kemp et al. 1977b). The small amount of nitrite that is absorbed is either oxidized to nitrate or eliminated directly; it is short-lived in the systemic circulation (Kiese 1974; Vertregt 1977). During intoxications, there is a 100-fold or more differential in nitrate/nitrite balance, favoring the former. This portends a greater difference than the 2.5-to-10-fold increase in toxicity reported for nitrite over nitrate (Williams & Hines 1940). In vitro, nitrite is stable in plasma but not in blood (Watts et al. 1969). The half-lives of nitrate and nitrite are 4.2 hours and 0.5 hours, respectively, in sheep (Schneider & Yeary 1975). Nitrate half-life in cattle appears to be appreciably longer, about 9 hours (Johnson et al. 1992). Very little of the nitrate that is absorbed into the systemic circulation is reduced to nitrite. In the fetus the half-life of nitrate may be longer than 24 hours (Johnson et al. 1992).

- nitrite, oxidation of hemoglobin to methemoglobin, MHb incapable of oxygen transport

Under unusual circumstances and associated with high nitrate levels in plants, the conversion of nitrate to nitrite may exceed the rate of nitrite conversion to ammonia in the rumen, resulting in substantial nitrite in the blood. The nitrite ion is the culpable toxicant, causing vasodilation, and then, if blood concentrations become sufficiently high, there is direct oxidation of hemoglobin (Hb) to methemoglobin (MHb) (Haldane et al. 1897; Emerick 1965; Kiese 1974; Vertregt 1977). Appreciation of the nitrate-nitrite-MHb relationship as a problem in livestock is largely the result of the pioneering studies by Bradley and co-workers (1939, 1940). A dark discoloration of blood is noticeable when approximately 20% of the hemoglobin is oxidized to MHb (Kemp et al. 1977b). This sign is due to the color imparted directly by the presence of MHb rather than to a lack of oxygenation. MHb is not capable of oxygen transport in blood, and the clinical signs reflect the deficiency of tissue oxygenation. Signs of uneasiness and apprehension are reported for swine at 20% MHb (London et al. 1967) However, in most instances with cattle, tissue anoxia results in readily observed signs when MHb concentrations approach or exceed 50%. Although MHb% is used as a guide to severity, it alone is not entirely accurate in this respect, because it reflects only one portion of the overall sequence. The total functional Hb remaining is a truer measure (Dollahite & Holt 1969). Vasodilation is an early response to the nitrite ion, but it does not seem to contribute significantly to the eventual overall intoxication response (Asbury & Rhode 1964; Holtenius 1967).

If nitrite accumulation is not lethal, the MHb is slowly reduced to Hb; a naturally occurring NADH-dependent diaphorase or reductase system in red blood cells mediates this reaction. The half-life of methemoglobinemia in sheep is about 1.5 hours, and thus baseline values reappear in 12–24 hours (Crawford et al. 1966; Watts et al. 1969; Schneider & Yeary 1975; Froslie 1976). Therapeutically, a similar but 5 times more active NADPH-dependent system can be utilized by administration of dyes such as methylene blue, which serve as electron carriers. Their administration results in prompt reduction of MHb levels and restoration of vital tissue oxygenation (Bradley et al. 1939; Bishop & Surgenor 1964; Kiese 1974; Burrows 1979). Fortunately, the effectiveness of the catalytic acceleration of MHb reduction by methylene blue parallels the increased tendency for susceptibility to MHb formation in ruminants (Kiese 1974). Methylene blue was originally identified as an antidote for methemoglobinemia (Steele & Spink 1933; Williams & Challis 1933), and it was later confirmed as appropriate for treatment of nitrate/nitrite intoxication in most animal species; it is most effective in humans and ruminants (Campbell & Morgan 1939; Wendel 1939; Bradley et al. 1940; Scott 1941; Kiese 1974).

- MHb slowly reduced to Hb by natural system; reduction rate greatly increased by methylene blue via an alternative pathway

Administration of methylene blue results in an almost immediate sharp drop in MHb in even the most severe cases (Crawford et al. 1966; Burrows 1979; Bhikane et al. 1990). The dosage may range from 1–2 mg/kg b.w. up to 10 mg/kg in particularly severe cases. The risk of toxicity with methylene blue is low because the LD_{50} is about 42 mg/kg b.w. in sheep, or approximately 10 times the typical dose (Burrows

1984). Its half-life in serum is about 1.5 hours; therefore most is cleared in 7–8 hours, and the dosage can be repeated. However, much of the dye seems to be accumulated in tissues, and prolonged withdrawal times of 30 days or more should be adhered to following treatment.

Other related dyes such as tolonium chloride (tolonium blue), which facilitate MHb reduction to Hb, are effective but have a much narrower therapeutic index (Burrows 1979; Gupta et al. 1992; Cudd et al. 1996). There is little to recommend their use instead of methylene blue. Experimentally, oxygen has been shown to provide an added benefit to methylene blue in mice, but it is not likely to be available or economically appropriate in most instances (Sheehy & Way 1974). Alternative agents such as ascorbic acid and menadione may exert a slight beneficial effect on reducing MHb to Hb, but they are much too slow in action to reverse an acute life-threatening episode (Kemp et al. 1976; Van Dijk et al. 1983; Besong et al. 1994). Sodium selenite also facilitates MHb reduction, but only very slowly (Masukawa & Iwata 1979).

- reproductive effects, acute intoxication, abortions; chronic exposure, effects unknown, weak calves

An additional complication of nitrate intoxication may be abortion. There is little doubt that abortion is a serious problem with acute nitrate intoxication in pregnant cows. The effects are likely nonspecific and probably occur with all stages of gestation, although they seem to be most common in the last trimester, when fetal oxygen demands are highest. Abortion may be expected several days to weeks after the appearance of clinical signs in the dam. The risk of abortion in the absence of disease in the dam is probably also possible if there are appreciable nitrate levels in the plants or if the circulating levels of MHb are high, i.e., about 50% or close to that, causing clinical signs (Winter & Hokanson 1964; Abbitt 1982; Johnson et al. 1994).

The significance of ingestion of low levels of nitrates as a cause of disease is not clear. Much has been said about this possibility, and indeed in some instances there has appeared to be a strong relationship between low levels of nitrates and reproductive effects such as abortion (Sund et al. 1957; Simon et al. 1958, 1959a,b; Kahler et al. 1975). Other effects suggested include interference with vitamin A and decreased milk and meat production (Case 1957). However, the bulk of the evidence as yet fails to confirm the reality of low nitrate consumption as a clear reproductive risk (Davison et al. 1964, 1965; Winter & Hokanson 1964; Emerick 1974; O'Hara & Fraser 1975; Bruning-Fann & Kaneene 1993). In addition, there seem not to be any clinical effects on vitamin A, the thyroid, or animal productivity. Concentrations of 1.24% in oat hay given for 35 days did not produce abortion or reduce weight gains in pregnant heifers (Crawford et al. 1966). In spite of the lack of experimental evidence and the skepticism about nitrate as a cause of disease, there continue to be reports of sublethal effects such as depression of newborn calves and decreased milk production in cows (Johnson et al. 1994). Furthermore, weak calves with elevated ocular nitrate levels but no increase in MHb respond to methylene blue treatment (Johnson et al. 1983).

The longer half-life of nitrate in cattle than in sheep indicates the possibility of cumulative effects in the former (Johnson et al. 1992). Cumulative effects may occur because of nitrate ingestion throughout the day, but even with ingestion restricted to brief periods, cumulative actions on MHb formation are apparent (Kemp et al. 1976). There is also the possibility of cumulative effects even at a dosage not associated with appreciable MHb formation, because effects other than via MHb formation may also be of importance (Case 1957; Setchell & Williams 1962). The likelihood of effects on the fetus are underscored by the persistence of nitrate in the body beyond 24 hours (Johnson et al. 1994), even though Malestein and co-workers (1980) reported MHb higher in maternal than fetal blood and low oxygen transfer across the placenta. Other studies indicate that the low PO_2 in the fetus results from a marked reduction in oxygen transfer from maternal blood due to high levels of MHb and low perfusion pressures rather than from the minimal increases of MHb in fetal blood (Van't Klooster et al. 1990).

Clinical Signs—When 40–50% of the circulating Hb has been oxidized to MHb, mild clinical signs such as discoloration of the mucous membranes, depression, and a slight increase in respiratory rate may be seen upon close observation. It is generally not until MHb is increased to about 60% that distinct signs such as severe incoordination, rapid respiration, apprehension, and sometimes belligerence are seen. The animals may remain recumbent unless approached; they may be unable to rise or may only do so with much difficulty. Further increases in MHb are accompanied by a worsening of the signs until death ensues at approximately 75–80% MHb or more.

■ depression, increased respiration rate, ataxia, apprehension, recumbency

Diagnosis may be based on evaluation of tissues in the animal or surmised from the signs and determination of nitrate concentrations in the plants. The toxicants can be measured in body fluids. Determination of nitrate level is generally most appropriate because nitrate has a longer half-life and remains elevated for a longer time than nitrite and is therefore easier to detect (Garner et al. 1960; Schneider & Yeary 1975). This is especially true in cattle, in which nitrate is detectable in blood for 24 hours, whereas nitrite is negligible after 2 hours. In many instances ocular fluid (aqueous humor) is the most appropriate sample because it is easy to obtain and evaluate. Typically nitrate concentrations are several ppm in body fluids to 100 ppm or more in the blood in the more obvious instances of intoxication. Nitrate concentrations in ocular fluid are about 65% of those in blood, declining gradually by about 50% over 48 hours (Boermans 1990; Johnson et al. 1994). If refrigerated, they remain high for about 1 week, and if frozen, they are high for about 1 month (Boermans 1990). Nitrate values in excess of 10 ppm in ocular fluid are suspicious and in excess of 20 ppm are indicative of toxicity. Newborn calves may present a special diagnostic problem because their levels of ocular fluid nitrate are typically increased in the first 24 hours after birth, in contrast to levels in the fetus or adult (Johnson et al. 1994). Nitrite in blood increases from nondetectable before exposure to nitrate up to nearly 1 ppm when signs of intoxication are observed.

■ diagnosis, determination of nitrate, in body fluids, ocular; in plants; testing with diphenylamine

Several approaches are used to test for the presence of nitrate and/or nitrite on a qualitative rather than quantitative basis. The most commonly used is the diphenylamine (DPA) test (table 58.13), which has been widely used for screening urine and ocular fluid samples. It is also useful for plasma and for pericardial and peritoneal fluids (Bhikane & Singh 1990). An easier method is the use of commercial

Table 58.13. The diphenylamine test

1. Preparation of stock solution of the reagent
   a. Dissolve 0.5 g of diphenylamine in 20 ml of water.
   b. Add concentrated sulfuric acid to make 100 ml.
2. Testing plant stems
   a. Use stock solution directly, or dilute to 0.25 or 0.1%.
   b. Apply 1 drop to cut surface of lower stem.
3. Testing body fluids
   a. Immediately prior to use, dilute 5 ml of stock solution with 40 ml of concentrated sulfuric acid to yield 0.063% solution.
   b. Add 1 drop of a body fluid to 3 drops of diluted reagent.
4. Results and interpretation
   a. Immediate formation of a dark blue color indicates the presence of toxic levels of nitrate—10,000 ppm in plant stem and 10 ppm in body fluid.
   b. The test does not distinguish between nitrate and nitrite. Chlorates, bromides, iodates, and selenites give a false positive test.

Sources: Hoffer 1926; Helwig & Setchell 1960; Householder et al. 1966; Feigl et al. 1972.

reagent test strips commonly used to determine pH, glucose, and nitrite produced by bacteria (Rodriguez et al. 1992; Montgomery & Hum 1995). This is a use not indicated on the label, but it appears to be useful in evaluating the presence of nitrites and indicative of exposure to toxic amounts of nitrate or nitrite. It may be used with urine or other body fluids, including aqueous humor. Another test that has been used is the diazotization test, which employs sulfanilic acid/α-naphthylamine in a color reaction for nitrite (Diven et al. 1962; Osweiler et al. 1985). This test is not as widely used as the diphenylamine test, but it can be used as either a qualitative or a quantitative test (Bhikane & Singh 1991). The presence of MHb can also be determined, but because it is labile, the test is not often done. The MHb in blood samples can be stabilized for 48 hours by diluting 1:20 or 1:100 in 1/60 M phosphate buffer (Sleight & Sinha 1968; Watts et al. 1969). The diphenylamine test can also be applied to forage by placing a drop of the reagent directly on the stems.

- DPA test, reagent concentration must be proportional to amount of nitrate present because excess required for 2-step reaction

The DPA test is a two-step reaction that first involves nitrate oxidation of DPA to diphenylbenzidine. In the second step, additional nitrate oxidizes the colorless diphenylbenzidine to the blue quinone imonium ion (Feigl et al. 1972). Development of color requires an excess of nitrate to ensure completion of both reaction steps. The concentration of the reagent is different for stems and fluids in order to be roughly equivalent to the amount of nitrate present; otherwise an excess of DPA would use up the nitrate, leaving insufficient for the second step. Under such conditions little or no blue quinone imonium ion will be generated (Monier-Williams 1931). Thus, high concentrations of DPA are required for the high nitrate levels in plants, whereas more-dilute DPA solutions are appropriate for the low concentrations in body fluids. The sensitivity can be increased 10-fold by using diphenylbenzidine directly, but in most instances this is not desirable, because sensitivity needs to be reduced in order to have positive reactions reflect only those individuals with significant nitrate concentrations.

- few lesions, perhaps dark-colored tissues

Pathology—Few distinctive lesions are evident at necropsy. The blood and tissues may be dark- or chocolate-colored, but this is not a consistent feature. There also may be scattered, small splotchy hemorrhages and increased foamy fluid in the terminal airways of the lungs, but these changes are not usually extensive.

- methylene blue

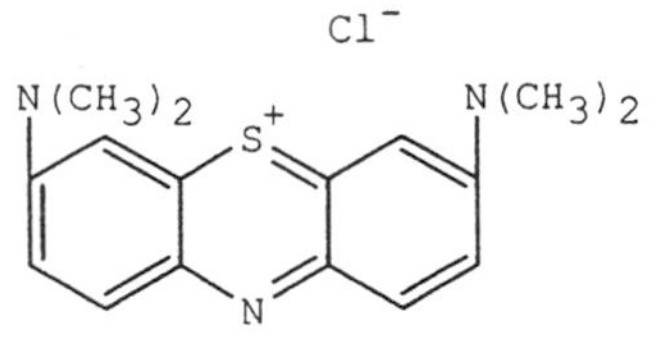

Cl^-

$N(CH_3)_2$ S^+ $N(CH_3)_2$ N

Methylene blue

Treatment—When distinct clinical signs are evident, immediate treatment with methylene blue is imperative because the levels of MHb are very close to being lethal. The response to i.v. administration of a 2% solution of methylene blue at a dosage of 20 ml/100 kg b.w. results in alleviation of the clinical signs in several minutes. In severe cases, treatment with methylene blue can be repeated. The rapid development of signs and death several hours after feeding often results in animals' being found dead without observation of premonitory signs and with little opportunity for treatment.

- management may reduce disease risk; careful nitrogen fertilizer application, harvest under appropriate conditions; corn supplements, *Proprionibacterium* supplements to promote adaptation of ruminal microflora; excess feed to allow animals to eat leafy portions preferentially; ensiling may reduce nitrate levels

A number of approaches are available to reduce the risk of nitrate intoxication (Geurink et al. 1982). They can be grouped into three categories: those directed toward reducing nitrate accumulation in the plants, those minimizing the likelihood of effects in the animal being fed high-nitrate forage, and the possibility of reducing the levels of nitrate already present in the forage. With respect to forage production, several recommendations have been made. One is the judicious use of nitrogen fertilizer appropriate for the available soil moisture. This requires several applications throughout the growing season as appropriate for the moisture levels rather than application of a single large amount early in the season. For example, with the sorghums, less than 100 lb of nitrogen per acre should be applied if less than 30 inches of rainfall is anticipated (Clay et al. 1976). A second recommendation is that the harvest of forages should be done when soil moisture is good and temperatures are warm. If necessary, hay should be cut late in the day. After a prolonged dry period, harvest should be delayed for 1 week after the first good rain if possible (Strickland et al. 1995). All of these recommendations are associated with effective nitrate assimilation by the leafy plants rather than accumulation in the stems. In addition, the higher stem concentrations can be avoided somewhat by raising the cutter bar above 6 inches. Nitrate concentrations are decreased in the ensiling process but may still remain in the toxic range (Wilson 1943; Mizukami et al. 1997).

In some years, hay bales, particularly those of sorghum, often contain 15,000–20,000 ppm nitrate, with some bales having 30,000–40,000 ppm and a few possibly as high as 100,000 ppm (Haliburton & Edwards 1978). The toxicity level is approximately 6,000–10,000 ppm. Avoid the use of forages with greater than 25,000 ppm nitrate as the sole roughage; if necessary, dilute with other forages.

As noted above, the disease cannot be controlled effectively by treatment, because of its abrupt onset and the short time period before death. However, the possibility of intoxication with forages with less than 25,000 ppm can be reduced. The toxicity potential can be reduced by supplementing high-nitrate forage with approximately 3 or more kg of corn per day per adult cow. Other approaches include dilution of the problem forage with hay containing low nitrate levels and/or feeding excessive amounts so that the animals are able to select the upper leafy stems, which typically contain less nitrate than the lower stems. In order to avoid significant changes in the ruminal populations of microorganisms, hay should be tested for nitrate content and fed in such a manner as to avoid abrupt changes in nitrate concentrations. When testing hay bales, sampling should comprise 1 core from each of 20 small square bales; 2 cores from each of 10 large square bales; or 2 cores from the curved side of each of 10 large round bales (Strickland et al. 1995). Another approach to reducing the risk of intoxication is through manipulation of the ruminal microflora. The ability of ruminal bacteria to adapt by increasing their capacity to reduce nitrite as well as nitrate can be augmented by feeding supplements containing *Proprionibacterium* or by administration of the same as an oral ruminal inoculum paste (Bova-pro, FarMore Biochemical, Milwaukee, Wisconsin) (Swartzlander et al. 1993a,b). This supplementation should begin 1–2 weeks prior to the nitrate challenge.

It should be noted that many of the same environmental factors are involved with nitrate toxicity as with cyanogenesis; however, the sites of accumulation and animal management practices differ. These differences are reflected in the recommendations for avoiding intoxication. With cyanide, high stocking rates are recommended, after test feeding with a few individuals, to force the animals to eat all portions of the plants, whereas with nitrate, light stocking rates are used to allow animals to selectively eat the leafy portions and avoid the more toxic stems.

Cyanogenesis

Disease Problems—Cyanide intoxication is characterized by the abrupt onset of severe neurologic signs of a peracute nature; they may begin within minutes of ingestion and become severe in a few more minutes. In cattle, the signs may occur 1–2 hours after the animals begin grazing a new *Sorghum* pasture or after weather changes such as a freeze or rain after a drought period. Many of the factors increasing the potential for intoxication due to cyanide such as rate of growth, stage of growth, moisture, and temperature were known in the 1800s (Leeman 1935).

■ abrupt onset of rapidly fatal disease

Disease Genesis—The cause of this syndrome is a cyanogenic glucoside, dhurrin, which is accompanied in the plant by a hydrolytic β-glucosidase (Dunstan & Henry

- disease of ruminants due to a cyanogenic glycoside, dhurrin; high levels in young plants, variable risk depending on type and variety of *Sorghum*

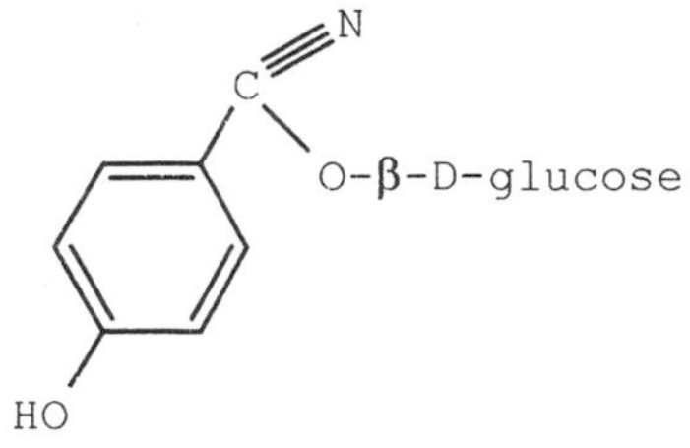

Dhurrin

1902; Erb et al. 1981). The glucoside and enzyme are separated in the plant under normal conditions; 90% of the dhurrin occurs in the vacuoles of the epidermal cells of the leaf blade, and the enzyme occurs in the cells of mesophyll (Saunders & Conn 1978; Kojima et al. 1979; Conn 1991). When the plant is macerated or the tissues otherwise damaged, the glucosidase comes in contact with and hydrolyzes the glucoside to an α-hydroxynitrile aglycone and glucose. The aglycone further dissociates to *p*-hydroxybenzaldehyde and the toxicant hydrocyanic, or prussic, acid (HCN). The latter reaction is not necessarily enzyme dependent. Dhurrin is present mainly as the monoglucoside, plus a small amount of diglucoside (Selmar et al. 1996). There is little or no free HCN in the plant. It is only after ingestion that HCN becomes available for absorption, following hydrolysis due to the glycosidase activity of either the plant or the ruminal microflora. The onset of clinical effects may occur as soon as 10–15 minutes after ingestion, with a similar duration of signs before death in instances of ingestion of plants with especially high concentrations of dhurrin. There is considerable variation in dhurrin concentration between the various varieties or hybrids. In general, grain sorghums are highest in HCN potential, Sudangrass lowest, and the hybrids intermediate (table 58.14) (Menaul & Dowell 1920).

Table 58.14. Examples of the cyanide potential of sorghums

Low Potential
Sudangrasses, common, 'Wheeler,' 'Piper'
Some Sudangrass x sorghum hybrids, 'Haygrazer,' 'Trudan,' 'Rancher,' 'Horizon SP-110'
Intermediate Potential
Sudangrass x sorghum hybrids
Sweet Sudans, common, 'Lahoma,' 'Greenleaf,' various hybrids
Forage sorghums, 'Amber,' 'Black Amber,' 'Dakota Amber,' 'Leoti,' 'Orange'
High Potential
Perennial grass sorghums, Columbusgrass, Johnsongrass
Sweet and grain sorghums, 'Chiltex,' feterita, hegari

Sources: Collison 1919; Couch et al. 1939; Nelson 1953; Gangstad 1959; Jung et al. 1964; Harrington 1966; Barnett & Caviness 1968; Easty et al. 1971; Harms & Tucker 1973; Eck & Hageman 1974; Gorz et al. 1977; Hedges et al. 1978.

- cyanogenic potential of *Sorghums* may be reduced with selective breeding

In general, the low-potential types seldom exceed 50–100 ppm d.w. HCN potential under any circumstances. Those of intermediate potential seldom exceed 200 ppm, but those of high potential may readily exceed 200 ppm in the appropriate conditions. The mean d.w. HCN potential of 11 grain sorghum cultivars 27 days after planting was 4,607 ppm (Eck 1976). In contrast, the HCN potential of Piper, a Sudangrass, was seldom greater than 50 ppm and never exceeded 200 ppm d.w. Piper and Rancher appear to be safe under almost all conditions, while Columbusgrass and Johnsongrass are often dangerous under similar conditions (Crawford 1906; Franzke 1945; Jung et al. 1964). Although some varieties and lines are generally of low risk, in the appropriate circumstances, as described in the following paragraphs, many of the sorghums may accumulate toxic concentrations of the glucosides. Conversely, even the more toxic varieties may be safe under most situations.

The genetic influence on HCN potential allows for selective breeding to reduce the problem; hybrids are generally intermediate to the parents in toxicity (Hogg & Ahlgren 1943; Eck et al. 1975). However, there are impediments to this approach because there may be a protective benefit against disease and predation afforded by the presence of dhurrin (Wheeler 1980). The genetic benefits are most pronounced in the young, high-glucoside, active growth stages (Eck et al. 1975). It is of interest that the intravarietal variation in HCN potential with respect to plants grown in different geographic locations is considerable and is possibly due to differences in climate, soil, altitude, and other environmental factors (Vinall 1921; Leemann 1935; Couch et al. 1939).

Essentially all sorghums contain at least traces of dhurrin. The range of HCN potential is from 10 to almost 8,000 ppm d.w., depending somewhat on which of the numerous assay methods is employed (Haskins et al. 1988). The danger breakpoint for cyanogenic glycosides in general has been variously estimated at between 75 and 200 ppm (Hindmarsh 1930; Couch et al. 1939; Moran 1954; Singh et al. 1983). Above these levels, there is a risk but not necessarily a likely problem, because of the rapidity with which HCN is detoxified in the body. More specifically for the sorghums, based on field studies, a higher risk scale of up to 500 ppm d.w. as generally safe and greater than 750 ppm as dangerous has been well accepted (Boyd et al. 1938). The toxic dose of cyanide in cattle is estimated at 0.6–1 g (Vinall 1921; Hindmarsh 1930; Boyd et al. 1938). Thus, for typical HCN potentials for sorghum of 164 ppm and Sudangrass of 66 ppm, it is estimated that 2–4.5 kg and 4.5–9 kg, respectively, are required for toxicity. However, the seriousness of the problem is underscored by the observation that under certain adverse circumstances, cattle are able to consume lethal quantities of *S. halepense* in as little as 15 minutes (Mathews 1932).

■ >750 ppm dhurrin dangerous; cattle may ingest lethal amount of plant in 15–30 min

Although genetic influences are important factors in the plant, numerous other factors are of equal or greater importance. Perhaps the most important is stage of growth. Peak dhurrin concentrations occur during the first week after caryopsis germination (Hogg & Ahlgren 1943; Akazawa et al. 1960; Loyd & Gray 1970; Singh et al. 1983). Because of the exceptionally high cyanide content of sprouts, serious risk is associated with their consumption by either humans or livestock (Panasiuk & Bills 1984; Ikediobi et al. 1988). The concentrations decline quickly over the next several weeks and then more gradually thereafter. This phenomenon occurs in both high- and low-dhurrin cultivars. The initially high levels in the 2- or 3-leaf stage decline to yield more even distribution between leaves and stems (Kim & Voigtlaender 1985). Even grain sorghums exceeding 4,000 ppm d.w. early in growth may decline to less than 200 ppm 4 months after planting (Eck 1976). When plants are 2.5–15 cm tall, the risk is very high, but it declines markedly as height exceeds 20 cm (Vinall 1921; Mathews 1932; Boyd et al. 1938). The danger is low when Sudangrass has gone to seed.

■ dhurrin concentrations very high in early growth; low when plants are >20 cm tall; very low when in seed

Management factors, especially fertilization practices, are also important in determining cyanogenic potential. Numerous studies confirm the effects of nitrogen on increasing HCN potential, but studies of the action of phosphorus to decrease it are less consistent (Boyd et al. 1938; Nelson 1953; Patel & Wright 1958; Harms & Tucker 1973; Wheeler et al. 1980). Generally, promotion of growth increases the risk of high HCN potential. Thus, large amounts of phosphorus may be required to decrease HCN potential (Kriedman 1964). The risk of increasing HCN potential is markedly increased if nitrogen is applied when phosphorus is limited. Although reports are conflicting, second growth after cutting seems to present a greater cyanogenic risk (Hogg & Ahlgren 1943). Herbicides are probably of little consequence in altering dhurrin concentrations (Wheeler 1980).

■ risk increased with high nitrogen in soil and with moisture stress, frosts, insect damage

risk low with hay or silage

The HCN potential may be reduced in some instances in hay and silage, but this is not a consistent change (Vinall 1921; Boyd et al. 1938; Harrington 1966). Furthermore, the HCN potential may actually increase in the first few days after drying (Harrington 1966). Generally, HCN potential decreases appreciably with slow, thorough drying (Dowell 1919a,b; Harms & Tucker 1973). Rapid or incomplete drying and baling are more likely to result in the retention of considerable HCN potential (Pammel 1919). Perhaps the most important factor is that *Sorghum* forage should not be cut for hay until its height is more than 45–60 cm, which minimizes the HCN potential. All other things being equal, the risk of toxicity is much reduced by delaying grazing until forages are 60 cm high.

It was noted very early, before the specific causes of intoxication were known, that environmental factors are very important. This is especially true for moisture stress. Periods of drought that wilt plants or delay growth so that they remain in the short, highly toxic stage may result in marked increases in HCN potential (Vinall 1921; Mathews 1932). Increased HCN potential is particularly evident when young plants are growing well and then become stunted when a lack of moisture interrupts continued development (Peters et al. 1903). On the other hand, long-term drought when plants grow poorly, if at all, may result in lower HCN potential (Rogers & Boyd 1936). Although high environmental temperature is typically associated with drought, it alone may lower plant cyanide (Kim & Voigtlaender 1985).

■ light frost, marked increase in HCN potential for several weeks

heavy frost, moderate increase in HCN potential for a few days

Similarly, the effects of early frosts on *Sorghum* toxicity differ considerably. The effects are inconsistent and in some situations, as with *S. halepense,* seem to be almost nil (Mathews 1932). Light frosts in which the plant tops are frozen but the plants survive and regrow produce a serious increase in toxicity potential similar to that seen in newly germinated seedlings (Boyd et al. 1938; Wattenbarger et al. 1968). It may be several weeks before the HCN potential subsides. In instances of heavy frosts in which the plants are killed, the risk of increased HCN potential may be apparent while the plants remain frozen, but it is very short-lived, typically several days after the plants thaw and die, or up to a week in some cultivars (Rogers & Boyd 1936; Harrington 1966; Haskins et al. 1984).

A third factor in changes in HCN potential is insect damage. It is not a consistent problem, but both aphid damage and especially locust or grasshopper infestations have been associated with increased risk (Wheeler 1980). It is not clear whether higher HCN levels are a response by the plant to increase the concentration of protective toxicants or whether it is secondary to leaf structural changes and breakdown of the compartmentalized separation of the glycoside and glycosidase.

All animal species are susceptible to the toxic effects of cyanide. However, ruminants appear to be more susceptible to cyanogenic glycosides in plants because the favorable conditions prevailing in the rumen facilitate rapid hydrolysis and absorption (Majak et al. 1990). On the other hand, preformed HCN or cyanide salts may be more toxic to monogastrics; for example, the lethal dose of HCN in horses is reported to be about 1 mg/kg b.w., or about half the toxic dose required for cattle and sheep (Couch 1932; Clawson et al. 1934).

Although it has not been confirmed experimentally, there have been suggestions of differences among cattle breeds in susceptibility to cyanogenic pastures. This may be due in part to animal aggressiveness in consuming the grasses. Brahman cattle are said to be more prone to intoxication, and Herefords the least (Jones 1952).

In addition to the acute toxic effects of cyanide, there may be long-term nutritional effects due to the possible decreased palatability of dhurrin-containing forage for sheep and cattle (Hedges et al. 1978). A loss of animal condition due to a possible deficiency of cystine, because of its role in detoxification of cyanide, and a subsequent decrease in tissue synthesis are also possible (Wheeler 1980). However, these effects are minimized when animals consume taller plants with the lower dhurrin concentrations.

Additional discussion of cyanogenesis is presented in the treatment of the Rosaceae (see chapter 64).

Clinical Signs—There is abrupt onset of apprehension and distress, which occur within a few minutes of grazing the toxic *Sorghum* plants. These signs are quickly followed by weakness, incoordination, and labored respiration. If the intoxication is severe, the animal may be unable to stand and may lay in lateral recumbency with periodic paddling and tetanic seizures. The entire sequence of signs may occur over a 5-to-15-minute period, with either death or recovery at the end of this time.

- very abrupt onset; apprehension, distress, weakness, ataxia, labored respiration, collapse, seizures

Pathology—There are few distinctive changes. For the most part, the changes are limited to congestion of the abdominal viscera and scattered, small splotchy hemorrhages.

- no lesions

Treatment—The disease usually occurs and progresses so rapidly that treatment is seldom available soon enough to be effective. Toxicosis due to marginally lethal dosages may be responsive to general supportive therapy, but more-specific treatment is much preferred. The primary antidote for ruminants is sodium thiosulfate given i.v. at a dose of 0.25–0.5 g/kg b.w. (Burrows 1981). Sodium nitrite, 10 mg/kg b.w. i.v., can be used to augment the effects of the sodium thiosulfate, but it is not required for effective treatment. Additional discussion is presented in the treatment of the Rosaceae (see chapter 64).

- sodium thiosulfate, sodium nitrite

Stipa L.

Taxonomy and Morphology—Distributed from the tropics almost to the Arctic, *Stipa,* a member of the subfamily Pooideae and the tribe Stipeae, comprises approximately 300 cool-season species. The generic name is from the Greek word *stupe,* for "tow" or "oakum," alluding to the feathery awns of some species. Grazed chiefly when young, several North American species are valuable forage plants. The sharp bases of the mature florets, however, cause mechanical injury around the mouth, especially in sheep. In the Old World, species of the genus are used in the manufacture of paper and cordage. Prized both for their robust growth and attractive inflorescences, some are cultivated as ornamentals. With greatest abundance in the West, the genus is represented in North America by some 27–35 species. Some taxonomists have reduced its circumscription and placed some of its species in the genera *Nassella* and *Piptochaetium.* Representative of the genus and of toxicologic interest are the following:

- *S. eminens*
  S. lobata
  S. robusta
  S. viridula

  needlegrass
  sleepygrass

S. eminens Cav.	southwestern needlegrass
S. lobata Swallen	little-awn needlegrass
S. robusta (Vasey) Scribn. (= *S. vaseyi* Scribn.)	sleepygrass
S. viridula Trin. (= *Nassella viridula* [Trin.] Barkworth)	green needlegrass

- perennials; leaves primarily basal; inflorescences panicles; spikelets with 1 floret; glumes longer than floret; floret round, hard, awned, base elongate, sharp

Plants perennials; cespitose. **Culms** 10–150 cm tall; erect; internodes hollow or solid. **Leaves** primarily basal. **Blades** involute or flat; 0.7–5 mm wide. **Ligules** membranous or ciliolate-membranous. **Auricles** absent. **Sheaths** glabrous or indumented; margins free or fused. **Inflorescences** panicles; open or contracted. **Disarticulation** above the glumes. **Spikelets** all alike; lanceolate; compression lateral or terete; 3–12 mm long. **Florets** 1; callus elongate, sharp, densely indumented. **Glumes** 2; equal or subequal; longer than body of lemma; elliptic; membranous to hyaline; apices attenuate; awns absent. **Lemmas** terete; involute, completely enclosing paleas and caryopses; indurate;

Stipa eminens

Stipa lobata

Stipa robusta

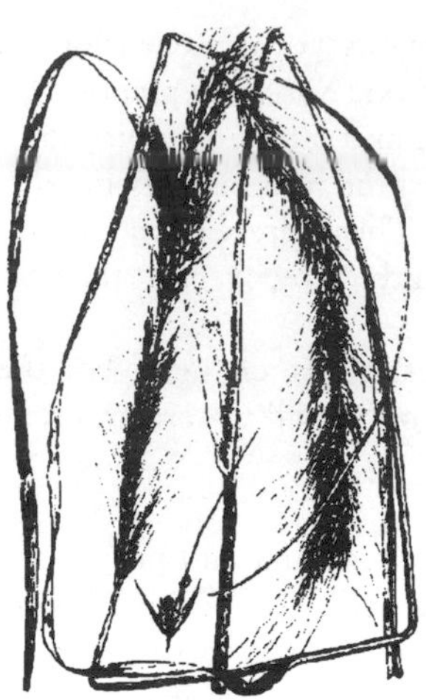

Stipa viridula

Stipa eminens

Stipa lobata

■ neurologic narcotic effects, horses, cattle, become docile, up to several days' duration

awns 1, apical, elongate, twisted proximally, twice geniculate. **Paleas** convex or flat; hyaline or coriaceous; nerves 2; keels absent.

Distribution and Habitat—Species of *Stipa* are present in both the Old World and the New World. In North America, greatest abundance is in the western half. In the northern tallgrass prairies, one of the upland communities is dominated by the genus. In contrast to the other species, *S. robusta* occurs only at high elevations.

Disease Problems—In the late 1800s, there was considerable interest in the narcotic effect of *S. robusta,* widely known as sleepygrass (Havard 1891; Bailey 1903). Although there are 27–35 species of *Stipa* in North America, only *S. robusta* and *S. viridula* have been recognized as causing disease (Bews 1929). An Asian species, *Stipa inebrians* (reported as *Achnatherum inebrians*), and possibly others, which were encountered by missionaries in Mongolia in the 1870s, are also known to have a narcotic effect (Hance 1876). The forage of *S. robusta,* either fresh or in hay, was recognized as a particular problem in horses, seemingly almost exclusively in the Sacramento and Sierra Blanca Mountains of New Mexico, while *S. viridula* was troublesome in the mountains of the state of Coahuila in northern Mexico (Palmer 1887; Marsh & Clawson 1929). Experimental results on plants of these species collected in other areas were generally negative. The narcotic effect has apparently been used to advantage by "experienced" or unscrupulous horse traders to attenuate unruly behavior of unbroken or otherwise intractable horses (Green 1967). The animals remain docile or sleepy for several days, with all their movements reduced to "slow motion." In some instances with inadvertent animal exposure to the plants, riders not only were exposed to dangers while riding but also were unable to escape pursuers when in hostile country (Smalley & Crookshank 1976).

Eaten primarily in the late fall and early spring, *S. robusta,* when present, prevents full utilization of some mountain grazing ranges. Horses are at greatest risk, but the effects are seen also in cattle and to a lesser extent in sheep. The grass seems to be equally toxic whether green or in hay (Marsh & Clawson 1929). Dosage of 1% b.w. of hay in horses and cattle and 4% b.w. in sheep produces a narcotic effect. An observation common to all reports on the toxicity of sleepygrass is that animals will eat it only once and then avoid it thereafter (Palmer 1887; Smalley & Crookshank 1976). The effects may last several days to a week in horses. Although *S. robusta* may temporarily increase in abundance with overgrazing of more-palatable forages, in the long run it does not compete well with other grasses in an area (Wolfe 1979). It is generally of little importance as a toxic plant because it is limited in distribution to higher ranges and is of low palatability. Sleepygrass is now more of a toxicologic curiosity than a significant disease problem.

Stipa robusta

Disease Genesis—The cause of the narcosis was originally thought to be extractable by either methanol or ethanol and to possibly be diacetone alcohol (Epstein et al. 1964; Smalley & Crookshank 1976). Experimentally, diacetone alcohol (1.2% d.w.), found in the leaves, causes a narcotic effect and postural abnormalities in rats and mice. However, a more likely cause of the problem are toxins produced by an endophytic fungus of the family Clavicepitaceae that have been isolated from several species of *Stipa* and consistently from *S. eminens* and *S. robusta. Neotyphodium chisosum* (formerly *Acremonium chisosum*), a fungus similar to that present in fescue and perennial ryegrass, has been proposed as a cause of the narcosis (White & Morgan-Jones 1987; White & Halisky 1992; Glenn et al. 1996). Similar-appearing fungi also have been observed in *S. lobata* and *S. viridula,* but less consistently. *Neotyphodium chisosum* produces a series of closely related ergot alkaloids such as lysergic acid amide, isolysergic acid amide, and ergonovine. Additional toxins include *N*-formyl-loline, chanoclavine, harman, and norharman, but these compounds are in such low concentrations as to be of little significance (Powell & Petroski 1992; TePaske et al. 1993). The effects of the dominant lysergic and isolysergic acid amides are consistent with the sedative effects of *S. robusta* in livestock (Fanchamps 1978; Petroski et al. 1992).

Stipa viridula

Another clavicepitaceous endophyte has been identified as the probable cause of the similar intoxication problem of livestock in Asia that consume *Stipa inebrians,* commonly known as drunken horse grass (Bruehl et al. 1994). Infection of the grass with this *Neotyphodium*-like fungus results in very high levels of ergonovine and lysergic acid in the plant tissues (Miles et al. 1996).

- cause is probably fungal endophytes that produce ergot alkaloids, lysergic acid amide, ergonovine

Chanoclavine I

N-Formyl-loline

Ergonovine

Clinical Signs and Pathology—Several hours after ingesting *Stipa,* the animals become quite docile. All movements are slowed, and the animals drag their feet when walking. Horses are especially prone to the effects and are sleepy and difficult to arouse. Their heads are drooped low, the limbs are crossed, and if coaxed to move, they stagger weakly. With forced movement, sweating may be particularly noticeable. The narcosis may last several days in horses, and usually 24 hours or less in cattle and sheep. Signs in sheep are quite mild. Deaths are rare and probably most likely due to misadventure.

- animals become docile, sleepy, difficult to arouse, lasting up to several days

Treatment—Because cattle and sheep would eat the forage only once, Smalley and Crookshank (1976) recommended feeding hay containing plants of the genus before taking animals to higher-elevation ranges, e.g., while they are in pens awaiting release onto the range.

- transient effects

Triticum L.

■ *T. aestivum*
bread wheat
common wheat
wheat

Taxonomy and Morphology—Native to southern Europe and western Asia, *Triticum,* a member of the subfamily Pooideae and the tribe Triticeae, is the principal cereal of temperate regions. It occupies more acreage of farmland than any other crop plant. Domesticated in 8,000–7,500 B.C., it comprises 10–20 cool-season species and countless cultivars and races. A classic example of diversification via polyploidy, its three ploidy levels, generally referred to as the einkorn, durum, and bread wheats, consist of wild and domesticated species. In North America, primarily hexaploid bread wheat is grown for its grain and to provide winter forage for livestock. The species is this:

T. aestivum L. wheat, bread wheat, common wheat

■ erect, cespitose annuals; ligules membranous; inflorescences 2-sided spikes; spikelets laterally compressed, with 2–5 florets; glumes and lemmas asymmetrical

Triticum aestivum

■ cultivated crop

■ disease problems—
tetany
nitrate accumulation
ARDS
photosensitization
sudden death

Plants annuals; cespitose. **Culms** 50–120 cm tall; erect; not branched above; nodes glabrous; internodes hollow. **Leaves** both basal and cauline. **Blades** flat; 10–20 mm wide. **Ligules** membranous. **Auricles** present. **Sheaths** glabrous; margins free, overlapping. **Inflorescences** terminal 2-sided spikes; spikelets 1 per node; attachment lateral (margins of lemmas against rachises); rachises continuous or disarticulating. **Disarticulation** above the glumes; mature caryopses dropping from florets. **Spikelets** all alike; compression lateral; 10–15 mm long excluding glumes. **Florets** 2–5. **Glumes** 2; unequal; shorter or longer than lemmas; asymmetrical; ovate; thick; indurate; nerves 3 or 5 or 7; keels present; awns absent or present. **Lemmas** asymmetrical; broadly ovate; keels present; awns absent or present, 1, apical. **Paleas** hyaline; nerves 2; keels 2.

Distribution and Habitat—In North America, *T. aestivum* is typically planted in the fall. A C_3 photosynthetic species, it grows through the winter, and in the southern portion of North America it commonly provides forage for cattle before flowering in the late spring or summer. Waifs are found along roadsides and in fallow fields but do not persist.

Disease Problems and Genesis—Several disease problems have been encountered in cattle on wheat pasture. These include grass tetany, nitrate toxicosis, acute respiratory distress syndrome (ARDS), and photosensitization, as described at the beginning of this chapter. In addition, sudden deaths unrelated to the foregoing problems occur occasionally (Hollis 1984). These deaths are associated with exceptionally high potassium levels in the serum.

The photosensitization appears to be similar to that seen with *Avena* and other grasses and is not unusual when animals are grazing lush winter wheat pasture (Schmidt 1931). Because the disease is not typically accompanied by icterus and other evidences of hepatic dysfunction, it is assumed to be a primary or direct photosensitization. The specific photodynamic agent or the factors involved in its formation are unknown.

Like other cereal grain forages, *Triticum* is capable of accumulating excess concentrations of nitrates under certain circumstances. Abundant nitrogen fertilization to promote high-grain yields is an important factor. Fresh forage is much less of a problem than hay or straw because its high moisture content makes its rate of intake slower than that of dry forage. Additional discussion of nitrate as a toxicant is presented in the treatment of *Sorghum,* in this chapter.

Dhurrin, a cyanogenic glucoside, may be present in *Triticum,* and under some conditions its concentrations may be elevated, but the toxicologic significance is not clear (Erb et al. 1981).

Zea L.

Taxonomy and Morphology—Native to Mexico and Central America, *Zea,* a member of the subfamily Panicoideae and the tribe Andropogoneae, comprises 1–3 warm-season species. The generic name is derived from the Greek *zea* or *zeia,* for a kind of grain. Apparently domesticated some 8,000–10,000 years ago, *Z. mays* is one of the most important and widely grown cereals. Innumerable varieties and races have been developed and are grown for grain, forage, fodder, oil, syrup, and alcohol. In North America, all species are present. Of toxicologic significance is the following:

- *Z. mays*
  corn
  Indian corn
  maize

Z. mays L. maize, corn, Indian corn

Although the common names *corn* and *maize* are quite familiar to us here in North America, it must be remembered that elsewhere in the world they may be applied to other grasses. *Corn* may be used in a general sense for the predominant cereal grown in an area; for example, *Triticum aestivum* (wheat) is known as corn in England. Likewise, *Avena sativa* (oats) is called corn in Scotland. The name *maize* is used by some individuals for cultivars of *Sorghum bicolor* (sorghum).

Plants robust annuals; culms solitary; monoecious. **Culms** 150–250 cm tall; erect; producing prop roots at lower nodes; not branched above; nodes glabrous; internodes solid. **Leaves** cauline. **Blades** flat; 30–70 mm wide; 35–60 cm long. **Ligules** membranous or ciliate-membranous. **Auricles** absent. **Sheaths** glabrous or rarely indumented; margins free, overlapping. **Inflorescences** of 2 types, staminate and pistillate different; staminate inflorescences (tassels) terminal panicles of rames; pistillate inflorescences (ears) comprising 8–30 rows of paired spikelets borne on a thickened axis (cob)and subtended by bracts (husk). **Disarticulation** not occurring. **Spikelets** of 2 forms, staminate and pistillate different; compression dorsal. **Staminate Spikelets** sessile or short pedicelled; lanceolate to elliptic; 7–9 mm long; florets 2, staminate; glumes 2, equal, different, longer than lemmas, membranous, nerves 5 or 7; lemmas hyaline, awns absent; paleas hyaline. **Pistillate Spikelets** sessile; florets 2, of 2 forms, lower floret neuter or rarely perfect, upper floret perfect; lemmas hyaline; paleas hyaline; styles elongate, ends protruding from husks at anthesis; caryopses large, thick, protruding from florets at maturity.

- robust annuals; staminate spikelets borne in terminal tassels; pistillate spikelets borne on axillary ears

Zea mays

Distribution and Habitat—Although unable to exist without cultivation, *Z. mays* is unrivaled among the cereals in terms of distributional and ecological range. It is grown throughout the world in a plethora of environmental conditions. Caryopses dropped from grain trucks account for its occasional appearance along roadsides or in waste areas.

- crop plant

Disease Problems—Several potential toxicologic problems involve *Zea mays.* One is cornstalk disease, a problem apparently caused by nitrate and possibly cyanide. Other problems are caused by mycotoxins.

- disease due to nitrate, cyanide, or mycotoxins

Nitrate Accumulation/Cyanogenesis

Disease Problems and Genesis—Cornstalk disease has long been recognized as a risk when cattle are turned into cornfields after harvest. The animals may sicken and die abruptly in 1–2 days (Price 1904). The cause of these abrupt deaths has been attributed to either cyanogenesis or nitrate accumulation by the plants. Even though corn leaves are weakly cyanogenic and contain the appropriate glucosidase enzymes, the

- cyanogenesis and high risk of nitrate accumulation, especially with high nitrogen fertilization, high temperature, low moisture

problem is more consistent with nitrate as a cause. Nitrate concentrations vary in *Z. mays* in the same fashion as in *Sorghum,* and most of the same factors are involved. Increased nitrate concentrations, especially in seedlings, are observed with increased temperature and decreased light and soil moisture (Younis et al. 1965). Concentrations in the stalks are 5-to-10-fold greater than those in leaves (Brady et al. 1955). Under certain conditions, especially in times of drought, corn represents a serious toxicologic risk as a nitrate accumulator. During the droughts of the 1890s, concentrations of 200,000 ppm KNO_3 or more were noted in cornstalks associated with cattle deaths (Mayo 1895). Crystals formed on the cut surfaces, and the stalks burned with a sizzle in much the same manner as a fuse. Similar episodes have occurred in subsequent times of severe moisture stress. Numerous deaths of cattle occurred in the 1950s in Missouri when nitrate concentrations of up to 80,000 ppm KNO_3 were recorded (Brady et al. 1955; Muhrer et al. 1955; Pfander 1955) and in Oklahoma in 1977 when KNO_3 concentrations reached 100,000 ppm (Haliburton & Edwards 1978). In Missouri, where the levels in the stalks were 5-to-10-fold greater than in the leaves, the problem was alleviated somewhat by ensiling the stalks, which decreased KNO_3 levels by about 50%. The rate of nitrogen fertilization has a marked influence on nitrate accumulation in *Z. mays.* In an experiment carried out in 1985–1986, postharvest NO_3 accumulation in lower stalks of corn left standing in the field varied from 6,000 ppm (with 100 lb of nitrogen applied per acre) to 18,000 ppm (with 200 lb/acre) and 28,000 ppm (with 300 lb/acre) (Johnson et al. 1992). During the winter the concentrations ranged from 100 to 60% of those at postharvest. By early spring the levels ranged from 70 to 50%. It was hypothesized that thawing caused soil water to move the NO_3 up the stalk, even long after senescence. Thus, high rates of nitrogen fertilization pose a serious risk of causing toxic levels of nitrate to accumulate.

Mycotoxicoses

- mycotoxicoses due to *Fusarium, Aspergillus,* infection of corn

  leukoencephalomalacia of horses eating moldy corn; also a tremorgenic disease of cattle eating unharvested corn

Disease Problems—Corn has been implicated as one of the primary hosts for various fungi, including *Fusarium* and *Aspergillus,* which under the appropriate circumstances produce mycotoxins. Infection of the caryopses occurs either in the field or in storage after harvest. First reported in the late 1800s, a serious neurologic problem of horses has been attributed to moldy corn (Schwarte 1938). Originally called moldy corn poisoning, it was later more specifically called equine leukoencephalomalacia. It was reproduced experimentally with administration of 1.5–1.75 lb of damaged corn daily for 5 weeks to a horse, which subsequently exhibited development of the typical softening of the white matter of the brain (Butler 1902).

A less common disease is associated with allowing cattle to feed in unharvested cornfields in which the overwintered stalks have become lodged. In this instance a tremorgenic or staggers syndrome results.

Disease Genesis—The fungus *Fusarium moniliforme* is now well accepted as the cause of equine leukoencephalomalacia, and presumably the fumonisins are the specific toxins involved.

- *Fusarium moniliforme* produces fumonisins, which cause leukoencephalomalacia

  Aspergillus flavus produces aflatrem, a tremorgen

In the case of the less common staggers syndrome, the cause appears to be infection of *Z. mays* with *Aspergillus flavus* (Gallagher et al. 1980a). In this instance the toxins are tremorgens, aflatrem, and other alkaloids such as paspalinine. They are produced in a different part of the fungal life cycle than is aflatoxin, the well-known hepatotoxin (Cole & Dorner 1986). Except for an apparent shorter duration of action, as shown experimentally in mice, aflatrem is very similar in effect and structure to lolitrem from *Lolium* (Gallagher & Hawkes 1986). High-conductance Ca^{2+}-activated K^{+} channels are inhibited, but this does not seem to be associated with tremors (Knaus et al. 1994). The signs and treatment are those of the other grass staggers syndromes, such as those produced by *Lolium* and *Paspalum* (treated in this chapter).

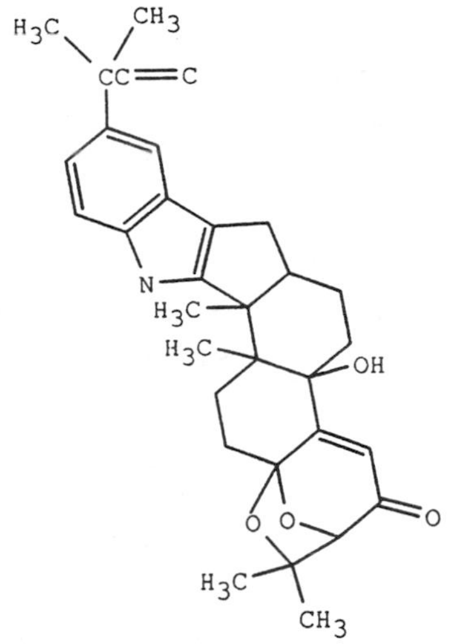

Aflatrem

Genera of Questionable Toxicity

Brachiaria (Trin.) Griseb.

- *B. decumbens*
  Surinam grass
- tropical pasture grass, not typically grown in North America
- photosensitization, crystalloid liver disease; sapogenin isomers; calcium salts of sapogenin glucuronides; retention of phylloerythrin

Comprising about 100 warm-season grasses, *Brachiaria,* commonly known as signal grass, is a member of the subfamily Panicoideae and the tribe Paniceae. It is most abundant in the tropics of the Old World. In North America, the genus is represented by 20 native and introduced species. Toxicity has been specifically associated with only 1 tropical species, *B. decumbens* Stapf (Surinam grass). This is an African grass grown for pasture in many areas of the world but not typically in North America. It is the cause of several problems, including hepatogenous photosensitization, kidney disease, and decreased rumen motility (Abas Mazni et al. 1983; Salam Abdullah et al. 1988, 1989). Similar to species of *Panicum, Tribulus, Agave,* and *Nolina,* it causes crystal-associated or crystalloid cholangiohepatopathy (Graydon et al. 1991). Neurotoxic effects including incoordination, head pressing, and circling, are seen occasionally, and are probably secondary to the liver problems (Salam Abdullah et al. 1989).

There has been some uncertainty concerning the cause of the liver disease due to this otherwise useful tropical forage (Mullenax 1991). Sporidesmin from the fungus *Pithomyces chartarum* has been suggested as a cause, and its presence has been confirmed in rare field cases (Mullenax 1991; Alessi et al. 1994). In most instances, however, sporidesmin has not been present in samples taken from forages causing liver disease. Furthermore, the crystalloid cholangiohepatopathy observed with *Brachiaria* differs from the hepatocellular injury and necrotizing, obliterative cholangitis and pericholangitis caused by sporidesmin (Smith et al. 1993; Meagher et al. 1996).

As do other genera causing crystalloid cholangiohepatopathy, *B. decumbens* contains large concentrations of a mixture of 3-spirostanol sapogenin isomers, mainly diosgenin and yamogenin (Meagher et al. 1996). These compounds are metabolized in the rumen to epitype genins—diosgenin to epismilagenin and yamogenin to episarsapogenin (Salam Abdullah et al. 1992; Lajis et al. 1993; Meagher et al. 1996). Ruminal fluid from intoxicated sheep, but not the actual plants, caused liver and kidney degeneration in cattle (Noordin et al. 1989). Similarly, extracts of sheep ruminal fluid containing these sapogenins caused degenerative liver changes when administered intraperitoneally to mice (Salam Abdullah et al. 1992). It may be expected that biliary crystals associated with the disease are calcium salts of sapogenin glucuronides, as discovered in the crystalloid cholangiohepatopathy produced by *Panicum* and *Tribulus.* Saponins also may be a factor in causing poor performance of weaner cattle grazing regrowth of *B. decumbens* after it is burned (Low et al. 1994). As a result of the effects of the crystalloid material on the bile ducts, there presumably is a relative inability to excrete phylloerythrin. The latter is formed in the digestive tract during microbial degradation of chlorophyll and normally eliminated by the liver via the biliary system and in urine (Ford & Gopinath 1974, 1976). Phylloerythrin is a photodynamic compound, which, when subjected to ultraviolet light penetration in the skin, causes the tissue necrosis.

The clinical signs, pathology, and treatment are typical of other types of photosensitization, as described at the beginning of this chapter.

Hordeum L.

■ *H. vulgare*
cultivated barley

■ photosensitization; nitrate accumulation; also alkaloids of unknown risk are present

Distributed in temperate regions throughout the world, *Hordeum*, commonly known as barley, is a member of the subfamily Pooideae and the tribe Triticeae and comprises some 40 cool-season species. Its members include the economically important cereal *H. vulgare* L. (cultivated barley), good forage species in the West, and a number of weedy species distributed across the continent.

Adverse effects, typical of the grasses in general, have been reported for *H. vulgare* and other species of the genus. They include mechanical irritation caused by the long bristles or awns of the spikelets, photosensitization, and nitrate accumulation. The latter two problems are relatively rare and generally of little economic importance.

In addition to nitrate and the toxicants that cause photosensitization, concentrations of gramine up to 1,400 ppm and hordenine up to 350 ppm d.w. occur in *Hordeum*. The levels of these two alkaloids vary considerably with genotype or cultivar, but environmental factors are the most important determinants; e.g., there is a marked increase of hordenine with decreased light levels (Lovett et al. 1994). However, there appears to be little risk of intoxication. In horses, i.v. administration of hordenine produces only transient, mild increases in respiratory and heart rates (Frank et al. 1990). In addition, its absorption is slow, and its elimination is rapid, as indicated by its half-life of 35 minutes; thus systemic levels are negligible.

In some instances, species of *Hordeum*, such as the European *H. bogdani* Wilensky and *H. brevisubulatum* (Trin.) Link, may be infected with endophytic fungi that produce small amounts of ergovaline and/or other similar alkaloids (TePaske et al. 1993). The toxicologic significance of these alkaloids is not known.

References

Abas Mazni O, Mohd Khusahry Y, Sheikh-Omar AR: Jaundice and photosensitization in indigenous sheep of Malaysia grazing on *Brachiaria decumbens*. *Malays Vet J* 7;254–263, 1983

Abbitt B: A case of nitrite-induced abortion in cattle. *Southwest Vet* 35;12, 1982

Abney LK, Oliver JW, Reinemeyer CR: Vasoconstrictive effects of tall fescue alkaloids on equine vasculature. *J Equine Vet Sci* 13;334–340, 1993

Acton KS, Boermans HJ, Bray TM: The role of prostaglandin H synthase in 3-methylindole–induced pneumotoxicity in goats. *Comp Biochem Physiol* 101;101–108, 1992

Adams LG, Dollahite JW, Romane WM, Bullard TL, Bridges CH: Cystitis and ataxia associated with sorghum ingestion by horses. *J Am Vet Med Assoc* 155;518–524, 1969

Aiken GE, Bransby DI, McCall CA: Growth of yearling horses compared to steers on high- and low-endophyte infected tall fescue. *J Equine Vet Sci* 13;26–28, 1993

Akazawa T, Miljanich P, Conn EE: Studies on cyanogenic glycoside of *Sorghum vulgare*. *Plant Physiol* 35;535–538, 1960

Aldrich CG, Paterson JA, Tate JL, Kerley MS: The effects of endophyte-infected tall fescue consumption on diet utilization and thermal regulation in cattle. *J Anim Sci* 71;164–170, 1993a

Aldrich CG, Rhodes MT, Miner JL, Kerley MS, Paterson JA: The effects of endophyte-infected tall fescue consumption and use of a dopamine antagonist on intake, digestibility, body temperature, and blood constituents in sheep. *J Anim Sci* 71;158–163, 1993b

Alessi AC, Fagliari JJ, Okuda HT, Passipieri M: Natural poisoning of cattle by sporidesmin mycotoxin. 4. Study on liver pathological changes. *Arq Bras Med Vet Zootec* 46;319–328, 1994

Allcroft R, Burns KN: Hypomagnesemia in cattle. *N Z Vet J* 16;109–128, 1968

Allen VG, Robinson DL: Occurrence of Al and Mn in grass tetany cases and their effects on the solubility of Ca and Mg in vitro. *Agron J* 72;957–960, 1980

Allison MJ, Reddy CA: Adaptations of gastrointestinal bacteria in response to changes in dietary oxalate and nitrate. In *Current Perspectives in Microbial Ecology,* Klug MJ, Reddy CA eds, American Society for Microbiology, Washington, DC, pp248–256, 1984

Anas K, Cross DL, Poling R, Redmond LM, Campbell CE: A survey concerning the equine fescue toxicosis malady. *J Equine Vet Sci* 18;631–637, 1998

Anderton N, Cockrum PA, Walker DW, Edgar JA: Identification of a toxin suspected of causing sudden death in livestock grazing phalaris pastures. In *Plant-Associated Toxins: Agricultural, Phytochemical, and Ecological Aspects,* Colegate SM, Dorling PR eds, CAB International, Wallingford, UK, pp269–274, 1994

Anderton N, Cockrum PA, Colegate SM, Edgar JA: Assessment of potential for toxicity of *Phalaris* spp. via alkaloid content determination: *P. coerulescens,* a case example. *Phytochem Anal* 10;113–118, 1999

Antoine D, Thejomoorthy P, Napolean RE, Prabakar TG, Appaji Rao VN, Padmanabhan PN, Kathaperumal V: Nitrite poisoning in cattle. *Indian Vet J* 70;1172–1173, 1993

Appleton ML, Larson DN, Skiles GL, Nichols WK, Yost GS: Toxicity of 3-methylene-oxindole, a proposed reactive intermediate in the metabolism of 3-methylindole. *Adv Exp Med Biol* 283;245–248, 1991

Asbury AC, Rhode EA: Nitrite intoxication in cattle: the effects of lethal doses of nitrite on blood pressure. *Am J Vet Res* 25;1010–1013, 1964

Bacon CW: Toxic endophyte-infected tall fescue and range grasses: historic perspectives. *J Anim Sci* 73;861–870, 1995

Bacon CW, De Battista J: Endophytic fungi of grasses. In *Handbook of Applied Mycology,* vol 1, *Soil and Plants,* Arora DK, Rai B, Mukerji KG, Knudsen GR eds, Marcel Dekker, New York, pp231–256, 1991

Bacon CW, Siegel MR: Endophyte parasitism of tall fescue. *J Prod Agric* 1;45–55, 1988

Bacon CW, Porter JK, Robbins JD: Toxicity and occurrence of *Balansia* on grasses from toxic fescue pastures. *Appl Microbiol* 29;553–556, 1975

Bacon CW, Porter JK, Robbins JD, Luttrell ES: *Epichloe typhina* from tall fescue grasses. *Appl Environ Microbiol* 34;576–581, 1977

Bailey EM Jr: *Panicum* spp. toxicity in southern USA. In *Plant Toxicology, Proceedings of the Australia-USA Poisonous Plants Symposium,* Seawright AA, Hegarty MP, James LF, Keeler RF eds, Animal Research Institute, Yeerongpilly, Brisbane, Australia, pp495–498, 1985

Bailey V: Sleepy grass and its effect on horses. *Science* 17;392–393, 1903

Ball DM, Hoveland CS, Lacefield GD: *Southern Forages.* Potash & Phosphate Institute/Foundation for Agronomic Research, Atlanta, Ga, pp169–174, 1991

Ball DM, Schmidt S, Lacefield G, Hoveland CS, Young WC: *Tall Fescue/Endophyte/ Animal Relationships.* Oregon Tall Fescue Commission, Salem, 1995

Bankowski RA, Wichmann RW, Stuart EE: Stomatitis of cattle and horses due to yellow bristle grass *(Setaria lutescens). J Am Vet Med Assoc* 129;149–152, 1956

Barker JR: Blood plasma changes and variations in the female bovine toxaemias. *Vet Rec* 51;575–580, 1939

Barnett RD, Caviness CE: Inheritance of hydrocyanic acid production in two sorghum × Sudangrass crosses. *Crop Sci* 8;89–91, 1968

Bartol JM, Thompson LJ, Minnier SM, Divers TJ: Hemorrhagic diathesis, mesenteric hematoma, and colic associated with ingestion of sweet vernal grass in a cow. *J Am Vet Med Assoc* 216;1605–1608, 2000

Barton NJ, McOrist S, McQueen DS, O'Conner PF: Poisoning of cattle by *Glyceria maxima. Aust Vet J* 60;220–221, 1983

Beconi MG, Howard MD, Forbes TDA, Muntifering RB, Bradley NW, Ford MJ: Growth and subsequent feedlot performance of estradiol-implanted vs. nonimplanted steers grazing fall-accumulated endophyte-infested or low-endophyte tall fescue. *J Anim Sci* 73;1576–1584, 1995

Bedotti DO, Busetti MR, Suarez VH, Fort MC: Hepatogenic photosensitization in sheep caused by ingestion of millet (*Panicum miliaceum* L.). *Vet Argent* 8;109–112, 1991

Begg H, Whiteford WA: Acute interstitial pulmonary emphysema of bovines. *Vet Rec* 60;135, 1948

Bennett-Wimbush K, Loch WE: The efficacy of fluphenazine as a treatment for fescue toxicosis in gravid pony mares. *Proc Equine Nutr Physiol Symp* 15;281–282, 1997

Berde B, Sturmer E: Introduction to the pharmacology of ergot alkaloids and related compounds as a basis of their therapeutic application. In *Handbook of Experimental Pharmacology,* vol 49, *Ergot Alkaloids and Related Compounds,* Berde B, Schild HO eds, Springer-Verlag, Berlin, pp1–28, 1978

Berg RT, McElroy LW: Effect of 2,4-D on the nitrate content of forage crops and weeds. *Can J Agric Sci* 33;354–358, 1953

Besong SA, Jackson JA, Hemken RW: Influence of dietary supplementation of vitamin E or L-ascorbic acid on the detoxification of nitrate toxicity in steers. *J Anim Sci* 72 (Suppl 1);187, 1994

Bews JW: *The World's Grasses.* Russell & Russell, New York, 1929

Bhikane AU, Singh B: Diphenylamine blue test for diagnosis of nitrite poisoning in crossbred calves. *Indian Vet J* 67;808–812, 1990

Bhikane AU, Singh B: Diazotization test for diagnosis of experimental nitrite poisoning in cattle. *Indian J Anim Sci* 61;61–62, 1991

Bhikane AU, Singh B, Salabat Ali M: Therapeutic efficacy of methylene blue against experimental nitrite poisoning in crossbred calves. *Indian Vet J* 67;459–462, 1990

Bicudo PL: Nitrate poisoning in cattle: investigation of possible inhibitors of nitrate reduction in vitro and in vivo in cattle. PhD dissertation, Univ Minnesota, 1986

Bishop C, Surgenor DM: *The Red Blood Cell: A Comprehensive Treatise.* Academic Press, New York, 1964

Blake JT, Thomas DW: Acute bovine pulmonary emphysema in Utah. *J Am Vet Med Assoc* 158;2047–2052, 1971

Blakemore F, Stewart J: Some observations on several outbreaks of so-called lactation tetany in cattle. *Third Rep Inst Anim Pathol, Univ Cambridge,* 1932/1933;159–168, 1933

Blakemore F, Stewart J: Studies on the magnesium content of the blood of cows in lactation tetany districts. *4th Rept Inst An Pathol, Univ Cambridge,* 1934/1935;103–110, 1935

Blohm H: *Poisonous Plants of Venezuela.* Harvard Press, Cambridge, 1962

Boermans HJ: Diagnosis of nitrate toxicosis in cattle, using biological fluids and a rapid ion chromatographic method. *Am J Vet Res* 51;491–495, 1990

Bohman VR, Lesperance AL, Harding GD, Grunes DL: Induction of experimental tetany in cattle. *J Anim Sci* 29;99–102, 1969

Bohman VR, Horn FP, Littledike ET, Hurst JG, Griffin D: Wheat pasture poisoning. 2. Tissue composition of cattle grazing cereal forages and related tetany. *J Anim Sci* 57;1364–1373, 1983a

Bohman VR, Horn FP, Stewart BA, Mathers AC, Grunes DL: Wheat pasture poisoning. 1. An evaluation of cereal pastures as related to tetany in beef cows. *J Anim Sci* 57;1352–1363, 1983b

Bolt DJ, Bond J: Effects in pregnant beef heifers grazing fungus-infected tall fescue on birth weight, milk yield, and calf growth. *Nutr Rep Int* 40;487–494, 1989

Bond J, Lynch GP, Bolt DJ, Hawk HW, Jackson C Jr, Wall RJ: Reproductive performance and lamb weight gains for ewes grazing fungus-infected tall fescue. *Nutr Rep Int* 37;1099–1115, 1988

Boosinger TR, Brendemuehl JP, Bransby DL, Wright JC, Kemppainen RJ, Kee DD: Prolonged gestation, decreased triiodothyronine concentration, and thyroid gland histomorphologic features in newborn foals and mares grazing *Acremonium coenophialum*–infected fescue. *Am J Vet Res* 56;66–69, 1995

Bourke CA: The evidence against the existence of so-called convulsive ergotism in ruminants. In *Plant-Associated Toxins: Agricultural, Phytochemical, and Ecological Aspects,* Colegate SM, Dorling PR eds, CAB International, Wallingford, UK, pp387–392, 1994

Bourke CA, Carrigan MJ: Mechanisms underlying *Phalaris aquatica* "sudden death" syndrome in sheep. *Aust Vet J* 69;165–167, 1992
Bourke CA, Carrigan MJ, Dixon RJ: Experimental evidence that tryptamine alkaloids do not cause *Phalaris aquatica* sudden death syndrome in sheep. *Aust Vet J* 65;218–220, 1988
Bourke CA, Carrigan MJ, Dixon RJ: The pathogenesis of the nervous syndrome of *Phalaris aquatica* toxicity in sheep. *Aust Vet J* 67;356–358, 1990
Bove FJ: *The Story of Ergot.* Karger, Basel, Switzerland, 1970
Boyd FT, Aamodt OS, Bohstedt G, Truog E: Sudan grass management for control of cyanide poisoning. *J Am Soc Agron* 30;569–582, 1938
Bradley BJ, Carlson JR: Ultrastructural pulmonary changes induced by intravenously administered 3-methylindole in goats. *Am J Pathol* 99;551–559, 1980
Bradley GA, Metcalf HC, Reggiardo C, Noon TH, Bicknell EJ, Lozano-Alarcon F, Reed RE, Riggs MW: Neuroaxonal degeneration in sheep grazing *Sorghum* pastures. *J Vet Lab Diagn Invest* 7;229–236, 1995
Bradley WB, Epson HF, Beath OA: Nitrate as the cause of oat hay poisoning. *J Am Vet Med Assoc* 94;541–542, 1939
Bradley WB, Epson HF, Beath OA: Methylene blue as an antidote for poisoning by oat hay and other plants containing nitrates. *J Am Vet Med Assoc* 96;41–42, 1940
Brady DE, Comfort JE, Lasley JF, Pfander WH: *Forage Poisoning in Missouri Due to Excessive Amounts of Nitrate.* Mo Agric Exp Stn Bull 652, pp19–20, 1955
Bransby DI: Steer weight gain responses to ivermectin when grazing fescue. *Large Anim Pract* 18;16–19, 1997
Bransby DI, Schmidt SP, Griffey W, Eason JT: Heavy grazing is best for infected fescue. *Highlights Agric Res, Ala Agric Exp Stn* 35(4);12, 1988
Bray TM, Carlson JR: Role of mixed-function oxidase in 3-methylindole–induced acute pulmonary edema in goats. *Am J Vet Res* 40;1268–1272, 1979
Bray TM, Kirkland JB: The metabolic basis of 3-methylindole-induced pneumotoxicity. *Pharmacol Ther* 46;105–118, 1990
Bridges CH, Camp BJ, Livingston CW, Bailey EM: Kleingrass (*Panicum coloratum* L.) poisoning in sheep. *Vet Pathol* 24;525–531, 1987
Brooks HV, Cahill JJ: The susceptibility of Canadian wapiti to ryegrass staggers [letter]. *N Z Vet J* 33;126, 1985
Brown CM, Burrows GE, Edwards WC: Nitrate intoxication [letter]. *Vet Hum Toxicol* 32;481–482, 1990
Brown HB, Ranck EM: *Forage Poisoning Due to* Claviceps paspali *on* Paspalum. Miss Agric Exp Stn Tech Bull 6, 1915
Brownie CF, Prasad RD: Suspected convulsive ergotism in beef calves on overgrown Dallis grass *(Paspalum dilatatum)* pasture. *Vet Hum Toxicol* 29;257–258, 1987
Browning R Jr: Physiological responses of Brahman and Hereford steers to an acute ergotamine challenge. *J Anim Sci* 78;124–130, 2000
Browning R Jr, Gissendanner SJ, Wakefield T Jr: Ergotamine alters plasma concentrations of glucagon, insulin, cortisol, and triiodothyronine in cows. *J Anim Sci* 78;690–698, 2000
Bruehl GW, Kaiser WJ, Klein RE: An endophyte of *Achnatherum inebrians,* an intoxicating grass of northwest China. *Mycologia* 86;773–776, 1994
Bruning-Fann CS, Kaneene JB: The effects of nitrate, nitrite, and *N*-nitroso compounds on animal health. *Vet Hum Toxicol* 35;237–253, 1993
Brunnich JC: Hydrocyanic acid in fodder-plants. *J Chem Soc* 83;788–796, 1903
Burau R, Stout PR: Trans-aconitic acid in range grasses in early spring. *Science* 150;766–767, 1965
Burfening PJ: Ergotism. *J Am Vet Med Assoc* 163;1288–1290, 1973
Burrows GE: Methylene blue or tolonium chloride antagonism of sodium nitrite induced methemoglobinemia. *J Vet Pharmacol Ther* 2;81–86, 1979
Burrows GE: Nitrate intoxication. *J Am Vet Med Assoc* 177;82–83, 1980
Burrows GE: Cyanide intoxication in sheep: therapeutics. *Vet Hum Toxicol* 23;22–28, 1981

Burrows GE: Methylene blue: effects and disposition in sheep. *J Vet Pharmacol Ther* 7;225–231, 1984

Burrows GE, Horn GW, McNew RW, Croy LI, Keeton RD, Kyle J: The prophylactic effect of corn supplementation on experimental nitrate intoxication in cattle. *J Anim Sci* 64;1682–1689, 1987

Burt AWA, Thomas DC: Dietary citrate and hypomagnesemia in the ruminant. *Nature* 192;1193, 1961

Bush LP, Cornelius PL, Buckner RC, Varney DR, Chapman RA, Burrus PB II, Kennedy CW, Jones TA, Saunders MJ: Association of *N*-acetyl loline and *N*-formyl loline with *Epichloe typhina* in tall fescue. *Crop Sci* 22;941–943, 1982

Butler T: Notes on a feeding experiment to produce leucoencephalitis in a horse, with positive results. *Am Vet Rev* 26;748–751, 1902

Button C, Paynter DI, Shiel MJ, Colson AR, Paterson PJ, Lyford RL: Crystal-associated cholangiohepatopathy and photosensitization in lambs. *Aust Vet J* 64;176–180, 1987

Camp BJ, Dollahite JW, Schwartz WL: Biochemical changes in sheep given transaconitic acid. *Am J Vet Res* 29;2009–2011, 1968

Campbell CW, Burfening PJ: Effects of ergot on reproductive performance in mice and gilts. *Can J Anim Sci* 52;567–569, 1972

Campbell D, Morgan TN: Cyanosis caused by sulphonamide compounds. *Lancet* 237;123–127, 1939

Carlson JR, Dickinson EO: Tryptophan-induced pulmonary edema and emphysema in ruminants. In *Effects of Poisonous Plants on Livestock,* Keeler RF, van Kampen KR, James LF eds, Academic Press, New York, pp261–272, 1978

Carlson JR, Dyer IA, Johnson RJ: Tryptophan-induced interstitial pulmonary emphysema in cattle. *Am J Vet Res* 29;1983–1989, 1968

Carlson JR, Yokoyama MT, Dickinson EO: Induction of pulmonary edema and emphysema in cattle and goats with 3-methylindole. *Science* 176;298–299, 1972

Carlson JR, Dickinson EO, Yokoyama MT, Bradley BJ: Pulmonary edema and emphysema in cattle after intraruminal and intravenous administration of 3-methylindole. *Am J Vet Res* 36;1341–1347, 1975

Carlton MW, Kunkel DB: St. Anthony's fire revisited. *Am Fam Physician* 52;1681, 1995

Case AA: Some aspects of nitrate intoxication in livestock. *J Am Vet Med Assoc* 130;323–329, 1957

Chen G, Janzen EG, Bray TM: Identification of 3-MI-derived *N*-centered radicals obtained from incubation of 3-MI with microsomal-NADPH system by EPR-HPLC spin trapping. *Free Radic Biol Med* 17;19–25, 1994

Clapp JB: *Pulmonary Emphysema Project—Tranquille Meadows, Kamloops.* Livestock Branch, British Columbia, Department of Agriculture, Victoria, 1962

Clare NT: Photosensitization in animals. *Adv Vet Sci* 2;182–211, 1955

Clark BJ, Chu D, Aellig WH: Actions on the heart and circulation. In *Handbook of Experimental Pharmacology,* vol 49, *Ergot Alkaloids and Related Compounds,* Berde B, Schild HO eds, Springer-Verlag, Berlin, pp321–420, 1978

Clark DA: Beating ryegrass staggers. In *Proceedings of the 44th Ruakura Farmer's Conference,* pp87–92, 1992

Clawson AB, Bunyea H, Couch JF: Remedies for cyanide poisoning in sheep and cattle. *J Wash Acad Sci* 24;369–385, 1934

Clay BR, Edwards WC, Peterson DR: Toxic nitrate accumulation in sorghums. *Bovine Pract* 11;28–32, 1976

Clegg FG: A convulsive syndrome in sheep. *Vet Rec* 71;824–827, 1959

Coetzer JAW, Kellerman TS, Sadler W, Bath GF: Photosensitivity in South Africa. 5. A comparative study of the pathology of the ovine hepatogenous photosensitivity diseases, facial eczema, and geeldikkop *(Tribulus ovis),* with special reference to their pathogenesis. *Onderstepoort J Vet Res* 50;59–71, 1983

Coffey KP, Moyer JL, Lomas LW, Smith JE, La Rue DC, Brazle FK: Implant and copper oxide needles for steers grazing *Acremonium coenophialum*–infected tall fescue pastures: effects on grazing and subsequent feedlot performance and serum constituents. *J Anim Sci* 70;3203–3214, 1992

Cole RJ, Dorner JW: Role of fungal tremorgens in animal disease. In *Mycotoxins and Phycotoxins,* Steyn PS, Vleggaar R eds, Elsevier Sci, Amsterdam, pp501–511, 1986

Colegate SM, Anderton N, Edgar J, Bourke CA, Oram RN: Suspected blue canary grass *(Phalaris coerulescens)* poisoning of a horse. *Aust Vet J* 77;537–538, 1999

Collison SE: *Prussic Acid in* Sorghum. Fla Agric Exp Stn Bull 155, pp49–54, 1919

Conn EE: The metabolism of a natural product: lessons learned from cyanogenic glycosides. *Planta Med* 57 (Suppl 1);S1–S9, 1991

Cook GM, Wells JE, Russell JB: Ability of *Acidaminococcus fermentans* to oxidize trans-aconitate and decrease the accumulation of tricarballylate, a toxic end product of ruminal fermentation. *Appl Environ Microbiol* 60;2533–2537, 1994

Coppock RW, Mostrom MS, Simon J, McKenna DJ, Jacobson B, Szlachta HL: Cutaneous ergotism in a herd of dairy cows. *J Am Vet Med Assoc* 194;549–551, 1989

Cornick JL, Carter GK, Bridges CH: Kleingrass-associated hepatotoxicosis in horses. *J Am Vet Med Assoc* 193;932–935, 1988

Cotrau M, Proca M, Zanoschi V: Seasonal variation in hydrogen cyanide in *Glyceria aquatica:* Lucrari Stiintifice. *Ser Zooteh Med Vet* 1977;107–108, 1977

Couch JF: *Poisoning of Livestock by Plants That Produce Hydrocyanic Acid.* USDA Leaflet 88, 1932 (rev 1940)

Couch JF, Briese RR, Martin JH: Hydrocyanic acid content of sorghum varieties. *J Wash Acad Sci* 29;146–161, 1939

Coulman BE, Woods DL, Clarke KW: Distribution within the plant, variation with maturity, and heritability of gramine and hordenine in reed canary grass. *Can J Plant Sci* 57;771–777, 1977

Crawford AC: *The Poisonous Action of Johnson Grass.* USDA Bur Plant Ind Bull 90, part 4, pp3–6, 1906

Crawford RF, Kennedy WK, Davison KL: Factors influencing the toxicity of forages that contain nitrate when fed to cattle. *Cornell Vet* 56;3–17, 1966

Crawford RJ Jr, Forwood JR, Belyea RL, Garner GB: Relationship between level of endophyte infection and cattle gains on tall fescue. *J Prod Agric* 2;147–153, 1989

Crawford RJ, Massie MD, Sleper DA, Mayland HF: Use of an experimental high-magnesium tall fescue to reduce grass tetany in cattle. *J Prod Agric* 11;491–496, 1998

Crookshank HR, Sims FH: Serum values in wheat pasture poisoning cases. *J Anim Sci* 14;964–969, 1955

Cross DL, Redmond LM, Strickland JR: Equine fescue toxicosis: signs and solutions. *J Anim Sci* 73;899–908, 1995

Cruz N, Araya O, Zaror L, Del Campo CH: Reproductive problems related to fescue toxicosis in mares. *Arch Med Vet* 29;167–169, 1997

Cudd LA, Burrows GE, Clarke CR: Pharmacokinetics and toxicity of tolonium chloride in sheep. *Vet Hum Toxicol* 38;329–332, 1996

Culvenor CCJ, Dal Bon R, Smith LW: The occurrence of indolealkylamine alkaloids in *Phalaris tuberosa* L. and *P. arundinacea* L. *Aust J Chem* 17;1301–1304, 1964

Custer FD: Tetany in cattle on winter rations. In *Proceedings Symposium on Magnesium and Agriculture,* West Virginia Univ, Morgantown, pp159–164, 1959

Davies EG, Ashton WM: Coumarin and related compounds of *Anthoxanthum puelii* and *Melilotus alba* and dicoumarol formation in spoilt sweet vernal and sweet clover hay. *J Sci Food Agric* 15;733–738, 1964

Davis CB, Camp BJ: The vasoactive potential of halostachine, an alkaloid of tall fescue (*Festuca arundinacea* Schreb.) in cattle. *Vet Hum Toxicol* 25;408–411, 1983

Davison KL, Hansel W, Krook L, McEntee K, Wright MJ: Nitrate toxicity in dairy heifers. 1. Effects on reproduction, growth, lactation, and vitamin A nutrition. *J Dairy Sci* 47;1065–1073, 1964

Davison KL, McEntee K, Wright MJ: Responses in pregnant ewes fed forages containing various levels of nitrate. *J Dairy Sci* 48;968–977, 1965

Debessai W, Luick BR, Cheeke PR: Effects of feeding endophyte-infected tall fescue seed on lamb performance and serum prolactin [abstract]. *J Anim Sci* 71 (Suppl 1);200, 1993
Deeb BS, Sloan KW: *Nitrate, Nitrites, and Health.* Ill Agric Exp Stn Bull 750, 1975
Dillon BE: Acute ergot poisoning in cattle. *J Am Vet Med Assoc* 126;136, 1955
Diven RH, Pistor WJ, Reed RE, Trautman RJ, Watts RE: The determination of serum or plasma nitrate and nitrite. *Am J Vet Res* 23;497–499, 1962
Doggett H: *Sorghum,* 2d ed. Longman Scientific & Technical, Essex, England, 1988
Dollahite JW, Holt EC: Nitrate poisoning. *Southwest Vet* 23;23–28, 1969
Dollahite JW, Rowe LD: Nitrate and nitrite intoxication in rabbits and cattle. *Southwest Vet* 27;246–248, 1974
Dollahite JW, Younger RL, Jones LP: Photosensitization in lambs grazing Kleingrass. *J Am Vet Med Assoc* 171;1264–1265, 1977
Dougherty CT, Lauriault LM, Bradley NW, Gay N, Cornelius PL: Induction of tall fescue toxicosis in heat-stressed cattle and its alleviation with thiamin. *J Anim Sci* 69; 1008–1018, 1991
Dowell CT: Cyanogenesis in *Andropogon sorghum. J Agric Res* 16;175–181, 1919a
Dowell CT: *A Study of Cyanogenesis in* Sorghum vulgare. Okla Agric Exp Stn Bull 122, 1919b
Dunstan WR, Henry TA: VII. Cyanogenesis in plants. Part II. The great millet, *Sorghum vulgare. Roy Soc Lond Phil Trans A* 199;399–410, 1902
Dyer DC: Evidence that ergovaline acts on serotonin receptors. *Life Sci* 53;PL223–228, 1993
East NE, Higgins RJ: Canary grass (*Phalaris* sp.) toxicosis in sheep in California. *J Am Vet Med Assoc* 192;667–669, 1988
Easty DB, Blaedel WJ, Anderson L: Continuous electrochemical determination of cyanide: application to cyanogenic glycosides in sudan grass. *Anal Chem* 43;509–514, 1971
Eck HV: Hydrocyanic acid potentials in leaf blade tissue of eleven grain sorghum hybrids. *Agron J* 68;349–351, 1976
Eck HV, Hageman RH: Nitrate reductase activity in Sudangrass cultivars. *Crop Sci* 14; 283–287, 1974
Eck HV, Gilmore EC, Ferguson DB, Wilson GC: Heritability of nitrate reductase and cyanide levels in seedlings of grain sorghum cultivars. *Crop Sci* 15;421–424, 1975
Elsasser TH, Bolt DJ: Dopaminergic-like activity in toxic fescue alters prolactin but not growth hormone or thyroid stimulating hormone in ewes. *Domest Anim Endocrinol* 4;259–269, 1987
Emerick RJ: Rate of formation and reduction of nitrite-induced methemoglobin in vitro and in vivo as influenced by diet of sheep and age of swine. *J Anim Sci* 24;221–230, 1965
Emerick RJ: Consequences of high nitrate levels in feed and water. *Fed Proc* 33;1183–1187, 1974
Epstein W, Gerber K, Karler R: The hypnotic constituent of *Stipa vaseyi* sleepy grass. *Experientia* 20;390, 1964
Erb N, Zinsmeister HD, Nahrstedt A: The cyanogenic glycosides of *Triticum, Secale,* and *Sorghum. Planta Med* 41;84–89, 1981
Everist SL: *Poisonous Plants of Australia,* rev ed. Angus & Robertson, London, pp305–307, 1981
Fanchamps A: Some compounds with hallucinogenic activity. In *Handbook of Experimental Pharmacology,* vol 49, *Ergot Alkaloids and Related Compounds,* Berde B, Schild HO eds, Springer-Verlag, Berlin, pp567–614, 1978
Feigl F, Anger W, Oesper RE: *Spot Tests in Inorganic Analysis,* 6th ed., English. Elsevier, Amsterdam pp359–361, 1972
Fichte BE: Bermuda grass tremors. *Farm Q* 1972 (March);62–63, 1972
Filipov Nm, Thompson FN, Hill NS, Dawe DL, Stuedemann JA, Price JC, Smith CK: Vaccination against ergot alkaloids and the effect of endophyte-infected fescue seed–based diets on rabbits. *J Anim Sci* 76;2456–2463, 1998

Finnie JW: Pathogenesis of corynetoxin poisoning. In *Plant-Associated Toxins: Agricultural, Phytochemical, and Ecological Aspects,* Colegate SM, Dorling PR eds, CAB International, Wallingford, UK, pp405–409, 1994

Finnie JW, O'Shea JD: Pathological and pathogenetic changes in the central nervous system of guinea pigs given tunicamycin. *Acta Neuropathol* 75;411–421, 1988

Flaoyen A, Smith BL, Miles CO: An attempt to reproduce crystal-associated cholangitis in lambs by the experimental dosing of sarsasapogenin or diosgenin alone and in combination with sporidesmin. *N Z Vet J* 41;171–174, 1993

Fletcher LR, Barrell GK: Reduced liveweight gains and serum prolactin levels in hoggets grazing ryegrasses containing *Lolium* endophyte. *N Z Vet J* 32;139–140, 1984

Fletcher LR, Harvey IC: An association of a *Lolium* endophyte with ryegrass staggers. *N Z Vet J* 29;185–186, 1981

Fletcher LR, Hume DE, Latch GCM, Easton HS: Grazing ryegrass/endophyte associations and their effect on animal health and performance. In *Proceedings of the Second International Symposium on Acremonium-Grass Interactions: Plenary Papers,* pp115–120, 1993

Fluckiger E, del Pozo E: Influence on the endocrine system. In *Handbook of Experimental Pharmacology,* vol 49, *Ergot Alkaloids and Related Compounds,* Berde B, Schild HO eds, Springer-Verlag, Berlin, pp615–690, 1978

Fontenot JB, Allen VG, Bunce GE, Goff JP: Factors influencing magnesium absorption and metabolism in ruminants. *J Anim Sci* 67;3445–3455, 1989

Foote JZ, Woodburn OJ, Walsh JR, Heazlewood PG: Responses in grazing sheep to toxins from perennial ryegrass/endophyte associations. In *Plant-Associated Toxins: Agricultural, Phytochemical, and Ecological Aspects,* Colegate SM, Dorling PR eds, CAB International, Wallingford, UK, pp375–380, 1994

Ford EJH, Gopinath C: The excretion of phylloerythrin and bilirubin by the horse. *Res Vet Sci* 16;186–198, 1974

Ford EJH, Gopinath C: The excretion of phylloerythrin and bilirubin by calves and sheep. *Res Vet Sci* 21;12–18, 1976

Frahn JL, O'Keefe DF: The occurrence of tetrahydro-ß-carboline alkaloids in *Phalaris tuberosa* (Gramineae). *Aust J Chem* 24;2189–2192, 1971

Frakes RA, Sharma RP, Willhite CC: Developmental toxicity of the cyanogenic glycoside linamarin in the golden hamster. *Teratology* 31;241–246, 1985

Francis CK: *The Poisoning of Livestock While Feeding on Plants of the Sorghum Group.* Okla Agric Exp Stn Circ 38, 1915

Frank M, Weckman TJ, Wood T, Woods WE, Tai CL, Chang SL, Ewing A, Blake JW, Tobin T: Hordenine: pharmacology, pharmacokinetics and behavioural effects in the horse. *Equine Vet J* 22;437–441, 1990

Franzke CJ: *Rancher: A Low Hydrocyanic Acid Forage Sorghum.* S D Agric Exp Stn Circ 57, 1945

Fraser DM, Dorling PR: Suspected ergotism in two heifers. *Aust Vet J* 60;303–305, 1983

Froslie A: Methemoglobin reduction and NADH-dependent methemoglobin reductase activity following DNBP- and nitrite-induced methemoglobinemia in sheep. *Acta Pharmacol Toxicol* 38;17–23, 1976

Galey FD, Tracy ML, Craigmill AL, Barr BC, Markegard G, Peterson R, O'Conner MO: Staggers induced by consumption of perennial ryegrass in cattle and sheep from northern California. *J Am Vet Med Assoc* 199;466–470, 1991

Gallagher CH, Koch JH, Moore RM, Steel JD: Toxicity of *Phalaris tuberosa* for sheep. *Nature* 204;542–545, 1964

Gallagher CH, Koch JH, Hoffman H: Diseases of sheep due to ingestion of *Phalaris tuberosa. Aust Vet J* 42;279–286, 1966

Gallagher CH, Koch JH, Hoffman H: Deaths of ruminants grazing *Phalaris tuberosa* in Australia. *Aust Vet J* 43;495–500, 1967a

Gallagher CH, Kock JH, Hoffman H: Electro-myographic studies on sheep injected with the *N,N*-dimethylated tryptamine alkaloids of *Phalaris tuberosa. Int J Neuropharmacol* 6;223–228, 1967b

Gallagher RT, Hawkes AD: The potent tremorgenic neurotoxins lolitrem B and aflatrem: a comparison of the tremor response in mice. *Experientia* 42;823–825, 1986
Gallagher RT, Clardy J, Wilson BJ: Aflatrem, a tremorgenic toxin from *Aspergillus flavus. Tetrahedron Lett* 21,239–242, 1980a
Gallagher RT, Finer J, Clardy J, Leutwiler A, Weibel F, Acklin W, Arigoni D: Paspalinine, a tremorgenic metabolite from *Claviceps paspali* Stevens et Hall. *Tetrahedron Lett* 21;235–238, 1980b
Gallagher RT, White EP, Mortimer PH: Ryegrass staggers: isolation of potent neurotoxins lolitrem A and lolitrem B from staggers-producing pastures. *N Z Vet J* 29;189–190, 1981
Gallagher RT, Campbell AG, Hawkes AD, Holland PT, McGaveston DA, Pansier EA, Harvey IC: Ryegrass staggers: the presence of lolitrem neurotoxins in perennial ryegrass seed. *N Z Vet J* 30;183–184, 1982
Gallagher RT, Hawkes AD, Steyn PS, Vleggaar R: Tremorgenic neurotoxins from perennial ryegrass causing ryegrass staggers disorder of livestock: structure elucidation of lolitrem B. *J Chem Soc Commun* 1984;614–616, 1984
Gander JE, Marum P, Martin GC, Hovin AW: The occurrence of 2-methyl-1,2,3,4-tetrahydro-beta-carboline and variation in alkaloids in *Phalaris arundinacea. Phytochemistry* 15;737–738, 1976
Gangstad EO: *Composition, Yield, and Grazing Studies of Sudan and Related Sorghums.* Tex Res Found Bull 7, 1959
Gant DB, Cole RJ, Valdes JJ, Eldefrawi ME, Eldefrawi AT: Action of tremorgenic mycotoxins on GABAa receptor. *Life Sci* 41;2207–2214, 1987
Garner GB, Cornell CN, Moyer JH, Muhrer ME, Pfander WH: Nitrate levels in bovine and porcine serum. *J Anim Sci* 19;1322, 1960
Geurink JH, Malestein A, Kemp A, van't Klooster ATh: Nitrate poisoning in cattle. 3. The relationship between nitrate intake with hay or fresh roughage and the speed of intake on the formation of methemoglobin. *Neth J Agron Sci* 27;268–276, 1979
Geurink JH, Malestein A, Kemp A, Korzeniowski A, van't Klooster ATh: Nitrate poisoning in cattle. 7. Prevention. *Neth J Agric Sci* 30;105–113, 1982
Ghorbani GR, Jackson JA, Hemken R: Performance of Holstein calves fed varying additions of endophyte-infected tall fescue. *Nutr Rep Int* 40;931–938, 1989
Gibbons WJ: "Downer" or Bermuda grass poisoning. *North Am Vet* 34;433–434, 1953
Gibbs RD: *Chemotaxonomy of Flowering Plants,* vol 1. McGill-Queens Univ Press, Montreal, pp57–69,629–635, 1974a
Gibbs RD: *Chemotaxonomy of Flowering Plants,* vol 3. McGill-Queens Univ Press, Montreal, pp1621–1623, 1974b
Glenn AE, Bacon CW, Price R, Hamlin RT: Molecular phylogeny of *Acremonium* and its taxonomic implications. *Mycologia* 88;369–383, 1996
Goodwin DE: Ergot poisoning of cattle grazing Dallisgrass. *J Am Vet Med Assoc* 151; 204–205, 1967
Gooneratne SR, Smith BL, Towers NR: Zinc, molybdenum and sulfur for prevention of facial eczema in sheep. In *Plant-Associated Toxins: Agricultural, Phytochemical, and Ecological Aspects,* Colegate SM, Dorling PR eds, CAB International, Wallingford, UK, pp507–511, 1994
Gorz HJ, Haag WL, Specht JE, Haskins FA: Assay of para-hydroxy-benzaldehyde as a measure of hydrocyanic acid potential in sorghums. *Crop Sci* 17;578–582, 1977
Grauwiler J, Schon H: Teratological experiments with ergotamine in mice, rats, and rabbits. *Teratology* 7;227–236, 1973
Graydon RJ, Hamid H, Zahari P, Gardiner C: Photosensitization and crystal-associated cholangiohepatopathy in sheep grazing *Brachiaria decumbens. Aust Vet J* 68;234–236, 1991
Greatorex JC, Mantle PG: Experimental ergotism in sheep. *Res Vet Sci* 15;337–346, 1973
Greatorex JC, Mantle PG: Effect of rye ergot on the pregnant sheep. *J Reprod Fert* 37;33–41, 1974

Green BK: *Horse Tradin'*. AA Knopf, New York, pp281–304, 1967
Green LW, Baker JF, Hardt PF: Use of animal breeds and breeding to overcome the incidence of grass tetany: a review. *J Anim Sci* 67;3463–3469, 1989
Griffith RW, Grauwiler J, Hodel Ch, Leist KH, Matter B: Toxicologic considerations. In *Handbook of Experimental Pharmacology,* vol 49, *Ergot Alkaloids and Related Compounds,* Berde B, Schild HO eds, Springer-Verlag, Berlin, pp805–851, 1978
Groger D, Tyler VE Jr, Dusenberry JE: Investigation of the alkaloids of *Paspalum* ergot. *Lloydia* 24;97–102, 1961
Grunes DL, Welch RM: Plant contents of magnesium, calcium, and potassium in relation to ruminant nutrition. *J Anim Sci* 67;3485–3494, 1989
Grunes DL, Stout PR, Brownell JR: Grass tetany of ruminants. *Adv Agron* 22;331–374, 1970
Guerin P: The probable causes of the poisonous effects of the darnel (*Lolium temulentum* L.). *Bot Gaz* 28;136–137, 1899
Gupta D, Singh B, Bhikane AU, Rajguru DN: Treatment of experimental nitrate poisoning with tolonium chloride in buffalo calves. *Indian J Anim Sci* 62;1180–1182, 1992
Haldane J, Makgill RH, Mavrogordato AE: The actions as poisons of nitrites and other physiologically related substances. *J Physiol* 21;160–189, 1897
Haliburton JC, Edwards WC: Nitrate poisoning in Oklahoma cattle during the winter of 1977–1978. *Vet Hum Toxicol* 20;401–403, 1978
Hammond AC, Carlson JR: Inhibition of ruminal degradation of L-tryptophan to 3-methylindole in vitro. *J Anim Sci* 51;207–214, 1980
Hammond AC, Bradley BJ, Yokayama MT, Carlson JR, Dickinson EO: 3-Methylindole and naturally occurring acute bovine pulmonary edema and emphysema. *Am J Vet Res* 40;1398–1401, 1979
Hammond AC, Carlson JR, Breeze RG: Prevention of tryptophan-induced acute bovine pulmonary oedema and emphysema (fog fever). *Vet Rec* 107;322–325, 1980
Hanafy MS, Bogan JA: Pharmacological modulation of the pneumotoxicity of 3-methylindole. *Biochem Pharmacol* 31;1765–1771, 1982
Hance HF: On a Mongolian grass producing intoxication in cattle. *J Bot* 14;210–212, 1876
Hansen DE, McCoy RD, Hedstrom OR, Snyder SP, Ballerstedt PB: Photosensitization associated with exposure to *Pithomyces chartarum* in lambs. *J Am Vet Med Assoc* 204;1668–1671, 1994
Hanway JJ, Englehorn AJ: Nitrate accumulation in some Iowa crop plants. *Agron J* 50;331–334, 1958
Hanway JJ, Herrick JB, Willrich TL, Bennett PC, McCall JT: *The Nitrate Problem.* Iowa Ext Ser Spec Rep 34, 20pp, 1963
Harlan JR, de Wet JMJ: A simplified classification of cultivated *Sorghum. Crop Sci* 12;172–176, 1972
Harms CL, Tucker BB: Influence of nitrogen fertilization and other factors on yield, prussic acid, nitrate, and total nitrogen concentrations of sudangrass cultivars. *Agron J* 65;21–26, 1973
Harrington JD: *Hydrocyanic Acid Content of Piper, Trudan I, and Six Sorghum-Sudangrass Hybrids.* Pa Agric Exp Stn Bull 735, 1966
Hartley WJ: Chronic phalaris poisoning or phalaris staggers. In *Effects of Poisonous Plants on Livestock,* Keeler RF, Van Kampen KR, James LF eds, Academic Press, New York, pp391–393, 1978
Haskins FA, Gorz HJ, Vogel KP: Cyanogenesis in Indiangrass seedlings. *Crop Sci* 19;761–765, 1979
Haskins FA, Gorz HJ, Hill RM, Youngquist JB: Effect of freezing on the hydrocyanic acid potential of field-grown sorghum tillers. *Crop Sci* 24;1183–1186, 1984
Haskins FA, Gorz HJ, Hill RM: Colorimetric determination of cyanide in enzyme-hydrolysed extracts of dried sorghum leaves. *J Agric Food Chem* 36;775–778, 1988
Havard V: The sleepy grass. *Gard Forest* 4;111, 1891

Hedges DA, Wheeler JL, Mulcahy C, Vincent MS: Composition and acceptability to sheep of twelve summer forage crops. *Aust J Exp Agric Anim Husb* 18;520–526, 1978

Helwig DM, Setchell BP: Observations on the diagnosis of nitrite poisoning in sheep. *Aust Vet J* 36;14–17, 1960

Hemken RW, Bull LS, Boling JA, Kane E, Bush LP, Buckner RC: Summer fescue toxicosis in lactating dairy cows and sheep fed experimental strains of ryegrass-tall fescue hybrids. *J Anim Sci* 49;641–646, 1979

Hemken RW, Boling JA, Bull LS, Hatton RH, Buckner RC, Bush LP: Interaction of environmental temperature and anti-quality factors on the severity of summer fescue toxicosis. *J Anim Sci* 52;710–714, 1981

Hemken RW, Jackson JA Jr, Boling JA: Toxic factors in tall fescue. *J Anim Sci* 58; 1011–1016, 1984

Hibbs CM, Stencel EL, Hill RM: Nitrate toxicosis in cattle. *Vet Hum Toxicol* 20;1–2, 1978

Hicks DR, Peterson RH: Defoliation and fertilizer nitrogen effects on nitrate-nitrogen profiles in maize. *Agron J* 68;476–478, 1976

Hignight KW, Muilenburg GA, van Wijk AJP: A clearing technique for detecting the fungal endophyte *Acremonium* sp. in grasses. *Biotech Histochem* 68;87–90, 1993

Hill NS, Stringer WC, Rottinghaus GE, Belesky DP, Parrott WA, Pope DD: Growth, morphological, and chemical component responses of tall fescue to *Acremonium coenophialum*. *Crop Sci* 30;156–161, 1990

Hill NS, Thompson FN, Dawe DL, Stuedemann JA: Antibody binding of circulating ergot alkaloids in cattle grazing tall fescue. *Am J Vet Res* 55;419–424, 1994

Hindmarsh WL: The lethal dose of hydrocyanic acid for ruminants. *J Counc Sci Ind Res Aust* 3;12–13, 1930

Hinebauch TD: *Rheumatism in Horses.* N D Agric Exp Stn Bull 7, 1892

Hinebauch TD: *Feeding of Millet to Horses.* N D Agric Exp Stn Bull 26, pp87–105, 1896

Hjerpe CA: Grass tetany in California cattle. *J Am Vet Med Assoc* 144;1406, 1964

Hoffer GN: A simple test for detecting the nutrient needs of corn [abstract]. *J Am Soc Agron* 18;29–31, 1926

Hogg PG, Ahlgren HL: Environmental, breeding, and inheritance studies of hydrocyanic acid in *Sorghum vulgare* var. *sudanense*. *J Agric Res* 67;195–210, 1943

Hogg RA: Poisoning of cattle fed ergotized silage. *Vet Rec* 129;313–314, 1991

Holland PT, Miles CO, Mortimer PH, Wilkins AL, Hawkes AD, Smith BL: Isolation of the steroidal sapogenin epismilagenin from the bile of sheep affected by *Panicum dichotomiflorum* toxicosis. *J Agric Food Chem* 39;1963–1965, 1991

Hollis LC: Observations on deaths of stocker cattle on wheat pasture. In *Proceedings of the National Wheat Pasture Symposium,* Okla Agric Exp Stn Misc Publ 115, pp199–204, 1984

Holtenius P: Nitrite poisoning in sheep, with special reference to the detoxification of nitrite in the rumen. *Acta Agric Scand* 7;113–163, 1967

Honeyfield DC, Carlson JR, Nocerini MR, Breeze RG: Duration of inhibition of 3-methylindole production by monensin. *J Anim Sci* 60;226–231, 1985

Horn G, Burrows GE, Lusby KS: Preliminary studies on the effect of yeast culture and supplementation on nitrate/nitrite induced methemoglobinemia in lambs and steers. *Vet Hum Toxicol* 26;309–313, 1984

Horn GA, Burrows GE: Primary photosensitization in cattle [letter]. *Vet Hum Toxicol* 32;331–332, 1990

Horvath DJ: So-called "grass tetany" in West Virginia: survey and laboratory findings. In *Proceedings Symposium on Magnesium and Agriculture,* West Virginia, Morgantown, pp197–214, 1959

Householder GT, Dollahite JW, Hulse R: Diphenylamine for the diagnosis of nitrate intoxication. *J Am Vet Med Assoc* 148;662–665, 1966

Hoveland CS: Importance and economic significance of the *Acremonium* endophytes to performance of animals and grass plant. *Agric Ecosyst Environ* 44;3–12, 1993

Howarth JA: Sudan grass as a photosensitizing agent causing dermatitis in sheep. *North Am Vet* 12;29–33, 1931

Huang TW, Carlson JR, Bray TM, Bradley BJ: 3-methylindole–induced pulmonary injury in goats. *Am J Pathol* 87;647–666, 1977

Humble JA: The biological production unit for food animal veterinarians. *J Am Vet Med Assoc* 209;1698–1699, 1996

Hunt LD, Blythe L, Holtan DW: Ryegrass staggers in ponies fed processed ryegrass straw. *J Am Vet Med Assoc* 182;285–286, 1983

Hyslop NStG: Bovine pulmonary emphysema. *Can Vet J* 10;251–257, 1969

Ikediobi CO, Olugboji O, Okoh PN: Cyanide profile of component parts of sorghum (*Sorghum bicolor* L. Moench) sprouts. *Food Chem* 27;167–175, 1988

Ireland FA, Loch WE, Worthy K, Anthony RV: Effects of bromocriptine and perphenazine on prolactin and progesterone concentrations in pregnant pony mares during late gestation. *J Reprod Fert* 92;179–186, 1991

Ishigami K, Inoue K: Metabolism of nitrate and methemoglobinemia in ruminants. *Res Bull Obihiro Univ* 10;45–55, 1976

Jackson JA, Hemken RW, Boling JA, Harmon RJ, Buckner RC, Bush LP: Loline alkaloids in tall fescue hay and seed and their relationship to summer fescue toxicosis in cattle. *J Dairy Sci* 67;104–109, 1984

Jackson JA, Hemken RW, Boling JA, Buckner RC: Effect of feeding *Acremonium coenophialum* infected tall fescue to primiparous dairy heifers and feeding an *Acremonium* free selection to dairy calves on the expression of fescue toxicosis. *Nutr Rep Int* 37;1265–1274, 1988

Jacobs BL, Klemfuss H: Brain stem and spinal cord mediation of a serotonergic behavioral syndrome. *Brain Res* 100;450–457, 1975

Jacobson DR, Miller WM, Seath DM, Yates SG, Tookey HL, Wolff IA: Nature of fescue toxicity and progress toward identification of the toxic entity. *J Dairy Sci* 46;416–422, 1963

Jessup TM, Dent CHR, Kemp JB, Christie B, Ahrens PJ, Burgess LW, Bryden WL: Bovine idiopathic hyperthermia. *Aust Vet J* 64;353–354, 1987

Johnson JL, Schneider NR, Grotelueschen DM: Nitrate exposure in perinatal beef calves. *Proc Annu Meet Am Assoc Vet Lab Diagn* 26;167–177, 1983

Johnson JL, Hergert GW, Schneider NR, Grabouski P: Post-harvest change in cornstalk nitrate and its relationship to bovine fetal nitrite/nitrate exposure. In *Poisonous Plants: Proceedings of the Third International Symposium,* James LF, Keeler RF, Bailey EM Jr, Cheeke PR, Hegarty MP eds, Iowa State Univ Press, Ames, 423–430, 1992

Johnson JL, Grotelueschen D, Knott M: Evaluation of bovine perinatal nitrate accumulation in western Nebraska. *Vet Hum Toxicol* 36;467–471, 1994

Johnson MC, Pirone TP, Siegel MR, Varney DR: Detection of *Epichloe typhina* in tall fescue by means of enzyme-linked immunosorbant assay. *Phytopathology* 72;647–650, 1982

Johnson MC, Dahlman DL, Siegel MR, Bush LP, Latch GCM, Potter DA, Varney DR: Insect feeding deterrents in endophyte-infected tall fescue. *Appl Environ Microbiol* 49;568–571, 1985

Jones TJ: Cyanide poisoning in cattle [letter]. *North Am Vet* 33;258, 1952

Jung GA, Lilly B, Shih SC, Reid RL: Studies with Sudangrass. 1. Effect of growth stage and level of nitrogen fertilizer upon yield of dry matter; estimated digestibility of energy, dry matter and protein; amino acid composition; and prussic acid potential. *Agron J* 56;533–537, 1964

Kahler LW, Jorgensen NA, Satter LD, Tyler WJ, Crowley JW, Finner MF: Effect of nitrate in drinking water on reproductive efficiency of dairy cattle. *J Dairy Sci* 58;771, 1975

Kappel LC, Youngberg H, Ingraham RH, Hembry FG, Robinson DL, Cherney JH: Effects of dietary aluminum on magnesium status of cows. *Am J Vet Res* 44;770–773, 1983

Kemp A, Geurink JH, Haalstra RT, Malestein A: Nitrate poisoning in cattle. 1. Discolouration of the vaginal mucous membrane as aid in the prevention of nitrate poisoning in cattle. *Stikstof* 19;40–48, 1976

Kemp A, Geurink JH, Haalstra RT, Malestein J: Nitrate contents of fresh forage, hay, and silage and nitrite poisoning of cattle. *Wirtsch Futter* 23;53–59, 1977a

Kemp A, Geurink JH, Haalstra RT, Malestein A: Nitrate poisoning in cattle. 2. Changes in nitrite in rumen fluid and methemoglobin formation in blood after high nitrate intake. *Neth J Agric Sci* 25;51–62, 1977b

Kennedy DJ, Cregan PD, Glastonbury JRW, Golland DT, Day DG: Poisoning of cattle grazing a low-alkaloid cultivar of *Phalaris aquatica,* Sirolan. *Aust Vet J* 63;88–89, 1986

Kennedy GS: Trans-aconitate utilization by sheep. *Aust J Biol Sci* 21;529–538, 1968

Kerr DR: Rapid death of cattle grazing recently irrigated *Phalaris tuberosa. Aust Vet J* 48;421, 1972

Kerr LA, Linnabary RD: A review of interstitial pneumonia in cattle. *Vet Hum Toxicol* 31;247–254, 1989

Kerr LA, McCoy CP, Boyle CR, Essig HW: Effects of ammoniation of endophyte fungus-infected fescue hay on serum prolactin concentration and rectal temperature in beef cattle. *Am J Vet Res* 51;76–78, 1990

Kidder RW, Beardsley DW, Erwin TC: Photosensitization in cattle grazing bermuda grass [abstract]. *Proc Assoc South Agric Work* 48;80, 1951

Kidder RW, Beardsley DW, Erwin TC: *Photosensitization in Cattle Grazing Frosted Common Bermudagrass.* Fla Agric Exp Stn Bull 630, 1961

Kiese M: *Methemoglobinemia: A Comprehensive Treatise.* CRC Press, Cleveland, 1974

Kim JG, Voigtlaender G: Studies on synthesis and accumulation pattern of cyanogenic glycosides in sorghum plants. *J Korean Soc Grassl Sci* 5;121–126, 1985

King B: Outbreak of ergotism in Wollo, Ethiopia. *Lancet* 1;1411, 1979

Kirkland JB, Bray TM: The effect of 3-methylindole on phospholipid synthesis in goat lung tissue slices. *Proc Soc Exp Biol Med* 175;30–34, 1984

Kirkland JB, Bray TM: Impaired surfactant function in 3-methylindole–induced lung injury in goats. *Comp Biochem Physiol C Comp Pharmacol Toxicol* 94;591–593, 1989

Knaus HG, McManus OB, Lee SH, Schmalhofer WA, Garcia-Calvo M, Helms LM, Sanchez M, Giangiacomo K, Reuben JP, Smith AB III et al: Tremorgenic indole alkaloids potently inhibit smooth muscle high-conductance calcium-activated potassium channels. *Biochemistry* 33:5819–5828, 1994

Knight PR: Equine cystitis and ataxia associated with grazing of pastures dominated by *Sorghum* species. *Aust Vet J* 44;257, 1968

Kojima M, Poulton JE, Thayer SS, Conn EE: Tissue distributions of dhurrin and of enzymes involved in its metabolism in leaves of *Sorghum bicolor. Plant Physiol* 63;1022–1028, 1979

Korzeniowski A, Geurink JH, Kemp A: Nitrate poisoning in cattle. 5. The effect of tungsten on nitrite formation by rumen microbes. *Neth J Agric Sci* 28;16–19, 1980

Korzeniowski A, Geurink JH, Kemp A: Nitrate poisoning in cattle. 6. Tungsten (Wolfram) as a prophylactic against nitrate-nitrite intoxication in ruminants. *Neth J Agric Sci* 29;37–47, 1981

Kosuge T, Gilchrist D: Chemistry of coumarin and related compounds. In *Mycotoxic Fungi, Mycotoxins, Mycotoxicosis: An Encyclopedic Handbook,* Wyllie TD, Morehouse LG eds, Marcel Dekker, New York, pp239–282, 1977

Kralovec J, Frycek A: The influence of high rates of nitrogen on the quality of fodder and on the health of cattle. *Vaxtodling* 29;139–144, 1974

Kriedman PE: Cyanide formation in *Sorghum almum* in relation to nitrogen and phosphorus nutrition. *Aust J Exp Agric Anim Husb* 4;15–16, 1964

Krishnamachari KAVR, Bhat RV: Poisoning by ergoty bajra (pearl millet) in man. *Indian J Med Res* 64;1624–1628, 1976

Lajis NH, Salam Abdullah A, Salim SJS, Bremner JB, Khan MN: Epi-sarsapogenin and epi-smilagenin: two sapogenins isolated from the rumen content of sheep intoxicated by *Brachiaria decumbens. Steroids* 58;387–389, 1993

Lancaster MJ, Vit I, Lyford RL: Analysis of bile crystals from sheep grazing *Panicum schinzii* (sweet grass). *Aust Vet J* 68;281, 1991

Larson BT, Sullivan DM, Samford MD, Kerley MS, Paterson JA, Turner JT: D2 dopamine receptor response to endophyte-infected tall fescue and an antagonist in the rat. *J Anim Sci* 72;2905–2910, 1994
Lean IJ, Anderson M, Kerfoot MG, Marten GC: Tryptamine alkaloid toxicosis in feedlot sheep. *J Am Vet Med Assoc* 195;768–771, 1989
Lee HJ, Kuchel RE: The aetiology of phalaris staggers in sheep. 1. Preliminary observations on the preventive role of cobalt. *Aust J Agric Res* 4;88–99, 1953
Lee HJ, Kuchel RE, Trowbridge RF: The aetiology of phalaris staggers in sheep II. The toxicity to sheep of three types of pasture containing *Phalaris tuberosa. Aust J Agric Res* 7;333–344, 1956
Leemann AC: A short summary on our botanical knowledge of *Lolium temulentum* L. *Onderstepoort J Vet Sci Anim Ind* 1;213–218, 1933
Leemann AC: Hydrocyanic acid in grasses. *Onderstepoort J Vet Sci Anim Ind* 5;97–136, 1935
Leffel EC, Mason KR: A study of winter tetany in Maryland. In *Proceedings Symposium on Magnesium and Agriculture,* West Virginia Univ, Morgantown, pp182–184, 1959
Lemberger L: The pathology of ergots: past and present. *Fed Proc* 37;2176–2180, 1978
Leuchtmann A: Systematics, distribution, and host specificity of grass endophytes. *Nat Toxins* 1;150–162, 1992
Lewis D: The metabolism of nitrate and nitrite in the sheep. 1. The reduction of nitrate in the rumen of the sheep. *Biochem J* 48;175–180, 1951a
Lewis D: The metabolism of nitrate and nitrite in the sheep. 2. Hydrogen donators in nitrate reduction by rumen microorganisms in vitro. *Biochem J* 49;149–153, 1951b
Lewis DC, Sparrow LA: Implications of soil type, pasture composition, and mineral content of pasture components for the incidence of grass tetany in the southeast of South Australia. *Aust J Exp Agric* 31;609–615, 1991
Linnabary RD, Tarrier MP: Acute bovine pulmonary emphysema caused by the fungus *Fusarium semitectum. Vet Hum Toxicol* 30;255–256, 1988
Lipham LB, Thompson FN, Stuedemann JA, Sartin JL: Effects of metoclopramide on steers grazing endophyte-infected fescue. *J Anim Sci* 67;1090–1097, 1989
Loew DM, van Deusen EB, Meier-Ruge W: Effects on the central nervous system. In *Handbook of Experimental Pharmacology,* vol 49, *Ergot Alkaloids and Related Compounds,* Berde B, Schild HO eds, Springer-Verlag, Berlin, pp421–531, 1978
London WT, Henderson W, Cross RF: An attempt to produce chronic nitrite toxicosis in swine. *J Am Vet Med Assoc* 150;398–402, 1967
Long HC: Poisonous plants on the farm—3. *J Minist Agric (London)* 33;1022–1031, 1927
Lopez TA, Odriozola ER, Mutti GL: Characteristics of intoxication of cattle with *Paspalum dilitatum* contaminated with *Claviceps paspali. Vet Argent* 3;863–871, 1986
Lopez TA, Campero CM, Chayer R, Hoyos M de: Ergotism and photosensitization in swine produced by the combined ingestion of *Claviceps purpurea* sclerotia and *Ammi majus* seeds. *J Vet Diagn Invest* 9;68–71, 1997
Lovett JV, Hoult AHC, Christen O: Biologically active secondary metabolites of barley. 4. Hordenine production by different barley lines. *J Ecol Chem* 20;1945–1954, 1994
Low SG, Jephcott SB, Bryden WL: Weaner illthrift of cattle grazing signal grass *(Brachiaria decumbens)* in Papua New Guinea. In *Plant-Associated Toxins: Agricultural, Phytochemical, and Ecological Aspects,* Colegate SM, Dorling PR eds, CAB International, Wallingford, UK, pp567–571, 1994
Loyd RC, Gray E: Amount and distribution of hydrocyanic acid potential during the life cycle of plants of three sorghum cultivars. *Agron J* 62;394–397, 1970
Lyons PC, Plattner RD, Bacon CW: Occurrence of peptide and clavine ergot alkaloids in tall fescue grass. *Science* 232;487–489, 1986
Mackey DR: Acute pulmonary emphysema in cattle. *Southwest Vet* 5;342–344, 1952
Mackintosh CG, Orr MB, Gallagher RT, Harvey IC: Ryegrass staggers in Canadian wapiti deer [letter]. *N Z Vet J* 30;106–107, 1982
Majak W, McDiarmid RE, van Ryswyk AL, Broersma K, Bonin SG: Alkaloid levels in

reed canarygrass grown on wet meadows in British Columbia. *J Range Manage* 32; 322–326, 1979

Majak W, McDiarmid RE, Hall JW, Cheng K-J: Factors that determine rates of cyanogenesis in bovine ruminal fluid in vitro. *J Anim Sci* 68;1648–1655, 1990

Malestein A, Geurink JH, Schuyt G, Schotman AJH, Kemp A, van't Klooster ATh: Nitrate poisoning in cattle. 4. The effect of nitrite dosing during parturition on the oxygen capacity of maternal blood and the oxygen supply to the unborn calf. *Vet Q* 2;149–159, 1980

Malone P: Monensin sodium toxicity in cattle. *Vet Rec* 103;477–478, 1978

Mantle PG: Interruption of early pregnancy in mice by oral administration of agroclavine and sclerotia of *Claviceps fusiformis* (Loveless). *J Reprod Fert* 18;81–88, 1969

Mantle PG: Metabolism and elimination of tremorgenic mycotoxins. In *Mycotoxins and Phycotoxins,* Steyn PS, Vleggaar R eds, Elsevier Science, Amsterdam, pp399–408, 1986

Mantle PG, Gunner DE: Abortions associated with ergotized pastures. *Vet Rec* 77; 885–886, 1965

Mantle PG, Penny RHC: Tremorgenic mycotoxins and neurological disorders—a review. *Vet Ann* 21;51–62, 1981

Mantle PG, Mortimer PH, White EP: Mycotoxic tremorgens of *Claviceps paspali* and *Penicillium cyclopium:* a comparative study of effects on sheep and cattle in relation to natural staggers syndromes. *Res Vet Sci* 24;49–56, 1977

Marsh CD, Clawson AB: *Sleepy Grass* (Stipa vaseyi) *as a Stock-Poisoning Plant.* USDA Tech Bull 114, 1929

Marshak, 1959: The role of magnesium in the tetany syndromes of cattle. In *Proceedings Symposium on Magnesium and Agriculture,* West Virginia Univ, Morgantown, pp169–178, 1959

Marten GC, Barnes RF, Simons AB, Wooding FJ: Alkaloids and palatability of *Phalaris arundinacea* L. grown in diverse environments. *Agron J* 65;199–201, 1973

Marten GC, Jordan RM, Hovin AW: Improved lamb performance associated with breeding for alkaloid reduction in reed canarygrass. *Crop Sci* 21;295–298, 1981

Mason RJ, Dobbs LG, Greenleaf RD, Williams MC: Alveolar type II cells. *Fed Proc* 36;2697–2702, 1977

Mason RW: Axis cylinder degeneration associated with ryegrass staggers in sheep and cattle. *Aust Vet J* 44;428, 1968

Masukawa T, Iwata H: Protective effect of selenite on nitrite toxicity. *Experientia* 35; 1360–1361, 1979

Mathews FP: Johnson grass *(Sorghum halepense)* poisoning. *J Am Vet Med Assoc* 81; 663–666, 1932

Mathews FP: *Poisonous Plants in the Davis Mountains Area.* Tex Agric Exp Stn Ann Rep 51, pp13–14, 1938

Mayland HF, Grunes DL: Soil-climate plant relationships in the etiology of grass tetany. In *Grass Tetany,* Rendig VV, Grunes DL eds, American Society of Agronomy, Crop Science Society of America, Soil Science Society of America, Madison, Wisc, pp123–175, 1979

Mayland HF, Wilkinson SR: Soil factors affecting magnesium availability in plant-animal systems: a review. *J Anim Sci* 67;3437–3444, 1989

Mayland HF, Grunes DL, Stuart DM: Chemical composition of *Agropyron desertorum* as related to grass tetany. *Agron J* 66;441–446, 1974

Mayland HF, Grunes DL, Lazar VA: Grass tetany hazard of cereal forages based upon chemical composition. *Agron J* 68;665–667, 1976

Mayo NS: *Cattle Poisoning by Nitrate of P.* Kans Agric Exp Stn Bull 49, 1895

McCann JS, Heusner GL, Amos HE: Concentrate and endophyte infected tall fescue hay diets: digestibility and effects on yearling horse growth rate. *Proc Equine Nutr Physiol Symp* 12;69, 1991

McDonald IW: A "staggers" syndrome in sheep and cattle associated with grazing on *Phalaris tuberosa. Aust Vet J* 18;182–189, 1942

McGeer PL, Eccles JC, McGeer EG: *Molecular Neurobiology of the Mammalian Brain.* Plenum Press, New York, pp295–320,491, 1978

McIntosh GH, Thomas MR: Toxicity of parasitized Wimmera ryegrass, *Lolium rigidum,* for sheep and cattle. *Aust Vet J* 43;349–353, 1967

McKenzie RA, McMicking LI: Ataxia and urinary incontinence in cattle grazing sorghum. *Aust Vet J* 53;496–497, 1977

Meagher LP, Wilkins AL, Miles CO, Collin RG, Fagliari JJ: Hepatogenous photosensitization of ruminants by *Brachiaria decumbens* and *Panicum dichotomiflorum* in the absence of sporidesmin: lithogenic saponins may be responsible. *Vet Hum Toxicol* 38;271–274, 1996

Menaul P, Dowell CT: Cyanogenesis in sudan grass: A modification of the Francis-Connell method of determining hydrocyanic acid. *J Agric Res* 18;447–450, 1920

Metson AJ, Saunders WMH, Collie TW, Graham VW: Chemical composition of pastures in relation to grass tetany in beef breeding cows. *N Z J Agric Res* 9;410–436, 1966

Miles CO, Munday SC, Holland PT, Smith BL, Embling PP, Wilkins AL: Identification of a sapogenin glucuronide in the bile of sheep affected by *Panicum dichotomiflorum* toxicosis. *N Z Vet J* 39;150–152, 1991

Miles CO, Munday SC, Holland PT, Lancaster MJ, Wilkins AL: Further analysis of bile crystals from sheep grazing *Panicum schinzii* (sweet grass). *Aust Vet J* 69;34, 1992a

Miles CO, Wilkins AL, Gallagher RT, Hawkes AD, Munday SC, Towers NR: Synthesis and tremorgenicity of paxitriols and lolitriol: possible biosynthetic precursors of lolitrem B. *J Agric Food Chem* 40;234–238, 1992b

Miles CO, Wilkins AL, Munday SC, Holland PT, Smith BL, Lancaster MJ, Embling PP: Identification of the calcium salt of epismilagenin -D-glucuronide in the bile crystals of sheep affected by *Panicum dichotomiflorum* and *Panicum schinzii* toxicoses. *J Agric Food Chem* 40;1606–1609, 1992c

Miles CO, Wilkins AL, Munday SC, Flaoyen A, Holland PT, Smith BL: Identification of insoluble salts of the β-D-glucuronides of episarsasapogenin and epismilagenin in the bile of lambs with alveld and examination of *Narthecium ossifragum, Tribulus terrestris,* and *Panicum miliaceum* for sapogenins. *J Agric Food Chem* 41;914–917, 1993

Miles CO, Lane GA, di Menna ME, Garthwaite I, Piper EL, Ball OJ-P, Latch GCM, Allen JM, Hunt MB, Bush LP, Min FK, Fletcher I, Harris PS: High levels of ergonovine and lysergic acid amide in toxic *Achnatherum inebrians* accompany infection by an *Acremonium*-like endophytic fungus. *J Agric Food Chem* 44;1285–1290, 1996

Mislevy P, Hodges EM, Martin FG: Hydrocyanic acid potential in stargrass (*Cynodon* spp.). In *Proceedings of XIV International Grasslands Congress,* Smith JA, Hays VW eds, Westview Press, Boulder, Colo, pp732–735, 1983

Miyazaki A, Kawashima R: Reduction pattern of nitrate in the rumen of sheep given young soiling rye. *Jpn J Zootechnol Sci* 47;158–165, 1976

Mizinga KM, Thompson FN, Stuedemann JA, Edwards GL: Neural dopamine D2 receptors in rats fed endophyte-infected fescue seed. *Drug Chem Toxicol* 16;307–319, 1993

Mizukami Y, Kanbe M, Inami S, Fukaya K: Changes in the nitrate content of millets and Sudangrass. *Res Bull Aichi-ken Agric Res Cent* 29;71–76, 1997

Moate PJ, Schneider KM, Leaver DD, Morris DC: Effect of 1,25 dihydroxyvitamin D3 on the calcium and magnesium metabolism of lactating cows. *Aust Vet J* 64;73–75, 1987

Monier-Williams GW: The nitrate test for the detection of added water in milk. *Analyst* 56;397–398, 1931

Monlux AW, Glenn BL, Panciera RJ, Corcoran JB: Bovine hepatogenous photosensitivity associated with feeding of alfalfa hay. *J Am Vet Med Assoc* 142;989–994, 1963

Monroe JL, Cross DL, Hudson LW, Hendricks DM, Kennedy SW, Bridges WC Jr: Effect of selenium and endophyte-contaminated fescue on performance and reproduction in mares. *Equine Vet Sci* 8;148–153, 1988

Monson WJ: The significance of magnesium in nutrition. In *Proceedings Nutrition Council American Feed Manufacturers Association,* Arlington, Va, 1963

Montgomery JF, Hum S: Field diagnosis of nitrite poisoning in cattle by testing aqueous humour samples with urine test strips. *Vet Rec* 137;593–594, 1995

Montigny C de, Aghajanian GK: Preferential action of 5-methoxytryptamine and 5-methoxydimethyltryptamine on presynaptic serotonin receptors: a comparative iontophoretic study with LSD and serotonin. *Neuropharmacology* 16;811–818, 1977

Moran EA: Some new cyanogenic plants. *J Wash Acad Sci* 30;237–239, 1940

Moran EA: Cyanogenic compounds in plants and their significance in animal industry. *Am J Vet Res* 15;171–176, 1954

Morgan SE, Johnson B, Brewer B, Walker J: Sorghum cystitis ataxia syndrome in horses [letter]. *Vet Hum Toxicol* 32;582, 1990

Mortimer PH, Ronaldson JW: Fungal-toxin–induced photosensitization. In *Handbook of Natural Toxins,* vol 1, *Plant and Fungal Toxins,* Keeler RF, Tu AT eds, Marcel Dekker, New York, pp361–419, 1983

Muchiri DJ, Bridges CH, Ueckert DN, Bailey EM: Photosensitization of sheep on Kleingrass pasture. *J Am Vet Med Assoc* 177;353–354, 1980

Muhrer ME, Case AA, Garner GB, Pfander WH: Toxic forages produced in a drought area [abstract]. *J Anim Sci* 14;1251, 1955

Mullenax CH: *Brachiaria decumbens* toxicity grass or fungus [letter]. *Vet Hum Toxicol* 33;464–465, 1991

Munday BL, Mason RW: Lesions in ryegrass staggers in sheep. *Aust Vet J* 43;598–599, 1967

Munday BL, Monkhouse IM, Gallagher RT: Intoxication of horses by lolitrem B in ryegrass seed cleanings. *Aust Vet J* 62;207, 1985

Munday SC, Wilkins AL, Miles CO, Holland PT: Isolation and structure elucidation of dichotomin, a furostanol saponin implicated in hepatogenous photosensitization of sheep grazing *Panicum dichotomiflorum. J Agric Food Chem* 41;267–271, 1993

Munday-Finch SC, Miles CO, Wilkins AL, Hawkes AD: Isolation and structural elucidation of lolitrem A, a tremorgenic mycotoxin from perennial ryegrass infected with *Acremonium lolii. J Agric Food Chem* 43;1283–1288, 1995

Murty BR, Govil JN: Description of 70 groups in genus *Sorghum* based on a modified Snowden's classification. *Indian J Genet Plant Breed* 27;75–91, 1967

Murty BR, Arunachalam V, Saxena MBL: Classification and catalogue of a world collection of *Sorghum. Indian J Genet Plant Breed* 27;1–74, 1967

Muth OH, Haag JR: Disease of Oregon cattle associated with hypomagnesemia and hypocalcemia. *North Am Vet* 26;216–219, 1945

Nakamura Y, Tada Y, Shibuya H, Yoshida J, Nakamura R: The influence of concentrates on the nitrate metabolism of microorganisms in the rumen of sheep. *Jpn J Zootech Sci* 50;782–789, 1979

Nason A: Symposium on metabolism of inorganic compounds. *Bacteriol Rev* 26;16–41, 1962

Neill JC: The endophyte of ryegrass *(Lolium perenne). N Z J Sci Technol* 21A;280–291, 1940

Nelson CE: Hydrocyanic acid content of certain sorghums under irrigation as affected by nitrogen fertilizer and soil moisture stress. *Agron J* 45;615–617, 1953

Newsom IE, Stout EN, Thorp F Jr, Barber CW, Groth AH: Oat hay poisoning. *J Am Vet Med Assoc* 90;66–75, 1937

Nichols WK, Larson DN, Yost GS: Bioactivation of 3-methylindole by isolated rabbit lung cells. *Toxicol Appl Pharmacol* 105;264–270, 1990

Nicholson SS: Tremorgenic syndromes in livestock. *Vet Clin North Am* 5(2);291–300, 1989

Nicholson SS, Olcott BM, Usenik EA, Casey HW, Brown CC, Urbatsch LE, Turnquist SE, Moore SC: Delayed phalaris grass toxicosis in sheep and cattle. *J Am Vet Med Assoc* 195;345–346, 1989

Noble JW: Suspected poisoning of cattle by *Claviceps* spp. on water couch. *Aust Vet J* 62;432–433, 1985

Nocerini MR, Carlson JR, Yost GS: Electrophilic metabolites of 3-methylindole as toxic intermediates in pulmonary oedema. *Xenobiotica* 14;561–564, 1984

Nocerini MR, Carlson JR, Yost GS: Glutathione adduct formation with microsomally activated metabolites of the pulmonary alkylating and cytotoxic agent, 3-methylindole. *Toxicol Appl Pharmacol* 81;75–84, 1985a

Nocerini MR, Honeyfield DC, Carlson JR, Breeze RG: Reduction of 3-methylindole production and prevention of acute bovine pulmonary edema and emphysema with lasalocid. *J Anim Sci* 60;232–238, 1985b

Nocerini MR, Yost GS, Carlson JR, Liberato DJ, Breeze RG: Structure of the glutathione adduct of activated 3-methylindole indicates that an imine methide is the electrophilic intermediate. *Drug Metab Dispos* 13;690–694, 1985c

Noordin MM, Salam Abdullah A, Rajion MA: Experimental *Brachiaria decumbens* toxicity in cattle. *Vet Res Commun* 13;491–494, 1989

Nordskog AW, Clark RT: Ergotism in pregnant sows, female rats, and guinea pigs. *Am J Vet Res* 6;107–116, 1945

Odriozola E, Campero C, Lopez T, Marin R, Casaro G, Andrada M: Neuropathological effects and deaths of cattle and sheep in Argentina from *Phalaris angusta. Vet Hum Toxicol* 33;465–467, 1991

O'Hara PJ, Fraser AJ: Nitrate poisoning in cattle grazing crops. *N Z Vet J* 23;45–53, 1975

Oliver JW, Abney LK, Strickland JR, Linnabary RD: Vasoconstriction in bovine vasculature induced by the tall fescue alkaloid lysergamide. *J Anim Sci* 71;2708–2713, 1993

Oliver JW, Schultze AE, Rohrbach BW, Fribourg HA, Ingle T, Waller JC: Alterations in hemograms and serum biochemical analytes of steers after prolonged consumption of endophyte-infected tall fescue. *J Anim Sci* 78;1029–1035, 2000

Olson OE, Whitehead E: Nitrate content of some South Dakota plants. *S D Acad Sci Proc* 20;95–101, 1940

Oram RN: Genetic and environmental control of the amount and composition of toxins in *Phalaris tuberosa* L. In *Proceedings of XI International Grasslands Congress,* Norman MJT ed, Univ Queensland Press, Australia, pp785–788, 1970

Osweiler GD, Carson TL, Buck WB, Van Gelder GA: *Clinical and Diagnostic Veterinary Toxicology,* 3d ed. Kendall/Hunt, Dubuque, Iowa, 1985

Palmer E: Special uses and properties of some Mexican grasses. *Bull Torrey Bot Club* 14;98–100, 1887

Pammel LH: *Manual of Poisonous Plants.* Torch Press, Cedar Rapids, Iowa, pp336–367, 1911

Pammel LH: Poisoning from sorghum and sudan grass. *Am J Vet Med* 14;30–31, 1919

Pammel LH: Department of livestock poisoning. *Am J Vet Med* 15;489–491, 1920

Panasiuk O, Bills DD: Cyanide content of sorghum sprouts. *J Food Sci* 49;791–793, 1984

Parmar SS, Brink VC: Tryptamine levels in pasturage implicated in bovine pulmonary emphysema. *Can J Plant Sci* 56;175–184, 1976

Patamalai B: The identification of hepatotoxins in Kleingrass and a study of their toxicity. PhD dissertation, Texas A&M Univ (*Diss Abstr Int B* 49;2080, 1988)

Patamalai B, Bridges CH, Hill DW, Camp BJ: The isolation and identification of steroidal sapogenins in Kleingrass. *Vet Hum Toxicol* 32;314–318, 1990

Patel CJ, Wright MJ: The effect of certain nutrients upon the hydrocyanic acid content of Sudangrass grown in nutrient solution. *Agron J* 50;645–647, 1958

Paterson J, Forcherio C, Larson B, Samford M, Kerley M: The effects of fescue toxicosis on beef cattle productivity. *J Anim Sci* 73;889–898, 1995

Pearson EG, Andreasen CB, Blythe LL, Craig AM: Atypical pneumonia associated with ryegrass staggers in calves. *J Am Vet Med Assoc* 209;1137–1142, 1996

Peckham JC, Mitchell FE, Jones OH, Doupnik B Jr: Atypical interstitial pneumonia in cattle fed moldy sweet potatoes. *J Am Vet Med Assoc* 160;169–172, 1972

Peet RL, McCarthy MR, Barbetti MJ: Hyperthermia and death in feedlot cattle associated with ingestion of *Claviceps purpurea. Aust Vet J* 68;121, 1991

Percival JC, Thornton RH: Relationship between the presence of fungal spores and a test for hepatotoxic grass. *Nature* 182;1095–1096, 1958

Perrin DD: Photosensitivity diseases in New Zealand. 15. A chemical procedure for the detection of facial eczema toxicity in pasture. *N Z J Agric Res* 2;266–273, 1959

Peters AT, Slade HB, Avery S: *Poisoning of Cattle by Common Sorghum and Kafir Corn* (Sorghum vulgare). Neb Agric Exp Stn Bull 77, 1903

Peterson DE, Schultheiss PC: Ulcerative stomatitis in horses due to ingestion of hay containing plant awns. *Vet Med/Small Anim Clin* 79;1509–1511, 1984

Petritz DC, Lechtenberg VL, Smith WH: Performance and economic returns of beef cows and calves grazing grass-legume herbage. *Agron J* 72;581–584, 1980

Petroski RJ, Powell RG, Clay K: Alkaloids of *Stipa robusta* (sleepygrass) infected with an *Acremonium* endophyte. *Nat Toxins* 1;84–88, 1992

Pfander WH: *Forage Poisoning in Missouri Due to Excessive Amounts of Nitrate.* Mo Agric Exp Stn Bull 652, pp19–20, 1955

Phillips WA, Von Tungeln D: Acute pulmonary edema and emphysema in steers fed Old-World bluestem hay. *Mod Vet Pract* 67;252–253, 1986

Pickrell JA, Oehme FW, Hickman SR: Drought increases forage nitrate and cyanide. *Vet Hum Toxicol* 33;247–251, 1991

Piening LJ: Effects of leaf rust on nitrate in rye. *Can J Plant Sci* 52;842–843, 1972

Pierson RE, Kainer RA: Clinical classification of pneumonias in cattle. *Bovine Pract* 15;73–79, 1980

Poppenga RH, Mostrom MS, Haschek WM, Lock TF, Buck WB, Beasley VR: Mare agalactia, placental thickening, and high foal mortality associated with the grazing of tall fescue: a case report. *Proc Am Assoc Vet Lab Diagn* 27;325–336, 1984

Porter JK: Analysis of endophyte toxins: fescue and other grasses toxic to livestock. *J Anim Sci* 73;871–880, 1995

Porter JK, Thompson FN: Effects of fescue toxicosis on reproduction in livestock. *J Anim Sci* 70;1594–1603, 1992

Porter JK, Bacon CW, Robbins JD: Major alkaloids of a *Claviceps* isolated from toxic Bermuda grass. *J Agric Food Chem* 22;838–841, 1974

Porter JK, Bacon CW, Robbins JD, Betowski D: Ergot alkaloid identification in Clavicepitaceae systemic fungi of pasture grasses. *J Agric Food Chem* 29;653–657, 1981

Porter JK, Stuedemann JA, Thompson FN, Buchanan BA, Tucker HA: Melatonin and pineal neurochemicals in steers grazed on endophyte-infected tall fescue: effects of metoclopramide. *J Anim Sci* 71;1526–1531, 1993

Potchoiba MJ, Carlson JR, Nocerini MR, Breeze RG: Effect of monensin and supplemental hay on ruminal 3-methylindole formation in adult cows after abrupt change to lush pasture. *Am J Vet Res* 53;129–133, 1992

Powell DG: Endophyte toxicity: a cause of foal mortality. *Equine Dis Q* 5(1);2, 1996

Powell RG, Petroski RJ: Alkaloid toxins in endophyte-infected grass. *Nat Toxins* 1;163–170, 1992

Price TM: Enzymes in cornstalks and their relation to cornstalk disease. *USDA Bur Anim Ind* 21;66–75, 1904

Prichard JT, Voss JL: Fetal ankylosis in horses associated with hybrid sudan pasture. *J Am Vet Med Assoc* 150;871–873, 1967

Pritchard DG, Markson LM, Brush PJ, Sawtell JAA, Bloxham PA: Haemorrhagic syndrome of cattle associated with the feeding of sweet vernal *(Anthoxanthum)* hay containing dicoumarol. *Vet Rec* 113;78–84, 1983

Puls R, Newschwander FP, Greenway JA: Cyanide poisoning from *Glyceria grandis* S.Wats. ex Gray (tall mannagrass) in a British Columbia beef herd. *Can Vet J* 19; 264–265, 1978

Puoli JR, Reid RL, Belesky DP: Photosensitization in lambs grazing switchgrass. *Agron J* 84;1077–1080, 1992

Purseglove JW: *Tropical Crops: Monocotyledons.* Longman Scientific & Technical, Essex, England, 1988

Putnam MR, Qualls CW, Rice LE, Dawson LJ: Hepatic enzyme changes in bovine hepatogenous photosensitivity caused by water-damaged alfalfa. *J Am Vet Med Assoc* 189;77–82, 1986

Putnam MR, Bransby DI, Schumacher J, Boosinger TR, Bush L, Shelby RA, Vaughan JT, Ball D, Brendemuehl JP: Effects of the fungal endophyte *Acremonium coenophialum* in fescue on pregnant mares and foal viability. *Am J Vet Res* 52;2071–2074, 1991

Rainey JW: Hydrocyanic acid poisoning from *Poa aquatica. Aust Vet J* 22;187, 1946

Rebhun WC, Georgi MA, Robbins JA, Hyppio PA, Sponenberg P: Stomatitis in a stable of horses caused by *Setaria lutescens. Equine Pract* 2;37–39, 1980

Redmond LM, Cross DL, Jenkins TC, Kennedy SW: The effect of *Acremonium coenophialum* on intake and digestibility of tall fescue hay in horses. *Equine Vet Sci* 11;215–219, 1991

Redmond LM, Cross DL, Strickland JR, Kennedy SW: Efficacy of dromperidone and sulpiride as treatments for equine fescue toxicosis in gravid mares grazing endophyte-infected tall fescue [abstract]. *J Anim Sci* 70 (Suppl 1);21, 1992

Restall DJ: Blood splash in lambs—a preliminary study using the one-stage prothrombin time test. *Meat Sci* 5;125–129, 1980–1981

Reynard GB, Norton JBS: *Poisonous Plants of Maryland in Relation to Livestock.* Md Agric Exp Stn Tech Bull A10, 1942

Rhodes MT, Paterson JA, Kerley MS, Garner HE, Laughlin MH: Reduced blood flow to peripheral and core body tissues in sheep and cattle induced by endophyte-infected tall fescue. *J Anim Sci* 69;2033–2043, 1991

Riley IT, Ophel KM: *Clavibacter toxicus* sp. nov., the bacterium responsible for annual ryegrass toxicity in Australia. *Int J Syst Bact* 42;64–68, 1992

Ritter RJ, Boling JA, Gay N: Labile magnesium reserves in beef cows subjected to different prepasture supplementation regimens. *J Anim Sci* 59;197–203, 1984

Robinson DL, Kappel LC, Boling JA: Management practices to overcome the incidence of grass tetany. *J Anim Sci* 67;3470–3484, 1989

Rodriguez FS, Santiyan MPM, Zamorano JDP: Evaluation of reagent strips for the rapid diagnosis of nitrite poisoning. *J Anal Toxicol* 16;63–65, 1992

Rogers CF, Boyd WL: Sudan grass and other cyanophoric plants as animal intoxicants. *J Am Vet Med Assoc* 88;489–499, 1936

Rohrbach BW, Green EM, Oliver JW, Schneider JF: Aggregate risk study of exposure to endophyte-infected *(Acremonium coenophialum)* tall fescue as a risk factor for laminitis in horses. *Am J Vet Res* 56;22–26, 1995

Romane WM, Adams LG, Bullard TL, Dollahite JW: Cystitis syndrome of the equine. *Southwest Vet* 19;95–99, 1966

Romero RR, Gonzalez SG, Garza AMN, Tavera FJT: Enfisema y edema pulmonar agudo en bovines. *Vet Mex* 24;43–46, 1993

Ross AD, Bryden WL, Bakau W, Burgess LW: Induction of heat stress in beef cattle feeding on the ergots of *Claviceps purpurea. Aust Vet J* 66;247–249, 1989

Rottinghaus GE, Garner GB, Cornell CN, Ellis JL: HPLC method for quantifying ergovaline in endophyte-infested tall fescue: seasonal variation of ergovaline levels in stems with leaf sheaths, leaf blades, and seed heads. *J Agric Food Chem* 39;112–115, 1991

Rowan DD: Lolitrems, peramine, and paxilline: mycotoxins of the ryegrass/endophyte interaction. *Agric Ecosyst Environ* 44;103–122, 1993

Rowe LD: Photosensitization problems in livestock. *Vet Clin North Am Food Anim Pract* 5(2);301–323, 1989

Ruangyuttikarn W, Skiles GL, Yost GS: Identification of a cysteinyl adduct of oxidized 3-methylindole from goat lung and human liver microsomal proteins. *Chem Res Toxicol* 5;713–719, 1992

Rudert CP, Oliver J: The effect of fertilizer nitrogen on the content of hydrocyanic acid, nitrate, and some minerals in star grass (*Cynodon aethiopicus,* Clayton & Harlan). *Rhod J Agric Res* 16;23–29, 1978

Russell JB: Enrichment and isolation of rumen bacteria that reduce *trans*-aconitic acid to tricarballylic acid. *Appl Environ Microbiol* 49;120–126, 1985

Russell JB, Van Soest PJ: In vitro ruminal fermentation of organic acids common in forage. *Appl Environ Microbiol* 47;155–159, 1984

Rutschmann J, Stadler PA: Chemical background. In *Handbook of Experimental Pharmacology,* vol 49, *Ergot Alkaloids and Related Compounds,* Berde B, Schild HO eds, Springer-Verlag, Berlin, pp29–85, 1978
Saffron M, Prado JL: Inhibition of aconitase by *trans*-aconitate. *J Biol Chem* 180;1301–1309, 1949
Saha DC, Jackson MA, Johnson-Cicalese JM: A rapid staining method for detection of endophytic fungi in turf and forage grasses. *Phytopathology* 78;237–239, 1988
Salam Abdullah A, Noordin MM, Rajion MA: Signal grass *(Brachiaria decumbens)* toxicity in sheep: changes in motility and pH of reticulo-rumen. *Vet Hum Toxicol* 30;256–258, 1988
Salam Abdullah A, Noordin MM, Rajion MA: Neurological disorders in sheep during signal grass *(Brachiaria decumbens)* toxicity. *Vet Hum Toxicol* 31;128–129, 1989
Salam Abdullah A, Lajis NH, Bremner JB, Davies NW, Mustapha W, Rajion MA: Hepatotoxic constituents in the rumen of *Brachiaria decumbens* intoxicated sheep. *Vet Hum Toxicol* 34;154–155, 1992
Salmon DE: Enzootics of ergotism. *Rep Comm Agric USDA* 1884;212–252, 1884
Samford-Grigsby MD, Larson BT, Forcherio JC, Lucas DM, Paterson JA, Kerley MS: Injection of a dopamine antagonist into Holstein steers to relieve symptoms of fescue toxicosis. *J Anim Sci* 75;1026–1031, 1997
Sandos J, Clare NT, White EP: Photosensitivity diseases in New Zealand. 16. Improved procedure for the beaker test for facial-eczema toxicity. *N Z J Agric Res* 2;623–626, 1959
Sapiro ML, Hoflund S, Clark R, Quin JI: Studies on the alimentary tract of the Merino sheep in South Africa. 16. The fate of nitrate in ruminal ingesta as studied in vitro. *Onderstepoort J Vet Sci Anim Ind* 22;357–372, 1949
Saunders JA, Conn EE: Presence of the cyanogenic glucoside dhurrin in isolated vacuoles from *Sorghum. Plant Physiol* 61;154–157, 1978
Saupe SG: Cyanogenic compounds and angiosperm phylogeny. In *Phytochemistry and Angiosperm Phylogeny,* Young DA, Seigler DS eds, Praeger, New York, pp80–116, 1981
Schillo KK, Leshin LS, Boling JA, Gay N: Effects of endophyte-infected fescue on concentrations of prolactin in blood sera and the anterior pituitary and concentrations of dopamine and dopamine metabolites in brains of steers. *J Anim Sci* 66;713–718, 1988
Schmidt H: *Swellhead in Sheep and Goats.* Tex Agric Exp Stn Annu Rep 44, p11, 1931
Schmidt SP, Hoveland CS, Clark EM, Davis ND, Smith LA, Grimes HW, Holliman JL: Association of an endophyte fungus with fescue toxicity in steers fed Kentucky 31 tall fescue seed or hay. *J Anim Sci* 55;1259–1263, 1982
Schneider DJ, Miles CO, Garthwaite I, Van Halderen A, First report of field outbreaks of ergot-alkaloid toxicity in South Africa. *Onderstepoort J Vet Res* 63;97–108, 1996
Schneider NR, Yeary RA: Nitrite and nitrate pharmacokinetics in the dog, sheep, and pony. *Am J Vet Res* 36;941–947, 1975
Schofield FW: Acute pulmonary emphysema of cattle. *J Am Vet Med Assoc* 112;254–259, 1948
Schroder VN: Hydrogen cyanide from forage plants. *Proc Soil Crop Soc Fla* 36;195–197, 1977
Schwarte LH: Moldy corn poisoning in horses. *J Am Vet Med Assoc* 92;152–158, 1938
Scott MIR: Methylene blue as an antidote for nitrite poisoning in sheep. *Aust Vet J* 17;71–72, 1941
Seaman JT, Smeal MG, Wright JC: The possible association of a sorghum *(Sorghum sudanese)* hybrid as a cause of developmental defects in calves. *Aust Vet J* 57;351–352, 1981
Seawright AA: *Animal Health in Australia.* Vol 2, *Chemical and Plant Poisons.* Australian Government Publ Serv, Canberra, pp91–94, 1989
Seigler DS: Plants of Oklahoma and Texas capable of producing cyanogenic compounds. *Proc Okla Acad Sci* 56;95–100, 1976a

Seigler DS: Plants of the northeastern United States that produce cyanogenic compounds. *Econ Bot* 30;395–407, 1976b

Seigler DS: The naturally occurring cyanogenic glycosides. In *Progress in Phytochemistry,* Reinhold L, Harbone JB, Swain T eds, Pergamon Press, Oxford, pp83–120, 1977

Seigler DS: Cyanogenic glycosides and lipids: structural types and distribution. In *Cyanide in Biology,* Vennesland B, Conn EE, Knowles CJ, Westley J, Wissing F eds, Academic Press, London, pp133–143, 1981

Selk GE, Wagner DG, Strickland GL, Buchanan DS: Nitrate concentrations in hybrid Sudangrass and pearl millet hays. *Prof Anim Sci* 11;20–25, 1995

Selman IE, Allan EM, Gibbs HA, Wiseman A, Dalgleish R, Young WB: The effect of anti-prostaglandin therapy in an acute respiratory distress syndrome induced in experimental cattle by the oral administration of 3-methylindole. *Bovine Pract* 20;124–126, 1985

Selmar D, Irandoost Z, Wray V: Dhurrin-6'-glucoside, a cyanogenic diglucoside from *Sorghum bicolor. Phytochemistry* 43;569–572, 1996

Setchell BP, Williams AJ: Plasma nitrate and nitrite concentration in chronic and acute nitrate poisoning in sheep. *Aust Vet J* 38;58–62, 1962

Sharman JR: Cyanide poisoning of cattle grazing "reed sweet-grass." *N Z Vet J* 15;7–8, 1967

Sharman JR: Seasonal variations in the cyanide content of *Poa aquatica* [letter]. *N Z Vet J* 16;85, 1968

Shaw JN, Muth OH: Some types of forage poisoning in Oregon cattle and sheep. *J Am Vet Med Assoc* 114;315–317, 1949

Sheehy MH, Way JL: Nitrite intoxication: protection with methylene blue and oxygen. *Toxicol Appl Pharmacol* 30;221–226, 1974

Shewmaker GE, Mayland HF, Hansen SB: Cattle grazing preference among eight endophyte-free tall fescue cultivars. *Agron J* 89;695–701, 1997

Siegel MR, Johnson MC, Varney DR, Nesmith WC, Buckner RC, Bush LP, Burrus PB II, Jones TA, Boling JA: A fungal endophyte in tall fescue: incidence and dissemination. *Phytopathology* 74;932–937, 1984a

Siegel MR, Varney DR, Johnson MC, Nesmith WC, Buckner RC, Bush LP, Burrus PB II, Hardison JR: A fungal endophyte of tall fescue: evaluation of control methods. *Phytopathology* 74;937–941, 1984b

Siegel MR, Latch GCM, Johnson MC: *Acremonium* fungal endophytes of tall fescue and perennial ryegrass: significance and control. *Plant Dis* 69;179–183, 1985

Sikula J: Poisonous plants. In *Veterinary Toxicology,* Bartik M, Piskac A eds, Elsevier Scientific, Amsterdam, pp291–292, 1981

Simon J, Sund JM, Wright MJ, Winter A, Douglas FD: Pathological changes associated with the lowland abortion syndrome in Wisconsin. *J Am Vet Med Assoc* 132;164–169, 1958

Simon J, Sund JM, Douglas FD, Wright MJ, Kowalczyk T: The effect of nitrate or nitrite when placed in the rumens of pregnant dairy cattle. *J Am Vet Med Assoc* 135;311–314, 1959a

Simon J, Sund JM, Wright MJ, Douglas FD: Prevention of noninfectious abortion in cattle by weed control and fertilization practices on lowland pastures. *J Am Vet Med Assoc* 135;315–317, 1959b

Simons AB, Marten GC: Relationship of indole alkaloids to palatability of *Phalaris arundinacea* L. *Agron J* 63;915–919, 1971

Sims FH, Crookshank HR: *Wheat Pasture Poisoning.* Tex Agric Exp Stn Bull 842, 1956

Singer RH, Grainger RB, Baker FH: *Investigations on a Complicated Grass Tetany Syndrome in ruminants of Kentucky. 1. Preliminary Observations.* Ky Agric Exp Stn Bull 658, 1958

Singh V, Singh D, Shinde DA, Namdeo KN: HCN content of varieties of fodder sorghum in relation to yield and nutritional value. *Indian J Agric Sci* 53;431–434, 1983

Sjollema B: Nutritional and metabolic disorders in cattle. *Nutr Abst Rev* 1;621–632, 1932

Skerritt JH, Guihot SL, McDonald SE, Culvenor RA: Development of immunoassays for tyramine and tryptamine toxins of *Phalaris aquatica* L. *J Agric Food Chem* 48;27–32, 2000

Skordos KW, Laycock JD, Yost GS: Thioether adducts of a new imine reactive intermediate of the pneumotoxin 3-methylindole. *Chem Res Toxicol* 11;1326–1331, 1998a

Skordos KW, Skiles GL, Laycock JD, Lanza DL, Yost GS: Evidence supporting the formation of 2,3-epoxy-3-methylindole: a reactive intermediate of the pneumotoxin 3-methylindole. *Chem Res Toxicol* 11;741–749, 1998b

Slade HB: Prussic acid in sorghum. *J Am Chem Soc* 25;55–59, 1903

Sleight SD, Sinha DP: Prevention of methemoglobin reduction in blood samples. *J Am Vet Med Assoc* 152;1521–1525, 1968

Sleper DA: Plant breeding, selection, and species in relation to grass tetany. In *Grass Tetany,* Rendig VV, Grunes DL eds, Am Soc Agron, Crop Sci Soc Am, Soil Sci Soc Am, Madison, Wisc, pp63–92, 1979

Smalley HE, Crookshank HR: Toxicity studies on sleepy grass, *Stipa robusta* (Vasey) Scribn. *Southwest Vet* 29;35–39, 1976

Smith BL: Mycotoxicoses of sheep in Australia and New Zealand. In *Veterinary Clinical Toxicology Proceedings 103,* Post-graduate Committee in Veterinary Science, Univ of Sydney, pp279–286, 1987

Smith BL, McLeay LM: Effects of paxalline, lolitrem B and penitrem on skeletal and smooth muscle electromyographic activity in sheep. In *Toxic plants and Other Natural Toxicants*, Garland T, Barr AC eds, CAB International, New York, pp 397–401, 1998

Smith BL, Towers NR: Pithomycotoxicosis (facial eczema) in New Zealand and the use of zinc salts for its prevention. In *Plant Toxicology, Proceedings of the Australia-USA Poisonous Plants Symposium,* Seawright AA, Hegarty MP, James LF, Keeler RF eds, Animal Research Institute, Yeerongpilly, Brisbane, Australia, pp70–79, 1985

Smith BL, McLeay LM, Embeling PP: Effect of the mycotoxins penitrem, paxilline, and lolitrem B on the electromyographic activity of skeletal and gastrointestinal smooth muscle of sheep. *Res Vet Sci* 62;111–116, 1997

Smith BL, Towers NR, Munday R, Morris CA, Collin RG: Control of the myotoxic hepatogenous photosensitization, facial eczema, in New Zealand. In *Toxic Plants and Other Natural Toxicants,* Garland T, Barr AC eds, CAB International, New York, pp469–473, 1998

Smith DJ, Skiles GL, Appleton ML, Carlson JR, Yost GS: Identification of goat and mouse urinary metabolites of the pneumotoxin, 3-methylindole. *Xenobiotica* 23;1025–1044, 1993

Smith RA, Suleiman A: Nitrite intoxication from large round bales. *Vet Hum Toxicol* 33;349–350, 1991

Smith TA: Tryptamine and related compounds in plants. *Phytochemistry* 16;171–175, 1977

Solomons RN, Oliver JW, Linnabary RD: Reactivity of dorsal pedal vein of cattle to selected alkaloids associated with *Acremonium coenophialum*–infected fescue grass. *Am J Vet Res* 50;235–238, 1989

Sousa RS, Irigoyen LF: Experimental poisoning by *Phalaris angusta* (Gramineae) in cattle. *Pesq Vet Bras* 19;116–122, 1999

Sperry OE, Turk RD, Hoffman GO, Stroud FB: *Photosensitization of Cattle in Texas.* Tex Agric Exp Stn Bull 812, 1955

Sprague R: *Diseases of Cereals and Grasses in North America.* Ronald Press, New York, pp59–66, 1950

Stalker M: Ergotism again. Iowa Agric Exp Stn Bull 17, pp453–456, 1892

Stamm MM, DelCurto T, Horney MR, Brandyberry SD, Barton RK: Influence of alkaloid concentration of tall fescue straw on the nutrition, physiology, and subsequent performance of beef steers. *J Anim Sci* 72;1068–1075, 1994

Steele CW, Spink WW: Methylene blue in the treatment of poisoning associated with methemoglobinemia. *N Engl J Med* 208;1152–1153, 1933

Stewart J: Hypomagnesemia and tetany of cattle and sheep. *Scott Agric* 34;68–73, 1954
Steyn DG: Fungi in relation to health in man and animals. *Onderstepoort J Vet Sci Anim Husb* 1;183–212, 1933a
Steyn DG: Poisoning of human beings by weeds contained in cereals (bread poisoning). *Onderstepoort J Vet Sci Anim Husb* 1;219–266, 1933b
Steyn DG: *The Toxicology of Plants in South Africa.* Central News Agency, South Africa, 1934
Stone CW: *Yeast Culture Research Briefs.* Diamond V Mills, Cedar Rapids, Iowa, 1974
Stone CW: *Yeast Culture Research Briefs.* Diamond V Mills Inc. Cedar Rapids, Iowa, 1976
Stout PR, Brownell J, Burau RG: Occurrences of trans-aconitate in range forage species. *Agric J* 59;21–24, 1967
Strain GM, Seger CL, Flory W: Toxic Bermuda grass tremor in a goat: an electroencephalographic study. *Am J Vet Res* 43;158–162, 1982
Strickland G, Selk G, Allen E, Thedford T: *Nitrate Toxicity in Livestock.* Okla State Univ Ext Facts, F-2903, 1995
Strickland JR, Oliver JW, Cross DL: Fescue toxicosis and its impact on animal agriculture. *Vet Hum Toxicol* 35;454–464, 1993
Stuedemann JA, Rumsey TS, Bond J, Wilkinson SR, Bush LP, Williams DJ, Caudle AB: Association of blood cholesterol with occurrence of fat necrosis in cows and tall fescue summer toxicosis in steers. *Am J Vet Res* 46;1990–1995, 1985
Sund JM, Wright MJ, Simon J: Weeds containing nitrates cause abortion in cattle. *Agron J* 49;278–279, 1957
Swartzlander JH, Rehberger TG, Hibberd CA: Selection and ruminal establishment of a denitrifying *Proprionibacterium* strain in beef cattle [abstract]. *J Anim Sci* 71 (Suppl 1);31, 1993a
Swartzlander JH, Rehberger TG, Hibberd CA: In vivo denitrification by a selected strain of *Proprionibacterium acidiproprionici* in beef cattle [abstract]. *J Anim Sci* 71 (Suppl 1);277, 1993b
Taylor MC, Loch WF, Ellersieck M: Toxicity in pregnant pony mares grazing Kentucky-31 fescue pastures. *Nutr Rep Int* 31;787–795, 1985
TePaske MR, Powell RG, Clement SL: Analyses of selected endophyte-infected grasses for the presence of loline-type and ergot-type alkaloids. *J Agric Food Chem* 41;2299–2303, 1993
Thompson FN, Porter JK: Tall fescue toxicosis in cattle: could there be a public health problem here? *Vet Hum Toxicol* 32 (Suppl);51–57, 1990
Thompson FN, Stuedemann JA, Sartin JL, Belesky DP, Devine OJ: Selected hormonal changes with summer fescue toxicosis. *J Anim Sci* 65;727–733, 1987
Thompson RW, Fribourg HA, Waller JC, Sanders WL, Reynolds JH, Phillips JM, Schmidt SP, Crawford RJ Jr, Allen VG, Faulkner DB, Hoveland CS, Fontenot JP, Carlisle RJ, Hunter PP: Combined analysis of tall fescue steer grazing studies in the eastern United States. *J Anim Sci* 71;1940–1946, 1993
Thornton-Manning JR, Nichols WK, Manning BW, Skiles GL, Yost GS: Metabolism and bioactivation of 3-methylindole by Clara cells, alveolar macrophages, and subcellular fractions from rabbit lungs. *Toxicol Appl Pharmacol* 122;182–190, 1993
Thorp F: Further observations on oat hay poisoning. *J Am Vet Med Assoc* 92;159–170, 1938
Tillman AD, Sheriha GM, Sirny RJ: Nitrate reduction studies with sheep. *J Anim Sci* 24;1140–1146, 1965
Tor-Agbidye J, Blythe LL, Craig AM: Correlation of quantities of ergovaline and lolitrem B toxins to clinical cases of tall fescue toxicosis and perennial ryegrass staggers. In *Plant-Associated Toxins: Agricultural, Phytochemical, and Ecological Aspects,* Colegate SM, Dorling PR eds, CAB International, Wallingford, UK, pp369–374, 1994
Tucker JO, Maki LR: Acute pulmonary emphysema of cattle. 1. Experimental production. *Am J Vet Res* 23;821–823, 1962

Turk MAM, Thomas DE: Effects of phenobarbital treatment on 3-methylindole toxicosis in ponies. *Am J Vet Res* 47;901–905, 1986

Tyler JW, Shelby RA, Sartin EA, Wolfe DF, Steiss JE, Sorjonen DC, Powe TA, Spano JA: Naturally occurring neurologic disease in calves fed *Claviceps* sp. infected Dallis grass hay and pasture. *Prog Vet Neurol* 3;101–106, 1992

Ueno K, Giam CS, Taber WA: Absence of sporidesmin in a Texas isolate of *Pithomyces chartarum. Mycologia* 66;360–362, 1974

Van der Walt SJ, Steyn DG: The hydrocyanic acid content of *Cynodon plectostachyum* (giant star grass) and its suitability as a pasture grass. *Onderstepoort J Vet Sci Anim Ind* 17;191–199, 1941

Van Dijk S, Lobsteyn AJH, Wensing T, Breukink HJ: Treatment of nitrate intoxication in a cow. *Vet Rec* 112;272–274, 1983

van Essen GJ, Blom M, Fink Gremmels-Gehrmann J: Ryegrass cramps in horses. *Tijdschr Diergeneeskd* 120;710–711, 1995

Van Kampen KR: Sudan grass and sorghum poisoning of horses: a possible lathyrogenic disease. *J Am Vet Med Assoc* 156;629–630, 1970

van Rensburg SJ, Altenkirk B: *Claviceps purpurea*—Ergotism. In *Mycotoxins,* Elsevier, Amsterdam, pp69–96, 1974

van Santen E: Animal preference of tall fescue during reproductive growth in the spring. *Agron J* 84;979–982, 1992

van't Klooster ATh, Taverne MAM, Malestein A, Akkersdijk EM: On the pathogenesis of abortion in acute nitrite toxicosis of pregnant dairy cows. *Theriogenology* 33;1075–1089, 1990

Vertregt N: The formation of methemoglobin by the action of nitrite on bovine blood. *Neth J Agric Sci* 25;243–254, 1977

Vinall HN: A study of the literature concerning poisoning of cattle by the prussic acid in sorghum, Sudan grass, and Johnson grass. *J Am Soc Agron* 13;267–280, 1921

Vogel KP, Haskins FA, Gorz HJ: Potential for hydrocyanic acid poisoning of livestock by Indiangrass. *J Range Manage* 40;506–509, 1987

Wallner BM, Booth NH, Robbins JD, Bacon CW, Porter JK, Kiser TE, Wilson R, Johnson B: Effect of an endophyte fungus isolated from toxic pasture grass on serum prolactin concentrations in the lactating cow. *Am J Vet Res* 44;1317–1322, 1983

Wang LC, Garcia-Rivera J, Burris RH: Metabolism of nitrate by cattle. *Biochem J* 81;237–242, 1961

Watt JM, Breyer-Brandwijk MG: *The Medicinal and Poisonous Plants of Southern and Eastern Africa.* E & S Livingston, Edinburgh, pp475–477, 1962

Wattenbarger DW, Gray E, Rice JS, Reynolds JH: Effects of frost and freezing on hydrocyanic acid potential of sorghum plants. *Crop Sci* 8;526–528, 1968

Watts H, Webster M, Chappel A, Leaver DD: Laboratory diagnosis of nitrite poisoning in sheep and cattle. *Aust Vet J* 45;492, 1969

Welty RE, Milbrath GM, Faulkenberry D, Azevedo MD, Meek L, Hall K: Endophyte detection in tall fescue seed by staining and ELISA. *Seed Sci Technol* 14;105–116, 1986

Wendel WB: The control of methemoglobinemia with methylene blue. *J Clin Invest* 18; 179–185, 1939

Wheeler JL: Increasing animal production from sorghum forage. *World Anim Rev* 35; 13–22, 1980

Wheeler JL, Hedges DA, Archer KA, Hamilton BA: Effect of nitrogen, sulphur, and phosphorus on the production, mineral content, and cyanide potential of forage sorghum. *Aust J Exp Agric Anim Husb* 20;330–338, 1980

White JF Jr: Widespread distribution of endophytes in the Poaceae. *Plant Dis* 71;340–342, 1987

White JF, Halisky PM: Association of systemic fungal endophytes with stock-poisoning grasses. In *Poisonous Plants: Proceedings of the Third International Symposium,* James LF, Keeler RF, Bailey EM Jr, Cheeke PR, Hegarty MP eds, Iowa State Univ Press, Ames, pp574–578, 1992

White JF, Morgan-Jones G: Endophyte-host associations in forage grasses. 7. *Acremonium chisosum*, a new species isolated from *Stipa eminens* in Texas. *Mycotaxon* 28;179–189, 1987

Whitehair CK, Young HC, Gibson ME, Short GE: A nervous disturbance in cattle caused by a toxic substance associated with mature Bermuda grass. Okla Agric Exp Stn Misc Publ MP–22, pp57–59, 1951

Whitehead EI, Moxon AL: *Nitrate Poisoning.* S D Agric Exp Stn Bull 424, 1952

Wilcox EW: *Poisonous Plants of Montana.* Mont Agric Exp Stn Bull 22, 1899

Williams CH, Hines HJG: The toxic properties of *Salvia reflexa. Aust Vet J* 16;14–20, 1940

Williams GD, Hatkin J, Jones LP: Acute respiratory distress syndrome occurring in Texas pastured cattle. *Proc Am Assoc Vet Lab Diagn* 20;327–338, 1977

Williams JR, Challis FE: Methylene blue as an antidote for aniline dye poisoning. *J Lab Clin Med* 19;166–171, 1933

Wilson AD, Gay CC, Fransen SC: An *Acremonium* endophyte of *Lolium perenne* associated with hyperthermia of cattle in Pacific County, Washington. *Plant Dis* 76;212, 1992

Wilson JK: Nitrate in plants: its relation to fertilizer injury, changes during silage making, and indirect toxicity to animals. *J Am Soc Agron* 35;279–290, 1943

Winter AJ: Studies on nitrate metabolism in cattle. *Am J Vet Res* 23;500–505, 1962

Winter AJ, Hokanson JF: Effects of long-term feeding of nitrate, nitrite, or hydroxylamine on pregnant dairy heifers. *Am J Vet Res* 25;353–361, 1964

Wittenberg KM, Duynisveld GW, Tosi HR: Comparison of alkaloid content and nutritive value for tryptamine- and beta-carboline-free cultivars of reed canarygrass (*Phalaris arundinacea* L.). *Can J Anim Sci* 72;903–909, 1992

Witters WL, Wilms RA, Hood RD: Prenatal effects of elymoclavine administration and temperature stress. *J Anim Sci* 42;1700–1705, 1975

Wolfe MH: More notes on sleepygrass. *Rangelands* 1;144, 1979

Woods AJ, Bradley Jones J, Mantle PG: An outbreak of gangrenous ergotism in cattle. *Vet Rec* 78;742–749, 1966

Woods LW, Wilson DW, Schiedt MJ, Giri SN: Structural and biochemical changes in lungs of 3-methylindole-treated rats. *Am J Pathol* 142;129–138, 1993

Wright MJ, Davison KL: Nitrate accumulation in crops and nitrate poisoning in animals. *Adv Agron* 16;197–247, 1964

Yates SG, Powell RG: Analysis of ergopeptine alkaloids in endophyte-infected tall fescue. *J Agric Food Chem* 36;337–340, 1988

Yates SG, Plattner RD, Garner GB: Detection of ergopeptine alkaloids in endophyte infected, toxic Ky-31 tall fescue. *J Agric Food Chem* 33;719–722, 1985

Yokoyama MT, Carlson JR: Dissimilation of tryptophan and related indolic compounds by ruminal microorganisms in vitro. *Appl Microbiol* 27;540–548, 1974

Younis MA, Pauli AW, Mitchell HL, Stickler FC: Temperature and its interaction with light and moisture in nitrogen metabolism of corn (*Zea mays* L). *Crop Sci* 5;321–326, 1965

Zanzalari KP, Heitmann RN, McLaren JB, Fribourg HA: Effects of endophyte-infected fescue and cimetidine on respiration rates, rectal temperatures, and hepatic mixed function oxidase activity as measured by hepatic antipyrine metabolism in sheep. *J Anim Sci* 67;3370–3378, 1989

59 Polygonaceae Juss.

- Buckwheat Family

  Fagopyrum
  Polygonum
  Rheum
  Rumex

Commonly known as the buckwheat family, the Polygonaceae comprises 30–51 genera and 750–1,150 species distributed primarily in the temperate region of the Northern Hemisphere. Not of great economic importance, the family contains a few food species, such as *Fagopyrum* (buckwheat) and *Rheum* (rhubarb); a few ornamentals, such as *Antigonon* (mountain rose vine) and *Polygonum aubertii* (silver lace vine); and several noxious weeds, such as *Polygonum aviculare* (prostrate knotweed). The achenes of several taxa are eaten by wildlife. In North America, approximately 29 genera and 450 species, both native and introduced, are present. Intoxication problems have been associated with 4 genera.

- herbs, vines, subshrubs; stems jointed, nodes swollen; leaves simple, alternate; stipules sheathing stems; petals absent; fruits achenes

Plants herbs or herbaceous vines or subshrubs; perennials or biennials or annuals; from fleshy roots or rhizomes or caudices or crowns; terrestrial or emergent or floating-leaved aquatics; bearing perfect flowers or dioecious or monoecious. **Stems** jointed; nodes typically swollen. **Leaves** simple; alternate or rarely opposite or whorled; petiolate or sessile; venation pinnate or a single vein or not apparent; stipules present, fused, sheathing stems (ocrea), or absent. **Inflorescences** simple or compound cymes or spikes or racemes or clusters or glomerules or solitary flowers; bracts absent or present. **Flowers** perfect or imperfect, similar; perianths in 1-series; radially symmetrical. **Sepals** 2-6; in 1 or 2 whorls; fused or free; sepaloid or petaloid; of various colors except blue. **Petals** absent. **Stamens** 2–9. **Pistils** 1; compound, carpels 2–4; stigmas 2–4; styles 1–4, free or fused; ovaries superior; locules 1; placentation basal. **Fruits** achenes; trigonous or lenticular. **Seeds** 1.

Fagopyrum Mill.

- *F. esculentum*
  F. tataricum

  buckwheat
  common buckwheat
  India wheat
  Kangra buckwheat

Taxonomy and Morphology—Comprising 6–15 species native to Eurasia, *Fagopyrum,* commonly known as buckwheat, is represented in North America by 2 naturalized species. Its name is derived from the Latin root *fagus,* meaning "beech," and the Greek root *pyros,* meaning "wheat," which allude to the fruits' being shaped like those of beech trees and yielding an edible flour. Likewise, its common name reflects this; *buche* is German for "beech." The buckwheat flour of commerce is obtained from *F. esculentum.* Wild buckwheat is derived from *Eriogonum,* another genus of the family. The following are the 2 introduced species:

F. esculentum Moench (= *F. sagittatum* Gilib.)	buckwheat, common buckwheat
F. tataricum (L.) Gaertn.	India wheat, Kangra buckwheat

- erect annual herbs; leaves simple, alternate; flowers borne in racemes; sepals 5, petal-like

Plants herbs; annuals; bearing perfect flowers. **Stems** erect; 20–60 cm tall; longitudinal lines of pubescence present. **Leaves** simple; alternate; long petiolate; blades broadly triangular-hastate to cordate; margins entire; ocreae obliquely truncate. **Inflorescences** clusters of flowers borne in racemes arising from axils of short bracts. **Flowers** perfect. **Sepals** 5; in 1 whorl; fused near base; unequal in width; petaloid; greenish or greenish white to pink. **Petals** absent. **Stamens** 8. **Pistils** with 3 carpels; stigmas 2, capitellate; styles 3. **Achenes** trigonous; shiny or dull; smooth or rough; longer than persistent perianths.

Fagopyrum esculentum

1. Racemes short; borne in dense terminal clusters. Sepals greenish white to greenish pink. Achenes shining; margins smooth . . . . *F. esculentum*
1. Racemes elongate; borne in leaf axils. Sepals greenish. Achenes dull; margins roughened . . . . *F. tataricum*

Distribution and Habitat—Apparently native to central Asia, both species are naturalized in North America and elsewhere. They typically occur in disturbed sites, although they are infrequently encountered or not long persistent. They are sometimes planted as summer cover crops to prevent erosion, in addition to being cultivated for their grain. Both are good bee plants for the commercial production of honey.

■ scattered occurrence across North America

Disease Problems—Both the herbage and intact achenes of *F. esculentum* have long been associated with photosensitization in a variety of animal species (Wender 1946). Flour made from the seed, however, poses no risk. This is a disease mainly of historical interest, because buckwheat has, for the most part, been supplanted by other forages for livestock.

■ herbage and achenes, photosensitization, of little concern for livestock

Disease Genesis—Called fagopyrism, the disease is due to the presence of naphthodianthrone derivatives, the most abundant of which are fagopyrin and protofagopyrin. These complex or extended quinone-type compounds are derivatives of hypericin and pseudohypericin, respectively, and are found in the flowers and achene pericarps. They were originally described as differing from hypericin and pseudohypericin by the presence of $C_6H_{11}ON$ side chains on each of the two methyl groups (Brockmann 1957). This interpretation of the structure was later revised to the nitrogen-containing rings positioned between the two hydroxyl groups on each side at the opposite end of the molecule from the CH_3 groups (Brockmann & Lachner 1979). These compounds may well be the same precursors found in *Hypericum*. In both *Polygonum* and *Hypericum*, the parent compound is hydrolyzed to the active dianthrone. It is eliminated from the body only slowly and therefore has prolonged toxic effects. The dianthrone is activated by light wavelengths of 540–610 nm in the presence of oxygen to produce singlet oxygen as described for hypericin (Sheard et al. 1928). Further discussion of this phenomenon is presented in the treatment of *Hypericum* (see chapter 24).

■ dianthrones, fagopyrin, protofagopyrin; photoactive, by UV light, causing skin lesions

Fagopyrin

Clinical Signs—As in cases involving *Hypericum,* the signs begin within a day or so after consumption of *Fagopyrum*. In cattle and most other species, the most striking effects are severe pruritis, reddening, swelling, edema, and sloughing of the skin of the teats, udder, escutcheon, and the less pigmented and the lightly haired areas. There is often a very striking demarcation in distribution of the lesions; they coincide exactly with the lighter-colored areas of the skin and extend right up to the edge of the darker areas. These obvious skin changes may be accompanied by signs of diarrhea, restlessness, agitation, loss of appetite, and increased respiratory rate. Occasionally, violent behavior during the onset of the disease may be a serious problem. In sheep,

■ less pigmented areas of skin, pruritis, reddening, swelling, edema, sloughing

because of the protective effects of wool, the skin lesions may be limited to ears, eyelids, face, lips, and coronets. The signs may last 2 weeks, and subsequent healing may require 2 or more months.

- gross pathology, skin thickened, crusty ulcerative areas

  microscopic, skin edema, necrosis, neutrophilic (PMN) infiltration

Pathology—Skin changes typically are limited to the lesser pigmented areas or to the head in sheep. Depending upon the amount of plant material eaten, the changes range from minimal reddening and blisterlike effects to extensive areas of sloughing. In more-severe cases, the skin will be thickened with a gelatinous appearance extending down into the deeper dermiss. Crusty and ulcerative or proliferative areas may be present, especially around the face. Microscopically, the edema of the skin will be accompanied by necrosis and a neutrophilic infiltrate in the deeper dermis.

- slow recovery when animals removed from plants

Treatment—Prevention of animal access to the plants is usually sufficient for recovery to begin within a few days, albeit the signs may disappear slowly as reepithelialization of the denuded areas occurs. In some cases relief of pain may be helpful, in addition to soothing skin medications.

Polygonum L.

- *P. aviculare*
  P. convolvulus
  P. cuspidatum
  P. hydropiper
  P. pensylvanicum
  P. persicaria

  bindweed
  black bindweed
  chivalry-grass
  cornbind
  dishwater grass
  doorweed
  dullseed
  fleeceflower
  heart's-ease
  heartweed
  Japanese bamboo
  knot bindweed
  knotweed
  lady's thumb
  marshpepper
  Mexican bamboo
  pinkweed
  smartweed
  water pepper
  wild buckwheat
  wireweed

Taxonomy and Morphology—Comprising 150–300 species, *Polygonum* is native to both the Old World and the New World. Its name is derived from the Greek roots *polys* and *gonu,* meaning "many knees" or "many joints," an allusion to the conspicuously swollen nodes. A morphologically and ecologically diverse genus distributed primarily in temperate regions, it has been divided into several genera by some taxonomists. Most, however, recognize 1 genus with several sections, as is done in this treatise. Some species of the genus are cultivated as ornamentals, others are invasive weeds, and still others are important food sources for aquatic birds and mammals. In North America, approximately 75 native and introduced species and innumerable varieties are present. Representative of the genus and implicated in toxicological problems are the following:

P. aviculare L.	prostrate knotweed, wireweed, doorweed, chivalry-grass, dishwater grass
P. convolvulus L.	wild buckwheat, black bindweed, ivy bindweed, knot bindweed, dullseed, cornbind
P. cuspidatum Siebold & Zucc.	Japanese knotweed, fleeceflower, Mexican bamboo, Japanese bamboo
P. hydropiper L.	marshpepper, smartweed, water pepper, water smartweed
P. pensylvanicum (L.) Small	smartweed, pink smartweed, Pennsylvania smartweed, pinkweed
P. persicaria L.	lady's thumb, spotted smartweed, heart's-ease, heartweed

- herbs, vines, subshrubs; ocreae conspicuous; flowers in spicate or racemose panicles or axillary clusters; sepals 4–6, petaloid; achenes shiny, black, or brown

Plants herbs or herbaceous vines or rarely subshrubs; perennials or annuals; terrestrial or emergent aquatics. **Stems** prostrate to erect; 20–150 cm tall. **Leaves** alternate; petiolate or occasionally sessile; margins entire; ocreae conspicuous. **Inflorescences** terminal spicate or racemose panicles, or axillary clusters. **Flowers** perfect or imperfect, similar; perianths funnelform or campanulate. **Sepals** 4–6; in 1 whorl; fused; petaloid; green or white or cream or pink or red. **Petals** absent. **Stamens** 3–9. **Pistils**

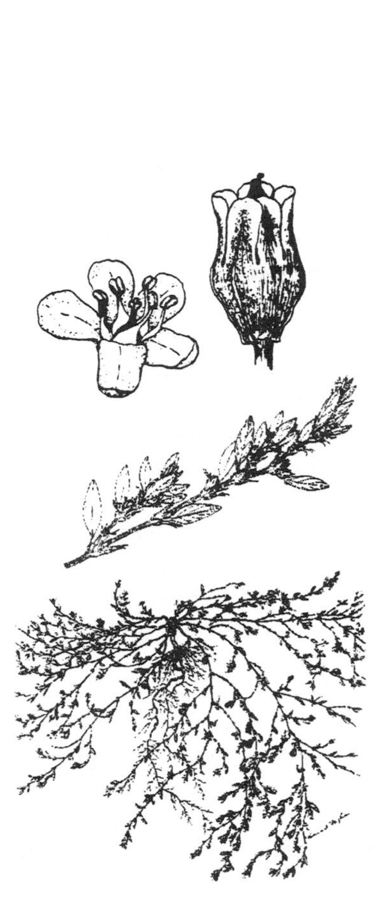

Polygonum aviculare

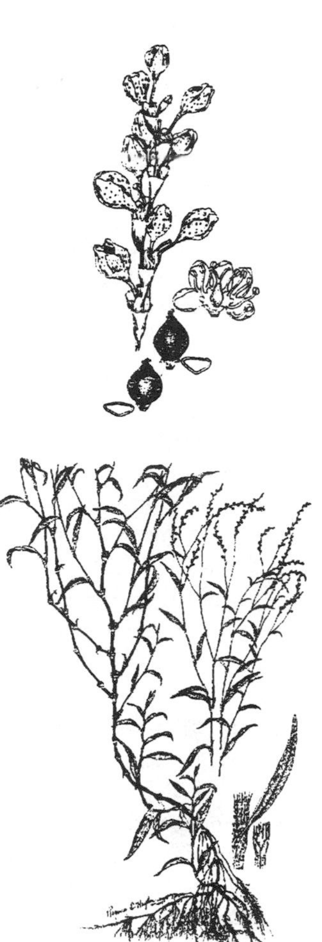

Polygonum hydropiper

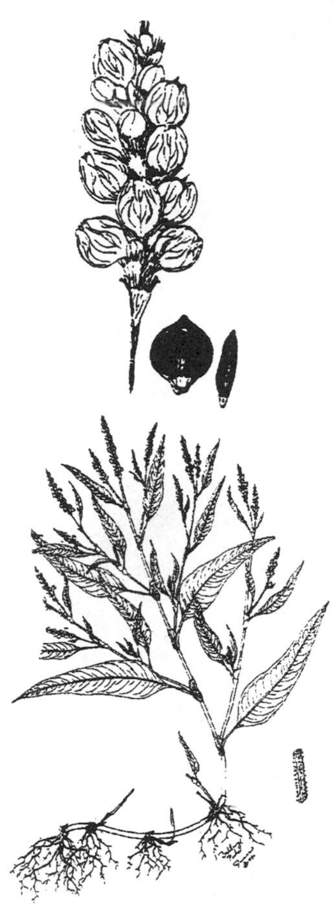

Polygonum pensylvanicum

with 2 or 3 or rarely 4 carpels; stigmas 2 or 3 or rarely 4, capitellate; styles 2 or 3 or rarely 4. **Achenes** trigonous or lenticular; shiny; black or brown; shorter or longer than persistent perianths.

Because of the large number of species in the genus, a taxonomic key for identification is not presented here. If identification is needed, floristic treatments for the reader's locality should be used.

Polygonum persicaria

Distribution and Habitat—Occupying diverse habitats, the species of *Polygonum* are adapted to a variety of environmental conditions. Some are aquatic and semiaquatic, whereas others flourish in dry, disturbed sites. Many are quite weedy and readily form dense populations. *Polygonum aviculare,* for example, is a native annual commonly encountered as a weed in lawns across the continent. It is very tolerant of trampling and also occupies waste areas and weedy sites. *Polygonum persicaria, P. hydropiper, and P. convolvulus* are introduced annuals that are also found throughout North America. They typically occupy damp, disturbed soils. *Polygonum cuspidatum* is an introduction from Japan that is cultivated, but it occasionally escapes and is found along roadsides.

Disease Problems and Genesis—Although species of *Polygonum* are not commonly regarded as causes of intoxication, they have been suspected of causing an acute dis-

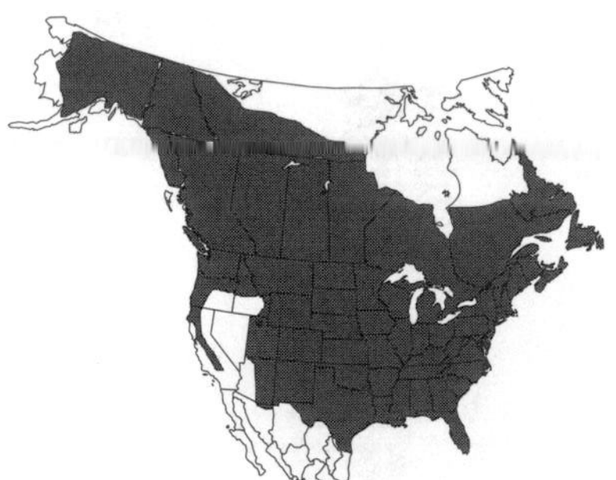
Polygonum aviculare

Polygonum cuspidatum

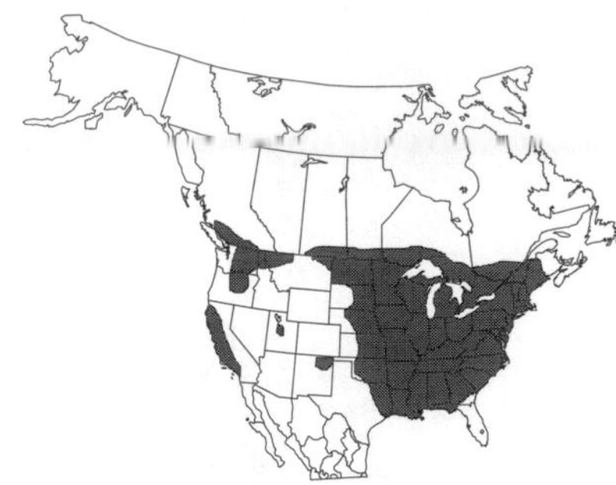
Polygonum hydropiper

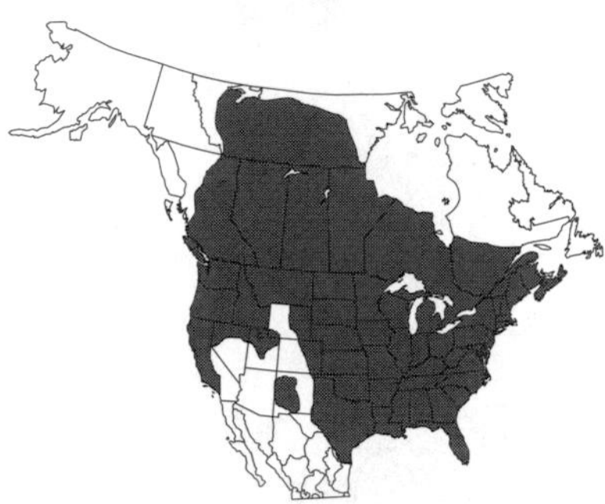
Polygonum persicaria

■ some aquatic, others weedy lawn invaders

■ abrupt deaths; more likely photosensitization

ease ending in death or less acute photosensitization. In Australia, several horses that ate large amounts of *P. aviculare* became recumbent and died a few hours later (Knight 1979). Nitrite intoxication was suspected because species of the genus accumulate nitrates (Case 1957), but the clinical signs and pathologic changes were not consistent with this diagnosis. There also is a report of *P. cuspidatum* causing the death of a goat in the United Kingdom (Cooper & Johnson 1984). In Argentina, after 160 cows were moved into a field containing *P. convolvulus,* 10 were found dead within 24 hours (Cantero et al. 1990). Although the cattle were removed from the field after the first deaths, 20 more died in the following 5 days. Tonic convulsions, irritability, liver necrosis, and pulmonary edema were observed. One to 2 months later, 10 cows developed photosensitization.

In other instances of *Polygonum* toxicity, only skin changes similar to those caused by *Fagopyrum* are produced. In the Canadian prairie provinces, fresh plants of *Polygonum* eaten in the pasture are credited with causing photosensitization (Looman 1983). The disease is locally termed yellows or bighead. Plants eaten as hay do not produce the adverse effects. In these provinces, plants growing in water and cut after the areas dry up are considered to make nutritious hay.

Experiments to confirm the toxicity of *Polygonum* have yielded at best only increased suspicion. Large amounts *P. hydropiper* and *P. persicaria* were fed without apparent adverse effects (Bruce 1917; McKenzie et al. 1988). Smaller amounts of *P. aviculare* and *P. punctatum* (reported as *P. acre*) also failed to produce changes. Some species are said to contain an acrid, irritating sap—hence the common name *smartweed.*

■ similar to *Fagopyrum*

Clinical Signs, Pathology, and Treatment—The clinical signs, pathology, and treatment for photosensitization are the same as for *Fagopyrum.*

Rheum L.

■ *R. rhabarbarum*

garden rhubarb
pie plant
rhubarb
wine plant

Taxonomy and Morphology—Native to temperate and subtropical Asia, *Rheum,* commonly known as rhubarb, comprises about 50 species. Its name is derived from *rha,* the ancient Greek word for rhubarb. In addition to the species cultivated for food and wine, several others are propagated for their striking foliage and tall, showy inflorescences. A few Chinese species are used medicinally. Most commonly encountered and of toxicologic interest is this species:

R. rhabarbarum L. (= *R. rhaponticum* L.) — rhubarb, garden rhubarb, pie plant, wine plant

Some taxonomists contend that the cultivated rhubarb is an ancient hybrid and use the binomial *R.* x*cultorum.* The common name *rhubarb* also is applied to several species of *Rumex;* e.g., *R. hymenosepalus* is called wild rhubarb and *R. alpinus* is known as mountain rhubarb.

Plants robust herbs; perennials; from thick roots. **Stems** erect; 50–150 cm tall; hollow. **Leaves** large; primarily basal; petioles long, stout, flat above; blades suborbicular to broadly ovate; margins entire to lobed; ocreae prominent. **Inflorescences** terminal panicles; branches appressed; pedicels jointed; flowers borne in fascicles at nodes; bracts absent. **Flowers** perfect. **Sepals** 6; in 2 whorls; free; petaloid; white or greenish white. **Petals** absent. **Stamens** 6 or 9. **Pistils** with carpels, stigmas, and styles 3. **Achenes** trigonous; winged; brown; longer than perianths.

- robust, perennial herbs; leaves large, primarily basal, on long stout petioles; flowers borne in terminal panicles; sepals 6, in 2 whorls; achenes trigonous, brown

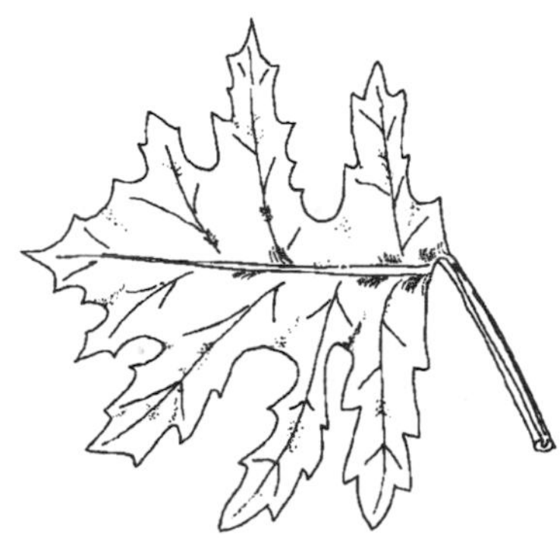

Rheum rhabarbarum

Distribution and Habitat—*Rheum rhabarbarum* is cultivated throughout North America. In some areas it has escaped cultivation and become sparingly established. It also may persist in abandoned gardens but apparently does not spread.

- cultivated food plant

Disease Problems—Rhubarb has somewhat of a reputation for being highly toxic to both humans and animals. During World War I, because of dwindling food supplies, efforts were made in England to utilize alternative foods. Rhubarb leaves were tried as a spinach substitute, and several intoxications occurred. Perhaps the most noteworthy case involved the death of a minister who had eaten the leaves ("Poisoning," 1917). Because oxalic acid (not crystals of calcium oxalate) was found in his tissues, it was thought that its presence was due to consumption of the rhubarb. Based upon the soluble oxalate concentrations in *Rheum,* this conclusion is questionable. The man would have had to have eaten a kilogram or more of the cooked leaves to cause intoxication. At about the same time, the death of a woman who had eaten fried rhubarb leaves and boiled stalks was reported in the United States (Robb & Sippy 1919). The disease involved considerable colic and vomiting, and her blood failed to clot terminally. These and other reports fostered an atmosphere of considerable concern, and warnings were expressed about the dire consequences of eating rhubarb leaves (Jeghers & Murphy 1945). It is now apparent that there is little risk in eating *Rheum* leaves.

- humans and livestock, leaves not substantiated as a cause of intoxications

Several cases also have been reported for animals. For example, a sow that was observed to have eaten heartily of rhubarb leaves suffered a loss of appetite and was found dead 2 days later with her stomach still full of them (Buchanan 1933). There is, however, little or no direct experimental evidence to substantiate the toxic reputation of *Rheum* other than reports of oxalate concentrations in the plants.

Disease Genesis—All portions of the plant, but especially the leaves, contain soluble oxalates, probably acid (potassium) oxalate, as occurs in *Rumex* (James 1972). In an examination of 78 rhubarb cultivars, concentrations of total oxalate in the petioles or stalks were 3.4–9.5% dry matter, with an average of 5.8% d.w. or 0.4% wet weight (Libert & Creed 1985). The range for soluble oxalates was 1.8–5.7% (Libert & Franceschi 1987). If concentrations are higher in the leaf blades, then indeed

- oxalates, risk for livestock; minimal risk for humans; possible adverse effects on digestive tract due to presence of emodin, aloe-emodin, rhein

OH O OH

HO CH3

O

Emodin

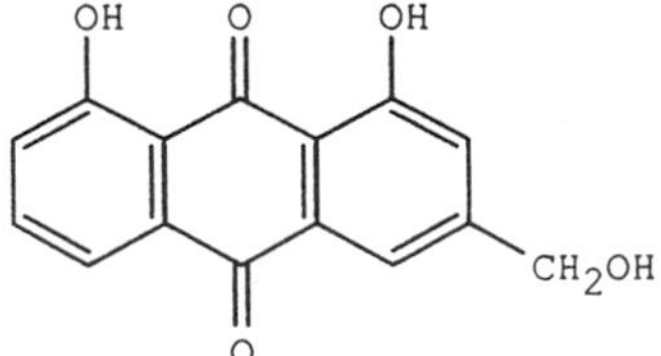

Aloe-emodin

OH O OH

COOH

O

Rhein

oxalates may pose a danger to livestock. Although there is little doubt that livestock consuming large amounts of forage are at risk with oxalate-accumulating plants, there is some doubt about the role of oxalate in human intoxications, because of the dosage required. Fassett (1973) has reviewed some of the reported serious human intoxications, and although the clinical signs are somewhat consistent with the irritating effects of oxalic acid, a role for other toxicants such as anthraquinones cannot be discounted (Cooper & Johnson 1984). Emodin, aloe-emodin, and rhein are anthraquinones that are present in the tissues of *Rheum* as reduced anthrone and bianthrone forms and as the more toxic glycosides (Thomson 1971; Robinson 1991). These toxicants are clearly capable of causing marked effects on the digestive tract.

■ oxalate
digestive disturbance

Clinical Signs, Pathology, and Treatment—Aspects of these facets of the oxalate-associated disease are presented in the treatment of *Rumex,* which follows in this chapter. *Rheum* also may produce signs of mild to moderate irritation of the digestive tract, manifested as diarrhea. These are distinct from problems associated with oxalate salts.

Rumex L.

■ *R. acetosa*
R. acetosella
R. crispus
R. venosus
dock
lengua de vaca
sorrel
sour grass
sour green
sourweed
wild begonia
wild hydrangea

Taxonomy and Morphology—Commonly known as dock or sorrel, *Rumex* comprises 125–200 species distributed primarily in the northern temperate regions. Its name was used by Pliny, the Roman natural historian. The leaves of some species are valued for their tart taste and are eaten raw in salads or cooked like spinach. Other species are considered pernicious weeds, and a few are grown as ornamentals for the bronze red color of their autumn foliage. In North America, about 50 introduced and native species and named hybrids are present. Exact identification of individual plants is sometimes troublesome because of the frequent hybridization. Representative of the genus and implicated in toxicologic problems are the following (see plate 59.1):

R. acetosa L.	green sorrel, tall sorrel, garden sorrel, meadow sorrel, sour dock
R. acetosella L.	red sorrel, sheep sorrel, common sorrel, cow sorrel, horse sorrel, mountain sorrel, field sorrel, sour grass, sourweed
R. crispus L.	curly dock, curly leaf dock, yellow dock, sour dock, lengua de vaca
R. venosus L.	veined dock, wild begonia, wild hydrangea, sour green, winged sorrel

The common name *sorrel* also is applied to species of *Oxalis* in the Oxalidaceae(see chapter 54), a notorious oxalate accumulator, as is implied by its genus name.

■ herbs, often reddish green at maturity; ocreae prominent; flowers in terminal panicles or racemes; sepals 6, in 2 whorls, outer 3 small, inner 3 small; achenes trigonous

Plants herbs; perennials or biennials or annuals; from taproots or fibrous roots or rhizomes; herbage typically becoming reddish green with maturity; bearing perfect flowers or dioecious or monoecious. **Stems** erect; 10–150 cm tall; typically persisting through winter. **Leaves** alternate; both basal and cauline; petiolate; blades flat or crisped; apices acute or obtuse to rounded; margins entire or undulate or shallowly toothed; ocreae prominent, often shredding with maturity. **Inflorescences** terminal panicles or racemes; flowers borne in verticillate clusters at nodes and typically persisting through winter. **Flowers** perfect or imperfect, similar. **Sepals** 6; in 2 whorls, outer whorl small and becoming reflexed, inner whorl enlarging, each typically bearing a corky tubercule at center or base; fused; sepaloid; green or reddish brown. **Petals**

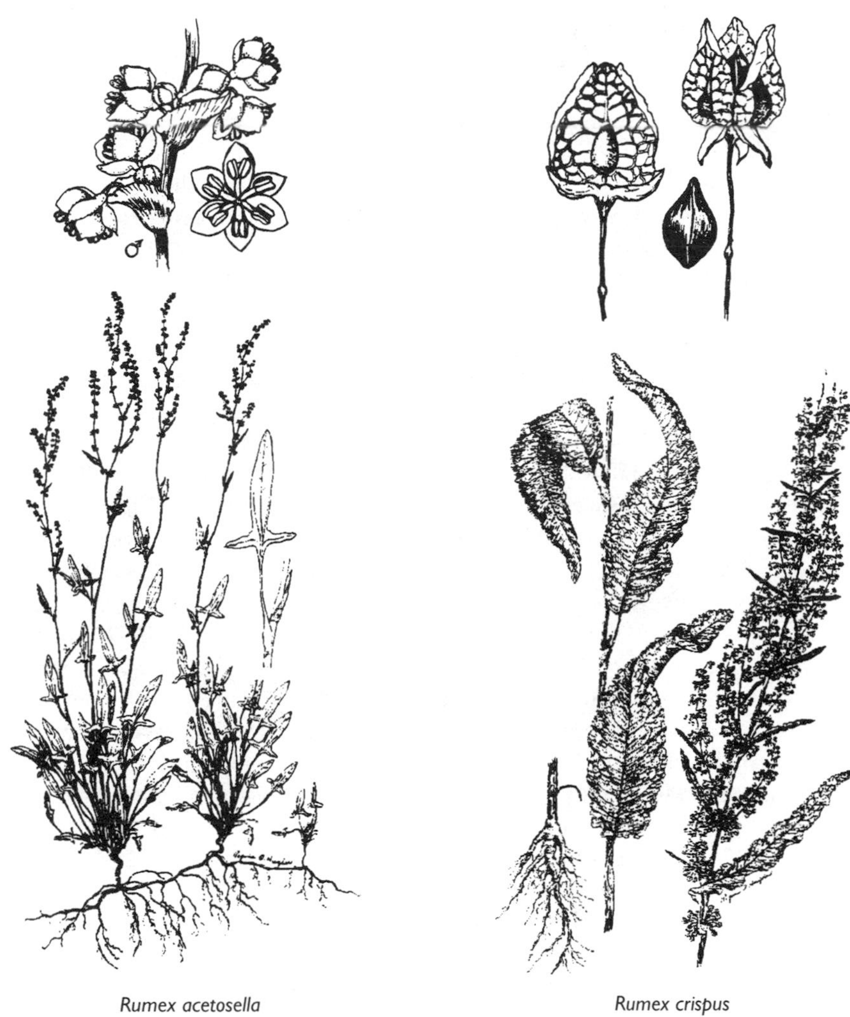

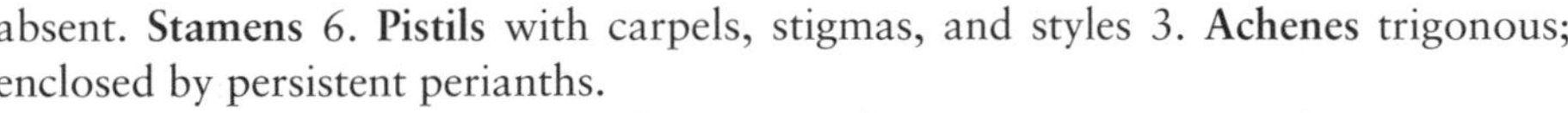
Rumex acetosella Rumex crispus

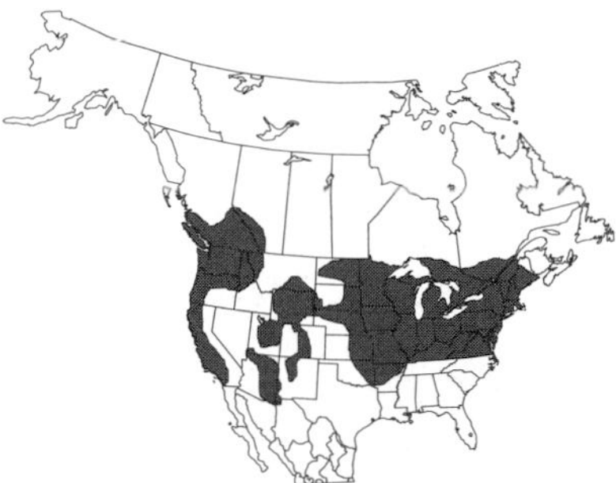
Rumex acetosella

Rumex crispus

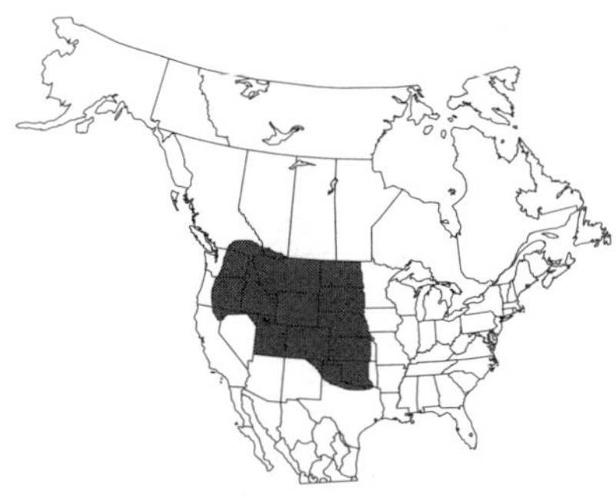
Rumex venosus

absent. **Stamens** 6. **Pistils** with carpels, stigmas, and styles 3. **Achenes** trigonous; enclosed by persistent perianths.

Because of the large number of species in the genus, a taxonomic key is not presented here. If identification is needed, floristic treatments for reader's locality should be used.

Distribution and Habitat—The species of *Rumex* occupy a variety of habitats and environmental conditions in North America. Native to Eurasia but now naturalized throughout most of North America, *R. acetosa* and *R. acetosella* are found along roadsides and in lawns, waste areas, and open woodlands. Also introduced from Europe, *R. crispus* occupies similar disturbed habitats, especially wet sites. Native to the sand dunes and riverbanks of the grasslands and deserts of western North America, *R. venosus* is adventive in the Midwest and Northeast.

■ waste area invaders; damp or wet sites

Disease Problems—Species of *Rumex* have had a reputation for toxicity in the Old World. Losses of sheep and horses due to *R. acetosella* was reported by Cornevin and by Muller (Craig & Kehoe 1921). This prompted an investigation of the toxicity of another species, *R. acetosa,* a perennial widely distributed in Europe. Most animals would not eat the plants. However, one bullock ate it readily, consuming 2.7–4.5 kg

■ intoxications uncommon; metabolic and renal disease

daily for 3 weeks without adverse effect (Craig & Kehoe 1921). This is in contrast to a later report of a flock of 90 ewes and lambs readily eating the species, followed by illness of 12 ewes and death of 5 (Coward 1949). In Colorado, cattle subsisting extensively on *R. venosus* developed severe renal disease with copious fluid in the body cavities and oxalate crystals in the kidneys (Dickie et al. 1978). Similarly, abundant *R. crispus* was the apparent cause of disease in hungry sheep that had been held overnight in a small pasture with limited other forage (Panciera et al. 1990). In this case, the initial clinical signs were suggestive of hypocalcemia, eventually accompanied by severe tubular nephrosis.

Species of *Rumex* are eaten commonly by humans throughout the world as potherbs and in rare instances may cause intoxications, accompanied by oxalate crystals in blood vessels and the urinary system and by liver disease (Farre et al. 1989).

■ acid/potassium oxalate

$H^+\ {}^-OOC—COO^-\ H^+$

$H^+\ {}^-OOC—COO^-\ Na^+$

$Na^+\ {}^-OOC—COO^-\ Na^+$

$\longrightarrow H^+\ {}^-OOC—COO^-\ K^+$

$K^+\ {}^-OOC—COO^-\ K^+$

$Ca^{++}\ {}^-OOC—COO^-\ Ca^{++}$

COO^- ^-OOC

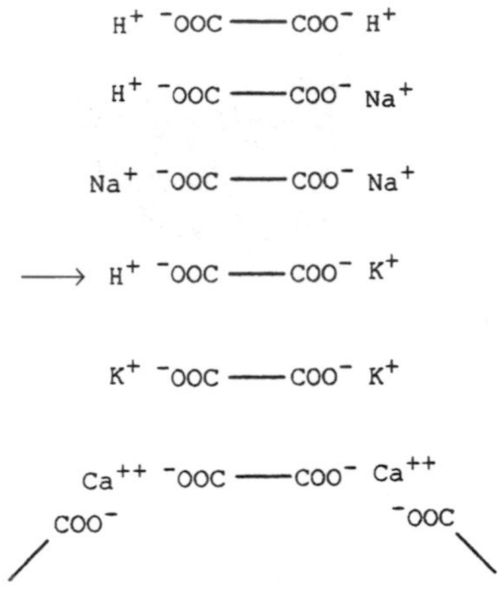

Oxalic acid and its salts

Disease Genesis—As its tart taste attests, the sap of some species of *Rumex* contains soluble acid/potassium oxalate at an acidic pH of approximately 2 (Oke 1969; James 1972, 1978). Oxalate concentrations may be less than 1%, and it has been suggested that greater than 10% d.w. soluble oxalate is the danger level when large amounts of the plants are eaten (James 1972, 1978; Guil et al. 1996). Indeed, in incidents of intoxication, oxalate concentrations for *R. venosus* in Colorado were reported to range from 9.2% d.w. in April to 13.9% d.w. in June (Dickie et al. 1978), and for *R. crispus* the concentration of oxalic acid ranged from 6.6 to 11.1% d.w. (Panciera et al. 1990). In other situations not necessarily associated with toxicosis, concentrations in *R. acetosa* have been reported to range up to 12.9% d.w. (Libert & Franceschi 1987). In Mexico, cultivars with exceptionally high concentrations (to 44% d.w.) have been reported (Mejia et al. 1985). Interestingly, in the case in which *R. acetosa* was seemingly eaten with impunity, oxalate concentrations of only 0.16% were reported (Craig & Kehoe 1921).

In addition, nitrates and anthraquinone glycosides are suggested to be of possible concern (Cooper & Johnson 1984). Species of *Rumex* contain several anthraquinones—e.g., rhein and aloe-emodin—as potent glycosides that may cause digestive tract problems related to the cathartic actions of these types of compounds (Thomson 1971; Robinson 1991). A more detailed discussion of these types of compounds is presented in the treatment of *Aloe* (see chapter 4).

■ anthraquinones of possible significance; oxalates seldom a problem because plants not eaten or else eaten slowly

It appears that only rarely are conditions appropriate for intoxication due to *Rumex* to occur. The concentrations of oxalate in the plants and the likelihood that large enough amounts of plant will be eaten must be high. Large amounts must be eaten in a relatively short period in order to exceed the capacity of the ruminal microflora to metabolize the oxalate (Morris & Garcia-Rivera 1955; James et al. 1967). In situations in which animals, especially goats, eat small amounts of the plant continuously, it may be difficult to exceed this capacity, because microbial adaptation will occur (Allison et al. 1977; Duncan et al. 1997). Thus conditions for intoxication will seldom prevail. In most cases of acute oxalate intoxication, hungry animals have been placed in a new environment where they become rather nonselective in their eating. Although it has not been determined in *Rumex*, species of other genera tend to accumulate oxalates under conditions of high levels of nitrogen and potassium and low phosphorus (Libert & Franceschi 1987). Thus plants growing in areas subject to high nitrogenous runoff may pose a special risk.

■ abrupt onset, depression, excess salivation, tremors, difficulty walking, collapse, labored respiration, elevated BUN, nitrogen, creatinine

Clinical Signs—The onset is abrupt, perhaps 8–12 hours after animals begin grazing a new pasture. Typically there is depression, excess salivation, a staggering gait, and tremors of the head, neck, and legs. When forced to walk, they do so only with difficulty, repeatedly falling and struggling to rise. Respiration is labored, and prostration comes quickly. Animals may die within a few hours of the onset of signs if not treated. Initially blood calcium concentrations are markedly decreased, while blood urea nitrogen and creatinine are elevated. The latter two concentrations increase in magnitude as the disease progresses.

Pathology—Grossly, the kidneys are pale and swollen, and excess fluid is present in the body cavities. There may also be edema of various tissues, including the lungs. Microscopically, there is renal tubular dilation and flattened degenerated tubular epithelium. Oxalate crystals are prominent in the renal tubules and may also be seen in the walls of small arteries in the rumen and less often in the brain.

- gross pathology, kidneys pale, swollen; excess fluid in body cavities

  microscopic, renal tubular dilation, oxalate crystals in kidney and ruminal vessels

Treatment—Prompt administration of parenteral calcium solutions will generally result in immediate relief of many of the signs, and the animals will become ambulatory within a few hours. Relapses may occur, and some animals will require treatment for acute renal failure.

- parenteral calcium

References

Allison MJ, Littledike ET, James LF: Changes in ruminal degradation rates associated with adaptation to oxalate ingestion. *J Anim Sci* 45;1173–1179, 1977

Brockmann H: Photodynamically active plant pigments. *Proc Chem Soc Lond* 1957; 304–312, 1957

Brockmann H, Lachner H: Zur konstitution des fagopyrin. *Tetrahedron Lett* 18;1575–1578, 1979

Bruce EA: Fagopyrismus (buckwheat poisoning) and similar affections. *J Am Vet Med Assoc* 52;189–194, 1917

Buchanan JM: Three cases of poisoning plants. *Vet Rec* 13;927, 1933

Cantero JJ, Giraudo J, Mena C: Intoxication of cattle with *Polygonum convolvulus*. *Rev Med Vet* 71;208–212, 1990

Case AA: Some aspects of nitrate intoxication in livestock. *J Am Vet Med Assoc* 130;323–329, 1957

Cooper MR, Johnson AW: *Poisonous Plants in Britain and Their Effects on Animals and Man*. Her Majesty's Stationery Office, London, pp186–191, 1984

Coward TG: Acute, fatal poisoning in sheep due to ingestion of common sorrel *(Rumex acetosa)*. *Vet Rec* 61;765–766, 1949

Craig JF, Kehoe D: Investigations as to the poisonous nature of common sorrel (*Rumex acetosa* Linn.) for cattle. *J Comp Pathol Ther* 34;27–35, 1921

Dickie CW, Hamann MH, Carroll WD, Chow FH: Oxalate *(Rumex venosus)* poisoning in cattle. *J Am Vet Med Assoc* 173;73–74, 1978

Duncan AJ, Frutos P, Young SA: Rates of oxalic acid degradation in the rumen of sheep and goats in response to different levels of oxalic acid administration. *Anim Sci* 65;451–455, 1997

Farre M, Xirgu J, Salgado A, Peracaula R, Reig R, Sanz P: Fatal oxalic acid poisoning from sorrel soup. *Lancet* 2;1527, 1989

Fassett DW: Oxalates. In *Toxicants Occurring Naturally in Foods*, 2d ed, National Research Council, National Academy of Sciences, Washington, DC, pp346–362, 1973

Guil JL, Torija ME, Gimenez JJ, Rodriguez-Garcia I, Gimenez A: Oxalic acid and calcium determination in wild edible plants. *J Agric Food Chem* 44;1821–1823, 1996

James LF: Oxalate toxicosis. *Clin Toxicol* 5;231–243, 1972

James LF: Oxalate poisoning in livestock. In *Effects of Poisonous Plants on Livestock*, Keeler RF, Van Kampen KR, James LF eds, Academic Press, New York, pp139–145, 1978

James LF, Street JC, Butcher JE: In vitro degradation of oxalate and cellulose by rumen ingesta from sheep fed *Halogeton glomeratus*. *J Anim Sci* 26;1438–1444, 1967

Jeghers H, Murphy R: Practical aspects of oxalate metabolism. *N Engl J Med* 233;208–215, 1945

Knight PR: Suspected nitrite toxicity in horses associated with the ingestion of wireweed *(Polygonum aviculare)*. *Aust Vet Pract* 9;175–177, 1979

Libert B, Creed C: Oxalate content of seventy-eight rhubarb cultivars and its relation to some other characters. *J Hortic Sci* 60;257–261, 1985

Libert B, Franceschi VR: Oxalate in crop plants. *J Agric Food Chem* 35;926–938, 1987

Looman J: *111 Range and Forage Plants of the Canadian Prairies*. Agriculture Canada Publ 1751, 1983

McKenzie RA, Dunster PJ, Burchill JC: Smartweeds (*Polygonum* spp.) and photosensitization of cattle. *Aust Vet J* 65;128, 1988

Mejia MR, Montalvo VR, Martinez RR: Levels of oxalates in wild forages coming from the states of Hidalgo, Guanajuato, Mexico, Tlaxcala, and the Federal District. *Vet Mex* 16;21–25, 1985

Morris MP, Garcia-Rivera J: The destruction of oxalates by the rumen contents of cows. *J Dairy Sci* 38;1169, 1955

Oke Ol: Oxalic acid in plants and in nutrition. *World Rev Nutr Diet* 10;262–303, 1969

Panciera RJ, Martin T, Burrows GE, Taylor DS, Rice LE: Acute oxalate poisoning attributable to ingestion of curly dock *(Rumex crispus)* in sheep. *J Am Vet Med Assoc* 196;1981–1984, 1990

Poisoning by rhubarb leaves. *Lancet* 1;847, 1947

Robb HF, Sippy JJ: Death from rhubarb leaves due to oxalic acid poisoning. *J Am Med Assoc* 73;627–628, 1919

Robinson T: *The Organic Constituents of Higher Plants,* 6th ed. Cordus Press, North Amherst, Mass, 1991

Sheard C, Caylor HD, Schlotthauer C: Photosensitization of animals after the ingestion of buckwheat. *J Exp Med* 47;1013–1028, 1928

Thomson RH: *Naturally Occurring Quinones,* 2d ed. Academic Press, London, 1971

Wender SH: The action of photosensitizing agents isolated from buckwheat. *Am J Vet Res* 7;486–489, 1946

Primulaceae Vent. 60

Comprising 25–30 genera and 600–1,000 species, the Primulaceae, or primrose family, is cosmopolitan in distribution but exhibits greatest diversity in the temperate and cold regions of the Northern Hemisphere. The family is economically important only for ornamentals—e.g., *Primula* (primula), *Dodecatheon* (shooting star), and *Cyclamen* (cyclamen)—which are grown in rock gardens or as potted plants. In North America, approximately 10 genera and 100 species are present. Intoxication problems are associated with 3 genera.

- Primrose Family

  Anagallis
  Cyclamen
  Primula

Plants herbs; annuals or perennials; terrestrial or emergent aquatics; caulescent or acaulescent. **Leaves** all alike or of 2 forms; simple; alternate or opposite or whorled; forming a basal rosette or cauline; venation pinnate; surfaces often gland-dotted or farinose; stipules absent. **Inflorescences** solitary flowers or racemes or panicles or umbels or verticillate clusters; bracts present. **Flowers** perfect; perianths in 2-series. **Sepals** 5, or rarely 4 or 6; fused. **Corollas** radially symmetrical. **Petals** 5, or rarely 4 or 6; fused; of various colors. **Stamens** 5, or rarely 4 or 6; opposite the petals; epipetalous; staminodia present or absent. **Pistils** 1; compound, carpels 5; stigmas 1, capitate; styles 1; ovaries superior or inferior; locules 1; placentation free-central. **Fruits** capsules; valvate or circumscissile. **Seeds** numerous.

- annual or perennial herbs; sepals and petals fused; fruits capsules with numerous seeds

Anagallis L.

Taxonomy and Morphology—Widespread throughout the world, *Anagallis*, commonly known as pimpernel, comprises 20–28 species. Its name, used by the ancient Greeks, is derived from *anagelao*, which means "to laugh" or "to delight." Plants were fabled to disperse sadness. In North America, 1 introduced species is of toxicologic interest:

- *A. arvensis*

  poison chickweed
  poor-man's-weatherglass
  red chickweed
  scarlet pimpernel
  shepherd's-clock
  shepherd's-weatherglass

A. arvensis L. — scarlet pimpernel, poison chickweed, red chickweed, shepherd's-weatherglass, poor-man's-weatherglass, shepherd's-clock

The common names alluding to the weather or time reflect the tendency of the flowers to be open only during fair weather. Darkness and cold temperatures cause them to close.

Plants annuals; caulescent; from taproots; somewhat mat-forming; herbage glabrous. **Stems** prostrate or decumbent or occasionally ascending, often rooting at nodes; 10–40 cm long; profusely branched; 4-angled. **Leaves** opposite; cauline; sessile; blades ovate to lanceolate; surfaces gland-dotted; margins entire, scarious; stipules absent. **Inflorescences** solitary flowers borne in axils of upper leaves; pedicels slender,

- annuals from taproots; leaves opposite, sessile, gland-dotted; sepals 5, nearly divided; petals 5, scarlet, white, or blue; capsules globose

Anagallis arvensis

■ weedy invader

ascending, recurved in fruit. **Flowers** rotate; variable in size. **Sepals** 5; divided nearly to base. **Petals** 5; convolute; lobes ovate to elliptic, margins denticulate or crenulate; scarlet or white or blue. **Stamens** 5; filaments pubescent. **Pistils** with styles filiform; ovaries superior. **Capsules** circumscissile; globose. **Seeds** numerous; minute; papillose; brown.

Distribution and Habitat—Introduced from Europe, *A. arvensis* is now widely distributed across North America, especially along roadsides and in disturbed sites. It has become an especially noxious weed in wheat fields and other croplands.

Anagallis arvensis

■ variable disease, digestive disturbance, renal disease, sheep and cattle

Disease Problems—Although *Anagallis* is no longer used for these purposes, it was employed therapeutically for centuries, its recorded use beginning with the ancient Greeks (Millspaugh 1974). Thus far it has not been clearly identified as a disease risk in North America, albeit it was suspected to be the culprit in the deaths of horses in California and calves in Pennsylvania (Reynard & Norton 1942; Fuller & McClintock 1986). It has, however, been considered a cause of kidney disease in sheep and cattle in Australia since the early 1900s, and the disease has been experimentally reproduced (Pullar 1939; Schneider 1978). In one episode involving 90 sheep grazing a pasture consisting of about 90% *A. arvensis*, 12 animals died and about half of the remainder became ill (Pullar 1939). Experimentally, as little as 624–680 g of fresh and/or dry plant material was lethal in 48 hours. More typical of the disease in the field, 2% b.w. per day for 1–2 weeks produced severe kidney disease (Schneider 1978).

In Brazil, marked weight loss, bloody diarrhea, and high mortality due to renal failure have been observed in cattle and sheep after ingestion of *A. arvensis* for several days or weeks (Riet-Correa et al. 1998).

Plants of the species, under some circumstances, seem to be toxic for only a short period. Experimentally, in one test 907 g produced only a transient diarrhea, while in another, 2,155 g fed in a 10-day period caused no untoward effects (Pullar 1939). This variability in toxicity is further demonstrated by observations that plant material was no longer toxic a few weeks after a clinical kidney disease problem had occurred (Rothwell & Marshall 1986). The disease seems to be more of a problem following excessive summer rainfall.

■ unknown toxin or toxins; possibly terpenoid glycosides, anagalligenins, and oxalates

Anagalligenin B

Disease Genesis—The diversity of clinical signs ranging from those indicative of a digestive tract disturbance to kidney problems is consistent with more than one toxin. A possible cause may be terpenoid saponins. These include the pentacyclic glycosides of anagalligenins A and B, such as anagalloside A, and of priverogenin B, such as anagalloside C. Both are acetates and pentaglycosides at C-3. Also present are arvenin I and arvenin II, glycosides of the quadricyclic cucurbitacin D (Connolly & Hill 1991; Shoji et al. 1994).

The crude saponin extract, when given to a dog, caused digestive tract irritation as well as other systemic signs (Millspaugh 1974). In addition to the irritation, the saponins have other effects, including increased intestinal spasticity and cardiac contractility, as revealed by administration of plant extracts to laboratory animals (Hilal et al. 1987). Oxalate concentrations, in the range of 9–13% oxalic acid, have been reported in association with acute tubular nephrosis (Sadekar et al. 1995). However, although oxalate crystals were found in the microvasculature of the brain, they were not present in the renal tubules. This type of effect is consistent with the disease as reported by others (Schneider 1978). In *A. arvensis* there may be a combination of effects from the two types of toxicants present, and these effects are perhaps dependent upon the amount of plant material eaten. Plant material associated with toxicosis in sheep was found to be negative for cyanogenesis (Pullar 1939).

Experimentally in rats, an ethanolic extract of 1–2 mg of the saponins per kilogram of animal b.w. caused impaired growth and bone ossification, soft-tissue abnormalities, and fetal death and resorptions (El-Sayed et al. 1987).

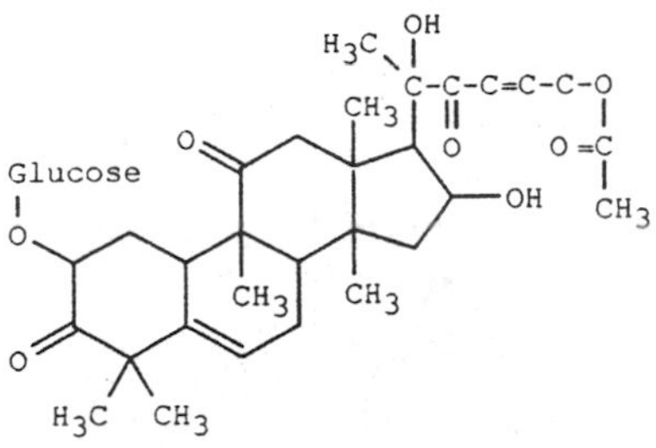

Arvenin I

Clinical Signs—The signs of *A. arvensis* intoxication are of rapid onset and initially include weakness, diarrhea, and incoordination. Weakness and depression become of increasing severity, with the head held low and the ears drooping, and eventually the animals are unable to stand. The respiratory rate may increase noticeably. There may also be subcutaneous edema in the perineal area and along the ventral abdomen. Changes in serum chemistry may be diagnostic; blood urea nitrogen increased markedly (>74 mg/dl), magnesium increased, and calcium decreased. Death occurs 1–2 days after onset of the signs.

In humans there may be headache, trembling, shivering, and rheumatic pains of a temporary nature.

■ animals, rapid onset, weakness, diarrhea, ataxia, depression; decrease in blood calcium, marked increase in BUN

humans, headache, shivering, rheumatic pains

Pathology—The lesions are rather inconspicuous. Grossly, in addition to the reddened mucosa of the stomach and intestines, the lungs are congested, the liver is pale and friable, and the kidneys are small and pale. There may be scattered, small splotchy hemorrhages in the kidneys and a few on the heart. The renal effects are accompanied by subcutaneous, perirenal, and lung edema, with free fluid in the abdomen and thorax. Microscopically, there is degeneration and necrosis of the proximal convoluted tubules of the kidney. This damage is less severe in the distal tubules. Intratubular hemorrhage may also be present in the kidneys.

■ gross pathology, reddened gut mucosa, edema and free fluid abdomen and thorax

microscopic, necrosis of renal tubules

Cyclamen L.

Taxonomy and Morphology—A well-known and popular florist's plant, *Cyclamen* comprises 15–20 species native primarily to the Mediterranean region. Its name is classical Greek and derived from *kyklos,* meaning "circular," which alludes to the spiraling of the scapes after flowering in some species. Although several species are cultivated horticulturally, the most commonly encountered plants are those derived by repeated selections of the following species:

■ *C. persicum*

alpine violet
cyclamen
florist's cyclamen
Persian violet
sowbread

C. persicum Mill.	florist's cyclamen, cyclamen, Persian violet, alpine violet, sowbread

Plants perennials; acaulescent; from depressed globose tubers. **Leaves** basal; folded when young; blades cordate or reniform to suborbicular, conspicuously variegated, greens and silver-grays; petioles of varying lengths. **Scapes** 5–20 cm long; straight after flowering. **Inflorescences** solitary flowers; nodding. **Flowers** showy; not fragrant. **Sepals** 5. **Petals** 5; reflexed 90–180°; tubes short; white or pale pink to deep carmine red, often with dark purple blotches at base. **Stamens** 5; rarely exserted beyond corolla tubes. **Pistils** with ovaries superior. **Capsules** dehiscent by 5 valves. **Seeds** numerous.

■ perennials from tubers; leaves basal, variegated; flowers showy, solitary, nodding; sepals 5; petals 5, reflexed, white, pale pink, or deep carmine red, often purple-splotched at base

Distribution and Habitat—In their natural habitats, species of *Cyclamen* occupy rocky terrain in semiarid locations. Native to the eastern end of the Mediterranean region, *C. persicum* is extensively cultivated as a houseplant in North America.

■ ornamental houseplants

Cyclamen persicum

■ digestive disturbance; irritant terpenoid glycosides of cyclamigenins, cyclamiretins

Disease Problems and Genesis—The introduced species of *Cyclamen* are not likely to be a serious toxicologic risk, but they occasionally may cause digestive tract and other disturbances (Spoerke et al. 1987). Like other members of the family, the genus contains a series of irritating, pentacyclic terpenoid saponins (Connolly & Hill 1991). They include cyclamin, isocyclamin, cyclaminorin, cyclacoumin, deglucocyclamin, and mirabilin. These saponins are mainly glycosides of cyclamiretins and cyclamigenins. They are especially abundant in the tubers and are the cause of the purgative effects (Frohne & Pfander 1984; Leroux 1986; Spoerke & Smolinske 1990). Cyclamin is piscicidal and a potent hemolytic agent (Hostettmann & Marston 1995). These steroidal saponins have cardiac activity similar to that of the cardiac glycosides, decreasing the rate of and augmenting the strength of contraction, as shown for the Eurasian *C. ibericum* (Gladkikh 1966). In vitro studies also indicate that they are uterocontractile (Calis et al. 1997). Whether amounts sufficient to cause systemic types of effects are ever absorbed is uncertain.

Cyclamigenin C

Cyclamiretin A

■ diarrhea, vomiting, colic
reddened gut mucosa
treatment not generally needed

Clinical Signs, Pathology, and Treatment—The primary effects are related to irritation of the digestive tract, including diarrhea with or without blood, vomiting, and colic. There may also be reddish discoloration of the urine due to hemoglobinuria and, with more-severe intoxication, cardiac arrhythmias and seizures. Irritation of the digestive tract will be manifested as reddening of the mucosa of the stomach and small intestine. The disease is not likely to be fatal. Treatment is based upon relief of the diarrhea, but in most instances will not be necessary.

Primula L.

■ *P. auricula*
P. obconica
auricula
bear's-ears
primrose

Taxonomy and Morphology—Comprising 200–500 species, *Primula,* commonly known as primrose, is native to both the Old World and the New World. Alluding to its winter and early spring flowering, its name is the diminutive of the Latin *primus,* meaning "first." It exhibits considerable diversity of form and several infrageneric classifications have been published. At present, 37 subgenera and sections are recognized (Fenderson 1986). Numerous species with a multitude of cultivars are popular ornamentals as border, window box, greenhouse, and rockery plants. Approximately 25 species are present in North America. Representative of the genus and of toxicologic interest are 2 introduced species:

P. auricula Balf.	auricula, primrose, bear's-ears
P. obconica Hance	German primrose, poison primrose, top primrose

Plants perennials; acaulescent; typically from short rhizomes. **Leaves** forming a basal rosette; petioled or sessile; blades of various shapes; surfaces often farinose; margins entire or toothed or lobed. **Scapes** 15–30 cm tall. **Inflorescences** umbels or heads or verticillate clusters or rarely solitary; bracts present. **Flowers** showy; funnelform or salverform; often winter-flowering. **Sepals** 5; calyces often angled or ridged. **Petals** 5; lobes spreading or rarely erect, entire or 2-lobed; yellow or orange or purple or white or rose; often bicolor. **Stamens** 5; included within corollas; filaments short. **Pistils** with styles filiform; ovaries superior. **Capsules** globose or cylindric; dehiscent by 5 or 10 valves. **Seeds** numerous.

- perennials; leaves in basal rosettes; flowers showy; sepals 5; petals 5, spreading, yellow, orange, purple, white, rose, or bicolored

Because of the great diversity in morphology in the genus, a taxonomic key is not presented here. If exact identification is needed, horticultural manuals or appropriate floristic treatments should be consulted.

Primula auricula

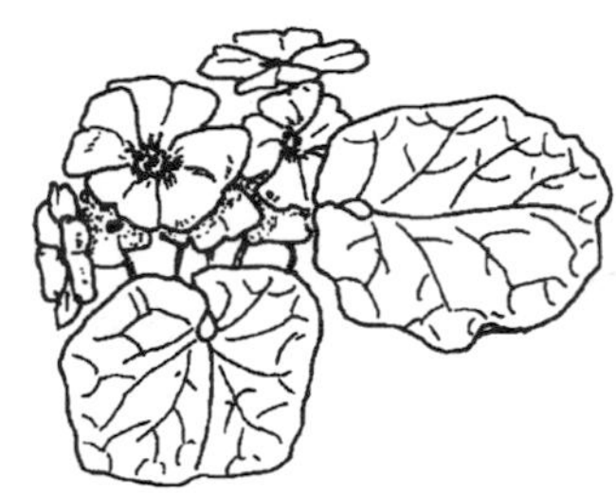

Primula obconica

Distribution and Habitat—Occurring in a variety of soil types, native species of *Primula* are characteristically found in rock crevices, on talus, along stream banks, or in moist meadows primarily in the northern half of the continent and at higher elevations in the mountains. *Primula auricula* is native to the Alps of Europe, and *P. obconica* to China.

- scattered in higher elevations, ornamentals

Disease Problems—The leaves of species of *Primula* bear glandular hairs that exude a noxious secretion that causes a contact dermatitis and eczematous reaction on the hands and face of humans; it is sometimes severe (Harshberger 1920). Actual intoxications are rare. In one of the few cases, *P. auricula* is reported to have caused vomiting in a child (O'Leary 1964). The role of *Primula* in animal intoxications is not clear, but studies of the compounds in plants of the genus indicate a strong potential for adverse effects.

- skin irritation, possibly digestive disturbance

Disease Genesis—The dermatitis is caused by a benzoquinone, primin, which acts as a contact irritant or allergen (Mitchell & Rook 1979). Of potentially greater significance are the pentacyclic terpenoid saponins present in the foliage of various species of *Primula* and other genera of the family (Connolly & Hill 1991). The roots of Eurasian species used for medicinal purposes contain 5–10% saponin (Hostettman & Marston 1995). In Old World species the terpenoids include glycosides of primulagenin and priverogenin, and the pridentigenins A–E (Usmanghani 1992). Some *Primula* species also contain glycosides of cyclamiretins similar to those of *Cyclamen*. Pridentigenin E is the same as cyclamiretin E, and pridentigenin C is a derivative of cyclamiretin C. Because of the presence of these terpenoid saponins, effects on the digestive tract may be seen following ingestion of appreciable amounts of foliage. However, severe or lethal effects have not been reported.

- skin irritation due to benzoquinone primin; digestive disturbance due to terpenoid glycosides of primulagenin, priverogenin

Primin

Clinical Signs—Following contact with the hairs and exudate, there is rapid onset of skin irritation. The affected area is pruritic and becomes reddened, swollen, and

Primulagenin A

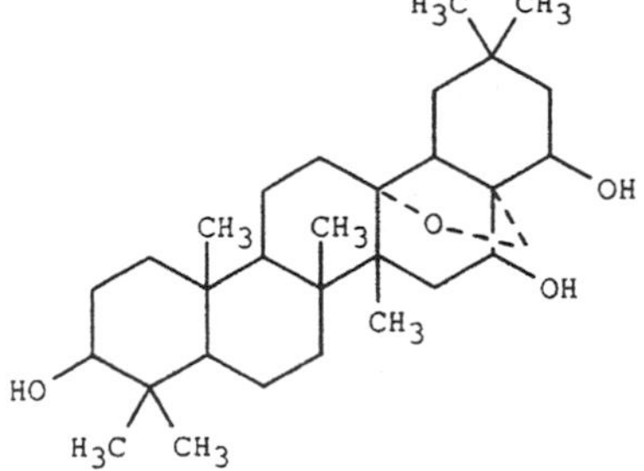

Priverogenin

Pridentigenin

- skin, rapid onset, reddened, swollen, itch, blisters, reddened conjunctiva, cornea
- application of ammonia to skin

blistered with blotchy or linear discolored streaks. Contact with the eyes causes reddening of the conjunctiva, the cornea, and even the iris in severe cases.

Treatment—The dermatologic response may be alleviated by the immediate application (within 15 minutes) of a 25% solution of ammonia (Mitchell & Rook 1979).

References

Calis I, Santana ME, Yuruker A, Kelican P, Demirdamar R, Alacam R, Tanker N, Ruegger H, Sticher O: Triterpene saponins from *Cyclamen mirabile* and their biological activities. *J Nat Prod* 60;315–318, 1997

Connolly JD, Hill RA: *Dictionary of Terpenoids.* Vol 2, *Di- and Higher Terpenoids.* Chapman & Hall, London, pp1174–1312, 1991

El-Sayed MGA, Youssef SAH, Fahmy MFM, Kadri M: Embryotoxic effect and histopathological changes induced by *Anagallis arvensis* in rats. *J Egypt Vet Med Assoc* 47; 427–437, 1987

Fenderson GK: *A Synoptic Guide to the Genus Primula.* Allen Press, Lawrence, Kans, 1986

Frohne D, Pfander HJ: *A Colour Atlas of Poisonous Plants,* 2d ed. Wolfe Science, London, pp149–151, 1984

Fuller TC, McClintock E: *Poisonous Plants of California.* Univ California Press, Berkeley, pp210–211, 1986

Gladkikh AS: Action of saponin from *Cyclamen ibericum* on some condition indicators in animal cardiovascular systems. *Farmakol Toksikol* 29;559–562, 1966

Harshberger JW: *Pastoral and Agricultural Botany.* Blakiston's Son & Co, Philadelphia, pp95–96, 1920

Hilal SH, Youssef SAH, El-Sayed MGA, Shalaby MA, Aboutabl EA: A study of some pharmacodynamic actions of *Anagallis arvensis. Assiut Vet Med J* 18;146–155, 1987

Hostettmann K, Marston A: *Saponins.* Cambridge Univ Press, Cambridge, 1995

Leroux V: Intoxications des animaux de compagnie par les plantes d'appartement. *Le Point Vet* 18;45–55, 1986

Millspaugh CF: *American Medicinal Plants.* Dover, New York, pp423–426, 1974 (original 1892)

Mitchell J, Rook A: *Botanical Dermatology: Plants and Plant Products Injurious to the Skin.* Greengrass, Vancouver, BC, pp546–567, 1979

O'Leary SB: Poisoning in man from eating poisonous plants. *Arch Environ Health* 9;216–242, 1964

Pullar EM: Studies on five suspected poisonous plants. *Aust Vet J* 15;19–23, 1939

Reynard GB, Norton JBS: *Poisonous Plants of Maryland in Relation to Livestock.* Md Agric Exp Stn Tech Bull A10, pp249–312, 1942

Riet-Correa F, Rivero R, Dutra F, Timm CD, Medez M de Carmen: Recently encountered poisonous plants of Rio Grande do Sul and Uruguay. In *Toxic Plants and Other Natural Toxicants,* Garland T, Barr AC eds, CAB International, pp1–5, 1998

Rothwell JT, Marshall DJ: Suspected poisoning of sheep by *Anagallis arvensis* (scarlet pimpernel). *Aust Vet J* 63;316, 1986

Sadekar RD, Bhandarkar AG, Udasi SD, Deshmukh PN, Phadnaik BS: Toxicity of a winter crop weed *Anagallis arvensis* (blue pimpernel) in cattle and buffaloes. *Indian Vet J* 72;1151–1153, 1995

Schneider DJ: Fatal ovine nephrosis caused by *Anagallis arvensis* L. *J S Afr Vet Assoc* 49;321–324, 1978

Shoji N, Umeyawa A, Yoshikawa K, Arihara S: Triterpenoid glycosides from *Anagallis arvensis. Phytochemistry* 37;1397–1402, 1994

Spoerke DG, Smolinske SC: *Toxicity of Houseplants.* CRC Press, Boca Raton, Fla, pp113–114, 1990

Spoerke DG, Spoerke SE, Hall A, Rumack BH: Toxicity of *Cyclamen persicum* (Mill.). *Vet Hum Toxicol* 29;250–251, 1987

Usmanghani K: Characterization of chemical constituents and toxicity evaluation of some poisonous plants. In *Poisonous Plants: Proceedings of the Third International Symposium,* James LF, Keeler RF, Bailey EM Jr, Cheeke PR, Hegarty MP eds, Iowa State Univ Press, Ames, pp314–326, 1992

61 Pteridaceae Rchb.

■ Maidenhair Fern Family

Astrolepis
Cheilanthes

Cosmopolitan in distribution, the Pteridaceae, or maidenhair fern family, comprises about 40 genera and 1,000 species. Taxonomists differ in their opinions as to whether 1 large taxon or 5 smaller ones should be recognized (Tyron & Tyron 1982). There is also disagreement about generic circumscriptions. In many publications, the family name Adiantaceae has been used rather than Pteridaceae. Thirteen genera and 90 species are present in North America (Windham 1993). Two genera are of toxicologic interest.

■ perennial herbs from rhizomes; fronds pinnately compound; sori borne on revolute margins or on abaxial surfaces of fronds

Plants herbs; perennials; from rhizomes; producing sporangia in sori on abaxial surfaces of fronds. **Fronds** all alike; vernation circinate or erect; 1-to-5-pinnately compound; stipitate; abaxial surfaces green or white or yellow, mealy or not mealy. **Sori** borne on revolute margins or on abaxial surfaces; false indusia formed by margins. **Sporangia** all alike; annuli present, vertical. **Spores** all alike; tetrahedral to globose. **Gametophytes** all alike; green; obcordate to reniform.

Astrolepis D.M.Benham & Windham

■ *A. cochisensis*
A. integerrima
A. sinuata

cloak fern
jimmy fern

Taxonomy and Morphology—A New World genus, *Astrolepis,* commonly known as star-scale cloak fern, comprises 8 species, of which 4 are present in North America (Windham 1993). Indicative of the starlike scales on the abaxial frond surfaces, its name is derived from the Greek roots *astro* for "star," and *lepis* for "scale." Its species traditionally were placed in *Notholaena* or *Cheilanthes* but are segregated into their own genus on the basis of biosystematic data (Benham & Windham 1992). Three species are of possible toxicologic interest:

A. cochisensis (Goodd.) D.M.Benham & Windham (= *Notholaena cochisensis* Goodd.) (= *N. sinuata* var. *cochisensis* [Goodd.] Weath.)	jimmy fern
A. integerrima (Hook.) D.M.Benham & Windham (= *N. integerrima* [Hook.] Hevly) (= *N. sinuata* var. *integerrima* Hook.) (= *Cheilanthes integerrima* [Hook.] Mickel)	
A. sinuata (Lag. ex Sw.) D.M.Benham & Windham (= *N. sinuata* [Lag. ex Sw.] Kaulf.) (= *C. sinuata* [Lag. ex Sw.] Domin)	wavy cloak fern, long cloak fern, bulb cloak fern

Astrolepis cochisensis comprises 3 subspecies, and *A. sinuata* 2 subspecies. Believed to be a hybrid between *A. cochisensis* and a Mexican species, *A. integerrima* has 2 morphological forms that are not formally recognized as subspecies (Windham 1993). *Astrolepis cochisensis* and *A. integerrima* have long been treated as varieties of *N. sinuata*.

- evergreen herbs from short rhizomes; fronds densely clustered, leathery, waxy; sori separate, submarginal

Plants evergreen; from rhizomes. **Rhizomes** short; erect or ascending; sparsely branched; scales linear-attenuate, tan to dark brown, concolor or bicolor. **Fronds** densely clustered; erect to spreading; 7–130 cm long; 1-pinnately compound; blades linear-oblong to ovate, leathery, adaxial and abaxial surfaces covered with stellate or ciliate-margined scales; waxy; pinnae stalked to subsessile, ovate or oblong or elongate-deltoid; vein tips free; stipes stiff; dark. **Sori** separate; submarginal, often clustered near notches between pinna lobes; borne on vein ends; oblong to suborbicular; false indusia absent.

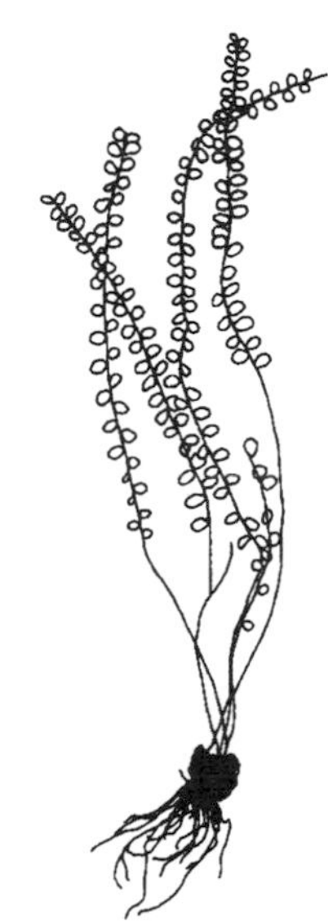

Astrolepis cochisensis

1. Largest pinnae 4–7 mm long. Scales peltate.
   Abaxial scales ovate . . . . *A. cochisensis*
1. Largest pinnae 7–35 mm long. Scales attached at bases.
   Abaxial scales lanceolate.
   2. Pinnae margins symmetrically lobed. Adaxial scales sparse . . *A. sinuata*
   2. Pinnae margins entire or asymmetrically lobed.
      Adaxial scales dense . . . . *A. integerrima*

Distribution and Habitat—*Astrolepis cochisensis* is a species of slopes and summits of limestone hills and mountains of the Southwest. *Astrolepis integerrima* occupies similar habitats and has a similar distributional range but extends farther north into Oklahoma and Nevada and farther south to central Mexico. *Astrolepis sinuata* ranges from central Texas to California and southward through northern Mexico. It also occurs in the West Indies and Central and South America.

- summits of limestone hills and mountains

Astrolepis cochisensis

Disease Problems—*Astrolepis cochisensis* (reported as *N. sinuata* var. *cochisensis*) causes a neurologic disease, commonly called jimmies, primarily in sheep and to a lesser extent in goats and cattle. It typically occurs in the winter, especially late winter, when both severity and frequency are increased. The morbidity may be quite high, but the case-fatality rate is typically low, especially if care is given to avoid undue animal stress (Mathews 1942). When fed to sheep or goats, *A. cochisensis* at 0.8–3% b.w. for 2–3 days proved to be toxic, whereas 6.5–12.7% *N. sinuata* (reported as *N. sinuata* var. *sinuata*) fed for 6–9 days did not produce adverse effects in either species (Mathews 1945). Thus it appears that risk is associated specifically with *A. cochisensis*. Dosage requirements for the other species may exceed that likely in clinical situations.

- neurologic disease, jimmies, mainly in sheep; goats and cattle less risk

Disease Genesis—Although many constituents of *Astrolepis* (reported as *Notholaena*) are known, the specific cause of the disease remains obscure (Murakami & Tanaka 1988). The toxicant appears to be eliminated in milk (Mathews 1942). Species do contain notholaenic acid, which affects electron flow in plants and inhibits photosynthesis in some situations (Gorham & Coughlan 1980). However, its effect in animals has not been reported.

- toxin unknown

Clinical Signs—As its name suggests, the disease is characterized by violent trembling of the entire body. The animal is without symptoms while quietly grazing, but when moved to water, a corral, or another location, or when the animal simply attempts

- when forced to move, abrupt onset, arched back, stilted gait, violent trembling, seizures; recovery in a few minutes, but tendency for seizures remains for 1–2 weeks

to keep pace with other animals as they move about, signs begin abruptly. At first there is an arched back, a few stilted steps, several steps backward, and then the seizure begins. The animal may die suddenly at the onset of the trembling seizure, or the signs may subside after a few minutes rest, only to reappear if the animal is forced to move for a prolonged period. Recovery from this tendency to have seizures occurs in 1–2 weeks.

■ no lesions

general nursing care

Pathology and Treatment—The disease is of such sudden onset and short course that there are no lesions except for a few hemorrhages in the lungs and some foam in the air passages. There is no specific treatment, only supportive general nursing care.

Cheilanthes Sw.

■ *C. feei*
C. gracillima
C. lanosa
C. tomentosa

lace fern
lip fern

Taxonomy and Morphology—Adapted to xeric habitats, *Cheilanthes,* commonly known as lip fern, comprises about 150 species distributed primarily in the Western Hemisphere. Its name is derived from the Greek roots *cheilos* and *anthus,* meaning "margin" and "flowering," which allude to the sporangia's being borne at the margins of the fronds. Morphologically quite diverse and closely related to *Notholaena* and *Pellaea,* its circumscription has been revised several times (Windham 1993). In North America, 28 species are present. Representative and most widely distributed are the following:

C. feei T.Moore	slender lip fern
C. gracillima D.C.Eaton	lace fern
C. lanosa (Michx.) D.C.Eaton	hairy lip fern
C. tomentosa Link	woolly lip fern

■ evergreen herbs from branched dark-colored rhizomes; fronds clustered, pinnately compound; sori borne on vein ends

Plants evergreen; from rhizomes. **Rhizomes** short or elongate; ascending or horizontal; usually branched; scales linear-subulate to ovate-lanceolate, brown to black or bicolor. **Fronds** clustered; erect to spreading; 4–60 cm long; 1-to-4-pinnately compound; blades linear-oblong to ovate; abaxial surfaces green, not mealy, pubescent or rarely glabrous; vein tips free; stipes slender, wiry, dark, shining. **Sori** contiguous or separate; borne on vein ends; roundish; false indusia formed by margins of ultimate segments, greenish to whitish, narrow, often concealing sporangia.

Because the species listed are merely representative of the genus, a taxonomic key is not presented here. If exact identification is need, floristic manuals specific for the reader's area should be consulted.

Cheilanthes feei

■ rocky sites

Distribution and Habitat—*Cheilanthes* is distributed across the continent. Some species, such as the four listed, have relatively wide distributions; others are much more restricted. Plants of the genus are typically encountered in rock crevices of bluffs, outcrops, and ledges and generally exposed to intense sunlight.

■ disease unknown; monogastric animals, neurologic

ruminants, hemorrhagic syndromes similar to diseases caused by *Pteridium;* risk is low because of limited availability of plant material

Disease Problems—*Cheilanthes* in North America has not been implicated in intoxications. Elsewhere, it is well recognized as a cause of disease problems. Perhaps best known is the Australasian species *C. seiberi,* which produces four syndromes: (1) a thiamine deficiency in sheep and probably monogastric species; (2) an acute hemorrhagic or radiomimetic syndrome in cattle (Clark & Dimmock 1971; Chick et al. 1985); (3) cystic hematuria or enzootic hematuria in cattle (McKenzie 1978); and (4) neoplasia of the bladder associated with the chronic cystitis. All of these problems also are produced by *Pteridium aquilinum,* in the Dennstaedtiaceae, and are described in depth in its treatment (see chapter 32). As with *Pteridium,* problems in horses are most likely to occur when they are fed contaminated hay.

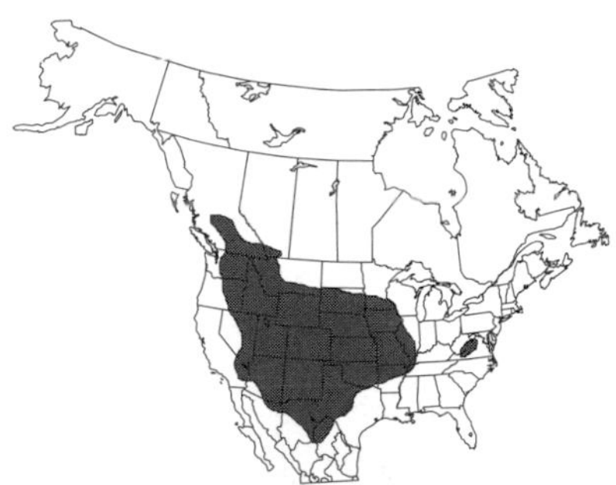

Cheilanthes feei

Disease Genesis—Both thiaminase I and ptaquiloside are present in species of *Cheilanthes.* Thiaminase activity is high, whereas ptaquiloside is about 0.3–0.4 mg/g, a concentration about half that in *Pteridium* (Chick et al. 1989; Smith et al. 1989; Agnew & Lauren 1991).

- neurologic thiaminase

  hemorrhagic ptaquiloside

Clinical Signs—The signs associated with thiaminase activity in the horse are those of a neurologic disorder, including incoordination and unsteadiness, especially when the animal turns. There may be yawning or stretching of the mouth and difficulty chewing, with the food falling from the mouth after a brief period of ineffectual chewing motions. Apparently blind or oblivious of its environment, the horse may walk in circles. After a few days it may lie down. At first it may rise with difficulty but soon is unable to do so. Death preceded by paddling seizures may occur shortly thereafter.

The other syndromes reported for the Australasian species—hemorrhagic, hematuria, and neoplastic—are discussed under *Pteridium aquilinum* in the Dennstaedtiaceae and are described in depth in its treatment (see chapter 32). For the acute hemorrhagic syndrome, following ingestion of plants by cattle for 2–4 weeks, platelets and neutrophils decline markedly, bleeding time increases, and capillary fragility is increased. Sheep seem to be little affected.

- neurologic, unsteadiness, ataxia, yawning, difficulty chewing, oblivious of surroundings

  hemorrhage, from nose, mouth or internal

Pathology—In horses that have died as a result of thiaminase activity, the gross changes are not distinctive, but there may be moderate swelling of the brain. Microscopically, there will be extensive edema of the brain, with numerous hemorrhages, degenerating neurons, and an influx of glia.

- microscopic, brain edema, loss of neurons

Treatment—As for other thiaminase-containing plants, treatment should include thiamine given parenterally twice a day for several days. Grain supplementation may be somewhat protective.

- neurologic, thiamine i.v., i.m., or i.p.

  hemorrhage, blood transfusion

References

Agnew MP, Lauren DR: Determination of ptaquiloside in bracken fern *(Pteridium esculentum). J Chromatogr* 538;462–468, 1991

Benham DM, Windham MD: Generic affinities of the star-scaled cloak ferns. *Am Fern J* 82:47–58, 1992

Chick BP, Quinn C, McCleary BV: Pteridophyte intoxication of livestock in Australia. In *Plant Toxicology, Proceedings Australia-USA Poisonous Plant Symposium,* Seawright AA, Hegarty MP, James LF, Keeler RF eds, Animal Research Institute, Brisbane, Yeerongpilly, Australia, pp453–464, 1985

Chick BF, McCleary BV, Beckett RJ: Thiaminases. In *Toxicants of Plant Origin,* vol 3, *Proteins and Amino Acids,* Cheeke PR ed, CRC Press, Boca Raton, Fla, pp73–91, 1989

Clark IA, Dimmock CK: The toxicity of *Cheilanthes seiberi* to cattle and sheep. *Aust Vet J* 47;149–152, 1971

Gorham J, Coughlan SJ: Inhibition of photosynthesis by stilbenoids. *Phytochemistry* 19;2059–2064, 1980

Mathews FP: *Fern* (Notholaena sinuata, *var.* crenata) *Poisoning in Sheep, Goats, and Cattle—The So-called "Jimmies" of the Trans-Pecos.* Tex Agric Exp Stn Bull 611, 1942

Mathews FP: A comparison of the toxicity of *Notholaena sinuata* and *N. sinuata* var. *cochisensis. Rhodora* 47;393–395, 1945

McKenzie RA: Bovine enzootic haematuria in Queensland. *Aust Vet J* 54;61–64, 1978

Murakami T, Tanaka N: Occurrence, structure, and taxonomic implications of fern constituents. In *Progress in the Chemistry of Organic Natural Products* 54, Herz W, Grisebach H, Kirby GW, Tamm C eds, Springer Verlag, Vienna, pp1–353, 1988

Smith BL, Embling PP, Lauren DR, Agnew MP, Ross AD, Greentree PL: Carcinogen in rock fern *(Cheilanthes seiberi)* from New Zealand and Australia. *Aust Vet J* 66;154–155, 1989

Tyron RM, Tyron AF: *Ferns and Allied Plants, with Special Reference to Tropical America.* Springer-Verlag, New York, 1982

Windham MD: Pteridaceae. In *Flora of North America,* vol 2, *Pteridophytes and Gymnosperms,* Flora of North America Editorial Committee eds, Oxford Univ Press, New York, pp122–185, 1993

Ranunculaceae Juss. 62

Cosmopolitan, but distributed primarily in the boreal and temperate regions of both the Old World and the New World, the Ranunculaceae comprises 46–60 genera and 1,800–2,500 species. In North America, 22 genera and 284 species, both native and introduced, are present (Whittemore & Parfitt 1997). A morphologically diverse but natural taxon, it is considered to be one of the most primitive families of flowering plants. Traditionally, 2–6 subfamilies and several tribes are recognized. Often quite showy, many genera are popular garden ornamentals, e.g., *Delphinium* (larkspur), *Clematis* (virgin's bower), *Aquilegia* (columbine), and *Anemone* (windflower). Some, such as *Hydrastis* (goldenseal), are used in herbal medicines, and others, such as *Aconitum* (monkshood), have long been recognized as extremely poisonous.

Members of the Ranunculaceae produce a variety of toxic effects. Two genera, *Adonis* and *Helleborus,* contain cardioactive glycosides with different structures but similar effects. Other genera produce toxicants of entirely different structure and action, e.g., the neurotoxic diterpenes of *Aconitum* and *Delphinium,* the vesicants of *Ranunculus,* and the extensive array of alkaloids of *Hydrastis* and *Thalictrum* that also are present in the Fumariaceae and Papaveraceae (see chapters 39 and 55, respectively).

■ Crowfoot Family

Aconitum
Actaea
Adonis
Caltha
Delphinium
Helleborus
Ranunculus

QUESTIONABLE
Anemone
Clematis
Hydrastis
Thalictrum

Plants herbs or vines or rarely shrubs; annuals or perennials; terrestrial or emergent aquatics; caulescent or acaulescent; bearing perfect flowers, or rarely dioecious or polygamodioecious. **Leaves** simple or palmately or 1-to-3-pinnately compound; alternate, or rarely opposite or whorled; basal or cauline or both; venation pinnate or palmate; petioles often swollen and stipular-like; stipules absent, or present and vestigial. **Inflorescences** solitary flowers or cymes or racemes or panicles or umbels; terminal or axillary; bracts and bracteoles present or absent. **Flowers** perfect or rarely imperfect, similar; perianths in 1- or 2-series; of various colors; receptacles typically elongate. **Perianths in 1-Series:** Parts radially symmetrical; 3–20; spiraled or whorled; free; sepaloid or petaloid. **Perianths in 2-Series:** Sepals 5 or 3 or 4; free; spurred or not spurred; sepaloid or petaloid. Corollas radially or bilaterally symmetrical. Petals 0–26; free or rarely fused; spurred or not spurred. **Stamens** 5 to numerous; spiraled or whorled; staminodia present or absent. **Pistils** 1 to numerous; spiraled or whorled; free; simple (compound in 1 genus), carpels 1; stigmas 1; styles 1; ovaries superior; locules 1; placentation parietal. **Nectaries** absent or present; petaliferous or sepaliferous. **Fruits** achenes or follicles or berries (capsule in 1 genus). **Seeds** 1 to numerous.

■ herbs, vines, rarely shrubs; flowers with perianths in 1- or 2-series; stamens 5 to many; pistils typically many

Aconitum L.

Taxonomy and Morphology—A popular garden ornamental and long known as poisonous, *Aconitum* comprises approximately 100 species. Its name, used by Theophrastus and Dioscorides, is derived from the Greek *akoniton,* meaning "leopard's

- *A. columbianum*
  A. delphiniifolium
  A. maximum
  A. napellus
  A. reclinatum
  A. uncinatum
  aconite
  aconitum
  helmetflower
  monkshood
  wolfsbane

bane" or "wolfsbane," which are two of its common names. Also used are *aconite* and *monkshood,* which reflect the distinctive cowl-like appearance of the flower. In North America, 5 native and 5 introduced species are present (Brink and Woods 1997). The following are representative taxa (see plate 62.1):

A. columbianum Nutt.	monkshood, western monkshood
A. delphiniifolium DC.	delphinium-leaved monkshood
A. maximum Pall. ex DC.	Kamachatka aconite, Kamachatka monkshood
A. napellus L.	wolfsbane, aconitum, monkshood, helmetflower, garden aconite
A. reclinatum A.Gray	white monkshood, trailing wolfsbane
A. uncinatum L.	southern monkshood, wild monkshood, clambering monkshood

- perennial herbs; leaves simple, alternate, palmately lobed; flowers in terminal racemes; sepals 5, hoodlike, various colors; petals 2 or 5, concealed in hood; fruits oblong, beaked follicles

Plants herbs; perennials; terrestrial; caulescent; from tubers or fascicled roots. **Stems** erect or reclining or twining; 10–250 cm tall. **Leaves** simple; alternate; basal and cauline; blades palmately lobed or cleft, lobes 3 or 5 or 7. **Inflorescences** racemes; terminal, sometimes also axillary; bracts foliaceous. **Flowers** perfect; perianths in 2-series; bilaterally symmetrical. **Sepals** 5; conspicuous; not spurred; petaloid; white or yellow or greenish to blue, violet, or purple; uppermost large, arched, concave, hoodlike, often beaked. **Petals** 2 or 5; free; concealed in hood; clawed; spurred. **Stamens** 25–50; filament bases expanded; staminodia absent. **Pistils** 3–5. **Fruits** follicles; oblong; beaked. **Seeds** numerous; deltoid.

Aconitum columbianum

1. Plants from elongate, fascicled roots. Hoods conic to cylindrical. Flowers white or cream .... *A. reclinatum*
1. Plants from tubers. Hoods conic-hemispheric to crescent-shaped. Flowers blue to purple or white or yellow.
   2. Plants of Alaska and northwestern Canada.
      3. Blades divided to bases. Racemes elongate. Tubers contiguous .... *A. delphiniifolium*
      3. Blades not divided to bases. Racemes capitate. Tubers separated .... *A. maximum*
   2. Plants of eastern and western United States.
      4. Blades divided to bases; lobes linear .... *A. napellus*
      4. Blades not divided to bases; lobes not linear.
         5. Tubers contiguous or nearly so .... *A. columbianum*
         5. Tubers separated .... *A. uncinatum*

- along streams, wooded areas

Distribution and Habitat—Circumboreal in distribution, *Aconitum* extends southward, via high-elevation habitats in mountain ranges, into northern Mexico and northern Africa. Plants typically are encountered in small populations along streams or in wooded areas. *Aconitum columbianum* is the most widespread species and occurs in moist sites along streams and in brushy or wooded areas to elevations of 2,900 m. It is western, ranging from Mexico to southern British Columbia, but disjunct populations occur in the east. As indicated in the taxonomic key, *A. delphiniifolium* and *A. maximum* occur in the far northwest corner of the continent and are found in meadows, thickets, forests, and tundra. Inhabiting rich woods, *A. reclinatum* and *A. uncinatum* have similar ranges in the eastern half of North America from Pennsylvania to the Carolinas, the latter occurring as far south as southern Georgia.

A European introduction, *A. napellus* is a cultivated ornamental species that may persist in old gardens or occasionally escape; it is most commonly encountered in New England.

Disease Problems—Potions of *Aconitum* have a history of use as a poison for humans and beasts that dates to Greek mythology. Similar to those of *Conium maculatum* (poison hemlock), decoctions of *Aconitum* were used to terminate life in Aegean societies (Baskin 1967). It was such use and subsequent abuse that led the Romans to declare cultivation of the plants a capital offense in imperial Rome. In later centuries the roots were used as an arrow poison and for homicidal and medicinal purposes in Europe and Asia. The common name *wolfsbane* is a holdover from earlier times when decoctions were used to poison predators and rodents (Benn & Jacyno 1983). The genus was also widely respected as a poison and a natural drug in oriental medicine and remains in common use today in Asian communities (Natori et al. 1981). Attendant to its common use, intoxications are also a continuing problem (Chan 1994).

The most consistent features of *Aconitum* intoxication are numbness and tingling of the mouth and limbs, muscle weakness, problems of the digestive tract, and cardiac disturbances (Chan 1994). These effects, especially the cardiac problems and the prospect of ventricular fibrillation and cardiac failure, can be very serious and life threatening (Chan et al. 1994a,b; Pui-Hay et al. 1994). In contrast to *Delphinium*, with which it has much in common, *Aconitum* is seldom eaten, and very few instances of intoxication have been reported in North America, in either people or livestock (Knight & Pfister 1997). Fleming and coworkers (1920) noted signs of only slight toxicosis from 1 kg of green plant material fed to sheep. Thus, although the genus is well known to be toxic, there seems to be little hazard.

Disease Genesis—Species of *Aconitum* contain 19-C diterpene alkaloids mainly of the aconitine type, with small and probably not significant amounts of lycoctonine, pyrodelphinine, and heteratisine types also present (Olsen & Manners 1989). Crude preparations of aconitine known as Bushi (Japanese) or Bachnag, Caowu, Chuanwu, and Fuzi (Chinese) may contain up to 0.6% of these 19-C alkaloids (Chan 1994; Chan et al. 1994b). The aconitine types, including aconitine and its derivatives mesaconitine and delphinine, have Na^+ channel and cardiac activity. Aconitine shares affinity for the type-2 Na^+ channel receptor site with veratridine (Catterall 1988). The resting Na^+ permeability of excitable cells is increased, resulting in incomplete inactivation of Na^+ channels in nerves (persistent activation) and a decrease in ion selectivity (Strichartz et al. 1987). This produces depolarization of resting nerve and muscle and an increase in repetitive discharge. With low dosage there is stimulation of medullary vagal centers, causing decreased heart rate and blood pressure (Stern 1954). Higher dosage predisposes to a variety of arrhythmias, including atrial flutter and fibrillation, due in part apparently to the complex effects of aconitine on the

Aconitine

Delphinine

Aconitum columbianum

Aconitum delphiniifolium

Aconitum reclinatum

- well-known topical anesthetic; variety of effects, neurologic, digestive disturbance, cardiac, but low risk
- diterpene alkaloids, aconitine type; bind to Na^+ channel and interfere

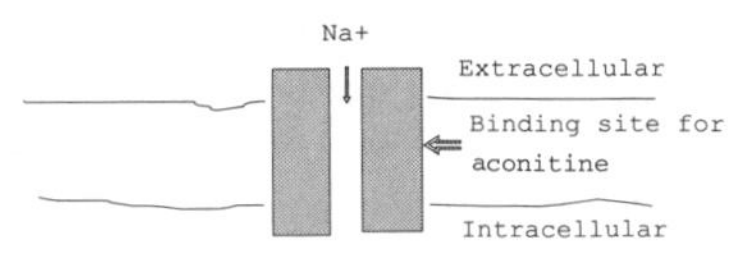

Sodium channel
(after Catterall 1986)

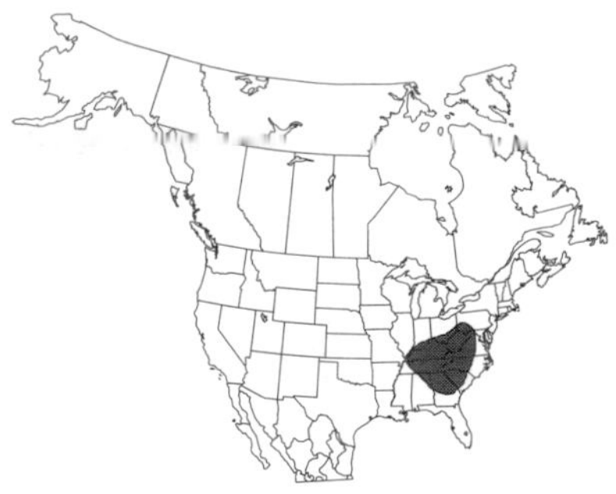

Aconitum uncinatum

central nervous system, which in turn affect the adrenergic and cholinergic systems (Scherf 1947; Benn & Jaycno 1983). The lethal dose of aconitine in humans is estimated to be about 2–3 mg (Jacyno 1996).

As indicated by its early use as a topical anesthetic, aconitine is absorbed through the intact skin or mucous membranes, affecting sensory perception and resulting in numbness, intense tingling, itching, and a feeling of warmth or burning (Stern 1954; Katz 1990). Other derivatives of aconitine, such as jesaconitine and lappaconitine, also have analgesic activity, with the substituents at the C-3 and C-8 positions being important in determining activity (Ono & Satoh 1990; Mori et al. 1991). The effects on the mouth are said to impart a feeling of constriction (Baskin 1967). An additional cardioactive alkaloidal component, higenamine, has been reported to occur in *Aconitum* (Natori et al. 1981).

■ numbness, tingling of tongue and mouth; tingling and prickling sensations of body; vomiting, diarrhea; later, weakness, labored respiration, irregular heart rate, respiratory failure

Clinical Signs—*Aconitine* causes numbness and tingling of the end of the tongue, a burning sensation in the mouth, constriction of the throat, decreased respiration, tingling and prickling over the body, vomiting, and diarrhea. Following these early signs, additional manifestations include muscle weakness, dyspnea, irregular heart rate, and respiratory failure. Death may occur due to heart block and ventricular fibrillation or to respiratory failure. The signs appear within minutes or hours after ingestion of the plants and are of short duration in most instances.

■ no lesions

control of vomiting, diarrhea, and rarely cardiac effects

Pathology and Treatment—*Aconitum* intoxication is an acute disease involving physiologic changes rather than anatomic changes; thus there are no distinctive gross or microscopic lesions. Treatment is symptomatic to control the vomiting and diarrhea and, in more severe intoxications, to control the cardiac arrhythmias or respiratory problems caused by neuromuscular blockade. Additional comments are presented in the treatment of *Delphinium,* in this chapter.

Actaea L.

■ *A. pachypoda*
A. rubra
A. spicata

baneberry
doll's-eyes
herb Christopher
necklaceweed
toadroot
white cohosh

Taxonomy and Morphology—Occurring in the temperate forests of both the Old World and the New World, *Actaea* comprises 8 species, of which 3 occur in North America. Its name is derived from the Greek *aktea,* a name that was originally used for *Sambucus* (elderberry) in the Caprifoliaceae and possibly reflects the similarity in leaves of the 2 genera. Present are the following species:

A. pachypoda Elliott (= *A. alba* [L.] Mill.)	white baneberry, toadroot, white cohosh, doll's-eyes, necklaceweed
A. rubra (Aiton) Willd.	red baneberry
A. spicata	black baneberry, European baneberry, herb Christopher

Taxonomists have differed in their opinions as to whether *A. pachypoda* or *A. alba* should be used for white baneberry in North America. It appears that *A. alba* is based on an illustration of the Eurasian species *A. spicata* and is not applicable to North American plants; thus *A. pachypoda* is the correct name for our plants (Ford 1997a).

■ perennial, erect herbs, herbage odoriferous; leaves usually 2, alternate, 2-to-3-pinnately compound; sepals 3–5; petals 4–10, clawed, cream-colored; fruits berries, red or white

Plants herbs; perennials; terrestrial; caulescent; from vertical rootstocks; herbage odoriferous. **Stems** erect; 40–90 cm tall; unbranched below. **Leaves** usually 2; large; 2-to-3-pinnately (ternately) compound; alternate, cauline; leaflets toothed; long petiolate. **Inflorescences** racemes; terminal or axillary; bracts and bracteoles present.

Flowers perfect; perianths in 2-series. **Sepals** 3–5; caducous; petaloid; whitish green; obovate to orbicular. **Corollas** radially symmetrical. **Petals** 4–10; free; deciduous; spathulate to obovate; clawed; cream-colored. **Stamens** 15–50; filaments widened distally. **Pistils** 1. **Fruits** berries; broadly ellipsoidal to globose; red or white; smooth; beaked. **Seeds** 9–16; dark brown or reddish brown; obconic.

Actaea spicata

1. Fruiting pedicels bright red; 0.9–3 mm in diameter. Stigmas 1.5–2.8 mm in diameter. Petals truncate or cleft. Berries white . . . . *A. pachypoda*
1. Fruiting pedicels dull green or brown; 0.3–0.7 mm in diameter. Stigmas 0.7–1.2 mm in diameter. Petals acute or obtuse. Berries black or red.
   2. Berries black . . . . *A. spicata*
   2. Berries red . . . . *A. rubra*

Although berry color is typically white in *A. pachypoda* and red in *A. rubra,* it must be noted that *A. pachypoda* sometimes has red fruits and *A. rubra* white. *Actaea spicata* and *A. rubra* may be conspecific; systematic work is needed to elucidate their relationship.

Distribution and Habitat—Native to North America, *A. pachypoda* is found in moist soils of woods throughout the eastern half of North America. Also native, *A. rubra* is widely distributed in moist woods across the northern half of the continent. Occasionally cultivated, *A. spicata* is native to Eurasia.

■ moist soils of woodlands

Actaea pachypoda

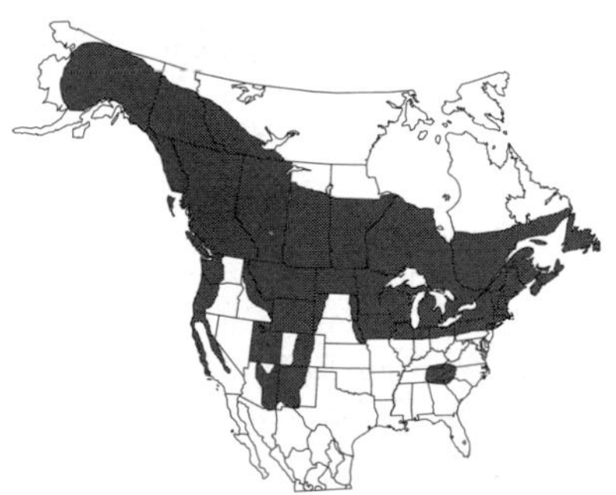

Actaea rubra

Disease Problems and Genesis—There is little question about the noxious nature of the foliage and fruits of *Actaea.* When the foliage is chewed, there is initially little discomfort, but a bitter and nauseating aftertaste quickly follows (Chesnut & Wilcox 1901). In a self-administered trial of toxicity, Bacon (1903) reported the berries to have a nauseous and bitter taste and to cause a burning sensation in his stomach. Large doses produced vague neurologic signs such as confusion, dizziness, and visions, in addition to colic, pain, and an increased and irregular heart rate. *Actaea rubra,* which is moderately toxic, is generally regarded as the most potent species (Jean-Blain 1981).

Actaea reportedly contains berberin, but more likely the adverse effects are caused by irritants similar to those of other members of the family (Pammel 1911).

■ foliage or fruit; digestive disturbance, neurologic and cardiac effects

toxicant unknown

Clinical Signs, Pathology, and Treatment—*Actaea* produces clinical signs and pathologic changes similar to those caused by *Ranunculus,* and a complete discussion of them is presented in the treatment of that genus, in this chapter.

■ digestive disturbance similar to that with *Ranunculus*

Adonis L.

■ *A. aestivalis*
A. annua
A. vernalis

adonis
bird's-eye
pheasant's-eye
yellow oxeye

Taxonomy and Morphology—Native to Europe and temperate Asia, *Adonis*, commonly known as pheasant's-eye or simply as adonis, comprises 20–35 species, of which 3 are present in North America. Its name comes from Greek mythology—the tale of Adonis, who was killed by a boar and changed by his lover Aphrodite into a plant with blood red flowers. Because of their showy foliage and flowers, some species of the genus have been brought into cultivation as border or understory plants. Typically encountered in North America are these species:

A. aestivalis L.	summer adonis, summer pheasant's-eye
A. annua L. (= *A. autumnalis* L.)	autumn adonis, flos adonis, bird's-eye
A. vernalis L.	pheasant's-eye, yellow oxeye, spring adonis

■ erect herbs; leaves simple, alternate, pinnately dissected; flowers solitary, terminal; sepals 5–8; petals 5–20, yellow, red, or white; fruits achenes

Plants herbs; annuals or perennials; terrestrial; caulescent; from taproots or rhizomes. **Stems** erect; 5–100 cm tall. **Leaves** simple; alternate; basal and cauline; blades 2 or 3 times pinnately dissected into linear segments. **Inflorescences** solitary flowers; terminal; bracts absent. **Flowers** perfect; perianths in 1-series. **Sepals** 5–8; deciduous; petaloid; colorless or green; obovate; apices more or less erose. **Corollas** radially symmetrical. **Petals** 5–20; free; oblanceolate; yellow to red or rarely white; with or without basal dark spot. **Stamens** 15–80; anthers yellow or purple-black; filaments filiform; staminodia absent. **Pistils** 20–50; spiraled; simple. **Fruits** achenes; aggregated; sessile; nearly globose; beaked. **Seeds** 1.

Adonis is morphologically very similar to *Ranunculus* (buttercup) but lacks nectaries at the petal bases.

1. Plants perennials. Rhizomes present. Flowers 4–8 cm in diameter. Petals 10–20. Anthers yellow. Achenes pubescent . . . . *A. vernalis*
1. Plants annuals. Rhizomes absent. Flowers 1.5–3 cm in diameter. Petals 5–8. Anthers purple-black. Achenes glabrous.
   2. Bodies of achenes with transverse ridge. Abaxial keels of achenes without tooth . . . . *A. aestivalis*
   2. Bodies of achenes without transverse ridge. Abaxial keels of achenes with 1 tooth . . . . *A. annua*

Adonis aestivalis

■ ornamentals, rarely meadows, open woods

Distribution and Habitat—Inhabiting mountain meadows and steppes, these Eurasian species were introduced into North America as garden plants; they are especially popular in rock gardens. All three have escaped cultivation. *Adonis aestivalis* has naturalized and is abundant in disturbed sites and open forests in the western United States. *Adonis vernalis* and *A. annua* occasionally escape cultivation in the eastern and south-central to southeastern portions of the continent.

■ digestive disturbance, low risk

Disease Problems—Considered unpalatable and seldom found in circumstances where livestock have access to them, plants of *Adonis* rarely cause intoxications, and cases have not been reported. In other areas of the world, severe digestive disturbances have been described in horses and sheep (Everist 1981). In sheep, death was caused by a dose of 450 g of plant material, whereas in cattle a dosage of up to 2.7 kg for several days was without consequence. Seeds are also of concern because pea screenings containing seeds of the European *A. microcarpa* caused complete feed refusal and vomiting when fed for several days to pigs (Davies & Whyte 1989).

Disease Genesis—Species of *Adonis* contain a series of potent 23-C cardenolides—adonitoxin (a mannoside of adonitoxigenin; = 19-oxogitoxigenin), cymarin, and K-strophanthin (a mixture of glycosides of strophanthidin); and vernadigin (a glycoside of strophadogenin; = 16-hydroxystrophanthidin) (Shoppee 1964; Elkiey et al. 1967; Singh 1970; Joubert 1989). The concentrations of these cardenolides are highest in the leaves and flowers, reaching levels of 1 mg or more of glycosides/g d.w. of plant at the time of fruiting (Elkiey et al. 1967). Concentrations are lower in the fruits and lowest in stems and roots. Because the primary intoxication problem is enteritis and accompanying diarrhea, the role of other toxins cannot be excluded; e.g., protoanemonin has been reported to occur in *A. annua* (Cooper & Johnson 1984).

- mixture of cardenolide glycosides of adonitoxigenin, strophanthidin, strophadogenin; leaves and flowers

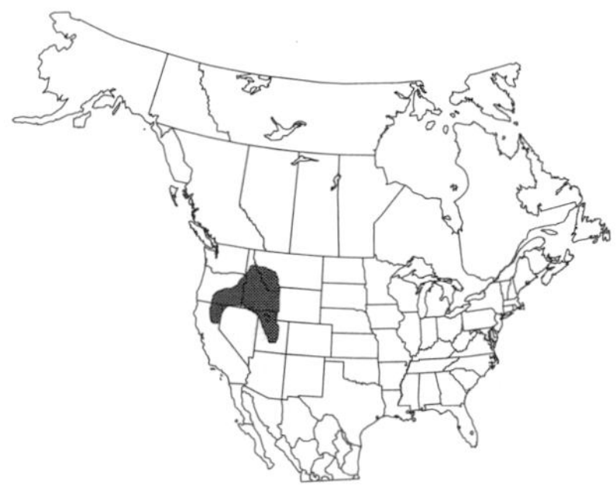

Adonis aestivalis

Adonis annua

Adonitoxin (adonitoxigenin)

Strophanthin (strophanthidin)

Strophandogenin

Clinical Signs, Pathology, and Treatment—The signs of *Adonis* intoxication are those related primarily to digestive disturbance: diarrhea with colic and vomiting. Gross necropsy changes are limited to reddening of the mucosa of the stomach and small intestine. Treatment is directed toward symptomatic relief of the digestive disturbance.

- diarrhea, colic, vomiting
  reddened gut mucosa
  treat for digestive disturbance

Caltha L.

Taxonomy and Morphology—Commonly known as marsh marigold, *Caltha* comprises 10–20 species present in both the Northern and Southern Hemispheres and in the Old World and the New World (Smit 1973). Its name was used by Virgil and Pliny. In North America, 3 species are present:

C. leptosepala DC. (= *C. biflora* DC.)	white marsh marigold, twin-flowered marsh marigold, broad-leaved marsh marigold, elkslip

- *C. leptosepala*
  C. natans
  C. palustris

bull's-eyes, cowflock, cowslip, elkslip, kingcup, populage des marais, marsh marigold

C. *natans* Pall.	
C. *palustris* L. (= C. *asarifolia* DC.)	marsh marigold, cowslip, kingcup, bull's-eyes, cowflock, populage des marais

Both C. *leptosepala* and C. *palustris* are now recognized to be morphologically and ecologically quite variable. Included within them are taxa such as C. *biflora* and C. *asarifolia,* which were formerly treated as distinct species (Ford 1997). The common name *cowslip* for C. *palustris* is used only in North America. In Europe, the same name is applied to *Primula veris,* in the Primulaceae (treated in chapter 60).

■ perennial herbs; leaves simple, alternate; sepals 5–12, white, yellow, pinkish, or orange; petals absent; fruits beaked follicles

Plants herbs; perennials; terrestrial or emergent aquatics; caulescent or acaulescent. **Stems** decumbent to erect; floating or creeping; 10–80 cm tall or long; rooting or not rooting at nodes. **Leaves** simple; alternate; basal or cauline or both; blades cordate to oblong or orbicular-reniform; margins entire to crenate or dentate. **Inflorescences** solitary flowers or cymes; terminal or axillary; bracts present, foliaceous. **Flowers** perfect; perianths in 1-series. **Calyces** radially symmetrical. **Sepals** 5–12; deciduous; petaloid; white or yellow or pinkish or orange; oval-orbicular to obovate. **Petals** absent. **Stamens** 10–40; filaments filiform; staminodia absent. **Pistils** 5–55. **Fruits** follicles; aggregated; sessile or stipitate; linear-oblong to ellipsoid; beaked. **Seeds** numerous; elliptic; brown; rugulose.

1. Stem leaves 1 or absent .... C. *leptosepala*
1. Stem leaves 2 to many.
   2. Stems erect. Sepals yellow or orange. Follicles 5–15 per flower; 8–15 mm long .... C. *palustris*
   2. Stems creeping or floating. Sepals white or pinkish. Follicles 20–55 per flower; 3.2–6.5 mm long .... C. *natans*

Caltha leptosepala

Caltha natans

Caltha palustris

Caltha leptosepala

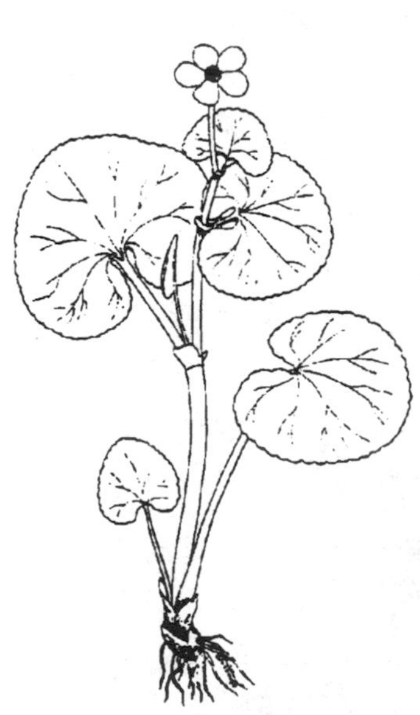

Caltha palustris

Distribution and Habitat—Emergent aquatics, plants of *Caltha* are characteristically encountered in open wet woods, meadows, marshes, swamps, bogs, ditches, and other shallow-water habitats. Greatest abundance is in the northern half of the continent and at higher elevations farther south.

■ moist woods, meadows, bogs, swamps

Disease Problems—*Caltha* has a reputation as both a potherb and a toxic plant in Europe and is reputed to be toxic in hay in North America (Pammel 1911). However, in most situations it is of limited hazard at worst. Fleming and coworkers (1920) fed 0.5 kg of green plants to a sheep without causing adverse effects. *Caltha palustris* is reputed to cause digestive tract problems, agalactia, and possibly bloody urine (Harshberger 1920).

■ digestive disturbance

Disease Genesis—*Caltha leptosepala* (including plants formerly called *C. biflora*) contains small concentrations of senecionine (0.001–0.005%) and an aporphine alkaloid, probably magnaflorine, as well (Stermitz & Adamovics 1977). *Caltha palustris* contains small amounts of protoanemonin, less than 1 ppm, presumably as the inactive glycosidic precursor (Poulsson 1917; Bruni et al. 1986). The concentrations of these compounds are quite low relative to those in *Helleborus* and *Ranunculus*. These small concentrations probably should be taken as an indication of the questionable toxicity of this genus or that some other toxicant is responsible. The report of toxicity when dried further suggests involvement of other toxicants (Pammel 1911).

■ toxicant unknown, low levels of protoanemonin

Clinical Signs, Pathology, and Treatment—A complete discussion of the clinical signs, pathology, and treatment is presented in the section on *Ranunculus*.

■ signs similar to those of *Ranunculus*

Delphinium L.

Taxonomy and Morphology—Comprising 200–300 species and innumerable varieties and cultivars, *Delphinium*, commonly known as larkspur or delphinium, is distributed primarily in the Northern Hemisphere. Only a few species in Africa occur south of the equator. In North America, 61 species are recognized (Warnock 1997). The genus name is a Latinization of the Greek word *delphinion*, which comes from *delphis*, meaning "dolphin," a possible allusion to the resemblance of the flower buds of some species to the animal. It is a genus of considerable horticultural interest, with cultivation and hybridization of its species beginning in the early 1800s, albeit some individuals assert that work actually began two centuries earlier. Interestingly, *Delphinium* is one of the few taxa in which all three primary colors are represented, as well as a plethora of other colors due to horticultural hybridization and selection.

The taxonomy of the genus is troublesome because of confusion about the differentiation of species and varieties and the misapplication of many names. For example, more than 200 names have been applied to the species native to North America (Ewan 1945). Some taxonomists recognize the segregate genera *Consolida* and *Delphinastrum*. This treatment of *Delphinium* by Warnock (1997) employs 2 sections and 9 subsections to classify the species.

Species are often informally classified as "tall," "intermediate," or "low" larkspurs (Marsh et al. 1916). In addition to the implied differences in height, these designations also reflect differences in general growth form, the root systems, and the seasonal pattern of growth. The so-called "tall" species are more robust, are often taller, and have shoots that persist throughout the summer growing season. In contrast, "low" species are typically smaller in stature. They begin growth early in May, and the aerial portions are generally gone by early July. As would be expected,

■ taxonomic problems, nomenclature changes confusing; informally divided into 3 types based upon plant size—low, intermediate, and tall

■ *D. andersonii*
D. barbeyi
D. bicolor
D. californicum
D. carolinianum
D. exaltatum
D. geyeri
D. glaucescens
D. glaucum
D. gypsophilum
D. hesperium
D. menziesii
D. nuttallianum
D. parryi
D. patens
D. recurvatum
D. robustum
D. sapellonis
D. scaposum
D. scopulorum
D. treleasei
D. tricorne
D. trolliifolium
D. variegatum

cow poison
larkspur
poisonweed

Delphinium barbeyi

Delphinium bicolor

"intermediate" species exhibit traits of both "tall" and "low" taxa. Representative toxic species of each type are as follows (see plate 62.2):

Low (1 m or less)

D. andersonii A.Gray	desert larkspur, Anderson's larkspur
D. bicolor Nutt.	little larkspur, Montana larkspur, flathead larkspur
D. gypsophilum Ewan	gypsum larkspur
D. hesperium A.Gray	western larkspur
D. menziesii DC.	spring larkspur, Menzies larkspur
D. nuttallianum Pritz. ex Walp. (= *D. nelsonii* Greene)	upland larkspur, blue larkspur, Nelson larkspur, dwarf larkspur, meadow larkspur, slim larkspur
D. parryi A.Gray	Parry's larkspur, maritime larkspur
D. patens Benth.	smooth larkspur
D. recurvatum Greene	alkali larkspur, recurved larkspur, valley larkspur
D. scaposum Greene	
D. treleasei Bush ex K.C.Davis	Trelease's larkspur
D. tricorne Michx.	dwarf larkspur
D. variegatum Torr. & A.Gray	royal larkspur

Intermediate

D. carolinianum Walter (= *D. virescens* Nutt.)	Carolina larkspur, blue larkspur, plains larkspur, prairie larkspur, white larkspur
D. geyeri Greene	plains larkspur, Geyer's larkspur, poisonweed

Tall (about 2 m tall)

D. barbeyi (Huth) Huth (= *D.* ×*occidentale* [S.Watson] S.Watson) (= *D. cucullatum* A.Nelson)	tall larkspur, dunce-cap larkspur, western larkspur
D. californicum Torr. & A.Gray	coast larkspur, California larkspur
D. exaltatum Aiton	tall larkspur
D. glaucescens Rydb.	waxy larkspur
D. glaucum S.Watson (= *D. brownii* Rydb.) (= *D. scopulorum* in part)	mountain larkspur, pale larkspur, mountain larkspur, Brown's larkspur, Hooker's larkspur, giant larkspur, western larkspur
D. robustum Rydb.	robust larkspur
D. sapellonis Cockerell	Sapello Canyon larkspur
D. scopulorum A.Gray (= *D. macrophyllum* Wooton)	Rocky Mountain larkspur
D. trolliifolium A.Gray	thin-leaf larkspur, Columbia larkspur, cow poison

Plants herbs; perennials; terrestrial; from fasciculate roots or rhizomes. **Stems** generally erect. **Leaves** simple; alternate; basal or cauline or both; basal leaves generally larger than cauline; blades orbicular to pentagonal or reniform, deeply palmately divided into 3 or 5 or 7 major lobes, each lobe often divided again; petioles of basal leaves long, of distal leaves short. **Inflorescences** terminal racemes or rarely panicles; bracts and bracteoles present. **Flowers** perfect; perianths in 2-series; bilaterally symmetrical. **Sepals** 5; deciduous; petaloid; ovate to elliptic; 1 spurred. **Petals** 4; paired; of 2 forms; upper 2 spurred, enclosed in spurred sepal; lower 2 not spurred, orbicular to ovate or elliptic, clawed; of various colors. **Stamens** 25–40; filament bases dilated. **Pistils** 3, or rarely 1 or 5. **Nectaries** present in petal spurs. **Fruits** follicles; aggregated; sessile; cylindrical; curved. **Seeds** 8–20; dark brown to black; rectangular to pyramidal.

- perennial herbs; leaves simple, alternate, palmately divided; flowers in terminal racemes or panicles, conspicuously spurred; petals of various colors; fruits curved, cylindrical follicles

Delphinium geyeri

Delphinium glaucum

Delphinium menziesii

Delphinium scopulorum

Because of the tremendous array of species in North America and the complexity of the variation pattern, a key is not presented here. If identification of a particular species is needed, taxonomic floras specific for the reader's locale should be used.

Distribution and Habitat—The numerous species of *Delphinium* are distributed across the continent, but greatest diversity and abundance are in the western mountain ranges. Some taxa form large stands and are locally abundant, whereas others are infrequently encountered as small, scattered populations. In general, the "tall" types are found at higher elevations in the mountains, especially in sheltered areas where snow accumulates. In contrast, the "low" species tend to occur lower in the hills and plains, albeit sometimes at fairly high elevations (table 62.1).

Disease Problems—Although *Delphinium* was recognized in ancient Greece for its toxic and medicinal properties, it was not generally appreciated as a highly toxic

■ great variety of habitats

Delphinium andersonii

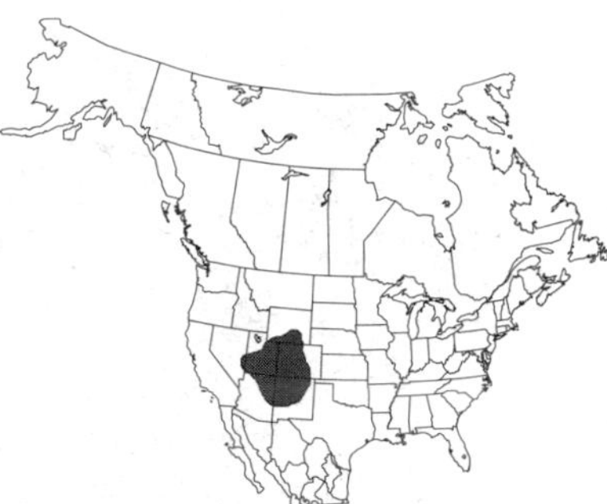

Delphinium barbeyi

Delphinium bicolor

Table 62.1. General habitats and distribution of toxic species of *Delphinium*

Low	
D. andersonii	desert flats and talus in basins and mountain valleys
D. bicolor	edges of dry meadows and open woods of the northern Great Plains
D. gypsophilum	gypsum-rich soils of grasslands, open oak woods, and chaparral of California
D. hesperium	grassy slopes and clearings in open oak woods, grasslands, open pine woods, and coastal chaparral of California
D. menziesii	often abundant in early spring on open hillsides and clearings throughout the Pacific Northwest
D. nuttallianum	hillsides of prairies, open woodlands, or open pine woods
D. parryi	locally abundant in oak woodlands, chaparral, and lower montane woods and on sand dunes in California and Baja California
D. patens	near streams of grasslands and open forests and at edges of thickets, chaparral, and woodlands of central California south into Mexico
D. recurvatum	open fields, grasslands, scrublands, and brushy areas of California
D. scaposum	brushy ravines, grasslands, and pine clearings of Arizona, Utah, Colorado, and New Mexico
D. treleasei	locally abundant on rocky slopes and in juniper glades on calcareous soils in Missouri and Arkansas
D. tricorne	moist soils of prairies, woods, bluffs, and ravines in eastern half of continent as far west as Oklahoma
D. variegatum	open grasslands, chaparral, and oak woodlands of central and coastal California
Intermediate	
D. carolinianum	open sites in prairies, at meadow edges, in forest openings, and on rocky outcrops of eastern half of continent from Manitoba south to northern Mexico
D. geyeri	open plains of grasslands or shrublands and grassy or brushy slopes
Tall	
D. barbeyi	subalpine stream banks and timberline thickets of the Rocky Mountains
D. californicum	forming dense stands in moist thickets, seeps, and creek bottoms in chaparral of coastal mountains in California
D. exaltatum	primarily rocky calcareous slopes in open forests and barrens of east-central portion of continent as far west as Missouri
D. glaucescens	uncommon, on rocky slopes and slides of subalpine coniferous forests in Idaho, Wyoming, and Montana
D. glaucum	meadows, bogs, wet thickets, stream bottoms, and forest edges
D. robustum	uncommon, in shaded pine forests, subalpine meadows, and along edges of aspen groves in Colorado and New Mexico
D. sapellonis	moist sites in open pine forest, in subalpine meadows, and along stream banks in New Mexico
D. scopulorum	scattered populations in riparian and forests
D. trolliifolium	moist sites in woods, along stream banks, and at pasture edges from California to Washington

Source: Ewan 1945.

Delphinium californicum

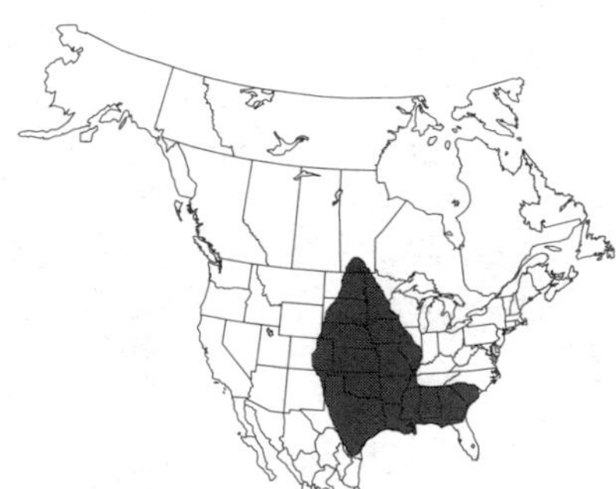

Delphinium carolinianum

Delphinium geyeri

genus in the Old World. In North America it was only when the ranges of the western states began to be extensively used for grazing in the late 1800s that *Delphinium* emerged as a serious intoxication problem. In the Front Range of Colorado, larkspur was considered second only to locoweed as a dangerous plant in the 1870s and thereafter (Glover 1906; Cook & Redente 1993). It continues to be recognized as one the most serious hazards to cattle on rangelands of the mountainous west (Cronin 1971; Cronin et al. 1988; Nielsen & Ralphs 1988; Pfister et al. 1999). In the spring and early summer the "tall" larkspurs may lie protected under the snow in drift sites, ready to burst forth in a flush of growth. Subsequent grazing by cattle is associated with appreciable losses. In 1915–1916, it was estimated that 90% of plant-associated cattle deaths in national forests were due to "tall" larkspurs (Aldous 1917). Estimated cattle losses exceeded 5,000 per year, with nearly 2,000 deaths in Colorado, more than 1,000 in Utah, and almost 1,000 in Idaho. More recently, on U.S. Forest Service allotments in the western mountain states, death losses have ranged as high as 15% (Cronin et al. 1976; Pfister et al. 1993). Although in many instances they are not preferred forages, the *Delphinium*, if not eaten in excess, are nutritious range plants with no apparent adverse effects on growth and production (Pfister et al. 1989).

- major problem, mainly higher mountain ranges of western North America

rapid onset, neuromuscular incapacitation of cattle; sheep and horses lower risk

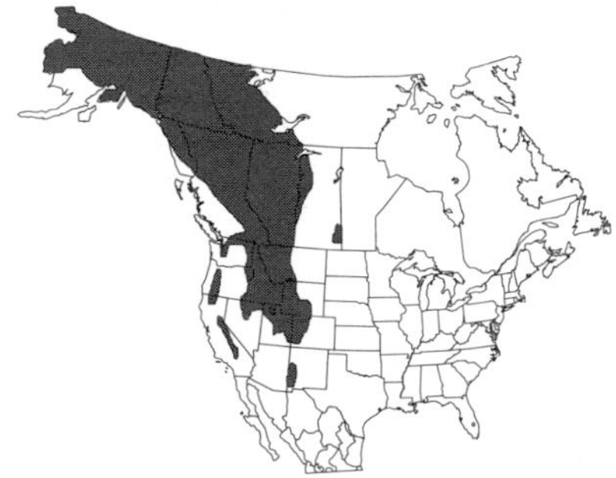

Delphinium glaucum

In the 1890s and early twentieth century, losses were reported for both cattle and sheep, especially the latter when *D. menziesii* was eaten almost exclusively for several days early in the season before flowering (Wilcox 1897; Chesnut & Wilcox 1901; Marsh et al. 1916). Later it appeared that sheep were much less susceptible to neuromuscular incapacitation, as revealed by observations of animals fed *D. andersonii* and other species (Marsh et al. 1916; Fleming et al. 1923). Nelson (1898) fed a sheep 11.25 kg b.w. of leaves, flowers, and fruits of *D. menziesii* over a 5-day period with no adverse consequences. It was eventually shown that the LD_{50} for *D. barbeyi* in the prebud and bud stage was 4 or more times greater for sheep than for cattle (Olsen 1978). These observations seem to confirm the use of sheep as a management tool, because they readily eat some species of *Delphinium* up to the bud stage and yet seldom show signs of intoxication (Ralphs et al. 1991). It should be pointed out, however, that resistance of sheep to intoxication is not apparent when the plant extract is administered subcutaneously (Olsen & Sisson 1991b). Horses are unlikely to be affected, because they are somewhat less susceptible and rarely eat the plants (Marsh et al. 1916). Laboratory animals are susceptible, showing signs similar to those seen in other species (Olsen & Sisson 1991b). Regarding these species, the mouse has been suggested as most representative of the effects in cattle.

Delphinium gypsophilum

Plant parts such as the roots may be consumed by humans because of their reputation for producing "highs" (Tomassoni et al. 1996). For example, a 17-year-old male ate a small potion of immature *Delphinium* root called monk root. In about 4 hours he experienced seizures, ventricular fibrillation, and cardiac arrest requiring cardiopulmonary resuscitation. He recovered, but 10 days later he described a floating feeling, which was suggested to be similar to older accounts of witches who used these toxins to "fly."

Delphinium menziesii

Disease Genesis—The principal toxins are diterpenoid alkaloids, although some species may contain small amounts of other alkaloids such as magnaflorine, an aporphine (Aiyar et al. 1979). The range of alkaloids present is extensive; e.g., some 27 compounds are found in *D. nuttallianum* (Kulanthaivel & Benn 1985; Kulanthaivel et al. 1986; Liang et al. 1991; Bai et al. 1992, 1995). These alkaloids can be grouped into three classes: the 19-C diterpenes, which are the most important toxicologically;

- 19-C diterpene alkaloids, lycoctonine types; MLA, nudicauline

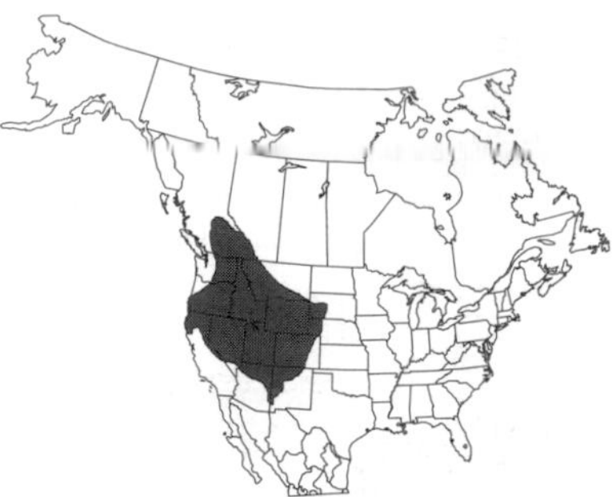

Delphinium nuttallianum

Delphinium recurvatum

Delphinium scopulorum

- alkaloids inhibit ACh nicotinic postsynaptic receptors; curare-like competitive neuromuscular blockade

20-C diterpenes, of only limited importance; and bis-diterpenes, of least risk (Olsen & Manners 1989). The 19-C diterpenes can be divided into four types, of which only the aconitine and lycoctonine types are represented in *Delphinium*. Intoxications are accounted for mainly by lycoctonine type *N*-(methylsuccinimido)anthranoyllycoctonine (MSAL) alkaloids such as methyllycaconitine (MLA), nudicauline, and 14-deacetylnudicauline. Others of lesser potency include ajacine, browniine, deltaline, and two unique compounds, glaucerine and glaucenine, esterified with methyl butanoyl at C-14 of the 5-member ring (Pelletier et al. 1989; Manners et al. 1991, 1993, 1995). The MSAL alkaloids are esterified at C-18 with *N*-(methylsuccinyl)anthranilic acid.

The lycoctonine types exert curare-like competitive neuromuscular blocking effects, inhibiting acetylcholine at nicotinic postsynaptic receptor sites (Aiyar et al. 1979; Nation et al. 1982). In contrast to nicotine, these alkaloids produce a nondepolarizing blockade that leads predominantly to loss of motor function, particularly with respect to the diaphragm and esophagus. Although there are prominent effects on the postsynaptic sites, actions on other nicotinic receptor sites, including presynaptic sites, cannot be excluded as important contributors to intoxication. There is considerable diversity among the nicotinic acetylcholine receptors of muscle, ganglia, brain (neuronal and nonneuronal), and autonomic neurons (Lukas & Bencherif 1992; Sargent 1993). These receptors differ in subunit composition and binding requirements. There are strong indications that the lycoctonine alkaloids affect many of these receptor types, some with high affinity (Ward et al. 1990; Alkondon et al. 1992; Drasdo et al. 1992). Thus, there is the possibility for an array of effects from these alkaloids, some perhaps producing important sublethal impairments in addition to or rather than acute neuromuscular dysfunction. Several structural prerequisites for toxicity have been proposed (Manners et al. 1995). These include a tertiary alkaloidal nitrogen and an anthranilic acid ester (ortho-imide-substituted C-18 benzoate ester). The tertiary nitrogen is suggested to be the primary binding site with the nicotinic receptor. Additional binding effects may occur with specific functional groups at C-14 and to a lesser extent at C-1 and C-16, decreasing in order of potency as follows: acetate > hydroxyl > methoxyl > carbonyl.

The lycoctonine alkaloids are especially potent on insect nicotinic cholinergic receptors (Jennings et al. 1986). Experiments with insect receptors confirm the special potency of alkaloids possessing the C-18 succinimide group, e.g., MLA and glaudelsine (Kukel & Jennings 1994). The most potent of these alkaloids, methyllycaconitine, has about half the potency of tubocurarine in inhibiting cholinergic neurotransmission. Indeed, around the beginning of the twentieth century, alkaloidal extracts of species such as *D. bicolor, D. menziesii,* and *D. nuttallianum* were used as substitutes for curare in laboratory experiments (Crawford 1907). A few species, especially the "low" types, contain small amounts of aconitine-type 19-C diterpenes such as delphidine and delphinine, but these compounds seem to be of little significance toxicologically. The toxic effects represent the sum of the effects of all the alkaloids. The subcutaneous LD_{50} in mice for MLA and 14-deacetylnudicauline is approximately 32 mg/kg b.w., whereas it is 175–200 or more for barbinine, browniine, deltaline, deltamine, and lycoctonine (Olsen & Manners 1989; Manners et al. 1993). Similarly, the mouse i.v. LD_{50}s of MLA, nudicauline, 14-acetylnudicauline, and grandiflorine are 7.5, 2.7, 4.0, and 8.0 mg/kg b.w., respectively, in contrast to 200 mg/kg for many of the other types of alkaloids (Manners et al. 1995, 1998). The orally administered dose of MLA and 14-deacetylnudicauline (the alkaloids in *D. barbeyi*) that causes sternal recumbency in cattle is 21 mg/kg b.w. (Pfister et al. 1994a, 1997b). The estimated LD_{50} is 40 mg/kg b.w. However, in spite of these differences in toxicity, species containing only small amounts of MLA are occasional problems in the field. This may reflect the differing potencies of the alkaloids on the various nicotinic receptors in the body.

In general, toxicity is considered to decrease only slightly with plant growth through flowering for the "low" and "intermediate" species, although there are some differences among studies; e.g., a marked decrease in toxicity of *D. carolinianum* (= *D. virescens*) with maturity is reported (Crawford 1907; Marsh et al. 1916; Beath 1919; Fleming et al. 1923; Marsh 1929; Majak 1993; Majak et al. 1999; Pfister & Gardner 1999). *Delphinium nuttallianum* contains up to 1.2% total alkaloids in its seeds and follicles, 0.87% in flowers, 0.38% in stems, and 0.07% in leaves (Bai et al. 1992). Of the specific alkaloids there may be as high as 0.3% d.w. MLA and smaller amounts of nudicauline in the flowers and fruits and half as much or less in the leaves and stems (Majak et al. 1987; Majak & Engelsjord 1988). Pfister and Gardner (1999) reported concentrations of 0.16% d.w. MLA and 0.07% nudicauline in whole plants of the same species. *Delphinium bicolor,* another "low" species, contains at least eight diterpene alkaloids, including methyllycaconitine (Kulanthaivel et al. 1986).

The toxicity of the "tall" species seems to be subject to greater variation during the growing season than that of the "low" species. Toxicity is initially quite high, declines somewhat through flowering, and then dramatically drops thereafter (Marsh et al. 1916; Beath 1919; Williams & Cronin 1966). *Delphinium barbeyi,* an especially toxic species, contains at least 16 alkaloids, including MLA (Pelletier et al. 1989). During its growing season, alkaloid concentrations in the leaves declined from 3% d.w. in early stages to 1.6% at flowering and to 1% or less in fruit. The content of flowers and follicles was also approximately 1% (Pfister et al. 1988a). Other studies with *D. glaucum, D. barbeyi, D. glaucescens,* and plants of *D. barbeyi* reported as *D. occidentale* confirm the decline in toxicity with advancing stages of plant maturity (Manners et al. 1991; Ralphs et al. 1997b). The relative degree of decline varied among the species and was most pronounced in *D. glaucum*, declining from 2% total alkaloids in the whole plant to less than 0.5% in the late pod stage. Similar findings are reported for the individual alkaloids present (Manners & Pfister 1993; Majak et al. 2000). There is considerable variation in alkaloid concentrations among geographical sites, with an apparent increase at higher elevations. Not only does the total alkaloid content decline during the growing season, but the relative toxicity of the array of alkaloids present declines as well, as shown for *D. barbeyi* (reported as *D. occidentale*) (Olsen 1983).

Alkaloid concentrations are not dependent on light, so there is no diurnal variation, but they may be higher with moisture deficiency, at least for "low" larkspur (Ralphs et al. 1998a; Majak et al. 1999). Ralphs and coworkers (1998a) have hypothesized that synthesis of the alkaloids occurs mainly in early growth and slows with maturity and that environmental influences on alkaloid levels are limited to the early growth period. They further speculate that alkaloids are probably translocated from vegetative to reproductive portions of the plant in later growth stages to account for the higher levels in these portions.

The "tall" larkspurs, such as *D. barbeyi* (including *D. occidentale*), *D. glaucum,* and *D. trolliifolium,* are generally considered to be of most concern, but appreciable problems have also been attributed to "low" and "intermediate" species, such as *D. andersonii, D. bicolor, D. geyeri, D. nuttallianum,* and *D. virescens* (now *D. carolinianum*) (Wilcox 1897; Marsh et al. 1916; Beath 1919, 1925; Fleming et al. 1923; Marsh 1929). It is clear that the "low" types represent a hazard because they may contain a potent array of alkaloids and 9–11 kg of leaves and flowers of *D. andersonii* is reported to be lethal to a 450 kg cow (Fleming et al. 1923). However, for those species with an appreciable content of MLA (ca. 1%), 2 kg may be lethal in an

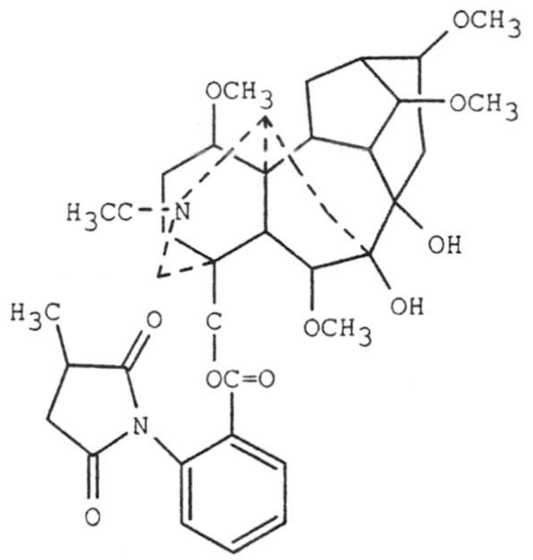

Methyllycaconitine

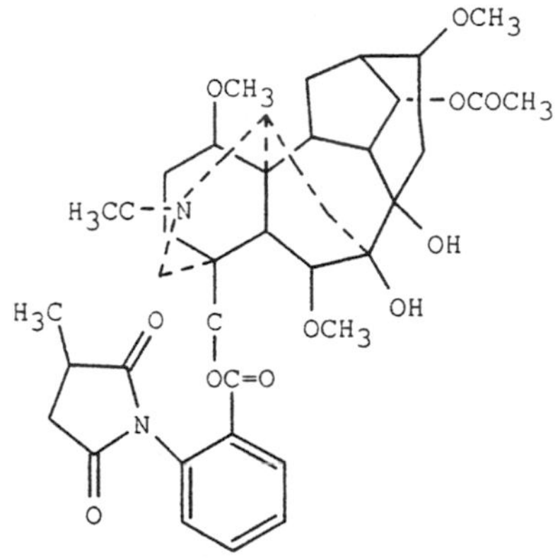

Nudicauline

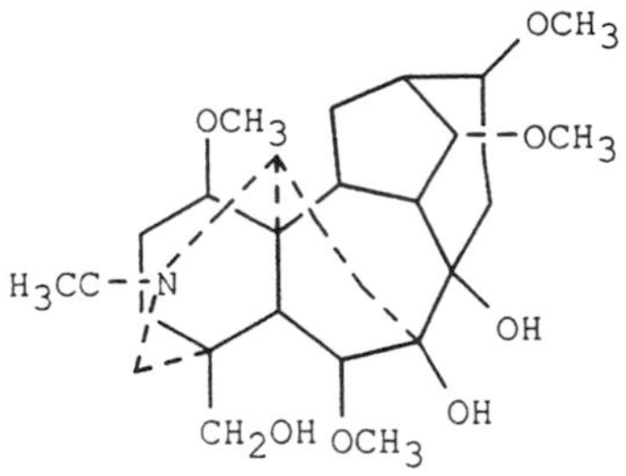

Ajacine

- risk mainly during flowering and early fruit formation when plants are both toxic and palatable

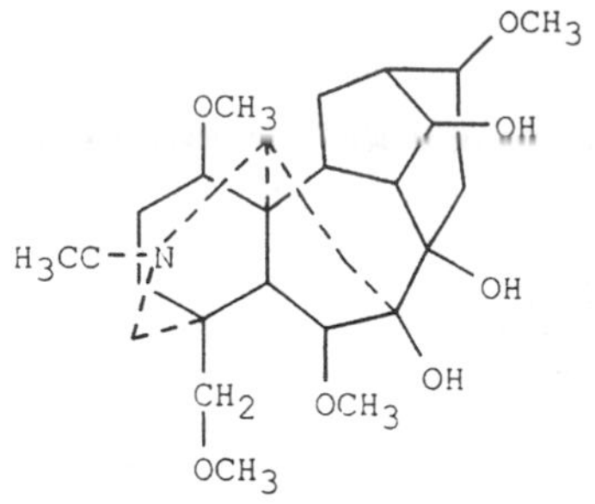

Browniine

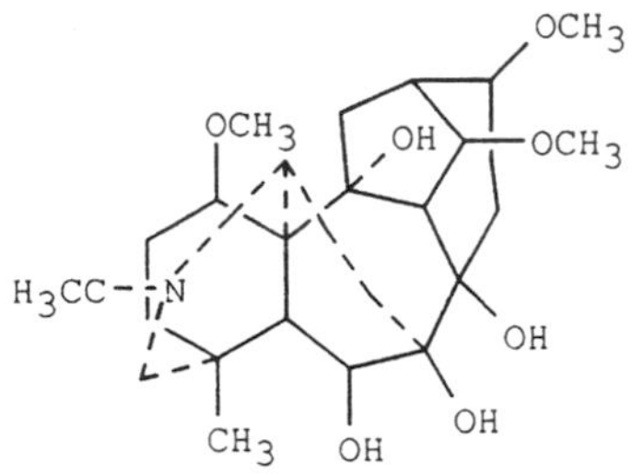

Deltaline

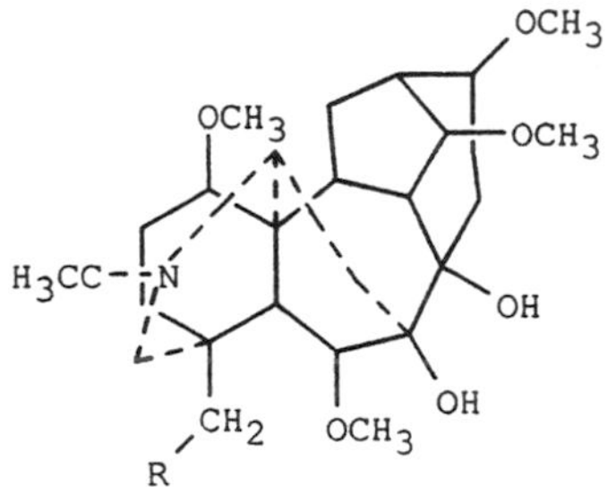

Lycoctonine

animal of the same size (Pfister et al. 1993). It has been estimated that an animal must eat about 20% of its diet as "low" larkspur to become intoxicated (Pfister & Gardner 1999). The dietary percentage is even lower for some of the "tall" larkspurs. Typically, toxicity declines after flowering in all larkspurs. Perhaps the most important factor is that the likelihood of an animal's eating a toxic dose is greater for the "tall" types because of the density of populations, the size of the plants, and the longer persistence of the plants through the growing season (Glover 1906). Many of the "low" and "intermediate" types are rather infrequently encountered, and those that are abundant are short-lived (Beath 1919; Fleming et al. 1923).

In addition to toxicity fluctuations, the question of palatability has always been paramount in determining when animals are at highest risk of intoxication. Palatability seems to be low early in plant growth, increasing during early flowering and then again during fruiting (Majak & Engelsjord 1988; Pfister et al. 1988a; Pfister & Gardner 1999). In a study of mountain rangelands, cattle ate little or no *D. barbeyi* (including *D. occidentale*) until the flowering stems had elongated (Pfister et al. 1997b). Thus, it is proposed that there is a toxic period (or window) of a few days to 1–2 weeks for "low" larkspurs and 4–5 weeks for "tall" larkspurs, during the flowering and early fruit stages, when the plants are both highly toxic and readily eaten by cattle (Pfister et al. 1988a; Sizelove et al. 1992; Pfister et al. 1997b). The risk for cattle on larkspur ranges is much reduced before and after the toxic period (Pfister et al. 1997a). The toxic period occurs early in the grazing season—late April to May for *D. carolinianum* (including *D. virescens*) in the southern plains and May and June for *D. bicolor* as far north as Alberta and Saskatchewan (Looman 1975). Both of these species are readily eaten by cattle; hence, they represent serious problems when they are abundant and constitute a significant portion of the available forage. Ingestion by sheep does not vary as much, because they seem more prone to eating larkspur species in all stages of growth (Ralphs and Olsen 1990).

Intoxications in cattle are most likely with consumption of plants on successive days, because there is a buildup in severity of signs with repetitive doses each day up to 4 days (Marsh et al. 1916; Olsen & Sisson 1991a). This, coupled with the prospect of development of short-term aversion due to negative digestive consequences when increasing amounts of plant material are eaten, may lead to cyclical changes in intoxication risk (Pfister et al. 1990). There is some indication that glutinous feeding may be precipitated by cold rain showers or other types of stormy weather (Wilcox 1899; Glover 1906; Marsh et al. 1916; Pfister et al. 1988b; Ralphs et al. 1988, 1994). Experimentally, neither mineral supplementation nor chronic administration of carbachol influenced larkspur consumption by cattle, but cattle, however, eat much less of the plant during drought (Pfister & Manners 1995).

■ biological control, larkspur mirid decreases palatability of larkspur

As with many other toxic range plants, biological control holds promise for reduction or elimination of problems due to larkspur infestations. Studies with the larkspur mirid *(Hopplomachus affiguratus)*, which breeds exclusively on "tall" larkspurs, indicate that it serves as a deterrent to foraging by cattle (Ralphs et al. 1997a; Jones et al. 1998). Plant alkaloid levels are slightly reduced with infestation by the mirid, but most importantly, the feeding nymphs terminate reproductive development of the plant and decrease palatability (Jones et al. 1998; Ralphs et al. 1998b).

■ signs occur within 3–4 hours, maximal at 5–8 hours

6 stages of disease—
1, slight tremors
2, lies down
3, unable to stand
4, sternal
5, lateral
6, death

Clinical Signs—The disease produced by consumption of *Delphinium* is generally described to encompass 6 stages. The animal in stage 1 exhibits slight tremors; in 2, cannot stand for long periods and thus periodically lies down; in 3, can lift body but cannot stand; in 4, remains sternal; in 5, is in lateral recumbency and unable to rise to sternal; and in 6, dies (Olsen & Sisson 1991a,b). In this scenario, clinical signs may appear 3–4 hours after ingestion and become fully developed in 5–8 hours. There is rapid onset of uneasiness, a wide straddle-legged stance, a stiff and staggering gait, repeated falling with kicking, tremors, ataxia, constipation, recumbency, and bloat, especially when the animal is down. There is usually an increase in heart rate. Respiratory difficulties are especially conspicuous, and death is eventually due to respiratory failure. This is an episodic weakness; the animal falls, struggles, rests, rises, and falls again. Signs develop rapidly and subside almost as quickly (within a few hours) on some occasions but may build up to longer periods of incapacitation with continued ingestion of the plants. Bloat is a common but not inevitable consequence of intoxication. Because animals are more likely to fall when locomotor problems occur on uneven ground, bloat will develop rapidly if they are lying and facing downhill. Indeed, if they can be moved to face uphill, recovery seems to be the rule.

Pathology—Although there are no distinctive gross or microscopic changes associated with *Delphinium* intoxication, diagnosis may be aided by discovery of plants or plant parts in the rumen or assay of the ruminal fluid for the alkaloids.

■ no lesions

Treatment—Because this is a disease primarily of cattle on rangelands, treatment is an infrequent option. Atropine has been used with some success; however, physostigmine is a more effective antidote. A mixture of 65 mg physostigmine salicylate, 130 mg pilocarpine, and 32 mg strychnine per 250 kg b.w. given s.c. to cattle has been recommended (Marsh et al. 1916). Physostigmine at a dose of 0.04–0.08 mg/kg b.w. given i.p. (when given i.v., use great care) can reverse the signs in cattle even when the animals are in lateral recumbency (Pfister et al. 1994b). The dosage may need to be repeated in a few hours. Animals that fall facing downhill and are unable to rise should be turned to an uphill direction to facilitate movement of gases from the rumen and reduce the incidence of fatal bloat.

■ treat with physostigmine; when animal is facing downhill, fatal bloat may occur quickly; turn animals to face uphill

larkspur ingestion while grazing may be prevented by previous aversion training, but protection may be lost under some conditions

Cattle may be conditioned to avoid eating larkspur through the use of noxious agents such as lithium chloride. This aversion may carry over from one season to another but is subject to interruption when averted cattle graze with nonaverted herd mates (Lane et al. 1990; Ralphs & Olsen 1990). A single training exposure of 200 mg/kg b.w. of lithium chloride together with some plant material will provide an aversion lasting several years (Ralphs 1997; Ralphs & Olsen 1998). Commercially available mineral salt supplements have been used in hopes of reducing larkspur consumption, but as yet there is no experimental evidence to support the effectiveness of this practice (Knowles 1974; Pfister & Manners 1991; Pfister et al. 1998). There also is no evidence to show that mineral-deprived cattle eat more larkspur (Knight & Pfister 1997). With appropriate management, the apparent resistance of sheep may provide a means of protecting cattle from larkspur intoxication, because sheep may eat some species in early growth (Ralphs & Olsen 1992). However, even this technique is not without risk, because the sheep may instead reduce overall availability of forage (Aldous 1917).

Helleborus L.

■ *H. foetidus*
H. niger
H. orientalis
H. viridis

bear's-foot, black hellebore, Christmas rose, Easter rose, green hellebore, Lenten rose, setterwort, stinking hellebore, stinkwort

Taxonomy and Morphology—Comprising 20–25 species, *Helleborus,* commonly known as hellebore, is an Old World genus. The name is Greek and was used by Theophrastus, although whether he was referring to this taxon or another is ambiguous. The common name *hellebore* is also used for *Veratrum* in the Liliaceae. Of toxicologic interest and encountered in cultivation or in the wild are the following species:

H. foetidus L.	stinking hellebore, stinkwort, setterwort
H. niger L.	Christmas rose, Easter rose, black hellebore
H. orientalis Lam.	Lenten rose
H. viridis L.	green hellebore, bear's-foot

■ perennial herbs or subshrubs; leaves simple or palmately compound, alternate; sepals 5 or 6, larger than petals, white, green, purple, red, or yellowish; petals modified into nectaries, dropping quickly; fruits follicles

Plants herbs or subshrubs; perennials; terrestrial; caulescent or acaulescent; from short rhizomes or fibrous roots. **Stems** absent or present; erect; 12–60 cm tall. **Leaves** simple or palmately compound; alternate; basal or cauline or both; blades orbicular to reniform; lobes 3–11, long; margins entire or toothed; petioles long, bases sheathing. **Inflorescences** solitary flowers or small cymes; bracts present, often foliaceous. **Flowers** large; perfect; perianths in 2-series. **Calyces** radially symmetrical. **Sepals** 5 or 6; larger than petals; persistent; petaloid; white or green to purple or red or yellowish, turning green after anthesis. **Petals** 5–15; caducous; free; funnel-shaped or tubular; modified into nectaries; clawed; green or yellow or pink or dark purple to black. **Stamens** numerous; longer than petals; filaments filiform. **Pistils** 3–10. **Fruits** follicles; aggregated; sessile; coriaceous; beaked. **Seeds** numerous; oblong to spherical; dark brown.

Helleborus viridis

1. Plants subshrubs. Stems persistent.
   Leaves strictly cauline. Rhizomes absent .... *H. foetidus*
1. Plants herbs. Stems dying back.
   Leaves both cauline and basal. Rhizomes present.
   2. Bract margins entire. Flowers white to pink or purple .... *H. niger*
   2. Bract margins divided. Flowers green or cream-colored to greenish yellow-brown.
      3. Follicles free; stipitate. Flowers cream to greenish yellow-brown. Leaves overwintering .... *H. orientalis*
      3. Follicles fused at bases; not stipitate. Flowers green. Leaves deciduous .... *H. viridis*

■ ornamentals; edges of woods and thickets

Distribution and Habitat—*Helleborus* is native to Europe and Asia. Several species have been planted in North America as ornamentals, but only *H. viridis* appears to persist after cultivation and is infrequently encountered in waste areas, on shaded roadsides, and along the edges of woods and thickets in the eastern states.

■ digestive and cardiac disturbances

Disease Problems—Used medicinally for centuries, *Helleborus* is well known for its cardiac actions (Chen et al. 1950). Superstitions and myths also are associated with the genus, and it is reputed to be lethal (Baskin 1967). Like other cardenolide-containing plants, the dry hellebore retains its toxicity. Although the plant is apparently distasteful, its potency is such that a relatively small amount of plant material is quite dangerous, with approximately 0.5 kg causing the death of several steers (Johnson & Routledge 1971; Holliman & Milton 1990).

■ bufadienolides, glycosides of hellebrigenin and protoanemonin

Disease Genesis—In contrast to *Adonis,* which contains the more typical 23-C cardenolides with a 5-member ring, *Helleborus* contains 24-C bufadienolides with a 6-member ring (Shoppee 1964; Bartik & Piskac 1981; Joubert 1989). It is of interest that hellebrigenin—the aglycone of hellebrin, helleborein, and other compounds—is more potent than the intact parent monosides or biosides (Chen et al. 1950). Interestingly, hellebrigenin is also found in several species of toads, including the common European toad (Shoppee 1964). The roots of *Helleborus* are particularly toxic and are the source of most isolations. The lack of consistency in bufadienolide content is an important consideration. Although the toxin has been readily isolated and identified from *H. niger* and other species, in at least one investigation *H. niger* seemed to be devoid of the toxin (Karrer 1943; Wissner & Kating 1971). Concentrations up to 1% are reported in the dry rhizomes of some species (Wissner & Kating 1971). *Helleborus niger* also contains protoanemonin, sometimes in high concentrations, i.e., up to 0.58% wet weight (Bonara et al. 1985). Thus, the effects on the digestive system may be due to a combination of the irritant effects of protoanemonin and those of the bufadienolides.

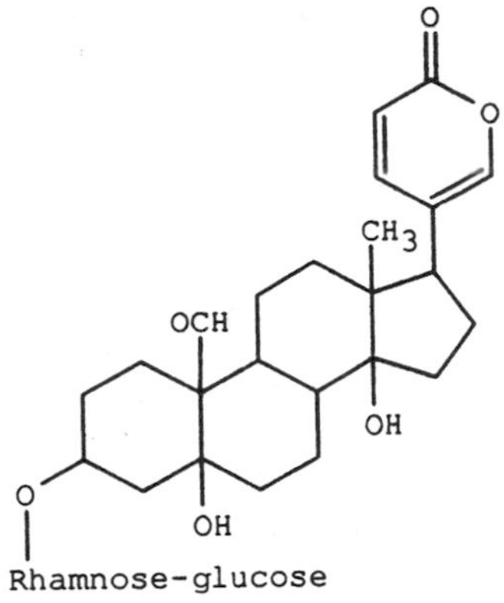

Rhamnose-glucose

Hellebrin

Clinical Signs—The signs produced by *Helleborus* are similar to those for *Digitalis,* with severe manifestations 2–3 days following consumption of the plants. The earliest signs are those associated with the gastrointestinal tract and include diarrhea (often with blood), vomiting (in pigs), and colic. The pain produced seems to be such that the animals are reluctant to move and may even resist standing. With complete anorexia, less severely affected animals may not die for several days. Effects on the heart range from premature beats and prolonged PR interval to various degrees of atrioventricular blockade. Although a slow irregular pulse is typical, there may be a secondary tachycardia. Death is due to ventricular fibrillation and cardiac failure.

■ diarrhea, vomiting, colic, irregular heart rate

Pathology—In most animals the most distinctive change will be reddening of the mucosa of the stomach and small intestine. There also may be subepicardial and subendocardial ecchymotic hemorrhages and petechia in other tissues. Because of the duration of effects, microscopic changes in the heart may be anticipated as described for *Digitalis* (see chapter 68).

■ reddened gut mucosa

Treatment—Treatment is mainly directed toward relief of the digestive disturbances. In the event of appearance of cardiotoxic effects, treatment will include activated charcoal and atropine, similar to that for the cardenolides in general (cf. *Digitalis,* in chapter 68).

■ relief of digestive disturbance; activated charcoal, atropine

Ranunculus L.

Taxonomy and Morphology—Almost cosmopolitan in distribution but with greatest diversity in the cold-temperate regions, *Ranunculus,* commonly known as buttercup or crowfoot, comprises 300–400 species. Used by the natural historian Pliny, the name is derived from the Latin roots *rana* and *unculus,* meaning "frog" and "little," possibly alluding to the wet habitats typically occupied by plants of the genus. A few species are cultivated as garden ornamentals, whereas others are weeds of wet or moist rights-of-way, other disturbed sites, and meadows. Large populations often develop and offer showy displays in the early spring.

Exhibiting considerable morphological and ecological diversity, the genus is taxonomically troublesome. Taxonomists differ in their opinions as to its boundaries and infrageneric relationships. For example, the 9 subgenera recognized by Benson (1948) have been treated as separate genera by other workers. In addition, many species are quite polymorphic and numerous varieties have been described. Hybridization also plays a role in the evolution of the genus. In North America, 77 native and introduced species occur (Whittemore 1997). Representatives are the following taxa:

■ *R. abortivus*
R. acris
R. arvensis
R. bulbosus
R. cymbalaria
R. flammula
R. hispidus
R. parviflorus
R. recurvatus
R. repens
R. sceleratus
R. testiculatus

blisterwort
buttercup
crowfoot
hungerweed
spearwort

Native Species

R. abortivus L.	smallflower buttercup, littleleaf buttercup, early wood buttercup, kidney-leaved buttercup
R. acris L.	tall buttercup, tall crowfoot, meadow buttercup, common buttercup

- annual or perennial herbs; leaves simple or palmately compound, alternate; sepals 3–5, yellow or white; petals 0–22, yellow, white, red, or green; fruits achenes

R. cymbalaria Pursh	shore buttercup, seaside buttercup, marsh buttercup, desert crowfoot
R. flammula L.	spearwort, creeping spearwort, lesser spearwort, creeping buttercup
R. hispidus Michx. (= *R. septentrionalis* Poir.)	creeping buttercup
R. recurvatus Poir.	hooked buttercup, blisterwort
R. sceleratus L.	celery-leaf buttercup, blister buttercup, ditch crowfoot, cursed crowfoot, celery-leaved crowfoot, bitter buttercup

Ranunculus abortivus

Naturalized Species

R. arvensis L.	corn buttercup, corn crowfoot, field buttercup, hungerweed
R. bulbosus L.	bulbous buttercup
R. parviflorus L.	small-flowered buttercup, stickseed crowfoot
R. repens L.	creeping buttercup
R. testiculatus Crantz	testiculate buttercup, bur buttercup, hornseed buttercup

Ranunculus acris

Plants herbs; annuals or perennials; terrestrial or aquatic; caulescent; from tuberous roots, caudices, rhizomes, or stolons. **Stems** erect to decumbent. **Leaves** simple or palmately compound; alternate; basal or cauline or both; basal leaves long petiolate; cauline leaves short petiolate or sessile; blades reniform to linear; margins entire or toothed. **Inflorescences** solitary flowers or cymes; terminal or axillary; bracts absent or present, large or small, foliaceous. **Flowers** perfect; perianths in 1- or 2-series, radially symmetrical or asymmetrical; inconspicuous or showy. **Sepals** 3–5 or rarely 6; deciduous or rarely persistent in fruit; green or sometimes purple, yellow, or white. **Petals** typically 5 (0–22); free; yellow or rarely white, red, or green; linear to orbicular. **Stamens** 5 to many; filaments filiform. **Pistils** 4–250. **Nectaries** present on abaxial surfaces of petals. **Fruits** achenes or rarely utricles; aggregated; sessile; lenticular or globose or obovoid or cylindrical; beaked or not beaked. **Seeds** 1.

Because of the large number of species present in North America and their extensive morphological variability, a key for the identification of taxa is not presented. The reader is encouraged to use floras specific for his or her locale.

Ranunculus bulbosus

Ranunculus hispidus

Ranunculus parviflorus

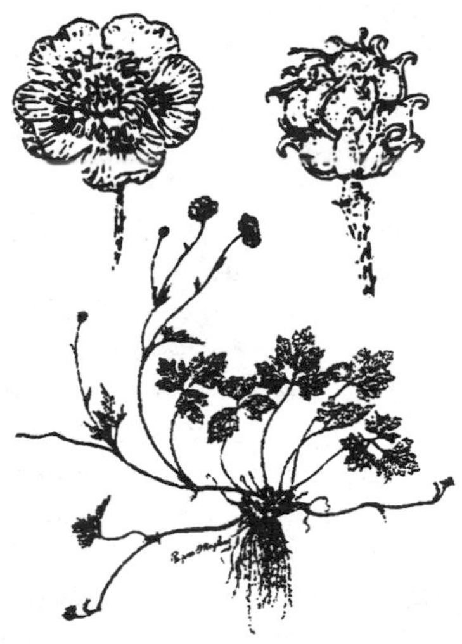

Ranunculus repens

Ranunculus sceleratus

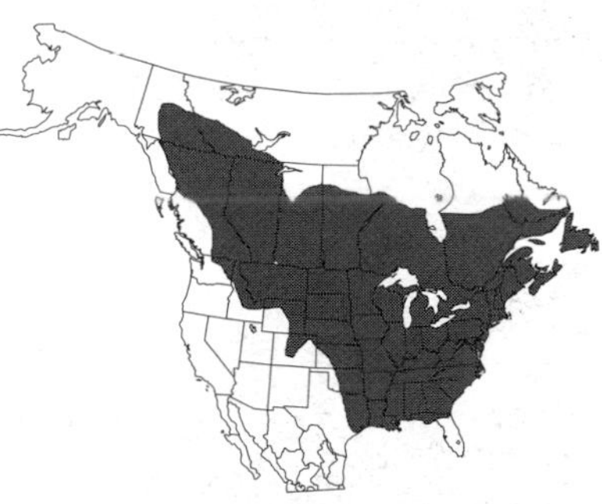

Ranunculus abortivus

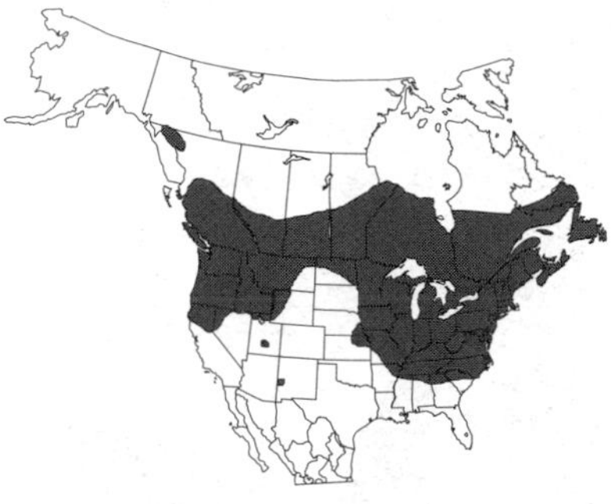

Ranunculus acris

■ variety of habitats, often in or around water, wet or moist soils

Distribution and Habitat—Species of *Ranunculus* occur across the continent in a variety of habitats. Moist or wet soils are typical, but some taxa occur in relatively dry sites while others are truly aquatic. Some species—e.g., *R. acris* and *R. abortivus*—have broad distributions, whereas others are restricted to small geographical areas.

Disease Problems—With a long history of various medicinal uses, species of *Ranunculus* are now known to have antibacterial, antifungal, and antimutagenic effects (Seegal & Holden 1945; Misra & Dixit 1980; Minakata et al. 1983). Although irritation of the digestive tract associated with various species (e.g., *R. acris, R. flammula,* and

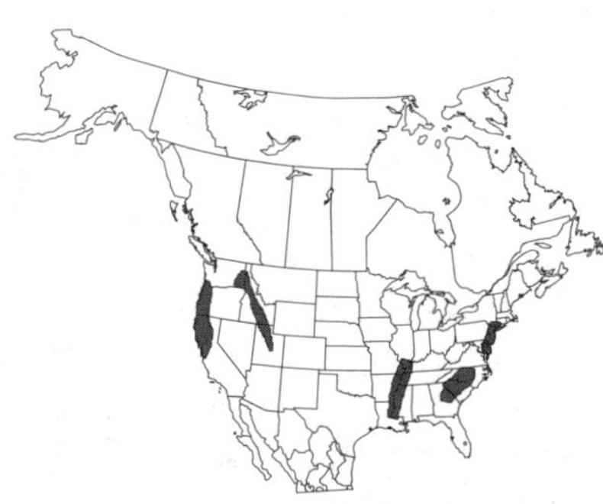

Ranunculus arvensis

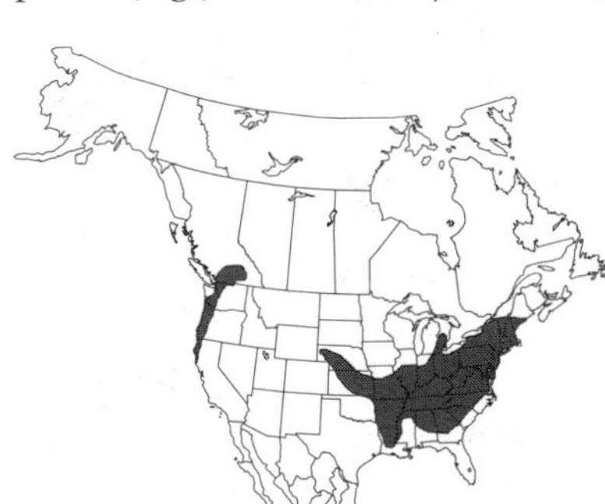

Ranunculus bulbosus

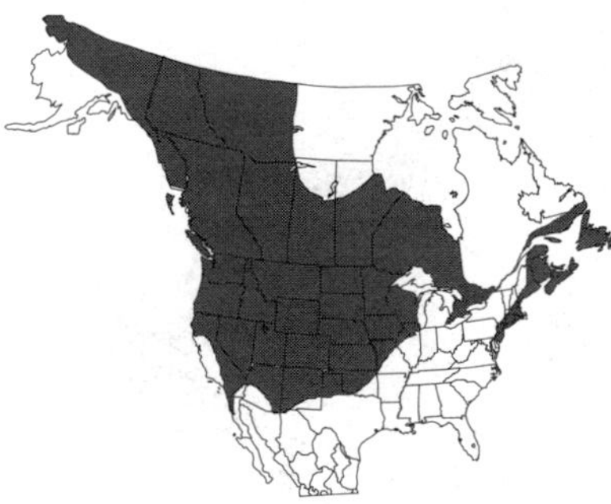

Ranunculus cymbalaria

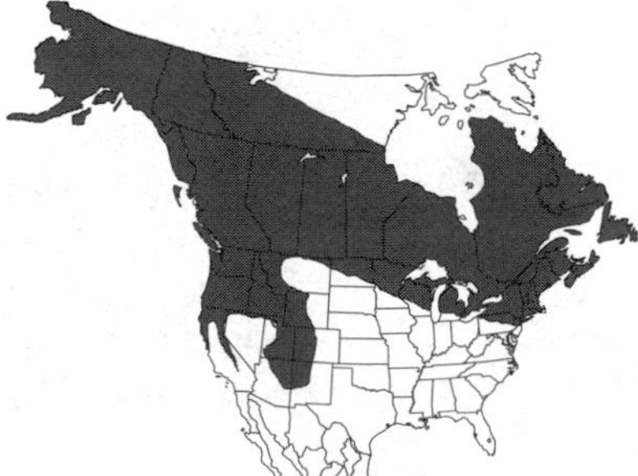

Ranunculus flammula

Ranunculus hispidus

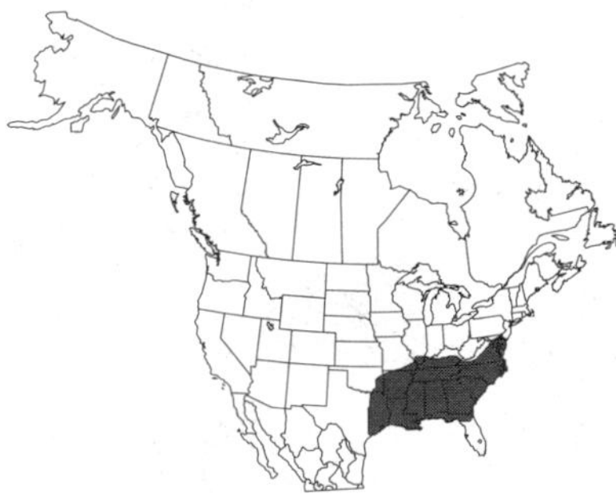

Ranunculus parviflorus

- irritant digestive disturbance, sometimes severe; rarely neurologic problems

Ranunculus recurvatus

Ranunculus repens

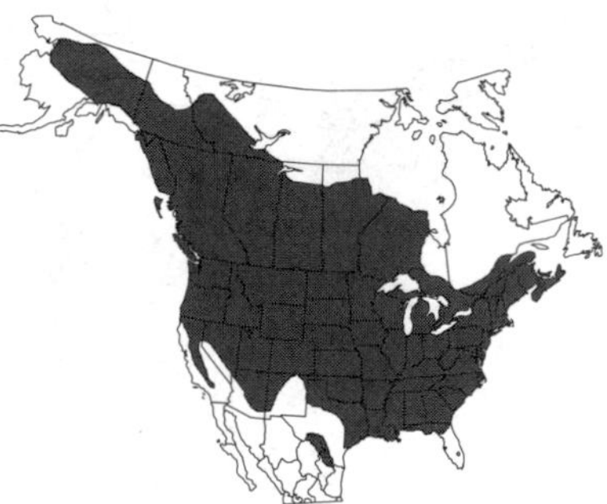

Ranunculus sceleratus

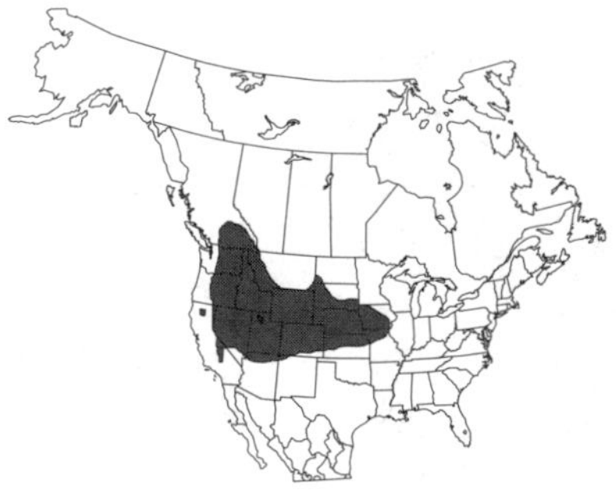

Ranunculus testiculatus

R. sceleratus) has been reported, risk due to these early-season plants seems to be overrated for most of them, at least with respect to lethality (Shearer 1938; Harvey et al. 1944). Experimentally, in laboratory animals, *R. repens* seems to be rather innocuous, and *R. acris* is toxic only with very large dosage (Nicholson 1933). *Ranunculus bulbosus* and *R. sceleratus* are of considerably greater toxic potential, producing digestive tract problems as well as neurologic effects with large dosage (Shearer 1938; Gunning 1949; Piekarz 1981). Other effects on the urinary system (hematuria) and mammary gland (reddish and bitter milk) have also been reported. A rather unusual episode that differs from these problems was the reaction occurring in a dog seen eating *R. acris* and from which plant parts were subsequently identified in its vomitus (Winters et al. 1976). Swelling of the face and head was very severe and accompanied by a few urticarial wheals. With conservative treatment, these effects were resolved within 24 hours without further problems. In another incident, *Ranunculus* was suspected as a cause of photosensitization (Kelch et al. 1992). In perhaps the most revealing study, *R. acris* at various stages of growth, including flowering, was fed as green chop or via strip grazing to cattle and sheep in very large amounts, with no apparent adverse effects other than decreased weight gains (Therrien et al. 1962; Hidiroglou & Knutti 1963). In contrast, in another instance involving the death of 2 horses on pasture, plant parts of *R. acris* were identified in the stomach contents (Griess & Rech 1997). In spite of the difficulties in substantiating disease experimentally, periodic clinical reports seem to confirm the capability for some *Ranunculus* species, when eaten in substantial amounts, to produce occasionally severe irritating effects on the digestive system.

In most situations only a few animals are seriously affected. Occasionally, large numbers of deaths are seen, as happened with the loss of 150 sheep in a flock of 800 after they ate *R. testiculatus* (Olsen et al. 1982, 1983). In this case, hungry sheep were grazed in a pasture composed mainly of *R. testiculatus*. Over the next 3–4 days, severe watery diarrhea, weakness, dyspnea, recumbency, and death developed. Experimentally, the LD_{50} wet weight of green plants was shown to be 1.1% b.w. Deaths occurred 2–48 hours after administration. Unfortunately, plants of the species in the early stages of their growth are readily eaten by sheep, and although not the animals' first choice, the plants are eaten even when mature and fruiting. There are suggestions that some species of *Ranunculus (R. acris* and *R. reprens)* produce abortifacient effects, but these have not been confirmed (Pammel 1911; Morales 1989).

Disease Genesis—Although alkaloids and low concentrations of cyanogenic glycosides are present in some species, toxicity appears to be due mainly to ranunculin, a glycoside. Ranunculin is hydrolyzed to protoanemonin when the plant tissues are macerated and then subsequently polymerized to form the inactive anemonin (Hill & Van Heyningen 1951). The identity of the glycosidic precursor to protoanemonin has been questioned, with ranunculin suggested to be only an artifact, but there is little question of the ultimate vesicant (Tschesche et al. 1972). Protoanemonin is a potent vesicant, but anemonin, which forms when the plant is dried, is inactive in this respect. Plants are said to be particularly irritating when flowering, with protoanemonin levels up to 1–2% d.w. (Hill & Van Heyningen 1951; Kingsbury 1964). The variation in toxicity among the various species of *Ranunculus* is in large measure due to the great range in protoanemonin concentrations. Example concentrations of protoanemonin, as reported by Shearer (1938), are as follows: R. acris, 0.26% d.w./1.45% wet weight; *R. arvensis,* 0.54%/1.74%; *R. bulbosus,* 0.26%/1.45%; *R. flammula,* 0.54%/2.29%; *R. parviflorus,* 0.37%/2.0%; *R. repens,* 0.05%/0.27%; *R. sceleratus,* 0.38%/2.5%. These values are in contrast to those reported by others: 0.78% wet weight for *R. bulbosus* and 0.13% for *R. sceleratus,* both species of more consistent vesicant effect, and 0.013% for *R. repens,* a species seldom, if ever, associated with toxicity (Minakata et al. 1983; Bonora et al. 1985). Concentrations of ranunculin in *R. testiculatus* of 0.5% wet weight or 2.3% d.w. at flowering decrease by about half with increasing maturity (Nachman & Olsen 1983). The presence of glycosides such as ranuncoside is of unknown significance (Tschesche et al. 1972).

A complicating factor in this genus is the possibility that cyanogenic glycosides are present in some species. The cyanogenic glucoside triglochin is reported in *R. arvensis* and *R. repens* in their native Old World ranges (Tjon Sie Fat 1979).

- glycoside ranunculin hydrolyzed to irritant protoanemonin, which when dried forms nonirritant anemonin

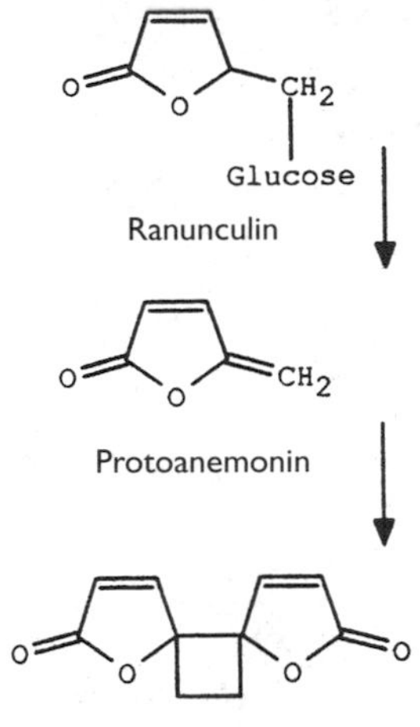

CH_2
Glucose
Ranunculin

CH_2
Protoanemonin

Anemonin

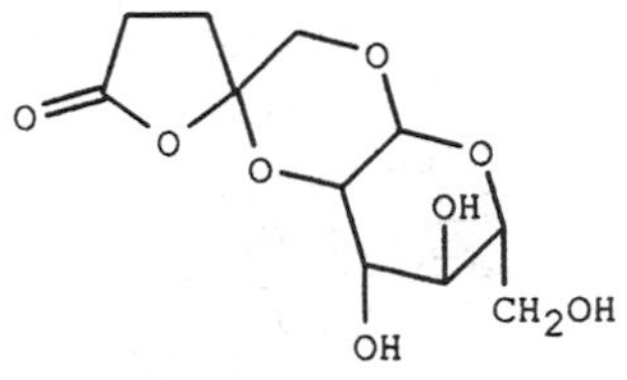

OH
CH_2OH
OH
Ranuncoside

Clinical Signs—In the unusual situation when clinical signs develop, they are typically those associated with blistering of the skin, the mouth, and the digestive system. Thus, erythema and swelling of the muzzle, nose, lips, and face may be seen with or without further effects. In most animals the signs are related to the digestive system and include nausea, vomiting, diarrhea, and colic. However, neurologic signs such as tremors, seizures, persistent weakness, and paralysis are reported to be produced by *R. acris* and *R. bulbosus* (Nicholson 1933; Cooper & Johnson 1984; Fuller & McClintock 1986). Signs with *R. testiculatus* include weakness, depression, diarrhea, and dyspnea.

Diagnosis may be aided by identification of plant fragments in the ingesta. Griess and Rech (1997) have described the microscopic identification of *R. acris, R. arvensis, R. flammula, R. repens,* and *R. sceleratus* using such fragments.

- erythema and swelling of muzzle, nose, lips, and face; vomiting, diarrhea, colic; rarely tremors, weakness, seizures

Pathology—There is edema of the serosa and reddening of the mucosa of the abomasum and small intestines, moderate excess fluid in the chest and abdomen, congestion of the abdominal viscera, and a few splotchy hemorrhages on the serosa of the intestines and endocardium.

- reddened gut mucosa

Treatment—This is primarily symptomatic relief of the irritation of the digestive tract and any accompanying neurologic or other signs. Inasmuch as the allelopathic (suppression of competitive plant growth) effects of protoanemonin are attenuated by thiols like British Anti-Lewisite (BAL), there may be a role for these compounds therapeutically (Thimann & Bonner 1949).

- relief of irritant effects on digestive system

Genera of Questionable Toxicity

Anemone L.

Comprising 120–150 species distributed primarily in north temperate and arctic regions of both the Old World and the New World, *Anemone,* commonly known as windflower, is represented in North America by 25–30 native and introduced taxa. Many have broad distributions and occur in the eastern or western half of the continent. Plants of the genus are perennials from rhizomes, caudices, or tubers with petiolate, basal, simple or palmately compound leaves. The showy, perfect flowers are borne singly or in 2-to-9-flowered umbels or cymes and are usually subtended by a leafy involucre. The radially symmetrical perianths are in 1-series, with the petaloid sepals ranging from white to purple, blue, green, yellow, pink, or red. Shorter than the sepals, the numerous stamens are borne in spirals on the typically elongated receptacle, as are the numerous pistils, which give rise to achenes.

- numerous species

windflower

Little is known of the toxicity potential of this genus, although it had some reputation as such and was used as an astringent by Native Americans. The foliage is very

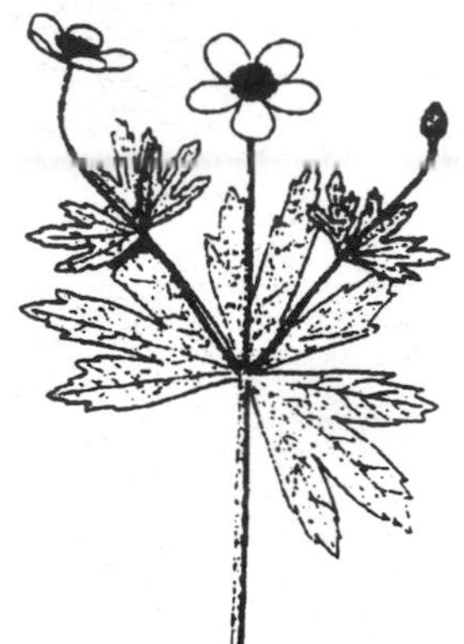

Anemone canadensis

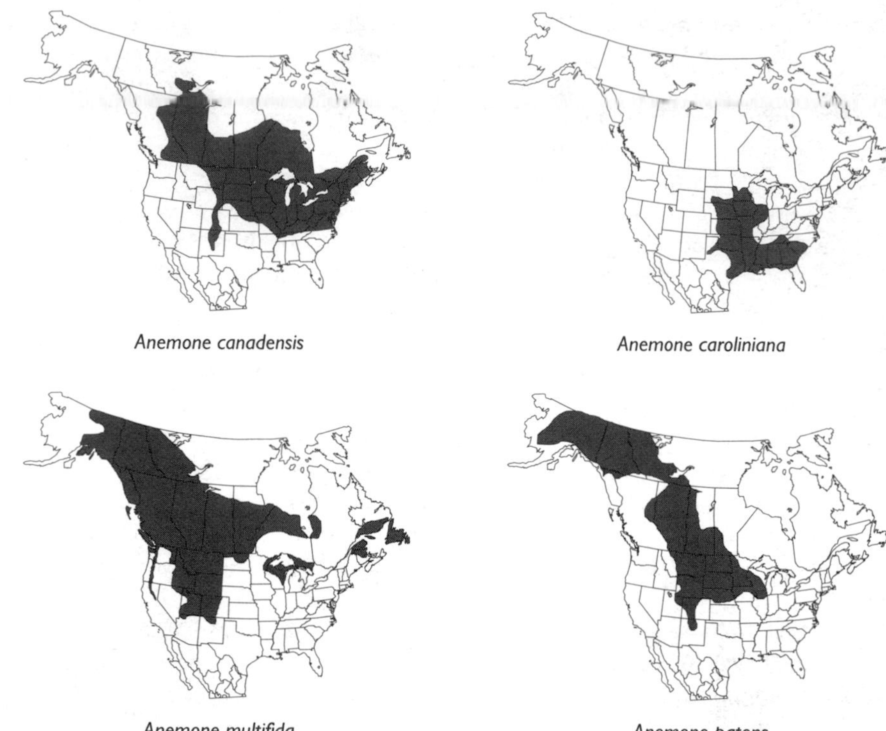

Anemone canadensis *Anemone caroliniana* *Anemone multifida* *Anemone patens*

Anemone caroliniana

- possibly irritant effects due to protoanemonin

acrid to the taste and is rarely eaten (Chesnut & Wilcox 1901). Most of the species are reported to be very irritating, in a manner similar to that of *Ranunculus,* causing blistering of the skin or of the digestive tract, resulting in colic (Pammmel 1911; Harvey et al. 1944). However, studies with species found in other areas of the world indicate protoanemonin concentrations to be low, in the range of 0.02%–0.05% wet weight (Minakata et al. 1983; Bonora et al. 1985). Thus, both the toxicity and/or mechanism of effect are in question. Plants also contain glycoside derivatives of oleanolic acid terpenoids (Connolly & Hill 1991). In the event of ingestion of large amounts of plant material, the signs, pathology, and treatment may be similar to those of *Ranunculus.*

- numerous species

clematis
leatherflower
virgin's bower

Clematis L.

One of the most prized garden ornamentals, *Clematis,* commonly known as virgin's bower, leatherflower, or simply clematis, comprises 200–300 species of both the Old World and the New World; 32–38 native and introduced taxa are present in North America. The genus has long been cultivated in eastern Asia. In Europe in the middle 1500s, species were being introduced and cultivated. The first hybrid was produced in 1835, and in the mid-1800s, numerous species were being exhibited at flower shows. The generic name is derived from the Greek root *clema,* meaning "vine" or "tendril," which reflects the genus's characteristic growth form.

Plants of *Clematis* are deciduous or evergreen, more or less woody perennial vines climbing by means of tendril-like petioles and rachises, or are erect herbs from elongated rhizomes. The cauline, opposite or rarely alternate leaves are simple or compound, with either entire or toothed margins. Perfect or imperfect, the radially symmetrical flowers are borne singly or in many-flowered cymes or panicles or fascicles. Bracts may be present or absent. The petals are absent and the 4 or 5 petaloid

Clematis virginiana

sepals are white, blue, violet, red, yellow, or greenish. They may be widely spreading, or erect and forming a campanulate or urceolate perianth. The numerous stamens have filiform or flattened filaments. Numbering 5–150, the pistils give rise to lenticular or terete achenes usually bearing long, persistent, feathery styles. The large-flowered taxa of gardens are typically hybrids and cultivars. Representative of the native species of the genus is *C. virginiana* L., commonly known as virgin's bower or devil's darning needle, which is the most abundant and widespread species of the genus in the eastern half of North America.

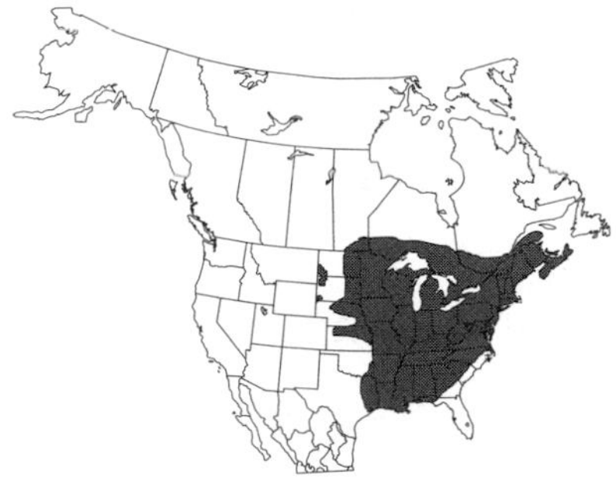

Clematis virginiana

■ possibly irritant effects due to protoanemonin

In spite of the apparent potential for causing problems, few actual episodes have been described. In one of the few reported cases, involving an Old World species, a large amount of plant material was found in the rumen of a cow that died suddenly (Moore 1971). The disease was of short duration, with few signs other than scattered ulcerations of the muzzle and colic but no diarrhea. In spite of the lack of diarrhea, there was inflammation of the wall of the rumen, abomasum, and small intestine, and free fluid in the gut lumen. There was also pulmonary edema and congestion and excess fluid (with blood) in the thoracic cavity. Experimentally, 1 g of spring-growth leaves of a cultivated variety of *Clematis* caused intermittent vomiting in 2 budgerigars (Shropshire et al. 1992). The attributes ascribed to various species of *Clematis* in Africa also substantiate its reputation as an irritant, although specific experiments failed to confirm toxicity (Watt & Breyer-Brandwijk 1962). Thus, the disease problem appears to be similar to that produced by other genera of the family that cause irritant effects (Pammel 1911).

The potent vesicant actions are similar to those of *Ranunculus*, but in the few species studied, concentrations of protoanemonin were low: 0.05% wet weight or less (Pammel 1911; Everist 1981; Bonora et al. 1985). Glycosidic derivatives of oleanolic acid terpenoids also are present in the plants (Connolly & Hill 1991). In the event of ingestion of large amounts of plant material, the signs, pathology, and treatment may be similar to those of *Ranunculus*.

Hydrastis L.

■ *H. canadensis*

goldenseal
ground raspberry
Indian dye
orangeroot
sceau d'or
tumeric root
yellow puccoon
yellowroot
wild curcima

Known by a variety of common names—goldenseal, yellow puccoon, orangeroot, yellowroot, wild curcima, tumeric root, Indian dye, sceau d'or, or ground raspberry—*Hydrastis* is represented in eastern North America by a single species, *H. canadensis* L. One other species occurs in similar habitats in eastern China, albeit some taxonomists contend that the 2 are the same taxon. Taxonomists also differ in their opinions as to whether the genus should be placed in the Ranunculaceae, as is done here, or segregated in its own family intermediate between the Ranunculaceae and the Berberidaceae, with which it shares several attributes. Flowering in the early spring, plants occur in mesic, shaded forest sites, often in clay soils from New England southwestward to Arkansas and Oklahoma. They are perennial herbs arising from thick, knotted, yellow rhizomes. The erect stems are unbranched and pubescent, bearing 2, alternate, cordate to orbicular, palmately 3-to-9-lobed leaves with serrate margins. Long-petioled basal leaves are present and resemble the cauline leaves. Solitary, perfect flowers terminate the stems. The radially symmetrical perianth is in 1-series and comprises 3, greenish white to creamy white sepals. Petals are absent. Fifty to 75 stamens surround 5–15 pistils. Each bearing 1 or 2 black seeds, the dark red berries are aggregated into a dense cluster.

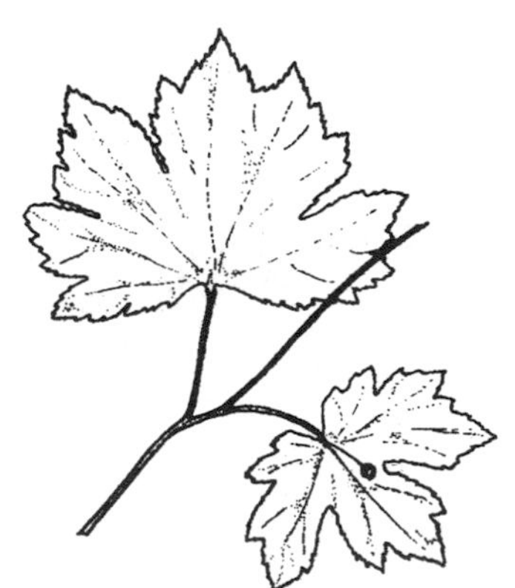

Hydrastis canadensis

An ingredient in a number of herbal remedies, *H. canadensis* is an important drug plant. Its use began with American Indians who used it medicinally to treat a variety of ailments. They also used it as a dye plant. Although the species is grown com-

Hydrastis canadensis

- alkaloids, berberine, canadine, hydrastine; possible digestive or neurologic disturbances

mercially, the rhizomes of wild plants are still avidly sought. Wanton collecting has resulted in its extinction in many areas.

Although the genus is not known as a cause of intoxications, *H. canadensis* contains a number of alkaloids similar to those present in members of the Fumariaceae and Papaveraceae. Perhaps the most noteworthy of these is hydrastine, a potent GABAa receptor competitive antagonist (Pammel 1911; Huang & Johnston 1990). With concentrations of hydrastine up to 2%, *H. canadensis* has the potential to produce striking convulsant effects like some of the *Corydalis* species. Other alkaloids present include the protoberberines such as berberine, canadaline, canadine, and corypalmine (Henry 1949; Manske & Ashford 1954; Bhakuni & Jain 1986). The quaternary berberine acts to augment tone and contraction of uterine muscles in some animals and also inhibits acetylcholinesterase. Primary concern is for irritation of the digestive tract, but based on the presence of biologically active alkaloids, neurologic and other systemic signs may also be expected.

Hydrastine

Canadine

Thalictrum L.

- numerous species

  meadow rue

Thalictrum fendleri

- numerous alkaloids, similar to those of Fumariaceae and Papaveraceae, possible digestive or neurologic disturbances, hypotensive

Comprising 120–200 species, *Thalictrum,* commonly known as meadow rue is almost worldwide in distribution but with greatest abundance in the northern temperate zones. It is represented in North America by 22 native species and 1 introduced species that occasionally escapes cultivation. Occurring in a variety of habitats, some species have broad distributions, whereas others are quite restricted. Plants of the genus are perennial herbs arising from woody rhizomes, caudices, or tuberous roots. The short to tall stems bear alternate, 1-to-4-ternately or pinnately compound leaves, the distal ones being sessile and the proximal ones petiolate. Basal leaves are also present. Perfect or imperfect and borne singly or in terminal panicles, racemes, corymbs, or umbels, the flowers are radially symmetrical with the perianth in 1-series. The 4–10 petaloid sepals are whitish to greenish yellow or purplish. Petals are absent. The 7–30 stamens surround 1–16 simple pistils, which give rise to achenes.

Plants of the genus have a reputation for use in herbal remedies. When used dried and smoked or sprinkled on a fire and inhaled, they effect purgation, diuresis, alleviation of stomach pains, and relief of headaches. The genus also has antibacterial activity against *Mycobacterium* (Wu et al. 1977).

Although not known for their toxicity, species of *Thalictrum* are exceptionally rich in alkaloids, rivaling genera of the Fumariaceae and Papaveraceae, with whom they share many compounds in common. Dozens of alkaloids are present in this genus, with a least some present in most species examined (Cieszynski & Borkowski 1965; Jeffs 1967; Bhakuni & Jain 1986). There are for example, at least a dozen alkaloids in *T. revolutum* DC., especially in the roots (Wu et al. 1977). They include benzylisoquinolines, protoberberines, and aporphines (magnaflorine = thalictrine). Because of the complexity of the various possible alkaloid combinations, it is difficult to ascribe particular syndromes to the various species. Berberine, common to most species, weakly modifies digestive system activity, enhances uterine muscular activity, and depresses cardiac function, and magnaflorine is hypotensive through ganglionic block (Schiff & Doskotch 1970; Bhakuni & Jain 1986). Six of the alkaloids from *T. revolutum* are powerful hypotensives, moderated presumably through smooth muscle effects (Patil et al. 1963). They also depress the nervous system and have various effects on the uterus, depending on the animal species. Adding to the toxicological complexity are reports that several Old World species contain the cyanogenic glucoside triglochinin (Ettlinger & Eyjolfsson 1972; Tjon Sie Fat 1979).

With such a rich array of alkaloids/toxicants, plants of *Thalictrum* are undoubtedly toxic, but they are apparently seldom, if ever, eaten in sufficient quantity to cause intoxication. This is apparently due to low palatability and/or to the limited amount of plant material available in a given location. This lack of a significant problem is reminiscent of the situation with the Fumareaceae and Papaveraceae. As with other members of the Ranunculaceae, vesicant effects similar to the effects of protoanemonin are ascribed to this genus. Whether these effects are due to some of the alkaloids listed above or other compounds is not clear. Although no disease has been described, the toxicants identified indicate possible digestive, neurologic, or cardiac effects.

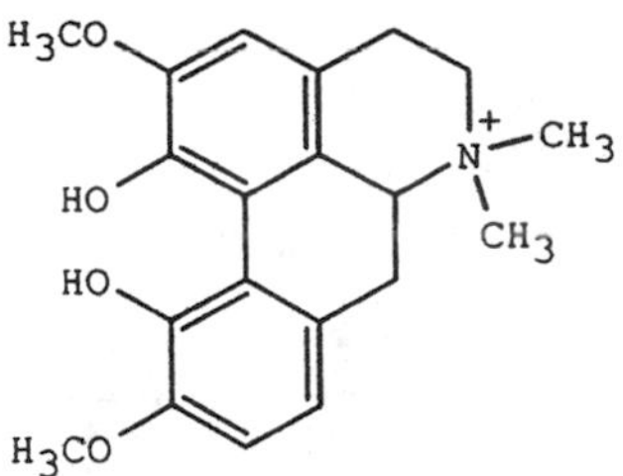

Magnaflorine

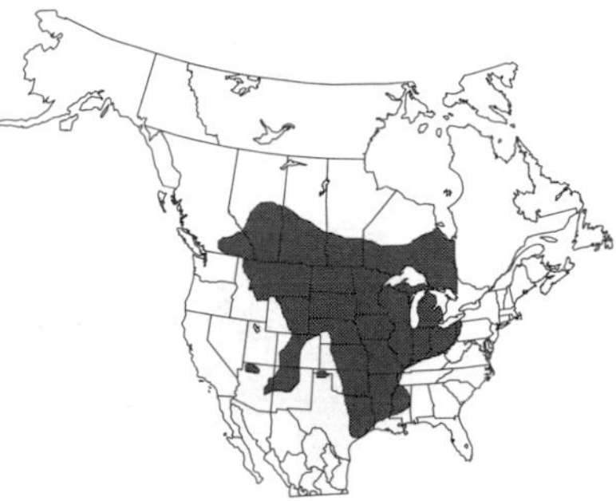
Thalictrum dasycarpum

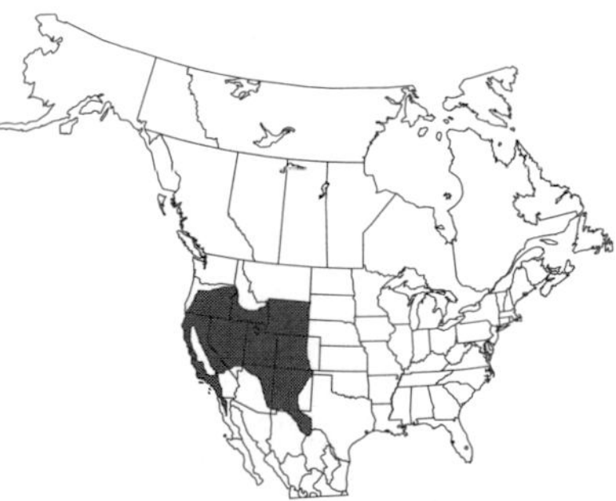
Thalictrum fendleri

Thalictrum revolutum

References

Aiyar VN, Benn MH, Hanna T, Jacyno J, Roth SH, Wilkens JL: The principal toxin of *Delphinium brownii* Rydb., and its mode of action. *Experientia* 35;1367–1368, 1979

Aldous AE: *Eradicating Tall Larkspur on Cattle Ranges in the National Forests.* USDA Farmers Bull 826, 1917

Alkondon M, Periera EFR, Wonnacott S, Albuqueque EX: Blockade of nicotinic currents in hippocampal neurons defines methyllycaconitine as a potent and specific receptor antagonist. *Mol Pharmacol* 41;802–808, 1992

Bacon AE: An experiment with the fruit of red baneberry. *Rhodora* 5;77–79, 1903

Bai Y, Benn MH, Majak W: The alkaloids of *Delphinium nuttallianum:* the cattle-poisoning low larkspur of interior British Columbia. In *Poisonous Plants: Proceedings of the Third International Symposium,* James LF, Keeler RF, Bailey EM Jr, Cheeke PR, Hegarty MP eds, Iowa State Univ Press, Ames, pp304–308, 1992

Bai Y, Desai HK, Pelletier SW: *N*-Oxides of some norditerpenoid alkaloids. *J Nat Prod* 58;929–933, 1995

Bartik M, Piskac A: *Veterinary Toxicology.* Elsevier, Amsterdam, p224, 1981

Baskin E: *The Poppy and Other Deadly Plants.* Delacorte Press, New York, 1967

Beath OA: *The Chemical Examination of Three Species of Larkspurs.* Wyo Agric Exp Stn Bull 120, 1919

Beath OA: *Chemical Examination of Three Delphiniums.* Wyo Agric Exp Stn Bull 143, 1925

Benn MH, Jacyno JM: The toxicology and pharmacology of diterpenoid alkaloids. In *Alkaloids: Chemical and Biological Perspectives,* Pelletier SW ed, Wiley & Sons, New York, pp153–210, 1983

Benson LD: A treatise on the North American *Ranunculi. Am Midl Nat* 40:1–261, 1948.

Bhakuni DS, Jain S: Protoberberine alkaloids. In *The Alkaloids,* vol 28, Brossi A ed, Academic Press, Orlando, Fla, pp95–181, 1986

Bonora A, Dall'Olio G, Bruni A: Separation and quantitation of protoanemonin in Ranunculaceae by normal- and reversed phase HPLC. *Planta Med* 51;364–367, 1985

Brink DE, Woods JA: *Aconitum.* In *Flora of North America,* vol 3, *Magnoliophyta: Magnoliidae and Hamamelidae.* Flora of North America Editorial Committee eds, Oxford Univ Press, New York, pp191–195, 1997

Bruni A, Bonora A, Dall'Olio G: Protoanemonin detection in *Caltha palustris. J Nat Prod* 49;1172–1173, 1986

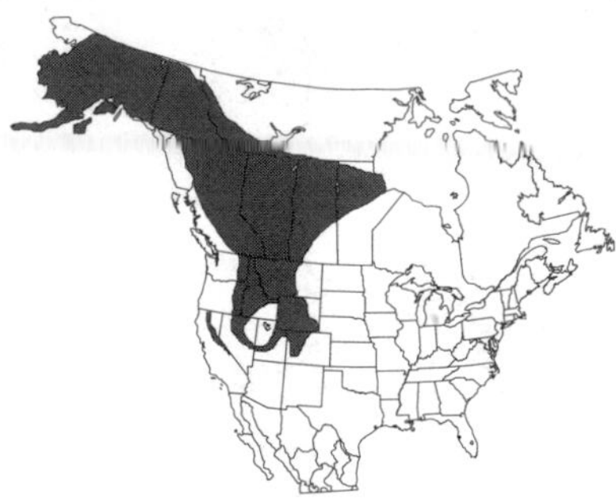

Thalictrum sparsiflorum

Thalictrum venulosum

Catterall WA: Structure and function of voltage-sensitive ion channels. *Science* 242;50–61, 1988

Chan TYK: Aconitine poisoning: a global perspective. *Vet Hum Toxicol* 36;326–328, 1994

Chan TYK, Chan JCN, Tomlinson B, Critchley JAJH: Poisoning by Chinese herbal medicines in Hong Kong: a hospital-based study. *Vet Hum Toxicol* 36;546–547, 1994a

Chan TYK, Tse LKK, Chan JCN, Chan WWM, Critchley JAJH: Aconitine poisoning due to Chinese herbal medicines: a review. *Vet Hum Toxicol* 36;452–455, 1994b

Chen KK, Henderson FG, Anderson RC: The cardiac action of *Helleborus* glycosides and their aglycones. *J Pharm Exp Ther* 99;395–400, 1950

Chesnut VK, Wilcox EV: *The stock-poisoning plants of Montana.* US Div Bot Bull 26, pp65–80, 1901

Cieszynski T, Borkowski B: Alkaloids of the genus *Thalictrum* L. *Acta Pol Pharm* 22;181–190, 1965

Connolly JD, Hill RA: *Dictionary of Terpenoids.* Vol 2, *Di- and Higher Terpenoids,* Chapman & Hall, London, pp655–1460, 1991

Cook CW, Redente EF: Development of the ranching industry in Colorado. *Rangelands* 15;204–207, 1993

Cooper MR, Johnson AW: *Poisonous Plants in Britain and Their Effects on Animals and Man.* Ministry of Agriculture, Fisheries and Food Ref Book 161, Her Majesties Stationery Office, London, pp192–202, 1984

Crawford AC: *The larkspurs as poisonous plants.* USDA Bur Plant Ind Bull 111, pp5–12, 1907

Cronin EH: Tall larkspur: some reasons for its continuing preeminance as a poisonous plant. *J Range Manage* 24;258–263, 1971

Cronin EH, Nielsen DB, Madson N: Cattle losses, tall larkspur, and their control. *J Range Manage* 29,364–367, 1976

Cronin EH, Olsen JD, Laycock WA: Ecological considerations of the larkspurs. In *The Ecology and Economic Impact of Poisonous Plants on Livestock Production,* James LF, Ralphs MH, Nielsen DB eds, Westview, Boulder, Colo, pp107–118, 1988

Davies RL, Whyte PBD: *Adonis microcarpa* (pheasants eye) toxicity in pigs fed field pea screenings. *Aust Vet J* 66;141–143, 1989

Drasdo A, Caulfield M, Bertrand D, Bertrand S, Wonnacott S: Methyllycaconitine: a novel nicotinic antagonist. *Mol Cell Neurosci* 3;237–243, 1992

Elkiey MA, Darwish Sayed M, Abdel Wahab SM, Soliman FM: Estimation of the cardiac glycosidal-contents of *Adonis autumnalis* Linn and *Adonis dentata* Del. *Planta Med* 15;201–204, 1967

Ettlinger M, Eyjolfsson R: Revision of the structure of the cyanogenic glucoside triglochinin. *J Chem Soc Chem Commun* 1972;572–573, 1972

Everist SL: *Poisonous Plants of Australia,* 2d ed. Angus & Robertson, Sydney, Australia, pp596–603, 1981

Ewan J: A synopsis of the North American species of *Delphinium. Univ Colo Stud,* ser D, 2;55–244, 1945

Fleming CE, Wright LH, Miller MR: *Poisonous Plants.* Nev Agric Exp Stn Annu Rep 1919, pp39–43, 1920

Fleming CE, Miller MR, Vawter LR: *The low larkspur* (Delphinium andersoni). Nev Agric Exp Stn Bull 105, 1923

Ford BA: *Actaea.* In *Flora of North America,* vol 3, *Magnoliophyta: Magnoliidae and Hamamelidae.* Flora of North America Editorial Committee eds, Oxford Univ Press, New York, pp181–183, 1997a

Ford BA: *Caltha.* In *Flora of North America,* vol 3, *Magnoliophyta: Magnoliidae and Hamamelidae.* Flora of North America Editorial Committee eds, Oxford Univ Press, New York, pp187–189, 1997b

Fuller TC, McClintock E: *Poisonous Plants of California.* Univ Calif Press, Berkeley, p433, 1986

Glover GH: *Larkspur and Other Poisonous Plants.* Colo Agric Exp Stn Bull 113, 1906
Griess D, Rech J: Diagnosis of acute poisoning by Rannunculi (*Ranunculus acris* L. and *Ficaria ranunculoides* L.) in horses. *Rev Med Vet* 148;55–60, 1997
Gunning OV: Suspected buttercup poisoning in a Jersey cow. *Br Vet J* 105;393, 1949
Harshberger JW: *Pastoral and Agricultural Botany.* P. Blakiston's Sons, Philadelphia, 1920
Harvey RB, Larson AH, Landon RH, Boyd WL, Erickson LC: *Weeds Poisonous to Livestock.* Minn Agric Exp Stn Bull 388, 1944
Henry TA: *The Plant Alkaloids,* 4th ed. Blakiston, Philadelphia, 1949
Hidiroglou M, Knutti HJ: The effects of green tall buttercup in roughage on the growth and health of beef cattle and sheep. *Can J Anim Sci* 43;68–71, 1963
Hill R, Van Heyningen R: Ranunculin: the precursor of the vesicant substance in buttercup. *Biochem J* 49;332–335, 1951
Holliman A, Milton D: *Helleborus foetidus* poisoning of cattle. *Vet Rec* 127:339–340, 1990
Huang J-H, Johnston GAR: (+)-Hydrastine, a potent competetive antagonist at mammalian GABAa receptors. *Br J Pharmacol* 99;727–730,1990
Jacyno JM: Chemistry and toxicology of the diterpenoid alkaloids. In *Chemistry and Toxicology of Diverse Classes of Alkaloids,* Blum MS ed, Alaken, Fort Collins, Colo, pp301–336, 1996
Jean-Blain C: Tableau de baiesnet fruits bacciformes les plus couramment recontres avec indication de leur toxicite. *Notes Toxicol Vet* 4;67–78, 1981
Jeffs PW: The protoberberine alkaloids. In *The Alkaloids,* vol 9, Manske RHF ed, Academic Press, New York, pp41–115, 1967
Jennings KR, Brown DG, Wright DPjr: Methyllycaconitine, a naturally occurring insecticide with a high affinity for the insect cholinergic receptor. *Experientia* 42;611–613, 1986
Johnson CT, Routledge JK: Suspected *Helleborus viridis* poisoning of cattle. *Vet Rec* 89;202, 1971
Jones WA, Ralphs MH, James LF: Use of a native insect to deter grazing and prevent poisoning in livestock. In *Toxic Plants and Other Natural Toxicants,* Garland T, Barr AC eds, CAB International, New York, pp23–28, 1998
Joubert JPJ: Cardiac glycosides. In *Toxicants of Plant Origin,* vol 2, *Glycosides,* Cheeke PR ed, CRC Press, Boca Raton, Fla, pp61–96, 1989
Karrer W: Uber helleborin, ein krystallisiertes glykosid aus radix *Hellebori nigri. Helv Chim Acta* 26;1353–1367, 1943
Katz A: Detection of diterpenoid alkaloids in aphids feeding on *Aconitum napellus* and *Aconitum paniculatum. J Nat Prod* 53;204–206, 1990
Kelch WJ, Kerr LA, Adair HS, Boyd GD: Suspected buttercup *(Ranunculus bulbosus)* toxicosis with secondary photosensitization in a Charolais heifer. *Vet Hum Toxicol* 34;238–239, 1992
Kingsbury JM: *Poisonous Plants of the United States and Canada.* Prentice Hall, Englewood Cliffs, NJ, 1964
Knight AP, Pfister JA: Larkspur poisoning in livestock: myths and misconceptions. *Rangelands* 19;10–13, 1997
Knowles K: *Prevention of larkspur intoxication in cattle.* Master's thesis, Univ Idaho, 1974
Kukel CF, Jennings KR: *Delphinium* alkaloids as inhibitors of alpha-bungarotoxin binding to rat and insect neural membranes. Can *J Physiol Pharmacol* 72;104–107, 1994
Kulanthaivel P, Benn M: Diterpenoid alkaloids from *Delphinium nudicaule* Torr. and Gray. *Heterocycles* 23;2515–2520, 1985
Kulanthaivel P, Benn M, Majak W: The C19-diterpenoid alkaloids of *Delphinium bicolor. Phytochemistry* 25;1511–1513, 1986
Liang X, Ross SA, Sohni YR, Sayed HM, Desai HK, Joshi BS, Pelletier SW: Norditerpenoid alkaloids from the stems and leaves of *Delphinium ajacis. J Nat Prod* 54;1283–1287, 1991
Lane MA, Ralphs MH, Olsen JD, Provenza FD, Pfister JA: Conditioned taste aversion: potential for reducing cattle loss to larkspur. *J Range Manage* 43;127–131, 1990

Looman J: Biological flora of the Canadian prairie provinces. 3. *Delphinium bicolor* Nutt. *Can J Plant Sci* 55;605–617, 1975

Lukas RJ, Bencherif M: Heterogeneity and regulation of nicotinic acetylcholine receptors. *Int Rev Neurobiol* 34;25–131, 1992

Majak W: Alkaloid levels in a species of low larkspur and their stability in rumen fluid. *J Range Manage* 46;100–104, 1993

Majak W, Engelsjord M: Levels of a neurotoxic alkaloid in a species of low larkspur. *J Range Manage* 41;224–226, 1988

Majak W, McDiarmid RE: Alkaloid levels of a tall larkspur species in southwestern Alberta. *J Range Manage* 53;207–210, 2000

Majak W, McDiarmid RE, Benn MH: Isolation and HPLC determination methyllycaconitine in a species of low larkspur *(Delphinium nuttallianum)*. *J Agric Food Chem* 35;800–802, 1987

Majak W, Van Ryswyk AL, Hall JW: Soil moisture influences low larkspur and death camas alkaloid levels. *J Range Manage* 52;127–131, 1999

Manners GD, Pfister JA: Normal phase liquid chromatographic analysis of toxic norditerpene alakloids. *Phytochem Anal* 4;14–18, 1993

Manners GD, Pfister JA, Ralphs MH, Panter KE, Olsen JD: Larkspur chemistry: toxic alkaloids in tall larkspurs. *J Range Manage* 45;63–67, 1991

Manners GD, Panter KE, Ralphs MH, Pfister JA, Olsen JD, James LF: Toxicity and chemical phenology of norditerpenoid alkaloids in the tall larkspurs (*Delphinium* species). *J Agric Food Chem* 41;96–110, 1993

Manners GD, Panter KE, Pelletier SW: Sructure-activity relationships of norditerpenoid alkaloids occurring in toxic larkspur *(Delphinium)* species. *J Nat Prod* 58;863–869, 1995

Manners GD, Panter KE, Pfister JA, Ralphs MH, James LF: The characterization and structure-activity evaluation of toxic norditerpenoid alkaloids from two *Delphinium* species. *J Nat Prod* 61;1086–1089, 1998

Manske RHF, Ashford WR: The protoberberine alkaloids. In *The Alkaloids*, vol 4, Manske RHF, Holmes HL eds, Academic Press, New York, pp77–118, 1954

Marsh CD: *Stock-Poisoning Plants of the Range*. USDA Bull 1245, pp11–21, 1929

Marsh CD, Clawson AB, Marsh H: *Larkspur Poisoning of Livestock*. USDA Bull 365, 1916

Minakata H, Komura H, Nakanishi K, Kada T: Protoanemonin, an antimitogen isolated from plants. *Mutat Res* 116;317–322, 1983

Misra SB, Dixit SN: Antifungal principle of *Ranunculus sceleratus*. *Econ Bot* 34;362–367, 1980

Moore RHS: Poisoning by old man's beard *(Clematis vitalba)*? *Vet Rec* 89;569–570, 1971

Morales H: Abortions in a dairy herd in the VIII Region of Chile attributed to the consumption of creeping buttercup (*Ranunculus repens* L.). *Arch Med Vet* 21;163–166, 1989

Mori T, Murayama M, Bando H, Kawahara N: Studies on the constituents of *Aconitum* species. 12. Synthesis of jesaconitine derivatives and their analgesic and toxic activities. *Chem Pharm Bull* 39;379–383, 1991

Nachman RJ, Olsen JD: Ranunculin: a toxic constituent of the poisonous range plant bur buttercup *(Ceratocephalus testiculatus)*. *J Agric Food Chem* 31;1358–1360, 1983

Nation PN, Benn MH, Roth SH, Wilkens JL: Clinical signs and studies of the site of action of purified larkspur alkaloid, methyllycaconitine, administered parenterally to calves. *Can Vet J* 23;264–266, 1982

Natori S, Ikekawa N, Suzuki M: *Advances in Natural Products Chemistry*. Halstead Press, New York, pp357–367, 1981

Nelson SB: Feeding wild plants to sheep. *Annu Rep Bur Anim Ind* 15;421–425, 1898

Nicholson JA: Section of comparative medicine. *Proc Roy Soc Med* 26;1273–1276, 1933

Nielsen DB, Ralphs MH: Larkspur: economic considerations. In *The Ecology and Economic Impact of Poisonous Plants on Livestock Production*, James LF, Ralphs MH, Nielsen DB eds, Westview Press, Boulder, Colo, pp119–129, 1988

Olsen JD: Tall larkspur poisoning in cattle and sheep. *J Am Vet Med Assoc* 173;762–765, 1978

Olsen JD: Relationship of relative total alkaloid concentration and toxicity of duncecap larkspur during growth. *J Range Manage* 36;550–552, 1983

Olsen JD, Manners GD: Toxicology of diterpenoid alkaloids in rangeland larkspur (*Delphinium* spp.). In *Toxicants of Plant Origin,* vol 1, Cheeke PR ed, CRC Press, Boca Raton, Fla, pp291–326, 1989

Olsen JD, Sisson DV: Description of a scale for rating the clinical response of cattle poisoned by larkspur. *Am J Vet Res* 52;488–493, 1991a

Olsen JD, Sisson DV: Toxicity of extracts of tall larkspur *(Delphinium barbeyi)* in mice, hamsters, rats, and sheep. *Toxicol Lett* 56;33–41, 1991b

Olsen JD, Anderson TE, Madsen G: Bur buttercup. *Utah Sci* 43;11–13, 1982

Olsen JD, Anderson TE, Murphy JC, Madsen G: Bur buttercup poisoning of sheep. *J Am Vet Med Assoc* 183;538–543, 1983

Ono M, Satoh T: Pharmacological studies of lappaconitine: analgesia produced by intracerebroventricular, intracisternal, and intrathecal injections. *J Pharmacobio-Dyn* 13; 374–377, 1990

Pammel LH: *A Manual of Poisonous Plants.* Torch Press, Cedar Rapids, Iowa, pp447–469, 1911

Patil PN, Tye A, Nelson JM, Beal JL: A study of the alkaloids of *Thalictrum.* 2. Pharmacology of an extract of *Thalictrum revolutum. Lloydia* 26;229–235, 1963

Pelletier SW, Kulanthaivel P, Olsen JD: Alakloids of *Delphinium barbeyi. Phytochemistry* 28;1521–1525, 1989

Pfister JA, Gardner DR: Consumption of low larkspur (*Delphinium nuttallianum*) by cattle. *J Range Manage* 52;378–383, 1999

Pfister JA, Manners GD: Mineral salt supplementation of cattle grazing tall larkspur–infested rangeland during drought. *J Range Manage* 44;105–11, 1991

Pfister JA, Manners GD: Effects of carbachol administration in cattle grazing tall larkspur–infested range. *J Range Manage* 48;343–349, 1995

Pfister JA, Manners GD, Ralphs MH, Hong ZX, Lane MA: Effects of phenology, site, and rumen fill on tall larkspur consumption by cattle. *J Range Manage* 41:509–514, 1988a

Pfister JA, Ralphs MH, Manners GD: Cattle grazing tall larkspur on Utah mountain rangeland. *J Range Manage* 41;118–122, 1988b

Pfister JA, Adams DC, Arambel MJ, Olsen JD, James LF: Sublethal levels of toxic larkspur: effects on intake and rumen dynamics in cattle. *Nutr Rep Int* 40;629–636, 1989

Pfister JA, Provenza FD, Manners GD: Ingestion of tall larkspur by cattle: separating effects of flavor from postingestive consequences. *J Chem Ecol* 16;1697–1705, 1990

Pfister JA, Ralphs MH, Manners GD, Panter KE, James LF, Stegelmeier BL, Gardner DR: Tall larkspur poisoning in cattle: current research and recommendations. *Rangelands* 15;157–160, 1993

Pfister JA, Panter KE, Manners GD: Effective dose in cattle of toxic alkaloids from tall larkspur *(Delphinium barbeyi). Vet Hum Toxicol* 36;10–11, 1994a

Pfister JA, Panter KE, Manners GD, Cheney CD: Reversal of tall larkspur *(Delphinium barbeyi)* poisoning in cattle with physostigmine. *Vet Hum Toxicol* 36;511–514, 1994b

Pfister JA, Gardner DR, Price KW: Grazing risk on tall larkspur–infested ranges. *Rangelands* 19;12–15, 1997a

Pfister JA, Ralphs MH, Manners GD, Gardner DR, Price KW, James LF: Early season grazing by cattle of tall larkspur–(*Delphinium* spp.) infested rangeland. *J Range Manage* 50;391–398, 1997b

Pfister JA, Cheney CD, Gardner DR, Manners GD: Mineral-salt supplement does not attenuate tall larkspur (*Delphinium barbeyi*) toxicosis in cattle. *J Range Manage* 51;566–569, 1998

Pfister JA, Gardner DR, Panter KE, Manners GD, Ralphs MH, Stegelmeier BL, Schoch TK: Larkspur (*Delphinium* spp.) poisoning in livestock. *J Nat Toxins* 8;81–94, 1999

Piekarz J: Buttercup poisoning in a horse. *Med Weter* 37;658, 1981

Poulsson E: Untersuchungen über *Caltha palustris. Arch Exp Pathol Pharmakol* 80;173–182, 1917

Pui-Hay P, Tai Y-T, Young K: Three fatal cases of herbal aconite poisoning. *Vet Hum Toxicol* 36;212–215, 1994

Ralphs MH: Persistence of aversions to larkspur in naive and native cattle. *J Range Manage* 50;367–370, 1997

Ralphs MH, Olsen JD: Adverse influence of social facilitation and learning context in training cattle to avoid eating larkspur. *J Anim Sci* 68;1944–1952, 1990

Ralphs MH, Olsen JD: Prior consumption by sheep reduces waxy larkspur consumption by cattle: an observation. *J Range Manage* 45;136–139, 1992

Ralphs MH, Olsen JD: Conditioned food aversion: a management tool to prevent livestock poisoning. In *Toxic Plants and Other Natural Toxicants,* Garland T, Barr AC eds, CAB International, New York, pp227–232, 1998

Ralphs MH, Olsen JD, Pfister JA, Manners GD: Plant-animal interactions in larkspur poisoning in cattle. *J Anim Sci* 66;2334–2342, 1988

Ralphs MH, Bowns JE, Manners GD: Utilization of larkspur by sheep. *J Range Manage* 44;619–622, 1991

Ralphs MH, Jensen DT, Pfister JA, Nielsen DB, James LF: Storms influence cattle to graze larkspur: an observation. *J Range Manage* 47;275–278, 1994

Ralphs MH, Jones WA, Pfister JA: Damage from the larkspur mirid deters cattle grazing of larkspur. *J Range Manage* 50;371–373, 1997a

Ralphs MH, Manners GD, Pfister JA, Gardner DR, James LF: Toxic alkaloid concentration in tall larkspur species in the western U.S. *J Range Manage* 50;497–502, 1997b

Ralphs MH, Manners GD, Gardner DR: Influence of light and photosynthesis on alkaloid concentration in larkspur. *J Chem Ecol* 24;167–182, 1998a

Ralphs MH, Gardner DR, Jones WA, Manners GD: Diterpenoid alkaloid concentration in tall larkspur plants damaged by larkspur mirid. *J Chem Ecol* 24;829–840, 1998b

Sargent PB: The diversity of neuronal nicotinic acetylcholine receptors. *Annu Rev Neurosci* 16;403–443, 1993

Scherf D: Studies on auricular tachycardia caused by aconitine administration. *Proc Soc Exp Biol Med* 64;233–239, 1947

Schiff PL Jr, Doskotch RW: *Thalictrum* alkaloids. *Lloydia* 33;403– 452,1970

Seegal BC, Holden M: The antibiotic activity of extracts of Ranunculaceae. *Science* 101;413–414, 1945

Shearer GD: Some observations on the poisonous properties of buttercups. *Vet J* 94;22–32, 1938

Shoppee CW: *Chemistry of the Steroids,* 2d ed. Butterworths, Washington, DC, pp326–397, 1964

Shropshire CM, Stauber E, Arai M: Evaluation of selected plants for acute toxicosis in budgerigars. *J Am Vet Med Assoc* 200;936–939, 1992

Singh B, Rastogi RP: Cardenolides—glycosides and genins. *Phytochemistry* 9;315–331, 1970

Sizelove W, McIlaim EH, Burrows GE: Possible larkspur intoxications responsible for acute deaths in cattle [letter]. *Vet Hum Toxicol* 34;458–459, 1992

Smit PG: A revision of *Caltha* (Ranunculaceae). *Blumea* 29:119–150, 1973

Stermitz FR, Adamovics JA: Alkaloids of *Caltha leptosepala* and *Caltha biflora. Phytochemistry* 16;500, 1977

Stern ES: The *Aconitum* and *Delphinium* alkaloids. In *The Alkaloids,* vol 4, Manske RHF, Holmes HL eds, Academic Press, New York, pp275–333, 1954

Strichartz G, Rando T, Wang GK: An integrated view of the molecular toxinology of sodium channel gating in excitable cells. *Annu Rev Neurosci* 10;237–267, 1987

Therrien HP, Hidiroglou M, Charette LA: Note on the toxicity of tall buttercup (*Ranunculus acris* L.) to cattle. *Can J Anim Sci* 42;123–124, 1962

Thimann KV, Bonner WDjr: Inhibition of plant growth by protoanemonin and coumarin and its prevention by BAL. *Proc Nat Acad Sci* 35;272–276, 1949

Tjon Sie Fat L: Contribution to the knowledge of cyanogenesis in angiosperms. 14.

Cyanogenesis in Ranunculaceae. *Proc Ned Akad Wet,* ser C, 82;197–209, 1979 (*Chem Abstr* 91:71718e,1979)

Tomassoni AJ, Snook CP, McConville BJ, Siegel EG: Recreational use of *Dephinium*—an ancient poison revisited [abstract]. *J Toxicol Clin Toxicol* 34;598, 1996

Tschesche R, Welmar K, Wulff E, Snatzke G: On glycosides with lactone-forming aglycones. 6. About subsequent products of a still unknown genuine precursor of ranunculin in Ranunculaceae. *Chem Ber* 105;290–300, 1972

Ward JM, Cockroft VB, Lunt GG, Smilie FS, Wonnacott S: Methyllycaconitine: a selective probe for neuronal α-bungarotoxin binding sites. *FEBS Lett* 270;45–48, 1990

Warnock MJ: *Delphinium.* In *Flora of North America,* vol 3, *Magnoliophyta: Magnoliidae and Hamamelidae,* Flora of North America Editorial Committee eds, Oxford Univ Press, New York, pp196–240, 1997

Watt JM, Breyer-Brandwijk MG: *The Medicinal and Poisonous Plants of Southern and Eastern Africa,* 2d ed, E & S Livingston, Edinburgh, pp877–882, 1962

Whittemore AT: *Ranunculus.* In *Flora of North America,* vol 3, *Magnoliophyta: Magnoliidae and Hamamelidae,* Flora of North America Editorial Committee eds, Oxford Univ Press, New York, pp88–135, 1997

Whittemore AT, Parfitt BD: Ranunculaceae. In *Flora of North America,* vol 3, *Magnoliophyta: Magnoliidae and Hamamelidae,* Flora of North America Editorial Committee eds, Oxford Univ Press, New York, pp85–271, 1997

Wilcox EV: *Larkspur Poisoning in Sheep.* Mont Agric Exp Stn Bull 15, 1897

Wilcox EV: *Cattle poisoning by the tall larkspur.* Mont Agric Exp Stn Bull 22, pp45–47, 1899

Williams MC, Cronin EH: Five poisonous range weeds—when and why they are dangerous. *J Range Manage* 19;274–279, 1966

Winters JB: Severe urticarial reaction in a dog following ingestion of tall field buttercup. *Vet Med/Small Anim Clin* 71;307,1976

Wissner W, Kating H: Untersuchungen uber die hellebrinfuhrung der unterirdischen organe von *Helleborus-arten. Planta Med* 20;344–349, 1971

Wu WN, Beal JL, Doskotch RW: Alkaloids of *Thalictrum.* 22. Isolation of alkaloids with hypotensive and antimicrobial activity from *Thalictrum revolutum. Lloydia* 40;508–514, 1977

63 Rhamnaceae Juss.

■ Buckthorn Family

Karwinskia
Rhamnus

Cosmopolitan in distribution but exhibiting greatest diversity in the tropics and subtropics, the Rhamnaceae, commonly known as the buckthorn family, comprises 45–58 genera and about 900 species. It is a source of dyes, purgatives and other medicines, soap substitutes, and edible fruits, such as the jujube, or Chinese date. Species of the American genus *Ceanothus* (California lilac) are prized ornamentals. In addition, the flowers are good sources of nectar, and the fruits are an important winter food for thrushes and small mammals. In North America, approximately 17 genera and 120 species and hybrids are present. Intoxication problems are associated with 2 genera.

■ woody vines, shrubs, or trees; leaves simple, alternate; petals typically clawed and hooded

Plants trees or shrubs or woody vines; armed or not armed with thorns; bearing perfect flowers or polygamodioecious. **Stems** erect or twining or climbing by hooks or tendrils. **Leaves** simple; alternate; venation pinnate or palmate; stipules minute or spinose, caducous. **Inflorescences** simple cymes or panicles or corymbs or umbels or solitary flowers; terminal or axillary. **Flowers** perfect or imperfect, similar; perianths in 2- or rarely 1-series. **Sepals** 5 or 4; free. **Corollas** radially symmetrical. **Petals** 5 or 4 or rarely absent; free; clawed or not clawed; spathulate or cucullate. **Stamens** 5 or 4; opposite the petals; anthers typically partially enclosed by petal hoods; epipetalous. **Pistils** 1; compound, carpels 2–4; stigmas 1, 2-to-4-lobed; styles 1; ovaries superior or inferior; locules 2–4; placentation axile; ovules 1 or 2 per locule. **Hypanthia** absent or present; tubular or cup-shaped or annular. **Fruits** drupes or capsular schizocarps or septicidal capsules. **Seeds** 3–10; indurate or cartilaginous; often with dorsal groove.

Karwinskia Zucc.

■ *K. humboldtiana*
K. parvifolia

buckthorn	gallita bush
cacachila	margarita
cacatsi	negrito
callotio	palo negrito
capulín tullidor	pimientillo
capulincillo	riventdore
cayote	tanglefoot
coyotillo	tullidora
coyotio	wild cherry

■ shrubs or small trees; leaves with venation pinnate, small translucent glands; sepals 5; petals 5, clawed; fruits drupes, purple to black, with 2 or 3 stones

Taxonomy and Morphology—Native to the New World, *Karwinskia* comprises 11 or 12 species (Johnston 1966). Two species are present in our range:

K. humboldtiana (Roem. ex Schult.) Zucc.	tullidora, coyotillo, buckthorn, capulín tullidor, capulincillo, wild cherry, cacatsi, callotio, coyotio, cayote, riventdore, margarita, cacachila, gallita bush, tanglefoot, negrito, pimientillo, palo negrito
K. parvifolia Rose	coyotillo

Plants shrubs or small trees; not armed with thorns. **Stems** erect 100–300 cm tall. **Leaves** alternate or subopposite; sessile or short petiolate; venation pinnate; blades ovate to oblong-elliptic or oblong, with small translucent glands, abaxial veins with dark and light flecks; margins entire or rarely crenate. **Inflorescences** axillary cymes or umbels or rarely solitary flowers. **Flowers** perfect. **Sepals** 5; deltoid; deciduous or persistent in fruit. **Petals** 5; clawed; cucullate. **Stamens** 5; filaments subulate. **Pistils** with carpels 2 or 3; ovaries superior; locules 2 or 3; ovules 2 per locule. **Hypanthia**

turbinate or cup-shaped. **Fruits** drupes; ovoid to subglobose; purple to black; stones 2 or 3. **Seeds** 1 per stone.

1. Leaf blades 4–8 cm long; apices acute to acuminate; surfaces glabrous. Sepals persistent. Drupes 7–10 mm long . . . . . . *K. humboldtiana*
1. Leaf blades 1–4 cm long; apices obtuse or rounded; surfaces puberulent to pubescent. Sepals deciduous. Drupes 4–6 mm long . . . . *K. parvifolia*

Taxonomists question whether *K. humboldtiana* and *K. parvifolia* are distinct species (Johnston 1966). Morphologically the 2 taxa appear to intergrade. Following Johnston's treatment, they are maintained as discrete taxa here.

Distribution and Habitat—The most widely distributed species of the genus, *K. humboldtiana* occurs in arroyos, bajadas, and rocky hillsides in two disjunct regions (Johnston 1966). A geographical gap separates populations occurring in northeastern and northwestern Mexico. Partially occupying the gap in the distribution of *K. humboldtiana, K. parvifolia* occurs on dry hillsides. This distributional pattern is believed by some taxonomists to be further evidence for a possible relationship with *K. humboldtiana.*

Karwinskia humboldtiana

Karwinskia parvifolia

- plains and brushy areas
- seeds neurotoxic, progressive neuropathy with ascending paralysis, flaccid quadriplegia, all animal species but mainly a problem in children

Disease Problems—The toxicity of *K. humboldtiana* has been recognized for centuries, and some Indians of Mexico are reported to have made a practice of eating the sweet fruits while carefully avoiding the very toxic seeds (Marsh et al. 1928; Weller et al. 1980). Furthermore, it was recognized that the signs of intoxication were delayed in onset, occurring several or more days after consumption of the fruit. The resulting paralysis, although not always permanent, was nevertheless a much feared debilitating disease. The name *tullidora*, which means "crippled" or "maimed," refers to its overall effects (Weller et al. 1980). Interestingly, an extract from the roots is reputed to counteract the paralysis if given immediately upon onset of signs of intoxication (Pesman 1962).

In Mexico, *K. humboldtiana* continues to be a very serious threat for humans, mainly because the fruits are sweet and very attractive to children who, unlike knowledgeable adults, eat the seeds with the fruit (Carrada-Bravo et al. 1983; Ascherio et al. 1992). Most cases are reported in the states of Nuevo León, San Luis Potosí, and Tamaulipas, where there may be 8–10 cases annually (Bermudez et al. 1986; Bermudez et al. 1995). A related species, *K. calderoni,* is the cause of similar disease in Central America.

The disease is a progressive symmetric polyneuropathy with an ascending flaccid paralysis similar to the polyradiculoneuritis of the Guillain-Barré syndrome. The end result varies from death by respiratory arrest or bulbar paralysis to slow recovery (Bermudez et al. 1986). In one case a 5-year-old girl exhibited gradual onset of flaccid quadriplegia and respiratory insufficiency several weeks after eating the fruit and seeds (Calderon-Gonzalez & Rizzi-Hernandez 1967). Even though there was loss of peripheral segmental myelin, suggestive of a primary Schwann cell effect, there was gradual recovery of limb function. In an apparent outbreak type of problem in Nicaragua, 10 children from rural areas developed signs of ascending symmetric paralysis despite being in general good health (Ascherio et al. 1992). Poliomyelitis was the initial diagnosis but was not confirmed by tests. Subsequently it was discovered that all of the children had eaten the fruit of *K. calderoni* 1–3 weeks earlier. In experimental studies with goats using fruits, dosage at 0.025–0.15% b.w. produced, in addition to obvious neurologic effects, microscopic and serum enzymatic evidence of

Karwinskia humboldtiana

Karwinskia parvifolia

- series of complex anthracenones T-514, T-544; progressive degeneration of large motor peripheral nerves; dying-back syndrome
- disruption of axoplasmal transport, degeneration of distal axonal segment initially, followed by myelin degeneration

Tullidonol (T-544)

widespread muscular degeneration, including heart and all major skeletal muscle groups (Dewan et al. 1965; Dollahite and Henson 1965). In rare cases in which large amounts of the plants are eaten, death occurs without the development of neurologic signs. The toxin appears to affect any animal eating the seeds—humans, goats, cattle, sheep, cats, rats, chickens, and monkeys (Marsh et al. 1928; Joiner et al. 1975).

Disease Genesis—The cause of the disease was initially isolated as a polyhydroxy aldehyde with a molecular weight of 544 (Kim & Camp 1972). It is now known that a series of four or more complex anthracenones is found principally in the seeds, especially when the fruit is immature (Dreyer et al. 1975). Based upon their molecular weights, the toxins have been designated T-496, T-514, T-516, and T-544 (tullidinol). T-544 is generally the most abundant, with concentrations of 1% or more, and T-514 is the most toxic, several times more so than the others (Bermudez et al. 1986; Flores-Otero et al. 1987; Guerrero et al. 1987). T-544 is a mixture of four isomers, one of which is karwinol A (Kim & Stipanovic 1998).

The progressive paralysis characteristic of the disease is associated with degenerative changes in peripheral nerves. In an extensive series of experiments involving goats, segmental myelin and Wallerian degeneration were demonstrated in large and medium motor nerves, especially distal sites of long axons (Charlton & Pierce 1970a–c; Charlton et al. 1970, 1971). It was postulated that Schwann cell swelling and degeneration occurred first, leading to demyelination and axonal degeneration, although even at that time stasis of axonal flow was suggested. It was also noted that metabolic deficits, with mitochondrial degeneration and possibly a decrease in ATP production, were apparent, in addition to many obvious intra-axonal structural changes and decreases in neuronal conduction velocity. Further studies were supportive of the role of mitochondrial respiration deficits (with uncoupling of oxidative phosphorylation) and the preeminence of Schwann cell injury (Wheeler & Camp 1971; Mitchell et al. 1978). However, in other studies it became apparent that axonal changes were not necessarily coupled to or secondary to Schwann cell injury.

A decrease in axoplasmal transport was suggested as an alternative mechanism that caused axonal degeneration (Munoz-Martinez & Chavez 1979). Experiments in cats demonstrated a 50% or more reduction in myelin protein in motor nerves and a 25% reduction in cutaneous nerves (Aoki & Munoz-Martinez 1981). Maximal reduction occurred in the distal portion of the sciatic nerve. It was concluded that the toxin entered through the motor nerve terminus, and degeneration began distally. The effects on neuronal synthesis and axonal transport of proteins in long axons were demonstrated in these and further studies (Aoki & Munoz-Martinez 1981; Munoz-Martinez et al. 1981; Heath et al. 1982; Munoz-Martinez et al. 1984). In mice exposed to T-496 and T-544 for 48 hours to 14 days, the sequence of effects began with axonal changes, including focal swelling, axoplasmal disruption, Wallerian degeneration, and later myelin degeneration (Heath et al. 1982). Eventually axons were shown to be affected first and myelin secondarily, with axons and Schwann cells perhaps affected independently (Hernandez-Cruz & Munoz-Martinez 1983; Munoz-Martinez et al. 1983). Degeneration of the axon begins distally at the myoneural junction and extends retrograde to the proximal portion. The more highly branched, large motor nerves of the pelvic limbs seem to be more susceptible to the axoplasmal transport deficit. The small branched nerves of the diaphragm and those to the eye muscles are affected little if at all. In some cases of long duration, there is impulse transmission presumably between degenerated/demyelinated axons, resulting in multiple discharges and spontaneous firing (Hernandez-Cruz & Munoz-Martinez 1984). Axonal regeneration and remyelination occur in some animals, apparently accounting for the reported recoveries from the disease (Charlton & Pierce 1970a; Joiner et al. 1975).

The kinetics of T-514, as determined in rabbits, shows a moderately long half-time in serum of 3.5 hours, a volume of distribution of 5.61 l/kg b.w., and a clearance rate of 18.7 ml/min per kg b.w. (Adame-Reyna et al. 1994). The relatively rapid clearance suggests that either some of the toxin is strongly bound in the neurons or that damage to the neuron is quickly initiated but takes several weeks for sufficient degeneration to occur to cause signs of intoxication. Higher dosage then yields more-rapid effects on other tissues where signs of degeneration occur more quickly.

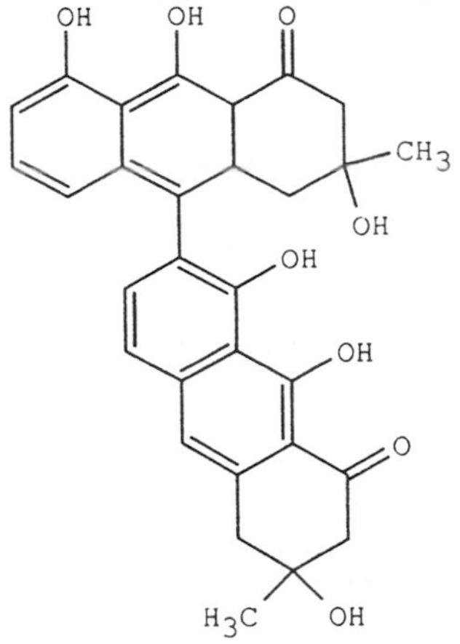

OH OH O CH3 OH OH OH O H3C OH

T-514

Although it was originally thought that the seeds were singularly toxic, it is now known that all plant parts are so—leaves the least, fruit pulp intermediate, and seeds the most (Marsh et al. 1928). In addition, there may be considerable seasonal and geographical variation in toxicant concentration (Guerrero et al. 1987). Interestingly, toxicity of the seeds appears to decrease several-fold with maturation. This occurs in spite of the relatively constant levels of T-514 and T-544 (Bermudez et al. 1986).

The toxins are cumulative in effect, although a single dose of 0.1–0.2% b.w. of fruits and seeds may be lethal in cattle and sheep (Marsh et al. 1928). Goats, guinea pigs, and chickens require slightly higher dosage. The leaves, being less toxic, must be eaten daily for several weeks at approximately 1% b.w. to produce severe illness or death. Whether by single or repeated dosage, the onset of signs of intoxication typically occurs 1–2 weeks after initial ingestion. Although the disease is primarily of neurologic importance because of axonal degeneration, there are other neurologic changes and important extraneural effects (Ortiz et al. 1992; Pineyro-Lopez et al. 1994). The toxins have marked effects on the liver and lungs in vitro and in vivo, and acute deaths following large doses may be due to hepatic or respiratory effects (Bermudez et al. 1986, 1992; Garza-Ocanas et al. 1992). Both T-514 and T-544 cause hemorrhage and necrosis of the liver and death in a few days at dosages of the fruit of 2–5 mg/kg b.w. in mice (Bermudez et al. 1986).

The toxins also have been shown to be teratogenic and embryotoxic in mice (Martinez de Villarreal et al. 1990). One-half of the LD_{50} of T-544, but not T-514, given at day 8 of gestation resulted in an increased incidence of fetal resorptions and malformations in mice (Valadez-Alvizo et al. 1993).

Clinical Signs—There is gradual onset of signs several days to weeks after initial ingestion of *Karwinskia,* beginning with increased alertness, hypersensitivity to touch and sounds, and tremors. These are followed by weakness of the legs with a stumbling/staggering gait, then dragging of the pelvic limbs, and finally loss of motor control and recumbency. In some animals there may be spasmodic leg movements and a stiff/stilted gait with exaggerated leg actions. Although they may be paralyzed, sensory and autonomic functions are not compromised, and thus eating, drinking, rumination, urination, and defecation are not adversely affected. In monkeys there is abnormal high-stepping involving the pelvic limbs and rapid progression to flaccid paralysis (Joiner et al. 1975). Cerebrospinal fluid analysis results will be within normal values for protein and cells.

- animals, gradual onset, days to weeks; increased alertness, hypersensitivity, tremors, weakness of limbs, loss of motor control, collapse

humans, ascending paralysis, tendon reflexes absent, sensory normal, decreased motor conduction, normal CSF

In humans, mainly young children, there is gradual onset of ascending systemic paralysis beginning with the pelvic limbs and ascending to the upper body over 3–7 days. Tendon reflexes are absent in the affected areas, but there are no sensory deficits and no cranial nerve involvement. Motor nerve conduction velocities will be markedly reduced. Cerebrospinal fluid examination will reveal normal values for glucose and proteins and a lack of cells. Diagnosis can be aided by peripheral nerve biopsy and especially by analysis of blood, using a thin-layer chromatographic method for detection of the toxins (T-514) (Bermudez et al. 1995; Martinez et al. 1998).

■ gross pathology, few lesions

microscopic, Wallerian degeneration, demyelination, Schwann cell swelling and degeneration of large motor axons

Pathology—There are few gross pathologic lesions other than scattered petechia and congestion of lungs, liver, and kidneys. Microscopically, there is mild renal tubular degeneration, myocardial degeneration, pulmonary congestion, and hepatocellular necrosis. The lesions of primary interest are those of the peripheral nervous system. The large motor axons (sciatic) are preferentially affected; Wallerian degeneration, demyelination, and swelling and degeneration of Schwann cells are the most pronounced changes. Swollen axons of the white and gray areas of the spinal cord and swollen Purkinje's cells of the cerebellum are also seen. There may also be evidence of remyelination and axonal regeneration.

When large doses of *Karwinskia* are consumed and death occurs quickly, hepatic necrosis and pulmonary congestion are the most consistent findings.

■ general nursing care

Treatment—Little can be done to counteract the effects of the toxins, although daily administration of thiamine hydrochloride is reported to be of some benefit (Weller et al. 1980). The primary hope is to nurse the individuals through the paralytic period to allow nerve regeneration and/or compensatory adjustment for a persisting neurologic deficit.

Rhamnus L.

■ *R. californica*
R. caroliniana
R. cathartica
R. frangula
R. purshiana

buckthorn
cascara
cascara sagrada
chittam bark
coffeeberry
hartshorn
Indian cherry
pigeonberry
sacred bark

Taxonomy and Morphology—Native to both the Old World and the New World, *Rhamnus,* commonly known as buckthorn, comprises more than 125 species divided into 2 well-marked subgenera (Johnston & Johnston 1978). Some taxonomists treat the 2 subgenera as distinct genera. Members of the genus are planted occasionally as ornamentals and have been the source of dyes, medicine, and charcoal. The fruits are eaten by birds and small mammals. In North America, some 17 native and introduced species are present. Representative and commonly encountered are the following:

R. californica Eschsch.	California buckthorn, coffeeberry, pigeonberry
R. caroliniana Walter	Carolina buckthorn, Indian cherry
R. cathartica L.	common buckthorn, hartshorn, purging buckthorn
R. frangula L. (= *Frangula alnus* Mill.)	alder buckthorn, glossy buckthorn
R. purshiana DC. (= *Frangula purshiana* [DC.] Cooper)	cascara, cascara sagrada, cascara buckthorn, chittam bark, sacred bark

Specific epithets ending in *-us* rather than *-a* are encountered in some taxonomic works. Adhering to a strict interpretation of the rule of botanical nomenclature, their authors contend that the *-us* should be used to make the specific epithet and generic name agree in gender. The rules, however, provide for exceptional endings in *Rhamnus* and several other genera of familiar trees.

■ shrubs or trees; leaf venation pinnate; petals clawed, greenish or yellowish white; fruits drupes, brown, orange-red, purplish black, with 2–4 stones

Plants trees or shrubs; deciduous or evergreen; armed or not armed with thorns; bearing perfect flowers or polygamodioecious. **Stems** erect; 1–10 m tall; branchlets glabrous or indumented; buds with or without bud scales. **Leaves** alternate or rarely

Rhamnus californica

Rhamnus caroliniana

■ chaparral, woodlands, and forests

Rhamnus californica

Rhamnus caroliniana

Rhamnus cathartica

subopposite; petiolate; venation pinnate; blades ovate to oblong or elliptic to obovate; margins entire or serrate or dentate; stipules small; deciduous. **Inflorescences** umbels or solitary flowers; axillary; sessile or peduncled. **Flowers** perfect or imperfect; perianths in 2- or 1-series. **Sepals** 5 or 4; triangular. **Petals** 5 or 4 or absent; clawed; cucullate; greenish or yellowish white. **Stamens** 5 or 4. **Pistils** with carpels 2–4; styles 1, not divided or 2–4 divided; ovaries superior; locules 2–4; ovules 1 per locule. **Hypanthia** turbinate or cup-shaped. **Fruits** drupes; oblong or globose; brown to orange-red to purplish black; stones 2–4. **Seeds** 1 per stone.

Because the species cited are merely representative of the genus, a taxonomic identification key is not presented. If identification is needed, floristic treatments for the reader's area should be consulted.

Distribution and Habitat—*Rhamnus cathartica* and *R. frangula,* of Eurasian origin, are encountered mainly as ornamentals, although both have escaped cultivation and become established in wet soils of the northeastern states and Canada. *Rhamnus cathartica* occurs almost as far west as the eastern Great Plains. *Rhamnus purshiana* is found up to elevations of 2,000 m in coniferous forests and chaparral, and *R. californica* is present in chaparral, woodlands, and coniferous forests, both along the Pacific Coast. Typically occurring as a forest understory species, *R. caroliniana* is distributed primarily in the southeastern quarter of the continent.

■ digestive disturbance, bark mainly, fruit and leaves less potent

Disease Problems—Species of *Rhamnus* are known mainly for their purgative effects. Their bark is an especially rich source of purgatives. Perhaps best known is *R. purshiana,* which yields cascara sagrada, or sacred bark, and has long been used as a cathartic for horses. The fruits and, to a lesser extent, the leaves also have laxative properties. Because such a large amount of leaves is required to produce purgation, they appear to be of little hazard.

- glycosidic anthrones, anthraquinones, hydrolyzed to aloe-emodin; cascara sagrada

Disease Genesis—The purgative action of *Rhamnus* is due to anthrones/anthraquinones (Thomson 1971). The parent cascaroside glycosides are hydrolyzed to yield the active aglycone aloe-emodin and other compounds (Thomson 1971; Cooper & Johnson 1984). Little is known of the toxicity potential of the other species of the genus.

- nausea, vomiting, colic, diarrhea

Clinical Signs—The signs produced by *Rhamnus* are those expected with potent purgatives and include nausea, vomiting, colic, and watery/bloody diarrhea, which occur within a few hours after the plants are eaten.

- reddened gut mucosa

treat with activated charcoal, fluids, electrolytes

Pathology and Treatment—In the rare event that the genus causes death, reddened mucosa of the gastrointestinal tract would be the main lesion expected. Treatment is symptomatic and includes the use of activated charcoal and demulcents orally and fluids and electrolytes orally and/or parenterally.

Rhamnus frangula

Rhamnus purshiana

OH O OH

CH_2OH

O-Sugars

Cascarosides

OH O OH

HO OH

O

Emodin

References

Adame-Reyna M, Guerrero-Olazaran M, Belmontes-Hernandez R, Viader-Salvado JM: Pharmacokinetic design for use in preclinical phase: application to the study of anthracenone. *Rev Invest Clin* 46;53–58, 1994

Aoki K, Munoz-Martinez EJ: Quantitative changes in myelin proteins in a peripheral neuropathy caused by tullidora *(Karwinskia humboldtiana)*. *J Neurochem* 36;1–8, 1981

Ascherio A, Bermudez CS, Garcia D: Outbreak of buckthorn paralysis in Nicaragua. *J Trop Pediatr* 38;87–89, 1992

Bermudez MV, Gonzalez-Spencer D, Guerrero M, Waksman N, Pineyro A: Experimental intoxication with fruit and purified toxins of buckthorn *(Karwinskia humboldtiana)*. *Toxicon* 24;11–12, 1986

Bermudez MV, Martinez FJ, Salazar ME, Waksman N, Pineyro A: Experimental acute intoxication with ripe fruit of *Karwinskia humboldtiana* (tullidora) in rat, guinea-pig, hamster, and dog. *Toxicon* 30;1493–1496, 1992

Bermudez MV, Lozano-Melendez FE, Tamez-Rodriguez VA, Diaz-Cuello G, Pineyro-Lopez A: Frequency of intoxication with *Karwinskia humboldtiana* in Mexico. *Salud Publica Mex* 37;57–62, 1995

Calderon-Gonzalez R, Rizzi-Hernandez H: Buckthorn polyneuropathy. *N Engl J Med* 277;69–71, 1967

Carrada-Bravo T, Lopez LH, Vasquez AG, Ley-Lopez A: Brote epidemico de poliradiculoneuritis por tullidora *Karwinskia humboldtiana*. *Bol Med Hosp Infant Mex* 40;139–147, 1983

Charlton KM, Pierce KR: A neuropathy in goats caused by experimental coyotillo *(Karwinskia humboldtiana)* poisoning. 2. Lesions in the peripheral nervous system. *Pathol Vet* 7;385–407, 1970a

Charlton KM, Pierce KR: A neuropathy in goats caused by experimental coyotillo *(Karwinskia humboldtiana)* poisoning. 3. Distribution of lesions in peripheral nerves. *Pathol Vet* 7;408–419, 1970b

Charlton KM, Pierce KR: A neuropathy in goats caused by experimental coyotillo *(Karwinskia humboldtiana)* poisoning. 4. Light and electron microscopic lesions in peripheral nerves. *Pathol Vet* 7;420–434, 1970c

Charlton KM, Pierce KR, Storts RW, Bridges CH: A neuropathy in goats caused by experimental coyotillo *(Karwinskia humboldtiana)* poisoning. 5. Lesions in the central nervous system. *Pathol Vet* 7;435–447, 1970

Charlton KM, Claborn LD, Pierce KR: A neuropathy in goats caused by experimental coyotillo *(Karwinskia humboldtiana)* poisoning: clinical and neurophysiologic studies. *Am J Vet Res* 32;1381–1389, 1971

Cooper MR, Johnson AW: *Poisonous Plants in Britain and Their Effects on Animals and Man.* Ref Book 161. Her Majesty's Stationery Office, London, pp203–204, 1984

Dewan ML, Henson JB, Dollahite JW, Bridges CH: Toxic myodegeneration in goats produced by feeding mature fruits from the coyotillo plant *(Karwinskia humboldtiana). Am J Pathol* 46;215–226, 1965

Dollahite JW, Henson JB: Toxic plants as the etiologic agent of myopathies in animals. *Am J Vet Res* 26;749–752, 1965

Dreyer DL, Arai I, Bachman CD, Anderson WR Jr, Smith RG, Daves GD Jr: Toxins causing noninflammatory paralytic neuronopathy: isolation and structure elucidation. *J Am Chem Soc* 97;4985– 4990, 1975

Flores-Otero G, Cueva J, Munoz-Martinez EJ, Rubio-Franchini C: Spectrophotometric and chromatographic detection of *Karwinskia humboldtiana* (tullidora) toxin in rat serum after tullidora ingestion. *Toxicon* 25;419–426, 1987

Garza-Ocanas L, Hsieh GC, Acosta D, Torres-Alanis O: Toxicity assessment of toxins T-514 and T-544 of buckthorn *(Karwinskia humboldtiana)* in primary skin and liver cell cultures. *Toxicology* 73;191–201, 1992

Guerrero M, Pineyro A, Waksman N: Extraction and quantification of toxins from *Karwinskia humboldtiana* (tullidora). *Toxicon* 25;565–568, 1987

Heath JW, Ueda S, Bornstein MB, Daves GD Jr, Raine CS: Buckthorn neuropathy in vitro: evidence for a primary neuronal effect. *J Neuropathol Exp Neurol* 41;204–220, 1982

Hernandez-Cruz A, Munoz-Martinez EJ: Distal reduction of the conduction velocity of alpha-axons in tullidora (buckthorn) neuropathy. *Exp Neurol* 82;335–343, 1983

Hernandez-Cruz A, Munoz-Martinez EJ: Axon-to-axon transmission in tullidora (buckthorn) neuropathy. *Exp Neurol* 84;533–548, 1984

Johnston MC: Systematic studies in the plant genus *Karwinskia* in Mexico and Central America. *Year Book Am Philos Soc* 1966;351–357, 1966

Johnston MC, Johnston LVA: *Rhamnus.* In *Flora Neotropica,* Monograph 20, Organization for Flora Neotropica, New York, 1978

Joiner GN, Russell LH, Bush DE, Gleiser CA, Johnston TD, Fedigan L, Fedigan L: A spontaneous neuropathy of free-ranging Japanese macaques. *Lab Anim Sci* 25;232–237, 1975

Kim HL, Camp BJ: Isolation of a neurotoxic substance from *Karwinskia humboldtiana* Zucc. (Rhamnaceae). *Toxicon* 10;83–84, 1972

Kim HL, Stipanovic RD: Isolation of karwinol A from coyotillo *(Karwinskia humboldtiana)* fruits. In *Toxic Plants and Other Natural Toxicants,* Garland T, Barr AC eds, CAB International, New York, pp440–446, 1998

Marsh CD, Clawson AB, Roe GC: *Coyotillo* (Karwinskia humboldtiana) *as a poisonous plant.* USDA Tech Bull 29, 1928

Martinez HR, Bermudez MV, Rangel-Guerra RA, Flores LL: Clinical diagnosis in *Karwinskia humboldtiana* polyneuropathy. *J Neurol Sci* 154;49–54, 1998

Martinez de Villarreal L, Velazco-Campos R, Pineyro Lopes A, Gonzalez Alanis R: Effects of toxin T-544 from the *Karwinskia humboldtiana* (buckthorn) plant upon mouse embryos explanted at 11 days. *Toxicon* 28;449–452, 1990

Mitchell J, Weller RO, Evans H, Arai I, Daves GD Jr: Buckthorn neuropathy: effects of intraneural injection of *Karwinskia humboldtiana* toxins. *Neuropathol Appl Neurobiol* 4;85–97,1978

Munoz-Martinez EJ, Chavez B: Conduction block and functional denervation caused by tullidora *(Karwinskia humboldtiana). Exp Neurol* 65;255–270, 1979

Munoz-Martinez EJ, Nunez R, Sanderson A: Axonal transport: a quantitative study of retained and transported protein fraction in the cat. *J Neurobiol* 12;15–26, 1981

Munoz-Martinez EJ, Cueva J, Joseph-Nathan P: Denervation caused by tullidora *(Karwinskia humboldtiana). Neuropathol Appl Neurobiol* 9;121–134, 1983

Munoz-Martinez EJ, Massieu D, Ochs S: Depression of fast axonal transport produced by tullidora. *J Neurobiol* 15;375–392,1984

Ortiz GG, Gonzalez-Burgos I, Feria-Velasco A: Structural study of the acute effect of *Karwinskia humboldtiana* on cerebral cortex, hippocampus, and caudate nucleus of the rat. *Gen Pharmacol* 23;543–547, 1992

Pesman MW: *Meet Flora Mexicana.* DS King, Globe, Ariz, 1962
Pineyro-Lopez A, Martinez de Villarreal L, Gonzalez-Alanis R: In vitro selective toxicity of toxin T-514 from *Karwinskia humboldtiana* (buckthorn) plant on various human tumor cell lines. *Toxicology* 92,217–227, 1994
Thomson RH: *Naturally Occurring Quinones,* 2d ed. Academic Press, London, 1971
Valadez-Alvizo R, Rodriguez-Alvarado A, Martinez de Villarreal L, Gonzalez-Alanis R, Pineyro-Lopez A: Effect of toxins 544 and 514 from *Karwinskia humboldtiana* (buckthorn) plant upon fetal development of the mouse. *Toxicon* 31;1329–1332, 1993
Weller RO, Mitchell J, Daves GD: Buckthorn *(Karwinskia humboldtiana)* toxins. In *Experimental and Clinical Neurotoxicology,* Spencer PS, Schaumburg HH eds, Williams & Wilkins, Baltimore, pp336–347, 1980
Wheeler MH, Camp BJ: Inhibitory and uncoupling actions of extracts from *Karwinskia humboldtiana* on respiration and oxidative phosphorylation. *Life Sci* 10(2);41–51, 1971

Rosaceae Juss. 64

Cosmopolitan in distribution but with greatest diversity in the temperate and subtropical regions of eastern Asia, Europe, and North America, the Rosaceae, or rose family, is one of the larger families, comprising 100–122 genera and 3,000–3,400 species. In North America, 62–68 genera, 870–907 species, numerous hybrids, and a plethora of cultivars are present. The members of Rosaceae have conspicuously similar flowers. Floral structure varies primarily in the position and number of ovaries and types of fruits. Genera are traditionally grouped by taxonomists into 4–6 subfamilies and 12 or more tribes (table 64.1).

- Rose Family

Amelanchier
Cercocarpus
Cotoneaster
Heteromeles
Malus
Photinia
Prunus
Pyracantha
Rhodotypos

QUESTIONABLE
Coleogyne
Spiraea

Table 64.1. Major subfamilies of the Rosaceae and their principal distinguishing features

Subfamily	Representative Taxa	Distinguishing Features
Amygdaloideae (including Prunoideae)	cherries, plums, peaches, apricots	leaves simple, ovaries superior, fruits drupes
Maloideae (including Pomoideae)	apples, pears, hawthorns, loquats	leaves simple or compound, ovaries inferior, fruits pomes
Rosoideae	roses, blackberries, strawberries, raspberries	leaves compound, ovaries superior, fruits aggregations of achenes or drupelets
Spiraeoideae	spiraeas, ninebarks	leaves simple or compound, ovaries superior, fruits aggregations of follicles

Source: Cronquist 1981.

Plants trees or shrubs or herbs; annuals or perennials; armed or not armed with thorns or prickles. **Stems** erect or arching. **Leaves** basal to cauline; simple or 1-pinnately compound or palmately compound; alternate or rarely opposite; venation pinnate or palmate; leaflets of compound leaves 3–15, terminal one present; stipules present or rarely absent. **Inflorescences** solitary flowers or simple cymes or racemes or spikes or panicles or clusters; terminal or axillary. **Flowers** produced before or simultaneously with or after leaves; fragrant or not fragrant; perfect or rarely imperfect; perianths in 2- or 1-series. **Sepals** typically 5 or rarely 4; free or fused; herbaceous or rarely petaloid. **Corollas** radially symmetrical. **Petals** typically 5, or rarely 4 or 10 or more; free; of various colors but predominantly white or pink or yellow. **Stamens** numerous or rarely 1; in 1 or more whorls; arising from receptacles or hypanthia. **Pistils** numerous or 1; simple or compound, carpels 1–5; stigmas 1–5; styles 1–5, free; ovaries superior or inferior; locules 1–5; placentation axile or basal. **Hypanthia** present or absent; cup-shaped or saucer-shaped or tubular. **Nectaries** absent or present; hypanthial. **Fruits** achenes or drupes or follicles or pomes or hips, or aggregates of drupelets or achenes. **Seeds** 1 to numerous.

- erect herbs, shrubs or trees; corollas radially symmetrical; petals free; stamens typically numerous

Disease Problems—The Rosaceae is of considerable economic importance. Many of its genera are cultivated as ornamentals, and others are prized for their edible fruits

- foliage, seeds, neurotoxic; fruits, digestive disturbance

or are valued as browse for wild ruminants and livestock (Dayton 1931). Cyanogenesis, however, is characteristic of many genera and is the predominant disease problem in the family. The potential toxicity of some species consumed as food also is known. For example, the poisonous nature of peach stones/seeds has been recognized for centuries. Blyth in 1895 indicated that the toxicity of the seeds was well known to the Egyptians (Herbert 1979). Subsequently, they were used by the Romans as a suicide poison. Extracts were known as the "penalty of the peach." In more-recent years the use of apricot and peach seeds as natural flavorings has come into vogue with health food faddists. Unfortunately, there have been serious intoxications due to these products. In addition, the use of laetrile for treatment of various ailments, especially neoplasia, resulted in serious intoxications (Herbert 1979; Chandler et al. 1984). In those instances, the signs are often of slow onset and persistent over several days. Although not confirmed as a risk, the fruits of species of *Crataegus* (hawthorn) have been suggested to cause cardiotoxic effects (Turner & Szczawinski 1979).

In addition to chemical intoxications, there is a potential for other problems from ingestion of large amounts of the fruit of some members of the Rosaceae. Mechanical problems and colic due to bowel obstruction similar to that caused by sand have been reported for horses that ingested large amounts of the fruits of *Crataegus crus-galli* (Rook et al. 1991). Overeating of ripe fruits may also cause other types of digestive disturbance.

■ cyanogenesis characteristic of many members of family

Disease Genesis—Cyanogenesis is a consistent feature among many genera of the family. Although there are numerous examples of cyanogenic taxa (table 64.2), few are serious risks to livestock. In the Rosaceae, *Prunus laurocerasus, P. serotina,* and *P. virginiana* are the most consistent cyanogenic threats. Most of the other cyanogenic taxa do not contain HCN potential of a hazardous magnitude or maintain it throughout the year. Representative species of those genera more specifically associated with

Table 64.2. Genera and species of the Rosaceae for which cyanogenic potential is reported

Genus and Species	Compound and Plant Part	Common Name(s)
Adenostoma		
A. fasciculatum	lvs	chamise, greasewood
A. sparsifolium	lvs	red shank
Amelanchier		
A. alnifolia	P, all/A, frs	Saskatoon serviceberry
A. arborea	?	Juneberry, shadberry
A. bartramiana	lvs	mountain Juneberry
A. canadensis	P, all/A, frs	shadblow serviceberry
A. humilis	sds	low serviceberry
A. laevis	sds, lvs	Allegheny serviceberry
Aronia		
A. arbutifolia	lvs	chokeberry
A. melanocarpa	lvs	black chokeberry
A. ×*prunifolia*	lvs	purple-fruited chokeberry
Aruncus		
A. dioicus		
(= *Spiraea aruncus*)	lvs	goatsbeard
Cercocarpus		
C. intricatus	lvs	dwarf mountain mahogany
C. ledifolius	lvs	curl-leaf mountain mahogany
C. montanus		
(numerous varieties)	lvs	mountain mahogany
C. traskiae	lvs	Catalina Island mountain mahogany
Chaenomeles		
C. speciosa		
(= *C. lagenaria*)	lvs, sds	flowering quince

Table 64.2. (Continued)

Genus and Species	Compound and Plant Part	Common Name(s)
Cotoneaster		
C. acutifolius	lvs	Peking cotoneaster
C. adpressus	?	creeping cotoneaster
C. apiculatus	?	cranberry cotoneaster
C. bullatus	P/A	
C. buxifolius	sds, lvs	
C. congestus	P	Pyrenees cotoneaster
C. dammeri	P	bearberry cotoneaster
C. dielsianus	P/A	
C. divaricatus	?	spreading cotoneaster
C. franchetii	sds, lvs	
C. horizontalis	sds, lvs	rock cotoneaster
C. 'Hybridus Pendulus'	P	
C. integerrimus	P/A	
C. lucidus	?	hedge cotoneaster
C. microphyllus	P, lvs	littleleaf cotoneaster
C. multiflorus	P	many-flowered cotoneaster
C. 'Nan Shan' (= *C. praecox*)	P/A	
C. pannosus	sds, lvs	
C. roseus	?	
C. zabelii	?	
Coleogyne		
C. ramosissima	lvs	blackbrush
Crataegus		
C. chrysocarpa (= *C. irrasa*)	lvs	fireberry hawthorn
C. coccinea (= *C. pedicellata*)	lvs	scarlet hawthorn
C. crus-galli	lvs	cockspur hawthorn
C. curvisepala	sds, lvs	hawthorn
C. flabellata (= *C. iracunda*)	lvs	fanleaf hawthorn
C. intricata	sds, lvs	thicket hawthorn
C. monogyna	frs	European hawthorn
C. phaenopyrum	lvs	Washington hawthorn
C. pruinosa	lvs	frosty hawthorn
Cydonia		
C. oblonga	P, lvs, sds/A	quince
Docynia		
D. delavayii	lvs	
Eriobotrya		
E. japonica	A	loquat, Japanese plum
Exochorda		
E. giraldii	lvs	redbud pearlbush
E. serratifolia	lvs	
Heteromeles		
H. arbutifolia	P, all	California holly
Kerria		
K. japonica	lvs	Japanese kerria
Malus		
M. angustifolia	?	southern crabapple
M. baccata	sds	Siberian crab
M. coronaria	sds	sweet crabapple
M. pumila	A	common apple
M. sylvestris	A	common crabapple
Neviusia		
N. alabamensis	lvs	snow wreath
Oemleria cerasiformis (= *Nuttallia cerasiformis*)	lvs	Indian plum, osoberry
Peraphyllum		
P. ramosissimum	lvs	squaw apple
Photinia		
P. davidiana (= *Stranvaesia davidiana*)	P, lvs	

Table 64.2. (Continued)

Genus and Species	Compound and Plant Part	Common Name(s)
P. davidsoniae	lvs	
P. prionophylla	lvs	
P. serratifolia		
(= *P. serrulata*)	A/lvs	Chinese photinia
P. villosa	sds	Oriental photinia
Porteranthus		
P. trifoliatus		
(= *Gillenia trifoliata*)	P	Bowman's root
Prunus		
P. alleghaniensis	?	Allegheny plum
P. armeniaca	P, lvs/A	apricot
P. avium	P, lvs	sweet cherry, gean
P. brachybotrya	lvs, frs	aguacatillo
P. caroliniana	lvs	laurel cherry
P. cerasus	P/A	sour pie cherry
P. domestica	lvs	European plum
P. dulcis		
(= *P. amygdalus*)	A	almond
P. laurocerasus	P, lvs/A	cherry laurel
P. lusitanica	lvs	laurel, Portugal laurel
P. lyonii	lvs	Catalina cherry
P. macrophylla	P, lvs	
P. myrtifolia	sd	
P. nigra	shs	wild plum
P. occidentalis	sds	
P. padus	P, lvs/A	European bird cherry
P. pensylvanica	?	bird or pin cherry
P. persica	P/A	peach
P. pumila	lvs	sand or dwarf cherry
P. serotina	P, lvs	black cherry
P. serrulata	P	flowering cherry
P. spinosa	lvs	blackthorn, sloe
P. tomentosa	lvs	bush cherry
P. virginiana	?	chokecherry
Pyracantha		
P. coccinea	P	pyracantha, firethorn
Pyrus sensu lato		
innominate species	A	crabapple, pear
Rhodotypos		
R. scandens	lvs	jetbead
Sorbaria		
S. kirilowii		
(= *S. arborea*)	lvs	false spiraea
S. sorbifolia	lvs	Ural false spiraea
Sorbus		
S. americana	?	American Mountain ash
S. aucuparia	P, shs/A	mountain ash, rowan tree
Spiraea		
S. japonica	?	Japanese spiraea
S. prunifolia	lvs	bridal wreath

Sources: Kingsbury 1964; Gibbs 1974a,b; Seigler 1976a,b, 1977; Fikenscher et al. 1981; Saupe 1981; Barnea et al. 1993.
Note: A, amygdalin; *P,* prunasin; *lvs,* leaves; *sds,* seeds; *frs,* fruits; *shs,* shoots; *?,* unknown; *all,* all plant parts.

serious toxicity problems are presented in the discussion of individual genera. Unless otherwise noted, the genera discussed individually can be considered to be cyanogenic.

■ cyanogenic glycosides, hydrolyzed by glycosidases in plant tissues to HCN or cyanide

Cyanogenesis is an important feature of numerous genera in other families as well, mainly those of Berberidaceae, Caprifoliaceae, Euphorbiaceae, Juncaginaceae, Poaceae, Leguminosae, and Linaceae (Seigler 1976a,b). The ubiquitousness of the phenomenon in flowering plants is exemplified by the numbers of cyanogenic taxa given in various reports of its occurrence. For example, Seigler (1976b) identified 135 species in 46 families; Gibbs (1974a) observed a positive test for cyanide in 232 species of 135 genera and 58 families; and Saupe (1981) estimated that about 2,000 species from 110 families were cyanogenic. Very low HCN potential has even been reported for the forages of barley, rice, and wheat (Conn 1979). Numerous reviews of this important toxicologic problem have been published, e.g., Seigler (1977, 1981), Poulton (1983), Tewe and Iyayi (1989).

Cyanide in plants occurs mainly as O-β-glycosides of α-hydroxynitriles, of which about 24 are synthesized by the plants from amino acids (Seigler 1977). Most are monoglycosides of mandelonitrile or its hydroxy derivative, including dhurrin, prunasin, sambunigrin, taxiphyllin, and zierin (table 64.3). Production or release of cyanide is typically a two-step process. Initially there is hydrolysis by β-glycosidases (emulsin in the Rosaceae) to yield a sugar and an aliphatic or aromatic α-hydroxy-nitrile aglycone (cyanohydrin). Subsequently there is formation of an aldehyde or a ketone—typically benzaldehyde—and the toxic hydrogen cyanide (HCN), or prussic acid (Conn 1979). The latter reaction may proceed with or without the enzymatic action of hydroxynitrile lyase. In the case of amygdalin and vicianin, diglycosides of mandelonitrile, there is a three-step enzymatic pathway because of the necessity of hydrolysis of the second sugar to form the monoglycoside. Many of the β-glycosidases are specific for the individual glycosides with which they co-occur (Hosel 1981). In a few instances plant species or specific plant parts may contain the cyanogenic glycosides but lack the appropriate glycosidase; these taxa are generally of minor importance toxicologically. In other situations the glycosidase may be present only during certain stages of growth (Swain et al. 1992). In addition to cyanogenic glycosides, there are cyanide-containing lipids from Sapindaceae, but these compounds do not represent a source of HCN for animals.

Table 64.3. Representative cyanogenic glycosides

Glycoside	Examples of Plant Sources
acacipetalin	*Acacia*
amygdalin	Rosaceae only
dhurrin	*Sorghum, Juncus*
holocalin	*Sambucus*
p-hydroxy-mandelonitrile	*Nandina, Thalictrum*
linamarin + lotaustralin	*Phaseolus, Manihot, Linum, Thalictrum*
lotaustralin + linamarin	*Trifolium*
prunasin	Rosaceae, *Eremophila, Eucalyptus, Linaria, Sambucus*
sambunigrin	*Sambucus*
taxiphyllin	*Taxus, Phyllanthus*
triglochinin	*Triglochin*
vicianin	*Vicia*
zierin	*Zieria, Sambucus*

Sources: McIlroy 1951; Gibbs 1974a; Seigler 1977.

■ Rosaceae, diglucoside amygdalin, monoglucoside prunasin

Cyanogenesis in the Rosaceae is associated almost exclusively with amygdalin and prunasin; these glycosides are used as taxonomic characters in subdividing the family into subfamilies (Saupe 1981). Both are derived from phenylalanine and differ only in size; amygdalin is a diglucoside, and prunasin is a monoglucoside. Amygdalin is found only in the Rosaceae, where it is present in the seeds of some species (Seigler 1981). Prunasin is found in other families as well as the Rosaceae. Typically, cyanogenesis from ingestion of foliage is due to prunasin and from seeds is due either to amygdalin or prunasin. Exceptions to the exclusiveness of these glycosides for Rosaceae are the leucine-derived (*S*)-cardiospermin-*p*-hydroxybenzoate present in *Sorbaria kirilowii* and the methylbut-2-ene nitriles of *Oemleria cerasiformis* (Saupe

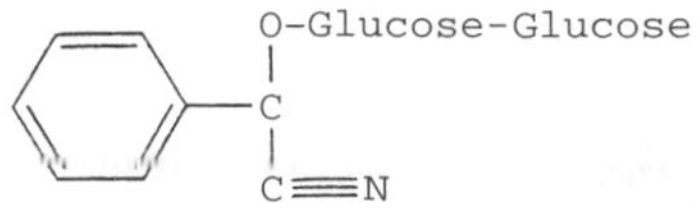

O-Glucose-Glucose
C
C≡N

Amygdalin

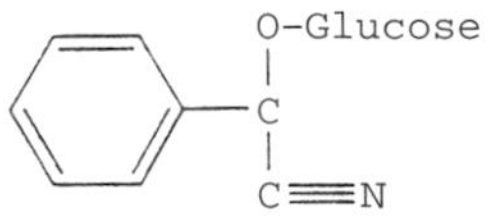

O-Glucose
C
C≡N

Prunasin

- cyanogenic glycosides and hydrolytic enzymes in outer portions of plant tissues, activated by chewing and other damage

1981; Lechtenberg et al. 1994). For an extensive discussion of the chemotaxonomic relationships of the Rosaceae, see Hegnauer (1961, 1973, 1986, 1990).

Under normal growing conditions, the glycoside and the hydrolytic enzymes are in separate sites in the plant. This is clearly demonstrated in studies with *Sorghum*, in the Poaceae, and *Phaseolus*, in the Fabaceae (Saunders & Conn 1978; Kojima et al. 1979; Conn 1991). The specific sites are probably not the same for all cyanogenic plant species, but in general the glycosides are located in the epidermis. The tendency for the highest concentrations of glycosides to occur in the outer portions is true for leaves, roots, stems, or tree trunks (Gomez et al. 1980; Reilly & Okie 1985). This distribution is exemplified by the marked difference in glycoside concentrations in different layers of cassava root. Concentrations are high in cortical tissue (50+ ppm in the peel) and low (10 ppm) in the middle of the tuber (Cooke 1978). This differential seems to be more so in cassava cultivars of low overall cyanogenic potential (Dufour 1988). In varieties with high potential, this difference is small. Distribution of the compounds in the outer tissues is consistent with the hypothesis that cyanogenesis represents a defense against herbivory, even though it may not be consistently effective against all insect herbivores (Jones et al. 1978; Spencer 1988). Presumably when the forage is damaged by maceration or crushing, the barriers separating the storage sites are broken down, and the glycosides are exposed to the hydrolytic action of the glycosidases. Both HCN potential and glycosidase activity are polymorphic in nature, subject to genetic manipulation, as shown for several genera, including *Sorbus* of the Rosaceae (Jones et al. 1978; Conn 1979).

There is little or no free HCN in plant tissues (Conn 1973; Cooke 1978). Therefore, results of tests generating and measuring HCN from plant material are traditionally referred to as "HCN potential." Alternatively, the amounts of the glycosides or aglycones may be measured and converted to HCN equivalents. In addition, the concentrations may be expressed as dry or wet weight. Glycoside concentrations may be as high as 6% prunasin d.w. in the leaves of *Prunus virginiana* (Majak et al. 1981) or 8% in *Amelanchier alnifolia* var. *cusickii* (Brooke et al. 1988). HCN potential in the range of 2,000+ ppm is typical of the immature leaves of either *P. serotina* or *P. virginiana* (Smeathers et al. 1973; Majak et al. 1981). Cyanogenic glycoside content may vary considerably among varieties of a species, as shown by the peak concentrations of 5–8% prunasin in var. *cusickii* and 1–3% in var. *alnifolia* of *A. alnifolia* (Brooke et al. 1988).

- all plant parts may have cyanogenic potential, typically lowest in flesh of fruit; greatest risk for livestock eating foliage is during early growing season

All parts of plants may have cyanogenic potential, even bark or fruit in some species. Prunasin content of buds, flowers, and fruits are 2.6, 2.5, and 0.9% d.w. for *A. alnifolia* and 3.6, 2.6, and 1.2% d.w. for *P. virginiana*, respectively (Majak et al. 1981). *Prunus persica* leaves are generally of little concern because they are of low HCN potential (100 ppm wet weight), whereas seeds may be in excess of 450 ppm (Machel & Dorsett 1970; Herbert 1979). Prunasin content of 0.7–0.9% in twigs and 0.2–0.3% in bark are reported for *P. persica* (Elberta peach) (Reilly & Okie 1985). Nevertheless, intoxications are reported occasionally for animals eating peach leaves as well as seeds (Hansen 1928). The relative concentrations vary not only between but also within tissues. The contrast between concentrations in the flesh and in the seed of fruits is marked, generally being quite low in the former.

Cyanogenic glycoside concentrations in the plant are subject to many endogenous and exogenous influences, most of which have been well recognized since the early 1900s (Brunnich 1903). Growth stage or season of the year is very important. In some instances, concentrations are much higher in immature than mature leaves. In *Prunus serotina*, concentrations expressed as HCN potential ranged from 2,500 ppm in newly emerged leaves, down to less than 500 ppm by 6 weeks, and then gradually down to about 40 ppm after several months (Smeathers et al. 1973). The drop in cyanide concentrations may not be as abrupt in *P. virginiana,* but the levels at the end of the growing season are nevertheless markedly reduced such that by late summer or early fall the amount of leaf material required for a toxic dose may have increased by 4–5 times (Fleming et al. 1926). This is about the time when the fruit is dark purple. However, a word of caution is in order because changes due to maturation are subject to environmental influences, and other investigators have noted the presence of reduced but still toxic concentrations of prunasin even at this late stage of growth (Majak et al. 1981). Deaths from ingestion of the leaves of *P. serotina* occasionally occur even in the latter part of summer (Gough 1995). The variation among and between plant species and genera with respect to fluctuation of concentrations is exemplified by the differences between *P. virginiana* and *Amelanchier alnifolia.* In the former the concentrations of prunasin declined from a maximum 6.1% d.w. of leaves early in the season to 2.9% at the time of fruit ripening (much less than the drop of 4–5 times above). In the latter species, the decline was from a maximum of 3.2% d.w. early, down to 0.3% at maturity. Thus, in the one case there is substantial risk throughout the season, and in the other the risk is limited to 30–45 days early in the growing season (Majak et al. 1981). *Amelanchier alnifolia* may be useful forage later in the season. *Prunus brachybotrya* also has sustained cyanogenic potential, 1.6% in leaves in early March, up to 2% in May, and down to 1.2% in July (Lopez et al. 1992). Levels in fruits are more variable, ranging from a maximum of 1.7% early to nil at maturity. Moisture levels in the forage may be a significant factor as well. During dry periods, cyanide potential may increase. This has been shown for *P. serotina, P. virginiana,* and to a lesser extent *A. alnifolia* (Smeathers et al. 1973; Majak et al. 1981). However, a word of caution: this increase applies to dry periods within a season. When the entire year has been exceptionally dry, there may be diminished growth and development of foliage, with less overall glycoside available.

- typically cyanogenic potential is lost with drying but may be retained in plant trimmings in cool or moist weather

Because HCN is volatile, once it is formed following hydrolysis of a glycoside, there is a likelihood of continued loss with time. Air-dried leaves of *A. alnifolia* lose much of their HCN potential in 24 and 48 hours, dropping from 2,500 ppm to less than 200 ppm and 40 ppm, respectively. This is in contrast to the loss of only 15% of HCN potential from leaves or twigs, chopped and stored at 4°C for 24 hours (Majak et al. 1980). Occasionally, toxicity is retained, and livestock losses occur when plant trimmings are discarded in a pasture in a cool or moist environment where the leaves remain green (Gill & McGregor 1928). Peach seeds from opened pits also lose considerable HCN with time (Machel & Dorsett 1970). There is generally marked loss of HCN potential following ensiling (Conn 1973).

- 200 ppm HCN potential in plant tissue hazardous, especially to ruminants

It is difficult to establish a specific concentration that is toxic because of differences among the glycosides, their rates of hydrolysis, and their rates of ingestion, but a plant tissue concentration of 200 ppm HCN potential has been suggested as the lower limit of hazard (Hindmarsh 1930; Couch, as quoted by Moran 1954). This level has continued to be the standard for evaluating toxicity potential (Kingsbury 1964), although a higher level of 500 ppm has been used more specifically for the sorghums. Plant concentrations of 2–3% d.w. of prunasin could readily result in a lethal dose of cyanide for cattle (Majak et al. 1980).

All animal species are susceptible to the toxic effects of cyanide. However, ruminants appear to be more susceptible to cyanogenic glycosides in plants because conditions prevailing in the rumen facilitate rapid hydrolysis and absorption. On the other hand, preformed HCN or cyanide salts may be more toxic to monogastrics; the lethal dose of HCN for the horse is reported to be about 1 mg/kg b.w. or about half of the toxic dose required for cattle and sheep (Couch 1932; Clawson et al. 1934).

The lethal dose for HCN in humans is estimated to be 0.5–3.5 mg/kg b.w. orally (Gettler & Bain 1938; Conn 1973; Way 1981). In mice the lethal dose of KCN i.p. is in the range of 6–8 mg/kg b.w. (Norris et al. 1986; Moore et al. 1991), whereas in rabbits the LD_{50} of KCN is 3–3.3 mg/kg i.m. (Ballantyne et al. 1972)

The unusual susceptibility of ruminants to the effects of the glycosides is shown by the similarity in the lethal doses for cyanide salts and cyanide from glycosides (Leemann 1935). The lethal dose of cyanide when given as prunasin in plant material from *A. alnifolia* is 5 mg HCN/kg b.w. in cattle (Majak et al. 1980). When given directly as a salt, the dose is quite similar, 3–4 mg/kg b.w. as the sodium salt or 2.2 mg/kg as HCN (Hindmarsh 1930; Steyn 1934; Burrows & Way 1977). The toxic dose of plant material varies with the particular plant species. For *Prunus* and *Amelanchier,* the lethal dose for sheep or cattle ranges between 1 and 4 g plant material/kg b.w., depending on its prunasin content (Fleming et al. 1926; Majak et al. 1980). Neither tolerance nor immunity develops. The effects are not cumulative, and one-half the lethal dose can be given repeatedly during the course of a day such that a total dose of 4–5 times the single lethal dose can be tolerated. Because of the rapid disposition of cyanide, the minimum lethal dose is variable, depending on the rate of intake. Whereas the single minimum lethal dose is 2 and 2.3 mg/kg b.w. for cattle and sheep, respectively, the daily grazing minimum dose is estimated at 15–50 mg/kg b.w. (Clawson et al. 1934; Coop & Blakley 1950).

■ wild ruminants such as deer are susceptible; birds also at risk, especially from preformed sources of cyanide

Wild ruminants are also susceptible. A dose of 1 kg wet weight of *A. alnifolia* per day was fatal to mature mule deer (Majak et al. 1978). It is estimated that when the diet of deer is composed of 35% or more *A. alnifolia,* there is risk of toxicity (Quinton 1985). This is of concern because at times deer forage heavily on this species; the danger is amplified if they consume new growth. Occasionally, intoxications from plant glycosides as well as from preformed sources of HCN are reported for horses and pigs (Hansen 1928; Enge 1939; Prodjoharjono 1974). In a few instances cyanogenic glycoside plants have been suspected as causes of death in dogs or cats (Stauffer 1970; Clay 1977; Bradley et al. 1988). Birds also are susceptible to cyanide, and the flesh-eating species especially so. The LD_{50} of cyanide in vultures, kestrels, and owls is in the range of 4–9 mg/kg b.w., whereas in seed-eating quail, chickens, and starlings it is 9–21 mg/kg b.w. (Wiemeyer et al. 1986). Cyanogenic glycosides may be responsible occasionally for wildfowl mortalities, as illustrated by an incident in British Columbia with gizzard tissue concentrations of 15–200 ppm HCN in dead birds (Cameron 1972).

■ rapid glycoside hydrolysis may occur in rumen because of moderate pH and hydrolytic glycosidases from bacteria; hydrolysis slower in gut of monogastric animals

Prunasin is readily absorbed intact without hydrolytic cleavage of the glucose when given orally in rats (>50%), but amygdalin is essentially not absorbed until it is hydrolyzed to prunasin (Rauws et al. 1982, 1983; Strugala et al. 1995). Linamarin is also poorly absorbed (Barrett et al. 1977). The increased susceptibility of ruminants to plant cyanogenic glycosides is due to the relatively high pH of rumen fluids and the presence of microbial-derived glycosidases capable of hydrolyzing the parent glycoside. Hydrolysis of amygdalin or prunasin proceeds in dilute acid only at temperatures exceeding 60°C (Conn 1979). Thus, free hydrolysis is of limited importance at the low pH characteristic of the monogastric stomach. More importantly, the glycosides are rapidly hydrolyzed by β-glycosidases from a variety of ruminal bacteria; prunasin is hydrolyzed faster than linamarin, which in turn is hydrolyzed faster than amygdalin (Coop & Blakley 1950; Majak & Cheng 1984, 1987). There is also considerable, but slower and somewhat delayed, glycoside hydrolysis in the gut of humans and hamsters and lesser amounts in mice, rats, guinea pigs, and monkeys (Stavric & Klassen 1980; Adewusi & Oke 1985; Frakes et al. 1986). This hydrolytic activity is much influenced by diet and the presence of an active gut microflora (Carter et al.

1980). The activity of the microbial enzymatic degradation pathway, and the hydrolysis of the resulting cyanohydrin, are strongly pH dependent. Both increase as the pH rises above 6 (Majak et al. 1990). Thus, ruminal pH and the rate of hydrolysis are lowest immediately after eating, rising slowly thereafter, especially when the animal is feeding on fresh forages. High-grain diets attenuated the postprandial increase, suggesting a possible approach to reducing the intoxication potential of cyanogenic forages (Majak et al. 1990). Ruminal hydrolysis is also dependent upon the presence of sufficient fluid. In unusual situations with sheep on range where water is limited, hydrolysis to yield the toxic HCN appears to be delayed until the animal drinks deeply. Thereafter, signs of intoxication appear quickly, or in a few instances the signs and resolution are slow and protracted (Fleming & Dill 1928). Except for taxiphyllin, these glycosides are not especially heat-labile. Interestingly, the glycosidase specificity for glycosides by the microflora in hamsters differs from that of most other species, including ruminants, by being highest for amygdalin (Frakes et al. 1986).

- cyanide causes inhibition of many enzymes in body; effects mainly due to inhibition of cytochrome oxidase pathway, especially in brain; action is quick and rapidly lethal; rarely delayed action; chronic effects restricted mainly to humans from long-term ingestion of foods with high cyanogenic potential

The effect of HCN is, predominantly, reversible inhibition of cytochrome oxidase in the mitochondrial matrix, which prevents electron flow through the respiratory chain to oxygen (Keilin 1930; Warburg 1931; Hogness 1939; Albaum et al. 1946). The block occurs mainly with the oxidized form (Fe^{3+}) between cytochromes a and a_3 of complex IV of the respiratory chain and results in histotoxic anoxia. However, many other enzymes, including those of the metabolic pathways, are inhibited as well (Solomonson 1981). Cytochrome oxidase is inhibited in cells throughout the body, but this is especially important in the brain because there is very little rhodanese in this tissue. The effects on other tissues, including the heart, may contribute to the signs observed, but most of the nonneurologic effects are of little overall consequence, because of the overwhelming effects on the brain (Way 1984). Acute, subacute, and chronic neurologic effects can be reproduced experimentally or are associated with cyanide in clinical diseases of humans. Repeated exposure seems to predispose to basal ganglia disorders that cause Parkinsonian-like signs and dopaminergic interference effects (Uitti et al. 1985; Carella et al. 1988; Cassel & Persson 1992; Kanthasamy et al. 1994). Many of the neurologic effects involve alterations in calcium content or movement (Johnson et al. 1986). The wide variety of neurologic effects is not surprising, because of the array of neurons affected, many with differing characteristics and functions (Rosenberg et al. 1989; Patel et al. 1992; Valenzuela et al. 1992).

Intoxication with HCN in animals is usually peracute, but there are exceptions. Under some conditions of limited ruminal fluid content, the disease onset is slow and prolonged, apparently owing to slow hydrolysis of the glycosides (Fleming & Dill 1928). An unusual intoxication involving *Prunus laurocerasus* was reported in which, instead of the typical abrupt onset and rapid resolution of clinical signs, there was a delay in onset and then repeated episodes in the animal, even in the absence of continued plant consumption (Robb & Campbell 1941; Wilson & Gordon 1941). In some instances in humans there is apparently slow hydrolysis and absorption such that the signs occur over a more protracted time. Also in humans, chronic intoxication due to long-term ingestion of cyanogenic tropical foods is a serious problem. Several chronic neurologic syndromes are associated with HCN, including tropical ataxic neuropathy, tropical amblyopia, West Indian amblyopia, ataxic neuropathy of Nigeria, and tobacco amblyopia (Montgomery 1965; MacKenzie & Phillips 1968; Osuntokun 1968; Howlett et al. 1990). The foods most commonly associated with these problems are cassava and to a lesser extent tropical lima beans. The result is bilateral optic atrophy and nerve deafness, posterior column myelopathy (spinal), and sometimes polyneuropathy. There may be severe sensory deficit in the legs. It is not clear whether the neurologic effects are due solely to direct effects of cyanide or to depletion of cysteine and other sulfur compounds in the tissues or are complications caused by nutritional deficiencies.

The role of cyanide as a cause of chronic intoxication in livestock is not clear. Such effects have not heretofore been recognized in the field as associated with cyanide, with the possible exception of the ataxia-urinary incontinence syndrome in horses on sorghum pastures. In light of the various neurologic problems, including spinal effects in humans, the role of cyanide as a chronic neurotoxicant in animals merits reevaluation.

A complicating factor is the tendency of thiocyanate to accumulate in the body because of its long half-life—about 2.7 days in humans (Schulz 1984). It is a goitrogen, and long-term cyanide exposure under some conditions may cause thyroxin deficiency and compensatory thyroid enlargement.

■ possible but not confirmed cyanogenic glycoside teratogenic potential

An additional problem of importance is the potential for reproductive effects, mainly teratogenesis. Laetrile is suspected to present a teratogenic risk in humans. This suspicion is supported by a report of limb deformities and agenesis of the tail and anus in pigs, apparently resulting from the dams' eating the leaves and bark of *P. serotina* (Selby et al. 1971). This potential has subsequently been confirmed by production of limb malformations and microcephaly when cassava is fed to rats (Singh 1981), and anterior neural tube and axial skeletal defects (encephaloceles and exencephaly) when amygdalin and to a lesser extent linamarin are given to hamsters (Willhite 1982; Frakes et al. 1985). However, with administration of linamarin, developmental effects were produced only with dosage that caused overt signs of intoxication in the dams. Sodium cyanide produced the same effects as amygdalin, and these effects were prevented by prophylaxis with sodium thiosulfate, which strongly implicates cyanide as a causative factor (Doherty et al. 1982). However, the potential for cyanide to cause these types of problems in the field is mitigated by the need for high dosage to produce effects, dosage typically of such magnitude as to cause signs and possibly lethality in the dams.

An interesting interaction of questionable significance has been noted between cyanide and selenium. There seems to be a mutually protective action against intoxication between the two (Palmer 1981). Possibly, intoxication from cyanogenic plants may be less of a problem in areas of high soil selenium.

■ cyanide is rapidly absorbed from gut and appreciable levels are present in blood; it is rapidly detoxified; eliminated in urine and exhaled air and excreted as thiocyanate

Following hydrolysis, there is rapid turnover and absorption of HCN, beginning in seconds or minutes (Coop & Blakley 1950; Majak et al. 1980). HCN readily diffuses across mucous membranes and epithelial barriers and is widely distributed throughout body tissues (Halstrom & Moller 1945). In humans, about 98–99% of the HCN in blood is bound to methemoglobin (Schulz 1984). Because of this, blood is the preferred tissue for analysis. Concentrations in liver and other tissues are lower and more variable (Gettler & Baine 1938; Egekeze & Oehme 1979). However, even when there is evidence of intoxication, a large amount of cyanide is likely to remain in the digestive tract (Gettler & Baine 1938; Finck 1969). Blood HCN concentrations in lethal intoxications in humans range from 0.3 to more than 3 ppm, but in most cases are 1 ppm or more (Gettler & Baine 1938; Sunshine & Finkle 1964). A dose of 6–10 mg/kg b.w. of KCN resulted in blood concentrations of 2 ppm at 12 hours and 1.6 ppm at 22 hours, while ingestion of 20–40 apricot seeds produced peak blood HCN concentrations of 3.2 ppm (Graham et al. 1977; Rubino & Davidoff 1979). In the case of large lethal doses of cyanide salts in humans, concentrations may be in the range of 7–8 ppm in blood and 150–200 ppm in the stomach (Finck 1969). Elevations in blood cyanide are not always as markedly increased, as, for example, in the case of an 11-month-old girl who ingested laetrile (amygdalin) and eventually died in spite of therapy. She had an early blood HCN of only 0.29 ppm (Humbert et al. 1977). Lethal blood concentrations in other species are similar—approximately 1 ppm in dogs, 1.5–2.6+ ppm for mice and rats, and 4.5 ppm for rabbits (Gettler & Baine 1938; Ballantyne et al. 1972; Egekeze & Oehme 1979; Carter et al. 1980; Moore et al. 1991). In the dog, it is estimated that 1 ppm in blood will cause respiratory arrest (Christel et al. 1977). Lethal blood concentrations in sheep are similar, 0.8–1.3 ppm (Burrows 1981). With storage of blood under some conditions, small amounts of cyanide may be formed without actual exposure to the toxin (Ballantyne 1977).

As is indicated by the large daily dose that is tolerated, cyanide is rapidly detoxified in the body. Small amounts are eliminated directly as CO_2 and/or formate metabolites in the animal's exhaled air or urine. The majority of the cyanide is detoxified to thiocyanate and excreted slowly in urine. Urinary elimination is slowed because of tubular reabsorption. Formation of thiocyanate is the result of the action of sulfurtransferases. These include 3-mercaptopyruvate sulfurtransferase, which results in the formation of thiocyanate directly from cysteine, or from thiosulfate as a sulfane sulfur. The other and the best known sulfurtransferase is rhodanese, a mitochondrial enzyme found throughout the body, which catalyzes the formation of thiocyanate from thiosulfate (Westley 1981). The availability of sulfane sulfur is the rate-limiting factor in the body, and administration of exogenous thiosulfate markedly accelerates the formation of thiocyanate from cyanide (Frankenberg & Sorbo 1975). It has been suggested that HCN is detoxified to formic acid and ammonium chloride in the acidic environment of the monogastric stomach (Boyd et al. 1938).

Because of the assumed slow action of the enzymatic thiosulfate-induced detoxification of cyanide, methemoglobin (MHb) formers, most notably nitrites, have been used for their initial beneficial effects of providing large amounts of available ferric iron (in MHb) for binding cyanide directly. However, the long-held assumption that nitrite acts primarily as a MHb former appears to be unjustified (Way et al. 1987). Although the rapidly acting MHb formers such as dimethylaminophenol and para-aminopropiophenone (Kiese & Weger 1969; Burrows 1981) are effective antidotes, some vasodilators such as chlorpromazine and phenoxybenzamine, when combined with sodium thiosulfate, also provide positive effects (Burrows & Way 1976; Way & Burrows 1976; Burrows 1981; Way et al. 1987). Thus, the effects of nitrite are not definitely or possibly totally due to MHb formation. Perhaps deficits in blood flow to the brain due to HCN are subject to alteration by some of these compounds (Wheatley 1947).

Clinical Signs—Signs of HCN intoxication typically occur quickly, in many instances within 15–20 minutes after ingestion of the plants. There is a rapid progression of signs, beginning with apprehension and distress and quickly followed by severe weakness, ataxia, and rapid, labored respiration. The animal is barely able to stand, or it becomes recumbent and exhibits kicking and tetanic-type seizures. Prior to recumbency, attempts to urinate are common. Cessation of respiration and very slow heart rate with or without arrhythmias may be apparent at least for an early brief period. Cyanosis is not an early sign; on the contrary, red, well-oxygenated venous blood is present because oxygen is used minimally by the tissues (Manges 1935). The animal eventually becomes comatose, and, terminally, there is a brief period of paddling convulsions. The entire sequence of signs may occur in 10–15 minutes or, in the instance of lower dosage, may take 45–60 minutes. In these situations, if the animal survives the first hour, the prognosis for recovery is usually favorable. Less commonly, signs develop more slowly, and resolution takes hours to occur. The development of labored respiration and seizures portends a grave prognosis.

■ animals, rapid onset, signs within 15–20 minutes; quickly progress through apprehension, distress, weakness, ataxia, labored respiration, collapse, seizures, death; if death does not occur in first hour, recovery is likely

humans, headache, rapid heart rate, vomiting, flushing of skin, problems of vision and sensory perception

In humans, the signs caused by cyanogenic sources include headache, abdominal discomfort, rapid heart rate, vomiting, dizziness, vision problems, flushing of the skin, and strange sensory perceptions. In serious intoxications, collapse and other signs seen in animals will be apparent.

Signs in birds are somewhat similar to those in mammals. In chickens there is panting, rapid eye blinking, excess salivation, and lethargy of abrupt onset and rapid development (Wiemeyer et al. 1986). Signs in vultures are similar, with ataxia, head bowing, wing droop, and terminal prostration and seizures accompanied by tail fanning, opisthotonus, and deep gasping respiration.

The dose of cyanide that merely produces signs is very close to the lethal dose. Therefore, the presence of signs of intoxication should be viewed with concern and treatment promptly administered.

- diagnosis, based on history of exposure to toxic plants, rapid onset of signs, and response to treatment; animal tissue (blood) analysis difficult and often unreliable, not commonly carried out for livestock deaths; plant tests for cyanogenic potential commonly applied, Guignard (picrate) and other tests

Diagnosis of cyanide intoxication is usually based on knowledge of the animal's access to cyanogenic plants, clinical signs, and response to treatment. There may be an odor of almond extract or benzaldehyde in the rumen or intestine. Confirmation by quantitative determination of cyanide in plant or animal tissue is difficult and tedious. If it is attempted, blood is much preferred to plasma (Baar 1966; Ballantyne et al. 1972; Way 1984; Wiemeyer et al. 1986). Concentrations of 1 ppm or more are consistent with severe intoxication in mammals and birds. Proper handling of sample material is very important because of the volatile nature of HCN, and results are often otherwise questionable. There seems to be negligible loss of HCN from tissues with putrefaction (Gettler & Bain 1938), but blood should be kept in airtight containers at 4°C (Egekeze & Oehme 1979). Maintaining tissues at a pH of 10 or more may also be beneficial in preserving the cyanide. Storage of chopped leaves or twigs for 24 hours even at 4°C resulted in the loss of 15% of the HCN potential (Majak et al. 1980). Plant tissues can be preserved in a 1% solution of mercuric chloride for months with minimal loss of HCN potential (Briese & Couch 1941).

Confirmation of the cyanogenic potential of suspect plants may be helpful to ascertain the hazard of grazing or to incriminate them in the event of an intoxication. The Guignard (1906) or picrate test for cyanide can be performed in the field to determine this potential (table 64.4). Although it lacks absolute specificity, it remains a very useful presumptive test. The test can also be used to identify the presence of

Table 64.4. The picrate test for the presence or liberation of cyanide

Preparation of Paper Test Strips

1. Cut filter paper into strips measuring 1 cm x 5 cm.
2. Dissolve 0.5 g picric acid and 5 g sodium bicarbonate in 100 ml of distilled water.
3. Dampen one-half of each strip in the solution and allow to dry.
4. Store the strips in a dark, airtight container, and keep container cool. The strips should be usable for a year or more when stored in these conditions.

The Test

1. Cut or shred enough leaf material to fill the bottom one-third of a 15–30 ml test tube.
2. Add a few drops each of chloroform and water.
3. Suspend a strip of the prepared filter paper from the top of the tube so that it does not touch the leaves, and stopper the tube tightly.
4. Hold the tube vertically in the hand or place it in warm (37°C) water for up to 2 hours.

Interpretation of Results

1. If HCN is released from the leaves, the yellow filter paper will turn brick red.
2. The rapidity of the change is indicative of the intoxication potential.
3. The potential is high if the color change occurs within 10 minutes of the beginning of the test.
4. The potential is minimal if little or no color change is apparent within 60–120 minutes.

Source: Modified from Guignard 1906.

cyanide in the stomach or rumen if done soon after death. Under some conditions it may be necessary to add a β-glycosidase to facilitate release of the cyanide from the plant material. The picrate test, and its various modifications, including the use of chloroform or toluene, has been described in numerous reports (Mirande 1909; Nowosad & MacVicar 1940; Hogg & Ahlgren 1942; Burnside 1954; Bergsten 1964). A note of caution: a solution of pure picric acid should not be allowed to crystallize and dry, because the crystals are unstable under some conditions and may detonate.

This test has been modified to permit quantitative determination of the cyanogenic glycosides by aqueous extraction of the reacted picrate/filter paper strips and spectrophotometric determination of the isopurpurine produced by the interaction of cyanide with picrate (Indira & Sinha 1969; Lucas & Sotelo 1984).

Other spot tests have been developed that are not necessarily exclusive for cyanide. One such test employs 1% copper ethylacetoacetate and 4,4'-tetramethyldiaminodiphenylmethane (tetra base) (Feigl & Anger 1966). The reagents, dissolved in chloroform, are spotted on filter paper, which is then suspended over the material to be tested. A blue color indicates the presumed presence of cyanide at levels of 1 µg or more.

Pathology—For the acute disease, there are no clearly incriminating pathologic changes, but only a few scattered, small splotchy hemorrhages scattered on the heart and other viscera. Mild congestion and edema of the lungs with froth in the airways are not uncommon. Microscopically, there may be degeneration and necrosis of the nervous system and heart in cases with more-prolonged signs, but the changes are neither distinctive nor diagnostic. In the subacute/chronic disease of humans or laboratory animals, there are distinctive microscopic changes in the optic and auditory nerves and spinal pathways.

- no specific lesions

Treatment—The disease usually occurs and progresses so rapidly that treatment is seldom available soon enough to be effective. This is not because the treatment itself is ineffective but because the disease is so rapidly fatal. Toxicosis due to marginally lethal dosages may be responsive to general supportive therapy, but specific treatment using nitrite to form methemoglobin and thiosulfate as a sulfur donor is much preferred. This approach has been recognized as effective since the early 1930s. It remains the mainstay of treatment for humans, dogs, and most other animal species (Chen et al. 1933; Bunyea et al. 1934; Clawson et al. 1934; Couch et al. 1935; Chen & Rose 1952; Berlin 1970; Burrows et al. 1973; Burrows 1981). The antidotes are administered i.v. Sodium thiosulfate given orally is of little or no value, because it is not absorbed and therefore is not utilized as a sulfur donor for rhodanese (Schulz 1984). When administered i.v., thiosulfate is oxidized to sulfur or eliminated unchanged, with a half-life of 15–20 minutes, as measured in humans (Schulz 1984). Oral therapy with glucose, molasses, or glyceraldehyde may provide some benefit probably by cyanhydrin formation (Haustein et al. 1969). However, a note of caution: this approach provides slow antagonism and should not be relied upon as a primary treatment. It should be used in conjunction with the traditional parenteral antidotes or possibly as a short-duration preventive (Fleming et al. 1926; Marsh et al. 1929).

- ruminants, sodium thiosulfate i.v., sodium nitrite may provide added benefit; other animals, combination of sodium nitrite and thiosulfate; other compounds and oxygen may also provide some benefit; dietary changes may be beneficial

In contrast to other animals, sheep and possibly other ruminants can be treated effectively with thiosulfate alone. They should be treated with a 30–40% solution of sodium thiosulfate i.v. at a dose of approximately 25–50 g/100 kg b.w. Early studies indicated a favorable response with doses as low as 2 g (20 ml of 10% solution), but the efficacy is much improved with the larger dosage. It is often stated that sodium thiosulfate is essentially nontoxic, but this is not necessarily true. Dosage of 3 g/kg b.w. given i.v. has proven lethal in dogs, in large measure because of hypernatremia and hyperosmolality (Dennis & Fletcher 1966). However, dosage of 1.5 g/kg was not accompanied by apparent adverse effects. Because of the large volume required, it is unlikely that toxic doses will be given in livestock. Sodium nitrite can be given concurrently, but it adds an additional risk that is probably not justified by the small increase in effectiveness. In humans, dogs, and probably most other animals, it is necessary to use 10–20 mg/kg b.w. of sodium nitrate in combination with sodium thiosulfate (Burrows et al. 1973; Baskin et al. 1992). In some cases, treatment may be effective several hours after intoxication, even in ruminants (Shaw 1986). This is especially true when plants with marginally lethal levels of the cyanogenic glycosides are eaten.

The use of other therapeutic agents has been the subject of several reviews (Burrows 1981; Way 1984; Salkowski & Penney 1994), and a number of such agents have been evaluated, including atmospheric or hyperbaric oxygen (Way et al. 1972; Burrows & Way 1977), chlorpromazine (Levine & Klein 1959; Way & Burrows 1976; Kong et al. 1983), naloxone (Leung et al. 1986), pyruvate (Cittadini et al. 1972), cobaltous chloride and various vasoactive drugs (Burrows & Way 1976; Burrows et al. 1977), and α-ketoglutaric acid (Hume et al. 1995). Although not all have been thoroughly evaluated, some increase the effectiveness of sodium thiosulfate or the classic nitrite/thiosulfate combination. However, in most instances the increase in effectiveness is not sufficient to justify their use, or, as with cobaltous chloride and the various MHb producers, there is an increased risk of toxicity (Way et al. 1972; Burrows & Way 1977). Methylene blue is an inappropriate antidote because it is not an effective producer of MHb (Couch et al. 1935; Manges 1935; Stossel & Jennings 1966).

Although little has been done to formulate a preventive approach to cyanide intoxication, the possibility of doing so remains. One approach would be the use of a cobalt or other heavy metal bolus in the rumen to inactivate the cyanide as it is formed. Another, more recent development involves the use of infused erythrocytes as carriers of the enzyme rhodanese in combination with butanethiosufonate, an organic, lipid-soluble sulfur donor. This approach has been shown experimentally to be effective (Petrikovics et al. 1995). Another approach to reducing the risk with cyanogenic forages involves dietary manipulation. Various dietary alterations have been suggested but not experimentally confirmed. This includes the use of starchy diets and/or manipulation to increase glucose content of the rumen (Peters et al. 1903; Vinall 1921; Couch 1932). Drenching with corn syrup or molasses may thus be useful as a preventive or in marginal intoxications (Peters et al. 1903). The role of pH on the rate of hydrolysis of the cyanogenic glycosides and the effect of grain supplementation on these factors may provide a management approach for reducing risk from cyanogenic forages. This is especially true when animals are newly introduced to pastures. Animals put on new grazing areas should be fully fed, preferably with a grain supplement, because both factors will provide protection by decreasing ruminal pH (Majak et al. 1990). Sulfur supplementation in salt to increase the availability of sulfur donors for the rhodanese conversion of cyanide to thiocyanate has been suggested as a way to provide for an increase in the rate of detoxication and resistance (Steyn 1934; Wheeler 1980).

■ *A. alnifolia*
A. canadensis
Juneberry
serviceberry
shadbush
sugar pear

Amelanchier Medik.

Taxonomy and Morphology—Comprising 10–20 species distributed in the Northern Hemisphere of both the Old World and the New World, *Amelanchier* is commonly known as serviceberry, shadbush, or Juneberry. Its name is believe to be a Latinization of an old French name. Hybridization, polyploidy, and apomixis are rampant, with intergradation of species common. Currently, 18 species and numerous varieties are recognized to be present in North America. Representative and of toxicologic interest are the following:

A. alnifolia Nutt.	western serviceberry, Saskatoon serviceberry
A. canadensis (L.) Medik.	shadbush, eastern serviceberry, Juneberry, sugar pear, shadblow serviceberry

Plants shrubs or trees; deciduous; not armed with thorns or prickles. **Stems** erect. **Leaves** simple; alternate; margins serrate or almost entire; stipules linear-lanceolate, caducous. **Inflorescences** short racemes at ends of branches; bracts present. **Flowers** produced before or simultaneously with leaves; perfect; perianths in 2-series. **Sepals** 5; erect to reflexed; persistent. **Petals** 5; white to pinkish. **Stamens** 10 or 15 or 20; shorter than petals. **Pistils** 1; compound, carpels 2–5; styles 2–5; free or fused; ovaries inferior; locules 2–5, typically divided by false septa; placentation axile. **Hypanthia** campanulate to urceolate. **Fruits** pomes; purple-black. **Seeds** 4–10.

■ shrubs or trees; leaves simple, alternate, stipules quickly dropping; petals 5, white to pinkish; stamens 5, 10, or 20; fruits pomes, purple-black

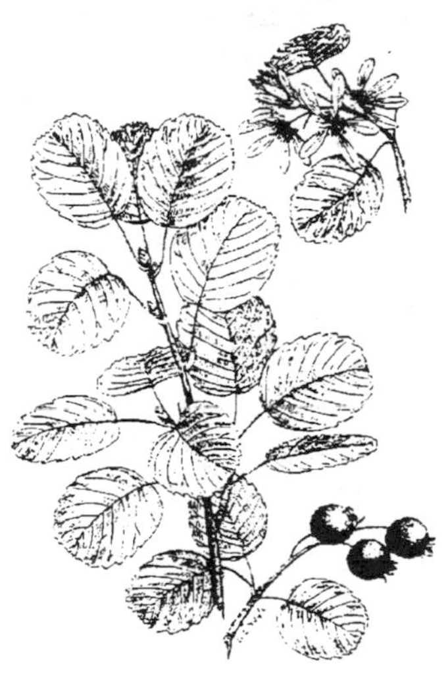

Amelanchier alnifolia

Amelanchier canadensis

Amelanchier alnifolia

Amelanchier canadensis

1. Ovary apices glabrous. Plants of eastern North America . . . . *A. canadensis*
1. Ovary apices tomentose. Plants of western North America . . . . *A. alnifolia*

Distribution and Habitat—Species of *Amelanchier* are distributed across the continent. A shrub or small tree, *A. alnifolia* occurs in meadows, brushlands, prairies, and open woodlands from Alaska to New Mexico and northeastward. *Amelanchier canadensis* is a shrub of swamps, water edges, and moist woodlands of the East, especially on the coastal plain.

■ meadows, prairies, woodlands

Disease Problems and Genesis—Members of *Amelanchier* are important browse for livestock and deer (Dayton 1931; Quinton 1985). However, under some circumstances cyanogenesis may represent a serious hazard. The primary toxicant is prunasin in all plant parts, with amygdalin additionally present in the fruits (Majak et al. 1978). Prunasin concentrations are typically 2–3% d.w., but early in the growing season may be as high as 4% (Majak et al. 1980). A level of 2–3% d.w. prunasin is equivalent to about 200–250 mg HCN/100 g of plant material.

■ browse for livestock; cyanogenesis risk early in growing season

Cercocarpus Kunth

- *C. montanus*
  birchleaf mountain mahogany

Taxonomy and Morphology—Native to western North America and northern Mexico, *Cercocarpus*, commonly known as mountain mahogany, comprises 10–13 species. Its name is derived from the Greek roots *kerkos* and *carpos*, for "tail" and "fruit," which allude to the long, feathery styles that persist on the mature fruits. Of toxicologic significance is the following:

C. montanus Raf. (= *C. breviflorus* A.Gray)	birchleaf, alder-leaf mountain mahogany, hairy mountain mahogany

- shrubs or trees, bark smooth, grayish to brown; leaves simple, alternate, clustered; sepals 5, yellowish, villous; petals absent; fruits achenes, with plumose style

Plants shrubs or small trees; not armed with thorns or prickles. **Stems** erect; 1–7 m tall; spur branches present; bark smooth, grayish to brown. **Leaves** simple; alternate; typically clustered; blades lanceolate or oblanceolate to subrotund; margins entire proximally, serrate or dentate distally; stipules deciduous. **Inflorescences** solitary or clusters of flowers in axils of leaves at ends of spur branches. **Flowers** produced after leaves; perfect; perianths in 1-series. **Sepals** 5; yellowish; recurved; silky-villous or tomentose; deciduous. **Petals** absent. **Stamens** 20–40; arising from hypanthia; anthers pubescent. **Pistils** 1; simple, carpels 1; styles 1, exserted beyond perianth, plumose, persistent; ovaries superior; locules 1. **Hypanthia** salverform. **Fruits** achenes; terete; pilose-villous; bearing elongate, plumose style. **Seeds** 1.

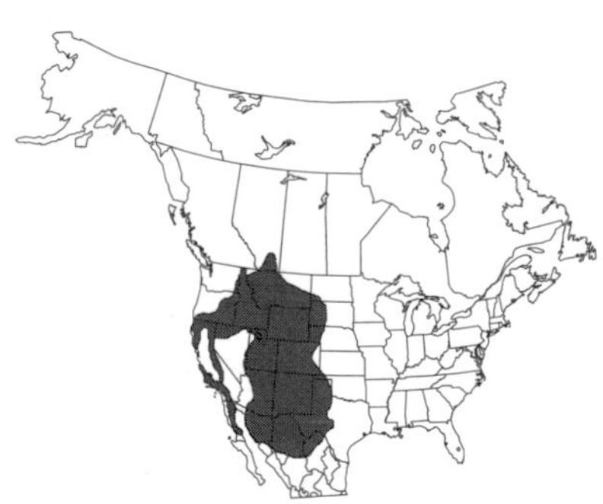

Cercocarpus montanus

Cercocarpus montanus

- grasslands, shrublands, and pine woodlands

Distribution and Habitat—Occurring on dry rocky hillsides, cliffs, and mesas, *C. montanus* is a species of grasslands, sagebrush lands, chaparral, and open pine woodlands to 2,500 m elevation throughout the western half of North America, including Mexico.

- valuable browse; rarely a cyanogenic risk, such as after first frosts in the fall

Disease Problems—*Cercocarpus montanus* is ordinarily considered to be a palatable and valuable browse for livestock, especially for sheep and goats (Dayton 1931; Chapline 1936). It also is extensively browsed by mule deer in winter, spring, and fall in some locations (Krausman et al. 1997). However, under some conditions it represents a cyanogenic risk, especially in the fall after the first frosts.

Cotoneaster Medik.

Taxonomy and Morphology—An Old World genus of the Northern Hemisphere comprising 50–70 species, *Cotoneaster* is a horticulturally important genus. It is grown for its brilliant foliage in autumn, colorful fruits in winter, and profusion of small flowers in spring. Quite diverse in habit and hardiness, numerous cultivars have been developed. The generic name is derived from the Latin *cotoneum,* for "quince," and the suffix *aster,* meaning "inferior," which indicate the similarity of its leaves to those of quince. *Cotoneaster* is closely related to *Crataegus* and is distinguished by the absence of thorns and by entire margins. Representative of the genus and possibly the most popular species are the following:

- *C. divaricatus*
  C. horizontalis

  cotoneaster
  rock cotoneaster
  rock spray
  spreading cotoneaster

C. divaricatus Rehder & E.H.Wilson	spreading cotoneaster
C. horizontalis Decne.	rock cotoneaster, rock spray

Plants shrubs or small trees; deciduous or evergreen; not armed with thorns or prickles. **Stems** erect or spreading. **Leaves** simple; alternate; blades typically small; margins entire; stipules narrow, caducous. **Inflorescences** solitary flowers or corymbose clusters. **Flowers** small; numerous; produced after leaves; perfect; perianths in 2-series. **Sepals** 5; persistent. **Petals** 5; white or rarely light pink. **Stamens** 10 or 20. **Pistils** 1; compound, carpels 2–5; styles 1–5; ovaries inferior; placentation axile. **Hypanthia** cup-shaped. **Fruits** pomes; drupelike; red or black or rarely yellow. **Seeds** 2–5.

- shrubs or small trees; leaves simple, alternate, small, stipules quickly dropping; flowers small; petals 5, white or light pink; stamens 10 or 20; fruits pomes, red, black, or yellow

A taxonomic key is not presented here, because the cultivars are so numerous and the listed species are only representative. If exact identification is needed, appropriate horticultural treatments should be consulted. Taxonomists have differed in their use of masculine or feminine endings for the specific epithets of *Cotoneaster.* Because the genus name is masculine, the specific epithets are treated as masculine in this work.

Distribution and Habitat—The majority of species of *Cotoneaster* introduced into North America are from China. Their native habitats are typically rocky, calcareous soils. Tolerant of atmospheric pollution and a variety of soils, they are cultivated here across the continent. In California, plants of *C. franchetii* and *C. pannosus* are encountered occasionally in disturbed sites near dwellings and in mixed-evergreen forests.

- ornamentals

Disease Problems and Genesis—HCN potentials vary considerably in *Cotoneaster.* For example, the potential of *C. zabelii* was 12 times higher than that of *C. roseus,* which had the lowest potential in a comparison of several species (Tidwell et al. 1970). Cyanide concentrations were about 1.5 mg/100 g in fruits and 17.5 mg/100 g in the leaves of *C. zabelii* and *C. divaricatus.* These levels are almost at the threshold of toxicity of 20 mg/100 g stated by Kingsbury (1964). Dosage of up to 6 g fruit/kg b.w. in the cat and 10 g fruit/kg b.w. in the dog were not toxic, whereas the LD_{50} of the fruit in rats was 3 g/kg b.w. However, even considering the low toxic dose for rats as representative of the toxic dose in humans, a dosage of 0.5 g/kg b.w., which is equivalent to 78 fruits for a child weighing 15 kg, would not be toxic.

- possibly cyanogenic, but risk is low

Heteromeles M.Roem.

Taxonomy and Morphology—A monotypic genus native to California and Baja California, *Heteromeles* is generally regarded as one of the state's most popular native shrubs. Bearing bright red fruits, it is widely planted as an ornamental and used at Christmas for decoration. Its name is derived from the Greek roots *hetero* and *mela,* for "different" and "apple," alluding to its dissimilarity to the apple. Closely related

- *H. arbutifolia*

  Christmas berry
  Christmas holly
  tollon
  toyon

to *Photinia,* it is distinguished by having 10 stamens rather than 20. The single species is the following:

H. arbutifolia (Lindl.) M.Roem. Christmas berry, toyon, California holly, tollon

■ evergreen shrubs or trees, bark grayish; leaves simple, alternate, thick, leathery, stipules deciduous; flowers in large flat-topped panicles; petals 5, white, spreading; stamens 10, borne in pairs; fruits pomes, bright red, ovoid

Plants shrubs or trees; evergreen; not armed with thorns or prickles. **Stems** erect; to 9 m tall; bark grayish; branchlets puberulent. **Leaves** simple; alternate; thick; leathery; blades elliptic to oblong, upper surfaces shiny dark green; margins serrate-dentate; stipules deciduous. **Inflorescences** large panicles; terminal; flat-topped. **Flowers** perfect; perianths in 2-series. **Sepals** 5; persistent. **Petals** 5; white; spreading. **Stamens** 10; borne in pairs opposite sepals. **Pistils** 1; compound, carpels 2 or 3; styles 2 or 3; ovaries inferior; locules 2 or 3. **Hypanthia** obconic. **Fruits** pomes; ovoid; bright red; pulp mealy; persistent sepals inflexed. **Seeds** 3–6; flattened; brown.

Heteromeles arbutifolia

■ chaparral, oak and conifer woodlands

Distribution and Habitat—*Heteromeles arbutifolia* occurs in chaparral, oak woodlands, and mixed-evergreen forests up to 1,300 m elevation. It is widely planted as an ornamental in warmer climates. It is not very tolerant of environmental stress.

■ cyanogenic risk mainly winter and spring

Heteromeles arbutifolia

Disease Problems and Genesis—*Heteromeles arbutifolia* is definitely cyanogenic (Dement & Mooney 1974; Dahlman & Johnson 1980). As is typical of other genera of the family, the cyanogenic potential is highest in the newly developing leaves. Peak concentrations of prunasin equivalents as converted from HCN potential are from 1.5–6+% d.w. of plant in the early spring, April–May. These levels decline somewhat during the summer months and then drop markedly in the fall. The levels rise again after the fall rains. The cyanogenic potential of the fruits is an interesting feature of this species. The pulp, but not the seeds, of immature fruits exhibits low HCN potential. But at maturation, there is little or no potential in the pulp, and concentrations in the seeds increase (Dement & Mooney 1974). Because of its evergreen nature, this species is a distinct hazard to grazing livestock during the winter and spring, when its cyanogenic potential is high and there is likely to be limited alternative forage. *Heteromeles* may be more of a problem in wet years because of decreased foliage development in dry years. It is apparently not always hazardous, because its presence on Isle Guadalupe, off the west coast of Baja California, has almost been eliminated by foraging feral goats (Wiggins 1980).

Malus Mill.

Taxonomy and Morphology—Comprising 25–35 species distributed in both the Old World and the New World, *Malus* is commonly known as apple or crabapple. Its name comes from the classical Latin word for "apple." Treated as a subgenus of *Pyrus,* or pear, by some taxonomists, it differs in having thorns, fused styles, and folded leaves in the buds and lacking stone cells in the fruit pulp. The distinction between an apple and a crabapple is nontechnical; it is simply based on fruit diameter, with the latter typically being less than 5 cm. Species of *Malus* are prized as ornamentals as well as for their edible fruits. The apple of commerce, now with numerous pomological varieties, is derived from the Eurasian *M. pumila* and/or its hybrids with *M. sylvestris,* which is also Eurasian. Its origins have been traced to the Caucasus region as early as 6,500 B.C. In North America, about 17 native, hybrid, and introduced species occur. The following are representative:

- *M. angustifolia*
- *M. coronaria*
- *M. ioensis*
- *M. pumila*
- *M. sylvestris*

apple
crabapple

M. angustifolia (Aiton) Michx.	southern crab
M. coronaria L.	sweet crab
M. ioensis (Wood) Britton	prairie crab
M. pumila Mill. (= *M. domestica* [Borkh.] Borkh.)	common apple, orchard apple
M. sylvestris Mill. (= *Pyrus malus* L.)	apple, common crabapple

Plants trees or shrubs; deciduous; usually armed with thorns. **Stems** erect; much branched; spur branches present. **Leaves** simple; alternate; blades ovate to oblong or elliptic, rolled or folded in bud; margins serrate or occasionally lobed; stipules deciduous. **Inflorescences** corymbs at ends of spur branches. **Flowers** showy; produced before or simultaneously with leaves; fragrant; perfect; perianths in 2-series. **Sepals** 5; persistent. **Petals** 5; white or pink or carmine red; obovate to orbicular. **Stamens** 20 or rarely 15 or more; arising from hypanthia. **Pistils** 1; compound, carpels 2–5; styles 2–5, fused at bases; ovaries inferior; locules 2–5; placentation axile. **Hypanthia** ovoid to globose or urceolate. **Fruits** pomes; red or yellow or green and numerous intermediate shades. **Seeds** 2–10.

- deciduous shrubs or trees, spur branches; leaves simple, alternate, stipules deciduous; flowers in corymbs at ends of spur branches; petals 5, white, pink, or carmine red; stamens 20; fruits pomes, red, yellow, or green

A taxonomic key is not presented here, because the cultivars are so numerous and the listed species are only representative. If exact identification is needed, appropriate horticultural treatments should be consulted.

Malus

Distribution and Habitat—Species of *Malus* occur across the continent. The native species occur principally in the woods of the eastern portion. The introduced species occasionally escape and persist.

- woodlands, cultivated species

Disease Problems and Genesis—Consumption of excess apples directly from the tree or as windfalls may produce a severe disease in about 24 hours. Animals give the appearance of being drunk. There is a marked reduction in milk production, digestive disturbance, and incoordination (Merrill 1952). The disease seems to worsen if animals are allowed to drink water after overeating of the apples, and it may be fatal in a few instances. Ruminal acidosis may be a factor with this or other disease manifestations (Teli et al. 1986).

- digestive disturbance, ataxia; cyanogenesis limited to low risk with seeds

Cyanogenesis is typically not a problem with *Malus* except for the seeds, which are of little risk. The specific toxicant causing the problems associated with excessive fruit consumption is not known. However, the seeds, fruits, and leaves contain phloretin as its glucoside, phlorizin, at 1% in leaves and 300–400 ppm in the fruits and seeds (Singleton & Kratzer 1973). Phlorizin blocks active transport of glucose in the kidney and intestine. It prevents transtubular reabsorption of glucose from urine to blood and small intestinal uptake, thereby producing pronounced glucosuria. Phlorizin is very potent, but because of hydrolysis and poor oral absorption, both it and phloretin are relatively inactive orally. Nevertheless, these compounds may play some role when large numbers of apples are eaten.

- loss of appetite and milk production, diarrhea, weakness, colic, ataxia

  treatment, nursing care, laxatives

Clinical Signs and Treatment—There is a marked reduction in milk production, loss of appetite, diarrhea, weakness, paresis, colic, and incoordination. Eventually coma and death follow in more-severe cases. The specific cause is unknown, and treatment is symptomatic, including the use of laxatives.

Photinia Lindl.

- *P. glabra*
  P. serratifolia
  P. villosa

  Chinese Christmas berry
  Japanese Christmas berry
  oriental Christmas berry
  photinia

Taxonomy and Morphology—Native to eastern and southeastern Asia, *Photinia,* commonly known as Christmas berry, comprises about 60 species. Indicative of its glossy leaves, its name is derived from the Greek root *photeinos,* for "shining." Plants are prized for their attractive foliage—typically bright scarlet or bronze to coppery red—in the spring and their showy red fruits, which persist in the winter. They are grown as wall shrubs or as lawn specimens. Most commonly encountered in North America are the following:

P. glabra (Thunb.) Maxim.	Japanese photinia, Japanese Christmas berry
P. serratifolia (Desf.) Kalkman (= *P. serrulata* Lindl.)	Chinese photinia, Chinese Christmas berry
P. villosa (Thunb.) DC.	oriental photinia, oriental Christmas berry

- shrubs or trees; leaves simple, alternate, leathery, dark glossy green, reddish when young; flowers in showy panicles; petals 5, white; fruits red pomes

Plants trees or shrubs; deciduous or evergreen; not armed with thorns or prickles. **Stems** 3–12 m tall; short lateral branches present; branchlets glabrous or villous. **Leaves** simple; alternate; leathery; red when young becoming dark glossy green; blades ovate to elliptic to obovate; margins serrate; stipules foliaceous. **Inflorescences** showy terminal corymbose panicles. **Flowers** perfect; perianths in 2-series. **Sepals** 5; persistent. **Petals** 5; white or creamy white; spreading. **Stamens** about 20. **Pistils** 1; compound, carpels 2–5; styles 2 or rarely 1 or 3–5, fused at base; ovaries inferior; locules 2–4; placentation axile. **Hypanthia** obconic or globose. **Fruits** pomes; globose or ellipsoid or ovoid; red. **Seeds** 2–4.

Photonia serratifolia

1. Plants deciduous. Branchlets villous when young. Panicles 2.5–5 cm in diameter. Pedicels with warty lenticels. Pomes ellipsoid . . . . *P. villosa*
1. Plants evergreen. Branchlets glabrous. Panicles 5–15 cm in diameter. Pedicels without warty lenticels. Pomes globose.
   2. Leaves 10–20 cm long. Petals glabrous . . . . *P. serratifolia*
   2. Leaves 5–9 cm long. Petals villous at base . . . . *P. glabra*

- ornamentals

Distribution and Habitat—Native to eastern Asia, all three species are widely cultivated ornamentals found throughout our range except in the warmest and coldest portions. *Photinia villosa* is the most hardy in the North.

Disease Problems and Genesis—As is characteristic of other members of the family, the leaves and seeds of different species of *Photinia* exhibit cyanogenic potential. Both prunasin and amygdalin have been detected.

- low risk of cyanogenesis

Prunus L.

Taxonomy and Morphology—With greatest abundance in the temperate regions of the Northern Hemisphere, *Prunus* comprises 400–450 species, a number of named hybrids, and innumerable varieties and cultivars. A few taxonomists contend there are only about 200 species. Its name is an ancient Latin one for the plum and cherry. Its members are prized as ornamentals for their leaf color, bark color, and spring displays of showy flowers. Its wood is used for furniture and carved objects, and 1 species yields sloe. Of greatest importance are the stone fruits such as cherries, peaches, plums, nectarines, almonds, and apricots. Some taxonomists divide the genus into several genera, and thus cherries are placed in *Padus* and *Cerasus,* almonds in *Amygdalus,* apricots in *Armeniaca,* and peaches in *Persica.* More commonly, as is done in this treatment, a broad circumscription of *Prunus* with subgenera is used. In North America approximately 50 native and introduced species are present. Of particular toxicologic concern are these:

P. armeniaca L.	apricot
P. brachybotrya Zucc.	aguacatillo, cerezo
P. caroliniana (Mill.) Aiton	laurel cherry, mock orange
P. dulcis (Mill.) D.A.Webber	almond
P. fasciculata (Torr.) A.Gray	California desert almond, desert peach
P. laurocerasus L.	cherry laurel
P. macrophylla Siebold & Zucc.	
P. pensylvanica L.f.	bird cherry, pin cherry, wild red cherry
P. persica (L.) Batsch	peach
P. serotina Ehrh.	black cherry, wild cherry, rum cherry
P. virginiana L. (= *P. demissa* [Nutt.] Walp.)	chokecherry

- *P. armeniaca*
  P. brachybotrya
  P. caroliniana
  P. dulcis
  P. fasciculata
  P. laurocerasus
  P. macrophylla
  P. pensylvanica
  P. persica
  P. serotina
  P. virginiana

aguacatillo
almond
apricot
bird cherry
black cherry
California desert almond
cerezo
cherry
cherry laurel
chokecherry
desert peach
laurel cherry
mock orange
peach
pin cherry

Plants trees or shrubs; armed with thorns; may or may not form thickets via root sprouts. **Stems** spur branches typically present; bark smooth or breaking into smooth plates; lenticels conspicuous, horizontal. **Leaves** simple; alternate; blades conduplicate in bud, of various shapes, typically glossy above; margins entire or serrate; 1 or 2 glands present at base of blades or on petioles; stipules deciduous. **Inflorescences** racemes or umbellate clusters or rarely solitary flowers; borne at ends of branches or on spurs. **Flowers** produced before or simultaneously with or after leaves; perfect; perianths in 2-series. **Sepals** 5; spreading. **Petals** 5 or 10; spreading; white to pink or red or greenish; deciduous. **Stamens** 15 or 20; in whorls; arising from hypanthia. **Pistils** 1; simple, carpels 1; styles 1; ovaries superior; locules 1. **Hypanthia** cup-shaped or obconic or urceolate. **Fruits** fleshy drupes; endocarp hard. **Seeds** 1 per stone; round or flattened.

- shrubs or trees, spur branches, bark smooth, horizontal lenticels conspicuous; leaves simple, alternate, glands 1 or 2 at blade bases or on petioles; petals 5 or 10, spreading, white, pink, red, or greenish; fruits fleshy drupes

Because the listed species are only representative and there are so many cultivars, a taxonomic key is not presented here. If exact identification is needed, appropriate horticultural treatments should be consulted.

Prunus caroliniana

Prunus serotina

Prunus virginiana

■ cultivated; grasslands, deserts, woodland forests

Prunus caroliniana

Distribution and Habitat—Species of *Prunus* occur across the continent in a variety of habitats, ranging from the rich moist soils of forests and floodplains in the East to the sandy or rocky soils of the prairies and deserts. Some, such as *P. serotina,* have relatively broad geographical and ecological distributions, whereas others are more restricted, e.g., *P. fasciculata.* Introduced Old World species often persist at old homesites or occasionally escape and appear to naturalize somewhat.

Disease Problems and Genesis—During certain times of the year and in some localities, the foliage of *Prunus* species is useful browse of moderate palatability (Dayton 1931). Many species are cyanogenic, but most are not a significant hazard. The foliage of *P. brachybotrya, P. laurocerasus, P. serotina,* and *P. virginiana* probably pose the most significant risks. They also are reasonably palatable (Dayton 1931; Chapline 1936). Although instances of intoxication in monogastric species due to *Prunus* are rare, the young or wilted foliage of the wild cherries poses some risk for dogs and horses (Hansen 1930; Stauffer 1970; Clay 1977). In some instances the seeds within

Prunus fasciculata

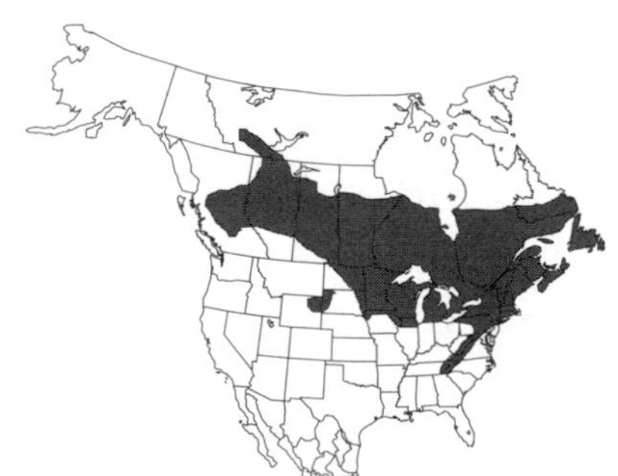

Prunus pensylvanica

Prunus serotina

the pits are a greater risk than the leaves. This is especially true for apricots and peaches; in contrast the seeds of cultivated cherries seem to have little cyanogenic potential at maturity (Swain et al. 1992). Apricot seeds, which are sold in health food stores oftentimes without appropriate warnings, are a health risk. In one instance, a woman ingested 20–30 raw apricot seeds and within minutes became deathly ill (Wallace et al. 1996). She was revived by prompt treatment with sodium nitrite and sodium thiosulfate. The pulp of the fruit is not toxic, but the seeds have about 4 times more cyanide than the leaves, 40+ mg/100 g and 10 mg/kg, respectively, whether an Elberta or Dixired type of peach (Machel & Dorsett 1970). In the case of the wild or bitter almonds, i.e., *P. fasciculata,* the leaves as well as the seeds may have high cyanogenic potential, up to 1,700 ppm (Moran et al. 1940). Wild black cherries *(P. serotina)* also pose some risk for humans, as do chokecherries *(P. virginiana)*. Several cases of intoxication have been reported in which individuals experienced sudden onset of peracute neurologic dysfunction and death following ingestion of very large amounts of chokecherries, including the seeds (Pijoan 1942). In addition to the neurologic and respiratory effects, epigastric pain and vomiting were commonly seen as early signs.

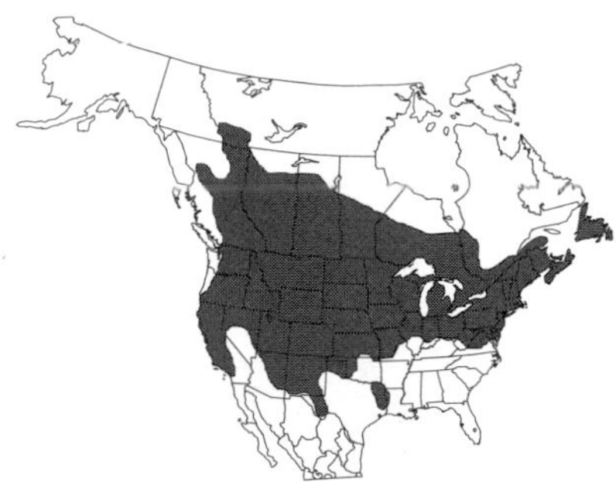

Prunus virginiana

- cyanogenesis potential from foliage mainly in spring with *P. brachybotrya, P. laurocerasus, P. serotina, P. virginiana*

  seeds also pose a risk, especially apricots

Pyracantha Roem.

Taxonomy and Morphology—Commonly known as firethorn, *Pyracantha* comprises 6 or 7 Old World species native to southeastern Europe and Asia. Its name is derived from the Greek roots *pyr* and *acantha,* for "fire" and "thorn." Closely related to *Cotoneaster* and *Crataegus,* it is a popular ornamental used for hedging or screening, and numerous cultivars have been developed. Perhaps most commonly encountered is this species:

- *P. coccinea*

  buisson ardent
  everlasting thorn
  fiery thorn
  firethorn
  pyracantha

P. coccinea Roem.	firethorn, pyracantha, fiery thorn, buisson ardent, everlasting thorn

Plants shrubs; evergreen; armed with thorns. **Stems** to 2 m tall; short lateral branches present; thorns leafy; young branchlets grayish pubescent. **Leaves** simple; alternate; blades narrowly elliptic, dark glossy green; apices acute; margins crenulate or serrulate; stipules small, deciduous. **Inflorescences** corymbs borne at ends of lateral branches; peduncles and pedicels pubescent. **Flowers** perfect; perianths in 2-series. **Sepals** 5; persistent. **Petals** 5; white; orbicular. **Stamens** 20; filaments fused at bases. **Pistils** 1; compound, carpels 5; styles 5; ovaries half inferior; locules 5. **Hypanthia** obconic to globose. **Fruits** pomes; globose or subglobose; orange to red or scarlet. **Seeds** 5.

- evergreen shrubs, armed with thorns; leaves simple, alternate, glossy green; flowers in corymbs; petals 5, white; fruits pomes, orange, red, or scarlet

Distribution and Habitat—Species of *Pyracantha* are cultivated as ornamentals throughout much of North America. *P. coccinea* occasionally escapes cultivation.

- ornamentals

Disease Problems and Genesis—Prunasin has been detected in the tissues of *P. coccinea.*

- possibly cyanogenic but risk very low

Rhodotypos Siebold & Zucc.

Taxonomy and Morphology—A monotypic genus native to Japan and China, *Rhodotypos* is an unusual member of the Rosaceae in that it has opposite leaves and 4-merous flowers. Its name is derived from the Greek roots *rhodon* and *typos,* for

- *R. scandens*

  jetbead

"rose" and "type," which reflect the resemblance of its flowers to those of *Rosa*. Plants are cultivated for their prolific production of showy flowers from spring to midsummer and for its shining jet black, pea-size fruits that remain on the plant in winter. The single species is the following:

R. scandens (Thunb.) Makino — jetbead
(= *R. tetrapetala* [Siebold] Makino)
(= *Kerria tetrapetala* Siebold)

■ deciduous shrubs; leaves simple, opposite; flowers solitary, showy; sepals 4, fused; petals 4, white; pistils 4, simple; fruits drupes, shiny jet black

Plants shrubs; deciduous; not armed with thorns or prickles. **Stems** 1–5 m tall; bark brown becoming reddish, glabrous. **Leaves** simple; opposite; blades ovate to oblong, rugose above; apices acuminate; margins doubly serrate; bases rounded; stipules linear, free, membranous, pubescent. **Inflorescences** solitary flowers at ends of leafy shoots. **Flowers** showy; perfect; perianths in 2-series; epicalyces present, bracts 4, linear. **Sepals** 4; fused, tubes short; persistent. **Petals** 4; white; orbicular. **Stamens** numerous. **Pistils** 4; simple, carpels 1; styles 1; ovaries superior. **Hypanthia** cup-shaped. **Fruits** drupes; slightly ellipsoid; shiny; jet black. **Seeds** 1.

■ ornamentals

Distribution and Habitat—Introduced from Asia, *R. scandens* is cultivated as a border plant and in woodland gardens. It is tolerant of a wide range of soil conditions, shade levels, and temperatures.

■ low risk of cyanogenesis

Disease Problems—Like other members of the Rosaceae, the leaves of *R. scandens* exhibit cyanogenic potential.

Genera of Questionable Toxicity

Coleogyne Torr.

■ *C. ramosissima*
blackbrush

■ thorny shrub; leaves in opposite fascicles; flowers at stem tips, yellow; sepals 4; petals absent; stamens numerous; fruits achenes

A desert shrub, *C. ramosissima* Torr., commonly known as blackbrush, is the sole species of *Coleogyne*. Its thorny, much branched, more or less strigose stems are 30–200 cm tall and bear opposite fascicles of thick, linear-oblanceolate leaves with entire margins. Borne at the stem tips and subtended by a linear 2-lobed bract, the solitary, generally yellow flowers consist of a campanulate, leathery hypanthium with

Coleogyne ramosissima

Coleogyne ramosissima

a sheath at the top that encloses the single pistil. The 4 sepals are elliptic, erect, and persistent, and their outer surfaces are often reddish. The petals are absent, and there are 30–40 stamens. The pistil has a superior ovary and a persistent, laterally arising style bearing long hairs. The fruit is a brown, glabrous, more or less crescent-shaped achene.

- high tannin content presents nutritional problem; very low cyanogenic risk

Coleogyne ramosissima is typically considered to be poor browse for livestock, but under some conditions it is somewhat useful for deer (Dayton 1931). The current season's growth is of limited palatability, and older growth is preferred by animals. This preference may be the result of the concentration of condensed tannins being 2–4 times higher in the new growth (Provenza et al. 1990). Such high tannin content limits the nutrient value of the plants and represents a potential nutritional problem. Although cyanogenesis has been reported for the species, it is not likely to be a significant risk.

Spiraea L.

- *Spiraea*

  bridal wreath
  spiraea

Native to both the Old World and the New World, *Spiraea,* commonly known as spiraea or bridal wreath, comprises 50–70 species of deciduous shrubs, many of which are considered valuable ornamentals. Its name is derived from the Greek root *speira,* for "coil," which is believed to reflect its use in garlands. The common name *spiraea* is also used for species of the genera *Aruncus* and *Filipendula* in this family and *Astilbe* in the Saxifragaceae.

- shrubs, leaves serrate; petals white, pink, or carmine; pistils 5, simple, enclosed in hypanthium; fruits follicles

Species of *Spiraea* vary considerably in general growth form, but all have simple, alternate leaves, typically with serrate or dentate margins. Stipules are usually absent. The perfect, white, pink, or carmine flowers are borne in racemes, panicles, or corymbs. Borne on the rim of the cupulate or campanulate or obconic hypanthium are 5 persistent sepals; 4 or 5 petals (sometimes more in some cultivars); and 15–60 stamens. Within the hypanthium are 5 simple pistils (rarely 1–4) with superior ovaries and beaked styles that give rise to 2-to-10-seeded, indurate follicles.

- diterpenoid alkaloids, unknown toxicity potential; cyanogenic risk very low

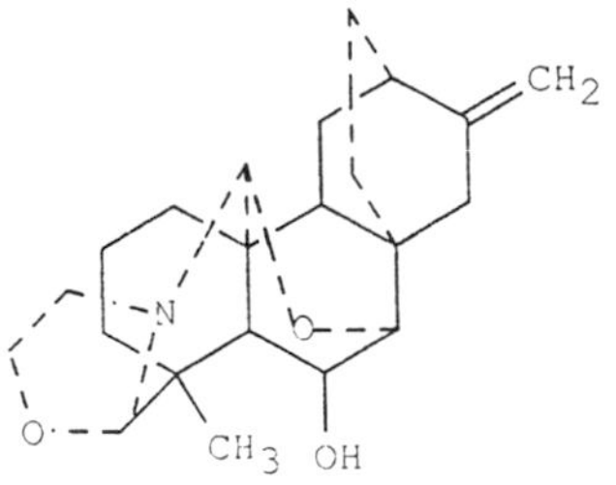

In contrast to other genera in the family, *Spiraea* contains a series of diterpenoid alkaloids in addition to cyanogenic glycosides. Found in *S. japonica* and *S. koreana,* these alkaloids include spiradine, spiramine, spirasine, and spirajin (Yunusov 1991, 1993). They have analgesic and antinociceptive activity similar to that of the alkaloids in *Aconitum,* in the Ranunculaceae (see chapter 62), but their role as toxicants is not known.

Spiradine G

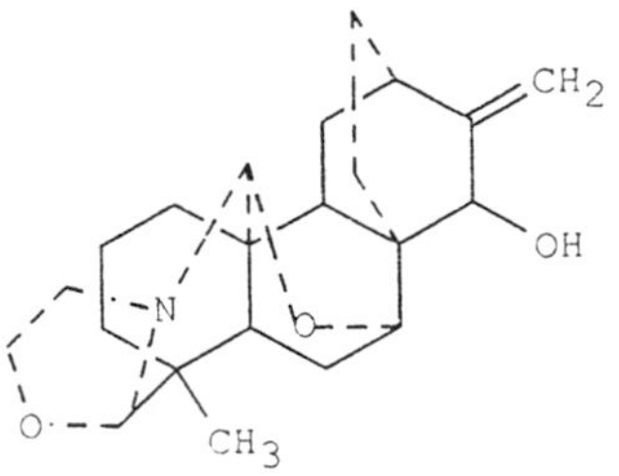

Spiramine C

References

Adewusi SRA, Oke OL: On the metabolism of amygdalin. 2. The distribution of β-glucosidase activity and orally administered amygdalin in rats. *Can J Physiol* 63;1084–1087, 1985

Albaum HG, Tepperman J, Bodansky O: The in vivo inactivation by cyanide of brain cytochrome oxidase and its effect on glycolysis and on the high energy phosphorus compounds in brain. *J Biol Chem* 164;45–51, 1946

Baar S: The micro determination of cyanide: its application to the analysis of whole blood. *Analyst* 91;268–272, 1966

Ballantyne B: In vitro production of cyanide in normal human blood and the influence of thiocyanate and storage temperature. *Clin Toxicol* 11;173–193, 1977

Ballantyne B, Bright J, Swanson DW, Williams P: Toxicity and distribution of free cyanides given intramuscularly. *Med Sci Law* 12;209–219, 1972

Barnea A, Harborne JB, Pannell C: What parts of fleshy fruits contain secondary compounds toxic to birds and why? *Biochem Syst Ecol* 21;421–429, 1993

Barrett MD, Hill DC, Alexander JC, Zitnak A: Fate of orally dosed linamarin in the rat. *Can J Physiol Pharmacol* 55;134–136, 1977

Baskin SI, Horowitz AM, Nealley EW: The antidotal action of sodium nitrite and sodium thiosulfate against cyanide poisoning. *J Clin Pharmacol* 32;368–375, 1992

Bergsten ML: Testing for prussic acid. *Vet Med/Small Anim Clin* 59;720–721, 1964

Berlin CM: The treatment of cyanide poisoning in children. *Pediatrics* 46;794–796, 1970

Boyd FT, Aamodt OS, Bohstedt G, Truog E: Sudan grass management for control of cyanide poisoning. *J Am Soc Agron* 30;569–582, 1938

Bradley M, Neiman LJ, Burrows GE: Seizures in a puppy [letter]. *Vet Hum Toxicol* 30;121, 1988

Briese RR, Couch JF: Mercuric chloride as a preservative of cyanogenetic plants for chemical analysis. *J Agric Res* 62;493–507, 1941

Brooke BM, McDiarmid RE, Majak W: The cyanide potential in two varieties of *Amelanchier alnifolia. Can J Plant Sci* 68;543–547, 1988

Brunnich JC: Hydrocyanic acid in fodder-plants. *J Chem Soc* 83;788–796, 1903

Bunyea H, Couch JF, Clawson AB: The nitrite-thiosulphate combination as a remedy for cyanide poisoning in sheep. *J Wash Acad Sci* 24;528–532, 1934

Burnside JE: Animal poisoning and diagnosis. *Vet Med* 49;136–146, 1954

Burrows GE: Cyanide intoxication in sheep: therapeutics. *Vet Hum Toxicol* 23;22–28, 1981

Burrows GE, Way JL: Antagonism of cyanide toxicity by phenoxybenzamine [abstract]. *Fed Proc* 35;533, 1976

Burrows GE, Way JL: Cyanide intoxication in sheep: therapeutic value of oxygen or cobalt. *Am J Vet Res* 38;223–227, 1977

Burrows GE, Liu DHW, Way JL: Effect of oxygen on cyanide intoxication. 5. Physiologic effects. *J Pharmacol Exp Ther* 184;739–748, 1973

Burrows GE, King L, Tarr S, Way JL: The protective effect of vasoactive compounds as cyanide antagonists [abstract]. *Proc Int Congr Toxicol* 1;50, 1977

Cameron JF: Natural substances suspected of killing birds in British Columbia. *Biol Conserv* 4;223, 1972

Carella F, Grassi MP, Savoiardo M, Contri P, Rapuzzi B, Mangoni A: Dystonic-Parkinsonian syndrome after cyanide poisoning: clinical and MRI findings. *J Neurol Neurosurg Psychiatry* 51;1345–1348, 1988

Carter JH, McLafferty MA, Goldman P: Role of the gastrointestinal microflora in amygdalin (laetrile)-induced cyanide toxicity. *Biochem Pharmacol* 29;301–304, 1980

Cassel G, Persson SA: Effects of acute lethal cyanide intoxication on central dopaminergic pathways. *Pharmacol Toxicol* 70;148–151, 1992

Chandler RF, Phillipson JD, Anderson LA: Controversial laetrile. *J Pharm* 232;330–332, 1984

Chapline WR: *Observation of Feeding Habits of Sheep on the Range.* N M Agric Exp Stn Bull 241, pp41–53, 1936

Chen KK, Rose CL: Nitrite and thiosulfate therapy in cyanide poisoning. *J Am Med Assoc* 149;113–119, 1952

Chen KK, Rose CL, Clowes GHA: Methylene blue, nitrites, and sodium thiosulfate against cyanide poisoning. *Proc Soc Exp Biol Med* 31;250–252, 1933

Christel D, Eyer P, Hegemann M, Kiese M, Lorcher W, Weger N: Pharmacokinetics of cyanide in poisoning of dogs, and the effect of 4-dimethylaminophenol or thiosulfate. *Arch Toxicol* 38;177–189, 1977

Cittadini A, Caprino L, Terranova T: Effect of pyruvate on the acute cyanide poisoning in mice. *Experientia* 28;943–944, 1972

Clawson AB, Bunyea H, Couch JF: Remedies for cyanide poisoning in sheep and cattle. *J Wash Acad Sci* 24;369–385, 1934

Clay BR: Poisoning and injury by plants. In *Current Veterinary Therapy*, vol. 6, *Small Animal Practice,* Kirk RW ed, Saunders, Philadelphia, pp179–184, 1977

Conn EE: Cyanogenetic glycosides. In *Toxicants Occurring Naturally in Foods,* Committee on Food Protection, National Research Council, National Academy of Sciences, Washington, DC, pp299–308, 1973

Conn EE: Cyanogenic glycosides. *Int Rev Biochem, Biochem Nutr 1A* 27;21–43, 1979

Conn EE: The metabolism of a natural product: lessons learned from cyanogenic glycosides. *Planta Med* 57 (Suppl 1);S1–S9, 1991

Cooke RD: An enzymatic assay for the total cyanide content of cassava (*Manihot esculenta* Crantz). *J Sci Food Agric* 29;345–352, 1978

Coop IE, Blakley RL: The metabolism and toxicity of cyanides and cyanogenic glycosides in sheep. 3. The toxicity of cyanides and cyanogenic glycosides. *N Z J Sci Technol* 31;44–58, 1950

Couch JF: *Poisoning of Livestock by Plants That Produce Hydrocyanic Acid.* USDA Leaflet 88, 1932 (rev 1940)

Couch JF, Bunyea H, Clawson AB: The relationship between time of administration and effectiveness of remedies for cyanide poisoning. *J Wash Acad Sci* 25;57–59, 1935

Cronquist, A: *An Integrated System of Classification of Flowering Plants.* Columbia Univ Press, New York, 1981

Dahlman DL, Johnson V: *Heteromeles arbutifolia* (Rosaceae: Pomoideae) found toxic to insects. *Entomol News* 91;141–142, 1980

Dayton WA: *Important Western Browse Plants.* USDA Misc Publ 101, 1931

Dement WA, Mooney HA: Seasonal variation in the production of tannins and cyanogenic glucosides in the chaparral shrub, *Heteromeles arbutifolia. Oecologia* 15;65–76, 1974

Dennis DL, Fletcher WS: Toxicity of sodium thiosulfate (NSC-45624), a nitrogen mustard antagonist, in the dog. *Cancer Chemother Rep* 50;255–257, 1966

Doherty PA, Ferm VH, Smith RP: Congenital malformations induced by infusion of sodium cyanide in the golden hamster. *Toxicol Appl Pharmacol* 64;456–464, 1982

Dufour DL: Cyanide content of cassava (*Manihot esculenta,* Euphorbiaceae) cultivars used by Tukanoan Indians in northwest Amazonia. *Econ Bot* 42;255–266, 1988

Egekeze JO, Oehme FW: Blood and liver cyanide concentrations in rats poisoned with oral doses of potassium cyanide. *Toxicol Lett* 3;243–247, 1979

Enge EH: Wild-cherry poisoning in pigs. *J Am Vet Med Assoc* 94;123–124, 1939

Feigl F, Anger V: Replacement of benzidine by copper ethylacetoacetate and tetra base as spot-test for hydrogen cyanide and cyanogen. *Analyst* 91;282–284, 1966

Fikenscher LH, Hegnauer R, Ruijgrok HWL: Distribution of hydrocyanic acid in Cormophyta. 15. New observations on cyanogenesis in Rosaceae. *Planta Med* 41; 313–327, 1981

Finck PA: Postmortem distribution studies of cyanide. *Med Ann DC* 38;357–358, 1969

Fleming CE, Dill R: *The Poisoning of Sheep on Mountain Grazing Ranges in Nevada by the Western Chokecherry.* Nev Agric Exp Stn Bull 110, 1928

Fleming CE, Miller MR, Vawter LR: *The Common Chokecherry as a Plant Poisonous to Sheep and Cattle.* Nev Agric Exp Stn Bull 109, 1926

Frakes RA, Sharma RP, Willhite CC: Developmental toxicity of the cyanogenic glycoside linamarin in the golden hamster. *Teratology* 31;241–246, 1985

Frakes RA, Sharma RP, Willhite CC: Comparative metabolism of linamarin and amygdalin in hamsters. *Food Chem Toxicol* 24;417–420, 1986

Frankenberg L, Sorbo B: Effect of cyanide antidotes on the metabolic conversion of cyanide to thiocyanate. *Arch Toxicol* 33;81–89, 1975

Gettler AO, Baine JO: The toxicology of cyanide. *Am J Med Sci* 195;182–198, 1938

Gibbs RD: *Chemotaxonomy of Flowering Plants,* vol 1. McGill-Queens Univ Press, Montreal, pp57–69, 629–635, 1974a

Gibbs RD: *Chemotaxonomy of Flowering Plants,* vol 3. McGill-Queens Univ Press, Montreal, pp1621–1623, 1974b

Gill DA, McGregor P: Laurel poisoning in stock. *N Z J Agric* 37;407, 1928

Gomez G, de la Cuesta D, Valdivieso M, Kawano K: Total and free cyanide content of the parenchyma and peel of ten promising cassava genotypes. *Turrialba* 30;361–365, 1980

Gough JF: Black cherry poisoning in an Angora goat. *Can Vet J* 36;45, 1995

Graham DL, Laman D, Theodore J, Robin ED: Acute cyanide poisoning complicated by lactic acidosis and pulmonary edema. *Arch Intern Med* 137;1051–1055, 1977

Guignard L: Le haricot á acide cyanhydrique. Etude historique, botanique et chimique; Nouveau procede pour deceler l'acide cyanhydrique. *Bull Sci Pharmacol* 13;415–417, 1906

Halstrom F, Moller KO: The content of cyanide in human organs from cases of cyanide taken by mouth. *Acta Pharmacol* 1;18–28, 1945

Hansen AA: Stock poisoning plants. *North Am Vet* 9;24–27, 1928

Hansen AA: *Indiana Plants Injurious to Livestock*. Ind Agric Exp Stn Circ 175, 1930

Haustein J, Gabsch HC, Lohs KH, Detoxification of cyanide by DL-glyceraldehyde. *Pharmazie* 24;482–483, 1969

Hegnauer R: Taxonomic value of cyanogenesis in higher plants. *Rec Adv Bot* 1(2);82–86, 1961

Hegnauer R: *Chemotaxonomie der Pflanzen,* vol 6. Birkhauser Verlag, Basel, pp727–731, 1973

Hegnauer R: *Chemotaxonomie der Pflanzen,* vol 7. Birkhauser Verlag, Basel, pp345–373, 1986

Hegnauer R: *Chemotaxonomie der Pflanzen,* vol 9. Birkhauser Verlag, Basel, pp369–396, 1990

Herbert V: Laetrile: the cult of cyanide poisoning. Promoting poison for profit. *Am J Clin Nutr* 32;1121–1158, 1979

Hindmarsh WL: The lethal dose of hydrocyanic acid for ruminants. *J Counc Sci Ind Res Aust* 3;12–13, 1930

Hogg PG, Ahlgren HL: A rapid method for determining hydrocyanic acid content of single plants of sudan grass. *J Am Soc Agron* 34;199–200, 1942

Hogness TR: Cytochrome oxidase. *Symp Quant Biol* 7;121–127, 1939

Hosel W: The enzymatic hydrolysis of cyanogenic glycosides. In *Cyanide in Biology,* Vennesland B, Conn EE, Knowles CJ, Westley J, Wissing F eds, Academic Press, London, pp217–232, 1981

Howlett WP, Brubaker GR, Mlingi N, Rosling H: Konzo, an epidemic upper motor neuron disease studied in Tanzania. *Brain* 113;223–235, 1990

Humbert JR, Tress JH, Braico KT: Fatal cyanide poisoning: accidental ingestion of amygdalin [letter]. *J Am Med Assoc* 238;482, 1977

Hume AS, Mozingo JR, McIntyre B, Ho IK: Antidotal efficacy of alpha-ketoglutaric acid and sodium thiosulfate in cyanide poisoning. *Clin Toxicol* 33;721–724, 1995

Indira P, Sinha SK: Colorimetric method for determination of HCN in tubers and leaves of cassava (*Manihot esculenta* Crantz). *Indian J Agric Sci* 39;1021–1023, 1969

Johnson JD, Meisenheimer TL, Isom GE: Cyanide-induced neurotoxicity: role of neuronal calcium. *Toxicol Appl Pharmacol* 84;464–469, 1986

Jones DA, Keymer RJ, Ellis WM: Cyanogenesis in plants and animal feeding. In *Biochemical Aspects of Plant and Animal Coevolution,* Harbone JB ed, Academic Press, London, pp21–34, 1978

Kanthasamy AG, Borowitz JL, Pavlakovic G, Isom GE: Dopaminergic neurotoxicity of cyanide: neurochemical, histological, and behavioral characterization. *Toxicol Appl Pharmacol* 126;156–163, 1994

Keilin D: Cytochrome and intracellular oxidase. *Proc Roy Soc,* ser B, 106;418–444, 1930

Kiese M, Weger N: Formation of ferrihaemoglobin with aminophenols in the human for the treatment of cyanide poisoning. *Eur J Pharmacol* 7;97–105, 1969

Kingsbury JM: *Poisonous Plants of the United States and Canada.* Prentice-Hall, Englewood Cliffs, NJ, 1964

Kojima M, Poulton JE, Thayer SS, Conn EE: Tissue distributions of dhurrin and of enzymes involved in its metabolism in leaves of *Sorghum bicolor. Plant Physiol* 63;1022–1028, 1979

Kong A, Shen A, Burrows GE, Sylvester D, Isom GE, Way JL: Effect of chlorpromazine on cyanide intoxication. *Toxicol Appl Pharmacol* 71;407–413, 1983

Krausman PR, Kuenze AJ, Etchberger RC, Rautenstrauch KR, Ordway LL, Hervert JJ: Diets of desert mule deer. *J Range Manage* 50;513–522, 1997

Lechtenberg M, Nahrstadt A, Wray V, Fronczek FR: Cyanoglucosides from *Osmaronia cerasiformis* (Rosaceae). *Phytochemistry* 37;1039–1043, 1994

Leemann AC: Hydrocyanic acid in grasses. *Onderstepoort J Vet Sci Anim Ind* 5;97–136, 1935

Leung P, Sylvester DM, Chiou F, Way LL, Way EL, Way JL: Stereospecific effect of naloxone hydrochloride on cyanide intoxication. *Toxicol Appl Pharmacol* 83;525–530, 1986

Levine S, Klein M: Effect of chlorpromazine on cyanide intoxication. *Proc Soc Exp Biol Med* 102;192–194, 1959

Lopez JAP, Martinez FG, Martinez RR, Horta JM: Variacion del contenido de cianuro en *Prunus brachybotrya* en la segunda mitad de su desarrollo. *Vet Mex* 23;131–133, 1992

Lucas B, Sotelo A: A simplified test for the quantitation of cyanogenic glucosides in wild and cultivated seeds. *Nutr Rep Int* 29;711–719, 1984

Machel AR, Dorsett CI: Cyanide analysis of peaches. *Econ Bot* 24;51–52, 1970

MacKenzie AD, Phillips CI: West Indian amblyopia. *Brain* 91;249–260, 1968

Majak W, Cheng K-J: Cyanogenesis in bovine rumen fluid and pure cultures of rumen bacteria. *J Anim Sci* 59;784–790, 1984

Majak W, Cheng K-J: Hydrolysis of the cyanogenic glycosides amygdalin, prunasin and linamarin by ruminal microorganisms. *Can J Anim Sci* 67;1133–1137, 1987

Majak W, Bose RJ, Quinton DA: Prunasin, the cyanogenic glycoside in *Amelanchier alnifolia. Phytochemistry* 17;803, 1978

Majak W, Udenberg T, Clark LJ, McLean A: Toxicity of Saskatoon serviceberry to cattle. *Can Vet J* 21;74–76, 1980

Majak W, McDiarmid RE, Hall JW: The cyanide potential of Saskatoon serviceberry *(Amelanchier alnifolia)* and chokecherry *(Prunus virginiana)*. *Can J Anim Sci* 61;681–686, 1981

Majak W, McDiarmid RE, Hall JW, Cheng K-J: Factors that determine rates of cyanogenesis in bovine ruminal fluid in vitro. *J Anim Sci* 68;1648–1655, 1990

Manges JD: Cyanide poisoning. *Vet Med* 30;347–349, 1935

Marsh CD, Clawson AB, Roe GC: *Arrow grass* (Triglochin maritima) *as a stock-poisoning plant.* USDA Tech Bull 113, 1929

McIlroy RJ: *The Plant Glycosides.* Edward Arnold, London, pp20–23, 1951

Merrill SD: Apple poisoning in dairy cows. *Vet Med* 47;405–406, 1952

Mirande M: Influence exercee par certaines vapeurs sur la cyanogenese vegetale. Procede rapide pour la recherche des plants a acide cyanhydrique. *C R* 149;140–142, 1909

Montgomery RD: The medical significance of cyanogen in plant foodstuffs. *Am J Clin Nutr* 17;103–113, 1965

Moore SJ, Ho IK, Hume AS: Severe hypoxia produced by concomitant intoxication with sublethal doses of carbon monoxide and cyanide. *Toxicol Appl Pharmacol* 109;412–420, 1991

Moran EA: Some new cyanogenic plants. *J Wash Acad Sci* 30;237–239, 1940

Moran EA: Cyanogenetic compounds in plants and their significance in animal industry. *Am J Vet Res* 15;171–176, 1954

Norris JC, Moore SJ, Hume AS: Synergistic lethality induced by the combination of carbon monoxide and cyanide. *Toxicol* 40;121–129, 1986

Nowosad FS, MacVicar RM: Adaptation of the "picric acid test" method for selecting HCN free lines in sudan grass. *Sci Agric* 20;566–569, 1940

Osuntokun BO: An ataxic neuropathy in Nigeria: a clinical, biochemical, and electrophysiological study. *Brain* 91;215–248, 1968

Palmer IS: Interactions between selenium and cyanogenic glycosides or cyanide. In *Selenium in Biology and Medicine,* Spallholz JE, Martin JL, Ganther HE eds, AVI Publishing, Westport, Conn, pp336–342, 1981

Patel MN, Yim GKW, Isom GE: Blockade of *N*-methyl-D-aspartate receptors prevents cyanide-induced neuronal injury in primary hippocampal culture. *Toxicol Appl Pharmacol* 115;124–129, 1992

Peters AT, Slade HB, Avery S: *Poisoning of Cattle by Common Sorghum and Kafir Corn* (Sorghum vulgare). Neb Agric Exp Stn Bull 77, 1903

Petrikovics I, Cannon EP, McGuinn WD, Pei L, Pu L, Lindner LE, Way JL: Cyanide antagonism with carrier erythrocytes and organic thiosulfonates. *Fundam Appl Toxicol* 24;86–93, 1995

Pijoan M: Cyanide poisoning from choke cherry seed. *Am J Med Sci* 204;550–553, 1942

Poulton JE: Cyanogenic compounds in plants and their toxic effects. In *Handbook of Natural Toxins,* vol 1, Keeler RF, Tu AT eds, Marcel Dekker, New York, pp117–157, 1983

Prodjoharjono S: Cyanide poisoning in horses. *Vet Med/Small Anim Clin* 69;1166–1167, 1974

Provenza FD, Burritt EA, Clausen TP, Bryant JP, Reichardt PB, Distel RA: Conditioned flavor aversion: a mechanism for goats to avoid condensed tannins in blackbrush. *Am Nat* 136;810–828, 1990

Quinton DA: Saskatoon serviceberry toxic to deer. *J Wildl Manage* 49;362–364, 1985

Rauws AG, Olling M, Timmerman A: The pharmacokinetics of amygdalin. *Arch Toxicol* 49;311–319, 1982

Rauws AG, Olling M, Timmerman A: The pharmacokinetics of prunasin, a metabolite of amygdalin. *J Toxicol Clin Toxicol* 19;851–856, 1983

Reilly CC, Okie WR: Method for quantifying the cyanogenic glucoside, prunasin in peach trees. *Hortic Sci* 20;905–907, 1985

Robb W, Campbell D: Poisoning of sheep by the consumption of laurel leaves. *Vet Rec* 53;93–95, 1941

Rook JS, Marteniuk JV, Arden W, Hammer R, Poppy G: An outbreak of impaction colic due to ingestion of cockspur hawthorn fruit. *Equine Pract* 13;28–32, 1991

Rosenberg NL, Myers JA, Martin WR: Cyanide-induced parkinsonism: Clinical, MRI, and 6-fluorodopa PET studies. *Neurology* 38;142–144, 1989

Rubino MJ, Davidoff F: Cyanide poisoning from apricot seeds. *J Am Med Assoc* 241;359, 1979

Salkowski AA, Penney DG: Cyanide poisoning in animals and humans: a review. *Vet Hum Toxicol* 36;455–466, 1994

Saunders JA, Conn EE: Presence of the cyanogenic glycoside dhurrin in isolated vacuoles from *Sorghum. Plant Physiol* 61;154–157, 1978

Saupe SG: Cyanogenic compounds and angiosperm phylogeny. In *Phytochemistry and Angiosperm Phylogeny,* Young DA, Seigler DS eds, Praeger, New York, pp80–116, 1981

Schulz V: Clinical pharmacokinetics of nitroprusside, cyanide, thiosulphate, and thiocyanate. *Clin Pharmacokinet* 9;239–251, 1984

Seigler DS: Plants of the northeastern United States that produce cyanogenic compounds. *Econ Bot* 30;395–407, 1976a

Seigler DS: Plants of Oklahoma and Texas capable of producing cyanogenic compounds. *Proc Okla Acad Sci* 56;95–100, 1976b

Seigler DS: The naturally occurring cyanogenic glycosides. In *Progress in Phytochemistry,* Reinhold L, Harbone JB, Swain T eds, Pergamon Press, Oxford, pp83–120, 1977

Seigler DS: Cyanogenic glycosides and lipids: structural types and distribution. In *Cyanide in Biology,* Vennesland B, Conn EE, Knowles CJ, Westley J, Wissing F eds, Academic Press, London, pp133–143, 1981

Selby LA, Manges RW, Houser EC, Flatt RE, Case AA: Outbreak of swine malformations associated with the wild black cherry, *Prunus serotina. Arch Environ Health* 22;496–501, 1971

Shaw JM: Suspected cyanide poisoning in two goats caused by ingestion of crab apple leaves and fruits. *Vet Rec* 119;242–243, 1986

Singh JD: The teratogenic effects of dietary cassava on the pregnant albino rat: a preliminary report. *Teratology* 24;289–291, 1981

Singleton VL, Kratzer FH: Plant phenolics. In *Toxicants Occurring Naturally in Foods,*

2d ed, Committee on Food Protection, National Research Council, National Academy of Sciences, Washington, DC, pp309–345, 1973

Smeathers DM, Gray E, James JH: Hydrocyanic acid potential of black cherry leaves as influenced by aging and drying. *Agron J* 65;775–777, 1973

Solomonson LP: Cyanide as a metabolic inhibitor. In *Cyanide in Biology,* Vennesland B, Conn EE, Knowles CJ, Westley J, Wissing F eds, Academic Press, London, pp11–28, 1981

Spencer KC: Glycosides: the interface between plant secondary and insect primary metabolism. In *Biologically Active Natural Products,* Cutler HG ed, American Chemical Society, Washington DC, pp401–416, 1988

Stauffer VD: Hydrocyanic acid poisoning from choke cherry leaves. *J Am Vet Med Assoc* 157;1324, 1970

Stavric B, Klassen R: Enzymatic hydrolysis of amygdalin by fecal samples and some foods. In *Natural Toxins: 6th International Symposium on Animal, Plant, and Microbial Toxins,* Eaker D, Wadstrom T eds, Pergamon Press, New York, pp655–660, 1980

Steyn DG: *The Toxicology of Plants in South Africa.* South Africa: Central News Agency, 1934

Stossel TP, Jennings RB: Failure of methylene blue to produce methemoglobinemia in vitro. *Am J Clin Pathol* 45;600–604, 1966

Strugala GJ, Stahl R, Elsenhans B, Rauws AG, Forth W: Small-intestinal transfer mechanism of prunasin, the primary metabolite of the cyanogenic glycoside amygdalin. *Hum Exp Toxicol* 14;895–901, 1995

Sunshine I, Finkle B: The necessity for tissue studies in fatal cyanide poisoning. *Int Arch Gewerbepathol* 20;558–561, 1964

Swain E, Li CP, Poulton JE: Development of the potential for cyanogenesis in maturing black cherry (*Prunus serotina* Ehrh.) fruits. *Plant Physiol* 98;1423–1428, 1992

Teli AA, Chauhan HVS, Gupta BS, Srivastava JP: Acidosis in ewes caused by feeding of damaged apple *(Malus sylvestris)* diet. *Indian Vet J* 63;591–593, 1986

Tewe OO, Iyayi EA: Cyanogenic glycosides. In *Toxicants of Plant Origin,* vol 2, *Glycosides,* CRC Press, Boca Raton, Fla, pp43–60, 1989

Tidwell RH, Beal JL, Patel DG, Tye A, Patil PN: A study of the cyanogenetic content and toxicity of the fruit of selected species of *Cotoneaster. Econ Bot* 24;47–50, 1970

Turner NJ, Szczawinski AF: Edible wild fruits and nuts of Canada. In *Edible Wild Plants of Canada,* no 3, National Museum of Natural Sciences, National Museums of Canada, Ottawa, Ontario, 1979

Uitti RJ, Rajput AH, Ashenhrust EM, Rozdilsky B: Cyanide-induced parkinsonism: a clinicopathological report. *Neurology* 35;921–925, 1985

Valenzuela R, Court J, Godoy J: Delayed cyanide-induced dystonia. *J Neurol Neurosurg Psychiatry* 55;198–199, 1992

Vinall HN: A study of the literature concerning poisoning of cattle by the prussic acid in sorghum, sudan grass, and Johnson grass. *J Am Soc Agron* 13;267–280, 1921

Wallace KL, Gerkin R, Mitchell RB: Acute cyanide poisoning due to apricot ingestion [abstract]. *J Toxicol Clin Toxicol* 34;599, 1996

Warburg O: Über nicht-hemmung der zellatmung durch blausaure. *Biochem Zentralbl* 231;493–497, 1931

Way JL: Pharmacologic aspects of cyanide and its antagonism. In *Cyanide in Biology,* Vennesland B, Conn EE, Knowles J, Westley J, Wissing F eds, Academic Press, London, pp28–49, 1981

Way JL: Cyanide intoxication and its mechanism of antagonism. *Annu Rev Pharmacol Toxicol* 24;451–481, 1984

Way JL, Burrows GE: Cyanide intoxication: protection with chlorpromazine. *Toxicol Appl Pharmacol* 36;93–97, 1976

Way JL, End E, Sheehy MH, de Miranda P, Feitknecht UF, Bachand R, Gibbon SL, Burrows GE: Effect of oxygen on cyanide intoxication. 4. Hyperbaric oxygen. *Toxicol Appl Pharmacol* 22;415–421, 1972

Way JL, Leung P, Sylvester DM, Burrows GE, Way JL, Tamulinas C: Methaemoglobin formation in the treatment of acute cyanide intoxication. In *Clinical and Experimental Toxicology of Cyanides,* Ballantyne B, Marrs TC eds, Wright, Bristol, UK, pp402–412, 1987

Westley J: Cyanide and sulfane sulfur. In *Cyanide in Biology,* Vennesland B, Conn EE, Knowles CJ, Westley J, Wissing F eds, Academic Press, London, pp61–76, 1981

Wheatley MD: Mechanism of cerebral death with doses of NaCN from which the heart recovers. *J Neuropath Exp Neurol* 6;295–298, 1947

Wheeler JL: Increasing animal production from sorghum forage. *World Anim Rev* 35; 13–22, 1980

Wiemeyer SN, Hill EF, Carpenter JW, Krynitsky AJ: Acute oral toxicity of sodium cyanide in birds. *J Wildl Dis* 22;538–546, 1986

Wiggins IL: *Flora of Baja California.* Stanford Univ Press, Stanford, Calif, 1980

Willhite CC: Congenital malformations induced by laetrile. *Science* 215;1513–1515, 1982

Wilson DR, Gordon WS: Laurel poisoning in sheep. *Vet Rec* 53;95–96, 1941

Yunusov MS: Diterpenoid alkaloids from *Spiraea japonica. Nat Prod Rep* 8;499–526, 1991

Yunusov MS: Diterpenoid alkaloids from *Spiraea japonica. Nat Prod Rep* 10;471–486, 1993

Rubiaceae Juss.

65

One of the largest families of flowering plants, the Rubiaceae, or madder family, comprises 450–500 genera and 6,500–7,000 species. Many of the genera are monotypic or quite small. Two or 3 subfamilies and 29 tribes are generally recognized. The majority of species are trees, shrubs, or woody vines in tropical and subtropical regions, whereas herbaceous species are more common in temperate areas. The family is economically quite significant. Coffee is the most important product. In addition, it is the source of quinine, ipecac, and dyes. Familiar ornamentals include gardenia, madder, and ixora. Known as bedstraw or cleavers, some species of *Galium* are troublesome weeds. In North America, about 63 genera and 320 species, both native and introduced, are present.

- Madder Family

  Cephalanthus

Plants herbs or shrubs; annuals or perennials; from rhizomes or caudices or taproots. **Leaves** simple; opposite or whorled; venation pinnate or a single vein; stipules leaflike or triangular, or sheathing and bristly, free or fused to petioles. **Inflorescences** solitary flowers or cymes or glomerules or heads; axillary or terminal. **Flowers** perfect; perianths in 1-series or 2-series. **Sepals** 4 or 5 or absent; free. **Corollas** radially symmetrical; salverform or funnelform or tubular or rotate. **Petals** 4 or 5; fused; of various colors. **Stamens** 4 or 5; epipetalous. **Pistils** 1; compound, carpels 2; stigmas 1–4, linear or capitate; styles 1; ovaries inferior; locules 2; placentation axile or basal or apical. **Fruits** capsules or nutlets or schizocarps or berries. **Seeds** 2 to numerous.

- herbs, shrubs; leaves opposite or whorled, interpetiolar stipules; corollas radially symmetrical; ovaries inferior; stamens isomerous, epipetalous

With the exception of the Caprifoliaceae, the Rubiaceae is readily distinguished from other families by its opposite or whorled leaves, interpetiolar stipules, radially symmetrical corollas, inferior ovaries, and epipetalous, isomerous stamens. The presence of stipules separates it from the Caprifoliaceae.

- bearers of economically important alkaloids; *Coffea,* methylxanthines; *Cinchona,* quinine type; *Cephaelis,* ipecac type

Large and diverse, the Rubiaceae contains several genera of great economic and medicinal significance. Biologically active alkaloids of many structural types are present (Raffauf 1996). Foremost among them are the ubiquitous coffee products and methylxanthine alkaloids such as caffeine, obtained from species of *Coffea.* Also of considerable significance are the cinchona (quinine-type) alkaloids obtained from a few genera, but mainly species of *Cinchona* and *Remijia* native to the eastern slopes of the Andes (Turner & Woodward 1953). Found in a number of genera, such as *Cephaelis,* the ipecac alkaloids, such as emetine, have been widely used as emergency drugs to induce vomiting (Janot 1953). These alkaloids, in addition to their emetic activity, have a wide range of effects on other organs, including the heart (Gibbs 1974; Fujii & Ohba 1983). Elsewhere in the world, the family also contains a number of important toxic plants. In North America, only one genus is occasionally responsible for intoxications.

Cephalanthus L.

- *C. occidentalis*

  buttonbush
  button willow
  globeflowers
  honeyballs

Taxonomy and Morphology—Comprising 7–17 species native to both the Old World and the New World, *Cephalanthus,* commonly known as buttonbush, is one of the few woody taxa of the family present in temperate regions. Its name is derived from the Greek *kephale,* for "head," and *anthos,* for "flower," which reflects the dense, globose inflorescence. In North America, 2 species are present. Only 1 has been implicated in toxicologic problems:

C. *occidentalis* L. buttonbush, button willow, honeyballs, globeflowers

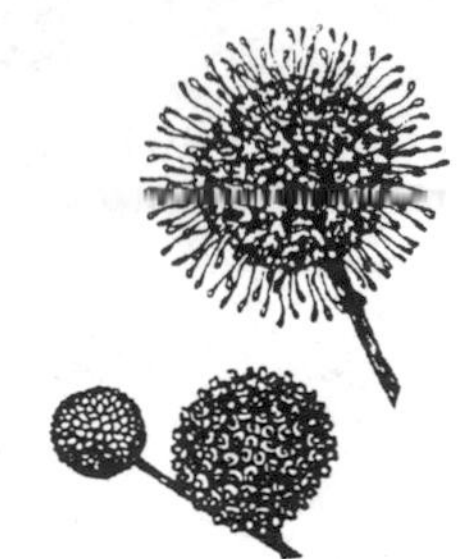

Cephalanthus occidentalis

- shrubs; flowers in globose heads; petals 4, white or cream-colored
- edges of ponds, lakes, creeks, streams

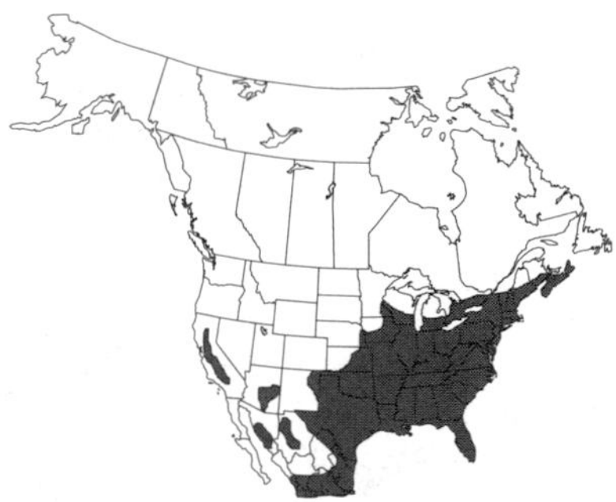

Cephalanthus occidentalis

- neurologic effects; weakness, seizures; cattle mainly; toxin unknown

Plants shrubs or rarely small, irregularly shaped trees; deciduous. **Stems** erect to spreading; 1–5 m tall; branchlets terete or 4-angled; grayish brown; glabrous. **Leaves** whorled and/or opposite; petiolate; glossy green; blades elliptic or lanceolate-oblong; surfaces glabrous; apices acuminate; margins entire; stipules deltoid. **Inflorescences** globose heads borne at ends of long, often branched peduncles. **Flowers** with perianths in 2-series; funnelform. **Sepals** 4. **Petals** 4; white or cream-colored. **Stamens** 4. **Pistils** with stigmas 1, capitate; styles 1, exserted well beyond corollas. **Fruits** schizocarps; obconic; brown; dehiscent from base upward; mericarps 2–4. **Seeds** 1 per mericarp.

Distribution and Habitat—*Cephalanthus occidentalis* is distributed throughout much of the eastern half of North America. It is a classic indicator of wetlands, being found at the edges of ponds, lakes, marshes, creeks, and intermittent streams. Capable of withstanding complete submergence for several weeks at a time, it also occurs at the edges of reservoirs with fluctuating water levels.

Disease Problems and Genesis—Intoxication due to *C. occidentalis* is an uncommon problem that occurs mainly in cattle. It causes neurologic signs such as muscular weakness and convulsive seizures (Cary et al. 1924; Standley 1926; Miller et al. 1980). Although the plant is described as toxic in several reviews, few actual clinical cases are reported. In one of the few, Hansen (1930) reported illness in a horse but not death. The species has been reported to contain the bitter glucosides cephalanthin and cephalin, but these compounds have not been characterized (Pammel 1911; Cary et al. 1924).

Although it may be prudent to accept *C. occidentalis* as a toxic plant, important questions remain as to the type of toxin present and its primary actions. In instances such as this, other toxic members of the family may serve for comparison, but in this case no other members of the family are causes of intoxication in North America. However, it is of interest that important intoxication problems are caused by genera of the Rubiaceae in South America and South Africa. There are two in particular that cause serious problems; both of them have cumulative effects, cause sudden death precipitated by exercise, and yield cardiac lesions. It would be interesting to determine if there are similarities between the diseases caused by these plants and the intoxication produced by *C. occidentalis.* A brief description of these diseases follows.

Species of *Palicourea,* which are among the most important toxic plants in Brazil, produce sudden death of unknown cause in cattle, sheep, and goats (Tokarnia et al. 1981, 1983, 1986, 1991). Clinical signs, which are precipitated by exercise, include tremors, weakness, stiffness, recumbency with paddling, and opisthotonus, suggestive of either neurologic or cardiac actions. Histopathologic findings of pulmonary edema and myocardial necrosis and fibrosis are consistent with death due to cardiac problems. Fluoroacetic acid has been isolated from *P. marcgravii* but only in very low concentrations (Krebs et al. 1994). In rats, this compound causes signs similar to those caused by the plant extracts, and the monofluoroacetic acid antagonist acetamide is protective against *P. marcgravii*–induced intoxication in rats (Gorniak et al. 1994; de Moraes Moreau et al. 1995). It is also of interest that the effects of the plants are cumulative and result in the possibility of sudden death of animals during forced exercise.

The second group of plants of interest is species of *Fadogia* and *Pachystigma,* which cause "gousiekte," a form of acute heart failure, which is of considerable importance in South Africa (Kellerman et al. 1988). This disease of ruminants occurs with sudden onset 4–8 weeks after ingestion of plants. Animals may die following onset of signs typical of heart failure or without signs but with lesions of myocardial necrosis and fibrosis apparent at necropsy. In sheep that were experimentally dosed with *P. pygmaeum,* cardiac changes typical of congestive heart failure—dilation, murmurs, and hemodynamic changes—appeared (Pretorius et al. 1988). A marked increase in serum AST appears to be a useful diagnostic parameter (Fourie et al. 1989). The specific toxin has not been identified.

Clinical Signs, Pathology, and Treatment—The disease produced by *C. occidentalis* has not been well characterized. Based on the few descriptions from clinical cases, there is abrupt onset of weakness or paresis accompanied by seizures and eventually recumbency and perhaps paralysis. The disease may be fatal, but pathologic lesions are not described. Because little is known of the specific intoxication problem, treatment is general supportive care.

■ abrupt onset, weakness, seizures, paresis, collapse

References

Cary CA, Miller ER, Johnstone GR: *Poisonous Plants of Alabama.* Ala Agric Exp Stn Circ 71, 1924

de Moraes Moreau RL, Haraguchi M, Morita H, Palermo Neto J: Chemical and biological demonstration of the presence of monofluoroacetate in the leaves of *Palicourea marcgravii* St.Hil. *Braz J Med Biol Res* 28;685–692, 1995

Fourie N, Schultz RA, Prozesky L, Kellerman TS, Labuschagne L: Clinical pathological changes in gousiekte, a plant-induced cardiotoxicosis of ruminants. *Onderstepoort J Vet Res* 56;73–80, 1989

Fujii T, Ohba M: Ipecac alkaloids and β-carboline congeners. In *The Alkaloids: Chemistry and Physiology,* vol 22, Manske RHF ed, Academic Press, New York, pp1–50, 1983

Gibbs RD: *Chemotaxonomy of Flowering Plants,* vol 3. McGill-Queens Univ Press, Montreal, pp1323–1334, 1974

Gorniak SL, Palermo NJ, Spinosa HS: Effects of acetamide on experimentally-induced *Palicourea marcgravii* (St Hill) poisoning in rats. *Vet Hum Toxicol* 36;101–102, 1994

Hansen AA: *Indiana Plants Injurious to Livestock.* Ind Agric Exp Stn Circ 175, 1930

Janot M-M: The ipecac alkaloids. In *The Alkaloids: Chemistry and Physiology,* vol 3, Manske RHF, Holmes HL eds, Academic Press, New York, pp363–394, 1953

Kellerman TS, Coetzer JAW, Naude TW: *Plant Poisonings and Mycotoxicoses of Livestock in Southern Africa.* Oxford Univ Press, Cape Town, South Africa, 1988

Krebs HC, Kemmerling W, Habermehl G: Qualitative and quantitative determination of fluorometric acid in *Arrabidea bilabiata* and *Palicourea marcgravii* by 19F-NMR spectroscopy. *Toxicon* 32;909–913, 1994

Miller JF, Davis DE, Kates AH, McCormack J: *Poisonous Plants of the Southern United States.* Ga Coop Ext Serv, 1980

Pammel LH: *Manual of Poisonous Plants.* Torch Press, Cedar Rapids, Iowa, 1911

Pretorius PJ, Van Rooyen JM, Van Ryssen JCJ, Minnaar JP: Experimental gallop rhythm in sheep with gousiekte: correlation of changes in amplitude with haemodynamic parameters. *Onderstepoort J Vet Res* 55;221–225, 1988

Raffauf RF: *Plant Alkaloids: A Guide to Their Discovery and Distribution.* Food Products Press, New York, 1996

Standley PC: Trees and shrubs of Mexico. *Contrib U S Natl Herb* 23;1–1721, 1920–1926

Tokarnia CH, Dobereiner J, Silva MF: Intoxication by *Palicourea grandiflora* (Rubiaceae) of cattle in territory of Rondonia, Brazil. *Pesq Vet Bras* 1;85–94, 1981

Tokarnia CH, Dobereiner J, Couceiro JEM, Cordeiro Silva AC: Intoxication by *Palicourea aenofusca* (Rubiaceae), a cause of "sudden death" of cattle in the forest region of Pernambuco, Brazil. *Pesq Vet Bras* 3;75–79, 1983

Tokarnia CH, Peixoto PV, Dobereiner J: Experimental intoxication by *Palicourea marcgravii* (Rubiaceae) in sheep. *Pesq Vet Bras* 6;121–131, 1986

Tokarnia CH, Peixoto PV, Dobereiner J: Experimental intoxication by *Palicourea marcgravii* (Rubiaceae) in goats. *Pesq Vet Bras* 11;65–70, 1991

Turner RB, Woodward RB: The chemistry of the cinchona alkaloids. In *The Alkaloids: Chemistry and Physiology,* vol 3, Manske RHF, Holmes HL eds, Academic Press, New York, pp1–63, 1953

Rutaceae Juss. 66

Nearly cosmopolitan in distribution, but mostly tropical and subtropical with greatest abundance in South Africa and Australia, the Rutaceae, commonly known as the citrus or rue family, comprises 140–150 genera and 900–1,500 species. It is of major economic importance. Most familiar are members of *Citrus,* the citrus fruits such as oranges, grapefruits, lemons, limes, and tangerines, which are prized for their flavor and high content of vitamin C. Commercially, citrus fruits are grown in the "citrus belt" of the southern United States, Mexico, South America, the Mediterranean region, South Africa, and Australia. Lesser known but quite tasty is the kumquat, produced by the genus *Fortunella.* In addition, oils produced by several genera are used in perfumes and medicines, and other species are ornamentals. Genera of the family are classified into 4 subfamilies. In North America, approximately 23 genera and 50 species, both native and introduced, are present. Toxicologic problems are associated with 3 genera.

- Citrus family
  Rue Family

  Ruta
  Thamnosma
  Zanthoxylum

Plants trees or shrubs or rarely herbs; deciduous or evergreen; strongly aromatic; armed or not armed with spines or thorns or prickles; bearing perfect flowers or polygamous or dioecious or polygamodioecious. **Leaves** simple or 1-pinnately compound; alternate or opposite; conspicuously glandular punctate; terminal leaflet present; stipules absent. **Inflorescences** solitary flowers or simple cymes or clusters or panicles; terminal or axillary. **Flowers** perfect or imperfect, similar, perianths in 2-series. **Sepals** 3–5; free or fused. **Corollas** radially or slightly bilaterally symmetrical. **Petals** 3–5; free or fused. **Stamens** 3–60; in 1 or 2 whorls; filaments free or fused at bases. **Pistils** 1–5; compound or simple; carpels, stigmas, styles, and locules 1 to many; ovaries superior; placentation axile. **Fruits** follicles or capsules or samaras or hesperidia. **Seeds** 1 to many.

- aromatic herbs, shrubs, trees; petals 3–5

- tropical species; citrus fruits, sources of pilocarpine

This family is well known for the numerous taxa containing photodynamic linear 6,7-furanocoumarins, especially *Citrus* and *Ruta* (Pathak et al. 1962; Gibbs 1974). Its members are second only to those of the Apiaceae, or carrot family, as sources of furanocoumarins. These compounds, especially linear types such as 8-methoxypsoralen (xanthotoxin), are activated by long-wavelength UV light (320–400 nm) and undergo type I anaerobic reactions with DNA bases (Dodge & Knox 1986). The interaction between furanocoumarins and DNA in the cells of the animal may follow from either contact between plants and skin (photodermatitis) or from ingestion and a subsequent primary photosensitization-type response. In either situation sunlight as a source of UV light is required, and the skin lesions develop following initial exposure. Photoactivation results in tissue destruction from alterations in the lipid portion of cell membranes. A more extensive discussion of the role of furanocoumarins is given in the treatment of genera in the family Apiaceae that cause photosensitization (see chapter 7). The effects of the furanocoumarins differ from the contact type of those allergens found in the Anacardiaceae; there is no immunologic component

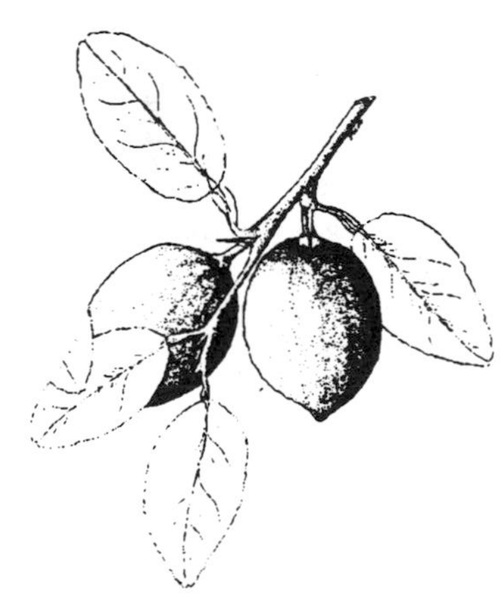

Citrus aurantifolia

to the response. Skin lesions may develop following initial exposure. Because of their content of furanocoumarins, the young foliage and fruit skins of various species of citrus such as lime, lemon, orange, and grapefruit are occasional causes of photodermatitis in humans (McCloud et al. 1992; Nigg et al. 1993).

In addition, the Rutaceae family includes two South American shrubs, *Pilocarpus alvaradoi* and *P. jaborandi,* which are particularly noteworthy as sources of the well-known cholinergic agonist pilocarpine.

Ruta L.

- *R. chalepensis*
  R. graveolens

  common rue
  herb of grace
  rue

Taxonomy and Morphology—Comprising 50–60 species, *Ruta is* native to the Mediterranean region and southwestern Asia. Its name is the classical Latin one for the common rue, a species prized for its supposed prophylactic properties. Mentioned in the writings of Shakespeare and Milton, rue was thought to protect against poisoning by serpents and insects and to ward off evil spells. Brushes made of rue were commonly used to sprinkle holy water in Catholic masses—hence the common name *herb of grace.* In North America, 2 introduced species are of toxicologic interest:

R. chalepensis L.	rue
R. graveolens L.	rue, common rue, herb of grace

- perennial, malodorous herbs or subshrubs; leaves simple, alternate, pinnately divided; petals 4 or 5, yellow or greenish yellow; fruits capsules
- ornamentals; naturalized in some areas

Plants herbs, often woody at bases; perennials; herbage glaucous, malodorous when crushed; bearing perfect flowers. **Stems** erect; 30–80 cm tall; profusely branched at bases. **Leaves** simple; alternate; blades 2- or 3-pinnately divided; lobes oblanceolate to obovate. **Inflorescences** terminal cymes; corymbose or paniculate; bracts present. **Flowers** perfect. **Sepals** 4 or 5. **Petals** 4 or 5; cymbiform; yellow or greenish yellow. **Stamens** 8 or 10. **Pistils** 1; compound; carpels 4 or 5; styles 1; locules 4 or 5. **Fruits** capsules; 4- or 5-lobed. **Seeds** numerous.

Ruta graveolens

Ruta chalepensis

Ruta graveolens

1. Cymes dense. Petal margins fimbriate.
   Capsule lobes pointed . . . . *R. chalepensis*
1. Cymes open. Petal margins not fimbriate.
   Capsule lobes rounded . . . . *R. graveolens*

Distribution and Habitat—Native to the Balkan Peninsula, *R. graveolens* is cultivated in old-fashioned gardens and occasionally is found as an escape along roadsides and in waste areas. Native to southern Europe and northern Africa, *R. chalepensis* is naturalized in disturbed sites, principally in Texas and California.

Disease Problems and Genesis—Species of *Ruta* have been known from ancient times for inducing primary photodermatitis following initial contact between humans and the leaves or stems. The response is variable between individuals, and even with significant contact there may be little or no reaction. The causes are furanocoumarins similar to those of the Apiaceae and include chalepin, chalepensin, imperitorin, and xanthotoxin (Gibbs 1974). The majority of the furanocoumarins are found on the surface and oil ducts of leaves, rendering the plants potent contact photosensitizers, especially the green leaves (Zobel 1991; Zobel & Brown 1990, 1991). The dried leaves are much less likely to cause detrimental effects. Household pets, but not livestock, appear to be at some risk.

■ photodermatitis, furanocoumarins, imperitorin

Imperitorin

Clinical Signs—Within 48 hours after initial and subsequent exposures, there is painful reddening and blistering of the skin. This may last for several days, and in severe cases residual hyperpigmentation may develop at the site and persist for several months (Mitchell & Rook 1977).

■ reddened, blistered skin

Thamnosma Torr. & Frem.

Taxonomy and Morphology—Exhibiting a rather unusual distribution that includes the southwestern portion of North America, the Arabian Peninsula, Socotra Island, and South Africa, *Thamnosma,* commonly known as turpentine broom, comprises 4–9 species. Its name is derived from the Greek roots *thamnos* and *osme,* meaning "shrub" and "odor," which reflect the strong odor when the foliage is crushed or bruised. Two species occur in North America:

■ *T. montana*
T. texana

blisterweed
Dutchman's-breeches
ruda del monte
Texas desert rue
turpentine broom

T. montana Torr. & Frem.	turpentine broom
T. texana (A.Gray) Torr.	blisterweed, Texas desert rue, ruda del monte, Dutchman's-breeches

■ herbs, subshrubs, shrubs; leaves simple, alternate, with blisterlike glands; petals 4, blue, purple, or yellow; fruits leathery capsules

Plants shrubs or subshrubs or herbs; perennials; bearing perfect flowers. **Stems** erect; 10–60 cm tall. **Leaves** simple; alternate; quickly deciduous or persistent; with conspicuous blisterlike glands; blades linear to filiform or spathulate, somewhat fleshy, margins entire. **Inflorescences** racemose cymes or clusters; bracts absent. **Flowers** perfect; urceolate or cylindric or campanulate. **Sepals** 4. **Petals** 4; blue or purple or yellow. **Stamens** 8; in 2 whorls. **Pistils** 1; compound; carpels 2; styles 1, filiform; ovaries deeply 2-lobed, stipitate; locules 2. **Fruits** capsules; leathery; 2-lobed; dehiscent at lobe apices. **Seeds** 8 or 9 per lobe; smooth or rugose or tuberculate.

Thamnosma montana

1. Plants shrubs. Herbage yellowish green.
   Petals dark blue to purplish; 8–14 mm long . . . . *T. montana*
1. Plants herbs or subshrubs. Herbage bluish green.
   Petals yellow or purplish with yellow centers; 3–5 mm long . . . . *T. texana*

Although the common name Dutchman's-breeches is sometimes used for *T. texana,* it is more commonly applied to *Dicentra cucullaria,* in the Fumariaceae (see chapter 39).

Thamnosma texana

Distribution and Habitat—Both species of *Thamnosma* are native to dry, rocky slopes in the southwestern part of the continent. *Thamnosma montana* is distributed from southeastern California to New Mexico and south into northern Mexico. More eastern, *T. texana* occurs from Texas to Arizona and northern Mexico.

■ dry rocky slopes

■ photodermatitis, photosensitization, contact or consumption

furanocoumarins, heraclenin, isopimpinellin, thamnosmin

Disease Problems and Genesis—As the colloquial name *blisterweed* implies, a severe primary photosensitization causing blistering and epithelial sloughing results from either contact with or consumption of the plants. A dosage of 1% b.w./day resulted in disease within 1 or 2 days, whereas lower dosages produced less severe signs after 3 or 4 days. Toxicity, as shown for *T. texana,* is apparently little reduced by drying of the plants, and in spite of the blistering effect they are somewhat palatable (Oertli et al. 1983).

Photosensitization is due to a series of linear furanocoumarins similar to those of Apiaceae. The toxicants of *T. texana* include bergapten, heraclenin, imperatorin, oxypeucedanin, psoralen, 8-methoxypsoralen, isopimpinellin, and alloimperatorin methyl ether (Oertli et al. 1984). *Thamnosma montana* contains the latter two plus several other 8-methoxypsoralen derivatives and a novel coumarin epoxide, thamnosmin (Bennett & Bonner 1953; Dreyer 1953; Kutney et al. 1972; Chang et al. 1976).

Thamnosma montana

Thamnosma texana

Heraclenin

Isopimpinellin

Thamnosmin

■ erythema, edema; muzzle, ears, udder, teats; blisters; cloudy corneas

Clinical Signs—As with other furanocoumarin photosensitizers, the signs in sheep include erythema and edema of the skin of the muzzle, ears, udder, teats, and vulva and subsequently an exudative dermatitis. In other animal species, the skin involvement will include the less pigmented areas. In some cases, the lesions may be limited to areas of direct contact with the plants; in others, they may be more widely distributed in less protected and minimally pigmented skin areas. Blisters may form in the skin, with exudation and subsequent ulceration. Involvement of the eyes may also be seen, with increased sensitivity to light, cloudy corneas, and reddening of the edges of the cornea and conjunctiva.

■ skin lesions

■ remove from plants

Pathology and Treatment—The disease is self-limiting, and treatment is not generally required. Prevention of animal access to the plants is sufficient to allow recovery. However, recovery in severe cases may require several weeks. In the meantime, nursing offspring may require supplemental feeding.

Zanthoxylum L.

Taxonomy and Morphology—Comprising 200–250 species, *Zanthoxylum,* commonly known as prickly ash, is native to the warmer regions of both the Old World and the New World. Sometimes incorrectly spelled *Xanthoxylum,* its name is derived from the Greek roots *xanthos* and *xylon*, meaning "yellow" and "wood." Some species are grown as border or specimen plants for their attractive, aromatic foliage. Representative of the genus in North America are the following:

Z. americanum Mill.	toothache tree, northern prickly ash
Z. clava-herculis L.	Hercules'-club, southern prickly ash, sting tongue, chew bark, pepperwood, toothache tree, sea ash, wait-a-bit
Z. fagara (L.) Sarg.	wild lime, colima, una de gato, correosa
Z. flavum Vahl	yellowheart, satinwood, yellowwood, yellowheart prickly ash

■ *Z. americanum*
Z. clava-herculis
Z. fagara
Z. flavum

chew bark
colima
correosa
Hercules'-club
pepperwood
prickly ash
satinwood
sea ash
sting tongue
toothache tree
wait-a-bit
yellowheart
yellowwood

Plants trees or shrubs; deciduous or evergreen; aromatic; foliage armed with paired prickles; dioecious or polygamodioecious. **Stems** 1–17 m tall; erect or sometimes spreading; trunks sometimes armed with curved prickles. **Leaves** 1-pinnately compound; alternate; conspicuously glandular-punctate; leaflets 5–19, sessile or subsessile, margins entire or dentate or serrate or crenate. **Inflorescences** small paniculate cymes; terminal or axillary. **Flowers** small; perfect or imperfect. **Sepals** 3–5 or absent; free or fused. **Petals** 4 or 5 or absent; yellow-green. **Stamens** 3–5. **Pistils** 2–5; simple; styles coherent. **Fruits** follicles; dry or somewhat fleshy. **Seeds** 1 or 2.

■ aromatic, shrubs or trees; leaves pinnately compound, alternate, glandular-punctate; flowers in small paniculate cymes; petals 4 or 5 or absent, yellow-green

1. Sepals absent. Flowers appearing before leaves .... *Z. americanum*
1. Sepals present. Flowers appearing after leaves.
   2. Inflorescences axillary. Flowers 4-merous .... *Z. fagara*
   2. Inflorescences terminal. Flowers 5-merous.
      3. Sepals free; deciduous; ovate to lanceolate .... *Z. clava-herculis*
      3. Sepals fused; persistent; triangular or deltoid .... *Z. flavum*

Zanthoxylum americanum

Distribution and Habitat—Occurring in moist woods and thickets, *Z. americanum* and *Z. clava-herculis* are distributed throughout much of the eastern half of North America. As their common names indicate, the former is more northern and the latter more southern. *Zanthoxylum fagara* and *Z. flavum* are found in coastal hammocks of the Florida Keys. *Zanthoxylum fagara* also occurs along the Gulf Coast to Texas and Mexico.

■ moist woods, thickets

Disease Problems—As the common names suggest, there has been considerable interest in the use of various species of *Zanthoxylum* for medicinal purposes, mainly as a toothache relief due to the paralyzing sensation of the mouth when the bark is chewed (Jacobson 1948). Millspaugh (1974) used the bark of *Z. americanum* to treat himself for tooth pain and confirmed its effectiveness, but he also indicated that it was very irritating, suggesting the name *pepperwood.* He also cited the use of bark for relief of colic and rheumatism (Millspaugh 1974). The berries are said to provide an effective remedy for typhoid-type diseases. The use of the bark of *Z. clava-herculis* also was apparently well known to members of Indian tribes of eastern North America (Fielder 1975). A species of prickly ash is used in traditional Chinese medicine as a local anesthetic (G. Helman, pers. comm., 1999). Highly regarded and commonly used by people in Sichuan Province, the seeds cause a definite numbing sensation in the mouth.

■ medicinal use of bark for relief of tooth pain; bark also toxic, neurologic effects, blindness

In contrast to other members of the family, contact photodermatitis has not been a problem with *Zanthoxylum.* Although the irritant effects of the bark suggest the possibility of the occurrence of adverse effects, there have been few reports of intoxications

Zanthoxylum americanum

Zanthoxylum clava-herculis

in livestock. Hansen (1930) suggested that the leaves of *Z. americanum* were the likely cause of death of sheep in Indiana. An illness in beef cattle in Georgia was ascribed to their consumption of the bark from *Z. clava-herculis* (Bowen et al. 1996). In a herd of 30 animals, 15 became ill and 8 died. The signs were mainly neurologic in nature, including apparent blindness and inability to drink and swallow. There were no distinctive lesions in the animals that died.

- toxin unknown, possibly neuromuscular or ganglionic blocker; also an alkaloid, nitidine, inhibitor of brain Na^+,K^+-ATPase, possible cardiotoxin

O O OCH3 H3CN+ OCH3

Nitidine

$CH_3(CH_2)_2C{=}C(CH_2)_4C{=}CCNCC(CH_3)_2$ (with C=O)

Herculin

Disease Genesis—The specific toxin in the bark or foliage is not known. Bowen and coworkers (1996) subjected organic bark extracts to various in vitro and in vivo tests and concluded that the toxin may be a neuromuscular and ganglionic blocker perhaps similar in action to quaternary ammonium compounds. Some aspects of the inhibition of in vitro contractile responses were overcome by increased calcium in the bathing solution or by addition of neostigmine.

Species of *Zanthoxylum* also contain several alkaloids, of which the benzophenanthridine nitidine is common to many (Phillips & Castle 1981; Fish et al. 1985). Nitidine has a wide variety of biological effects. It has been studied for its antitumor activity, which is presumably due to inhibition of human topoisomerase I and II (Gatto et al. 1996). Difficulties in administration due to lack of aqueous solubility and precipitation in the peritoneal cavity have decreased interest in its use as an antitumor agent. The capacity of the alkaloid to inhibit guinea pig brain Na^+,K^+-ATPase is of interest and may be associated with its clinical effects (Cohen et al. 1978). The apparent low activity may have been biased by the difficulties in solubilizing the drug. Inhibition of Na^+,K^+-ATPase activity suggests the possibility of cardiovascular effects similar to those of sanguinarine found in *Argemone* in the Papaveraceae, or poppy family (see chapter 55). However, the cardiovascular effects of nitidine as tested by i.v. administration in dogs appear to be minor (Hamlin et al. 1976).

Other bioactive compounds also have been isolated from the bark of *Z. clava-herculis:* herculin, an isobutylamide of a 12-C unsaturated acid with insecticidal activity similar to that of pyrethrin (Jacobsen 1948); and the fish toxins asarinin and sesamin (Rao & Davies 1986). Herculin, placed on the tongue, caused a burning sensation and then numbing of the tongue and mucous membranes of the mouth, consistent with the use of the bark as a toothache remedy (Jacobsen 1948). However, the toxicity of this compound is low; rats fed 500 mg/kg b.w. experienced only an increase in salivation.

- blindness, high-stepping gait, inability to swallow and drink, recumbency; most recover

  no lesions

  general nursing care, possibly neostigmine or calcium

Clinical Signs, Pathology, and Treatment—The disease begins as a neurologic problem with apparent blindness, a high-stepping gait, inability to drink and swallow, and eventually recumbency with struggling. Some animals may die in the next few days, but most will recover over the following weeks. Distinctive gross or microscopic lesions have not been observed.

Treatment is supportive nursing care. The possibility that this is a disease related to neuromuscular blockade suggests the use of calcium preparations, acetylcholinesterase inhibitors such as neostigmine, or perhaps even low dosage of neuromuscular/ganglionic agonists (Bowen et al. 1996).

References

Bennett EL, Bonner J: Isolation of plant growth inhibitors from *Thamnosma montana. Am J Bot* 40;29–33, 1953

Bowen JM, Cole RJ, Bedell D, Schabdach D: Neuromuscular effects of toxins from southern prickly ash *(Zanthoxylum clava-herculis)* bark. *Am J Vet Res* 57;1239–1244, 1996

Chang PTO, Cordell GA, Aynilian GH, Fong HHS, Farnsworth NR: Alkaloids and coumarins of *Thamnosma montana. Lloydia* 39;134–140, 1976

Cohen HG, Seifen EE, Straub KD, Tiefenbach CX, Stermitz FR: Structural specificity of the NaK-ATPase inhibition by sanguinarine, an isoquinoline benzophenanthridine alkaloid. *Biochem Pharmacol* 27;2555–2558, 1978

Dodge AD, Knox JP: Photosensitizers from plants. *Pestic Sci* 17;579–586, 1986

Dreyer DL: Constituents of *Thamnosma montana* Torr. and Frem. *Tetrahedron* 22;2923–2927, 1953

Fielder M: *Plant Medicine and Folklore.* Winchester Press, New York, 1975

Fish F, Gray AI, Waterman PG, Donachie F: Alkaloids and coumarins from North American *Zanthoxylum* species. *Lloydia* 38;268–270, 1975

Gatto B, Sanders MM, Yu C, Wu HY, Makhey D, LaVoie EJ, Liu LF: Identification of topoisomerase I as the cytotoxic target of the protoberberine alkaloid coralyne. *Cancer Res* 56;2795–2800, 1996

Gibbs RD: *Chemotaxonomy of Flowering Plants,* vol 1. McGill-Queens Univ Press, Montreal, 1974

Hamlin RL, Pipers FS, Nguyen K, Milkalko P, Folk RM: *Acute Cardiovascular Effects of Nitidine Chloride Dihydrate (NSC-146397) following Continuous Intravenous Infusion to Anesthetized Beaglehounds.* US NTIS PB Rep PB-261 267, 1976

Hansen AA: *Indiana Plants Injurious to Livestock.* Ind Agric Exp Stn Circ 175, 1930

Jacobson M: Herculin, a pungent insecticidal constituent of southern prickly ash bark. *J Am Chem Soc* 70;4234–4237, 1948

Kutney JP, Verma AK, Young RN: Studies on constituents of *Thamnosma montana* Torr. and Frem.: the structure of thamnosmin, a novel coumarin epoxide. *Tetrahedron* 28;5091–5104, 1972

McCloud ES, Berenbaum MR, Tuveson RW: Furanocoumarin content and phototoxicity of rough lemon *(Citrus jambhiri)* foliage exposed to enhanced ultraviolet-B (UVB) irradiation. *J Chem Ecol* 18;1125–1137, 1992

Millspaugh CF: *American Medicinal Plants.* Dover Press, New York, 1974 (reprint from 1892)

Mitchell JC, Rook AJ: Diagnosis of contact dermatitis from plants. *Int J Dermatol* 16;257–266, 1977

Nigg HN, Nordby HE, Beier RC, Dillman A, Macias C, Hansen RC: Phototoxic coumarins in limes. *Food Chem Toxicol* 31;331–335, 1993

Oertli EH, Rowe LD, Lovering SL, Ivie GW, Bailey EM: Phototoxic effect of *Thamnosma texana* (Dutchmans breeches) in sheep. *Am J Vet Res* 44;1126–1129, 1983

Oertli EH, Beier RC, Ivie GW, Rowe LD: Linear furanocoumarins and other constituents from *Thamnosma texana. Phytochemistry* 23;439–441, 1984

Pathak MA, Daniels F Jr, Fitzpatrick TB: The presently known distribution of furocoumarins (psoralens) in plants. *J Invest Dermatol* 39;225–239, 1962

Phillips SD, Castle RN: A review of the chemistry of the antitumor benzo[c]phenanthridine alkaloids nitidine and fagaronine and of the related antitumor alkaloid coralyne. *J Heterocyclic Chem* 18;223–232, 1981

Rao KV, Davies R: The ichthyotoxic principles of *Zanthoxylum clava-herculis. J Nat Prod* 49;340–342, 1986

Zobel AM: Comparison of furanocoumarin concentrations of greenhouse-grown *Ruta chalepensis* with outdoor plants later transferred to a greenhouse. *J Chem Ecol* 17;21–27, 1991

Zobel AM, Brown SA: Dermatitis-inducing furanocoumarins on the leaf surfaces of eight species rutaceous and umbelliferous plants. *J Chem Ecol* 16;693–700, 1990

Zobel AM, Brown SA: Psoralens in senescing leaves of *Ruta graveolens. J Chem Ecol* 17;1801–1809, 1991

67 Sapindaceae Juss.

- Soapberry Family

  Blighia
  Sapindus

Commonly known as the soapberry family, the Sapindaceae comprises 130–150 genera and 1,500–2,000 species widely distributed in tropical and subtropical regions, especially in Asia and the Americas. Relatively few species occur in temperate regions. The family is of minor economic importance, with a few ornamentals, most notably *Koelreuteria* (golden-rain tree), *Melicocca* (Spanish lime), *Ungnadia* (Mexican buckeye), and *Xanthoceras* (xanthoceras). *Litchi chinensis* (lychee) and *Blighia sapida* (akee) are prized for the soft, fleshy, edible arils surrounding their seeds. Most commonly encountered in the extreme southern part of the continent, 18 genera and 33 species, both native and introduced, are present in North America. Intoxication problems are associated with 2 genera.

- shrubs, trees, vines; leaves pinnately compound; petals 4 or 5 or rarely 3

Plants trees or shrubs or vines; perennials or annuals; vines with tendrils; polygamodioecious. **Stems** with terminal buds absent in woody species. **Leaves** 1- or 2-pinnately compound; alternate; stipules absent or present in viny species. **Inflorescences** compound cymes; racemose or paniculate or corymbose; terminal or axillary. **Flowers** perfect or imperfect, similar; perianths in 2-series. **Sepals** 4 or 5; all alike or of 2 forms; free or fused. **Corollas** radially or bilaterally symmetrical **Petals** 4 or 5 or rarely 3; with an adaxial scalelike or petaloid, indumented appendage at bases; free or fused; clawed or not clawed. **Stamens** 4–10, typically 8; borne on a circular or 1-sided disk; in 2 whorls; free; longer in staminate flowers. **Pistils** 1; compound; carpels 2–6, typically 3; stigmas 1–3; styles 1 or absent; ovaries superior; locules 3; placentation axile. **Fruits** berries or drupes or capsules. **Seeds** 1–3; arils present or absent.

Blighia Koenig

- *B. sapida*

  ackee
  akee

Taxonomy and Morphology—Native to West Africa, *Blighia,* commonly known as akee or ackee, comprises 4–7 species. Its name honors Captain William Bligh, famous commander of *HMS Bounty.* One species causes toxicologic problems:

B. sapida Koenig (= *Cupania sapida* Voigt.)	ackee, akee

- trees; leaves 1-pinnately compound, leaflets 6-10; flowers small, fragrant; petals 5, greenish white; fruits ovoid capsules, 3-lobed; seeds 3, shiny black, arils fleshy, creamy white

Plants trees; to 15 m tall; branches stiff; crowns typically asymmetrical. **Leaves** 1-pinnately compound; leaflets 6–10, subsessile, blades elliptic or ovate or obovate, margins entire. **Inflorescences** racemose; axillary; pendulous. **Flowers** small; fragrant. **Sepals** 5; fused; imbricate or valvate. **Corollas** radially symmetrical. **Petals** 5; greenish white. **Stamens** 8 or 10. **Pistils** with 3-lobed stigmas; ovules 1 per locule. **Fruits** capsules; ovoid; 3-lobed; pericarps coriaceous; stramineous-yellow to magenta-red at maturity. **Seeds** 3; large; shiny black; arils fleshy, creamy white.

Distribution and Habitat—Native to tropical West Africa, *B. sapida* was introduced into the West Indies in 1793 by Captain Bligh. It is now found also in southern Florida and parts of Mexico.

■ mainly imported fruit

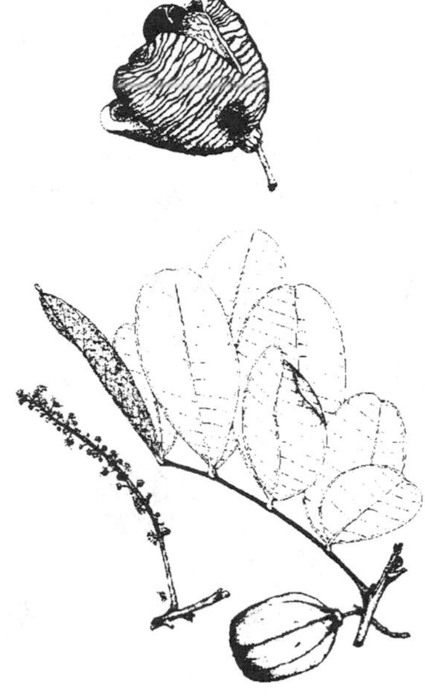

Blighia sapida

Disease Problems—Ackee is a staple Jamaican food. The fleshy aril is the part of the fruit that is eaten when ripe, either fresh or preserved (Morton 1982). Ackees are also canned to be exported and to be served in restaurants. When they are cooked, the water should be discarded to avoid the adverse effects—abrupt onset of vomiting and a precipitous decline in blood sugar (McTague & Forney 1994). The danger lies in ingestion of spoiled or unripe fruits, those not opening on the tree (Jordan & Burrows 1937). Blohm (1962) reported that the fruit is said to be safe to eat if it is open, the aril is firm, and the outer portion is bright red. To be certain, the flesh should be boiled, and the water discarded.

■ humans, vomiting sickness, unripe fruit, fresh or canned

The toxic effects of *B. sapida* have been shown to be the same as those of a disease long known as vomiting sickness of Jamaica and now referred to as akee poisoning (Jordan & Burrows 1937). It was noted as a serious problem in the late 1800s, mainly in children. The incidence increased during the winter when other foods were scarce and the ackee fruit not yet ripe (Bressler 1976). Literally hundreds and perhaps thousands of deaths are attributed to this cause over the years, although this is difficult to substantiate. Prior to 1952, the case-fatality rate was 80% (Bressler 1976). Although the importance of the disease has waned since the 1930s, episodes still occur; 28 cases, including 6 deaths, were reported from January 1989 to July 1991 in Jamaica (Figeroa et al. 1992). In the same report, vomiting occurred in 77% of the cases as the major and most common sign, with coma and seizures in about 25%. There seems to be less of a problem in areas where the tree grows other than Jamaica, perhaps because the fruit is not eaten as much in these other areas. It is not clear if any cases have occurred in Florida (Morton 1982). However, even though it is not legal to import the fruit to the United States, several cases have occurred here with fresh or canned fruit.

As suggested previously, liver injury is a possible sequela to long-term, low-level toxin ingestion (Tanaka et al. 1972). This is shown by a report of liver disease following chronic ingestion (months) of canned ackee fruit (Larson et al. 1994). There was marked abdominal pain, diarrhea, pruritis, and jaundice. The case was not fatal, but biopsy revealed zonal liver cell necrosis and bile stasis. The toxins of *B. sapida* are also suspected as a cause of fatal encephalopathy in children 2–6 years of age (Meda et al. 1999).

Disease Genesis—The toxins are amino acids, i.e., hypoglycins A and B, 2-methylenecyclopropane-alanine and its γ-glutamyl derivative, respectively, which are most abundant in the seeds of unripe fruit (Hassal & Reyle 1955; Tanaka et al. 1976). The level of the toxins is 2–4 times greater in the seeds of unripe fruit. Chase and co-workers (1990) examined the toxin levels in fruit grown in Florida and found that concentrations were 939, 711, and 41.6 mg/100 g in the seeds, aril, and husk, respectively, of unripe fruit, and 269, <1.2, and 41.6 in ripe fruits. This relationship compares with the levels reported previously of 111 and 8 mg/100 g for the unripe and ripe fruits, respectively (Ellington 1975). Hypoglycin A but not B occurs in the fleshy aril of the fruit, but its presence is inconsistent and, further, is much reduced in concentration during ripening. The toxins are water soluble, are not heat labile, and are retained in cooked/canned fruits unless the water of preparation is discarded. Unfortunately it appears that canned products, in some instances, may contain unripe

■ toxic amino acids; hypoglycins, mainly in unripe seeds and arils; biotransformed in liver to a metabolic toxin

fruit. This is shown by the appreciable concentrations of toxin found in some canned products, up to a range of 10–26 mg/100 g of fruit (Scott et al. 1974; Chase et al. 1989; Sarwar & Botting 1994). To reduce the likelihood of toxic effects from canned products, the liquid portion, which may contain appreciable amounts of the water-soluble toxins, should be drained from the fruit.

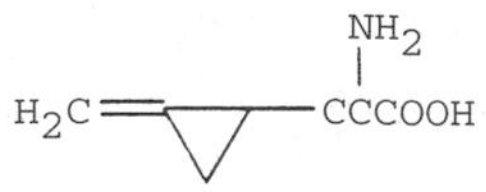

NH_2
H_2C= CCCOOH

Hypoglycin A

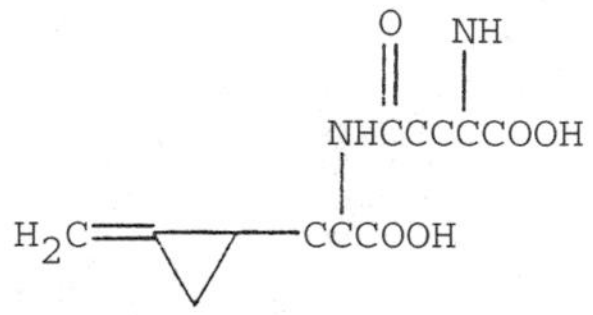

O NH
NHCCCCCOOH
H_2C= CCCOOH

Hypoglycin B

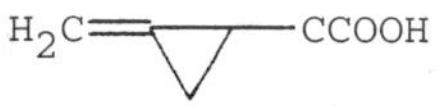

H_2C= CCOOH

Methylenecyclopropylacetic acid

Hypoglycin A, the most potent, is biotransformed to methylenecyclopropylacetic acid, which has a profound inhibitory effect on many metabolic enzymes and on mitochondrial metabolism. The overall effect is a complex series of events involving inhibition of numerous enzymes, not clearly known, but resulting eventually in profound hypoglycemia, as reviewed by Sherrat (1986) and Kean (1989). The effects include inhibition of acyl-CoA (especially short- and medium-chain) and butyryl-CoA dehydrogenases, which limits β-oxidation of fatty acids (Sherrat 1986; Ikeda & Tanaka 1990). Oxidation of fatty acids occurs only as far as butyryl units, leading to their buildup and an increase in ketone bodies. There is an overall block in the transport of long-chained fatty acids into mitochondria (Tanaka et al. 1972, 1976). This limits the availability of cofactors CoA and carnitine (due to formation of nonmetabolizable acyl-CoA esters) and, together with inhibition of pyruvate carboxylase, leads to decreased hepatic gluconeogenesis and subsequently to a profound decrease in blood glucose, to less than 10 mg% (Bressler 1976; Tanaka et al. 1976; Sherratt 1986). There is also marked inhibition of glucose-6-phosphatase. The name for the toxin, hypoglycin, derives from the profound hypoglycemia. Inhibition of amino acid dehydrogenases results in accumulation of butyrate and related products. The overall effect is a complex series of events involving inhibition of numerous enzymes, not clearly known, but resulting in profound hypoglycemia (Sherratt 1986). The hypoglycemia appears to be due to the combined impairment of glucose utilization and gluconeogenesis (Kean 1989). There may be direct neurologic effects of the toxic metabolites, as shown by the prevention of the early vomiting by ablation of the medullary chemoreceptor trigger zone in cats, but this did not affect the overall outcome of later vomiting and coma (Borison et al. 1974).

Experimentally, hypoglycin A causes a variety of mild malformations in rats but not chicks (Persaud & Kaplan 1970). Although hypoglycin A may be considered a leucine analog, the teratogenic effects were reduced only by riboflavin and not by leucine or carnitine (Persaud 1974).

- rapid onset, severe vomiting, headache; eventually seizures, coma

  severe hypoglycemia, increased urinary short-chained fatty acids

Clinical Signs—There is usually moderately rapid onset of severe vomiting and headache, sometimes within minutes but more typically in a few hours after eating the fruit. Following these early signs there is a period of apparent remission with drowsiness and then eventually seizures and coma. Rarely is there a significant diarrhea. Nausea and numbness of the trunk and extremities have also been reported. These signs are accompanied in the latter stages by a marked decrease in blood glucose (to as low as 10 mg/dl or less), which may account for the neurologic signs. Less commonly there is rapid onset of seizures and collapse without the early vomiting.

Serum chemistry changes include high levels of short-chained fatty (adipic) acids and dicarboxylic acids (Tanaka et al. 1976). The increased urinary excretion of the short-chained fatty acids can be used to aid diagnosis, together with a blood glucose less than 10 mg/dl.

- microscopic, fatty degeneration, liver

Pathology—There are few distinctive changes, only a few scattered splotchy hemorrhages and reddening of the mucosa of the intestines. Microscopically, there is a characteristic fatty degeneration of the liver, kidneys, and other organs, with fine cytoplasmic lipid droplets present. There is also degeneration of the renal tubular epithelium.

Treatment—Fluids and electrolytes are needed to offset losses due to vomiting, and glucose is needed to counteract the effects of the toxins. Oral glycine to promote conjugation of hypoglycin may have a useful role in reversing the metabolic effects (Sherratt 1986). Other antidotes that have been suggested, and some used with apparent success in laboratory animals, include riboflavin, carnitine, and clofibrate.

■ fluids, electrolytes; glycine

Sapindus L.

Taxonomy and Morphology—Comprising 13–15 species, *Sapindus,* commonly known as soapberry or soap nut, is native to both the Old World and the New World. Its name is derived from the Latin roots *sapo* and *indicus,* meaning "soap" and "Indian." Both the common and scientific names reflect use of the berries as a soap substitute because their high saponin content causes them to lather freely in water. Necklaces, bracelets, and buttons are made from the seeds. Two species occur in North America:

■ *S. drummondii*
S. saponaria

Chinaberry
false dogwood
jaboncillo
soapberry
wild China tree

S. drummondii Hook. & Arn.	western soapberry, Chinaberry, wild China tree
S. saponaria L. (= *S. marginatus* Willd.)	wingleaf soapberry, jaboncillo, false dogwood

Plants trees; solitary or colonial; to 18 m tall; branchlets gray-green; crowns rounded. **Bark** light gray-brown; becoming rough and coarsely furrowed. **Leaves** 1-pinnately compound; leaflets 7–18, subsessile, blades elliptic or lanceolate, margins entire or undulate-sinuate; rachises winged or not winged. **Inflorescences** paniculate; terminal; large. **Flowers** small. **Sepals** 4 or 5; free. **Corollas** radially or bilaterally symmetrical. **Petals** 4 or 5; white or greenish white. **Stamens** 8 or 10. **Pistils** with stigmas 3; styles 1. **Fruits** large berries; nearly translucent; amber when mature, turning black or brown; persistent on branches. **Seeds** 1; large; round; dark brown; arils absent.

■ trees; leaves 1-pinnately compound, leaflets 7–18; flowers small; petals 4 or 5, white or greenish white; fruits large berries, nearly translucent, amber, turning black or brown

1. Plants deciduous. Leaf rachises not winged. Leaflets 8–18; lanceolate to elliptic-lanceolate; margins not undulate-sinuate. Berries black at maturity .... *S. drummondii*
1. Plants evergreen. Leaf rachises winged. Leaflets 7–9; elliptic to oblong-lanceolate; margins undulate-sinuate. Berries brown at maturity .... *S. saponaria*

Taxonomists differ in their opinions as to whether the two species are distinct. Many treat *S. drummondii* as a variety of *S. saponaria,* but it seems appropriate to maintain them as separate species because of their differences in habitat, geographic distribution, and leaf and berry morphology. Although *S. drummondii* is often called Chinaberry, this common name is more properly used for *Melia azedarach* in the Meliaceae (see chapter 51).

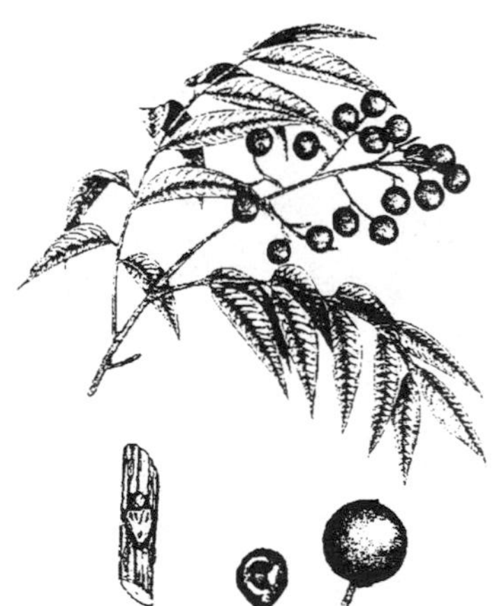

Sapindus drummondii

Distribution and Habitat—Common and widely distributed in tropical America, *S. saponaria* occurs in southern Florida. *Sapindus drummondii* is found along the edges of woods, streams, or fencerows.

■ edges of woods and streams

Disease Problems and Genesis—It is not uncommon to find numerous berries of *S. drummondii* in the rumen of deer, goats, or other ruminants without indications of

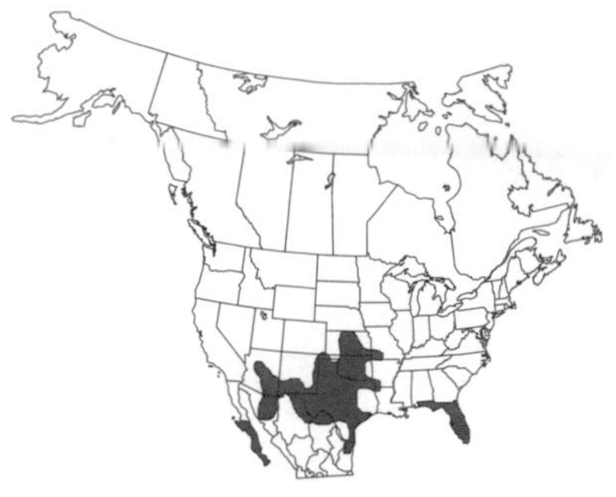

Sapindus drumondii

- possible digestive disturbance; irritant terpenoid glycoside saponins in berries; sapindosides

Sapindoside A

- mild diarrhea

any adverse effects. However, the presence of significant amounts (up to 10%) of a series of pentacyclic terpenoid glycoside saponins in the berries indicates the potential for irritant effects (George 1965). These saponins, sapindosides A–E, are glycosides of hederagenin as found in *Hedera helix,* except for different combinations of sugars (Connolly & Hill 1991). Experimentally, alcoholic extracts of the fruits of an Asian species, *S. trifoliata,* inhibited oocyte implantation in rats (Bodhankar et al. 1974).

Clinical Signs and Treatment—If signs occur, they will be related to irritation of the digestive tract and include diarrhea and possibly excess salivation. Treatment with fluids and electrolytes, if needed, will suffice in most instances.

References

Blohm H: *Poisonous Plants of Venezuela.* Harvard Univ Press, Cambridge, 1962

Bodhankar SL, Garg SK, Mathur VS: Antifertility screening of plants, part 9: effect of five indigenous plants on early pregnancy in female albino rats. *Indian J Med Res* 62;831–837, 1974

Borison HL, Pendleton J Jr, McCarthy LE: Central vs. systemic neurotoxicity of hypoglycin (ackee toxin) and related substances in the cat. *J Pharmacol Exp Ther* 190;327–333, 1974

Bressler R: The unripe akee—forbidden fruit. *N Engl J Med* 295;500–501, 1976

Chase GW Jr, Landen WO Jr, Gelbaum LT, Soliman A-GM: Ion-exchange chromatographic determination of hypoglycin A in canned ackee fruit. *J Assoc Off Anal Chem* 72;374–377, 1989

Chase GW Jr, Landen WO Jr, Soliman A-GM: Hypoglycin A content in the aril, seeds, and husks of ackee fruit at various stages of ripeness. *J Assoc Off Anal Chem* 73;318–3319, 1990

Connolly JD, Hill RA: *Dictionary of Terpenoids.* Vol 2, *Di- and Higher Terpenoids.* Chapman & Hall, London, pp1277–1279, 1991

Ellington EV: Hypoglycin A: a contribution to the problem of vomiting sickness. In *Hypoglycin,* Kean EA ed, Academic Press, New York, pp1–2, 1975

Figeroa P, Nembhard O, Coleman A, Betton M: Toxic hypoglycemic syndrome—Jamaica 1989–1991. *Morb Mortal Wkly Rep* 41;53–54, 1992

George AJ: Legal status and toxicity of saponins. *Food Cosmet Toxicol* 3;85–91, 1965

Hassal CH, Reyle K: Hypoglycin A and B, two biologically active polypeptides from *Blighia sapida. Biochem J* 60;334–339, 1955

Ikeda Y, Tanaka K: Selective inactivation of various acyl coenzyme A dehydrogenases by (methylenecyclopropyl)acetyl coenzyme A. *Biochim Biophys Acta* 1038;216–221, 1990

Jordan EO, Burrows W: The vomiting sickness of Jamaica, B.W.I., and its relation to akee poisoning. *Am J Hyg* 25;520–545, 1937

Kean EA: Hypoglycin. In *Toxicants of Plant Origin,* vol 3, *Proteins and Amino Acids,* Cheeke PR ed, CRC Press, Boca Raton, Fla, pp229–262, 1989

Larson J, Vender R, Camuto P: Cholestatic jaundice due to ackee fruit poisoning. *Am J Gastroenterol* 89;1577–1578, 1994

McTague JA, Forney R Jr: Jamaican vomiting sickness in Toledo, Ohio. *Ann Emerg Med* 23;1116–1118, 1994

Meda HA, Diallo B, Buchet JP, Lison D, Barennes H, Ouangre A, Sanou M, Cousens S, Tall F, Van de Perre P: Epidemic of fatal encephalopathy in preschool children in Burkina Faso and consumption of unripe ackee *(Blighia sapida)* fruit. *Lancet* 353;536–540, 1999

Morton JF: *Plants Poisonous to People in Florida and Other Warm Areas.* Morton, Miami, pp63–64, 1982

Persaud TVN: Mechanism of teratogenic action of hypoglycin-A. *Experientia* 27;414–415, 1974

Persaud TVN, Kaplan S: The effects of hypoglycin-A, a leucine analogue, on the development of rat and chick embryos. *Life Sci* 9;1305–1313, 1970

Sarwar G, Botting HG: Reversed-phase liquid chromatographic determination of hypoglycin A (HG-A) in canned ackee fruit samples. *J AOAC Int* 77;1175–1179, 1994

Scott PM, Botting HG, Kennedy PC, Knipfel JE: The determination of hypoglycin A in ackee. *J Food Sci* 39;1057–1058, 1974

Sherratt HSA: Hypoglycin, the famous toxin of the unripe Jamaican ackee fruit. *Trends Pharmacol Sci* 7;186–191, 1986

Tanaka K, Isselbacher KJ, Shih V: Isovaleric and α-methylbutyric acidemias induced by hypoglycin A: mechanism of Jamaican vomiting sickness. *Science* 175;69–71, 1972

Tanaka K, Kean EA, Johnson B: Jamaican vomiting sickness: biochemical investigation of two cases. *N Engl J Med* 295;461–467, 1976

68 Scrophulariaceae Juss.

- Figwort Family
  Snapdragon Family
  Digitalis
  QUESTIONABLE
  Castilleja
  Linaria
  Pedicularis
  Penstemon
  Scrophularia
  Verbascum
  Veronica

Cosmopolitan but most abundant in temperate regions, the Scrophulariaceae, commonly known as the figwort or snapdragon family, comprises 190–222 genera and 3,000–4,450 species. More than half belong to 8 familiar genera: *Castilleja* (Indian paintbrush), *Penstemon* (beardtongue), *Calceolaria* (slipperwort), *Pedicularis* (lousewort), *Verbascum* (mullein), *Linaria* (toadflax), *Scrophularia* (figwort), and *Veronica* (speedwell). In North America, 73–81 genera, about 860 species, numerous named hybrids, and a plethora of varieties are present. Introduced as ornamentals, several genera are native to other continents but have naturalized and become quite widespread in North America.

The Scrophulariaceae is morphologically quite diverse, especially with respect to corolla structure and color. Traditionally, 3 subfamilies are recognized on the basis of leaf arrangement and corolla aestivation. Its taxa are sometimes confused with members of the related families Lamiaceae, Bignoniaceae, Acanthaceae, and Orobanchaceae, which have bilabiate corollas. The last family is submerged in the Scrophulariaceae by some taxonomists.

- many ornamentals, such as snapdragon

The family is of economic importance because of its many garden ornamentals such as *Antirrhinum* (snapdragon), *Mimulus* (monkey flower), *Verbascum, Calceolaria, and Penstemon.* Species of some 64 genera are in horticultural usage. In addition, species of genera such as *Castilleja* and *Penstemon* are beautiful roadside wildflowers, and a few species, e.g., *Striga asiatica* (witchweed), are noxious weeds in cereal crops. Historically, American Indians and early Europeans used various species medicinally, albeit the most familiar medicinal plant of the family is the introduced, Old World *Digitalis* (foxglove), the source of digitalin, digitoxin, and digoxin. Because of the notoriety of the digitalis glycosides, the Scrophulariaceae is often associated with cardenolide activity. This association, however, is misleading because genera in the family contain several other types of toxins. Several genera contain alkaloids instead of the typical cardenolides. The cardenolides of medicinal or toxicologic importance are only in *Digitalis.*

- herbs; leaves simple; flowers showy; sepals and petals fused; corollas typically 2-lipped; fruits septicidal capsules

Plants herbs or rarely shrubs; annuals or perennials or biennials; autophytic or semiparasitic on roots of hosts; terrestrial or emergent or floating aquatics; caulescent or acaulescent. **Leaves** simple; alternate or opposite, or alternate above and opposite below, or rarely whorled; cauline, basal or forming a basal rosette; venation pinnate; margins entire to pinnately lobed; stipules absent. **Inflorescences** racemes or panicles or solitary flowers or spikes or thyrses or simple cymes; axillary or terminal; bracts absent or present; bracteoles absent or present. **Flowers** perfect; perianths in 2-series. **Calyces** bilaterally symmetrical. **Sepals** 4 or 5; fused. **Corollas** bilaterally or rarely radially symmetrical; bilabiate or campanulate or salverform or rotate. **Petals** 5; fused, 5- or 4-lobed; of various colors. **Stamens** 2 or 4 or rarely 5; of equal lengths or didynamous; epipetalous; staminodia absent or present, 1, filamentous or reduced to a scale. **Pistils** 1; compound, carpels 2; stigmas 1 or 2, not lobed or 2-lobed; styles 1; ovaries superior; locules 2; placentation axile. **Nectaries** absent or present; receptacular; annular. **Fruits** septicidal capsules. **Seeds** numerous.

Digitalis L.

Taxonomy and Morphology—An Old World genus from Europe, northwest Africa, and west-central Asia, *Digitalis,* commonly known as foxglove, comprises 20–25 species. Its name is derived from the Latin *digitus,* alluding to the shape of the corolla. In North America, 7 species have been introduced as ornamentals. Three have escaped, of which 2 are occasionally encountered, and 1 is naturalized. Of toxicologic interest are the following species (see plate 68.1):

■ *D. lanata*
D. lutea
D. purpurea
foxglove

D. lanata Ehrh.	Grecian foxglove
D. lutea L.	straw foxglove
D. purpurea L.	purple foxglove, common foxglove

Plants herbs; perennials or biennials; caulescent; herbage glabrous or tomentose or villous. **Stems** erect; 30–180 cm tall; rarely branched. **Leaves** alternate; lower often crowded; blades lanceolate to oblong or ovate; margins entire to dentate. **Inflorescences** elongate racemes; terminal; usually secund; flowers pendant; bracts present. **Flowers** showy; tubular-campanulate. **Sepals** 5; fused only at bases. **Petals** 5; lobes 4; upper lip shorter than lower; purple or pink or bronze-brown or yellow or white; throat typically spotted or streaked. **Stamens** 4; didynamous. **Pistils** with stigmas 2; flattened. **Capsules** ovoid to conical.

■ erect, rarely branched herbs; leaves alternate, lower crowded; flowers in elongate terminal racemes, showy, tubular-campanulate; petals 5, 4 lobes, purple, pink, bronze-brown, yellow, or white, throat spotted or streaked

1. Corollas 4–5.5 cm long . . . . *D. purpurea*
1. Corollas 2–3 cm long.
   2. Lower middle lobe of corollas conspicuously longer than lateral lobes. Rachises and pedicels glandular-villous . . . *D. lanata*
   2. Lower middle lobe of corollas not conspicuously longer than lateral lobes. Rachises and pedicels glabrous . . . . *D. lutea*

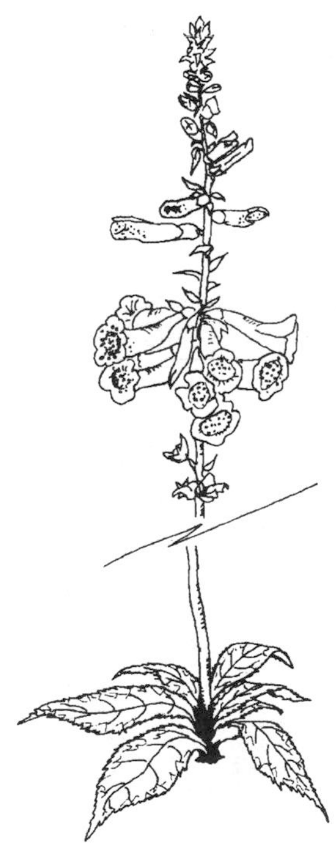

Digitalis purpurea

Distribution and Habitat—Originally introduced from Europe, *D. purpurea* has naturalized locally in the Pacific Northwest. Plants prefer moist areas along forest edges and roadsides, and in open pastures. They are frequently found along the perimeter of logged-over areas of Douglas fir. The species also may be found occasionally eastward. It is a common cultivated garden ornamental throughout North America but generally flourishes where summer temperatures are mild to moderate and moisture is assured. *Digitalis lanata* and *D. lutea* also are European introductions but are encountered only rarely in the northeastern corner of the continent.

■ moist woodlands and open logged-over areas

Disease Problems—*Digitalis* has long been known to have special curative powers. It has been used for hundreds of years for numerous ailments, particularly various types of edema (Hoffman & Bigger 1990). Although not the first plant to be used for cardiac disease, it is certainly the most well known, beginning in 1785 with the published work of William Withering (1941). Although the genus had been used by the populace of rural England for centuries, it was Withering who developed its great medicinal potential (Estes & White 1965; Rahimtoola 1975). In the ensuing years it became widely used as a herbal remedy for cardiac disease and remains a mainstay in the treatment of cardiac disease to the present day, albeit as a source of the

Digitalis purpurea

- cardiotoxic, long used to treat cardiac disease

purified individual cardenolides rather than a component of herbal preparations. Of course during this lengthy period of therapeutic use, cardiotoxicity problems have been discovered; e.g., repeated large dosage of the leaves sometimes causes a distortion of vision in which yellows are accentuated (Estes & White 1965). This discovery has led to the speculation that Vincent Van Gogh's infatuation with yellows in his paintings may have resulted from the possible use of digitalis as a treatment for his epilepsy (Lee 1981).

- steroidal 23-C cardenolide glycosides

Disease Genesis—Both the medicinal and toxic effects of *Digitalis* are due primarily to a series, perhaps several dozen, of steroidal 23-C, C-17β-lactone, C-5β-cardenolides (ring A/B cis). These cardenolides or glycosides contain one or more sugars linked at the C-3 position of the aglycone or genin. Much of the variety in these cardenolides is due to the variety of the different combinations of sugars with specific genins. The sugars are responsible in large measure for the distribution characteristics after absorption and the overall potency of the cardenolides (Shoppee 1964; Hoffman & Bigger 1990). In general, monosides are more potent than biosides or triosides. The best known of the cardenolides of *Digitalis* are digoxin and digitoxin and their respective genins, digoxigenin and digitoxigenin. Other genins of the genus include gitoxigenin, gitaloxigenin, and the numerous but inactive digitenolides. These same active cardenolides are well represented—albeit in differing sugar combinations—among the members of the Apocynaceae (dogbane family) and to a lesser extent the Asclepiadaceae (milkweed family) (Shoppee 1964; Coffey 1970; Singh & Rastogi 1970).

- inhibition of Na^+,K^+-ATPase, followed by increased intracellular calcium, resulting in conduction problems, AV block, ventricular fibrillation

Because these cardenolides bind to the exterior cell surface, their primary effect is a conformational change in the Na^+,K^+-ATPase on the internal surface and subsequent alteration of Na^+ and K^+ movement and concentration balance across the cell membrane. This fosters transmembranal movement and an increase in intracellular Ca^{2+} (Hoffman & Bigger 1990). At therapeutic dosages there is improvement of myocardial function. However, higher dosages produce toxic effects, primarily attributable to alterations in cardiac conduction and other changes in electrical activity. AV block is one of the most common manifestations of toxicity, and death typically occurs because of ventricular fibrillation (Detweiler & Knight 1977; Hoffman & Bigger 1990).

- all plant parts, fresh or dry; all animal species, including ruminants

Cardenolides are present in all parts of the plants, especially the seeds, and toxicity is retained with drying (Everist 1981; Fuller & McClintock 1986). Concentrations of the toxins are highest in the fruits, flowers, and immature leaves, with up to 480 μg cardenolide/g d.w. in the mature capsules (Evans & Cowley 1972). Mature leaves of flowering plants typically have low cardenolide content. Although the cardenolides and their genins are of limited solubility in water, intoxications may nonetheless result when children or pets drink water containing cut stems of the plants. The cardenolides also pose a risk in herbal teas brewed from *Digitalis* intentionally or accidentally (Dickstein & Kunkel 1980). As with other cardenolide-containing species, palatability is low, and rather unique conditions exist when intoxications occur. More likely to be eaten when in hay, the biennial plants may also be less objectionable in the first year's rosette stage. However in one case, cows were reported to have eaten preflowering *Digitalis* in preference to good green grass (Thomas et al. 1987). A change in feed might also encourage consumption, as happened in another case in which turkeys ate plants of the genus after a mash was substituted for pelleted feed (Parker 1951).

Digit-digit-digitoxose

Digoxin (digioxygenin)

Ruminants have long been considered to be somewhat resistant to the effects of digitalis, and therefore presumably to those of cardenolides, because compounds from different taxa are structurally very similar (Craig & Kehoe 1925; Brander et al. 1991). Indeed, experiments with concentrated digitalis preparations in vitro indicate the ability of ruminal microbes to reduce cardenolide activity (Westermarck 1959). This however, has not apparently been inevitably true for in vivo experiments with other cardenolides (Wilson 1909). In addition, reports of intoxication in ruminants are as numerous as those in monogastric species. Thus, this possible resistance cannot be relied upon. An additional factor of some relevance is the reported protective effects of estrogens against digitalis, thus reducing the vulnerability of the female animal to the toxins (Grinnell & Smith 1957; Grinnell et al. 1961). These observations have been substantiated for other cardenolides as well (Schwartz et al. 1974). In addition to the above factors that affect sensitivity to the cardenolides, there is considerable species variation. Rats and mice, especially some *Peromyscus* species, are quite resistant to cardenolide effects (Glendinning 1992).

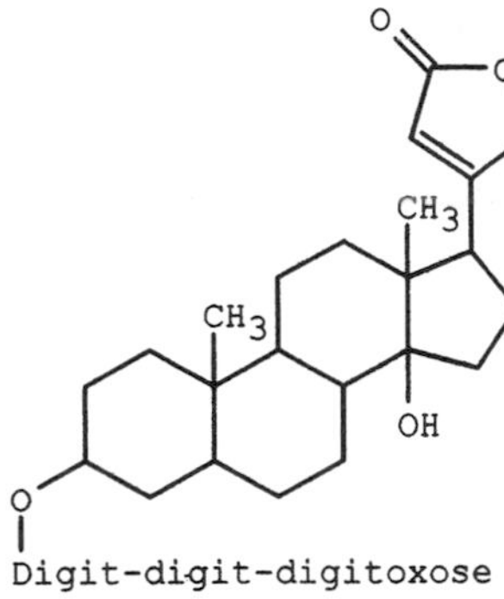

Digit-digit-digitoxose

Digitoxin (digitoxigenin)

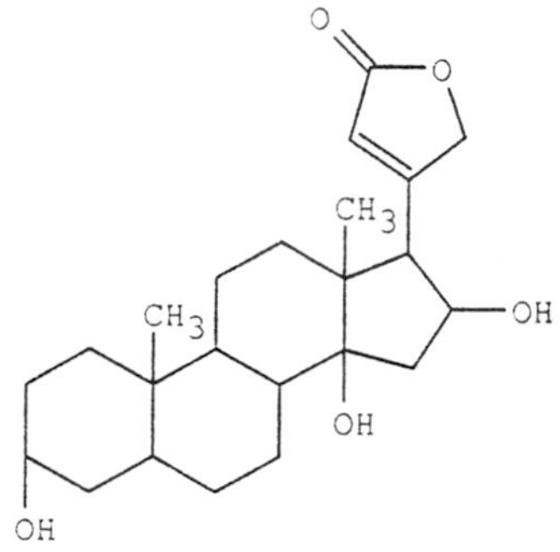

Gitoxigenin

Gitaloxigenin

Clinical Signs—Probably most animal species are susceptible to these toxins; cardiotoxic effects typically predominate. The signs produced in general mimic those of the cardenolides of other families (Benson et al. 1979). The earliest signs, which may not be apparent until 12 or more hours after ingestion, include marked weakness and stupor, which are consistent among the various animal species. In dogs, vomiting and diarrhea are especially prominent (Detweiler & Knight 1977). Diarrhea (sometimes bloody) is common in most other animals as well. During this early period, weakness may be of such severity that a horse may almost be unable to raise its head from the ground while standing (Craig & Kehoe 1925). In sheep and possibly cattle and deer, colic and labored respiration with a peculiar forced expiratory sound may be quite noticeable (Maclean 1966; Benson et al. 1979). During the period of initial signs, the heartbeat may be loud and strong, but as the disease progresses during the following hours, it will slow and become irregular and almost imperceptible (Craig & Kehoe 1925; Maclean 1966; Corrigal et al. 1978; Bartik & Piskac 1981). Occasionally, the heartbeat may be very loud but irregular. Arrhythmias that may be noted include: sinus bradycardia, ventricular extrasystoles, and supraventricular tachycardia, as well as heart block. In the horse, marked subcutaneous edema of head, lips, and eyelids may accompany the drooping of the head and cardiac failure (Craig & Kehoe 1925). Other accompanying signs include increased frequency of urination, grinding of the teeth, and cyanotic/congested mucous membranes. It is interesting to note that in a report of intoxication in dairy cows there was no residual adverse effect on milk production in spite of the occurrence of severe disease (Thomas et al. 1987). Death typically ensues in several days, resulting from ventricular fibrillation. In the event that small quantities of the plants are eaten daily, signs may be limited to poor body condition and weight loss.

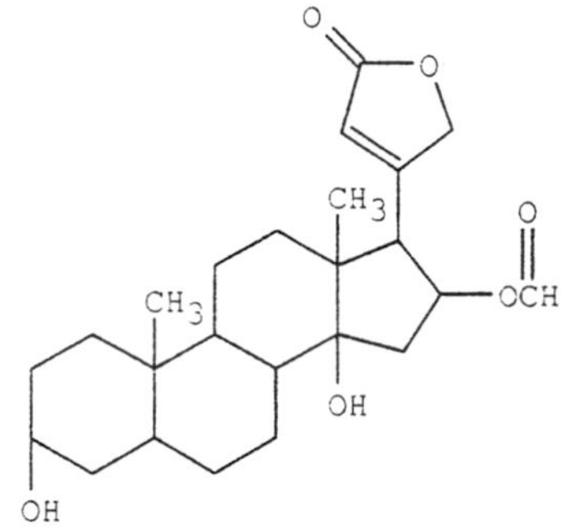

■ marked weakness, stupor, vomiting, diarrhea, labored respiration, slow irregular heartbeat

Although cardiac effects are the typical manifestations, some animals may respond with neurologic signs predominating, because of either high cardiotoxin dosage or the vagaries of certain animal species (Gold et al. 1947). This phenomenon is similar to the effects observed in other cardenolide-containing genera, and it may be that mice and rats generally are more likely to display neurologic signs (Chen & Henderson 1965; Szabuniewicz & Schwartz 1972). In turkeys, signs included drowsiness, inappetence, crop stasis, and seizures just prior to death (Parker 1951). Although the birds were not eating, their crops remained full but not impacted for several days. Poultry appear to be relatively resistant to the cardiotoxic effect of the cardenolides (Ogden et al. 1992).

Diagnosis may be facilitated by measurement of ATPase activity in tissues such as RBCs or by radioimmunoassay for digitoxin in serum or other tissues, although false positives may occur (Smith 1970; Aronson 1986).

Pathology—In many instances the disease will be of such brief duration that distinctive gross lesions will be absent. In such cases there may be congestion of the brain

- gross pathology, reddened gut mucosa

  microscopic, swollen, fragmented muscle cells; leaf fragments in stomach

and abdominal viscera, and scattered, small splotchy hemorrhages, especially subepicardial and subendocardial. Typically, there will be reddening of the mucosa of the stomach and intestine and increased fluidity of their sometimes bloody contents (Craig & Kehoe 1925; Gold et al. 1947; Maclean 1966; Corrigal et al. 1978). The heart appearance is quite variable, but most likely it will be dilated.

Surprisingly, microscopic myocardial degenerative changes may be quite evident in less acute cases. These changes range from mild myocardial degeneration to swollen and hydropic muscle cells, especially in the papillary muscle of the left ventricle; areas of hemorrhage with granular/lytic muscle fiber fragments; and leucocytic and lymphocytic infiltrates. There also may be areas of interstitial edema; proliferation of histiocytes and fibroblasts, which replace degenerative muscle cells; and foci of fragmented partly hyalinized muscle cells. Vascular changes in the heart include edema of walls, contraction of arteries, and dilation of veins (Hueper & Ichniowski 1941). Similar changes have been ascribed to the cardenolides of other plant families (Newsholme & Coetzer 1984; Ogden et al. 1992).

The leaves are readily digested and thus are difficult to recognize in the stomach. However, the upper leaf epidermis is less readily degraded and may be identified microscopically (Corrigal et al. 1978).

- activated charcoal, atropine, phenytoin, propanalol, Fab fragments

Treatment—Similar to that in other acute cardenolide intoxications, treatment may be directed toward prevention of further absorption of the toxins in the digestive tract by oral administration of activated charcoal (2 g/kg b.w.) (Joubert & Schultz 1982a,b). If carried out early in the disease in humans, it may be quite effective. In a trial of 150 patients, there was a marked positive response in 90% (with resolution of all signs in 80%) and improvement in 10% (Smith 1991). To alleviate worrisome cardiac problems, the charcoal may be augmented by parenteral use of atropine to effect for the decreased heart rate associated with sinus bradycardia and AV block or with phenytoin, which is quite effective against the more severe ventricular arrhythmias induced by digitalis (Hoffman & Bigger 1990). Propranolol may also be used in some instances. Wijnberg and co-workers (1999) successfully treated a pony intoxicated from *Digitalis,* using phenytoin at 7.5 mg/kg b.w. i.v. and 10 mg/kg b.w. orally. In particularly severe intoxications, digoxin-specific Fab fragment antibodies can be given to rapidly inactivate cardenolides in serum (Schmidt & Butler 1971; Smith et al. 1976, 1982; Zucker et al. 1982). The half-life and clearance for 50,000 dalton FAB in a 60–70 kg b.w. adult are 14–16 hours and 18–20 ml/minute, respectively (Schaumann et al. 1986; Grene-Lerouge et al. 1996).

Genera of Questionable Toxicity

Castilleja Mutis ex L.f.

- *C. flava*
  C. miniata
  C. rhexifolia
  C. sulphurea

  paintbrush
  painted cups
  prairie fire

- semiparasitic herbs; leaves alternate; flowers in terminal spikes, bilabiate, subtended by brightly colored bracts, yellow, orange, red, scarlet, or maroon

A genus of about 200 species occurring primarily in western North America, *Castilleja* was named to honor Domingo Castillejo, an eighteenth-century Spanish botanist. Typically semiparasitic on the roots of grasses, plants of the genus are annual or perennial herbs with decumbent to erect stems bearing alternate, sessile leaves with entire, pinnatifid or lobed margins. The small, strongly bilabiate flowers are borne in terminal spikes and subtended by showy, conspicuous, brightly colored bracts whose appearance is reflected in the vernacular names of *Indian paintbrush, painted cups,* and *prairie fire.* Colors range from yellow to orange, red, scarlet, and maroon. Species occupy a variety of habitats—dry woodlands, prairies, overgrazed pastures, cold deserts, alpine meadows, tundra, and stream banks.

Species of *Castilleja* are the only members of the Scrophulariaceae typically eaten by livestock, mainly sheep and less so by cattle (Looman 1983). Only a few of the them have been examined for potential toxicants. The four for which specific information is available are *C. flava* S.Watson (yellow paintbrush), *C. miniata* Douglas (scarlet or common paintbrush), *C. rhexifolia* Rydb. (rhexia-leaved paintbrush), and *C. sulphurea* Rydb. (sulfur paintbrush). These species are particularly common in moist meadows and along streams, primarily in the Great Plains and the mountainous West.

Known previously as selenium accumulators, these four may contain potentially troublesome alkaloids (Kingsbury 1964; Stermitz & Suess 1978; McCoy & Stermitz 1983; Roby & Stermitz 1984). *Castilleja rhexifolia* may contain several pyrrolizidine alkaloids, including senecionine, 40% of which is present as the *N*-oxide (Roby & Stermitz 1984). It also may contain rhexifoline, a pyridine monoterpene alkaloid. *Castilleja flava* and *C. sulphurea* have been examined and found negative for pyrrolizidine alkaloids (Stermitz & Suess 1978). *Castilleja miniata* may contain dipiperidine and quinolizidine alkaloids (McCoy & Stermitz 1983). The presence of toxic alkaloids in these semiparasites appears to be the result of their growth on, and accumulation of constituents from, the roots of their host plants—pyrrolizidine or quinolizidine alkaloids, from species of *Senecio* in the Asteraceae and of *Lupinus* or *Thermopsis* in the Fabaceae, respectively (Stermitz & Harris 1987). With the exception of senecionine, the role of the other alkaloids as disease agents is as yet undefined. It is doubtful that *Castilleja* as a source of toxic alkaloids represents a serious risk by itself for livestock. However, as Stermitz and coworkers (1992) have pointed out, the possibility of alkaloid accumulation under some conditions poses a definite risk when these plants are used as herbal preparations for humans. A more complete discussion of the pyrrolizidine alkaloids is given in the treatment of *Senecio* (see chapter 13).

Castilleja

- possibly contain toxic alkaloids from adjacent plants

Linaria Mill.

Comprising 100–150 primarily Eurasian species, *Linaria*, commonly called toadflax or butter-and-eggs, is represented in North America by 15 introduced species and 1 or possibly 2 native species (Alex 1962). Its name is derived from the Greek root

- *L. dalmatica*
  L. genistifolia
  L. purpurea
  L. vulgaris

  butter-and-eggs
  toadflax
  wild snapdragon

- herbs; leaves alternate above; flowers in terminal spikes or racemes; corollas bilabiate, upper lip 2-lobed, lower lip 3-lobed, with long tubular spur; petals blue, lavender, purple, or yellow

Linaria vulgaris

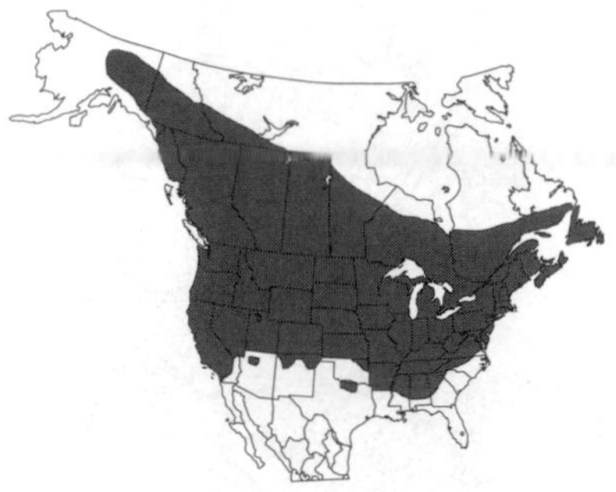

Linaria vulgaris

- possible risk from quinazoline alkaloids

Vasicine

Vasicinone

linon, meaning "flax," an allusion to the flaxlike leaves of some species. The North American taxa are segregated in the genus *Nuttallanthus* by some taxonomists. Closely related to *Antirrhinum* (snapdragon), plants of *Linaria* are annual or perennial herbs with branched or unbranched stems 30–100 cm tall. The sessile, linear or oblong leaves are typically opposite or whorled below and alternate above. Bracteate spikes or racemes terminate the erect stems; rarely are the flowers solitary in the leaf axils. The most distinctive features of the genus are the long, tubular spur arising from a conspicuously bilabiate corolla with a 2-lobed upper lip; a 3-lobed lower lip; and a basal palate that closes the tube. Four stamens are included within the blue, lavender, purple, or yellow corollas, and the capsules are globose. The introduced species typically are encountered in disturbed soils of roadsides, waste areas, and overgrazed pastures across the continent. Also across the continent, but most common in the eastern half and the Pacific Northwest, the native species *L. canadensis* (L.) Dum.Cours. occurs in a variety of soil and vegetation types.

Representative species for which toxicologic information is available include *L. dalmatica* Mill. (Dalmatian toadflax), *L. genistifolia* (L.) Mill. (brown-leaved toadflax); *L. purpurea* (L.) Mill. (purple toadflax); and *L. vulgaris* Mill. (butter-and-eggs, wild snapdragon, or yellow toadflax). These species contain a series of quinazoline alkaloids; vasicine (peganine), vasicinone, and deoxyvasicinone (Openshaw 1953; Groeger & Johne 1965; Johne 1986). They also contain several flavonoid glycosides of unknown toxicologic importance, including linarin and linarasin (Pethes et al. 1975; Bartik & Piskac 1981). Intoxication problems have not been reported; however, vasicine causes bronchodilation, hypotension, and uterine stimulation (Johne 1986). Furthermore, extracts of *L. vulgaris* increase respiration rate; decrease heart rate, blood pressure, and intestinal tone; and alter EEG waveforms in cats, dogs, and rabbits (Karimova et al. 1967). Thus, should plants be consumed in sufficient quantity, signs attendant to the above effects may be possible.

Pedicularis L.

- *P. bracteosa*
  P. crenulata
  P. procera
  P. racemosa

  elephant's-head
  Indian warrior
  lousewort
  wood betony

- semiparasitic erect herbs; leaves simple; flowers in spikes or racemes; corollas bilabiate, upper lip helmet-shaped, lower lip 3-lobed; petals white, yellow, pink, red, or purple

- possibly contain toxic alkaloids from adjacent plants

Native to the northern temperate and boreal regions of both the Old World and the New World and comprising 350–500 species, *Pedicularis* (commonly known as lousewort, wood betony, Indian warrior, or elephant's-head) is widespread across North America, where approximately 35 species occur. Its name comes from the Latin *pediculus,* for "louse," alluding to the superstition that livestock eating these plants will become infested with lice. Plants of this herbaceous, semiparasitic genus are erect annuals, biennials, or perennials from short rhizomes or fibrous or tuberous roots. Alternate or opposite, the simple, toothed to bipinnatifid leaves are nearly all basal to all cauline or both. Borne in bracteate racemes or spikes, the perfect flowers are bilabiate, with the upper lip helmet-shaped and often laterally compressed and extended into a beak or projection—hence the vernacular name *elephant's-head.* The lower lip is 3-lobed. Petal color varies from white to yellow, pink, red, or purple. There are 4 stamens, and the capsule is ovate to oblong, flattened, and often asymmetrical. Species are typically encountered in moist shady sites, open forests, streamsides, moist alpine meadows, and tundra.

Species of *Pedicularis* are reported to contain alkaloids such as *N*-methylcytisine, indicaine, pedicularine, and plantagonine (Ubaev et al. 1963; Lumfullin et al. 1965; Raffauf 1970). The accumulation of these alkaloids appears to be the result of the growth of these species as semiparasites on the roots of other genera containing toxic compounds (Stermitz & Harris 1987). Stermitz and coworkers (1992) reported the presence of senecionine in *P. bracteosa* Benth. (meadow lousewort) growing near

Senecio triangularis; anagyrine in *P. crenulata* Benth. (scalloped-leaf lousewort) growing near *Thermopsis montana;* methylcytisine in *P. procera* A.Gray (gray lousewort; reported as *P. grayi*), growing near *Thermopsis rhombifolia* (reported as *T. divaricarpa*); and quinolizidine alkaloids in *P. racemosa* Douglas ex Hook. (leafy lousewort) growing in association with *Lupinus argenteus.* The toxicologic importance of alkaloids in these and other species of *Pedicularis* in North America is unknown because there are no reports of disease in livestock. However, as Stermitz and coworkers (1992) pointed out, the use of these plants in herbal medicaments for humans may be dangerous. In Europe, digestive tract disorders and anemia are reported for *P. palustris* L. (Bartik & Piskac 1981). However, in another report from Europe, no toxic effects were observed for *P. comosa* L. (Karimova et al. 1967).

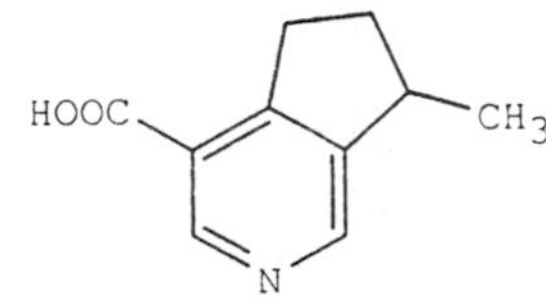

Plantagonine

Pedicularis sp.

Penstemon Schmidel

With many species prized as garden ornamentals and others often offering spectacular displays when flowering in the wild, *Penstemon,* commonly known as beardtongue, comprises about 250 species almost exclusively North American—1 species occurs in northeastern Asia. The generic name is sometimes incorrectly spelled *Pentstemon.* Exhibiting a diversity of growth forms, plants of the genus are herbaceous perennials, subshrubs, or small shrubs with deciduous or evergreen, opposite or whorled leaves, of which the lower are typically petiolate and the upper sessile and alternate. The inflorescences are terminal racemes or panicles or rarely solitary flowers. Showy and perfect, the flowers have a 5-parted calyx and a tubular to campanulate, bilabiate corolla with a 2-lobed upper lip and a 3-lobed lower lip. Petal color is most commonly shades of blue, ranging from purple to lilac, but may also be white to yellow. The most distinctive feature of the genus is the presence of a conspicuous staminodium, typically hirsute, in addition to the 4 fertile stamens—hence the genus's vernacular name. The style is elongate and terminates in a capitate stigma. The majority of species occur in the western half of the continent and occupy a wide range of habitats in prairies, coniferous forests, deserts, and alpine tundra.

Little is known of *Penstemon*'s toxicity potential, because there have been no reports of problems. However, some species, including *P. whippleanus* A.Gray (Whipple's penstemon) of the Rocky Mountains, are reported to contain several pyridine monoterpene alkaloids (McCoy & Stermitz 1983; Connolly & Hill 1991a).

- *P. whippleanus*

  beardtongue
  Whipple's penstemon

- perennial herbs to shrubs; leaves opposite or whorled; flowers showy; corollas bilabiate, upper lip 2-lobed, lower lip 3-lobed; petals blue, purple, lilac, white, or yellow; conspicuous hirsute staminodium

- contain pyridine monoterpene alkaloids

Scrophularia L.

Scrophularia, commonly known as figwort, comprises 150–300 species, with greatest abundance in Europe and Asia. About 16 species occur in North America in a variety of vegetation types and habitats. The genus name is derived from the medieval Latin *scrofula,* the name of a disease reputedly cured by treatment with some of its species, Plants of the genus are herbaceous perennials or shrubs with square stems 50–200 cm tall that bear opposite, petiolate, toothed leaves. Borne in terminal, paniculate, bracteolate inflorescences, the smallish flowers have a saucer-shaped calyx with 5 fused sepals; a bilabiate corolla with a forward-directed 2-lobed upper lip and

- figwort

- perennial herbs or shrubs, square stems; leaves opposite, toothed; flowers in terminal panicles; corollas bilabiate, upper lip 2-lobed, lower lip 3-lobed; petals dull reddish brown or greenish purple a 3-lobed lower lip with the middle lobe deflexed; 4 stamens; and a slender style that terminates in a capitate stigma. Petal color is typically dull reddish brown to greenish purple.

- possibly contain alkaloids

Species of *Scrophularia* contain several flavonoid glycosides of unknown significance and probably alkaloids as well, inasmuch as they produce effects similar to those of *Linaria vulgaris* (Bartik & Piskac 1981; Karimova et al. 1967; Pethes et al. 1975). An Old World species, *S. nodosa* L., taxonomically similar to *S. marilandica* L. of eastern North America, is reported to cause emesis, excess salivation, diarrhea, and nephropathy (Bartik & Piskac 1981). There are no reports of intoxications in North America.

Verbascum L.

- *V. thapsus*

  common mullein
  flannel mullein
  mullein

- biennial herbs; leaves in basal rosettes first year, woolly appearance; corollas almost radially symmetrical, rotate, 5-lobed

- contain oleanane terpenoids

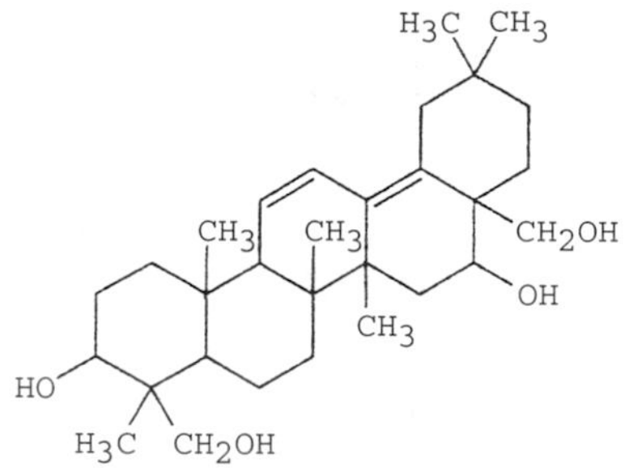

H_3C CH_3 CH_3 CH_3 CH_2OH OH CH_3 HO H_3C CH_2OH

Saikogenin

A popular garden ornamental and used in numerous herbal remedies, *Verbascum,* generally called mullein, is a European and temperate Asian genus of 250–360 species. In North America some 14 species have naturalized, typically as weeds in disturbed sites. Mostly biennial herbs, species of the genus produce large rosettes of leaves the first year and, in the second year, erect stems 50–200 cm tall that bear alternate leaves and terminal, bracteate spikes, racemes, or panicles. The herbage generally is densely woolly-tomentose, and the plants thus have a greenish gray appearance. Unlike most other members of the family, this genus has flowers that are more or less radially symmetrical, with a deeply 5-lobed calyx, a rotate 5-lobed corolla, and 5 stamens of 2 forms. The globose capsule subsequently produced contains numerous, long-lived seeds.

The most commonly encountered species of the genus is *V. thapsus* L. (common or flannel mullein), which is naturalized throughout much of North America. Although two pentacyclic oleanane terpenoids, saikogenin and smithiandienol, are present in this species, there is little reason to ascribe any toxicity potential to the genus, inasmuch as there are no reports of intoxications, and in studies of an Old World species no ill effects were produced (Karimova et al. 1967; Connolly & Hill 1991b).

Veronica L.

- speedwell

- herbs or subshrubs; leaves opposite; corollas almost radially symmetrical, 4- or 5-lobed; stamens 2; capsules typically obcordate

Simultaneously prized for its ornamentals and detested for its weeds, *Veronica,* commonly known as speedwell, comprises 250–300 species distributed primarily in the Old World. In North America, about 35 native and introduced species are present. They are annual and perennial herbs or subshrubs with prostrate to erect stems that bear opposite leaves, although those immediately below the inflorescences may be whorled or alternate. The inflorescences are solitary flowers borne in the leaf axils or bracteate, terminal or axillary racemes. The almost radially symmetrical flowers have a 4- or 5-parted calyx, a rotate to campanulate 4- or 5-lobed corolla, 2 exserted stamens, and a generally obcordate capsule flattened perpendicular to its septum. The introduced species typically are weeds in disturbed sites, whereas the native species are found in a variety of habitats and vegetation types across the continent, generally at higher latitudes.

In North America, species of *Veronica* have not been implicated in intoxication problems. However, an Old World species, *V. nigrum* is reported to slow the frog heart and increase amplitude of contractions in vitro (Karimova et al. 1967). These effects are probably due to compounds other than the flavonoid glycosides and monoterpenes reported to be present in the genus (Pethes et al. 1975; Connolly & Hill 1991a).

References

Alex JF: The taxonomy, history, and distribution of *Linaria dalmatica. Can J Bot* 40;295–307, 1962

Aronson JK: Glycosides of plants and men. *Med J Aust* 144;505–506, 1986

Bartik M, Piskac A: *Veterinary Toxicology.* Elsevier Scientific, Amsterdam, pp276–280, 1981

Benson JM, Seiber JN, Bagley CV, Keeler RF, Johnson AE, Young S: Effects on sheep of the milkweeds *Asclepias eriocarpa* and *A. labriformis* and of cardiac glycoside-containing derivative material. *Toxicon* 17;155–165, 1979

Brander GC, Pugh DM, Bywater RJ, Jenkins WL: *Veterinary Applied Pharmacology and Therapeutics,* 5th ed. Bailliere Tindall, London, 1991

Chen KK, Henderson FG: Pharmacology of sixty-four cardiac glycosides and aglycones. *J Pharm Exp Ther* 111;365–383, 1965

Coffey S: *Rodd's Chemistry of Carbon Compounds,* 2d ed. Vol 2, part D, Alicyclic compounds. Elsevier, Amsterdam, pp360–421, 1970

Connolly JD, Hill RA: *Dictionary of Terpenoids,* vol 1. Chapman & Hall, London, 1991a

Connolly JD, Hill RA: *Dictionary of Terpenoids,* vol 2. Chapman & Hall, London, 1991b

Corrigal W, Moody RR, Forbes JC: Foxglove *(Digitalis purpurea)* poisoning in farmed red deer *(Cervus elaphus). Vet Rec* 102;119–122, 1978

Craig JF, Kehoe D: Plant poisoning. *Vet Rec* 5;795–816, 1925

Detweiler DK, Knight DH: Congestive heart failure in dogs: therapeutic concepts. *J Am Vet Med Assoc* 171;106–114, 1977

Dickstein ES, Kunkel FW: Foxglove tea poisoning. *Am J Med* 69;167–169, 1980

Estes JW, White PD: William Withering and the purple foxglove. *Sci Am* 212;110–119, 1965

Evans FJ, Cowley PS: Cardenolides and spirostanols in *Digitalis purpurea* at various stages of development. *Phytochemistry* 11;2971–2975, 1972

Everist SL: *Poisonous Plants of Australia,* 2d ed. Angus & Robertson, Sydney, Australia, pp620–622, 1981

Fuller TC, McClintock E: *Poisonous Plants of California.* Univ Calif Press, Berkeley, pp230–231, 1986

Glendinning JI: Effectiveness of cardenolides as feeding deterrents to *Peromyscus* mice. *J Chem Ecol* 18;1559–1575, 1992

Gold H, Modell W, Cattell M, Benton JG, Cotlove EW: Action of digitalis glycosides on the central nervous system with special reference to the convulsant action of red squill. *J Pharm Exp Ther* 91;15–30, 1947

Grene-Lerouge NAM, Bazin-Redureau MI, Debray M, Scherrmann JMG: Interspecies scaling of clearance and volume of distribution for digoxin-specific Fab. *Toxicol Appl Pharmacol* 138;84–89, 1996

Grinnell EH, Smith PW: Effect of estrogens on myocardial sensitivity to toxic effects of digoxin. *Proc Soc Exp Biol Med* 94;524–526, 1957

Grinnell EH, Johnson JR, Rhone JR, Tillotson A, Noffsinger J, Huffman MN: Oestrogen protection against acute digitalis toxicity in dogs. *Nature* 190;1117–1118, 1961

Groeger D, Johne S: Occurrence of peganine in *Linaria* species. *Planta Med* 13;182–188, 1965

Hoffman BF, Bigger JT Jr: Digitalis and allied cardiac glycosides. In *Goodman and Gilman's The Pharmacological Basis of Therapeutics,* 8th ed, Gilman AG, Rall TW, Nies AS, Taylor P eds, McGraw-Hill, New York, pp814–839, 1990

Hueper WC, Ichniowski CT: Experimental studies in cardiovascular pathology. 2. Pathologic lesions in organs of cats, guinea pigs, and frogs produced by digitalis poisoning. *J Lab Clin Med* 26;1565–1574, 1941

Johne S: Quinazoline alkaloids. In *The Alkaloids,* vol 29, Brossi A ed, Academic Press, New York, pp99–140, 1986

Joubert JPJ, Schultz RA: The minimal effective dose of activated charcoal in the treatment of sheep poisoned with cardiac glycoside containing plant *Moraea polystachia* (Thunb.) Ker-Gawl. *J S Afr Vet Assoc* 53;265–266, 1982a

Joubert JPJ, Schultz RA: The treatment of *Moraea polystachya* (Thunb.) Ker-Gawl. (cardiac glycoside) poisoning in sheep and cattle with activated charcoal and potassium chloride. *J S Afr Vet Assoc* 53;249–253, 1982b

Karimova SG, Smirnova SG, Nasyrov KhM: Chemical composition and pharmacology of the Scrophulariaceae. *Chem Abstr* 67;20384q, 1967

Kingsbury JM: *Poisonous Plants of the United States and Canada.* Prentice Hall, Englewood Cliffs, NJ, pp300–303, 1964

Lee TC: Van Gogh's vision—digitalis intoxication? *J Am Med Assoc* 245;727–729, 1981

Looman J: *111 Range and Forage Plants of the Canadian Prairies.* Agric Can Publ 1751, 1983

Lumfullin KL, Yuldashev PKh, Yunusov SYu: Investigation of the alkaloids of *Pedicularis olgae*—the structure of plantagonine and indicaine. *Chem Nat Compd* 1;287–288, 1965

Maclean A: Suspected foxglove poisoning in sheep [letter]. *Vet Rec* 79;817–818, 1966

McCoy JW, Stermitz FR: Alkaloids from *Castilleja miniata* and *Penstemon whippleanus,* two host species for the plume moth, *Amblyptilia (Platyptilia) pica. J Nat Prod* 46;902–907, 1983

Newsholme SJ, Coetzer JAW: Myocardial pathology of domestic ruminants in Southern Africa. *J S Afr Vet Assoc* 55;89–96, 1984

Ogden L, Burrows GE, Tyrl RJ, Gorham SL: Comparison of *Asclepias* species based on their toxic effects on chickens. In *Poisonous Plants: Proceedings of the Third International Symposium,* James LF, Keeler RF, Bailey EM Jr, Cheeke PR, Hegarty MP eds, Iowa State Univ Press, Ames, pp500–505, 1992

Openshaw HT: The quinazoline alkaloids. In *The Alkaloids,* vol 3, Manske RHF, Holmes HL eds, Academic Press, New York, pp101–118, 1953

Parker WH: Foxglove *(Digitalis purpurea)* poisoning in turkeys. *Vet Rec* 63;416, 1951

Pethes E, Marczal G, Kery A, Petho M: Flavonoids as biologically active agents and their occurrence in the Scrophulariaceae family. *Acta Pharm Hung* 44 (Suppl);83–90, 1974 (*Chem Abstr* 82;28547a, 1975)

Raffauf RF: *A Handbook of Alkaloids and Alkaloid-Containing Plants.* Wiley-Interscience, New York, 1970

Rahimtoola SH: Digitalis and William Withering, the clinical investigator: editorial. *Circulation* 52;969–971, 1975

Roby MR, Stermitz FR: Chemistry of the Scrophulariaceae. 3. Pyrrolizidine and pyridine monoterpene alkaloids from two *Castilleja* plant hosts of the plume moth, *Platyptilia pica. J Nat Prod* 47;846–853, 1984

Schaumann W, Kaufmann B, Neubert P, Smolarze A: Kinetics of Fab fragments of digoxin antibodies and of bound digoxin in patients with severe digoxin intoxication. *Eur J Clin Pharmacol* 30;527–533, 1986

Schmidt DH, Butler VP Jr: Reversal of digoxin toxicity with specific antibodies. *J Clin Invest* 50;1738–1744, 1971

Schwartz WL, Bay WW, Dollahite JW, Storts RW, Russell LH: Toxicity of *Nerium oleander* in the monkey *(Cebus apella). Vet Path* 11;259–277, 1974

Shoppee CW: *Chemistry of the Steroids,* 2d ed. Butterworths, Washington, DC, pp326–397, 1964

Singh B, Rastogi RP: Cardenolides—glycosides and genins. *Phytochemistry* 9;315–331, 1970

Smith TW: Radioimmunoassay for serum digitoxin concentration: methodology and clinical experience. *J Pharm Exp Ther* 175;352–360, 1970

Smith TW: Review of clinical experience with digoxin immune Fab (ovine). *Am J Emerg Med* 9;1–6, 1991

Smith TW, Haber E, Yeatman L, Butler VP Jr: Reversal of advanced digoxin intoxication with FAB fragments of digoxin-specific antibodies. *N Engl J Med* 294;797–800, 1976

Smith TW, Butler VP Jr, Haber E, Fozzard H, Marcus FI, Bremner WF, Schulman IC, Phillips A: Treatment of life-threatening digitalis intoxication with digoxin-specific Fab antibody fragments. *N Engl J Med* 307;1357–1362, 1982

Stermitz FR, Harris GH: Transfer of pyrrolizidine and quinolizidine alkaloids to *Castilleja* (Scrophulariaceae) hemiparasites from composite and legume host plants. *J Chem Ecol* 13;1917–1925, 1987

Stermitz FR, Suess TR: Pyrrolizidine alkaloids in *Castilleja rhexifolia* (Scrophulariaceae). *Phytochemistry* 17;2142, 1978

Stermitz FR, Schneider MJ, Schell LD, McGregor S: Toxic pyrrolizidine, quinolizidine, and piperidine alkaloids of root parasitic plants used in folk medicine. In *Poisonous Plants: Proceedings of the Third International Symposium,* James LF, Keeler RF, Bailey EM Jr, Cheeke PR, Hegarty MP eds, Iowa State Univ Press, Ames, pp204–207, 1992

Szabuniewicz M, Schwartz WL, McCrady JD, Russell LH, Camp BJ: Experimental oleander poisoning and treatment. *Southwest Vet* 25;105–114, 1972

Thomas DL, Quick MP, Morgan RP: Suspected foxglove *(Digitalis purpurea)* poisoning in a dairy cow [letter]. *Vet Rec* 120;300–301, 1987

Ubaev Kh, Yuldashev PKh, Yunusov SYu: *Pedicularis olgae* alkaloids. *Uzb Khim Zh* 7;33–36, 1963 (*Chem Abst* 59;15602a, 1963)

Westermarck H: Digitalis inactivation in vitro due to rumen microbes from sheep. *Acta Vet Scand* 1;67–73, 1959

Wijnberg ID, van der Kolk JH, Hiddink EG; Use of phenytoin to treat digitalis-induced cardiac arrhythmias in a miniature Shetland pony. *Vet Rec* 144;259–261, 1999

Wilson FW: Oleander poisoning of livestock. *Ariz Agric Exp Stn Bull* 59;381–397, 1909

Withering W: An account of the foxglove and some of its medical uses, from 1785. In *Cardiac Classics,* Willius FA, Keys TE eds, Mosby, St Louis, pp225–252, 1941

Zucker AR, Lacina SJ, DasGupta DS, Fozzard HA, Mehlman D, Butler VP, Haber E, Smith TW: Fab fragments of digoxin-specific antibodies used to reverse ventricular fibrillation induced by digoxin ingestion in a child. *Pediatrics* 70;468–471, 1982

69 Solanaceae Juss.

■ Nightshade Family
Potato Family

Brugmansia
Brunfelsia
Capsicum
Cestrum
Datura
Hyoscyamus
Lycium
Lycopersicon
Nicotiana
Solandra
Solanum

QUESTIONABLE
Nicandra
Physalis

■ herbs, shrubs, or trees; leaves simple; sepals 5, fused; petals 5

Almost cosmopolitan in distribution but with greatest diversity and abundance in South and Central America, the Solanaceae, commonly known as the potato or nightshade family, comprises 76–90 genera and 2,000–2,900 species. Traditionally, 5 subfamilies and 10 tribes are recognized. The family is of considerable economic importance; it includes food plants such as potatoes, tomatoes, eggplants, bell peppers, and cayenne peppers; ornamentals such as petunias, night-blooming jessamine, Chinese lantern, and thorn apple; and drug plants such as tobacco, henbane, and belladonna. In North America, 33–35 genera and 199–204 species, both native and introduced, are present. Toxicologic problems have been associated with 11 genera.

Plants herbs or shrubs or trees; perennials or annuals; evergreen or deciduous; armed or not armed with thorns and/or prickles. **Leaves** simple; alternate, or occasionally subopposite or fascicled; petiolate or sessile; venation pinnate or a single vein; margins entire or variously dissected or lobed; stipules absent. **Inflorescences** simple or compound cymes or solitary flowers. **Flowers** perfect; perianths in 2-series. **Sepals** 5; fused; may be enlarged in fruit. **Corollas** radially or slightly bilaterally symmetrical; plicate or rarely valvate; rotate or funnelform or salverform. **Petals** 5; fused; yellow or white or blue to purple. **Stamens** 5; epipetalous; free, often connivent; anthers dehiscing poricidally or longitudinally. **Pistils** 1; compound, carpels 2; stigmas 1, not lobed or rarely 2-lobed; styles 1; ovaries superior; locules 2 or rarely 3 to 5; placentation axile. **Nectaries** absent or present; receptacular. **Fruits** berries or capsules. **Seeds** numerous.

■ members of family well known for medicinal, hallucinogenic, poisonous effects; mainly risk of abuse for humans; *Atropa belladonna*

Atropa belladonna

Disease Problems—The medicinal, hallucinogenic, and poisonous properties of members of the Solanaceae have long been known and used by humans (Muller 1998). Indeed, almost mystical properties have been attributed to the family, in large measure due to the reputations of the genera *Hyoscyamus* (henbane), *Mandragora* (mandrake), *Atropa* (belladonna), and *Datura* (jimsonweed). *Hyoscyamus* was well known as a narcotic to the ancient Greeks, including Dioscorides and Pliny, and has since been cultivated for its medicinal value (Hocking 1947). In medieval Europe, *Mandragora officinarum* was known for its many actions and used for a variety of purposes. It was known to have potent narcotic activity, capable of anesthetizing an individual or producing a deathlike trance in those being executed. Citing it specifically, Shakespeare underscored its use for these purposes in *Romeo and Juliet, Henry VI,* and *Anthony and Cleopatra* (Tippo & Stern 1977). This species was also likely the source of the knockout drops used in the first Mickey Finn. Somewhat resembling the legs and genitalia of a man, the forking root system of mandrake and the rituals associated with digging plants further embellished the mysticism attributed to both it and the Solanaceae. The European *Atropa belladonna,* also known as deadly nightshade, likewise has strongly contributed to the family's reputation, although in most instances it is not as deadly as other taxa and is now more widely recognized for its hallucinogenic actions (Schneider et al. 1996). It continues to be one of the most common causes of severe intoxications in humans in Europe (Jaspersen-Schib et al. 1996). *Datura* produces hallucinations; was of historical significance in colonial America, as described below; and has been used ritualistically by American Indians and others. With both anecdotal and documented properties attributed to these genera, the ready acceptance of the tales of poisonous qualities of other members of the family is understandable; e.g., until the middle 1800s the common garden tomato was thought in the United States to be dangerous to eat.

Several types of toxicants are represented in the Solanaceae, the two most common being tropane alkaloids and steroidal glycoalkaloids. The tropane types seem to be present primarily in members of the subfamily Atropoideae and the tribe Atropeae. They produce severe nervous derangement in any species of animal. Among them, the calystegins are of particular interest. The steroidal glycoalkaloids are present in other members of the family, especially the genus *Solanum*. These compounds also are neurotoxins, and all animal species are susceptible to their effects. Complete discussions of these two types of toxicants are presented in the treatments of *Datura* and *Solanum*. Kaurene glycosides, a unique third type of toxicants that produce hepatotoxic effects, have been isolated from a single species—*Cestrum parqui*—in the family.

■ neurotoxic tropane alkaloids; steroidal glycoalkaloids; hepatotoxic kaurene glycosides

Brugmansia Pers.

■ *B. aurea*
B. xcandida
B. sanguinea
B. suaveolens
B. versicolor

angel's-trumpet
floripondio

Taxonomy and Morphology—Native to the Andes Mountains, *Brugmansia*, comprises 5 species, 3 named hybrids, and several cultivars. Its name commemorates Sebald Justinus Brugmans, a late-eighteenth-century Dutch professor of natural history. Its species may be encountered as ornamentals in the southern parts of the continent. They are quite similar to and were originally included in *Datura*. The large, pendulous, trumpetlike, five-toothed flowers are reflected in the common name for the genus, *angel's-trumpet*. The following taxa are of toxicologic interest:

B. aurea Lagerh.	golden angel's-trumpet
B. x*candida* Pers. (= *B. aurea* x *B. versicolor*) (= *Datura candida* [Pers.] Saff.)	angel's-trumpet, floripondio
B. sanguinea (Ruiz & Pav.) D.Don	red angel's-trumpet
B. suaveolens (Humb. & Bonpl. ex Willd.) Bercht. & Presl. (= *Datura suaveolens* Humb. & Bonpl. ex Willd.)	angel's-trumpet
B. versicolor Lagerh.	

■ evergreen shrubs or trees; flowers solitary, large, showy; corollas funnelform, yellow, white, peach-pink, or red; fruits berries

Plants shrubs or trees; evergreen. **Stems** erect or spreading; to 11 m tall. **Leaves** ovate to elliptic or oblong; margins entire to dentate. **Inflorescences** solitary flowers; pendulous; stipitate. **Flowers** showy; large, to 50 cm long; fragrant or not fragrant. **Calyces** green; spathe-like or 2-to-5-dentate; enlarging in fruit. **Corollas** radially symmetrical; 5-lobed; yellow or white or peach-pink or red; funnelform; ribbed; plicate in bud. **Petal Lobes** reflexed; narrowly acute. **Stamens** 5; with anthers free or fused in ring about style. **Fruits** berries; subglobose to cylindrical or fusiform.

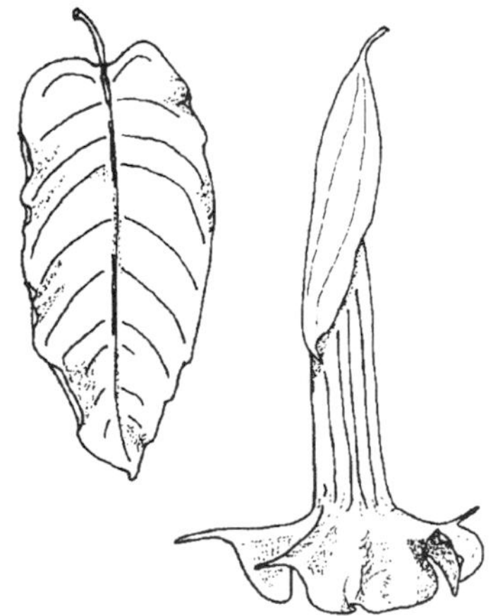

Brugmansia aurea

1. Calyces spathe-like.
   2. Leaf margins undulate. Berries to 45 cm long . . . . *B. versicolor*
   2. Leaf margins not undulate. Berries to 15 cm long . . . . *B. xcandida*
1. Calyces toothed, not spathe-like.
   3. Corollas red or orange-red . . . . *B. sanguinea*
   3. Corollas white to yellow.
      4. Calyx lobes 4. Corolla lobes to 6 cm long; recurved. Anthers not connivent. Berries ovoid to cylindric . . . . *B. aurea*
      4. Calyx lobes 5. Corolla lobes to 2.5 cm long; spreading. Anthers connivent. Berries fusiform . . . . . . *B. suaveolens*

■ ornamentals

Distribution and Habitat—The species of *Brugmansia* are all ornamentals in our area and will be found in gardens in the warmer, southern areas or as tub plants elsewhere. *Brugmansia xcandida* occurs as an escape in the Mexican states of Sinaloa and San Louis Potosí, in southern Florida, and perhaps along the Gulf coast of the United States. *Brugmansia sanguinea* and *B. suaveolens,* also found in southern Florida, are widely cultivated in Mexico. These species normally do not bear fruit in the environmental conditions of most of North America.

■ neurotoxic, tropane alkaloids; risk of abuse in humans

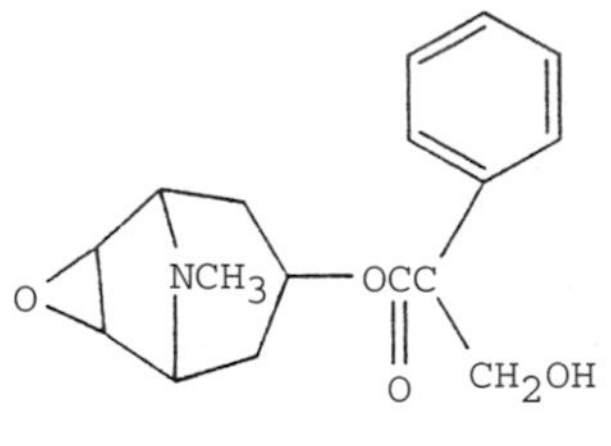

Scopolamine

Disease Problems and Genesis—The flowers and leaves of species of *Brugmansia* contain various tropane alkaloids, including hyoscine (scopolamine), norhyoscine, and hyoscyamine (Evans et al. 1965; Evans & Major 1966; El-Dabbas & Evans 1982; Roses et al. 1987, 1988). The concentrations of hyoscine in *B. sanguinea* subsp. *vulcanicola* are 0.83% in the flowers, 0.74% in the berries, 0.4% in leaves, and 0.32% in the bark (Rivera et al. 1989). Other members of the genus seem to be similarly endowed, with up to 3 mg total alkaloids per flower (Niess et al. 1999). Even nicotine has been identified in the flowers of *B. xcandida* (Gambaro & Roses 1989). Because of the presence of these alkaloids and their reputation as hallucinogens, there is a considerable risk for abuse, especially in adolescent males (Hall et al. 1977; Francis & Clarke 1999; Niess et al. 1999). Green and co-workers (1996) reported a 10-fold increase in reported ingestions in Florida in 1994. Typically there are 6–8 cases per year. The researchers described 3 cases involving young males who exhibited pronounced hallucinogenic effects. In most instances of intoxication, the outcome is satisfactory with conservative treatment, but occasionally the patients may experience severe life-threatening seizures or death or they may fall victim to accidents secondary to the hallucinations, such as drowning or self-inflicted gunshot wounds (McHenry & Hall 1978; Niess et al. 1999). For a more complete discussion, the reader is referred to the treatment of *Datura,* in this chapter. The intense and exquisite fragrance associated with the blossoms is said to be intoxicating. It is disagreeable to some people and may cause headaches. However, chemical examination has not confirmed the presence of tropane alkaloids or other toxins in the fragrance (Kite & Leon 1995).

■ humans; restlessness, pupil dilation, dry mouth, excitement, increased heart rate, labored respiration

Clinical Signs—A problem mainly in humans, the signs produced by *Brugmansia* are similar to those caused by *Datura* and include restlessness, confusion, pupil dilation, increased heart rate, labored respiration, excitement, bizarre behavior, dry mouth, bloat, and rarely seizures and/or death.

■ no lesions

treatment generally not needed; protect from intense light

Pathology and Treatment—Seldom are lethal effects encountered with ingestion of plants of this genus, and few gross or microscopic lesions are associated with tropane alkaloids. Treatment is generally not required, other than conservative approaches to control any serious problems. Because of the dilation of the pupils, individuals should be protected from bright light. A more complete discussion is given in the treatment of *Datura*.

Brunfelsia L.

Taxonomy and Morphology—Native to South and Central America and the West Indies, *Brunfelsia* comprises 30–40 species of primarily winter-flowering shrubs or trees, a few of which are grown for their fragrant or showy blooms. Its name honors the noted sixteenth-century German herbalist Otto Brunfels. Within a day or so of opening, the flowers typically exhibit a rapid color change from dark blue-purple to pale blue to white. This fading is reflected in the genus's common names of *morning-noon-and-night* or *yesterday-today-and-tomorrow*. Representative species include the following:

■ *B. americana*
B. australis
B. grandiflora
B. latifolia
B. nitida
B. pauciflora
B. undulata
B. uniflora

Franciscan rain tree
kiss-me-quick
lady-of-the-night
manacá
morning-noon-and-night
Paraguayan jasmine
rain tree
vegetable mercury
yesterday-and-today
yesterday-today-tomorrow

B. americana L.	lady-of-the-night, Franciscan rain tree
B. australis Benth.	yesterday-today-tomorrow, morning-noon-and-night, Paraguayan jasmine
B. grandiflora D.Don	
B. latifolia (Pohl) Benth.	kiss-me-quick
B. nitida Benth.	
B. pauciflora (Cham. & Schltdl.) Benth. (= *B. calycina* Benth.)	yesterday-today-tomorrow, morning-noon-and-night, yesterday-and-today
B. undulata Sw.	rain tree
B. uniflora (Pohl) D.Don (= *B. hopeana* [Hook.] Benth.)	manacá, vegetable mercury

■ evergreen shrubs or small trees; leaves thick, leathery; flowers showy; corollas 5-lobed, violet, purple, or blue, changing to white or yellowish

Plants shrubs or small trees; evergreen. **Stems** with branchlets glabrous or pubescent. **Leaves** thick; leathery; upper surfaces lustrous; blades ovate to elliptic or oblong or obovate; margins entire. **Inflorescences** of solitary or clustered flowers at branch ends. **Flowers** showy; fragrant or not fragrant. **Calyces** tubular to campanulate; 5-toothed or 5-lobed. **Corollas** radially symmetrical; 5-lobed; violet or purple or blue to white or yellowish; funnelform or salverform. **Petal Lobes** broad; blunt; spreading. **Stamens** 4; didynamous. **Fruits** capsules; dry or rarely fleshy.

Because the species listed are merely representative of the genus, a taxonomic key is not presented here. If exact identification is required, appropriate horticultural manuals should be consulted.

Brunfelsia americana

Distribution and Habitat—All species of *Brunfelsia* are ornamentals in North America and will be found in gardens in the southern, warmer regions or as tub plants elsewhere. Native to the West Indies, *B. americana* grows wild in Key West and on the mainland of extreme southern Florida.

■ ornamentals

Disease Problems—*Brunfelsia* has somewhat of a reputation for hallucinogenic properties, but its effects are probably exaggerated because it was most often used to supplement other, more potent sources (Schultes 1969; Flores & Lewis 1978). Many of its species are also well accepted for medicinal purposes in areas where they are indigenous. Plants of this genus are probably toxic for all animal species as has been reported for cattle (Tokarnia et al. 1991). Dogs seem to have a particular propensity

- neurotoxic, reputation as hallucinogen; risk for humans; also toxic to animals, especially dogs

for eating the fruits and developing a neurotoxicosis (McBarron & de Sarem 1975; Neilson & Burren 1983; Banton et al. 1989; Spainhour et al. 1990).

- several possible toxicants— scopoletin neurodepressant hopeanine neuroexcitant brunfelsamidine

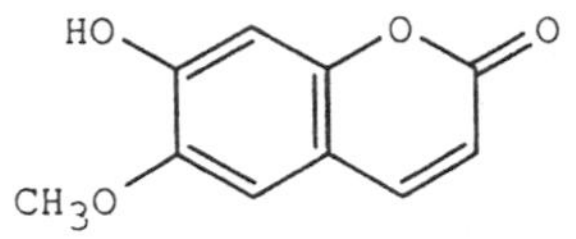

HO, CH_3O

Scopoletin

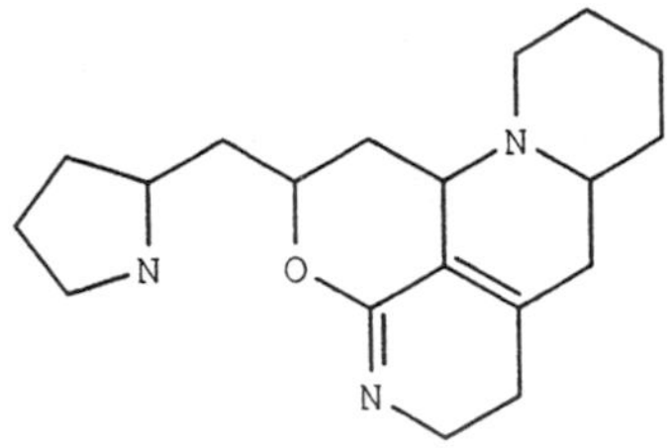

N, O, N

Hopeanine

NH, C, NH_2, N

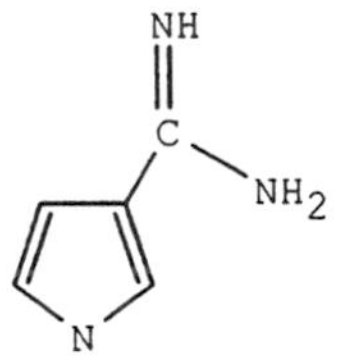

Brunfelsamidine

Disease Genesis—Several biologically active compounds have been isolated from this genus. The first compound was scopoletin (6-methoxy-7-hydroxycoumarin), or gelseminic acid, a compound subsequently shown to be common in many families, in some cases serving as a phytoallexin to reduce competition from other plants (Mors & Ribeiro 1957; Zeringue 1984). Scopoletin is a smooth muscle relaxant and a hypotensive (Jarboe et al. 1967; Ojewole & Adesina 1983b). It also produces a neuromuscular block and negative effects on cardiac rate and strength of contraction, actions not reversed by atropine (Ojewole & Adesina 1983a). Perhaps of even greater interest are its prostaglandin synthetase and anti-inflammatory properties, several times more potent than aspirin (Farah & Samuelsson 1992). However, despite these actions, scopoletin does not seem to account for the toxic effects of this genus.

Two other compounds have emerged as more likely toxicants: hopeanine from *B. uniflora* (reported as *B. hopeana*) and brunfelsamidine from *B. grandiflora* (Iyer et al. 1977; Iyer 1978; Lloyd et al. 1985). Both compounds are convulsants in mice or rats. Hopeanine causes decreased activity, paralysis mixed with seizures, and hypersensitivity, whereas brunfelsamidine produces excitement, clonic and tonic seizures, and death. Thus, it appears that there are some species differences, because the first compound is a neurodepressant and the latter is a neuroexcitant. At present, brunfelsamidine appears to be the most likely toxicant, given that its effects more closely simulate those caused by ingestion of the plants. Furthermore, most of the species for which intoxications are reported produce similar symptomatology.

- anxiety, excitement, coughing, sneezing, nystagmus, tremors, seizures

Clinical Signs—The onset of signs may vary from minutes to several hours or more and begin with anxiety and excitement. Persistent coughing, sneezing and gagging, and less often nystagmus are also reported as early signs. After a few hours, tremors begin and then the most characteristic signs—repeated extensor rigidity-type seizures—and occasional paddling seizures appear. The seizures are precipitated by external stimuli and bear considerable resemblance to those caused by strychnine. The period of tremors and seizures may last a few hours or days, and complete recovery can take as long as a week or more.

- no lesions

Pathology—Few changes are seen at necropsy. Some of the less specific changes include mild congestion of the lungs and lymph nodes and mild reddening of the mucosa of the stomach and intestines. There are no consistent serum chemistry or hematologic changes.

- oral-activated charcoal, sedation to control seizures

Treatment—Treatment is mainly symptomatic and will include sedation of the animal with pentobarbital or diazepam to control seizures and administration of fluids to provide maintenance during the prolonged time of incapacity. Oral-activated charcoal is quite helpful if given early, but emetics must be used with care or not at all, because they may precipitate seizures.

Capsicum L.

Taxonomy and Morphology—Native to tropical America and introduced in Europe by Columbus, *Capsicum,* commonly known as green pepper or chili pepper, comprises 10–50 species. Closely related to *Solanum,* its name is derived from the Greek root *kapto,* meaning "to bite," which reflects the hot, pungent taste of the fruits. The genus is cultivated for food, condiments, and muscle balms and as ornamentals. The fruits are eaten raw or cooked and are the source of paprika, chili powder, and cayenne pepper. Despite their common name, members of *Capsicum* are not the source of black or white pepper, which comes from *Piper nigrum*, in the quite unrelated family Piperaceae. Most of the cultivated peppers, of which there are many, are varieties of *C. annuum*. The following are representative of the genus:

C. annuum L.	sweet pepper, hot pepper, capsicum, spice pepper, ornamental pepper, red pepper, cayenne pepper
C. frutescens L.	tabasco pepper, spur pepper, hot pepper

- *C. annuum*
  C. frutescens

  capsicum
  cayenne pepper
  hot pepper
  ornamental pepper
  red pepper
  spice pepper
  spur pepper
  sweet pepper
  tabasco pepper

Plants subshrubs or shrubs; herbage glabrous or rarely pubescent. **Stems** erect or spreading; typically much branched. **Leaves** dark glossy green; blades narrowly lanceolate to elliptic to broadly ovate; margins entire. **Inflorescences** solitary flowers or clusters of 2 or 3. **Flowers** showy; erect or declined. **Calyces** tubular to campanulate; 5-toothed; fleshy; enlarging slightly in fruit. **Corollas** radially symmetrical; deeply 5-lobed; white, sometimes greenish- or purple-tinged; rotate. **Petal Lobes** broad; blunt; spreading. **Stamens** 5; equal; protruding; anthers connivent, blue-tinged. **Fruits** berries; shiny; of various colors and shapes and sizes; locules 2–6; pungent to the taste.

- erect, spreading, branched subshrubs or shrubs; leaves dark glossy green; flowers showy; corollas 5-lobed, white; fruits berries

Capsicum annum

Distribution and Habitat—Although commonly grown as annuals for food or decoration in North America, species of *Capsicum* are actually subshrubs or shrubs; thus, plants may persist in the warmer frost-free regions of the South. Commonly known as bird pepper or chillipiquin, *C. annuum* var. *minus* occurs in thickets and groves from Texas to Arizona and in southern Florida.

- cultivated crops

Disease Problems—The use of *Capsicum* fruits or chiles for food or flavoring dates back to about 7,000 B.C. and their presence in ancient burial sites in Central and South America (Govindarajan 1985; Cordell & Araujo 1993). Chiles have since been used rather extensively in the medical treatment of various arthritic syndromes and to control pain and headaches. As a consequence of their pungency, they have the potential to cause irritation of the mucous membranes of the mouth and eyes. *Capsicum* remains one of the most common causes of inquiries to poison information centers in the United States (Krenzelok & Provost 1995). Because of the intense burning and stinging sensation of the lips, tongue, and mouth that follows contact with plants, exposure causes symptoms more commonly than with most other causes of inquiries to poison control centers (Borys et al. 1987). However, it is rare for a person to have other than transient effects on the mucous membranes or skin where contact occurs. This is because the toxic principles are poorly absorbed. Other than direct irritation, the toxicity of these plants is rather low. Experimentally, mice given up to 10% *C. annuum* (ground red chili) in the diet for a prolonged period experienced no apparent adverse effects and only slight pathologic changes (Jang et al. 1992). However, there may be indirect effects on other compounds in the diet because chile powder slows gastric emptying and hastens intestinal transit of ingesta (Horowitz et al. 1992).

- potent irritant effects, especially on mouth and eyes; very large dosage neurotoxic

Disease Genesis—The pungent and toxic principles in peppers are capsaicinoids (Nelson & Dawson 1923). Capsaicin, a vanillyl amide of 8-methyl-6-nonenoyl vanillylamide, is the major constituent (Nelson & Dawson 1923). There are lesser amounts of dihydrocapsaicin and even smaller amounts of homocapsaicin, nordihydrocapsaicin,

- capsaicinoids, especially in mature fruits

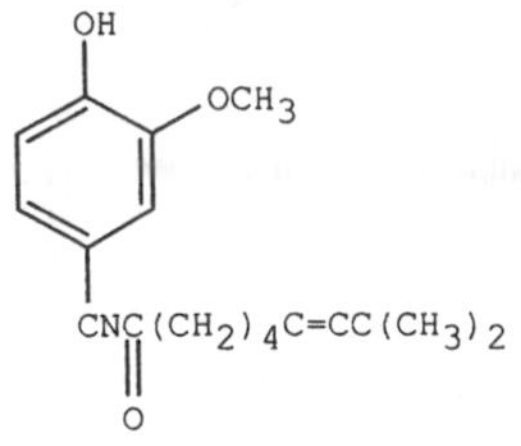

Capsaicin

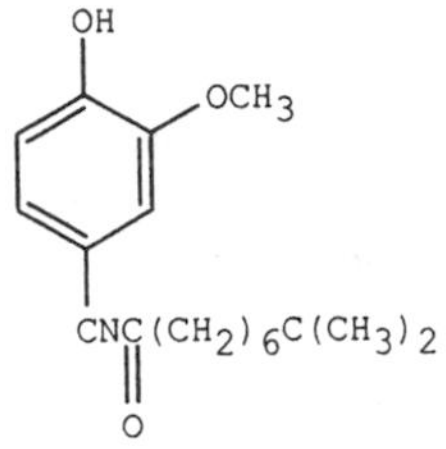

Dihydrocapsaicin

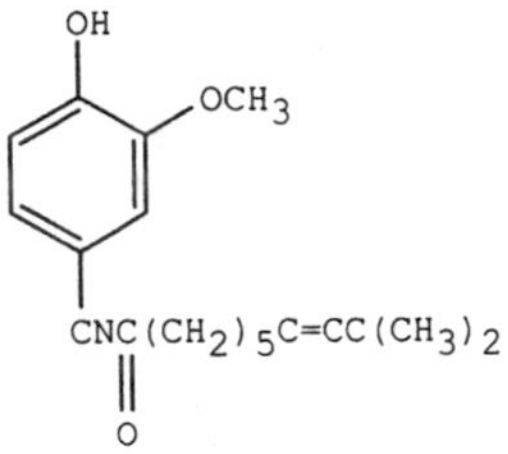

Homocapsaicin

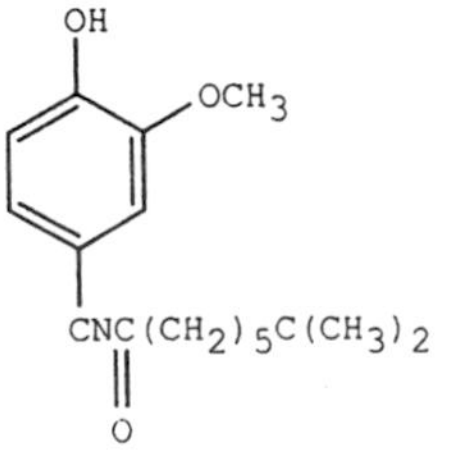

Nordihydrocapsaicin

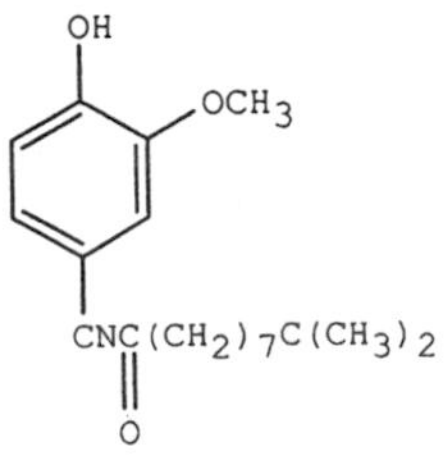

Homodihydrocapsaicin

- act on afferent, sensory neurons; stimulation and then blockade; depletion of substance P; high dosage causes other actions via effects on ion channels

and homodihydrocapsaicin (Masada et al. 1971; Iwai et al. 1979). The concentrations of these compounds are variable but are typically highest in the mature fruits 5–6 weeks after flowering. A key factor in the activity of capsaicin is the presence of homovanillic acid ester.

Capsaicinoids are biotransformed by the liver and excreted as glucuronides (Kawada et al. 1984, 1986). Not only are they metabolized, but they may modify their own hepatic biotransformation or that of other compounds, inhibiting initially and inducing after several days of ingestion (Miller et al. 1983; Oi et al. 1992). In rats pretreated for 1 week with 25 mg/kg capsaicin daily, antipyrine clearance was decreased and the half-life increased from 1.6 to 2.1 hours, but there was no effect on disposition of quinine or theophylline (Wanwimolruk et al. 1993).

In spite of the low toxicity of the plants, the capsaicinoids are capable of causing remarkable effects. Experimentally, in small doses capsaicin acts specifically on primary afferent neurons, i.e., thin, mainly unmyelinated sensory nerves (C fibers). The result is stimulation and then receptor blockade, producing desensitization to further stimuli (Buck & Burks 1986; Holzer 1991). The nociceptive effects are due to interactions with substance P. Peripheral pain nociceptors, when activated, transmit signals via release of substance P in afferent nerves, eventually reaching the spinal cord and higher centers (Gamse 1979). The neurotransmitter, substance P, may be depleted from sensory neurons, and the effect may extend up through the lamina (I, II, V) of the superficial dorsal horn to higher brain centers, resulting in functional and possibly morphologic sensory nerve impairment. The effects appear to be related to actions on cation channels.

Capsaicin acts on ion channels insensitive to conventional blockers to increase Na^+ and Ca^{2+} and decrease H^+ entry, causing depolarization and release of neurotransmitters (O'Neill 1991; Dray 1992). There is also inhibition of anterograde and especially retrograde axonal transport (Otten et al. 1983; Buck & Burks 1986). Inhibition of K^+ movement is associated with prolongation of ventricular action potential (Castle 1992). Larger doses produce nonselective effects on a variety of other cells concomitant with inhibition of voltage-gated Na^+ and K^+ channels. The overall effect is to stabilize membranes and is exemplified by decreased activity of cardiac and visceral smooth muscle. Contrasting excitation of thermosensitive neurons of the hypothalamus and vascular smooth muscle may also occur. The effects on the heart and vessels result in variable blood pressure changes (Toda et al. 1972). The ion channel effects are associated with an interference of reflex activity in a variety of systems. Perhaps the most important is stimulation of the Bezold-Jarisch reflex, which results in respiratory distress, a serious side effect of therapeutic use of the capsaicinoids (Szolcsanyi et al. 1991). The capsaicinoids have been exploited medicinally, especially for use against syndromes involving pain, which is possibly mediated through depletion of substance P and the resulting desensitization (Cordell & Araujo 1993).

Although in high dosage capsaicin can cause lethal convulsions, this condition is quite unlikely because it would require an adult person to eat 2 kg d.w. of fruits, as extrapolated from the toxic oral dose in mice (Glinsukon et al. 1980). Birds are much less sensitive to the effects of capsaicin than are mammals. The irritant effects of capsaicin have been exploited for use in repellents (Weidner 1980; Houpt et al. 1984).

Other toxicants found in *Capsicum* include solanine and scopoletin. For a discussion of scopoletin, see *Brunfelsia*. It is of interest that analogs of greatly increased potency have been isolated from other plants or are being synthesized. One of these is resiniferatoxin, from species of *Euphorbia* in the family Euphorbiaceae (see chapter 36), which also contains the key functional homovanillic acid group (Szallasi & Blumberg 1989).

Clinical Signs—The signs are directly related to the biting irritation at the site of contact with the peppers, whether it be the mouth, eyes, or skin. When large amounts are ingested, vomiting and diarrhea may occur. The effects are temporary—in almost all cases they are reversible. In cases of prolonged contact or when the protective epithelial barrier is damaged, there may be more-severe effects, including blistering. Hypothalamic actions may lead to hypothermia except when the exposure is chronic.

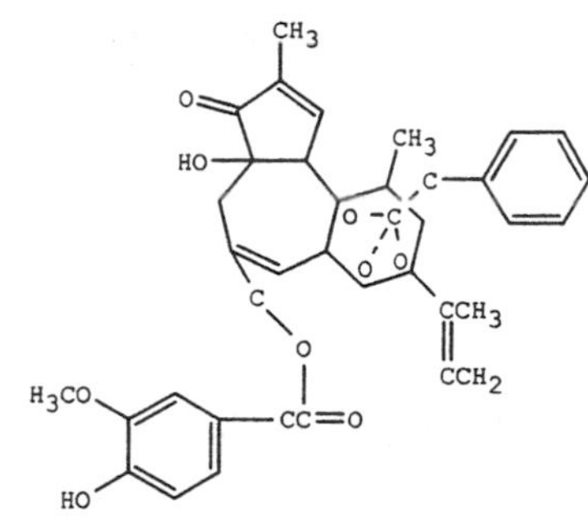

Resiniferatoxin

Pathology—*Capsicum* is not likely to cause a lethal disease problem. There may be reddening of the mucosa of the digestive tract, but there are generally no neurologic lesions except for the spinal changes produced experimentally.

- pain in mouth, skin, eyes; rarely blistering, vomiting, diarrhea
- reddening of areas of contact

Treatment—Topical relief of the burning irritating effects is the normal treatment. Cool-water irrigation provides adequate relief for most cases of skin or eye irritation. Mineral or vegetable oils will provide longer duration benefit. In only rare cases will local or topical anesthetics be required.

- topical relief of irritant effects, cool water, mineral oil

Cestrum L.

Taxonomy and Morphology—Comprising 150–175 species native to tropical and subtropical America, *Cestrum* is now cultivated throughout the world for its clusters of frequently fragrant flowers. The common names *jessamine* and *jasmine* are used for some of its species, but they also are applied to members of *Jasminum* in the Oleaceae and *Gelsemium* in the Loganiaceae. Species of importance in North America include the following (see plate 69.1):

- *C. aurantiacum*, *C. diurnum*, *C. elegans*, *C. fasciculatum*, *C. nocturnum*, *C. parqui*

bastard jasmine
cestrum
Chinese inkberry
day jessamine
huele de noche
jessamine
night-blooming jessamine
poisonberry
wild jasmine
willow-leaved jessamine

C. aurantiacum Lindl.	orange-flowered cestrum
C. diurnum L.	day jessamine, day cestrum, wild jasmine, Chinese inkberry
C. elegans (Brongn.) Schltdl. (= *C. purpureum* [Lindl.] Standl.)	jessamine, bastard jasmine
C. fasciculatum (Schltdl.) Miers	
C. nocturnum L.	night-blooming jessamine, poisonberry, huele de noche
C. parqui L'Hér.	willow-leaved jessamine, green cestrum

- shrubs or small trees; leaves narrow; flowers in clusters; corollas mostly funnelform, 5-lobed, yellow, white, red, or greenish; fruits berries

Plants shrubs or small trees; evergreen or deciduous. **Stems** erect or spreading or climbing. **Leaves** narrow; lanceolate to elliptic or oval or oblong; margins entire. **Inflorescences** clusters of flowers; terminal or axillary. **Flowers** fragrant or not fragrant. **Calyces** short; tubular to campanulate; 5-toothed or 5-lobed. **Corollas** radially symmetrical; 5-lobed; yellow or red or white or greenish; salverform or funnelform. **Petal Lobes** acute; erect to spreading or reflexed. **Stamens** 5; equal or unequal; included within corollas. **Fruits** small berries; globose to ellipsoid; whitish or red or black or violet-brown; locules 2.

Cestrum diurnum

1. Corollas red or purple-red; throats abruptly constricted.
   2. Abaxial surfaces of corollas glabrous. Inflorescences open; nodding . . . . *C. elegans*
   2. Abaxial surfaces of corollas pubescent. Inflorescences dense; erect . . . . *C. fasciculatum*

■ ornamentals, occasional escapes

Cestrum parqui

1. Corollas white or yellow or orange-yellow; throats tapered.
   3. Corolla lobes becoming reflexed. Flowers fragrant during day.
      4. Corollas orange-yellow. Leaves ovate to lanceolate, malodorous when crushed. Stamens unequal ...... *C. aurantiacum*
      4. Corollas white or greenish white. Leaves oblong to elliptic; not malodorous when crushed. Stamens equal .. *C. diurnum*
   3. Corolla lobes erect to spreading. Flowers fragrant at night.
      5. Leaves lanceolate; apices long acute. Berries violet-brown .... *C. parqui*
      5. Leaves ovate-oblong to elliptic; apices short acuminate. Berries white .... *C. nocturnum*

Distribution and Habitat—Species of *Cestrum* are showy ornamentals that have been cultivated in warmer areas of the world. They are restricted to the warm South in North America, except when used as tub plants to be taken indoors in winter. Native to Chile, *C. parqui* is the hardiest and has the broadest range; it is found in the woods and thickets of the coastal plain from Florida to Texas. *Cestrum elegans* and *C. fasciculatum* from Mexico are also somewhat hardy. The other species are limited to southern Florida, parts of the Gulf Coast, California, and Mexico. They occasionally escape in areas where they are used ornamentally. For example, *C. diurnum,* native to the West Indies, is found in waste areas in southern Florida, including the Florida Keys, southern Texas, and Sinaloa, Mexico. Also native to the West Indies, *C. nocturnum* ranges from southern Texas and Coahuila south to central Mexico.

■ some *(C. diurnum, C. nocturnum),* dystrophic calcification and chronic debilitation in livestock; others *(C. aurantiacum, C. parqui),* hepatotoxic and possibly cardiotoxic and neurotoxic

Disease Problems—In areas where livestock have access to plants of *Cestrum,* intoxications are not uncommon. Following the identification of *Solanum glaucophyllum* (= *S. malacoxylon*) as a cause of dystrophic calcification in livestock in South America (Worker & Carrillo 1967), other members of the Solanaceae have been shown to have a similar propensity, including *C. diurnum* and *C. nocturnum* (Krook et al. 1975b). The excess calcification results in chronic debilitation, weight loss, and lameness. Horses as well as cattle and sheep are affected and possibly other species such as pigs to a lesser extent (Krook et al. 1975a,b; Done et al. 1976).

Other *Cestrum* species such as *C. aurantiacum* and *C. parqui* cause very different effects, being for the most part hepatotoxic (Shone & Drummond 1965; McLennan & Kelly 1984; Kudo et al. 1985; Riet-Correa et al. 1986; Kellerman et al. 1988; Lugt et al. 1992). Their effects are consistent with those of *C. laevigatum* in South Africa, which is well known as a cause of hepatic necrosis in cattle and sheep with repeated ingestion of the foliage (Lugt et al. 1991, 1992). *Cestrum parqui* also appears to have neurologic and cardiotoxic effects (Gonzalez et al. 1988). As with *Solanum,* the immature fruits of *Cestrum* are typically more toxic than the ripe ones (Shone & Drummond 1965). The ripe fruits of *C. diurnum* are eaten by birds but, like those of *C. nocturnum,* are considered to be toxic to humans (Morton 1974).

1,25-Dihydroxycholecalciferol

■ dystrophic calcification, glycoside of vitamin D; hepatotoxicity, kaurene glycosides similar to metabolic toxin in *Xanthium*

Disease Genesis—The toxin responsible for the dystrophic calcification caused by *C. diurnum* and *C. nocturnum* is a glycoside of 1,25-dihydroxycholecalciferol (Weissenberg 1989). The glycoside is hydrolyzed to yield the active vitamin D_3 (Mello & Habermehl 1992). There are about 30,000 IU D_3 equivalents/kg plant; the resultant effects are increased calcium absorption and increased calcium-binding protein (Wasserman et al. 1975, 1976; Hughes et al. 1977). The glycoside in *C. diurnum* is much less water soluble than that in *Solanum glaucophyllum* and causes less extensive and less severe calcification of soft tissues. These two species of *Cestrum* also are reported to contain cardioactive saponins that produce effects similar to those of ouabain, but these seem to be of little consequence (Roy & Chatterjee 1968).

The hepatotoxic effects are due to kaurene glycosides, such as carboxyparquin and the less toxic parquin, which have been isolated from *C. parqui* (Pearce et al. 1992). Carboxyparquin is highly toxic in mice; a dose of 4.3 mg/kg b.w. causes severe liver and moderate kidney disease. The compound is very similar structurally to carboxyatractyloside, the toxicant of *Xanthium* (cockleburs), differing only in the sugar substituents. The role of solasonine, a solasodine-based glycoalkaloid that has been reported for *C. parqui,* in its toxicity is not clear (Schreiber 1968). *Cestrum parqui* also is reported to contain small amounts of cardenolides based upon digitogenin, gitogenin, gitoxigenin, and tigogenin (Lopez et al. 1984). The small amounts of nicotine and nornicotine reported in *C. diurnum* and *C. nocturnum* are probably of little consequence toxicologically (Halim et al. 1971).

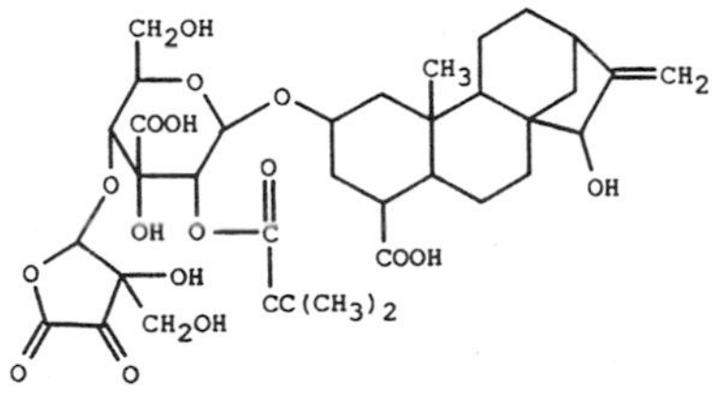

Parquin

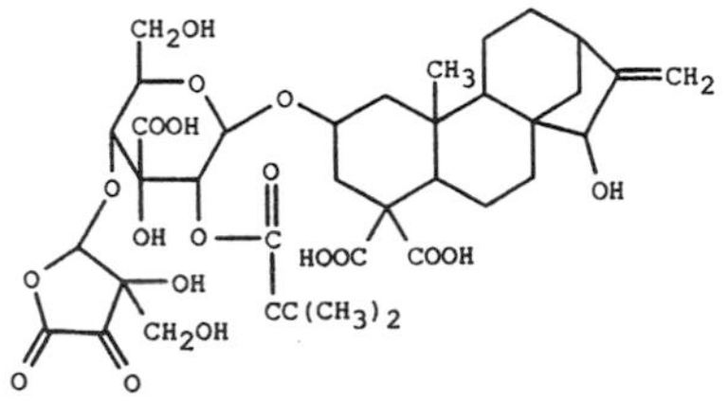

Carboxyparquin

Clinical Signs—The calcinogenic species, *C. diurnum* and *C. nocturnum,* produce signs of chronic debilitation, weight loss, and progressive lameness. Serum calcium is usually distinctly elevated (>13 mg/dl). The resulting dystrophic calcification is especially noticeable in the horse, as manifested by bony prominences, tense abdominal muscles, stiffness with a short choppy walk, reluctance to move, and eventually recumbency. There may be pain on palpation of tendons and ligaments. The syndrome may be slowly progressive over several years or develop more rapidly to become severe within a few months. Large amounts of plant material may cause a more acute disease with signs related to irritation of the digestive tract or cardiac dysfunction, presumably due to excess calcium.

Signs associated with hepatic necrosis include depression, incoordination, seizures, and collapse/coma. In this case, serum chemistry changes indicative of hepatic damage—marked increase in AST, LDH, GGT activity, or prothrombin time—will be noteworthy. This is typically an acute, rapidly developing disease.

■ calcification, debilitation, weight loss, lameness; horses, bony prominences, tense abdomen, stiffness, reluctance to move, pain when tendons and ligaments palpated

hepatotoxicosis, depression, ataxia, seizures, collapse, elevated serum hepatic enzymes

Pathology—The chronic wasting disease associated with dystrophic calcification is characterized by calcification of elastic tissues, including tendons, ligaments, and arteries. There are bony changes as well, including primary osteonecrosis and secondary osteopetrosis.

Grossly, in the hepatotoxicosis caused by *C. parqui* in sheep, the liver has a nutmeg color or reticular pattern, and the gallbladder wall is thick and edematous. There also may be excess serofibrinous peritoneal fluid and small splotchy hemorrhages in various tissues, including the heart. Microscopically, there will be centrilobular hepatocellular necrosis and moderate renal tubular necrosis.

■ calcification of tendons, ligaments, arteries, bony changes

hepatotoxicosis, gross pathology, liver nutmeg color, gallbladder wall thick, peritoneal fluid

microscopic, hepatic necrosis, moderate renal tubular necrosis

Treatment—When dystrophic calcification is present, calcitonin concentrations are usually already elevated, and treatment is mainly directed toward relief of the discomfort. Disease associated with hepatocellular necrosis is best treated with good nursing care and a high-quality diet.

■ calcification, relief of discomfort

hepatotoxicity, general nursing care, good diet

Datura L.

Taxonomy and Morphology—Commonly known as thorn apple, *Datura* comprises 8–25 species native to the tropics and subtropics. Its name is apparently derived from an East Indian or Arabic vernacular name. Woody species with pendulous flowers formerly included in the genus are now segregated in their own genus, *Brugmansia,*

■ *D. discolor*
D. ferox
D. inoxia
D. metel
D. quercifolia
D. stramonium
D. wrightii

datura
devil's trumpet
horn-of-plenty
Indian apple
Jamestown weed
jimsonweed
thorn apple
tolguacha

which is also treated in this chapter. In North America, 7 native and introduced species are present. All species are of toxicologic interest:

D. discolor Bernh.	
D. ferox L.	
D. inoxia Mill. (= *D. meteloides* DC. ex Dunal)	large-flowered thorn apple, sacred datura, Indian apple
D. metel L. (= *D. fastuosa* L.)	black datura, devil's trumpet, Hindu datura, horn-of-plenty, downy thorn apple, hairy thorn apple
D. quercifolia Kunth	oak-leaf thorn apple
D. stramonium L.	jimsonweed, Jamestown weed, thorn apple
D. wrightii Regel	sacred datura, Indian apple, tolguacha

■ branching herbs or subshrubs; leaves short petiolate; flowers large, showy, fragrant, erect; corollas tubular to funnelform, white, yellow, or violet to purple; fruits prickly capsules

Plants herbs or subshrubs; annuals or perennials; herbage typically malodorous, glabrous or pubescent. **Stems** erect; 50–150 cm tall; branching profuse, candelabriform. **Leaves** short petiolate; blades ovate to elliptic or rhomboidal or triangular; margins entire to coarsely toothed or lobed, plane or undulate. **Inflorescences** extra-axillary; flowers solitary in branch forks. **Flowers** large; showy; fragrant; short-lived; erect. **Calyces** tubular or angled; 5-toothed; circumscissile, distal halves deciduous, proximal halves persistent and enlarging in fruit. **Corollas** radially symmetrical; 5- or 10-lobed; white or yellow or violet to purple; tubular to funnelform. **Petal Lobes** long or short; acute to acuminate. **Stamens** 5; equal or unequal; included within corollas or exserted. **Fruits** capsules; globose to ovoid or ellipsoid; prickles present; locules 4. **Seeds** flat; black or brown or tan.

Datura stramonium

1. Capsules erect.
   2. Capsule prickles 5–10 mm long; subequal . . . . *D. stramonium*
   2. Capsule prickles 10–30 mm long; conspicuously unequal.
      3. Plants less than 60 cm tall. Prickles numerous; stiff . . . . *D. ferox*
      3. Plants 100–150 cm tall. Prickles sparse; flexible . . . . . *D. quercifolia*
1. Capsules nodding.
   4. Stems densely villous or glandular villous.
      5. Corollas 10-toothed. Seeds black . . . . *D. discolor*
      5. Corollas 5-toothed. Seeds brown . . . . *D. inoxia*
   4. Stems glabrous or puberulent.
      6. Corollas 10-toothed. Calyces 5–6.5 cm long. Capsules ovoid; 4–6 cm long. Prickles more than 1 cm long . . . . *D. metel*
      6. Corollas 5-toothed. Calyces 7–12 cm long. Capsules globose; 3–4 cm in diameter. Prickles less than 1 cm long . . . . *D. wrightii*

■ ornamentals and weedy invaders

Distribution and Habitat—Native to both the Old World and the New World, species of *Datura* are now widely naturalized, and the geographic origins of some are debated by taxonomists. Illustrative of this is the uncertainty whether *D. stramonium,* which is widespread throughout the world in disturbed sites, is native to Asia or tropical America. Flowering for long periods in the summer and fall, some species are cultivated as ornamentals, whereas others are weeds of waste areas, feedlots, croplands, and roadsides.

Disease Problems—Species of *Datura* are of considerably more importance as risks for humans rather than for livestock or other animals. Plants of this genus, especially their seeds, have been used for visionary purposes by many societies for centuries (Baskin 1967). The common names *Jamestown weed* and its derivative, *jimsonweed*, date to an incident in Virginia in 1676 when soldiers sent to Jamestown by the governor to quell an insurrection promoted by Captain Nathaniel Bacon accidentally consumed *Datura* leaves in a salad and became intoxicated. They were incapacitated for 11 days and then recovered. Other episodes involving large numbers of individuals have since been reported; each intoxication resulting in bizarre behavior and incapacitation of up to several days duration (Hughes & Clark 1939). The seeds in small quantities are thought to be hallucinogenic and are used ritualistically (especially *D. inoxia*) by American Indians and others. Their use for this purpose is not without danger, because in large amounts, severe derangement of the parasympathetic nervous system may occur and present serious problems. Smoke from burning the leaves and seeds has been used therapeutically to alleviate respiratory distress. The potential for serious problems arises when inhalant asthma powders are taken orally, because of the lack of control over the dose. Commercially available thorn apple tea preparations are also a potential problem (Coremans et al. 1994). Numerous poisonings have been reported, especially in children (Jennings 1935; Mitchell & Mitchell 1955). Hallucinatory episodes are reported in many cases, some individuals exhibiting bizarre (wandering naked) and/or aggressive and violent behavior (Dean 1963; Gowdy 1972; Klein-Schwartz & Oderda 1984).

■ neurotoxic, mainly in humans, seeds considered hallucinogenic; used for relief of respiratory distress

Datura inoxia

Datura metel

Datura quercifolia

Datura seeds, roots, and extracts continue to be misused in a variety of ways throughout North America (Shervette et al. 1970; Mahler 1975; Klein-Schwartz & Oderda 1984; Guharoy & Barajas 1991; Hanna et al. 1992; Rodgers & Von Kanel 1993; Perrotta et al. 1995; Dewitt et al. 1997). Although most intoxications are resolved without serious long-term consequences, occasionally there are more serious outcomes, even deaths (Perrotta et al. 1995). Toxicity in humans accrues from small amounts of plant material, possibly about 0.1% b.w.

The poor value of *D. stramonium* as feed for livestock is reflected in the reference to the "lowly jimsonweed" in Gene Autry's theme song, "Back in the Saddle Again." Disease in livestock from consumption of fresh plants seems rather limited in importance, because it is generally unpalatable and the development of early signs diminishes the animals' appetite. The plants should be considered most toxic when mature, and they remain so after drying. This is important because, although fresh plants are quite unpalatable, dried material in hay may be eaten inadvertently. This is particularly true for ruminants. Up to 1% b.w./day of leaves or fruits fed to sheep or goats produced clinical signs after 2 or more days, but it was not lethal unless fed for several weeks or longer (El Dirdiri et al. 1981). Such a situation of prolonged daily ingestion is unlikely because of the plant's effects on appetite. This is illustrated by an episode in which linseed meal contaminated with seeds of *D. stramonium* caused only decreased appetite and weight loss (Janssens & Wilde 1989). However, the species is capable of causing serious problems in some circumstances, as is indicated by the death of calves reported by Case (1956). When dried whole plants were administered to sows at 0.12% b.w./day, severe signs developed within a few days, and 1 sow died after 4 days (Keeler 1981). In rare instances, mass intoxications may occur; e.g., in Czechoslovakia, of 510 cows at risk from eating *D. stramonium*, 44 died and 22 more had to be destroyed (Ofukany et al. 1983). Intoxications also occur in wildlife, as is documented by a case in which *D. inoxia*, introduced into South Africa, was suspected to be the cause of disease in springbok (Lindeque & Scheepers 1992).

■ generally little problem in livestock, unpalatable; small amounts cause reduced motility of gut and decreased appetite

Datura stramonium

Datura wrightii

Congenital arthrogryposis in pigs has been reported to be caused by *D. stramonium* (Leipold et al. 1973). However, this has not been confirmed experimentally, even when plant material has been given for prolonged periods ranging from 20 to 90 days of gestation at a dosage sufficient to produce prominent signs of intoxication in the dams and death in 1 sow (Keeler 1981).

Disease Genesis—Most of the toxic effects of species of *Datura* are attributable to potent anticholinergic tropane alkaloids such as L-hyoscyamine and scopolamine (L-hyoscine), an epoxy derivative of hyoscyamine. Atropine is a mixture of D- and L-hyoscyamine, the racemization occurring as an artifact of extraction. The toxicity problems are similar to those of the well-known deadly nightshade of European origin, *Atropa belladonna,* which contains mainly hyoscyamine (0.3–0.4%) and little scopolamine (Oshima et al. 1989). In *D. metel,* alkaloid concentrations increase with plant maturity, up to 1.5% at fruiting (Afsharypuor et al. 1995). Typically, scopolamine is present in leaves and flowers at levels of 0.2–0.27%, whereas hyoscyamine is at 0.06–0.13% (Oshima et al. 1989; Afsharypuor et al. 1995). Concentrations of tropane alkaloids in *D. stramonium* seeds range from 0.2 to 0.6%, with a ratio of hyoscyamine to scopolamine of 4:1 to 5:1 (Worthington et al. 1981; Nelson et al. 1982; Dugan et al. 1989; Friedman & Levine 1989). Hyoscyamine also predominates in seeds of *D. inoxia* (0.14% hyoscyamine and 0.02% scopolamine), but the reverse is true for leaves, albeit with lower concentrations (<0.07% scopolamine and traces of hyoscyamine) (Duez et al. 1985).

- anticholinergic tropane alkaloids, hyoscyamine, scopolamine, competitive antagonists of ACh at muscarinic cholinergic receptors

Tropane alkaloids are competitive antagonists of acetylcholine at muscarinic cholinergic receptors and are capable of producing severe effects in most animal species. The affected receptors are of several subtypes and are found mainly in the autonomic nervous system, central nervous system, heart and smooth muscle, and secretory tissues (Brown 1990). The most prominent effects common to both types of tropanes are (1) blockage of the contraction of the muscles of the iris and lens, thus dilating the pupil and abolishing the reflex constriction of the pupil to light (cycloplegia); (2) decreased digestive tract motility due to inhibition of smooth muscles; and (3) decreased salivary secretions (Brown 1990). There is also increased vagal activity, with an increase in respiration and heart rates. Atropine or hyoscyamine effects are mainly peripheral because of limited penetration of the blood-brain barrier. Apoatropine, the anhydride of atropine, has greatly increased neurotoxic potential (Krantz et al. 1954). Scopolamine causes marked neurologic alterations, including depression, drowsiness, sleep alterations, and euphoria.

- prominent effects—
  dilation of pupils
  decreased digestive tract motility
  decreased salivary secretion

There is evidence that some tropane alkaloids are teratogenic. Experimentally, a high proportion of early embryonic deaths and malformed embryos were seen in chicks when eggs were inoculated at day 2 or 4 with 10 μg or more of a 98:2 mixture of scopolamine and L-hyoscyamine (Magras et al. 1993). Whether similar effects are caused by other proportions of these alkaloids remains to be determined. In contrast, neither atropine nor scopolamine were teratogenic, based upon limited studies using the frog embryo teratogenesis assay (FETAX) (Friedman et al. 1992). At present, the risk of teratogenesis is not clear, but species containing high proportions of scopolamine should be of concern as potential teratogens.

- tropane alkaloids possibly teratogenic

The highest concentrations of alkaloids are found in the seeds, and numerous instances of accidental or deliberate poisoning in humans have been attributed to their ingestion or to plant decoctions. The problems are similar to those caused by *Atropa belladona,* long recognized as a serious hazard for children and animals (Cooper & Johnson 1984). In most situations the severity of the disease in animals is limited by the onset of the effects (especially decreased appetite), which preclude

- contamination of feed with seeds poses some risk for livestock; pigs, 1–2%, or 1,300–2,600 seeds/kg feed, only if ground, not whole seeds; horses, 0.5%, or 595 seeds/kg feed; cattle, 4,408 seeds/kg feed; chickens, 3% ground seeds in feed

continued ingestion of the plants or feed containing the seeds. The high concentrations of alkaloids in the seeds and their relative insensitivity to heat are of special concern because of possible contamination of cereal grains and have led to evaluation of the effects of seed contamination of feeds for various animal species. Rats appear to eat and tolerate reasonably well up to 5% seeds in their diet for 90 days (Dugan et al. 1989). There were some effects with 0.5% or more, but they were not serious, although there were decreases in body and some tissue (liver) weights. In pigs in which the toxic dose of alkaloid is 1.2–2 mg/kg b.w., 1–2% ground seed by weight in feed is estimated as the toxic contamination level (Worthington et al. 1981). This is equivalent to approximately 1,300–2,600 seeds/kg feed. However, whole seeds in the diet of pigs are of little consequence because they may be rejected or, if consumed, pass undigested in feces. In horses, 595 seeds/kg of feed, or 0.5% (by weight), produced typical signs of intoxication: anorexia, increased heart rate, dilated pupils, diarrhea, and increased drinking and urination (Williams & Scott 1984). This level of contamination represented a dose of 0.075 mg seeds/kg b.w. per day for 10 days. Schulman and Bolton (1998) describe severe digestive disturbances in 2 horses fed a feed supplement contaminated with seeds of a species of *Datura*. One horse died of acute gastric dilation, and the other was euthanized because of a nonresponsive ileus. In cattle, 4,408 seeds/kg feed caused depressed feed intake and periodically more-severe signs of intoxication (Nelson et al. 1982). Contamination with 1,000 seeds/kg feed is cause for concern in adult cattle. In chickens, there is a decrease in feed intake at 3% of seeds ground in meal, and the no-effect level is 1% by weight (Day & Dilworth 1984; Flunker et al. 1987). Concern over the potential problems with seeds of *Datura*, especially of *D. ferox*, in feed for chickens prompted a study of the effects of scopolamine and hyoscyamine in a ratio of 98:2 added to the feed (Kovatsis et al. 1993). With up to 150 ppm alkaloids in feed, the only adverse effects were a decrease in egg production and a temporary decrease in body weight. There were no signs or pathologic changes indicative of intoxication. There were no adverse effects up to 75 ppm in feed.

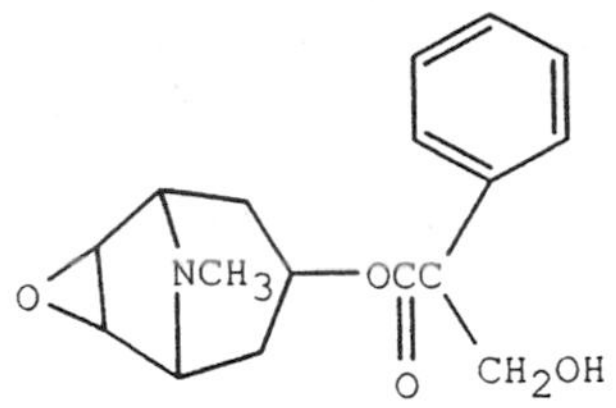

NCH$_3$ —OCC(=O) CH$_2$OH

Scopolamine

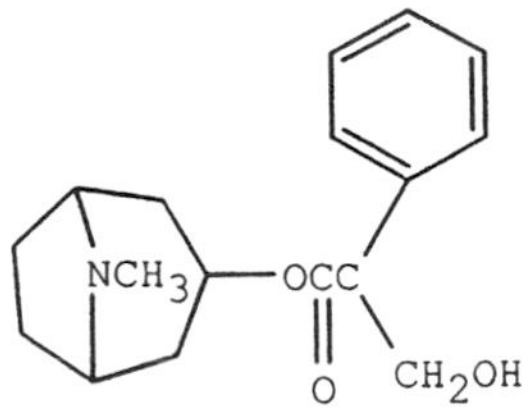

NCH$_3$ —OCC(=O) CH$_2$OH

Hyoscyamine

The wide margin of safety seen with most animals is not apparent for humans. The lethal dose of *D. stramonium* for children has been estimated as 4–5 g of crude leaf or seed (Mitchell & Mitchell 1955). Because of the apparent high sensitivity to atropine, contamination of food products with several seeds/kg of food may pose a serious risk (Kellerman et al. 1988). Species of *Datura* are also reported to be a serious hazard for ostrich chicks (Steyn 1934).

There is considerable variation between and within different animal species in ability to hydrolyze the ester bond and to further metabolize atropine and scopolamine (Bernheim & Bernheim 1938; Wada et al. 1991). Metabolic polymorphism is especially noteworthy in the rabbit (Sawin & Glick 1943; Margolis & Feilgelson 1963).

■ calystegin B_2, low levels in leaves

It is of interest that calystegin B_2 has been isolated from the leaves of *D. wrightii* (Nash et al. 1993), but concentrations in *D. stramonium* are quite low (Drager et al. 1995). Although no disease has been associated with this compound in *Datura*, it is suspected to be the cause of a cerebellar degenerative disease, crazy-cow disease, associated with consumption of *Solanum dimidiatum*. This may portend a potential for adverse effects with long-term ingestion of some species of *Datura*.

A few species such as *D. quercifolia* and especially the Old World *D. ferox* contain 28-C steroid lactone with anolides, but these are probably primarily of allelopathic rather than toxicologic interest (Evans et al. 1984).

Clinical Signs—The early signs of intoxication due to *Datura* reflect its effects on the parasympathetic nervous system: intermittent depression, restlessness, irritability,

- depression, restlessness, irritability, weakness, tachycardia, dilated pupils, photophobia, dry mouth, constipation; rarely ataxia, delirium; plus in humans fever and flushed red appearance

  confirm alkaloid presence by placing urine in eye of test animal

bizarre behavior, weakness, sinus tachycardia, marked and persistent mydriasis and photophobia, dryness of mucous membranes, and constipation. If severe intoxication occurs, incoordination and paresis may develop, followed by depression and respiratory failure. In humans, fever and a flushed red appearance may accompany the other signs. High dosage in humans may result in delirium, fever, or rarely arrhythmias and seizures but is seldom fatal. Deaths are generally due to misadventure such as drowning while the individual is impaired.

The tropane alkaloids may be detected in urine or stomach contents, or it is asserted that a drop of urine from the intoxicated animal, placed on the eye of a test animal, will cause the rapid onset of mydriasis (Smith et al. 1956). The test administration of parasympathomimetic drugs will not cause increases in salivation, lacrimation, or gastrointestinal activity in intoxicated animals.

- no lesions

Pathology—There are no distinct or specific lesions. Grossly, scattered hemorrhages in the liver and renal medulla, pulmonary congestion, and peritoneal and pericardial effusions have been reported.

- oral activated charcoal; most cases no treatment, rarely cholinergic physostigmine

Treatment—Activated charcoal given orally as an absorbent, and laxatives, may have some value, but the gastrointestinal inhibitory effects of the alkaloids limit the effectiveness of this approach, especially in ruminants. Control of seizures should be accomplished with drugs such as diazepam rather than phenothiazine-type tranquilizers, to avoid the anticholinergic side effects of the latter.

If necessary, the disease can be controlled by the use of cholinergic drugs until the toxicant is eliminated. Physostigmine is preferred because it is a tertiary amine and crosses the blood-brain barrier (Duvoisin & Katz 1968; Rumack 1973). However, it is of short duration, and the dose (0.5 mg total up to 2 mg/100 kg parenterally) may need to be repeated hourly. If given i.v., it must be done slowly. That this approach is not required in most human intoxications is shown by a study in which 29 cases were treated conservatively without physostigmine, and all individuals recovered without problems (Rodgers & Von Kanel 1993). Some clinicians prefer a conservative and supportive approach because physostigmine is not very effective against the central nervous system actions of the tropane alkaloids (Schneider et al. 1996).

Hyoscyamus L.

- *H. niger*

  fetid nightshade
  henbane
  hog's-bean
  insane root

Taxonomy and Morphology—An Old World genus of some 15 species centered in the Mediterranean region, *Hyoscyamus,* commonly known as henbane, is one of the best-known poisonous plants. Its name, used by Dioscorides, is derived from the Greek roots *hyos* and *kyamous,* for "hog" and "bean," and alludes to ingestion of the fruits by swine. One species is naturalized in North America:

H. niger L.	black henbane, henbane, hog's-bean, fetid nightshade, insane root

- erect, malodorous, annual or biennial herbs; basal rosettes of leaves first year; upper leaves clasping; flowers in terminal spikes or racemes; corollas 5-lobed, funnelform, greenish yellow with purple veins; fruits capsules, enclosed in calyces

Plants herbs; annuals or biennials; from fusiform roots; herbage malodorous, conspicuously villous. **Stems** erect; 30–100 cm tall. **Leaves** in large rosettes the first year; lower cauline leaves petiolate; upper sessile, clasping; blades broadly ovate to oblong; margins coarsely pinnately lobed or toothed, lobes triangular. **Inflorescences** racemes or spikes; terminal; somewhat 1-sided; bracts large, leafy. **Flowers** showy. **Calyces** urceolate or campanulate; reticulate veined; 5-toothed; villous; enlarging in fruit. **Corollas** slightly bilaterally symmetrical; 5-lobed; greenish yellow with purple veins and throat; funnelform. **Petal Lobes** spreading; broad; rounded. **Stamens** 5; unequal; mostly exserted. **Fruits** circumscissile capsules; 5-lobed; completely enclosed by calyces.

Hyoscyamine

- naturalized, weedy

Distribution and Habitat—*Hyoscyamus niger* was introduced into North America in the early 1600s and cultivated for its medicinal properties. It escaped and naturalized by the late 1600s and now occurs as a weed of roadsides and waste areas across the continent in southern Canada and the northern United States.

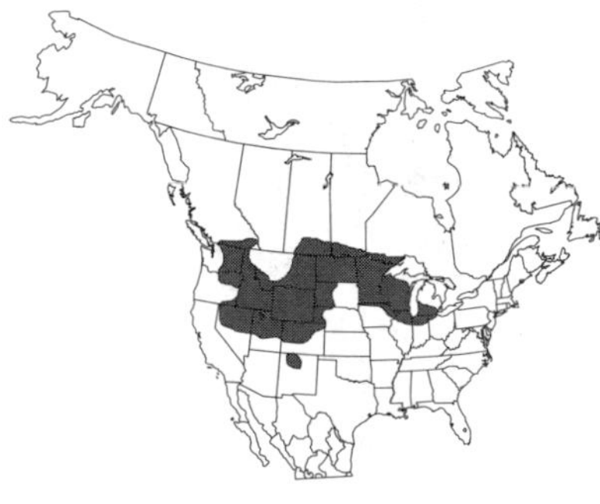

Hyoscyamus niger

Disease Problems—*Hyoscyamus* has proven to be much more of a problem in humans than in animals. During a 10-year period in Iran, more than 900 cases of intoxication were reported to be caused by *H. niger,* mainly in the spring in male children (Daneshvar et al. 1992). Moderate atropinic signs were noted, although there was considerable nervous stimulation with delirium, hallucinations, and seizures. There were no fatalities. Similar descriptions are noted in Europe, where the plant is strongly associated with sorcery (Cooper & Johnson 1984). There are many tales and traditions of its medicinal and other uses (Baskin 1967). There are also reports of toxic honey from bees that fed on nectar and tainting of milk from cows that ate the plants (Cooper & Johnson 1984).

- neurotoxic, mainly in humans, abuse, delirium, hallucinations

Disease Genesis—Hyoscyamine and scopolamine occur in about equal proportions in the leaves, with total concentrations of 0.06–0.13% (Hocking 1947; Oshima et al. 1989). Scopolamine undergoes hepatic biotransformation, but there is considerable variation in extent among animal species (Wada et al. 1991). Rabbits and to a lesser extent guinea pigs have appreciable esterase activity toward scopolamine, as with atropine.

Hyoscyamine

Scopolamine

- tropane alkaloids, hyoscyamine, scopolamine, low levels of calystegin B_2

As with some other members of the family, calystegin B_2 has been identified in *H. niger* in concentrations of 25–50 μg/g in the upper leaves and roots (Drager et al. 1995). The significance of this as a risk for livestock is not known, but this toxin is associated with neurotoxic effects, as seen with *Solanum dimidiatum.*

- restlessness, pupillary dilation, dry mouth, excitement, increased heart rate, labored respiration

Clinical Signs—The signs produced by *H. niger* are similar to those caused by *Datura* and include restlessness, pupillary dilation, increased heart rate, labored respiration, excitement, dry mouth, bloat, and more rarely seizures.

- no lesions; specific treatment generally not needed

Pathology and Treatment—Lethal effects are seldom produced by this genus, and few gross or microscopic lesions are associated with the action of the tropane alkaloids. Treatment is generally not required, other than conservative approaches to control any serious problems. In animals in which cholinergic signs are particularly prominent, acetylcholinesterase inhibitors such as physostigmine may be useful. A more complete discussion is presented in the treatment of *Datura,* in this chapter.

Lycium L.

- *L. barbarum*

  boxthorn
  Chinese boxthorn
  Christmas berry
  Duke of Argyll's tea tree
  matrimony vine

- deciduous, thorny shrubs; leaves alternate or fascicled; flowers in leaf axils, showy; corollas funnelform, lavender to purple; fruits bright red berries

Taxonomy and Morphology—Native to both the Old World and the New World, with almost equal numbers of species in each, *Lycium* comprises 90–100 species of deciduous or evergreen shrubs in temperate, subtropical, and tropical regions. Its name is derived from the Greek *lykion,* for the ancient country Lycia in southwestern Asia. Common names for the genus are numerous. *Boxthorn* and *matrimony vine* are most commonly used, but in the Southwest, desertthorn, wolfberry, squawberry, and cilindrillo also are used. In North America, approximately 21 species are present, but only 1 introduced species is reported to be of toxicologic significance:

L. barbarum L. (= *L. halimifolium* Mill.) (= *L. chinense* Mill.)	matrimony vine, boxthorn, Christmas berry, Chinese boxthorn, Duke of Argyll's tea tree

Lycium barbarum

Plants shrubs; deciduous; typically armed with leafy thorns; herbage glabrous. **Stems** arched or climbing; 1–6 m long. **Leaves** alternate or fascicled; short petiolate; blades elliptic to lanceolate or ovate or obovate; margins entire. **Inflorescences** of 1–4 flowers in leaf axils; pedicels long or short. **Flowers** showy. **Calyces** campanulate to tubular; 2- or 3-lobed. **Corollas** radially symmetrical; 4- or 5-lobed; lavender to purple; funnelform. **Petal Lobes** spreading; broad. **Stamens** 5; subequal; exserted. **Fruits** berries; ellipsoid or ovoid; bright red.

- ornamental, naturalized in Northeast

Distribution and Habitat—Native to southern Europe and southwestern Asia, *L. barbarum* is cultivated as an ornamental but has escaped to become naturalized and scattered in waste areas and thickets in much of North America, especially the northeast quarter.

- neurotoxic, low risk, problems similar to those with *Solanum*

Disease Problems and Genesis—Only a few instances of intoxication are reported to be caused by *L. barbarum.* In 2 separate cases, calves and sheep were affected when they grazed plants growing around houses. In both cases a large amount of material—1.8 kg or more—was eaten by the calves (Hansen 1927). The calves developed neurologic signs, including seizures, and the sheep exhibited signs of severe irritation of the small intestine. Thus, *Lycium* should be considered a toxic plant; unknown is whether both fruit and foliage are toxic (Pammel 1911; Watt & Breyer-Brandwijk 1962). The types of clinical signs produced are suggestive of glycoalkaloids (Kingsbury 1964). Refer to the treatment of *Solanum* for a complete discussion of the mode of action, clinical signs, pathology, and treatment associated with these compounds.

Lycopersicon Mill.

- *L. esculentum*

  love apple
  tomato

Taxonomy and Morphology—Believed to be native to northwestern South America and comprising 6 or 7 annual or perennial species of herbs, *Lycopersicon,* commonly known as tomato, is closely related to *Solanum.* The 2 genera differ primarily in their anthers. *Lycopersicon* has anthers that are elongate, fused, and open by longitudinal slits, whereas those of *Solanum* are short, connivent, and open by apical pores. *Lycopersicon*'s name is derived from the Greek roots *lycos* for "wolf" and *persikon* for "peach," possibly alluding to the early misconception that its berries were poisonous in contrast to the edible peach. Its name is sometimes misspelled *Lycopersicum.* In North America, 2 species have been reported to occur, but only 1 is typically encountered:

L. esculentum Mill. (= *L. lycopersicum* [L.] Karst.) — tomato, cultivated tomato, love apple

- aromatic, perennial herbs; leaves simple to pinnately compound; flowers extra-axillary; corollas yellow; fruits juicy berries

Plants herbs; perennials; herbage clammy-glandular, aromatic. **Stems** erect or becoming decumbent and sprawling; often rooting at nodes; often fleshy. **Leaves** simple to 1- or 2-pinnately compound; blades of various shapes, often ovate or lanceolate; margins irregularly toothed or lobed. **Inflorescences** extra-axillary cymes; racemose; with 3–12 flowers; pedicels jointed, reflexed in fruit. **Flowers** often nodding. **Calyces** green; deeply 5- or 6-lobed; reflexed in fruit. **Corollas** radially symmetrical; yellow; rotate. **Petal Lobes** reflexed when mature; narrowly acute to acuminate. **Stamens** 5 or 6; anthers fused in conelike ring about style. **Fruits** juicy berries; subglobose to ellipsoid; yellow or orange or red when mature; lobed or not lobed.

Lycopersicon esculentum

- cultivated

Distribution and Habitat—Native to the Andes of South America, *L. esculentum*, with numerous varieties and cultivars, is now cultivated throughout the world. Although a perennial, it is typically grown as an annual. Because the berries are so readily consumed by both animals and humans, the seeds are often widely dispersed. As a result, plants appear in waste areas and abandoned fields and along roadsides but generally do not persist. In extreme southern Texas, the variety *cerasiforme* may be encountered in palm groves, thickets, and open woodlands.

- foliage, neurologic, digestive disturbances; rare occurrence in livestock

- glycoalkaloids, similar to those of *Solanum;* but lower risk, possibly teratogenic

Disease Problems—On occasion, horses, cattle, or other livestock gain access to tomato foliage in gardens. In some instances this results in the appearance of acute neurologic and/or digestive disturbances, sometimes lethal. Often the link between the plant and disease is not firmly established, and thus few cases are reported. Although there are anecdotal reports of toxicity due to the foliage, the vines have been fed in large amounts to cattle experimentally for a sustained period without adverse effects other than weight loss (Shlosberg et al. 1996). However, there continue to be field reports of intoxications, and tomato foliage should be considered of some risk, especially during early growth and with ingestion of large amounts without previous adaptation (Pammel 1911). Case (1956) reported that even the mature foliage in the fall was a cause of deaths in pigs and calves.

Tomatidine

Tomatine

Disease Genesis—Tomatoes contain α-tomatine, a trisaccharide glycoalkaloid of the aglycone tomatidine. Tomatine is found throughout the plant but especially in flowers and leaves (up to 5%) (Jadhav et al. 1981). Concentrations are high in young green fruits but decline markedly as a result of metabolism during maturation and ripening (Hostettmann & Marston 1995). Friedman and Levine (1995) reported α-tomatine concentrations of 450–750 mg/100 g of dried material or 50–130 mg/100 g of fresh plants. Concentrations in plant parts arranged from highest to lowest are as follows: flowers > small stems > leaves > calyces > small immature fruits. Very little tomatine was found in mature red fruits.

Tomatine shares many features with solanine and chaconine, including weak to moderate cardiotoxicity and cholinesterase inhibition (Roddick 1989). However, it is even more poorly absorbed orally than these related glycoalkaloids and is seldom a hazard. Extrapolation from mice suggests a toxic dosage, even with concentrations of 5% in leaves, to be about 1 kg leaves/100 kg b.w. Thus, the hazard in most situations is low.

The potential for teratogenicity is difficult to assess because tomatine has not produced terata in laboratory animals, but it is strongly embryotoxic and teratogenic when tested in frog embryos (*Xenopus*) (Keeler et al. 1991b; Friedman et al. 1992). Mainly lethal malformations were produced.

■ similar to *Solanum*

Clinical Signs, Pathology, and Treatment—For signs, pathology, and treatment see the treatment of *Solanum* (in this chapter) and its discussion of glycoalkaloids.

Nicotiana L.

■ *N. attenuata*
N. glauca
N. longiflora
N. quadrivalvis
N. repanda
N. tabacum
N. trigonophylla

buena moza
Don Juan
gigante
mustard tree
tabaco cimarrón
tabaquillo
tobacco

■ aromatic herbs, shrubs or small trees; lower leaves often large; flowers in terminal racemes or panicles, showy; corollas 5-lobed, funnelform, white, yellow, greenish, pink, red, or purple; fruits capsules

Nicotiana attenuata

Taxonomy and Morphology—Native to the Americas, Australia, and Polynesia, *Nicotiana*, commonly known as tobacco, comprises 60–67 species. Its name honors Jean Nicot, the Frenchman who is said to have introduced the genus to the royal courts of Portugal and France in the late 1500s. In addition to being cultivated commercially for its leaves, the genus also is grown as a garden or border ornamental for its fragrant, delicately colored flowers. In North America approximately 15 native and introduced species are present. Representative of the genus are the following (see plate 69.2):

N. attenuata Torr. ex S.Watson	coyote tobacco
N. glauca Graham	tree tobacco, mustard tree, buena moza, Don Juan, gigante
N. longiflora Cav.	longflower tobacco
N. quadrivalvis Pursh (= *N. bigelovii* [Torr.] S.Watson)	western tobacco
N. repanda Willd. ex Lehm.	fiddleleaf tobacco, wild tobacco, tabaco cimarrón
N. tabacum L.	cultivated tobacco, burley tobacco
N. trigonophylla Dunal (= *N. palmeri* [A.Gray] M.E.Jones)	desert tobacco, tabaquillo

Plants herbs or shrubs or small trees; annuals or perennials; deciduous or evergreen; herbage typically clammy-viscid, aromatic. **Stems** erect. **Leaves** sessile or subsessile or petiolate; lower leaves often quite large; blades lanceolate or ovate to oblong or panduriform; margins entire or undulate. **Inflorescences** terminal panicles or racemes. **Flowers** showy; opening in evenings and quite fragrant. **Calyces** green; tubular to globose or campanulate; 5-lobed or 5-toothed; enlarging in fruit. **Corollas** radially symmetrical; 5-lobed; white or yellow or greenish or pink or purple or red; salverform or funnelform. **Petal Lobes** spreading; broad. **Stamens** 5; equal or 1 shorter; included within or exserted beyond corollas. **Fruits** capsules; globose to ovoid or ellipsoid.

1. Plants shrubs or trees. Herbage glabrous; glaucous .... *N. glauca*
1. Plants herbs or subshrubs. Herbage clammy-viscid; not glaucous.
   2. Leaves panduriform. Margins conspicuously undulate ..... *N. repanda*
   2. Leaves lanceolate or ovate to oblong. Margins entire or inconspicuously undulate.
      3. Corollas funnelform .... *N. tabacum*
      3. Corollas salverform.
         4. Corolla tubes 40–120 mm long; pale yellow tinged with gray or purple .... *N. longiflora*

4. Corolla tubes 12–40 mm long; white or cream or greenish.
   5. Middle cauline leaves petiolate . . . . *N. attenuata*
   5. Middle cauline leaves sessile.
      6. Plants perennials from woody bases. Corolla tubes 12–23 mm long . . . . *N. trigonophylla*
      6. Plants annuals from herbaceous bases. Corolla tubes 25–40 mm long . . . . *N. quadrivalvis*

Nicotiana glauca

■ cultivated, some weedy invaders, dry open areas to thickets

■ neurologic and muscular disturbances, teratogenic; livestock, main risk with cultivated tobacco low risk with wild species; pets, high risk exposure to tobacco products

Distribution and Habitat—*Nicotiana* is mainly a western genus, with its species occupying a variety of habitats. Introduced from South America, *N. glauca* is now naturalized in much of the southwestern United States and northern Mexico. *Nicotiana attenuata* is found in dry, often disturbed sites of deserts and montane forests up to 3,000 m elevation. *Nicotiana longiflora* is a cultivated species from South America that has escaped in some areas such as southeastern Texas. At one time extensively cultivated by American Indians in the West, *N. quadrivalvis* occurs in open areas of washes and on slopes up to 1,500 m elevation. *Nicotiana repanda* is of restricted distribution in thickets and wooded draws in southern Texas and northern Mexico. *Nicotiana trigonophylla* is a subshrub of the deserts. With numerous cultivars and native to tropical America, *N. tabacum* is the tobacco of commerce and is grown primarily in the southeastern quarter of the continent. Plants occasionally escape cultivation.

Disease Problems—Well known for its chronic toxic and addictive effects in humans, *N. tabacum* also presents problems for pets and livestock. Intoxications with neurologic and muscular disturbances are typically acute, and in the case of exposure in animals in early pregnancy, teratogenic effects may be seen (Panter et al. 1999). Pets gain access to the plant material mainly by ingestion of commercial tobacco products, whereas livestock may encounter and eat the cultivated and wild species. A pack of cigarettes ingested by a small dog may be lethal (Kaplan 1968). Similarly, sweetened chewing tobacco also represents an attractive danger for pets (Vig 1990). Acute intoxications of livestock are rarely seen with wild tobaccos, because the plants are seldom eaten in sufficient quantity. On the other hand, problems with cultivated tobacco are not uncommon in pigs allowed to eat the stalks after the leaves are harvested (Crowe & Pike 1973). The fetal arthrogryposis that results may occur without signs of intoxication in the dams and without fetal deaths or abortions, but the deformed piglets usually die shortly after birth (Crowe & Swerczek 1974; Crowe 1978). However, when *N. glauca* is eaten, at least experimentally, fetal deformities are preceded by the appearance of signs of acute toxicity in the dam (Keeler et al. 1981a). The teratogenic potential of other species of tobacco is not known. Species of *Nicotiana* have been a particular problem for ostrich chicks in South Africa (Steyn 1934).

Nicotiana attenuata

Nicotiana glauca

Disease Genesis—The toxicants of *Nicotiana* are alkaloids, the best known of which is the pyridine nicotine. Nornicotine is also an important pyridine in some species. In *N. attenuata, N. bigelovii,* and *N. trigonophylla,* nicotine is the predominant alkaloid (Marsh et al. 1927; Bush & Crowe 1989). In contrast, in *N. glauca,* anabasine, a substituted piperidine akin to coniine, accounts for almost the total alkaloid content (Smith 1935; Rindl & Sapiro 1949; Keeler et al. 1981a). Alkaloid concentrations, which vary considerably because of environmental factors, may exceed 1% in leaves and outer portions of stems for nicotine, may be almost that high for anabasine, and may be in excess of 2% for nornicotine, depending on the species (Bush & Crowe 1989). Nicotine is a rapidly acting sympathetic and parasympathetic ganglionic depolarizer; low dosage causes stimulation and high dosage causes blockade.

- alkaloids, nicotine, anabasine; ganglionic and neuromuscular stimulation and blockade; CNS stimulation; many body systems and functions affected; rapid absorption of nicotine through skin; extent of CNS stimulation varies with plant species
- fetal effects, anabasine, anatabine; arthrogryposis

Nicotiana quadrivalvis

Nicotiana repanda

Nicotiana trigonophylla

- rapid elimination of alkaloids, actions generally of short duration

It has similar effects on the neuromuscular endplate and also stimulates the central nervous system (Taylor 1990). There is a rapid response via ligand-gated ion channels, increasing Na^+ and K^+ permeability in the process of depolarization and excitation. It, thus, has very complex actions because of the diversity of the systems affected. Nicotine causes a marked decrease in heart and respiration rates and blood pressure. The toxicity of nicotine is exemplified by the appearance of "green tobacco sickness," which afflicts workers handling wet or fresh-cut leaves (McKnight et al. 1994). The signs, in addition to digestive tract problems, include headache, dizziness, respiratory difficulty, abdominal cramps, and fluctuations in blood pressure and heart rate. The effects are consistent with dermal absorption of nicotine and its subsequent actions on a variety of cholinergic receptors.

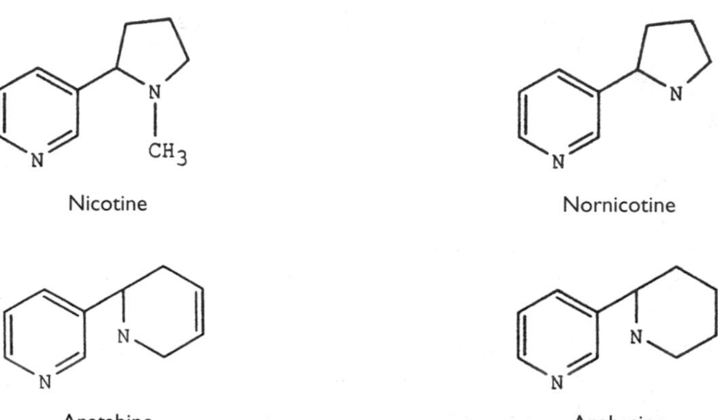

Anabasine appears to be about half as toxic as nicotine in producing acute nicotinic effects but is much less potent as a central nervous system stimulant (Upshall 1972; Sloan et al. 1988). Anabasine is also capable of reducing heart and respiration rates and blood pressure. Both are potent teratogens in the chick embryo model (Landauer 1960; Upshall 1972). However, in mammals, anabasine and perhaps anatabine (as in *N. tabacum*) appear to account for the fetal deformities (Keeler et al. 1984; Keeler & Crowe 1985). In pigs, 2.6 mg/kg b.w. of anabasine given 2 times per day for several days is highly teratogenic, with the specific effects dependent upon stage of pregnancy (Keeler et al. 1984).

Within the limitations imposed by the varying concentrations of alkaloids in plant collections, toxic dosages can be proposed. A dosage of 0.07% b.w. of dry plant material or 0.75% b.w. of fresh plants in cows produces signs of intoxication, sometimes within minutes of ingestion (Marsh et al. 1927; Keeler et al. 1981a). Dosage of fresh plant in excess of 2% b.w. is lethal. In pigs and sheep dosages of 0.13% b.w. of dry *N. glauca* produced prominent clinical signs, whereas a slightly higher dosage of 0.17% b.w. was teratogenic (Keeler et al. 1981a; Keeler & Crowe 1984; Keeler et al. 1984). Thus, it is apparent that teratogenic dosages will typically cause distinctive but reversible signs of intoxication in the dams (Keeler et al. 1984). Fetal malformations characterized by arthrogryposis occur following continued daily ingestion between 45 and 75 days of gestation for cattle, between 30 and 60 days for sheep, and between 18 and 68 days for pigs (Keeler et al. 1981a,b; Keeler & Crowe 1984).

Because it is a weak base ($pK_{a1} = 6.2$, $pK_{a2} = 11$), nicotine is absorbed from the alkaline environment of the small intestine. It has a high extraction rate by liver, and thus its removal is dependent upon blood flow in that organ. It is therefore subject to a marked first-pass effect following absorption from the intestine, with oral bioavailability limited to approximately 30% (Benowitz & Jacob 1987). In humans and dogs, the acute effects of nicotine and presumably anabasine are of short duration; the alkaloids are rapidly eliminated in urine as both the parent compound and its metabolites (Larson & Haag 1942; Monji & Castro 1979; Benowitz & Jacob 1987). Urine pH provides further control over elimination rate, given that these alkaloids are best eliminated in an acid environment.

Clinical Signs—Typical signs of *Nicotiana* intoxication in any species include urination, defecation, excess salivation, mydriasis, rapid and labored respiration, incoordination, tremors, and severe weakness. Because of nicotine's effect as a depolarizing ganglionic blocker, early effects are excitatory and related to ganglionic depolarization, whereas the later effects related to ganglionic blockade are depressant. Muscle twitching or trembling over the shoulders and then the entire body, a peculiar stiff-legged walk, and protrusion of the third eyelid are other distinctive findings. In most instances the signs probably last 5–6 hours or less. However, death, due to respiratory failure and then cardiac arrest, may occur within a few hours. Because of the relatively rapid elimination of the alkaloids, survival beyond 12 hours portends a more favorable prognosis. After 24 hours the condition should be much improved.

■ excitement, salivation, tremors, third-eyelid protrusion; later depression, weakness, ataxia, labored respiration

fetal, flexure and rotation of joints of legs, spinal deformities, cleft palate

Deformities in the newborn tend to be severe flexure and lateral rotation of the carpal joints, moderate rotation of the fetlocks, and less commonly spinal malformations resulting in lordosis or scoliosis and cleft palate. In many cases, the limb deformities become less of a problem as the animals grow and mature.

Pathology—Grossly, congestion of abdominal viscera and lungs and scattered splotchy hemorrhages and foam in the trachea may be seen, but there are no distinctive pathologic changes. As an aid to diagnosis, urine or tissues may be analyzed for alkaloid content using GC/MS (Plumlee et al. 1993).

■ no lesions

Treatment—There are no specific antidotes for nicotinic effects. Care must be used when treating the early excitement phase, because sedatives may worsen the later depressant phase. A symptomatic approach is necessary and may possibly require respiratory ventilation assistance in very severe intoxications. Induction of vomiting, gastric lavage, and/or activated charcoal to limit absorption may be useful.

■ oral activated charcoal, general nursing care

late stages, respiratory assistance

Solandra (L.) Sw.

Taxonomy and Morphology—Comprising 8–10 species native to tropical America, *Solandra*'s name honors Daniel Carlsson Solander, a noted eighteenth-century Swedish botanist who worked with Linneaus, Joseph Banks, and William Aiton. Species of the genus are shrubby climbers with large, showy white or yellow, goblet-shaped flowers that are reflected in the common names *chalice vine, golden cup, copa de oro (cup of gold),* and *trumpet flower*. The following are introduced species representative of the genus:

■ *S. guttata*
S. grandiflora
S. guerrerense
S. longiflora
S. maxima

bolsa de Judas
chalice vine
copa de oro
cup of gold
milkcup
trumpet plant

S. guttata D.Don	
S. grandiflora Sw.	
S. guerrerense Martinez	
S. longiflora Tussac	trumpet plant
S. maxima (Sesse & Moc.) P.S.Green	cup of gold, copa de oro, milkcup, chalice vine, bolsa de Judas

- climbing, evergreen shrubs; leaves shiny, leathery; flowers solitary, large, showy; corollas 5-lobed, long, narrow, broadened distally, yellow, white or purple-blue; fruits berries, enclosed by calyces

Plants shrubs; evergreen. **Stems** climbing; to 5 m tall; branched; bark ridged. **Leaves** shiny; leathery; blades ovate to elliptic to obovate; margins entire. **Inflorescences** solitary flowers borne at branch ends. **Flowers** showy; large, to 25 cm long; fragrant at night. **Calyces** long tubular; 2-to-5-toothed or lobed. **Corollas** radially symmetrical; 5-lobed; yellow or white or purple-blue; basal tube long, narrow; distal portion broadly campanulate. **Petal Lobes** reflexed; overlapping in bud. **Stamens** 5; included within corollas. **Fruits** berries; conical; 2-loculed; enclosed by calyxes.

Because cultivation of *Solandra* is limited in North America, a taxonomic key to its species is not presented in this treatment. If identification is needed, appropriate horticultural references should be consulted.

Solandra maxima

- ornamentals

Distribution and Habitat—The species of *Solandra* are indigenous to tropical America and have had limited use as ornamentals in the warmer, southernmost regions of the continent. Plants of *S. maxima,* which is native to Mexico, Central America, and extreme northern South America, may be encountered occasionally as a winter-blooming vine in southern Florida.

- neurotoxic, tropane alkaloids, similar to *Datura*; problems with abuse in humans

Disease Problems and Genesis—Species of *Solandra* have a long history of use for medicinal and religious purposes. As an example, *S. maxima* was used in traditional medicine long before the arrival of the Spanish in Mexico (Lozoya et al. 1988). Extracts of the flowers have marked anticholinergic activity. As with species of *Datura,* the principal toxins are the neurotoxic tropane alkaloids. Hyoscyamine is synthesized in the roots and transformed in the leaves to atropine, noratropine (norhyoscyamine), and tropine, the major alkaloids in most species (Willaman & Schubert 1961; Ghani 1986). In contrast, scopolamine is the main alkaloid of *S. maxima,* perhaps an indication of the species' greater neurologic activity. The reported effects with most species—pupillary dilation, incoordination, and hallucinations—are consistent with the effects of tropane alkaloids and seem to have led to intentional use of the flowers for recreational abuse (Morton 1982). A more complete discussion of these compounds is presented in the treatment of *Datura,* in this chapter.

Hyoscyamine

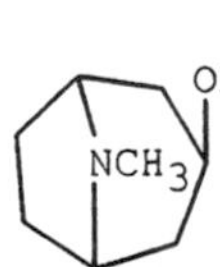

Tropine

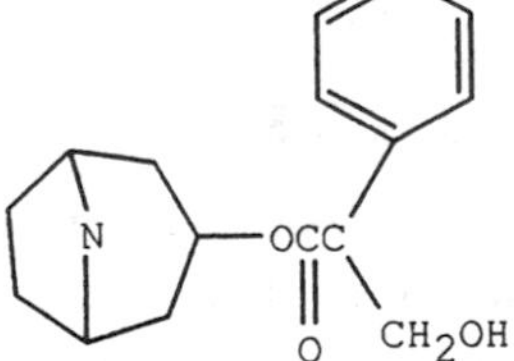

Norhyoscyamine (noratropine)

- similar to signs of *Datura,* pupillary dilation, dry mouth, hallucinations

Clinical Signs—The signs produced by *Solandra* are very similar to those of *Datura* and include restlessness, pupillary dilation, increased heart rate, labored respiration, excitement, dry mouth, and bloat. There also may be hallucinations and more rarely convulsive seizures and cessation of respiration.

- no lesions

treatment generally not needed

Pathology and Treatment—Seldom are lethal effects encountered with these plants, and few gross or microscopic lesions are associated with the tropane alkaloids. Treatment generally is not required, other than conservative approaches to control any serious neurologic problems. In instances in which cholinergic signs are particularly prominent, acetylcholinesterase inhibitors such as physostigmine may be useful.

Solanum L.

Taxonomy and Morphology—Cosmopolitan in distribution, but with greatest diversity in the American tropics, *Solanum,* commonly known as nightshade, comprises 1,500–1,700 species and is one of the largest genera of flowering plants as well as the largest genus in the Solanaceae. In North America, approximately 76 native and introduced species are present. The genus name, used by the natural historian Pliny, is derived from the Latin root *solamen,* which means "to quiet" and alludes to the sedative effect of its chemical constituents. Horticulturally quite important, some of its species provide foods, the most important being potato and eggplant; others are noxious weeds, such as silverleaf nightshade; and still others are grown as ornamentals for both their foliage and flowers. Toxic species are also a significant problem. Representative toxic species of the genus include the following (see plate 69.3):

Non-ornamentals

S. americanum Mill. (= *S. nodiflorum* Jacq.)	black nightshade, common nightshade, stubbleberry, yerba mora negra, chichiquelite
S. carolinense L.	horse nettle, bull nettle, stickerweed, tread-softly, Carolina horse nettle, ball nightshade, ball nettle, apple-of-Sodom
S. dimidiatum Raf. (= *S. torreyi* A.Gray)	western horse nettle
S. dulcamara L.	bitter nightshade, bittersweet, felonwort, blue bindweed, climbing nightshade
S. elaeagnifolium Cav.	silverleaf nightshade, white horse nettle, trompillo, meloncillo del campo, tomatillo, tomato weed, sand brier, revienta caballa, devilbush, desert nightshade, western horse nettle
S. nigrum L.	black nightshade, poisonberry, morelle noire, yerba mora, duscle
S. physalifolium Rusby (= *S. sarrachoides* auct. non Sendt.)	hairy nightshade, viscid nightshade, morelle poilue
S. ptychanthum Dunal (= *S. americanum* auct. non Mill.)	eastern black nightshade, black nightshade, morelle noire de l'est
S. rostratum Dunal	buffalo bur, Kansas thistle, duraznillo, Texas thistle, mala mujer, beaked nightshade, Colorado bur
S. triflorum Nutt.	cutleaf nightshade, three-flowered nightshade
S. tuberosum L.	Irish potato, white potato

Ornamentals

S. jasminoides Paxton	potato vine
S. pseudocapsicum L.	Jerusalem cherry, Christmas cherry
S. seaforthianum Andrews	potato creeper, glycine, Brazilian nightshade, Italian jasmine
S. wendlandii Hook.f.	Costa Rican nightshade, marriage vine, divorce vine, giant potato creeper

■ *S. americanum*
S. carolinense
S. dimidiatum
S. dulcamara
S. elaeagnifolium
S. jasminoides
S. nigrum
S. physalifolium
S. pseudocapsicum
S. ptychanthum
S. rostratum
S. seaforthianum
S. triflorum
S. tuberosum
S. wendlandii

apple-of-Sodom
ball nettle
bittersweet
blue bindweed
buffalo bur
bull nettle
Carolina horse nettle
chichiquelite
Christmas cherry
Colorado bur
devilbush
divorce vine
duraznillo
duscle
felonwort
giant potato creeper
glycine
horse nettle
Irish potato
Italian jasmine
Jerusalem cherry
Kansas thistle
mala mujer
marriage vine
meloncillo del campo
morelle noire
morelle noire de l'est
morelle poilue
nightshade
poisonberry
potato creeper
potato vine
revienta caballa
sand brier
stickerweed
stubbleberry
Texas thistle
tomatillo
tomato weed
tread-softly
trompillo
western horse nettle
white horse nettle
white potato
yerba mora
yerba mora negra

- taxonomic confusion with *S. nigrum*; in most reports, this name is most likely misapplied; should be other species

- herbs, shrubs, or small trees; leaves alternate or paired; flowers often showy; corollas 5-lobed, yellow, white, blue, or purple; stamens 5, connivent, opening by apical pores; fruits globose berries

There has been and continues to be considerable taxonomic confusion and differences of opinion among taxonomists about the identity of North American members of the *S. nigrum* group, a worldwide polyploid complex (Rogers & Ogg 1981; Schilling 1981). The binomials *S. nigrum, S. americanum, S. ptycanthum, S. douglasii, S. nodiflorum,* and *S. sarrachoides* have been used in North America. However, there is now almost general agreement that *S. nigrum* sensu stricto is a Eurasian species sporadically introduced into North America and that the plants of most earlier reports are actually members of the other species listed (Correll & Johnston 1970; Cronquist et al. 1984; Gleason & Cronquist 1991). Because of this confusion, much of the information in the toxicologic literature reported for *S. nigrum* must be regarded with some reservation. It is most likely that the data are for the other species. The binomial *S. sarrachoides* has been used in numerous toxicological papers, but taxonomic work suggests that it is strictly a South American species and that the name *S. physalifolium* should be used for the North American plants. An additional point of confusion is that members of the complex are occasionally called deadly nightshades. This is unfortunate because this common name is more appropriately applied to the more dangerous *Atropa belladonna.*

Solanum americanum

Plants herbs or shrubs or small trees; annuals or perennials; armed or not armed with prickles; herbage glabrous to densely pubescent. **Stems** decumbent to erect. **Leaves** alternate or paired with 1 leaf smaller; blades of various shapes; margins entire to bipinnatifid. **Inflorescences** solitary flowers or racemose cymes; lateral; extra-

Solanum dulcamara

Solanum carolinense

Solanum elaeagnifolium

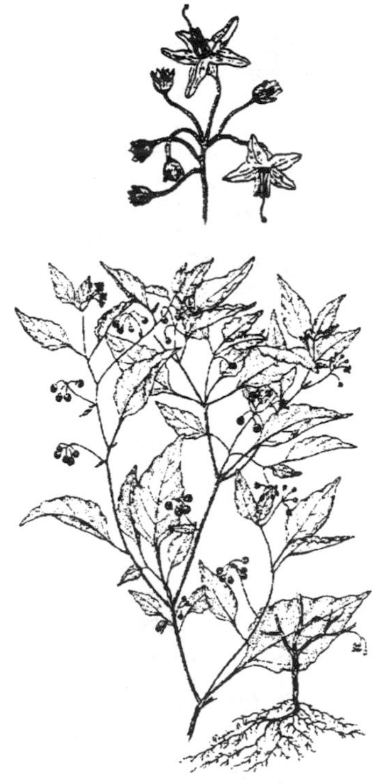

Solanum nigrum

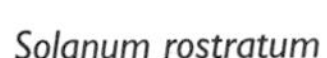

Solanum rostratum

Solanum triflorum

axillary. **Flowers** of various sizes; often showy. **Calyces** rotate to campanulate; 5-lobed; enlarging or not enlarging in fruit. **Corollas** radially symmetrical; 5-lobed or cleft or angled; yellow or white or blue to purple; rotate or broadly campanulate; plaited in bud. **Petal Lobes** spreading or reflexed; long or short; acute. **Stamens** 5; connivent; filaments short; anthers opening by apical pores. **Fruits** berries; globose; fleshy or coriaceous.

Distribution and Habitat—Species of *Solanum* are widespread across North America and especially abundant in the southern Great Plains. A variety of habitats are occupied, albeit most exhibit disturbance—waste areas, construction sites, croplands, feedlots, overgrazed pastures, and so on. Most of the non-ornamental species are considered weeds or indicators of disturbance. *Solanum elaeagnifolium,* for example, is of considerable agronomic concern because it is a noxious weed and its range seems to be expanding as a result of its ease of dispersal via seeds and rhizome fragments (Davis et al. 1945; Boyd et al. 1984).

- wide range of habitats, often waste area invaders

Solanum americanum

Solanum carolinense

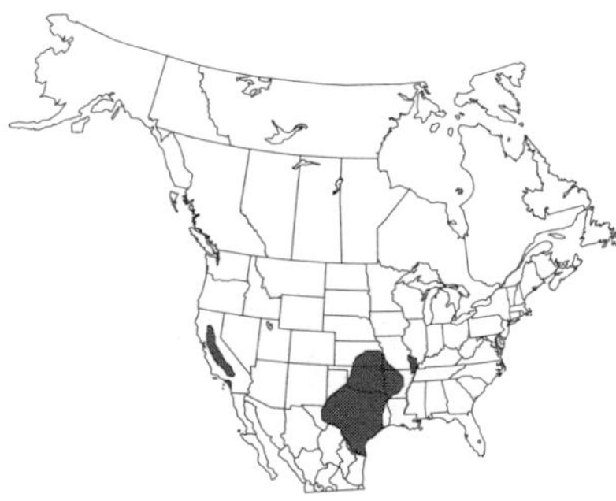

Solanum dimidiatum

Disease Problems—Even though several of its species are edible, *Solanum* is a genus that has long been associated with toxic effects. Its adverse effects are mainly related to irritation of the digestive tract. Surprisingly, extracts of *S. nigrum* are considered in some societies to be protective against gastric ulcers, and *S. incanum,* an Asian species, is used as a protection against chemically induced liver injury (Akhtar & Munir 1989; Chiu et al. 1989). However, cases of intoxication in humans are not common. Most interest at present is focused on the possible detrimental effects of certain potato varieties and/or conditions that might increase toxicant concentrations in them. This is partly due to an earlier reported relationship between potato consumption (especially of blighted potatoes) by pregnant women and fetal deformities such as spinal bifida (Renwick 1972). The validity of this relationship has been much studied, and deformities in laboratory rats due to consumption of potatoes or their toxins have been difficult to reproduce experimentally (Ruddick et al. 1974). Hopkins (1995) reviewed the status of the potato threat and concluded, based upon on experimental animal models and human epidemiological data, that there does not appear to be a reproductive risk. However, there may be other risks.

- digestive disturbance; humans, concern for possible teratogenic effects from damaged potatoes, suspect as a cause of spinal bifida

  livestock, possible problems from sprouted potatoes or the foliage, low risk

It is well recognized that the toxins in potatoes are a risk to livestock, especially in potatoes that have sprouted. Episodes of intoxication occasionally occur when cattle are allowed to graze fields from which potatoes have been recently harvested (SAC Veterinary Services 1998). However, young dairy cattle experimentally fed 6.8–13.6 kg sprouted potatoes or 0.9–2.7 kg of sprouts daily for 6–7 weeks failed to develop adverse effects (Dice 1931). There was also no milk taint. The concern about the

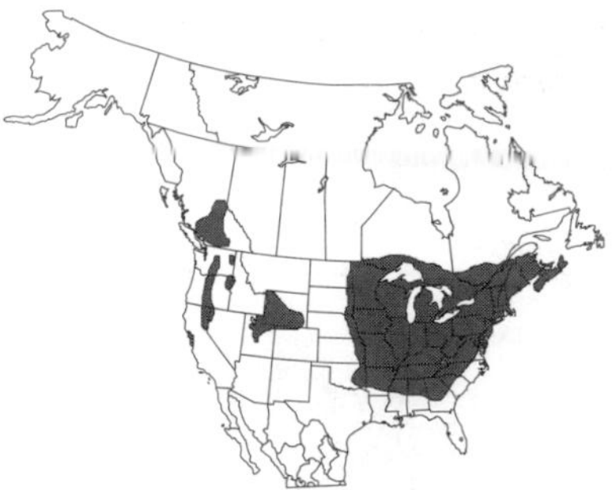

Solanum dulcamara

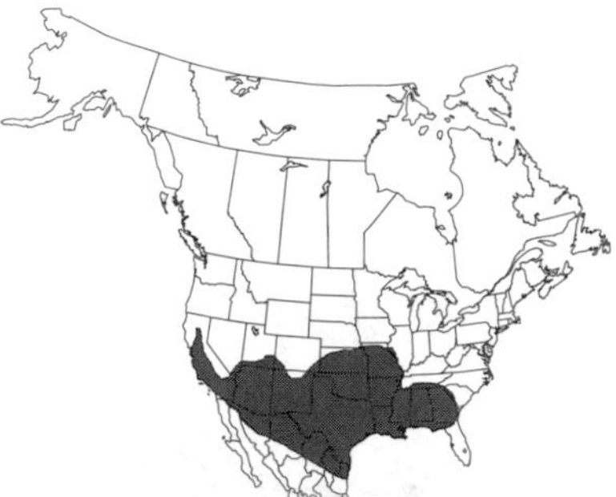

Solanum elaeagnifolium

■ nightshades, including *S. nigrum* complex; livestock, mainly digestive disturbances; possibly neurotoxic

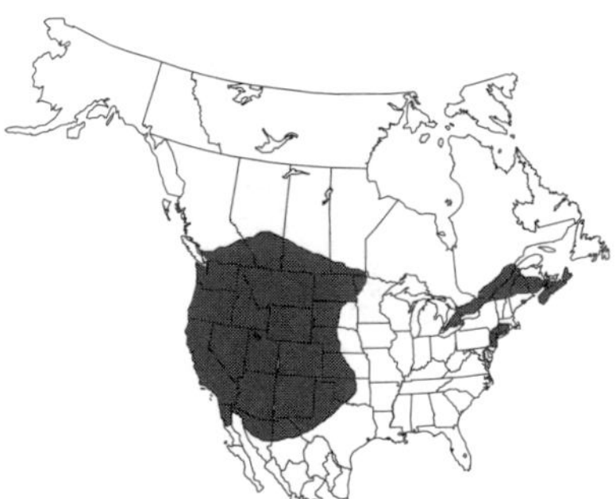

Solanum nigrum

Solanum physalifolium

toxicity potential of potatoes extends to the ingestion of the foliage and stems, because of the economic desire to make use of this potential forage, which has reasonable crude protein for livestock (15–18% in young plants of the Kennebec variety), and because of the consumption of potato greens by certain ethnic groups (Domek et al. 1995; Phillips et al. 1996). However, in neither case does the risk appear to be serious.

The fruits and/or foliage of many species of *Solanum* are eaten by livestock, and intoxications are reported for a variety of animal species, including horses, cattle, sheep, ducks, and chickens (Pammel 1921; Craig & Kehoe 1925; Hansen 1927; Casselberry 1939; Hubbs 1947; Gunning 1949; Carey 1955; Buck et al. 1960). Even the heavily prickled and formidable *S. rostratum* may be ingested by pigs and cause digestive tract disturbances, including reddening, ulceration, and hemorrhage (Simic 1943; Case 1956).

There has been particular concern about the toxicity potential of the species of *Solanum* commonly called nightshades, but the hazard with the various species is quite erratic, and evidence is conflicting as to the actual overall risk (Bassett & Munro 1985). In some instances, investigators have been unable to produce intoxication, e.g., using the berries of *S. trifolium* (Stevens 1933). In other instances, digestive disturbances are reported (Case 1956), but cardiac effects such as bradycardia and arrhythmias, including various degrees of heart block, are also of concern as being caused by *S. pseudocapsicum* (Watt et al. 1932). Boughton and Hardy (1939) fed the ripe fruit of *S. elaeagnifolium* to sheep and a calf; a dosage of 0.5–0.7% b.w. produced severe respiratory difficulty and the death of 1 of 2 sheep. Two doses of 0.31% b.w. given 6 hours apart caused severe respiratory difficulty in the calf. There was no problem in sheep fed green leaves. However, in another incident, 2 horses that were fed grass hay consisting of perhaps 50% or more of the fruits and foliage of *S. elaeagnifolium* developed signs of severe colic, became recumbent, and exhibited epileptiform seizures (L Westall, pers. comm., 1996). The animal that ate the most fruits died within 48 hours.

An interesting interaction apparently occurs between the anthelmintic ivermectin and *S. elaeagnifolium* in horses (Garland et al. 1998). Dosage of ivermectin that otherwise would not be expected to cause adverse effects produced severe depression and death of at least 2 horses, in animals exposed to the plants in hay or pasture. Brain concentrations of ivermectin were markedly elevated. The findings, which were experimentally reproduced in rabbits, suggest the possibility of interactions between species of *Solanum* and other potentially neurotoxic plants.

The role of plants of the *S. nigrum* complex in causing disease is uncertain. Bonner (1938) described an intoxication episode in cattle attributed to *S. nigrum* in which the signs were mainly neurologic but there were no pathologic changes. Others also have ascribed toxicologic effects to *S. nigrum;* however, the taxonomic difficulties in distinguishing among *S. americanum, S. nigrum,* and *S. ptychanthum,* as discussed above, make it difficult to make definitive statements about clinical cases. In addition, the fruits of some species, such as the garden huckleberry *(S. melanocerasum; = S. nigrum* var. *guineense),* are eaten and used in preserves. In farming operations, the main problems with species of the *S. nigrum* complex include interference with harvesting by the sticky, tangled masses of foliage; staining of soybeans; and contamination and devaluation of harvested peas (Bassett & Munro 1985). There is a continuing problem because plants of the complex are prolific seed producers, and the seeds are long-lived (Weller & Phipps 1978–1979).

There is little question that most species of *Solanum,* under some circumstances, are capable of causing digestive tract and less commonly neurologic problems in a variety of animal species (Case 1956; Dollahite & Allen 1960; Baker et al. 1987, 1989; Drew & Fowler 1991). There is some evidence to indicate that ruminants are less susceptible than other animals, and based on observations of a small number of animals, sheep and especially goats appear to be less susceptible than cattle, at least for *S. elaeagnifolium* (Case 1956; Buck et al. 1960; King & McQueen 1981). Similar results are reported for *S. tuberosum* (Craig & Kehoe 1925; Hansen 1928; Milligan 1941; Ogilvie 1943; Owen 1985).

Solanum ptychanthum

The fruits of various nightshade species are a concern because of their possible attractiveness to children. However, evaluation of cases of ingestions of the berries of *S. dulcamara* over several years as reported to a poison control center indicated a low risk (Borys et al. 1992). Only in 17 cases of 235 ingestions by children less than 5 years of age did signs such as drowsiness (9 cases), vomiting (4), and diarrhea (4) appear. There was no correlation between the occurrence of signs and the maturity of the berries. That there is a risk, however, is illustrated by a case of acute cholinergic crisis with lethargy, pupillary dilation, and elevated heart rate and body temperature in a 4-year-old child following ingestion of approximately 50 berries of *S. dulcamara* (Ceha et al. 1997). The child responded favorably to gastric lavage and physostigmine i.v. For livestock, the ripe berries generally are less of a risk than the more toxic green or immature berries (Bonner 1938; Tehon et al. 1946). Three or four unripe nightshade berries given to mice caused lethargy, labored respiration, and severe erosions of the gastric mucosa, whereas ripe berries were nontoxic at similar dosage (Hornfeldt & Collins 1990). An exception to the above occurs in *S. elaeagnifolium,* which has ripe berries that are as toxic (0.1–0.2% b.w.) or more so than green ones (Buck et al. 1960; Dollahite & Allen 1960). There are other species in which both the unripe and ripe berries are toxic (Steyn 1931). Thus, caution is warranted when considering the toxicity potential for the berries of any species. However, this warning is tempered by the fact that it is sometimes difficult to produce intoxication with leaves, stems, or berries. Furthermore, it is reported that potato leaves form part of the diet of members of Bangladeshi communities in the United Kingdom (Phillips et al. 1996).

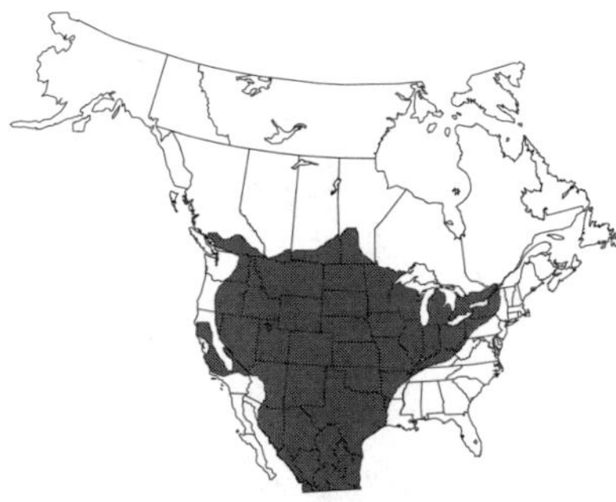

Solanum rostratum

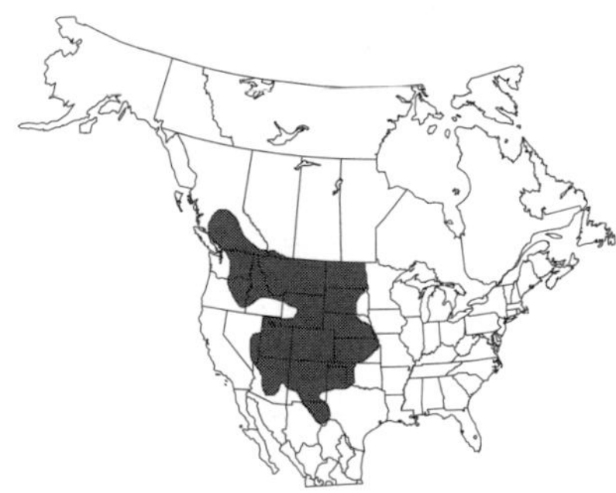

Solanum triflorum

- nightshade fruits, humans, occasional serious problem with ingestion of fruits by children

  livestock, mainly with ingestion of immature fruits

A rather uncommon disease due to ingestion of *S. dimidiatum* for several months has been reported in central Texas. Known as crazy-cow syndrome since the 1920s, the disease is manifested by animals that appear normal until they are disturbed, at which time they fall and struggle to stand without success until they rest for a few minutes (Menzies et al. 1979). The condition is not lethal but remains a permanent disability inasmuch as there is degeneration and loss of Purkinje cells and axons associated with cerebellar nuclei. Similar diseases are associated with long-term ingestion of *S. kwebense* in South Africa (Pienaar et al. 1976), *S. bonariensis* and *S. fastigiatum* in South America (Riet-Correa et al. 1983), and *S. cinereum* and possibly *S. esuriale* in Australia (McMeniman 1976; O'Sullivan 1976; Bourke 1997). The disease appears to be unique to ruminants, because in at least one study, it could not be reproduced in laboratory animals (Santos Zambrano et al. 1985). In South Africa the disease is also suspected but not confirmed to occur in horses, donkeys, and goats (Pienaar et al. 1976)

- *S. dimidiatum,* crazy-cow syndrome, cerebellar problem with long-term ingestion of plants

There has been some concern about possible problems with feed-grain contamination by seeds of the plants of the *S. nigrum* complex, because it is a very common weed pest. However, experimentally, concentrations of seeds as high as 32% in feed produced only transient depression of feed consumption and weight gain in rats (Dugan & Gumbmann 1990).

Disease Genesis—Species of *Solanum* contain steroid glycosides of the spirostanol type such as soladulcosides in *S. dulcamara* and furostanol types such as melongosides in *S. melongena.* However, toxicity of most species is attributed to an extensive series of 27-C steroidal alkaloid derivatives of cholestane. Sometimes called pseudoalkaloids,

- assortment of steroid alkaloids, in varying proportions of glycosides and aglycones

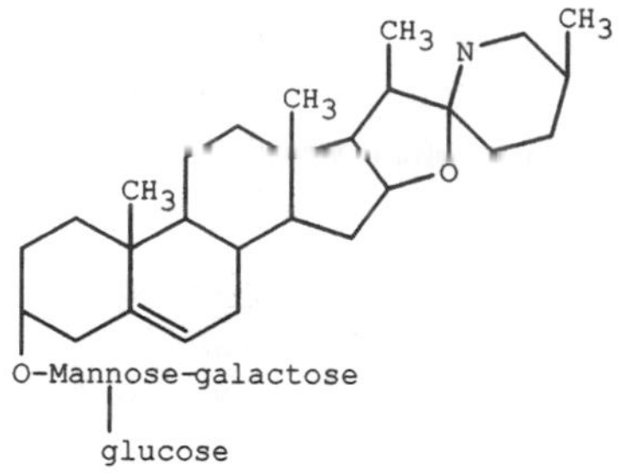

Solasonine (solasodine genin)

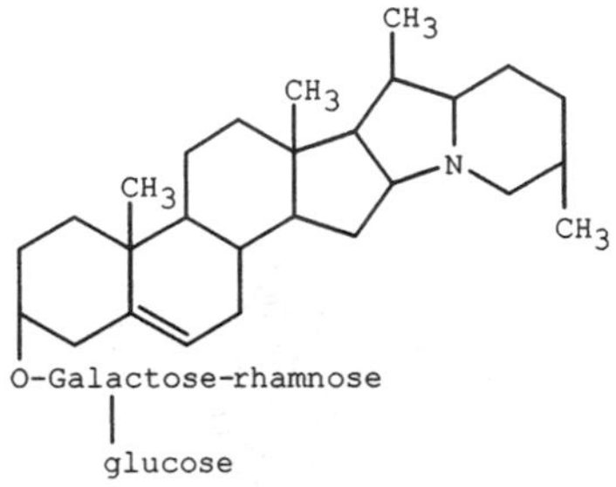

Solanine (solanidine genin)

■ of special interest in potatoes are solanidines α-chaconine and α-solanine; highest levels in fruits and sprouts, appreciable amounts in foliage at time of flowering

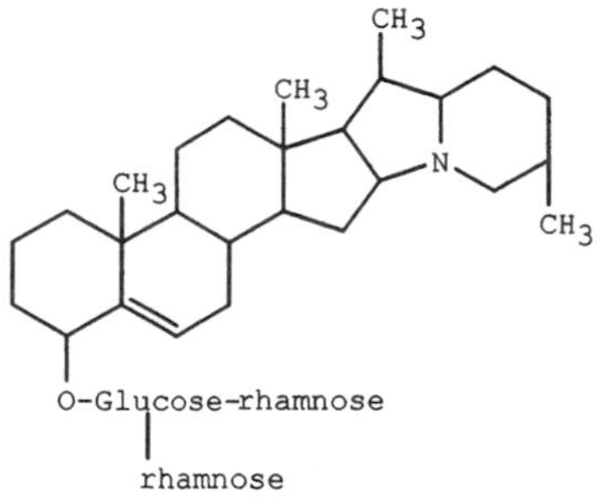

Chaconine

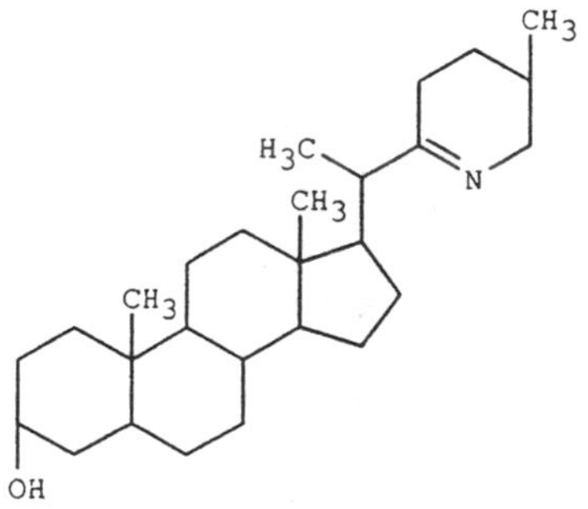

Solacongestidine

but more often referred to as glycoalkaloids (GAs), the aglycones can be divided among two major groups; spirosolanes (solasodine, tomatidine) and solanidanes (solanidine), and 3 minor groups; epiminocholestanes (solacongestidine), solanocapsines (solanocapsine), and 3-aminospirostanes (jurubidine) (Schreiber 1968; Ripperger & Schreiber 1981). These alkaloids are widely represented in the genus (Vieira & Freire de Carvalho 1993). They occur mainly as glycosides complexed in various linkages with up to five sugars. For example, the fruits of *S. dulcamara, S. elaeagnifolium, S. melongena, S. ptychanthum,* and *S. sarrachoides* contain only glycoalkaloids, whereas in *S. tuberosum* only 67% of the alkaloids are present as the glycoforms (Keeler et al. 1990). An alkaloid aglycone such as solanidine may exist in a variety of oligosaccharide complexes with varying toxicity potentials, but generally the aglycones are less toxic than the glycosides.

Glycoalkaloids of solasodine (solasonine, solamargine, solasurine) are common to many species of *Solanum* (Schreiber 1968; Ripperger & Schreiber 1981; Gaffield et al. 1991). The toxicity of these glycoalkaloids is illustrated by the effects of *S. elaeagnifolium* and *S. dulcamara,* which contain primarily glycosides of solasodine and produce severe gastric and intestinal necrosis in hamsters (Baker et al. 1989). Solasodines also predominate in the mature fruits of *S. ptychanthum*; 10.6 mg/g solasodines is present, and only 1.5 mg/g of solanidines (Eldridge & Hockridge 1983).

Most interest has been focused on the more toxic solanidines α-chaconine and α-solanine, because they are present in potatoes. Based upon a trout embryo toxicity assay procedure, α-chaconine is about 4.5 times more toxic than either α-solanine or solanidine (Crawford & Kocan 1993). In addition to being toxic, chaconine and solanine, which are not heat labile, are thought to impart a bitterness to potatoes when the concentrations exceed 11 mg% (Sinden 1987). Chaconine and solanine have been negative in the Ames test for mutagenicity and negative in the micronucleus test for clastogenicity (Slanina 1990; Friedman & Henika 1992).

Typical concentrations of total glycoalkaloids (more than 50% chaconine) in potato tubers are in the range of 5–10 mg/100 g (Jadhav et al. 1981). Concentrations range from highest to lowest as follows: flowers > sprouts > peels and eyes > shoot tips. Solanidine is typically 2 times higher in fruits than foliage, in the range of 0.4–3% d.w. (Vieira & Freire de Carvalho 1993). Glycoalkaloid synthesis is highly heritable, but concentrations can also be increased by trauma and stress such as cuts, bruises, high temperature, and greening due to exposure to light (Sinden 1987). Although glycoalkaloids increase with greening, there does not seem to be a direct link between chlorophyll and alkaloid accumulation (Edwards et al. 1998). Concentrations of glycoalkaloids may be as high as 400 mg/100 g in sprouts and up to 100 mg/100 g in peels and leaves, but generally decrease with plant maturity, especially in the foliage. Concentrations in foliage are high at the time of flowering and fruiting and decline thereafter (Dijkstra & Reestman 1943). Friedman and Dao (1992) found that plants with tuber concentrations of 14.7 mg/100 g chaconine/solanine had much higher levels in the foliage—145.1 mg/100 g. Similar findings are reported by Phillips and co-workers (1996), with chaconine levels several times higher than those of solanine. Glycoalkaloid concentrations in tubers is dependent upon environmental influences; molybdenum applications during growth may reduce the levels by 50% (Mondy & Munshi 1993).

In most species the toxins in the fruits are metabolized to nitrogen-free neutral saponins and decline with maturation (Hostettmann & Marston 1995). Thus, immature berries are typically higher in glycoalkaloids and more toxic than when ripe (Steyn 1934; Watt & Breyer-Brandewijk 1962). However, even ripe berries may contain substantial concentrations of glycoalkaloids (Eldridge & Hockridge 1983).

Humans appear to be the most susceptible species; glycoalkaloid concentrations in excess of 20 mg/100 g (200 ppm) are considered toxic and, when in excess of 40 mg/100 g, are associated with illness (Sinden 1987). When expressed as dosage, the threshold has been estimated at about 1 mg/kg, or in the range of 2–5 mg/kg b.w. of potato glycoalkaloid (Morris & Lee 1984; Hopkins 1995). This is in contrast to the 225 mg/kg toxic dose in sheep reported for solanine (Jadhav et al. 1981). Similarly, oral dosage of 5–50 mg/kg b.w. with a 3:1 ratio of chaconine to solanine for 3 days was without adverse effects (Phillips et al. 1996). Intoxication, which is associated mainly with a saponin-like effect on the digestive system and disruption of cellular membranes due to cholesterol binding, may be caused by consumption of immature foliage or berries from many of the *Solanum* species or under very unusual circumstances from parts of the potato tuber.

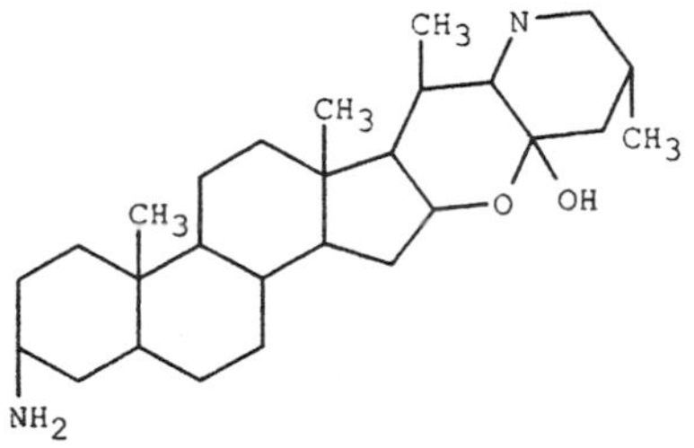

Solanocapsine

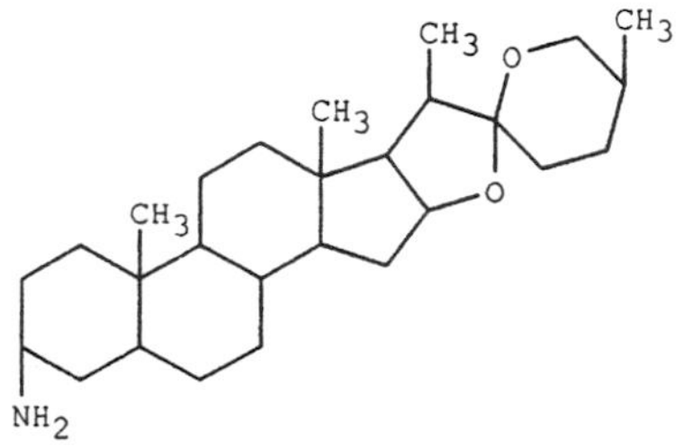

The glycoalkaloids are poorly absorbed from the digestive tract and are much more toxic when given by injection (Nishie et al. 1971). The kinetics of glycoalkaloid disposition have been studied in humans, based upon oral ingestion of mashed potatoes containing glycoalkaloid concentrations equivalent to 1 mg/kg b.w., with a ratio of 4:6 α-solanine to α-chaconine (Hellenas et al. 1992). The half-lives in serum were 10 and 20 hours, respectively, for solanine and chaconine, based upon the median values for 7 adult males.

Jurubidine

■ other toxic effects; cardiac, neurologic, hepatic, inhibition of ACh

The glycoalkaloids produce a variety of effects in addition to those on the digestive system, including cardenolide-like effects on cardiac activity and rhythm; retinal hemorrhage; and liver damage and neurotoxicosis (Nishie et al. 1971; Dalvi 1985; Baker et al. 1991; Caldwell et al. 1991). Slight to moderate inhibition of human plasma acetylcholinesterase activity is apparent with the crude glycoalkaloids, in some instances about 20 times less than inhibition with neostigmine (Orgell 1963). The purified glycoalkaloids are more potent than aglycones, but even the most potent is 15-to-30-fold less potent than typical organophosphate or carbamate insecticides (Bushway et al. 1987). Chaconine and solanine, which are of similar potency, are between 10 and 100 times less potent than physostigmine, and tomatine is even slightly less toxic (Roddick 1989). Although atropine is apparently somewhat protective (Patil et al. 1972), acetylcholinesterase inhibition, which may contribute to signs such as salivation, is not typically a major factor in toxicity (Bushway et al. 1987). However, in a case involving an individual swallowing about 50 fruits of *S. dulcamara*, there were signs suggestive of marked anticholinergic activity, and the patient was responsive to treatment with physostigmine (Ceha et al. 1997).

Even though leaves contain much higher (10 times) concentrations of solanidine than tubers, they have been safely fed to laboratory animals and used for feeding cattle and sheep (Lawson et al. 1992; Phillips et al. 1996). This result may be due in part to hydrolysis of the glycoalkaloid in the rumen to the less toxic aglycone and thence to dihydro analogs (King & McQueen 1981). Although some difficulties are encountered in making silage from potato foliage, the resulting product has been safely fed for several weeks to months to sheep and cattle (Dijkstra & Reestman 1943; Dijkstra 1945; Watson & Nash 1960; Nicholson et al. 1978). Other investigators have not been as confident about the protection afforded by the ensiling process (Weller & Phipps 1980).

■ possible teratogenic effects such as neural tube defects, difficult to confirm experimentally

The hazard of the occurrence of terata in animals exposed to various glycoalkaloids remains in question. Because of the early linking of spinal bifida in infants to potato consumption in Ireland by Renwick (1972), considerable interest in understanding the relationship has developed. Subsequent studies with high dosage of both sprouts and purified glycoalkaloids in rats showed no adverse effects, but in hamsters there was a low but significant increase in neural-tube malformations (Ruddick et al. 1974; Renwick et al. 1984). Sprouts were the most potent, but effects were also seen with α-chaconine and to a lesser extent with α-solanine. Using hamsters, Keeler and co-workers (1990) were able to produce deformities in 10–20% of the litters when dams were fed various species of *Solanum* containing glycosides of solasodine or solanidine. However, they were unable to cause deformities with the solasodine glycosides even at dosages lethal to 50% of the treated dams. α-Chaconine and to a lesser extent α-solanine and solasonine have also been shown to be teratogenic, using the frog embryo assay (FETAX) (Friedman et al. 1992). Although it has been possible to reproduce the reproductive effects only in a select few species (Keeler et al. 1991b), there seems to be justification to continue to suspect glycoalkaloids as teratogens, albeit probably at high dosages. The presence of additional toxicants, such as the calystegins discussed below, amplifies the potential risks.

Configurational requirements relative to unhindered nitrogen electron pair accessibility for involvement in active site binding in the embryo have been proposed but not confirmed (Brown & Keeler 1978; Keeler et al. 1991b). Chaconine and solanine, the principal glycoalkaloids of potatoes, in typical form do not meet the proposed configurational requirements, but there is evidence to incriminate them as teratogens in some laboratory animal species when given in high dosage, as in sprouts (Gaffield et al. 1992). Furthermore, in the frog embryo *(Xenopus)* test, these and other glycoalkaloids are clearly teratogenic without respect to orientation of unshared nitrogen electrons (Friedman et al. 1991; 1992). The presence of unsaturation at C-5/C-6 appears to be the most important determinant of teratogenicity in hamsters, more than the molecular configuration at C-22 (Gaffield & Keeler 1993, 1994). However, the embryo effects are enhanced by the presence of sugars and their structure and order or arrangement, and a basic nitrogen in ring F is required. Solasodine glycoside, a spirosolane that meets proposed configurational requirements, also causes terata in laboratory animals, as shown by results with *S. elaeagnifolium* and *S. dulcamara* (Keeler et al. 1991a). The incidence of neural tube defects in hamsters with *S. dimidiatum* is quite low; with a high dosage of ripe fruit at 1.52 mg/g b.w. of the dam, defects occurred in only 3 fetuses from 1 litter of 12 litters evaluated (Casteel et al. 1989). Thus, it appears that many of the *Solanum* species may be suspected to be teratogens under certain circumstances.

■ tropane alkaloids, calystegins present in potatoes, many plant parts of other species of *Solanum*; calystegins, potent enzyme inhibitors causing neurologic problems; crazy-cow syndrome

Other compounds of interest in species of *Solanum* include traces of the benzodiazepines diazepam, lorazepam, and delorazepam, which have been isolated from potatoes and presumably synthesized in the plant (Wildmann et al. 1988). However, the concentrations of these compounds are too low to be of significance. Glycosides of 1,25-dihydrocholecalciferol present in South American species of *Solanum* such as *S. glaucophyllum* are not present in North American species (Skliar et al. 1992). Of greater interest are the calystegins that have been isolated from the tubers of potatoes and the roots, leaves, and fruits of several other species (Tepfer et al. 1988). The calystegins are a series of novel bicyclic, tropane-type alkaloids that appear to serve as carbon and nitrogen sources for certain soil bacteria and may play an additional biological role as allelochemicals secreted by roots (Molyneux et al 1993; Goldmann et al. 1996). Calystegin B_1 has been shown to be a competitive inhibitor of almond β-glucosidase and bovine liver β-galactosidase, whereas the more potent B_2 inhibits β-glucosidase and coffee bean α-galactosidase (Asano et al. 1995). B_1 and B_2 are particularly potent inhibitors, comparable in potency with swainsonine inhibition of α-mannosidase activity, and thus have effects analogous to those of the indolizidine alkaloid swainsonine in species of *Astragalus* in the Fabaceae (see chapter 37) (Molyneux et al. 1993). The other calystegins are of lesser potency. The neurologic disease associated specifically with *S. dimidiatum* is a cerebellar degenerative disease that has many similarities to locoism caused by ingestion of species of *Astragalus.* Although *S. dimidiatum* also contains solanine and chaconine (Casteel et al. 1989), the neurologic disease known as crazy-cow syndrome appears to be caused by calystegin B_2 present in the leaves and fruit (Nash et al. 1993; Molyneux et al. 1994). This appears to be another disease associated with inhibition of the biodegradation of complex sugars and their subsequent detrimental accumulation in lysosomes, leading to cerebellar and Purkinje cell degeneration. Calystegin B_2 is also present in *S. kwebense,* causing a similar neurologic disease. Other species of *Solanum* that contain calystegin B_2 include the leaves of *S. dulcamara, S. melongena,* and *S. tuberosum* (Nash et al. 1993). This is particularly of concern with potatoes because concentrations of 0.01% in the peel and 0.001% in the tuber of calystegins mainly A_3 and B_2 in a 1:2 ratio have been identified in a commercial cultivar (Nash et al. 1993). In addition to their metabolic effects, calystegins given in water to canaries caused diarrhea. Other species that contain calystegins include *Atropa belladonna, Datura wrightii, Hyoscyamus niger,* and *Physalis alkekengi* in this family and *Convolvulus arvensis* and *Calystegia sepium* of the Convolvulaceae (see chapter 25) (Tepfer et al. 1988; Nash et al. 1993; Asano et al. 1995; Drager et al. 1995).

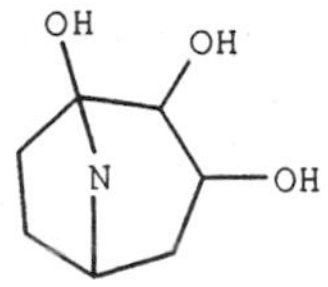

OH OH N OH A_3

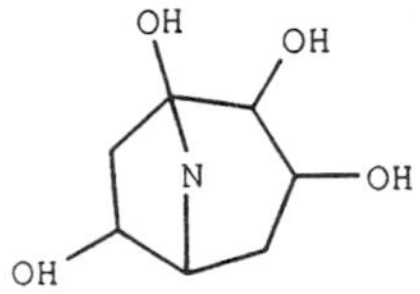

OH OH N OH OH B_1

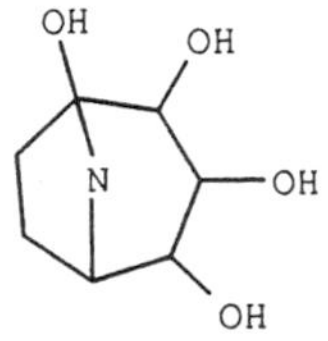

OH OH N OH OH B_2

Calystegins

Clinical Signs—Substantial dosage in most species, including humans, causes severe irritation of the digestive tract following a latent period of several to 24 hours. Typically this is manifested as vomiting, diarrhea, and abdominal pain. These signs may be accompanied by depression, drowsiness, weakness, depressed respiration, pupillary dilation, and alterations in vision. There may also be cardiac arrhythmias and other accompanying abnormalities.

■ humans, vomiting, diarrhea, abdominal pain, pupillary dilation, depression, weakness, drowsiness, alterations in vision

In cattle, ingestion of the foliage and/or berries of *Solanum* may cause a response similar to that of milk fever. It may begin suddenly or follow several days of depression and inappetence. Signs include drowsiness, incoordination, trembling weakness, grinding the teeth, rapid/labored respiration, and constipation early but evolving to a fetid diarrhea in some cases. In calves, edema, local ventral subcutaneous swelling, emaciation, and abdominal fluid accumulation may be prominent with ingestion of *S. carolinense* fruits (Basset & Munro 1986). Pigs display similar signs and also vomiting, whereas in goats there may be vomiting/regurgitation, respiratory difficulties, and neurologic signs. Pupillary dilation is common to all animal species, but it is not as consistent as with the tropane alkaloids of *Datura.* In horses, colic may result from eating large amounts of either foliage or fruits. In the rare severe case, there may also be muscular twitching, collapse, nystagmus, epileptiform seizures, and death within a few days.

■ cattle, depression, loss of appetite, drowsiness, ataxia, weakness, labored respiration

calves, edema, ventral swelling, abdominal fluid accumulation

The neurologic signs associated with *S. dimidiatum* occur following some type of stimulus, such as exercise, loading, or fright. There quickly follows a loss of balance, collapse, nystagmus, and struggling to stand. After a brief rest, the animal will again be able to stand. In less severely affected animals there may be only ataxia, head tilt, and tremors of the shoulders and flanks. Some animals may have hypermetria or dysmetria and appear as if stargazing. Epileptiform seizures may also occur. A diagnostic test may be used that is known as the head raising test (HRT), which involves forcibly holding the animal's head up for 1 minute and then suddenly releasing it. If the animal is affected by *S. dimidiatum,* there will be development of signs such as loss of balance and falling. Animals seldom die of this disease; more likely they are victims of misadventure such as drowning.

■ neurologic, "crazy cow" following an appropriate stimulus; tremors, ataxia, collapse, nystagmus, struggle to stand; able to rise after brief rest; test for by raising head for 1 minute and suddenly releasing, animal may collapse

Pathology—In cases with obvious digestive tract involvement, reddening and necrosis of the mucosa of the stomach and intestines are to be expected. Grossly, in livestock,

- digestive disturbance, fluid in body cavities, reddened mucosa of stomach and intestines, rumen flaccid, distended with nightshade berries

  neurologic, microscopic, Purkinje cell loss

fluid accumulates in the various body cavities, especially the abdomen. The rumen may also be flaccid and distended with partially digested nightshade berries. In swine, fluid accumulation around the kidney is described, a type of perirenal edema, but it is not certain whether this is due to *S. nigrum* as well as *Atropa belladonna* (Hofferd 1937; Quin 1938; Breed 1945). Diagnosis is aided by placing a drop of the urine in the eye of a cat; prompt pupillary dilation in the treated eye represents a positive test (Smith et al. 1956).

Although the neurological incapacitation due to *S. dimidiatum* is not typically lethal, there are distinctive pathologic changes. Grossly, there may be congestion of the intestinal mucosa and ulceration of the mucosa of the abomasum and jejunum. Microscopically, consistently seen are Purkinje cell swelling, chromatolysis, and vacuolation and sometimes complete loss of the Purkinje cell layer of the cerebellum.

- alleviation of signs, possibly atropine

Treatment—If presenting signs such as salivation are suggestive of cholinergic activity, then atropine may be at least partially protective (Patil et al. 1972). Otherwise, treatment is directed toward alleviation of the signs of intoxication and in some instances may even include the use cholinergic drugs such as physostigmine (Ceha et al. 1997).

Genera of Questionable Toxicity

Nicandra Adans.

- *N. physalodes*

  apple of Peru
  shoofly plant

- succulent, hollow-stemmed annuals; flowers solitary in leaf axils; corollas campanulate; petals blue

Native to Peru and Chile, *Nicandra* consists of the single species *N. physalodes* (L.) Gaertn., commonly known as apple of Peru or shoofly plant. The second common name reflects its purported value in repelling flies. Long cultivated as a garden plant, especially in the South, the species has escaped and naturalized in fields and waste places in the southeastern quarter of the continent and sporadically elsewhere. Closely related to *Physalis,* or ground-cherry, it is a glabrous annual with a stout, succulent, hollow, 5-ribbed, branching stem 20–100 cm tall. The leaves are broadly lanceolate with lacerate-dentate margins. Borne singly in the leaf axils, the blue, broadly campanulate flowers have essentially separate, sagittate sepals that enlarge and surround the berry.

- glycoalkaloids, but low if any risk

Nicandra physalodes is sometimes considered to be of some hazard because of the presence of glycoalkaloids, but there is little or no reason at present to consider it a risk (Pammel 1911; Perkins & Payne 1981). In a reported case, 2 sheep grazing plants developed severe bloat, and although treated, both died (Cohen 1970). The possibility of bloat secondary to tropane alkaloid effects cannot be excluded.

Physalis L.

- *P. angulata*

  Chinese lantern
  cutleaf ground-cherry
  ground-cherry

- annual or perennial herbs; flowers solitary in leaf axils; fruits berries enclosed in a papery envelope

Taxonomy and Morphology—Native to both the Old World and the New World and comprising 80–100 species, *Physalis,* commonly known as ground-cherry or Chinese lantern, is readily recognized by its greatly inflated calyces that enclose the edible berries. Derived from the Greek *physa* for "bladder," its name reflects this trait. Approximately 30 native and introduced species are present throughout North America but are most common in the southern regions. Plants are annual or perennial, glabrous or pubescent herbs with stems to 10–100 cm tall that bear alternate or seemingly opposite leaves with entire to pinnately lobed margins. Solitary in the leaf axils, the flowers are white, blue to purple, yellow, or greenish yellow with 5-toothed calyces that enlarge, become papery, and completely enclose the spherical, greenish or yellow fruits. The rotate to campanulate corollas are 5-lobed or toothed, and stamen number is 5.

Although the fruits of some species of this genus are widely used as foods—e.g., *P. philadelphica* Lam. (tomatillo) and *P. peruviana* L. (cape gooseberry)—the genus has been of some interest toxicologically (Watt & Breyer-Brandwijk 1962), but there is little evidence of toxicity. Two cases involving *P. angulata* L. (cutleaf ground-cherry) in hay or forage are described. In dairy cattle there were signs of a transient mastitis, and 11 sheep in a flock of 100 died and upon necropsy exhibited liver disease and hemorrhagic lung lesions (Fuller & McClintock 1986). However, toxicity of the plants has not been experimentally confirmed. There is little reason to consider them toxic other than that they may contain solanine-type glycoalkaloids. Calystegins, including B_2, have been isolated from *P. alkekengi* L. (Asano et al. 1995).

- glycoalkaloids but toxicity not confirmed experimentally

Physalis spp

References

Afsharypuor S, Mostajeran A, Mokhtary R: Variation of scopolamine and atropine in different parts of *Datura* during development [letter]. *Planta Med* 61;383–384, 1995

Akhtar MS, Munir M: Evaluation of the gastric antiulcerogenic effects of *Solanum nigrum, Brassica oleracea,* and *Ocimum basilicum* in rats. *J Ethnopharmacol* 27;163–176, 1989

Asano N, Kato A, Oseki K, Kizu H, Matsui K: Calystegins of *Physalis alkekengi* var. *francheti* (Solanaceae) structure determination and their glycosidase inhibitory activities. *Eur J Biochem* 229;369–376, 1995

Baker D, Keeler R, Gaffield W: Lesions of potato sprout and extracted potato sproat alkaloid toxicity in Syrian Hamsters. *Clin Toxicol* 25;199–208, 1987

Baker DC, Keeler RF, Gaffield W: Pathology in hamsters administered *Solanum* plant species that contain steroidal alkaloids. *Toxicon* 27;1331–1337, 1989

Baker DC, Keeler RF, Gaffield W: Toxicosis from steroidal alkaloids of *Solanum* species. In *Toxicology of Plant and Fungal Components: Handbook of Natural Toxins,* vol 6, Keeler RF, Tu AT eds, Marcel Dekker, New York, pp71–82, 1991

Banton MI, Renegar KR, Nicholson SS: *Brunfelsia pauciflora* (yesterday, today, and tomorrow) poisoning in a dog [letter]. *Vet Hum Toxicol* 31;496–497, 1989

Baskin E: *The Poppy and Other Deadly Plants.* Delacorte Press, New York, 1967

Bassett IJ, Munro DB: The biology of Canadian weeds. 67. *Solanum ptycanthum* Dun., *S. nigrum* L. and *S. sarrachoides* Sendt. Can *J Plant Sci* 65;401–414, 1985

Bassett IJ, Munro DB: The biology of Canadian weeds. 78. *Solanum carolinense* L. and *Solanum rostratum* Dunal. *Can J Plant Sci* 66;977–991, 1986

Benowitz NL, Jacob P III: Metabolism, pharmacokinetics, and pharmacodynamics of nicotine in man. In *Tobacco Smoking and Nicotine: A Neurobiological Approach,* Martin WR, Van Loon GR, Iwamoto ET, Davis L eds, Plenum Press, New York, pp357–373, 1987

Bernheim F, Bernheim MLC: The hydrolysis of homatropine and atropine by various tissues. *J Pharmacol Exp Ther* 64;209–215, 1938

Bonner WG: Poisoning of cattle. Black (or garden) nightshade. *N Z J Agric* 57;99–100, 1938

Borys DJ, Setzer SC, Hornfeldt CS: A retrospective review of plant exposures as reported to the Hennepin Regional Poison Center in 1985. *Vet Hum Toxicol* 29;83–85, 1987

Borys DJ, Herrick ME, Krenzelok EP, Reisdorf JJ, Spillar HA, Bobbink SR, Guevara DS: The clinical effects of *Solanum dulcamara* (deadly nightshade) in pediatric patients [abstract]. *Vet Hum Toxicol* 34;351, 1992

Boughton IW, Hardy WT: Feeding trials of suspected plants. Tex Agric Exp Stn Annu Rep 52, pp239–240, 1939

Bourke CA: Cerebellar degeneration in goats grazing *Solanum cinereum* (Narrawa burr). *Aust Vet J* 75;363–365, 1997

Boyd JW, Murray DS, Tyrl RJ: Silverleaf nightshade, *Solanum elaeagnifolium,* origin, distribution, and relation to man. *Econ Bot* 38;210–217, 1984
Breed F: Plant poisoning in swine. *Norden News* 19;6, 1945
Brown D, Keeler RF: Structure-activity relation of steroid teratogens. 3. Solanidan epimers. *J Agric Food Chem* 26;566–569, 1978
Brown JH: Atropine, scopolamine and related antimuscarinic drugs. In *Goodman and Gilman's The Pharmacological Basis of Therapeutics,* 8th ed, Gilman AG, Rall TW, Nies AS, Taylor P eds, McGraw-Hill, New York, pp150–165, 1990
Buck SH, Burks TF: The pharmacology of capsaicin: review of some recent observations. *Pharmacol Rev* 38;179–226, 1986
Buck WB, Dollahite JW, Allen TJ: *Solanum eleagnifolium,* silver-leaved nightshade poisoning in livestock. *J Am Vet Med Assoc* 137;348–351, 1960
Bush LP, Crowe MW: *Nicotiana* alkaloids. In *Toxicants of Plant Origin,* vol 1, *Alkaloids,* Cheeke PR ed, CRC Press, Boca Raton, Fla, pp87–107, 1989
Bushway RJ, Savage SA, Ferguson BS: Inhibition of acetylcholinesterase by solanaceous glycoalkaloids and alkaloids. *Am Potato J* 64;409–413, 1987
Caldwell KA, Grosjean OK, Henika PR, Friedman M: Hepatic ornithine decarboxylase induction by potato glycoalkaloids in rats. *Food Chem Toxicol* 29;531–535, 1991
Carey JC: Black nightshade poisoning in swine. *North Am Vet* 36;446, 1955
Case AA: Nightshade poisoning. *Southwest Vet* 9;140–143, 1956
Casselberry NH: Nightshade poisoning of swine. *Vet Med* 34;444–445, 1939
Casteel SW, Bailey EM, Camp BJ, Bridges CH, Hejtmancik E, Ueckert DN: The developmental toxicity of an isolate from the plant *Solanum dimidiatum* (potato-weed) in Syrian golden hamsters. *Toxicon* 27;757–762, 1989
Castle NA: Differential inhibition of potassium currents in rat ventricular myocytes by capsaicin. *Cardiovasc Res* 26;1137–1144, 1992
Ceha L, Presperin C, Young E, Allswede M, Erickson T: Anticholinergic toxicity from nightshade berry poisoning responsive to physostigmine. *J Emerg Med* 15;65–69, 1997
Chiu H-F, Lin C-C, Yang C-C, Yang F: The pharmacological and pathological studies on several hepatic protective crude drugs from Taiwan (2). *Am J Chin Med* 17;17–23, 1989
Cohen RDH: Bloat in sheep grazing wild gooseberry, *Nicandra physalodes* [letter]. *Aust Vet J* 46;559, 1970
Cooper MR, Johnson AW: *Poisonous Plants in Britain and Their Effects in Man and Animals.* Her Majesty's Stationery Office, London, pp211–223, 1984
Cordell GA, Araujo OE: Capsaicin: identification, nomenclature, and pharmacotherapy. *Ann Pharmacother* 27;330–336, 1993
Coremans P, Lambrecht G, Schepens P, Vanwelden J, Verhaegen H: Anticholinergic intoxication with commercially available thorn apple tea. *J Toxicol Clin Toxicol* 32;589–592, 1994
Correll DS, Johnston MC: *Manual of the Vascular Plants of Texas.* Texas Research Foundation, Renner, Tex, 1970
Craig JF, Kehoe D: Plant poisoning. *Vet Rec* 37;795–816, 1925
Crawford L, Kocan RM: Steroidal alkaloid toxicity to fish embryos. *Toxicol Lett* 66;175–181, 1993
Cronquist A, Holmgren AH, Holmgren NH, Reveal JL, Holmgren PK: *Intermountain Flora.* Vol 4, *Subclass Asteridae.* New York Botanical Garden, Bronx, New York, 1984
Crowe MW: Tobacco—a cause of congenital arthrogryposis. Inn *Effects of Poisonous Plants on Livestock,* Keeler RF, Van Kampen KR, James LF eds, Academic Press, New York, pp419–427, 1978
Crowe MW, Pike HT: Congenital arthrogryposis associated with ingestion of tobacco stalks by pregnant sows. *J Am Vet Med Assoc* 162;453–455, 1973
Crowe MW, Swerczek TW: Congenital arthrogryposis in offspring of sows fed tobacco *(Nicotiana tabacum). Am J Vet Res* 35;1071–1073, 1974
Dalvi RR: Comparative assessment of the effect of solanine administered orally and intraperitoneally on hepatic function in male rats. *Jpn J Vet Sci* 47;657–659, 1985

Daneshvar S, Mirhossaini ME, Balali M: *Hyoscyamus* poisoning in Mashhad [abstract]. *Toxicon* 30;501, 1992
Davis CH, Smith TJ, Hawkins RS: *Eradication of the White Horsenettle in Southern Arizona.* Ariz Agric Exp Stn Bull 195, 1945
Day EJ, Dilworth BC: Toxicity of jimson weed seed and cocoa shell meal to broilers. *Poult Sci* 63;466–468, 1984
Dean ES: Self-induced stramonium intoxication. *J Am Med Assoc* 185;882, 1963
Dewitt MS, Swain R, Gibson LB Jr: The dangers of jimson weed and its abuse by teenagers in the Kanawha Valley of West Virginia. *W Va Med J* 93;182–185, 1997
Dice JR: *Potatoes for Dairy Cows.* N D Agric Exp Stn Bull 249, 1931
Dijkstra ND: Voederproef met geënsileered aardappelloof. *Versl Landbouwkd Onderz* 51;91–129, 1945
Dijkstra ND, Reestman AJ: Het gebruik van aardappelloof als veevoeder. *Landbouwkd Tijdschr* 55;191–210, 1943
Dollahite JW, Allen TJ: *Silverleaf Nightshade Poisoning in Livestock.* Tex Agric Exp Stn Prog Rep 2146, 1960
Domek JM, Cantelo WW, Wagner RM, Li BW, Miller-Ihli NJ: Nutritional composition of potato foliage. *J Agric Food Chem* 43;1512–1515, 1995
Done SH, Tokarnia CH, Dammrich K, Dobereiner J: *Solanum malacoxylon* poisoning in pigs. *Res Vet Sci* 20;217–219, 1976
Drager B, van Almsick A, Mrachatz G: Distribution of calystegins in several Solanaceae. *Planta Med* 61;577–579, 1995
Dray A: Mechanism of action of capsaicin-like molecules on sensory neurons. *Life Sci* 51;1759–1765, 1992
Drew ML, Fowler ME: Poisoning of black and white ruffed lemurs *(Varecia variegata)* by hairy nightshade *(Solanum sarrachoides). J Zoo Wildl Med* 22;494–496, 1991
Duez P, Chamart S, Hanocq M, Molle L, Vanhaelen M, Vanhaelen-Fastre R: Comparison between thin-layer chromatography-densitometry and high-performance liquid chromatography for the determination of hyoscyamine and hyoscyne in leaves, fruit, and seeds of datura (*Datura* spp.) *J Chromatogr* 329;415–421, 1985
Dugan GM, Gumbmann MR: Toxicological evaluation of sicklepod and black nightshade seeds in short-term feeding studies in rats. *Food Chem Toxicol* 28;101–107, 1990
Dugan GM, Gumbmann MR, Friedman M: Toxicological evaluation of jimson weed *(Datura stramonium)* seed. *Food Chem Toxicol* 27;501–510, 1989
Duvoisin RC, Katz R: Reversal of central anticholinergic syndrome in man by physostigmine. *J Am Med Assoc* 206;1963–1965, 1968
Edwards EJ, Saint RE, Cobb AH: Is there a link between greening and light-enhanced glycoalkaloid accumulation in potato (*Solanum tuberosum* L.) tubers. *J Sci Food Agric* 76;327–333, 1998
El-Dabbas SW, Evans WC: Alkaloids of the genus *Datura,* section Brugmansia. *Planta Med* 44;184–185, 1982
El Dirdiri NI, Wasfi IA, Adam SEI, Edds GT: Toxicity of *Datura stramonium* to sheep and goats. *Vet Hum Toxicol* 23,241–246, 1981
Eldridge AC, Hockridge ME: High-performance liquid chromatographic separation of eastern black nightshade *(Solanum ptycanthum)* glycoalkaloids. *J Agric Food Chem* 31;1218–1220, 1983
Evans WC, Major VA: The alkaloids of the genus *Datura,* section *Brugmansia.* 4. New alkaloids of *Datura sanguinea* R. and P. *J Chem Soc* (*Perkin Trans* 1) 18;1621–1623, 1966
Evans WC, Major VA, Than MP: The alkaloids of the genus *Datura,* section *Brugmansia.* 3. *Datura sanguinea* R. and P. *Planta Med* 13;353–358, 1965
Evans WC, Grout RJ, Mensah MLK: Withanolides of *Datura* spp. and hybrids. *Phytochemistry* 23;1717–1720, 1984
Farah MH, Samuelsson G: Pharmacologically active phenylpropanoids from *Senra incana. Planta Med* 58,14–18, 1992
Flores FA, Lewis WH: Drinking the South American hallucinogenic ayahuasca. *Econ Bot* 32;154–156, 1978

Flunker LK, Damron BL, Sundlof SF: Jimsonweed seed contamination of broiler chick and White Leghorn hen diets. *Nutr Rep Int* 36;551–556, 1987

Francis PD, Clarke CF: Angel trumpet lily poisoning in five adolescents: clinical findings and management. *J Paediatr Child Health* 35;93–95, 1999

Friedman M, Dao L: Distribution of glycoalkaloids in potato plants and commercial potato products. *J Agric Food Chem* 40;419–423, 1992

Friedman M, Henika PR: Absence of genotoxicity of potato alkaloids α-chaconine, α-solanine and solanidine in the Ames *Salmonella* and adult and foetal erythrocyte micronucleus assay. *Food Chem Toxicol* 30;689–694, 1992

Friedman M, Levine CE: Composition of jimson weed *(Datura stramonium)* seeds. *J Agric Food Chem* 37;998–1005, 1989

Friedman M, Levine CE: α-Tomatine content in tomato and tomato products determined by HPLC with pulsed amperometric detection. *J Agric Food Chem* 43;1507–1511, 1995

Friedman M, Rayburn JR, Bantle JA: Developmental toxicology of potato alkaloids in the frog embryo teratogenesis assay—*Xenopus* (FETAX). *Food Chem Toxicol* 29;537–547, 1991

Friedman M, Rayburn JR, Bantle JA: Structural relationships and developmental toxicity of *Solanum* alkaloids in the frog embryo teratogenesis assay-*Xenopus. J Agric Food Chem* 40;1617–1624, 1992

Fuller TC, McClintock E: *Poisonous Plants of California.* Univ Calif Press, Berkeley, pp232–251, 1986

Gaffield W, Keeler RF: Implication of C-5, C-6 unsaturation as a key structural factor in steroidal alkaloid-induced mammalian teratogenesis. *Experientia* 49;922–924, 1993

Gaffield W, Keeler RF: Structure-activity relations of teratogenic natural products. *Pure Appl Chem* 66;2407–2410, 1994

Gaffield W, Keeler RF, Baker DC: *Solanum* glycoalkaloids: plant toxins possessing disparate physiologically active structural entities. In *Toxicology of Plant and Fungal Components: Handbook of Natural Toxins,* vol 6, Keeler RF, Tu AT eds, Marcel Dekker, New York, pp135–158, 1991

Gaffield W, Keeler RF, Baker DC, Stafford AE: Studies on the *Solanum tuberosum* sprout teratogen. In *Poisonous Plants: Proceedings of the Third International Symposium,* James LF, Keeler RF, Bailey EM Jr, Cheeke PR, Hegarty MP, eds, Iowa State Univ Press, Ames, pp418–422, 1992

Gambaro VE, Roses OE: Presence of nicotine in extracts and decoctions from flowers of *Brugmansia candida* Pers. *Acta Farm Bonaerense* 8;17–22, 1989

Gamse R, Molnar A, Lembeck F: Substance P release from spinal cord slices by capsaicin. *Life Sci* 25;629–636, 1979

Garland T, Bailey EM Jr, Reagor JC, Binford E: Probable interaction between *Solanum eleagnifolium* and ivermectin in horses. In *Toxic Plants and Other Natural Toxicants*, Garland T, Barr AC eds, CAB International, New York, pp423–427, 1998

Ghani A: The site of synthesis and secondary transformation of hyoscyamine in *Solandra grandiflora. Phytochemistry* 25;617–619, 1986

Gleason HA, Cronquist A: *Manual of Vascular Plants of Northeastern United States and Adjacent Canada,* 2d ed. New York Botanical Garden, Bronx, New York, 1991

Glinsukon T, Stitmunnaithum V, Toskulkao C, Buranawuti T, Tangkrisanavinont V: Acute toxicity of capsaicin in several animal species. *Toxicon* 18;215–220, 1980

Goldmann A, Message B, Tepfer D, Molyneux RJ, Duclos O, Boyer F-D, Pan YT, Elbein AD: Biological activities of the nortropane alkaloid, calystegin B_2, and analogs: structure-function relationships. *J Nat Prod* 59;1137–1142, 1996

Gonzalez JE, Garcia Ovando H, Gonzalez Quintana H, Maffrand C, Mortara L: Clinical and pathological study of dogs poisoned experimentally with *Cestrum parqui* l'Herit. *Pet's Cienc* 4;207–213, 1988

Govindarajan VS: *Capsicum* production, technology, chemistry, and quality. Part 1. History, botany, cultivation, and primary processing. *CRC Crit Rev Food Sci Nutr* 22;109–176, 1985

Gowdy JM: *Stramonium* intoxication, review of symptomatology in 212 cases. *J Am Med Assoc* 221;585–587, 1972

Greene GS, Patterson SG, Warner E: Ingestion of angel's trumpet: an increasingly common source of toxicity. *South Med J* 89;365–369, 1996

Guharoy SR, Barajas M: Atropine intoxication from the ingestion and smoking of jimson weed *(Datura stramonium). Vet Hum Toxicol* 33;588–589, 1991

Gunning OV: Poisoning in goats by black nightshade *(Solanum nigrum). Br Vet J* 105; 473–474, 1949

Halim AF, Collins RP, Berigari MS: Alkaloids produced by *Cestrum nocturnum* and *Cestrum diurnum. Planta Med* 20;44–49, 1971

Hall RC, Popkin MK, McHenry LE: Angel's trumpet psychosis: a central nervous system anticholinergic syndrome. *Am J Psychiatry* 134;312–314, 1977

Hanna JP, Schmidley JW, Braselton WE Jr: *Datura* delirium. *Clin Neuropharmacol* 15;109–113, 1992

Hansen AA: Stock poisoning by plants in the nightshade family. *J Am Vet Med Assoc* 71;221–227, 1927

Hansen AA: Potato poisoning. *North Am Vet* 9;31–34, 1928

Hellenas K-E, Nyman A, Slanina P, Loof L, Gabrielsson J: Determination of potato glycoalkaloids and their aglycone in blood serum by high-performance liquid chromatography. Application to pharmacokinetic studies in humans. *J Chromatogr Biomed Appl* 573;69–78, 1992

Hocking GM: Henbane—healing herb of Hercules and of Apollo. *Econ Bot* 1;306–316, 1947

Hofferd RM: Belladonna poisoning in swine. *Bio-Chem Rev* 8;21, 1937

Holzer P: Capsaicin: cellular targets, mechanisms of action, and selectivity for thin sensory neurons. *Pharmacol Rev* 43;143–201, 1991

Hopkins J: The glycoalkaloids: naturally of interest (but a hot potato?). *Food Chem Toxicol* 33;323–328, 1995

Hornfeldt CS, Collins JE: Toxicity of nightshade berries *(Solanum dulcamara)* in mice. *Clin Toxicol* 28;185–192, 1990

Horowitz M, Wishart J, Maddox A, Russo A: The effect of chilli on gastrointestinal transit. *J Gastroenterol Hepatol* 7;52–56, 1992

Hostettmann K, Marston A: *Saponins.* Cambridge Univ Press, New York, 1995

Houpt K, Zgoda JC, Stahlbaum CC: Use of taste repellants and emetics to prevent accidental poisoning of dogs. *Am J Vet Res* 45;1501–1503, 1984

Hubbs JC: Belladonna poisoning in pigs. *Vet Med* 42;428–429, 1947

Hughes JD, Clark JA: *Datura stramonium* poisoning. *J Am Med Assoc* 112;2500–2502, 1939

Hughes MR, McCain TA, Chang SY, Haussler MR, Villareale M, Wasserman RH: Presence of 1,25-dihydroxyvitamin D_3-glycoside in the calcinogenic plant *Cestrum diurnum. Nature* 268;347–349, 1977

Iwai K, Suzuki T, Fujiwake H: Formation and accumulation of pungent principle of hot pepper fruits, capsaicin and its analogues, in *Capsicum annuun* var. *annuun* cv. Karayatsubusa at different growth stages after flowering. *Agric Biol Chem* 43;2493–2498, 1979

Iyer RP: *Brunfelsia hopeana*—pharmacologic screening: isolation and characterization of hopeanine. PhD thesis, Univ Pacific, 1978 ((*Diss Abstr Int* 39B;761, 1978)

Iyer RP, Brown JK, Chaubal MG, Malone MH: *Brunfelsia hopeana.* 1. Hippocratic screening and antiinflammatory evaluation. *Lloydia* 40;356–360, 1977

Jadhav SJ, Sharma RP, Salunkhe DK: Naturally occurring toxic alkaloids in foods. *CRC Crit Rev Toxicol* 9;21–104, 1981

Jang JJ, Devor DE, Logsdon DL, Ward JM: A four-week feeding study of ground red chilli *(Capsicum annuum)* in male B6C3F1 mice. *Food Chem Toxicol* 30;783–787, 1992

Janssens G. Wilde R de: Toxicity of thornapple *(Datura stramonium* and/or *D. ferox)* seed present in pig feed. *Vlaam Diergeneesk Tijdschr* 58;84–86, 1989

Jarboe CH, Zirvi KA, Nicholson JA, Schmidt CM: Scopoletin, an antispasmodic component of *Viburnum opulus* and *prunifolium*. *J Med Chem* 10;488–489, 1967

Jaspersen-Schib R, Theus L, Guirguis-Oeschger M, Gossweiler B, Meier-Abt PJ: Serious plant poisonings in Switzerland 1966–1994. *Schweiz Med Wochenschr* 126;1085–1098, 1996

Jennings RE: Stramonium poisoning. *J Pediatr* 6;657–664, 1935

Jessell TM, Iversen LL, Cuello AC: Capsaicin-induced depletion of substance P from primary sensory neurons. *Brain Res* 152;183–188, 1978

Kaplan B: Acute nicotine poisoning in a dog. *Vet Med/Small Anim Clin* 63;1033–1034, 1968

Kawada T, Iwai K: In vivo and in vitro metabolism of dihydrocapsaicin, a pungent principle of hot pepper, in rats. *Agric Biol Chem* 49;441–448, 1985

Kawada T, Suzuki T, Takahashi M, Iwai K: Gastrointestinal absorption and metabolism of capsaicin in rats. *Toxicol Appl Pharmacol* 72;449–456, 1984

Kawada T, Watanabe T, Takaishi T, Tanaka T, Iwai K: Capsaicin-induced β-adrenergic action on energy metabolism in rats: influence of capsaicin on oxygen consumption, the respiratory quotient, and substrate utilization. *Proc Soc Exp Biol Med* 183;250–256, 1986

Keeler RF: Absence of arthrogryposis in newborn Hampshire pigs from sows ingesting toxic levels of jimsonweed during gestation. *Vet Hum Toxicol* 23;413–415, 1981

Keeler RF, Crowe MW: Teratogenicity and toxicity of wild tree tobacco, *Nicotiana glauca* in sheep. *Cornell Vet* 74;50–59, 1984

Keeler RF, Crowe MW: Anabasine, a teratogen from the *Nicotiana* genus. In *Plant Toxicology, Proceedings of the Australia-USA Poisonous Plants Symposium,* Seawright AA, Hegarty MP, James LF, Keeler RF eds, Animal Research Institute, Yeerongpilly, Australia, pp324–333, 1985

Keeler RF, Balls LD, Panter K: Teratogenic effects of *Nicotiana glauca* and concentration of anabasine, the suspect teratogen in plant parts. *Cornell Vet* 71;47–53, 1981a

Keeler RF, Shupe JL, Crowe MW, Olson A, Balls LD: *Nicotiana glauca*–induced congenital deformities in calves: clinical and pathological aspects. *Am J Vet Res* 42;1231–1234, 1981b

Keeler RF, Crowe MW, Lambert EA: Teratogenicity in swine of the alkaloid anabasine isolated from *Nicotiana glauca*. *Teratology* 30;61–69, 1984

Keeler RF, Baker DC, Gaffield W: Spirosolane-containing *Solanum* species and induction of congenital craniofacial malformations. *Toxicon* 28;873–884, 1990

Keeler RF, Baker DC, Gaffield W: *Solanum* alkaloids. In *Mycotoxins and Phytoalexins,* Scharma RP, Salunkhe DK, eds, CRC Press, Boca Raton, Florida, p607–636, 1991a

Keeler RF, Baker DC, Gaffield W: Teratogenic *Solanum* species and the responsible teratogens. In *Toxicology of Plant and Fungal Components, Handbook of Natural Toxins* Vol 6, Keeler Rf, Tu AT eds, Marcel Dekker, New York, pp83–99, 1991b

Kellerman TS, Coetzer JAW, Naude TW: *Plant Poisoning and Mycotoxicoses of Livestock in Southern Africa.* Oxford Univ Press, Cape Town, South Africa, pp13–15, 1988

King RR, McQueen RE: Transformations of potato alkaloids by rumen microorganisms. *J Agric Food Chem* 29;1101–1103, 1981

Kingsbury JM: *Poisonous Plants of the United States and Canada.* Prentice Hall, Englewood Cliffs, NJ, pp275–294, 1964

Kite GC, Leon C: Volatile compounds emitted from flowers and leaves of *Brugmansia* ×*candida* (Solanaceae). *Phytochemistry* 40;1093–1095, 1995

Klein-Schwartz W, Oderda GM: Jimsonweed intoxication in adolescents and young adults. *Am J Dis Child* 138;737–739, 1984

Kovatsis A, Flaskos J, Nikolaidis E, Kotaki-Kovatsi VP: Toxicity study of the main alkaloids of *Datura* in broilers. *Food Chem Toxicol* 31;841–845, 1993

Krantz JC Jr, Forrest JW, Heisse CK: Contribution to the pharmacology of apoatropine and its methyl bromide. *Proc Soc Exp Biol Med* 86;511–512, 1954

Krenzelok E, Provost FJ: The ten most common plant exposures reported to the poison information centers in the United States. *J Nat Toxins* 4;195–202, 1995

Krook L, Wasserman RH, McEntee K, Brokken TD, Teigland MB: *Cestrum diurnum* poisoning in Florida cattle. *Cornell Vet* 65;557–575, 1975a

Krook L, Wasserman RH, Shively JN, Tashjian AH Jr, Brokken TD, Morton JF: Hypercalcemia and calcinosis in Florida horses: implication of the shrub, *Cestrum diurnum,* as the causative agent. *Cornell Vet* 65;26–56, 1975b

Kudo K, Kelly WR, Oelrichs PB: Experimental poisoning of mice and sheep with *Cestrum parqui.* In *Plant Toxicology, Proceedings of the Australia-USA Poisonous Plants Symposium,* Seawright AA, Hegarty MP, James LF, Keeler RF eds, Animal Research Institute, Yeerongpilly, Australia, pp533–540, 1985

Landauer W: Nicotine-induced malformations of chicken embryos and their bearing on the phenocopy problem. *J Exp Zool* 143;107–122, 1960

Larson PS, Haag HB: Studies on the fate of nicotine in the body. 2. On the fate of nicotine in the dog. *J Pharmacol Exp Ther* 76;240–244, 1942

Lawson DR, Erb WA, Miller AR: Analysis of *Solanum* alkaloids using internal standardization and capillary gas chromatography. *J Agric Food Chem* 40;2186–2191, 1992

Leipold HW, Oehme FW, Cook JE: Congenital arthrogryposis associated with ingestion of jimsonweed by pregnant sows. *J Am Vet Med Assoc* 162;1059–1060, 1973

Lindeque M, Scheepers JL: Use of *Datura inoxia* by ungulates in the Hoanib River, Namibia. *S Afr J Wildl Res* 22;45–48, 1992

Lloyd HA, Fales HM, Goldman ME, Jerina DM, Plowman T, Schultes RE: Brunfelsamidine: a novel convulsant from the medicinal plant *Brunfelsia grandiflora. Tetrahedron Lett* 26;2623–2624, 1985

Lopez T, Keeler RF, Sharma RP, Shupe JL: Some substances related to the toxicity of *Cestrum parqui.* L'Heritier. *Vet Argent* 1;966–967, 1984

Lozoya X, Meckes-Lozoya M, Chavez MA, Becerril G: Estudio farmacologico y fitoquimico de dos especies de *Solandra* nativas de Mexico [Pharmacological and phytochemical study of two species of *Solandra* native to Mexico]. *Ars Pharm* 29;35–42, 1988

Lugt JJ van der, Nel PW, Kitching JP: The pathology of *Cestrum laevigatum* (Schlechtd.) poisoning in cattle. *Onderstepoort J Vet Res* 58;211–221, 1991

Lugt JJ van der, Nel PW, Kitching JP: Experimentally-induced *Cestrum laevigatum* (Schlechtd.) poisoning in sheep. *Onderstepoort J Vet Res* 59;135–144, 1992

Magras IN, Kotsaki-Kovatsi VP, Kovatsis A, Adamidou L: Teratogenic effects of a mixture of scopolamine and hyoscyamine in chick embryos. *Vet Hum Toxicol* 35;434–435, 1993

Mahler DA: The jimson-weed high [letter]. *J Am Med Assoc* 231;138, 1975

Margolis F, Feilgelson P: Purification and characterization of a genetically determined serum esterase. *J Biol Chem* 238;2620–2627, 1963

Marsh CD, Clawson AB, Roe GC: *Wild Tobaccos* (Nicotiana trigonophylla *Dunal* and *Nicotiana attenuata* Torrey) *as Stock-Poisoning Plants.* USDA Tech Bull 22, 1927

Masada Y, Hashimoto K, Inoue T, Suz M: Analysis of the pungent principles of *Capsicum annuum* by combined gas chromatography–mass spectrometry. *J Food Sci* 36;858–8860, 1971

McBarron EJ, de Sarem W: Poisoning of dogs by the fruits of the garden shrub *Brunfelsia bonodora* [letter}. *Aust Vet J* 51;280, 1975

McHenry LE, Hall RC: Angel's trumpet: lethal and psychogenic aspects. *J Fla Med Assoc* 65;192–196, 1978

McLennan MW, Kelly WR: *Cestrum parqui* (green cestrum) poisoning in cattle. *Aust Vet J* 61;289–291, 1984

McMeniman NP: *Solanum esuriale,* a possible cause of humpy back in sheep. *Aust Vet J* 52;432–433, 1976

McNight RH, Levine EJ, Rodgers GC: Detection of green tobacco sickness by a regional poison center. *Vet Hum Toxicol* 36;505–510, 1994

Mello JR, Habermehl GG: Calcinogenic plants and the incubation effect of rumen fluid. *Dtsch Tierarztl Wochenschr* 99;371–376, 1992

Menzies JS, Bridges CH, Bailey EM Jr: A neurological disease associated with *Solanum dimidiatum. Southwest Vet* 32;45–49, 1979

Miller MS, Brendel K, Burks TF, Sipes IG: Interaction of capsaicinoids with drug-metabolizing systems: relationship to toxicity. *Biochem Pharmacol* 32;547–551, 1983

Milligan JB: Suspected potato poisoning in cattle. *Vet Rec* 53;512, 1941

Mitchell JE, Mitchell FN: Jimson weed *(Datura stramonium)* poisoning in childhood. *J Pediatr* 47;227–230, 1955

Molyneux RJ, Pan YT, Goldmann A, Tepfer DA, Elbein AD: Calystegins, a novel class of alkaloid glycosidase inhibitors. *Arch Biochem Biophys* 304;81–88, 1993

Molyneux RJ, James LF, Ralphs MH, Pfister JA, Panter KE, Nash RJ: Polyhydroxy alkaloid glycosidase inhibitors from poisonous plants of global distribution: analysis and identification. In *Plant-Associated Toxins: Agricultural, Phytochemical, and Ecological Aspects,* Colegate SM, Dorling PR eds, CAB International, Wallingford, UK, pp107–112, 1994

Mondy NI, Munshi CB: Effect of soil and foliar application of molybdenum on the glycoalkaloid and nitrate concentration of potatoes. *J Agric Food Chem* 41;256–258, 1993

Monji N, Castro A: Plasma nicotine pharmacokinetics in dogs after intravenous administration: Determination by radioimmunoassay. *Res Comm Chem Pathol Pharmacol* 23;267–277, 1979

Morris SC, Lee TH: The toxicity and teratogenicity of Solanaceae glycoalkaloids, particularly those of the potato *(Solanum tuberosum):* a review. *Food Technol Aust* 36;118–124, 1984

Mors WB, Ribeiro O: Occurrence of scopoletin in the genus *Brunfelsia. J Org Chem* 22;978–979, 1957

Morton JF: *500 Plants of South Florida.* EA Seeman, Miami, 1974

Morton JF: *Plants Poisonous to People in Florida.* JF Morton, Miami, pp19–21, 1982

Muller JL: Love potions and the ointment of witches: historical aspects of the nightshade alkaloids. *J Toxicol Clin Toxicol* 36;617–624, 1998

Nash RJ, Rothschild M, Porter EA, Watson AA, Waigh RD, Waterman PG: Calystegines in *Solanum* and *Datura* species and the deaths-head hawk moth *(Acherontia atropus). Phytochemistry* 34;1281–1283, 1993

Neilson J, Burren V: Intoxication of two dogs by fruit of *Brunfelsia australis. Aust Vet J* 60;379–380, 1983

Nelson EK, Dawson LE: The constitution of capsaicin, the pungent principle of *Capsicum.* 3. *J Am Chem Soc* 45;2179–2181, 1923

Nelson PD, Mercer HD, Essig HW, Minyard JP: Jimson weed seed toxicity in cattle. *Vet Hum Toxicol* 24;321–325, 1982

Nicholson JWG, Young DA, McQueen RE, De Jong H, Wood FA: The feeding value potential of potato vines. *Can J Anim Sci* 58;559–569, 1978

Niess C, Schnabel A, Kauert G: Angel trumpet: a poisonous garden plant as a new addictive drug? *Dtsch Med Wochenschr* 124;1444–1447, 1999

Nishie K, Gumbmann MR, Keyl AC: Pharmacology of solanine. *Toxicol Appl Pharmacol* 19;81–92, 1971

Ofukany L, Frantova E, Turon J: Mass poisoning of cattle with *Datura stramonium. Veterinarstvi* 33;316–317, 1983

Ogilvie DD: *Solanum* poisoning in pigs. *Vet Rec* 55;249, 1943

Oi Y, Kawada T, Watanabe T, Iwai K: Induction of capsaicin-hydrolysing enzyme activity in rat liver by continuous oral administration of capsaicin. *J Agric Food Chem* 40;467–470, 1992

Ojewole JAO, Adesina SK: Cardiovascular and neuromuscular actions of scopoletin from fruit of *Tetrapleura tetraptera. Planta Med* 49;99–102, 1983a

Ojewole JAO, Adesina SK: Mechanism of the hypotensive effect of scopoletin isolated from the fruit of *Tetrapleura tetraptera. Planta Med* 49;46–50, 1983b

O'Neill TP: Mechanism of capsaicin action: recent learnings. *Respir Med* 85 (Suppl A);35–41, 1991

Orgell WH: Inhibition of human plasma cholinesterase in vitro by alkaloids, glycosides, and other natural substances. *Lloydia* 26;36–43, 1963

Oshima T, Sagara K, Tong YY, Zhang G, Chen YH: Application of ion-pair high performance liquid chromatography for analysis of hyoscyamine and scopolamine in solanaceous crude drugs. *Chem Pharm Bull* 37;2456–2458, 1989

O'Sullivan BM: Humpy back of sheep. Clinical and pathological observations. *Aust Vet J* 52;414–418, 1976

Otten U, Lorez HP, Businger F: Nerve growth factor antagonizes the neurotoxic action of capsaicin on primary sensory neurons. *Nature* 301;515–517, 1983

Owen Rh ap Rh: Potato poisoning in a horse. *Vet Rec* 117;246, 1985

Pammel LH: *Manual of Poisonous Plants.* Torch Press, Cedar Rapids, Iowa, pp713–733, 1911

Pammel LH: Three-flowered nightshade poisonous. *Vet Med* 16;46–47, 1921

Panter KE, James LF, Gardner DR: Lupines, poison-hemlock, and *Nicotiana* spp.: toxicity and teratogenicity in livestock. *J Nat Toxins* 8;117–133, 1999

Patil BC, Sharma RP, Salunkhe DK, Salunkhe K: Evaluation of solanine toxicity. *Food Cosmetic Toxicol* 10;395–398, 1972

Pearce CM, Skelton NJ, Naylor S, Kanaan R, Kelland J, Oelrichs PB, Sanders JKM, Williams DH: Parquin and carboxyparquin, toxic kaurene glycosides from the shrub *Cestrum parqui. J Chem Soc (Perkin Trans* 1) 1992;593–600, 1992

Perkins KD, Payne WW: *Guide to the Poisonous Plants of Florida.* Fla Coop Ext Serv Circ 441, 1981

Perrotta DM, Nickey LN, Raid M, Caraccio T, Mofenson HC, Waters C, Morse D, Osorio AM, Hoshiko S, Rutherford GW III: Jimson weed poisoning—Texas, New York, and California, 1994. *Morb Mortal Wkly Rep* 44;41–44, 1995

Phillips BJ, Hughes JA, Phillips JC, Walters DG, Anderson D, Tahourdin CSM: A study of the toxic hazard that might be associated with the consumption of green potato tops. *Food Chem Toxicol* 34;439–448, 1996

Pienaar JG, Kellerman TS, Basson PA, Jenkins WL, Vahrmeijer J: Maldronksiekte in cattle: A neuropathy caused by *Solanum kwebense* N.E.Br. *Onderstepoort J Vet Res* 43;67–74, 1976

Plumlee KH, Holstege DM, Blanchard PC, Fiser KM, Galey FD: *Nicotiana glauca* toxicosis of cattle. *J Vet Diagn Invest* 5;498–499, 1993

Quin AH: Diagnostic point in nightshade poisoning in swine. *Bio-Chem Rev* 10;18, 1938

Renwick JH: Spina bifida and the potato. *New Sci* 56;277–279, 1972

Renwick JH, Claringbold WDB, Earthy ME, Few JD, McLean ACS: Neural-tube defects produced in Syrian hamsters by potato glycoalkaloids. *Teratology* 30;371–381, 1984

Riet-Correa F, Carmen Mendez M del, Schild AL, Summers BA, Oliveira JA: Intoxication by *Solanum fastigiatum* var. *fastigiatum* as a cause of cerebellar degeneration in cattle. *Cornell Vet* 73;240–256, 1983

Riet-Correa F, Schild AL, Mendez M del C, Pinheiro MH: Intoxication with *Cestrum parqui* (Solanaceae) in cattle in Rio Grande do Sul. *Pesq Vet Bras* 6;111–115, 1986

Rindl M, Sapiro ML: A chemical investigation of the constituents of *Nicotiana glauca* R.Grah. (Solanaceae) (wild tobacco). *Onderstepoort J Vet Sci Anim Ind* 22;301–312, 1949

Ripperger H, Schreiber K: *Solanum* steroid alkaloids. In *The Alkaloids,* vol 19, Rodrigo RGA ed, Academic Press, New York, pp81–192, 1981

Rivera A, Calderon E, Gonzalez MA, Valbuena S, Joseph-Nathan P: *Brugmansia sanguinea* subsp. *vulcanicola,* a good source of scopolamine. *Fitoterapia* 60;542–544, 1989

Roddick JG: The acetylcholinesterase-inhibitory activity of steroidal glycosides and their aglycones. *Phytochemistry* 28;2631–2634, 1989

Rodgers GC, Von Kanel RL: Conservative treatment of jimsonweed ingestion. *Vet Hum Toxicol* 35;32–33, 1993

Rogers BS, Ogg AG Jr: *Biology of Weeds of the* Solanum nigrum *complex* (Solanum *section* Solanum) *in North America.* USDA Agric Rev Man ARM-W-23, 1981

Roses OE, Lopez CM, Garcia Fernandez JC: Isolation and identification of tropane alkaloids in species of the genus *Brugmansia* (Solanaceae). *Acta Farm Bonaerense* 6;167–174, 1987

Roses OE, Gambaro VE, Rofi R: Presence of nor-hyoscine and hyoscine in *Brugmansia candida* Pers. flowers as possibly characteristic of its origin. *Acta Farm Bonaerense* 7;85–90, 1988

Roy AR, Chatterjee ML: Pharmacological observations on saponins from *Cestrum diurnum* and *Cestrum nocturnum. Indian J Exp Biol* 6;160–162, 1968

Ruddick JA, Harwig J, Scott PM: Nonteratogenicity in rats of blighted potatoes and compounds contained in them. *Teratology* 9;165–168, 1974

Rumack BH: Anticholinergic poisoning: treatment with physostigmine. *Pediatrics* 52; 449–451, 1973

SAC Veterinary Services: Solanine poisoning causes deaths in Scottish cattle. *Vet Rec* 142;125, 1998

Sampson AW, Malmsten HE: *Stock-Poisoning Plants of California.* Calif Agric Exp Bull 593, 1942 (rev)

Santos Zambrano Mdos, Riet Correa F, Schild AL, Carmen Mendez M del: Intoxication with *Solanum fastigiatum* var. *fastigiatum*: evolution and reversibility of the lesions in cattle, and susceptibility of sheep, rabbits, guinea pigs, and rats. *Pesq Vet Bras* 5:133–141, 1985

Sawin PB, Glick D: Atropinase, a genetically determined enzyme in the rabbit. *Proc Am Acad Sci* 29;55–59, 1943

Schilling EE: Systematics of *Solanum* section *Solanum* (Solanaceae) in North America. *Syst Bot* 6;172–185, 1981

Schneider F, Lutun P, Kintz P, Astruc D, Flesch F, Tempe J-D: Plasma and urine concentrations of atropine after the ingestion of cooked deadly nightshade berries. *Clin Toxicol* 34;113–117, 1996

Schreiber K: Steroid alkaloids: the *Solanum* group. In *The Alkaloids,* vol 10, Manske RHF ed, Academic Press, New York, pp1–192, 1968

Schulman ML, Bolton LA: Datura seed intoxication in two horses. *J S Afr Vet Assoc* 69; 27–29, 1998

Schultes RE: Hallucinogens of plant origin. *Science* 163;245–254, 1969

Shervette RE III, Schydlower M, Lampe RM, Fearnow R: Jimson "loco" weed abuse in adolescents. *Pediatrics* 63;520–523, 1970

Shlosberg A, Bellaiche M, Hanji V, Ershov E, Bogin E, Avidar Y, Perl S, Shore L, Shemesh M: The effect of feeding dried tomato vines to beef cattle. *Vet Hum Toxicol* 38;135–136, 1996

Shone DK, Drummond RB: Poisonous plants of Rhodesia, *Cestrum aurantiacum* Lindl. *Rhod Agric J* 62;44–45, 1965

Simic WJ: Solanine poisoning in swine. *Vet Med* 38;353–354, 1943

Sinden SL: Potato glycoalkaloids. *Acta Hortic* 207;41–47, 1987

Skliar MI, Boland RL, Mourino A, Tojo G: Isolation and identification of vitamin D_3, 25-hydroxyvitamin D_3, 1,25-dihydroxyvitamin D_3, and 1,24,25-trihydroxyvitamin D_3 in *Solanum malacoxylon* incubated in ruminal fluid. *J Steroid Biochem Mol Biol* 43;677–682, 1992

Slanina P: Solanine (glycoalkaloids) in potatoes: toxicological evaluation. *Food Chem Toxicol* 28;759–761, 1990

Sloan JW, Martin WR, Bostwick M, Hook R, Wala E: The comparative binding characteristics of nicotine ligands. *Pharmacol Biochem Behav* 30;255–267, 1988

Smith CR: Occurrence of anabasine in *Nicotiana glauca* R.Grah. (Solanaceae). *J Am Chem Soc* 57;959–960, 1935

Smith HC, Tausig RA, Peterson PC: Deadly nightshade poisoning in swine. *J Am Vet Med Assoc* 129;116–117, 1956

Spainhour CB, Fiske RA, Flory W, Reagor JC: A toxicological investigation of the garden shrub *Brunfelsia calycina* var. *floribunda* (yesterday-today-and-tomorrow) in three species. *J Vet Diagn Invest* 2;3–8, 1990

Stevens OA: *Poisonous Plants and Plant Products.* N D Agric Exp Stn Bull 265, 1933

Steyn DG: Recent investigations into the toxicity of plants. *J S Afr Med Vet Assoc* 2;115–117, 1931

Steyn DG: *The Toxicology of Plants in South Africa.* Central News Agency, South Africa, 1934

Szallasi A, Blumberg PM: Resiniferatoxin, a phorbol-related diterpene, acts as an ultrapotent analog of capsaicin, the irritant constituent in red pepper. *Neuroscience* 30;515–520, 1989

Szolcsanyi J, Szallasi A, Szallasi Z, Joo F, Blumberg PM: Resiniferatoxin, an ultrapotent neurotoxin of capsaicin-sensitive primary afferent neurons. *Ann N Y Acad Sci* 632;473–474, 1991

Taylor P: Agents acting at the neuromuscular junction and autonomic ganglia. In *Goodman and Gilman's The Pharmacological Basis of Therapeutics,* 8th ed, Gillman AG, Rall TW, Nies A, Taylor P eds, McGraw-Hill, New York, pp166–186, 1990

Tehon LR, Morrill CC, Graham R: *Illinois Plants Poisonous to Livestock.* Ill Agric Exp Stn Serv Circ 599, 1946

Tepfer D, Goldmann A, Pamboukdjian N, Maille M, Lepingle A, Chevalier D, Denarie J, Rosenberg C: A plasmid of *Rhizobium meliloti* 41 encodes catabolism of two compounds from root exudate of *Calystegium sepium. J Bacteriol* 170;1153–1161, 1988

Tippo O, Stern WL: *Humanistic Botany.* WW Norton, New York, 1977

Toda N, Usui H, Nishino N, Fujiwara M: Cardiovascular effects of capsaicin in dogs and rabbits. *J Pharmacol Exp Ther* 181;512–521, 1972

Tokarnia CH, Gava A, Stolf L, Peixoto PV: Experimental poisoning in cattle by *Brunfelsia pauciflora* (Solanaceae). *Pesq Vet Bras* 11;9–12, 1991

Upshall DG: Correlation of chick embryo teratogenicity with the nicotinic activity of a series of tetrahydropyrimidines. *Teratology* 5;287–294, 1972

Vieira RF, Freire de Carvalho LDA: Espécies do gênero *Solanum* productoras de alcalóides esteroidais. *Rev Bras Farm* 74;97–111, 1993

Vig MM: Nicotine poisoning in a dog. *Vet Hum Toxicol* 32;573–575, 1990

Wada S, Yoshimitsu T, Koga N, Yamada H, Oguri K, Yoshimura H: Metabolism in vivo of the tropane alkaloid, scopolamine, in several mammalian species. *Xenobiotica* 21;1289–1300, 1991

Wanwimolruk S, Nyika S, Kepple M, Ferry DG, Clark CR: Effects of capsaicin on the pharmacokinetics of antipyrine, theophylline, and quinine in rats. *J Pharm Pharmacol* 45;618–621, 1993

Wasserman RH, Corradino RA, Krook LP: *Cestrum diurnum:* a domestic plant with 1,25-dihydroxycholecalciferol-like activity. *Biochem Biophys Res Commun* 62;85–91, 1975

Wasserman RH, Corradino RA, Krook L, Hughes MR, Haussler MR: Studies on the 1α,25-dihydroxycholecalciferol-like activity in a calcinogenic plant, *Cestrum diurnum,* in the chick. *J Nutr* 106;457–465, 1976

Watson SJ, Nash MJ: *The Conservation of Grass and Forage Crops.* Oliver & Boyd, Edinburgh, 1960

Watt JM, Breyer-Brandwijk MG: *The Medicinal and Poisonous Plants of Southern and Eastern Africa.* E & S Livingston, Edinburgh, pp942–1012, 1962

Watt JM, Heimann HL, Epstein E: Solanocarpine—a new alkaloid with cardiac action. *Q J Pharm Pharmacol* 5;649–656, 1932

Weidner J: Possible harmful effects from a capsaicin base aerosol dog repellent. *Vet Hum Toxicol* 22;18–19, 1980

Weissenberg M: Calcinogenic glycosides. In *Toxicants of Plant Origin,* vol 2, *Glycosides,* Cheeke PR ed, CRC Press, Boca Raton, Fla, pp201–238, 1989

Weller RF, Phipps RH: A review of black nightshade (*Solanum nigrum* L.). *Prot Ecol* 1;121–139, 1978–1979

Weller RF, Phipps RH: Herbicidal control of black nightshade (*Solanum nigrum* L.) in forage maize. *Prot Ecol* 2;127–133, 1980

Wildmann J, Vetter W, Ranalder UB, Schmidt K, Maurer R, Mohler H: Occurrence of pharmacologically active benzodiazepines in trace amounts in wheat and potato. *Biochem Pharmacol* 37;3549–3559, 1988

Willaman JJ, Schubert BG: *Alkaloid-Bearing Plants and Their Contained Alkaloids.* USDA Tech Bull 1234, pp216–230, 1961

Williams S, Scott P: The toxicity of *Datura stramonium* (thorn apple) to horses. *N Z Vet J* 32;47, 1984
Worker BA, Carrillo BJ: "Enteque Seco" calcification and wasting in grazing animals in the Argentine. *Nature* 215;72–74, 1967
Worthington TR, Nelson EP, Bryant MJ: Toxicity of thornapple (*Datura stramonium* L.) seeds to the pig. *Vet Rec* 108;208–211, 1981
Zeringue HJ Jr: The accumulation of five fluorescent compounds in the cotton leaf induced by cell-free extracts of *Aspergillus flavus*. *Phytochemistry* 23;2501–2503, 1984

Taxaceae A.Gray

70

Distributed primarily in the Northern Hemisphere, the Taxaceae, or yew family, comprises 5 genera and 17–20 species. In North America, 2 genera—*Taxus* and *Torreya*—and 7–12 species, both native and introduced, are present. Both genera are widely cultivated as ornamental shrubs. The hard but easily worked wood of *Taxus* is used for cabinets, novelty toys, and archery bows, and the Asian species of *Torreya* yield edible seeds and cooking oil. Only *Taxus* is of toxicologic significance.

■ Yew Family

Taxus

Plants trees or shrubs; evergreen; resinous or not resinous; aromatic or not aromatic; producing pollen cones and seed cones; dioecious or monoecious. **Stems** with lateral branches well developed. **Bark** scaly or fissured. **Leaves** needlelike; persistent for several years; simple; alternate; spirally arranged but often twisted and appearing 2-ranked; linear to linear-lanceolate; decurrent. **Pollen Cones** produced annually; axillary; solitary or clustered; globose to ovoid; microsporophylls peltate; microsporangia 2–16 per microsporophyll. **Seed Cones** reduced to 1 or 2 ovules arising from inconspicuous decussate bracts; produced annually; maturing in 1–2 years; axillary. **Seeds** 1; nutlike; erect; partially or wholly enclosed by fleshy or leathery, green or red-orange to scarlet-red aril.

■ evergreen trees or shrubs, producing pollen and seed cones; leaves needlelike, simple, alternate, sometimes appearing 2-ranked; seeds enclosed by arils

Taxus L.

Taxonomy and Morphology—Its name an ancient Latin one for "yew," *Taxus* comprises 6–10 morphologically quite similar but geographically separated species. Because of their similarity, some taxonomists have even treated them all as subspecies of a single species. In North America, the genus is represented by 3 natives, several introduced ornamentals, hybrids, and numerous cultivars. The following are representative of the native and commonly encountered introduced taxa:

■ *T. baccata*
T. brevifolia
T. canadensis
T. chinensis
T. cuspidata
T. floridana
T. xmedia

ground hemlock
yew

Native Species

T. brevifolia Nutt.	western yew, Pacific yew
T. canadensis Marshall	Canada yew, ground hemlock, American yew
T. floridana Nutt. ex Chapm.	Florida yew

Introduced Ornamental Species

T. baccata L.	English yew
T. chinensis (Pilg.) Rehder	Chinese yew
T. cuspidata Siebold & Zucc.	Japanese yew
T. x*media* Rehder (*T. baccata* x *T. cuspidata*)	Hicks yew

■ trees or shrubs; leaves flexible; pollen cones globose, yellowish; seeds partially enclosed in red-orange to scarlet-red, fleshy cup-shaped arils

Plants trees or shrubs; not resinous; not aromatic. **Stems** 2–25 m tall; branches ascending to drooping; twigs irregularly alternate; green or yellow-green when young; reddish brown when old. **Bark** scaly; reddish brown. **Leaves** flexible; stomates abaxial, in 2 broad, pale bands; apices mucronate; resin canals absent. **Pollen Cones** globose; yellowish. **Seeds** maturing in 1 year; brown; partially enclosed by arils; arils cup-shaped, fleshy, red-orange to scarlet-red.

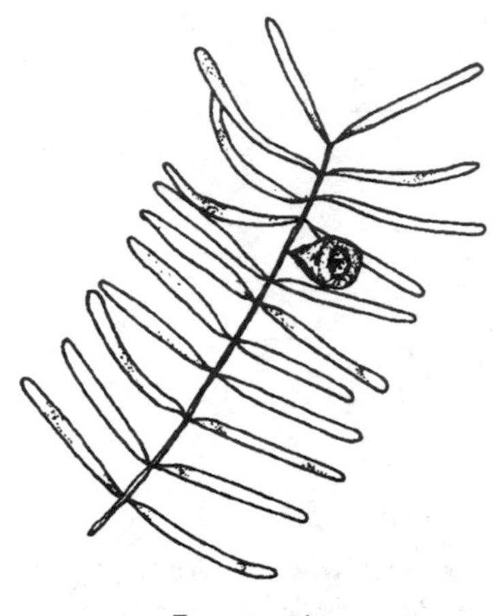

Taxus cuspidata

1. Leaf apices acuminate. Scales of winter buds not keeled.
   2. Plants trees . . . . *T. baccata*
   2. Plants shrubs . . . . *T. chinensis*
1. Leaf apices mucronate. Scales of winter buds keeled.
   3. Leaves not borne in one plane, but rather forming V-shaped trough.
      4. Mature branchlets reddish brown. Apices of winter bud scales acute . . . . *T. cuspidata*
      4. Mature branchlets olive-green. Apices of winter bud scales obtuse . . . . *T.* x*media*
   3. Leaves borne in one plane.
      5. Plants of northeastern North America; low shrubs; monoecious. Stems spreading to prostrate . . . . *T. canadensis*
      5. Plants of western North America or Florida; tall shrubs or small trees; dioecious. Stems erect or ascending.
         6. Plants of western North America. Upper leaf surfaces yellow-green. Seeds ovoid; 2-to-4-angled . . . *T. brevifolia*
         6. Plants of northwestern Florida. Upper leaf surfaces dark green. Seeds ellipsoid; not angled . . . . *T. floridana*

Torreya, the other genus of the family in North America, is distinguished from *Taxus* by its stiff, spine-tipped leaves, which have resin canals, and leathery, resinous arils, which are green or green with purple streaks and completely enclose the seeds.

■ ornamentals; forests and woodlands

Distribution and Habitat—Yews are often planted near fences or buildings that provide some shelter and shade. They are widely planted over North America except perhaps for the desert Southwest. In more northerly areas they are commonly used as hedges or screens. *Taxus cuspidata* is more winter-hardy and is more common in the north, whereas *T. baccata* and *T.* x*media* are typically encountered in the south. *Taxus brevifolia* grows to about 20 m tall and typically is found as an understory tree of forests in the Northwest from British Columbia to Montana and California. *Taxus canadensis* is a straggly shrub of woodlands. It is very cold-hardy and ranges from the Ohio River valley to the far north of eastern Canada. *Taxus floridana* is a small tree of localized distribution along the Apalachicola River in Florida.

■ considered a magical, mystical plant; toxic to most animal species, especially horses

Disease Problems—Long treasured for making bows, arrows, and other weapons, the yew is perhaps more associated with superstition than any other tree. It was well known as a magical, sacred, and toxic plant in antiquity. The Egyptians, Greeks, and Romans used it as a symbol of mourning; its wood was used in funeral pyres. In Europe, beginning with the Druids, it was used in sacred rites and thereafter was traditionally planted in cemeteries and churchyards, practices that further linked it with the morbid and mystical. Even its shade was considered dangerous to anyone sleeping underneath it (Voliotis 1986). It has been the subject of graphic lines of poetry (Nicholson 1992).

Having a notorious reputation in Europe for causing sudden death in cattle and horses (Masheter 1937; Persson et al. 1988), yew is reputed to be the most dangerous shrub in England and probably is among the most hazardous in North America as well. Many intoxications have occurred following the disposal of shrubbery trimmings in pastures where they were subsequently eaten by livestock. In an example case, 18 cows and calves escaped through a fence into a house yard and stripped the leaves from several large shrubs of *T. cuspidata* as high as they could reach. One cow and 2 calves died the same morning; subsequently 1 cow died each of the next 3 mornings, for a total of 6 deaths. There were no premonitory signs and no pathologic abnormalities, only large amounts of the leaves in the ruminal ingesta. The poisoning occurred in February, and the cattle apparently were attracted to the evergreen foliage, which they ate avidly.

Approximately 0.1% b.w. of leaves of *T. cuspidata* may be lethal in a mature horse and several times this amount in adult cattle (Lowe et al. 1970; Beal 1975; Alden et al. 1977). Even less of *T. baccata,* perhaps half of the above amounts, may be lethal (Humphries 1988). Panter and co-workers (1993) estimated the toxic dose of *T. baccata* for cattle to be 0.36–0.7 g fresh plant/kg b.w., or 0.036–0.07% b.w. In adult humans, doses of 50–100 g of yew needles are considered lethal (Frohne & Pribilla 1965). Fresh plant material is sometimes considered to be of low palatability because of the volatile irritant oil present, but animals frequently eat the leaves readily, especially in the winter, a time when the evergreen foliage is presumably more attractive to livestock (Edwards 1949; Kerr & Edwards 1981; Karns 1983). Dried or stored leaves remain toxic for at least several months and represent a serious hazard when discarded where they can be eaten by livestock (Alden et al. 1977; Thomsom & Barker 1978; Veatch et al. 1988).

■ leaves, fresh or dried, livestock, 0.05–0.7 g/kg b.w.; humans 50–100 g; seeds toxic but not fruits

Because of the small amounts of plant material required for intoxication, dogs, cats and other house pets may be at risk by chewing the branches (Evans & Cook 1991). Clay (1977) reported a case in which shreds of the bark of *T. cuspidata* were found in the stomach and considered to be the cause of death of a 1-year-old dog. Experimentally, 0.5 g of fruits and spring or fall growth of leaves of *T.* x*media* caused retching, vomiting, and labored respiration in 4 budgerigars (Shropshire et al. 1992). Death occurred in 1 of 2 given the fall growth leaves. Reproductive effects, mainly increased fetal resorptions, also have been noted experimentally in rats given small dosage of *T. baccata* leaf extracts (Garg 1972).

There have been a number of human intoxications, mainly suicides (Frohne & Pribilla 1965; von Dach & Streuli 1988; Sinn & Porterfield 1991; Van Ingen et al. 1992; Musshoff et al. 1993). In most cases the victims are found dead and the diagnosis is based on the discovery of leaf fragments in the digestive tract. In an unusual case involving a 3-year-old boy, death occurred about 4 hours after ingestion of yew berries (Thompson 1868). In contrast to several other boys who ate the berries but were unaffected, the deceased failed to spit out the toxic seeds. Rarely are clinical signs observed and electrocardiographic (ECG) evaluations carried out. However, the most distinctive ECG change is a marked distortion and widening of the QRS complex (von Dach & Streuli 1988; Van Ingen et al. 1992). In spite of the potency of the toxins in *Taxus*, the great majority of incidental ingestions by humans involve insufficient plant material to cause clinical signs. Krenzelok and co-workers (1998) assessed data from U.S. poison control centers for 1985–1994 and found only 30 moderate and 4 major intoxications from among more than 7,000 cases.

Disease Genesis—*Taxus* contains a wide variety of biologically active constituents, including taxane derivatives, cyanogenic glycosides, lignins, flavonoids, molting hormones, ephedrine, and a volatile irritant oil (Khan & Parveen 1987). Of these, only the taxane derivatives are of concern toxicologically. Cyanide is not consistently

■ toxicity due to taxines; taxols, cyanogenic glycoside also present

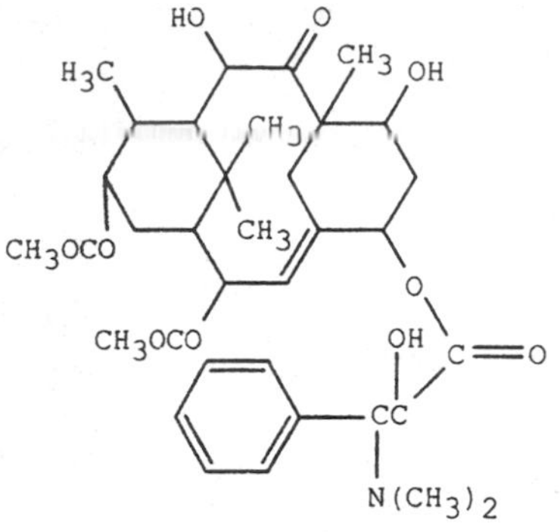

Taxine A

- taxine effects mainly on heart; interferes with cardiac conduction; heart block

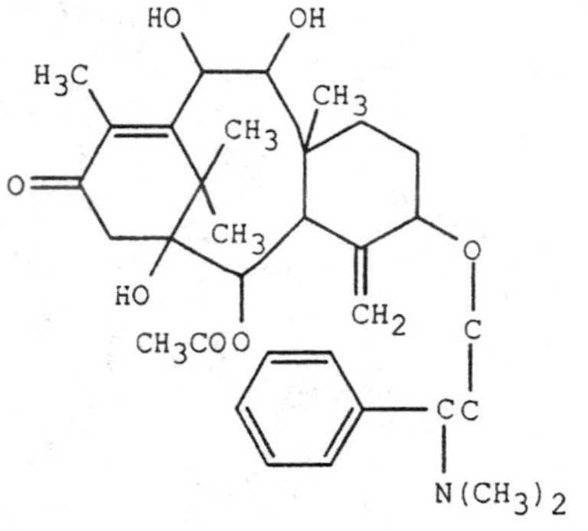

Taxine B

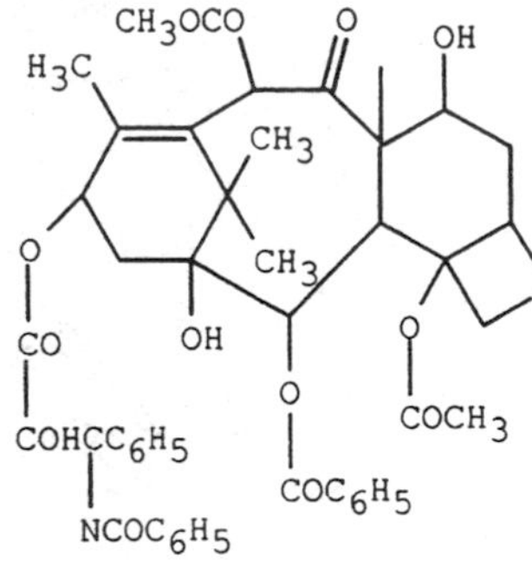

Taxol

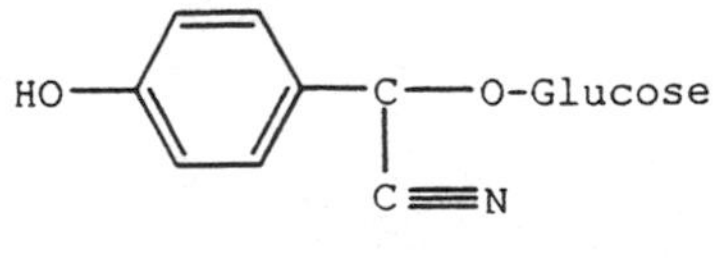

Taxiphyllin

present and does not seem to play a prominent role in intoxications. The most important of the extensive series of taxane derivatives are the taxines and taxols; both are nitrogenous ester alkaloids, sometimes called pseudoalkaloids. Concentrations of the taxols are quite low, and intoxications are due to the cardiotoxic taxines. Originally, the structures of taxine I and II were hypothesized from the structures of more-stable cinnaminic acid derivatives (Lythgoe 1968). Eventually, 2 of the 11 or more taxines were specifically identified, their structures were determined, and they were named taxines A and B (Graf 1956; Graf et al. 1986; Ettouati et al. 1991; Appendino et al. 1993). Taxine B appears to be the equivalent of the hypothesized taxine I (Graf et al. 1986). Taxines occur in differing proportions in the various species of the genus, with taxine B and isotaxine B (acetyl group on C-9 rather than C-10) reported as the predominant forms in *T. baccata* but not necessarily in *T. cuspidata* (Lythgoe 1968; Appling 1984; Graf et al. 1986; Jenniskens et al. 1996; Adeline et al. 1997; Kite et al. 2000).

Taxine B appears to be much more toxic than taxine A (Bauereis & Steiert 1959). Little is known of other closely related taxines that have been identified in *T. baccata* (Appendino 1993; Kite et al. 2000). Many of the other taxane derivatives are non-nitrogenous (Khan & Parveen 1987). In vitro studies indicate that taxine B and its cinnamate derivative cause decreased cardiac contractility, maximal rate of depolarization, and coronary blood flow (Alloatti et al. 1996). In vivo, taxines markedly slow atrial and especially ventricular rates, the ventricles eventually stopping in diastole at death (Bryan-Brown 1932; Tekol 1985). Thus, the effects seem to be mainly on AV conduction possibly via inhibition of sodium and calcium currents in cardiac cells and/or possible potassium channel effects (Tekol & Kameyama 1987). This seems to be borne out by the marked slowing of the heart and the distinctive widening of the QRS complexes (Frohne & Pribilla 1965; Schulte 1975; von Dach & Streuli 1988). In addition, in vitro studies indicate that taxine has selective negative inotropic and chronotropic Ca^{2+} channel inhibitory actions on the heart similar to but less potent than those of verapamil (Tekol & Gogusten 1999).

Although taxines are present in all species of *Taxus*, they are quite low in *T. canadensis* and probably of little hazard. In addition, both total concentration and relative proportions of the taxines may vary considerably in other species, especially in *T. cuspidata* (Appling 1984). All parts of the plant are toxic except the fleshy aril of the fruit. Taxine is quite toxic, the lethal dose ranging from 4 to 20 mg/kg in various laboratory animals (Bryan-Brown 1932; Tekol 1991).

Unfortunately there has been a confusing proliferation of nomenclature for these complex three-dimensional caged structures. In addition to the names *taxine* and *taxol*, names such as *taxicin, taxacin, taxagifine, taxinin, taxusin, taxanine, taxicatin, taxinol,* and *taxininol* have been applied to various isolations (Miller 1980; Khan & Parveen 1987). Some of these compounds may be extraction artifacts. Taxicins are the nonesterified cage structures of taxines, and taxinines and taxinols (or taxininol) are breakdown products of taxines. Other active and inactive taxane derivatives have been isolated (some in cell culture) in the search for antineoplastic compounds akin to taxol (Yoshizaki et al. 1988; Beutler et al. 1991; Ma et al. 1994).

The taxols and their precursor baccatins are of great medical interest because of their antineoplastic effects. These compounds inhibit mitosis by promoting microtubular assembly and stabilizing the polymerized microtubules (Schiff et al. 1979; Schiff & Horwitz 1980). Thus, the microtubules cannot dissociate into tubulin subunits. As might be expected, this is a beneficial effect when attempting to control neoplasms, but at high dosage these compounds are neurotoxic because they interfere with microtubular axonal transport (Anthony & Graham 1991). Their overall effect is similar to, but the mechanism is opposite, that of the other mitotic inhibitors such as colchicine, podophyllin, vinblastine, and vincristine. Because of its uniqueness, the mechanism is exploited in studies of microtubules (Vallee 1986). Cells of a broad range of organisms from mammals to fungi are affected (Young et al. 1992).

■ taxol effects, mitosis inhibition, antineoplastic, toxic via long-term action on axonal structure

Taxol has been extracted primarily from the bark of *T. brevifolia,* but it and its baccatin precursor also are found in most of the other species of *Taxus* as well (Khan & Parveen 1987; Witherup et al. 1990). Concentrations are quite low even in the bark, ranging from 0.015 to 0.076% d.w. (Vidensek et al. 1990; Kelsey & Vance 1992). Of interest are the slightly lower concentrations (0.001–0.01%) in the leaves of species such as *T.* x*media, T. canadensis,* and *T. cuspidata* (Witherup et al. 1990).

It is of interest that although the genus appears to be highly toxic to many animal species, including livestock and poultry (Cooper & Johnson 1984), wild ruminants, such as moose and deer, browse extensively on some species of *Taxus* without apparent ill effects (Risenhoover & Maass 1987; Conover 1989). In the case of moose, the species browsed is *T. canadensis,* which as pointed out above, is of little hazard to any animal. Deer, however, during the winter often avidly browse the new shoots of other species, especially the cultivated ones, typically without adverse effects. This may reflect the generally lower concentrations of the various alkaloids in newer leaves than in older leaves and possibly the development of tolerance as well (Bryan-Brown 1932; Appling 1984; Kelsey & Vance 1992).

Alkaloids of *Taxus* are biotransformed in the liver and, at least in the case of taxol, are excreted in bile as the parent compound or one of numerous metabolites (Monsarrat et al. 1990). These alkaloids also apparently increase some hepatic biotransformation pathways, given that mixed-function oxidase activity is increased in rats fed *T. cuspidata* (Appling 1984). That they may also influence the biotransformation of xenobiotics is shown by the increase in toxicity of pyrethroid insecticides for weevils feeding on *T.* x*media* when compared with those feeding on other plants (Shanks & Chamberlain 1988). In addition, rats fed *Taxus* for several days sleep longer than otherwise when pentobarbital is administered (Appling 1984). *Taxus* therefore seems to have possible inhibitory effects on hepatic biotransformation.

Clinical Signs—In most cases, few premonitory signs are noted prior to sudden death of the animal. Even when the animal is closely observed, the signs may exist only for several minutes. Horses typically collapse and die in a matter of minutes, preceded only by a few minutes of trembling and quivering. Often they appear to drop suddenly as if they had been shot. Signs likely to be seen in cattle include depression, weakness, nervousness, tremors, ataxia, hypothermia, bradycardia, jugular distension and pulsation, and dyspnea with an expiratory grunt. The stress of forced movement or restraint may precipitate sudden collapse and death (Masheter 1937; Knowles 1949; Alden et al. 1977). The lethal dose is so small that the animal is able to readily consume an amount much greater than that needed to cause death. Death may occur a few minutes after eating the plant or up to several days later (Knowles 1949; Alden et al. 1977; Karns 1983; Ogden 1988). The presence of clinical signs, however, is not always an indicator of impending death, because in a few cases prominent signs were seen without subsequent development of more-severe signs or death (Masheter 1937; Knowles 1949; Casteel & Cook 1985).

■ few signs; horses may tremble, then quickly collapse and die; cattle if closely observed may show vague signs but are most commonly found dead

humans experience dizziness, weakness, slow heart rate, collapse

Signs in humans include dizziness, unconsciousness, marked pupillary dilation, reddening of the lips, slow pulse, and circulatory problems that result in weakness, especially of the legs (Frohne & Pribilla 1965; von Dach & Streuli 1988).

In contrast to the preponderance of cases in which signs of cardiotoxicity were reported, signs suggestive of central nervous system involvement, including aggressiveness, were noted in a case in which a dog was thought to have chewed the leaves of *T. cuspidata* (Evans & Cook 1991).

■ no lesions; leaf fragments in rumen; analysis of stomach contents for taxine or other compounds

Pathology—There are few lesions other than nonspecific organ congestion, including moderate pulmonary congestion and edema, which are changes typical of circulatory disturbances. In animals in which death is delayed for 1 or more days, reddening and edema of the mucosa of the stomach and small intestine may be present. Perhaps the most useful diagnostic techniques are examination of the stomach and/or intestinal contents for leaf fragments and/or extraction and analysis of stomach contents for taxine fragments by gas chromatography/mass spectrometry (GC/MS) analysis with or without acetylation or by using liquid chromatography/mass spectrometry (LC/MS) (Smith 1989; Lang et al. 1997; Kite et al. 2000). Alternatively, 3,5-dimethoxyphenol, the aglycone of taxicatine, has been suggested as a marker in stomach contents or blood as an indicator of yew ingestion (Musshoff et al. 1993). The leaf fragments may not be identifiable in the finely chewed ingesta of the horse, and extraction and analysis may be required (Rook 1994).

■ atropine, activated charcoal; in humans, lidocaine or external cardiac pacing

Treatment—Because of the severe problem with cardiac conduction, death appears to be due to peracute cardiac failure. Although no antidote is recognized, cardiac conduction enhancers such as atropine may be helpful in cases in which ingestion is observed but signs have not yet occurred. Because the lethal effects appear to be related to ventricular rate and rhythm abnormalities, lidocaine has been used with success in a human (von Dach & Streuli 1988). This approach may not be as useful in livestock, because it would require continuous monitoring and repeated i.v. administration of lidocaine. External cardiac pacing has been used with apparent success in humans to overcome the conduction inhibition (Cummins et al. 1990). In livestock, the amount of *Taxus* ingested may be quite large, and once signs occur, it is generally too late for therapy, although activated charcoal or rumenotomy may be of value in some cases. Occasionally, digestive tract irritation may occur in cattle because of the volatile oil, in which case treatment is symptomatic.

References

Adeline MT, Wang XP, Poupat C, Ahond A, Potier P: Evaluation of taxoids from *Taxus* sp. crude extracts by high performance liquid chromatography. *J Liq Chromatogr Relat Technol* 20;3135–3145, 1997

Alden CL, Fosnaugh CJ, Smith JB, Mohan R: Japanese yew poisoning of large domestic animals in the Midwest. *J Am Vet Med Assoc* 170;314–316, 1977

Alloatti G, Penna C, Levi RC, Gallo MP, Appendino G, Fenoglio I: Effects of yew alkaloids and related compounds on guinea-pig isolated perfused heart and papillary muscle. *Life Sci* 58;845–854, 1996

Anthony DC, Graham DG: Toxic responses of the nervous system. In *Casarett and Doull's Toxicology,* 4th ed, Amdur MO, Doull J, Klaassen CD eds, Pergamon Press, New York, pp407–429, 1991

Appendino G, Ozen HC, Fanoglis I, Gariboldi P, Gabetta B, Bombardelli E: Pseudoalkaloid taxanes from *Taxus baccata. Phytochemistry* 33;1521–1523, 1993

Appling JW: Factors affecting the outcome of animal poisoning by species of *Taxus,* the yew. PhD thesis, Cornell Univ, 1984

Bauereis R, Steiert W: Pharmacological properties of taxine A and B. *Arzneim-Forsch* 9;77–79, 1959

Beal J: Poisonous properties of *Taxus.* Paper presented at International Taxus Symposium, Ohio Agricultural Research and Development Center, Wooster, 1975

Beutler JA, Chmurny GM, Look SA, Witherup KM: Taxinine M, a new tetracyclic taxane from *Taxus brevifolia. J Nat Prod* 54;893–897, 1991

Bryan-Brown T: The pharmacological actions of taxine. *Q J Pharm Pharmacol* 5;203–219, 1932

Casteel SW, Cook WO: Japanese yew poisoning in ruminants. *Mod Vet Pract* 66;875–877, 1985

Clay BR: Poisoning and injury by plants. In *Current Veterinary Therapy,* vol 6, *Small Animal Practice,* Kirk RW ed, Saunders, Philadelphia, pp179–184, 1977

Conover MR: Relationships between characteristics of nurseries and deer browsing. *Wildl Soc Bull* 17;414–418, 1989

Cooper MR, Johnson AW: *Poisonous Plants of Britain and Their Effects on Animals and Man.* Ref Book 161. Her Majesty's Stationery Office, London, pp223–226, 1984

Cummins RO, Haulman J, Quan L, Reid-Graves J, Peterson D, Horan S: Near-fatal yew berry intoxication treated with external cardiac pacing and digoxin-specific Fab antibody fragments. *Ann Emerg Med* 19;38–43, 1990

Edwards CM: Some observations on plant poisoning in grazing animals. *Vet Rec* 61; 864–865, 1949

Ettouati L, Ahond A, Poupat C, Potier P: Révision structurale de la taxine B, alcaloïde majoritaire des feuilles de l'if d'Europe, *Taxus baccata. J Nat Prod* 54;1455–1458, 1991

Evans KL, Cook JR: Japanese yew poisoning in a dog. *J Am Anim Hosp Assoc* 27;300–302, 1991

Frohne D, Pribilla O: Tödliche vergiftung mit *Taxus baccata. Arch Toxikol* 21;150–162, 1965

Garg SK: Antifertility screening of plants. 8. Investigation on *Taxus baccata* Linn. leaves. *Indian J Med Res* 60;159–163, 1972

Graf E: Zur chemie des taxins. *Angew Chem* 68;249–250, 1956

Graf E, Weinandy S, Koch B, Breitmaier E: ^{13}C-NMR-untersuchung von taxin B aus *Taxus baccata* L. *Liebigs Ann Chem* 1986;1147–1151, 1986

Humphries DJ: *Veterinary Toxicology,* 3d ed. Bailliere Tindall, London, pp276–277, 1988

Jenniskens LHD, van Rozendaal ELM, van Beek TA, Wiegerinck PHG, Scheeren HW: Identification of six taxine alkaloids from *Taxus baccata* needles. *J Nat Prod* 59;117–123, 1996

Karns PA: Intoxication in horses due to ingestion of Japanese yew *(Taxus cuspidata). Equine Pract* 5;12–15, 1983

Kelsey RG, Vance NC: Taxol and cephalomannine concentrations in the foliage and bark of shade-grown sun-exposed *Taxus brevifolia* trees. *J Nat Prod* 55;912–917, 1992

Kerr LA, Edwards WC: Japanese yew: a toxic ornamental shrub. *Vet Med/Small Anim Clin* 76;1339–1340, 1981

Khan NUD, Parveen N: The constituents of the genus *Taxus. J Sci Indus Res* 46;512–516, 1987

Kite GC, Lawrence TJ, Dauncey EA: Detecting *Taxus* poisoning in horses by using liquid chromatography/mass spectrometry. *Vet Human Toxicol* 42;151–154, 2000

Knowles JW: From my case book. *Vet Rec* 61;421–422, 1949

Krenzelok EP, Jacobsen TD, Aronis J: Is the yew really poisonous to you? *J Toxicol Clin Toxicol* 36;219–223, 1998

Lang DG, Smith RA, Miller RE: Detecting *Taxus* poisoning using GC/MS [letter]. *Vet Hum Toxicol* 39;314, 1997

Lowe JE, Hintz HF, Schryver HF, Kingsbury JM: *Taxus cuspidata* (Japanese yew) poisoning in horses. *Cornell Vet* 60;36–39, 1970

Lythgoe B: The *Taxus* alkaloids. In *The Alkaloids,* Manske RHF ed, Academic Press, New York, pp597–626, 1968

Ma W, Park GL, Gomez GA, Nieder MH, Adams TL, Aynsley JS, Sahai OP, Smith RJ, Stahlhut RW, Hylands PJ, Bitsch F, Shackleton C: New bioactive taxoids from cell cultures of *Taxus baccata. J Nat Prod* 57;116–122, 1994

Masheter JWH: Yew poisoning in cattle [letter]. *Vet Rec* 49;265–266, 1937

Miller RW: A brief survey of *Taxus* alkaloids and other taxane derivatives. *J Nat Prod* 43;425–437, 1980

Monsarrat B, Mariel F, Cros S, Gares M, Guenard D, Gueritte-Voegelein F, Wright M: Taxol metabolism: isolation and identification of three major metabolites of taxol in rat bile. *Drug Metab Dispos* 18;895–901, 1990

Musshoff F, Jacob B, Fowinkel C, Daldrup T: Suicidal yew leaf ingestion—phloroglucin-dimethylether (3,5-dimethoxyphenol) as a marker for poisoning from *Taxus baccata*. *Int J Leg Med* 106;45–50, 1993

Nicholson R: Death and taxus. *Nat Hist* 101;20–23, 1992

Ogden L: *Taxus* (yews)—a highly toxic plant [letter]. *Vet Hum Toxicol* 30;563–564, 1988

Panter KE, Molyneux RJ, Smart RA, Mitchell L, Hansen S: English yew poisoning in 43 cattle. *J Am Vet Med Assoc* 202;1476–1477, 1993

Persson L, Thornberg CG, Vesterlund S: Yew poisoning: sudden death of horses and cattle. *Sven Veterinartidn* 40;27–28, 1988

Risenhoover KL, Maass SA: The influence of moose on the composition and structure of Isle Royale forests. *Can J Forest Res* 17;357–364, 1987

Rook JS: Japanese yew toxicity. *Vet Med* 89;950–951, 1994

Schiff PB, Horwitz SB: Taxol stabilizes microtubules in mouse fibroblast cells. *Proc Nat Acad Sci* 77;1561–1565, 1980

Schiff PB, Fant J, Horwitz SB: Promotion of microtubule assembly in vitro by taxol. *Nature* 277;665–667, 1979

Schulte T: Lethal intoxication with leaves of the yew tree *(Taxus baccata)*. *Arch Toxicol* 34;153–158, 1975

Shanks CH Jr, Chamberlain JD: Effect of *Taxus* foliage and extract on the toxicity of some pyrethroid insecticides to adult black vine weevil (*Coleoptera:* Curculionidae). *J Econ Entomol* 81;98–101, 1988

Shropshire CM, Stauber E, Arai M: Evaluation of selected plants for acute toxicosis in budgerigars. *J Am Vet Med Assoc* 200;936–939, 1992

Sinn LE, Porterfield JF: Fatal taxine poisoning from yew leaf ingestion. *J Forensic Sci* 36;599–601, 1991

Smith RA: Comments on diagnosis of intoxication due to *Taxus* [letter]. *Vet Hum Toxicol* 31;177, 1989

Tekol Y: Negative chronotropic and atrioventricular blocking effects of taxine on isolated frog heart and its acute toxicity in mice. *Planta Med* 51;357–360, 1985

Tekol Y: Acute toxicity of taxine in mice and rats. *Vet Hum Toxicol* 33;337–338, 1991

Tekol Y, Gogusten B: Comparative determination of the cardioselectivity of taxine and verapamil in the isolated aorta, atrium, and jejunum preparations of rabbits. *Arzneimittelforschung* 49;673–678, 1999

Tekol Y, Kameyama M: Electrophysiological studies of the mechanism of action of the yew toxin taxine on the heart. *Arzneim-Forsch* 37;428–431, 1987

Thompson J: Poisoning by yew-berries [letter]. *Lancet* 2;530, 1868

Thomson GW, Barker IK: Japanese yew *(Taxus cuspidata)* poisoning in cattle. *Can Vet J* 19;320–321,1978

Vallee RB: Purification of brain microtubules and microtubule-associated protein 1 using taxol. *Methods Enzymol* 134;104–115, 1986

Van Ingen G, Visser R, Peltenburg H, Van Der Ark AM, Voortman M: Sudden unexpected death due to *Taxus* poisoning: a report of five cases, with review of the literature. *Forensic Sci Int* 56;81–87, 1992

Veatch JK, Reid FM, Kennedy GA: Differentiating yew poisoning from other toxicoses. *Vet Med* 83;298–300, 1988

Vidensek N, Campbell A, Carlson C: Taxol content in bark, wood, root, leaf, twig, and seedling from several *Taxus* species. *J Nat Prod* 53;1609–1610, 1990

Voliotis D: Historical and environmental significance of the yew (*Taxus baccata* L.). *Isr J Bot* 35;47–52,1986 von Dach B, Streuli RA: Lidocainbehandlung einer vergiftung mit eibennadeln (*Taxus baccata* L.). *Schweiz Med Wochenschr* 118;1113–1116, 1988

Witherup KM, Look SA, Stasko MW, Ghiorzi TJ, Muschik GM: *Taxus* spp. needles contain certain amounts of taxol comparable to the bark of *Taxus brevifolia:* analysis and isolation. *J Nat Prod* 53;1249–1255, 1990

Yoshizaki F, Fukuda M, Hisamichi S, Ishida T, In Y: Structures of taxane diterpenoids from the seeds of Japanese yew, *Taxus cuspidata. Chem Pharm Bull* 36;2098–2102, 1988

Young DH, Michelotti EL, Swindell CS, Krauss NE: Antifungal properties of taxol and various analogues. *Experientia* 48;882–885, 1992

71 Thymelaeaceae Juss.

- Mezereum Family

  Daphne
  Dirca
  Pimelea

Cosmopolitan in distribution but exhibiting greatest diversity in Australia, tropical Africa, and the Mediterranean region, the Thymelaeaceae, commonly known as the mezereum family, comprises 45–50 genera and about 500 species divided into 4 distinctive subfamilies. It is of little economic importance, although some species are popular ornamental shrubs and in some areas the bark is used for fiber. In North America, 6 or 7 genera and 20–25 species, most introduced, are present. Toxicologic problems are associated with 3 genera.

- shrubs or trees; leaves simple, venation pinnate; petals absent

Plants shrubs or trees or rarely herbs; evergreen or deciduous; bearing perfect flowers or dioecious or polygamous. **Leaves** simple; alternate or opposite; venation pinnate; margins entire; stipules absent. **Inflorescences** capitate clusters or umbels; terminal or axillary or extra-axillary; bracts present or absent. **Flowers** perfect or imperfect; perianths in 1-series. **Calyces** radially or slightly bilaterally symmetrical. **Sepals** 4 or 5; small; petaloid. **Petals** absent. **Stamens** 8 or 10; borne at summit of hypanthium. **Pistils** 1; compound, carpels 2–5; stigmas 1, capitate; styles 1 or absent; ovaries superior; locules 1 or rarely 2; placentation apical. **Hypanthia** campanulate or tubular or urceolate; petaloid; colored. **Fruits** drupes or loculicidal capsules or achenes. **Seeds** 1.

Daphne L.

- *D. cneorum*
  D. genkwa
  D. laureola
  D. mezereum
  D. odora

  copse laurel
  daphne
  dwarf bay
  dwarf laurel
  flax olive
  genkwa
  mezereum
  spurge laurel
  wild pepper
  wood laurel

Taxonomy and Morphology—An Old World genus comprising 50–70 species and numerous cultivars, *Daphne* is grown extensively throughout the world for its fragrant and showy flowers (Brickell & Mathew 1976). Its name is an ancient Greek one once used for the unrelated sweet bay *(Laurus nobilis)*. Of toxicologic interest are the following species:

D. cneorum L.	garland daphne
D. genkwa Siebold & Zucc.	genkwa
D. laureola L.	wood laurel, spurge laurel, copse laurel
D. mezereum L.	mezereum, spurge laurel, spurge olive, flax olive, dwarf bay, dwarf laurel, February daphne, wild pepper
D. odora Thunb.	winter daphne

- erect shrubs; leaves glossy green, tough; sepals 4, spreading, white, purple, yellow, orange, or greenish white; fruits leathery or fleshy drupes, red, orange, yellow, or black

Plants shrubs; evergreen or deciduous. **Stems** erect; to 1 m tall; branches flexible, tough. **Leaves** alternate or opposite; short petiolate or sessile; glossy green; tough; blades ovate to oblong or elliptic to oblanceolate; surfaces glabrous or densely indumented. **Inflorescences** capitate clusters; terminal or axillary. **Flowers** perfect. **Sepals**

4; spreading; white or pink or purple or yellow or orange or greenish white. **Stamens** in 2 whorls. **Pistils** with styles short or absent. **Hypanthia** tubular; not constricted above ovary; colors same as sepals. **Fruits** drupes; globose to ovoid; leathery or fleshy; red or orange or yellow or black.

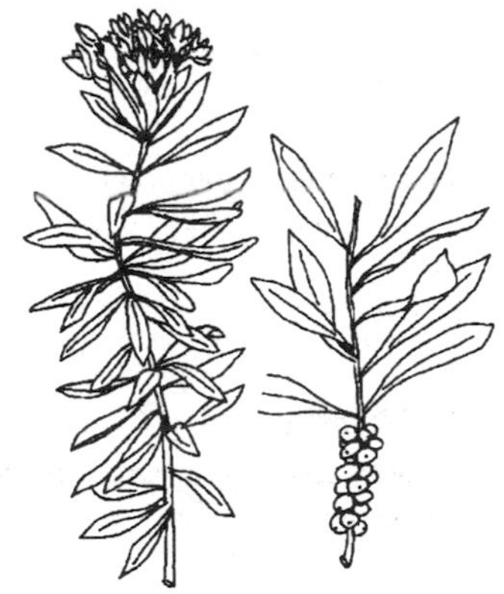

Daphne mezereum

1. Inflorescences terminal.
   2. Leaves 1–2.5 cm long. Abaxial surfaces of hypanthial tube densely pilose . . . . *D. cneorum*
   2. Leaves 2.5–7.5 cm long. Abaxial surfaces of hypanthial tube glabrous or sparsely pilose . . . . *D. odora*
1. Inflorescences axillary.
   3. Leaves evergreen . . . . *D. laureola*
   3. Leaves deciduous.
      4. Leaves alternate . . . . *D. mezereum*
      4. Leaves opposite . . . . *D. genkwa*

■ ornamentals

Distribution and Habitat—*Daphne* exhibits greatest diversity in Europe, northern Africa, and temperate and subtropical Asia. Of the several species cultivated in North America, only *D. mezereum* has escaped and naturalized in the forests of the northeastern corner of the continent.

Daphne mezereum

■ digestive disturbance, dermatitis; neurologic effects in children

Disease Problems—*Daphne* has long been a problem in Europe, for livestock and humans, especially children. The bitter berries and other plant parts are very irritating, and ingestion of only several drupes is thought to present a problem, mainly of the digestive tract (Pammel 1911; Kingsbury 1964). Cases are described in children, in whom seizures, tremors, and death have occurred in addition to the typical digestive tract problems produced (Kingsbury 1964; Cooper & Johnson 1984). Although ingestion of the drupes by children purportedly causes a mortality rate of 30% in Europe (Habermehl 1992), experiences in North America indicate that severe intoxications are rare. This in large measure is due to the bitter, acrid sap, which limits ingestion to a few drupes. In most instances, livestock also eat little of the foliage or drupes. Contact with the foliage can produce a severe dermatitis (Mitchell & Rook 1979).

■ numerous irritant daphnane and tigliane diterpenes

Disease Genesis—Foliage prunings, bark, and berries of species such as *D. mezereum* and *D. odora* contain tricyclic daphnane and tigliane diterpenes, mezerein (an ester of daphnetoxin), odoracin, thymelein, gnidizin, and mezerenol, which are intensely irritating (Evans & Soper 1978; Tyler & Howden 1985; Connolly & Hill 1991). The irritant properties are retained even upon drying. There are also coumarin glycosides present that may contribute to the general unpalatability and possibly to the toxicity (Kingsbury 1964). Because of their taste, only a few berries are eaten by children in

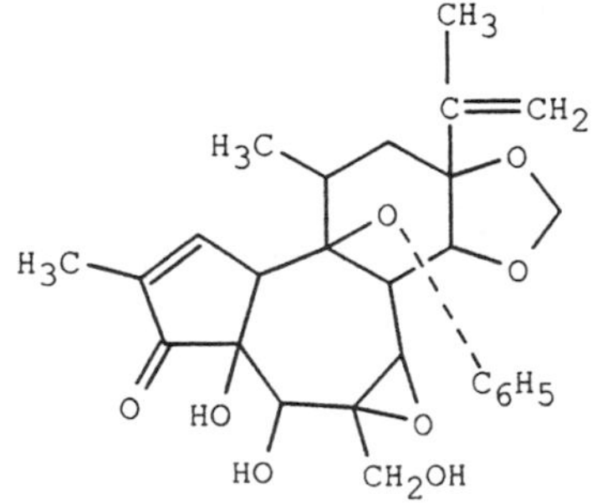

Daphnetoxin

Mezerein

any episode. It is doubtful that intoxications in humans will be of such severity as to be lethal.

- blistering of lips, mouth, and tongue; thirst, colic, nausea, vomiting, diarrhea

Clinical Signs—Wherever contact occurs, such as the mucous membranes of the mouth or eyes or the lower digestive tract, intense reddening and edema occur. There may be blistering of the lips, mouth, and tongue, as well as thirst, colic, nausea, and vomiting, and perhaps diarrhea, if sufficient plant material is ingested and swallowed.

- reddened gut mucosa

treat with activated charcoal, fluids, electrolytes

Pathology and Treatment—The changes are reflective of the severe irritating effects of the diterpenes, i.e., marked reddening and inflammation of the stomach and intestinal mucosa. There may be free blood in the lumina as well. Treatment involves control of the vomiting and diarrhea, oral administration of activated charcoal, and fluid and electrolyte therapy as needed.

Dirca L.

- *D. occidentalis*
D. palustris

American mezereon
leatherwood
moosewood
rope bark
swampwood
thong bark
wicopy

Taxonomy and Morphology—Native to North America and comprising 2 species, *Dirca* is commonly known as leatherwood because of its tough, flexible bark. Its name comes from *dirke,* which was the name of the fountain at Thebes and of mythological significance. The 2 species are both of toxicological significance:

D. occidentalis A.Gray	western leatherwood
D. palustris L.	leatherwood, moosewood, American mezereon, wicopy, rope bark, swampwood, thong bark

- erect, deciduous shrubs, branches yellow-tinged; leaves alternate, green and blue-tinged; flowers clustered, appearing before leaves; sepals 4, minute, pale yellow; fruits ovoid drupes, pale green, yellowish- or reddish-tinged

Plants shrubs; deciduous. **Stems** erect; 100–200 cm tall; profusely branched; branches tough, flexible, yellow-tinged; lateral buds small, enclosed by petiole bases; nodes swollen, socket-jointed by circular scars at start of annual growth. **Leaves** alternate; short-petiolate; green and blue-tinged; blades elliptic-oblong to obovate or ovate; apices acute; bases cuneate. **Inflorescences** capitate clusters; terminal or axillary; appearing before leaves. **Flowers** 2–4 per cluster; sessile or subsessile; perfect. **Sepals** 4; minute; pale yellow. **Stamens** filiform; exserted beyond hypanthia. **Pistils** with styles exserted, filiform. **Hypanthia** funnelform; pale yellow. **Fruits** drupes; ovoid; pale green with yellowish or reddish tinge.

Dirca palustris

1. Flowers subsessile. Bud scales sparsely pubescent.
   Plants of the eastern United States and Canada .... *D. palustris*
1. Flowers sessile. Bud scales densely pubescent.
   Plants of California .... *D. occidentalis*

■ shade, moist soils

Distribution and Habitat—*Dirca occidentalis* occurs in partial shade on moist slopes in the hills around San Francisco Bay in California. *Dirca palustris* is found in rich, moist forests in the eastern half of the continent.

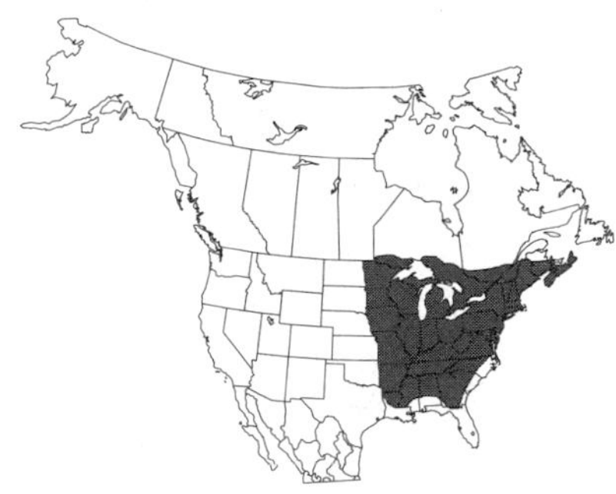

Dirca palustris

Disease Problems and Genesis—Plants of *Dirca* have the potential to cause severe irritation of the digestive tract and urinary bladder very similar to the effects produced by *Daphne*. Consumption of the fruit is associated with dizziness, stupor, and loss of sensory perception (Millspaugh 1974). The bark when chewed may cause excess salivation, paresis of the tongue, and severe irritation of the mucous membrane surfaces of the mouth, esophagus, and stomach.

Although not as well known as a toxic genus as other members of the family are, *Dirca* contains irritant daphnane ester diterpenoids such as dircin, linimacrin B, and *Pimelea* factor P_2 (Badawi et al. 1983; Tyler & Howden 1985). It thus presents a similar but probably smaller risk than *Daphne*, producing similar clinical and pathological signs and requiring the same treatment. Novel tricyclic phenolic glycosides of unknown toxicologic significance are also present (Ramsewak et al. 1999).

■ severe irritation of mouth and lower digestive tract; fruits, neurologic effects

irritant diterpenoids

Pimelea Banks & Sol. ex Gaertn.

Taxonomy and Morphology—Native to Australia, New Zealand, and Timor, *Pimelea*, commonly known as rice flower, comprises 108 species. Its name is derived from the Greek root *pimele*, meaning "fat" and alluding to the fleshy seeds. Representative of the genus are the following species:

P. ferruginea Labill.	pimelea
P. ligustrina Labill.	bunjong, banjine
P. prostrata (J.R.Forst. & G.Forst.) Willd.	prostrate pimelea, creeping pimelea

■ *P. ferruginea*
P. ligustrina
P. prostrata

banjine
bunjong
pimelea
rice flower

Plants compact shrubs; evergreen; polygamous or dioecious. **Stems** erect or prostrate and mat-forming; 1–8 m tall. **Leaves** opposite or rarely alternate; typically decussate; short-petiolate or sessile; blades ovate to oblong. **Inflorescences** capitate clusters; terminal; bracts present, green or colored. **Flowers** perfect or imperfect. **Sepals** 4; white or pink or red or yellow. **Stamens** 2. **Pistils** with stigmas papillose or indumented; styles long. **Hypanthia** tubular; not constricted above ovary; colors same as sepals. **Fruits** drupes; ovoid; green or red or black.

■ compact, evergreen shrubs; leaves opposite; flowers in capitate clusters; sepals 4, white, pink, red, or yellow; fruits ovoid drupes, green, red, or black

1. Stems prostrate and mat-forming .... *P. prostrata*
1. Stems erect and not mat-forming.
   2. Hypanthia white; circumscissile.
      Ovaries distally indumented .... *P. ligustrina*
   2. Hypanthia pink; not circumscissile. Ovaries glabrous .... *P. ferruginea*

Distribution and Habitat—In North America, several species of the genus are occasionally cultivated as outdoor ornamentals in the Southwest, where mild winters and hot dry summers are the norm. These plants are uncommon.

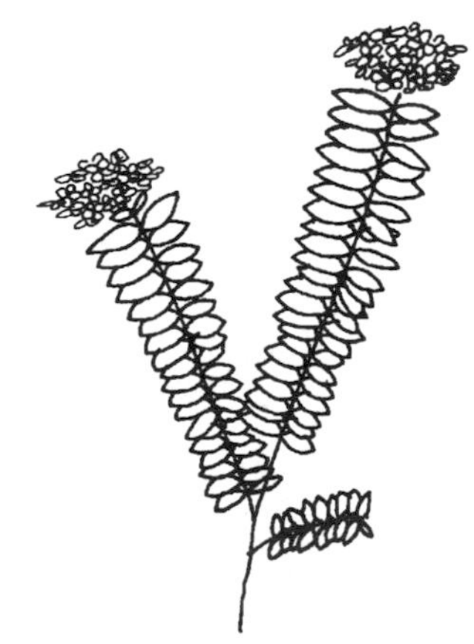

Pimelia prostrata

Disease Problems and Genesis—Acute intoxication due to these species of *Pimelea* is manifested by marked disturbance of the digestive tract (Seawright 1989). This may

- ornamentals
- digestive disturbance, all plant parts; dried trimmings are a special risk
- irritant daphnane and tigliane diterpenoids

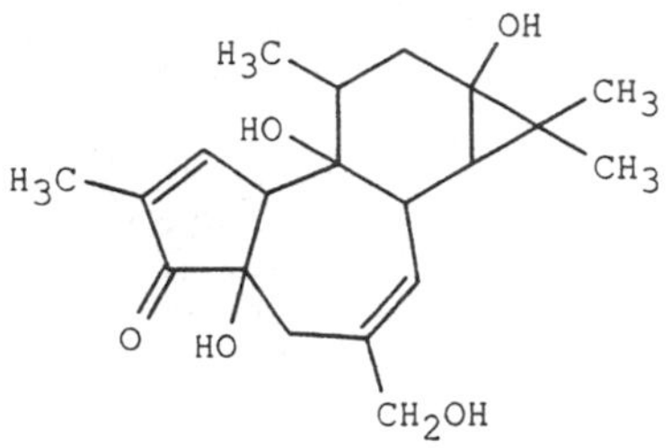

Prostratin

- vomiting, diarrhea

reddened gut mucosa

treat with activated charcoal, fluids, electrolytes

result from ingestion of fresh green foliage or more likely from more-palatable, dried trimmings of foliage discarded into a pasture where they can be eaten by livestock. Probably all species of the genus are capable of causing the problem.

Most species of *Pimelea* contain diterpenoids of the daphnane and tigliane types such as prostratin, gnidilatin, and linimacrin (Tyler & Howden 1985; Connolly & Hill 1991). All animal species are susceptible.

In Australia, a chronic cardiopulmonary disease known as St. George disease is caused by *Pimelea.* It is not a risk in North America, because the species causing it are not grown here as ornamentals. Intoxication also requires sustained ingestion of plant material for days or weeks, amounts unlikely to be available even of the species cultivated in North America. Simplexin, the cause of St. George disease, is apparently not present in the plant species typically cultivated in North America.

Clinical Signs, Pathology, and Treatment—*Pimelea* produces signs similar to those associated with other members of the family. They include severe vomiting and bloody diarrhea. This is likely to result in fluid loss and dehydration. The pathologic changes reflect the severe irritant effects of the diterpenes—marked reddening of the mucosa of the stomach and intestinal mucosa. There may be free blood in the lumina as well. In addition to control of the vomiting and diarrhea, activated charcoal and fluid and electrolyte therapy may be needed.

References

Badawi MM, Handa SS, Kinghorn AD, Cordell GA, Farnsworth NR: Plant anticancer agents. 27. Antileukemic and cytotoxic constituents of *Dirca occidentalis* (Thymelaeaceae). *J Pharm Sci* 72;1285–1287, 1983

Brickell CD, Mathew B: *Daphne: The Genus in the Wild and in Cultivation.* Alpine Garden Society, Woking, Surrey, England, 1976

Connolly JD, Hill RA: *Dictionary of Terpenoids.* Vol 2, *Di- and Higher Terpenoids.* Chapman & Hall, London, pp655–1460, 1991

Cooper MR, Johnson AW: *Poisonous Plants in Britain and Their Effects on Animals and Man.* Ministry of Agriculture, Fisheries and Food, Her Majesty's Stationery Office, London, pp226–227, 1984

Evans FJ, Soper CJ: The tigliane, daphnane, and ingenane diterpenes, their chemistry, distribution, and biological activities: a review. *Lloydia* 41;193–216, 1978

Habermehl GG: Poisonous plants of Europe. In *Poisonous Plants: Proceedings of the Third International Symposium,* James LF, Keeler RF, Bailey EM Jr, Cheeke PR, Hegarty MP eds, Iowa State Univ Press, Ames, pp74–83, 1992

Kingsbury JM: *Poisonous Plants of the United States and Canada.* Prentice-Hall, Englewood Cliffs, NJ, 1964

Millspaugh CF: *American Medicinal Plants.* Dover, New York, 1974 (reprint from 1892)

Mitchell J, Rook A: *Botanical Dermatology: Plants and Plant Products Injurious to the Skin.* Greengrass, Vancouver, BC, pp672–673, 1979

Pammel LH: *A Manual of Poisonous Plants.* Torch Press, Cedar Rapids, Iowa, pp63–64, 1911

Ramsewak RS, Nair MG, DeWitt DL, Mattson WG, Zasada J: Phenolic glycosides from *Dirca palustris. J Nat Prod* 62;1558–1561, 1999

Seawright AA: *Animal Health in Australia.* Vol 2, 2d ed, *Chemical and Plant Poisons.* Australian Government Publ Serv, Canberra, 1989

Tyler MI, Howden MEH: Antitumor and irritant diterpenoid esters of Thymelaeaceae species. In *Plant Toxicology, Proceedings of the Australian-USA Poisonous Plants Symposium,* Seawright AA, Hegarty MP, James LF, Keeler RF eds, Animal Research Institute, Yeerongpilly, Brisbane, Australia, 367–374, 1985

Urticaceae Juss.

72

Widely distributed in tropical and subtropical regions, the Urticaceae, or nettle family, comprises 39–45 genera and 700–1,000 species. It is closely related to the Moraceae, Cannabaceae, and Cecropiaceae; and the 4 taxa are treated as subfamilies of 1 large family by some taxonomists (Zomlefer 1994). Differences among them, however, are conspicuous and permit easy recognition; thus they are maintained as distinct families in this treatment, as is traditional. The family is of little economic importance; fiber used in textile production, and forage is obtained from some species. In North America, contact irritant dermatitis is associated primarily with 2 genera.

- Nettle Family

  Laportea
  Urtica

Plants herbs or shrubs; annuals or perennials; armed or not armed with stinging hairs; monoecious or dioecious or rarely polygamomonoecious. **Leaves** simple; alternate or opposite; venation pinnate or pinnipalmate; surfaces glabrous or pubescent; apices acuminate to obtuse; margins serrate or dentate or entire; bases cuneate or cordate or rounded; stipules present or absent. **Inflorescences** simple cymes or panicles or glomerules; axillary or rarely terminal; bracts present, minute. **Flowers** small; imperfect; staminate and pistillate different; perianths in 1-series or absent; radially symmetrical. **Staminate Flowers** with sepals 3–5, free or fused, green; petals absent; stamens 3–5. **Pistillate Flowers** with sepals 2–5 or 0, fused, green; pistils 1, compound, carpels 2, appearing to be 1; stigmas 1, capitate or linear; styles 1 or 0; ovaries superior; locules 1; placentation basal. **Fruits** achenes; sometimes enclosed in persistent calyces. **Seeds** 1.

- herbs or shrubs, some with stinging hairs; leaves simple; flowers small, imperfect, staminate, pistillate, different; petals absent

The presence of stinging hairs is an attribute characteristic of many genera of the family, including *Urtica* and *Laportea.* These hairs, sometimes termed emergences, are derived from epidermal and subepidermal cells (Thurston 1974). Solitary and very conspicuous or scattered among other surface hairs, each consists of a slender, siliceous, hollow shaft arising from a multicellular, bulbous, sheathing supporting pedestal, which may be up to one-third the length of the hair. The length of the hair and its base may exceed 1 mm. The slightly swollen tip of the hair is easily fractured and typically shears off upon contact. The resulting sharp-pointed, hypodermic-like shaft penetrates the skin, thus allowing extrusion of the cell's chemical contents into the skin and producing a contact irritant dermatitis (Thurston & Lersten 1969; Thurston 1974).

- stinging hairs, bulbous supporting pedestal, hollow shaft, easily fractured, swollen tip that when broken may penetrate and extrude its chemical contents into skin

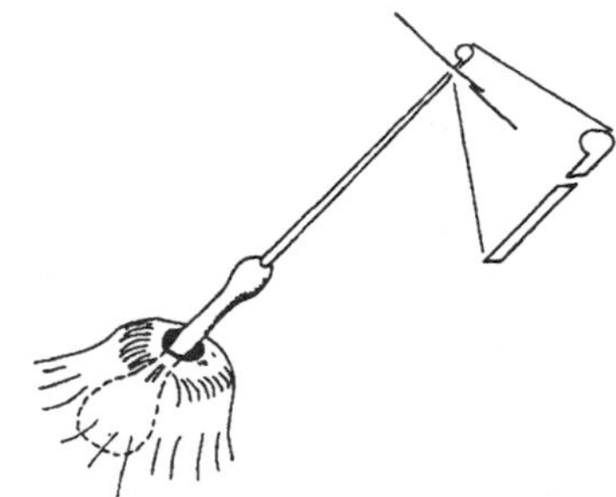

Stinging hair

Most animals appear to be susceptible to the toxins present in the hairs of *Urtica* and *Laportea,* although some animal species such as cattle rarely show any response. Cattle, sheep, and rabbits may consume the plants as forage (Pollard & Briggs 1984). The most serious problems of contact dermatitis occur in dogs, particularly the hunting breeds (Davis & Clark 1958; Peterson 1968; Louisiana Animal Health Notes 1982; Edwards & Remer 1983). Late winter and early spring are problem times in southeastern states, when dogs are used in field trials and for hunting in the open woods and bottomlands. In some of these areas the disease has been referred to as

- mainly in humans, dogs; contents appear to have histamine-like activity

bull nettle syndrome (Davis & Clark 1958). As the dogs move through areas with high *Urtica chamaedryoides* density, there is repeated contact with the plants. Contact probably occurs over the animal's entire body as well as around the mouth and tongue; there is possible aspiration as well. Contact with an occasional plant is not expected to produce severe signs.

At various times, the contents of the stinging hairs of the Urticaceae have been reported to be: formic acid, enzymes, tartaric acid, glycosides, alkaloids, histamine, acetylcholine, and 5-OH-tryptamine (serotonin). The exact nature of the toxicants is still not certain, although pharmacologic and chromatographic data suggest histaminic, acetylcholinergic, and serotoninergic compounds (Emmelin & Feldberg 1947, 1949; Collier & Chesher 1956). The clinical syndrome also is consistent with the toxic effects produced by histamine and acetylcholine (Davis & Clark 1958). In an Australian species of *Laportea*, a toxin has been identified that has a molecular weight greater than 1,000; is not heat labile, dialyzable, susceptible to proteolytic enzymes, or precipitated by trichloroacetic acid; and suffers no appreciable loss of activity with drying or preservation for many years (MacFarlane 1963; Oelrichs & Robertson 1970).

- site of contact with skin, erythema, swelling, pain, pruritis persisting for a few minutes or rarely a few hours; with massive exposure in dogs, salivation, licking of paws, scratching of body, retching, vomiting, labored respiration, muscle fasciculation, depression, may last a few hours or a day or more; rarely fatal

Clinical Signs—Contact with the stinging hairs of *Urtica* and *Laportea* characteristically, and especially in humans, produces erythema, swelling, pain, and pruritis at the contact site. These signs persist for minutes to hours. There may be systemic signs with massive or repeated contact. Signs may occur in any animal, although they are rarely seen in cattle. Horses occasionally react markedly at contact sites on their limbs and abdomen. Signs are readily observed in dogs. Within minutes or up to an hour after exposure to the plants, dogs begin salivating profusely, licking their paws, and scratching at their nose, face, and other body areas. Retching and vomiting may also occur. In more-severe cases the signs include rapid and/or labored respiration, persistent vomiting, hemorrhage from the nostrils, muscle fasciculations, and ataxia that may progress to partial paresis of the pelvic limbs. The signs subside in most cases within an hour or in more-serious cases last up to several days. Depression may be quite noticeable during this period. The disease is usually self-limiting but rarely may be lethal.

- atropine, antihistamines

Treatment—If administered early, atropine and antihistamines may be of value. Otherwise, treatment is symptomatic.

Laportea Gaudich.

- *L. canadensis*

  Canada wood nettle

Taxonomy and Morphology—Comprising approximately 25–75 species in both the Old World and the New World, *Laportea* is commonly known as wood nettle. Its name honors F. L. de Laporte, a nineteenth-century French entomologist. In North America, 1 species is native and of toxicologic significance:

L. canadensis (L.) Wedd. Canada wood nettle

- erect, perennial herbs armed with stinging hairs, monoecious; leaves alternate, long petioles; flowers axillary

Plants herbs; perennials; from fibrous roots or rhizomes; armed with stinging hairs; monoecious. **Stems** erect; 50–100 cm tall; stinging hairs dense or sparse. **Leaves** alternate; long-petiolate; blades elliptic to broadly ovate; midribs occasionally bearing stinging hairs; apices acuminate; margins coarsely serrate; bases subcordate or rounded; stipules lanceolate, chartaceous, brownish, deciduous. **Inflorescences** axillary cymes or glomerules; staminate in lower axils, pistillate in upper axils, open,

Laportea canadensis

Laportea canadensis

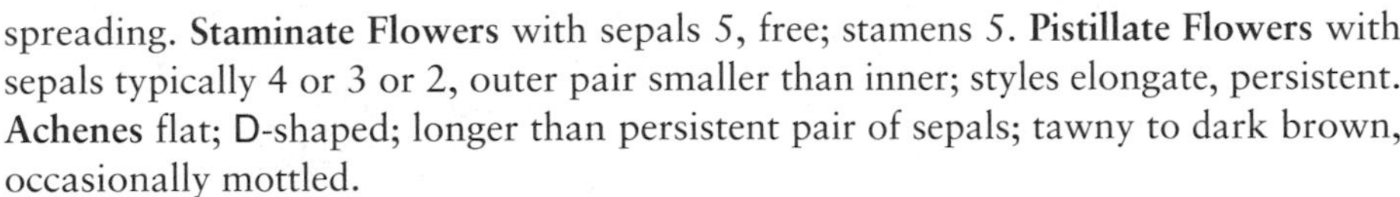

spreading. **Staminate Flowers** with sepals 5, free; stamens 5. **Pistillate Flowers** with sepals typically 4 or 3 or 2, outer pair smaller than inner; styles elongate, persistent. **Achenes** flat; D-shaped; longer than persistent pair of sepals; tawny to dark brown, occasionally mottled.

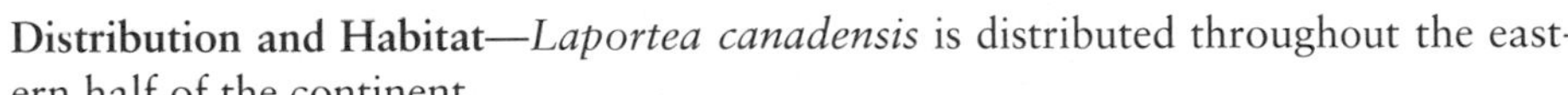

Distribution and Habitat—*Laportea canadensis* is distributed throughout the eastern half of the continent.

Urtica L.

Taxonomy and Morphology—Generally thought of as one of the most troublesome genera because of its stinging hairs, *Urtica,* commonly known as nettle, comprises 25–100 species distributed principally in temperate regions. Used by Pliny, the Roman natural historian, its name is derived from the Latin root *uro,* which means "to burn" and reflects the painful effects of the stinging hairs. The density of these hairs may vary considerably, and hairs may be almost lacking in some populations (Pollard & Briggs 1982). Other features are likewise quite variable, and numerous species have been described. Systematic work by Woodland (1982) indicated that 4 species are present in North America:

U. chamaedryoides Pursh	heartleaf nettle, weak nettle, ortiguilla
U. dioica L.	American stinging nettle, tall nettle, slim nettle, hoary nettle
U. gracilenta Greene	mountain nettle
U. urens L.	dwarf nettle, dog nettle, burning nettle

■ *U. chamaedryoides*
U. dioica
U. gracilenta
U. urens

American stinging nettle
burning nettle
dog nettle
dwarf nettle
heartleaf nettle
hoary nettle
mountain nettle
ortiguilla
tall nettle
weak nettle

Plants herbs; annuals or perennials; armed with stinging hairs; monoecious or dioecious. **Stems** erect; 50–200 cm tall; branched or unbranched; terete or angled; stinging hairs dense or sparse; other hairs of various types. **Leaves** opposite; long- or short-petiolate; blades lanceolate to ovate or cordate; venation pinnipalmate; stinging hairs dense or sparse; other hairs of various types; apices acuminate; margins serrate or dentate; bases cuneate to cordate; stipules present. **Inflorescences** axillary panicles or glomerules. **Staminate Flowers** with sepals 4, fused, hispid; stamens 4, anthers explosively dehiscent. **Pistillate Flowers** with sepals 4, fused, outer 2 smaller; styles absent. **Achenes** ovate to deltoid; lenticular; enclosed by 2 inner sepal lobes.

■ erect, annual or perennial herbs, armed with stinging hairs; leaves opposite, venation pinnipalmate, stipules present; flowers axillary

1. Plants perennials. Rhizomes present. Stipules 5–15 mm long, erect. Inflorescences longer than petioles . . . . *U. dioica*
1. Plants annuals. Rhizomes absent. Stipules 1–4 mm long, spreading or reflexed. Inflorescences shorter than petioles.
    2. Staminate and pistillate flowers in separate inflorescences . . . . *U. gracilenta*

Urtica chamaedryoides

Urtica dioica

Urtica urens

2. Staminate and pistillate flowers mixed in same inflorescences.
   3. Upper leaves smaller than lower leaves. Inflorescences subglobose .... *U. chamaedryoides*
   3. Upper leaves not smaller than lower. Inflorescences elongate .... *U. urens*

■ moist sites, bottom-lands, waste areas

Distribution and Habitat—Species of *Urtica* occur across the continent and thrive in disturbed and nitrogen-rich soils of bottomlands, moist open woods, stream banks, and waste areas. Comprising three subspecies, of which two are native and one naturalized from Europe, perennial *U. dioica* exhibits the greatest range from Alaska to the Atlantic Coast. Native to Europe, annual *U. urens* also occurs from coast to coast, but its populations are scattered and less commonly encountered than those of *U. dioica*. *Urtica chamaedryoides* is distributed primarily in the southeastern quarter of the continent. Exhibiting the most restricted distribution, *U. gracilenta* is found in the Southwest.

Urtica chamaedryoides

Urtica dioica

References

Collier HOJ, Chesher GB: Identification of 5-hydroxytryptamine in the sting of the nettle *(Urtica dioica)*. *Br J Pharmacol* 11;186–189, 1956

Davis DE, Clark CH: Questions and answers: "bull nettle syndrome" in hunting dogs. *Mod Vet Pract* 39;50–52, 1958

Edwards WC, Remer JC: Nettle poisoning in dogs. *Vet Med/Small Anim Clin* 78;347,350, 1983

Emmelin N, Feldberg W: The mechanism of the sting of the common nettle *(Urtica urens)*. *J Physiol* 106;440–455, 1947

Emmelin N, Feldberg W: Distribution of acetylcholine and histamine in nettle plants. *New Phytol* 48;143–148, 1949

Louisiana Animal Health Notes: Stinging nettle (*Urtica* sp.) and dogs. *Vet Hum Toxicol* 24;247, 1982

MacFarlane WV: The stinging properties of *Laportea*. *Econ Bot* 17;303–311, 1963

Oelrichs PB, Robertson PA: Purification of pain-producing substances from *Dendrocnide (Laportea) moroides*. *Toxicon* 8;89–90, 1970

Peterson DR: Nettle poisoning in the dog. *Okla Vet* 20;4–6, 1968

Pollard AJ, Briggs D: Genecological studies of *Urtica dioica* L. 1. The nature of intraspecific variation in *U. dioica*. *New Phytol* 92;453–470, 1982

Pollard AJ, Briggs D: Genecological studies of *Urtica dioica* L. 3. Stinging hairs and plant-herbivore interactions. *New Phytol* 97;507–522, 1984

Thurston EL: Morphology, fine structure, and ontogeny of the stinging emergence of *Urtica dioica*. *Am J Bot* 61;809–817, 1974

Thurston EL, Lersten NR: The morphology and toxicology of plant stinging hairs. *Bot Rev* 35;393–412, 1969

Woodland DW: Biosystematics of the perennial North American taxa of *Urtica*. 2. Taxonomy. *Syst Bot* 7;282–290, 1982

Zomlefer WB: *Guide to Flowering Plant Families*. Univ North Carolina Press, Chapel Hill, 1994

Urtica gracilenta

Urtica urens

73 Verbenaceae J.St.-Hil.

■ Verbena Family
Vervain Family
Aloysia
Duranta
Lantana

Primarily pantropical in distribution and with only a few genera in temperate regions, the Verbenaceae, commonly known as the verbena or vervain family, comprises 75–100 genera and 2,600–3,000 species. In North America, approximately 28 genera and 160 species, both native and introduced, are present. The family is diverse in habit and gynoecial structure, and its circumscription is vigorously debated by taxonomists. Some individuals segregate a number of its genera as distinct families; others position approximately half of its taxa in the Lamiaceae; and a few place two-thirds of its taxa in the Lamiaceae. The traditional, broad interpretation with 4 subfamilies based on inflorescence type and fruit structure is maintained in this treatment. Although the family is unquestionably closely related to the Lamiaceae, its members can be distinguished by their undivided ovary, apical style, and usually nonverticillate inflorescences.

Economically the family is noted for its ornamentals, e.g., *Verbena, Lantana,* and *Vitex* (chaste tree). It also is the source of teak, zither wood, and valuable essential oils. Several species were used medicinally by American Indians and settlers and today are constituents of some herbal medicines. Toxicologic problems are associated with 3 genera.

■ herbs, shrubs, or trees, often aromatic; sepals fused; petals fused

Plants herbs or shrubs or trees; annuals or perennials; often strongly aromatic. **Stems** 4-sided or terete. **Leaves** simple or palmately compound or pinnately compound; opposite or whorled or rarely alternate; stipules absent. **Inflorescences** spikes or compound cymes or panicles or racemes; terminal or axillary; bracts present. **Flowers** perfect or rarely imperfect; perianths in 2-series. **Sepals** 5 or 4 or rarely 8; fused; radially or bilaterally symmetrical. **Corollas** bilaterally or rarely radially symmetrical; salverform or funnelform or bilabiate. **Petals** 5 or 4 or rarely 8; fused; of various colors. **Stamens** 4 or 2 or rarely 5; didynamous or of equal lengths; epipetalous. **Pistils** 1; compound, carpels 2; stigmas 1, 2-lobed; styles 1, apical; ovaries superior; terete or lobed, locules 4 or 2 or 1; placentation axile. **Nectaries** present; receptacular; annular. **Fruits** nutlets or achenes or drupes or capsules. **Seeds** 1–8.

Aloysia Juss.

■ *A. gratissima*
angel
bee-bush
cedron
palo amarillo
white brush
white bush

Taxonomy and Morphology—Native to the Americas, *Aloysia,* commonly known as bee-brush, comprises 35–40 species of sweet-aromatic shrubs. Its name honors Maria Louisa, wife of King Charles IV of Spain. Grown primarily for its attractive, strongly aromatic foliage, which is dried and used in cooking, teas, potpourri, and sachets, *A. triphylla,* or lemon verbena, is the best-known member of the genus. Three species are present in North America, and toxicologic problems are associated with 1:

A. gratissima (Gilles & Hook.) Tronc.	white brush, white bush, bee-bush,
(= *Lippia lycioides* [Cham.] Steud.)	cedron, palo amarillo, angel
(= *Lippia ligustrina* of auth.)	

■ deciduous shrubs, strongly aromatic; leaves simple; flowers in spikes or racemes, vanilla odor, 2-lipped; sepals 4; petals 5, white; fruits 2 nutlets

Plants shrubs; deciduous; herbage strongly aromatic. **Stems** erect; 1–3 m tall; much branched; branches stiff, gray, grayish pubescent; spinescent; wood yellow. **Leaves** simple; opposite or in fascicles; blades lanceolate to oblong or elliptic or obovate; lower surfaces densely pubescent or resinous punctate; margins entire. **Inflorescences** dense spikes or racemes; borne well above leaves; bractlets deciduous. **Flowers** with strong vanilla odor; perfect. **Sepals** 4; tubular-campanulate. **Corollas** bilaterally symmetrical; salverform, 2-lipped. **Petals** 5; white or tinged with violet; throats villous. **Stamens** 4; didynamous. **Pistils** lobed; locules 2. **Fruits** nutlets; 2; thin-walled. **Seeds** 1 per nutlet.

Aloysia gratissima

Distribution and Habitat—*Aloysia gratissima* is found in a variety of sandy or rocky sites in open grasslands, creek beds, washes, woods, and thickets in Texas, New Mexico, and Mexico.

■ sandy or rocky sites

Aloysia gratissima

Disease Problems and Genesis—*Aloysia gratissima* appears to be the cause of a rickets-like disease in horses (Mathews 1942). Experimentally, horses pastured exclusively on the species lost weight, became incoordinated after 1 month, and in a few more weeks were unable to stand and then died. During this time the animals exhibited conspicuous weakness and a propensity to sweat.

A toxin has not been identified. An apparent antimetabolite of some type is present, however, as demonstrated experimentally in rats (Mathews 1942). In addition, an African species, *Lippia rehmannii,* contains toxic triterpenes very similar in structure and activity to the lantadenes of *Lantana* (Barton & de Mayo 1954a,b).

■ horses, rickets-like disease; toxin unknown

Clinical Signs—Initially there is a loss of stamina and a tendency to sweat excessively, as well as a tendency for the animal to fall when turned while running. Later, there is difficulty walking and ataxia due to the loss of tone of the leg muscles, followed by terminal prostration and apparent paralysis.

■ loss of stamina, sweating, ataxia

Pathology and Treatment—There may be some edema of the meninges and congestion of the liver and kidneys, but overall there is a lack of distinctive lesions either grossly or microscopically. No specific treatment is known, except for good nursing care and provision of high-quality feeds.

■ no lesions

nursing care

Duranta L.

Taxonomy and Morphology—Named in honor of Castor Durantes, a sixteenth-century botanist and physician in Rome, *Duranta,* commonly known as golden dewdrop, comprises 30–36 species and is native from Mexico to South America and in the West Indies. One species is of toxicologic interest:

D. erecta L.	golden dewdrop, yellow hat tree, pigeonberry,
(= *D. plumieri* Jacq.)	adonis morado, espina de palma, cuenta de oro,
(= *D. repens* L.)	garbancillo, coralillo, sky flower

■ *D. erecta*

adonis morado
coralillo
cuenta de oro
espina de palma
garbancillo
golden dewdrop
pigeonberry
sky flower
yellow hat tree

Plants shrubs or small trees; evergreen; armed or not armed with thorns. **Stems** erect; to 6 m tall; branches drooping or trailing. **Leaves** simple; opposite; blades obovate

- evergreen shrubs or trees; leaves simple, opposite; flowers in racemes; sepals 5; petals 5, white, blue, lavender, or purple; fruits drupes, berrylike, yellow, stones 8

to ovate-elliptic; apices acuminate or obtuse; margins entire or serrate; bases cuneate. **Inflorescences** racemes; loosely flowered; terminal or axillary; erect or recurved; bractlets caducous. **Flowers** perfect. **Sepals** 5; tubular; angled; persistent; yellowish and enlarged in fruit. **Corollas** salverform. **Petals** 5; white or blue to lavender or purple. **Stamens** 4; didynamous. **Pistils** globose. **Fruits** drupes; berrylike; yellow; enclosed by inflated beaked calyx; with 8 stones. **Seeds** 1 per stone.

Duranta erecta

- Florida, Texas, ornamental elsewhere

Distribution and Habitat—*Duranta erecta* is native to southern Florida, the Caribbean, and Mexico. It is cultivated widely in the South and has escaped and naturalized in some areas such as the Rio Grande Valley and plains of Texas.

- possibly digestive disturbance, neurologic problems

  toxin unknown, possibly monoterpenoids

Disease Problems and Genesis—Intoxication problems caused by *D. erecta* are not well documented, but digestive tract and neurologic effects have been reported (DerMarderosian & Liberti 1988). The toxicants are not known, but the species contains several monoterpenoids, durantosides I, II, and III (Connolly & Hill 1991). Similar to lamiide found in species of *Lamium,* these compounds may be of some toxicologic importance.

Durantoside I

- vomiting, diarrhea, edema of face, somnolence, seizures

  no lesions

  supportive treatment

Clinical Signs, Pathology, and Treatment—There may be evidence of digestive tract irritation such as vomiting and diarrhea, edema of the lips and eyelids, and systemic effects such as somnolence, mydriasis, increased heart rate, and seizures. The disease is unlikely to be lethal, and no specific lesions are associated with it. Treatment is directed toward relief of the specific signs.

Lantana L.

- *L. camara*
  L. montevidensis

  lantana
  red sage
  yellow sage

Taxonomy and Morphology—Native to the tropics and subtropics of both the Old World and the New World, *Lantana* comprises 150–160 species and more than 75 named varieties and cultivars. Widely grown for the showy, dense clusters of flowers that are produced from spring to fall, it is used in window boxes and hanging baskets and as ground cover or a border plant. The numerous cultivars and hybrids exhibit a wide range of colors. Most striking are the changes in petal color as the flowers mature—yellow to red, white to pink, yellow to pink, and red to different shades. In the tropics, some species are widespread weeds. Of the approximately 13 native and introduced species in North America, only 2 species are of toxicologic significance (see plate 73.1):

L. camara L.	lantana, red sage, yellow sage
L. montevidensis (Spreng.) Briq. (= *L. sellowiana* Link & Otto)	creeping lantana, trailing lantana, weeping lantana

- perennial herbs or shrubs, malodorous, stems 4-sided; leaves simple, opposite or whorled; flowers in dense spikes or heads; sepals 4 or 5; petals 4 or 5, white, red, yellow, orange, blue, or purple; fruits glossy black drupes

Plants herbs or shrubs; perennials; foliage typically malodorous when crushed. **Stems** erect or trailing and mat-forming; to 2 m tall or 1 m long; 4-sided; armed or not armed with prickles. **Leaves** simple; opposite or whorled; blades typically rugose; margins toothed. **Inflorescences** dense spikes or heads; typically flat-topped or hemispheric; terminal or axillary; peduncles often long; bractlets present. **Flowers** perfect. **Sepals** 4 or 5; small. **Corollas** bilaterally symmetrical; salverform. **Petals** 4 or 5; white or red or yellow or orange or blue or purple or of intermediate colors. **Stamens** 4; didynamous. **Pistils** with 2 locules. **Fruits** fleshy drupes with 2 stones; globose; glossy black. **Seeds** 1 per stone.

The flowers of *Lantana* resemble those of *Verbena* but differ by having very short calyces and drupes rather than long calyces and nutlets.

1. Stems erect; prickly. Bracts oblong to lanceolate.
   Petals yellow, becoming red to orange .... *L. camara*
1. Stems trailing; not prickly. Bracts ovate.
   Petals rose-red to purple .... *L. montevidensis*

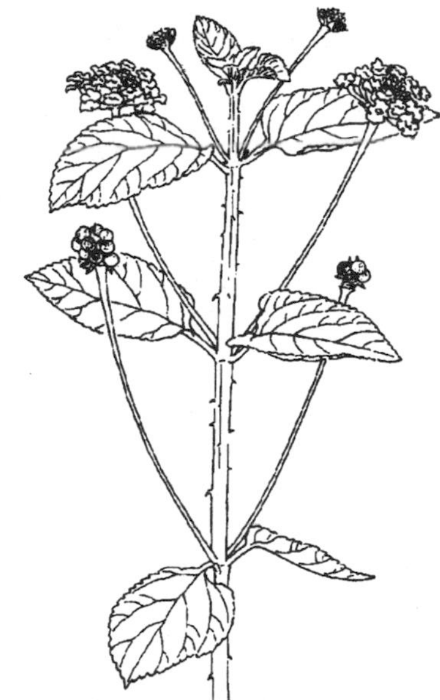

Lantana camara

■ ornamentals, weedy in southern North America

Distribution and Habitat—Presumably native to Central America, *L. camara* is now widely distributed throughout the tropics of the world. It occurs as a weed along roadsides and in other disturbed sites across the southern United States and in Mexico. Not winter-hardy, it is restricted to the South. Farther north, it is commonly grown as a tub plant on patios and then taken indoors during the colder months. Rapid-growing, it is also used in temporary ornamental plantings established from hothouse cuttings in the spring. Introduced from South America, *L. montevidensis* also has escaped cultivation and naturalized in disturbed sites in the southern states.

Lantana camara

■ leaves, digestive disturbance, liver disease, photosensitization; green, unripe fruits, children, digestive disturbance

Disease Problems—Throughout most of its current range in the world, *Lantana* has been reported to cause intoxications in livestock. In the southern parts of North America, it produces a severe digestive tract irritation and liver disease with accompanying photosensitization (Sanders 1946; de Aluja 1970; Sperry et al. 1977). Morton (1994) recounted numerous cases of intoxication in dogs and horses that ate the leaves and in children who consumed the green fruits.

There are numerous cultivars and hybrids, and toxicity seems to vary considerably among them. Several studies have been conducted on cultivars in Australia, and the common pink-flowered type has been consistently low in lantadene A and B and low in toxicity (Seawright 1965a; Hart et al. 1976 a,b). The red-flowered types had high levels of lantadene A and B. Other cultivars such as the orange-flowered types were intermediate in toxicity. The potency of the red-flowered cultivars has been confirmed in other studies in New Zealand and India (Black & Carter 1985; Sharma et al. 1991). Unfortunately, it is difficult to correlate the Australian varieties with those present in North America. In general, based upon the foregoing, the types with red flowers or yellow flowers changing to red should be regarded as the most toxic. The types with white-to-pink or yellow-to-pink flowers are probably of low risk, although caution is warranted in relying unduly on flower color as a reliable indicator of toxicity. The lower toxicity for a pink-flowered species *(L. velutina)* and a white-flowered species *(L. achyranthifolia)* has been substantiated in Mexico (de Aluja & Skewes 1971).

■ red- or yellow-flowered types high risk; all animal species, mainly ruminants; leaves fresh or dried; mature fruits no problem

Intoxication seems to be a problem mainly in ruminants, but most animal species, including common laboratory species, dogs, horses, wildlife, and nonruminant zoo animals such as kangaroos, are also at risk (Sharma et al. 1988, 1989; Morton 1994; Johnson & Jensen 1998). Goats, which are browsers often thought to be more tolerant of noxious weeds, are quite susceptible to lantana. Obwolo and co-workers (1991) described a case in goats in Zimbabwe that is very typical except for the lack of photosensitization and the presence of marked renal tubular degeneration. Similar findings are reported for Boer goats in South Africa (Ide & Tutt 1998).

The leaves are toxic, and dosage of 5–10 g of dried leaves/kg b.w. produces severe and often fatal intoxication in sheep and cattle. In Brazil, toxicity varies considerably among locations. In the high-risk areas, toxicity in cattle ranges from 30 to 50 g of fresh leaves/kg b.w. of single or daily dosage over 5 days (Tokarnia et al. 1984). Sanders (1946) reported that in Florida, 454 g of dried leaves was sufficient to intoxicate a 182 kg calf. An adult sheep may be intoxicated by eating approximately 0.5 kg of the plant in 1 or 2 days (Gopinath & Ford 1969). The ripe fruits and seeds, however, are not toxic and apparently are relished by sheep and goats (Lall et al. 1983b; Sharma 1988). The fruits have been fed in substantial amounts for several months to sheep with no apparent adverse effects as measured by hematologic parameters or serum bilirubin and hepatic enzymes (Lall et al. 1983a,b). The ripe fruits also are consumed by humans, especially children, without apparent adverse effects (Standley 1926; Sharma 1994). In contrast to the ripe fruits, green ones have caused severe disease in children (Wolfson & Solomons 1964; Morton 1994). The effects of the green fruits seem to be of a different nature from those of the leaves, given that there is no apparent liver damage (Sharma 1988).

Intoxication can be acute but is more often subacute after consumption of the plant for 1 or 2 days. Experimentally, the rate of ingestion determines the speed of onset of signs and doses may be low enough that the only sign is a transient loss of appetite (Gopinath & Ford 1969). The skin sloughing is secondary to liver disease and is presumably due to phylloerythrin, which has been shown to be present in the peripheral circulation. In addition, serum bilirubin and serum hepatic enzymes (SDH, AST) are elevated (Gopinath & Ford 1969). Clearance of BSP is markedly reduced.

There has been a suggestion of breed differences in susceptibility to intoxication, with exotic breeds derived from *Bos indicus* less likely to develop problems. However, this has not been shown to be true except for a general increase in resistance to environmental stresses for these types of breeds (Frisch et al. 1984).

Although an uncommon cause of intoxication, *Lantana* is clearly a risk, because it has been responsible for significant losses in areas where is grows wild, e.g., parts of Mexico (de Aluja 1970). Mortality is high in severe cases, but with more-moderate intoxications the response to treatment may be more favorable (Hari et al. 1973).

■ triterpenes, lantadenes, irritant action, direct effect on digestive tract, excreted metabolites cause obstructive cholangitis, retention of bilirubin and phylloerythrin

Disease Genesis—*Lantana* contains a number of biologically active triterpenes, including some of novel structure (Barre et al. 1997). The poisonous principles are hepatotoxic pentacyclic triterpenoid lantadenes, the most toxic and abundant of which are lantadene A and its reduced form. Originally termed lantanin and shown to cause photosensitization, it was later renamed lantadene (Louw 1943; 1948). The isolation and toxicity of these compounds have subsequently been reconfirmed (Seawright & Hrdlicka 1977). Lantadenes are very similar in structure and activity to the toxic triterpenes from *Lippia rehmannii* (Barton & de Mayo 1954a,b). Other lantadenes of more limited importance include B and reduced B. Leaves are the source of the toxins. Flowers, fruits, and shoots lack lantadene A, and they do not cause hepatic or bilirubin changes in guinea pigs (Sharma et al. 1980). However, the shoots cause hair loss unrelated to liver changes. The unripe or green fruits may have other toxins in them to account for the reported effects in children in which liver injury is absent.

We are indebted to workers in Australia who in an extensive series of studies provided detailed descriptions of the pathology and pathophysiology of *Lantana* intoxication in sheep and cattle (Seawright 1964, 1965a,b; Seawright & Allen 1972; Pass

1986). The primary problem caused by the genus seems to be obstructive cholangitis, as indicated by the elevation of conjugated bilirubin and supported by ultrastructural studies (Pass et al. 1978a; Sharma et al. 1988). The lantadenes are slowly absorbed, mainly from the small intestine and secondarily from the stomachs, including the rumen (Pass et al. 1981; McSweeney & Pass 1983). Absorption is not only slow but also incomplete, as is shown by the 20-to-30-fold difference between oral and i.v. toxicity (Pass & Stewart 1992). The initial signs of intoxication are the result of absorption of a small amount of the toxin. It is the prolonged, continuous, low-level absorption from the digestive tract that provides a continuing source of toxins to cause a pronounced insult to the liver tissues (Pass 1986). The lantadenes (including the reduced forms) are then biotransformed in the liver and excreted in bile; in the liver they damage bile canalicular membranes and microvilli, causing decreased ATPase activity, as well as closure and blockage or backup of bile flow (Pass et al. 1978b, 1985a; Pass & Goosem 1983). There is hepatocyte damage, but the ability to conjugate bilirubin is at least partially retained (Pass et al. 1978b). This eventually results in retention of conjugated bilirubin and phylloerythrin and elevation of these constituents in blood (Agarwala et al. 1962; Gopinath & Ford 1969). As has been shown in guinea pigs, some biotransformation of the lantadenes in the lower intestinal tract may occur because of the large amount of these compounds that remains unabsorbed (Sharma et al. 2000).

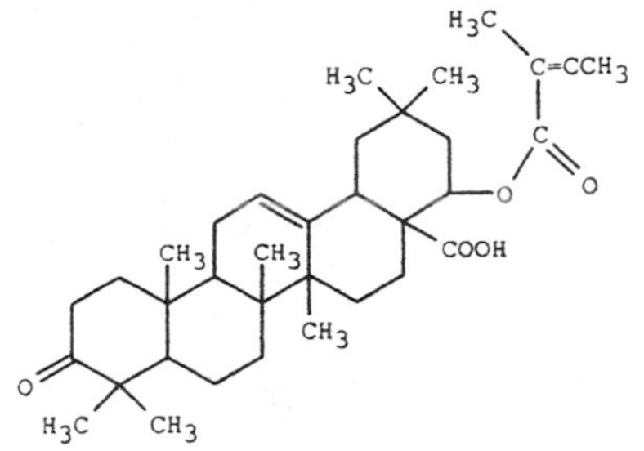

Lantadene A

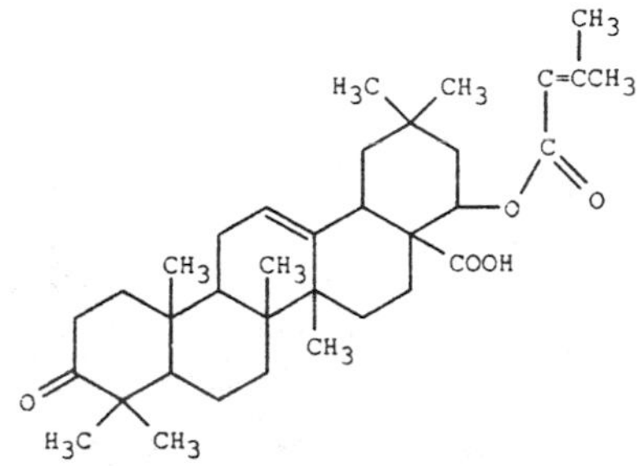

Lantadene B

Because of the continued insult to the liver, lantadenes have cumulative effects. Tokarnia and co-workers (1999) have shown in cattle that although severe intoxication and death were caused by single doses of 20 g and 40 g/kg b.w., respectively, a similar outcome could be produced by the same total dosage in divided doses given daily over 4–5 days (5 g and 10 g/kg b.w.). The lantadenes are also quite irritating to the mucosa of the digestive tract, and depending on the situation, either constipation or dark brown to black diarrhea may be seen (Fourie et al. 1987). Death is apparently due to a combination of liver and kidney injury, dehydration and electrolyte imbalance, and metabolic acidosis (Pass et al. 1985b). The toxins do not directly alter intestinal motility, but absorption from the digestive tract and worsening of the disease are facilitated by ruminal stasis secondary to liver disease (Pass et al. 1985b; Pass 1986, 1992). Although not affecting motility, the toxins do markedly reduce ruminal microflora activity and cause an increase in rumen pH and ammonia nitrogen (Mandial & Randhawa 1998). The biliary effects are exemplified by the paralytic effect of lantadenes on the gallbladder, causing its dilation (Pass & Heath 1977). In some instances and in individual animals, hepatocellular necrosis may predominate over the biliary and other effects on the liver.

Clinical Signs—With acute intoxication, signs appear several hours after plant ingestion and begin with depression, loss of appetite, marked decrease in ruminal motility, and severe constipation. By the second or third day, icterus and, in some cases, skin changes of hepatogenous photosensitization are apparent, leading to eventual severe sloughing of the epithelium and mucous membranes in the animal's light-colored areas. As the dose increases, the likelihood of diarrhea as an early sign increases. There is marked elevation of serum bilirubin, especially the conjugated form. Death, which may occur in several days to a few weeks, is usually due to a combination of hepatic disease and renal failure.

- acute disease, depression, loss of appetite, ruminal atony, constipation, icterus, photosensitization, later diarrhea

chronic disease, mainly photosensitization

humans, mainly children, green fruits, weakness, vomiting, diarrhea

The chronic disease develops over several weeks and is manifested mainly as photosensitization. There is cracking and desquamation of the skin, causing it to peel off in shreds. This is accompanied by itching, inflammation of the sclera, clouding of the cornea, and increased sensitivity of the eyes to bright light.

Ingestion of the green, unripe fruits of *Lantana* by children may produce in a few hours weakness, lethargy, vomiting, diarrhea, labored respiration, and mydriasis. In most cases the signs will be mild to moderate, and recovery uncomplicated.

- gross pathology, liver swollen, bile stained, gallbladder enlarged, kidneys swollen, perirenal edema

  microscopic, liver centrilobular necrosis, portal fibrosis, bile duct hyperplasia; edema of gallbladder wall; renal tubular necrosis

Pathology—Grossly, the effects of lantadene as a cholestatic are prominent. The liver is swollen and often orange-colored because of bile staining. The gallbladder typically is much enlarged. The kidneys will also be swollen, and there may be perirenal edema. Microscopically, there is periportal degeneration, portal fibrosis, bile duct hyperplasia, and sometimes mild to moderate centrilobular necrosis. Enlarged hepatocytes and those with enlarged or multiple nuclei may be apparent in cattle (Seawright & Allen 1972). There is edema of the wall of the gallbladder and tubular necrosis with casts in the kidneys. In a few animals, myocardial degeneration and fibrosis and pulmonary congestion and edema may be evident. Because of the irritating nature of *Lantana,* reddening and edema of the mucosa of the digestive tract may be severe in acute cases following large doses. The lesions in the small intestines may be suggestive of salmonellosis.

- oral activated charcoal, fluids, electrolytes, general supportive care

  humans, gastric lavage, supportive care as above

Treatment—Control of absorption of the toxins from the digestive tract is very helpful. This may be accomplished using activated charcoal or bentonite (McSweeney et al. 1985; McLennan & Amos 1989; McKenzie 1991). Supportive care with fluids, glucose, and electrolytes given i.v. is indicated. Sodium thiosulfate, 0.5 g/kg b.w. i.v., also has been suggested as an antidote (Forman & Kidder 1997). Protective immunization has been suggested, but vaccination of cattle i.m. or sheep intradermally using lantadene A or B or derivatives thereof conjugated with bovine serum albumin or hemocyanin provided only limited protection (Stewart et al. 1988; Pass & Stewart 1992). However, the results were thought nonetheless to be promising. Skin involvement will require symptomatic care of the affected areas and providing shade. Maintenance of an animal with severe hepatic and renal disease is difficult in livestock and generally unrewarding, so evaluation using serum hepatic enzymes, blood urea nitrogen, and/or creatinine is advisable.

In humans, treatment is mainly conservative and may include gastric lavage and general supportive care as above. Rarely will systemic signs be present that require attention.

Various methods are available for elimination of *Lantana* infestations of pastures and rangeland and include both herbicides and biological control (Sharma 1988). The plant is quite susceptible to herbicides such as 2,4-D.

References

Agarwala ON, Negi SS, Mahadevan V: Serum bilirubin and icteric index values in cattle and sheep in experimental lantana poisoning. *Curr Sci* 31;506–507, 1962

Barre JT, Bowden BF, Coll JC, Jesus J de, Fuente VE de la, Janairo GC, Ragasa CY: A bioactive triterpene from *Lantana camara. Phytochemistry* 45;321–324, 1997

Barton DHR, de Mayo P: Triterpenoids. Part 15. The constitution of icterogenin, a physiologically active triterpenoid. *J Chem Soc* 1954;887–900, 1954a

Barton DHR, de Mayo P: Triterpenoids. Part 16. The constitution of rehmannic acid. *J Chem Soc* 1954;900–903, 1954b

Black H, Carter RG: Lantana poisoning of cattle and sheep in New Zealand. *N Z Vet J* 33;136–137, 1985

Connolly JD, Hill RA: *Dictionary of Terpenoids,* vol 1. Chapman & Hall, London, 1991

de Aluja AS: *Lantana camara* poisoning in cattle in Mexico [letter]. *Vet Rec* 86;628, 1970 de Aluja AS, Skewes HR: Further investigation regarding the toxicity of members of the genus *Lantana* in Mexico. *19th World Vet Congr* 1;327–331, 1971

DerMarderosian A, Liberti L: *Natural Product Medicine.* GF Stickley, Philadelphia, 1988

Forman CR, Kidder RW: Sodium thiosulfate as a specific effective treatment for otherwise frequently fatal photosensitization in cattle. Paper presented at Fifth International Symposium on Poisonous Plants, San Angelo, Texas, May 1997

Fourie N, Van Der Lugt JJ, Newsholme SJ, Nel PW: Acute *Lantana camara* toxicity in cattle. *J S Afr Vet Assoc* 58;173–178, 1987

Frisch JE, O'Neill CJ, Burrow HM: The incidence and effect of poisoning with *Lantana camara* in different cattle breeds. *J Agric Sci* 102;191–195, 1984

Gopinath C, Ford EJH: The effect of *Lantana camara* on the liver of sheep. *J Pathol* 99;75–85, 1969

Hari R, Shivnani GA, Joshi HC: Therapeutic efficacy in lantana poisoning in buffalo calves in relation to clinical and haematological studies. *Indian Vet J* 50;764–770, 1973

Hart NK, Lamberton JA, Sioumis AA, Suares H: New triterpenes of *Lantana camara:* a comparative study of the constituents of several taxa. *Aust J Chem* 29;655–671, 1976a

Hart NK, Lamberton JA, Sioumis AA, Suares H, Seawright AA: Triterpenes of toxic and non-toxic taxa of *Lantana camara. Experientia* 32;412–413, 1976b

Ide A, Tutt CLC: Acute *Lantana camara* poisoning in a Boer goat kid. *J S Afr Vet Assoc* 69;30–32, 1998

Johnson JH, Jensen JM: Hepatotoxicity and secondary photosensitization in a red kangaroo *(Megaleia rufus)* due to ingestion of *Lantana camara. J Zoo Wildl Med* 29;203–207, 1998

Lall D, Lohan OP, Vaid J, Sharma OP, Negi SS: Can lantana seeds be utilized in lamb rations? *Anim Feed Sci Technol* 9;29–36, 1983a

Lall D, Sharma OP, Negi SS: Nutritive value of lantana *(Lantana camara)* seeds for rams. *Indian J Anim Sci* 53;1034–1036, 1983b

Louw PGJ: Lantanin, the active principle of *Lantana camara* L. Part 1. Isolation and preliminary results on the determination of its constitution. *Onderstepoort J Vet Sci Anim Ind* 18;197–202, 1943

Louw PGJ: Lantadene A, the active principle of *Lantana camara* L. Part 2. Isolation of lantadene B, and the oxygen functions of lantadene A and lantadene B. *Onderstepoort J Vet Sci Anim Ind* 23;233–238, 1948

Mandial RK, Randhawa SNS: Physicochemical and microbial alterations in rumen liquor during *Lantana camara* toxicity in buffalo calves *(Bubalus bubalis). Indian J Anim Sci* 68;1191–1192, 1998

Mathews FP: Whitebrush *(Lippia ligustrina)* poisoning in horses. *J Am Vet Med Assoc* 101;35–38, 1942

McKenzie RA: Bentonite as therapy for *Lantana camara* poisoning of cattle. *Aust Vet J* 68;146–148, 1991

McLennan, Amos ML: Treatment of lantana poisoning in cattle. *Aust Vet J* 66;93–94, 1989

McSweeney CS, Pass MA: The role of the rumen in absorption of lantana toxins in sheep. *Toxicon* (Suppl) 3;285–288, 1983

McSweeney CS, Stewart C, Pass MA: Treatment of lantana poisoning of cattle and sheep. In *Plant Toxicology, Proceedings of the Australia-USA Poisonous Plants Symposium,* Animal Research Institute, Yeeongpilly, Brisbane, Australia, pp61–69, 1985

Morton JF: Lantana, or red sage (*Lantana camara* L., [Verbenaceae]), notorious weed and popular garden flower: some cases of poisoning in Florida. *Econ Bot* 48;259–270, 1994

Obwolo MJ, Basudde CDK, Odiawo GO, Goedegebuure SA: Clinicopathological features of experimental acute *Lantana camara* poisoning in indigenous Zimbabwean goats. *Bull Anim Health Prod Afr* 39;339–346, 1991

Pass MA: Current ideas on the pathophysiology and treatment of lantana poisoning of ruminants. *Aust Vet J* 63;169–171, 1986

Pass MA: Toxins and mechanisms of intoxication. In *Poisonous Plants: Proceedings of the Third International Symposium,* James LF, Keeler RF, Bailey EM Jr, Cheeke PR, Hegarty MP eds, Iowa State Univ Press, Ames, pp405–413, 1992

Pass MA, Goosem MW: Bile canalicular injury in lantana poisoning. *Toxicon* (Suppl) 3;337–340, 1983

Pass MA, Heath T: Gallbladder paralysis in sheep during lantana poisoning. *J Comp Pathol* 87;301–306, 1977

Pass MA, Stewart C: Immunization against lantana toxins. In *Poisonous Plants: Proceedings of the Third International Symposium,* James LF, Keeler RF, Bailey EM Jr, Cheeke PR, Hegarty MP eds, Iowa State Univ Press, Ames, pp443–447, 1992

Pass MA, Gemmell RT, Heath TJ: Effect of lantana on the ultrastructure of the liver of sheep. *Toxicol Appl Pharmacol* 43;589–596, 1978a

Pass MA, Seawright AA, Heath TJ, Gemmell RT: Lantana poisoning: a cholestatic disease of cattle and sheep. In *Effects of Poisonous Plants on Livestock,* Keeler RF, Van Kampen KR, James LF eds, Academic Press, New York, pp229–237, 1978b

Pass MA, McSweeney CS, Reynoldson JA: Absorption of the toxins of *Lantana camara* L. from the digestive system of sheep. *J Appl Toxicol* 1;38–41, 1981

Pass MA, Goosem MW, Pollott S: A relationship between hepatic metabolism of reduced lantadene A and its toxicity in rats and sheep. *Comp Biochem Physiol* 82C;457–461, 1985a

Pass MA, Pollitt S, Goosem MW, McSweeney CS: The pathogenesis of lantana poisoning, in plant toxicology. In *Proceedings of the Australia-USA Poisonous Plants Symposium,* Animal Research Institute, Yeerongpilly, Brisbane, Australia, pp487–494, 1985b

Sanders DA: Lantana poisoning in cattle. *J Am Vet Med Assoc* 109;139–141, 1946

Seawright AA: Studies on the pathology of experimental lantana (*Lantana camara* L.) poisoning of sheep. *Path Vet* 1;504–529, 1964

Seawright AA: Electron microscopic observations of the hepatocytes of sheep in lantana poisoning. *Path Vet* 2;175–196, 1965a

Seawright AA: Toxicity of *Lantana* spp in Queensland. *Aust Vet J* 41;235–238, 1965b

Seawright AA, Allen JG: Pathology of the liver and kidney in lantana poisoning of cattle. *Aust Vet J* 48;323–331, 1972

Seawright AA, Hrdlicka J: The oral toxicity for sheep of triterpene acids isolated from *Lantana camara. Aust Vet J* 53;230–235, 1977

Sharma OmP: How to combat lantana (*Lantana camara* L.) menace? A current perspective. *J Sci Ind Res* 47;611–616, 1988

Sharma OmP, Dawra RK, Krishna L, Makkar HPS: Toxicity of lantana (*Lantana camara* L.) leaves and isolated toxins to rabbits. *Vet Hum Toxicol* 30;214–218, 1988

Sharma OmP, Dawra RK, Makkar HPS: Toxicity of isolated lantana (*Lantana camara* L.) constituents to male and female guinea pigs. *Vet Hum Toxicol* 31;10–13, 1989

Sharma OmP, Vaid J, Sharma PD: Comparison of lantadenes content and toxicity of different taxa of the lantana plant. *J Chem Ecol* 17;2283–2291, 1991

Sharma OP: Plant toxicoses in north-western India. In *Plant-Associated Toxins: Agricultural, Phytochemical, and Ecological Aspects,* Colegate SM, Dorling PR eds, CAB International, Wallingford, UK, pp19–24, 1994

Sharma OP, Makkar HPS, Pal RN, Negi SS: Lantadene A content and toxicity of the lantana plant (*Lantana camara,* Linn.) to guinea pigs. *Toxicon* 18;485–488, 1980

Sharma S, Sharma OP, Singh B, Bhat TK: Biotransformation of lantadenes, the pentacyclic triterpenoid hepatotoxins of lantana plant, in guinea pigs. *Toxicon* 38;1191–1202, 2000

Sperry OE, Dollahite JW, Hoffman GO, Camp BJ: *Texas Plants Poisonous to Livestock.* Tex Agric Exp Stn Publ B-1028, 1977 (reprint)

Standley PC: Trees and shrubs of Mexico. *Contrib U S Natl Herb* 23;1–1721, 1920–1926

Stewart C, Lamberton JA, Fairclough RJ, Pass MA: Vaccination as a possible means of preventing lantana poisoning. *Aust Vet J* 65;349–352, 1988

Tokarnia CH, Dobereiner J, Lazzari AA, Peixoto PV: Poisoning of cattle by species of *Lantana* in Mato Grosso and Rio de Janeiro states. *Pesq Vet Bras* 4;129–141, 1984
Tokarnia CH, Armien AG, Barros SS de, Peixoto PV, Dobereiner J: Complementary studies on the toxicity of *Lantana camara* (Verbenaceae) in cattle. *Pesq Vet Bras* 19;128–132, 1999
Wolfson SL, Solomons TWG: Poisoning by fruit of *Lantana camara*. *Am J Dis Child* 107;173–176, 1964

74 Viscaceae Miers

- Christmas Mistletoe Family
  Phoradendron

Comprising 7–11 genera and 350–450 species of semiparasites on tree branches, the Viscaceae, or Christmas mistletoe family, is cosmopolitan in distribution but with greatest abundance in the tropics. Although traditionally combined with the Loranthaceae, or showy mistletoe family, and treated as a subfamily, it is now recognized as distinct on the basis of embryological studies. Chlorophyllous and photosynthesizing, members of the family occur on a variety of tree hosts that provide water and inorganic nutrients taken up by haustoria, which penetrate the host's vascular system. Dispersal from one tree to another is facilitated by the mucilaginous seeds, which adhere to birds eating the fruits and are subsequently scraped off on another tree.

The family is of some economic importance. Species of the genus *Arceuthobium*, commonly known as dwarf mistletoe, are destructive branch parasites of native conifers, causing them to form "witches' brooms" and often to die. Although also semiparasitic, species of *Phoradendron* are less destructive and are the source of decorative mistletoe at Christmas time. In Europe, the Old World genus *Viscum* is used instead. Mistletoe was a plant revered by the Druids and used in their ceremonies to celebrate the winter solstice. In North America, 4 genera and 43 species are present. Toxicologic problems are associated with 1 genus.

- evergreen, semiparasitic, woody epiphytes, roots penetrating branches of woody hosts; leaves simple, opposite, well developed, or scalelike; flowers imperfect; fruits berries

Plants woody epiphytes; evergreen; semiparasitic; monoecious or dioecious. **Root Systems** haustorial; penetrating branches of woody hosts. **Stems** dichotomously branched; nodes jointed; chlorophyllous; brittle. **Leaves** well developed or scalelike; simple; opposite; decussate; stipules absent. **Inflorescences** spicate cymes; axillary. **Flowers** imperfect, staminate and pistillate similar; perianths in 1-series; radially symmetrical. **Perianth Parts** 3, or rarely 2 or 4; free or fused; scalelike; yellowish green. **Stamens** 3 or rarely 2 or 4; borne at bases of perianth parts; anthers dehiscing transversely; filaments absent; androecial rudiments absent in pistillate flowers. **Pistils** 1; compound, carpels 3 or 4; stigmas 1, capitate; styles 1; ovaries inferior; locules 1; ovules not differentiated from placentae; gynoecial rudiments present in staminate flowers. **Fruits** berries; shining; mesocarp mucilaginous; explosive in 1 genus. **Seeds** 1.

- *P. bolleanum* *P. macrophyllum*
  P. californicum *P. rubrum*
  P. juniperinum *P. tomentosum*
  P. leucarpum *P. villosum*

  acacia mistletoe, hairy mistletoe, American mistletoe, injerto, bigleaf mistletoe, juniper mistletoe, cedar mistletoe, legume mistletoe, Christmas mistletoe, oak mistletoe, dense mistletoe, rough mistletoe, desert mistletoe, scaly juniper mistletoe, eastern mistletoe, West Indian mistletoe

Phoradendron Nutt.

Taxonomy and Morphology—A New World genus most abundant in the tropics, *Phoradendron* comprises about 200 species. Aptly named because of its semiparasitic habit, its Greek roots *phor* and *dendron* mean "tree thief." Species relationships and the names to be applied have been the subject of considerable taxonomic debate. The treatment of Wiens (1964) is followed here. Four sections with 19–20 species are present in North America. Of toxicologic interest are the following:

P. bolleanum (Seem.) Eichler	dense mistletoe, rough mistletoe
P. californicum Nutt.	acacia mistletoe, desert mistletoe, legume mistletoe
P. juniperinum A.Gray	juniper mistletoe, cedar mistletoe, scaly juniper mistletoe
P. leucarpum (Raf.) Reveal & M.C.Johnst. (= *P. serotinum* [Raf.] M.C.Johnst.) (= *P. flavescens* Nutt. ex Engelm.)	eastern mistletoe, Christmas mistletoe
P. macrophyllum (Engelm.) Cockerell	bigleaf mistletoe
P. rubrum (L.) Griseb.	West Indian mistletoe
P. tomentosum (DC.)Engelm. ex A.Gray	American mistletoe, hairy mistletoe, injerto
P. villosum (Nutt.) Nutt. (= *P. coryae* Trel.)	oak mistletoe

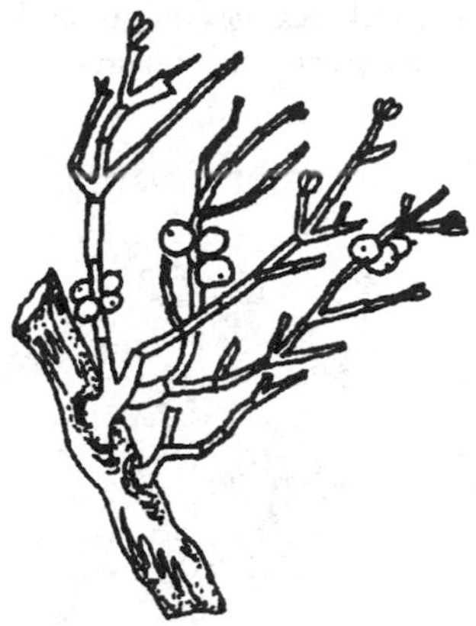

Phoradendron juniperinum

■ nomenclatural confusion about name for eastern mistletoe

The correct name for the mistletoes of the eastern half of the North American continent is perhaps the most confusing issue in the taxonomy of the genus. Initially plants of the Southeast were called *P. flavescens.* It was determined to be an illegitimate name, and the name *P. serotinum* was used instead (Johnston 1957). Some taxonomists, however, merge *P. serotinum* with the more western *P. tomentosum* (which was also called *P. flavescens* in some publications) and use the latter binomial. Others continue to recognize 2 species, as is done here. A further complication is that an earlier name, *P. leucarpum,* exists and now must be used in place of the familiar name *P. serotinum* (Reveal & Johnston 1989).

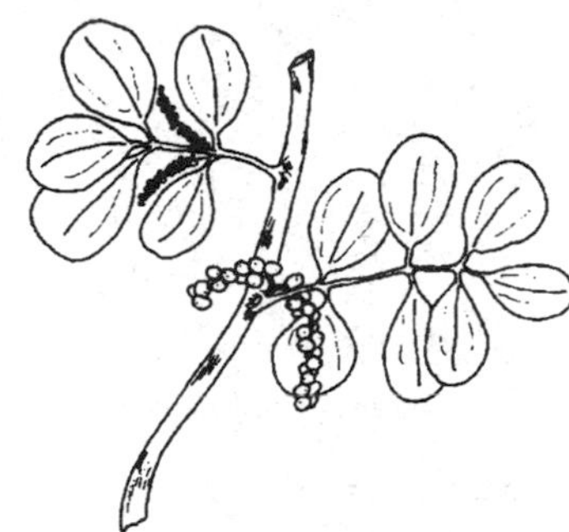

Phoradendron leucarpum

■ shrublike, forming dense masses; stems dark green; berries globose, white or dark red

Plants shrublike; forming dense masses. **Stems** dark green; readily disarticulating at nodes. **Leaves** well developed or scalelike; coriaceous to succulent; short-petiolate; blades elliptic or obovate to oval; venation pinnate; apices obtuse to rounded; margins entire. **Flowers** inconspicuous. **Perianth Parts** free; deltoid; persistent. **Berries** globose or ellipsoid-globose; white or dark red; not explosive.

Because of the nomenclatural ambiguity in the genus and localized host specificity, a taxonomic key is not presented in this treatment. If exact identification of an unknown mistletoe is needed, manuals specific for the reader's locality should be consulted.

Phoradendron villosum

■ wide distribution, especially southern North America

Distribution and Habitat—Species of *Phoradendron* occur across North America. Some, such as *P. leucarpum, P. juniperinum,* and *P. tomentosum,* are widely distributed, whereas others are more restricted. Greatest abundance is in the southern half of the continent, and many are restricted to Mexico.

Tree species serving as hosts vary among the species of *Phoradendron* (Wiens 1964). For example, *P. juniperinum* parasitizes *Juniperus* exclusively, and *P. californicum* is found on leguminous shrubs of the Sonoran deserts of the southwestern United States and Mexico. In contrast, *P. tomentosum* is parasitic on many deciduous trees, including oaks, willows, and poplars. Likewise, *P. leucarpum* parasitizes a variety of woody plants of the eastern deciduous forest.

Disease Problems—The mistletoes are somewhat of a toxicological enigma. Mystical qualities have been ascribed to the European *Viscum album* since the time of its use in religious rites of the Druids (Crawford 1911). It also was used widely for an assort-

■ medicinal uses, unknown toxicity, weak pressor action on uterus

Phoradendron bolleanum

Phoradendron californicum

■ animals, few cases reported humans, rarely digestive disturbance

Phoradendron juniperinum

ment of medicinal purposes but at the same time was occasionally rumored to be toxic. The latter quality seemed of little importance, given that it had been fed to livestock and dogs without problem and had even been used as a tobacco substitute. Some of the lore attached to the Old World species of *Viscum* was eventually extrapolated to the New World species of *Phoradendron.* Although an extract of *P. leucarpum* (reported as *P. serotinum*) was apparently of some popularity in the early 1900s for its oxytocic activity, anecdotal implications of toxicity from the plant soon surfaced. Claims of toxicity with vomiting, purgation, collapse, and a pounding pulse were occasionally presented, and medicinal use of the plant eventually declined (Crawford 1911).

There remains considerable uncertainty as to the actual toxicity potential of *Phoradendron* and whether the type of toxicity and clinical signs produced are the same among the different species in North America. Members of western American Indian tribes frequently used the berries of various species of mistletoe for food or drink (Ebeling 1986). It is of interest, however, that they were very selective about the host plant from which the mistletoe was collected, because the host was considered to be a guide as to whether the mistletoe was toxic (Chesnut 1902). The role of the host plant in influencing the toxicity potential has subsequently been shown for many parasitic plants, including Old World *Viscum* (Atsatt 1977; Fukunaga et al. 1989; Bloor & Molloy 1991).

Although species of *Phoradendron* are often listed as toxic, few cases in animals have been reported. In reality, the species form part of the typical diet of deer, elk, and several small mammals (Atsatt 1977). In a California case, 13 cows eating *P. villosum* (growing on oaks) over a 3-month period died, each within 12 hours after becoming ill (Fuller & McClintock 1986). However, the pathology was not reported, and the possibility of oak toxicity or other causes cannot be excluded. Surprisingly little information is available on the toxicity of the whole plant given orally, because most studies involve extracts or isolated compounds given parenterally. In one study, foliage and berries of *P. leucarpum* (reported as *P. serotinum*) fed to pigs failed to produce disease (Cary et al. 1924). A tea made from *P. latifolium,* a tropical species, given daily for 30 days caused hepatic and renal degeneration in rats (Queiroz-Neto et al. 1985). An extract of *P. tomentosum* given orally resulted in marginal effects on pulse rate and uterine activity (Hanzlik & French 1924). In studies carried out by the authors, 3% b.w. of the foliage of *P. tomentosum* given orally to a sheep did not cause any apparent adverse effects. There seems to be little reason for undue concern regarding serious disease potential to livestock of species of *Phoradendron.*

Although there are few reports of ingestions in animals, numerous incidents in humans have been reported. A woman in South Carolina is reported to have brewed and drunk an unknown amount of mistletoe tea made presumably with locally available *P. leucarpum* (Moore 1963). Within 1–2 hours she became abruptly ill, with severe abdominal pain, vomiting, and diarrhea. She was eventually hospitalized with cardiovascular collapse and profound hypotension and died 8–10 hours after becoming ill. In contrast, Hall and co-workers (1986) examined 14 cases of ingestion of mistletoe, 11 of berries and 3 of leaves. There were no adverse effects noted for any of the ingestions. They also reviewed 318 cases reported to poison control centers from 1978 to 1984 and found that few resulted in the development of signs of intoxication. There were, however, 2 fatalities, possibly due to tea brewed from the berries or other plant parts. These investigators concluded that ingestion of 1–3 berries or leaves was unlikely to cause serious intoxication. Similar findings have been reported by others. Krenzelok and co-workers (1997) reviewed 1,754 mistletoe exposures reported to poison information centers over an 8-year period (1985–1992). There were no adverse signs in more than 99% of the exposures, no fatalities, and only 2 cases represented serious problems. Spiller and co-workers (1996) reviewed 92 cases of mistletoe exposure (some data may overlap with those of Krenzelok) from 4 poison control centers in individuals ranging in age from 4 months to 42 years. Of these, 11 developed signs of illness; 6, digestive tract problems; 2, drowsiness; 1 (infant), seizures; 1, eye problems; and 1, incoordination. The researchers concluded that ingestion of even up to 20 berries or 5 leaves would rarely cause serious problems.

■ various compounds present; glycoprotein lectins, proteins, alkaloids, amines

Disease Genesis—A number of compounds of potential toxicologic importance have been isolated from various Old and New World species of mistletoe, including protein lectins (thionens), small basic proteins, alkaloids, flavonoids, choline, γ-aminobutyric acid (GABA), tyramine and other amines, and high concentrations of potassium. The compounds present in the European and New World genera are quite similar.

Glycoprotein lectins composed of chains with molecular weights of 31,000 and 38,000 have been isolated from *P. californicum*. These lectins inactivate 60S ribosomal subunits and inhibit protein synthesis in a manner similar to the action of ricin and abrin but with much lower potency (Thunberg & Samuelsson 1982; Thunberg 1983; Endo et al. 1989; Tonevitsky et al. 1991). Similar to viscumin from *Viscum album,* the lectins from *P. californicum* are, however, much less active, requiring 100 times greater dose to produce intoxication (Luther et al. 1980; Stirpe et al. 1980, 1982; Olsnes et al. 1982). Viscumin (= lectin I) also inhibits protein synthesis in whole cells in a manner similar to that of ricin; the facilitator chain of viscumin interacts with different surface receptors (Olsnes et al. 1982; Stirpe et al. 1982). Studies on these toxic lectins and the smaller viscotoxins from *V. album* have led to speculation that these compounds exist as glycoconjugates with alkaloids (Khwaja et al. 1986). If true for *Phoradendron* proteins, this characteristic may play a role in enhancing the likelihood of intoxication by ingestion. Otherwise, it is doubtful that either type of protein would be an important factor in disease, because the conditions of the digestive system would lead to their destruction. In addition, antiserum to viscumin does not prevent circulatory collapse (Olsnes et al. 1982).

Phoradendron leucarpum

Small basic proteins—called viscotoxins A2, A3, and B—and phoratoxins have been isolated from *V. album* and *P. tomentosum,* respectively (Samuelsson 1973). These proteins, composed of 46 amino acids with 3 disulfide bridges and a *N*-terminal lysine, have negative inotropic effects on the heart, causing hypotension and decreased cardiac rate and force of contraction (Rosell & Samuelsson 1966). Like the larger lectins, phoratoxins are less potent than the viscotoxins (Samuelsson & Ekblad 1967; Mellstrand & Samuelsson 1973). Similar protein toxins, such as ligatoxin A from *P. liga,* have been shown to occur in other tropical species of *Phoradendron* (Samuelsson et al. 1981; Thunberg & Samuelsson 1982). Interestingly, phoratoxin B has a depolarizing action on frog skeletal muscle, possibly due to a detergent-like action on the membrane phospholipid layer (Sauviat 1990). Other compounds contributing to the hypotensive effects reported especially for tropical phoradendrons include γ-aminobutyric acid and flavones (Durand et al. 1962; Queiroz-Neto et al. 1986; Fukunaga et al. 1989). *Phoradendron rubrum* is reported to contain a hypotensive alkaloid called rubrine C (West & Feng 1967). Some of the early studies on *P. tomentosum* extracts indicated that instead of being hypotensive, they were vasoconstrictive and increased uterine smooth muscle activity (Hanzlik & French 1924). These effects were ascribed to tyramine and other phenethylamine-type agents (*p*-OH-phenethylamine), compounds that have subsequently been found in many species of *Phoradendron* (Crawford & Watanabe 1914, 1916; Graziano et al. 1967; Smith 1977; Queiroz-Neto et al. 1986). This finding also prompted

Phoradendron macrophyllum

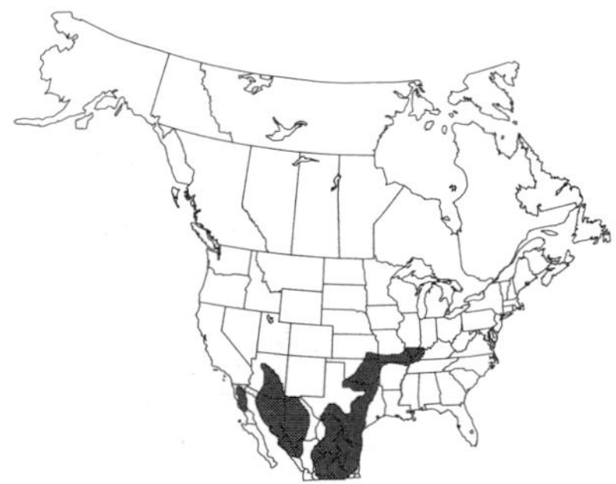
Phoradendron tomentosum

concerns about toxicity. However, all of the above toxicants are effective when given parenterally but not necessarily when given orally.

■ possible adverse effects of herbal preparations, hypotension or hypertension

The pharmacologic effects of the various North American species have not been thoroughly studied, although *P. juniperinum* and *P. leucarpum* are reported to be hypertensive and *P. villosum* hypotensive (Crawford 1911; Hanzlik & French 1924). In some instances there appear to be disparate actions on smooth muscle of various tissues; there may be relaxation of vascular smooth muscle and hypotension together with stimulation in other tissues such as the uterus (Hanzlik & French 1924; Queiroz-Neto & Melito 1990). Detailed studies on the South American mistletoe *P. latifolium* revealed that there is initial hypertension of short duration followed by a more sustained hypotension. It was concluded that the hypotension was associated with amino compounds and was not due to effects on muscarinic or adrenergic receptors (Cortez et al. 1988; Queiroz-Neto et al. 1988). The potential for hypotensive effects associated with teas made from species of *Phoradendron* is of most concern. Problems are not likely from ingestion of the fresh berries or foliage directly.

Phoradendron villosum

Another protein in *Phoradendron* is viscin, the mucilaginous seed covering, which is important in the dispersal of the seeds by birds and allows the seeds to become attached to host plants in new sites. Viscin is composed of proteins and neutral sugars (Gedalovich et al. 1988; Gedalovich-Shedletzky et al. 1989). The proteins are of unknown toxicologic significance.

■ vomiting, possibly weakness due to hypotension

treat with fluids, electrolytes

Clinical Signs, Pathology, and Treatment—The most likely signs of *Phoradendron* intoxication are vomiting and perhaps some diarrhea, although nonlethal systemic effects, especially hypotension, are possible. With respect to treatment, the principal concern is correction of fluid and electrolyte deficits that may develop. In rare instances, the use of sympathomimetics to control the hypotension may be indicated.

References

Atsatt PR: The insect herbivore as a predictive model in parasitic seed plant biology [letter]. *Am Nat* 3;579–586, 1977

Bloor SJ, Molloy BPJ: Cytotoxic norditerpene lactones from *Ileostylus micranthus*. *J Nat Prod* 54;1326–1330, 1991

Cary CA, Miller ER, Johnstone GR: *Poisonous Plants of Alabama*. Ala Polytech Inst Ext Serv Circ 71, p10, 1924

Chesnut VK: Plants used by the Indians of Mendocino County, California. *Contrib U S Natl Herb* 7;295–408, 1902

Cortez DAG, Queiroz-Neto A, Hubner DV, Melito I: Contribution to the phytochemical study of *Phoradendron latifolium* (SW) Griseb. (erva-de-passarinho). *Acta Amazonica* (Suppl) 18;433–438, 1988

Crawford AC: The pressor action of an American mistletoe. *J Am Med Assoc* 57;865–868, 1911

Crawford AC, Watanabe WK: Parahydroxyphenylethylamine, a pressor compound in American mistletoe. *J Biol Chem* 19;303–304, 1914

Crawford AC, Watanabe WK: The occurrence of *p*-hydroxyphenethylamine in various mistletoes. *J Biol Chem* 24;169–172, 1916

Durand E, Ellington EV, Feng PC, Haynes LJ, Magnus KE, Philip N: Simple hypotensive and hypertensive principles from some West Indian medicinal plants. *J Pharm Pharmacol* 14;562–566, 1962

Ebeling W: *Handbook of Indian Foods and Fibers of Arid America*. Univ California Press, Berkeley, 1986

Endo Y, Oka T, Tsurugi K, Franz H: The mechanism of action of the cytotoxic lectin from *Phoradendron californicum*: the RNA *N*-glycosidase activity of the protein. *FEBS Lett* 248;115–118, 1989

Fukunaga T, Nishiya K, Kajikawa I, Takeya K, Itokawa H: Studies on the constituents of Japanese mistletoes from different host trees, and their antimicrobial and hypotensive properties. *Chem Pharm Bull* 37;1543–1546, 1989

Fuller TC, McClintock E: *Poisonous Plants of California*. Univ Calif Press, Berkeley, 1986

Gedalovich E, Kuijt J, Carpita NC: Chemical composition of viscin, an adhesive involved in dispersal of the parasite *Phoradendron californicum* (Viscaceae). *Physiol Mol Plant Pathol* 32;61–76, 1988

Gedalovich-Shedletzky E, Delmer DP, Kuijt J: Chemical composition of viscin mucilage from three mistletoe species—a comparison. *Ann Bot* 64;249–252, 1989

Graziano MN, Widmer GA, Coussio JD: Isolation of tyramine from five species of Loranthaceae. *Lloydia* 30;242–244, 1967

Hall AH, Spoerke DG, Rumack BH: Assessing mistletoe toxicity. *Ann Emerg Med* 15;1320–1323, 1986

Hanzlik PJ, French WO: The pharmacology of *Phoradendron flavescens* (American mistletoe). *J Pharmacol Exp Ther* 23;269–306, 1924

Johnston MC: *Phoradendron serotinum* for *P. flavescens* (Loranthaceae): nomenclatural corrections. *Southwest Nat* 2:45–47, 1957

Khwaja TA, Dias CB, Pentecost S: Recent studies on the anticancer activities of mistletoe *(Viscum album)* and its alkaloids. *Oncology* 43 (Suppl 1);42–50, 1986

Krenzelok EP, Jacobsen TD, Aronis JM: American mistletoe exposures. *Am J Emerg Med* 15;516–520, 1997

Luther P, Theise H, Chatterjee B, Karduck D, Uhlenbruck G: The lectin from *Viscum album* L.—isolation, characterization, properties, and structure. *Int J Biochem* 11; 429–435, 1980

Mellstrand ST, Samuelsson G: Phoratoxin, a toxic protein from the mistletoe *Phoradendron tomentosum* subsp. *macrophyllum* (Loranthaceae). *Eur J Biochem* 32; 143–147, 1973

Moore HW: Mistletoe poisoning. *J S C Med Assoc* 59;269–271, 1963

Olsnes S, Stirpe F, Sandvig K, Pihl A: Isolation and characterization of viscumin, a toxic lectin from *Viscum album* L. (mistletoe). *J Biol Chem* 257;13263–13270, 1982

Queiroz-Neto A, Melito I: Changes in sensitivity of the isolated guinea-pig vas deferens induced by a lyophilized *Phoradendron latifolium* leaf infusion. *J Ethnopharmacol* 28;183–189, 1990

Queiroz-Neto A, Hubner DV, Cortez DAG, Alessi AC, Melito I: Toxicological evaluation of mistletoe tea *(Phoradendron latifolium)*. *Ars Vet* 1;9–16, 1985

Queiroz-Neto A, Cortez DA, Prado WA, Bracht A: Identification of γ-aminobutyric acid in *Phoradendron latifolium* (SW) Griseb. *Ars Vet* 2;185–193, 1986

Queiroz-Neto A, Cortez DAG, Hubner DV, Melito I: Pharmacological evaluation of the active principles present in the tea of the mistletoe *Phoradendron latifolium* (SW) Griseb. *Acta Amazonica* (Suppl) 18;197–202, 1988

Reveal JL, Johnston MC: A new combination in *Phoradendron* (Viscaceae). *Taxon* 38:107–108, 1989

Rosell S, Samuelsson G: Effect of mistletoe viscotoxin and phoratoxin on blood circulation. *Toxicon* 4;107–110, 1966

Samuelsson G: Mistletoe toxins. *Syst Zool* 22;566–569, 1973

Samuelsson G, Ekblad M: Isolation and properties of phoratoxin, a toxic protein from *Phoradendron serotinum* (Loranthaceae). *Acta Chem Scand* 21;849–856, 1967

Samuelsson G, Borsub L, Jayawardine AL, Falk L, Ziemilis S: Screening of plants of the families Loranthaceae and Viscaceae for toxic proteins. *Acta Pharm Suec* 18;179–184, 1981

Sauviat M-P: Effect of phoratoxin B, a toxin isolated from mistletoe, on frog skeletal muscle fibers. *Toxicon* 28;83–89, 1990

Smith TA: Phenethylamine and related compounds in plants. *Phytochemistry* 16;9–18, 1977

Spiller HA, Willias DB, Gorman SE, Sanftleban J: Retrospective study of mistletoe ingestion. *J Toxicol Clin Toxicol* 34;405–408, 1996

Stirpe F, Legg RF, Onyon LJ, Ziska P, Franz H: Inhibition of protein synthesis by a toxic lectin from *Viscum album* L. (mistletoe). *Biochem J* 190;843–845, 1980
Stirpe F, Sandvig K, Olsnes S, Pihl A: Action of viscumin, a toxic lectin from mistletoe, on cells in culture. *J Biol Chem* 257;13271–13277, 1982
Thunberg E: Phoratoxin B, a toxic protein from the mistletoe *Phoradendron tomentosum* ssp. *macrophyllum*. *Acta Pharm Suec* 20;115–122, 1983
Thunberg E, Samuelsson G: Isolation and properties of ligatoxin A, a toxic protein from the mistletoe *Phoradendron liga*. *Acta Pharm Suec* 19;285–292, 1982
Tonevitsky AG, Toptygin AY, Pfuller U, Bushueva TL, Ershova GV, Gelbin M, Pfuller K, Agapov II, Franz H: Immunotoxin with mistletoe lectin I A-chain and ricin A-chain directed against CD5 antigen of human T-lymphocytes: comparison of efficiency and specificity. *Int J Immunopharmacol* 13;1037–1041, 1991
West ME, Feng PC: Rubrine C, a pharmacologically active alkaloid from *Phoradendron rubrum*. *J Pharm Pharmacol* 19;197–198, 1967
Wiens D: Revision of the acataphyllous species of *Phoradendron*. *Brittonia* 16;11–54, 1964

Zamiaceae Horan.

75

- Sago-palm Family

Ceratozamia
Dioon
Zamia

Comprising 8 genera and 80–100 species in the tropics and subtropics of both the Old World and the New World, the Zamiaceae, or sago-palm family, is the largest family of cycads. Its taxa were originally classified in the broadly circumscribed Cycadaceae but were segregated because of differences in mode of leaflet development, appearance of seed cones, and other attributes (Johnson 1959; Stevenson 1992). Like the Cycadaceae, the Zamiaceae has a fossil history extending to the Paleozoic, and its members dominated the world's vegetation in the Jurassic and early Cretaceous periods, a time known as the Age of the Cycads, or the Age of the Dinosaurs (Norstog & Nicholls 1997). In North America, 3 genera are present.

- primitive seed-bearing plants with motile sperm

As the common name of the family implies, these cycads are superficially similar to palms in the family Arecaceae, and they are often mistaken for them. However, they have few features in common. The leaves of several genera are used in funerals and as decoration in places of worship in both Central America and Australasia (Jones 1993). Among the most primitive of the seed-bearing plants, species of the family have motile sperm that move to fertilize the ovule in a manner analogous to fertilization in animals.

- treelike or herbs, perennial, evergreen, pollen and seed cones; foliage leaves large, pinnately compound, clustered at stem apices; cones erect, terminal or axillary; seeds large, drupelike

Plants treelike or herbs; perennials; evergreen; caulescent or acaulescent; producing sporangia borne in pollen and seed cones; dioecious. **Stems** subterranean or aerial; erect; columnar; 2–10 m tall; not branched or rarely so; bases of old leaves present or absent. **Leaves** of 2 forms, foliage and scale (cataphylls) different; foliage leaves large, 1-pinnately compound, spirally clustered at stem apices, petiolate, blades oblong, leaflets straight or rarely circinate in bud, coriaceous to papery, bases articulated or not articulated; petiole bases sheathing, persistent or deciduous; cataphylls alternating with foliage leaves. **Pollen Cones** terminal or axillary; erect; cylindrical to elliptic; microsporophylls numerous, spirally arranged or in vertical rows; microsporangia numerous; pollen with motile sperm. **Seed Cones** terminal or axillary; erect; megasporophylls numerous, consisting of basal linear stalk and expanded apical lobe; ovules 2 or 3 per megasporophyll, attached to stalk. **Seeds** large; drupelike; globose to subcylindrical; variously angled; seed coat differentiated into indurate inner and fleshy outer layers.

- seeds, leaves, pulp, mainly hepatotoxic, less often neurotoxic

Disease Problems—It appears that all genera of the Zamiaceae cause essentially the same toxicologic problems, which are also similar to those produced by *Cycas* (see chapter 30). Ingestion of the seeds, leaves, or pulp typically produces hepatotoxic effects and occasionally neurologic disease in livestock (Hall & McGavin 1968; Gabbedy et al. 1975; Hooper 1978). In Australia the disease is sometimes called zamia staggers (Seawright 1989). There also have been several cases of hepatotoxicity in dogs that consumed the leaves, stems, or seeds of *Dioon, Zamia,* or *Macrozamia* (Senior et al. 1985; Botha et al. 1991; Mills et al. 1996). Dogs that chewed the stems developed marked effects within 1 hour and exhibited evidence of liver involvement. They recovered with only conservative treatment. In another incident, dogs that were fed roasted seeds of *Macrozamia* developed severe digestive disturbance followed by intractable abdominal pain 3 days later. The pain was accompanied by anemia, a clotting disorder, and elevation of serum hepatic enzymes (Mills et al. 1996). In some instances, both hepatotoxic and neurotoxic effects appear in the same group of animals. Cattle in Puerto Rico that ate the leaves and nuts of *Z. portoricensis* Urb. (reported as *Z. puertoriquensis*) developed severe liver cell necrosis and brain and spinal lesions (Reams et al. 1993). The type of disease manifested may be a function of the plant part eaten, although this has not been confirmed experimentally.

■ glycosides cycasin and macrozamin hydrolyze to MAM, which alkylates RNA and DNA of liver and may be neurotoxic; L-BMAA also may have neurologic effects as a weak excitotoxin

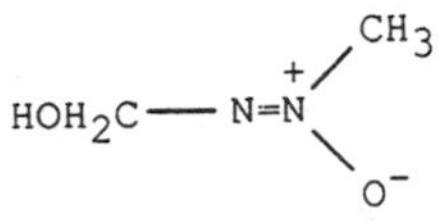

Methylazoxymethanol (MAM)

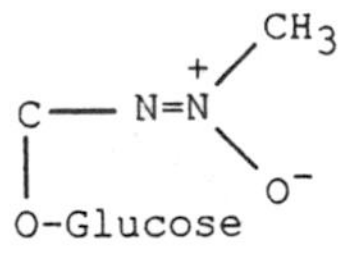

Cycasin

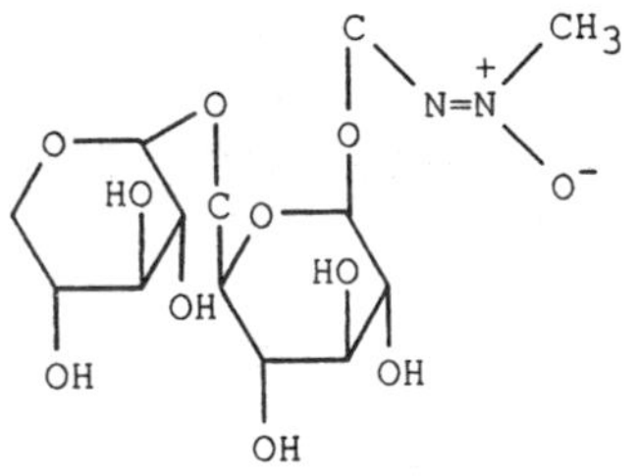

Macrozamin

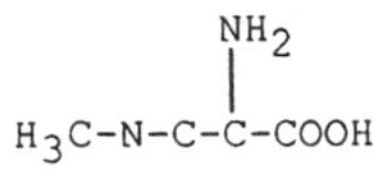

BMAA

Disease Genesis—The genera of the Zamiaceae contain two different types of toxins—glycosides of methylazoxymethanol (MAM) and the neurotoxic amino acid β-*N*-methylamino-L-alanine (L-BMAA). The principal glycosides of MAM present are the β-glucoside known as cycasin and the β-primeveroside known as macrozamin. Several minor glycosides, neocycasins, also are present (Yagi & Tadera 1987). The glycosides are subject to hydrolysis by β-glycosidases in the plant or more importantly by bacterial enzymes in the digestive tract of the animal to yield the aglycone and sugars (Spatz et al. 1967). The aglycone MAM is the toxic portion, alkylating DNA and RNA and causing hepatotoxicosis (Laqueur & Spatz 1968). MAM also is carcinogenic, mutagenic, neurotoxic, and teratogenic; whether it causes significant problems in animals via these actions is unknown (Smith 1966; Spatz et al. 1967; Tustin 1983). The teratogenic potential may be of particular concern (Keeler 1972). The glycosides are found in all parts of the plants. The highest concentrations of macrozamin are present in the seeds (Yagi & Tadera 1987).

Present in the seeds, β-*N*-methylamino-L-alanine (L-BMAA) is the second type of toxicant present in members of the Zamiaceae. It is an unusual neurotoxic amino acid similar in structure and action to the neurolathyrogens except that the glutamate receptor of major concern is the *N*-methyl-D-aspartate, or NMDA, type (Ross et al. 1987). L-BMAA, which acts as a weak excitotoxin in some circumstances, also affects the kainate and quisqualate types of glutamate receptors (Weiss et al. 1989). It has been suggested that BMAA itself may not be the culpable toxin but rather that, in the presence of bicarbonate, the α-carbamate form of the L-α-amino acid forms a more specific structural analog of NMDA and interactant at the receptor (Nunn et al. 1991). L-BMAA is subject to active uptake transport across the blood-brain barrier via the neutral amino acid carrier (Smith et al. 1992). Although BMAA is less potent than the neurolathyrogen L-BOAA, it is nevertheless an excitatory neurotoxicant; however, it takes hours rather than minutes to produce the presumably indirect actions (Nunn et al. 1987). Additional information about the roles of MAM and BMAA in the neurotoxic effects of cycad toxicity is presented in the treatment of the Cycadaceae (see chapter 30).

■ dogs, eating seeds; persistent vomiting, anorexia, constipation, icterus, increased serum hepatic enzymes and bilirubin

similar hepatic disease changes in sheep

neurologic effects in cattle, stumbling, staggering, ataxia

Clinical Signs—Because taxa of the Zamiaceae are so limited in distribution and abundance in North America, intoxications are most likely to involve pets that eat the seeds. In dogs, the initial sign is vomiting, which may begin within minutes of ingestion and may persist, with periodic episodes, for hours. There may also be increased salivation and increased water consumption. During the next few days anorexia, constipation, and icterus will be observed in conjunction with a marked decrease in blood platelets and an increase in serum bilirubin (especially conjugated), hepatic enzymes, urea nitrogen, and creatinine. At this point the prognosis is guarded because the liver involvement may be life-threatening.

A similar disease occurs in sheep but may, on occasion, be subacute to almost chronic. Emaciation, pale to icteric mucus membranes, and other evidence of liver disease are observed. The animals will be lethargic, but there is seldom diarrhea. In cattle, the neurologic disease occurs after consumption of the leaves for several weeks or more. The signs are those related to a proprioceptive deficit and progressive weakness in the pelvic limbs, as indicated by stumbling, staggering, incoordination, and sometimes goose-stepping.

Pathology—In dogs, sheep, and cattle displaying both liver and brain involvement, the characteristic gross change is an enlarged, congested, nutmeg liver. Microscopically, there is hepatic necrosis, which may be focal centrilobular or midzonal. In more chronic cases fibrosis, bile duct proliferation, bile stasis, and moderately enlarged hepatocytes may be noted. Additionally there may be scattered ecchymotic hemorrhages, visceral edema, and mild tubular nephrosis.

- hepatic form, gross pathology, enlarged, nutmeg liver

  microscopic, hepatic necrosis, fibrosis, bile duct proliferation

  neurologic form, microscopic, spinal axonal and myelin degeneration

There are no distinctive gross changes in cattle in instances in which the disease in exclusively neurologic, but there are numerous microscopic lesions. Bilaterally symmetrical degeneration in the ventral, lateral, and dorsal columns is present through the length of the cord but is especially evident in the cervical portion (Reams et al. 1993). Axonal and myelin degeneration is particularly noticeable in the gracile fascicle and the dorsal spinocerebellar tracts.

Treatment—Supportive care as for hepatic disease in general is indicated. There are no specific treatments for either the hepatic or the neurologic syndromes.

- general nursing care

Ceratozamia Brongn.

Taxonomy and Morphology—A New World genus comprising 11–12 species, *Ceratozamia* is morphologically quite similar to larger members of the genus *Zamia* (Nortstog & Nicholls 1997). Its name reflects this relationship and the presence of paired hornlike projections on the sporophylls. The Greek root *ceras* means "horn." In North America, only 1 species is of toxicologic interest:

- *C. mexicana*

  horned cycad
  horned zamia

C. mexicana Brongn. horned zamia, horned cycad

Plants herbs. **Stems** partially subterranean; erect; columnar; 10–50 cm tall; bases of old leaves absent. **Leaves** 10–20; arching; 1–3 m long; leaflets 14–150, articulated, dark green, glossy, blades lanceolate to linear, apices acute; petioles with prickles. **Pollen Cones** cylindrical to elliptic; 20–60 cm long; microsporophylls numerous, each with pair of recurved horns. **Seed Cones** cylindrical; 30–40 cm long; megasporophylls 200 or more, each with pair of recurved horns. **Seeds** yellow-white.

- erect columnar herbs; leaves arching; sporophylls of seed cones each bearing a pair of recurved horns

Distribution and Habitat—*Ceratozamia* is indigenous to central Mexico, Guatemala, and Belize. Plants are typically found in cloud or mist forests in the mountains. *Ceratozamia mexicana* occurs primarily in the state of Veracruz, in eastern Mexico. It is cultivated as an ornamental in other warm areas or as a tub plant. Unfortunately, populations have been damaged by poachers who collect plants to sell. All species of the genus are listed in Appendix 1 of CITES (Jones 1993).

- mountain forests of eastern Mexico

Dioon Lindl.

Taxonomy and Morphology—Comprising 10 New World species, *Dioon* is distinguished from other genera of the family by its nonarticulated leaflets and stalked ovules. Its name is derived from the Greek roots *di* and *oon,* for "two" and "egg,"

■ *D. edule*
D. spinulosum
chamal
coyolillo
jango
palma de la virgen
virgin's palm

reflecting the paired seeds borne on each megasporophyll. Two species are occasionally encountered in North America:

D. edule Lindl.	chamal, coyolillo, jango, palma de la virgen, virgin's palm
D. spinulosum Dyer	

■ treelike or herbs; leaves arching or straight; sporophylls leaflike; seeds paired, whitish

Plants treelike or herbs. **Stems** partially subterranean or aerial; erect; columnar; 3–6 m tall; bases of old leaves present or absent. **Leaves** 15–20; arching or straight; 1–3 m long; leaflets 120–240, not articulated, light to dark green, glossy, blades linear to lanceolate, apices acute, margins entire or spinulose, bases decurrent; petioles with or without prickles. **Pollen Cones** cylindrical; 20–55 cm long; microsporophylls numerous, leaflike, peltate. **Seed Cones** ovoid; 20–90 cm long; woolly; megasporophylls numerous, leaflike, peltate; ovules paired, short-stalked. **Seeds** whitish or yellowish white.

1. Petioles with prickles. Leaflet margins spinulose.
   Leaflets 14–20 mm wide . . . . *D. spinulosum*
1. Petioles without prickles. Leaflet margins entire.
   Leaflets 6–9 mm wide . . . . *D. edule*

Dioon edule

■ poor, rocky soils of mountain slopes in Mexico

Distribution and Habitat—With the exception of one species in Honduras, *Dioon* is strictly a Mexican genus with the nine species widely separated in undisturbed mountain habitats. Plants typically grow in poor, rocky soils of steep slopes and ravines from sea level to about 2,500 m elevation (Jones 1993; Norstog & Nicholls 1997). They are rarely cultivated as ornamentals in other warm areas or as tub plants elsewhere.

Zamia L.

■ *Z. integrifolia*
comptie
coontie
Florida arrowroot
guayiga
Seminole bread

Taxonomy and Morphology—Native to the tropics of both the Old World and the New World, *Zamia* comprises 30–60 species and is ecologically and morphologically the most diverse genus of the cycads (Norstog & Nicholls 1997). Its taxonomy is yet to be resolved, with taxonomists expressing numerous opinions about species relationships. There also is disagreement about whether its name is derived from the Latin *azania,* meaning a type of pinecone, or from *zamia,* meaning "loss." In perhaps the earliest description, the Roman natural historian Pliny described the withered appearance of the pollen cones. In North America, a single species occurs naturally in Florida and Georgia:

Z. integrifolia L.f. (= *Z. floridana* A.DC.)	coontie, Seminole bread, guayiga, Florida arrowroot, comptie

■ herbs; leaves straight, leaflets stiff; cones red to reddish brown; seeds orange or orange-red

Plants herbs. **Stems** subterranean or only apices exposed; bases of old leaves present or absent. **Leaves** 4–10; straight; erect or ascending; 0.2–1.5 m long; leaflets 20–80; articulated, dark green, glossy, blades linear, stiff; margins entire or serrulate or denticulate distally, often revolute; petioles without prickles. **Pollen Cones** narrowly cylindrical; 5–16 cm long; red to reddish brown; indumented; sporophylls numerous, hexagonal. **Seed Cones** cylindrical-ellipsoidal; 5–15 cm long; red to reddish brown; indumented; megasporophylls numerous, hexagonal; ovules sessile. **Seeds** orange or orange-red.

Distribution and Habitat—A widely distributed species, *Z. integrifolia* occurs from southeastern Georgia and Florida to the Bahamas, Cuba, and Puerto Rico. Plants occupy a variety of habitats, including grasslands, dunes, and woodlands. Habitat destruction is a threat. They are grown as tub plants.

Zamia integrifolia

■ grasslands, woodlands, ornamentals

Zamia integrifolia

References

Botha CJ, Naude TW, Swan GE, Ashton MM, Van Der Wateren JF: Suspected cycad *(Cycas revoluta)* intoxication in dogs. *J S Afr Vet Assoc* 62;189–190, 1991

Gabbedy BJ, Meyer EP, Dickson J: Zamia palm *(Macrozamia reidlei)* poisoning of sheep. *Aust Vet J* 51;303–305, 1975

Hall WTK, McGavin MD: Clinical and neuropathological changes in cattle eating the leaves of *Macrozamia lucida* or *Bowenia serrulata* (family Zamiaceae). *Pathol Vet* 5;26–34, 1968

Hooper PT: Cycad poisoning in Australia—etiology and pathology. In *Effects of Poisonous Plants on Livestock,* Keeler RF, Van Kampen KR, James LF eds, Academic Press, New York, pp337–347, 1978

Johnson LAS: The families of cycads and Zamiaceae of Australia. *Proc Linn Soc N S W* 84;64–117, 1959

Jones DL: *Cycads of the World.* Smithsonian Institution Press, Washington, DC, 1993

Keeler RF: Known and suspected teratogenic hazards in range plants. *Clin Toxicol* 5;529–565, 1972

Laqueur GL, Spatz M: Toxicology of cycasin. *Cancer Res* 28;2262–2267, 1968

Mills JN, Lawley MJ, Thomas J: *Macrozamia* toxicosis in a dog. *Aust Vet J* 73;69–72, 1996

Norstog KJ, Nicholls TJ: *The Biology of the Cycads.* Cornell Univ Press, Ithaca, NY, 1997

Nunn PB, Seelig M, Zagoren JC, Spencer PS: Stereospecific acute neuronotoxicity of "uncommon" plant amino acids linked to human motor-system diseases. *Brain Res* 410;375–379, 1987

Nunn PB, Davis AJ, O'Brien P: Carbamate formation and the neurotoxicity of L-α-amino acids. *Science* 251;1619, 1991

Reams RY, Janovitz EB, Robinson FR, Sullivan JM, Casanova CR, Mas E: Cycad *(Zamia puertoriquensis)* toxicosis in a group of dairy heifers in Puerto Rico. *J Vet Diagn Invest* 5;488–494, 1993

Ross SM, Seelig M, Spencer PS: Specific antagonism of excitotoxic action of "uncommon" amino acids assayed in organotypic mouse cultures. *Brain Res* 425;120–127, 1987

Seawright AA: *Animal Health in Australia.* Vol 2, 2d ed, *Chemical and Plant Poisons.* Australian Government Publ Serv, Canberra, pp46–48, 1989

Senior DF, Sundlof SF, Buergelt CD, Hines SA, O'Neil-Foil CS, Meyer DJ: Cycad poisoning in the dog. *J Am Anim Hosp Assoc* 21;103–109, 1985

Smith DWE: Mutagenicity of cycasin aglycone (methylazoxy-methanol), a naturally occurring carcinogen. *Science* 152;1273–1274, 1966

Smith QR, Nagura H, Takada Y, Duncan MW: Facilitated transport of the neurotoxin, β-*N*-methylamino-L-alanine, across the blood-brain barrier. *J Neurochem* 58;1330–1337, 1992

Spatz M, Smith DWE, McDaniel EG, Laqueur GL: Role of intestinal microorganisms in determining cycasin toxicity. *Proc Soc Exp Biol Med* 124;691–697, 1967

Stevenson D: A formal classification of the extant cycads. *Brittonia* 44;220–223, 1992

Tustin RC: Notes on the toxicity and carcinogenicity of some South African cycad species with special reference to that of *Encephalartos lanatus. J S Afr Vet Assoc* 54;33–42, 1983

Weiss JH, Koh J-Y, Choi DW: Neurotoxicity of β-*N*-methylamino-L-alanine (BMAA) and β-*N*-oxalylamino-L-alanine (BOAA) on cultured cortical neurons. *Brain Res* 497;64–71, 1989

Yagi F, Tadera K: Azoxyglycoside contents in seeds of several cycad species and various parts of Japanese cycad. *Agric Biol Chem* 51;1719–1721, 1987

76 Zygophyllaceae R.Br.

■ Caltrop Family
Creosote Bush Family

Kallstroemia
Larrea
Peganum
Tribulus

Commonly known as the caltrop or creosote bush family, the Zygophyllaceae comprises 25–30 genera and 240–250 species. Its distribution is primarily tropical and subtropical; species are typical of dry and/or saline sites. Creosote bush, for example, dominates considerable portions of the warm deserts of both North and South America. Economically, the family is a source of timber, medicinal resins, some edible fruits, essential oils used in perfumes, and dyes (turkey red). *Tribulus,* or puncture vine, is notorious for its small, hard, spiny fruits. In North America, 7 genera and 18 species are present. Species of 4 genera have been associated with intoxication problems.

■ herbs, shrubs, or subshrubs; leaves 1-pinnately compound, with stipules; flowers axillary; fruits capsular schizocarps

Plants herbs or shrubs or subshrubs; annuals or perennials. **Leaves** 1-pinnately compound; opposite or alternate; often fleshy or coriaceous; terminal leaflet absent; stipules present. **Inflorescences** solitary or paired flowers or cymes; axillary. **Flowers** perfect; perianths in 2-series. **Sepals** 4 or 5; free; persistent or caducous. **Corollas** radially symmetrical; imbricate. **Petals** 4 or 5; free; yellow or orange or white. **Stamens** 10–15; in 2 or 3 whorls, of 2 different lengths, outer epipetalous. **Pistils** 1; compound, carpels 2–5; stigmas 1–5; styles 1; persistent; ovaries superior, lobes 2–5 or 10–12; locules 2–5 or 10; placentation axile. **Fruits** capsular schizocarps; rugose to tuberculate or with spines. **Seeds** 1 per mericarp.

Kallstroemia Scop.

■ *K. californica*
K. grandiflora
K. hirsutissima
K. maxima
K. parviflora
K. perennans
K. pubescens

caltrop
carpetweed
desert poppy
golondrina

Taxonomy and Morphology—The largest and taxonomically most troublesome genus of the family, *Kallstroemia* (caltrop), comprises 17 species native to the Americas (Porter 1969). It is represented in North America by 7 native and introduced species:

K. californica (S.Watson) Vail	California caltrop, golondrina
K. grandiflora Torr.	orange caltrop, desert poppy
K. hirsutissima Vail	hairy caltrop, carpetweed
K. maxima (L.) Hook. & Arnold	
K. parviflora Norton	warty caltrop
K. perennans B.L.Turner	perennial caltrop
K. pubescens (G.Don) Dandy	

■ herbs or subshrubs, stems prostrate, diffusely branched; leaves opposite, 1 of each pair smaller or absent; flowers solitary; sepals 5; petals 5, white to bright orange; schizocarps long-beaked

Plants herbs or subshrubs; annuals or perennials; from taproots. **Stems** prostrate or decumbent; to 100 cm long or 10–20 cm high; diffusely branched. **Leaves** opposite, with 1 of each pair typically smaller than other or absent; leaflets 2–10 pairs; blades elliptic to oblong or obovate; bases asymmetrical. **Inflorescences** solitary flowers;

pedicels 0.2–10 cm long. **Sepals** 5; persistent or deciduous; hirsute or strigose. **Petals** 5; elliptic to obovate; white to bright orange. **Stamens** 10–12; in 2 whorls. **Pistils** with stigmas capitate; ovaries 10-to-12-lobed. **Schizocarps** ovoid or conical or pyramidal; long-beaked; 10-to-12-lobed; glabrous or indumented; mericarps wedge-shaped, tuberculate.

Kallstroemia hirsutissima

- deserts, grasslands, open disturbed sites

1. Leaves obovate in outline. Terminal leaflets larger than middle leaflets.
   2. Sepals deciduous. Schizocarps with elongate tubercules . . *K. californica*
   2. Sepals persistent. Schizocarps rugose or tuberculate, but tubercules not elongate.
      3. Ovaries and schizocarps glabrous . . . . *K. maxima*
      3. Ovaries and schizocarps pubescent.
         4. Sepals appressed. Schizocarps 6–8 mm wide; beaks hirsute . . . . *K. hirsutissima*
         4. Sepals spreading. Schizocarps 3–6 mm wide; beaks appressed short pilose . . . . *K. pubescens*
1. Leaves elliptic in outline. Terminal leaflets smaller than middle leaflets.
   5. Plants perennials; from woody rootstocks . . . . *K. perennans*
   5. Plants annuals; from taproots.
      6. Pedicels shorter than subtending leaves. Petals 3–5 mm long. Sepals deciduous . . . . *K. californica*
      6. Pedicels longer than subtending leaves. Petals 6–35 mm long. Sepals persistent.
         7. Flowers 2.5–6 cm in diameter. Pedicels 4–10 cm long . . . . *K. grandiflora*
         7. Flowers 1–2.5 cm in diameter. Pedicels 1–4 cm long . . . . *K. parviflora*

Kallstroemia hirsutissima

Distribution and Habitat—Populations of *Kallstroemia* are most commonly encountered in the deserts and grasslands of Mexico and the adjacent southwestern part of the United States. *Kallstroemia maxima* occurs along the Atlantic and Gulf coasts of Florida. The most widely distributed species, *K. parviflora* (warty caltrop), ranges from central Mexico northward to Illinois. Plants are typically encountered in open, disturbed habitats, both natural and anthropogenic. Early-day ecologists used species of the genus as indicators of overgrazing in the arid grasslands of the Southwest, and agronomists noted how quickly *K. parviflora* became a troublesome weed in southern Kansas soon after the conversion of the grasslands into cultivated fields.

Kallstroemia parviflora

Disease Problems—Causing neurologic problems in cattle, sheep, and goats, *Kallstroemia* is of concern mainly in late summer when formerly cultivated areas are used for grazing and animals gain access to large populations of the plants that have established themselves in the fallow soils. Because toxicity varies considerably from year to year, location to location, and growth stage to growth stage, disease problems are typically erratic in appearance, only being of serious concern every few years. Animals must eat plants as a major portion of their diet (about 90%) for 1 or more weeks before clinical signs become apparent (Mathews 1944; Dollahite 1975).

- neurotoxic, late summer, when eaten in large amounts by sheep, goats, or cattle

Disease Genesis—The cause of the disease is yet unknown. However, the similarity in neurologic signs produced by *Kallstroemia* to those produced by other members of the family has led to speculation that the toxicants are indole alkaloids such as harman and harmaline (Bourke et al. 1990).

- toxin unknown, possibly indole alkaloids as with some other genera of the family

■ knuckling over of fetlocks of pelvic limbs, stand or walk on carpal joints, weakness, ataxia

Clinical Signs—Typically there is knuckling over of the fetlocks of the pelvic limbs in cattle, sheep, and goats. In some cases, this effect may persist for months, even though plants are no longer being eaten. Some animals, especially sheep and goats, may stand and walk on their carpal joints. Additional signs include weakness, pelvic limb incoordination, and flexure of the hocks. In more-severe cases, animals are unable to stand but may continue to eat and drink. If they continue to eat the plants, the disease ends in seizures and death.

■ no lesions

supportive nursing care

Pathology and Treatment—*Kallstroemia* does not cause distinctive lesions, only congestion and scattered small splotchy hemorrhages. Prevention of continued access to the plants, and supportive nursing care is the only treatment approach available to animals that are exhibiting toxicity caused by *Kallstroemia.*

Larrea Cav.

■ *L. divaricata*

creosote bush
gobernadora
hediondilla

Taxonomy & Morphology—*Larrea* is a New World genus with 4 species in South America and 1 in North America. Its name honors Don Anthony Hernandez de Larrea, an eighteenth-century Spaniard who was dean of Saragossa and a patron of science. The North American species is the following:

L. divaricata Cav. (= *L. tridentata* [Moc. & Sessé ex DC.] Coville)	creosote bush, gobernadora, hediondilla

■ evergreen, perennial shrubs, resinous aromatic; leaves opposite, leaflets 2, leathery, shiny-varnished appearance; flowers solitary; sepals 5; petals 5, short-clawed, bright yellow; schizocarps silvery white to red, densely woolly

Taxonomists differ in their interpretations of the relationship between the South and North American populations of creosote bush. Some believe them to be different species and use the binomial *L. tridentata* for the North American plants. Other taxonomists believe that a single species is present on both continents and use the binomial *L. divaricata,* as is done here.

Larrea divaricata

Plants shrubs; perennials; evergreen; from a large crown or short rhizomes; herbage resinous aromatic. **Stems** numerous; spreading; 1–3 m tall; radiating from base; flexuous; jointed; bark gray; nodes conspicuous, swollen, darkened. **Leaves** opposite; subsessile to short-petiolate; leaflets 2; leathery; glossy dark green or more often yellowish green; blades ovate to oblong or obovate; surfaces with resinous coating giving them a shiny-varnished appearance; apices acute; bases fused. **Inflorescences** solitary flowers; borne on pedicels that arise from ends of short lateral branches. **Sepals** 5; unequal; deciduous. **Petals** 5; obovate; concave; short-clawed; bright yellow. **Stamens** 10; filaments winged. **Pistils** with ovaries 1-to-5-lobed. **Schizocarps** globose; 5-lobed; silvery white to red; densely woolly.

■ deserts

Larrea divaricata

Distribution and Habitat—*Larrea divaricata* is the most common and widely distributed shrub of the North American deserts, with plants occurring from sea level to about 1,300 m elevation in the foothills of the western mountain ranges. It is especially common in the lower Sonoran and Chihuahuan deserts, sometimes forming extensive, pure stands. Plants occur in hot, dry sites in a variety of soil types. Flowering may occur throughout the year but is generally most abundant from February through April.

Disease Problems—Because of the widespread distribution of *L. divaricata* in the arid regions of both North and South America, many attempts have been made to use it for food and medicinal purposes (Timmermann 1977). It is used as a contraceptive in Mexico and possibly has uterorelaxant activity (Obermeyer et al. 1995). The leaves are commonly used as herbal supplements and for making chaparral tea (Brent 1999). However, in a few instances, these latter uses have resulted in the development of acute toxic hepatitis in persons consuming *L. divaricata*, usually for a month or more (Katz & Saibil 1990; CDC Report, 1992; Alderman et al. 1994; Batchelor et al. 1995). The hepatitis, accompanied by marked elevations of serum hepatic enzymes (AST, ALT, GGT), resolved favorably in most patients several weeks or more after the use of the tea or herbal product was discontinued. However, in a few instances the toxicosis was severe, resulting in cirrhosis and/or liver failure and requiring transplantation (Gordon et al. 1995; Sheikh et al. 1997).

- humans, mild liver disease when used as tea or herbal supplement

livestock, toxicity unknown

The toxicity potential for livestock is not clear. This unpalatable shrub is seldom eaten except when it is the only green forage available on a range. When fed the apparently nutritious leaves, treated with sodium hydroxide to improve palatability, cattle performed well, without any apparent problems (Timmermann 1977). However, it also is suspected of causing losses in cattle (Fuller & McClintock 1986). The death of 6 of 10 calves was associated with the presence of large amounts of the leaves and fruits of *L. divaricata* in the stomachs.

Disease Genesis—The foliage of *L. divaricata* contains numerous polyphenolic types of compounds such as flavones and their glycosides (Mabry et al. 1977). There is also a high content (5–10% d.w.) of the lignan nordihydroguaiaretic acid (NDGA), which has been used as an antioxidant in foods. A recognized lipoxygenase inhibitor, NDGA extracted from *L. divaricata* is commonly used experimentally as an inhibitor of the cyclooxygenase-lipoxygenase pathway (Tappel & Marr 1954; Hamberg 1976; Thomson 1984). It is converted to a quinone metabolite in the intestine. In long-term feeding studies, only minor toxic effects such as cystic enlargement of the paracecal lymph nodes have been consistently observed (Lehman et al. 1951). The oral LD_{50} ranges from 830 mg/kg b.w for guinea pigs up to 4–5 g/kg in mice and rats. Grice et al. (1968) fed rats up to 1% NDGA in their diet for 74 weeks and produced only mild kidney tubular vacuolation and cystic lymphoid changes. A level of 2% caused the same lesions in shorter time. They also found little NDGA in the tissues, most having been converted to the O-quinone metabolite in the ileum and cecum prior to absorption. Only the quinone was found in rat tissues. The role of NDGA in acute toxicosis is not clear, but high concentrations of reactive quinones are plausible causes of either hepatic or renal lesions. In addition, in vitro studies indicate that NDGA is a photoactive compound capable of causing photosensitization in livestock (Downum & Rodriguez 1986; Downum et al. 1989). The structural similarity of the *Larrea* lignan NDGA, and its metabolites, to diethylstilbesterol suggest the possibility of hepatotoxic effects similar to those caused by estrogens (Obermeyer et al. 1995).

- toxin unknown, possibly NDGA

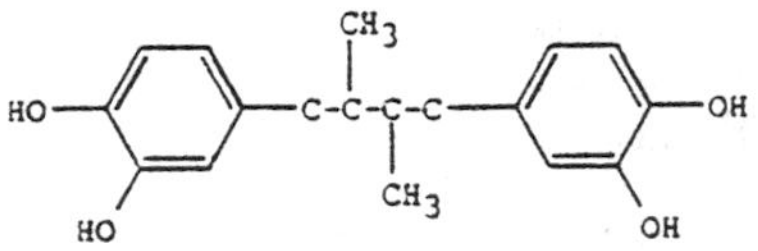

HO, HO, CH_3, C-C-C-C, CH_3, OH, OH

Nordihydroguaiaretic acid (NDGA)

The foliage of *L. divaricata* also contains numerous polyphenolic types of compounds such as flavones and their glycosides and appreciable concentrations of saponins (10–15% d.w.) such as larreagenin A and larreic acid (Mabry et al. 1977). A role for polyphenolics as contributing factors cannot be dismissed inasmuch as their levels are quite high, in the range of 16–20% d.w. or more of the foliage (Gisvold & Thaker 1974). Although no evidence indicates a role for saponins in this disease, these types of compounds are recognized as important factors in the development of liver disease, as is discussed in the treatment of *Tribulus* (in this chapter).

Clinical Signs—In humans the signs, typical of hepatic involvement, include nausea, loss of appetite, fatigue, pruritis, icterus, abdominal pain, dark-colored urine, and an enlarged liver. Changes in serum chemistry include elevated bilirubin and markedly

- humans, loss of appetite, icterus, abdominal pain, dark-colored urine, elevated hepatic enzymes in blood elevated hepatic enzymes such as AST, LDH, ALT, and GGT. Alkaline phosphatase increases only slightly. Because the disease has not been clearly identified in animals, the signs, pathology, and treatment are at present unknown. If, however, the disease affects the liver, as is the case in humans, then signs of and treatment directed toward liver insufficiency or failure may be expected.

- microscopic, mild to moderate hepatic necrosis

  treatment not usually needed; stop use of herbal supplement

Pathology and Treatment—Microscopically, the hepatic lesions associated with *Larrea* intoxication include cholestasis and mild to moderate periportal necrosis, often without a pronounced inflammatory reaction. The lesions and signs in most instances are not particularly severe, and specific treatment is not required. They appear to regress readily within a few weeks upon discontinuance of the herbal supplement or tea. Less commonly the necrosis may be severe and require treatment for liver failure.

Peganum L.

- *P. harmala*
  P. mexicanum

  African rue
  garbancillo
  harmal
  harmal peganum
  limoncillo
  Mexican peganum

Taxonomy and Morphology—A genus of 3 species occurring in both the Old World and the New World, *Peganum* is represented in North America by 2 species, 1 introduced and 1 native:

P. harmala L.	harmal, African rue, harmal peganum
P. mexicanum A.Gray	Mexican peganum, garbancillo, limoncillo

- perennial herbs, profusely branched, dense foliage; leaves alternate, fleshy; flowers solitary; petals 4 or 5, subequal, white to yellow

Plants herbs; perennials; from woody rootstocks; foliage dense. **Stems** prostrate to ascending; to 90 cm long and 30 cm tall; profusely branched; pubescent or glabrous. **Leaves** alternate; somewhat fleshy; margins irregularly pinnatifid with linear lobes. **Inflorescence** solitary flowers; pedicels present. **Sepals** 4 or 5; foliaceous; margins entire or pinnatifid; persistent. **Petals** 4 or 5; subequal; white to yellow. **Stamens** 12–15; filaments linear; bases dilated. **Pistils** 2-to-4-lobed. **Schizocarps** subglobose; beak as long as or longer than body; mericarps triangular, dark.

1. Flowering pedicels shorter than petals. Fruiting pedicels recurved; shorter than schizocarps . . . . *P. mexicanum*
1. Flowering pedicels longer than petals. Fruiting pedicels erect; longer than schizocarps . . . . *P. harmala*

Peganum harmala

- semiarid rangelands

Distribution and Habitat—Both species are inhabitants of semiarid rangelands. Native to the Chihuahuan Desert, *P. mexicanum* is infrequently encountered in alluvial and gypsum soils of the deserts of the northern half of Mexico. Introduced from the Mediterranean region in the 1930s, *P. harmala* is a weed in the western United States. Its seeds are the source of the dye turkey red.

- neurotoxic, large amounts, seldom eaten because not palatable

Disease Problems—Although *P. harmala* has long been recognized as a toxic plant in North Africa, it has not evoked great concern in North America, because of its limited range. However, growing awareness of it as a noxious weed and toxic plant has resulted in laws making its cultivation illegal in some states, including California (Fuller & McClintock 1986). The seeds at a dose of 0.16% b.w. in a single feeding were lethal in a guinea pig, causing weakness, loss of muscular control of the pelvic limbs, and shaking several hours before death. Young leaves also caused clinical signs of disease but required higher doses, whereas signs could not be induced with old leaves (Moran et al. 1940). Mild disease symptoms were seen in a heifer given 0.31% b.w. of seeds and leaves as d.w. The fresh foliage is extremely bitter and disagreeable to the taste, apparently because of a volatile oil. Thus, it is very difficult to induce an animal to eat enough plant material to produce toxic effects (Mathews 1941). *Peganum harmala* is eaten more readily by sheep if it is allowed to dry for several days (Black & Parker 1936). The lack of early concern about this genus was in part due to the negative results of experiments, even though the dosages used were quite low. Dosages of 44.5 g seeds/day for 4 days and 136.2 g young leaves/day for 20 days in sheep of 27–40 kg b.w. were not sufficient to cause adverse effects. Similarly, in cattle of 355–400 kg b.w., 250 g seeds/day for 29 days and 406 g young leaves/day for 15 days did not cause intoxication (Black & Parker 1936).

Peganum harmala

Disease signs have been reported for camels forced to eat the plant (El Bahri & Chemli 1991). However, in this study, signs related to the digestive system were seen, such as excess salivation, vomiting, and diarrhea, in addition to the predominant neurologic signs. Indeed the signs due to *P. harmala* are said to resemble closely the cholinergic effects caused by species of *Hyoscyamus* in the Solanaceae(see chapter 69) (El Bahri et al. 1996).

Early work by Boughton and Hardy (1938) indicated *P. mexicanum* to be nontoxic, but confirmation of this is needed because of the accumulation of more-recent information about *P. harmala.*

Peganum mexicanum

■ indole, β-carboline alkaloids

Disease Genesis—*Peganum harmala* contains a series of related β-carboline indole alkaloids: harman, norharman, tetrahydroharmane, harmine, harmaline, harmol, and harmalol (Marion 1962; Kamel et al. 1970b; Puzii et al. 1980; Bourke et al. 1990). Alkaloid concentrations are highest in mature seeds (approximately 4%) and lowest in leaves and stems (1%), with considerable seasonal variation. The methoxy-containing harmine and its dihydro derivative harmaline account for 90% or more of the plants' alkaloid content. In Egypt the ratios of harmine to harmaline are reported to vary from 1:2 in winter to 2:1 in the summer dry period (Kamel et al. 1970a). A decrease in total alkaloid concentration, in concert with a change in proportions favoring harmine, may result in a decrease in toxicity because harmaline is twice as toxic as harmine.

■ alkaloids inhibit MAO, increase serotonin activity, excite inferior olivary neurons, causing tremors and spasticity

The effects of these β-carboline alkaloids are the result of a combination of short-duration competitive inhibition of monoamine oxidase (MAO); interference with amine uptake, leading to increased serotonin activity; and excitation of the inferior olives, inducing tremors (Udenfriend et al. 1958; Lessin et al. 1967). The two compounds are hallucinogenic in humans (McGeer et al. 1978). Inhibition of MAO possibly augments the dopaminergic and serotonergic activity in the nervous system. Hence, it is not surprising that both of these alkaloids cause spastic activity in rats at low parenteral dosages (0.01–0.02% b.w. s.c.) (Gunn & Simonart 1930; Gunn & MacKeith 1931; Puzii et al. 1980). However, the most intriguing mechanism involved with harmaline is its action on inferior olivary neurons. These neurons are part of the extrapyramidal system. Located in the lower medullary region near the pyramids, they connect with the cerebellum via olivocerebellar fibers. The climbing fibers allow interaction of the spinal cord, red nucleus, cerebellum, and basal ganglia with Purkinje cells of the cerebellum. The overall effect is on skilled, coordinated movements and involves increase tone and rigidity in the extremities. Harmaline causes a very characteristic generalized tremor by a central action transmitted through the inferior olive and olivocerebellar pathway to the Purkinje cells in the cerebellar vermis (Lamarre & Mercier 1971; Llinás & Volkind 1973; Knowles & Phillips 1980). The olivary effects are due to hyperpolarization of neurons and facilitation of low-threshold conductance (Llinás 1991). It has been suggested that harmaline causes selective degeneration of Purkinje cells, especially of the olivocerebellar climbing fiber projections

Harmine

Harmaline

Harmalol

via excitatory amino acid release and NMDA receptors (O'Hearn & Molliver 1993; Du & Harvey 1997).

The other alkaloids produced by *Peganum* contribute little toward toxicity because their concentrations are low and some such as harmol and harmalol are not convulsants (Gunn & Simonart 1930; Gunn & MacKeith 1931). The seeds also contain vasicine (= peganine), which seems not to be of toxicological concern (Openshaw 1953). Because of its possible dopaminergic actions, harmalol has been tried, but with only limited success, in the treatment of parkinsonism. Alkaloidal extracts of *P. harmala,* however, are used i.v. for treatment of theileriosis in cattle and sheep with few adverse effects (Cooper & Gunn 1931; Puzii & Serov 1983). Much of the extract is eliminated in urine in the first 24 hours, but some remains in the animal's tissues for a week or more.

■ tremors, ataxia, excitability, narcosis, weakness

no lesions

treat with supportive nursing care

Clinical Signs, Pathology, and Treatment—In most cases, the signs of *Peganum* toxicity will be mild and of short duration. They include tremors and mild incoordination, especially when the animals are forced to move. In more-severe cases, the animal may be excitable at first and then have alternating periods of narcosis and obvious weakness of the pelvic limbs. There may be accompanying effects on the digestive system as well, including diarrhea and increased salivation. There may be congestion of some organs and a few scattered small hemorrhages, but there are no distinctive lesions. General supportive care is the most appropriate treatment.

Tribulus L.

■ *T. cistoides*
T. terrestris

bullhead
burnut
goathead
Jamaican feverplant
Mexican sandbur
puncture vine
tacks

Taxonomy and Morphology—Comprising 25–30 species, *Tribulus* is a genus of the Old World deserts. Its name is derived from the Greek root *tribolos,* meaning "three-pointed," which alludes to the three spines of the fruits. In North America, 2 introduced species are present:

T. cistoides L.	Jamaican feverplant
T. terrestris L.	puncture vine, caltrop, burnut, goathead, bullhead, tacks, Mexican sandbur

■ herbs, diffusely branched; leaves opposite, 1 of pair smaller or absent; flowers solitary; petals 5, yellow, orange, or white; schizocarps 5-angled, each side with 2–4 stout horns or spines

Plants herbs; annuals or perennials; from taproots or woody rootstocks. **Stems** prostrate to ascending; to 100 cm long; diffusely branched; sericeous and hirsute. **Leaves** opposite; 1 of each pair typically smaller than other or rarely absent; leaflets 3–9 pairs; blades elliptic to oblong; bases asymmetrical. **Inflorescences** solitary flowers; borne in axils of smaller leaves; pedicels present. **Sepals** 5; caducous. **Petals** 5; obovate; yellow or orange or rarely white. **Stamens** 10; in 2 whorls. **Pistils** with stigmas 5; ovaries 5-lobed, hirsute-pilose. **Schizocarps** subglobose; short-beaked; 5-angled; each side bearing 2–4 stout horns or spines; tubercules and small spines often present.

1. Plants annuals; from taproots. Flowers 5–10 mm in diameter. Pedicels shorter than subtending leaves .... *T. terrestris*
1. Plants perennials; from woody rootstocks. Flowers 20–40 mm in diameter. Pedicels longer than subtending leaves ...... *T. cistoides*

Tribulus cistoides

Distribution and Habitat—Both species are native to the Old World and are now widely distributed as weeds in disturbed soils throughout the southern half of the continent. As the common names imply, the stout spines of the fruits readily penetrate skin and tires, causing painful injury and flat tires. The seeds are long-lived and once plants are established in an area, they are difficult to eradicate.

Disease Problems—After being suspected as a toxic plant for many years, *T. terrestris* was finally definitely incriminated as producing photosensitization due to a class of toxicants associated with an inability of the liver to clear the blood of phylloerythrin (Rimington & Quin 1934; Blum 1941). Phylloerythrin, derived from the breakdown of chlorophyll in the digestive tract, has been identified as a photodynamic compound. Eventually, in what is now considered a classical treatise, this type was termed hepatogenous photosensitization (Clare 1955). Considerable effort was expended in an attempt to elucidate the mechanism of *Tribulus*'s effects on the liver (Brown 1968). Eventually, the possible roles of various cofactors, such as fungal toxins, became apparent (Kellerman et al. 1980). It had long been recognized that the liver lesions included crystalloid structures in the bile ducts and other sites, and it is now apparent that the disease caused by *T. terrestris* is essentially the same as the crystal-associated or crystalloid cholangiohepatopathy (Button et al. 1987) caused by several other plants, including *Agave lechuguilla, Nolina,* and various species of *Panicum.* Photosensitization associated with crystalloid material in the liver and biliary microliths has also been observed in goats not eating known toxic plants but grazing green oats *(Avena sativa)* infected with the fungus *Drechslera campanulata* (Collett & Spicket 1989).

Hepatogenous photosensitization had been recognized for many years in South Africa as a problem associated with *T. terrestris* and had been experimentally produced on several occasions, even using frozen plant material (Bath et al. 1978). Eventually it was recognized as a problem in Australia, each case exhibiting the same prominent biliary crystalloid deposits and high case-fatality rate (Bourke 1983; Glastonbury et al. 1984; Jacob & Peet 1987). It also appears to be an occasional problem in South and North America, having been implicated in several instances of sheep intoxication in New Mexico and Colorado (Hershey 1945; Durrell et al. 1950; Tapia et al. 1994). In these last cases, *T. terrestris* may have been the only forage available, and the disease was perhaps precipitated by driving the flocks hard during warm weather. It is important to note that *T. terrestris* is palatable and is readily eaten by sheep (Kellerman et al. 1980).

Neurotoxic effects also are ascribed to *T. terrestris* in Australia; sheep grazing extensively on plants for months during drought developed an irreversible, progressively asymmetrical weakness of the pelvic limbs (Bourke 1984). They stood with legs apart and walked awkwardly. After a few weeks, the changes became obviously asymmetrical as the sheep walked or ran sideways. Eventually, weakness of the thoracic limbs developed and then permanent recumbency. Death occurred mainly from thirst or misadventure. The same disease has also been observed with ingestion of another species, *T. micrococcus* (Bourke & Macfarlane 1985).

Other effects, including debilitation, anemia, and death, are reported to occur as a consequence of continuous feeding of large amounts of *T. terrestris* to sheep (Pachalag & Patil 1977). Field observations in Australia suggested possible decreased survival of lambs borne of ewes grazing *Tribulus.* However, experiments carried out on a limited number of sheep failed to confirm any obvious reproductive effects except for a decrease in fetal breathing movements while dams were feeding on plants during the last trimester of gestation (Walker et al. 1992).

Disease Genesis—Saponins have long been suspected as possible factors in the development of the liver lesions caused by *Tribulus* and other photosensitizing plants, and

Tribulus terrestris

- weedy in disturbed sites
- hepatogenous photosensitization, crystalloid cholangiohepatopathy
- neurotoxic but not confirmed in North America

Tribulus cistoides

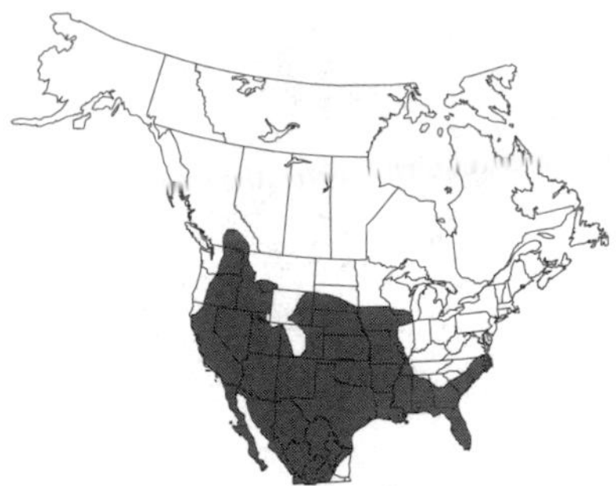

Tribulus terrestris

- photosensitization possibly due to blocking of bile ducts by calcium salts of sapogenic glucuronides; associated with high levels of lithogenic diosgenin and yamogenin in some *Tribulus;* phylloerythrin accumulates because of inability of liver to remove it from blood

several have been identified, including diosgenin (DeKock & Enslin 1958; Hsu et al. 1966; Kellerman et al. 1991; Miles et al. 1993). *Tribulus* typically contains a mixture of genins such as diosgenin, yamogenin, epismilagenin, tigogenin, neotigogenin, gitogenin, and neogitogenin (Miles et al. 1994a). Although gitogenin and neogitogenin are highest in concentration in some plants, diosgenin and yamogenin are the saponins of primary toxicologic interest. The disease has been reproduced using a crude saponin mixture extract from the genus (Kellerman et al. 1991; Miles et al. 1994b). As with the diseases caused by *Narthecium* and *Panicum,* the crystalloid biliary deposits produced by *Tribulus* have been identified as insoluble calcium salts of sapogenic glucuronides (Lancaster et al. 1991; Miles et al. 1991, 1992a,b, 1993; Miles et al. 1994b). In experimentally induced disease, the calcium salts of the β-D-glucuronides of epismilagenin and episarsapogenin have been recovered from the bile of a sheep in a 6:1 mixture (Miles et al. 1994b). A proposed mechanism of intoxication is that the saponins in the plant undergo hydrolysis in the digestive tract to form free sapogenins. Diosgenin and yamogenin are further reduced and epimerized to the more stable epismilagenin and episarsapogenin, also in the rumen (Miles et al. 1994a). The epimerized sapogenins are then glucuronidated in the liver, with subsequent precipitation as the insoluble calcium salts at sites of high concentration such as bile or renal tubules (Miles et al. 1991, 1994b). This results in occlusion of the bile ducts mechanically or via irritation of the bile canalicular membranes (Coetzer et al. 1983). Only diosgenin and yamogenin are involved in the disease; the other sapogenins in the crude plant extracts are not. However, it is not clear at present whether the saponins alone are always responsible for the disease or whether in some instances additional toxicants are involved. In some studies, investigators have been unable to produce the liver disease in lambs dosed daily intraruminally with 9 g of sarsapogenin or diosgenin (Flåøyen et al. 1993). However, either of these sapogenins were hepatotoxic when administered with a small amount of sporidesmin. In spite of this, the lesions in crystalloid-associated photosensitization are quite different from those induced by the mycotoxin sporidesmin (Coetzer et al. 1983). The association of toxicity solely with diosgenin and yamogenin is of considerable importance. As pointed out by Miles and co-workers (1994a), this could account for the lack of occurrence of disease in many areas where *Tribulus* is common. This diversity in occurrence has been noted in Australia and South Africa and seems to correlate with plants high in the lithogenic diosgenin and yamogenin in areas with disease and those with high levels of nonlithogenic tigogenin and neotigogenin in areas lacking disease (Miles et al. 1994a;

Diosgenin

Epismilagenin

Ca^{2+} salt of epismilagenin β-D-glucuronide

Phylloerythrin

Wilkins et al. 1994, 1996). Although this variation has not been confirmed in North America, it may account in part for the relative lack of problems on this continent.

The toxicant directly responsible for photosensitivity is the porphyrin phylloerythrin. It is formed in the digestive tract during microbial degradation of chlorophyll and normally eliminated by the liver via the biliary system and in urine (Ford & Gopinath 1974, 1976). In the event of liver dysfunction due to either hepatocellular or biliary tract disease, phylloerythrin accumulates in the circulation in sufficient quantities (10 μg/dl in serum) to function as a photodynamic agent. Light at wavelengths of 320–400 nm excites the porphyrin, resulting in the subsequent generation of singlet oxygen. In the absence of quenching agents (carotenoids in plants), singlet oxygen reacts at the site of its formation (in capillaries of skin penetrable by light) with amino acids, guanine, and lipids. The resulting lipid peroxidation is in large measure responsible for membrane disruption, cellular degradation, and tissue necrosis (Dodge & Knox 1986). For additional discussion of photosensitization, see Blaauboer and van Graft (1983), Johnson (1983), and Rowe (1989). Sunburn, which in some cases may appear similar, is caused by wavelengths of 290–320 nm, and its effects are seen very soon after exposure. However, in both situations the lesser pigmented areas are at greatest risk.

In Australia, *T. terrestris* has been shown to contain the β-carboline indole alkaloids harman and norharman, which are also found in *Peganum* (Bourke et al. 1992). However, the more toxic harmaline, responsible mainly for the toxic effects of *Peganum*, is not present in *Tribulus* (Allen & Holmstedt 1980). Experimentally, these alkaloids and other closely related indoles have been shown to produce signs consistent with those of the naturally occurring neurologic disease (Bourke et al. 1990, 1992). These alkaloids may accumulate in neurons and cause functional changes because of their ability to inhibit monoamine oxidase and possibly enhance either dopamine or serotonin activity (McGeer et al. 1978). The clinical signs in the presence of normal spinal reflex arcs have been interpreted as indicative of a primary nigrostriatal dopaminergic disorder. In affected animals there was a moderate decrease in dopamine and 3,4-dihydroxyphenylacetic acid in the striatum (Bourke 1987). It is hypothesized that there is malfunction of striatal presynaptic receptors, with slowly developing, progressive upper motor inhibition of the lower motor γ-neurons of the pelvic limb extensor muscle spindles. Thus, there is a slowly developing limb weakness and a disease very similar to that caused by *Kallstroemia* and *Peganum*.

- neurotoxic, indole, β-carboline alkaloids; nigrostriatal disorder

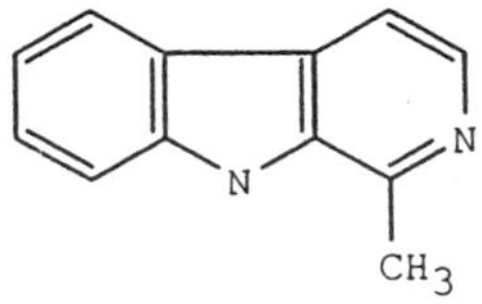

N N CH_3

Harman

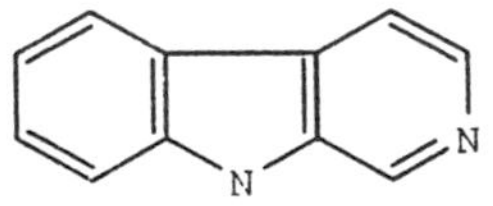

N N

Norharman

Clinical Signs—In the early stages of *Tribulus*-induced liver disease, depression, reduced appetite, and reddening and swelling of the less pigmented areas of the skin are prominent. In sheep, the ears, lips, eyelids, and face around the nose are most affected. Icterus, or yellowing of the mucous membranes, is variable. The skin over the affected areas becomes covered with serous exudate and then cracked, culminating in necrosis and sloughing of large patches. There may be obvious loss of body weight and reddening of the sclera and cornea, resulting in temporary vision problems, including increased sensitivity to light. In sheep that have recovered, the ears may be absent, and occasionally eyes may be obliterated because of the severe swelling and sloughing of and around them.

- photosensitization, depression, loss of appetite; skin lesions in less pigmented areas, reddened, swollen, cracked, necrotic

neurologic, gradual onset, weakness, pelvic limbs, ataxia, fetlocks of pelvic limbs knuckle over, walk with great difficulty

The neurologic disease is a slowly progressive, irreversible, weakness of the pelvic limbs. Initially there is incoordination with an awkward walk, and then after a few weeks the fetlock joints of the pelvic limbs begin to knuckle over and the hock joints are kept flexed. Gradually, the effects on the pelvic limbs become more severe, and affected animals begin to walk at almost a squat or with a dragging of the limbs while moving on a diagonal. The thoracic limbs may become so weak that the animal goes down in permanent recumbency. Reflex arcs are not diminished.

- photosensitization, gross pathology, tissues dark-colored, liver yellowish brown, bile duct walls thickened

  microscopic, hepatic necrosis, portal fibrosis, bile duct proliferation, crystals or clefts in bile ducts

  neurologic microscopic, segmental demyelination of peripheral nerves

Pathology—Grossly, with photosensitization, the tissues are often icteric, and the liver and kidneys somewhat swollen. The liver is yellowish brown, with edema and thickening of bile duct walls. A chalky white sediment may be discernible in the bile. The kidneys are mottled brown. Microscopically, there is necrosis of some hepatocytes and moderate portal fibrosis and bile duct proliferation. The most noteworthy change is the presence of polarizing crystals or needle-shaped to lenticular clefts in and around the bile ducts, hepatocytes, Kupffer cells, and renal tubules. Crystalloid material will be present in the bile and may occlude the lumens of some ducts. Moderate peribiliary edema may be present around affected bile ducts. The renal tubular crystals may be accompanied by tubular nephrosis.

In the case of the neurologic disease caused by *Tribulus,* there are mild to moderate changes in the spinal cord and peripheral nerves. There may be mild segmental demyelination of the larger peripheral nerves. Mild to moderate Wallerian degeneration is seen at all levels of the cord, especially in the lumbar white matter.

- supportive nursing care

Treatment—In the case of photosensitization, treatment is limited to supportive care. Although it is often not feasible, animals may be protected from sunlight by moving them to shaded areas or indoors. Because the disease is associated with a metabolic product of chlorophyll, it seems appropriate to reduce the load of green plant material in the diet. There may be some cases where even a small reduction of phylloerythrin overload is of some benefit, but it may not always be efficacious, given that skin effects are reported to persist even when the animals are being fed only concentrate rations (Clare 1955).

References

Alderman S, Kailas S, Goldfarb S, Singaram C, Malone DG: Cholestatic hepatitis after ingestion of chaparral leaf: confirmation by endoscopic retrograde cholangiopancreatography and liver biopsy. *J Clin Gastroenterol* 19;242–247, 1994

Allen JRF, Holmstedt BR: The simple β-carboline alkaloids. *Phytochemistry* 19;1573–1582, 1980

Batchelor WB, Heathcote J, Wanless IR: Chaparral-induced hepatic injury. *Am J Gastroenterol* 90;831–833, 1995

Bath GF, Van Tonder EM, Basson PA: Geeldikkop: preservation of toxic material. *J S Afr Vet Assoc* 49;23–25, 1978

Blaauboer BJ, van Graft M: Photosensitization in ruminants: porphyrins and phylloerythrin. In *Veterinary Pharmacology and Toxicology,* Ruckebusch Y, Toutain PL, Koritz GD eds, AVI Publishing, Westport, Conn, pp671–680, 1983

Black WL, Parker KW: *Toxicity Tests on African Rue* (Peganum harmala *L.*). N M Agric Exp Stn Bull 240, 1936

Blum HF: *Photodynamic Action and Disease Caused by Light.* Am Chem Soc Mono Ser No 85. Reinhold Publ, New York, 1941

Boughton IB, Hardy WT: *Suspected Plants.* Tex Agric Exp Stn Annu Rep 51, p220, 1938

Bourke CA: Hepatopathy in sheep associated with *Tribulus terrestris. Aust Vet J* 60;189, 1983

Bourke CA: Staggers in sheep associated with the ingestion of *Tribulus terrestris. Aust Vet J* 61;360–363, 1984

Bourke CA: A novel nigrostriatal dopaminergic disorder in sheep affected by *Tribulus terrestris* staggers. *Res Vet Sci* 43;347–350, 1987

Bourke CA, Macfarlane JA: A transient ataxia of sheep associated with the ingestion of *Tribulus micrococcus* (yellow vine). *Aust Vet J* 62;282, 1985

Bourke CA, Carrigan MJ, Dixon RJ: Upper motor neuron effects in sheep of some β-carboline alkaloids identified in zygophyllaceous plants. *Aust Vet J* 67;248–251, 1990

Bourke CA, Stevens GR, Carrigan MJ: Locomotor effects in sheep of alkaloids identified in Australian *Tribulus terrestris. Aust Vet J* 69;163–165, 1992

Brent J: Three new herbal hepatotoxic syndromes. *Clin Toxicol* 37;715–719, 1999

Brown JMM: Biochemical studies on geeldikkop and enzootic icterus. *Onderstepoort J Vet Res* 35;319–376, 1968

Button C, Paynter DI, Shiel MJ, Colson AR, Paterson PJ, Lyford RL: Crystal-associated cholangiohepatopathy and photosensitization in lambs. *Aust Vet J* 64;176–180, 1987

CDC Report: Chaparral-induced toxic hepatitis—California and Texas, 1992. *J Am Med Assoc* 268;3295–3298, 1992

Clare NT: Photosensitization in animals. *Adv Vet Sci* 2;182–211, 1955

Coetzer JAW, Kellerman TS, Sadler W, Bath GF: Photosensitivity in South Africa. 5. A comparative study of the pathology of the ovine hepatogenous photosensitivity diseases, facial eczema, and geeldikkop *(Tribulosis ovis),* with special reference to their pathogenesis. *Onderstepoort J Vet Res* 50;59–71, 1983

Collett MG, Spickett AM: Unusual hepatic parenchymal crystalloid material and biliary microliths in goats. *J S Afr Vet Assoc* 60;134–138, 1989

Cooper HA, Gunn JA: Harmalol hydrochloride in the treatment of parkinsonism. *Lancet* 221;901–902, 1931

DeKock WT, Enslin PR: Chemical investigations of photosensitization diseases of domestic animals. Part 1. Isolation and characterization of steroidal sapogenins from *Tribulus terrestris. J S Afr Chem Inst* 11;33–36, 1958

Dodge AD, Knox JP: Photosensitizers from plants. *Pestic Sci* 17;579–586, 1986

Dollahite JW: Toxicity of *Kallstroemia parviflora* (warty caltrop) to sheep, goats, and rabbits. *Southwest Vet* 28;135–139, 1975

Downum KR, Rodriguez E: Toxicological action and ecological importance of plant photosensitizers. *J Chem Ecol* 12;823–834, 1986

Downum KR, Villegas S, Rodriguez E, Keil DJ: Plant photosensitizers: a survey of their occurrence in arid and semiarid plants from North America. *J Chem Ecol* 15;345–355, 1989

Du W, Harvey JA: Harmaline-induced tremor and impairment of learning are both blocked by dizocilpine in the rabbit. *Brain Res* 745;183–188, 1997

Durrell LW, Jensen R, Klinger B: *Poisonous and Injurious Plants in Colorado.* Colo Agric Exp Stn Bull 412A, 1950

El Bahri L, Chemli R: *Peganum harmala* L.: a poisonous plant of North Africa. *Vet Hum Toxicol* 33;276–277, 1991

El Bahri L, Belguith J, Ben Youssef S, Bellil H: *Hyoscyamus falezlez* C.: a poisonous plant of North Africa. *Vet Hum Toxicol* 38;378–379, 1996

Flåøyen A, Smith BL, Miles CO: An attempt to reproduce crystal-associated cholangitis in lambs by the experimental dosing of sarsapogenin or diosgenin alone or in combination with sporidesmin. *N Z Vet J* 41;171–174, 1993

Ford EJH, Gopinath C: The excretion of phylloerythrin and bilirubin by the horse. *Res Vet Sci* 16;186–198, 1974

Ford EJH, Gopinath C: The excretion of phylloerythrin and bilirubin by calves and sheep. *Res Vet Sci* 21;12–18, 1976

Fuller TC, McClintock E: *Poisonous Plants of California.* Univ Calif Press, Berkeley, pp262–264, 1986

Gisvold O, Thaker E: Lignans from *Larrea divaricata. J Pharm Sci* 63;1905–1907, 1974

Glastonbury JRW, Doughty FR, Whitaker SJ, Sergeant E: A syndrome of hepatogenous photosensitization, resembling geeldikkop, in sheep grazing *Tribulus terrestris. Aust Vet J* 61;314–316, 1984

Gordon DW, Rosenthal G, Hart J, Sirota R, Baker AL: Chaparral ingestion: the broadening spectrum of liver injury caused by herbal medications. *J Am Med Assoc* 273;489–490, 1995

Grice HC, Becking G, Goodman T: Toxic properties of nordihydroguaiaretic acid. *Food Cosmet Toxicol* 6;155–161, 1968

Gunn JA, MacKeith RC: The pharmacological actions of harmol. *Q J Pharm Pharmacol* 4;33–51, 1931

Gunn JA, Simonart AJL: The pharmacological actions of harmalol. *Q J Pharm Pharmacol* 3;218–237, 1930

Hamberg M: On the formation of thromboxane B_2 and 12L-hydroxy-5,8,10,14-eicosatetraenoic acid (12ho-20:4) in tissues from the guinea pig. *Biochim Biophys Acta* 431;651–654, 1976

Hershey AL: *Some Poisonous Plant Problems in New Mexico.* N M Agric Exp Stn Bull 322, 1945

Hsu CC, Odell G, Williams T: Characterization of the saponin fraction of *Tribulus terrestris. Proc Okla Acad Sci* 47;21–24, 1966

Jacob RH, Peet RL: Poisoning of sheep and goats by *Tribulus terrestris* (caltrop). *Aust Vet J* 64;288–289, 1987

Johnson AE: Photosensitizing toxins from plants and their biological effects. In *Plant and Fungal Toxins: Handbook of Natural Toxins,* vol 1, Keeler RF, Tu AT eds, Marcel Dekker, New York, p345–359, 1983

Kamel SH, Ibrahim TM, Afifi AA, Hamza SM: Estimation of toxic doses of harmine-harmaline hydrochloride 2:1 mixture, harmine and harmaline hydrochlorides in white rats. *U A R J Vet Sci* 7;87–107, 1970a

Kamel SH, Ibrahim TM, Afifi AA, Hamza SM: Major alkaloidal constituents of the Egyptian plant, *Peganum harmala. U A R J Vet Sci* 7;71–86, 1970b

Katz M, Saibil F: Herbal hepatitis: subacute hepatic necrosis secondary to chaparral leaf. *J Clin Gastroenterol* 12;203–206, 1990

Kellerman TS, Van Der Westhuizen GCA, Coetzer JAW, Roux C, Marasas WFO, Minne JA, Bath GF, Basson PA: Photosensitivity in South Africa. 2. The experimental production of the ovine hepatogenous photosensitivity disease geeldikkop *(Tribulosis ovis)* by the simultaneous ingestion of *Tribulus terrestris* plants and cultures of *Pithomyces chartarum* containing the mycotoxin sporidesmin. *Onderstepoort J Vet Res* 47;231–261, 1980

Kellerman TS, Erasmus GL, Coetzer JAW, Brown JMM, Maartens BP: Photosensitivity in South Africa. 6. The experimental induction of geeldikkop in sheep with crude steroidal saponins from *Tribulus terrestris. Onderstepoort J Vet Res* 58;47–53, 1991

Knowles WD, Phillips MI: Neurophysiological and behavioral maturation of cerebellar function studied with tremorogenic drugs. *Neuropharmacology* 19;745–756, 1980

Lamarre Y, Mercier L-A: Neurophysiological studies of harmaline-induced tremor in the cat. *Can J Physiol Pharmacol* 49;1049–1058, 1971

Lancaster MJ, Vit I, Lyford RL: Analysis of bile crystals from sheep grazing *Panicum schinzii* (sweet grass). *Aust Vet J* 68;281, 1991

Lehman AJ, Fitzhugh OG, Nelson AA, Woodard G: The pharmacological evaluation of antioxidants. *Adv Food Res* 3;197–208, 1951

Lessin AW, Long RF, Parkes MW: The central stimulant properties of some substituted indolylalkylamines and β-carbolines and their activities as inhibitors of monoamine oxidase and the uptake of 5-hydroxytryptamine. *Br J Pharmacol Chemother* 29;70–79, 1967

Llinás RR: The noncontinuous nature of movement execution. In *Motor Control: Concepts and Issues,* Humphrey DR, Freund H-J eds, Wiley, Chichester, UK, 223–242, 1991

Llinás R, Volkind RA: The olivo-cerebellar system: functional properties as revealed by harmaline-induced tremor. *Exp Brain Res* 18;69–87, 1973

Mabry TJ, DiFeo DR Jr, Sakakibara M, Bohnsredt CF Jr, Seigler D: The natural products chemistry of *Larrea.* In *Creosote Bush—Biology and Chemistry of* Larrea *in New World*

Deserts, Mabry TJ, Hunziker JH, DiFrio DR Jr eds, Dowden, Hutchinson & Ross, Stroudsburg, Pa, pp115–134, 1977

Marion L: The indole alkaloids. In *The Alkaloids,* vol 2, Manske RHF, Holmes HL eds, Academic Press, New York, pp369–390, 1962

Mathews FP: *Poisonous Plants in the Davis Mountains.* Tex Agric Exp Stn Annu Rep 54, p93, 1941

Mathews FP: The toxicity of *Kallstroemia hirsutissima* (carpet weed) for cattle, sheep, and goats. *J Am Vet Med Assoc* 105;152–155, 1944

McGeer PL, Eccles JC, McGeer EG: *Molecular Neurobiology of the Mammalian Brain.* Plenum Press, New York, pp295–320, 1978

Miles CO, Munday SC, Holland PT, Smith BL, Embling PP, Wilkins AL: Identification of a sapogenin glucuronide in the bile of sheep affected by *Panicum dichotomiflorum* toxicosis. *N Z Vet J* 39;150–152, 1991

Miles CO, Munday SC, Holland PT, Lancaster MJ, Wilkins AL: Further analysis of bile crystals from sheep grazing *Panicum schinzii* (sweet grass). *Aust Vet J* 69;34, 1992a

Miles CO, Wilkins AL, Munday SC, Holland PT, Smith BL, Lancaster MJ, Embling PP: Identification of the calcium salt of epismilagenin beta-D-glucuronide in the bile crystals of sheep affected by *Panicum dichotomiflorum* and *Panicum schinzii* toxicoses. *J Agric Food Chem* 40;1606–1609, 1992b

Miles CO, Wilkins AL, Munday SC, Flåøyen A, Holland PT, Smith BL: Identification of insoluble salts of the β-D-glucuronides of episarsapogenin and epismilagenin in the bile of lambs with Alveld and examination of *Narthecium ossifragum, Tribulus terrestris,* and *Panicum miliaceum* for sapogenins. *J Agric Food Chem* 41;914–917, 1993

Miles CO, Wilkins AL, Erasmus GL, Kellerman TS: Photosensitivity in South Africa. 8. Ovine metabolism of *Tribulus terrestris* saponins during experimentally induced geeldikkop. *Onderstepoort J Vet Res* 61;351–359, 1994a

Miles CO, Wilkins AL, Erasmus GL, Kellerman TS, Coetzer JAW: Photosensitivity in South Africa. 7. Chemical composition of biliary crystals from a sheep with experimentally induced geeldikkop. *Onderstepoort J Vet Res* 61;215–222, 1994b

Moran EA, Couch JF, Clawson AB: *Peganum harmala,* a poisonous plant in the Southwest. *Vet Med* 35;234–235, 1940

Obermeyer WR, Musser SM, Betz JM, Casey RE, Pohland AE, Page SW: Chemical studies of phytoestrogens and related compounds in dietary supplements: flax and chaparral. *Proc Soc Exp Biol Med* 208;6–12, 1995

O'Hearn E, Molliver ME: Degeneration of Purkinje cells in parasagittal zones of the cerebellar vermis after treatment with ibogain or harmaline. *Neuroscience* 55;303–310, 1993

Openshaw HT: The quinazoline alkaloids. In *The Alkaloids,* vol 3, Manske RHF, Holmes HL eds, Academic Press, New York, pp101–118, 1953

Pachalag SV, Patil BD: Effect of gokharu (*Tribulus terrestris* Linn.) as sole feed to weaner lambs. *Indian Vet J* 54;586–587, 1977

Porter DM: The genus *Kallstroemia* (Zygophyllaceae). *Contrib Gray Herb* 198;41–153, 1969

Puzii AD, Serov VM: Persistence of pegarmin (a preparation of the alkaloids of *Peganum harmala*) in the body (of cattle). *Veterinariya (Moscow)* 5;62–64, 1983 (*Vet Bull* 53;6698, 1983)

Puzii AD, Vecherkin SS, Tribunskii MP, Romakhov VG: Toxicity of the combined alkaloids of harmel (*Peganum harmale,* Zygophyllaceae). *Veterinariya (Moscow)* 4;57–58, 1980 (*Vet Bull* 50;7693, 1980)

Rimington C, Quin JI: Studies on the photosensitization of animals in South Africa. 7. The nature of the photosensitizing agent in geeldikkop. *Onderstepoort J Vet Sci Anim Ind* 3;137–157, 1934

Rowe LD: Photosensitization problems in livestock. *Vet Clin North Am Food Anim Pract* 5(2);301–323, 1989

Sheikh NM, Philen RM, Love LA: Chaparral-associated hepatotoxicity. *Arch Intern Med* 157;913–919, 1997

Tapia MO, Giordano MA, Gueper HG: An outbreak of hepatogenous photosensitization in sheep grazing *Tribulus terrestris* in Argentina. *Vet Hum Toxicol* 36;311–313, 1994

Tappel AL, Marr AG: Effect of α-tocopherol, propyl gallate, and NDGA on enzymatic reactions. *J Agric Food Chem* 2;554–558, 1954

Thomson DMP: Various authentic chemoattractants mediating leukocyte adherence inhibition. *J Natl Cancer Inst* 73;595–605, 1984

Timmermann BN: Practical uses of *Larrea*. In *Creosote Bush—Biology and Chemistry of Larrea in New World Deserts,* Mabry TJ, Hunziker JH, DiFeo DR eds, Dowden, Hutchinson & Ross, Stroudsburg, Pa, pp252–256, 1977

Udenfriend S, Witkop B, Redfield BG, Weissbach H: Studies with reversible inhibitors of monoamine oxidase: harmaline and related compounds. *Biochem Pharmacol* 1;160–165, 1958

Walker D, Bird A, Flora T, O'Sullivan B: Some effects of feeding *Tribulus terrestris, Ipomoea lonchophylla* and the seed of *Abelmoschus ficulneus* on fetal development and the outcome of pregnancy in sheep. *Reprod Fertil Dev* 4;135–144, 1992

Wilkins AL, Miles CO, Smith BL, Meagher LP, Ede RM: GC/MS method for analysis of plant and animal samples associated with ovine photosensitization. In *Poisonous Plants of the World: Agricultural, Phytochemical, and Ecological Aspects,* Colegate SM, Dorling PR eds, CAB International, Wallingford, UK, pp263–268, 1994

Wilkins AL, Miles CO, De Kock WT, Erasmus GL, Basson AT, Kellerman TS: Photosensitization in South Africa. 9. Structure elucidation of a β-glucosidase-treated saponin from *Tribulus terrestris,* and the identification of saponin chemotypes of South African *T. terrestris. Onderstepoort J Vet Res* 63;327–334, 1996

Families of Questionable Toxicity or Significance

77

1. Aizoaceae F.Rudolphi

Commonly known as the fig marigold family, the Aizoaceae comprises 12–143 genera and 2,300–2,500 species, some of which are popular garden plants, e.g., *Mesembryanthemum* (ice plant) and *Lampranthus*. The family's primary centers of diversity are South Africa and Australia. Known for their propensity to accumulate oxalate salts, 4 genera in North America have the potential to cause toxicity problems. They have not as yet been implicated, but some have caused disease elsewhere. The lack of problems in North America may be due in part to their limited availability to livestock. These 4 genera have been reported to contain significant and hazardous concentrations of oxalates, up to 10–12% soluble and 15% total (Mathams & Sutherland 1952; Williams 1979; Jacob & Peet 1989; Jacob et al. 1994). The pH of the cell sap of *M. nodiflorum* L. is about 6 (Jacob et al. 1994). This suggests the presence of soluble sodium oxalate salts such as are found in taxa of the Chenopodiaceae (Oke 1969). Thus, as happens in cases of *Halogeton* intoxication, onset of disease may be rapid, and the animal may be less likely to respond to calcium therapy. The concentrations of oxalates increase with plant maturity, being highest when the plants become senescent and dry. Animals newly introduced into an area containing members of the Aizoaceae are at highest risk.

The disease caused by oxalates is of abrupt onset, with muscle twitching, recumbency, and coma. Terminally, there may be bloat. Serum chemistry changes include a decline in calcium and an increase in magnesium. Grossly, there is edema and congestion of the lungs, pale and swollen kidneys, increased fluid in the pericardial sac, hemorrhages on the surfaces of the heart, and reddening of the mucosa of the rumen. Microscopically, the renal tubules will be filled with proteinaceous casts and oxalate crystals, and the epithelium flattened and necrotic. Oxalate crystals may also be present in the walls of microvessels in the submucosa of the rumen.

Prompt administration of parenteral calcium solutions generally will produce temporary relief of the signs of oxalate intoxication. The animals may become ambulatory within a few hours, but the treatment is not curative and relapses can be expected. Some animals also may eventually die of renal failure.

- Fig Marigold Family

  Galenia
  Mesembryanthemum
  Tetragonia
  Trianthema

- oxalates, probably mainly soluble sodium salts, highest levels when plants are mature and dry

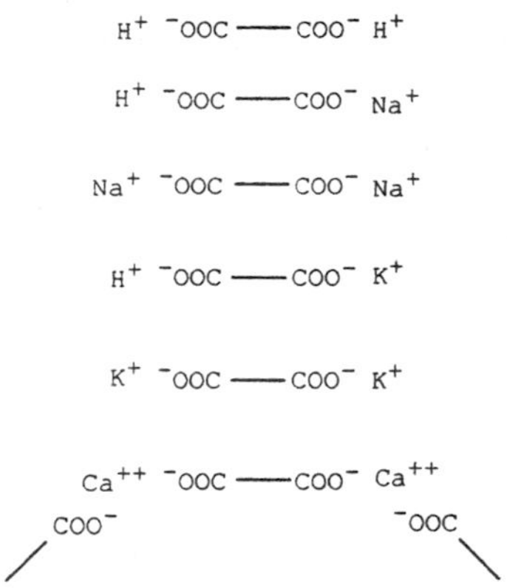

$H^+ \ {}^-OOC—COO^- \ H^+$

$H^+ \ {}^-OOC—COO^- \ Na^+$

$Na^+ \ {}^-OOC—COO^- \ Na^+$

$H^+ \ {}^-OOC—COO^- \ K^+$

$K^+ \ {}^-OOC—COO^- \ K^+$

$Ca^{++} \ {}^-OOC—COO^- \ Ca^{++}$
$COO^- \quad {}^-OOC$

Oxalic acid and its salts

- abrupt onset, muscle twitching, collapse, coma

  gross pathology, kidneys edematous

  microscopic, renal tubules dilated with casts and crystals, epithelium flattened

- treat with parenteral calcium solutions

Galenia L.

Comprising some 20–27 species, *Galenia* is an African genus with greatest diversity in South Africa. In North America, 2 species have been introduced: *G. pubescens* (Eckl. & Zeyh.) Druce and *G. secunda* (L.f.) Sond. These prostrate or decumbent herbs form dense stands 3 m or more in diameter. Their leaves are simple, alternate, entire, and somewhat fleshy. Lacking petals, the flowers are white to pink or yellow. Introduced into California from South Africa as a ground cover and for control of

- *G. pubescens*
  G. secunda

  ornamentals; evergreen, perennial ground covers

roadside erosion, *Galenia* was also considered a candidate for use as a fire retardant on hillsides because it is an evergreen perennial, grows well in the winter in areas with mild climates, and is drought and salt tolerant (Kleinkopf et al. 1976). A good assimilator of soil nitrogen, it develops leaf protein levels of 15–18%. Its use as a ground cover was discontinued after it was discovered to be a potential intoxication risk.

■ nitrate and oxalate accumulation

Although *G. pubescens* has many desirable traits as a ground cover, it may be toxic under some circumstances. Experimentally, sheep given 900–1,100 g/day of plant material died 4–6 hours after dosing or on the next day (Williams 1979). A related species in South Africa, *G. africana* L., when eaten by sheep for several weeks in times of drought, is associated with eventual cardiac failure (van der Lugt et al. 1988). Experimentally, plants of *G. pubescens* grown in the field contained 2.55% KNO_3, whereas those grown in nutrient solution had 6%, in addition to significant amounts of oxalates (Williams 1979). Oxalate concentrations were high in the leaves and low in the stems; in contrast, the nitrate levels were the reverse. Toxicity seems to be ascribable to both compounds because both methemoglobinemia and oxalate crystalluria were apparent in sheep dosed with the plant (Williams 1979). In chicks fed the plant, MHb levels in the blood were as much as 37% of the total hemoglobin, and serum calcium was down to 5.1 mg/dl. Organic nitrotoxins or cyanogenic compounds were not detected. There is rapid onset of distress and weakness, with death occurring in a few hours. The blood may be dark, MHb levels will be elevated markedly, serum calcium will be low, and magnesium may be elevated. Grossly, the tissues are dark-colored, and there is excess fluid in body cavities. Microscopically, there may be pulmonary congestion and edema, as well as renal lesions typical of oxalate intoxication. Prompt administration of parenteral calcium solutions will produce temporary relief of the signs of oxalate intoxication. Nitrate intoxication should be treated with methylene blue, 2–4 mg/kg b.w. i.v.

Mesembryanthemum L.

■ *M. crystallinum*
M. nodiflorum

crystalline ice plant
ice plant
molina
sea fig
slender-leafed ice plant

ornamental ground covers

Originally comprising some 2,000 species, *Mesembryanthemum* was divided into a number of smaller genera on the basis of differences in fruits and now has only about 74 species. Numerous species have been transported throughout the world as ornamentals, the most familiar being ice plant. In North America, 2 species, *M. crystallinum* L. (crystalline ice plant, sea fig, or molina) and *M. nodiflorum* L. (slender-leafed ice plant), have escaped from cultivation and naturalized along the southwestern coast. Annuals or biennials, plants of *Mesembryanthemum* have prostrate to ascending stems, and their alternate or opposite leaves are flat or terete, turning reddish as they mature or when stressed. The small, white flowers have numerous free linear petals.

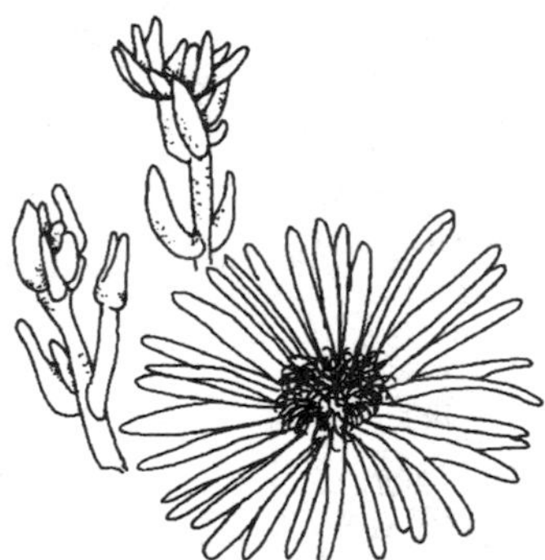

Mesembryanthemum crystallinum

Mesembryanthemum crystallinum

Mesembryanthemum nodiflorum

As do the other members of the Aizoaceae, plants of *Mesembryanthemum* accumulate oxalates. They have caused acute intoxication of sheep in Australia (Jacob & Peet 1989).

■ oxalate accumulation

Tetragonia L.

■ *T. tetragonioides*
New Zealand spinach
cultivated potherb

Comprising approximately 50 species distributed primarily in the Southern Hemisphere, *Tetragonia* is represented in North America by a single species: *T. tetragonioides* (Pall.) Kuntze (= *T. expansa* Murr), commonly known as New Zealand spinach. This annual has spreading, prostrate stems that bear foliage with small crystalline papillae. The alternate leaves have winged petioles. The small flowers lack petals. Cultivated as a perennial in warmer regions or as an annual elsewhere, *T. tetragonioides* has escaped and naturalized in areas with mild winters.

Tetragonia tetragonioides

Tetragonia tetragonioides

Like other members of the family, *T. tetragonioides* is an oxalate accumulator, and intoxication problems have been reported in Australia (Mathams & Sutherland 1952). In North America, no cases of poisoning have been reported, perhaps because of its very low palatability. The levels of oxalates are highest in the early stages of growth, decreasing as the plants mature. Sheep can be induced to eat only older growth.

■ oxalate accumulation, especially young plants

Trianthema L.

■ *T. portulacastrum*
horse purslane
verdolaga blanca
ground cover

Comprising approximately 20 species distributed primarily in tropics and subtropics, and especially Australia, *Trianthema* is represented in North America by a single species: *T. portulacastrum* L. (horse purslane, or verdolaga blanca). Like other members of the family, this annual has prostrate or decumbent stems. The opposite leaves are in unequal pairs, with their petioles sheathing the stems. The pink to rose-purple, inconspicuous flowers are sessile in the leaf axils. *Trianthema portulacastrum* is a widespread tropical weed distributed both in the Old World and the New World. In North America, it occurs in sandy sites across the southern edge of the continent from coast to coast.

Trianthema portulacastrum

As do other members of the family, species of *Trianthema* accumulate oxalates. The leaves of *T. portulacastrum* contain hazardous concentrations of oxalic acid (Bharadwaj & Chandra 1987). In India the species (reported as *T. monogyna* L.) was found to contain up to 9.2% total oxalates and 2,530 ppm NO_3-N in the stems and caused renal failure when fed to goats for a week or more (Gupta et al. 1983).

■ oxalate and possibly nitrate accumulation

2. Annonaceae Juss.

■ Custard-apple Family
Sweet & Soursop Family
Asimina

Commonly known as the custard-apple or sweet & soursop family, the Annonaceae comprises 120–130 genera and 2,000–2,300 species of primarily trees and shrubs. Distribution is chiefly tropical, especially the lowland evergreen forests of both the Old World and the New World. In North America, 7 genera are present, including the common *Asimina* (pawpaw) of the eastern deciduous forests and *Annona reticulata* Shuttlew. ex Chapm. (custard apple, or bullock's-heart) of southern Florida. The common names for the family are related to the edible, fleshy fruits. Intoxication problems are limited to the dermatitis problems associated with species of *Asimina.*

Asimina Adans.

■ *A. parviflora*
A. triloba

American custard apple
dwarf pawpaw
false banana
pawpaw
poor man's banana

deciduous shrubs or small trees; aromatic leaves; pulpy fruits

Asimina triloba

■ fruit, allergic dermatitis, digestive disturbance

■ acetogenins, seeds and bark, possible contact allergens

alkaloids, unknown significance

In North America, the genus *Asimina* is represented by 8 species and 1 named hybrid. The most widespread and commonly encountered are *A. parviflora* (Michx.) Dunal (dwarf pawpaw) and *A. triloba* (L.) Dunal (common pawpaw, poor man's banana, false banana, or American custard apple). These 2 species are small deciduous trees or shrubs. Their oblong to obovate leaves are aromatic when crushed. Produced before or with the leaves, the flowers are green at first, gradually turning brown or lurid purple at maturity. Eaten only after the first frost, the oblong, pulpy fruits are berries with several large flattened seeds. The custardlike flesh, though sweet, is characterized as delicious by some and insipid by others. *Asimina parviflora* occurs in pine or oak woods along the Atlantic and Gulf coast from Virginia to Florida and Texas. In contrast, *A. triloba* is found in rich, moist soils of the forests of eastern North America.

Although generally considered to produce an edible fruit, *A. triloba* may rarely cause dermatitis after the fruit is handled or irritation of the digestive tract after it is eaten (Mitchell & Rook 1979). Barber (1905) reported his own urticarial response to eating ripe pawpaw. Although he had eaten pawpaw previously without harm, a severe response ensued about 6 hours after he ate ground-ripened fruit on one occasion. The fruit eaten by another person at the same time evoked no adverse effects. Thereafter, Barber suffered the same response each time he ate pawpaw, i.e., severe nausea and hives that lasted for several hours. It is assumed that the same potential for dermatitis exists for *A. parviflora.*

The irritation of the digestive tract was perhaps the basis for the medicinal use of pawpaw at one time. A tincture made from the seeds was used as an emetic (Millspaugh 1974).

The specific cause of the dermatitis caused by *Asimina* is not known. However, studies of chemical constituents reveal the presence of two types—acetogenins and alkaloids—that may have toxic effects. *Asimina* contains numerous bioactive compounds that share structural similarities with those of the Anacardiaceae (sumac family), well known for its contact allergens such as those in poison ivy. Chief among these compounds in the Annonaceae are the cytotoxic acetogenins, long-chain 35-to-38-C compounds found in the seeds and bark of various genera and species. The acetogenins can be subdivided into several groups based upon the arrangement of the tetrahydrofuran (THF) substituents. Arranged in order of increasing potency, they include mono-THF, nonadjacent bis-THF, and adjacent bis-THF (Gu et al. 1995; Alfonso et al. 1996). There also are a few tri-THF compounds. These cytotoxic compounds may be responsible for the allergic dermatitis experienced by some individuals after becoming sensitized to the fruit.

Some acetogenins, called annonins, are also noteworthy for their potential as cytotoxic antineoplastic agents, especially the extensive series found in *Annona bullata*

A.Rich., native to Cuba (Rupprecht et al. 1990; Hui et al. 1992; Gu et al. 1994; Zhao et al. 1994, 1996). The principal mechanism of their action is inhibition of mitochondrial electron transport and oxidative phosphorylation, specifically through competitive inhibition of NADH:ubiquinone oxidoreductase of complex I (Londershausen et al. 1991). Because complex II is not affected, succinate has been proposed as a likely antidote should intoxications occur (Gu et al. 1995). A secondary toxic effect is inhibition of NADH oxidase in the plasma membrane. The acetogenins of *A. triloba*'s bark and seeds—asimin, asimicin, asiminacin, bullatacin, trilobin, and trilobacin—are potent adjacent bis-THF types (Rupprecht et al. 1990; Zhao et al. 1992; Gu et al. 1995). Bullatacin is generally accepted as the compound with which the others are compared. It markedly inhibits Complex I (Alfonso et al. 1996). Trilobin and asiminacin are of equal or greater potency. Albeit their roles in disease are not clear, acetogenins produce a range of other effects; e.g., asimicin is cytotoxic, antimalarial, and pesticidal (Rupprecht et al. 1986).

Asimina triloba

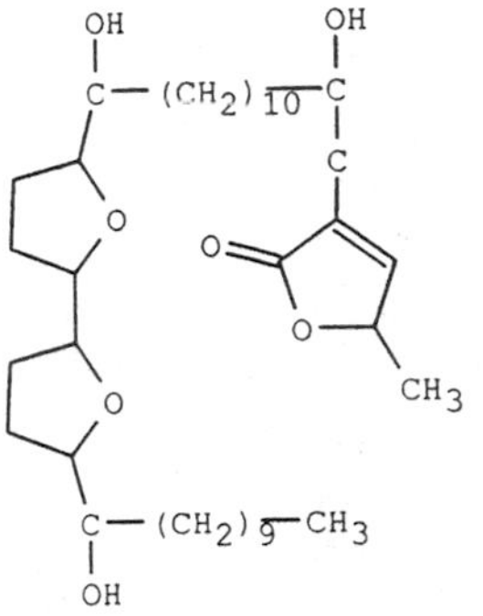

Asimicin

Asimina also produces alkaloids: asimicilone in *A. parviflora*, asimilobine in *A. triloba*, and others such as isocorydine, mainly in the woody portions of the plants (Tomita & Kosuka 1965; Ratnayake et al. 1992). The crude seed-oil extract from *A. triloba* is acutely toxic to rats; it can be rendered nontoxic by refining but not by heating (Matsui 1980).

Intoxication due to *Asimina* is apparently a problem only in humans. In sensitized individuals, 3–12 hours after ingestion of the fruit, there is rapid onset of urticaria, or hives, on the face, and it then spreads to other areas. This is accompanied by nausea, vomiting, and chills. Complete recovery usually occurs in 24–48 hours. Intoxication in some instances may be characterized by marked overall weakness, and treatment may be directed toward counteracting this effect, possibly with a parenteral source of succinate.

- sensitized humans, urticaria (hives), possibly nausea, vomiting, 24–48 hours duration

3. Arecaceae Schultz Sch.

- Palm Family

  Caryota

Commonly known as the palm family, the Arecaceae comprises 200–212 genera and 2,780–3,000 species of tall, slender trees or stout shrubs with unbranched stems bearing evergreen leaves crowded in terminal clusters. Nearly all are restricted to tropical or warm-tropical regions. An alternative family name used primarily by European taxonomists is Palmae. Of considerable economic importance, the family includes the coconut, date, sugar, and betel-nut palms, as well as those commonly planted as ornamentals, such as royal, fishtail, and queen palms. In North America, approximately 20 genera and 45 species are present; 1 genus has the potential to cause toxicologic problems.

Caryota L.

- *C. mitis*

  C. ochlandra

  C. urens

  fishtail palm
  sago palm
  wine palm

  ornamentals, frost-sensitive

Native to tropical Asia and Australia, *Caryota* (fishtail palm) comprises 12 species, of which 3—*C. mitis* Lour. (tufted or clustered fishtail palm), *C. ochlandra* Hance (Canton fishtail palm), and *C. urens* L. (wine or sago palm)—are frequently cultivated for their feathery foliage. Their solitary or clustered stems are partially covered by the persistent, fibrous sheaths of previously present leaves, which are typically 2-pinnately compound with obdeltoid to triangular, irregularly toothed leaflets that give the genus its characteristic appearance, as indicated by its vernacular name. Very frost sensitive, these ornamentals are planted as specimen plantings in Mexico, southern Florida, and California or as tub plants elsewhere.

- fruits, irritant calcium oxalate raphides

  burning sensation, lips and tongue, itching of skin

The mature fruits of *Caryota* cause contact irritation because of the presence of calcium oxalate raphides made up of monohydrate crystals approximately 250–400 μm long and 6–9 μm in diameter (Allen 1943; Snyder et al. 1979). In contrast to the raphides of the Araceae (arum family), electron microscopy reveals no evidence of barbs or grooves on these crystals. The irritant effects appear to be due solely to mechanical action, given that washing the crystals or treating them with enzyme inhibitors failed to change the irritant response, whereas mechanical disruption of the crystals prevented the response (Snyder et al. 1979).

Depending on the site of contact, there is an intense burning sensation of the mouth, lips, and tongue or itching of the skin with redness and swelling of a few hours' duration. Problems can be avoided or reduced by washing the fruits vigorously to remove the crystals.

- Balsam Family
  Touch-me-not Family

  Impatiens

4. Balsaminaceae A.Rich.

Native to both the Old World and the New World, Balsaminaceae (balsam or touch-me-not family) comprises 2–4 genera and 450–600 species of subsucculent annual or perennial herbs or rarely subshrubs bearing conspicuously bilaterally symmetrical flowers. Typically 1 of the 3 sepals is petaloid and long-spurred. Of minor economic importance, plants of the family are cultivated as garden ornamentals or houseplants. In North America, only 1 genus that may cause toxicologic problems is present.

- *I. balsamina*
  I. capensis
  I. pallida

  impatiens
  garden balsam
  jewelweed
  lady's slipper
  orange touch-me-not
  pale touch-me-not
  touch-me-not
  yellow touch-me-not

  moist soils, woods and meadows; ornamentals

- irritant naphthoquinone, lawsone

O
OH
O

Lawsone

Impatiens L.

The largest genus of the family, with 450–500 species, *Impatiens* is commonly known as touch-me-not because of the explosive dehiscence of its capsules. In North America, 11 native and introduced species are present, including *I. balsamina* L. (garden balsam, or lady's slipper), *I. capensis* Meerb. (jewelweed, or orange touch-me-not), and *I. pallida* Nutt. (pale or yellow touch-me-not). Plants of these species are annual herbs with a mucilaginous sap, hollow internodes, and swollen nodes. The stems bear simple, alternate or rarely opposite leaves and axillary, scorpioid cymes of 1–4 chasmogamous or cleistogamous flowers with perianths in 2-series. Bilaterally symmetrical, the calyces consist of 3 sepals of 2 forms. The lateral 2 are small, whereas the middle or lower is petaloid, large, recurved, free, spurred, and yellow or orange or white, often with red or reddish brown markings. The corollas are likewise bilaterally symmetrical, with 5 petals appearing to be 3 because of fusion, and of 2 forms, 1 free and 4 fused in 2 pairs, each pair resembling a single bilobed petal. Moist soils of woods or meadows, often adjacent to water, are typical habitats.

Although the crushed leaves and stems of several species of *Impatiens* are reputed to help prevent a reaction to poison ivy and to promote healing of minor burns, they, and the flowers, contain the naphthoquinone lawsone (naphthalenic acid), the yellow pigment of the dye henna. It is chemically very similar to jugalone and may cause mild to moderate irritation of the digestive tract (Thomson 1971).

- Catalpa Family
  Trumpet-creeper Family

  Bignonia
  Campsis
  Crescentia

5. Bignoniaceae Juss.

Widespread, but with greatest diversity in tropical America, the Bignoniaceae, the trumpet-creeper or catalpa family, comprises 100–120 genera and 650–800 species of trees and woody vines. Its members characteristically produce large, showy flowers, and several of its species are prized as ornamentals. Familiar street trees in tropical cities are African tulip tree, jacaranda, calabash, and empress tree. In temperate cities both catalpa and trumpet creeper are often encountered. In North America, 18–20 genera and 28–30 species, both native and introduced, are present. Plants of 3 genera are of questionable toxicity.

Bignonia L.

Although originally circumscribed to include numerous species, *Bignonia,* commonly known as cross vine or quarter vine, is now considered to have only 1 species, *B. capreolata* L. (= *Anisostichus capreolatus* [L.] Bureau). The other taxa formerly in this genus are placed in other genera. It is an evergreen, perennial woody climber with stems up to 20 m long, bearing opposite, pinnately compound leaves with 1 pair of leaflets and a 3-branched tendril with adhesive disks at the branch ends. Borne in axillary cymes, the large, campanulate to funnelform flowers are deep orange to reddish. As is characteristic of the family, the slender capsules are elongate.

■ *B. capreolata*

cross vine
quarter vine

evergreen, perennial, woody climber

Native to moist woods of the southeastern quarter of North America, *B. capreolata* was long thought to be toxic, but there are no authenticated cases or reports (Pammel 1911).

■ risk unknown

Campsis Lour.

Comprising 2 species native to eastern North America and eastern Asia, *Campsis* is commonly known as trumpet creeper and is widely grown for its dense, dark green foliage and showy flowers. Native to moist woods in the eastern half of the continent is *C. radicans* Lour., which may be an aggressive weed in fencerows and along roadsides. It is cultivated elsewhere around the world. Synonyms appearing in the older literature are *Tecoma radicans* (L.) Juss. and *Bignonia radicans* L. Plants of the species are woody vines that climb either by trailing or by aerial rootlets along the stems. The opposite, pinnately compound leaves, with 5–13, lanceolate to ovate leaflets, are serrate and acuminate. Large, orange-red, and showy, the flowers are borne in clusters at the ends of the branches from midsummer until fall. The long, spindle-shaped capsules bear numerous winged seeds.

■ *C. radicans*

trumpet creeper

weedy vine, ornamental; flowers showy, orange-red

As with *Bignonia* and other members of the family, there are suspicions of *Campsis* toxicity but a lack of actual reports (Pammel 1911). It is doubtful that toxicity is an important feature for this species, because it is cultivated as an ornamental throughout the world and there are no reports of problems other than a minor problem of dermatitis in some individuals who come in contact with the leaves and flowers (O'Leary 1964; Morton 1982).

■ risk unknown, possibly dermatitis

Crescentia L.

Its name honoring the thirteenth-century Italian agricultural writer Pietro de Crescenzi, *Crescentia* comprises 3–5 species native to tropical America, especially the West Indies, of which *C. cujete* L. (calabash tree, cujete, or camasa) yields the calabash, which is used as a water gourd. Plants of the species are trees 6–12 m tall, with large, horizontally spreading branches bearing clusters of simple, oblanceolate leaves. The solitary, pendulous, yellowish purple flowers are tubular and produce a nearly globose, smooth, many-seeded berry 45–55 cm in diameter and with a hard rind.

■ *C. cujete*

calabash tree
camasa
cujete

ornamental trees

The fruits have been used for a variety of medicinal purposes and are considered to be edible (Watt & Breyer-Brandwijk 1962). Under some circumstances and for some animals, their pulp is considered to be slightly toxic, with purgative and possibly abortifacient actions (Blohm 1962). There are few reports of serious problems.

■ fruit pulp, possibly digestive disturbance

6. Buxaceae Dumort.

■ Boxwood Family

Buxus

Widely distributed in both temperate and tropical regions of the Old World and the New World, the Buxaceae, or boxwood family, comprises 4 or 5 genera and 60–100 species of evergreen shrubs or trees or perennial herbs. Several taxa are cultivated as ornamentals, and the hard, fine-grained wood of some species is valuable and used for drafting tools, musical instruments, pipes, inlay, and furniture. In North America, 3 genera and 8 species are present; members of 1 genus have been associated with intoxication problems.

Buxus L.

■ *B. harlandii*

B. microphylla

B. sempervirens

box

boxwood

ornamentals, evergreen shrubs or trees

Prized as hedging and as specimen plantings for its dense growth and attractive foliage, *Buxus*, commonly known as boxwood or box, comprises 35–70 species and numerous cultivars. *Buxus sempervirens* L. (English or common boxwood) was one of the first plants brought to the American colonies. Other commonly encountered species are *B. harlandii* Hance (Korean boxwood) and *B. microphylla* Siebold & Zucc. (Japanese boxwood). Plants of the genus are profusely branched evergreen shrubs or trees bearing opposite, short-petiolate or sessile, coriaceous, simple leaves. The tiny, apetalous, imperfect flowers are borne in axillary or terminal, racemose to capitate clusters. Globose to ovoid, the 3-horned, coriaceous capsules contain glossy black seeds.

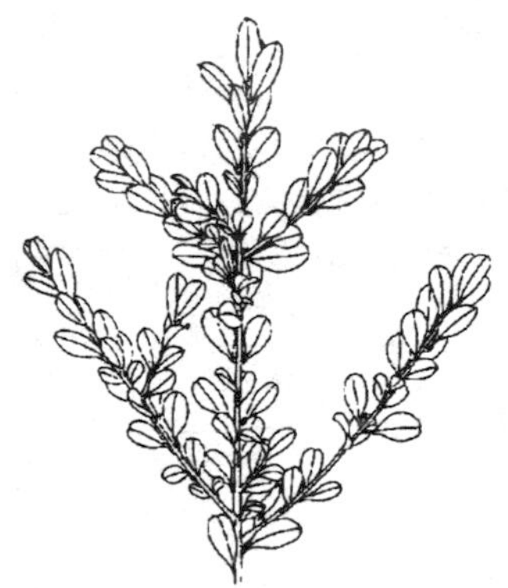

Buxus microphylla

Boxwood has long been employed for a variety of medicinal uses (Cordell 1981). Ingestion of large amounts of its foliage or bark is an uncommon cause of serious digestive tract disease in livestock but is suspected of causing death in some instances (Harshberger 1920; West & Emmel 1952; van Soest et al. 1965; Camy et al. 1986). Livestock affected include cattle, horses, and pigs. The foliage contains numerous alkaloids, as well as glycosides and flavonoids (Atta-ur-Rahman et al. 1989, 1991). A dozen or more steroid alkaloids have been isolated from *B. microphylla*, and two dozen from *B. sempervirens* (Cerny & Sorm 1967; Tomko & Voticky 1973). Examples include buxidine, cyclobuxines, cyclobuxamines, and cyclobuxidines.

■ irritant steroid alkaloids, digestive disturbance

Cyclobuxine D

Cyclobuxamine A

■ diarrhea, colic, excess salivation

reddened gut mucosa

treat with activated charcoal, fluids, electrolytes

Within a few hours of ingestion, the bitter alkaloids cause profuse diarrhea with tenesmus, severe colic, excess salivation, dehydration, and apparent thirst. There may also be systemic signs such as tremors, seizures, and respiratory difficulty. In the rare fatality, there will be severe inflammation of the mucosa of the stomach and small intestine. Treatment involves rehydration with fluids and electrolytes, and this can be augmented with activated charcoal to absorb any toxins remaining in the digestive tract.

7. Capparaceae Juss.

Closely related to the Brassicaceae, the Capparaceae, or caper family, is distributed primarily in the tropics and subtropics of both the Old World and the New World, particularly Africa. It comprises 40–50 genera and 700–800 species of trees, shrubs, and herbs. The spelling Capparaceae has been conserved over the traditional Capparidaceae in the International Code of Botanical Nomenclature. A few of its members are cultivated as ornamentals, and capers, the pickled flower buds of *Capparis spinosa* L., are used for seasoning and garnish. Of the 9 genera and approximately 45 species present in North America, the members of 2 genera may pose potential toxicity problems.

■ Caper Family

Cleome
Wislizenia

Cleome L.

Comprising 150–200 species distributed chiefly in Africa and the Americas, *Cleome* is known by a variety of common names—*bee plant, stinking flower, stinking clover, clammy weed, spider flower*—each of which is indicative of a feature of the genus. In North America some 18 species are present. Perhaps most commonly encountered are *C. lutea* Hook. and *C. serrulata* Pursh, which occur in a variety of habitats in the western two-thirds of the continent. Plants of the genus are glabrous to glandular-pubescent, annual or perennial herbs (perennial in North America) whose alternate, palmately compound leaves have 3–7, entire to serrulate leaflets. Borne in terminal, bracteate racemes, the showy, bilaterally symmetrical flowers have 4 sepals and 4 petals, 6 to many stamens, and a 2-carpellate, superior ovary with 1 locule. Typically long-clawed, the petals are purplish, white, yellow, or greenish. The fruits are capsules containing several to many seeds.

■ *C. lutea*
C. serrulata

bee plant
clammy weed
spider flower
stinking clover
stinking flower

herbs; leaves palmately compound; 3–7 leaflets; petals long-clawed

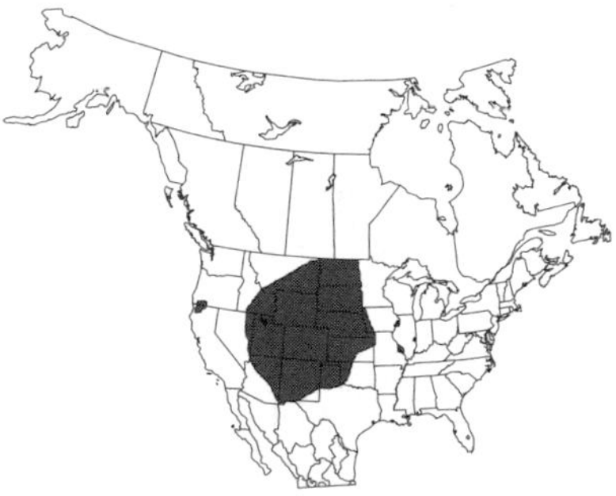

Cleome serrulata

■ potherbs; possible nitrate, selenium accumulation

Cleome serrulata

Cleome serrulata was an important food plant for western Indian tribes. Its seeds were used to make a meal, and the foliage was used as a potherb but required several changes of the cooking water to eliminate the plant's unpleasant odor (Harrington 1967). The species was so highly regarded that it was immortalized in song by the Hopi and the Tewa (Ebeling 1986). Both *C. lutea* and *C. serrulata* were utilized as sources of dyes. Species of *Cleome* are commonly listed among the plants that accumulate nitrate and selenium (Tucker et al. 1961; Kingsbury 1964), but it is doubtful that actual problems are caused by the presence of either of these toxicants in *Cleome*.

■ *W. refracta*
jackass clover
foul-smelling herbs

■ potherb; risk unknown

Wislizenia Engelm.

A monotypic genus of the southwestern corner of the continent, *Wislizenia,* commonly known as jackass clover, occurs in desert washes, alkaline flats, and fields and along roadsides. The single species is *W. refracta* Engelm. Rather foul-smelling, plants of this species are glabrous to puberulent annuals or perennials with profusely branched stems to 2 m tall. The alternate compound leaves, resembling those of clover, typically have 3 leaflets. The small, yellow flowers are borne in dense, terminal racemes. The exserted stamens also are yellow. Borne on geniculate pedicels, the capsules are conspicuously 2-lobed.

Wislizenia refracta was used as a potherb in a manner similar to *Cleome,* and a meal was made from the ground seed (Ebeling 1986). It also is valued as a source of good honey. However, the foliage is thought to be lethal at less than 1% of body weight, but plants are quite unpalatable (Kingsbury 1964; Fuller & McClintock 1986). Actual case reports are lacking.

Wislizenia refracta

8. Commelinaceae R.Br.

■ Spiderwort Family
Tradescantia

Widespread in tropical and warm-temperate regions of the world, the Commelinaceae, or spiderwort family, comprises 34–50 genera and 500–700 species, some of which are cultivated as garden ornamentals or houseplants, such as wandering Jew, boat-lily, spiderwort, and dayflower. Others are somewhat weedy and thrive in disturbed sites. Taxa of the family are typically semisucculent, mucilaginous perennials or annuals that are acaulescent or caulescent, with jointed stems and alternate, simple leaves that are somewhat dilated at their bases into inflated sheaths. Borne in terminal or axillary cymes and subtended by a spathe, their showy, ephemeral, radially or bilaterally symmetrical flowers are 3-merous, with differentiated sepals and petals. The superior ovary gives rise to a capsule.

Some members of the family are used medicinally or eaten (Heywood 1978). For example, stem and leaf extracts of African and Asian species are used as a laxative or as a treatment for inflammation of the eyes, and the rhizomes and young shoots of some American taxa were eaten by Americans Indians. In North America, approximately 13 genera and 58 species, native and introduced, are present. Two species of 1 genus may cause problems.

Tradescantia L.

■ *T. pallida*
T. spathacea
boat-lily
Moses-in-his-cradle
purple heart
purple queen
spiderwort
ornamentals, weedy invaders

Honoring John Tradescant, a seventeenth-century English naturalist and gardener, *Tradescantia* (spiderwort) comprises 50–70 New World species. It differs from other members of the family by having umbellate cymes subtended by paired foliaceous bracts, radially symmetrical flowers, and 6 fertile stamens with villous filaments. Flower color varies from white to pink or blue or purple. With some brought into cultivation as garden ornamentals, species of the genus naturally occur in the eastern half of the continent in prairie and forest habitats. A few are weedy and are present in highway and railroad rights-of-way. The 2 species implicated in problems are *T. pallida* (Rose) D.R.Hunt (purple queen, or purple heart), which is native to eastern Mexico, and *T. spathacea* Sw. (boat-lily, or Moses-in-his-cradle), which is native to Central America and called *Rhoeo spathacea* (Sw.) Stearn by some taxonomists.

■ irritant, stinging, burning sensation of skin, eyes, or mouth

The sap from the foliage of these taxa has been used medicinally for astringent effects, e.g., to promote clotting of blood, but it also may cause contact irritation consisting of a stinging or burning sensation of the skin, eyes, and mouth (Morton 1982).

The skin becomes reddened temporarily, and there may be some transient respiratory difficulty. The sclera of the eye may become inflamed and if plant material is eaten, the mouth, lips, and tongue may be reddened and swollen. There will also be excess salivation.

9. Cornaceae Dumort.

Distributed primarily in the north temperate regions, the Cornaceae, or dogwood family, comprises 11–13 genera and 90–100 species of evergreen or deciduous trees and shrubs. Only a few are perennial herbs from woody rhizomes. Prized as ornamentals, the best-known genera are *Aucuba* (spotted laurel) and *Cornus* (dogwood). The fruits of 1 species of *Cornus* are made into preserves and an alcoholic beverage, and the wood of others was used at one time for weaving shuttles and bobbins, because its tight grain did not snag the textile fibers. In addition to its woody habit, the family is characterized by small, 4- or 5-merous flowers that are subtended by large, showy bracts in some taxa. The 2-carpellate pistils have inferior ovaries and give rise to drupes. In North America, only *Cornus* is native. Possible toxicologic problems have been associated with it and the introduced *Aucuba.*

■ Dogwood Family

Aucuba
Cornus

Aucuba Thunb.

Comprising 3 or 4 species, *Aucuba* is native to eastern and central Asia. Only *A. japonica* Thunb., commonly known as spotted laurel, gold-dust tree, or Japanese aucuba, is widely cultivated. Tolerating a variety of environmental conditions, including industrial pollution, it is one of the most common garden evergreens in England. It is cultivated as an outdoor ornamental in the southern half of North America and as an indoor tub plant elsewhere. Used as screening and hedging, plants are stout, dioecious shrubs 1–5 m tall with green stems and branches that bear simple, opposite, glossy, green and yellow to white-spotted, coriaceous leaves that have elliptic to narrowly ovate blades and serrate margins. The staminate plants are cultivated for their foliage, and the pistillate are cultivated both for foliage and the small, 4-merous, star-shaped, purple flowers, borne in terminal panicles that give rise to bright scarlet, oblong to ovoid drupes.

The leaves and fruits of *Aucuba* are occasionally the cause of a slow onset of mild purgative actions when ingested by pets (Inouye et al. 1974; Leroux 1986). This effect is probably due to the presence of aucubin and asperuloside, monohydrate, iridoid glucosides of unstable furan derivative aglycones (Trim & Hill 1952). All aerial portions of the plants should be considered toxic, but the effects are generally mild. Triterpenoids may also contribute to the irritant effects. There may be mild vomiting and perhaps diarrhea, but typically treatment will not be required. This is usually a problem of limited duration and severity.

■ *A. japonica*

gold-dust tree
Japanese aucuba
spotted laurel

evergreen ornamental

■ irritant iridoid glucosides; mild digestive disturbance, vomiting

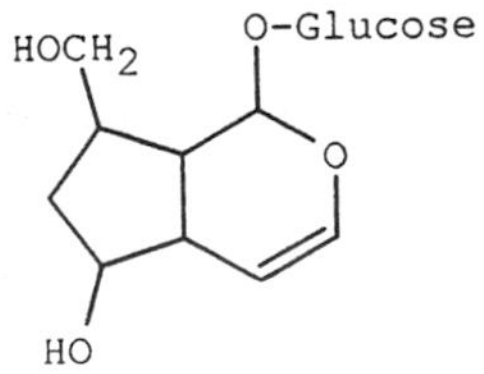

$HOCH_2$ O-Glucose O HO

Aucubin

Cornus L.

Simultaneously prized for the showy involucre of many of its species in the spring and the brilliant color of its autumn foliage, *Cornus,* known as dogwood or cornel, comprises 45–55 species. In North America, 20 species, numerous named hybrids, and many cultivars are present. The genus has been divided by some taxonomists into 3–6 smaller genera—e.g., *Swida, Cynoxylon,* and *Chamaepericlymenum*—on the basis of differences in habit, phyllotaxy, and involucre. Representative of the genus are *C. florida* L. (flowering dogwood, cornel, dog tree, or bitter redberry), *C. rugosa*

■ *C. florida*
C. rugosa
C. sericea

cornel
dog tree
dogwood
green osier
swamp sassafras

shrubs or trees; showy spring flowers and autumn foliage

eastern North America, some ornamentals

Lam. (round-leaved dogwood, alder dogwood, green osier, or swamp sassafras), and *C. sericea* L. (swamp dogwood, silky cornel, blueberry cornel, or female dogwood). Plants of the genus are deciduous or rarely evergreen trees, shrubs, or rarely herbs, with opposite, entire leaves, except for 1 species in which they are alternate. Borne in terminal glomerules or umbellate to paniculate cymes, the small flowers are 4-merous and produce large or small drupes of various colors. Perhaps the most distinctive feature is the presence in many species of an involucre of large, spreading, petaloid bracts, often mistaken for petals. Their appearance in the woods of the eastern half of the continent is traditionally taken as a harbinger of spring.

■ bark, irritant, digestive disturbance

Species of *Cornus* generally are not considered to be toxic, but the bark has effects on the heart and, when chewed and swallowed, may cause signs indicative of irritation of the digestive tract (Millspaugh 1974). Signs include nausea, headache, vomiting, abdominal pain, diarrhea, sweating, and sedation. It is not known whether all species of the genus cause these effects, but they are consistent among *C. florida, C. rugosa,* and *C. sericea.* Probably all taxa should be suspected.

10. Corynocarpaceae Engl.

■ Karaka Family
New Zealand Laurel Family

Corynocarpus

Commonly known as the New Zealand laurel or karaka family, the Corynocarpaceae consists of a single genus, *Corynocarpus,* with 5 species of trees and shrubs native to New Zealand, Australia, and Polynesia.

Corynocarpus J.R.Forst. & G.Forst.

■ *C. laevigatus*

karaka nut
evergreen trees

ornamental, moderate climates

The genus is represented in North America by a single species, *C. laevigatus* J.R.Forst. & G.Forst. (karaka nut). Not winter-hardy, it is grown as an ornamental only in the more moderate climates of California and Mexico and along the Gulf Coast. Attaining heights of 16 m, trees of this species have spreading crowns of stout branches that bear simple, alternate, evergreen, glossy leaves with obovate-oblong blades and entire margins. The small, perfect, 5-merous flowers are borne in terminal panicles and produce orange, ovoid drupes.

■ seeds, heat-labile nitro-compound, karakin

nausea, vomiting, rarely neurotoxic effects

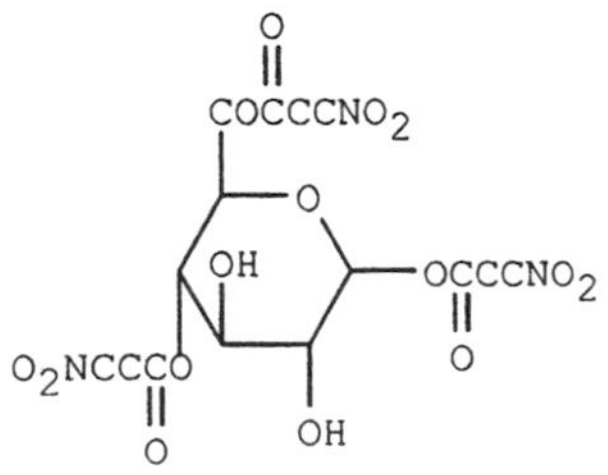

$COCCCNO_2$ (O)
OH
$OCCCNO_2$ (O)
O_2NCCCO (O)
OH

Karakin

The flesh of the drupes of *C. laevigatus* is edible, but the bitter seeds have the reputation of causing violent convulsive seizures in humans and paralysis of the pelvic limbs of animals (Bell 1974). Both seeds and fruits were eaten by the Maoris, but only after extensive preparation involving heating and washing. The effects are attributable to karakin, a nitropropanoyl glucopyranose that is identical to endecaphyllin B from *Indigofera spicata* in the Fabaceae (Carter & McChesney 1949; Finnegan & Stephani 1970; Bell 1974). Other glucopyranoyls present in the species include corollin, coronarian, cibarian, and corynocarpin (Moyer et al. 1979). The convulsive seizures can be reproduced with either aqueous extracts of the seeds, karakin, or nitropropionic acid (3-NPA) (Bell 1974). The nectar is reported to be toxic to bees (McBarron 1976; Fuller & McClintock 1986). In the rare instance of intoxication there will be nausea and vomiting and rarely spasms and paralysis. There are no distinctive pathological changes. Sedation to control the seizures is the primary treatment.

11. Cuscutaceae Dumort.

■ Dodder Family

Cuscuta

Apparently derived from the Convolvulaceae, or morning-glory family, the Cuscutaceae, commonly known as the dodder family, consists of a single genus, *Cuscuta,* with 150–170 species of worldwide distribution but with greatest diversity in the New World. Approximately 50 species are present in North America.

Cuscuta L.

Known by a variety of common names, including *dodder, strangle vine, love vine, witches' shoelaces, devil's gut,* or *angel's hair,* members of *Cuscuta* are rootless, leafless parasites. They produce orange or white, sticky, filamentous stems that form tangled masses on and between their host plants. Seeds in the soil germinate to produce filaments that elongate until they come in contact with an appropriate host plant. The filaments coil tightly about the host stem and produce small extensions of tissue, interpreted to be haustoria, that penetrate the host tissues and through which the dodder plant receives its nutrients. Proliferation of new filaments occurs, and eventually small, 5-merous flowers and capsules are produced. Species of the genus parasitize a wide range of woody and herbaceous taxa; some species are very host specific, whereas others are generalists. Representative species include *C. pentagona* Engelm. (field dodder), *C. gronovii* Willd. & Schult. (common dodder), *C. indecora* Choisy (collared dodder), and *C. coryli* (hazel dodder).

■ *C. coryli*
C. gronovii
C. indecora
C. pentagona

angel's hair
devil's gut
dodder
love vine
strangle vine
witches' shoelaces

rootless, leafless parasites with filamentous stems forming tangled masses on host plants

Some species of *Cuscuta* are suspected to cause purgative effects (Watt & Breyer-Brandwijk 1962). *Cuscuta pentagona* has been associated with severe digestive tract irritation when hay infested with up to 50% dodder was fed to cattle for a month (Movsesyan & Azaryan 1971). There were also signs of liver disease, which was substantiated by experiments in horses, cattle, and rabbits (Movsesyan & Azaryan 1973). Coupled with the appearance of neurologic signs, this suggests that hepatoencephalopathy may be a complicating factor. European species are also reputed to be neurotoxic when the filaments are fresh, but not when dried (Bartik & Piskac 1981). Horses seem to be especially prone to digestive tract problems when fed substantial amounts of hay infested with *Cuscuta* (Pergat & Stolyarova 1961).

■ irritant effects, digestive disturbance

If the range of toxic effects produced by *Cuscuta* seems to be unusually diverse, the cause may be due to the wide range of hosts parasitized by the genus. As with other parasitic plants, such as *Phoradendron* (mistletoe) in the Viscaceae, there seem to be significant differences in the chemical content of the dodder plants depending on the host plant as indicated by the reaction of aphids feeding on *Cuscuta* filaments. When growing on host plants such as *Allium* or *Amaranthus, Cuscuta* becomes unsuitable for aphid growth (Harvey 1966). Species of *Cuscuta* seem to be particularly prone to accumulate alkaloids from a variety of host plants (Czygan et al. 1988). Although it is very unlikely for *Cuscuta* to be eaten in sufficient quantity to cause disease, hepatotoxic and neurotoxic effects are possible, depending on the host plant alkaloids. The most likely signs are those related to digestive tract irritation, mainly diarrhea. Treatment should be directed toward relief of these effects.

Cuscuta pentagona

12. Cyperaceae Juss.

■ Sedge Family

Scirpus

Morphologically very similar to the Poaceae, or grass family, and often mistaken for it, the Cyperaceae, or sedge family, comprises 70–146 genera and 4,000–5,300 species. Approximately two-thirds of the species belong to 6 genera. Cosmopolitan in distribution, the family is most abundant in subarctic and temperate regions; its taxa are characteristically encountered in wet or damp habitats. In contrast to the grass family, the Cyperaceae is of minor economic importance. The stems and leaves of some taxa provide mediocre forage for cattle or are used for basketry and matting. Papyrus is derived from the stems of *Cyperus*. Other taxa are used as pot and water-garden ornamentals, and a few are used medicinally or eaten. Members of the family are perennial or annual, grasslike herbs with solid, typically 3-sided stems bearing 3-ranked leaves that have fused sheath margins. Wind-pollinated and inconspicuous, the flowers are borne in spikes, and each is subtended by a scarious bract. A stem may bear 1 spike or many spikes clustered together. The fruit is an achene. In North America, 24–26 genera and about 1,000 species are present. Intoxication problems are associated with 1 genus.

Scirpus L.

■ *S. americanus*

bulrush
swordgrass
three-square bulrush

herbs; leaves basal, long linear; in or near water

Typically encountered in marshes, bogs, ponds, or waterways and often forming dense stands, *Scirpus,* commonly known as bulrush, consists of 100–300 species. Some taxonomists divide the genus into 5 smaller genera. The erect stems are 3-sided or cylindrical and bear basal or cauline, linear leaves that are sometimes vestigial and scalelike. Solitary or borne in paniculate or capitate clusters, the spikes are turbinate with spiraled bracts. Each flower has 6 rough bristles (perianth parts), 3 stamens, and a 3-branched styles. The smooth achenes are trigonous or lenticular. In North America, approximately 60 native and introduced species are present. Some are used as ornamentals, and others provide habitat for aquatic wildlife.

■ suspect as a cause of ARDS in cattle

Elsewhere in the world, *Scirpus* is utilized in a variety of ways (Heywood 1978). In Hindu medicine the roots of some species are used as a purgative and as a treatment for diarrhea. In Japan and China, the tubers of another species are eaten as a vegetable. The stems are used for matting, baskets, canoes, and roof thatching. In North America, however, *S. americanus* Pers. (three-square bulrush, or swordgrass) is suspected to be a cause of acute respiratory distress syndrome (ARDS) in cattle following their transfer from a poor pasture to one with good growth in which the species was abundant (Beath et al. 1953). In this situation, the cattle appeared to have been eaten the bulrushes readily, in contrast to the animals' typical reluctance to eat them. The disease appears to be the same as that caused by members of the Poaceae (see chapter 58), presumably a result of the sudden increase in levels of ruminal tryptophan, predisposing formation of excess 3-methylindole. The original descriptions of the disease were prompted mainly by the continuing problems occurring in cattle when they were first allowed to graze rape or kale in the fall of the year (Schofield 1948).

■ labored respiration, expiratory grunt; severe pulmonary emphysema, edema

In the rare instance of intoxication, signs will occur 7–10 days after the animals begin to graze *Scirpus.* The animals will be in great distress, with labored, open-mouth breathing with the head extended and a noticeable expiratory grunt. The heart and respiratory rates are increased, and there may be either constipation or fetid diarrhea. At necropsy, the lungs will be uncollapsed, wet, heavy, and rubbery, with pockets of air in the interlobular septa. The tissue will be dark red and firm, with numerous irregular sections divided by interlobular emphysema. Edema and congestion throughout the lung and frothy exudate in the trachea and bronchi also can be anticipated. The liver may be pale, friable, and mottled; the gallbladder distended with bile; and the kidneys pale. Microscopically, there is rupture, collapse, and filling of the alveoli with serous exudate, and edema and air in the interlobular connective tissue. There may be extensive areas of hepatocellular degeneration and necrosis around the central veins of the liver, presumably due to hypoxia.

It is very important to minimize stress during the period of severe respiratory distress. Numerous drugs have been used, including corticosteroids, antihistamines, nonsteroidal anti-inflammatories, atropine, diuretics, and antibiotics, but therapy appears to be of limited value. Often, handling the animal for treatment causes stress and death due to acute respiratory embarrassment.

13. Droseraceae Salisb.

■ Sundew Family

Drosera

Capturing insects by means of glandular hairs or hinged leaf blades, members of the Droseraceae, or sundew family, are perennial or annual herbs or subshrubs, distributed in temperate and tropical regions throughout the world. Of no economic significance other than as botanical or ornamental curiosities, 4 genera, 3 of which are monotypic, constitute the family. The most famous is *Dionaea* (Venus flytrap), native to the southeastern United States. One genus has been reported to be toxic.

Drosera L.

Comprising about 100 species, *Drosera*, commonly known as sundew, is distributed in both the Old World and the New World. Nine species and named hybrids are present in North America. Representative is circumboreal *D. rotundifolia* L., generally called round-leaved sundew, red rot, moorgrass, or youth-root. Typically encountered in moist or wet soils, plants of the genus are acaulescent, with a rosette of basal leaves, the blades of which bear long, gland-tipped hairs. Insects become entrapped in the sticky mucilage of these hairs and are digested by proteolytic enzymes and ribonucleases secreted by the plant. With white, pink, or purple petals, the typically 5-merous, radially symmetrical flowers are borne in cymes at the ends of elongate scapes. Commonly encountered in swamps, bogs, and wet sites, species of the genus occur across the continent.

- *D. rotundifolia*

  moorgrass
  red rot
  round-leaved sundew
  sundew
  youth-root

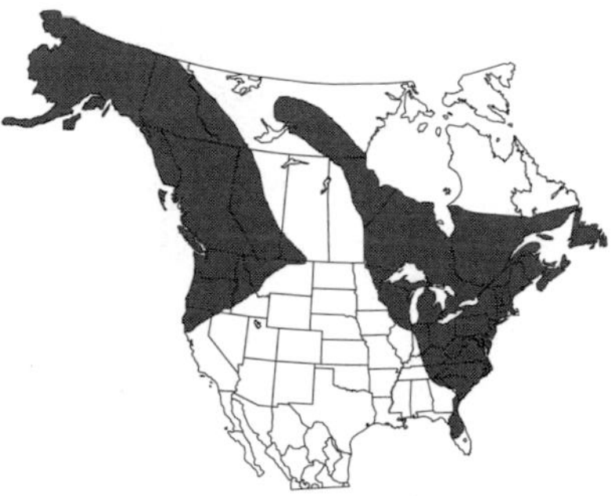

Drosera rotundifolia

Drosera rotundifolia

There seems to be little doubt that *D. rotundifolia* is toxic, but few if any actual problems have been associated with it. It is reported to cause irritation of the digestive tract in sheep, and a tincture made from the foliage and flowers causes respiratory tract changes, including increased secretions (Millspaugh 1974). An herbal diuretic is prepared from the entire plant (Olin 1989). These effects may be due to plumbagin (2-methyljugalone) or its isomer 7-methyljugalone, the simple quinone found in species of *Plumbago* (Pammel 1911; Thomson 1971). This quinone is an irritant and vesicant on the skin and digestive tract. Some Australian species of the genus are cyanogenic (Everist 1981).

- irritant quinones, possible digestive disturbance

14. Dryopteridaceae Herter

Comprising about 60 genera and more than 3,000 species, the Dryopteridaceae, commonly known as the wood fern family, includes some of the most commonly encountered and widely distributed fern genera in North America. Taxonomists differ in their circumscription of the family, some individuals dividing it into 7 small families. When the family is broadly defined, as is done here, its members are rhizomatous,

- Wood Fern Family

  Dryopteris
  Onoclea

perennial herbs that produce sporangia in sori on abaxial surfaces of fronds, which lack needlelike hairs. The fronds are all alike or of 2 forms, with the sterile and fertile quite different. They exhibit circinate vernation and are simple, 1 pinnately, or 2 pinnately compound. Borne on the veins, the sori are separate or contiguous; are orbicular, oblong, or U-shaped; and have their indusia attached beneath them, at their centers, or at 1 side. In North America, 18 genera and 79 species are present.

Although disease problems due to ingestion of ferns are generally associated with *Pteridium* (bracken fern) in the Dennstaedtiaceae (see chapter 32), the potential to cause problems is not exclusive to that genus. A feature common to many fern species is their capacity to cause thiaminase-induced neurointoxication (Chick et al. 1989). However, an appreciable amount of plant material must be consumed for a sustained period, thus limiting the likelihood of disease. In any cases of neurologic disease in horses where ferns are present, the possibility of thiaminase activity should be considered.

■ some eaten by humans, occasional problems

Although potentially toxic, ferns have been and continue to be used as food sources by some cultures. For example, the croziers, or fiddleheads, of *Matteucia struthiopteris* (L.) Tod. (ostrich fern) have long been deemed edible as spring vegetables by members of American Indian tribes and early settlers in northeastern North America (CDC 1995). This species is found across Canada from Alaska to the northeastern United States and Great Lakes region. The croziers are available fresh, frozen, or canned. Served in restaurants, they are generally lightly cooked or sautéed prior to serving. Occasionally problems have been associated with eating this fern (CDC 1995). Problems are also linked to 2 other genera.

Dryopteris Adans.

■ *D. filix-mas*
D. fragrans
D. marginalis

male fern
shield fern
wood fern

shaded, moist sites

Commonly known as shield fern, male fern, and wood fern, *Dryopteris* comprises about 250 New World and Old World species, of which 14 occur in North America (Montgomery & Wagner 1993). Numerous hybrids also are present and species identification can be difficult. Widespread, commonly encountered representatives are *D. filix-mas* (L.) Schott (male fern), *D. fragrans* (L.) Schott (fragrant wood fern), and *D. marginalis* (L.) A.Gray (marginal wood fern). In addition to the morphological features of the family, the genus exhibits thick, erect or short creeping rhizomes from which clusters of usually evergreen, semileathery fronds typically arise. The blades are deltoid-ovate to lanceolate, are simple (pinnatifid) to 3-pinnately compound, and bear sori, with orbicular to reniform indusia, in 1 row between margins and midribs. Habitats occupied by species of *Dryopteris* include moist or swampy woods, thickets, stream banks, shaded cliffs, and talus slopes.

■ *D. filix-mas,* buds of rhizomes, digestive disturbance, neurologic effects, blindness; *D. fragrans, D. marginalis,* toxicity unknown

Only *D. filix-mas* is definitely known to be toxic, and reports are mainly from Europe, where episodes can occur in the spring, fall, or winter when other forage is limited. Acute intoxication follows ingestion of the apical buds of the rhizomes and is associated with severe digestive disturbance and blindness, which may be permanent in some animals (Edgar & Thin 1968). In one case in the United Kingdom, 3 of 40 yearling Simmental cattle developed abrupt onset of ataxia, depression, and bilateral blindness (Mitchell & Wain 1983). The pupils were dilated and unresponsive. One animal subsequently recovered its sight, whereas the others were permanently blind. In another incident, 8 of 61 animals initially affected remained blind (Rosen et al. 1970). Interestingly, affected blind animals are often found standing or lying in water (Macleod et al. 1978).

Dryopteris filix-mas

An extract from the frond bases and apical buds of *D. filix-mas* has long been used as a taeniacide to combat tapeworms in dogs. Toxicity from the extract is consistent with that seen in clinical cases in cattle, with digestive disturbance and neurologic problems, including blindness that is rarely permanent (Grant 1962). Although experimental evidence is lacking, both *D. fragrans* and *D. marginalis* have been considered to be of some risk (Pammel 1911).

Toxicity and anthelmintic activity appear to be due primarily to phloroglucinol derivatives or a mixture of acylphloroglucinols, which is referred to as crude filicin (Widen et al. 1980; Richter et al. 1987; Murakami & Tanaka 1988). Also present in other members of the family, they include albaspidins, aspidins, and filixic acids. These compounds, which fluoresce, are localized in the intracellular spaces of rhizomes and leaf bases (Franchi & Ferri 1988). Crude filicin is toxic, and several ounces of the concentrated extract caused the death of a 300 kg cow (Harvey et al. 1944). The cause of the blindness is unknown, but the disease seems to be quite different in origin from that caused by *Pteridium,* because hematologic parameters are within normal limits and the disease is not responsive to vitamin administration.

With respect to clinical signs, there is acute depression, decreased rumination, dry/dark feces, painful-appearing seizures, and rapid/weak pulse. There is little effect on the animal's appetite. The blindness, of varying duration, is a very prominent feature. Ocular examination in cases of acute disease reveals papilledema with solitary to multiple focal hemorrhages on and/or around the optic disk and on the retina (Rosen et al. 1970). In more-chronic cases the optic disk is shrunken and light gray, the vessels of the disk and retina are thin, and the tapetal reflex is coarse and fragmented.

Gross pathologic changes are limited to reddening of the mucosa of the abomasum and small intestine. Microscopically, there is mild focal edema of the gray matter of the cerebrum and cerebellum and some intramyelinic and periaxonal edema. Blindness is associated with acute bilateral retrobulbar optic neuropathy with edema and atrophy of the optic nerve and degeneration of axons and myelin. There is no specific treatment other than supportive general nursing care.

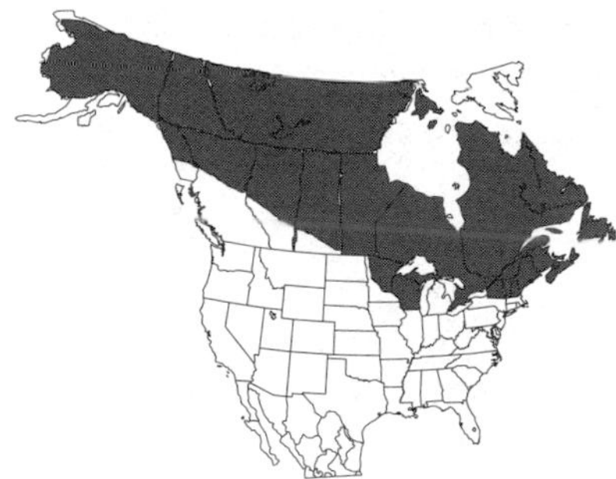

Dryopteris fragrans

Dryopteris marginalis

- phloroglucinols, irritants; cause of blindness unknown

  depression, dry/dark feces, seizures, blindness; changes in optic disk

- gross pathology, reddened gut mucosa

  microscopic, cerebral edema, optic nerve edema, atrophy

Onoclea L.

A monotypic genus native to the temperate regions of eastern North America and eastern Asia, *Onoclea* is commonly known as sensitive fern. Its name is derived from the Greek roots *onos* and *kleio,* for "vessel" and "to close," alluding to the enclosure of its sori by the revolute frond margins. The common name alludes to its sensitivity to frost, which quickly turns its foliage black. The single species is *O. sensibilis* L., which often forms dense stands in swamps, marshes, and moist woods. It is characterized by its dimorphic fronds, pinnatifid to 1-pinnately compound sterile fronds, areolate venation, 2-pinnately compound fertile fronds, and sori enclosed in small, globose, indurate pinnules.

- *O. sensibilis*

  sensitive fern

  fern of moist woods

Onoclea sensibilis is of limited concern toxicologically because it is seldom in sufficient quantity in hay to be a problem. However, when it makes up 10–20% of the hay, it produces a disease of horses closely akin to that caused by *Pteridium* (Waller et al. 1944). The specific cause has not been confirmed, but the supposition is that disease is most likely due to thiaminase activity similar to that for other ferns and *Equisetum.* Because the habitats for this fern may also be occupied by these other taxa, there may be superimposition of thiaminase effects.

Onoclea sensibilis

The clinical signs are those of a neurologic disorder and include incoordination and unsteadiness, especially when the animal is turned. There may be yawning or stretching of the mouth and difficulty in chewing, with the food falling from the mouth after a brief period of ineffectual chewing motions. The horse may walk in circles,

- possible thiaminase activity; neurologic disorder in horses; ataxia, seizures, blindness

■ microscopic, brain, edema and hemorrhage

treat with parenteral thiamine

apparently blind or oblivious to its environment. After a few days it may lie down. At first it may rise with difficulty, but soon it is unable to rise at all. Death, preceded by paddling seizures, may occur shortly thereafter.

The pathologic changes are not striking, but there may be icterus and a slightly swollen liver and moderate edema of the brain. Microscopically, there will be extensive edema of the brain, with a few small hemorrhages, degenerating neurons, and gliosis. Liver changes are mild and include cloudy swelling and fatty change, with some loss of hepatocellular nuclei.

Treatment is the same as for other thiaminase-containing plants and should include thiamine given parenterally twice a day for several days. Grain supplementation may be somewhat protective.

15. Ephedraceae Dumort.

■ Mormon-tea Family
Joint-fir Family

Ephedra

Classified in the division Gnetophyta and morphologically quite different from other gymnosperms, the Ephedraceae, commonly known as the Mormon-tea or joint-fir family, comprises a single genus, *Ephedra,* with 42–60 species in the temperate and warm regions of the world except for Australia.

Ephedra L.

■ *E. nevadensis*
E. viridis

Indian-tea
Mexican-tea
Mormon-tea

shrubs or vines, pollen and seed cones; mainly dry arid sites

The genus *Ephedra* is represented in North America by 12 species (Cutler 1939). Representative and known to cause adverse effects on the digestive system is *E. viridis* Coville. Superficially resembling *Equisetum* (horsetail), it and other species of the genus are dioecious shrubs or vines that produce pollen and seed cones. Their stems are jointed and bear whorled, fascicled, or opposite, terete, photosynthetic branches. Scalelike and typically caducous, the leaves are opposite with basal sheaths. The pollen cones are elliptic, with 5–8 pairs of opposite microsporophylls. The seed cones also are elliptic, maturing in 1 year and bearing 4–6 pairs of opposite ovuliferous scales. There are 1 or 2 seeds per cone, and they are conspicuously exserted, 3- or 4-angled, dark brown, smooth, and wingless. Species of the genus inhabit arid or alkaline sites, dry riverbeds, and other habitats in the deserts and mountainous areas of the southwestern United States and Mexico.

Ephedra nevadensis

Ephedra

■ used for food and drink; sap has irritant effects

Species of *Ephedra,* especially those of the Old World, have been used medicinally for centuries, as an antipyretic, antisyphilitic, circulatory stimulant, antihistamine, and cough suppressant (Stevenson 1993). In North America, *E. nevadensis* S.Watson and *E. viridis* were popular among the western Indian tribes as a source of a tea—hence the common names *Indian-tea, Mexican-tea,* and *Mormon-tea.* However, *E. viridis* was reputed to make the best tea, and although it was commonly used, its extensive use was thought to be detrimental to health (Ebeling 1986). The seeds were used to make a flour for bread and cakes. The sap from these plants also is a cause of contact irritation of the skin (Mitchell & Rook 1979).

Ephedra viridis

■ ephedrine in Old World species but not species of North America; ruminal impaction

Because of the extensive use of Old World species for their profound medicinal effects, the species in North America were suspect as causes of toxicity. However, this has not been borne out, and some species are considered to be important fall and winter browse for livestock (Dayton 1931). *Ephedra nevadensis* is of moderate palatability and is readily eaten by sheep with no apparent adverse effects, even with dosage as high as 1.75% b.w. given for 50 days (Keeler 1989). Although *E. viridis* is considered to be useful forage, it is not eaten as readily, and if eaten in sufficient quantity, it can produce serious adverse effects, primarily related to digestive tract dysfunction. Experimentally, as little as 0.5% b.w. daily for several weeks in cattle caused adverse effects. Up to 2.1% b.w. per day caused severe ruminal impaction in sheep, with a few deaths (Keeler 1989). After consumption of the plants for several weeks or more, there is gradual onset of regurgitation, loss of appetite and weight, bloat, and, in a few animals, diarrhea. Sheep seem to be particularly prone to ruminal impaction, and even with low dosage of *E. viridis,* the disease may be fatal. There are no lesions other than the impaction of the rumen. Treatment is nonspecific other than directed toward relief of the signs, including the impactions. Neither *E. nevadensis* nor *E. viridis* cause adverse effects on the fetus.

Old World species such as *E. equisetina* Bunge, *E. sinica* Stapf, and *E. vulgaris* Rich. contain 1% or more of ephedrine and related compounds. However, species in North America, in spite of their purported medicinal use as teas, contain no detectable amounts of these alkaloids (Nielsen et al. 1927; Terry 1927; Nielsen & McCausland 1928; Reti 1953). They do contain appreciable amounts of tannins, especially *E. viridis,* which has 4 times as much as *E. nevadensis* (Keeler 1989).

16. Gentianaceae Juss.

■ Gentian Family

Centaurium

Cosmopolitan but most common in temperate and subtropical regions, especially mountain habitats, the Gentianaceae, commonly known as the gentian family, comprises 75–83 genera and 900–1,000 species. Its taxa typically are herbs, often mycorrhizal, with simple, opposite, decussate, sessile, estipulate leaves. Radially symmetrical and usually perfect, the 4- or 5-merous flowers are borne in cymes or rarely are solitary. The convolute, brightly colored, fused petals exhibit an array of hues from white to pink, purple, or blue. Corollas are rotate, salverform, funnelform, or campanulate. The superior ovaries give rise to capsules or rarely berries. Species of more than 15 genera are cultivated as garden ornamentals, and others yield bitter glucosides used in flavorings and medicines. In North America, 17 genera and 117–120 species are present. Species of 1 genus have caused toxicologic problems.

Centaurium Hill

Comprising 40–50 species in both the Old World and the New World, *Centaurium,* commonly known as centaury, is represented in North America by about 17 species

- *C. beyrichii*
  C. calycosum
  centaury
  mountain pink
  rosita
  herbs of dry, open sites

of annual or biennial herbs with numerous small, pink to rose-purple or white flowers in terminal corymbose or spicate cymes. Exserted beyond the funnelform or salverform, 4- or 5-parted corollas are the stamens, with their distinctive spirally twisted anthers. Representative of the genus and causing toxicologic problems are *C. beyrichii* (Torr. & A.Gray) B.L.Rob. (mountain pink, or rock centaury) and *C. calycosum* (Buckley) Fernald (rosita, or Buckley's centaury). The former occurs in limestone and granitic soils, whereas the latter is widespread in the prairies and meadows.

Centaurium beyrichii

Centaurium beyrichii

Centaurium calycosum

- digestive disturbance, colic, diarrhea, loss of appetite, depression, weight loss
  reddened gut mucosa, with ulceration and hemorrhage

Centaurium beyrichii, which is unpalatable and eaten only under unusual circumstances, was incriminated in a toxicity problem in Mexico that also involved *Lobelia berlandieri* in the Campanulaceae (see chapter 18) (Dollahite & Allen 1962). Subsequent experiments in goats confirmed its toxicity potential. Feeding 2.8% b.w. caused death by the third day and 1% fed for 4 days caused death on the fifth day. Dosage as low as 0.5% b.w. was toxic to goats. Experiments with sheep were not conclusive, because of the low dosage fed. *Centaurium calycosum* was also toxic to a goat, but a slightly higher dosage was required. Ingestion of the plants caused digestive tract disturbance, especially pain. These digestive problems led to colic, diarrhea, decreased appetite, depression, marked weight loss, and sometimes increased frequency of urination. Grossly, there is severe reddening and edema of the mucosa of the rumen and abomasum, accompanied by ulcerations and hemorrhages. There is also congestion of the kidneys, a friable liver, and distension of the gallbladder.

17. Geraniaceae Juss.

- Geranium Family
  Erodium

Widely distributed in temperate and subtropical regions throughout the world, the Geraniaceae, commonly known as the geranium family, comprises 11–14 genera and 700–775 species. It is of considerable economic importance. Geraniums—actually members of the South African genus *Pelargonium*—are extensively grown as ornamentals, as are species of *Geranium* (crane's-bill) and *Erodium* (stork's-bill). Used in perfumes and lotions, geranium oil is extracted from species of *Pelargonium.* Members of the family typically are aromatic herbs or shrubs with simple or palmately or pinnately compound, stipulate leaves. The radially symmetrical, 5-merous flowers are borne singly or in cymes. Their brightly colored petals surround 1–3 whorls of stamens, which usually have their filaments fused at the base, and a superior ovary that gives rise to a schizocarp that dehisces elastically from a central column into 5, 1-seeded, curled mericarps. In North America, 3 genera and about 60 native and introduced species are present. Species of 1 genus may be a potential hazard.

Erodium L'Hér. ex Aiton

Simultaneously prized as a border or rockery ornamental and cursed as a noxious lawn weed, *Erodium*, commonly known as stork's-bill or heron's-bill, comprises 75–90 species native to both the Old World and the New World. An introduction from Europe, which is now widely naturalized throughout North America, is *E. cicutarium* (L.) L'Hér. ex Aiton (alfilerillo, California filaree, stork's-bill, red-stem filaree, crowfoot, pin grass). It is a winter annual or biennial with prostrate or ascending stems radiating from a rosette of leaves. Both the rosette and cauline leaves are fernlike, with narrow pinnatifid to pinnately compound divisions. Borne in umbellate cymes, the small, radially symmetrical flowers are 5-merous and have purplish pink petals.

■ *E. cicutarium*
alfilerillo
crowfoot
filaree
heron's-bill
pin grass
stork's-bill

leaves in basal rosette, fernlike; weedy invaders

Erodium cicutarium

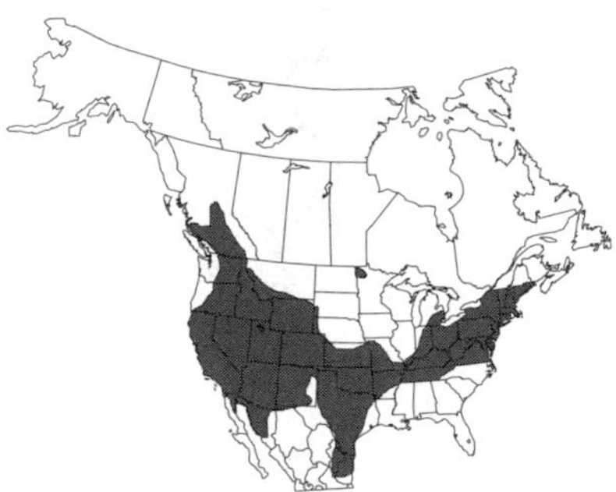

Erodium cicutarium

Most species of *Erodium* are eaten without harm, and some were prepared by members of American Indian tribes for use either raw when very young or as a potherb (Ebeling 1986). Occasionally, the fruit causes contact irritation. The foliage also is reported as a cause of primary photosensitization in calves (Ford 1965). Some species are cyanogenic under certain conditions, such as after a light frost (Fuller & McClintock 1986).

■ potherb, possibly photosensitization, cyanogenic

18. Haemodoraceae R.Br.

Commonly known as the bloodwort or bloodroot family, because of the reddish orange sap of the rhizomes and roots, the Haemodoraceae is a family of 14–16 genera and about 100 species of tropical and warm-temperate, liliaceous herbs distributed primarily in South Africa and Australia. Of little economic significance, a few species are cultivated as ornamentals, the most famous of which is *Anigozanthos* (kangaroo paw), the state emblem of Western Australia. Species of other genera are sources of red dyes and herbal medicines. Three genera, each with one species, are present in North America. One has been linked to toxicologic problems.

■ Bloodroot Family
Bloodwort Family

Lachnanthes

- *L. caroliniana*

dandy
dye-root
paint-root
redroot

perennial herb from red rhizomes

wet, wooded sites

- possible porphyria-like photosensitization

Lachnanthes caroliniana

Lachnanthes Elliott

A monotypic genus occurring in acid soils of wet habitats such as swamps, bogs, and wet pine forests along the East Coast from Nova Scotia to Florida, *Lachnanthes* is commonly known as dandy, redroot, paint-root, or dye-root. The single species is *L. caroliniana* (Lam.) Dandy. Synonyms appearing in the older literature are *L. tinctoria* (J.F.Gmel.) Salisb. and *Gyrotheca tinctoria* (J.F.Gmel.) Elliott. With erect stems 30–90 cm tall, bearing mostly basal, equitant, linear, glabrous leaves, it is a perennial from red rhizomes. The upper stems and inflorescence axes are densely tomentose. Each forming a helicoid cyme, the corymbose branches bear bracts and 3-merous, dull yellow flowers. The inferior ovary gives rise to a capsule.

Lachnanthes caroliniana has rarely been associated with disease problems. Harshberger (1920) described it as a cause of a porphyria-like photosensitization, with affected animals exhibiting reddish pink bones. Blum (1941) also reported, perhaps using the same source as Harshberger, that the species was reputed by early Virginia farmers to be the cause of pink bones, skin lesions, and sloughing of the hooves in light-colored pigs. This reputation of toxicity for all plant parts has persisted for decades but has not been substantiated since the appearance of the initial reports (Fernald & Kinsey 1943; Hardin 1973). The species contains lachnanthocarpone, a phenalenone, but its role in toxicity is not known (Forte et al. 1976).

19. Hydrangeaceae Dumort.

- Hydrangea Family

Hydrangea

Comprising about 17 genera and 170–250 species of woody plants with simple, opposite, estipulate leaves and 4- or 5-merous flowers borne in compound cymes, the Hydrangeaceae, or hydrangea family, is primarily a family of the Northern Hemisphere. It has long been treated as a subfamily of the Saxifragaceae, but molecular and cladistic studies indicate considerable differences between the 2 taxa. The family is economically important for its ornamentals, most notably hydrangea and mock orange. In North America, 11 genera and 49–52 native and introduced species are present. Species of 1 genus are of possible toxicologic significance.

Hydrangea L.

- *H. arborescens*
H. macrophylla
H. quercifolia

hydrangea
wild hydrangea

shrubs with very showy flowers

wooded sites, ornamentals

Species of *Hydrangea,* numbering about 80, are characterized by having flowers of 2 quite different forms—perfect and neutral. The perfect, radially symmetrical flowers are relatively inconspicuous and have free, white or pink or greenish white petals,

Hydrangea macrophylla

Hydrangea quercifolia

8-numerous stamens in 2 or more whorls, and ovaries that are wholly or partially inferior. They produce many-seeded capsules. In contrast, the neutral flowers have radially or slightly bilaterally symmetrical calyces consisting of 3–5, persistent, large, petaloid sepals that are white or greenish white. Their petals are absent, the 8–10 stamens are sterile, and the pistils are abortive. In North America, about 7 native and introduced species are present. Representative of the genus are *H. arborescens* L. (American or wild hydrangea) and *H. quercifolia* Bartram (oak-leaved hydrangea), native species of the eastern half of the continent, and the prized ornamental *H. macrophylla* (Thunb.) Ser., a native of Japan.

Hydrangea arborescens

Hydrangea quercifolia

Intoxication problems involving *Hydrangea* are rare. Bruce (1920) reported that a horse, after eating about 1.8 kg of the foliage of *H. macrophylla* var. *otaksa,* developed a watery and bloody diarrhea and colic but recovered overnight. A similar problem also was noted in a cow. Cary (1924) reported that although the foliage may cause purgative effects and dizziness, *H. arborescens* is seldom if ever eaten by livestock. Consumption of the leaves and buds also causes digestive tract disturbance in humans (West & Emmel 1952). Experimentally, guinea pigs given an alkaloid type of extract of the leaves and flowers developed severe diarrhea and inflammation of the mucosa of the stomach and intestines.

The young leaves are thought to be cyanogenic, but the clinical signs that generally appear are not consistent with cyanogenesis. Other bioactive constituents such as hydramacroside, a secoiridoid, have been isolated from the foliage of hydrangeas, but their toxicologic significance is not known (Yoshikawa et al. 1994). Another constituent, the isocoumarin hydrangenol, is a strong contact sensitizer and a rare cause of allergic contact dermatitis in humans with occupational exposure to the plants (Hausen et al. 1982; Hausen 1991; Kuligowski et al. 1992). Concentrations range from 0.04% in leaves to 0.1% in the flowers; however, some cultivars are devoid of the compound.

- diarrhea
- digestive disturbance; contact skin irritation, dermatitis, isocoumarin hydrangenol

Hydrangenol

There may be nausea, vomiting, diarrhea with or without blood, colic, and lethargy. Treatment may include activated charcoal and other medicaments aimed at relief of the digestive tract irritation.

20. Hydrophyllaceae R.Br.

- Waterleaf Family

Phacelia

Almost cosmopolitan in distribution, the Hydrophyllaceae, or waterleaf family, comprises 18–20 genera and about 250 species of herbs or small shrubs usually with alternate leaves and 5-merous, radially symmetrical flowers producing capsules borne in cymes or helicoid cymes. In North America, 16 genera and approximately 230 species are present. Despite its common name, its members are most commonly encountered in dry habitats of the western United States. The family is of little economic importance, species of *Phacelia* (phacelia) and *Nemophila* (baby-blue-eyes) are cultivated as garden ornamentals. One species of *Phacelia* causes toxicologic problems.

Phacelia Juss.

■ *P. crenulata*
desert heliotrope
false heliotrope
phacelia
scorpion weed

A New World genus with greatest species diversity in the western United States, *Phacelia,* commonly known as phacelia or scorpion weed, comprises 150–175 species of annual, biennial, and perennial herbs that are typically downy to glandular pubescent. The leaves are simple to 2-pinnately compound and generally reduced upward. Borne in 1-sided, coiled cymes, the flowers have tubular to funnelform corollas with the 5 limbs spreading. Their color ranges from white to blue to violet to purple.

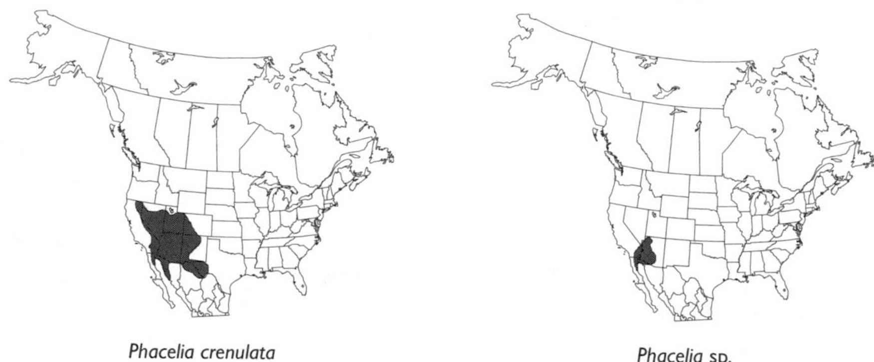

Phacelia crenulata *Phacelia* sp.

■ contact allergic dermatitis

The viscid glandular hairs of some of these arid-habitat species, such as *P. crenulata* Torr. ex S.Watson (desert heliotrope, or false heliotrope), cause a contact allergic dermatitis that is similar to that caused by poison ivy and lasts about 10 days (Munz 1932; Fuller & McClintock 1986).

21. Juncaceae Juss.

■ Rush Family
Juncus

One of the 3 graminoid families, the Juncaceae comprises 8 or 9 genera and 300–400 species of worldwide distribution, but primarily in wet habitats of cold-temperate or montane regions. Only 2 genera, *Juncus* (rush) and *Luzula* (wood rush), occur in the Northern Hemisphere. The family is of minor economic importance; the stems are used in basketry and chair caning and for matting. A few species are cultivated as water ornamentals. Members of the family are perennial or annual, grasslike herbs with erect or decumbent, generally unbranched stems and simple, alternate leaves with basal sheaths and linear, flat or cylindrical, frequently septate blades. The inflorescences are cymose or capitate or spicate clusters. Subtended by small bracts, the perfect, radially symmetrical, 3-merous flowers have 6, free, membranous or scarious perianth parts that are green, reddish brown, black, brown, or stramineous. There are 6 or 3 stamens, and the single pistil, with a superior ovary and 3 styles, produces a capsule. In North America, only the 2 northern genera and about 130 species are present. Intoxication problems are associated with 1 genus.

Juncus L.

■ *J. effusus*
J. inflexus
J. phaeocephalus
J. squarrosus
rush
leaves used to make baskets

Its name an ancient Latin one, the genus *Juncus,* or rush, comprises about 225 species, of which 107–120 species and named hybrids are present in North America. In addition, there are numerous varieties. The common name *rush* is applied to other taxa, e.g., *Scirpus* (bulrush) and *Eleocharis* (spike rush) in the Cyperaceae. Members of the genus exhibit the features of the family and are distinguished by their 3-loculed, many-seeded capsules and sheaths with overlapping margins. Species occur across the continent, some with wide distributions and others with restricted distributions.

The rushes have a long history of use by humans. Because of their delicate, sweet odor, the floors of parlors and churches were strewn at one time with their foliage during the summer and especially for holy days (Baskin 1967). Some species were used extensively for basketry and raft making by some American Indian tribes (Ebeling 1986).

There are only a few isolated instances of intoxication that can be attributed to the genus. In California, 2 dairy heifers died abruptly on a pasture in which new growth of *J. phaeocephalus* Engelm. (brown-headed rush) was one of the few plants available to be eaten (Fuller & McClintock 1986). Some species, such as *J. squarrosus* L., are apparently readily eaten by cattle, whereas others, such as *J. inflexus* L., are eaten only under unusual circumstances and in some instances cause diarrhea, nervousness, loss of weight, and later seizures (Cooper & Johnson 1984). There are few lesions other than some hemorrhages in the brain. Blindness is one of the neurologic signs reported, but it seems to disappear within a few weeks, perhaps with resolution of blood clots in the brain. Some caution is warranted in allowing access to *J. inflexus,* because animals may develop an appetite for the plants (Cooper & Johnson 1984). *Juncus effusus* L., a cosmopolitan species quite common in North America, is suspected as a possible cause of disease associated with swampy areas in South Africa (Mogg 1927).

■ possible neurologic problems, blindness; possibly cyanogenic under some circumstances

In a few instances the cyanogenic glycoside dhurrin has been incriminated as a cause of intoxication involving species of *Juncus* (Seigler 1976). A species in Australia, *J. holoschoenus* R.Br., associated with the deaths of 3 calves, tested positive for cyanide, especially the growth in young preflowering stages (Albiston 1937). In a California case, the same problem was associated with new growth during a period following frosts (Fuller & McClintock 1986). An extract of *J. inflexus* also was reputed to have a narcotic effect (Baskin 1967).

22. Lythraceae J.St.-Hil.

■ Loosestrife Family

Lagerstroemia

Comprising 22–25 genera and 450–500 species, the Lythraceae, commonly known as the loosestrife family, is widespread in tropical regions, but with relatively few taxa—mostly herbs in damp habitats—in temperate regions. Economically, the family is somewhat important as a source of commercial timber; several ornamentals, including crape myrtle and pomegranate; and a few noxious weeds. Henna, a popular orange-red dye, is extracted from the leaves of 1 genus. Members of the family are trees, shrubs, and herbs with opposite, entire-margined leaves. Borne in spicate, racemose, paniculate, or fasciculate cymes, the radially or bilaterally symmetrical, perfect flowers are 4- or 6-merous. Most distinctive is the well-developed, tubular to campanulate hypanthium. The fruits are capsules. In North America, 11 or 12 genera and 36 or 37 native and introduced species are present. One genus is of possible toxicologic interest.

The Lythraceae contains several important cultivated ornamentals, some of which are known to contain biologically active compounds. However, genera of the family are not definitely recognized as toxic; thus the actual risk is difficult to establish. There is an array of similar alkaloids represented in the family. For example, those present in *Heimia salicifolia* (Kunth) Link include the *trans*-fused compounds lythrine, lyfoline, sinine, nesodine, and dehydrodecodine and the *cis*-fused humidine, sinicuichine, and vertine; and heimine, of unknown structure (Golebiewski & Wrobel 1981; Southon & Buckingham 1989). Similar types are found in *Lagerstroemia* and other genera. The role of these alkaloid types as toxicants is not known, although some are biologically active; for example, nesodine and vertine have sedative effects, vertine also is hypotensive, and decinine has diuretic activity (Southon & Buckingham 1989).

Lagerstroemia L.

- *L. indica*
  L. speciosa

  crape myrtle
  pride-of-India

  shrubs or trees, ornamentals

Native to tropical Asia and Australia, *Lagerstroemia,* commonly known as crape myrtle, comprises 30–50 species, several of which are prized as ornamentals. Members of the genus are deciduous or evergreen shrubs or trees typically bearing opposite, distichous leaves with obovate to elliptic or lanceolate blades and entire margins. The axillary or terminal panicles bear perfect, radially symmetrical flowers with 6–9 sepals and commonly 6, clawed petals with fringed or crinkled blades. Flower colors vary from white to red or purple or lavender-pink. There are 2 introduced species in North America; one is *L. indica* L. (crape myrtle), a native of China and somewhat naturalized in the southern United States. The other is *L. speciosa* (L.) Pers. (queen crape myrtle, giant crape myrtle, or pride-of-India), which is indigenous from India to Australia.

Lagerstroemia indica

Although ingestion of the roots, leaves, and flowers of *L. indica* may cause a purgative effect, the plants are unlikely to represent a significant problem (Watt & Breyer-Brandwijk 1962). Organic solvent and ethanol extracts are reputed to be narcotic, and experimentally, they cause sedation and decreased social aggression in male mice when given i.p. (Al-Maliki et al. 1988). Alkaloids present include the *trans*-fused decinine; the *cis*-fused decamine, dihydroverticillatine, and lagerstroemine (= indicamine); and the biphenyl ether lagerine (Golebiewski & Wrobel 1981). Similar alkaloids are present in other species (Southon & Buckingham 1989).

- possible digestive disturbance, alkaloids present; vomiting, diarrhea

Lagerstroemine

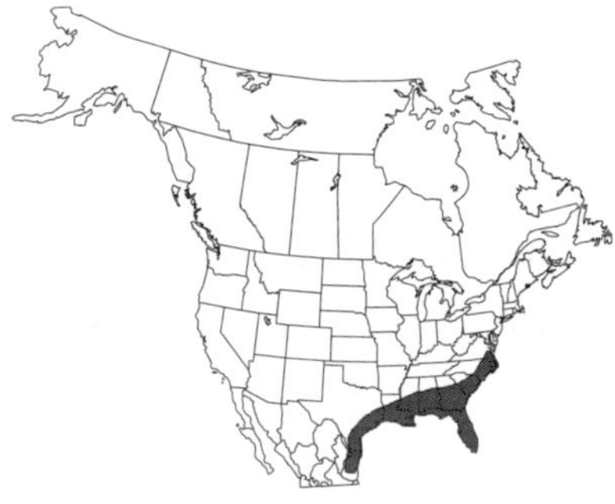

Lagerstroemia indica

The effects of these compounds are most likely to be evidenced by mild digestive tract problems, including retching, vomiting, and diarrhea of short duration. Orally administered charcoal may be helpful; otherwise treatment, if needed, should be directed toward relief of the digestive tract effects.

23. Melianthaceae Link

- Melianthus Family

  Melianthus

Native to South Africa and believed to be closely related to the Sapindaceae, the Melianthaceae, or melianthus family, has 2 genera and 8 species of trees and shrubs with pinnately compound, alternate, stipulate leaves and perfect, 5-merous, resupinate flowers borne in terminal racemes. The family is of little economic use; a few species are cultivated as ornamentals, a few provide timber, and a few are used in herbal medicines. In North America, only 1 genus is occasionally cultivated and may pose a toxicologic risk.

Melianthus L.

- *M. comosus*
  M. major

  honeybush
  honeyflower

Comprising 6 species, *Melianthus* is known as honeybush or honeyflower because the calyces are typically filled with nectar and quite attractive to bees. Only rarely cultivated in the extreme southern portions of North America are *M. comosus* Vahl and *M. major* L. They are aromatic, evergreen shrubs to 2.5 m tall with toothed leaflets. Their bilaterally symmetrical flowers are brown-red to brick-red, with 5 petals fused into a short spur and hooded tube. Containing many, shiny, black seeds, the papery capsules are 4-lobed and 4-winged.

- aromatic evergreen ornamental shrubs
- cardiotoxic bufadienolides; possible digestive disturbance

These 2 species are apparently very disagreeable to the taste and seldom consumed by livestock except in unusual circumstances. They contain several cardiotoxic bufadienolides, including hellebrigenin-3-acetate, which are compounds of some importance medicinally (Watt & Breyer-Brandwijk 1962; Kellerman et al. 1988). Reports from areas where the plants are indigenous indicate the primary problem to be severe irritation of the digestive tract. The clinical signs include diarrhea with colic and, in the most severe cases, weak pulse, cyanosis, and respiratory difficulty (Watt & Breyer-Brandwijk 1962). Pathologic changes are limited to reddening of the mucosa of the small intestine; scattered splotchy hemorrhages, especially on the heart surfaces; and perhaps some lung edema (Kellerman et al. 1988). Treatment, including the use of atropine, is mainly directed toward the cardiac effects.

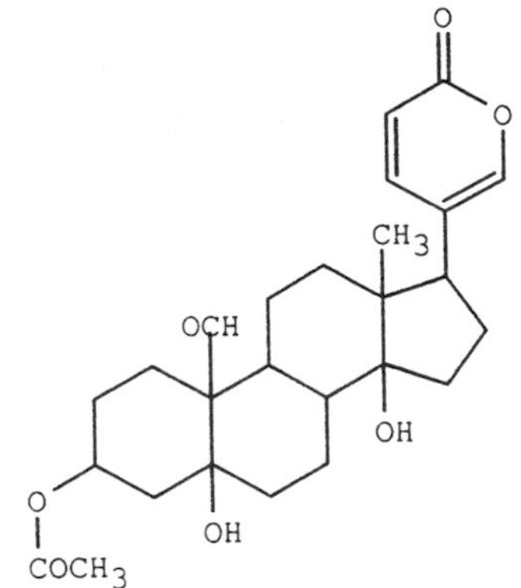

Hellebrigenin-3-acetate

24. Menispermaceae Juss.

- Moonseed Family

  Cocculus
  Menispermum

Comprising 65–70 genera and 350–400 species, the Menispermaceae, commonly known as the moonseed family, is widespread in the tropics and subtropics but has relative few taxa in temperate regions. Members of the family typically are dioecious, woody or herbaceous vines or scandent shrubs, with simple, alternate leaves that characteristically have palmate venation and entire or palmately lobed margins. Their small, radially symmetrical, imperfect, 3-merous flowers are borne in axillary or extra-axillary cymes. The sepals, petals, and stamens typically are each in 2 whorls, and the fruits are drupes with horseshoe-shaped seeds.

Although the family is not of major economic significance, several of its tropical taxa are used in medicine—both formal and herbal—or as fish poisons, parasiticides, and curare-type arrow poisons (Heywood 1978; Zomlefer 1994). Members of the family have a considerable array of alkaloids, especially the protoberberines (Bhakuni & Jain 1986). In North America, 5 genera and 8–10 species are present. Two genera are of some toxicologic concern.

Cocculus L.

- *C. carolinus*
  C. diversifolius

  coral beads
  correhuela
  moonseed
  orient-vine
  snailseed

  herbaceous vines with bright red drupes

Comprising 10–12 species, of which 3 occur in North America, *Cocculus* is commonly known as moonseed, snailseed, or coral beads. Most commonly encountered are *C. carolinus* (L.) DC., distributed throughout the southeastern quarter of the

Cocculus carolinus

Cocculus carolinus

continent, and *C. diversifolius* DC., generally known as correhuela or orient-vine, which is found from Texas to Arizona and southward into Mexico. Both have the attributes of the family. In addition, their flowers are borne in racemose or paniculate cymes and produce bright red drupes.

■ biologically active alkaloids

Although there are few, if any, reports of intoxications due to species of *Cocculus*, they contain several biologically active alkaloids, such as cocculine and trilobine (Southon & Buckingham 1989) and even palmatine in *C. carolinus* (Bhakuni & Jain 1986). Thus, there is a basis for considering these species as minor toxicologic risks if they are eaten in sufficient quantity. Cocculine, a representative alkaloid, is a weak ganglionic blocker, causing reduced blood pressure. The clinical signs, pathology, and treatment would most likely be similar to those associated with *Menispermum*.

Cocculine

Menispermum L.

■ *M. canadense*

maple vine
moonseed
yellow parilla
yellow sarsaparilla

herbaceous vines with bluish black drupes

Also known as moonseed, *Menispermum* consists of 2 species, 1 in eastern North America and the other in eastern Asia. Occurring in moist woods and thickets, the North American species is *M. canadense* L., often called yellow parilla, maple vine, yellow sarsaparilla, or Canadian moonseed. It is not uncommon but is seldom encountered in abundance. In addition to the characters of the family, it has peltate leaves and bluish black drupes.

A tincture made from the roots of *M. canadense* was at one time used as a tonic (Millspaugh 1974). The drupes were purportedly the cause of neurotoxicosis and death of humans prior to 1911 (O'Leary 1964). However, the lack of reports of intoxications since that time makes this seem less likely. The scarcity of vines in any one location minimizes the likelihood of livestock intoxications; however, the attractive fruits may be eaten by children and may possibly result in their intoxication.

Menispermum canadense

■ biologically active alkaloids; possible neurologic, digestive problems

Menispermum canadense contains in its rootstocks numerous alkaloids: dauricine, daurinoline, desmethyldauricine, acutumidine, magnaflorine, and two protoberberines, one of which is dehydrocheilanthifoline (Willaman & Schubert 1961; Doskotch & Knapp 1971; Bhakuni & Jain 1986). Acutumine is present in the aerial portions (Doskotch & Knapp 1971) and possibly picrotoxin in the wood (Pammel 1911). The various protoberberines are known to cause decreased intestinal peristalsis, increased

Acutumidine

Dauricine

uterine contraction, sedative and other neurologic effects, cardiovascular changes, and even inhibition of acetylcholinesterase (Bhakuni & Jain 1986). However, because actual intoxications are so poorly documented, it is difficult to ascribe specific adverse effects to the compounds present.

The signs of intoxication range from transient digestive disturbance to sedation or perhaps even seizures, but undue alarm is generally not warranted. Treatment with emetics and/or activated charcoal should suffice in most instances. Otherwise, treatment should be directed toward relief of the presenting signs.

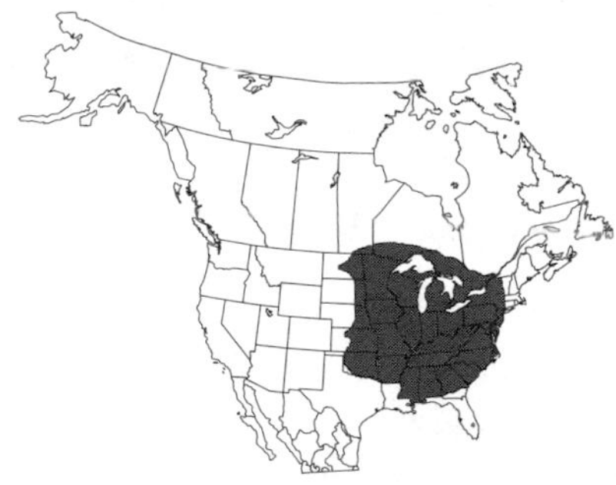

Menispermum canadense

25. Menyanthaceae Dumort.

■ Bogbean Family
Buckbean Family

Menyanthes

Closely related to the Gentianaceae and included within it by some taxonomists, the Menyanthaceae, commonly known as the buckbean or bogbean family, comprises 5 genera and 30–40 species of aquatic and semiaquatic herbs with alternate, simple or trifoliolate leaves, and radially symmetrical, 5-merous, white or pink or yellow flowers. Distribution of the family is almost cosmopolitan. A few species are cultivated as aquatic ornamentals, a few have local use as food or medicine, and several are aquatic weeds (Heywood 1978). In North America, 3 genera and 9 species are present. Toxicologic problems are associated with 1 species.

Menyanthes L.

■ *M. trifoliata*

bitterroot
bogbean
buckbean
marsh trefoil
water shamrock

leaves basal, palmately compound, with 3 leaflets

shallow, quiet water

■ monoterpenes, alkaloids; possible digestive, neuromuscular disturbances

The single species of the genus *Menyanthes, M. trifoliata* L., is commonly known as buckbean, bogbean, marsh trefoil, bitterroot, or water shamrock. It is circumboreal in distribution in quiet, shallow water. A glabrous, rhizomatous perennial, its basal leaves have sheathing petiole bases and palmately compound leaves with 3, elliptic to obovate leaflets. Borne in bracteate racemes on scapes 15–35 cm long, the perfect flowers are heterostylous, with deeply divided calyces and conspicuously fringed, recurved petal lobes. The globose, corky capsules rupture irregularly to release many, shiny, yellowish brown seeds.

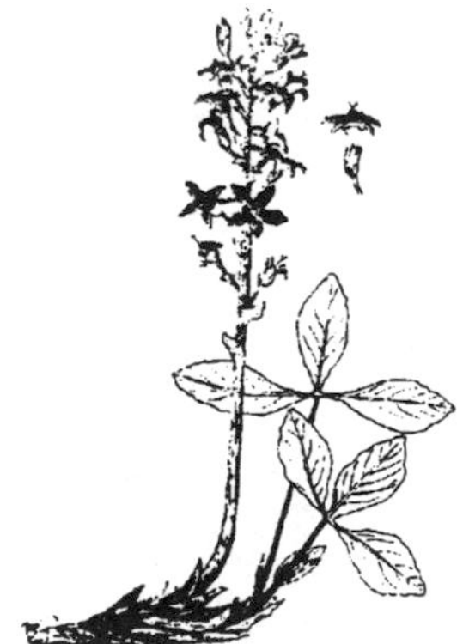

Menyanthes trifoliata

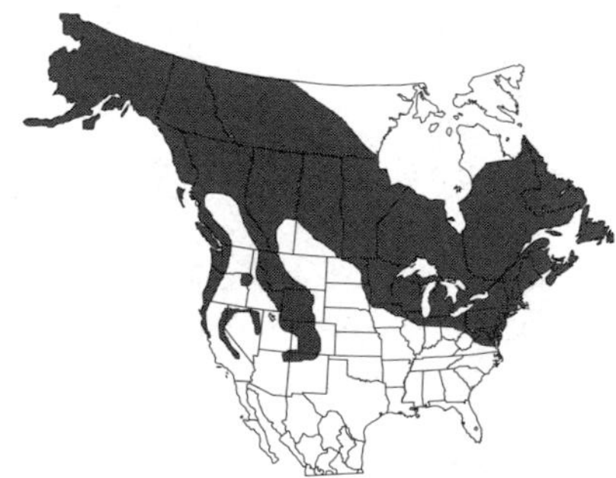

Menyanthes trifoliata

Because of its habitat and habit, *M. trifoliata* seems an unlikely cause of disease. However, ingestion of the rhizomes by humans was considered to cause profuse vomiting, diarrhea, and neurologic effects (Millspaugh 1974). The effects of consumption of other plant parts are not clear, but they also should probably be suspected to cause digestive disturbances (Pammel 1911) because they contain several potential

Menthifolin

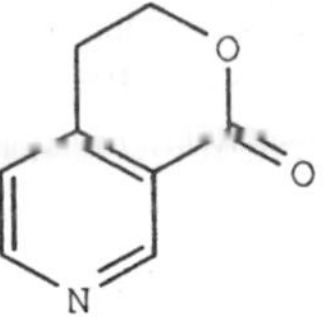

Gentianidine

toxicants, including monoterpenes and alkaloids. The monoterpenes, including foliamenthin, dihydrofoliamenthin, menthifolin, and onikulactone, may account for the digestive tract irritation (Connolly & Hill 1991). The alkaloids gentianidine and gentiatibetine have biological effects, such as muscle relaxation and hypotension, but are of unknown toxicologic significance (Southon & Buckingham 1989).

26. Moraceae Link

■ Mulberry Family
Maclura

Centered in the tropics, but with numerous taxa in temperate regions, the Moraceae, commonly known as the mulberry family, comprises 40–53 genera and 1,000–1,400 species. Taxonomists differ in their interpretations of its boundaries; some place it in the Urticaceae, whereas others maintain it as distinct but include the Cannabaceae within it. The preponderance of systematic evidence now favors a family distinct from both the Urticaceae and Cannabaceae, an interpretation followed here. The family is economically quite important for its fruits, such as figs, mulberries, and breadfruits, and for its ornamentals such as India rubber plant. Members of the family are dioecious or monoecious trees or shrubs or herbs, typically with a viscous, milky sap. The simple, alternate or rarely opposite leaves bear stipules, and the inflorescences are catkins, heads, or clusters of flowers on flattened or in hollowed receptacles. The staminate and pistillate, 4-merous flowers are similar or different and have their perianths in 1, radially symmetrical series. The fruits are achenes or multiple syncarps with fleshy calyces partially or entirely enclosing the achenes on frequently fleshy rachises. In North America, 14 genera and approximately 30 species, both native and introduced, are present. One genus is of toxicologic significance.

Maclura Nutt.

■ *M. pomifera*
bodark
bois d'arc
bowwood
hedge apple
horse apple
Osage orange

thorny tree; fruits large, fleshy, yellow or green, with rugose surface

A genus of some 12 species distributed in the warmer regions of the Americas, Asia, and Africa, *Maclura* has a variety of common names, including Osage orange, bodark, bowwood, hedge apple, horse apple, and bois d'arc. It is represented in North America

Maclura pomifera

by 1 species, *M. pomifera* (Raf.) C.K.Schneid., whose original distribution was the Red River drainage of Oklahoma and Texas. It is now widely planted and naturalized across the continent. Plants of this species are trees, often with several trunks and branches armed with thorns, bearing ovate, acuminate, entire leaves on spurs or solitary at the nodes. Typically 10–20 cm in diameter, the fleshy multiple fruit is yellow-green or bright green with a rugose surface and persistent filiform styles.

Maclura pomifera generally has not been considered to be a disease problem for livestock, but in one instance, a sheep that ate green leaves from a fallen branch developed abrupt onset of an offensive diarrhea (Pullar 1939). Experimentally, sheep given 454 g of fresh green or partly dried leaves for 2 days died on the second or third day. The only pathologic change was a slight congestion of the mucosa of the abomasum and small intestine. Leaves dried for several months produced no adverse effects. *Maclura pomifera* contains at least two pentacyclic triterpenoids, such as monogynol, but not as the glycosides (Connolly & Hill 1991). Their role as inducers of disease is uncertain.

Monogynol A

■ terpenoids, possible digestive disturbance

27. Moringaceae Dumort.

Commonly known as the moringa family, the Moringaceae consists of 1 genus and 10–14 species of deciduous, xerophytic trees of acacia-like habit occurring from Africa and Madagascar through the Middle East to India. Of minor economic importance, the family is the source of nondrying ben oil, which is used as a lubricant, in perfumery, and as a salad oil. Some species are cultivated as ornamentals.

■ Moringa Family

Moringa

Moringa Adans.

The genus is represented in North America by a single species, *M. oleifera* Lam. (= *M. pterygosperma* Gaertn.), which is native to India and now naturalized in southern Florida, the West Indies, and other parts of tropical America. Known as horseradish tree, ben, or oil-of-ben tree, this species attains heights of 6–10 m and has thick trunks, smooth corky bark, and brittle branches bearing large, alternate 2-to-3-pinnately compound leaves with numerous ovate leaflets. The creamy white, slightly bilaterally symmetrical, 5-merous flowers are borne in axillary panicles and produce podlike capsules containing winged seeds.

■ *M. oleifera*

ben
horseradish tree
oil-of-ben tree

tree, smooth, corky bark

Moringa oleifera has a long history of medical and cosmetic uses, especially in its native India (LePoole 1996). Similar to sesame and olive oils, the oil from its seeds, and those of other species of *Moringa,* was held in high esteem by many ancient Mideastern cultures. As its common name suggests, its roots smell and taste like horseradish and are used in the same way. The shoots, leaves, flowers, and long, green immature fruits are also cooked and eaten as vegetables. However, the mature seeds are reportedly highly toxic when fed to rats (Grant et al. 1995). Extracts of the seeds have antispasmodic activity via inhibition of acetylcholine-induced contraction of the small intestine (Caceres et al. 1992).

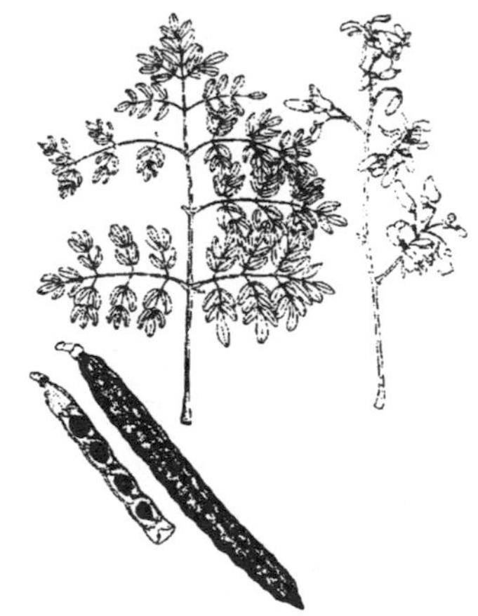

Moringa oleifera

Consistent with the species' reputation as an abortifacient plant in India, reproductive effects have been shown experimentally in rats fed *M. oleifera.* Aqueous extracts of the leaves given daily from days 5–10 of pregnancy at a dosage equivalent to 175 mg/kg b.w. caused resorption of all fetuses in 7 rats while only reducing weight gain in the dams (Sethi et al. 1988; Nath et al. 1992). Extracts of the roots cause other types of reproductive alterations, mainly but not exclusively estrogenic in nature (Shukla et al. 1988). The popularity of *M. oleifera* for cosmetic purposes lies with the oil, which is extracted from the mature seeds. Called behen or ben oil, it is composed primarily of oleic acid (71%); the remaining portion is mainly

■ seed oil, cosmetic, medicinal uses; other plant parts cooked and eaten; seeds, toxicity unknown; leaves possible abortifacient effects

Moringa oleifera

■ seeds, fatty acids, mainly oleic; leaves, mustard-oil glycosides and other biologically active compounds

saturated fatty acids, such as stearic, arachidic, and lignoceric (LePoole 1996). It is very stable and resistant to oxidation and is widely appreciated for use as a base in cosmetics, perfumes, lubricants, and ointments and for a variety of religious purposes.

The foliage contains several thiocarbamate mustard-oil glycosides—niazinins A and B, niazimicin, and niaziminins A and B (Faizi et al. 1992). These compounds cause a lowering of blood pressure. In addition, a hypotensive isothiocyanate and two nitrile glycosides, niazirin and niazirinin, which are not active in lowering of blood pressure, have also been isolated (Faizi et al. 1994). The cause of the toxic effects of the seeds in rats has not been specifically identified, but it does not appear to be toxic lectins (Grant et al. 1995).

S

C-N-C-O-CH_3

O-Rhamnose

Niazinin A

Dimeric cationic flocculating proteins are present in the seeds, which have been exploited for treating poor-quality water and wastewater in developing countries (Gupta & Chaudhuri 1992; Gassenschmidt et al. 1995; Ndabigengesere et al. 1995).

28. Myrtaceae Juss.

■ Myrtle Family
Eucalyptus

Comprising 100–144 genera and about 3,000 species of evergreen shrubs and trees, the Myrtaceae, or myrtle family, is mostly tropical and subtropical in distribution, with Australia and the Americas being centers of species diversity. The family is ecologically and economically quite important. Species of *Eucalyptus* dominate the hardwood forests of Australia and provide timber and oil. Other genera provide spices such as cloves, allspice, and pimento and oils such as oil of bay rum and cajeput oil. The fruits of the American and West Indian guava and jaboticaba are eaten. Taxa cultivated as ornamentals are common in the family as well. A few species are troublesome weeds in the South. In North America the family is represented by 17–20 genera and 80–125 native and introduced species. Species of 1 genus are of toxicologic significance.

Eucalyptus L'Hér.

■ *E. cladocalyx*
E. viminalis

eucalyptus
gum tree
manna gum
sugar gum

aromatic trees

Ranging in habit from small shrubs to trees more than 100 m tall, *Eucalyptus* is commonly known as eucalyptus or gum tree. With the exception of 2 or 3 species in Indo-Malaysia, it is native to Australia and Tasmania and comprises about 500 species adapted to a diversity of habitats. Aromatic and oily, the bark typically exfoliates in scales or strings, leaving the trunk smooth. The generally evergreen leaves are heterophyllous, with the juvenile ones typically opposite and sessile or paired perfoliate. In contrast the adult leaves are typically alternate, petiolate, lanceolate, and falcate. Borne in umbellate or paniculate cymes, the generally perfect, radially symmetrical flowers are 4- or 5-merous and have a well-developed hypanthium. The biseriate perianth is caplike and is shed at anthesis. The capsules are woody, flat, and open at the apices. Many species of the genus are cultivated in California, and a few have naturalized.

Of the numerous species of this genus, 2—*E. cladocalyx* R.Br. (sugar gum) and *E. viminalis* Labill. (manna gum)—have caused disease and abrupt death in sheep and cattle that were allowed to eat foliage from recently felled trees or branches or regrowth from stumps (Everist 1981). In one case, 20 Angora goats in a flock of 50 were affected within a few hours after being seen eating the leaves from a sugar gum felled only hours before (Webber et al. 1985). Within 2 hours of felling the tree, 9 goats had died, and by the next day, 24 of the 50 were dead. The cause of death by these species of *Eucalyptus* is due to cyanide, present as prunasin in these species (McBarron 1976). Presumably the glycosidase is also present to aid in hydrolysis of the glucoside. In the instance of the intoxication of the Angora goats, cyanide was implicated on the basis of a strong odor of bitter almonds from the rumen and a strong positive picrate test on rumen contents (Webber et al. 1985). Guinea fowl that ate the leaves of *E. cladocalyx* died of apparent cyanide intoxication, as indicated by the discovery of 10 ppm cyanide in the livers of the dead birds (Reece 1982).

- foliage cyanogenic, prunasin; abrupt onset, neurologic disturbance, rapidly fatal

29. Nyctaginaceae Juss.

Native primarily to the tropics and subtropics of both the Old World and the New World, but with greatest diversity in the latter, the Nyctaginaceae, commonly known as the four-o'clock family, comprises 30 genera and 290–300 species. Most familiar are the ornamental bougainvillea and garden four-o'clock. Plants of the family also are eaten as vegetables and incorporated in herbal medicines to treat a variety of ailments. Members are herbs, shrubs, or trees with simple, alternate or opposite, estipulate leaves. The inflorescences are cymose and bear perfect or imperfect, 5-merous flowers with uniseriate perianths that are typically subtended by showy involucres. The calyx is generally petaloid and tubular. The petals are absent, and the fruits are achenes. In North America, 16 genera and 119–124 native and introduced species are present. One genus has been associated with a toxicologic problem.

- Four-o'clock Family

  Mirabilis

Mirabilis L.

An American genus of 50–60 species distributed primarily in the warmer regions, *Mirabilis* is commonly known as four-o'clock or umbrellawort. Although there are numerous species, only *M. jalapa* L. (garden four-o'clock, marvel-of-Peru, or beauty-of-the-night) has been associated with adverse effects. Typically grown as an annual, it is a glabrous to sparsely pubescent perennial herb 40–100 cm tall from large fleshy or woody roots. Its opposite, dark green, ovate to cordate leaves are petiolate below and sessile above. The flowers are borne in terminal clusters and are subtended by a calyx-involucre. Petaloid and funnelform, the calyx opens in the late afternoon—

- *M. jalapa*

  beauty-of-the-night
  four-o'clock
  garden four-o'clock
  marvel-of-Peru
  umbrellawort

  terminal flower clusters open in afternoon

- irritants, possible digestive disturbance

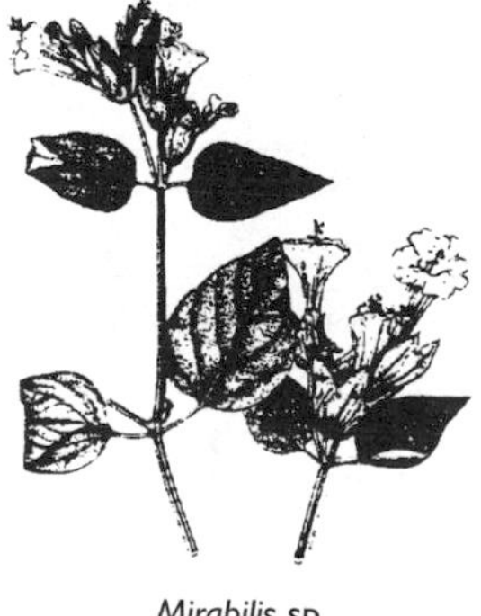

Mirabilis sp.

Trigonelline

hence the common names—and may be white, yellow, red, purple, or variegated. The achenes are 5-angled.

Ingestion of the achenes of *M. jalapa* is reported to have caused toxicity in a child (Watt & Breyer-Brandwijk 1962; O'Leary 1964). They and the roots contain the alkaloid trigonelline, which is an irritant of the skin and the digestive tract.

30. Pittosporaceae R.Br.

■ Pittosporum Family
Pittosporum

Native to the Old World tropics and warm-temperate regions and especially abundant in Australia, the Pittosporaceae, or pittosporum family, comprises 9 genera and 200–240 species of evergreen trees, shrubs, or woody vines with attractive leathery leaves. Members of the family are cultivated as ornamentals. In North America, 3 genera are cultivated, and 1 of them has the potential to cause toxicologic problems.

Pittosporum Banks ex Sol.

■ *P. crassifolium*
P. eugenioides
P. phillyraeoides
P. tenuifolium
P. tobira
P. undulatum

karo
lemonwood
mock orange
pittosporum
tarata
tawhiwhi
Victorian box

shrubs or small trees; fruits capsules with sticky seeds

■ irritant saponins, possible digestive disturbance

Comprising 150–200 species, *Pittosporum,* commonly known as Victorian box or mock orange, is widely cultivated in southern California, where several species have escaped and become established in disturbed urban areas. Representative of the genus are *P. crassifolium* Banks & Sol. ex A.Cunn. (karo), *P. eugenioides* A.Cunn. (lemonwood, or tarata), *P. phillyraeoides* DC. (willow-leaved pittosporum), *P. tenuifolium* Banks & Sol. ex Gaertn. (tawhiwhi), *P. tobira* (Thunb.) W.T.Aiton (Japanese pittosporum), and *P. undulatum* Vent. (Victorian box). Plants of the genus are shrubs or small trees with simple, alternate, estipulate leaves that are clustered at the branch ends. Borne in terminal or axillary panicles or corymbose clusters, the perfect or imperfect flowers are 5-merous and radially symmetrical, with white, yellow, pink to dark red, or greenish petals. The resinous, woody or leathery capsule is 2- or 3-valved and contains sticky seeds—hence the generic name, which is derived from the Greek roots *pitta,* for "pitch," and *sporos,* for "seed."

Ingestion of large numbers of the capsules of *Pittosporum* has caused irritation of the digestive tract (Fuller & McClintock 1986). The fruit and foliage also contain saponins, which may cause digestive tract disturbance. These saponins are glycosides of various genins, including the pentaterpenoid oleananes such as barringtogenol and 16-deoxybarringtogenol; the pentaterpenoid taraxastanes such as taraxastanoic acid; and the sesquiterpenoid aromadendranes and germacranes such as pittosporanoside(Connolly & Hill 1991). In cases of intoxication there will be nausea, vomiting, colic, and lethargy.

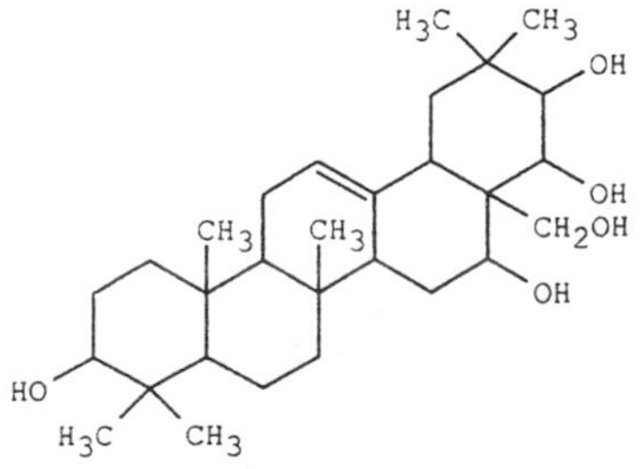

Barringtogenol C

31. Plumbaginaceae Juss.

Cosmopolitan in distribution, especially in dry or saline habitats of the Old World, the Plumbaginaceae, commonly known as the leadwort family, comprises 10–22 genera and 440–500 species of herbs and shrubs, several of which are valued as ornamentals or are used medicinally. In North America, 5 genera and approximately 30 species are present; most are introductions and cultivated. One genus is of toxicologic interest.

Plumbago L.

■ *P. auriculata*
leadwort
herbs and shrubs

Native to the tropical and warm-temperate regions of both the Old World and the New World, *Plumbago,* commonly known as leadwort, comprises 12–20 species of shrubs and annual or perennial herbs with simple, alternate leaves that have entire margins and auriculate or clasping bases. Borne in terminal, bracteate spikes or racemes, the perfect, radially symmetrical, 5-merous flowers have a salverform corolla that ranges in color from white to blue or purplish red or scarlet. The fruits are circumscissile capsules. Representative of the genus is *P. auriculata* Lam. (cape leadwort), the species most commonly cultivated.

■ irritant quinone, plumbagin; possible dermatitis, digestive disturbance, reproductive effects

Species of *Plumbago* have been widely used folk medicaments in Asian countries (Itoigawa et al. 1991). However, some portions of the plants may cause adverse effects. Direct contact with the sap causes irritation of the skin, which may require several days to subside (Fuller & McClintock 1986). Large dosage of an extract of the root of Asian *P. indica* L. (reported as *P. rosea* L.) caused severe diarrhea in rats and mice (Solomon et al. 1993). A Brazilian species, *P. scandens* L., is lethal to cattle at 1% b.w. (Tokarnia & Dobereiner 1982). The signs of intoxication in experimental animals indicated severe irritation of the digestive tract. They included excessive salivation, degeneration of the buccal mucosa, submandibular edema, loss of appetite, ruminal inactivity, bloat, and reddish brown urine. Diarrhea was not a feature of the disease.

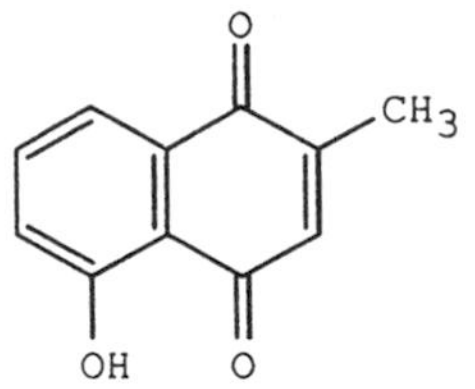

The roots, leaves, and flowers of *Plumbago* contain plumbagin, an irritant and vesicant quinone (Thomson 1971). The quinone is particularly irritating to smooth muscle and the uterus (Bhatia & Lal 1933). Plants containing it have been used for a variety of medicinal purposes, including induction of abortion, promotion of digestion, and a multitude of other uses with both oral and topical application. Experimentally, oral administration of plumbagin at a dose of 10 mg/kg b.w. caused significant reproductive effects, altering estrus, inhibiting mating behavior, preventing implantation, and causing abortion (Premakumari et al. 1977). Effects also were noted in rabbits. Long-term administration caused testicular degeneration in dogs (Bhargava 1984). The toxic effects of 40 mg/kg b.w. given orally to mice included loss of righting ability, drowsiness, tremors, and increased respiration and heart rates within a few hours. The oral LD_{50} of plumbagin was 40 mg/kg in mice and 65 mg/kg in rats (Premakumari et al. 1977).

O
CH_3
OH O

Plumbagin

In vitro experiments using heart tissue of guinea pigs showed positive inotropic effects for plumbagin (Itoigawa et al. 1991). These effects appear to be due to intracellular Ca^{2+} accumulation secondary to inhibition of the sarcoplasmic reticulum Ca^{2+} pump (Floreani et al. 1996).

The signs of intoxication are mainly those associated with irritation of the digestive tract such as excess salivation, vomiting, and perhaps diarrhea. Treatment, if necessary, should be directed toward relief of the signs.

32. Polygalaceae R.Br.

■ Milkwort Family
Polygala

Nearly cosmopolitan in distribution except for the Arctic and New Zealand, the Polygalaceae, commonly known as the milkwort family, comprises 12–17 genera and

750–1,000 species of herbs, shrubs, woody vines, and small trees. Its most notable attribute is the resemblance of its flowers to those of the Fabaceae, albeit the similar-appearing parts are not homologous, and novice botanists often misidentify its taxa. The family is of little economic importance; a few species are used in folk medicines, and a few are cultivated as ornamentals. In North America, 3 genera and 64 native and introduced species are present. One genus may pose a potential toxicologic problem.

Polygala L.

- *P. senega*

  Seneca snakeroot
  senga root

  herbs and shrubs

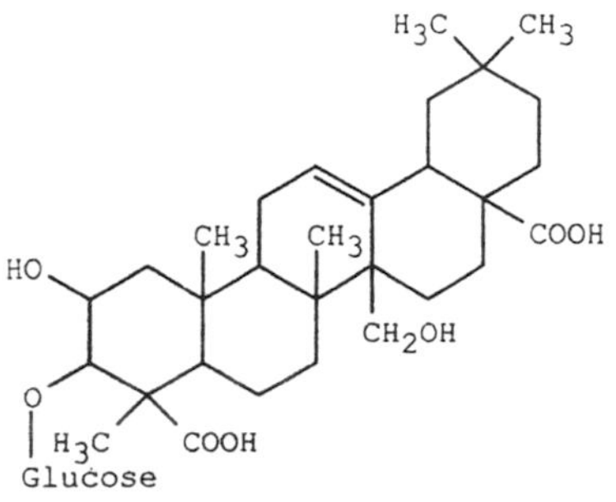

Tenuifolin

- irritant terpenoids; possible digestive disturbance

Comprising 500–600 species, *Polygala* is the largest genus of the family. Its name, used by Dioscorides, is derived from the Greek roots *poly* and *gala,* for "much milk," alluding to its reputation for stimulating milk secretion. Plants of the genus are herbs, subshrubs, and shrubs with simple, alternate or opposite or whorled, stipulate leaves with entire or toothed margins. Borne in bracteate racemes or spikes, the perfect flowers are bilaterally symmetrical, with their perianths in 2-series. There are 5, deciduous or persistent sepals of 3 forms in 2 whorls. The outer 3 are smaller and herbaceous or partially petaloid, and the inner 2 are larger, petaloid, winglike, and free. The petals number 3, or rarely 5, but if 5, 2 are minute. They are fused at their bases; white or whitish green or pale to deep pink or rose or magenta or bicolored; and of 2 forms. The lower is keeled or bowl-shaped, and usually fringed or lobed at the apices, whereas the upper 2 are ligulate and not fringed or lobed at apices. The capsular fruits bear seeds with arils. Species of the genus are found in a variety of habitats across the continent.

Although *Polygala* has not been shown to be definitely toxic in North America (Pammel 1929), a South American species, *P. klotzschii,* causes digestive tract and neurologic problems in cattle, and the disease has been experimentally reproduced in rabbits (Rezende et al. 1981). The pathologic changes were consistent with a toxin that irritated the digestive tract. Because many species of *Polygala* contain numerous pentaterpenoid oleananes such as arjunolic acid, polygalic acid, presenegenin, tenuifolic acid, tenuifolin, and others (Connolly & Hill 1991), they are considered likely causes of the digestive tract irritation. The roots of *P. senega* L. (senga root, Seneca snakeroot) are a source of senega, which was formerly used as an irritant drug (Kingsbury 1964).

33. Portulacaceae Juss.

- Purslane Family

  Claytonia
  Portulaca

Commonly known as the purslane family, the Portulacaceae comprises 19 or 20 genera and 350–500 species of more or less succulent herbs and subshrubs. It has cosmopolitan distribution but exhibits greatest diversification in western North America, the Andes Mountains of South America, and South Africa. Of taxonomic interest are its flowers, with perianths seemingly in 2-series. Anatomical studies have revealed that the "sepals" are bracts, and the "petals" are petaloid sepals; true petals are absent. The family is noted for several popular garden and rock garden ornamentals such as rose moss, flameflower, bitterroot, and rock purslane. A few species are eaten as potherbs, and a few are weedy. In North America, 8 genera and 108 species are present. Species of 2 genera are of toxicologic interest.

Claytonia L.

- *C. virginica*

  spring beauty

  herbs with 2, opposite cauline leaves; sepals 2; petals 5

Comprising 20–28 species distributed primarily in North America, *Claytonia,* commonly known as spring beauty, honors John Clayton, a colonial American botanist.

Plants of the genus are glabrous annual or perennial herbs from rhizomes, tubers, or taproots. The mostly basal, simple leaves are entire and fleshy. Typically there are 2, opposite cauline leaves. Borne in terminal, 1-sided racemes, the radially symmetrical flowers have 2 "sepals"; 5 free "petals," with white or pink with pink veins; and a superior ovary with a 3-lobed style and free-central placentation that gives rise to a capsule. Species of the genus occur across the continent in a variety of habitats; some, such as *C. virginica* L., are widely distributed.

Not known to be toxic, species of *Claytonia* were used as potherbs by members of western American Indian tribes (Ebeling 1986). The corms/roots were cooked and eaten, and the foliage of *C. perfoliata* Donn ex Willd. (miner's lettuce) was eaten raw as well. However, there may be appreciable amounts of soluble oxalates similar to those found in species of *Portulaca*.

■ oxalate accumulation

Portulaca L.

Both prized for its ornamental species and cursed for its weedy ones, *Portulaca*, commonly known as purslane, comprises 100–200 species distributed primarily in warm regions. In North America, approximately 17 native and introduced species are present. They are annual herbs with prostrate to erect stems that bear alternate or opposite leaves with flat or terete blades. The uppermost 2-5 leaves form a leafy involucre that subtends the often showy, radially symmetrical flowers with 2 "sepals," 4–6 "petals," 4–20 stamens, and an inferior ovary. Flower colors are yellow, orange, pink, red, magenta-purple, or variously striped. The capsule is circumscissile. Representative of the genus and most commonly encountered are *P. grandiflora* Hook. (rock moss), a popular ornamental; and *P. oleracea* L. (common purslane), a troublesome weed.

■ *P. grandiflora*
P. oleracea

common purslane
rock moss

annual herbs; sepals 2; petals 4–6

Portulaca oleracea

Portulaca oleracea has been used for food and medicinal purposes for centuries (Nuez & Hernandez Bermejo 1994). It was used in salads and as a winter vegetable and was even thought to be an aphrodisiac. The foliage was prepared and eaten as a potherb by members of western American Indian tribes (Ebeling 1986). The seeds were also eaten. In spite of the ready acceptance of these plants as a reliable source of food, intoxications have occasionally occurred in livestock. Typically, problems develop when hungry animals are put in a pasture or lot where *Portulaca* is one of the few forages available. When large amounts are eaten in a few hours, most commonly overnight, an acute intoxication may occur. The onset of signs is abrupt, typically involving only a small number of the more voracious eaters, and deaths occur in a few hours or several days.

■ eaten as potherb by humans; oxalate accumulation, possible metabolic and renal problems in livestock

Aerial portions of *P. oleracea* are reported to contain 7.2–10.3% soluble and 9.7–17% total oxalates (Libert & Franceschi 1987). Even higher oxalate concentrations are reported in plants from some Mexican states (Mejia et al. 1985). Australia seems to be the country where *Portulaca* has been most commonly associated with toxicity problems. Oxalate concentrations of *P. oleracea* at various locations in Queensland ranged from 4.3 to 9.4% soluble and from 3.5 to 13.3% total (Mathams & Sutherland 1952). In specific instances where problems developed, the levels were 4.6–7.4% soluble and 7.7–9.3% total oxalates.

The signs, pathology, and treatment are similar to the acute intoxication associated with other oxalate-accumulating plants. The onset of signs is acute, perhaps 8–12 hours after animals graze a new pasture. They may die within a few hours of the onset of signs if not treated. Prompt administration of parenteral calcium solutions will generally result in immediate relief of many of the signs, and the animals will become ambulatory within a few hours. Relapses may occur, and some individuals may eventually die of acute renal failure.

34. Simaroubaceae DC.

■ Quassia Family
Ailanthus
Quassia

Pantropical in distribution, with only a few species in warm-temperate regions, Simaroubaceae, commonly known as the quassia family, comprises 20–25 genera and 120–150 species of trees and shrubs, mostly with very bitter-tasting bark, wood, and seeds. The family includes the ornamental genus *Ailanthus* and the medicinal genus *Quassia*, both of which are of toxicologic significance in North America.

Ailanthus Desf.

■ *A. altissima*
Chinese sumac
copal tree
tree-of-heaven

deciduous tree; leaves large, 1-pinnately compound; malodorous

Ailanthus altissima

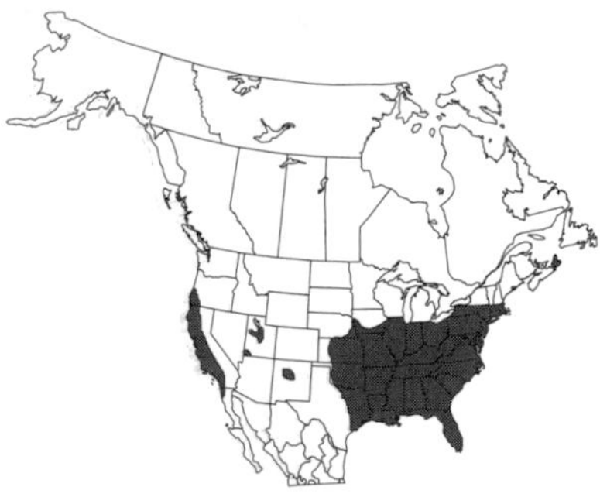

Ailanthus altissima

■ irritant compounds; dermatitis, systemic effects unknown

An Old World genus of about 10 species, *Ailanthus* is represented in North America by *A. altissima* (Mill.) Swingle (tree-of-heaven), a native of eastern Asia widely planted as a street shade tree but escaping and naturalizing in various areas. Also known as copal tree and Chinese sumac, it is a rapidly growing, deciduous tree to 20 m tall with large, 1-pinnately compound, glandular leaves up to 1 m long. Borne in large, terminal panicles, the malodorous, imperfect, 5-merous flowers are greenish or yellowish and give rise to schizocarps of samaras.

Ailanthus altissima is suspected of being toxic in humans; the noxious odor of the foliage is associated with headaches and nausea, and contact may result in dermatitis with vesicular eruptions (Mitchell & Rook 1979). Although the noxious odor reduces consumption of this species by livestock, intoxications occasionally occur. Sheep eating the foliage developed weakness, increased respiratory rate, and paralysis of the pelvic limbs. Water from the flowers caused irritation of the skin and digestive tract (Fuller & McClintock 1986). In humans, a tincture made from the foliage and blossoms caused nausea, vomiting, and muscular relaxation (Millspaugh 1974).

The intoxication potential has not been borne out by experimental efforts to reproduce disease (Bourke 1996). Goats fed 75 kg of fresh leafy stems as the sole ration over a 3.5-month period remained in good health and maintained their body weight. Similarly, crude methanol extracts of the bark fed to goats at the equivalent of 191 g fresh bark per day for 56 days failed to cause adverse effects.

The leaves, bark, and flowers contain ailanthone, which is an antimalarial and a phytoalexin, and biologically active quassinoids such as ailantinols A and B (Bray et al. 1987; Lin et al. 1995; Kubota et al. 1996). A quinone irritant, 2,6-dimethoxybenzoquinone, also is present (Thomson 1971).

Ailanthone

2,6-Dimethoxybenzoquinone

Quassia L.

■ *Quassia*
bitterwood

shrub or tree not present in North America

Commonly known as bitterwood and used to treat fevers, to kill insects, or to substitute for hops in beer, *Quassia* consists of 1–40 species, depending upon the taxonomic treatment used. If only 1 species is recognized, it is called *Q. amara* L. A shrub or small tree to 8 m tall, it has alternate, 1-pinnately compound leaves with 3 or 5 or 7 leaflets borne on winged rachises. The bright red flowers are borne in terminal racemes or panicles.

Quassia amara is not found in North America, but shavings of its wood have been imported for use as animal bedding. They have been implicated, however, as a cause of contact dermatitis in horses (Atwater 1993). The result was vesicular eruptions and skin sloughing of the faces of show horses. The wood contains an irritant and allergenic quinone, 2,6-dimethoxybenzoquinone which may account for the effects on the skin of horses (Polonsky & Lederer 1959).

- wood shavings imported for animal bedding; irritant effects in horses

35. Typhaceae Juss.

Commonly known as the cattail family, the Typhaceae consists of a single genus, *Typha,* with 10–15 species found in marshes and wetlands throughout the world.

- Cattail Family

  Typha

Typha L.

The genus *Typha* is represented in North America by 3 species and 1 named hybrid. Distributed across the continent, it is important in wetland succession. It provides cover and food for wildlife, and various parts of the plants are eaten by humans. The plants are vigorously rhizomatous, emergent aquatic herbs with long, linear, 2-ranked, equitant leaves. The erect, unbranched stems terminate in dense, cylindrical spikes with the staminate flowers borne above the pistillate. The flowers have perianth parts of numerous bristles. Borne on the wind and easily dispersed in great numbers, the fruits are follicles.

- *Typha*

  cattail

  aquatic herbs from rhizomes; flowers borne in terminal cylindrical spikes on long, erect unbranched stems

Plants of *Typha* are renowned as edible and are said to represent an outdoor pantry (Harrington 1967; Morton 1975). The starchy rootstocks are used like cucumbers in salads, cooked and eaten, dried to make a flour, and squeezed to make a juice drink (Fernald & Kinsey 1943). Other parts of the plants are also relished: the young shoots are eaten like bamboo shoots, the pollen is used to make pancakes, and the immature spikes are boiled, steamed, or coated and baked (Fernald & Kinsey 1943; Harrington 1967). In spite of the recognized value of these plants, intoxication of livestock has been suspected in a few instances. Signs are primarily related to digestive tract problems, but one case involved stiffness, tremors, and sweating in horses (Hansen 1930). Experimentally, the leaves have not proven toxic to a calf or to sheep (Morton 1975). They are unlikely to be a significant livestock risk.

- eaten by humans; possible digestive disturbance in livestock

36. Vitaceae Juss.

Comprising 11–13 genera and 700–735 species, the Vitaceae, commonly known as the grape family, exhibits greatest diversity in the tropics and subtropics, with relatively few taxa in temperate regions. Members of the family are primarily woody vines that climb by means of tendrils opposite the leaves. A few are shrubs or small trees. The family is of considerable economic importance because of *Vitis vinifera,* the wine grape. It and other species of the genus provide raisins, currants, and juice as well as wine. Some taxa of the family are sometimes cultivated as ornamentals. In North America, 5 genera and approximately 45 native and introduced species are present. One genus has been implicated in toxicologic problems.

- Grape Family

  Parthenocissus

Parthenocissus Planch.

Native to North America and Asia, *Parthenocissus,* commonly known as Virginia creeper, comprises 10–15 species, of which *P. quinquefolia* (L.) Planch. is the most

- *P. quinquefolia*

  Virginia creeper

- climbing vine with tendrils; leaves palmately compound, with 5 leaflets

Parthenocissus quinquefolia

- possible digestive disturbance

commonly encountered throughout the eastern half of the continent. Often it is the "ivy" that cloaks buildings, especially those on university campuses. Climbing in trees or on walls by branched tendrils with terminal adhesive pads, it has alternate, palmately compound leaves with 5 (3–7) leaflets with serrate margins that typically turn a brilliant scarlet-red in the autumn. Borne in paniculate cymes, the small, greenish, 4- or 5-merous flowers produce dark blue or black berries with 1–4 seeds.

There are several reports of intoxication due to *P. quinquefolia,* including lethal effects in children and experimental reproduction of disease in guinea pigs. Kingsbury (1964), however, questioned whether the plants causing the problem were properly identified. Experimentally, 1 g of fall-growth leaves caused depression and intermittent vomiting of 5 hours' duration in budgerigars (Shropshire et al. 1992). Spring-growth leaves did not produce adverse effects. However, the foliage and berries may cause digestive tract disturbance in humans. In one instance 3 young children who ate the leaves experienced mild stomachache (O'Leary 1964). In another case, 2 children, 2.5 and 5 years of age, chewed the leaves and swallowed the juice from them. They developed vomiting and diarrhea with straining, collapsed with profuse sweating, and then lapsed into a deep sleep of short duration (Millspaugh 1974).

References

Albiston HE: The joint leaf rush *(Juncus holoschoenus):* a cyanogenetic plant. *Aust Vet J* 13;200, 1937

Alfonso D, Johnson HA, Colman-Saizarbitoria T, Presley CP, McCabe GP, McLaughlin JL: SARs of annonaceous acetogenins in rat liver mitochondria. *Nat Toxins* 4;181–188, 1996

Allen PH: Poisonous and injurious plants of Panama. *Am J Trop Med* 23 (Suppl);1–76, 1943

Al-Maliki SJ, Hatam NA, Ibrahem DK, Yousif NJ: The influence of various extracts of *Lagerstroemia indica* on social aggression and behavioural responses in male mice. *J Biol Sci Res* 19;53–64, 1988

Atta-ur-Rahman, Asif E, Ahmed D, Sener B, Turkoz S: New Alkaloids from *Buxus sempervirens. J Nat Prod* 52;1319–1322, 1989

Atta-ur-Rahman, Ahmed D, Asif E, Ahmad S, Sener B, Turkoz S: Chemical constituents of *Buxus sempervirens. J Nat Prod* 54;79–82, 1991

Atwater C: Tracking a toxin. *Equus* 194;13–14, 1993

Barber MA: Poisoning due to the papaw *(Asimina triloba). J Am Med Assoc* 45;2013–2014, 1905

Bartik M, Piskac A: *Veterinary Toxicology.* Elsevier Scientific, Amsterdam, p270, 1981

Baskin E: *The Poppy and Other Deadly Plants.* Delacorte Press, New York, 1967

Beath AO, Gilbert CS, Eppson HF, Rosenfeld I: *Poisonous Plants and Livestock Poisoning.* Wyo Agric Exp Stn Bull 324, 1953

Bell ME: Toxicology of karaka kernel, karakin, and β-nitropropionic acid. *N Z J Sci* 17;327–334, 1974

Bhakuni DS, Jain S: Protoberberine alkaloids. In *The Alkaloids: Chemistry and Pharmacology,* vol 28, Brossi A ed, Academic Press, Orlando, Fla, pp95–181, 1986

Bharadwaj K, Chandra V: *Trianthema portulacastrum* Linn.: natural resource of oxalic acid. *Res Ind* 32;117–123, 1987

Bhargava SK: Effects of plumbagin on reproductive function of male dog. *Indian J Exp Biol* 22;153–156, 1984

Bhatia RB, Lal S: The pharmacological actions of *Plumbago zeylanica* and its active principles (plumbagin). *Indian J Med Res* 23;777–780, 1933

Blohm H: *Poisonous Plants of Venezuela.* Harvard Univ Press, Cambridge, 1962

Blum HF: *Photodynamic Action and Diseases Caused by Light.* Am Chem Soc Monogr Ser 85, Reinhold, New York, 1941

Bourke CA: Lack of toxicity of *Ailanthus altissima* (tree-of-heaven) for goats. *Aust Vet J* 74;465, 1996

Bray DH, Boardman P, O'Neill MJ, Chan KL, Phillipson JP, Warhurst DC, Suffness M: Plants as a source of antimalarial drugs. 5. Activities of *Ailanthus altissima* stem constituents and of some related quassinoids. *Phytother Res* 1;22–24, 1987

Bruce EA: Hydrangea poisoning. *J Am Vet Med Assoc* 58;313–315, 1920

Caceres A, Saravia A, Rizzo S, Zabala L, Leon E, Nave F: Pharmacologic properties of *Moringa oleifera*. 2. Screening for antispasmodic, antiinflammatory, and diuretic activity. *J Ethnopharmacol* 36;233–237, 1992

Camy G, Leveille JL, Nevers B: Case report: box *(Buxus sempervirens)* poisoning in cattle. *Point Vet* 18;97,203–204, 1986 (*Vet Bull* 56;7273, 1986)

Carter CL, McChesney WJ: Hiptagenic acid identified as β-nitropropionic acid. *Nature* 164;575–576, 1949

Cary CA, Miller ER, Johnstone GR: *Poisonous Plants of Alabama.* Ala Coop Ext Serv Circ 71, 1924

CDC Editorial Note: Ostrich fern poisoning—New York and western Canada, 1994. *J Am Med Assoc* 273;912–913, 1995

Cerny V, Sorm F: Steroid alkaloids: alkaloids of Apocynaceae and Buxaceae. In *The Alkaloids: Chemistry and Physiology,* vol 9, Manske RHF ed, Academic Press, New York, pp305–426, 1967

Chick BF, McCleary BV, Beckett RJ: Thiaminases. In *Toxicants of Plant Origin,* vol 3, *Proteins and Amino Acids,* Cheeke PR ed, CRC Press, Boca Raton, Fla, pp73–91, 1989

Connolly JD, Hill RA: *Dictionary of Terpenoids.* Vol 2, *Di- and Higher Terpenoids,* Chapman & Hall, London, pp655–1460, 1991

Cooper MR, Johnson AW: *Poisonous Plants in Britain and Their Effects on Animals and Man.* Ministry of Agriculture, Fisheries and Food, Her Majesty's Stationery Office, London, pp130–131, 1984

Cordell GA: *Introduction to Alkaloids: A Biogenetic Approach.* John Wiley & Sons, New York, pp907–910, 1981

Cutler HC: Monograph of the North American species of the genus *Ephedra. Ann Mo Bot Gard* 26;373–424, 1939

Czygan F-C, Wessinger B, Warmuth K: *Cuscuta* and its property to take up and accumulate alkaloids of the host plant. *Biochem Physiol Pflanz* 183;495–501, 1988

Dayton WA: *Important Western Browse Plants.* USDA Misc Publ 101, 1931

Dollahite JW, Allen TJ: Poisoning of cattle, sheep, and goats with *Lobelia* and *Centaurium* species. *Southwest Vet* 15;126–130, 1962

Doskotch RW, Knapp JE: Alkaloids from *Menispermum canadense. Lloydia* 34;292–300, 1971

Ebeling W: *Handbook of Indian Foods and Fibers of Arid America.* Univ California Press, Berkeley, 1986

Edgar JT, Thin IM: Plant poisoning involving male fern [letter]. *Vet Rec* 82;33–34, 1968

Everist SL: *Poisonous Plants of Australia,* 2d ed. Angus & Robertson, London, pp546–548, 1981

Faizi S, Siddiqui BS, Saleem R, Siddiqui S, Aftab K, Gilani A-ul-H: Isolation and structure elucidation of novel hypotensive agents, niazinin A, niazinin B, niazimicin and niaziminin A + B from *Moringa oleifera:* the first naturally occurring thiocarbamates. *J Chem Soc (Perkin Trans)* 1992;3237–3241, 1992

Faizi S, Siddiqui BS, Saleem R, Siddiqui S, Aftab K: Isolation and structure elucidation of new nitrile and mustard oil glycosides from *Moringa oleifera* and their effect on blood pressure. *J Nat Prod* 57;1256–1261, 1994

Fernald ML, Kinsey AC: *Edible Wild Plants of Eastern North America.* Idlewild Press, Cornwall-on-Hudson, New York, 1943

Finnegan RA, Stephani RA: The structure of karakin. *Lloydia* 33;491–492, 1970

Floreani M, Forlin A, Pandolfo L, Petrone M, Bellin S: Mechanisms of plumbagin action on guinea pig isolated atria. *J Pharmacol Exp Ther* 278;763–770, 1996

Ford GE: Photosensitivity due to *Erodium* spp. [letter]. *Aust Vet J* 41;56, 1965

Forte GJ, Zito JA, Edwards JM: Synthesis of lachnanthocarpone. *Lloydia* 39;192–196, 1976

Franchi GG, Ferri S: Localization of the active principles of the male fern (*Dryopteris filix-mas* [L.] Schott.) by fluorescence microscopy. *Pharmacol Res Commun* 20 (Suppl 5);135–138, 1988

Fuller TC, McClintock E: *Poisonous Plants of California.* Univ California Press, Berkeley, 1986

Gassenschmidt U, Jany KD, Tauscher B, Niebergall H: Isolation and characterization of a flocculating protein from *Moringa oleifera* Lam. *Biochim Biophys Acta* 1243;477–481, 1995

Golebiewski WM, Wrobel JT: The Lythraceae alkaloids. In *The Alkaloids*, vol 18, Rodrigo REA ed, Academic Press, New York, pp263–322, 1981

Grant G, More LJ, McKenzie NH, Dorward PM, Buchan WC, Telek L, Pusztai A: Nutritional and haemagglutination properties of several tropical seeds. *J Agric Sci* 124;437–445, 1995

Grant WM: *Toxicology of the Eye.* CC Thomas, Springfield, Ill p236, 1962

Gu Z-M, Fang X-P, Hui Y-H, McLaughlin JL: 10-, 12-, and 29-hydroxybullatacinones: new cytotoxic annonaceous acetogenins from *Annona bullata* Rich. (Annonaceae). *Nat Toxins* 2;49–55, 1994

Gu Z-M, Zhao G-X, Oberlies NH, Zeng L, McLaughlin JL: Annonaceous acetogenins, potent mitochondrial inhibitors with diverse applications. In *Recent Advances in Phytochemistry,* vol 29, pp249–310, 1995

Gupta A, Chaudhuri M: Domestic water purification for developing countries. *Aqua* 41;290–298, 1992

Gupta BK, Gupta SC, Thind IS: Effect of feeding "itsit" (*Trianthema monogyna* L.) to goats. *J Res Punjab Agric Univ* 20;539–543, 1983

Hansen AA: *Indiana Plants Injurious to Livestock.* Ind Agric Exp Stn Circ 175, 1930

Hardin JW: *Stock-Poisoning Plants of North Carolina.* N C Agric Exp Stn Bull 414, 1973

Harrington HD: *Edible Native Plants of the Rocky Mountains.* Univ New Mexico Press, Albuquerque, 1967

Harshberger JW: *Pastoral and Agricultural Botany.* Blakiston's Son & Co, Philadelphia, p79, 1920

Harvey RB, Larson AH, Landon RH, Boyd WL, Erickson LC: *Weeds Poisonous to Livestock.* Minn Agric Exp Stn Bull 388, pp3–5, 1944

Harvey TL: Aphids, dodder *(Cuscuta campestris),* and dodder-host plant interactions. *Ann Entomol Soc Am* 59;1276–1282, 1966

Hausen BM: Hydrangenol: a strong contact sensitizer found in hydrangea, *Hydrangea* spp., Hydrangeaceae. *Contact Dermatitis* 24;233–235, 1991

Hausen BM, Baurle G, Schmalle HW: Hydrangea-allergy. *Aktuel Dermatol* 8;141–145, 1982

Heywood VH: *Flowering Plants of the World.* Mayflower Books, New York, 1978

Hui Y-H, Wood KV, McLaughlin JL: Bullatencin, 4-deoxyasimicin, and the uvariamicins: additional bioactive annonaceous acetogenins from *Annona bullata* Rich. (Annonaceae). *Nat Toxins* 1;4–14, 1992

Inouye H, Takeda Y, Uobe K, Yamauchi K, Yabuuchi N, Kuwano S: Purgative activities of iridoid glucosides. *Planta Med* 25;285–288, 1974

Itoigawa M, Takeya K, Furukawa H: Cardiotonic action of plumbagin on guinea-pig papillary muscle. *Planta Med* 57;317–319, 1991

Jacob RH, Peet RL: Acute oxalate toxicity of sheep associated with slender iceplant *(Mesembryanthemum nodiflorum). Aust Vet J* 66;91–92, 1989

Jacob RH, Shaw T, Penny SR: Seasonal oxalate concentrations of six plant species in the eastern wheatbelt of western Australia. In *Plant-Associated Toxins: Agricultural, Phytochemical, and Ecological Aspects,* Colegate SM, Dorling PR eds, CAB International, Wallingford, UK, pp66–70, 1994

Keeler RF: Investigation of maternal and embryo/fetal toxicity of *Ephedra viridis* and *Ephedra nevadensis* in sheep and cattle. *J Range Manage* 42;31–35, 1989

Kellerman TS, Coetzer JAW, Naude TW: *Plant-Poisonings and Mycotoxicoses of Livestock in Southern Africa.* Oxford Univ Press, Cape Town, South Africa, 1988
Kingsbury JM: *Poisonous Plants of the United States and Canada.* Prentice-Hall, Englewood Cliffs, NJ, 1964
Kleinkopf GE, Wallace A, Hartsock TL: *Galenia pubescens*—salt-tolerant, drought-tolerant potential source of leaf protein. *Plant Sci Lett* 7;313–320, 1976
Kubota K, Fukamuja N, Hamada T, Okano M, Tagahara K, Lee K-H: Two new quassinoids, ailantinols A and B, and related compounds from *Ailanthus altissima. J Nat Prod* 59;683–686, 1996
Kuligowski ME, Chang A, Leemreize JHM: Allergic contact hand dermatitis from hydrangea: report of a 10th case. *Contact Dermatitis* 26;269–270, 1992
LePoole HA: Behen oil: classic oil for modern cosmetics. *Cosmet Toiletries* 111;77–80, 1996
Leroux V: Poisoning of pets by house plants. *Point Vet* 18;45–55, 1986
Libert B, Franceshi VR: Oxalate in crop plants. *J Agric Food Chem* 35;926–938, 1987
Lin L-J, Peiser G, Ying B-P, Mathias K, Karasina F, Wang Z, Itatani J, Green L, Hwang Y-S: Identification of plant growth inhibitory principles in *Ailanthus altissima* and *Castela tortuosa. J Agric Food Chem* 43;1708–1711, 1995
Londershausen M, Leicht W, Lieb F, Moeschler H, Weiss H: Molecular mode of action of annonins. *Pestic Sci* 33;427–438, 1991
Macleod NSM, Greig A, Bonn JM, Angus KW: Poisoning in cattle associated with *Dryopteris filix-mas* and *D. borreri. Vet Rec* 102;239–240, 1978
Mathams RH, Sutherland AK: The oxalate content of some Queensland pasture plants. *Queensl J Agric Sci* 9;317–334, 1952
Matsui T: Studies on the utilization of pawpaw seeds a source of seed oil. *Meiji Daigaku Kogakubu Kenkyu Hokoku* 52;43–53, 1980 (*Chem Abstr* 93;237238, 1980)
McBarron EJ: *Medical and Veterinary Aspects of Plant Poisons in New South Wales.* Department of Agriculture, New South Wales, Glenfield, Australia, 1976
Mejia MR, Montalvo VR, Martinez RR: Levels of oxalates in wild forages coming from the states of Hidalgo, Guanajuato, Mexico, Tlaxcala, and the Federal District. *Vet Mex* 16;21–25, 1985
Millspaugh CF: *American Medicinal Plants.* Dover, New York, 1974 (reprint from 1892)
Mitchell GBB, Wain EB: Suspected male fern poisoning in cattle. *Vet Rec* 113;188, 1983
Mitchell J, Rook A: *Botanical Dermatology—Plants and Plant Products Injurious to the Skin.*. Greengrass, Vancouver, BC, 1979
Mogg AOD: Vlei poisoning. *S Afr J Sci* 26;269–277, 1927
Montgomery JD, Wagner WH: *Dryopteris.* In *Flora of North America,* vol 2, *Pteridophytes and Gymnosperms,* Flora of North America Editorial Committee eds, Oxford Univ Press, New York, pp280–288, 1993
Morton JF: Cattails (*Typha* spp.)—weed problem or potential crop? *Econ Bot* 29;7–29, 1975
Morton JF: *Plants Poisonous to People in Florida and Other Warm Areas.* Morton, Miami, 1982
Movsesyan TB, Azaryan KhA: Pathological changes in cattle poisoned by the dodder *Cuscuta campestris* Yuneker. *Biol Zh Arm* 24;67–70, 1971 (*Vet Bull* 42;2086, 1972)
Movsesyan TB, Azaryan KhA: Poisoning produced by feeding dodder *(Cuscuta campestris)* to animals. *Veterinariya (Moscow)* 6;92, 1973 (*Vet Bull* 44;330, 1974)
Moyer BG, Pfeffer PE, Valentine KM, Gustine DL: 3-Nitropropanoyl-D-glucopyranoses of *Corynocarpus laevigatus. Phytochemistry* 18;111–113, 1979
Munz PA: Dermatitis produced by *Phacelia* (Hydrophyllaceae). *Science* 76;194, 1932
Murakami T, Tanaka N: Occurrence, structure, and taxonomic implications of fern constituents. In *Progress in the Chemistry of Organic Natural Products* 54, Herz W, Grisebach H, Kirby GW, Tamm C eds, Springer-Verlag, Vienna, pp1–353, 1988
Nath D, Sethi N, Singh RK, Jain AK: Commonly used Indian abortifacient plants with special reference to their teratological effects in rats. *J Ethnopharmacol* 36;147–154, 1992

Ndabigengesere A, Narasiah KS, Talbot BG: Active agents and mechanism of coagulation of turbid waters using *Moringa oleifera. Water Res* 29;703–710, 1995

Nielsen C, McCausland H: The occurrence and alkaloidal content of various *Ephedra* species. 2. *J Am Pharm Assoc* 17;427–430, 1928

Nielsen C, McCausland H, Spruth HC: The occurrence and alkaloidal content of various *Ephedra* species. *J Am Pharm Assoc* 16;288–294, 1927

Nuez F, Hernandez Bermejo JE: Neglected horticultural crops. In *Neglected Crops: 1492 from a Different Perspective,* Hernandez Bermejo JE ed, Food and Agriculture Organization, United Nations, Rome, pp303–332, 1994

Oke OL: Oxalic acid in plants and nutrition. *World Rev Nutr Diet* 10;262–302, 1969

O'Leary SB: Poisoning in man from eating poisonous plants. *Arch Environ Health* 9;216–242, 1964

Olin BR: Herbal diuretics. In *The Lawrence Review of Natural Products,* Lippincott, St Louis, May 1989

Pammel LH: *Manual of Poisonous Plants.* Torch Press, Cedar Rapids, Iowa, 1911

Pammel LH: Horses and milkwort. *Vet Med* 24;514, 1929

Pergat FF, Stolyarova AG: Clinical and pathological changes in horses poisoned with peppery cuscus, *Cuscuta breviflora. Tr Uzbek Inst Vet* 14;239–247, 1961

Polonsky J, Lederer E: Note sur l'isolement de la dimethoxy-2,6 benzoquinone des ecorces et du bois quelques Simarubacees et Meliacees. *Bull Soc Chem Fr* 1959;1157–1158, 1959

Premakumari P, Rathinam K, Santhakumari G: Antifertility activity of plumbagin. *Indian J Med Res* 65;829–838, 1977

Pullar EM: Studies on five suspected poisonous plants. *Aust Vet J* 15;19–23, 1939

Ratnayake S, Fang XP, Anderson JE, McLaughlin JL, Evert DR: Bioactive constituents from the twigs of *Asimina parviflora. J Nat Prod* 55;1462–1467, 1992

Reece RL: Observations on the accidental poisoning of birds by organophosphate insecticides and other toxic substances. *Vet Rec* 111;453–455, 1982

Reti L: Ephedra bases. In *The Alkaloids: Chemistry and Physiology,* vol 3, Manske RHF, Holmes HL eds, Academic Press, New York, pp339–362, 1953

Rezende AML de, Dobereiner J, Tokarnia CH: Experimental poisoning of rabbits with *Polygala klotzchii* (Polygalaceae). *Pesq Vet Bras* 1;61–64, 1981

Richter WJ, Raschdorf F, Euw Jv, Reichstein T, Widen C-J: Field-desorption mass spectra of fern phloroglucinols containing three to six ring constituents. *Helv Chim Acta* 70;881–893, 1987

Rosen ES, Edgar JT, Smith JLS: Male fern retrobulbar neuropathy in cattle. *J Small Anim Pract* 10;619–625, 1970

Rupprecht JK, Chang CJ, Cassady JM, McLaughlin JL, Mikolajczak KL, Weisleder D: Asimicin, a new cytotoxic and pesticidal acetogenin from the pawpaw, *Asimina triloba. Heterocycles* 24;1197–1201, 1986

Rupprecht JK, Hui Y-H, McLaughlin JL: Annonaceous acetogenins: a review. *J Nat Prod* 53;237–278, 1990

Schofield FW: Acute pulmonary emphysema of cattle. *J Am Vet Med Assoc* 112;254–259, 1948

Seigler DS: Plants of the Northeastern United States that produce cyanogenic compounds. *Econ Bot* 30;395–407, 1976

Sethi N, Nath D, Shukla SC, Dyal R: Abortifacient activity of a medicinal plant "*Moringa oleifera*" in rats. *Ancient Sci Life* 7;172–174, 1988

Shropshire CM, Stauber E, Arai M: Evaluation of selected plants for acute toxicosis in budgerigars. *J Am Vet Med Assoc* 200;936–939, 1992

Shukla S, Mathur R, Prakesh AO: Antifertility profile of the aqueous extract of *Moringa oleifera* roots. *J Ethnopharmacol* 22;51–62, 1988

Snyder DS, Hatfield GM, Lampe KF: Examination of the itch response from the raphides of the fishtail palm *Caryota mitis. Toxicol Appl Pharmacol* 48;287–292, 1979

Solomon FE, Sharada AC, Devi PU: Toxic effects of crude root extract of *Plumbago rosea* (Rakta chitraka) on mice and rats. *J Ethnopharmacol* 38;79–84, 1993

Southon IW, Buckingham J: *Dictionary of Alkaloids.* Chapman & Hall, London, 1989
Stevenson DM: Ephedraceae. In *Flora of North America,* vol 2, *Pteridophytes and Gymnosperms,* Flora of North America Editorial Committee eds, Oxford Univ Press, New York, pp428–434, 1993
Terry RE: A study of *Ephedra nevadensis. J Am Pharm Assoc* 16;397–407, 1927
Thomson RH: *Naturally Occurring Quinones,* 2d ed. Academic Press, London, 1971
Tokarnia CH, Dobereiner J: Experimental poisoning of cattle with *Plumbago scandens* (Plumbaginaceae). *Pesq Vet Bras* 2;105–112, 1982
Tomita M, Kosuka M: Alkaloids of *Asimina triloba. Yakugaku-Zasshi* 85;77–82, 1965 (*Chem Abstr* 62;11863d, 1965)
Tomko J, Voticky Z: Steroid alkaloids: the *Veratrum* and *Buxus* groups. In *The Alkaloids: Chemistry and Physiology,* vol 14, Manske RHF ed, Academic Press, New York, pp1–82, 1973
Trim AR, Hill R: The preparation and properties of aucubin, asperuloside, and some related glycosides. *Biochem J* 50;310–318, 1952
Tucker JM, Cordy DR, Berry LJ, Harvey WA, Fuller TC: *Nitrate Poisoning in Livestock.* Calif Agric Exp Stn Circ 506, 1961
van der Lugt JJ, Fourie N, Anitra Schultz R: *Galenia africana* L. (Aizoaceae) poisoning in sheep. *J S Afr Vet Assoc* 59;100, 1988
van Soest H, Gotink WM, van der Vooren LJ: Poisoning in pigs and cows by box-tree leaves *(Buxus sempervirens). Tijdschr Diergeneesk* 90;387–389, 1965
Waller EF, Prince FS, Hodgdon AR, Colovos NF: *Sensitive-fern Poisoning of Horses.* N H Agric Exp Stn Tech Bull 83, 1944
Watt JM, Breyer-Brandwijk MG: *The Medicinal and Poisonous Plants of Southern and Eastern Africa,* 2d ed. Livingston, Edinburgh, 1962
Webber JJ, Roycroft CR, Callinan JD: Cyanide poisoning of goats from sugar gums *(Eucalyptus cladocalyx). Aust Vet J* 62;28, 1985
West E, Emmel MW: *Poisonous Plants in Florida.* Fla Agric Exp Stn Bull 510, 1952
Widen C-J, Pyysalo H, Salovaara P: Separation of naturally occurring acylphloroglucinols by high-performance liquid chromatography. *J Chromatogr* 188;213–220, 1980
Willaman JJ, Schubert BG: *Alkaloid Bearing Plants and Their Contained Alkaloids.* USDA Tech Bull 1234, 1961
Williams MC: Toxicological investigations on *Galenia pubescens. Weed Sci* 27;506–508, 1979
Yoshikawa M, Ueda T, Matsuda H, Yamahara J, Murakami N: Absolute stereostructures of hydramacrosides A and B, new bioactive secoiridoid glucoside complexes from the leaves of *Hydrangea macrophylla* Seringe var. *thunbergii* Makino. *Chem Pharm Bull* 42;1691–1693, 1994
Zhao G, Hui Y, Rupprecht JK, McLaughlin, Wood KV: Additional bioactive compounds and trilobacin, a novel highly cytotoxic acetogenin from the bark of *Asimina triloba. J Nat Prod* 55;347–356, 1992
Zhao G-X, Miesbauer LR, Smith DL, McLaughlin JL: Asimin, asiminacin, and asiminecin: novel highly cytotoxic asimicin isomers from *Asimina triloba. J Med Chem* 37;1971–1976, 1994
Zhao G-X, Chao J-F, McLaughlin JL: (2,4-*cis*)-Asimicinone and (2,4-*trans*)-asimicinone: two novel bioactive ketolactone acetogenins from *Asimina triloba* (Annonaceae). *Nat Toxins* 4;128–134, 1996
Zomlefer WB: *Guide to Flowering Plant Families.* Univ North Carolina Press, Chapel Hill, 1994

78 Identification of Toxic Plants

When plants are suspected to be the cause of disease or death in animals or humans, it is imperative that they be quickly and correctly identified. The easiest way is disarmingly simple—ask someone who knows. This approach works well when an expert is near at hand, ready to name the unknown plants. A second approach is to compare the unknown plant with photographs or illustrations in field guides specific for the locality. The major drawbacks in using these popularized publications are that they typically illustrate only the showy-flowered species and may not include all of the species of a genus present in the area. The ideal way to identify an unknown plant is to use a taxonomic key. Taxonomists have been writing and using them for centuries as they have inventoried the world's flora. Such keys appear in many of the genus treatments in this book.

Floristic Treatments—Taxonomic keys for identification of plant families, genera, and species typically are incorporated in floristic treatments known as floras or manuals. These works are designed to facilitate identification of the plants in an area and generally comprise the keys, descriptions of the morphology of each taxon, and abbreviated comments about each taxon's distribution, ecology, flowering time, and taxonomic relationships.

When attempting to identify an unknown, possibly toxic plant, one should use, whenever possible, the keys appearing in a modern treatment written specifically for the area or state. Representatives of such works are presented in table 78.1. If a local manual is not available, a regional or continental treatment can be used, such as those listed in table 78.2. Recommendations about the best manual(s) to use in an area can be obtained from taxonomists working at area universities.

Table 78.1. Representative taxonomic treatments for states and small regions

Diggs GM, O'Kennon RJ, Lipscomb BL: *Shinners and Mahler's Illustrated Flora of North Central Texas*. Botanical Research Institute of Texas, Fort Worth, 1999

Hickman JC ed: *The Jepson Manual: Higher Plants of California*. Univ California Press, Berkeley, 1993

Richardson A: *Plants of the Rio Grande Delta,* rev ed. Univ Texas Press, Austin, 1995

Smith EB: *Keys to the Flora of Arkansas*. Univ Arkansas Press, Fayetteville, 1994

Yatskievych G: *Steyermark's Flora of Missouri,* vol 1. Missouri Department of Conservation and Missouri Botanical Garden Press, St Louis, 1998

Use of a Key—Although a taxonomic key looks intimidating at first, its use is quite easy. For individuals who have not used one before, the following comments are offered.

A key simply presents the user with a progressive series of choices between pairs of alternative features. For example, following examination of an unknown plant, the user might be required to choose between *plants trailing or climbing vines* versus

Table 78.2. Representative taxonomic treatments of regional or continental scope

Cronquist A, Holmgren AH, Holmgren NH, Reveal JL, Holmgren PK: *Intermountain Flora: Vascular Plants of the Intermountain West, USA,* vols 3–6. Columbia Univ Press, New York, 1977–1997
Flora of North America Editorial Committee eds: *Flora of North America North of Mexico,* vols 2, 3, 22. Oxford Univ Press, New York, 1993–1999
Gleason HA, Cronquist A: *Manual of Vascular Plants of Northeastern United States and Adjacent Canada,* 2d ed. New York Botanical Garden, Bronx, 1991
Great Plains Flora Association: *Flora of the Great Plains.* Univ Kansas Press, Lawrence, 1986
Hitchcock CL, Cronquist A: *Flora of the Pacific Northwest: An Illustrated Manual.* Univ Washington Press, Seattle, 1973

plants erect or prostrate, but not trailing or climbing vines. Selection of the most applicable alternative leads the user to other pairs of alternatives and ultimately to a scientific name. Using a key is thus analogous to following a path along which each junction is a Y. To reach the proper destination, the traveler must take the correct fork in the road at each junction.

The pair of alternative features is referred to as a couplet, and the alternatives of a single couplet are called leads. To facilitate use of the key, the couplets typically are successively indented to the right, with both leads of a single couplet equally indented and numbered. A person commences keying at couplet 1 by reading both leads, observing the plant's features, and making a decision as to which lead applies. After one of the two leads has been selected, the user proceeds to the first indented couplet immediately under the selected lead. The couplets under the nonselected lead are disregarded. Successive indented couplets are followed until a name is reached. For example, a key to five families might be as follows:

1. Leaf venation palmate or dichotomous.
    2. Leaves flabellate. Venation dichotomous. Flowers absent. Pollen cones present. Ovules exposed; borne at broadened apices of branched peduncles . . . . GINKGOACEAE
    2. Leaves not flabellate. Venation palmate. Flowers present. Pollen cones absent. Ovules enclosed; borne in ovaries. . . . . ROSACEAE
1. Leaf venation pinnate.
    3. Flowers perfect. Fruits capsules . . . . ALOACEAE
    3. Flowers imperfect. Fruits nuts or achenes.
        4. Plants monoecious. Pistillate flowers solitary or in clusters of 2 or 3. Fruits nuts; subtended individually or in clusters of 3 by spiny or muricate or caplike involucral cupule . . . . FAGACEAE
        4. Plants dioecious. Pistillate flowers in heads. Fruits achenes; not subtended by cupule . . . . ASTERACEAE

Sometimes the leads of a single couplet will be separated by numerous other couplets; the numbers thus help in locating the contrasting leads. The user should always read both leads of a couplet and observe the plant carefully before making a decision. Although the first choice may be applicable, the second may be better. In the key given above, note that the different characters in the leads are separated by periods, whereas semicolons generally are used to separate aspects of one character.

Should a couplet be encountered for which the selection of a lead is tenuous, the user should proceed both ways. Subsequent couplets posing inapplicable alternatives generally will indicate which lead is the correct one.

When a taxon's name—family, genus, or species—is reached in the key, the user should compare the taxon's morphological description with the features of the unknown plant. If they match, identification is accomplished. If they don't match, the user should reexamine the features of the unknown plant and begin keying again, being sure to read carefully both leads of each couplet before selecting one of the two.

Diagnostic Characters of Families—As one becomes familiar with a variety of species, it becomes possible to recognize plant families and immediately relate unknown plants to them. Identification of unknown plants via keying is accomplished much faster if one can begin keying at the family level.

Many families are quite distinctive and can be recognized by two or three easily observable characters. For example, members of the Asteraceae, or sunflower family, have flowers borne on a common receptacle in involucrate heads and produce achenes, whereas members of the Lamiaceae, or mint family, have square stems; simple, opposite, aromatic leaves; 5-merous, bilabiate flowers with gynobasic styles and deeply 4-lobed ovaries that give rise to 4 nutlets.

In the following paragraphs, the diagnostic characters of 24 families commonly encountered are presented. They are presented in the hope that if the users of this book are attempting to identify the family of an unknown plant suspected of being toxic, a quick comparison of the plant's features with the synopses will aid identification. Because not all families are summarized and there are morphologically unusual genera in each, the synopses must be used with caution. They are intended only to be a quick guide and merely a supplement to the taxonomic manuals available for the user's area.

Aceraceae—shrubs or trees; simple, opposite, palmately lobed, palmately veined leaves; radially symmetrical flowers; fruits 2-winged schizocarps (double samaras). (See illustration of *Acer rubrum* on page 8.)

Apiaceae—aromatic herbs with hollow stems; compound, alternate leaves with sheathing petiole bases; 5-merous, 2-carpellate, epigynous flowers in umbels; fruits schizocarps. (See illustration of *Conium maculatum* on page 54; cf. Araliaceae, chapter 11.)

Araceae—herbs or vines; simple or compound, alternate, typically fleshy leaves; flowers in spadices on sheathing petiole bases subtended by conspicuous, often brightly colored spathes. (See illustration of *Spathiphyllum* sp. on page 106.)

Asclepiadaceae—herbs or vines with milky sap; simple, opposite or whorled leaves; 5-merous, radially symmetrical flowers with petaloid coronas, anthers fused to styles to form a central column; fruits follicles with numerous, comose seeds. (See illustration of *Asclepias* flower on page 126; cf. Apocynaceae, chapter 8.)

Asteraceae—herbs or shrubs; 5-merous, epigynous flowers borne on a common receptacle in involucrate heads; fruits achenes. (See illustration of composite head on page 148.)

Boraginaceae—bristly herbs with round stems; simple, alternate leaves; radially symmetrical, 5-merous flowers in curled, 1-sided inflorescences that appear to unroll from the ends (helicoid cymes); fruits nutlets in clusters of 4. (See illustration of *Amsinckia intermedia* on page 265.)

Brassicaceae—herbs with acrid sap; simple, alternate leaves; racemes of radially symmetrical flowers with 4 free sepals and petals (as a cross), 4 long and 2 short stamens, and a 2-carpellate pistil with superior ovary; fruits siliques or silicles. (See illustration of *Raphanus sativus* on page 301.)

Caryophyllaceae—herbs typically with swollen nodes; simple, opposite leaves connected at bases with a transverse line; 4- or 5-merous, radially symmetrical flowers with perianth parts in 1- or 2-series and superior 2-to-5-carpellate ovaries with free-central or basal placentation; fruits capsules. (See illustration of *Saponaria officinalis* on page 331.)

Convolvulaceae—climbing or trailing vines often with milky sap; simple or palmately compound, alternate leaves; 5-merous, radially symmetrical flowers with tubular, pleated corollas, epipetalous stamens, and 2-carpellate pistils; fruits capsules. (See illustration of *Ipomoea hederacea* on page 375.)

Cucurbitaceae—tendril-bearing, trailing or climbing vines; simple or palmately compound, alternate leaves; imperfect, 5-merous, radially symmetrical, epigynous flowers; fruits pepos. (See illustration of *Cucurbita foetidissima* on page 394.)

Cupressaceae—evergreen, coniferous shrubs or trees; 2- or 4-ranked, scalelike leaves; pollen cones and fleshy berrylike seed cones. (See illustration of *Juniperus* sp. on page 402.)

Equisetaceae—rhizomatous herbs with jointed, longitudinally ribbed, hollow stems; whorled, scalelike leaves fused to form toothed sheaths at the nodes; sporangia in strobili at stem tips. (See illustration of *Equisetum arvense* on page 435.)

Ericaceae—shrubs or small trees; simple, alternate, usually evergreen leaves that bear urn-, pitcher- or bell-shaped flowers with anthers opening by terminal pores. (See illustration of *Kalmia latifolia* on page 446.)

Euphorbiaceae—monoecious or dioecious herbs, shrubs or trees often with milky sap; simple leaves; imperfect, radially symmetrical flowers with 3-carpellate ovaries; fruits 3-seeded capsular schizocarps. (See illustration of *Euphorbia esula* on page 468.)

Fabaceae—herbs, shrubs, or trees; compound, alternate, stipulate leaves; 5-merous, radially or bilaterally symmetrical (papilionaceous) flowers with 10 or 5 stamens and a simple pistil; fruits legumes. (See illustration of *Crotalaria spectabilis* on page 545.)

Fagaceae—monoecious shrubs or trees; simple, alternate leaves; staminate flowers in pendulous catkins and pistillate flowers singly or in clusters of 2–4; fruits nuts partially subtended by an involucral cap (acorn) or enclosed in spiny or prickly, dehiscent husks. (See illustration of *Quercus marilandica* on page 688.)

Hippocastanaceae—shrubs or trees; palmately compound, opposite leaves; panicles of 5-merous, bilaterally symmetrical flowers. (See illustration of *Aesculus glabra* on page 715.)

Lamiaceae—aromatic herbs or shrubs with square stems; simple, opposite leaves; 5-merous, 2-lipped flowers with deeply 4-lobed ovary and gynobasic style; fruits 4 nutlets. (See illustration of *Salvia reflexa* on page 739; cf. Verbenaceae, chapter 73, and Scrophulariaceae, chapter 68.)

Liliaceae—perennial herbs; simple leaves; showy, 3-merous, radially symmetrical flowers with petaloid perianth parts and 6 stamens. (See illustration of *Lilium pardalinum* on page 772.)

Malvaceae—herbs or shrubs with stellate pubescence; simple, alternate, palmately veined leaves; 5-merous, radially symmetrical flowers with numerous stamens fused by filaments into a tube and 5-to-many-carpellate pistils. (See illustration of *Malva neglecta* on page 822.)

Poaceae—herbs; linear, 2-ranked leaves with ligules and overlapping sheath margins; flowers borne in spikelets. (See illustration of *Festuca arundinacea* on page 896; cf. Cyperaceae and Juncaceae in chapter 77.)

Ranunculaceae—herbs; leaves with sheathing bases and compound or divided blades; typically radially symmetrical flowers with numerous, spirally arranged stamens and pistils. (See illustration of *Ranunculus repens* on page 1021.)

Rosaceae—herbs, shrubs, or trees; simple or compound, alternate, stipulate leaves; 5-merous, radially symmetrical flowers with numerous stamens and free petals arising from the rim of a hypanthium. (See illustration of *Prunus virginiana* on page 1064.)

Scrophulariaceae—herbs; 5-merous, bilaterally symmetrical, frequently 2-lipped flowers bearing 2 or 4 stamens; fruits many-seeded capsules. (See illustration of *Digitalis purpurea* on page 1093; cf. Lamiaceae, chapter 45, and Solanaceae, chapter 69.)

Plant Taxa Listed by Their Principal Adverse Effects

Anemia – Hemolytic

Acer
Allium
Brassica

Blindness

Brassica
Cheilanthes
Dryopteris
Equisetum
Hemerocallis
Pteridium
Zanthoxylum

Cystitis

Cheilanthes
Pteridium
Sorghum

Globus Pallidus/Substantia Nigra Necrosis

Centaurea

Hair Loss – Severe

Crotalaria
Leucaena
Narcissus

Hallucinations

Brugmansia
Datura
Hyoscyamus
Ipomoea
Solandra
Sophora

Head Pressing/Pacing/Aimless Walking

Boraginaceae
Centaurea
Crotalaria
Erechtites
Indigofera
Kochia
Lupinus
Senecio
Trifolium

Hemorrhage

Anthoxanthum
Cheilanthes
Melilotus
Pteridium

Hepatic – Crystalloid Cholangiohepatopathy

Agave
Nolina
Panicum
Tribulus

Hepatic – Necrosis/Fibrosis

Boraginaceae
Ceratozamia
Cestrum
Crotalaria
Cycas
Dioon
Erechtites
Eupatorium
Gossypium
Lantana
Myoporum
Phyllanthus
Psathyrotes
Reverchonia
Senecio
Tetradymia
Xanthium
Zamia

Lameness/Laminitis/Limb Edema

Berteroa
Cestrum
Juglans
Lathyrus
Robinia
Setaria

Lip or Jaw Movement/Eating/Drinking – Abnormal

Astragalus
Centaurea
Cheilanthes
Descurainia
Onoclea
Oxytropis
Prosopis
Pteridium
Solidago
Zanthoxylum

Mastitis

Persea

Mouth Dry/Pupillary Dilation

Brugmansia
Datura
Hyoscyamus
Solandra

Rectal Straining/Tenesmus

Boraginaceae
Castilleja
Crotalaria
Erechtites
Senecio

Renal

Amaranthus
Anagallis

Chenopodiaceae
Lilium
Oxalis
Quercus
Rheum
Rumex

Reproductive – Abortion/Weak Neonate/Stillbirths

Astragalus
Cupressus
Gutierrezia
Iva
Juniperus
Pinus

Reproductive – Decreased Fertility

Linum
Medicago
Trifolium

Reproductive – Teratogenesis

Astragalus
Baptisia
Conium
Lupinus
Nicotiana
Sophora
Veratrum

Respiratory Difficulty – Severe

Astragalus
Brassica
Galega
Ipomoea
Medicago
Perilla
Poaceae
Scirpus
Sphenosciadium
Verbesina

Salivation – Profuse

Amianthium
Ericaceae
Veratrum
Zigadenus

Stringhalt-like Symptoms

Hypochaeris
Lathyrus

Vomiting/Retching – Persistent

Blighia
Colchicum
Ericaceae

Plants of Concern for Dogs, Cats, and Other Pets

Plants of Concern for Dogs, Cats, and Other Pets

Genus	Common Name	System Affected
Abrus	rosary pea	digestive
Allium	onion	blood
*Aloe**	aloe	digestive
*Amaryllis**	belladonna lily, naked lady	digestive
*Brugmansia**	angel's-trumpet	nervous, ocular
Brunfelsia	yesterday-today-tomorrow	nervous
Cannabis	marijuana, hemp	nervous
*Cycas**	sago palm	hepatic
*Cyclamen**	cyclamen, sowbread	digestive
Daphne	mezerum, spurge laurel	digestive
Datura	jimsonweed	nervous, ocular
Digitalis	foxglove	digestive, cardiac
*Dieffenbachia**	dumb cane	digestive
*Epipremnum**	ivy arum, devil's ivy, pothos	digestive
Erythrina	coral bean	neuromuscular
Gladiolus	gladiolus	digestive
*Gloriosa**	glory lily	digestive
Hedera	English ivy	digestive
Hyacinthus	hyacinth	digestive
Ipomoea	morning glory	nervous
Iris	iris, flag	digestive
*Kalanchoe**	mother-of-millions	digestive, cardiac
Kalmia	mountain laurel	digestive, cardiac
Laburnum	golden chain tree	nervous
*Lilium**	lily	renal
*Narcissus**	daffodil	digestive
Nerium	oleander	digestive, cardiac
Nicotiana	tobacco	neuromuscular
*Philodendron**	philodendron	digestive, renal
Pieris	lily-of-the-valley bush, pieris	digestive, cardiac
*Rhododendron**	rhododendron	digestive, cardiac
Ricinus	castor bean	digestive
Setaria	bristlegrass	mechanical injury
*Solandra**	chalice vine, cup of gold	nervous, ocular
Taxus	yew	cardiac
Thevetia	yellow oleander	digestive, cardiac
*Tulipa**	tulip	digestive
Urginea	squill	digestive, cardiac
Urtica	American stinging nettle	cardiovascular
Zamia	coontie, Seminole bread	hepatic
Zantedeschia	calla lily	digestive

*May be present as a houseplant.

Glossary

Botanical Terms

abaxial—(1) with side of lateral organ away from axis; (2) pertaining to underside of leaf or outside of petal.
abortion—arrested development of an organ.
acaulescent—stemless or apparently stemless; cf. *caulescent.*
accrescent—increasing in size with age; e.g., as a calyx after flowering.
achene—small, dry, indehiscent, 1-seeded fruit with ovary wall free from the seed except at funiculus.
achlorophyllous—without chlorophyll.
acicular—needle-shaped.
acorn—fruit comprising nut and basal cupule of fused bracts (cap); characteristic of *Quercus* (oak).
acuminate—tapering to a sharp point with concave sides.
acute—tapering to pointed apex, sides more or less straight.
adaxial—(1) with side of lateral organ next to axis; (2) pertaining to upper side of leaf or inner side of petal.
adherent—different organs, usually separate, sticking together but not fused; e.g., as anthers to the style.
adnate—with unlike parts fused.
aerial—pertaining to portion of plant above ground or water surface.
aggregate—collected together.
alternate—(1) arrangement of organs such as leaves, branches, flowers, or fruits 1 per node; (2) arrangement of organs in 1 floral whorl between organs of adjacent floral whorl; e.g., petals alternate with sepals.
alveolate—pertaining to surface with pits separated by partitions; honeycombed.
ament—synonym of *catkin.*
amplexicaul—clasping or embracing stem; e.g., leaf base or stipules; synonym of *clasping.*
anastomosis—union of 1 vein with another to form a net.
androecium (pl. *androecia*)—collective term for stamens of a flower.
androgynophore—stalk supporting androecium and gynoecium.
andromonoecious—having both staminate and perfect flowers on same plant.
annual—plant completing its life cycle in 1 growing season.
annular—arrangement of organs in circle or ring.
annulus (pl. *annuli*)—(1) ring of elastic cells in sporangia of ferns that functions in dehiscence; (2) incompletely developed sheath below strobilus in *Equisetum* (horsetail); (3) corona or rim of corolla in *Asclepias* (milkweed).
anther—pollen-bearing portion of stamen.
antheridium (pl. *antheridia*)—in lower plants, multicellular organ producing male gametes.
anthesis—act of flowering; used to designate flowering period.
antrorse—directed upward or forward or toward apex; cf. *retrorse.*
apex (pl. *apices*)—morphological tip of an organ.
apiculate—ending abruptly in minute, pointed tip.
appendage—secondary part attached to main structure.
appressed—lying flat against an organ.
aquatic—growing naturally in water; either completely or partially submerged.
aril—fleshy appendage of funiculus or hilum, partially or entirely enveloping the seed.
aristate—with awn or bristle at tip.
armed—bearing thorns or spines or prickles or sharp projections.
articulate—jointed or appearing so, with joints sometimes separating at maturity.
ascending—rising or extending upward at an oblique angle.
asymmetrical—(1) not divisible into mirror-image halves in any plane; (2) irregular or unequal in shape or outline; cf. *bilateral* and *radial.*
attenuate—tapering gradually to narrow tip.
auricle—earlike lobe or appendage, typically at base of organ.
autophyte—plant capable of photosynthesis.
awl-shaped—narrowly wedge-shaped and tapering to sharp point.
awn—bristlelike appendage extending beyond body of organ, usually at tip or from dorsal surface.
axil—upper angle formed between axis and any organ arising from it; e.g., between leaf and stem.

axile—positioned on or pertaining to the axis.
axillary—borne in axil of leaf.
axis—principal or central line of development of plant or organ; main stem.
banner—in Fabaceae (legumes), uppermost (adaxial) petal in papilionaceous corolla; synonym of *standard.*
barb—point projecting backward, like barb of fishhook.
barbellate—finely barbed.
bark—tissues of woody plant stem or root external to vascular cambium.
basal—attached or situated at base of plant or organ.
basifixed—attached by base; e.g., anther attached to filament at base rather than side; cf. *dorsifixed* and *versatile.*
beak—long, pointed, terminal projection of fruit or seed.
berry—fleshy, indehiscent fruit derived from 1 pistil and having a pulpy or fleshy pericarp; e.g., *Vitis* (grape) or *Lycopersicon* (tomato).
bi- —prefix meaning two, twice, or double.
biennial—plant completing its life cycle in 2 seasons, generally with vegetative growth the first season and reproductive growth the second.
bifid—tip deeply 2-cleft or forked to form 2 lobes.
bilabiate—2-lipped.
bilateral—divisible into mirror-image halves in only 1 plane; cf. *asymmetrical* and *radial.*
biserrate—with marginal teeth that are themselves toothed.
bladder—membranous sac filled with air or liquid.
blade—thin, expanded portion of leaf or petal; synonym of *lamina.*
bloom—white or waxy covering of some fruits, leaves, and stems.
bract—modified leaf subtending flower or an inflorescence.
bracteole—(1) small bract; (2) small bract borne on pedicel and not fully subtending flower.
branch—lateral division of stem or floral axis; axis of growth.
branchlet—ultimate divisions of branch; small branch.
bristle—stiff hair.
bud—undeveloped, terminal or axillary organ or shoot; meristem typically enclosed in bud scales.
bulb—short, thick, subterranean stem bearing fleshy leaves; e.g., *Allium* (onion).
bur—fruit or fruiting inflorescence with a rough or prickly covering derived from pericarp, persistent calyx, or involucre.
bush—low, dense shrub profusely branched at ground level.
caducous—falling off early; e.g., sepals in some plants; cf. *persistent.*
calyx (pl. *calyces*)—collective term for sepals.
campanulate—bell-shaped.
capillary—very slender and hairlike.
capitate—headlike or in a head-shaped cluster.
capsule—dry, dehiscent fruit derived from a compound pistil and splitting along 2 or more lines of suture.
carpel—megasporophyll bearing ovules on inner surface; characteristic of Magnoliophyta (flowering plants)
cartilaginous—tough and firm but elastic and flexible; like cartilage.
caryopsis—in Poaceae (grasses), dry, indehiscent, 1-seeded fruit with seed coat fused to pericarp; synonym of *grain.*
catkin—spicate or racemose inflorescence bearing apetalous, unisexual flowers, the entire inflorescence typically dropping as unit; synonym of *ament.*
caudate—with a tail-like, basal or terminal appendage.
caudex (pl. *caudices*)—short, vertical, typically woody base of herbaceous perennial plant.
caulescent—producing an aerial stem with distinct nodes and internodes; cf. *acaulescent.*
cauline—pertaining to or arising from aerial stem.
cespitose—growing in dense clumps or multistemmed tufts.
chaff—in Asteraceae (sunflowers), collective term for receptacular bracts.
chartaceous—having the texture of writing paper; usually not green.
chasmogamous—with pollination occurring while flower is open; cf. *cleistogamous.*
ciliate—with marginal fringe of hairs.
circinate—coiled from tip downward, with apex in center; e.g., young fern fronds or leaflets.
circumscissile—form of dehiscence in which capsule opens along a transverse line, allowing top to come off like a lid.
cladophyll—branch or stem with form and functions of leaf.
clasping—foliaceous structure partly or wholly surrounding the stem; synonym of *amplexicaul.*
clathrate—pierced with openings; latticelike in appearance.
clavate—(1) club-shaped; (2) gradually thickening toward apex.
claw—long, narrow, petiole-like base of petals or sepals in some flowers.
cleft—divided into lobes separated by narrow or acute sinuses that extend more than halfway to midrib.
cleistogamous—with self-pollination occurring while flower is closed; cf. *chasmogamous.*
coherent—sticking together but not fused.
collateral—situated side-by-side; e.g., axillary and accessory buds.
colonial—forming colonies, typically asexually via rhizomes, stolons, or root sprouts.
column—in Malvaceae (mallows) tubular structure formed by fusion of staminal filaments.

coma—tuft of hairs at end of structure, especially seeds.
compound—organ divided into 2 or more similar parts; e.g., compound leaves or compound pistils.
conduplicate—folded lengthwise usually into 2 equal, appressed halves with adaxial surface inside.
cone—(1) dense cluster of overlapping sporophylls on an axis; strobilus; (2) inflorescence or fruit with overlapping bracts.
confluent—with merger of one part and another; passing by degrees of one part into another.
conical—cone-shaped; tapering evenly from base to apex in 3 dimensions; e.g., *Daucus* (carrot).
connate—united; applied to similar structures when fused into a single structure.
connective—portion of stamen connecting pollen sacs of anther.
connivent—coming together or converging, but not fused.
conserved—pertaining to a scientific name whose continued use has been authorized by International Code of Botanical nomenclature, even though it is illegitimate according to the rules (*nominia conservanda*).
convex—rounded or curved outward on surface.
convolute—(1) longitudinally rolled or twisted together; (2) pertaining to leaf blades or sepals or petals in which edge of one overlaps that of next one; like shingles on a roof.
cordate—heart-shaped.
coriaceous—leathery or tough but smooth and pliable.
corm—short, solid, thickened underground stem with or without small, papery leaves; e.g., *Crocus* (crocus).
corolla—collective term for petals.
corona—crown or cuplike appendage or ring of appendages arising from petals or stamens in *Narcissus* (daffodil) or *Asclepias* (milkweed); synonym of *crown*.
coroniform—crown-shaped.
corymb—short and broad, more or less flat-topped indeterminate inflorescence in which outer flowers open first and lower pedicels are longer.
costa (pl. *costae*)—(1) prominent midvein or midrib; (2) in ferns, midrib of pinna or pinnule.
cotyledon—seed leaf; primary leaf or leaves in embryo.
creeping—running along ground and rooting at the nodes.
crenate—margin with shallow, rounded or blunt teeth.
crenulate—finely crenate.
crown—(1) base of tufted, herbaceous perennial where shoot and root systems merge; (2) overall appearance or form of branches of tree or shrub; (3) synonym of *corona*.
crozier—young, circinately coiled frond of most ferns.
cruciform—cross-shaped.
cucullate—hooded or hood-shaped.
culm—aerial stem of Poaceae (grasses), Cyperaceae (sedges), and Juncaceae (rushes).
cuneate—triangular or tapering to narrow end at point of attachment; wedge-shaped.
cupule—involucre composed of hardened bracts that are fused, at least at bases, partially or completely enclosing fruit or fruits.
cuspidate—abruptly tipped with sharp, rigid point.
cyathium (pl. *cyathia*)—modified inflorescence consisting of a pedicelled pistillate flower arising from the center of a cup-like involucre bearing 1 or more glands on its rim and staminate flowers with 1 stamen on its inner surface; characteristic only of *Euphorbia* (spurge) and a few related genera.
cymbiform—boat-shaped.
cyme—broad, more or less flat-topped determinate inflorescence, with central or terminal flowers opening first.
deciduous—(1) not persistent, falling off when no longer functional; e.g., leaves falling in autumn or floral parts after anthesis; (2) pertaining to a plant that drops all of its leaves annually.
decumbent—lying horizontally on ground, but with terminal portion ascending.
decurrent—extending downward as wing or ridge from point of insertion.
deflexed—bent or turned abruptly downward.
dehiscent—splitting open spontaneously along suture lines when mature to release contents; e.g., anthers or fruits; cf. *indehiscent*.
deliquescent—dissolving or melting away; e.g., petals of *Tradescantia* (spiderwort).
deltoid—triangular; shaped like the Greek letter delta.
dentate—pertaining to a margin with sharp teeth pointing outward, rather than forward.
denticulate—minutely or finely dentate.
diadelphous—stamens united by filaments into 2 groups.
dichotomous—branching by repeatedly forking into pairs.
didynamous—with 2 pairs of stamens of unequal filament length, as in members of the Lamiaceae (mints).
dimorphic—occurring in 2 forms.
dioecious—with pistillate and staminate flowers borne on different plants; cf. *monoecious*.
discoid—(1) flattened and circular; (2) disk-shaped; disciform.
disk—(1) more or less fleshy and/or raised outgrowth of receptacle; (2) flattened adhesive tip of tendril; (3) in Asteraceae (sunflowers), central portion of receptacle bearing disk florets.
dissected—deeply divided, or cut into segments.
distal—away from point of origin or attachment or center; cf. *proximal*.
divergent—spreading away from center or each other.
dorsal—pertaining to or located on back; (2) abaxial surface of organ; cf. *ventral*.

dorsifixed—pertaining to anther attached to filament along dorsal edge; cf. *basifixed* and *versatile.*

drupe—fleshy, 1-seeded, indehiscent fruit with seed enclosed in hard or papery endocarp called the pit or stone; e.g., *Prunus persica* (peach).

drupelet—small drupe that is part of aggregate fruit; e.g., *Rubus* (blackberry).

e- —prefix meaning part or parts missing.

ebracteate—without bracts.

echinate—armed with blunt prickles or spines.

elater—club-shaped, hygroscopic band attached to spores and functioning in their dispersal; e.g., *Equisetum* (horsetail).

ellipsoid—solid body elliptic in longitudinal section and circular in cross section.

elliptic—ellipse-shaped; broadest at middle and narrowed at rounded ends.

elongate—lengthened; stretched; extended.

emarginate—with shallow notch at apex.

embryo—immature plant within seed, comprising radicle, plumule, cotyledons, and possibly endosperm.

emergent—(1) growing from surface; (2) raised above and out of water.

endocarp—inner layer of fruit wall or pericarp that is differentiated into 3 layers.

endogenous—growing from, or originating from, within.

ensiform—sword-shaped; e.g., *Iris* (iris) leaves.

entire—pertaining to a continuous margin, without teeth or divisions.

ephemeral—lasting a very short time.

epi- —prefix meaning upon; e.g., epipetalous, or borne on petals.

epicalyx (pl. *epicalyces*)—set of bracts immediately below and resembling calyx.

epiphyte—plant that grows on other plants but is not parasitic; e.g., *Tillandsia* (Spanish moss).

equitant—with folded leaves overlapping lengthwise in 2 ranks; e.g., *Iris* (iris).

erect—upright or perpendicular to surface; vertically oriented.

erose—with margin irregularly toothed as if gnawed, torn, or eroded.

evergreen—(1) plant having foliage that remains green for more than 1 growing season and not falling en masse; (2) never lacking green leaves; (3) remaining green through winter.

excrescence—small, warty outgrowth.

exfoliating—coming off in thin layers or shreds or plates; e.g., flaking bark of *Platanus* (sycamore).

exocarp—typically thin, outermost layer of fruit wall or pericarp that is differentiated into 3 layers.

exserted—extending beyond or protruding from surrounding parts.

extrafloral—outside the flower.

extrorse—facing or opening outward, away from axis; e.g., dehiscence of anther; cf. *introrse.*

falcate—sickle-shaped; strongly curved sideways.

farinose—having a mealy, granular texture.

fascicle—tight cluster or bundle of organs, appearing to arise from a common point.

fertile—(1) capable of producing viable seeds; (2) capable of producing functional pollen except in Poaceae (grasses); (3) in Poaceae, pertaining to floret or spikelet producing a caryopsis.

fibrous—(1) having numerous, woody fibers, e.g., mesocarp of *Cocos* (coconut); (2) having all roots and their branches approximately the same diameter, as in Poaceae (grasses).

filament—(1) part of stamen that supports anther; (2) threadlike structure.

filiferous—with filaments or filament-like appendages.

filiform—filamentous, long and very slender.

fimbriate—margin with fringe of fine hairs or hairlike structures.

flabellate—fan-shaped; broadly wedge-shaped, sometimes plaited.

flaccid—weak and limp; flabby; lax.

fleshy—succulent; firm and juicy.

floating—resting on or at the surface of water.

floret—(1) in Poaceae (grasses), flower and 2 associated bracts, the lemma and palea; subunit of grass spikelet; (2) in Asteraceae (sunflowers), 1 flower of a composite head with 5 fused petals, a highly modified calyx, 5 stamens fused by their anthers, and an inferior ovary; 3 types recognized—disk, ray, and ligulate.

flower—determinate shoot axis bearing sepals, petals, stamens, and carpels in spiral or whorls; reproductive organ of flowering plants; site of sporogenesis, gametogenesis, and seed formation.

fluted—marked by alternating ridges and groovelike depressions.

foliaceous—leaflike; e.g., bracts and sepals that look like leaves in texture, size, and/or color.

foliage—collective term for leaves of plant.

follicle—dry, dehiscent fruit derived from simple pistil and dehiscing along 1 line of suture.

forb—(1) a broadleaf herbaceous plant (dicot); (2) any herbaceous plant other than members of the Poaceae (grasses), Cyperaceae (sedges), and Juncaceae (rushes); (3) any herbaceous plant other than a member of the Poaceae.

free—separate, not fused or attached to another organ.

fringed—with marginal hairs, bristles, or hairlike structures.
frond—(1) leaf of ferns; (2) large, compound leaf of palms.
fruit—mature ovary or ovaries, plus any other structures maturing with them after fertilization.
funiculus (pl. *funiculi*)—stalk or threadlike structure connecting ovule or seed to placenta.
funnelform—funnel-shaped; with corolla tube widening gradually upward toward apex; e.g., many flowers of *Convolvulus* (morning glory).
furrow—longitudinal channel or groove.
fused—parts or organs joined to one another.
fusiform—spindle-shaped; broadest near middle and tapering toward ends.
gametophyte—haploid (1*n*) gamete-producing generation of plants with a sporic reproductive cycle; in flowering plants, the embryo sac and pollen grain.
gibbous—(1) swollen on 1 side; (2) with pouchlike enlargement of base of organ.
glabrous—devoid of pubescence or hairs of any form.
gland—(1) appendage, protuberance, or depression on surface of organ or at end of hair that secretes usually sticky fluid; (2) protuberance that may not be secretory, e.g., warty swelling at base of leaf blade in *Prunus* (plum).
glaucous—covered with fine, waxy substance that may be rubbed off and that imparts a whitish or bluish cast to surface; e.g., "bloom" of *Vitis* (grape).
globose—spherical or nearly so; globular.
glomerule—compact, cymose cluster of flowers or heads subtended by involucre.
glume—in Poaceae (grasses), 1 of a pair of bracts at base of spikelet, typically dry and membranous.
glutinous—covered with sticky exudation.
gymnosperm—general term for a plant that bears naked seeds, usually in cones.
gynobase—enlargement or elongation of receptacle bearing the ovary.
gynoecium (pl. *gynoecia*)—collective term for pistil or pistils of flower.
gynomonoecious—bearing pistillate and perfect flowers on same plant.
habit—general appearance of plant; may refer to single character or collectively to all characters.
habitat—environmental conditions or type of area normally occupied by a plant taxon; e.g., borrow ditch, tallgrass prairie, or waste area.
hastate—arrowhead-shaped, with 2 basal lobes turned outward; cf. *sagittate.*
haustorium (pl. *haustoria*)—specialized, rootlike organ of parasitic plant that penetrates tissues of host plant and through which nutrients pass.
head—(1) dense spherical, hemispheric, or flat-topped inflorescence of sessile or subsessile flowers; (2) in Asteraceae (sunflowers), dense spherical, hemispheric, or flat-topped inflorescence of sessile flowers borne on a common receptacle and subtended by involucre.
helicoid—coiled in a spiral or helix, as in some l-sided cymose inflorescences.
herb—plant with nonwoody aerial stems that die at end of growing season.
herbaceous—(l) pertaining to a herb; (2) having the texture and/or color of a foliage leaf; (3) lacking woody tissues.
herbage—collective term for leaves and young branches of plant.
hesperidium (pl. *hesperidia*)—modified berry, with tough rind and septate juicy flesh; e.g., *Citrus* (lemon).
heterosporous—producing spores of 2 types; e.g., microspores and megaspores.
heterostyly—the state of having styles of different lengths in flowers of same species.
hip—berrylike, urceolate hypanthium enclosing numerous achenes; characteristic of *Rosa* (rose).
hirsute—with stiff, long, straight hairs.
hispid—with rigid, long, tapered, straight hairs; bristly.
hood—hollow, arched covering often enclosing other organs; e.g., corona of *Asclepias* (milkweed).
horn—tapering projection resembling the horn of a cow.
host—plant that provides nutrients to a parasite.
husk—covering, generally derived from perianth or involucre, that encloses fruit.
hyaline—thin and translucent or transparent.
hypanthium (pl. *hypanthia*)—cup-shaped to elongate tubular extension of floral axis that surrounds gynoecium, produced by fusion of basal parts of perianth and androecium or occasionally as expansion of receptacle.
imbricate—partial overlapping of organs so as to resemble shingles on a roof.
imperfect—pertaining to a flower that lacks either stamens or pistil or both.
incised—cut sharply, deeply, and usually irregularly.
included—not protruding beyond the surrounding organ or organs; included within; not exserted.
indehiscent—not splitting open to release contents when mature; cf. *dehiscent.*
indumentum—surface covering of hairs, scales, or scurf; pubescence.
indurate—hardened.
indusium (pl. *indusia*)—thin epidermal outgrowth that covers or surrounds sorus on frond of many ferns.
inferior—(1) pertaining to an organ lower than another; (2) pertaining to an ovary below apparent point of attachment of other floral organs; cf. *superior.*
inflexed—bent or turned toward main axis or center.

inflorescence—arrangement of flowers and accessory parts on an axis.
inserted—attached to or growing out of.
internode—portion of stem between successive nodes.
intrastaminal—within the androecium.
introrse—facing or opening inward toward axis; e.g., dehiscence of anther; cf. *extrorse.*
involucel—small, secondary involucre, as in secondary umbels of Apiaceae (carrots).
involucre—(1) collective term for a dense cluster of bracts subtending a flower or an inflorescence; (2) collective term for a single bract or a pair or whorl of bracts subtending a flower or an inflorescence.
involute—rolled inward from edges so that adaxial (upper) surface is partially concealed; cf. *revolute.*
jointed—with distinct nodes or points of actual or apparent separation.
keel—(1) prominent longitudinal ridge, analogous to keel of boat; (2) in Fabaceae (legumes), the 2, fused, abaxial (lower) petals in the papilionaceous corolla.
labium (pl. *labia*)—lower lip of a bilabiate flower.
lacerate—appearing irregularly cut or cleft, as if torn.
laciniate—deeply cut into narrow and unusually irregular lobes.
lamina—blade or flat expanded portion of leaf or petal.
lanate—with soft, long, dense, tangled and matted hairs; woolly.
lanceolate—lance-shaped; rather narrow, tapering to both ends with broadest part below middle.
lanose—synonym of *lanate.*
lanulose—diminutive of *lanose,* meaning minutely woolly.
lateral—on or at side of axis or organ.
leaf (pl. *leaves*)—lateral appendage arising from stem at node and typically subtending axillary bud; comprising blade, petiole, and stipules; and having primary functions of photosynthesis and transpiration.
leaflet—single segment of compound leaf.
legume—in Fabaceae (legumes), dry, dehiscent, 1-locular, 2-valved fruit derived from a simple pistil and opening along 2 sutures.
lemma—in Poaceae (grasses), outer and generally large bract of 2 subtending flowers; 1 of 2 bracts of a floret.
lenticel—elliptical, slightly raised, somewhat corky area or pore on surface of young bark.
lenticular—lens-shaped; elliptical, with both sides convex.
ligneous—woody.
ligule—(1) tongue- or strap-shaped organ; (2) in Asteraceae (sunflowers), strap-shaped limb of ray or ligulate floret; (3) in Poaceae (grasses), membranous or hairy appendage on adaxial surface of leaf at junction of sheath and blade.
limb—broadened or flattened part of organ extending from narrower base; e.g., the expanded portion of a fused corolla.
linear—long and narrow, with margins parallel or nearly so.
lip—1 of the 2 projections or segments of a bilabiate corolla or calyx.
lobe—segment or portion of organ separated from adjacent segment by sinuses or clefts, with division about halfway to middle or less.
locule—cavity or chamber in ovary, anther, or fruit.
loculicidal—dehiscing longitudinally and dorsally directly into locule of capsule; cf. *septicidal.*
lodicule—in Poaceae (grasses), small, hyaline appendage of tissue at base of ovary; modified sepal.
loment—flat legume constricted between seeds and disarticulating at constrictions into 1-seeded joints when mature.
longitudinal—extending along long axis of an organ.
lunate—crescent-shaped.
malodorous—with disagreeable odor.
mammillate—with nipplelike protuberances.
margin—edge or border of organ.
mealy—dry, powdery or granular covering; with consistency of fine meal.
megasporangium (pl. *megasporangia*)—multicellular structure producing megaspores; in flowering plants, the nucellus within the ovule.
megaspore—larger spore of heterosporous plants that gives rise to megagametophyte or embryo sac in flowering plants.
membranous—thin, pliable, and somewhat translucent; having a thin membrane.
mericarp—1-seeded, 1-carpellate segment of a schizocarp; characteristic of Apiaceae (carrots) and Geraniaceae (geraniums).
-merous—suffix denoting number of parts in each whorl of a flower; e.g., 3-merous, indicating sepals, petals, stamens, and carpels in 3s or multiples of 3.
mesocarp—middle layer of fruit wall or pericarp that is differentiated into 3 layers.
microsporangium (pl. *microsporangia*)—multicellular structure producing microspores; in flowering plants, the tapetum of the anther.
microspore—smaller spore of heterosporous plants, giving rise to microgametophyte or pollen grain in flowering plants.
microsporophyll—modified leaf bearing microsporangia; a stamen in flowering plants.
monadelphous—with stamens united by filaments or anthers into 1 group, usually forming tube or column.
monoecious—with pistillate and staminate flowers borne on same plant; cf. *dioecious.*

monotypic—(1) pertaining to a genus with a single species; (2) having only 1 component.
mucilage—substance that becomes slimy or gelatinous when wet.
mucro—short, sharp, abrupt point, usually at tip of leaf or other organ.
mucronulate—tipped with very small mucro.
multiple—(1) having several, or more than 1; (2) pertaining to fruit produced by ripened ovaries of several to many flowers, e.g., *Morus* (mulberry).
muricate—having a rough surface with short, hard points.
mycotrophic—forming symbiotic relationship with fungus, usually mycorrhizal.
necrotic—having dead cells or tissues.
nectar—secretion from glands or nectaries in flowers and certain leaves.
nectary—organ or tissue in which nectar is secreted.
needle—stiff, linear, general evergreen leaf, characteristic of conifers; e.g., *Pinus* (pine).
nerve—prominent longitudinal vein or slender rib.
net-veined—having anastomosing veins that form an irregular network; reticulate.
neuter—flower that lacks stamens and pistils; cf. *neutral.*
neutral—pertaining to a flower that lacks functional stamens and pistils—organs may be present but sterile and do not produce spores and gametophytes; cf. *neuter.*
node—point on stem where leaves, branches, or flowers originate.
nut—dry indehiscent, usually 1-loculed, 1-seeded fruit derived from compound ovary with bony, woody, leathery, or papery pericarp and partially or wholly enclosed in involucre.
nutlet—(1) dry, indehiscent, 1-seeded fruit derived from half of carpel; characteristic of Lamiaceae (mints), Boraginaceae (borages), and Verbenaceae (vervains); (2) small nut.
ob- —prefix of adjectives that indicates inversion of shape; e.g., *obcordate* describes a cordate shape, but with sinus at apex rather than base.
oblique—slanted with unequal sides.
oblong—rectangular but with rounded ends; 2-3 times longer than broad and with sides nearly parallel.
obsolete—organ or structure so reduced as to be absent or barely detectable; cf. *rudiment* and *vestigial.*
obtuse—blunt or rounded at end, with sides forming an angle of more than 90 degrees.
ocrea—sheath around stem formed by fusion of 2 stipules; characteristic of Polygonaceae (knotweeds).
opposite—(1) arrangement of organs such as leaves, branches, flowers, or fruits 2 per node; (2) arrangement of organs in 1 floral whorl superposed with those of adjacent floral whorl, e.g., stamens opposite the petals.
orbicular—essentially circular in outline; applied to flat organs.
oval—broadly elliptic, the width more than half the length.
ovary—basal portion of pistil containing ovules and maturing into fruit after fertilization.
ovate—egg-shaped in 2 dimensions and attached at broader end; applied to plane surfaces.
ovoid—egg-shaped in 3 dimensions; applied to solids.
ovulate—pertaining to ovule, or possessing ovules.
ovule—multicellular structure containing megagametophyte with egg enclosed within nucellus and 1 or 2 integuments; becomes seed after fertilization.
pale—in Asteraceae (sunflowers), membranous or hyaline scale or bristle arising from receptacle and subtending disk floret; synonym of *receptacular bract.*
palea—in Poaceae (grasses), inner and generally smaller bract of 2 that subtend flower; 1 of 2 bracts of floret.
palmate—with 2 or more lobes or segments radiating from the same basal point; resembling fingers spreading from palm of hand.
palmatifid—cut in palmate fashion nearly to petioles.
paludal—pertaining to a marsh; swampy.
pan- —prefix meaning whole or all; e.g., pantropical distribution.
panicle—indeterminate, branched inflorescence comprising rachis, 1 to several series of branches, and pedicels bearing flowers.
papilionaceous—in Fabaceae (legumes), corolla shape comprising 5 petals differentiated into a standard (banner), wings, and keel; butterfly-like.
papilla (pl. *papillae*)—small, rounded projection or excrescence.
pappus—in Asteraceae (sunflowers), highly modified calyx of florets, usually in form of capillary bristles, plumose bristles, scales, awns, or short crown.
parallel—pertaining to veins extended in same direction, but equally distant at every point.
parallel-convergent—pertaining to parallel veins extending from base to apex of leaf in curved manner.
parasite—plant deriving nutrients from another plant, the host, to which it is attached.
pedicel—(1) stalk bearing an individual flower, or the ultimate axis of compound inflorescence; (2) in Poaceae (grasses), stalk bearing a spikelet.
peduncle—primary axis or stalk terminating in inflorescence; axis from node to lowest flower or branch of inflorescence.
peltate—shield-shaped; pertaining to an organ or a structure with supporting stalk attached inside margin rather than at edge; e.g., leaf blade with petiole attached in center.

pendulous—hanging or drooping downward.
pepo—modified berry derived from inferior ovary, with 1 locule, many seeds, and tough rind; characteristic of Cucurbitaceae (gourds).
perennial—plant whose life cycle takes 3 to many growing seasons to complete.
perfect—pertaining to flowers with both stamens and pistils present and functional.
perfoliate—pertaining to sessile leaf with basal portion fused around stem, so that it appears to pass through blade.
perianth—collective term for calyx and corolla; if both are present, perianth is said to be in 2-series; if calyx or corolla is absent, perianth is said to be in 1-series.
pericarp—wall of fruit or mature ovary; may be dry or fleshy, and undifferentiated or differentiated into 2 or 3 layers.
persistent—remaining attached; not falling off; cf. *caducous*.
personate—bilabiate corolla with upper lip arched and lower lip closing throat by prominent projection (palate).
petal—1 member of second whorl of floral organs; typically colored and showy.
petaliferous—(1) bearing petals; (2) bearing nectar glands at bases of petals.
petaloid—petal-like in color and/or shape and/or texture.
petiole—stalk of leaf connecting blade and stem node.
petiolule—stalk of leaflet of compound leaf.
phreatophyte—plant with root system typically in soil saturated with water.
phyllary—in Asteraceae (sunflowers), involucral bract of head subtending florets.
pilose—with soft hairs.
pinna (pl. *pinnae*)—(1) primary division of pinnately compound leaf; (2) primary division of fern frond.
pinnate—featherlike; with closely arranged secondary veins radiating from midvein (primary) in 2 rows.
pinnately—descriptive of a compound leaf with leaflets borne in 2 rows along an axis; if 1-compound, leaflets are attached to rachis; if 2- or 3-compound, leaflets are attached to secondary or tertiary axes (rachillas).
pinnatifid—pinnately cleft into narrow lobes almost to midrib.
pinnatisect—pinnately cut to midrib into very narrow divisions but not leaflets.
pinnipalmate—with venation intermediate between pinnate and palmate, as in leaf with first pair of veins at base of blade larger and more distinctive than others arising from midvein.
pinnule—(1) leaflet of 2-pinnately compound leaf; secondary pinna; (2) secondary division of fern frond.
pistil—innermost floral organ; normally differentiated into ovary, style, and stigma; female organ; when simple, comprising 1 carpel; when compound, comprising 2 or more carpels.
pistillate—pertaining to flower that bears pistil or pistils but lacks functional stamens.
pith—centermost tissue of stem, usually soft or spongy.
placenta (pl. *placentae*)—ovule-bearing tissue in ovary.
placentation—arrangement of placentae within ovary; 5 types recognized: (1) apical—ovule(s) attached to apex; (2) axile—ovules attached to central axis and separated by septa; (3) free-central—ovules attached to central axis and not separated by septa; (4) basal—ovule(s) attached at base; (5) parietal—ovules attached in 1 or more sets to ovary wall. wall.
plicate—plaited or folded longitudinally, as a folding fan.
plumose—with indumentum simulating feather or plume.
pollen—mature microspores or developing microgametophytes in seed plants.
pollinium (pl. *pollinia*)—coherent mass of pollen; characteristic of Orchidaceae (orchids) and Asclepiadaceae (milkweeds).
polygamodioecious—bearing perfect and staminate flowers or perfect and pistillate flowers on separate plants.
polygamomonoecious—bearing perfect, staminate, and pistillate flowers on same plant.
polygamous—bearing perfect and imperfect flowers on same plant.
pome—fleshy, indehiscent fruit derived from compound pistil with inferior ovary surrounded by receptacular tissue and enclosing seeds within papery or cartilaginous endocarps; e.g., Malus (apple).
pore—orifice or opening, as in anthers or fruits.
poricidal—opening by pores; e.g., anthers of Solanum (nightshade).
prehensile—clasping or grasping, as tendrils.
prickle—small, spinelike outgrowth of bark or epidermis.
prostrate—lying flat on ground.
proximal—nearest to point of origin or attachment or center of organ; cf. distal.
pseudo- —prefix meaning false or closely resembling.
puberulent—minutely pubescent.
pubescent—(1) general term pertaining to presence of any type of hairs; (2) covered with short, soft hairs; down.
pulverulent—appearing dusty or powdery.
pulvinus (pl. pulvini)—(1) swollen base of petiole or petiolule, especially conspicuous in some legumes; (2) in Poaceae (grasses), swelling of culm or sheath just above node.
punctate—marked with pits, colored dots, or translucent sunken glands.
pustule—small blisterlike or pimplelike area on surface elevations.

pyramidal—pyramid-shaped; tetrahedral.

raceme—indeterminate inflorescence consisting of a rachis bearing a number of pedicelled flowers.

rachilla—(1) secondary axis of rachis; (2) in Poaceae (grasses) and Cyperaceae (sedges), axis of spikelet.

rachis (pl. rachises)—central axis of compound leaf or compound inflorescence, bearing flowers or leaflets.

radial—(1) denoting radius; (2) developing uniformly on all sides; (3) divisible into mirror-image halves in multiple planes.

rame—inflorescence consisting of dorsally compressed paired spikelets, 1 sessile and 1 pedicelled, borne on a disarticulating rachis; characteristic of the Andropogoneae (bluestem tribe) of the Poaceae (grasses).

rank—a vertical row along an axis.

ray—(1) in Asteraceae (sunflowers), pistillate or neutral floret with strap-shaped corolla; (2) 1 of radiating stalks arising from apex of peduncle of compound umbel.

receptacle—expanded portion of stem that bears floral organs.

receptacular bract—in Asteraceae (sunflowers), membranous or hyaline scale or bristle arising from receptacle and subtending disk floret; synonym of *pale.*

recurved—gradually bent or turned backward.

reflexed—abruptly bent downward or backward.

reniform—kidney-shaped.

resin—yellowish to dark, sticky, organic exudate.

reticulate—forming network of anastomosing veins; net-veined.

retrorse—directed downward or backward or toward base; cf. *antrorse.*

retuse—with shallow notch in round or blunt apex.

revolute—rolled downward from edges so that abaxial (lower) surface is partially concealed; cf. *involute.*

rhizome—horizontal underground stem producing aerial stems and leaves along its length or end, with scale leaves present or absent.

rhomboidal—diamond-shaped.

ribbed—with prominent ribs or veins.

ridge—long, relatively narrow, raised strip with sloping sides.

root—descending axis of plant, growing opposite stem, without nodes and leaves, typically developing underground and having primary functions of anchorage and absorption of water and nutrients.

rootlet—small, slender root or secondary root.

rosette—cluster of leaves radiating from common center or crown; usually at or near ground level.

rotate—pertaining to circular corolla shape with short tube and wide limb(s) spreading at right angles to tube.

rotund—nearly circular in outline; intermediate between orbicular and oblong.

rudiment—incompletely developed organ or structure; cf. *obsolete* and *vestigial.*

rugose—irregularly and coarsely wrinkled.

runcinate—oblanceolate, with retrorse teeth or incised lobes.

saccate—bag-shaped or pouchlike.

sagittate—arrowhead-shaped, with 2 basal lobes turned downward or toward axis; cf. *hastate.*

salverform—pertaining to trumpet-shaped corolla with slender tube and abruptly expanded limb, at right angles to tube; e.g., *Phlox* (phlox).

samara—winged, dry, indehiscent, 1-seeded, achene-like fruit; e.g., *Ulmus* (elm) and *Fraxinus* (ash).

sap—aqueous fluid circulated throughout a plant.

saprophyte—plant deriving all of its nutrients from dead or decaying organic matter and usually lacking chlorophyll.

saxicolous—dwelling or growing among or on rocks.

scaberulous—minutely roughened.

scabrous—having surface rough to the touch because of epidermal projections, scales, or short stiff hairs.

scale—(1) small, thin, usually scarious leaf or bract; (2) small leaf modified for protection of bud meristem; (3) small, scarious or coriaceous appendage of tissue within perianth; (4) small bract subtending individual flower in spike of members of Cyperaceae (sedges).

scandent—climbing.

scape—erect or ascending, leafless peduncle arising from ground, bearing 1 to many flowers, and possibly bearing scales or bracts but not foliage leaves.

scar—(1) mark or indentation left on stem by separation of leaf or bud scale; (2) mark or indentation on seed when it detaches.

scarious—thin, dry, and membranous; not green; semi-translucent.

schizocarp—dry, dehiscent fruit derived from a compound pistil that splits into separate 1-seeded, 1-carpellate mericarps at maturity.

scorpioid—like a scorpion's tail; pertaining to a coiled cyme with branches or flowers borne in 2 opposite ranks.

scrambling—climbing upon or over a support such as a rock, a fence, or a plant; cf. *twining.*

scurfy—covered with small branlike scales.

secund—with flowers or branches arranged on 1 side of an axis.

seed—multicellular structure developing from ovule after fertilization and typically comprising embryo, seed coat, and storage tissue such as endosperm.

semiparasite—plant capable of carrying on photosynthesis but requiring some nutrients from host plant.

sepal—1 member of outer most whorl of floral organs, typically green and sometimes leaflike or petaloid.

sepaliferous—(1) bearing sepals; (2) bearing nectary glands at bases of sepals.

sepaloid—sepal-like in color and/or shape and/or texture.

septicidal—with capsules dehiscing through the septa between the locules; cf. *loculicidal.*

septum (pl. *septa*)—partition or wall; e.g., separating adjacent locules in ovary.

sericeous—silky, with long, soft, slender, generally appressed hairs.

series—a number of structures standing in order and related in some fashion; e.g., calyx and corolla are each a series of floral parts.

serrate—with sharp teeth pointing forward on margin of organ.

serrulate—serrate with minute teeth.

sessile—stalkless; attached directly at base, as leaf without petiole.

setaceous—bristlelike; bearing bristles; synonym of *setiform.*

setose—covered with sharply pointed bristles.

sheath—long, more or less tubular structure partly or wholly surrounding another organ or part.

shoot—(1) collective term for stem and leaves; (2) young stems or branches arising from roots or older stems or branches.

shrub—woody plant with multiple stems from base but lacking a single trunk.

silicle—dry, dehiscent, 2-carpellate fruit of the Brassicaceae (mustards), less than 2 times longer than wide and with 2 valves separating from persistent septum (replum) and placentae.

silique—dry, dehiscent, 2-carpellate fruit of the Brassicaceae (mustards), more than 2 times longer than wide and with 2 valves separating from persistent septum (replum) and placentae.

simple—pertaining to an organ not divided into distinct parts or segments; not compound; e.g., leaf with undivided blade or pistil with 1 carpel; cf. *compound.*

sinuate—with wavy margin in horizontal plane alternatively concave and convex (in and out); cf. *undulate.*

sinus—space or indentation between 2 lobes or segments of an expanded organ such as leaf or petal.

solitary—single; 1-flowered.

sorus (pl. *sori*)—cluster of sporangia in ferns.

spadix (pl. *spadices*)—spike bearing small flowers crowded on fleshy axis; characteristic of the Araceae (arums).

spathe—bract or pair of bracts surrounding or subtending solitary flower, spadix, or other inflorescence.

spathulate—spatula-shaped, obovate or oblong apically but attenuated at base; synonym of *spatulate.*

spicate—spikelike; inflorescence with appearance of spike but flowers not truly sessile.

spike—inflorescence consisting of central rachis bearing 1 or more sessile flowers.

spikelet—in Poaceae (grasses), highly modified inflorescence typically consisting of 2 glumes, 1 or more florets, and rachilla.

spine—(l) sharp, stiff projection from stem; a modified leaf or stipule; (2) structure with appearance of spine.

spinescent—ending in a spine or spinelike sharp point.

spiral—coiled or wound around an axis and gradually and continuously ascending or descending.

sporangium (pl. *sporangia*)—multicellular organ producing spores.

spore—reproductive cell capable of developing into an adult without fusion with another cell, and, in flowering plants, giving rise to gametophytes.

sporocarp—specialized structure containing sporangia.

sporophyll—sporangium-bearing leaf, scale, or frond.

sporophyte—diploid ($2n$), spore-producing generation of plants with a sporic reproductive cycle; in flowering plants, the familiar green plant.

spreading—extending more or less horizontally outward.

spur—(1) short, compact branch with little or no internodal development and often bearing fascicles of leaves or flowers; (2) tubular or saclike, basal extension of sepal or petal.

stalk—general term for any supporting or connecting structure such as petiole, peduncle, pedicel, filament, or stipe.

stamen—floral organ producing pollen and normally consisting of filament and anther.

staminate—pertaining to a flower that bears stamens but lacks a functional pistil.

staminodium (pl. *staminodia*)—sterile stamen with vestigial or absent anther and sometimes petal-like or otherwise highly modified.

standard—synonym of *banner.*

stellate—(1) star-shaped, radiating like points of a star; (2) pertaining to hairs with several branches radiating from 1 point.

stem—ascending axis of plant bearing and supporting leaves, flowers, and fruits; growing opposite root; having nodes; and typically developing aboveground.

sterile—(1) not producing viable seeds; (2) not producing pollen; (3) in Poaceae (grasses), floret or spikelet not producing a caryopsis.

stigma—portion of pistil that receives pollen, typically at end of style.

stipe—(1) stalk of pistil or other small organ; (2) petiole of fern frond.

stipel—1 of pair of appendages borne at base of leaflet.

stipule—1 of pair of appendages borne at base of leaf.

stolon—elongate, horizontal, aboveground stem typically rooting at nodes and giving rise to new shoots.

stoloniferous—producing stolons.
stone—hard, woody endocarp of drupe; pit.
stramineous—strawlike in color or texture.
striate—with narrow, usually parallel lines or grooves.
strigose—with stiff, long, sharp, straight or curved, appressed hairs with bulbous bases.
strobilus (pl. *strobili*)—small cone, or dense conelike cluster of sporophylls.
style—slender stalk typically connecting stigma to ovary.
stylopodium (pl. *stylopodia*)—enlargement or disklike expansion at base of style, as in Apiaceae (carrots) and Cyperaceae (sedges).
sub- —prefix meaning somewhat, nearly, or slightly less; e.g., *subsessile* and *suboval.*
subtend—to enclose or extend beyond structure in its axil; e.g., as bract below flower or leaf below bud.
subulate—awl-shaped; tapering from base to fine, sharp point.
succulent—fleshy and juicy.
suffrutescent—obscurely or somewhat shrubby, slightly woody at base, and with aerial portions dying back to woody base.
sulcate—with 1 or more longitudinal grooves.
superior—(1) pertaining to an organ higher than another; (2) pertaining to ovary attached at summit or center of receptacle and above point of attachment of other floral organs; cf. *inferior.*
suture—junction or seam of fusion; line of opening or dehiscence of fruit or anther.
symmetry—condition whereby half of organ or structure is mirror image of other half; correspondence in size, shape, and relative position of parts of 2 halves.
syncarp—multiple fruit derived from fused pistils of 2 or more flowers; e.g., *Morus* (mulberry).
taproot—persistent, well-developed primary root, generally larger than secondary roots that arise from it.
taxon (pl. *taxa*)—general taxonomic term for any group of any rank.
tendril—rotating or twisting threadlike process or extension by which a plant grasps an object for support; may be a modified stem, leaf, leaflet, or stipule.
terete—cylindrical and circular in cross section.
terminal—tip or apex of stem, summit of axis; e.g., terminal leaflet of 1-pinnately compound leaf.
terrestrial—growing on land, not in water.
tetradynamous—having 4 long stamens and 2 short ones, as in Brassicaceae (mustards).
tetrahedral—4-sided, each side triangular; pyramidal.
thallus—plant body not differentiated into roots, stems, and leaves; e.g., *Marchantia* (liverwort).
theca (pl. *thecae*)—half of anther containing 2 pollen sacs or microsporangia.
thicket—thick, dense growth of shrubs.
thorn—stiff, woody, modified stem with sharp point.
thyrse—compact, cylindrical, or ovate panicle with indeterminate main axis and determinate cymose subaxes.
tomentose—with soft, short, dense, matted or tangled hairs; woolly.
toothed—general term encompassing margin with serrations, dentations, or crenations.
torulose—irregularly swollen and constricted cylinder.
trailing—prostrate or creeping but not rooting.
translator—connecting structure between pollinia of adjacent anthers in Asclepiadaceae (milkweeds).
translucent—semitransparent; allowing passage of light but not permitting a clear view of any object.
tree—perennial woody plant of considerable stature at maturity with 1 main stem (trunk), generally branching well aboveground and exhibiting well-developed crown.
trichome—any hairlike outgrowth of epidermis.
trifid—divided into 3 parts.
trigonous—with 3 angles and sides.
truncate—ending abruptly, with base or apex transversely straight or nearly so, as if cut off.
tube—hollow, cylindrical structure, as constricted basal portion of some fused corollas.
tuber—short, thickened branch of subterranean rhizome bearing nodes and buds, serving in food storage; e.g., *Solanum tuberosum* (potato).
tubercule—small swelling or projection.
tubular—having the form of tube or cylinder.
tuft—dense cluster of organs or structures.
tunica—loose, dry, papery, or membranous or fibrous covering; e.g., outer layer or skins of *Allium* (onion).
turbinate—top-shaped; inversely conical but contracted near apex.
twig—small shoot or branch of woody plant.
twiner—climbing plant that twines around a support.
umbel—indeterminate, flat-topped or hemispheric inflorescence consisting of flowers borne on pedicels arising from common point of attachment; in simple umbels, pedicels arise from end of peduncle; in compound umbels, pedicels arise from end of rays that arise from end of peduncle.
umbellet—1 of ultimate umbellate clusters of compound umbel.
undulate—with wavy margin in vertical plane; alternatively convex and concave (up and down); cf. *sinuate.*
uni- —prefix meaning one; as in unilabiate or unisexual.
urceolate—pertaining to urn-shaped or pitcher-like corolla; globose to cylindrical but contracted at or just below mouth.
utricle—small, thin walled, l-seeded, bladderlike, fruit; e.g., *Atriplex* (saltbush).

valvate—(1) edges of adjacent organs meeting but not overlapping; (2) opening by valves breaking away from septum or septa.
valve—(1) 1 of several segments into which a dehiscent fruit naturally separates at maturity; (2) in poricidal anthers, portion of anther wall covering pore.
vascular bundle scar—mark within leaf scar left by vascular bundle at leaf separation.
vein—vascular bundle visible externally; primary veins are largest, occurring singly as midvein or series; secondary veins are next size class, arising from primary veins.
venation—arrangement of veins on surface of organ.
ventral—(1) pertaining to or located on front; (2) pertaining to adaxial surface of organ; cf. *dorsal.*
vernation—arrangement of leaves within bud.
verrucose—surface covered with wartlike projections; warty.
versatile—attached near middle rather than at 1 end, as in some anthers; cf. *basifixed* and *dorsifixed.*
verticil—whorl of leaves or flowers.
vesicle—small bladderlike sac or cavity.
vestigial—organ much reduced and likely nonfunctional as a result of evolutionary reduction; cf. *obsolete* and *rudimentary.*
villose—with long, soft, wavy hairs; not matted.
vine—climbing or scrambling plant with elongate, flexible, non-self-supporting stems.
viscid—covered with sticky or gelatinous exudate; gummy to the touch.
viscous—having sticky or glutinous consistency.
waxy—resembling wax in appearance, consistency, or adhesive qualities; e.g., surfaces of leaves and fruit in Myricaceae (bayberries).
weed—plant species that aggressively colonizes disturbed habitats and cultivated lands.
whorl—(1) arrangement of organs such as leaves, branches, flowers, or fruits 3 or more per node; (2) 2 or more cyclic groups of sepals or petals or stamens.
wing—(1) thin, flat membranous expansion of pericarp in samara; (2) lateral petal of papilionaceous flower; (3) cortical eruptions of bark forming ridges on branches of some woody plants; (4) thin, flat, membranous expansion forming ridge or flange on organ.
xerophyte—plant adapted to dry conditions of air or soil; desert plant.

General Terms

abomasum—glandular, fourth, or true stomach of ruminants, including cattle, goats, and sheep.
abortifacient—capable of causing abortion.
acidosis—state of low blood pH indicating low blood bicarbonate.
acinar—pertaining to lobules of a gland.
adipose—fat or fatty.
adrenolytic—capable of inhibiting the adrenergic system or action of epinephrine and norepinephrine.
agalactia—lack of milk secretion.
agglutinate—to cause to aggregate.
agonist—agent that stimulates a receptor, causes positive effects.
allelopathy—prevention of growth of other plants.
ALT—alanine transaminase, an enzyme derived from liver and present in blood in increasing amounts with liver damage.
alveolar congestion—accumulation of blood in alveoli.
alveolitis—inflammation of alveoli such as air sacs of lungs.
amblyopia—loss of vision without obvious structural changes of the eye.
AMPA—α-amino-3-hydroxy-5-methyl-isoxazole-4-proprionate.
anaerobic—growing or chemically active in absence of oxygen.
analog—compound structurally very similar to another.
anorexia—loss of appetite.
antagonist—agent that acts opposite of another.
anthelmintic—agent that kills worms or helminths.
antiepinephrine—antagonist to action of epinephrine.
antigen—substance that stimulates formation of antibody.
anti-inflammatory—capable of reducing or suppressing inflammation.
antileukemic—reduction of effects of or cause of leukemia.
antimicrobial—with action against microbes.
antineoplastic—pertaining to reduction of effects of or cause of neoplasia or cancer.
antioxidant—substance that inhibits or prevents oxidation in cells.
anxiolytic—capable of reducing anxiety.
AP—alkaline phosphatase, an enzyme derived from various tissues and present in blood in increasing amounts with damage to those tissues.
apnea—cessation of or lack of breathing.
arrhythmia—irregularity of rhythm, such as with heart.
arthrogryposis—persistent flexure or contracture of a joint.
ascites—accumulation of fluid in abdomen or peritoneal cavity.
aspiny—without spines.
AST—aspartate transaminase, an enzyme derived from liver and present in blood in increasing amounts with liver damage.
ataxia—incoordination.
atony—without muscular tone.
atrioventricular block—interference with electrical conduc-

tion through pathway of AV (atrioventricular) node of heart.
atrium (pl. *atria*)—1 of 2 upper heart chambers.
atrophy—diminishing of size of organ or tissue.
autonomic—referring to autonomic nervous system, the involuntary nervous system that regulates function without conscious control, composed of sympathetic and parasympathetic portions.
AV block—see *atrioventricular block*.
axonal—pertaining to the extension of a neuron essential for conduction of an electrical impulse to other nerves or excitable tissues.
basal ganglia—portion of cerebral hemispheres of central nervous system responsible for motor performance.
Betz cells—giant pyramidal neurons of primary motor cortex with axons that pass via corticospinal tract, providing fine movement input to distal muscles.
bilirubin—breakdown product of hemoglobin, found in bile, blood, and urine.
birefringent—doubly refractile, often referring to crystals.
blepharospasm—spasm of muscles of eyelids.
bradycardia—slow heart rate.
BSP—bromosulfophthalein or sulfobromophthalein.
b.w.—body weight.
C-3—carbon number 3 of compound.
3-C—with 3 carbon atoms.
cardenolides—23-carbon steroidal compounds with a 5-member constituent ring, acting on the heart.
cardiac insufficiency—cardiac function not sufficient to maintain normalcy.
cardiodepressor—agent causing diminished cardiac function.
cardiomyopathy—degenerative disease of cardiac muscle.
carpal—pertaining to carpus or wrist.
catabolism—breakdown or metabolism of a substance.
cathartic—compound increasing evacuation of intestines; purgative.
caudate nucleus—nucleus of the basal ganglia of the central nervous system.
caudoputamen—neostriatum or corpus striatum, 2 partially fused nuclei with similar cell types and responsible for receiving most of the afferent input from cerebral cortex and thalamus to basal ganglia.
caustic—burning or corrosive; compound with burning or corrosive action.
centrilobular—in center of a hepatic lobule in region of central vein.
cerebellum—posterior portion of brain, associated with coordination of movement.
cerebrocortical—pertaining to outer portion or cortex of cerebrum.
cholangitis—inflammation of bile ducts.
cholinergic receptor—receptor activated by acetylcholine, mainly parasympathetic.
cholinesterase inhibitor—substance that inhibits enzymes responsible for breakdown of and termination of acetylcholine activity.
cholinomimetic—chemical that mimics the action of acetylcholine.
chorea—involuntary muscular activity, often a fine tremor or rapid, jerky twitching.
cirrhotic—affected with increased connective tissue or fibrosis of liver, nodular regeneration, and bile duct hyperplasia.
CNS—central nervous system.
colloid—gelatinous or gluelike substance.
conjunctiva—membrane lining eyelids and covering eye.
conjunctivitis—inflammation of conjunctiva.
contact dermatitis—inflammation of skin caused by direct contact with an irritant.
corneal—pertaining to cornea of eye.
corticosteroid—compound with actions similar to those caused by steroids from adrenal cortex.
creatine phosphokinase—an enzyme in muscle responsible for increasing metabolic energy and present in blood in increasing amounts with muscle damage.
crystalloid—crystal-like.
crystalluria—condition of having crystals in urine.
cyanogenic—capable of release or production of cyanide.
cyanogenic glycosides—sugar-linked compounds with cyanogenic potential.
cyanosis—bluish discoloration of skin.
cystitis—inflammation of urinary bladder.
cytolytic—capable of lysis of cells.
cytotoxic—capable of causing adverse effects on cells.
debilitation—loss of strength or vitality.
decoction—potion prepared by boiling.
degranulator—agent causing loss of granules from cells.
delirium—transient mental disturbance, with hallucinations and incoherence.
demyelination—loss of myelin sheath around axons.
denaturing—change in structure such as with proteins.
dendrite—process of a neuron that receives stimuli.
dendritogenesis—formation of multiple new dendritic processes.
depolarization—elimination of polarity of cell by transmission of electrical charge.
dermatitis—inflammation of skin.
diazepam—drug causing sedation.
diuresis—increased secretion of urine.
dopaminergic pathways—pathways of nervous system in which dopamine is the transmission link.

d.w.—dry weight.
dyskeratosis—abnormal or imperfect keratinization of skin.
dyspnea—difficult or labored respiration or breathing.
dystonia—abnormal muscle tonicity.
dystrophic—defective or faulty.
ecchymotic—splotchy foci of hemorrhage.
ectopic—displaced or malpositioned.
edema—intercellular accumulation of fluid.
electrocardiograph—instrument for recording electrical activity of heart muscle.
electroencephalograph—instrument for recording electrical activity of brain.
electrophilic—electron-seeking chemical reactants.
ELISA—enzyme-linked immunosorbent assay.
emetic—substance causing vomiting.
emphysema—distension of tissues caused by trapped air.
endocardial—pertaining to lining of heart chambers (endocardium).
endoplasmic reticulum—cell organelle; smooth endoplasmic reticulum (ER) is responsible for metabolism of steroids and other compounds, and rough ER for synthesis of cell proteins intended for secretion.
endothelium—cells lining blood vessels.
enkephalin—a group of 5-amino-acid peptides that serve as endogenous analgesic endorphins.
enterohepatic—within liver or between digestive tract and liver.
eosinophilic—pinkish coloration with eosin dye, indicating presence of proteins.
epicardial—on external surface of heart.
epimer—compound of similar structure but with altered configuration around 1 specific atom.
epithelium—cells lining various surfaces of body.
epoxide—tricyclic moiety with 2 carbons and 1 oxygen.
erythrocyte—red blood cell.
erythrophagia—engulfment of red blood cells by other cells such as macrophages.
erythropoiesis—production of red blood cells.
ester—compound formed by interaction of an acid and an alcohol, with loss of a water.
estrogenic—compound with feminizing action or causing estrus.
eversion—turning inside out.
excitotoxicity—adverse effect due to increased activity of excitatory neuronal pathways in brain.
excoriation—loss of superficial layers of skin.
extrahepatic—outside the liver.
fasciculations—spontaneous contractions of muscle fiber bundles.
fibrinolytic—capable of liquefying fibrin.
fibrosis—formation of fibrous connective tissue.
fistula—deep penetrating ulcer or tract, often leading to an internal organn.
flexor—muscle causing flexion of joint or limb.
foci (sing. *focus*)—multiple points/locations.
folia—convolutions or gyri of cerebellar surface.
follicular—pertaining to a small secretory sac or gland.
fundic—pertaining to base of an organ.
GABA—gamma-aminobutyric acid.
ganglion—a controlling mass of neurons.
ganglioplegic—inhibition of nerve impulses through a ganglion.
ganglioside—galactose-containing compound in brain.
GGT—gamma glutamyl transaminase, an enzyme derived from bile ducts of liver and present in blood in increasing amounts with liver damage.
globus pallidus—a nucleus of the basal ganglia.
glomerular—pertaining to glomerulus of kidney.
glucuronide—compound conjugated with glucuronic acid.
glutaminergic—compound that stimulates glutamine-sensitive receptors.
goitrogenesis—process of development of enlarged thyroids.
Golgi apparatus—cell organelle that processes proteins and lipids from rough endoplasmic reticulum to form compounds such as glycoproteins for secretion.
GOT (or *SGOT*)—older term for the enzyme AST.
Heinz body—aggregate of denatured hemoglobin in red blood cells.
hematuria—blood in urine.
hemiplegia—paralysis of 1 side of body.
hemodialysis—circulation of blood through artificial kidney-type apparatus.
hemoglobinuria—condition of having hemoglobin in urine.
hemolysis—lysis of red blood cells.
hemoperfusion—circulation of blood through an instrument to remove undesirable substances.
hemorrhagic—causing bleeding.
hemosiderin—intracellular iron containing pigment.
hepatocellular—pertaining to liver cells.
hepatoencephalopathy—liver disease that causes signs of nervous derangement.
hepatogenous—produced in liver.
hepatohyperplasia—proliferation of liver cells.
hepatotoxic—causing adverse effects on liver.
herbivore—animal that subsists on plants for food.
heterocyclic—pertaining to compounds composed of closed chains containing different elements.
heterodimeric—composed of 2 dissimilar parts.
hippocampus—structure of middle horn of lateral ventricle of cerebral cortex.
histamine—potent vasodilator and gastric secretion stimulant.

hydrolysis—decomposition of a compound due to addition of water.
hydrothorax—fluid in chest or pleural cavity.
hyperesthesia—excess sensitivity of skin or other sensory organs.
hyperexcitability—increased excitability.
hyperflexion—overflexion of joint or limb.
hypermetria—excessive range of limb movement.
hyperplasia—increase in number of cells.
hyperthermia—high body temperature.
hypertonicity—excess tone or activity of muscle.
hypertrophy—increase in size of an organ due to increased size of individual cells.
hypocalcemic—low blood calcium.
hypocholesterolemic—low blood cholesterol.
hypoglycemia—low blood glucose or sugar.
hypoplasia—incomplete development.
hypotensive—causing low blood pressure.
hypothalamus—center for regulation of emotions and various activities such as thirst, hunger, and body temperature.
hypoxia—lack of oxygen.
icterus—jaundice or discoloration of blood and tissues, usually due to bilirubin.
inotropic—affecting strength of muscular contraction.
intraperitoneal—within peritoneal or abdominal cavity.
intravenous—within a vein.
in utero—within the uterus.
in vitro—outside the body in an observable system such as a test tube.
in vivo—within the body.
i.p.—intraperitoneal.
i.v.—intravenous.
karyomegaly—enlargement of cell nuclei.
keratosis—thickening of keratin layer of skin.
ketosis—excessive formation of ketones in body.
Kupffer cells—macrophages of sinusoids of liver.
lacrimation—tear secretion.
lactogenesis—production of milk.
laminitis—inflammation of lamina of hoof.
laryngeal—pertaining to the larynx.
laxative—mild cathartic.
leaching—act of washing in a liquid such as water.
lectin—compound capable of lysing red blood cells.
leukemia—disease with abnormal increase in white blood cells.
leukopenia—deficiency of white blood cells.
libido—sexual desire.
ligation—act of tying off, or occlusion of, a vessel or structure.
lithogenic—capable of causing formation of calculi or concretions in tissues.
liver enzymes—ALT, AST, GGT, SDH.
locomotor—associated with movement.
luminal cells—cells lining a tubular or hollow organ.
luteal—pertaining to corpus luteum of ovary.
lymphoma—neoplasm or tumor of lymphoid tissue.
lysis—dissolution or destruction.
lysosomes—cell organelles that contain acid hydrolases for breakdown and digestion of various compounds present in cells.
macrophage—activated phagocytic monocyte outside the bloodstream.
malacia—degenerative softening of tissues.
mannosidosis—accumulation of mannose-containing oligosaccharides in cells.
medullary—pertaining to central portion of an organ.
megalocytosis—enlargement of cell size.
meganeuritis—inflammation of large nerves.
meninges—the 3 membranes that envelop brain and spinal cord—dura mater, pia mater, and arachnoid.
metabolites—breakdown products of a substance.
methemoglobin—hemoglobin containing oxidized heme, dark-colored and unable to release oxygen to tissues.
methemoglobinemia—excess methemoglobin in blood.
MHb—methemoglobin.
MHbemia—methemoglobinemia.
microcornea—abnormal smallness of cornea.
microphthalmia—abnormal smallness of eye.
microtubule—subunit of tubulin structure in cells.
mitochondrial—pertaining to the cellular organelle responsible for production of energy to drive cellular reactions.
molluscicidal—capable of killing mollusks or snails.
monogastric—simple or single-stomached animal.
morbidity—rate of illness; ratio of sick to well individuals.
motility—spontaneous movement.
mucilage—viscid or gumlike substance.
mucopurulent—composed of pus and mucus.
mucosa—mucous membrane lining.
muscarinic acetylcholine receptors—receptors of parasympathetic nervous system affecting glandular secretion, heart, eye, and digestive tract.
mutagenicity—ability to induce genetic mutation.
mycotoxicosis—intoxication due to fungal toxins.
mydriasis—dilation of pupils of eye.
myelin—fatty substance forming sheath around nerves.
myocarditis—inflammation of heart muscle.
myocardium—heart muscle.
myodegeneration—degeneration of muscle.
myoglobinuria—presence of myoglobin in urine from muscle breakdown.
narcosis—deep unconsciousness.
necropsy—postmortem examination; autopsy.
necrosis—cell death.

neonate—newborn.
neoplastic—tumorous or cancerous.
nephrosis—degenerative disease of kidney.
nephrotoxic—causing toxic effects on kidney.
neural crest cells—embryonic precursor cells from dorsal region of neural tube that form neurons, Schwann cells, and others cells such as melanocytes.
neurotoxic—causing toxic effects on nervous system.
nicotinic acetylcholine receptors—receptors of parasympathetic nervous system that affect skeletal muscle.
NMDA—*N*-methyl-D-aspartate.
nonsteroidal—of a different class of drugs from corticosteroids.
NPA (or 3-NPA)—nitropropionic acid.
nystagmus—rapid, involuntary movement of eyeball.
oliguria—secretion of reduced amount of urine.
omentum—membranous saclike covering of stomach and intestines.
opioid—opium-like drug.
opisthotonus—spasm in which head is bent backward and body is bowed forward.
organelle—intracellular organ.
osmotic—pertaining to movement of fluid from compartments of lower to higher solute concentrations.
osteodystrophy—defective bone formation.
oxidant—compound causing oxidation.
palatable—readily eaten.
parakeratotic—abnormality of keratin layer of skin.
parasympathetic—portion of autonomic nervous system.
parenterally—other than by oral route.
paresis—slight or incomplete paralysis.
paroxysmal tachycardia—recurring episodes of rapid heart rate.
pars reticularis—synonym of *pars reticulata,* a zone of the substantia nigra of the basal ganglia.
parturition—act of giving birth.
PCV—packed cell volume.
peracute—excessively acute or sharp, occurring in minutes to hours.
pericardial—around heart.
pericholangitis—inflammation of tissues around bile ducts.
perirenal—around kidney.
peristalsis—wavelike contractions of digestive tract, propelling its contents.
peroxidation—oxidation due to peroxides.
petechia (sing. *petechium*)—small pinpoint hemorrhages.
phagocytic—pertaining to ingestion of materials by cells.
phosphorylation—introduction or addition of phosphorus group on an organic molecule.
photophobia—intolerance to light.
photosensitization—condition of increased sensitivity to ultraviolet light, commonly manifest by skin changes.
phototoxic—substance capable of causing adverse effects in combination with exposure to light.
phylloerythrin—product of chlorophyll metabolism in digestive tract.
phytobezoar—digestive tract concretion derived from vegetable or plant matter.
piloerection—erection of hair.
piscicide—substance capable of killing fish.
pituitary—portion of central nervous system under control of hypothalamus and regulating various hormones.
pleural—pertaining to membranous lining on surfaces of thoracic cavity and lungs.
pneumocytes—cells of alveoli in the lungs.
pneumotoxic—causing adverse effects on lungs.
polioencephalomalacia—degenerative softening of gray matter of brain.
poliomalacia—degenerative softening of gray matter of spinal cord.
polyphenolic—compound with benzene rings containing multiple hydroxyl groups.
polypnea—rapid respiration.
progesterone—female hormone responsible for maintenance of pregnancy.
proprioceptive—pertaining to sensory input concerning position and movement of body.
prostaglandin—regulatory compound in body derived locally from arachidonic acid.
proteolytic—capable of digestion or hydrolysis of proteins.
prothrombin—precursor to thrombin subject to activation by Ca^{2+}.
protoplasmic—pertaining to the viscid material that fills cells.
protraction—protrusion of jaw.
pruritis—intense itching.
psychedelic—causing extreme changes in consciousness, stimulating auditory and visual responses such as delusions and hallucinations.
pulmonary alveoli—air sacs or dilations in lungs where oxygen exchange occurs.
pulmonary edema—accumulation of fluid in lungs.
purgative—cathartic causing evacuation of bowels.
Purkinje's cells—large branched neurons of middle layer of cerebellum.
putamen—outer part of lentiform nucleus of brain.
pyelonephritis—inflammation of kidney, including pelvic region.
pyknotic—pertaining to degeneration of a cell in which the nucleus shrinks.
pyogranulomatous—pertaining to a form of chronic inflammation in which pus predominates.
regurgitation—backward flowing, such as of undigested food from stomach to mouth.

retinopathy—noninflammatory disease or degeneration of retina of eye.
retrograde—moving backward.
ring cleavage—chemical reactive disruption of a ring structure.
rumen—main fermentative stomach of ruminants.
rumenotomy—cutting open or making an opening in the rumen.
ruminant—herbivore with rumen as 1 of 4 digestive compartments.
saponins—glycosidic compounds from plants which form stable foam in solution and lyse red blood cells.
sarcoplasm—cytoplasm of muscle cells.
SDH—sorbitol dehydrogenase, an enzyme derived from liver and present in blood in increasing amounts with liver damage.
seleniferous—capable of causing accumulation of selenium in plants or animals.
selenosis—intoxication caused by selenium.
septal cell—cells of the supporting septum.
serofibrinous—fluid with appearance of serum and containing clumps of fibrin.
serotonin—synonym of *5-hydroxytryptamine,* a compound in the body with a variety of actions on smooth muscle and neurotransmission in the central nervous system.
Sertoli cells—cells of tubules of testes to which spermatids attach while maturing to spermatozoa.
sinoatrial—pertaining to area of sinus venosus and atrium of heart.
spermatocyte—precursor cell to spermatid.
spermatogenesis—process of formation of spermatozoa.
steroids—compounds containing a specific 5-membered ring structure.
striatal—pertaining to corpus striatum, neostriatum, or caudoputamen in brain.
stupor—disorder of consciousness with reduced responsiveness.
subcutaneous—beneath the skin.
subendocardial—under the membranous lining of the inner heart surface.
subepicardial—under the membranous lining of the outer heart surface.
sublingual—below the tongue.
submandibular—below the mandible of the jaw.
substantia nigra—portion of basal ganglia that, with globus pallidus, forms major output nuclei.
sulci (sing. *sulcus*)—furrows or grooves on surface of brain.
surfactant—compound that alters surface tension of liquid.
sympathetic—division of autonomic nervous system.
synapse—junction between 2 neurons.
syndrome—set of signs that occur together.
tardive dyskinesia—muscular movement disorder of basal ganglia involving dopaminergic pathways.
tarsal—pertaining to tarsus of rear, or pelvic, limbs.
T **cell**—thymus-dependent lymphocyte involved in cell-mediated immunity.
tenesmus—rectal or intestinal straining.
teratogenic—capable of causing fetal deformities or literally genesis of monsters.
tetanic—muscular spasms of the tonic type or continuous in appearance.
thalamic—pertaining to the thalamus, the relay center responsible for processing sensory and motor input to cerebral cortex.
thermolabile—susceptible to destruction by heat.
thromboplastic—accelerating blood clot formation.
thrombosis—formation of blood clot.
tincture—alcoholic solution.
toxicant—toxin.
toxin—substance causing adverse effects on body.
tremorgenic—capable of causing muscle tremors.
urticaria—vascular eruptions visible as smooth, elevated patches of skin.
vacuole—membrane-bound space formed in protoplasm of cell.
vagus nerve—tenth cranial nerve.
vasoconstriction—constriction of blood vessels.
vasodilation—dilation of blood vessels.
vasospasm—spasm of contraction of blood vessels.
ventricular fibrillation—ineffective convulsive, contractile movements of ventricles of heart.
ventricular tachycardia—excessively rapid beating of ventricles of heart.
vermifuge—substance for killing worms.
vesicant—substance causing blistering of skin or mucous membranes.
viscera—internal body organs such as heart, lungs, and liver.
vitiligo—skin disease with smooth, light-colored or unpigmented areas.
Wallerian degeneration—a lesion of the nervous system with degeneration of axons, resulting in impaired or interrupted nerve transmission.

Index